Principles of GENETICS

Fifth Edition

International Student Version

D. Peter Snustad
University of Minnesota

Michael J. Simmons
University of Minnesota

John Wiley & Sons, Inc.

Cover image from © Ovidiu lordachi/shutterstock.com

Contributing Subject Matter Expert: Ashley Hagler, M.S., M.A.T., Gaston College

ISBN: 978-0-470-39842-5

Printed in Asia

10 9 8 7 6 5 4 3 2 1

Dedications

To Judy, my wife and best friend.
D.P.S.

To my family, especially to Benjamin.
M.J.S.

About the Authors

D. Peter Snustad is a professor in the Department of Plant Biology at the University of Minnesota, Twin Cities. He received his B.S. degree in science specialization from the University of Minnesota and his M.S. and Ph.D. degrees in genetics from the University of California, Davis. During his 43 years as a member of the faculty at Minnesota, he has taught courses at all levels from general biology to advanced biochemical genetics. For 20 years, his research focused on bacteriophage T4 morphogenesis and the interaction between T4 and its host, *Escherichia coli*. For the past 23 years, his research group has studied the genetic control of the cytoskeleton in *Arabidopsis thaliana* and the glutamine synthetase gene family in *Zea mays*. He has served on the National Institutes of Health Molecular Cytology Study Section and as the Program Chairperson for the Annual Meeting of the Genetics Society of America. His honors include the Morse-Amoco and Stanley Dagley Memorial teaching awards. A lifelong love of the Canadian wilderness has kept him in nearby Minnesota.

Michael J. Simmons is a Professor in the Department of Genetics, Cell Biology and Development at the University of Minnesota, Twin Cities. He received his B.A. degree in biology from St. Vincent College in Latrobe, Pennsylvania, and his M.S. and Ph.D. degrees in genetics from the University of Wisconsin, Madison. Dr. Simmons has taught a variety of courses, including genetics and population genetics. He has also mentored many students on research projects in his laboratory. Early in his career he received the Morse-Amoco teaching award from the University of Minnesota in recognition of his contributions to undergraduate education. Dr. Simmons's research focuses on the genetic significance of transposable elements in the genome of *Drosophila melanogaster*. He has served on advisory committees at the National Institutes of Health and he has been a member of the Editorial Board of the journal *Genetics* for 20 years. One of his favorite activities, figure skating, is especially compatible with the Minnesota climate.

Preface

In recent years the science of genetics has been undergoing a sea change. The DNA of genomes, even large ones, can now be analyzed in great detail; the functions of individual genes can be studied with an impressive array of new techniques; and organisms can be changed genetically by introducing alien or altered genes into their genomes. All of these developments have placed genetics in the center of a technological revolution that is affecting agriculture, medicine, and society. At a more basic level, genetics has become a key science—some would say *the* key science—in all of biology. Genetic information and genetic analysis are now crucial to research in almost every biological discipline. We have created the fifth edition of *Principles of Genetics* to provide a comprehensive, up-to-date introduction to this important science.

Goals

This edition of *Principles of Genetics* continues our efforts to offer a book that balances new information with fundamental principles. As in previous editions, we have been guided by four overarching goals:

- **To focus on the basic principles of genetics** by presenting the important concepts of classical, molecular, and population genetics carefully and thoroughly. We believe that an understanding of current advances in genetics and an appreciation for their practical significance must be based on a strong foundation. Furthermore, we believe that the breadth and depth of coverage in the different areas of genetics—classical, molecular, and populational—must be balanced, and that the ever-growing mass of information in genetics must be organized by a sturdy—but flexible—framework of key concepts.

- **To focus on the scientific process** by showing how scientific concepts develop from observation and experimentation. Our book provides numerous examples to show how genetic principles have emerged from the work of different scientists. We emphasize that genetics is an ongoing process of observation, experimentation, and discovery. For example, each chapter contains a *Milestone in Genetics*, which focuses on an important advance in genetics and discusses how that advance came about. We also present the experimental evidence for important genetic principles.

- **To focus on human genetics** by incorporating human examples and showing the relevance of genetics to societal issues. Experience has shown us that students are keenly interested in the genetics of their own species. Because of this interest, they find it easier to comprehend complex concepts when these concepts are illustrated with human examples. Consequently, we have used human examples to illustrate genetic principles wherever possible. We have also included discussions of the Human Genome Project, human gene mapping, genetic disorders, gene therapy, and genetic counseling throughout the text. Issues such as genetic screening, DNA fingerprinting, genetic engineering, cloning, stem cell research, and gene therapy have sparked vigorous debates about the social, legal, and ethical ramifications of genetics. We believe that it is important to involve students in discussions about these issues, and we hope that this textbook will provide students with the background to engage in such discussions thoughtfully.

- **To focus on developing critical thinking skills** by emphasizing the analysis of experimental data and problems. Genetics has always been a bit different from other disciplines in biology because of its heavy emphasis on problem solving. In this text, we have fleshed out the analytical nature of genetics in many ways—in the development of principles in classical genetics, in the discussion of experiments in molecular genetics, and in the presentation of calculations in population genetics. Throughout the book we have emphasized the integration of observational and experimental evidence with logical analysis in the development of key concepts. Each chapter has two sets of worked-out problems—the *Basic Exercises* section, which contains simple problems that illustrate basic genetic analysis, and the *Testing Your Knowledge* section, which contains more complex problems that integrate different concepts and techniques. A set of *Questions and Problems* follows the worked-out problems so that students can enhance their understanding of the concepts in the chapter and develop their analytical skills. In each chapter in this edition we have also incorporated a *Focus on Problem Solving*. This feature poses a problem, lists the pertinent facts and concepts, and then analyzes the problem and presents a solution. Finally, we have added a new feature, *Genomics on the Web*, which poses questions that can be answered by going to the National Center for Biotechnology Information web site. With this feature, students can learn to use the vast repository of genetic information that is accessible via that web site, and they can apply it to specific problems.

Content and Organization of the Fifth Edition

The organization of this edition of *Principles of Genetics* is similar to that of the previous edition. However, some of the material has been consolidated and many passages have been rewritten to reflect how the science of genetics has changed in the past few years. The art program has also been completely redesigned. There are many new photographs and the illustrations have been redrawn to provide a clear, consistent style throughout the text.

The text comprises 25 chapters—two less than the previous edition. Chapters 1–2 introduce the science of genetics, basic features of cellular reproduction, and some of the model genetic organisms; Chapters 3–8 present the concepts of classical genetics and the basic procedures for the genetic analysis of microorganisms; Chapters 9–14 present the topics of molecular genetics, including DNA replication, transcription, translation, mutation, and definitions of the gene; Chapters 15–18 cover more advanced topics in molecular genetics and genomics; Chapters 19–22 deal with the regulation of gene expression and the genetic bases of development, immunity, and cancer; Chapters 23–25 present the concepts of quantitative, population, and evolutionary genetics. In selecting material to be included in this edition of *Principles of Genetics*, we have tried to be comprehensive but not encyclopedic. To keep the length of the text reasonable, we had to make difficult decisions about what to include and exclude, and we had to streamline some of the older material to make room for emerging developments.

As in previous editions, we have tried to create a text that can be adapted to different course formats. Many instructors prefer to present the topics in much the same way as we have, starting with classical genetics, progressing into molecular genetics, and finishing with quantitative, population, and evolutionary genetics. However this text is constructed so that teachers can present topics in different orders. They may, for example, begin with basic molecular genetics (Chapters 9–14), then present classical genetics (Chapters 3–8), progress to more advanced topics in molecular genetics (Chapters 15–22), and finish the course with quantitative, population, and evolutionary genetics (Chapters 23–25). Alternatively, they may wish to insert quantitative and population genetics between classical and molecular genetics.

Pedagogy of the Fifth Edition

This text includes special features designed to emphasize the relevance of the topics discussed, to facilitate the comprehension of important concepts, and to assist students in evaluating their grasp of these concepts.

- **Chapter-Opening Vignette.** Each chapter opens with a brief story that highlights the significance of the topics discussed in the chapter.
- **Chapter Outline.** The main sections of each chapter are conveniently listed on the chapter's first page.
- **Section Summary.** The content of each major section of text is briefly summarized at the beginning of that section. These opening summaries serve to focus attention on the main ideas developed in a chapter.
- **Key Points.** These learning aids appear at the end of each major section in a chapter. They are designed to help students review for exams and to recapitulate the main ideas of the chapter.

Topical Focus Boxes. Throughout the text, special topics are presented in separate *Focus* boxes. The material in these boxes supports or develops concepts, techniques, or skills that have been introduced in the text of the chapter.

Focus on Problem Solving. Each chapter contains a box that guides the student through the analysis and solution of a representative problem. We have chosen a problem that involves important material in the chapter. The box lists the facts and concepts that are relevant to the problem, and then explains how to obtain the solution. Ramifications of the problem are discussed in the online feature Wiley*PLUS*.

A Milestone in Genetics. Each *Milestone* explores a key development in genetics—usually an experiment or a discovery. We cite the original papers that pertain to the subject of the *Milestone*, and we include two *Questions for Discussion* to provide students with an opportunity to investigate the current significance of the subject. These questions are suitable for cooperative learning activities in the classroom, or for reflective writing exercises that go beyond the technical aspects of genetic analysis.

- **Basic Exercises.** At the end of each chapter we present several worked-out problems to reinforce each of the fundamental concepts developed in the chapter. These simple, one-step exercises are designed to illustrate basic genetic analysis or to emphasize important information.

- **Testing Your Knowledge.** Each chapter also has more complicated worked-out problems to help students hone their analytical and problem-solving skills. The problems in this section are designed to integrate different concepts and techniques. In the analysis of each problem, we walk the students through the solution step by step.

- **Questions and Problems.** Each chapter ends with a set of questions and problems of varying difficulty organized according to the sequence of topics in the chapter. The more difficult questions and problems have been designated with orange numbers. These sets of questions and problems provide students with the opportunity to enhance their understanding of the concepts covered in the chapter and to develop their analytical skills. Also, some of the questions and problems—called GO problems—have been selected for interactive solutions in Wiley*PLUS*. The GO problems are designated with a special icon.

- **Genomics on the Web.** Information about genomes, genes, DNA sequences, mutant organisms, polypeptide sequences, biochemical pathways and evolutionary relationships is now freely available on an assortment of web sites. Researchers routinely access this information, and we believe that students should become familiar with it. To this end, we have incorporated a set of questions at the end of each chapter that can be answered by using the National Center for Biotechnology Information (NCBI) web site, which is sponsored by the U.S. National Institutes of Health.

- **Glossary**. This section of the book defines important terms. Students find it useful in clarifying topics and in preparing for exams.

- **Answers**. Answers to the odd-numbered questions and problems are given at the end of the text.

SUPPLEMENTS

Wiley *PLUS*

A powerful online tool that provides instructors and students with an integrated suite of teaching and learning resources in one easy-to-use web site.

These include:

TEST BANK

The test bank, by Ashley Hagler, Gaston College, is available on both the instructor companion site and within Wiley*PLUS*. The test bank contains approximately 50 test questions per chapter and is available online as MS Word files and a computerized test bank. This easy-to-use test-generation program fully supports graphics, print tests, student answer sheets, and answer keys. The software's advanced features allow you to produce an exam to your exact specifications.

POWERPOINT PRESENTATIONS

Laurie Russell, Saint Louis University, designed these presentations to be highly visual and to convey key text concepts illustrated by imbedded text art. The presentations may be accessed on either the instructor companion site or within Wiley*PLUS*.

PRE AND POST LECTURE ASSESSMENT

This assessment tool, created specifically for Wiley*PLUS* by Pamela Marshall, Arizona State University, allows instructors to assign a quiz prior to lecture to assess student understanding and encourage reading and following lecture to gauge improvement and weak areas. Two quizzes are provided for every chapter.

PERSONAL RESPONSE SYSTEM QUESTIONS

These questions by Carolyn Beam, Emory University, are designed to provide readymade pop quizzes and to foster student discussion and debate in class. Available on both the instructor companion site and within Wiley*PLUS*.

ANIMATIONS

Located within Wiley*PLUS*, these animations illustrate key concepts from the text and aid students in grasping some of the most difficult concepts in genetics. Also included are animations that will give students a refresher in basic biology.

GO PROBLEMS

Dubear Kroening, University of Wisconsin—Fox Valley, has implemented select End of Chapter Questions and Problems within Wiley*PLUS* in a guided tutorial format. GO Problems enhance interactivity and hone problem solving skills to give students the confidence they need to tackle complex problems in genetics.

FOCUS ON PROBLEM SOLVING DISCUSSION

Further discussion of each Focus on Problem Solving box is provided within Wiley*PLUS*. Students may link to this discussion directly from the electronic book as they are reading the chapter. Further discussion allows students to more fully understand the problem and put it in context.

ANSWERS TO QUESTIONS AND PROBLEMS

Answers to odd numbered Questions and Problems are located at the end of the text for easy access for students. Answers to all Questions and Problems in the text are available only to instructors on the instructor companion site and within Wiley*PLUS*.

ILLUSTRATIONS AND PHOTOS

All line illustrations and photos from *Principles of Genetics, 5th Edition*, are available on the instructor companion site and within Wiley*PLUS* in both jpeg files and PowerPoint format. Line illustrations are enhanced to provide the best presentation experience.

BOOK COMPANION WEB SITE

(www.wiley.com/go/global/snustad)
This text-specific web site provides students with additional resources and extends the chapters of the text to the resources of the World Wide Web.

Resources include:
For Students: web quizzes, by Carolyn Beam, covering key concepts for each chapter of text, flashcards, and the Biology NewsFinder.
For Instructors: Test Bank, PowerPoint Presentations, line art and photos in jpeg and PowerPoint formats, personal response system questions, and all answers to end of chapter Questions and Problems.

Acknowledgments

As with previous editions, this edition of *Principles of Genetics* has been influenced by the genetics courses we teach. We thank our students for their constructive feedback on both content and pedagogy, and we thank our colleagues at the University of Minnesota for sharing their knowledge and expertise. Genetics professors at other institutions also provided many helpful suggestions. In particular, we acknowledge the help of the following reviewers and focus group participants:

5th Edition Reviewers
Scott Baird, Wright State University; Jay Brewster, Pepperdine University; Jeff DeJong, University of Texas, Dallas; Charles B. Fenster, University of Maryland; Gary Kuleck, Loyola Marymount University; Paul F. Lurquin, Washington State University; Mark Meade, Jacksonville State University; Jessica L. Moore, University of South Florida; David H. Reed, University of Mississippi; Valery Soyfer, George Mason University; Mark Sturtevant, University of Michigan, Flint; Ted Weinert, University of Arizona

5th Edition Focus Group Participants
Aaron Cassill, University of Texas, San Antonio; Nestor DeOcampo, Michigan State University; Robert G. Fowler, San Jose State University; Maria Gallo, University of Florida; Michael Gilchrist, University of Tennessee, Knoxville; Adam W. Hrincevich, Louisiana State University; David Kass, Eastern Michigan University; Gregory J. Podgorski, Utah State University; Inder M. Saxena, University of Texas, Austin

Reviewers of Previous Editions
Colleen Belk, University of Minnesota, Duluth; John Belote, Syracuse University; Paul Bottino, University of Maryland; Jay Brewster, Pepperdine University; Joan Burnside, University of Delaware; Pat Calie, Eastern Kentucky University; David Carroll, Florida Institute of Technology; Glen Collier, University of Tulsa; Jon Coren, Elizabethtown College; Stephen J. D'Surney, University of Mississippi; David S. Durica, University of Oklahoma; Larry Eckroat, Pennsylvania State University—Erie; Bert Ely, University of South Carolina; David W. Foltz, Louisiana State University; Matthew Gilg, University of North Florida; Patrick Guilfoile, Bemidji State University; Linda Hensel, Mercer University; Tim Ho, University of Wisconsin, Eau Claire; Margaret Hollingsworth, SUNY at Buffalo; Robert Karn, Butler University; Jeffrey M. Marcus, Western Kentucky University; Deb McDonough, University of New England; Michael Polymenis, Texas A & M University; Michael Shintaku, University of Hawaii; Lisa Timmons, University of Kansas; Xiaofei Wang, Tennessee State University; Dan Wells, University of Houston Marvin Whiteley, University of Oklahoma; Mark S. Wilson, Humboldt State University

Many people contributed to the development and production of this edition. Kevin Witt, Senior Editor initiated the project and provided ideas about some of the text's features. Merillat Staat, Associate Editor, worked tirelessly—and enthusiastically—at all stages of the project. We deeply appreciate her guidance and input. Alissa Etrheim, Editorial Assistant, helped with many of the logistical details, while Teri Stratford, Photo Researcher, and Hilary Newman, Photo Manager, researched and obtained many new photographs for this edition. Kathleen Naylor, Developmental Art Editor, heroically analyzed the art in the previous edition, expertly laid out specifications for new illustrations, and made many suggestions for new photographs. We are very grateful for all their contributions. We thank Kevin Murphy, Senior Designer, for creating a fresh text layout and Sigmund Malinowski for executing the new, complex illustration program. Elizabeth Swain, Senior Production Editor superbly coordinated the production of this edition, Betty Pessagno faithfully copyedited the manuscript, Julie Nemer did the final proofreading, and Steve Ingle prepared the index. We greatly appreciate the excellent work of all these people. We also thank Clay Stone, Marketing Manager, for helping to get this edition into the hands of prospective users. With an eye toward the next edition, we encourage students, teaching assistants, instructors, and other readers to send us comments on this edition in care of Merillat Staat at John Wiley & Sons, Inc., 111 River Street, 6-01, Hoboken, NJ, 07030.

Wiley Publishers would like to thank Ashley Hagler, M.S., M.A.T., Gaston College, for her contribution to this International Student Version.

Contents

Chapter 1
The Science of Genetics

CHAPTER OUTLINE

- An Invitation
- Three Great Milestones in Genetics
- DNA as the Genetic Material
- Genetics and Evolution
- Levels of Genetic Analysis
- Genetics in the World: Applications of Genetics to Human Endeavors

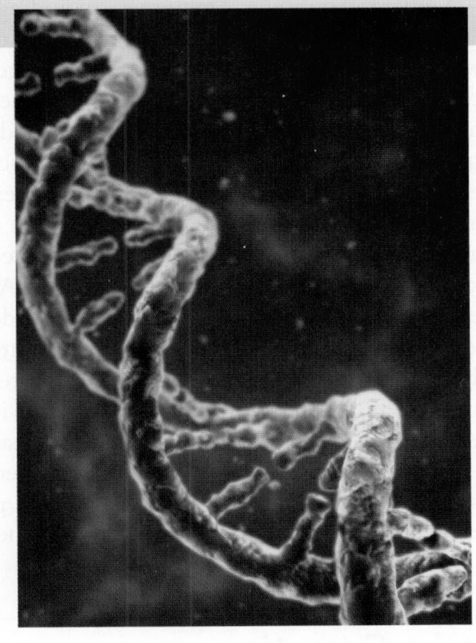

Computer artwork of deoxyribonucleic acid (DNA).

The Personal Genome

Each of us is composed of trillions of cells, and each of those cells contains very thin fibers a few centimeters long that play a major role in who we are, as human beings and as persons. These all-important intracellular fibers are made of DNA. Every time a cell divides, its DNA is replicated and apportioned equally to two daughter cells. The DNA content of these cells—what we call the genome—is thereby conserved. This genome is a master set of instructions, in fact a whole library of information, that cells use to maintain the living state. Ultimately, all the activities of a cell depend on it. To know the DNA is therefore to know the cell and, in a larger sense, to know the organism to which that cell belongs.

Given the importance of the DNA, it should come as no surprise that great efforts have been expended to study it, down to the finest details. In fact, in the last decade of the twentieth century a worldwide campaign, the Human Genome Project, took shape, and in 2001 it produced a comprehensive analysis of human DNA samples that had been collected from a small number of anonymous donors. This work—stunning in scope and significance—laid the foundation for all future research on the human genome. Then, in 2007, the analysis of human DNA took a new turn. Two of the architects of the Human Genome Project had their own DNA decoded. It would, of course, be an exaggeration to say that the analysis of a person's genome is now easy or affordable. Considerable work must be done to obtain genome information, and the cost of this work is still high. But as the techniques for studying DNA become more efficient, and as the cost of such studies comes down, it might become possible for each of us to have our own genome analyzed—a truly astounding prospect. Given the rapid progress that has already occurred in this field of science, personal genome analysis may soon become commonplace.

An Invitation

This book is about genetics, the science that deals with DNA. Genetics is also one of the sciences that has a profound impact on us. Through applications in agriculture and medicine, it helps to feed us and keep us healthy. It also provides insights into what makes us human and into what distinguishes each of us as individuals. Genetics is a relatively young science—it emerged only at the beginning of the twentieth century, but it has grown in scope and significance, so much so that it now has a prominent, and some would say commanding, position in all of biology.

Genetics began with the study of how the characteristics of organisms are passed from parents to offspring—that is, how they are inherited. Until the middle of the twentieth century, no one knew for sure what the hereditary material was. However, geneticists recognized that this material had to fulfill three requirements. First, it had to replicate so that copies could be transmitted from parents to offspring. Second, it had to encode information to guide the development, functioning, and behavior of cells and the organisms to which they belong. Third, it had to change, even if only once in a great while, to account for the differences that exist among individuals. For several decades, geneticists wondered what the hereditary material could be. Then in 1953 the structure of DNA was elucidated and genetics had its great clarifying moment. In a relatively short time, researchers discovered how DNA functions as the hereditary material—that is, how it replicates, how it encodes and expresses information, and how it changes. These discoveries ushered in a new phase of genetics in which phenomena could be explained at the molecular level. In time, geneticists learned how to analyze the DNA of whole genomes, including our own. This progress—from studies of heredity to studies of whole genomes—has been amazing.

As practicing geneticists and as teachers, we have written this book to explain the science of genetics to you. As its title indicates, this book is designed to convey the principles of genetics, and to do so in sufficient detail for you to understand them clearly. We invite you to read each chapter, to study its illustrations, and to wrestle with the questions and problems at the chapter's end. We all know that learning—and research, teaching, and writing, too—takes effort. As authors, we hope your effort studying this book will be rewarded with a good understanding of genetics.

This introductory chapter provides an overview of what we will explain in more detail in the chapters to come. For some of you, it will be a review of knowledge gained from studying basic biology and chemistry. For others, it will be new fare. Our advice is to read the chapter without dwelling on the details. The emphasis here is on the grand themes that run through genetics. The many details of genetics theory and practice will come later.

Three Great Milestones in Genetics

Genetics is rooted in the research of Gregor Mendel, a monk who discovered how traits are inherited. The molecular basis of heredity was revealed when James Watson and Francis Crick elucidated the structure of DNA. The Human Genome Project is currently engaged in the detailed analysis of human DNA.

Scientific knowledge and understanding usually advance incrementally. In this book we will examine the advances that have occurred in genetics during its short history—barely a hundred years. Three great milestones stand out in this history: (1) the discovery of rules governing the inheritance of traits in organisms; (2) the identification of the material responsible for this inheritance and the elucidation of its structure; and (3) the comprehensive analysis of the hereditary material in human beings and other organisms.

MENDEL: GENES AND THE RULES OF INHERITANCE

Although genetics developed during the twentieth century, its origin is rooted in the work of *Gregor Mendel* (**FIGURE 1.1**), a Moravian monk who lived in the nineteenth century. Mendel carried out his pathbreaking research in relative obscurity. He studied the inheritance of different traits in peas, which he grew in the monastery garden. His method involved interbreeding plants that showed different traits—for example, short plants were bred with tall plants—to see how the traits were inherited by the offspring. Mendel's careful analysis enabled him to discern patterns, which led him to postulate the existence of hereditary factors responsible for the traits he studied. We now call these factors **genes.**

Mendel studied several genes in the garden pea. Each of the genes was associated with a different trait—for example, plant height, or flower color, or seed texture. He discovered that these genes exist in different forms, which we now call **alleles.** One form of the gene for height, for example, allows pea plants to grow more than 2 meters tall; another form of this gene limits their growth to about half a meter.

Mendel proposed that pea plants carry two copies of each gene. These copies may be the same or different. During reproduction, one of the copies is randomly incorporated into each sex cell or gamete. The female gametes (eggs) unite with the male gametes (sperm) at fertilization to produce single cells, called zygotes, which then develop into new plants. The reduction in gene copies from two to one during gamete formation and the subsequent restoration of two copies during fertilization underlie the rules of inheritance that Mendel discovered.

Mendel emphasized that the hereditary factors—that is, the genes—are discrete entities. Different alleles of a gene can be brought together in the same plant through hybridization and

Figure 1.1 ▶ Gregor Mendel.

can then be separated from each other during the production of gametes. The coexistence of alleles in a plant therefore does not compromise their integrity. Mendel also found that alleles of different genes are inherited independently of each other.

These discoveries were published in 1866 in the proceedings of the Natural History Society of Brünn, the journal of the scientific society in the city where Mendel lived and worked. The article was not much noticed, and Mendel went on to do other things. In 1900, sixteen years after he died, the paper finally came to light, and the science of genetics was born. In short order, the type of analysis that Mendel pioneered was applied to many kinds of organisms, and with notable success. Of course, not every result fit exactly with Mendel's principles. Exceptions were encountered, and when they were investigated more fully, new insights into the behavior and properties of genes emerged. We shall delve into Mendel's research and its applications to the study of inheritance, including heredity in human beings, in Chapter 3, and we shall explore some ramifications of Mendel's ideas in Chapter 4. In Chapters 5, 6, and 7 we shall see how Mendel's principles of inheritance are related to the behavior of chromosomes—the cellular structures where genes reside.

WATSON AND CRICK: THE STRUCTURE OF DNA

The rediscovery of Mendel's paper launched a plethora of studies on inheritance in plants, animals, and microorganisms. The big question on everyone's mind was "What is a gene?" In the middle of the twentieth century, this question was finally answered. Genes were shown to consist of complex molecules called **nucleic acids.**

Nucleic acids are made of elementary building blocks called **nucleotides** (**FIGURE 1.2**). Each nucleotide has three components: (1) a sugar molecule; (2) a phosphate molecule, which has acidic chemical properties; and (3) a nitrogen-containing molecule, which has slightly basic chemical properties. In **ribonucleic acid,** or RNA, the constituent sugar is ribose; in **deoxyribonucleic acid,** or **DNA,** it is deoxyribose. Within RNA or DNA, one nucleotide is distinguished from another by its nitrogen-containing base. In RNA, the four kinds of bases are adenine (A), guanine (G), cytosine (C), and uracil (U); in DNA, they are A, G, C, and thymine (T). Thus, in both DNA and RNA there are four kinds of nucleotides, and three of them are shared by both types of nucleic acid molecules.

The big breakthrough in the study of nucleic acids came in 1953 when *James Watson* and *Francis Crick* (**FIGURE 1.3**) deduced how nucleotides are organized within DNA. Watson and Crick knew that the nucleotides are linked, one to another, in a chain. The linkages are formed by chemical interactions between the phosphate of one nucleotide and the sugar of another nucleotide. The nitrogen-containing bases are not involved in these interactions. Thus, a chain of nucleotides consists of a phosphate-sugar backbone to which bases are attached, one base to each sugar in the backbone. From one end of the chain to the other, the bases form a linear sequence characteristic of that particular chain. This sequence of bases is what distinguishes one gene from another. Watson and Crick proposed that DNA molecules consist of two chains of nucleotides (**FIGURE 1.4*a***). These chains are held together by weak chemical attractions between particular pairs of bases; A pairs with T, and G pairs with C. Because of these base-pairing rules, the sequence of one nucleotide chain in a double-stranded DNA molecule can

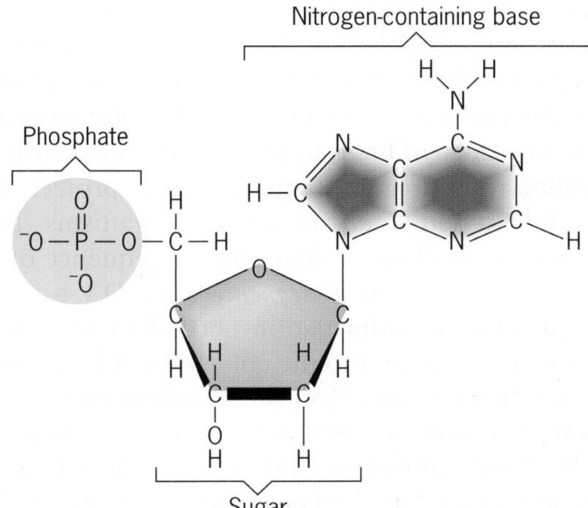

Figure 1.2 ▶ Structure of a nucleotide. The molecule has three components: a phosphate group, a sugar (in this case deoxyribose), and a nitrogen-containing base (in this case adenine).

Figure 1.3 ▶ Francis Crick and James Watson.

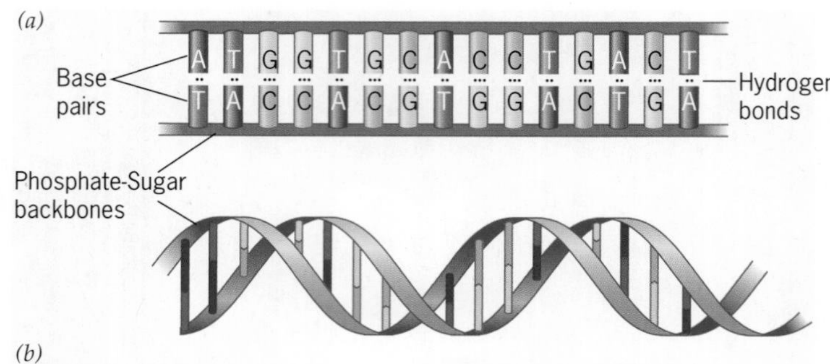

Figure 1.4 ▶ DNA, a double-stranded molecule held together by hydrogen bonding between paired bases. (*a*) Two-dimensional representation of the structure of a DNA molecule composed of complementary nucleotide chains. (*b*) A DNA molecule shown as a double helix.

be predicted from that of the other. In this sense, then, the two chains of a DNA molecule are complementary.

A double-stranded DNA molecule is often called a duplex. Watson and Crick discovered that the two strands of a DNA duplex are wound round each other in a helical configuration (**FIGURE 1.4*b***). These helical molecules can be extraordinarily large. Some contain hundreds of millions of nucleotide pairs, and their end-to-end length exceeds 10 centimeters. Were it not for their extraordinary thinness (about a hundred-millionth of a centimeter), we would be able to see them with the unaided eye.

RNA, like DNA, consists of nucleotides linked one to another in a chain. However, unlike DNA, RNA molecules are usually single-stranded. The genes of most organisms are composed of DNA, although in some viruses they are made of RNA. We will examine the structures of DNA and RNA in detail in Chapter 9, and we will investigate the genetic significance of these macromolecules in Chapters 10, 11, and 12.

THE HUMAN GENOME PROJECT: SEQUENCING DNA AND CATALOGUING GENES

If geneticists in the first half of the twentieth century dreamed about identifying the stuff that genes are made of, geneticists in the second half of that century dreamed about ways of determining the sequence of bases in DNA molecules. Near the end of the century, their dreams became reality as projects to determine DNA base sequences in several organisms, including human beings, took shape. Obtaining the sequence of bases in an organism's DNA—that is, *sequencing the DNA*—should, in principle, provide the information needed to analyze all that organism's genes. We refer to the collection of DNA molecules that is characteristic of an organism as its **genome.** Sequencing the genome is therefore tantamount to sequencing all the organism's genes—and more, for we now know that some of the DNA does not comprise genes. The function of this nongenic DNA is not always clear; however, it is present in many genomes, and sometimes it is abundant. A Milestone in Genetics: ΦX174, the First DNA Genome Sequenced, at the end of this chapter describes how genome sequencing got started.

The paragon of all the sequencing programs is the **Human Genome Project,** a worldwide effort to determine the sequence of approximately 3 billion nucleotide pairs in human DNA. As initially conceived, the Human Genome Project was to involve collaborations among researchers in many different countries, and much of the work was to be funded by their governments. However, a privately funded project initiated by Craig Venter, a scientist and an entrepreneur, soon developed alongside the publicly funded project. In 2001 all these efforts culminated in the publication of two lengthy articles about the human genome. The articles reported that 2.7 billion nucleotide pairs of human DNA had been sequenced. Computer analysis of this DNA suggested that the human genome contained between 30,000 and 40,000 genes. More recent analyses have revised the human gene number downward, to between 20,000 and 25,000. These genes have been catalogued by location, structure, and potential function. Efforts are now focused on studying how they influence the myriad characteristics of human beings.

The genomes of many other organisms—bacteria, fungi, plants, protists, and animals—have also been sequenced. Much of this work has been done under the auspices of the Human Genome Project, or under projects closely allied to it. Initially, the sequencing efforts focused on organisms that are especially favorable for genetic research. We discuss some of these model organisms in Chapter 2, and we explore ways in which researchers have used them to advance genetic knowledge in many places in this book. Current sequencing projects have moved beyond the model organisms to diverse plants, animals, and microbes. For example, the genomes of the mosquito and the malaria parasite that it carries have both been sequenced, as have the genomes of the honeybee, the poplar tree, and the sea squirt. Some of the targets of these sequencing projects have a medical, agricultural, or commercial significance; others simply help us to understand how genomes are organized and how they have diversified during the history of life on Earth.

Figure 1.5 ► A researcher preparing samples for an automated DNA sequencer.

All the DNA sequencing projects have transformed genetics in a fundamental way. Genes can now be studied at the molecular level with relative ease, and vast numbers of genes can be studied simultaneously. This approach to genetics, rooted in the analysis of the DNA sequences that make up a genome, is called **genomics.** It has been made possible by advances in DNA sequencing technology, robotics, and computer science (**FIGURE 1.5**). Researchers are now able to construct and scan enormous databases containing DNA sequences to address questions about genetics. Although a large number of useful databases are currently available, we will focus on the databases assembled by the *National Center for Biotechnology Information (NCBI)*, maintained by the U.S. National Institutes of Health. The NCBI databases—available free on the web at http://www.ncbi.nih.gov/—are invaluable repositories of information about genes, proteins, genomes, publications, and other important data in the fields of genetics, biochemistry, and molecular biology. They contain the complete nucleotide sequences of all genomes that have been sequenced to date, and they are continually updated. In addition, the NCBI web site contains tools that can be used to search for specific items of interest—gene and protein sequences, research articles, and so on. In Chapter 16, we will introduce you to some of these tools, and throughout this book, we will encourage you to visit the NCBI web site at the end of each chapter to answer specific questions.

KEY POINTS

- Gregor Mendel postulated the existence of particulate factors—now called genes—to explain how traits are inherited.
- Alleles, the alternate forms of genes, account for heritable differences among individuals.
- James Watson and Francis Crick elucidated the structure of DNA, a macromolecule composed of two complementary chains of nucleotides.
- DNA is the hereditary material in all life forms except some types of viruses, in which RNA is the hereditary material.
- The Human Genome Project determined the sequence of nucleotides in the DNA of the human genome.
- Sequencing the DNA of a genome provides the data to identify and catalogue all the genes of an organism.

DNA as the Genetic Material

In biology information flows from DNA to RNA to protein.

In all cellular organisms, the genetic material is DNA. This material must be able to *replicate* so that copies can be transmitted from cell to cell and from parents to offspring; it must contain *information* to direct cellular activities and to guide the development, functioning, and behavior of organisms; and it must be able to *change* so that, over time, groups of organisms can adapt to different circumstances.

DNA REPLICATION: PROPAGATING GENETIC INFORMATION

The genetic material of an organism is transmitted from a mother cell to its daughters during cell division. It is also transmitted from parents to their offspring during reproduction. The faithful transmission of genetic material from one cell or organism to another is based on the ability of double-stranded DNA molecules to be replicated. DNA replication is extraordinarily exact. Molecules consisting of hundreds of millions of nucleotide pairs are duplicated with few, if any, mistakes.

The process of DNA replication is based on the complementary nature of the strands that make up duplex DNA molecules (**FIGURE 1.6**). These strands are held together by hydrogen bonds between specific base pairs—A paired with T, and G paired with C. When these bonds are broken, the separated strands can serve as templates for the synthesis of new partner strands. The new strands are assembled by the stepwise incorporation of nucleotides opposite to nucleotides in the template strands. This incorporation conforms to the base-pairing rules. Thus, the sequence of nucleotides in a strand being synthesized is dictated by the sequence of nucleotides in the template strand. At the end of the replication process, each template strand is paired with a newly synthesized partner strand. Thus, two identical DNA duplexes are created from one original duplex.

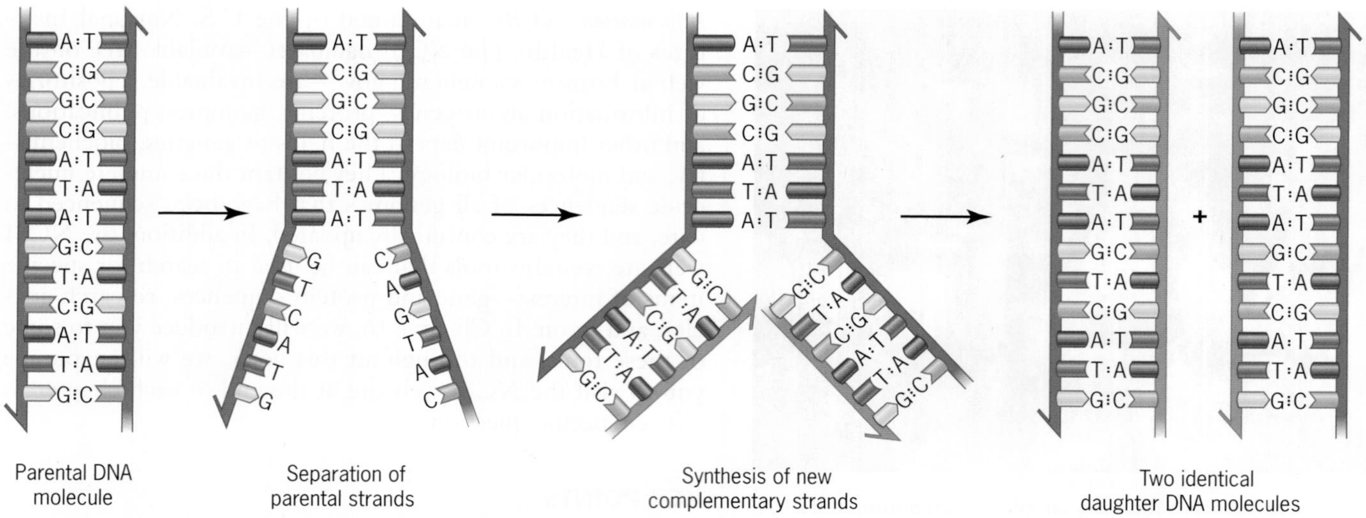

Figure 1.6 ▶ DNA replication. The two strands in the parental molecule are oriented in opposite directions (see arrows). These strands separate, and new strands are synthesized using the parental strands as templates. When replication is completed, two identical double-stranded DNA molecules have been produced.

The process of DNA replication does not occur spontaneously. Like most biochemical processes, it is catalyzed by enzymes. We shall explore the details of DNA replication, including the roles played by different enzymes, in Chapter 10.

GENE EXPRESSION: USING GENETIC INFORMATION

DNA molecules contain information to direct the activities of cells and to guide the development, functioning, and behavior of the organisms that comprise these cells. This information is encoded in sequences of nucleotides within the DNA molecules of the genome. Among cellular organisms, the smallest known genome is that of *Mycoplasma genitalium*: 580,070 nucleotide pairs. By contrast, the human genome consists of 3.2 billion nucleotide pairs. In these and all other genomes, the information contained within the DNA is organized into the units we call genes. An *M. genitalium* has 482 genes, whereas a human sperm cell has between 20,000 and 25,000. Each gene is a stretch of nucleotide pairs along the length of a DNA molecule. A particular DNA molecule may contain thousands of different genes. In an *M. genitalium* cell, all the genes are situated on one DNA molecule—the single chromosome of this organism. In a human sperm cell, the genes are situated on 23 different DNA molecules corresponding to the 23 chromosomes in the cell. Most of the DNA in *M. genitalium* comprises genes, whereas most of the DNA in human beings does not—that is, most of the human DNA is nongenic. We shall investigate the genic and nongenic composition of genomes in many places in this book, especially in Chapter 16.

How is the information within individual genes organized and expressed? This question is central in genetics, and we will turn our attention to it in Chapters 11, 12, and 14. Here, suffice it to say that most genes contain the instructions for the synthesis of proteins. Each protein consists of one or more polypeptides, which are chains of amino acids. The 20 different kinds of amino acids that occur naturally can be combined in myriad ways to form polypeptides. Each polypeptide has a characteristic sequence of amino acids. Some polypeptides are short—just a few amino acids long—whereas others are enormous—thousands of amino acids long.

The sequence of amino acids in a polypeptide is specified by a sequence of elementary coding units within a gene. These elementary coding units, called **codons,** are triplets of adjacent nucleotides. A typical gene may contain hundreds or even thousands of codons. Each codon specifies the incorporation of an amino acid into a polypeptide. Thus, the information encoded within a gene is used to direct the synthesis of a polypeptide, which is often referred to as the gene's product. Sometimes, depending on how the coding information is utilized, a gene may encode several polypeptides; however, these polypeptides are usually all related by sharing some common sequence of amino acids.

The expression of genetic information to form a polypeptide is a two-stage process (**FIGURE 1.7**). First, the information contained in a gene's DNA is copied into a molecule of RNA. The RNA is assembled in stepwise fashion along one of the strands of the DNA duplex. During this assembly process, A in the RNA pairs with T in the DNA, G in the RNA pairs with C in the DNA, C in the RNA pairs with G in the DNA, and U in the RNA pairs with A in the DNA. Thus, the nucleotide sequence of the RNA is determined by the nucleotide sequence of a strand of DNA in the gene. The process that produces this RNA molecule is called **transcription,** and the RNA itself is called a **transcript.** The RNA transcript eventually separates from its DNA template and, in some organisms, is altered by the addition, deletion, or modification of nucleotides. The finished molecule,

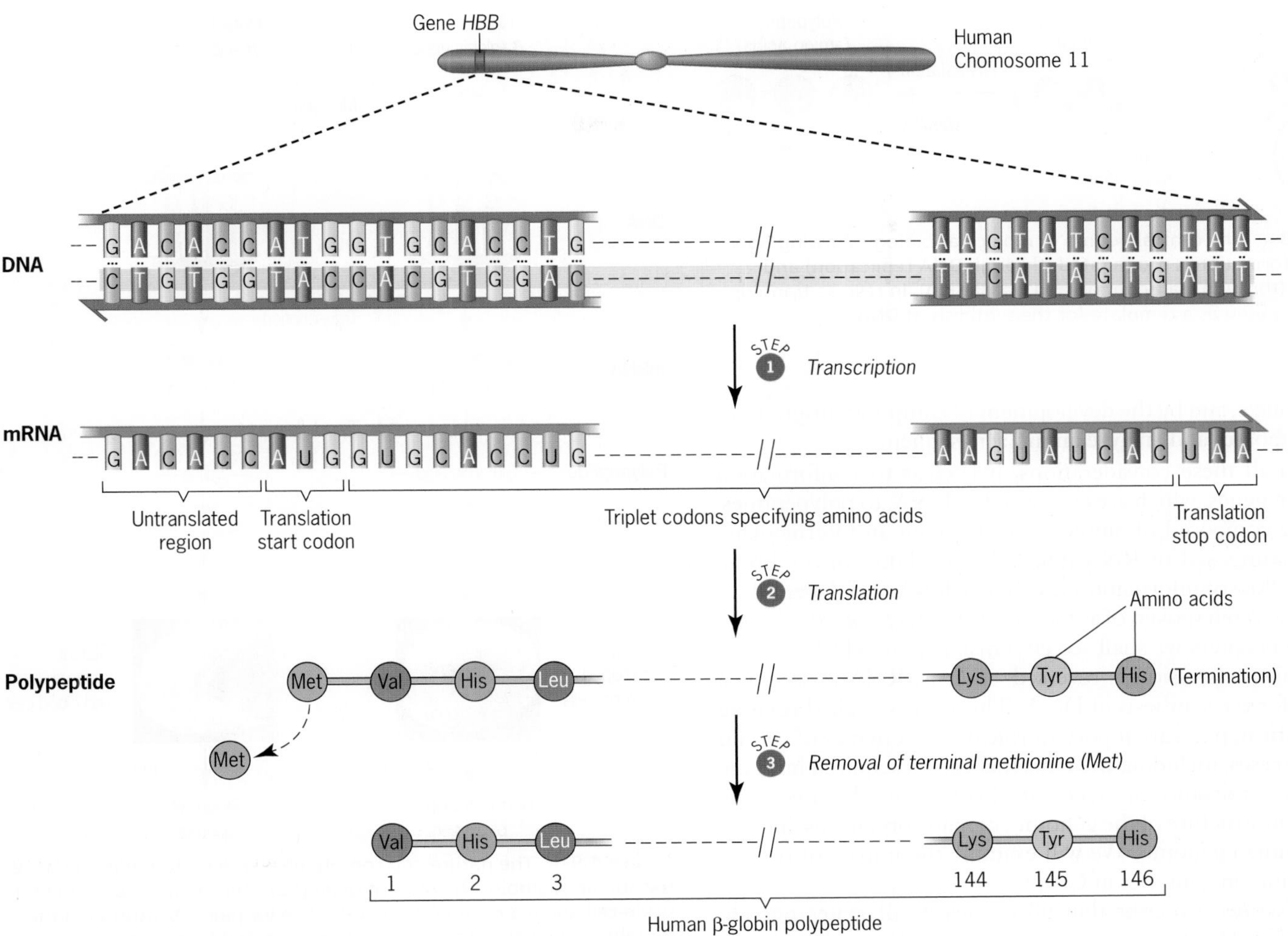

Figure 1.7 ▶ Expression of the human gene *HBB* coding for the β-globin polypeptide of hemoglobin. During transcription (step 1), one strand of the *HBB* DNA (here the bottom strand shown highlighted) serves as a template for the synthesis of a complementary strand of RNA. After undergoing modifications, the resulting mRNA (messenger RNA) is used as a template to synthesize the β-globin polypeptide. This process is called translation (step 2). During translation, each triplet codon in the mRNA specifies the incorporation of an amino acid in the polypeptide chain. Translation is initiated by a start codon, which specifies the incorporation of the amino acid methionine (Met), and it is terminated by a stop codon, which does not specify the incorporation of any amino acid. After translation is completed, the initial methionine is removed (step 3) to produce the mature β-globin polypeptide.

called the **messenger RNA** or simply **mRNA,** contains all the information needed for the synthesis of a polypeptide.

The second stage in the expression of a gene's information is called **translation.** At this stage, the gene's mRNA acts as a template for the synthesis of a polypeptide. Each of the gene's codons, now present within the sequence of the mRNA, specifies the incorporation of a particular amino acid into the polypeptide chain. One amino acid is added at a time. Thus, the polypeptide is synthesized stepwise by reading the codons in order. When the polypeptide is finished, it dissociates from the mRNA, folds into a precise three-dimensional shape, and then carries out its role in the cell. Some polypeptides are altered by the removal of the first amino acid, which is usually methionine, in the sequence.

We refer to the collection of all the different proteins in an organism as its **proteome.** Humans, with between 20,000 and 25,000 genes, may have hundreds of thousands of different proteins in their proteome. One reason for the large size of the human proteome is that a particular gene may encode several different, but related, polypeptides, and these polypeptides may combine in complex ways to produce different proteins. Another reason is that proteins may be produced by combining polypeptides encoded by different genes. If the number of genes in the human genome is large, the number of proteins in the human proteome is truly enormous.

The study of all the proteins in cells—their composition, the sequences of amino acids in their constituent polypeptides, the interactions among these polypeptides and among different proteins, and, of course, the functions of these complex molecules—is called **proteomics.** Like genomics, proteomics has been made possible by advances in the technologies used to study genes and

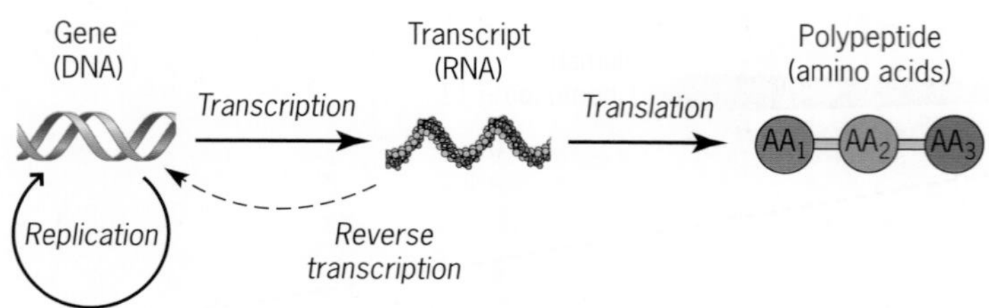

Figure 1.8 ▶ The central dogma of molecular biology showing how genetic information is propagated (through DNA replication) and expressed (through transcription and translation). In reverse transcription, RNA is used as a template for the synthesis of DNA.

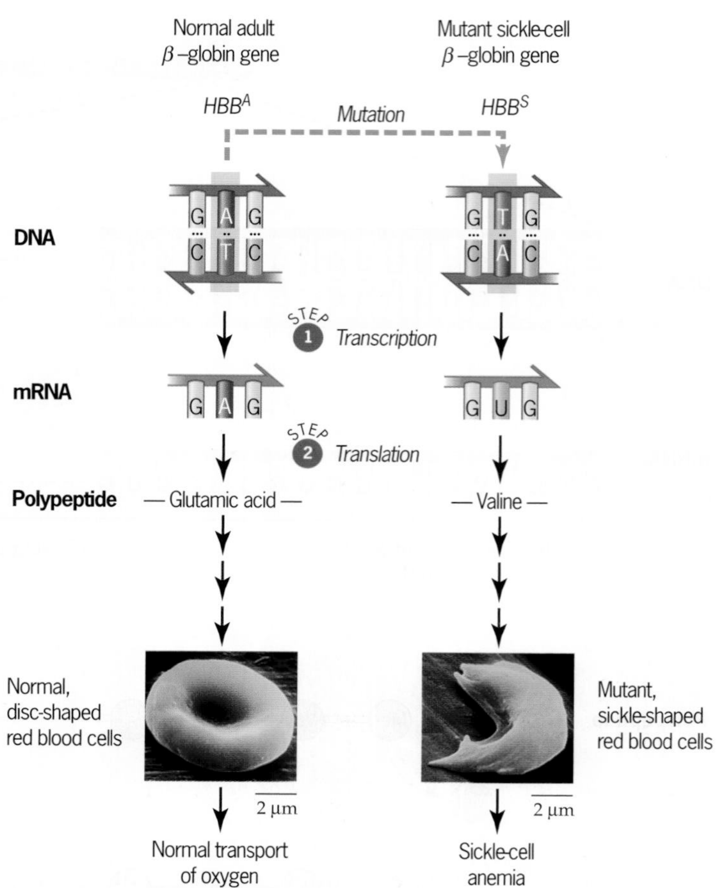

Figure 1.9 ▶ The nature and consequence of a mutation in the gene for human β-globin. The mutant gene (*HBB^S* top right) responsible for sickle-cell anemia resulted from a single base-pair substitution in the β-globin gene (*HBB^A* top left). Transcription and translation of the mutant gene produce a β-globin polypeptide containing the amino acid valine (center right) at the position where normal β-globin contains glutamic acid (center left). This single amino acid change results in the formation of sickle-shaped red blood cells (bottom right) rather than the normal disc-shaped cells (bottom left). The sickle-shaped cells cause a severe form of anemia.

gene products, and by the development of computer programs to search databases and analyze amino acid sequences.

From all these considerations, it is clear that information flows from genes, which are composed of DNA, to polypeptides, which are composed of amino acids, through an intermediate, which is composed of RNA (**FIGURE 1.8**). Thus, in the broad sense, the flow of information is DNA → RNA → polypeptide, a progression often spoken of as the *central dogma of molecular biology*. In several chapters we shall see circumstances in which the first part of this progression is reversed—that is, RNA is used as a template for the synthesis of DNA. This process, called **reverse transcription,** plays an important role in the activities of certain types of viruses, including the virus that causes acquired immune deficiency syndrome, or AIDS; it also profoundly affects the content and structure of the genomes of many organisms, including the human genome. We will examine the impact of reverse transcription on genomes in Chapter 18.

It was once thought that all or nearly all genes encode polypeptides. However, recent research has shown this idea to be incorrect. Many genes do not encode polypeptides; instead, their end products are RNA molecules that play important roles within cells. We will explore these RNAs and the genes that produce them in Chapters 11 and 20.

MUTATION: CHANGING GENETIC INFORMATION

DNA replication is an extraordinarily accurate process, but it is not perfect. At a low but measurable frequency, nucleotides are incorporated incorrectly into growing DNA chains. Such changes have the potential to alter or disrupt the information encoded in genes. DNA molecules are also sometimes damaged by electromagnetic radiation or by chemicals. Although the damage induced by these agents may be repaired, the repair processes often leave scars. Stretches of nucleotides may be deleted or duplicated, or they may be rearranged within the overall structure of the DNA molecule. We call all these types of changes **mutations.** Genes that are altered by the occurrence of mutations are called mutant genes.

Often mutant genes cause different traits in organisms (**FIGURE 1.9**). For example, one of the genes in the human genome encodes the polypeptide known as β-globin. This polypeptide, 146 amino acids long, is a constituent of hemoglobin, the protein that transports oxygen in the blood. The 146 amino acids in β-globin correspond to 146 codons in the β-globin gene. The sixth of these codons specifies the incorporation of glutamic acid into the polypeptide. Countless generations ago, in the germ line of some nameless individual, the middle nucleotide pair in this codon was changed from A:T to T:A, and the resulting mutation was passed on to the individual's descendants. This mutation, now widespread in some human populations, altered the sixth codon so that it specifies the incorporation of valine into the β-globin polypeptide. This seemingly insignificant change has a deleterious effect on the structure of the cells that make and store hemoglobin—the red blood cells. People who carry two copies of the mutant version of the β-globin gene have sickle-shaped red blood cells, whereas people who carry two copies of the nonmutant version of this gene have disc-shaped red blood cells. The sickle-shaped cells do not transport oxygen efficiently through the body. Consequently, people with sickle-shaped red blood cells develop a serious disease, so serious in fact that they eventually die from it. This disease, called sickle-cell anemia, is therefore traceable to a mutation in the β-globin gene. We shall investigate the nature and causes of mutations like this one in Chapter 13.

The process of mutation has another aspect—it introduces variability into the genetic material of organisms. Over time, the mutant genes created by mutation may spread through a population. For example, you might wonder why the mutant β-globin gene is relatively common in some human populations. It turns out that people who carry both a mutant and a nonmutant allele of this gene are less susceptible to infection by the blood parasite that causes malaria. These people therefore have a better chance of surviving in environments where malaria is a threat. Because of this enhanced survival, they produce more children than other people, and the mutant allele that they carry can spread. This example shows how the genetic makeup of a population—in this case, the human population—changes over time.

KEY POINTS

- When DNA replicates, each strand of a duplex molecule serves as the template for the synthesis of a complementary strand.
- When genetic information is expressed, one strand of a gene's DNA duplex is used as a template for the synthesis of a complementary strand of RNA.
- For most genes, RNA synthesis (transcription) generates a molecule (the RNA transcript) that becomes a messenger RNA (mRNA).
- Coded information in an mRNA is translated into a sequence of amino acids in a polypeptide.
- Mutations can alter the DNA sequence of a gene.
- The genetic variability created by mutation is the basis for biological evolution.

Genetics and Evolution

Genetics has much to contribute to the scientific study of evolution.

As mutations accumulate in the DNA over many generations, we see their effects as differences among organisms. Mendel's strains of peas carried different mutant genes, and so do people from different ethnic groups. In almost any species, at least some of the observable variation has an underlying genetic basis. In the middle of the nineteenth century, *Charles Darwin* and *Alfred Wallace* (**FIGURE 1.10**), both contemporaries of Mendel, proposed that this variation makes it possible for species to change—that is, to evolve—over time.

(*a*)

(*b*)

Figure 1.10 ▶ (*a*) Charles Darwin. (*b*) Alfred Wallace.

The ideas of Darwin and Wallace revolutionized scientific thought. They introduced an historical perspective into biology and gave credence to the concept that all living things are related by virtue of descent from a common ancestor. However, when these ideas were proposed, Mendel's work on heredity was still in progress and the science of genetics had not yet been born. Research on biological evolution was stimulated when Mendel's discoveries came to light at the beginning of the twentieth century, and it took a new turn when DNA sequencing techniques emerged at the century's end. With DNA sequencing we can see similarities and differences in the genetic material of diverse organisms. On the assumption that sequences of nucleotides in the DNA are the result of historical processes, it is possible to interpret these similarities and differences in a temporal framework. Organisms with very similar DNA sequences are descended from a recent common ancestor; organisms with less similar DNA sequences are descended from a more remote common ancestor. Using this logic, researchers can establish the historical relationships among organisms (**FIGURE 1.11**). We call these relationships a phylogenetic tree, or more simply, a **phylogeny,** from Greek words meaning "the origin of tribes."

Today the construction of phylogenetic trees is an important part of the study of evolution. Biologists use the burgeoning DNA sequence data from the genome projects and other research ventures, such as the United States National Science Foundation's "Tree of Life" program, in combination with anatomical data collected from living and fossilized organisms to discern the evolutionary relationships among species. We shall explore the genetic basis of evolution in Chapters 24 and 25.

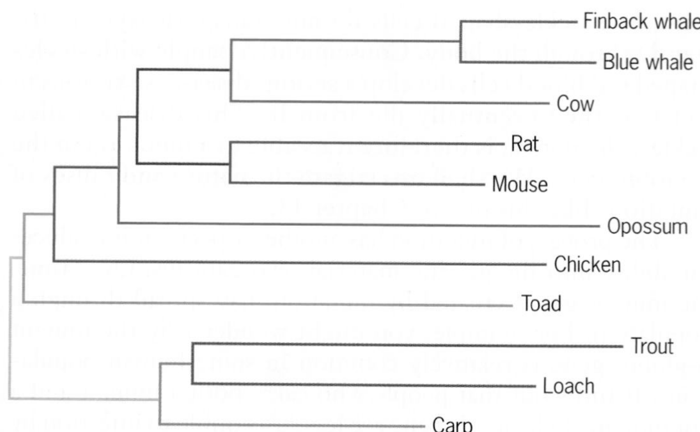

Figure 1.11 ▶ Phylogenetic tree showing the evolutionary relationships among 11 different vertebrates. This tree was constructed by comparing the sequences of the gene for cytochrome b, which is carried by the DNA found in the mitochondria of these animals. The 11 different animals have been positioned in the tree according to the similarity of their cytochrome b gene sequences. This tree is consistent with other information (e.g., data obtained from the study of fossils), except for the positions of the three fish species. The loach is actually more closely related to the carp than it is to the trout. This discrepancy points out the need to interpret the results of DNA sequence comparisons carefully.

KEY POINTS

- Evolution depends on the occurrence, transmission, and spread of mutant genes in groups of organisms.
- DNA sequence data provide a way of studying the historical process of evolution.

Levels of Genetic Analysis

Geneticists approach their science from different points of view—from that of a gene, a DNA molecule, or a population of organisms.

Genetic analysis is practiced at different levels. The oldest type of genetic analysis follows in Mendel's footsteps by focusing on how traits are inherited when different strains of organisms are hybridized. Another type of genetic analysis follows in the footsteps of Watson and Crick and the army of people who have worked on the various genome projects by focusing on the molecular makeup of the genetic material. Still another type of genetic analysis imitates Darwin and Wallace by focusing on entire populations of organisms. All these levels of genetic analysis are routinely used in research today. Although we shall encounter them in many different places in this book, we provide brief descriptions of them here.

CLASSICAL GENETICS

The period prior to the discovery of the structure of DNA is often spoken of as the era of *classical genetics*. During this time, geneticists pursued their science by analyzing the outcomes of crosses between different strains of organisms, much as Mendel had done in his work with peas. In this type of analysis, genes are identified by studying the inheritance of trait differences—tall pea plants versus short pea plants, for example—in the offspring of crosses. The trait differences are due to the alternate forms of genes. Sometimes more than one gene influences a trait, and sometimes environmental conditions—for example, temperature and nutrition—exert an effect. These complications can make the analysis of inheritance difficult.

The classical approach to the study of genes can also be coordinated with studies on the structure and behavior of chromosomes, which are the cellular entities that contain the genes. By analyzing patterns of inheritance, geneticists can localize genes to specific chromosomes. More detailed analyses allow them to localize genes to specific positions within chromosomes—a practice called chromosome mapping.

Because these studies emphasize the transmission of genes and chromosomes from one generation to the next, they are often referred to as exercises in *transmission genetics*. However, classical genetics is not limited to the analysis of gene and chromosome transmission. It also studies the nature of the genetic material—how it controls traits and how it mutates. We present the essential features of classical genetics in Chapters 3–8.

MOLECULAR GENETICS

With the discovery of the structure of DNA, genetics entered a new phase. The replication, expression, and mutation of genes could now be studied at the molecular level. This approach to genetic analysis was raised to a new level when it became possible to sequence DNA molecules easily. Molecular genetic analysis is rooted in the study of DNA sequences. Knowledge of a DNA sequence and comparisons to other DNA sequences allow a geneticist to define a gene chemically. The gene's internal components—coding sequences, regulatory sequences, and noncoding sequences—can be identified, and the nature of the polypeptide encoded by the gene can be predicted.

But the molecular approach to genetic analysis is much more than the study of DNA sequences. Geneticists have learned to cut DNA molecules at specific sites. Whole genes, or pieces of genes, can be excised from one DNA molecule and inserted into another DNA molecule. These "recombinant" DNA molecules can be replicated in bacterial cells or even in test tubes that have been supplied with appropriate enzymes. Milligram quantities of a particular gene can be generated in the laboratory in an afternoon. In short, geneticists have learned how to manipulate genes more or less at will. This artful manipulation has allowed researchers to study genetic phenomena in great detail. They have even learned how to transfer genes from one organism to another. We present examples of molecular genetic analysis in many chapters in this book.

POPULATION GENETICS

Genetics can also be studied at the level of an entire population of organisms. Individuals within a population may carry different alleles of a gene; perhaps they carry different alleles of many genes. These differences make individuals genetically distinct, possibly even unique. In other words, the members of a population vary in their genetic makeup. Geneticists seek to document this variability and to understand its significance. Their most basic approach is to determine the frequencies of specific alleles in a population and then to ascertain if these frequencies change over time. If they do, the population is evolving. The assessment of genetic variability in a population is therefore a foundation for the study of biological evolution. It is also useful in the effort to understand the inheritance of complex traits, such as body size or disease susceptibility. Often, complex traits are of considerable interest because they have an agricultural or a medical significance. We discuss genetic analysis at the population level in Chapters 23, 24, and 25.

KEY POINTS

- In classical genetic analysis, genes are studied by following the inheritance of traits in crosses between different strains of an organism.
- In molecular genetic analysis, genes are studied by isolating, sequencing, and manipulating DNA and by examining the products of gene expression.
- In population genetic analysis, genes are studied by assessing the variability among individuals in a group of organisms.

Genetics in the World: Applications of Genetics to Human Endeavors

Genetics is relevant in many venues outside the research laboratory.

Genetics began in a European monastic enclosure; today, it is a worldwide enterprise. The significance and international scope of genetics are evident in today's scientific journals, which showcase the work of geneticists from many different countries. They are also evident in the myriad ways in which genetics is applied in agriculture, medicine, and many other human endeavors all over the world. We shall consider some of these applications in Chapters 15, 16, 17, 24, and 25. Some of the highlights are introduced in this section.

GENETICS IN AGRICULTURE

By the time the first civilizations appeared, humans had already learned to cultivate crop plants and to rear livestock. They had also learned to improve their crops and livestock by selective breeding. This pre-Mendelian application of genetic principles had telling effects. Over thousands of generations, domesticated plant and animal species came to be quite different from their wild ancestors. For example, cattle were changed in appearance and behavior (**FIGURE 1.12**), and corn, which is descended from a wild grass called teosinte (**FIGURE 1.13**), was changed so much that it could no longer grow without human cultivation.

Selective breeding programs—now informed by genetic theory—continue to play important roles in agriculture. High-yielding varieties of wheat, corn, rice, and many other plants have been developed by breeders to feed a growing human population. Selective breeding techniques have also been

Angus

Beef master

Simmental

Charolais

Figure 1.12 ▶ Breeds of beef cattle.

applied to animals such as beef and dairy cattle, swine, and sheep, and to horticultural plants such as shade trees, turf grass, and garden flowers.

Beginning in the 1980s, classical approaches to crop and livestock improvement were supplemented—and in some cases, supplanted—by approaches from molecular genetics. Detailed genetic maps of the chromosomes of several species were constructed to pinpoint genes of agricultural significance. By locating genes for traits such as grain yield or disease resistance, breeders could now design schemes to incorporate particular alleles into agricultural varieties. These mapping projects have been carried on relentlessly and for a few species have culminated in the complete sequencing of the genome. Other crop and livestock genome sequencing projects are still in progress. All sorts of potentially useful genes are being identified and studied in these projects.

Plant and animal breeders are also employing the techniques of molecular genetics to introduce genes from other species into crop plants and livestock. This process of changing the genetic makeup of an organism was initially developed using test species such as fruit flies. Today it is widely used to augment the genetic material of many kinds of creatures. Plants and animals that have been altered by the introduction of foreign genes are called **GMOs—genetically modified organisms.** BT corn is an example. Many corn varieties now grown in the United States carry a gene from the bacterium *Bacillus thuringiensis*. This gene encodes a protein that is toxic to many insects. Corn strains that carry the gene for BT toxin are resistant to attacks by the European corn borer, an insect that has caused enormous damage in the past (**FIGURE 1.14**). Thus, BT corn plants produce their own insecticide.

The development and use of GMOs has stirred up controversy worldwide. For example, African and European countries have been reluctant to grow BT corn or to purchase BT corn grown in the United States. Their reluctance is due to several factors, including the conflicting interests of small farmers and large agricultural corporations, and concerns about the safety of consuming genetically modified food. There is also a concern that BT corn might kill nonpest species of insects such as butterflies and honeybees. Advances in molecular genetics have provided the tools and the materials to change agriculture profoundly. Today, policy makers are wrestling with the implications of these new technologies.

(a)

(b)

6.5 cm

Figure 1.13 ▶ Ears of corn (*a*) and its ancestor, teosinte (*b*).

(a)

(b)

Figure 1.14 ▶ Use of a genetically modified plant in agriculture. (*a*) European corn borer eating away the stalk of a corn plant. (*b*) Side-by-side comparison of corn stalks from plants that are resistant (top) and susceptible (bottom) to the corn borer. The resistant plant is expressing a gene for an insecticidal protein derived from *Bacillus thuringiensis*.

GENETICS IN MEDICINE

Classical genetics has provided physicians with a long list of diseases that are caused by mutant genes. The study of these diseases began shortly after Mendel's work was rediscovered. In 1909 Sir Archibald Garrod, a British physician and biochemist, published a book entitled *Inborn Errors of Metabolism*. In this book, Garrod documented how metabolic abnormalities can be traced to mutant alleles. His research was seminal, and in the next several decades, a large number of inherited human disorders were identified and catalogued. From this work, physicians have learned to diagnose genetic diseases, to trace them through families, and to predict the chances that particular individuals might inherit them. Today some hospitals have professionals known as **genetic counselors** who are trained to advise people about the risks of inheriting or transmitting genetic diseases. We shall discuss some aspects of genetic counseling in Chapter 3.

Genetic diseases like the ones that Garrod studied are individually rather rare in most human populations. For example, among newborns, the incidence of phenylketonuria, a disorder of amino acid metabolism, is only one in 10,000. However, mutant genes also contribute to more prevalent human maladies—heart disease and cancer, for example. In Chapter 23 we shall explore ways of assessing genetic risks for complex traits such as the susceptibility to heart disease, and in Chapter 22 we shall investigate the genetic basis of cancer.

Advances in molecular genetics are providing new ways of detecting mutant genes in individuals. Diagnostic tests based on analysis of DNA are now readily available. For example, a hospital lab can test a blood sample or a cheek swab for the presence of a mutant allele of the *BRCA1* gene, which strongly predisposes its carriers to develop breast cancer. If a woman carries the mutant allele, she may be advised to undergo a mastectomy to prevent breast cancer from occurring. The application of these new molecular genetic technologies therefore often raises difficult issues for the people involved.

Molecular genetics is also providing new ways to treat diseases. For decades diabetics had to be given insulin obtained from animals—usually pigs. Today, perfect human insulin is manufactured in bacterial cells that carry the human insulin gene. Vats of these cells are grown to produce the insulin polypeptide on an industrial scale (**FIGURE 1.15**). Human growth hormone, previously isolated from cadavers, is also manufactured in bacterial cells. This hormone is used to treat children who cannot make sufficient amounts of the hormone themselves because they carry a mutant allele of the growth hormone gene. Without the added hormone, these children would be affected with dwarfism. Many other medically important proteins are now routinely produced in bacterial cells that have been transformed with the appropriate human gene. The large-scale production of such proteins is one facet of the burgeoning biotechnology industry. We shall explore ways of producing human proteins in bacterial cells in Chapter 17.

Human gene therapy is another way in which molecular genetic technologies are used to treat diseases. The strategy

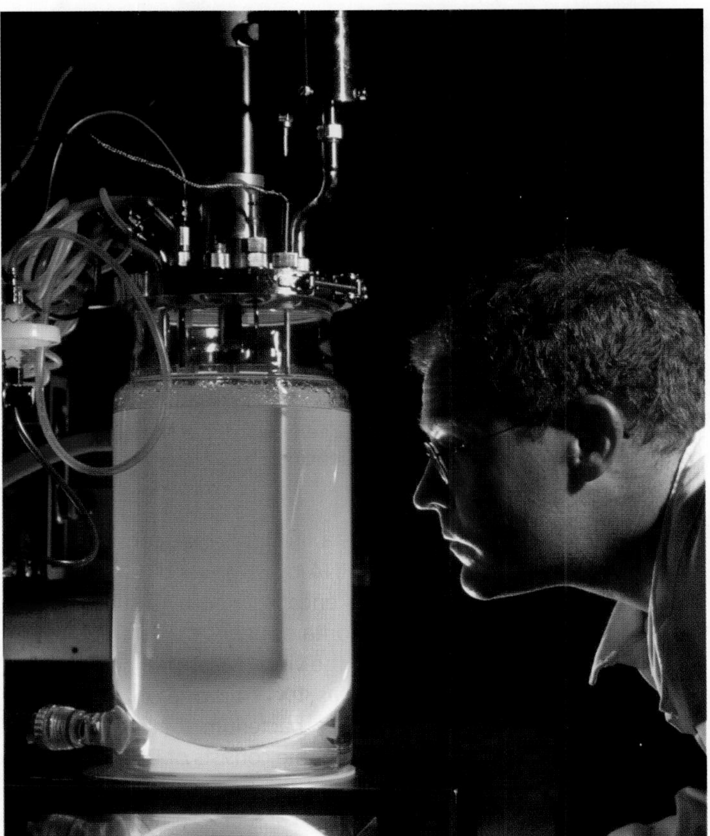

Figure 1.15 ▶ A researcher examines a fermenter growing *E. coli* to produce the painkilling drug hydroxymorphine.

in this type of therapy is to insert a healthy, functional copy of a particular gene into the cells of an individual who carries only mutant copies of that gene. The inserted gene can then compensate for the faulty genes that the individual inherited. To date, human gene therapy has had mixed results. Efforts to cure individuals with cystic fibrosis (CF), a serious respiratory disorder, by introducing copies of the normal *CF* gene into lung cells have not been successful. However, medical geneticists have had some success in treating immune system and blood cell disorders by introducing the appropriate normal genes into bone marrow cells, which later differentiate into immune cells and blood cells. We shall discuss the emerging technologies for human gene therapy and some of the risks involved in Chapter 17.

GENETICS IN SOCIETY

Modern societies depend heavily on the technology that emerges from research in the basic sciences. Our manufacturing and service industries are built on technologies for mass production, instantaneous communication, and prodigious information processing. Our lifestyles also depend on these technologies. At a more fundamental level, modern societies rely on technology to provide food and health care. We have already seen how genetics is contributing to these important needs. However, genetics impacts society in other ways too.

One way is economic. Discoveries from genetic research have initiated countless business ventures in the biotechnology industry. Companies that market pharmaceuticals and diagnostic

▶ A MILESTONE IN GENETICS: ΦX174, the First DNA Genome Sequenced

Figure 1 ▶ Frederick Sanger, developer of techniques to sequence DNA molecules. For his work, Sanger shared the Nobel Prize in Chemistry in 1980. Previously, Sanger had been awarded a Nobel Prize for his pioneering work on protein sequencing.

Sometimes life is simple. A virus is a life form that consists of little more than a nucleic acid—either RNA or DNA—and a few kinds of protein. The nucleic acid, which is encapsulated by a subset of the proteins, is the virus's genome. It contains genes that are needed for the formation and propagation of the virus. Most viral genomes are small, and that of the virus known as ΦX174 is one of the smallest.

ΦX174 infects *Escherichia coli,* a bacterium that naturally lives inside the human gut. The ΦX174 genome consists of DNA, but unlike the DNA of *E. coli* and its human host, the DNA of ΦX174 is single-stranded. This single-stranded molecule can act as a template for the synthesis of a partner strand. It does so after entering an *E. coli* cell, and it also does so *in vitro*—that is, in a test tube—as long as a researcher provides each of the four nucleotides, a DNA-synthesizing enzyme, and a short strand of DNA that is complementary to a segment on the ΦX174 DNA molecule. This short strand of DNA serves as a primer to get the process of DNA synthesis on the ΦX174 template started.

In the 1970s chemists and biochemists were trying to develop ways to sequence viral genomes. Many different strategies were employed, and the ΦX174 genome had become a test case. For several years, Frederick Sanger (**FIGURE 1**) and his colleagues at the MRC Laboratory of Molecular Biology at Cambridge University in the United Kingdom worked on the problem of sequencing the ΦX174 genome. Their approach was to synthesize DNA *in vitro* using viral DNA as a template and then analyze the products for the positions of individual nucleotides. It was demanding work. In February 1977 they published a nearly complete sequence of 5375 nucleotides based on their results with a technique they called the "plus and minus method"—because it involved adding or subtracting individual nucleotides from the *in vitro* reaction mixture. In a reflective passage about their work, Sanger and his colleagues wrote:

> As with other methods of sequencing nucleic acids, the plus and minus technique used by itself cannot be regarded as a completely reliable system and occasional errors may occur. Such errors and uncertainties can only be eliminated by more laborious experiments, and, although much of the sequence has been so confirmed, it would probably be a long time before the complete sequence could be established. We are not certain that there is any scientific justification for establishing every detail and, as it is felt that the results may be useful to other workers, it has been decided to publish the sequence in its present form.[1]

[1]Sanger, F., G. M. Air, B. G. Barrell, N. L. Brown, A. R. Coulson, J. C. Fields, C. A. Hutchison III, P. M. Slocombe, and M. Smith. 1977. Nucleotide sequence of bacteriophage ΦX174 DNA. *Nature* 265:687–695.

tests, or that provide services such as DNA profiling, have contributed to worldwide economic growth. Another way is legal. DNA sequences differ among individuals, and by analyzing these differences, people can be identified uniquely. Such analyses are now routinely used in many situations—to test for paternity, to convict the guilty and to exonerate the innocent of crimes for which they are accused, to authenticate claims to inheritances, and to identify the dead. Evidence based on analysis of DNA is now commonplace in courtrooms all over the world.

But the impact of genetics goes beyond the material, commercial, and legal aspects of our societies. It strikes the very core of our existence, because, after all, DNA—the subject of genetics—is a crucial part of us. Discoveries from genetics raise deep, difficult, and sometimes disturbing existential questions. Who are we? Where do we come from? Does our genetic makeup determine our nature? our talents? our ability to learn? our behavior? Does it play a role in setting our customs? Does it affect the ways we organize our societies? Does it influence our attitudes toward other people? Will knowledge about our genes and how they influence us affect our ideas about morality and justice, innocence and guilt, freedom and responsibility? Will this knowledge change how we think about what it means to be human? Whether we like it or not, these and other probing questions await us in the not-so-distant future.

KEY POINTS

- Discoveries in genetics are changing procedures and practices in agriculture and medicine.
- Advances in genetics are raising ethical, legal, political, social, and philosophical questions.

And so they did. The sequence data were presented on two pages, which because of the way in which the article was printed, came right in the midst of the passage quoted above. Thus, the data were embedded within an apology for their own shortcomings.

Despite these shortcomings, the ΦX174 DNA data were more extensive than any sequence data that had ever been collected. They revealed the detailed structure of nine genes in the virus's genome, and they showed that nearly all of ΦX174's DNA was genic. As the authors noted, "The most striking feature of the ΦX174 DNA sequence is the way in which the various functions of the genome are compressed within the 5,375 nucleotides."[2] In fact, two of the nine genes were found to reside wholly within two other genes. Thus, in the ΦX174 genome there were genes within genes—a phenomenon that no one had ever seen before. We shall discuss this discovery in Chapter 14 when we consider the ways in which genes can be defined.

In December 1977 Sanger and two of his colleagues reported a new technique for sequencing DNA.[3] It relied on a way of terminating *in vitro* DNA synthesis at specific nucleotides, for instance, at places where an A should be incorporated into the growing DNA chain, or at places where a T should be incorporated. By analyzing the lengths of the chains produced under these conditions, Sanger and his colleagues could identify the locations of all the A's, all the T's, and so on. The chain-terminating technique was a big improvement over all the previous methods that had been used to sequence DNA, including the plus and minus method that had provided most of the ΦX174 sequence data. Sanger and his colleagues immediately used the new method to clear up the uncertainties left over from their earlier work. In 1978 they published the complete sequence of the ΦX174 genome—5386 nucleotides in all.[4] By this time, however, readers were less interested in the actual sequence than in the technique that had been used to obtain it. Sanger's chain-terminating technique became the standard method to sequence DNA; it is still used today. In Chapter 15 we shall discuss this method of sequencing DNA, and in Chapter 16 we shall see how it has been applied to sequence genomes enormously larger than the ΦX174 genome. Without this technique, the various genome projects, including, of course, the Human Genome Project, would never have materialized.

QUESTIONS FOR DISCUSSION

1. Scientific advances often depend on the development of key techniques. Sanger's chain-terminating technique for sequencing DNA is one example. Can you think of other examples?
2. Viruses may appear to be simple, but their genetic organization can be complex. ΦX174 has ten genes—not nine as Sanger's team originally thought—and three of those genes are nested completely within other genes. From what you know about the way genetic information is encoded and expressed, how could one gene be contained wholly within another gene? What challenges might the gene-within-a-gene arrangement pose for the evolution of a virus like ΦX174?

[2]Ibid.

[3]Sanger, F., S. Nicklen, and A. R. Coulson. 1977. DNA sequencing with chain-terminating inhibitors. *Proc. Natl. Acad. Sci. USA* 74:5463–5467.

[4]Sanger, F., A. R. Coulson, T. Friedmann, G. M. Air, B. G. Barrell, N. L. Brown, J. C. Fields, C. A. Hutchison III, P. M. Slocombe, and M. Smith. 1978. The nucleotide sequence of bacteriophage ΦX174. *J. Molec. Biol.* 125:225–246.

▶ Basic Exercises

ILLUSTRATE BASIC GENETIC ANALYSIS

1. How is genetic information expressed in cells?

Answer: The genetic information is encoded in sequences in the DNA. Initially, these sequences are used to synthesize RNA complementary to them—a process called transcription—and then the RNA is used as a template to specify the incorporation of amino acids in the sequence of a polypeptide—a process called translation. Each amino acid in the polypeptide corresponds to a sequence of three nucleotides in the DNA. The triplets of nucleotides that encode the different amino acids are called codons.

2. What is the evolutionary significance of mutation?

Answer: Mutation creates variation in the DNA sequences of genes (and in the nongenic components of genomes as well). This variation accumulates in populations of organisms over time and may eventually produce observable differences among the organisms. One population may come to differ from another according to the kinds of mutations that have accumulated over time. Thus, mutation provides the input for different evolutionary outcomes at the population level.

▶ Testing Your Knowledge

INTEGRATE DIFFERENT CONCEPTS AND TECHNIQUES

1. Suppose a gene contains 10 codons. How many coding nucleotides does the gene contain? How many amino acids are expected to be present in its polypeptide product? Among all possible genes composed of 10 codons, how many different polypeptides could be produced?

Answer: The gene possesses 30 coding nucleotides. Its polypeptide product is expected to contain 10 amino acids, each corresponding to one of the codons in the gene. If each codon can specify one of 20 naturally occurring amino acids, among all possible gene sequences 10 codons long, we can imagine a total of 20^{10} polypeptide products—a truly enormous number!

▶ Questions and Problems

ENHANCE UNDERSTANDING AND DEVELOP ANALYTICAL SKILLS

1.1 Which bases are present in DNA? Which bases are present in RNA? Which sugars are present in each of these nucleic acids?

1.2 What is the difference between transcription and translation?

1.3 In a few sentences, what were Mendel's key ideas about inheritance?

1.4 GO A gene contains 138 codons. How many nucleotides are present in the gene's coding sequence? How many amino acids are expected to be present in the polypeptide encoded by this gene?

1.5 The template strand of a gene being transcribed is TAGCTTAGT. What will be the sequence of the RNA made from this template?

1.6 Both DNA and RNA are composed of nucleotides. What molecules combine to form a nucleotide?

1.7 The sequence of a strand of DNA is TAAGCCTGC. If this strand serves as the template for DNA synthesis, what will be the sequence of the newly synthesized strand?

1.8 GO The gene for α-globin is present in all vertebrate species. Over millions of years, the DNA sequence of this gene has changed in the lineage of each species. Consequently, the amino acid sequence of α-globin has also changed in these lineages. Among the 141 amino acid positions in this polypeptide, human α-gobin differs from shark α-globin in 79 positions; it differs from carp α-globin in 68 and from cow α-globin in 17. Do these data suggest an evolutionary phylogeny for these vertebrate species?

1.9 Sickle-cell anemia is caused by a mutation in one of the codons in the gene for β-globin; because of this mutation, the sixth amino acid in the β-globin polypeptide is a valine instead of a glutamic acid. A less severe type of anemia is caused by a mutation that changes this same codon to one specifying lysine as the sixth amino acid in the β-globin polypeptide. What word is used to describe the two mutant forms of this gene? Do you think that an individual carrying these two mutant forms of the β-globin gene would suffer from anemia? Explain.

1.10 Hemophilia is an inherited disorder in which the blood-clotting mechanism is defective. Because of this defect, people with hemophilia may die from cuts or bruises, especially if internal organs such as the liver, lungs, or kidneys have been damaged. One method of treatment involves injecting a blood-clotting factor that has been purified from blood donations. This factor is a protein encoded by a human gene. Suggest a way in which modern genetic technology could be used to produce this factor on an industrial scale. Is there a way in which the inborn error of hemophilia could be corrected by human gene therapy?

1.11 RNA is synthesized using DNA as a template. Is DNA ever synthesized using RNA as a template? Explain.

Genomics on the Web

at http://www.ncbi.nlm.nih.gov/

You might enjoy using the NCBI web site to explore the Human Genome Project. In the sidebar, click on Education, then under Genomes and Genetics click on Human Genome Project, and finally click on All About the Human Genome Project.

Chapter 2 Cellular Reproduction and Model Genetic Organisms

CHAPTER OUTLINE

- Cells and Chromosomes
- Mitosis
- Meiosis
- Genetics in the Laboratory: An Introduction to Some Model Research Organisms

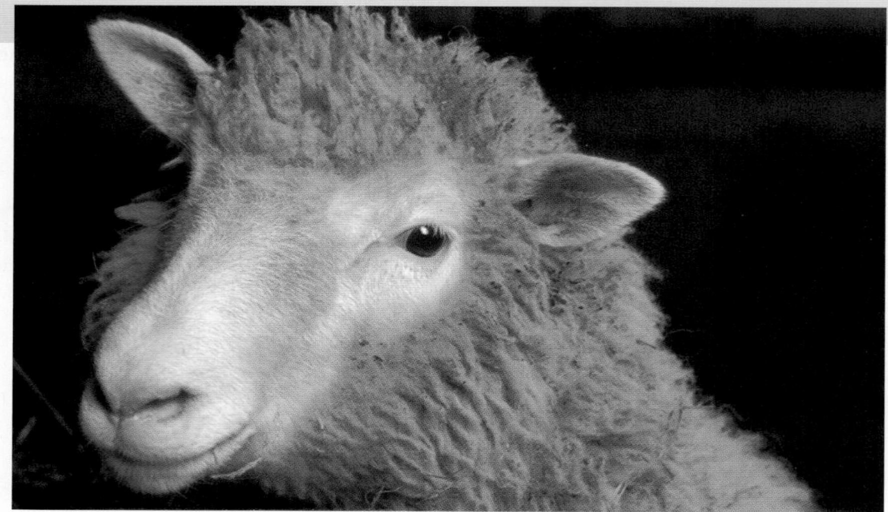

Dolly, the first cloned mammal. The photo below shows the cloning process.

Dolly

Sheep have grazed on the hard-scrabble landscape of Scotland for centuries. Finn Dorsets and Scottish Blackfaces are some of the breeds raised by shepherds there. Every spring, the lambs that were conceived during the fall are born. They grow quickly and take their places in flocks—or in butcher shops. Early in 1997, a lamb unlike any other came into the world. This lamb, named Dolly, did not have a father, but she did have three mothers; furthermore, her genes were identical to those of one of her mothers. In a word, Dolly was a clone.

Scientists at the Roslin Institute near Edinburgh, Scotland, produced Dolly by fusing an egg from a Blackface ewe (the egg cell mother) with a cell from the udder of a Finn Dorset ewe (the genetic mother). The genetic material in the Blackface ewe's egg had been removed prior to fusing the egg with the udder cell. Subsequently, the newly endowed egg was stimulated to divide. It produced an embryo, which was implanted in the uterus of another Blackface ewe (the gestational or surrogate mother). This embryo grew and developed, and when the surrogate mother's pregnancy came to term, Dolly was born.

The technology that produced Dolly emerged from a century of basic research on the cellular basis of reproduction. In the ordinary course of events, an egg cell from a female is fertilized by a sperm cell from a male, and the resulting zygote divides to produce genetically identical cells. These cells then divide many times to produce a multicellular organism. Within that organism, a particular group of cells embarks on a different mode of division to produce specialized reproductive cells—either eggs or sperm. An egg from one such organism then unites with a sperm from another such organism to produce a new offspring. The offspring grows up and the cycle continues, generation after generation. But among the sheep that graze on Scotland's pastures, Dolly did not conform to this standard pattern of reproduction. Clearly, her creators had something else in mind.

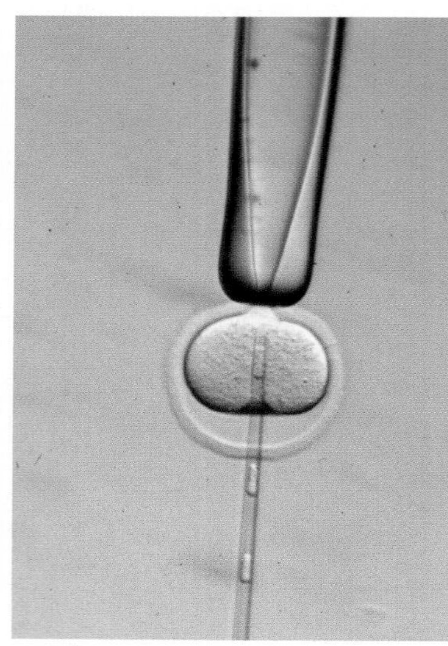

A nucleus inside a long, thin micropipette is being injected into an enucleated egg that is being held in place by a wider pipette.

Cells and Chromosomes

In both prokaryotic and eukaryotic cells, the genetic material is organized into chromosomes.

In the early part of the nineteenth century, a few decades before Gregor Mendel carried out his experiments with peas, biologists established the principle that living things are composed of cells. Some organisms consist of just a single cell; others consist of trillions of cells. Each cell is a complicated assemblage of molecules that can acquire materials, recruit and store energy, and carry out diverse activities, including reproduction. The simplest life forms, viruses, are not composed of cells. However, viruses must enter cells in order to function. Thus, all life has a cellular basis. As preparation for our journey through the science of genetics, we now review the biology of cells. We also discuss chromosomes—the cellular structures in which genes reside.

THE CELLULAR ENVIRONMENT

Living cells are made of many different kinds of molecules. The most abundant is water. Small molecules—for example, salts, sugars, amino acids, and certain vitamins—readily dissolve in water, and some larger molecules interact favorably with it. All these sorts of substances are said to be hydrophilic. Other kinds of molecules do not interact well with water. They are said to be hydrophobic. The inside of a cell, called the **cytoplasm,** contains both hydrophilic and hydrophobic substances.

The molecules that make up cells are diverse in structure and function. **Carbohydrates** such as starch and glycogen store chemical energy for work within cells. These molecules are composed of glucose, a simple sugar. The glucose subunits are attached one to another to form long chains, or polymers. Cells obtain energy when glucose molecules released from these chains are chemically degraded into simpler compounds—ultimately, to carbon dioxide and water. Cells also possess an assortment of **lipids.** These molecules are formed by chemical interactions between glycerol, a small organic compound, and larger organic compounds called fatty acids. Lipids are important constituents of many structures within cells. They also serve as energy sources. **Proteins** are the most diverse molecules within cells. Each protein consists of one or more polypeptides, which are chains of amino acids. Often a protein consists of two polypeptides—this protein is a dimer; sometimes a protein consists of many polypeptides—that is, it is a multimer. Within cells, proteins are components of many different structures. They also catalyze chemical reactions. We call these catalytic proteins **enzymes.** Cells also contain nucleic acids—DNA and RNA—which, as already described in Chapter 1, are central to life.

Cells are surrounded by a thin layer called a **membrane.** Many different types of molecules make up cell membranes; however, the primary constituents are lipids and proteins. Membranes are also present inside cells. These internal membranes may divide a cell into compartments, or they may help to form specialized structures called **organelles.** Membranes are fluid and flexible. Many of the molecules within a membrane are not rigidly held in place by strong chemical forces. Consequently, they are able to slip by one another in what amounts to an ever-changing molecular sea. Some kinds of cells are surrounded by tough, rigid walls, which are external to the membrane. Plant cell walls are composed of cellulose, a complex carbohydrate. Bacterial cell walls are composed of a different kind of material called murein.

Walls and membranes separate the contents of a cell from the outside world. However, they do not seal it off. These structures are porous to some materials, and they selectively allow other materials to pass through them via channels and gates. The transport of materials in and through walls and membranes is an important activity of cells. Cell membranes also contain molecules that interact with materials in a cell's external environment. Such molecules provide a cell with vital information about conditions in the environment, and they also mediate important cellular activities.

PROKARYOTIC AND EUKARYOTIC CELLS

When we survey the living world, we find two basic kinds of cells: prokaryotic and eukaryotic (**FIGURE 2.1**). **Prokaryotic cells** are usually less than a thousandth of a millimeter long, and they typically lack a complicated system of internal membranes and membranous organelles. Their hereditary material—that is, the DNA—is not isolated in a special subcellular compartment. Organisms with this kind of cellular organization are called prokaryotes. Examples include the bacteria, which are the most abundant life forms on earth, and the archaea, which are found in extreme environments such as salt lakes, hot springs, and deep-sea volcanic vents. All other organisms—plants, animals, protists, and fungi—are eukaryotes.

Eukaryotic cells (**FIGURE 2.2**) are larger than prokaryotic cells, usually at least 10 times bigger, and they possess complicated systems of internal membranes, some of which are associated with conspicuous organelles. For example, eukaryotic cells typically contain one or more **mitochondria** (singular, mitochondrion), which are ellipsoidal organelles dedicated to the recruitment of energy from foodstuffs. Algal and plant cells contain another kind of energy-recruiting organelle called the **chloroplast,** which captures solar energy and converts it into chemical energy. Both mitochondria and chloroplasts are surrounded by membranes.

The hallmark of all eukaryotic cells is that their hereditary material is contained within a large, membrane-bounded structure called the **nucleus.** The nuclei of eukaryotic cells provide a safe haven for the DNA, which is organized into discrete structures called **chromosomes.** Individual chromosomes become visible during cell division, when they condense and thicken. In prokaryotic cells, the DNA is usually not housed

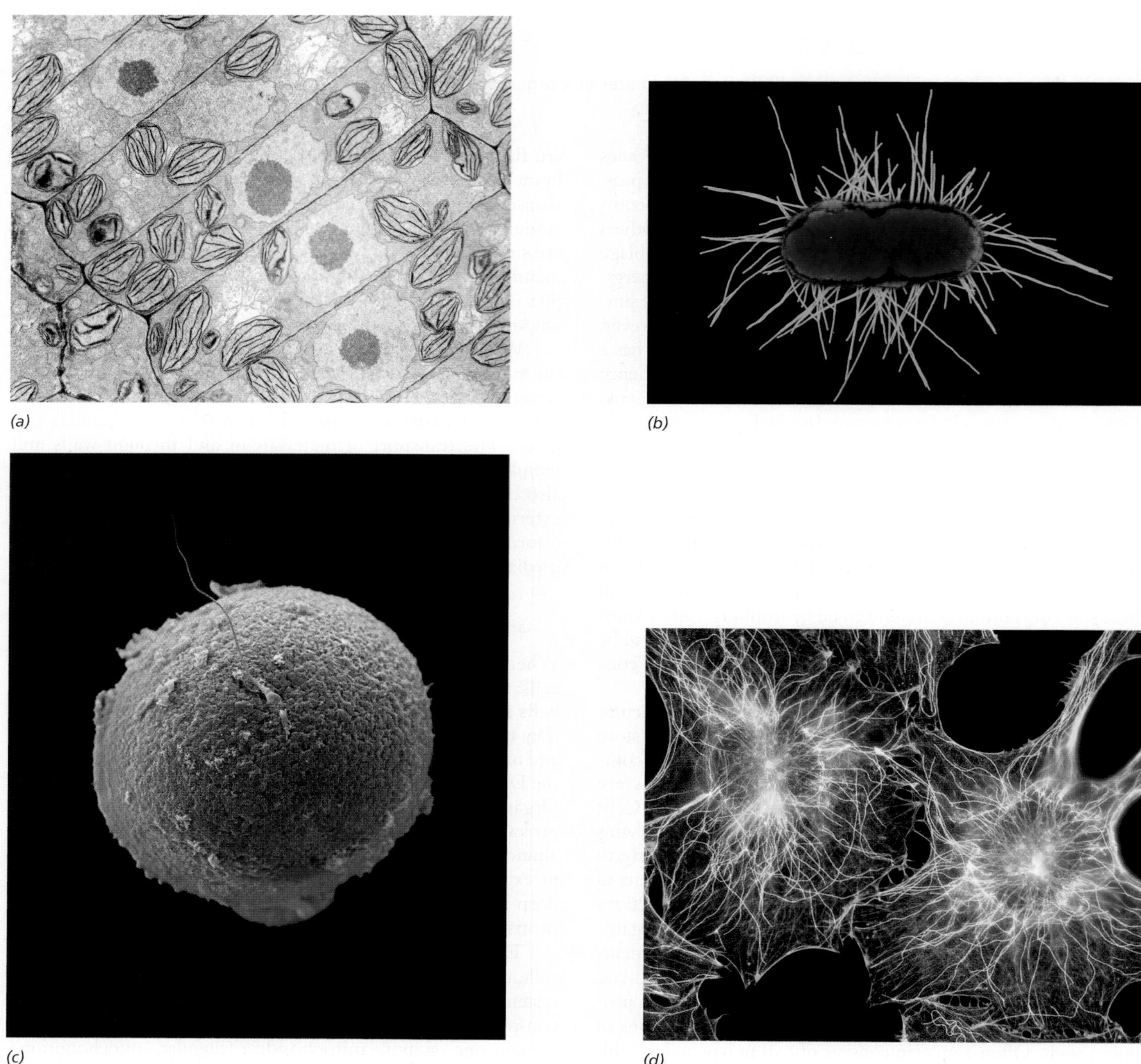

Figure 2.1 ▶ Various types of cells. (*a*) Color-enhanced transmission electron micrograph of duckweed (*Spirodela oligorrhiza*) root cells, showing cell walls, centralized nuclei, and numerous chloroplasts. (*b*) Color-enhanced scanning electron micrograph of a bacterial cell (*Escherichia coli*). (*c*) Color-enhanced scanning electron micrograph of a human egg with sperm. (*d*) Fluorescent light micrograph of two fibroblast cells, showing their nuclei (green) and cytoskeleton.

within a well-defined nucleus. We shall investigate the ways in which chromosomal DNA is organized in prokaryotic and eukaryotic cells in Chapter 9.

Some of the DNA within a eukaryotic cell is not situated within the nucleus. This extranuclear DNA is located in the mitochondria and chloroplasts. We shall examine its structure and function in Chapter 16.

Both prokaryotic and eukaryotic cells possess numerous **ribosomes,** which are small organelles involved in the synthesis of proteins, a process that we shall investigate in Chapter 12.

Animal cell

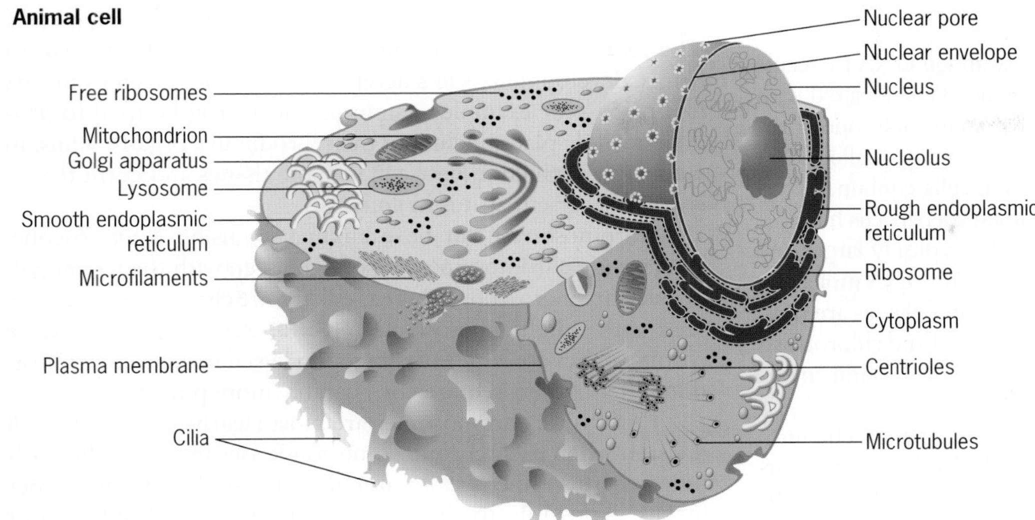

(*a*)

Plant cell

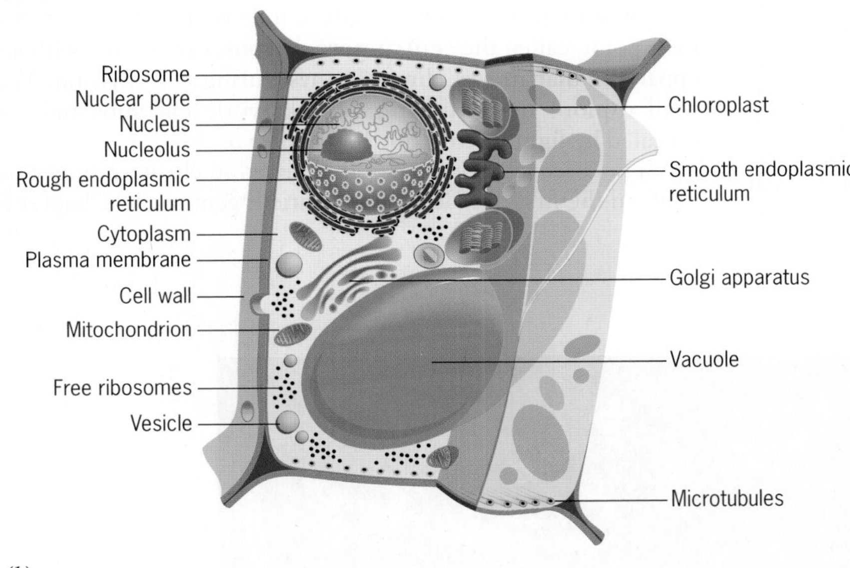

(*b*)

Figure 2.2 ▶ The structures of animal cells (*a*) and plant cells (*b*).

Ribosomes are found throughout the cytoplasm. Although ribosomes are not composed of membranes, in eukaryotic cells they are often associated with a system of membranes called the **endoplasmic reticulum.** The reticulum may be connected to the **Golgi complex,** a set of membranous sacs and vesicles that are involved in the chemical modification and transport of substances within cells. Other small, membrane-bound organelles may also be found in eukaryotic cells. In animal cells, **lysosomes** are produced by the Golgi complex. These organelles contain different kinds of digestive enzymes that would harm the cell if they were released into the cytoplasm. Both plant and animal cells may contain **peroxisomes,** which are small organelles dedicated to the metabolism of substances such as fats and amino acids. The internal membranes and oganelles of eukaryotic cells create a system of subcellular compartments that vary in chemical conditions such as pH and salt content. This variation provides cells with different internal environments that are adapted to the many processes that cells carry out.

The shapes and activities of eukaryotic cells are influenced by a system of filaments, fibers, and associated molecules that collectively form the **cytoskeleton.** These materials give form to cells and enable some types of cells to move through their environment—a phenomenon referred to as cell motility. The cytoskeleton holds organelles in place, and it plays a major role in moving materials to specific locations within cells—a phenomenon called **trafficking.**

CHROMOSOMES: WHERE GENES ARE LOCATED

Each chromosome consists of one double-stranded DNA molecule plus an assortment of proteins; RNA may also be associated with chromosomes. Prokaryotic cells typically contain only one chromosome, although sometimes they also possess many smaller DNA molecules called **plasmids.** Most eukaryotic cells contain several different chromosomes—for example, human sperm cells have 23. The chromosomes of eukaryotic cells are also typically larger and more complex than those of prokaryotic cells. The DNA molecules in prokaryotic chromosomes and plasmids are circular, as are most of the DNA molecules found in the mitochondrial and chloroplasts of eukaryotic cells. However, the DNA molecules found in the chromosomes in the nuclei of eukaryotic cells are linear.

Many eukaryotic cells possess two copies of each chromosome. This condition, referred to as the **diploid** state, is characteristic of the cells in the body of a eukaryote—that is, the **somatic** cells. By contrast, the sex cells, or **gametes,** usually possess only one copy of each chromosome, a condition referred to as the **haploid** state. Gametes are produced from diploid cells located in the **germ line,** which is the reproductive tissue of an organism. In some creatures, such as plants, the germ line produces both sperm and eggs. In other creatures, such as human beings, it produces one kind of gamete or the other. When a male and a female gamete unite during fertilization, the diploid state is reestablished, and the resulting zygote develops into a new organism. During animal development, a small number of cells are set aside to form the germ line. All the gametes that will ever be produced are derived from these few cells. The remaining cells form the somatic tissues of the animal. In plants, development is less determinate. Tissues taken from part of a plant—for example, a stem or a leaf—can be used to produce a whole plant, including the reproductive organs. Thus, in plants the distinction between somatic tissues and germ tissues is not as clear-cut as it is in animals.

Chromosomes can be examined by using a microscope. Prokaryotic chromosomes can only be seen with the techniques of electron microscopy, whereas eukaryotic chromosomes can be seen with a light microscope (**FIGURE 2.3**). Some eukaryotic chromosomes are large enough to be viewed with low magnification (20×); others require considerably more power (>500×).

Eukaryotic chromosomes are most clearly seen during cell division when each chromosome condenses into a smaller volume. At this time the greater density of the chromosomes makes it possible to discern certain structural features. For example, each chromosome may appear to consist of two parallel rods held together at a common point (**FIGURE 2.3*b***). Each of the rods is an identical copy of the chromosome created during a duplication process that precedes condensation, and the common point, called the **centromere,** becomes associated with an apparatus that moves chromosomes during cell division. We shall explore the structures of eukaryotic chromosomes as revealed by light microscopy in Chapter 6.

The discovery that genes are located in chromosomes was made in the first decade of the twentieth century. In Chapter 5

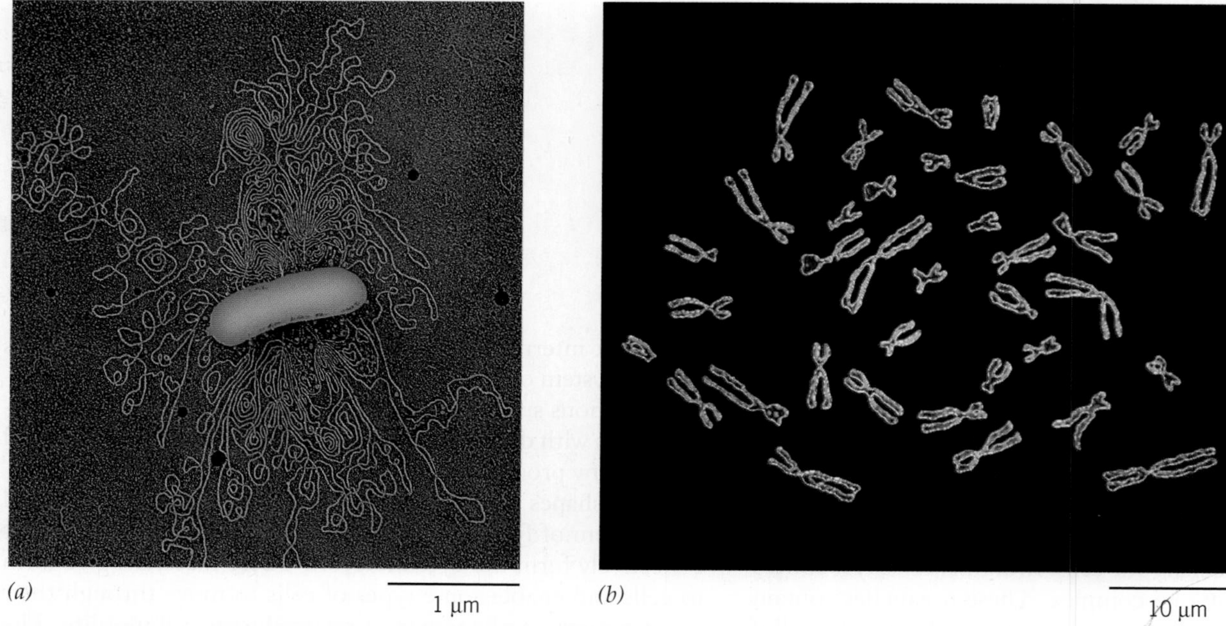

Figure 2.3 ▶ (*a*) Electron micrograph showing a bacterial chromosome extruded from a cell. (*b*) Light micrograph of human chromosomes during cell division. The constriction in each of the duplicated chromosomes is the centromere, the point at which spindle fibers attach to move the chromosome during cell division.

we shall examine the experimental evidence for this discovery, and in Chapters 7 and 8 we shall study some of the techniques for locating genes within chromosomes.

CELL DIVISION

Among the many activities carried out by living cells, division is the most astonishing. A cell can divide into two cells, each of which can also divide into two, and so on through time, to create a population of cells called a **clone.** Barring errors, all the cells within a clone are genetically identical. Cell division is an integral part of the growth of multicellular organisms, and it is also the basis of reproduction.

A cell that is about to divide is called a **mother cell,** and the products of division are called **daughter cells.** When prokaryotic cells divide, the contents of the mother cell are more or less equally apportioned between the two daughter cells. This process is called **fission.** The mother cell's chromosome is duplicated prior to fission, and copies of it are bequeathed to each of the daughter cells. Under optimal conditions, a prokaryote such as the intestinal bacterium *Escherichia coli* divides every 20 to 30 minutes. At this rate, a single *E. coli* cell could form a clone of approximately 2^{50} cells—more than a quadrillion—in just one day. In reality, of course, *E. coli* cells do not sustain this high rate of division. As cells accumulate, the rate of division declines because nutrients are exhausted and waste products pile up. Nevertheless, a single *E. coli* cell can produce enough progeny in a single day to form a mass visible to the unaided eye. We call such a mass of cells a **colony.**

The division of eukaryotic cells is a more elaborate process than the division of prokaryotic cells. Typically, many chromosomes must be duplicated, and the duplicates must be distributed equally and exactly to the daughter cells. Organelles—mitochondria, chloroplasts, endoplasmic reticulum, Golgi complex, and so on—must also be distributed to the daughter cells. However, for these entities the distribution process is not equal and exact. Mitochondria and chloroplasts are randomly apportioned to the daughter cells. The endoplasmic reticulum and the Golgi complex are fragmented at the time of division and later are re-formed in the daughter cells.

Each time a eukaryotic cell divides, it goes through a series of phases that collectively form the **cell cycle** (**FIGURE 2.4**). The progression of phases is $G_1 \rightarrow S \rightarrow G_2 \rightarrow M$. In this progression, S is the period in which the chromosomes are duplicated—an event that requires DNA *synthesis*, to which the label "S" refers. The M phase in the cell cycle is the time when the mother cell actually divides. This phase usually has two components: (1) **mitosis,** which is the process that distributes the duplicated chromosomes equally and exactly to the daughter cells, and (2) **cytokinesis,** which is the process that physically separates the two daughter cells from each other. The label "M" refers to the term *mitosis*, which is derived from a Greek word for thread; during mitosis, the chromosomes appear as threadlike bodies inside cells. The G_1 and G_2 phases are "gaps" between the S and M phases.

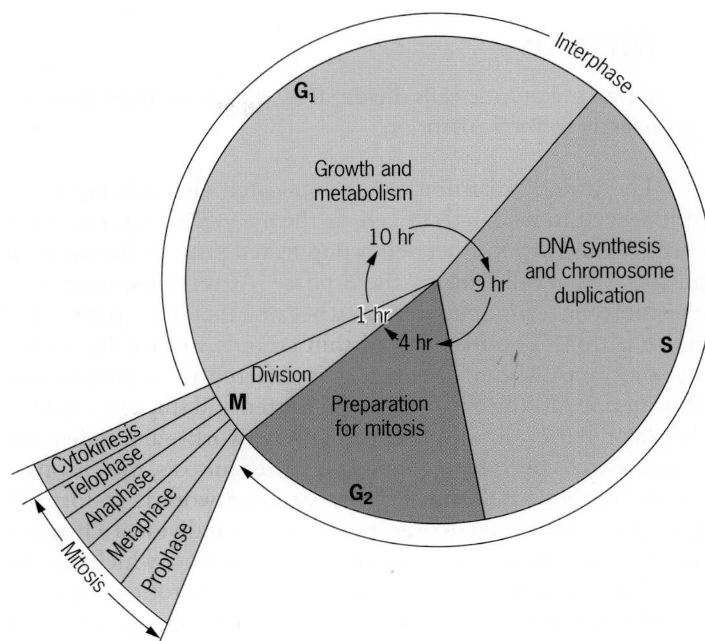

Figure 2.4 ▶ The cycle of an animal cell. This cycle is 24 hours long. The duration of the cycle varies among different types of eukaryotic cells.

The length of the cell cycle varies among different types of cells. In embryos, where growth is rapid, the cycle may be as short as 30 minutes. In slow-growing adult tissues, it may last several months. Some cells, such as those in nerve and muscle tissues, cease to divide once they have acquired their specialized functions. The progression of eukaryotic cells through their cycle is tightly controlled by different types of proteins. When the activities of these proteins are disrupted, cells divide in an unregulated fashion. This deregulation of cell division may lead to cancer, which is a major cause of death among people today. In Chapter 22 we shall investigate the genetic basis of cancer.

KEY POINTS

- Cells, the basic units of all living things, are enclosed by membranes.
- Chromosomes, the cellular structures that carry the genes, are composed of DNA, RNA, and protein.
- In eukaryotes, chromosomes are contained within a membrane-bounded nucleus; in prokaryotes they are not.
- Eukaryotic cells possess complex systems of internal membranes as well as membranous organelles such as mitochondria, chloroplasts, and the endoplasmic reticulum.
- Haploid eukaryotic cells possess one copy of each chromosome; diploid cells possess two copies.
- Prokaryotic cells divide by fission; eukaryotic cells divide by mitosis and cytokinesis.
- Eukaryotic chromosomes duplicate when a cell's DNA is synthesized; this event, which precedes mitosis, is characteristic of the S phase of the cell cycle.

Mitosis

When eukaryotic cells divide, they distribute their genetic material equally and exactly to their offspring.

The orderly distribution of duplicated chromosomes in a mother cell to its daughter cells is the essence of mitosis. Each chromosome in a mother cell is duplicated prior to the onset of mitosis, specifically during the S phase. At this time individual chromosomes cannot be identified because they are too extended and too thin. The network of thin strands formed by all the chromosomes within the nucleus is referred to as **chromatin.** During mitosis, the chromosomes shorten and thicken—that is, they "condense" out of the chromatin network—and individual chromosomes become recognizable. After mitosis, the chromosomes "decondense" and the chromatin network is re-formed. Biologists often refer to the period when individual chromosomes cannot be seen as **interphase.** This period, which may be quite lengthy, is the time between successive mitotic events.

When mitosis begins, each chromosome has already been duplicated. The duplicates, called **sister chromatids,** remain intimately associated with each other and are joined at the chromosome's centromere. The term *sister* is something of a misnomer because these chromatids are copies of the original chromosome; therefore, they are more closely related than sisters. Perhaps the word "twin" would describe the situation better. However, "sister" is commonly used, and we shall use it here.

The distribution of duplicated chromosomes to the daughter cells is organized and executed by **microtubules,** which are components of the cytoskeleton. These fibers, composed of proteins called tubulins, attach to the chromosomes and move them about within the dividing mother cell. During mitosis the microtubules assemble into a complex array called the **spindle** (**FIGURE 2.5*a***). The formation of the spindle is associated with **microtubule organizing centers (MTOCs),** which are found in the cytoplasm of eukaryotic cells, usually near the nucleus. In animal cells, the MTOCs are differentiated into small organelles called **centrosomes;** these organelles are not present in plant cells. Each centrosome contains two barrel-shaped **centrioles,** which are aligned at right angles to each other (**FIGURE 2.5*b***). The centrioles are surrounded by a diffuse matrix called the pericentriolar material, which initiates the formation of the microtubules that will make up the mitotic spindle. The single centrosome that exists in an animal cell is duplicated during interphase. As the cell enters mitosis, microtubules develop around each of the daughter centrosomes to form a sunburst pattern called an **aster.** These centrosomes then move around the nucleus to opposite positions in the cell, where they establish the axis of the upcoming mitotic division. The final positions of the centrosomes define the poles of the dividing mother cell. In plant cells, MTOCs that do not have distinct centrosomes define these poles and establish the mitotic spindle.

The initiation of spindle formation and the condensation of duplicated chromosomes from the diffuse network of chromatin are hallmarks of the first stage of mitosis, called **prophase** (**FIGURE 2.6**). Formation of the spindle is accompanied by fragmentation of many intracellular organelles—for instance, the endoplasmic reticulum and the Golgi complex. The **nucleolus,** a dense body involved in RNA synthesis within the nucleus, also disappears; however, other types of organelles such as mitochondria and chloroplasts remain intact. Concomitant with the fragmentation of the endoplasmic reticulum, the nuclear membrane (also known as the nuclear envelope) breaks up into many small vesicles, and microtubules formed within the cytoplasm invade the nuclear space. Some of these

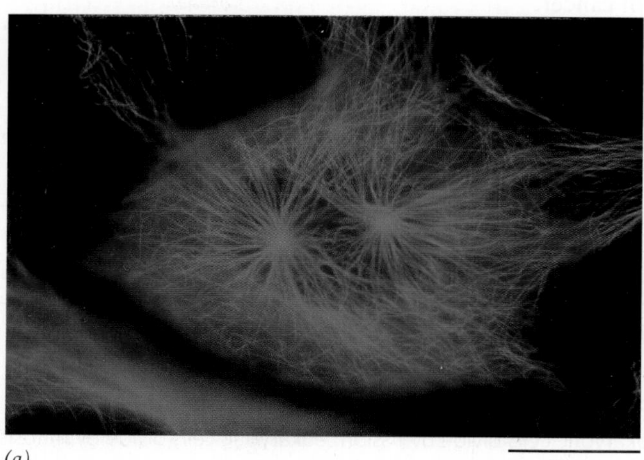

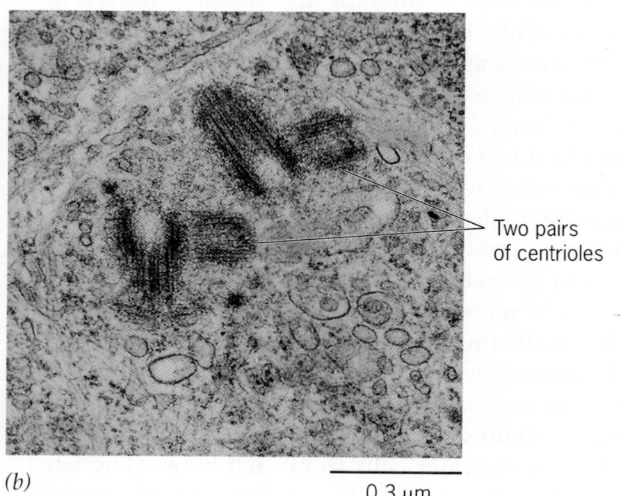

Figure 2.5 ▶ (*a*) The mitotic spindle in a cultured animal cell, which has been stained to show the microtubules (green) emanating from the two asters. (*b*) Electron micrograph showing two pairs of centrioles.

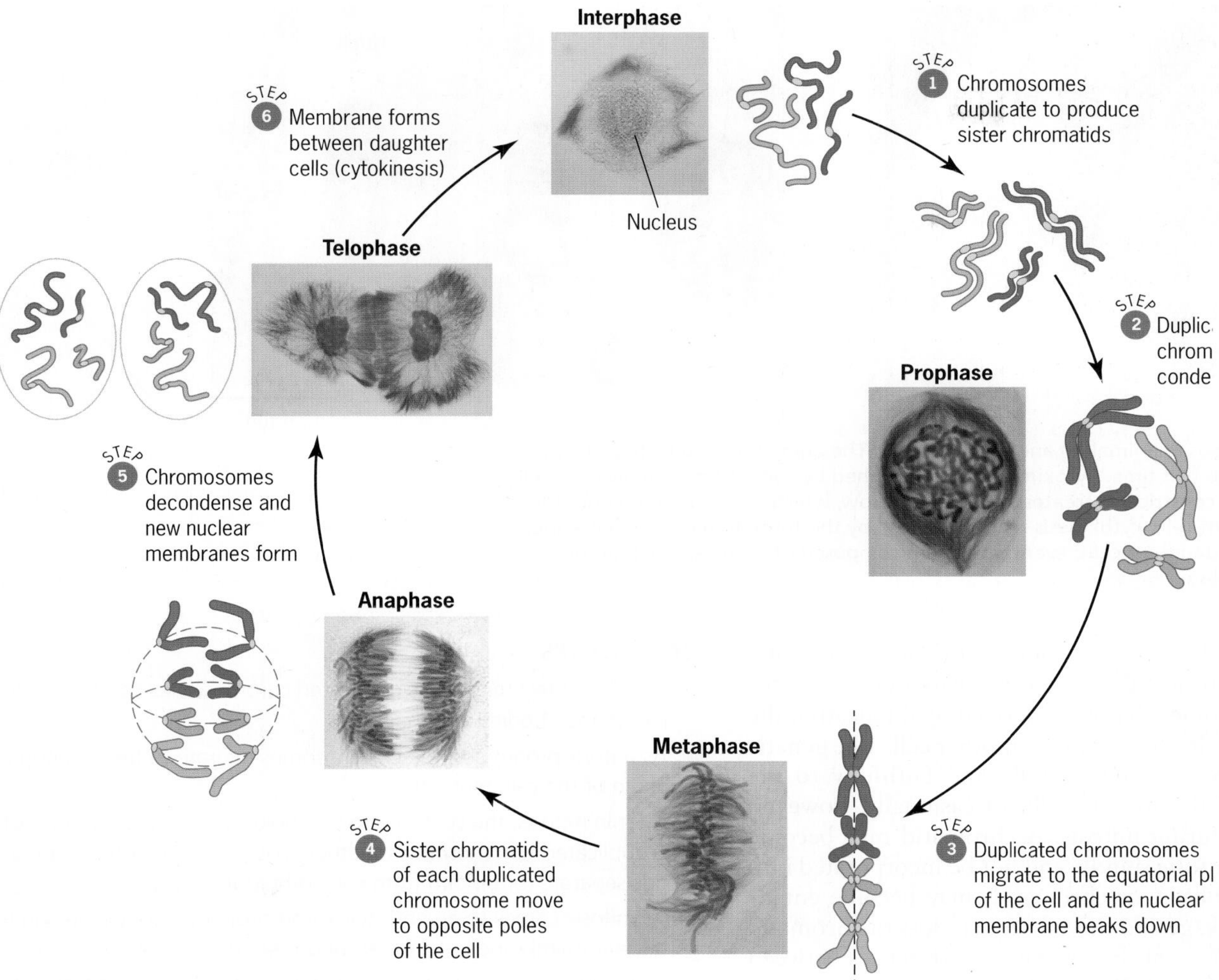
Interphase
STEP 1 Chromosomes duplicate to produce sister chromatids
Nucleus
STEP 6 Membrane forms between daughter cells (cytokinesis)
Telophase
STEP 2 Duplic
chrom
conde
Prophase
STEP 5 Chromosomes decondense and new nuclear membranes form
Anaphase
Metaphase
STEP 4 Sister chromatids of each duplicated chromosome move to opposite poles of the cell
STEP 3 Duplicated chromosomes migrate to the equatorial pl
of the cell and the nuclear membrane beaks down

Figure 2.6 ► Mitosis in the blood lily *Haemanthus.*

microtubules attach to the **kinetochores,** which are proteinaceous structures associated with the centromeres of the duplicated chromosomes. Attachment of spindle microtubules to the kinetochores indicates that the cell is entering the **metaphase** of mitosis.

During metaphase the duplicated chromosomes move to positions midway between the spindle poles. This movement is leveraged by changes in the length of the spindle microtubules and by the action of force-generating motor proteins that work near the kinetochores. The spindle apparatus also contains microtubules that are not attached to kinetochores. These additional microtubules appear to stabilize the spindle apparatus. Through the operation of the spindle apparatus, the duplicated chromosomes come to lie in a single plane in the middle of the cell. This equatorial plane is called the *metaphase plate.* At this stage, each sister chromatid of a duplicated chromosome is connected to a different pole via microtubules attached to its kinetochore. This polar alignment of the sister chromatids is crucial for the equal and exact distribution of genetic material to the daughter cells.

The sister chromatids of duplicated chromosomes are separated from each other during the **anaphase** of mitosis. This separation is accomplished by shortening the microtubules attached to the kinetochores and by degrading materials that hold the sister chromatids together. As the microtubules shorten, the sister chromatids are pulled to opposite poles of the cell. The separated sister chromatids are now referred to as chromosomes. While the chromosomes are moving toward the poles, the poles themselves also begin to move apart. This double movement cleanly separates the two sets of chromosomes into distinct spaces within the dividing cell. Once this separation has been achieved, the chromosomes decondense into a network of chromatin fibers, and the organelles that were lost at the onset of mitosis re-form. Each set of chromosomes becomes enclosed by a nuclear membrane. The decondensation of the chromosomes and the restoration of the internal organelles are characteristic of the **telophase** of mitosis. When mitosis is complete, the two daughter cells are separated by the formation of membranes between them. In plants, a wall is also laid down between the daughter cells. This physical separation of the daughter cells is called cytokinesis (**FIGURE 2.7**).

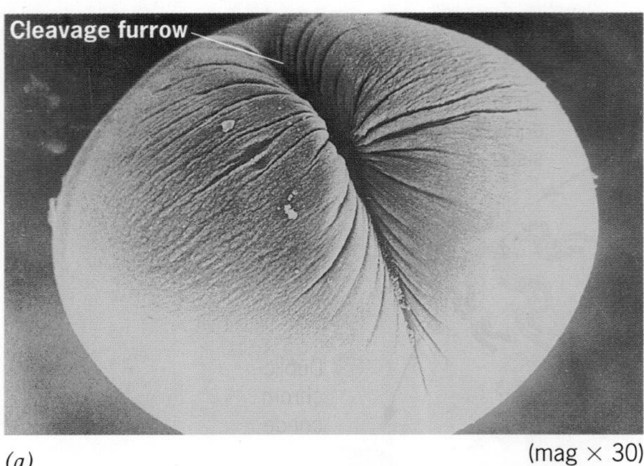

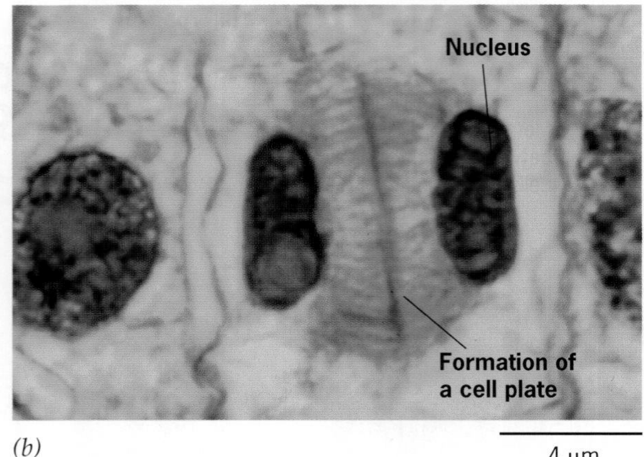

(a) (mag × 30) (b) 4 μm

Figure 2.7 ▶ Cytokinesis in animal (*a*) and plant (*b*) cells. The animal cell is a fertilized egg, which is dividing for the first time. Cytokinesis is accomplished by constricting the dividing cell around its middle. This constriction creates a cleavage furrow, which is seen here on one side of the dividing cell. In plant cells, cytokinesis is accomplished by the formation of a membranous cell plate between the daughter cells; eventually, walls composed of cellulose are built on either side of the cell plate.

The daughter cells that are produced by the division of a mother cell are genetically identical. Each daughter has a complete set of chromosomes that were derived by duplicating the chromosomes originally present in the mother cell. The genetic material is therefore transmitted fully and faithfully to the daughter cells from the mother cell. Occasionally, however, mistakes are made during mitosis. A chromatid may become detached from the mitotic spindle and not be incorporated into one of the daughter cells, or chromatids may become entangled, leading to breakage and the subsequent loss of chromatid parts. These types of events cause genetic differences between the daughter cells. We shall consider some of their consequences in Chapter 6 and again in Chapter 22.

KEY POINTS

- As a cell enters mitosis, its duplicated chromosomes condense into rod-shaped bodies (prophase).
- As mitosis progresses, the chromosomes migrate to the equatorial plane of the cell (metaphase).
- Later in mitosis, the centromere that holds the sister chromatids of a duplicated chromosome together splits, and the sister chromatids separate (or disjoin) from each other (anaphase).
- As mitosis comes to an end, the chromosomes decondense and a nuclear membrane re-forms around them (telophase).
- Each daughter cell produced by mitosis and cytokinesis has the same set of chromosomes; thus, daughter cells are genetically identical.

Meiosis

Sexual reproduction involves a mechanism that reduces the number of chromosomes by half.

If we denote the number of chromosomes in a gamete by the letter n, then the zygote produced by the union of two gametes has $2n$ chromosomes. We refer to the n chromosomes of a gamete as the haploid state, and the $2n$ chromosomes of the zygote as the diploid state. **Meiosis**—from a Greek word meaning "diminution"—is the process that reduces the diploid state to the haploid state—that is, it reduces the number of chromosomes in a cell by half. The resulting haploid cells either directly become gametes or divide to produce cells that later become gametes. Meiosis therefore plays a key role in reproduction among eukaryotes. Without it, organisms would double their chromosome number every generation—a situation that would quickly become unsupportable given the obvious limitations on the size and metabolic capacity of cells.

If we look at the chromosomes in a diploid cell, we find that they come in pairs (**FIGURE 2.8**). For example, somatic human cells have 23 pairs of chromosomes. Each pair is distinct. Different pairs of chromosomes carry different sets of genes. The members of a pair are called homologous chromosomes, or simply **homologues,** from a Greek word meaning "in agreement with." Homologues carry the same set of genes, although as we shall see in Chapter 5, they may carry different alleles of these genes. Chromosomes from different pairs are called **heterologues.** During meiosis, homologues associate intimately with each other. This association is the basis of an orderly process that ultimately reduces the chromosome number to the haploid state. The reduction in chromosome number occurs in such a way that each of the resulting haploid cells receives exactly one member of each chromosome pair.

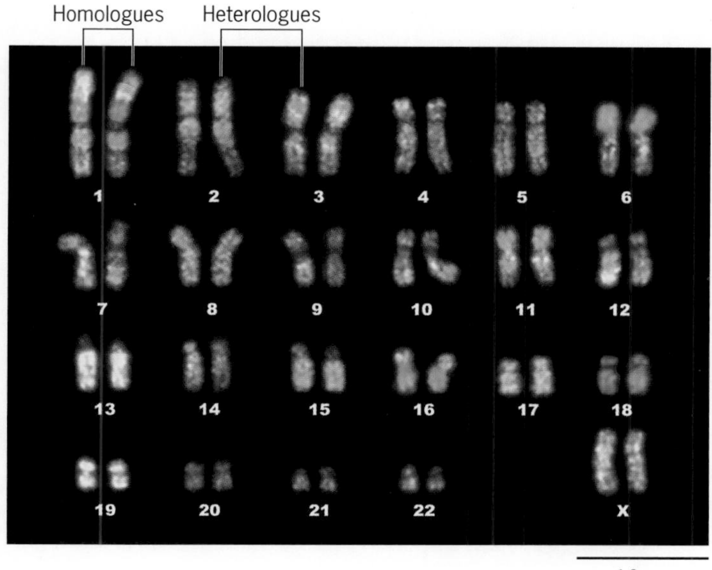

Figure 2.8 The 23 pairs of homologous chromosomes found in human cells.

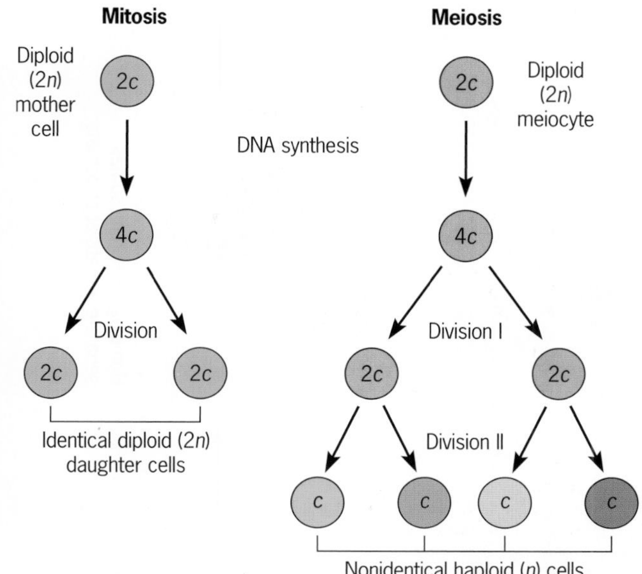

Figure 2.9 Comparison between mitosis and meiosis; *c* is the haploid amount of DNA in the genome.

The process of meiosis involves two cell divisions (**FIGURE 2.9**). Chromosome duplication, which is associated with DNA synthesis, occurs prior to the first of these divisions. It does not occur between the two divisions. Thus, the progression of events is: chromosome duplication → meiotic division I → meiotic division II. If we denote the amount of DNA in a haploid set of chromosomes as c, in sequence these events double the amount of DNA (from $2c$ to $4c$), cut it in half (from $4c$ to $2c$), then cut it in half again (from $2c$ to c). The overall effect is to reduce the diploid chromosome number ($2n$) to the haploid chromosome number (n).

MEIOSIS I

The events in the two meiotic divisions are illustrated in **FIGURE 2.10**. The first meiotic division is complicated and protracted. When it begins, the chromosomes have already been duplicated; consequently, each of them consists of two sister chromatids. The prophase of meiosis I—or simply, **prophase I**—is divided into five stages, each denoted by a Greek term. These terms convey key features about the appearance or behavior of the chromosomes.

Leptonema, from Greek words meaning "thin threads," is the earliest stage of prophase I. During leptonema (also referred to as the leptotene stage), the duplicated chromosomes condense out of the diffuse chromatin network. With a light microscope, individual chromosomes can barely be seen. With an electron microscope, each of the chromosomes appears to consist of two sister chromatids. As chromosome condensation continues, the cell progresses into **zygonema** (from Greek words meaning "paired threads"). During zygonema (also the zygotene stage), homologous chromosomes come together intimately. This process of pairing between homologues is called **synapsis.** In some species, synapsis begins at the ends of chromosomes and then spreads toward their middle regions. Synapsis is usually accompanied by the formation of a proteinaceous structure between the pairing chromosomes (**FIGURE 2.11**). This structure, called the **synaptonemal complex,** consists of three parallel rods—one associated with each of the chromosomes (called the lateral elements) and one located midway between them (called the central element)—and a large number of ladderlike rungs connecting the lateral elements with the central element. The role of the synaptonemal complex in chromosome pairing and in subsequent meiotic events is not fully understood. In some types of meiotic cells it does not even appear. Thus, it may not be absolutely essential for pairing during prophase I. The process by which homologues find each other in prophase I also is not well understood. Recent studies suggest that homologues may actually begin to pair early in meiosis I, during leptonema. This pairing may be facilitated by a tendency for homologous chromosomes to remain in the same region of the nucleus during interphase. Thus, homologues may not have far to go to find each other.

As synapsis progresses, the duplicated chromosomes continue to condense into smaller volumes. The thickened chromosomes that result from this process are characteristic of **pachynema** (from Greek words for "thick threads"). At pachynema (also the pachytene stage), paired chromosomes can easily be seen with a light microscope. Each pair consists of two duplicated homologues, which themselves consist of two sister chromatids. If we count homologues, the pair is referred to as a **bivalent** of chromosomes, whereas if we count strands, it is referred to as a **tetrad** of chromatids. During

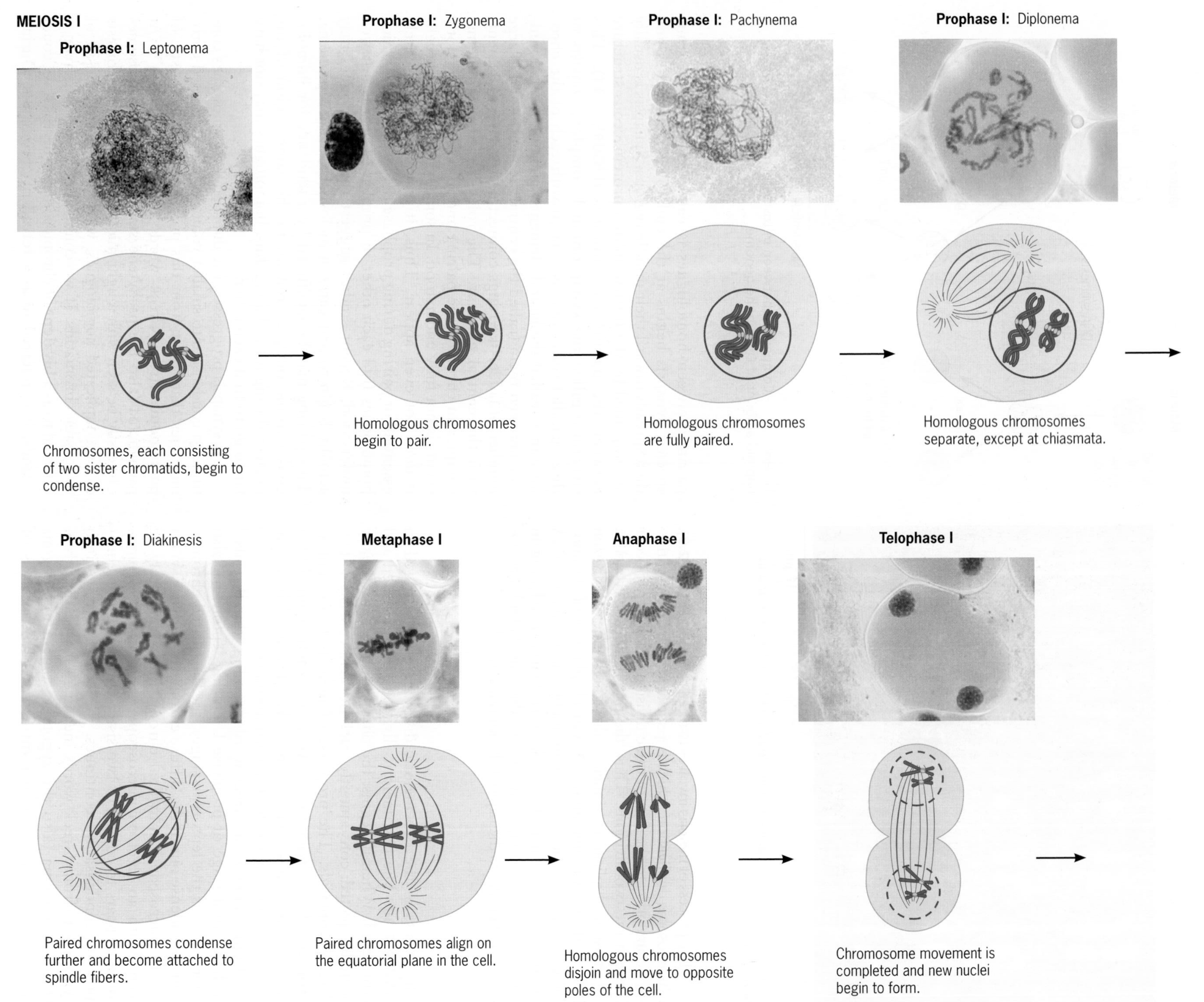
MEIOSIS I
Prophase I: Leptonema
Chromosomes, each consisting of two sister chromatids, begin to condense.
Prophase I: Zygonema
Homologous chromosomes begin to pair.
Prophase I: Pachynema
Homologous chromosomes are fully paired.
Prophase I: Diplonema
Homologous chromosomes separate, except at chiasmata.
Prophase I: Diakinesis
Paired chromosomes condense further and become attached to spindle fibers.
Metaphase I
Paired chromosomes align on the equatorial plane in the cell.
Anaphase I
Homologous chromosomes disjoin and move to opposite poles of the cell.
Telophase I
Chromosome movement is completed and new nuclei begin to form.

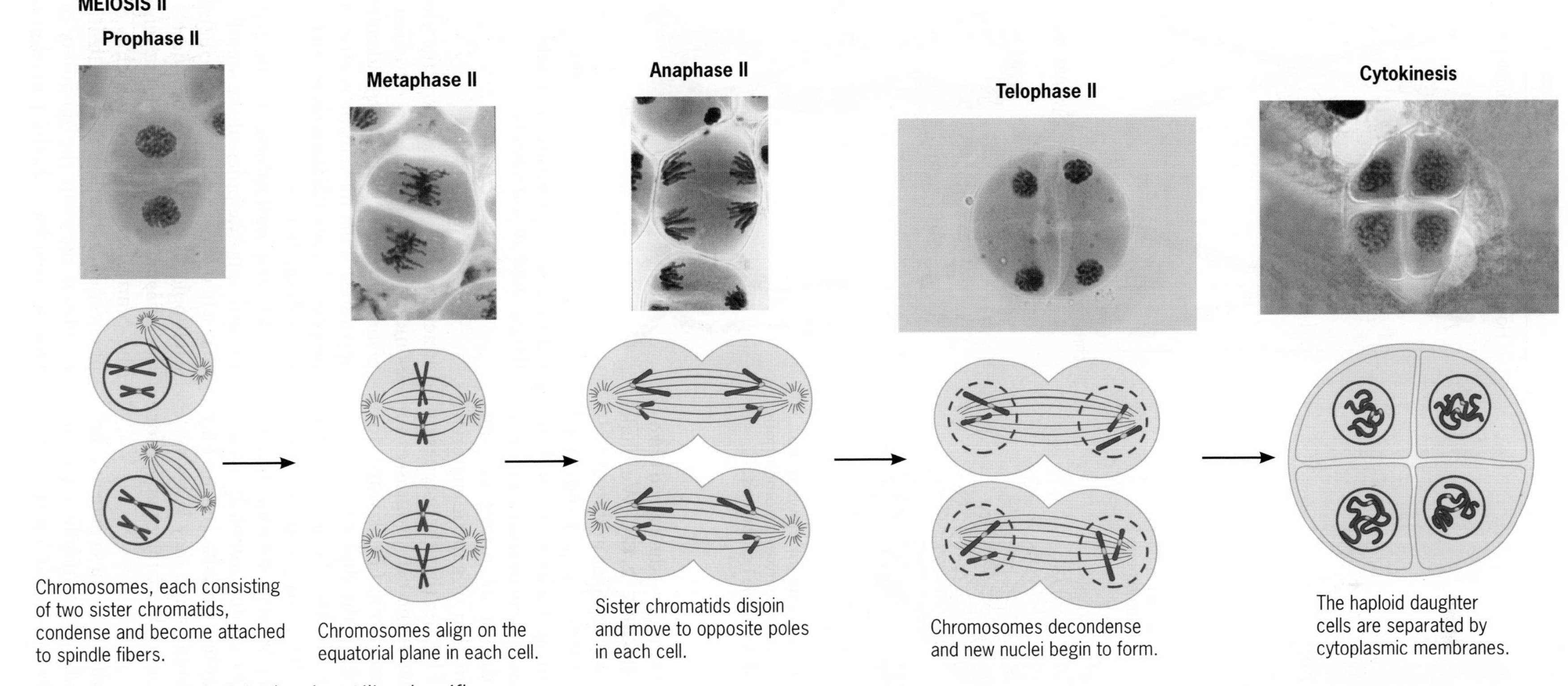
MEIOSIS II
Prophase II
Metaphase II
Anaphase II
Telophase II
Cytokinesis
Chromosomes, each consisting of two sister chromatids, condense and become attached to spindle fibers.
Chromosomes align on the equatorial plane in each cell.
Sister chromatids disjoin and move to opposite poles in each cell.
Chromosomes decondense and new nuclei begin to form.
The haploid daughter cells are separated by cytoplasmic membranes.

Figure 2.10 ▶ Meiosis in the plant *Lilium longiflorum*.

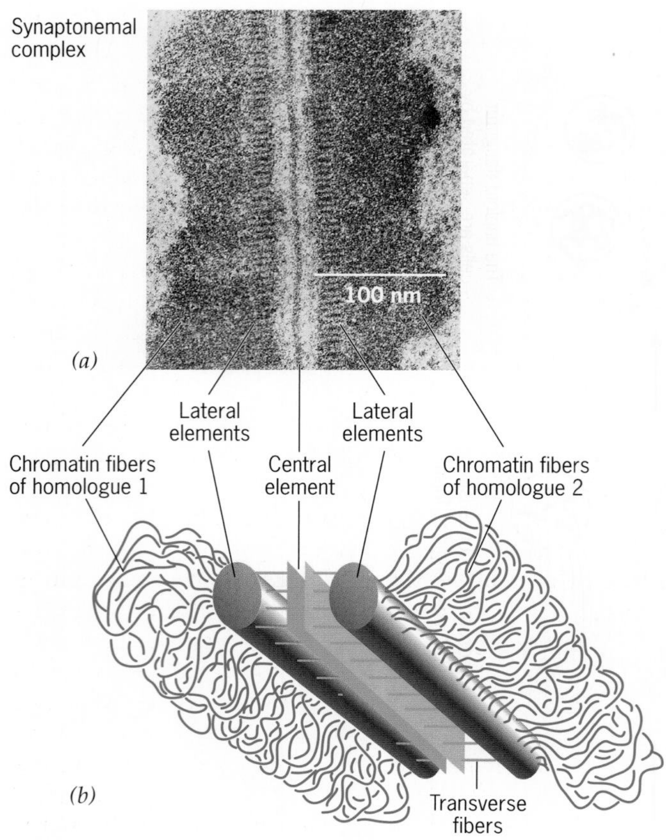

Figure 2.11 ▶ Electron micrograph (*a*) and diagram (*b*) showing the structure of the synaptonemal complex that forms between homologous chromosomes during the zygotene stage of prophase I of meiosis.

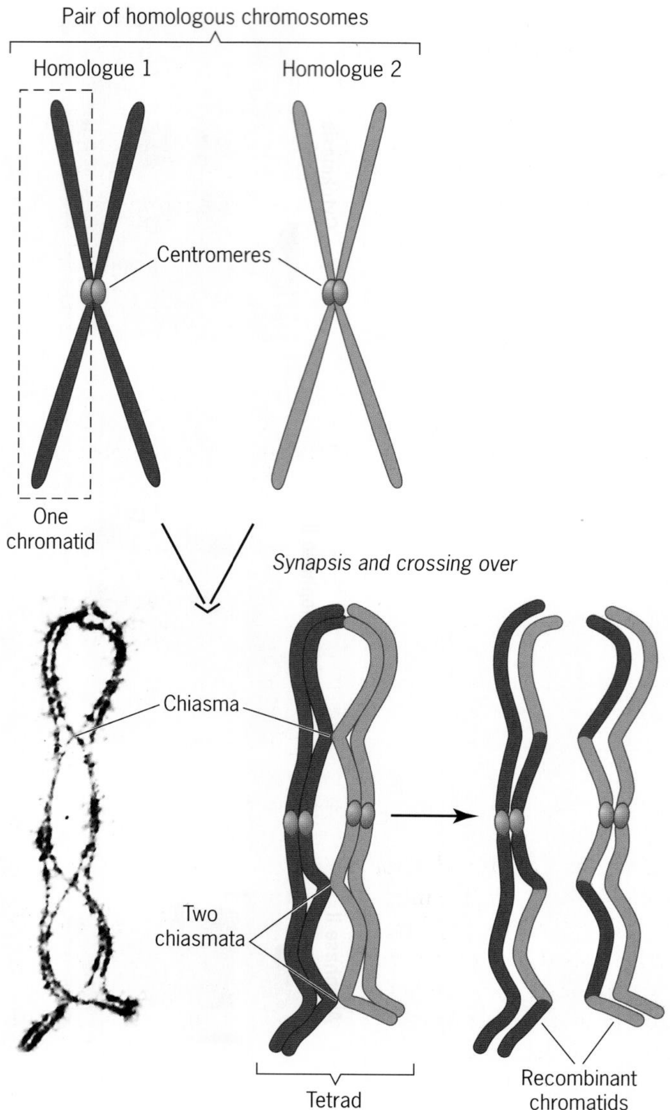

Figure 2.12 ▶ Chiasmata in a bivalent of homologous chromosomes during prophase I of meiosis.

pachynema—or perhaps a bit before or after—the paired chromosomes may exchange material (**FIGURE 2.12**). We shall explore this phenomenon, called **crossing over,** and its consequences in Chapter 7. Here, suffice it to say that individual sister chromatids may be broken during pachynema, and the broken pieces may be swapped between chromatids within a tetrad. The breakage and reunion that occur during crossing over may therefore lead to recombination of genetic material between the paired chromosomes. The fact that these types of exchanges have occurred can be seen as the cell progresses to the next stage of meiosis I, **diplonema** (from Greek words for "two threads"). During diplonema (also the diplotene stage), the paired chromosomes separate slightly. However, they remain in close contact where they have crossed over. These contact points are called **chiasmata** (singular, chiasma, from a Greek word meaning "cross"). Close examination of the chiasmata indicates that each of them involves only two of the four chromatids in the tetrad. The diplotene stage may last a very long time. In human females, for example, it may persist for more than 40 years.

Near the end of prophase I, the chromosomes condense further, the nuclear membrane fragments, and a spindle apparatus forms. Spindle microtubules penetrate into the nuclear space and attach to the kinetochores of the chromosomes. The chromosomes, still held together by the chiasmata, then move to a central plane of the cell that is perpendicular to the axis of the spindle apparatus. This movement is characteristic of the last stage of prophase I, called **diakinesis** (from Greek words meaning "movement through").

During **metaphase I,** the paired chromosomes orient toward opposite poles of the spindle. This orientation ensures that when the cell divides, one member of each pair will go to each pole. At the end of prophase I and during metaphase I, the chiasmata that hold the bivalents together slip away from the centromeres toward the ends of the chromosomes. This phenomenon, called terminalization, reflects the growing repulsion between the members of each chromosome pair. During **anaphase I,** the paired chromosomes separate from each other definitively. This separation, called *chromosome disjunction*, is mediated by the spindle apparatus acting on each of the bivalents in the cell. As the separating chromosomes gather at opposite poles, the first meiotic division comes to an end. During the next stage, called **telophase I,** the spindle apparatus is disassembled, the daughter cells are separated from each other by membranes, the chromosomes are decondensed, and a nucleus is formed around the chromosomes in each daughter cell. In some species, chromosome decondensation is incomplete, the daughter nuclei do not form, and the daughter cells proceed immediately into the second meiotic division. The cells produced by meiosis I contain the haploid number of chromosomes; however, each chromosome still consists of two sister chromatids, which may not be genetically identical because they might have exchanged material with their pairing partners during prophase I.

MEIOSIS II AND THE OUTCOMES OF MEIOSIS

During meiosis II, the chromosomes condense and become attached to a new spindle apparatus **(prophase II).** They then move to positions in the equatorial plane of the cell **(metaphase II),** and their centromeres split to allow the constituent sister chromatids to move to opposite poles **(anaphase II),** a phenomenon called *chromatid disjunction.* During **telophase II,** the separated chromatids—now called chromosomes—gather at the poles and daughter nuclei form around them. Each daughter nucleus contains a haploid set of chromosomes. Mechanistically, meiosis II is therefore much like mitosis. However, its products are haploid, and unlike the products of mitosis, the cells that emerge from meiosis II are not genetically identical.

One reason these cells differ is that homologous chromosomes pair and disjoin from each other during meiosis I. Within each pair of chromosomes, one homologue was inherited from the organism's mother, and the other was inherited from its father. During meiosis I, the maternally and paternally inherited homologues come together and synapse. They are positioned on the meiotic spindle and become oriented randomly with respect to the spindle's poles. Then they disjoin. Half the daughter cells produced by the first meiotic division receive the maternally inherited homologue, and the other half receive the paternally inherited homologue. Thus, from the end of the first meiotic division, the products of meiosis are destined to be different. These differences are compounded by the number of chromosome pairs that disjoin during meiosis I. Each of the pairs disjoins independently. Thus, if there are 23 pairs of chromosomes, as there are in humans, meiosis I can produce 2^{23} chromosomally different daughter cells—that is, more than 8 million possibilities.

Another reason the cells that emerge from meiosis differ is that during meiosis I, homologous chromosomes exchange material by crossing over. This process can create countless different combinations of genes. When we superimpose the variability created by crossing over on the variability created by the random disjunction of homologues, it is easy to see that no two products of meiosis are likely to be the same.

Overall, the two meiotic divisions produce four cells, each with the haploid number of chromosomes. These cells have different fates in different types of organisms. In baker's yeast, *Saccharomyces cerevisiae*, they divide mitotically to produce populations of haploid cells, whereas in other fungi, such as the bread mold *Neurospora crassa*, they divide mitotically to produce filaments of cells called hyphae (singular, hypha), which form the body of the fungus. Under appropriate environmental conditions, these haploid organisms can produce sexual cells, which are effectively their gametes (**FIGURE 2.13**). When two different sexual cells from these organisms unite, they produce a diploid zygote, which usually proceeds into meiosis without any intervening mitotic growth. Thus, the haploid state predominates in the life cycle.

None of these simple organisms has differentiated male and female sexes. Rather, they have mating types. In *Saccharomyces*, the two mating types are denoted *a* and alpha, and in *Neurospora*, they are denoted *A* and *a*. Fertilization always involves the union of sexual cells from opposite mating types.

In lower plants, such as the mosses, the haploid cells derived from meiosis divide mitotically to produce branched filaments that eventually differentiate into tissues such as stems and leaves. When mature, these haploid plants produce gametes—either eggs or sperm. Consequently, they are called **gametophytes.** The gametes unite at fertilization to form diploid zygotes, which divide mitotically and develop into structures called **sporophytes.** In higher plants, the gametophyte is much reduced in size—it consists of just a few haploid cells. Thus, the sporophyte is the conspicuous part of the life cycle.

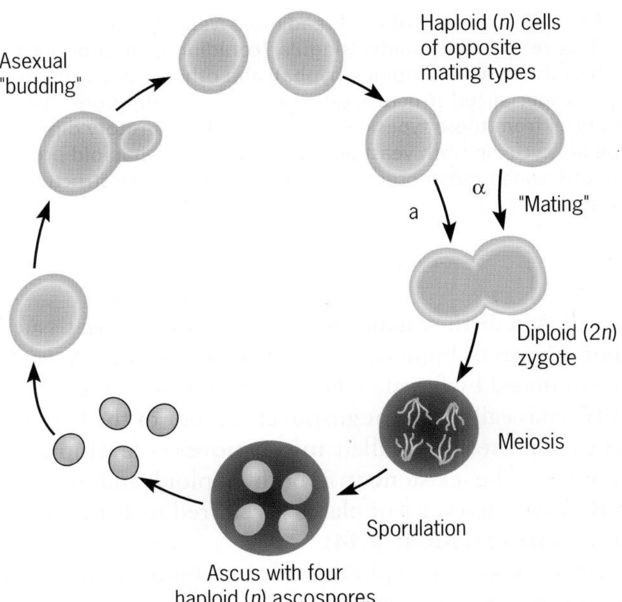

Figure 2.13 ► (*a*) Life cycle of the yeast *Saccharomyces cerevisiae* (n = 16).

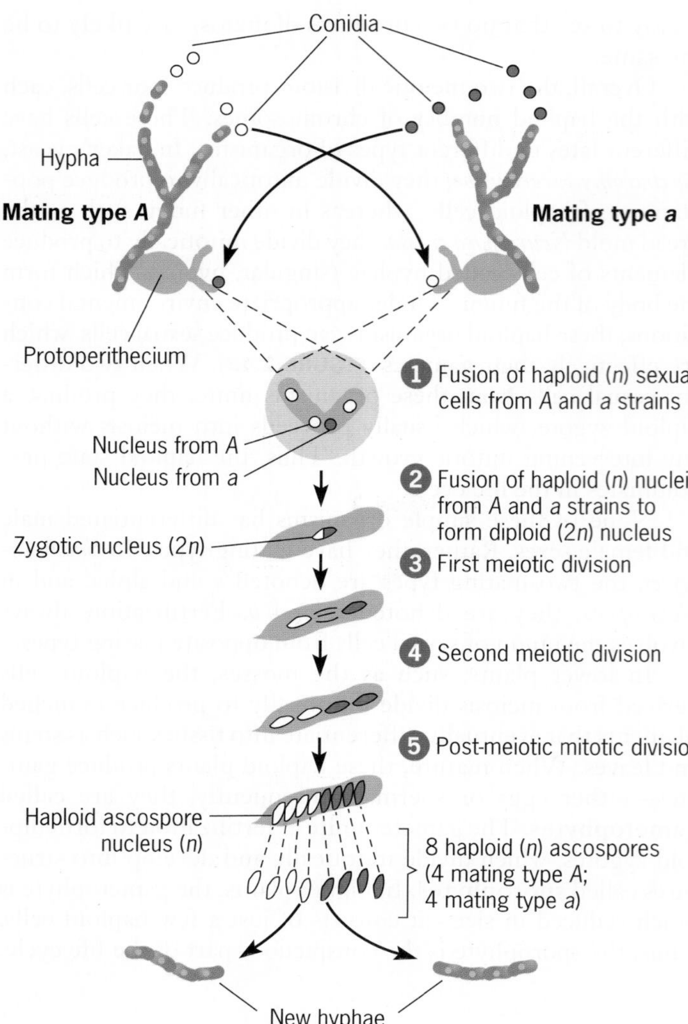

Figure 2.13 ▶ (*b*) Life cycle of the bread mold *Neurospora crassa* ($n = 7$). Sexual cells, called conidia (singular, conidium), are produced at the ends of the hyphae (singular, hypha), and other sexual cells are produced in a specialized structure called the protoperithecium. The fusion of nuclei from these types of cells creates the transitory diploid part of the life cycle. In both yeast and *Neurospora*, the haploid products of meiosis, called ascospores, are contained in a saclike structure called the ascus.

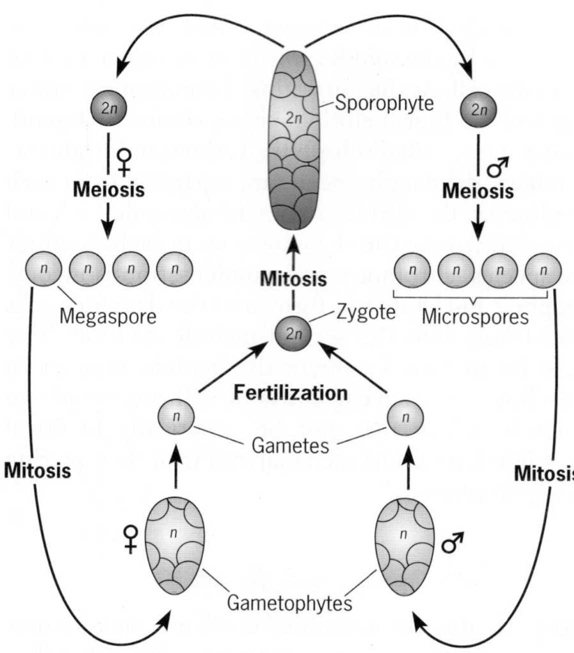

Figure 2.14 ▶ Alternation of generations between haploid gametophytes and diploid sporophytes in plants. In the lower plants, the gametophyte is dominant, whereas in the higher plants, the sporophyte is dominant.

Meiosis occurs in distinct male and female reproductive tissues in the sporophytes of higher plants. Only one of the four haploid cells produced by female meiosis develops into a gametophyte; this cell is called the **megaspore.** All four of the haploid cells from male meiosis—called **microspores**—develop into gametophytes. The existence of both haploid and diploid organisms in the life cycles of plants is referred to as the *alternation of generations* (**FIGURE 2.14**).

Among animals, the haploid products of meiosis develop directly into gametes (**FIGURE 2.15**). Usually, only one of the four haploid cells from female meiosis becomes an egg, or ovum; the other three cells, called polar bodies, degenerate. By contrast, all four of the haploid cells from male meiosis develop into sperm. The formation of gametes, a process called **gametogenesis,** occurs in the gonads of animals. Oogenesis, the formation of eggs, occurs in the ovaries, which are the female gonads, and spermatogenesis, the formation of sperm, occurs in the testes, which are the male gonads. These processes begin when undifferentiated diploid cells, called oogonia or spermatogonia, undergo meiosis to produce haploid cells. The haploid cells then differentiate into mature gametes.

To assess your understanding of the events of meiosis, work through the Focus on Problem Solving: Counting Chromosomes and Chromatids.

KEY POINTS

- Diploid eukaryotic cells form haploid cells by meiosis, a process involving one round of chromosome duplication followed by two cell divisions (meiosis I and meiosis II).
- During meiosis I, homologous chromosomes pair (synapse), exchange material (cross over), and separate (disjoin) from each other.
- During meiosis II, chromatids disjoin from each other.
- In many organisms, the haploid products of meiosis develop directly into gametes.
- In plants, the products of meiosis divide mitotically to form haploid gametophytes.
- The gametophytic phase of a plant's life cycle alternates with the sporophytic phase, which is diploid; meiosis occurs in the sporophyte.

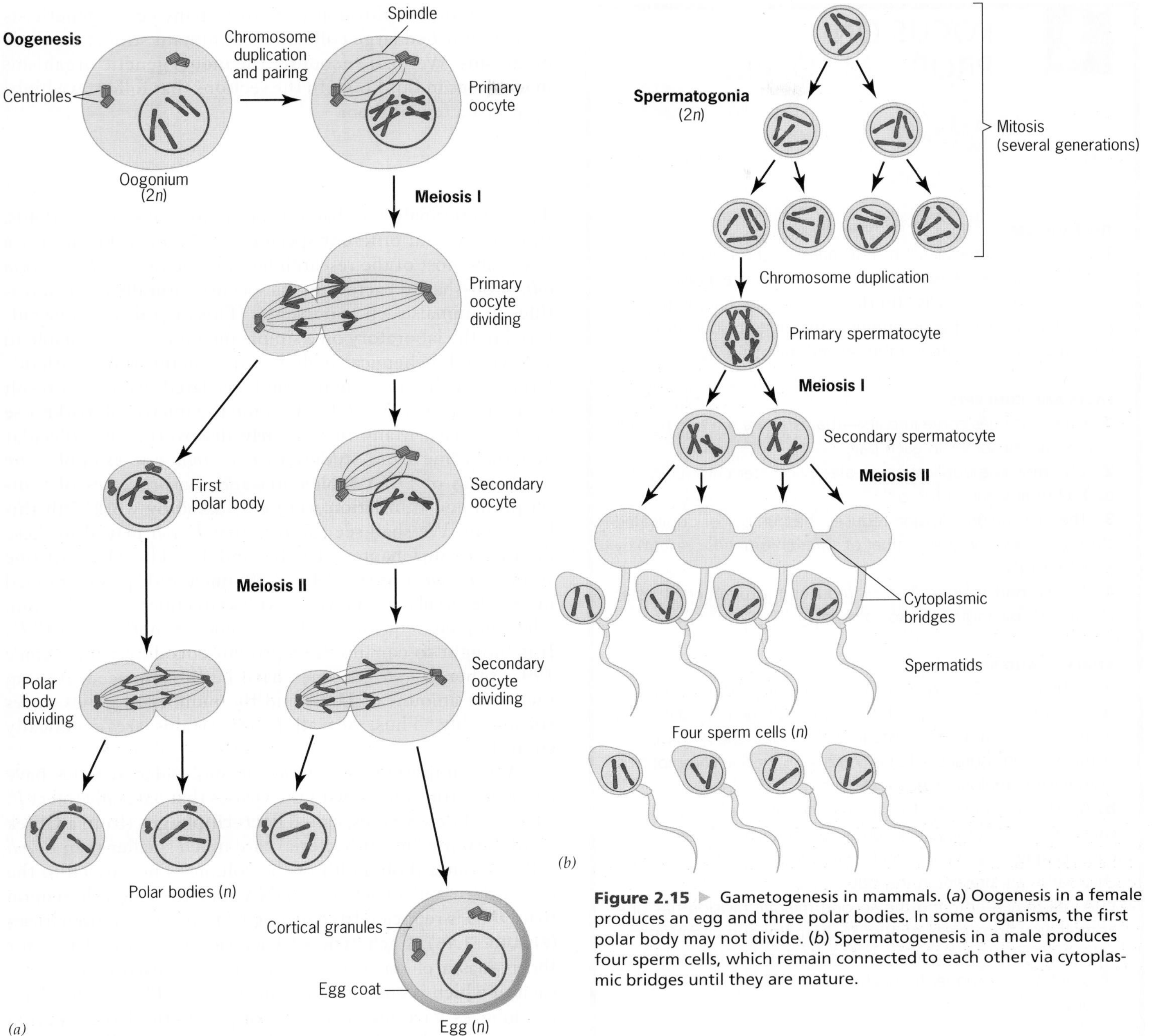

Figure 2.15 ▶ Gametogenesis in mammals. (*a*) Oogenesis in a female produces an egg and three polar bodies. In some organisms, the first polar body may not divide. (*b*) Spermatogenesis in a male produces four sperm cells, which remain connected to each other via cytoplasmic bridges until they are mature.

Genetics in the Laboratory: An Introduction to Some Model Research Organisms

Geneticists focus their research on microorganisms, plants, and animals well suited to experimentation.

When genetics began, the organisms that were used for research were the ones that came to hand from the garden or the barnyard. Some early geneticists branched out to study inheritance in other types of creatures—moths and canaries, for example—and as genetics progressed, research became focused on organisms that were well suited for controlled

▶ FOCUS ON PROBLEM SOLVING Counting Chromosomes and Chromatids

THE PROBLEM

The cat (*Felis domesticus*) has 36 pairs of chromosomes in its somatic cells. (a) How many chromosomes are present in a cat's mature sperm cells? (b) How many sister chromatids are present in a cell that is entering the first meiotic division? (c) In a cell that is entering the second meiotic division?

FACTS AND CONCEPTS

1. Chromosomes come in pairs—that is, there are *two* homologous chromosomes in each pair.
2. Chromosome duplication creates *two* sister chromatids for each chromosome in the cell.
3. The first meiotic division reduces the number of duplicated chromosomes (and the number of sister chromatids present) by a factor of *two*.
4. The second meiotic division reduces the number of sister chromatids by another factor of *two*.

ANALYSIS AND SOLUTION

a. If the cat has 36 pairs of chromosomes in its diploid somatic cells—that is, $2 \times 36 = 72$ chromosomes altogether—a haploid sperm cell, which is an end product of meiosis, should have half as many chromosomes—that is, $72/2 = 36$, or one chromosome from each homologous pair.
b. A cell that is entering the first meiotic division has just duplicated its 72 chromosomes. Because each chromosome now consists of two sister chromatids, altogether $72 \times 2 = 144$ sister chromatids are present in this cell.
c. A cell that is entering the second meiotic division has one homologue from each of the 36 homologous chromosome pairs, and each of these homologues consists of two sister chromatids. Consequently, such a cell has $36 \times 2 = 72$ sister chromatids.

For further discussion go to your *WileyPLUS* course.

experimentation in laboratories or field plots. Today a select group of microorganisms, plants, and animals are favored in genetic research. These creatures, often called **model organisms,** lend themselves well to genetic analysis. For the most part, they are easily cultured in the laboratory, their life cycles are relatively short, and they are genetically variable. In addition, through work over many years, geneticists have established large collections of mutant strains for these organisms. We shall encounter the model genetic organisms many times in this book. In the sections that follow, we highlight just a few of them.

ESCHERICHIA COLI, A BACTERIUM

The genetic analysis of bacteria began in earnest in the 1940s. Initially, several different species were studied, but in just a few years most of the research became focused on *Escherichia coli*, a rod-shaped bacterium that occurs naturally in the intestines of animals such as ourselves. This organism can be cultured in the laboratory on a simple medium, it is amenable to all sorts of biochemical analyses, and mutant strains with different growth requirements can be isolated easily. As a result of these features, *E. coli* has become the microbial workhorse of geneticists. Many of the early discoveries in molecular genetics came from studies with *E. coli*. For example, the mechanism of DNA replication and the processes of transcription and translation were elucidated by work with this bacterium. We shall see the role that *E. coli* played in these discoveries in Chapters 10, 11, and 12. The DNA of one strain of *E. coli* consists of 4.6×10^6 nucleotide pairs arranged in a single circular chromosome with a circumference of 1.4 mm. The complete sequence of this genome was obtained in 1997. It is thought to contain 4288 protein-encoding genes. Since 1997, other *E. coli* genomes have been sequenced. Among them, the amount of DNA and the number of genes varies considerably. Thus, not all *E. coli* strains are genetically equivalent.

Many important discoveries in molecular genetics have also come from studies with the viruses that naturally infect *E. coli* cells. These viruses, called **bacteriophages** (from a Greek word meaning "to eat bacteria") are much smaller than *E. coli* cells—about a thousandth of the volume. They attach to the cell surface and inject their DNA into the cell, whereupon that DNA is replicated and expressed to produce more viruses (**FIGURE 2.16**). Each type of bacteriophage—or phage, for short—has a characteristic structure composed of proteins, each of which is encoded by a phage gene. The smallest bacteriophages have just a handful of genes; the largest have a couple of hundred. Because of their relative simplicity, bacteriophages are beautifully suited for studying basic genetic processes.

E. coli cells and the viruses that infect them are tiny creatures that can be cultured in the laboratory to produce tens of billions of their own kind in a short period of time. These large population sizes allow researchers to screen efficiently for rare events such as the occurrence of a particular kind of mutation. Obviously, research with such organisms can advance more rapidly than research with large organisms with a limited reproductive ability. We shall explore ways of analyzing the genes of *E. coli* and its viruses in Chapter 8.

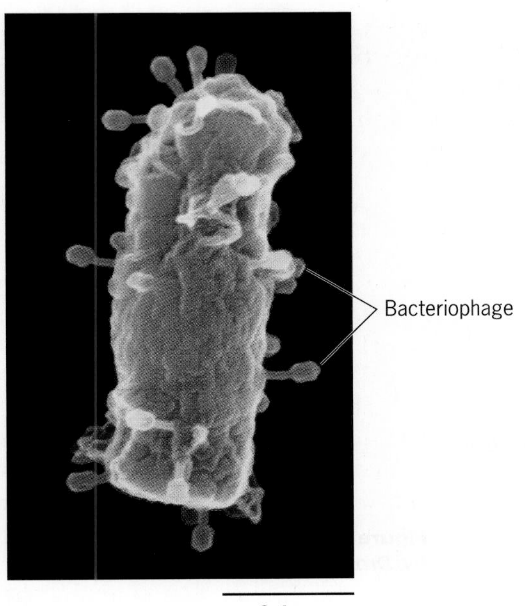

Figure 2.16 ▶ Bacteriophage attaching to the surface of an *E. coli* cell.

SACCHAROMYCES CEREVISIAE, BAKER'S YEAST

Baker's yeast came into genetics research about the same time as *E. coli*. However, long before it was commonplace in genetics laboratories, this organism was used in kitchens as a leavening agent for making bread. Yeast is a unicellular fungus, although under some conditions, its cells divide to form long filaments, which resemble the hyphae of other fungi. Like *E. coli*, yeast cells can be cultured on simple media in the laboratory, and large numbers of cells can be obtained from a single mother cell in just a few days. In addition, mutant strains with different growth characteristics can be readily isolated.

The great difference between yeast and *E. coli* is that yeast is a eukaryote. Its 16 chromosomes are contained within a defined nucleus, and each chromosome consists of a single DNA molecule combined with a large number of protein molecules. Paradoxically, some of these DNA molecules are smaller than the DNA molecule that makes up the *E. coli* chromosome. Yeast cells possess mitochondria, which also contain DNA molecules—although they are much smaller than the ones found in the nucleus, and they have a complex system of intracellular membranes. The yeast genome consists of 12×10^6 nucleotide pairs, and it is thought to comprise 6268 genes. Its sequence was reported in 1996—the first eukaryotic genome to be analyzed completely.

S. cerevisiae reproduces both sexually and asexually (**FIGURE 2.13**). Asexual reproduction occurs by a process called budding, which involves a mitotic division of the haploid nucleus. After this division, one daughter nucleus moves into a small "bud" or progeny cell. Eventually, the bud is separated from the mother cell by cytokinesis. Sexual reproduction in *S. cerevisiae* occurs when haploid cells of opposite mating types (denoted *a* and alpha) come together—an event referred to as mating—to fuse and form a diploid cell, which then undergoes meiosis. The four haploid products of meiosis are created in a sac called the ascus (plural, asci), and each of the products is called an ascospore. By dissecting this sac, a researcher can isolate each meiotic product and place it in a culture dish to start a new yeast colony. We shall see how ascospores from yeast and other fungi are analyzed in Chapter 7.

INVERTEBRATE ANIMALS: *DROSOPHILA MELANOGASTER*, A FRUIT FLY, AND *CAENORHABDITIS ELEGANS*, A ROUND WORM

With untold millions of species, insects are the most diverse group of animals on earth. It therefore seems appropriate that one of the model organisms for genetic analysis is an insect. The fruit fly *Drosophila melanogaster* came on the genetics scene in 1909, decades before bacteria and fungi. It appealed to researchers because it is easy to rear in the laboratory, it has a relatively short life cycle (about 10 days at 25°C), and it reproduces prolifically. These features continue to make it an attractive animal for genetic studies.

The adult fly is about 2 mm long. It possesses a complex nervous system and many specialized tissues and organs. Like all insects, an adult *Drosophila* has three pairs of legs. However, unlike most insects, it has only one pair of wings; the second pair of wings has been modified into small appendages called halteres, which help the animal to balance during flight. The surface of the adult body is covered with sensory hairs and bristles, which are connected to the nervous system. Other prominent sensory organs—the eyes and the antennae—are located on the head. The reproductive organs are located in the abdomen.

A female *Drosophila* can produce hundreds of eggs. When fertilized, each egg becomes an embryo, which hatches out of the eggshell to become a wormlike *larva* (plural, larvae) (**FIGURE 2.17**). The larva feeds voraciously for about a week. At intervals during this period it sheds its skin to allow for increases in body size. Each larval stage between these molts is called an *instar*. The third larval instar is the last stage before metamorphosis into the *imago*, or adult fly. The skin of the third instar larva hardens into a case, and the animal becomes a *pupa* (plural, pupae), a stage during which the larva's tissues are radically reorganized. Packets of cells called **imaginal discs** grow and differentiate into adult structures such as eyes, wings, and legs. In about four days, an adult fly emerges from the pupal case.

Thousands of mutant strains of *Drosophila* have been obtained in the nearly hundred years this animal has been studied. Some show unusual eye or body colors; others have an abnormal anatomy—short wings, extra legs, misplaced bristles, and so on. In one strain, the halteres are converted into wings. The anatomical complexity of *Drosophila* provides geneticists with opportunities to study how genes control the formation of body parts. It also allows them to investigate the processes by which a fertilized egg develops into an embryo and eventually into an adult.

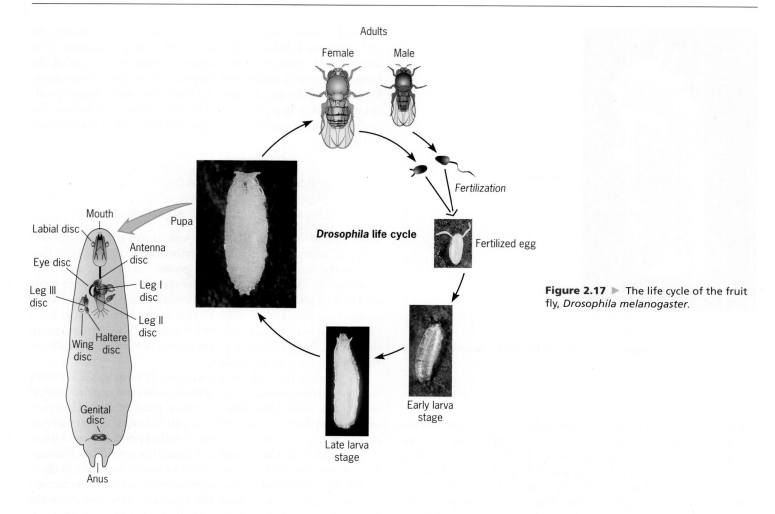

Figure 2.17 ▶ The life cycle of the fruit fly, *Drosophila melanogaster*.

Another invertebrate, the round worm *Caenorhabditis elegans* (**FIGURE 2.18**)—known to geneticists simply as *C. elegans*—also has played a major role in research. In the 1960s, Sydney Brenner, who initially worked with bacteriophages, chose this small, free-living soil nematode as a model for studying animal development. Adult *C. elegans* are about 1 mm long, roughly the same size as a *Drosophila* egg. They reproduce quickly and prolifically and can be reared easily on agar plates that have been seeded with *E. coli* bacteria as food. Under optimal conditions, the life cycle—egg, larval stages, and adult—is completed in a mere three days.

C. elegans is a hermaphroditic species—that is, one in which individual organisms are capable of producing both sperm and eggs. Because there is no system of self-incompatibility, sperm can fertilize eggs from the same animal. *C. elegans* populations also contain some animals that are strictly male, and these males mate with hermphrodites to produce offspring. One of the great advantages of *C. elegans* as an experimental animal is that it is transparent. Researchers can watch cells dividing and moving from the time an egg is fertilized. Thus, they can trace the formation of tissues and organs over the course of development.

Both the *Drosophila* and *C. elegans* genomes have been sequenced. These genomes are much larger than the yeast genome—100×10^6 nucleotide pairs for *C. elegans* and 170×10^6

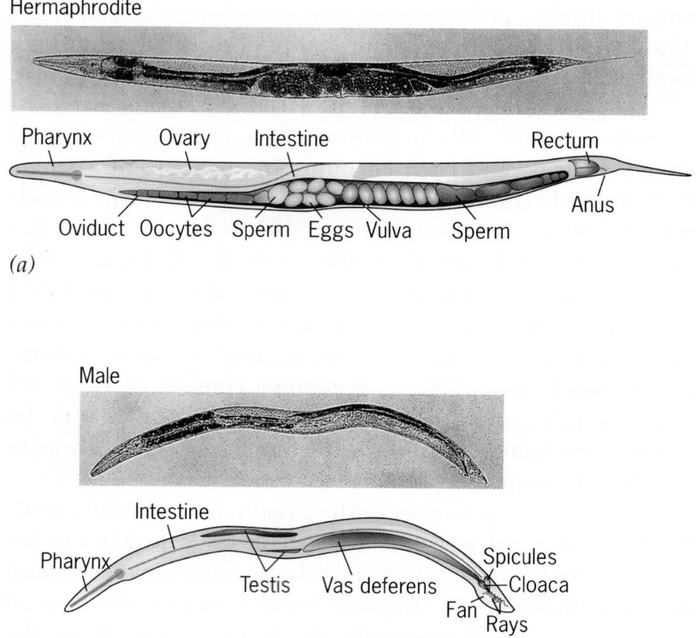

Figure 2.18 ▶ Hermaphrodite (*a*) and male (*b*) *Caenorhabditis elegans*.

(a) (b)

Figure 2.19 ▶ The mouse, *Mus musculus*. (*a*) Wild-type mouse. (*b*) Mutant nude mouse.

nucleotide pairs for *Drosophila*—and they contain more genes—an estimated 20,516 in *C. elegans* and 13,792 in *Drosophila*. However, the number of genes in these multicellular animals is only two to three times greater than the number of genes in yeast. Geneticists are trying to reconcile this relatively modest increase in gene number with the dramatic increase in biological complexity.

VERTEBRATE ANIMALS: *MUS MUSCULUS*, THE MOUSE, AND *DANIO RERIO*, THE ZEBRA FISH

The analysis of human DNA has been aided by parallel studies on the DNA of other vertebrate species, including preeminently the mouse *Mus musculus* (**FIGURE 2.19**) and the zebra fish *Danio rerio* (**FIGURE 2.20**). These two model organisms can be subjected to genetic experimentation, and many of the discoveries that have come from work with them are relevant to genetic investigations of our own species.

The mouse has been especially important because of its widespread use in many areas of biomedical research. Mice have been used to study health and disease, and they have been the subjects of innumerable projects to ascertain the effects of drugs, chemicals, foods, and other materials relevant to human health. Mouse genetics began early in the twentieth century with studies on the inheritance of coat color. Since those days, it has developed into an impressive enterprise. Special strains have been produced by intensive inbreeding to provide researchers with animals that are genetically uniform, and thousands of mutant strains have been collected to identify genes and map chromosomes. Sequencing the mouse genome has provided a vantage point from which researchers can assess the significance of sequences in the human genome. The lineages that gave rise to mice and humans diverged from a common ancestor about 80 million years ago. If a sequence has been conserved in these two genomes over all this evolutionary time, it must have an important function. Thus, comparing the genomes of mice and humans—the practice of comparative genomics—provides a way of ascertaining which sequences are functionally important.

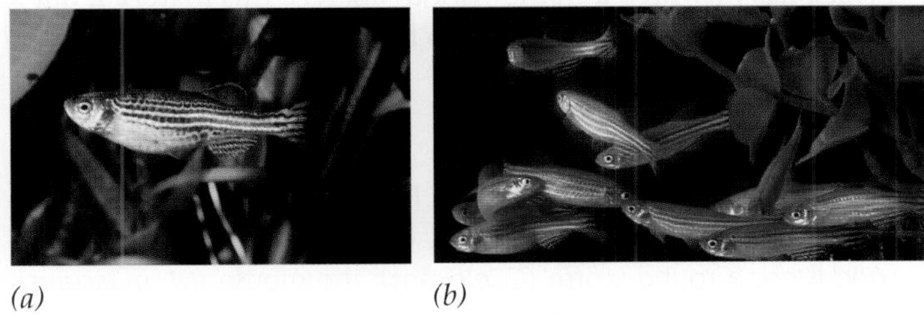

(a) (b)

Figure 2.20 ▶ The zebra fish, *Danio rerio*. Fish from the wild (*a*), and fish that have been genetically transformed with a gene from a sea anemone (*b*). The anemone gene produces a protein that fluoresces red when illuminated.

The zebra fish was developed as a model genetic organism in the late 1960s and early 1970s by George Streisinger, another bacteriophage geneticist who switched to the study of eukaryotes. This small, freshwater fish is native to southern Asia and is a popular choice for household aquariums. Zebra fish are easily bred, and under optimal conditions, they can complete a generation in five to six months. Streisinger believed that the zebra fish could be used as a model for the study of vertebrate development, especially because the eggs are transparent and because they are fertilized externally. His belief was realized in the 1990s when researchers in several countries began projects to identify genes in this organism and to investigate how they control development.

The mouse and zebra fish genomes have both been sequenced. The mouse genome contains about 2.9×10^9 nucleotide pairs of DNA and comprises an estimated 25,396 genes. The zebra fish genome contains approximately 1.6×10^9 nucleotide pairs of DNA and is thought to comprise 23,524 genes.

ARABIDOPSIS THALIANA, A FAST-GROWING PLANT

By capturing solar energy and storing it in simple sugars such as glucose, plants play an indispensable role in the web of life. As a by-product of this process, they also generate oxygen, which we and all other nonphotosynthetic organisms breathe. Garden plants were the first organisms to be studied genetically. Today geneticists focus their attention on *Arabidopsis thaliana*, a weed sometimes called the mouse ear cress. This member of the crucifer family is related to food plants such as radish, cabbage, and canola. However, it has no agronomic or horticultural value.

Arabidopsis is small, and it has a relatively short generation time (about five weeks). Like Mendel's garden peas, *Arabidopsis* is a self-fertilizing species; however, different strains can be cross-fertilized in the laboratory to produce hybrids. The male gametes or microspores are produced from microspore mother cells by meiosis in **anthers,** which are atop the **stamens** within *Arabidopsis* flowers (**FIGURES 2.21, 2.22**). Each microspore then undergoes mitosis to produce a mature pollen grain, which contains two generative or sperm cells located within a vegetative cell; each of these cells contains a haploid nucleus. The female gametes, or megaspores, are produced from megaspore mother cells by meiosis in the **ovary,** which is located within the **pistil** at the center of the flower. Meiosis in this female tissue produces four cells; however, three subsequently degenerate, leaving only one functional megaspore. The haploid nucleus in each megaspore undergoes three mitotic divisions to produce an immature embryo sac containing eight haploid nuclei. Cytokinesis then occurs, creating three antipodal cells, two synergid cells, and an egg cell. Two polar nuclei remain in the large central cell of the embryo sac. These polar nuclei subsequently fuse to form a diploid secondary endosperm nucleus, and the three antipodal cells degenerate.

When a mature pollen grain lands on the stigma atop the pistil, a pollen tube grows down through the style to an egg cell within the ovary. In plants, fertilization involves two events. (1) The diploid zygote, which will grow into an embryo, is formed when one sperm cell from the pollen tube fuses with the egg cell of the female gametophyte. (2) The triploid endosperm nucleus is formed when the other sperm cell nucleus combines with the diploid secondary endosperm nucleus of the female gametophyte. The triploid endosperm tissue, which develops from this second fertilization event, feeds the embryo while the seed is germinating.

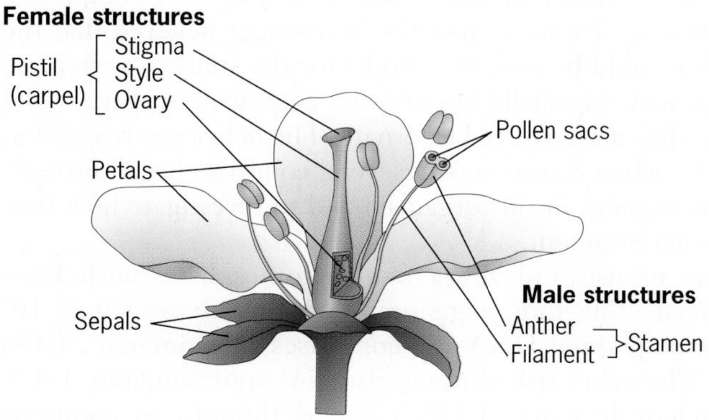

Figure 2.21 ▶ Male and female reproductive organs in a typical flower.

The genome of *Arabidopsis* consists of 157×10^6 nucleotide pairs of DNA, distributed over five chromosomes, and it is thought to contain 27,706 genes. Many of these genes are present in other plant species, including agriculturally significant crops such as maize, rice, and wheat. Thus, the genetic analysis of *Arabidopsis* is providing a framework to understand how genes function in important food plants. This analysis has already yielded a wealth of information about plant development, disease resistance, photosynthesis, cold tolerance, and many other phenomena.

HOMO SAPIENS, OUR OWN SPECIES

The study of heredity in human beings began in the nineteenth century with statistical analyses of traits such as height and weight in parents and offspring. In 1900, when Mendel's discoveries came to light, the emphasis shifted to studying the inheritance of simple, discrete traits in families. Often these traits were diseases such as phenylketonuria, which arises from the faulty metabolism of amino acids, or hemophilia, which arises from a failure of the blood-clotting mechanism. With this approach, many human genes were identified, but progress was understandably slow. Human beings cannot be subjected to genetic experimentation like *Drosophila* or yeast. In this respect, *Homo sapiens* is not a model organism. However, the desire to learn about the genetic material of our own species is very strong, and great efforts have been expended to satisfy this desire in spite of many obstacles.

One breakthrough came in the middle of the twentieth century, when biologists learned to grow human cells in culture; see A Milestone in Genetics: Culturing Human Cells. Such cells can be manipulated in all sorts of experiments. Another breakthrough came in the last quarter of the twentieth century, when researchers learned how to isolate segments of human DNA and propagate them inside *E. coli* cells—a technique called *DNA cloning*. Ultimately, this technology paved the way for the Human Genome Project. The human genome is not the largest or most complex genome on earth. Its 3.2×10^9 nucleotide pairs contain 20,000 to 25,000 genes, as well as lots of nongenic DNA. The Human Genome Project, allied to other genome sequencing projects, has provided a framework to analyze this material, but a great deal of effort is still needed to fill in all the details.

KEY POINTS

- The bacterium *E. coli* is the premier prokaryote for genetic analysis.
- Model eukaryotes include yeast *(S. cerevisiae)*, a fruit fly *(D. melanogaster)*, a round worm *(C. elegans)*, the mouse *(M. musculus)*, the zebra fish *(D. rerio)*, and a fast-growing plant *(A. thaliana)*.
- Techniques such as cell culture and DNA cloning have made it possible to study the genetic material of human beings and many other organisms.

A MILESTONE IN GENETICS: **Culturing Human Cells**

The model genetic organisms are small and have short life cycles. As a result, they can be reared cheaply in large numbers, and multiple generations—parents, offspring, grand-offspring, and so on—can be examined in weeks or months rather than in years. The easy husbandry and the ability to study members from different generations in a reasonable period of time greatly facilitate progress in genetic research.

Unlike the model organisms, our own species is large and has a long life cycle. Anybody inclined to perform genetic research on human beings would have to become reconciled to the difficulties and expenses associated with caring for human subjects, as well as the long time frames needed to obtain results. The prospects for success in this kind of research are not very good, and besides, not too many people are favorably disposed toward it—either as researchers or as research subjects. Research involving human subjects is regulated by moral and ethical principles. Sometimes scientists step outside the boundaries established by these principles, and when they do, we judge them to be aberrant, demented, or evil. Fortunately, scientists have been able to devise ways to investigate genetic phenomena outside the human organism. These *in vitro* approaches use human cells growing in culture.

Cell culture techniques were developed during the first half of the twentieth century. Over several decades, researchers learned to grow cells from a variety of organisms, including mice, chickens, and other vertebrates, in culture tubes. Complex media that contained blood serum or fluid extracted from embryos, together with an assortment of salts, were required to grow these cells *in vitro*. Many attempts were also made to culture human cells, but success was limited. The best results were obtained when the cells were derived from human cancers. For instance, in 1951 George Gey, a professor at the Johns Hopkins School of Medicine, established a culture from the cancerous cervical cells of a 30-year-old woman named Henrietta Lacks. These cells flourished in the culture medium that Gey used. Samples from the culture were taken to establish subcultures, and in just a few years, the HeLa cell line (**FIGURE 1**) was being grown in laboratories all over the world. As Gey put it in a lecture he delivered in 1955: "[T]he HeLa carcinoma strain grows abundantly and has been maintained for over four years. . . . It has been grown in our laboratory and in many others throughout the world in a variety of diluted sera with and without synthetic supplements. Its wide use has opened up a number of new opportunities in the study of living human cancer cells and their responses to various agents."[1]

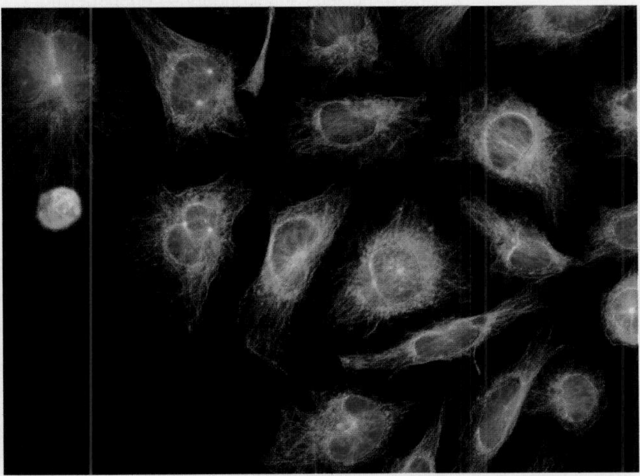

Figure 1 ▶ HeLa cells growing in culture.

Many other types of human cell lines have since been established—most, like the HeLa cell line, being derived from cancerous cells. However, it is also possible to culture noncancerous cells *in vitro*, although these cells may grow for only a limited time. Both cancerous and noncancerous human cells have been used in countless experiments to study cell structure, metabolism, growth, division, and communication, as well as the processes of differentiation and development. In genetics, they have been used to investigate the structure of chromosomes, to localize genes on chromosomes, and to analyze the expression of genes and the functions of their products. In effect, geneticists have used cultured human cells to study genetic phenomena just as they have used microorganisms such as *E. coli* and yeast.

Recently, researchers have begun to focus their attention on ways to isolate and culture human stem cells. These special cells can be obtained from certain kinds of adult tissues as well as from embryos. With appropriate stimulation, they can be induced to differentiate in culture into specialized cell types, such as muscle fibers, lymphocytes, or neurons. Because of their ability to form diverse kinds of cells, stem cells may have the potential to produce replacements for worn-out tissues and body parts. Thus, they are much in the news. However, the way in which these cells are obtained has provoked controversy. Many people object to the destruction of human embryos to obtain stem cells. Others believe that it is legitimate to procure stem cells from early human embryos. They have no quarrel with the procedure, and furthermore, they believe that public funds should be made available to pay for research with stem cells of embryonic origin. Recently, scientists in Japan and the United States have had some success in altering differentiated human cells so that they behave like stem cells. This technical advance may ultimately provide a way to sidestep the reservations that many people have about using embryonic stem cells in research and medicine.

QUESTIONS FOR DISCUSSION

1. Noncancerous cells can be grown in culture for 50 to 100 generations. Then they stop dividing, and eventually they die. Cancerous cells can often be grown indefinitely in culture. What might explain the different growth characteristics of cancerous and noncancerous cells? What does the finite life span of noncancerous cultured cells imply for the phenomenon of aging?
2. What are some of the potential therapeutic uses of stem cells? What is the evidence that some of these potential therapies might actually work?

[1]Gey, G. O. 1954–1955. Some aspects of the constitution and behavior of normal and malignant cells maintained in continuous culture. *Harvey Lectures* 50:154–229.

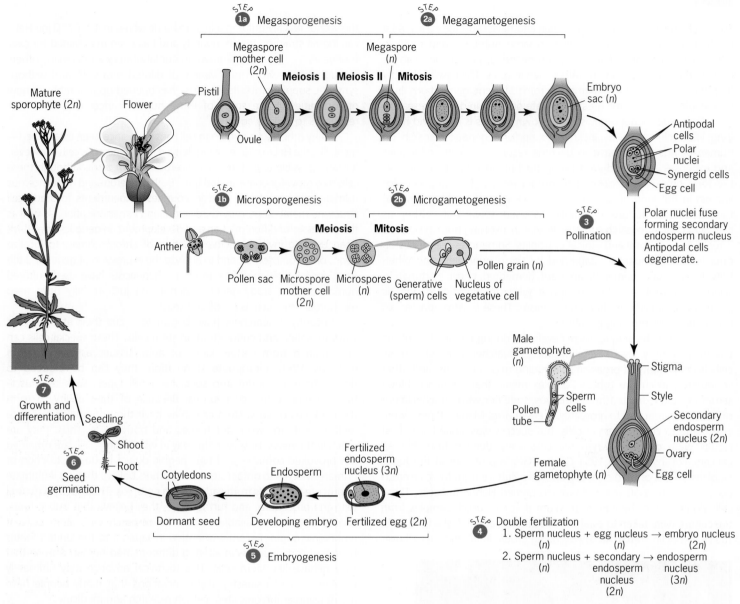

Figure 2.22 ► The life cycle of the model plant, *Arabidopsis thaliana*.

► Basic Exercises

ILLUSTRATE BASIC GENETIC ANALYSIS

1. Among the model organisms discussed in this chapter, which are prokaryotes and which are eukaryotes? Do viruses fit into this classification scheme?

Answer: *E. coli* is a prokaryote; *S. cerevisiae*, *D. melanogaster*, *C. elegans*, *M. musculus*, *D. rerio*, *A. thaliana*, and *H. sapiens* are all eukaryotes. Viruses are found in both prokaryotic and eukaryotic cells; however, by themselves, viruses are neither prokaryotes nor eukaryotes.

2. Identify the stages of mitosis in the following drawings.

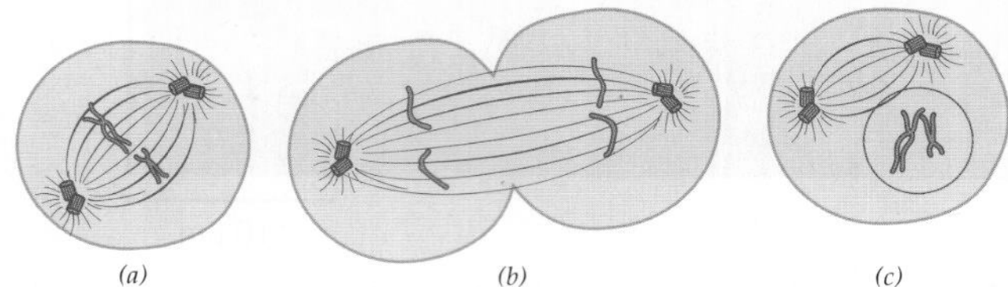

Answer: (*a*) metaphase; (*b*) anaphase; (*c*) prophase

3. Why does a diploid mother cell that undergoes meiosis produce four haploid cells?

Answer: During meiosis, chromosome duplication precedes two division events. If the number of chromosomes in the diploid mother cell is $2n$, then after duplication the cell contains $4n$ chromatids. During the first meiotic division, homologous chromosomes pair and then are separated into different daughter cells, each of which receives $2n$ chromatids. During the second meiotic division, the centromere that holds the two chromatids of each chromosome together splits and the chromatids are separated into different daughter cells. Each of the four cells resulting from these successive meiotic divisions therefore contains n chromatids (now called chromosomes). Thus, the diploid state of the mother cell is reduced to the haploid state in the four cells that emerge from meiosis.

4. Identify the stages of prophase I of meiosis in the following drawings.

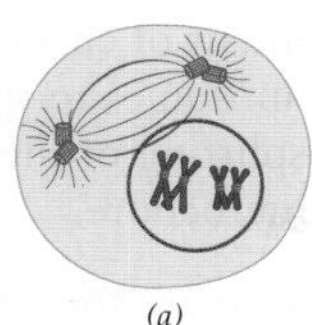
(a)

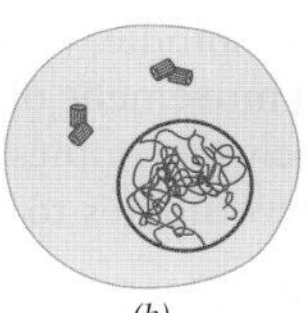
(b)

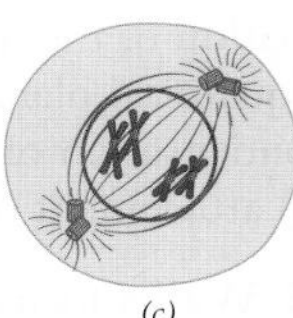
(c)

Answer: (*a*) diplonema; (*b*) leptonema; (*c*) diakinesis

5. Twenty pairs of chromosomes are present in a somatic cell of the mouse. How many sister chromatids are present in (a) a primary oocyte, (b) a secondary spermatocyte, (c) a mature sperm cell?

Answer: (a) 80, because each of the 40 chromosomes (20 pairs × 2 chromosomes/pair) had been duplicated prior to the cell's entry into meiosis I; (b) 40, because homologous chromosomes (each still consisting of two sister chromatids) were apportioned to different cells during the first meiotic division; (c) 20, the haploid chromosome number.

► Testing Your Knowledge

INTEGRATE DIFFERENT CONCEPTS AND TECHNIQUES

1. What are the principal differences between mitosis and meiosis?

Answer: In mitosis, one division event follows one round of chromosome duplication. In meiosis, two division events follow one round of chromosome duplication. Furthermore, during the first meiotic division, homologous chromosomes pair with each other. This homology-based pairing does not normally occur during mitosis. The two cells produced by a mitotic division are identical to each other and to the mother cell from which they were derived. The four cells produced by the two successive meiotic divisions are not identical to each other or to the mother cell from which they were derived. When a diploid cell undergoes mitosis, the two cells derived from it will also be diploid. When a diploid cell undergoes meiosis, the four cells derived from it will be haploid.

2. *C. elegans* hermaphrodites have five pairs of chromosomes. How many chromosomes are present (a) in a sperm cell from a hermaphrodite? (b) in a fertilized egg from a hermaphrodite? How many sister chromatids are present in a hermaphrodite's cell (c) that is entering the first meiotic division? (d) that is entering the second meiotic division? (e) that has completed the second meiotic division?

Answer: (a) 5, because sperm are haploid. (b) 10, because a fertilized egg contains chromosomes from the egg and the sperm that fertilized it. (c) 20, because each of the 10 chromosomes in a cell entering meiosis I has been duplicated to produce two sister chromatids. (d) 10, because homologous chromosomes have been apportioned to different cells during the first meiotic division; however, the sister chromatids of each homologue are still held together by a common centromere. (e) 5, because the end products of meiosis are haploid.

3. A human sperm cell contains about 3.2×10^9 nucleotide pairs of DNA. How much DNA is present in each of the following: (a) a primary human spermatocyte; (b) a secondary human spermatocyte; (c) the first polar body produced by division of a primary oocyte?

Answer: (a) $4 \times 3.2 \times 10^9 = 12.8 \times 10^9$ nucleotide pairs because a primary spermatocyte contains the $4c$ amount of DNA; (b) $2 \times 3.2 \times 10^9 = 6.4 \times 10^9$ nucleotide pairs because a secondary spermatocyte contains the $2c$ amount of DNA; (c) $2 \times 3.2 \times 10^9 = 6.4 \times 10^9$ nucleotide pairs because a first polar body contains the $2c$ amount of DNA.

► Questions and Problems

ENHANCE UNDERSTANDING AND DEVELOP ANALYTICAL SKILLS

2.1 Compare the sizes and structures of prokaryotic and eukaryotic chromosomes.

2.2 Distinguish between the haploid and diploid states. What types of cells are haploid? What types of cells are diploid?

2.3 What are the principal differences between prokaryotic and eukaryotic cells?

2.4 In what way do the microtubule organizing centers of plant and animal cells differ?

2.5 Match the stages of mitosis with the events they encompass: **Stages:** (1) anaphase, (2) metaphase, (3) prophase, (4) telophase. **Events:** (a) re-formation of the nucleolus, (b) disappearance of the nuclear membrane, (c) condensation of the chromosomes,

(d) formation of the mitotic spindle, (e) movement of chromosomes to the equatorial plane, (f) movement of chromosomes to the poles, (g) decondensation of the chromosomes, (h) splitting of the centromere, (i) attachment of microtubules to the kinetochore.

2.6 With a focus on the chromosomes, what are the key events during interphase and M phase in the eukaryotic cell cycle?

2.7 Which typically lasts longer, interphase or M phase? Can you explain why one of these phases lasts longer than the other?

2.8 Arrange the following events in the correct temporal sequence during eukaryotic cell division, starting with the earliest: (a) condensation of the chromosomes, (b) movement of chromosomes to the poles, (c) duplication of the chromosomes, (d) formation of the nuclear membrane, (e) attachment of microtubules to the kinetochores, (f) migration of centrosomes to positions on opposite sides of the nucleus.

2.9 Does crossing over occur before or after chromosome duplication in cells going through meiosis?

2.10 What visible characteristics of chromosomes indicate that they have undergone crossing over during meiosis?

2.11 During meiosis, when does *chromosome* disjunction occur? When does *chromatid* disjunction occur?

2.12 In flowering plants, is sporophytic tissue haploid or diploid? How many nuclei are present in the female gametophyte? How many are present in the male gametophyte? Are these nuclei haploid or diploid?

2.13 In human beings, the gene for β-globin is located on chromosome 11, and the gene for α-globin, which is another component of the hemoglobin protein, is located on chromosome 16. Would these two chromosomes be expected to pair with each other during meiosis? Explain your answer.

2.14 GO A *human* sperm cell contains 23 chromosomes. How many *chromosomes* would be present in a spermatogonial cell about to enter meiosis? How many *chromatids* would be present in a spermatogonial cell at metaphase I of meiosis? How many would be present at metaphase II?

2.15 In terms of DNA content, are yeast chromosomes larger or smaller than the *E. coli* chromosome? What is the significance of your answer?

2.16 T1 bacteriophages infect *E. coli* cells and destroy them. The infection process is so efficient that with equal numbers of phage and bacteria in a culture tube, all the bacterial cells are killed in less than 12 hours. Suppose that a mixture of large numbers of phage and bacteria is incubated for 12 hours and then is spread over the surface of a culture plate to determine whether any bacterial cells have survived. The next day, a single colony of cells is visible on the surface of the plate. How would you interpret this result?

2.17 From the information given in this chapter, is there a relationship between genome size (measured in base pairs of DNA) and gene number? Explain.

2.18 GO The *Drosophila* haploid genome contains about 1.2×10^8 nucleotide pairs of DNA. How many nucleotide pairs of DNA are present in each of the following cells: (a) somatic cell, (b) sperm cell, (c) fertilized egg, (d) primary oocyte, (e) fist polar body, (f) secondary spermatocyte?

2.19 In flowering plants, two nuclei from the pollen grain participate in the events of fertilization. With which nuclei from the female gametophyte do these nuclei combine? What tissues are formed from the fertilization events?

2.20 Given the way that chromosomes behave during meiosis, is there any advantage for an organism to have an even number of chromosome pairs (such as *Drosophila* does), as opposed to an odd number of chromosome pairs (such as human beings do)?

2.21 *Neurospora* plants have 14 chromosomes (7 pairs) in their somatic cells. How many chromosomes are present in each of the following: (a) ascus nucleus in the female gametophyte, (b) ascus nucleus in a male gametophyte, (c) fertilized nucleus?

▶ Genomics on the Web

at http://www.ncbi.nlm.nih.gov/

1. The genomes of the model organisms described in this chapter have all been sequenced. Using the NCBI web site, find out when the sequence of each of these genomes was first released.

Hint: On the web site, click on Genomic Biology, and then on the appropriate taxonomic group to bring up a page that lists all the organisms with sequenced genomes in that group. The release dates for these sequences are given in the table.

2. Use the links on the NCBI web site to locate web sites dedicated to each of the eukaryotic model organisms described in this chapter: SGD (Saccharomyces Genome Database), Flybase, Wormbase, ZIRC (Zebrafish International Resource Center), MGI (Mouse Genomic Informatics), and TAIR (The Arabidopsis Information Resource).

Chapter 3
Mendelism: The Basic Principles of Inheritance

CHAPTER OUTLINE

- Mendel's Study of Heredity
- Applications of Mendel's Principles
- Testing Genetic Hypotheses
- Mendelian Principles in Human Genetics

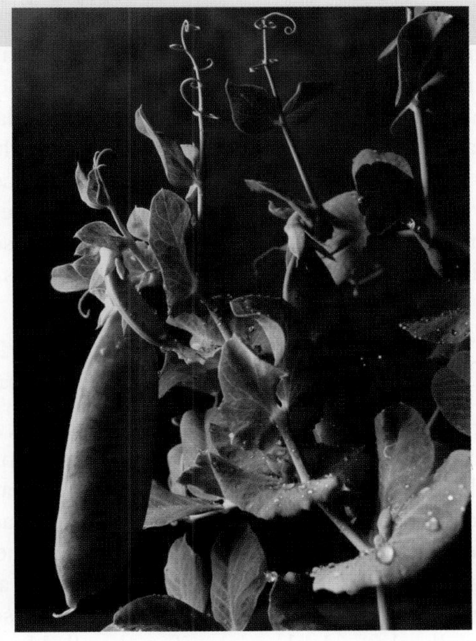

Pisum sativum, the subject of Gregor Mendel's experiments.

The Birth of Genetics: A Scientific Revolution

Science is a complex endeavor involving the careful observation of natural phenomena, reflective thinking about these phenomena, and the formulation of testable ideas about their causes and effects. Progress in science often depends on the work of a single insightful individual. Consider, for example, the effect that Nicolaus Copernicus had on astronomy, that Isaac Newton had on physics, or that Charles Darwin had on biology. Each of these individuals altered the course of his scientific discipline by introducing radically new ideas. In effect, they began scientific revolutions.

In the middle of the nineteenth century, the Austrian monk Gregor Mendel, a contemporary of Darwin, laid the foundation for another revolution in biology, one that eventually produced an entirely new science—genetics. Mendel's ideas, published in 1866 under the title "Experiments with Plant Hybrids," endeavored to explain how the characteristics of organisms are inherited. Many people had attempted such an explanation previously but without much success. Indeed, Mendel commented on their failures in the opening paragraphs of his article:

> *To this object, numerous careful observers, such as Kölreuter, Gärtner, Herbert, Lecoq, Wichura and others, have devoted a part of their lives with inexhaustible perseverance....*
>
> *[However], Those who survey the work in this department will arrive at the conviction that among all the numerous experiments made, not one has been carried out to such an extent and in such a way as to make it possible to determine the number of different forms under which the offspring of the hybrids appear, or to arrange these forms with certainty according to their separate generations, or definitely to ascertain their statistical relations.*[1]

He then described his own efforts to elucidate the mechanism of heredity:

> *It requires indeed some courage to undertake a labor of such far-reaching extent; this appears, however, to be the only right way by which we can finally reach the solution of a question the importance of which cannot be overestimated in connection with the history of the evolution of organic forms.*
>
> *The paper now presented records the results of such a detailed experiment. This experiment was practically confined to a small plant group, and is now, after eight years' pursuit, concluded in all essentials. Whether the plan upon which the separate experiments were conducted and carried out was the best suited to attain the desired end is left to the friendly decision of the reader.*[2]

[1,2]Peters, J. A., ed. 1959. *Classic Papers in Genetics*. Prentice-Hall, Englewood Cliffs, NJ.

Mendel's Study of Heredity

Gregor Mendel's experiments with peas elucidated how traits are inherited.

The life of Gregor Johann Mendel (1822–1884) spanned the middle of the nineteenth century. His parents were farmers in Moravia, then a part of the Hapsburg Empire in Central Europe. A rural upbringing taught him plant and animal husbandry and inspired an interest in nature. At the age of 21, Mendel left the farm and entered a Catholic monastery in the city of Brünn (today, Brno in the Czech Republic). In 1847 he was ordained a priest, adopting the clerical name Gregor. He subsequently taught at the local high school, taking time out between 1851 and 1853 to study at the University of Vienna. After returning to Brünn, he resumed his life as a teaching monk and began the genetic experiments that eventually made him famous.

Mendel performed experiments with several species of garden plants, and he even tried some experiments with honeybees. His greatest success, however, was with peas. He completed his experiments with peas in 1864. In 1865, Mendel presented the results before the local Natural History Society, and the following year, he published a detailed report in the society's proceedings (see A Milestone in Genetics: Mendel's 1866 Paper later in this chapter). Unfortunately, this paper languished in obscurity until 1900, when it was rediscovered by three botanists—Hugo de Vries in Holland, Carl Correns in Germany, and Eric von Tschermak-Seysenegg in Austria. As these men searched the scientific literature for data supporting their own theories of heredity, each found that Mendel had performed a detailed and careful analysis 35 years earlier. Mendel's ideas quickly gained acceptance, especially through the promotional efforts of a British biologist, William Bateson. This champion of Mendel's discoveries coined a new term to describe the study of heredity: genetics, from the Greek word meaning "to generate."

MENDEL'S EXPERIMENTAL ORGANISM, THE GARDEN PEA

One reason for Mendel's success is that he chose his experimental material astutely. The garden pea, *Pisum sativum*, is a dicot, a type of plant that sprouts two leaves, or cotyledons, from a germinating seed. Peas are easily grown in experimental gardens or in pots in a greenhouse.

One peculiarity of pea reproduction is that the petals of the flower close down tightly, preventing pollen grains from entering or leaving. This enforces a system of self-fertilization, in which the male and female gametes from the same flower unite with each other to produce seeds. As a result, individual pea strains are highly inbred, displaying little if any genetic variation from one generation to the next. Because of this uniformity, we say that such strains are *true-breeding*.

At the outset, Mendel obtained many different true-breeding varieties of peas, each distinguished by a particular characteristic. In one strain, the plants were 2 meters high, whereas in another they measured only a half meter. Another variety produced green seeds, and still another produced yellow seeds. Mendel took advantage of these contrasting traits to determine how the characteristics of pea plants are inherited. His focus on these singular differences between pea strains allowed him to study the inheritance of one trait at a time—for example, plant height. Other biologists had attempted to follow the inheritance of many traits simultaneously, but because the results of such experiments were complex, they were unable to discover any fundamental principles about heredity. Mendel succeeded where these biologists had failed because he focused his attention on contrasting differences between plants that were otherwise the same—tall versus short, green seeds versus yellow seeds, and so forth. In addition, he kept careful records of the experiments that he performed.

MONOHYBRID CROSSES: THE PRINCIPLES OF DOMINANCE AND SEGREGATION

In one experiment, Mendel *cross-fertilized*—or, simply, crossed—tall and dwarf pea plants to investigate how height was inherited (**FIGURE 3.1**). He carefully removed the anthers from one variety before its pollen had matured and then applied pollen from the other variety to the stigma, a sticky organ on top of the pistil that leads to the ovary. The seeds that resulted from these cross-fertilizations were sown the next year, yielding hybrids that were uniformly tall. Mendel obtained tall plants regardless of the way he performed the cross (tall male with dwarf female or dwarf male with tall female); thus, the two reciprocal crosses gave the same results. Even more significantly, however, Mendel noted that the dwarf characteristic seemed to have disappeared in the progeny of the cross, for all the hybrid plants were tall. To explore the hereditary makeup of these tall hybrids, Mendel allowed them to undergo self-fertilization—the natural course of events in peas. When he examined the progeny, he found that they consisted of both tall and dwarf plants. In fact, among 1064 progeny that Mendel cultivated in his garden, 787 were tall and 277 were dwarf—a ratio of approximately 3:1.

Mendel was struck by the reappearance of the dwarf characteristic. Clearly, the hybrids that he had made by crossing tall and dwarf varieties had the ability to produce dwarf progeny even though they themselves were tall. Mendel inferred that these hybrids carried a latent genetic factor for dwarfness, one that was masked by the expression of another factor for tallness. He said that the latent factor was **recessive** and that the expressed factor was **dominant.** He also inferred that these recessive and dominant factors separated from each other when the hybrid plants reproduced. This enabled him to explain the reappearance of the dwarf characteristic in the next generation.

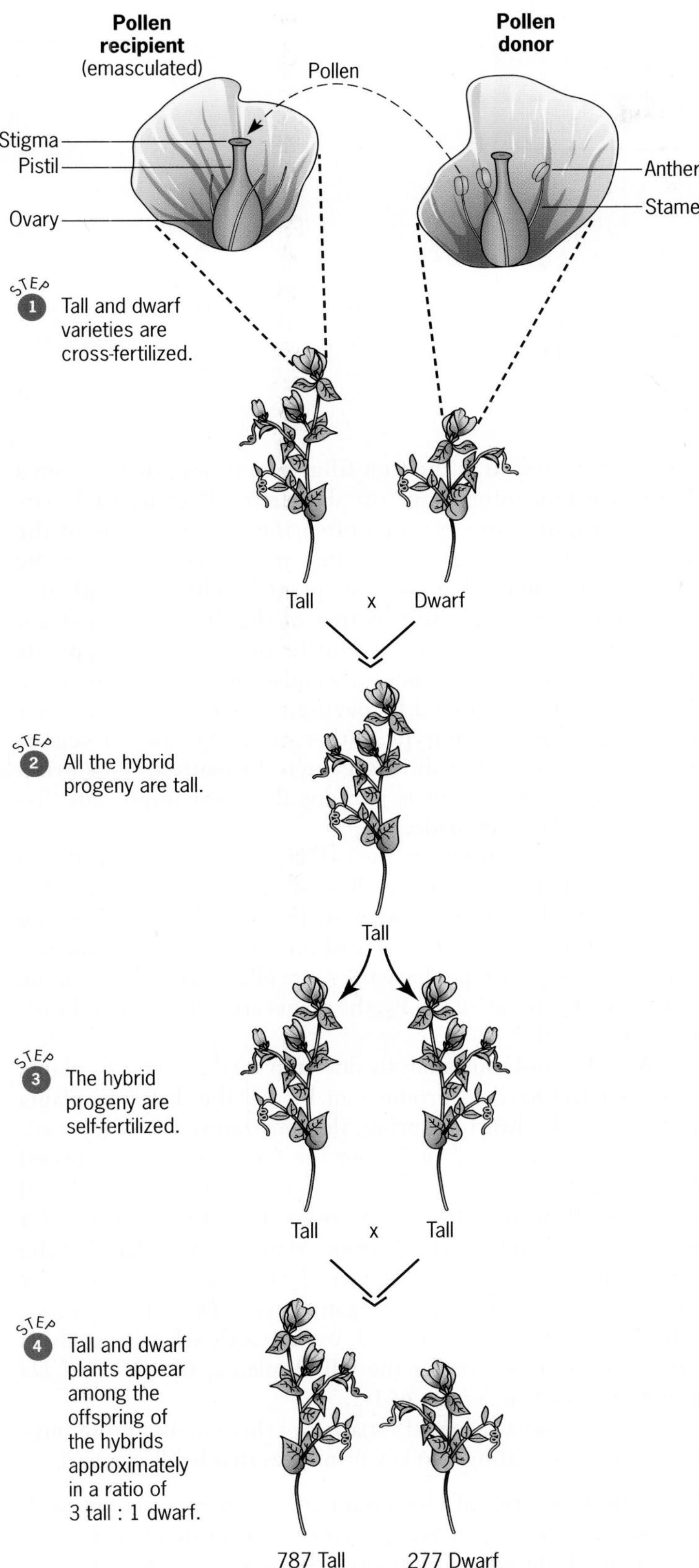
Pollen recipient (emasculated)
Pollen
Pollen donor
Stigma
Pistil
Ovary
Anther
Stamen
STEP 1 Tall and dwarf varieties are cross-fertilized.
Tall x Dwarf
STEP 2 All the hybrid progeny are tall.
Tall
STEP 3 The hybrid progeny are self-fertilized.
Tall x Tall
STEP 4 Tall and dwarf plants appear among the offspring of the hybrids approximately in a ratio of 3 tall : 1 dwarf.
787 Tall
277 Dwarf

Figure 3.1 ▶ Mendel's crosses involving tall and dwarf varieties of peas.

Mendel performed similar experiments to study the inheritance of six other traits: seed texture, seed color, pod shape, pod color, flower color, and flower position (**TABLE 3.1**). In each experiment—called a **monohybrid cross** because a single trait was being studied—Mendel observed that only one of the two contrasting characteristics appeared in the hybrids and that when these hybrids were self-fertilized, they produced two types of progeny, each resembling one of the plants in the original crosses. Furthermore, he found that these progeny consistently appeared in a ratio of 3:1. Thus, each trait that Mendel studied seemed to be controlled by a heritable factor that existed in two forms, one dominant, the other recessive. These factors are now called **genes,** a word coined by the Danish plant breeder Wilhelm Johannsen in 1909; their dominant and recessive forms are called **alleles**—from the Greek word meaning "of one another." Alleles are alternate forms of a gene.

The regular numerical relationships that Mendel observed in these crosses led him to another important conclusion: that genes come in pairs. Mendel proposed that each of the parental strains that he used in his experiments carried two identical copies of a gene—in modern terminology, they are diploid and **homozygous**. However, during the production of gametes, Mendel proposed that these two copies are reduced to one; that is, the gametes that emerge from meiosis carry a single copy of a gene—in modern terminology, they are haploid.

Mendel recognized that the diploid gene number would be restored when sperm and egg unite to form a zygote. Furthermore, he understood that if the sperm and egg came from genetically different plants—as they did in his crosses—the hybrid zygote would inherit two different alleles, one from the mother and one from the father. Such an offspring is said to be **heterozygous.** Mendel realized that the different alleles that are present in a heterozygote must coexist even though one is dominant and the other recessive, and that each of these alleles would have an equal chance of entering a gamete when the heterozygote reproduces. Furthermore, he realized that random fertilizations with a mixed population of gametes—half carrying the dominant allele and half carrying the recessive allele—would produce some zygotes in which both alleles were recessive. Thus, he could explain the reappearance of the recessive characteristic in the progeny of the hybrid plants.

Mendel used symbols to represent the hereditary factors that he postulated—a methodological breakthrough. With symbols, he could describe hereditary phenomena clearly and concisely, and he could analyze the results of crosses mathematically. He could even make predictions about the outcome of future crosses. Although the practice of using symbols to analyze genetic problems has been much refined since Mendel's time, the basic principles remain the same. The symbols stand for genes (or, more precisely, for their alleles), and they are manipulated according to the rules of inheritance that Mendel discovered. These manipulations are the essence of formal genetic analysis. As an introduction to this subject, let us consider the symbolic representation of the cross between tall and dwarf peas (**FIGURE 3.2**).

▶ **TABLE 3.1**

Results of Mendel's Monohybrid Crosses

Parental Strains	F_2 Progeny	Ratio
Tall plants × dwarf plants	787 tall, 277 dwarf	2.84:1
Round seeds × wrinkled seeds	5474 round, 1850 wrinkled	2.96:1
Yellow seeds × green seeds	6022 yellow, 2001 green	3.01:1
Violet flowers × white flowers	705 violet, 224 white	3.15:1
Inflated pods × constricted pods	882 inflated, 299 constricted	2.95:1
Green pods × yellow pods	428 green, 152 yellow	2.82:1
Axial flowers × terminal flowers	651 axial, 207 terminal	3.14:1

The two true-breeding varieties, tall and dwarf, are homozygous for different alleles of a gene controlling plant height. The allele for dwarfness, being recessive, is symbolized by a lowercase letter *d*; the allele for tallness, being dominant, is symbolized by the corresponding uppercase letter *D*. In genetics, the letter that is chosen to denote the alleles of a gene is usually taken from the word that describes the recessive trait (*d*, for *d*warfness). Thus, the tall and dwarf pea strains are symbolized by *DD* and *dd*, respectively. The allelic constitution of each strain is said to be its **genotype.** By contrast, the physical appearance of each strain—the tall or dwarf characteristic—is said to be its **phenotype.**

As the **parental** strains, the tall and dwarf pea plants form the **P** generation of the experiment. Their hybrid progeny are referred to as the first **filial** generation, or $\mathbf{F_1}$, from a Latin word meaning "son" or "daughter." Because each parent contributes equally to its offspring, the genotype of the F_1 plants must be *Dd*; that is, they are heterozygous for the alleles of the gene that controls plant height. Their phenotype, however, is the same as that of the *DD* parental strain because *D* is dominant over *d*. During meiosis, these F_1 plants produce two kinds of gametes, *D* and *d*, in equal proportions. Neither allele is changed by having coexisted with the other in a heterozygous genotype; rather, they separate, or **segregate,** from each other during gamete formation. This process of allele segregation is perhaps the most important discovery that Mendel made.

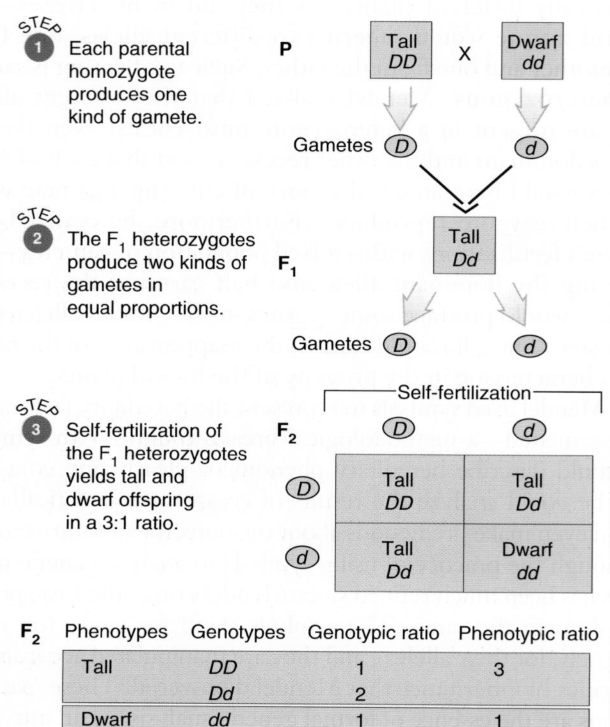

Figure 3.2 ▶ Symbolic representation of the cross between tall and dwarf peas.

Upon self-fertilization, the two kinds of gametes produced by heterozygotes can unite in all possible ways. Thus, they produce four kinds of zygotes (we write the contribution of the egg first): *DD*, *Dd*, *dD*, and *dd*. However, because of dominance, three of these genotypes have the same phenotype. Thus, in the next generation, called the $\mathbf{F_2}$, the plants are either tall or dwarf, in a ratio of 3:1.

Mendel took this analysis one step further. The F_2 plants were self-fertilized to produce an F_3. All the dwarf F_2 plants produced only dwarf offspring, demonstrating that they were homozygous for the *d* allele, but the tall F_2 plants comprised two categories. Approximately one-third of them produced only tall offspring, whereas the other two-thirds produced a mixture of tall and dwarf offspring. Mendel concluded that the third that were true-breeding were *DD* homozygotes and that the two-thirds that were segregating were *Dd* heterozygotes. These proportions, 1/3 and 2/3, were exactly what his analysis predicted because, among the tall F_2 plants, the *DD* and *Dd* genotypes occur in a ratio of 1:2.

We summarize Mendel's analysis of this and other monohybrid crosses by stating two key principles that he discovered:

1. **The Principle of Dominance:** *In a heterozygote, one allele may conceal the presence of another.* This principle is a statement about genetic function. Some alleles evidently control the phenotype even when they are present in a single copy. We consider the physiological explanation for this phenomenon in later chapters.

2. **The Principle of Segregation:** *In a heterozygote, two different alleles segregate from each other during the formation of gametes.* This principle is a statement about genetic transmission. An allele is transmitted faithfully to the next generation, even if it was present with a different allele in a heterozygote. The biological basis for this phenomenon is the pairing and subsequent separation of homologous chromosomes during meiosis, a process we discussed in Chapter 2. We will consider the experiments that led to this chromosome theory of heredity in Chapter 5.

DIHYBRID CROSSES: THE PRINCIPLE OF INDEPENDENT ASSORTMENT

Mendel also performed experiments with plants that differed in two traits (**FIGURE 3.3**). He crossed plants that produced yellow, round seeds with plants that produced green, wrinkled seeds. The purpose of the experiments was to see if the two seed traits, color and texture, were inherited independently. Because the F_1 seeds were all yellow and round, the alleles for these two characteristics were dominant. Mendel grew plants from these seeds and allowed them to self-fertilize. He then classified the F_2 seeds and counted them by phenotype.

The four phenotypic classes in the F_2 represented all possible combinations of the color and texture traits. Two classes—yellow, round seeds and green, wrinkled seeds—resembled the parental strains. The other two—green, round seeds and yellow, wrinkled seeds—showed new combinations of traits. The four classes had an approximate ratio of 9 yellow, round:3 green, round:3 yellow, wrinkled:1 green, wrinkled (**FIGURE 3.3**). To Mendel's insightful mind, these numerical relationships suggested a simple explanation: Each trait was controlled by a different gene segregating two alleles, and the two genes were inherited independently.

Let's analyze the results of this two-factor, or **dihybrid cross,** using Mendel's methods. We denote each gene with a letter, using lowercase for the recessive allele and uppercase for the dominant (**FIGURE 3.4**). For the seed color gene, the two alleles are *g* (for green) and *G* (for yellow), and for the seed texture gene, they are *w* (for wrinkled) and *W* (for round). The parental strains, which were true-breeding, must have been doubly homozygous; the yellow, round plants were *GG WW* and the green, wrinkled plants were *gg ww*. Such two-gene genotypes are customarily written by separating pairs of alleles with a space.

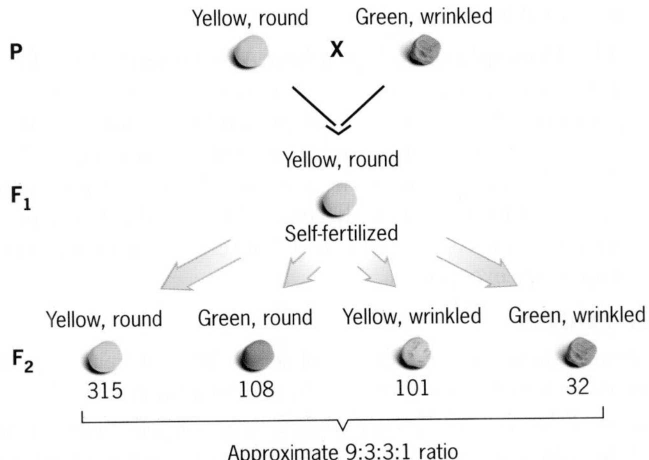

Figure 3.3 ▶ Mendel's crosses between peas with yellow, round seeds and peas with green, wrinkled seeds.

The haploid gametes produced by a diploid plant contain one copy of each gene. Gametes from *GG WW* plants therefore contain one copy of the seed color gene (the *G* allele) and one copy of the seed texture gene (the *W* allele). Such gametes are symbolized by *G W*. By similar reasoning, the gametes from *gg ww* plants are written *g w*. Cross-fertilization of these two types of gametes produces F_1 hybrids that are doubly heterozygous, symbolized by *Gg Ww*, and their yellow, round phenotype indicates that the *G* and *W* alleles are dominant.

The Principle of Segregation predicts that the F_1 hybrids will produce four different gametic genotypes: (1) *G W*, (2) *G w*, (3) *g W*, and (4) *g w*. If each gene segregates its alleles independently, these four types will be equally frequent; that is, each will be 25 percent of the total. On this assumption, self-fertilization in the F_1 will produce an array of 16 equally frequent zygotic genotypes. We obtain the zygotic array by systematically combining the gametes, as shown in Figure 3.4. We then obtain the phenotypes of these F_2 genotypes by noting that *G* and *W* are the dominant alleles. Altogether, there are four distinguishable phenotypes, with relative frequencies indicated by the number of positions occupied in the array. For absolute frequencies, we divide each number by the total, 16:

yellow, round	9/16
yellow, wrinkled	3/16
green, round	3/16
green, wrinkled	1/16

This analysis is predicated on two assumptions: (1) that each gene segregates its alleles, and (2) that these segregations are independent of each other. The second assumption implies that there is no connection or linkage between the segregation events of the two genes. For example, a gamete that receives *W* through the segregation of the texture gene is just as likely to receive *G* as it is to receive *g* through the segregation of the color gene.

Do the experimental data fit with the predictions of our analysis? **FIGURE 3.5** compares the predicted and observed frequencies of the four F_2 phenotypes in two ways—by proportions and by numerical frequencies. For the numerical frequencies, we calculate the predicted numbers by multiplying the predicted proportion by the total number of F_2 seeds examined. With either method, there is obviously good agreement between the observations and the predictions. Thus, the assumptions on which we have built our analysis—independent segregation of the seed color and seed texture genes—are consistent with the observed data.

Mendel conducted similar experiments with other combinations of traits and in each case observed that the genes

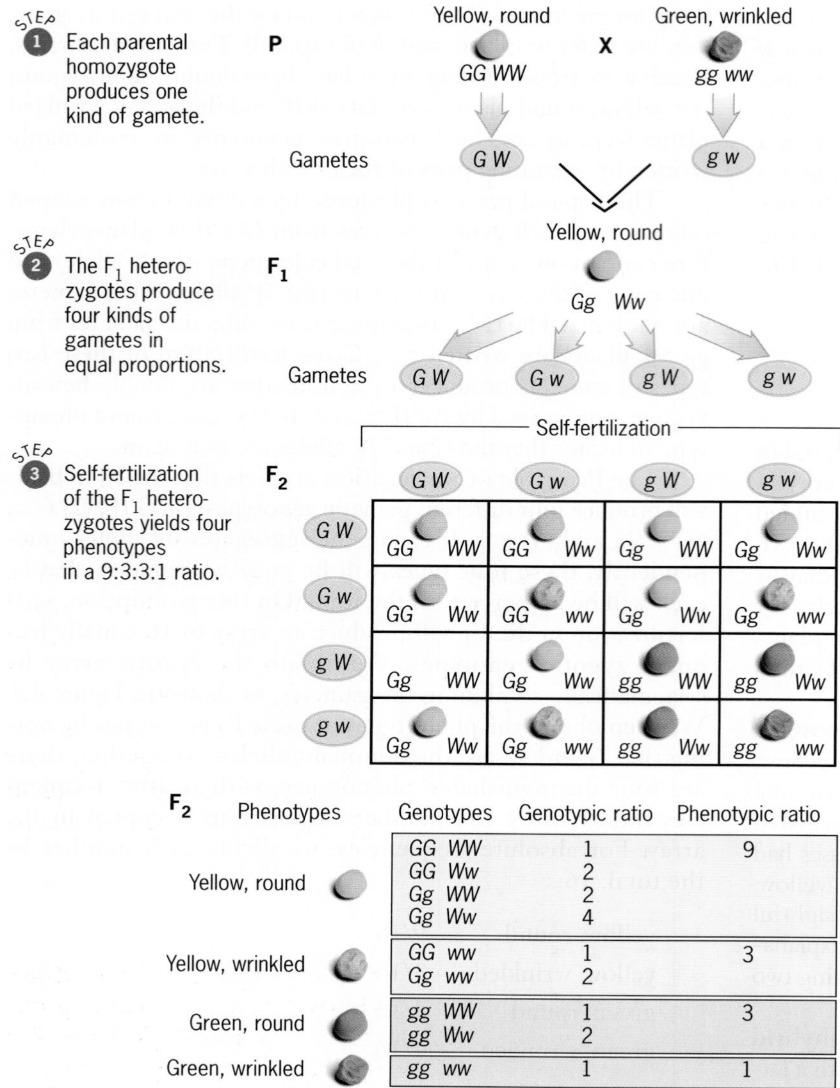

Figure 3.4 ▶ Symbolic representation of Mendel's dihybrid cross.

F_2 phenotypes	Observed Number	Observed Proportion	Expected Number	Expected Proportion
Yellow, round	315	0.567	313	0.563
Green, round	108	0.194	104	0.187
Yellow, wrinkled	101	0.182	104	0.187
Green, wrinkled	32	0.057	35	0.063
Total	**556**	**1.000**	**556**	**1.000**

Figure 3.5 ▶ Comparing the observed and expected results of Mendel's dihybrid cross.

segregated independently. The results of these experiments led him to a third key principle:

3. **The Principle of Independent Assortment:** *The alleles of different genes segregate, or as we sometimes say, assort, independently of each other.* This principle is another rule of genetic transmission, based, as we will see in Chapter 5, on the behavior of different pairs of chromosomes during meiosis. However, not all genes abide by the Principle of Independent Assortment. In Chapter 7 we consider some important exceptions.

KEY POINTS

- Mendel studied the inheritance of seven different traits in garden peas, each trait being controlled by a different gene.
- Mendel's research led him to formulate three principles of inheritance: (1) the alleles of a gene are either dominant or recessive, (2) different alleles of a gene segregate from each other during the formation of gametes, and (3) the alleles of different genes assort independently.

Applications of Mendel's Principles

Mendel's principles can be used to predict the outcomes of crosses between different strains of organisms.

If the genetic basis of a trait is known, Mendel's principles can be used to predict the outcome of crosses. There are three general procedures, two relying on the systematic enumeration of all the zygotic genotypes or phenotypes and one relying on mathematical insight.

THE PUNNETT SQUARE METHOD

For situations involving one or two genes, it is possible to write down all the gametes and combine them systematically to generate the array of zygotic genotypes. Once these have been obtained, the Principle of Dominance can be used to determine the associated phenotypes. This procedure, called the *Punnett square method* after the British geneticist R. C. Punnett, is a straightforward way of predicting the outcome of crosses. We have used it to analyze the zygotic output of the cross with Mendel's yellow, round F_1 hybrids—a type of mating commonly called an **intercross** (**FIGURE 3.4**). However, in more complicated situations, like those involving more than two genes, the Punnett square method is unwieldy. We shall see in **FIGURE 3.8** how the Punnett square method is related to an approach to genetic problems that uses the concept of probability.

THE FORKED-LINE METHOD

Another procedure for predicting the outcome of a cross involving two or more genes is the *forked-line method*. However, instead of enumerating the progeny in a square by combining the gametes systematically, we tally them in a diagram of branching lines. As an example, let us consider an intercross between peas that are heterozygous for three independently assorting genes—one controlling plant height, one controlling seed color, and one controlling seed texture. This is a trihybrid cross—*Dd Gg Ww* × *Dd Gg Ww*—that can be partitioned into three monohybrid crosses—*Dd* × *Dd*, *Gg* × *Gg*, and *Ww* × *Ww*—because all the genes assort independently. For each gene, we expect the phenotypes to appear in a 3:1 ratio. Thus, for example, *Dd* × *Dd* will produce a ratio of 3 tall plants:1 dwarf plant. Using the forked-line method (**FIGURE 3.6**), we can combine these separate ratios into an overall phenotypic ratio for the offspring of the cross.

We can also use this method to analyze the results of a cross between multiply heterozygous individuals and multiply homozygous individuals. This type of cross is called a **testcross.** For example, if *Dd Gg Ww* pea plants are crossed with *dd gg ww* pea plants, we can predict the phenotypes of the progeny by noting that each of the three genes in the heterozygous parent segregates dominant and recessive alleles in a 1:1 ratio, and that the homozygous parent transmits only recessive alleles of these genes. Thus, the genotypes—and ultimately the phenotypes—of the offspring of this cross depend on which alleles the heterozygous parent transmits (**FIGURE 3.7**).

THE PROBABILITY METHOD

An alternative method to the Punnett square and forked-line methods—and a quicker one—is based on the principle of **probability** (see the Focus on the Rules of Probability). Mendelian segregation is like a coin toss; when a heterozygote produces gametes, half contain one allele and half contain the other. If two segregating heterozygotes are crossed, their gametes are combined randomly, producing the zygotic genotypes (**FIGURE 3.8**).

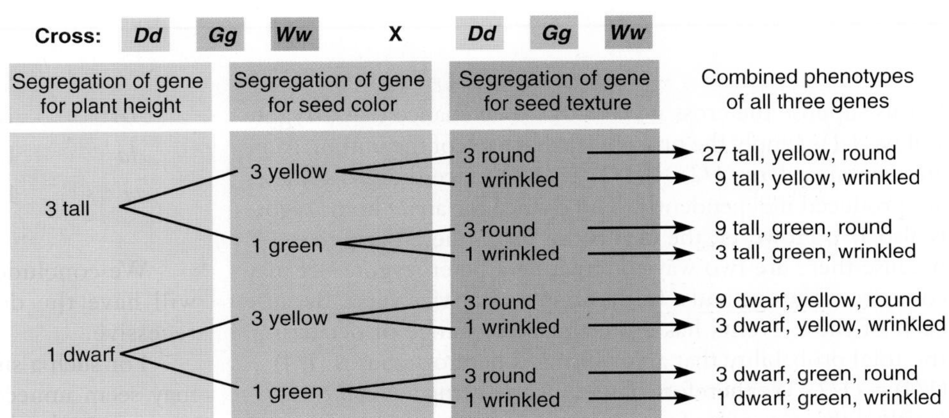

Figure 3.6 ► The forked-line method for predicting the outcome of an intercross involving three independently assorting genes in peas.

▶ FOCUS ON The Rules of Probability

Probability theory accounts for the frequency of events—for example, the chance of getting a head on a coin toss, drawing an ace from a deck of cards, or obtaining a dominant homozygote from a mating between two heterozygotes. In each case, the event is the outcome of a process—tossing a coin, drawing a card, producing an offspring. To determine the probability of a particular event, we must consider all possible outcomes of the process. The collection of all events is called the *sample space*. For a coin toss, the sample space contains two events, head and tail; for drawing a card, it contains 52, one for each card; and for heterozygotes producing an offspring, it contains three, *GG*, *Gg*, and *gg*. *The probability of an event is the frequency of that event in the sample space.* For example, the probabilities associated with each of the progeny from a mating between two heterozygotes are 1/4 (for *GG*), 1/2 (for *Gg*), and 1/4 (for *gg*).

Two kinds of questions often arise in problems involving probabilities: (1) What is the probability that two events, A and B, will occur together? (2) What is the probability that at least one of two events, A or B, will occur at all? The first question specifies the joint occurrence of two events—A *and* B must occur together to satisfy this question. The second question is less stringent—if *either* A *or* B occurs, the question will be satisfied. A simple diagram can help to explain the different meanings of these two questions.

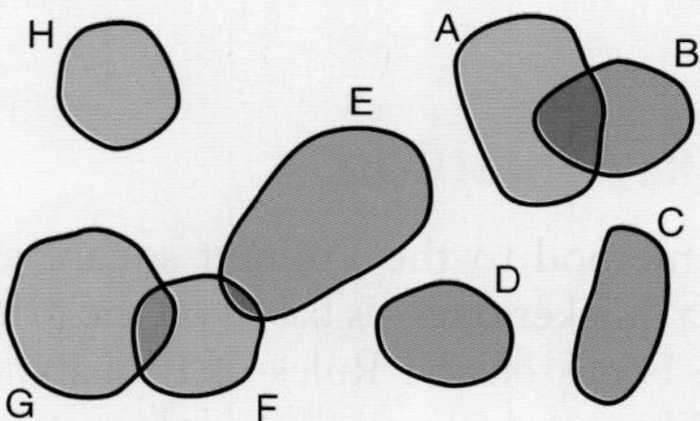

The shapes in the diagram represent events in the sample space, and the sizes of the shapes reflect their relative frequencies. Overlaps between shapes indicate the joint occurrence of two events. If the events do not overlap, then they can never occur together. The first question seeks the probability that both A and B will occur; this probability is represented by the size of the overlap between the two events. The second question seeks the probability that either A or B will occur; this probability is represented by the combined shapes of the two events, including, of course, the overlap between them.

The Multiplicative Rule: If the events A and B are independent, the probability that they occur together, denoted P(A and B), is P(A) × P(B).

Here P(A) and P(B) are the probabilities of the individual events. Note that independent does not mean that they do not overlap in the sample space. In fact, nonoverlapping, or disjoint, events are not independent, for if one occurs, then the other cannot. In probability theory, independent means that one event provides no information about the other. For example, if a card drawn from a deck turns out to be an ace, we have no clue about the card's suit. Thus, drawing the ace of hearts represents the joint occurrence of two independent events—the card is an ace (A) and it is a heart (H). According to the Multiplicative Rule, P(A and H) = P(A) × P(H), and because P(A) = 4/52 and P(H) = 1/4, P(A and H) = (4/52) × (1/4) = 1/52.

The Additive Rule: If the events A and B are independent, the probability that at least one of them occurs, denoted P(A or B), is P(A) + P(B) – [P(A) × P(B)].

Here the term P(A) × P(B), which is the probability that A and B occur together, is subtracted from the sum of the probabilities, P(A) + P(B), because the straight sum includes this term twice. As an example, suppose we seek the probability that a card drawn from a deck is either an ace or a heart. According to the Additive Rule, P(A or H) = P(A) + P(H) – [P(A) × P(H)] = (4/52) + (1/4) – [(4/52) × (1/4)] = 16/52.

If the two events do not overlap in the sample space, the Additive Rule reduces to a simpler expression: P(A or B) = P(A) + P(B). For example, suppose we seek the probability that a card drawn from a deck is either an ace or a king (K). These two events do not overlap in the sample space; they are said to be mutually exclusive. Thus, P(A or K) = P(A) + P(K) = (4/52) + (4/52) = 8/52.

Let us suppose the cross is *Aa* × *Aa*. The chance that a zygote will be *AA* is simply the probability that each of the uniting gametes contains *A*, or (1/2) × (1/2) = (1/4), since the two gametes are produced independently. The chance for an *aa* homozygote is also 1/4. However, the chance for an *Aa* heterozygote is 1/2 because there are two ways of creating a heterozygote—*A* may come from the egg and *a* from the sperm, or vice versa. Because each of these events has a one-quarter chance of occurring, the total probability that an offspring is heterozygous is (1/4) + (1/4) = (1/2). We therefore obtain the following *probability distribution* of the genotypes from the mating *Aa* × *Aa*:

AA	1/4
Aa	1/2
aa	1/4

We conclude that (1/4) + (1/2) = (3/4) of the progeny will have the dominant phenotype and 1/4 will have the recessive.

For such a simple situation, use of the probability method may seem unnecessary. However, in more complicated situations, it is clearly the most practical approach to predict the

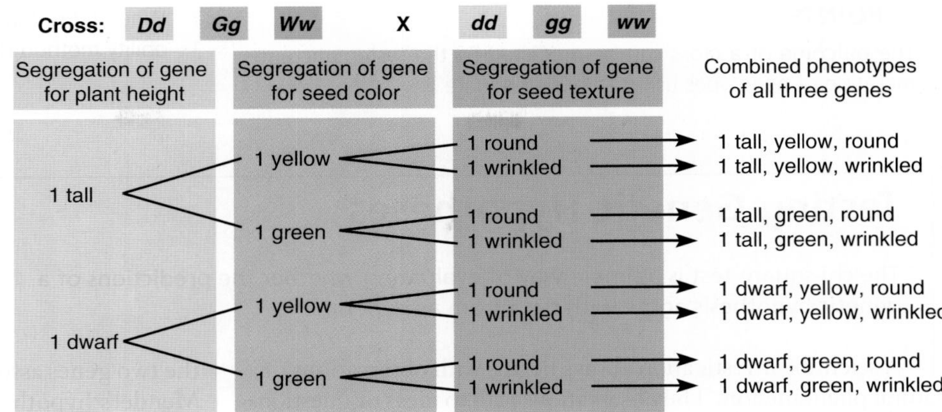

Figure 3.7 ▶ The forked-line method for predicting the outcome of a testcross involving three independently assorting genes in peas.

outcome of crosses. Consider, for example, a cross between plants heterozygous for four different genes, each assorting independently. What fraction of the progeny will be homozygous for all four recessive alleles? To answer this question, we consider the genes one at a time. For the first gene, the fraction of offspring that will be recessive homozygotes is 1/4, as it will be for the second, third, and fourth genes. Therefore, by the Principle of Independent Assortment, the fraction of offspring that will be quadruple recessive homozygotes is $(1/4) \times (1/4) \times (1/4) \times (1/4) = (1/256)$. Surely, using the probability method is a better approach than diagramming a Punnett square with 256 entries!

Now let's consider an even more difficult question. What fraction of the offspring will be homozygous for all four genes? Before computing any probabilities, we must first decide what genotypes satisfy the question. For each gene there are two types of homozygotes, the dominant and the recessive, and together they constitute half the progeny. The fraction of progeny that will be homozygous for all four genes will therefore be $(1/2) \times (1/2) \times (1/2) \times (1/2) = (1/16)$.

To see the full power of the probability method, we need to consider one more question. Suppose the cross is *Aa Bb* × *Aa Bb* and we want to know what fraction of the progeny will show the recessive phenotype for at least one gene (**FIGURE 3.9**). Three kinds of genotypes would satisfy this condition: (1) *A- bb* (the dash stands for either *A* or *a*), (2) *aa B-*, and (3) *aa bb*. The answer to the question must therefore be the sum of the probabilities corresponding to each of these genotypes. The probability for *A- bb* is $(3/4) \times (1/4) = (3/16)$, that for *aa B-* is $(1/4) \times (3/4) = (3/16)$, and that for *aa bb* is $(1/4) \times (1/4) = (1/16)$. Adding these together, we find that the answer is 7/16.

Cross: *Aa* X *Aa*

Male gametes ♂

Female gametes ♀	*A* (1/2)	*a* (1/2)
A (1/2)	*AA* (1/4)	*Aa* (1/4)
a (1/2)	*aA* (1/4)	*aa* (1/4)

Progeny:

Genotype	Frequency	Phenotype	Frequency
AA	1/4	Dominant	3/4
Aa	1/2		
aa	1/4	Recessive	1/4

Figure 3.8 ▶ An intercross showing the probability method in the context of a Punnett square. The frequency of each genotype from the cross is obtained from the frequencies in the Punnett square, which are, in turn, obtained by multiplying the frequencies of the two types of gametes produced by the heterozygous parents.

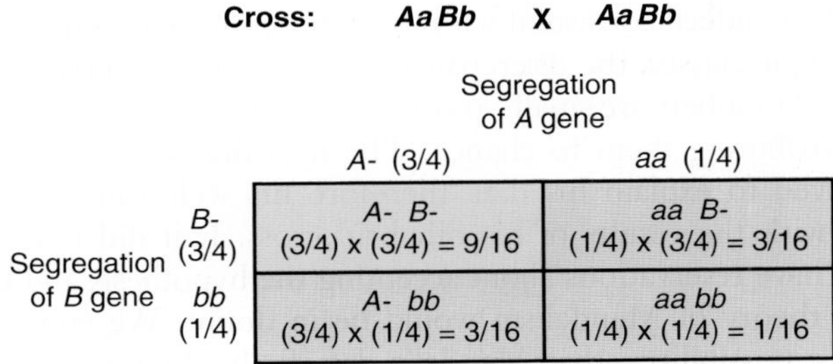

Cross: Aa Bb X Aa Bb
Segregation of A gene
A- (3/4)
aa (1/4)
Segregation of B gene
B- (3/4)
bb (1/4)
A- B- (3/4) x (3/4) = 9/16
aa B- (1/4) x (3/4) = 3/16
A- bb (3/4) x (1/4) = 3/16
aa bb (1/4) x (1/4) = 1/16

Progeny:

Genotype	Frequency	Phenotype	Frequency
A- B-	9/16	Dominant for both genes	9/16
aa B-	3/16	Recessive for at least one gene	7/16
A- bb	3/16		
aa bb	1/16		

Figure 3.9 ▶ Application of the probability method to an intercross involving two genes. In this cross, each gene segregates dominant and recessive phenotypes, with probabilities 3/4 and 1/4, respectively. Because the segregations occur independently, the frequencies of the combined phenotypes within the square are obtained by multiplying the marginal probabilities. The frequency of progeny showing the recessive phenotype for at least one of the genes is obtained by adding the frequencies in the relevant cells (tan color).

KEY POINTS

- The outcome of a cross can be predicted by the systematic enumeration of genotypes using a Punnett square.
- When more than two genes are involved, the forked-line or probability method is used to predict the outcome of a cross.

Testing Genetic Hypotheses

The chi-square test is a simple way of evaluating whether the predictions of a genetic hypothesis agree with data from an experiment.

A scientific investigation always begins with observations of a natural phenomenon. The observations lead to ideas or questions about the phenomenon, and these ideas or questions are explored more fully by conducting further observations or by performing experiments. A well-formulated scientific idea is called a **hypothesis.** Data collected from observations or from experimentation enable scientists to test hypotheses—that is, to determine if a particular hypothesis should be accepted or rejected.

In genetics, we are usually interested in deciding whether or not the results of a cross are consistent with a hypothesis. As an example, let's consider the data that Mendel obtained from his dihybrid cross involving the color and texture of peas. In the F_2, 556 peas were examined and sorted into four phenotypic classes (**FIGURE 3.3**). From the data, Mendel hypothesized that the pea color and texture were controlled by different genes, that each of the genes segregated two alleles—one dominant, the other recessive—and that the two genes assorted independently. Are the data from the experiment actually consistent with this hypothesis? To answer this question, we need to compare the results of the experiment with the predictions of the hypothesis. The comparison laid out in **FIGURE 3.5** suggests that the experimental results are indeed consistent with the hypothesis. Across the four phenotypic classes, the discrepancies between the observed and expected numbers are small, so small in fact that we are comfortable attributing them to chance. The hypothesis that Mendel conceived to explain his data therefore fits well—almost too well—with the results of his dihybrid cross. If it did not, we would have reservations about accepting the hypothesis and the whole theory of Mendelism would be in doubt. We consider another possibility—that Mendel's data fit his hypothesis too well—in the Milestone in Genetics at the end of this chapter.

Unfortunately, the results of a genetic experiment do not always agree with the predictions of a hypothesis as clearly as Mendel's did. Take, for example, data obtained by Hugo DeVries, one of the rediscoverers of Mendel's work. DeVries crossed different varieties of the campion, a plant that grew in his experimental garden. One variety had red flowers and hairy foliage; the other had white flowers and smooth foliage. The F_1 plants all had red flowers and hairy foliage, and when intercrossed, they produced F_2 plants that sorted into four phenotypic classes (**FIGURE 3.10**). To explain the results of these crosses, DeVries proposed that flower color and foliage type were controlled by two different genes, that each gene segregated two alleles—one dominant, the other recessive—and that the two genes assorted independently; that is, he simply applied Mendel's hypothesis to the campion. However, when we compare DeVries's data with the predictions of the Mendelian hypothesis, we find some disturbing discrepancies. Are these discrepancies large enough to raise questions about the experiment or the hypothesis?

THE CHI-SQUARE TEST

With DeVries's data, and with other genetic data as well, we need an objective procedure to compare the results of the experiment with the predictions of the underlying hypothesis. This

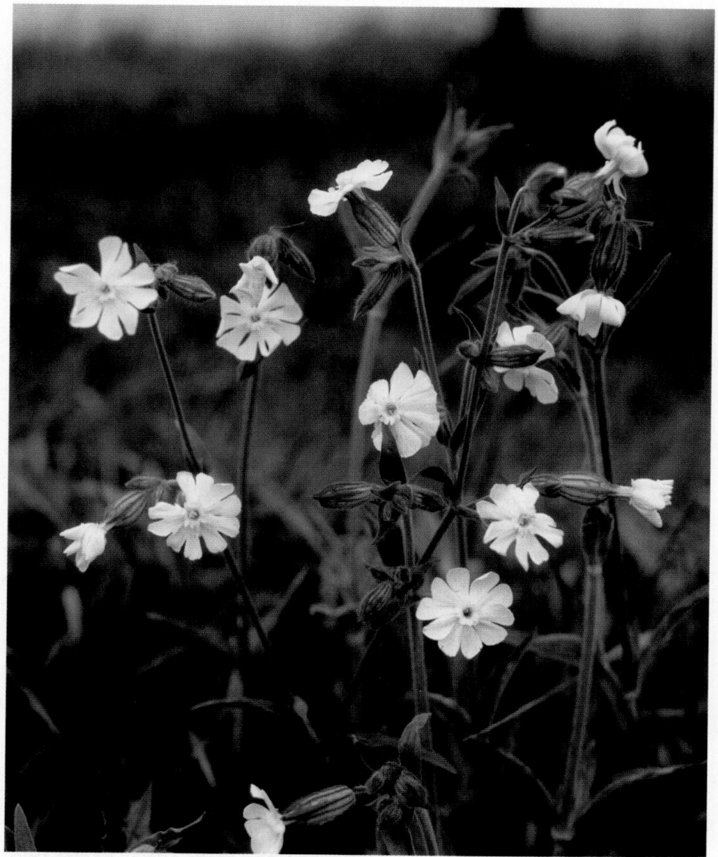

Figure 3.10 ▶ (*a*) The white campion.

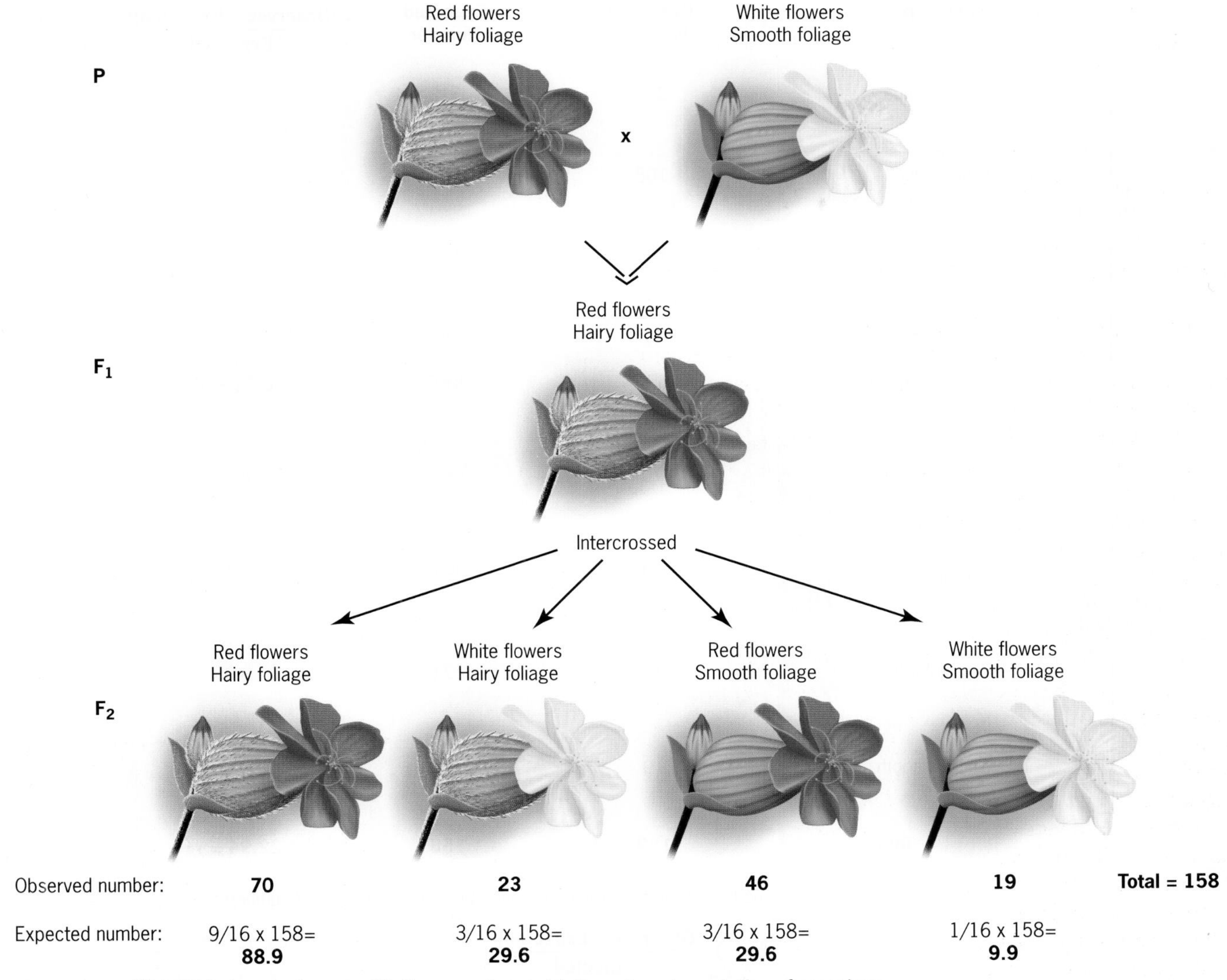

Figure 3.10 ▶ (*b*) DeVries's experiment with flower color and foliage type in varieties of campions.

procedure has to take into account how chance might affect the outcome of the experiment. Even if the hypothesis is correct, we do not anticipate that the results of the experiment will exactly match the predictions of the hypothesis. If they deviate a bit, as Mendel's data did, we would ascribe the deviations to chance variation in the outcome of the experiment. However, if they deviate grossly, we would suspect that something was amiss. The experiment might have been executed poorly—for example, the crosses might have been improperly carried out, or the data might have been incorrectly recorded—or, perhaps, the hypothesis is simply wrong. The possible discrepancies between observations and expectations obviously lie on a continuum from small to large, and we must decide how large they need to be for us to entertain doubts about the execution of the experiment or the acceptability of the hypothesis.

One procedure for assessing these discrepancies uses a statistic called **chi-square** (χ^2). A *statistic* is a number calculated from data—for example, the mean of a set of examination scores. The χ^2 statistic allows a researcher to compare data, such as the numbers we get from a breeding experiment, with their predicted values. If the data are not in line with the predicted values, the χ^2 statistic will exceed a critical number and we will decide either to reevaluate the experiment—that is, look for a mistake in technique—or reject the underlying hypothesis. If the χ^2 statistic is below this number, we tentatively conclude that the results of the experiment are consistent with the predictions of the hypothesis. The χ^2 statistic therefore reduces hypothesis testing to a simple, objective procedure.

As an example, let's consider the data from the experiments of Mendel and DeVries. Mendel's F_2 data seemed to be consistent with the underlying hypothesis, whereas DeVries's F_2 data showed some troubling discrepancies. **FIGURE 3.11** outlines the calculations.

	F_2 Phenotype	Observed Number	Expected Number	$\frac{(\text{Observed} - \text{Expected})^2}{\text{Expected}}$
Mendel's dihybrid cross	Yellow, round	315	313	0.01
	Green, round	108	104	0.15
	Yellow, wrinkled	101	104	0.09
	Green, wrinkled	32	35	0.26
	Total:	**556**	**556**	**0.51** = χ^2
DeVries's dihybrid cross	Red, hairy	 70	88.9	4.02
	White, hairy	 23	29.6	1.47
	Red, smooth	 46	29.6	9.09
	White, smooth	 19	9.9	8.36
	Total:	**158**	**158**	**22.94** = χ^2

Formula for chi-square statistic to test for agreement between observed and expected numbers:

$$\chi^2 = \sum \frac{(\text{Observed} - \text{Expected})^2}{\text{Expected}}$$

Figure 3.11 ▶ Calculating χ^2 for Mendel's and DeVries's F_2 data.

For each phenotypic class in the F_2, we compute the difference between the observed and expected numbers of offspring and square these differences. The squaring operation eliminates the canceling effects of positive and negative values among the four phenotypic classes. Then we divide each squared difference by the corresponding expected number of offspring. This operation scales each squared difference by the size of the expected number. If two classes have the same squared difference, the one with the smaller expected number contributes relatively more in the calculation. Finally, we sum all the terms to obtain the χ^2 statistic. For Mendel's data, the χ^2 statistic is 0.51 and for DeVries's data it is 22.94. These statistics summarize the discrepancies between the observed and expected numbers across the four phenotypic classes in each experiment. If the observed and expected numbers are in basic agreement with each other, the χ^2 statistic will be small, as it happens to be with Mendel's data. If they are in serious disagreement, it will be large, as it happens to be with DeVries's data. Clearly, we must decide what value of χ^2 on the continuum between small and large casts doubt on the experiment or the hypothesis. This **critical value** is the point where the discrepancies between observed and expected numbers are not likely to be due to chance.

To determine the critical value, we need to know how chance affects the χ^2 statistic. Assume for the moment that the underlying genetic hypothesis is true. Now imagine carrying out the experiment—carefully and correctly—many times, and each time, calculating a χ^2 statistic. All these statistics can be compiled into a graph that shows how often each value occurs. We call such a graph a *frequency distribution*. Fortunately, the χ^2 frequency distribution is known from statistical theory (**FIGURE 3.12**)—so we don't actually need to carry out many

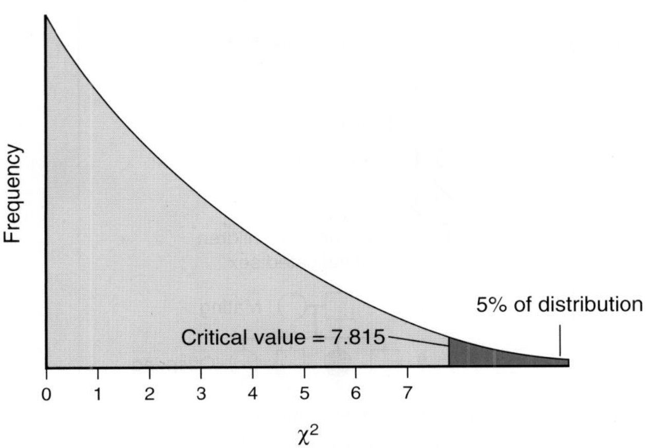

Figure 3.12 ▶ Distribution of a χ^2 statistic.

replications of the experiment to get it. The critical value is the point that cuts off the upper 5 percent of the distribution. By chance alone, the χ^2 statistic will exceed this value 5 percent of the time. Thus, if we perform an experiment once, compute a χ^2 statistic, and find that the statistic is greater than the critical value, we have either observed a rather unlikely set of results—something that happens less than 5 percent of the time—or there is a problem with the way the experiment was executed or with the appropriateness of the hypothesis. Assuming that the experiment was done properly, we are inclined to reject the hypothesis. Of course, we must realize that with this procedure we will reject a true hypothesis 5 percent of the time.

Thus, as long as we know the critical value, the χ^2 testing procedure leads us to a decision about the fate of the hypothesis. However, this critical value—and the shape of the associated frequency distribution—depend on the number of phenotypic classes in the experiment. Statisticians have tabulated critical values according to the **degrees of freedom** associated with the χ^2 statistic (**TABLE 3.2**). This index to the set of χ^2 distributions is determined by subtracting one from the number of phenotypic classes. In each of our examples, there are $4 - 1 = 3$ degrees of freedom. The critical value for the χ^2 distribution with 3 degrees of freedom is 7.815. For Mendel's data, the calculated χ^2 statistic is 0.51, much less than the critical value and

▶ **TABLE 3.2**

Table of Chi-Square (χ^2) 5% Critical Values[a]

Degress of Freedom	5% Critical Value
1	3.841
2	5.991
3	7.815
4	9.488
5	11.070
6	12.592
7	14.067
8	15.507
9	16.919
10	18.307
15	24.996
20	31.410
25	37.652
30	43.773

[a]Selected entries from R. A. Fisher and Yates, 1943, *Statistical Table for Biological, Agricultural, and Medical Research.* Oliver and Boyd, London.

therefore no threat to the hypothesis being tested. However, for DeVries's data the calculated χ^2 statistic is 22.94, very much greater than the critical value. Thus, the observed data do not fit with the genetic hypothesis. Ironically, when DeVries presented these data in 1905, he judged them to be consistent with the genetic hypothesis. Unfortunately, he did not perform a χ^2 test. DeVries also argued that his data provided further evidence for the correctness and widespread applicability of Mendel's ideas—not the only time that a scientist has come to the right conclusion for the wrong reason.

KEY POINTS

- The chi-square statistic is calculated as $\chi^2 = \Sigma$ (observed number − expected number)^2/expected number, with the sum computed over all categories comprising the data.
- Each chi-square statistic is associated with an index, the degrees of freedom, which is equal to the number of data categories minus one.

Mendelian Principles in Human Genetics

Mendel's principles can be applied to study the inheritance of traits in humans.

The application of Mendelian principles to human genetics began soon after the rediscovery of Mendel's paper in 1900. However, because it is not possible to make controlled crosses with human beings, progress was obviously slow. The analysis of human heredity depends on family records, which are often incomplete. In addition, human beings—unlike experimental organisms—do not produce many progeny, making it difficult to discern Mendelian ratios, and humans are not maintained and observed in a controlled environment. For these and other reasons, human genetic analysis has been a difficult endeavor. Nonetheless, the drive to understand human heredity has been very strong, and today, despite all the obstacles, we have learned about thousands of human genes. **TABLE 3.3** lists some of the conditions they control. We discuss many of these conditions in later chapters of this book.

▶ **TABLE 3.3**

Inherited Condition in Human Beings

Dominant Traits
Achondroplasia (dwarfism)
Brachydactyly (short fingers)
Congenital night blindness
Ehler-Danlos syndrome (a connective tissue disorder)
Huntington's disease (a neurological disorder)
Marfan syndrome (tall, gangly stature)
Neurofibromatosis (tumorlike growths on the body)
Phenylthiocarbamide (PTC) tasting
Widow's peak
Woolly hair
Recessive Traits
Albinism (lack of pigment)
Alkaptonuria (a disorder of amino acid metabolism)
Ataxia telangiectasia (a neurological disorder)
Cystic fibrosis (a respiratory disorder)
Duchenne muscular dystrophy
Galactosemia (a disorder of carbohydrate metabolism)
Glycogen storage disease
Phenylketonuria (a disorder of amino acid metabolism)
Sickle-cell anemia (a hemoglobin disorder)
Tay-Sachs disease (a lipid storage disorder)

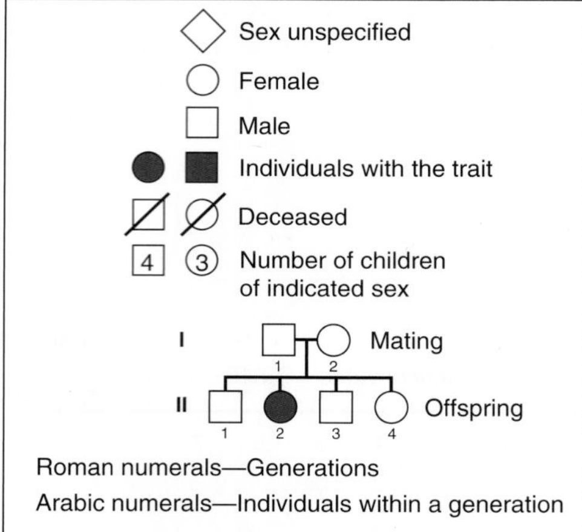

(a) Pedigree conventions

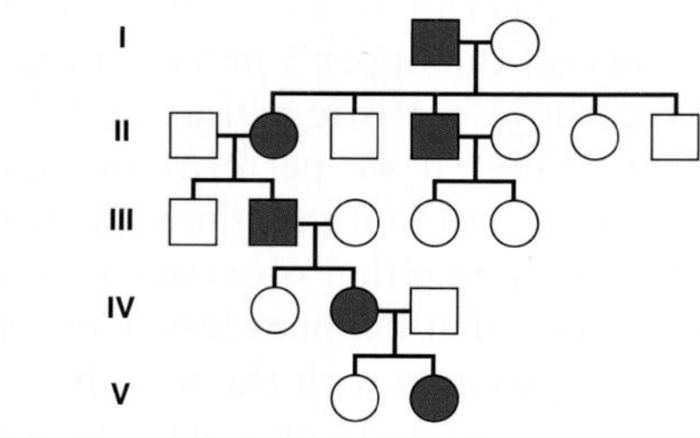

(b) Dominant trait

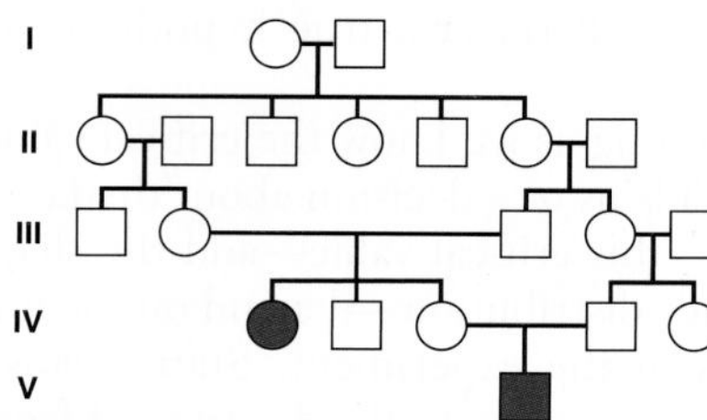

(c) Recessive trait

Figure 3.13 ▶ Mendelian inheritance in human pedigress. (*a*) Pedigree conventions. (*b*) Inheritance of a dominant trait. The trait appears in each generation. (*c*) Inheritance of a recessive trait. The two affected individuals are the offspring of relatives.

PEDIGREES

Pedigrees are diagrams that show the relationships among the members of a family (**FIGURE 3.13a**). It is customary to represent males as squares and females as circles. A horizontal line connecting a circle and a square represents a mating. The offspring of the mating are shown beneath the mates, starting with the first born at the left and proceeding through the birth order to the right. Individuals that have a genetic condition are indicated by coloring or shading. The generations in a pedigree are usually denoted by Roman numerals, and particular individuals within a generation are referred to by Arabic numerals following the Roman numeral.

Traits caused by dominant alleles are the easiest to identify. Usually, every individual who carries the dominant allele manifests the trait, making it possible to trace the transmission of the dominant allele through the pedigree (**FIGURE 3.13b**). Every affected individual is expected to have at least one affected parent, unless, of course, the dominant allele has just appeared in the family as a result of a new mutation—a change in the gene itself. However, the frequency of most new mutations is very low—on the order of one in a million; consequently, the spontaneous appearance of a dominant condition is an extremely rare event. Dominant traits that are associated with reduced viability or fertility never become frequent in a population. Thus, most of the people who show such traits are heterozygous for the dominant allele. If their spouses do not have the trait, half their children should inherit the condition.

Recessive traits are not so easy to identify because they may occur in individuals whose parents are not affected. Sometimes several generations of pedigree data are needed to trace the transmission of a recessive allele (**FIGURE 3.13c**). Nevertheless, a large number of recessive traits have been observed in human beings—at last count, over 4000. Rare recessive traits are more likely to appear in a pedigree when spouses are related to each other—for example, when they are first cousins. This increased incidence occurs because relatives share alleles by virtue of their common ancestry. Siblings share one-half their alleles, half siblings one-fourth their alleles, and first cousins one-eighth their alleles. Thus, when such relatives mate, they have a greater chance of producing a child who is homozygous for a particular recessive allele than do unrelated parents. Many of the classical studies in human genetics have relied on the analysis of matings between relatives, principally first cousins. We will consider this subject in more detail in Chapter 4.

MENDELIAN SEGREGATION IN HUMAN FAMILIES

In human beings, the number of children produced by a couple is typically small. Today in the United States, the average is around two. In developing countries, it is six to seven. Such numbers provide nothing close to the statistical power that Mendel had in his experiments with peas. Consequently, phenotypic ratios in human families often deviate significantly from their Mendelian expectations.

As an example, let's consider a couple who are each heterozygous for a recessive allele that, in homozygous condition, causes cystic fibrosis, a serious disease in which breathing is impaired by an accumulation of mucus in the lungs and respiratory tract. If the couple were to have four children, would we expect exactly three to be unaffected and one to be affected by cystic fibrosis? The answer is no. Although this is a possible outcome, it is not the only one. There are, in fact, five distinct possibilities:

1. Four unaffected, none affected.
2. Three unaffected, one affected.
3. Two unaffected, two affected.
4. One unaffected, three affected.
5. None unaffected, four affected.

Intuitively, the second outcome seems to be the most likely, since it conforms to Mendel's 3:1 ratio. We can calculate the probability of this outcome, and of each of the others, by using Mendel's principles and by treating each birth as an independent event (**FIGURE 3.14**).

For a particular birth, the chance that the child will be unaffected is 3/4. The probability that all four children will be unaffected is therefore $(3/4) \times (3/4) \times (3/4) \times (3/4) = (3/4)^4 = 81/256$. Similarly, the chance that a particular child will be affected is 1/4; thus, the probability that all four will be affected is $(1/4)^4 = 1/256$. To find the probabilities for the three other outcomes, we need to recognize that each actually represents a collection of distinct events. The outcome of three unaffected children and one affected child, for instance, comprises four distinct events; if we let U symbolize an unaffected child and A an affected child, and if we write the children in their order of birth, we can represent these events as

UUUA, UUAU, UAUU, and AUUU

Because each has probability $(3/4)^3 \times (1/4)$, the total probability for three unaffected children and one affected, regardless of birth order, is $4 \times (3/4)^3 \times (1/4)$. The coefficient 4 is the number of ways in which three children could be unaffected and one could be affected in a family with four children. Similarly, the probability for two unaffected children and two affected is $6 \times (3/4)^2 \times (1/4)^2$, since in this case there are six distinct events. The probability for one unaffected child and three affected is $4 \times (3/4) \times (1/4)^3$, since

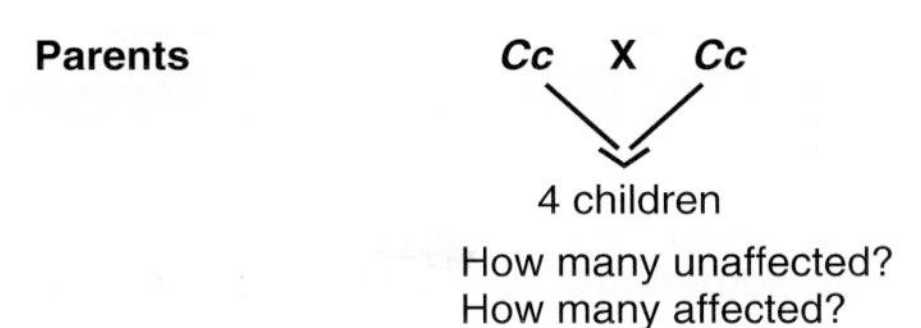

Number of children that are:

Unaffected	Affected	Probability
4	0	1 x (3/4) x (3/4) x (3/4) x (3/4) = 81/256
3	1	4 x (3/4) x (3/4) x (3/4) x (1/4) = 108/256
2	2	6 x (3/4) x (3/4) x (1/4) x (1/4) = 54/256
1	3	4 x (3/4) x (1/4) x (1/4) x (1/4) = 12/256
0	4	1 x (1/4) x (1/4) x (1/4) x (1/4) = 1/256

Probability distribution:

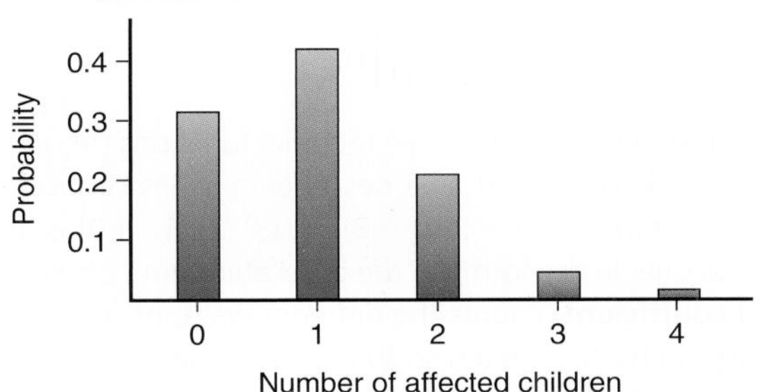

Figure 3.14 ▶ Probability distribution for families with four children segregating a recessive trait.

in this case there are four distinct events. **FIGURE 3.14** summarizes the calculations in the form of a probability distribution. As anticipated, three unaffected children and one affected child is the most probable outcome (probability 108/256).

In this example, the children fall into two possible phenotypic classes. Because there are only two classes, the probabilities associated with the various outcomes are called **binomial probabilities.** The Focus on Binomial Probabilities generalizes the analysis of this example to other situations involving two phenotypic classes.

GENETIC COUNSELING

The diagnosis of genetic conditions is often a difficult process. Typically, diagnoses are made by physicians who have been trained in genetics. The study of these conditions requires a great deal of careful research, including examining patients, interviewing relatives, and sifting through vital statistics on births, deaths, and marriages. The accumulated data provide the basis for defining the condition clinically and for determining its mode of inheritance.

Prospective parents may want to know whether their children are at risk to inherit a particular condition, especially if other family members have been affected. It is the responsibility of the genetic counselor to assess such risks and to explain them to the prospective parents. Risk assessment requires familiarity with probability and statistics, as well as a thorough knowledge of genetics.

As an example, let's consider a pedigree showing the inheritance of **nonpolypoid colorectal cancer** (**FIGURE 3.15**). This

▶ FOCUS ON **Binomial Probabilities**

The progeny of crosses sometimes segregate into two distinct classes—for example, male or female, healthy or diseased, normal or mutant, dominant phenotype or recessive phenotype. To be general, we can refer to these two kinds of progeny as P and Q, and note that for any individual offspring, the probability of being P is p and the probability of being Q is q. Because there are only two classes, $q = 1 - p$. Suppose that the total number of progeny is n and that each one is produced independently. We can calculate the **binomial probability** that exactly x of the progeny will fall into one class and y into the other:

$$\text{Probability of } x \text{ in class P and } y \text{ in class Q} = \left[\frac{n!}{x!\,y!}\right] p^x q^y$$

The bracketed term contains three factorial functions ($n!$, $x!$, and $y!$), each of which is computed as a descending series of products. For example, $n! = n(n-1)(n-2)(n-3) \ldots (3)(2)(1)$. If 0! is needed, it is defined as one. In the formula, the bracketed term, often called the **binomial coefficient,** counts the different ways, or orders, in which n offspring can be segregated so that x fall in the P class and y fall in the Q class. The other term, $p^x q^y$, gives the probability of obtaining a particular way or order. Because each of the orders is equally likely, multiplying this term by the bracketed term gives the probability of obtaining x progeny in the P class and y in the Q class, regardless of the order of occurrence.

If, for fixed values of n, p, and q, we systematically vary x and y, we can calculate a whole set of probabilities. This set constitutes a binomial probability distribution. With the distribution, we can answer questions such as "What is the probability that x will exceed a particular value?" or "What is the probability that x will lie between two particular values?" For example, let's consider a family with six children. What is the probability that at least four will be girls? To answer this question, we note that for any given child, the probability that it will be a girl (p) is 1/2 and the probability that it will be a boy (q) is also 1/2. The probability that exactly four children in a family will be girls (and two will be boys) is therefore $[(6!)/(4!\,2!)](1/2)^4\,(1/2)^2 = 15/64$, which is one of the terms in the binomial distribution. However, the probability that at least four will be girls (and that no more than two will be boys) is the sum of three terms from this distribution:

Event	**Binomial Formula**	**Probability**
4 girls and 2 boys	$[(6!)/(4!\,2!)] \times (1/2)^4\,(1/2)^2 =$	15/64
5 girls and 1 boy	$[(6!)/(5!\,1!)] \times (1/2)^5\,(1/2)^1 =$	6/64
6 girls and 0 boys	$[(6!)/(6!\,0!)] \times (1/2)^6\,(1/2)^0 =$	1/64

Therefore, the answer is $(15/64) + (6/64) + (1/64) = 22/64$.

The binomial distribution also provides answers to other kinds of questions. For example, what is the probability that at least one but no more than four of the children will be girls? Here the answer is the sum of four terms:

Event	**Binomial Formula**	**Probability**
1 girl and 5 boys	$[(6!)/(1!\,5!)] \times 6/64$	$(1/2)^1\,(1/2)^5 =$
2 girls and 4 boys	$[(6!)/(2!\,4!)] \times 15/64$	$(1/2)^2\,(1/2)^4 =$
3 girls and 3 boys	$[(6!)/(3!\,3!)] \times 20/64$	$(1/2)^3\,(1/2)^3 =$
4 girls and 2 boys	$[(6!)/(4!\,2!)] \times 15/64$	$(1/2)^4\,(1/2)^2 =$

Summing up, we find that the answer is 56/64.

Let's now consider the example discussed in the section on Mendelian Segregation in Human Families. A man and a woman, who are both heterozygous for the recessive mutant allele that causes cystic fibrosis, plan to have four children. What is the chance that one of these children will have cystic fibrosis and the other three will not? We have already seen by enumeration that the answer to this question is 108/256 (see **FIGURE 3.14**). However, this answer could also be obtained by using the binomial formula. The probability that a particular child will be affected is $p = 1/4$, and the probability that it will not be affected is $q = 3/4$. The total number of children is $n = 4$, the number of affected children is $x = 1$, and the number of unaffected children is $y = 3$. Putting all this together, we can calculate the probability that exactly one of the couple's four children will have cystic fibrosis as

$$[4!/(1!3!)]\,(1/4)^1\,(3/4)^3 = 4 \times (1/4) \times (27/64) = 108/256$$

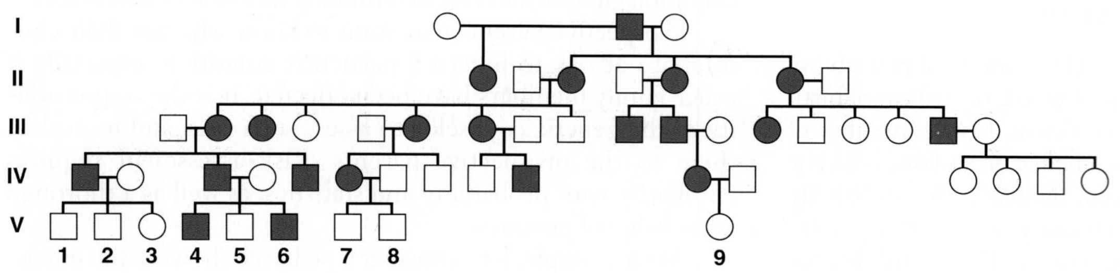

Figure 3.15 ▶ Pedigree showing the inheritance of hereditary non-polypoid colorectal cancer.

disease is one of several types of cancer that are inherited. It is due to a dominant mutation that affects about 1 in 500 individuals in the general population. The median age when hereditary nonpolypoid colorectal cancer appears in an individual who carries the mutation is 42. In the pedigree, we see that the cancer is manifested in at least one individual in each generation and that every affected individual has an affected parent. These facts are consistent with the dominant mode of inheritance of this disease.

The counseling issue arises in generation V. Among the nine individuals shown, two are affected and seven are not. Yet each of the seven unaffected individuals had one affected parent who must have been heterozygous for the cancer-causing mutation. Some of these seven unaffected individuals may therefore have inherited the mutation and would be at risk to develop nonpolypoid colorectal cancer later in life. Only time will tell. As the unaffected individuals age, those who carry the mutation will be at increased risk to develop the disease. Thus, the longer they remain unaffected, the greater the probability that they are actually not carriers. In this situation, the risk is a function of an individual's age and must be ascertained empirically from data on the age of onset of the disease among individuals from the same population, if possible from the same family. Each of the seven unaffected individuals will, of course, have to live with the anxiety of being a possible carrier of the cancer-causing mutation. Furthermore, at some point they will have to decide if they wish to reproduce and risk transmitting the mutation to their children. We shall discuss other inherited cancers and related counseling issues in Chapter 22.

As another example, consider the situation shown in **FIGURE 3.16.** A couple, denoted R and S in Figure 3.16a, is concerned about the possibility that they will have a child (T) with **albinism,** a recessive condition characterized by a complete

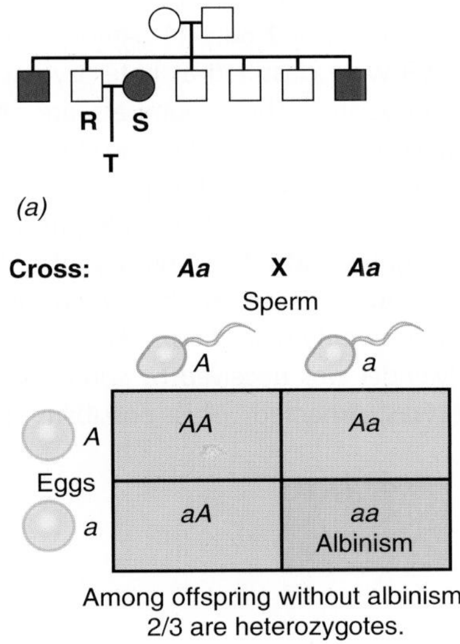

Figure 3.16 ▶ Genetic counseling in a family with albinism. (*a*) Pedigree showing the inheritance of albinism. (*b*) Punnett square showing that, among offspring without albinism, the frequency of heterozygotes is 2/3.

▶ FOCUS ON PROBLEM SOLVING
Making Predictions from Pedigrees

THE PROBLEM

This pedigree shows the inheritance of a recessive trait in humans. Individuals that have the trait are homozygous for a recessive allele *a*. If H and I, who happen to be first cousins, marry and have a child, what is the chance that this child will have the recessive trait?

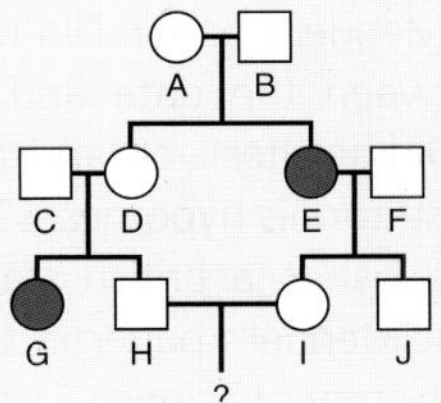

FACTS AND CONCEPTS

1. The child can show a recessive trait only if both of its parents carry the recessive allele.
2. One parent (H) has a sister (G) with the trait.
3. The other parent (I) has a mother (E) with the trait.
4. The chance that a heterozygote will transmit a recessive allele to its offspring is 1/2.
5. In a mating between two heterozygotes, 2/3 of the offspring that do not show the trait are expected to be heterozygotes (see **FIGURE 3.16b**).

ANALYSIS AND SOLUTION

I must be a heterozygous carrier of the recessive allele because her mother E is homozygous for it but she herself does not show the trait. I therefore has a 1/2 chance of transmitting the recessive allele to her child. Because H's sister has the trait, both of her parents must be heterozygotes. H, who does not show the trait, therefore has a 2/3 chance of being a heterozygote, and if he is, there is a 1/2 chance that he will transmit the recessive allele to his child. Putting all these factors together, we calculate the chance that the child of H and I will show the trait as 1/2 (the chance that I transmits the recessive allele) × 2/3 (the chance that H is a heterozygote) × 1/2 (the chance that H transmits the recessive allele assuming that he is a heterozygote) = 1/6, which is a fairly substantial risk.

For further discussion go to your *WileyPLUS* course.

▶ A MILESTONE IN GENETICS: **Mendel's 1866 Paper**

The paper that launched the science of genetics had the title "Versuche über Pflanzenhybriden"—which translates from the German as "Experiments with Plant-Hybrids." This paper detailed Gregor Mendel's studies on inheritance in peas. It was published in 1866 in the proceedings of the Natural History Society of Brünn.[1] Early in the previous year, Mendel had presented the results of his studies in lectures at two of the Society's meetings.

Not many scientific papers have had the impact of Mendel's has had. Initially, the paper was ignored. However, when its significance was recognized, it became required reading for anyone interested in the study of heredity. Scrutiny by generations of readers has raised many questions about Mendel's paper. Did Mendel report his results literally? Is the fit between the data and the predictions of his hypotheses too good? Did he alter—or fabricate—the data to present the most compelling case for his hypotheses?

In 1936 Ronald A. Fisher, a British statistician and geneticist, presented an analysis of Mendel's paper in the *Annals of Science,* a journal devoted to the history of science.[2] Fisher carefully attempted to reconstruct what Mendel actually did and when he did it. Mendel's research seems to have begun with the cultivation of stocks of peas in 1857. The first hybridizations between different varieties were apparently performed in 1858; other hybridizations were performed in 1859. Mendel followed the progeny of these crosses for as many as six generations. Fisher conjectures that Mendel began his dihybrid and trihybrid crosses in 1861. In that year, he also apparently began testcrosses to determine gametic ratios from heterozygous plants. Altogether, Mendel's experiments with peas spanned eight years—from 1857 to 1864. The magnitude of these experiments is impressive. In some years Mendel grew more than 5000 pea plants in the monastery garden.

Fisher considers the question of whether or not Mendel's account of the experiments with peas can be taken literally:

> Mendel's paper is, as has been frequently noted, a model in respect of the order and lucidity with which the successive relevant facts are presented, and such orderly presentation would be much facilitated had the author felt himself at liberty to ignore the particular crosses and years to which the plants contributing to any special result might belong. Mendel was an experienced and successful teacher, and might well have adopted a style of presentation suitable for the lecture-room without feeling under any obligation to complicate his story by unessential details. The style of didactic presentation, with its conventional simplifications, represents, as is well known, a tradition far more ancient among scientific writers than the more literal narratives in which experiments are now habitually presented.[3]

After examining the evidence in the paper, Fisher concludes that Mendel's account of his experiments should be taken literally: "His experiments were carried out in just the way and much in the order that they are recounted."[4]

Fisher also considers the question of whether or not Mendel's data agree too well with the predictions of his hypotheses. For example, using the data Mendel obtained in 1863, Fisher calculates a χ^2 statistic to test for the goodness of fit between the observations and the expectations. The result, 15.54, is less than half the expected value of the χ^2 distribution with 41 degrees of freedom, and Fisher says "that so low a value could scarcely occur by chance once in 2000 trials."[5] Thus, he concludes, "There can be no doubt that the data from the later years of the experiment have been biased strongly in the direction of agreement with expectation."[6]

The bias in favor of the expected results is most apparent in the experiments Mendel performed to determine if F_2 plants with a dominant phenotype were homozygous or heterozygous for the dominant allele. His procedure was to allow the plants to self-fertilize; then he examined 10 of the offspring. If any of the 10 showed the recessive phenotype, the parent was classified as a heterozygote. If none of the 10 showed this phenotype, it was classified as a homozygote. Fisher notes that with this procedure, some heterozygotes will incorrectly be classified as homozygotes simply by chance. The probability of this error is $(0.75)^{10} = 0.0563$. Thus, the expected ratio of segregating to nonsegregating parents is not 2:1 but $2 - 2 \times 0.0563$: $1 + 2 \times 0.0563$, or 1.88:1.11. Among the 600 plants that Mendel tested, 399 were classified as heterozygotes and 201 were classified as homozygotes. These numbers are very close to the expectations based on a 2:1 ratio, but not so close to the expectations based on the ratio corrected for the probability of misclassifying heterozygotes as homozygotes. With this ratio, the expected numbers are 377.5 heterozygotes and 222.5 homozygotes. Fisher notes that the deviation of Mendel's data from these predicted numbers "is to be expected once in twenty-nine trials."[7] For this discrepancy, Fisher suggests "that Mendel was deceived by some assistant who knew too well what was expected. This possibility is supported by

[1]Mendel, G. 1866. Versuche über Pflanzenhybriden. *Verhandlungen Naturforshender Vereines in Brünn* 10.

[2]Fisher, R. A. 1936. Has Mendel's work been rediscovered? *Annals of Science* 1:115–137.

[3]Ibid.

[4]Ibid.

[5]Ibid.

[6]Ibid.

[7]Ibid.

independent evidence that the data of most, if not all, of the experiments have been falsified so as to agree closely with Mendel's expectations."[8]

In 2004 the respected geneticist Edward Novitski published an article about Fisher's criticism of Mendel's results.[9] Novitski concedes that overall, these results "conform more closely with the ratios . . . theoretically expected than one might reasonably expect to obtain on a chance basis," and that "Particularly troubling are those two groups of experiments in which Mendel's results are in close agreement with ratios that Mendel may have considered appropriate, but which were, according to Fisher (1936), incorrect."[10] However, Novitski argues that Fisher's criticism of Mendel—especially his suggestion that some of the experimental data might have been falsified—is unfounded. For one thing, Mendel's procedure of ascertaining whether an F_2 plant with a dominant phenotype was homozygous or heterozygous for the dominant allele may have been affected by an error in the opposite direction of the one that Fisher described. Mendel based his decision about the genotype of a plant on the phenotypes of 10 of its progeny. Novitski proposes that for some of the plants, Mendel could not obtain the required 10 progeny. The failure rate for seed germination may have been as high as 2 percent. If fewer than 10 progeny were examined, and at least one of them had the recessive phenotype, then the parent plant could have been correctly classified as a heterozygote. However, if fewer than 10 progeny were examined—say only 8 or 9—and none of them showed the recessive phenotype, what was Mendel to do? Novitski conjectures that Mendel would not have classified the parent plant as a homozygote for the dominant allele; rather, he would have excluded that plant from consideration and replaced it with another plant that had been held in reserve. The excluded plant was most likely a dominant homozygote, and the reserve plant was most likely a heterozygote. Thus, in an effort to adhere to his rule of classifying a plant as a dominant homozygote only after counting 10 dominant progeny, Mendel may have skewed his results in favor of heterozygous plants. This error would counterbalance the one that Fisher described and move the observed ratio of heterozygotes to dominant homozygotes toward 2:1, which is what Mendel reported.

Novitski also suggests that Mendel may have repeated some of the experimental runs in which the results he obtained "did not appear to bear out his expectations, not with any intent to deceive, but to make certain for his own benefit that those runs were in fact not *bona fide* cases of exceptions to his rules. Having obtained additional data, he could have either replaced the earlier deviating data with the 'better' numbers or combined the two sets of data, which would usually obscure the extent of the deviations in the first set. . . . It might be observed that such procedures probably exist in preparing data for publication even today."[11]

As for the possibility that Mendel altered or manipulated his data, Novitski says that if he did, his alterations "would stem not from any desire to mislead, but as a concession to his ill-prepared audience. We can imagine that at the time of writing for oral presentation, Mendel changed, for didactic purposes, some specific results that might have distracted his audience from the main theme of the article because of their seemingly aberrant nature. Surely words like fraud or dishonesty should be used with caution. Perhaps his situation can be compared to that of the competent high school science teacher who, in explaining the structure of the atom to his students, falls back on the simple Bohr model, well aware that while it is not the correct picture, it is appropriate for the audience for which it is intended."[12]

QUESTIONS FOR DISCUSSION

1. Today many scientific papers contain four main sections of text—Introduction, Materials and Methods, Results, and Discussion—and they are usually preceded by an abstract (or summary) and followed by a bibliography of references cited. How does this format vary among journals that publish papers on genetics (e.g., *Cell, Genetics, Proceedings of the National Academy of Sciences, Nature,* and *Science*)? How does it differ from the format of papers in other disciplines such as sociology, law, history, and literature?

2. Gregor Mendel seems to have done his research to satisfy his own curiosity. He did not benefit from the research financially or professionally, and, except for space in the monastery garden and perhaps the help of a few assistants, his work did not require any other type of support, say, for example, a government grant. What do these circumstances say about the kind of person Mendel was? How does he compare with professional scientists working on research projects today?

[8]Ibid.

[9]Novitski, E. 2004. On Fisher's criticism of Mendel's results with the garden pea. *Genetics* 166:1133–1136.

[10]Ibid.

[11]Ibid.

[12]Ibid.

absence of melanin pigment in the skin, eyes, and hair. S, the prospective mother, has albinism, and R, the prospective father, has two siblings with albinism. It would therefore seem that the child has some risk of being born with albinism.

This risk depends on two factors: (1) the probability that R is a heterozygous carrier of the albinism allele (*a*), and (2) the probability that he will transmit this allele to T if he actually is a carrier. S, who is obviously homozygous for the albinism allele, must transmit this allele to her offspring.

To determine the first probability, we need to consider the possible genotypes for R. One of these, that he is homozygous for the recessive allele (*aa*), is excluded because we know that he does not have albinism himself. However, the other two genotypes, *AA* and *Aa*, remain distinct possibilities. To calculate the probabilities associated with each of these, we note that both of R's parents must be heterozygotes, because they have had two children with albinism. The mating that produced R was therefore *Aa* × *Aa*, and from such a mating we would expect 2/3 of the offspring without albinism to be *Aa* and 1/3 to be *AA* (**FIGURE 3.16b**). Thus, the probability that R is a heterozygous carrier of the albinism allele is 2/3. To determine the probability that he will transmit this allele to his child, we simply note that *a* will be present in half of his gametes.

In summary, the risk that T will be *aa*

= [Probability that R is *Aa*] × [Probability that R transmits *a*, assuming that R is *Aa*]

= (2/3) × (1/2) = (1/3)

The example in Figure 3.16 illustrates a simple counseling situation in which the risk can be determined precisely. Often the circumstances are much more complicated, making the task of risk assessment quite difficult. The genetic counselor's responsibility is to analyze the pedigree information and determine the risk as precisely as possible.

For practice in calculating genetic risks, work through the example in the Focus on Problem Solving.

KEY POINTS

- Pedigrees are used to identify dominant and recessive traits in human families.
- The analysis of pedigrees allows genetic counselors to assess the risk that an individual will inherit a particular trait.

▸ Basic Exercises

ILLUSTRATE BASIC GENETIC ANALYSIS

1. Two highly inbred strains of mice, one with black fur and the other with gray fur, were crossed, and all of the offspring had black fur. Predict the outcome of intercrossing the offspring.

Answer: The two strains of mice are evidently homozygous for different alleles of a gene that controls fur color: *G* for black fur and *g* for gray fur; the *G* allele is dominant because all the F_1 animals are black. When these mice, genotypically *Gg*, are intercrossed, the *G* and *g* alleles will segregate from each other to produce an F_2 population consisting of three genotypes, *GG*, *Gg*, and *gg*, in the ratio 1:2:1. However, because of the dominance of the *G* allele, the *GG* and *Gg* genotypes will have the same phenotype (black fur); thus, the phenotypic ratio in the F_2 will be 3 black:1 gray.

2. A plant heterozygous for three independently assorting genes, *Aa Bb Cc*, is self-fertilized. Among the offspring, predict the frequency of (a) *AA BB CC* individuals, (b) *aa bb cc* individuals, (c) individuals that are either *AA BB CC* or *aa bb cc*, (d) *Aa Bb Cc* individuals, (e) individuals that are not heterozygous for all three genes.

Answer: Because the genes assort independently, we can analyze them one at a time to obtain the answers to each of the questions. (a) When *Aa* individuals are selfed, 1/4 of the offspring will be *AA*; likewise, for the *B* and *C* genes, 1/4 of the individuals will be *BB* and 1/4 will be *CC*. Thus, we can calculate the frequency (that is, the probability) of *AA BB CC* offspring as (1/4) × (1/4) × (1/4) = 1/64. (b) The frequency of *aa bb cc* individuals can be obtained using similar reasoning. For each gene the frequency of recessive homozygotes among the offspring is 1/4. Thus, the frequency of triple recessive homozygotes is (1/4) × (1/4) × (1/4) = 1/64. (c) To obtain the frequency of offspring that are either triple dominant homozygotes or triple recessive homozygotes—these are mutually exclusive events—we sum the results of (a) and (b): 1/64 + 1/64 = 2/64 = 1/32. (d) To obtain the frequency of offspring that are triple heterozygotes, again we multiply probabilities. For each gene, the frequency of heterozygous offspring is 1/2; thus, the frequency of triple heterozygotes should be (1/2) × (1/2) × (1/2) = 1/8. (e) Offspring that are not heterozygous for all three genes occur with a frequency that is one minus the frequency calculated in (d). Thus, the answer is 1 − 1/8 = 7/8.

3. Two true-breeding strains of peas, one with tall vines and violet flowers and the other with dwarf vines and white flowers, were crossed. All the F_1 plants were tall and produced violet flowers. When these plants were backcrossed to the dwarf, white parent strain, the following offspring were obtained: 53 tall, violet; 48 tall, white; 47 dwarf, violet; 52 dwarf, white. Do the genes that control vine length and flower color assort independently?

Answer: The hypothesis of independent assortment of the vine length and flower color genes must be evaluated by calculating a chi-square test statistic from the experimental results. To obtain this statistic, the results must be compared to the predictions of the genetic hypothesis. Under the assumption that the two genes assort independently, the four phenotypic classes in the F_2 should each be 25 percent of the total (200); that is, each should contain 50 individuals. To compute the chi-square statistic, we must obtain the difference between each observation and its predicted value, square these differences, divide each squared difference by the predicted value, and then sum the results:

$$\chi^2 = (53 - 50)^2/50 + (48 - 50)^2/50 + (47 - 50)^2/50 + (52 - 50)^2/50 = 0.52$$

This statistic must then be compared to the critical value of the chi-square frequency distribution for 3 degrees of freedom (calculated as the number of phenotypic classes minus one). Because the computed value of the chi-square statistic (0.52) is much less than the critical value (7.815; see Table 3.2), there is no evidence to reject the hypothesis of independent assortment of the vine length and flower color genes. Thus, we may tentatively accept the idea that these genes assort independently.

4. Is the trait that is segregating in the following pedigree due to a dominant or a recessive allele?

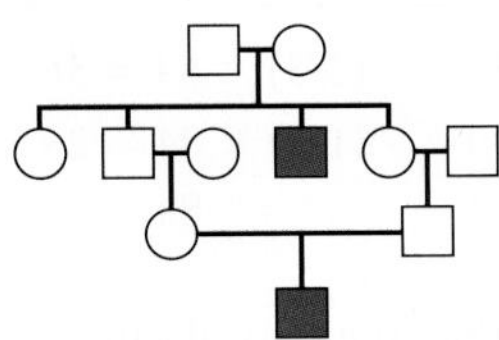

Answer: Both affected individuals have two unaffected parents, which is inconsistent with the hypothesis that the trait is due to a dominant allele. Thus, the trait appears to be due to a recessive allele.

5. In a family with three children, what is the probability that two are boys and one is a girl?

Answer: To answer this question, we must apply the theory of binomial probabilities. For any one child, the probability that it is a boy is 1/2 and the probability that it is a girl is 1/2. Each child is produced independently. Thus, the probability of two boys and one girl is $(1/2)^3$ times the number of ways in which two boys and one girl can appear in the birth order. By enumerating all the possible birth orders—BBG, BGB, and GBB—we find that the number of ways is 3. Thus, the final answer is $3 \times (1/2)^3 = 3/8$.

Testing Your Knowledge

INTEGRATE DIFFERENT CONCEPTS AND TECHNIQUES

1. Phenylketonuria, a metabolic disease in humans, is caused by a recessive allele, k. If two heterozygous carriers of the allele marry and plan a family of five children: (a) What is the chance that all their children will be unaffected? (b) What is the chance that four children will be unaffected and one affected with phenylketonuria? (c) What is the chance that at least three children will be unaffected? (d) What is the chance that the first child will be an unaffected girl?

Answer: Before answering each of the questions, note that from a mating between two heterozygotes, the probability that a particular child will be unaffected is 3/4, and the probability that a particular child will be affected is 1/4. Furthermore, for any one child born, the chance that it will be a boy is 1/2 and the chance that it will be a girl is 1/2.

(a) To calculate the chance that all five children will be unaffected, use the Multiplicative Rule. For each child, the chance that it will be unaffected is 3/4, and all five children are independent. Consequently, the probability of five unaffected children is $(3/4)^5 = 0.237$. This is the first term of the binomial probability distribution with $p = 3/4$ and $q = 1/4$.

(b) To calculate the chance that four children will be unaffected and one affected, compute the second term of the binomial distribution:

$$= [5!/(4!\ 1!)] \times (3/4)^4 \times (1/4)^1 = 5 \times (81/1024) = 0.399$$

(c) To find the probability that at least three children will be unaffected, sum the first three terms of the binomial distribution:

Event	Binomial Formula	Probability
5 unaffected, 0 affected	$[(5!)/(5!\ 0!)] \times (3/4)^5 (1/4)^0 =$	0.237
4 unaffected, 1 affected	$[(5!)/(4!\ 1!)] \times (3/4)^4 (1/4)^1 =$	0.399
3 unaffected, 2 affected	$[(5!)/(3!\ 2!)] \times (3/4)^3 (1/4)^2 =$	0.264
	Total 0.900	

(d) To determine the probability that the first child will be an unaffected girl, use the Multiplicative Rule: P(unaffected child and girl) = P(unaffected child) × P(girl) = (3/4) × (1/2) = (3/8).

2. Mice from wild populations typically have gray-brown (or *agouti*) fur, but in one laboratory strain, some of the mice have yellow fur. A single yellow male is mated to several agouti females. Altogether, the matings produce 40 progeny, 22 with agouti fur and 18 with yellow fur. The agouti F_1 animals are then intercrossed with each other to produce an F_2, all of which are agouti. Similarly, the yellow F_1 animals are intercrossed with each other, but their F_2 progeny segregate into two classes; 30 are agouti and 54 are yellow. Subsequent crosses between yellow F_2 animals also segregate yellow and agouti progeny. What is the genetic basis of these coat color differences?

Answer: We note that the cross agouti × agouti produces only agouti animals and that the cross yellow × yellow produces a mixture of yellow and agouti. Thus, a reasonable hypothesis is that yellow fur is caused by a dominant allele, A, and that agouti fur is caused by a recessive allele, a. According to this hypothesis, the agouti females used in the initial cross would be aa and their yellow mate would be Aa. We hypothesize that the male was heterozygous because he produced approximately equal numbers of agouti and yellow F_1 offspring. Among these, the agouti animals should be aa and the yellow animals Aa. These genotypic assignments are borne out by the F_2 data, which show that the F_1 agouti mice have bred true and the F_1 yellow mice have segregated. However, the segregation ratio of yellow to agouti (54:30) seems to be out of line with the Mendelian expectation of 3:1. Is this lack of fit serious enough to reject the hypothesis?

We can use the χ^2 procedure to test for disagreement between the data and the predictions of the hypothesis. According to the hypothesis, 3/4 of the F_2 progeny from the yellow × yellow intercross should be yellow and 1/4 should be agouti. Using these proportions, we can calculate the expected numbers of progeny in each class and then calculate a χ^2 statistic with $2 - 1 = 1$ degree of freedom.

F_2 Phenotype	Obs	Exp	(Obs-Exp)^2/Exp
yellow (*AA* and *Aa*)	54	$(3/4) \times 84 = 63$	1.286
agouti (*aa*)	30	$(1/4) \times 84 = 21$	3.857
Total	84	84	5.143

The χ^2 statistic (5.143) is much greater than the critical value (3.841) for a χ^2 distribution with 1 degree of freedom. Consequently, we reject the hypothesis that the coat colors are segregating in a 3:1 Mendelian fashion.

What might account for the failure of the coat colors to segregate as hypothesized? We obtain a clue by noting that subsequent yellow × yellow crosses failed to establish a true-breeding yellow strain. This suggests that the yellow animals are all *Aa* heterozygotes and that the *AA* homozygotes produced by matings between heterozygotes do not survive to the adult stage. Embryonic death is, in fact, why the yellow mice are underrepresented in the F_2 data. Examination of the uteruses of pregnant females reveals that about 1/4 of the embryos are dead. These dead embryos must be genotypically *AA*. Thus, a single copy of the *A* allele produces a visible phenotypic effect (yellow fur), but two copies cause death. Taking this embryonic mortality into account, we can modify the hypothesis and predict that 2/3 of the live-born F_2 progeny should be yellow (*Aa*) and 1/3 should be agouti (*aa*). We can then use the χ^2 procedure to test this modified hypothesis for consistency with the data.

F_2 Phenotype	Obs	Exp	(Obs-Exp)^2/Exp
yellow (*Aa*)	54	$(2/3) \times 84 = 56$	0.071
agouti (*aa*)	30	$(1/3) \times 84 = 28$	0.143
Total	84	84	0.214

This χ^2 statistic is less than the critical value for a χ^2 distribution with 1 degree of freedom. Thus, the data are in agreement with the predictions of the modified hypothesis.

▶ Questions and Problems

ENHANCE UNDERSTANDING AND DEVELOP ANALYTICAL SKILLS

3.1 On the basis of Mendel's observations, predict the results from the following crosses with peas: (a) a tall (dominant and homozygous) variety crossed with a dwarf variety; (b) the progeny of (a) self-fertilized; (c) the progeny from (a) crossed with the original tall parent; (d) the progeny of (a) crossed with the original dwarf parent.

3.2 A woman has a rare abnormality of the eyelids called ptosis, which prevents her from opening her eyes completely. This condition is caused by a dominant allele, *P*. The woman's father had ptosis, but her mother had normal eyelids. Her father's mother had normal eyelids.

(a) What are the genotypes of the woman, her father, and her mother?

(b) What proportion of the woman's children will have ptosis if she marries a man with normal eyelids?

3.3 In pigeons, a dominant allele *C* causes a checkered pattern in the feathers; its recessive allele *c* produces a plain pattern. Feather coloration is controlled by an independently assorting gene; the dominant allele *B* produces red feathers, and the recessive allele *b* produces brown feathers. Birds from a true-breeding checkered, red variety are crossed to birds from a true-breeding plain, brown variety.

(a) Predict the phenotype of their progeny.

(b) If these progeny are intercrossed, what phenotypes will appear in the F_2, and in what proportions?

3.4 In shorthorn cattle, the genotype *RR* causes a red coat, the genotype *rr* causes a white coat, and the genotype *Rr* causes a roan coat. A breeder has red, white, and roan cows and bulls. What phenotypes might be expected from the following matings, and in what proportions?

(a) red × red

(b) red × roan

(c) red × white

(d) roan × roan

3.5 How many different kinds of F_1 gametes, F_2 genotypes, and F_2 phenotypes would be expected from the following crosses?

(a) *AA* × *aa*

(b) *AA BB* × *aa bb*

(c) *AA BB CC* × *aa bb cc*

(d) What general formulas are suggested by these answers?

3.6 GO In mice, the allele *F* for colored fur is dominant over the allele *f* for white fur, and the allele *B* for normal behavior is dominant over the allele *b* for waltzing behavior, a form of discoordination. Give the genotypes of the parents in each of the following crosses:

(a) Colored, normal mice mated with white, normal mice produced 31 colored, normal and 8 colored, waltzing progeny.

(b) Colored, normal mice mated with colored, normal mice produced 39 colored, normal; 14 colored, waltzing; 12 white, normal; and 3 white; waltzing progeny.

(c) Colored, normal mice mated with white, waltzing mice produced 8 colored, normal; 7 colored, waltzing; 9 white, normal; and 6 white, waltzing progeny.

3.7 In rabbits, the dominant allele *B* causes black fur and the recessive allele *b* causes brown fur; for an independently assorting gene, the dominant allele *R* causes long fur and the recessive allele *r* (for *rex*) causes short fur. A homozygous rabbit with long, black fur is crossed with a rabbit with short, brown fur, and the offspring are intercrossed. In the F_2, what proportion of the rabbits with long, black fur will be homozygous for both genes?

3.8 GO A researcher studied six independently assorting genes in a plant. Each gene has a dominant and a recessive allele: *R* black stem, *r* red stem; *D* tall plant, *d* dwarf plant; *C* full pods, *c* constricted pods; *O* round fruit, *o* oval fruit; *H* hairless leaves, *h* hairy leaves; *W* purple flower, *w* white flower. From the cross (P1) *Rr Dd cc Oo Hh Ww* × (P2) *Rr dd Cc oo Hh ww*,

(a) How many kinds of gametes can be formed by P1?
(b) How many genotypes are possible among the progeny of this cross?
(c) How many phenotypes are possible among the progeny?
(d) What is the probability of obtaining the *Rr Dd cc Oo hh ww* genotype in the progeny?
(e) What is the probability of obtaining a black, dwarf, constricted, oval, hairy, purple phenotype in the progeny?

3.9 Albinism in humans is caused by a recessive allele *a*. From marriages between people known to be carriers (*Aa*) and people with albinism (*aa*), what proportion of the children would be expected to have albinism? Among three children, what is the chance of one with albinism and two without albinism?

3.10 If both husband and wife are known to be carriers of the allele for albinism, what is the chance of the following combinations in a family of five children: (a) all five unaffected; (b) three unaffected and two affected; (c) two unaffected and three affected; (d) one unaffected and four affected?

3.11 If a man and a woman are heterozygous for a gene, and if they have four children, what is the chance that all four will also be heterozygous?

3.12 If eight babies are born on a given day: (a) What is the chance that four will be boys and four girls? (b) What is the chance that all eight will be girls? (c) What combination of boys and girls among eight babies is most likely? (d) What is the chance that at least one baby will be a girl?

3.13 In a family of five children, what is the chance that at least two are girls?

3.14 GO Mendel testcrossed pea plants grown from yellow, round F_1 seeds to plants grown from green, wrinkled seeds and obtained the following results: 30 yellow, round; 35 green, round; 28 yellow, wrinkled; and 27 green, wrinkled. Are these results consistent with the hypothesis that seed color and seed texture are controlled by independently assorting genes, each segregating two alleles?

3.15 In humans, cataracts in the eyes and fragility of the bones are caused by dominant alleles that assort independently. A man with cataracts and normal bones marries a woman without cataracts but with fragile bones. The man's father had normal eyes, and the woman's father had normal bones. What is the probability that the first child of this couple will (a) be free from both abnormalities; (b) have cataracts but not have fragile bones; (c) have fragile bones but not have cataracts; (d) have both cataracts and fragile bones?

3.16 In generation V in the pedigree in **Figure 3.15**, what is the probability of observing seven children without the cancer-causing mutation and two children with this mutation among a total of nine children?

3.17 For each of the following situations, determine the degrees of freedom associated with the χ^2 statistic and decide whether or not the observed χ^2 value warrants acceptance or rejection of the hypothesized genetic ratio.

	Hypothesized Ratio	Observed χ^2
(a)	3:1	8.0
(b)	1:2:1	12.0
(c)	1:1:1:1	6.0
(d)	9:3:3:1	2.0

3.18 The following pedigree shows the inheritance of a dominant trait. What is the chance that the offspring of the following matings will show the trait: (a) III-1 × III-3; (b) III-2 × III-4?

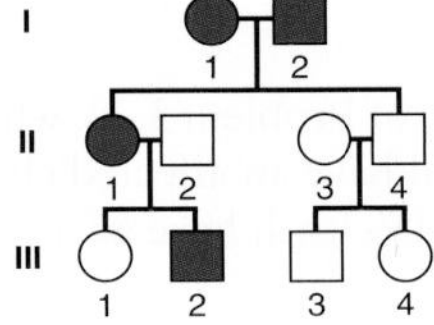

3.19 Perform a chi-square test to determine if an observed ratio of 30 tall: 20 dwarf pea plants is consistent with an expected ratio of 1:1 from the cross *Dd* × *dd*.

3.20 GO Peas heterozygous for three independently assorting genes were intercrossed.

(a) What proportion of the offspring will be homozygous for all three recessive alleles?
(b) What proportion of the offspring will be homozygous for all three genes?
(c) What proportion of the offspring will be homozygous for one gene and heterozygous for the other two?
(d) What proportion of the offspring will be homozygous for the recessive allele of at least one gene?

3.21 The following pedigree shows the inheritance of a recessive trait. Unless there is evidence to the contrary, assume that the individuals who have married into the family do not carry the recessive allele. What is the chance that the offspring of the following matings will show the trait: (a) III-1 × III-12; (b) II-4 × III-14; (c) III-6 × III-13; (d) IV-1 × IV-2?

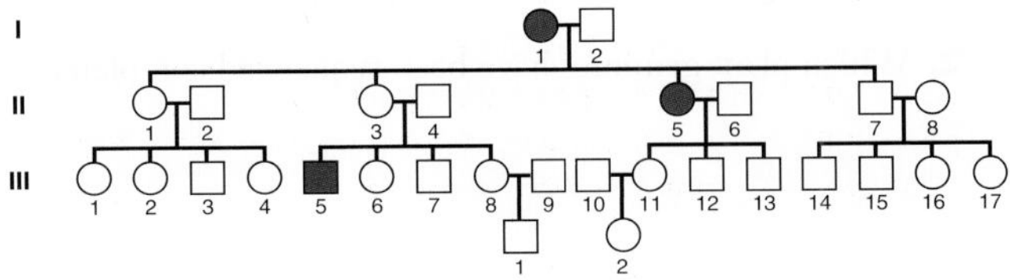

3.22 In the following pedigrees, determine whether the trait is more likely to be due to a dominant or a recessive allele. Assume the trait is rare in the population.

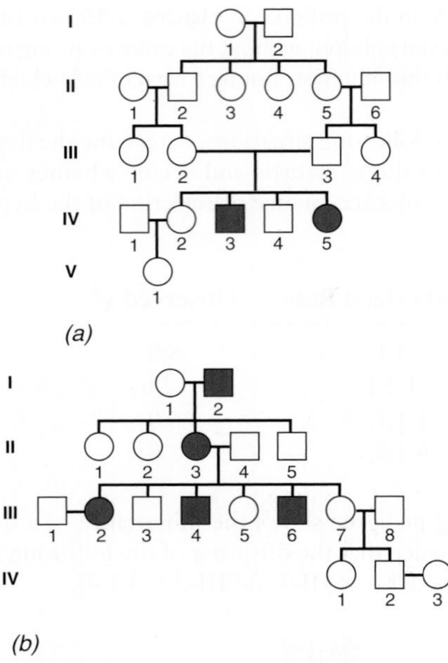

3.23 In pedigree (*b*) of Problem 3.22, what is the chance that the couple III-1 and III-2 will have an affected child? What is the chance that the couple IV-2 and IV-3 will have an affected child?

3.24 GO A geneticist crosses tall pea plants with short pea plants. All the F_1 plants are tall. The F_1 plants are then allowed to self-fertilize, and the F_2 plants are classified by height: 62 tall and 26 short. From these results, the geneticist concludes that shortness in peas is due to a recessive allele (*s*) and that tallness is due to a dominant allele (*S*). On this hypothesis, 2/3 of the tall F_2 plants should be heterozygous *Ss*. To test this prediction, the geneticist uses pollen from each of the 62 tall plants to fertilize the ovules of emasculated flowers on short pea plants. The next year, three seeds from each of the 62 crosses are sown in the garden, and the resulting plants are grown to maturity. If none of the three plants from a cross is short, the male parent is classified as having been homozygous *SS*; if at least one of the three plants from a cross is short, the male parent is classified as having been heterozygous *Ss*. Using this system of progeny testing, the geneticist concludes that 29 of the 62 tall F_2 plants were homozygous *SS* and that 33 of these plants were heterozygous *Ss*.

(a) Using the chi-square procedure, evaluate these results for goodness of fit to the prediction that 2/3 of the tall F2 plants should be heterozygous.

(b) Explain why the geneticist's procedure for classifying tall F2 plants by genotype is not definitive.

(c) Adjust for the uncertainty in the geneticist's classification procedure and calculate the expected frequencies of homozygotes and heterozygotes among the tall F_2 plants.

(d) Evaluate the predictions obtained in (c) using the chi-square procedure.

3.25 The following pedigree shows the inheritance of a recessive trait. What is the chance that the couple III-3 and III-4 will have an affected child?

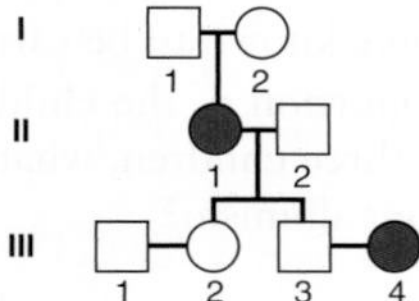

3.26 A researcher who has been studying albinism has identified a large group of families with four children in which at least one child shows albinism. None of the parents in this group of families shows albinism. Among the children, the ratio of those without albinism to those with albinism is 1.7:1. The researcher is surprised by this result because he thought that a 3:1 ratio would be expected on the basis of Mendel's Principle of Segregation. Can you explain the apparently non-Mendelian segregation ratio in the researcher's data?

▶ Genomics on the Web

at http://www.ncbi.nlm.nih.gov/

1. Gregor Mendel worked out the rules of inheritance by performing experiments with peas (*Pisum sativum*). Has the genome of this organism been sequenced, or is it currently being sequenced?

2. Which plant genomes have been sequenced completely?

3. What is the scientific or agricultural significance of the plants whose genomes have been sequenced completely?

Hint*:* At the web site, Genomic Biology → Entrez Genome → Plant Genomes Central.

Chapter 4
Extensions of Mendelism

CHAPTER OUTLINE

- Allelic Variation and Gene Function
- Gene Action: From Genotype to Phenotype
- Inbreeding: Another Look at Pedigrees

Diverse species of plants growing in a garden. Experiments with many different plants extended Mendel's Principles of Dominance, Segregation, and Independent Assortment.

Genetics Grows Beyond Mendel's Monastery Garden

In 1902, enthused by what he read in Mendel's paper, the British biologist William Bateson published an English translation of Mendel's German text and appended to it a brief account of what he called "Mendelism—the Principles of Dominance, Segregation, and Independent Assortment." Later, in 1909, he published *Mendel's Principles of Heredity,* in which he summarized all the evidence then available to support Mendel's findings. This book was remarkable for two reasons. First, it examined the results of breeding experiments with many different plants and animals and in each case demonstrated that Mendel's principles applied. Second, it considered the implications of these experiments and raised questions about the fundamental nature of genes, or, as Bateson called them, "unit-characters." At the time Bateson's book was published, the word "gene" had not yet been coined.

Bateson's book played a crucial role in spreading the principles of Mendelism to the scientific world. Botanists, zoologists, naturalists, horticulturalists, and animal breeders got the message in plain and simple language: Mendel's principles—tested by experiments with peas, beans, sunflowers, cotton, wheat, barley, tomatoes, maize, and assorted ornamental plants, as well as with cattle, sheep, cats, mice, rabbits, guinea pigs, chickens, pigeons, canaries, and moths—were universal. In the preface to his book, Bateson remarked that "The study of heredity thus becomes an organized branch of physiological science, already abundant in results, and in promise unsurpassed."[1]

[1]Bateson, W. 1909. *Mendel's Principles of Heredity.* University Press, Cambridge, England.

Allelic Variation and Gene Function

The diverse kinds of genes affect phenotypes in different ways.

Mendel's experiments established that genes can exist in alternate forms. For each of the seven traits that he studied—seed color, seed texture, plant height, flower color, flower position, pod shape, and pod color—Mendel identified two alleles, one dominant, the other recessive. This discovery suggested a simple functional dichotomy between alleles, as if one allele did nothing and the other did everything to determine the phenotype. However, research early in the twentieth century demonstrated this to be an oversimplification. Genes can exist in more than two allelic states, and each allele can have a different effect on the phenotype.

INCOMPLETE DOMINANCE AND CODOMINANCE

An allele is dominant if it has the same phenotypic effect in heterozygotes as in homozygotes—that is, the genotypes *Aa* and *AA* are phenotypically indistinguishable. Sometimes, however, a heterozygote has a phenotype different from that of either of its associated homozygotes. Flower color in the snapdragon, *Antirrhinum majus*, is an example. White and red varieties are homozygous for different alleles of a color-determining gene; when crossed, they produce heterozygotes that have pink flowers. The allele for red color (*W*) is therefore said to be **incompletely,** or **partially, dominant** over the allele for white color (*w*). The most likely explanation is that the intensity of pigmentation in this species depends on the amount of a product specified by the color gene (**FIGURE 4.1**). If the *W* allele specifies this product and the *w* allele does not, *WW* homozygotes will have twice as much of the product as *Ww* heterozygotes do and will therefore show deeper color. When the heterozygote's phenotype is midway between the phenotypes of the two homozygotes, as it is here, the partially dominant allele is sometimes said to be **semidominant** (from the Latin word for "half"—thus half-dominant).

Another exception to the principle of simple dominance arises when a heterozygote shows characteristics found in each of the associated homozygotes. This occurs with human blood types, which are identified by testing for special cellular products called *antigens.* An antigen is detected by its ability to react with factors obtained from the serum portion of the blood. These factors, which are produced by the immune system, recognize antigens quite specifically. Thus, for example, one serum, called anti-M, recognizes only the M antigen on human blood cells; another serum, called anti-N, recognizes only the N antigen on these cells (**FIGURE 4.2**). When one of these sera detects its specific antigen in a blood-typing test, the cells clump together in a reaction called *agglutination.* Thus, by testing cells for agglutination with different sera, a medical technologist can identify which antigens are present and thereby determine the blood type.

The ability to produce the M and N antigens is determined by a gene with two alleles. One allele allows the M antigen to be produced; the other allows the N antigen to be produced. Homozygotes for the M allele produce only the M antigen, and homozygotes for the N allele produce only the N antigen. However, heterozygotes for these two alleles produce both kinds of antigens. Because the two alleles appear to contribute independently to the phenotype of the heterozygotes, they are said to be **codominant.** Codominance implies that there is an independence of allele function. Neither allele is dominant, or even partially dominant, over the other. It would therefore be inappropriate to distinguish the alleles by upper and lowercase letters, as we have in all previous examples. Instead, codominant alleles are represented by superscripts on the symbol for the gene, which in this case is the letter *L*—a tribute to Karl Landsteiner, the discoverer of blood-typing. Thus, the M allele is L^M and the N allele is L^N. **FIGURE 4.2** shows the three possible genotypes formed by the L^M and L^N alleles, and their associated phenotypes.

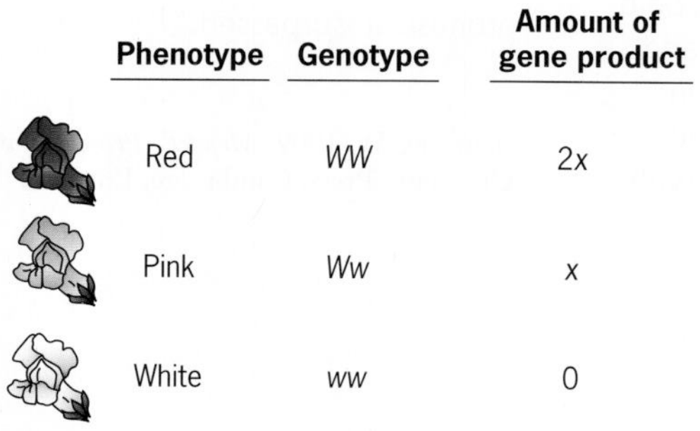

Phenotype	Genotype	Amount of gene product
Red	*WW*	2x
Pink	*Ww*	x
White	*ww*	0

Figure 4.1 ▶ Genetic basis of flower color in snapdragons. The allele *W* is incompletely dominant over *w*. Differences among the phenotypes could be due to differences in the amount of the product specified by the *W* allele.

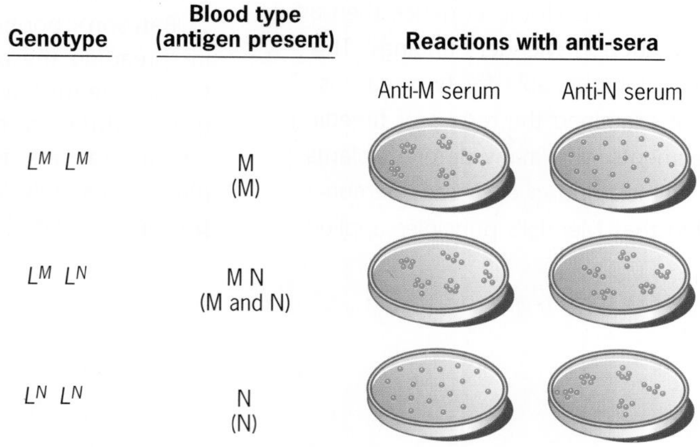

Genotype	Blood type (antigen present)	Reactions with anti-sera: Anti-M serum	Anti-N serum
$L^M L^M$	M (M)		
$L^M L^N$	M N (M and N)		
$L^N L^N$	N (N)		

Figure 4.2 ▶ Detection of the M and N antigens on blood cells by agglutination with specific anti-sera. With the anti-M and anti-N sera, three blood types can be identified.

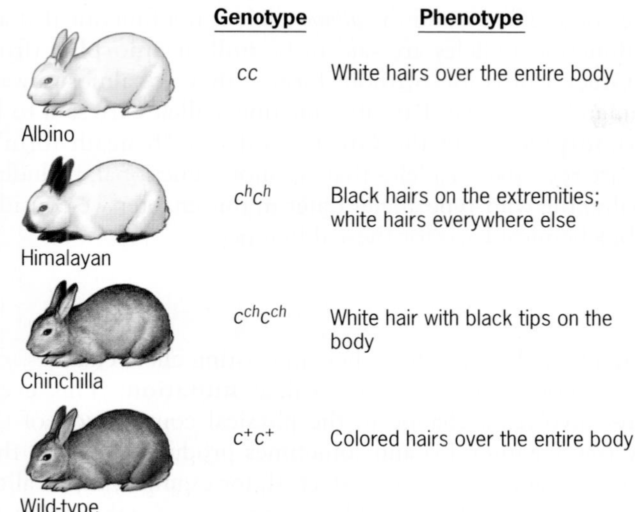

Genotype
Phenotype
Albino
cc
White hairs over the entire body
Himalayan
c^hc^h
Black hairs on the extremities; white hairs everywhere else
Chinchilla
$c^{ch}c^{ch}$
White hair with black tips on the body
Wild-type
c^+c^+
Colored hairs over the entire body

Figure 4.3 ▶ Coat colors in rabbits. The different phenotypes are caused by four different alleles of the *c* gene.

MULTIPLE ALLELES

The Mendelian concept that genes exist in no more than two allelic states had to be modified when genes with three, four, or more alleles were discovered. A classic example of a gene with **multiple alleles** is the one that controls coat color in rabbits (**FIGURE 4.3**). The color-determining gene, denoted by the lowercase letter *c*, has four alleles, three of which are distinguished by a superscript: *c (albino)*, c^h *(himalayan)*, c^{ch} *(chinchilla)*, and c^+ (wild-type). In homozygous condition, each allele has a characteristic effect on the coat color. Because most rabbits in wild populations are homozygous for the c^+ allele, this allele is called the **wild-type.** In genetics it is customary to represent wild-type alleles by a superscript plus sign after the letter for the gene. When the context is clear, the letter is sometimes omitted and only the plus sign is used; thus, c^+ may be abbreviated simply as +.

The other alleles of the *c* gene are **mutants**—altered forms of the wild-type allele that must have arisen sometime during the evolution of the rabbit. The *himalayan* and *chinchilla* alleles are denoted by superscripts, but the *albino* allele is denoted simply by the letter *c* (for colorless, another word for the albino condition). This notation reflects another custom in genetics nomenclature: genes are often named for a mutant allele, usually the allele associated with the most abnormal phenotype. The convention of naming a gene for a mutant allele is generally consistent with the convention we discussed in Chapter 3—that of naming genes for a recessive allele—because most mutant alleles are recessive. However, sometimes a mutant allele is dominant, in which case the gene is named after its associated phenotype. For example, a gene in mice controls the length of the tail. The first mutant allele of this gene that was discovered caused a shortening of the tail in heterozygotes. This dominant mutant was therefore symbolized by *T*, for *t*ail-length. All other alleles of this gene—and there are many—have been denoted by an uppercase or lowercase letter "t," depending on whether they are dominant or recessive; different alleles are distinguished from each other by superscripts.

Another example of multiple alleles comes from the study of human blood types. The A, B, AB, and O blood types, like the M, N, and MN blood types discussed previously, are identified by testing a blood sample with different sera. One serum detects the A antigen, another the B antigen. When only the A antigen is present on the cells, the blood is type A; when only the B antigen is present, the blood is type B. When both antigens are present, the blood is type AB, and when neither antigen is present, it is type O. Blood-typing for the A and B antigens is completely independent of blood-typing for the M and N antigens.

The gene responsible for producing the A and B antigens is denoted by the letter *I*. It has three alleles: I^A, I^B, and *i*. The I^A allele specifies the production of the A antigen, and the I^B allele specifies the production of the B antigen. However, the *i* allele does not specify an antigen. Among the six possible genotypes, there are four distinguishable phenotypes—the A, B, AB, and O blood types (**TABLE 4.1**). In this system, the I^A and I^B alleles are codominant, since each is expressed equally in the I^A I^B heterozygotes, and the *i* allele is recessive to both the

▶ **TABLE 4.1**

Genotypes, Phenotypes, and Frequencies in the ABO Blood-Typing System

Frequency in Genotype	Blood Type	A Antigen Present	B Antigen Present	U.S. White Population (%)
I^AI^A *or* I^Ai	A	+	–	41
I^BI^B *or* I^Bi	B	–	+	11
I^AI^B	AB	+	+	4
ii	O	–	–	44

I^A and I^B alleles. All three alleles are found at appreciable frequencies in human populations; thus, the *I* gene is said to be **polymorphic**, from the Greek words for "having many forms." We consider the population and evolutionary significance of genetic polymorphisms in Chapter 25.

ALLELIC SERIES

The functional relationships among the members of a series of multiple alleles can be studied by making heterozygous combinations through crosses between homozygotes. For example, the four alleles of the *c* gene in rabbits can be combined with each other to make six different kinds of heterozygotes: $c^h c$, $c^{ch} c$, $c^+ c$, $c^{ch} c^h$, $c^+ c^h$, and $c^+ c^{ch}$. These heterozygotes allow the dominance relations among the alleles to be studied (**FIGURE 4.4**). The wild-type allele is completely dominant over all the other alleles in the series; the *chinchilla* allele is partially dominant over the *himalayan* and *albino* alleles, and the *himalayan* allele is completely dominant over the *albino* allele. These dominance relations can be summarized as $c^+ > c^{ch} > c^h > c$.

Notice that the dominance hierarchy parallels the effects that the alleles have on coat color. A plausible explanation is that the *c* gene controls a step in the formation of black pigment in the fur. The wild-type allele is fully functional in this process, producing colored hairs throughout the body. The *chinchilla* and *himalayan* alleles are only partially functional, producing some colored hairs, and the *albino* allele is not functional at all. Nonfunctional alleles are said to be **null** or **amorphic** (from the Greek words for "without form"); they are almost always completely recessive. Partially functional alleles are said to be **hypomorphic** (from the Greek words for "beneath form"); they are recessive to alleles that are more functional, including (usually) the wild-type allele. Later in this chapter we consider the biochemical basis for these differences.

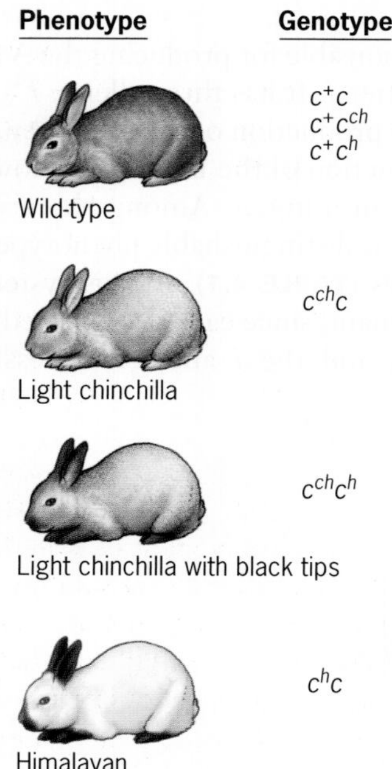

Figure 4.4 ▶ Phenotypes of different combinations of *c* alleles in rabbits. The alleles form a series, with the wild-type allele, c^+, dominant over all the other alleles and the null allele, *c* (*albino*), recessive to all the other alleles; one hypomorphic allele, c^{ch} (*chinchilla*), is partially dominant over the other, c^{ch} (*himalayan*).

TESTING GENE MUTATIONS FOR ALLELISM

A mutant allele is created when an existing allele changes to a new genetic state—a process called **mutation**. This event always involves a change in the physical composition of the gene (see Chapter 13) and sometimes produces an allele that has a detectable phenotypic effect. If, for example, the c^+ allele mutated to a null allele, a rabbit homozygous for this mutation would have the albino phenotype. However, it is not always possible to assign a new mutation to a gene on the basis of its phenotypic effect. In rabbits, for example, several genes determine coat color, and a mutation in any one of them could reduce, alter, or abolish pigmentation in the hairs. Thus, if a new coat color appears in a population of rabbits, it is not immediately clear which gene has been mutated.

A simple test can be used to determine the allelic identity of a new mutation, providing that the new mutation is recessive. The procedure involves crosses to combine the new recessive mutation with recessive mutations of known genes (**FIGURE 4.5**). If the hybrid progeny show a mutant phenotype, then the new mutation and the tester mutation *are* alleles of the same gene. If the hybrid progeny show a wild phenotype, then the new mutation and the tester mutation *are not* alleles of the same gene. This test is based on the principle that mutations of the same gene impair the same genetic function. If two such mutations are combined, the organism will be abnormal for this function and will show a mutant phenotype, even if the two mutations had an independent origin.

It is important to remember that this test applies only to recessive mutations. Dominant mutations cannot be tested in this way because they exert their effects even if a wild-type copy of the gene is present.

As an example, let's consider the analysis of two recessive mutations affecting eye color in the fruit fly, *Drosophila*

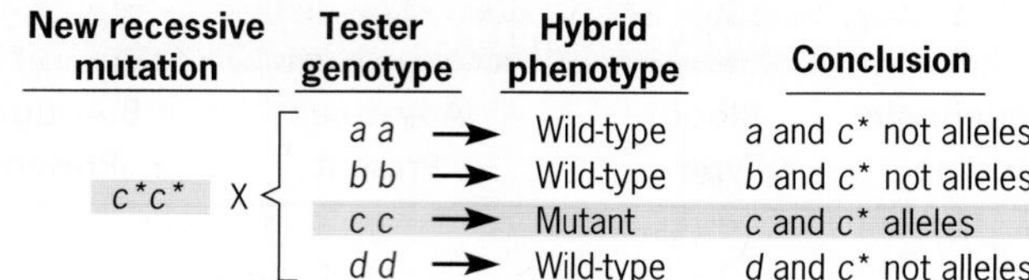

New recessive mutation
Tester genotype
Hybrid phenotype
Conclusion
c*c* X
a a → Wild-type
a and c* not alleles
b b → Wild-type
b and c* not alleles
c c → Mutant
c and c* alleles
d d → Wild-type
d and c* not alleles

Figure 4.5 ▶ A general scheme to test recessive mutations for allelism. Two mutations are alleles if a hybrid that contains both of them has the mutant phenotype.

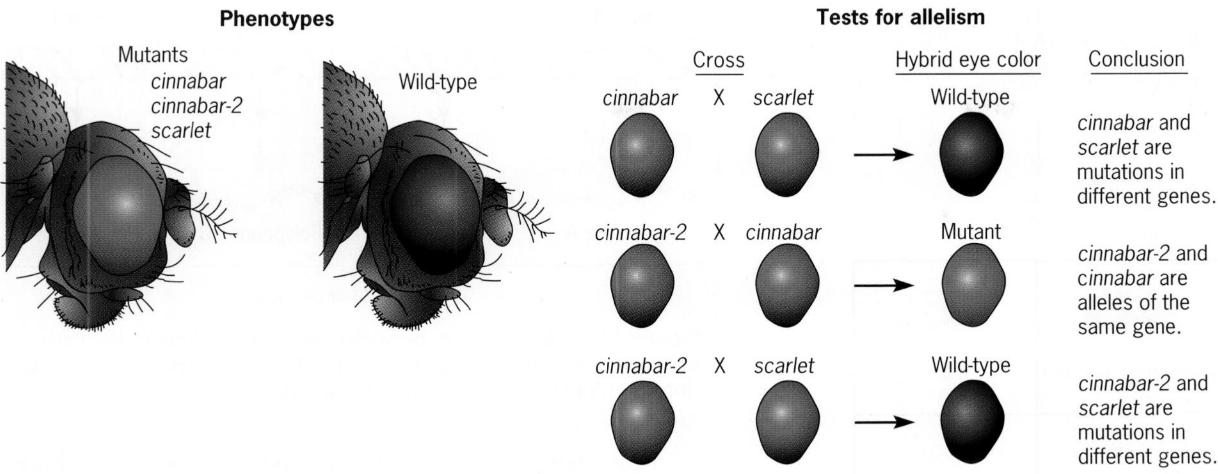

Figure 4.6 ▶ A test for allelism involving recessive eye color mutations in *Drosophila*. Three phenotypically identical mutations, *cinnabar, scarlet,* and *cinnabar-2,* are tested for allelism by making pairwise crosses between flies homozygous for different mutations. The phenotypes of the hybrids show that the *cinnabar* and *cinnabar-2* mutations are alleles of a single gene and that the *scarlet* mutation is not an allele of this gene.

melanogaster (**FIGURE 4.6**). Geneticists have investigated this organism for nearly a century, and a great many different mutations have been identified. Two independently isolated recessive mutations, called *cinnabar* and *scarlet*, are phenotypically indistinguishable, each causing the eyes to be bright red. In wild-type flies, the eyes are dark red. We wish to know whether the *cinnabar* and *scarlet* mutations are alleles of a single color-determining gene or if they are mutations in two different genes. To find the answer, we must cross the homozygous mutant strains with each other to produce hybrid progeny. If the hybrids have bright red eyes, we will conclude that *cinnabar* and *scarlet* are alleles of the same gene. If they have dark red eyes, we will conclude that they are mutations in different genes.

The hybrid progeny turn out to have dark red eyes; that is, they are wild-type rather than mutant. Thus, *cinnabar* and *scarlet* are not alleles of the same gene but, rather, mutations in two different genes, each apparently involved in the control of eye pigmentation. When we test a third mutation, called *cinnabar-2*, for allelism with the *cinnabar* and *scarlet* mutations, we find that the hybrid combination of *cinnabar-2* and *cinnabar* has the mutant phenotype (bright red eyes) and that the hybrid combination of *cinnabar-2* and *scarlet* has the wild phenotype (dark red eyes). These results tell us that the mutations *cinnabar* and *cinnabar-2* are alleles of one color-determining gene and that the *scarlet* mutation is not an allele of this gene. Rather, the *scarlet* mutation defines another color-determining gene.

The test to determine whether mutations are alleles of a particular gene is based on the phenotypic effect of combining the mutations in the same individual. If the hybrid combination is mutant, we conclude that the mutations are alleles; if it is wild-type, we conclude that they are not alleles. Chapter 14 discusses how this test—called the *complementation test* in modern terminology—enables geneticists to define the functions of individual genes.

VARIATION AMONG THE EFFECTS OF MUTATIONS

Genes are identified by mutations that alter the phenotype in some conspicuous way. For instance, a mutation may change the color or shape of the eyes, alter a behavior, or cause sterility or even death. The tremendous variation among the effects of individual mutations suggests that each organism carries many different kinds of genes and that each of these can mutate in different ways. In nature, mutations provide the raw material for evolution (see Chapter 25).

Mutations that alter some aspect of morphology, such as seed texture or color, are called *visible mutations*. Most visible mutations are recessive, but a few are dominant. Geneticists have learned much about genes by analyzing the properties of these mutations. We will encounter many examples of this analysis throughout this textbook. Mutations that limit reproduction are called *sterile mutations*. Some sterile mutations affect both sexes, but most affect either males *or* females. As with visible mutations, steriles can be either dominant or recessive. Some steriles completely prevent reproduction, whereas others only impair it slightly.

Mutations that interfere with necessary vital functions are called *lethal mutations*. Their phenotypic effect is death. We know that many genes are capable of mutating to the lethal state. Thus, each of these genes is absolutely essential for life. Dominant lethals that act early in life are lost one generation after they occur because the individuals that carry them die; however, dominant lethals that act later in life, after reproduction, can be passed on to the next generation. Recessive lethals may linger a long time in a population because they can be hidden in heterozygous condition by a wild-type allele. Recessive lethal mutations are detected by observing unusual segregation ratios in the progeny of heterozygous carriers. An example is the *yellow-lethal* mutation, A^Y, in the mouse

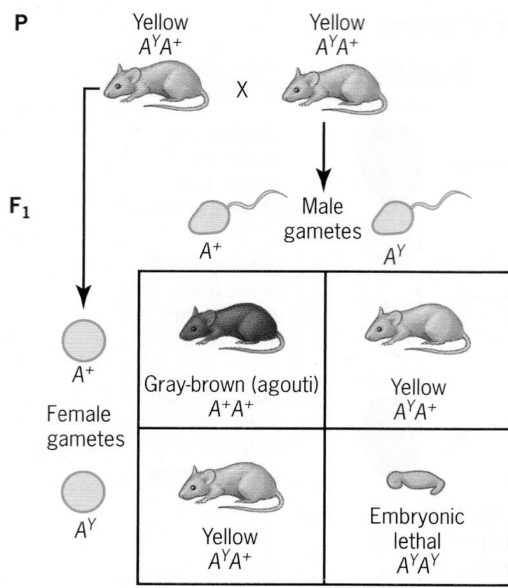

Figure 4.7 ▸ A^Y, the *yellow-lethal* mutation in mice: a dominant visible that is also a recessive lethal. A cross between carriers of this mutation produces yellow heterozygotes and gray-brown (agouti) homozygotes in a ratio of 2:1. The yellow homozygotes die as embryos.

(**FIGURE 4.7**). This mutation is a dominant visible, causing the fur to be yellow instead of gray-brown (the wild-type color, also known as *agouti*, which is determined by the allele A^+). In addition, the A^Y mutation is a recessive lethal, killing $A^Y A^Y$ homozygotes early in their development. A cross between $A^Y A^+$ heterozygotes produces two kinds of viable progeny, yellow ($A^Y A^+$) and gray-brown ($A^+ A^+$) in a ratio of 2:1. The $A^Y A^Y$ homozygotes die during embryonic development.

Geneticists have used different conventions to symbolize genes and their mutations. Mendel began the practice of using letters to denote genes. However, he simply started with the letter *A* and proceeded through the alphabet as symbols were needed to represent genes in his crosses. William Bateson was the first person to use letters mnemonically to symbolize genes. For the symbol, Bateson chose the first letter of the word that described the gene's phenotypic effect—thus, *B* for a gene causing *b*lue flowers, *L* for a gene causing *l*ong pollen grains. As the number of known genes grew, it became necessary to use two or more letters to represent newly discovered genes. Variations in the procedures for naming and symbolizing genes are discussed in the Focus on Genetic Symbols.

GENES FUNCTION TO PRODUCE POLYPEPTIDES

The extensive variation revealed by mutations indicates that organisms contain many different genes and that these genes can exist in multiple allelic states. However, it does not tell us how genes actually affect the phenotype. What is it about a gene that enables it to influence a trait such as eye color, seed texture, or plant height?

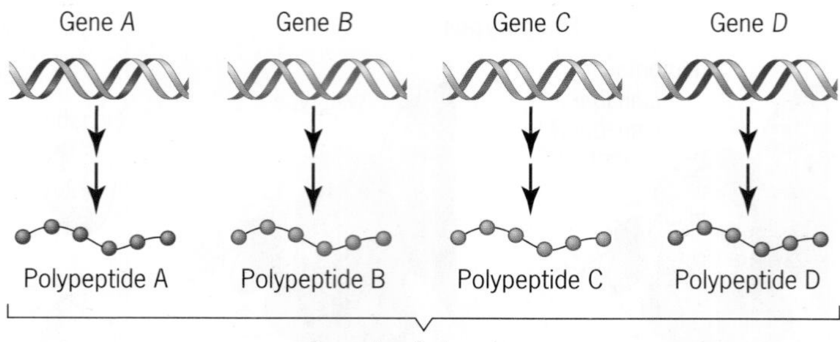

Figure 4.8 ▸ Relationship between genes and polypeptides. Each gene specifies a different polypeptide. These polypeptides then function to influence the organism's phenotype.

The early geneticists had no answer to this question. However, today it is clear that most genes specify a product that subsequently affects the phenotype. This idea, which was discussed in Bateson's book and which was supported by the research of many scientists, including, most notably, the British physician Sir Archibald Garrod (see A Milestone in Genetics: Garrod's Inborn Errors of Metabolism later in this chapter), was forcefully brought out in the middle of the twentieth century when George Beadle and Edward Tatum discovered that the products of genes are *polypeptides* (**FIGURE 4.8**).

Polypeptides are macromolecules built of a linear chain of *amino acids*. Every organism makes thousands of different polypeptides, each characterized by a specific amino acid sequence. These polypeptides are the fundamental constituents of *proteins*. Two or more polypeptides may combine to form a protein. Some proteins, called *enzymes*, function as catalysts in biochemical reactions; others form the structural components of cells; and still others are responsible for transporting substances within and between cells. Beadle and Tatum proposed that each gene is responsible for the synthesis of a particular polypeptide. When a gene is mutated, its polypeptide product either is not made or is altered in such a way that its role in the organism is changed. Mutations that eliminate or alter a polypeptide are often associated with a phenotypic effect. Whether this effect is dominant or recessive depends on the nature of the mutation. In Chapter 12 we consider the details of how genes produce polypeptides, and in Chapter 13 we discuss the molecular basis of mutation.

WHY ARE SOME MUTATIONS DOMINANT AND OTHERS RECESSIVE?

The discovery that genes specify polypeptides provides insight into the nature of dominant and recessive mutations. Dominant mutations have phenotypic effects in heterozygotes as well as in homozygotes, whereas recessive mutations have these effects only in homozygotes. What accounts for this striking difference in expression?

Recessive mutations often involve a loss of gene function, that is, when the gene no longer specifies a polypeptide or when

▶ FOCUS ON Genetic Symbols

William Bateson started the practice of choosing gene symbols mnemonically. In discussing Mendel's work, for example, he symbolized the dominant allele for tall pea plants as *T* and the recessive allele for short plants as *t*. Later, when it became customary to choose allele symbols based on the mutant trait, these symbols were changed to *D* (for tall) and *d* (for *d*warf). This convention provided a simple and consistent notation in which the dominant and recessive alleles of a particular gene were represented by a single letter, and that letter was mnemonic for the trait influenced by the gene. Bateson also coined the words *genetics, alleleomorph* (which was later shortened to *allele*), *homozygote,* and *heterozygote,* and he introduced the practice of denoting the generations in a breeding scheme as P, F_1, F_2, and so forth.

The gene-naming system that Bateson developed worked well until the number of genes that had been identified exceeded the capacity of the English alphabet; thereupon it became necessary to use two or more letters to symbolize a gene. For example, a particular mutant allele in *Drosophila* causes the eyes to be carmine instead of red. When this allele was discovered, it was given the symbol *cm* because the single letter *c* had already been used to represent a mutant allele that causes the wings to be curved instead of straight. Today, with thousands of genes identified, it is often necessary to use three or four letters, or combinations of letters and numbers, to symbolize genes. For example, mutations in the *cmp* gene of *Drosophila* cause the wings to be *crump*led, and mutations in the *Sh1* and *Sh2* genes of maize cause the kernels to be *sh*runken.

The discovery of multiple alleles made genetic notation even more complicated; because upper- and lowercase letters were no longer adequate to distinguish among alleles, geneticists began to combine a basic gene symbol with an identification symbol. *Drosophila* geneticists were the first to apply this procedure. They made the identification symbol a superscript on the basic gene symbol. Usually, both the gene symbol and the superscript had some mnemonic significance. Thus, for example, cn^2 was used to symbolize the *second* *cn*nabar eye color allele that was discovered in *Drosophila*; and ey^D was used to symbolize a *d*ominant allele that causes *Drosophila* to be *ey*eless. This convention was extended to other experimental animals, such as rabbits and mice. For example, c^{ch} was used to symbolize the *ch*inchilla allele of the gene that determines whether a rabbit's fur is colored or colorless. Plant geneticists adopted a variation of this practice. They use hyphenated symbols to identify mutant alleles; for example, *sh2–6801* represents a mutant allele of the *Sh2* gene that was discovered in 1968.

As genetic nomenclature developed, it became necessary to use a special symbol to represent the wild-type allele. The early *Drosophila* geneticists proposed using a plus sign (+), sometimes written as a superscript on the basic gene symbol (for example, c^+). This simple notation conveys the idea that the wild-type allele is the standard, or normal, allele of the gene and is widely used today. However, other gene-naming practices persist. Plant geneticists tend to use the gene symbol itself to represent the wild-type allele, but to make it stand out, they capitalize the first letter. Thus, *Sh2* is the wild-type allele of the second shrunken gene discovered in maize, whereas *sh2* is a mutant allele.

Genetic nomenclature has been further complicated by the discovery of genes through the polypeptides they specify. These discoveries have introduced gene symbols that are mnemonic for polypeptide gene products. For example, the human gene that specifies the polypeptide *h*ypoxanthine-guanine *p*hospho*r*ibosyl *t*ransferase is symbolized by *HPRT*, and the plant gene that specifies the polypeptide *a*lcohol *deh*ydrogenase is symbolized by *Adh*. Whether uppercase letters are used throughout the gene symbol or only for the first letter depends on the organism.

Today there are many specialized systems for symbolizing genes and alleles. Researchers who work with different organisms—*Drosophila*, mice, plants, or humans—speak slightly different languages. Later, we will see that still other genetic dialects have been created to describe the genes of viruses, bacteria, and fungi. These different systems of nomenclature indicate that the symbols in genetics have evolved in response to new discoveries—visible evidence of growth in a dynamic, young science.

it specifies a nonfunctional or underfunctional polypeptide (**FIGURE 4.9**). Recessive mutations are therefore typically **loss-of-function** alleles. Such alleles have little or no discernible effect in heterozygous condition with a wild-type allele because the wild-type allele specifies a functional polypeptide that will carry out its normal role in the organism. The phenotype of a mutant/wild heterozygote will therefore be the same, or essentially the same, as that of a wild-type homozygote. The *cinnabar* mutation in *Drosophila* is an example of a recessive loss-of-function allele. The wild-type allele of the *cinnabar* gene produces a polypeptide that functions as an enzyme in the synthesis of the brown pigment that is deposited in *Drosophila* eyes. Flies that are homozygous for a loss-of-function mutation in the *cinnabar* gene cannot produce this enzyme, and consequently, they do not synthesize any brown pigment in their eyes. The phenotype of homozygous *cinnabar* mutants is bright red—the color of the mineral cinnabar, for which the gene is named. However, flies that are heterozygous for the *cinnabar* mutation and its wild-type allele have dark red eyes; that is, they are phenotypically identical to wild-type. In these flies, the loss-of-function allele is recessive to the wild-type allele because the latter produces enough enzyme to synthesize normal amounts of brown pigment. The *scarlet* mutation mentioned earlier in this chapter is also an example of a recessive loss-of-function allele. The wild-type allele of the *scarlet* gene produces a different enzyme than the wild-type allele of the *cinnabar* gene. Both enzymes—and therefore both wild-type alleles—are necessary for the synthesis of brown pigment in *Drosophila* eyes. If either enzyme is

Wild-type allele produces a functional polypeptide.

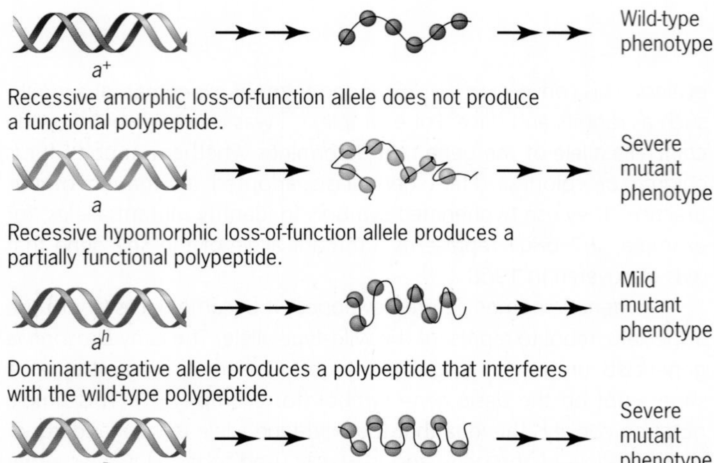

a^+
Wild-type phenotype
Recessive amorphic loss-of-function allele does not produce a functional polypeptide.
a
Severe mutant phenotype
Recessive hypomorphic loss-of-function allele produces a partially functional polypeptide.
a^h
Mild mutant phenotype
Dominant-negative allele produces a polypeptide that interferes with the wild-type polypeptide.
a^D
Severe mutant phenotype

(*a*)

Genotype	Polypeptides present	Phenotype	Nature of mutant allele
a^+ a		Wild-type	Recessive
a^+ a^h		Wild-type	Recessive
a^+ a^D		Mutant	Dominant

(*b*)

Figure 4.9 ▶ Differences between recessive loss-of-function mutations and dominant gain-of-function mutations. (*a*) Polypeptide products of recessive and dominant mutations. (*b*) Phenotypes of heterozygotes carrying a wild-type allele and different types of mutant alleles.

missing, the eyes are bright red rather than dark red because they lack this brown pigment.

Some recessive mutations result in a partial loss of gene function. For example, the *himalayan* allele of the coat color gene in mammals such as rabbits and cats specifies a polypeptide that functions only in the parts of the body where the temperature is reduced. This partial loss of function explains why animals homozygous for the *himalayan* allele have pigmented hair on their extremities—tail, legs, ears, and tip of the nose—but not on the rest of their bodies. In the extremities, the polypeptide specified by this allele is functional, whereas in the rest of the body, it is not. The expression of the *himalayan* allele is therefore temperature-sensitive.

Some dominant mutations may also involve a loss of gene function. If the phenotype controlled by a gene is sensitive to the amount of gene product, a loss-of-function mutation can evoke a mutant phenotype in heterozygous condition with a wild-type allele. In such cases, the wild-type allele, by itself, is not able to supply enough gene product to provide full, normal function. In effect, the loss-of-function mutation reduces the level of gene product below the level that is needed for the wild phenotype.

Other dominant mutations actually interfere with the function of the wild-type allele by specifying polypeptides that inhibit, antagonize, or limit the activity of the wild-type polypeptide (**FIGURE 4.9**). Such mutations are called *dominant-negative* mutations. Some of the mutations of the *T* gene in the mouse are examples of dominant-negative mutations. We have already seen that in heterozygous condition, these mutations cause a shortening of the tail. In homozygous condition, they are lethal. The wild-type allele of the *T* gene is therefore essential for life. At the cellular level, the polypeptide product of this allele regulates important events during embryological development. Dominant-negative *T* alleles produce slightly shorter polypeptides than the wild-type *T* allele. In heterozygotes, these shorter polypeptides interfere with the function of the wild-type polypeptide. The result is a completely tailless mouse.

Some dominant mutations cause a mutant phenotype in heterozygous condition with a wild-type allele because they enhance the function of the gene product. The enhanced function may arise because the mutation specifies a novel polypeptide or because it causes the wild-type polypeptide to be produced where or when it should not be. Dominant mutations that work in these ways are called **gain-of-function mutations.** In *Drosophila*, the mutation known as *Antennapedia* (*Antp*) is a dominant gain-of-function mutation. In heterozygous condition with a wild-type allele, *Antp* causes legs to develop in place of the antennae on the head of the fly. The reason for this bizarre anatomical transformation is that the *Antp* mutation causes the polypeptide product of the *Antennapedia* gene to be produced in the head, where, ordinarily, it is not produced; the *Antennapedia* gene product has therefore expanded the domain of its function.

We should note that not all genes produce polypeptides as the work of Beadle and Tatum implied. Modern research has identified many genes whose end products are RNA molecules rather than polypeptides. We shall explore these kinds of genes later in this book.

KEY POINTS

- Genes often have multiple alleles.
- Mutant alleles may be dominant, recessive, incompletely dominant, or codominant.
- If a hybrid that inherited a recessive mutation from each of its parents has a mutant phenotype, then the recessive mutations are alleles of the same gene; if the hybrid has a wild phenotype, then the recessive mutations are alleles of different genes.
- Most genes encode polypeptides.
- In homozygous condition, recessive mutations often abolish or diminish polypeptide activity.
- Some dominant mutations produce a polypeptide that interferes with the activity of the polypeptide encoded by the wild-type allele of a gene.

Gene Action: From Genotype to Phenotype

Phenotypes depend on both environmental and genetic factors.

At the beginning of the twentieth century, geneticists had imprecise ideas about how genes evoke particular phenotypes. They knew nothing about the chemistry of gene structure or function, nor had they developed the techniques to study it. Everything that they proposed about the nature of gene action was inferred from the analysis of phenotypes. These analyses showed that genes do not act in isolation. Rather, they act in the context of an environment and in concert with other genes. These analyses also showed that a particular gene can influence many different traits.

INFLUENCE OF THE ENVIRONMENT

A gene must function in the context of both a biological and a physical environment. The factors in the physical environment are easier to study, for particular genotypes can be reared in the laboratory under controlled conditions, allowing an assessment of the effects of temperature, light, nutrition, and humidity. As an example, let's consider the *Drosophila* mutation known as *shibire*. At the normal culturing temperature, 25°C, *shibire* flies are viable and fertile, but are extremely sensitive to a sudden shock. When a *shibire* culture is shaken, the flies—temporarily paralyzed—fall to the bottom of the culture. Indeed, *shibire* is the Japanese word for "paralysis." However, if a culture of *shibire* flies is placed at a slightly higher temperature, 29°C, all the flies fall to the bottom and die, even without a shock. Thus, the phenotype of the *shibire* mutation is temperature-sensitive. At 25°C, the mutation is viable, but at 29°C it is lethal. A plausible explanation is that at 25°C the mutant gene makes a partially functional protein, but at 29°C this protein is totally nonfunctional.

ENVIRONMENTAL EFFECTS ON THE EXPRESSION OF HUMAN GENES

Human genetic research provides an example of how the physical environment can influence a phenotype. **Phenylketonuria** (PKU) is a recessive disorder of amino acid metabolism. Infants homozygous for the mutant allele accumulate toxic substances in their brains that can impair mental ability by affecting the brain's development. The harmful aspects of PKU are traceable to a particular amino acid, phenylalanine, which is ingested in the diet. Though not toxic itself, phenylalanine is metabolized into other substances that are. Infants with PKU who are fed normal diets ingest enough phenylalanine to bring out the worst manifestations of the disease. However, infants who are fed low-phenylalanine diets usually mature without serious mental impairment. Because PKU can be diagnosed in newborn babies, the clinical impact of this disease can be reduced if infants that are PKU homozygotes are placed on a low-phenylalanine diet shortly after birth. This example illustrates how an environmental factor—diet—can be manipulated to modify a phenotype that would otherwise become a personal tragedy.

The biological environment can also influence the phenotypic expression of genes. **Pattern baldness** in humans is a well-known example. Here the relevant biological factor is gender. Premature pattern baldness is due to an allele that is expressed differently in the two sexes. In males, both homozygotes and heterozygotes for this allele develop bald patches, whereas in females, only the homozygotes show a tendency to become bald, and this is usually limited to general thinning of the hair. The expression of this allele is probably triggered by the male hormone **testosterone.** Females produce much less of this hormone and are therefore seldom at risk to develop bald patches. The sex-influenced nature of pattern baldness shows that biological factors can control the expression of genes.

PENETRANCE AND EXPRESSIVITY

When individuals do not show a trait even though they have the appropriate genotype, the trait is said to exhibit *incomplete penetrance*. An example of incomplete penetrance in humans is **polydactyly**—the presence of extra fingers and toes (**FIGURE 4.10*a***). This condition is due to a dominant mutation, *P*, that is manifested in some of its carriers. In the pedigree in **FIGURE 4.10*b***, the individual denoted III-2 must be a carrier even though he does not have extra fingers or toes. The reason is that both his mother and three of his children are polydactylous—an indication of the transmission of the mutation through III-2. Incomplete penetrance can be a serious problem in pedi-

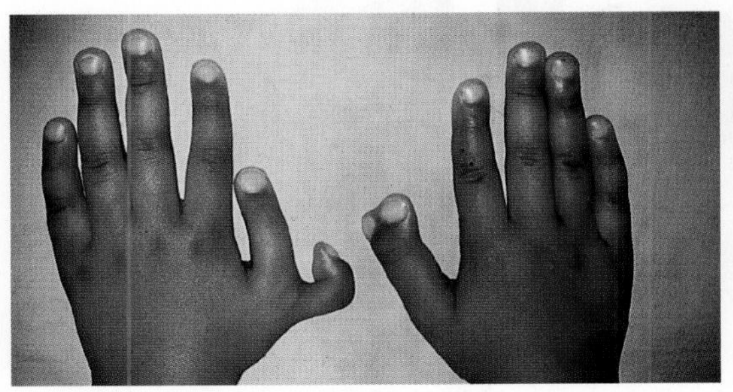
(*a*)

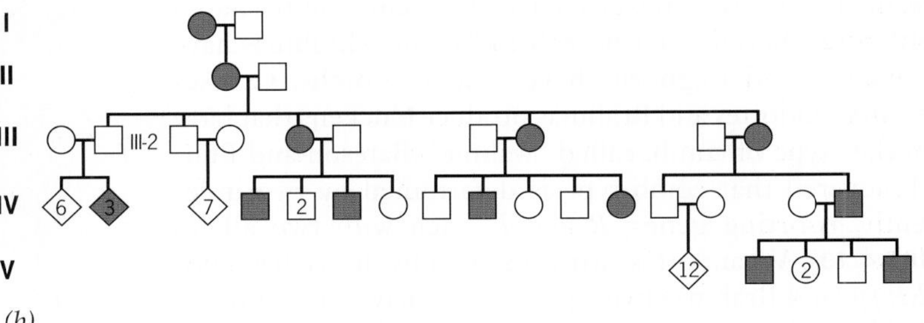

(*b*)

Figure 4.10 ▶ Polydactyly in human beings. (*a*) Phenotype showing extra fingers. (*b*) Pedigree showing the inheritance of this incompletely penetrant dominant trait.

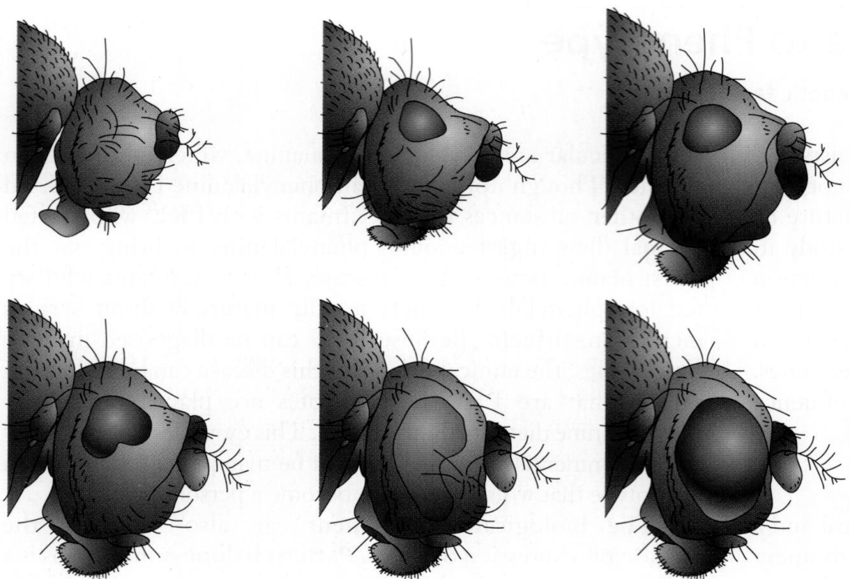

Figure 4.11 ▶ Variable expressivity of the *Lobe* mutation in *Drosophila.* Each fly is heterozygous for this dominant mutation; however, the phenotypes vary from complete absence of the eye to a nearly wild-type eye.

gree analysis because it can lead to the incorrect assignment of genotypes.

The term *expressivity* is used if a trait is not manifested uniformly among the individuals that show it. The dominant *Lobe* eye mutation (**FIGURE 4.11**) in *Drosophila* is an example. The phenotype associated with this mutation is extremely variable. Some heterozygous flies have tiny compound eyes, whereas others have large, lobulated eyes; between these extremes, there is a full range of phenotypes. The *Lobe* mutation is therefore said to have *variable expressivity.*

Incomplete penetrance and variable expressivity indicate that the pathway between a genotype and its phenotypes is subject to considerable modulation. Geneticists know that some of this modulation is due to environmental factors, but some is also due to factors in the genetic background. Clear-cut evidence for such factors comes from breeding experiments showing that two or more genes can affect a particular trait.

GENE INTERACTIONS

Some of the first evidence that a trait can be influenced by more than one gene was obtained by Bateson and Punnett from breeding experiments with chickens. Their work was carried out shortly after the rediscovery of Mendel's paper. Domestic breeds of chickens have different comb shapes (**FIGURE 4.12**): Wyandottes have "rose" combs, Brahmas have "pea" combs, and Leghorns have "single" combs. Crosses between Wyandottes and Brahmas produce chickens that have yet another type of comb, called "walnut." Bateson and Punnett discovered that comb type is determined by two independently assorting genes, *R* and *P*, each with two alleles (**FIGURE 4.13**). Wyandottes (with rose combs) have the genotype *RR pp*, and Brahmas (with pea combs) have the genotype *rr PP*. The F_1 hybrids between these two varieties are therefore *Rr Pp*, and phenotypically they have walnut combs. If these hybrids are intercrossed with each other, all four types

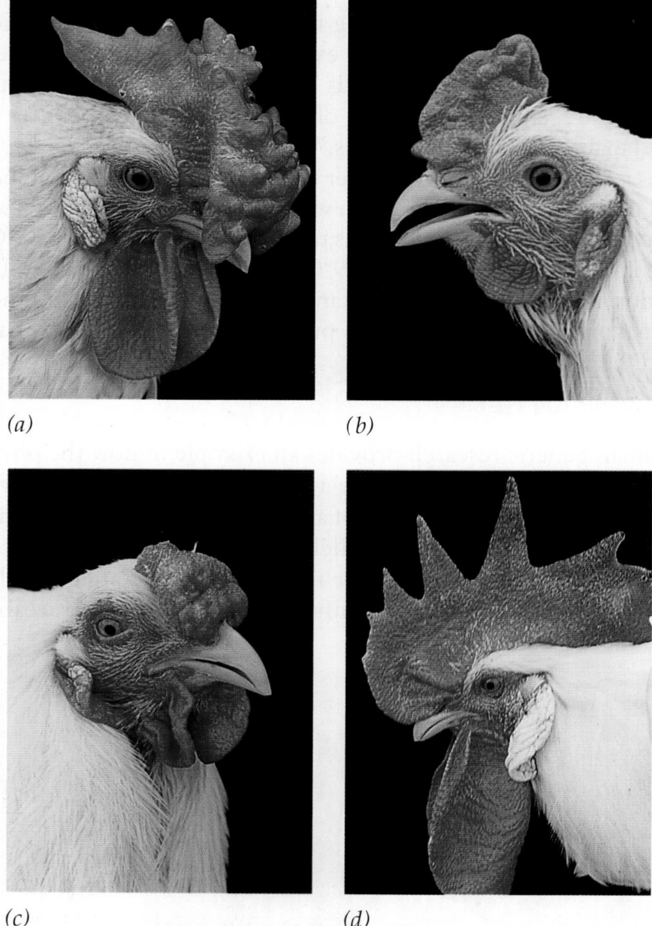

Figure 4.12 ▶ Comb shapes in chickens of different breeds. (*a*) Rose, Wyandottes; (*b*) pea, Brahmas; (*c*) walnut, hybrid from cross between chickens with rose and pea combs; (*d*) single, Leghorns.

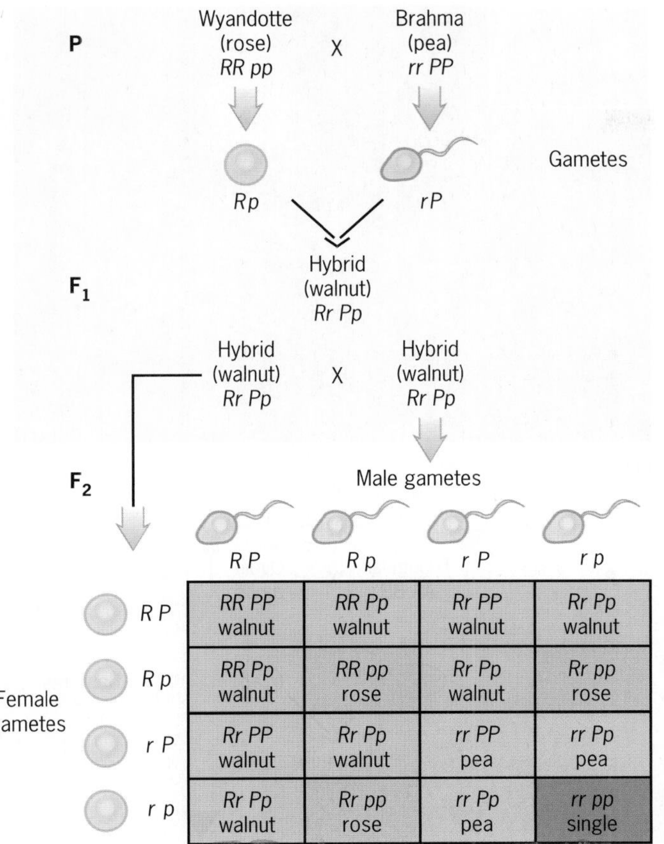

Figure 4.13 ▶ Bateson and Punnett's experiment on comb shape in chickens. The intercross in the F_1 produces four phenotypes, each highlighted by a different color in the Punnett square, in a 9:3:3:1 ratio.

of combs appear in the progeny: 9/16 walnut (*R- P-*), 3/16 rose (*R- pp*), 3/16 pea (*rr P-*), and 1/16 single (*rr pp*). The Leghorn breed, which has the single-comb type, must therefore be homozygous for both of the recessive alleles.

The work of Bateson and Punnett demonstrated that two independently assorting genes can affect a trait. Different combinations of alleles from the two genes resulted in different phenotypes, presumably because of interactions between their products at the biochemical or cellular level.

EPISTASIS

When two or more genes influence a trait, an allele of one of them may have an overriding effect on the phenotype. When an allele has such an overriding effect, it is said to be epistatic to the other genes that are involved; the term **epistasis** comes from a Greek word meaning to "stand above." For example, we know that eye pigmentation in *Drosophila* involves a large number of genes. If a fly is homozygous for a null allele in any one of these genes, the pigment-synthesizing pathway can be blocked, and an abnormal eye color will be produced. This allele essentially nullifies the work of all the other genes, masking their contributions to the phenotype.

A mutant allele of one gene is epistatic to a mutant allele of another gene if it conceals the latter's presence in the genotype. We have already seen that a recessive mutation in the *cinnabar* gene of *Drosophila* causes the eyes of the fly to be bright red. A recessive mutation in a different gene causes the eyes to be white. When both of these mutations are made homozygous in the same fly, the eye color is white. Thus, the *white* mutation is epistatic to the *cinnabar* mutation.

What physiological mechanism makes the *white* mutation epistatic to the *cinnabar* mutation? The polypeptide product of the wild-type allele of the *white* gene transports pigment into the *Drosophila* eye. When this gene is mutated, the transporter polypeptide is not made. Flies that are homozygous for the *cinnabar* mutation cannot synthesize brown pigment, but they can synthesize red pigment. When these flies are also homozygous for the *white* mutation, the red pigment cannot be transported into the eyes. Consequently, flies that are homozygous for both the *cinnabar* and *white* mutations have white eyes.

The analysis of epistatic relationships such as the one between *cinnabar* and *white* can suggest ways in which genes control a phenotype. A classic example of this analysis is again from the work of Bateson and Punnett, who studied the genetic control of flower color in the sweet pea, *Lathyrus odoratus* (**FIGURE 4.14*a***). The flowers in this plant are either purple or white—purple if they contain anthocyanin pigment and white if they do not. Bateson and Punnett crossed two different varieties with white flowers to obtain hybrids, which all had purple flowers. When these hybrids were intercrossed, Bateson and Punnett obtained a ratio of 9 purple: 7 white plants in the F_2 They explained the results by proposing that two independently assorting genes, *C* and *P*, are involved in anthocyanin synthesis and that each gene has a recessive allele that abolishes pigment production (**FIGURE 4.14*b***).

Given this hypothesis, the parental varieties must have had complementary genotypes: *cc PP* and *CC pp*. When the two varieties were crossed, they produced *Cc Pp* double heterozygotes that had purple flowers. In this system, a dominant allele from each gene is necessary for the synthesis of anthocyanin pigment. In the F_2, 9/16 of the plants are *C- P-* and have purple flowers; the remaining 7/16 are homozygous for at least one of the recessive alleles and have white flowers. Notice that the double recessive homozygotes, *cc pp*, are not phenotypically different from either of the single recessive homozygotes. Bateson and Punnett's work established that each of the recessive alleles is epistatic over the dominant allele of the other gene. A plausible explanation is that each dominant allele produces an enzyme that controls a step in the synthesis of anthocyanin from a biochemical precursor. If a dominant allele is not present, its step in the biosynthetic pathway is blocked and anthocyanin is not produced:

Gene		*C*		*P*	
	Precursor	→	Intermediate	→	Anthocyanin
Genotype					
C– P–	+		+		+
cc P–	+		–		–
C– pp	+		+		–
cc pp	+		–		–

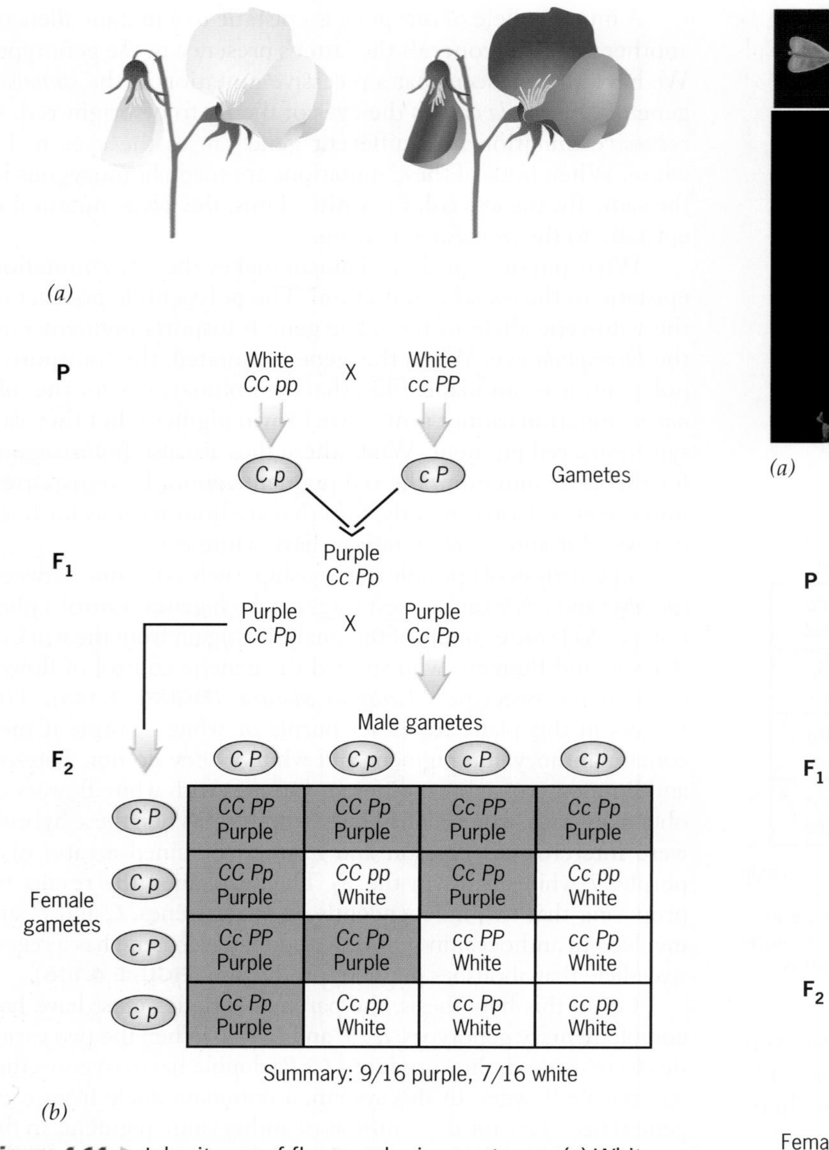

Figure 4.14 ▶ Inheritance of flower color in sweet peas. (*a*) White and purple flowers of the sweet pea. (*b*) Bateson and Punnett's experiment on the genetic control of flower color in sweet peas.

Notice that Bateson's and Punnett's first cross was a test for allelism between two white-flowered strains of the sweet pea. Each strain was homozygous for a recessive mutation in a gene involved in the production of purple pigment. When the two white strains were crossed, the F_1 plants had purple flowers. This result tells us that the white strains were homozygous for mutations in different genes involved in the synthesis of purple pigment.

Another classic study of epistasis was performed by George Shull using a weedy plant called the shepherd's purse, *Bursa bursa-pastoris* (**FIGURE 4.15*a***). The seed capsules of this plant are either triangular or ovoid in shape. Ovoid capsules are produced only if a plant is homozygous for the recessive alleles of two genes—that is, if it has the genotype *aa bb*. If the dominant allele of either gene is present, the plant produces triangular capsules. The evidence for this conclusion comes from

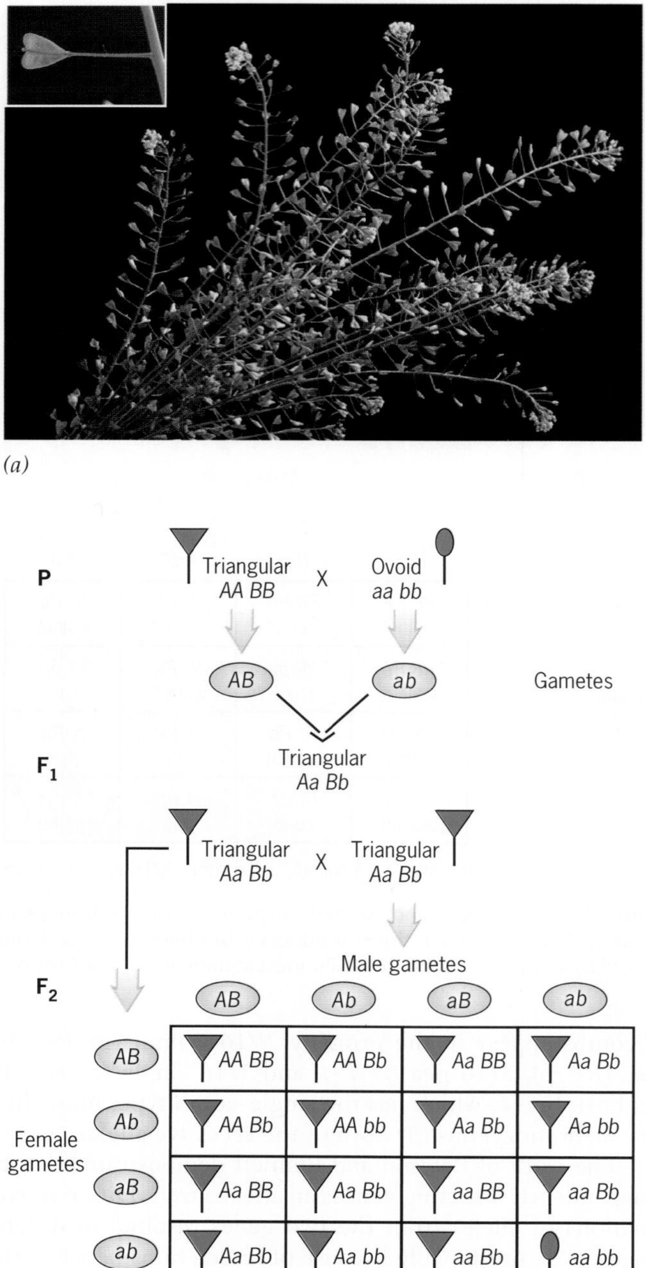

Figure 4.15 ▶ (*a*) The shepherd's purse, *Bursa bursa-pastoris.* The inset shows a close-up of the seed capsule. (*b*) Crosses showing duplicate gene control of seed capsule shape in the shepherd's purse.

crosses between doubly heterozygous plants (**FIGURE 4.15*b***). Such crosses produce progeny in a ratio of 15 triangular:1 ovoid, indicating that the dominant allele of one gene is epistatic over the recessive allele of the other. The data suggest that capsule shape is determined by duplicate developmental pathways, either of which can produce a triangular capsule. One pathway involves the dominant allele of the *A* gene, and the other the dominant allele of the *B* gene. A precursor substance can be converted into a product that leads to a triangular seed capsule through either of these pathways. Only when both pathways are blocked by homozygous recessive alleles is the triangular phenotype suppressed and an ovoid capsule produced:

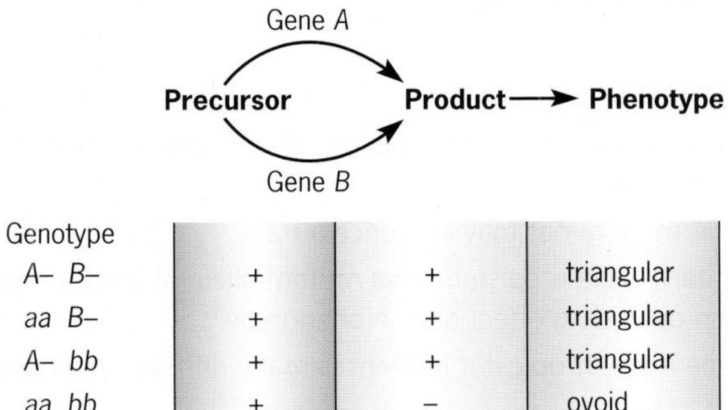

In other cases of epistasis, the product of one gene may inhibit the expression of another gene. Consider, for example, the inheritance of fruit color in summer squash plants. Plants that carry the dominant allele *C* produce white fruit, whereas plants that are homozygous for the recessive allele *c* produce colored fruit. If a squash plant is also homozygous for the recessive allele *g* of an independently assorting gene, the fruit will be green. However, if it carries the dominant allele *G* of this gene, the fruit will be yellow. These observations suggest that the two genes control steps in the synthesis of green pigment. The first step converts a colorless precursor into a yellow pigment, and the second step converts this yellow pigment into a green pigment. If the first step is blocked (by the presence of the *C* allele), neither of the pigments is produced and the fruit will be white. If only the second step is blocked (by the presence of the *G* allele), the yellow pigment cannot be converted into the green pigment and the fruit will be yellow. We can summarize these ideas with a diagram that shows the genetic control of pigment synthesis in this biochemical pathway:

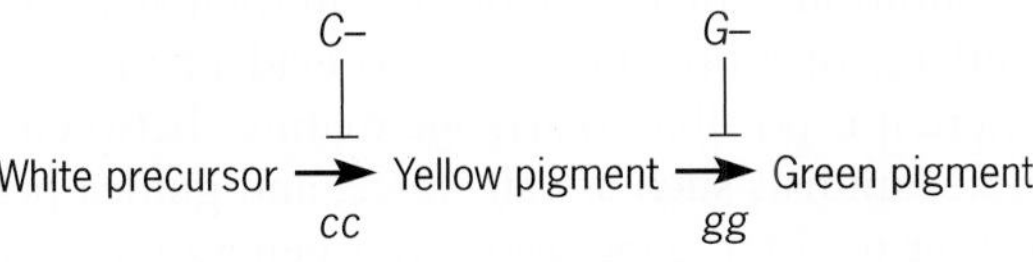

Genotype				Phenotype
C– G–	+	–	–	white
C– gg	+	–	–	white
cc G–	+	+	–	yellow
cc gg	+	+	+	green

The arrows in the diagram show the steps in the pathway. The genotype below an arrow allows that step to occur, whereas the genotype above an arrow inhibits that step from occurring. It is customary in genetics to symbolize the inhibitory effect of a genotype by drawing a blunted arrow (—|) from the genotype to the relevant step in the pathway. In this example, the *C* allele inhibits the first step and the *G* allele inhibits the second step.

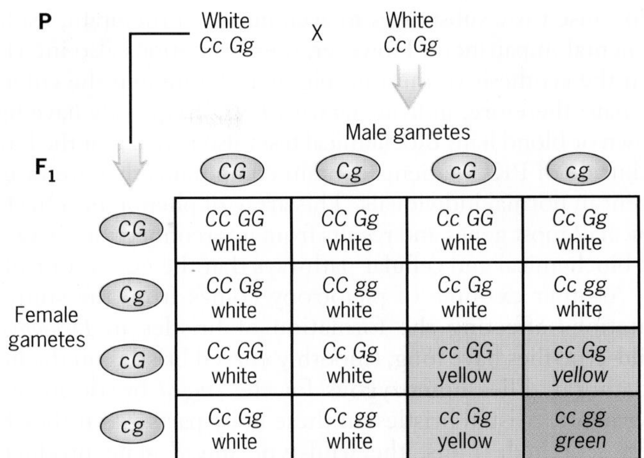

Summary: 12/16 white, 3/16 yellow, 1/16 green

Figure 4.16 Segregation in the offspring of a cross between summer squash plants heterozygous for two genes controlling fruit color.

Because of its role as an inhibitor of the first step, the *C* allele is epistatic to both of the alleles of the other gene. No matter which of the alleles of this other gene is present in a plant, the *C* allele will cause that plant to produce white fruit.

FIGURE 4.16 shows the outcome of a cross between plants heterozygous for both of the fruit-color-determining genes. When *Cc Gg* plants are intercrossed, they produce progeny that sort into three phenotypic classes: white, yellow, and green. The offspring with green fruit are homozygous for the recessive alleles of both genes; that is, they are *cc gg*, and their frequency is 1/16. The offspring with yellow fruit are homozygous for *c*, and they carry at least one copy of *G*; their frequency is 3/16. The offspring with white fruit carry at least one copy of *C*; the rest of the genotype does not matter. The frequency of the white-fruited plants is 12/16. To test your ability to make genetic predictions from a biochemical pathway, work through the exercise in the Focus on Problem Solving: Going from Pathways to Phenotypic Ratios.

These examples indicate that a particular phenotype is often the result of a process controlled by more than one gene. Each gene governs a step in a pathway that is part of the process. When a gene is mutated to a nonfunctional or partially functional state, the process can be disrupted, leading to a mutant phenotype. Much of modern genetic analysis is devoted to the investigation of pathways involved in important biological processes such as metabolism and development. Studying the epistatic relationships among genes can help to sort out the role that each gene plays in these processes.

PLEIOTROPY

Not only is it true that a phenotype can be influenced by many genes, but it is also true that a gene can influence many phenotypes. When a gene affects many aspects of the phenotype, it is said to be **pleiotropic,** from the Greek words for "to take many turns." The gene for phenylketonuria in human beings is an example. The primary effect of recessive mutations in this gene is to cause toxic substances to accumulate in the brain, leading to mental impairment. However, these mutations also interfere with the synthesis of melanin pigment, lightening the color of the hair; therefore, individuals with PKU frequently have light brown or blond hair. Biochemical tests also reveal that the blood and urine of PKU patients contain compounds that are rare or absent in normal individuals. This array of phenotypic effects is typical of most genes and results from interconnections between the biochemical and cellular pathways that the genes control.

Another example of pleiotropy comes from the study of mutations affecting the formation of bristles in *Drosophila.* Wild-type flies have long, smoothly curved bristles on the head and thorax. Flies homozygous for the *singed* bristle muation have short, twisted bristles on these body parts—as if they had been scorched. Thus, the wild-type *singed* gene product is needed for the proper formation of bristles. It is also needed for the production of healthy, fertile eggs. We know this fact because females that are homozygous for certain *singed* mutations are completely sterile; they lay flimsy, ill-formed eggs that never hatch. However, these mutations have no adverse effect on male fertility. Thus, the *singed* gene pleiotropically controls the formation of both bristles and eggs in females and the formation of bristles in males.

KEY POINTS

- Gene action is affected by biological and physical factors in the environment.
- Two or more genes may influence a trait.
- A mutant allele is epistatic to a mutant allele of another gene if it has an overriding effect on the phenotype.
- A gene is pleiotropic if it influences many different phenotypes.

Inbreeding: Another Look at Pedigrees

Geneticists use a simple statistic, the inbreeding coefficient, to analyze the effects of matings between relatives.

Geneticists have always been interested in the phenomenon of inbreeding, whether to make true-breeding strains or to reveal the homozygous effects of recessive alleles. In addition, when inbreeding occurs in nature, it can affect the character of plant and animal populations. In this section we consider ways to analyze the effects of inbreeding. We also introduce the techniques needed to study common ancestry in pedigrees.

THE EFFECTS OF INBREEDING

Inbreeding occurs when mates are related to each other by virtue of common ancestry. A mating between relatives is often referred to as a **consanguineous mating,** from Latin words meaning "of the same blood." In human populations, these types of matings are rare, with the incidence depending on cultural and ethnic traditions and on geography. In many cultures, marriages between close relatives—for example, between siblings or half siblings—are expressly forbidden, and marriages between more distant relatives, though allowed, must be approved by civil or religious authorities before they can occur. These restrictions exist because inbreeding tends to produce more diseased and debilitated children than matings between unrelated individuals. This tendency, as we now know, arises from an increased chance for the children of a consanguineous mating to be homozygous for a harmful recessive allele. In some cultures, however, consanguineous matings have been accepted and even encouraged. In ancient Egypt, for example, the royal line was perpetuated by brother–sister marriages, presumably to preserve the "purity" of the royal blood. Similar practices existed in Polynesia until relatively recent times.

The occurrence of consanguineous matings in human populations has helped in the analysis of genetic conditions caused by recessive alleles. In fact, the very first gene to be identified in humans was brought to light by observing a greater frequency of recessive homozygotes in the children of first cousins; for more information, see A Milestone in Genetics: Garrod's Inborn Errors of Metabolism later in this chapter. Many of the classic studies in human genetics were based on the analysis of consanguineous matings in socially closed groups—for example, the Amish, a religious sect scattered in small communities in the eastern and midwestern United States. **FIGURE 4.17** shows an Amish pedigree in which 10 individuals have albinism. The affected individuals are all descendants of two people (I-1 and I-2) who had immigrated from Europe. The consanguineous matings in the pedigree are indicated by double lines connecting the mates. All the affected individuals come from such matings. Thus, this pedigree shows how inbreeding brings out a recessive condition, which geneticists can then analyze.

The effects of inbreeding are also evident in experimental species where it is possible to arrange matings between relatives. For example, animals such as rats, mice, and guinea pigs can be mated brother to sister, generation after generation, to create an **inbred line.** Although these lines are genetically quite pure—that is, they do not segregate different alleles of particular genes—they often are less vigorous than lines maintained by matings between unrelated individuals. We refer to this loss of vigor as **inbreeding depression.** In plants where self-fertilization is possible, very highly inbred lines can be created by repeated self-fertilization over several generations. Each line would be expected to be homozygous for different alleles that were present in the founding population of plants. **FIGURE 4.18** shows the result of this process in maize. The inbred plants are short and produce small ears with few kernels. By comparison, the plants generated by crossing the two inbred strains are tall and produce large ears with many kernels. These plants are expected to be heterozygous for many genes. Their robustness is a phenomenon called hybrid

▶ FOCUS ON PROBLEM SOLVING
Going from Pathways to Phenotypic Ratios

THE PROBLEM

Flower color in a plant is determined by two independently assorting genes, *B* and *D*. The dominant allele *B* allows a pigment precursor to be converted into blue pigment. In homozygous condition, the recessive allele of this gene, *b*, blocks this conversion, and without blue pigment, the flowers are white. The dominant allele of the other gene, *D*, causes the blue pigment to degrade, whereas the recessive allele of this gene, *d*, has no effect. True-breeding blue and white strains of the plant were crossed, and all the F_1 plants had white flowers. (a) What was the genotype of the F_1 plants? (b) What were the genotypes of the plants used in the initial cross? (c) If the F_1 plants are self-fertilized, what phenotypes will appear in the F_2, and in what proportions?

FACTS AND CONCEPTS

1. The dominant allele (*D*) of one gene is epistatic to both alleles (*B* and *b*) of the other gene.
2. Plants with blue flowers must carry at least one *B* allele, but they cannot carry even one *D* allele.
3. Plants with white flowers can be *bb*, or they can be *BB* or *Bb* as long as they also carry at least one *D* allele.
4. True-breeding strains are homozygous for their genes.
5. When genes assort independently, we multiply the probabilities associated with the components of the complete genotype.

ANALYSIS AND SOLUTION

A good place to start the analysis is to diagram the biochemical pathway—that is, to transform the "word problem" into a diagram that will guide our search for a solution.

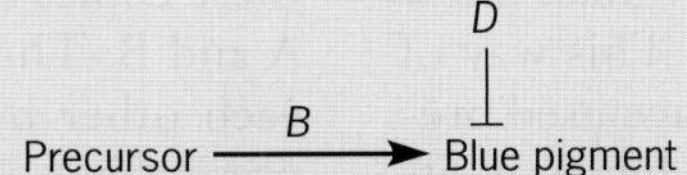

The positive action of the *B* allele is required for the synthesis of blue pigment. The negative action of the dominant allele *D* is indicated by a blunted arrow pointed at this pigment. Now we can address the questions in the problem.

a. The key observation is that the flowers of the F_1 plants are white. Because these plants had a true-breeding blue parent, they must carry the *B* allele, but the blue pigment produced through the action of this allele must be degraded. The F_1 plants must therefore also carry the *D* allele. However, they cannot be homozygous for it because their blue parent could not have carried it. Thus, the F_1 plants must be heterozygous for the *D* allele. Genotypically, they are either *BB Dd* or *Bb Dd*. From the information given in the problem, we cannot distinguish between these two possibilities.

b. The blue plants used in the cross must have been *BB dd*. The white plants could have been either *BB DD* or *bb DD*—we cannot be certain which of these genotypes they were.

c. If the F_1 plants are *BB Dd*, then when they are selfed only the *D* and *d* alleles will segregate, and 1/4 of their offspring will be blue (*BB dd*) and 3/4 will be white (*BB DD* or *BB Dd*). If the F_1 plants are *Bb Dd*, then when they are selfed, both genes will segregate dominant and recessive alleles. Among the offspring, those that are *BB dd* or *Bb dd* will be blue. This phenotypic class will constitute $(3/4) \times (1/4) = 3/16$ of the total. All the other offspring—$1 - 3/16 = 13/16$ of the total—will be white.

For further discussion go to your *WileyPLUS* course.

vigor, or **heterosis**. This term was introduced in 1914 by George Shull, a pioneering plant breeder who began the practice of crossing inbred strains to produce uniformly high-yielding, heterozygous offspring. Shull's technique has since become standard in the plant breeding industry.

GENETIC ANALYSIS OF INBREEDING

Matings between full siblings, between half siblings, and between first cousins are all examples of inbreeding. When such matings occur, we speak of the offspring as being *inbred*. Inbred individuals differ from the offspring of unrelated parents in one important way: the two copies of a gene they carry may be identical to each other by virtue of common ancestry—that is, because the genes have descended from a gene that was present in an ancestor of the inbred individual. To understand this concept, let's consider a simple pedigree that illustrates a mating between half siblings.

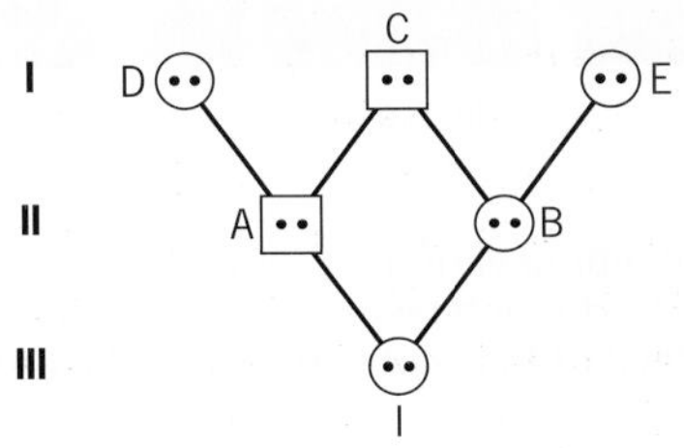

I D C E
II A B
III I

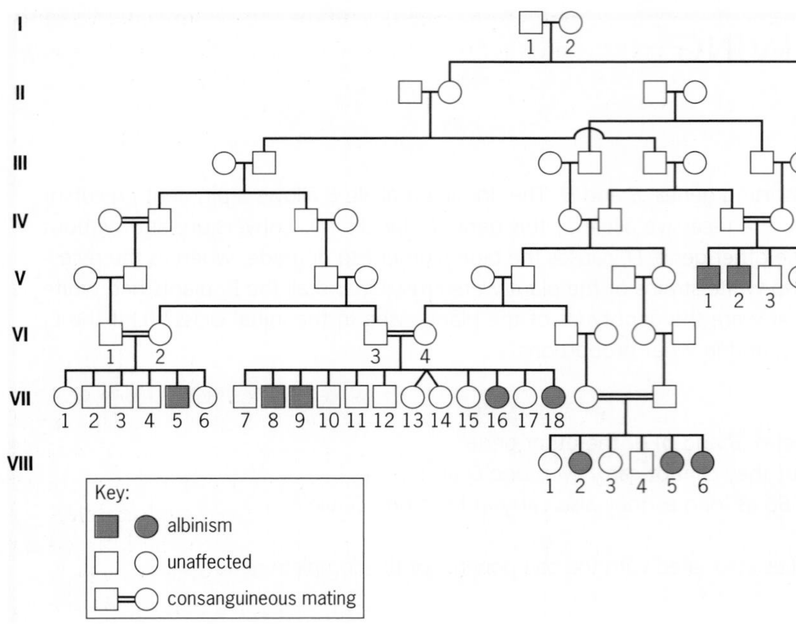

Figure 4.17 ▶ Albinism in the offspring of consanguineous marriages in an Amish community from the Midwestern United States. Consanguineous marriages are indicated by double lines between the mates. The individuals with albinism, who are homozygous for a recessive allele, all come from consanguineous matings.

The two dots in each individual represent the two copies of a particular gene, and the lines that connect individuals show how genes have passed from parent to offspring. This way of drawing a pedigree is different from the one we have used previously. It clarifies how each parent contributes genes to its offspring, and it allows us to trace the descent of a particular gene through multiple generations.

The two individuals in Generation II, labeled A and B, are half siblings. These individuals had a common father, C, but different mothers (D and E). The mating between A and B produced an offspring, I, who is inbred. Notice that I inherits one gene copy from A and one copy from B. However, both of these copies might have originated in C, the common father of A and B. Thus, the two gene copies in I might be identical to each other by descent from one of the gene copies that was present in C. This possibility of *identity by descent* is the important consequence of inbreeding. Any individual whose gene copies are identical by descent must be homozygous for a particular allele of that gene. Thus, consanguineous matings are expected to produce relatively more homozygotes than matings between unrelated individuals, which, as we have seen, is one of the conspicuous effects of inbreeding.

Figure 4.18 ▶ (*a*) Inbred varieties of maize and the hybrid produced by crossing them. The inbred plants are shorter and less robust than the hybrid plant. (*b*) Cobs from inbred plants are considerably smaller than cobs from hybrid plants.

In the pedigree we are considering, C is referred to as the *common ancestor* of I because two paths of descent from C converge in I, the inbred individual. The two paths are C → A → I and C → B → I, and together they form what geneticists call an *inbreeding loop*. This loop shows how a particular gene copy in C can be passed down both sides of the pedigree to produce two identical gene copies in I.

The fundamental determination in any analysis of inbreeding is to calculate the probability that two gene copies in an individual are identical by descent. Intuitively, this probability should increase with the intensity of inbreeding. Thus, the offspring of a mating between full siblings should have a greater probability of identity by descent than the offspring of a mating between half siblings. The effort to measure inbreeding intensity began with the pioneering work of the American geneticist Sewall Wright. In 1921 Wright discovered a mathematical quantity he called the **inbreeding coefficient**. Wright's investigations—too complicated to be discussed here—involved an analysis of correlations between the individuals in a pedigree. In these investigations, he discovered how to calculate the inbreeding coefficient and used it to measure the intensity of inbreeding. Then, in the 1940s, another American, Charles Cotterman, showed that Wright's inbreeding coefficient was equivalent to the probability of identity by descent. Thus, we can define the inbreeding coefficient, symbolized by the letter F, as the probability that two gene copies in an individual are identical by descent from a common ancestor.

To calculate the inbreeding coefficient, we follow the procedures developed by Wright and Cotterman. First, we identify the common ancestor(s) of the inbred individual. A common ancestor is connected to the inbred individual through both of that individual's parents. In the pedigree we are considering, I has only one common ancestor; however, in other types of pedigrees, an inbred individual might have more than one common ancestor. For example, the offspring of a mating between full siblings has two common ancestors:

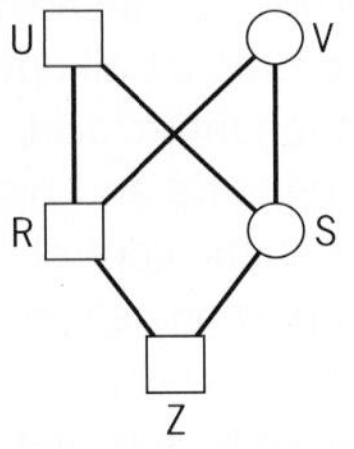

In this case, both of Z's grandparents (U and V) are common ancestors. Two genetic paths descend from each grandparent and converge in Z. Thus, the pedigree for full-sib mating has two distinct inbreeding loops:

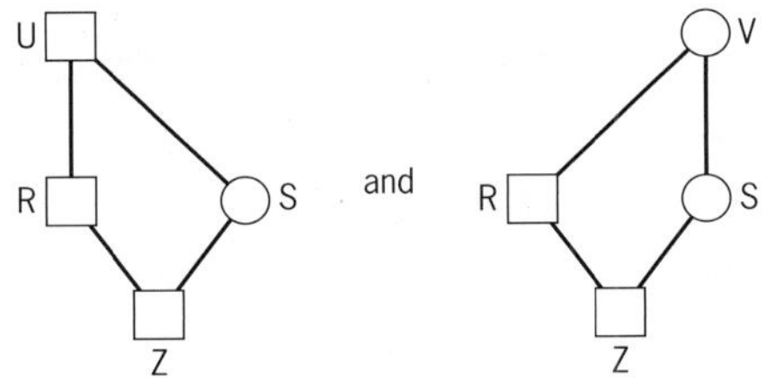

The second step in calculating the inbreeding coefficient is to count the number of individuals (n) in each inbreeding loop defined by a common ancestor. In the pedigree for mating between half siblings, there is one inbreeding loop and it has three individuals. (We do not count the inbred individual itself.) Thus, for the pedigree with half-sib mating, $n = 3$. In the pedigree for full-sib mating there are two inbreeding loops, each with three individuals; thus, for each of these loops, $n = 3$.

The third step in the procedure to calculate the inbreeding coefficient is to compute the quantity $(1/2)^n$ for each inbreeding loop and then sum the results. The sum we obtain is the inbreeding coefficient, F, of the inbred individual—that is, the probability that its two gene copies are identical to each other by descent from a common ancestor. For the offspring of a mating between half siblings, we obtain $F = (1/2)^3 = 1/8$. For the offspring of a mating between full siblings, we obtain $F = (1/2)^3 + (1/2)^3 = 1/4$. Thus, the inbreeding coefficient of the offspring of full-sib mating is greater than the inbreeding coefficient of the offspring of half-sib mating, as expected.

The factor $(1/2)^n$ that we compute for each inbreeding loop is the probability that *either* of the two gene copies in the common ancestor of that loop produces two identical gene copies in the inbred individual. To understand this probability, let's focus on the mating between half siblings. We must consider two cases, labeled 1 and 2 in the following illustration.

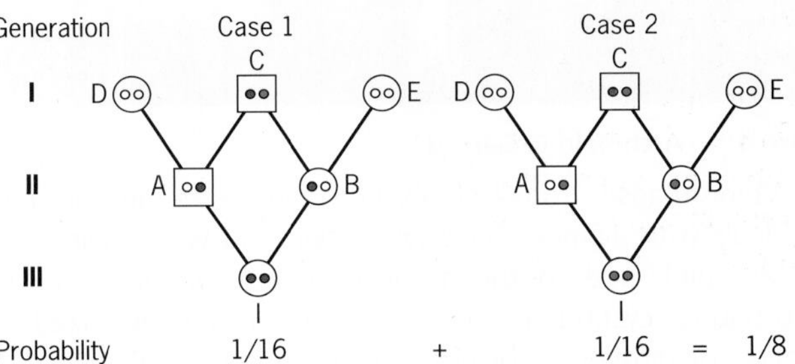

In Case 1, the chance that the gene copy on the left (shown in red) in the common ancestor C is transmitted to the daughter A is 1/2; once in A, the chance that this gene copy is transmitted to I is 1/2. Thus, the probability that the "left" gene copy in C makes its way down to I through A is $(1/2) \times (1/2) = 1/4$. Similarly, the chance that the "left" gene copy makes its way down to I through B is $(1/2) \times (1/2) = 1/4$. Altogether, then, the probability that the "left" gene copy in C produces two identical gene copies in I, one transmitted through A and the other through B, is $(1/4) \times (1/4) = 1/16$. By similar reasoning in Case 2, we find the probability that the "right" gene copy (shown in blue) in C produces two identical gene copies in I to be 1/16. Thus, the probability that *either* the "left" or the "right" gene copies in C will produce two identical gene copies in I is $(1/16) + (1/16) = 1/8$, which, as we have seen, is $(1/2)^3$. The procedure of calculating the factor $(1/2)^n$ is therefore a shortcut to find the probability that either of the gene copies in a particular common ancestor will give rise to two identical gene copies in the inbred individual.

This method of calculating inbreeding coefficients works for most pedigrees. However, when a common ancestor is itself

▶ A MILESTONE IN GENETICS: Garrod's Inborn Errors of Metabolism

At the beginning of the twentieth century, the British physician Archibald E. Garrod (**FIGURE 1**) saw connections between the emerging scientific disciplines of genetics and biochemistry. Garrod was fascinated by what he called the "inborn errors of metabolism"—by which he meant inherited conditions that "result from failure of some step or other in the series of chemical changes which constitute metabolism."[1] Garrod focused his attention on four of these conditions: albinism, alkaptonuria, cystinuria, and pentosuria.

Figure 1 ▶ Archibald E. Garrod.

Albinism arises from an inability to synthesize melanin pigment. Individuals with albinism have white skin and white hair, and their eyes are pink due to the absence of any pigment in the iris. Alkaptonuria, cystinuria, and pentosuria are characterized by the excretion of different substances in the urine. In alkaptonuria, the substance is homogentisic acid, which in Garrod's day was called alkapton. This substance turns black upon exposure to air. Thus, alkaptonuria is easily diagnosed—even in babies, where a darkened diaper is a sure sign of the condition. In cystinuria, the amino acid cystine is excreted in the urine. If the cystine becomes too concentrated, it crystallizes and forms calculi—"kidney stones"—which are painful and may damage the kidneys. In pentosuria, the sugar xylulose is excreted in the urine. Unlike cystinuria, neither alkaptonuria nor pentosuria is associated with a serious pathology.

[1]Garrod, A. E. 1909. *Inborn Errors of Metabolism.* In *Garrod's Inborn Errors of Metabolism*, 1963, by H. Harris, Oxford University Press.

In 1902, Garrod published a short paper in the British medical journal *The Lancet.*[2] This paper focused on studies of alkaptonurics, which Garrod found came disproportionately from parents who were first cousins. Discussions with William Bateson, the great champion of Mendelism, provided Garrod with a plausible explanation for this observation. Garrod hypothesized that alkaptonurics were homozygous for a recessive mutant allele. Though rare in the overall population, this allele was more likely to become homozygous in the offspring of relatives than in the offspring of unrelated individuals. In a book based on lectures he presented to the Royal College of Physicians in 1908, Garrod summarized his ideas this way:

> It was pointed out by Bateson . . . that the mode of incidence of alkaptonuria finds a ready explanation if the anomaly in question be regarded as a rare recessive character in the Mendelian sense. Mendel's law asserts that as regards two mutually exclusive characters, one of which tends to be dominant and the other recessive, crossbred individuals will tend to manifest the dominant character, but when they interbreed the offspring of the hybrids will exhibit one or other of the characters and will consist of dominants and recessives in definite proportions. . . . Only when two recessive gametes meet in fertilization will the resulting individual show the recessive character.
>
> If the recessive character be a rare one many generations may elapse before the union of two such gametes occurs, for the families in which they are produced will be few in number and the chance that in any given marriage both parents will contribute such gametes will be very small. When, however, intermarriage occurs between two members of such a family the chance will be much greater, and of the offspring of such a marriage several are likely to exhibit the peculiarity.[3]

Thus, the first recessive allele in humans was identified. What is more, this allele was shown to be associated with a metabolic abnormality. Garrod supposed that alkaptonuria was caused by an inability to metabolize homogentisic acid, which is then excreted in the urine. This merger of genetics and biochemistry led to the idea that metabolism is under genetic control. Garrod also had evidence that albinism occurs disproportionately among the offspring of first cousins and is therefore caused by a recessive allele, but for cystinuria and pentosuria, he could only speculate about a genetic basis.

[2]Garrod, A. E. 1902. The incidence of alkaptonuria: a study of chemical individuality. *Lancet* ii, 1616–1620.

[3]*Garrod, Inborn Errors of Metabolism.*

Nevertheless, based on his detailed knowledge of alkaptonuria, Garrod was convinced that genes influence the chemical makeup of individuals.

We now know that albinism, alkaptonuria, and another inborn error, phenylketonuria, are caused by blocks in the biochemical pathways that metabolize two amino acids, phenylalanine and tyrosine, which are derived from protein ingested in the food (**FIGURE 2**). In phenylketonuria, the amino acid phenylalanine cannot be converted into tyrosine because of a defect in the enzyme phenylalanine hydroxylase. Instead, phenylalanine is converted into another substance, phenylpyruvic acid, which accumulates to levels that will damage some tissues, especially those in the brain. If not treated by reducing phenylalanine intake in the diet, this metabolic abnormality can impair brain development. In albinism, the pathway that converts tyrosine to melanin pigment is blocked by a defect in the enzyme tyrosinase. Consequently, melanin pigment cannot be made. In alkaptonuria, the metabolic block is in a reaction that converts homogentisic acid into malylacetoacetic acid, which is, in turn, converted by other reactions into fumaric acid and acetoacetic acid. This block, as Garrod correctly proposed, causes homogentisic acid to accumulate and be excreted in the urine. Modern research has shown that pentosuria is due to a block in a pathway that breaks down the sugar xylulose. However, cystinuria, the most complex of the conditions that Garrod studied, is not due to a block in the metabolism of cystine. Rather, it stems from an inability of the renal tubules to reabsorb cystine from the filtered liquid in the kidneys.

Garrod's work forged a critical link between genetics and biochemistry. However, it took more than three decades for its full significance to be recognized. The clarifying event came in 1941, when George Beadle and Edward Tatum demonstrated that genes encode the enzymes that catalyze biochemical reactions. With Beadle and Tatum's discovery, the era of biochemical genetics was off and running.

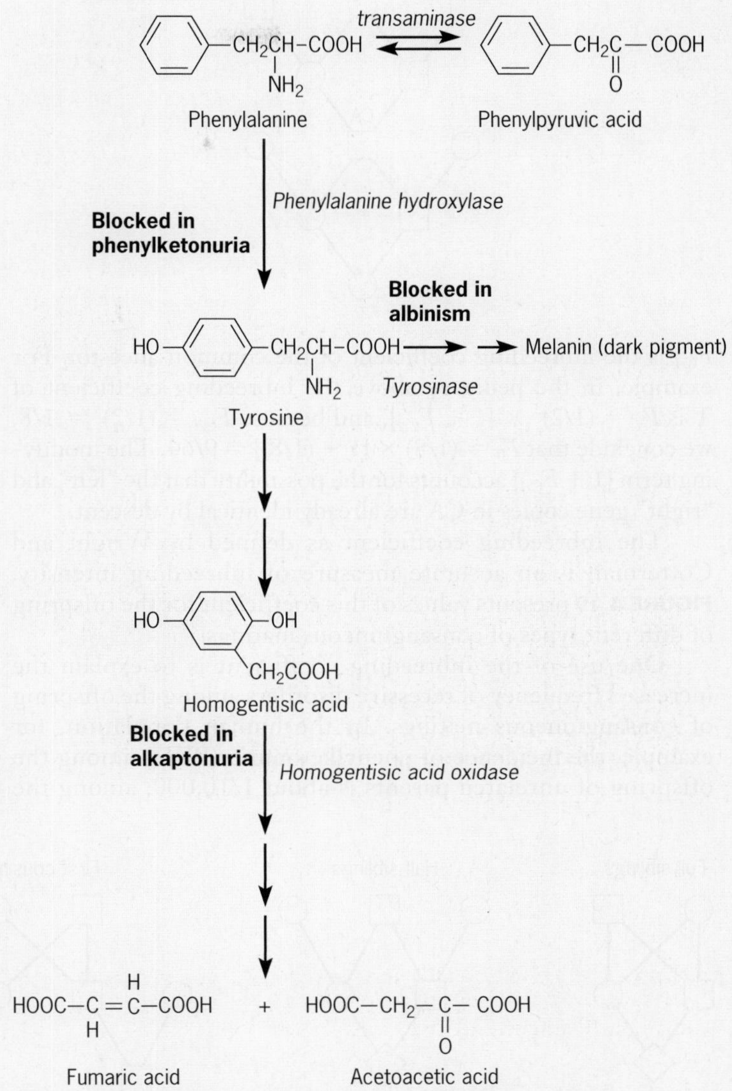

Figure 2 ▶ Biochemical pathways involving the breakdown of the amino acids phenylalanine and tyrosine.

QUESTIONS FOR DISCUSSION

1. Garrod's work exemplifies an interdisciplinary approach to research. Can you find current examples of interdisciplinary research projects? Are there special difficulties associated with these types of projects?
2. Many of the data obtained by Garrod and his colleagues came from experiments with human subjects. Chemicals were administered to these subjects to study aspects of metabolism. Other types of data came from studies with animals such as rabbits and dogs. What ethical issues are involved in research on human subjects and animals? How is this research regulated today?

inbred, the method needs to be modified. We multiply the factor $(1/2)^n$ for the common ancestor by the term $[1 + F_{CA}]$, where

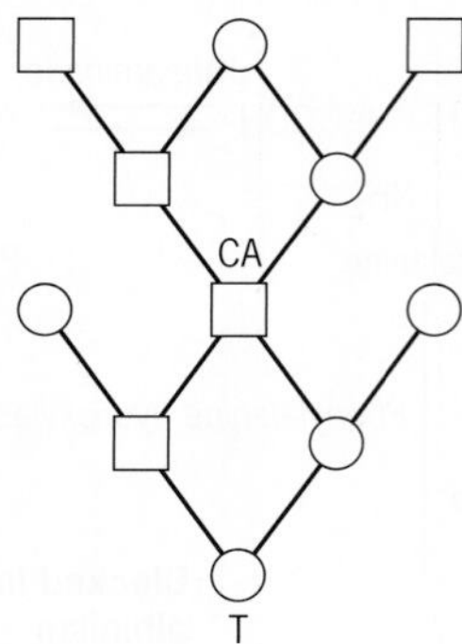

F_{CA} is the inbreeding coefficient of the common ancestor. For example, in the pedigree above the inbreeding coefficient of T is $F_T = (1/2)^3 \times [1 + F_{CA}]$, and because $F_{CA} = (1/2)^3 = 1/8$, we conclude that $F_T = (1/8) \times [1 + (1/8)] = 9/64$. The modifying term $[1 + F_{CA}]$ accounts for the possibility that the "left" and "right" gene copies in CA are already identical by descent.

The inbreeding coefficient as defined by Wright and Cotterman is an accurate measure of inbreeding intensity. **FIGURE 4.19** presents values of this coefficient for the offspring of different types of consanguineous matings.

One use of the inbreeding coefficient is to explain the increased frequency of recessive disorders among the offspring of consanguineous matings. In the human population, for example, the incidence of phenylketonuria (PKU) among the offspring of unrelated parents is about 1/10,000; among the offspring of first-cousin marriages, it is about 7/10,000. The difference between these frequencies, 6/10,000, is the effect of inbreeding with $F = 1/16$. For the offspring of closer relatives, we would expect a greater difference in the frequencies of PKU. For example, the offspring of half siblings have an inbreeding coefficient of 1/8, twice that of the offspring of first cousins. Because the effect of inbreeding is proportional to F, we would expect the incidence of PKU among the offspring of half siblings to be twice the inbreeding effect seen with the offspring of first cousins, plus the incidence of PKU in the general population. Thus, the predicted frequency of PKU among the offspring of half siblings is $2 \times (0.0006) + 0.0001 = 0.0013$. Among the offspring of full siblings, the predicted frequency is $4 \times (0.0006) + 0.0001 = 0.0025$ (because they have an inbreeding coefficient four times that of the offspring of first cousins).

Another use of the inbreeding coefficient is to measure the decline in a complex phenotype, such as plant height or crop yield. Such traits are influenced by large numbers of genes. **FIGURE 4.20** shows data collected from inbred strains of maize that were obtained through a program of repeated self-fertilization. Seed was saved at each stage of the inbreeding process, and at the end, maize plants were grown from the seed in test plots to study two traits, plant height and crop yield. As **FIGURE 4.20** shows, both of these traits declined linearly as a function of the inbreeding coefficient. The simplest explanation for this linear decline is that recessive alleles of different genes were made homozygous as the inbreeding proceeded—that is, in proportion to the value of F—and that

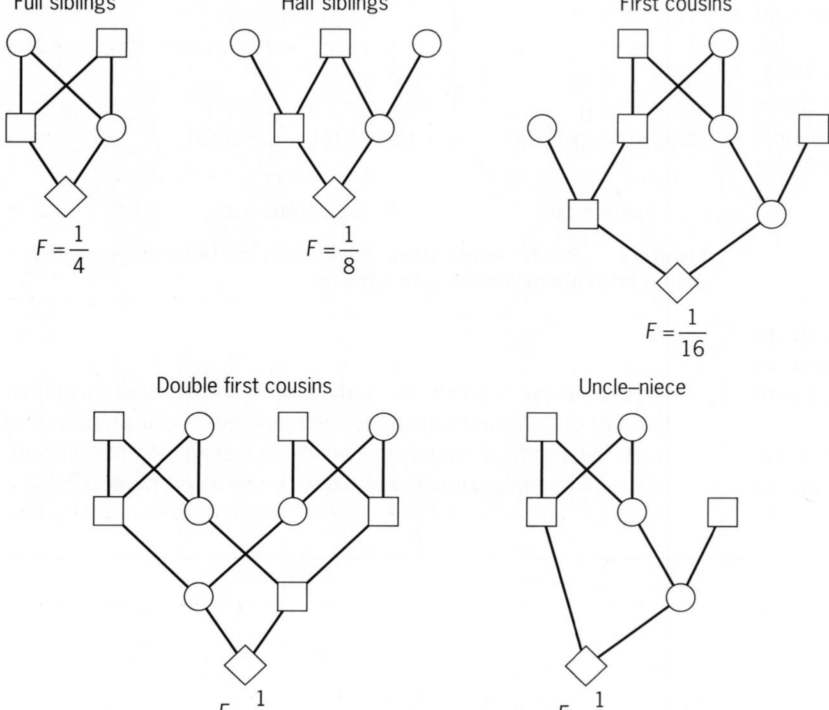

Figure 4.19 ▶ Values of the inbreeding coefficient, *F*, for different pedigrees.

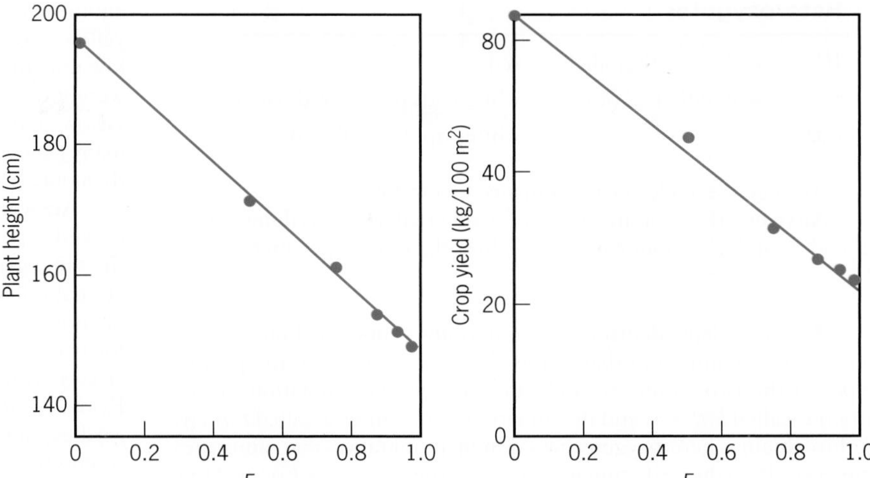

Figure 4.20 ▶ Inbreeding decline in plant height and crop yield in maize. The intensity of inbreeding is measured by the inbreeding coefficient, *F*.

these homozygotes manifested lower values for the traits. Thus, an increase in the incidence of deleterious recessive homozygotes is the basis for inbreeding depression.

MEASURING GENETIC RELATIONSHIPS

The inbreeding coefficient can also be used to measure the closeness of genetic relationships. Full siblings are obviously more closely related than half siblings. Are uncle and niece more closely related than half siblings? Half siblings are more closely related than first cousins. Are they more closely related than double first cousins? To answer these questions, we must determine the fraction of genes that two relatives share by virtue of common ancestry.

For regular relatives—that is, relatives that are not themselves inbred—we can calculate the fraction of genes that are shared by imagining that the relatives have mated and produced an offspring. Obviously, because this offspring is inbred, we can calculate its inbreeding coefficient according to the usual procedure. Then, to determine the fraction of genes that the two relatives share, we simply multiply the offspring's inbreeding coefficient by 2. The result is sometimes called the **coefficient of relationship.** For full siblings, the inbreeding coefficient of an imaginary offspring is 1/4; thus, the coefficient of relationship of full siblings (or the fraction of genes they share) is $2 \times (1/4) = 1/2$. By similar reasoning, the coefficient of relationship of half siblings is 1/4, that of first cousins is 1/8, and that of double first cousins is 1/4. For uncle and niece, the coefficient of relationship is 1/4. Thus, half siblings, double first cousins, and uncle and niece are equivalently related because each shares the same fraction of their genes, 1/4. Siblings, by comparison, are more closely related because they share half their genes, and single first cousins are less closely related because they share only one-eighth of their genes.

KEY POINTS

- Inbreeding increases the frequency of homozygotes and decreases the frequency of heterozygotes.
- The effects of inbreeding are proportional to the inbreeding coefficient, which is the probability that two gene copies in an individual are identical by descent from a common ancestor.
- The coefficient of relationship is the fraction of genes that two individuals share by virtue of common ancestry.

▶ Basic Exercises

ILLUSTRATE BASIC GENETIC ANALYSIS

1. A researcher has discovered a new blood-typing system for human beings. The system involves two antigens, P and Q, each determined by a different allele of a gene named *N*. The alleles for these antigens are about equally frequent in the general population. If the N^P and N^Q alleles are codominant, what antigens should be detected in the blood of N^PN^Q heterozygotes?

Answer: Both the P and the Q antigens should be detected because codominance implies that both of the alleles will be expressed in heterozygotes.

2. Flower color in a garden plant is under the control of a gene with multiple alleles. The phenotypes of the homozygotes and heterozygotes of this gene are as follows:

Homozygotes

WW	red
ww	pure white
w^sw^s	white stippled with red
w^pw^p	white with regular red patches

Heterozygotes

W	with any other allele	red
w^p	with either w^s or w	white with regular red patches
$w^s w$		white stippled with red

Arrange the alleles in a dominance hierarchy.

Answer: W is dominant to all the other alleles, w^p is dominant to w^s and w, and w^s is dominant to w. Thus, the dominance hierarchy is $W > w^p > w^s > w$.

3. Two independently discovered strains of mice are homozygous for a recessive mutation that causes the eyes to be small; the phenotypes of the two strains are indistinguishable. The mutation in one strain is called *little eye*, and the mutation in the other is called *tiny eye*. A third strain is heterozygous for a dominant mutation that eliminates the eyes altogether; the mutation in this strain is called *Eyeless*. How would you determine if the *little eye*, *tiny eye*, and *Eyeless* mutations are alleles of the same gene?

Answer: The procedure to determine if two recessive mutations are alleles of the same gene is to cross their respective homozygotes to obtain hybrid progeny and then evaluate the phenotype of the hybrids. If the phenotype is mutant, the mutations are alleles of the same gene; if it is wild-type, they are not alleles. In this case, we should therefore cross *little eye* mice with *tiny eye* mice and look at their offspring. If the offspring have small eyes, the two mutations are alleles of the same gene; if they have eyes of normal size, the two mutations are alleles of different genes. For a dominant mutation such as *Eyeless*, no test of allelism is possible. Thus, we cannot determine if *Eyeless* is an allele of either the *little eye* or the *tiny eye* mutation.

4. Distinguish between incomplete penetrance and variable expressivity.

Answer: Incomplete penetrance occurs when an individual with the genotype for a trait does not express that trait at all. Variable expressivity occurs when a trait is manifested to different degrees in a set of individuals with the genotype for that trait.

5. In a species of fly, the wild-type eye color is red. In a mutant strain homozygous for the w mutation, the eye color is pure white; in another mutant strain homozygous for the y mutation, the eye color is yellow. Homozygous white mutants were crossed to homozygous yellow mutants, and the offspring all had red eyes. When these offspring were intercrossed, they produced three classes of progeny: 92 red, 33 yellow, and 41 pure white. (a) From the results of these crosses, how many genes control eye color? Explain. (b) If the answer to (a) is greater than one, is any one mutant gene epistatic to any other mutant gene?

Answer: To answer (a), we note that the F_1 flies all had red—that is, wild-type—eyes. The w and y mutations are therefore not alleles of the same gene, and we conclude that at least two genes must control eye color in this species. To answer (b), we note that in the F_2 flies, the phenotypic segregation ratio departs from the 9:3:3:1 ratio expected for two genes assorting independently. The F_2 consists of only three classes, which, moreover, appear in the ratio of 9 red:4 white:3 yellow. Evidently, the ww homozygotes cause the flies to have white eyes regardless of what alleles of the y gene are present. Thus, the w mutant should be considered epistatic to the y mutant.

6. Sewall Wright, the discoverer of the inbreeding coefficient, was the offspring of a marriage between first cousins. Draw the pedigree of Dr. Wright's family and identify his common ancestors and the inbreeding loops they define. Then calculate Dr. Wright's inbreeding coefficient.

Answer: A pedigree for a first-cousin marriage is:

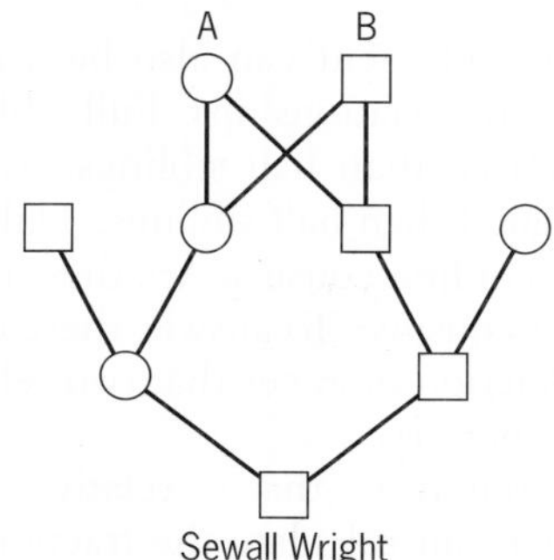

In it there are two common ancestors, A and B, each defining an inbreeding loop that terminates in the inbred individual. One loop is on the left side of the pedigree, the other on the right. Not counting the inbred individual, each of the loops contains five people. Thus, assuming that the common ancestors are not affected by prior inbreeding, the inbreeding coefficient of the offspring of the first-cousin marriage (Dr. Wright) is $(1/2)^5 + (1/2)^5 = 1/16$.

▶ Testing Your Knowledge

INTEGRATE DIFFERENT CONCEPTS AND TECHNIQUES

1. A geneticist has obtained two true-breeding strains of mice, each homozygous for an independently discovered recessive mutation that prevents the formation of hair on the body. One mutant strain is called *naked*, and the other is called *hairless*. To determine whether the two mutations are alleles, the geneticist crosses *naked* and *hairless* mice with each other. All the offspring are phenotypically wild-type; that is, they have hairs all over their bodies. After intercrossing these F_1 mice, the geneticist observes 115 wild-type mice and 85 mutant mice in the F_2 Are the *naked* and *hairless* mutations alleles? How would you explain the segregation of wild-type and mutant mice in the F_2?

Answer: The *naked* and *hairless* mutations are not alleles because the F_1 hybrids are phenotypically wild-type. Thus, *naked* and *hairless* are mutations of two different genes. To explain the phenotypic ratio in the F_2 let's first adopt symbols for these mutations and their dominant wild-type alleles:

n = naked mutation, N = wild-type allele
h = hairless mutation, H = wild-type allele

With these symbols, the genotypes of the true-breeding parental strains are $nn\ HH$ (*naked*) and $NN\ hh$ (*hairless*). The F_1 hybrids produced by crossing these strains are therefore *Nn Hh*. When these hybrids are intercrossed, we expect many different genotypes to appear in the offspring. However, each recessive allele, when homozygous, prevents the formation of hair on the body. Thus, only mice that are genotypically *N- H-* will develop hair; all the others—homozygous *nn* or homozygous *hh*, or homozygous for both recessive alleles—will fail to develop body hair. We can predict the frequencies of the wild and mutant phenotypes if we assume that the naked and hairless genes assort independently. The frequency of mice that will be *N- H-* is $(3/4) \times (3/4) = 9/16 = 0.56$ (by the Multiplicative Rule of Probability), and the frequency of mice that will be either *nn* or *hh* (or both) is $(1/4) + (1/4) - [(1/4) \times (1/4)] = 7/16 = 0.44$ (by the Additive Rule of Probability). Thus, in a sample of 200 F_2 progeny, we expect $200 \times 0.56 = 112$ to be wild-type and $200 \times 0.44 = 88$ to be mutant. The observed frequencies of 115 wild-type and 85 mutant mice are close to these expected numbers, suggesting that the hypothesis of two independently assorting genes for body hair is indeed correct.

2. In fruit flies a recessive mutation, *w*, causes the eyes to be white, another recessive mutation, *v*, causes them to be vermilion, and a third recessive mutation, *bw*, causes them to be brown. The wild-type eye color is dark red. Hybrids produced by crossing any two homozygous mutants have dark red eyes, and all the doubly homozygous mutant combinations have white eyes. How many genes do these three mutations define? If the dark red color of wild-type eyes is due to the accumulation of two different pigments, one red and the other brown, which gene controls the expression of which pigment? Can the genes be ordered into a pathway for pigment accumulation?

Answer: The three mutations define three different genes because when any two homozygous mutations are crossed, the offspring have wild-type eye color. The *w* mutation prevents the expression of all pigment because flies homozygous for it have neither red nor brown pigment in their eyes; the *v* mutation prevents the expression of brown pigment because flies homozygous for it have vermilion (bright red) eyes; and the *bw* mutation prevents the expression of red pigment because flies that are homozygous for it have brown eyes. Thus, the wild-type *v* gene controls the expression of brown pigment, the wild-type *bw* gene controls the expression of red pigment, and the wild-type *w* gene is necessary for the expression of both pigments. We can summarize these findings by proposing that each pigment is expressed in a different pathway and that the functioning of these pathways depends on the wild-type *w* gene.

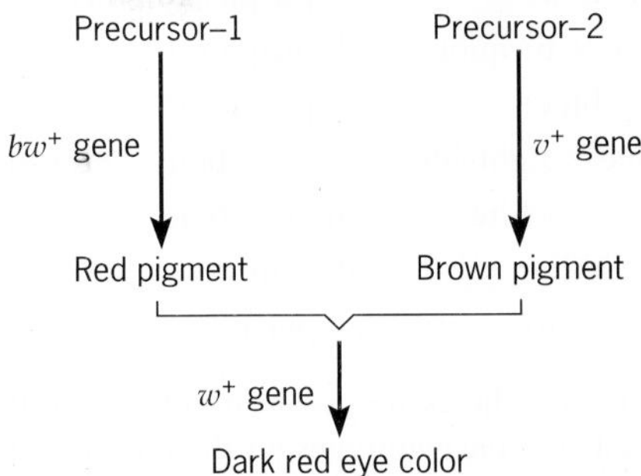

3. In the following pedigree, calculate the inbreeding coefficient of M.

Answer: M has three common ancestors, B, C, and D, because two lines of descent from each of these individuals ultimately converge in M. There are four distinct inbreeding loops (common ancestor underlined):

1. A B <u>C</u> D E $(n = 5)$
2. A D <u>C</u> B E $(n = 5)$
3. A <u>B</u> E $(n = 3)$
4. A <u>D</u> E $(n = 3)$

To calculate the inbreeding coefficient of M, F_M, we raise 1/2 to the power *n* for each of the loops and sum the results:

$$F_M = (1/2)^5 + (1/2)^5 + (1/2)^3 + (1/2)^3 = 5/16$$

Questions and Problems

ENHANCE UNDERSTANDING AND DEVELOP ANALYTICAL SKILLS

4.1 In mice, a series of five alleles determines fur color. In order of dominance, these alleles are: A^Y, yellow fur but homozygous lethal; A^L, agouti with light belly; A^+, agouti (wild-type); a^t, black and tan; and *a*, black. For each of the following crosses, give the coat color of the parents and the phenotypic ratios expected among the progeny: (a) $A^YA^L \times A^YA^L$; (b) $A^Ya \times A^La^t$; (c) $a^ta \times A^Ya$; (d) $A^La^t \times A^LA^L$; (e) $A^LA^L \times A^YA^+$; (f) $A^+a^t \times a^ta$; (g) $a^ta \times aa$; (h) $A^YA^L \times A^+a^t$; and (i) $A^Ya^L \times A^YA^+$

4.2 In rabbits, coloration of the fur depends on alleles of the gene *c*. From information given in the chapter, what phenotypes and proportions would be expected from the following crosses: (a) $c^+c^+ \times cc$; (b) $c^+c \times c^+c$; (c) $c^+c^h \times c^+c^{ch}$ (d) $cc^{ch} \times cc$; (e) $c^+c^h \times c^+c$; (f) $c^hc \times cc$?

4.3 What blood types could be observed in children born to a woman who has blood type N and a man who has blood type M?

4.4 GO The flower colors of plants in a particular population may be blue, purple, turquoise, light-blue, or white. A series of crosses between different members of the population produced the following results:

Cross	Parents	Progeny
1	purple × blue	all purple
2	purple × purple	76 purple, 25 turquoise
3	blue × blue	86 blue, 29 turquoise

4	purple × turquoise	49 purple, 52 turquoise
5	purple × purple	69 purple, 22 blue
6	purple × blue	50 purple, 51 blue
7	purple × blue	54 purple, 26 blue, 25 turquoise
8	turquoise × turquoise	all turquoise
9	purple × blue	49 purple, 25 blue, 23 light-blue
10	light-blue × light-blue	60 light-blue, 29 turquoise, 31 white
11	turquoise × white	all light-blue
12	white × white	all white
13	purple × white	all purple

How many genes and alleles are involved in the inheritance of flower color? Indicate all possible genotypes for the following phenotypes: (a) purple; (b) blue; (c) turquoise; (d) light-blue; (e) white.

4.5 A woman with type AB blood gave birth to a baby with type B blood. Two different men claim to be the father. One has type A blood, the other type O blood. Can the genetic evidence decide in favor of either?

4.6 A woman with type A blood gave birth to a baby, with type O blood. The woman stated that a man with type AB blood was the father of the baby. Is there any merit to her statement?

4.7 From information in the chapter about the ABO blood types, what phenotypes and ratios are expected from the following matings: (a) $I^Ai \times I^BI^B$; (b) $I^BI^B \times ii$; (c) $I^Ai \times I^Bi$; and (d) $I^Ai \times ii$?

4.8 In the fruit fly, recessive mutations in either of two independently assorting genes, *brown* and *purple*, prevent the synthesis of red pigment in the eyes. Thus, homozygotes for either of these mutations have brownish-purple eyes. However, heterozygotes for both of these mutations have dark red, that is, wild-type eyes. If such double heterozygotes are intercrossed, what kinds of progeny will be produced, and in what proportions?

4.9 A woman who has blood type AB and blood type M marries a man who has blood type O and blood type MN. If we assume that the genes for the A-B-O and M-N blood-typing systems assort independently, what blood types might the children of this couple have, and in what proportions?

4.10 From information given in the chapter, explain why mice with yellow coat color are not true-breeding.

4.11 A couple has four children. Neither the father nor the mother is bald; one of the two sons is bald, but neither of the daughters is bald.

(a) If one of the daughters marries a bald man and they have a son, what is the chance that the son will become bald as an adult?

(b) If the couple has a daughter, what is the chance that she will not be bald as an adult?

4.12 A Japanese strain of mice has a peculiar, uncoordinated gait called waltzing, which is due to a recessive allele, *v*. The dominant allele *V* causes mice to move in a coordinated fashion. A mouse geneticist has recently isolated another recessive mutation that causes uncoordinated movement. This mutation, called *tango*, could be an allele of the *waltzing* gene, or it could be a mutation in an entirely different gene. Propose a test to determine whether the *waltzing* and *tango* mutations are alleles, and if they are, propose symbols to denote them.

4.13 Congenital deafness in human beings is inherited as a recessive condition. In the following pedigree, two deaf individuals, each presumably homozygous for a recessive mutation, have married and produced four children with normal hearing. Propose an explanation.

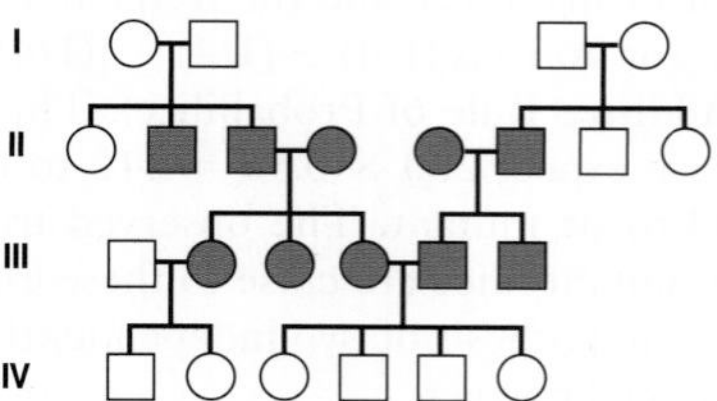

4.14 The following pedigree shows the inheritance of ataxia, a rare neurological disorder characterized by uncoordinated movements. Is ataxia caused by a dominant or a recessive allele? Explain.

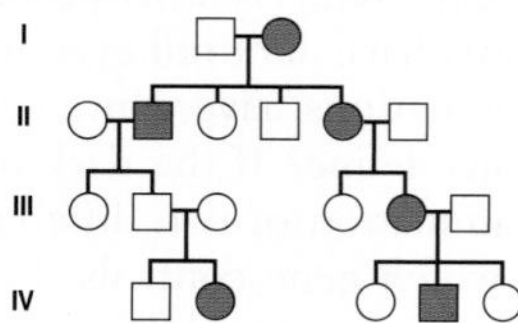

4.15 Summer squash plants with the dominant allele *C* bear white fruit, whereas plants homozygous for the recessive allele *c* bear colored fruit. When the fruit is colored, the dominant allele *G* causes it to be yellow; in the absence of this allele (that is, with genotype *gg*), the fruit color is green. What are the F_2 phenotypes and proportions expected from intercrossing the progeny of *CC GG* and *cc gg* plants? Assume that the *C* and *G* genes assort independently.

4.16 Rose-comb chickens mated with walnut-comb chickens produced 20 walnut-, 10 rose-, 4 pea-, and 8 single-comb chicks. Determine the genotypes of the parents.

4.17 Chickens that carry both the alleles for rose comb (*R*) and pea comb (*P*) have walnut combs, whereas chickens that lack both of these alleles (that is, they are genotypically *rr pp*) have single combs. From the information about interactions between these two genes given in the chapter, determine the phenotypes and proportions expected from the following crosses:

(a) *RR Pp* × *rr Pp*

(b) *rr PP* × *Rr Pp*

(c) *Rr Pp* × *Rr pp*

(d) *Rr pp* × *rr pp*

4.18 Consider the following hypothetical scheme of determination of coat color in a mammal. Gene *A* controls the conversion of a white pigment P_0 into a gray pigment P_1; the dominant allele *A* produces the enzyme necessary for this conversion, and the recessive allele *a* produces an enzyme without biochemical activity. Gene *B* controls the conversion of the gray pigment P_1 into a black pigment P_2; the dominant allele *B* produces the active enzyme for this conversion, and the recessive allele b produces an enzyme without activity. The dominant allele C of a third gene produces a polypeptide that completely inhibits the activity of the enzyme produced by gene A; that is, it prevents the reaction $P_0 \rightarrow P_1$. Allele c of this gene produces a defective polypeptide that does not inhibit the reaction $P_0 \rightarrow P_1$. Genes A, B, and C assort independently, and no other genes are involved. In the F_2 of the cross $AA\ bb\ CC \times aa\ BB\ cc$, what is the expected phenotypic segregation ratio?

4.19 Two plants with white flowers, each from true-breeding strains, were crossed. All the F_1 plants had red flowers. When these F_1 plants were intercrossed, they produced an F_2 consisting of 177 plants with red flowers and 142 with white flowers. (a) Propose an explanation for the inheritance of flower color in this plant species. (b) Propose a biochemical pathway for flower pigmentation and indicate which genes control which steps in this pathway.

4.20 The white Leghorn breed of chickens is homozygous for the dominant allele C, which produces colored feathers. However, this breed is also homozygous for the dominant allele I of an independently assorting gene that inhibits coloration of the feathers. Consequently, Leghorn chickens have white feathers. The white Wyandotte breed of chickens has neither the allele for color nor the inhibitor of color; it is therefore genotypically $cc\ ii$. What are the F_2 phenotypes and proportions expected from intercrossing the progeny of a white Leghorn hen and a white Wyandotte rooster?

4.21 Fruit flies homozygous for the recessive mutation *scarlet* have bright red eyes because they cannot synthesize brown pigment. Fruit flies homozygous for the recessive mutation *brown* have brownish-purple eyes because they cannot synthesize red pigment. Fruit flies homozygous for both of these mutations have white eyes because they cannot synthesize either type of pigment. The *brown* and *scarlet* mutations assort independently. If fruit flies that are heterozygous for both of these mutations are intercrossed, what kinds of progeny will they produce, and in what proportions?

4.22 Multiple crosses were made between true-breeding lines of black and yellow Labrador retrievers. All the progeny were black. When these progeny were intercrossed, they produced an F_2 consisting of 92 black, 40 yellow, and 28 chocolate. (a) Propose an explanation for the inheritance of coat color in Labrador retrievers. (b) Propose a biochemical pathway for coat color determination and indicate how the relevant genes control coat coloration.

4.23 Mabel and Frank are half siblings, as are Tina and Tim. However, these two pairs of half sibs do not have any common ancestors. If Mabel marries Tim and Frank marries Tina and each couple has a child, what fraction of their genes will these children share by virtue of common ancestry? Will the children be more or less closely related than first cousins?

4.24 GO The Micronesian Kingfisher, *Halcyon cinnamomina*, has a cinnamon-colored face. In some birds, the color continues onto the chest, producing one of three patterns: a circle, a shield, or a triangle; in other birds, there is no color on the chest. A male with a colored triangle was crossed with a female that had no color on her chest, and all their offspring had a colored shield on the chest. When these offspring were intercrossed, they produced an F_2 with a phenotypic ratio of 3 circle:6 shield:3 triangle:4 no color. (a) Determine the mode of inheritance for this trait and indicate the genotypes of the birds in all three generations. (b) If a male without color on his chest is mated to a female with a colored shield on her chest and the F_1 segregate in the ratio of 1 circle:2 shield:1 triangle, what are the genotypes of the parents and their progeny?

4.25 In the following pedigrees, what are the inbreeding coefficients of A, B, and C?

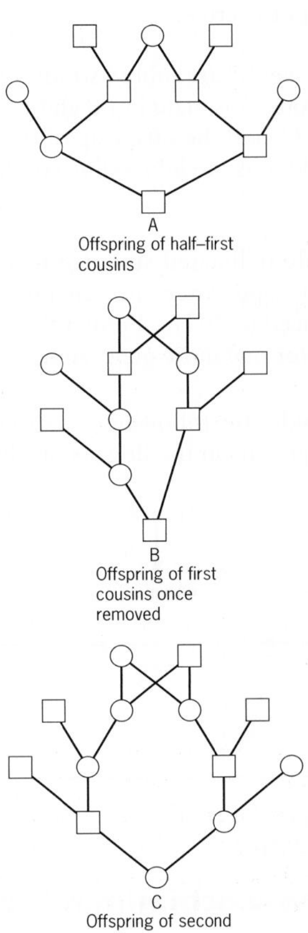

4.26 Suppose that the inbreeding coefficient of I in the following pedigree is 0.25. What is the inbreeding coefficient of I's common ancestor, C?

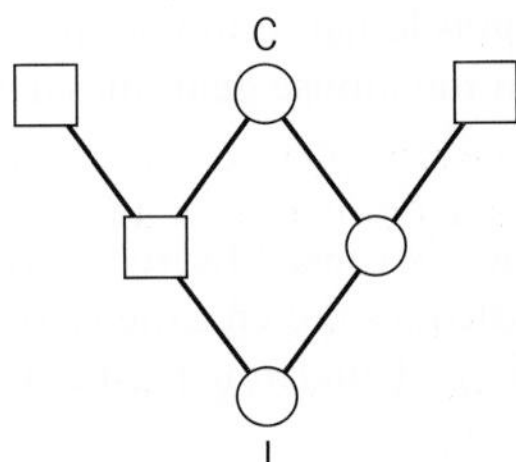

4.27 In a species of tree, seed color is determined by four independently assorting genes: A, B, C, and D. The recessive alleles of each of these genes (a, b, c, and d) produce abnormal enzymes that cannot catalyze a reaction in the biosynthetic pathway for seed pigment. This pathway is diagrammed as follows:

White precursor $\xrightarrow{A}$ Yellow $\xrightarrow{B}$ Orange $\xrightarrow{C}$ Red

Orange $\xrightarrow{D}$ Blue

When both red and blue pigments are present, the seeds are purple. Trees with the genotypes *Aa Bb Cc Dd* and *Aa Bb Cc dd* were crossed.

(a) What color are the seeds in these two parental genotypes?

(b) What proportion of the offspring from the cross will have white seeds?

(c) Determine the relative proportions of red, white, and blue offspring from the cross.

4.28 GO A, B, and C are inbred strains of mice, assumed to be completely homozygous. A is mated to B and B to C. Then the A × B hybrids are mated to C, and the offspring of this mating are mated to the B × C hybrids. What is the inbreeding coefficient of the offspring of this last mating?

4.29 A randomly pollinated strain of maize produces ears that are 24 cm long, on average. After one generation of self-fertilization, the ear length is reduced to 20 cm. Predict the ear length if self-fertilization is continued for one more generation.

4.30 GO Consider the following genetically controlled biosynthetic pathway for pigments in the flowers of a hypothetical plant:

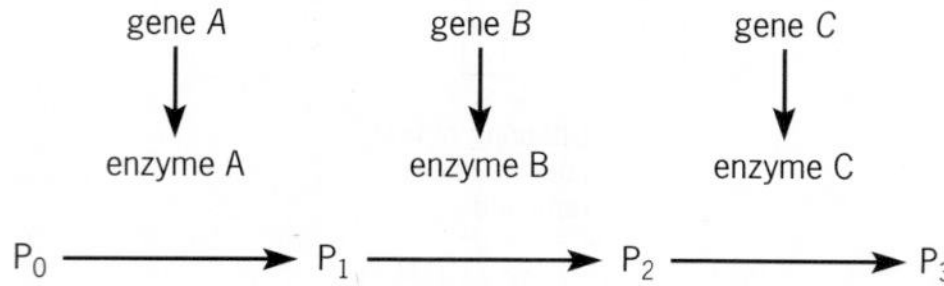

Assume that gene *A* controls the conversion of a white pigment, P_0, into another white pigment, P_1; the dominant allele *A* specifies an enzyme necessary for this conversion, and the recessive allele *a* specifies a defective enzyme without biochemical function. Gene *B* controls the conversion of the white pigment, P_1, into a pink pigment, P_2; the dominant allele, *B*, produces the enzyme necessary for this conversion, and the recessive allele, *b*, produces a defective enzyme. The dominant allele, *C*, of the third gene specifies an enzyme that converts the pink pigment, P_2, into a red pigment, P_3; its recessive allele, *c*, produces an altered enzyme that cannot carry out this conversion. The dominant allele, *D*, of a fourth gene produces a polypeptide that completely inhibits the function of enzyme *C*; that is, it blocks the reaction $P_2 \rightarrow P_3$. Its recessive allele, *d*, produces a defective polypeptide that does not block this reaction. Assume that flower color is determined solely by these four genes and that they assort independently. In the F_2 of a cross between plants of the genotype *AA bb CC DD* and plants of the genotype *aa BB cc dd*, what proportion of the plants will have (a) red flowers? (b) pink flowers? (c) white flowers?

▶ Genomics on the Web

at http://www.ncbi.nlm.nih.gov/

Coat color in mammals is controlled by many different genes.

1. In the mouse, the A^Y mutation, a dominant allele of the *a* gene, makes the coat yellow instead of agouti; in homozygous condition, this mutation is lethal. Can you find a description of the *a* gene and its A^Y allele in the mouse genomics database? What is the official name of this gene?

2. Albinism in mice is caused by recessive mutations in a gene called *Tyr*. This gene encodes the enzyme tyrosinase, which catalyzes a step in the production of melanin pigment from the amino acid tyrosine. What other symbol is used to denote this gene? Do you suspect that this gene is related, in an evolutionary sense, to the gene that, when mutant, causes albinism in rabbits?

3. Do humans have a gene related to the *Tyr* gene of mice? If they do, with what condition might this gene, when mutant, be associated?

Hint: At the web site, Genomic Biology → Mouse Genome Resources → Mouse Genome Informatics → Search for A <Y> or for Tyr; then try searching on the Human Genome Resources page for Tyr.

Chapter 5
The Chromosomal Basis of Mendelism

CHAPTER OUTLINE

- **Chromosomes**
- **The Chromosome Theory of Heredity**
- **Sex-Linked Genes in Human Beings**
- **Sex Chromosomes and Sex Determination**
- **Dosage Compensation of X-Linked Genes**

The fruit fly, Drosophila melanogaster.

Sex, Chromosomes, and Genes

What causes organisms to develop as males or females? Why are there only two sexual phenotypes? Is the sex of an organism determined by its genes? These and related questions have intrigued geneticists since the rediscovery of Mendel's work at the beginning of the twentieth century.

The discovery that genes play a role in the determination of sex emerged from a fusion between two previously distinct scientific disciplines, genetics—the study of heredity—and cytology—the study of cells. Early in the twentieth century, these disciplines were brought together through a friendship between two remarkable American scientists, Thomas Hunt Morgan and Edmund Beecher Wilson. Morgan was the geneticist and Wilson the cytologist.

As the cytologist, Wilson was interested in the behavior of chromosomes. These structures would prove to be important for sex determination in many species, including our own. Wilson was one of the first to investigate differences in the chromosomes of the two sexes. Through careful study, he and his colleagues showed that these differences were confined to a special pair of chromosomes called sex chromosomes. Wilson found that the behavior of these chromosomes during meiosis could account for the inheritance of sex.

As the geneticist, Morgan was interested in the identification of genes. He focused his research on the fruit fly, *Drosophila melanogaster,* and rather quickly discovered a gene that gave different phenotypic ratios in males and females. Morgan hypothesized that this gene was located on one of the sex chromosomes, and one of his students, Calvin Bridges, eventually proved this hypothesis to be correct. Morgan's discovery that genes reside on chromosomes was a great achievement. The abstract genetic factors postulated by Mendel were finally localized on visible structures within cells. Geneticists could now explain the Principles of Segregation and Independent Assortment in terms of meiotic chromosome behavior.

The discovery that specific genes determine the sex of an organism came much later, only after another scientific discipline, molecular biology, had joined forces with genetics and cytology. Through their combined efforts, cytologists, geneticists, and molecular biologists identified specific sex-determining genes by studying rare individuals in which the sexual phenotype was inconsistent with the sex chromosomes that were present. Today, researchers in all three fields are earnestly trying to figure out how these genes control sexual development.

Chromosomes

Each species has a characteristic set of chromosomes.

Chromosomes were discovered in the second half of the nineteenth century by a German cytologist, W. Waldeyer. Subsequent investigations with many different organisms established that chromosomes are characteristic of the nuclei of all cells. They are best seen by applying dyes to dividing cells; during division, the material in a chromosome is packed into a small volume, giving it the appearance of a tightly organized cylinder. During the interphase between cell divisions, chromosomes are not so easily seen, even with the best of dyes. Interphase chromosomes are loosely coiled, forming thin threads that are distributed throughout the nucleus. Consequently, when dyes are applied, the whole nucleus is stained and individual chromosomes cannot be identified. This diffuse network of threads is called **chromatin.** Some regions of the chromatin stain more darkly than others, suggesting an underlying difference in organization. The light regions are called the **euchromatin** (from the Greek word for "true") and the dark regions are called the **heterochromatin** (from the Greek word for "different"). We shall explore the functional significance of these different types of chromatin in Chapter 20.

CHROMOSOME NUMBER

Within a species, the number of chromosomes is almost always an even multiple of a basic number. In human beings, for example, the basic number is 23; mature eggs and sperm have this number of chromosomes. Most other types of human cells have twice as many (46), although a few kinds, such as some liver cells, have four times (92) the basic number.

The **haploid,** or basic, chromosome number **(n)** defines a set of chromosomes called the *haploid genome.* Most somatic cells contain two of each of the chromosomes in this set and are therefore **diploid (2n).** Cells with four of each chromosome are **tetraploid (4n),** those with eight of each are **octoploid (8n),** and so on.

The basic number of chromosomes varies among species. Chromosome number is unrelated to the size or biological complexity of an organism, with most species containing between 10 and 40 chromosomes in their genomes (**TABLE 5.1**). The muntjac, a tiny Asian deer, has only three chromosomes in its genome, whereas some species of ferns have many hundreds.

▶ **TABLE 5.1**

Chromosome Number in Different Organisms

Organism	Haploid Chromosome Number
Simple Eukaryotes	
Baker's yeast (*Saccharomyces cerevisiae*)	16
Bread mold (*Neurospora crassa*)	7
Unicellular green alga (*Chlamydomonas reinhardtii*)	17
Plants	
Maize (*Zea mays*)	10
Bread wheat (*Triticum aestivum*)	21
Tomato (*Lycopersicon esculentum*)	12
Broad bean (*Vicia faba*)	6
Giant sequoia (*Sequoia sempervirens*)	11
Crucifer (*Arabidopsis thaliana*)	5
Invertebrate Animals	
Fruit fly (*Drosophila melanogaster*)	4
Mosquito (*Anopheles culicifacies*)	3
Starfish (*Asterias forbesi*)	18
Nematode (*Caenorhabditis elegans*)	6
Mussel (*Mytilus edulis*)	14
Vertebrate Animals	
Human being (*Homo sapiens*)	23
Chimpanzee (*Pan troglodytes*)	24
Cat (*Felis domesticus*)	36
Mouse (*Mus musculus*)	20
Chicken (*Gallus domesticus*)	39
Toad (*Xenopus laevis*)	17
Fish (*Esox lucius*)	25

SEX CHROMOSOMES

In some animal species such as grasshoppers, females have one more chromosome than males (**FIGURE 5.1*a***). This extra chromosome, originally observed in other insects, is called the **X chromosome.** Females of these species have two X chromosomes, and males have only one; thus, females are cytologically XX and males are XO, where the "O" denotes the absence of a chromosome. During meiosis in the female, the two X chromosomes pair and then separate, producing eggs that contain a single X chromosome. During meiosis in the male, the solitary X chromosome moves independently of all the other chromosomes and is incorporated into half the sperm; the other half receive no X chromosome. Thus, when sperm and eggs unite, two kinds of zygotes are produced: XX, which develop into females, and XO, which develop into males. Because each of these types is equally likely, the reproductive mechanism preserves a 1:1 ratio of males to females in these species.

In many other animals, including human beings, males and females have the same number of chromosomes (**FIGURE 5.1*b***). This numerical equality is due to the presence of a chromosome in the male, called the **Y chromosome,** which pairs with the X during meiosis. The Y chromosome is morphologically distinguishable from the X chromosome. In humans, for example, the Y is much shorter than the X, and its centromere is

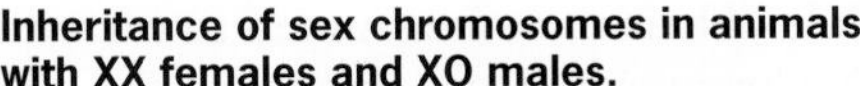

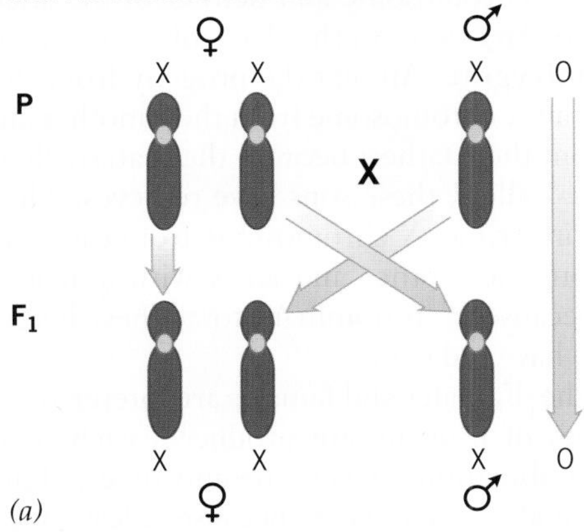

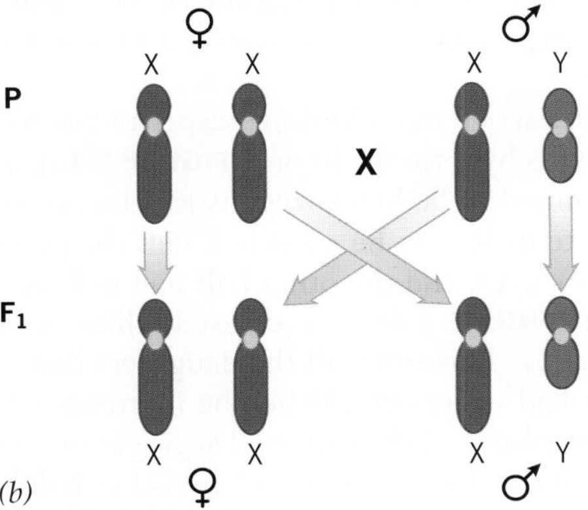

Figure 5.1 ▶ Inheritance of sex chromosomes in animals. (*a*) XX female/XO male animals, such as some grasshoppers. (*b*) XX female/XY male animals, such as human beings and *Drosophila*.

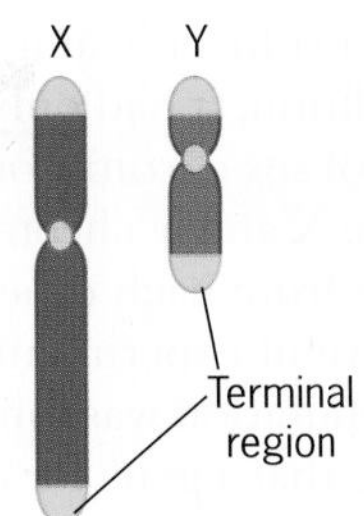

Figure 5.2 ▶ Human X and Y chromosomes. The terminal regions are common to both sex chromosomes.

located closer to one of the ends (**FIGURE 5.2**). The material common to the human X and Y chromosomes is limited, consisting mainly of short segments near the ends of the chromosomes. During meiosis in the male, the X and Y chromosomes separate from each other, producing two kinds of sperm, X-bearing and Y-bearing; the frequencies of the two types are approximately equal. XX females produce only one kind of egg, which is X-bearing. If fertilization were to occur randomly, approximately half the zygotes would be XX and the other half would be XY, leading to a 1:1 sex ratio at conception. However, in human beings, Y-bearing sperm have a fertilization advantage because they are lighter and move faster, and the zygotic sex ratio is about 1.3:1. During development, the excess of males is diminished by differential viability of XX and XY embryos, and at birth, males are only slightly more numerous than females (sex ratio 1.07:1). By the age of reproduction, the excess of males is essentially eliminated and the sex ratio is very close to 1:1.

The X and Y chromosomes are called **sex chromosomes.** All the other chromosomes in the genome are called **autosomes.** Sex chromosomes were discovered in the first few years of the twentieth century through the work of the American cytologists C. E. McClung, N. M. Stevens, W. S. Sutton, and E. B. Wilson. This discovery coincided closely with the emergence of Mendelism and stimulated research on the possible relationships between Mendel's principles and the meiotic behavior of chromosomes.

KEY POINTS

- Individual chromosomes become visible during cell division; between divisions they form a diffuse network of fibers called chromatin.
- Diploid somatic cells have twice as many chromosomes as haploid gametes.
- Sex chromosomes are different between the two sexes, whereas autosomes are the same.

The Chromosome Theory of Heredity

Studies on the inheritance of a sex-linked trait in *Drosophila* provided the first evidence that the meiotic behavior of chromosomes is the basis for Mendel's Principles of Segregation and Independent Assortment.

By 1910 many biologists suspected that genes were situated on chromosomes, but they did not have definitive proof. Researchers needed to find a gene that could be unambiguously linked to a chromosome. This goal required that the gene be defined by a mutant allele and that the chromosome be morphologically distinguishable. Furthermore, the pattern of gene transmission had to reflect the chromosome's behavior during reproduction. All these requirements were fulfilled when the American biologist Thomas H. Morgan discovered a particular eye color mutation in the fruit fly, *Drosophila melanogaster*. Morgan began experimentation with this species of fly about 1909. It was ideally suited for genetics research because it reproduced quickly and prolifically and was inexpensive to rear in the laboratory. In addition, it had only four pairs of chromosomes, one being a pair of sex chromosomes—XX in the female and XY in the male. The X and Y chromosomes were morphologically distinguishable from each other and from each of the autosomes. Through careful experiments, Morgan was able to show that the eye color mutation was inherited along with the X chromosome, suggesting that a gene for eye color was physically situated on that chromosome. Later, one of his students, Calvin B. Bridges, obtained definitive proof for this Chromosome Theory of Heredity.

EXPERIMENTAL EVIDENCE LINKING THE INHERITANCE OF GENES TO CHROMOSOMES

Morgan's experiments commenced with his discovery of a mutant male fly that had white eyes instead of the red eyes of wild-type flies. When this male was crossed to wild-type females, all the progeny had red eyes, indicating that white was recessive to red. When these progeny were intercrossed with each other, Morgan observed a peculiar segregation pattern: all of the daughters, but only half of the sons, had red eyes; the other half of the sons had white eyes. This pattern suggested that the inheritance of eye color was linked to the sex chromosomes. Morgan proposed that a gene for eye color was present on the X chromosome, but not on the Y, and that the white and red phenotypes were due to two different alleles, a mutant allele denoted w and a wild-type allele denoted w^+.

Morgan's hypothesis is diagrammed in **FIGURE 5.3**. The wild-type females in the first cross are assumed to be homozygous for the w^+ allele. Their mate is assumed to carry the mutant w allele on its X chromosome and neither of the alleles on its Y chromosome. An organism that has only one copy of a gene is called a **hemizygote.** Among the progeny from the cross, the sons inherit an X chromosome from their mother and a Y chromosome from their father; because the maternally inherited X carries the w^+ allele, these sons have red eyes. The daughters, in contrast, inherit an X chromosome from each parent—an X with w^+ from the mother and an X with w from the father. However, because w^+ is dominant to w, these heterozygous F_1 females also have red eyes.

When the F_1 males and females are intercrossed, four genotypic classes of progeny are produced, each representing a different combination of sex chromosomes. The XX flies, which are female, have red eyes because at least one w^+ allele is present. The XY flies, which are male, have either red or white eyes, depending on which X chromosome is inherited from the heterozygous F_1 females. Segregation of the w and w^+ alleles in these females is therefore the reason half the F_2 males have white eyes.

Morgan carried out additional experiments to confirm the elements of his hypothesis. In one (**FIGURE 5.4*a***), he crossed F_1 females assumed to be heterozygous for the eye color gene to mutant white males. As he expected, half the progeny of each sex had white eyes, and the other half had red eyes. In another experiment (**FIGURE 5.4*b***), he crossed white-eyed females to red-eyed males. This time, all the daughters had red eyes, and all the sons had white eyes. When he intercrossed these progeny, Morgan observed the expected segregation: half the progeny of each sex had white eyes, and the other half had red eyes. Thus, Morgan's hypothesis that the gene for eye color was linked to the X chromosome withstood additional experimental testing.

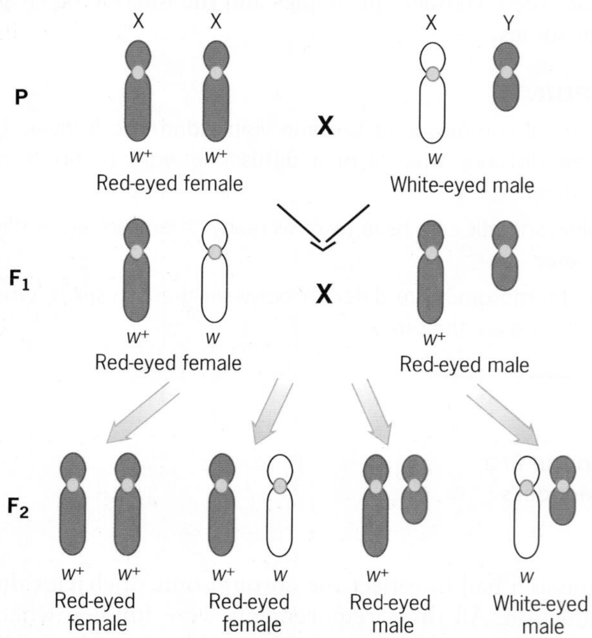

Figure 5.3 ▶ Morgan's experiment studying the inheritance of white eyes in *Drosophila*. The transmission of the mutant condition in association with sex suggested that the gene for eye color was present on the X chromosome but not on the Y chromosome.

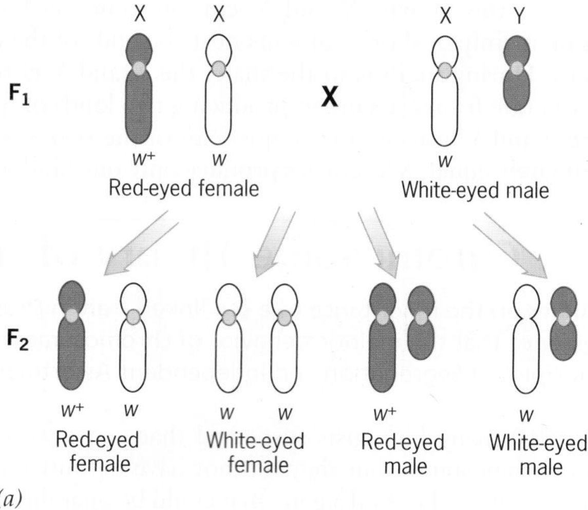

Figure 5.4 ▶ Experimental tests of Morgan's hypothesis that the gene for eye color in *Drosophila* is X-linked. (*a*) Experiment in which heterozygous females were crossed to white-eyed males.

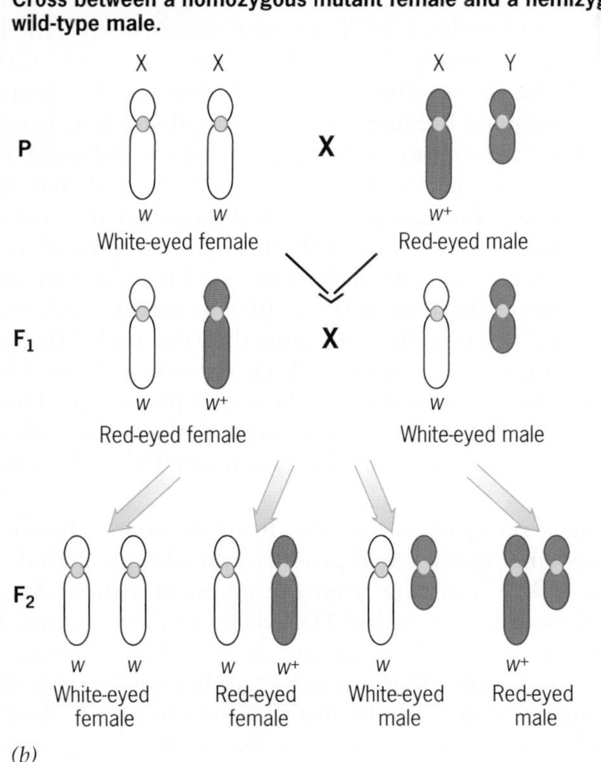

Figure 5.4 ▶ (*b*) Experiment in which white-eyed females were crossed to wild-type males.

CHROMOSOMES AS ARRAYS OF GENES

Morgan and his students soon identified other X-linked genes in *Drosophila*. In each case, simple breeding experiments demonstrated that recessive mutations of these genes were transmitted along with the X chromosome. As the evidence accumulated, it became clear that many genes were located on the X chromosome. However, Morgan's research group also identified genes that were not on the X chromosome. These genes followed the Mendelian Principle of Segregation, but they did not segregate with sex, as the gene for eye color did. Morgan correctly concluded that such genes were located on one of the three autosomes in the *Drosophila* genome. Thus, each *Drosophila* chromosome appeared to contain a different set of genes.

Morgan's laboratory then attempted to determine the relationships among the genes on a particular chromosome. They proceeded on the assumption that the genes were arranged in a linear array—an idea inspired by cytological evidence that the chromosome was a long, thin thread. In just a few years, Morgan's students were able to show that genes were indeed situated at different sites, or loci (from the Latin word for "place"; singular; locus), on a linear structure. This analysis, which we will discuss in Chapter 7, produced the world's first genetic maps—diagrams showing the positions of genes and the relative distances between them (**FIGURE 5.5**). Morgan's laboratory pioneered the methods for genetic mapmaking and laid the foundation for subsequent research on the physical structure of chromosomes. Eventually, the linearity of chromosomes was connected to the linear structure of DNA (see Chapter 9).

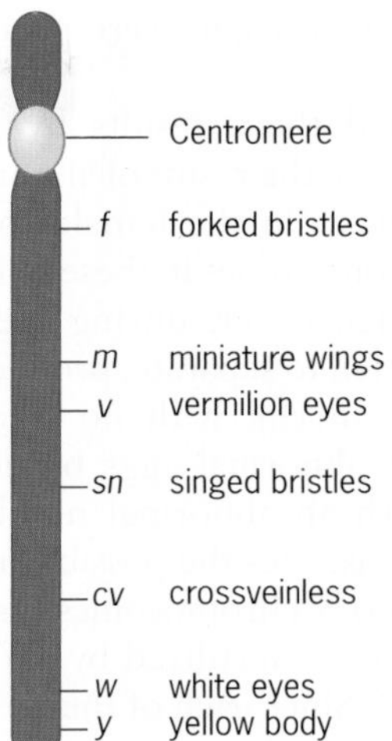

Figure 5.5 ▶ A map of genes on the X chromosome of *Drosophila*.

These early studies with *Drosophila*—primarily the work of Morgan and his students (see A Milestone in Genetics: Morgan's Fly Room later in this chapter)—greatly strengthened the view that all genes were located on chromosomes and that Mendel's principles could be explained by the transmissional properties of chromosomes during reproduction. This idea, called the **Chromosome Theory of Heredity,** stands as one of the most important achievements in biology. Since its formulation in the early part of the twentieth century, the Chromosome Theory of Heredity has provided a unifying framework for all studies of inheritance.

NONDISJUNCTION AS PROOF OF THE CHROMOSOME THEORY

Morgan showed that a gene for eye color was on the X chromosome of *Drosophila* by correlating the inheritance of that gene with the transmission of the X chromosome during reproduction. However, as noted earlier, it was one of his students, C. B. Bridges, who secured proof of the chromosome theory by showing that exceptions to the rules of inheritance could also be explained by chromosome behavior.

Bridges performed one of Morgan's experiments on a larger scale. He crossed white-eyed female *Drosophila* to red-eyed males and examined many F_1 progeny. Although as expected, nearly all the F_1 flies were either red-eyed females or white-eyed males, Bridges found a few exceptional flies—white-eyed females and red-eyed males. He crossed these exceptions to determine how they might have arisen. The exceptional males all proved to be sterile; however, the exceptional females were fertile, and when crossed to normal red-eyed males, they produced many progeny, including large numbers of white-eyed daughters and red-eyed sons. Thus, the exceptional F_1 females, though rare in their own right, were prone to produce many exceptional progeny.

Bridges explained these results by proposing that the exceptional F_1 flies were the result of abnormal X chromosome behavior during meiosis in the females of the P generation. Ordinarily, the X chromosomes in these females should **disjoin,** or separate from each other, during meiosis. Occasionally, however, they might fail to separate, producing an egg with two X chromosomes or an egg with no X chromosome at all. Fertilization of such abnormal eggs by normal sperm would produce zygotes with an abnormal number of sex chromosomes. **FIGURE 5.6** illustrates the possibilities.

If an egg with two X chromosomes (usually called a diplo-X egg; genotype X^wX^w) is fertilized by a Y-bearing sperm, the zygote will be X^wX^wY. Since each of the X chromosomes in this zygote carries a mutant w allele, the resulting fly will have white eyes. If an egg without an X chromosome (usually called a nullo-X egg) is fertilized by an X-bearing sperm (X^+), the zygote will be X^+O. (Once again, "O" denotes the absence of a chromosome.) Because the single X in this zygote carries a w^+ allele, the zygote will develop into a red-eyed fly. Bridges inferred that XXY flies were female and that XO flies were male. The exceptional white-eyed females that he observed were therefore X^wX^wY, and the exceptional red-eyed males were X^+O. Bridges confirmed the chromosome constitutions of these exceptional flies by direct cytological observation. Because the XO animals were male, Bridges concluded that in *Drosophila* the Y chromosome has nothing to do with the determination of the sexual phenotype. However, because the XO males were always sterile, he realized that this chromosome must be important for male sexual function.

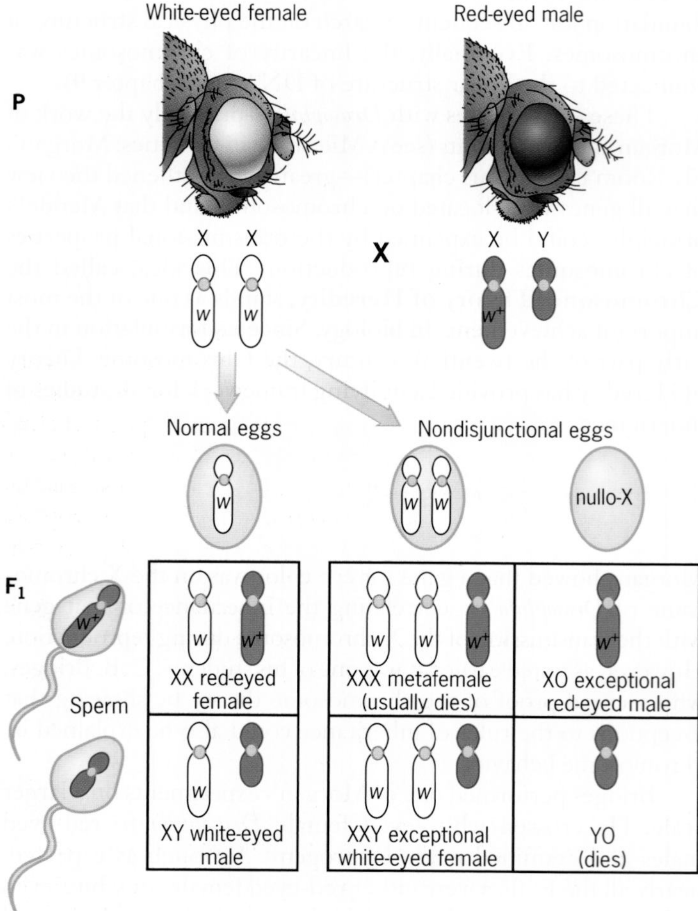

Figure 5.6 ▶ X chromosome nondisjunction is responsible for the exceptional progeny that appeared in Bridges' experiment. Nondisjunctional eggs that contain either two X chromosomes or no X chromosome unite with normal sperm that contain either an X chromosome or a Y chromosome to produce four types of zygotes. The XXY zygotes develop into white-eyed females, the XO zygotes develop into red-eyed, sterile males, and the XXX and YO zygotes die.

Bridges recognized that the fertilization of abnormal eggs by normal sperm could produce two additional kinds of zygotes: $X^wX^wX^+$, arising from the union of a diplo-X egg and an X-bearing sperm, and YO, arising from the union of a nullo-X egg and a Y-bearing sperm. The $X^wX^wX^+$ zygotes develop into females that are red-eyed, but weak and sickly. These "metafemales" can be distinguished from XX females by a syndrome of anatomical abnormalities, including ragged wings and etched abdomens. Generations of geneticists have inappropriately called them "superfemales"—a term coined by Bridges—even though there is nothing super about them. The YO zygotes turn out to be completely inviable; that is, they die. In *Drosophila,* as in most other organisms with sex chromosomes, at least one X chromosome is needed for viability.

Bridges' ability to explain the exceptional progeny that came from these crosses showed the power of the chromosome theory. Each of the exceptions was due to anomalous chromosome behavior during meiosis. Bridges called the anomaly **nondisjunction** because it involved a failure of the chromosomes to disjoin during one of the meiotic divisions. This failure could result from faulty chromosome movement, imprecise or incomplete pairing, or centromere malfunction. From Bridges' data, it is impossible to specify the exact cause. However, Bridges did note that the exceptional XXY females go on to produce a high frequency of exceptional progeny, presumably because their sex chromosomes can disjoin in different ways: the X chromosomes can disjoin from each other, or either X can disjoin from the Y. In the latter case, a diplo- or nullo-X egg is produced because the X that does not disjoin from the Y is free to move to either pole during the first meiotic division. When fertilized by normal sperm, these abnormal eggs will produce exceptional zygotes.

Bridges observed the effects of chromosome nondisjunction that had occurred during meiosis in females. We should note, however, that with appropriate experiments the effects of nondisjunction during meiosis in males can also be studied.

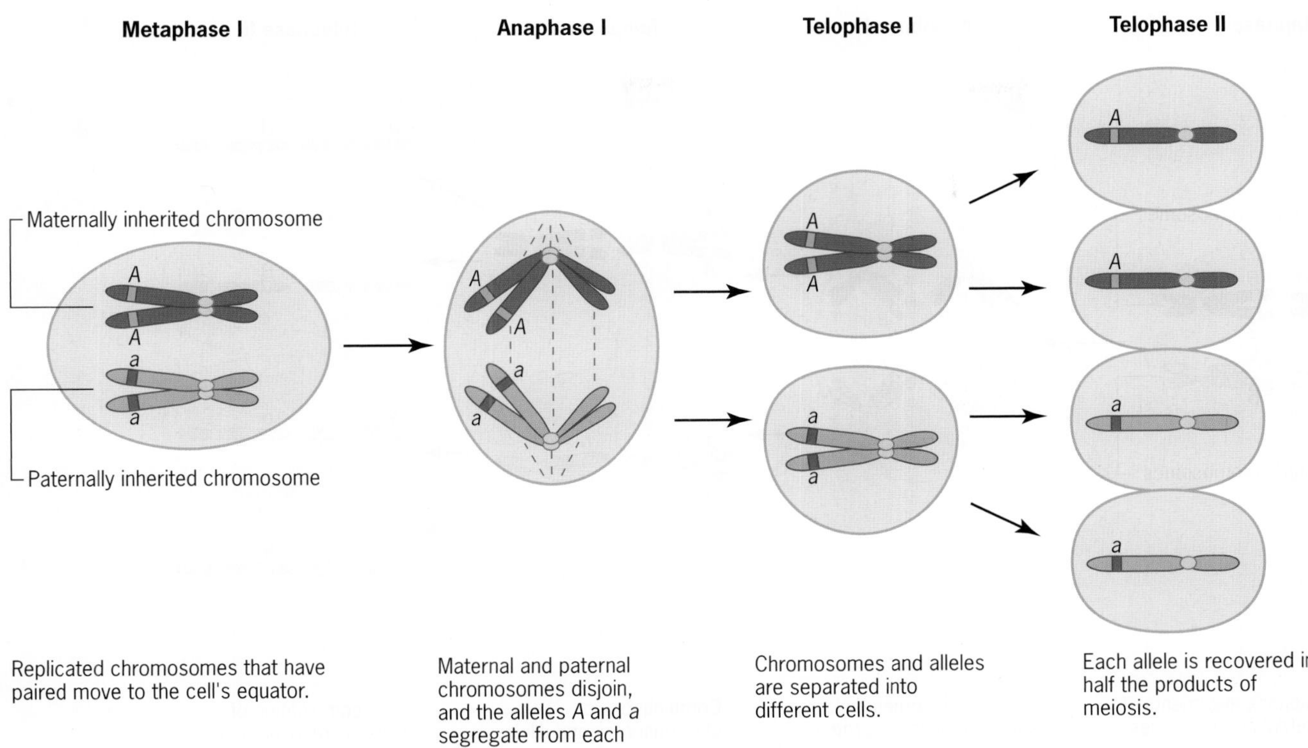
Metaphase I
Anaphase I
Telophase I
Telophase II
Maternally inherited chromosome
Paternally inherited chromosome
A
A
a
a
Replicated chromosomes that have paired move to the cell's equator.
Maternal and paternal chromosomes disjoin, and the alleles A and a segregate from each other.
Chromosomes and alleles are separated into different cells.
Each allele is recovered in half the products of meiosis.

Figure 5.7 ▶ Mendel's Principle of Segregation and meiotic chromosome behavior. The segregation of alleles corresponds to the disjunction of paired chromosomes in the anaphase of the first meiotic division.

THE CHROMOSOMAL BASIS OF MENDEL'S PRINCIPLES OF SEGREGATION AND INDEPENDENT ASSORTMENT

Mendel established two principles of genetic transmission: (1) the alleles of a gene segregate from each other, and (2) the alleles of different genes assort independently. The finding that genes are located on chromosomes made it possible to explain these principles (as well as exceptions to them) in terms of the meiotic behavior of chromosomes.

The Principle of Segregation

During the first meiotic division, homologous chromosomes pair. One of the homologues comes from the mother, the other from the father. If the mother was homozygous for an allele, *A*, of a gene on this chromosome, and the father was homozygous for a different allele, *a*, of the same gene, the offspring must be heterozygous, that is, *Aa*. In the anaphase of the first meiotic division, the paired chromosomes separate and move to opposite poles of the cell. One carries allele *A* and the other allele *a*. This physical separation of the two chromosomes segregates the alleles from each other; eventually, they will reside in different daughter cells. Mendel's Principle of Segregation (see **FIGURE 5.7**) is therefore based on the separation of homologous chromosomes during the anaphase of the first meiotic division.

The Principle of Independent Assortment

The Principle of Independent Assortment (**FIGURE 5.8**) is also based on this anaphase separation. To understand the relationship, we need to consider genes on two different pairs of chromosomes. Suppose that a heterozygote *Aa Bb* was produced by mating an *AA BB* female to an *aa bb* male; also, suppose that the two genes are on different chromosomes. During the prophase of meiosis I, the chromosomes with the *A* and *a* alleles will pair, as will the chromosomes with the *B* and *b* alleles. At metaphase, the two pairs will take up positions on the meiotic spindle in preparation for the upcoming anaphase separation. Because there are two pairs of chromosomes, there are two distinguishable metaphase alignments:

$$\frac{A}{a}\,\frac{B}{b} \quad \text{or} \quad \frac{A}{a}\,\frac{b}{B}$$

Each of these alignments is equally likely. Here the space separates different pairs of chromosomes, and the bar separates the homologous members of each pair. During anaphase, the alleles above the bars will move to one pole, and the alleles below them will move to the other. When disjunction occurs, there is therefore a 50 percent chance that the *A* and *B* alleles will move together to the same pole and a 50 percent chance that they will move to opposite poles. Similarly, there is a 50 percent chance that the *a* and *b* alleles will move to the same pole and a 50 percent chance that they will move to opposite poles. At the end of

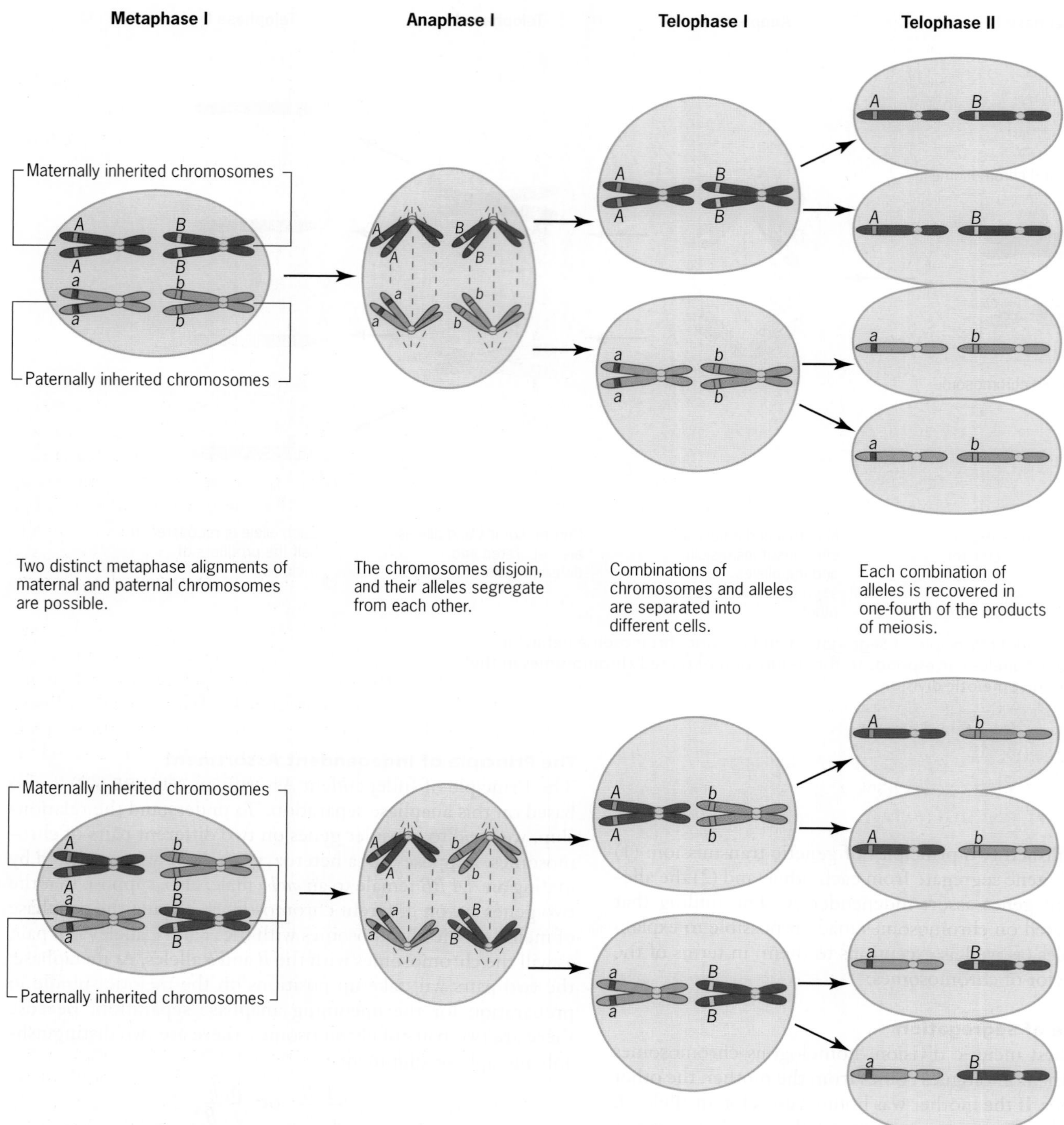

Metaphase I
Anaphase I
Telophase I
Telophase II
Maternally inherited chromosomes
Paternally inherited chromosomes
Two distinct metaphase alignments of maternal and paternal chromosomes are possible.
The chromosomes disjoin, and their alleles segregate from each other.
Combinations of chromosomes and alleles are separated into different cells.
Each combination of alleles is recovered in one-fourth of the products of meiosis.
Maternally inherited chromosomes
Paternally inherited chromosomes

Figure 5.8 ▶ Mendel's Principle of Independent Assortment and meiotic chromosome behavior. Alleles on different pairs of chromosomes assort independently in the anaphase of the first meiotic division because maternally and paternally inherited chromosomes have aligned randomly on the cell's equator.

meiosis, when the chromosome number is finally reduced, half the gametes should contain a parental combination of alleles (*A B* or *a b*), and half should contain a new combination (*A b* or *a B*). Altogether, there will be four types of gametes, each one-fourth of the total. This equality of gamete frequencies is a result of the independent behavior of the two pairs of chromosomes during the first meiotic division. Mendel's Principle of Independent Assortment is therefore a statement about the random alignment of different pairs of chromosomes at metaphase. In Chapter 7, we will see that genes on the same pair of chromosomes do not assort independently. Instead, because they are physically linked to each other, they tend to travel together through meiosis, violating the Principle of Independent Assortment. To test your understanding of the chromosomal basis of independent assortment, work through the Focus on Problem Solving: Tracking X-Linked and Autosomal Inheritance.

KEY POINTS

- Genes are located on chromosomes.
- The disjunction of chromosomes during meiosis is responsible for the segregation and independent assortment of genes.
- Nondisjunction during meiosis leads to abnormal numbers of chromosomes in gametes, and ultimately, in zygotes.

Sex-Linked Genes in Human Beings

X- and Y-linked genes have been studied in humans.

The development of the chromosome theory depended on the discovery of the *white* eye mutation in *Drosophila*. Subsequent analysis demonstrated that this mutation was a recessive allele of an X-linked gene. Although some of us might credit this important episode in the history of genetics to extraordinarily good luck, Morgan's discovery of the *white* eye mutation was not so remarkable. Such mutations are among the easiest to detect because they show up immediately in hemizygous males. In contrast, autosomal recessive mutations show up only after two mutant alleles have been brought together in a homozygote—a much more unlikely event.

In human beings too, recessive X-linked traits are much more easily identified than are recessive autosomal traits. A male needs only to inherit one recessive allele to show an X-linked trait; however, a female needs to inherit two—one from each of her parents. Thus, the preponderance of people who show X-linked traits are male.

HEMOPHILIA, AN X-LINKED BLOOD-CLOTTING DISORDER

People with **hemophilia** are unable to produce a factor needed for blood clotting; the cuts, bruises, and wounds of hemophiliacs continue to bleed and, if not stopped by transfusion with clotting factor, can cause death. The principal type of hemophilia in humans is due to a recessive X-linked mutation, and nearly all the individuals who have it are male. These males have inherited the mutation from their heterozygous mothers. If they reproduce, they transmit the mutation to their daughters, who usually do not develop hemophilia because they inherit a wild-type allele from their mothers. Affected males never transmit the mutant allele to their sons. Other blood-clotting disorders are found in both males and females because they are due to mutations in autosomal genes.

The most famous case of X-linked hemophilia occurred in the Russian imperial family at the beginning of the twentieth century (**FIGURE 5.9**). Czar Nicholas and Czarina Alexandra had four daughters and one son. The son, Alexis, suffered from hemophilia. The X-linked mutation responsible for Alexis's disease was transmitted to him by his mother, who was a heterozygous carrier. Czarina Alexandra was a granddaughter of Queen Victoria of Great Britain, who was also a carrier. Pedigree records show that Victoria transmitted the mutant allele to three of her nine children: Alice, who was Alexandra's mother, Beatrice, who had two sons with the disease, and Leopold, who had the disease himself. The allele that Victoria carried evidently arose as a new mutation in her germ cells, or in those of her mother, father, or a more distant maternal ancestor.

Throughout history hemophilia has been a fatal disease. Most of the people who have had it have died before the age of 20. Today, due to the availability of effective and relatively inexpensive treatments, hemophiliacs live long, healthy lives. The therapies that are now used to treat hemophilia were developed by applying modern genetic technologies. The Focus on Hemophilia discusses these medical advances.

COLOR BLINDNESS, AN X-LINKED VISION DISORDER

In human beings, color perception is mediated by light-absorbing proteins in the specialized cone cells of the retina in the eye. Three such proteins have been identified—one to absorb blue light, one to absorb green light, and one to absorb red light. Color blindness may be caused by an abnormality in any of these receptor proteins. The classic type of color blindness, involving faulty perception of red and green light, follows an X-linked pattern of inheritance. About 5 to 10 percent of human males are red-green color blind; however, a much smaller fraction of females, less than 1 percent, has this disability, suggesting that the mutant alleles are recessive. Molecular studies have shown that there are two distinct genes for color perception on the X chromosome; one encodes the receptor for green light, and the other encodes the receptor for red light. Detailed analyses have demonstrated that these two receptors are structurally very similar, probably because the genes encoding them evolved from an ancestral color-receptor gene. A third gene for color perception, the one encoding the receptor for blue light, is located on an autosome.

(a)

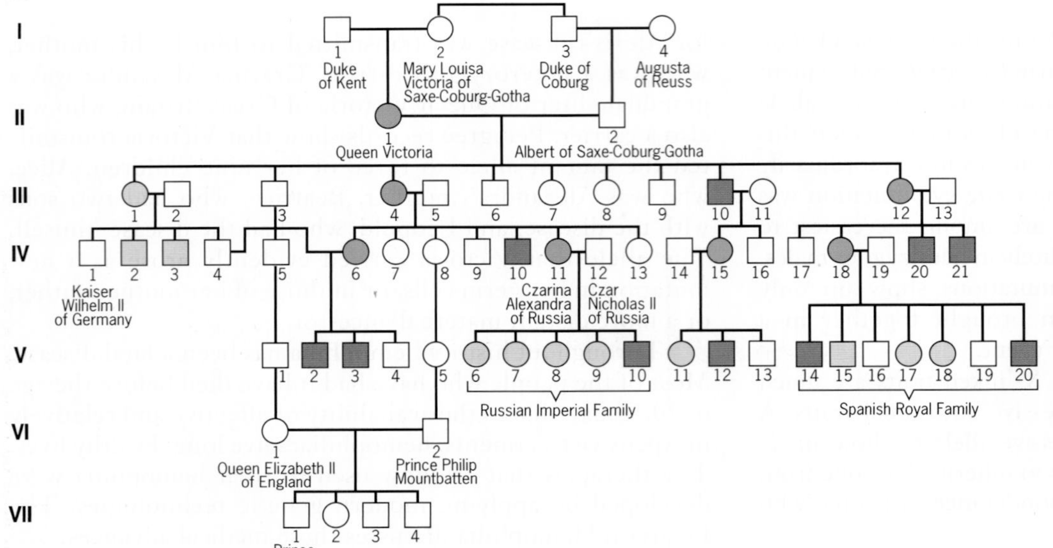

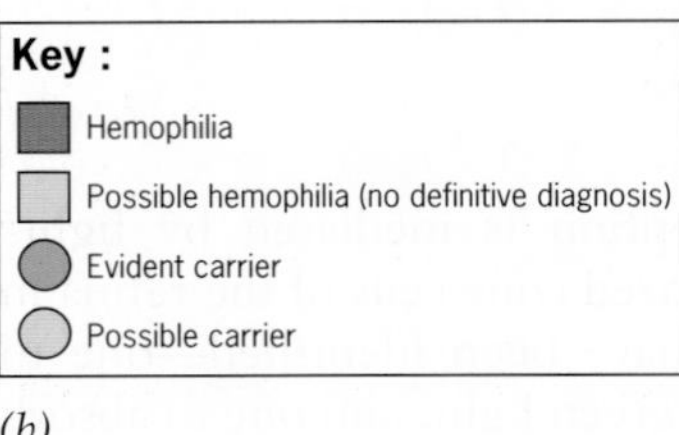

(b)

Figure 5.9 ▶ Royal hemophilia. (*a*) The Russian imperial family of Czar Nicholas II. (*b*) X-linked hemophilia in the royal families of Europe. Through intermarriage, the mutant allele for hemophilia was transmitted from the British royal family to the German, Russian, and Spanish royal families.

In **FIGURE 5.10** color blindness is used to illustrate the procedures for calculating the risk of inheriting a recessive X-linked condition. A heterozygous carrier, such as III-4 in the figure, has a 1/2 chance of transmitting the mutant allele to her children. However, the risk that a particular child will be color blind is only 1/4 since the child must be a male in order to manifest the trait. The female labeled IV-2 in the pedigree could be a carrier of the mutant allele for color blindness because her mother was. This uncertainty about the genotype of IV-2 introduces another factor of 1/2 in the risk of having a color-blind child; thus, the risk for her child is $1/4 \times 1/2 = 1/8$.

GENES ON THE HUMAN Y CHROMOSOME

The Human Genome Project has identified 307 genes on the human Y chromosome. By comparison, it has identified more

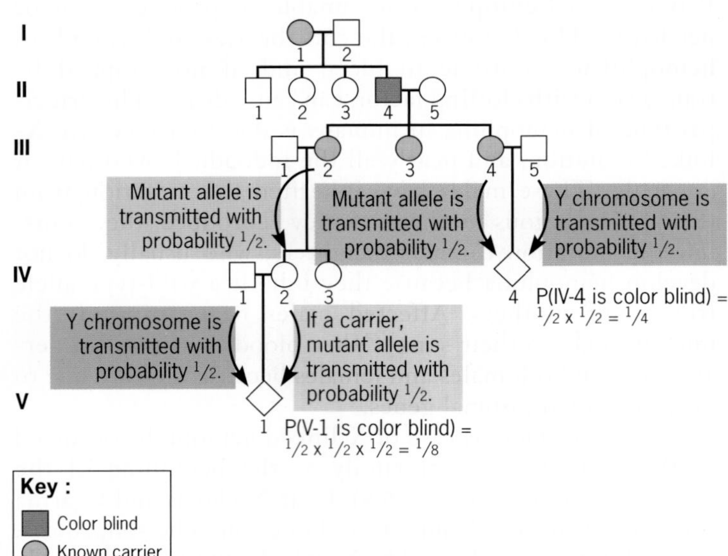

Figure 5.10 ▶ Analysis of a pedigree showing the segregation of X-linked color blindness.

► FOCUS ON PROBLEM SOLVING
Tracking X-Linked and Autosomal Inheritance

THE PROBLEM

In *Drosophila,* one of the genes controlling wing length is located on the X chromosome. A recessive mutant allele of this gene makes the wings miniature—hence, its symbol *m*; the wild-type allele of this gene, m^+, makes the wings long. One of the genes controlling eye color is located on an autosome. A recessive mutant allele of this gene makes the eyes brown—hence, its symbol *bw*; the wild-type allele of this gene, bw^+, makes the eyes red. Miniature-winged, red-eyed females from one true-breeding strain were crossed to normal-winged, brown-eyed males from another true-breeding strain.(a) Predict the phenotypes of the F_1 flies. (b) If these flies are intercrossed with one another, what phenotypes will appear in the F_2, and in what proportions?

FACTS AND CONCEPTS

1. Male and female offspring from a cross may show different phenotypes if the trait is X-linked.

2. A male inherits its X chromosome from its mother, whereas a female inherits its X chromosome from its father.

3. X-linked and autosomal genes assort independently.

4. When genes assort independently, we multiply the probabilities associated with the components of the complete genotype.

ANALYSIS AND SOLUTION

a. The parents in the initial cross were m/m; bw^+/bw^+ females and m^+/Y; bw/bw males. In the F_1, the females will be m/m^+; bw/bw^+ and because both mutant alleles are recessive, they will have long wings and red eyes. The F_1 males will be m/Y; bw/bw^+, and because they are hemizygous for the recessive X-linked mutation, they will have miniature wings; however, because they carry the dominant autosomal allele bw^+, they will have red eyes.

b. To obtain the F_2 phenotypes and their proportions, let's subdivide the problem into two parts: an X-linked part and an autosomal part. For the X-linked part, crossing the F_1 m/m^+ females to their m/Y brothers will produce four classes of offspring—(1) m/m females with miniature wings, (2) m/m^+ females with long wings, (3) m/Y males with miniature wings, and (4) m^+/Y males with long wings, and each class should be 1/4 of the total. For the autosomal part, crossing the F_1 bw/bw^+ females to their bw/bw^+ brothers will produce three classes of offspring—(1) bw^+/bw^+ flies with red eyes, (2) bw/bw^+ flies with red eyes, and (3) bw/bw flies with brown eyes, and the phenotypic ratio will be 3 red: 1 brown. To combine the results of the X-linked and autosomal parts of the problem, we construct a 2 x 4 table of phenotypic frequencies. The two autosomal phenotypes and the four X-linked phenotypes define the rows and columns of the table, and the values within the cells are the frequencies of the combined phenotypes, obtained by multiplying the frequencies in the margins.

		X-Linked Phenotypes			
		Miniature female (1/4)	**Normal female (1/4)**	**Miniature male (1/4)**	**Normal male (1/4)**
Autosomal Phenotypes	**Red (3/4)**	3/16	3/16	3/16	3/16
	Brown (1/4)	1/16	1/16	1/16	1/16

For further discussion go to your *WileyPLUS* course

than 1000 genes on the human X chromosome. Prior to the work of the Human Genome Project, little was known about the genetic makeup of the Y chromosome. Only a handful of Y-linked traits had been detected, even though transmission from father to son should make such traits easy to identify in conventional pedigree analysis. The results of the Human Genome Project have provided one possible explanation for the apparent lack of Y-linked traits. Several of the genes on the human Y chromosome seem to be required for male fertility. Obviously, a mutation in such a gene will interfere with a man's ability to reproduce; hence, that mutation will have little or no chance of being transmitted to the next generation.

GENES ON BOTH THE X AND Y CHROMOSOMES

Some genes are present on both the X and Y chromosomes, mostly near the ends of the short arms (see **FIGURE 5.2**). Alleles of these genes do not follow a distinct X- or Y-linked pattern of inheritance. Instead, they are transmitted from mothers and fathers to sons and daughters alike, mimicking the inheritance of an autosomal gene. Such genes are therefore called **pseudoautosomal genes.** In males, the regions that contain these genes seem to mediate pairing between the X and Y chromosomes.

KEY POINTS

- Disorders such as hemophilia and color blindness, which are caused by recessive X-linked mutations, are more common in males than in females.
- In humans the Y chromosome carries fewer genes than the X chromosome.
- In humans pseudoautosomal genes are located on both the X and Y chromosomes.

▶ FOCUS ON **Hemophilia**

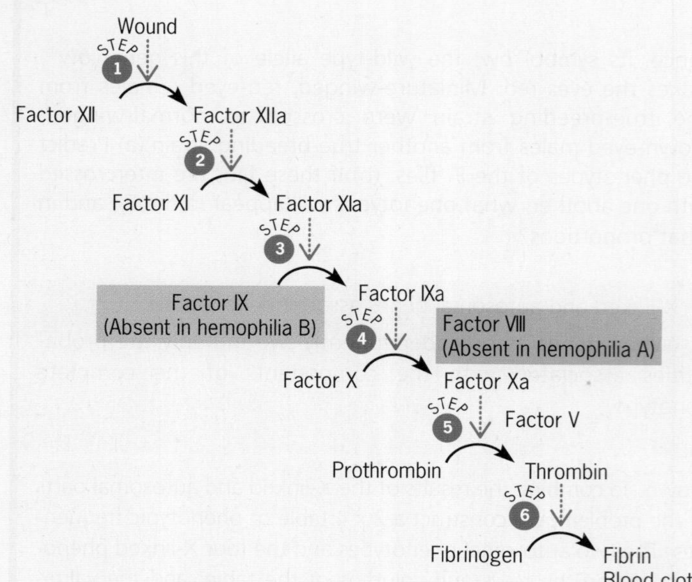

4 μm

Figure 1 ▶ Simplified pathway of blood coagulation in humans. The cascade is activated by cell damage at a wound site. Each factor is the product of a specific gene. At each step in the pathway, a factor is converted to its active form (denoted by the letter *a*), which then activates the next factor in the pathway. Some factors, such as factor VIII, participate in the activation of other factors without prior activation. In the last two steps in the pathway, the factors prothrombin and fibrinogen are converted to their active forms thrombin and fibrin, respectively. Long strands of fibrin stabilize blood clots by trapping cells called platelets (see photograph above). The absence of factors VIII and IX in individuals with hemophilia A and B, respectively, blocks the clotting process prior to the formation of fibrin, which results in excessive bleeding or hemorrhaging from a cut or bruise.

There are two major types of hemophilia. The classic form of the disease is hemophilia A, sometimes called royal hemophilia because it occurred in the descendants of Queen Victoria. The other form is hemophilia B, sometimes called Christmas disease because it was first observed in a patient named Stephen Christmas. Although both types of hemophilia are X-linked, they are caused by mutations in different genes on the X chromosome. Hemophilia A accounts for about 80 percent of all cases of X-linked hemophilia.

Both types of hemophilia result from defects in blood coagulation—the cascade of reactions that causes blood to clot at the site of a wound. A simplified version of part of this pathway is shown in **FIGURE 1**. People with hemophilia A are deficient in a blood-clotting protein called factor VIII; people with hemophilia B lack a different blood-clotting protein, factor IX. Each of these proteins is encoded by a different gene. When either gene is mutated, the encoded protein is missing or defective, and the blood does not clot normally.

The research that revealed the biochemical basis of hemophilia suggested that hemophiliacs could be treated by transfusions with the missing blood-clotting factors. Beginning in the 1960s, the clotting proteins were purified from blood obtained from large numbers of donors and administered to hemophiliacs in concentrated form. This process was expensive, and the concentrated factors often were not available for hemophiliacs in many countries. Fortunately, advances in genetic engineering provided simpler, cheaper, and more reliable ways of obtaining the crucial clotting factors. The normal genes that encode factors VIII and IX were isolated from DNA samples, and each gene was introduced into cells that could be cultured in the laboratory. Cell lines capable of synthesizing large amounts of the clotting factors were then developed from these cultures to permit the industrial production of each factor. As a result of this work, factors VIII and IX are now readily available to treat people who suffer from hemophilia.

The use of clotting factors produced in cultured cells has also made the treatment of hemophilia safer than it used to be. During the 1980s many people with hemophilia developed Acquired Immune Deficiency Syndrome—AIDS—because they were given clotting factors that had been collected from donated blood. The blood donations had been pooled to obtain the clotting factors in quantity, and alas, some of the blood was contaminated with the human immunodeficiency virus (HIV), which causes AIDS. Hemophilia patients who were given clotting factors obtained from contaminated blood became infected with HIV and developed AIDS. Most of them died. Today, because clotting factors are produced in cultured cells, people with hemophilia have essentially no risk of being infected with HIV when they receive their treatments.

Sex Chromosomes and Sex Determination

In some organisms, chromosomes—in particular, the sex chromosomes—determine male and female phenotypes.

In the animal kingdom, sex is perhaps the most conspicuous phenotype. Animals with distinct males and females are sexually dimorphic. Sometimes this dimorphism is established by environmental factors. In one species of turtles, for example, sex is determined by temperature. Eggs that have been incubated above 30°C hatch into females, whereas eggs that have been incubated at a lower temperature hatch into males. In many other species, sexual dimorphism is established by genetic factors, often involving a pair of sex chromosomes.

SEX DETERMINATION IN HUMAN BEINGS

The discovery that human females are XX and that human males are XY suggested that sex might be determined by the number of X chromosomes or by the presence or absence of a Y chromosome. As we now know, the second hypothesis is correct. In humans and other placental mammals, maleness is due to a dominant effect of the Y chromosome (**FIGURE 5.11**). The evidence for this fact comes from the study of individuals with an abnormal number of sex chromosomes. XO animals develop as females, and XXY animals develop as males. The dominant effect of the Y chromosome is manifested early in development, when it directs the primordial gonads to develop into testes. Once the testes have formed, they secrete testosterone, a hormone that stimulates the development of male secondary sexual characteristics.

Researchers have shown that the **testis-determining factor (TDF)** is the product of a gene called *SRY* (for **sex-determining region *Y***), which is located just outside the pseudoautosomal region in the short arm of the Y chromosome. The discovery of *SRY* was made possible by the identification of unusual individuals whose sex was inconsistent with their

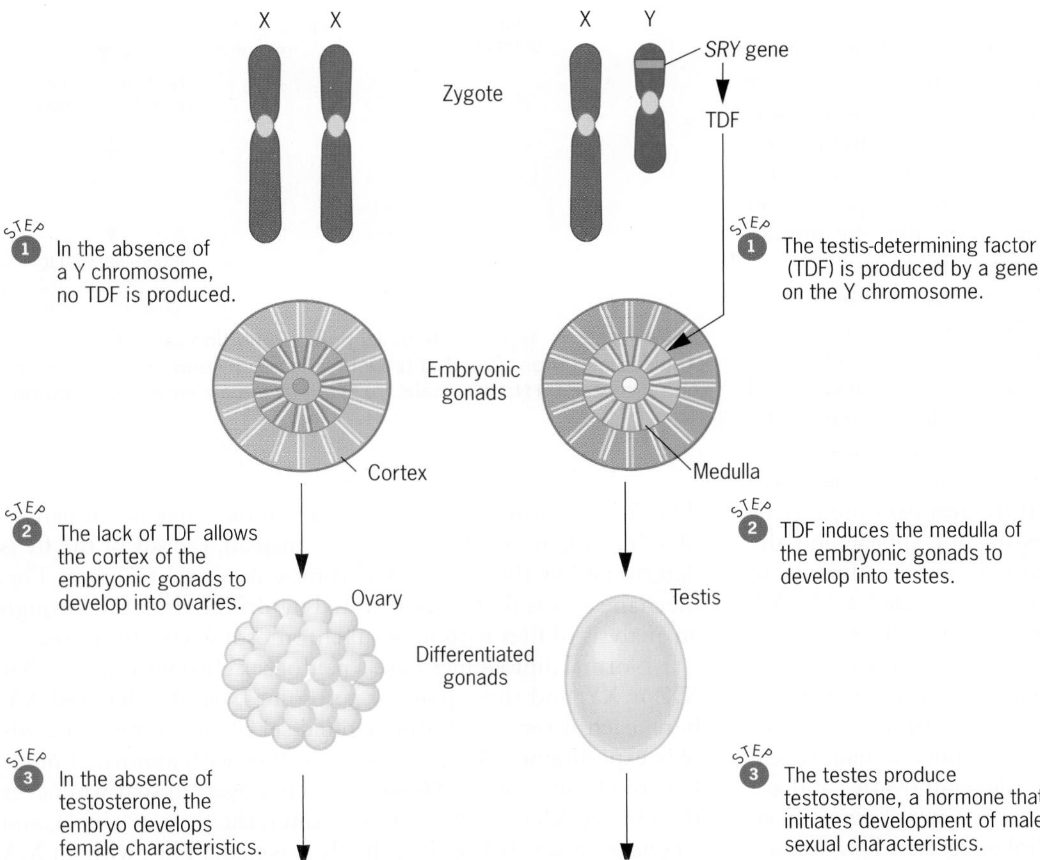

Figure 5.11 ▶ The process of sex determination in human beings. Male sexual development depends on the production of the testis-determining factor (TDF) by a gene on the Y chromosome. In the absence of this factor, the embryo develops as a female.

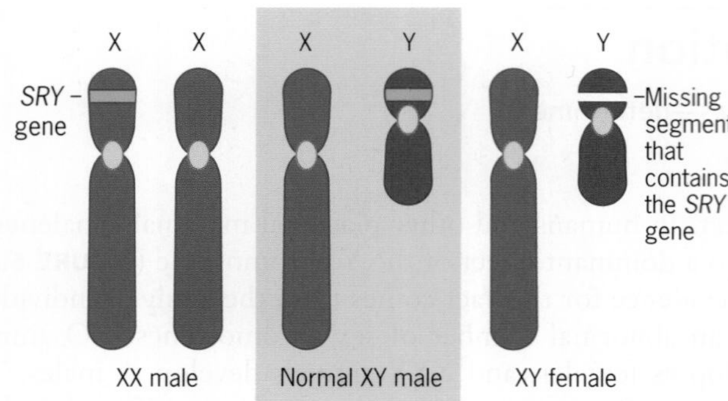

Figure 5.12 ▶ Evidence localizing the gene for the testis-determining factor (TDF) to the short arm of the Y chromosome in normal males. The TDF is the product of the *SRY* gene. In XX males, a small region containing this gene has been inserted into one of the X chromosomes, and in XY females, it has been deleted from the Y chromosome.

chromosome constitution—XX males and XY females (**FIGURE 5.12**). Some of the XX males were found to carry a small piece of the Y chromosome inserted into one of the X chromosomes. This piece evidently carried a gene responsible for maleness. Some of the XY females were found to carry an incomplete Y chromosome. The part of the Y chromosome that was missing corresponded to the piece that was present in the XX males; its absence in the XY females apparently prevented them from developing testes. These complementary lines of evidence showed that a particular segment of the Y chromosome was needed for male development. Molecular analyses subsequently identified the *SRY* gene in this male-determining segment. Additional research has shown that an *SRY* gene is present on the Y chromosome of the mouse, and that—like the human *SRY* gene—it triggers male development.

After the testes have formed, testosterone secretion initiates the development of male sexual characteristics. Testosterone is a hormone that binds to receptors in many kinds of cells. Once bound, the hormone–receptor complex transmits a signal to the nucleus, instructing the cell in how to differentiate. The concerted differentiation of many types of cells leads to the development of distinctly male characteristics such as heavy musculature, beard, and deep voice. If the testosterone signaling system fails, these characteristics do not appear and the individual develops as a female. One reason for failure is an inability to make the testosterone receptor (**FIGURE 5.13**). XY individuals with this biochemical deficiency initially develop as males—testes are formed and testosterone is produced. However, the testosterone has no effect because it cannot transmit the developmental signal inside its target cells. Individuals lacking the testosterone receptor therefore acquire female sexual characteristics. They do not, however, develop ovaries and are therefore sterile. This syndrome, called *testicular feminization*, results from a mutation in an X-linked gene, *Tfm*, which encodes the testosterone receptor. The *tfm* mutation is transmitted from mothers to their hemizygous XY offspring (who are phenotypically female) in a typical X-linked pattern.

Normal male with the wild-type *Tfm* gene.

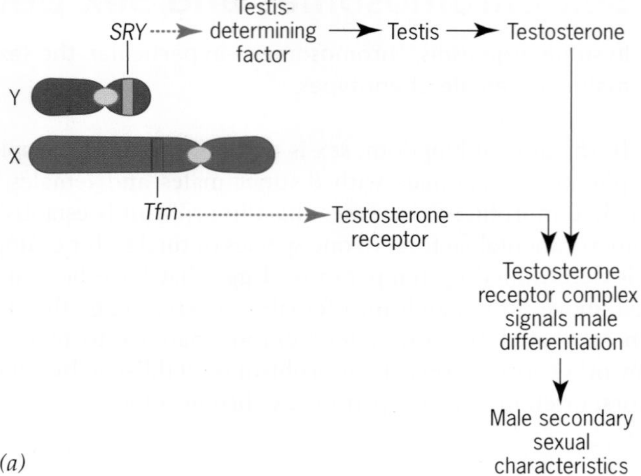

Male with the *tfm* mutation and testicular feminization.

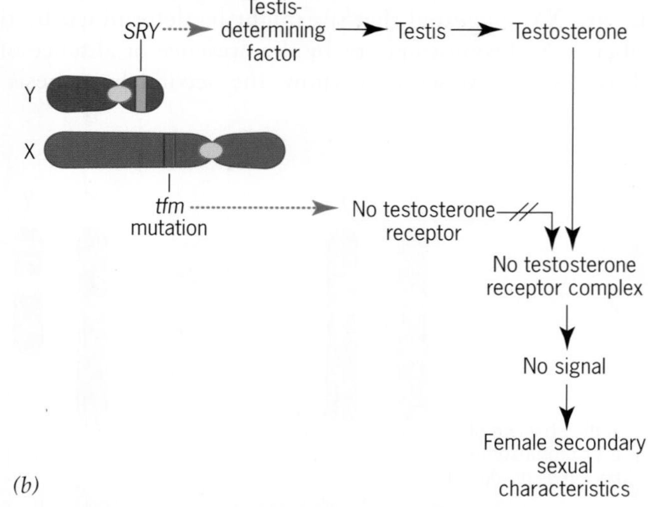

Figure 5.13 ▶ Testicular feminization, a condition caused by an X-linked mutation, *tfm*, that prevents the production of the testosterone receptor. (*a*) Normal male. (*b*) Feminized male with *tfm* mutation.

SEX DETERMINATION IN *DROSOPHILA*

The Y chromosome in *Drosophila*—unlike that in humans—plays no role in sex determination. Instead, the sex of the fly is determined by the ratio of X chromosomes to autosomes. This mechanism was first demonstrated by Bridges in 1921 through an analysis of flies with unusual chromosome constitutions.

Normal diploid flies have a pair of sex chromosomes, either XX or XY, and three pairs of autosomes, usually denoted AA; here, each A represents one haploid set of autosomes. In complex experiments, Bridges contrived flies with abnormal numbers of chromosomes (**TABLE 5.2**). He observed that whenever the ratio of X's to A's was 1.0 or greater, the fly was female, and whenever it was 0.5 or less, the fly was male. Flies with an X:A ratio between 0.5 and 1.0 developed characteristics of both sexes; thus, Bridges called them *intersexes*. In none of these flies did the Y chromosome have any effect on the sexual phenotype.

TABLE 5.2

Ratio of X Chromosomes to Autosomes and the Corresponding Phenotype in *Drosophila*

X Chromosomes (X) and Sets of Autosomes (A)	X:A Ratio	Phenotype
1X 2A	0.5	Male
2X 2A	1.0	Female
3X 2A	1.5	Metafemale
4X 3A	1.33	Metafemale
4X 4A	1.0	Tetraploid female
3X 3A	1.0	Triploid female
3X 4A	0.75	Intersex
2X 3A	0.67	Intersex
2X 4A	0.5	Tetraploid male
1X 3A	0.33	Metamale

It was, however, required for male fertility. The molecular details of sex determination in *Drosophila* are discussed in Chapters 20 and 21.

SEX DETERMINATION IN OTHER ANIMALS

In both *Drosophila* and human beings, males produce two kinds of gametes, X-bearing and Y-bearing. For this reason, they are referred to as the **heterogametic sex;** in these species females are the **homogametic sex.** In birds, butterflies, and some reptiles, this situation is reversed (**FIGURE 5.14**). Males are homogametic (usually denoted ZZ) and females are heterogametic (ZW). However, little is known about the mechanism of sex determination in the Z–W sex chromosome system.

In honeybees, sex is determined by whether the animal is haploid or diploid (**FIGURE 5.15**). Diploid embryos, which develop from fertilized eggs, become females; haploid embryos, which develop from unfertilized eggs, become males. Whether or not a given female will mature into a reproductive form (queen) depends on how she was nourished as a larva. In this system, a queen can control the ratio of males to females by regulating the proportion of unfertilized eggs that she lays. Because this number is small, most of the progeny are female, albeit sterile, and serve as workers for the hive. In a haplo-diplo system of sex determination, eggs are produced through meiosis in the queen, and sperm are produced through mitosis in the male. This system ensures that fertilized eggs will have the diploid chromosome number and that unfertilized eggs will have the haploid number.

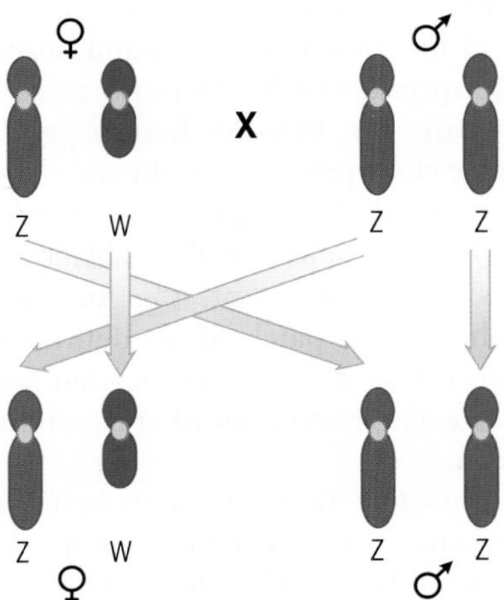

Figure 5.14 Sex determination in birds. The female is heterogametic (ZW), and the male is homogametic (ZZ). The sex of the offspring is determined by which of the sex chromosomes, Z or W, is transmitted by the female.

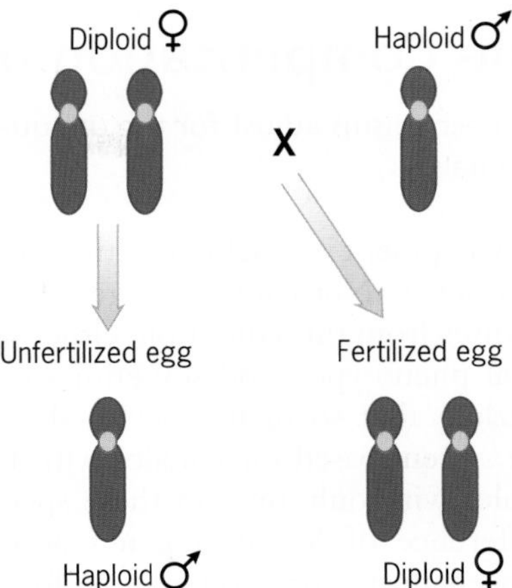

Figure 5.15 Sex determination in honeybees. Females, which are derived from fertilized eggs, are diploid, and males, which are derived from unfertilized eggs, are haploid.

Some wasps also have a haplo-diplo method of sex determination. In these species diploid males are sometimes produced, but they are always sterile. Detailed genetic analysis in one species, *Bracon hebetor*, has indicated that the diploid males are homozygous for a sex-determining locus, called *X*; diploid females are always heterozygous for this locus. Evidently, the sex locus in *Bracon* has many alleles; crosses between unrelated males and females therefore almost always produce heterozygous diploid females. However, when the mates are related, there is an appreciable chance that their offspring will be homozygous for the sex locus, in which case they develop into sterile males.

KEY POINTS

- In humans sex is determined by a dominant effect of the *SRY* gene on the Y chromosome; the product of this gene, the testis-determining factor (TDF), causes a human embryo to develop into a male.
- In *Drosophila*, sex is determined by the ratio of X chromosomes to sets of autosomes (X:A); for X:A $\leq$ 0.5, the fly develops as a male, for X:A $\geq$ 1.0, it develops as a female, and for 0.5 $<$ X:A $<$ 1.0, it develops as an intersex.
- In honeybees, sex is determined by the number of chromosome sets; haploid embryos develop into males and diploid embryos develop into females.

Dosage Compensation of X-Linked Genes

Different mechanisms adjust for the unequal dosage of X-linked genes in male and female animals.

Animal development is usually sensitive to an imbalance in the number of genes. Normally, each gene is present in two copies. Departures from this condition, either up or down, can cause abnormal phenotypes, and sometimes even death. It is therefore puzzling that so many species should have a sex-determination system based on females with two X chromosomes and males with only one. In these species, how is the numerical difference of X-linked genes accommodated? *A priori*, three mechanisms may compensate for this difference: (1) each X-linked gene could work twice as hard in males as it does in females, or (2) one copy of each X-linked gene could be inactivated in females, or (3) each X-linked gene could work half as hard in females as it does in males. Extensive research has shown that all three mechanisms are utilized, the first in *Drosophila*, the second in mammals, and the third in the nematode *Caenorhabditis elegans*. These mechanisms are discussed in detail in Chapter 20; here we provide brief descriptions of the dosage compensations systems in *Drosophila* and mammals.

HYPERACTIVATION OF X-LINKED GENES IN MALE *DROSOPHILA*

In *Drosophila*, dosage compensation of X-linked genes is achieved by an increase in the activity of these genes in males. This phenomenon, called *hyperactivation*, involves a complex of different proteins that binds to many sites on the X chromosome in males and triggers a doubling of gene activity (see Chapter 20). When this protein complex does not bind, as is the case in females, hyperactivation of X-linked genes does not occur. In this way, total X-linked gene activity in males and females is approximately equalized.

INACTIVATION OF X-LINKED GENES IN FEMALE MAMMALS

In placental mammals, dosage compensation of X-linked genes is achieved by the *inactivation* of one of the female's X chromosomes (**FIGURE 5.16**). This mechanism was first proposed in 1961 by the British geneticist Mary Lyon, who inferred it from studies on mice. Subsequent research by Lyon and others has shown that the inactivation event occurs when the mouse embryo consists of a few thousand cells. At this time, each cell makes an independent decision to silence one of its X chromosomes. The chromosome to be inactivated is chosen at random; once chosen, however, it remains inactivated in all the descendants of that cell. Thus, female mammals are *genetic mosaics* containing two types of cell lineages; the maternally inherited X chromosome is inactivated in roughly half of these cells, and the paternally inherited X is inactivated in the other half. A female that is heterozygous for an X-linked gene is therefore able to show two different phenotypes.

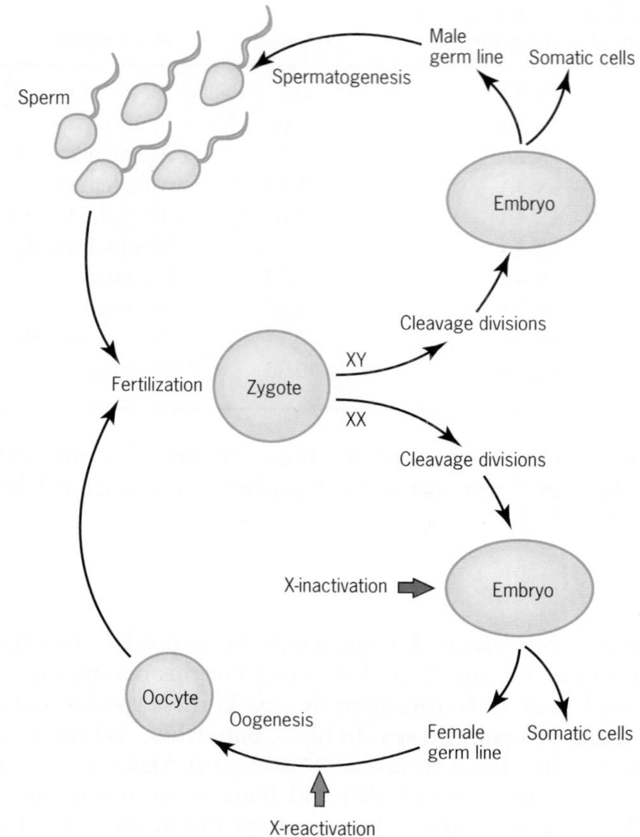

Figure 5.16 ▶ X chromosome inactivation in mammals. One of the X chromosomes in XX females is inactivated in each cell of the early embryo. In the germ line, the inactivated X chromosomes are subsequently reactivated during oogenesis.

One of the best examples of this phenotypic mosaicism comes from the study of fur coloration in cats and mice (**FIGURE 5.17**). In both of these species, the X chromosome carries a gene for pigmentation of the fur. Females heterozygous for different alleles of this gene show patches of light and dark fur. The light patches express one allele, and the dark patches express the other. In cats, where one allele produces black pigment and the other produces orange pigment, this patchy phenotype is called tortoiseshell. Each patch of fur defines a clone of pigment-producing cells, or melanocytes, that were derived by mitosis from a precursor cell present at the time of X chromosome inactivation.

An X chromosome that has been inactivated does not look or act like other chromosomes. Chemical analyses show that its DNA is modified by the addition of numerous methyl groups. In addition, it condenses into a darkly staining structure called a **Barr body** (**FIGURE 5.18**), after the Canadian geneticist Murray Barr, who first observed it. This structure becomes attached to the inner surface of the nuclear

A MILESTONE IN GENETICS: **Morgan's Fly Room**

The genus *Drosophila* (from the Greek words meaning "lover of dew") comprises a large number of species, the most intensively studied being *D. melanogaster*. This species was described in the middle of the nineteenth century under the name *D. ampelophilia,* a name meaning "lover of grape vines." After being established as an experimental animal, *D. ampelophilia* was renamed *D. melanogaster,* which means "black belly."

C. W. Woodworth seems to have been the first person to culture *Drosophila* in the laboratory. It is from Woodworth that W. E. Castle, a professor of zoology at Harvard University, learned of the advantages of using this animal as an experimental organism. Castle then recommended it to T. H. Morgan, who began to culture *Drosophila* in 1909. From its intensive work with *Drosophila,* Morgan's laboratory at Columbia University became known as "The Fly Room" (see **FIGURE 1**). Numerous students worked in this laboratory from 1910 until 1926, the year Morgan moved his research to the California Institute of Technology. The most famous of Morgan's students were Calvin Bridges, Alfred Sturtevant, and Hermann Muller. Bridges provided the proof for the Chromosome Theory of Heredity, Sturtevant produced the world's first chromosome map, and Muller discovered that mutations could be induced by X-irradiation. These and other *"Drosophila* workers" were instrumental in developing much of classical genetic analysis. Indeed, William Bateson commented that "not even the most skeptical of readers can go through the *Drosophila* work unmoved by a sense of admiration for the zeal and penetration with which it has been conducted, and for the great extension of genetic knowledge to which it has led—greater far than has been made in any one line of work since Mendel's own experiments."[1]

In 1939 Morgan reminisced about the early days in the Fly Room:

> It was not unusual for the six of us to carry on in this small room, the only space at our disposal. These were the days when bananas were used as fly food and in one corner of the room a bunch of bananas was generally on hand—an adjunct to our researches which interested other members of the laboratory in a different way. As there were no incubators, a bookcase and a wallcase were rigged up with electric bulbs and a cheap thermostat, which behaved badly at times, with consequent loss of cultures. The use of milk bottles came into the program at an early date, but where they came from was not known, or at least not mentioned. . . . Our proximity to each other led to cooperation in everything that went on.[2]

Figure 1 ▶ T. H. Morgan in his laboratory at Columbia University.

Later, Morgan's student, Alfred Sturtevant, noted that "There was a give-and-take atmosphere in the fly room. As each new result or new idea came along, it was discussed freely by the group."[3] The excitement of using *Drosophila* as an experimental organism and the intellectual camaraderie created by the close quarters of the Fly Room combined to stimulate research. The rapid isolation and analysis of new mutations turned *Drosophila* into the premier organism for experimental genetics.

Since the days of the Fly Room, research on *Drosophila* has grown into a worldwide endeavor. Hundreds of different laboratories are currently investigating the genes and chromosomes of this animal. In the United States, one large stock center maintains cultures of mutant flies for distribution to interested researchers, and every year *Drosophila* workers gather for an international five-day meeting to discuss the results of their investigations; typically, more than a thousand people attend this annual meeting. In 1992, at the meeting in Philadelphia, a compendium of 80 years of *Drosophila* research was unveiled. This 1133-page volume, entitled *The Genome of* Drosophila melanogaster,[4] describes more than 4000 genes. Information about *Drosophila* genes is now available from a regularly updated electronic database, called FlyBase, which is accessible on the Internet (http://www.flybase.org/). Then in 2000, at the meeting in Pittsburgh, the completion of a project to obtain the DNA sequence for most of the *Drosophila* genome was announced; 120 million base pairs comprising an estimated 13,601 genes were analyzed—a legacy that would certainly have made T. H. Morgan happy.

QUESTIONS FOR DISCUSSION

1. *Drosophila melanogaster* was the first species to become established as a "model" organism for genetic research. Other organisms from different biological kingdoms have since been elevated to this special status. What are the advantages and disadvantages of concentrating research on "model" organisms?
2. Morgan's Fly Room demonstrated the power of teamwork in scientific research, and it established a pattern for laboratory organization that many geneticists have followed—students working under a "principal investigator" to study different aspects of a research problem. What are the benefits and drawbacks of doing research this way? What are the responsibilities of a principal investigator and his or her students on a research team?

[1]Bateson, W. 1916. The mechanism of Mendelian heredity (a review). *Science* 44:536–543.

[2]Morgan, T. H. 1939. "Personal recollections of Calvin B. Bridges." *Journal of Heredity* 30:355.

[3]Sturtevant, A. H. 1965. *A History of Genetics.* Harper and Row, New York, p. 49.

[4]Lindsley, D. L., and G. G. Zimm. 1992. *The Genome of* Drosophila melanogaster. Academic Press, New York.

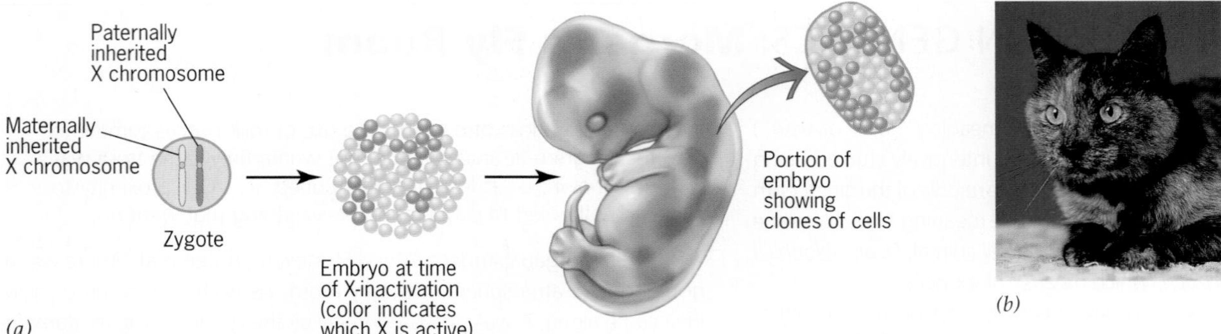

Figure 5.17 ▶ Color mosaics resulting from X chromosome inactivation in female mammals. (*a*) Formation of clones of cells in a cat embryo that produce different patches of fur in the adult. (*b*) A tortoiseshell cat. This female is heterozygous for an X-linked coat color gene. The orange and black patches are due to inactivation of different alleles in the pigment-producing cells of the body.

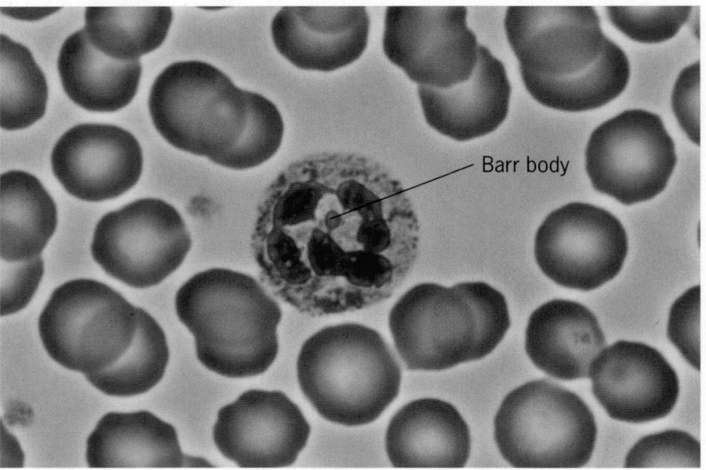

Figure 5.18 ▶ Barr body in a human female cell.

membrane, where it replicates out of step with the other chromosomes in the cell. The inactivated X chromosome remains in this altered state in all the somatic tissues. However, in the germ tissues it is reactivated, perhaps because two copies of some X-linked genes are needed for the successful completion of oogenesis. The molecular mechanism of X-inactivation is discussed in Chapter 20.

Cytological studies have identified human beings with more than two X chromosomes (see Chapter 6). For the most part, these people are phenotypically normal females, apparently because all but one of their X chromosomes is inactivated. Often all the inactivated X's congeal into a single Barr body. These observations suggest that cells may have a limited amount of some factor needed to prevent X-inactivation. Once this factor has been used to keep one X chromosome active, all the others quietly succumb to the inactivation process.

KEY POINTS

- In *Drosophila,* dosage compensation for X-linked genes is achieved by hyperactivating the single X chromosome in males.
- In mammals, dosage compensation for X-linked genes is achieved by inactivating one of the two X chromosomes in females.

▶ Basic Exercises

ILLUSTRATE BASIC GENETIC ANALYSIS

1. A mutant *Drosophila* male with prune-colored eyes was crossed to a wild-type female with red eyes. All the F_1 offspring of both sexes had red eyes. When these offspring were intercrossed, three different classes of F_2 flies were produced: females with red eyes, males with red eyes, and males with prune eyes. The males and females were equally frequent in the F_2, and among the males, the two eye color classes were equally frequent. Do these results suggest that the *prune* mutation is on the X chromosome?

Answer: The results of these crosses are consistent with the hypothesis that the *prune* mutation is on the X chromosome. According to this hypothesis, the male in the first cross must have been hemizygous for the *prune* mutation; his mate must have been homozygous for the wild-type allele of the *prune* gene. Among the F_1, the daughters must have been heterozygous for the mutation and the wild-type allele, and the sons must have been hemizygous for the wild-type allele. When the F_1 flies were intercrossed, they produced daughters that inherited the wild-type allele from their fathers—these flies must therefore have had red eyes—and they produced sons that inherited either the mutant allele or the wild-type allele from their mothers, with each of these possibilities being equally likely. Thus, according to the hypothesis, among the F_2, all the daughters and half the sons should have red eyes, and half the sons should have prune eyes, which is what was observed.

2. The following pedigree shows the inheritance of hemophilia in a human family. (a) What is the probability that II-2 is a carrier of the allele for hemophilia? (b) What is the probability that III-1 will be affected with hemophilia?

Answer: (a) II-2 has an affected brother, which indicates that her mother was a carrier. Her chance of also being a carrier is therefore simply the probability that her mother transmitted the mutant allele to her, which is 1/2. (b) The chance that III-1 will be affected depends on three events: (1) that II-2 is a carrier, (2) that II-2 transmits the mutant allele, if she carries it, and (3) that II-3 transmits a Y chromosome. Each of these events is associated with a probability of 1/2. Thus, the probability that III-1 will be affected is $(1/2) \times (1/2) \times (1/2) = 1/8$.

3. How do the chromosomal mechanisms of sex determination differ between humans and *Drosophila?*

Answer: In humans, sex is determined by a dominant effect of the Y chromosome. In the absence of a Y chromosome, the individual develops as a female; in its presence, the individual develops as a male. In *Drosophila*, sex is determined by the ratio of X chromosomes to sets of autosomes. When the X:A ratio is greater than or equal to one, the individual develops as a female; when the X:A ratio is less than or equal to 0.5, it develops as a male; in between these limits, the individual develops as an intersex.

4. How do the mechanisms that compensate for different doses of the X chromosome in the two sexes differ between humans and *Drosophila*?

Answer: In humans, one of the two X chromosomes in an XX female is inactivated in the somatic cells early in development. In *Drosophila*, the single X chromosome in a male is hyperactivated so that its genes are as active as the double dose of X-linked genes present in an XX female.

▶ Testing Your Knowledge

INTEGRATE DIFFERENT CONCEPTS AND TECHNIQUES

1. The Lesch-Nyhan syndrome is a serious metabolic disorder affecting about one in 50,000 males in the population of the United States. A class of molecules called purines, which are biochemical precursors of DNA, accumulate in the nervous tissues and joints of people with the Lesch-Nyhan syndrome. This biochemical abnormality is caused by a deficiency for the enzyme hypoxanthine phosphoribosyltransferase (HPRT), which is encoded by a gene located on the X chromosome. Individuals deficient for this enzyme are unable to control their movements and unwillingly engage in self-destructive behavior such as biting and scratching themselves. The males labeled IV-5 and IV-6 in the following pedigree have the Lesch-Nyhan syndrome. What are the risks that V-1 and V-2 will inherit this disorder?

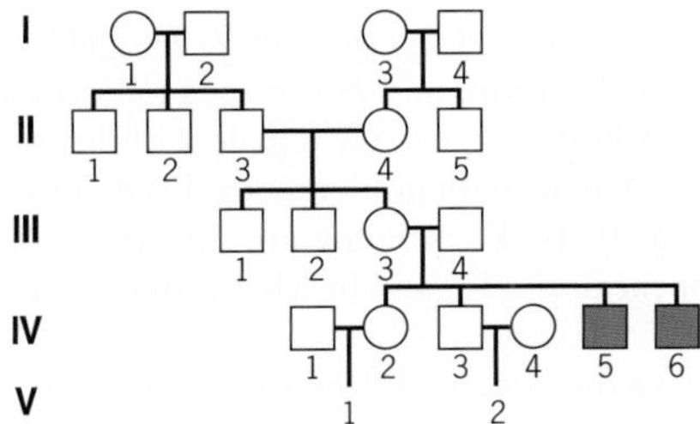

Answer: We know that III-3 must be a heterozygous carrier of the mutant allele (*h*) because two of her sons are affected. However, because she herself does not show the mutant phenotype, we know that her other X chromosome must carry the wild-type allele (*H*). Given that III-3 is genotypically *Hh*, there is a one-half chance that she passed the mutant allele to her daughter (IV-2). If she did, there is a one-half chance that IV-2 will transmit this allele to her child (V-1), and there is a one-half chance that this child will be a male. Thus, the risk that V-1 will have the Lesch-Nyhan syndrome is $(1/2) \times (1/2) \times (1/2) = 1/8$. For V-2, the risk of inheriting the Lesch-Nyhan syndrome is essentially zero. This child's father (IV-3) does not have the mutant allele, and even if he did, he would not transmit it to a son. The child's mother comes from outside the family and is very unlikely to be a carrier because the trait is rare in the general population. Thus, V-2 has virtually no chance of suffering from the Lesch-Nyhan syndrome.

2. A geneticist crossed *Drosophila* females that had white eyes and ebony bodies to wild-type males, which had red eyes and gray bodies. Among the F_1, all the daughters had red eyes and gray bodies, and all the sons had white eyes and gray bodies. These flies were intercrossed to produce F_2 progeny, which were classified for eye and body color and then counted. Among 384 total progeny, the geneticist obtained the following results:

Phenotypes			
Eye Color	**Body Color**	**Males**	**Females**
white	ebony	20	21
white	gray	70	73
red	ebony	28	25
red	gray	76	71

How would you explain the inheritance of eye color and body color?

Answer: The results in the F_1 tell us that both mutant phenotypes are caused by recessive alleles. Furthermore, because the males and females have different eye color phenotypes, we know that the eye color gene is X-linked and that the body color gene is autosomal. In the F_2, the two genes assort independently, as we would expect for genes located on different chromosomes. In the following table, we show the genotypes of the different classes of flies in this experiment, using *w* for the *white* mutation and *e* for the *ebony* mutation; the wild-type alleles are denoted by plus signs. Following the convention of *Drosophila* geneticists, we write the sex chromosomes (X and Y) on the left and the autosomes on the right. A question mark in a genotype indicates that either the wild-type or mutant alleles could be present.

Phenotypes		Genotypes	
Eye Color	**Body Color**	**Males**	**Females**
white	ebony	*w*/*Y* *e*/*e*	*w*/*w* *e*/*e*
white	gray	*w*/*Y* +/?	*w*/*w* +/?
red	ebony	+/*Y* *e*/*e*	+/*w* *e*/*e*
red	gray	+/*Y* +/?	+/*w* +/?

3. In 1906 the British biologists L. Doncaster and G. H. Raynor reported the results of breeding experiments with the currant moth, *Abraxas*. This moth exists in two color forms in Great Britain. One, called grossulariata, has large black spots on its wings; the other, called lacticolor, has much smaller black spots. Doncaster and Raynor crossed lacticolor females with grossulariata males and found that all the F_1 progeny were grossulariata. They then intercrossed the F_1 moths to produce an F_2, which consisted of two types of females (grossulariata and lacticolor) and one type of males (grossulariata). Doncaster and Raynor also testcrossed the F_1 moths. Grossulariata F_1 females crossed to lacticolor males produced lacticolor females and grossulariata males—the first grossulariata males ever seen; and grossulariata F_1 males crossed to lacticolor females produced four kinds of offspring: grossulariata males, grossulariata females, lacticolor males, and lacticolor females. Propose an explanation for the results of these experiments.

Answer: The inheritance of the grossulariata and lacticolor phenotypes is obviously linked to sex. In moths, however, females are heterogametic (ZW) and males are homogametic (ZZ). Thus, we can hypothesize that lacticolor females are hemizygous for a recessive allele (*l*) on the Z chromosome and that grossulariata males are homozygous for a dominant allele (*L*) on this chromosome. When the two types of moths are crossed, they produce grossulariata females that are hemizygous for the dominant allele (*L*) and grossulariata males that are heterozygous for the two alleles (*Ll*). An intercross between these F_1 moths produces grossulariata (*L*) and lacticolor (*l*) females, each hemizygous for a different allele, and grossulariata males that are either homozygous *LL* or heterozygous *Ll*. The hypothesis that the spotting pattern in *Abraxas* is controlled by a gene on the Z chromosome also explains the results of the testcrosses with the F_1 grossulariata animals. Grossulariata F_1 females, which are hemizygous for the dominant allele *L*, when crossed to homozygous *ll* lacticolor males produce hemizygous *l* lacticolor females and heterozygous *Ll* grossulariata males. Grossulariata F_1 males, which are *Ll* heterozygotes, when crossed to hemizygous *l* lacticolor females produce heterozygous *Ll* grossulariata males, hemizygous *L* grossulariata females, homozygous *ll* lacticolor males, and hemizygous *l* lacticolor females. Unfortunately, at the time Doncaster and Raynor reported their work, the sex chromosome constitution of *Abraxas* was not known. Consequently, they did not make the conceptual link between the inheritance of wing spotting and transmission of the sex chromosomes. Had they done so, T. H. Morgan's demonstration of sex linkage in *Drosophila* might today appear to be an afterthought.

▶ Questions and Problems

ENHANCE UNDERSTANDING AND DEVELOP ANALYTICAL SKILLS

5.1 In grasshoppers, rosy body color is caused by a recessive mutation; the wild-type body color is green. If the gene for body color is on the X chromosome, what kind of progeny would be obtained from a mating between a homozygous rosy female and a hemizygous wild-type male? (In grasshoppers, females are XX and males are XO.)

5.2 In human beings, a recessive X-linked mutation, *g*, causes green-defective color vision; the wild-type allele, *G*, causes normal color vision. A man (a) and a woman (b), both with normal vision, have three children, all married to people with normal vision: a color-defective son (c), who has a daughter with normal vision (f); a daughter with normal vision (d), who has one color-defective son (g) and two normal sons (h); and a daughter with normal vision (e), who has six normal sons (i). Give the most likely genotypes for the individuals (a to i) in this family.

5.3 What are the sexual phenotypes of the following genotypes in *Drosophila*: XX, XY, XXY, XXX, XO?

5.4 In the mosquito *Anopheles culicifacies*, *golden* body (*go*) is a recessive X-linked mutation, and *brown* eyes (*bw*) is a recessive autosomal mutation. A homozygous XX female with brown eyes is mated to a hemizygous XY male with golden body. Predict the phenotypes of their F_1 offspring. If the F_1 progeny are intercrossed, what kinds of progeny will appear in the F_2, and in what proportions?

5.5 What are the genetic differences between male- and female-determining sperm in animals with heterogametic males?

5.6 A normal woman, whose father had hemophilia, marries a hemophiliac man. What is the chance that their first child will have hemophilia?

5.7 A *Drosophila* female homozygous for a recessive X-linked mutation that causes vermilion eyes is mated to a wild-type male with red eyes. Among their progeny, all the sons have vermilion eyes, and nearly all the daughters have red eyes; however, a few daughters have vermilion eyes. Explain the origin of these vermilion-eyed daughters.

5.8 GO In *Drosophila*, the gene for *bobbed* bristles (recessive allele *bb*, bobbed bristles; wild-type allele+, normal bristles) is located on the X chromosome and on a homologous segment of the Y chromosome. Give the genotypes and phenotypes of the offspring from the following crosses:

(a) $X^{bb} X^{bb} \times X^{bb} Y^{+}$
(b) $X^{bb} X^{bb} \times X^{bb} Y^{+}$
(c) $X^{+} X^{bb} \times X^{+} Y^{bb}$
(d) $X^{+} X^{bb} \times X^{bb} Y^{+}$

5.9 In *Drosophila*, vermilion eye color is due to a recessive allele (*v*) located on the X chromosome. Curved wings is due to a recessive allele (*cu*) located on one autosome, and ebony body is due to a recessive allele (*e*) located on another autosome. A vermilion male is mated to a curved, ebony female, and the F_1 males are phenotypically wild-type. If these males were backcrossed to curved, ebony females, what proportion of the F_2 offspring will be wild-type males?

5.10 GO A *Drosophila* female heterozygous for the recessive X-linked mutation *w* (for *white* eyes) and its wild-type allele w^{+} is mated to a male with white eyes. Among the sons, half have white eyes and half have red eyes. Among the daughters, some have red eyes; and some have white eyes. Explain the origin of these white-eyed daughters.

5.11 A man with X-linked color blindness marries a woman with no history of color blindness in her family. The daughter of this couple marries a normal man, and their daughter also marries a normal man. What is the chance that this last couple will have a child with color blindness? If this couple has already had a child with color blindness, what is the chance that their next child will be color blind?

5.12 A man who has color blindness and type A blood has children with a woman who has normal color vision and type O blood. The woman's father had color blindness. Color blindness is determined by an X-linked gene, and blood type is determined by an autosomal gene.

(a) What are the genotypes of the man and the woman?
(b) What proportion of their children will have color blindness and type B blood?
(c) What proportion of their children will have color blindness and type A blood?
(d) What proportion of their children will be color blind and have type AB blood?

5.13 Would a human with two X chromosomes and a Y chromosome be male or female?

5.14 A woman carries the testicular feminization mutation (*tfm*) on one of her X chromosomes; the other X carries the wild-type allele (*Tfm*). If the woman marries a normal man, what fraction of her children will be phenotypically female? Of these, what fraction will be fertile?

5.15 Predict the sex of *Drosophila* with the following chromosome compositions (A = haploid set of autosomes):

(a) 4X 4A
(b) 3X 4A
(c) 4X 5A
(d) 1X 3A
(e) 6X 6A
(f) 1X 2A

5.16 GO In *Drosophila*, a recessive mutation called *chocolate* (*c*) causes the eyes to be darkly pigmented. The mutant phenotype is indistinguishable from that of an autosomal recessive mutation called *brown* (*bw*). A cross of chocolate-eyed females to homozygous brown males yielded wild-type F_1 females and darkly pigmented F_1 males. If the F_1 flies are intercrossed, what types of progeny are expected, and in what proportions? (Assume the double mutant combination has the same phenotype as either of the single mutants alone.)

5.17 Suppose that a mutation occurred in the *SRY* gene on the human Y chromosome, knocking out its ability to produce the testis-determining factor. Predict the phenotype of an individual who carried this mutation and a normal X chromosome.

5.18 In chickens, the absence of barred feathers is due to a recessive allele. A barred rooster was mated with a nonbarred hen, and all the offspring were barred. These F_1 chickens were intercrossed to produce F_2 progeny, among which all the males were barred; half the females were barred and half were nonbarred. Are these results consistent with the hypothesis that the gene for barred feathers is located on one of the sex chromosomes?

5.19 A *Drosophila* male carrying a recessive X-linked mutation for yellow body is mated to a homozygous wild-type female with gray body. The daughters of this mating all have uniformly gray bodies. Why aren't their bodies a mosaic of yellow and gray patches?

5.20 GO A breeder of sun conures (a type of bird) has obtained two true-breeding strains, A and B, which have red eyes instead of the normal brown found in natural populations. In Cross 1, a male from strain A was mated to a female from strain B, and the male and female offspring all had brown eyes. In Cross 2, a female from strain A was mated to a male from strain B, and the male offspring had brown eyes and the female offspring had red eyes. When the F_1 birds from each cross were mated brother to sister, the breeder obtained the following results:

Phenotype	Proportion in F_2 of Cross 1	Proportion in F_2 of Cross 2
Brown male	6/16	3/16
Red male	2/16	5/16
Brown female	3/16	3/16
Red female	5/16	5/16

Provide a genetic explanation for these results.

5.21 Males in a certain species of deer have two nonhomologous X chromosomes, denoted X_1 and X_2, and a Y chromosome. Each X chromosome is about half as large as the Y chromosome, and its centromere is located near one of the ends; the centromere of the Y chromosome is located in the middle. Females in this species have two copies of each of the X chromosomes and lack a Y chromosome. How would you predict the X and Y chromosomes to pair and disjoin during spermatogenesis to produce equal numbers of male- and female-determining sperm?

5.22 What is the maximum number of Barr bodies in the nuclei of human cells with the following chromosome compositions?

(a) XY
(b) XX
(c) XXY
(d) XXXX
(e) XXXXX
(f) XXYY

5.23 In 1908 F. M. Durham and D. C. E. Marryat reported the results of breeding experiments with canaries. Cinnamon canaries have pink eyes when they first hatch, whereas green canaries have black eyes. Durham and Marryat crossed cinnamon females with green males and observed that all the F_1 progeny had black eyes, just like those of the green strain. When the F_1 males were crossed to green females, all the male progeny had black eyes, whereas all the female progeny had either black or pink eyes, in about equal proportions. When the F_1 males were crossed to cinnamon females, four classes of progeny were obtained: females with black eyes, females with pink eyes, males with black eyes, and males with pink eyes—all in approximately equal proportions. Propose an explanation for these findings.

▶ Genomics on the Web

at http://www.ncbi.nlm.nih.gov/

Both humans and mice have X and Y sex chromosomes. In each species the Y is smaller than the X and has fewer genes.

1. What are the sizes of the human X and Y chromosomes in nucleotide pairs? How many genes does each of these chromosomes contain?

2. How do the sizes of the mouse sex chromosomes compare with those of the human sex chromosomes?

3. The *SRY* gene responsible for sex determination in humans is located in the short arm of the Y chromosome, near but not in the pseudoautosomal region. Can you find its homologue, *Sry*, on the Y chromosome of the mouse?

Hint: Using the Map Viewer feature on the web site, click on the species whose genome you want to see, and then click on one of the sex chromosomes. Use the Search function to find the *Sry* gene on the mouse's Y chromosome.

Chapter 6
Variation in Chromosome Number and Structure

CHAPTER OUTLINE

- Cytological Techniques
- Polyploidy
- Aneuploidy
- Rearrangements of Chromosome Structure

Chromosomes, Agriculture, and Civilization

The cultivation of wheat originated some 10,000 years ago in the Middle East. Today, it is the principal food crop for more than a billion people. Wheat is grown in diverse environments, from Norway to Argentina. More than 17,000 varieties have been developed, each adapted to a different locality. The total wheat production of the world is 60 million metric tons annually, accounting for more than 20 percent of the food calories consumed by the entire human population. Wheat is clearly an important agricultural crop and, some would argue, a mainstay of civilization.

Wheat field in Italy.

Modern cultivated wheat, *Triticum aestivum,* is a hybrid of at least three different species. Its progenitors were low-yielding grasses that grew in Syria, Iran, Iraq, and Turkey. Some of these grasses appear to have been cultivated by the ancient peoples of this region. Although we do not know the exact course of events, two of the grasses apparently interbred, producing a species that excelled as a crop plant. Through human cultivation, this hybrid species was selectively improved, and then it, too, interbred with a third species, yielding a triple hybrid that was even better suited for agriculture. Modern wheat is descended from these triple hybrid plants.

What made the triple hybrid wheats so superior to their ancestors? They had larger grains, they were more easily harvested, and they grew in a wider range of conditions. We now understand the chromosomal basis for these improvements. Triple-hybrid wheat contains the chromosomes of each of its progenitors. Genetically, it is an amalgamation of the genomes of three different species.

Cytological Techniques

Geneticists use stains to identify specific chromosomes and to analyze their structures.

Geneticists study chromosome number and structure by staining dividing cells with certain dyes and then examining them with a microscope. The analysis of stained chromosomes is the main activity of the discipline called **cytogenetics.**

Cytogenetics had its roots in the research of several nineteenth-century European biologists who discovered chromosomes and observed their behavior during mitosis, meiosis, and fertilization. This research blossomed during the twentieth century, as microscopes improved and better procedures for preparing and staining chromosomes were developed. The demonstration that genes reside on chromosomes boosted interest in this research and led to important studies on chromosome number and structure. Today, cytogenetics has significant applied aspects, especially in medicine, where it is used to determine whether disease conditions are associated with chromosome abnormalities.

ANALYSIS OF MITOTIC CHROMOSOMES

Researchers perform most cytological analyses on dividing cells, usually cells in the middle of mitosis. To enrich for cells at this stage, they have traditionally used rapidly growing material such as animal embryos and plant root tips. However, the development of cell-culturing techniques has made it possible to study chromosomes in other types of cells (**FIGURE 6.1**). For example, human white blood cells can be collected from peripheral blood, separated from the nondividing red blood cells, and put into culture. The white cells are then stimulated to divide by chemical treatment, and midway through division a sample of the cells is prepared for cytological analysis. The usual procedure is to treat the dividing cells with a chemical that disables the mitotic spindle. The effect of this interference is to trap the chromosomes in mitosis, when they are most easily seen. Mitotically arrested cells are then swollen by immersion in a hypotonic solution that causes the cells to take up water by osmosis. The contents of each cell are diluted by the additional water, so that when the cells are squashed on a microscope slide, the chromosomes are spread out in an uncluttered fashion. This technique greatly facilitates subsequent analysis, especially if the chromosome number is large. For many years it was erroneously thought that human cells contained 48 chromosomes. The correct number, 46, was determined only after the swelling technique was used to separate the chromosomes within individual mitotic cells. See the Milestone in Genetics: Tjio and Levan Count Human Chromosomes Correctly later in this chapter.

Until the late 1960s and early 1970s, chromosome spreads were usually stained with Feulgen's reagent, a purple dye that reacts with the sugar molecules in DNA, or with aceto-carmine, a deep red dye. Because these types of dyes stain the chromosomes uniformly, they do not allow a researcher to distinguish one chromosome from another unless the chromosomes are very different in size or in the positions of their centromeres. Today, cytogeneticists use dyes that stain chromosomes differentially along their lengths. *Quinacrine,* a chemical relative of the antimalarial drug quinine, was one of the first of these more discriminating reagents. Chromosomes that have been stained with quinacrine show a characteristic pattern of bright bands on a darker background. However, because quinacrine is a fluorescent compound, the bands appear only when the chromosomes are exposed to ultraviolet (UV) light. Ultraviolet irradiation causes some of the quinacrine molecules that have inserted into the chromosome to emit energy. Parts of the chromosome shine brightly, whereas other parts remain dark. This bright-dark banding pattern is highly reproducible and is also specific for each chromosome

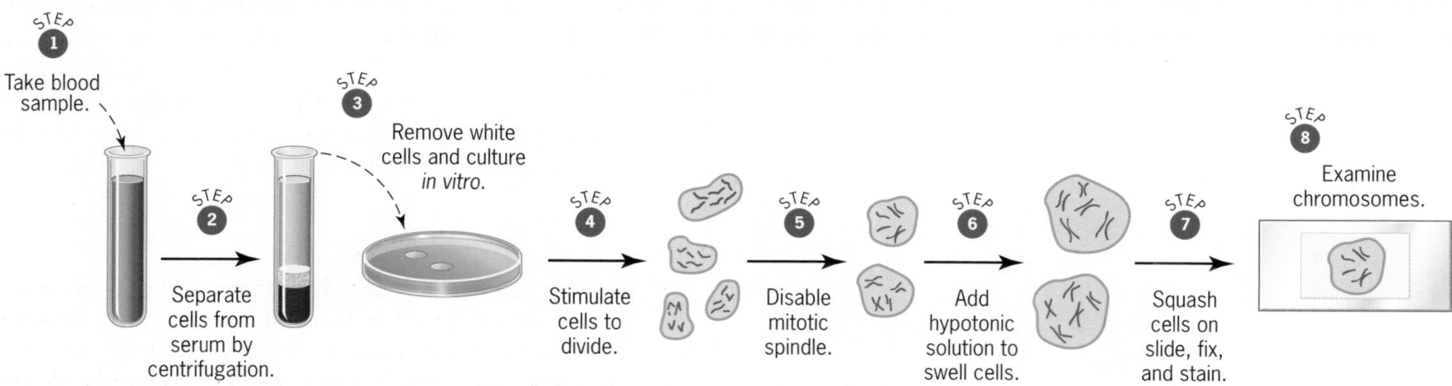

Figure 6.1 ▶ Preparation of cells for cytological analysis.

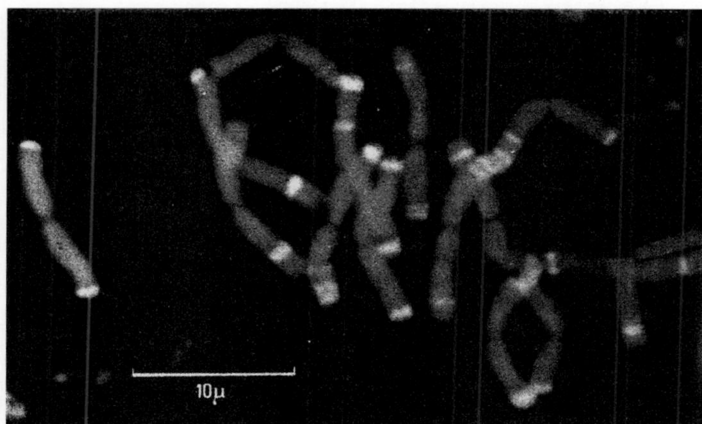

Figure 6.2 ► Metaphase chromosomes of the plant *Allium carinatum,* stained with quinacrine.

(**FIGURE 6.2**). Thus with quinacrine banding, cytogeneticists can identify particular chromosomes in a cell, and they can also determine if a chromosome is structurally abnormal—for example, if it is missing certain bands.

Excellent nonfluorescent staining techniques have also been developed. The most popular of these uses *Giemsa* stain, a mixture of dyes named after its inventor, Gustav Giemsa. Like quinacrine, Giemsa creates a reproducible pattern of bands on each chromosome (**FIGURE 6.3**). It is still not clear why chromosomes show bands when they are stained with quinacrine or Giemsa. It may be that these types of dyes react preferentially with certain DNA sequences, or with the proteins associated with them, and that these target DNA sequences are distributed in a characteristic way within each chromosome.

The most advanced technique used by cytogeneticists today is called **chromosome painting**. With this technique, colorful chromosome images are created by treating chromosome spreads with fluorescently labeled DNA fragments that have been isolated and characterized in the laboratory. Such a fragment may, for instance, come from a particular gene. The DNA fragment is chemically labeled with a fluorescent dye in the laboratory and then applied to chromosomes that have been spread on a glass slide. Under the right conditions, the DNA fragment will bind to chromosomal DNA that is complementary to it in sequence. This binding, in effect, labels the chromosomal DNA with the fluorescent

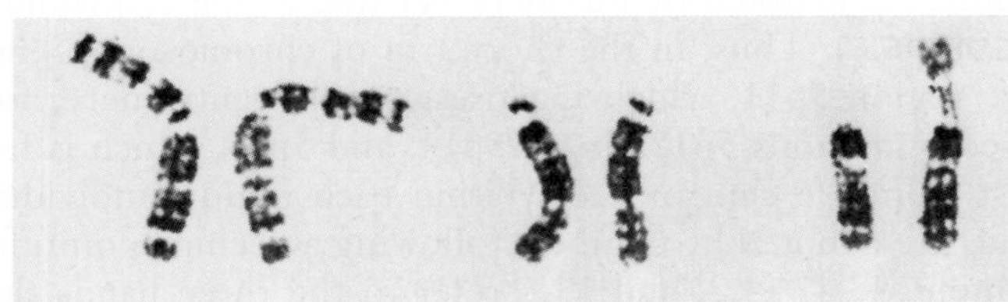

Figure 6.3 ► Metaphase chromosomes of the deerlike Asian muntjak stained with Giemsa.

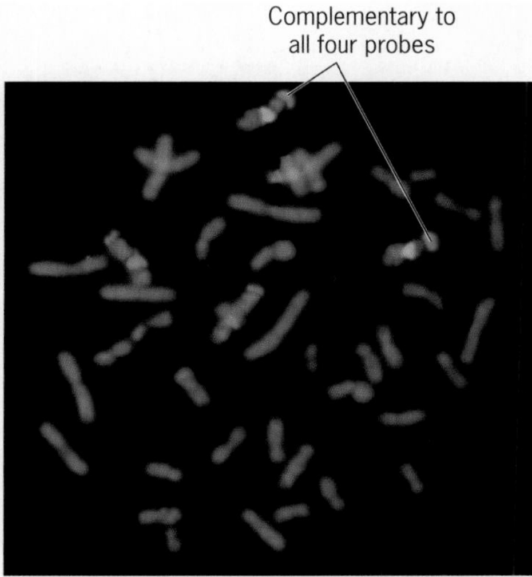

Figure 6.4 ► Chromosome painting. Probes made from human DNA have been applied to the chromosomes of the gibbon, *Hylobates concolor*. Each probe has been labeled with a dye that fluoresces a different color (orange, green, pink, and yellow). A pair of gibbon chromosomes that contain sequences complementary to all four probes is noted at the top right of the micrograph.

dye that is present in the DNA fragment. Because of the specific nature of the interaction between the DNA fragment and the complementary DNA in the chromosomes, we often call the DNA fragment a *probe*. It seeks out and binds to its complement in the large mass of chromosomal DNA in a cell. After the probe has bound, the chromosome spreads are irradiated with light of an appropriate wavelength. The resulting bands or dots of color reveal where the complementary DNA sequence—the target of the probe—is located in the chromosomes. **FIGURE 6.4** shows the chromosomes of the gibbon that have been analyzed with this technique. The gibbon chromosomes were simultaneously painted with four different human DNA fragments, each labeled with a dye that fluoresces a different color. The positions of the colored bands show which of the gibbon chromosomes carries sequences that are complementary to the human DNA fragments—that is, they show the positions of human-like DNA sequences in the chromosomes of this nonhuman primate. **FIGURE 2.9** shows human chromosomes that have been painted with a panel of probes made from human DNA fragments. Each of the pairs of chromosomes has a characteristic pattern of colored bands. Thus, each pair can be uniquely identified using this technique.

THE HUMAN KARYOTYPE

Diploid human cells contain 46 chromosomes—44 autosomes and two sex chromosomes, which are XX in females and XY in males. At mitotic metaphase, each of the 46 chromosomes consists of two identical sister chromatids. When stained appropriately, each of the duplicated chromosomes

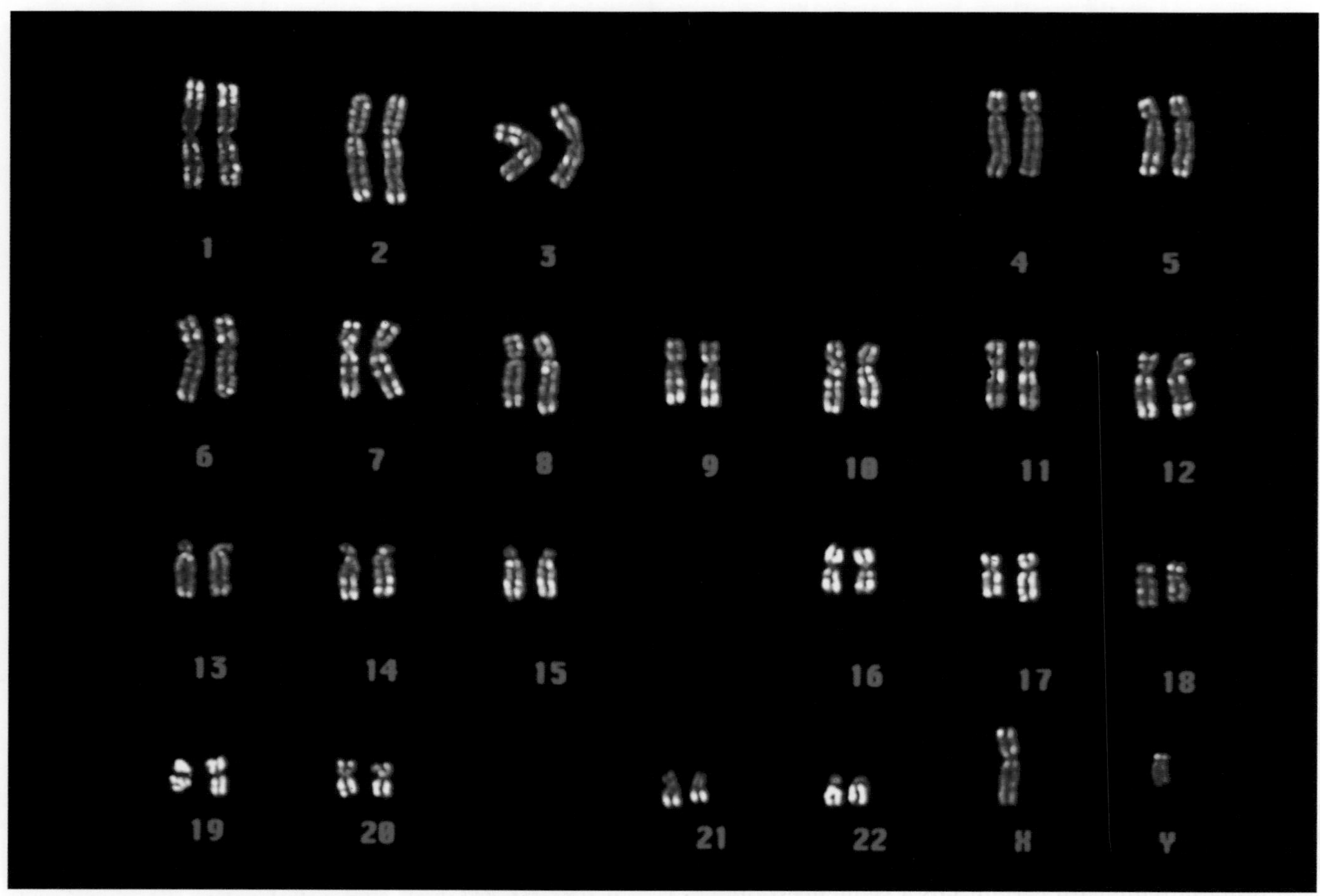

Figure 6.5 ▶ The karyotype of a human male stained to reveal bands on each of the chromosomes. The autosomes are numbered from 1 to 22. The X and Y are the sex chromosomes.

can be recognized by its size, shape, and banding pattern. For cytological analysis, well-stained metaphase spreads are photographed, and then each of the chromosome images is cut out of the picture, matched with its partner to form homologous pairs, and arranged from largest to smallest on a chart (**FIGURE 6.5**). The largest autosome is number 1, and the smallest is number 21. (For historical reasons, the second smallest chromosome has been designated number 22.) The X chromosome is intermediate in size, and the Y chromosome is about the same size as chromosome 22. This chart of chromosome cutouts is called a **karyotype** (from the Greek word meaning "kernel," a reference to the contents of the nucleus). A skilled researcher can use a karyotype to identify abnormalities in chromosome number and structure.

Before the banding and painting techniques were available, it was difficult to distinguish one human chromosome from another. Cytogeneticists could only arrange the chromosomes into groups according to size, classifying the largest as group A, the next largest as group B, and so forth. Although they could recognize seven different groups, within these groups it was nearly impossible to identify a particular chromosome. Today—as a result of the banding and painting techniques—we can routinely identify each of the chromosomes. The banding and painting techniques also make it possible to distinguish each arm of a chromosome and to investigate specific regions within them. The centromere divides each chromosome into long and short arms. The short arm is denoted by the letter *p* (from the French word *petite*, meaning "small") and the long arm by the letter *q* (because it follows "p" in the alphabet). Thus, for example, a cytogeneticist can refer specifically to the short arm of chromosome 5 simply by writing "5p." Within each arm, specific regions are denoted by numbers, starting at the centromere (**FIGURE 6.6**). Thus, in the short arm of chromosome 5, we have region 5p11, which is closest to the centromere, followed by regions 5p12, 5p13, 5p14, and 5p15, which is farthest from the centromere. Within each region, individual bands are denoted by numbers following a decimal point; for example, 13.1, 13.2, and 13.3 refer to the three bands that make up region 5p13. The pattern of bands within the chromosome is called an *ideogram*.

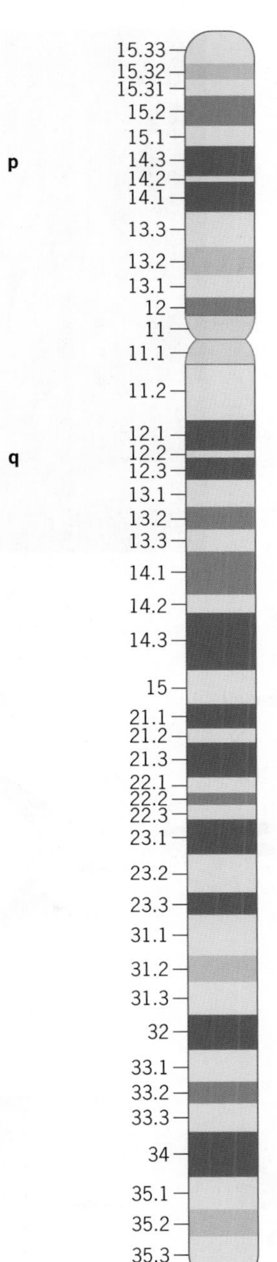

Figure 6.6 ► The ideogram of human chromosome 5. Regions within each arm are numbered consecutively starting at the centromere. Bands within each region are denoted by numbers following a decimal point.

CYTOGENETIC VARIATION: AN OVERVIEW

The phenotypes of many organisms are affected by changes in the number of chromosomes in their cells; sometimes even changes in part of a chromosome can be significant. These numerical changes are usually described as variations in the *ploidy* of the organism (from the Greek word meaning "fold," as in "twofold"). Organisms with complete, or normal, sets of chromosomes are said to be *euploid* (from the Greek words meaning "good" and "fold"). Organisms that carry extra sets of chromosomes are said to be *polyploid* (from the Greek words meaning "many" and "fold"), and the level of polyploidy is described by referring to a basic chromosome number, usually denoted n. Thus, diploids, with two basic chromosome sets, have $2n$ chromosomes; triploids, with three sets, have $3n$; tetraploids, with four sets, have $4n$; and so forth. Organisms in which a particular chromosome, or chromosome segment, is under- or overrepresented are said to be *aneuploid* (from the Greek words meaning "not," "good," and "fold"). These organisms therefore suffer from a specific genetic imbalance. The distinction between aneuploidy and polyploidy is that aneuploidy refers to a numerical change in part of the genome, usually just a single chromosome, whereas polyploidy refers to a numerical change in a whole set of chromosomes. Aneuploidy implies a genetic imbalance, but polyploidy does not.

Cytogeneticists have also catalogued various types of structural changes in the chromosomes of organisms. For example, a piece of one chromosome may be fused to another chromosome, or a segment within a chromosome may be inverted with respect to the rest of that chromosome. These structural changes are called *rearrangements.* Because some rearrangements segregate irregularly during meiosis, they can be associated with aneuploidy. In the sections that follow, we consider all these cytogenetic variations—polyploidy, aneuploidy, and chromosome rearrangements.

KEY POINTS

- Cytogenetic analysis usually focuses on chromosomes in dividing cells.
- Dyes such as quinacrine and Giemsa create banding patterns that are useful in identifying individual chromosomes within a cell.
- A karyotype shows the duplicated chromosomes of a cell arranged for cytogenetic analysis.

Polyploidy

Extra sets of chromosomes in an organism can affect the organism's appearance and fertility.

Polyploidy, the presence of extra chromosome sets, is fairly common in plants but very rare in animals. One-half of all known plant genera contain polyploid species, and about two-thirds of all grasses are polyploids. Many of these species reproduce asexually. In animals, where reproduction is primarily by sexual means, polyploidy is rare, probably because it interferes with the sex-determination mechanism.

One general effect of polyploidy is that cell size is increased, presumably because there are more chromosomes in the nucleus. Often this increase in size is correlated with an overall

Figure 6.7 ▶ Polyploid plants with agricultural or horticultural significance: (*a*) Chrysanthemum (tetraploid), (*b*) strawberry (octoploid), (*c*) cotton (tetraploid), (*d*) banana (triploid).

increase in the size of the organism. Polyploid species tend to be larger and more robust than their diploid counterparts. These characteristics have a practical significance for human beings, who depend on many polyploid plant species for food. These species tend to produce larger seeds and fruits, and therefore provide greater yields in agriculture. Wheat, coffee, potatoes, bananas, strawberries, and cotton are all polyploid crop plants. Many ornamental garden plants, including roses, chrysanthemums, and tulips, are also polyploid (**FIGURE 6.7**).

STERILE POLYPLOIDS

In spite of their robust physical appearance, many polyploid species are sterile. Extra sets of chromosomes segregate irregularly in meiosis, leading to grossly unbalanced (that is, aneuploid) gametes. If such gametes unite in fertilization, the resulting zygotes almost always die. This inviability among the zygotes explains why many polyploid species are sterile.

As an example, let us consider a triploid species with three identical sets of n chromosomes. The total number of chromosomes is therefore $3n$. When meiosis occurs, each chromosome will try to pair with its homologues (**FIGURE 6.8**). One possibility is that two homologues will pair completely along their length, leaving the third without a partner; this solitary chromosome is called a **univalent.** Another possibility is that all three homologues will synapse, forming a **trivalent** in which each member is partially paired with each of the others. In either case, it is difficult to predict how the chromosomes will move during anaphase of the first meiotic division. The more likely event is that two of the homologues will move to one pole, and one homologue will move to the other, yielding gametes with one or two copies of the chromosome. However, all three homologues might move to one pole, producing gametes with zero or three copies of the chromosome. Because this segregational uncertainty applies to each trio of chromosomes in the cell, the total number of chromosomes in a gamete can vary from zero to $3n$.

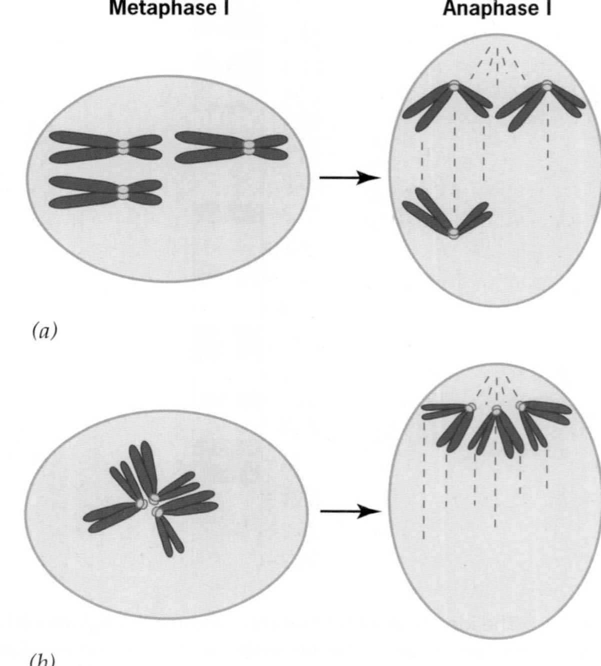

Figure 6.8 ▶ Meiosis in a triploid. (*a*) Univalent formation. Two of the three homologues synapse, leaving a univalent free to move to either pole during anaphase. (*b*) Trivalent formation. All three homologues synapse, forming a trivalent, which may move as a unit to one pole during anaphase. However, other anaphase disjunctions are possible.

Zygotes formed by fertilization with such gametes are almost certain to die; thus, most triploids are completely sterile.

In agriculture and horticulture, this sterility is circumvented by propagating the species asexually. The many methods of asexual propagation include cultivation from cuttings (bananas), grafts (Winesap, Gravenstein, and Baldwin apples), and bulbs (tulips). In nature, polyploid plants can also reproduce asexually. One mechanism is **apomixis,** which involves a modified meiosis that produces unreduced eggs; these eggs then form seeds that germinate into new plants. The dandelion, a highly successful polyploid weed, reproduces in this way.

FERTILE POLYPLOIDS

The meiotic uncertainties that occur in triploids also occur in tetraploids with four identical chromosome sets. Such tetraploids are therefore also sterile. However, some tetraploids are able to produce viable progeny. Close examination shows that these species contain two distinct sets of chromosomes and that each set has been duplicated. Thus, fertile tetraploids seem to have arisen by chromosome duplication in a hybrid that was produced by a cross of two different, but related, diploid species; most often these species have the same or very similar chromosome numbers. **FIGURE 6.9** shows a plausible mechanism for the origin of such a tetraploid. Two diploids, denoted A and B, are crossed to produce a hybrid that receives one set of chromosomes from each of the parental species. Such a hybrid will probably be sterile because the A and B chromosomes cannot pair with each other. However, if the chromosomes in this hybrid are duplicated, meiosis will proceed in reasonably good order. Each of the A and B chromosomes will be able to pair with a perfectly homologous partner. Meiotic segregation can therefore produce gametes with a complete set of A and B chromosomes. In fertilization, these "diploid" gametes will unite to form tetraploid zygotes, which will survive because each of the parental sets of chromosomes will be balanced.

This scenario of hybridization between different but related species followed by chromosome doubling has evidently occurred many times during plant evolution. In some cases, the process has occurred repeatedly, generating complex polyploids with distinct chromosome sets. One of the best examples is modern bread wheat, *Triticum aestivum* (**FIGURE 6.10**). This important crop species is a hexaploid containing three different chromosome sets, each of which has been duplicated. There are seven chromosomes in each set, for a total of 21 in the gametes and 42 in the somatic cells. Thus, as we noted at the beginning of this chapter, modern wheat seems to have been formed by two hybridization events. The first involved two diploid species that combined to form a tetraploid, and the second involved a combination between this tetraploid and another diploid, to produce a hexaploid. Cytogeneticists have identified primitive cereal plants in the Middle East that may have participated in this evolutionary process.

Because chromosomes from different species are less likely to interfere with each other's segregation during meiosis, polyploids arising from hybridizations between different species

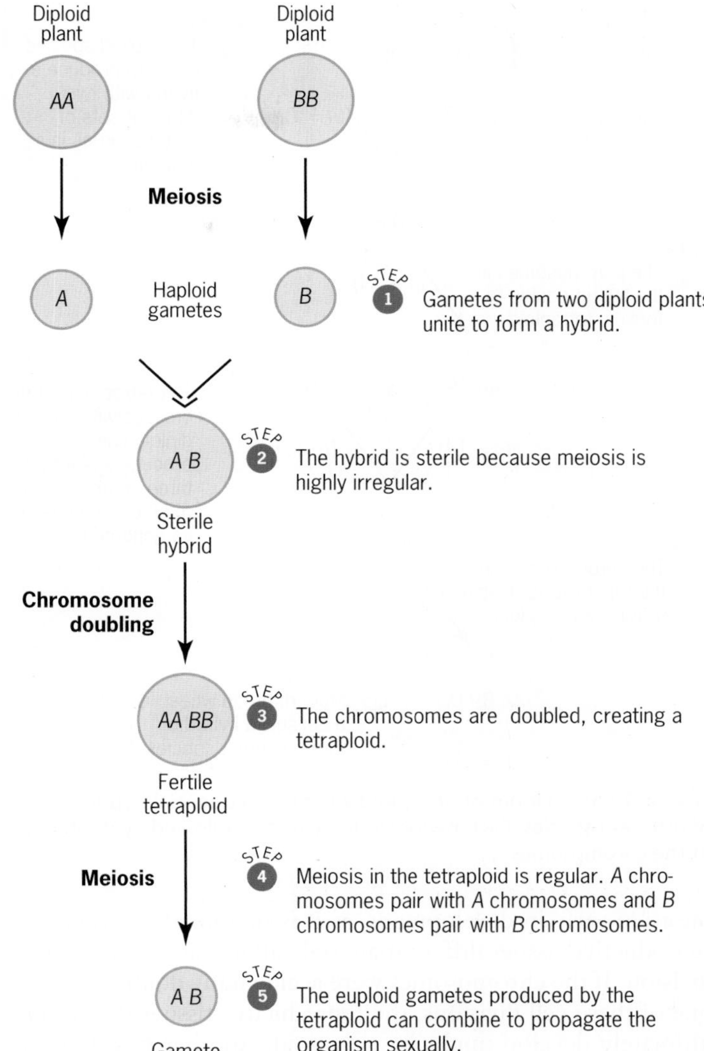

Figure 6.9 ► Origin of a fertile tetraploid by hybridization between two diploids and subsequent doubling of the chromosomes.

have a much greater chance of being fertile than do polyploids arising from the duplication of chromosomes in a single species. Polyploids created by hybridization between different species are called ***allo*polyploids** (from the Greek prefix for "other"); in these polyploids, the contributing genomes are qualitatively different. Polyploids created by chromosome duplication within a species are called ***auto*polyploids** (from the Greek prefix for "self"); in these polyploids, a single genome has been multiplied to create extra chromosome sets.

Chromosome doubling is a key event in the formation of polyploids. One possible mechanism for this event is for a cell to go through mitosis without going through cytokinesis. Such a cell will have twice the usual number of chromosomes. Through subsequent divisions, it could then give rise to a polyploid clone of cells, which might contribute either to the asexual propagation of the organism or to the formation of gametes. In plants it must be remembered that the germ line is not

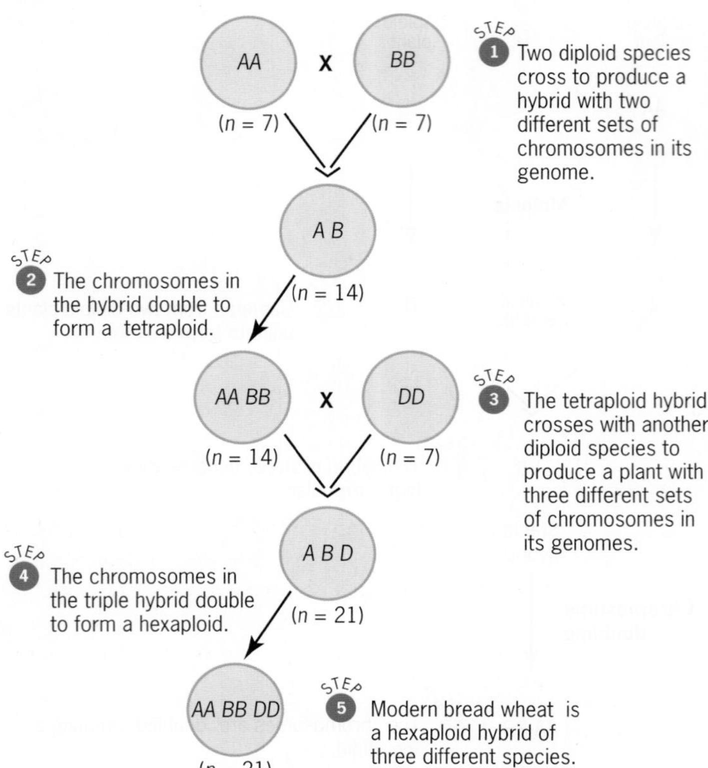

Figure 6.10 ▶ Origin of hexaploid wheat by sequential hybridization of different species. Each hybridization event is followed by doubling of the chromosomes.

set aside early in development, as it is in animals. Rather, the reproductive tissues differentiate only after many cycles of cell division. If the chromosomes were accidentally doubled during one of these cell divisions, the reproductive tissues that would ultimately develop might be polyploid. Another possibility is for meiosis to be altered in such a way that unreduced gametes (with twice the normal number of chromosomes) are produced. If such gametes participate in fertilization, polyploidy zygotes will be formed. These zygotes may then develop into mature organisms, which, depending on the nature of the polyploidy, may be able to produce gametes themselves.

TISSUE-SPECIFIC POLYPLOIDY AND POLYTENY

In some organisms, certain tissues become polyploid during development. This polyploidization is probably a response to the need for multiple copies of each chromosome and the genes it carries. The process that produces such polyploid cells, called **endomitosis,** involves chromosome duplication, followed by separation of the resulting sister chromatids. However, because there is no accompanying cell division, extra chromosome sets accumulate within a single nucleus. In the human liver and kidney, for example, one round of endomitosis produces tetraploid cells.

Sometimes polyploidization occurs without the separation of sister chromatids. In these cases, the duplicated chromosomes pile up next to each other, forming a bundle of strands that are aligned in parallel. The resulting chromosomes are said to be **polytene,** from the Greek words meaning "many threads." The most spectacular examples of polytene chromosomes are found in the salivary glands of *Drosophila* larvae. Each chromosome undergoes about nine rounds of replication, producing a total of about 500 copies in each cell. All the copies pair tightly, forming a thick bundle of chromatin fibers. This bundle is so large that it can be seen under low magnification with a dissecting microscope. Differential coiling along the length of the bundle causes variation in the density of the chromatin. When dyes are applied to these chromosomes, the denser chromatin stains more deeply, creating a pattern of dark and light bands (**FIGURE 6.11**). This pattern is highly reproducible, permitting detailed analysis of chromosome structure.

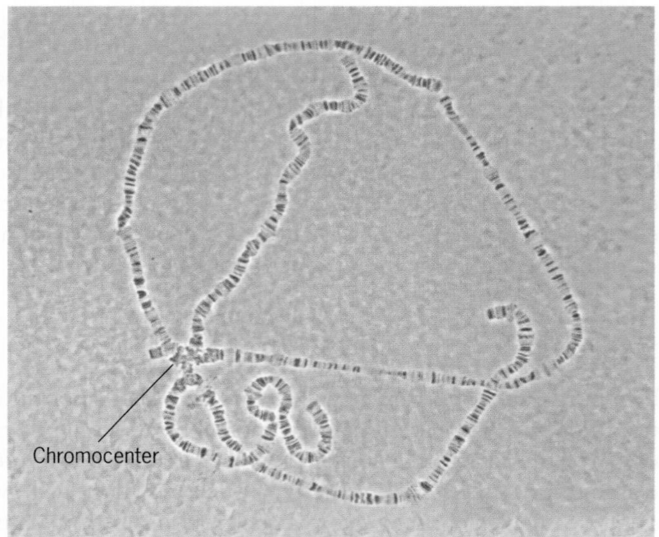

Figure 6.11 ▶ Polytene chromosomes of *Drosophila.*

The polytene chromosomes of *Drosophila* show two additional features:

1. **Homologous Polytene Chromosomes Pair:** Ordinarily, we think of pairing as a property of meiotic chromosomes; however, in many insect species the somatic chromosomes also pair—probably as a way of organizing the chromosomes within the nucleus. When *Drosophila* polytene chromosomes pair, the large chromatin bundles become even larger. Because this pairing is precise—point-for-point along the length of the chromosome—the two homologues come into perfect alignment. Thus, the banding patterns of each are exactly in register, so much so that it is almost impossible to distinguish the individual members of a pair.
2. **All the Centromeres of Drosophila Polytene Chromosomes Congeal Into a Body Called the Chromocenter:** Material flanking the centromeres is also drawn into this mass. The result is that the chromosome arms seem to emanate out of the chromocenter. These arms, which are banded, consist of euchromatin, that portion of the chromosome that contains most of the genes; the chromocenter consists

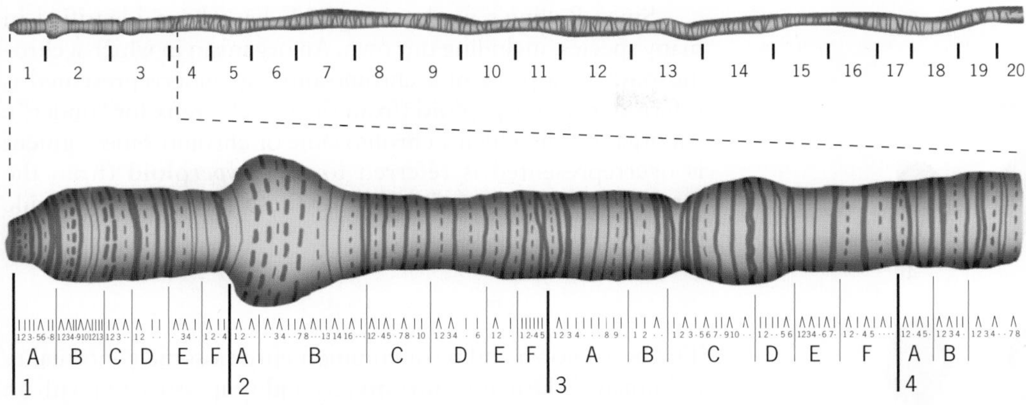

Figure 6.12 ▶ Bridges' polytene chromosome maps. (*Top*) Banding pattern of the polytene X chromosome. The chromosome is divided into 20 numbered sections. (*Bottom*) Detailed view of the left end of the polytene X chromosome showing Bridges' system for denoting individual bands.

of heterochromatin, a gene-poor material that surrounds the centromere. Unlike the euchromatic chromosome arms, this centric heterochromatin does not become polytene. Thus, compared to the euchromatin, it is vastly underreplicated.

In the 1930s C. B. Bridges published detailed drawings of the polytene chromosomes (**FIGURE 6.12**). Bridges arbitrarily divided each of the chromosomes into sections, which he numbered; each section was then divided into subsections, which were designated by the letters *A* to *F*. Within each subsection, Bridges enumerated all the dark bands, creating an alphanumeric directory of sites along the length of each chromosome. Bridges' alphanumeric system is still used today to describe the features of these remarkable chromosomes.

The polytene chromosomes of *Drosophila* are trapped in the interphase of the cell cycle. Thus, although most cytological analyses are performed on mitotic chromosomes, the most thorough and detailed analyses are performed on polytenized interphase chromosomes. Such chromosomes are found in many species within the insect order Diptera, including flies and mosquitoes. Unfortunately, human beings do not have polytene chromosomes; thus, the high-resolution structural analysis that is possible for *Drosophila* is not possible for our own species.

KEY POINTS

- Polyploids contain extra sets of chromosomes.
- Many polyploids are sterile because their multiple sets of chromosomes segregate irregularly in meiosis.
- Polyploids produced by chromosome doubling in interspecific hybrids may be fertile if their constituent genomes segregate independently.
- In some somatic tissues—for example, the salivary glands of *Drosophila* larvae—successive rounds of chromosome replication occur without intervening cell divisions and produce large polytene chromosomes that are ideal for cytogenetic analysis.

Aneuploidy

The under- or overrepresentation of a chromosome or a chromosome segment can affect a phenotype.

Aneuploidy describes a numerical change in part of the genome, usually a change in the dosage of a single chromosome. Individuals that have an extra chromosome, are missing a chromosome, or have a combination of these anomalies are said to be aneuploid. This definition also includes pieces of chromosomes. Thus, an individual in which a chromosome arm has been deleted is also considered to be aneuploid.

Aneuploidy was originally studied in plants, where it was shown that a chromosome imbalance usually has a phenotypic effect. The classic study was one by Albert Blakeslee and John Belling, who analyzed chromosome anomalies in Jimson weed, *Datura stramonium*. This diploid species has 12 pairs of chromosomes, for a total of 24 in the somatic cells. Blakeslee collected plants with altered phenotypes and discovered that in some cases the phenotypes were inherited in an irregular way. These peculiar mutants were apparently caused by dominant factors that were transmitted primarily through the female. By examining the chromosomes of the mutant plants, Belling found that in every case an extra chromosome was present. Detailed analysis established that the extra chromosome was different in each mutant strain. Altogether there were 12 different mutants, each corresponding to a triplication of one of the *Datura* chromosomes (**FIGURE 6.13**). Such triplications are called **trisomies.** The transmissional irregularities of these mutants were due to anomalous chromosome behavior during meiosis.

Belling also discovered the reason for the preferential transmission of the trisomic phenotypes through the female.

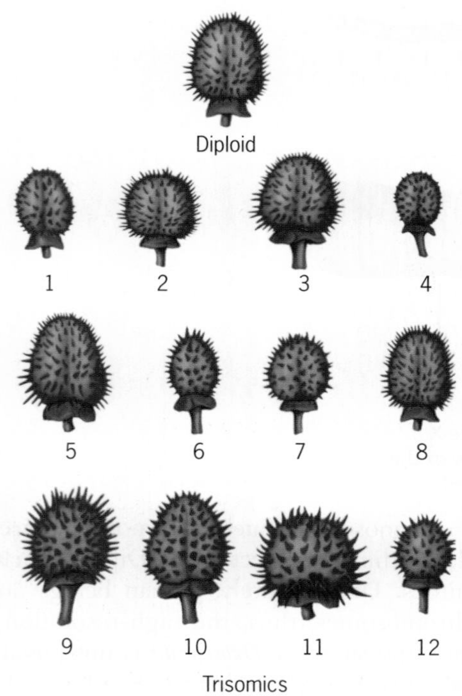

Figure 6.13 ▶ Seed capsules of normal and trisomic *Datura stramonium*. Each of the 12 trisomies is shown.

During pollen tube growth, aneuploid pollen—in particular, pollen with $n + 1$ chromosomes—does not compete well with euploid pollen. Consequently, trisomic plants almost always inherit their extra chromosome from the female parent. Belling's work with *Datura* demonstrated that each chromosome must be present in the proper dosage for normal growth and development.

Since Belling's work, aneuploids have been identified in many species, including our own. An organism in which a chromosome, or a piece of a chromosome, is underrepresented is referred to as a ***hypo*ploid** (from the Greek prefix for "under"). An organism in which a chromosome or chromosome segment is overrepresented is referred to as a ***hyper*ploid** (from the Greek prefix for "over"). Each of these terms covers a wide range of abnormalities.

TRISOMY IN HUMAN BEINGS

The best-known and most common chromosome abnormality in humans is **Down syndrome,** a condition associated with an extra chromosome 21 (**FIGURE 6.14*a***). This syndrome was first described in 1866 by a British physician, Langdon Down, but its chromosomal basis was not clearly understood until 1959. People with Down syndrome are typically short in stature and loose-jointed, particularly in the ankles; they have broad skulls, wide nostrils, large tongues with a distinctive furrowing, and stubby hands with a crease on the palm. Impaired mental abilities require that they be given special education and care. The life span of people with Down syndrome is much shorter than that of other people. Down syndrome individuals also almost invariably develop Alzheimer's disease, a form of dementia that is fairly common among the elderly. However, people with Down syndrome develop this disease in their fourth or fifth decade of life, much sooner than other people.

The extra chromosome 21 in Down syndrome is an example of a trisomy. **FIGURE 6.14*b*** shows the karyotype of a female Down patient. Altogether, there are 47 chromosomes, including two X chromosomes as well as the extra chromosome 21. The karyotype of this individual is therefore written 47, XX, +21.

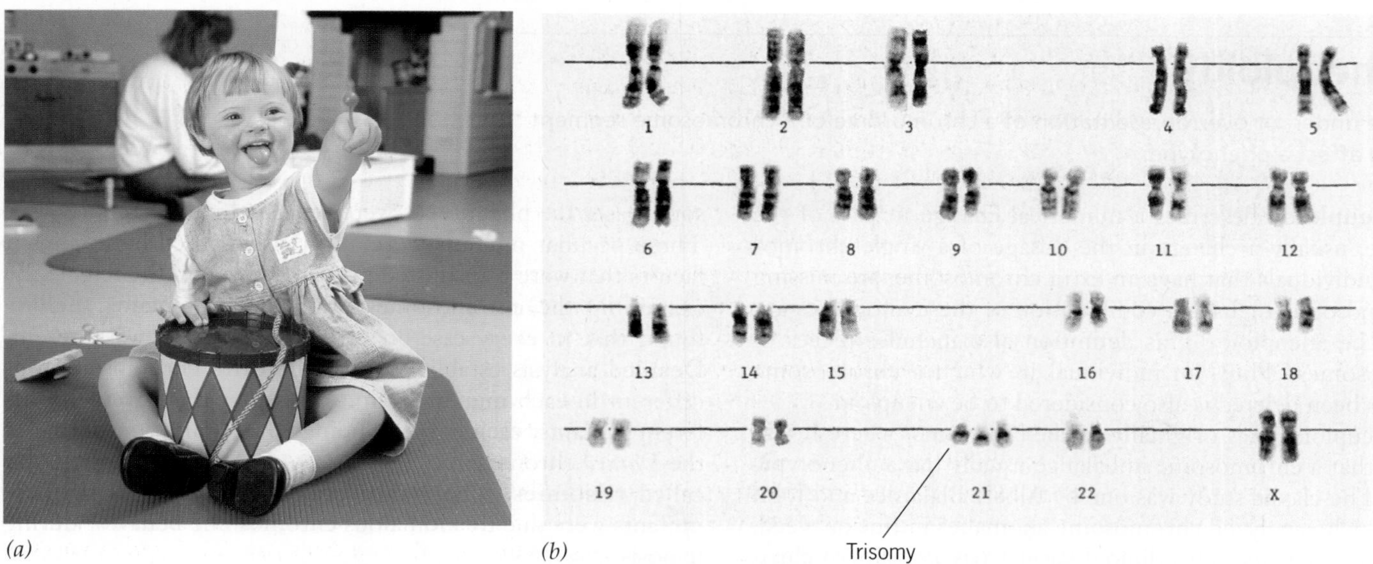

Figure 6.14 ▶ Down syndrome. (*a*) A young girl with Down syndrome. (*b*) Karyotype of a child with Down syndrome, showing trisomy for chromosome 21 (47, XX, +21).

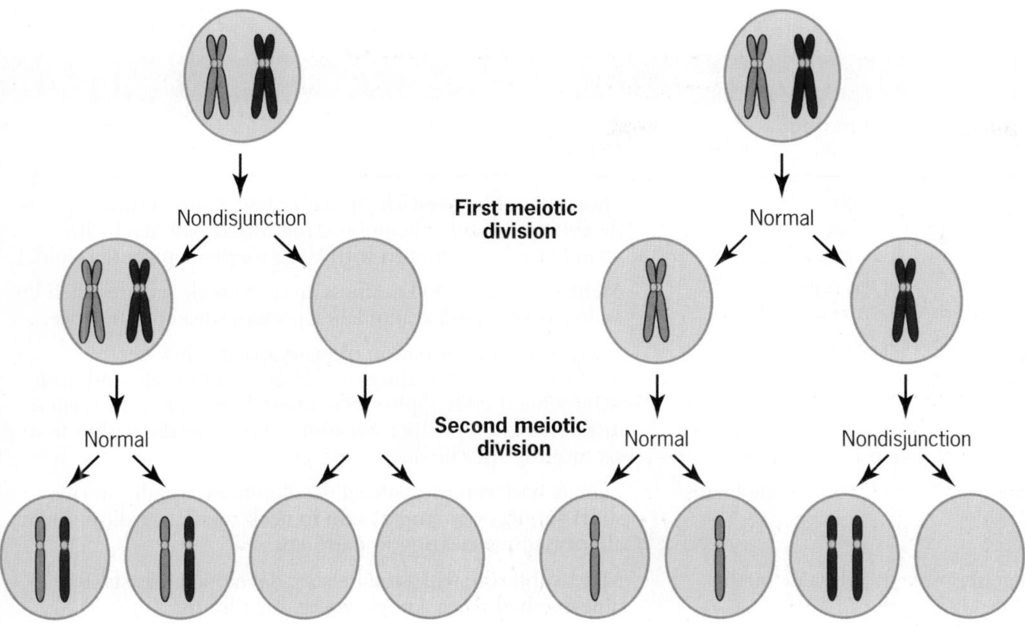

Figure 6.15 ▶ Meiotic nondisjunction of chromosome 21 and the origin of Down syndrome. Nondisjunction at meiosis I produces no normal gametes. Nondisjunction at meiosis II produces a gamete with two identical sister chromosomes, a gamete lacking chromosome 21, and two normal gametes.

Trisomy 21 can be caused by chromosome nondisjunction in one of the meiotic cell divisions (**FIGURE 6.15**). The nondisjunction event can occur in either parent, but it seems to be more likely in females. In addition, the frequency of nondisjunction increases with maternal age. Thus, among mothers younger than 25 years old, the risk of having a child with Down syndrome is about 1 in 1500, whereas among mothers 40 years old, it is 1 in 100. This increased risk is due to factors that adversely affect meiotic chromosome behavior as a woman ages. In human females, meiosis begins in the fetus, but it is not completed until after the egg is fertilized. During the long time prior to fertilization, the meiotic cells are arrested in the prophase of the first division. In this suspended state, the chromosomes may become unpaired. The longer the time in prophase, the greater the chance for unpairing and subsequent chromosome nondisjunction. Older females are therefore more likely than younger females to produce aneuploid eggs.

Trisomies for chromosomes 13 and 18 have also been reported. However, these are rare, and the affected individuals show serious phenotypic abnormalities and are short-lived, usually dying within the first few weeks after birth. Another viable trisomy that has been observed in human beings is the triplo-X karyotype, 47, XXX. These individuals survive because two of the three X chromosomes are inactivated, reducing the dosage of the X chromosome so that it approximates the normal level of one. Triplo-X individuals are female and are phenotypically normal, or nearly so; sometimes they exhibit a slight mental impairment and reduced fertility.

The 47, XXY karyotype is also a viable trisomy in human beings. These individuals have three sex chromosomes, two X's and one Y. Phenotypically, they are male, but they can show some female secondary sexual characteristics and are usually sterile. In 1942 H. F. Klinefelter described the abnormalities associated with this condition, now called **Klinefelter syndrome;** these include small testes, enlarged breasts, long limbs, knock-knees, and underdeveloped body hair. The XXY karyotype can originate by fertilization of an exceptional XX egg with a Y-bearing sperm or by fertilization of an X-bearing egg with an exceptional XY sperm. The XXY karyotype accounts for about three-fourths of all cases of Klinefelter syndrome. Other cases involve more complex karyotypes such as XXYY, XXXY, XXXYY, XXXXY, XXXXYY, and XXXXXY. All individuals with Klinefelter syndrome have one or more Barr bodies in their cells, and those with more than two X chromosomes usually have some degree of mental impairment.

The 47, XYY karyotype is another viable trisomy in human beings. These individuals are male, and except for a tendency to be taller than 46, XY men, they do not show a consistent syndrome of characteristics. All the other trisomies in human beings are embryonic lethals, demonstrating the importance of correct gene dosage. Unlike *Datura*, in which each of the possible trisomies is viable, human beings do not tolerate many types of chromosomal imbalance (see **TABLE 6.1**).

MONOSOMY

Monosomy occurs when one chromosome is missing in an otherwise diploid individual. In human beings, there is only one viable monosomic, the 45, X karyotype. These individuals have a single X chromosome as well as a diploid complement of autosomes. Phenotypically, they are female, but because their ovaries are rudimentary, they are almost always sterile. 45, X individuals are usually short in stature; they have webbed necks, hearing deficiencies, and significant cardiovascular abnormalities. Henry H. Turner first described the condition in 1938; thus, it is now called **Turner syndrome.** 45, X individuals can originate from eggs or sperm that lack a sex chromosome or from the loss of a sex chromosome in mitosis sometime after

▶ **TABLE 6.1**

Aneuploidy Resulting from Nondisjunction in Human Beings

Karyotype	Chromosome Formula	Clinical Syndrome	Estimated Frequency at Birth	Phenotype
47, +21	$2n + 1$	Down	1/700	Short, broad hands with palmar crease, short stature, hyperflexibility of joints, mental retardation, broad head with round face, open mouth with large tongue, epicanthal fold.
47, +13	$2n + 1$	Patau	1/20,000	Mental deficiency and deafness, minor muscle seizures, cleft lip and/or palate, cardiac anomalies, posterior heel prominence.
47, +18	$2n + 1$	Edward	1/8000	Congenital malformation of many organs, low-set, malformed ears, receding mandible, small mouth and nose with general elfin appearance, mental deficiency, horseshoe or double kidney, short sternum; 90 percent die within first six months after birth.
45, X	$2n - 1$	Turner	1/2500 female births	Female with retarded sexual development, usually sterile, short stature, webbing of skin in neck region, cardiovascular abnormalities, hearing impairment.
47, XXY	$2n + 1$	Klinefelter	1/500 male births	Male, subfertile with small testes, developed breasts, feminine-pitched voice, knock-knees, long limbs.
48, XXXY	$2n + 2$			
48, XXYY	$2n + 2$			
49, XXXXY	$2n + 3$			
50, XXXXXY	$2n + 4$			
47, XXX	$2n + 1$	Triplo-X	1/700	Female with usually normal genitalia and limited fertility, slight mental retardation.

fertilization (**FIGURE 6.16**). This latter possibility is supported by the finding that many Turner individuals are *somatic mosaics*. These people have two types of cells in their bodies; some are 45, X and others are 46, XX. This karyotypic mosaicism evidently arises when an X chromosome is lost during the development of a 46, XX zygote. All the descendants of the cell in which the loss occurred are 45, X. If the loss occurs early in development, an appreciable fraction of the body's cells will be aneuploid and the individual will show the features of Turner syndrome. If the loss occurs later, the aneuploid cell population will be smaller and the severity of the syndrome is likely to be reduced. For a discussion of procedures used to detect aneuploidy in human fetuses, see the Focus on Amniocentesis and Chorionic Biopsy.

XX/XO chromosome mosaics also occur in *Drosophila*, where they produce a curious phenotype. Because sex in this species is determined by the ratio of X chromosomes to autosomes, such flies are part female and part male. XX cells develop in the female direction, and XO cells develop in the male direction. Flies with both male and female structures are called **gynandromorphs** (from Greek words meaning "woman," "man," and "form").

People with the 45, X karyotype have no Barr bodies in their cells, indicating that the single X chromosome that is present is not inactivated. Why, then, should Turner patients, who have the same number of active X chromosomes as normal XX females, show any phenotypic abnormalities at all? The answer probably involves a small number of genes that remain active on both of the X chromosomes in normal 46, XX females. These noninactivated genes are apparently needed in double dose for proper growth and development. The finding that at least some of these special X-linked genes are also present on the Y chromosome would explain why XY males grow and develop normally. In addition, the X chromosome that has been inactivated in 46, XX females is reactivated during oogenesis, presumably because two copies of some X-linked genes are required for normal ovarian function. 45, X individuals, who have only one copy of these genes, cannot meet this quantitative requirement and are therefore sterile.

Origin of monosomy at fertilization.

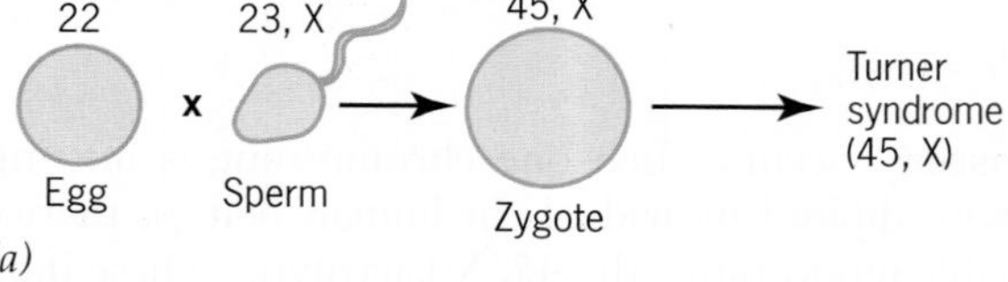

Origin of monosomy in the cleavage division following fertilization.

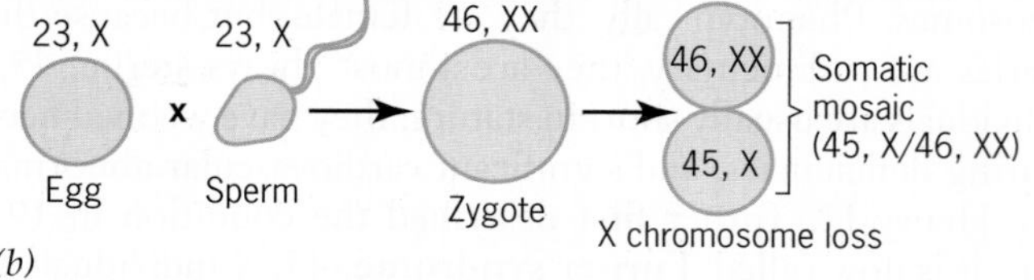

Figure 6.16 ▶ Origin of the Turner syndrome karyotype at fertilization (*a*) or at the cleavage division following fertilization (*b*).

► FOCUS ON **Amniocentesis and Chorionic Biopsy**

The Andersons, a couple living in Minneapolis, were expecting their first baby. Neither Donald nor Laura Anderson knew of any genetic abnormalities in their families, but because of Laura's age—38—they decided to have the fetus checked for aneuploidy. Laura's physician performed a procedure called **amniocentesis.** A small amount of fluid was removed from the cavity surrounding the developing fetus by inserting a needle into Laura's abdomen (**FIGURE 1**). This cavity, called the amniotic sac, is enclosed by a membrane. To prevent discomfort during the procedure, Laura was given a local anesthetic. The needle was guided into position by following an ultrasound scan, and some of the amniotic fluid was drawn out. Because this fluid contains nucleated cells sloughed off from the fetus, it is possible to determine the fetus's karyotype (**FIGURE 2**). Usually the fetal cells are purified from the amniotic fluid by centrifugation, and then the cells are cultured for several days to a few weeks. Cytological analysis of these cells will reveal if the fetus is aneuploid. Additional tests may be performed on the fluid recovered from the amniotic sac to detect other sorts of abnormalities, including neural tube defects and some kinds of mutations. The results of all these tests may take up to three weeks. In Laura's case, no abnormalities of any sort were detected, and 20 weeks after the amniocentesis, she gave birth to a healthy baby girl.

Chorionic biopsy provides another way of detecting chromosomal abnormalities in the fetus. The chorion is a fetal membrane that interdigitates with the uterine wall, eventually forming the placenta. The minute chorionic projections into the uterine tissue are called *villi* (singular, villus). At 10–11 weeks of gestation, before the placenta has developed, a sample of chorionic villi can be obtained by passing a hollow plastic tube into the uterus through the cervix. This tube can be guided by an ultrasound scan, and when the tube is in place, a tiny bit of material can be drawn up into the tube by aspiration. The recovered material usually consists of a mixture of maternal and fetal tissue. After these tissues are separated by dissection, the fetal cells can be analyzed for chromosomal abnormalities. Chorionic biopsy can be performed earlier than amniocentesis (10–11 weeks gestation versus 14–16 weeks), but it is not as reliable. In addition, it seems to be associated with a slightly greater chance of miscarriage than amniocentesis does, perhaps 2 to 3 percent. For these reasons, it tends to be used only in pregnancies where there is a strong reason to expect a genetic abnormality. In routine pregnancies, such as Laura Anderson's, amniocentesis is the preferred procedure.

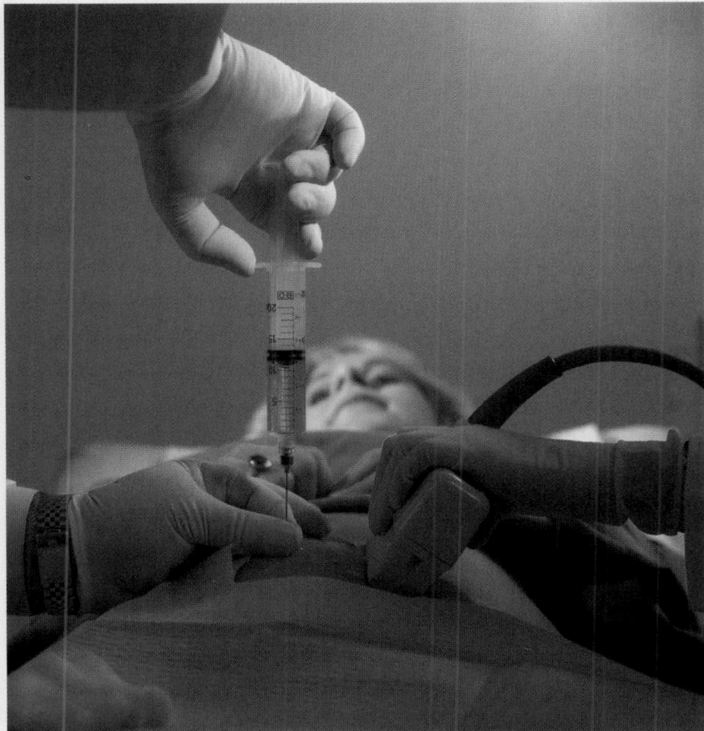

Figure 1 ► A physician taking a sample of fluid from the amniotic sac of a pregnant woman for prenatal diagnosis of a chromosomal or biochemical abnormality.

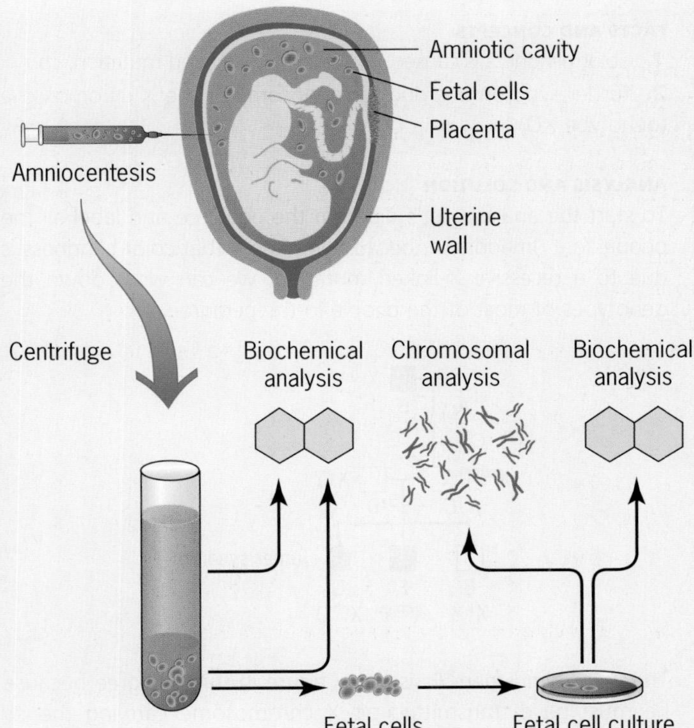

Figure 2 ► Amniocentesis and procedures for prenatal diagnosis of chromosomal and biochemical abnormalities.

Curiously, the cognate of the XO Turner karyotype in the mouse exhibits no anatomical abnormalities. This finding implies that the mouse homologues of the human genes that are involved in Turner syndrome need only be present in one copy for normal growth and development. To investigate the origin of the XO Turner karyotype, work through the Focus on Problem Solving: Tracing Sex Chromosome Nondisjunction.

DELETIONS AND DUPLICATIONS OF CHROMOSOME SEGMENTS

A missing chromosome segment is referred to either as a **deletion** or as a **deficiency.** Large deletions can be detected cytologically by studying the banding patterns in stained chromosomes, but small ones cannot. In a diploid organism, the deletion of a chromosome segment makes part of the genome hypoploid. This hypoploidy may be associated with a phenotypic effect, especially if the deletion is large. A classic example is the ***cri-du-chat* syndrome** (from the French words for "cry of the cat") in human beings (**FIGURE 6.17**). This condition is caused by a deletion in the short arm of chromosome 5. The size of the deletion varies. Individuals heterozygous for the deletion and a normal chromosome have the karyotype 46 del(5)(p14), where the terms in parentheses indicates that bands in region 14 of the short arm (p) of one of the chromosomes 5 is missing. These individuals may be severely impaired, mentally as well as physically; their plaintive, catlike crying during infancy gives the syndrome its name.

An extra chromosome segment is referred to as a **duplication.** The extra segment can be attached to one of the chromosomes, or it can exist as a new and separate chromosome, that is, as a "free duplication." In either case, the effect is the same: The organism is hyperploid for part of its genome. As with deletions, this hyperploidy can be associated with a phenotypic effect.

▶ FOCUS ON PROBLEM SOLVING
Tracing Sex Chromosome Nondisjunction

THE PROBLEM

A color-blind man married a normal woman. Their daughter, who was phenotypically normal, married a normal man and the couple produced three children: a normal boy, a color-blind boy, and a color-blind girl with Turner syndrome. Explain the origin of the color-blind girl with Turner syndrome.

FACTS AND CONCEPTS

1. Color blindness is caused by a recessive X-linked mutation, *cb*.

2. Turner syndrome is due to monosomy of the X chromosome (genotype XO).

3. Monosomy can arise from chromosome nondisjunction during mitosis or meiosis.

4. Mitotic nondisjunction in an XX individual can create a mosaic of XO and XX cells.

ANALYSIS AND SOLUTION

To start the analysis, let's diagram the pedigree and label all the people in it. In addition, because we know that color blindness is due to a recessive X-linked mutation, we can write down the genotypes of most of the people in the pedigree.

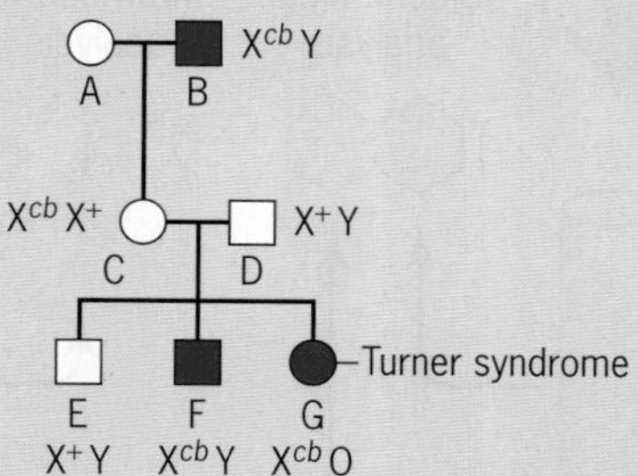

The color-blind man, B, is a key figure in this pedigree because he must have transmitted an X chromosome carrying the *cb* mutation to his daughter C, who is the mother of the child in question. C is not color blind herself, so she must be heterozygous for the mutant allele—that is, her genotype is X^{cb} X^{+}. Likewise, her husband, D, is not color blind, so he must have the genotype X^{+} Y. The genotypes of the couple's first two children are also known with certainty. The last child, G, has Turner syndrome, which implies that she has just one sex chromosome—an X. Because this girl is color blind, her genotype is presumably X^{cb} O. This genotype could have been created by fertilization of an egg containing the X^{cb} chromosome by a sperm that lacked a sex chromosome. In this scenario, there must have been nondisjunction of the sex chromosomes during meiosis in G's father. Another possibility is that the X^{cb}-bearing egg was fertilized by a sperm that carried an X chromosome and this chromosome was lost during one of the early divisions in the embryo. On this second hypothesis, G would be a somatic mosaic of XO and XX cells (see **FIGURE 6.16*b***). However, this explanation does not square with the observation that G is color blind, for if G were a somatic mosaic, her XX cells would have to be X^{cb} X^{+}, and some of these cells would be expected to have formed normal photoreceptor cells in her retinas, thereby giving her normal color vision. The fact that G is color blind indicates that she does not have X^{cb} X^{+} cells in her retinas—or probably anywhere else in her body. Sex chromosome nondisjunction during meiosis in G's father is therefore the more plausible explanation for her color-blind, Turner phenotype.

For further discussion go to your *WileyPLUS* course.

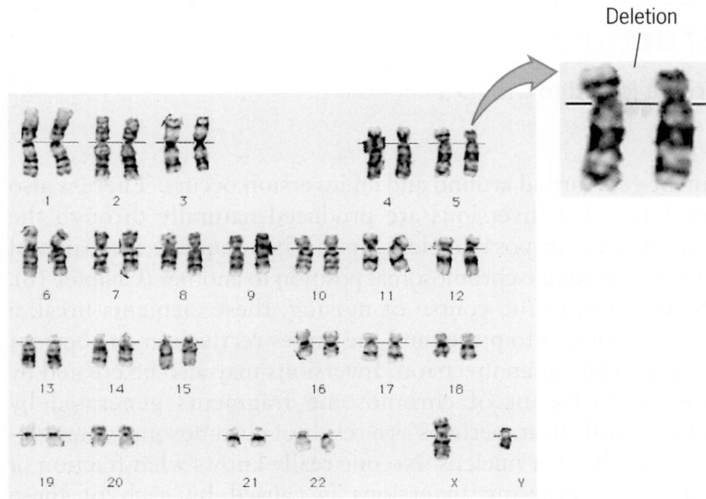

Figure 6.17 ► Karyotype of a male with The *cri-du-chat* syndrome, 46, XY del(5)(p14). There is a deletion in region 14 of the short arm (p) of chromosome 5.

Deletions and duplications are two types of aberrations in chromosome structure. Large aberrations can be detected by examination of mitotic chromosomes that have been stained with banding agents such as quinacrine or Giemsa. However, small aberrations are difficult to detect in this way and are usually identified by other genetic and molecular techniques. The best organism for studying deletions and duplications is *Drosophila*, where the polytene chromosomes afford an unparalleled opportunity for detailed cytological analysis. **FIGURE 6.18*b*** shows a deletion in one of two paired homologous chromosomes in a *Drosophila* salivary gland. Because the two chromosomes have separated slightly, we can see that a small region is missing in the lower one.

Duplicated segments can also be recognized in polytene chromosomes. **FIGURE 6.18c** shows a tandem duplication of a segment in the middle of the X chromosome of *Drosophila*. Because tandem copies of this segment pair with each other, the chromosome appears to have a knot in its middle. The *Bar* eye mutation in *Drosophila* is associated with a tandem duplication (**FIGURE 6.19**). This dominant X-linked mutation alters the size and shape of the compound eyes, transforming them from large, spherical structures into narrow bars. In the 1930s C. B. Bridges analyzed X chromosomes carrying the *Bar* mutation and found that the 16A region, which apparently contains a gene for eye shape, had been tandemly duplicated. Tandem triplications of 16A were also observed, and in these cases the compound eye was extremely small—a phenotype referred to as double-bar. The severity of the mutant eye phenotype is therefore related to the number of copies of the 16A region—clear evidence for the importance of gene dosage in determining a phenotype. Many other tandem duplications have been found in *Drosophila*, where polytene chromosome analysis makes their detection relatively easy. Today, molecular techniques have made it possible to detect very small tandem duplications in a wide variety of organisms. For example, the genes that encode the hemoglobin proteins have been tandemly duplicated in mammals (Chapter 20). Gene duplications appear to be relatively common and provide a significant source of variation for evolution.

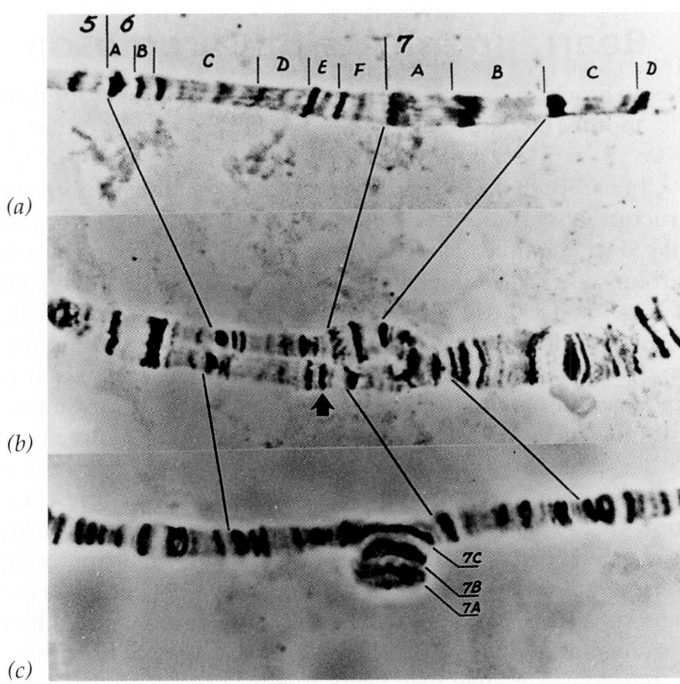

Figure 6.18 ► Polytene chromosomes showing (*a*) the normal structure of regions 6 and 7 in the middle of the *Drosophila* X chromosome, (*b*) a heterozygote with a deletion of region 6F-7C in one of the chromosomes (arrow), and (*c*) an X chromosome showing a reverse tandem duplication of region 6F–7C. In (*b*) the prominent bands in regions 7A and 7C are present in the upper chromosome but absent in the lower one, indicating that the lower chromosome has undergone a deletion. In (*c*) the duplicated sequence reads 7C, 7B, 7A, 7A, 7B, 7C from left to right.

KEY POINTS

- In a trisomy, such as Down Syndrome in humans, three copies of a chromosome are present; in a monosomy, such as Turner Syndrome in humans, only one copy of a chromosome is present.
- Aneuploidy may involve the deletion or duplication of a chromosome segment.

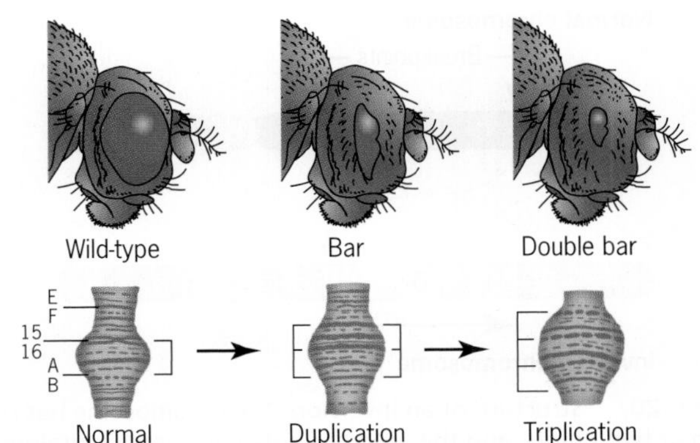

Figure 6.19 ► Effects of duplications for region 16A of the X chromosome on the size of the eyes in *Drosophila*.

Rearrangements of Chromosome Structure

A chromosome may become rearranged internally, or it may become joined to another chromosome.

In nature there is considerable variation in the number and structure of chromosomes, even among closely related organisms. For example, *Drosophila melanogaster* has four pairs of chromosomes, including a pair of sex chromosomes, two pairs of large, metacentric autosomes—chromosomes with the centromere in the middle—and a pair of small, dotlike autosomes. *Drosophila virilis*, which is not too distantly related, has a pair of sex chromosomes, four pairs of acrocentric autosomes—chromosomes with the centromere near one end—and a pair of dotlike autosomes. Thus, even in the same genus, species can have different chromosome arrangements. These differences imply that over evolutionary time, segments of the genome are rearranged. In fact, the observation that chromosome rearrangements can be found as variants within a single species suggests that the genome is continuously being reshaped. These rearrangements may change the position of a segment within a chromosome, or they may bring together segments from different chromosomes. In either case, the order of the genes is altered. Cytogeneticists have identified many kinds of chromosome rearrangements. Here we consider two types: inversions, which involve a switch in the orientation of a segment within a chromosome, and translocations, which involve the fusion of segments from different chromosomes. In humans, chromosome rearrangements have a medical significance because some of them are involved in predisposing individuals to develop certain types of cancer. We consider these kinds of rearrangements, and their connection to cancer, in Chapter 22.

INVERSIONS

An **inversion** occurs when a chromosome segment is detached, flipped around 180°, and reattached to the rest of the chromosome; as a result, the order of the segment's genes is reversed (**FIGURE 6.20**). Such rearrangements can be induced in the laboratory by X-irradiation, which breaks chromosomes into pieces. Sometimes the pieces reattach, but in the process a segment gets turned around and an inversion occurs. There is also evidence that inversions are produced naturally through the activity of transposable elements—DNA sequences capable of moving from one chromosomal position to another (Chapter 18). Sometimes, in the course of moving, these elements break a chromosome into pieces and the pieces reattach in an aberrant way, producing an inversion. Inversions may also be created by the reattachment of chromosome fragments generated by mechanical shear, perhaps as a result of chromosome entanglement within the nucleus. No one really knows what fraction of naturally occurring inversions is caused by each of these mechanisms.

Cytogeneticists distinguish between two types of inversions based on whether or not the inverted segment includes the chromosome's centromere (**FIGURE 6.21**). **Pericentric** inversions include the centromere, whereas **paracentric** inversions do not. The consequence is that a pericentric inversion may change the relative lengths of the two arms of the chromosome, whereas a paracentric inversion has no such effect. Thus, if an acrocentric chromosome acquires an inversion with a breakpoint in each of the chromosome's arms (that is, a pericentric inversion), it can be transformed into a metacentric chromosome. However, if an acrocentric chromosome acquires an inversion in which both of the breaks are in the chromosome's

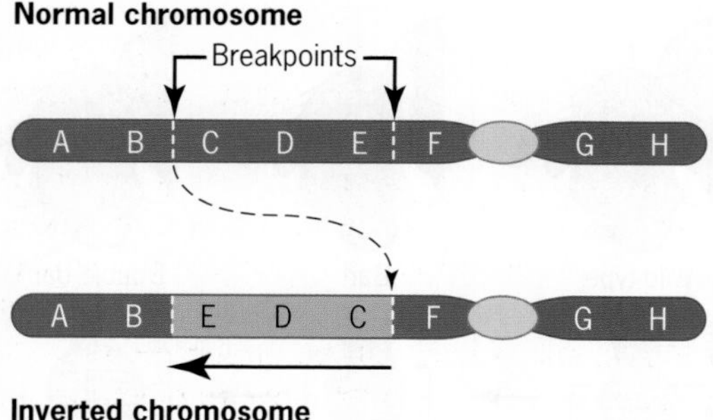

Figure 6.20 ▶ Structure of an inversion. The chromosome has been broken at two points, and the segment between them (containing regions C, D, and E) has been inverted.

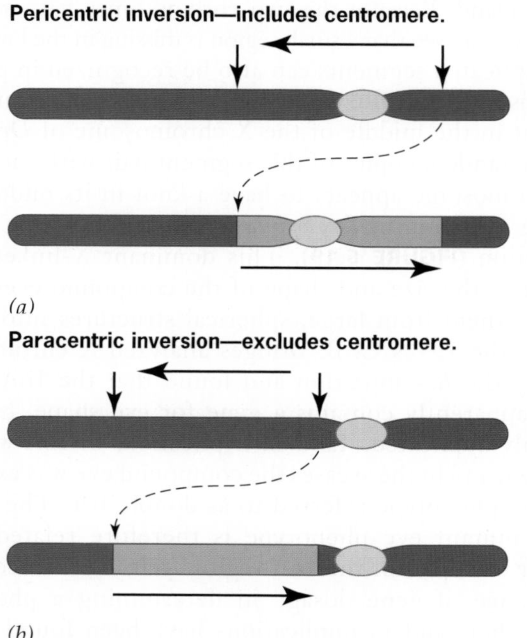

Figure 6.21 ▶ Pericentric and paracentric inversions. A pericentric inversion (*a*) changes the size of the chromosome arms because the centromere is included within the inversion. By contrast, a paracentric inversion (*b*) does not because it excludes the centromere.

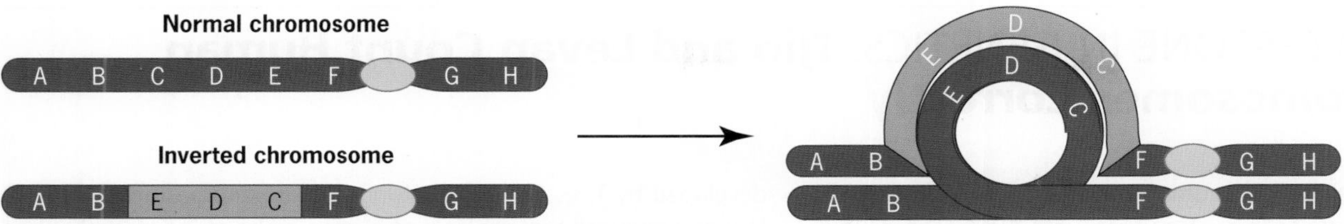

Figure 6.22 ▶ Pairing between normal and inverted chromosomes.

long arm (that is, a paracentric inversion), the morphology of the chromosome will not be changed. Hence, with the use of standard cytological methods, pericentric inversions are much easier to detect than paracentric inversions.

An individual in which one chromosome is inverted but its homologue is not is said to be an inversion heterozygote. During meiosis, the inverted and noninverted chromosomes pair point-for-point along their length. However, because of the inversion, the chromosomes must form a loop to allow for pairing in the region where their genes are in reversed order. **FIGURE 6.22** shows this pairing configuration; only one of the chromosomes is looped, and the other conforms around it. In practice, either the inverted or noninverted chromosome can form the loop to maximize pairing between them. However, near the ends of the inversion, the chromosomes are stretched, and there is a tendency for some de-synapsis. We consider the genetic consequences of inversion heterozygosity in Chapter 7.

TRANSLOCATIONS

A **translocation** occurs when a segment from one chromosome is detached and reattached to a different (that is, nonhomologous) chromosome. The genetic significance is that genes from one chromosome are transferred to another.

When pieces of two nonhomologous chromosomes are interchanged without any net loss of genetic material, the event is referred to as a *reciprocal translocation.* **FIGURE 6.23*a*** shows a reciprocal translocation between two large autosomes. These chromosomes have interchanged pieces of their right arms. During meiosis, these translocated chromosomes would be expected to pair with their untranslocated homologues in a cruciform, or crosslike, pattern (**FIGURE 6.23*b***). The two translocated chromosomes face each other opposite the center of the cross, and the two untranslocated chromosomes do likewise; to maximize pairing, the translocated and untranslocated chromosomes alternate with each other, forming the arms of the cross. This pairing configuration is diagnostic of a translocation heterozygote. Cells in which the translocated chromosomes are homozygous do not form a cruciform pattern. Instead, each of the translocated chromosomes pairs smoothly with its structurally identical partner.

Because cruciform pairing involves four centromeres, which may or may not be coordinately distributed to opposite poles in the first meiotic division, chromosome disjunction in translocation heterozygotes is a somewhat uncertain process, prone to produce aneuploid gametes. Altogether there are

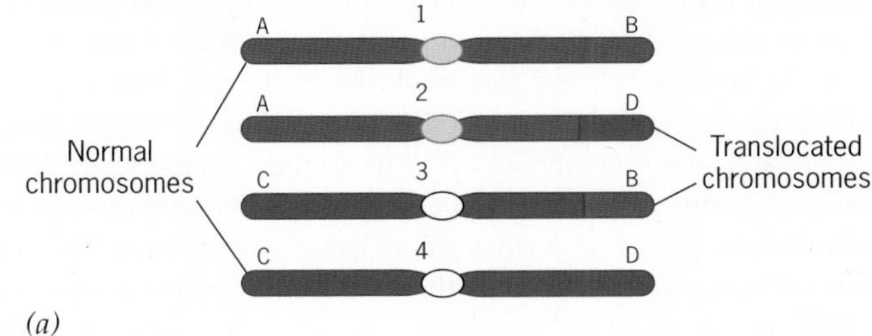

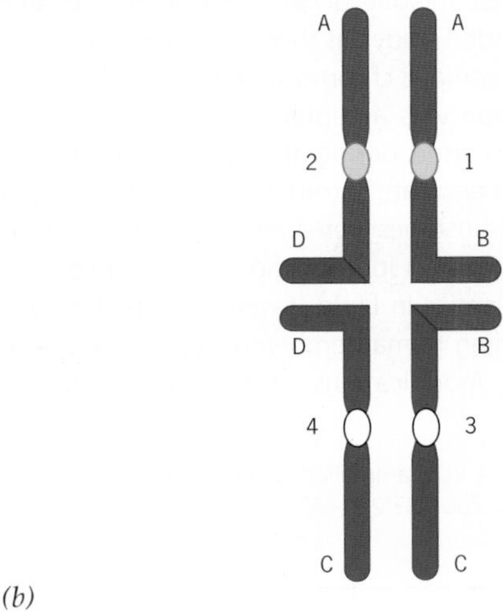

Figure 6.23 ▶ Structure (*a*) and pairing behavior (*b*) of a reciprocal translocation between chromosomes. In (*b*) pairing occurs during the prophase of meiosis I, after the chromosomes have been duplicated.

three possible disjunctional events, illustrated in **FIGURE 6.24**. This simplified figure shows only one of the two sister chromatids of each chromosome. In addition, each of the centromeres is labeled to keep track of chromosome movements; the two white centromeres are homologous (that is, derived from the same chromosome pair), as are the two gray centromeres.

If centromeres 2 and 4 move to the same pole, forcing 1 and 3 to the opposite pole, all the resulting gametes will be aneuploid—because some chromosome segments will be deficient for genes, and others will be duplicated (**FIGURE 6.24*a***). Similarly, if centromeres 1 and 2 move to one pole and 3 and 4 to the other, only aneuploid gametes will be produced (**FIGURE 6.24*b***). Each of these cases is referred to as *adjacent disjunction* because centromeres that were next to each other in

▶ A MILESTONE IN GENETICS: Tjio and Levan Count Human Chromosomes Correctly

Each diploid human cell has 46 chromosomes—22 pairs of autosomes and a pair of sex chromosomes. However, until 1956, the diploid chromosome number was thought to be 48. This incorrect estimate for the number of human chromosomes first entered the scientific literature in 1923 when Theophilus Painter, a gifted cytologist at the University of Texas, reported his analysis of dividing cells in testes that had been removed from three inmates of the Texas State Insane Asylum. In those days, inmates could be castrated if their sexual behavior was considered abnormal. The testicular tissue was preserved in alcohol and later prepared for cytological study. Painter's procedure involved embedding the dissected tissues in paraffin and then slicing off ultrathin sections with an instrument called a microtome. The cells in the sections were stained with iron haematoxylin, a dye that provides good contrast between the chromosomes and the surrounding cytoplasm. Painter studied spermatogonial cells, both in mitosis and in the first meiotic division. Based on his analysis, he concluded "that extended study has shown that 48 is, in all probability, the correct diploid or somatic chromosome number."[1]

Painter's conclusion was accepted for many years because the cytological techniques used during the first half of the twentieth century did not permit accurate chromosome counts, especially when large numbers of chromosomes were present. Painter's counting error was corrected in 1956 when Joe Hin Tjio and Albert Levan, working at the Institute of Genetics in Lund, Sweden, published analyses of chromosomes in dividing human embryonic cells. The previous year, Levan had tried out modifications of techniques that had been developed by T. C. Hsu for studying human chromosomes. In 1952 Hsu had reported that explants of human embryonic tissues that had accidentally been washed with a hypotonic solution yielded mitotic cells that could be analyzed easily.[2] The hypotonic solution caused the cells to take up water, swelling them so that when they were squashed on a microscope slide, their chromosomes were dispersed. Tjio and Levan adopted Hsu's techniques to study chromosomes in dividing fibroblasts from tissue explants obtained from four human embryos. The explants were cultured for a few days in fluid obtained from cow embryos, and then colchicine, a drug that poisons the mitotic spindle, was added to trap dividing cells in the metaphase of mitosis. After treatment with a hypotonic solution for one or two minutes, the cells were stained with acetic orcein, a dye derived from lichens, and then squashed on slides. With this technique, accurate chromosome counts could be obtained from many cells; **FIGURE 1** shows an example. Tjio and Levan reported that they "were surprised to find that the chromosome number 46 predominated in the tissue cultures from all

[1]Painter, T. 1923. Studies in mammalian spermatogenesis, II. The spermatogenesis of man. *J. Exp. Zool.* 37:291–338.

[2]Hsu. T. C. 1952. Mammalian chromosomes in vitro. I. The karyotype of man. *J. Hered.* 43:167–172. Hsu's paper had the following addendum: "It was found after this article had been sent to press that the well-spread metaphases and the seemingly C-mitotic anaphases were the results of an accident. Instead of being washed in isotonic saline, the cultures had been washed in hypotonic Tryode solution before fixation. Furthermore, it was found that Dr. Arthur Hughes of the Strangeways Research Laboratory, Cambridge, England, had been carrying on experiments on the effect of hypotonicity upon dividing cells and his findings were almost identical with ours. We owe our sincere thanks to Dr. Hughes for allowing us to read his original manuscript prior to its publication."

the cruciform pattern moved to the same pole. When the centromeres that move to the same pole are from different chromosomes (that is, they are *heterologous*), the disjunction is referred to as adjacent I (**FIGURE 6.24*a***); when the centromeres that move to the same pole are from the same chromosome (that is, they are *homologous*), the disjunction is referred to as adjacent II (**FIGURE 6.24*b***). Another possibility is that centromeres 1 and 4 move to the same pole, forcing 2 and 3 to the opposite pole. This case, called *alternate disjunction*, produces only euploid gametes, although half of them will carry only translocated chromosomes (**FIGURE 6.24*c***).

The production of aneuploid gametes by adjacent disjunction explains why translocation heterozygotes have reduced fertility. When such gametes fertilize a euploid gamete, the resulting zygote will be genetically unbalanced and therefore will be unlikely to survive. In plants, aneuploid gametes are themselves often inviable, especially on the male side, and fewer zygotes are produced. Translocation heterozygotes are therefore characterized by low fertility.

COMPOUND CHROMOSOMES AND ROBERTSONIAN TRANSLOCATIONS

Sometimes one chromosome fuses with its homologue, or two sister chromatids become attached to each other, forming a single genetic unit. A **compound chromosome** can exist stably in a cell as long as it has a single functional centromere; if there are two centromeres, each may move to a different pole during division, pulling the compound chromosome apart. A compound chromosome may also be formed by the union of homologous chromosome segments. For example, the right arms of the two second chromosomes in *Drosophila* might detach from their left arms and fuse at the centromere, creating a compound half-chromosome. Cytogeneticists sometimes

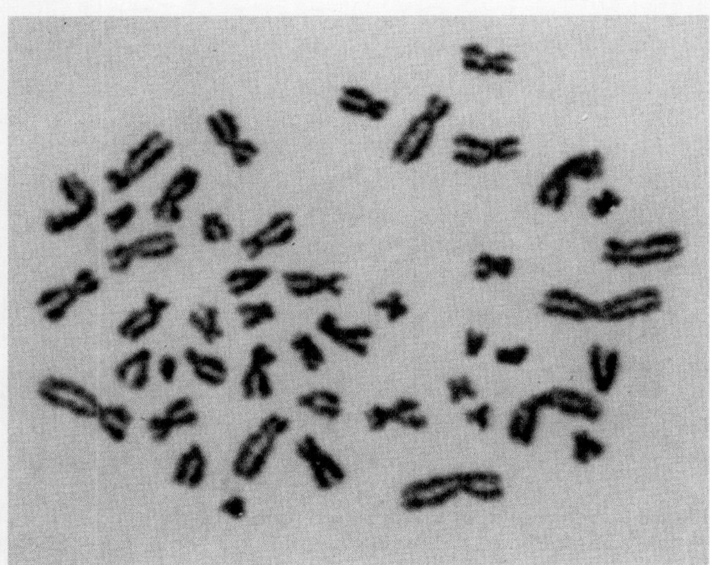

Figure 1 ▶ Metaphase chromosomes in a dividing human fibroblast cultured *in vitro*.

four embryos."[3] However, they were cautious in interpreting their results: "Before a renewed, careful control has been made of the chromosome number in spermatogonial mitoses of man we do not wish to generalize our present findings into a statement that the chromosome number of man is $2n = 46$, but it is hard to avoid the conclusion that this would be the most natural explanation of our observations."[4]

Frank H. Ruddle, a distinguished human geneticist at Yale University, has attempted to explain Painter's counting error.[5] Ruddle has speculated that Painter overestimated the number of chromosomes because in the material he studied, each arm of the largest human chromosome (denoted chromosome number 1) may have appeared to be separate and distinct chromosomes. Iron haematoxylin, the reagent that Painter used, does not stain a region in the middle of this large chromosome well. With the middle weakly stained, the end segments of this chromosome might appear to be unconnected to each other and would therefore be counted as two separate chromosomes. Thus, Painter's counting error might have been due to the peculiarities of haematoxylin staining.

QUESTIONS FOR DISCUSSION

1. Hsu's discovery that treatment with a hypotonic solution helps to spread chromosomes for cytological study was the result of an accident. Can you identify other scientific discoveries that involved "accidents"? Do you think that "luck" plays much of a role in scientific discoveries?
2. The cytological material that Painter studied came from men who had been castrated by official order, and the material that Tjio and Levan studied came from legally aborted embryos. Today castration and abortion are both controversial practices. How is current scientific research affected by moral and ethical controversies?

[3]Tjio, J. H., and A. Levan. 1956. The chromosome number of man. *Hereditas* 42:1–6.

[4]Ibid.

[5]Ruddle, F. H. 2004. Theophilus Painter: First steps toward an understanding of the human genome. *J. Exp. Zool.* 301:375–377.

call this structure an **isochromosome** (from the Greek prefix for "equal"), because its two arms are equivalent. Compound chromosomes differ from translocations in that they involve fusions of homologous chromosome segments. Translocations, by contrast, always involve fusions between nonhomologous chromosomes.

The first compound chromosome was discovered in 1922 by Lillian Morgan, the wife of T. H. Morgan. This compound was formed by fusing the two X chromosomes in *Drosophila*, creating double-X or *attached-X chromosomes*. The discovery was made through genetic experimentation rather than cytological analysis. Lillian Morgan crossed females homozygous for a recessive X-linked mutation to wild-type males. From such a cross, we would ordinarily expect all the daughters to be wild-type and all the sons to be mutant. However, Morgan observed just the opposite: all the daughters were mutant and all the sons were wild-type. Further work established that the X chromosomes in the mutant females had become attached to each other. **FIGURE 6.25** illustrates the genetic significance of this attachment. The attached-X females produce two kinds of eggs, diplo-X and nullo-X, and their mates produce two kinds of sperm, X-bearing and Y-bearing. The union of these gametes in all possible ways produces two kinds of viable progeny: mutant XXY females, which inherit the attached-X chromosomes from their mothers and a Y chromosome from their fathers; and phenotypically wild-type XO males, which inherit an X chromosome from their fathers and no sex chromosome from their mothers. Because the Y chromosome is needed for fertility, these XO males are sterile. Lillian Morgan was able to propagate the attached-X chromosomes by backcrossing XXY females to wild-type XY males from another stock. Because the sons of this cross inherited a Y chromosome from their mothers, they were fertile and could be crossed to their XXY sisters to establish a stock in which

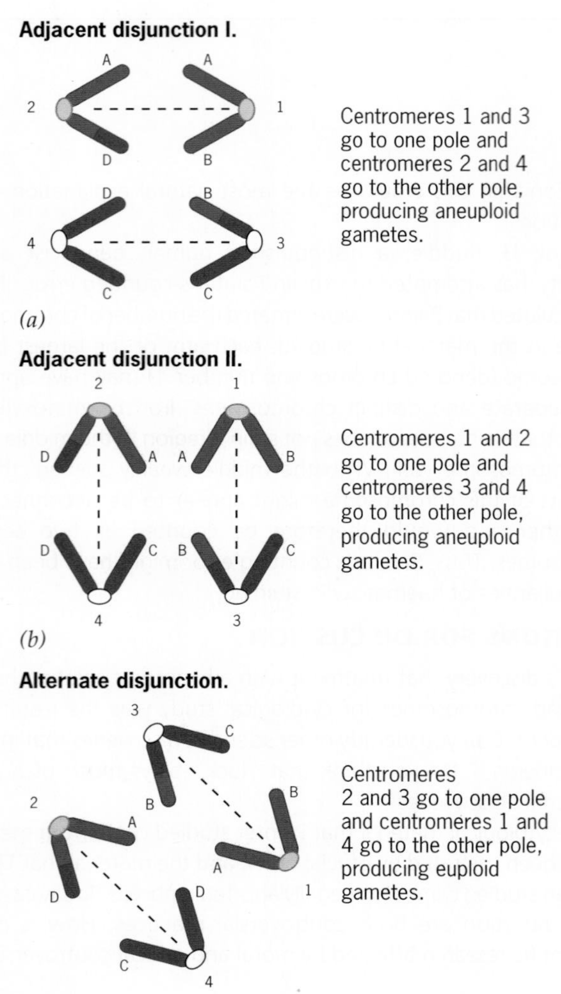

Figure 6.24 ▶ Types of disjunction in a translocation heterozygote during meiosis I. For simplicity, only one sister chromatid of each duplicated chromosome is shown. (*a*) One form of adjacent disjunction in which homologous centromeres go to opposite poles during anaphase. (*b*) Another form of adjacent disjunction in which homologous centromeres go to the same pole during anaphase. (*c*) Alternate disjunction in which homologous centromeres go to opposite poles during anaphase.

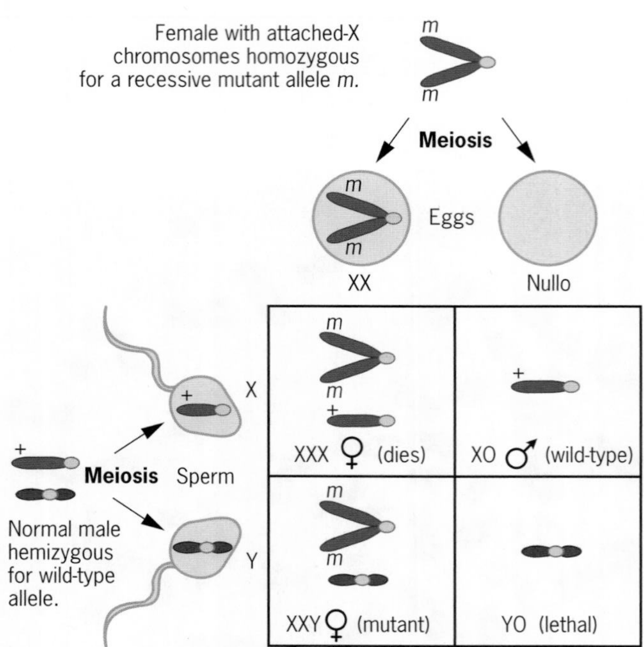

Figure 6.25 ▶ Results of a cross between a normal male and a female with attached-X chromosomes.

the attached-X chromosomes were permanently maintained in the female line.

Nonhomologous chromosomes can also fuse at their centromeres, creating a structure called a **Robertsonian translocation** (**FIGURE 6.26**), named for the cytologist F. W. Robertson. For example, if two acrocentric chromosomes fuse, they will produce a metacentric chromosome; the tiny short arms of the participating chromosomes are simply lost in this process. Apparently, such chromosome fusions have occurred quite often in the course of evolution.

Chromosomes can also fuse end-to-end to form a structure with two centromeres. If one of the centromeres is inactivated, the chromosome fusion will be stable. Such a fusion evidently occurred in the evolution of our own species. Human chromosome 2, which is a metacentric, has arms that correspond to two different acrocentric chromosomes in the genomes of the great apes. Detailed cytological analysis has shown that the ends of the short arms of these two chromosomes apparently fused to create human chromosome 2.

KEY POINTS

- An inversion reverses the order of genes in a segment of a chromosome.
- A translocation interchanges segments between two nonhomologous chromosomes.
- Compound chromosomes result from the fusion of homologous chromosomes, or from the fusion of the arms of homologous chromosomes.
- Robertsonian translocations result from the fusion of nonhomologous chromosomes.

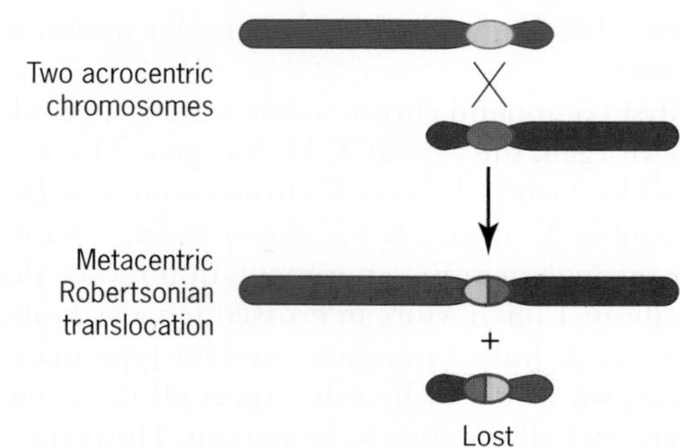

Figure 6.26 ▶ Formation of a metacentric Robertsonian translocation by exchange between two nonhomologous acrocentric chromosomes.

▶ Basic Exercises

ILLUSTRATE BASIC GENETIC ANALYSIS

1. A species has two pairs of chromosomes, one long and the other short. Draw the chromosomes at metaphase of mitosis. Show each chromatid. Are homologous chromosomes paired?

Answer: Mitotic metaphase in this species would look something like the picture below. Because each chromosome is duplicated, it consists of two sister chromatids. However, because the picture shows mitosis rather than meiosis, homologous chromosomes are not paired.

2. Plant species A shows 10 bivalents of chromosomes at metaphase of meiosis I; plant species B shows 14 bivalents at this stage. The two species are crossed, and the chromosomes in the offspring are doubled. (a) How many bivalents will be seen at metaphase of meiosis I in the offspring? (b) Is the offspring expected to be fertile or sterile?

Answer: (a) The offspring is a composite of the chromosomes of the two parents. In species A, the basic chromosome number is 10; in species B, it is 14. The basic chromosome number in the offspring is therefore 10 + 14 = 24, and with the chromosomes having been doubled, this is the number of bivalents that should be seen at metaphase of meiosis I. (b) The offspring is an allotetraploid and should therefore be fertile.

3. What are the karyotypes of (a) a female with Down syndrome, (b) a male with trisomy 13, (c) a female with Turner syndrome, (d) a male with Klinefelter syndrome, (e) a male with a deletion in the short arm of chromosome 11?

Answer: (a) 47, XX, +21; (b) 47, XY, +13; (c) 45, X; (d) 47, XXY; (e) 46, XY del (11p).

4. What kind of pairing configuration would be seen in prophase of meiosis I in (a) an inversion heterozygote, (b) a translocation heterozygote?

Answer: (a) Loop configuration, (b) cross configuration.

▶ Testing Your Knowledge

INTEGRATE DIFFERENT CONCEPTS AND TECHNIQUES

1. A *Drosophila* geneticist has obtained females that carry attached-X chromosomes homozygous for a recessive mutation (y) that causes the body to be yellow instead of gray. In one experiment, she crosses some of these females to ordinary wild-type males, and in another, she crosses these females to wild-type males that have their X and Y chromosomes attached to each other; that is, they carry a compound XY chromosome. Predict the phenotypes of the progeny from these two crosses and indicate which, if any, will be sterile.

Answer: To predict the phenotypes of the progeny, we need to know their genotypes. The easiest way to determine these genotypes is to diagram the kinds of zygotes produced by each cross.

First, we consider the cross between the yellow-bodied attached-X females and the ordinary wild-type males. The females produce two kinds of gametes, XX and nullo. The males also produce two kinds of gametes, X and Y. When these are combined in all possible ways, four types of zygotes are produced; however, only two types are viable. The XXY zygotes will develop into yellow-bodied females—like their mothers except that they carry a Y chromosome—and the XO zygotes will develop into gray-bodied males—like their fathers except that they lack a Y chromosome. The extra Y chromosome in the females will have no effect on fertility, but the missing Y chromosome in the males will cause them to be sterile.

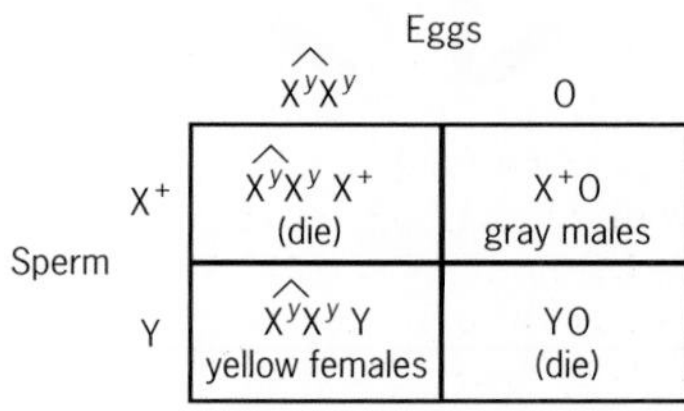

		Eggs	
		$\widehat{X^yX^y}$	0
Sperm	X^+	$\widehat{X^yX^y}\,X^+$ (die)	X^+O gray males
	Y	$\widehat{X^yX^y}\,Y$ yellow females	YO (die)

Now we consider the cross between the yellow-bodied attached-X females and the males with a compound XY chromosome. Both sexes produce two kinds of gametes—for the females, the same as above, and for the males, either XY or nullo. When these are united in all possible ways, we find that two types of zygotes will be viable: yellow-bodied females with attached-X chromosomes and gray-bodied males with a compound XY chromosome. Both types of these viable progeny will be fertile.

		Eggs	
		$\widehat{X^y\,X^y}$	0
Sperm	$\widehat{X^+Y}$	$\widehat{X^y\,X^y}\,\widehat{X^+Y}$ (die)	$\widehat{X^+Y}\,O$ gray males
	0	$\widehat{X^y\,X^y}\,O$ yellow females	O O (die)

2. A phenotypically normal man carries a translocated chromosome that contains the entire long arm of chromosome 14, part of the short arm of chromosome 14, and most of the long arm of chromosome 21:

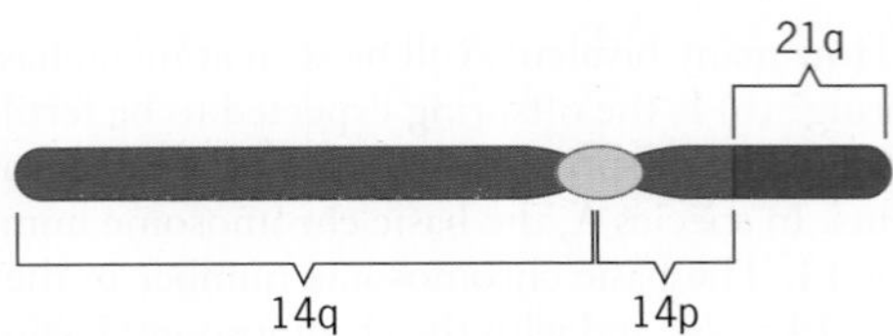

The man also carries a normal chromosome 14 and a normal chromosome 21. If he marries a cytologically (and phenotypically) normal woman, is there any chance that the couple will produce phenotypically abnormal children?

Answer: Yes, the couple could produce children with Down syndrome as a result of meiotic segregation in the cytologically abnormal man. During meiosis in this man, the translocated chromosome, T(14, 21), will synapse with the normal chromosomes 14 and 21, forming a trivalent. Disjunction from this trivalent will produce six different types of sperm, four of which are aneuploid.

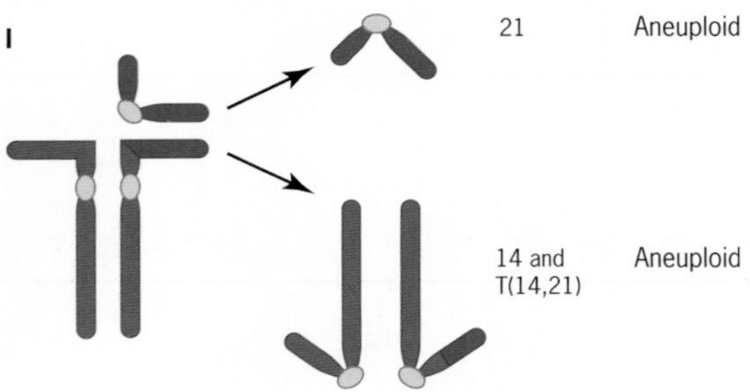

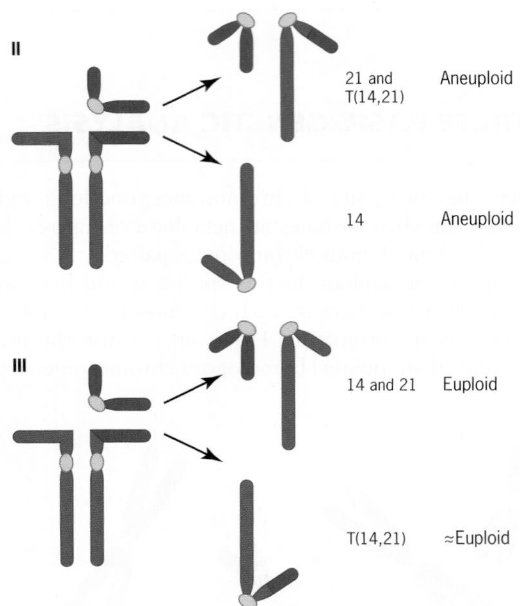

Fertilization of an egg containing one chromosome 14 and one chromosome 21 by any of the aneuploid sperm will produce an aneuploid zygote as shown in the following table. Although trisomy or monosomy for chromosome 14 and monosomy for chromosome 21 are all lethal conditions, trisomy for chromosome 21 is not. Thus, it is possible for the couple to give birth to a child with Down syndrome.

Disjunction	Sperm	Zygote	Condition	Outcome
I	21	14, 21, 21	monosomy 14	dies
	14, T(14, 21)	14, 14, T(14, 21), 21	trisomy 14	dies
II	14	14, 14, 21	monosomy 21	dies
	T(14, 21), 21	14, T(14, 21), 21, 21	trisomy 21	Down
III	14, 21	14, 14, 21, 21	euploid	normal
	T(14, 21)	14, T(14, 21), 21	≈euploid	normal

▶ Questions and Problems

ENHANCE UNDERSTANDING AND DEVELOP ANALYTICAL SKILLS

6.1 During meiosis, why do some tetraploids behave more regularly than triploids?

6.2 GO The following table presents chromosome data on four species of plants and their F_1 hybrids:

Meiosis I Metaphase

Species or F_1 Hybrid	Root Tip Chromosome Number	Number of Bivalents	Number of Univalents
A	20	10	0
B	20	10	0
C	10	5	0
D	10	5	0
A × B	20	0	20
A × C	15	5	5
A × D	15	5	5
C × D	10	0	10

(a) Deduce the chromosomal origin of species A.

(b) How many bivalents and univalents would you expect to observe at meiotic metaphase I in a hybrid between species C and species B?

(c) How many bivalents and univalents would you expect to observe at meiotic metaphase I in a hybrid between species D and species B?

6.3 Identify the sexual phenotypes of the following genotypes in human beings: XX, XY, XO, XXX, XXY, XYY.

6.4 In human beings, a cytologically abnormal chromosome 22, called the "Philadelphia" chromosome because of the city in which it was discovered, is associated with chronic leukemia. This chromosome is missing part of its long arm. How would you denote the karyotype of an individual who had 46 chromosomes in his somatic cells, including one normal 22 and one Philadelphia chromosome?

6.5 A *Drosophila* female homozygous for a recessive X-linked mutation causing yellow body was crossed to a wild-type male. Among the progeny, one fly had sectors of yellow pigment in an otherwise gray body. These yellow sectors were distinctly male, whereas the gray areas were female. Explain the peculiar phenotype of this fly.

6.6 GO In humans, Hunter syndrome is known to be an X-linked trait with complete penetrance. In family A, two phenotypically normal parents have produced a normal son, a *daughter* with Hunter and Turner syndromes, and a son with Hunter syndrome. In family B, two phenotypically normal parents have produced two phenotypically normal daughters and a *son* with Hunter and Klinefelter syndromes. In family C, two phenotypically normal parents have produced a phenotypically normal daughter, a *daughter* with Hunter syndrome, and a son with Hunter syndrome. For each family, explain the origin of the child indicated in italics.

6.7 A plant species A, which has seven chromosomes in its gametes, was crossed with a related species B, which has nine. The hybrids were sterile, and microscopic observation of their pollen mother cells showed no chromosome pairing. A section from one of the hybrids that grew vigorously was propagated vegetatively, producing a plant with 32 chromosomes in its somatic cells. This plant was fertile. Explain.

6.8 The *Drosophila* fourth chromosome is so small that flies monosomic or trisomic for it survive and are fertile. Several genes, including *eyeless* (*ey*), have been located on this chromosome. If a cytologically normal fly homozygous for a recessive eyeless mutation is crossed to a fly monosomic for a wild-type fourth chromosome, what kinds of progeny will be produced, and in what proportions?

6.9 A woman with X-linked color blindness and Turner syndrome had a color-blind father and a normal mother. In which of her parents did nondisjunction of the sex chromosomes occur?

6.10 If nondisjunction of chromosome 18 occurs in the division of a secondary oocyte in a human female, what is the chance that a mature egg derived from this division will receive three number 18 chromosomes?

6.11 Although XYY men are phenotypically normal, would they be expected to produce more children with sex chromosome abnormalities than XY men? Explain.

6.12 In a *Drosophila* salivary chromosome, the bands have a sequence of 1 2 3 4 5 6 7 8. The homologue with which this chromosome is synapsed has a sequence of 1 2 3 6 5 4 7 8. What kind of chromosome change has occurred? Draw the synapsed chromosomes.

6.13 Other chromosomes have sequences as follows: (a) 1 2 5 6 7 8; (b) 1 2 3 4 4 5 6 7 8; (c) 1 2 3 4 5 8 7 6. What kind of chromosome change is present in each? Illustrate how these chromosomes would pair with a chromosome whose sequence is 1 2 3 4 5 6 7 8.

6.14 In *Drosophila*, the genes *bw* and *st* are located on chromosomes 2 and 3, respectively. Flies homozygous for *bw* mutations have brown eyes, flies homozygous for *st* mutations have scarlet eyes, and flies homozygous for *bw* and *st* mutations have white eyes. Doubly heterozygous males were mated individually to homozygous *bw; st* females. All but one of the matings produced four classes of progeny: wild-type, and brown-, scarlet-, and white-eyed. The single exception produced only wild-type and white-eyed progeny. Explain the nature of this exception.

6.15 One chromosome in a plant has the sequence A B C D E F, and another has the sequence M N O P Q R. A reciprocal translocation between these chromosomes produced the following arrangement: A B C P Q R on one chromosome and M N O D E F on the other. Illustrate how these translocated chromosomes would pair with their normal counterparts in a heterozygous individual during meiosis.

6.16 A man has attached chromosomes 21. If his wife is cytologically normal, what is the chance their first child will have Down syndrome?

6.17 A yellow-bodied *Drosophila* female with attached-X chromosomes was crossed to a white-eyed male. Both of the parental phenotypes are caused by X-linked recessive mutations. Predict the phenotypes of the progeny.

6.18 Distinguish between a compound chromosome and a Robertsonian translocation.

6.19 A phenotypically normal boy has 45 chromosomes, but his sister, who has Down syndrome, has 46. Suggest an explanation for this paradox.

6.20 GO A male mouse that is heterozygous for a reciprocal translocation between the X chromosome and an autosome is crossed to a female mouse with a normal karyotype. The autosome involved in the translocation carries a gene responsible for coloration of the fur. The allele on the male's translocated autosome is wild-type, and the allele on its nontranslocated autosome is mutant; however, because the wild-type allele is dominant to the mutant allele, the male's fur is wild-type (dark in color). The female mouse has light color in her fur because she is homozygous for the mutant allele of the color-determining gene. When the offspring of the cross are examined, all the males have light fur and all the females have patches of light and dark fur. Explain these peculiar results.

6.21 Analysis of the polytene chromosomes of three populations of *Drosophila* has revealed three different banding sequences in a region of the second chromosome:

Population	Banding Sequence
P1	1 2 3 4 5 6 7 8 9 10
P2	1 2 3 9 8 7 6 5 4 10
P3	1 2 3 9 8 5 6 7 4 10

Explain the evolutionary relationships among these populations.

6.22 In *Drosophila*, the autosomal genes *cinnabar* (*cn*) and *brown* (*bw*) control the production of brown and red eye pigments, respectively. Flies homozygous for *cinnabar* mutations have bright red eyes, flies homozygous for *brown* mutations have brown eyes, and flies homozygous for mutations in both of these genes have white eyes. A male homozygous for mutations in the *cn* and *bw* genes has bright red eyes because a small duplication that carries the wild-type allele of *bw* (bw^+) is attached to the Y chromosome. If this male is mated to a karyotypically normal female that is homozygous for the *cn* and *bw* mutations, what types of progeny will be produced?

6.23 The following diagram shows two pairs of chromosomes in the karyotypes of a man, a woman, and their child. The man and the woman are phenotypically normal, but the child (a boy) suffers from a syndrome of abnormalities, including poor motor control and severe mental impairment. What is the genetic basis of the child's abnormal phenotype? Is the child hyperploid or hypoploid for a segment in one of his chromosomes?

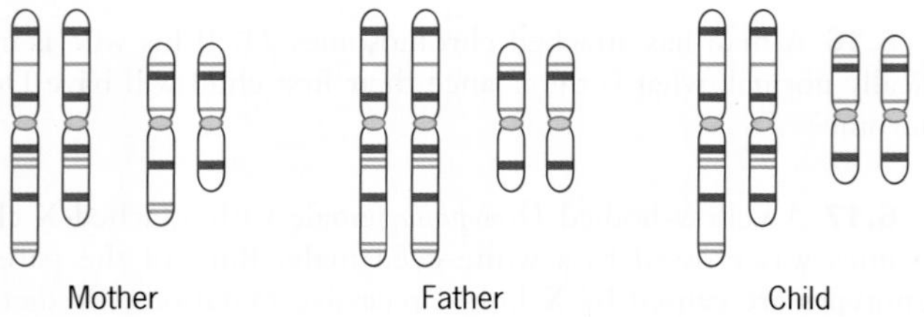

6.24 In *Drosophila*, vestigial wing (*vg*), hairy body (*h*), and eyeless (*ey*) are recessive mutations on chromosomes 2, 3, and 4, respectively. Wild-type males that had been irradiated with X rays were crossed to triply homozygous recessive females. The F_1 males (all phenotypically wild-type) were then testcrossed to triply homozygous recessive females. Most of the F_1 males produced eight classes of progeny in approximately equal proportions, as would be expected if the *vg*, *h*, and *ey* genes assort independently. However, one F_1 male produced only four classes of offspring, each approximately one-fourth of the total: (1) wild-type; (2) eyeless; (3) vestigial, hairy; and (4) vestigial, hairy, eyeless. What kind of chromosome aberration did the exceptional F_1 male carry, and which chromosomes were involved?

6.25 Cytological examination of the sex chromosomes in a man has revealed that he carries an insertional translocation. A small segment has been deleted from the Y chromosome and inserted into the short arm of the X chromosome; this segment contains the gene responsible for male differentiation (*SRY*). If this man marries a karyotypically normal woman, what types of progeny will the couple produce?

6.26 Each of six populations of *Drosophila* in different geographic regions had a specific arrangement of bands in one of the large autosomes:

(a) 12345678
(b) 12263478
(c) 15432678
(d) 14322678
(e) 16223478
(f) 154322678

Assume that arrangement (a) is the original one. In what order did the other arrangements most likely arise, and what type of chromosomal aberration is responsible for each change?

▸ Genomics on the Web

at http://www.ncbi.nlm.nih.gov/

1. Many crop plants are polyploid. What progress has been made in sequencing the polyploid genomes of soybean (*Glycine max*), wheat (*Triticum aestivum*), and potato (*Solanum tuberosum*)?

Hint: At the web site, under Genome Resources, click on Plants. Find each species and read about ongoing DNA sequencing efforts.

2. When triplicated, chromosome 21, the smallest of the autosomes in the human genome, causes Down syndrome. How many nucleotide pairs are present in this chromosome? How many genes does it contain?

Hint: Use Map Viewer to find chromosome 21 and then determine its size and gene content.

3. The gene for amyloid precursor protein is located on human chromosome 21. This protein appears to play an important role in the etiology of Alzheimer's disease. Locate the *APP* gene on the ideogram of human chromosome 21. In what band does it lie?

Hint: Search for APP using the "Find in This View" function. Click on the highlighted gene name to find more information about it.

4. Chromosome 21 as well as a few other chromosomes in the human genome have secondary constrictions as well as a primary constriction, which is situated at the centromere. The material distal to the secondary constriction—that is, going away from the centromere toward the nearest end of the chromosome—is called a satellite. Find the secondary constriction and the satellite on the ideogram of chromosome 21.

5. Secondary constrictions on some chromosomes contain genes for ribosomal RNA. Is this true for human chromosome 21?

Hint: Use the Map Viewer function to examine the ideogram of chromosome 21. Search for ribosomal RNA genes using the "Find in This View" function.

Chapter 7 Linkage, Crossing Over, and Chromosome Mapping in Eukaryotes

CHAPTER OUTLINE

- Linkage, Recombination, and Crossing Over
- Chromosome Mapping
- Cytogenetic Mapping
- Tetrad Analysis in Fungi
- Linkage Analysis in Humans
- Recombination and Evolution

Linkage between genes was first discovered in experiments with sweet peas.

The World's First Chromosome Map

The modern picture of chromosome organization emerged from a combination of genetic and cytological studies. T. H. Morgan laid the foundation for these studies when he demonstrated that the gene for *white* eyes in *Drosophila* was located on the X chromosome. Soon afterward Morgan's students showed that other genes were X-linked, and eventually they were able to locate each of these genes on a map of the chromosome. This map was a straight line, and each gene was situated at a particular point, or locus, on it (**FIGURE 7.1**). The structure of the map therefore implied that a chromosome was simply a linear array of genes.

The procedure for mapping chromosomes was invented by Alfred H. Sturtevant, an undergraduate working in Morgan's laboratory. One night in 1911 Sturtevant put aside his algebra homework in order to evaluate some experimental data. Before the sun rose the next day, he had constructed the world's first chromosome map. How was Sturtevant able to determine the map locations of individual genes? No microscope was powerful enough to see genes, nor was any measuring device accurate enough to obtain the distances between them. In fact, Sturtevant did not use any sophisticated instruments in his work. Instead, he relied completely on the analysis of data from experimental crosses with *Drosophila*. His method was simple and elegant, and exploited a phenomenon that regularly occurs during meiosis. This methodology laid the foundation for all subsequent efforts to study the organization of genes in chromosomes.

Linkage, Recombination, and Crossing Over

Genes that are on the same chromosome travel through meiosis together; however, alleles of chromosomally linked genes can be recombined by crossing over.

Sturtevant based his mapping procedure on the principle that genes on the same chromosome should be inherited together. Because such genes are physically attached to the same structure, they should travel as a unit through meiosis. This phenomenon is called **linkage.** The early geneticists were unsure about the nature of linkage, but some of them, including Morgan and his students, thought that genes were attached to one another much like beads on a string. Thus, these researchers clearly had a linear model of chromosome organization in mind.

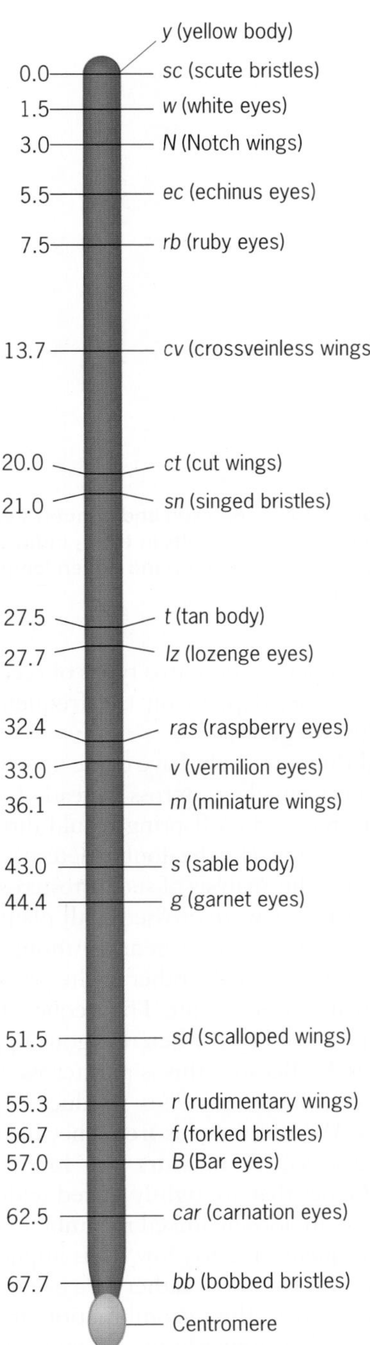

Figure 7.1 ► A map of genes on the X chromosome of *Drosophila melanogaster.*

The early geneticists also knew that linkage was not absolute. Their experimental data demonstrated that genes on the same chromosome could be separated as they went through meiosis and that new combinations of genes could be formed. However, this phenomenon, called **recombination,** was difficult to explain by simple genetic theory.

One hypothesis was that during meiosis, when homologous chromosomes paired, a physical exchange of material separated and recombined genes. This idea was inspired by the cytological observation that chromosomes could be seen in pairing configurations that suggested they had switched pieces with each other. At the switch points, the two homologues were crossed over, as if each had been broken and then reattached to its partner. A crossover point was called a **chiasma** (plural, **chiasmata**), from the Greek word meaning "cross." Geneticists began to use the term *crossing over* to describe the process that created the chiasmata—that is, the actual process of exchange between paired chromosomes. They considered recombination—the separation of linked genes and the formation of new gene combinations—to be a result of the physical event of crossing over.

EARLY EVIDENCE FOR LINKAGE AND RECOMBINATION

Some of the first evidence for linkage came from experiments performed by W. Bateson and R. C. Punnett (**FIGURE 7.2**). These researchers crossed varieties of sweet peas that differed in two traits, flower color and pollen length. Plants with red flowers and long pollen grains were crossed to plants with white flowers and short pollen grains. All the F_1 plants had red flowers and long pollen grains, indicating that the alleles for these two phenotypes were dominant. When the F_1 plants were self-fertilized, Bateson and Punnett observed a peculiar distribution of phenotypes among the offspring. Instead of the 9:3:3:1 ratio expected for two independently assorting genes, they obtained a ratio of 24.3:1.1:1:7.1. We can see the extent of the disagreement between the observed results and the expected results at the bottom of **FIGURE 7.2**. Among the 803 F_2 plants that were examined, the classes that resembled the original parents (called the parental classes) are significantly overrepresented and the two other (nonparental) classes are significantly underrepresented. For such obvious discrepancies, it hardly seems necessary to calculate a chi-square statistic to test the

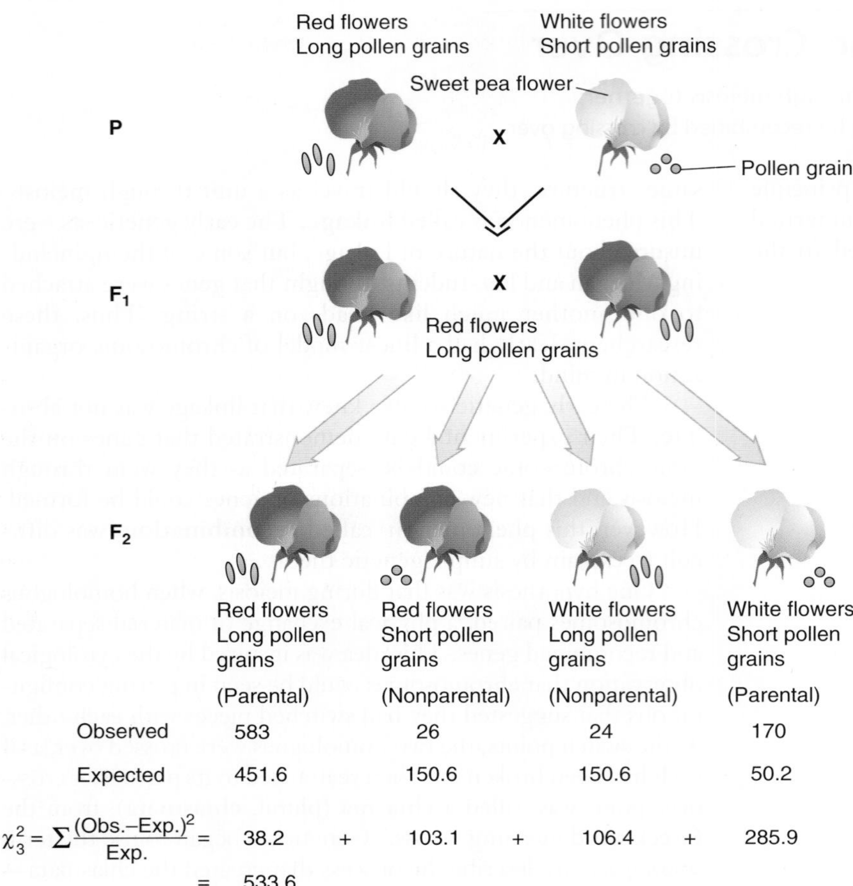

Figure 7.2 ▶ Bateson and Punnett's experiment with sweet peas. The results in the F_2 indicate that the genes for flower color and pollen length do not assort independently.

hypothesis that the two traits, flower color and pollen grain length, have assorted independently. Clearly they have not. Nevertheless, we have included the chi-square calculation in **FIGURE 7.2** just to show how much the observed results are out of line with the expected results. The chi-square value is enormous—much greater than 7.8, which is the critical value for a chi-square distribution with three degrees of freedom (see **TABLE 3.2**). Consequently, we must reject the hypothesis that the genes for flower color and pollen grain length have assorted independently.

Bateson and Punnett devised a complicated explanation for their results, but it turned out to be wrong. The correct explanation for the lack of independent assortment in the data is that the genes for flower color and pollen length are located on the same chromosome—that is, they are linked. This explanation is diagrammed in **FIGURE 7.3**. The alleles of the flower color gene are *R* (red) and *r* (white), and the alleles of the pollen length gene are *L* (long) and *l* (short); the *R* and *L* alleles are dominant. (Note here that for historical reasons, the allele symbols are derived from the dominant rather than the recessive phenotypes.) Because the flower color and pollen length genes are linked, we expect the doubly heterozygous F_1 plants to produce two kinds of gametes, *R L* and *r l*. However, once in a while a crossover will occur between the two genes and their alleles will be recombined, producing two other kinds of gametes, *R l* and *r L*. The frequency of these two types of recombinant gametes should, of course, depend on the frequency of crossing over between the two genes.

Bateson and Punnett might have come up with this explanation if they had performed a testcross instead of an intercross in the F_1. With a testcross the offspring would directly reveal the types of gametes produced by the doubly heterozygous F_1 plants. **FIGURE 7.4** presents the analysis of such a testcross. Doubly heterozygous F_1 sweet peas were crossed with plants homozygous for the recessive alleles of both genes. Among 1000 progeny scored, 920 resemble one or the other of the parental strains and the remaining 80 are recombinant. The frequency of the recombinant progeny produced by the heterozygous F_1 plants is therefore 80/1000 = 0.08. Because this is a testcross, 0.08 is also the frequency of recombinant gametes produced by the heterozygous F_1 plants. We can use this frequency, usually called the *recombination frequency*, to measure the intensity of linkage between genes. Genes that are tightly linked seldom recombine, whereas genes that are loosely linked recombine often. Here the recombination frequency is fairly low. This implies that crossing over between the two genes is a rather rare event.

For any two genes, the recombination frequency never exceeds 50 percent. This upper limit is obtained when genes are on different chromosomes; 50 percent recombination is, in fact, what we mean when we say that the genes assort independently.

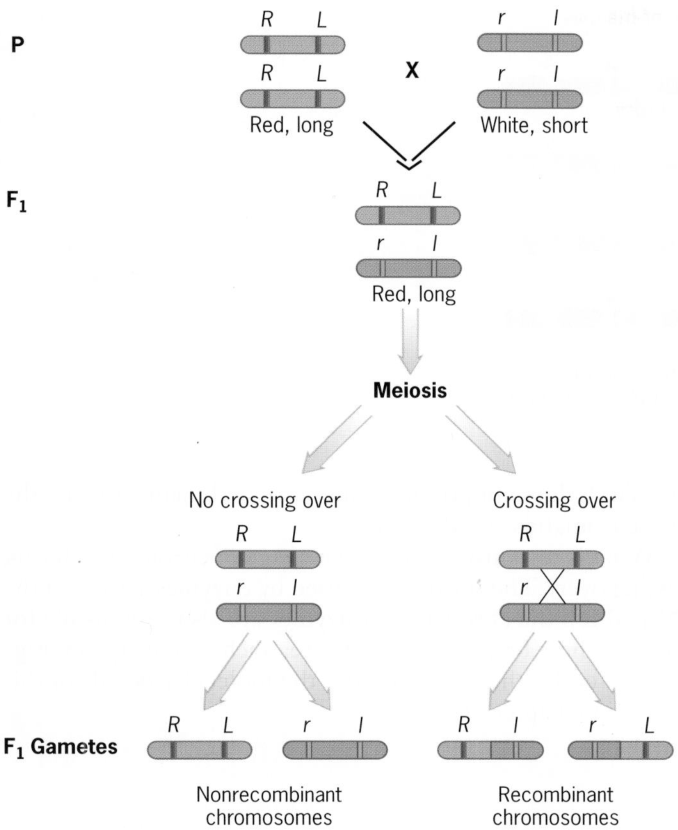

Figure 7.3 ▶ Hypothesis of linkage between the genes for flower color and pollen length in sweet peas. In the F_1 plants the two dominant alleles, *R* and *L*, of the genes are situated on the same chromosome; their recessive alleles, *r* and *l*, are situated on the homologous chromosome.

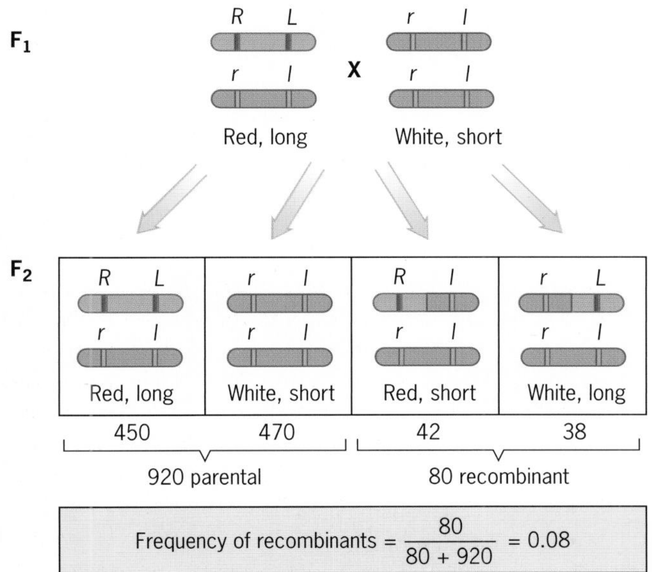

Figure 7.4 ▶ A testcross for linkage between genes in sweet peas. Because the recombinant progeny in the F_2 are 8 percent of the total, the genes for flower color and pollen length are rather tightly linked.

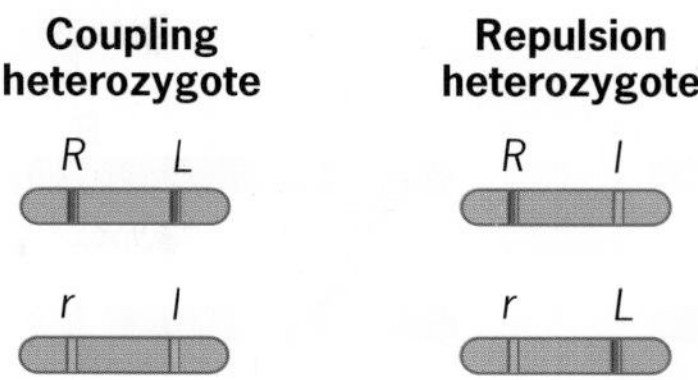

Figure 7.5 ▶ Coupling and repulsion linkage phases in double heterozygotes.

For example, let's assume that genes *A* and *B* are on different chromosomes and that an *AA BB* individual is crossed to an *aa bb* individual. From this cross, the *Aa Bb* offspring are then testcrossed to the double recessive parent. Because the *A* and *B* genes assort independently, the F_2 will consist of two classes (*Aa Bb* and *aa bb*) that are phenotypically like the parents in the original cross, and two classes (*Aa bb* and *aa Bb*) that are phenotypically recombinant. Furthermore, each F_2 class will occur with a frequency of 25 percent (see **FIGURE 5.8**). Thus, the total frequency of recombinant progeny from a testcross involving two genes on different chromosomes will be 50 percent. A frequency of recombination less than 50 percent implies that the genes are linked on the same chromosome.

Crosses involving linked genes are usually diagrammed to show the *linkage phase*—the way in which the alleles are arranged in heterozygous individuals (**FIGURE 7.5**). In Bateson and Punnett's sweet pea experiment, the heterozygous F_1 plants received two dominant alleles, *R* and *L*, from one parent and two recessive alleles, *r* and *l*, from the other. Thus, we write the genotype of these plants *R L*/*r l*, where the slash (/) separates alleles inherited from different parents. Another way of interpreting this symbolism is to say that the alleles on the left and right of the slash entered the genotype on different homologous chromosomes, one from each parent. Whenever the dominant alleles are all on one side of the slash, as in this example, the genotype has the *coupling* linkage phase. When the dominant and recessive alleles are split on both sides of the slash, as in *R l*/*r L*, the genotype has the *repulsion* linkage phase. These terms provide us with a way of distinguishing between the two kinds of double heterozygotes.

CROSSING OVER AS THE PHYSICAL BASIS OF RECOMBINATION

Recombinant gametes are produced as a result of crossing over between homologous chromosomes. This process involves a physical exchange between the chromosomes, as diagrammed in **FIGURE 7.6**. The exchange event occurs during the prophase of the first meiotic division, when duplicated chromosomes have paired. Although four homologous chromatids are present, forming what is called a **tetrad,** only two chromatids cross over at any one point. Each of these chromatids breaks at the site of the crossover, and the resulting pieces reattach to produce the recombinants. The other two chromatids are not recombinant at this site. Each crossover event therefore produces two recombinant chromatids among a total of four.

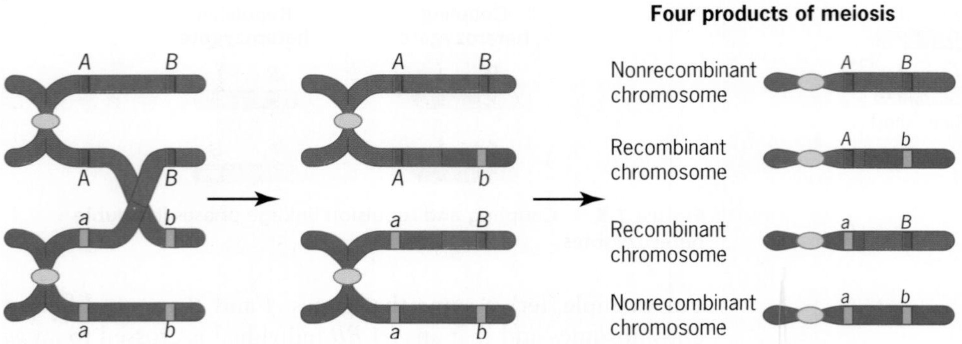

Four products of meiosis

Nonrecombinant chromosome — A B

Recombinant chromosome — A b

Recombinant chromosome — a B

Nonrecombinant chromosome — a b

Figure 7.6 ▶ Crossing over as the basis of recombination between genes. An exchange between paired chromosomes during meiosis produces recombinant chromosomes at the end of meiosis.

Although only two chromatids are involved in an exchange at any one point, the other two chromatids may cross over at a different point. Thus, there is a possibility for multiple exchanges in a tetrad of chromatids (**FIGURE 7.7**). There may, for example, be two, three, or even four separate exchanges—customarily called double, triple, or quadruple crossovers. (We consider the genetic significance of these in a later section of this chapter.) Note, however, that an exchange between sister chromatids does not produce genetic recombinants because the sister chromatids are identical.

What is responsible for the breakage of chromatids during crossing over? The breaks are caused by enzymes acting on the DNA within the chromatids. Enzymes are also responsible for repairing these breaks—that is, for reattaching chromatid fragments to each other. We consider the molecular details of this process in Chapter 13.

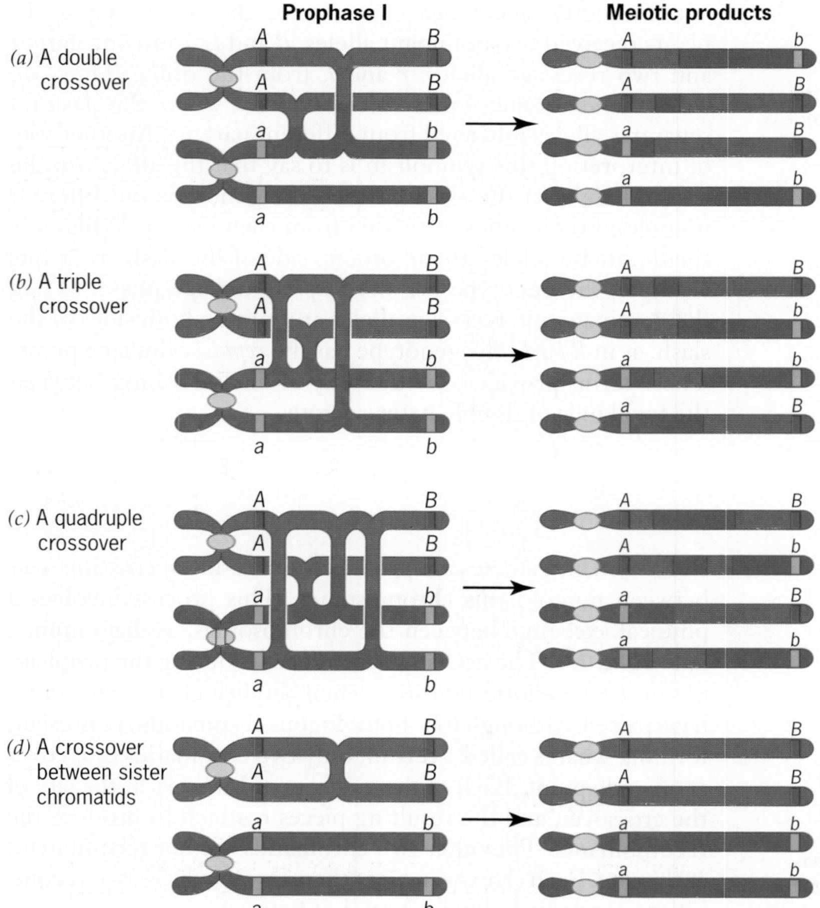

Figure 7.7 ▶ Consequences of multiple exchanges between chromosomes and exchange between sister chromatids during prophase I of meiosis.

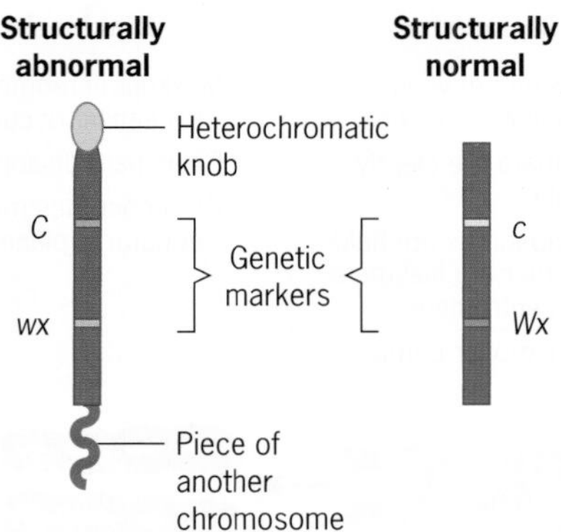

Figure 7.8 ▶ Two forms of chromosome 9 in maize used in the experiments of Creighton and McClintock.

EVIDENCE THAT CROSSING OVER CAUSES RECOMBINATION

In 1931 Harriet Creighton and Barbara McClintock obtained evidence that genetic recombination was associated with a material exchange between chromosomes. Creighton and McClintock studied homologous chromosomes in maize that were morphologically distinguishable. The goal was to determine whether physical exchange between these homologues was correlated with recombination between some of the genes they carried.

Two forms of chromosome 9 were available for analysis; one was normal, and the other had cytological aberrations at each end—a heterochromatic knob at one end and a piece of a different chromosome at the other (**FIGURE 7.8**). These two forms of chromosome 9 were also genetically marked to detect recombination. One marker gene controlled kernel color (*C*, colored; *c*, colorless), and the other controlled kernel texture (*Wx*, starchy; *wx*, waxy). Creighton and McClintock performed the following testcross:

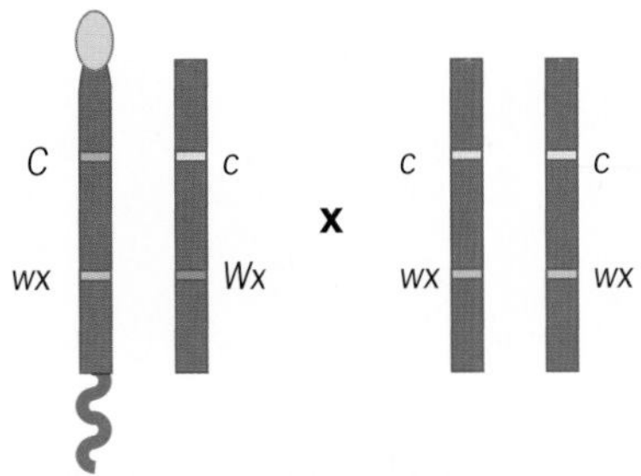

They then examined the recombinant progeny for evidence of exchange between the two different forms of chromosome 9. Their results showed that the *C Wx* and *c wx* recombinants carried a chromosome with only one of the abnormal cytological markers; the other abnormal marker had evidently been lost through an exchange with the normal chromosome 9 in the previous generation:

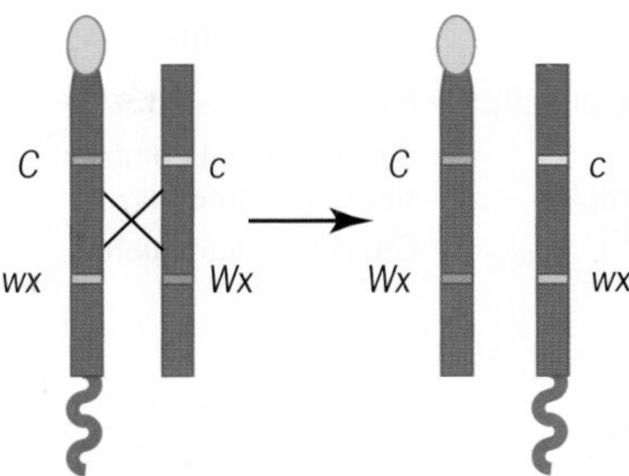

These findings strongly argued that recombination was caused by a physical exchange between paired chromosomes.

CHIASMATA AND THE TIME OF CROSSING OVER

The cytological evidence for crossing over can be seen during late prophase of the first meiotic division when the chiasmata become clearly visible. At this time paired chromosomes repel each other slightly, maintaining close contact only at the centromere and at each chiasma (**FIGURE 7.9**). This partial separation makes it possible to count the chiasmata accurately. As we might expect, large chromosomes typically have more chiasmata than small chromosomes. Thus, the number of chiasmata is roughly proportional to chromosome length.

The appearance of chiasmata late in the first meiotic prophase might imply that it is then that crossing over occurs. However, evidence from several different experiments suggests that it occurs earlier. Some of these experiments used heat shocks to alter the frequency of recombination. When the heat shocks were administered late in prophase, there was little effect, but when they were given earlier, the recombination frequency was changed. Thus, the event responsible for recombination, namely, crossing over, occurs rather early in the meiotic prophase. Additional evidence comes from molecular studies on

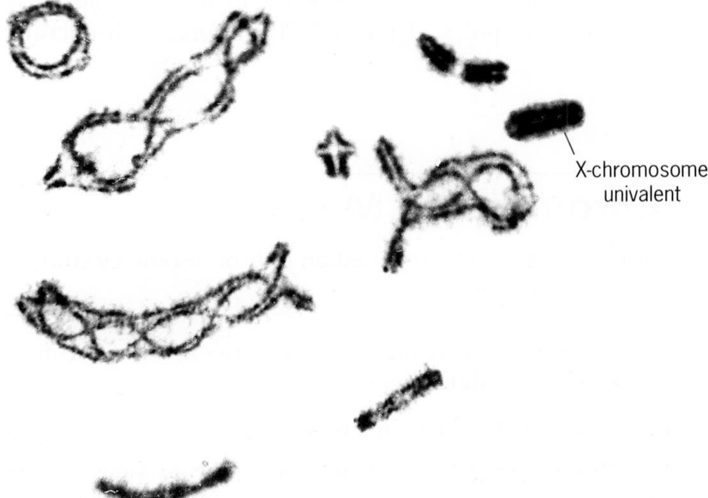

Figure 7.9 ▶ Diplonema of male meiosis in the grasshopper *Chorthippus parallelus*. There are eight autosomal bivalents and an X-chromosome univalent. The four smaller bivalents each have one chiasma. The remaining bivalents have two to five chiasmata.

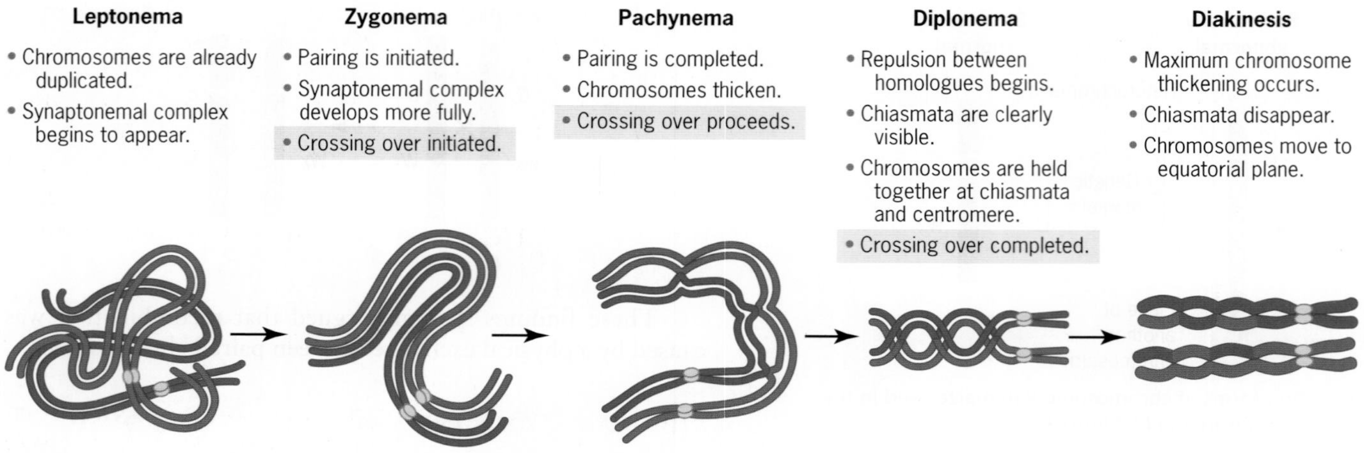

Leptonema
- Chromosomes are already duplicated.
- Synaptonemal complex begins to appear.

Zygonema
- Pairing is initiated.
- Synaptonemal complex develops more fully.
- Crossing over initiated.

Pachynema
- Pairing is completed.
- Chromosomes thicken.
- Crossing over proceeds.

Diplonema
- Repulsion between homologues begins.
- Chiasmata are clearly visible.
- Chromosomes are held together at chiasmata and centromere.
- Crossing over completed.

Diakinesis
- Maximum chromosome thickening occurs.
- Chiasmata disappear.
- Chromosomes move to equatorial plane.

Figure 7.10 ▶ Five stages of prophase of meiosis I. The synaptonemal complex is a structure involved in chromosome pairing (see Chapter 2). Crossing over occurs in early to mid-prophase. Chiasmata become visible in diplonema and disappear during diakinesis.

the time of DNA synthesis. Although almost all the DNA is synthesized during the interphase that precedes the onset of meiosis, a small amount is made during the first meiotic prophase. This limited DNA synthesis has been interpreted as part of a process to repair broken chromatids, which, as we have discussed, is thought to be associated with crossing over. Careful timing experiments have shown that this DNA synthesis occurs in early to mid-prophase, but not later. The accumulated evidence therefore suggests that crossing over occurs in early to mid-prophase, long before the chiasmata can be seen.

What, then, are chiasmata, and what do they mean? Most geneticists believe that the chiasmata are merely vestiges of the actual exchange process. Chromatids that have experienced an exchange probably remain entangled with each other during most of prophase. Eventually, these entanglements are resolved, and the chromatids are separated by the meiotic spindle apparatus to opposite poles of the cell. Therefore, each chiasma probably represents an entanglement that was created by a crossover event earlier in prophase (**FIGURE 7.10**).

KEY POINTS

- Linkage between genes is detected as a deviation from expectations based on Mendel's Principle of Independent Assortment.
- The frequency of recombination measures the intensity of linkage. In the absence of linkage, this frequency is 50 percent; for very tight linkage, it is close to zero.
- Recombination is caused by a physical exchange between paired homologous chromosomes early in prophase of the first meiotic division after chromosomes have duplicated.
- At any one point along a chromosome, the process of exchange (crossing over) involves only two of the four chromatids in a meiotic tetrad.
- Late in prophase I, crossovers become visible as chiasmata.

Chromosome Mapping

Linked genes can be mapped on a chromosome by studying how often their alleles recombine.

Crossing over during the prophase of the first meiotic division has two observable outcomes:

1. Formation of chiasmata in late prophase.
2. Recombination between genes on opposite sides of the crossover point.

However, the second outcome can only be seen in the next generation, when the genes on the recombinant chromosomes are expressed.

Geneticists construct chromosome maps by counting the number of crossovers that occur during meiosis. However, because the actual crossover events cannot be seen, they cannot count them directly. Instead, they must estimate how many crossovers have taken place by counting either chiasmata or recombinant chromosomes. Chiasmata are counted through cytological analysis, whereas recombinant chromosomes are counted through genetic analysis. Before we consider either of these procedures, we must define what we mean by distance on a chromosome map.

CROSSING OVER AS A MEASURE OF GENETIC DISTANCE

Sturtevant's fundamental insight was to estimate the distance between points on a chromosome by counting the number of crossovers between them. Points that are far apart should have more crossovers between them than points that are close together. However, the number of crossovers must be understood in a statistical sense. In any particular cell, the chance that a crossover will occur between two points may be low, but in a large population of cells, this crossover will probably occur several times simply because there are so many independent opportunities for it. Thus, the quantity that we really need to measure is the *average* number of crossovers in a particular chromosome region. Genetic map distances are, in fact, based on such averages. This idea is sufficiently important to justify a formal definition: *The distance between two points on the genetic map of a chromosome is the average number of crossovers between them.*

One way for us to understand this definition is to consider 100 oogonia going through meiosis (**FIGURE 7.11**). In some cells, no crossovers will occur between sites *A* and *B*; in others, one, two, or more crossovers will occur between these loci. At the end of meiosis, there will be 100 gametes, each containing a chromosome with either zero, one, two, or more crossovers between *A* and *B*. We estimate the genetic map distance between these loci by calculating the average number of crossovers in this sample of chromosomes. The result is 0.42.

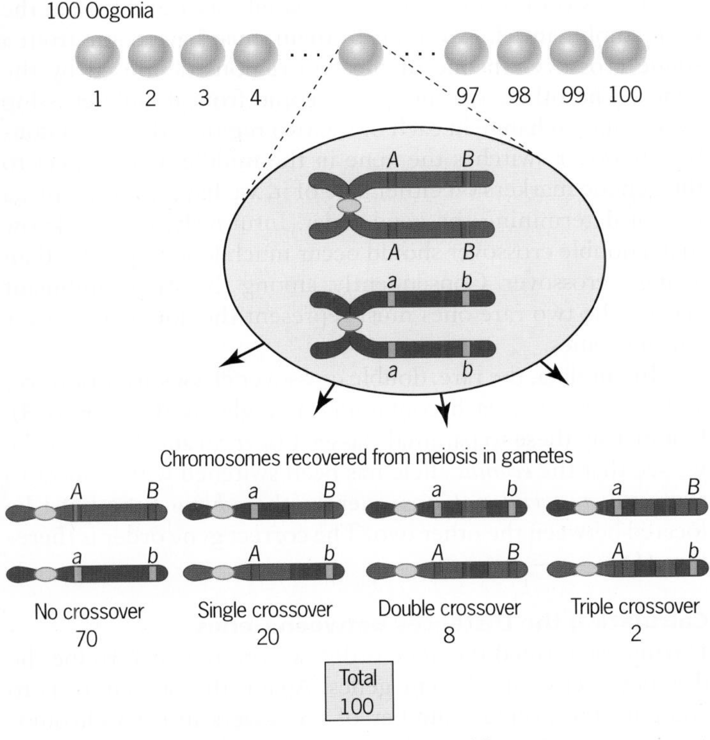

$$0\times\left(\frac{70}{100}\right) + 1\times\left(\frac{20}{100}\right) + 2\times\left(\frac{8}{100}\right) + 3\times\left(\frac{2}{100}\right) = 0.42$$

Figure 7.11 ▶ Calculating the average number of crossovers between genes on chromosomes recovered from meiosis.

In practice, we cannot "see" each of the exchange points on the chromosomes coming out of meiosis. Instead, we infer their existence by observing the recombination of the alleles that flank them. A chromosome in which alleles have recombined must have arisen by crossing over. Counting recombinant chromosomes therefore provides a way of counting crossover exchange points.

RECOMBINATION MAPPING WITH A TWO-POINT TESTCROSS

To illustrate the mapping procedure, let's consider the two-point testcross in **FIGURE 7.12**. Wild-type *Drosophila* females were mated to males homozygous for two autosomal mutations—*vestigial* (*vg*), which produces short wings, and *black* (*b*), which produces a black body. All the F_1 flies had long wings and gray bodies; thus, the wild-type alleles (vg^+ and b^+) are dominant. The F_1 females were then testcrossed to vestigial, black males, and the F_2 progeny were sorted by phenotype and counted. As the data show, there were four phenotypic classes, two abundant and two rare. The abundant classes had the same phenotypes as the original parents, and the rare classes had recombinant phenotypes. We know that the *vestigial* and *black* genes are linked because the recombinants are much fewer than 50 percent of the total progeny counted. These genes must therefore be on the same chromosome. To determine the distance between them, we must estimate the average number of crossovers in the gametes of the doubly heterozygous F_1 females. We can do this by calculating the frequency of recombinant F_2 flies and noting that each such fly inherited a chromosome that had crossed over once between *vg* and *b*. The average number of crossovers in the whole sample of progeny is therefore

$$\underset{\text{nonrecombinants}}{(0) \times 0.82} + \underset{\text{recombinants}}{(1) \times 0.18} = 0.18$$

In this expression, the number of crossovers for each class of flies is placed in parentheses; the other number is the frequency of that class. The nonrecombinant progeny obviously do not add any crossover chromosomes to the data, but we include them in the calculation to emphasize that we must calculate the average number of crossovers by using all the data, not just those from the recombinants.

This simple analysis indicates that, on average, 18 out of 100 chromosomes recovered from meiosis had a crossover between *vg* and *b*. Thus, *vg* and *b* are separated by 18 *units* on the genetic map. Sometimes geneticists call a map unit a **centiMorgan,** abbreviated cM, in honor of T. H. Morgan; 100 centiMorgans equal one Morgan (M). We can therefore say that *vg* and *b* are 18 cM (or 0.18 M) apart. Notice that the map distance is equal to the frequency of recombination, written as a percentage. Later we will see that when the frequency of recombination approaches 0.5, it underestimates the map distance.

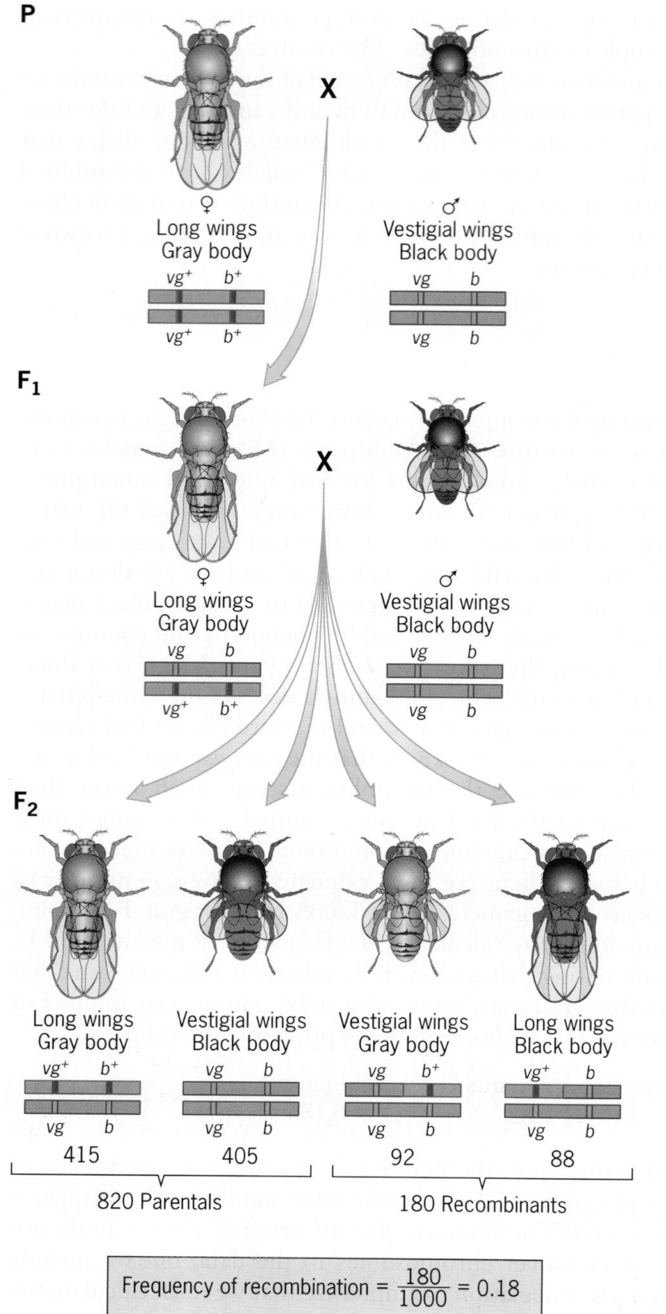

Figure 7.12 ▶ An experiment involving two linked genes, *vg* (*vestigial* wings) and *b* (*black* body), in *Drosophila*.

RECOMBINATION MAPPING WITH A THREE-POINT TESTCROSS

We can also use the recombination mapping procedure with data from testcrosses involving more than two genes. **FIGURE 7.13** illustrates an experiment by C. B. Bridges and T. M. Olbrycht, who crossed wild-type *Drosophila* males to females homozygous for three recessive X-linked mutations—*scute* (*sc*) bristles, *echinus* (*ec*) eyes, and *crossveinless* (*cv*) wings. They then intercrossed the F_1 progeny to produce F_2 flies, which they classified and counted. We note that the F_1 females in this intercross carried the three recessive mutations on one of their X chromosomes and the wild-type alleles of these mutations on the other X chromosome. Furthermore, the F_1 males carried the three recessive mutations on their single X chromosome. Thus, this intercross was equivalent to a testcross with all three genes in the F_1 females present in the coupling configuration.

The F_2 flies from the intercross comprised eight phenotypically distinct classes, two of them parental and six recombinant. The parental classes were by far the most numerous. The less numerous recombinant classes each represented a different kind of crossover chromosome. To figure out which crossovers were involved in producing each type of recombinant, we must first determine how the genes are ordered on the chromosome.

Determining the Gene Order

There are three possible gene orders:

1. *sc* — *ec* — *cv*

2. *ec* — *sc* — *cv*

3. *ec* — *cv* — *sc*

Other possibilities, such as *cv* – *ec* – *sc*, are the same as one of these because the left and right ends of the chromosome cannot be distinguished. Which of the orders is correct?

To answer this question, we must take a careful look at the six recombinant classes. Four of them must have come from a single crossover in one of the two regions delimited by the genes. The other two must have come from double crossing over—one exchange in each of the two regions. Because a double crossover switches the gene in the middle with respect to the genetic markers on either side of it, we have, in principle, a way of determining the gene order. Intuitively, we also know that a double crossover should occur much less frequently than a single crossover. Consequently, among the six recombinant classes, the two rare ones must represent the double crossover chromosomes.

In our data, the rare, double crossover classes are 7 ($sc\ ec^+\ cv$) and 8 ($sc^+\ ec\ cv^+$), each containing a single fly (**FIGURE 7.13**). Comparing these to parental classes 1 ($sc\ ec\ cv$) and 2 ($sc^+\ ec^+\ cv^+$), we see that the *echinus* allele has been switched with respect to *scute* and *crossveinless*. Consequently, the *echinus* gene must be located between the other two. The correct gene order is therefore **(1)** *sc* – *ec* – *cv*.

Calculating the Distances Between Genes

Having established the gene order, we can now determine the distances between adjacent genes. Again, the procedure is to compute the average number of crossovers in each chromosomal region (**FIGURE 7.14**).

We can obtain the length of the region between *sc* and *ec* by identifying the recombinant classes that involved a crossover between these genes. There are four such classes: 3 ($sc\ ec^+\ cv^+$), 4 ($sc^+\ ec\ cv$), 7 ($sc\ ec^+\ cv$), and 8 ($sc^+\ ec\ cv^+$). Classes 3 and 4 involved

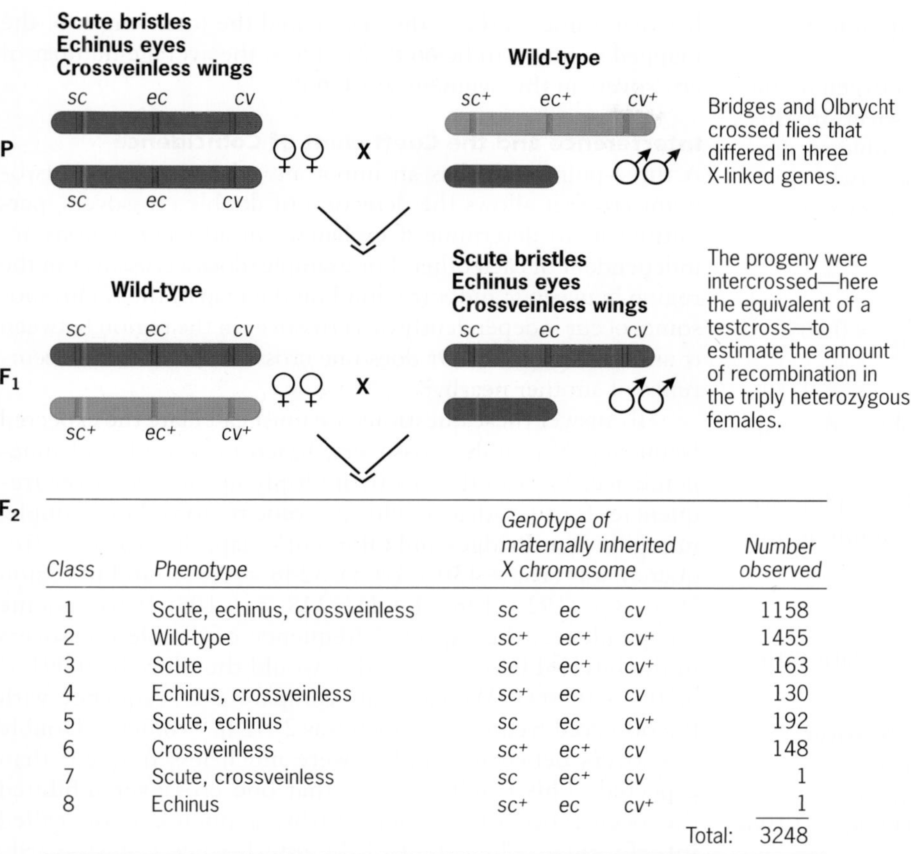

Class	Phenotype	Genotype of maternally inherited X chromosome			Number observed
1	Scute, echinus, crossveinless	*sc*	*ec*	*cv*	1158
2	Wild-type	*sc*+	*ec*+	*cv*+	1455
3	Scute	*sc*	*ec*+	*cv*+	163
4	Echinus, crossveinless	*sc*+	*ec*	*cv*	130
5	Scute, echinus	*sc*	*ec*	*cv*+	192
6	Crossveinless	*sc*+	*ec*+	*cv*	148
7	Scute, crossveinless	*sc*	*ec*+	*cv*	1
8	Echinus	*sc*+	*ec*	*cv*+	1
				Total:	3248

Figure 7.13 ► Bridges and Olbrycht's three-point cross with the X-linked genes *sc* (*scute* bristles), *ec* (*echinus* eyes), and *cv* (*crossveinless* wings) in *Drosophila*.

a single crossover between *sc* and *ec*, and classes 7 and 8 involved two crossovers, one between *sc* and *ec* and the other between *ec* and *cv*. We can therefore use the frequencies of these four classes to estimate the average number of crossovers between *sc* and *ec*:

$$\frac{\overset{\text{Class 3}}{163} + \overset{\text{Class 4}}{130} + \overset{\text{Class 7}}{1} + \overset{\text{Class 8}}{1}}{\text{Total}} = \frac{295}{3248} = 0.091$$

Thus, in every 100 chromosomes coming from meiosis in the F_1 females, 9.1 had a crossover between *sc* and *ec*. The

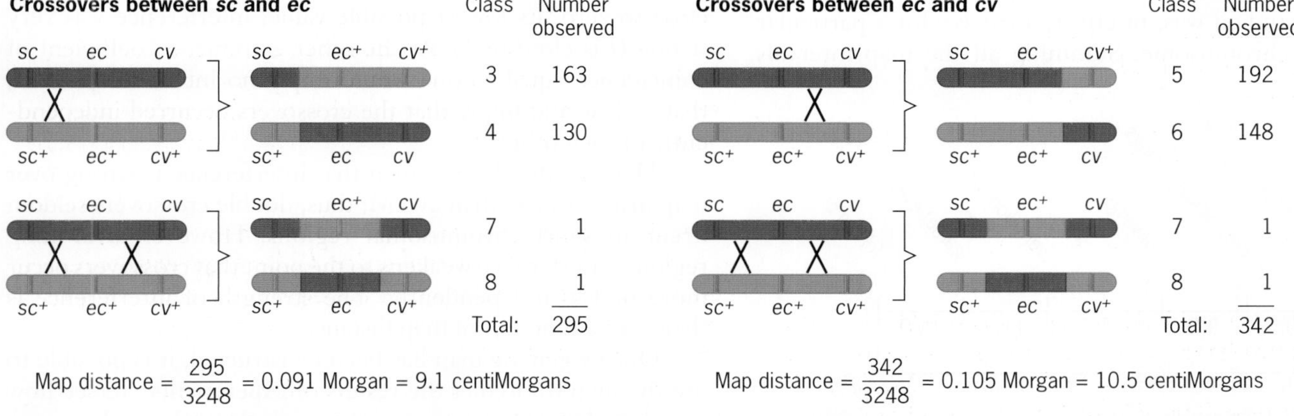

Figure 7.14 ► Calculation of genetic map distances from Bridges and Olbrycht's data. The distance between each pair of genes is obtained by estimating the average number of crossovers.

distance between these genes is therefore 9.1 map units (or, if you prefer, 9.1 centiMorgans).

In a similar way, we can obtain the distance between *ec* and *cv*. Four recombinant classes involved a crossover in this region: 5 ($sc\ ec\ cv^+$), 6 ($sc^+ ec^+ cv$), 7, and 8. The double recombinants are also included here because one of their two crossovers was between *ec* and *cv*. The combined frequency of these four classes is:

$$\frac{\begin{array}{ccccccc}\text{Class 5} & & \text{Class 6} & & \text{Class 7} & & \text{Class 8}\\ 192 & + & 148 & + & 1 & + & 1\end{array}}{\text{Total}} = \frac{342}{3248} = 0.105$$

Consequently, *ec* and *cv* are 10.5 map units apart.

Combining the data for the two regions, we obtain the map

sc—9.1—*ec*—10.5—*cv*

Map distances computed in this way are additive. Thus, we can estimate the distance between *sc* and *cv* by summing the lengths of the two map intervals between them:

$$9.1\ \text{cM} + 10.5\ \text{cM} = 19.6\ \text{cM}$$

We can also obtain this estimate by directly calculating the average number of crossovers between these genes:

Non–crossover classes 1 and 2		Single crossover classes 3, 4, 5, and 6		Double crossover classes 7 and 8
(0) × 0.805	+	(1) × 0.195	+	(2) × 0.0006 = 0.196

Here the number of crossovers is given in parentheses, and its multiplier is the combined frequency of the classes with that many crossovers. In other words, each recombinant class contributes to the map distance according to the product of its frequency and the number of crossovers it represents.

Bridges and Olbrycht actually studied seven X-linked genes in their recombination experiment: *sc*, *ec*, *cv*, *ct* (*cut* wings), *v* (*vermilion* eyes), *g* (*garnet* eyes), and *f* (*forked* bristles). By calculating recombination frequencies between each pair of adjacent genes, they were able to construct a map of a large segment of the X chromosome (**FIGURE 7.15**); *sc* was at one end, and *f* was at the other. Each of the seven genes that Bridges and Olbrycht studied was, in effect, a *marker* for a particular site on the X chromosome. Summing all the map intervals between these markers, they estimated the total length of the mapped segment to be 66.8 cM. Thus, the average number of crossovers in this segment was 0.668.

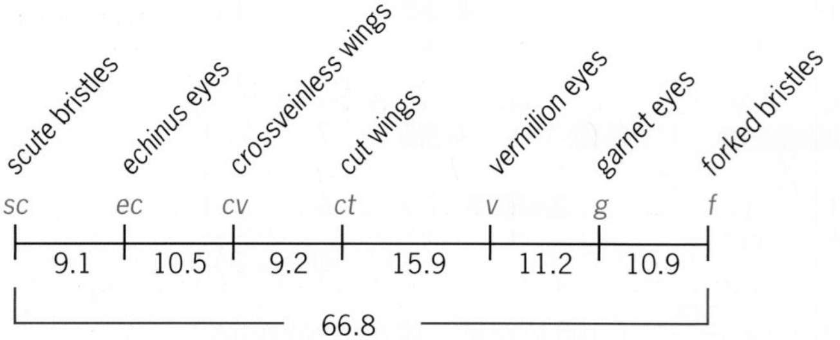

Figure 7.15 ▶ Bridges and Olbrycht's map of seven X-linked genes in *Drosophila*. Distances are given in centiMorgans.

Interference and the Coefficient of Coincidence

A three-point cross has an important advantage over a two-point cross: it allows the detection of double crossovers, permitting us to determine if exchanges in adjacent regions are independent of each other. For example, does a crossover in the region between *sc* and *ec* (region I on the map of the X chromosome) occur independently of a crossover in the region between *ec* and *cv* (region II)? Or does one crossover inhibit the occurrence of another nearby?

To answer these questions, we must calculate the expected frequency of double crossovers, based on the idea of independence. We can do this by multiplying the crossover frequencies for two adjacent chromosome regions. For example, in region I on Bridges and Olbrycht's map, the crossover frequency was $(163 + 130 + 1 + 1)/3248 = 0.091$, and in region II, it was $(192 + 148 + 1 + 1)/3248 = 0.105$. If we assume independence, the expected frequency of double crossovers in the interval between *sc* and *cv* would therefore be $0.091 \times 0.105 = 0.0095$. We can now compare this frequency with the observed frequency, which was $2/3248 = 0.0006$. Double crossovers between *sc* and *cv* were much less frequent than expected. This result suggests that one crossover inhibited the occurrence of another nearby, a phenomenon called **interference.** The extent of the interference is customarily measured by the **coefficient of coincidence,** *c*, which is the ratio of the observed frequency of double crossovers to the expected frequency:

$$c = \frac{\text{observed frequency of double crossovers}}{\text{expected frequency of double crossovers}} = \frac{0.0006}{0.0095} = 0.063$$

The level of interference, symbolized *I*, is calculated as $I = 1 - c = 0.937$.

Because in this example the coefficient of coincidence is close to zero, its lowest possible value, interference was very strong (*I* is close to 1). At the other extreme, a coefficient of coincidence equal to one would imply no interference at all; that is, it would imply that the crossovers occurred independently of each other.

Many studies have shown that interference is strong over map distances less than 20 cM; thus, double crossovers seldom occur in short chromosomal regions. However, over long regions, interference weakens to the point that crossovers occur more or less independently. The strength of interference is therefore a function of map distance.

Once a genetic map has been constructed, it is possible to use the map to predict the results of experiments. To see how map-based predictions are made, work through the Focus on Problem Solving: Using a Genetic Map to Predict the Outcome of a Cross.

▶ FOCUS ON PROBLEM SOLVING Using a Genetic Map to Predict the Outcome of a Cross

THE PROBLEM

The genes *r, s,* and *t* reside in the middle of the *Drosophila* X chromosome; *r* is 15 cM to the left of *s*, and *t* is 20 cM to the right of *s*. In this region, the coefficient of coincidence (*c*) is 0.2. A geneticist wishes to create an X chromosome that carries the recessive mutant alleles of all three genes. One stock is homozygous for *r* and *t,* and another stock is homozygous for *s*. By crossing the two stocks, the geneticist obtains females that are triple heterozygotes, $r\ s^+ t/r^+ s\ t^+$. These females are then crossed to wild-type males. If the geneticist examines 10,000 sons from these females, how many of them will be triple mutants, *r s t*?

FACTS AND CONCEPTS

1. For small map intervals (<20 cM), the map distance equals the frequency of a single crossover in the interval.
2. The coefficient of coincidence equals the observed frequency of double crossovers/expected frequency of double crossovers.
3. The expected frequency of double crossovers is calculated on the assumption that the two crossovers occur independently.
4. Males inherit their X chromosome from their mothers.

ANALYSIS AND SOLUTION

Triple mutant males will be produced only if a double crossover occurs in the $r\ s^+ t/r^+ s\ t^+$ females that were crossed to wild-type males. The frequency of such double crossovers is a function of the two map distances (15 cM and 20 cM) and the level of interference, which is measured by the coefficient of coincidence (here $c = 0.2$). Because c = observed frequency of double crossovers/expected frequency of double crossovers, we can solve for the observed frequency of double crossovers after a simple algebraic rearrangement: observed frequency of double crossovers = $c \times$ expected frequency of double crossovers. The expected frequency of double crossovers is calculated from the map distances assuming that crossovers in adjacent map intervals occur independently: $0.15 \times 0.20 = 0.03$. Thus, among 10,000 sons, 300 should carry an X chromosome that had one crossover between the *r* and *s* genes and another crossover between the *s* and *t* genes. However, only half of these 300 sons—that is, 150—will carry the triply mutant X chromosome; the other 150 will be triply wild-type.

For further discussion go to your *WileyPLUS* course.

RECOMBINATION FREQUENCY AND GENETIC MAP DISTANCE

In the preceding sections, we have considered how to construct chromosome maps from data on the recombination of genetic markers. These data allow us to infer where crossovers have occurred in a sample of chromosomes. By localizing and counting these crossovers, we can estimate the distances between genes and then place the genes on a chromosome map.

This method works well as long as the genes are fairly close together. However, when they are far apart, the frequency of recombination may not reflect the true map distance (**FIGURE 7.16**). As an example, let us consider the genes at the ends of Bridges and Olbrycht's map of the X chromosome; *sc*, at the left end, was 66.8 cM away from *f*, at the right end. However, the frequency of recombination between *sc* and *f* was 50 percent—the maximum possible value. Using this frequency to estimate map distance, we would conclude that *sc* and *f* were 50 map units apart. Of course, the distance obtained by summing the lengths of the intervening regions on the map, 66.8 cM, is much greater.

This example shows that the true genetic distance, which depends on the average number of crossovers on a chromosome,

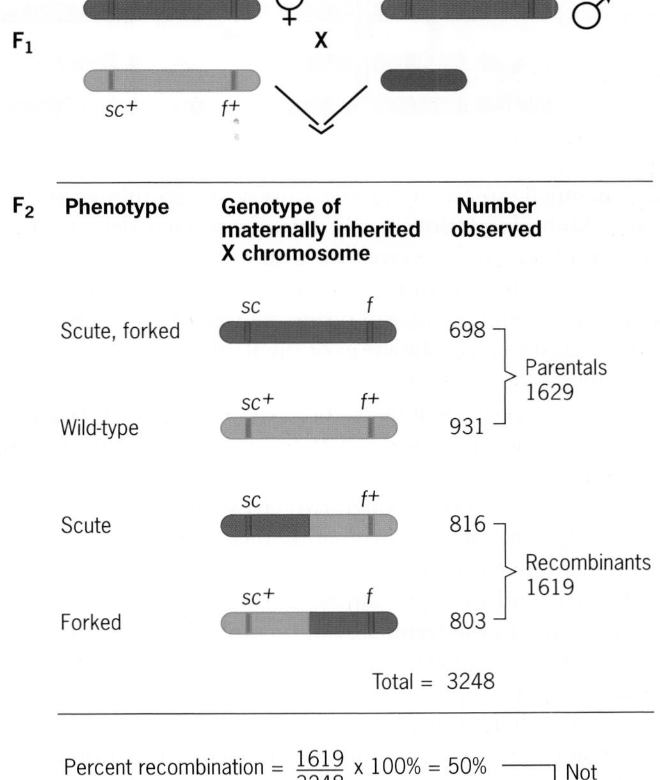

Figure 7.16 ▶ A discrepancy between map distance and percent recombination. The map distance between the genes *sc* and *f* is greater than the observed percent recombination between them.

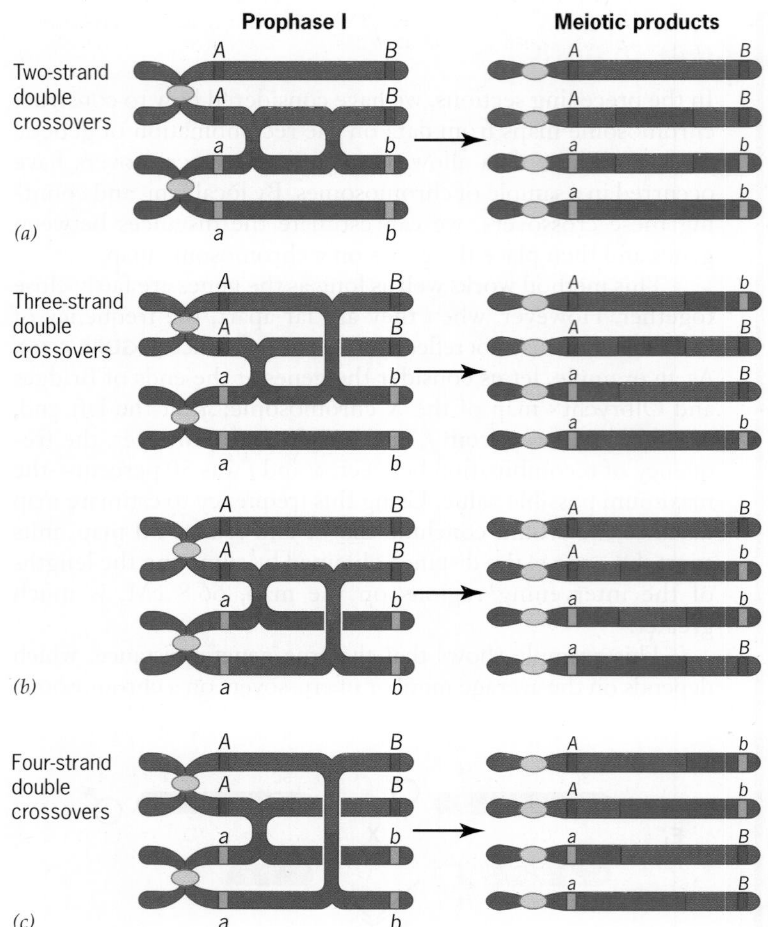

Figure 7.17 ▶ Consequences of double crossing over between two loci. (*a*) Two-strand double crossovers produce only nonrecombinant chromosomes. (*b*) Three-strand double crossovers produce half recombinant and half nonrecombinant chromosomes. (*c*) Four-strand double crossovers produce only recombinant chromosomes.

may be much greater than the observed recombination frequency. Multiple crossovers may occur between widely separated genes, and some of these crossovers may not produce genetically recombinant chromosomes (**FIGURE 7.17**). To see this, let us assume that a single crossover occurs between two chromatids in a tetrad, causing recombination of the flanking genetic markers. If another crossover occurs between these same two chromatids, the flanking markers will be restored to their original configuration; the second crossover essentially cancels the effect of the first, converting the recombinant chromatids back into nonrecombinants. Thus, even though two crossovers have occurred in this tetrad, none of the chromatids that come from it will be recombinant for the flanking markers.

This second example shows that a double crossover may not contribute to the frequency of recombination, even though it contributes to the average number of exchanges on a chromosome. A quadruple crossover would have the same effect. These and other multiple exchanges are responsible for the discrepancy between recombination frequency and genetic map distance. In practice, this discrepancy is small for distances less than 20 cM. Over such distances, interference is strong enough to suppress almost all multiple exchanges, and the recombination frequency is a good estimator of the true genetic distance. For values greater than 20 cM, these two quantities diverge, principally because multiple exchanges become much more likely. **FIGURE 7.18** shows the mathematical relationship between recombination frequency and genetic map distance.

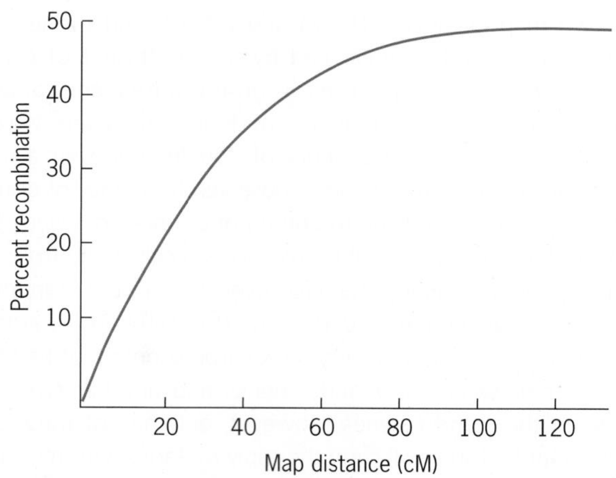

Figure 7.18 ▶ Relationship between frequency of recombination and genetic map distance. For values less than 20 cM, there is approximately a linear relationship between percent recombination and map distance; for values greater than 20 cM, the percent recombination underestimates the map distance.

CHIASMA FREQUENCY AND GENETIC MAP DISTANCE

Recombination between genetic markers is one outcome of crossing over. The other is chiasma formation during prophase of the first meiotic division. Each chiasma is thought to represent the resolution of a crossover that occurred earlier in prophase. Thus, by counting chiasmata, we should be able to estimate the average number of crossovers occurring on a chromosome, which we can then use as an estimate of genetic map length.

Let's suppose, for example, that in a group of 100 cells going through meiosis, 5 cells show five chiasmata in a particular pair of chromosomes, 15 show four chiasmata in this pair, 15 show three, 30 show two, 25 show one, and 10 show none (**FIGURE 7.19**). Per cell, the average number of chiasmata in this pair of chromosomes is 2.15. This translates into an average of 1.07 chiasmata per chromatid because each chiasma affects only two of the four chromatids in the chromosome pair. Thus, at the completion of meiosis, the average number of crossovers per chromatid will be 1.07. This implies that the genetic length of this chromosome is 1.07 M, or 107 cM. If an average of 2.15 chiasmata are equivalent to 107 cM, then an average of one chiasma is equivalent to 107/2.15 = 50 cM. In general, we can conclude that on average each chiasma corresponds to 50 cM on the genetic map.

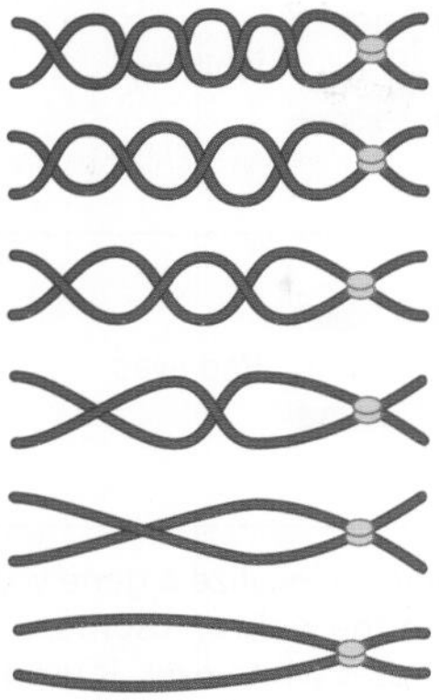

Meiotic chromosomes	Number of chiasmata (A)	Number of cells observed (B)	*Product* (A × B)
	5	5	25
	4	15	60
	3	15	45
	2	30	60
	1	25	25
	0	10	0
	Total:	100	215

Average number of chiasmata per cell $= \frac{215}{100} = 2.15$

Average number of chiasmata per chromatid $= \frac{2.15}{2} = 1.07$

Genetic length of chromosome = 1.07 Morgans = 107 centiMorgans

Relationship between genetic length and average number of chiasmata $= \frac{107 \text{ cM}}{2.15 \text{ chiasmata}} = 50 \text{ cM/chiasma}$

Figure 7.19 ▶ Computing map length from chiasmata frequencies in meiosis. The average number of chiasmata per chromatid estimates the genetic length of the chromosome.

KEY POINTS

- The genetic maps of chromosomes are based on the average number of crossovers that occur during meiosis.
- Genetic map distances are estimated by calculating the frequency of recombination between genes in experimental crosses.
- Recombination frequencies less than 20 percent estimate map distance directly; however, recombination frequencies greater than 20 percent underestimate map distance because multiple crossover events do not always produce recombinant chromosomes.
- An average of one chiasma during meiosis is equivalent to 50 centiMorgans of genetic map distance.

Cytogenetic Mapping

Geneticists have developed techniques to localize genes on the cytological maps of chromosomes.

Recombination mapping allows us to determine the relative positions of genes by using the frequency of crossing over as a measure of distance. However, it does not allow us to localize genes with respect to cytological landmarks, such as bands, on chromosomes. This kind of localization requires a different procedure that involves studying the phenotypic effects of chromosome rearrangements, such as deletions and duplications. Because these types of rearrangements can be recognized cytologically, their phenotypic effects can be correlated with particular regions along the length of a chromosome. If these phenotypic effects can be associated with genes that have already been positioned on a recombination map, then the map positions of those genes can be tied to locations on the cytological map of a chromosome. This process, called *cytogenetic mapping*, has been most thoroughly developed in *Drosophila* genetics, where the large, banded polytene chromosomes provide researchers with extraordinarily detailed cytological maps.

LOCALIZING GENES USING DELETIONS AND DUPLICATIONS

As an example of cytogenetic mapping, let's consider ways of localizing the X-linked *white* gene of *Drosophila*, a wild-type copy of which is required for pigmentation in the eyes. This gene is situated at map position 1.5 near one end of the X chromosome.

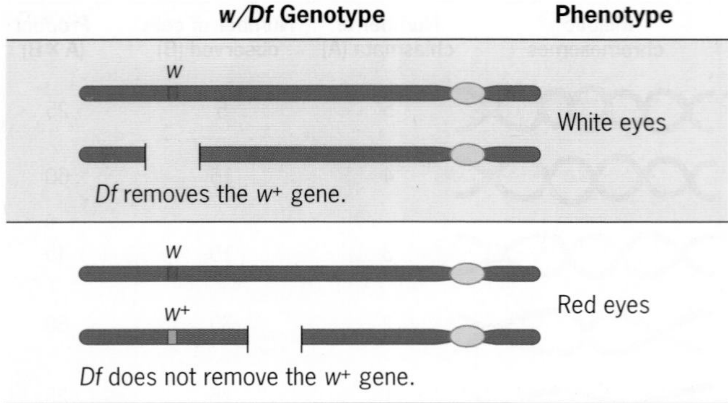

Figure 7.20 ▶ Principles of deletion mapping to localize a gene within a *Drosophila* chromosome. The *white* gene on the X chromosome, defined by the recessive mutation *w* which causes white eyes, is used as an example.

But which of the two ends is it near, and how far is it, in cytological terms, from that end? To answer these questions, we need to find the position of the *white* gene on the cytological map of the polytene X chromosome.

One procedure is to produce flies that are heterozygous for a recessive null mutation of the *white* gene (*w*) and a cytologically defined deletion (or deficiency, usually symbolized *Df*) for part of the X chromosome (**FIGURE 7.20**). These *w/Df* heterozygotes provide a functional test for the location of *white* relative to the deficiency. If the *white* gene has been deleted from the *Df* chromosome, then the *w/Df* heterozygotes will not be able to make eye pigment because they will not have a functional copy of the *white* gene on either of their X chromosomes. The eyes of the *w/Df* heterozygotes will therefore be white (the mutant phenotype). If, however, the *white* gene has not been deleted from the *Df* chromosome, then the *w/Df* heterozgyotes will have a functional *white* gene somewhere on that chromosome, and their eyes will be red (the wild phenotype). By looking at the eyes of the *w/Df* heterozygotes, we can therefore determine whether or not a specific deficiency has deleted the *white* gene. If it has, *white* must be located within the boundaries of that deficiency.

Different X chromosome deficiencies have allowed researchers to locate the *white* gene to a position near the left end of the X chromosome (**FIGURE 7.21**). Each deficiency was combined with a recessive *white* mutation, but only one of the deficiencies, $Df(1)w^{rJ1}$, produced white eyes. Because this deficiency "uncovers" the *white* mutation, we know that the *white* gene must be located within the segment of the chromosome that it deletes, that is, somewhere between polytene chromosome bands 3A1 and 3C2. With smaller deficiencies, the *white* gene has been localized to polytene chromosome band 3C2, near the right boundary of $Df(1)w^{rJ1}$.

We can also use duplications to determine the cytological locations of genes. The procedure is similar to the one using deletions, except that we look for a duplication that masks the phenotype of a recessive mutation. **FIGURE 7.22** shows an example utilizing duplications for small segments of the X chromosome that have been translocated to another chromosome. Only one of these duplications, *Dp2*, masks—or, as geneticists like to say, "covers"—the *white* mutation; thus, a wild-type copy

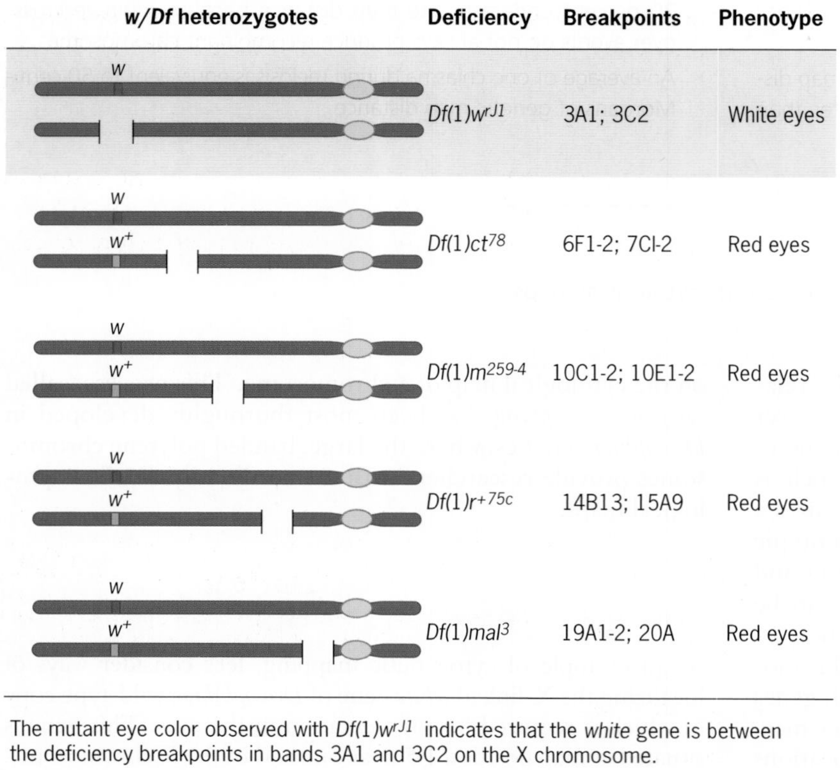

***w/Df* heterozygotes**	**Deficiency**	**Breakpoints**	**Phenotype**
w / (deletion)	$Df(1)w^{rJ1}$	3A1; 3C2	White eyes
w / w+	$Df(1)ct^{78}$	6F1-2; 7Cl-2	Red eyes
w / w+	$Df(1)m^{259\text{-}4}$	10C1-2; 10E1-2	Red eyes
w / w+	$Df(1)r^{+75c}$	14B13; 15A9	Red eyes
w / w+	$Df(1)mal^{3}$	19A1-2; 20A	Red eyes

The mutant eye color observed with $Df(1)w^{rJ1}$ indicates that the *white* gene is between the deficiency breakpoints in bands 3A1 and 3C2 on the X chromosome.

Figure 7.21 ▶ Localization of the *white* gene in the *Drosophila* X chromosome by deletion mapping. The deficiency breakpoints are presented using the coordinates of Bridges' cytological map of the polytene X chromosome.

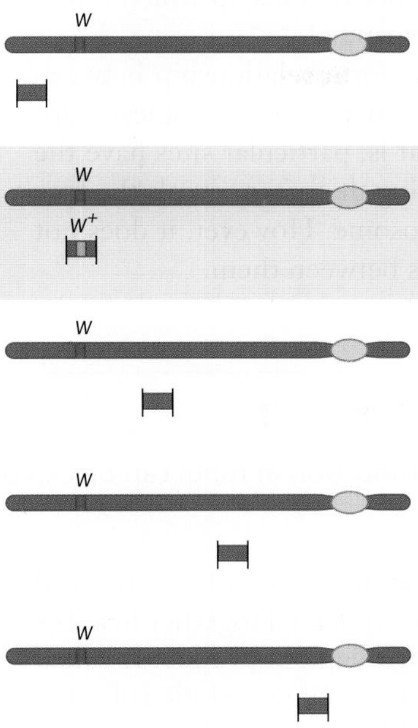

w/Dp combinations	Duplication	Breakpoints	Phenotype
w	*Dp1*	tip; 1E2-4	White eyes
w / w^+	*Dp2*	2D; 3D	Red eyes
w	*Dp3*	6E2; 7C4-6	White eyes
w	*Dp4*	9F3; 10E3-4	White eyes
w	*Dp5*	14B13; 15A9	White eyes

The wild-type eye color observed with *Dp2* indicates that the *white* gene is between the duplication breakpoints in regions 2D and 3D on the X chromosome.

Figure 7.22 ▶ Localization of the *white* gene in the *Drosophila* X chromosome by duplication mapping. Each duplication is a segment of the X chromosome that has been translocated to another chromosome. For simplicity, however, the other chromosome is not shown. The duplication breakpoints are presented using the coordinates of Bridges' cytological map of the polytene X chromosome.

of *white* must be present within it. This localizes the *white* gene somewhere between sections 2D and 3D on the polytene X chromosome, which is consistent with the results of the deletion tests already discussed.

Deletions and duplications have been extraordinarily useful in locating genes on the cytological maps of *Drosophila* chromosomes. The basic principle in *deletion mapping* is that a deletion that *uncovers* a recessive mutation must lack a wild-type copy of the mutant gene. This fact localizes that gene within the boundaries of the deletion. The basic principle in *duplication mapping* is that a duplication that *covers* a recessive mutation must contain a wild-type copy of the mutant gene. This fact localizes that gene within the boundaries of the duplication.

GENETIC DISTANCE AND PHYSICAL DISTANCE

The procedures for measuring genetic distance and for constructing recombination maps are based on the incidence of crossing over between paired chromosomes. Intuitively, we expect that long chromosomes should have more crossovers than short ones and that this relationship will be reflected in the lengths of their genetic maps. For the most part, our assumption is true; however, within a chromosome some regions are more prone to crossing over than others. Thus, distances on the genetic map do not correspond exactly to physical distances along the chromosome's cytological map (**FIGURE 7.23**). Crossing over is less likely to occur near the ends of a chromosome

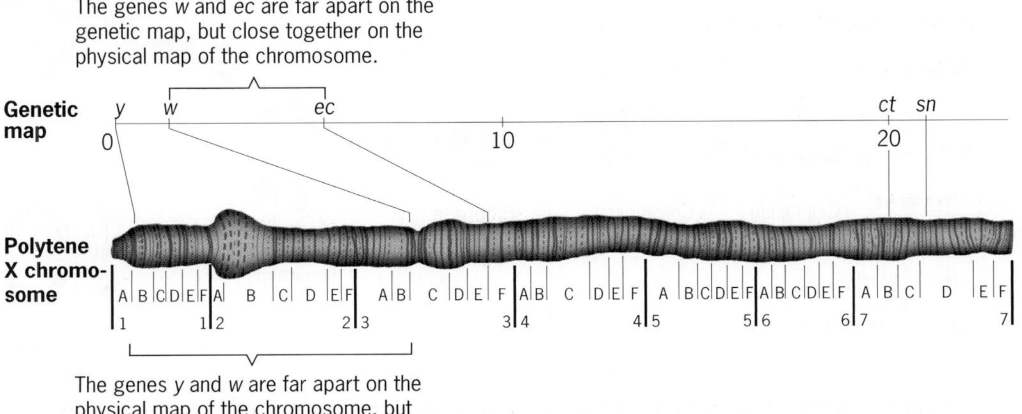

Figure 7.23 ▶ Left end of the polytene X chromosome of *Drosophila* and the corresponding portion of the genetic map showing the genes for *yellow* body (*y*), *white* eyes (*w*), *echinus* eyes (*ec*), *cut* wings (*ct*), and *singed* bristles (*sn*).

and also around the centromere; consequently, these regions are condensed on the genetic map. Other regions, in which crossovers occur more frequently, are expanded.

Even though there is not a uniform relationship between genetic and physical distance, the genetic and cytological maps of a chromosome are colinear; that is, particular sites have the same order. Recombination mapping therefore reveals the true order of the genes along a chromosome. However, it does not tell us the actual physical distances between them.

KEY POINTS

- In *Drosophila,* genes can be localized on maps of the polytene chromosomes by combining recessive mutations with cytologically defined deletions and duplications.
- A deletion will reveal the phenotype of a recessive mutation located between its endpoints, whereas a duplication will conceal the mutant phenotype.
- Genetic and cytological maps are colinear; however, genetic distances are not proportional to cytological distances.

Tetrad Analysis in Fungi

Unique features of sexual reproduction in fungi can be exploited to map genes and chromosomes.

Among fungi in the class Ascomycetes, all the haploid products of a single meiotic cell are recovered together in a saclike structure called the **ascus.** Each haploid product is called an **ascospore.** The ascospores from a single meiotic cell provide a geneticist with the opportunity to analyze the course of the preceding meiotic divisions in detail. As we shall see, they also provide a powerful way to construct genetic maps.

The yeast *Saccharomyces cerevisiae* is the preeminent fungus—indeed, the preeminent microbial eukaryote—used in genetic analysis today. This organism has 16 chromosomes. It came on the genetics scene in the 1940s and has since been used in innumerable experiments to elucidate the structure and function of genes. Haploid yeast cells of opposite mating types (a and α) fuse to form a diploid cell, which then undergoes meiosis to produce four haploid ascospores inside an ascus. Each ascospore receives one of the four chromatids that were present in each chromosome pair during prophase of the first meiotic division (**FIGURE 7.24**). If the ascospores are removed from the ascus carefully, each can be induced to germinate and divide mitotically to form a colony of identical, haploid cells. Phenotypic analysis of these colonies can provide information about the genotypes of the ascospores that produced them. For instance, if a colony grows on medium supplemented with leucine, but does not grow on a minimal medium lacking leucine, it must be mutant for a gene involved in leucine biosynthesis. The coordinated analysis of colonies derived from the four ascospores therefore allows a researcher to determine the genotype of each product of a single meiotic cell. Because these colonies are studied in groups of four, this method is called **tetrad analysis.**

The ascospores in a yeast ascus are arranged in no particular order. We say that they are **unordered tetrads** of ascospores. Another fungus, the bread mold *Neurospora crassa*, produces asci in which the ascospores are arranged in the order in which they were created during the meiotic divisions. *Neurospora* is a multicellular fungus. It grows as a branching mass of haploid filaments called a mycelium. Each cell in the mycelium contains seven different chromosomes. Specialized cells from different

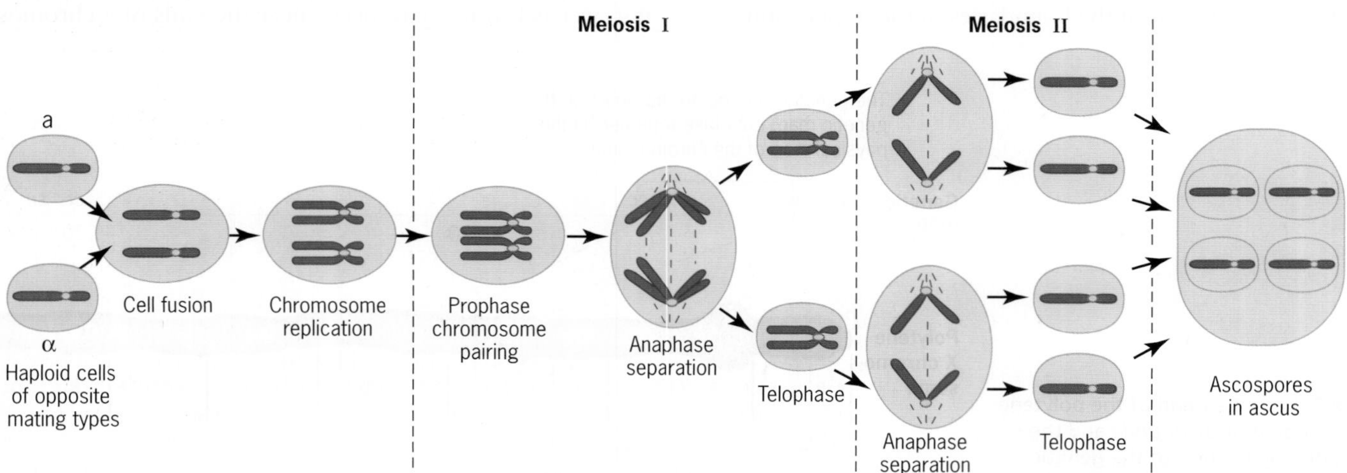

Figure 7.24 ▶ Formation of tetrads in *Saccharomyces cerevisiae*, or baker's yeast. All four products of meiosis are recovered in the ascus.

Neurospora strains can fuse—if they have opposite mating types—to create a diploid cell, which then undergoes meiosis to produce four haploid nuclei within a long, narrow ascus (**FIGURE 2.13*b***). The shape of the ascus constrains the nuclei to develop in a single plane. The four postmeiotic nuclei, stacked one on top of another, then divide mitotically to create four pairs of nuclei. Because the members of each pair were produced by mitosis, they are genetically identical. Furthermore, because of physical constraints, these twin nuclei remain next to each other in the ascus. The eight postmitotic nuclei are then partitioned into separate cells, which develop into ascospores capable of germinating into new organisms. As with yeast, *Neurospora* ascospores can be dissected from the ascus and analyzed to determine their genotypes. If the ascospores are removed carefully—from one end of the ascus to the other—a researcher can analyze them in the order in which they were created. Thus, *Neurospora* provides **ordered tetrads** of spore pairs, which, as we shall see, are more informative than the unordered tetrads of spores obtained from yeast.

DETECTING LINKAGE AND MAPPING GENES IN YEAST

We can detect linkage in yeast by analyzing tetrads of ascospores. A cross between two yeast strains with mutations on different chromosomes produces different classes of tetrads (**FIGURE 7.25**). We distinguish between alleles by using upper- and lowercase letters; the lowercase letters represent mutants, and the uppercase letters represent wild-type. The segregation and independent

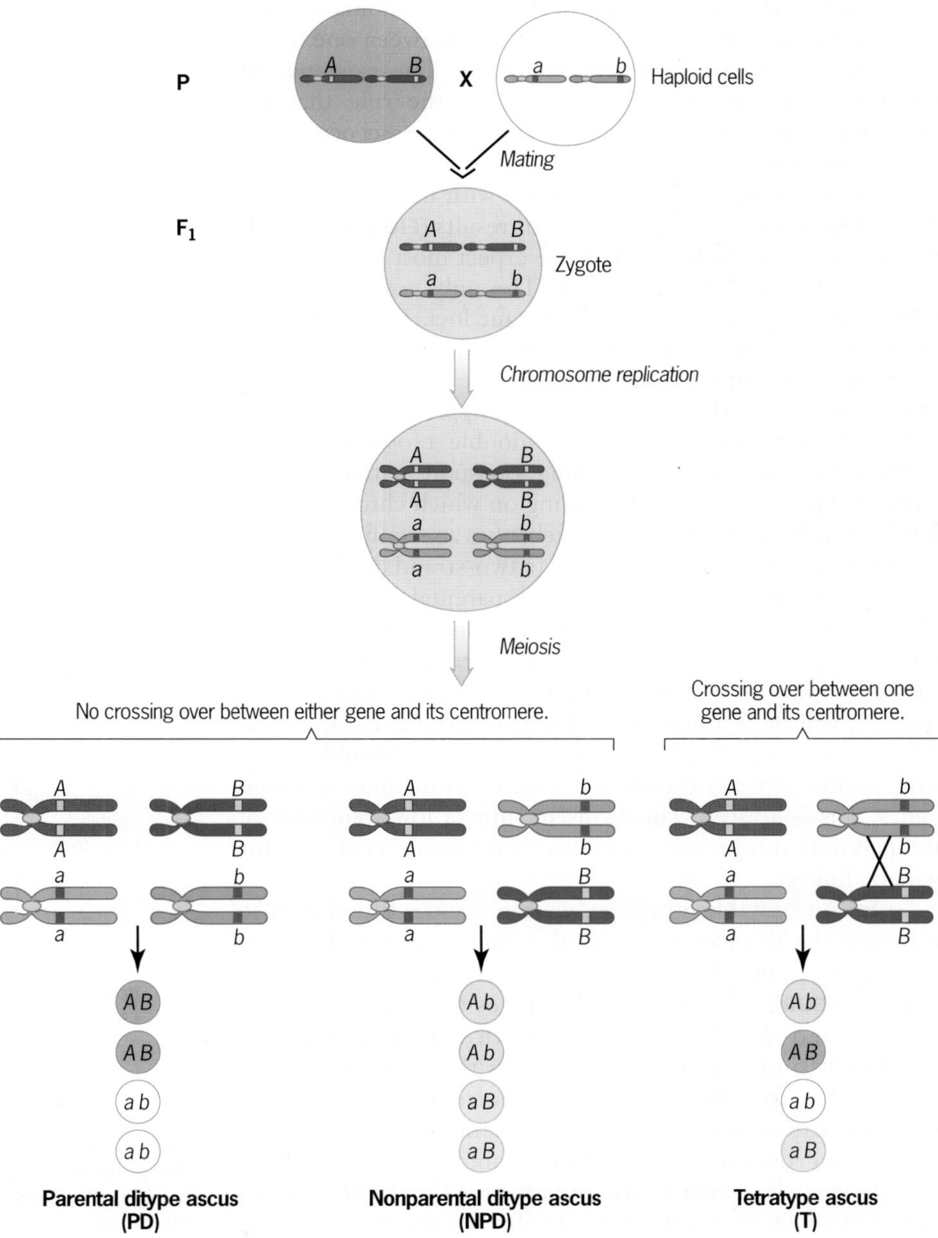

Figure 7.25 ▶ Cross with unlinked yeast mutations. The frequencies of parental and nonparental ditype asci are expected to be equal because the two metaphase I alignments of the *A/a* and *B/b* chromosomes are equally likely.

assortment of mutations on different chromosomes produce two main types of asci. In one type of ascus, two of the ascospores are genotypically like one of the parents and two are like the other parent. An ascus with this pattern is called a **parental ditype** ascus. In the other type, all the spores are recombinant; two of the spores show one genotype, and two show the reciprocal genotype. Neither of the two genotypes is like the parental genotypes; thus, the ascus is called a **nonparental ditype** ascus. Because different chromosomes assort independently during the first meiotic division, a cross involving unlinked genes produces approximately equal frequencies of parental and nonparental ditype asci.

The segregation and independent assortment of unlinked genes can also produce a third type of ascus, called a **tetratype.** In this type there are four kinds of ascospores, one that is like each of the parents and two that are recombinant. Such asci occur whenever there is crossing over between one of the genes and its centromere (**FIGURE 7.25**). The existence of tetratype asci does not, however, break the rule that for unlinked genes, parental and nonparental ditype asci occur with equal frequency.

A cross between two yeast strains with mutations on the same chromosome produces different results (**FIGURE 7.26**). If the mutations are tightly linked, we expect most of the asci to show the parental ditype pattern. The only exceptions arise from crossing over between the mutant loci. For example, a single crossover between these loci will produce an ascus with four kinds of ascospores: one spore like one parent, one like the other, and two with recombinant genotypes. Such an ascus shows the tetratype pattern. If a double crossover occurs between the two loci, the resulting ascospore pattern will be either ditype or tetratype, depending on which chromatids are involved in the exchanges. If the exchanges are confined to only two of the chromatids in the tetrad (a two-strand double crossover), the resulting ascus will have the parental ditype pattern. If the exchanges affect three of the chromatids (a three-strand double crossover), the ascus will have a tetratype pattern, and if they affect four of the chromatids (a four-strand double crossover), it will have a nonparental ditype pattern. Nonparental ditype asci are produced only when a four-strand double crossover occurs between the mutant loci. Even for loosely linked genes, this is a rare event. Consequently, a low frequency of nonparental ditype asci is considered strong evidence for genetic linkage.

Once we establish linkage, we can use the tetrad data to estimate the distance between the genes. A cross between two yeast strains illustrates how genetic distances are estimated (**FIGURE 7.27**). One strain carries two recessive mutations, *py* and *th*, which prevent growth unless the vitamins pyridoxine and thiamin are added to the medium; the *py* mutation causes the pyridoxine requirement, and the *th* mutation causes the thiamin requirement. The other strain, which can grow without any vitamin supplements, carries the wild-type alleles of these two mutations, *PY* and *TH*. Diploid cells formed by crossing the two haploid strains were induced

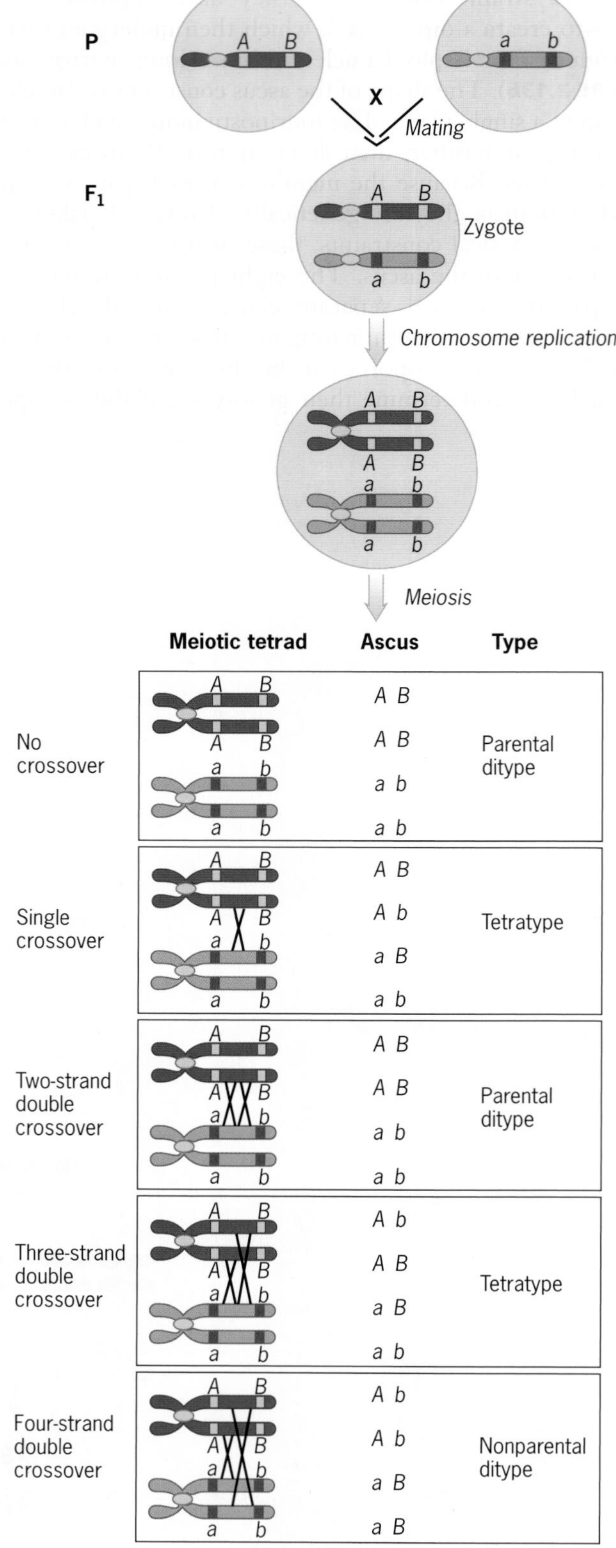

Figure 7.26 ▶ Cross with linked yeast mutations. The frequencies of parental and nonparental ditype asci are expected to be unequal. The nonparental ditype asci, which are produced by four-strand double crossovers, should be much less numerous than the parental ditype asci.

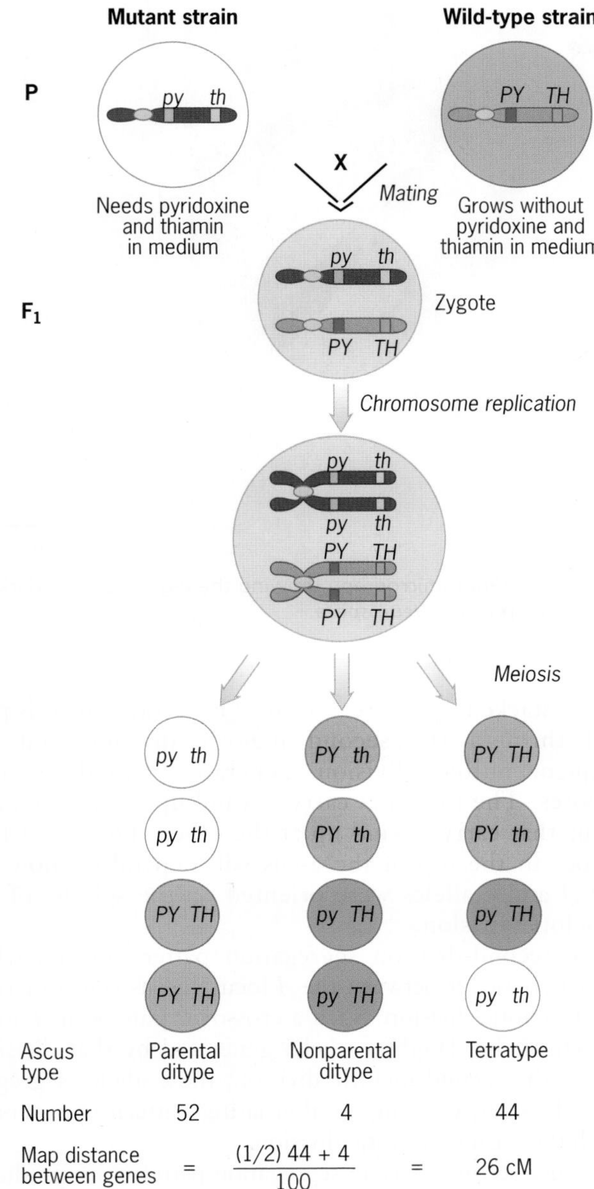

Figure 7.27 ▶ Tetrad analysis with two linked yeast mutations, *py* (pyridoxine requirement) and *th* (thiamin requirement).

to go through meiosis, and the ascospores from 100 of the resulting tetrads were isolated and grown into independent yeast colonies. Cells from each colony were then separately tested for their ability to grow in the absence of pyridoxine or thiamin, thereby revealing the genotype of the spore from which they came. When all the tests were finished, each tetrad of spores could be classified into one of the three possible types.

The data indicate that the two genes are linked because nonparental ditype tetrads occur much less frequently than do parental ditype tetrads. To calculate the distance between the genes, we must estimate the average number of crossovers between them on a chromosome. A simple estimate is the frequency of recombination, which we obtain from the formula

$$\text{Recombination frequency} = [(1/2)\text{T} + \text{NPD}]/\text{total asci}$$

Here T and NPD are the observed numbers of tetratype and nonparental ditype tetrads, respectively. We multiply T by one-half because only half the chromatids in tetratype tetrads are genetically recombinant. However, all the chromatids in a nonparental ditype tetrad are recombinant; thus, we count these tetrads fully in the formula. We do not count the parental ditype tetrads because none of their chromatids is genetically recombinant.

Applying the formula, we find that the recombination frequency = [(1/2)44 + 4]/100 = 0.26, or a distance of 26 centiMorgans.

This mapping procedure usually underestimates the true distance between genes, albeit slightly. Two-strand double crossovers, which produce parental ditype tetrads, are not counted at all, and three-strand double crossovers, which produce tetratype tetrads, are not counted correctly. In these latter tetrads, the average number of exchanges per chromatid is 3/4 rather than 1/2; consequently, the coefficient 1/2 in the formula is too small. To compensate for these shortcomings, some researchers prefer to estimate the map distance between genes with the formula

$$\text{Map distance} = [(1/2)\text{T} + 3\ \text{NPD}]/\text{total asci}$$

This formula corrects for the failure to count the two- and three-strand double crossovers properly by inflating the contribution of the nonparental ditype tetrads. In this case,

$$\text{Map distance} = [(1/2)44 + 3 \times 4]/100 = 34\ \text{centiMorgans}$$

Tetrad analysis is also possible in several other experimental organisms, including the unicellular green alga *Chlamydomonas reinhardii*.

MAPPING CENTROMERES USING ORDERED TETRAD DATA

The ordered tetrads of spore pairs obtained from crosses between different strains of *Neurospora* can be analyzed for linkage between genes just as we have analyzed the unordered tetrads of yeast. However, *Neurospora*'s ordered tetrads have one great advantage over unordered tetrads: they readily allow us to locate a chromosome's centromere on a genetic map. Centromeres are the points at which spindle fibers attach to the chromosomes to move them during meiosis. In *Neurospora* we can map these important loci by studying the segregation patterns of nearby genes. **FIGURE 7.28** illustrates the principles.

To map a chromosome's centromere, let's assume that the *A* locus is a short distance away from the centromere and that we can distinguish its two alleles, *A* and *a*, in haploid *Neurospora*. (They might, for example, produce different pigments in the ascospores; see **FIGURE 7.29.**) A heterozygous *Neurospora* cell going through meiosis can segregate these alleles in two different ways. One type of segregation produces an ascus in which the ascospores with the *A* allele are all at one end

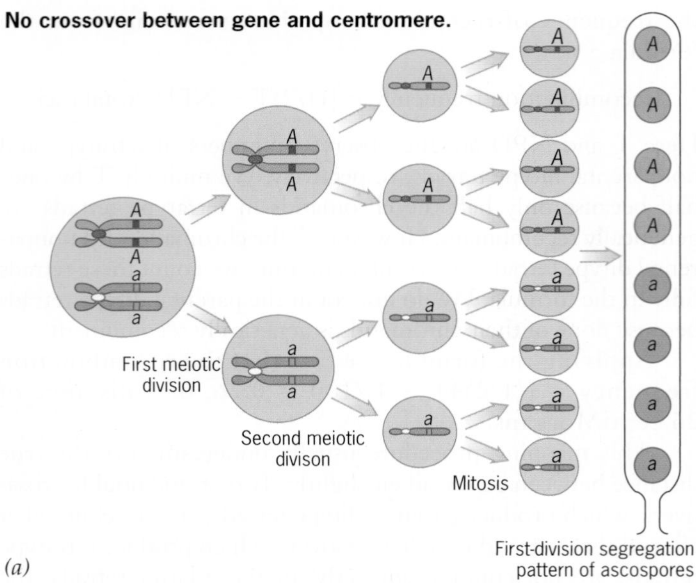

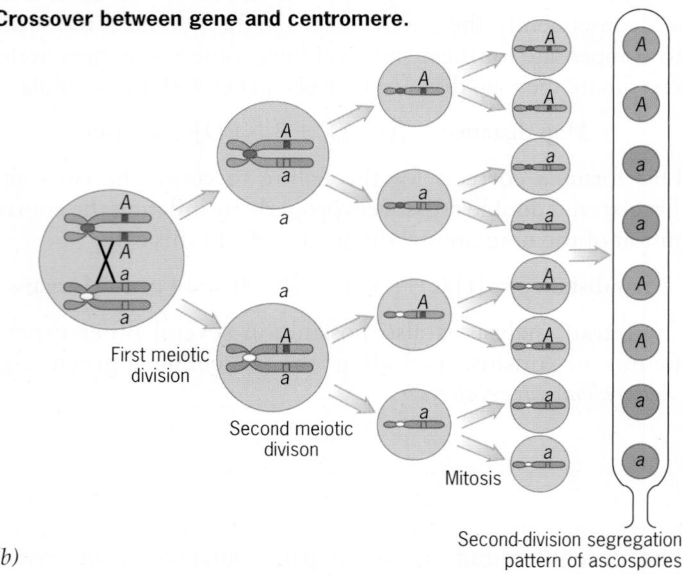

Figure 7.28 ▶ Segregation patterns in *Neurospora* asci. (*a*) First-division segregation, resulting from the absence of a crossover between a gene and its centromere. (*b*) Second-division segregation, resulting from a crossover between a gene and its centromere.

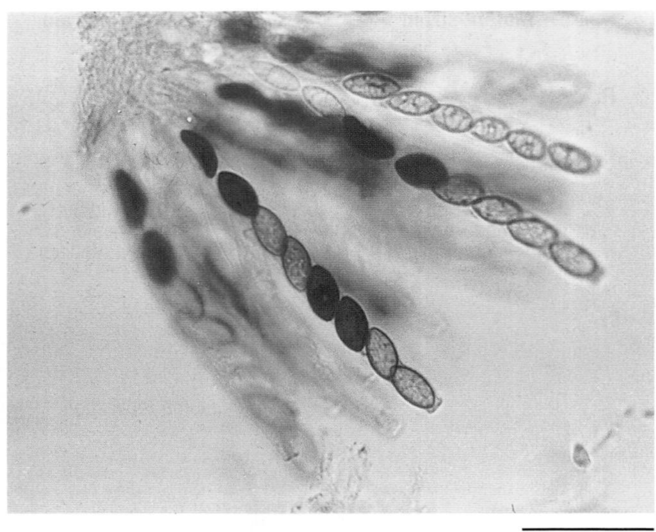

Figure 7.29 ▶ Photomicrograph showing the segregation of dark and light ascospores in *Neurospora*.

and those with the *a* allele are at the other; this is called the *first-division segregation* pattern. The other type produces an alternating array of ascospores; this is called the *second-division segregation* pattern.

The genetic basis for these two patterns is simple. The first-division segregation pattern occurs when the *A* and *a* alleles separate in the first meiotic division; that is, there is no crossing over. The sister chromatids with the *A* allele move to one pole, and the sister chromatids with the *a* allele move to the other. Because this positioning takes place within the narrow confines of the developing ascus, the alleles become "stacked" in a linear array. This array, which is preserved through the second meiotic division and the subsequent mitotic division, generates a total of eight ascospores. The four that carry *A* wind up at one end, and the four that carry *a* wind up at the other. The set of four that goes to the top of the ascus will depend on how the paired *A* and *a* alleles were oriented on the spindle of the first meiotic division.

The second-division segregation pattern occurs when there is a crossover between the *A* locus and its centromere in the first meiotic division. Such a crossover puts both *A* and *a* into each of the daughter nuclei generated by that division. Then, in the second meiotic division, these alleles segregate from each other, creating an alternating pattern that persists through the ensuing mitotic division.

Because second-division segregation patterns are produced by crossing over between the gene and its centromere, their frequency allows us to estimate the distance between these two loci. The gene-to-centromere distance is simply $(1/2) \times$ frequency of second-division segregation asci. The factor 1/2 enters in this expression because only half the chromatids in a second-division segregation tetrad have crossed over between the gene and the centromere.

As an example of ordered tetrad analysis, let's consider data from a cross performed with different strains of *Neurospora* by Mary Houlahan, George Beadle, and Hermione Calhoun (**FIGURE 7.30**). These researchers mapped a large number of *Neurospora* genes. In one experiment, they crossed a mutant strain (*thi*) that could not grow in the absence of the vitamin thiamin with a wild-type strain (thi^+); 132 asci were analyzed for first- and second-division segregation patterns. From the results, Houlahan, Beadle, and Calhoun inferred that the *thi* gene was $(1/2) \times (28/132) = 10.6$ cM from its centromere.

KEY POINTS

- The genotypes of all four products of meiosis—the tetrad—can be analyzed in Ascomycetes such as the yeast *Saccharomyces cerevisiae* and the bread mold *Neurospora crassa*.
- The unordered tetrads of yeast can be used to detect linkage and map genes.
- The ordered tetrads of *Neurospora crassa* can be used to map the positions of both centromeres and genes on chromosomes.

Linkage Analysis in Humans

Pedigree analysis provides ways of localizing genes on human chromosomes.

To detect and analyze linkage in humans, geneticists must collect data from pedigrees. Often these data are limited or incomplete, or the information they provide is ambiguous. The task of constructing human linkage maps therefore confronts researchers with many challenging problems. Classical studies of linkage in humans focused on pedigrees in which it was possible to follow the inheritance of two or more genes simultaneously. Today, modern molecular methods permit researchers to analyze the inheritance of dozens of different markers in the same set of pedigrees. This multi-locus analysis has greatly increased the ability to detect linkage and to construct detailed chromosome maps. The linkage relationships that are easiest to study in human beings are those between genes on the X chromosome. Such genes follow a pattern of inheritance that is readily identified. If two genes show this pattern, they must be linked. Determining linkage between autosomal genes is much more difficult. The human genome has 22 different autosomes, and a gene that does not show X-linkage could be on any one of them. Which autosome is the gene on, what other genes are linked to it, and what are the map positions of these genes? These are challenging questions for the human geneticist.

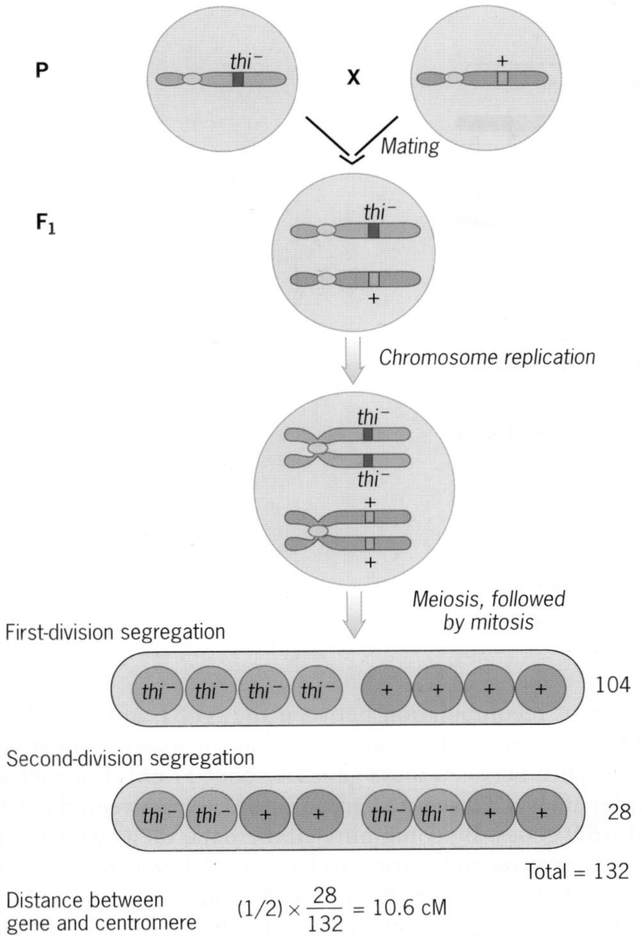

To see how linkage is detected in human pedigrees, let's examine some of the work of J. H. Renwick and S. D. Lawler. In 1955 these researchers reported evidence for linkage between the gene controlling the ABO blood groups (see Chapter 4) and a dominant mutation responsible for a rare, autosomal disorder called the nail-patella syndrome. People with this syndrome have abnormal nails and kneecaps. A portion of one of the pedigrees that Renwick and Lawler studied is shown in **FIGURE 7.31*a*.** Each individual in this pedigree was characterized for the presence or absence of the mutation for the nail-patella syndrome, denoted *NPS1*; in addition, most of the individuals were typed for the ABO blood group.

The woman in generation II must represent a new occurrence of the *NPS1* mutation. Neither of her parents nor any of her 11 siblings showed the nail-patella phenotype. Among the five individuals who showed the nail-patella syndrome in this pedigree, all but one (III-6) of them had blood type B. This observation suggests that the *NPS1* mutation is genetically linked to the *B* allele of the *ABO* blood group locus. If we assume this inference to be correct, then the woman in generation II must have the genotype *NPS1 B/+ O;* that is, she is a repulsion heterozygote. Her husband's genotype is clearly *+O/+O*.

FIGURE 7.31*b* illustrates the genetic phenomena underlying this pedigree and suggests a strategy to estimate, albeit crudely, the distance between the *NPS1* and *ABO* loci. The mating indicated in **FIGURE 7.24*b*** is essentially a testcross. The woman II-1 can produce four different kinds of gametes, two carrying recombinant chromosomes and two carrying nonrecombinant chromosomes. When these gametes are combined with the single type of gamete (+*O*) produced by the man II-2, four different genotypes can result. As the pedigree in **FIGURE 7.31*a*** shows, II-1 and II-2 produced all four

$$(1/2) \times \frac{28}{132} = 10.6 \text{ cM}$$

Figure 7.30 ► Centromere mapping in *Neurospora*. The mutant strain, thi^-, requires thiamin for growth.

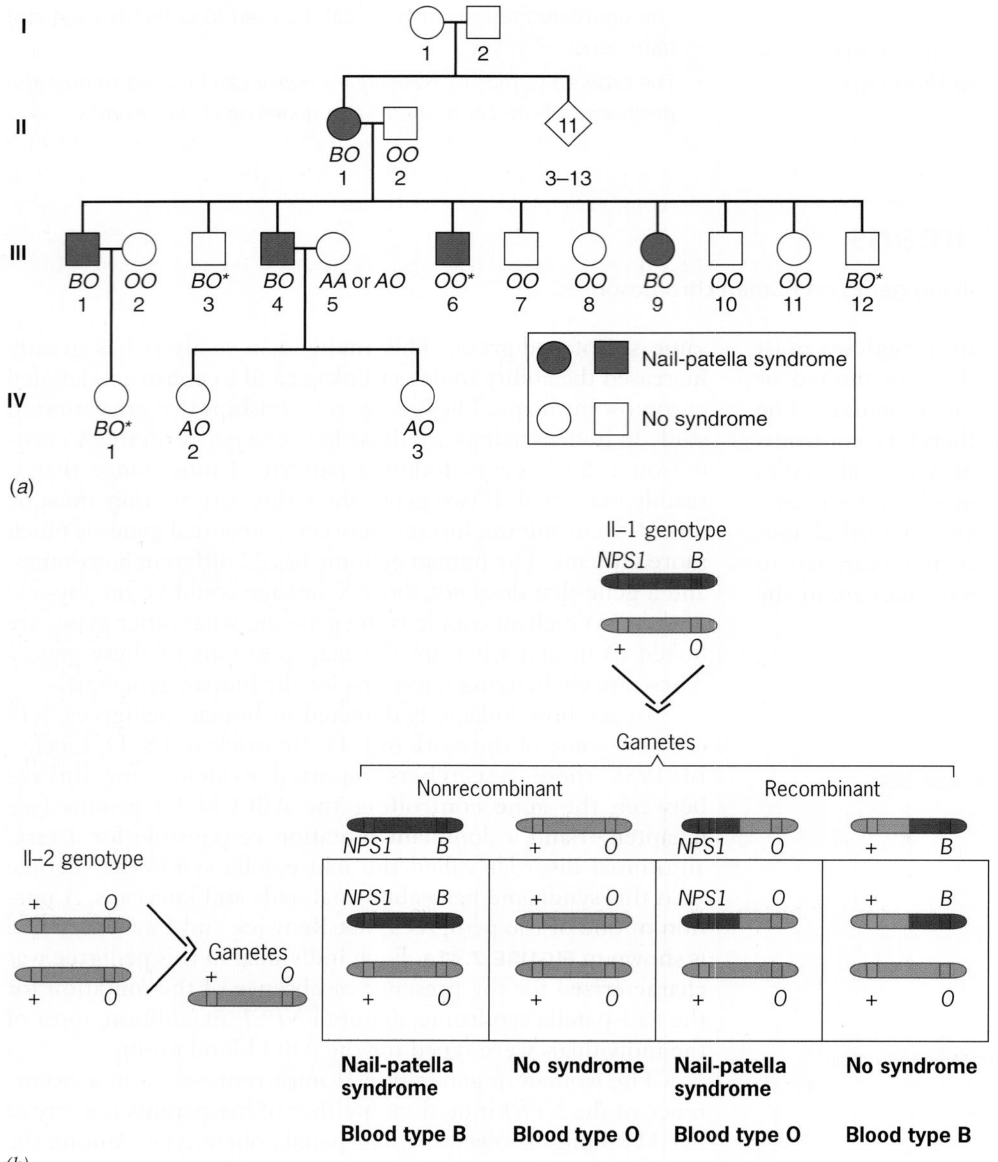

Figure 7.31 ▶ Linkage analysis in a human pedigree. (*a*) A portion of a pedigree showing linkage between the *ABO* and nail-patella loci. Individuals affected with the nail-patella syndrome are denoted by red symbols. Where known, the genotype of the *ABO* locus is given underneath each symbol. Asterisks denote recombinants. (*b*) A Punnett square showing the genotypes produced by the couple in generation II.

types of children. However, only 3 (III-3, III-6, III-12, indicated by asterisks in **FIGURE 7.31*a***) of their 10 children were recombinants; the other 7 were nonrecombinants. Thus, we can estimate the frequency of recombination between the *NPS1* and *ABO* loci as 3/10 = 30 percent. However, this estimate does not use all the information in the pedigree. To refine it, we can incorporate the information from the couples' three grandchildren, only one (IV-1) of whom was a recombinant. Altogether, then, 3 + 1 = 4 of the 10 + 3 = 13 offspring in the pedigree were recombinants. Thus, we conclude that the frequency of recombination between the *NPS1* and *ABO* loci is 4/13 = 31 percent. In terms of a linkage map, we estimate that the distance between these genes is about 31 cM. Renwick and Lawler analyzed other pedigrees for linkage between the *NPS1* and *ABO* genes. By combining all the data, they estimated the frequency of recombination to be about 10 percent. Thus, the distance between the *NPS1* and *ABO* genes is about 10 cM.

Renwick and Lawler's study of the *NPS1* and *ABO* loci established that these two genes are linked, but it could not identify the specific autosome that carried them. The first localization of a gene to a specific human autosome came in 1968, when R. P. Donahue and coworkers demonstrated that the Duffy blood group locus, denoted *FY*, is on chromosome 1. This demonstration hinged on the discovery of a variant of chromosome 1 that was longer than normal. Pedigree analysis showed that in a particular family, this long chromosome segregated with specific *FY* alleles. Thus, the *FY* locus was assigned to chromosome 1. Subsequent research has placed this locus at region 1p31 on that chromosome. By use of different techniques, the *NPS1* and *ABO* loci have been situated near the tip of the long arm of chromosome 9.

Until the early 1980s, progress in human gene mapping was extremely slow because it was difficult to find pedigrees that were segregating linked markers—say, for example, two different genetic diseases. In the 1980s, however, it became possible to identify genetic variants in the DNA itself. These variants result from differences in the DNA sequence in parts of chromosomes. For example, in one individual a particular sequence might be GAATTC on one of the DNA strands, and in another individual the corresponding DNA sequence might be GATTTC—a difference of just one nucleotide. Although we must defer to later chapters a discussion of the techniques that are used to reveal such molecular differences, here we can explore how they have helped to map human genes, including many that are involved in serious inherited diseases. If, in addition to the usual phenotypic analysis, the members of a pedigree are analyzed for the presence or absence of molecular markers in the DNA, a researcher can look for linkage between each marker and the gene under study. Then, with appropriate statistical techniques, he or she can estimate the distances between the gene and the markers that are linked to it.

This approach has allowed geneticists to map a large number of genes involved in human diseases. One of the most dramatic examples is the research that located the gene for Huntington's disease (*HD*), a debilitating and ultimately fatal neurological disorder, on chromosome 4. This effort, discussed in A Milestone in Genetics: Mapping the Gene for Huntington's Disease on the next page, analyzed large pedigrees for linkage between the *HD* gene and an array of molecular markers. Through painstaking work, the *HD* gene was mapped to within 4 cM of one of these markers. This precise localization laid the foundation for the isolation and molecular characterization of the *HD* gene itself.

Molecular markers have also made it possible to build up maps of human chromosomes from completely independent analyses. If gene *A* has been shown to be linked to marker *x* in one set of pedigrees, and gene *B* has been shown to be linked to marker *x* in another set of pedigrees, then gene *A* and gene *B* are obviously linked to each other. Thus, the analysis of these markers allows human geneticists to determine linkage relationships between genes that are not segregating in the same pedigrees.

The analysis of recombination data from pedigrees allows geneticists to construct linkage maps of chromosomes. However, except in the case of X linkage, this analysis does not tell us which chromosome is being mapped or where a particular gene resides on the physical image of that chromosome. These challenges have been addressed by applying cytological techniques such as chromosome banding and chromosome painting (Chapter 6).

KEY POINTS

- Linkage between human genes can be detected by analyzing pedigrees.
- Pedigree analysis also provides estimates of recombination frequencies to map genes on human chromosomes.

Recombination and Evolution

Recombination—or the lack of it—plays a key role in evolution.

Recombination is an essential feature of sexual reproduction. During meiosis, when chromosomes come together and cross over, there is an opportunity to create new combinations of alleles. Some of these may benefit the organism by enhancing survival or reproductive ability. Over time, such beneficial combinations would be expected to spread through a population and become standard features of the genetic makeup of the species. Meiotic recombination is therefore a way of shuffling genetic variation to potentiate evolutionary change.

EVOLUTIONARY SIGNIFICANCE OF RECOMBINATION

We can appreciate the evolutionary advantage of recombination by comparing two species, one capable of reproducing sexually and the other not. Let's suppose that a beneficial mutation has arisen in each species. Over time, we would expect these mutations to spread. Let's also suppose that while they are spreading, another beneficial mutation occurs in a nonmutant individual within each species. In the asexual organism, there is no possibility that this second mutation will be recombined with the first, but in the sexual organism, the two mutations can be recombined to produce a strain that is better than either of the single mutants by itself. This recombinant strain will be able to spread through the whole species population. In evolutionary terms, recombination can allow favorable alleles of different genes to come together in the same organism.

SUPPRESSION OF RECOMBINATION BY INVERSIONS

The gene-shuffling effect of recombination can be thwarted by chromosome rearrangements. Crossing over is usually inhibited near the breakpoints of a rearrangement in heterozygous

▶ A MILESTONE IN GENETICS: **Mapping the Gene for Huntington's Disease**

Figure 1 ▶ Dr. Nancy Wexler, one of the leading researchers on Huntington's disease, is Professor of Neuropsychology at Columbia University.

Huntington's disease (HD) is a neurodegenerative disorder that appears in people during their fourth or fifth decade of life. It is rare in most populations, with an incidence usually less than 1 in 10,000. The symptoms of HD are disquieting. People with HD gradually, but inexorably, lose motor control and mental function. The involuntary writhing and swaying they exhibit is described in medical language as "chorea," from the Greek word for "dance." The symptoms of HD worsen with age. Often severe depression sets in, the patient becomes bedridden, and eventually he or she dies.

George Huntington, an American physician, described this disease near the end of the nineteenth century. Early in the twentieth century, pedigree analysis revealed that HD is inherited as an autosomal dominant condition. Near the end of the twentieth century, the *HD* gene was isolated and its DNA sequence was determined.

The isolation and molecular characterization of the *HD* gene was made possible by research that mapped the gene to a small region of chromosome 4. James Gusella, Nancy Wexler, and collaborators analyzed pedigrees from the United States and Venezuela for linkage between *HD* and an assortment of genetic markers defined by fragments of DNA that had been characterized previously.[1] Each of the DNA fragments was able to identify small differences between the DNA sequences of a particular locus in the genome. These differences are, in effect, alleles of that locus.

One DNA fragment known as G8 defines a locus called *D4S10*. When G8 is used to analyze DNA extracted from individuals, it identifies four different alleles of the *D4S10* locus: *A, B, C,* and *D*. Gusella, Wexler, and coworkers discovered that in a pedigree from the United States, *HD* was tightly linked to the *A* allele of *D4S10*. In another pedigree from Venezuela, it was tightly linked to the *C* allele. The *DS410* locus was then mapped by cytological techniques to chromosome 4. Subsequently, it was localized more precisely to a region near the tip of the short arm of chromosome 4.

The researchers who mapped *HD* to chromosome 4 remarked on the significance of their finding:

> *The discovery of a DNA marker genetically linked to the Huntington's disease locus has profound implications both for investigations of the basic gene defect and for clinical care. . . . The discovery of a marker linked to the Huntington's disease gene makes it feasible to attempt the cloning and characterization of the abnormal gene on the basis of its map location. Understanding the nature of the genetic defect may ultimately lead to the development of improved treatments. . . . It is likely that Huntington's disease is only the first of many hereditary autosomal diseases for which a DNA marker will provide the initial indication of chromosomal location of the gene defect.*[2]

Ten years after linkage between *HD* and *D4S10* was established, the Huntington's Disease Collaborative Research Group reported the isolation and basic molecular characterization of the *HD* gene. This achievement was the culmination of much work by many people, and one of them, Dr. Nancy Wexler, had a personal as well as a professional

[1] Gusella, James F., et al. 1983. A polymorphic DNA marker genetically linked to Huntington's disease. *Nature* 306:234–238.

[2] Ibid.

condition, probably because the rearrangement disrupts chromosome pairing. Many rearrangements are therefore associated with a reduction in the frequency of recombination. This effect is most pronounced in inversion heterozygotes because the inhibition of crossing over that occurs near the breakpoints of the inversion is compounded by the selective loss of chromosomes that have undergone crossing over within the inverted region.

To see this recombination-suppressing effect, we consider an inversion in the long arm of a chromosome (**FIGURE 7.32**). If a crossover occurs between inverted and noninverted chromatids within the tetrad, it will produce two recombinant chromatids; however, both of these chromatids are likely to be lost during or after meiosis. One of the chromatids lacks a centromere—it is an *acentric* fragment—and will therefore be interest in it. Dr. Wexler's mother died of Huntington's disease.[3] So did her grandfather and three of her uncles. Because of the mode of inheritance, both she and her sister have a 50 percent chance of developing HD. Here is how Dr. Wexler describes living with the knowledge of this risk:

> *Our first reaction upon learning that we were at risk for inheriting Huntington's disease was one of horror. It's not anything you would ever expect. It feels like a total invasion of everything that you are. It feels like it's taking away everything that you wanted to be in the future. When you are in high school or college, the thought of getting seriously ill or of dying . . . well, that's not on your radar. It's not what you think about. Suddenly, you feel as if you have been turned into a very old person. At this point, my sister and I decided not to have children.*
>
> *Even though we both knew that this was a disease that tended to come on in your thirties and forties, we suddenly became extremely aware of every single little movement that we would make. I was at a Senate hearing on Huntington's disease, and I was really nervous. As I started the program, I spilled a glass of water all over everything, everybody. This was it. I was not only mortified and felt stupid, but in the back of my mind I was saying that now I have Huntington's and this is the first symptom and everyone knows it and is looking at me. So I become hyper-vigilant and very self-conscious. A person goes through a process of mourning. I know I did. I was depressed and irritable through this phase. People just can't pretend that nothing has happened.*
>
> *When people have a crisis in their life, if they don't allow themselves to mourn, they don't allow themselves to really recognize what has happened to them. It lasts much longer, and it creeps out in other ways. For me, the turning-around point was when I started to fight it, started getting involved with it. At this point in time, there was not an awful lot going on in the science of Huntington's disease. I started meeting with other families with Huntington's, and my father, who is a psychologist and really my role model in life, started the Hereditary Disease Foundation. So right away we started interacting with scientists. The science was so enthralling, so fascinating, so intriguing; and the people I met were and still are my closest friends; but this personal and scientific involvement in Huntington's disease was a lifesaver for me. It turned everything around.*

[3]The Huntington's Disease Collaborative Research Group. 1993. A novel gene containing a trinucleotide repeat that is expanded and unstable on Huntington's disease chromosomes. *Cell* 72:971–983.

QUESTIONS FOR DISCUSSION

1. A DNA marker tightly linked to a disease-causing mutation can help to identify individuals at risk to develop the disease. However, testing individuals for such markers, and divulging the resulting information—to the individuals or to others—can be a sensitive matter. In fact, in the description of one of the pedigrees they analyzed, Gusella, Wexler, and coworkers state, "Although a number of younger at-risk individuals were also analysed as part of this study, for the sake of these family members the data are not shown due to their predictive nature." And Nancy Wexler herself has decided not to have her DNA tested for the presence of the *HD* mutation. What are the pros and cons of testing an individual for DNA markers that could reveal the presence of a disease-causing mutation? If you were at risk to develop Huntington's disease, would you want to be tested?
2. The ultimate isolation and characterization of the *HD* gene began with efforts to localize the gene on a chromosome map. Current research is now focused on understanding how the protein product of this gene functions in cells. Someday this research may suggest ways of treating, or even curing, people with Huntington's disease. This approach—mapping a gene, isolating the gene's DNA, and then studying its protein product—is now standard in genetic research. Can you find examples where this approach has led to treatments or cures for inherited human diseases?

unable to move to its proper place during anaphase of the first meiotic division. The other chromatid has two centromeres and will therefore be pulled in opposite directions, forming a *dicentric chromatid bridge*. Eventually, this bridge will break and split the chromatid into pieces. Even if the acentric and dicentric chromatids produced by crossing over within the inversion survive meiosis, they are not likely to form viable zygotes. Both of these chromatids are aneuploid—duplicate for some genes and deficient for others—and such aneuploidy is usually lethal. These chromatids will therefore be eliminated by natural selection in the next generation. The net effect of this chromatid loss is to suppress recombination between inverted and noninverted chromosomes in heterozygotes.

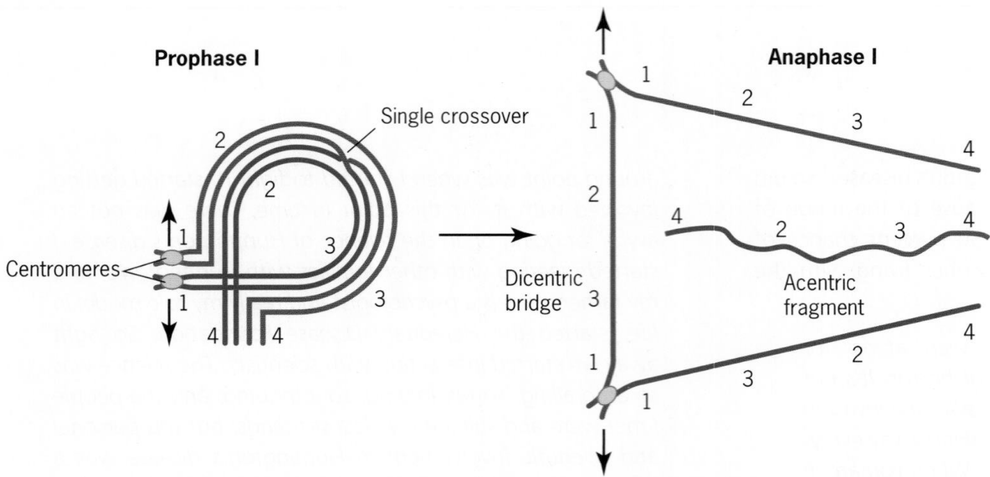

Figure 7.32 ▶ Suppression of recombination in an inversion heterozygote. The dicentric (1 2 3 1) and acentric (4 3 2 4) chromosomes formed from the crossover chromatids are aneuploid and will cause inviability in the next generation. Consequently, the products of crossing over between the inverted and noninverted chromosomes are not recovered.

Sometimes, however, euploid products result from crossing over between inverted and noninverted chromatids; for example, when two crossovers occur within the inverted region (**FIGURE 7.33**). Both of the crossovers must involve the same two chromatids—a so-called two-strand double exchange. If they involve different chromatids, the products of the exchanges will be aneuploid.

Geneticists have exploited the recombination-suppressing properties of inversions to keep alleles of different genes together on the same chromosome. Let's assume, for example, that a chromosome that is structurally normal carries the recessive alleles *a*, *b*, *c*, *d*, and *e*. If this chromosome is paired with another structurally normal chromosome that carries the corresponding wild-type alleles a^+, b^+, c^+, d^+, and e^+, the recessive and wild-type alleles will be scrambled by recombination. To prevent this scrambling, the chromosome with the recessive alleles can be paired with a wild-type chromosome that has an inversion. Unless double crossovers occur within the inverted region, this structural heterozygosity will suppress recombination. The multiply mutant chromosome can then be transmitted to the progeny as an intact genetic unit.

This recombination-suppressing technique has often been used in experiments with *Drosophila*, where the inverted chromosome usually carries a dominant mutation that permits it to be tracked through a whole series of crosses without cytological examination. Such marked inversion chromosomes are called **balancers** because they allow a chromosome of interest to be kept in heterozygous condition without recombinational breakup.

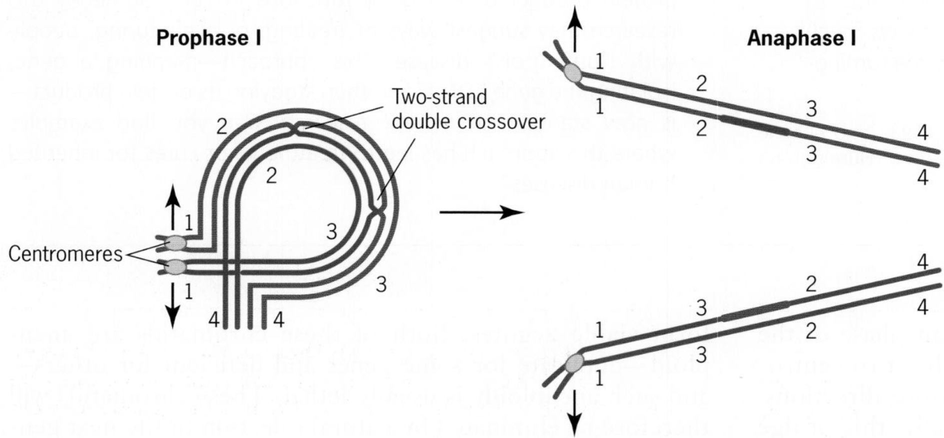

Figure 7.33 ▶ Double crossing over in an inversion heterozygote. None of the chromatids from a two-strand double crossover tetrad is aneuploid; consequently, they will be recovered in the next generation.

Suppression of recombination by inversions seems to have played an important role in the evolution of the sex chromosomes in mammals. The evidence comes from analyses by Bruce Lahn and David Page, who studied 19 genes that are present on both the human X and Y chromosomes. These shared genes occupy different postions on the X and Y chromosomes—a finding indicating that inversions have rearranged them relative to one another during the course of evolution. In addition, the DNA sequences of the X- and Y-linked copies of these shared genes have diverged from one another to different extents. By analyzing variation in the extent of divergence, Lahn and Page have discerned four "evolutionary strata" in the human sex chromosomes—regions in which recombination has been suppressed for different lengths of evolutionary time. Lahn and Page conjecture that the X and Y chromosomes originated from a pair of autosomes sometime after the mammalian evolutionary line diverged from the line of ancient reptiles that led to dinosaurs, crocodiles, and birds. Between 240 and 320 million years ago, an inversion in what was to become the Y chromosome led to regional suppression of recombination between the X and the Y. In the lineage that ultimately led to humans, at least three additional inversions occurred, two of them sometime between 80 and 130 million years ago, and one of them between 30 and 50 million years ago. The net effect of these inversions has been to suppress recombination between most of the regions on the X and Y chromosomes. Through natural selection, functional genes have been retained on the X chromosome, but on the Y chromosome most of the genes have degenerated through the accumulation of random mutations. Thus, today the Y chromosome has many fewer functional genes than the X chromosome, and the ones that remain are arranged in a different order (**FIGURE 7.34**).

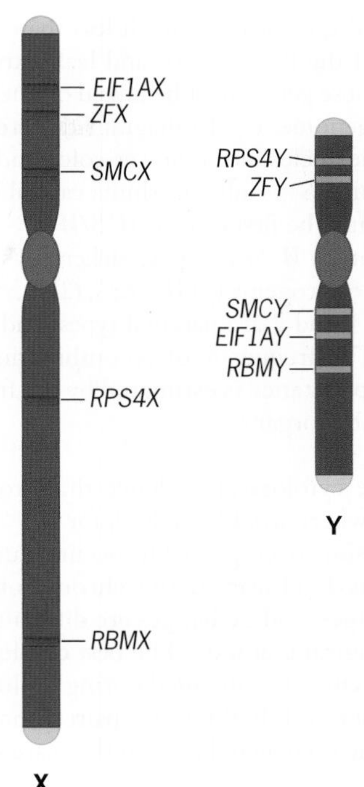

Figure 7.34 ▶ Order of shared genes outside the pseudoautosomal regions on the human X and Y chromosomes.

GENETIC CONTROL OF RECOMBINATION

It is not surprising that a process as important as recombination should be under genetic control. Studies with several organisms, including yeast and *Drosophila*, have demonstrated that recombination involves the products of many genes. Some of these gene products play a role in chromosome pairing, others catalyze the process of exchange, and still others help to rejoin broken chromatid segments. We will consider some of these activities in greater detail in Chapter 13.

One curious phenomenon, which no one has yet explained, is that there is no crossing over in *Drosophila* males. In this regard, *Drosophila* is different from most species, including our own, where crossing over occurs in both sexes. In addition, we know that the amount of recombination varies among species. Perhaps the events that lead to recombination are themselves subject to evolutionary change.

KEY POINTS

- Recombination can bring favorable mutations together.
- Chromosome rearrangements, especially inversions, can suppress recombination.
- Recombination is under genetic control.

▶ Basic Exercises

ILLUSTRATE BASIC GENETIC ANALYSIS

1. An inbred strain of snapdragons with violet flowers and dull leaves was crossed to another inbred strain with white flowers and shiny leaves. The F_1 plants, which all had violet flowers and dull leaves, were backcrossed to the strain with white flowers and shiny leaves, and the following F_2 plants were obtained: 50 violet, dull; 46 white, shiny; 12 violet, shiny; and 10 white, dull. (a) Which of the four classes in the F_2 are recombinants? (b) What is the evidence that the genes for flower color and leaf texture are linked? (c) Diagram the crosses of this experiment. (d) What is the frequency of recombination between the flower color and leaf texture genes? (e) What is the genetic map distance between these genes?

Answer: (a) The last two classes—violet, shiny and white, dull—in the F_2 are recombinants. Neither of these combinations of phenotypes was present in the strains used in the initial cross. (b) The recombinants are 18.6 percent of the F_2 plants—much less than the 50 percent that would be expected if the flower color and leaf texture genes were unlinked. Therefore, these genes must be linked on the same chromosome in the snapdragon genome. (c) To diagram the crosses, we must first assign symbols to the alleles of the flower color and leaf texture genes: W = violet, w = white; S = dull, s = shiny; capital letters indicate that the allele is dominant. The first cross is $W\,S/W\,S \times w\,s/w\,s$, yielding F_1 plants with the genotype $W\,S/w\,s$. The backcross is $W\,S/w\,s \times w\,s/w\,s$, yielding four classes of progeny: (1) $W\,S/w\,s$, (2) $w\,s/w\,s$, (3) $W\,s/w\,s$, and (4) $w\,S/w\,s$. Classes 1 and 2 are parental types, and classes 3 and 4 are recombinants. (d) The frequency of recombination is 18.6 percent. (e) The genetic map distance is estimated by the frequency of recombination as 18.6 centiMorgans.

2. What is the cytological evidence that crossing over has occurred? When and where would you look for it?

Answer: Crossing over probably occurs during early to mid-prophase of meiosis I. However, the chromosomes are not easily analyzed in these stages, and exchanges are difficult, if not impossible, to identify by cytological methods. The best cytological evidence that crossing over has occurred is obtained during diplonema near the end of the prophase of meiosis I. In this stage, paired homologues repel each other slightly, and the exchanges between them are seen as chiasmata.

3. A geneticist has estimated the number of exchanges that occurred during meiosis on each of 100 chromatids that were recovered in gametes. The data are as follows:

Number of Exchanges	Frequency
0	18
1	20
2	40
3	16
4	6

What is the genetic length in centiMorgans of the chromosome analyzed in this study?

Answer: The genetic length of a chromosome is the average number of exchanges on a chromatid at the end of meiosis. For the data at hand, the average is $0 \times (18/100) + 1 \times (20/100) + 2 \times (40/100) + 3 \times (16/100) + 4 \times (6/100) = 1.72$ Morgans or 172 centiMorgans.

4. *Drosophila* females heterozygous for three recessive X-linked markers, y (yellow body), ct (*cut* wings), and m (*miniature* wings), and their wild-type alleles were crossed to $y\ ct\ m$ males. The following progeny were obtained:

Phenotypic Class	Number
1. yellow, cut, miniature	30
2. wild-type	33
3. yellow	10
4. cut, miniature	12
5. miniature	8
6. yellow, cut	5
7. yellow, miniature	1
8. cut	1
	Total: 100

(a) Which classes are parental types? (b) Which classes represent double crossovers? (c) Which gene is in the middle of the other two? (d) What was the genotype of the heterozygous females used in the cross? (Show the correct linkage phase as well as the correct order of the markers along the chromosome.)

Answer: (a) The parental classes are the most numerous; therefore, in these data, classes 1 and 2 are parental types. (b) The double crossover classes are the least numerous; therefore, in these data, classes 7 and 8 are the double crossover classes. (c) The parental classes tell us that all three mutant alleles entered the heterozygous females on the same X chromosome; the other X chromosome in these females must have carried all three wild-type alleles. The double crossover classes tell us which of the three genes is in the middle because the middle marker will be separated from each of the flanking markers by the double exchange process. In these data, the ct allele is separated from y and m in the double crossover classes; therefore, the ct gene must lie between the y and m genes. (d) The genotype of the heterozygous females used in the cross must have been $y\ ct\ m/+\,+\,+$.

5. A *Drosophila* geneticist has conducted experiments to localize the *singed* (*sn*) bristle gene on the cytological map of the X chromosome. Males hemizygous for a recessive *sn* mutation were mated to females that carried various deficiencies (symbolized *Df*) in the X chromosome balanced over a multiply inverted X chromosome marked with the semidominant mutation for *Bar* (*B*) eyes. Thus, the crossing scheme was *sn/Y* males × *Df/B* females. The results of crosses with four different deficiencies are as follows:

Deficiency	Breakpoints	Phenotype of Non-Bar Daughters
1	2F; 3C	wild-type
2	4D; 5C	wild-type
3	6F; 7E	singed
4	7C; 8C	singed

The cytological map of the X chromosome is divided into 20 numbered sections, each subdivided into subsections A–F. Where is the *singed* gene on this cytological map?

Answer: The non-Bar daughters that were examined for the singed phenotype were genotypically *Df/sn*. The *singed* mutation was "uncovered" by two of the deficiencies, 3 and 4; thus, it must lie in the deleted region on the X chromosome that is common to both—that is, in region 7C-7E.

6. A yeast geneticist has analyzed 100 tetrads in a cross involving two genes; 60 of the tetrads are parental ditypes, 8 are nonparental ditypes, and 32 are tetratypes. (a) What fact establishes that the two genes are linked? (b) What is the map distance between these genes?

Answer: (a) For unlinked genes, the frequencies of the parental ditype and nonparental ditype asci should be approximately equal. Their obvious inequality in these data indicates that the genes are linked. (b) The frequency of recombination between the genes is estimated by the formula [(1/2)T + NPD]/total, where T and NPD are the number of tetratype and nonparental ditype asci observed. The recombination frequency is therefore 0.24. We can use this number to estimate the genetic distance in centiMorgans—that is, 24 centiMorgans—or we can adjust the number upward to account for multiple exchanges by the formula [(1/2)T + 3 NPD]/total, which yields a map distance of 40 centiMorgans.

7. The following pedigree shows four generations of a family described in 1928 by M. Madlener. The great-grandfather, I-1, has both color blindness and hemophilia. Letting *c* represent the allele for color blindness and *h* represent the allele for hemophilia, what are the genotypes of the man's five grandchildren? Do any of the individuals in the pedigree provide evidence of recombination between the genes for color blindness and hemophilia?

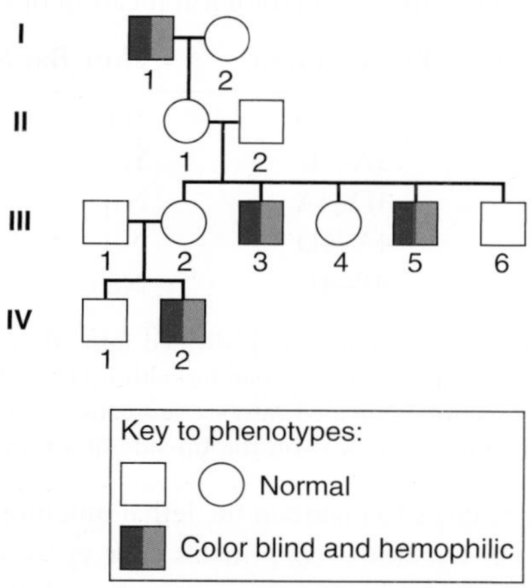

Answer: The genes for color blindness and hemophilia are X-linked. Because I-1 has both color blindness and hemophilia, his genotype must be *c h*. His daughter, II-1, is phenotypically normal and must therefore carry the nonmutant alleles, *C* and *H*, of these two X-linked genes. Moreover, because II-1 inherited both *c* and *h* from her father, the two nonmutant alleles that she carries must be present on the X chromosome she inherited from her mother. II-1's genotype is therefore *C H*/*c h*—that is, she is a coupling heterozygote for the two loci. III-2, the first granddaughter of I-1, is also a coupling heterozygote. We infer that she has this genotype because her son has both color blindness and hemophilia (*c h*), and her father is phenotypically normal (*C H*). Evidently, III-2 inherited the *c h* chromosome from her mother. Among the grandsons of I-1, two (III-3 and III-5) of them have both hemophilia and color blindness; thus, these grandsons are genotypically *c h*. The other grandson (III-6) is neither color blind nor hemophilic; his genotype is therefore *C H*. The genotype of the remaining granddaughter (III-4) is uncertain. This woman inherited a *C H* chromosome from her father. However, the chromosome she inherited from her mother could be *C H*, *c h*, *C h*, or *c H*. The pedigree does not allow us to determine which of these chromosomes she received. The most we can say about III-4's genotype is that she carries a chromosome with the *C* and *H* alleles.

None of the four grandchildren to whom we can assign genotypes provides evidence of recombination between the genes for color blindness and hemophilia. Neither do the two great-grandchildren shown in generation IV. One of these great-grandchildren is genotypically *C H*; the other is genotypically *c h*. Thus, in the pedigree as a whole there is no evidence for recombination between the *C* and *H* genes.

► Testing Your Knowledge

INTEGRATE DIFFERENT CONCEPTS AND TECHNIQUES

1. R. K. Sakai, K. Akhtar, and C. J. Dubash (1985, *J. Hered.* 76:140–141) reported data from a set of testcrosses with the mosquito *Anopheles culicifacies*, a vector for malaria in southern Asia. The data involved three mutations: *bw* (*brown* eyes), *c* (*colorless* eyes), and *Blk* (*black* body). In each cross, repulsion heterozygotes were mated to mosquitoes homozygous for the recessive alleles of the genes, and the progeny were scored as having either a parental or a recombinant genotype. Are any of the three genes studied in these crosses linked? If so, construct a map of the linkage relationships.

Cross	Repulsion Heterozygote	Progeny Parental	Progeny Recombinant	Percent Recombination
1	*bw* +/+ *c*	850	503	37.2
2	*bw* +/+ *Blk*	750	237	24.0
3	*c* +/+ *Blk*	629	183	22.5

Answer: In each cross, the frequency of recombination is less than 50 percent, so all three loci are linked. To place them on a linkage map, we estimate the distances between each pair of genes from the observed recombination frequencies:

bw—24.0—*Blk*—22.5—*c*

37.2

Notice that the recombination frequency between *bw* and *c* (37.2 percent, from Cross 1) is substantially less than the actual distance between these genes (46.5). This shows that for widely separated genes, the recombination frequency underestimates the true map distance.

2. Singed bristles (*sn*), crossveinless wings (*cv*), and vermilion eye color (*v*) are due to recessive mutant alleles of three X-linked genes in *Drosophila melanogaster*. When a female heterozygous for each of the three genes was testcrossed with a singed, crossveinless, vermilion male, the following progeny were obtained:

Class	Phenotype	Number
1	singed, crossveinless, vermilion	3
2	crossveinless, vermilion	392
3	vermilion	34
4	crossveinless	61
5	singed, crossveinless	32
6	singed, vermilion	65
7	singed	410
8	wild-type	3
		Total: 1000

What is the correct order of these three genes on the X chromosome? What are the genetic map distances between *sn* and *cv*, *sn* and *v*, and *cv* and *v*? What is the coefficient of coincidence?

Answer: Before attempting to analyze these data, we must establish the genotype of the heterozygous female that produced the eight classes of offspring. We do this by identifying the two parental classes (2 and 7), which are the most numerous in the data. These classes tell us that the heterozygous female had the *cv* and *v* mutations on one of her X chromosomes and the *sn* mutation on the other. Her genotype was therefore (*cv* + *v*)/(+*sn*+), with the parentheses indicating uncertainty about the gene order.

To determine the gene order, we must identify the double crossover classes among the six types of recombinant progeny. These are classes 1 and 8—the least numerous. They tell us that the *singed* gene is between *crossveinless* and *vermilion*. We can verify this by investigating the effect of a double crossover in a female with the genotype

$$\frac{cv + v}{+ sn +}$$

Two exchanges in this genotype will produce gametes that are either *cv sn v* or + + +, which correspond to classes 1 and 8, the observed double crossovers. Thus, the proposed gene order—*cv sn v*—is correct.

Having established the gene order, we can now determine which recombinant classes represent crossovers between *cv* and *sn*, and which represent crossovers between *sn* and *v*.

Crossovers between *cv* and *sn:*

Class:	3		5		1		8	
Number:	34	+	32	+	3	+	3	= 72

Crossovers between *sn* and *v:*

Class:	4		6		1		8	
Number:	61	+	65	+	3	+	3	= 132

We determine the distances between these pairs of genes by calculating the average number of crossovers. Between *cv* and *sn*, the distance is 72/1000 = 7.2 cM, and between *sn* and *v* it is 132/1000 = 13.2 cM. We can estimate the distance between *cv* and *v* as the sum of these values: 7.2 + 13.2 = 20.4 cM. The linkage map of these three genes is therefore:

cv—7.2—*sn*—13.2—*v*

To calculate the coefficient of coincidence, we use the observed and expected frequencies of double crossovers:

$$c = \frac{\text{observed frequency of double crossovers}}{\text{expected frequency of double crossovers}} = \frac{0.006}{0.072 \times 0.132} = 0.63$$

which indicates only moderate interference.

3. A *Drosophila* geneticist is studying a recessive lethal mutation, *l(1)r13*, located on the X chromosome. This mutation is maintained in a stock with a balancer X chromosome marked with a semidominant mutation for *Bar* eyes (*B*). In homozygous and hemizygous condition, the *B* mutation reduces the eyes to narrow bars. In heterozygous condition, it causes the eyes to be kidney-shaped. Flies that are homozygous or hemizygous for the wild-type allele of *B* have large, spherical eyes. To maintain the *l(1)r13* mutation in stock, for each generation the geneticist crosses *B* males to *l(1)r13/B* females and selects daughters with kidney-shaped eyes for crosses with their Bar-eyed brothers. The geneticist wishes to determine the cytological location of *l(1)r13*. To accomplish this goal, she crosses *l(1)r13/B* females to various males that carry duplications for short segments of the X chromosome in their genomes. Each duplication is attached to the Y chromosome. Thus, the genotype of the males used in these crosses can be represented as *X/Y-Dp*. The geneticist screens the progeny of each cross for the presence of non-Bar sons. From the results shown in the following table, determine the cytological location of *l(1)r13*.

Dp Name	Dp Segment*	Non-Bar Sons Present
1	2D–3D	Yes
2	3A–3E	Yes
3	3D–4A	No
4	4A–4D	No
5	4B–4E	No

*The long arm of the X chromosome is divided into 20 numbered sections, starting with section 1 at the tip and ending with section 20 near the centromere. Each section is divided into six subsections, ordered alphabetically A through F. Subsection A is on the tip-side of a numbered section.

Answer: The cross to maintain the lethal mutation in stock is *B/Y* males × *l(1)r*13/*B* females → *B/Y* males (Bar eyes), *l(1)r*13/*Y* males (die), *l(1)r*13/*B* females (kidney-shaped eyes), and *B/B* females (Bar eyes). Each generation, the *B/Y* males and the *l(1)r*13/*B* females are selected for crosses to perpetuate the lethal mutation. A cross to determine the cytological location of the lethal mutation can be represented as *l(1)r*13/*B* females × *X/Y-Dp* males → *l(1)r*13/*Y-Dp* males (if viable, non-Bar eyes), *B/Y-Dp* males (Bar eyes), *l(1)r*13/*X* females (non-Bar eyes), and *B/X* females (kidney-shaped eyes). The first class of flies—males with non-Bar eyes—provides the data on whether or not a specific duplication "covers" the lethal mutation. If it does, these males will appear among the progeny in the culture. If it does not, they will not appear. From the data, we see that two duplications, *Dp 1* and *Dp 2*, cover the lethal mutation. Thus, the mutation must lie within the boundaries of these duplications—that is, somewhere between 2D and 3E. We can refine the lethal mutation's location by noting that the two duplications overlap from subsection 3A to subsection 3D. The mutation must therefore lie within the 3A–3D region of the X chromosome.

4. Mary Houlahan, George Beadle, and Hermione Calhoun (*Genetics* 34:493–507, 1949) studied two mutant strains of *Neurospora crassa*. One strain (*pdx*) could not grow without pyridoxine, and the other (*pan*) could not grow without pantothenic acid. Tetrads of ascospores from a cross between these strains were dissected and analyzed. From the results, determine the linkage relationships between the *pdx* and *pan* genes, and between each gene and its centromere. (A+ represents the wild-type allele of a gene; thus, for example, *pdx* + = *pdx* pan^+ is the genotype of a spore that cannot grow without pyridoxine but that can grow without pantothenic acid.)

Tetrad Classes

1	2	3	4	5	6
pdx +	*pdx pan*	*pdx* +	*pdx* +	*pdx* +	*pdx* +
pdx +	*pdx pan*	*pdx pan*	+ +	+ *pan*	+ *pan*
+ *pan*	+ +	+ +	*pdx pan*	*pdx* +	*pdx pan*
+ *pan*	+ +	+ *pan*	+ *pan*	+ *pan*	+ +
Number observed					
15	1	17	1	13	2

Answer: Before analyzing the data, we note that the cross was between strains with the genotypes *pdx* + and + *pan*; these genotypes are indicative of parental-type ascospores in the progeny. Nonparental (that is, recombinant) ascospores are either *pdx pan* or + +. Using this nomenclature, we can classify each tetrad as parental ditype (PD), nonparental ditype (NPD), or tetratype (T). Furthermore, because *Neurospora* tetrads are ordered, we can classify them as having first-(F) or second-(S) division segregation patterns with respect to each marker:

Tetrad class	1	2	3	4	5	6
Number observed	15	1	17	1	13	2
Type	PD	NPD	T	T	PD	T
Segregation with respect to *pdx*	F	F	F	S	S	S
Segregation with respect to *pan*	F	F	S	F	S	S

The analysis of the data involves several steps. First, we note that the NPD tetrads (1) are much less numerous than the PD tetrads (15 + 13 = 28). This observation indicates that *pdx* and *pan* are linked. Second, we use the standard mapping formula to estimate the distance between *pdx* and *pan*:

$$[(1/2)\text{T} + \text{NPD}]/\text{total asci} = [(1/2)(17 + 1 + 2) + 1]/49 = 22.4 \text{ cM}$$

Third, we estimate the distance between *pdx* and its centromere by calculating the frequency of second-division segregation patterns:

$$[(1/2)\text{S}]/\text{total asci} = [(1/2)(1 + 13 + 2)/49 = 16.3 \text{ cM}$$

Similarly, we estimate the distance between *pan* and its centromere:

$$[(1/2)\text{S}]/\text{total asci} = [(1/2)(17 + 13 + 2)/49 = 32.7 \text{ cM}$$

The combined results tell us that *pdx* and *pan* are on the same side of the centromere and that *pan* is farther away. We summarize our analysis in a linkage map:

centromere—16.3 cM—*pdx*—22.4 cM—*pan*

Notice that the frequency of second-division segregation patterns underestimates the distance between *pan* and the centromere because it overlooks a few double exchanges. We can obtain a better estimate by summing the distances between the centromere and *pdx* and between *pdx* and *pan*: 16.3 + 22.4 = 38.7 cM.

5. A woman has two dominant traits, each caused by a mutation in a different gene: cataract (an eye abnormality), which she inherited from her father, and polydactyly (an extra finger), which she inherited from her mother. Her husband has neither trait. If the genes for these two traits are 15 cM apart on the same chromosome, what is the chance that the first child of this couple will have both cataract and polydactyly?

Answer: To calculate the chance that the child will have both traits, we first need to determine the linkage phase of the mutant alleles in the woman's genotype. Because she inherited the cataract mutation from her father and the polydactyly mutation from her mother, the mutant alleles must be on opposite chromosomes, that is, in the repulsion linkage phase:

$$\frac{C\,+}{+\,P}$$

For a child to inherit both mutant alleles, the woman would have to produce an egg that carried a recombinant chromosome, *C P*. We can estimate the probability of this event from the distance between the two genes, 15 cM, which, because of interference, should be equivalent to 15 percent recombination. However, only half the recombinants will be *C P*. Thus, the chance that the child will inherit both mutant alleles is (15/2) percent = 7.5 percent.

▶ Questions and Problems

ENHANCE UNDERSTANDING AND DEVELOP ANALYTICAL SKILLS

7.1 Two yeast strains differing in three linked genes were crossed: *A B C* × *a b c*. Among the tetrads that were analyzed, one contained the following spores: *A B C*, *A b C*, *a B c*, and *a b c*. How did this tetrad originate?

7.2 If two loci are 13 cM apart, what proportion of the cells in prophase of the first meiotic division will contain a single crossover in the region between them?

7.3 Genes *a* and *b* are 20 cM apart. An $a^+\ b^+/a^+\ b^+$ individual was mated with an *a b*/*a b* individual.

(a) Diagram the cross and show the gametes produced by each parent and the genotype of the F_1.

(b) What gametes can the F_1 produce, and in what proportions?

(c) If the F_1 was crossed to *a b*/*a b* individuals, what offspring would be expected, and in what proportions?

(d) Is this an example of the coupling or repulsion linkage phase?

(e) If the F_1 were intercrossed, what offspring would be expected and in what proportions?

7.4 From a cross between individuals with the genotypes *Cc Dd Ee* × *cc dd ee*, 1000 offspring were produced. The class that was *C- D- ee* included 400 individuals. Are the genes *c*, *d*, and *e* on the same or different chromosomes? Explain.

7.5 If *a* is linked to *b*, and *b* to *c*, and *c* to *d*, does it follow that a recombination experiment would detect linkage between *a* and *d*? Explain.

7.6 A homozygous variety of maize with red leaves and normal seeds was crossed with another homozygous variety with green leaves and tassel seeds. The hybrids were then backcrossed to the green, tassel-seeded variety, and the following offspring were obtained: red, normal 201; red, tassel 211; green, normal 210; green, tassel 199. Are the genes for plant color and seed type linked? Explain.

7.7 If the recombination frequency in the previous two problems were 30 percent instead of 20 percent, what change would occur in the proportions of gametes and testcross progeny?

7.8 Another phenotypically wild-type female fruit fly heterozygous for the two genes mentioned in the previous problem was crossed to a homozygous black, vestigial male. The cross produced the following progeny: gray body, normal wings 15; gray body, vestigial wings 200; black body, normal wings 225; black body, vestigial wings 11. Do these data indicate linkage? What is the frequency of recombination? Diagram the cross, showing the arrangement of the genetic markers on the chromosomes.

7.9 A phenotypically wild-type female fruit fly that was heterozygous for genes controlling body color and wing length was crossed to a homozygous mutant male with black body (allele *b*) and vestigial wings (allele *vg*). The cross produced the following progeny: gray body, normal wings 140; gray body, vestigial wings 39; black body, normal wings 42; black body, vestigial wings 47. Do these data indicate linkage between the genes for body color and wing length? What is the frequency of recombination? Diagram the cross, showing the arrangement of the genetic markers on the chromosomes.

7.10 Answer questions (a)–(e) in question 7.3 under the assumption that the original cross was $a^+\ b/a^+\ b \times a\ b^+/a\ b^+$.

7.11 In rabbits, the dominant allele *C* is required for colored fur; the recessive allele *c* makes the fur colorless (albino). In the presence of at least one *C* allele, another gene determines whether the fur is black (*B*, dominant) or brown (*b*, recessive). A homozygous strain of brown rabbits was crossed with a homozygous strain of albinos. The F_1 were then crossed to homozygous double recessive rabbits, yielding the following results: black 21; brown 58; albino 98. Are the genes *b* and *c* linked? What is the frequency of recombination? Diagram the crosses, showing the arrangement of the genetic markers on the chromosomes.

7.12 GO In *Drosophila*, genes *a* and *b* are located at positions 22.0 and 42.0 on chromosome 2, and genes *c* and *d* are located at positions 10.0 and 25.0 on chromosome 3. A fly homozygous for the wild-type alleles of these four genes was crossed with a fly homozygous for the recessive alleles, and the F_1 daughters were backcrossed to their quadruply recessive fathers. What offspring would you expect from this backcross, and in what proportions?

7.13 In *Drosophila*, the genes *sr* (*stripe* thorax) and *e* (*ebony* body) are located at 62 and 70 cM, respectively, from the left end of chromosome 3. A striped female homozygous for e^+ was mated with an ebony male homozygous for sr^+. All the offspring were phenotypically wild-type (gray body and unstriped).

(a) What kind of gametes will be produced by the F_1 females, and in what proportions?

(b) What kind of gametes will be produced by the F_1 males, and in what proportions?

(c) If the F_1 females are mated with striped, ebony males, what offspring are expected, and in what proportions?

(d) If the F_1 males and females are intercrossed, what offspring would you expect from this intercross, and in what proportions?

7.14 In *Drosophila*, the genes *st* (*scarlet* eyes), *ss* (*spineless* bristles), and *e* (*ebony* body) are located on chromosome 3, with map positions as indicated:

st	*ss*	*e*
30	55	76

Each of these mutations is recessive to its wild-type allele (st^+, dark red eyes; ss^+, smooth bristles; e^+, gray body). Phenotypically wild-type females with the genotype $st\ ss\ e^+/st^+ ss^+ e$ were crossed with triply recessive males. Predict the phenotypes of the progeny and the frequencies with which they will occur assuming (a) no interference and (b) complete interference.

7.15 The *Drosophila* genes *vg* (*vestigial* wings) and *cn* (*cinnabar* eyes) are located at 67.0 and 57.0, respectively, on chromosome 2. A female from a homozygous strain of vestigial flies was crossed with a male from a homozygous strain of cinnabar flies. The F_1 hybrids were phenotypically wild-type (long wings and dark red eyes).

(a) How many different kinds of gametes could the F_1 females produce, and in what proportions?

(b) If these females are mated with cinnabar, vestigial males, what kinds of progeny would you expect, and in what proportions?

7.16 In tomatoes, tall vine (*D*) is dominant over dwarf (*d*), and spherical fruit shape (*P*) is dominant over pear shape (*p*). The genes for vine height and fruit shape are linked with 20 percent recombination between them. One tall plant (I) with spherical fruit was crossed with a dwarf, pear-fruited plant. The cross produced the following results: tall, spherical 81; dwarf, pear 79; tall, pear 22; dwarf spherical 17. Another tall plant with spherical fruit (II) was crossed with the dwarf, pear-fruited plant, and the following results were obtained: tall, pear 21; dwarf, spherical 18; tall, spherical 5; dwarf, pear 4. Diagram these two crosses, showing the genetic markers on the chromosomes. If the two tall plants with spherical fruit were crossed with each other, that is, I × II, what phenotypic classes would you expect from the cross, and in what proportions?

7.17 A *Drosophila* geneticist made a cross between females homozygous for three X-linked recessive mutations (*y*, *yellow* body; *ec*, *echinus* eye shape; *w*, *white* eye color) and wild-type males. He then mated the F_1 females to triply mutant males and obtained the following results:

Females	Males	Number
+ + +/*y ec w*	+ + +	475
y ec w/*y ec w*	*y ec w*	469
y + +/*y ec w*	*y* + +	8
+ *ec w*/*y ec w*	+ *ec w*	7
y + *w*/*y ec w*	*y* + *w*	18
+ *ec* +/*y ec w*	+ *ec* +	23
+ + *w*/*y ec w*	+ + *w*	0
y ec +/*y ec w*	*y ec* +	0

Determine the order of the three loci *y*, *ec*, and *w*, and estimate the distances between them on the linkage map of the X chromosome.

7.18 GO In maize, the genes *Tu*, *j2*, and *gl3* are located on chromosome 4 at map positions 90, 97, and 107, respectively. If plants homozygous for the recessive alleles of these genes are crossed with plants homozygous for the dominant alleles, and the F_1 plants are testcrossed to triply recessive plants, what genotypes would you expect, and in what proportions? Assume that interference is complete over this map interval.

7.19 Female *Drosophila* heterozygous for three recessive mutations *e* (*ebony* body), *st* (*scarlet* eyes), and *ss* (*spineless* bristles) were testcrossed, and the following progeny were obtained:

Phenotype	Number
wild-type	67
ebony	8
ebony, scarlet	68
ebony, spineless	347
ebony, scarlet, spineless	78
scarlet	368
scarlet, spineless	10
spineless	54

(a) What indicates that the genes are linked?
(b) What was the genotype of the original heterozygous females?
(c) What is the order of the genes?
(d) What is the map distance between *e* and *st*?
(e) Between *e* and *ss*?
(f) What is the coefficient of coincidence?
(g) Diagram the crosses in this experiment.

7.20 A *Drosophila* geneticist crossed females homozygous for three X-linked mutations (*y*, *yellow* body; *B*, *bar* eye shape; *v*, *vermilion* eye color) to wild-type males. The F_1 females, which had gray bodies and bar eyes with dark red pigment, were then crossed to $y\ B^+ v$ males, yielding the following results:

Phenotype	Number
yellow, bar, vermilion / wild-type	581
yellow / bar, vermilion	200
yellow, vermilion / bar	173
yellow, bar / vermilion	46

Determine the order of these three loci on the X chromosome and estimate the distances between them.

7.21 In the nematode *Caenorhabditis elegans*, the linked genes *dpy* (*dumpy* body) and *unc* (*uncoordinated* behavior) recombine with a frequency *P*. If a repulsion heterozygote carrying recessive mutations in these genes is self-fertilized, what fraction of the offspring will be both dumpy and uncoordinated?

7.22 GO Assume that in *Drosophila* there are three genes *x*, *y*, and *z*, with each mutant allele recessive to the wild-type allele. A cross between females heterozygous for these three loci and wild-type males yielded the following progeny:

Females	+ + +	1515
Males	+ + +	55
	+ + *z*	630
	+ *y z*	65
	x + +	43
	x y +	640
	x y z	52
	Total:	3000

Using these data, construct a linkage map of the three genes and calculate the coefficient of coincidence.

7.23 In *Drosophila*, the X-linked recessive mutations *prune* (*pn*) and *garnet* (*g*) recombine with a frequency of 0.4. Both of these mutations cause the eyes to be brown instead of dark red. Females homozygous for the *pn* mutation were crossed to males hemizygous for the *g* mutation, and the F_1 daughters, all with dark red eyes, were crossed with their brown-eyed brothers. Predict the frequency of sons from this last cross that will have dark red eyes.

7.24 In the following testcross, genes *a* and *b* are 30 cM apart, and genes *b* and *c* are 20 cM apart: *a* + *c*/+ *b* + × *a b c*/*a b c*. If the coefficient of coincidence is 0.5 over this interval on the linkage map, how many triply homozygous recessive individuals are expected among 1000 progeny?

7.25 *Drosophila* females heterozygous for three recessive mutations, *a*, *b*, and *c*, were crossed to males homozygous for all three mutations. The cross yielded the following results:

Phenotype	Number
+ + +	75
+ + *c*	348
+ *b c*	96
a + +	110
a b +	306
a b c	65

Construct a linkage map showing the correct order of these genes and estimate the distances between them.

7.26 GO Consider a female *Drosophila* with the following X chromosome genotype:

$$\frac{w\ dor^+}{w^+\ dor}$$

The recessive alleles *w* and *dor* cause mutant eye colors (white and deep orange, respectively). However, *w* is epistatic over *dor*; that is, the genotypes *w dor*/*Y* and *w dor*/*w dor* have white eyes. If there is 28 percent recombination between *w* and *dor*, what proportion of the sons from this heterozygous female will show a mutant phenotype? What proportion will have either red or deep orange eyes?

7.27 The total map length of the *Drosophila* genome is about 270 cM. On the average, how many chiasmata will occur in an oocyte going through meiosis?

7.28 A chromosome is 150 cM long. On the average, how many chiasmata will occur when it goes through meiosis?

7.29 A geneticist obtained the following ordered tetrad data from a cross with *Neurospora:*

Spore Pairs				
Top of Ascus		Bottom of Ascus		Number of Tetrads
(1,2)	(3,4)	(5,6)	(7,8)	
A	*A*	*a*	*a*	61
a	*a*	*A*	*A*	55
a	*A*	*a*	*A*	40
A	*a*	*A*	*a*	44
			Total =	200

What is the distance between the *A* locus and the centromere on the chromosome?

7.30 In yeast, the *ad* mutation requires adenine for growth, and the *in* mutation requires inositol for growth. The wild-type alleles of these two mutations are *AD* and *IN*. C. C. Lindegren analyzed 48 tetrads from the cross *AD IN* × *ad in*; among these tetrads, 20 were parental ditype, 7 were nonparental ditype, and 21 were tetratype. Are the *AD* and *IN* genes linked? If so, what is the distance between them?

7.31 A *Drosophila* second chromosome that carried a recessive lethal mutation, *l(2)g14*, was maintained in a stock with a balancer chromosome marked with a dominant mutation for curly wings. This latter mutation, denoted *Cy*, is also associated with a recessive lethal effect—but this effect is different from that of *l(2)g14*. Thus, *l(2)g14/Cy* flies survive, and they have curly wings. Flies without the *Cy* mutation have straight wings. A researcher crossed *l(2)g14/Cy* females to males that carried second chromosomes with different deletions (all homozygous lethal) balanced over the *Cy* chromosome (genotype *Df/Cy*). Each cross was scored for the presence or absence of progeny with straight wings.

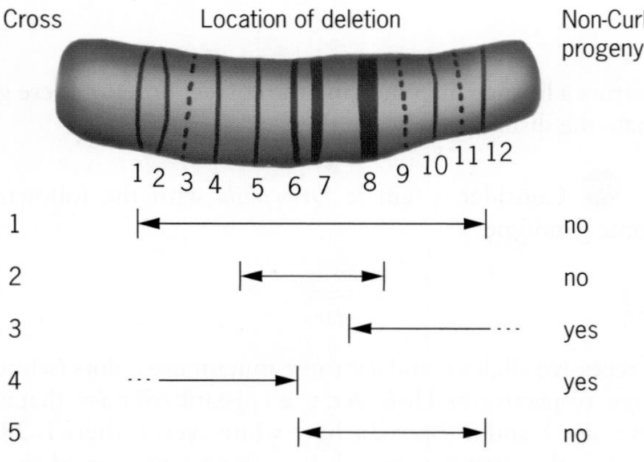

In which band is the lethal mutation *l(2)g14* located?

7.32 The following tetrad data were obtained from the cross *A B* × *a b* in *Neurospora*:

Spore Pairs				
Top of Ascus		Bottom of Ascus		Number of Tetrads
(1,2)	(3,4)	(5,6)	(7,8)	
A B	*A B*	*a b*	*a b*	755
A B	*a B*	*A b*	*a b*	230
A B	*A b*	*a B*	*a b*	15
			Total =	1000

Use these data to construct a linkage map of the two genes and the centromere on their chromosome.

7.33 Analysis of unordered yeast tetrads from the cross + + + × *a b c* yielded the following data:

Tetrad Class	Spores				Number of Asci
1	*a b c*	*a b c*	+ + +	+ + +	36
2	*a b c*	*a* + *c*	+ *b* +	+ + +	14
3	*a* + +	*a* + +	+ *b c*	+ *b c*	32
4	*a b* +	*a* + +	+ *b c*	+ + *c*	16
5	*a b* +	*a b* +	+ + *c*	+ + *c*	2

Which of these genes are linked, and what is the map distance between them?

7.34 The following tetrads were produced by a cross of a *Neurospora* strain that had white spores (*w*) and a nutritional requirement for the amino acid arginine (*arg*) with a strain that had dark spores and no arginine requirement:

Spore Pairs	Tetrads					
	1	2	3	4	5	6
1–2	*w arg*	*w arg*	*w arg*	*w arg*	*w* +	*w* +
3–4	*w arg*	*w* +	+ *arg*	+ +	*w* +	+ +
5–6	+ +	+ *arg*	+ +	+ *arg*	+ *arg*	*w arg*
7–8	+ +	+ +	*w* +	*w* +	+ *arg*	+ *arg*
No.	58	14	15	2	1	10
					Total =	100

(a) What is the map distance between the *arg* locus and the centromere? (b) What is the map distance between the *w* and *arg* loci? (c) Are the *arg* and *w* loci found on the same or different arms of the chromosome?

7.35 The order of three genes and the centromere on one chromosome of *Neurospora* is centromere —*x*—*y*—*z*. A cross between + + + and *x y z* produced an ascus with the following ordered array of ascospores (only one member of each spore pair is shown): (+ + *z*) (+ *y z*) (*x* + +) (*x y* +).

(a) Is this ascus most likely the result of a meiotic event in which 0, 1, 2, or 3 crossovers occurred?
(b) In what interval(s) did the crossover(s) most likely occur?
(c) If double or triple crossovers were involved, were they two-strand, three-strand, or four-strand multiple crossovers?

7.36 In *Neurospora*, the mutations *arg*, *thi*, and *leu* block the synthesis of arginine, thiamine, and leucine, respectively. The following ordered tetrad data come from two crosses with these mutations:

Cross	Spore Pairs				Number of Tetrads
	(1,2)	(3,4)	(5,6)	(7,8)	
arg × *thi*	*arg* +	*arg* +	+ *thi*	+ *thi*	46
	arg thi	*arg thi*	+ +	+ +	56
					Total = 100
arg × *leu*	*arg* +	*arg* +	+ *leu*	+ *leu*	155
	arg +	*arg leu*	+ +	+ *leu*	44
	arg +	+ *leu*	*arg* +	+ *leu*	1
					Total = 200

Using these data, construct the linkage map(s) for these genes. Show the position of the centromere(s).

7.37 A *Drosophila* geneticist has identified a strain of flies with a large inversion in the left arm of chromosome 3. This inversion includes two mutations, *e* (*ebony* body) and *cd* (*cardinal* eyes), and is flanked by two other mutations, *sr* (*stripe* thorax) on the right and *ro* (*rough* eyes) on the left. The geneticist wishes to replace the *e* and *cd* mutations inside the inversion with their wild-type alleles; he plans to accomplish this by recombining the multiply mutant, inverted chromosome with a wild-type, inversion-free chromosome. What event is the geneticist counting on to achieve his objective? Explain.

7.38 Two strains of maize, M1 and M2, are homozygous for four recessive mutations, *a*, *b*, *c*, and *d*, on one of the large chromosomes in the genome. Strain W1 is homozygous for the dominant alleles of these mutations. Hybrids produced by crossing M1 and W1 yield many different classes of recombinants, whereas hybrids produced by crossing M2 and W1 do not yield any recombinants at all. What is the difference between M1 and M2?

7.39 The following pedigree, described in 1937 by C. L. Birch, shows the inheritance of X-linked color blindness and hemophilia in a family. What is the genotype of II-2? Do any of her children provide evidence for recombination between the genes for color blindness and hemophilia?

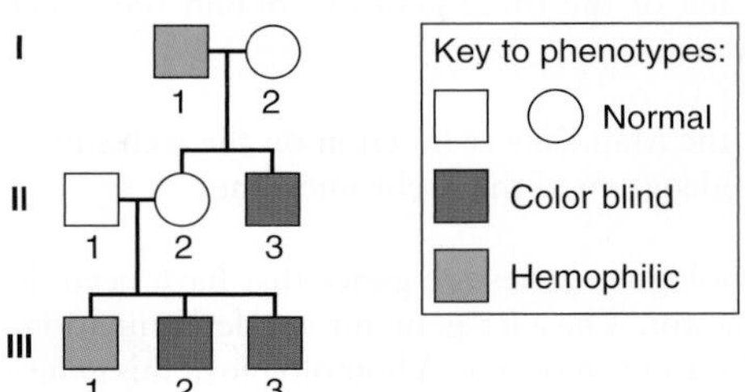

7.40 The following pedigree, described in 1938 by B. Rath, shows the inheritance of X-linked color blindness and hemophilia in a family. What are the possible genotypes of II-1? For each possible genotype, evaluate the children of II-1 for evidence of recombination between the color blindness and hemophilia genes.

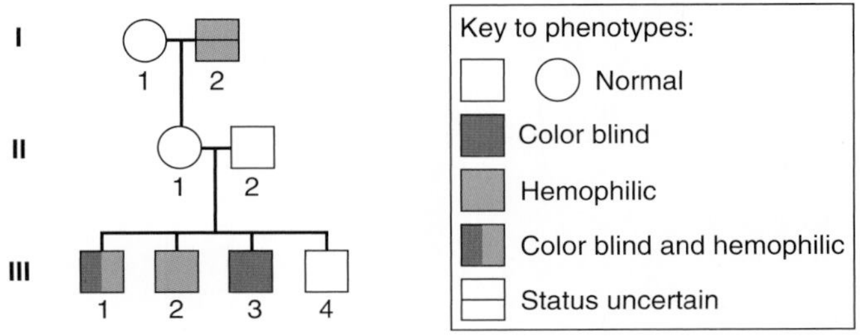

7.41 A normal woman with a color-blind father married a normal man, and their first child, a boy, had hemophilia. Both color blindness and hemophilia are due to X-linked recessive mutations, and the relevant genes are separated by 10 cM. This couple plans to have a second child. What is the probability that it will have hemophilia? color blindness? both hemophilia and color blindness? neither hemophilia nor color blindness?

▶ Genomics on the Web

at http://www.ncbi.nlm.nih.gov/

Chromosome maps were first developed by T. H. Morgan and his students, who used *Drosophila* as an experimental organism.

1. Find the genetic map positions of the genes *w* (white eyes), *m* (miniature wings), and *f* (forked bristles) on the X chromosome (also denoted as chromosome 1) of *Drosophila melanogaster*.

2. Find the positions of these three genes on the cytogenetic map of the X chromosome of *D. melanogaster*.

Hint: At the web site, click on Genomic Biology and then under Genome resources, click on Insects. From there, open the page on *Drosophila melanogaster*, and then under Related Resources in the

Resource Links sidebar, click on FlyBase, which is the database for genomic information about *Drosophila*. On the FlyBase main page, search for each of the three genes to obtain the genetic and cytological locations.

3. Use the Map Viewer function on the web site to locate *w*, *m*, and *f* on the ideogram of the X chromosome.

4. Homologous genes are genes that have been derived from a common ancestor. The *SRY* gene for sex determination in humans is located on the Y chromosome. A homologue of this gene, called *SOX3*, is located on the X chromosome. Find these two genes on the ideograms of the human sex chromosomes. In what bands do they lie?

5. *RBMX* and *RBMY* are another pair of homologous genes on the human X and Y chromosomes. Locate these two genes relative to *SOX3* and *SRY*. Considering the evolutionary history of the X and Y chromosomes, what might account for the positions of these two pairs of genes on the sex chromosomes?

Hint: Search using the "Find in This View" function on the Map Viewer page of the web site.

Chapter 8
The Genetics of Bacteria and Their Viruses

CHAPTER OUTLINE

- Viruses and Bacteria in Genetics
- The Genetics of Viruses
- The Genetics of Bacteria
- Mechanisms of Genetic Exchange in Bacteria
- The Evolutionary Significance of Genetic Exchange in Bacteria

Multi-Drug-Resistant Bacteria: A Ticking Timebomb?

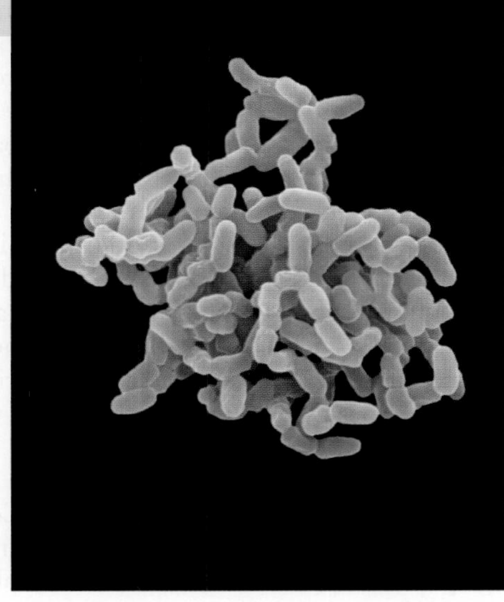

Mycobacterium tuberculosis, the bacterium that causes tuberculosis in humans.

Oscar Peterson was a happy child, the son of Norwegian immigrants who moved to the Minnesota frontier at the end of the nineteenth century. However, his happy childhood was short-lived. His mother soon became very ill, with incessant coughing, chest pains, and high fevers. She had tuberculosis (TB), a dreaded disease caused by the bacterium *Mycobacterium tuberculosis.* TB is highly contagious because *M. tuberculosis* is transmitted via aerosolized droplets produced when an infected person coughs or sneezes. The disease was often fatal because there was no effective treatment at the time. Fresh air was prescribed, so the Peterson family slept with the windows open, even during the cold winter months. Because TB is so contagious, families with the disease lived in almost total isolation. Their friends were afraid to visit for fear of contracting the disease. When Oscar was 14 years old, his mother died, and his life changed immediately. He quit school so that he could take care of his three younger brothers while his dad worked.

Thousands of frontier families like the Petersons fought to survive the scourge of TB in the first part of the twentieth century. Then, Alexander Fleming discovered penicillin, and a revolution in the treatment of bacterial diseases followed. During the 1940s and 1950s, scientists discovered an arsenal of highly effective antibiotics. As a result, the incidence of TB decreased sharply in the United States during the 1970s. Indeed, many physicians thought that TB might be totally eliminated. Unfortunately, they were wrong!

On November 16, 1991, a headline in the *New York Times* stated "A Drug-Resistant TB Results in 13 Deaths in New York Prisons." Then, a prison guard in Syracuse was killed by the same drug-resistant strain of *M. tuberculosis* as the prisoners, and, unfortunately, this drug-resistant strain was just the tip of the iceberg. Today, many strains of *M. tuberculosis* are resistant to a whole battery of drugs and antibiotics. The most resistant of these bacteria are called extensively drug-resistant (XDR-TB) strains. These drug-resistant strains are of two types: multi-drug-resistant (MDR) strains—those resistant to most normally prescribed antibiotics, and extensively drug-resistant (XDR) strains—those also resistant to the antibiotics used to treat MDR-TB. MDR and XDR strains of *M. tuberculosis* are present throughout the world, with especially high frequencies in prisons from New York to Siberia. The genetic basis of this multi-drug resistant TB is discussed later in this chapter (see Plasmids and Episomes in the section Mechanisms of Genetic Exchange in Bacteria) and in Chapter 18.

How serious a threat does the evolution of MDR and XDR bacteria pose to human health? Dr. Lee Reichman, one of the world's leading experts on TB, has referred to MDR-*M. tuberculosis* as a "timebomb." Worldwide, 2 billion people (15 million in the United States) are infected with latent *M. tuberculosis.* Of these, 8.4 million develop active TB and 2 million die every year. Although most of these cases of TB are currently treatable with antibiotics, the World Health Organization estimates that 450,000 people develop MDR-TB and XDR-TB each year, and they have concluded that an effective global response plan could save the lives of hundreds of thousands of MDR-TB and XDR-TB patients in the next few years. Perhaps we should initiate steps to confront the crisis of MDR-TB now—before the "timebomb" explodes.

Viruses and Bacteria in Genetics

Bacteria and viruses have made important contributions to the science of genetics.

We live in a world along with countless bacteria and viruses. Some bacteria, like *M. tuberculosis*, are harmful; others, like those we use to make yogurt, are helpful. Bacteria play important roles in the earth's ecosytems. They erode rock, capture energy from materials in their environments, fix atmospheric nitrogen into compounds that other organisms can use, and break down the bodies of organisms that have died. If bacteria did not carry out these functions, life as we know it would not be possible. These tiny organisms enable large, multicellular organism like us to survive.

Geneticists began to study bacteria and their viruses in the middle of the twentieth century, years after Mendel's Principles and the Chromosome Theory of Heredity had been firmly established. To the first bacterial and viral geneticists, these tiny organisms seemed to offer the possibility of extending genetic analysis to a deeper, biochemical level—indeed, to the very molecules that make up genes and chromosomes. As we shall see in this and succeeding chapters, this exciting prospect was realized. The genetic analysis of bacteria and viruses has allowed researchers to probe the chemical nature of genes and their products. All that we now call molecular biology has been founded on the study of bacteria and viruses.

For a research scientist, bacteria and viruses have several advantages compared to creatures like maize or *Drosophila*. First, they are small, reproduce quickly, and form large populations in just a matter of days. An experimenter can grow 10^{10} bacteria in a small culture tube; 10^{10} *Drosophila*, by contrast, would fill a 14 ft × 14 ft × 14 ft room. Second, bacteria and viruses can be grown on biochemically defined culture media. Because the constituents of the culture medium can be changed as desired, a researcher can identify the chemical needs of the organism and investigate how it processes these chemicals during its metabolism. Drugs such as antibiotics can also be added to the medium to kill bacteria selectively. This type of treatment allows a researcher to identify resistant and sensitive strains of a bacterial species—for example, to determine whether *M. tuberculosis* cultured from a patient is resistant to a particular antibiotic. Third, bacteria and viruses have relatively simple structures and physiology. They are therefore ideal for studying fundamental biological processes. Finally, genetic variability is easy to detect among these tiny microorganisms. If we examine bacteria or viruses, we almost always find that they manifest different phenotypes and that these differences are heritable. For example, some strains of a bacterial species can grow on a biochemically defined medium containing lactose as the only energy source, whereas other strains cannot. Strains that are not able to grow on this type of medium are mutant with respect to the metabolism of lactose. The ability to obtain mutant strains of bacteria and viruses has allowed geneticists to dissect complex phenomena such as energy recruitment, protein synthesis, and cell division at the molecular level.

The advances in molecular biology during the last few decades have provided a wealth of information about the genomes of many bacteria and viruses. Today, we know the complete nucleotide sequences of the genomes of a large number of viruses and bacteria. These sequences are providing detailed information about the genetic control of metabolism in diverse microbial species and, especially, about their evolutionary relationships. We will examine some of this information in Chapter 16 (see Comparative Genomics).

In this chapter we will concentrate on a few bacteria and viruses that have played major roles in genetic analysis. These tiny organisms include the bacterium *Escherichia coli* and two viruses that infect it. We will begin our investigation with the simplest microorganisms—the viruses that infect bacteria such as *E. coli*.

KEY POINTS

- Their small size, short generation time, and simple structures have made bacteria and viruses valuable model systems for genetic studies.
- Many basic concepts of genetics were first deduced from studies of bacteria and viruses.

The Genetics of Viruses

Viruses can only reproduce by infecting living host cells. Bacteriophages are viruses that infect bacteria. Several important genetic concepts have been discovered through studies of bacteriophages.

Viruses straddle the line between the living and the nonliving. Consider, for example, a virus that causes discoloration on the leaves of tobacco plants, a condition called tobacco mosaic disease. The tobacco mosaic virus (TMV) can be crystallized and stored on a shelf for years. In this state, it exhibits none of the properties normally associated with living systems: it does not reproduce; it does not grow or develop; it does not utilize energy; and it does not respond to environmental stimuli.

However, if a liquid suspension containing TMV is rubbed onto the leaf of a tobacco plant, the viruses in the suspension infect the cells, reproduce, utilize energy supplied by the plant cells, and respond to cellular signals. Clearly, they exhibit the properties of living systems.

Indeed, it is the simplicity of viruses that has made them ideal research tools for genetic analysis. Questions that have been difficult to answer using more complicated eukaryote systems have often been addressed using viruses. In Chapter 9, we will discuss experiments that used viruses to demonstrate that genetic information is stored in DNA and RNA. In Chapters 10, 11, and 12, we will discuss experiments that used viruses to elucidate the mechanisms of DNA replication, transcription, and translation. In this chapter, we will focus on viruses that infect bacteria: we will discuss the organization of their genomes and the methods that geneticists have developed to analyze them.

BACTERIOPHAGES T4 AND LAMBDA

Viruses that infect bacteria are called **bacteriophages** (from the Greek "to eat bacteria"). Among the many bacteriophages that have been identified, two have played especially important roles in the elucidation of genetic concepts. Both of these viruses infect the colon bacillus *Escherichia coli.* Bacteriophages can be categorized into two types—virulent and temperate—based on their lifestyles in infected cells. Bacteriophage T4 (phage T4) is a virulent phage; it uses the metabolic machinery of the host cell to produce progeny viruses and kills the host in the process. Bacteriophage lambda (λ), a temperate phage, is another coliphage (phage that infects *E. coli*); however, this phage can either kill the host cell like phage T4, or it can enter into a special association with the host and replicate its genome along with the host cell's genome during each cell duplication. The results of studies performed on bacteriophages T4 and lambda have established genetic paradigms that are relevant to understanding other types of viruses, such as the human immunodeficiency virus, HIV (see Chapter 18 for a discussion of HIV).

Bacteriophage T4

Bacteriophage T4 is a large virus that stores its genetic information in a double-stranded DNA molecule packaged inside a proteinaceous head (**FIGURE 8.1**). The virus is composed almost entirely of proteins and DNA—approximately half of each (**FIGURE 8.2**). The T4 chromosome is approximately 168,800 base pairs long and contains about 150 characterized genes and an equal number of uncharacterized sequences thought to be genes. The tail of the virus contains several important components. Its central hollow core provides the channel through which the phage DNA is injected into the bacterium. The tail sheath functions as a small muscle that contracts and pushes the tail core through the bacterial cell wall. The six tail fibers are used to locate receptors on the host cell, and the tail pins on the baseplate then attach firmly to these receptors. All of these components must function correctly for the phage to infect an *E. coli* cell successfully.

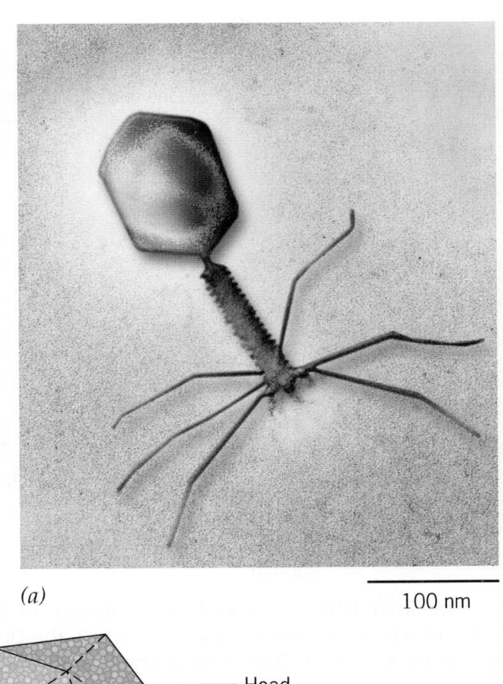

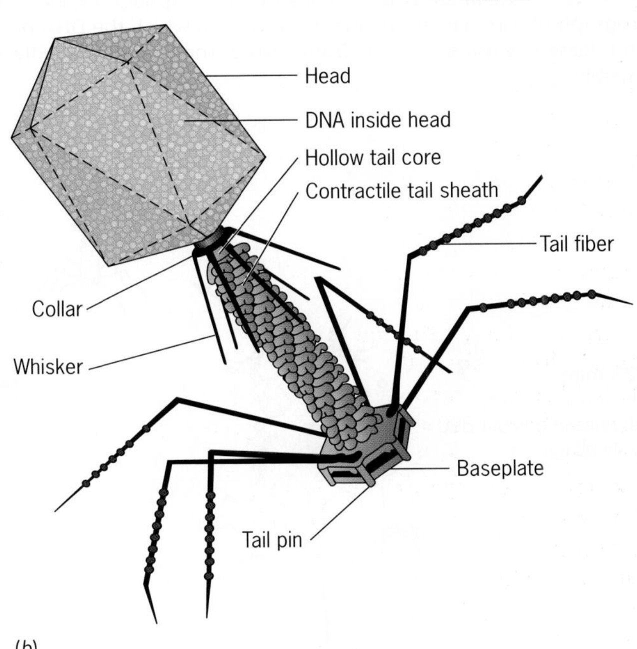

Figure 8.1 ▸ Bacteriophage T4. Electron micrograph (*a*) and diagram (*b*) showing the structure of bacteriophage T4.

Bacteriophage T4 is a **lytic phage;** when it infects a bacterium, it replicates and kills the host, producing about 300 progeny viruses per infected cell (**FIGURE 8.3**). After the phage DNA is injected into the host bacterium, it quickly (within 2 minutes) directs the synthesis of proteins that shut off the transcription, translation, and replication of bacterial genes, allowing the virus to take control of the metabolic machinery of the host. Some of the phage genes encode nucleases that degrade the host DNA. Other phage proteins initiate the replication of

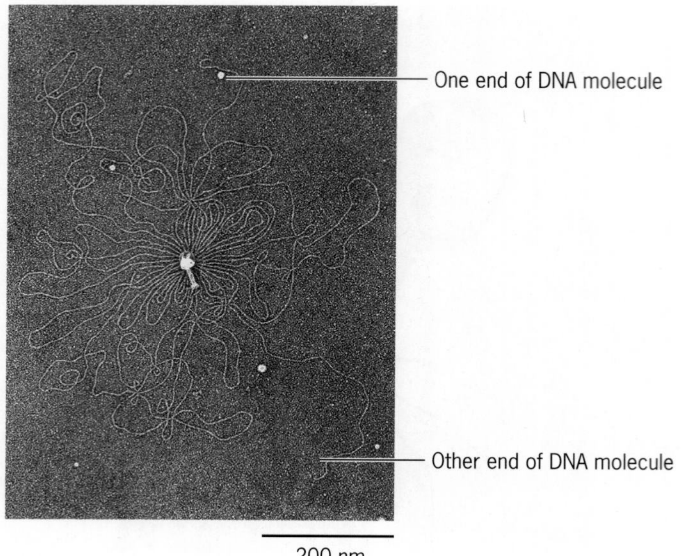
One end of DNA molecule
Other end of DNA molecule
200 nm

Figure 8.2 ▶ The linear DNA genome of bacteriophage T4. Electron micrograph of a T4 bacteriophage (center) from which the DNA has been released by osmotic shock. Both ends of the linear DNA molecule are visible.

phage DNA. Somewhat later, the genes that encode the structural components of the virus are expressed. Thereafter, the assembly of progeny phage begins; infectious progeny phage start to accumulate in the host cell at about 17 minutes after infection. At about 25 minutes after infection, a phage-encoded enzyme called *lysozyme* degrades the bacterial cell wall and ruptures the host bacterium, releasing about 300 progeny phage per infected cell.

As mentioned earlier, T4 encodes nucleases that degrade the host DNA. The degradation products are then used in the synthesis of phage DNA. But how do these enzymes degrade host DNA without destroying the DNA of the virus? The answer is that T4 DNA contains an unusual base—5-hydroxymethylcytosine (HMC; cytosine with a —CH_2OH group attached to one of the atoms in the cytosine molecule)—instead of cytosine. In addition, derivatives of glucose molecules are attached to the HMC. These modifications protect T4 DNA from degradation by the nucleases that it uses to degrade the DNA of the host cell.

Bacteriophage Lambda

Bacteriophage lambda (λ) is another coliphage that has made large contributions to genetics. Lambda is smaller than T4;

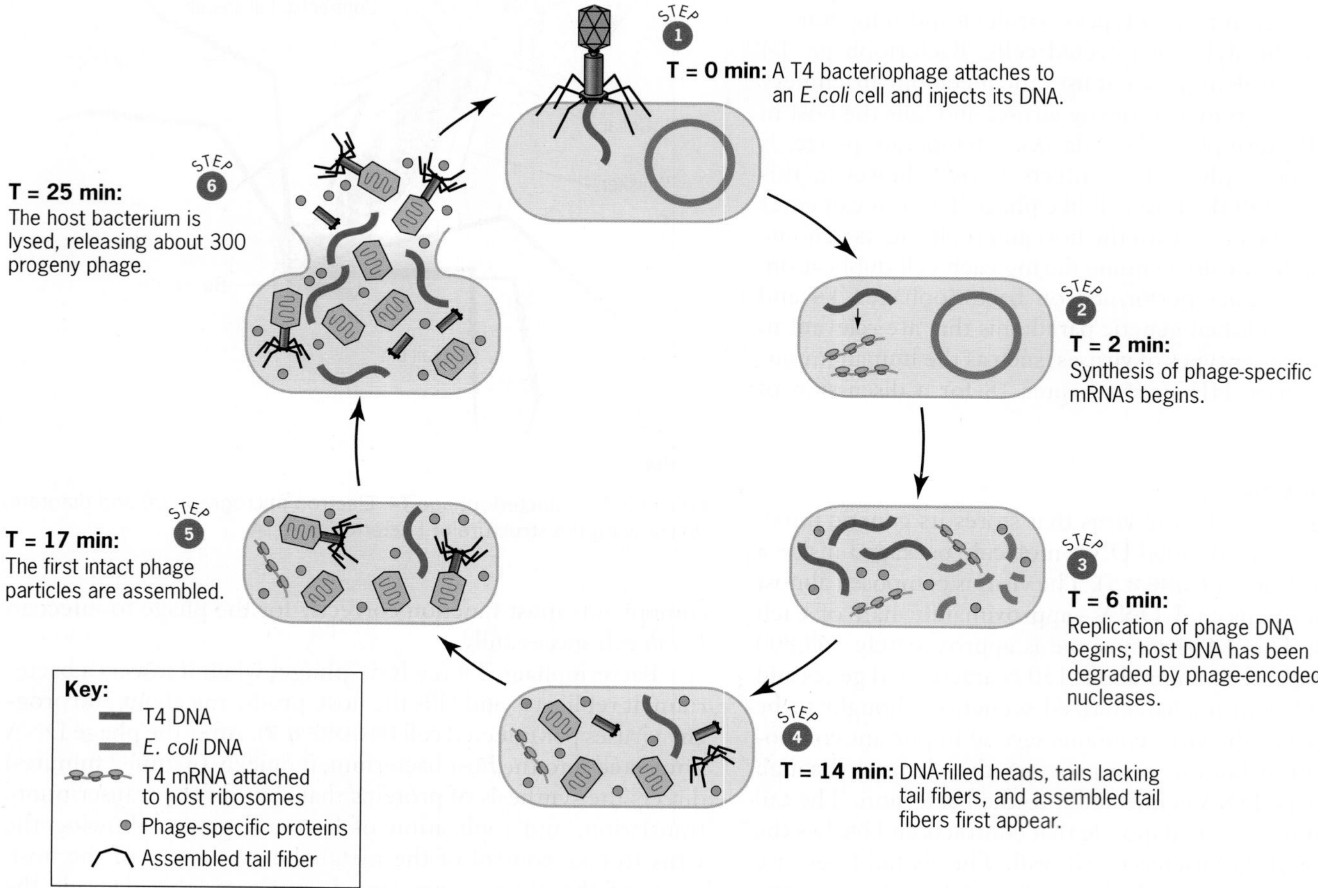
STEP 1
T = 0 min: A T4 bacteriophage attaches to an E.coli cell and injects its DNA.
STEP 2
T = 2 min:
Synthesis of phage-specific mRNAs begins.
STEP 3
T = 6 min:
Replication of phage DNA begins; host DNA has been degraded by phage-encoded nucleases.
STEP 4
T = 14 min: DNA-filled heads, tails lacking tail fibers, and assembled tail fibers first appear.
STEP 5
T = 17 min:
The first intact phage particles are assembled.
STEP 6
T = 25 min:
The host bacterium is lysed, releasing about 300 progeny phage.
Key:
T4 DNA
E. coli DNA
T4 mRNA attached to host ribosomes
Phage-specific proteins
Assembled tail fiber

Figure 8.3 ▶ The life cycle of bacteriophage T4.

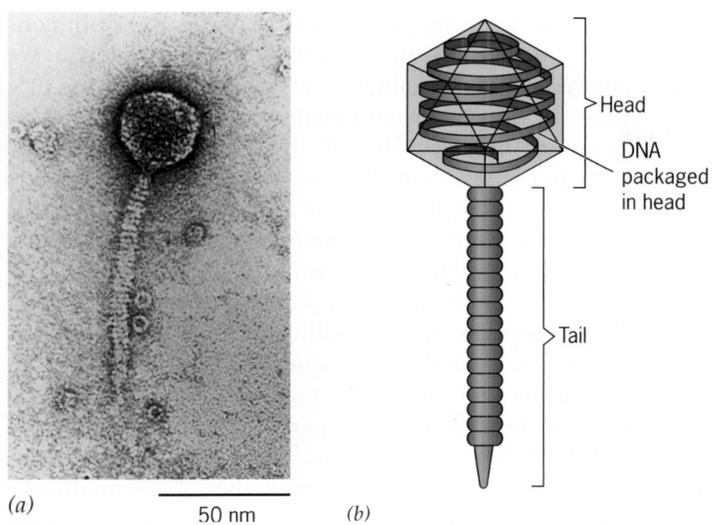

Figure 8.4 ► Bacteriophage λ. Electron micrograph (*a*) and diagram (*b*) showing the structure of bacteriophage λ.

however, its life cycle is more complex. The lambda genome contains about 50 genes in a double-stranded DNA molecule 48,502 base pairs long. This linear DNA molecule is packaged in the λ head (**FIGURE 8.4**). Soon after it is injected into an *E. coli* cell, the λ DNA molecule is converted to a circular form, which participates in all subsequent intracellular events.

Inside the cell, the circular λ chromosome can proceed down either of two pathways (**FIGURE 8.5**). It can enter a lytic cycle, during which it reproduces and encodes enzymes that lyse the host cell, just like phage T4. Or, it can enter a **lysogenic** pathway, during which it is inserted into the chromosome of the host bacterium and thereafter is replicated along with that chromosome. In this integrated state, the λ chromosome is called a **prophage.** For this state to continue, the genes of the prophage that encode products involved in the lytic pathway—for example, enzymes involved in the replication of phage DNA, structural proteins required for phage morphogenesis, and the lysozyme that catalyzes cell lysis—must not be expressed. The mechanism by which the expression of these genes is repressed is discussed in Chapter 19.

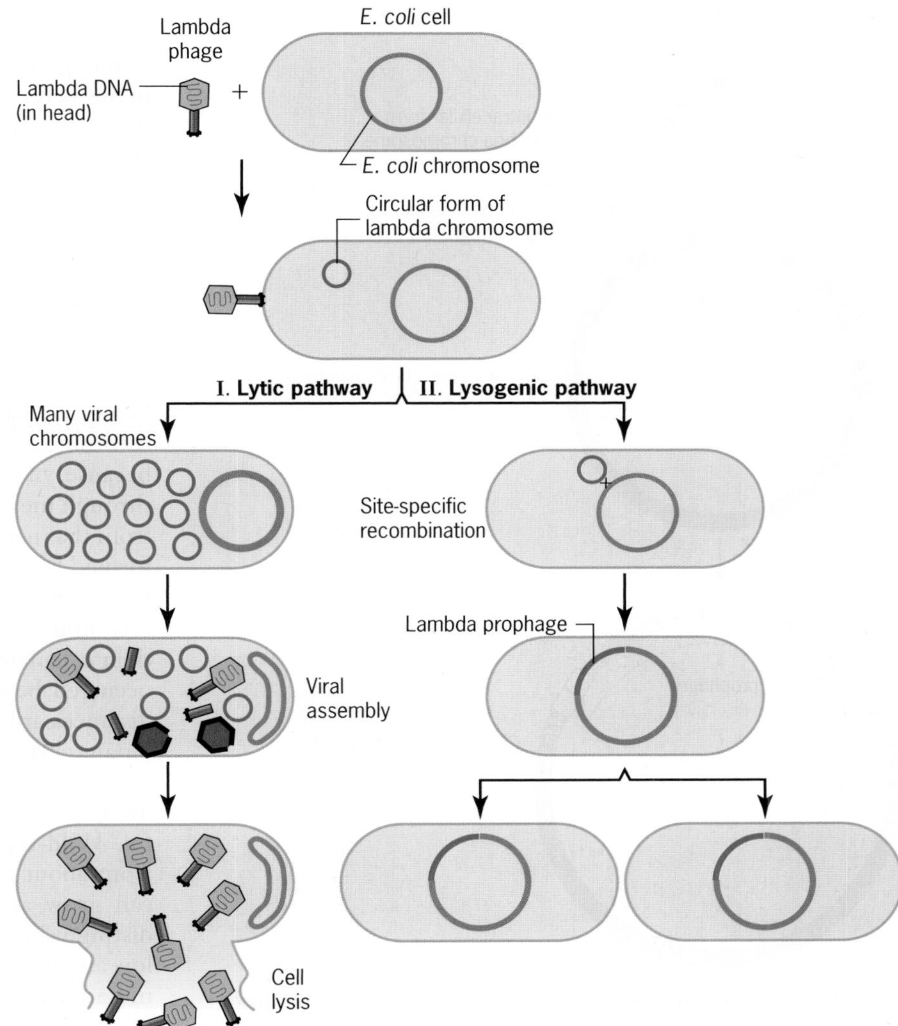

Figure 8.5 ► The life cycle of bacteriophage λ. The two intracellular states of bacteriophage lambda: lytic growth and lysogeny.

Integration of the λ chromosome occurs by a site-specific recombination event between the circular λ DNA and the circular *E. coli* chromosome (**FIGURE 8.6**). This recombination occurs at specific attachment sites—*attP* on the λ chromosome and *attB* on the bacterial chromosome—and is mediated by the product of the λ *int* gene, the λ integrase. It covalently inserts the λ DNA into the chromosome of the host cell. The site-specific recombination occurs in the central region of the attachment sites where both *attP* and *attB* have the same sequence of 15 nucleotide pairs:

GCTTTTTTATACTAA
CGAAAAAATATGATT

With the exception of this core sequence, *attP* and *attB* have quite different sequences. Because recombination occurs within this core sequence during integration, the resulting *attB/P* and *attP/B* sites that flank the integrated prophage also both contain the 15-nucleotide-pair sequence. These structures are important because they facilitate excision of the prophage by a very similar site-specific recombination event.

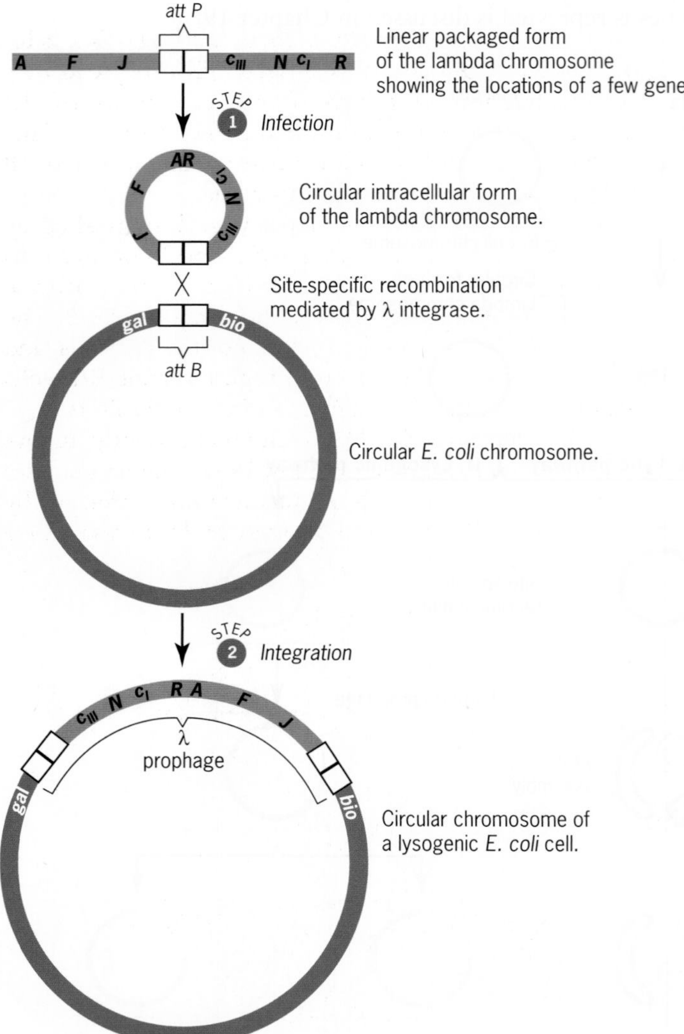

Figure 8.6 ▶ Integration of the λ DNA molecule into the chromosome of *E. coli*.

About once in every 10^5 cell divisions, the λ prophage spontaneously excises from the host chromosome and enters the lytic pathway. This phenomenon is the reason the prophage is said to be in a *lysogenic* state, that is, one capable of causing lysis, albeit at low frequency. Excision of the λ prophage can also be induced, for example, by irradiation with ultraviolet light. The excision process is usually precise, with site-specific recombination between the core sequences in *attB/P* and *attP/B*. It produces an autonomous λ chromosome that has the original pre-integration form. Excision requires the λ integrase and the product of the λ *xis* gene, λ excisase. These two enzymes mediate a site-specific recombination event that is essentially the reverse of the integration event. Occasionally, excision occurs anomalously, and bacterial DNA is excised along with phage DNA. When this occurs, the resulting virus can transfer bacterial genes from one host bacterium to another. We will discuss this process later in this chapter (see the section Mechanisms of Genetic Exchange in Bacteria).

Studies on phage λ have contributed much to our understanding of genetic phenomena. We will discuss the replication of the λ chromosome in Chapter 9 and the elegant genetic switch that controls the entry into either the lysogenic state or the lytic state in Chapter 19. The discovery of the λ prophage (for which André Lwoff was awarded a share of the 1965 Nobel Prize in Physiology or Medicine) provided the paradigm for the proviral states of the human immunodeficiency virus (HIV) (Chapter 18) and various vertebrate RNA tumor viruses (Chapter 22).

MAPPING GENES IN BACTERIOPHAGE

Genes on bacteriophage chromosomes can be mapped using recombination frequencies, just as in eukaryotes. However, because viruses have a single chromosome that does not go through meiosis, the mapping procedure is somewhat different from that used for an organism like *Drosophila*. Crosses are performed by simultaneously infecting host bacteria with two different types of phage and then screening the progeny phage for recombinant genotypes. Map distances, in centiMorgans, are then calculated as the average number of crossovers that have occurred between genetic markers. For short distances, map distances are approximately equal to the percentage of recombinant chromosomes among the progeny.

There are many different kinds of mutant alleles in phage. Temperature-sensitive (*ts*) mutations are among the most useful. Wild-type coliphages can grow at temperatures ranging from about 25° to over 42°C, whereas heat-sensitive mutants can grow at 25° but not at 42°C. Thus, *ts* mutants can be distinguished from wild-type phage by culturing the phage at low and high temperatures. We will discuss these heat-sensitive mutations in Chapter 13. For now, let's examine some other types of mutant phage.

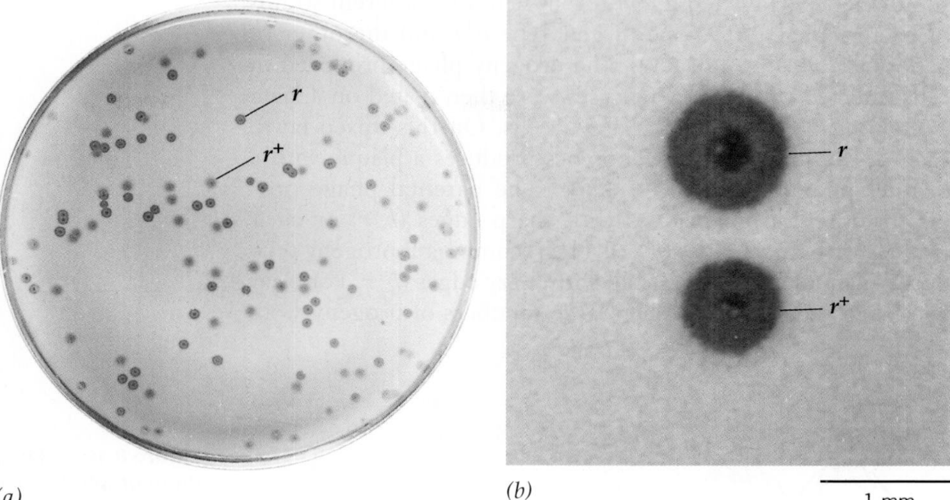

Figure 8.7 ▶ Bacteriophage T4 plaque morphology. (*a*) Plaques formed by T4 wild-type (r^+) and rapid lysis (r) mutants on a confluent lawn of *E. coli* strain B cells. (*b*) Individual r^+ and r plaques shown at higher magnification.

If you spread *E. coli* cells on the surface of semisolid (agar-containing) growth medium in a sterile petri dish and incubate the dish at 37°C overnight, the cells will grow, divide, and produce a confluent "lawn" of bacteria on the surface of the medium. If you add one T4 phage to the mixture, it will infect one bacterium, lyse it, and release about 300 progeny phage about 25 minutes later. Each of the 300 progeny viruses will, in turn, infect bacteria and repeat the lytic cycle. After this cycle is repeated many times, all the bacteria in the vicinity of the original phage particle will be lysed, yielding a clear spot, or hole, in the lawn of bacteria. This clear zone of lysed bacteria is called a **plaque**. In the case of bacteriophage T4, or its close relative T2, each overnight plaque will contain approximately 10^8 viruses.

Some of the first mutant phage that were studied exhibited altered plaque morphology. The *rapid lysis* (r) mutants of phage T4 are the most intensely studied of the plaque morphology mutants. They produce large plaques with sharp margins, which are distinct from the small plaques with fuzzy margins produced by wild-type T4 (**FIGURE 8.7**). When a T4 phage attaches to an *E. coli* cell that is already infected with wild-type T4, it triggers the synthesis of new cell wall material and delays lysis for up to 2 hours. This phenomenon, called *lysis inhibition*, is responsible for the fuzzy margins of wild-type plaques. They contain a mixture of lysed cells and lysis-inhibited cells. Lysis inhibition does not occur in bacteria infected with r mutants; thus, the cells infected with r mutants all lyse rapidly, yielding plaques with sharp, or clearly defined, edges.

Another type of mutant used in early studies altered the ability of the phage to infect different host strains. Such phage are said to be *host range* mutants. For example, *E. coli* strain B cells can be infected by all wild-type T-even (T2, T4, and T6) bacteriophages. However, *E. coli* strain B/2, a mutant derivative of *E. coli* B, is resistant to infection by phage T2. *E. coli* B/2 cells harbor a mutation that alters the phage T2 receptor on the bacterial surface so that T2 phage cannot attach to them. However, a mutant strain of T2, T2*h*, carries a host range mutation (*h*) that allows the virus to infect both *E. coli* B and B/2 cells. This alternation of resistance and susceptibility may continue with bacterial mutations that make *E. coli* resistant to *both* T2 and T2*h*, and with viral mutations that allow infection of the new resistant strains. Similar host range mutants occur in all other bacterial viruses that have been studied.

T2 wild-type and T2*h* mutants can be distinguished by growing them on a mixed lawn of *E. coli* B and *E. coli* B/2 cells (**FIGURE 8.8**). The *h* mutants produce clear plaques because they infect and lyse all host cells—whether strain B or B/2. The wild-type viruses infect only the B cells, not the B/2 cells, and thus produce turbid plaques. The phage-resistant B/2 cells continue to grow within the plaque, causing the turbidity.

Alfred Hershey and Max Delbrück independently discovered genetic recombination in phage in 1946. The first crosses between host range and rapid lysis mutants were performed by Hershey and Raquel Rotman shortly thereafter. These researchers

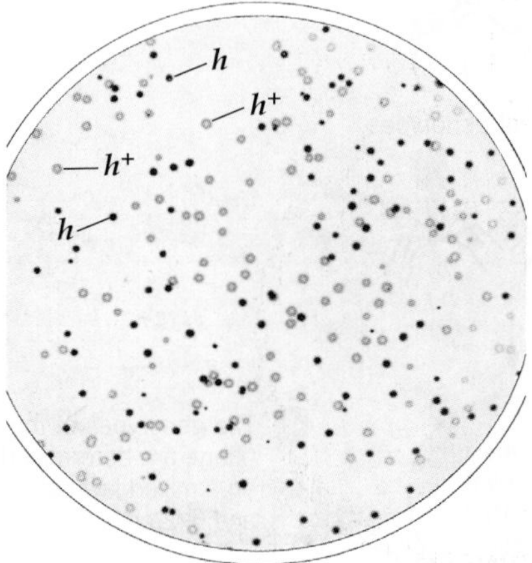

Figure 8.8 ▶ Bacteriophage T2 plaque morphology. Plaques formed by T2 wild-type (h^+) and host range (h) mutants when grown on a mixed lawn of *E. coli* B and B/2 cells.

simultaneously infected *E. coli* B cells with two different strains of bacteriophage T2—one of genotype h^+r and the other of genotype $h\ r^+$ (**FIGURE 8.9**). The progeny phage produced in cells infected with the two viruses were then plated on a lawn containing both *E. coli* B and *B*/2 cells. On this mixed lawn, each of the four possible genotypes produces a plaque with a distinct phenotype (**FIGURE 8.10**). The parental phage produced either turbid plaques with sharp edges (h^+r) or clear plaques with fuzzy edges ($h\ r^+$). The recombinant progeny produced plaques that were turbid with fuzzy edges (h^+r^+) or clear with sharp edges ($h\ r$). When large numbers of progeny were analyzed, about 2 percent had recombinant genotypes. On the basis of these results, the distance between the *r* and *h* genes was estimated to be about 2 centiMorgans. The results also indicated that phage recombination is a reciprocal process, because the two recombinant genotypes (h^+r^+ and $h\ r$) were present among the progeny with approximately the same frequency.

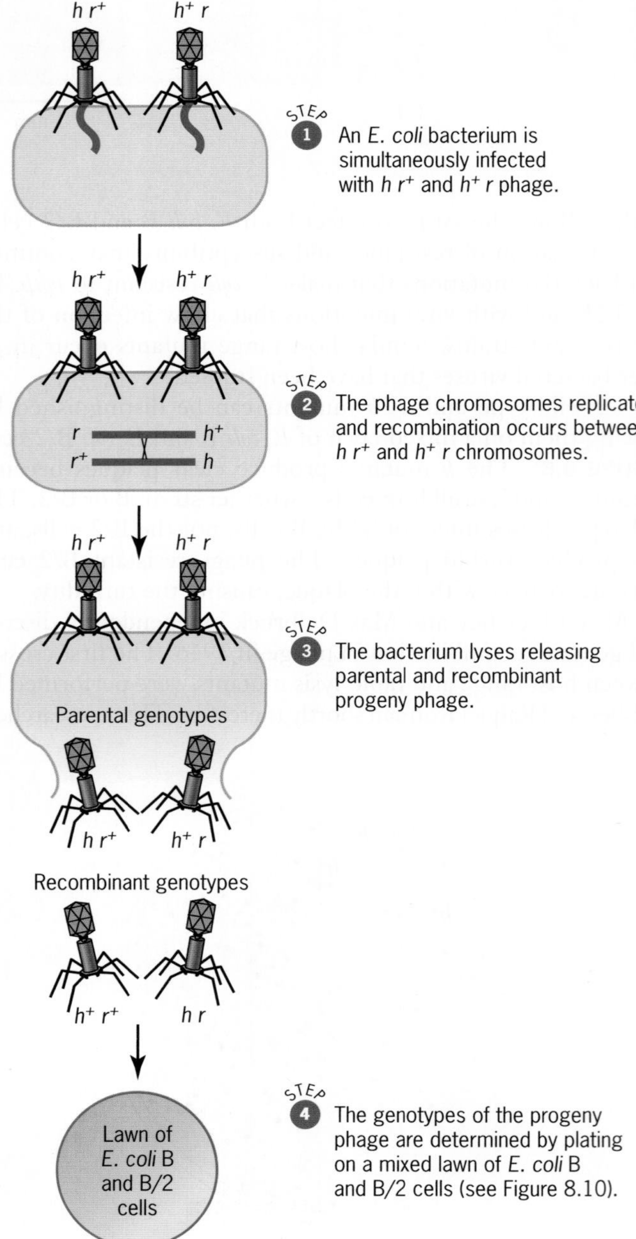

h r+
h+ r
STEP 1 An E. coli bacterium is simultaneously infected with h r+ and h+ r phage.
r
r+
h+
h
STEP 2 The phage chromosomes replicate and recombination occurs between h r+ and h+ r chromosomes.
STEP 3 The bacterium lyses releasing parental and recombinant progeny phage.
Parental genotypes
Recombinant genotypes
h+ r+
h r
Lawn of E. coli B and B/2 cells
STEP 4 The genotypes of the progeny phage are determined by plating on a mixed lawn of E. coli B and B/2 cells (see Figure 8.10).

Figure 8.9 ▶ A phage cross in which *E. coli* B cells are simultaneously infected with T2 $h\ r^+$ and h^+r phage.

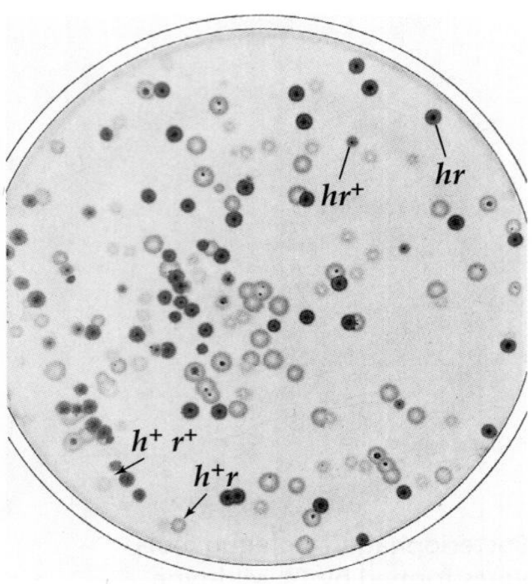

Figure 8.10 ▶ Phage T2 host range and rapid-lysis plaque morphology. Photograph showing the types of plaques formed by T2 h^+r^+, $h\ r$, h^+r, $h\ r^+$ phage when grown on a lawn containing both *E. coli* B and *B*/2 cells.

One of the major advantages of studying recombination with bacteriophage is that large numbers of progeny can be analyzed with relative ease, allowing researchers to detect and study rare events. Indeed, in Chapter 14, we will discuss Seymour Benzer's analysis of the *rII* locus of phage T4—the most detailed genetic map constructed to date.

BACTERIOPHAGE T4: A LINEAR CHROMOSOME AND A CIRCULAR GENETIC MAP

All of the genetic information of phage T4 is stored in a single linear molecule of DNA (see **FIGURE 8.2**). Thus, the T4 linkage map was expected to be linear. However, the results of early genetic crosses exhibited some unexpected, but repeatable, inconsistencies. The results of two-factor crosses indicated that the mutations *h*42, *ac*41 (*ac*ridine-resistant mutant 41), and *r*67 were linked with the order *h*42—*ac*41—*r*67. But the results of three-factor crosses contradicted the two-factor data, indicating that the order was *ac*41—*h*42—*r*67. The contradiction could be resolved by making the T4 genetic map circular (**FIGURE 8.11**). With a circular map, *h*42 maps between *ac*41 and *r*67 starting at *ac*41 and moving counterclockwise. However, given the circular map and moving clockwise from *h*42 yields the sequence *h*42—*ac*41—*r*67. So, a circular genetic map resolved the paradox. But how can a circular genetic map be generated from a linear chromosome?

In the mid-1960s, George Streisinger proposed that the T4 chromosome was both **terminally redundant** and **circularly permuted.** Initially, both geneticists and virologists were

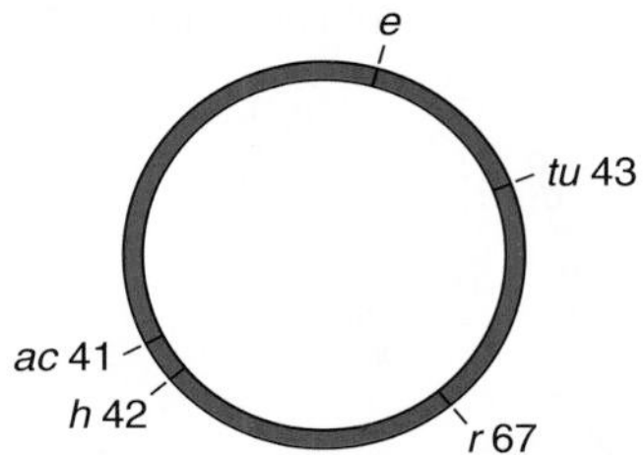

Figure 8.11 ► An early circular map of the bacteriophage T4 chromosome. The mutant alleles are *e* (endolysin = lysozyme), *tu* (turbid plaque), *r* (rapid lysis), *h* (host range), and *ac* (acridine resistance).

skeptical of his proposal. But both components of the hypothesis were soon proven correct.

Terminally redundant DNA molecules contain the same nucleotide sequence at both ends of a linear molecule, as in the sequence AAGGCCTTGACTA............TACGTAAGGCCTT. This DNA strand is terminally redundant; the sequence AAGGCCTT is present at both ends.

Circularly permuted sequences are obtained by breaking a circular structure containing a linear sequence (with the ends joined) of markers at random to obtain a collection of linear structures, as shown in the following:

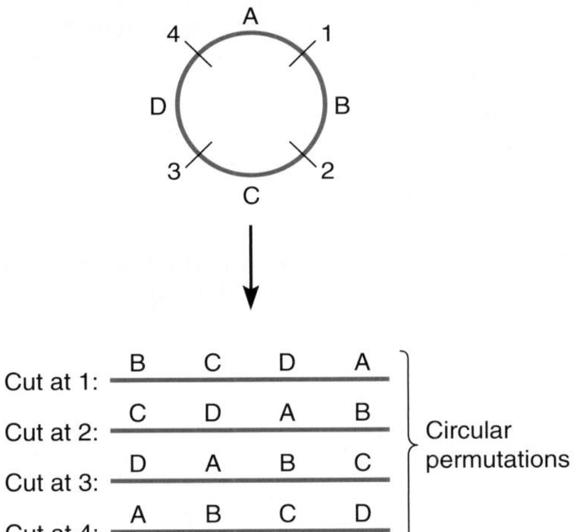

The ends of a T4 chromosome contain the same genes in the same order; that is, they are terminally redundant (*abcdefg* ....*wxyzabc*). Note that the long terminally redundant nucleotide sequences of the phage chromosome are denoted *abc* here for brevity. However, the endpoints for each chromosome are different; that is, one chromosome may have the sequence *abcdef*.........*xyzabc*, whereas another will have the sequence *defghi*..........*xyzabcdef*, and a third will have the sequence *fghijk*......... *xyzabcdefgh*. Thus, a population of T4 chromosomes consists of a set of circularly permutated DNA molecules that are also terminally redundant. Sometimes segment *d* is within the terminally redundant ends of the molecule, and sometimes it is in the middle of the DNA molecule.

The combination of these two features—circular permutations and terminal redundancy—results in a circular genetic map from a population of linear chromosomes. But how can a population of circularly permuted, terminally redundant progeny phage chromosomes be generated in a cell infected with a single T4 virus particle? That is, how can a parental phage that is redundant for *abc* at its ends produce progeny that are redundant for *cde*, *def*, *efg*, and so forth? The key to the answer is the way that T4 DNA replicates and is packaged into phage heads. During replication of T4 DNA molecules, recombination occurs between the terminally redundant ends of these molecules and generates long DNA molecules called **concatamers** (**FIGURE 8.12**). Each concatamer contains many copies of the phage T4 genome joined end-to-end.

During the formation of progeny viruses, the T4 head proteins condense around these concatameric DNA molecules until the head is full. The amount of DNA required to fill a head is *slightly more* than one complete set of T4 genes, usually symbolized *abc*..........*xyz*. Because there is still room left in the head after one complete set of genes (*abc*.........*xyz*) has entered, a few additional genes are added, yielding a phage head containing gene sequence *abc*.........*xyzabc*. The head is now full, and the DNA is cut. This particular virus is redundant for genes *a*, *b*, and *c*. The next virus particle begins adding DNA starting at *d*, has packaged a complete chromosome when *c* is reached, and then adds the redundant region *def*. The subsequent virus begins packaging at *g* and is redundant for the *ghi*

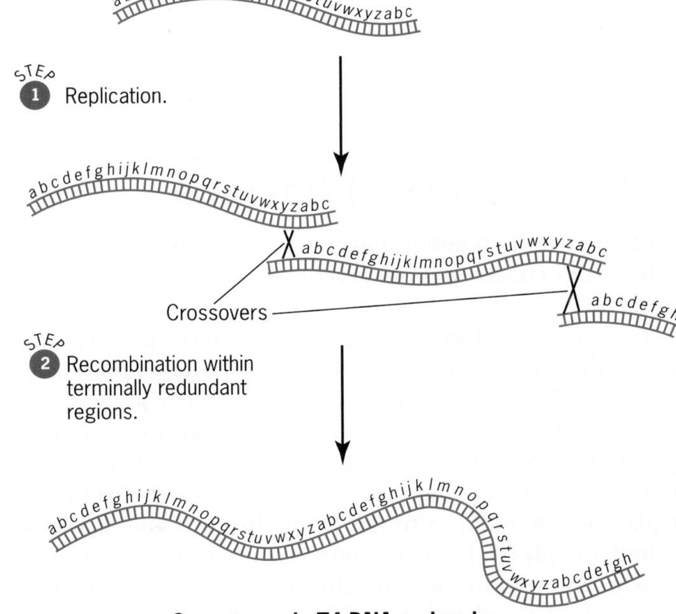

Figure 8.12 ► Phage T4 DNA concatamers. The formation of concatameric replicative DNA molecules by "head-to-tail" recombination within the terminally redundant regions of T4 chromosomes.

segment, and so on. This mode of DNA packaging is called the *headful mechanism;* it produces T4 chromosomes that are both terminally redundant and circularly permuted (**FIGURE 8.13**). However, the progeny chromosomes are redundant for different genes.

Phage T4 is not unique in packaging DNA by the headful mechanism. Many bacteriophages, including T2, T6, and P1, use the headful mechanism of DNA packaging.

The circularly permuted chromosomes of phage T4 will yield a circular genetic map, even though each phage contains a linear molecule of DNA. Map distances are based on average recombination frequencies, which are population parameters. Although two closely linked genes may be located at opposite (terminally redundant) ends of an individual phage chromosome, they will be close together in the internal, nonredundant region of most of the chromosomes in the T4 progeny population.

KEY POINTS

- Viruses are obligate parasites that can reproduce only by infecting living host cells.
- Bacteriophages are viruses that infect bacteria.
- Bacteriophage T4 is a lytic phage that infects *E. coli*, reproduces, and lyses the host cell.
- Bacteriophage lambda (λ) can enter a lytic pathway, like T4, or it can enter a lysogenic pathway, during which its chromosome is inserted into the chromosome of the bacterium.
- In its integrated state, the λ chromosome is called a prophage, and its lytic genes are kept turned off.
- Phage T4 has a circular genetic map but a linear chromosome. The circular map is produced because T4 chromosomes are terminally redundant and circularly permuted.

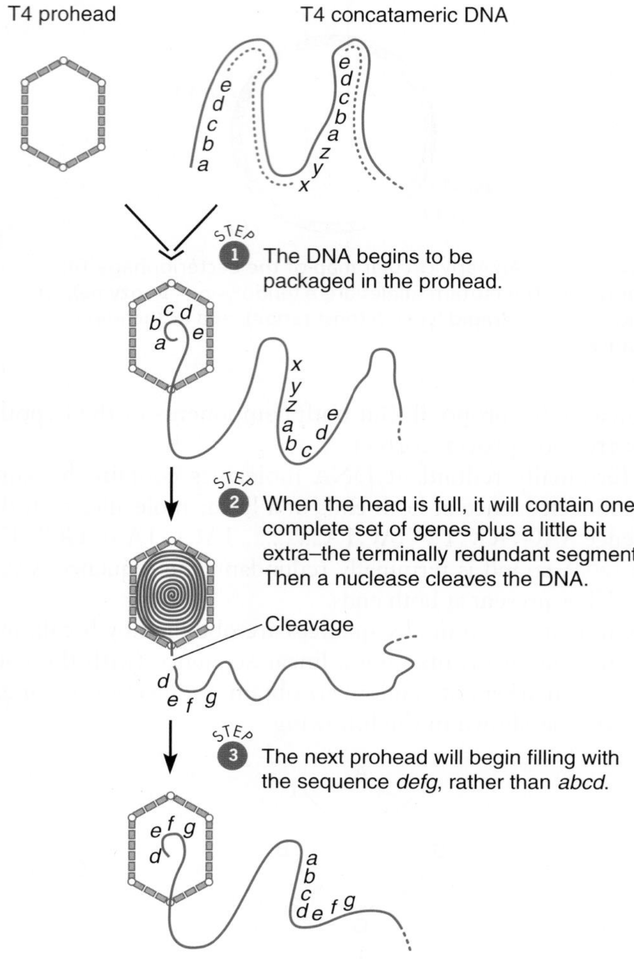

Figure 8.13 ▶ Generation of circularly permuted, terminally redundant T4 chromosomes by the "headful" mechanism.

The Genetics of Bacteria

Bacteria contain genes that mutate to produce altered phenotypes. Gene transfer in bacteria is unidirectional—from donor cells to recipient cells.

The genetic information of most bacteria is stored in a single main chromosome carrying a few thousand genes and a variable number of "mini-chromosomes" called plasmids and episomes. Plasmids are autonomously replicating, circular DNA molecules that carry anywhere from three genes to several hundred genes. Some bacteria contain as many as 11 different plasmids in addition to the main chromosome. Episomes are similar to plasmids, but episomes can replicate either autonomously or as part of the main chromosome—in an integrated state like the λ prophage.

Bacteria reproduce asexually by simple fission, with each daughter cell receiving one copy of the chromosome. They are monoploid but "multinucleate"; that is, the cell usually contains two or more identical copies of the chromosome. The chromosomes of bacteria do not go through the mitotic and meiotic condensation cycles that occur during cell division and gametogenesis in eukaryotes. Therefore, the recombination events—independent assortment and meiotic crossing over—that occur during sexual reproduction in eukaryotes do not occur in bacteria.

Nevertheless, recombination has been just as important in the evolution of bacteria as it has been in the evolution of eukaryotes. Indeed, processes that are akin to sexual reproduction—parasexual processes—occur in bacteria. We will consider these processes after discussing some of the types of mutants used in bacterial genetics and the unidirectional nature of gene transfer between bacteria.

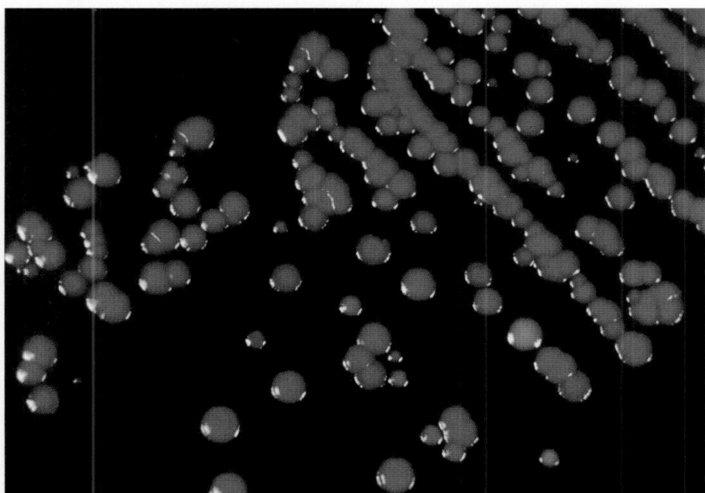

Figure 8.14 Bacterial colonies. Photograph showing colonies of the bacterium *Serratia marcescens* growing on agar-containing medium. The distinctive color of the colonies results from the red pigment produced by this species.

MUTANT GENES IN BACTERIA

Bacteria will grow in liquid medium, often requiring aeration, or on the surface of semisolid medium containing agar. If grown on semisolid medium, each bacterium will divide and grow exponentially, producing a visible colony on the surface of the medium. The number of colonies that appear on a culture plate can be used to estimate the number of bacteria that were originally present in the suspension applied to the plate.

Each bacterial species produces colonies with a specific color and morphology. *Serratia marcescens*, for example, produces a red pigment that results in distinctive red colonies (**FIGURE 8.14**). Mutations in bacterial genes can change both colony color and morphology. Moreover, any mutation that slows the growth rate of the bacterium will produce small or petite colonies. Some mutations alter the morphology of the bacterium without changing colony morphology. Besides these colony color and morphology mutants, other types of mutants have been useful in genetic studies of bacteria.

Mutants Blocked in Their Ability to Utilize Specific Energy Sources

Wild-type *E. coli* can use almost any sugar as an energy source. However, some mutants are unable to grow on the milk sugar lactose. They grow well on other sugars but cannot grow on medium containing lactose as the sole energy source. Other mutants are unable to grow on galactose, and still others are unable to grow on arabinose. The standard nomenclature for describing these and other types of mutants in bacteria is to use three-letter abbreviations with appropriate superscripts. For phenotypes, the first letter is capitalized; for genotypes, all three-letters are lowercase and italicized. Therefore, wild-type *E. coli* is phenotypically Lac^{+} (able to use lactose as an energy source) and genotypically lac^{+}. Mutants that are unable to utilize lactose as an energy source are phenotypically Lac^{-} and genotypically lac^{-} (or sometimes just *lac*).

Mutants Unable to Synthesize an Essential Metabolite

Wild-type *E. coli* can grow on medium (minimal medium) containing an energy source and some inorganic salts. These cells can synthesize all of the metabolites—amino acids, vitamins, purines, pyrimidines, and so on—they need from these substances. These wild-type bacteria are called **prototrophs.** When a mutation occurs in a gene encoding an enzyme required for the synthesis of an essential metabolite, the bacterium carrying that mutation will have a new growth requirement. It will grow if the metabolite is added to the medium, but it will not grow in the absence of the metabolite. Such mutants are called **auxotrophs;** they require auxiliary nutrients for growth. As an example, wild-type *E. coli* can synthesize tryptophan *de novo*; these cells are phenotypically Trp^{+} and genotypically trp^{+}. Tryptophan auxotrophs are Trp^{-} and trp^{-}.

Mutants Resistant to Drugs and Antibiotics

Wild-type *E. coli* cells are killed by antibiotics such as ampicillin and tetracycline. Phenotypically, they are Amp^{s} and Tet^{s}. The mutant alleles that make *E. coli* resistant to these antibiotics are designated amp^{r} and tet^{r}, respectively. Bacteria that contain these mutant alleles can grow on medium containing the antibiotics, whereas wild-type bacteria cannot. Thus, antibiotics can be used to select bacteria that carry genes for resistance. The resistance genes function as dominant selectable markers.

Bacteria divide rapidly and produce large populations of cells for genetic studies. Moreover, media that select specific bacterial genotypes (selective media) are relatively easy to prepare. As a result, bacteria have been used to study rare events such as mutation within genes and recombination between closely linked genes. We will discuss some of these studies in Chapters 13 and 14.

UNIDIRECTIONAL GENE TRANSFER IN BACTERIA

The recombination events that occur in bacteria involve transfers of genes from one bacterium to another, rather than the reciprocal exchanges of genes that occur during meiosis in eukaryotes. Thus, gene transfer is *unidirectional* rather than bidirectional. Recombination events in bacteria usually occur between a fragment of one chromosome (from a **donor cell**) and a complete chromosome (in a **recipient cell**), rather than between two complete chromosomes as in eukaryotes. With rare exceptions, the recipient cells become partial diploids, containing a linear piece of the donor chromosome and a complete circular recipient chromosome. As a result, *crossovers must occur in pairs* and must insert a segment of the donor chromosome into the recipient chromosome (**FIGURE 8.15*a***). If a single crossover (or any odd number of crossovers) occurs, it will destroy the integrity of the recipient chromosome, producing a nonviable linear DNA molecule (**FIGURE 8.15*b***).

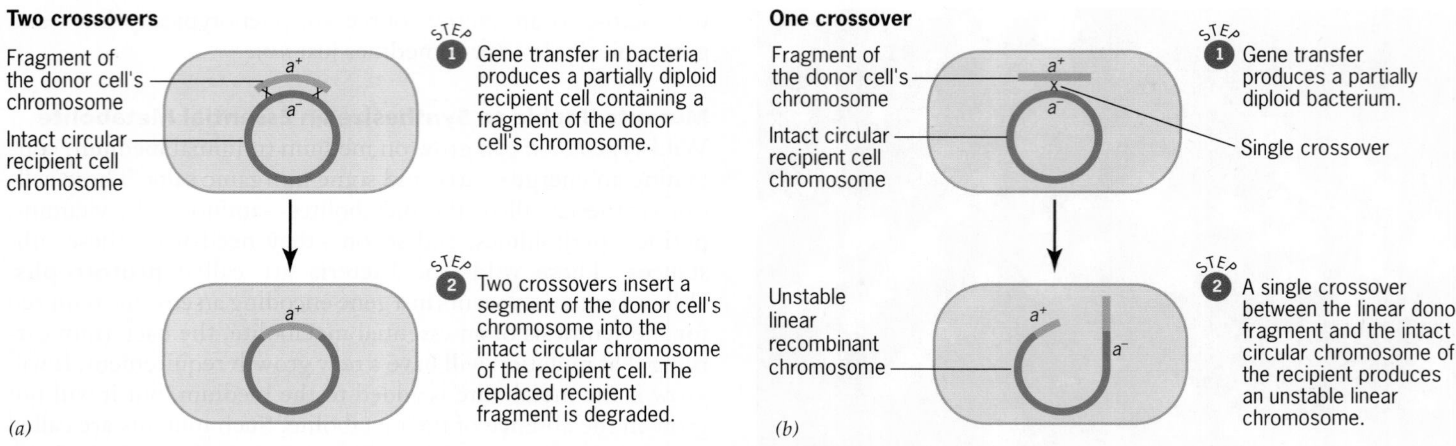
Two crossovers
Fragment of the donor cell's chromosome
Intact circular recipient cell chromosome
a^+
a^-
STEP 1 Gene transfer in bacteria produces a partially diploid recipient cell containing a fragment of the donor cell's chromosome.
STEP 2 Two crossovers insert a segment of the donor cell's chromosome into the intact circular chromosome of the recipient cell. The replaced recipient fragment is degraded.
(a)
One crossover
Fragment of the donor cell's chromosome
Intact circular recipient cell chromosome
STEP 1 Gene transfer produces a partially diploid bacterium.
Single crossover
Unstable linear recombinant chromosome
STEP 2 A single crossover between the linear donor fragment and the intact circular chromosome of the recipient produces an unstable linear chromosome.
(b)

Figure 8.15 ▶ Recombination in bacteria. The parasexual processes that occur in bacteria produce partial diploids containing linear fragments of the donor cell's chromosome and intact circular chromosomes of the recipient cells. (*a*) To maintain the integrity of the circular chromosomes, crossovers must occur in pairs, inserting segments of the donor chromosomes into the chromosomes of the recipient. (*b*) A single crossover between a fragment of a donor chromosome and a circular recipient chromosome destroys the integrity of the circular chromosome, producing a linear DNA molecule that is unable to replicate and is subsequently degraded.

KEY POINTS

- Bacteria usually contain one main chromosome.
- Wild-type bacteria are prototrophs; they can synthesize everything they need to grow and reproduce given an energy source and some inorganic molecules.
- Auxotrophic mutant bacteria require additional metabolites for growth.
- Gene transfer in bacteria is unidirectional; genes from a donor cell are transferred to a recipient cell, with no transfer from recipient to donor.

Mechanisms of Genetic Exchange in Bacteria

Bacteria exchange genetic material through three different parasexual processes.

Three distinct parasexual processes occur in bacteria. The most obvious difference between these three processes is the mechanism by which DNA is transferred from one cell to another (**FIGURE 8.16**). **Transformation** involves the uptake of free DNA molecules released from one bacterium (the donor cell) by another bacterium (the recipient cell). **Conjugation** involves the direct transfer of DNA from a donor cell to a recipient cell. **Transduction** occurs when bacterial genes are carried from a donor cell to a recipient cell by a bacteriophage.

The three parasexual processes of gene transfer—transformation, conjugation, and transduction—in bacteria can be distinguished by two simple criteria (**TABLE 8.1**). (1) Does the process require cell contact? (2) Is the process sensitive to deoxyribonuclease (DNase), an enzyme that degrades DNA? These two criteria can be tested experimentally quite easily. Sensitivity to DNase is determined simply by adding the enzyme to the medium in which the bacteria are growing. If gene transfer no longer occurs, the process involves transformation. The protein coats of bacteriophages and the walls and membranes of bacterial

Transformation: uptake of free DNA.

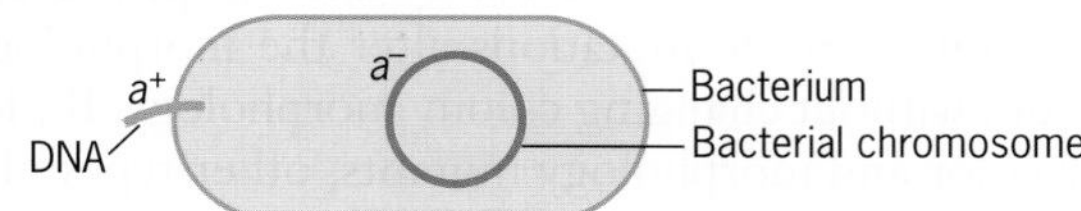
a^+
a^-
DNA
Bacterium
Bacterial chromosome

Conjugation: direct transfer of DNA from one bacterium to another.

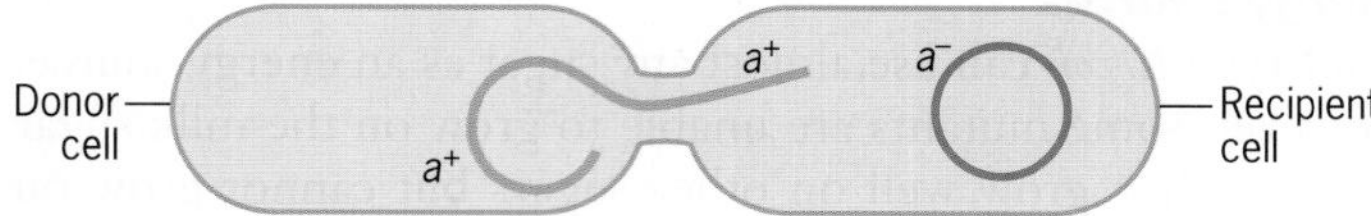
Donor cell
a^+
a^+
a^-
Recipient cell

Transduction: transfer of bacterial DNA by a bacteriophage.

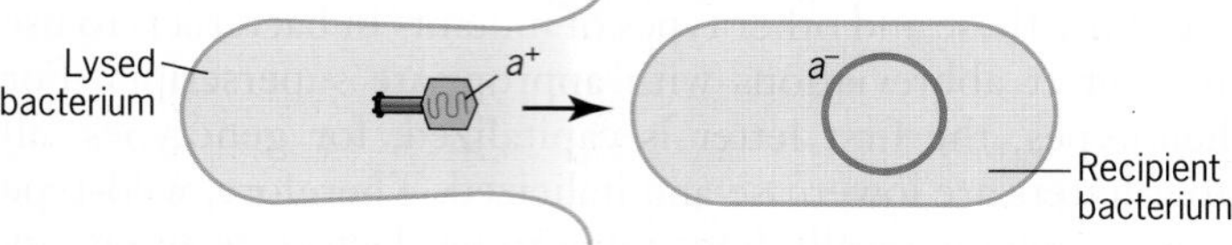
Lysed bacterium
a^+
a^-
Recipient bacterium

Figure 8.16 ▶ The three types of gene transfer in bacteria.

► **TABLE 8.1**

Distinguishing between the Three Parasexual Processes in Bacteria

	Criterion	
Recombination Process	Cell Contact Required?	Sensitive to DNase?
Transformation	no	yes
Conjugation	yes	no
Transduction	no	no

cells protect the donor DNA from degradation by DNase during transduction and conjugation, respectively.

A simple experiment can determine whether or not cell contact is required for bacterial gene transfer. In this experiment, bacteria with different genotypes are placed in opposite arms of a U-shaped culture tube (**FIGURE 8.17**). The two arms are separated by a glass filter that has pores large enough to allow DNA molecules and viruses, but not bacteria, to pass through it. If gene transfer occurs between the bacteria growing in opposite arms of the U-tube, the process cannot be conjugation, which requires direct contact between donor and recipient cells. If the observed gene transfer occurs in the presence of DNase and in the absence of cell contact, it must involve transduction.

All three parasexual processes do not occur in all bacterial species; in fact, transduction probably is the only process that occurs in all bacteria. Whether or not transformation or conjugation occurs in a species depends on whether the required genes and metabolic machinery have evolved in that species. *E. coli*, for example, does not contain genes that encode the proteins required to take up free DNA. Thus, transformation does not occur in *E. coli* growing under natural conditions. Only conjugation and transduction occur in *E. coli* cells growing in natural habitats. However, scientists have discovered how to transform *E. coli* cells in the laboratory by using chemical or physical treatments that make them permeable to DNA. In Chapter 15, we will discuss the use of artificial transformation methods to "clone" (make many copies of) foreign genes in *E. coli* cells.

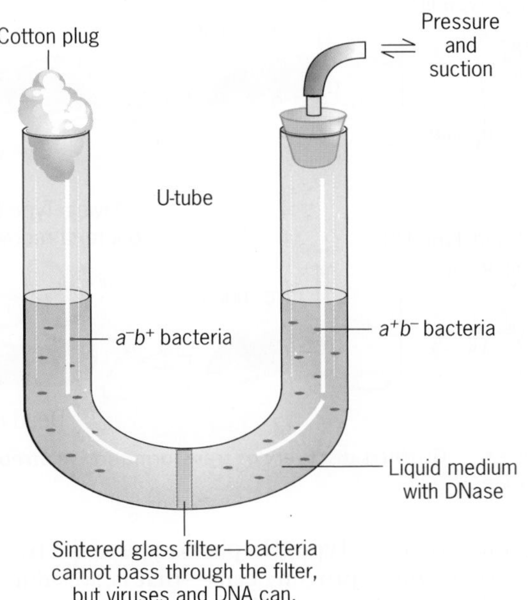

Figure 8.17 ► The U-tube experiment with bacteria. The U-tube is used to determine whether or not cell contact is required for recombination to occur. Bacteria of different genotypes are placed in separate arms of the tube, separated by a glass filter that prevents contact between them. If recombination occurs, it cannot be due to conjugation.

TRANSFORMATION

Frederick Griffith discovered transformation in *Streptococcus pneumoniae* (pneumococcus) in 1928. Pneumococci, like all other living organisms, exhibit genetic variability that can be recognized by the existence of different phenotypes (**TABLE 8.2**). The two phenotypic characteristics of importance in Griffith's demonstration of transformation are (1) the presence or absence of a polysaccharide (complex sugar polymer) capsule surrounding the bacterial cells, and (2) the type of capsule—that is, the specific molecular composition of the polysaccharides present in the capsule. When grown on blood agar medium in petri dishes, pneumococci with capsules form large, smooth colonies (**FIGURE 8.18**) and are thus designated Type S. Encapsulated pneumococci are virulent (pathogenic), causing pneumonia in mammals such as mice and humans. The virulent Type S pneumococci mutate to an avirulent (nonpathogenic) form that has no polysaccharide capsule at a frequency of about one per 10^7 cells. When grown on blood agar medium, such nonencapsulated, avirulent pneumococci produce small, rough-surfaced colonies (**FIGURE 8.18**) and are thus designated Type R. The polysaccharide capsule is required for virulence because it protects the bacterial cell from destruction by white blood cells. When a capsule is present, it may be of several different antigenic types (Type I, II, III, and so forth), depending on the specific molecular composition of the polysaccharides and, of course, ultimately depending on the genotype of the cell.

The different capsule types can be identified immunologically. If Type II cells are injected into the bloodstream of rabbits, the immune system of the rabbits will produce antibodies that react specifically with Type II cells. Such Type II antibodies will agglutinate Type II pneumococci but not Type I or Type III pneumococci.

▶ **TABLE 8.2**

Characteristics of *Streptococcus pneumoniae* Strains When Grown on Blood Agar Medium

	Colony Morphology				Reaction with Antiserum Prepared Against	
Type	Appearance	Size	Capsule	Virulence	Type IIS	Type IIIS
IIR[a]	Rough	Small	Absent	Avirulent	None	None
IIS	Smooth	Large	Present	Virulent	Agglutination	None
IIIR[a]	Rough	Small	Absent	Avirulent	None	None
IIIS	Smooth	Large	Present	Virulent	None	Agglutination

[a]Although Type R cells are nonencapsulated, they carry genes that would direct the synthesis of a specific kind (antigenic Type II or III) of capsule if the block in capsule formation were not present. When Type R cells mutate back to encapsulated Type S cells, the capsule Type (II or III) is determined by these genes. Thus, R cells derived from Type IIS cells are designated Type IIR. When these Type IIR cells mutate back to encapsulated Type S cells, the capsules are of Type II.

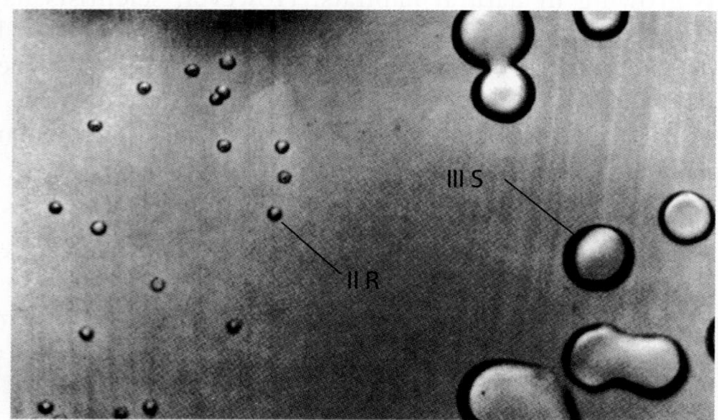

Figure 8.18 ▶ Colony phenotypes of the two strains of *Streptococcus pneumoniae* studied by Griffith in 1928.

Griffith's unexpected discovery was that if he injected heat-killed Type IIIS pneumococci (virulent when alive) plus live Type IIR pneumococci (avirulent) into mice, many of the mice succumbed to pneumonia, and live Type IIIS cells were recovered from the carcasses (**FIGURE 8.19**). When mice were injected with heat-killed Type IIIS pneumococci alone, none of the mice died. The observed virulence was therefore not due to a few Type IIIS cells that survived the heat treatment. The live pathogenic pneumococci recovered from the carcasses had Type III polysaccharide capsules. This result is important because nonencapsulated Type R cells can mutate back to encapsulated Type S cells. However, when such a mutation occurs in a Type IIR cell, the resulting cell will become Type IIS, not Type IIIS. Thus, the transformation of avirulent Type IIR cells to virulent Type IIIS cells cannot be explained by mutation. Instead, some component of the dead Type IIIS cells (the "transforming principle") must have converted living Type IIR cells to Type IIIS.

Subsequent experiments by Richard Sia and Martin Dawson in 1931 showed that the phenomenon described by Griffith, now called transformation, was not mediated in any way by a living host. The same phenomenon occurred in the test tube when live Type IIR cells were grown in the presence of heat-killed Type IIIS cells. Since Griffith's experiments

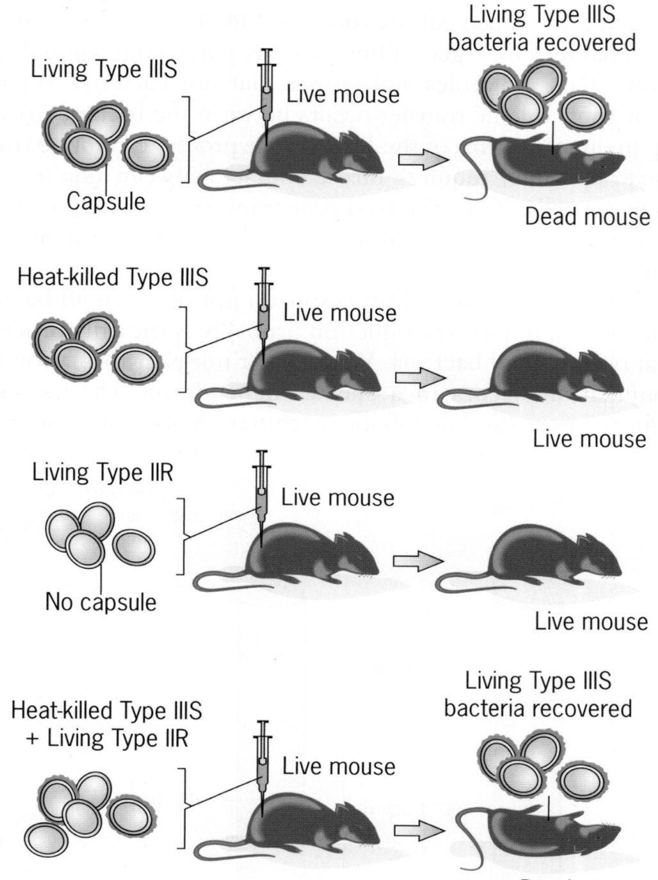

Figure 8.19 ▶ Griffith's discovery of transformation in *Streptococcus pneumoniae*.

demonstrated that the Type IIIS phenotype of the transformed cells was passed on to progeny cells—that is, was due to a permanent inherited change in the genotype of the cells—the demonstration of transformation set the stage for determining the chemical basis of heredity in pneumococcus. Indeed, the first proof that genetic information is stored in DNA rather than proteins was the 1944 demonstration by Oswald Avery,

Colin MacLeod, and Maclyn McCarty that DNA was responsible for transformation in pneumococci. Because of its pivotal role in establishing DNA as the genetic material, we will discuss this demonstration in Chapter 9.

The mechanism of transformation has been studied in considerable detail in *S. pneumoniae*, *Bacillus subtilis*, *Haemophilus influenzae*, and *Neisseria gonorrhoeae*. The basic process is similar in all four species; however, variations in the mechanism occur in each species. *S. pneumoniae* and *B. subtilis* will take up DNA from any source, whereas *H. influenzae* and *N. gonorrhoeae* will only take up their own DNA or DNA from closely related species. *H. influenzae* and *N. gonorrhoeae* will only take up DNA that contains a special short nucleotide-pair sequence (11 base pairs in *Haemophilus*; 10 in *Neisseria*) that is present in about 600 copies in their respective genomes.

Even in the bacterial species that have the ability to take up DNA from their environment, not all cells can do so. Indeed, only cells that are expressing the genes that encode proteins required for the process are capable of taking up DNA. These bacteria are said to be **competent,** and the proteins that mediate the transformation process are called **competence (Com) proteins.** Bacteria develop competence during the late phase of their growth cycle—when cell density is high but before cell division stops. The process by which cells become competent is understood best in *B. subtilis*, where small peptides called competence pheromones are secreted by cells and accumulate at high cell density. High concentrations of the pheromones induce the expression of the genes encoding proteins required for transformation to occur.

Let's focus on the mechanism of transformation in *B. subtilis* (**FIGURE 8.20**). The competence genes are located in clusters, and each cluster is designated by a letter, for example, *A*, *B*, *C*. The first gene in each cluster is designated *A*, the second *B*, and so on. Thus, the protein encoded by the first gene in the fifth cluster is designated ComEA. ComEA and ComG proteins bind double-stranded DNA to the surfaces of competent cells. As the bound DNA is pulled into the cell by the ComFA DNA translocase (an enzyme that moves or "translocates" DNA), one strand of DNA is degraded by a deoxyribonuclease (an enzyme that degrades DNA), and the other strand is protected from degradation by a coating of single-stranded DNA-binding protein and RecA protein (a protein required for recombination). With the aid of RecA and other proteins that mediate recombination, the single strand of transforming DNA invades the chromosome of the recipient cell, pairing with the complementary strand of DNA and replacing the equivalent strand. The replaced recipient

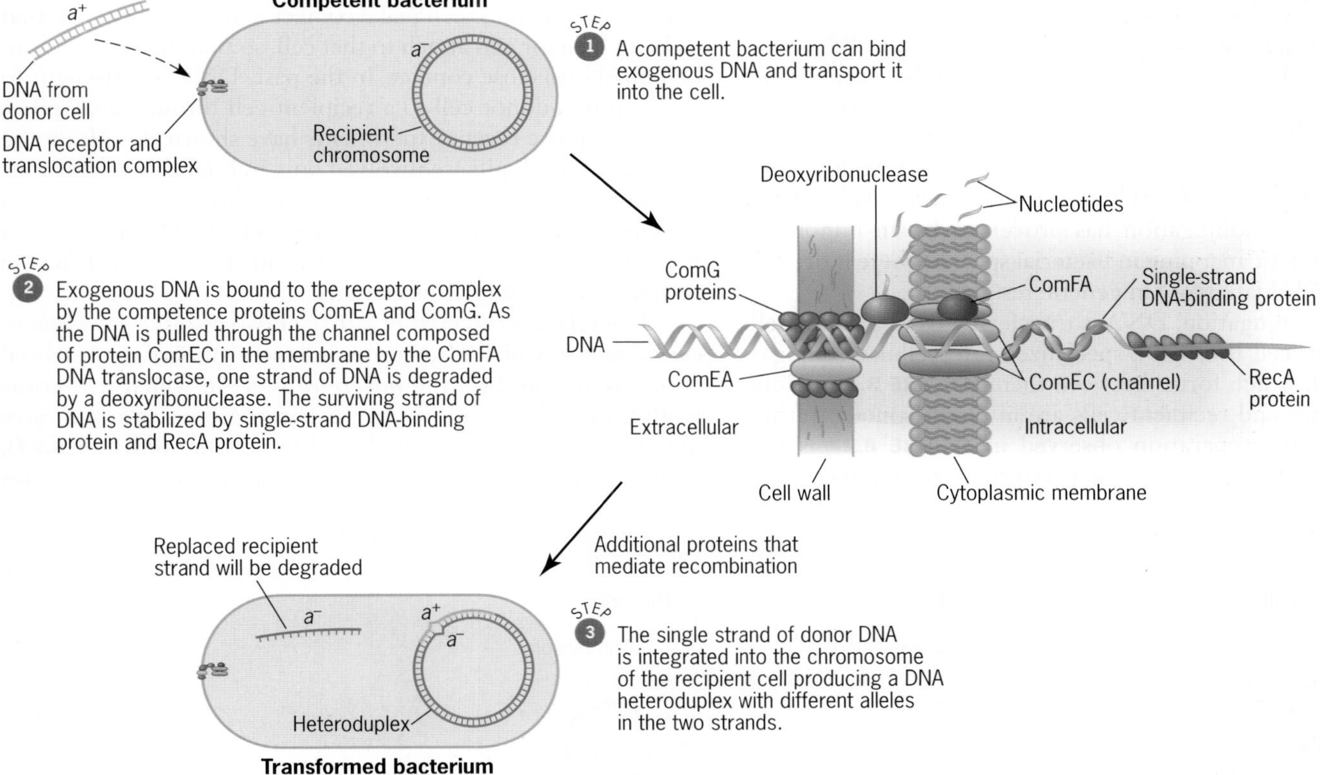

Figure 8.20 ▶ The mechanism of transformation in *Bacillus subtilis.* A competent bacterium contains a DNA receptor/translocation complex that can bind exogenous DNA and transport it into the cell, where it can recombine with chromosomal DNA of the recipient cell. ComEA, EC, FA, and G are competence proteins; they are synthesized only in competent cells. See the text for additional details.

strand is then degraded. If the donor and recipient cells carry different alleles of a gene, the resulting recombinant double helix will have one allele in one strand and the other allele in the second strand. A DNA double helix of this type is called a **heteroduplex** (a "heterozygous" double helix); it will segregate into two homoduplexes when it replicates.

The DNA molecules taken up by competent cells during transformation are usually only 0.2 to 0.5 percent of the complete chromosome. Therefore, unless two genes are quite close together, they will never be present on the same molecule of transforming DNA. Double transformants for two genes (say, a to a^+ and b to b^+, using an $a^+\ b^+$ donor and an $a\ b$ recipient) will require two independent transformation events (uptake and integration of one DNA molecule carrying a^+ and of another molecule carrying b^+). The probability of two such independent events occurring together will equal the product of the probability of each occurring alone. If, by contrast, two genes are closely linked, they may be carried on a single molecule of transforming DNA, and double transformants may be formed at a high frequency. The frequency with which two genetic markers are cotransformed can thus be used to estimate how far apart they are on the host chromosome.

CONJUGATION

Transformation does not occur in *E. coli*—the most intensely studied bacterial species—under natural conditions. Thus, we could ask, is there any kind of gene transfer between *E. coli* cells? The answer to this question is "yes." In 1946, Joshua Lederberg and Edward Tatum discovered that *E. coli* cells transfer genes by conjugation. We discuss their important discovery in A Milestone in Genetics: Conjugation in *Escherichia coli* later in this chapter. Conjugation has proven to be an important method of genetic mapping in bacterial species where it occurs, and it is an invaluable tool in genetic research.

During conjugation, DNA is transferred from a donor cell to a recipient cell through a specialized intercellular conjugation channel, which forms between them (**FIGURE 8.21**). Note that the donor and recipient cells are in direct contact during conjugation; the separation observed in **FIGURE 8.21** is the result of stretching forces during preparation for microscopy.

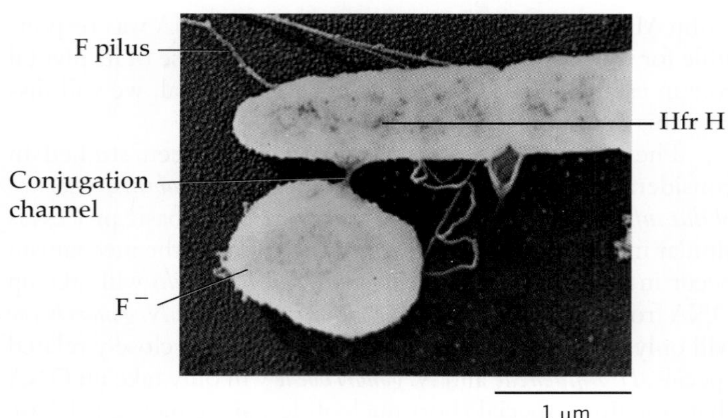

Figure 8.21 ▶ Conjugation in *E. coli*. This early electron micrograph by Thomas F. Anderson shows conjugation between an Hfr H cell and an F⁻ cell. Donor and recipient cells are actually in close juxtaposition during conjugation. The conjugation channel shown here has been stretched during preparation for microscopy.

Donor cells have cell-surface appendages called F pili (singular, F pilus). The synthesis of these F pili is controlled by genes present on a small circular molecule of DNA called an **F factor** (for *f*ertility *factor*). Most F factors are approximately 10^5 nucleotide pairs in size (see **FIGURE 8.30**). Bacteria that contain an F factor are able to transfer genes to other bacteria. The F pili of a donor cell make contact with a recipient cell that lacks an F factor and attach to that cell, so that the two cells can be pulled into close contact. In the past, DNA was thought to move from a donor cell to a recipient cell through an F pilus. However, more recent experiments have shown this idea to be incorrect. The F pili are involved only in establishing cell contact, not in DNA transfer. After the F pili bring a donor cell and a recipient cell together, a conjugation channel forms between the cells, and DNA is transferred from the donor cell to the recipient cell through this channel.

The F factor can exist in either of two states: (1) the *autonomous state*, in which it replicates independently of the bacterial chromosome, and (2) the *integrated state*, in which it is covalently inserted into the bacterial chromosome and replicates like any other segment of that chromosome (**FIGURE 8.22**). Genetic elements with these properties are called episomes

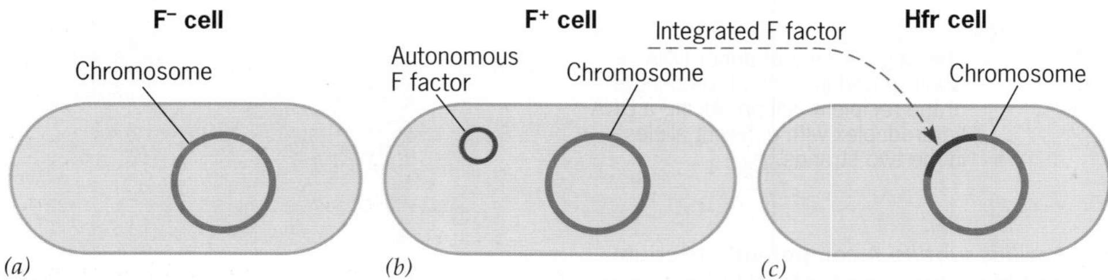

Figure 8.22 ▶ The F factor in *E. coli*: F⁻, F⁺, and Hfr cells. (*a*) An F⁻ cell has no F factor. (*b*) An F⁺ cell contains an F factor that replicates independently of the chromosome, and (*c*) an Hfr cell contains an F factor that is integrated—covalently inserted—in the chromosome.

(see PLASMIDS AND EPISOMES later in this chapter). A donor cell carrying an autonomous F factor is called an **F^+ cell.** A recipient cell lacking an F factor is called an F^- cell. When an F^+ cell conjugates (or "mates") with an F^- recipient cell, only the F factor is transferred. Both cells (donor and recipient) become F^+ cells because the F factor is replicated during transfer, and each cell receives a copy. Thus, if a population of F^+ cells is mixed with a population of F^- cells, virtually all of the cells will acquire an F factor.

The F factor can integrate into the bacterial chromosome by site-specific recombination events (**FIGURE 8.23**). The integration of the F factor is mediated by short DNA sequences that are present in multiple copies in both the F factor and the bacterial chromosome. Thus, an F factor can integrate at many different sites in the bacterial chromosome. A cell that carries an integrated F factor is called an **Hfr cell** (for *h*igh-*fr*equency *r*ecombination). In its integrated state, the F factor mediates the transfer of the chromosome from the Hfr cell to a recipient (F^-) cell during conjugation. Usually, the cells separate before chromosome transfer is complete; thus, only rarely will an entire chromosome be transferred from an Hfr cell to a recipient cell.

The mechanism that transfers DNA from a donor cell to a recipient cell during conjugation appears to be the same if just the F factor is being transferred, as in $F^+ \times F^-$ matings, or if the bacterial chromosome is being transferred, as in Hfr $\times$ F^- matings. Transfer is initiated at a special site called *oriT*—the *ori*gin of *t*ransfer—one of three sites on the F factor at which DNA replication can be initiated. The other two sites—*oriV* and *oriS*—are used to initiate replication during cell division, not during conjugation. *OriV* is the primary origin of replication during cell fission; *oriS* is a secondary origin that performs this function when *oriV* is absent or nonfunctional.

During conjugation, one strand of the circular DNA molecule is cut at *oriT* by an enzyme, and one end is transferred into the recipient cell through the channel that forms between the conjugating cells (**FIGURE 8.24**). The F factor or the Hfr chromosome containing the F factor replicates during transfer by a mechanism called *rolling-circle replication,* because the circular DNA molecule "rolls" during replication (see Chapter 10, **FIGURE 10.28**). During conjugation, one copy of the donor chromosome is synthesized in the donor cell, and the transferred strand of donor DNA is replicated in the recipient cell.

Because transfer is initiated within the integrated F factor, part of the F factor is transferred prior to the transfer of chromosomal genes in Hfr $\times$ F^- matings. The rest of the F factor is transferred after the chromosomal genes. Thus, the recipient cell acquires a complete F factor and is converted to an Hfr cell only in rare cases when an entire Hfr chromosome is transferred.

Several of the details of conjugation were worked out using one particular Hfr strain called Hfr H (for the English microbial geneticist William Hayes, who isolated it). In this strain, the F factor is integrated near the *thr* (threonine) and *leu* (leucine) loci, as shown in **FIGURE 8.23**. In 1957 Elie Wollman and François Jacob, working at the Pasteur Institute in Paris, provided new insight into the process of conjugation by crossing Hfr H cells of genotype thr^+ leu^+ azi^s ton^s lac^+ gal^+ str^s with F^- cells of genotype thr^- leu^- azi^r ton^r lac^- gal^- str^r. The *thr* gene and the *leu* gene are responsible for the syntheses of the amino acids threonine and leucine, respectively. Allele pairs azi^s/azi^r, ton^s/ton^r, and str^s/str^r control sensitivity (*s*) or resistance (*r*) to sodium azide, bacteriophage T1, and streptomycin, respectively. Alleles lac^+ and lac^- and alleles gal^+ and gal^- govern the ability (+) or inability (–) to utilize lactose and galactose, respectively, as energy sources.

At varying times after the Hfr H and F^- cells were mixed to initiate mating, samples were removed and agitated vigorously in a blender to break the conjugation bridges and separate the conjugating cells. These cells, whose mating had been so unceremoniously interrupted, were then plated on medium containing the antibiotic streptomycin but lacking the amino acids threonine and leucine. Only recombinant cells carrying the thr^+ and leu^+ genes of the Hfr H parent and the str^r gene of the F^- parent could grow on this *selective medium*. The Hfr H donor cells would be killed by the streptomycin, and the F^- recipient cells would not grow without threonine and leucine.

Colonies produced by thr^+ leu^+ str^r recombinants were then transferred (see Chapter 13, **FIGURE 13.2**) to a series of plates containing different selective media to determine which

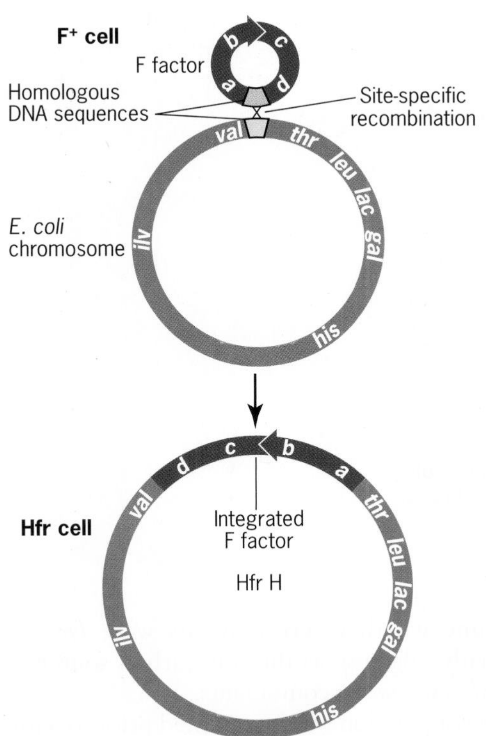

Figure 8.23 ▶ The formation of an Hfr cell by the integration of an autonomous F factor. The F factor is covalently inserted into the chromosome by site-specific recombination between homologous DNA sequences in the F factor and the chromosome.

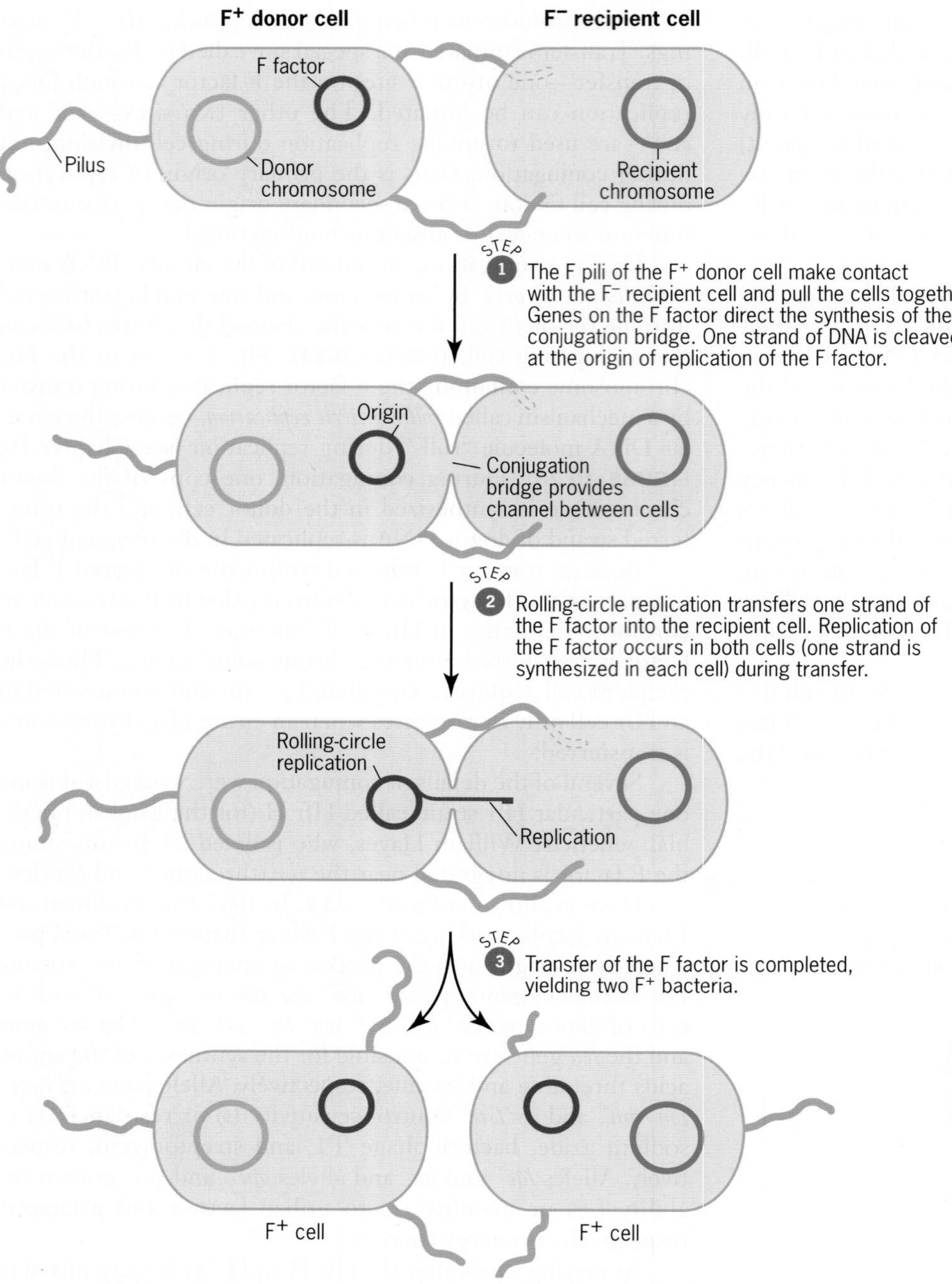
F⁺ donor cell
F⁻ recipient cell
F factor
Pilus
Donor chromosome
Recipient chromosome
STEP 1 The F pili of the F⁺ donor cell make contact with the F⁻ recipient cell and pull the cells together. Genes on the F factor direct the synthesis of the conjugation bridge. One strand of DNA is cleaved at the origin of replication of the F factor.
Origin
Conjugation bridge provides channel between cells
STEP 2 Rolling-circle replication transfers one strand of the F factor into the recipient cell. Replication of the F factor occurs in both cells (one strand is synthesized in each cell) during transfer.
Rolling-circle replication
Replication
STEP 3 Transfer of the F factor is completed, yielding two F⁺ bacteria.
F⁺ cell
F⁺ cell

Figure 8.24 ▶ Mating between an F^+ cell and an F^- cell. The F factor of the donor cell is replicated during transfer from an F^+ cell to an F^- cell. When the process is complete, each cell has a copy of the F factor.

of the other donor markers were present. The series of plates included medium containing specific supplements that allowed Wollman and Jacob to determine whether the recombinants contained donor or recipient alleles of each of the genes. Medium containing sodium azide was used to distinguish between azi^s and azi^r cells. Medium containing bacteriophage T1 was used to score recombinant bacteria as ton^s or ton^r. Medium containing lactose as the sole carbon source was used to determine whether recombinants were lac^+ or lac^-, and medium with galactose as the sole carbon source was used to identify gal^+ and gal^- recombinants.

When conjugation was interrupted prior to 8 minutes after mixing the Hfr H cells and the F^- cells, no thr^+ leu^+ str^r recombinants were detected. Recombinants (thr^+ leu^+ str^r) first appeared at about 8½ minutes after mixing the Hfr H and F^- cells and accumulated to a maximum frequency within a few

Summary of the results

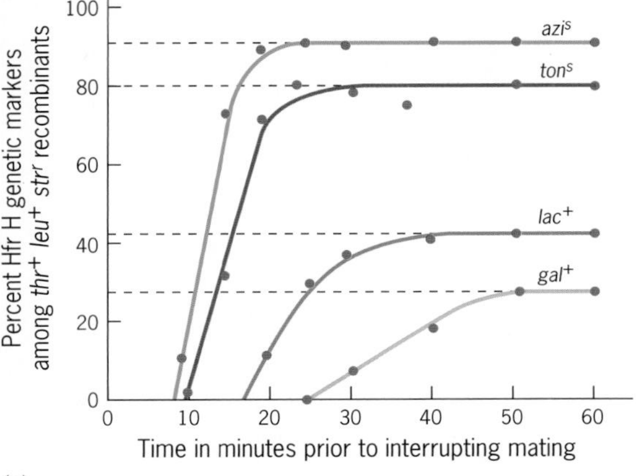

(a)

Interpretation of the results

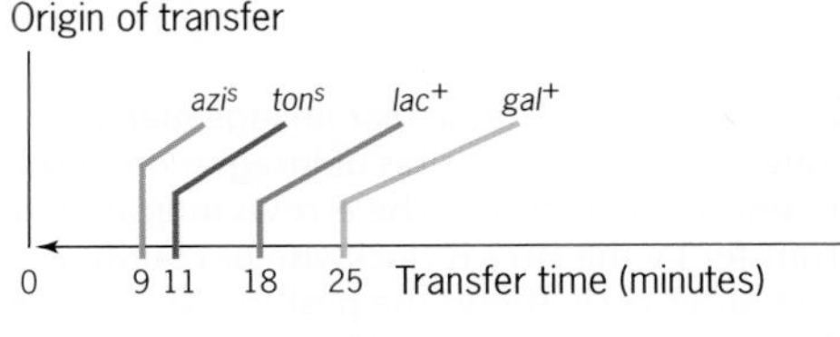

(b)

Figure 8.25 ▶ Wollman and Jacob's classic interrupted mating experiment. (a) The frequencies of the unselected donor alleles present in thr^+ leu^+ str^r recombinants are shown as a function of the time at which mating was interrupted. (b) Interpretation of the results based on the linear transfer of genes from the Hfr cell to the F^- cell. Transfer is initiated at the origin on the F factor, and the time at which a gene is transferred to the F^- cell depends on its distance from the F factor.

minutes. When the presence of the other donor markers was examined at varying times after mixing the donor and recipient cells, donor alleles were transferred to recipient cells in a specific temporal sequence (**FIGURE 8.25**). The Hfr H azi^s gene first appeared in recombinants at about 9 minutes after mixing the Hfr and F^- bacteria. The ton^s, lac^+, and gal^+ markers first appeared after 11, 18, and 25 minutes of mating, respectively. These results indicated that the genes from Hfr H were being transferred to the F^- cells in a specific temporal order, reflecting the order of the genes on the chromosome (**FIGURE 8.26**).

Subsequent studies with different Hfr strains revealed that gene transfer could be initiated at different sites on the chromosome. We now know that the F factor can integrate at many different sites in the *E. coli* chromosome and that the site of integration determines where gene transfer is initiated in each Hfr strain. Moreover, the orientation of F factor integration—either *d c b a* reading clockwise or *a b c d* reading clockwise (see **FIGURE 8.23**)—determines whether the transfer of genes is clockwise relative to the *E. coli* linkage map or counterclockwise (**FIGURE 8.27**).

The transfer of a complete chromosome from an Hfr to an F^- cell takes about 100 minutes, and transfer appears to proceed at a fairly constant rate. Thus, the time required for transfer of genes during conjugation can be used to map genes on bacterial chromosomes. A map distance of 1 minute corresponds to the length of a chromosomal segment transferred in

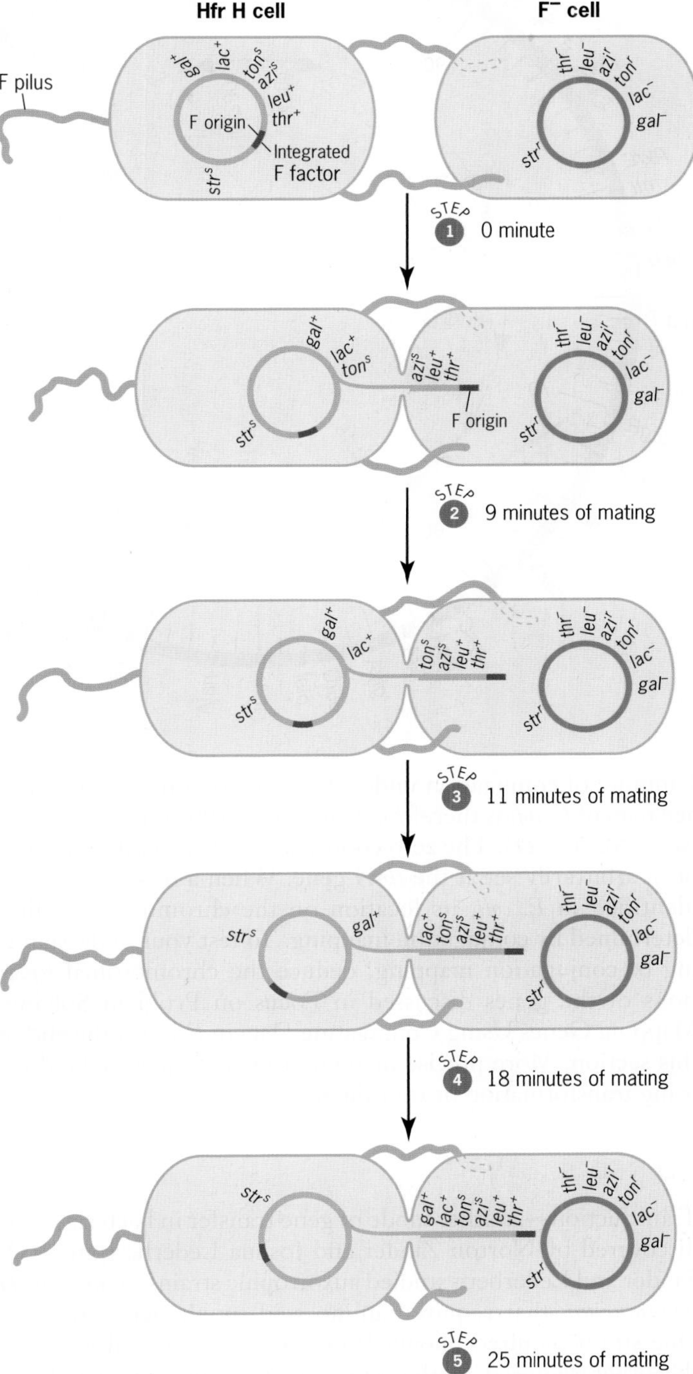

Figure 8.26 ▶ The interpretation of Wollman and Jacob's interrupted mating experiment. A linear transfer of genes occurs from the donor (Hfr H) cell to the recipient (F^-) cell. Transfer begins at the origin of replication on the integrated F factor and proceeds with the sequential transfer of genes based on their location on the chromosome. The chromosome replicates during the transfer process so that the Hfr and F^- cells both end up with a copy of the transferred DNA.

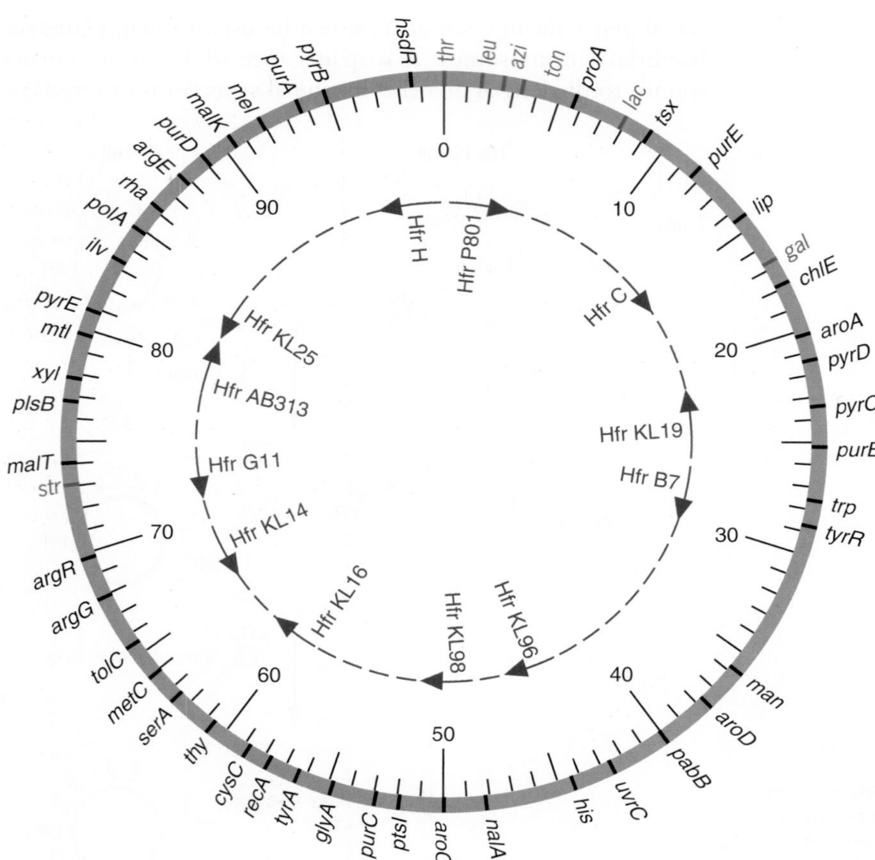

Figure 8.27 ▶ The circular linkage map of *E. coli*. The inner circle shows the sites of integration of the F factor in selected Hfr strains. The arrows indicate whether transfer by the Hfr's is clockwise or counterclockwise. The outer circle shows the position of selected genes. The map is divided into 100 units, where each unit is the length of DNA transferred during one minute of conjugation. The genes shown in red were used in Wollman and Jacob's famous interrupted mating experiment (see Figures 8.25 and 8.26).

1 minute of conjugation under standard conditions. The linkage map of *E. coli* is therefore divided into 100-minute intervals (see **FIGURE 8.27**). The zero coordinate of this circular map has been arbitrarily set at the *thrA* gene. When a new mutation is identified in *E. coli*, its location on the chromosome is first determined by conjugation mapping. To test your understanding of conjugation mapping, deduce the chromosomal locations of the genes discussed in Focus on Problem Solving: Mapping Genes Using Conjugating Data in *E. coli* at the end of this section. More precise mapping can subsequently be done using transformation or transduction.

TRANSDUCTION

Transduction—another mode of gene transfer in bacteria—was discovered by Norton Zinder and Joshua Lederberg in 1952. Zinder and Lederberg studied auxotrophic strains of *Salmonella typhimurium* that required amino acid supplements to grow. One strain required phenylalanine, tryptophan, and tyrosine; the other required methionine and histidine. Neither strain could grow on minimal medium lacking these amino acids. However, when Zinder and Lederberg grew the strains together, rare prototrophs were produced. Moreover, when they grew the strains in medium containing DNase, but separated them in the two arms of a U-tube (see **FIGURE 8.17**), prototrophic recombinants were still produced. The insensitivity to DNase ruled out transformation as the underlying mechanism, and the fact that cell contact was not required for the appearance of the prototrophs eliminated conjugation. Subsequent experiments showed that one of the strains was infected with a virus called bacteriophage P22 and that this virus was carrying genes from one cell (the donor) to another (the recipient). The rare prototrophs that Zinder and Lederberg detected were therefore due to recombination between bacterial DNA carried by the virus and DNA in the chromosome of the recipient cell.

Later studies revealed that there are two very different types of transduction. In **generalized transduction,** a random or nearly random fragment of bacterial DNA is packaged in the phage head in place of the phage chromosome. In **specialized transduction,** a recombination event occurs between the host chromosome and the phage chromosome, producing a phage chromosome that contains a piece of bacterial DNA. Phage particles that contain bacterial DNA are called *transducing particles*. Generalized transducing particles contain only bacterial DNA. Specialized transducing particles always contain both phage and bacterial DNA.

Generalized Transduction

Generalized transducing phages can transport any bacterial gene from one cell to another—thus, the name generalized transduction. The best-known generalized transducing phages are P22 in *S. typhimurium* and P1 in *E. coli*. Only about 1 to

2 percent of the phage particles produced by bacteria infected with P22 or P1 contain bacterial DNA, and only about 1 to 2 percent of the transferred DNA is incorporated into the chromosome of the recipient cell by recombination. Thus, the process is quite inefficient; the frequency of transduction for any given bacterial gene is about 1 per 10^6 phage particles.

Specialized Transduction

Specialized transduction is characteristic of viruses that transfer only certain genes between bacteria. Bacteriophage lambda (λ) is the best-known specialized transducing phage; λ carries only the *gal* (required for the utilization of galactose as an energy source) and *bio* (essential for the synthesis of biotin) genes from one *E. coli* cell to another. Earlier in this chapter, we discussed the site-specific insertion of the λ chromosome into the *E. coli* chromosome to establish a lysogenic state (see Bacteriophage Lambda). The insertion site is between the *gal* genes and the *bio* genes on the *E. coli* chromosome (see **FIGURE 8.6**), which explains why λ transduces these genes specifically.

The integrated λ chromosome—the λ prophage—in a lysogenic cell undergoes rare (about one in 10^5 cell divisions) spontaneous excision, whereupon it enters the lytic pathway. Prophage excision can also be induced, for example, by irradiating lysogenic cells with ultraviolet light. Normal excision is essentially the reverse of the site-specific integration process and yields intact circular phage and bacterial chromosomes (**FIGURE 8.28*a***). Occasionally, the excision is anomalous, with the crossover occurring at a site other than the original attachment site. When this happens, a portion of the bacterial chromosome is excised with the phage DNA and a portion of the phage chromosome is left in the host chromosome (**FIGURE 8.28*b***). These anomalous prophage excisions produce specialized transducing phage carrying either the *gal* or *bio* genes of the host. These transducing phage are denoted λ*dgal* (for λ *defective*

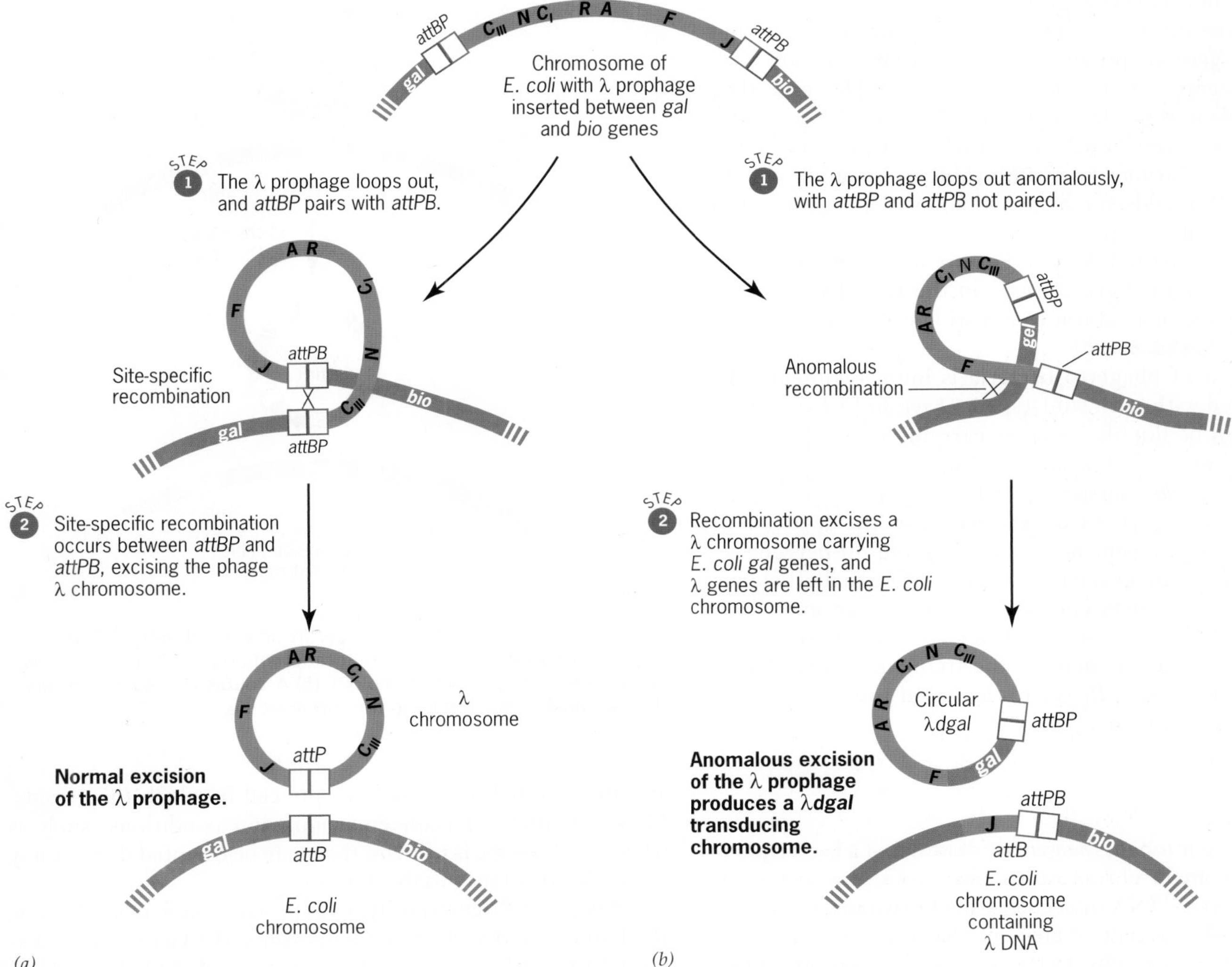

Figure 8.28 ▶ Lambda prophage excision. Comparison of (*a*) the normal excision of the λ prophage with (*b*) anomalous excision producing recombinant λ*dgal* transducing chromosomes.

phage carrying *gal* genes) and λ*dbio* (λ *d*efective phage carrying *bio* genes), respectively. They are defective phage particles because one or more genes required for lytic or lysogenic reproduction were left in the host chromosome.

Because of the small size of the phage head, only bacterial genes located close to the prophage can be excised with the phage DNA and packaged in phage heads. Another specialized transducing phage, Φ80, integrates near the *E. coli trp* genes (required for the synthesis of the amino acid tryptophan); this phage transduces *trp* markers. If specialized transducing particles are formed during prophage excision, as shown in **FIGURE 8.28*b***, they should be produced only when lysogenic cells enter the lytic pathway. Indeed, transducing particles are not present in lysates produced from primary lytic infections. The frequency of transducing particles in lysates produced by induction of lysogenic cells is about one in 10^6 progeny particles; therefore, these lysates are called *Lft* (*l*ow-*f*requency *t*ransduction) lysates.

The fate of the λ*dgal* and λ*dbio* DNA molecules after their injection into new host cells will depend on which λ genes are missing. If genes for lytic growth are missing, but an *att* site and *int* (integrase) gene are present, the defective chromosomes will be able to integrate into the host chromosome. However, they will not be able to reproduce lytically unless a wild-type λ, acting as a "helper" phage, is present. If the *int* gene is missing, the defective phage chromosome will be able to integrate only in the presence of a wild-type helper. If a λ*dgal*$^+$ phage infects a *gal*- recipient cell, integration of the λ*dgal*$^+$ will produce an unstable *gal*$^+$/*gal*$^-$ partial diploid (**FIGURE 8.29*a***), whereas rare recombination events between *gal*$^+$ in the transducing DNA and *gal*$^-$ in the recipient chromosome will produce a stable *gal*$^+$ transductant (**FIGURE 8.29*b***).

If the ratio of phage to bacteria is high, recipient cells will be infected with both wild-type λ *phage* and λ*dgal*$^+$; thus, these cells will be double lysogens carrying one wild-type λ prophage and one λ*dgal* prophage. The resulting transductants will be *gal*$^+$/*gal*$^-$ partial diploids. If the *gal*$^+$/*gal*$^-$ transductants are induced with ultraviolet light, the lysates will contain about 50 percent λ*dgal* particles and 50 percent λ$^+$ particles. Both prophages will replicate with equal efficiency using the gene products encoded by the λ$^+$ genome. Such lysates are called *Hft* (*h*igh-*f*requency *t*ransduction) lysates. *Hft* lysates dramatically increase the frequency of transduction events; therefore, *Hft* lysates are used preferentially in transduction experiments.

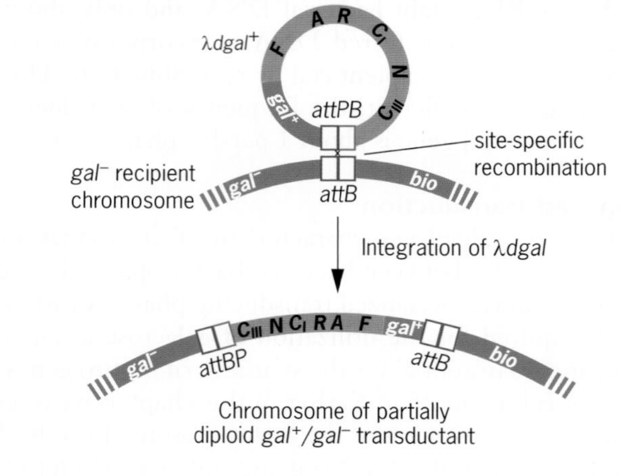

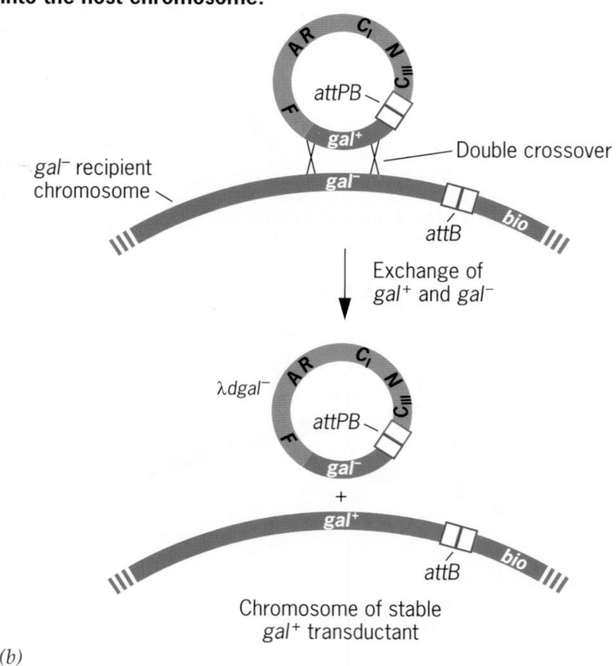

Figure 8.29 ▶ Recombination in *gal*$^-$ recipient cells infected with λ*dgal*$^+$ transducing phage. (*a*) Integration of λ*dgal*$^+$ at *attB* produces an unstable *gal*$^+$/*gal*$^-$ partial diploid. (*b*) A double crossover transfers the *gal*$^+$ allele from λ*dgal*$^+$ to the chromosome.

PLASMIDS AND EPISOMES

As previously mentioned, the genetic material of a bacterium is carried in one main chromosome plus from one to several extrachromosomal DNA molecules called plasmids. By definition, a **plasmid** is a genetic element that can replicate independently of the main chromosome in an extrachromosomal state. Most plasmids are dispensable to the host; that is, they are not required for survival of the cell in which they reside. However, under certain environmental conditions, such as when an antibiotic is present, they may be essential if they carry a gene for resistance to the antibiotic.

There are three major types of plasmids in *E. coli:* F factors, R plasmids, and Col plasmids. Fertility (F) factors were discussed earlier (see Conjugation). R plasmids (*r*esistance plasmids) carry genes that make host cells resistant to antibiotics

▶ FOCUS ON The Overuse of Antibiotics

Selman Waksman, a Ukrainian immigrant to the United States, discovered streptomycin in 1943. Later he named this class of antibacterial drugs *antibiotics*. The first documented treatment of a human with streptomycin involved a 21-year-old patient at the Mayo Clinic in Rochester, Minnesota. This woman had an advanced case of tuberculosis. She began receiving experimental injections of streptomycin in 1944, and to everyone's surprise her tuberculosis was cured. Streptomycin and other antibiotics quickly became "miracle drugs." Their application to people with bacterial infections saved many lives.

Humans soon began using large amounts of antibiotics. In 1950, the world used 10 tons of streptomycin. By 1955, worldwide use of streptomycin had increased to 50 tons, along with about 10 tons each of chloramphenicol and tetracycline.

However, the bacteria soon began to fight back. They evolved new genes encoding products that protected them from antibiotics. The evolution of antibiotic-resistant bacteria confirmed the power of natural selection. A bacterium without an antibiotic-resistance gene is killed by an antibiotic. A bacterium with an antibiotic-resistance gene grows, divides, and produces a population of bacteria, all resistant to the antibiotic. The result was inevitable: antibiotic-resistant bacteria spread.

Some of the first studies documenting the evolution of antibiotic and drug-resistant bacteria were performed in Japan on the four "species" of *Shigella—S. dysenteriae, S. flexneri, S. boydii, and S. sonnei,* which cause dysentery. Only 0.2 percent of the *Shigella* strains isolated from sewers and polluted rivers in 1953 were resistant to any of the antibiotics and drugs tested. Just 12 years later, the frequency of antibiotic- and drug-resistant *Shigella* strains isolated from the same places had increased to 58 percent. However, the really bad news was not that these strains were resistant to antibiotics; rather, it was that most of them were resistant to at least four of the six antibiotics and drugs—ampicillin, kanamycin, tetracycline, streptomycin, sulfanilamide, and chloramphenicol—tested. They were multi-drug-resistant (MDR) strains of *Shigella.* MDR strains of other bacteria, such as *M. tuberculosis,* also began appearing.

The genes that protect bacteria from antibiotics often are present on small DNA molecules called R-plasmids (see PLASMIDS AND EPISOMES). Many R-plasmids are self-transmissible; that is, they carry genes that mediate their own transfer from one cell to another and even from one species to another. In addition, the antibiotic-resistance genes are often present on genetic elements—transposable genetic elements or "jumping genes"—that can move from one DNA molecule to another (see Chapter 18). Thus, the genes on these R-plasmids can spread rapidly through bacterial populations.

One reason that MDR strains of bacteria have evolved so rapidly is that we overuse antibiotics. All too often, antibiotics are prescribed for viral infections such as the common cold and flu. Antibiotics have no antiviral activity and should not be used to treat viral infections. In addition, antibiotics are widely used in vast amounts as "growth-promoters" in animal feed, where they prevent bacterial infections that reduce growth rate. Indeed, almost half of the antibiotics produced in the United States are used as additives in animal feed. They are added at rates of 2 to 50 grams per ton of feed, and the inevitable happens—antibiotic-resistant bacteria evolve. These resistant bacteria are then transmitted to humans caring for the animals, working in the meat-packaging industry, or consuming undercooked meat products.

Given the widespread evolution of MDR strains of *M. tuberculosis, Staphylococcus aureus, Shigella dysenteriae,* and other pathogenic bacteria, perhaps we should restrict the use of some of our best antibiotics to the treatment of potentially fatal human diseases. Indeed, in 1986, Sweden passed a law banning all nontherapeutic use of antibiotics, including use as animal growth promoters. Actually, Denmark was the first to restrict the use of antibiotics in animal feed; it banned the use of penicillins and tetracyclines as growth promoters in the 1970s. Then, in 1998, Denmark initiated a voluntary ban on the use of all antibiotics as animal growth promoters and began requiring prescriptions for the therapeutic treatment of animals with antibiotics. The negative effects of banning the use of antibiotics in animal feed on productivity have been minimal. Moreover, in Sweden, the overall use of antibiotics has decreased by 55 percent since the ban on nontherapeutic uses began. Perhaps it is time for the United States and the rest of the world to follow Scandinavia's lead—to ban, or at least limit, the nontherapeutic use of antibiotics. Do we really need antibiotics in animal feed? And do we need them in our hand soap?

and other antibacterial drugs. Col plasmids (previously called colicinogenic factors) encode proteins that kill sensitive *E. coli* cells. There are a large number of distinct Col plasmids; however, they will not be discussed further here.

Some plasmids endow host cells with the ability to conjugate. All F^+ plasmids, many R plasmids, and some Col plasmids have this property; we say that they are conjugative plasmids. Other R and Col plasmids do not endow cells with the ability to conjugate; we say that they are nonconjugative. The conjugative nature of many R plasmids plays an important role in the rapid spread of antibiotic and drug resistance genes through populations of pathogenic bacteria, as discussed at the beginning of this chapter. The evolution of R plasmids that make host bacteria resistant to multiple antibiotics has become a serious medical problem, and the use of antibiotics for nontherapeutic purposes has contributed to the rapid evolution of multiple drug-resistant bacteria (see Focus on the Overuse of Antibiotics above).

In 1958 François Jacob and Elie Wollman recognized that the F factor and certain other genetic elements had unique properties. They defined this class of elements and called them episomes. According to Jacob and Wollman, an **episome** is a genetic element that is unessential to the host and that can replicate either autonomously or be integrated (covalently inserted) into the chromosome of the host bacterium. The terms *plasmid* and *episome* are not synonyms. Many plasmids do not exist in integrated states and are thus not episomes. Similarly, many lysogenic phage chromosomes, such as the phage λ genome, are episomes but are not plasmids.

The ability of episomes to insert themselves into chromosomes depends on the presence of short DNA sequences called insertion sequences (or IS elements). The IS elements are present in both episomes and bacterial chromosomes. These short sequences (from about 800 to about 1400 nucleotide pairs in length) are transposable; that is, they can move from one chromosome to a different chromosome (see Chapter 18). In addition, IS elements mediate recombination between otherwise nonhomologous genetic elements. The role of IS elements in mediating the integration of episomes is well documented in the case of the F factor in *E. coli*. Crossing over between IS elements in the F factor and the bacterial chromosome produces Hfr's with different origins and directions of transfer during conjugation (**FIGURE 8.30**).

F′ FACTORS AND SEXDUCTION

As previously discussed, phage λ transducing particles are produced by the anomalous excision of the λ prophage (see **FIGURE 8.28*b***). An Hfr strain is produced by the integration of an F factor into the chromosome (see **FIGURE 8.30**) by a mechanism similar to prophage formation. Moreover, rare F^+ cells are present in Hfr cultures, indicating that excision of the F factor also occurs, and, as would be expected, anomalous excision events produce autonomous F factors carrying bacterial genes (**FIGURE 8.31**). These modified F factors, called F′ ("F-prime") factors, were first identified by Edward Adelberg and Sarah Burns in 1959. Since F′ factors do not require packaging in a phage head like specialized transducing DNAs, the size of the segment of the bacteria chromosome carried by the F′ is not restricted. Thus, F′ factors range in size from those carrying a single bacterial gene to those carrying up to half the bacterial chromosome (**FIGURE 8.32**).

Transfer of F′ factors to recipient (F^-) cells is called **sexduction;** it occurs by the same mechanism as F factor transfer in $F^+ \times F^-$ matings (see **FIGURE 8.24**)—with one important difference: bacterial genes incorporated into F′ factors are transferred to recipient cells at a much higher frequency. The F′ factors are valuable tools for genetic studies; they can be used to produce partial diploids carrying two copies of any

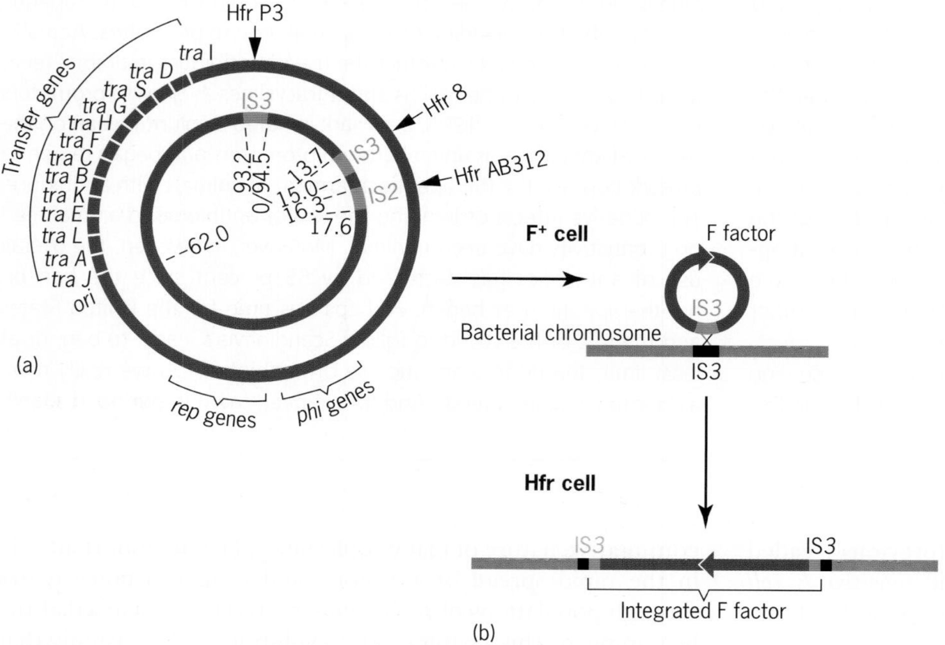

Figure 8.30 ▶ IS elements mediate the integration of the F factor. (*a*) An abbreviated map of the structure of the F factor in *E. coli* strain K12, with distances given in kilobases (1000 nucleotide pairs). The locations of genes required for conjugative transfer (*tra* genes), replication (*rep* genes), and the inhibition of phage growth (*phi* genes) are shown, along with the positions of three IS elements. The arrows denote the specific IS element that mediated the integration of the F factor during the formation of the indicated Hfr strains. (*b*) Recombination between IS elements inserts the F factor into the bacterial chromosome, producing an Hfr.

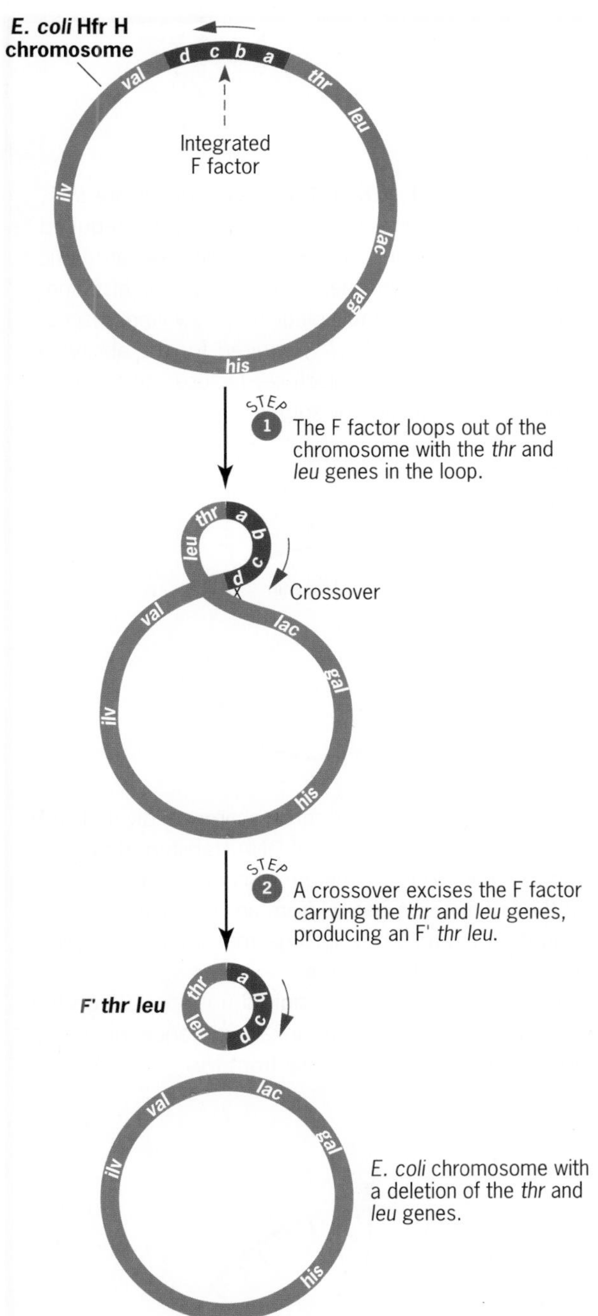

Figure 8.31 ▶ Formation of an F′. The anomalous excision of the F factor from an Hfr chromosome produces an F factor carrying the *E. coli thr* and *leu* genes and designated F′ *thr leu*.

gene or set of linked genes. Thus, sexduction can be used to determine dominance relationships between alleles and perform other genetic tests requiring two copies of a gene in the same cell.

Consider an F′ *thr*$^+$ *leu*$^+$ factor generated by anomalous excision of the F factor from Hfr H, as shown in **FIGURE 8.31**. Matings between F′ *thr*$^+$ *leu*$^+$ donor cells and *thr*$^-$ *leu*$^-$ recipient

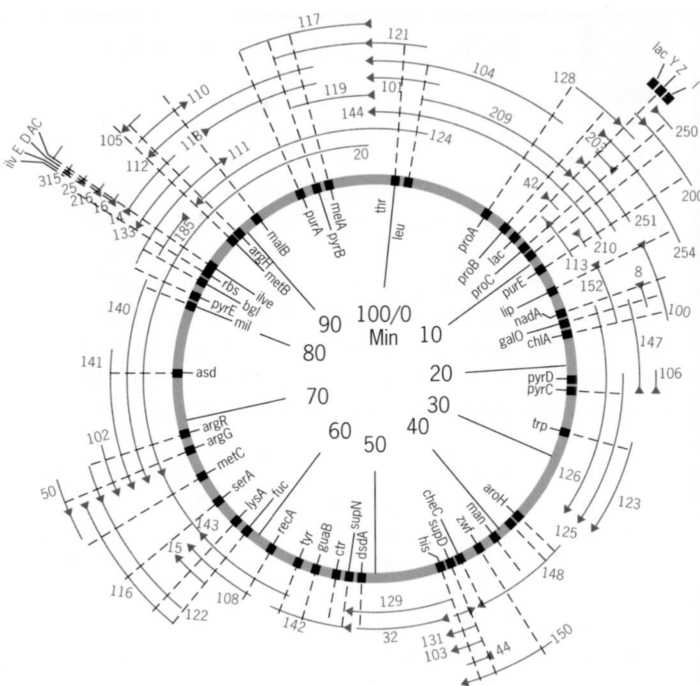

Figure 8.32 ▶ F′s in *E. coli*. Map of the chromosome of *E. coli* K12 showing the genes present in representative F′s. The F′s are drawn as linear structures in order to align them with the segments of the chromosome that they contain. In reality, they are circular DNA molecules—the structures formed by joining the two ends of each F′.

cells produce *thr*$^-$ *leu*$^-$/F′ *thr*$^+$*leu*$^+$ partial diploids. These partial diploids are unstable because the F′ factor may be lost, producing *thr*$^-$ *leu*$^-$ haploids, or recombination may occur between the chromosome and the F′, producing stable *thr*$^+$ *leu*$^+$ recombinants.

USING PARTIAL DIPLOIDS TO MAP CLOSELY LINKED GENES

The unambiguous mapping of closely linked genes in bacteria requires three-factor cross data, just as in eukaryotes (see Chapter 7). The rationale behind such mapping experiments with bacteria is essentially the same as for eukaryotes; namely, the more crossovers required to produce a recombinant genotype, the lower its frequency. However, recombination in bacteria usually involves crossovers between a linear fragment of the donor DNA and an intact circular recipient chromosome, that is, in partial diploids (see **FIGURE 8.15**). Thus, crossovers must occur in pairs. However, the rationale is the same—recombinant genotypes that require two pairs of crossovers will be less frequent than those that require only one pair, and so on.

Crosses between double-mutant and single-mutant bacteria can be performed in two ways: (1) single-mutant donor × double-mutant recipient and (2) double-mutant donor × single-mutant

▶ FOCUS ON PROBLEM SOLVING
Mapping Genes Using Conjugation Data in *E. coli*

THE PROBLEM

You have identified a mutant *E. coli* strain that cannot synthesize the amino acid tryptophan (Trp^-). To determine the location of the trp^- mutation on the *E. coli* chromosome, you have carried out interrupted mating experiments with four different Hfr strains. In all cases, the Hfr strains carried the dominant wild-type alleles of the marker genes, and the F^- strain carried the recessive mutant alleles of these genes. The following chart shows the time of entry in minutes (in parentheses) of the wild-type alleles of the marker genes into the Trp^- F^- strain. The marker genes are thr^+, aro^+, his^+, tyr^+, met^+, arg^+ and ilv^+ (encoding enzymes required for the synthesis of the amino acids threonine, the aromatic amino acids phenylalanine, tyrosine, and tryptophan, histidine, tyrosine, methionine, arginine, and isoleucine plus valine, respectively) and man^+, gal^+, lac^+, and xyl^+ (required for the ability to catabolize the sugars mannose, galactose, lactose, and xylose, respectively, and use them as energy sources).

Hfr A ——	man^+ (1)	trp^+ (9)	aro^+ (17)	gal^+ (20)	lac^+ (29)	thr^+ (37)
Hfr B ——	trp^+ (6)	*man* (14)	*his* (22)	*tyr* (34)	*met* (42)	*arg* (48)
Hfr C ——	*thr* (3)	ilv^+ (20)	xyl^+ (25)	arg^+ (33)	met^+ (39)	tyr^+ (47)
Hfr D ——	met^+ (2)	arg^+ (8)	xyl^+ (16)	ilv^+ (21)	thr^+ (38)	lac^+ (46)

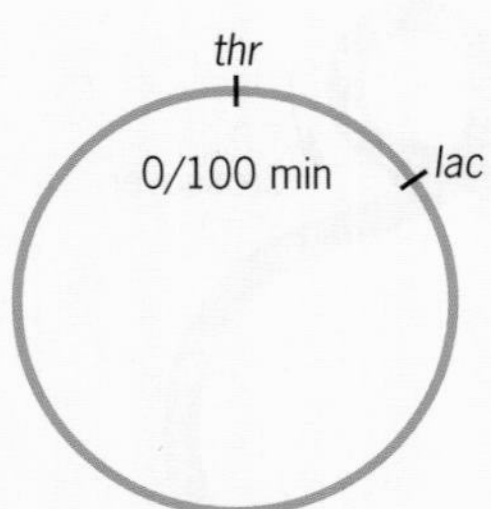

On the map of the circular *E. coli* chromosome shown on the right, indicate (1) the relative location of each gene, (2) the position where the F factor is integrated in each of the four Hfr's, and (3) the direction of chromosome transfer for each Hfr (clockwise or counterclockwise; indicate direction with an arrow).

FACTS AND CONCEPTS

1. The chromosome of *E. coli* contains a circular DNA molecule.

2. Chromosomal DNA is transferred from Hfr donor cells to F^- recipient cells by rolling-circle replication.

3. Rolling-circle replication, and thus transfer of chromosomal genes, is initiated at the origin of replication on the integrated F factor.

4. The direction of transfer (clockwise or counterclockwise) depends on the orientation of the F factor in the Hfr chromosome.

5. The F factor can integrate at many sites in the *E. coli* chromosome and in either orientation (clockwise or counterclockwise).

6. The genetic map of the *E. coli* chromosome is divided into minutes, where 1 minute is the length of DNA transferred from an Hfr strain to an F^- strain during 1 minute of conjugation.

7. Transfer of the entire chromosome from an Hfr cell to an F^- cell takes 100 minutes; therefore, the linkage map of the complete circular chromosome totals 100 minutes.

8. The *thr* locus has been arbitrarily assigned position "0" on the map of the *E. coli* chromosome, with linkage distance increasing from 0 to 100 minutes moving clockwise from *thr*.

ANALYSIS AND SOLUTION

If we examine the sequence in which genes are transferred from each Hfr strain to the F^- strain, we observe a linear sequence in each case. Moreover, note that regardless of the sequence in which genes are transferred by different Hfr strains, the distance between adjacent genes remains the same. The distance between *man* and *trp* is 8 minutes, for example, regardless of whether Hfr strain A or B is used in the experiment. Indeed, if we combine the results obtained using the four Hfr strains and place *thr* at position 0, the data yield the circular genetic map shown on the right. The circular map is a satisfying result given that we know that the chromosomal DNA of *E. coli* is also circular.

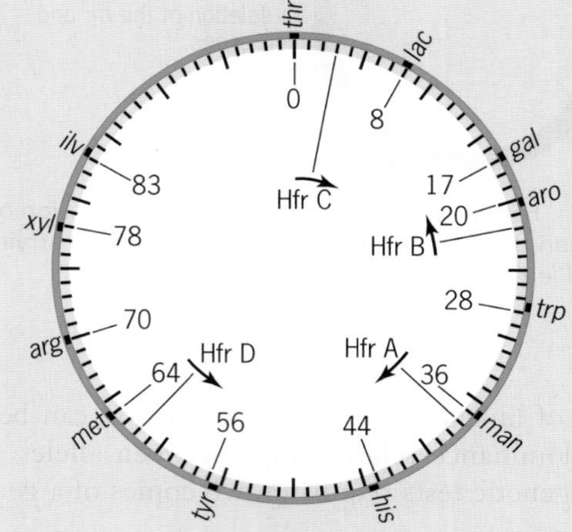

For further discussion go to your *WileyPLUS* course.

recipient. The progeny genotype frequencies of these reciprocal crosses are not necessarily the same. In fact, the results of such reciprocal crosses can be used to order the markers involved. Suppose that one wishes to order two genes (y and z) at one locus relative to a marker (x) at a nearby locus. The two reciprocal crosses are

1. $x^+ y^+ z^-$ donor X $x^- y^- z^+$ recipient, and
2. $x^- y^- z^+$ donor X $x^+ y^+ z^-$ recipient.

Note that the order of the loci $x\ y\ z$ is unknown; the crosses are being done to determine the actual order. Often, such crosses are done with auxotophic mutants so that selective media can be prepared on which only the prototrophic ($x^+y^+z^+$) recombinants can grow. The use of selective media greatly simplifies the identification and enumeration of recombinants.

Two orders are possible for the mutations y and z relative to the outside marker x: x-y-z and x-z-y. If the correct order is x-y-z, $x^+y^+z^+$, recombinants will occur in approximately the same frequency in both crosses. The formation of $x^+y^+z^+$ recombinants will require two crossovers in both crosses. However, if the order is x-z-y, $x^+y^+z^+$ recombinants will be much more frequent in cross 2 than in cross 1. If the order is x-z-y, the formation of $x^+y^+z^+$ recombinants will require four crossovers in cross 1, but only two crossovers in cross 2 (**FIGURE 8.33**).

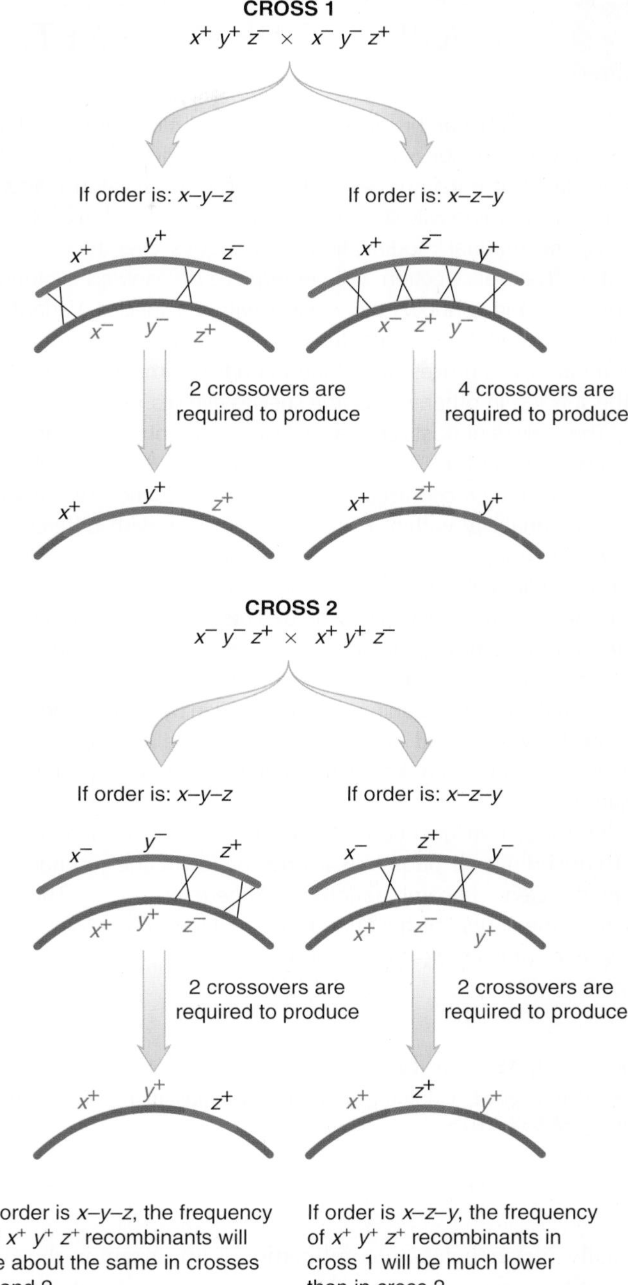

Figure 8.33 ► Mapping genes in bacteria. The diagram shows the rationale used to order closely linked genes by reciprocal three-factor crosses that produce partial diploids in bacteria.

KEY POINTS

- Three parasexual processes—transformation, conjugation, and transduction—occur in bacteria. These processes can be distinguished by two criteria: whether the gene transfer is inhibited by deoxyribonuclease and whether it requires cell contact.
- Transformation involves the uptake of free DNA by bacteria.
- Conjugation occurs when a donor cell makes contact with a recipient cell and then transfers DNA to the recipient cell.
- Transduction occurs when a virus carries bacterial genes from a donor cell to a recipient cell.
- Plasmids are self-replicating extrachromosomal genetic elements.
- Episomes can replicate autonomously or as integrated components of bacterial chromosomes.
- F factors that contain chromosomal genes (F′ factors) are transferred to F$^-$ cells by sexduction.
- Closely linked genes can be mapped in bacteria by three-factor crosses.

The Evolutionary Significance of Genetic Exchange in Bacteria

Genetic exchange is as important in bacteria as it is in other organisms.

Mutation is the source of new genetic variation, and recombination then produces new combinations of this variation—new assortments of genes responsible for the phenotypes acted on by natural selection. In eukaryotes, recombination events—independent assortment and crossing over—are an integral part of sexual reproduction. However, bacteria do not reproduce

▶ A MILESTONE IN GENETICS: Conjugation in *Escherichia coli*

Charles Darwin began his classic book on evolution with an extensive discussion of variation within and between species.[1] His account is rich in detail. It encompasses variation in domesticated plants and animals, as well as in wild species. Darwin understood that evolution—that is, the gradual modification of a species over time—requires variation. Natural selection, the centerpiece of Darwin's evolutionary theory, sifts and winnows the variants within a species. Variants that survive transmit their characteristics to the next generation. The repetition of this process over long periods of time leads to changes in the species. In other words, the species evolves.

The early geneticists embraced Darwin's emphasis on variation as the raw material for evolutionary change. With their knowledge of Mendelism, they recognized that sexual reproduction was a way to enhance variability within a species. Individuals with different traits can be crossed, and their genes can be recombined to produce offspring unlike either parent. Recombination, whether by independent assortment of unlinked genes or by crossing over between linked genes, produces new genotypes on which natural selection can act.

Although the variation-enhancing role of recombination was evident in the plant and animal species studied by the early geneticists, no one knew if it had any place in the world of microorganisms. Indeed, no one even knew if recombination occurred in these organisms.

In 1946, two geneticists settled the matter. Joshua Lederberg and Edward Tatum published a brief paper in the journal *Nature* entitled "Gene Recombination in *Escherichia coli*."[2] This paper changed the course of genetics. It also influenced the way in which the science of microbiology developed.

[1]Darwin, C. R. 1859. *The Origin of Species.* London, John Murray.

[2]Lederberg, J., and E. L. Tatum. 1946. Gene recombination in *Escherichia coli. Nature* 158:558–560.

Lederberg and Tatum employed a simple strategy to look for recombinants in cultures of *E. coli.* They grew two auxotrophic strains in the same culture vessel and then screened the culture for prototrophic cells. If prototrophs were produced, Lederberg and Tatum would have evidence that genes from the parental strains had recombined. The experiment was designed to rule out reverse mutation as a source of prototrophs. Each of the parental strains was a triple mutant. One parent, denoted Y-10, required the amino acids threonine and leucine plus the vitamin thiamine for growth. The other parent, denoted Y-24, required the amino acids phenylalanine and cysteine and the vitamin biotin. Reversion of either auxotrophic strain to the prototrophic condition would require three mutations in the same cell—an extraordinarily unlikely event. The two parental strains were grown together in a rich medium that met all their nutritional requirements. Then the cells were shifted to a minimal medium in which only prototrophs could grow. Lederberg and Tatum found the prototrophs they were looking for (**FIGURE 1**). Thus, *E. coli* cells were shown to enjoy the benefits of genetic recombination.

Subsequent experiments demonstrated that the process of recombination in *E. coli* involves the transfer of chromosomal material from one cell (the donor) to another cell (the recipient) through a conjugation tube. Crossovers within the recipient cell then insert fragments of the donor cell's chromosome into the recipient's chromosome. In the end, a genetically recombinant cell is produced. This parasexual process has played an important role in the genetic analysis of *E. coli* and other bacteria. It has also played an important role in the evolution of multi-drug-resistant pathogens, such as the strain of *Mycobacterium tuberculosis* discussed at the beginning of this chapter. Clearly, what is good for the bacteria is not always good for us.

QUESTIONS FOR DISCUSSION

1. Biologists often say that organisms belonging to different species are reproductively isolated from each other; that is,

sexually. Nevertheless, recombination is important in the evolution of bacteria, just as in eukaryotes. Thus, it is not surprising that mechanisms for exchanging genetic information and producing recombinant combinations of genes have evolved in bacteria. These parasexual mechanisms—transformation, conjugation, and transduction—generate new combinations of genes and allow bacteria to evolve and adapt to new environmental niches and to sudden changes in existing habitats.

Although parasexual processes are beneficial to bacteria, they create serious problems for humans who are attempting to combat bacterial diseases. These parasexual processes are partially responsible for the rapid evolution and spread of plasmids that confer antibiotic and drug resistance to bacteria. The extensive use of antibiotics in agriculture and medicine has resulted in the evolution of plasmids that carry a whole battery of antibiotic-resistance genes (see the Focus on the Overuse of Antibiotics). Indeed, the resurgence of tuberculosis in New York City in the 1990s has in large part been due to the evolution of a strain of *M. tuberculosis* that is resistant to seven different antibiotics. Moreover, the New York strain is closely related to the predominant strains of antibiotic-resistant *M. tuberculosis* in China and other regions of the world. Thus, what is beneficial to pathogenic bacteria is often harmful to humans.

KEY POINTS

- Parasexual recombination mechanisms produce new combinations of genes in bacteria.
- Parasexual mechanisms enhance the ability of bacteria to adapt to changes in the environment.

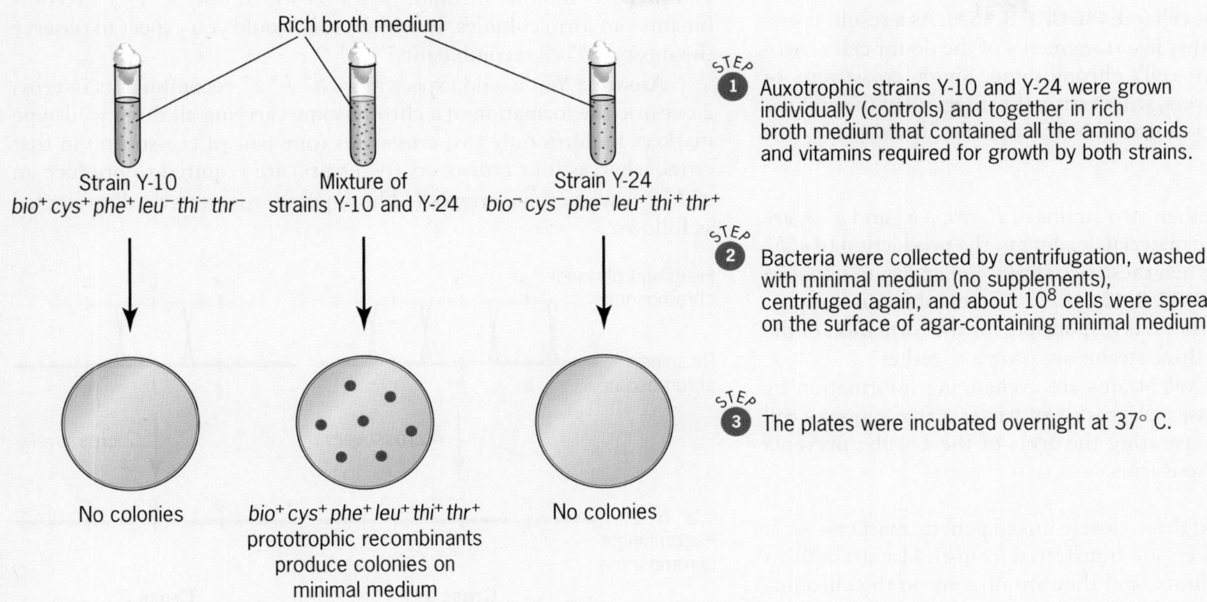
Rich broth medium
STEP 1 Auxotrophic strains Y-10 and Y-24 were grown individually (controls) and together in rich broth medium that contained all the amino acids and vitamins required for growth by both strains.
Strain Y-10
bio⁺ cys⁺ phe⁺ leu⁻ thi⁻ thr⁻
Mixture of strains Y-10 and Y-24
Strain Y-24
bio⁻ cys⁻ phe⁻ leu⁺ thi⁺ thr⁺
STEP 2 Bacteria were collected by centrifugation, washed with minimal medium (no supplements), centrifuged again, and about 10⁸ cells were spread on the surface of agar-containing minimal medium.
STEP 3 The plates were incubated overnight at 37° C.
No colonies
bio⁺ cys⁺ phe⁺ leu⁺ thi⁺ thr⁺ prototrophic recombinants produce colonies on minimal medium
No colonies

Figure 1 ▶ Lederberg and Tatum's demonstration of genetic recombination in *Escherichia coli*.

they do not exchange genes. However, conjugation allows genes to be transferred among bacteria as diverse as *E. coli*, *Salmonella typhimurium* (which causes food poisoning), *Proteus mirabilis* (which causes urinary tract infections), *Shigella dysenteriae* (which causes dysentery), and *Haemophilius influenzae* (which causes meningitis). What do these observations imply for the concept of a species in the world of bacteria?

2. Sensitive bacteria are killed by antibiotics; resistant bacteria are not. Genes that confer resistance to antibiotics provide scientists with dominant selectable markers that are extraordinarily useful in microbial research. However, these antibiotic-resistant genes are also found in multi-drug-resistant pathogens such as *Mycobacterium tuberculosis.* What precautions should be taken in handling antibiotic-resistant strains of bacteria in a research laboratory? What precautions should be taken with people infected by multi-drug-resistant pathogens?

▶ Basic Exercises

ILLUSTRATE BASIC GENETIC ANALYSIS

1. What advantages for genetic research do viruses have over cellular and multicellular organisms?

Answer: The two major advantages for genetic studies that viruses have over cellular and multicellular organisms are (1) their structural simplicity and (2) their short life cycle. Viruses usually contain a single chromosome with a relatively small number of genes, and they can complete their life cycle in from about 20 minutes to a few hours. Phage T4 crosses can be done in less than an hour, and the results can be analyzed the same day.

2. How is a cross performed with bacteriophage?

Answer: A phage cross is performed by simultaneously infecting the same host cells with two phage of interest (usually mutants) and screening for recombinants (usually wild-type) among the progeny. The ratio of phage to bacteria is kept high enough (about 5 to 10 phage of each genotype for each bacterium) so that most cells are infected with both parental phage. A major advantage of bacteriophage for such genetic studies is that large numbers of progeny (10^8 to 10^9) can easily be screened for rare recombinants.

3. What are the major differences between crossing over in bacteria and in eukaryotes?

Answer: In bacteria, crossing over usually occurs between a fragment of the chromosome from a donor cell and an intact circular chromosome in a recipient cell (see **FIGURE 8.15*a***). As a result, crossovers must occur in pairs that insert segments of the donor cell's chromosome into the recipient cell's chromosome. Single crossovers, or any odd number of crossovers, will destroy the integrity of the circular chromosome and yield a linear DNA molecule in its place (see **FIGURE 8.15*b***).

4. When grown together, two strains of *E. coli*, $a\ b^+$ and $a^+\ b$, are known to exchange genetic material, leading to the production of a^+b^+ recombinants. However, when these two strains are grown in opposite arms of a U-tube (see **FIGURE 8.17**), no $a^+\ b^+$ recombinants are produced. What parasexual process is responsible for the formation of the $a^+\ b^+$ recombinants when these strains are grown together?

Answer: The two *E. coli* strains are exchanging information by conjugation, the only parasexual process in bacteria that requires cell contact. The glass filter separating the arms of the U-tube prevents contact between cells in these arms.

5. You have identified three closely linked genetic markers—*a*, *b*, and *c*—in *E. coli*. The markers are transferred from an Hfr strain to an F^- strain in less than 1 minute, and they are present on the chromosome in the order *a*—*b*—*c*. You perform phage P1 transduction experiments using strains of genotype $a^+\ b\ c^+$ and $a\ b^+\ c$. In cross 1, the donor cells $a^+\ b\ c^+$ and the recipient cells are $a\ b^+\ c$. In cross 2, the donor cells $a\ b^+\ c$ and the recipient cells are $a^+\ b\ c^+$. For both crosses, you prepare minimal medium plates on which only $a^+\ b^+\ c^+$ recombinants can form colonies. In which cross would you expect to observe the most $a^+\ b^+\ c^+$ recombinants?

Answer: You would expect more $a^+\ b^+\ c^+$ recombinants in cross 2 because the formation of a chromosome carrying all three wild-type markers requires only two crossovers (one pair of crossovers) in that cross, whereas four crossovers (two pairs) are required to produce an $a^+\ b^+\ c^+$ chromosome in cross 1. The required crossovers are diagrammed as follows.

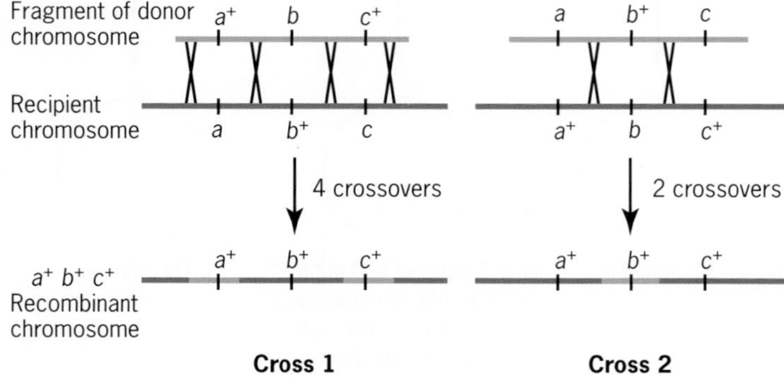

▶ Testing Your Knowledge

INTEGRATE DIFFERENT CONCEPTS AND TECHNIQUES

1. You have identified a mutant *E. coli* strain that cannot synthesize histidine (His⁻). To determine the location of the *his*⁻ mutation on the *E. coli* chromosome, you perform interrupted mating experiments with five different Hfr strains. The following chart shows the time of entry (minutes, in parentheses) of the wild-type alleles of the first five markers (mutant genes) into the His⁻ strain.

Hfr A	*bio* (4)	*glu* (20)	*his* (27)	*cys* (37)	*tyr* (45)
Hfr B	*xyl* (6)	*met* (18)	*tyr* (24)	*cys* (32)	*his* (42)
Hfr C	*his* (3)	*cys* (13)	*tyr* (21)	*met* (27)	*xyl* (39)
Hfr D	*xyl* (7)	*thr* (25)	*lac* (40)	*bio* (48)	*glu* (62)
Hfr E	*his* (4)	*glu* (11)	*bio* (27)	*lac* (35)	*thr* (50)

(a) On the following map of the circular *E. coli* chromosome, indicate (1) the relative location of each gene, (2) the position where the sex factor is integrated in each of the five Hfr's, and (3) the direction of chromosome transfer for each Hfr (indicate direction with an arrow).

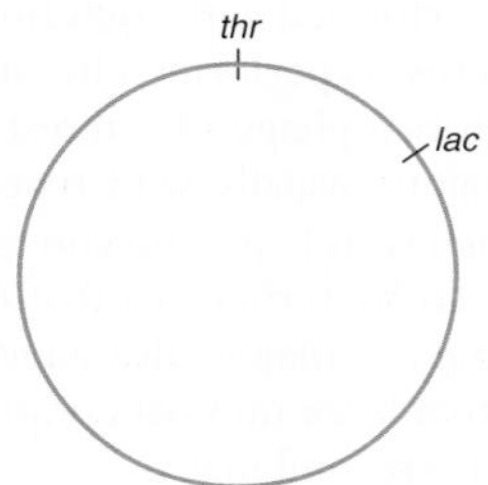

(b) To further define the location of the *his*⁻ mutation on the chromosome, you use the mutant strain as a recipient in a bacteriophage P1 transduction experiment. Which, if any, of the genes shown in the chart above would you expect to be cotransduced with the *his*⁺ allele of your *his*⁻ mutant gene, given that phage P1 can package about 1 percent of the *E. coli* chromosomal DNA molecule? Note that the *E. coli* chromosome contains 4.6 million nucleotide pairs and that transfer of the entire chromosome during conjugation takes 100 minutes. Explain your answer.

Answer: (a) The gene order is as shown on the following map, and the sites of F factor integration and direction of transfer for each of the Hfr's are indicated by the arrowheads labeled A through E. (b) None of the markers would be cotransduced with *his*⁺ because phage P1 can package only 1 percent of the *E. coli* chromosome, and none of the other genes is within 1 min of *his*.

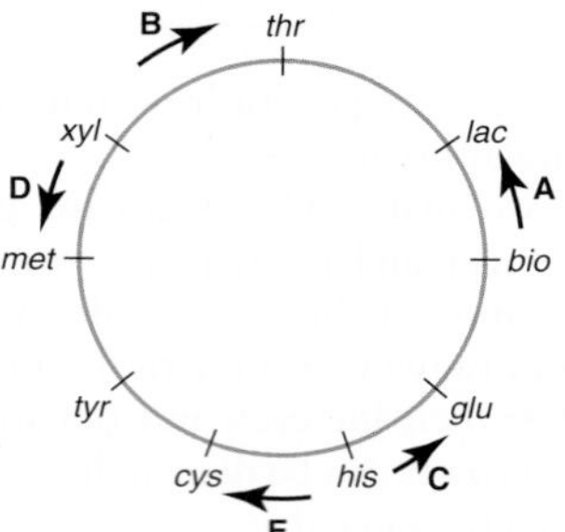

2. Reciprocal three-point transduction crosses were used to determine the order of two mutations, leu_1 and leu_2, in the *leuA* gene relative to the linked *thrA* gene of *E. coli*. In each cross, leu^+ recombinants were selected on minimal medium containing threonine but no leucine, and then tested for thr^+ or thr^- by replica plating onto plates containing no threonine. The results are given in the following table.

Cross				
Donor Markers	**Recipient Markers**	***thr* Allele in leu^+ Recombinants**	**Percent thr^+**
1. thr^+ leu_1	thr^- leu_2	350 thr^+ : 349 thr^-	50
2. thr^+ leu_2	thr^- leu_1	60 thr^+ : 300 thr^-	17

What is the order of leu_1 and leu_2 relative to the outside marker *thr*?

Answer: The two crosses are diagrammed in this question showing the two possible orders, with the dashed red lines marking the portions of the two chromosomes that must be present in thr^+-leu_1^+-leu_2^+ (+++) recombinants. Note that if order 1 is correct, the formation of +++ recombinants will require 4 crossovers (2 pairs of crossovers) in cross 1 and only 2 crossovers (1 pair) in cross 2, therefore predicting more +++ recombinants in cross 2 and fewer in cross 1. However, if order 2 is correct, there should be more +++ recombinants in cross 1 and fewer in cross 2. Since the second result was observed, the correct order is *thr*-leu_2-leu_1.

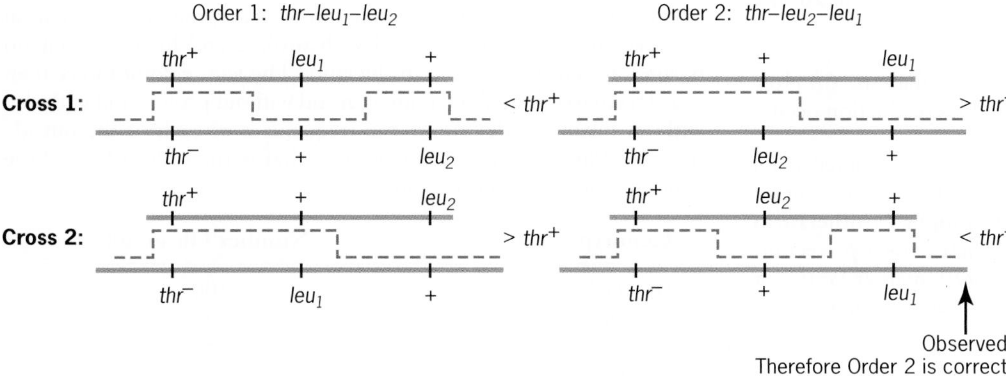

Questions and Problems

ENHANCE UNDERSTANDING AND DEVELOP ANALYTICAL SKILLS

8.1 In what ways do the life cycles of bacteriophages T4 and λ differ? In what respects are they the same?

8.2 How are genetic crosses performed with viruses? In what ways do these crosses differ from crosses in eukaryotes such as *Drosophila*?

8.3 Bacteriophage T4 has a linear chromosome but a circular genetic map. How can the two structures be reconciled?

8.4 How do bacteriophages differ from other viruses?

8.5 By what criteria are viruses living? nonliving?

8.6 Geneticists have used mutations that cause altered phenotypes such as white eyes in *Drosophila*, white flowers and wrinkled seeds in peas, and altered coat color in rabbits to determine the locations of genes on the chromosomes of these eukaryotes. What kinds of mutant phenotypes have been used to map genes in bacteria?

8.7 You have identified three mutations —*a*, *b*, and *c*—in *Streptococcus pneumoniae*. All three are recessive to their wild-type alleles a^+, b^+, and c^+. You prepare DNA from a wild-type donor strain and use it to transform a strain with genotype *a b c*. You observe a^+b^+ transformants and a^+c^+ transformants, but no b^+c^+ transformants. Are these mutations closely linked? If so, what is their order on the *Streptococcus* chromosome?

8.8 A nutritionally defective *E. coli* strain grows only on a medium containing thymine, whereas another nutritionally defective strain grows only on medium containing leucine. When these two strains were grown together, a few progeny were able to grow on a minimal medium with neither thymine or leucine. How can this result be explained?

8.9 In what way does the integration of the λ chromosome into the host chromosome during a lysogenic infection differ from crossing over between homologous chromosomes?

8.10 If a single T4 phage—with its linear chromosome—infects an *E. coli* cell and goes through the lytic cycle, the chromosomes of the progeny phage will be circularly permuted with different terminally redundant ends. How are these circularly permuted, terminally redundant molecules produced?

8.11 Assume that you have just demonstrated genetic recombination (e.g., when a strain of genotype $a\ b^+$ is present with a strain of genotype a^+b, some recombinant genotypes, $a^+\ b^+$ and $a\ b$, are formed) in a previously unstudied species of bacteria. How would you determine whether the observed recombination resulted from transformation, conjugation, or transduction?

8.12 What are the basic differences between generalized transduction and specialized transduction?

8.13 What does the term *cotransduction* mean? How can cotransduction frequencies be used to map genetic markers?

8.14 How can bacterial genes be mapped by interrupted mating experiments?

8.15 What roles do IS elements play in the integration of F factors?

8.16 (a) What are the genotypic differences between F^- cells, F^+ cells, and Hfr cells? (b) What are the phenotypic differences? (c) By what mechanism are F^- cells converted to F^+ cells? F^+ cells to Hfr cells? Hfr cells to F^+ cells?

8.17 (a) Of what use are F′ factors in genetic analysis? (b) How are F′ factors formed? (c) By what mechanism does sexduction occur?

8.18 GO The data in the following table were obtained from three-point transduction tests made to determine the order of mutant sites in the *A* gene encoding the α subunit of tryptophan synthetase in *E. coli*. *Anth* is a linked, unselected marker. In each cross, trp^+ recombinants were selected and then scored for the *anth* marker ($anth^+$ or $anth^-$). What is the linear order of *anth* and the three mutant alleles of the *A* gene indicated by the data in the table?

Cross	Donor Markers	Recipient Markers	*anth* Allele in trp^+ Recombinants	% $anth^+$
1	$anth^+$ — *A34*	$anth^-$ — *A223*	60 $anth^+$: 300 $anth^-$	17
2	$anth^+$ — *A46*	$anth^-$ — *A223*	168 $anth^+$: 200 $anth^-$	46
3	$anth^+$ — *A223*	$anth^-$ — *A34*	395 $anth^+$: 380 $anth^-$	51
4	$anth^+$ — *A223*	$anth^-$ — *A46*	50 $anth^+$: 270 $anth^-$	16

8.19 An F^+ strain, marked at 10 loci, gives rise spontaneously to Hfr progeny whenever the F factor becomes incorporated into the chromosome of the F^+ strain. The F factor can integrate into the circular chromosome at many points, so that the resulting Hfr strains transfer the genetic markers in different orders. For any Hfr strain, the order of markers entering a recipient cell can be determined by interrupted mating experiments. From the following data for several Hfr strains derived from the same F^+, determine the order of markers in the F^+ strain.

Hfr Strain	Markers Donated in Order
1	— Z-H-E-R →
2	— O-K-S-R →
3	— K-O-W-I →
4	— Z-T-I-W →
5	— H-Z-T-I →

8.20 GO Two additional mutations in the *trp A* gene of *E. coli*, *trp A58* and *trp A4487*, were ordered relative to *trp A223* and the outside marker *anth* by three-factor transduction crosses as described in Problem 8.18. The results of these crosses are summarized in the following table. What is the linear order of *anth* and the three mutant sites in the *trp A* gene?

Cross	Donor Markers	Recipient Markers	*anth* Allele in trp^+ Recombinants	% $anth^-$
1	$anth^+$ — *A487*	$anth^-$ — *A223*	60 $anth^+$: 300 $anth^-$	83
2	$anth^+$ — *A58*	$anth^-$ — *A223*	168 $anth^+$: 200 $anth^-$	54
3	$anth^+$ — *A223*	$anth^-$ — *A487*	395 $anth^+$: 380 $anth^-$	49
4	$anth^+$ — *A223*	$anth^-$ — *A58*	50 $anth^+$: 270 $anth^-$	84

8.21 Bacteriophage P1 mediates generalized transduction in *E. coli*. A P1 transducing lysate was prepared by growing P1 phage on pur^+ pro^- his^- bacteria. Genes *pur*, *pro*, and *his* encode enzymes required for the synthesis of purines, proline, and histidine, respectively. The phage and transducing particles in this lysate were then allowed to infect pur^- pro^+ his^+ cells. After incubating the infected bacteria for a period of time sufficient to allow transduction to occur, they were plated on minimal medium supplemented with proline and histidine, but no purines, to select for pur^+ transductants. The pur^+ colonies were then transferred to minimal medium with and without proline and with and without histine to determine the frequencies of each of the outside markers. Given the following results, what is the order of the three genes on the *E. coli* chromosome?

Genotype	Number Observed
pro^+ his^+	100
pro^- his^+	22
pro^+ his^-	150
pro^- his^-	1

8.22 GO In *E. coli*, the ability to utilize lactose as a carbon source requires the presence of the enzymes β-galactosidase and β-galactoside permease. These enzymes are encoded by two closely linked genes, *lacZ* and *lacY*, respectively. Another gene, *proC*, controls, in part, the ability of *E. coli* cells to synthesize the amino acid proline. The alleles str^r and str^s, respectively, control resistance and sensitivity to streptomycin. Hfr H is known to transfer the two *lac* genes, *proC*, and *str*, in that order, during conjugation.

A cross was made between Hfr H of genotype $lacZ^-$ $lacY^+$ $proC^+$ str^s and an F^- strain of genotype $lacZ^+$ $lacY^-$ $proC^-$ str^r. After about 2 hours, the mixture was diluted and plated out on medium containing streptomycin but no proline. When the resulting $proC^+$ str^r recombinant colonies were checked for their ability to grow on medium containing lactose as the sole carbon source, very few of them were capable of fermenting lactose. When the reciprocal cross (Hfr H $lacZ^+$ $lacY^-$ $proC^+$ str^s X F^- $lacZ^-$ $lacY^+$ $proC^-$ str^r) was done, many of the $proC^+$ str^r recombinants were able to grow on medium containing lactose as the sole carbon source. What is the order of the *lacZ* and *lacY* genes relative to *proC*?

8.23 You have identified a mutant *E. coli* strain that cannot synthesize histidine (His^-). To determine the location of the his^- mutation on the *E. coli* chromosome, you perform interrupted mating experiments with five different Hfr strains. The following chart shows the time of entry (minutes, in parentheses) of the wild-type alleles of the first five markers (mutant genes) into the His^- strain.

Hfr A ——	*his* (1)	*man* (9)	*gal* (28)	*lac* (37)	*thr* (45)
Hfr B ——	*man* (15)	*his* (23)	*cys* (38)	*ser* (42)	*arg* (49)
Hfr C ——	*thr* (3)	*lac* (11)	*gal* (20)	*man* (39)	*his* (47)

Hfr D ——	*cys* (3)	*his* (18)	*man* (26)	*gal* (45)	*lac* (54)
Hfr E ——	*thr* (4)	*rha* (18)	*arg* (36)	*ser* (43)	*cys* (47)

On the following map of the circular *E. coli* chromosome, indicate (1) the relative location of each gene relative to *thr* (located at 0/100 min), (2) the position where the sex factor is integrated in each of the five Hfr's, and (3) the direction of chromosome transfer for each Hfr (indicate direction with an arrow or arrowhead).

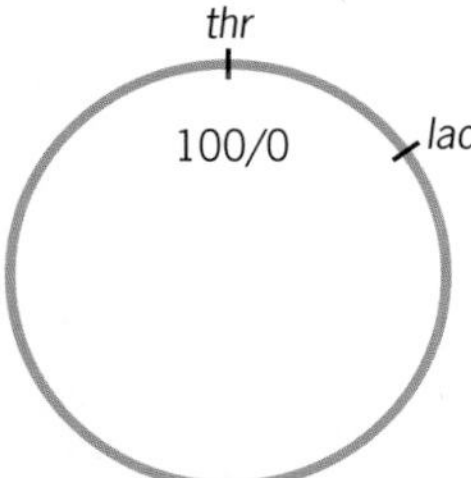

8.24 Mutations *nrd* 11 (gene *nrd B*, encoding the beta subunit of the enzyme ribonucleotide reductase), *am* M69 (gene 63, encoding a protein that aids tail-fiber attachment), and *nd* 28 (*denA*, encoding the enzyme endonuclease II) are known to be located between gene 31 and gene 32 on the bacteriophage T4 chromosome. Mutations *am* N54 and *am* A453 are located in genes 31 and 32, respectively. Given the three-factor cross data in the following table, what is the linear order of the five mutant sites?

Three-Factor Cross Data

Cross		% Recombination[a]
1	*am A*453—*am M*69 × *nrd* 11	2.6
2	*am A*453—*nrd* 11 × *am M*69	4.2
3	*am A*453—*am M*69 × *nd* 28	2.5
4	*am A*453—*nd* 28 × *am M*69	3.5
5	*am A*453—*nrd* 11 × *nd* 28	2.9
6	*am A*453—*nd* 28 × *nrd* 11	2.1
7	*am N*54—*am M*69 × *nrd* 11	3.5
8	*am N*54—*nrd* 11 × *am M*69	1.9
9	*am N*54—*nd* 28 × *am M*69	1.7
10	*am N*54—*am M*69 × *nd* 28	2.7
11	*am N*54—*nd* 28 × *nrd* 11	2.9
12	*am N*54—*nrd* 11 × *nd* 28	1.9

[a]All recombination frequencies are given as $\frac{2(\text{wild-type progeny})}{\text{total progeny}} \times 100$.

► Genomics on the Web

at http://www.ncbi.nlm.nih.gov/

The *E. coli* genome was one of the first bacterial genomes sequenced. The complete nucleotide sequence (4.6 million nucleotide pairs) of the genome of *E. coli* strain K12 was published in September 1997.

1. How many different strains of *E. coli* have had their genomes sequenced since 1997?

2. Are these genomes all about the same size? If not, how much variation in size is observed between the genomes of different *E. coli* strains?

3. Some *E. coli* strains, for example, 0157:H7, are more pathogenic to humans and other mammals than strains such as K12. Do these strains have larger or smaller genomes than K12. Might comparisons of the genes in the pathogenic and nonpathogenic strains provide hints as to why some strains are pathogenic and others are not?

Hint: At the NCBI web site, Genome Biology → Entrez Genome → Microbial Genomes → Complete Genomes → *Escherichia coli*

Chapter 9
DNA and the Molecular Structure of Chromosomes

CHAPTER OUTLINE

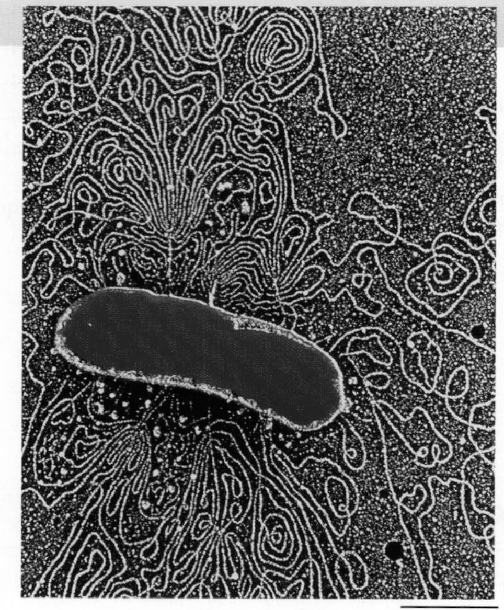

Color-enhanced transmission electron micrograph of a ruptured *E. coli* cell with much of its DNA extruded.

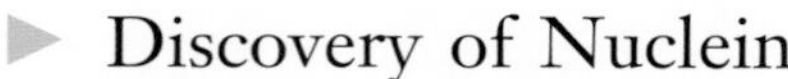

Discovery of Nuclein

In 1868, Johann Friedrich Miescher, a young Swiss medical student, became fascinated with an acidic substance that he isolated from pus cells obtained from bandages used to dress human wounds. He first separated the pus cells from the bandages and associated debris, and then treated the cells with pepsin, a proteolytic enzyme that he isolated from the stomachs of pigs. After the pepsin treatment, he recovered an acidic substance that he called "nuclein." Miescher's nuclein was unusual in that it contained large amounts of both nitrogen and phosphorus, two elements known at the time to coexist only in certain types of fat. Miescher wrote a paper describing his discovery of nuclein in human pus cells and submitted it for publication in 1869. However, the editor of the journal to which the paper was sent was skeptical of the results and decided to repeat the experiments himself. As a result, Miescher's paper describing nuclein was not published until 1871, two years after its submission.

At the time, the importance of the substance that Miescher called nuclein could not have been anticipated. The existence of polynucleotide chains, the key component of the acidic material in Miescher's nuclein, was not documented until the 1940s. The role of nucleic acids in storing and transmitting genetic information was not established until 1944, and the double-helix structure of DNA was not discovered until 1953. Even in 1953, many geneticists were reluctant to accept the idea that nucleic acids, rather than proteins, carried the genetic information because nucleic acids exhibited less structural variability than proteins.

Functions of the Genetic Material

The genetic material must replicate, control the growth and development of the organism, and allow the organism to adapt to changes in the environment.

In 1865, Mendel showed that "Merkmalen" (now "genes") transmitted genetic information, and in the first part of the twentieth century, their patterns of transmission were studied extensively. Although these classical genetic studies provided little insight into the molecular nature of genes, they did demonstrate that the genetic material must perform three essential functions:

1. The genotypic function, **replication.** The genetic material must store genetic information and accurately transmit that information from parents to offspring, generation after generation.
2. The phenotypic function, **gene expression.** The genetic material must control the development of the phenotype of the organism. That is, the genetic material must dictate the growth of the organism from the single-celled zygote to the mature adult.
3. The evolutionary function, **mutation.** The genetic material must undergo changes to produce variations that allow organisms to adapt to modifications in the environment so that evolution can occur.

Other early genetic studies established a precise correlation between the patterns of transmission of genes and the behavior of chromosomes during sexual reproduction, providing strong evidence that genes are usually located on chromosomes. Thus, further attempts to discover the chemical basis of heredity focused on the molecules present in chromosomes.

Chromosomes are composed of two types of large organic molecules (macromolecules) called **proteins** and **nucleic acids.** The nucleic acids are of two types: **deoxyribonucleic acid (DNA)** and **ribonucleic acid (RNA).** During the 1940s and early 1950s, the results of elegant experiments clearly established that the genetic information is stored in nucleic acids, not in proteins. In most organisms, the genetic information is encoded in the structure of DNA. However, in many small viruses, the genetic information is encoded in RNA.

KEY POINT

- The genetic material must perform three essential functions: the genotypic function—replication, the phenotypic function—gene expression, and the evolutionary function—mutation.

Proof That Genetic Information Is Stored in DNA

In most organisms, the genetic information is encoded in DNA. In some viruses, RNA is the genetic material. Viroids are infectious, naked RNA molecules, and prions are infectious, heritable proteins.

Several lines of indirect evidence suggested that DNA harbors the genetic information of living organisms. For example, most of the DNA of cells is located in the chromosomes, whereas RNA and proteins are also abundant in the cytoplasm. Also, a precise correlation exists between the amount of DNA per cell and the number of sets of chromosomes per cell. Most somatic cells of diploid organisms contain twice the amount of DNA as the haploid germ cells (gametes) of the same species. The molecular composition of the DNA is the same (with rare exceptions) in all the cells of an organism, whereas the composition of both RNA and proteins is highly variable from one cell type to another. DNA is more stable than RNA or proteins. Since the genetic material must store and transmit information from parents to offspring, we might expect it to be stable, like DNA. Although these correlations strongly suggest that DNA is the genetic material, they by no means prove it.

PROOF THAT DNA MEDIATES TRANSFORMATION

Frederick Griffith's discovery of transformation in *Streptococcus pneumoniae* was discussed in Chapter 8. When Griffith injected both heat-killed Type IIIS bacteria (virulent when alive) and live Type IIR bacteria (avirulent) into mice, many of the mice developed pneumonia and died, and live Type IIIS cells were recovered from their carcasses. Something from the heat-killed cells—the "transforming principle"—had converted the live Type IIR cells to Type IIIS. In 1931, Richard Sia and Martin Dawson performed the same experiment *in vitro*, showing that the mice played no role in the transformation process (**FIGURE 9.1**). Sia and Dawson's experiment set the stage for Oswald Avery, Colin MacLeod, and Maclyn McCarty's demonstration that the "transforming principle" in *S. pneumoniae* is DNA. Avery and colleagues showed that DNA is the only

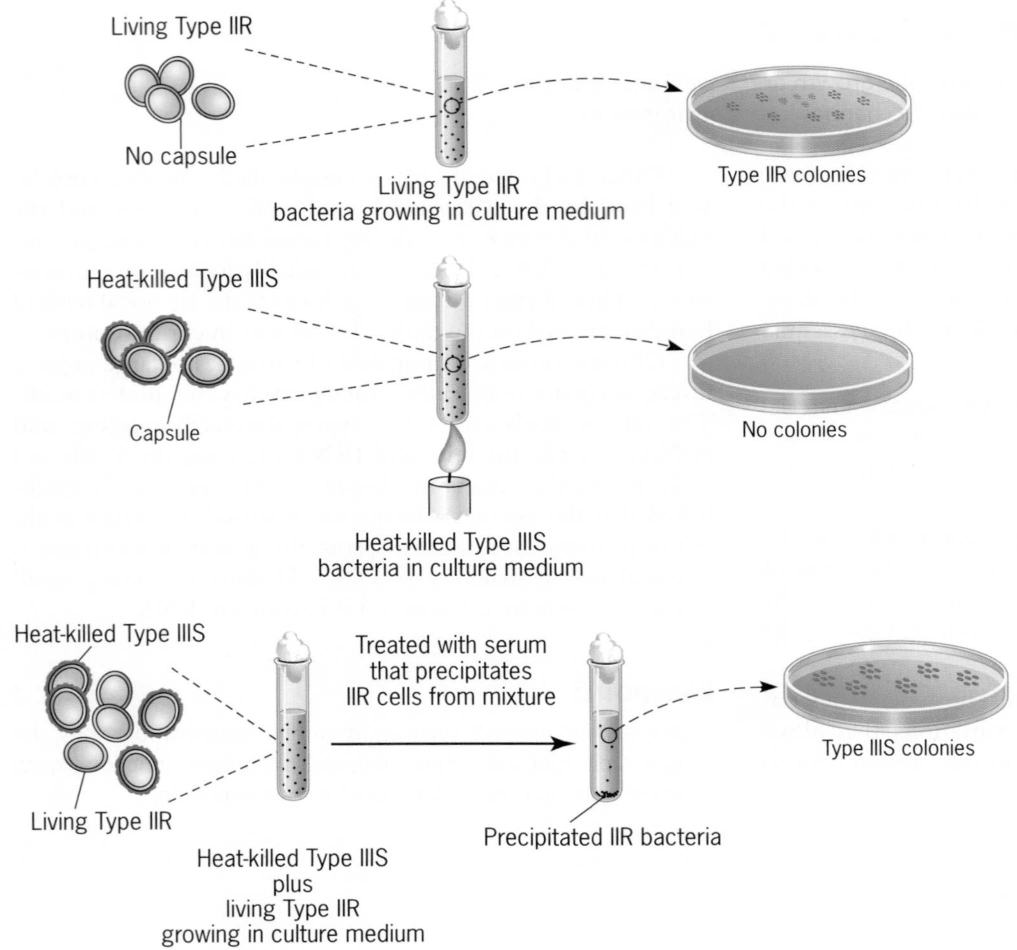

Figure 9.1 ▶ Sia and Dawson's demonstration of transformation in *Streptococcus pneumoniae in vitro.*

component of the Type IIIS cells required to transform Type IIR cells to Type IIIS (**FIGURE 9.2**).

But how could they be sure that the DNA was really pure? Proving the purity of any macromolecular substance is extremely difficult. Maybe the DNA preparation contained a few molecules of protein, and these contaminating proteins were responsible for the observed transformation. The most definitive experiments in Avery, MacLeod, and McCarty's proof that DNA was the transforming principle involved the use of enzymes that degrade DNA, RNA, or protein. In separate experiments, highly purified DNA from Type IIIS cells was treated with the enzymes (1) **deoxyribonuclease (DNase),** which degrades DNA; (2) **ribonuclease (RNase),** which degrades RNA; or (3) **proteases,** which degrade proteins; the DNA was then tested for its ability to transform Type IIR cells to Type IIIS. Only DNase treatment had any effect on the transforming activity of the DNA preparation—it eliminated all transforming activity (**FIGURE 9.2**).

Although the molecular mechanism by which transformation occurs remained unknown for many years, the results of Avery and coworkers clearly established that the genetic information in *Streptococcus* is present in DNA. Geneticists now know that the segment of DNA in the chromosome of *Streptococcus* that carries the genetic information specifying the synthesis of a Type III capsule is physically inserted into the chromosome of the Type IIR recipient cell during the transformation process.

PROOF THAT DNA CARRIES THE GENETIC INFORMATION IN BACTERIOPHAGE T2

Additional evidence demonstrating that DNA is the genetic material was published in 1952 by Alfred Hershey (1969 Nobel Prize winner) and Martha Chase. The results of their experiments showed that the genetic information of a particular bacterial virus (bacteriophage T2) was present in its DNA. Their results had a major impact on scientists' acceptance of DNA as the genetic material. This impact was the result of the simplicity of the Hershey-Chase experiment.

Viruses are the smallest living organisms; they are living, at least in the sense that their reproduction is controlled by genetic information stored in nucleic acids via the same processes as in cellular organisms (Chapter 8). However, viruses are acellular

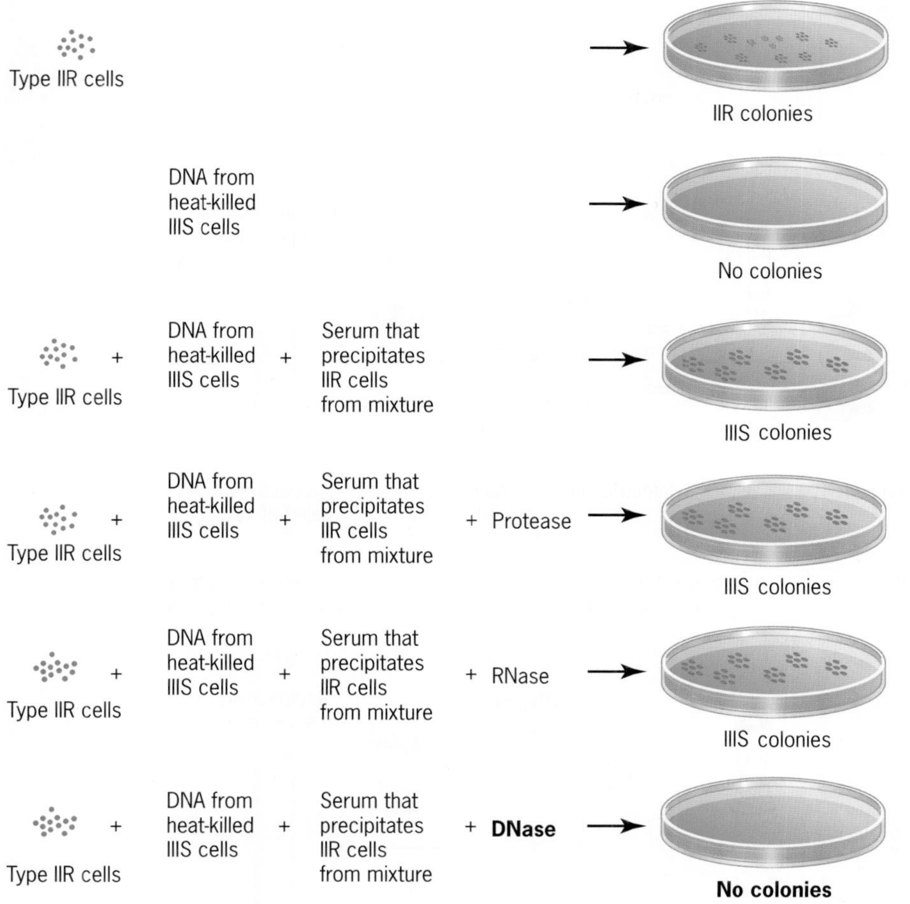

Figure 9.2 ▶ Avery, MacLeod, and McCarty's proof that the "transforming principle" is DNA.

parasites that can reproduce only in appropriate host cells. Their reproduction is totally dependent on the metabolic machinery (ribosomes, energy-generating systems, and other components) of the host. Viruses have been extremely useful in the study of many genetic processes because of their simple structure and chemical composition (many contain only proteins and nucleic acids) and their very rapid reproduction (15 to 20 minutes for some bacterial viruses under optimal conditions).

Bacteriophage T2, which infects the common colon bacillus *Escherichia coli*, is composed of about 50 percent DNA and about 50 percent protein (**FIGURE 9.3**). Experiments prior to 1952 had shown that all bacteriophage T2 reproduction takes place within *E. coli* cells. Therefore, when Hershey and Chase showed that the DNA of the virus particle entered the cell, whereas most of the protein of the virus remained adsorbed to the outside of the cell, the implication was that the genetic information necessary for viral reproduction was present in DNA. The basis for the Hershey-Chase experiment is that DNA contains phosphorus but no sulfur, whereas proteins contain sulfur but virtually no phosphorus. Thus, Hershey and Chase were able to label specifically either (1) the phage DNA by growth in a medium containing the radioactive isotope of phosphorus, ^{32}P, in place of the normal isotope, ^{31}P; or (2) the phage protein coats by growth in a medium containing radioactive sulfur, ^{35}S, in place of the normal isotope, ^{32}S (**FIGURE 9.3**).

When T2 phage particles labeled with ^{35}S were mixed with *E. coli* cells for a few minutes and the phage-infected cells were then subjected to shearing forces in a Waring blender, most of the radioactivity (and thus the proteins) could be removed from the cells without affecting progeny phage production. When T2 particles in which the DNA was labeled with ^{32}P were used, however, essentially all the radioactivity was found inside the cells; that is, the DNA was not subject to removal by shearing in a blender. The sheared-off phage coats were separated from the infected cells by low-speed centrifugation, which pellets (sediments) cells while leaving phage particles suspended. These results indicated that the DNA of the virus enters the host cell, whereas the protein coat remains outside the cell. Since progeny viruses are produced inside the cell, Hershey and Chase's results indicated that the genetic information directing the synthesis of both the DNA molecules and the protein coats of the progeny viruses must be present in the parental DNA. Moreover, the progeny particles were shown to contain some of the ^{32}P, but none of the ^{35}S, of the parental phage.

There was one problem with Hershey and Chase's proof that the genetic material of phage T2 is DNA. Their results

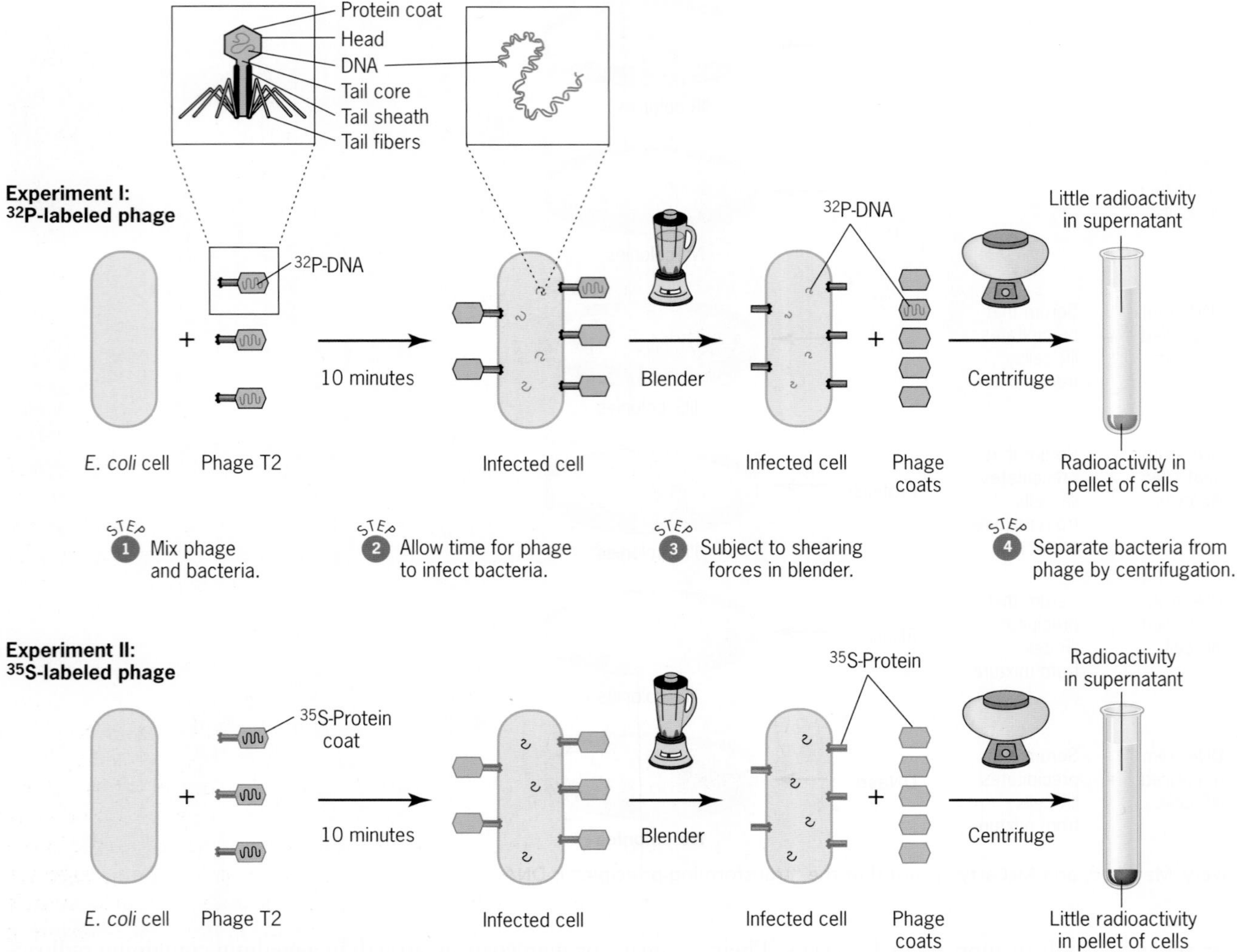

Figure 9.3 ▸ Hershey and Chase's demonstration that the genetic information of bacteriophage T2 resides in its DNA.

showed that a significant amount of ^{35}S (and thus protein) was injected into the host cells with the DNA. Thus, it could be argued that this small fraction of the phage proteins contained the genetic information. More recently, scientists have developed procedures by which protoplasts (cells with the walls removed) of *E. coli* can be infected with pure phage DNA. Normal infective progeny phage are produced in these experiments, called **transfection** experiments, proving that the genetic material of such bacterial viruses is DNA.

PROOF THAT RNA STORES THE GENETIC INFORMATION IN SOME VIRUSES

As more and more viruses were identified and studied, it became apparent that many of them contain RNA and proteins, but no DNA. In all cases studied to date, it is clear that these RNA viruses—like all other organisms—store their genetic information in nucleic acids rather than in proteins, although in these viruses the nucleic acid is RNA. One of the first experiments that established RNA as the genetic material in RNA viruses was the so-called reconstitution experiment of Heinz Fraenkel-Conrat and coworkers, published in 1957. Their simple, but definitive, experiment was done with tobacco mosaic virus (TMV), a small virus composed of a single molecule of RNA encapsulated in a protein coat. Different strains of TMV can be identified on the basis of differences in the chemical composition of their protein coats.

Fraenkel-Conrat and colleagues treated TMV particles of two different strains with chemicals that dissociate the protein coats of the viruses from the RNA molecules and separated the proteins from the RNA. Then they mixed the proteins from one strain with the RNA molecules from the other strain under conditions that result in the reconstitution of complete, infective viruses composed of proteins from one strain and RNA from the other strain. When tobacco leaves were infected with these reconstituted mixed viruses, the progeny viruses were always phenotypically and genotypically identical to the parent strain from which the RNA had been obtained (**FIGURE 9.4**). Thus, the genetic information of TMV is stored in RNA, not in protein.

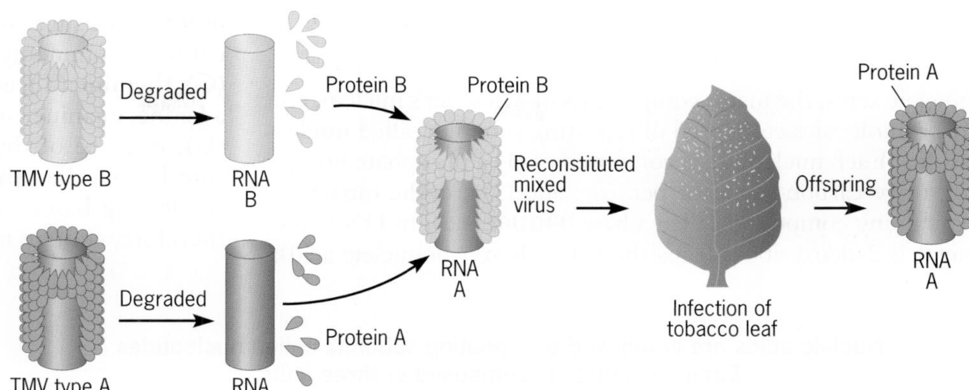

Figure 9.4 ► The genetic material of tobacco mosaic virus (TMV) is RNA, not protein. TMV contains no DNA; it is composed of just RNA and protein.

VIROIDS, HERITABLE INFECTIOUS NAKED RNA MOLECULES

Some infectious agents that cause diseases in plants, and others that are thought to cause diseases in animals, contain small circular molecules of RNA. However, unlike RNA viruses such as TMV, these RNA molecules are not packaged in protein coats. They were discovered by Theodore Diener and colleagues in the 1960s and were named **viroids**—meaning "virus-like." The potato spindle tuber viroid (PSTV) was one of the first viroids studied; it causes the tubers of infected potatoes to be long and pointed like spindles. PSTV is a circular RNA molecule consisting of 359 nucleotides; it replicates autonomously in potato tuber cells. PSTV has a rod-shaped structure because of base-pairing within the molecule. The pathogenic effects of viroids are thought to result from their ability to alter normal patterns of gene expression.

PRIONS, HERITABLE INFECTIOUS PROTEINS

Other transmissible infectious agents contain no nucleic acid of any kind, just protein. These proteinaceous agents were named **prions** (derived from *pro*tein and *in*fectious) in 1982 by Stanley Prusiner, who received the 1997 Nobel Prize in Physiology or Medicine for his research on these unique proteins. Prions are altered forms of normal cellular proteins in mammals. They are responsible for a group of fatal neurodegenerative diseases including Creutzfeldt-Jakob disease (CJD) and kuru in humans, bovine spongiform encephalopathy (BSE)—commonly called "mad cow disease"—in cattle, chronic wasting disease (CWD) in deer and elk, and scrapie disease in sheep. The importance of these diseases has been documented by the slaughter of entire herds of cattle exposed to BSE in Great Britain and of elk exposed to CWD in the United States during the last decade. Prion diseases are spread by the consumption of meat containing the infective proteins. In the case of kuru in the Fore people of New Guinea, the disease is thought to have been transmitted by their ritualistic cannibalism of the brains of deceased relatives. "Mad cow disease" is known to have spread when tissues from infected animals ended up in cattle feed supplements.

The prions are aberrant forms of proteins encoded by normal mammalian genes. Once the aberrant pathogenic form of the protein—the prion—has formed, it acts as a template that converts more of the normal cellular form of the protein to the infectious prion form. The aberrant prion proteins clump together and eventually kill the host cells. The brains of sick animals develop a spongelike morphology with empty spaces resulting from cell death. Once symptoms of the disease appear, neurodegeneration occurs quite rapidly, followed quickly by death.

KEY POINTS

- The genetic information of most living organisms is stored in deoxyribonucleic acid (DNA).
- In some viruses, the genetic information is present in ribonucleic acid (RNA).
- Viroids and prions are infectious naked molecules of RNA and protein, respectively.

The Structures of DNA and RNA

DNA is usually double-stranded, with adenine paired with thymine and guanine paired with cytosine. RNA is usually single-stranded and contains uracil in place of thymine.

The genetic information of all living organisms, except the RNA viruses, is stored in DNA. What is the structure of DNA, and in what form is the genetic information stored? What features of the structure of DNA facilitate the accurate transmission of genetic information from generation to generation? The answers to these questions are without doubt three of the most important facets of our understanding of the nature of life.

NATURE OF THE CHEMICAL SUBUNITS IN DNA AND RNA

Nucleic acids, the major components of Miescher's nuclein, are macromolecules composed of repeating subunits called **nucleotides.** Each nucleotide is composed of (1) a phosphate group; (2) a five-carbon sugar, or pentose; and (3) a cyclic nitrogen-containing compound called a base (**FIGURE 9.5**). In DNA, the sugar is 2-deoxyribose (thus the name deoxyribonucleic acid); in RNA, the sugar is ribose (thus ribonucleic acid). Four different bases commonly are found in DNA: **adenine (A), guanine (G), thymine (T), and cytosine (C).** RNA also usually contains adenine, guanine, and cytosine but has a different base, **uracil (U),** in place of thymine. Adenine and guanine are double-ring bases called **purines;** cytosine, thymine, and uracil are single-ring bases called **pyrimidines.** Both DNA and RNA, therefore, contain four different subunits, or nucleotides: two

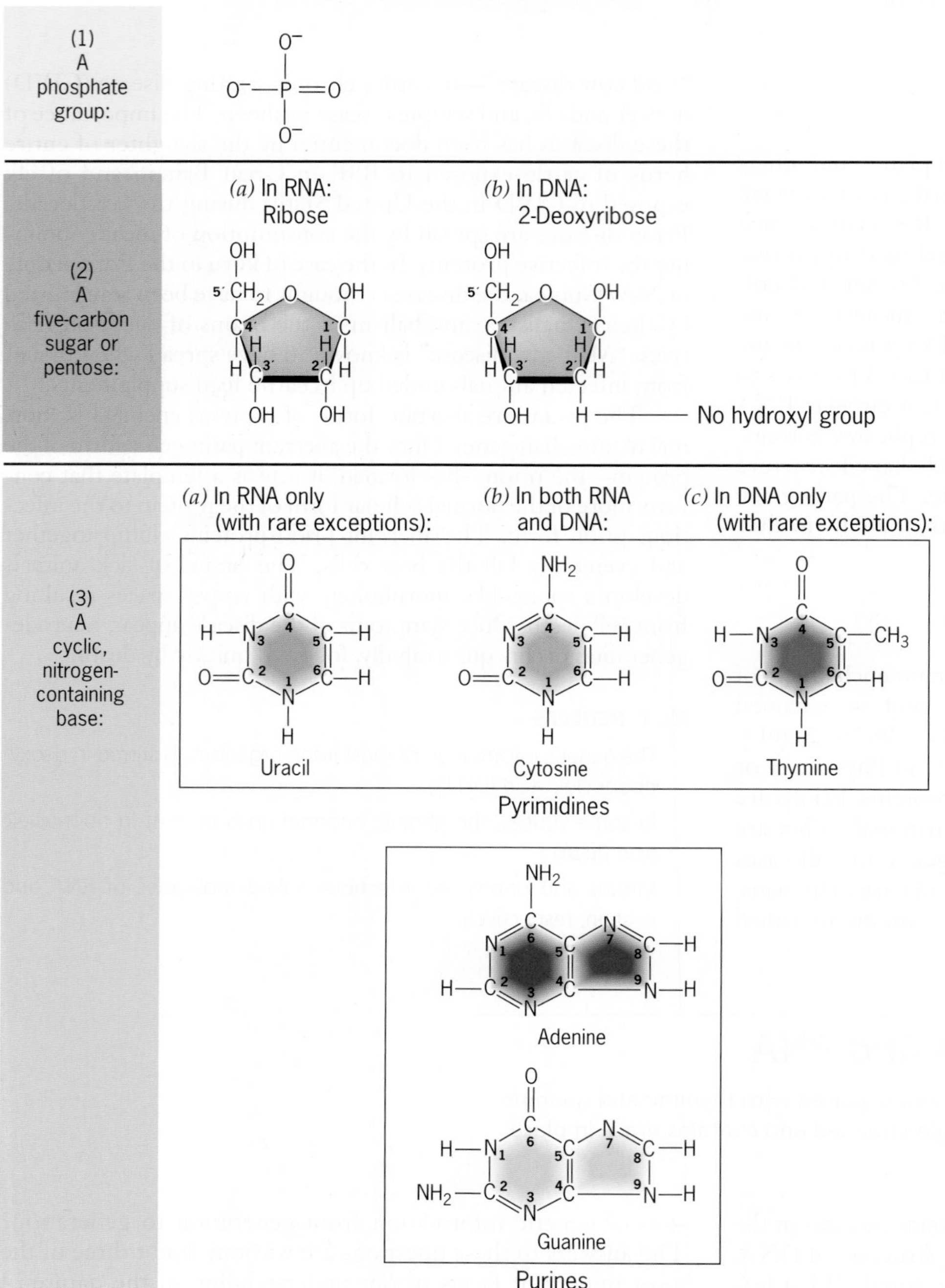

Figure 9.5 ▶ Structural components of nucleic acids. The standard numbering systems for the carbons in pentoses and the carbons and nitrogens in the ring structures of the bases are shown in (2) and (3), respectively. Single-ring bases are called pyrimidines, and double-ring bases are purines.

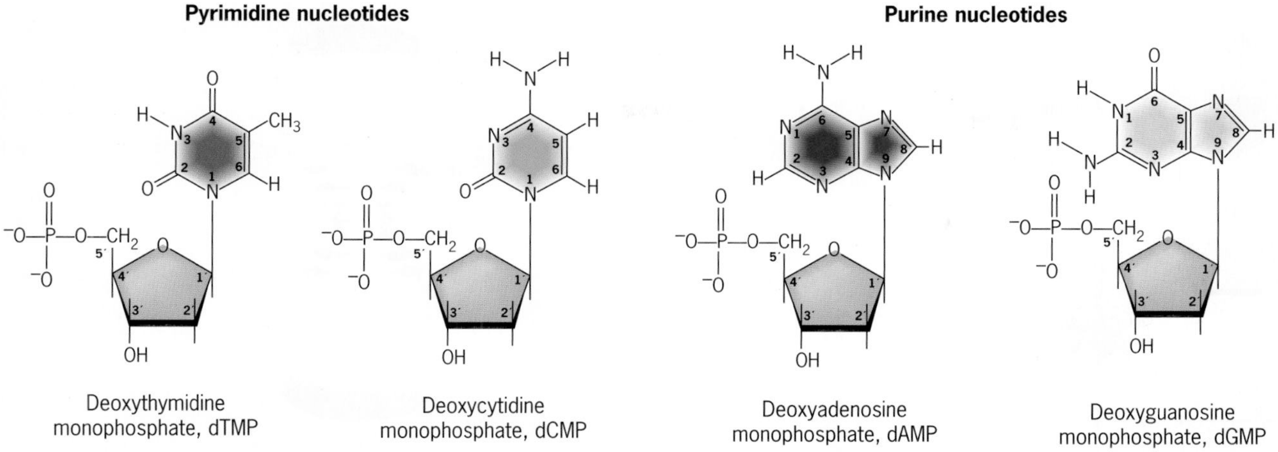

Pyrimidine nucleotides

Purine nucleotides

Deoxythymidine monophosphate, dTMP

Deoxycytidine monophosphate, dCMP

Deoxyadenosine monophosphate, dAMP

Deoxyguanosine monophosphate, dGMP

Figure 9.6 ▶ Structures of the four common deoxyribonucleotides present in DNA. The carbons and nitrogens in the rings of the bases are numbered 1 through 6 (pyrimidines) and 1 through 9 (purines). Therefore, the carbons in the sugars of nucleotides are numbered 1′ through 5′ to distinguish them from the carbons in the bases.

purine nucleotides and two pyrimidine nucleotides (**FIGURE 9.6**). In polynucleotides such as DNA and RNA, these subunits are joined together in long chains (**FIGURE 9.7**). RNA usually exists as a single-stranded polymer that is composed of a long sequence of nucleotides. DNA has one additional—and very important—level of organization: it is usually a double-stranded molecule.

DNA STRUCTURE: THE DOUBLE HELIX

One of the most exciting breakthroughs in the history of biology occurred in 1953 when James Watson and Francis Crick deduced the correct structure of DNA. Their double-helix model of the DNA molecule immediately suggested an elegant mechanism for the transmission of genetic information (see A Milestone in Genetics: The Double Helix later in this chapter). Watson and Crick's double-helix structure was based on two major kinds of evidence:

1. When Erwin Chargaff and colleagues analyzed the composition of DNA from many different organisms, they found that the concentration of thymine was always equal to the concentration of adenine and the concentration of cytosine was always equal to the concentration of guanine (**TABLE 9.1**). Their results strongly suggested that thymine and adenine as well as cytosine and guanine were present in DNA in

▶ **TABLE 9.1**

Base Composition of DNA from Various Organisms

Species	% Adenine	% Guanine	% Cytosine	% Thymine	Molar Ratios $\frac{A+G}{T+C}$	Molar Ratios $\frac{A+T}{G+C}$
I. Viruses						
Bacteriophage λ	26.0	23.8	24.3	25.8	0.99	1.08
Bacteriophage T2	32.6	18.1	16.6	32.6	1.03	1.88
Herpes simplex	13.8	37.7	35.6	12.8	1.06	0.36
II. Bacteria						
Escherichia coli	26.0	24.9	25.2	23.9	1.04	1.00
Micrococcus lysodeikticus	14.4	37.3	34.6	13.7	1.07	0.39
Ramibacterium ramosum	35.1	14.9	15.2	34.8	1.00	2.32
III. Eukaryotes						
Saccharomyces cerevisiae	31.7	18.3	17.4	32.6	1.00	1.80
Zea mays (corn)	25.6	24.5	24.6	25.3	1.00	1.04
Drosophila melanogaster	30.7	19.6	20.2	29.4	1.01	1.51
Homo sapiens (human)	30.2	19.9	19.6	30.3	1.01	1.53

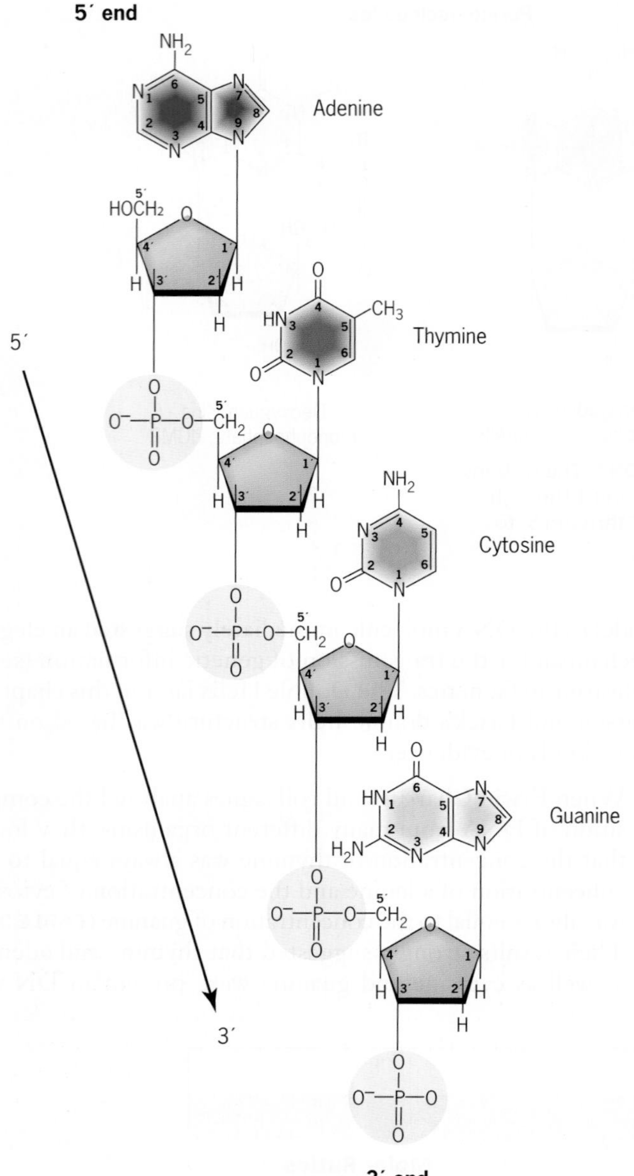

Figure 9.7 ▶ Structure of a polynucleotide chain. The tetranucleotide chain shown is a DNA chain containing the sugar 2′-deoxyribose. RNA chains contain the sugar ribose. The nucleotides in polynucleotide chains are joined by phosphodiester (C—O—P—O—C) linkages. Note that the polynucleotide shown has a 5′ (top) to 3′ (bottom) chemical polarity because each phosphodiester linkage joins the 5′ carbon of 2′-deoxyribose in one nucleotide to the 3′ carbon of 2′-deoxyribose in the adjacent nucleotide. Therefore, the chain has a 5′ carbon terminus at the top and a 3′ carbon terminus at the bottom.

some fixed interrelationship. Their data also showed that the total concentration of pyrimidines (thymine plus cytosine) was always equal to the total concentration of purines (adenine plus guanine; see **TABLE 9.1**).

2. When X rays are focused through fibers of purified molecules, the rays are deflected by the atoms of the molecules in specific patterns, called diffraction patterns, which provide information about the organization of the components of the molecules. These *X-ray diffraction patterns* can be recorded on X-ray-sensitive film just as patterns of light can be recorded with a camera and light-sensitive film. Watson and Crick used X-ray diffraction data on DNA structure (**FIGURE 9.8**) provided by Maurice Wilkins, Rosalind Franklin, and their coworkers. These data indicated that DNA was a highly ordered, two-stranded structure with repeating substructures spaced every 0.34 nanometer (1 nm = 10^{-9} meter) along the axis of the molecule.

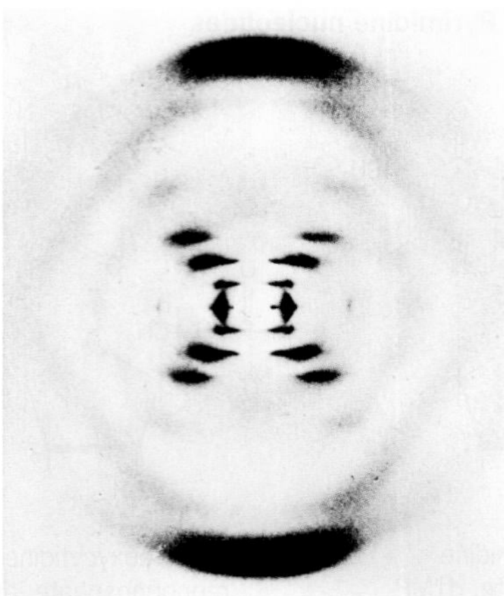

Figure 9.8 ▶ Photograph of the X-ray diffraction pattern obtained with DNA. The central cross-shaped pattern indicates that the DNA molecule has a helical structure, and the dark bands at the top and bottom indicate that the bases are stacked perpendicular to the axis of the molecule with a periodicity of 0.34 nm.

On the basis of Chargaff's chemical data, Wilkins' and Franklin's X-ray diffraction data, and inferences from model building, Watson and Crick proposed that DNA exists as a right-handed **double helix** in which the two polynucleotide chains are coiled about one another in a spiral (**FIGURE 9.9**). Each polynucleotide chain consists of a sequence of nucleotides linked together by phosphodiester bonds, joining adjacent deoxyribose moieties (**TABLE 9.2**). The two polynucleotide strands are held together in their helical configuration by hydrogen bonding (**TABLE 9.2**) between bases in opposing strands; the resulting base pairs are stacked between the two chains perpendicular to the axis of the molecule like the steps of a spiral staircase (**FIGURE 9.9**). The base-pairing is specific: adenine is always paired with thymine, and guanine is always paired with cytosine. Thus, all base pairs consist of one purine and one pyrimidine. The specificity of base-pairing results from the hydrogen-bonding capacities of the bases in their normal configurations (**FIGURE 9.10**). In their common structural configurations, adenine and thymine form two hydrogen bonds, and guanine and cytosine form three hydrogen bonds. Hydrogen bonding is not possible between cytosine and adenine or thymine and guanine when they exist in their common structural states.

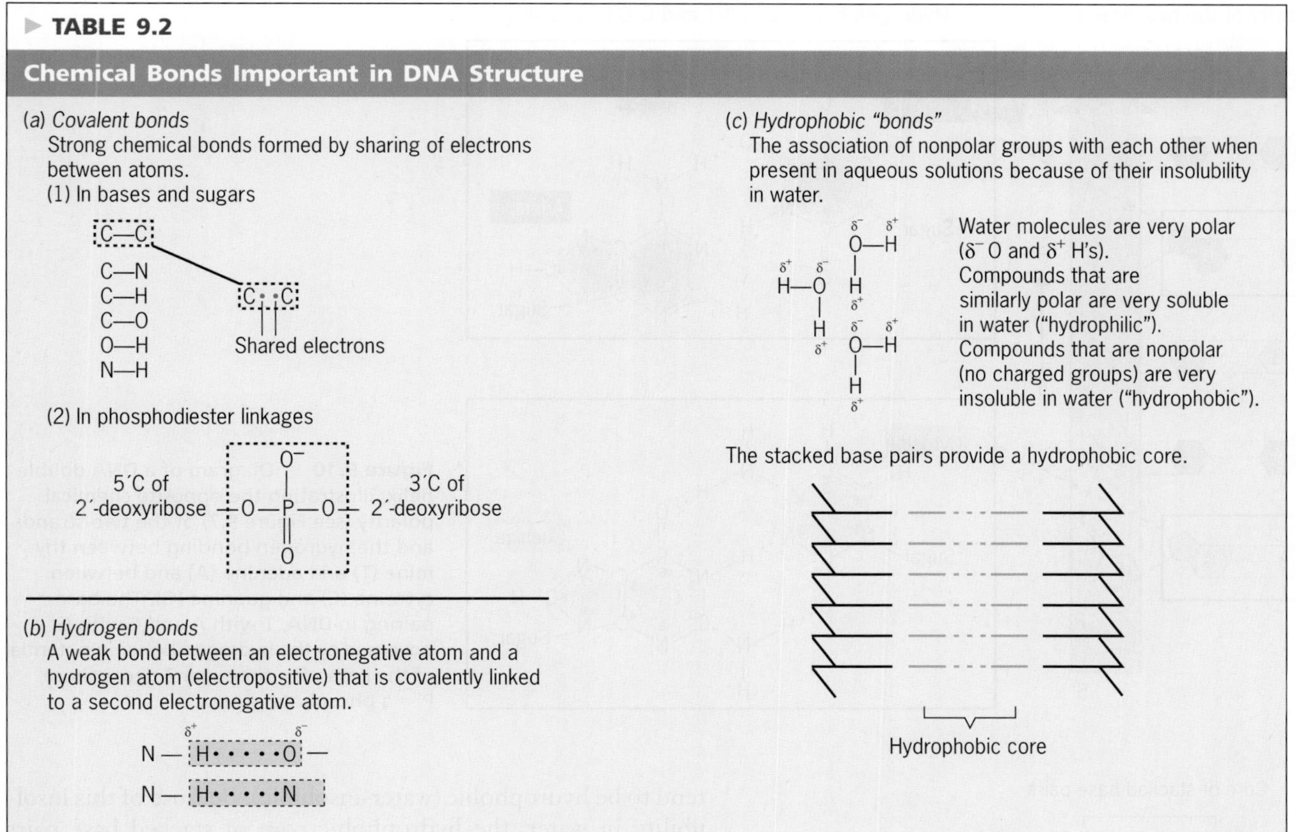

TABLE 9.2

Chemical Bonds Important in DNA Structure

(a) Covalent bonds
Strong chemical bonds formed by sharing of electrons between atoms.
(1) In bases and sugars

C—C
C—N
C—H
C—O
O—H
N—H

C : C
Shared electrons

(2) In phosphodiester linkages

5′C of 2′-deoxyribose —O—P—O— 3′C of 2′-deoxyribose
(O^- above P; O double-bonded below P)

(b) Hydrogen bonds
A weak bond between an electronegative atom and a hydrogen atom (electropositive) that is covalently linked to a second electronegative atom.

N — $H^{\delta+}$ • • • • • $O^{\delta-}$ —
N — H • • • • • N

(c) Hydrophobic "bonds"
The association of nonpolar groups with each other when present in aqueous solutions because of their insolubility in water.

Water molecules are very polar (δ^- O and δ^+ H's).
Compounds that are similarly polar are very soluble in water ("hydrophilic").
Compounds that are nonpolar (no charged groups) are very insoluble in water ("hydrophobic").

The stacked base pairs provide a hydrophobic core.

Hydrophobic core

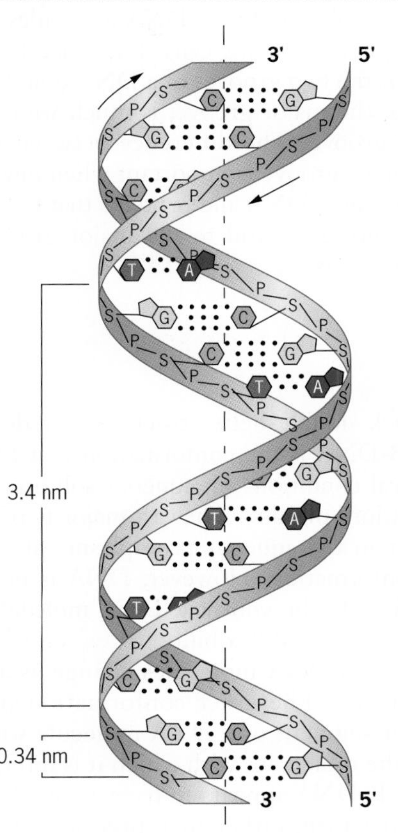

Figure 9.9 ▸ Diagram of the double-helix structure of DNA.

Once the sequence of bases in one strand of a DNA double helix is known, the sequence of bases in the other strand is also known because of the specific base-pairing (see Focus on Problem Solving: Calculating Base Content in DNA). The two strands of a DNA double helix are thus said to be complementary. *This property, the* **complementarity** *of the two strands of the double helix, makes DNA uniquely suited to store and transmit genetic information from generation to generation* (Chapter 10).

The base pairs in DNA are stacked about 0.34 nm apart, with 10 base pairs per turn (360°) of the double helix (**FIGURE 9.9**). The sugar-phosphate backbones of the two complementary strands are *antiparallel* (**FIGURE 9.10**). Unidirectionally along a DNA double helix, the phosphodiester bonds in one strand go from a 3′ carbon of one nucleotide to a 5′ carbon of the adjacent nucleotide, whereas those in the complementary strand go from a 5′ carbon to a 3′ carbon. This "opposite polarity" of the complementary strands of a DNA double helix plays an important role in DNA replication, transcription, and recombination.

The stability of DNA double helices results in part from the large number of hydrogen bonds between the base pairs (even though each hydrogen bond by itself is weak, much weaker than a covalent bond) and in part from the hydrophobic bonding (or stacking forces) between adjacent base pairs (**TABLE 9.2**). The stacked nature of the base pairs is best illustrated with a space-filling diagram of DNA structure (**FIGURE 9.11**). The planar sides of the base pairs are relatively nonpolar and thus

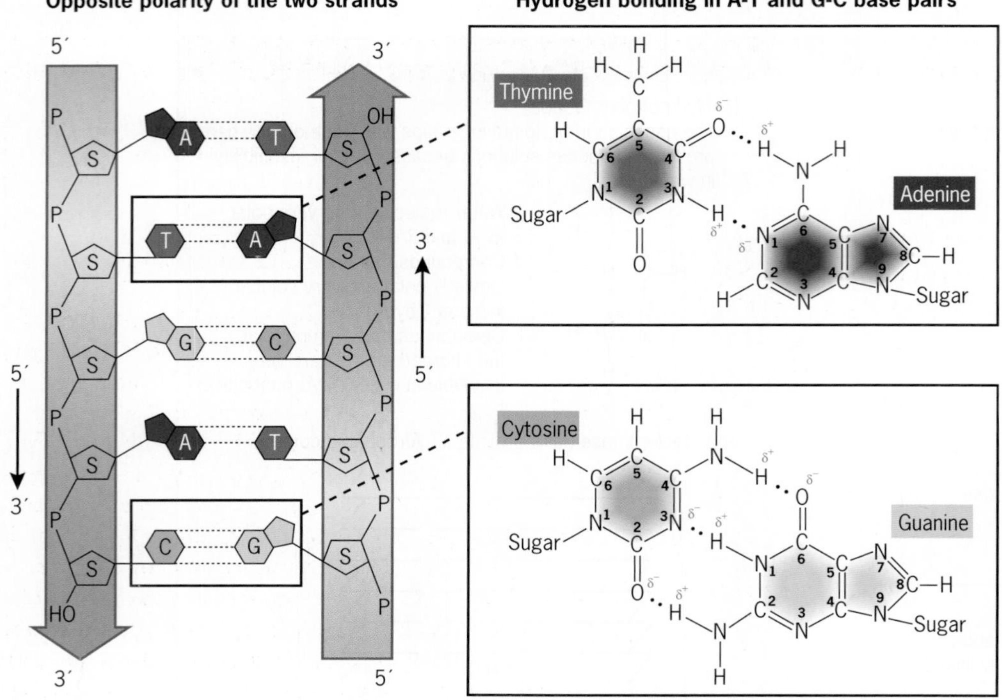

Figure 9.10 ▶ Diagram of a DNA double helix, illustrating the opposite chemical polarity (see Figure 9.7) of the two strands and the hydrogen bonding between thymine (T) and adenine (A) and between cytosine (C) and guanine (G). The base-pairing in DNA, T with A and C with G, is governed by the hydrogen-bonding potential of the bases. S = the sugar 2-deoxyribose; P = a phosphate group.

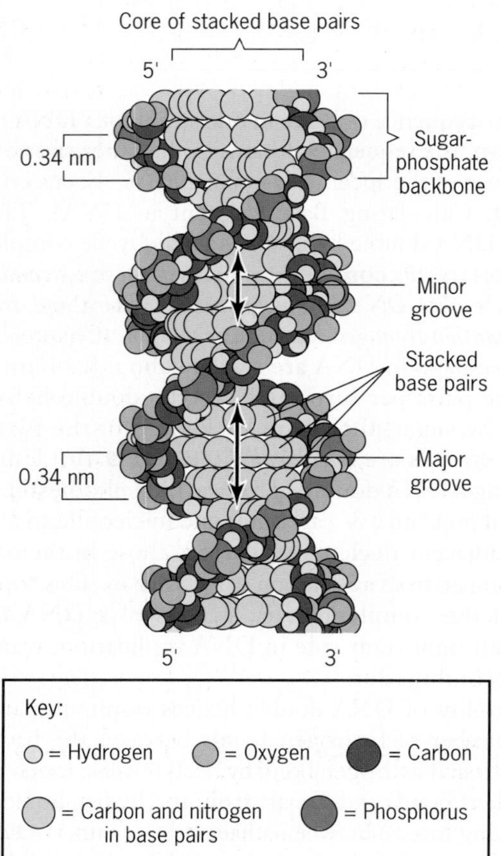

Figure 9.11 ▶ Space-filling diagram of a DNA double helix.

tend to be hydrophobic (water-insoluble). Because of this insolubility in water, the hydrophobic core of stacked base pairs contributes considerable stability to DNA molecules present in the aqueous protoplasms of living cells. The space-filling drawing also shows that the two grooves of a DNA double helix are not identical; one, the major groove, is much wider than the other, the minor groove. The difference between the major groove and the minor groove is important when one examines the interactions between DNA and proteins that regulate gene expression. Some proteins bind to the major groove; others bind to the minor groove.

DNA STRUCTURE: ALTERNATE FORMS OF THE DOUBLE HELIX

The Watson-Crick double-helix structure just described is called **B-DNA.** B-DNA is the conformation that DNA takes under physiological conditions (in aqueous solutions containing low concentrations of salts). The vast majority of the DNA molecules present in the aqueous protoplasms of living cells exist in the B conformation. However, DNA is not a static, invariant molecule. On the contrary, DNA molecules exhibit considerable conformational flexibility.

The structures of DNA molecules change as a function of their environment. The exact conformation of a given DNA molecule or segment of a DNA molecule will depend on the nature of the molecules with which it is interacting. In fact, intracellular B-DNA appears to have an average of 10.4 nucleotide pairs per turn, rather than precisely 10 as shown

► FOCUS ON PROBLEM SOLVING
Calculating Base Content in DNA

THE PROBLEM

Double-stranded genomic DNA was isolated from the bacterium *Mycobacterium tuberculosis,* and chemical analysis showed that 33 percent of the bases in the DNA were guanine residues. Based on this information, is it possible to determine what percent of the bases in the DNA of *M. tuberculosis* were adenine residues?

Single-stranded genomic DNA was isolated from bacteriophage ΦX174, and chemical analysis showed that 22 percent of the bases in the ΦX174 DNA were cytosines. Based on this information, is it possible to determine what percent of the bases in the DNA packaged in the ΦX174 virion were adenines?

FACTS AND CONCEPTS

1. In double-stranded DNA, adenine in one strand is always paired with thymine in the complementary strand and guanine in one strand is always paired with cytosine in the other strand.

2. In single-stranded DNA, there is no strict base-pairing. There is some base-pairing between bases within the single strands forming hairpin structures, but there is no strict A:T and G:C base-pairing as in double-stranded DNA.

ANALYSIS AND SOLUTION

In the double-stranded genomic DNA of *M. tuberculosis,* every A in one strand is hydrogen-bonded to a T in the complementary strand, and every G is hydrogen-bonded to a C in the complementary strand. Thus, if 33 percent of the bases are guanines, 33 percent of the bases are cytosines. That means that 66 percent of the bases are G's and C's and 34 percent (100% − 66%) of the bases are A's and T's. Since A always pairs with T, half are A's and half are T's. Therefore, 17 percent (34% × ½) of the bases in the DNA of *M. tuberculosis* are adenines.

In the single-stranded DNA of bacteriophage ΦX174, there is no strict base-pairing, but only the occasional pairing between bases within the single strand of DNA. As a result, one cannot predict the proportion of adenine residues in the DNA based on the proportion of cytosines. Indeed, one cannot even predict the percentage of adenines based on the percentage of thymines in single-stranded DNA, like the DNA packaged in the ΦX174 virion.

For further discussion go to your *WileyPLUS* course.

► TABLE 9.3

Alternate Forms of DNA

Helix Form	Helix Direction	Base Pairs per Turn	Helix Diameter
A	Right-handed	11	2.3 nm
B	Right-handed	10	1.9 nm
Z	Left-handed	12	1.8 nm

in **FIGURE 9.9.** In high concentrations of salts or in a partially dehydrated state, DNA exists as **A-DNA,** which is a right-handed helix like B-DNA, but with 11 nucleotide pairs per turn (**TABLE 9.3**). A-DNA is a shorter, thicker double helix with a diameter of 2.3 nm. DNA molecules almost certainly never exist as A-DNA *in vivo.* However, the A-DNA conformation is important because DNA-RNA heteroduplexes (double helices containing a DNA strand base-paired with a complementary RNA strand) or RNA-RNA duplexes exist in a very similar structure *in vivo.*

Certain DNA sequences have been shown to exist in a left-handed, double-helical form called **Z-DNA** (Z for the zig-zagged path of the sugar-phosphate backbones of the structure). Z-DNA was discovered by X-ray diffraction analysis of crystals formed by DNA oligomers containing alternating G:C and C:G base pairs. Z-DNA occurs in double helices that are G:C-rich and contain alternating purine and pyrimidine residues. In addition to its unique left-handed helical structure, Z-DNA (**TABLE 9.3**) differs from the A and B conformations in having 12 base pairs per turn, a diameter of 1.8 nm, and a single deep groove. The function of Z-DNA in living cells is still not clear.

DNA STRUCTURE: NEGATIVE SUPERCOILS *IN VIVO*

All the functional DNA molecules present in living cells display one other very important level of organization—they are supercoiled. **Supercoils** are introduced into a DNA molecule when one or both strands are cleaved and when the complementary strands at one end are rotated or twisted around each other with the other end held fixed in space—and thus not allowed to spin. This supercoiling causes a DNA molecule to collapse into a tightly coiled structure similar to a coiled telephone cord or twisted rubber band (**FIGURE 9.12**, lower right). Supercoils are introduced into and removed from DNA molecules by enzymes that play essential roles in DNA replication (Chapter 10) and other processes.

Supercoiling occurs only in DNA molecules with fixed ends, ends that are not free to rotate. Obviously, the ends of the circular DNA molecules (**FIGURE 9.12**) present in most prokaryotic chromosomes and in the chromosomes of eukaryotic organelles such as mitochondria are fixed. The large linear DNA molecules present in eukaryotic chromosomes are also fixed by their attachment at intervals and at the ends to non-DNA components of the chromosomes. These attachments allow enzymes to introduce supercoils into the linear DNA molecules present in eukaryotic chromosomes, just as they are

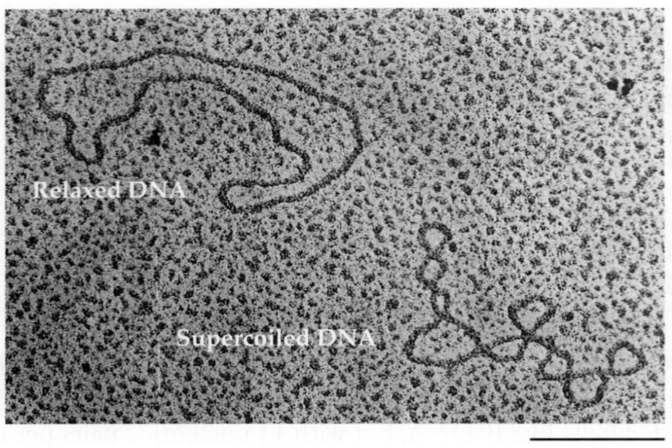

Figure 9.12 ▶ Comparison of the relaxed and negatively supercoiled structures of DNA. The relaxed structure is B-DNA with 10.4 base pairs per turn of the helix. The negatively supercoiled structure results when B-DNA is underwound, with less than one turn of the helix for every 10.4 base pairs.

incorporated into the circular DNA molecules present in most prokaryotic chromosomes.

We can perhaps visualize supercoiling most easily by considering a circular DNA molecule. If we cleave one strand of a covalently closed, circular double helix of DNA, and rotate one end of the cleaved strand a complete turn (360°) around the complementary strand while holding the other end fixed, we will introduce one supercoil into the molecule (**FIGURE 9.13**). If we rotate the free end in the same direction as the DNA double helix is wound (right-handed), a positive supercoil (overwound DNA) will be produced. If we rotate the free end in the opposite direction (left-handed), a negative supercoil (underwound DNA) will result. Although this is the simplest way to define supercoiling in DNA, it is not the mechanism by which supercoils are produced in DNA *in vivo*. That mechanism is discussed in Chapter 10.

The DNA molecules of almost all organisms, from the smallest viruses to the largest eukaryotes, exhibit **negative supercoiling** *in vivo*, and many of the biological functions of chromosomes can be carried out only when the participating DNA molecules are negatively supercoiled. (The DNA of some viruses that infect Archaea is positively supercoiled.) Considerable evidence indicates that negative supercoiling is involved in replication (Chapter 10), recombination, gene expression, and the regulation of gene expression. Similar amounts of negative supercoiling exist in the DNA molecules present in bacterial chromosomes and eukaryotic chromosomes.

KEY POINTS

- DNA usually exists as a double helix, with the two strands held together by hydrogen bonds between the complementary bases: adenine paired with thymine and guanine paired with cytosine.
- The complementarity of the two strands of a double helix makes DNA uniquely suited to store and transmit genetic information.
- The two strands of a DNA double helix have opposite chemical polarity.
- RNA usually exists as a single-stranded molecule containing uracil instead of thymine.
- The functional DNA molecules in cells are negatively supercoiled.

Chromosome Structure in Prokaryotes and Viruses

The DNA molecules of prokaryotes and viruses are organized into negatively supercoiled domains.

Much of the information about the structure of DNA has come from studies of prokaryotes, primarily because they are less complex, both genetically and biochemically, than eukaryotes. Prokaryotes are monoploid (*mono* = one); they have only one set of genes (one copy of the genome). ("Monoploid" should not be confused with "haploid," which refers specifically

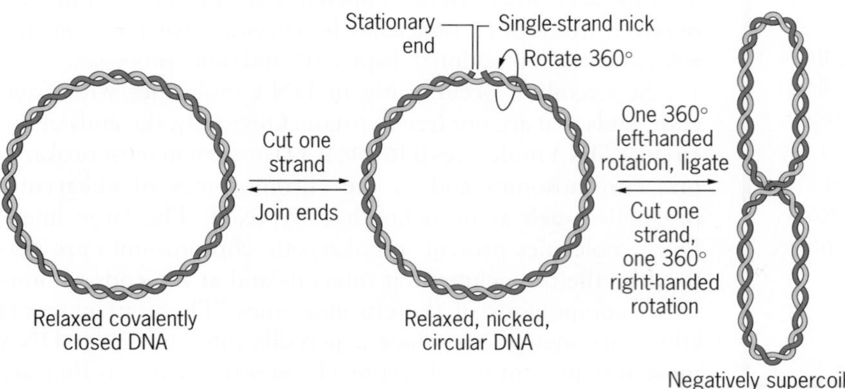

Figure 9.13 ▶ A visual definition of negatively supercoiled DNA. Although the structure of DNA supercoils is most clearly illustrated by the mechanism shown here, DNA supercoils are produced by a different mechanism *in vivo* (see Chapter 10).

to the reduced chromosome number in gametes.) In most viruses and prokaryotes, the single set of genes is stored in a single chromosome, which in turn contains a single molecule of nucleic acid (either RNA or DNA).

The smallest known RNA viruses have only three genes, and the complete nucleotide sequences of the genomes of many viruses are known. For example, the single RNA molecule in the genome of bacteriophage MS2 consists of 3569 nucleotides and contains 4 genes. The smallest known DNA viruses have only 9 to 11 genes. Again, the complete nucleotide sequences are known in several cases. For example, the genome of bacteriophage ϕX174 is a single DNA molecule 5386 nucleotides in length that contains 11 genes. The largest DNA viruses, like bacteriophage T2 and the animal pox viruses, contain about 150 genes. Bacteria like *E. coli* have 2500 to 3500 genes, most of which are present in a single molecule of DNA.

In the past, prokaryotic chromosomes were often characterized as "naked molecules of DNA," in contrast to eukaryotic chromosomes with their associated proteins and complex morphology. This misconception resulted in part because (1) most of the published pictures of prokaryotic "chromosomes" were electron micrographs of isolated DNA molecules, not metabolically active or functional chromosomes, and (2) most of the published photographs of eukaryotic chromosomes were of highly condensed meiotic or mitotic chromosomes—again, metabolically inactive chromosomal states. Functional prokaryotic chromosomes, or nucleoids (nucleoids rather than nuclei because they are not enclosed in a nuclear membrane) are now known to bear little resemblance to the isolated viral and bacterial DNA molecules seen in electron micrographs, just as the metabolically active interphase chromosomes of eukaryotes have little morphological resemblance to mitotic or meiotic metaphase chromosomes.

The contour length of the circular DNA molecule present in the chromosome of the bacterium *Escherichia coli* is about 1500 μm. Because an *E. coli* cell has a diameter of only 1 to 2 μm, the large DNA molecule present in each bacterium must exist in a highly condensed (folded or coiled) configuration. When *E. coli* chromosomes are isolated by gentle procedures in the absence of ionic detergents (commonly used to lyse cells) and are kept in the presence of a high concentration of cations such as polyamines (small basic or positively charged proteins) or 1 *M* salt to neutralize the negatively charged phosphate groups of DNA, the chromosomes remain in a highly condensed state comparable in size to the nucleoid *in vivo*. This structure, called the **folded genome,** is the functional state of a bacterial chromosome. Though smaller, the functional intracellular chromosomes of bacterial viruses are very similar to the folded genomes of bacteria.

Within the folded genome, the large DNA molecule in an *E. coli* chromosome is organized into 50 to 100 *domains* or loops, each of which is independently negatively supercoiled (**FIGURE 9.14**). RNA and protein are both components of the folded genome, which can be partially relaxed by treatment

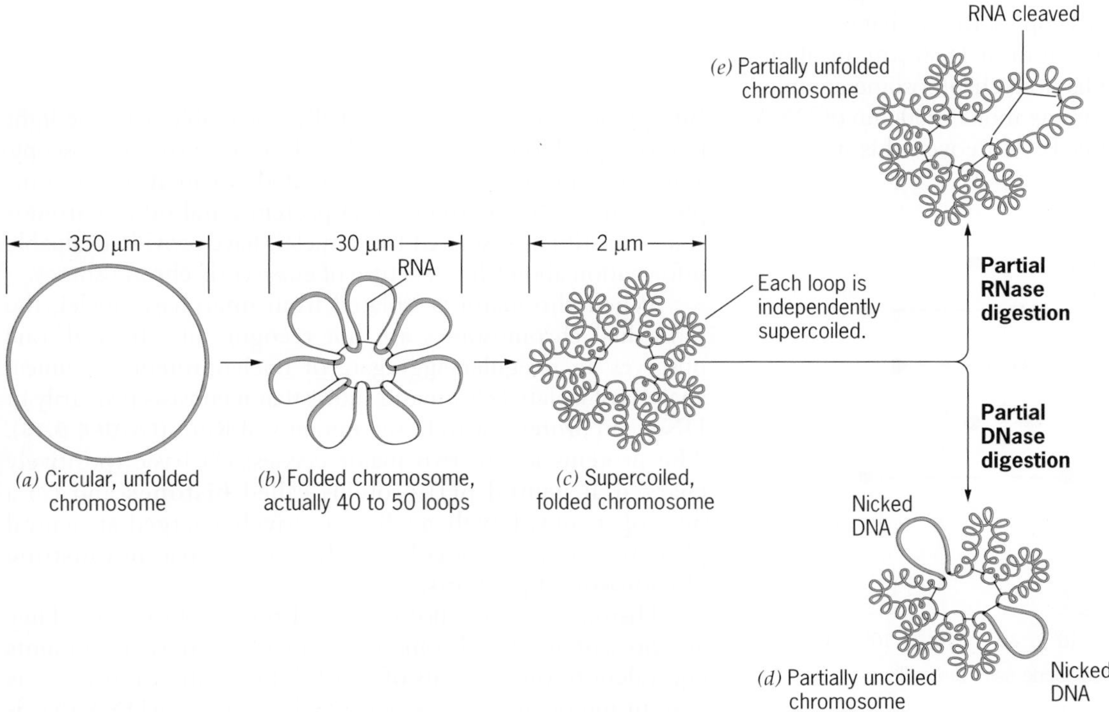

Figure 9.14 ► Diagram of the structure of the functional state of the *E. coli* chromosome.

with either deoxyribonuclease (DNase) or ribonuclease (RNase). Because each domain of the chromosome is independently supercoiled, the introduction of single-strand "nicks" in DNA by treatment of the chromosomes with a DNase that cleaves DNA at internal sites will relax the DNA only in the nicked domains, and all unnicked loops will remain supercoiled. Destruction of the RNA connectors by RNase will unfold the folded genome partially by eliminating the organization of the DNA molecule into 50 to 100 loops. However, RNase treatment will not affect the supercoiling of the domains of the chromosome.

KEY POINTS

- The DNA molecules in prokaryotic and viral chromosomes are organized into negatively supercoiled domains.
- Bacterial chromosomes contain circular molecules of DNA segregated into about 50 domains.

Chromosome Structure in Eukaryotes

Eukaryotic chromosomes contain huge molecules of DNA that are highly condensed during mitosis and meiosis. The centromeres and telomeres of eukaryotic chromosomes have unique structures.

Eukaryotic genomes contain levels of complexity that are not encountered in prokaryotes. In contrast to prokaryotes, most eukaryotes are diploid, having two complete sets of genes, one from each parent. As we discussed in Chapter 6, some flowering plants are polyploid; that is, they carry several copies of the genome. Although eukaryotes have only about 2 to 15 times as many genes as *E. coli*, they have orders of magnitude more DNA (**FIGURE 9.15**). Moreover, much of this DNA does not contain genes, at least not genes encoding proteins or RNA molecules.

Not only do most eukaryotes contain many times the amount of DNA in prokaryotes, but also this DNA is packaged into several chromosomes, and each chromosome is present in two (diploids) or more (polyploids) copies. Recall that the chromosome of *E. coli* has a contour length of 1500 μm, or about 1.5 mm. Now consider that the haploid chromosome complement, or genome, of a human contains about 1000 mm of DNA (or about 2000 mm per diploid cell). Moreover, this meter of DNA is subdivided among 23 chromosomes of variable size and shape, with each chromosome containing 15 to 85 mm of DNA. In the past, geneticists had little information as to how this DNA was arranged in the chromosomes. Is there one molecule of DNA per chromosome as in prokaryotes, or are there many? If many, how are the molecules arranged relative to each other? How does the 85 mm (85,000 μm) of DNA in the largest human chromosome get condensed into a mitotic metaphase structure that is about 0.5 μm in diameter and 10 μm long? What are the structures of the metabolically active interphase chromosomes? We consider the answers to some of these questions in the following sections.

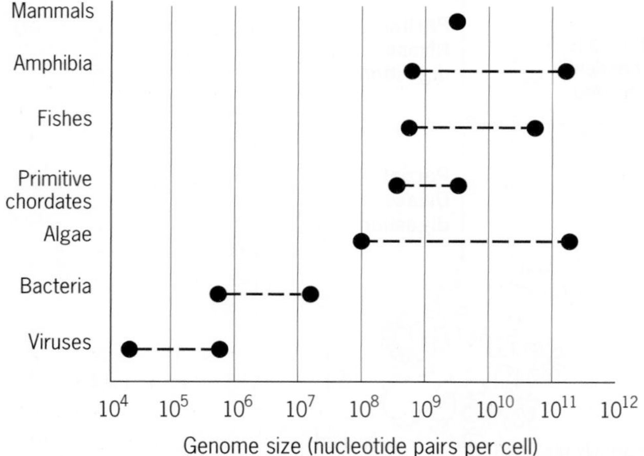

Figure 9.15 ▶ Increased genome size in organisms with increased developmental complexity.

CHEMICAL COMPOSITION OF EUKARYOTIC CHROMOSOMES

Interphase chromosomes are usually not visible with the light microscope. However, chemical analysis, electron microscopy, and X-ray diffraction studies of isolated **chromatin** (the complex of the DNA, chromosomal proteins, and other chromosome constituents isolated from nuclei) have provided valuable information about the structure of eukaryotic chromosomes.

When chromatin is isolated from interphase nuclei, the individual chromosomes are not recognizable. Instead, one observes an irregular aggregate of nucleoprotein. Chemical analysis of isolated chromatin shows that it consists primarily of DNA and proteins with lesser amounts of RNA (**FIGURE 9.16**). The proteins are of two major classes: (1) basic (positively charged at neutral pH) proteins called **histones** and (2) a heterogeneous, largely acidic (negatively charged at neutral pH) group of proteins collectively referred to as **nonhistone chromosomal proteins.**

Histones play a major structural role in chromatin. They are present in the chromatin of all eukaryotes in amounts equivalent to the amounts of DNA. This relationship suggests that an interaction occurs between histones and DNA that is conserved in eukaryotes. The histones of all plants and animals

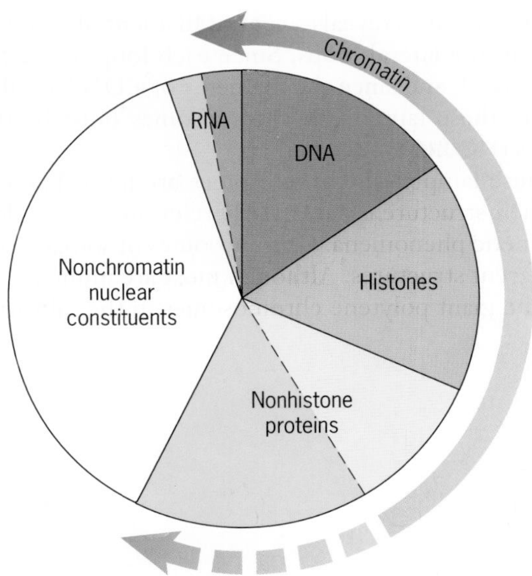

Figure 9.16 ▶ The chemical composition of chromatin as a function of the total nuclear content. The DNA and histone contents of chromatin are relatively constant, but the amount of nonhistone proteins present depends on the procedure used to isolate the chromatin (dashed arrow).

consist of five classes of proteins. These five major histone types, called *H1*, *H2a*, *H2b*, *H3*, and *H4*, are present in almost all cell types. A few exceptions exist, most notably some sperm, where the histones are replaced by another class of small basic proteins called **protamines.**

The five histone types are present in molar ratios of approximately 1 H1:2 H2a:2 H2b:2 H3:2 H4. Four of the five types of histones are specifically complexed with DNA to produce the basic structural subunits of chromatin, small (approximately 11 nm in diameter by 6.5 nm high) ellipsoidal beads called **nucleosomes.** The histones have been highly conserved during evolution—four of the five types of histone are similar in all eukaryotes.

Most of the 20 amino acids in proteins are neutral in charge; that is, they have no charge at pH 7. However, a few are basic and a few are acidic. The histones are basic because they contain 20 to 30 percent arginine and lysine, two positively charged amino acids (**FIGURE 9.17**). The exposed $-NH_3^+$ groups of arginine and lysine allow histones to act as polycations. The positively charged side groups on histones are important in their interaction with DNA, which is polyanionic because of the negatively charged phosphate groups.

The remarkable constancy of histones H2a, H2b, H3, and H4 in all cell types of an organism and even among widely divergent species is consistent with the idea that they are important in chromatin structure (DNA packaging) and are only nonspecifically involved in the regulation of gene expression. However, as will be discussed later, chemical modifications of histones can alter chromosome structure, which, in turn, can enhance or decrease the level of expression of genes located in the modified chromatin.

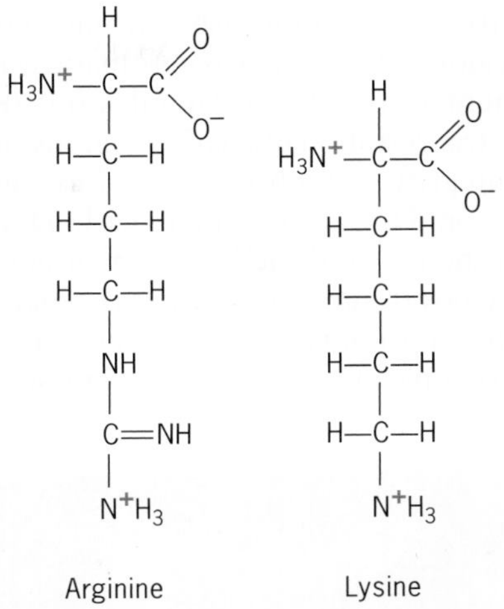

Arginine

Lysine

Figure 9.17 ▶ Structures of the amino acids arginine and lysine (at pH 7), which together account for 20 to 30 percent of the amino acids in histones.

In contrast, the nonhistone protein fraction of chromatin consists of a large number of heterogeneous proteins. Moreover, the composition of the nonhistone chromosomal protein fraction varies widely among different cell types of the same organism. Thus, the nonhistone chromosomal proteins probably do not play central roles in the packaging of DNA into chromosomes. Instead, they are likely candidates for roles in regulating the expression of specific genes or sets of genes.

ONE LARGE DNA MOLECULE PER CHROMOSOME

A typical eukaryotic chromosome contains 1 to 20 cm (10^4 to 2×10^5 μm) of DNA. During metaphase of meiosis and mitosis, this DNA is packaged in a chromosome with a length of only 1 to 10 μm. How is all of this DNA condensed into the compact chromosomes that are present during mitosis and meiosis? Do many DNA molecules run parallel throughout the chromosome—the *multineme* or "multistrand" model—or is there just one DNA double helix extending from one end of the chromosome to the other—the *unineme* or single-strand model? (Note that "strand" here refers to the DNA double helix, not the individual polynucleotide chains of DNA.) Considerable evidence now indicates that each chromosome contains a single, giant molecule of DNA that extends from one end through the centromere all the way to the other end of the chromosome.

Some of the strongest evidence supporting the unineme model of chromosome structure has come from studies of **lampbrush chromosomes** (so named because they resemble the brushes used to clean the mantles of oil lamps) present during prophase I of oogenesis in many vertebrates, particularly amphibians. Lampbrush chromosomes are up to 800 μm long, and their large size has allowed cytologists to carry out microscopic studies of chromosome structure that are not feasible with smaller chromosomes. The homologous chromosomes are paired, and each has duplicated to produce two chromatids. Each lampbrush chromosome contains a central axial region, where the two chromatids are highly condensed, and numerous pairs of lateral loops (**FIGURE 9.18**). The loops are transcriptionally active regions of single chromatids. The integrity of both the central axis and the lateral loops depends on DNA. Treatment with DNase produces breaks in both the axis and the loops. Treatment with RNase or proteases removes surrounding matrix material but does not destroy the continuity of either the axis or the loops. Electron microscopy of RNase- and protease-treated lampbrush chromosomes reveals a central filament of about 2 nm in diameter in the lateral loops. Since each loop is a segment of one chromatid, and since the diameter of a DNA double helix is 1.9 nm, these lampbrush chromosomes must be unineme structures (**FIGURE 9.18**).

Because lampbrush chromosomes are germ-line chromosomes, their structure is particularly relevant to an understanding of genetic phenomena. Chromosomes of somatic cells may have different structures. Although most are unineme, some—such as the giant polytene chromosomes in the salivary glands

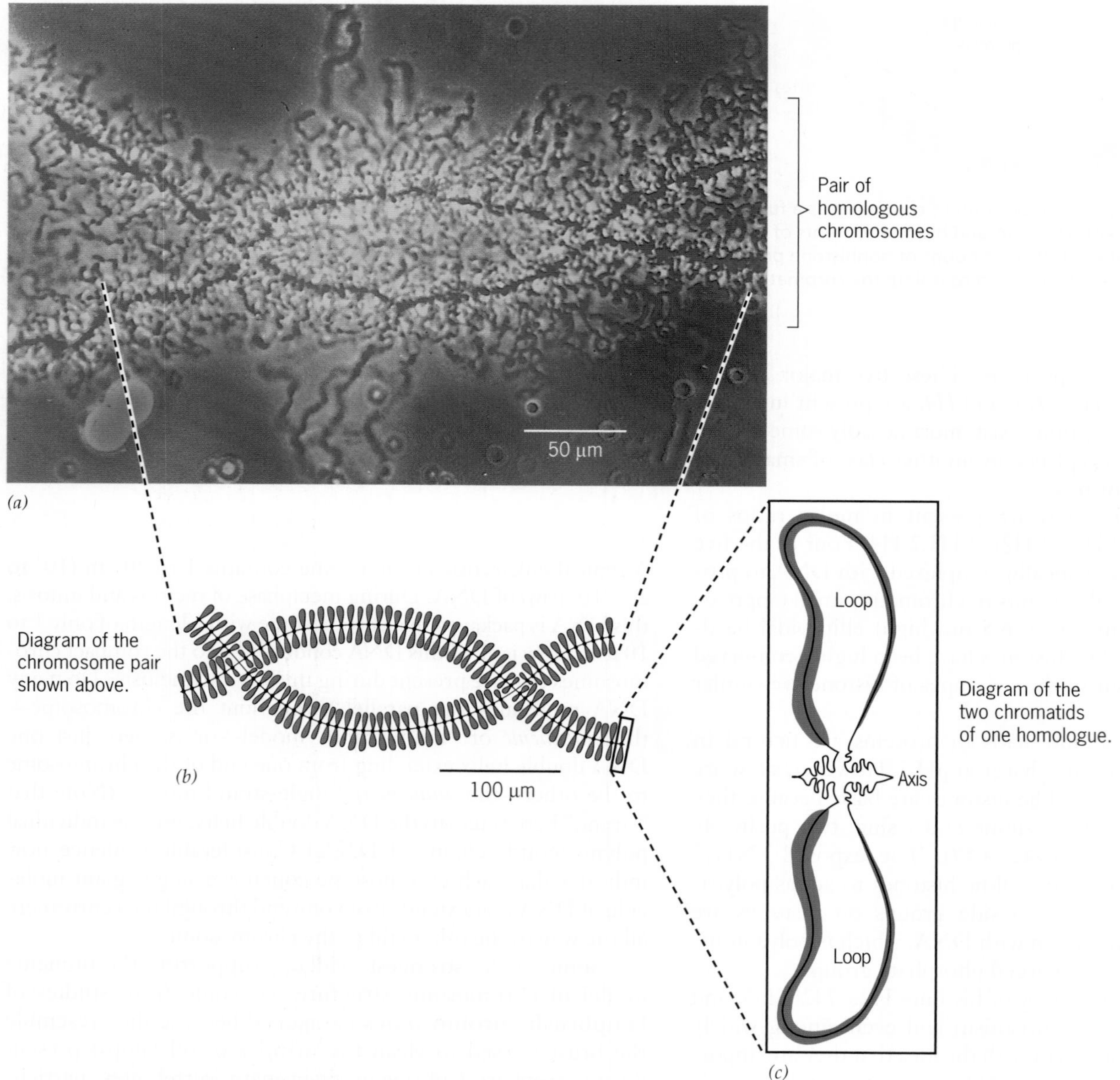

Figure 9.18 ▶ Phase contrast micrograph (*a*) and diagram (*b*) of a pair of lampbrush chromosomes in an oocyte of the newt, *Triturus viridescens*. The structures of axial and loop regions of the two chromatids of a single lampbrush chromosome are shown in (*c*). The central element in each chromatid (both axial regions and lateral loops) is a single molecule of DNA. The material surrounding the DNA molecule is primarily nascent RNA that is synthesized on the extended DNA in the loop regions.

of *Drosophila* (Chapter 6)—are known to be multineme structures composed of many identical DNA molecules.

The question of whether the unineme chromosomes of eukaryotes contain a single large molecule of DNA or many smaller molecules linked end-to-end has proven difficult to answer with rigorous experimental evidence. A centimeter-long molecule of DNA has a length-to-diameter ratio of 5 million to 1. Such a structure is extremely sensitive to shearing. If such a DNA molecule is in solution in a test tube, the slightest vibration will break the molecule into many fragments. For this and other reasons, accurate estimates of the sizes of eukaryotic DNAs cannot be obtained with the procedures used to analyze prokaryotic DNA molecules. However, by modifying old techniques and developing some new ones, scientists have obtained evidence indicating that each eukaryotic chromosome, no matter how large, contains one giant DNA double helix.

Some lower eukaryotic organisms, such as the mold *Neurospora crassa* and the yeast *Saccharomyces cerevisiae*, have relatively small chromosomes. With these organisms, a procedure called **pulsed-field gel electrophoresis** has been used to demonstrate that each chromosome contains a single molecule of DNA. The technique of gel electrophoresis is a powerful tool for separating macromolecules such as proteins and nucleic acids based on their size and charge (see Chapter 15). A semisolid gel (usually polyacrylamide or agarose) provides an inert matrix with pores in a given size range through which the macromolecules migrate when placed in an electric field. Positively charged molecules migrate toward the cathode (the negative electrode), and negatively charged molecules move toward the anode (the positive electrode). Proteins may be either positively or negatively charged, depending on their amino acid composition. Nucleic acids are negatively charged with one phosphate group per nucleotide. Thus, nucleic acids have an approximately constant charge per unit of mass, and all would migrate at the same rate in the absence of sieving. However, polyacrylamide gels have relatively small pores, and agarose gels have somewhat larger pores. These gels act as molecular sieves such that small molecules migrate faster than larger molecules with the same charge per unit of mass. As a result, the rate of migration of a nucleic acid during gel electrophoresis is almost exclusively a function of its size. Sometimes, conformation is a factor; for example, supercoiled DNAs migrate faster than relaxed molecules of the same size.

Pulsed-field gel electrophoresis, which is used to separate large DNA molecules, differs from standard gel electrophoresis in that instead of a single (one-dimensional), constant electric field, two electric fields offset by about 90° are applied across the gel in an alternating or pulsed manner. In standard gel electrophoresis, the DNA molecules pass through the gel in an end-first or snakelike fashion. In pulsed-field gel electrophoresis, the application of intermittent and alternating electric fields requires the molecules to reorient themselves before continuing to migrate through the gel. Larger molecules take longer to undergo these reorientation events and move more slowly. As a result, pulsed-field gel electrophoresis yields superior separation of very large DNA molecules. When this technique was used to separate intact DNA molecules from the fungi *N. crassa* and *S. cerevisiae*, the results showed that the number of different-sized DNA molecules was equal to the number of nonhomologous chromosomes in these species (**FIGURE 9.19**).

Unfortunately, the very large DNA molecules present in the chromosomes of higher eukaryotes such as *Drosophila* and humans cannot be separated even by pulsed-field gel electrophoresis. Researchers have used additional approaches in attempts to demonstrate that the large chromosomes of higher animals and plants each contain one molecule of DNA. Autoradiography and viscoelastometry are two approaches that have yielded important results.

Autoradiography is a method for detecting and localizing radioactive isotopes in cytological preparations or macromolecules by exposure to a photographic emulsion that is sensitive to low-energy radiation. The emulsion contains silver halides that produce tiny black spots—often called silver grains—when they are exposed to the charged particles emitted during the decay of radioactive isotopes. Autoradiography permits a researcher to prepare an image of the localization of radioactivity in macromolecules, cells, or tissues, just as photography permits us to make a picture of what we see. The difference is that the film used for autoradiography is sensitive to radioactivity, whereas the film we use in a camera is sensitive to visible light. Autoradiography is particularly useful in studying DNA

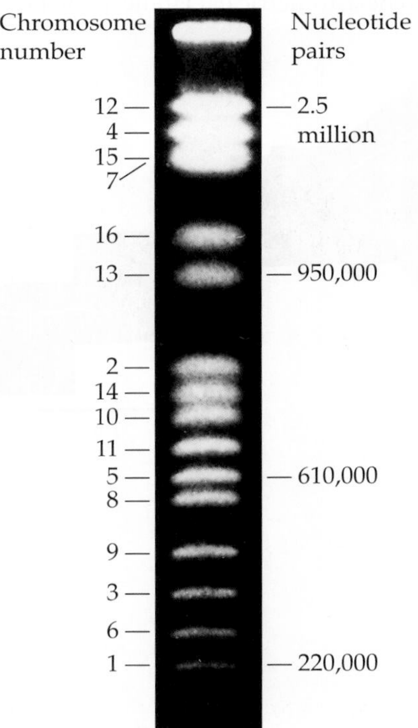

Figure 9.19 ▶ Separation of the chromosome-size DNA molecules of the yeast *Saccharomyces cerevisiae* by pulsed-field agarose gel electrophoresis. The large DNA molecules present in all 16 of the yeast chromosomes can be resolved by this procedure.

metabolism because DNA can be specifically labeled by growing cells on ^{3}H-thymidine, a deoxyribonucleoside of thymine that contains a radioactive isotope of hydrogen (tritium). Thymidine is incorporated almost exclusively into DNA; it is not present in any other major component of the cell.

In the early 1960s, Ruth Kavenoff, Lynn Klotz, and Bruno Zimm grew *Drosophila* cells in culture medium containing ^{3}H-thymidine for 24 hours, lysed the cells gently so as not to break the chromosomal DNA molecules, and carefully collected the DNA molecules on protein-coated glass slides. Then they covered the slides with emulsion sensitive to β-particles (the low-energy electrons emitted during decay of tritium) and stored them in the dark for a period of time to allow sufficient radioactive decays. The greatest challenge to Kavenoff and her coworkers was to spread out the molecules with no tangles or overlaps on the slides so that the entire length of a molecule would be visible. Their best autoradiographs showed DNA molecules with contour lengths of up to 1.2 cm (**FIGURE 9.20**). DNA molecules of this length would have a mass of about 3×10^{10} daltons (one dalton is the mass of one hydrogen atom) and would contain about two-thirds of the DNA known to be present in the largest chromosomes of *D. melanogaster*. Thus, these results provide support for the concept of chromosome-size DNA molecules in *Drosophila*.

Kavenoff and colleagues also used a technique called **viscoelastometry,** a procedure for analyzing the viscosity of molecules in solution, to determine the sizes of the DNA molecules in the largest *Drosophila* chromosomes. Kavenoff and coworkers' viscoelastometric data indicate that the largest DNA molecules in *Drosophila* have a mass of 4.1×10^{10} daltons. Because the largest chromosomes of *Drosophila* have been shown to contain about 4.3×10^{10} daltons of DNA (total, whether one molecule or many) by direct biochemical analysis, the viscoelastometric estimate of the size of the largest DNA molecules in *Drosophila* nuclei correlates almost exactly with the total amount of DNA present in the largest chromosome.

These and other results have provided strong, but not definitive, evidence that each eukaryotic chromosome contains one long double helix of DNA extending from one end of the chromosome through the centromere all the way to the other end. However, as we will discuss in the following section, this giant DNA molecule is highly condensed (coiled and folded) within the chromosome.

THREE LEVELS OF DNA PACKAGING IN EUKARYOTIC CHROMOSOMES

The largest chromosome in the human genome contains about 85 mm (85,000 μm, or 8.5×10^7 nm) of DNA that is believed to exist as one giant molecule. This DNA molecule somehow gets packaged into a metaphase structure that is about 0.5 μm in diameter and about 10 μm in length—a condensation of almost 10^4-fold in length from the naked DNA molecule to the metaphase chromosome. How does this condensation occur? What components of the chromosomes are involved in the packaging processes? Is there a universal packaging scheme? Are there different levels of packaging? Clearly, meiotic and mitotic chromosomes are more extensively condensed than

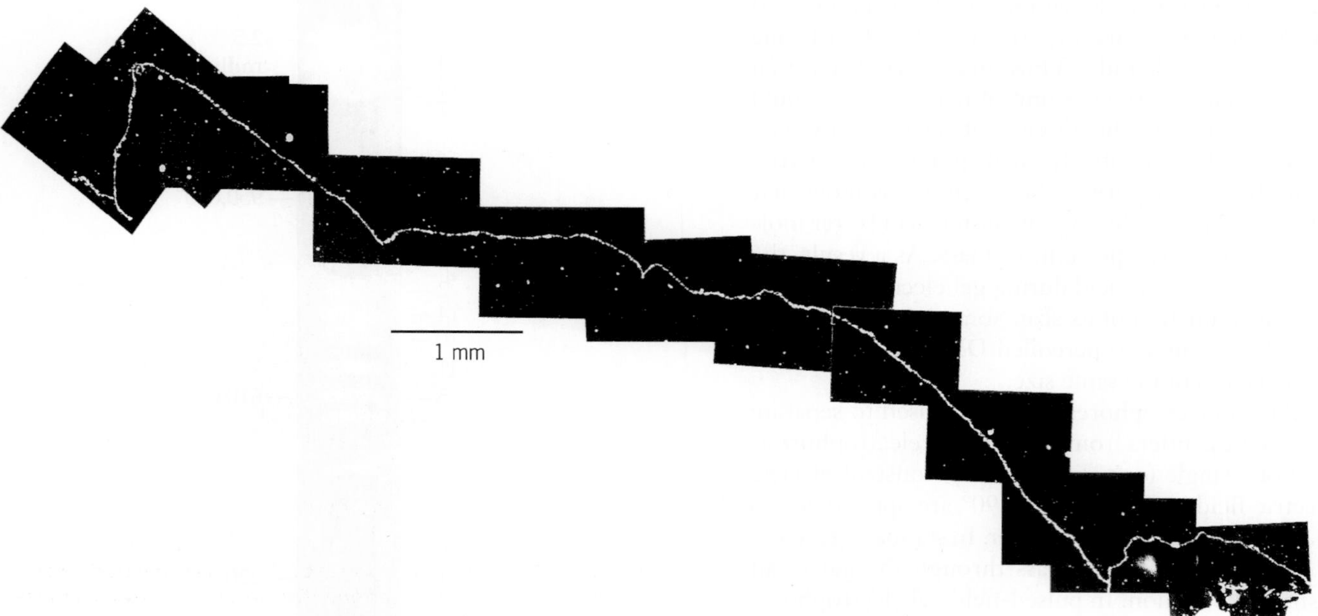

Figure 9.20 ▶ Visualization of a near-chromosome-size DNA molecule from *Drosophila melanogaster*. The largest *Drosophila* chromosome contains 4.3×10^{10} daltons of DNA. If this DNA is present in a single DNA molecule, it will be 1.8 cm in length. The molecule shown is 1.2 cm long, or two-thirds the total length of the DNA in the chromosome, whether present in one molecule or more than one molecule.

interphase chromosomes. What additional levels of condensation occur in these special structures that are designed to assure the proper segregation of the genetic material during cell divisions? Are DNA sequences of genes that are being expressed packaged differently from those of genes that are not being expressed? Let us investigate some of the evidence that establishes the existence of three different levels of packaging of DNA into chromosomes.

When isolated chromatin from interphase cells is examined by electron microscopy, it is found to consist of a series of ellipsoidal beads (about 11 nm in diameter and 6.5 nm high) joined by thin threads (**FIGURE 9.21*a***). Further evidence for a regular, periodic packaging of DNA has come from studies on the digestion of chromatin with various nucleases. Partial digestion of chromatin with these nucleases yielded fragments of DNA in a set of discrete sizes that were integral multiples of the smallest size fragment. These results are nicely explained if chromatin has a repeating structure, supposedly the bead seen by electron microscopy (**FIGURE 9.21*a***), within which the DNA is packaged in a nuclease-resistant form (**FIGURE 9.21*b***). This "bead" or chromatin subunit is called the **nucleosome.** According to the present concept of chromatin structure, the **linkers,** or interbead threads of DNA, are susceptible to nuclease attack.

After partial digestion of the DNA in chromatin with an endonuclease (an enzyme that cleaves DNA internally), DNA approximately 200 nucleotide pairs in length is associated with each nucleosome (produced by a cleavage in each linker region). After extensive nuclease digestion, a 146-nucleotide-pair-long segment of DNA remains present in each nucleosome. This nuclease-resistant structure is called the **nucleosome core.** Its structure—essentially invariant in eukaryotes—consists of a 146-nucleotide-pair length of DNA and two molecules each of histones H2a, H2b, H3, and H4. The histones protect the segment of DNA in the nucleosome core from cleavage by endonucleases.

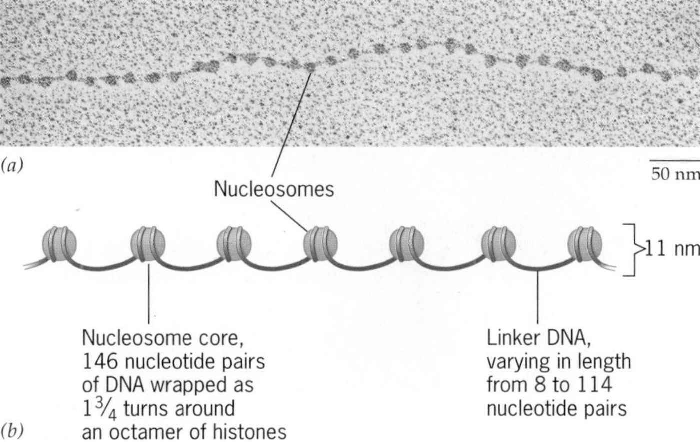

Figure 9.21 ▶ Electron micrograph (*a*) and low-resolution diagram of the beads-on-a-string nucleosome substructure of chromatin isolated from interphase nuclei. *In vivo*, the DNA linkers are probably wound between the nucleosomes forming a condensed 11 nm fiber.

Physical studies (X-ray diffraction and similar analyses) of nucleosome-core crystals have shown that the DNA is wound as 1.65 turns of a superhelix around the outside of the histone octamer (**FIGURE 9.22*a***).

The complete chromatin subunit consists of the nucleosome core, the linker DNA, and the associated nonhistone chromosomal proteins, all stabilized by the binding of one molecule of histone H1 to the outside of the structure (**FIGURE 9.22*b***). The size of the linker DNA varies from species to species and from one cell type to another. Linkers as short as eight nucleotide pairs and as long as 114 nucleotide pairs have been reported. Evidence suggests that the complete nucleosome (as opposed to the nucleosome core) contains two full turns of DNA superhelix (a 166-nucleotide-pair length of DNA) on the surface of the histone octamer, and the stabilization of this structure by the binding of one molecule of histone H1 (**FIGURE 9.22*b***).

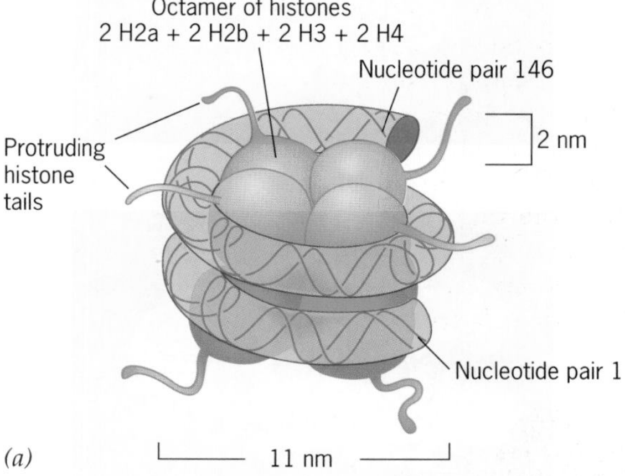

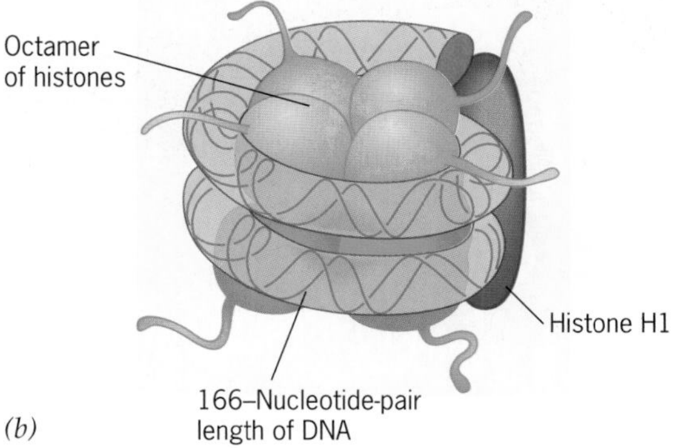

Figure 9.22 ▶ Diagrams of the gross structure of (a) the nucleosome core and (b) the complete nucleosome. The nucleosome core contains 146 nucleotide pairs wound as 1.65 turns of negatively supercoiled DNA around an octamer of histones—two molecules each of histones H2a, H2b, H3, and H4. The complete nucleosome contains 166 nucleotide pairs that form almost two superhelical turns of DNA around the histone octamer. One molecule of histone H1 is thought to stabilize the complete nucleosome.

The structure of the nucleosome core has been determined with resolution to 0.28 nm by X-ray diffraction studies. The resulting high-resolution map of the nucleosome core shows the precise location of all eight histone molecules and the 146 nucleotide pairs of negatively supercoiled DNA (**FIGURE 9.23*a*** and ***b***). Some of the terminal segments of the histones pass over and between the turns of the DNA superhelix to add stability to the nucleosome. The interactions between the various histone molecules and between the histones and DNA are seen most clearly in the structure of one-half of the nucleosome core (**FIGURE 9.23*c***), which contains only 73 nucleotide pairs of supercoiled DNA.

The basic structural component of eukaryotic chromatin is the nucleosome. But are the structures of all nucleosomes the same? What role(s), if any, does nucleosome structure play in gene expression and the regulation of gene expression? The structure of nucleosomes in transcriptionally active regions of chromatin is known to differ from that of nucleosomes in transcriptionally inactive regions. But what are the details of this structure–function relationship? The tails of some of the histone molecules protrude from the nucleosome and are accessible to enzymes that add and remove chemical groups such as methyl ($-CH_3$) and acetyl groups. The addition of these groups can change the level of expression of genes packaged in nucleosomes containing the modified histones (see Chapter 20).

Electron micrographs of isolated metaphase chromosomes show masses of tightly coiled or folded lumpy fibers (**FIGURE 9.24**). These **chromatin fibers** have an average diameter of 30 nm.

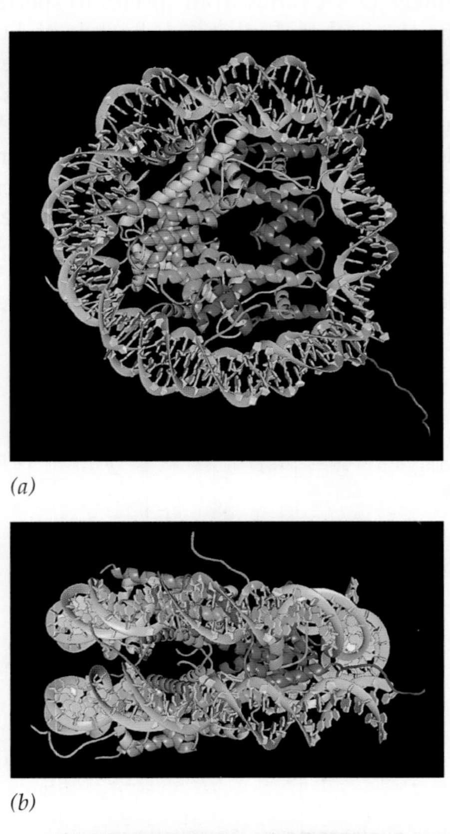

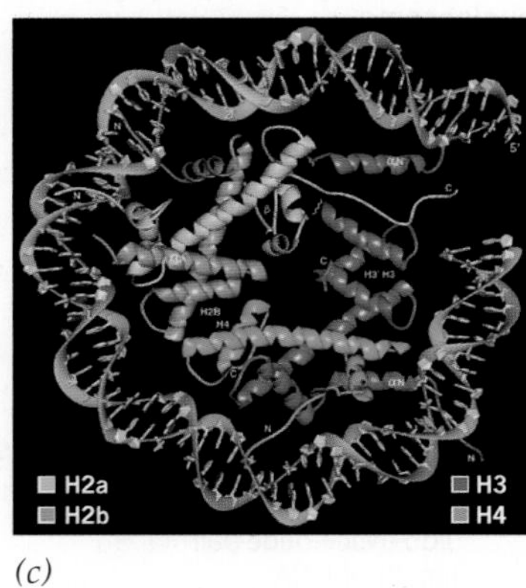

Figure 9.23 ▸ Structure of the nucleosome core based on X-ray diffraction studies with 0.28-nm resolution. The macromolecular composition of the nucleosome core is shown looking along (*a*) or perpendicular (*b*) to the axis of the superhelix. (*c*) Diagram of the structure of a half-nucleosome, which shows the relative positions of the DNA superhelix and the histones more clearly. The complementary strands of DNA are shown in brown and green, and histones H2a, H2b, H3, and H4 are shown in yellow, red, blue, and green, respectively.

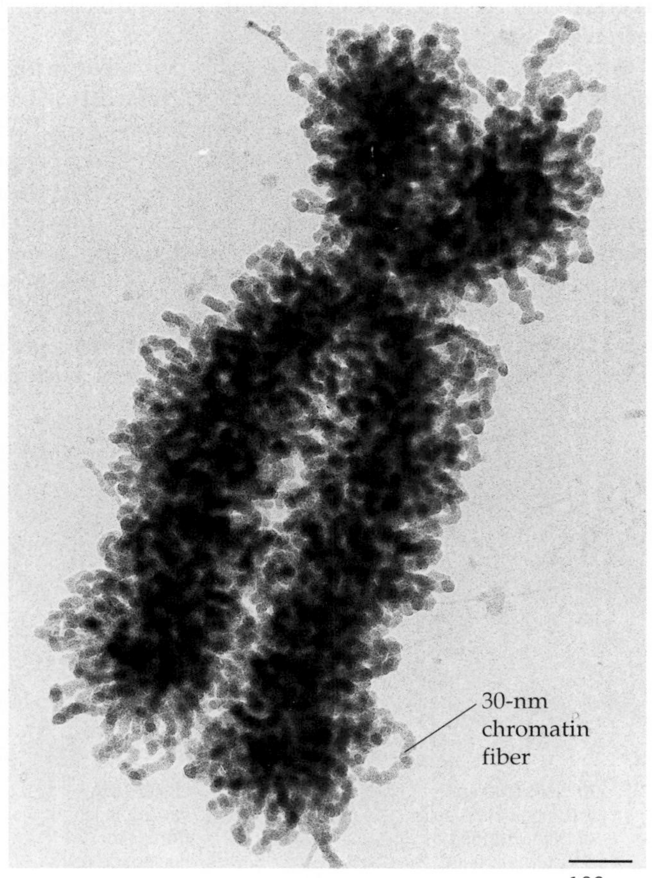

Figure 9.24 ▸ Electron micrograph of a human metaphase chromosome showing the presence of 30-nm chromatin fibers. The available evidence indicates that each chromatid contains one large, highly coiled or folded 30-nm fiber.

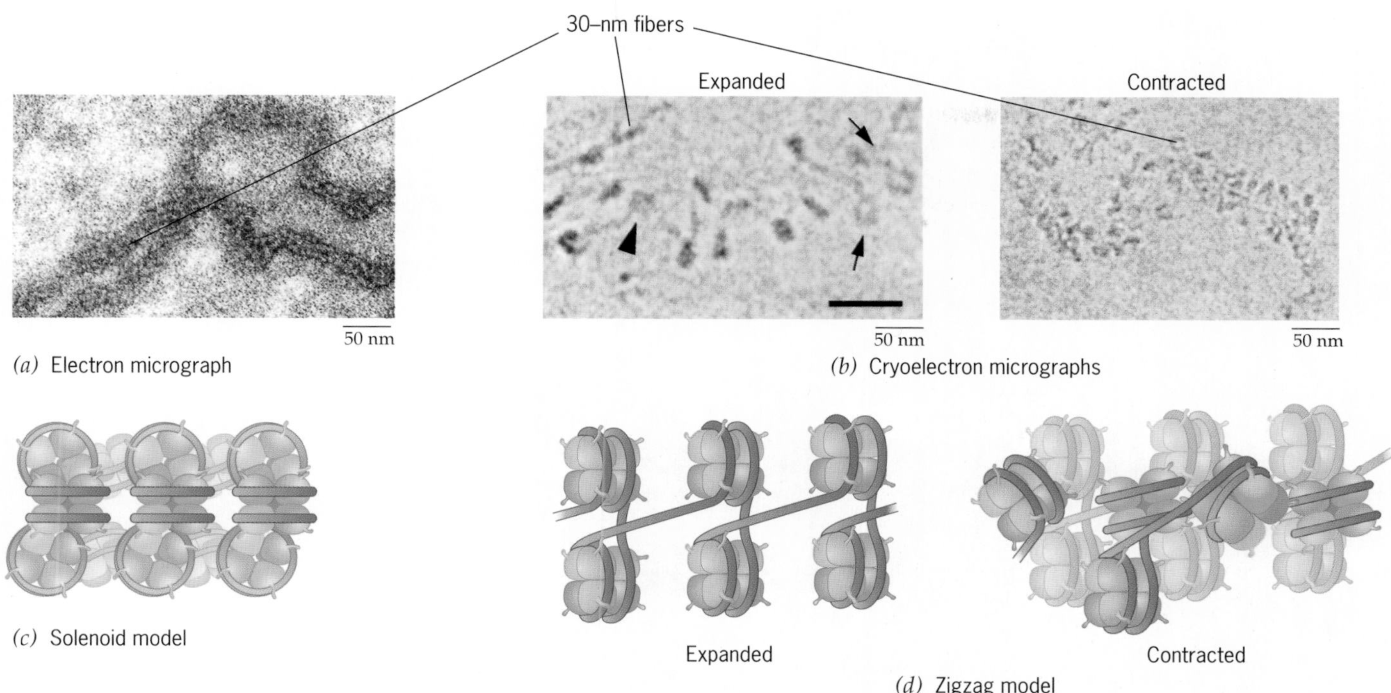

Figure 9.25 ▶ Electron micrograph (*a*) and cryoelectron micrographs (*b*) of the 30-nm chromatin fibers in eukaryotic chromosomes. The structure of 30-nm chromatin fibers seems to vary based on the procedures used to isolate and photograph them. (*c*) According to one popular model, the 30-nm fiber is produced by coiling the 11-nm nucleosome fiber into a solenoid structure with six nucleosomes per turn. (*d*) However, when chromatin is visualized after cryopreservation (quick freezing) without fixation, it exhibits a zigzag structure whose density—expanded versus relaxed—varies with ionic strength and with chemical modifications of the histone molecules.

When the structures seen by light and electron microscopy during earlier stages of meiosis are compared, it becomes clear that the light microscope simply permits one to see those regions where these 30-nm fibers are tightly packed or condensed. Indeed, when interphase chromatin is isolated using very gentle procedures, it also consists of 30-nm fibers (**FIGURE 9.25*a***). However, the structure of these fibers seems to be quite variable and depends on the procedures used. When observed by cryoelectron microscopy (microscopy using quickly frozen chromatin rather than fixed chromatin), the 30-nm fibers show less tightly packed "zigzag" structures (**FIGURE 9.25*b***).

What is the substructure of the 30-nm fiber seen in chromosomes? The two most popular models are the solenoid model (**FIGURE 9.25*c***) and the zigzag model (**FIGURE 9.25*d***). *In vivo*, the nucleosomes clearly interact with one another to condense the 11-nm nucleosomes into 30-nm chromatin fibers. Whether these have solenoid structures or zigzag structures or both, depending on the conditions, is still uncertain. What is certain is that chromatin structure is not static; chromatin can expand and contract in response to chemical modifications of histone H1 and the histone tails that protrude from the nucleosomes.

Metaphase chromosomes are the most condensed of normal eukaryotic chromosomes. Clearly, the role of these highly condensed chromosomes is to organize and package the giant DNA molecules of eukaryotic chromosomes into structures that will facilitate their segregation to daughter nuclei without the DNA molecules of different chromosomes becoming entangled and, as a result, being broken during the anaphase separation of the daughter chromosomes. As we noted in the preceding section, the basic structural unit of the metaphase chromosome is the 30-nm chromatin fiber. However, how are these 30-nm fibers further condensed into the observed metaphase structure? Unfortunately, there is still no clear answer to this question. There is evidence that the gross structure of metaphase chromosomes is not dependent on histones. Electron micrographs of isolated metaphase chromosomes from which the histones have been removed reveal a **scaffold,** or central core, which is surrounded by a huge pool or halo of DNA (**FIGURE 9.26**). This chromosome scaffold must be composed of nonhistone chromosomal proteins. Note the absence of any apparent ends of DNA molecules in the micrograph shown in **FIGURE 9.26**; this finding again supports the concept of one giant DNA molecule per chromosome.

In summary, at least three levels of condensation are required to package the 10^3 to 10^5 μm of DNA in a eukaryotic chromosome into a metaphase structure a few microns long (**FIGURE 9.27**).

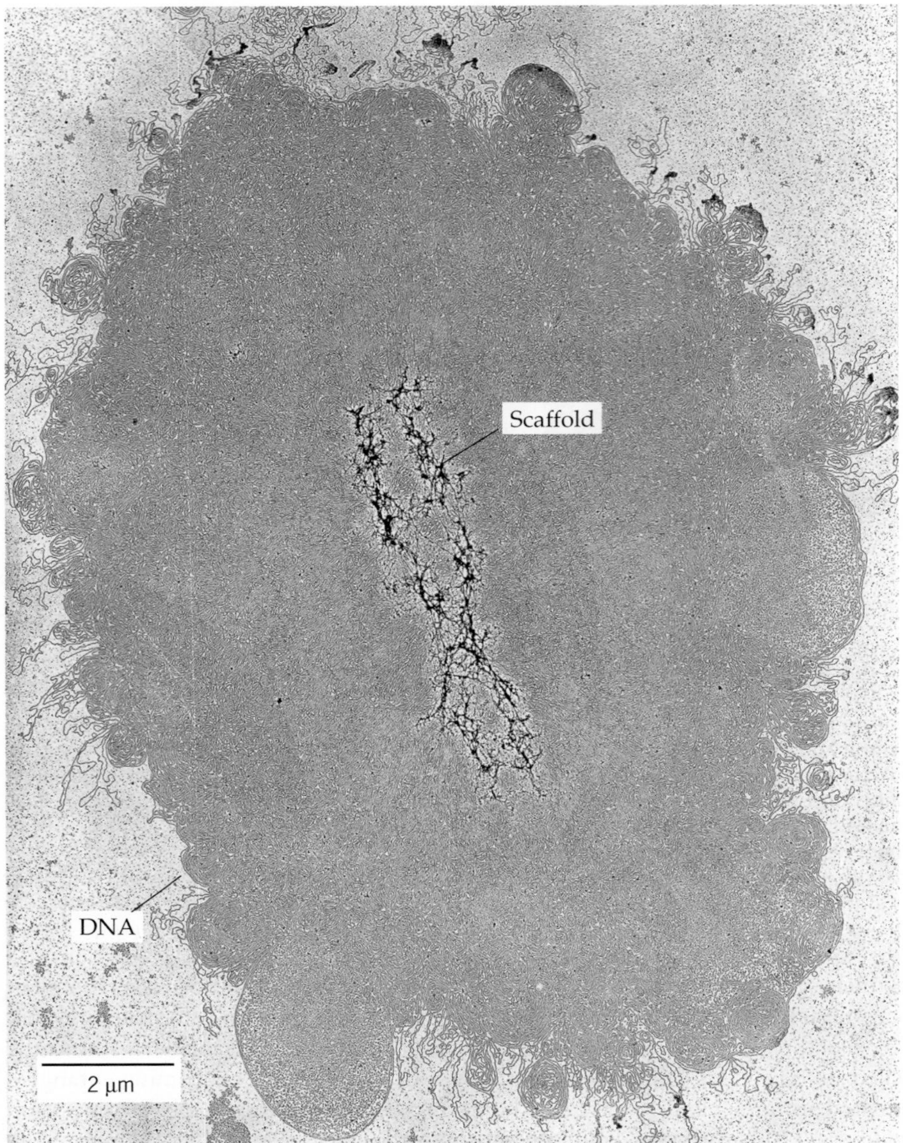

Figure 9.26 ▶ Electron micrograph of a human metaphase chromosome from which the histones have been removed. A huge pool of DNA surrounds a central "scaffold" composed of nonhistone chromosomal proteins. Note that the scaffold has roughly the same shape as the metaphase chromosome prior to removal of the histones. Also note the absence of ends of DNA molecules in the halo of DNA surrounding the scaffold.

1. The first level of condensation involves packaging DNA as a negative supercoil into nucleosomes to produce the 11-nm-diameter interphase chromatin fiber. This clearly involves an octamer of histone molecules, two each of histones H2a, H2b, H3, and H4.
2. The second level of condensation involves an additional folding or supercoiling of the 11-nm nucleosome fiber to produce the 30-nm chromatin fiber. Histone H1 is involved in this supercoiling of the 11-nm nucleosome fiber to produce the 30-nm chromatin fiber.
3. Finally, nonhistone chromosomal proteins form a scaffold that is involved in condensing the 30-nm chromatin fiber into the tightly packed metaphase chromosomes. This third level of condensation appears to involve the separation of segments of the giant DNA molecules present in eukaryotic chromosomes into independently supercoiled domains or loops. The mechanism by which this third level of condensation occurs is not known.

CENTROMERES AND TELOMERES

As we discussed in Chapter 2, the two homologous chromosomes (each containing two sister chromatids) of each chromosome pair separate to opposite poles of the meiotic spindle during anaphase I of meiosis. Similarly, during anaphase II of meiosis and the single anaphase of mitosis, the sister chromatids of each chromosome move to opposite spindle poles and become daughter chromosomes. These anaphase movements

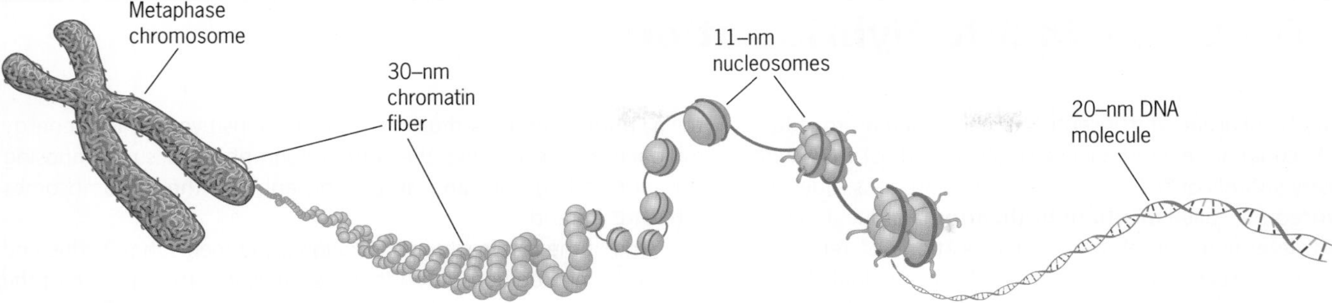

Figure 9.27 ► Diagram showing the different levels of DNA packaging in chromosomes. The 2-nm DNA molecule is first condensed into 11-nm nucleosomes, which are further condensed into 30-nm chromatin fibers. The 30-nm fibers are then segregated into supercoiled domains or loops via their attachment to chromosome scaffolds composed of nonhistone chromosomal proteins.

depend on the attachment of spindle microtubules to specific regions of the chromosomes, the centromeres. Because all centromeres perform the same basic function, it is not surprising that the centromeres of different chromosomes of a species contain similar structural components.

The centromere of a metaphase chromosome (**FIGURE 9.28**) can usually be recognized as a constricted region (see **FIGURE 9.24**). In fact, the production of two functional centromeres is a key step in the transition from metaphase to anaphase, and a functional centromere must be present on each daughter chromosome to avoid the deleterious effects of nondisjunction. Acentric chromosomal fragments are usually lost during mitotic and meiotic divisions.

The centromeres (*CEN* regions) of all the chromosomes of baker's yeast, *S. cerevisiae*, have been isolated and characterized. The *CEN* regions of different chromosomes are interchangeable: replacing the *CEN* region of one chromosome with the *CEN* region of another chromosome has no detectable effect on the host cell or its capacity to undergo a normal cell division. Molecular studies have shown that a functional *S. cerevisiae* centromere is 110 to 120 nucleotide pairs in length and has three essential regions (**FIGURE 9.29**). Regions I and III are short, conserved boundary sequences, and region II is an A:T-rich (>90 percent A:T) central segment about 90 nucleotide pairs long. The length and A:T-rich nature of region II are probably more important than its actual nucleotide sequence, whereas regions I and III contain specific sequences that serve as binding sites for proteins involved in spindle-fiber attachment.

The centromeres of multicellular eukaryotes are much larger and more complex than those of baker's yeast, but their actual structures are still uncertain. The centromeres of higher plants and animals contain large amounts of DNA sequences that are repeated many times, frequently in long tandem arrays. Other DNA sequences are often found embedded within these tandem arrays. Each centromere of human chromosomes, for example, contains 5000 to 15,000 copies of a 171 base-pair-long sequence called the alpha (sometimes "alphoid") satellite sequence (**FIGURE 9.30**). (Satellite sequences form distinct "satellite" bands during centrifugation in density gradients. See Focus on Centrifugation Techniques in Chapter 10.) Huntington Willard and colleagues

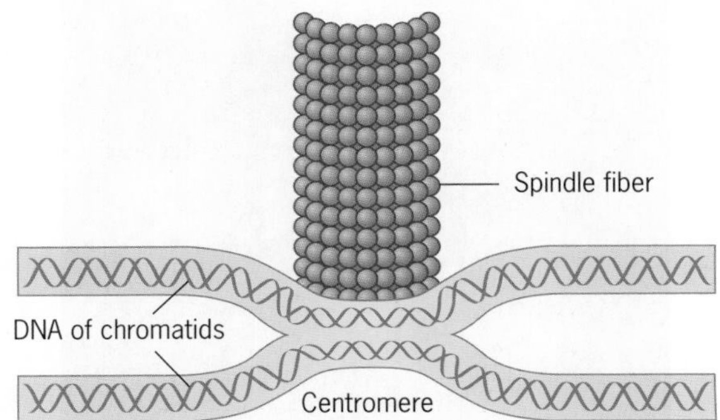

Figure 9.28 ► Model of centromere structure in a metaphase chromosome. The spindle fibers, which attach to centromeres, are responsible for the separation of homologous chromosomes during anaphase I of meiosis and progeny chromosomes (derived from chromatids) during anaphase II of meiosis and anaphase of mitosis (Chapter 2).

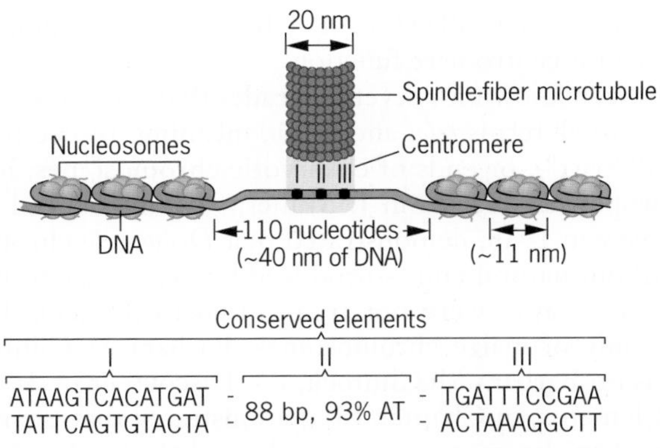

Figure 9.29 ► Diagram of the conserved structure of the centromeres in *Saccharomyces cerevisiae* (top) and the sequence of the *CEN* region of chromosome 3 of this species (bottom).

▶ FOCUS ON *In Situ* Hybridization

In 1969, Mary Lou Pardue and Joseph Gall developed a procedure by which they could anneal radioactive single strands of DNA with complementary strands of DNA in chromosomes on glass slides. By using this procedure, called ***in situ* hybridization**, Pardue and Gall were able to determine the chromosomal locations of repetitive DNA sequences. (The Latin term *in situ* means "in its original place"; hybridization is the formation of "hybrid" duplex molecules by the annealing of complementary or partially complementary strands of DNA or RNA.) Classical *in situ* hybridization involved spreading mitotic chromosomes on a glass slide (see **FIGURE 6.1**), denaturing the DNA in the chromosomes by exposure to alkali (0.07 *N* NaOH) for a few minutes, rinsing with buffer to remove the alkaline solution, incubating the slide in hybridization solution containing radioactive copies of the nucleotide sequence of interest, washing off the radioactive strands that have not hybridized with complementary sequences in the chromosomes, exposing the slide to a photographic emulsion that is sensitive to low-energy radioactivity, developing the autoradiograph, and superimposing the autoradiograph and a photograph of the chromosomes (**FIGURE 1*a*** and ***b***).

One of the first *in situ* hybridization experiments that Pardue and Gall performed demonstrated that the satellite DNA sequence of the mouse is located in heterochromatic regions that flank the centromeres of the mouse chromosomes (**FIGURE 1*b***). The mouse genome contains about 10^6 copies of this satellite DNA sequence, which is about 400 nucleotide pairs long and makes up about 10 percent of the mouse genome. Similar studies have subsequently been done with the satellite DNAs of several other species, and these repetitive

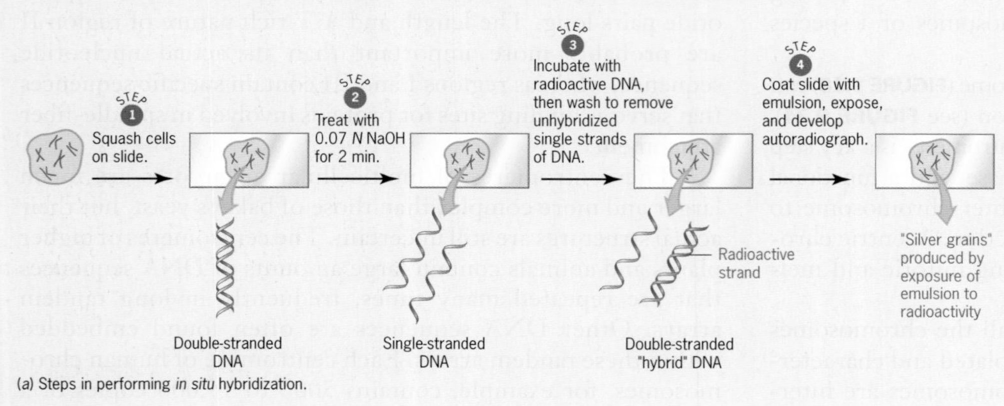
STEP 1
Squash cells on slide.
STEP 2
Treat with 0.07 N NaOH for 2 min.
STEP 3
Incubate with radioactive DNA, then wash to remove unhybridized single strands of DNA.
STEP 4
Coat slide with emulsion, expose, and develop autoradiograph.
Double-stranded DNA
Single-stranded DNA
Radioactive strand
Double-stranded "hybrid" DNA
"Silver grains" produced by exposure of emulsion to radioactivity

(a) Steps in performing *in situ* hybridization.

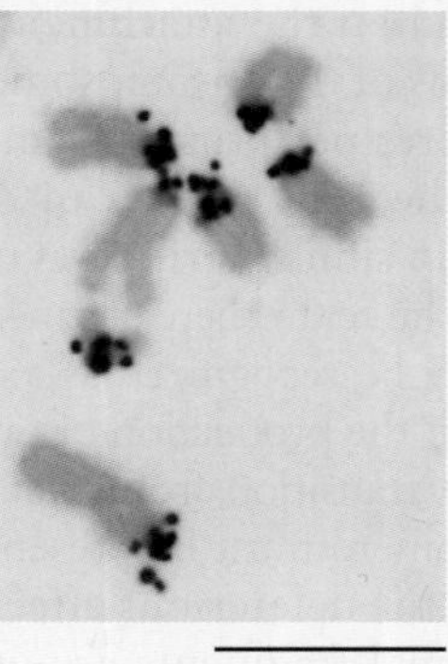
1 μm

(*b*) Autoradiograph showing chromosomal locations of mouse satellite DNA sequences.

have shown that a 450,000 base-pair segment of the centromere of the human X chromosome is sufficient for centromere function. This segment consists mostly of alpha satellite sequences but contains interspersed centromere protein (CENP) binding sites called CENP-B boxes. Both components are essential for centromere function.

It has been known for several decades that the **telomeres** (from the Greek terms *telos* and *meros*, meaning, respectively, "end" and "part"), or ends of eukaryotic chromosomes, have unique properties. Hermann J. Muller, who introduced the term *telomere* in 1938, demonstrated that *Drosophila* chromosomes without natural ends—produced by breaking chromosomes with X rays—were not transmitted to progeny. In a classical study of maize chromosomes, Barbara McClintock (see Focus on Barbara McClintock, the Discoverer of Transposable Elements, in Chapter 18) demonstrated that the new ends of broken chromosomes are sticky and tend to fuse with each other. In contrast, the natural ends of normal (unbroken) chromosomes are stable and show no tendency to fuse with

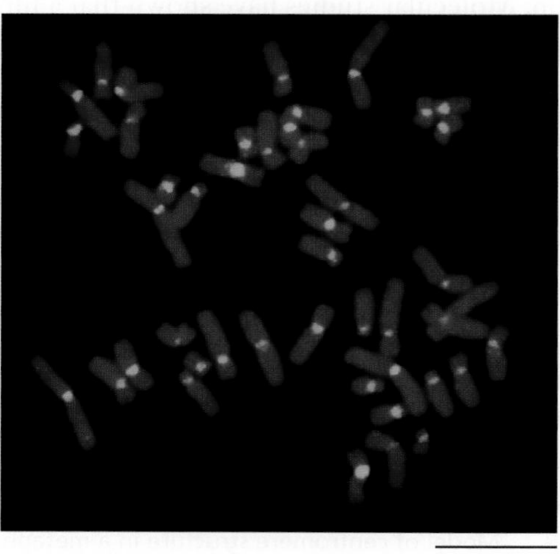
5 μm

Figure 9.30 ▶ The location of alpha satellite DNA sequences (yellow) in the centromeres of human chromosomes (red). See the Focus on *In Situ* Hybridization.

DNA sequences are usually located in centromeric heterochromatin or adjacent to telomeres. A repetitive DNA sequence can be identified as satellite DNA only if the sequence has a base composition sufficiently different from that of mainband DNA to produce a distinct band during density-gradient centrifugation. Therefore, centrifugation cannot be used to identify all repetitive DNA sequences. Satellite DNA sequences usually are not expressed; that is, they do not encode RNA or protein products.

Today, *in situ* hybridization experiments are often done by using hybridization probes that are linked to fluorescent dyes or antibodies tagged with fluorescent compounds (**FIGURE 1*c*** and ***d***). In one protocol, DNA or RNA hybridization probes are linked to the vitamin biotin, which is bound with high affinity by the egg protein avidin (**FIGURE 1*c***). By using avidin covalently linked to a fluorescent dye, the chromosomal location of the hybridized probe can be detected by the fluorescence of the dye. This procedure, called **FISH** (**F**luorescent ***I****n* ***S****itu* **H**ybridization), has been used to demonstrate the presence of the repetitive sequence TTAGGG in the telomeres of human chromosomes (**FIGURE 1*d***). The FISH procedure is very sensitive and can be used to detect the locations of single-copy sequences in human mitotic and interphase chromosomes.

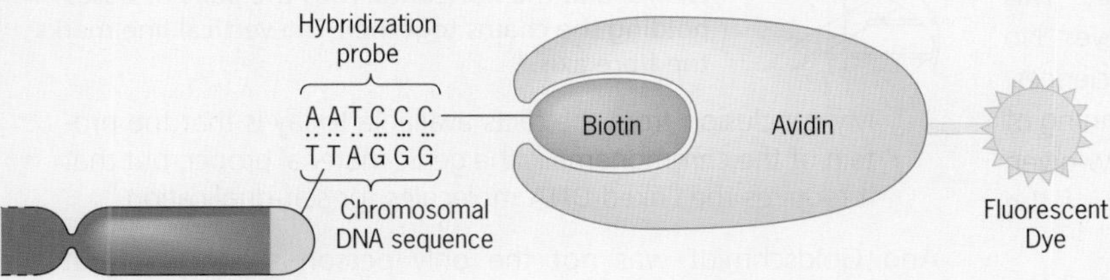

(*c*) Visualization of human telomeres by using fluorescent dyes and *in situ* hybridization.

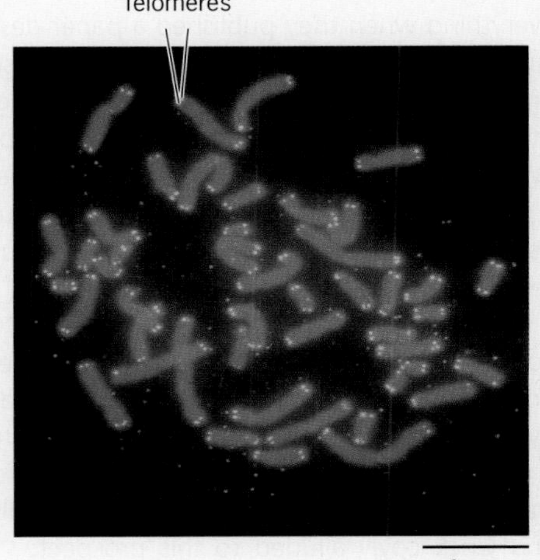

(*d*) Human telomeres visualized using fluorescent probes and *in situ* hybridization.

Figure 1 ▶ Localization of repeated DNA sequences in chromosomes by *in situ* hybridization performed with radioactive probes (*a* and *b*) or fluorescent probes (*c* and *d*). The *in situ* hybridization procedure developed by Pardue and Gall is shown in (*a*), and one of their autoradiographs demonstrating the presence of the mouse satellite DNA sequence in centromeric heterochromatin is shown in (*b*). Use of fluorescent dyes to localize the TTAGGG repeat sequence to the telomeres of human chromosomes is illustrated in (*c*), and a photomicrograph demonstrating its telomeric location is shown in (*d*).

other broken or native ends. McClintock's results indicated that telomeres must have special structures different from the ends produced by breakage of chromosomes.

Another reason for postulating that telomeres have unique structures is that the known mechanisms of replication of linear DNA molecules do not permit duplication of both strands of DNA at the ends of the molecules (Chapter 10). Thus, telomeres must have unique structures that facilitate their replication, or there must be some special replication enzyme that resolves this enigma. Whatever their structure, telomeres must provide at least three important functions. They must (1) prevent deoxyribonucleases from degrading the ends of the linear DNA molecules, (2) prevent fusion of the ends with other DNA molecules, and (3) facilitate replication of the ends of the linear DNA molecules without loss of material.

The telomeres of eukaryotic chromosomes have unique structures that include short nucleotide sequences present as tandem repeats. Although the sequences vary somewhat in different species, the basic repeat unit has the pattern $5'\ T_{1-4}A_{0-1}G_{1-8}\text{-}3'$ in all but a few species. For example, the repeat sequence in humans and other vertebrates is TTAGGG, that of the protozoan *Tetrahymena thermophila* is TTGGGG, and that of the plant *Arabidopsis thaliana* is TTTAGGG. In most species, additional repetitive DNA sequences are present adjacent to telomeres; these are referred to as telomere-associated sequences.

In vertebrates, the TTAGGG repeat is highly conserved; it has been identified in more than 100 species, including mammals, birds, reptiles, amphibians, and fishes. The number of copies of this basic repeat unit in telomeres varies from species to species, from chromosome to chromosome within a species, and even on the same chromosome in different cell types. In normal (noncancerous) human somatic cells, telomeres usually contain 500 to 3000 TTAGGG repeats and gradually shorten with age. In contrast, the telomeres of germ-line cells and cancer cells do not shorten with age (see Telomere Length and Aging in Humans in Chapter 10).

▶ A MILESTONE IN GENETICS: **The Double Helix**

The early geneticists told us where genes were—on chromosomes, that they controlled phenotypes, and that they could mutate to different forms, but they could not tell us what genes were made of. The first insights into the chemical nature of genes came in the middle of the twentieth century. In 1944, Avery, MacLeod, and McCarty showed that DNA is responsible for transformation in *S. pneumoniae,* and in 1952, Hershey and Chase demonstrated that the genetic information of phage T2 is present in DNA. Thus, genes seemed to be made of DNA. However, no one knew the structure of DNA or how DNA could store and transmit genetic information.

On April 25, 1953, James Watson and Francis Crick clarified everything when they published a paper describing the double-helix structure of DNA (**FIGURE 1**) in the British journal *Nature*.[1] This brief—slightly over one page—paper changed biology forever. No other publication has had a comparable impact on the life sciences. Watson and Crick's discovery of the double helix was the beginning of a revolution in molecular biology—the revolution that has now given us the sequences of entire genomes, including two drafts of the sequence of the human genome.

What was so striking about Watson and Crick's double helix? The complementarity of the two strands—adenine in one strand hydrogen-bonded to thymine in the other strand and guanine in one strand hydrogen-bonded to cytosine in the other strand—clearly suggested a mechanism by which the genetic material could be duplicated and transferred from parental cells to progeny cells. In their paper, Watson and Crick coyly alluded to this prospect: "It has not escaped our notice that the specific pairing we have postulated immediately suggests a possible copying mechanism for the genetic material."[2] Five weeks later, they presented their mechanism for the replication of DNA in another short paper in *Nature,*[3] and in 1958 their "template" mechanism of DNA replication was proven correct by Matthew Meselson and Franklin Stahl (see Chapter 10).

Nevertheless, the significance of Watson and Crick's discovery was not universally acknowledged at the time. Some geneticists thought that DNA was chemically too simple to account for all the phenotypic diversity evident in the world. They argued that proteins, which are chemically more complex, were a better fit to the observed phenotypic variability. Indeed, in 1955 (two years after the publication of Watson and Crick's paper), Richard Goldschmidt—Professor Emeritus at the University of California, Berkeley, and one of the leading geneticists of his generation—wrote in support of the idea that genes are made of protein in his book *Theoretical Genetics:*

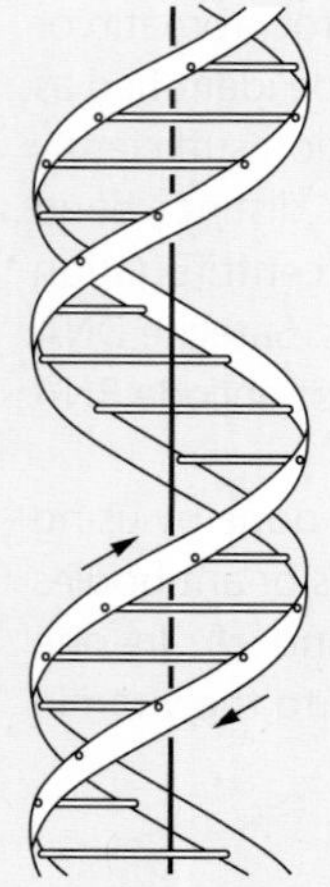

Figure 1 ▶ The diagram of DNA structure in Watson and Crick's Nobel Prize-winning paper. The original legend states: "This figure is purely diagrammatic. The two ribbons symbolize the two phosphate—sugar chains, and the horizontal rods the pairs of bases holding the chains together. The vertical line marks the fibre axis."

> My conclusion from the facts available today is that the protein of the chromonema is the genic material proper, but that it requires the linked DNA molecules for self-duplication.[4]

And Goldschmidt was not the only person skeptical about the significance of the double helix. In a special *Nature* supplement published on the jubilee anniversary of Watson and Crick's landmark paper, Robert Olby, author of *The Path to the Double Helix,*[5] wrote:

> Past discoveries usually become aggrandized in retrospect, especially in jubilee celebrations, and the double helix is no exception. The historical record reveals a muted response by the scientific community to the proposal of this structure in 1953. Indeed, it was only when the outlines appeared of a mechanism for DNA's involvement in protein synthesis that the biochemical community began to take a serious interest in the structure.[6]

However, by 1962, when Watson, Crick, and Maurice Wilkins shared the Nobel Prize in Physiology or Medicine, everyone recognized that the DNA double helix was perfectly designed to store and transmit genetic information.

Why did Wilkins share the Nobel Prize with Watson and Crick? Wilkins, Herbert Wilson, and Alec Stokes of King's College in London provided important X-ray diffraction data that helped Watson and Crick work out the double-helix structure of DNA. Their paper[7] was one of two X-ray studies on DNA published together with Watson

[1]Watson, J. D., and F. H. C. Crick. 1953. Molecular structure of nucleic acids. A structure for deoxyribose nucleic acid. *Nature* 171: 737–738.

[2]Ibid.

[3]Watson, J. D., and F. H. C. Crick. 1953. Genetical implications of the structure of deoxyribonucleic acid. *Nature* 171: 964–967.

[4]Goldschmidt, R. B. 1955. *Theoretical Genetics.* University of California Press, Berkeley, p. 57.

[5]Olby, R. C. 1974. *The Path to the Double Helix, The Discovey of DNA.* Macmillan Press, London.

[6]Olby, R. C. 1974. Quiet debut for the double helix. p. 402. In *The Double Helix—50 years. Nature* 421: 402–405.

[7]Wilkins, M. H. F., A. R. Stokes, and H. R. Wilson. 1953. Molecular structure of deoxypentose nucleic acids. *Nature* 171: 738–740.

Figure 2 ▶ The four major players—Francis Crick, Maurice Wilkins, James Watson, and Rosalind Franklin (clockwise from top left)—in the discovery of the double-helix structure of DNA.

and Crick's paper in the April 25th issue of *Nature*. The other paper[8] was by Rosalind Franklin and Raymond Gosling, also of King's College. Franklin was a rising young star in the field of X-ray crystallography. Her X-ray photographs of DNA crystals were the best available at the time; they provided evidence for the helical structure of DNA and helped to determine the spacing between base pairs.

Watson's autobiographical account of the discovery of the double helix[9] was somewhat dismissive of Franklin; however, some people have suggested that Franklin was close to deciphering the structure of DNA herself. For instance, her student Aaron Klug put it this way:

> In his book Watson wrote that Franklin's "instant acceptance" of the Watson-Crick model amazed him at first. But he went on to say that on further reflexion it was not so surprising to him. It is not in the least surprising when one studies her papers and notebooks and realizes how close she herself had come in the progress of her work—albeit in disconnected fashion at different times—to various features of the structure contained in the correct solution.[10]

Unfortunately, Franklin died of ovarian cancer in 1958 at the age of 37, and one of Alfred Nobel's rules in establishing the Nobel Prizes was that they could not be awarded posthumously. Therefore, Franklin was not eligible to share the award in 1962, even though her

[8]Franklin, R. E., and R. G. Gosling. 1953. Molecular configuration in sodium thymonucleate. *Nature* 171: 740–741.

[9]Watson, J. D. 1968. *The Double Helix: A Personal Account of the Discovery of the Structure of DNA.* Atheneum, New York.

[10]Klug, A. 1968. Rosalind Franklin and the discovery of the structure of DNA. p. 844 in *Nature* 219: 808–810, 843–844.

▶ A MILESTONE IN GENETICS: The Double Helix *(continued ...)*

contributions to Watson and Crick's double helix were equal to those of Wilkins. Nevertheless, the work of all four scientists—Watson, Crick, Wilkins, and Franklin (**FIGURE 2**)—contributed to the elucidation of the structure of DNA.

QUESTIONS FOR DISCUSSION

1. The fact that Rosalind Franklin did not share the Nobel Prize with Watson, Crick, and Wilkins is sometimes called an injustice. Some feminists consider her omission the result of bias against women scientists in a male-dominated discipline. Clearly, there were few women biophysicists in the 1950s, and there is undoubtedly some truth to this argument. Yet, the critical facts are that Franklin died in 1958—far too early in her short but distinguished career and four years before Watson, Crick, and Wilkins received the Nobel Prize, and the rules governing the Nobels exclude posthumous awards. Therefore, Franklin was not eligible to share the Nobel Prize awarded for the double-helix structure of DNA. What do you think the Nobel Selection Committee would have done had she still been alive in 1962?

2. Watson and Crick relied heavily on the unpublished X-ray diffraction patterns of Wilkins, Stokes, and Wilson and of Franklin and Gosling in working out the double-helix structure of DNA. Wilkins had shown Watson the most informative photograph produced by Franklin and Gosling some three months before their papers were published. A month later, Max Perutz showed Watson and Crick his copy of the Medical Research Council's report on ongoing research, including that of Wilkins, Franklin, and coworkers. The discovery of the double helix, therefore, documents the positive effects of sharing information on scientific progress. However, the free exchange of information also can make it difficult to assign appropriate credit to contributors. Thus, modern societies have copyright, trademark, and patent laws that protect an individual's or group's intellectual property. What are the pros and cons of the free exchange of data and ideas? What are the advantages and disadvantages of copyright, trademark, and patent laws? What is the proper balance between the free exchange of information and ideas and the protection of an individual's intellectual property?

The telomeres of a few species are not composed of short tandem repeats of the type described earlier. In *D. melanogaster*, for example, telomeres are composed of two specialized DNA sequences that can move from one location in the genome to other locations. Because of their mobility, such sequences are called *transposable genetic elements* (see Chapter 18).

Most telomeres terminate with a G-rich single-stranded region of the DNA strand with the 3′ end (a so-called 3′ overhang). These overhangs are short (12 to 16 bases) in ciliates such as *Tetrahymena*, but they are quite long (125 to 275 bases) in humans. The guanine-rich repeat sequences of telomeres have the ability to form hydrogen-bonded structures distinct from those produced by Watson-Crick base-pairing in DNA. Oligonucleotides that contain tandem telomere repeat sequences form these special structures in solution, but whether they exist *in vivo* remains unknown.

The telomeres of humans and a few other species have recently been shown to form structures called **t-loops,** in which the single strand at the 3′ terminus invades an upstream telomeric repeat and pairs with the complementary strand, displacing the equivalent strand (**FIGURE 9.31**). The displaced strand is protected from degradation by being coated with a protein called POT-1 (**P**rotection **O**f **T**elomeres-**1**). Note that the t-loop structure nicely protects the free end of the DNA molecule. Two telomere-specific protein complexes—TRF-1 and TRF-2—are associated with these t-loops. Both complexes contain several known proteins that are not identified here for the sake of brevity. One of these proteins is an enzyme that unwinds DNA and can disrupt the special hydrogen-bonded structures formed by G-rich telomere sequences, allowing the 3′ terminus to invade an upstream telomere region and form a t-loop. To date, t-loops have been identified in the telomeres of vertebrates, the ciliate *Oxytricha fallax*, the protozoan *Trypanosoma*

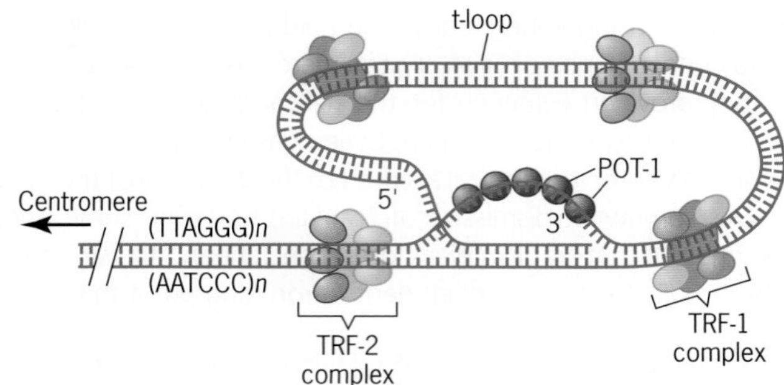

Figure 9.31 ▶ Model of a human telomere stabilized by the formation of a t-loop. The 3′-terminus forms a t-loop by invading an upstream telomere repeat and pairing with the complementary strand. TRF-1 and TRF-2 are telomere repeat-binding factors 1 and 2; both are complexes containing several proteins. Protein POT-1 (**P**rotection **O**f **T**elomeres protein **1**) coats the DNA single strand displaced by the invading 3′-terminus of the telomeric DNA.

brucei, and the plant *Pisum sativum* (peas). Thus, they may be important components of the telomeres of most species.

REPEATED DNA SEQUENCES

The centromeres and telomeres discussed above contain DNA sequences that are repeated many times. Indeed, the chromosomes of eukaryotes contain many DNA sequences that are repeated in the haploid chromosome complement, sometimes as many as a million times. DNA containing such repeated sequences, called **repetitive DNA,** is a major component (15 to 80 percent) of eukaryotic genomes.

The first evidence for repetitive DNA came from centrifugation studies of eukaryotic DNA. When the DNA of a prokaryote, such as *E. coli*, is isolated, fragmented, and centrifuged at high speeds for long periods of time in a 6*M* cesium chloride (CsCl) solution, the DNA will form a single band in the centrifuge tube at the position where its density is equal to the density of the CsCl solution (Chapter 10). For *E. coli*, this band will form at a position where the CsCl density is equal to the density of DNA containing about 50 percent A:T and 50 percent G:C base pairs. DNA density increases with increasing G:C content. The extra hydrogen bond in a G:C base pair results in a tighter association between the bases and thus a higher density than for A:T base pairs. The centrifugation of DNAs from eukaryotes to equilibrium conditions in such CsCl solutions usually reveals the presence of one large main band of DNA and one to several small bands. These small bands of DNA are called **satellite bands** (from the Latin word *satelles*, meaning "an attendant" or "subordinate") and the DNAs in these bands are often referred to as **satellite DNAs.** For example, the genome of *Drosophila virilis*, a distant relative of *Drosophila melanogaster*, contains three distinct satellite DNAs, each composed of a repeating sequence of seven base pairs. Other satellite DNAs in eukaryotes have long repetitive sequences.

Much of what we know about the types of repeated DNA sequences in the chromosomes of various eukaryotic species resulted from DNA renaturation experiments. The two strands of a DNA double helix are held together by a large number of relatively weak hydrogen bonds between complementary bases. When DNA molecules in aqueous solution are heated to near 100°C, these bonds are broken and the complementary strands of DNA separate. This process is called **denaturation.** If the complementary single strands of DNA are cooled slowly under the right conditions, the complementary base sequences will find each other and will re-form base-paired double helices. This re-formation of double helices from the complementary single strands of DNA is called **renaturation.**

If a DNA sequence is repeated many times, denaturation will yield a large number of complementary single strands that will renature rapidly, faster than the rate of renaturation of sequences that are present only once in the genome. Indeed, the rate of DNA renaturation is directly proportional to copy number (the number of copies of the sequence in the genome)—the higher the copy number, the faster the rate and the less time required for renaturation. Mathematical analyses of the rates of renaturation of DNA sequences in eukaryotic genomes provided strong evidence for the presence of different classes of repeated DNA sequences, or repetitive DNA, in eukaryotic chromosomes. The recent genome sequencing projects have provided additional information about the different types of repetitive DNA sequences in eukaryotic genomes. The locations of different DNA sequences in chromosomes can be determined directly by procedures similar to the renaturation experiments described here. With this procedure, called *in situ* hybridization, labeled strands of DNA form double helices with denatured DNA still present in chromosomes (see Focus on *In Situ* Hybridization).

The most highly repeated sequences in eukaryotic genomes do not encode proteins. Indeed, they are not even transcribed. Other less repetitive sequences encode proteins, such as ribosomal proteins and the muscle proteins actin and myosin that are needed in large amounts and are each encoded by several genes. The genes that specify ribosomal RNAs are also multicopy genes because cells need large amounts of ribosomal RNA to produce the ribosomes required for protein synthesis.

The most prevalent of the repeated DNA sequences are **transposable genetic elements,** DNA sequences that can move from one location in a chromosome to another or even to a different chromosome (Chapter 18), or inactive sequences derived from transposable elements. In *D. melanogaster*, about 90 different families of transposable elements have been characterized and given interesting names such as *hobo*, *pogo*, and *gypsy* that suggest their mobility. A much larger proportion—between 40 and 50 percent—of the human genome contains transposable elements or sequences derived from them. As much as 80 percent of the corn genome may consist of transposable genetic elements or their derivatives. These repetitive transposable elements are discussed in more detail in Chapters 16 and 18.

KEY POINTS

- Each eukaryotic chromosome contains one giant molecule of DNA packaged into 11-nm ellipsoidal beads called nucleosomes.
- The condensed chromosomes that are present in mitosis and meiosis and carefully isolated interphase chromosomes are composed of 30-nm chromatin fibers.
- At metaphase, the 30-nm fibers are segregated into domains by scaffolds composed of nonhistone chromosomal proteins.
- The centromeres (spindle-fiber-attachment regions) and telomeres (termini) of chromosomes have unique structures that facilitate their functions.
- Eukaryotic genomes contain repeated DNA sequences, with some sequences present a million times or more.

▶ Basic Exercises

ILLUSTRATE BASIC GENETIC ANALYSIS

1. What differences in the chemical structures of DNA and protein allow scientists to label one or the other of these macromolecules with a radioactive isotope?

Answer: DNA contains phosphorus (the common isotope is ^{31}P) but no sulfur; DNA can be labeled by growing cells on medium containing the radioactive isotope of phosphorus, ^{32}P. Proteins contain sulfur (the common isotope is ^{32}S) but usually little or no phosphorus; proteins can be labeled by growing cells on medium containing the radioactive isotope of sulfur, ^{35}S.

2. If the sequence of one strand of a DNA double helix is ATCG, what is the sequence of the other strand?

Answer: Because the two strands of a double helix are complementary—adenine always paired with thymine and guanine always paired with cytosine—the sequence of the second strand can be deduced from the sequence of the first strand. For ATCG, the double helix will have the following structure:

ATCG
TAGC

3. How should the sequence of the complementary strand in the double helix in exercise 2 be written as a single strand of DNA?

Answer: Remember that the two strands of a DNA double helix have opposite chemical polarity; one strand has $5' \rightarrow 3'$ polarity, and the other has $3' \rightarrow 5'$ polarity, when both are read in the same direction. Because the accepted convention is to write sequences starting with the 5′-terminus on the left and ending with the 3′-terminus on the right, the top strand of the double helix should be written 5′-ATCG-3′ and the complementary strand, 5′-CGAT-3′. The structure of the double helix should be written:

5′-ATCG-3′
3′-TAGC-5′

4. If a mixture of DNA and protein is shown to contain genetic information by some assay such as transformation in bacteria, how can a researcher determine whether that genetic information is present in the DNA or the protein component?

Answer: The biological specificity of enzymes provides a powerful tool for use in many investigations. The enzyme deoxyribonuclease (DNase) degrades DNA to mononucleotides, and proteases degrade proteins to smaller components. If the mixture of DNA and protein is treated with DNase and the genetic information is destroyed, it is stored in DNA. If the mixture is treated with protease and the genetic information is lost, it resides in the protein component of the mixture.

5. How can a researcher determine where a specific DNA sequence is located in a chromosome?

Answer: Under the appropriate conditions, complementary sequences of DNA pair with one another to form double helices. This process is called annealing or hybridization. If a specific DNA sequence is isolated and labeled with a radioactive isotope or a fluorescent dye, it can be used to locate complementary DNA sequences in chromosomes. The procedure is called *in situ* hybridization. The DNA in the chromosome is denatured by exposure to high temperature or pH and then allowed to hybridize to the labeled complementary strand of DNA or RNA (the "probe"). After hybridization has occurred, the labeled strand or probe in the chromosome is located by autoradiography or fluorescent microscopy (see Focus on *In Situ* Hybridization).

▶ Testing Your Knowledge

INTEGRATE DIFFERENT CONCEPTS AND TECHNIQUES

1. The red alga *Polyides rotundus* stores its genetic information in double-stranded DNA. When DNA was extracted from *P. rotundus* cells and analyzed, 32 percent of the bases were found to be guanine residues. From this information, can you determine what percentage of the bases in this DNA were thymine residues? If so, what percentage? If not, why not?

Answer: The two strands of a DNA double helix are complementary to each other, with guanine (G) in one strand always paired with cytosine (C) in the other strand and, similarly, adenine (A) always paired with thymine (T). Therefore, the concentrations of G and C are always equal, as are the concentrations of A and T. If 32 percent of the bases in double-stranded DNA are G residues, then another 32 percent are C residues. Together, G and C comprise 64 percent of the bases in *P. rotundus* DNA; thus, 36 percent of the bases are A's and T's. Since the concentration of A must equal the concentration of T, 18 percent (36% × ½) of the bases must be T residues.

2. The *E. coli* virus ΦX174 stores its genetic information in single-stranded DNA. When DNA was extracted from ΦX174 virus particles and analyzed, 21 percent of the bases were found to be G residues. From this information, can you determine what percentage of the bases in this DNA were thymine residues? If so, what percentage? If not, why not?

Answer: No! The A = T and G = C relationships occur only in double-stranded DNA molecules because of their complementary strands.

Since base-pairing does not occur or occurs only as limited intrastrand pairing in single-stranded nucleic acids, you cannot determine the percentage of any of the other three bases from the G content of the ΦX174 DNA.

3. If each G_1-stage human chromosome contains a single molecule of DNA, how many DNA molecules would be present in the chromosomes of the nucleus of (a) a human egg, (b) a human sperm, (c) a human diploid somatic cell in stage G_1, (d) a human diploid somatic cell in stage G_2, (e) a human primary oocyte?

Answer: A normal human haploid cell contains 23 chromosomes, and a normal human diploid cell contains 46 chromosomes, or 23 pairs of homologues. If prereplication chromosomes contain a single DNA molecule, postreplication chromosomes will contain two DNA molecules, one in each of the two chromatids. Thus, normal human eggs and sperm contain 23 chromosomal DNA molecules; diploid somatic cells contain 46 and 92 chromosomal DNA molecules at stages G_1 and G_2, respectively; and a primary oocyte contains 92 such DNA molecules.

▶ Questions and Problems

ENHANCE UNDERSTANDING AND DEVELOP ANALYTICAL SKILLS

9.1 (a) What was the objective of the experiment carried out by Hershey and Chase? (b) How was the objective accomplished? (c) What is the significance of this experiment?

9.2 (a) What background material did Watson and Crick have available for developing a model of DNA? (b) What was their contribution to building the model?

9.3 (a) How did the transformation experiments of Griffith differ from those of Avery and his associates? (b) What was the significant contribution of each? (c) Why was Griffith's work not evidence for DNA as the genetic material, whereas the experiments of Avery and coworkers provided direct proof that DNA carried the genetic information?

9.4 How did the reconstitution experiment of Fraenkel-Conrat and colleagues show that the genetic information of tobacco mosaic virus (TMV) is stored in its RNA rather than its protein?

9.5 (a) What distinguishes a viroid from an RNA virus? (b) What distinguishes a prion from a viroid? (c) In what ways are viroids and prions similar?

9.6 A cell-free extract is prepared from Type IIIS pneumococcal cells. What effect will treatment of this extract with (a) protease, (b) RNase, and (c) DNase have on its subsequent capacity to transform recipient Type IIR cells to Type IIIS? Why?

9.7 What are the differences between DNA and RNA?

9.8 (a) If a virus particle contained double-stranded DNA with 400,000 base pairs, how many nucleotides would be present? (b) How many complete spirals would occur on each strand? (c) How many atoms of phosphorus would be present? (d) What would be the length of the DNA configuration in the virus?

9.9 RNA was extracted from TMV (tobacco mosaic virus) particles and found to contain 35 percent cytosine (20 percent of the bases were cytosine). With this information, is it possible to predict what percentage of the bases in TMV are adenine? If so, what percentage? If not, why not?

9.10 GO DNA was extracted from cells of *Staphylococcus afermentans* and analyzed for base composition. It was found that 45 percent of the bases are cytosine. With this information, is it possible to predict what percentage of the bases are adenine? If so, what percentage? If not, why not?

9.11 (a) Why did Watson and Crick choose a double helix for their model of DNA structure? (b) Why were hydrogen bonds placed in the model to connect the bases?

9.12 If one strand of DNA in the Watson–Crick double helix has a base sequence of 5′-ACTGCACA-3′, what is the base sequence of the complementary strand?

9.13 Indicate whether each of the following statements about the structure of DNA is true or false. (Each letter is used to refer to the concentration of that base in DNA.)

(a) $A + T = G + C$
(b) $A = G$; $C = T$
(c) $A/T = C/G$
(d) $T/A = C/G$
(e) $A + G = C + T$
(f) $G/C = 1$
(g) $A = T$ within each single strand.
(h) Hydrogen bonding provides stability to the double helix in aqueous cytoplasms.
(i) Hydrophobic bonding provides stability to the double helix in aqueous cytoplasms.

(j) When separated, the two strands of a double helix are identical.
(k) Once the base sequence of one strand of a DNA double helix is known, the base sequence of the second strand can be deduced.
(l) The structure of a DNA double helix is invariant.
(m) Each nucleotide pair contains two phosphate groups, two deoxyribose molecules, and two bases.

9.14 The temperature at which one-half of a double-stranded DNA molecule has been denatured is called the melting temperature, T_m. Why does T_m depend directly on the GC content of the DNA?

9.15 Compare and contrast the structures of the A, B, and Z forms of DNA.

9.16 GO A diploid rye plant, *Secale cereale*, has $2n = 14$ chromosomes and approximately 6×10^8 bp of DNA. How much DNA is in a nucleus of a rye cell at (a) mitotic metaphase, (b) meiotic metaphase I, (c) mitotic telophase, and (d) meiotic telophase II?

9.17 The relationship between the melting T_m and GC content can be expressed, in its much simplified form, by the formula $T_m = 69 + 0.41$ (% GC). (a) Calculate the melting temperature of *E. coli* DNA that has about 40% GC. (b) Estimate the % GC of DNA from a human kidney cell where $T_m = 75°C$.

9.18 The nucleic acids from various viruses were extracted and examined to determine their base composition. Given the following results, what can you hypothesize about the physical nature of the nucleic acids from these viruses?
(a) 40% A, 40% T, 10% G, and 10% C
(b) 35% A, 15% T, 25% G, and 25% C
(c) 35% A, 30% U, 30% G, and 5% C

9.19 The satellite DNAs of *Drosophila virilis* can be isolated, essentially free of main-band DNA, by density-gradient centrifugation. If these satellite DNAs are sheared into approximately 40-nucleotide-pair-long fragments and are analyzed in denaturation–renaturation experiments, how would you expect their hybridization kinetics to compare with the renaturation kinetics observed using similarly sheared main-band DNA under the same conditions? Why?

9.20 GO A diploid nucleus of *Drosophila melanogaster* contains about 3.4×10^8 nucleotide pairs. Assume (1) that all the nuclear DNA is packaged in nucleosomes and (2) that an average internucleosome linker size is 60 nucleotide pairs. How many nucleosomes would be present in a diploid nucleus of *D. melanogaster*? How many molecules of histone H2a, H2b, H3, and H4 would be required?

9.21 Are eukaryotic chromosomes metabolically most active during prophase, metaphase, anaphase, telophase, or interphase?

9.22 Experimental evidence indicates that most highly repetitive DNA sequences in the chromosomes of eukaryotes do not produce any RNA or protein products. What does this indicate about the function of highly repetitive DNA?

9.23 The available evidence indicates that each eukaryotic chromosome (excluding polytene chromosomes) contains a single giant molecule of DNA. What different levels of organization of this DNA molecule are apparent in chromosomes of eukaryotes at various times during the cell cycle?

9.24 Are the scaffolds of eukaryotic chromosomes composed of histone or nonhistone chromosomal proteins? How has this been determined experimentally?

9.25 Of what special interest is the biophysical technique of viscoelastometry to geneticists?

9.26 (a) Which class of chromosomal proteins, histones or nonhistones, is the more highly conserved in different eukaryotic species? Why might this difference be expected? (b) If one compares the histone and nonhistone chromosomal proteins of chromatin isolated from different tissues or cell types of a given eukaryotic organism, which class of proteins will exhibit the greater heterogeneity? Why are both classes of proteins not expected to be equally homogeneous in chromosomes from different tissues or cell types?

9.27 When total DNA from the kangaroo rat (*Dipodomys ordii*) is analyzed by equilibium density-gradient centrifugation, a fraction of the DNA forms a distinct satellite band in the gradient. This satellite DNA contains a highly repeated DNA sequence that can be isolated from the gradient free of other DNA sequences. Many such highly repetitive DNA sequences are located in regions of heterochromatin adjacent to the centromeres of chromosomes ("centromeric heterochromatin"). How could a researcher determine where the satellite DNA sequence is located in the chromosomes of the kangaroo rat?

9.28 (a) If the haploid human genome contains 3×10^9 nucleotide pairs and the average molecular weight of a nucleotide pair is 660, how many copies of the human genome are present, on average, in 1 mg of human DNA? (b) What is the weight of one copy of the human genome? (c) If the haploid genome of the small plant *Arabidopsis thaliana* contains 7.7×10^7 nucleotide pairs, how many copies of the *A. thaliana* genome are present, on average, in 1 mg of *A. thaliana* DNA? (d) What is the weight of one copy of the *A. thaliana* genome? (e) Of what importance are calculations of the above type to geneticists?

9.29 How could it be demonstrated that the mixing of heat-killed Type III pneumococcus with live Type II resulted in a transfer of genetic material from Type III to Type II rather than a restoration of viability to Type III by Type II?

► Genomics on the Web

at http://www.ncbi.nlm.nih.gov/

The available evidence indicates that each eukaryotic chromosome contains one giant DNA double helix running from one end through the centromere all the way to the other end. Of course, those DNA molecules are highly condensed into nucleosomes, 30-nm fibers, and higher-order folding or coiling. Human cells contain 46 chromosomes. How large is the DNA molecule in the largest human chromosome?

1. Which human chromosome contains the largest DNA molecule? How large is it? How many genes does it contain?

2. Which human chromosome contains the smallest DNA molecule? How many base pairs does it contain? How many genes?

3. Which human chromosomes contain genes encoding H1 histones? Other histone genes? How many histone genes are present in the human genome?

Hint: At the NCBI web site, Map Viewer → *Homo sapiens* genome viewer (click on largest and smallest chromosomes shown) → Search (Q. 3).

Chapter 10
Replication of DNA and Chromosomes

CHAPTER OUTLINE

- Basic Features of DNA Replication *In Vivo*
- DNA Polymerases and DNA Synthesis *In Vitro*
- The Complex Replication Apparatus
- Unique Aspects of Eukaryotic Chromosome Replication

Seven pairs of identical twins. Although the twins were given no instructions regarding how to pose for the photograph, note the similar posture, hand placement, and facial expressions of both members of each pair of twins.

Monozygotic Twins: Are They Identical?

From the day of their birth, through childhood, adolescence, and adulthood, Merry and Sherry have been mistaken for one another. When they are apart, Merry is called Sherry about half of the time, and Sherry is misidentified as Merry with equal frequency. Even their parents have trouble distinguishing them. Merry and Sherry are monozygotic ("identical") twins; they both developed from a single fertilized egg. At an early cleavage stage, the embryo split into two cell masses, and both groups of cells developed into complete embryos. Both embryos developed normally, and on April 7, 1955, one newborn was named Merry, the other Sherry.

People often explain nearly identical phenotypes of monozygotic twins like Merry and Sherry by stating that "they contain the same genes." Of course, that is not true. To be accurate, the statement should be that identical twins contain progeny replicas of the same parental genes. But this simple colloquialism suggests that most people do, indeed, believe that the progeny replicas of a gene actually are identical. If the human genome contains 25,000 to 35,000 genes, are the progeny replicas of all these genes exactly the same in identical twins?

A human life emerges from a single cell, a tiny sphere about 0.1 mm in diameter. That cell gives rise to hundreds of billions of cells during fetal development. At maturity, a human of average size contains about 65 trillion (65,000,000,000,000) cells. With some exceptions, each of these 65 trillion cells contains a progeny replica of each of the 25,000 to 35,000 genes. Moreover, the cells of the body are not static; in some tissues, old cells are continuously being replaced by new cells. For example, in a healthy individual, the bone marrow cells produce about 2 million red blood cells per minute. Although not all of the progeny replicas of genes in the human body are identical, the process by which these genes are duplicated is very accurate. The human haploid genome contains about 3×10^9 nucleotide pairs of DNA, all of which must be duplicated during each cell division. In this chapter, we examine how DNA replicates, and we focus on the mechanisms that ensure the fidelity of this process.

Basic Features of DNA Replication *In Vivo*

DNA replication occurs semiconservatively, is initiated at unique origins, and usually proceeds bidirectionally from each origin of replication.

In humans, the synthesis of a new strand of DNA occurs at the rate of about 3000 nucleotides per minute. In bacteria, about 30,000 nucleotides are added to a nascent DNA chain per minute. Clearly, the cellular machinery responsible for DNA replication must work very fast, but, even more importantly, it must work with great precision. Indeed, the fidelity of DNA replication is amazing, with an average of only one mistake per billion nucleotides incorporated after synthesis and the correction of mistakes during and immediately after replication. Thus, the majority of the genes of identical twins are indeed identical, but some will have changed owing to replication errors and other types of mutations (Chapter 13). Most of the key features of the mechanism by which the rapid and accurate replication of DNA occurs are now known, although many molecular details remain to be elucidated.

The synthesis of DNA, like the synthesis of RNA (Chapter 11) and proteins (Chapter 12), involves three steps: (1) chain initiation, (2) chain extension or elongation, and (3) chain termination. In this and the following two chapters, we examine the mechanisms by which cells carry out each of the three steps in the synthesis of these important macromolecules. First, however, we consider some key features of DNA replication.

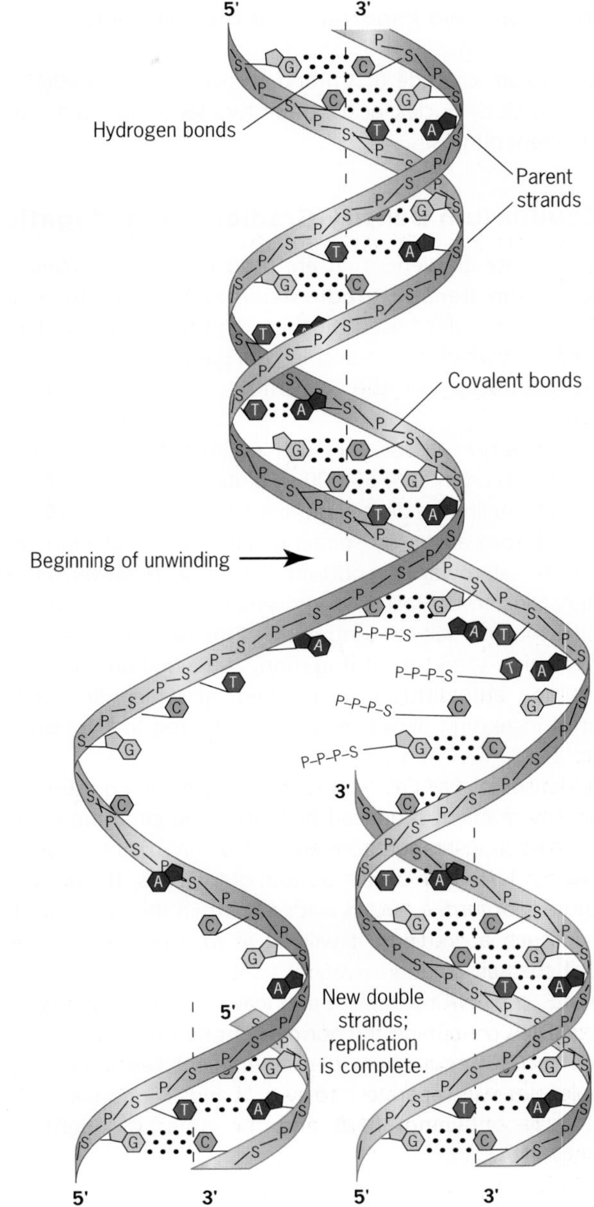

Figure 10.1 ► Semiconservative DNA replication. Watson and Crick first proposed this mechanism of DNA replication based on complementary base-pairing between the two strands of the double helix. Note that each of the parental strands is conserved and serves as a template for the synthesis of a new complementary strand; that is, the base sequence in each progeny strand is determined by the hydrogen-bonding potentials of the bases in the parental strand.

SEMICONSERVATIVE REPLICATION

When Watson and Crick deduced the double-helix structure of DNA with its complementary base-pairing, they immediately recognized that the base-pairing specificity could provide the basis for a simple mechanism for DNA duplication. If the two complementary strands of a double helix separated, by breaking the hydrogen bonds of each base pair, each parental strand could direct the synthesis of a new complementary strand because of the specific base-pairing requirements (**FIGURE 10.1**). That is, each parental strand could function as a template, a single strand of DNA that specifies the nucleotide sequence of a new complementary strand. Adenine, for example, in the parent strand would serve as a template via its hydrogen-bonding potential for the incorporation of thymine in the nascent complementary strand. This mechanism of DNA replication is called **semiconservative replication** because each of the complementary strands of the parental double helix is conserved (or the double helix is "half-conserved") during the process.

Watson and Crick's semiconservative mechanism of DNA replication was first shown to occur in the bacterium *Escherichia coli* by Matthew Meselson and Franklin Stahl in 1958. Their definitive study is discussed in detail in A Milestone in Genetics: DNA Replicates Semiconservatively. Meselson and Stahl used the heavy isotope of nitrogen, ^{15}N, in place of the normal light nitrogen, ^{14}N, to increase the density of DNA and then studied changes in density during replication by density-gradient centrifugation. See the Focus on Centrifugation Techniques for a discussion of the different types of density gradients and their use in isolating DNA molecules and studying their structure

▶ FOCUS ON Centrifugation Techniques

Cesium Chloride Equilibrium Density-Gradient Centrifugation and Sucrose Velocity Density-Gradient Centrifugation: Two Important and Distinct Tools

Although cesium chloride (CsCl) and sucrose density gradients are both used to study nucleic acids, these two techniques are used for totally different purposes.

CsCl Equilibrium Density-Gradient Centrifugation

The CsCl density gradients that are used to analyze nucleic acids are **equilibrium density gradients.** They are used to separate nucleic acid molecules based on their densities, which determine the positions at which the molecules will band in the linear density gradient produced in the centrifuge tube at equilibrium conditions.

When a heavy salt solution such as 6*M* CsCl is centrifuged at very high speeds (30,000 to 50,000 revolutions per minute) for 48 to 72 hours, an equilibrium density gradient is formed (**FIGURE 1**). The centrifugal force caused by spinning the solution at high speeds sediments the salt toward the bottom of the tube. However, diffusion, a countervailing force, results in movement of salt molecules back toward the top (low salt concentration) of the tube. After a sufficient period of high-speed centrifugation, an equilibrium between sedimentation and diffusion is reached, at which time a linear gradient of increasing density exists from the top to the bottom of the tube.

The density of 6*M* CsCl is about 1.7 g/cm^3; at equilibrium, the linear density gradient produced by high-speed centrifugation of a 6*M* CsCl solution will range from about 1.65 g/cm^3 at the top of the tube to about 1.75 g/cm^3 at the bottom of the tube. The densities of most naturally occurring nucleic acids fall within this range. If DNA is present in such a gradient, it will move to a position where the density of the salt solution is equal to its own density. Thus, if a mixture of *E. coli* DNA containing the heavy isotope of nitrogen, ^{15}N, and *E. coli* DNA containing the normal light nitrogen isotope, ^{14}N, is subjected to CsCl equilibrium density-gradient centrifugation, the DNA molecules will separate into two "bands," one consisting of "heavy" (^{15}N-containing) DNA and the other of "light" (^{14}N-containing) DNA.

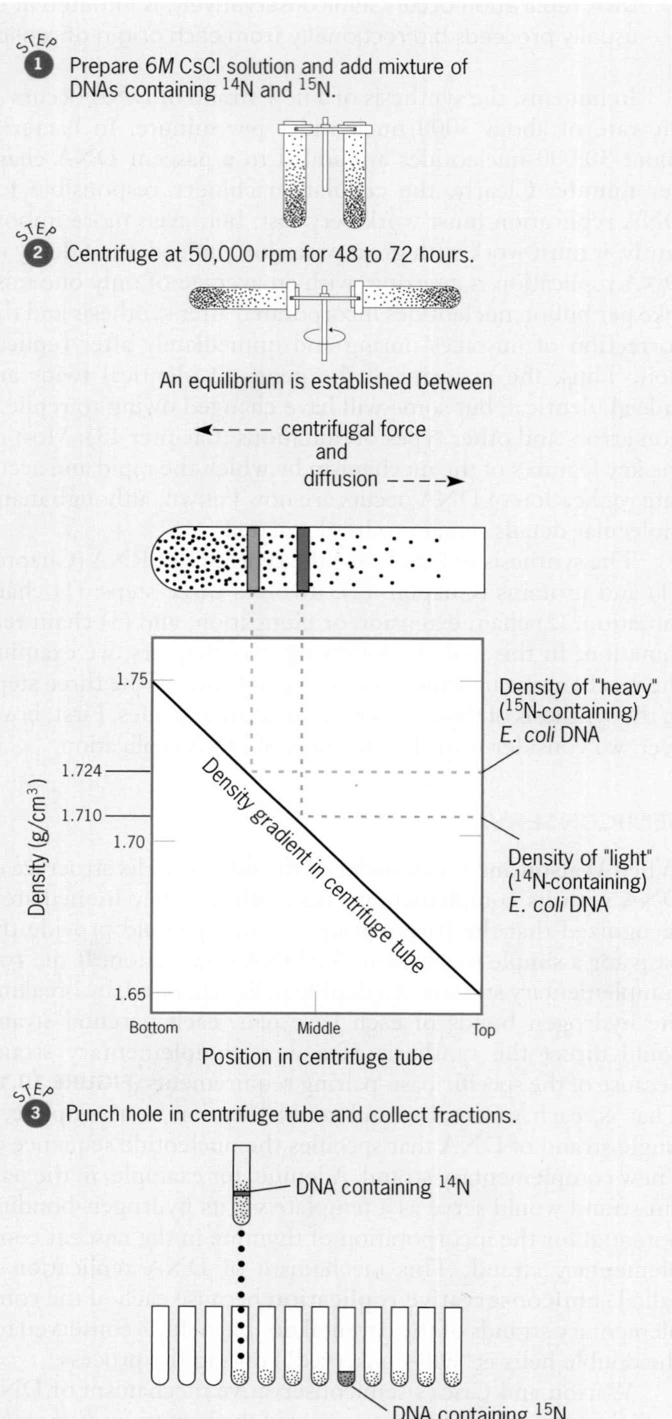

Figure 1 ▶ CsCl Equilibrium Density Gradient Centrifugation.

Sucrose Velocity Density-Gradient Centrifugation

The sucrose density gradients that are used to analyze nucleic acids are **velocity density gradients.** They are used to estimate the sizes of nucleic acid molecules based on their rates or velocities of movement through a preformed sucrose density gradient when exposed to a centrifugal force.

A gradient maker is used to prepare a preformed density gradient in a centrifuge tube (**FIGURE 2**). The gradient maker contains two chambers for sucrose solutions. One chamber is filled with a sucrose solution of the concentration desired at the bottom of the centrifuge tube; the other is filled with a sucrose solution of the concentration desired at the top of the tube. As the tube is filled, the high-concentration sucrose solution is progressively diluted with the low-concentration solution, producing a linear gradient in the tube.

Although any range of sucrose concentrations can be used, 5 to 20 percent and 10 to 40 percent sucrose gradients are widely used to separate nucleic acids. The solution containing the molecules to be analyzed is layered on the top of the sucrose gradient prior to centrifugation. The density of this solution must be lower than the density of the sucrose at the top of the centrifuge tube so that it will stay on the surface of the sucrose gradient. The tube is then placed in a swinging bucket rotor and centrifuged at high speed (20,000 to 30,000 revolutions per minute) for one to a few hours. The centrifugal force causes the macromolecules to migrate through the gradient, with larger molecules moving faster than smaller molecules with the same shape. If centrifugation is continued too long, all the macromolecules will accumulate at the bottom of the tube. If the dye ethidium bromide, which binds to DNA and fluoresces when exposed to ultraviolet (UV) light, is present, the bands of DNA in the gradient can be seen under UV illumination. After centrifugation, a small hole is punched through the bottom of the centrifuge tube, and drops of solution are collected in test tubes for subsequent analysis. The collected fractions represent successive layers of the sucrose gradient and any molecules therein at the time that centrifugation was terminated.

The velocity at which a macromolecule will move through such a gradient is determined primarily by its molecular weight and shape. A molecule's rate of movement through a solution of lower density during sedimentation is measured by its sedimentation coefficient (s), which equals velocity/centrifugal force. Most macromolecules have s values between 10^{-13} and 10^{-11} seconds. Thus, a sedimentation coefficient of 10^{-13} seconds has been designated one Svedberg (S) unit. It was named in honor of The Svedberg, the scientist who invented the ultracentrifuge. The Svedberg unit is the commonly used unit of sedimentation velocity in dealing with macromolecules or macromolecular complexes. For example, nucleic acids are sometimes referred to as 5S or 10S molecules, and ribosomal subunits of prokaryotes are frequently called 30S and 50S subunits.

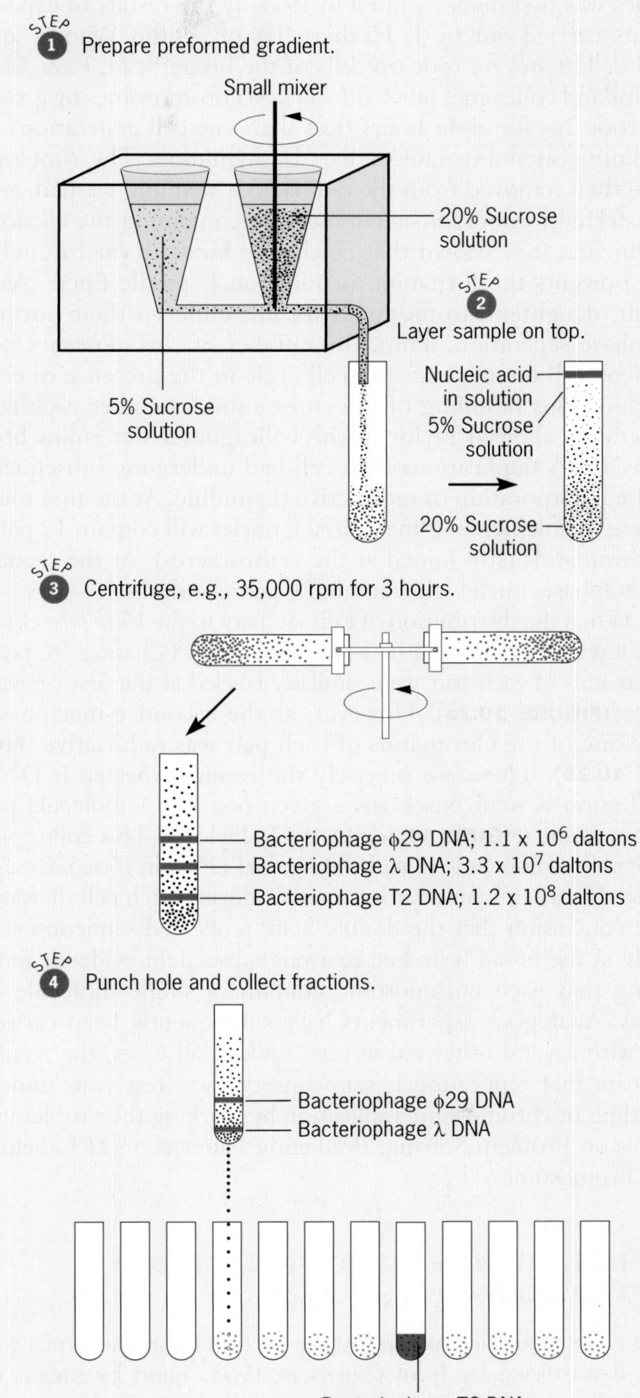

Figure 2 Sucrose Velocity Density Gradient Centrifugation.

and replication. DNA replication was subsequently shown to occur semiconservatively in several other microorganisms.

The semiconservative replication of eukaryotic chromosomes was first demonstrated in 1957 by the results of experiments carried out by J. Herbert Taylor, Philip Woods, and Walter Hughes on root-tip cells of the broad bean, *Vicia faba*. Taylor and colleagues labeled *Vicia faba* chromosomes by growing root tips for eight hours (less than one cell generation) in medium containing radioactive ^{3}H-thymidine. The root tips were then removed from the radioactive medium, washed, and transferred to nonradioactive medium containing the alkaloid colchicine. It is known that colchicine binds to microtubules and prevents the formation of functional spindle fibers. As a result, daughter chromosomes do not undergo their normal anaphase separation. Thus, the number of chromosomes per nucleus will double once per cell cycle in the presence of colchicine. This doubling of the chromosome number each cell generation allowed Taylor and his colleagues to determine how many DNA duplications each cell had undergone subsequent to the incorporation of radioactive thymidine. At the first metaphase in colchicine (c-metaphase), nuclei will contain 12 pairs of chromatids (still joined at the centromeres). At the second c-metaphase, nuclei will contain 24 pairs, and so on.

When the distribution of radioactivity in the *Vicia faba* chromosomes was examined by autoradiography (Chapter 9), both chromatids of each pair were similarly labeled at the first c-metaphase (**FIGURE 10.2*a***). However, at the second c-metaphase, only one of the chromatids of each pair was radioactive (**FIGURE 10.2*b***). These are precisely the results expected if DNA replication is semiconservative, given one DNA molecule per chromosome (**FIGURE 10.2*c***). In 1957, Taylor and his colleagues were able to conclude that chromosomal DNA in *Vicia faba* segregated in a semiconservative manner during each cell division. The conclusion that the double helix replicated semiconservatively in the broad bean had to await subsequent evidence indicating that each chromosome contains a single molecule of DNA. Analogous experiments have subsequently been carried out with several other eukaryotes, and, in all cases, the results indicate that replication is semiconservative. Test your understanding of chromosome replication by working the problem in Focus on Problem Solving: Predicting Patterns of ^{3}H Labeling in Chromosomes.

Autoradiographs of *Vicia faba* chromosomes

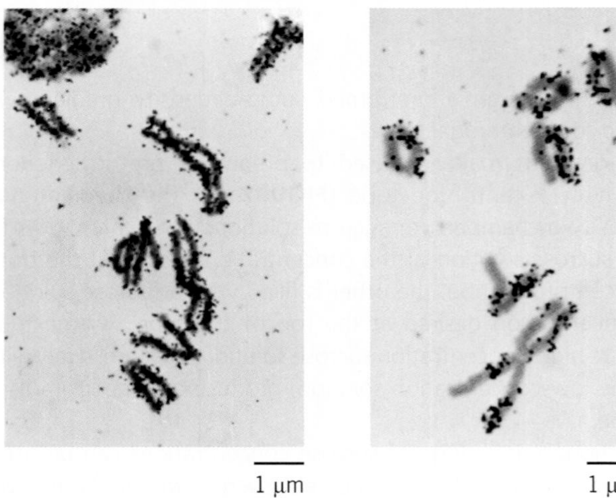

(*a*) First metaphase after replication in ^{3}H-thymidine.

(*b*) Second metaphase after an additional replication in ^{1}H-thymidine.

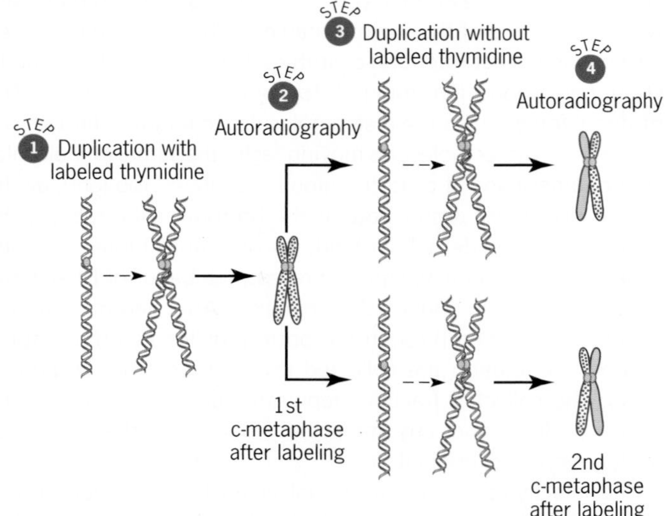

(*c*) Interpretation of the autoradiographs above in terms of semiconservative replication.

Figure 10.2 ▶ Proof of semiconservative replication of DNA in the broad bean, *Vicia faba*. The results obtained by Taylor, Woods, and Hughes (*a*, *b*) are predicted by the semiconservative replication of the DNA (*c*).

VISUALIZATION OF REPLICATION FORKS BY AUTORADIOGRAPHY

The gross structure of replicating bacterial chromosomes was first determined by John Cairns in 1963, again by means of autoradiography. Cairns grew *E. coli* cells in medium containing ^{3}H-thymidine for varying periods of time, lysed the cells gently so as not to break the chromosomes (long DNA molecules are sensitive to shearing), and carefully collected the chromosomes on membrane filters. These filters were affixed to glass slides, coated with emulsion sensitive to β-particles (the low-energy electrons emitted during decay of tritium), and stored in the dark for a period of time to allow sufficient radioactive decay. When the films were developed, the autoradiographs (**FIGURE 10.3*a***) showed that the chromosomes of *E. coli* are circular structures that exist as θ-shaped intermediates during replication. The autoradiographs further indicated that the unwinding of the two complementary parental strands (which is necessary for their separation) and their semiconservative replication occur simultaneously or are closely coupled. Since the parental double helix must rotate 360° to unwind each gyre of the helix, some kind of "swivel" must exist. Geneticists now

FOCUS ON PROBLEM SOLVING

Predicting Patterns of ^{3}H Labeling in Chromosomes

THE PROBLEM

Haplopappus gracilis is a diploid plant with two pairs of chromosomes ($2n = 4$). A G_1-stage cell of this plant, not previously exposed to radioactivity, was placed in culture medium containing ^{3}H-thymidine. After one generation of growth in this medium, the two progeny cells were washed with nonradioactive medium and transferred to medium containing ^{1}H-thymidine and colchicine. They were allowed to grow in this medium for one additional cell generation and on to metaphase of a second cell division. The chromosomes from each cell were then spread on a microscope slide, stained, photographed, and exposed to an emulsion sensitive to low-energy radiation. One of the daughter cells exhibited a metaphase plate with eight chromosomes, each with two daughter chromatids. Draw this metaphase plate showing the predicted distribution of radioactivity on the autoradiograph. Assume no crossing over!

FACTS AND CONCEPTS

1. Each G_1-stage (prereplicative) chromosome contains a single DNA double helix.
2. DNA replication is semiconservative.
3. Daughter chromatids remain attached to a single centromere at metaphase of mitosis.
4. The centromere duplicates prior to anaphase; at that time, each chromatid becomes a daughter chromosome.
5. Colchicine binds to the proteins that form the spindle fibers responsible for the separation of daughter chromosomes during anaphase and prevents the formation of functional spindles. As a result, chromosome number doubles during each cell generation in the presence of colchicine.

ANALYSIS AND SOLUTION

All four chromosomes will go through the same replication events. Therefore, we only need to follow one chromosome. The first replication in the presence of ^{3}H-thymidine, but no colchicine, is shown in the following diagram with radioactive strands in red.

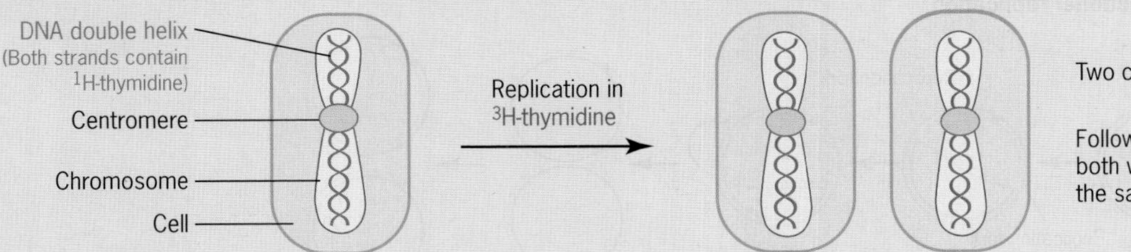

The second and third replications (in ^{1}H-thymidine and colchcine) are shown in the following diagram.

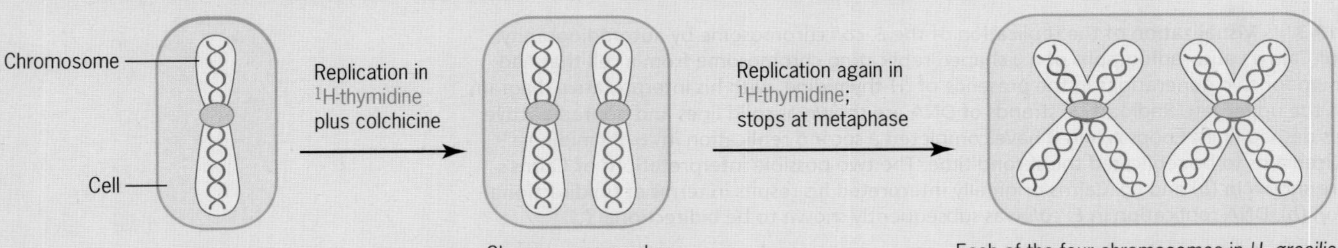

When the resulting metaphase chromosomes are subjected to autoradiography, the distribution of radioactivity (indicated by red dots) on the eight chromosomes will be as follows.

For further discussion go to your *WileyPLUS* course.

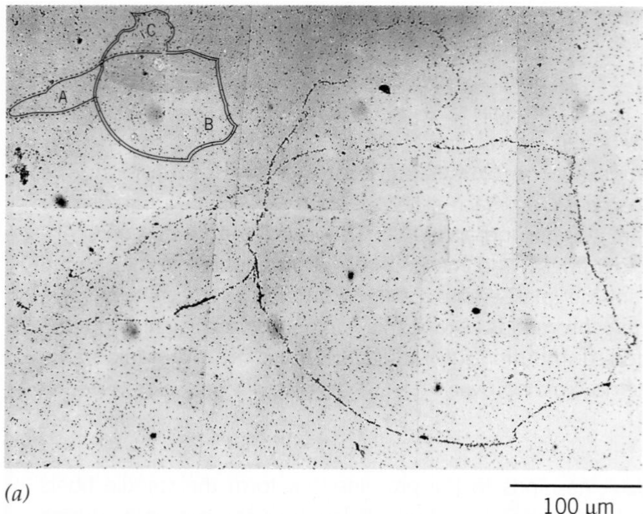

(a)

Original interpretation: unidirectional replication.

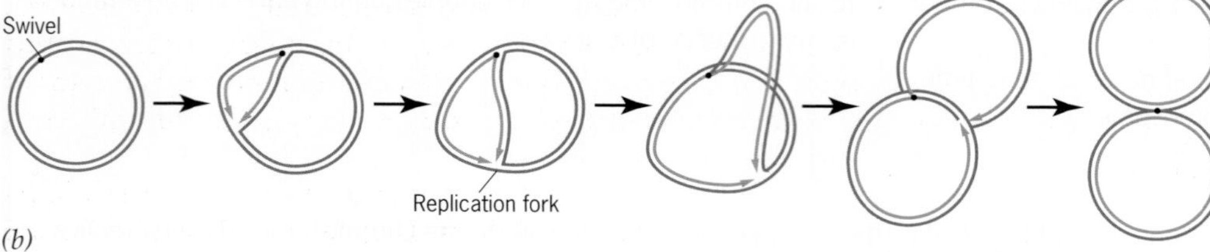

(b)

Correct interpretation: bidirectional replication.

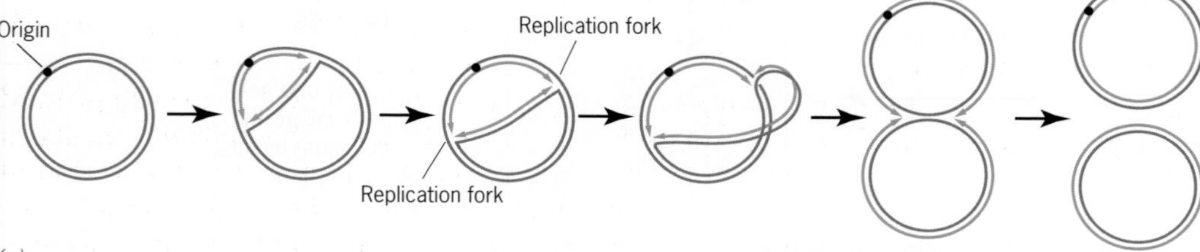

(c)

Figure 10.3 ▶ Visualization of the replication of the *E. coli* chromosome by autoradiography. (*a*) One of Cairns's autoradiographs of a θ-shaped replicating chromosome from a cell that had been grown for two generations in the presence of ^{3}H-thymidine, with his interpretative diagram shown at the upper left. Radioactive strands of DNA are shown as solid lines and nonradioactive strands as dashed lines. Loops A and B have completed a second replication in ^{3}H-thymidine; section C remains to be replicated the second time. The two possible interpretations of Cairns's results are shown in (*b*) and (*c*). Cairns originally interpreted his results in terms of unidirectional replication (*b*). DNA replication in *E. coli* was subsequently shown to be bidirectional (*c*).

know that the required swivel is a transient single-strand break (cleavage of one phosphodiester bond in one strand of the double helix) produced by the action of enzymes called topoisomerases.

Cairns's interpretation of the autoradiographs was that semiconservative replication of the *E. coli* chromosome started at a specific site, which he called the *origin*, and proceeded sequentially and unidirectionally around the circular structure (**FIGURE 10.3*b***). Subsequent evidence has shown his original interpretation to be incorrect on one point: replication of the *E. coli* chromosome actually proceeds bidirectionally, not unidirectionally. Each Y-shaped structure is a **replication fork,** and the two replication forks move in opposite directions sequentially around the circular chromosome (**FIGURE 10.3*c***).

The bidirectional replication of the circular *E. coli* chromosome just discussed occurs during cell division. It should not be confused with rolling-circle replication, which mediates the transfer of chromosomes from Hfr cells to F$^-$ cells (Chapter 8). Some viral chromosomes replicate by the rolling-circle mechanism; see the section ROLLING-CIRCLE REPLICATION later in this chapter.

UNIQUE ORIGINS OF REPLICATION

Cairns's results established the existence of a site of initiation or **origin** of replication on the circular chromosome of *E. coli* but provided no hint as to whether the origin was a unique site or occurred at randomly located sites in a population of replicating chromosomes. In bacterial and viral chromosomes, there is usually one unique origin per chromosome, and this single origin controls the replication of the entire chromosome. In the large chromosomes of eukaryotes, multiple origins collectively control the replication of the giant DNA molecule present in each chromosome. Current evidence indicates that these multiple replication origins in eukaryotic chromosomes also occur at specific sites. Each origin controls the replication of a unit of DNA called a *replicon;* thus, most prokaryotic chromosomes contain a single replicon, whereas eukaryotic chromosomes usually contain many replicons.

The single origin of replication, called *oriC,* in the *E. coli* chromosome has been characterized in considerable detail. *oriC* is 245 nucleotide pairs long and contains two different conserved repeat sequences (**FIGURE 10.4**). One 13-bp sequence is present as three tandem repeats. These three repeats are rich in A:T base pairs, facilitating the formation of a localized region of strand separation referred to as the *replication bubble.* Recall that A:T base pairs are held together by only two hydrogen bonds as opposed to three in G:C base pairs (Chapter 9). Thus, the two strands of AT-rich regions of DNA come apart more easily, that is, with the input of less energy. The formation of a localized zone of denaturation is an essential first step in the replication of all double-stranded DNAs. Another conserved component of *oriC* is a 9-bp sequence that is repeated four times and is interspersed with other sequences. These four sequences are binding sites for a protein that plays a key role in

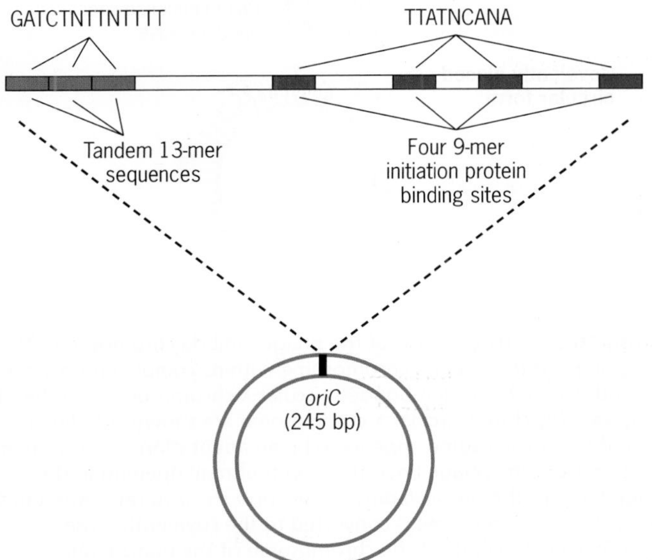

Figure 10.4 ▶ Structure of *oriC*, the single origin of replication in the *E. coli* chromosome.

the formation of the replication bubble. Later in this chapter we discuss additional details of the process of initiation of DNA synthesis at origins and the proteins that are involved.

The multiple origins of replication in eukaryotic chromosomes also appear to be specific DNA sequences. In the yeast *Saccharomyces cerevisiae*, segments of chromosomal DNA that allow a fragment of circularized DNA to replicate as an independent unit (autonomously), that is, as an extrachromosomal self-replicating unit, have been identified and characterized. These sequences are called *ARS* (for **A**utonomously **R**eplicating **S**equences) *elements.* Their frequency in the yeast genome corresponds well with the number of origins of replication, and some have been shown experimentally to function as origins. ARS elements are about 50 base pairs in length and include a core 11-bp AT-rich sequence,

ATTTATPuTTTA
TAAATAPyAAAT

(where Pu is either of the two purines and Py is either of the two pyrimidines) and additional imperfect copies of this sequence. The ability of ARS elements to function as origins of replication is abolished by base-pair changes within this conserved core sequence.

Attempts to characterize origins of replication in multicellular eukaryotes have been largely unsuccessful. Despite evidence that replication is initiated at specific sequences *in vivo* and the availability of the sequences of entire genomes, the components of a functional origin have remained elusive. There appear to be two major reasons for this failure to identify replication origins. First, the functional assays used in yeast—the ability of the origin to support the replication of a plasmid or artificial chromosome—do not yield reliable results in other eukaryotes. Sequences that support the replication of plasmids in mammalian cells, for example, often result in the initiation of replication at random or multiple sites. Second, considerable evidence now suggests that the initiation of replication involves relatively long DNA sequences—up to several thousand base pairs—in metazoans, making origins difficult to characterize.

One successful approach to characterizing origins of replication that function in higher eukaryotes has been to study the replication of DNA viruses. The replication of simian virus 40 (SV40) has been studied extensively. SV40 reproduces in primate cells growing in culture. The SV40 genome is a circular DNA double helix containing just 5243 nucleotide pairs, making it ideal for studies of DNA replication in eukaryotes. Indeed, most of the details of SV40 replication are known. A 64-nucleotide-pair sequence, the core origin (**FIGURE 10.5**), is all that is required to initiate DNA replication in infected cells. The core origin contains a 27-nucleotide-pair central element flanked by an AT-rich sequence and another sequence. Replication is initiated when a viral protein called the large T antigen binds to the core origin and, with the help of cellular proteins, initiates unwinding of DNA strands at the origin. Interestingly, some large DNA viruses such as Epstein-Barr contain three origins

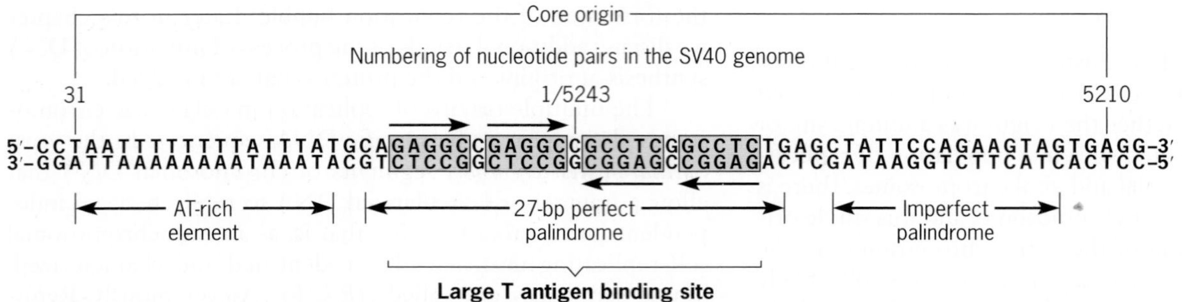

Figure 10.5 ▸ The core origin of replication in simian virus 40 (SV40). The core origin is the minimal sequence that can initiate replication *in vivo*. Palindromes are DNA sequences with the same nucleotide sequence when read in opposite directions from a central point of symmetry. They are often the binding sites for proteins that interact with DNA, such as the large T antigen of SV40, which is the first protein to bind to the core origin during the initiation of SV40 replication.

of replication. One origin is used for replication during latent infections, and the other two origins control replication in productive, or lytic, infections.

BIDIRECTIONAL REPLICATION

Cairns interpreted his autoradiograms of replicating *E. coli* chromosomes based on the prevailing view that replication started at a single origin and that a single replication fork moved unidirectionally around the circular chromosome (**FIGURE 10.3*b***). However, his results can be explained equally well by the formation of two forks at the single origin, with the two forks moving in opposite directions, or bidirectionally, around the circular chromosome (**FIGURE 10.3*c***). We now have definitive evidence showing that replication in *E. coli* and many other prokaryotes proceeds bidirectionally from a unique origin.

Bidirectional replication was first convincingly demonstrated by experiments with some of the small bacterial viruses that infect *E. coli*. Bacteriophage lambda (phage λ) contains a single linear molecule of DNA only 17.5 μm long. The phage λ chromosome is somewhat unusual in that it has a single-stranded region, 12 nucleotides long, at the 5′ end of each complementary strand (**FIGURE 10.6**). These single-stranded ends, called "cohesive" or "sticky" ends, are complementary to each other. The cohesive ends of a lambda chromosome can thus base-pair to form a hydrogen-bonded circular structure. One of the first events to occur after a lambda chromosome is injected into a host cell is its conversion to a covalently closed circular molecule (**FIGURE 10.6**). This conversion from the hydrogen-bonded circular form to the covalently closed circular form is catalyzed by *DNA ligase*, an important enzyme that seals single-strand breaks in DNA double helices. DNA ligase is required in all organisms for DNA replication, DNA repair, and recombination between DNA molecules. Like the *E. coli* chromosome, the lambda chromosome replicates in its circular form via θ-shaped intermediates.

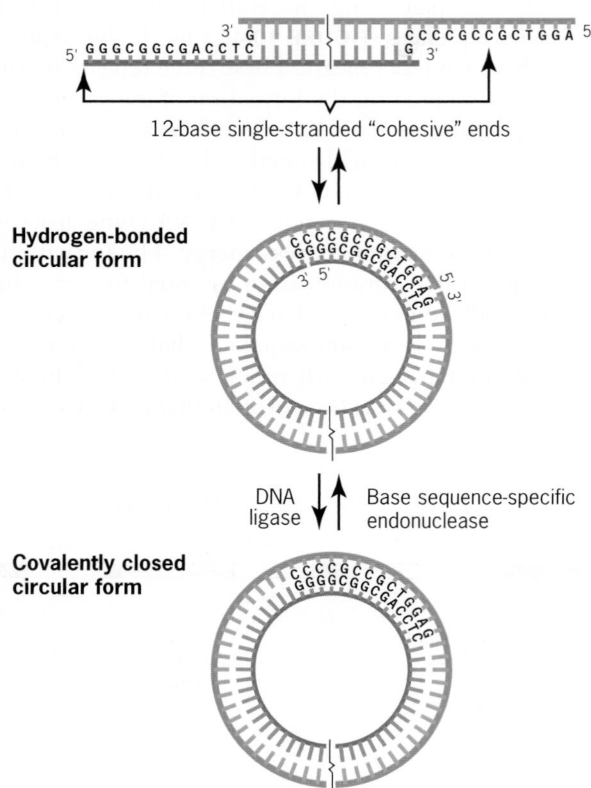

Figure 10.6 ▸ Three forms of the phage lambda chromosome. The conversions of the linear λ chromosome with its complementary cohesive ends to the hydrogen-bonded circular λ chromosome and then to the covalently closed circular λ chromosome are shown. The linear form of the chromosome appears to be an adaptation to facilitate its injection from the phage head through the small opening in the phage tail into the host cell during infection. Prior to replicating in the host cell, the chromosome is converted to the covalently closed circular form. Only the ends of the chromosome of the mature phage are shown; the jagged vertical line indicates that the central portion of the chromosome is not shown. The entire lambda chromosome is 48,502 nucleotide pairs long.

The feature of the lambda chromosome that facilitated the demonstration of bidirectional replication is its differentiation into regions containing high concentrations of adenine and thymine (AT-rich regions) and regions with large amounts of guanine and cytosine (GC-rich regions). In particular, it contains a few segments with high AT content (AT-rich clusters). In the late 1960s, Maria Schnös and Ross Inman used these AT-rich clusters as physical markers to demonstrate, by means of a technique called denaturation mapping, that replication of the lambda chromosome is initiated at a unique origin and proceeds bidirectionally rather than unidirectionally.

When DNA molecules are exposed to high temperature (100°C) or high pH (11.4), the hydrogen and hydrophobic bonds that hold the complementary strands together in the double-helix configuration are broken, and the two strands separate—a process called denaturation. Because AT base pairs are held together by only two hydrogen bonds, compared with three hydrogen bonds in GC base pairs, AT-rich molecules denature more easily (at lower pH or temperature) than GC-rich molecules. When lambda chromosomes are exposed to pH 11.05 for 10 minutes under the appropriate conditions, the AT-rich clusters denature to form denaturation bubbles, which are detectable by electron microscopy, whereas the GC-rich regions remain in the duplex state (**FIGURE 10.7**). These denaturation bubbles can be used as physical markers whether the lambda chromosome is in its mature linear form, its circular form, or its θ-shaped replicative intermediates. By examining the positions of the branch points (Y-shaped structures) relative to the positions of the denaturation bubbles in a large number of θ-shaped replicative intermediates, Schnös and Inman demonstrated that both branch points are replication forks that move in opposite directions around the circular chromosome. **FIGURE 10.8** shows the results expected in Schnös and Inman's experiment if replication is (*a*) unidirectional or (*b*) bidirectional. The results clearly demonstrated that replication of the lambda chromosome is bidirectional.

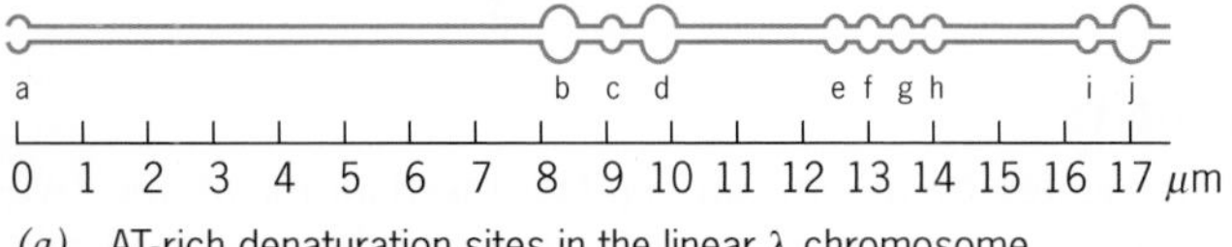

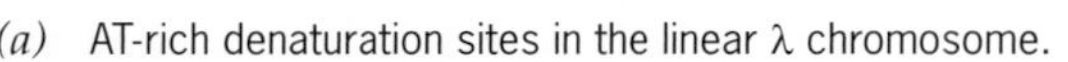

(*a*) AT-rich denaturation sites in the linear λ chromosome.

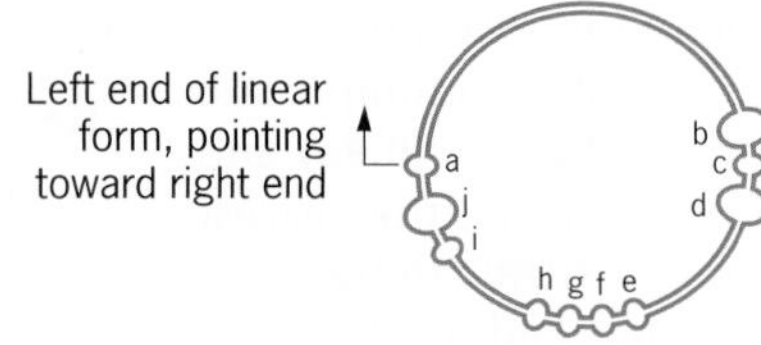

(*b*) AT-rich denaturation sites in the circular form of the λ chromosome.

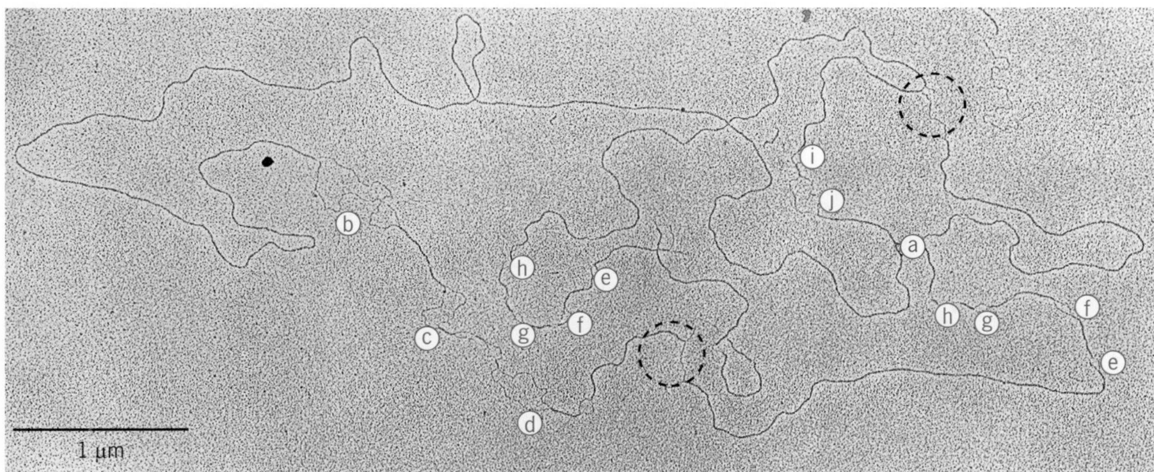

(*c*) AT-rich denaturation bubbles in a θ-shaped replicating λ chromosome.

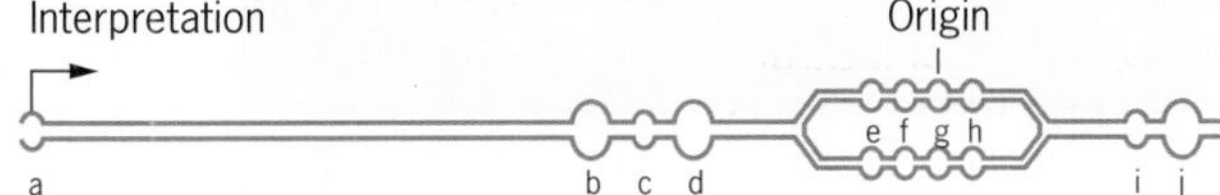

(*d*) Diagram in linear form of the λ replicative intermediate shown in (*c*).

Figure 10.7 ▶ The use of AT-rich denaturation sites as physical markers to prove that the phage λ chromosome replicates bidirectionally rather than unidirectionally. The positions of the AT-rich denaturation bubbles are shown for the linear (*a*) and circular (*b*) forms of the λ chromosome. The electron micrograph (*c*) shows the positions of denaturation bubbles (labeled *a-j*) and replication forks (circled) in a partially replicated λ chromosome. The structure of the partially replicated chromosome in (*c*) is diagrammed in (*d*).

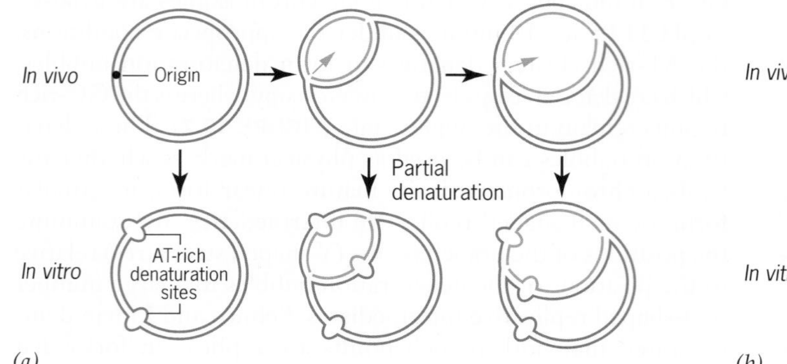

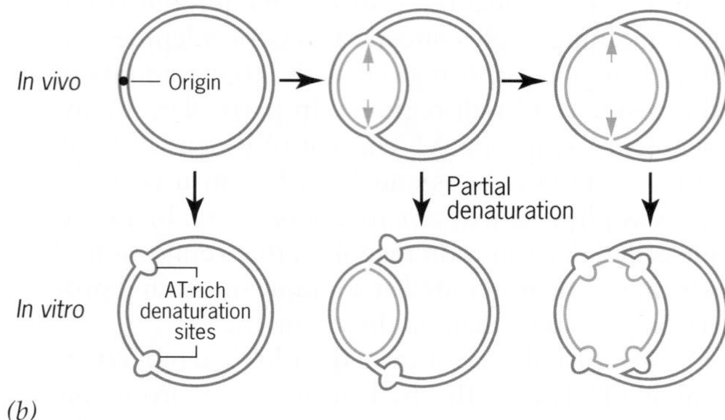

Figure 10.8 ▶ Rationale of the denaturation mapping procedure used by Schnös and Inman to distinguish between (*a*) unidirectional and (*b*) bidirectional modes of chromosome replication.

Bidirectional replication from a fixed origin has also been demonstrated for several organisms with chromosomes that replicate as linear structures. Replication of the chromosome of phage T7, another small bacteriophage, begins at a unique site near one end to form an "eye" structure (**FIGURE 10.9*a***) and then proceeds bidirectionally until one fork reaches the nearest end. Replication of the Y-shaped structure (**FIGURE 10.9*b***) continues until the second fork reaches the other end of the molecule, producing two progeny chromosomes.

Replication of chromosomal DNA in eukaryotes is also bidirectional in those cases where it has been investigated. However, bidirectional replication is not universal. The chromosome of coliphage P2, which replicates as a θ-shaped structure like the lambda chromosome, replicates unidirectionally from a unique origin.

KEY POINTS

- DNA replicates by a semiconservative mechanism: as the two complementary strands of a parental double helix unwind and separate, each serves as a template for the synthesis of a new complementary strand.
- The hydrogen-bonding potentials of the bases in the template strands specify complementary base sequences in the nascent DNA strands.
- Replication is initiated at unique origins and usually proceeds bidirectionally from each origin.

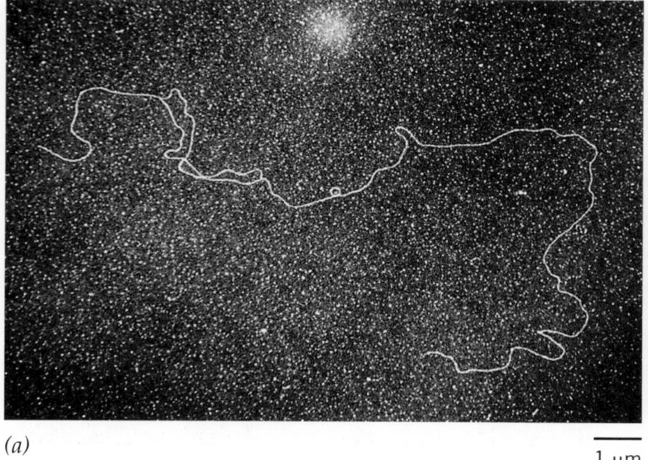

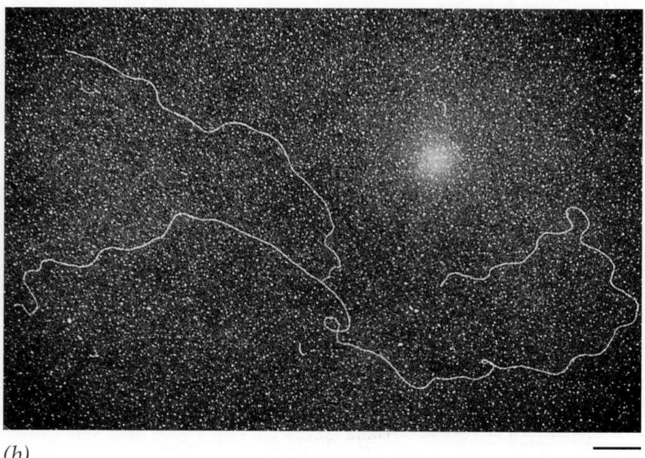

Figure 10.9 ▶ Electron micrographs of replicating bacteriophage T7 chromosomes. The phage T7 chromosome, unlike the *E. coli* and phage λ chromosomes, replicates as a linear structure. Its origin of replication is located 17 percent of the length of the chromosome from one end (the left end of the chromosomes shown). The chromosome in (*a*) illustrates the "eye" form (-○-) characteristic of early stages in the replication of linear DNA molecules. Replication proceeds bidirectionally from the origin until the fork moving in a leftward direction reaches the left end of the molecule, yielding a Y-shaped structure such as that shown in (*b*).

DNA Polymerases and DNA Synthesis *In Vitro*

Much of what we know about DNA synthesis was deduced from *in vitro* studies.

Much has been learned about the molecular mechanisms involved in biological processes by disrupting cells, separating the various organelles, macromolecules, and other components, and then reconstituting systems in the test tube, so-called *in vitro* systems that are capable of carrying out particular metabolic events. Such *in vitro* systems can be dissected biochemically much more easily than *in vivo* systems. Clearly, the information obtained from studies on *in vitro* systems has been invaluable. However, we should never assume that a phenomenon demonstrated *in vitro* occurs *in vivo*. Such an extrapolation should be made only when independent evidence from *in vivo* studies validates the *in vitro* studies.

DISCOVERY OF DNA POLYMERASE I IN *ESCHERICHIA COLI*

The *in vitro* synthesis of DNA was first accomplished by Arthur Kornberg and his coworkers in 1957. Kornberg, who received a Nobel Prize in 1959 for this work, isolated an enzyme from *E. coli* that catalyzes the covalent addition of nucleotides to preexisting DNA chains. Initially called DNA polymerase or "Kornberg's enzyme," it is now known as **DNA polymerase I.** The enzyme requires the 5′-triphosphates of each of the four deoxyribonucleosides—deoxyadenosine triphosphate (dATP), deoxythymidine triphosphate (dTTP), deoxyguanosine triphosphate (dGTP), and deoxycytidine triphosphate (dCTP)—and is active only in the presence of Mg^{2+} ions and preexisting DNA. This DNA must provide two essential components, one serving a primer function and the other a template function (**FIGURE 10.10**).

1. The *primer DNA* provides a terminus with a free 3′-OH to which nucleotides are added during DNA synthesis. DNA polymerase I cannot initiate the synthesis of DNA chains *de novo*. It has an absolute requirement for a free 3′-hydroxyl on a preexisting DNA chain. DNA polymerase I catalyzes the formation of a phosphodiester bridge between the 3′-OH at the end of the primer DNA chain and the 5′-phosphate of the incoming deoxyribonucleotide.

2. The *template DNA* provides the nucleotide sequence that specifies the complementary sequence of the nascent DNA chain. DNA polymerase I requires a DNA template whose base sequence dictates, by its base-pairing potential, the synthesis of a complementary base sequence in the strand being synthesized.

The reaction catalyzed by DNA polymerase I is a nucleophilic attack by the 3′-OH at the terminus of the primer strand on the nucleotidyl or interior phosphorus atom of the nucleoside triphosphate precursor with the elimination of pyrophosphate. This reaction mechanism explains the absolute requirement of DNA polymerase I for a free 3′-OH group on the primer DNA strand that is being covalently extended and dictates that *the direction of synthesis is always* 5′ → 3′ (**FIGURE 10.11**).

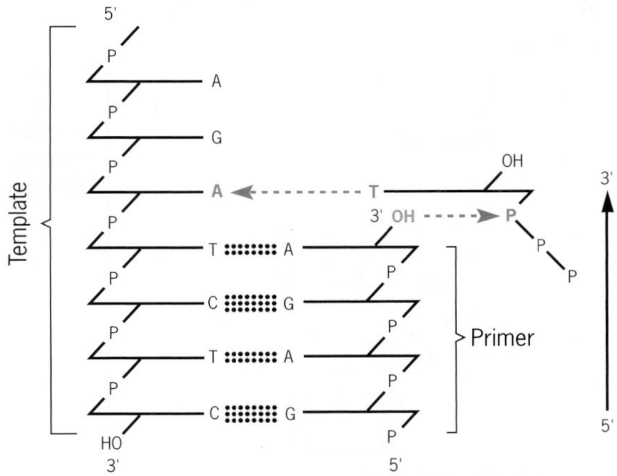

Figure 10.10 Template and primer requirements of DNA polymerases. The DNA molecule is shown here as a flattened "stick" diagram with one complementary strand on the left and the other on the right. "Stick" diagrams are especially useful in showing the 5′ → 3′ and 3′ → 5′ chemical polarity of DNA strands. All DNA polymerases require a primer strand (shown on the right) with a free 3′-hydroxyl. The primer strand is covalently extended by the addition of nucleotides (such as dTMP, derived from the incoming precursor dTTP shown). In addition, DNA polymerases require a template strand (shown on the left), which determines the base sequence of the strand being synthesized. The new strand will be complementary to the template strand.

DNA polymerase I is a single polypeptide with a molecular weight of 103,000 encoded by a gene called *polA*. However, subsequent research has shown that DNA polymerase I is not the true "DNA replicase" in *E. coli*. It does not catalyze the semiconservative replication of the *E. coli* chromosome; that function is performed by another enzyme. Nevertheless, DNA polymerase I does perform important functions in the *E. coli* cell, including playing a key role in chromosome replication and a central role in repairing damaged DNA. To understand how DNA polymerase I performs these functions, we must get to know this enzyme better.

In addition to the polymerase activity illustrated in **FIGURE 10.11**, DNA polymerase I has two other enzymatic activities, both exonuclease activities. A **nuclease** is an enzyme that degrades nucleic acids. An **exonuclease** degrades nucleic acids starting at one or both ends, whereas an **endonuclease** cleaves nucleic acids at internal sites. DNA polymerase I contains both 5′ → 3′ *exonuclease activity*, which cuts back DNA strands starting at 5′ termini, and 3′ → 5′ *exonuclease activity*, which cleaves off mononucleotides from the 3′ termini of DNA strands. The 5′ → 3′ exonuclease activity of DNA polymerase I usually excises small oligomers containing up to 10 nucleotides. Thus,

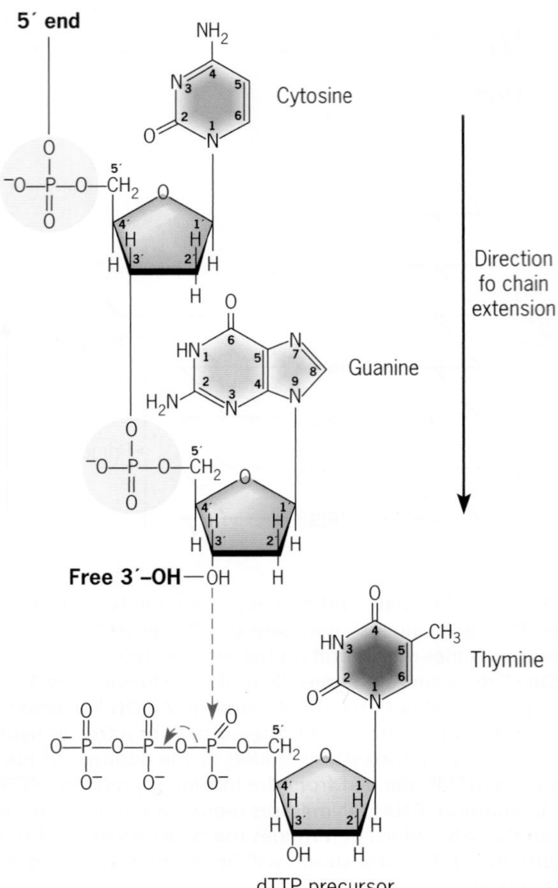

Figure 10.11 ▶ Mechanism of action of DNA polymerase I: covalent extension of a DNA primer strand in the 5′ → 3′ direction. The existing chain terminates at the 3′ end with the nucleotide deoxyguanylate (deoxyguanosine-5′-phosphate). The diagram shows the DNA polymerase-catalyzed addition of deoxythymidine monophosphate (from the precursor deoxythymidine triphosphate, dTTP) to the 3′ end of the chain with the release of pyrophosphate (P_2O_7).

DNA polymerase I has three different enzymatic activities: (1) a 5′ → 3′ polymerase activity, (2) a 5′ → 3′ exonuclease activity, and (3) a 3′ → 5′ exonuclease activity. These three activities are illustrated in **FIGURE 10.12.**

The first evidence that DNA polymerase I was not the true DNA replicase was published in 1969 by Paula DeLucia and John Cairns. They reported that DNA replication occurred in an *E. coli* strain lacking the 5′ → 3′ polymerase activity (but not the 5′ → 3′ exonuclease activity) of this enzyme owing to a mutation in the *polA* gene. Their results demonstrated that DNA replication in *E. coli* does not require DNA polymerase I activity, at least not the 5′ → 3′ polymerase activity of the enzyme. However, DeLucia and Cairns also discovered that this *polA1* mutant was extremely sensitive to ultraviolet light (UV). All three enzymatic activities of DNA polymerase I play important roles in the cell. The major function of DNA polymerase I in *E. coli* is to repair defects in DNA, such as those induced by UV (Chapter 13). However, as we will see later in this chapter, the 5′ → 3′ exonuclease activity of DNA polymerase I is also involved at one stage of chromosome replication.

MULTIPLE DNA POLYMERASES

If DNA polymerase I does not catalyze the semiconservative replication of the *E. coli* chromosome, another polymerase must carry out this function. In fact, there are at least four other DNA polymerases, **DNA polymerase II, DNA polymerase III, DNA polymerase IV,** and **DNA polymerase V,** in *E. coli*. Like DNA polymerase I, DNA polymerase II is a DNA repair enzyme; but it represents a small proportion of the polymerase activity in an *E. coli* cell. DNA polymerase II is a single polypeptide with 5′ → 3′ polymerase and 3′ → 5′ exonuclease activities. However, it has no 5′ → 3′ exonuclease activity. In contrast to DNA polymerases I and II, DNA polymerase III is a complex enzyme composed of many different subunits. Like DNA polymerase II, DNA polymerase III has 5′ → 3′ polymerase and 3′ → 5′ exonuclease activities; however, it has a 5′ → 3′ exonuclease that is active only on single-stranded DNA. The more recently characterized DNA polymerases IV and V, along with polymerase II, play important roles in the replication of damaged DNA, with the polymerase involved depending on the type of damage. Their functions are discussed further in Chapter 13.

Eukaryotic organisms are even more complex—with at least 13 different DNA polymerases having been identified so far. The eukaryotic DNA polymerases have been named α, β, γ, δ, ε, ζ, η, θ, ι, κ, λ, μ, and σ. Two or more of the DNA polymerases (α, δ, and/or ε) work together to carry out the semiconservative replication of nuclear DNA. DNA polymerase γ is responsible for the replication of DNA in mitochondria, and DNA polymerases β, ζ, η, θ, ι, κ, λ, μ, and σ are DNA repair enzymes or perform other metabolic functions. Some of the eukaryotic DNA polymerases lack the 3′ → 5′ exonuclease activity that is present in prokaryotic DNA polymerases.

All of the DNA polymerases studied to date, prokaryotic and eukaryotic, catalyze the same basic reaction: a nucleophilic attack by the free 3′-OH at the primer strand terminus on the nucleotidyl phosphorus of the nucleoside triphosphate precursor. Thus, all DNA polymerases have an absolute requirement for a free 3′-hydroxyl group on a preexisting primer strand; none of these DNA polymerases can initiate new DNA chains *de novo*, and all DNA synthesis occurs in the 5′ → 3′ direction.

The major replicative DNA polymerases are amazingly accurate, incorporating incorrect nucleotides with an initial frequency of 10^{-5} to 10^{-6}. (Some repair polymerases are error-prone—see Chapter 13.) Recent studies of the crystal structure of the complex formed by a monomeric DNA polymerase, a nucleoside triphosphate precursor, and a template-primer DNA have contributed to our understanding of the high fidelity of DNA synthesis. In these studies, published in 1998, Sylvie Doublié and colleagues determined the structure of the phage T7 polymerase, which is similar to DNA polymerase I of *E. coli*, with resolution to 0.22 nm. The results show that the polymerase is shaped like a little hand, with the incoming nucleoside

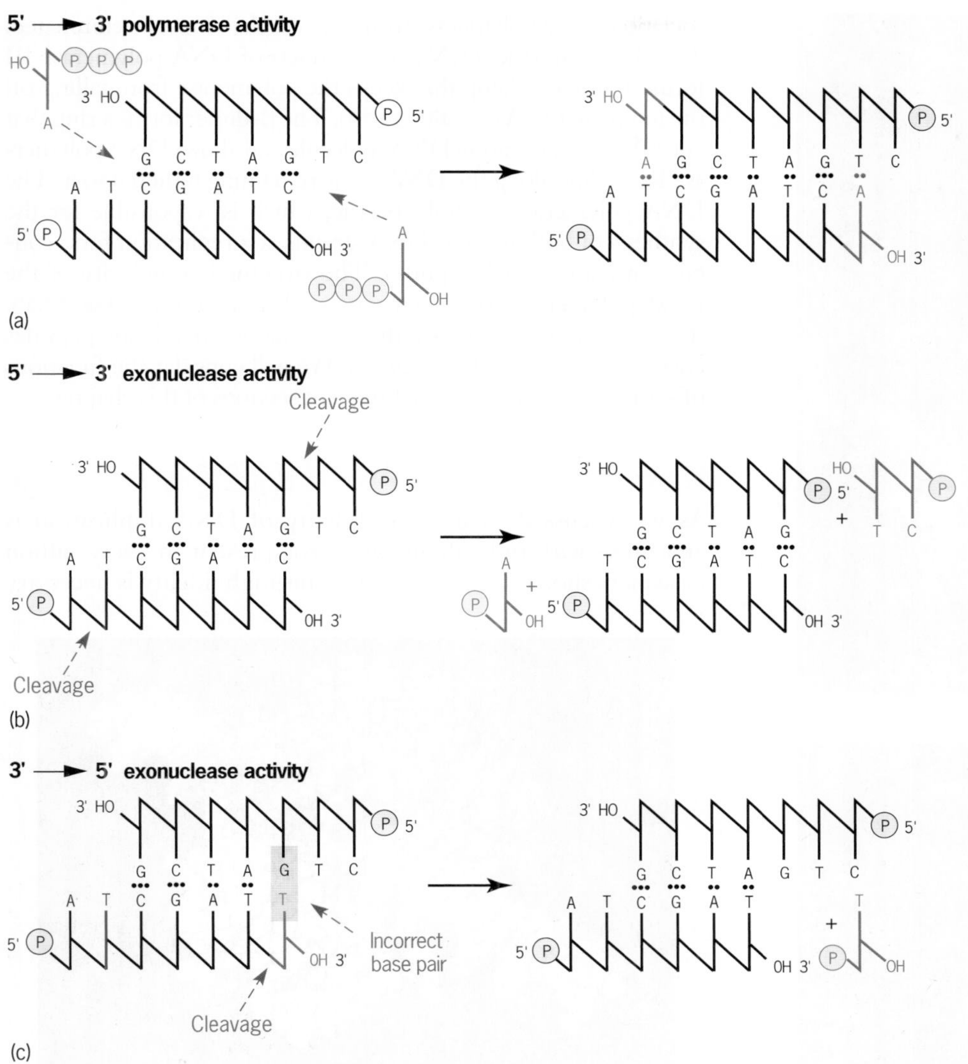

Figure 10.12 ▶ The three activities of DNA polymerases I in *E. coli*. As in Figure 10.10, the DNA molecules are shown in flattened "stick" form, which emphasizes the opposite chemical polarity (5′ → 3′ and 3′ → 5′) of the complementary strands. As is discussed in the text, all three activities—(*a*) 5′ → 3′polymerase activity, (*b*) 5′ → 3′ exonuclease activity, and (*c*) 3′ → 5′ exonuclease activity—play important roles in *E. coli* cells.

triphosphate, the template, and the primer terminus all tightly grasped between the thumb, the fingers, and the palm (**FIGURE 10.13**). The enzyme positions the incoming nucleoside triphosphate in juxtaposition with the terminus of the primer strand in a position to form hydrogen bonds with the first unpaired base in the template strand. Thus, the structure of this polymerase complex provides a simple explanation for the template-directed selection of incoming nucleotides during DNA synthesis.

DNA POLYMERASE III: THE REPLICASE IN *ESCHERICHIA COLI*

Evidence indicating that DNA polymerase III is the true DNA replicase responsible for the semiconservative replication of DNA in *E. coli* was first provided by isolating and characterizing a mutant strain with a mutation in a gene called *polC*, now renamed *dnaE*. This mutant strain produced active DNA polymerase III when grown at 25°C, but totally inactive polymerase III when grown at 43°C. When *dnaE* mutant cells growing at 25°C were shifted to 43°C, DNA replication stopped, indicating that the product of the *dnaE* gene is required for DNA synthesis. The *dnaE* gene was subsequently shown to encode the catalytic α subunit, the subunit with 5′ → 3′ polymerase activity, of DNA polymerase III.

DNA polymerase III is a multimeric enzyme (an enzyme with many subunits) with a molecular mass of about 900,000 daltons in its complete or **holoenzyme** form. The minimal core that has catalytic activity *in vitro* contains three subunits: α (the *dnaE* gene product), ε (the *dnaQ* product), and θ (the *holE* product). Addition of the τ subunit (the *dnaX* product) results in

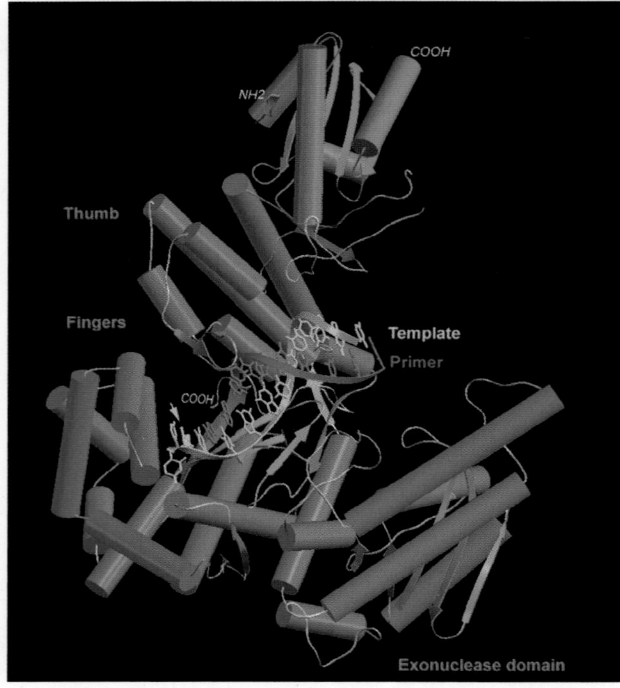

(a)

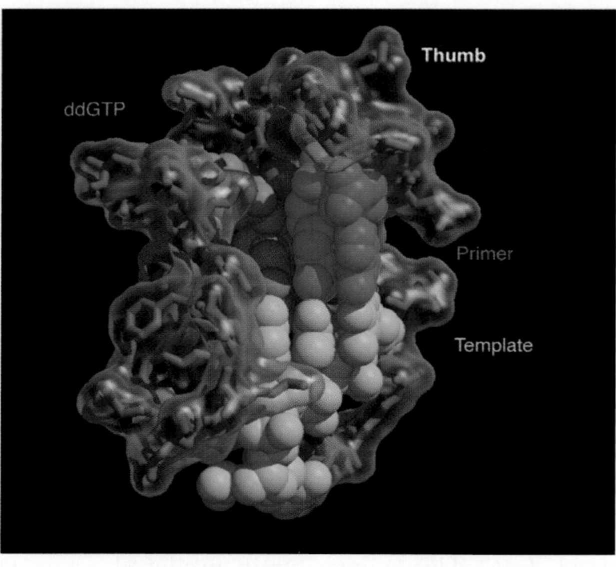

(b)

Figure 10.13 ▶ Schematic diagram (*a*) and space-filling model (*b*) of the structure of the complex between the phage T7 DNA polymerase, template-primer DNA, and a nucleoside triphosphate (ddGTP) precursor molecule. The template strand, primer strand, and nucleoside triphosphate are shown in yellow, magenta, and cyan, respectively. Protein components are shown in purple, green, orange, and gray. Note the tight juxtaposition between the nucleoside triphosphate, the primer terminus, and the template strand in (*b*).

dimerization of the catalytic core and increased activity. The catalytic core synthesizes rather short DNA strands because of its tendency to fall off the DNA template. In order to synthesize the long DNA molecules present in chromosomes, this frequent dissociation of the polymerase from the template must be eliminated. The β subunit (the *dnaN* gene product) of DNA polymerase III forms a dimeric clamp that keeps the polymerase from falling off the template DNA (**FIGURE 10.14**). The β-dimer forms a ring that encircles the replicating DNA molecule and allows DNA polymerase III to slide along the DNA while remaining tethered to it. The DNA polymerase III holoenzyme, which is responsible for the synthesis of both nascent DNA strands at a replication fork, contains at least 20 polypeptides. The structural complexity of the DNA polymerase III holoenzyme is illustrated in **FIGURE 10.15;** the diagram shows 16 of the best-characterized polypeptides encoded by seven different genes. We will consider the functions of some of the subunits in subsequent sections of this chapter.

PROOFREADING ACTIVITIES OF DNA POLYMERASES

As we discussed earlier, the fidelity of DNA duplication is amazing—with only about one error present in every billion base pairs shortly after synthesis. This high fidelity is necessary

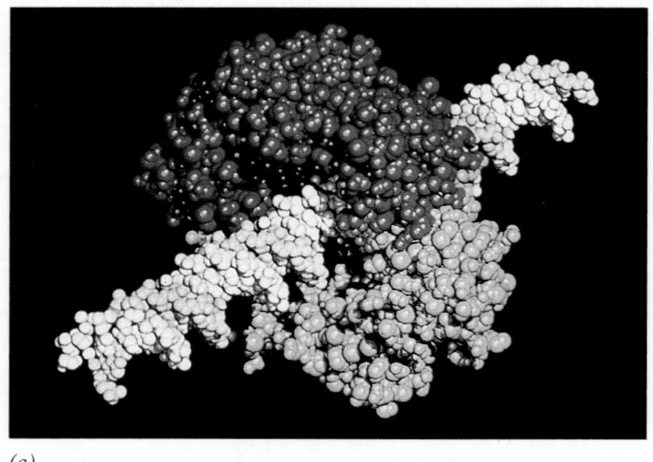

(a)

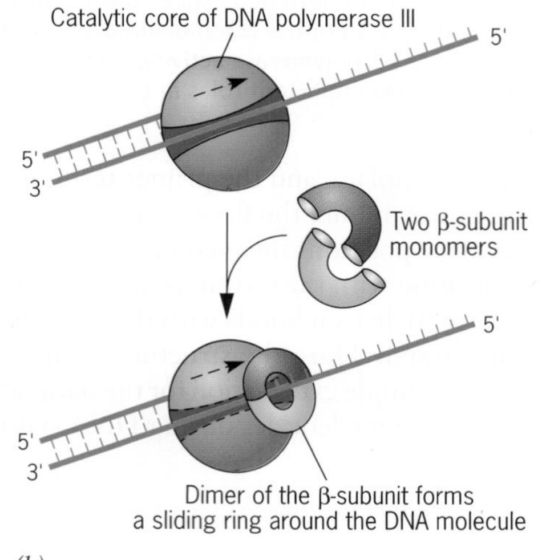

(b)

Figure 10.14 ▶ Space-filling model (*a*) and diagram (*b*) showing how two β-subunits (light and dark green) of DNA polymerase III clamp the enzyme to the DNA molecule (blue).

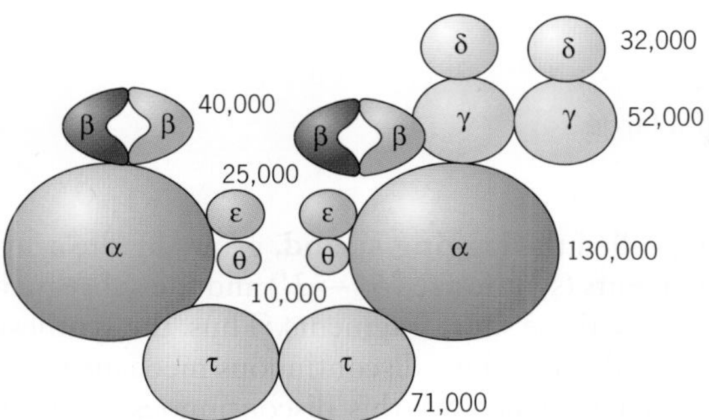

Figure 10.15 ▶ Structure of the *E. coli* DNA polymerase III holoenzyme. The numbers give the masses of the subunits in daltons.

to keep the mutation load at a tolerable level, especially in large genomes such as those of mammals, which contain 3×10^9 nucleotide pairs. Without the high fidelity of DNA replication, the monozygotic twins discussed at the beginning of this chapter would be less similar in phenotype. Indeed, based on the dynamic structures of the four nucleotides in DNA, the observed fidelity of DNA replication is much higher than expected. The thermodynamic changes in nucleotides that allow the formation of hydrogen-bonded base pairs other than A:T and G:C predict error rates of 10^{-5} to 10^{-4}, or one error per 10,000 to 100,000 incorporated nucleotides. The predicted error rate of 10,000 times the observed error rate raises the question of how this high fidelity of DNA replication can be achieved.

Living organisms have solved the potential problem of insufficient fidelity during DNA replication by evolving a mechanism for **proofreading** the nascent DNA chain as it is being synthesized. The proofreading process involves scanning the termini of nascent DNA chains for errors and correcting them. This process is carried out by the $3' \rightarrow 5'$ exonuclease activities of DNA polymerases. When a template-primer DNA has a terminal mismatch (an unpaired or incorrectly paired base or sequence of bases at the 3′ end of the primer), the $3' \rightarrow 5'$ exonuclease activity of the DNA polymerase clips off the unpaired base or bases (**FIGURE 10.16**). When an appropriately base-paired terminus is produced, the $5' \rightarrow 3'$ polymerase activity of the enzyme begins resynthesis by adding nucleotides to the 3′ end of the primer strand. In monomeric enzymes like DNA polymerase I of *E. coli*, this activity is built in. In multimeric enzymes, the $3' \rightarrow 5'$ proofreading exonuclease activity is often present on a separate subunit. In the case of DNA polymerase III of *E. coli*, this proofreading function is carried out by the ε subunit. DNA polymerase IV of *E. coli* contains no exonuclease activity. In eukaryotes, DNA polymerases γ, δ, and ε contain $3' \rightarrow 5'$ proofreading exonuclease activities, but polymerases α and β lack this activity. Given the importance of proofreading, we might speculate that accessory proteins must carry out the proofreading function for DNA polymerases α and β.

Without proofreading during DNA replication, Merry and Sherry, the twins discussed at the beginning of this chapter, would be less similar in appearance. Without proofreading, changes would have accumulated in their genes during the billions of cell divisions that occurred during their growth from small embryos to adults. Indeed, the identity of the genotypes of identical twins depends both on DNA proofreading during replication and on the activity of an army of DNA repair enzymes (Chapter 13). These enzymes continually scan DNA for various types of damage and make repairs before the alterations cause inherited genetic changes.

KEY POINTS

- DNA synthesis is catalyzed by enzymes called DNA polymerases.
- All DNA polymerases require a primer strand, which is extended, and a template strand, which is copied.
- All DNA polymerases have an absolute requirement for a free 3′-OH on the primer strand, and all DNA synthesis occurs in the 5′ to 3′ direction.
- The $3' \rightarrow 5'$ exonuclease activities of DNA polymerases proofread nascent strands as they are synthesized, removing any mispaired nucleotides at the 3′ termini of primer strands.

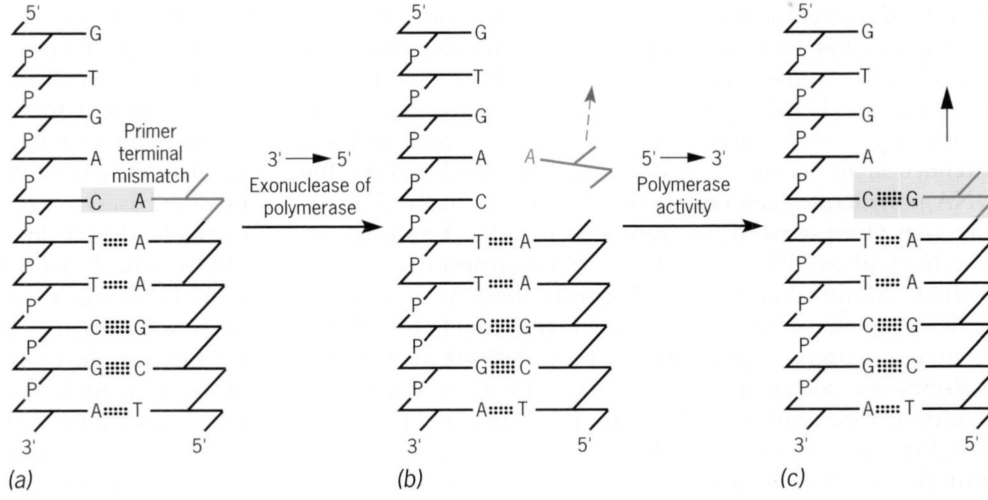

Figure 10.16 ▶ Proofreading by the $3' \rightarrow 5'$ exonuclease activity of DNA polymerases during DNA replication. As introduced in Figure 10.10, the DNA molecules are shown as "stick" diagrams. If DNA polymerase is presented with a template and primer containing a 3′ primer terminal mismatch (*a*), it will not catalyze covalent extension (polymerization). Instead, the $3' \rightarrow 5'$ exonuclease activity, an integral part of many DNA polymerases, will cleave off the mismatched terminal nucleotide (*b*). Then, presented with a correctly base-paired primer terminus, DNA polymerase will catalyze $5' \rightarrow 3'$ covalent extension of the primer strand (*c*).

The Complex Replication Apparatus

DNA replication is a complex process, requiring the concerted action of a large number of proteins.

The results of studies of DNA replication by autoradiography and electron microscopy indicate that the two progeny strands being synthesized at each replicating fork are being extended in the same overall direction at the macromolecular level. Because the complementary strands of a double helix have opposite polarity, synthesis is occurring at the 5′ end of one strand (3′ → 5′ extension) and the 3′ end of the other strand (5′ → 3′ extension). However, as we have previously discussed, all known polymerases have an absolute requirement for a free 3′-hydroxyl; they only carry out 5′ → 3′ synthesis. These apparently contradictory results created an interesting paradox. For many years biochemists searched for new polymerases that could catalyze 3′ → 5′ synthesis. No such polymerase was ever found. Instead, experimental evidence has shown that all DNA synthesis occurs in the 5′ → 3′ direction.

Clearly, the mechanism of DNA replication must be more complex than researchers originally thought (see **FIGURE 10.1**). In addition, given the absolute requirement of DNA polymerase for a free 3′-OH on the primer strand, this enzyme cannot begin the synthesis of a new strand *de novo*. How is the synthesis of a new DNA strand initiated? Since the two parental strands of DNA must be unwound, we have to deal with the need for a swivel or axis of rotation, especially for circular DNA molecules like the one present in the *E. coli* chromosome. Finally, how does the localized zone of strand separation or replication bubble form at the origin? These considerations and others indicate that DNA replication is more complicated than scientists thought when Watson and Crick proposed the semiconservative mechanism of replication in 1953.

CONTINUOUS SYNTHESIS OF ONE STRAND; DISCONTINUOUS SYNTHESIS OF THE OTHER STRAND

As discussed in the preceding section, the two nascent DNA strands being synthesized at each replicating fork are being extended in the same direction at the macromolecular level. Because the complementary strands of a DNA double helix have opposite chemical polarity, one strand is being extended in an overall 5′ → 3′ direction and the other strand is being extended in an overall 3′ → 5′ direction (**FIGURE 10.17*a***). But DNA polymerases can only catalyze synthesis in the 5′ → 3′ direction. This paradox was resolved with the demonstration that the synthesis of one strand of DNA is **continuous,** whereas synthesis of the other strand is **discontinuous.** At the molecular level, synthesis of the complementary strands of DNA is occurring in opposite physical directions (**FIGURE 10.17*b***), but both new strands are extended in the same 5′ → 3′ chemical direction. The synthesis of the strand being extended in the overall 5′ → 3′ direction, called the **leading strand,** is **continuous.** The strand being extended in the overall 3′ → 5′ direction, called the **lagging strand,** grows by the synthesis of short fragments (synthesized 5′ → 3′) and the subsequent covalent joining of these short fragments. Thus, the synthesis of the lagging strand occurs by a discontinuous mechanism.

The first evidence for this discontinuous mode of DNA replication came from studies in which intermediates in DNA synthesis were radioactively labeled by growing *E. coli* cells and bacteriophage T4-infected *E. coli* cells for very short periods of time in medium containing 3H-thymidine (pulse-labeling experiments). The labeled DNAs were then isolated, denatured, and characterized by measuring their velocity of sedimentation through sucrose gradients during high-speed centrifugation (see Focus on Centrifugation Techniques). When *E. coli* cells were pulse-labeled for 5, 10, or 30 seconds, for example, much of the label was found in small fragments of DNA, 1000 to 2000 nucleotides long (**FIGURE 10.17*c***). These small fragments of DNA have been named Okazaki fragments after Reiji Okazaki and Tuneko Okazaki, the scientists who discovered them in the late 1960s. In eukaryotes, the Okazaki fragments are only 100 to 200 nucleotides in length. When longer pulse-labeling periods are used, more of the label is recovered in large DNA molecules, presumably the size of *E. coli* or phage T4 chromosomes. If cells are pulse-labeled with ^{3}H-thymidine for a short period and then are transferred to nonradioactive medium for an extended period of growth (pulse-chase experiments), the labeled thymidine is present in chromosome-size DNA molecules. The results of these pulse-chase experiments are important because they indicate that the Okazaki fragments are true intermediates in DNA replication and not some type of metabolic by-product.

COVALENT CLOSURE OF NICKS IN DNA BY DNA LIGASE

If the lagging strand of DNA is synthesized discontinuously as described in the preceding section, a mechanism is needed to link the Okazaki fragments together to produce the large DNA strands present in mature chromosomes. This mechanism is provided by the enzyme **DNA ligase.** DNA ligase catalyzes the covalent closure of nicks (missing phosphodiester linkages; no missing bases) in DNA molecules by using energy from nicotinamide adenine dinucleotide (NAD) or adenosine triphosphate (ATP). The *E. coli* DNA ligase uses NAD as a cofactor, but some DNA ligases use ATP. The reaction catalyzed by DNA ligase is shown in **FIGURE 10.18.** First, AMP of the ligase-AMP intermediate forms a phosphoester linkage with the 5′-phosphate at the nick, and then a nucleophilic attack by the 3′-OH at the nick on the DNA-proximal phosphorous atom produces a phosphodiester linkage between the adjacent nucleotides at the site of the nick. DNA ligase alone has no activity at breaks

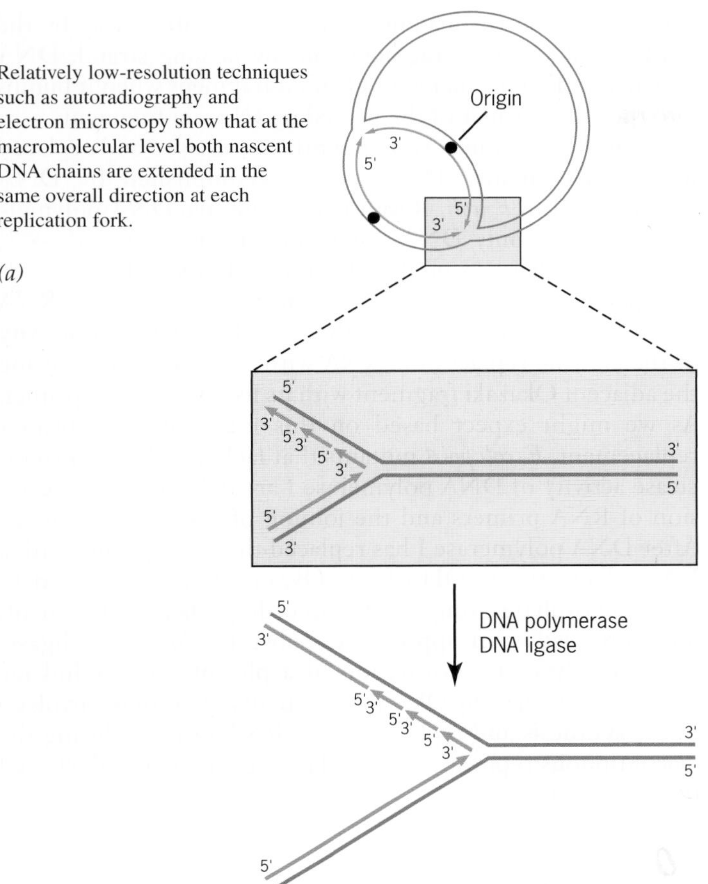

Relatively low-resolution techniques such as autoradiography and electron microscopy show that at the macromolecular level both nascent DNA chains are extended in the same overall direction at each replication fork.

(a)

(b) High-resolution biochemical techniques such as pulse-labeling and density-gradient analysis show that replication of the lagging strand is discontinuous—short fragments are synthesized in the 5'→ 3' direction and subsequently joined by DNA ligase.

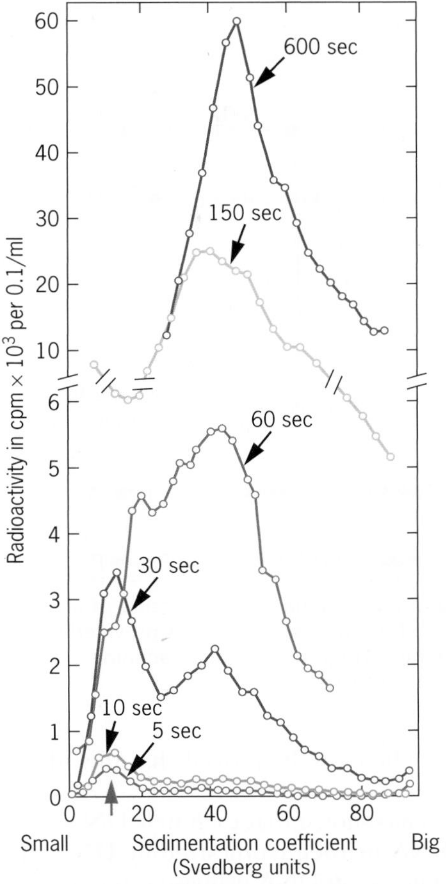

(c) Velocity sucrose gradient analysis of *E. coli* DNA pulse-labeled with ^{3}H-thymidine, extracted, and denatured during centrifugation.

Figure 10.17 ▶ Evidence for discontinuous synthesis of the lagging strand. (*a*) Although both strands of nascent DNA synthesized at a replication fork appear to be extended in the same direction, (*b*) at the molecular level, they are actually being synthesized in opposite directions. (*c*) The results of pulse-labeling experiments of Reiji and Tuneko Okazaki and colleagues showing that nascent DNA in *E. coli* exists in short fragments 1000 to 2000 nucleotides long. The red arrow shows the position of the "Okazaki fragments" in the gradient.

in DNA where one or more nucleotides are missing—so-called gaps. Gaps can be filled in and sealed only by the combined action of a DNA polymerase and DNA ligase. DNA ligase plays an essential role not only in DNA replication, but also in DNA repair and recombination (Chapter 13).

INITIATION OF DNA CHAINS WITH RNA PRIMERS

As discussed earlier, all known DNA polymerases have absolute requirements for a free 3′-OH on a DNA primer strand and an appropriate DNA template strand for activity. No known DNA polymerase can initiate the synthesis of a new strand of DNA. Thus, some special mechanism must exist to initiate or prime new DNA chains. Whereas the continuous synthesis of the leading strand requires the priming function only at the origin of replication, a priming event is required to initiate each Okazaki fragment during the discontinuous synthesis of the lagging strand. RNA polymerase, a complex enzyme that catalyzes the synthesis of RNA molecules from DNA templates, has long been known to be capable of initiating the synthesis of new RNA chains at specific sites on the DNA. When this occurs, an RNA-DNA hybrid is formed in which the nascent RNA is hydrogen-bonded to the DNA template. Because DNA polymerases are capable of extending polynucleotide chains containing a DNA or an RNA primer with a free 3′-OH, scientists began testing the idea that DNA synthesis might be initiated

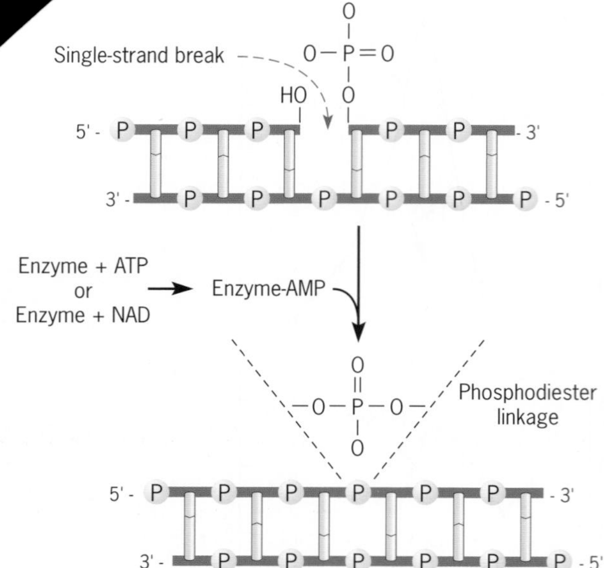

Figure 10.18 ▶ DNA ligase catalyzes the covalent closure of nicks in DNA. The energy required to form the ester linkage is provided by either adenosine triphosphate (ATP) or nicotinamide-adenine dinucleotide (NAD), depending on the species.

by using RNA primers. Their results proved that this idea is correct.

Subsequent research has shown that each new DNA chain is initiated by a short **RNA primer** synthesized by **DNA primase** (**FIGURE 10.19**). The *E. coli* DNA primase is the product of the *dnaG* gene. In prokaryotes, these RNA primers are 10 to 60 nucleotides long, whereas in eukaryotes they are shorter, only about 10 nucleotides long. The RNA primers provide the free 3′-OHs required for covalent extension of polynucleotide chains by DNA polymerases. In *E. coli*, deoxyribonucleotides are added to the RNA primers by DNA polymerase III, either

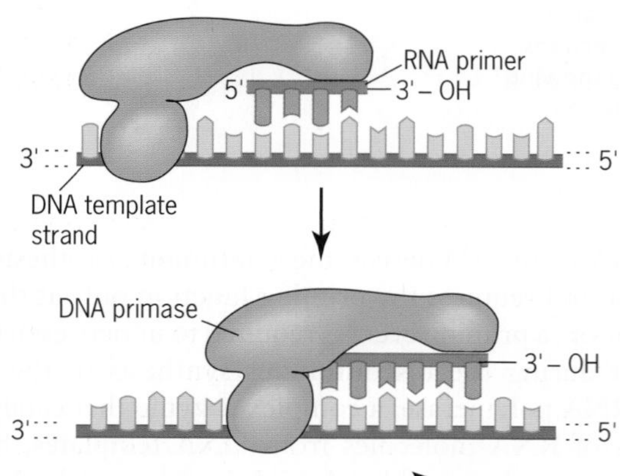

Figure 10.19 ▶ The initiation of DNA strands with RNA primers. The enzyme DNA primase catalyzes the synthesis of short (10 to 60 nucleotides long) RNA strands that are complementary to the template strands. DNA polymerase III then uses the free 3′-hydroxyls of the RNA primers to extend the chains by the addition of deoxyribonucleotides (see Figure 10.20).

continuously on the leading strand or discontinuously by the synthesis of Okazaki fragments on the lagging strand. DNA polymerase III terminates an Okazaki fragment when it bumps into the RNA primer of the preceding Okazaki fragment.

The RNA primers subsequently are excised and replaced with DNA chains. This step is accomplished by DNA polymerase I in *E. coli*. Recall that of the five DNA polymerases in *E. coli*, only DNA polymerase I possesses a $5' \rightarrow 3'$ exonuclease that acts on double-stranded DNA. The $5' \rightarrow 3'$ exonuclease activity of DNA polymerase I excises the RNA primer, and, at the same time, the $5' \rightarrow 3'$ polymerase activity of the enzyme replaces the RNA with a DNA chain by using the adjacent Okazaki fragment with its free 3′-OH as a primer. As we might expect based on this mechanism of primer replacement, *E. coli polA* mutants that lack the $5' \rightarrow 3'$ exonuclease activity of DNA polymerase I are defective in the excision of RNA primers and the joining of Okazaki fragments. After DNA polymerase I has replaced the RNA primer with a DNA chain, the 3′-OH of one Okazaki fragment is next to the 5′-phosphate group of the preceding Okazaki fragment. This product is an appropriate substrate for DNA ligase, which catalyzes the formation of a phosphodiester linkage between the adjacent Okazaki fragments. The steps involved in the synthesis and replacement of RNA primers during the discontinuous replication of the lagging strand are illustrated in **FIGURE 10.20.**

UNWINDING DNA WITH HELICASES, DNA-BINDING PROTEINS, AND TOPOISOMERASES

Semiconservative replication requires that the two strands of a parental DNA molecule be separated during the synthesis of new complementary strands. Since a DNA double helix contains two strands that cannot be separated without untwisting them turn by turn, DNA replication requires an unwinding mechanism. Given that each gyre, or turn, is about 10 nucleotide pairs long, a DNA molecule must be rotated 360° once for each 10 replicated base pairs. In *E. coli*, DNA replicates at a rate of about 30,000 nucleotides per minute. Thus, a replicating DNA molecule must spin at 3000 revolutions per minute to facilitate the unwinding of the parental DNA strands. The unwinding process (**FIGURE 10.21*a***) is catalyzed by enzymes called **DNA helicases.** The major replicative DNA helicase in *E. coli* is the product of the *dnaB* gene. DNA helicases unwind DNA molecules using energy derived from ATP.

Once the DNA strands are unwound by DNA helicase, they must be kept in an extended single-stranded form for replication. They are maintained in this state by a coating of **single-strand DNA-binding protein** (SSB protein) (**FIGURE 10.21*b***). The binding of SSB protein to single-stranded DNA is cooperative; that is, the binding of the first SSB monomer stimulates the binding of additional monomers at contiguous sites on the DNA chain. Because of the cooperativity of SSB protein

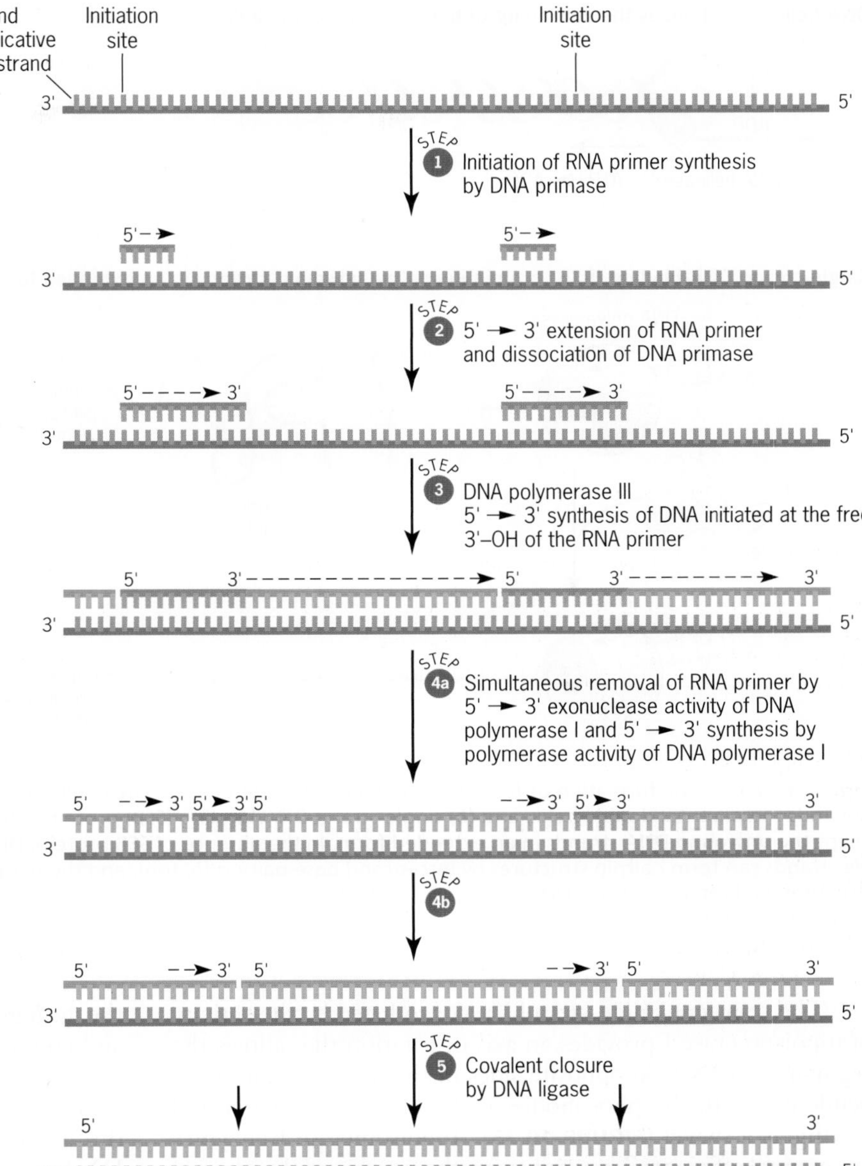

Figure 10.20 ▶ Synthesis and replacement of RNA primers during replication of the lagging strand of DNA. A short RNA strand is synthesized to provide a 3′-OH primer for DNA synthesis (see Figure 10.19). The RNA primer is subsequently removed and replaced with DNA by the dual 5′ → 3′ exonuclease and 5′ → 3′ polymerase activities built into DNA polymerase I. DNA ligase then covalently closes the nascent DNA chain, catalyzing the formation of phosphodiester linkages between adjacent 3′-hydroxyls and 5′-phosphates (see **FIGURE 10.18**).

binding, an entire single-stranded region of DNA is rapidly coated with SSB protein. Without the SSB protein coating, the complementary strands could renature or form intrastrand hairpin structures by hydrogen bonding between short segments of complementary or partially complementary nucleotide sequences. Such hairpin structures are known to impede the activity of DNA polymerases. In *E. coli*, the SSB protein is encoded by the *ssb* gene.

Recall that the *E. coli* chromosome contains a circular molecule of DNA. With the *E. coli* DNA spinning at 3000 revolutions per minute to allow the unwinding of the parental strands during replication (**FIGURE 10.22**), what provides the swivel or axis of rotation that prevents the DNA from becoming tangled (positively supercoiled) ahead of the replication fork? The required axes of rotation during the replication of circular DNA molecules are provided by enzymes called **DNA topoisomerases.** The topoisomerases catalyze transient breaks in DNA molecules but use covalent linkages to themselves to hold on to the cleaved molecules. The topoisomerases are of two types: (1) DNA topoisomerase I enzymes produce temporary single-strand breaks or nicks in DNA, and (2) DNA topoisomerase II enzymes produce transient double-strand breaks in DNA. An important result of this difference is that topoisomerase I activities remove supercoils from DNA one at a time, whereas topoisomerase II enzymes remove and introduce supercoils two at a time.

DNA helicase catalyzes the unwinding of the parental double helix.

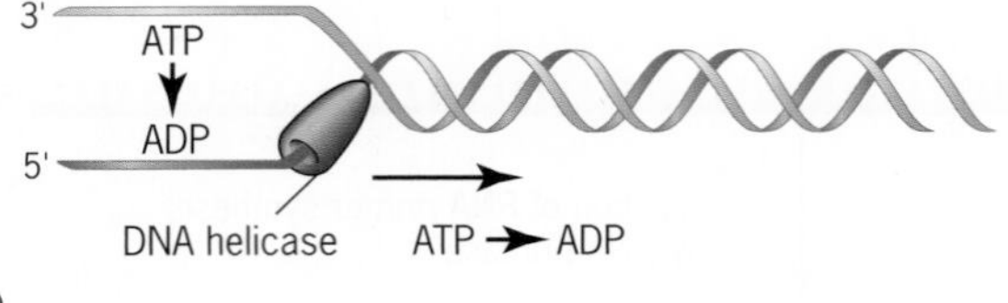

(*a*)

Single-strand DNA-binding (SSB) protein keeps the unwound strands in an extended form for replication.

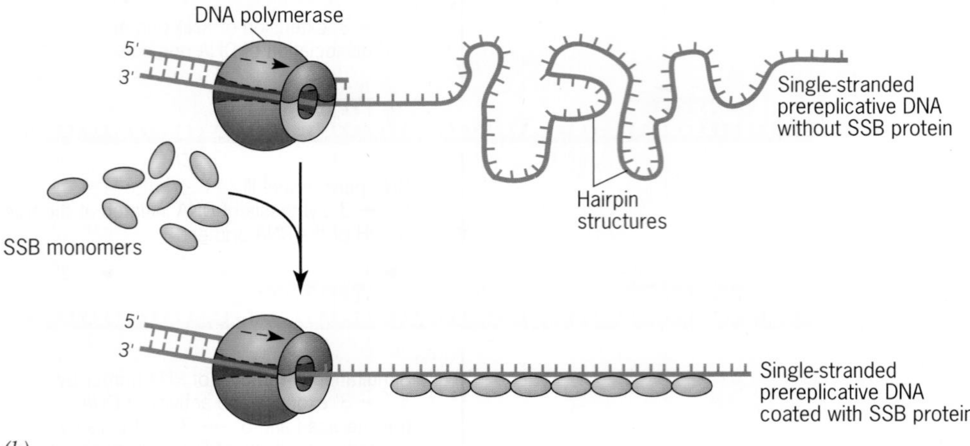

(*b*)

Figure 10.21 ▶ The formation of functional template DNA requires (*a*) DNA helicase, which unwinds the parental double helix, and (*b*) single-strand DNA-binding (SSB) protein, which keeps the unwound DNA strands in an extended form. In the absence of SSB protein, DNA single strands can form hairpin structures by intrastrand base-pairing (*b*, top), and the hairpin structures will retard or arrest DNA synthesis.

The transient single-strand break produced by the activity of topoisomerase I provides an axis of rotation that allows the segments of DNA on opposite sides of the break to spin independently, with the phosphodiester bond in the intact strand serving as a swivel (**FIGURE 10.23**). Topoisomerase I enzymes are energy-efficient. They conserve the energy of the cleaved phosphodiester linkages by storing it in covalent linkages between themselves and the phosphate groups at the cleavage sites; they then reuse this energy to reseal the breaks.

DNA topoisomerase II enzymes induce transient double-strand breaks and add negative supercoils or remove positive supercoils two at a time by an energy (ATP)-requiring mechanism. They carry out this process by cutting both strands of DNA, holding on to the ends at the cleavage site via covalent bonds, passing the intact double helix through the cut, and resealing the break (**FIGURE 10.24**). In addition to relaxing supercoiled DNA and introducing negative supercoils into DNA, topoisomerase II enzymes can separate interlocking circular molecules of DNA.

The best-characterized type II topoisomerase is an enzyme named **DNA gyrase** in *E. coli*. DNA gyrase is a tetramer with two α subunits encoded by the *gyrA* gene (originally *nalA*, for nalidixic acid) and two β subunits specified by the *gyrB* gene (formerly *cou*, for coumermycin). Nalidixic acid and coumermycin are antibiotics that block DNA replication in *E. coli* by inhibiting the activity of DNA gyrase. Nalidixic acid and coumermycin inhibit DNA synthesis by binding to the α and β subunits, respectively, of DNA gyrase. Thus, DNA gyrase activity is required for DNA replication to occur in *E. coli*.

Recall that chromosomal DNA is negatively supercoiled in *E. coli* (Chapter 9). The negative supercoils in bacterial chromosomes are introduced by DNA gyrase, with energy supplied by ATP. This activity of DNA gyrase provides another solution to the unwinding problem. Instead of creating positive supercoils ahead of the replication fork by unwinding the complementary strands of relaxed DNA, replication may produce relaxed DNA ahead of the fork by unwinding negatively supercoiled DNA. Because superhelical tension is reduced during unwinding—that is, strand separation is energetically favored—the negative supercoiling behind the fork may drive the unwinding process. If so, this mechanism nicely explains why DNA gyrase activity is required for DNA replication in bacteria. Alternatively, gyrase may simply remove positive supercoils that form ahead of the replication fork.

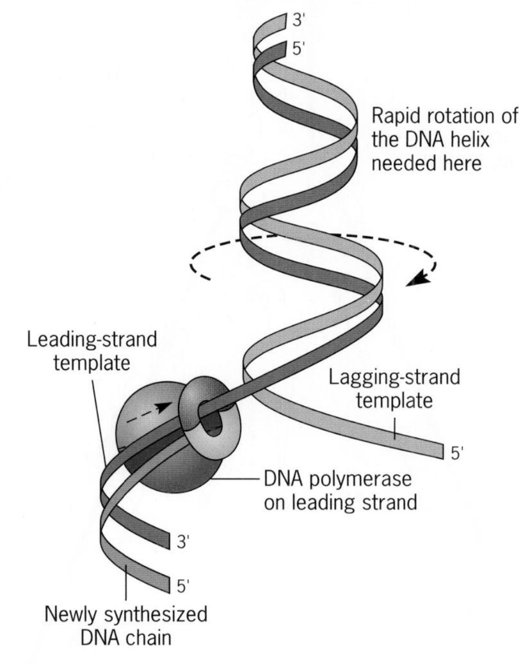

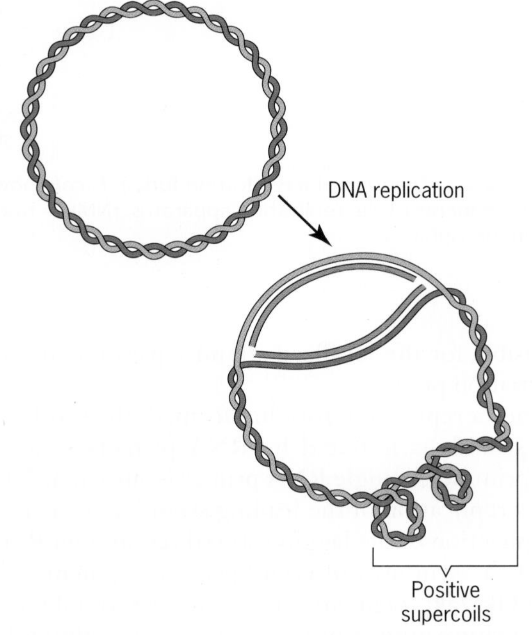

Figure 10.22 ► A swivel or axis of rotation is required during the replication of circular molecules of DNA like those in the *E. coli* or phage λ chromosomes. (*a*) During replication, the DNA in front of a replication fork must spin to allow the strands to be unwound by the helicase. (*b*) In the absence of an axis of rotation, unwinding will result in the production of positive supercoils in the DNA in front of a replication fork.

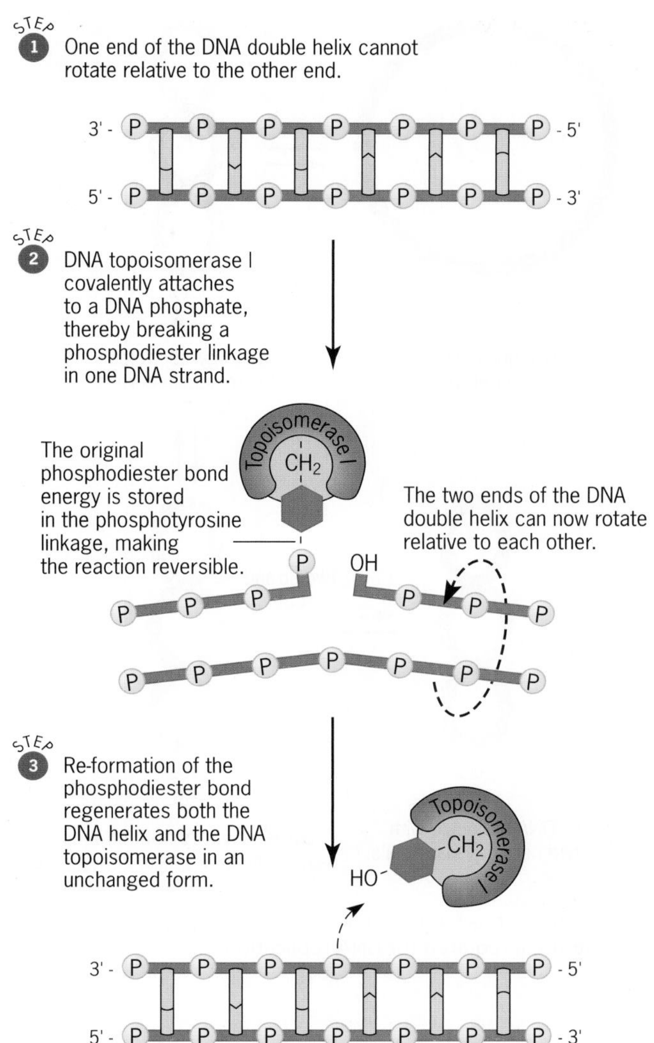

STEP 1 One end of the DNA double helix cannot rotate relative to the other end.
3' - P P P P P P P - 5'
5' - P P P P P P P - 3'
STEP 2 DNA topoisomerase I covalently attaches to a DNA phosphate, thereby breaking a phosphodiester linkage in one DNA strand.
Topoisomerase I
CH2
The original phosphodiester bond energy is stored in the phosphotyrosine linkage, making the reaction reversible.
The two ends of the DNA double helix can now rotate relative to each other.
OH
STEP 3 Re-formation of the phosphodiester bond regenerates both the DNA helix and the DNA topoisomerase in an unchanged form.
HO
3' - P P P P P P P - 5'
5' - P P P P P P P - 3'

Figure 10.23 ► DNA topoisomerase I produces transient single-strand breaks in DNA that act as axes of rotation or swivels during DNA replication.

THE REPLICATION APPARATUS: PREPRIMING PROTEINS, PRIMOSOMES, AND REPLISOMES

The basic features of DNA replication and most of the important components of the replication apparatus have been introduced in the preceding sections of this chapter. Now we will put the pieces together and look at the coordinated activities of these components during the replication of chromosomal DNA. To do so, we will consider the sequence of events that occur during the replication of the circular molecule of DNA in the *E. coli* chromosome. **FIGURE 10.25** shows the most important components of the replication apparatus in *E. coli* and indicates where they are located on a replication fork.

The replication of the *E. coli* chromosome begins at *oriC*, the unique sequence at which replication is initiated, with the formation of a localized region of strand separation called the

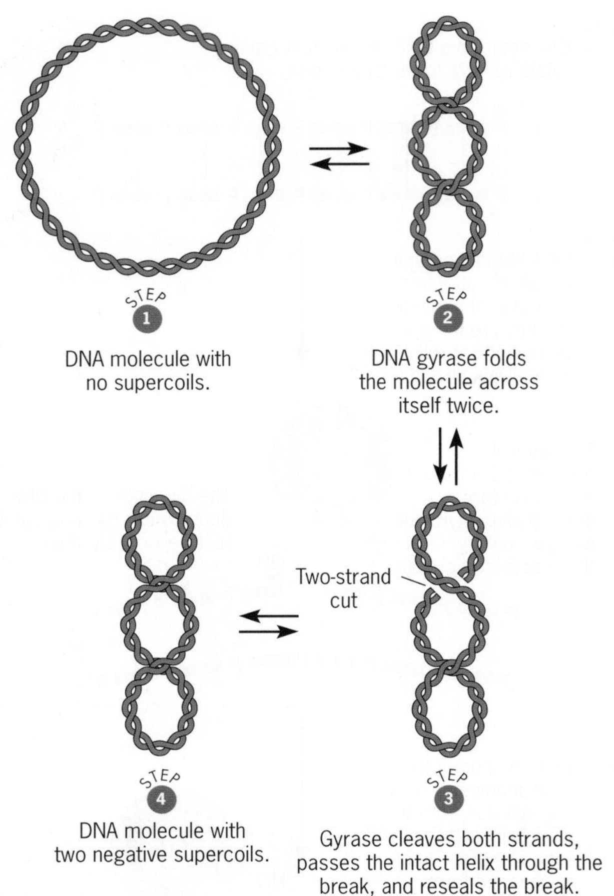

Figure 10.24 ▶ Mechanism of action of DNA gyrase, an *E. coli* DNA topoisomerase II required for DNA replication.

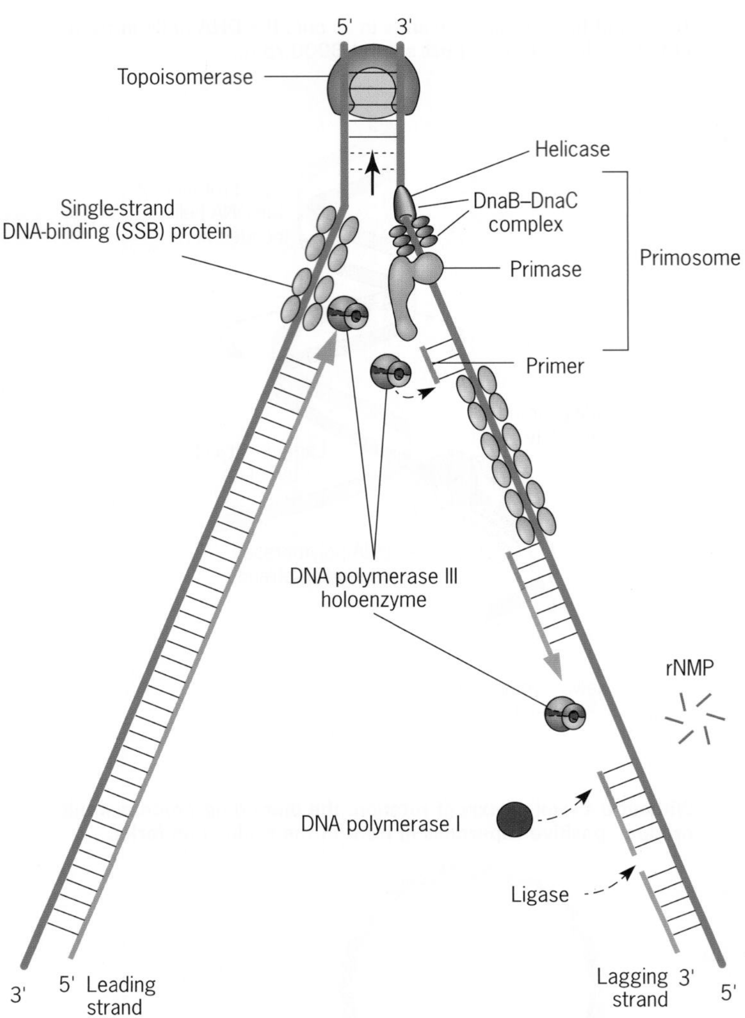

Figure 10.25 ▶ Diagram of a replication fork in *E. coli* showing the major components of the replication apparatus. rNMP = ribonucleoside monophosphates.

replication bubble. This replication bubble is formed by the interaction of *prepriming proteins* with *oriC* (**FIGURE 10.26**). The first step in prepriming appears to be the binding of four molecules of the *dnaA* gene product—DnaA protein—to the four 9-base-pair (bp) repeats in *oriC*. Next, DnaA proteins bind cooperatively to form a core of 20 to 40 polypeptides with *oriC* DNA wound on the surface of the protein complex. Strand separation begins within the three tandem 13-bp repeats in *oriC* and spreads until the replication bubble is created. A complex of DnaB protein (the hexameric DNA helicase) and DnaC protein (six molecules) joins the initiation complex and contributes to the formation of two bidirectional replication forks. The DnaT protein also is present in the prepriming protein complex, but its function is unknown. Other proteins associated with the initiation complex at *oriC* are DnaJ protein, DnaK protein, PriA protein, PriB protein, PriC protein, DNA-binding protein HU, DNA gyrase, and single-strand DNA-binding (SSB) protein. In some cases, however, their functional involvement in the prepriming process has not been established; in other cases, they are known to be involved, but their roles are unknown. The DnaA protein appears to be largely responsible for the localized strand separation at *oriC* during the initiation process.

Once a replication fork has formed, the synthesis of new DNA strands is initiated by RNA primers synthesized by DNA primase. A single RNA primer is sufficient for the continuous replication of the leading strand, but the discontinuous replication of the lagging strand requires an RNA primer to start the synthesis of each Okazaki fragment. The initiation of Okazaki fragments on the lagging strand is carried out by the **primosome,** a protein complex containing DNA primase and DNA helicase. The primosome moves along a DNA molecule, powered by energy from ATP; DNA helicase unwinds the double helix, and DNA primase synthesizes the RNA primers for successive Okazaki fragments. The RNA primers are covalently extended with deoxyribonucleotides by DNA polymerase III. DNA topoisomerases provide transient breaks in DNA that serve as swivels for DNA

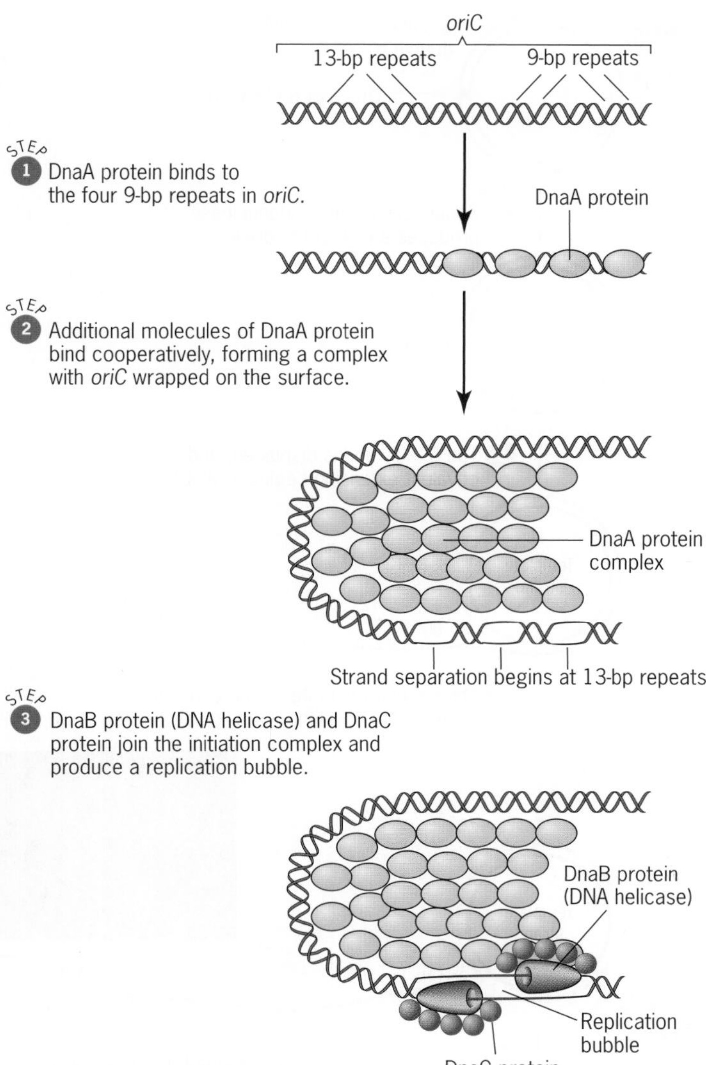

Figure 10.26 ▶ Prepriming of DNA replication at *oriC* in the *E. coli* chromosome.

unwinding and keep the DNA untangled. Single-strand DNA-binding protein coats the unwound prereplicative DNA and keeps it in an extended state for DNA polymerase III. The RNA primers are replaced with DNA by DNA polymerase I, and the single-strand breaks left by polymerase are sealed by DNA ligase. The DNA is then condensed into the nucleoid, or folded genome, of *E. coli*, in part through negative supercoiling introduced by DNA gyrase. All of these enzymes and DNA-binding proteins function in concert at each replication fork.

As a replication fork moves along a parental double helix, two DNA strands (the leading strand and the lagging strand) are replicated in the highly coordinated series of reactions described above. The complete replication apparatus moving along the DNA molecule at a replication fork is called the **replisome** (**FIGURE 10.27**). The replisome contains the DNA polymerase III holoenzyme; one catalytic core replicates the leading strand, the second catalytic core replicates the lagging strand, and the primosome unwinds the parental DNA molecule and sythesizes the RNA primers needed for the discontinuous synthesis of the lagging strand. In order for the two catalytic cores of the polymerase III holoenzyme to synthesize both the nascent leading and lagging strands, the lagging strand is thought to form a loop from the primosome to the second catalytic core of DNA polymerase III (**FIGURE 10.27**).

In *E. coli*, the termination of replication occurs at variable sites within regions called *terA* and *terB*, which block the movement of replication forks advancing in the counterclockwise and clockwise directions, respectively. DNA topoisomerases or special recombination enzymes then facilitate the separation of the nascent DNA molecules. At the beginning of this chapter, we noted the striking fidelity of DNA replication. Now that we have examined the cellular machinery responsible for DNA replication in living organisms, this fidelity seems less amazing. A very sophisticated apparatus, with built-in safeguards against malfunctions, has evolved to assure that the genetic

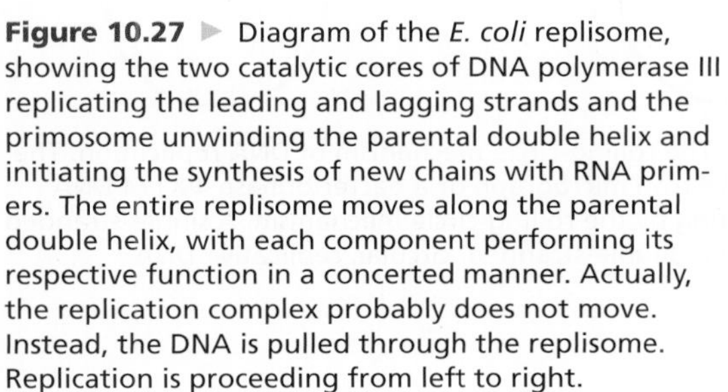
Figure 10.27 ▶ Diagram of the *E. coli* replisome, showing the two catalytic cores of DNA polymerase III replicating the leading and lagging strands and the primosome unwinding the parental double helix and initiating the synthesis of new chains with RNA primers. The entire replisome moves along the parental double helix, with each component performing its respective function in a concerted manner. Actually, the replication complex probably does not move. Instead, the DNA is pulled through the replisome. Replication is proceeding from left to right.

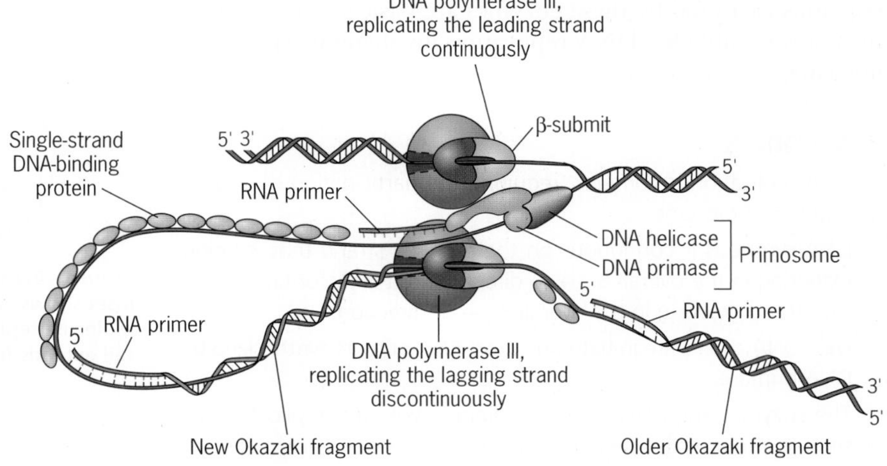

information of *E. coli* is transmitted accurately from generation to generation.

ROLLING-CIRCLE REPLICATION

In the preceding sections of this chapter, we have considered θ-shaped, eye-shaped, and Y-shaped replicating DNAs. We will now examine another important type of DNA replication called **rolling-circle replication.** Rolling-circle replication is used (1) by many viruses to duplicate their genomes, (2) in bacteria to transfer DNA from donor cells to recipient cells during one type of genetic exchange (Chapter 8), and (3) in amphibians to amplify extrachromosomal DNAs carrying clusters of ribosomal RNA genes during oogenesis (Chapter 20).

As the name implies, rolling-circle replication is a mechanism for replicating circular DNA molecules. The unique aspect of rolling-circle replication is that one parental circular DNA strand remains intact and rolls (thus the name rolling circle) or spins while serving as a template for the synthesis of a new complementary strand (**FIGURE 10.28**). Replication is initiated when a sequence-specific endonuclease cleaves one strand at the origin, producing 3′-OH and 5′-phosphate termini. The 5′ terminus is displaced from the circle as the intact template strand turns about its axis. Covalent extension occurs at the 3′-OH of the cleaved strand. Since the circular template DNA may turn 360° many times, with the synthesis of one complete or unit-length DNA strand during each turn, rolling-circle replication generates single-stranded tails longer than the contour length of the circular chromosome (**FIGURE 10.28**). Rolling-circle replication can produce either single-stranded or double-stranded progeny DNAs. Circular single-stranded progeny molecules are produced by site-specific cleavage of the single-stranded tails at the origins of replication and recircularization of the resulting unit-length molecules. To produce double-stranded progeny molecules, the single-stranded tails are used as templates for the discontinuous synthesis of complementary strands prior to cleavage and circularization. The enzymes involved in rolling-circle replication and the reactions catalyzed by these enzymes are basically the same as those responsible for DNA replication involving θ-type intermediates.

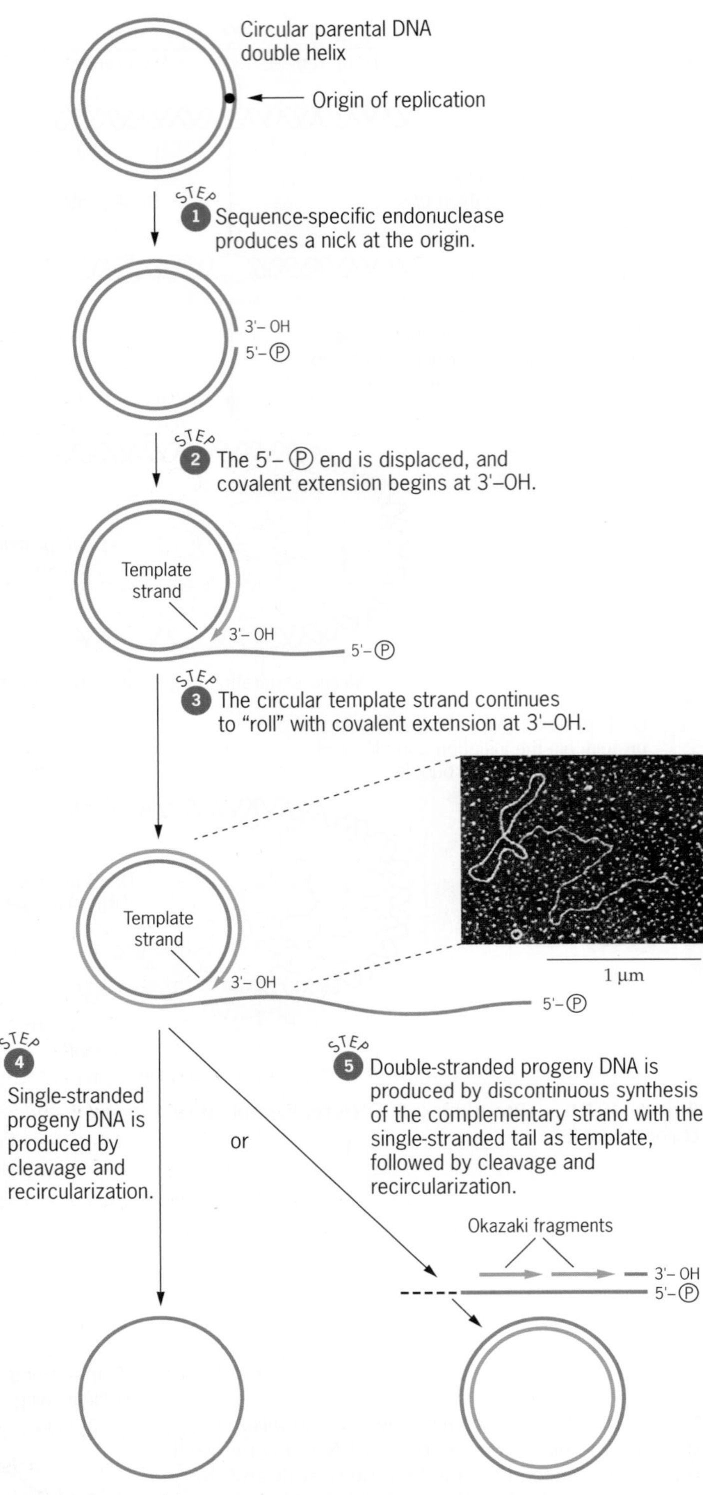
Circular parental DNA double helix
Origin of replication
STEP 1 Sequence-specific endonuclease produces a nick at the origin.
3'- OH
5'-Ⓟ
STEP 2 The 5'- Ⓟ end is displaced, and covalent extension begins at 3'–OH.
Template strand
3'- OH
5'-Ⓟ
STEP 3 The circular template strand continues to "roll" with covalent extension at 3'–OH.
Template strand
3'- OH
5'-Ⓟ
1 μm
STEP 4 Single-stranded progeny DNA is produced by cleavage and recircularization.
or
STEP 5 Double-stranded progeny DNA is produced by discontinuous synthesis of the complementary strand with the single-stranded tail as template, followed by cleavage and recircularization.
Okazaki fragments
3'- OH
5'-Ⓟ

Figure 10.28 ▶ The rolling-circle mechanism of DNA replication. The inset shows an electron micrograph of a bacteriophage ΦX174 DNA molecule replicating by the rolling-circle mechanism. A single-stranded tail extends from a double-stranded, circular, replicative DNA.

KEY POINTS

- DNA replication is complex, requiring the participation of a large number of proteins.
- DNA synthesis is continuous on the progeny strand that is being extended in the overall 5′ → 3′ direction, but is discontinuous on the strand growing in the overall 3′ → 5′ direction.
- New DNA chains are initiated by short RNA primers synthesized by DNA primase.
- The enzymes and DNA-binding proteins involved in replication assemble into a replisome at each replication fork and act in concert as the fork moves along the parental DNA molecule.

Unique Aspects of Eukaryotic Chromosome Replication

Although the main features of DNA replication are the same in all organisms, some processes occur only in eukaryotes.

Most of the information about DNA replication has resulted from studies of *E. coli* and some of its viruses. Less information is available about DNA replication in eukaryotic organisms. However, enough information is available to conclude that most aspects of DNA replication are similar in prokaryotes and eukaryotes, including humans. RNA primers and Okazaki fragments are shorter in eukaryotes than in prokaryotes, but the leading and lagging strands replicate by continuous and discontinuous mechanisms, respectively, in eukaryotes just as in prokaryotes. Nevertheless, a few aspects of eukaryotic DNA replication are unique to these structurally more complex species. For example, DNA synthesis takes place within a small portion of the cell cycle in eukaryotes, not continuously as in prokaryotes. The giant DNA molecules present in eukaryotic chromosomes would take much too long to replicate if each chromosome contained a single origin. Thus, eukaryotic chromosomes contain multiple origins of replication. Rather than using two catalytic complexes of one DNA polymerase to replicate the leading and lagging strands at each replication fork, eukaryotic organisms utilize two or more different polymerases.

As we discussed in Chapter 9, eukaryotic DNA is packaged in histone-containing structures called nucleosomes. Do these nucleosomes impede the movement of replication forks? If not, how does a replisome move past a nucleosome? Is the nucleosome completely or partially disassembled, or does the fork somehow slide past the nucleosome as the replisome duplicates the DNA molecule while it is still present on the surface of the nucleosome? Lastly, eukaryotic chromosomes contain linear DNA molecules, and the discontinuous replication of the ends of linear DNA molecules creates a special problem. We will address these aspects of chromatin replication in eukaryotes in the final sections of this chapter.

THE CELL CYCLE

When bacteria are growing on rich media, DNA replication occurs nonstop throughout the cell cycle. However, in eukaryotes, DNA replication is restricted to the S phase (for synthesis; Chapter 2). Recall that a normal eukaryotic cell cycle consists of G_1 phase (immediately following the completion of mitosis; G for gap), S phase, G_2 phase (preparation for mitosis), and M phase (mitosis) (see Chapter 2 for details). In rapidly dividing embryonic cells, G_1 and G_2 are very short or nonexistent. In all cells, decisions to continue on through the cell cycle occur at two points: (1) entry into S phase and (2) entry into mitosis. These checkpoints help to ensure that the DNA replicates once and only once during each cell division.

MULTIPLE REPLICONS PER CHROMOSOME

The giant DNA molecules in the largest chromosomes of *Drosophila melanogaster* contain about 6.5×10^7 nucleotide pairs. The rate of DNA replication in *Drosophila* is about 2600 nucleotide pairs per minute at 25°C. A single replication fork would therefore take about 17.5 days to replicate one of these giant DNA molecules. With two replication forks moving bidirectionally from a central origin, such a DNA molecule could be replicated in just over 8.5 days. Given that the chromosomes of *Drosophila* embryos replicate within 3 to 4 minutes and the nuclei divide once every 9 to 10 minutes during the early cleavage divisions, it is clear that each giant DNA molecule must contain many origins of replication. Indeed, the complete replication of the DNA of the largest *Drosophila* chromosome within 3.5 minutes would require over 7000 replication forks distributed at equal intervals along the molecule. Thus, multiple origins of replication are required to allow the very large DNA molecules in eukaryotic chromosomes to replicate within the observed cell division times.

The first evidence for multiple origins in eukaryotic chromosomes resulted from pulse-labeling experiments with Chinese hamster cells growing in culture. In 1968, when Joel Huberman and Arthur Riggs pulse-labeled cells with ^{3}H-thymidine for a few minutes, extracted the DNA, and performed autoradiographic analysis of the labeled DNA, they observed tandem arrays of exposed silver grains (**FIGURE 10.29*a***). The simplest interpretation of their results is that individual macromolecules of DNA contain multiple origins of replication. When the pulse-labeling period was followed by a short interval of growth in nonradioactive medium (pulse-chase experiments), the tandem arrays contained central regions of high-grain density with tails of decreasing grain density at *both* ends (**FIGURE 10.29*b***). This result indicates that replication in eukaryotes is bidirectional just as it is in most prokaryotes. The tails of decreasing grain density result from the gradual dilution of the intracellular pools of ^{3}H-thymidine by ^{1}H-thymidine as the replication forks move bidirectionally from central origins toward replication termini (**FIGURE 10.29*c***).

A segment of DNA whose replication is under the control of one origin and two termini is called a **replicon.** In prokaryotes, the entire chromosome is usually one replicon. The existence of multiple replicons per eukaryotic chromosome has been verified directly by autoradiography and electron microscopy in several different species. The genomes of humans and other mammals contain about 10,000 origins of replication distributed throughout the chromosomes at 30,000 to 300,000-base-pair intervals. Clearly, the number of replicons per chromosome is not fixed throughout the growth and development

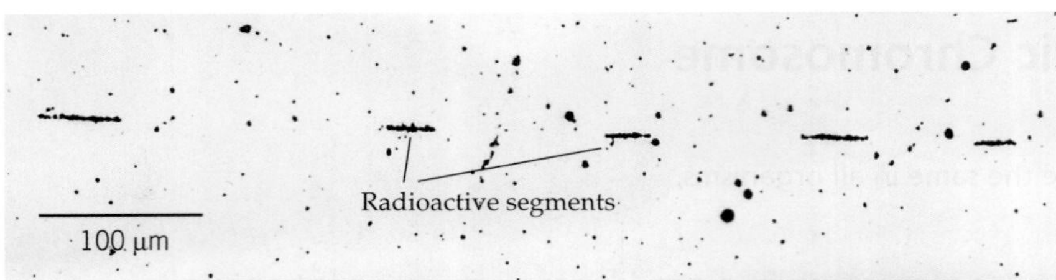

(*a*) Autoradiograph of a portion of a DNA molecule from a Chinese hamster cell that had been pulse-labeled with ^{3}H-thymidine.

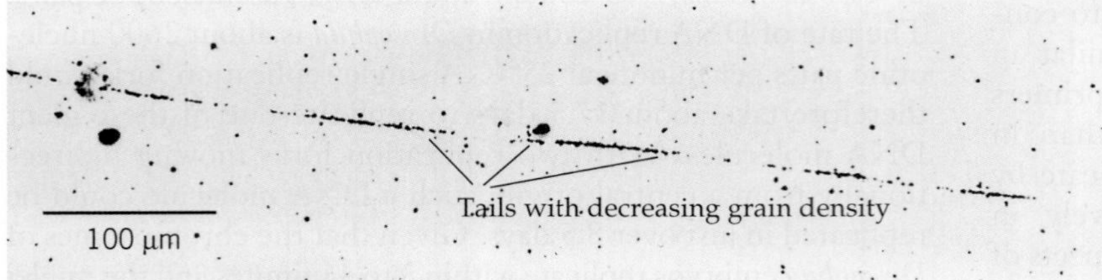

(*b*) Autoradiograph of a segment of a DNA molecule from a Chinese hamster cell that was pulse-labeled with ^{3}H-thymidine and then transferred to non-radioactive medium for an additional growth period.

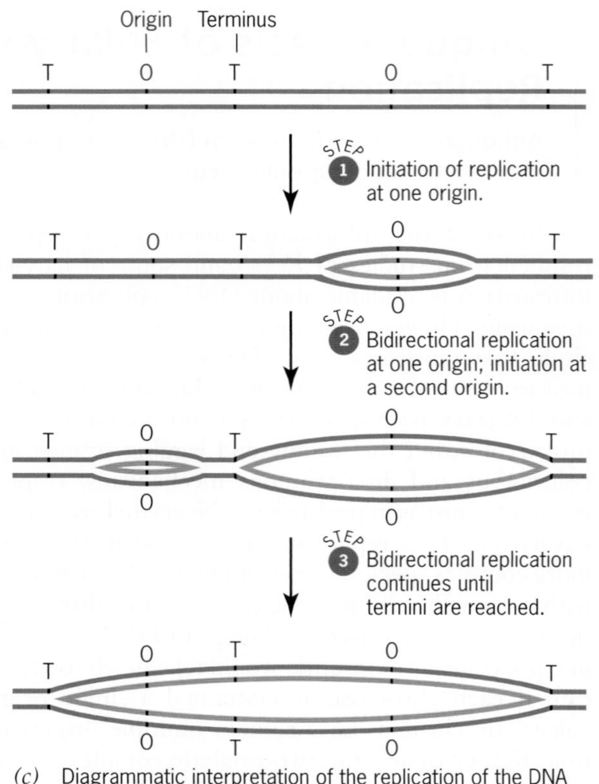

(*c*) Diagrammatic interpretation of the replication of the DNA molecules visualized in *(a)* and *(b)*.

Figure 10.29 ▶ Evidence for bidirectional replication of the multiple replicons in the giant DNA molecules of eukaryotes. The tandem arrays of radioactivity in (*a*) indicate that replication occurs at multiple origins; tails with decreasing grain density observed in (*b*) indicate that replication occurs bidirectionally from each origin (*c*).

of a multicellular eukaryote. Replication is initiated at more sites during the very rapid cell divisions of embryogenesis than during later stages of development. Unfortunately, geneticists don't know what factors determine which origins are operational at any given time or in a particular type of cell.

TWO OR MORE DNA POLYMERASES AT A SINGLE REPLICATION FORK

Given the complexity of the replisome in the simple bacterium *E. coli* (see **FIGURES 10.25** and **10.27**), it seems likely that the replication apparatus is even more complex in eukaryotes. Although knowledge of the structure of the replicative machinery in eukaryotes is still limited, many features of DNA replication are similar in eukaryotes and prokaryotes.

As in the case of prokaryotes, much of the information about DNA synthesis in eukaryotes has come from the development and dissection of *in vitro* DNA replication systems. Studies of the replication of DNA viruses of eukaryotes have proven informative, and of these viruses, Simian virus 40 (SV40) has proven particularly useful. The replication of SV40 is carried out almost entirely by the host cell's replication apparatus. Only one viral protein, the so-called T antigen, is required for replication of the SV40 chromosome.

As in prokaryotes, the unwinding of the parental DNA strands requires a DNA topoisomerase and a DNA helicase. The unwound strands are kept in the extended state by a single-strand DNA-binding protein called replication protein A (Rp-A). However, unlike the process in prokaryotes, the replication of chromosomal DNA in eukaryotes requires the activity of three different DNA polymerases—polymerase α (Pol α), polymerase δ (Pol δ), and polymerase ε (Pol ε). At least two polymerases, perhaps all three, are present in each replication fork (replisome), and each polymerase contains multiple subunits. Also, whereas the *E. coli* replisome contains 13 known proteins, the replisomes of yeast and mammals contain at least 27 different components.

In eukaryotes, Pol α is required for the initiation of replication at origins and for the priming of Okazaki fragments during the discontinuous synthesis of the lagging strand. Pol α exists in a stable complex with DNA primase; indeed, they copurify during isolation. The primase synthesizes the RNA primers, which are then extended with deoxyribonucleotides by Pol α to produce an RNA–DNA chain about 30 nucleotides in total length. These RNA–DNA primer chains are then extended by Pol δ (perhaps Pol ε in some cases). Pol δ is thought to catalyze most of the processive synthesis of chromosomal DNA. Pol δ must interact with proteins PCNA (proliferating cell nuclear antigen) and replication factor C (Rf-C) to be active (**FIGURE 10.30**). PCNA is the sliding clamp that tethers Pol δ to the DNA to allow processive replication (to prevent the polymerase from falling off the template); PCNA is equivalent to the β subunit of DNA polymerase III in *E. coli* (see **FIGURE 10.14**). Rf-C is required for PCNA to load onto

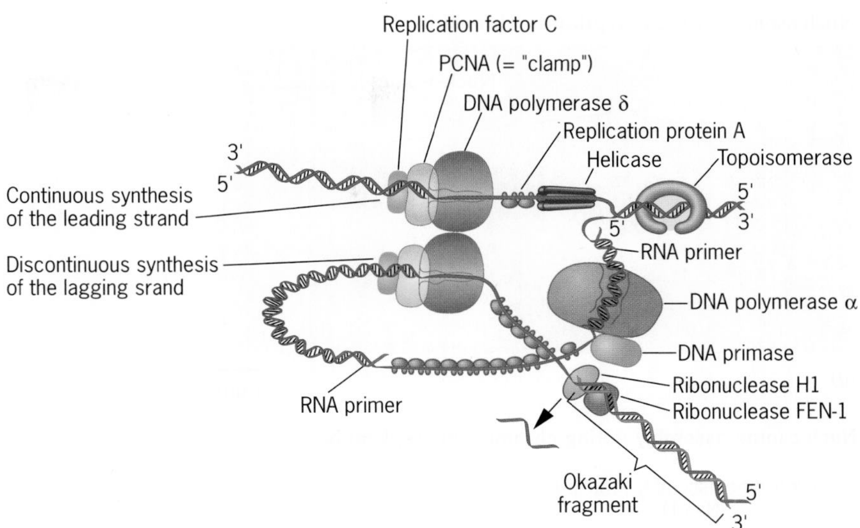

Figure 10.30 ▶ Some of the important components of a replisome in eukaryotes. Each replisome contains two different polymerases, α and δ (or, perhaps ε). The DNA polymerase α-DNA primase complex synthesizes the RNA primers and adds short segments of DNA. DNA polymerase δ (sometimes ε) then completes the synthesis of the Okazaki fragments in the lagging strand and catalyzes the continuous synthesis of the leading strand. PCNA (*p*roliferating *c*ell *n*uclear *a*ntigen) is equivalent to the β subunit of *E. coli* DNA polymerase III; it clamps polymerase δ to the DNA molecule facilitating the synthesis of long DNA chains. Ribonucleases H1 and FEN-1 (F1 nuclease 1) remove the RNA primers, polymerase δ fills in the gap, and DNA ligase (not shown) seals the nicks, just as in *E. coli* (see **FIGURE 10.20**).

DNA. PCNA is a trimeric protein that forms a closed ring; Rf-C induces a change in the conformation of PCNA that allows it to encircle DNA, providing the essential sliding clamp.

Polymerases δ and ε both contain the 3′ → 5′ exonuclease activity required for proofreading (see **FIGURE 10.16**). However, they do not have 5′ → 3′ exonuclease activity; thus, they cannot remove RNA primers like DNA polymerase I of *E. coli* does. Instead, the RNA primers are excised by two nucleases, ribonuclease H1 (which degrades RNA present in RNA–DNA duplexes) and ribonuclease FEN-1 (F1 nuclease 1). Pol δ then fills in the gaps, and DNA ligase seals the nicks, producing covalently closed progeny strands.

DNA polymerase ε is required for replication *in vivo* and for the repair of certain kinds of damage to DNA; however, its exact function(s) is uncertain. One proposal is that Pol δ is responsible for both leading and lagging strand synthesis, with Pol ε performing unknown specialized functions. Another model suggests that Pol δ replicates the lagging strand and Pol ε replicates the leading strand. A third model has their roles reversed. Clearly, more research will be required to determine the functions of the multiple polymerases and other proteins in the eukaryotic replisome.

As mentioned earlier, there are at least 13 different DNA polymerases—α, β, γ, δ, ε, ζ, η, θ, ι, κ, λ, μ, and σ—in eukaryotes. DNA polymerase E_i is responsible for the replication of DNA in mitochondria, and the other DNA polymerases have important roles in DNA repair and other pathways (Chapter 13).

DUPLICATION OF NUCLEOSOMES AT REPLICATION FORKS

As we discussed in Chapter 9, the DNA in eukaryotic interphase chromosomes is packaged in approximately 11-nm beads called nucleosomes. Each nucleosome contains 166 nucleotide pairs of DNA wound in two turns around an octamer of histone molecules. Given the size of nucleosomes and the large size of DNA replisomes, it seems unlikely that a replication fork can move past an intact nucleosome. Yet, electron micrographs of replicating chromatin in *Drosophila* clearly show nucleosomes with approximately normal structure and spacing on both sides of replication forks (**FIGURE 10.31*a***); that is, nucleosomes appear to have the same structure and spacing immediately behind a replication fork (postreplicative DNA) as they do in front of a replication fork (prereplicative DNA). This observation suggests that nucleosomes must be disassembled to let the replisome duplicate the DNA packaged in them and then be quickly reassembled; that is, DNA replication and nucleosome assembly must be tightly coupled.

Since the mass of the histones in nucleosomes is equivalent to that of the DNA, large quantities of histones must be synthesized during each cell generation in order for the nucleosomes to duplicate. Although histone synthesis occurs throughout the cell cycle, there is a burst of histone biosynthesis during S phase that generates enough histones for chromatin duplication. When density-transfer experiments were performed to examine the mode of nucleosome duplication, the nucleosomes on both progeny DNA molecules were found to contain both old (prereplicative) histone complexes and new (postreplicative) complexes. Thus, at the protein level, nucleosome duplication appears to occur by a dispersive mechanism.

A number of proteins are involved in the disassembly and assembly of nucleosomes during chromosome replication in eukaryotes. Two of the most important are *n*ucleosome *a*ssembly *p*rotein-*1* (Nap-1) and *c*hromatin *a*ssembly *f*actor-*1* (CAF-1). Nap-1 transports histones from their site of synthesis in the cytoplasm to the nucleus, and CAF-1 carries them to the chromosomal sites of nucleosome assembly (**FIGURE 10.31*b***). CAF-1 delivers histones to the sites of DNA replication by binding to PNCA (*p*roliferating *c*ell *n*uclear *a*ntigen)—the clamp that tethers DNA polymerase δ to the DNA template (see **FIGURE 10.30**). CAF-1 is an essential protein in *Drosophila*,

Nucleosome spacing in replicating chromatin.

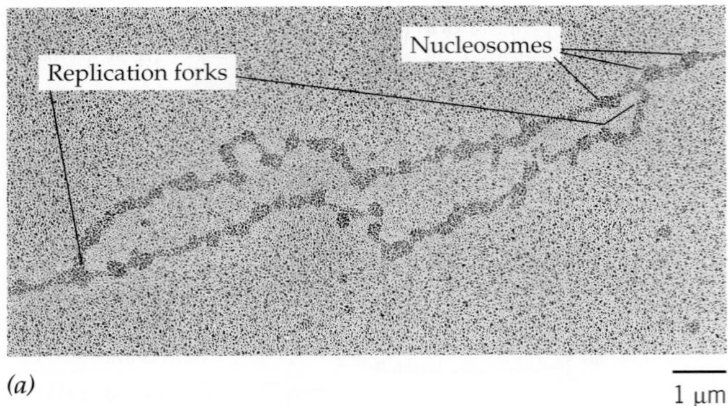

Nucleosome assembly during chromosome replication.

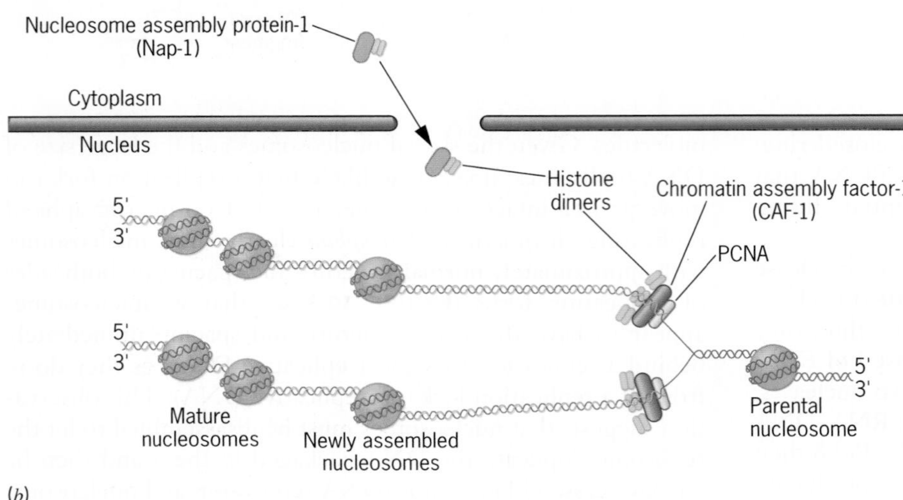

Figure 10.31 ▶ The disassembly and assembly of nucleosomes during the replication of chromosomes in eukaryotes. (*a*) An electron micrograph showing nucleosomes on both sides of two replication forks in *Drosophila*. Recall that DNA replication is bidirectional; thus, each branch point is a replication fork. (*b*) The assembly of new nucleosomes during chromosome replication requires proteins that transport histones from the cytoplasm to the nucleus and that concentrate them at the site of nucleosome assembly. PCNA = proliferating cell nuclear antigen (see **FIGURE 10.30**).

but not in yeast where other proteins can perform some of its functions.

Many other proteins affect nucleosome structure. Some are involved in chromatin remodeling—changing nucleosome structure in ways that activate or silence the expression of the genes packaged therein. Others modify nucleosome structure by adding methyl or acetyl groups to specific histones. In addition, eukaryotes contain several minor histones with structures slightly different from the major histones, and the incorporation of these minor histones into nucleosomes can change their structure. In *Drosophila*, for example, the incorporation of histone H3.3 into nucleosomes results in high levels of transcription of the genes therein. Thus, nucleosome structure is not invariant; to the contrary, it plays an important role in modulating gene expression (see Chapter 20).

TELOMERASE: REPLICATION OF CHROMOSOME TERMINI

We discussed the unique structures of telomeres, or chromosome ends, in Chapter 9. An early reason for thinking that telomeres must have special structures was that DNA polymerases cannot replicate the terminal DNA segment of the lagging strand of a linear chromosome. At the end of the DNA molecule being replicated discontinuously, there would be no DNA strand to provide a free 3′-OH (primer) for polymerization of deoxyribonucleotides after the RNA primer of the terminal Okazaki fragment has been excised (**FIGURE 10.32*a***). Either (1) the telomere must have a unique structure that facilitates its replication or (2) there must be a special enzyme that resolves this enigma of replicating the terminus of the lagging strand. Indeed, evidence has shown that both are correct. The special structure of telomeres provides a neat mechanism for the addition of telomeres by an RNA-containing enzyme called **telomerase.**

The telomeres of humans, which contain the tandemly repeated sequence TTAGGG, will be used to illustrate how telomerase adds ends to chromosomes (**FIGURE 10.32*b***). Telomerase recognizes the G-rich telomere sequence on the 3′ overhang and extends it 5′ → 3′ one repeat unit at a time. Telomerase does not fill in the gap opposite the 3′ end of the template strand; it simply extends the 3′ end of the template strand. The unique feature of telomerase is that it contains a built-in RNA template. After several telomere repeat units are added by telomerase, DNA polymerase catalyzes the synthesis of the complementary strand. Without telomerase activity, linear chromosomes would become progressively shorter. If the resulting terminal deletions extended into an essential gene or genes, this chromosome shortening would be lethal.

One change observed in many cancer cells is that the genes encoding telomerase are expressed, whereas they are not expressed in most somatic cells. Thus, one approach to cancer treatments has been to try to develop telomerase inhibitors, so that the chromosomes in cancer cells will lose their telomeres and die. However, other cancer cells do not contain active telomerase, making this approach problematic.

TELOMERE LENGTH AND AGING IN HUMANS

Unlike germ-line cells, most human somatic cells lack telomerase activity. When human somatic cells are grown in culture, they divide only a limited number of times (usually only 20 to 70 cell generations) before senescence and death occur. When telomere lengths are measured in various somatic-cell cultures, a correlation is observed between telomere length and the number of cell divisions preceding senescence and death. Cells with longer telomeres survive longer—go through more cell divisions—than cells with shorter telomeres. As would be expected in the absence of telomerase activity, telomere length decreases as the age of the cell culture increases. Occasionally, somatic cells are observed to acquire the ability to proliferate in culture indefinitely, and these immortal cells have been shown to contain telomerase activity, unlike their progenitors. Since the one common feature of all cancers is uncontrolled cell division or immortality, scientists have proposed that one way to combat human cancers would be to inhibit the telomerase activity in cancer cells.

Further evidence of a relationship between telomere length and aging in humans has come from studies of individuals with disorders called **progerias,** inherited diseases characterized by premature aging. In the most severe form of progeria, Hutchinson–Gilford syndrome (**FIGURE 10.33**), senescence—wrinkles, baldness, and other symptoms of aging—begins immediately after birth, and death usually occurs in the teens. This syndrome is caused by a dominant mutation in the gene encoding lamin A, a protein involved in the control of the shape of nuclei in cells. Why this mutation leads to premature aging

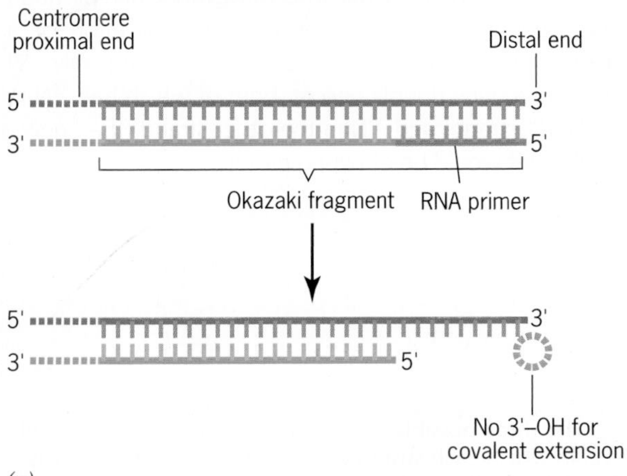

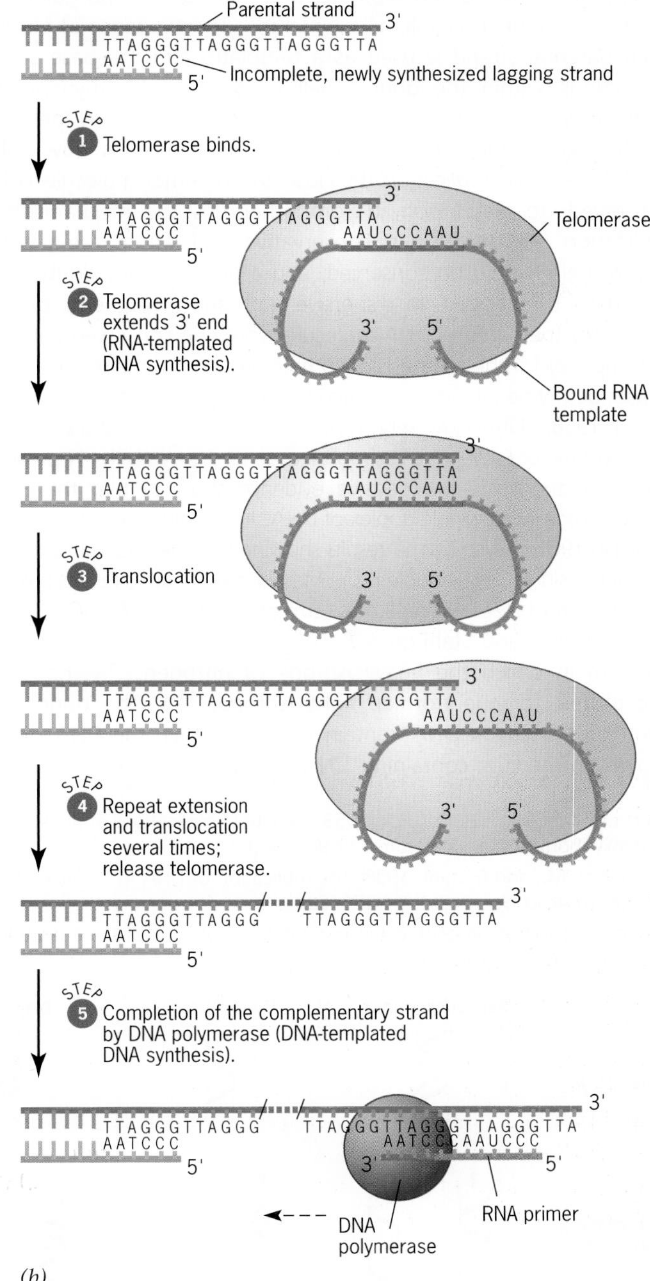

Figure 10.32 ▶ Replication of chromosome telomeres. (*a*) Because of the requirement for a free 3′-OH at the end of the primer strand, DNA polymerases cannot replace an RNA primer that initiates DNA synthesis close to or at the terminus of the lagging strand. (*b*) These termini of chromosomes are replicated by a special enzyme called telomerase, which prevents the ends of chromosomes from becoming shorter during each replication. The nucleotide sequence at the terminus of the lagging strand is specified by a short RNA molecule present as an essential component of telomerase. The telomere sequence shown is that of humans.

▶ A MILESTONE IN GENETICS: DNA Replicates Semiconservatively

Five weeks after the appearance of Watson and Crick's paper describing the double-helix structure of DNA, they published another paper[1] proposing a mechanism by which the double helix could replicate. They proposed that the two complementary strands of the double helix unwind and separate, and that each strand guides the synthesis of a new complementary strand. The sequence of bases in each parental strand is used as a template, and the base-pairing restrictions within the double helix—adenine with thymine and guanine with cytosine—dictate the sequence of bases in the newly synthesized strand. The Watson-Crick mechanism is now called *semiconservative replication* (because the parental molecule is half conserved) to distinguish it from other possible mechanisms of replication (**FIGURE 1**). In conservative replication, the parental double helix would be conserved, and a new progeny double helix would be synthesized. In dispersive replication, segments of both strands of the parental DNA molecule would be conserved and used as templates for the synthesis of complementary segments that would subsequently be joined to produce progeny DNA strands.

In 1958, Matthew Meselson and Franklin Stahl demonstrated that the chromosome of *Escherichia coli* replicates semiconservatively.[2] However, at the time, there was no decisive evidence showing that the *E. coli* chromosome was a single duplex of DNA. John Cairns provided that evidence in 1962.[3] Given Cairns' results showing that the *E. coli* chromosome contains a single DNA double helix, Meselson and Stahl's results provided definitive proof that DNA replicates semiconservatively in *E. coli*.

Meselson and Stahl grew *E. coli* cells for many generations in a medium in which the heavy isotope of nitrogen, ^{15}N, had been substituted for the normal, light isotope, ^{14}N. The purine and pyrimidine bases in DNA contain nitrogen. Thus, the DNA of cells grown on medium containing ^{15}N will have a greater density (mass per unit volume) than the DNA of cells grown on medium containing ^{14}N. Molecules with different densities can be separated by a procedure called **equilibrium density-gradient centrifugation** (see Focus on Centrifugation Techniques). As a result, Meselson and Stahl were able to distinguish between the three possible modes of DNA replication by following the changes in the density of DNA from cells grown on ^{15}N medium and then transferred to ^{14}N medium for various periods of time—so-called density-transfer experiments.

The density of most DNAs is about the same as the density of concentrated solutions of heavy salts such as cesium chloride (CsCl). For example, the density of 6*M* CsCl is about 1.7 g/cm^3. *E. coli* DNA containing ^{14}N has a density of 1.710 g/cm^3. Substitution of ^{15}N for ^{14}N increases the density of *E. coli* DNA to 1.724 g/cm^3. When a 6*M* CsCl solution is centrifuged at very high speeds for long periods of time, an equilibrium density gradient is formed (see Focus on Centrifugation Techniques). If DNA is present in such a gradient, it will move to a position where the density of the CsCl solution is equal to its own density.

Meselson and Stahl took cells that had been growing in medium containing ^{15}N for several generations (and thus contained "heavy" DNA), washed them to remove the medium containing ^{15}N, and transferred them to medium containing ^{14}N. After the cells were allowed to grow in the presence of ^{14}N for varying periods of time, the DNAs were extracted and analyzed in CsCl equilibrium density gradients. The results of their experiment (**FIGURE 2**) are consistent only with semiconservative replication, excluding both conservative and dispersive models of DNA synthesis. All the DNA isolated from cells after one generation of growth in medium containing ^{14}N had a density halfway between the densities of "heavy" DNA and "light" DNA. This intermediate density is usually referred to as "hybrid" density. After two generations of growth in medium containing ^{14}N, half of the DNA was of hybrid density and half was light. These results are precisely those predicted by the Watson and Crick semiconservative mode of replication (**FIGURE 2**). One generation of semiconservative replication of a parental double helix containing ^{15}N in medium containing only ^{14}N would produce two progeny double helices, both of which had ^{15}N in one strand (the "old" strand) and ^{14}N in the other strand (the "new" strand). Such molecules would be of hybrid density.

[1]Watson, J. D., and F. H. C. Crick. 1953. Genetical implications of the structure of deoxyribonucleic acid. *Nature* 171:964–969.

[2]Meselson, M., and F. Stahl. 1958. The replication of DNA in *Escherichia coli*. *Proc. Natl. Acad. Sci., U.S.A.* 44:671–682.

[3]Cairns, J. 1962. A proof that the replication of DNA involves separation of the strands. *Nature* 194: 1274.

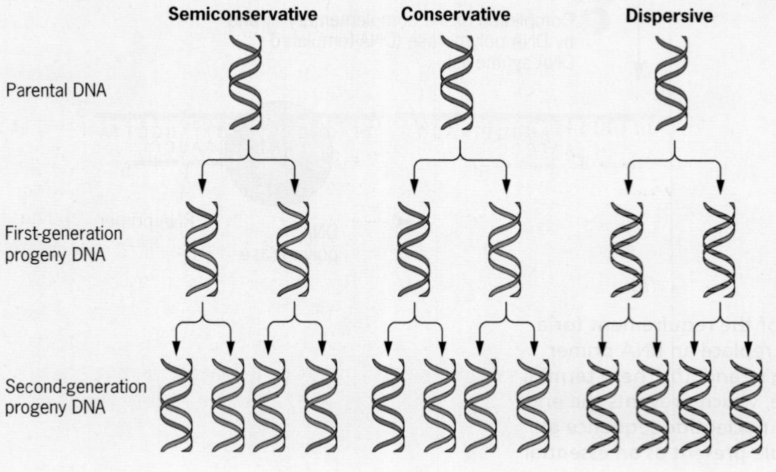

Figure 1 ▶ The three possible modes of DNA replication: (1) semiconservative, in which each strand of the parental double helix is conserved and directs the synthesis of a new complementary progeny strand; (2) conservative, in which the parental double helix is conserved and directs the synthesis of a new progeny double helix; and (3) dispersive, in which segments of each parental strand are conserved and direct the synthesis of new complementary strand segments that are subsequently joined to produce new progeny strands.

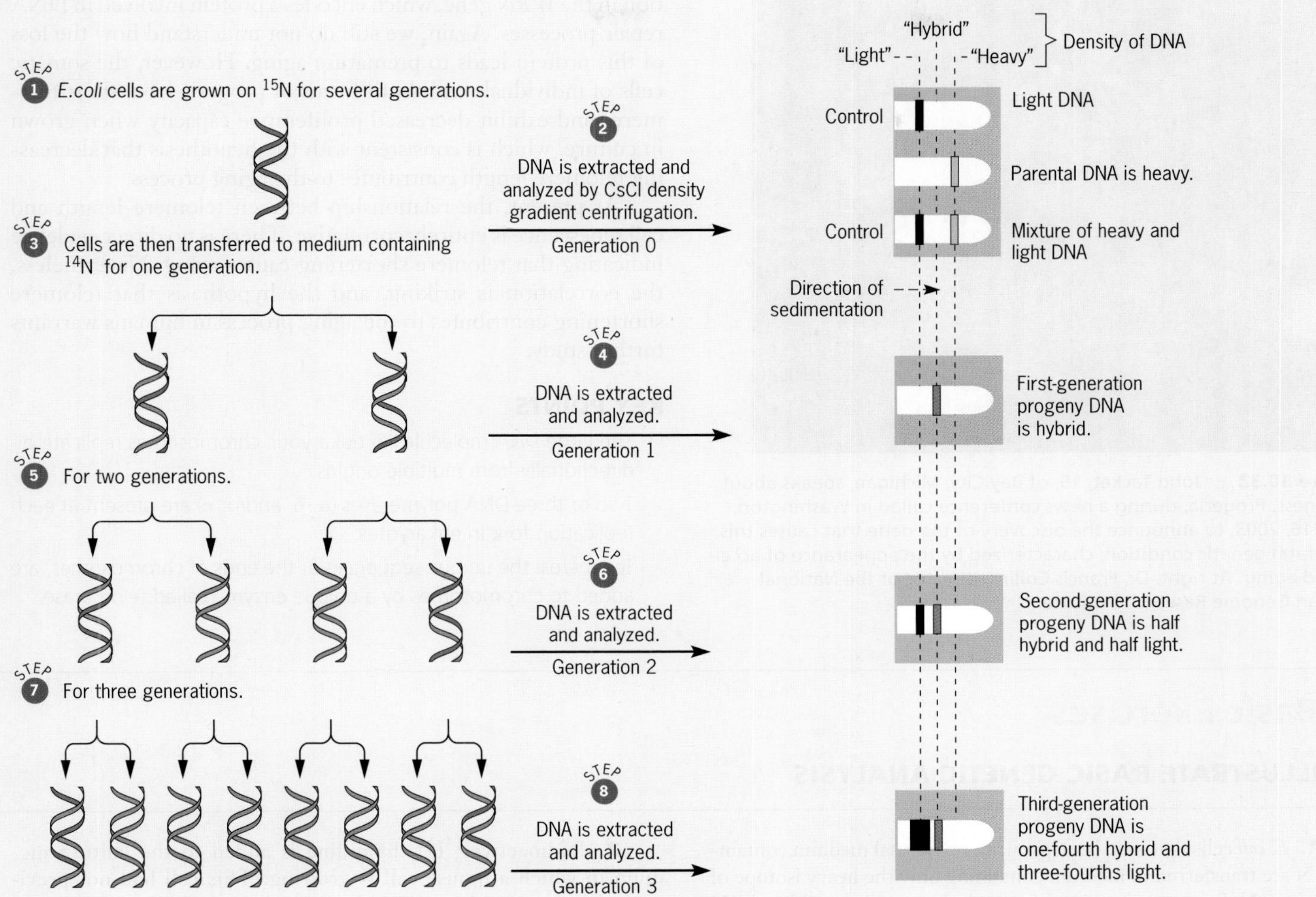

STEP 1 E.coli cells are grown on 15N for several generations.
STEP 2 DNA is extracted and analyzed by CsCl density gradient centrifugation.
Generation 0
STEP 3 Cells are then transferred to medium containing 14N for one generation.
STEP 4 DNA is extracted and analyzed.
Generation 1
STEP 5 For two generations.
STEP 6 DNA is extracted and analyzed.
Generation 2
STEP 7 For three generations.
STEP 8 DNA is extracted and analyzed.
Generation 3
"Hybrid"
"Light"
"Heavy"
Density of DNA
Control
Light DNA
Parental DNA is heavy.
Control
Mixture of heavy and light DNA
Direction of sedimentation
First-generation progeny DNA is hybrid.
Second-generation progeny DNA is half hybrid and half light.
Third-generation progeny DNA is one-fourth hybrid and three-fourths light.

Figure 2 ▶ **Meselson and Stahl's demonstration of semiconservative DNA replication in *E. coli*. The diagram shows that the results of their experiment are those expected if the *E. coli* chromosome replicates semiconservatively. Different results would have been obtained if DNA replication in *E. coli* were either conservative or dispersive (see FIGURE 1).**

Conservative replication would not produce any DNA molecules with hybrid density; after one generation of conservative replication of heavy DNA in light medium, half of the DNA still would be heavy and the other half would be light. If replication were dispersive, Meselson and Stahl would have observed a shift of the DNA from heavy toward light in each generation (that is, "half heavy" or hybrid after one generation, "quarter heavy" after two generations, and so forth). These possibilities are clearly inconsistent with the results of Meselson and Stahl's experiment.

QUESTIONS FOR DISCUSSION

1. DNA replication must be highly accurate to facilitate the transfer of genetic information from parents to offspring. However, it cannot be too accurate because new genetic variation is required for evolution to occur—to allow organisms to adapt to changes in the environment. What determines the proper balance between the accurate transmission of genetic material from parents to progeny and the generation of new variation—that is, the occurrence of new mutations?

2. One copy of the human genome contains about 3 billion nucleotide pairs. Only slightly over 1 percent of this DNA encodes proteins, whereas the noncoding portions of genes make up about 24 percent of the genome. Thus, almost 75 percent of the human genome is intergenic DNA. Given that it takes a lot of energy to replicate this noncoding DNA each time a cell divides, why is this DNA maintained in the genome? Does it perform important functions? What functions might noncoding DNA perform? Some of the intergenic sequences are highly conserved in related species. Does their conservation provide any information about their importance?

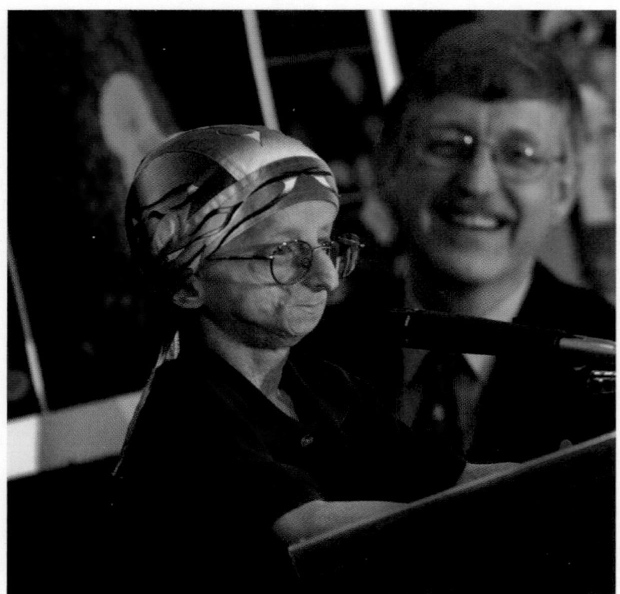

Figure 10.33 ▶ John Tacket, 15, of Bay City, Michigan, speaks about his illness, Progeria, during a news conference called in Washington, April 16, 2003, to announce the discovery of the gene that causes this rare, fatal genetic condition, characterized by the appearance of accelerated aging. At right, Dr. Francis Collins, director of the National Human Genome Research Institute.

is unknown. In a less severe form of progeria, Werner syndrome, senescence begins in the teenage years, with death usually occurring in the 40s. Werner syndrome is caused by a recessive mutation in the *WRN* gene, which encodes a protein involved in DNA repair processes. Again, we still do not understand how the loss of this protein leads to premature aging. However, the somatic cells of individuals with both forms of progeria have short telomeres and exhibit decreased proliferative capacity when grown in culture, which is consistent with the hypothesis that decreasing telomere length contributes to the aging process.

At present, the relationship between telomere length and cell senescence is entirely correlative. There is no direct evidence indicating that telomere shortening causes aging. Nevertheless, the correlation is striking, and the hypothesis that telomere shortening contributes to the aging process in humans warrants further study.

KEY POINTS

- The large DNA molecules in eukaryotic chromosomes replicate bidirectionally from multiple origins.
- Two or three DNA polymerases (α, δ, and/or ε) are present at each replication fork in eukaryotes.
- Telomeres, the unique sequences at the ends of chromosomes, are added to chromosomes by a unique enzyme called telomerase.

▶ Basic Exercises

ILLUSTRATE BASIC GENETIC ANALYSIS

1. *E. coli* cells that have been growing on normal medium containing ^{14}N are transferred to medium containing only the heavy isotope of nitrogen, ^{15}N, for one generation of growth. How will the ^{14}N and ^{15}N be distributed in the DNA of these bacteria after one generation?

Answer: Because DNA replicates semiconservatively, the parental strands of DNA containing ^{14}N will be conserved and used as templates to synthesize new complementary strands containing ^{15}N. Thus, each DNA double helix will contain one light strand and one heavy strand, as shown in the following diagram.

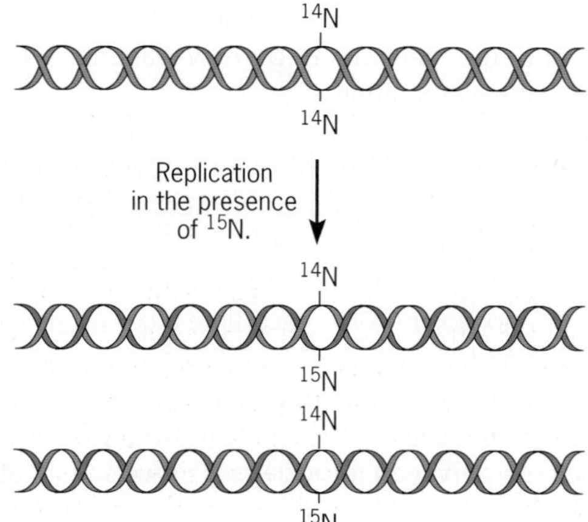

2. Radioactive (^3H) thymidine is added to the culture medium in which a mouse cell is growing. This cell had not previously been exposed to any radioactivity. If the cell is entering S phase at the time the ^3H-thymidine is added, what distribution of radioactivity will be present in the chromosomal DNA at the subsequent metaphase (the first metaphase after the addition of ^3H-thymidine)?

Answer: Remember that each prereplication chromosome contains a single giant DNA molecule extending from one end of the chromosome through the centromere all the way to the other end. This DNA molecule will replicate semiconservatively just like the DNA molecules in *E. coli* discussed in question 1. However, at metaphase, the two progeny double helices will be present in sister chromatids still joined at the centromere, as shown here:

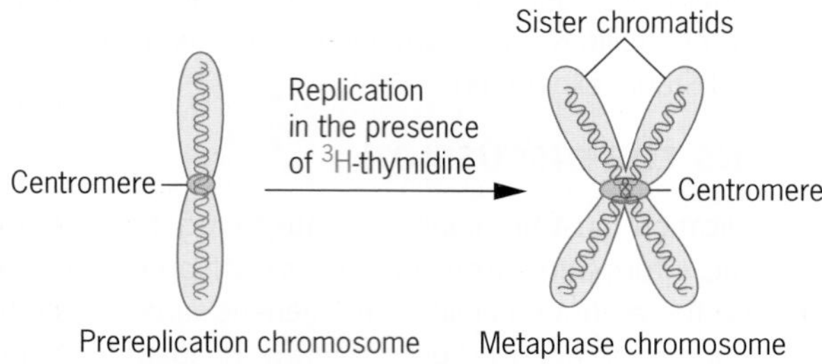

3. How can velocity sucrose gradient centrifugation be used to distinguish between (1) discontinuous synthesis of both nascent DNA strands at each replication fork and (2) discontinuous replication of one strand and continuous replication of the other strand at each fork?

Answer: Use ^{3}H-thymidine to pulse-label DNA in cells growing in culture, then extract the labeled DNA, denature it, and determine what proportion of the label is present in Okazaki fragments by velocity sucrose gradient centrifugation (see Focus on Centrifugation Techniques). If both strands are replicating discontinuously, 100 percent of the pulse-labeled DNA should be present in Okazaki fragments. If only one strand is replicating discontinuously, 50 percent of the label should be present in Okazaki fragments and the other 50 percent in chromosome-size DNA molecules. These two possibilities are diagrammed as follows:

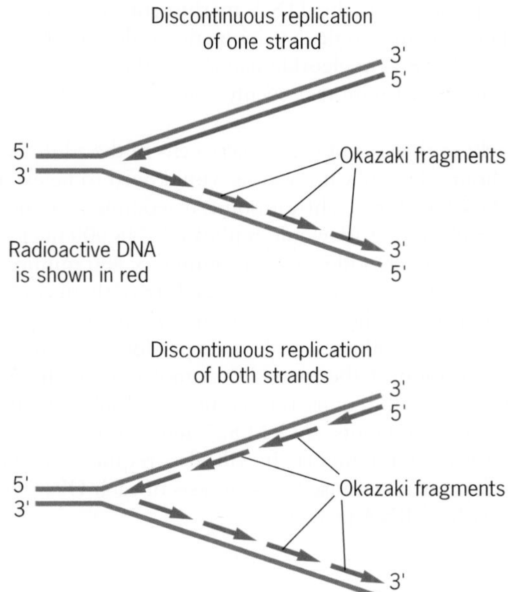

4. DNA polymerases are only able to synthesize DNA in the presence of both a template strand and a primer strand. Why? What are the functions of these two strands?

Answer: DNA polymerases can only extend DNA chains with a free 3′-OH because the mechanism of extension involves a nucleophilic attack by the 3′-OH on the interior phosphorus of the deoxyribonucleoside triphosphate precursor with the elimination of pyrophosphate. The strand with the 3′-OH is the primer strand; it is extended during synthesis. The template strand specifies the nucleotide sequence of the strand being synthesized; the new strand will be complementary to the template strand. These functions are illustrated as follows:

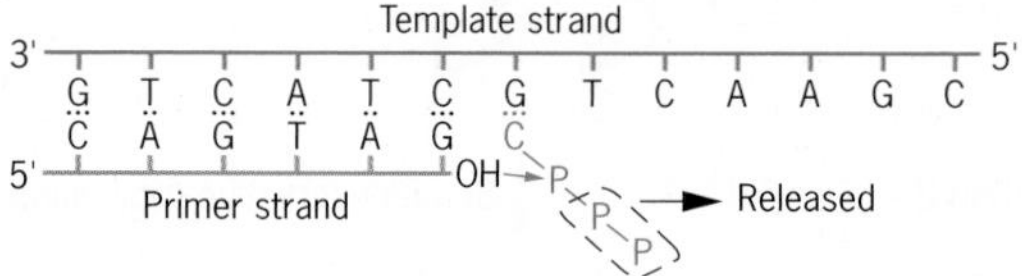

5. How can autoradiography be used to distinguish between uni- and bidirectional replication of DNA?

Answer: If cells are grown in medium containing ^{3}H-thymidine for a short period of time and are then transferred to nonradioactive medium for further growth (a pulse-chase experiment), uni- and bidirectional replication predict different labeling patterns, and these patterns can be distinguished by autoradiography, as shown here:

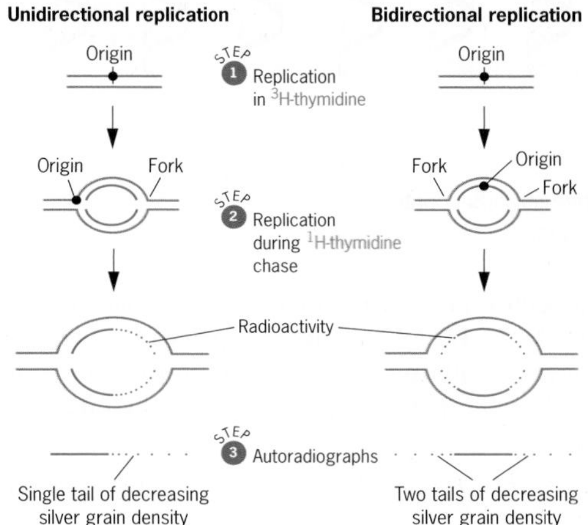

▶ Testing Your Knowledge

INTEGRATE DIFFERENT CONCEPTS AND TECHNIQUES

1. *Escherichia coli* cells were grown for many generations in a medium in which the only available nitrogen was the heavy isotope ^{15}N. The cells were then collected by centrifugation, washed with a buffer, and transferred to a medium containing ^{14}N (the normal light nitrogen isotope). After two generations of growth in the ^{14}N-containing medium, the cells were transferred back to ^{15}N-containing medium for one final generation of growth. After this final generation of growth in the presence of ^{15}N, the cells were collected by centrifugation. The DNA was then extracted from these cells and analyzed by CsCl equilibrium density-gradient centrifugation. How would you expect the DNA from these cells to be distributed in the gradient?

Answer: Meselson and Stahl demonstrated that DNA replication in *E. coli* is semiconservative. Their control experiments showed that DNA double helices with (1) ^{14}N in both strands, (2) ^{14}N in one strand and ^{15}N in the other strand, and (3) ^{15}N in both strands separated into three distinct bands in the gradient, called (1) the light band, (2) the hybrid band, and (3) the heavy band, respectively. If you start with a DNA double helix with ^{15}N in both strands, and replicate it semiconservatively for two generations in the presence of ^{14}N and then for one generation in the presence of ^{15}N, you will end up with eight DNA molecules, two with ^{15}N in both strands and six with ^{14}N in one strand and ^{15}N in the other strand, as shown in the following diagram.

Therefore, 75 percent (6/8) of the DNA will appear in the hybrid band, and 25 percent (2/8) will appear in the heavy band.

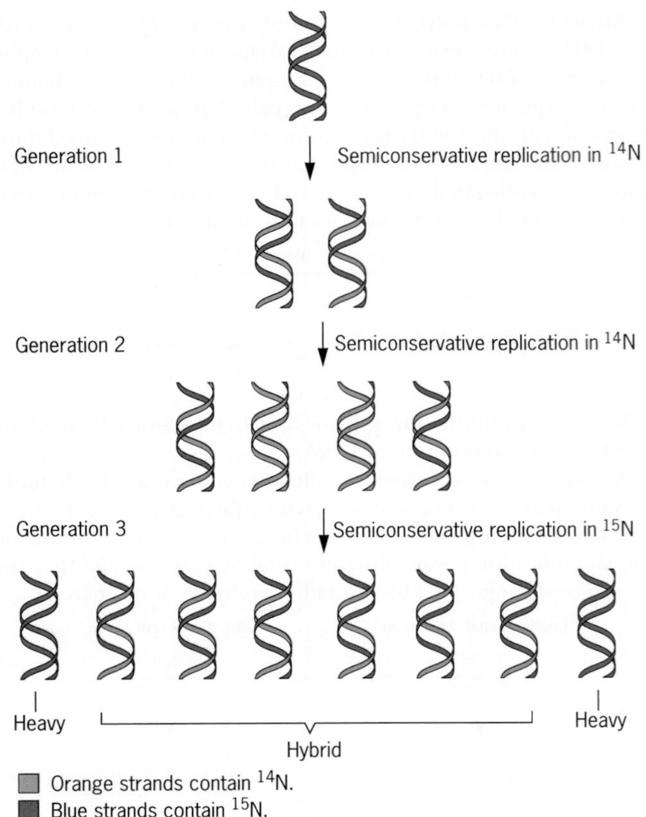

2. The X chromosome of *Drosophila melanogaster* contains a giant DNA molecule 22,422,827 nucleotide pairs long. During the early cleavage stages of embryonic development, nuclear division takes only 10 minutes. If each replication fork travels at the rate of 2,600 nucleotide pairs per minute, how many replication forks would be required to replicate the entire X chromosome in 10 minutes? Assume that these replication forks are evenly spaced along the DNA molecule.

Cell division occurs more slowly in the somatic cells of the adult fruit fly. If you are studying somatic cells with a generation time of 20 hours and an S phase of 8 hours, how many replication forks would be needed to complete the replication of the X chromosome during the S phase of mitosis?

If the average size of Okazaki fragments in *Drosophila* is 250 nucleotides, how many Okazaki fragments are synthesized during the replication of the X chromosome? How many RNA primers are needed?

Answer: If a replication fork moves at the rate of 2600 nucleotide pairs per minute, it will traverse 26,000 nucleotide pairs during 10 minutes and catalyze the synthesis of DNA chains 26,000 nucleotides long in each of the two daughter double helices. Given the presence of 22,422,827 nucleotide pairs in the X chromosome, and the replication of 26,000 nucleotide pairs by each replication fork in 10 minutes, the complete replication of the DNA in this chromosome during the cleavage stages of embryonic development would require 862 replication forks (22,422,827 nucleotide pairs/26,000 nucleotide pairs replicated per fork in 10 minutes) evenly spaced along the DNA molecule.

Similarly, in the case of the somatic cells of the adult fly with an S phase of 8 hours, 18 replication forks would need to be evenly spaced along the DNA in the X chromosome to complete replication in 8 hours. One replication fork would replicate 1,248,000 nucleotide pairs in 8 hours (2600 nucleotide pairs per minute x 480 minutes). Therefore, if replication forks were evenly spaced, 18 of them could replicate the DNA molecule in the X chromosome in 8 hours (22,422,827 nucleotide pairs/1,248,000 nucleotide pairs per fork per 8 hours).

The replication of the giant DNA molecule in the X chromosome of *Drosophila* will require the synthesis and subsequent joining of 89,691 Okazaki fragments (22,422,827 nucleotide pairs/250 nucleotides per Okazaki fragment). It will also require the synthesis of 89,691 RNA primers because the synthesis of each Okazaki fragment is initiated with an RNA primer.

▶ Questions and Problems

ENHANCE UNDERSTANDING AND DEVELOP ANALYTICAL SKILLS

10.1 How might continuous and discontinuous modes of DNA replication be distinguished experimentally?

10.2 A DNA template plus primer with the structure

3′ P —TGCGAATTAGCGACAT— P 5′
5′ P —ATCGGTACGACGCTTAAC—OH 3′

(where P = a phosphate group) is placed in an *in vitro* DNA synthesis system (Mg^{2+}, an excess of the four deoxyribonucleoside triphosphates, etc.) containing a mutant form of *E. coli* DNA polymerase I that lacks 5′ → 3′ exonuclease activity. The 5′ → 3′ polymerase and 3′ → 5′ exonuclease activities of this aberrant enzyme are identical to those of normal *E. coli* DNA polymerase I. It simply has no 5′ → 3′ exonuclease activity.

(a) What will be the structure of the final product?
(b) What will be the first step in the reaction sequence?

10.3 DNA polymerase I of *E. coli* is a single polypeptide of molecular weight 103,000.

(a) What enzymatic activities other than polymerase activity does this polypeptide possess?
(b) What are the *in vivo* functions of these activities?

(c) Are these activities of major importance to an *E. coli* cell? Why?

10.4 *E. coli* cells contain five different DNA polymerases—I, II, III, IV, and V. Which of these enzymes catalyzes the semiconservative replication of the bacterial chromosome during cell division? What are the functions of the other four DNA polymerases in *E. coli*?

10.5 A culture of bacteria is grown for many generations in a medium in which the only available nitrogen is the heavy isotope (^{15}N). The culture is then switched to a medium containing only ^{14}N for one generation of growth; it is then returned to a ^{15}N-containing medium for one final generation of growth. If the DNA from these bacteria is isolated and centrifuged to equilibrium in a CsCl density gradient, how would you predict the DNA to band in the gradient?

10.6 Arrange the following enzymes in the order of their action during DNA replication in *E. coli*: (1) DNA polymerase I, (2) DNA polymerase III, (3) DNA primase, (4) DNA gyrase, and (5) DNA helicase.

10.7 Thirteen distinct DNA polymerases—α, β, γ, δ, ε, ζ, η, θ, ι, κ, λ, μ and σ —have been characterized in mammals. What are the functions of these polymerases?

10.8 GO The *E. coli* chromosome contains approximately 4×10^6 nucleotide pairs and replicates as a single bidirectional replicon in approximately 40 minutes under a wide variety of growth conditions. The largest chromosome of *D. melanogaster* contains about 6×10^7 nucleotide pairs. (a) If this chromosome contains one giant molecule of DNA that replicates bidirectionally from a single origin located precisely in the middle of the DNA molecule, how long would it take to replicate the entire chromosome if replication in *Drosophila* occurred at the same rate as replication in *E. coli*? (b) Actually, replication rates are slower in eukaryotes than in prokaryotes. If each replication bubble grows at a rate of 5000 nucleotide pairs per minute in *Drosophila* and 100,000 nucleotide pairs per minute in *E. coli*, how long will it take to replicate the largest *Drosophila* chromosome if it contains a single bidirectional replicon as described in (a) above? (c) During the early cleavage divisions in *Drosophila* embryos, the nuclei divide every 9 to 10 minutes. Based on your calculations in (a) and (b) above, what do these rapid nuclear divisions indicate about the number of replicons per chromosome in *Drosophila*?

10.9 The Boston teaberry is an imaginary plant with a diploid chromosome number of 4, and Boston teaberry cells are easily grown in suspended cell cultures. ^{3}H-Thymidine was added to the culture medium in which a G_1-stage cell of this plant was growing. After one cell generation of growth in ^{3}H-thymidine-containing medium, colchicine was added to the culture medium. The medium now contained both ^{3}H-thymidine and colchicine. After two "generations" of growth in ^{3}H-thymidine-containing medium (the second "generation" occurring in the presence of colchicine as well), the two progeny cells (each now containing eight chromosomes) were transferred to culture medium containing nonradioactive thymidine (^{1}H-thymidine) plus colchicine. Note that a "generation" in the presence of colchicine consists of a normal cell cycle's chromosomal duplication but no cell division. The two progeny cells were allowed to continue to grow, proceeding through the "cell cycle," until each cell contained a set of metaphase chromosomes that looked like the following.

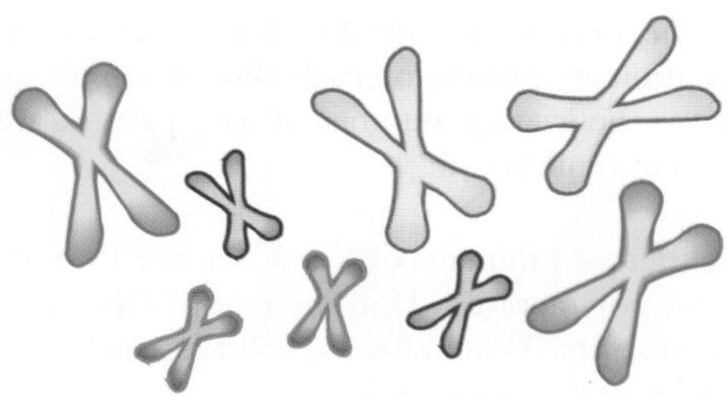

If autoradiography were carried out on these metaphase chromosomes (four large plus four small), what pattern of radioactivity (as indicated by silver grains on the autoradiograph) would be expected? (Assume no recombination between DNA molecules.)

10.10 Suppose that the experiment described in Problem 10.9 was carried out again, except this time replacing the ^{3}H-thymidine with nonradioactive thymidine at the same time that the colchicine was added (after one cell generation of growth in ^{3}H-thymidine-containing medium). The cells were then maintained in colchicine plus nonradioactive thymidine until the metaphase shown in Problem 10.7 occurred. What would the autoradiographs of these chromosomes look like?

10.11 What experimental techniques can be used to separate DNA molecules of mass 3×10^7 daltons isolated from lambda bacteriophage and DNA molecules of mass 1.3×10^8 daltons isolated from T2 bacteriophage?

10.12 The bacteriophage lambda chromosome has several AT-rich segments that denature when exposed to pH 11.05 for 10 minutes. After such partial denaturation, the linear packaged form of the lambda DNA molecule has the structure shown in **FIGURE 10.7a**. Following its injection into an *E. coli* cell, the lambda DNA molecule is converted to a covalently closed circular molecule by hydrogen bonding between its complementary single-stranded termini and the action of DNA ligase. It then replicates as a θ-shaped structure. The entire lambda chromosome is 17.5 μm long. It has a unique origin of replication located 14.3 μm from the left end of the linear form shown in **FIGURE 10.7a.** Draw the structure that would be observed by electron microscopy after both (1) replication of an approximately 6-μm-long segment of the lambda chromosomal DNA molecule (*in vivo*) and (2) exposure of this partially replicated DNA molecule to pH 11.05 for 10 minutes (*in vitro*), (a) *if* replication had proceeded bidirectionally from the origin, and (b) *if* replication had proceeded unidirectionally from the origin.

10.13 Why must each of the giant DNA molecules in eukaryotic chromosomes contain multiple origins of replication?

10.14 GO One species of tree has a very large genome consisting of 1.8×10^{12} base pairs of DNA.

(a) If this DNA was organized into a single linear molecule, how long (meters) would this molecule be?
(b) If the DNA is evenly distributed among 10 chromosomes and each chromosome has one origin of DNA replication, how long would it take to complete the S phase of the cell cycle, assuming that DNA polymerase can synthesize 2×10^4 bp of DNA per minute?
(c) An actively growing cell can complete the S phase of the cell cycle in approximately 300 minutes. Assuming that the origins

of replication are evenly distributed, how many origins of replication are present on each chromosome?
(d) What is the average number of base pairs between adjacent origins of replication?

10.15 Other *polA* mutants of *E. coli* lack the 3′ → 5′ exonuclease activity of DNA polymerase I. Will the rate of DNA synthesis be altered in these mutants? What effect(s) will these *polA* mutations have on the phenotype of the organism?

10.16 In *E. coli*, viable *polA* mutants have been isolated that produce a defective gene product with little or no 5′ → 3′ polymerase activity, but normal 5′ → 3′ exonuclease activity. However, no *polA* mutant has been identified that is completely deficient in the 5′ → 3′ exonuclease activity, while retaining 5′ → 3′ polymerase activity, of DNA polymerase I. How can these results be explained?

10.17 What enzyme activity catalyzes each of the following steps in the semiconservative replication of DNA in prokaryotes?
(a) The formation of negative supercoils in progeny DNA molecules.
(b) The synthesis of RNA primers.
(c) The removal of RNA primers.
(d) The covalent extension of DNA chains at the 3′-OH termini of primer strands.
(e) Proofreading of the nucleotides at the 3′-OH termini of DNA primer strands.

10.18 Many of the origins of replication that have been characterized contain AT-rich core sequences. Are these AT-rich cores of any functional significance? If so, what?

10.19 (a) Why isn't DNA primase activity required to initiate rolling-circle replication? (b) DNA primase is required for the discontinuous synthesis of the lagging strand, which occurs on the single-stranded tail of the rolling circle. Why?

10.20 How similar are the structures of DNA polymerase I and DNA polymerase III in *E. coli*? What is the structure of the DNA polymerase III holoenzyme? What is the function of the *dnaN* gene product in *E. coli*?

10.21 In *E. coli*, three different proteins are required to unwind the parental double helix and keep the unwound strands in an extended template form. What are these proteins, and what are their respective functions?

10.22 What is a primosome, and what are its functions? What essential enzymes are present in the primosome? What are the major components of the *E. coli* replisome? How can geneticists determine whether these components are required for DNA replication?

10.23 The *dnaA* gene product of *E. coli* is required for the initiation of DNA synthesis at *oriC*. What is its function? How do we know that the DnaA protein is essential to the initiation process?

10.24 DNA polymerase I is needed to remove RNA primers during chromosome replication in *E. coli*. However, DNA polymerase III is the true replicase in *E. coli*. Why doesn't DNA polymerase III remove the RNA primers?

10.25 In what ways does chromosomal DNA replication in eukaryotes differ from DNA replication in prokaryotes?

10.26 Two mutant strains of *E. coli* each have a temperature-sensitive mutation in a gene that encodes a product required for chromosome duplication. Both strains replicate their DNA and divide normally at 25°C, but are unable to replicate their DNA or divide at 42°C. When cells of one strain are shifted from growth at 25°C to growth at 42°C, DNA synthesis stops immediately. When cells of the other strain are subjected to the same temperature shift, DNA synthesis continues, albeit at a decreasing rate, for about a half hour. What can you conclude about the functions of the products of these two genes?

10.27 In the yeast *S. cerevisiae*, haploid cells carrying a mutation called *est1* (for *e*ver-*s*horter *t*elomeres) lose distal telomere sequences during each cell division. Predict the ultimate phenotypic effect of this mutation on the progeny of these cells.

10.28 GO (a) The chromosome of the bacterium *Salmonella typhimurium* contains about 4×10^6 nucleotide pairs. Approximately how many Okazaki fragments are produced during one complete replication of the *S. typhimurium* chromosome? (b) The largest chromosome of *D. melanogaster* contains approximately 6×10^7 nucleotide pairs. About how many Okazaki fragments are produced during the replication of this chromosome?

10.29 The chromosomal DNA of eukaryotes is packaged into nucleosomes during the S phase of the cell cycle. What obstacles do the size and complexity of both the replisome and the nucleosome present during the semiconservative replication of eukaryotic DNA? How might these obstacles be overcome?

10.30 Assume that the sequence of a double-stranded DNA shown in the following is present at one end of a large DNA molecule in a eukaryotic chromosome.

5′-(centromere sequence)-GATTCCCCGGGAAGCTTGGGGGGCCCATCTTCGTACGTCTTTGCA-3′
3′-(centromere sequence)-CTAAGGGGCCCTTCGAACCCCCCGGGTAGAAGCATGCAGAAACGT-5′

You have reconstituted a eukaryotic replisome that is active *in vitro*. However, it lacks telomerase activity. If you isolate the DNA molecule shown above and replicate it in your *in vitro* system, what products would you expect?

▶ Genomics on the Web

at http://www.ncbi.nlm.nih.gov/

1. DNA polymerase III catalyzes the semiconservative replication of the chromosome in *E. coli*. How many genes encode structural proteins of DNA polymerase III in *E. coli* strain K12? Which genes encode which subunits? Are these genes clustered to a specific region of the *E. coli* chromosome, or are they distributed throughout the chromosome? How large is the gene encoding the alpha subunit of DNA polymerase III in *E. coli* K12?

2. A single gene encodes DNA polymerase I in *E. coli*. What is the name of this gene? How large is the gene? Where is it located on the *E. coli* chromosome? What is the molecular weight of DNA polymerase I? How many amino acids does it contain?

Hint: At the NCBI web site, click Entrez Home → Entrez Gene → Search using protein name and organism, namely, DNA polymerase III AND *Escherichia K12*[orgn]. If you do not limit your search to strain K12, your search results will include the same genes for all of the other *E. coli* strains that have been sequenced. So, for the sake of simplicity, it is best to include K12 in the organism designator. In the search results, click Primary Source "Ecogene" for more information, including nucleotide coordinates and map position of the gene, protein size, and similar information.

Chapter 11 Transcription and RNA Processing

CHAPTER OUTLINE

- Transfer of Genetic Information: The Central Dogma
- The Process of Gene Expression
- Transcription in Prokaryotes
- Transcription and RNA Processing in Eukaryotes
- Interrupted Genes in Eukaryotes: Exons and Introns
- Removal of Intron Sequences by RNA Splicing

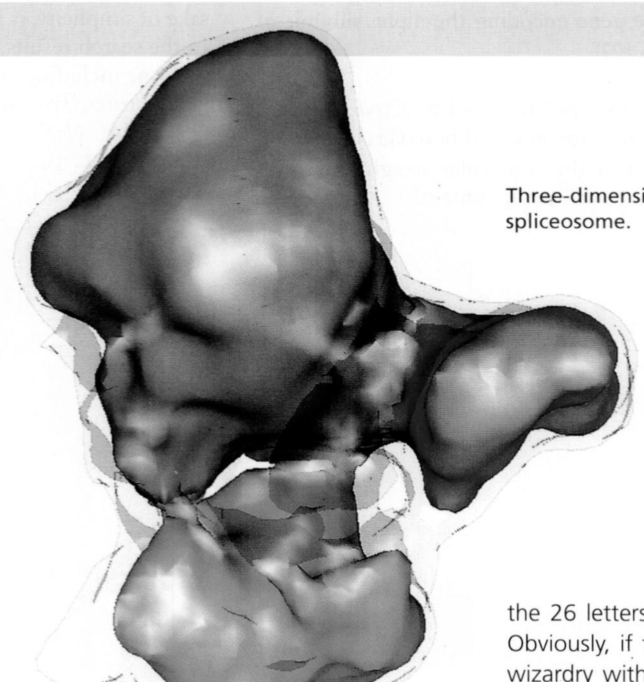

Three-dimensional structure of a mammalian spliceosome.

Storage and Transmission of Information with Simple Codes

We live in the age of the computer. It has an impact on virtually all aspects of our lives, from driving to work to watching spaceships land on the moon. These electronic wizards can store, retrieve, and analyze data with lightning-like speed. The "brain" of the computer is a small chip of silicon, the microprocessor, which contains a sophisticated and integrated array of electronic circuits capable of responding almost instantaneously to coded bursts of electrical energy. In carrying out its amazing feats, the computer uses a binary code, a language based on 0's and 1's. Thus, the alphabet used by computers is like that of the Morse code (dots and dashes) used in telegraphy. Both consist of only two symbols—in marked contrast to the 26 letters of the English alphabet. Obviously, if the computer can perform its wizardry with a binary alphabet, vast amounts of information can be stored and retrieved without using complex codes or lengthy alphabets. In this and the following chapter, we examine (1) how the genetic information of living creatures is written in an alphabet with just four letters, the four base pairs in DNA, and (2) how this genetic information is expressed during the growth and development of an organism. We shall see that RNA plays a key role in the process of gene expression.

Transfer of Genetic Information: The Central Dogma

The central dogma of biology is that information stored in DNA is transferred to RNA molecules during transcription and to proteins during translation.

According to the central dogma of molecular biology, genetic information usually flows (1) from DNA to DNA during its transmission from generation to generation and (2) from DNA to protein during its phenotypic expression in an organism (**FIGURE 11.1**). During the replication of RNA viruses, information is also transmitted from RNA to RNA. The transfer of genetic information from DNA to protein involves two steps: (1) **transcription,** the transfer of the genetic information from DNA to RNA, and (2) **translation,** the transfer of information from RNA to protein. In addition, genetic information flows from RNA to DNA during the conversion of the genomes of RNA tumor viruses to their DNA proviral forms (Chapter 18). Thus, the transfer of genetic information from DNA to RNA is sometimes reversible, whereas the transfer of information from RNA to protein is always irreversible.

TRANSCRIPTION AND TRANSLATION

As we just discussed, the expression of genetic information occurs in two steps: transcription and translation (**FIGURE 11.1**). During transcription, one strand of DNA of a gene is used as a template to synthesize a complementary strand of RNA, called the gene **transcript.** For example, in **FIGURE 11.1,** the DNA strand containing the nucleotide sequence AAA is used as a template to produce the complementary sequence UUU in the RNA transcript. During translation, the sequence of nucleotides in the RNA transcript is converted into the sequence of amino acids in the polypeptide gene product. This conversion is governed by the **genetic code,** the specification of amino acids by nucleotide triplets called **codons** in the gene transcript. For example, the UUU triplet in the RNA transcript shown in **FIGURE 11.1** specifies the amino acid phenylalanine

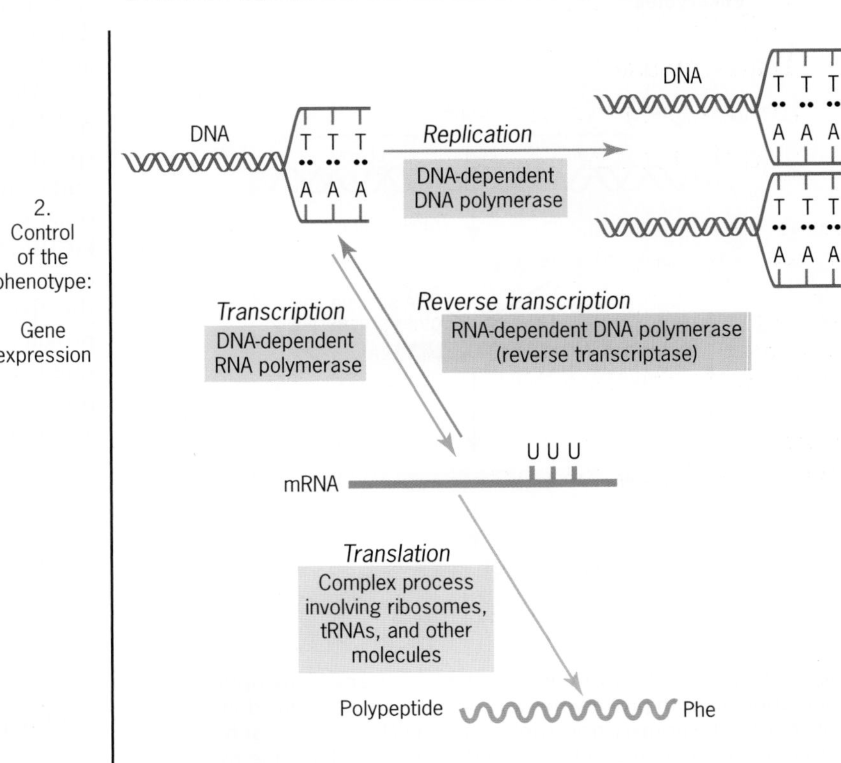

Figure 11.1 ► The flow of genetic information according to the central dogma of molecular biology. Replication, transcription, and translation occur in all organisms; reverse transcription occurs in cells infected with certain RNA viruses. Not shown is the transfer of information from RNA to RNA during the replication of RNA viruses.

(Phe) in the polypeptide gene product. Translation takes place on intricate macromolecular machines called **ribosomes,** which are composed of three to five RNA molecules and 50 to 90 different proteins. However, the process of translation also requires the participation of many other macromolecules. This chapter focuses on transcription; translation is the subject of Chapter 12.

The RNA molecules that are translated on ribosomes are called **messenger RNAs (mRNAs).** In prokaryotes, the product of transcription, the **primary transcript,** usually is equivalent to the mRNA molecule (**FIGURE 11.2*a***). In eukaryotes, primary transcripts often must be processed by the excision of specific sequences and the modification of both termini before they can be translated (**FIGURE 11.2*b***). Thus, in eukaryotes, primary transcripts usually are precursors to mRNAs and, as such, are called **pre-mRNAs.** Most of the nuclear genes in higher eukaryotes and some in lower eukaryotes contain noncoding sequences called *introns* that separate the expressed sequences or *exons* of these genes. The entire sequences of these *split genes* are transcribed into pre-mRNAs, and the noncoding intron sequences are subsequently removed by *splicing reactions* carried out on macromolecular structures called *spliceosomes.*

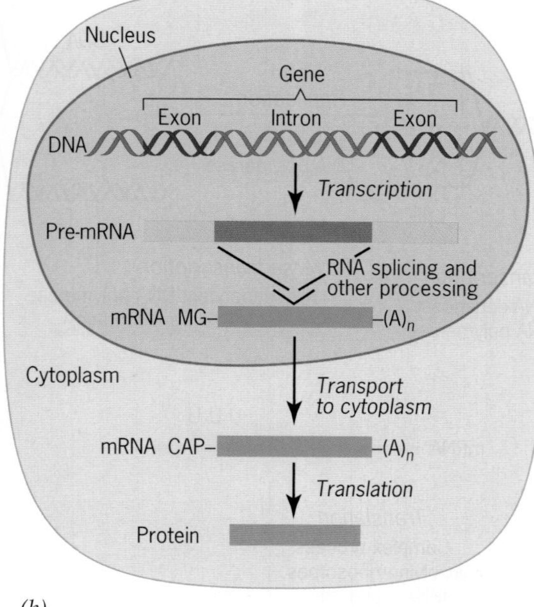

Figure 11.2 ▶ Protein synthesis involves two steps: transcription and translation, in both prokaryotes (*a*) and eukaryotes (*b*). In addition, in eukaryotes, the primary transcripts or pre-mRNAs often must be processed by the excision of introns and the addition of 5′ 7-methyl guanosine caps (CAP) and 3′ poly(A) tails [$(A)_n$].

FIVE TYPES OF RNA MOLECULES

Five different classes of RNA molecules play essential roles in gene expression. We have already discussed messenger RNAs, the intermediaries that carry genetic information from DNA to the ribosomes where proteins are synthesized. **Transfer RNAs (tRNAs)** are small RNA molecules that function as adaptors between amino acids and the codons in mRNA during translation. **Ribosomal RNAs (rRNAs)** are structural and catalytic components of the ribosomes, the intricate machines that translate nucleotide sequences of mRNAs into amino acid sequences of polypeptides. **Small nuclear RNAs (snRNAs)** are structural components of spliceosomes, the nuclear structures that excise introns from nuclear genes. **Micro RNAs (miRNAs)** are short 20 to 22–nucleotide single-stranded RNAs that are cleaved from small hairpin-shaped precursors and block the expression of complementary or partially complementary mRNAs by either causing their degradation or repressing their translation. The roles of mRNAs and snRNAs are discussed in this chapter. The structures and functions of tRNAs and rRNAs will be discussed in detail in Chapter 12. The mechanisms by which miRNAs regulate gene expression are discussed in Chapter 20.

All five types of RNA—mRNA, tRNA, rRNA, snRNA, and miRNA—are produced by transcription. Unlike mRNAs, which specify polypeptides, the final products of tRNA, rRNA, snRNA, and miRNA genes are RNA molecules. Transfer RNA, ribosomal RNA, snRNA, and miRNA molecules are not translated. **FIGURE 11.3** shows an overview of protein synthesis in eukaryotes, emphasizing the transcriptional origin and functions of the five types of RNA molecules. The process is similar in prokaryotes. However, in prokaryotes, the DNA is not separated from the ribosomes by a nuclear envelope. In addition, prokaryotic genes seldom contain noncoding sequences that are removed during RNA transcript processing.

KEY POINTS

- The central dogma of molecular biology is that genetic information flows from DNA to DNA during chromosome replication, from DNA to RNA during transcription, and from RNA to protein during translation.
- Transcription involves the synthesis of an RNA transcript complementary to one strand of DNA of a gene.
- Translation is the conversion of information stored in the sequence of nucleotides in the RNA transcript into the sequence of amino acids in the polypeptide gene product, according to the specifications of the genetic code.

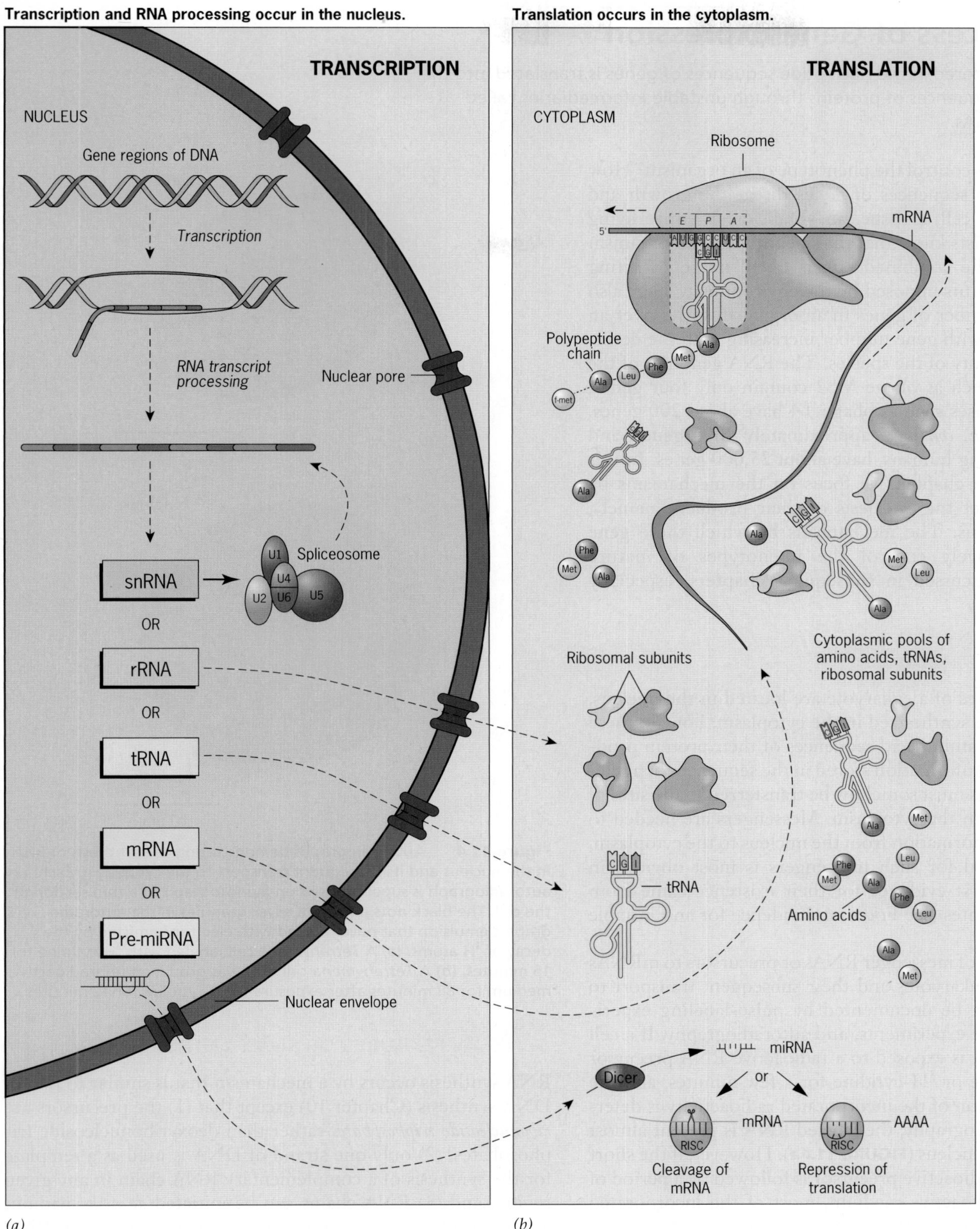

Figure 11.3 ► An overview of protein synthesis, emphasizing the transcriptional origin of miRNA, snRNA, tRNA, rRNA, and mRNA, the splicing function of snRNA, the regulation of gene expression by miRNA, and the translational roles of tRNA, rRNA, mRNA, and ribosomes. Dicer is a nuclease that processes the miRNA precursor into miRNA, and RISC is the *R*NA-*i*nduced *s*ilencing *c*omplex.

The Process of Gene Expression

Information stored in the nucleotide sequences of genes is translated into the amino acid sequences of proteins through unstable intermediaries called messenger RNAs.

How do genes control the phenotype of an organism? How do the nucleotide sequences of genes direct the growth and development of a cell, a tissue, an organ, or an entire living creature? Geneticists know that the phenotype of an organism is produced by the combined effects of all its genes acting within the constraints imposed by the environment. They also know that the number of genes in an organism varies over an enormous range, with gene number increasing with the developmental complexity of the species. The RNA genomes of the smallest viruses such as phage MS2 contain only four genes, whereas large viruses such as phage T4 have about 200 genes. Bacteria such as *E. coli* have approximately 4000 genes, and mammals, including humans, have about 25,000 genes. In this and the following chapter, we focus on the mechanisms by which genes direct the synthesis of their products, namely, RNAs and proteins. The mechanisms by which these gene products collectively control the phenotypes of mature organisms are discussed in subsequent chapters, especially Chapter 21.

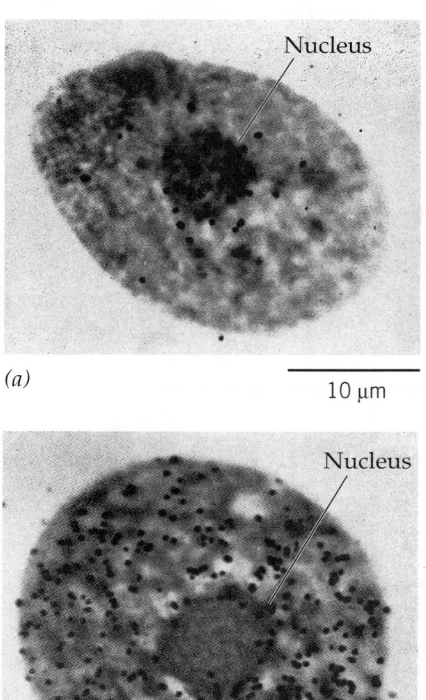

Figure 11.4 ▶ Autoradiographs demonstrating the synthesis of RNA in the nucleus and its subsequent transport to the cytoplasm. Each autoradiograph is superimposed on a photograph of a thin section of the cell. The black dots represent silver granules in the autoradiographic emulsion that have reacted with electrons emitted by the decay of ^{3}H atoms. (*a*) A *Tetrahymena* cell labeled with ^{3}H-cytidine for 15 minutes. (*b*) A *Tetrahymena* cell that was grown on nonradioactive medium for 88 minutes after exposure to ^{3}H-cytidine for 12 minutes.

AN mRNA INTERMEDIARY

If most of the genes of a eukaryote are located in the nucleus, and if proteins are synthesized in the cytoplasm, how do these genes control the amino acid sequences of their protein products? The genetic information stored in the sequences of nucleotide pairs in genes must somehow be transferred to the sites of protein synthesis in the cytoplasm. Messengers are needed to transfer genetic information from the nucleus to the cytoplasm. Although the need for such messengers is most obvious in eukaryotes, the first evidence for their existence came from studies of prokaryotes (see Focus on Evidence for an Unstable Messenger RNA).

The synthesis of messenger RNAs or precursors to mRNAs in the nuclei of eukaryotes and their subsequent transport to the cytoplasm can be documented by pulse-labeling experiments, pulse-chase experiments, and autoradiography. If a cell growing in culture is exposed to a radioactive RNA precursor such as ^{3}H-uridine or ^{3}H-cytidine for a few minutes, and the intracellular location of the incorporated radioactivity is determined by autoradiography, the labeled RNA is present almost exclusively in the nucleus (**FIGURE 11.4*a***). However, if the short exposure to the radioactive precursor is followed by a period of growth in nonradioactive medium, most of the incorporated radioactivity is present in the cytoplasm (**FIGURE 11.4*b***). Thus, the unstable RNA intermediaries are synthesized in the nucleus and are transported to the cytoplasm, where they direct the synthesis of proteins. In this chapter, we focus on the synthesis, processing, and transport of these messenger RNA molecules.

GENERAL FEATURES OF RNA SYNTHESIS

RNA synthesis occurs by a mechanism that is similar to that of DNA synthesis (Chapter 10) except that (1) the precursors are *ribonucleoside triphosphates* rather than deoxyribonucleoside triphosphates, (2) only one strand of DNA is used as a template for the synthesis of a complementary RNA chain in any given region, and (3) RNA chains can be initiated *de novo*, without any requirement for a preexisting primer strand. The RNA molecule produced will be complementary to the DNA **template strand** and identical, except that uridine residues replace thymidines, to the DNA **nontemplate strand** (**FIGURE 11.5**). If the RNA molecule is an mRNA, it will specify amino acids in

► FOCUS ON Evidence for an Unstable Messenger RNA

The first evidence for the existence of an RNA intermediary in protein synthesis came from studies by Elliot Volkin and Lawrence Astrachan on bacteria infected with bacterial viruses. Their results, published in 1956, suggested that the synthesis of viral proteins in infected bacteria involved unstable RNA molecules specified by viral DNA. Volkin and Astrachan observed a burst of RNA synthesis after infecting *E. coli* cells with bacteriophage T2. By labeling RNA with the radioactive isotope ^{32}P, they demonstrated that the newly synthesized RNA molecules were unstable, turning over with half-lives of only a few minutes. In addition, they showed that the nucleotide composition of the unstable RNAs was similar to the composition of T2 DNA and unlike that of *E. coli* DNA. Their results were soon extended by studies in other laboratories.

In 1961, Sol Spiegelman and coworkers reported that the unstable RNAs synthesized in phage T4-infected cells could form RNA–DNA duplexes with denatured T4 DNA, but not with denatured *E. coli* DNA. They pulse-labeled bacteria with ^{3}H-uridine at various times after infection with T4 phage, isolated total RNA from these cells, and determined whether the radioactive RNA molecules hybridized with *E. coli* DNA or phage T4 DNA. Their experiment is diagrammed in **FIGURE 1**.

Their results (**FIGURE 2**) demonstrated that most of the short-lived RNA molecules synthesized after infection were complementary to single strands of phage T4 DNA and noncomplementary to single strands of *E. coli* DNA, indicating that they were produced from phage T4 DNA templates, not from *E. coli* DNA templates.

In the same year that Spiegelman and colleagues published their results, Sydney Brenner, François Jacob, and Matthew Meselson demonstrated that phage T4 proteins were synthesized on *E. coli* ribosomes. Thus, the amino acid sequences of T4 proteins were not controlled by components of the ribosomes. Instead, the ribosomes provided the workbenches on which protein synthesis occured, but did not provide the specifications for individual proteins. These results strengthened the idea, first formally proposed by François Jacob and Jacques Monod in 1961, that unstable RNA molecules carried the specifications for the amino acid sequences of individual gene products from the genes to the ribosomes. Subsequent research firmly established the role of these unstable RNAs, now called messenger RNAs or mRNAs, in the transfer of genetic information from genes to the sites of protein synthesis in the cytoplasm.

STEP 1 Infect *E. coli* cells with bacteriophage T4.
Phage T4
Escherichia coli
STEP 2 Add ^{3}H-uridine to the medium at various times—2, 4, 6, 8, and 10 minutes—after infection, and incubate infected cells for one minute.
Radioactive RNA is synthesized in the bacteria.
^{3}H-uridine in medium and cells
STEP 3 Break open the bacteria and isolate the RNA.
RNA
STEP 4 Determine what proportions of the radioactive RNA hybridize to *E. coli* DNA and to phage T4 DNA.
All DNA is heat-denatured.
Nitrocellulose membranes containing:
Phage T4 DNA
E. coli DNA
No DNA
Hybridization solution containing radioactive RNA
STEP 5 Incubate at 65° C overnight. Remove and wash membranes extensively. Measure radioactivity on each membrane.
Radioactive RNA hybridized to phage T4 DNA
Background radioactivity

Figure 1 ► Spiegelman's experiment.

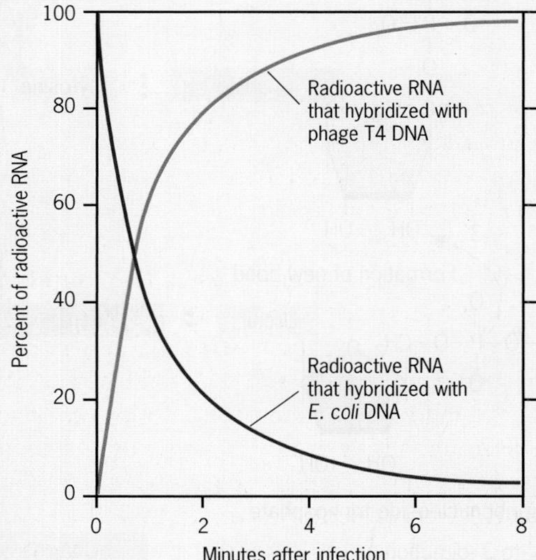

100
80
60
40
20
0
Percent of radioactive RNA
0
2
4
6
8
Minutes after infection
Radioactive RNA that hybridized with phage T4 DNA
Radioactive RNA that hybridized with *E. coli* DNA

Figure 2 ► Rapid switch from the transcription of *E. coli* genes to phage T4 genes in T4-infected bacteria.

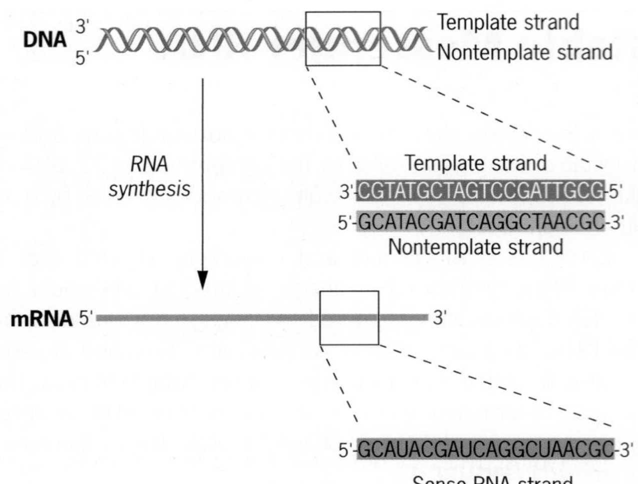

Figure 11.5 ▶ RNA synthesis utilizes only one DNA strand of a gene as template.

the protein gene product. Therefore, mRNA molecules are coding strands of RNA. They are also called **sense strands** of RNA because their nucleotide sequences "make sense" in that they specify sequences of amino acids in the protein gene products. An RNA molecule that is complementary to an mRNA is referred to as **antisense RNA.** This terminology is sometimes extended to the two strands of DNA. However, usage of the terms *sense* and *antisense* to denote DNA strands has been inconsistent. Thus, we will use *template strand* and *nontemplate strand* to refer to the transcribed and nontranscribed strands, respectively, of a gene.

The synthesis of RNA chains, like DNA chains, occurs in the $5' \rightarrow 3'$ direction, with the addition of ribonucleotides to the 3′-hydroxyl group at the end of the chain (**FIGURE 11.6**). The reaction involves a nucleophilic attack by the 3′-OH on the nucleotidyl (interior) phosphorus atom of the ribonucleoside triphosphate precursor with the elimination of pyrophosphate, just as in DNA synthesis. This reaction is catalyzed by enzymes called **RNA polymerases.** The overall reaction is as follows:

$$n(\text{RTP}) \xrightarrow[\text{RNA polymerase}]{\text{DNA template}} (\text{RMP})_n + n(\text{PP})$$

where n is the number of moles of ribonucleotide triphosphate (RTP) consumed, ribonucleotide monophosphate (RMP) incorporated into RNA, and pyrophosphate (PP) produced.

RNA polymerases bind to specific nucleotide sequences called **promoters** and, with the help of proteins called tran-

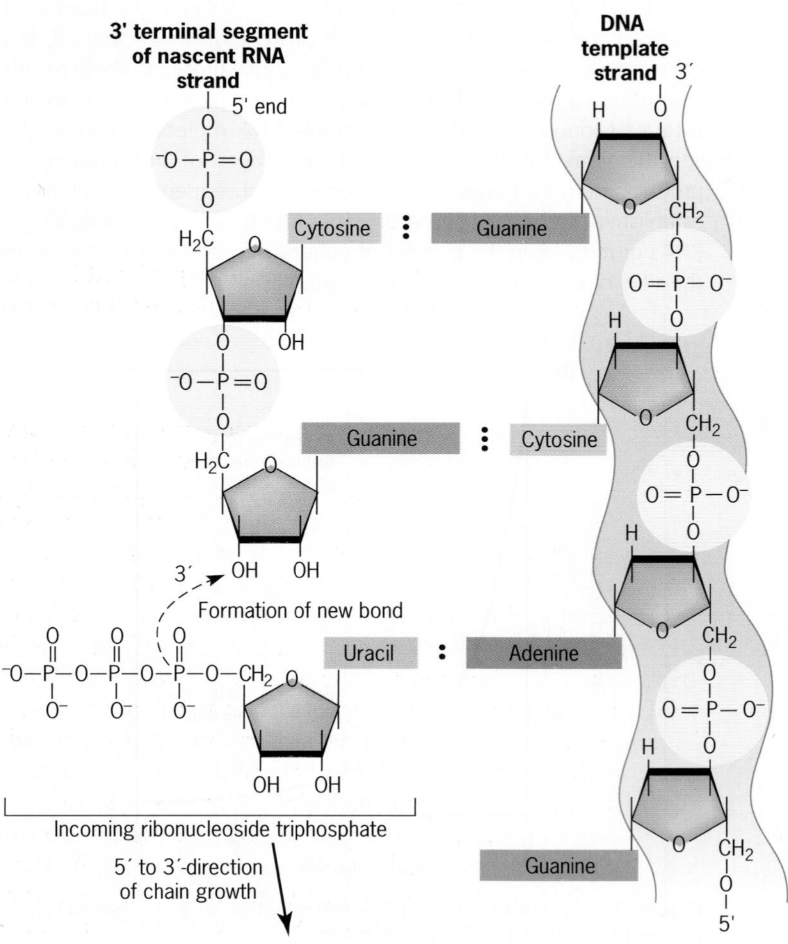

Figure 11.6 ▶ The RNA chain elongation reaction catalyzed by RNA polymerase.

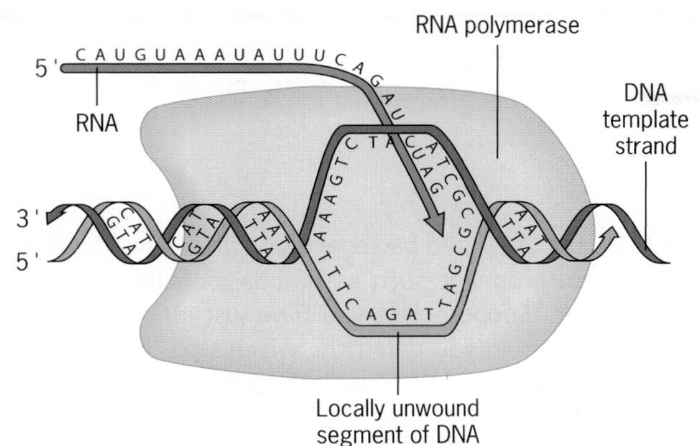

Figure 11.7 ▶ RNA synthesis occurs within a locally unwound segment of DNA. This *transcription bubble* allows a few nucleotides in the template strand to base-pair with the growing end of the RNA chain. The unwinding and rewinding of the DNA molecule are catalyzed by RNA polymerase.

scription factors, initiate the synthesis of RNA molecules at transcription start sites near the promoters. The promoters in eukaryotes are typically more complex than those of prokaryotes. A single RNA polymerase carries out all transcription in most prokaryotes, whereas three different RNA polymerases are present in eukaryotes, with each polymerase responsible for the synthesis of a distinct class of RNAs. RNA synthesis takes place within a locally unwound segment of DNA, sometimes called a **transcription bubble,** which is produced by RNA polymerase (**FIGURE 11.7**). The nucleotide sequence of an RNA molecule is complementary to that of its DNA template strand, and RNA synthesis is governed by the same base-pairing rules as DNA synthesis, but uracil replaces thymine. As a result, the origin of RNA transcripts can be determined by studying their hybridization to DNAs from different sources such as the chromosome(s) of the cell, viruses, and other infectious organisms (see Focus on Problem Solving: Distinguishing RNAs Transcribed from Viral and Host DNAs).

KEY POINTS

- In eukaryotes, genes are present in the nucleus, whereas polypeptides are synthesized in the cytoplasm.
- Messenger RNA molecules function as intermediaries that carry genetic information from DNA to the ribosomes, where proteins are synthesized.
- RNA synthesis, catalyzed by RNA polymerases, is similar to DNA synthesis in many respects.
- RNA synthesis occurs within a localized region of strand separation, and only one strand of DNA functions as a template for RNA synthesis.

Transcription in Prokaryotes

Transcription—the first step in gene expression—transfers the genetic information stored in DNA—genes—into messenger RNA molecules that carry the information to the ribosomes—the sites of protein synthesis—in the cytoplasm.

The basic features of transcription are the same in both prokaryotes and eukaryotes, but many of the details—such as the promoter sequences—are different. The RNA polymerase of *E. coli* has been studied in great detail and will be discussed here. It catalyzes all RNA synthesis in this species. The RNA polymerases of archaea have quite different structures; they will not be discussed here.

A segment of DNA that is transcribed to produce one RNA molecule is called a **transcription unit.** Transcription units may be equivalent to individual genes, or they may include several contiguous genes. Large transcripts that carry the coding sequences of several genes are common in bacteria. The process of transcription can be divided into three stages: (1) **initiation** of a new RNA chain, (2) **elongation** of the chain, and (3) **termination** of transcription and release of the nascent RNA molecule (**FIGURE 11.8**).

When discussing transcription, biologists often use the terms *upstream* and *downstream* to refer to regions located toward the 5′ end and the 3′ end, respectively, of the transcript from some site in the mRNA molecule. These terms are based on the fact that RNA synthesis always occurs in the 5′ to 3′ direction. Upstream and downstream regions of genes are the DNA sequences specifying the corresponding 5′ and 3′ segments of their transcripts relative to a specific reference point.

RNA POLYMERASES: COMPLEX ENZYMES

The RNA polymerases that catalyze transcription are complex, multimeric proteins. The *E. coli* RNA polymerase has a molecular weight of about 480,000 and consists of five polypeptides. Two of these are identical; thus, the enzyme contains four distinct polypeptides. The complete RNA polymerase molecule, the **holoenzyme,** has the composition $\alpha_2\ \beta\ \beta'\ \sigma$. The α subunits are involved in the assembly of the *tetrameric core* ($\alpha_2\ \beta\ \beta'$) of RNA polymerase. The β subunit contains the ribonucleoside triphosphate binding site, and the β' subunit harbors the DNA template binding region. One subunit, the **sigma (σ) factor,** is involved only in the initiation of transcription; it plays no role in chain elongation. After RNA chain initiation has occurred, the σ factor is released, and chain elongation (see **FIGURE 11.6**) is catalyzed by the core enzyme ($\alpha_2\ \beta\beta'$). The function of sigma is to recognize and bind RNA polymerase to the transcription initiation or promoter sites in DNA. The core enzyme (with no σ) will catalyze RNA synthesis from DNA

▶ FOCUS ON PROBLEM SOLVING
Distinguishing RNAs Transcribed from Viral and Host DNAs

THE PROBLEM

E. coli cells that have been infected with a virus present the opportunity for the cells to make two types of RNA transcripts: bacterial and viral. If the virus is a lytic bacteriophage such as T4, only viral transcripts are made; if it is a nonlytic bacteriophage such as M13, both viral and bacterial transcripts are made; and if it is a quiescent prophage such as lambda, only bacterial transcripts are made. Suppose that you have just identified a new DNA virus. How could you determine which types of RNA transcripts are made in cells infected with this virus?

FACTS AND CONCEPTS

1. During the first step in gene expression (transcription), one strand of DNA is used as a template for the synthesis of a complementary strand of RNA.
2. RNA can be labeled with ^3H by growing cells in medium containing ^3H-uridine.
3. DNA can be denatured—separated into its constituent single strands—by exposing it to high temperature or high pH.
4. Viral DNAs and host cell DNAs can both be purified, denatured, and bound to membranes for use in subsequent hybridization experiments (see Figure 1 in Focus on Evidence for an Unstable Messenger RNA).
5. Under the appropriate conditions, complementary single-stranded RNA and DNA molecules will form stable double helices *in vitro*.

ANALYSIS AND SOLUTION

The source of the RNA transcripts being synthesized in virus-infected cells can be determined by incubating the infected cells for a short period of time in medium containing ^3H-uridine, purifying the RNA from these cells, and then hybridizing it to single-stranded viral and bacterial DNAs (see Focus on Evidence for an Unstable Messenger RNA).

a. You should prepare one membrane with denatured viral DNA bound to it, a second membrane with denatured host DNA bound to it, and a third membrane with no DNA to serve as a control to measure nonspecific binding of ^3H-labeled RNA.
b. You should then prepare an appropriate hybridization solution and place the three membranes—one with viral DNA, one with host DNA, and one with no DNA—in this solution.
c. You next add a sample of the purified ^3H-labeled RNA and allow it to hybridize with the DNA on the membranes. Then you wash the membranes thoroughly to remove any nonhybridized RNA. The RNA that remains has either bound specifically to DNA on the membrane or it has bound nonspecifically to the membrane itself. The extent of the RNA binding can be determined by measuring how radioactive each membrane is.
d. Radioactivity on the membrane that had no DNA represents nonspecific "background" binding of RNA to the membrane. This radioactivity can be subtracted from the levels of radioactivity on the other two membranes to measure the specific binding of RNA to viral or bacterial DNA. The results will tell you whether the labeled transcripts were synthesized from viral DNA templates, bacterial DNA templates, or both. With phage T4-infected cells, phage M13-infected cells, and cells containing lambda prophages, the results might be summarized as follows. (The plus signs indicate the presence of RNA transcripts that hybridize specifically.)

	RNA Hybridized to Membrane Containing	
	E. coli **DNA**	**Phage DNA**
Phage T4-infected *E. coli* cells	–	+
Phage M13-infected *E. coli* cells	+	+
E. coli cells carrying lambda prophages	+	–

Which pattern do you observe in cells infected with the newly discovered virus?

For further discussion go to your *WileyPLUS* course.

templates *in vitro*, but, in so doing, it will initiate RNA chains at random sites on both strands of DNA. In contrast, the holoenzyme (σ present) initiates RNA chains *in vitro* only at sites used *in vivo*.

INITIATION OF RNA CHAINS

Initiation of RNA chains involves three steps: (1) the binding of the RNA polymerase holoenzyme to a promoter region in DNA; (2) the localized unwinding of the two strands of DNA by RNA polymerase, providing a template strand free to base-pair with incoming ribonucleotides; and (3) the formation of phosphodiester bonds between the first few ribonucleotides in the nascent RNA chain. The holoenzyme remains bound at the promoter region during the synthesis of the first eight or nine bonds; then the sigma factor is released, and the core enzyme begins the elongation phase of RNA synthesis. During initiation, short chains of two to nine ribonucleotides are synthe-

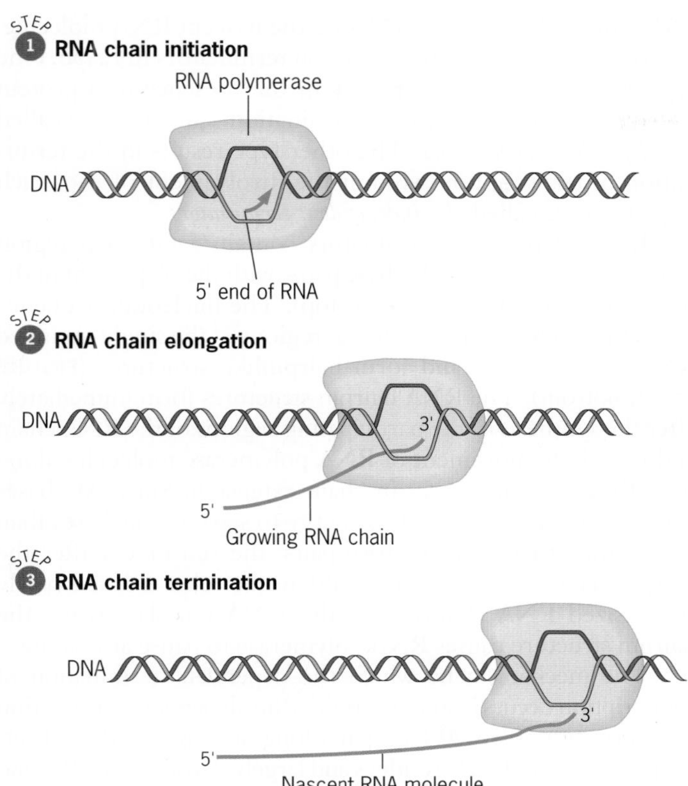

Figure 11.8 ▶ The three stages of transcription: initiation, elongation, and termination.

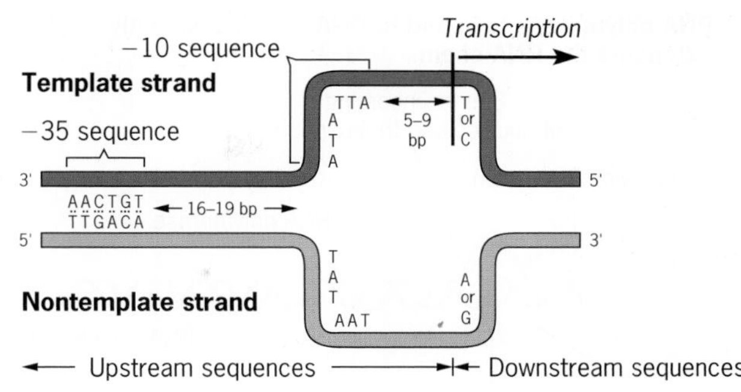

Figure 11.9 ▶ Structure of a typical promoter in *E. coli*. RNA polymerase binds to the –35 sequence of the promoter and initiates unwinding of the DNA strands at the AT-rich –10 sequence. Transcription begins within the transcription bubble at a site five to nine base pairs beyond the –10 sequence.

sized and released. This abortive synthesis stops once chains of 10 or more ribonucleotides have been synthesized and RNA polymerase has begun to move downstream from the promoter.

By convention, the nucleotide pairs or nucleotides within and adjacent to transcription units are numbered relative to the transcript initiation site (designated +1)—the nucleotide pair corresponding to the first (5′) nucleotide of the RNA transcript. Base pairs preceding the initiation site are given minus (−) prefixes; those following (relative to the direction of transcription) the initiation site are given plus (+) prefixes. Nucleotide sequences preceding the initiation site are referred to as **upstream sequences;** those following the initiation site are called **downstream sequences.**

As mentioned earlier, the sigma subunit of RNA polymerase mediates its binding to promoters in DNA. Hundreds of *E. coli* promoters have been sequenced and found to have surprisingly little in common. Two short sequences within these promoters are sufficiently conserved to be recognized, but even these are seldom identical in two different promoters. The midpoints of the two conserved sequences occur at about 10 and 35 nucleotide pairs, respectively, before the transcription-initiation site (**FIGURE 11.9**). Thus they are called the **−10 sequence** and the **−35 sequence,** respectively. Although these sequences vary slightly from gene to gene, some nucleotides are highly conserved. The nucleotide sequences that are present in such conserved genetic elements most often are called **consensus sequences.** The −10 consensus sequence in the nontemplate strand is TATAAT; the −35 consensus sequence is TTGACA. The sigma subunit initially recognizes and binds to the −35 sequence; thus, this sequence is sometimes called the **recognition sequence.** The AT-rich −10 sequence facilitates the localized unwinding of DNA, which is an essential prerequisite to the synthesis of a new RNA chain. The distance between the −35 and −10 sequences is highly conserved in *E. coli* promoters, never being less than 15 or more than 20 nucleotide pairs in length. In addition, the first or 5′ base in *E. coli* RNAs is usually (>90 percent) a purine.

ELONGATION OF RNA CHAINS

Elongation of RNA chains is catalyzed by the RNA polymerase core enzyme, after the release of the σ subunit. The covalent extension of RNA chains (see **FIGURE 11.6**) takes place within the transcription bubble, a locally unwound segment of DNA. The RNA polymerase molecule contains both DNA unwinding and DNA rewinding activities. RNA polymerase continuously unwinds the DNA double helix ahead of the polymerization site and rewinds the complementary DNA strands behind the polymerization site as it moves along the double helix (**FIGURE 11.10**). In *E. coli*, the average length of a transcription bubble is 18 nucleotide pairs, and about 40 ribonucleotides are incorporated into the growing RNA chain per second. The nascent RNA chain is displaced from the DNA template strand as RNA polymerase moves along the DNA molecule. The region of transient base-pairing between the growing chain and the DNA template strand is very short, perhaps only three base pairs in length. The stability of the transcription complex is maintained primarily by the binding of the DNA and the growing RNA chain to RNA polymerase, rather than by the base-pairing between the template strand of DNA and the nascent RNA.

TERMINATION OF RNA CHAINS

Termination of RNA chains occurs when RNA polymerase encounters a **termination signal.** When it does, the transcrip-

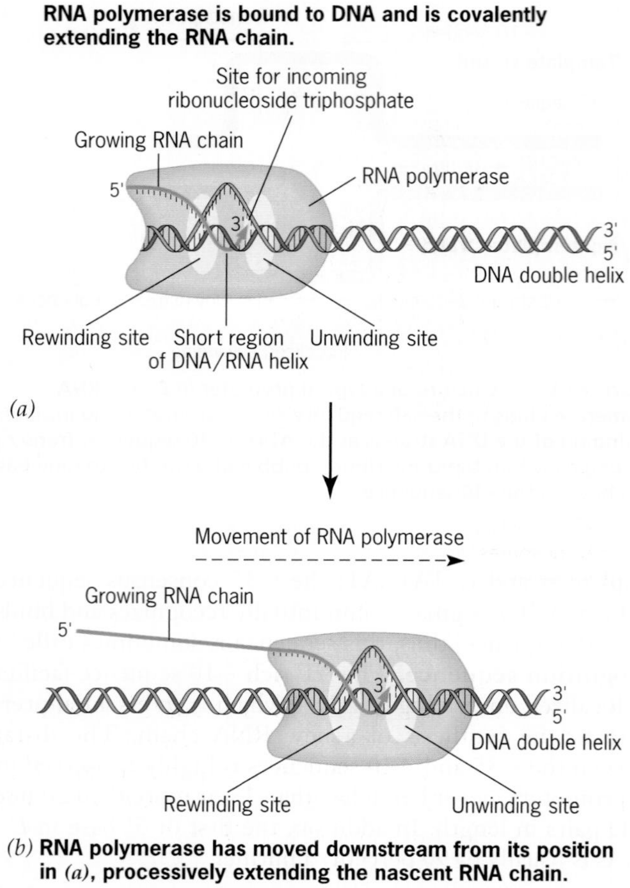

RNA polymerase is bound to DNA and is covalently extending the RNA chain.

Site for incoming ribonucleoside triphosphate

Growing RNA chain

RNA polymerase

5'

3'

3'

5'

DNA double helix

Rewinding site

Short region of DNA/RNA helix

Unwinding site

(a)

Movement of RNA polymerase

Growing RNA chain

5'

3'

3'

5'

DNA double helix

Rewinding site

Unwinding site

(b) **RNA polymerase has moved downstream from its position in *(a)*, processively extending the nascent RNA chain.**

Figure 11.10 ▶ Elongation of an RNA chain catalyzed by RNA polymerase in *E. coli*.

tion complex dissociates, releasing the nascent RNA molecule. There are two types of transcription terminators in *E. coli*. One type results in termination only in the presence of a protein called *rho* (ρ); therefore, such termination sequences are called *rho-dependent terminators*. The other type results in the termination of transcription without the involvement of rho; such sequences are called *rho-independent terminators*.

Rho-independent terminators contain a GC-rich region followed by six or more AT base pairs, with the A's present in the template strand (**FIGURE 11.11,** top). The nucleotide sequence of the GC-rich region is such that regions of the single-stranded RNA can base-pair and form hairpinlike structures (**FIGURE 11.11,** bottom). The RNA hairpin structures form immediately after the synthesis of the participating regions of the RNA chain and retard the movement of RNA polymerase molecules along the DNA, causing pauses in chain extension. Since AU base-pairing is weak, requiring less energy to separate the bases than any of the other standard base pairs, the run of U's after the hairpin region is thought to facilitate the release of the newly synthesized RNA chains from the DNA template when the hairpin structure causes RNA polymerase to pause at this site.

The mechanism by which rho-dependent termination of transcription occurs is still uncertain. Rho-dependent termination sequences are 50 to 90 base pairs long and specify RNA transcripts that are rich in C residues and largely devoid of G's. Beyond that, different rho-dependent termination signals have little in common. The rho protein binds to the growing RNA chain and moves 5′ to 3′ along the RNA, seeming to pursue the RNA polymerase molecule catalyzing the synthesis of the chain. When RNA polymerase slows down or pauses at the rho-dependent

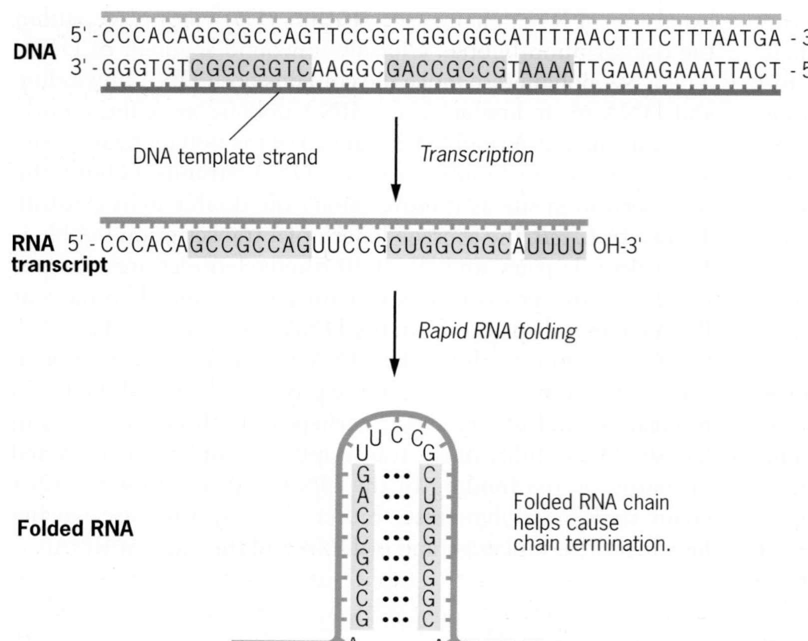

Figure 11.11 ▶ Structure of a rho-independent transcription terminator. Rho-independent terminator sequences contain a GC-rich region followed by at least six AT base pairs. Transcription of such a terminator sequence produces an RNA chain with GC-rich segments that base-pair with each other immediately after synthesis to form a hairpin-like structure. This structure retards the movement of RNA polymerase along the DNA molecule, which results in the termination of transcription in the adjacent AT tract and the release of the nascent RNA chain.

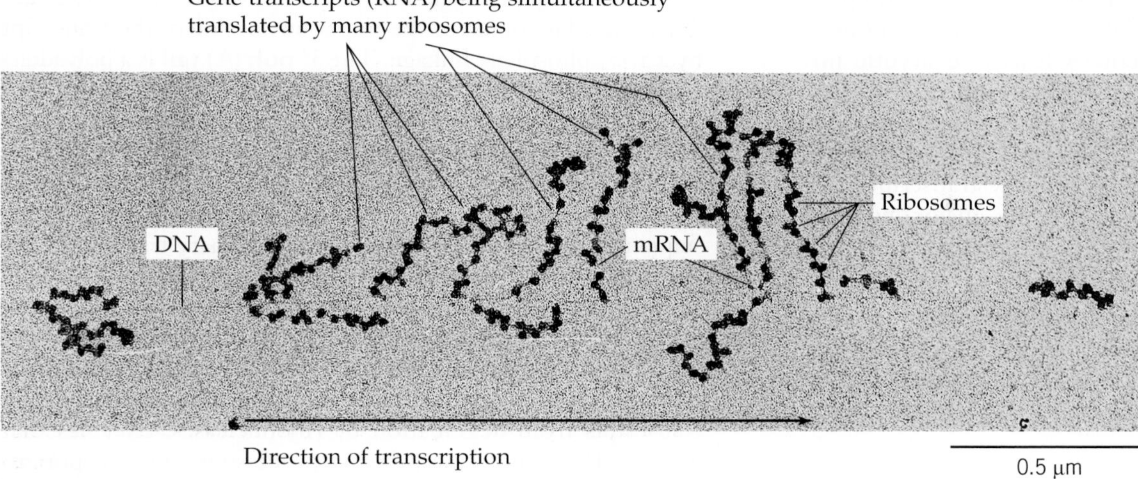

Figure 11.12 ▶ Electron micrograph prepared by Oscar Miller and Barbara Hamkalo showing the coupled transcription and translation of a gene in *E. coli*. DNA, mRNAs, and the ribosomes translating individual mRNA molecules are visible. The nascent polypeptide chains being synthesized on the ribosomes are not visible as they fold into their three-dimensional configuration during synthesis.

termination sequence, rho catches up with the polymerase and pulls the nascent RNA chain from the transcription bubble.

CONCURRENT TRANSCRIPTION, TRANSLATION, AND mRNA DEGRADATION

In prokaryotes, the translation and degradation of an mRNA molecule often begin before its synthesis (transcription) is complete. Since mRNA molecules are synthesized, translated, and degraded in the 5′ to 3′ direction, all three processes can occur simultaneously on the same RNA molecule. In prokaryotes, the polypeptide-synthesizing machinery is not separated by a nuclear envelope from the site of mRNA synthesis. Therefore, once the 5′ end of an mRNA has been synthesized, it can immediately be used as a template for polypeptide synthesis. Indeed, transcription and translation often are tightly coupled in prokaryotes. Oscar Miller, Barbara Hamkalo, and colleagues developed techniques that allowed them to visualize this coupling between transcription and translation in bacteria by electron microscopy. One of their photographs showing the coupled transcription of a gene and translation of its mRNA product in *E. coli* is reproduced in **FIGURE 11.12.**

KEY POINTS

- RNA synthesis occurs in three stages: (1) initiation, (2) elongation, and (3) termination.
- RNA polymerases—the enzymes that catalyze transcription—are complex multimeric proteins.
- The covalent extension of RNA chains occurs within locally unwound segments of DNA.
- Chain elongation stops when RNA polymerase encounters a transcription–termination signal.
- Transcription, translation, and degradation of mRNA molecules often occur simultaneously in prokaryotes.

Transcription and RNA Processing in Eukaryotes

Three different enzymes catalyze transcription in eukaryotes, and the resulting RNA transcripts undergo three important modifications, including the excision of noncoding sequences called introns. The nucleotide sequences of some RNA transcripts are modified posttranscriptionally by RNA editing.

Although the overall process of RNA synthesis is similar in prokaryotes and eukaryotes, the process is considerably more complex in eukaryotes. In eukaryotes, RNA is synthesized in the nucleus, and most RNAs that encode proteins must be transported to the cytoplasm for translation on ribosomes. However, there is evidence indicating that some translation in eukaryotes may occur in the nucleus. Prokaryotic mRNAs often contain the coding regions of two or more genes; such mRNAs are said to be multigenic. In contrast, many of the eukaryotic transcripts that have been characterized contain the coding region of a single gene (are monogenic). However, up to one-fourth of the transcription units in the small worm *Caenorhabditis elegans* may be multigenic. Clearly, eukaryotic mRNAs may be either monogenic or multigenic. Three different RNA polymerases are present in eukaryotes; each enzyme catalyzes the transcription of a specific class of genes. Moreover, in eukaryotes, the majority of the primary transcripts of genes that encode polypeptides undergo three major modifications prior to their transport to the cytoplasm for translation (**FIGURE 11.13**).

1. 7-Methyl guanosine caps are added to the 5′ ends of the primary transcripts.
2. Poly(A) tails are added to the 3′ ends of the transcripts, which are generated by cleavage rather than by termination of chain extension.
3. When present, intron sequences are spliced out of transcripts.

The **5′ cap** on most eukaryotic mRNAs is a 7-methyl guanosine residue joined to the initial nucleoside of the transcript by a 5′-5′ phosphate linkage. The 3′ **poly(A) tail** is a polyadenosine tract 20 to 200 nucleotides long.

In eukaryotes, the population of primary transcripts in a nucleus is called **heterogeneous nuclear RNA (hnRNA)** because of the large variation in the sizes of the RNA molecules present. Major portions of these hnRNAs are noncoding intron sequences, which are excised from the primary transcripts and degraded in the nucleus. Thus, much of the hnRNA actually consists of pre-mRNA molecules undergoing various processing events before leaving the nucleus. Also, in eukaryotes, RNA transcripts are coated with RNA-binding proteins during or immediately after their synthesis. These proteins protect gene transcripts from degradation by ribonucleases, enzymes that degrade RNA molecules, during processing and transport to the cytoplasm. The average half-life of a gene transcript in eukaryotes is about five hours, in contrast to an average half-

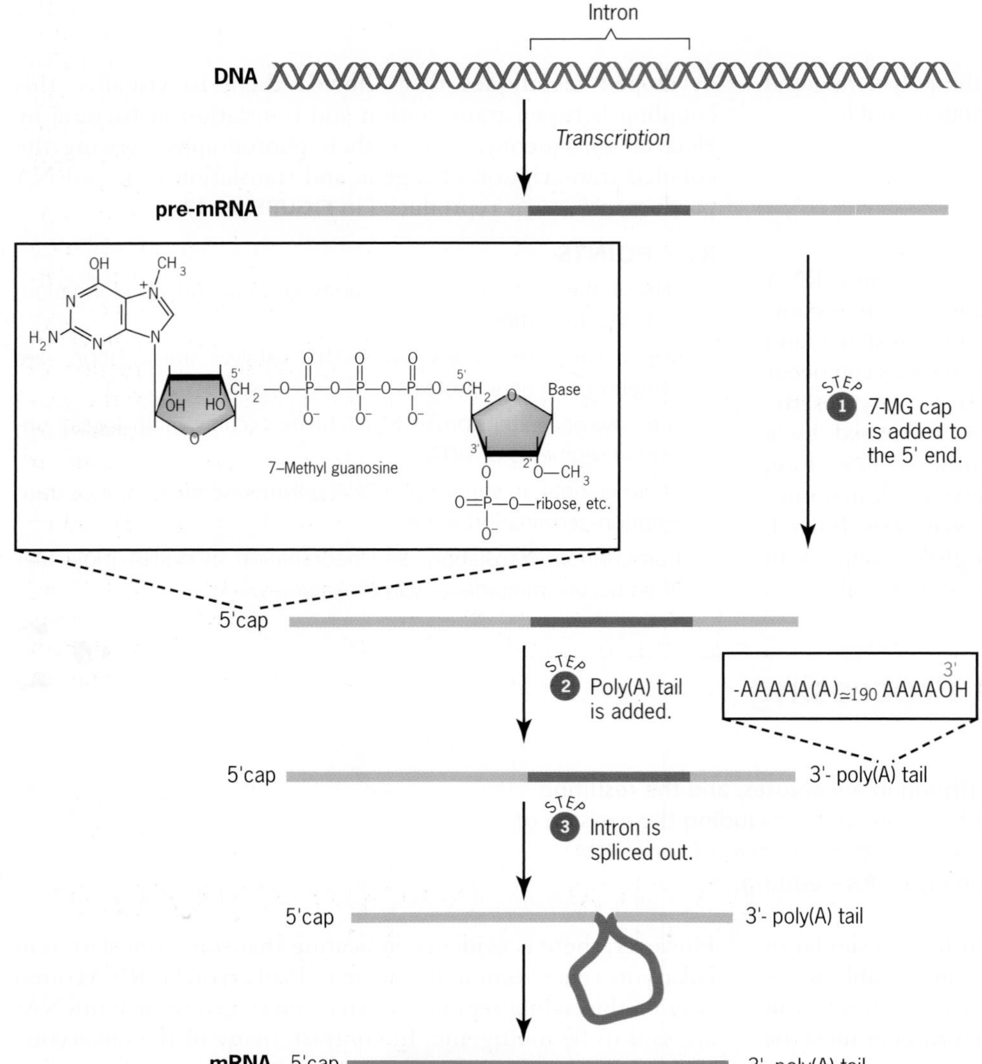

Figure 11.13 ▶ In eukaryotes, most gene transcripts undergo three different types of posttranscriptional processing.

▶ **TABLE 11.1**

Characteristics of the Three RNA Polymerases of Eukaryotes

Enzyme	Location	Products	Sensitivity to α-Amanitin
RNA polymerase I	Nucleolus	Ribosomal RNAs, excluding 5S rRNA	No sensitivity
RNA polymerase II	Nucleus	Nuclear pre-mRNAs	Complete sensitivity
RNA polymerase III	Nucleus	tRNAs, 5S rRNA, and other small nuclear RNAs	Intermediate sensitivity

life of less than five minutes in *E. coli*. This enhanced stability of gene transcripts in eukaryotes is provided, at least in part, by their presence in complexes with RNA-binding proteins.

THREE RNA POLYMERASES/THREE SETS OF GENES

Whereas a single RNA polymerase catalyzes all transcription in *E. coli*, eukaryotes ranging in complexity from the single-celled yeasts to humans contain three different RNA polymerases. All three eukaryotic enzymes, designated **RNA polymerases I, II,** and **III,** are more complex, with 10 or more subunits, than the *E. coli* RNA polymerase. Moreover, unlike the *E. coli* enzyme, all three eukaryotic RNA polymerases require the assistance of other proteins called **transcription factors** in order to initiate the synthesis of RNA chains.

The key features of the three eukaryotic RNA polymerases are summarized in **TABLE 11.1.** RNA polymerase I is located in the nucleolus, a distinct region of the nucleus where rRNAs are synthesized and combined with ribosomal proteins. RNA polymerase I catalyzes the synthesis of all ribosomal RNAs except the small 5S rRNA. RNA polymerase II transcribes nuclear genes that encode proteins and perhaps other genes specifying hnRNAs. RNA polymerase III catalyzes the synthesis of the transfer RNA molecules, the 5S rRNA molecules, and small nuclear RNAs.

The three RNA polymerases exhibit very different sensitivities to inhibition by α-amanitin, a metabolic poison produced by the mushroom *Amanita phalloides*. Whereas RNA polymerase I activity is insensitive to α-amanitin, RNA polymerase II activity is inhibited completely by low concentrations of α-amanitin, and RNA polymerase III exhibits an intermediate level of sensitivity to this drug. Thus, α-amanitin can be used to determine which RNA polymerase catalyzes the transcription of a particular gene.

INITIATION OF RNA CHAINS

Unlike their prokaryotic counterparts, eukaryotic RNA polymerases cannot initiate transcription by themselves. All three eukaryotic RNA polymerases require the assistance of protein transcription factors to start the synthesis of an RNA chain. Indeed, these transcription factors must bind to a promoter region in DNA and form an appropriate initiation complex before RNA polymerase will bind and initiate transcription. Different promoters and transcription factors are utilized by the RNA polymerases I, II, and III. In this section, we focus on the initiation of pre-mRNA synthesis by RNA polymerase II, which transcribes the vast majority of eukaryotic genes.

In all cases, the initiation of transcription involves the formation of a locally unwound segment of DNA, providing a DNA strand that is free to function as a template for the synthesis of a complementary strand of RNA (see **FIGURE 11.7**). The formation of the locally unwound segment of DNA required to initiate transcription involves the interaction of several transcription factors with specific sequences in the promoter for the transcription unit. The promoters recognized by RNA polymerase II consist of short conserved elements, or modules, located upstream from the transcription startpoint. The components of the promoter of the mouse thymidine kinase gene are shown in **FIGURE 11.14.** Other promoters that are recognized by RNA polymerase II contain some, but not all, of these components. The conserved element closest to the transcription start site (position +1) is called the **TATA box;** it has the consensus sequence TATAAAA (reading 5′ to 3′ on the

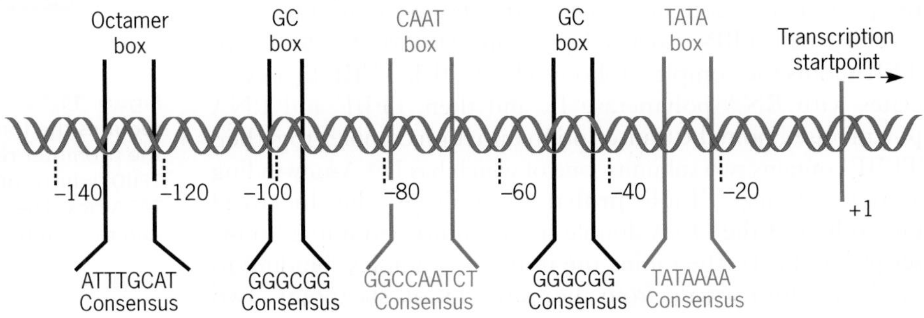

Figure 11.14 ▶ Structure of a promoter recognized by RNA polymerase II. The TATA and CAAT boxes are located at about the same positions in the promoters of most nuclear genes encoding proteins. The GC and octamer boxes may be present or absent; when present, they occur at many different locations, either singly or in multiple copies. The sequences shown here are the consensus sequences for each of the promoter elements. The conserved promoter elements are shown at their locations in the mouse thymidine kinase gene.

nontemplate strand) and is centered at about position −30. The TATA box plays an important role in positioning the transcription startpoint. The second conserved element is called the **CAAT box;** it usually occurs near position −80 and has the consensus sequence GGCCAATCT. Two other conserved elements, the *GC box*, consensus GGGCGG, and the *octamer box*, consensus ATTTGCAT, often are present in RNA polymerase II promoters; they influence the efficiency of a promoter in initiating transcription.

The initiation of transcription by RNA polymerase II requires the assistance of several **basal transcription factors.** Still other transcription factors and regulatory sequences called *enhancers* and *silencers* modulate the efficiency of initiation (Chapter 20). The basal transcription factors must interact with promoters in the correct sequence to initiate transcription effectively (**FIGURE 11.15**). Each basal transcription factor is denoted **TFIIX** (**T**ranscription **F**actor for polymerase **II,** where **X** is a letter identifying the individual factor).

The positions at which these transcription factors bind to the promoter DNA have been established in part by determining which segments of DNA are protected from degradation by nucleases when a specific factor has been added to the initiation complex. The DNA sequences that are protected by transcription factors or other DNA-binding proteins can be identified by DNase footprinting experiments. Homogeneous small DNA molecules that contain the potential binding site are isolated and labeled at one end with a radioactive isotope or a fluorescent dye. Half of the molecules are incubated with the DNA-binding protein of interest, and the other half are maintained protein-free. Both DNA samples are then treated briefly—just long enough so that most DNA molecules are cleaved once—with pancreatic DNase I, an enzyme that makes random internal cuts in DNA. The resulting cleavage products are then separated based on size by gel electrophoresis (see Chapter 9). The protein-free sample will contain DNA fragments of all possible lengths produced by cleavage between all adjacent nucleotide pairs in the molecule. The cleavage products produced from the DNA to which the protein was bound will show a "footprint" where there are no DNA fragment ends. The sequence to which the protein is bound will be protected from cleavage by DNase I. By comparing the radioactive DNA bands produced in the two samples—with and without the DNA-binding protein—a researcher can precisely define the nucleotide sequence to which the protein binds.

TFIID is the first basal transcription factor to interact with the promoter; it contains a TATA-binding protein (TBP) and several small TBP-associated proteins (**FIGURE 11.15**). Next, TFIIA joins the complex, followed by TFIIB. TFIIF first associates with RNA polymerase II, and then TFIIF and RNA polymerase II join the transcription initiation complex together. TFIIF contains two subunits, one of which has DNA-unwinding activity. Thus, TFIIF probably catalyzes the localized unwinding of the DNA double helix required to initiate transcription. TFIIE then joins the initiation complex, binding to the DNA downstream from the transcription startpoint. Two

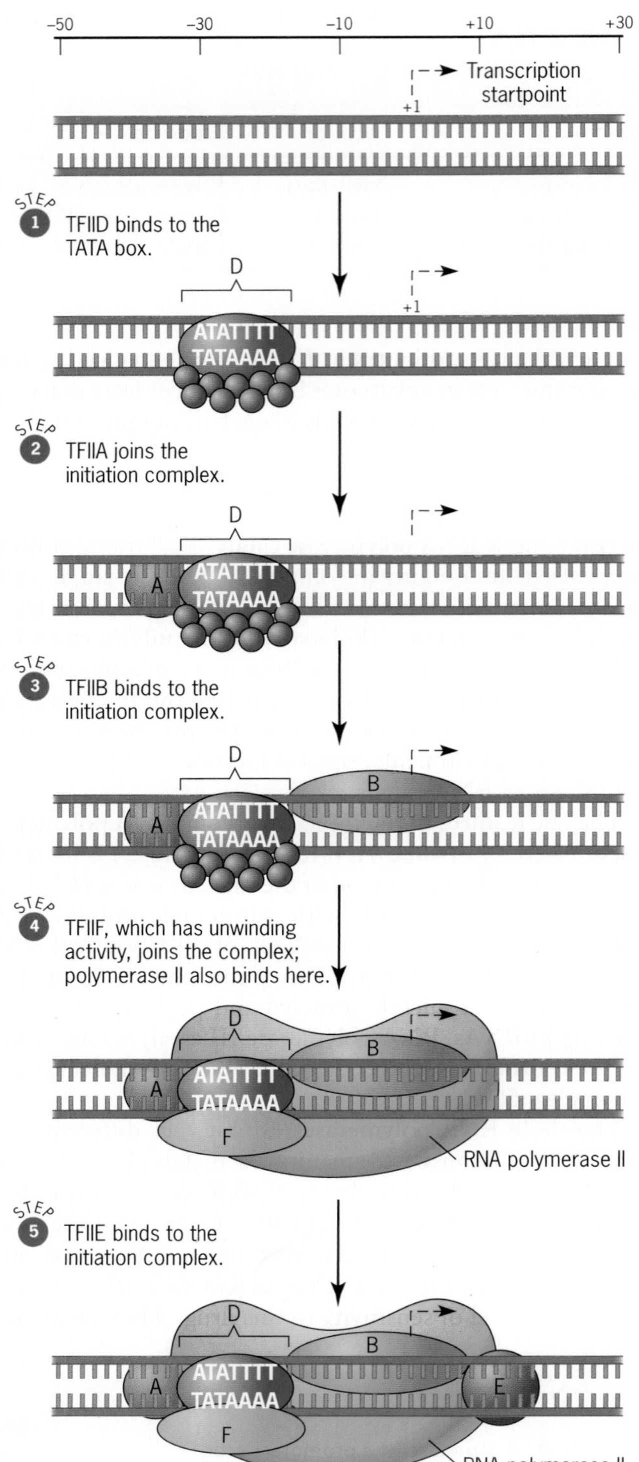

Figure 11.15 ▶ The initiation of transcription by RNA polymerase II requires the formation of a basal transcription initiation complex at the promoter region. The assembly of this complex begins when TFIID, which contains the TATA-binding protein (TBP), binds to the TATA box. The other transcription factors and RNA polymerase II join the complex in the sequence shown.

other factors, TFIIH and TFIIJ, join the complex after TFIIE, but their locations in the complex are unknown. TFIIH has helicase activity and travels with RNA polymerase II during elongation, unwinding the strands in the region of transcription (the "transcription bubble").

RNA polymerases I and III initiate transcription by processes that are similar, but somewhat simpler, than the one used by polymerase II. However, the promoters of genes transcribed by polymerases I and III are quite different from those utilized by polymerase II, even though they sometimes contain some of the same *cis*-acting elements. RNA polymerase I promoters are bipartite, with a core sequence extending from about −45 to +20, and an upstream control element extending from −180 to about −105. The two regions have similar sequences, and both are GC-rich. The core sequence is sufficient for initiation; however, the efficiency of initiation is strongly enhanced by the presence of the upstream control element.

Interestingly, the promoters of most of the genes transcribed by RNA polymerase III are located within the transcription units, downstream from the transcription startpoints, rather than upstream as in units transcribed by RNA polymerases I and II. The promoters of other genes transcribed by polymerase III are located upstream of the transcription start site, just as for polymerases I and II. Actually, polymerase III promoters can be divided into three classes, two of which have promoters located within the transcription unit.

RNA CHAIN ELONGATION AND THE ADDITION OF 5′ METHYL GUANOSINE CAPS

Once eukaryotic RNA polymerases have been released from their initiation complexes, they catalyze RNA chain elongation by the same mechanism as the RNA polymerases of prokaryotes (see **FIGURES 11.6** and **11.7**). Structural analysis of RNA polymerase II of *S. cerevisiae* reveals the presence of surface grooves thought to be DNA and RNA binding sites (**FIGURE 11.16**).

Early in the elongation process, the 5′ ends of eukaryotic pre-mRNAs are modified by the addition of 7-methyl guanosine (7-MG) caps. These 7-MG caps are added when the growing RNA chains are only about 30 nucleotides long (**FIGURE 11.17**). The 7-MG cap contains an unusual 5′-5′ triphosphate linkage (see **FIGURE 11.13**) and two or more methyl groups. These 5′ caps are added co-transcriptionally by the biosynthetic pathway shown in **FIGURE 11.17**. The 7-MG caps are recognized by protein factors involved in the initiation of translation (Chapter 12) and also help protect the growing RNA chains from degradation by nucleases.

Recall that eukaryotic genes are present in chromatin organized into nucleosomes (Chapter 9). How does RNA polymerase transcribe DNA packaged in nucleosomes? Does the nucleosome have to be disassembled before the DNA within it can be transcribed? Surprisingly, RNA polymerase II is able to

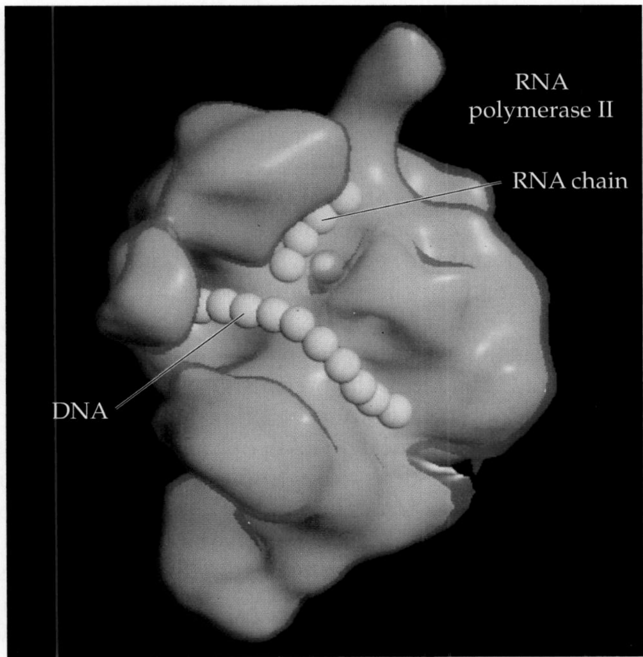

Figure 11.16 ► Model of the structure of RNA polymerase II in the yeast *S. cerevisiae*. Surface grooves on the enzyme are thought to be binding sites for DNA and the growing RNA chain. One groove about 2.5 nm wide and 1 nm deep is the putative binding site for DNA (the 10-beads-long chain); another groove about 1.5 nm wide and 2 nm deep could bind the growing RNA chain (the four-beads-long chain).

Early stage in the transcription of a gene by RNA polymerase II.

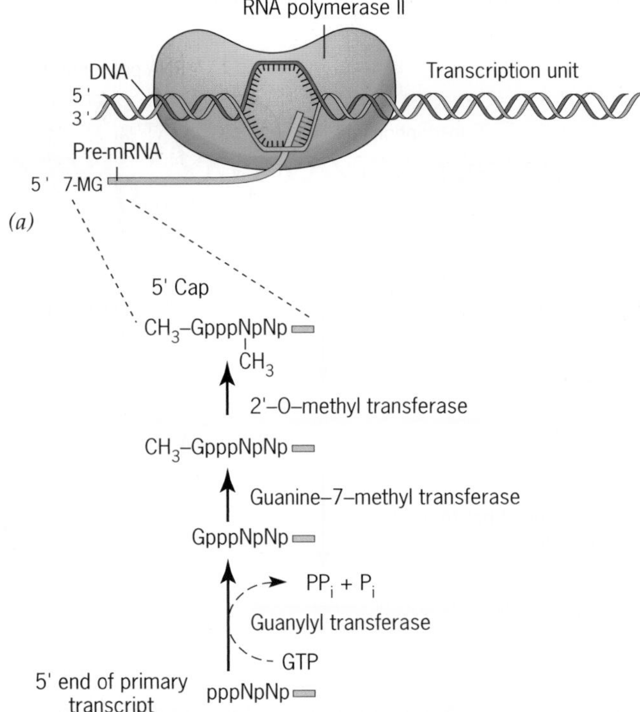

(*b*) **Pathway of biosynthesis of the 7-MG cap.**

Figure 11.17 ► 7-Methyl guanosine (7-MG) caps are added to the 5′ ends of pre-mRNAs shortly after the elongation process begins.

move past nucleosomes with the help of a protein complex called FACT (*fa*cilitates *c*hromatin *t*ranscription), which removes histone H2A/H2B dimers from the nucleosomes leaving histone "hexasomes." After polymerase II moves past the nucleosome, FACT and other accessory proteins help redeposit the histone dimers, restoring nucleosome structure. Also, we should note that chromatin containing genes that are being transcribed has a less compact structure than chromatin containing inactive genes. Chromatin in which active genes are packaged tends to contain histones with lots of acetyl groups (Chapter 9), whereas chromatin with inactive genes contains histones with fewer acetyl groups. These differences are discussed further in Chapter 20.

TERMINATION BY CHAIN CLEAVAGE AND THE ADDITION OF 3′ POLY(A) TAILS

The 3′ ends of RNA transcripts synthesized by RNA polymerase II are produced by endonucleolytic cleavage of the primary transcripts rather than by the termination of transcription (**FIGURE 11.18**). The actual transcription termination events often occur at multiple sites that are located 1000 to 2000 nucleotides downstream from the site that will become the 3′ end of the mature transcript. That is, transcription proceeds beyond the site that will become the 3′ terminus, and the distal segment is removed by endonucleolytic cleavage. The cleavage event that produces the 3′ end of a transcript usually occurs at a site 11 to 30 nucleotides downstream from a conserved sequence, consensus AAUAAA, and upstream from a GU-rich sequence located near the end of the transcription unit. After cleavage, the enzyme

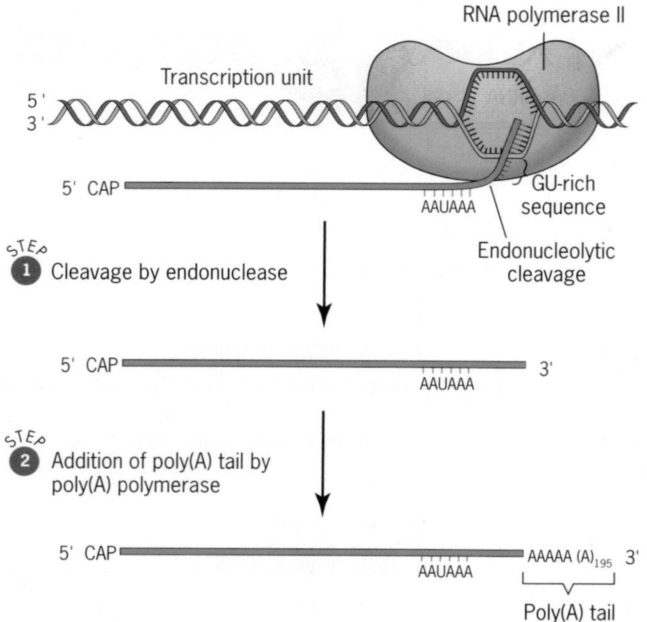

Figure 11.18 ▶ Poly(A) tails are added to the 3′ ends of transcripts by the enzyme poly(A) polymerase. The 3′ end substrates for poly(A) polymerase are produced by endonucleolytic cleavage of the transcript downstream from a polyadenylation signal, which has the consensus sequence AAUAAA.

poly(A) polymerase adds poly(A) tails, tracts of adenosine monophosphate residues about 200 nucleotides long, to the 3′ ends of the transcripts (**FIGURE 11.18**). The addition of poly(A) tails to eukaryotic mRNAs is called **polyadenylation.**

The formation of poly(A) tails on transcripts requires a specificity component that recognizes and binds to the AAUAAA sequence, a stimulatory factor that binds to the GU-rich sequence, an endonuclease, and the poly(A) polymerase. These proteins form a multimeric complex that carries out both the cleavage and the polyadenylation in tightly coupled reactions. The poly(A) tails of eukaryotic mRNAs enhance their stability and play an important role in their transport from the nucleus to the cytoplasm.

In contrast to RNA polymerase II, both RNA polymerase I and III respond to discrete termination signals. RNA polymerase I terminates transcription in response to an 18-nucleotide-long sequence that is recognized by an associated terminator protein. RNA polymerase III reponds to a termination signal that is similar to the rho-independent terminator in *E. coli* (see **FIGURE 11.11**).

RNA EDITING: ALTERING THE INFORMATION CONTENT OF mRNA MOLECULES

According to the central dogma of molecular biology, genetic information flows from DNA to RNA to protein during gene expression. Normally, the genetic information is not altered in the mRNA intermediary. However, the discovery of **RNA editing** has shown that exceptions do occur. RNA editing processes alter the information content of gene transcripts in two ways: (1) by changing the structures of individual bases and (2) by inserting or deleting uridine monophosphate residues.

The first type of RNA editing, which results in the substitution of one base for another base, is rare. This type of editing was discovered in studies of the apolipoprotein-B (*apo-B*) genes and mRNAs in rabbits and humans. Apolipoproteins are blood proteins that transport certain types of fat molecules in the circulatory system. In the liver, the *apo-B* mRNA encodes a large protein 4563 amino acids long. In the intestine, the *apo-B* mRNA directs the synthesis of a protein only 2153 amino acids long. Here, a C residue in the pre-mRNA is converted to a U, generating an internal UAA translation-termination codon, which results in the truncated apolipoprotein (**FIGURE 11.19**). UAA is one of three codons that terminates polypeptide chains during translation. If a UAA codon is produced within the coding region of an mRNA, it will prematurely terminate the polypeptide during translation, yielding an incomplete gene product. The C → U conversion is catalyzed by a sequence-specific RNA-binding protein with an activity that removes amino groups from cytosine residues. A similar example of RNA editing has been documented for an mRNA specifying a protein (the glutamate receptor) present in rat brain cells. More extensive mRNA editing of the C → U type occurs in the mitochondria of plants, where most of the gene transcripts are edited to some degree. Mitochondria have their own DNA

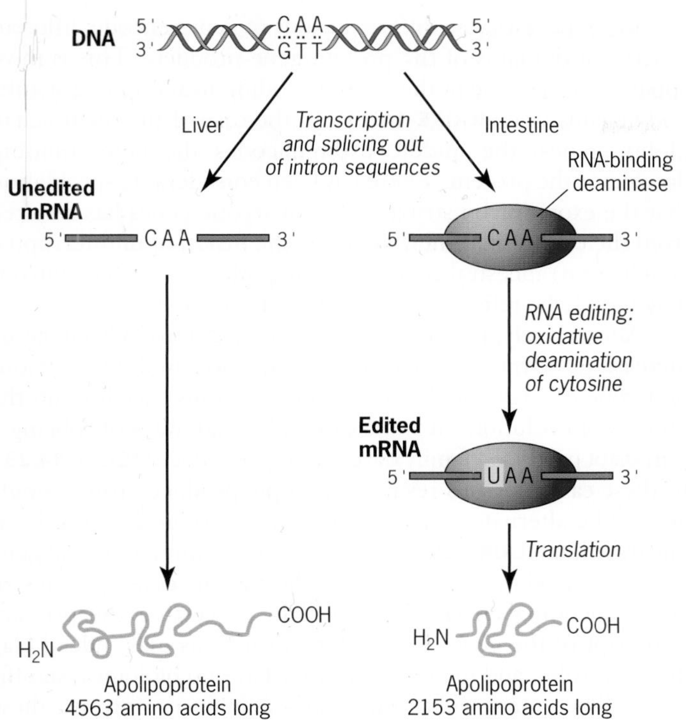

Figure 11.19 ▶ Editing of the apolipoprotein-B mRNA in the intestines of mammals.

genomes and protein-synthesizing machinery (Chapter 16). In some transcripts present in plant mitochondria, most of the C's are converted to U residues.

A second, more complex type of RNA editing occurs in the mitochondria of trypanosomes (a group of flagellated protozoa that causes sleeping sickness in humans). In this case, uridine monophosphate residues are inserted into (occasionally deleted from) gene transcripts, causing major changes in the polypeptides specified by the mRNA molecules. This RNA editing process is mediated by **guide RNAs** transcribed from distinct mitochondrial genes. The guide RNAs contain sequences that are partially complementary to the pre-mRNAs to be edited. Pairing between the guide RNAs and the pre-mRNAs results in gaps with unpaired A residues in the guide RNAs. The guide RNAs serve as templates for editing, as U's are inserted in the gaps in pre-mRNA molecules opposite the A's in the guide RNAs.

Why do these RNA editing processes occur? Why are the final nucleotide sequences of these mRNAs not specified by the sequences of the mitochondrial genes as they are in most nuclear genes? As yet, answers to these interesting questions are purely speculative. Trypanosomes are primitive single-celled eukaryotes that diverged from other eukaryotes early in evolution. Some evolutionists have speculated that RNA editing was common in ancient cells, where many reactions are thought to have been catalyzed by RNA molecules instead of proteins. Another view is that RNA editing is a primitive mechanism for altering patterns of gene expression. For whatever reason, RNA editing plays a major role in the expression of genes in the mitochondria of trypanosomes and plants.

KEY POINTS

- Three different RNA polymerases are present in eukaryotes, and each polymerase transcribes a distinct set of genes.
- Eukaryotic gene transcripts usually undergo three major modifications: (1) the addition of 7-methyl guanosine caps to 5′ termini, (2) the addition of poly(A) tails to 3′ ends, and (3) the excision of noncoding intron sequences.
- The information content of some eukaryotic transcripts is altered by RNA editing, which changes the nucleotide sequences of transcripts prior to their translation.

Interrupted Genes in Eukaryotes: Exons and Introns

Most eukaryotic genes contain noncoding sequences called introns that interrupt the coding sequences, or exons. The introns are excised from RNA transcripts prior to their transport to the cytoplasm.

Most of the well-characterized genes of prokaryotes consist of continuous sequences of nucleotide pairs, which specify colinear sequences of amino acids in the polypeptide gene products. However, in 1977, molecular analyses of three eukaryotic genes yielded a major surprise. Studies of mouse and rabbit β-globin (one of two different proteins in hemoglobin) genes and the chicken ovalbumin (an egg storage protein) gene revealed that they contain noncoding sequences intervening between coding sequences. They were subsequently found in the nontranslated regions of some genes. They are called **introns** (for **int**ervening sequences.) The sequences that remain present in mature mRNA molecules (both coding and noncoding sequences) are called **exons** (for **ex**pressed sequences). The discovery of introns is the subject of this chapter's Milestone (see A Milestone in Genetics: Introns). Geneticists now know that introns interrupt most, but not all, eukaryotic genes.

SOME VERY LARGE EUKARYOTIC GENES

Subsequent to the pioneering studies on the mammalian globin genes and the chicken ovalbumin gene (see Milestone), noncoding introns have been demonstrated in a large number of eukaryotic genes. In fact, interrupted genes are much more common than uninterrupted genes in higher animals and plants. For example, the *Xenopus laevis* gene that encodes vitellogenin A2 (which ends up as egg yolk protein) contains 33 introns, and the chicken 1α2 collagen gene contains at least 50 introns. The collagen gene spans 37,000 nucleotide pairs but gives rise to an mRNA molecule only about 4600 nucleotides long. Other genes contain relatively few introns, but some of the introns are very large. For example, the *Ultrabithorax* (*Ubx*) gene of *Drosophila* contains an intron that is approximately 70,000 nucleotide pairs in length. The largest gene characterized to date is the human *DMD* gene, which causes Duchenne muscular dystrophy when rendered nonfunctional by mutation. The *DMD* gene spans 2.5 million nucleotide pairs and contains 78 introns.

Although introns are present in most genes of higher animals and plants, they are not essential because not all such genes contain introns. The sea urchin histone genes and four *Drosophila* heat-shock genes were among the first animal genes shown to lack introns. We now know that many genes of higher animals and plants lack introns.

INTRONS: BIOLOGICAL SIGNIFICANCE?

At present, scientists know relatively little about the biological significance of the exon–intron structure of eukaryotic genes. Introns are highly variable in size, ranging from about 50 nucleotide pairs to thousands of nucleotide pairs in length. This fact has led to speculation that introns may play a role in regulating gene expression. The transcription of genes with large introns will take longer than the transcription of genes with small introns or no introns. Thus, the presence of large introns in genes would be expected to decrease the rate of transcript accumulation in a cell. The fact that introns accumulate new mutations much more rapidly than exons indicates that many of the specific nucleotide-pair sequences of introns, excluding the ends, are not very important.

In some cases, the different exons of genes encode different functional domains of the protein gene products. This is most apparent in the case of the genes encoding heavy and light antibody chains (see **FIGURE 14.24**). In the case of the mammalian globin genes, the middle exon encodes the heme-binding domain of the protein. There has been considerable speculation that the exon–intron structure of eukaryotic genes has resulted from the evolution of new genes by the fusion of uninterrupted (single exon) ancestral genes. If this hypothesis is correct, introns may merely be relics of the evolutionary process.

Alternatively, introns may provide a selective advantage by increasing the rate at which coding sequences in different exons of a gene can reassort by recombination, thus speeding up the process of evolution. In some cases, alternate ways of splicing a transcript produce a family of related proteins (see **FIGURE 14.23**). In these cases, introns result in multiple products from a single gene. The alternate splicing of the rat troponin T transcript is illustrated in **FIGURE 20.4**. In the case of the mitochondrial gene of yeast encoding cytochrome *b*, the introns contain exons of genes encoding enzymes involved in processing the primary transcript of the gene. Thus, different introns may indeed play different roles, and many introns may have no biological significance. Since many eukaryotic genes contain no introns, these noncoding regions are not required for normal gene expression.

KEY POINTS

- Most, but not all, eukaryotic genes are split into coding sequences called exons and noncoding sequences called introns.
- Some genes contain very large introns; others harbor large numbers of small introns.
- The biological significance of introns is still open to debate.

Removal of Intron Sequences by RNA Splicing

The noncoding introns are excised from gene transcripts by several different mechanisms.

Most nuclear genes that encode proteins in multicellular eukaryotes contain introns. Fewer, but still many, of the genes of unicellular eukaryotes such as the yeasts contain introns. Rare genes of archaea and of a few viruses of prokaryotes also contain introns. In the case of these "split" genes, the primary transcript contains the entire sequence of the gene, and the intron sequences are excised during RNA processing (see **FIGURE 11.13**).

For genes that encode proteins, the splicing mechanism must be precise; it must join exon sequences with accuracy to the single nucleotide to assure that codons in exons distal to introns are read correctly (**FIGURE 11.20**). Accuracy to this degree would seem to require precise splicing signals, presumably nucleotide sequences within introns and at the exon–intron junctions. However, in the primary transcripts of nuclear genes, the only completely conserved sequences of different introns are the dinucleotide sequences at the ends of introns, namely,

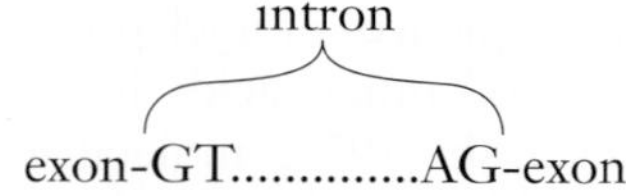

The sequences shown here are for the DNA nontemplate strand (equivalent to the RNA transcript, but with T rather than U). In addition, there are short consensus sequences at the exon–intron junctions. For nuclear genes, the consensus junctions are

exon — intron — exon

$A_{64}G_{73}G_{100}T_{100}A_{68}A_{68}G_{84}T_{63}$............$6Py_{74\text{-}87}N\ C_{65}A_{100}G_{100}$ N

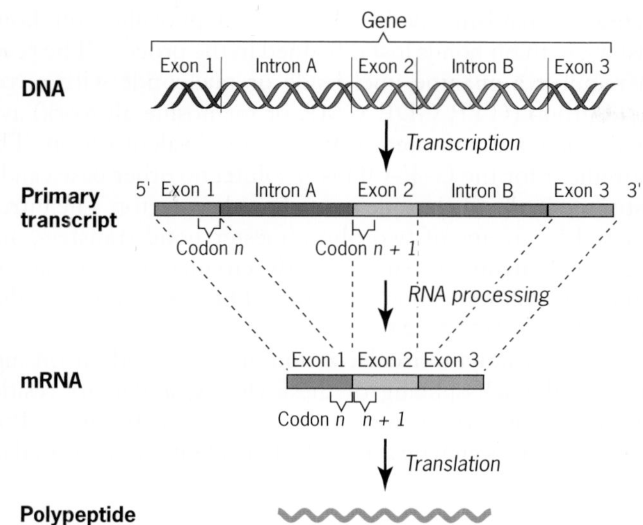

Figure 11.20 ▶ The excision of intron sequences from primary transcripts by RNA splicing. The splicing mechanism must be accurate to the single nucleotide to assure that codons in downstream exons are translated correctly to produce the right amino acid sequence in the polypeptide product.

The numerical subscripts indicate the percentage frequencies of the consensus bases at each position; thus, a 100 subscript indicates that a base is always present at that position. N and Py indicate that any of the four standard nucleotides or either pyrimidine, respectively, may be present at the indicated position. The exon–intron junctions are different for tRNA genes and structural genes in mitochondria and chloroplasts, which utilize different RNA splicing mechanisms (see the following section). There is only one short conserved sequence, the **TACTAAC box,** located about 30 nucleotides upstream from the 3′ splice site of introns in nuclear genes. The TACTAAC box is well conserved in baker's yeast but not in multicellular eukaryotes. The TACTAAC box does exhibit a strong preference for either a purine or a pyrimidine at each site, as follows:

$$Py_{80}\ N\ Py_{80}\ Py_{87}\ Pu_{75}\ A_{100}\ Py_{95}$$

The adenine residue at position six in the TACTAAC box is completely conserved and is known to play a key role in the splicing reaction. With the exception of the terminal dinucleotides and the TACTAAC box, the intron sequences of nuclear genes are highly divergent, apparently random sequences. The introns of genes of mitochondria and chloroplasts also contain conserved sequences, but they are different from those of nuclear genes.

The highly conserved nature of the 5′ and 3′ splice sites and the TACTAAC box indicates that they play an important role in the process of gene expression. Direct evidence for their importance has been provided by mutations at these sites that cause mutant phenotypes in many different eukaryotes. Indeed, such mutations are sometimes responsible for inherited diseases in humans, such as hemoglobin disorders.

The discovery of noncoding introns in genes stimulated intense interest in the mechanism(s) by which intron sequences are removed during gene expression. The early demonstration that the intron sequences in eukaryotic genes were transcribed along with the exon sequences focused research on the processing of primary gene transcripts. Just as *in vitro* systems provided important information about the mechanisms of transcription and translation, the key to understanding RNA splicing events was the development of *in vitro* splicing systems. By using these systems, researchers have shown that there are three distinct types of intron excision from RNA transcripts.

1. The introns of tRNA precursors are excised by precise endonucleolytic cleavage and ligation reactions catalyzed by special splicing endonuclease and ligase activities.
2. The introns of some rRNA precursors are removed autocatalytically in a unique reaction mediated by the RNA molecule itself. (No protein enzymatic activity is involved.)
3. The introns of nuclear pre-mRNA (hnRNA) transcripts are spliced out in two-step reactions carried out by complex ribonucleoprotein particles called spliceosomes.

These three mechanisms of intron excision are discussed in the following three sections. There are other mechanisms of intron excision, but for the sake of brevity they are not discussed here.

tRNA PRECURSOR SPLICING: UNIQUE NUCLEASE AND LIGASE ACTIVITIES

The tRNA precursor splicing reaction has been worked out in detail in the yeast *Saccharomyces cerevisiae*. Both *in vitro* splicing systems and temperature-sensitive splicing mutants have been used in dissecting the tRNA splicing mechanism in *S. cerevisiae*. The excision of introns from yeast tRNA precursors occurs in two stages. In stage I, a nuclear membrane-bound *splicing endonuclease* makes two cuts precisely at the ends of the intron. Then, in stage II, a *splicing ligase* joins the two halves of the tRNA to produce the mature form of the tRNA molecule. The specificity for these reactions resides in conserved three-dimensional features of the tRNA precursors, not in the nucleotide sequences per se.

AUTOCATALYTIC SPLICING

A general theme in biology is that metabolism occurs via sequences of enzyme-catalyzed reactions. These all-important enzymes are generally proteins, sometimes single polypeptides and sometimes complex heteromultimers. Occasionally, enzymes require nonprotein cofactors to perform their functions. When covalent bonds are being altered, it is usually assumed that the reaction is being catalyzed by an enzyme. Thus, the 1982 discovery by Thomas Cech and his coworkers that the intron in the rRNA precursor of *Tetrahymena thermophila* was excised without the involvement of any protein catalytic activity was quite surprising. However, it is now clearly established that the splicing activity that excises the intron from this rRNA precursor is intrinsic to the RNA molecule itself. Indeed,

Cech and Sidney Altman shared the 1989 Nobel Prize in Chemistry for their discovery of catalytic RNAs. Moreover, such *self-splicing* or *autocatalytic activity* has been shown to occur in rRNA precursors of several lower eukaryotes and in a large number of rRNA, tRNA, and mRNA precursors in mitochondria and chloroplasts of many different species. In the case of many of these introns, the self-splicing mechanism is the same as or very similar to that utilized by the *Tetrahymena* rRNA precursors (see **FIGURE 11.21**). For others, the self-splicing mechanism is similar to the splicing mechanism observed with nuclear mRNA precursors, but without the involvement of the spliceosome (see **FIGURE 11.22**).

The autocatalytic excision of the intron in the *Tetrahymena* rRNA precursor and certain other introns requires no external energy source and no protein catalytic activity. Instead, the splicing mechanism involves a series of phosphoester bond transfers, with no bonds lost or gained in the process. The reaction requires a guanine nucleoside or nucleotide with a free 3′-OH group (GTP, GDP, GMP, or guanosine all work) as a cofactor plus a monovalent cation and a divalent cation. The requirement for the G-3′-OH is absolute; no other base can be substituted in the nucleoside or nucleotide cofactor. The intron is excised by means of two phosphoester bond transfers, and the excised intron can subsequently circularize by means of another phosphoester bond transfer. These reactions are diagrammed in **FIGURE 11.21.**

The autocatalytic circularization of the excised intron suggests that the self-splicing of these rRNA precursors resides primarily, if not entirely, within the intron structure itself. Presumably, the autocatalytic activity is dependent on the secondary

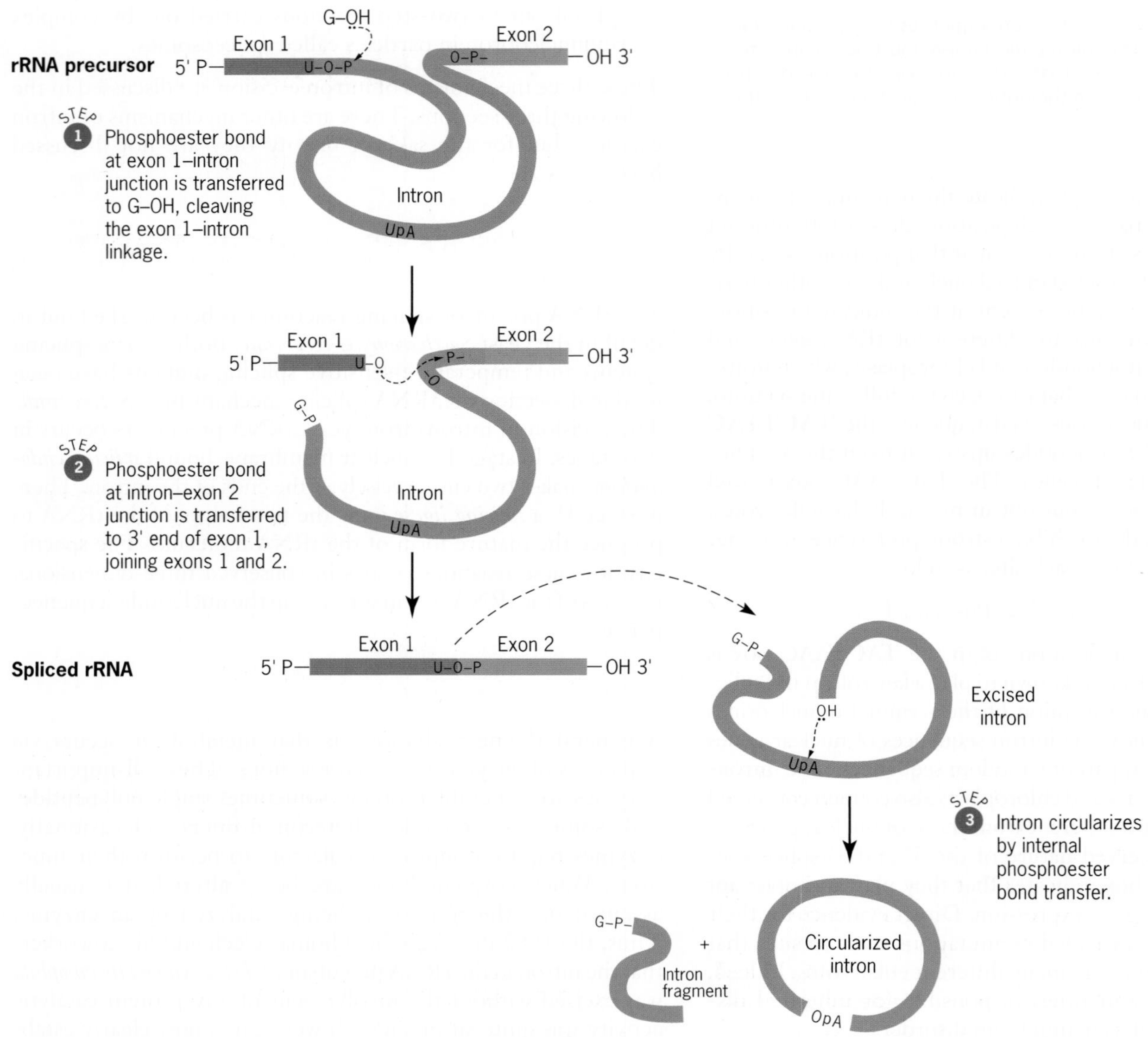

Figure 11.21 ▶ Diagram of the mechanism of self-splicing of the *Tetrahymena thermophila* rRNA precursor and the subsequent circularization of the excised intron.

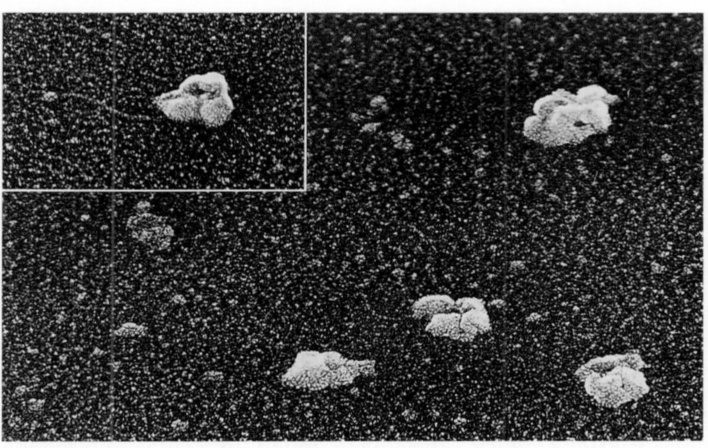

Figure 11.22 ▶ Electron microscope photographs of purified spliceosomes. Note the striking substructure of the particles, which have dimensions of 40 to 60 nanometers. The inset (top left) shows a particle with a thin filament and a smaller particle at the end of the filament.

structure of the intron or at least the secondary structure of the RNA precursor molecule. The secondary structures of these self-splicing RNAs must bring the reactive groups into close juxtaposition to allow the phosphoester bond transfers to occur. Since the self-splicing phosphoester bond transfers are potentially reversible reactions, rapid degradation of the excised introns or export of the spliced rRNAs to the cytoplasm may drive splicing in the forward direction.

Note that the autocatalytic splicing reactions are intramolecular in nature and thus are not dependent on concentration. Moreover, the RNA precursors are capable of forming an active center in which the guanosine-3′-OH cofactor binds. The autocatalytic splicing of these rRNA precursors demonstrates that catalytic sites are not restricted to proteins; however, there is no *trans* catalytic activity as for enzymes, only *cis* catalytic activity. Some scientists believe that autocatalytic RNA splicing may be a relic of an early RNA-based world.

PRE-mRNA SPLICING: snRNAs, snRNPs, AND THE SPLICEOSOME

The introns in nuclear pre-mRNAs are excised in two steps like the introns in yeast tRNA precursors and *Tetrahymena* rRNA precursors that were discussed in the preceding two sections. However, the introns are not excised by simple splicing nucleases and ligases or autocatalytically, and no guanosine cofactor is required. Instead, nuclear pre-mRNA splicing is carried out by complex RNA/protein structures called **spliceosomes** (**FIGURE 11.22**). These structures are in many ways like small ribosomes. They contain a set of small RNA molecules called snRNAs (small nuclear RNAs) and about 40 different proteins. The two stages in nuclear pre-mRNA splicing are known (**FIGURE 11.23**); however, some of the details of the splicing process are still uncertain.

Five snRNAs, called U1, U2, U4, U5, and U6, are involved in nuclear pre-mRNA splicing as components of the spliceosome. (snRNA U3 is localized in the nucleolus and probably is involved in the formation of ribosomes.) In mammals, these snRNAs range in size from 100 nucleotides (U6) to 215 nucleotides (U3). Some of the snRNAs in the yeast *S. cerevisiae* are much larger. These snRNAs do not exist as free RNA molecules. Instead, they are present in small nuclear RNA–protein complexes called **snRNPs** (**s**mall **n**uclear **r**ibo**n**ucleo**p**roteins). Spliceosomes are assembled from four different snRNPs and protein splicing factors during the splicing process. Characterization of snRNPs has been

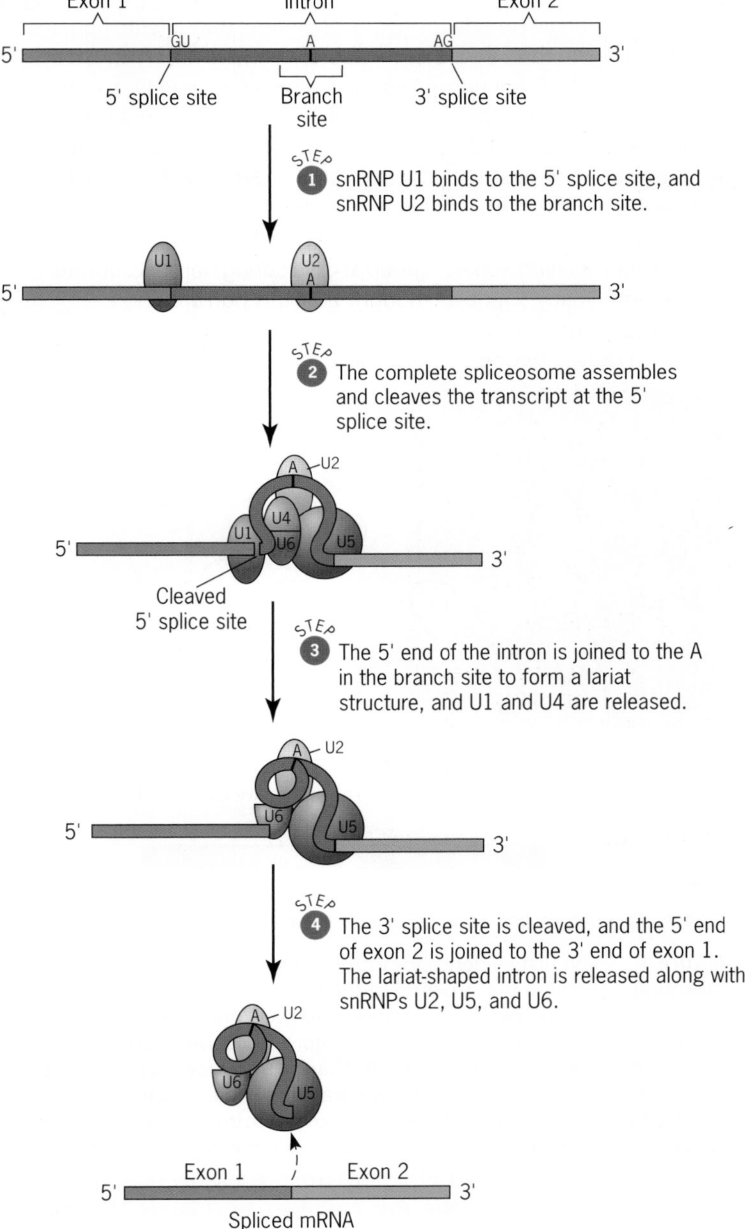

Figure 11.23 ▶ The postulated roles of the snRNA-containing snRNPs in nuclear pre-mRNA splicing.

▶ A MILESTONE IN GENETICS: **Introns**

During the 1960s and early 1970s, an elegantly simple picture of the gene emerged. The gene was a linear sequence of base pairs encoding a colinear sequence of amino acids in its polypeptide product. This view of the gene had been worked out almost exclusively by studying prokaryotes and their viruses. Of course, we also knew that only 1 to 2 percent of the DNA in mammals specified the sequences of amino acids in proteins. Nevertheless, everyone was taken by surprise in 1977 when noncoding sequences were discovered between the coding sequences of eukaryotic genes.

The first evidence that something unexpected was happening came from studies on adenovirus, a DNA virus that infects humans. When Susan Berget and Claire Moore, working in Philip Sharp's laboratory at MIT, hybridized purified adenovirus mRNA to single-stranded adenovirus DNA, they found that the mRNA did not hybridize to one region of the DNA as expected. Instead, the mRNA hybridized to three different regions of the DNA strand (**FIGURE 1**).[1] Somehow, different segments of the mRNA transcript had been spliced together. The same results were obtained independently by Louise Chow and coworkers in Richard Roberts' group at Cold Spring Harbor Laboratory.[2] At the time, many scientists thought that this odd splicing of mRNA segments might be something unique to viruses. However, that view was soon proven wrong.

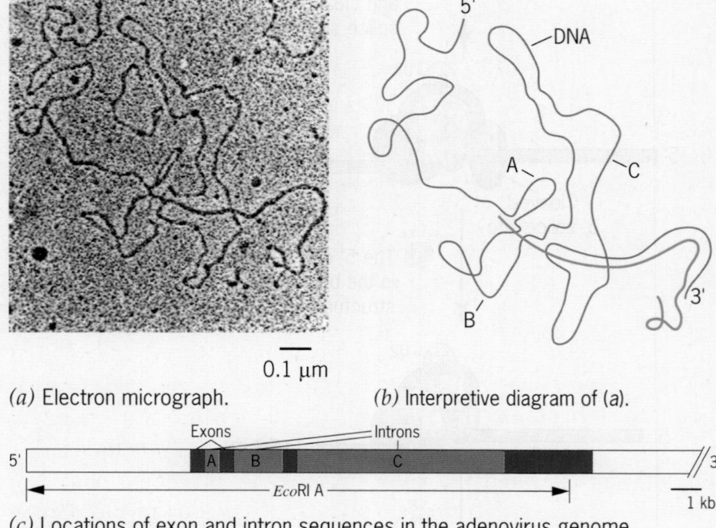

(a) Electron micrograph. *(b)* Interpretive diagram of (a).

(c) Locations of exon and intron sequences in the adenovirus genome.

Figure 1 ▶ **The first evidence for introns and RNA splicing. (*a*) Electron micrograph showing the structure of the heteroduplex formed by hybridizing the late mRNA of adenovirus to the complementary strand of DNA in the adenovirus chromosome (reproduced with permission from S. M. Berget, C. Moore, and P. A. Sharp. 1977. *Proc. Natl. Acad. Sci., U.S.A.* 74: 3171–3175). (*b*) Interpretive drawing of the structure shown in (*a*). (*c*) Diagram showing the locations of the introns (intervening sequences) and exons at the 5′ end of the adenovirus chromosome.**

[1]Berget, S. M., C. Moore, and P. A. Sharp. 1977. Spliced segments at the 5′ terminus of adenovirus 2 late RNA. *Proc. Natl. Acad. Sci. U.S.A.* 74:3171–3175.

[2]Chow, L. T., R. E. Gelinas, T. R. Broker, and R. J. Roberts. 1977. An amazing sequence arrangement at the 5′ ends of adenovirus 2 mRNA. *Cell* 12:1–9.

Research in four different laboratories demonstrated that intervening sequences were present in genes encoding different kinds of proteins in vertebrates. Work on the rabbit β-globin gene in Richard Flavell's lab,[3] the mouse β-globin gene in Philip Leder's lab,[4] and the chicken ovalbumin gene in both Pierre Chambon's lab[5] and Bert W. O'Malley's lab[6] independently documented the presence of noncoding intervening sequences, now called introns, in these structural genes. All four studies depended on the development of methods by which individual genes could be isolated ("cloned") and characterized (Chapter 15). When the genes were compared with their mRNA products, they were found to contain nucleotide sequences that were not present in the mRNAs.

Some of the earliest evidence for introns in mammalian β-globin genes resulted from the visualization of genomic DNA–mRNA hybrids by electron microscopy. DNA–RNA duplexes are more stable than DNA double helices. Thus, if partially denatured DNA double helices are incubated with homologous RNA molecules under the appropriate conditions, the RNA strands will hybridize with the complementary DNA strands, displacing the equivalent DNA strands (**FIGURE 2**). The resulting DNA–RNA hybrid structures will contain single-stranded regions of DNA called **R-loops,** where RNA molecules have displaced DNA strands to form DNA–RNA duplex regions. These R-loops can be visualized directly by electron microscopy. Thus, R-loop hybridization and electron microscopy provided a powerful tool for studies of gene structure.

When Shirley Tilghman, Philip Leder, and colleagues hybridized purified mouse β-globin mRNA to a DNA molecule that contained the mouse β-globin gene, they observed two R-loops separated by a loop of double-stranded DNA (**FIGURE 3*a***).[7] Their results demonstrated the presence of a sequence of nucleotide pairs in the middle of the β-globin gene that is not present in β-globin mRNA and, therefore, does not encode amino acids in the β-globin polypeptide. When Tilghman and coworkers repeated the R-loop experiments using purified β-globin gene transcripts isolated from nuclei and believed to be primary gene transcripts or pre-mRNA molecules, in place of cytoplasmic β-globin mRNA, they observed only one R-loop (**FIGURE 3*b***). This result indicated that the primary transcript contains the complete structural gene sequence, including both exons and introns. Together,

[3]Jeffreys, A. J., and R. A. Flavell. 1977. The rabbit β-globin gene contains a large insert in the coding sequence. *Cell* 12:1097–1108.

[4]Tilghman, S. M., D. C. Tiemeier, F. Polsky, M. H. Edgell, J. G. Seidman, A. Leder, L. W. Enquist, R. Norman, and P. Leder. 1977. Cloning specific segments of the mammalian genome: Bacteriophage λ containing mouse globin and surrounding gene sequences. *Proc. Natl. Acad. Sci. U.S.A.* 74:4406–4410.

[5]Breathnack, R., C. Benoist, K. O'Hare, F. Gannon, and P. Chambon 1978. Ovalbumin gene: Evidence for a leader sequence in mRNA and DNA sequences at the exon–intron boundaries. *Proc. Natl. Acad. Sci. U.S.A.* 75:4853–4857.

[6]Woo, S.L.C., A. Dugaiczyk, M-J. Tsai, E. C. Lai, J. F. Catterall, and B. W. O'Malley. 1978. The ovalbumin gene: Cloning of the natural gene. *Proc. Natl. Acad. Sci. U.S.A.* 75:3688–3692.

[7]Tilghman, S. M., D. C. Tiemeier, J. G. Seidman, B. M. Peterlin, M. Sullivan, J. V. Maizel, and P. Leder. 1978. Intervening sequence of DNA identified in the structural portion of a mouse β-globin gene. *Proc. Natl. Acad. Sci. U.S.A.* 75:725–729.

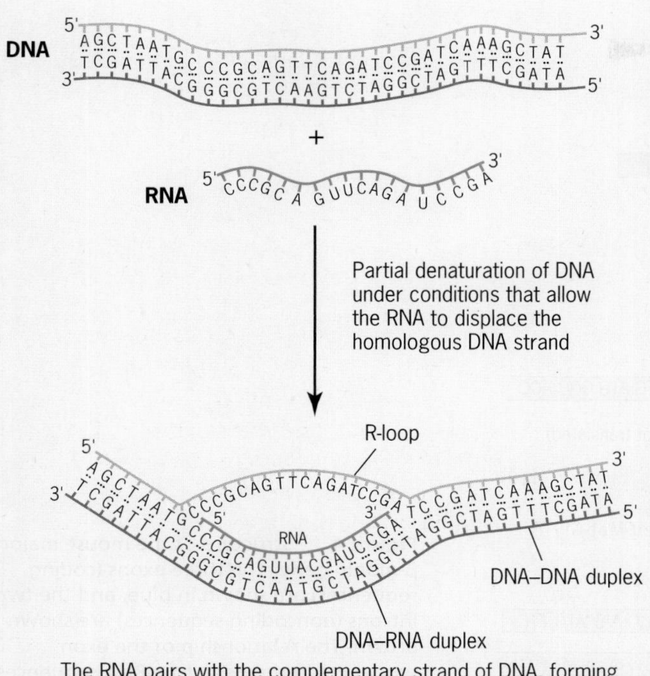

Figure 2 ▶ The technique of R-loop hybridization.

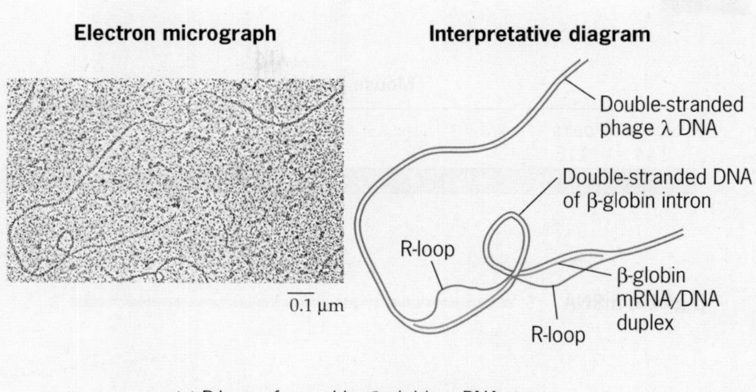

(*a*) R-loops formed by β-globin mRNA.

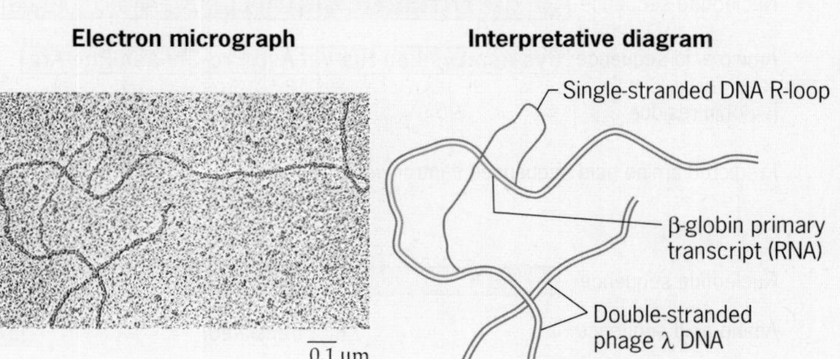

(*b*) R-loop formed by β-globin primary transcript (pre-mRNA).

Figure 3 ▶ **R-loop evidence for an intron in the mouse β-globin gene. (*a*) When mouse β-globin genes and mRNAs were hybridized under R-loop conditions, two R-loops were observed in the resulting DNA–RNA hybrids. (*b*) When primary transcripts or pre-mRNAs of mouse β-globin genes were used in the R-loop experiments, a single R-loop was observed. These results demonstrate that the intron sequence is present in the primary transcript but is removed during the processing of the primary transcript to produce the mature mRNA.**

the R-loop results obtained with cytoplasmic mRNA and nuclear pre-mRNA demonstrate that the intron sequence is excised and the exon sequences are spliced together during processing events that convert the primary transcript to the mature mRNA.

Tilghman and colleagues quickly verified their interpretation of the R-loop results by sequencing a segment of the mouse β-globin gene that spanned the junction between this large intron and the exon preceding it.[8] When they used the established codon assignments to compare the nucleotide sequence of the mouse β-globin gene with the known amino acid sequence of mouse β-globin, the presence of the noncoding sequence was clear (**FIGURE 4**). If there were no intron at this position in the mouse β-globin gene, the nucleotide sequence predicts that amino acid residues 105 through 110 would be Val-Ser-Leu-Met-Gly-Thr. However, the amino acid sequence of mouse β-globin has been determined experimentally, and amino acids 105 through 110 are Leu-Leu-Gly-Asn-Met-Ile. In a later paper, Tilghman, Leder, and coworkers sequenced the rest of the mouse β-globin gene and showed that amino acids 105 through 110 are encoded by the first six nucleotide-pair triplets in the exon distal to this intron. The large β-globin gene intron characterized in these early studies is 653 nucleotide pairs long. Shortly after the discovery of the large intron, Tilghman, Leder, and colleagues found that the mouse β-globin gene contains a second, smaller intron, 116 nucleotide pairs long (**FIGURE 4**). The results of Richard Flavell and coworkers showed that the structure of the rabbit β-globin gene is very similar to that of the mouse.

[8]Ibid.

Studies by Tom Maniatis, Richard Flavell, and others have shown that the structure of the human β-globin gene is also very similar to that of the mouse β-globin gene; both contain two introns that separate three exons. Each of the two human α-globin genes also contains two introns. In fact, the human genes encoding the embryonic α-like (ζ) and β-like (ε) globin chains, the fetal β-like (γ) globin chain, and the adult minor β-like (δ) globin chain, all contain two introns and three exons. Moreover, all of the human genes encoding α-like globins have the introns at the same positions (separating codon 31 from 32 and codon 99 from 100), as do all the human genes encoding β-like globins (separating codon 30 from 31 and codon 104 from 105). Thus, intron positions have been highly conserved during the evolution of the human globin genes.

While Philip Leder, Richard Flavell, and their coworkers were demonstrating the presence of two introns in mammalian β-globin genes, Pierre Chambon, Bert W. O'Malley, and their colleagues were performing similar experiments on the chicken ovalbumin gene. When the transcribed strand of the chicken ovalbumin gene was hybridized to the ovalbumin mRNA and visualized by electron

▶ A MILESTONE IN GENETICS: Introns (continued . . .)

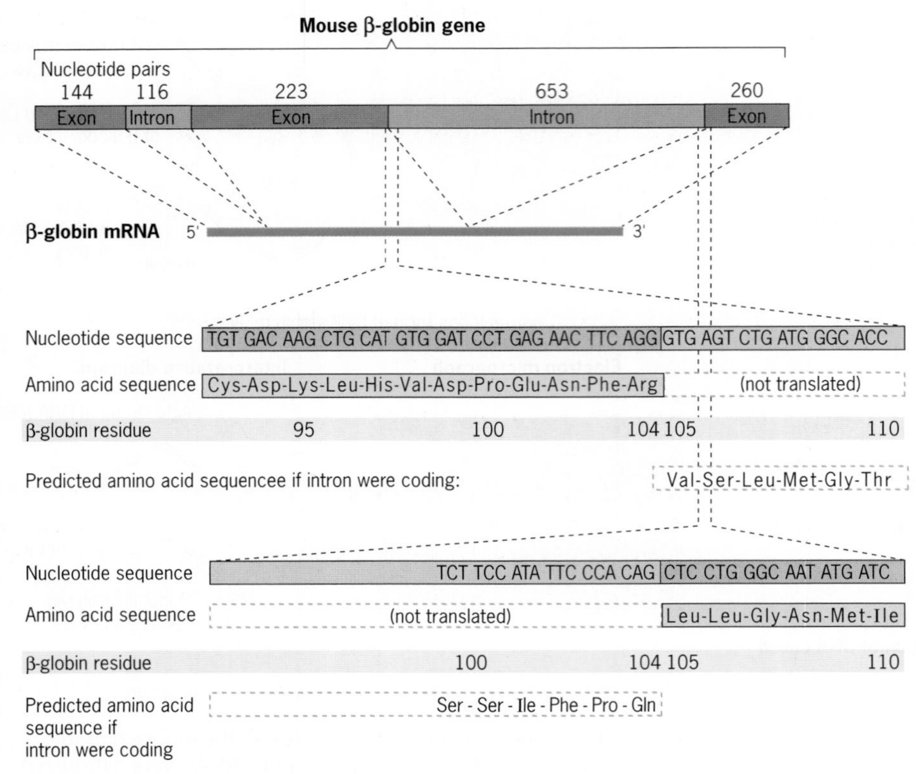

Figure 4 ▶ Structure of the mouse major β-globin gene. The three exons (coding sequences) are shown in blue, and the two introns (noncoding sequences) are shown in brown. The relationship of the exon sequences in the gene to mRNA sequences is indicated by the dashed diagonal lines. The nucleotide sequences of the nontranscribed DNA strand are shown for two exon–intron junctions. These nucleotide sequences are the same as the mRNA sequences, but T replaces U.

microscopy, seven single-stranded DNA loops were observed (**FIGURE 5**), suggesting that the ovalbumin gene contains seven introns. Later experiments, including direct comparisons of the sequences of the gene and the mRNA, showed conclusively that the chicken ovalbumin gene contains seven introns separating eight exons. Subsequent studies have shown that introns are a common feature of most eukaryotic genes. However, we still have a lot to learn about their functions.

QUESTIONS FOR DISCUSSION

1. Introns account for almost one-fourth of the DNA in the human genome. Are introns just "junk" sequences? Or do they play important roles in eukaryotic organisms? What functions can you think of that might be carried out by DNA sequences located in introns?
2. Very few prokaryotic genes contain introns, whereas most eukaryotic genes contain introns. Can you think of possible explanations for this difference? Which of your explanations do you think is most likely to be correct? Why?

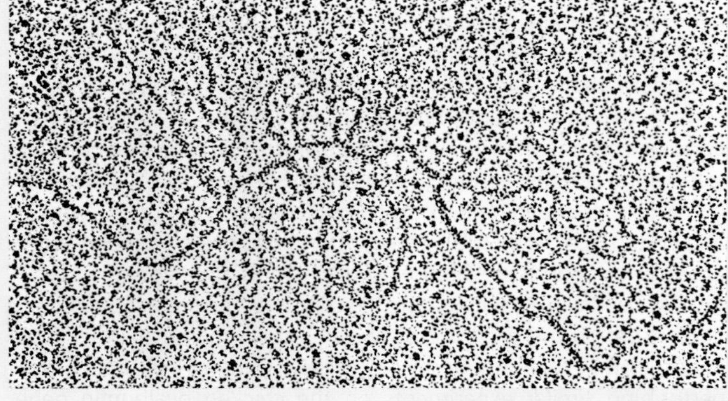

(a) Electron micrograph of ovalbumin DNA–mRNA heteroduplex.

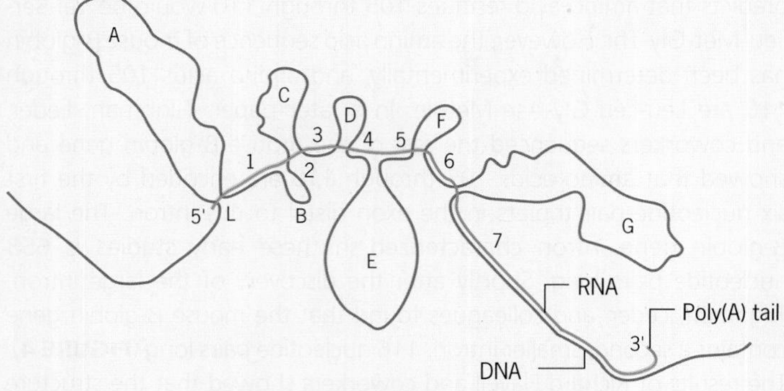

(b) Interpretative diagram of the DNA–mRNA heteroduplex.

Figure 5 ▶ Evidence for seven introns in the chicken ovalbumin gene. The mRNA molecule is represented by a green line; the DNA strand is shown as a blue line. Intron sequences are single-stranded DNA loops labeled A through G.

facilitated by the discovery that some patients with a sometimes fatal autoimmune disease called systemic lupus erythematosus produce antibodies that react with many of their own cellular components including snRNP proteins. These antibodies are called *autoantibodies* because they react with the patient's own proteins; normally, the human immune system will produce only antibodies that react with foreign proteins. In lupus patients, the autoantibodies cause inflammations of tissues and organs that can result in heart, kidney, or liver malfunction and even death. The autoantibodies from patients with systemic lupus erythematosus can be used to precipitate snRNPs; thus, they greatly facilitate the purification of snRNPs for structural and functional studies.

Each of the snRNAs U1, U2, and U5 is present by itself in a specific snRNP particle. snRNAs U4 and U6 are present together in a fourth snRNP; U4 and U6 snRNAs contain two regions of intermolecular complementarity that are base-paired in the U4/U6 snRNP. Each of the four types of snRNP particles contains a subset of seven well-characterized snRNP proteins plus one or more proteins unique to the particular type of snRNP particle. All four snRNP complexes are present in the isolated spliceosomes shown in **FIGURE 11.22**.

The first stage in nuclear pre-mRNA splicing involves cleavage at the 5′ intron splice site (↓GU-intron) and the formation of an intramolecular phosphodiester linkage between the 5′ carbon of the G at the cleavage site and the 2′ carbon of a conserved A residue near the 3′ end of the intron. This stage occurs on complete spliceosomes (**FIGURE 11.23**) and requires the hydrolysis of ATP. Evidence indicates that the U1 snRNP must bind at the 5′ splice site prior to the initial cleavage reaction. Recognition of the cleavage site at the 5′ end of the intron probably involves base-pairing between the consensus sequence at this site and a complementary sequence near the 5′ terminus of snRNA U1. However, the specificity of the binding of at least some of the snRNPs to intron consensus sequences involves both the snRNAs and specific snRNP proteins.

The second snRNP to be added to the splicing complex appears to be the U2 snRNP; it binds at the consensus sequence that contains the conserved A residue that forms the branch point in the lariat structure of the spliced intron. Thereafter, the U5 snRNP binds at the 3′ splice site, and the U4/U6 snRNP is added to the complex to yield the complete spliceosome (**FIGURES 11.22** and **11.23**). When the 5′ intron splice site is cleaved in step 1, the U4 snRNA is released from the spliceosome. During stage 2 of the splicing reaction, the 3′ splice site of the intron is cleaved, and the two exons are joined by a normal 5′ to 3′ phosphodiester linkage (**FIGURE 11.23**). The spliced mRNA is now ready for export to the cytoplasm and translation on ribosomes.

KEY POINTS

- Noncoding intron sequences are excised from RNA transcripts in the nucleus prior to their transport to the cytoplasm.
- Introns in tRNA precursors are removed by the concerted action of a splicing endonuclease and ligase, whereas introns in some rRNA precursors are spliced out autocatalytically—with no catalytic protein involved.
- The introns in nuclear pre-mRNAs are excised on complex ribonucleoprotein structures called spliceosomes.
- The intron excision process must be precise, with accuracy to the nucleotide level, to ensure that codons in exons distal to introns are read correctly during translation.

▶ Basic Exercises

ILLUSTRATE BASIC GENETIC ANALYSIS

1. If the template strand of a segment of a gene has the nucleotide sequence 3′-GCTAAGC-5′, what nucleotide sequence will be present in the RNA transcript specified by this gene segment?

Answer: The RNA transcript will be complementary to the template strand and will have the opposite chemical polarity, as in the following illustration:

DNA template strand:	3′-GCTAAGC-5′
RNA transcript:	5′-CGAUUCG-3′

2. If the nontemplate strand of a gene in *E. coli* had the sequence:

5′-TTGACA-(18 bases)-TATAAT-(8 bases)-GCCTTC-CAGTG-3′

what nucleotide sequence would be present in the RNA transcript of this gene?

Answer: The gene contains perfect –35 and –10 promoter sequences. Transcription should be initiated at a site five to nine bases downstream from the –10 TATAAT sequence, and the 5′-terminus of the transcript should contain a purine. The template strand and the 5′-end of the transcript should have the following structure:

DNA template strand:
(–35 sequence) (–10 sequence)
3′-AACTGT-(18 bases)-ATATTA-(8 bases)-CGGAAGGTCAC-5′
RNA transcript: 5′-GCCUUCCAGUG-3′

3. If the nontemplate strand shown in Exercise 2 were part of a gene in *Drosophila* rather than *E. coli*, would the same transcript be produced?

Answer: No, because the promoter sequences that control transcription in eukaryotes such as *Drosophila* are different from the promoters in prokaryotes such as *E. coli*. Therefore, the *E. coli* gene would probably not be transcribed if present in *Drosophila*.

4. A scientist is studying the transcription of a gene in mouse cells growing in culture. How can she determine whether the gene is transcribed by RNA polymerase I, RNA polymerase II, or RNA polymerase III?

Answer: She should determine what effect treatment of the cells with α-amanitin has on the transcription of the gene. If α-amanitin has no effect on transcription, the gene is transcribed by RNA polymerase I; if α-amanitin blocks transcription, the gene is transcribed by RNA polymerase II; if α-amanitin decreases, but does not completely block, transcription, the gene is transcribed by RNA polymerase III.

5. The primary transcript or pre-mRNA of a nuclear gene in a chimpanzee has the sequence:

5′-G—exon 1—AGGUAAGC—intron—CAGUC—exon 2—A-3′

After the intron has been excised, what is the most likely sequence of the mRNA?

Answer: Introns contain highly conserved dinucleotide termini: 5′-GT—AG-3′ in the DNA nontemplate strand or 5′-GU—AG-3′ in the RNA transcript. Thus, the intron sequence is almost certain to be 5′-GUUAAGC—intron—CAG-3′. With precise excision of the intron, the sequence of the mRNA will be:

5′-G—exon 1—AGUC—exon 2—A-3′

▶ Testing Your Knowledge

INTEGRATE DIFFERENT CONCEPTS AND TECHNIQUES

1. Certain medically important human proteins such as insulin and growth hormone are now being produced in bacteria. By using the tools of genetic engineering, DNA sequences encoding these proteins have been introduced into bacteria. You wish to introduce a human gene into *E. coli* and have that gene produce large amounts of the human gene product in the bacterial cells. Assuming that the human gene of interest can be isolated and introduced into *E. coli*, what problems might you encounter in attempting to achieve your goal?

Answer: The promoter sequences that are required to initiate transcription are very different in mammals and bacteria. Therefore, your gene will not be expressed in *E. coli* unless you first fuse its coding region to a bacterial promoter. In addition, your human gene probably will contain introns. Since *E. coli* cells do not contain spliceosomes or equivalent machinery with which to excise introns from RNA transcripts, your human gene will not be expressed correctly if it contains introns. As you can see, expressing eukaryotic genes in prokaryotic cells is not a trivial task.

2. A human β-globin gene has been purified and inserted into a linear bacteriophage lambda chromosome, producing the following DNA molecule:

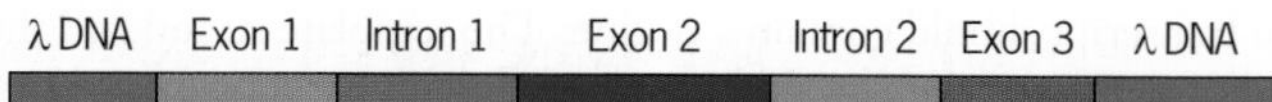

If this DNA molecule is hybridized to human β-globin mRNA using conditions that favor DNA:RNA duplexes over DNA:DNA duplexes (R-loop mapping conditions) and the product is visualized by electron microscopy, what nucleic acid structure would you expect to see?

Answer: The primary transcript of this human β-globin gene will contain both introns and all three exons. However, prior to its export to the cytoplasm, the intron sequences will be spliced out of the transcript. Thus, the mature mRNA molecule will contain the three exon sequences spliced together with no intron sequences present. Under R-loop conditions, the mRNA will anneal with the complementary strand of DNA, displacing the equivalent DNA strand. However, since the mRNA contains no intron sequences, the introns will remain as regions of double-stranded DNA as shown in the following diagram.

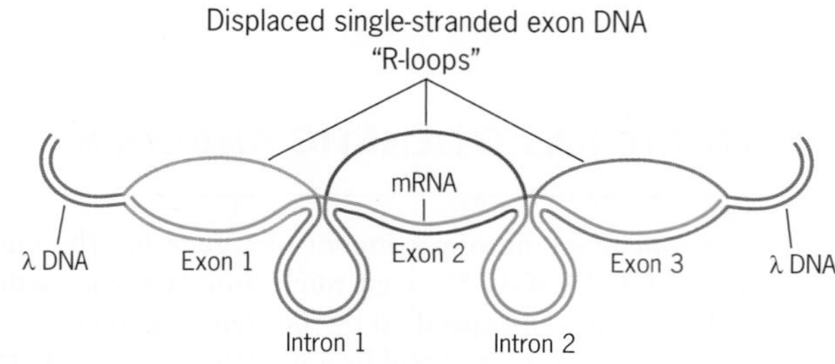

▶ Questions and Problems

ENHANCE UNDERSTANDING AND DEVELOP ANALYTICAL SKILLS

11.1 What bases in the transcribed strand of DNA would give rise to the following mRNA base sequence: 5′-AUGCU-3′?

11.2 List three ways in which the mRNAs of eukaryotes differ from the mRNAs of prokaryotes.

11.3 Distinguish between DNA and RNA (a) chemically, (b) functionally, and (c) by location in the cell.

11.4 What bases in the mRNA transcript would represent the following DNA template sequence: 5′-AGGTCAAC-3′?

11.5 What different types of RNA molecules are present in prokaryotic cells? in eukaryotic cells? What roles do these different classes of RNA molecules play in the cell?

11.6 What role(s) do spliceosomes play in pathways of gene expression? What is their macromolecular structure?

11.7 At what locations in a eukaryotic cell does protein synthesis occur?

11.8 Match one of the following terms with each of the descriptions given below. *Terms:* (1) sigma (σ) factor; (2) poly(A) tail; (3) TATAAT; (4) exons; (5) TATAAAA; (6) RNA polymerase III; (7) intron; (8) RNA polymerase II; (9) heterogeneous nuclear RNA (hnRNA); (10) snRNA; (11) RNA polymerase I; (12) TTGACA; (13) GGCCAATCT (CAAT box).

Descriptions:

(a) Intervening sequence found in many eukaryotic genes.
(b) A conserved nucleotide sequence (-30) in eukaryotic promoters involved in the initiation of transcription.
(c) Small RNA molecules that are located in the nuclei of eukaryotic cells, most as components of the spliceosome, that participate in the excision of introns from nuclear gene transcripts.
(d) A sequence (-10) in the nontemplate strand of the promoter of *E. coli* that facilitates the localized unwinding of DNA when complexed with RNA polymerase.
(e) The RNA polymerase in the nucleus that catalyzes the synthesis of all rRNAs except for the small 5S rRNA.
(f) The subunit of prokaryotic RNA polymerase that is responsible for the initiation of transcription at promoters.
(g) An *E. coli* promoter sequence located 35 nucleotides upstream from the transcription-initiation site; it serves as a recognition site for the sigma factor.
(h) The RNA polymerase in the nucleus that catalyzes the synthesis of the transfer RNA molecules and small nuclear RNAs.
(i) A polyadenosine tract 20 to 200 nucleotides long that is added to the 3′ end of most eukaryotic messenger RNAs.
(j) The RNA polymerase that transcribes nuclear genes that encode proteins.
(k) A conserved sequence in the nontemplate strand of eukaryotic promoters that is located about 80 nucleotides upstream from the transcription start site.
(l) Segments of a eukaryotic gene that correspond to the sequences in the final processed RNA transcript of the gene.
(m) The population of primary transcripts in the nucleus of a eukaryotic cell.

11.9 What components of the introns of nuclear genes that encode proteins in higher eukaryotes are conserved and required for the correct excision of intron sequences from primary transcripts by spliceosomes?

11.10 Many eukaryotic genes contain noncoding introns that separate the coding sequences or exons of these genes. At what stage during the expression of these split genes are the noncoding intron sequences removed?

11.11 For several decades, the dogma in biology has been that molecular reactions in living cells are catalyzed by enzymes composed of polypeptides. We now know that the introns of some precursor RNA molecules such as the rRNA precursors in *Tetrahymena* are removed autocatalytically ("self-spliced") with no involvement of any catalytic protein. What does the demonstration of autocatalytic splicing indicate about the dogma that biological reactions are always catalyzed by proteinaceous enzymes?

11.12 A segment of DNA in *E. coli* has the following sequence of nucleotide pairs:

When this segment of DNA is transcribed by RNA polymerase, what will be the sequence of nucleotides in the RNA transcript if the promoter is located to the left of the sequence shown?

11.13 A particular gene is inserted into the phage lambda chromosome and is shown to contain three introns. (*a*) The primary transcript of this gene is purified from isolated nuclei. When this primary transcript is hybridized under R-loop conditions with the recombinant lambda chromosome carrying the gene, what will the R-loop structure(s) look like? Label your diagram. (*b*) The mRNA produced from the primary transcript of this gene is then isolated from cytoplasmic polyribosomes and similarly examined by the R-loop hybridization procedure using the recombinant lambda chromosome carrying the gene. Diagram what the R-loop structure(s) will look like when the cytoplasmic mRNA is used. Again, label the components of your diagram.

11.14 A segment of DNA in *E. coli* has the following sequence of nucleotide pairs:

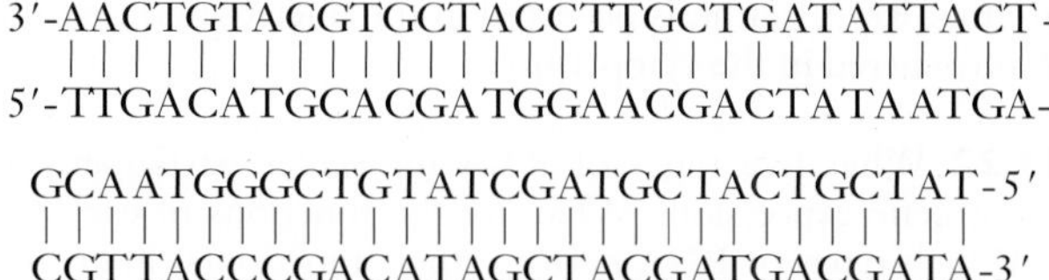

When this segment of DNA is transcribed by RNA polymerase, what will be the sequence of nucleotides in the RNA transcript?

11.15 (a) Which of the following nuclear pre-mRNA nucleotide sequences potentially contains an intron?

(1) 5′-UGACCAUGGCGCUAACACUGCCAAUUG-GCAAUACUGACCUGAUAGCAUCAGCCAA-3′

(2) 5′-UAGUCUCAUCUGUCCAUUGACUUC-GAAACUGAAUCGUAACUCCUACGUCUAUGGA-3′

(3) 5′-UAGCUGUUUGUCAUGACUGACUGGUCACU-AUCGUACUAACCUGUCAUGCAAUGUC-3′

(4) 5′-UAGCAGUUCUGUCGCCUCGUGGUGCU-GCUGGCCCUUCGUCGCUCGGGCUUAGCUA-3′

(5) 5′-UAGGUUCGCAUUGACGUACUUCUGAAAC-UACUAACUACUAACGCAUCGAGUCUCAA-3′

(b) One of the five pre-mRNAs shown in (a) may undergo RNA splicing to excise an intron sequence. What mRNA nucleotide sequence would be expected to result from this splicing event?

11.16 The genome of a human must store a tremendous amount of information using the four nucleotide pairs present in DNA. What do the Morse code and the language of computers tell us about the feasibility of storing large amounts of information using an alphabet composed of just four letters?

11.17 A segment of human DNA has the following sequence of nucleotide pairs:

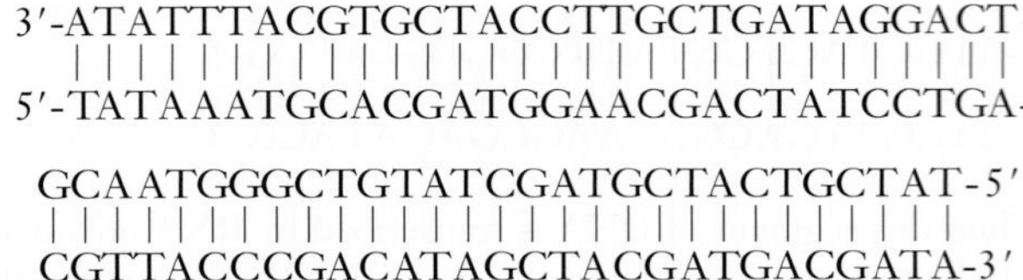

When this segment of DNA is transcribed by RNA polymerase, what will be the sequence of nucleotides in the RNA transcript?

11.18 The biosynthesis of metabolite X occurs via six steps catalyzed by six different enzymes. What is the minimal number of genes required for the genetic control of this metabolic pathway? Might more genes be involved? Why?

11.19 What do the processes of DNA synthesis, RNA synthesis, and polypeptide synthesis have in common?

11.20 What are the two stages of gene expression? Where do they occur in a eukaryotic cell? a prokaryotic cell?

11.21 Why was the need for an RNA intermediary in protein synthesis most obvious in eukaryotes? How did researchers first demonstrate that RNA synthesis occurred in the nucleus and that protein synthesis occurred in the cytoplasm?

11.22 What five types of RNA molecules participate in the process of gene expression? What are the functions of each type of RNA? Which types of RNA perform their function(s) in (a) the nucleus and (b) the cytoplasm?

11.23 What is the central dogma of molecular genetics? What impact did the discovery of RNA tumor viruses have on the central dogma?

11.24 Two eukaryotic genes encode two different polypeptides, each of which is 400 amino acids long. One gene contains a single exon; the other gene contains an intron 41,324 nucleotide pairs long. Which gene would you expect to be transcribed in the least amount of time? Why? When the mRNAs specified by these genes are translated, which mRNA would you expect to be translated in the least time? Why?

11.25 What two elements are almost always present in the promoters of eukaryotic genes that are transcribed by RNA polymerase II? Where are these elements located relative to the transcription start site? What are their functions?

11.26 Total RNA was isolated from human cells growing in culture. This RNA was mixed with nontemplate strands (single strands) of the human gene encoding the enzyme thymidine kinase, and the RNA–DNA mixture was incubated for 12 hours under renaturation conditions. Would you expect any RNA–DNA duplexes to be formed during the incubation? If so, why? If not, why not? The same experiment was then performed using the template strand of the thymidine kinase gene. Would you expect any RNA–DNA duplexes to be formed in this second experiment? If so, why? If not, why not?

11.27 Compare the structures of primary transcripts with those of mRNAs in prokaryotes and eukaryotes. On average, in which group of organisms do they differ the most?

11.28 In what ways are most eukaryotic gene transcripts modified? What are the functions of these posttranscriptional modifications?

11.29 What role did the human disease systemic lupus erythematosus play in the characterization of human snRNPs?

11.30 Transcription and translation are coupled in prokaryotes. Why is this not the case in eukaryotes?

11.31 How does RNA editing contribute to protein diversity in eukaryotes?

11.32 How do the mechanisms by which the introns of tRNA precursors, *Tetrahymena* rRNA precursors, and nuclear pre-mRNAs are excised differ? In which process are snRNAs involved? What role(s) do these snRNAs play?

11.33 Two preparations of RNA polymerase from *E. coli* are used in separate experiments to catalyze RNA synthesis *in vitro* using a purified fragment of DNA carrying the *argH* gene as template DNA. One preparation catalyzes the synthesis of RNA chains that are highly heterogeneous in size. The other preparation catalyzes the synthesis of RNA chains that are all the same length. What is the most likely difference in the composition of the RNA polymerases in the two preparations?

11.34 Total RNA was isolated from nuclei of human cells growing in culture. This RNA was mixed with a purified, denatured DNA fragment that carried a large intron of a housekeeping gene (a gene expressed in essentially all cells), and the RNA–DNA mixture was incubated for 12 hours under renaturation conditions. Would you expect any RNA–DNA duplexes to be formed during the incubation? If so, why? If not, why not? The same experiment was then performed using total cytoplasmic RNA from these cells. Would you expect any RNA–DNA duplexes to be formed in this second experiment? If so, why? If not, why not?

11.35 A mutation in an essential human gene changes the 5′ splice site of a large intron from GT to CC. Predict the phenotype of an individual homozygous for this mutation.

► Genomics on the Web

at http://www.ncbi.nlm.nih.gov/

Duchenne muscular dystrophy (DMD) is an X-linked recessive disease in humans that affects about one in 3300 newborn males. Individuals with DMD undergo progressive muscle degeneration starting early in life. They are usually confined to wheel chairs by their teens and commonly die in their late teens or early twenties. The disorder is caused by mutations in the human *DMD* gene, which encodes a protein called dystrophin. This protein is associated with the intracellular membranes of muscle cells. The *DMD* gene is one of the largest genes known and is composed of many exons and introns. Because of its medical importance, the ncbi web site contains a large amount of information on the *DMD* gene and its product dystrophin.

1. How large is the human *DMD* gene? How many exons and introns does it contain? How large is the *DMD* mRNA? The *DMD* protein coding sequence?

2. What is the largest exon in the human *DMD* gene? The smallest exon? Where are the mutations located that cause Duchenne muscular dystrophy? Some of the mutations in this gene cause a less severe form of muscular dystrophy called Becker muscular dystrophy. Where are these mutations located?

3. Do other species contain genes that are closely related to the human *DMD* gene and encode similar dystrophins? What species? How similar are these genes to each other and to the human *DMD* gene?

Hint: At the NCBI home page (click on each of the following) → Human genome resources → Gene Database → Search with query DMD AND human[orgn] → 1. DMD → Primary Source: HGNC:2928 → under Gene Symbol Links, click GENATLAS → DMD → See the exons. To view homologous genes in other organisms, go back to the results of your DMD gene search, and click on HomoloGene. Also, search the OMIM (the Online Medical Inheritance in Man) database for more information about Duchenne and Becker muscular dystrophies.

Chapter 12 Translation and the Genetic Code

CHAPTER OUTLINE

- Protein Structure
- Protein Synthesis: Translation
- The Genetic Code
- Codon-tRNA Interactions
- *In Vivo* Confirmation of the Nature of the Genetic Code

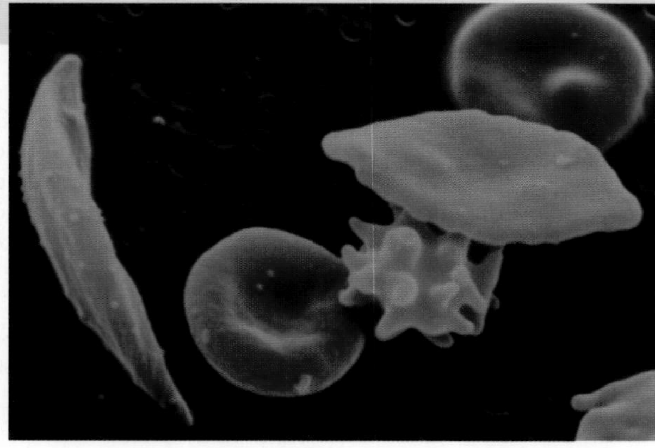

Scanning electron micrograph of sickle-shaped (left) and normal red blood cells.

Sickle-Cell Anemia: Devastating Effects of a Single Base-Pair Change

In 1904 James Herrick, a Chicago physician, and Ernest Irons, a medical intern working under Herrick's supervision, examined the blood cells of one of their patients. They noticed that many of the red blood cells of the young man were thin and elongated, in striking contrast to the round, donutlike red cells of their other patients. They obtained fresh blood samples and repeated their microscopic examinations several times, always with the same result. The blood of this patient always contained cells shaped like the sickles that farmers used to harvest grain at that time.

The patient was a 20-year-old college student who was experiencing periods of weakness and dizziness. In many respects, the patient seemed normal, both physically and mentally. His major problem was fatigue. However, a physical exam showed an enlarged heart and enlarged lymph nodes. His heart always seemed to be working too hard, even when he was resting. Blood tests showed that the patient was anemic; the hemoglobin content of his blood was about half the normal level. Hemoglobin is the complex protein that carries oxygen from the lungs to other tissues. Herrick charted this patient's symptoms for six years before publishing his observations in 1910. In his paper, Herrick emphasized the chronic nature of the anemia and the presence of the sickle-shaped red cells. In 1916, at age 32, the patient died from severe anemia and kidney damage.

James Herrick was the first to publish a description of sickle-cell anemia, the first inherited human disease to be understood at the molecular level. Hemoglobin contains four polypeptides—two α-globin chains and two β-globin chains—and an iron-containing heme group. In 1957, Vernon Ingram and colleagues demonstrated that the sixth amino acid of the β-chain of sickle-cell hemoglobin was valine, whereas glutamic acid was present at this position in normal adult human hemoglobin. This single amino acid change in a single polypeptide chain is responsible for all the symptoms of sickle-cell anemia.

How does the genetic information of an organism, stored in the sequence of nucleotide pairs in DNA, control the phenotype of the organism? How does a nucleotide-pair change in a gene—like the mutation that causes sickle-cell anemia—alter the structure of a protein, the emissary through which the gene acts? In Chapter 11, we discussed the transfer of genetic information stored in the sequences of nucleotide pairs in DNA to the sequences of nucleotides in mRNA molecules, which, in eukaryotes, carry that information from the nucleus to the sites of protein synthesis in the cytoplasm. The transfer of information from DNA to RNA (transcription) and RNA processing occur in the nucleus. In this chapter, we examine the process by which genetic information stored in sequences of nucleotides in mRNAs is used to specify the sequences of amino acids in polypeptide gene products. This process, **translation,** takes place in the cytoplasm on complex workbenches called ribosomes and requires the participation of many macromolecules.

Protein Structure

Proteins are complex macromolecules composed of 20 different amino acids.

Collectively, the proteins constitute about 15 percent of the wet weight of cells. Water molecules account for 70 percent of the total weight of living cells. With the exception of water, proteins are by far the most prevalent component of living organisms in terms of total mass. Not only are proteins major components in terms of cell mass, but they also play many roles vital to the lives of all cells. Before discussing the synthesis of proteins, we need to become more familiar with their structure.

POLYPEPTIDES: TWENTY DIFFERENT AMINO ACID SUBUNITS

Proteins are composed of polypeptides, and every polypeptide is encoded by a gene. Each polypeptide consists of a long sequence of amino acids linked together by covalent bonds. Twenty different amino acids are present in most proteins. Occasionally, one or more of the amino acids are chemically modified after a polypeptide is synthesized, yielding a novel amino acid in the mature protein. The structures of the 20 common amino acids are shown in **FIGURE 12.1**. All the amino acids except proline contain a *free amino group* and a *free carboxyl group*.

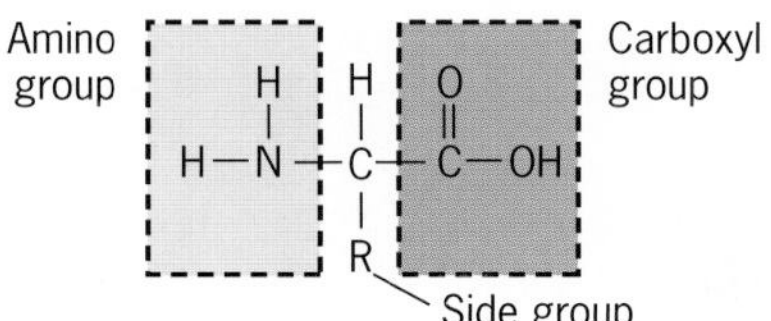

The amino acids differ from each other by the *side groups* (designated **R** for ***R***adical) that are present. The highly varied side groups provide the structural diversity of proteins. These side chains are of four types: (1) hydrophobic or nonpolar groups, (2) hydrophilic or polar groups, (3) acidic or negatively charged groups, and (4) basic or positively charged groups (**FIGURE 12.1**). The chemical diversity of the side groups of the amino acids is responsible for the enormous structural and functional versatility of proteins.

A peptide is a compound composed of two or more amino acids. Polypeptides are long sequences of amino acids, ranging in length from 51 amino acids in insulin to over 1000 amino acids in the silk protein fibroin. Given the 20 different amino acids commonly found in polypeptides, the number of different polypeptides that are possible is truly enormous. For example, the number of different amino acid sequences that can occur in a polypeptide containing 100 amino acids is 20^{100}. Since 20^{100} is too large to comprehend, let's consider a short peptide. There are 1.28 billion (20^7) different amino acid sequences possible in a peptide seven amino acids long. The amino acids in polypeptides are covalently joined by linkages called **peptide bonds.** Each peptide bond is formed by a reaction between the amino group of one amino acid and the carboxyl group of a second amino acid with the elimination of a water molecule (**FIGURE 12.2**).

PROTEINS: COMPLEX THREE-DIMENSIONAL STRUCTURES

Four different levels of organization—primary, secondary, tertiary, and quaternary—are distinguished in the complex three-dimensional structures of proteins. The *primary structure* of a polypeptide is its amino acid sequence, which is specified by the nucleotide sequence of a gene. The *secondary structure* of a polypeptide refers to the spatial interrelationships of the amino acids in segments of the polypeptide. The *tertiary structure* of a polypeptide refers to its overall folding in three-dimensional space, and the *quaternary structure* refers to the association of two or more polypeptides in a multimeric protein. Hemoglobin provides an excellent example of the complexity of proteins, exhibiting all four levels of structural organization (**FIGURE 12.3**).

Most polypeptides will fold spontaneously into specific conformations dictated by their primary structures. If denatured (unfolded) by treatment with appropriate solvents, most proteins will re-form their original conformations when the denaturing agent is removed. Thus, in most cases, all of the information required for shape determination resides in the primary structure of the protein. In some cases, protein folding involves interactions with proteins called **chaperones** that help nascent polypeptides form the proper three-dimensional structure.

The two most common types of secondary structure in proteins are *α helices* (see **FIGURE 12.3**) and *β sheets*. Both structures are maintained by hydrogen bonding between peptide bonds located in close proximity to one another. The α helix is a rigid cylinder in which each peptide bond is hydrogen bonded to the peptide bond between amino acids three and four residues away. Because of its rigid structure, proline cannot be present within an α helix. A β sheet occurs when a polypeptide folds back upon itself, sometimes repeatedly, and the parallel segments are held in place by hydrogen bonding between neighboring peptide bonds.

Whereas the spatial organization of adjacent amino acids and segments of a polypeptide determine its secondary structure, the overall folding of the complete polypeptide defines its tertiary structure, or *conformation*. In general, amino acids with hydrophilic side chains are located on the surfaces of proteins (in contact with the aqueous cytoplasm), whereas those with hydrophobic side chains interact with each other in the interior regions. The tertiary structure of a protein is maintained primarily by a large number of relatively weak noncovalent bonds. The only covalent bonds that play a significant role in protein conformation are disulfide (S—S) bridges that form between appropriately positioned cysteine residues (**FIGURE 12.4**). However, four different types of noncovalent interactions are involved: (1) ionic bonds, (2) hydrogen bonds, (3) hydrophobic interactions, and (4) Van der Waals interactions (**FIGURE 12.4**).

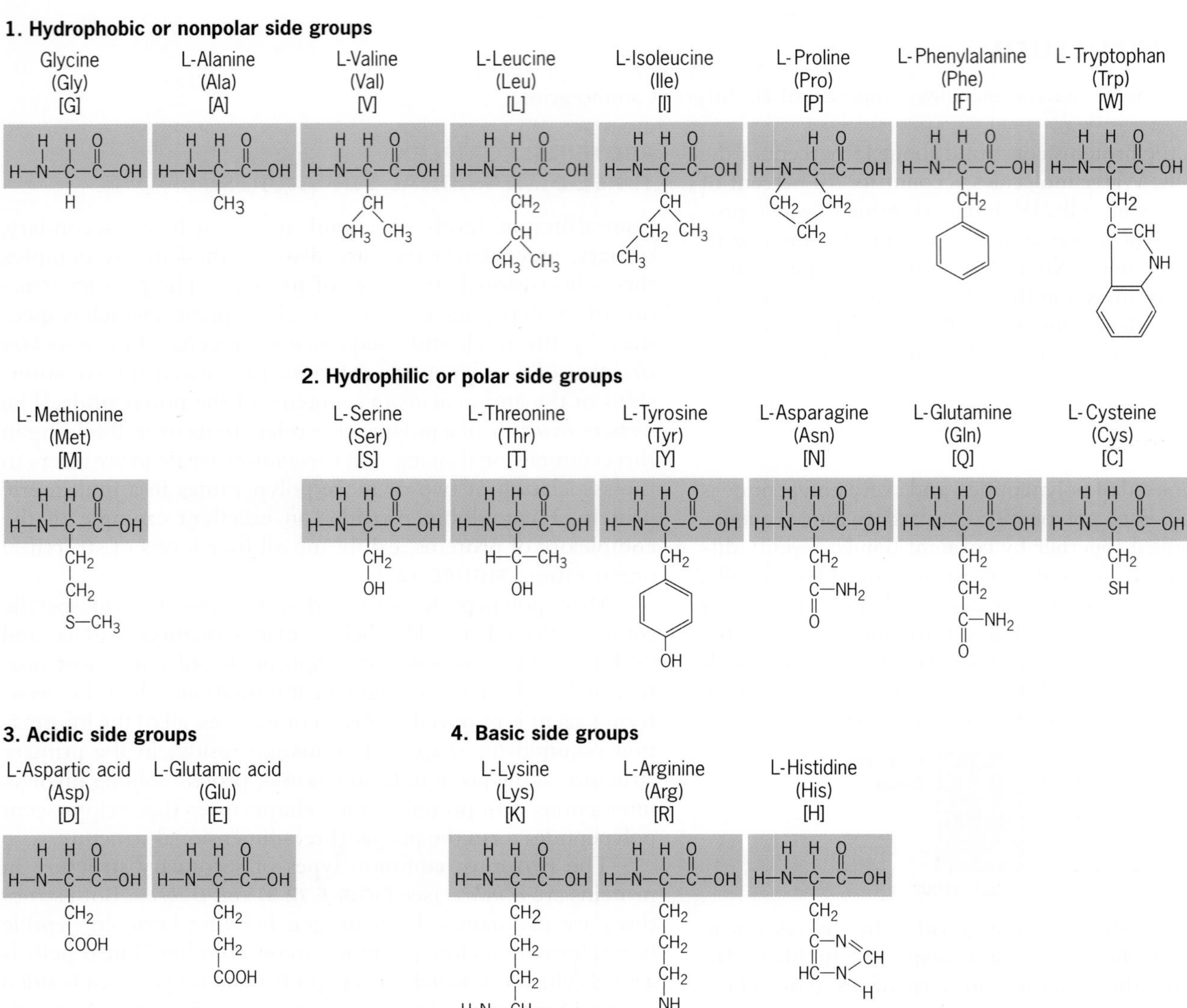

Figure 12.1 ▸ Structures of the 20 amino acids commonly found in proteins. The amino and carboxyl groups, which participate in peptide bond formation during protein synthesis, are shown in the shaded areas. The side groups, which are different for each amino acid, are shown below the shaded areas. The standard three-letter abbreviations are shown in parentheses. The one-letter symbol for each amino acid is given in brackets.

Ionic bonds occur between amino acid side chains with opposite charges—for example, the side groups of lysine and glutamic acid (see **FIGURE 12.1**). Ionic bonds are strong forces under some conditions, but they are relatively weak interactions in the aqueous interiors of living cells because the polar water molecules partially neutralize or shield the charged groups. **Hydrogen bonds** are weak interactions between electronegative atoms (which have a partial negative charge) and hydrogen atoms (which are electropositive) that are linked to other electronegative atoms. **Hydrophobic interactions** are associations of nonpolar groups with each other when present in aqueous

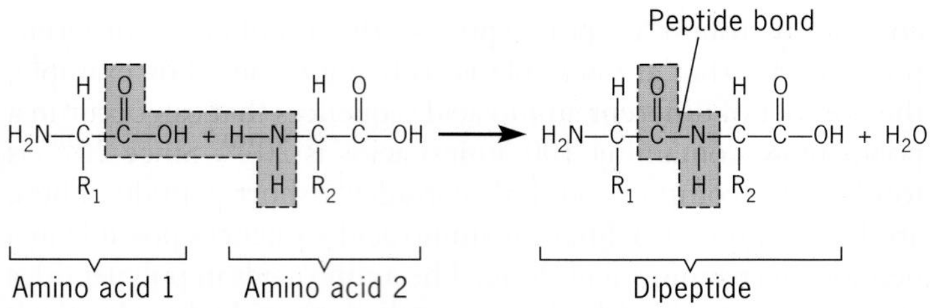

Peptide bond

$H_2N{-}CH(R_1){-}CO{-}OH + H{-}NH{-}CH(R_2){-}COOH \longrightarrow H_2N{-}CH(R_1){-}CO{-}NH{-}CH(R_2){-}COOH + H_2O$

Amino acid 1 — Amino acid 2 — Dipeptide

Figure 12.2 ▸ The formation of a peptide bond between two amino acids by the removal of water. Each peptide bond connects the amino group of one amino acid and the carboxyl group of the adjacent amino acid.

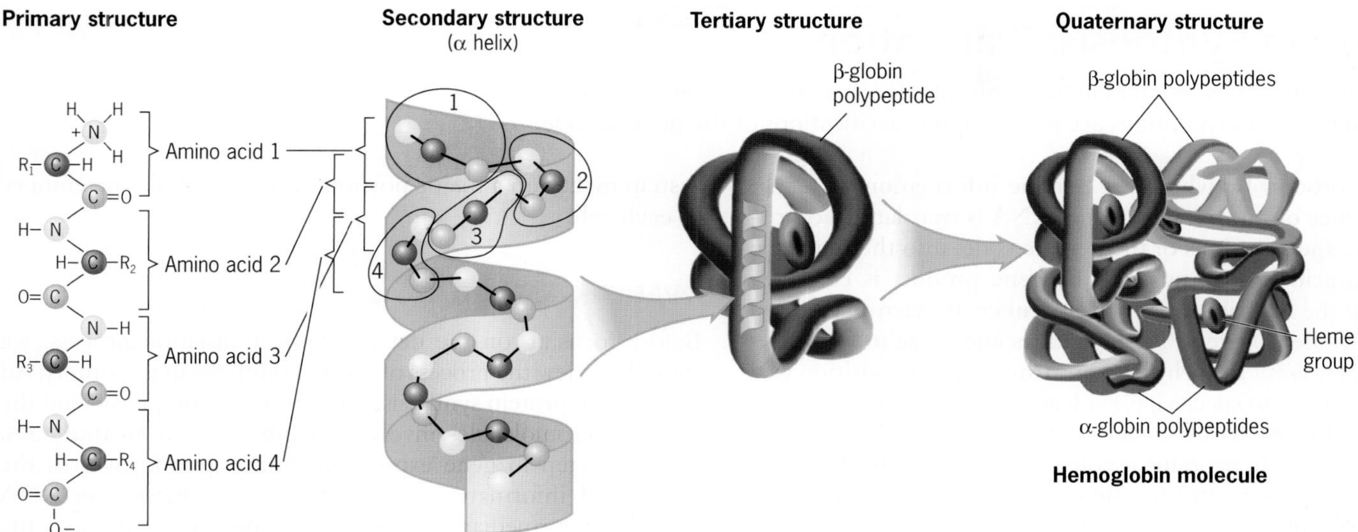

Figure 12.3 ► The four levels of organization in proteins—(1) primary, (2) secondary, (3) tertiary, and (4) quaternary structures—are illustrated using human hemoglobin as an example.

solutions because of their insolubility in water. Hydrogen bonds and hydrophobic interactions play important roles in DNA structure; thus, we have discussed them in some detail in Chapter 9 (see **TABLE 9.2**). **Van der Waals interactions** are weak attractions that occur between atoms when they are placed in close proximity to one another. Van der Waals forces are very weak, with about one one-thousandth of the strength of a covalent bond, but they play an important role in maintaining the conformations of closely aligned regions of macromolecules.

Quaternary structure exists only in proteins that contain more than one polypeptide. Hemoglobin provides a good illustration of quaternary structure, being a tetrameric molecule composed of two α-globin chains and two β-globin chains, plus four iron-containing heme groups (see **FIGURE 12.3**).

In a few cases, the primary translation products contain short amino acid sequences, called **inteins,** which excise themselves from the nascent polypeptides. Inteins occur in both eukaryotes and prokaryotes. For example, one of the first inteins discovered is in the RecA protein, which is involved in recombination and DNA repair in *Mycobacterium tuberculosis*, the bacterium that causes tuberculosis.

Since the secondary, tertiary, and quaternary structures of proteins and intein excision usually are determined by the primary structure(s) of the polypeptide(s) involved, in the rest of this chapter we will focus on the mechanisms by which genes control the primary structures of polypeptides.

KEY POINTS

- Most genes exert their effect(s) on the phenotype of an organism through proteins, which are large macromolecules composed of polypeptides.
- Each polypeptide is a chainlike polymer assembled from different amino acids.
- The amino acid sequence of each polypeptide is specified by the nucleotide sequence of a gene.
- The vast functional diversity of proteins results in part from their complex three-dimensional structures.

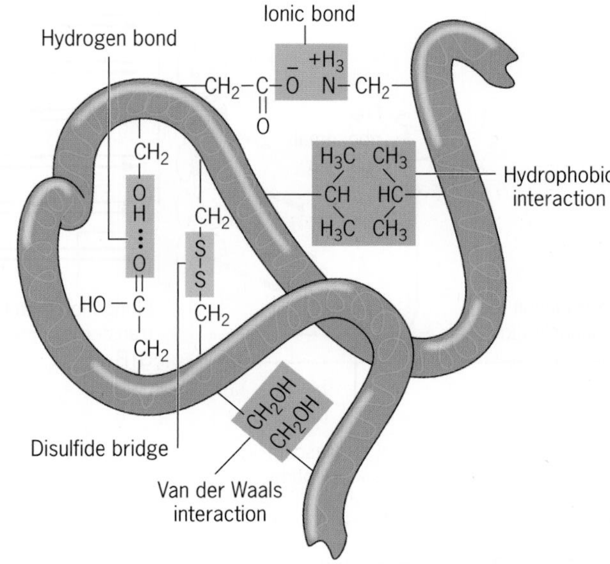

Figure 12.4 ► The five types of molecular interactions that determine the tertiary structure, or three-dimensional conformation, of a polypeptide. The disulfide bridge is a covalent bond; all other interactions are noncovalent.

Protein Synthesis: Translation

The genetic information in mRNA molecules is translated into the amino acid sequences of polypeptides according to the specifications of the genetic code.

The process by which the genetic information stored in the sequence of nucleotides in an mRNA is translated, according to the specifications of the genetic code, into the sequence of amino acids in the polypeptide gene product is complex, requiring the functions of a large number of macromolecules. These include (1) over 50 polypeptides and three to five RNA molecules present in each ribosome (the exact composition varies from species to species), (2) at least 20 amino acid-activating enzymes, (3) 40 to 60 different tRNA molecules, and (4) numerous soluble proteins involved in polypeptide chain initiation, elongation, and termination. Because many of these macromolecules, particularly the components of the ribosome, are present in large quantities in each cell, the translation system makes up a major portion of the metabolic machinery of each cell.

OVERVIEW OF PROTEIN SYNTHESIS

Before focusing on the details of the translation process, we should preview the process of protein synthesis in its entirety. An overview of protein synthesis, illustrating its complexity and the major macromolecules involved, is presented in **FIGURE 12.5.** The first step in gene expression, transcription, involves the transfer of information stored in genes to messenger RNA (mRNA) intermediaries, which carry that information to the sites of polypeptide synthesis in the cytoplasm. Transcription is

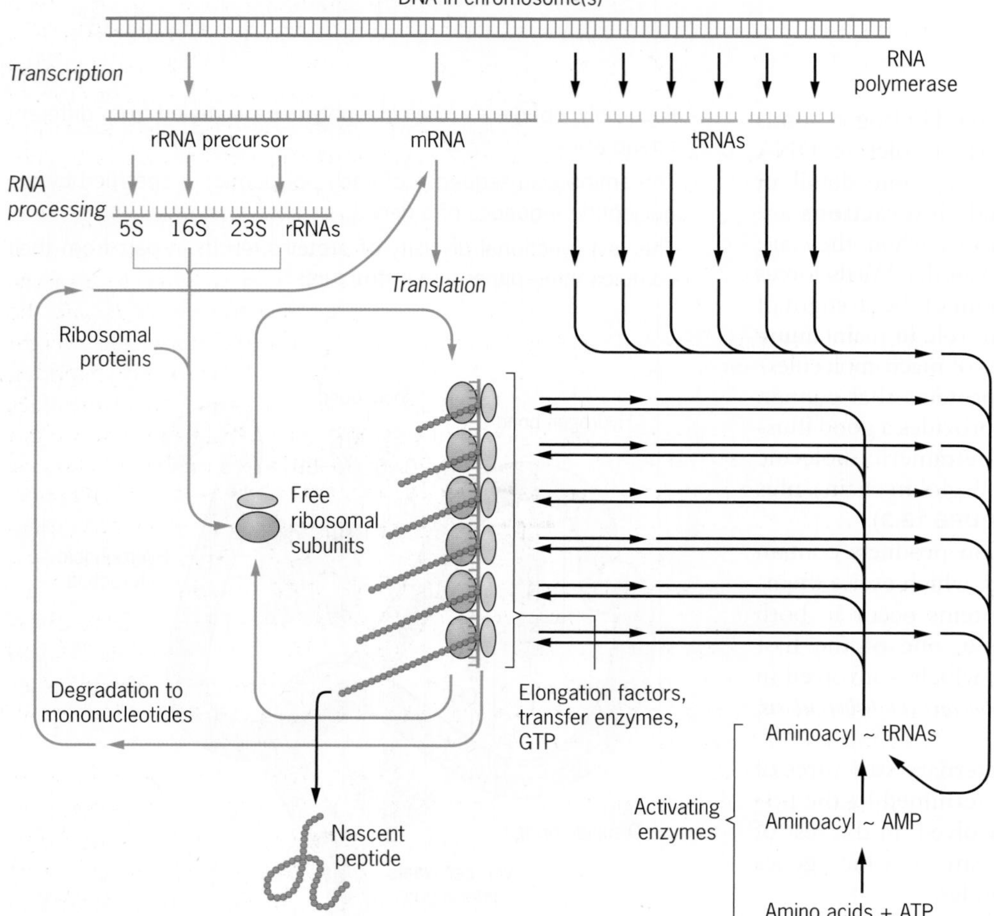

Figure 12.5 ▶ Overview of protein synthesis. The sizes of the rRNA molecules shown are correct for bacteria; larger rRNAs are present in eukaryotes. For simplicity, all RNA species have been transcribed from contiguous segments of a single DNA molecule. In reality, the various RNAs are transcripts of genes located at different positions on from one to many chromosomes. Details of the various stages of protein synthesis are discussed in subsequent sections of this chapter.

discussed in detail in Chapter 11. The second step, translation, involves the transfer of the information in mRNA molecules into the sequences of amino acids in polypeptide gene products.

Translation occurs on ribosomes, which are complex macromolecular structures located in the cytoplasm. Translation involves three types of RNA, all of which are transcribed from DNA templates (chromosomal genes). In addition to mRNAs, three to five RNA molecules (rRNA molecules) are present as part of the structure of each ribosome, and 40 to 60 small RNA molecules (tRNA molecules) function as adaptors by mediating the incorporation of the proper amino acids into polypeptides in response to specific nucleotide sequences in mRNAs. The amino acids are attached to the correct tRNA molecules by a set of activating enzymes called **aminoacyl-tRNA synthetases.**

The nucleotide sequence of an mRNA molecule is translated into the appropriate amino acid sequence according to the dictations of the genetic code. Each amino acid is specified by one or more codons, and each codon contains three nucleotides. Of the 64 possible nucleotide triplets, 61 specify amino acids—two of which also function as polypeptide chain initiation signals—and three specify polypeptide chain termination. The tRNA molecules contain nucleotide triplets called anticodons, which base-pair with the codons in mRNA during the translation process.

Some nascent polypeptides contain short amino acid sequences at the amino or carboxyl termini that function as signals for their transport into specific cellular compartments such as the endoplasmic reticulum, mitochondria, chloroplasts, or nuclei. Nascent secretory proteins, for example, contain a short *signal sequence* at the amino terminus that directs the emerging polypeptide to the membranes of the endoplasmic reticulum. Similar targeting sequences are present at the amino termini of proteins destined for import into mitochondria and chloroplasts. Some nuclear proteins contain targeting extensions at the carboxyl termini. In many cases, the targeting peptides are removed enzymatically by specific peptidases after transport of the protein into the appropriate cellular compartment.

The ribosomes may be thought of as workbenches, complete with machines and tools needed to make a polypeptide. They are nonspecific in the sense that they can synthesize any polypeptide (any amino acid sequence) encoded by a particular mRNA molecule, even an mRNA from a different species. Each mRNA molecule is simultaneously translated by several ribosomes, resulting in the formation of a polyribosome, or polysome. Given this brief overview of protein synthesis, we will now examine some of the more important components of the translation machinery more closely.

COMPONENTS REQUIRED FOR PROTEIN SYNTHESIS: RIBOSOMES

Living cells devote more energy to the synthesis of proteins than to any other aspect of metabolism. About one-third of the total dry mass of most cells consists of molecules that participate directly in the biosynthesis of proteins. In *E. coli*, the approximately 200,000 ribosomes account for 25 percent of the dry weight of each cell. This commitment of a major proportion of the metabolic machinery of cells to the process of protein synthesis documents its importance in the life forms that exist on our planet.

When the sites of protein synthesis were labeled in cells grown for short intervals in the presence of radioactive amino acids and were visualized by autoradiography, the results showed that proteins are synthesized on the ribosomes. In prokaryotes, ribosomes are distributed throughout cells; in eukaryotes, they are located in the cytoplasm, frequently on the extensive intracellular membrane network of the endoplasmic reticulum.

Ribosomes are approximately half protein and half RNA (**FIGURE 12.6**). They are composed of two subunits, one large and one small, which dissociate when the translation of an mRNA molecule is completed and reassociate during the initiation of translation. Each subunit contains a large, folded RNA molecule on which the ribosomal proteins assemble. Ribosome sizes are most frequently expressed in terms of their rates of sedimentation during centrifugation, in Svedberg units (see Focus on Centrifugation Techniques in Chapter 10). The *E. coli* ribosome, like the ribosomes of other prokaryotes, has a molecular weight of 2.5×10^6, a size of 70S, and dimensions of about 20 nm $\times$ 25 nm. The ribosomes of eukaryotes are larger (usually about 80S); however, size varies from species to species. The ribosomes present in the mitochondria and chloroplasts of eukaryotic cells are smaller (usually about 60S).

Although the size and macromolecular composition of ribosomes vary, the overall three-dimensional structure of the ribosome is basically the same in all organisms. In *E. coli*, the small (30S) ribosomal subunit contains a 16S (molecular weight about 6×10^5) RNA molecule plus 21 different polypeptides, and the large (50S) subunit contains two RNA molecules (5S, molecular weight about 4×10^4, and 23S, molecular weight about 1.2×10^6) plus 31 polypeptides. In mammalian ribosomes, the small subunit contains an 18S RNA molecule plus 33 polypeptides, and the large subunit contains three RNA molecules of sizes 5S, 5.8S, and 28S plus 49 polypeptides. In organelles, the corresponding rRNA sizes are 5S, 13S, and 21S.

Masayasu Nomura and his colleagues were able to disassemble the 30S ribosomal subunit of *E. coli* into the individual macromolecules and then reconstitute functional 30S subunits from the components. In this way, they studied the functions of individual rRNA and ribosomal protein molecules.

The ribosomal RNA molecules, like mRNA molecules, are transcribed from a DNA template. In eukaryotes, rRNA synthesis occurs in the nucleolus (**FIGURE 12.7**) and is catalyzed by RNA polymerase I. The nucleolus is a highly specialized component of the nucleus devoted exclusively to the synthesis of rRNAs and their assembly into ribosomes. The ribosomal RNA genes are present in tandemly duplicated arrays separated by intergenic spacer regions. The transcription of these tandem sets of rRNA genes can be visualized directly by electron microscopy (**FIGURE 12.8**).

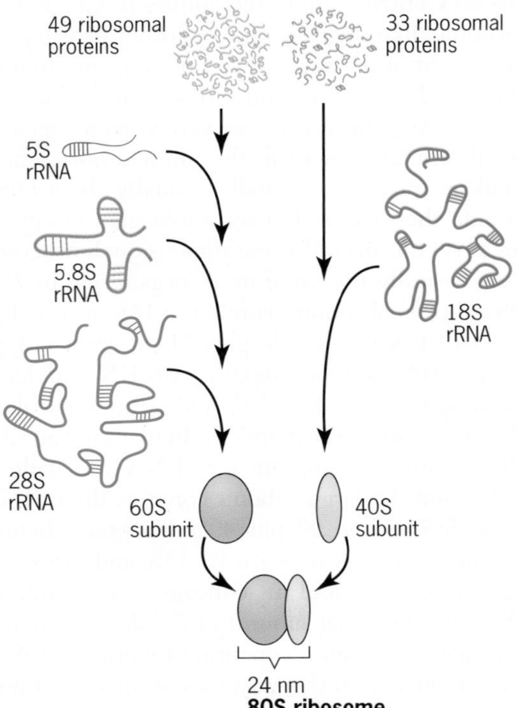

Figure 12.6 ▶ Macromolecular composition of prokaryotic and eukaryotic ribosomes.

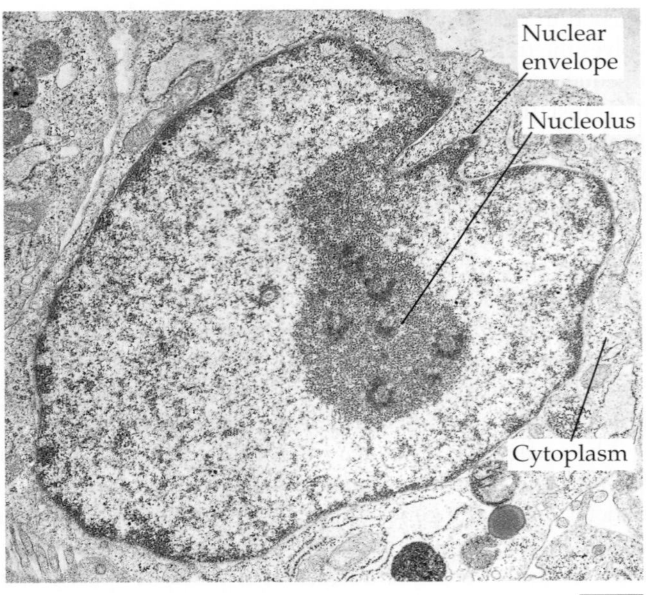

Figure 12.7 ▶ Electron micrograph of a human fibroblast, showing the nucleolus, the site of rRNA synthesis, and ribosome assembly.

The transcription of the rRNA genes produces precursors that are much larger than the RNA molecules found in ribosomes. These rRNA precursors undergo posttranscriptional processing to produce the mature rRNA molecules. In *E. coli*, the rRNA gene transcript is a 30S precursor, which undergoes endonucleolytic cleavages to produce the 5S, 16S, and 23S rRNAs plus one 4S transfer RNA molecule (**FIGURE 12.9*a***). In mammals, the 5.8S, 18S, and 28S rRNAs are cleaved from a 45S precursor (**FIGURE 12.9*b***), whereas the 5S rRNA is produced by posttranscriptional processing of a separate gene transcript. In addition to the posttranscriptional cleavages of rRNA precursors, many of the nucleotides in rRNAs are posttranscriptionally methylated. The methylation is thought to protect rRNA molecules from degradation by ribonucleases.

Multiple copies of the genes for rRNA are present in the genomes of all organisms that have been studied to date. This redundancy of rRNA genes is not surprising considering the large number of ribosomes present per cell. In *E. coli*, seven rRNA genes (*rrnA—rrnE*, *rrnG*, *rrnH*) are distributed among three distinct sites on the chromosome. In eukaryotes, the rRNA genes are present in hundreds to thousands of copies. The 5.8S-18S-28S rRNA genes of eukaryotes are present in tandem arrays in the **nucleolar organizer regions** of the chromosomes. In some eukaryotes, such as maize, there is a single pair of nucleolar organizers (on chromosome 6 in maize). In *Drosophila* and the South African clawed toad, *Xenopus laevis*, the sex chromosomes carry the nucleolar organizers. Humans have five pairs of nucleolar organizers located on the short arms of chromosomes 13, 14, 15, 21, and 22. The 5S rRNA genes in eukaryotes are not located in the nucleolar organizer regions. Instead, they are distributed over several chromosomes. However, the 5S rRNA genes are highly redundant, just as are the 5.8S-18S-28S rRNA genes.

COMPONENTS REQUIRED FOR PROTEIN SYNTHESIS: TRANSFER RNAs

Although the ribosomes provide many of the components required for protein synthesis, and the specifications for each

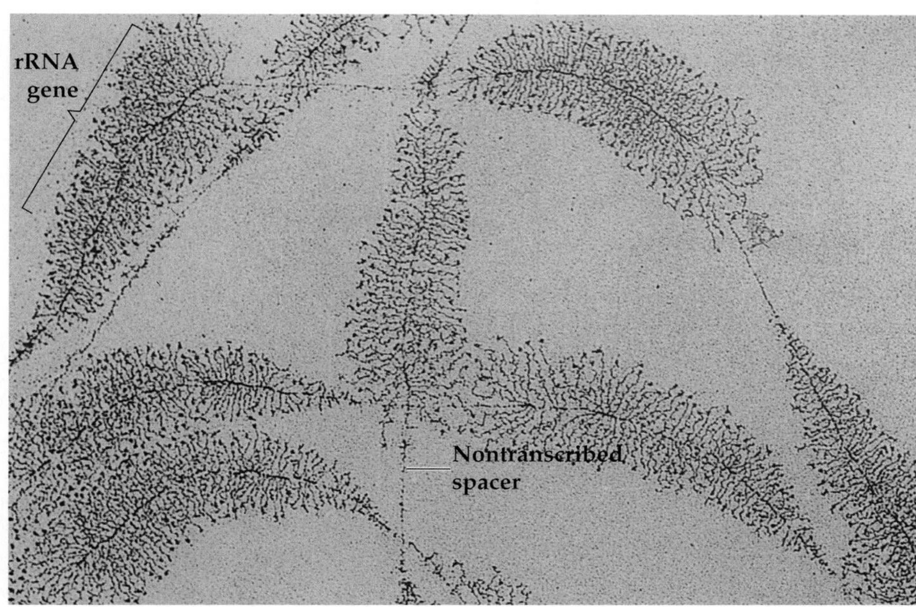

Figure 12.8 ► Electron micrograph showing the transcription of tandemly repeated rRNA genes in the nucleolus of the newt *Triturus viridescens*. A gradient of fibrils of increasing length is observed for each rRNA gene, and nontranscribed spacer regions separate the genes.

polypeptide are encoded in an mRNA molecule, the translation of a coded mRNA message into a sequence of amino acids in a polypeptide requires one additional class of RNA molecules, the transfer RNA (tRNA) molecules. Chemical considerations suggested that direct interactions between the amino acids and the nucleotide triplets or codons in mRNA were unlikely. Thus, in 1958, Francis Crick proposed that some kind of an adaptor molecule must mediate the specification of amino acids by codons in mRNAs during protein synthesis. The adaptor molecules were soon identified by other researchers and shown to be small (4S, 70–95 nucleotides long) RNA molecules. These molecules, first called soluble RNA (sRNA) molecules and subsequently transfer RNA (tRNA) molecules, contain a triplet nucleotide sequence, the anticodon, which is complementary to and base-pairs with the codon sequence in mRNA during translation. There are one to four tRNAs for each of the 20 amino acids.

The amino acids are attached to the tRNAs by high-energy (very reactive) bonds (symbolized ~) between the carboxyl groups of the amino acids and the 3′-hydroxyl termini of the tRNAs. The tRNAs are activated or charged with amino acids in a two-step process, with both reactions catalyzed by the same enzyme, aminoacyl-tRNA synthetase. There is at least one aminoacyl-tRNA synthetase for each of the 20 amino acids.

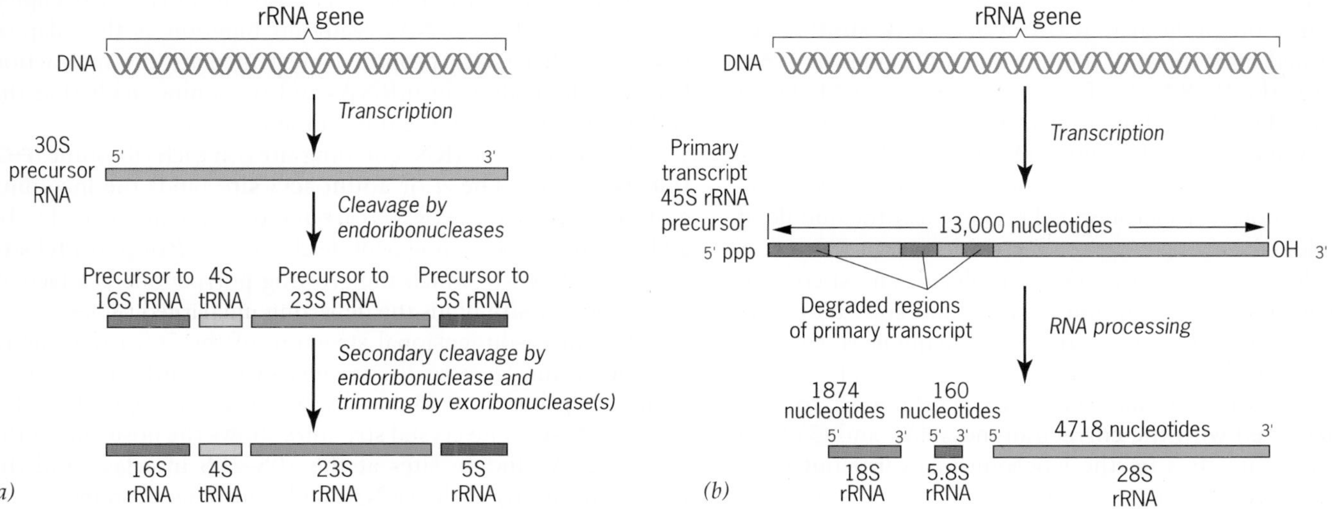

Figure 12.9 ► Synthesis and processing of (*a*) the 30S rRNA precursor in *E. coli* and (*b*) the 45S rRNA precursor in mammals.

The first step in aminoacyl-tRNA synthesis involves the activation of the amino acid using energy from adenosine triphosphate (ATP):

amino acid + ATP

aminoacyl-tRNA synthetase ↓

amino acid ~ AMP + Ⓟ ~ Ⓟ

The amino acid~AMP intermediate is not normally released from the enzyme before undergoing the second step in aminoacyl-tRNA synthesis, namely, the reaction with the appropriate tRNA:

amino acid ~ AMP + tRNA

aminoacyl-tRNA synthetase ↓

amino acid ~ tRNA + AMP

The aminoacyl~tRNAs are the substrates for polypeptide synthesis on ribosomes, with each activated tRNA recognizing the correct mRNA codon and presenting the amino acid in a steric configuration (three-dimensional structure) that facilitates peptide bond formation.

The tRNAs are transcribed from genes. As in the case of rRNAs, the tRNAs are transcribed in the form of larger precursor molecules that undergo posttranscriptional processing (cleavage, trimming, methylation, and so forth). The mature tRNA molecules contain several nucleosides that are not present in the primary tRNA gene transcripts. These unusual nucleosides, such as inosine, pseudouridine, dihydrouridine, 1-methyl guanosine, and several others, are produced by posttranscriptional, enzyme-catalyzed modifications of the four nucleosides incorporated into RNA during transcription.

Because of their small size (most are 70 to 95 nucleotides long), tRNAs have been more amenable to structural analysis than the other, larger molecules of RNA involved in protein synthesis. The complete nucleotide sequence and proposed cloverleaf structure of the alanine tRNA of yeast (**FIGURE 12.10**) were published by Robert W. Holley and colleagues in 1965; Holley shared the 1968 Nobel Prize in Physiology or Medicine for this work. The three-dimensional structure of the phenylalanine tRNA of yeast was determined by X-ray diffraction studies in 1974 (**FIGURE 12.11**). The anticodon of each tRNA occurs within a loop (nonhydrogen-bonded region) near the middle of the molecule.

It should be apparent that tRNA molecules must contain a great deal of specificity despite their small size. Not only must they (1) have the correct anticodon sequences, so as to respond to the right codons, but they also must (2) be recognized by the correct aminoacyl-tRNA synthetases, so that they are activated with the correct amino acids, and (3) bind to the appropriate sites on the ribosomes to carry out their adaptor functions.

François Chapeville and Günter von Ehrenstein and colleagues have proven, by means of a simple and direct experiment (**FIGURE 12.12**), that the specificity for codon recognition resides in the tRNA portion of an aminoacyl-tRNA, rather than in the amino acid. They treated cysteyl-tRNA^{Cys} (the cysteine tRNA activated with cysteine) with a strongly reducing nickel powder (Raney nickel), which converted (reduced) the cysteine to alanine while it was still attached to the cysteine tRNA. When this hybrid aminoacyl-tRNA, alanyl-tRNA^{Cys}, was used in *in vitro* protein-synthesizing systems, alanine was incorporated into polypeptides at positions normally occupied by cysteine. Thus, tRNAs really do function as the adaptor molecules that Crick proposed must mediate the interaction between the codons in mRNAs and the amino acids that the codons specify during the translation process.

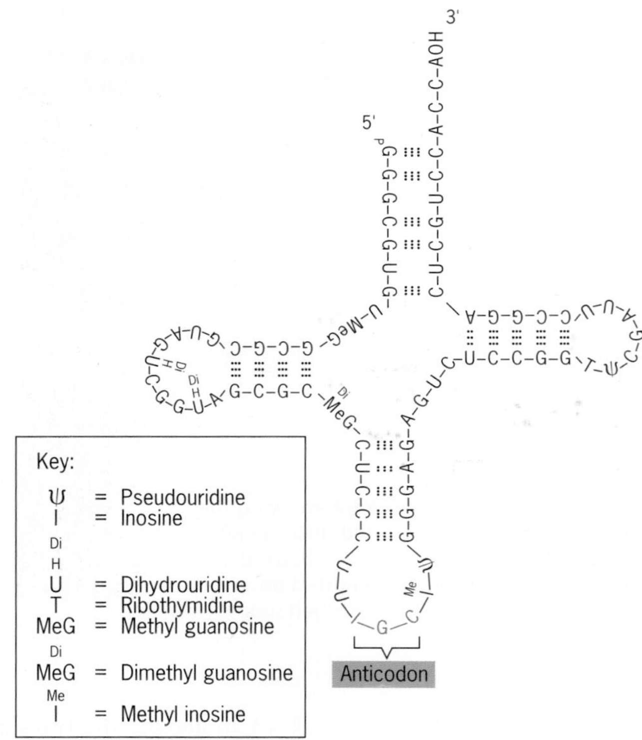

Figure 12.10 ▶ Nucleotide sequence and cloverleaf configuration of the alanine tRNA of *S. cerevisiae*. The names of the modified nucleosides present in the tRNA are shown in the inset.

There are three tRNA binding sites on each ribosome (**FIGURE 12.13*a*, *b***). The ***A*** or **aminoacyl site** binds the incoming aminoacyl-tRNA, the tRNA carrying the next amino acid to be added to the growing polypeptide chain. The ***P*** or **peptidyl site** binds the tRNA to which the growing polypeptide is attached. The ***E*** or **exit site** binds the departing uncharged tRNA.

The three-dimensional structure of the 70S ribosome of the bacterium *Thermus thermophilus* has recently been solved with resolution to 0.55 nm by X-ray crystallography (**FIGURE 12.14*a–c***). The crystal structure shows the positions of the three tRNA binding sites at the 50S–30S interface and the relative positions of the rRNAs and ribosomal proteins.

Although the aminoacyl-tRNA binding sites are located largely on the 50S subunit and the mRNA molecule is bound

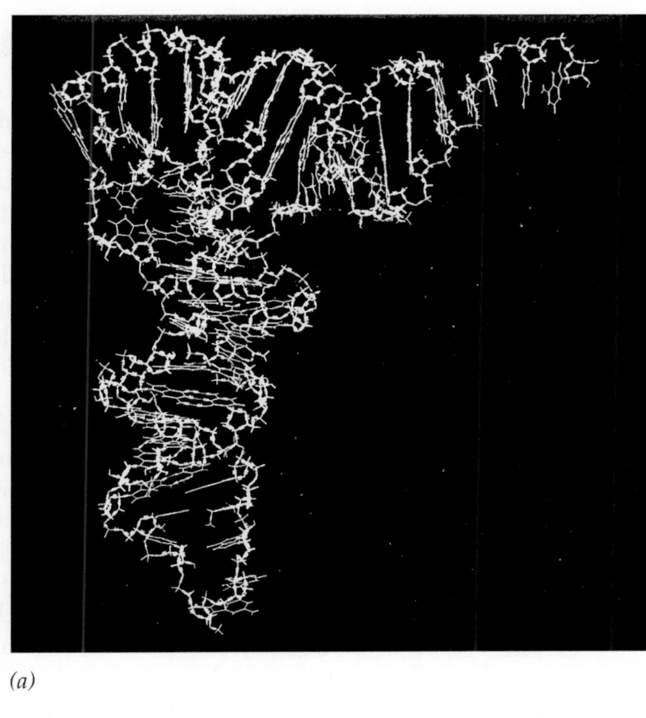

(*a*)

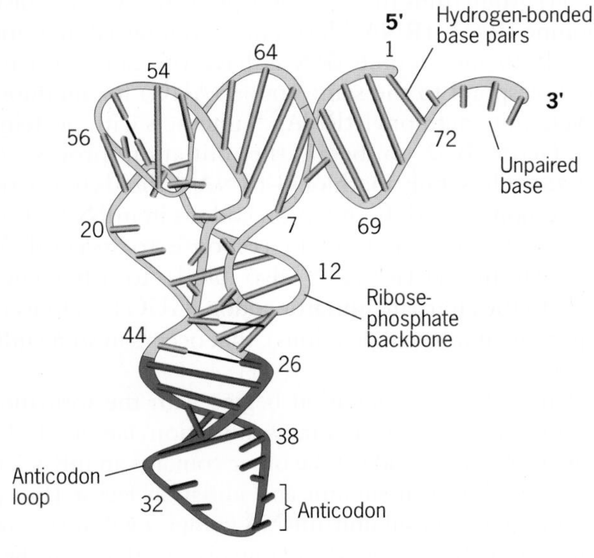

(*b*)

Figure 12.11 ▶ Photograph (*a*) and interpretative drawing (*b*) of a molecular model of the yeast phenylalanine tRNA based on X-ray diffraction data.

by the 30S subunit, the specificity for aminoacyl-tRNA binding in each site is provided by the mRNA codon that makes up part of the binding site (see **FIGURE 12.13*b***). As the ribosome moves along an mRNA (or as the mRNA is shuttled across the ribosome), the specificity for the aminoacyl-tRNA binding in the *A*, *P*, and *E* sites changes as different mRNA codons move into register in the binding sites. The ribosomal binding sites by themselves (minus mRNA) are thus capable of binding any aminoacyl-tRNA.

Anticodon

$tRNA^{Cys}$ + HO—C(=O)—C(H)(—NH_2)—CH_2—SH Cysteine

Cysteyl-$tRNA^{Cys}$ synthetase

O~C(=O)—C(H)(—NH_2)—CH_2—SH Cysteyl-$tRNA^{Cys}$

Raney nickel (strong reducing agent)

O~C(=O)—C(H)(—NH_2)—CH_3 Alanine

Alanyl-$tRNA^{Cys}$

In vitro protein-synthesizing system

Experiment I:

Using poly(UG) (UGUGU, etc., repeating UG copolymer) as an artificial messenger RNA. Alanine attached to $tRNA^{Cys}$ was incorporated, despite the fact that the alanine codons are GCU, GCC, GCA, and GCG. UGU = cysteine codon.

Experiment II:

Using the hemoglobin-synthesizing rabbit reticulocyte system containing native hemoglobin mRNAs. Demonstrated that alanine from alanyl-$tRNA^{Cys}$ was incorporated into positions in the rabbit globin chains normally occupied by cysteine.

Figure 12.12 ▶ Proof that the codon-recognizing specificity of an aminoacyl-tRNA complex resides in the tRNA rather than in the amino acid.

TRANSLATION: THE SYNTHESIS OF POLYPEPTIDES USING mRNA TEMPLATES

We now have reviewed all the major components of the protein-synthesizing system. The mRNA molecules provide the specifications for the amino acid sequences of the polypeptide gene products. The ribosomes provide many of the macromolecular components required for the translation process. The tRNAs provide the adaptor molecules needed to incorporate amino acids into polypeptides in response to codons in mRNAs. In addition, several soluble proteins participate in the process. The translation of the sequence of nucleotides in an mRNA molecule into the sequence of amino acids in its polypeptide product can be divided into three stages: (1) polypeptide chain initiation, (2) chain elongation, and (3) chain termination.

70S ribosome diagram

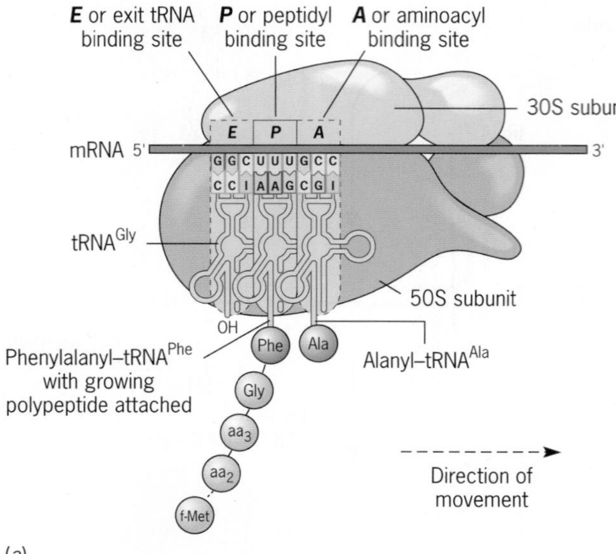

70S ribosome—cutaway view of model

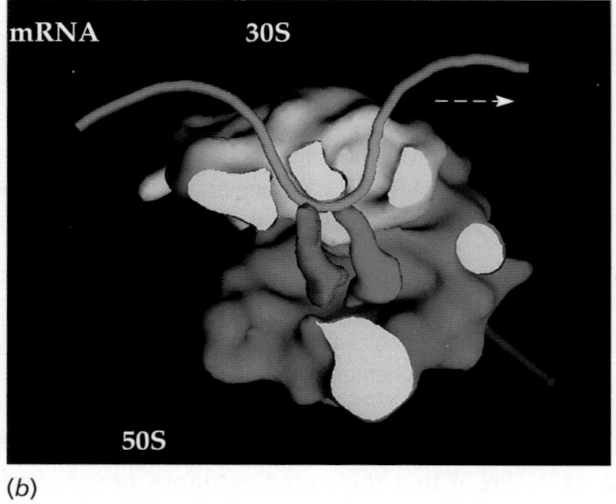

Figure 12.13 ▶ Ribosome structure in *E. coli*. (*a*) Each ribosome/mRNA complex contains three aminoacyl-tRNA binding sites. The *A* or aminoacyl-tRNA site is occupied by alanyl-$tRNA^{Ala}$. The *P* or peptidyl site is occupied by phenyl-alanyl-$tRNA^{Phe}$, with the growing polypeptide chain covalently linked to the phenylalanine tRNA. The *E* or exit site is occupied by $tRNA^{Gly}$ prior to its release from the ribosome. (*b*) An mRNA molecule (orange), which is attached to the 30S subunit (light green) of the ribosome, contributes specificity to the tRNA-binding sites, which are located largely on the 50S subunit (blue) of the ribosome. The aminoacyl-tRNAs located in the *P* and *A* sites are shown in red and dark green, respectively. The *E* site is unoccupied.

Translation: Polypeptide Chain Initiation

The **initiation** of translation includes all events that precede the formation of a peptide bond between the first two amino acids of the new polypeptide chain. Although several aspects of the initiation process are the same in prokaryotes and eukaryotes, some are different. Accordingly, we will first examine the initiation of polypeptide chains in *E. coli*, and we will then look at the unique aspects of translational initiation in eukaryotes.

In *E. coli*, the initiation process involves the 30S subunit of the ribosome, a special initiator tRNA, an mRNA molecule, three soluble protein **initiation factors: IF-1, IF-2,** and **IF-3,** and one molecule of GTP (**FIGURE 12.15**). Translation occurs on 70S ribosomes, but the ribosomes dissociate into their 30S and 50S subunits each time they complete the synthesis of a polypeptide chain. In the first stage of the initiation of translation, a free 30S subunit interacts with an mRNA molecule and the initiation factors. The 50S subunit joins the complex to form the 70S ribosome in the final step of the initiation process.

The synthesis of polypeptides is initiated by a special tRNA, designated $\mathbf{tRNA_f^{Met}}$, in response to a translation **initiation codon** (usually AUG, sometimes GUG). Therefore, all polypeptides begin with methionine during synthesis. The amino-terminal methionine is subsequently cleaved from many polypeptides. Thus, functional proteins need not have an amino-terminal methionine. The methionine on the initiator $tRNA_f^{Met}$ has the amino group blocked with a formyl ($\text{—}\overset{\text{O}}{\overset{\|}{\text{C}}}\text{—H}$) group (thus the "f" subscript in $tRNA_f^{Met}$). A distinct methionine tRNA, $\mathbf{tRNA^{Met}}$, responds to internal methionine codons. Both methionine tRNAs have the same anticodon, and both respond to the same codon (AUG) for methionine. However, only methionyl-$tRNA_f^{Met}$ interacts with protein initiation factor IF-2 to begin the initiation process (**FIGURE 12.15**). Thus, only methionyl-$tRNA_f^{Met}$ binds to the ribosome in response to AUG initiation codons in mRNAs, leaving methionyl-$tRNA^{Met}$ to bind in response to internal AUG codons. Methionyl-$tRNA_f^{Met}$ also binds to ribosomes in response to the alternate initiator codon, GUG (a valine codon when present at internal positions), that occurs in some mRNA molecules.

Polypeptide chain initiation begins with the formation of two complexes: (1) one contains initiation factor IF-2 and methionyl-$tRNA_f^{Met}$, and (2) the other contains an mRNA molecule, a 30S ribosomal subunit and initiation factor IF-3 (**FIGURE 12.15**). The 30S subunit/mRNA complex will form only in the presence of IF-3; thus, IF-3 controls the ability of the 30S subunit to begin the initiation process. The formation of the 30S subunit/mRNA complex depends in part on base-pairing between a nucleotide sequence near the 3′ end of the 16S rRNA and a sequence near the 5′ end of the mRNA molecule (**FIGURE 12.16**). Prokaryotic mRNAs contain a conserved polypurine tract, consensus AGGAGG, located about seven nucleotides upstream from the AUG initiation codon. This conserved hexamer, called the **Shine-Dalgarno sequence** after the scientists who discovered it, is complementary to a sequence near the 3′ terminus of the 16S ribosomal RNA. When the Shine-Dalgarno sequences of mRNAs are experimentally modified so that they can no longer base-pair with the 16S rRNA, the modified mRNAs either are not translated or are translated very inefficiently,

70S ribosome—crystal structure

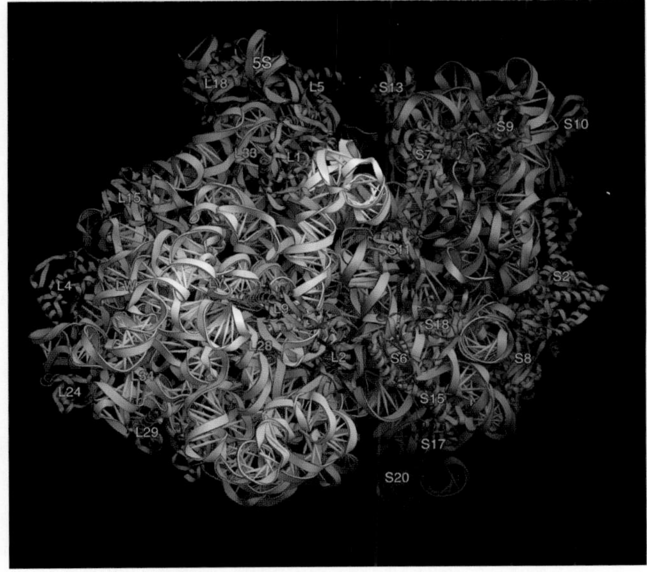

(a)

50S subunit—crystal structure

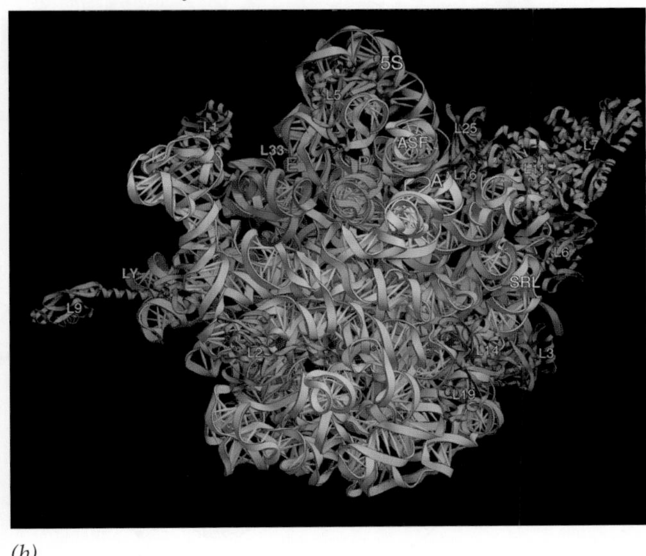

(b)

30S subunit—crystal structure

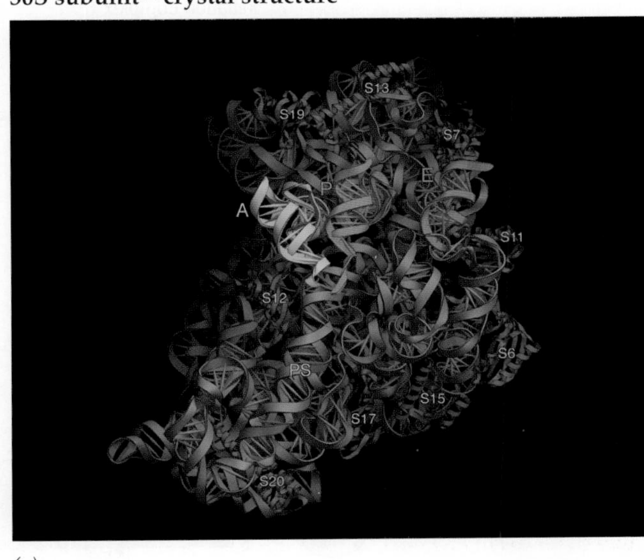

(c)

Figure 12.14 ▶ Ribosome structure in *Thermus thermophilus.* Crystal structure of the 70S ribosome with 0.55 nm resolution, showing the complete ribosome (*a*) and the interfaces of the 50S (*b*) and 30S (*c*) subunits. (*a*) 50S subunit on the left; 30S subunit on the right. (*b, c*) Interfaces of the 50S subunit and the 30S subunit obtained by rotating the structures shown in (*a*) 90° to the left (*b*) or to the right (*c*), respectively. The tRNAs in the *A, P,* and *E* sites are shown in gold, orange, and red, respectively. Components: 16S rRNA (cyan); 23S rRNA (gray); 5S rRNA (light blue); 30S subunit proteins (dark blue); and 50S subunit proteins (magenta). L1, large subunit protein 1; S7, small subunit protein 7.

indicating that this base-pairing plays an important role in translation.

The IF-2/methionyl-tRNA$_f^{Met}$ complex and the mRNA/30S subunit/IF-3 complex subsequently combine with each other and with initiation factor IF-1 and one molecule of GTP to form the complete 30S initiation complex. The final step in the initiation of translation is the addition of the 50S subunit to the 30S initiation complex to produce the complete 70S ribosome. Initiation factor IF-3 must be released from the complex before the 50S subunit can join the complex; IF-3 and the 50S subunit are never found to be associated with the 30S subunit at the same time. The addition of the 50S subunit requires energy from GTP and the release of initiation factors IF-1 and IF-2.

The addition of the 50S ribosomal subunit to the complex positions the initiator tRNA, methionyl-tRNA$_f^{Met}$, in the peptidyl (*P*) site with the anticodon of the tRNA aligned with the AUG initiation codon of the mRNA. Methionyl-tRNA$_f^{Met}$ is the only aminoacyl-tRNA that can enter the *P* site directly, without first passing through the aminoacyl (*A*) site. With the initiator AUG positioned in the *P* site, the second codon of the mRNA is in register with the *A* site, dictating the aminoacyl-tRNA binding specificity at that site

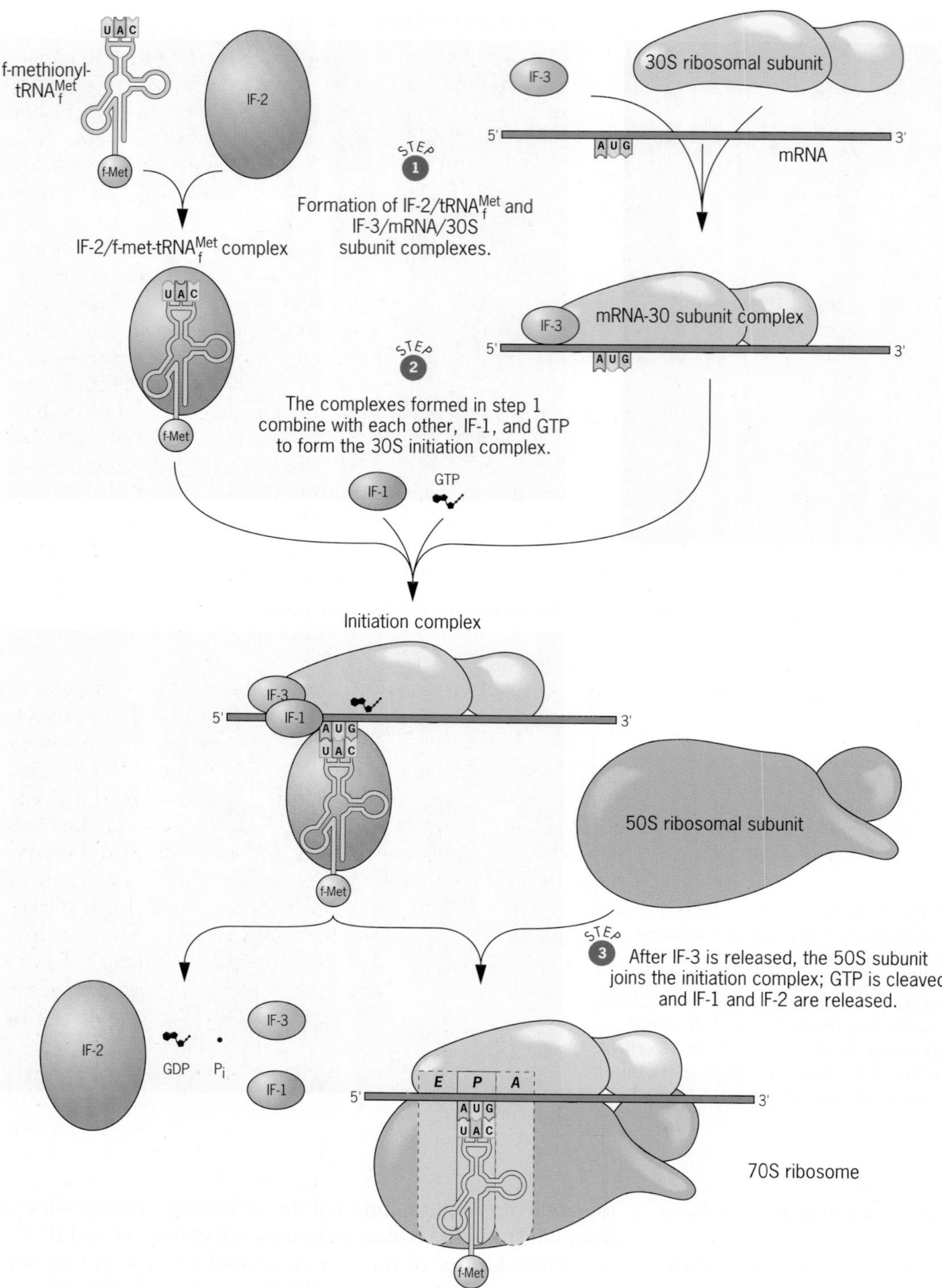

Figure 12.15 ▶ The initiation of translation in *E. coli*.

and setting the stage for the second phase in polypeptide synthesis, chain elongation.

The initiation of translation is more complex in eukaryotes, involving several soluble initiation factors. Nevertheless, the overall process is similar except for two features. (1) The amino group of the methionine on the initiator tRNA is not formylated as in prokaryotes. (2) The initiation complex forms at the 5′ terminus of the mRNA, not at the Shine-Dalgarno/AUG translation start site as in *E. coli*. In eukaryotes, the initiation complex scans the mRNA, starting at the 5′ end, searching for

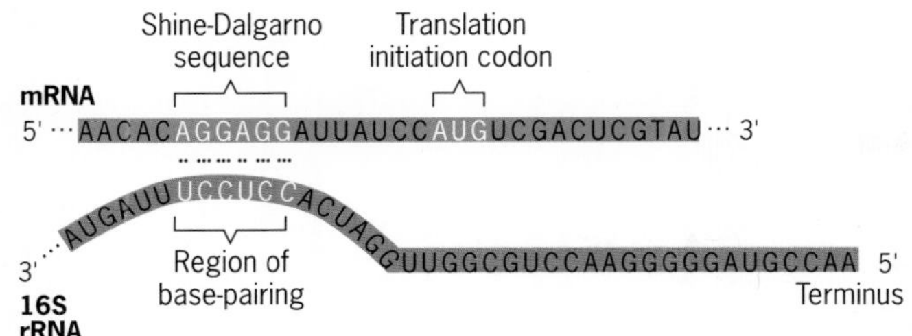

Figure 12.16 ► Base-pairing between the Shine-Dalgarno sequence in a prokaryotic mRNA and a complementary sequence near the 3′ terminus of the 16S rRNA is involved in the formation of the mRNA/30S ribosomal subunit initiation complex.

an AUG translation-initiation codon. Thus, in eukaryotes, translation frequently begins at the AUG closest to the 5′ terminus of the mRNA molecule, although the efficiency with which a given AUG is used to initiate translation depends on the contiguous nucleotide sequence. The optimal initiation sequence is 5′-GCC(A or G)CC**AUG**G-3′. The purine (A or G) three bases upstream from the **AUG** initiator codon and the G immediately following it are the most important—influencing initiation efficiency by tenfold or more. Changes of other bases in the sequence cause smaller decreases in initiation efficiency. These sequence requirements for optimal translation initiation in eukaryotes are called **Kozak's rules,** after Marilyn Kozak, who first proposed them.

Like prokaryotes, eukaryotes contain a special initiator tRNA, **$\text{tRNA}_i^{\text{Met}}$** ("i" for initiator), but the amino group of the methionyl-$\text{tRNA}_i^{\text{Met}}$ is not formylated. The initiator methionyl-$\text{tRNA}_i^{\text{Met}}$ interacts with a soluble initiation factor and enters the *P* site directly during the initiation process, just as in *E. coli*.

In eukaryotes, a cap-binding protein (CBP) binds to the 7-methyl guanosine cap at the 5′ terminus of the mRNA. Then, other initiation factors bind to the CBP-mRNA complex, followed by the small (40S) subunit of the ribosome. The entire initiation complex moves 5′ → 3′ along the mRNA molecule, searching for an AUG codon. When an AUG triplet is found, the initiation factors dissociate from the complex, and the large (60S) subunit binds to the methionyl-tRNA/mRNA/40S subunit complex, forming the complete (80S) ribosome. The 80S ribosome/mRNA/tRNA complex is ready to begin the second phase of translation, chain elongation.

Translation: Polypeptide Chain Elongation

The process of polypeptide chain **elongation** is basically the same in both prokaryotes and eukaryotes. The addition of each amino acid to the growing polypeptide occurs in three steps: (1) binding of an aminoacyl-tRNA to the *A* site of the ribosome, (2) transfer of the growing polypeptide chain from the tRNA in the *P* site to the tRNA in the *A* site by the formation of a new peptide bond, and (3) translocation of the ribosome along the mRNA to position the next codon in the *A* site (**FIGURE 12.17**). During step 3, the nascent polypeptide-tRNA and the uncharged tRNA are translocated from the *A* and *P* sites to the *P* and *E* sites, respectively. These three steps are repeated in a cyclic manner throughout the elongation process. The soluble factors involved in chain elongation in *E. coli* are described here. Similar factors participate in chain elongation in eukaryotes.

In the first step, an aminoacyl-tRNA enters and becomes bound to the *A* site of the ribosome, with the specificity provided by the mRNA codon in register with the *A* site (**FIGURE 12.17**). The three nucleotides in the anticodon of the incoming amino-acyl-tRNA must pair with the nucleotides of the mRNA codon present at the *A* site. This step requires **elongation factor Tu** carrying a molecule of GTP **(EF-Tu·GTP).** The GTP is required for aminoacyl-tRNA binding at the *A* site but is not cleaved until the peptide bond is formed. After the cleavage of GTP, EF-Tu·GDP is released from the ribosome. EF-Tu·GDP is inactive and will not bind to aminoacyl-tRNAs. EF-Tu·GDP is converted to the active EF-Tu·GTP form by **elongation factor Ts (EF-Ts),** which hydrolyzes one molecule of GTP in the process. EF-Tu interacts with all of the aminoacyl-tRNAs except methionyl-tRNA.

The second step in chain elongation is the formation of a peptide bond between the amino group of the aminoacyl-tRNA in the *A* site and the carboxyl terminus of the growing polypeptide chain attached to the tRNA in the *P* site. This uncouples the growing chain from the tRNA in the *P* site and covalently joins the chain to the tRNA in the *A* site (**FIGURE 12.17**). This key reaction is catalyzed by **peptidyl transferase,** an enzymatic activity built into the 50S subunit of the ribosome. We should note that the peptidyl transferase activity resides in the 23S rRNA molecule rather than in a ribosomal protein, perhaps another relic of an early RNA-based world. Peptide bond formation requires the hydrolysis of the molecule of GTP brought to the ribosome by EF-Tu in step 1.

During the third step in chain elongation, the peptidyl-tRNA present in the *A* site of the ribosome is translocated to the *P* site, and the uncharged tRNA in the *P* site is translocated to the *E* site, as the ribosome moves three nucleotides toward the 3′ end of the mRNA molecule. The translocation step requires GTP and **elongation factor G (EF-G).** The ribosome undergoes changes in conformation during the translocation process, suggesting that it may shuttle along the mRNA molecule. The energy for the movement of the ribosome is provided by the hydrolysis of GTP. The translocation of the peptidyl-tRNA from the *A* site to the *P* site leaves the *A* site unoccupied and the ribosome ready to begin the next cycle of chain elongation.

The elongation of one eukaryotic polypeptide, the silk protein fibroin, can be visualized with the electron microscope by using techniques developed by Oscar Miller,

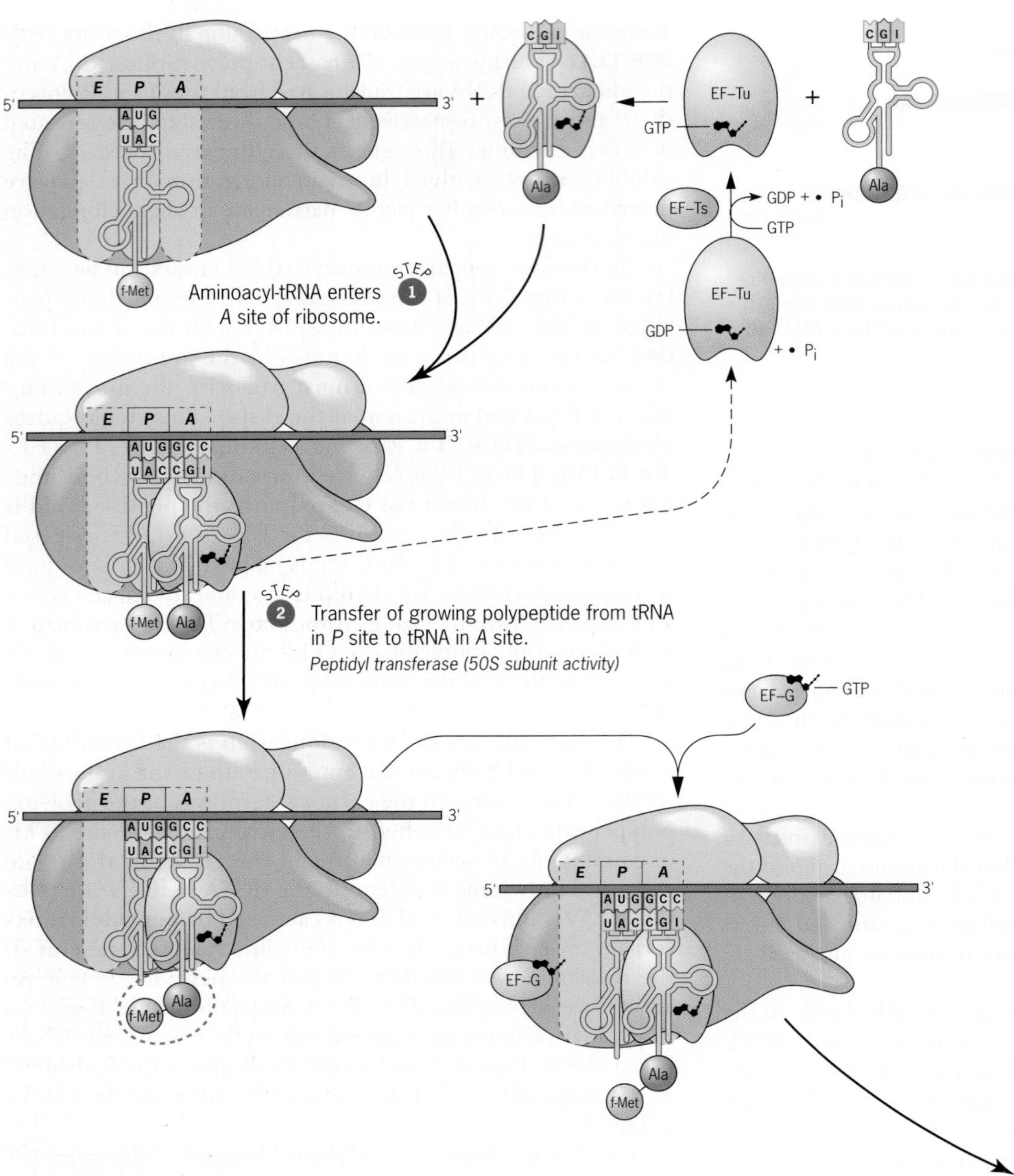

Figure 12.17 ▶ Polypeptide chain elongation in *E. coli*.

Barbara Hamkalo, and colleagues. Most proteins fold up on the surface of the ribosome during their synthesis. However, fibroin remains extended from the surface of the ribosome under the conditions used by Miller and coworkers. As a result, nascent polypeptide chains of increasing length can be seen attached to the ribosomes as they are scanned from the 5′ end of the mRNA to the 3′ end (**FIGURE 12.18**). Fibroin is a large protein with a mass of over 200,000 daltons; it is synthesized on large polyribosomes containing 50 to 80 ribosomes.

Polypeptide chain elongation proceeds rapidly. In *E. coli*, all three steps required to add one amino acid to the growing polypeptide chain occur in about 0.05 second. Thus, the synthesis of a polypeptide containing 300 amino acids takes only about 15 seconds. Given its complexity, the accuracy and efficiency of the translational apparatus are indeed amazing.

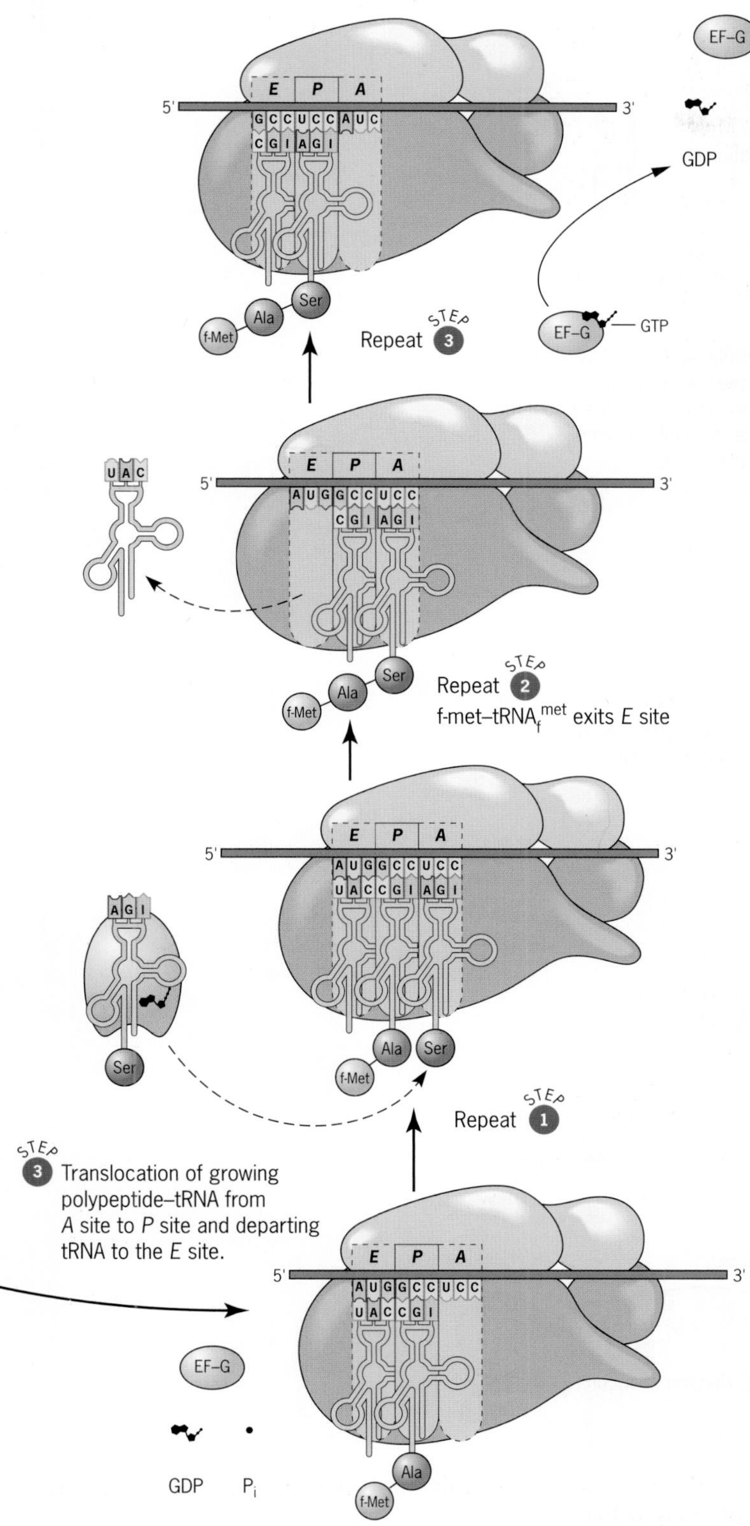

Figure 12.17 ► *(continued)*

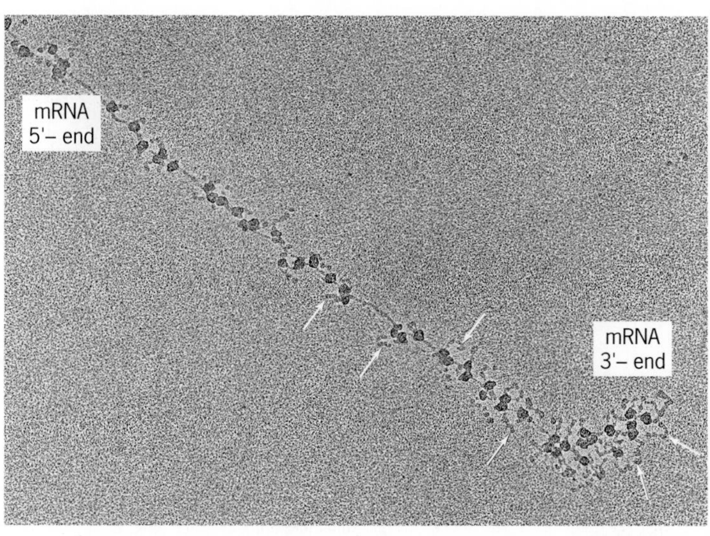

Figure 12.18 ► Visualization of the elongation of fibroin polypeptides in the posterior silk gland of the silkworm *Bombyx mori*. The arrows point to growing fibroin polypeptides. Note their increasing length as one approaches the 3′ end of the mRNA molecule.

Translation: Polypeptide Chain Termination

Polypeptide chain elongation undergoes **termination** when any of three **chain-termination codons** (UAA, UAG, or UGA) enters the *A* site on the ribosome (**FIGURE 12.19**). These three stop codons are recognized by soluble proteins called **release factors (RFs).** In *E. coli*, there are two release factors, RF-1 and RF-2. RF-1 recognizes termination codons UAA and UAG; RF-2 recognizes UAA and UGA. In eukaryotes, a single release factor **(eRF)** recognizes all three termination codons. The presence of a release factor in the *A* site alters the activity of peptidyl transferase such that it adds a water molecule to the carboxyl terminus of the nascent polypeptide. This reaction releases the polypeptide from the tRNA molecule in the *P* site and triggers the translocation of the free tRNA to the *E* site. Termination is completed by the release of the mRNA molecule from the ribosome and the dissociation of the ribosome into its subunits. The ribosomal subunits are then ready to initiate another round of protein synthesis, as previously described.

KEY POINTS

- Genetic information carried in the sequences of nucleotides in mRNA molecules is translated into sequences of amino acids in polypeptide gene products by intricate macromolecular machines called ribosomes.
- The translation process is complex, requiring the participation of many different RNA and protein molecules.
- Transfer RNA molecules serve as adaptors, mediating the interaction between amino acids and codons in mRNA.
- The process of translation involves the initiation, elongation, and termination of polypeptide chains and is governed by the specifications of the genetic code.

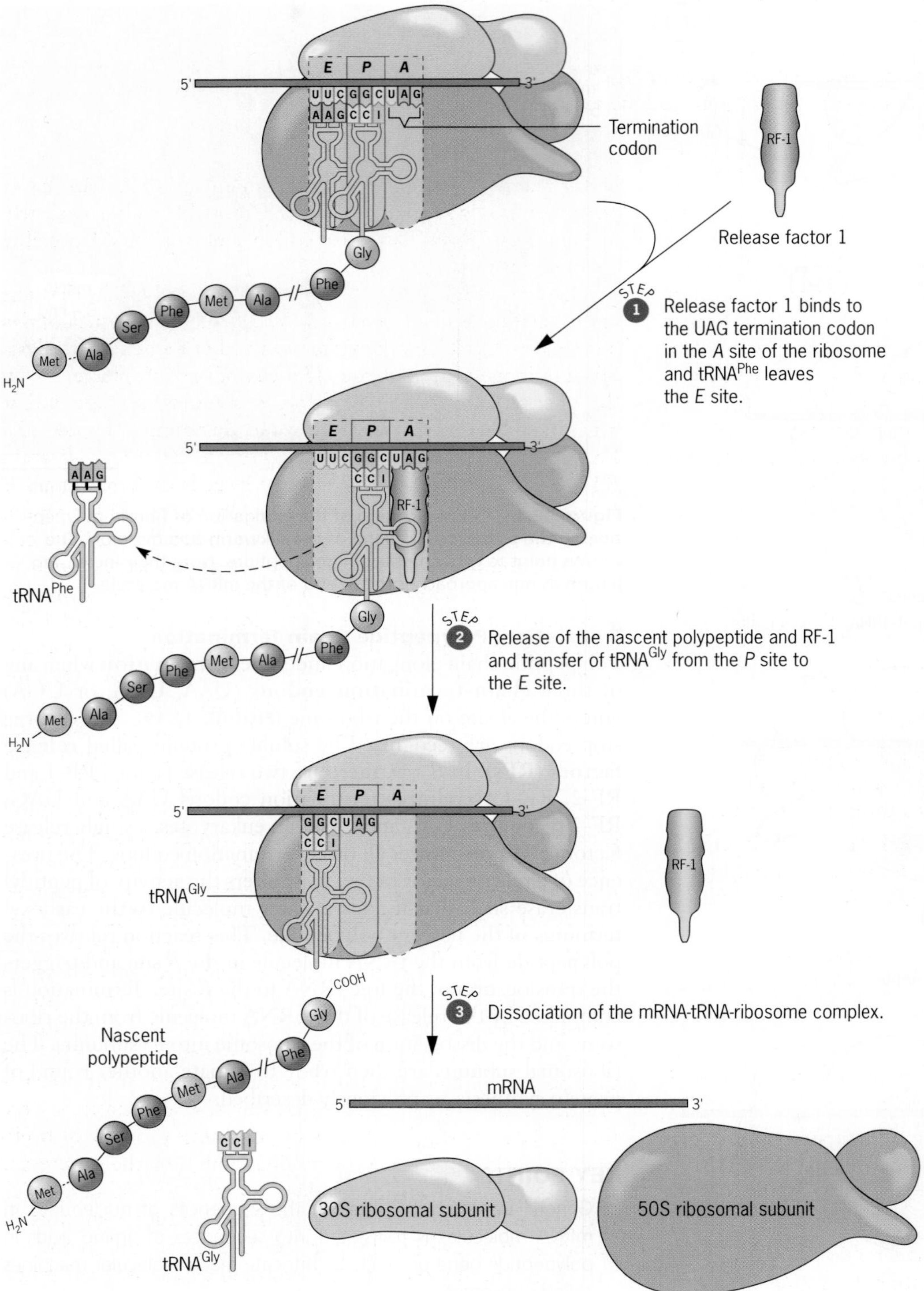

Figure 12.19 ▶ Polypeptide chain termination in *E. coli.* The formyl group of formylmethionine is removed during translation.

The Genetic Code

The genetic code is a nonoverlapping code, with each amino acid plus polypeptide initiation and termination specified by RNA codons composed of three nucleotides.

As it became evident that genes controlled the structure of polypeptides, attention focused on how the sequence of the four different nucleotides in DNA could control the sequence of the 20 amino acids present in proteins. With the discovery of the mRNA intermediary, the question became one of how the sequence of the four bases present in mRNA molecules could specify the amino acid sequence of a polypeptide. What is the nature of the genetic code relating mRNA base sequences to amino acid sequences? Clearly, the symbols or letters used in the code must be the bases; but what comprises a codon, the unit or word specifying one amino acid or, actually, one aminoacyl-tRNA?

PROPERTIES OF THE GENETIC CODE: AN OVERVIEW

The main features of the genetic code were worked out during the 1960s. Cracking the code was one of the most exciting events in the history of science, with new information reported almost daily. By the mid-1960s, the genetic code was largely solved. Before focusing on specific features of the code, let us consider its most important properties.

1. *The genetic code is composed of nucleotide triplets.* Three nucleotides in mRNA specify one amino acid in the polypeptide product; thus, each codon contains three nucleotides.
2. *The genetic code is nonoverlapping.* Each nucleotide in mRNA belongs to just one codon except in rare cases where genes overlap and a nucleotide sequence is read in two different reading frames.
3. *The genetic code is comma-free.* There are no commas or other forms of punctuation within the coding regions of mRNA molecules. During translation, the codons are read consecutively.
4. *The genetic code is degenerate.* All but two of the amino acids are specified by more than one codon.
5. *The genetic code is ordered.* Multiple codons for a given amino acid and codons for amino acids with similar chemical properties are closely related, usually differing by a single nucleotide.
6. *The genetic code contains start and stop codons.* Specific codons are used to initiate and to terminate polypeptide chains.
7. *The genetic code is nearly universal.* With minor exceptions, the codons have the same meaning in all living organisms, from viruses to humans.

THREE NUCLEOTIDES PER CODON

Twenty different amino acids are incorporated into polypeptides during translation. Thus, at least 20 different codons must be formed with the four bases available in mRNA. Two bases per codon would result in only 4^2 or 16 possible codons—clearly not enough. Three bases per codon yields 4^3 or 64 possible codons—an apparent excess.

In 1961, Francis Crick and colleagues published the first strong evidence in support of a *triplet code* (three nucleotides per codon). Crick and coworkers carried out a genetic analysis of mutations induced at the *r*II locus of bacteriophage T4 by the chemical proflavin. Proflavin is a mutagenic agent that causes single base-pair additions and deletions (Chapter 13). Phage T4 *r*II mutants are unable to grow in cells of *E. coli* strain K12, but grow like wild-type phage in cells of *E. coli* strain B (Chapter 8). Wild-type T4 grows equally well on either strain. Crick and coworkers isolated proflavin-induced revertants of a proflavin-induced mutation. These revertants were shown to result from the occurrence of additional mutations at nearby sites rather than reversion of the original mutation. Second-site mutations that restore the wild-type phenotype in a mutant organism are called **suppressor mutations** because they cancel, or suppress, the effect(s) of the original mutation.

Crick and colleagues reasoned that if the original mutation was a single base-pair addition or deletion, then the suppressor mutations must be single base-pair deletions or additions, respectively, occurring at a site or sites near the original mutation. If sequential nucleotide triplets in an mRNA specify amino acids, then every nucleotide sequence can be recognized or read during translation in three different ways. For example, the sequence AAAGGGCCCTTT can be read (1) AAA, GGG, CCC, TTT, (2) A, AAG, GGC, CCT, TT, or (3) AA, AGG, GCC, CTT, T. The **reading frame** of an mRNA is the series of nucleotide triplets that are read (positioned in the *A* site of the ribosome) during translation. A single base-pair addition or deletion will alter the reading frame of the gene and mRNA for that portion of the gene distal to the mutation. This effect is illustrated in **FIGURE 12.20*a***. The suppressor mutations were then isolated as single mutants by screening progeny of backcrosses to wild-type. Like the original mutation, the suppressor mutations were found to produce mutant phenotypes. Crick and colleagues next isolated proflavin-induced suppressor mutations of the original suppressor mutations, and so on.

Crick and colleagues then classified all the isolated mutations into two groups, plus (+) and minus (−) (for additions and deletions, although they had no idea which group was which), based on the reasoning that a (+) mutation would suppress a (−) mutation but not another (+) mutation, and vice versa (**FIGURE 12.20**). Then, Crick and coworkers constructed recombinants that carried various combinations of the (+) and the (−) mutations. Like the single mutants, recombinants with two (+) mutations or two (−) mutations always had the mutant phenotype.

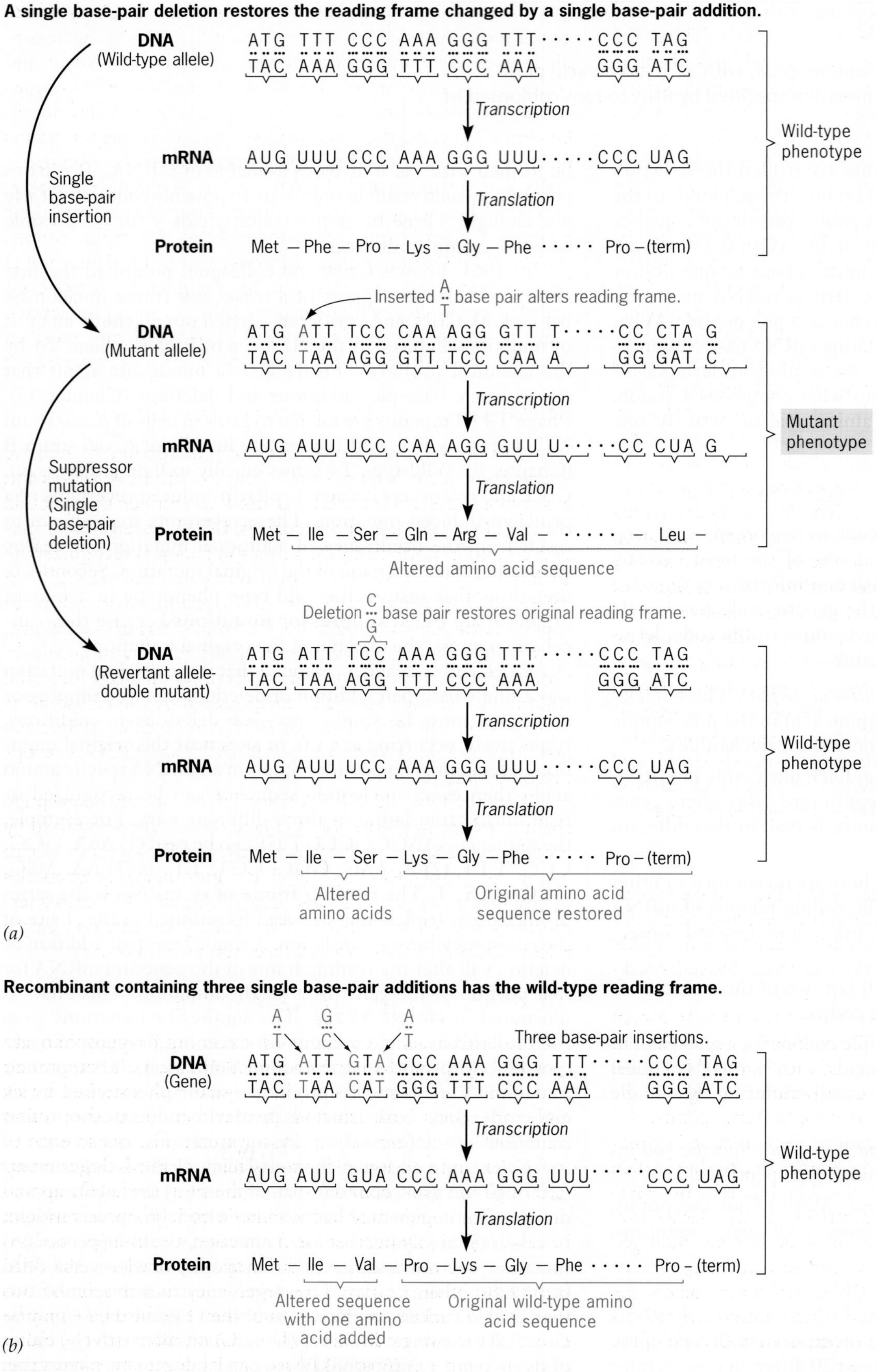

A single base-pair deletion restores the reading frame changed by a single base-pair addition.
DNA
(Wild-type allele)
ATG TTT CCC AAA GGG TTT·····CCC TAG
TAC AAA GGG TTT CCC AAA GGG ATC
Transcription
mRNA
AUG UUU CCC AAA GGG UUU·····CCC UAG
Translation
Protein
Met – Phe – Pro – Lys – Gly – Phe ····· Pro – (term)
Wild-type phenotype
Single base-pair insertion
Inserted A/T base pair alters reading frame.
DNA
(Mutant allele)
ATG ATT TCC CAA AGG GTT T·····CC CTA G
TAC TAA AGG GTT TCC CAA A GG GAT C
Transcription
mRNA
AUG AUU UCC CAA AGG GUU U·····CC CUA G
Translation
Protein
Met – Ile – Ser – Gln – Arg – Val – ····· – Leu –
Altered amino acid sequence
Mutant phenotype
Suppressor mutation (Single base-pair deletion)
Deletion C/G base pair restores original reading frame.
DNA
(Revertant allele-double mutant)
ATG ATT TCC AAA GGG TTT·····CCC TAG
TAC TAA AGG TTT CCC AAA GGG ATC
Transcription
mRNA
AUG AUU UCC AAA GGG UUU·····CCC UAG
Translation
Protein
Met – Ile – Ser – Lys – Gly – Phe ····· Pro – (term)
Altered amino acids
Original amino acid sequence restored
Wild-type phenotype
(a)
Recombinant containing three single base-pair additions has the wild-type reading frame.
A/T G/C A/T
Three base-pair insertions.
DNA
(Gene)
ATG ATT GTA CCC AAA GGG TTT·····CCC TAG
TAC TAA CAT GGG TTT CCC AAA GGG ATC
Transcription
mRNA
AUG AUU GUA CCC AAA GGG UUU·····CCC UAG
Translation
Protein
Met – Ile – Val – Pro – Lys – Gly – Phe ····· Pro – (term)
Altered sequence with one amino acid added
Original wild-type amino acid sequence
Wild-type phenotype
(b)

Figure 12.20 ▶ Early evidence that the genetic code is a triplet code. See the text for details.

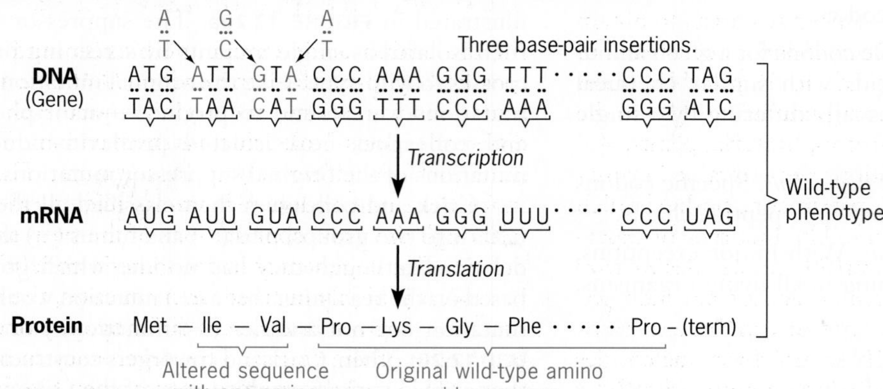

The critical result was that recombinants with three (+) mutations (**FIGURE 12.20*b***) or three (−) mutations often exhibited the wild-type phenotype. This indicated that the addition of three base pairs or the deletion of three base pairs left the distal portion of the gene with the wild-type reading frame. This result would be expected only if each codon contained three nucleotides.

Evidence from *in vitro* translation studies soon supported the results of Crick and colleagues and firmly established the triplet nature of the code. Some of the more important results follow: (1) Trinucleotides were sufficient to stimulate specific binding of aminoacyl-tRNAs to ribosomes. For example, 5′-UUC-3′ stimulated the binding of phenylalanyl-$tRNA^{Phe}$ to ribosomes. (2) Chemically synthesized mRNA molecules that contained repeating dinucleotide sequences directed the synthesis of copolymers (large chainlike molecules composed of two different subunits) with alternating amino acid sequences. For example, when $poly(UG)_n$ was used as an artificial mRNA in an *in vitro* translation system, the repeating copolymer $(cys\text{-}val)_m$ was synthesized. (The subscripts *n* and *m* refer to the number of nucleotides and amino acids in the respective polymers.) (3) In contrast, mRNAs with repeating trinucleotide sequences directed the synthesis of a mixture of three homopolymers (initiation being at random on such mRNAs in the *in vitro* systems). For example, $poly(UUG)_n$ directed the synthesis of a mixture of polyleucine, polycysteine, and polyvaline. These results are consistent only with a triplet code, with its three different reading frames. When $poly(UUG)_n$ is translated in reading frame 1, UUG, UUG, polyleucine is produced, whereas translation in reading frame 2, UGU, UGU, yields polycysteine, and translation in reading frame 3, GUU, GUU, produces polyvaline. Ultimately, the triplet nature of the code was definitively established by comparing the nucleotide sequences of genes and mRNAs with the amino acid sequences of their polypeptide products.

DECIPHERING THE CODE

The cracking of the genetic code in the 1960s took several years and involved intense competition between many different research laboratories. New information accumulated rapidly but sometimes was inconsistent with earlier data. Indeed, cracking the code proved to be a major challenge.

Deciphering the genetic code required scientists to obtain answers to several questions. (1) Which codons specify each of the 20 amino acids? (2) How many of the 64 possible triplet codons are utilized? (3) How is the code punctuated? (4) Do the codons have the same meaning in viruses, bacteria, plants, and animals? The answers to these questions were obtained primarily from the results of two types of experiments, both of which were performed with cell-free systems. The first type of experiment involved translating artificial mRNA molecules *in vitro* and determining which of the 20 amino acids were incorporated into proteins. In the second type of experiment, ribosomes were activated with mini-mRNAs just three nucleotides long. Then, researchers determined which aminoacyl-tRNAs were stimulated to bind to ribosomes activated with each of the trinucleotide messages.

The decade of the 1960s—the era of the cracking of the genetic code—was one of the most exciting times in the history of biology. Deciphering the genetic code was a difficult and laborious task, and progress came in a series of breakthroughs. We discuss these important developments in A Milestone in Genetics: Cracking the Genetic Code. By combining the results of *in vitro* translation experiments performed with synthetic mRNAs and trinucleotide binding assays, Marshall Nirenberg, Severo Ochoa, H. Ghobind Khorana, Philip Leder, and their colleagues worked out the meaning of all 64 triplet codons (**TABLE 12.1**). Nirenberg and Khorana shared the 1968 Nobel Prize in Physiology or Medicine for their work on the code with Robert Holley, who determined the complete nucleotide sequence of the yeast alanine tRNA. Ochoa had already received the 1959 Nobel Prize for his discovery of RNA polymerase.

INITIATION AND TERMINATION CODONS

The genetic code also provides for punctuation of genetic information at the level of translation. In both prokaryotes and eukaryotes, the codon AUG is used to initiate polypeptide chains (**TABLE 12.1**). In rare instances, GUG is used as an initiation codon. In both cases, the initiation codon is recognized by an initiator tRNA, $tRNA_f^{Met}$ in prokaryotes and $tRNA_i^{Met}$ in eukaryotes. In prokaryotes, an AUG codon must follow an appropriate nucleotide sequence, the Shine-Delgarno sequence, in the 5′ nontranslated segment of the mRNA molecule in order to serve as translation initiation codon. In eukaryotes, the codon must be the first AUG encountered by the ribosome as it scans from the 5′ end of the mRNA molecule. At internal positions, AUG is recognized by $tRNA^{Met}$, and GUG is recognized by a valine tRNA.

Three codons—UAG, UAA, and UGA—specify polypeptide chain termination (**TABLE 12.1**). These codons are recognized by protein release factors, rather than by tRNAs. Prokaryotes contain two release factors, RF-1 and RF-2. RF-1 terminates polypeptides in response to codons UAA and UAG, whereas RF-2 causes termination at UAA and UGA codons. Eukaryotes contain a single release factor that recognizes all three termination codons.

A DEGENERATE AND ORDERED CODE

All the amino acids except methionine and tryptophan are specified by more than one codon (**TABLE 12.1**). Three amino acids—leucine, serine, and arginine—are each specified by six different codons. Isoleucine has three codons. The other amino acids each has either two or four codons. The occurrence of more than one codon per amino acid is called **degeneracy** (although the usual connotations of the term are hardly appropriate). The degeneracy in the genetic code is not at random; instead, it is highly ordered. In most cases, the multiple codons specifying a given amino acid differ by only one base, the third or 3′ base of the codon. The degeneracy is primarily of two types. (1) Partial degeneracy occurs when the third base may be either of the two pyrimidines (U or C) or, alternatively, either of the two purines (A or G). With partial degeneracy, changing

▶ **TABLE 12.1**

The Genetic Code[a]

First (5′) letter	Second letter: U	Second letter: C	Second letter: A	Second letter: G	Third (3′) letter
U	UUU Phe (F)	UCU Ser (S)	UAU Tyr (Y)	UGU Cys (C)	U
U	UUC Phe (F)	UCC Ser (S)	UAC Tyr (Y)	UGC Cys (C)	C
U	UUA Leu (L)	UCA Ser (S)	UAA Stop (terminator)	UGA Stop (terminator)	A
U	UUG Leu (L)	UCG Ser (S)	UAG Stop (terminator)	UGG Trp (W)	G
C	CUU Leu (L)	CCU Pro (P)	CAU His (H)	CGU Arg (R)	U
C	CUC Leu (L)	CCC Pro (P)	CAC His (H)	CGC Arg (R)	C
C	CUA Leu (L)	CCA Pro (P)	CAA Gln (Q)	CGA Arg (R)	A
C	CUG Leu (L)	CCG Pro (P)	CAG Gln (Q)	CGG Arg (R)	G
A	AUU Ileu (I)	ACU Thr (T)	AAU Asn (N)	AGU Ser (S)	U
A	AUC Ileu (I)	ACC Thr (T)	AAC Asn (N)	AGC Ser (S)	C
A	AUA Ileu (I)	ACA Thr (T)	AAA Lys (K)	AGA Arg (R)	A
A	AUG *Met* (M) (initiator)	ACG Thr (T)	AAG Lys (K)	AGG Arg (R)	G
G	GUU Val (V)	GCU Ala (A)	GAU Asp (D)	GGU Gly (G)	U
G	GUC Val (V)	GCC Ala (A)	GAC Asp (D)	GGC Gly (G)	C
G	GUA Val (V)	GCA Ala (A)	GAA Glu (E)	GGA Gly (G)	A
G	GUG Val (V)	GCG Ala (A)	GAG Glu (E)	GGG Gly (G)	G

□ = Polypeptide chain initiation codon

■ = Polypeptide chain termination codon

[a]Each triplet nucleotide sequence or codon refers to the nucleotide sequence in **mRNA** (not DNA) that specifies the incorporation of the indicated amino acid or polypeptide chain termination. The one-letter symbols for the amino acids are given in parentheses after the standard three-letter abbreviations.

the third base from a purine to a pyrimidine, or vice versa, will change the amino acid specified by the codon. (2) In the case of complete degeneracy, any of the four bases may be present at the third position in the codon, and the codon will still specify the same amino acid. For example, valine is encoded by GUU, GUC, GUA, and GUG (**TABLE 12.1**).

Scientists have speculated that the **order** in the genetic code has evolved as a way of minimizing mutational lethality. Many base substitutions at the third position of codons do not change the amino acid specified by the codon. Moreover, amino acids with similar chemical properties (such as leucine, isoleucine, and valine) have codons that differ from each other by only one base. Thus, many single base-pair substitutions will result in the substitution of one amino acid for another amino acid with very similar chemical properties (for example, valine for isoleucine). In most cases, conservative substitutions of this type will yield active gene products, which minimizes the effects of mutations. Test your understanding of the genetic code by answering the question posed in Focus on Problem Solving: Predicting Amino Acid Substitutions Induced by Mutagens.

A NEARLY UNIVERSAL CODE

Vast quantities of information are now available from *in vitro* studies, from amino acid replacements due to mutations, and from correlated nucleic acid and polypeptide sequencing, which allow a comparison of the meaning of the 64 codons in different species. These data all indicate that the genetic code is nearly **universal;** that is, the codons have the same meaning, with minor exceptions, in all species.

The most important exceptions to the universality of the code occur in mitochondria of mammals, yeast, and several other species. Mitochondria have their own chromosomes and protein-synthesizing machinery (Chapter 16). Although the mitochondrial and cytoplasmic systems are similar, there are some differences. In the mitochondria of humans and other mammals, (1) UGA specifies tryptophan rather than chain termination, (2) AUA is a methionine codon, not an isoleucine codon, and (3) AGA and AGG are chain-termination codons rather than arginine codons. The other 60 codons have the same meaning in mammalian mitochondria as in nuclear mRNAs (**TABLE 12.1**). There are also rare differences

► FOCUS ON PROBLEM SOLVING

Predicting Amino Acid Substitutions Induced by Mutagens

THE PROBLEM

The chemical hydroxylamine (NH_2OH) transfers a hydroxyl (-OH) group to cytosine producing hydroxymethylcytosine (hmC), which, unlike cytosine, pairs with adenine. Therefore, hydroxylamine induces G:C to A:T base-pair substitutions in DNA. If you treat the double-stranded DNA of a virus such as phage T4 with hydroxylamine, what amino acid substitutions will be induced in the proteins encoded by the virus?

FACTS AND CONCEPTS

1. The nature of the genetic code—the meaning of the 64 triplet nucleotide sequences in mRNA—is shown in Table 12.1.
2. Complete degeneracy occurs when the first two nucleotides in an mRNA codon are sufficient to determine the amino acid in the polypeptide specified by the mRNA.
3. Partial degeneracy occurs when the same amino acid is specified if the base in the 3′ nucleotide of a codon is either of the two pyrimidines or either of the two purines.
4. Hydroxylamine will only alter codons specified by DNA base-pair triplets that contain G:C base pairs.
5. If the G:C base-pair occupies the third (3′) position of the triplet, hydroxylamine will induce amino acid substitutions only in cases where the genetic code is NOT degenerate, that is, where the base present as the 3′ nucleotide of the codon determines its meaning. Only two codons are not degenerate at the 3′ position; they are 5-AUG-3′ (methionine) and 5′-UGG-3′ (tryptophan).
6. For codons with complete or partial degeneracy at the 3′ position, hydroxylamine will not induce amino acid substitutions by modifying the base pair specifying the 3′ base in the codon. It will induce G:C → A:T and C:G → T:A substitutions (where the first base given is in the template strand). However, given the partial or complete degeneracy, the resulting codons will still specify the same amino acids. An AAG lysine codon, for example, could be changed to an AAA lysine codon, or a UUC phenylalanine codon could be changed to a UUU phenylalanine codon. But no amino acid substitution will occur in either case.

ANALYSIS AND SOLUTION

The answer to the question of which amino acid substitutions will be induced by hydroxylamine requires a careful analysis of the nature of the genetic code (**TABLE 12.1**). Potential targets of hydroxylamine mutagenesis are DNA triplets specifying mRNA codons containing C's and G's at the first (5′) and second positions in the codons and triplets specifying nondegenerate codons with G's or C's at the third (3′) position. Indeed, there are more potential targets than nontargets in genomes; 51 of the 64 DNA triplets contain G:C or C:G base pairs. Consider as an example the arginine codon 5′-AGA-3′; it will be transcribed from a DNA template strand with the sequence 3′-TCT-5′ (reversing the polarity to keep the bases in the same order). The C in this sequence can be hydroxymethylated, producing hmC, which will pair with adenine. After two semiconservative replications, the DNA template strand will contain the sequence 3′-TTT-5′ at this site, and transcription of this sequence will yield a 5′-AAA-3′ mRNA codon. Translation of the mRNA will result in the insertion of lysine in the resulting polypeptide because AAA is a lysine codon. Thus, one example of the effects of hydroxylamine will be the replacement of arginine residues with lysines. This process is diagramed on the right.

The only amino acids specified by codons with no targets of hydroxylamine-induced amino acid substitutions are phenylalanine (UUU and UUC), isoleucine (AUU, AUC, and AUA), tyrosine (UAU and UAC), asparagine (AAU and AAC), and lysine (AAA and AAG).

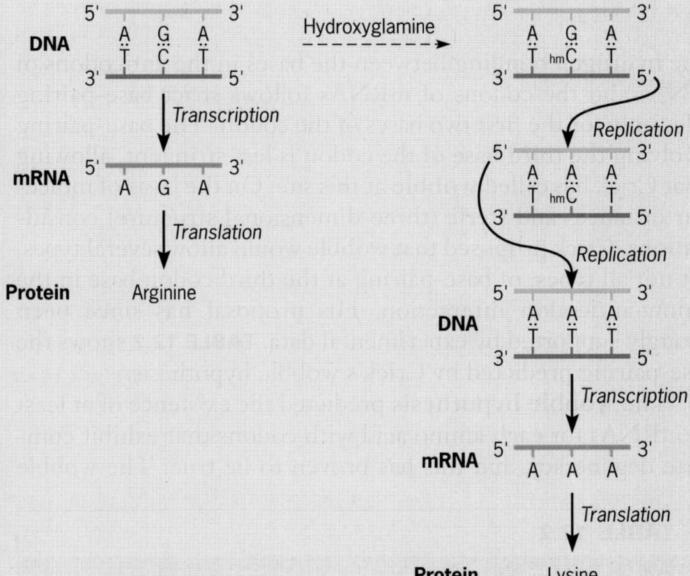

The other amino acids are all specified by DNA base-pair triplets that contain one or more G:C's, with the C's being potential targets of hydroxylamine mutagenesis.

For further discussion go to your *WileyPLUS* course.

in codon meaning in the mitochondria of other species and in nuclear transcripts of some protozoa. However, since these exceptions are rare, the genetic code should be considered nearly universal.

KEY POINTS

- Each of the 20 amino acids in proteins is specified by one or more nucleotide triplets in mRNA.
- Of the 64 possible triplets, given the four bases in mRNA, 61 specify amino acids and 3 signal chain termination.
- The code is nonoverlapping, with each nucleotide part of a single codon, degenerate, with most amino acids specified by two or four codons, and ordered, with similar amino acids specified by related codons.
- The genetic code is nearly universal; with minor exceptions, the 64 triplets have the same meaning in all organisms.

Codon-tRNA Interactions

Codons in mRNA molecules are recognized by aminoacyl-tRNAs during translation.

The translation of a sequence of nucleotides in mRNA into the correct sequence of amino acids in the polypeptide product requires the accurate recognition of codons by aminoacyl-tRNAs. Because of the degeneracy of the genetic code, either several different tRNAs must recognize the different codons specifying a given amino acid or the anticodon of a given tRNA must be able to base-pair with several different codons. Actually, both of these phenomena occur. Several tRNAs exist for certain amino acids, and some tRNAs recognize more than one codon.

RECOGNITION OF CODONS BY tRNAs: THE WOBBLE HYPOTHESIS

The hydrogen bonding between the bases in the anticodons of tRNAs and the codons of mRNAs follows strict base-pairing rules only for the first two bases of the codon. The base-pairing involving the third base of the codon is less stringent, allowing what Crick has called **wobble** at this site. On the basis of molecular distances and steric (three-dimensional structure) considerations, Crick proposed that wobble would allow several types, but not all types, of base-pairing at the third codon base in the codon–anticodon interaction. His proposal has since been strongly supported by experimental data. **TABLE 12.2** shows the base-pairing predicted by Crick's wobble hypothesis.

The **wobble hypothesis** predicted the existence of at least two tRNAs for each amino acid with codons that exhibit complete degeneracy, and this has proven to be true. The wobble hypothesis also predicted the occurrence of three tRNAs for the six serine codons. Three serine tRNAs have been characterized: (1) $tRNA^{Ser1}$ (anticodon AGG) binds to codons UCU and UCC, (2) $tRNA^{Ser2}$ (anticodon AGU) binds to codons UCA and UCG, and (3) $tRNA^{Ser3}$ (anticodon UCG) binds to codons AGU and AGC. These specificities were verified by the trinucleotide-stimulated binding of purified aminoacyl-tRNAs to ribosomes *in vitro*.

Finally, several tRNAs contain the base inosine, which is made from the purine hypoxanthine. Inosine is produced by a posttranscriptional modification of adenosine. Crick's wobble hypothesis predicted that when inosine is present at the 5′ end of an anticodon (the wobble position), it would base-pair with uracil, cytosine, or adenine in the codon. In fact, purified alanyl-tRNA containing inosine (I) at the 5′ position of the anticodon (see **FIGURE 12.10**) binds to ribosomes activated with GCU, GCC, or GCA trinucleotides (**FIGURE 12.21**). The same result has been obtained with other purified tRNAs with inosine at the 5′ position of the anticodon. Thus, Crick's wobble

TABLE 12.2

Base-Pairing Between the 5′ Base of the Anticodons of tRNAs and the 3′ Base of Codons of mRNAs According to the Wobble Hypothesis

Base in Anticodon	Base in Codon
G	U or C
C	G
A	U
U	A or G
I	A, U, or C

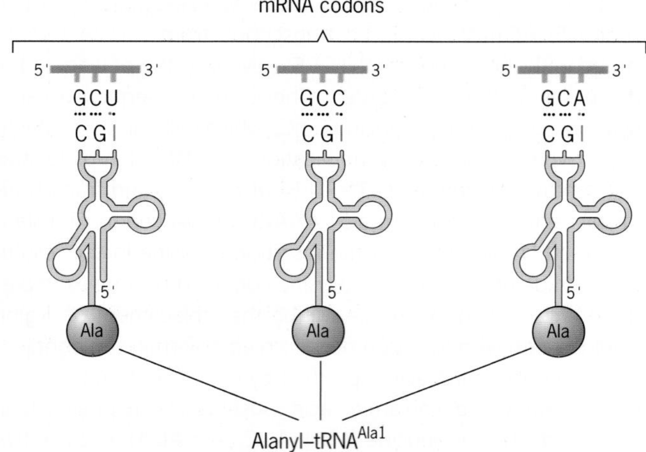

Figure 12.21 ▶ Base-pairing between the anticodon of alanyl-$tRNA^{Ala1}$ and mRNA codons GCU, GCC, and GCA according to Crick's wobble hypothesis. Trinucleotide-activated ribosome binding assays have shown that alanyl-$tRNA^{Ala1}$ does indeed base-pair with all three codons.

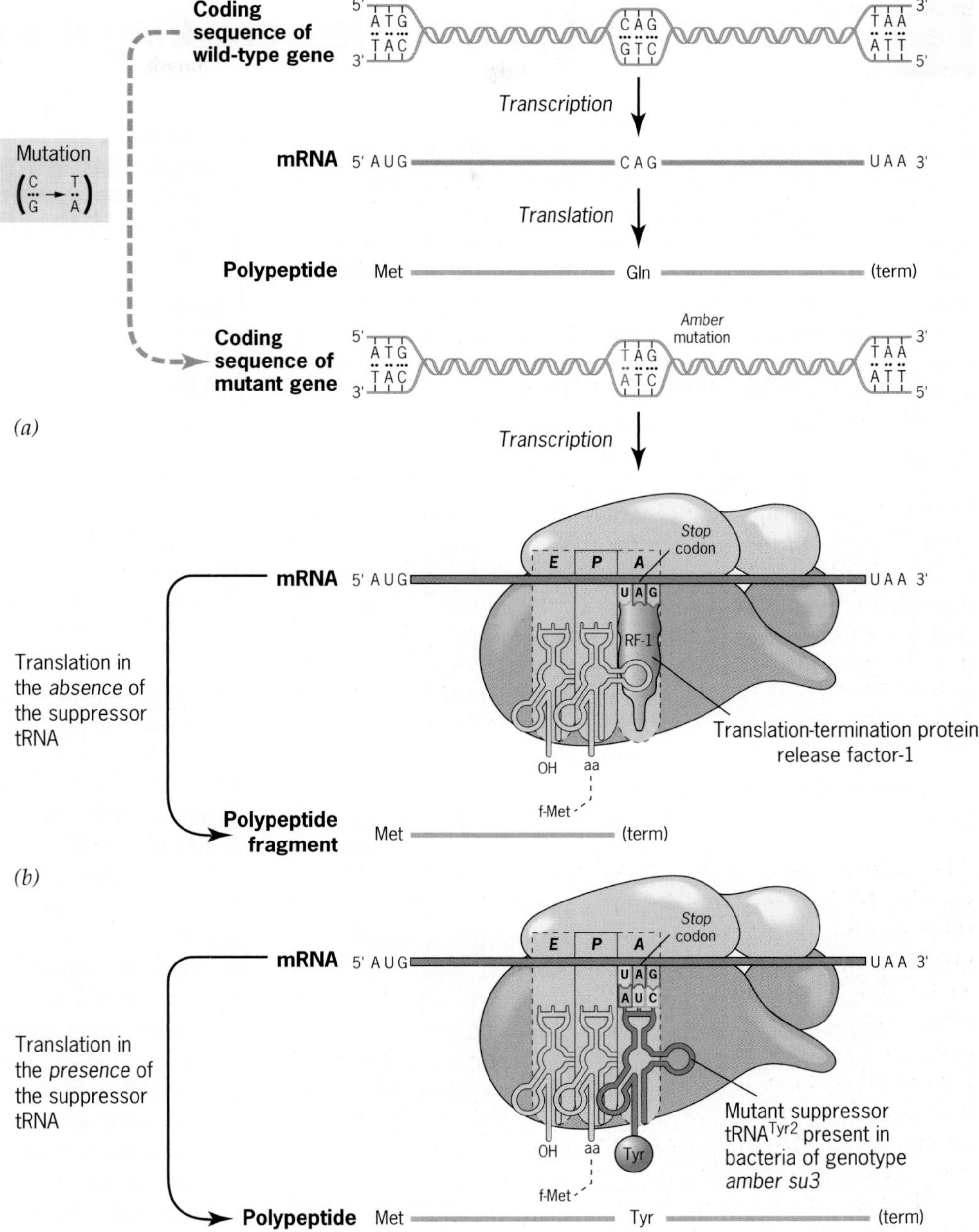

Figure 12.22 ▶ (*a*) The formation of an *amber* (UAG) chain-termination mutation. (*b*) Its effect on the polypeptide gene product in the absence of a suppressor tRNA, and (*c*) in the presence of a suppressor tRNA. The *amber* mutation shown changes a CAG glutamine (Gln) codon to a UAG chain-termination codon. The polypeptide containing the tyrosine inserted by the suppressor tRNA may or may not be functional; however, suppression of the mutant phenotype will occur only when the polypeptide is functional.

hypothesis nicely explains the relationships between tRNAs and codons given the degenerate, but ordered, genetic code.

SUPPRESSOR MUTATIONS THAT PRODUCE tRNAs WITH ALTERED CODON RECOGNITION

Even if we exclude the mitochondria, the genetic code is not absolutely universal. Minor variations in codon recognition and translation are well documented. In *E. coli* and yeast, for example, some mutations in tRNA genes alter the anticodons and thus the codons recognized by the mutant tRNAs. These mutations were initially detected as *suppressor mutations*, nucleotide substitutions that suppressed the effects of other mutations. The suppressor mutations were subsequently shown to occur in tRNA genes. Many of these suppressor mutations changed the anticodons of the altered tRNAs.

The best-known examples of suppressor mutations that alter tRNA specificity are those that suppress UAG chain-termination mutations within the coding sequences of genes. Such mutations, called *amber* mutations (after one of the

▶ A MILESTONE IN GENETICS: **Cracking the Genetic Code**

When scientists set out to decipher the genetic code in 1960, they knew that the four letters of the genetic alphabet—the four nucleotides in RNA—somehow had to specify the 20 amino acids in the polypeptide gene products plus the initiation and termination of polypeptide chains. A doublet code—two nucleotides per amino acid—would not provide enough information ($4^2 = 16$ codons). However, a triplet code seemed to provide too much information ($4^3 = 64$ codons). Then, in 1961, Francis Crick and colleagues provided strong evidence for a triplet code (see **FIGURE 12.20**). But which triplet codons specified which amino acids?

The first breakthrough came in 1961 when Marshall Nirenberg (1968 Nobel Prize recipient) and J. Heinrich Matthaei demonstrated that synthetic RNA molecules could be used as artificial mRNAs to direct *in vitro* protein synthesis.[1] When ribosomes, aminoacyl-tRNAs, and the soluble factors required for translation are purified free of natural mRNAs, these components can be combined *in vitro* and stimulated to synthesize polypeptides by the addition of chemically synthesized RNA molecules. If these synthetic mRNA molecules are of known nucleotide content, the amino acid composition of the resulting polypeptides can be used to deduce which codons specify which amino acids.

The first codon assignment (UUU for phenylalanine) was made when Nirenberg and Matthaei demonstrated that polyuridylic acid [poly(U) = $(U)_n$] directed the synthesis of polyphenylalanine [$(\text{phenylalanine})_m$]. They used radioactively labeled phenylalanine (^{14}C-phenylalanine) as a substrate in an *in vitro* translation system with poly(U), poly(C), and poly(A) as artificial messenger RNAs. The ^{14}C-phenylalanine was incorporated into polyphenylalanine only when poly(U) was used as the mRNA (**TABLE 1**). When each of 17 other labeled amino acids was used in the poly(U)-stimulated translation system, no radioactivity was incorporated into polypeptides. Given the triplet nature of the genetic code, Nirenberg and Matthaei concluded that UUU must be a codon for phenylalanine. Shortly thereafter, poly(A) and poly(C) were shown to encode polylysine and polyproline, respectively, allowing Nirenberg and colleagues to assign codon AAA to lysine and CCC to proline. Poly(G) does not function as an mRNA in the *in vitro* translation systems because base-pairing between the guanine residues results in complex three-stranded structures.

Researchers in the laboratories of Nirenberg and Severo Ochoa (1959 Nobel Prize recipient) extended the use of artificial mRNAs to synthetic copolymers with random sequences. For example, Nirenberg and coworkers synthesized a random copolymer containing approximately equal amounts of adenine and cytosine and used it as an artificial mRNA in their *in vitro* translation system.[2] A random AC copolymer composed of equal amounts of A and C will contain 12.5 percent (1/2 × 1/2 × 1/2 = 1/8) of each of the eight possible codons: AAA, AAC, ACA, CAA, CCA, CAC, ACC, and CCC. Nirenberg and colleagues observed that poly(AC) directed the incorporation of six amino acids—asparagine, glutamine, histidine, lysine, proline, and threonine—into polypeptides. Because they already knew that AAA and CCC were lysine and proline codons, their results indicated that codons composed of two A's plus one C and two C's plus one A specified asparagine, glutamine, histidine, and threonine. Approximately equal amounts of asparagine, glutamine, histidine, and lysine were incorporated, but threonine and proline were incorporated in about double the amount of each of the other four amino acids. This result indicated that random AC copolymers might contain twice as many threonine and proline codons as asparagine codons. Indeed, there are two threonine codons (ACC and ACA) and two proline codons (CCC and CCA) in random AC copolymers, but only one codon for each of the other four amino acids, for example, asparagine (AAC).

By varying the nucleotide composition of random copolymers, Ochoa and colleagues altered the relative frequencies of the eight codons and looked for correlations with the frequencies of the amino acids in the polypeptides synthesized in response to the copolymers.[3] For example, they synthesized random copolymers containing A and C in 5:1 and 1:5 ratios. When these copolymers

▶ **TABLE 1**

Incorporation of ^{14}C-Phenylalanine into Polyphenylalanine in an *In Vitro* Translation System Activated with Synthetic RNA Homopolymers[a]

Synthetic mRNA	Radioactivity Incorporated (counts per minute)
None	44
Poly (U)	39,800
Poly (A)	50
Poly (C)	38
Poly (I)[b]	57

[a]Data are from Nirenberg and Matthaei, 1961. *Proc. Natl. Acad. Sci. USA* 47:1588–1602.

[b]Poly (I) is polyinosinic acid, which contains the purine hypoxanthine. Hypoxanthine is like guanine in that it base-pairs with cytosine.

[1]Nirenberg, M. W., and J. H. Matthaei. 1961. The dependence of cell-free protein synthesis in *E. coli* upon naturally occurring or synthetic polyribonucleotides. Proc. Natl. *Acad. Sci. U.S.A.* 47: 1588–1602.

[2]Nirenberg, M. W., O. W. Jones, P. Leder, B. F. C. Clark, W. S. Sly, and S. Pestka. 1963. On the coding of genetic information. *Cold Spring Harbor Symp. Quant. Biol.* 28: 549–557.

[3]Speyer, J. F., P. Lengyel, C. Basilio, and S. Ochoa. 1962. Synthetic polynucleotides and the amino acid code, IV. *Proc. Natl. Acad. Sci. U.S.A.* 48: 441–448.

were used as mRNAs *in vitro,* the same six amino acids were incorporated as in Nirenberg's experiments with a 1A:1C random copolymer, but their relative frequencies in the polypeptide products were very different. With the 1A:5C copolymer, proline and lysine represented 60.8 percent and 1.2 percent, respectively, of the amino acid residues incorporated, in agreement with the earlier demonstration that CCC and AAA were proline and lysine codons. The values were reversed when the 5A:1C copolymer was used; 51 percent and 3.8 percent of the incorporated residues were lysine and proline, respectively. Ochoa's results indicated that the glutamine and asparagine codons contain 2 A's and 1 C, whereas the histidine codon contains 1 A and 2 C's, and so on.

A major breakthrough occurred when H. Ghobind Khorana (1968 Nobel Prize recipient) and colleagues developed a procedure by which copolymers with known repeating di-, tri- and tetranucleotide sequences could be synthesized.[4] Because the codon sequences in these repeating copolymers were fixed, their use as synthetic mRNAs yielded results that were more easily interpreted than those obtained with random copolymers. For example, an RNA molecule with a repeating UG dinucleotide sequence directed the synthesis of polypeptides containing alternating cysteine and valine residues, as shown in the following.

5′-UGU GUG UGU GUG UGU GUG-3′

H_2N—Cys—Val—Cys—Val—Cys—Val—COOH

Because the triplets UGU and GUG alternate in poly(UG)$_n$, these two codons must specify cysteine and valine, but the result does not tell us which codon specifies which amino acid. In contrast, a repeating UUG trinucleotide polymer directed the synthesis of a mixture of polyleucine, polycysteine, and polyvaline.

Synthetic mRNA

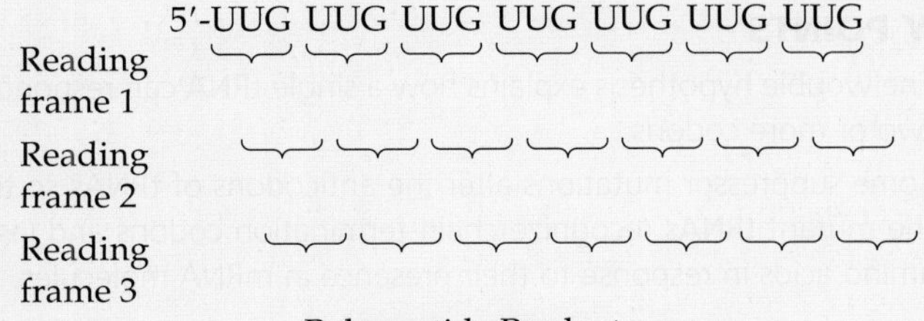

Polypeptide Products

Polyleucine Leu—Leu—Leu—Leu—Leu—Leu—Leu
Polycysteine Cys—Cys—Cys—Cys—Cys—Cys—Cys
Polyvaline Val—Val—Val—Val—Val—Val—Val

Because the initiation of translation occurs at random in these *in vitro* systems, some ribosomes will translate these polymers as UUG, UUG, UUG, and so on, whereas others will translate them as UGU, UGU, UGU. . . , and still others as GUU, GUU, GUU. . . . Thus, these three codons must specify leucine, cysteine, and valine. Because the poly(UG)$_n$ product was a cysteine-valine copolymer, one of the additional codons in poly(UUG)$_n$, either UUG or GUU, must be a leucine codon. By analyzing the amino acids incorporated in response to different repeating di-, tri-, and tetranucleotide copolymers, Khorana and coworkers were able to assign many of the codons to specific amino acids. For example, UGU specifies cysteine, UUG leucine, and GUU valine.

Additional information on the nature of the genetic code was obtained by assaying the binding of aminoacyl-tRNAs to ribosomes activated with small RNA oligomers. When an *in vitro* translation system is activated with poly(U), only one aminoacyl-tRNA, phenylalanyl-tRNA^{Phe}, binds to the ribosome. This specificity exists because the mRNA codon is part of the binding site, and binding involves base-pairing between the codon and the anticodon of the tRNA. In 1964, Nirenberg and Philip Leder developed an assay for aminoacyl-tRNA binding to ribosomes activated with *trinucleotides,* mini-mRNAs only three nucleotides long.[5] Nirenberg and Leder synthesized trinucleotides of known sequence and tested their ability to stimulate the binding of specific aminoacyl-tRNAs to ribosomes. They assayed the ability of each trinucleotide to serve as a mini-mRNA by using labeled amino acids to detect the formation of trinucleotide-aminoacyl-tRNA-ribosome complexes (**FIGURE 1**). For example, the trinucleotides 5′-UUU-3′ and 5′-UUC-3′ both stimulated the binding of phenylalanyl-tRNA^{Phe} to ribosomes, indicating that UUU and UUC are both phenylalanine codons.

By combining the results of trinucleotide binding assays and *in vitro* translation experiments performed with synthetic mRNAs, Nirenberg, Ochoa, Khorana, Leder, and others were able to decipher the meaning of all 64 triplet codons (see **TABLE 12.1**). These codon assignments are now firmly established, supported by definitive data from both *in vitro* and *in vivo* studies.

QUESTIONS FOR DISCUSSION

1. The genetic code is largely the same—that is, each triplet codon has the same meaning—in all organisms—viruses, bacteria, fungi, plants, and animals. Does this near-universality of the code have any implications about evolution and the origin of species? If so, what? If not, how can this near-universality of the code be explained?
2. If there are living organisms that store genetic information in nucleic acids and utilize proteins for catalysis on other planets, do you think that they would use a genetic code identical to ours? similar to ours? Why or why not?

[4]Khorana, H. G., H. Büchi, H. Ghosh, N. Gupta, T. M. Jacob, H. Kössel, R. Morgan, S. A. Narang, E. Ohtsuka, and R. D. Wells. 1966. Polynucleotide synthesis and the genetic code. *Cold Spring Harbor Symp. Quant. Biol.* 31: 39–49.

[5]Nirenberg, M., and P. Leder. 1964. RNA codewords and protein synthesis: The effect of trinucleotides upon the binding of sRNA to ribosomes. *Science* 145: 1399–1407.

▶ A MILESTONE IN GENETICS: **Cracking the Genetic Code** *(continued . . .)*

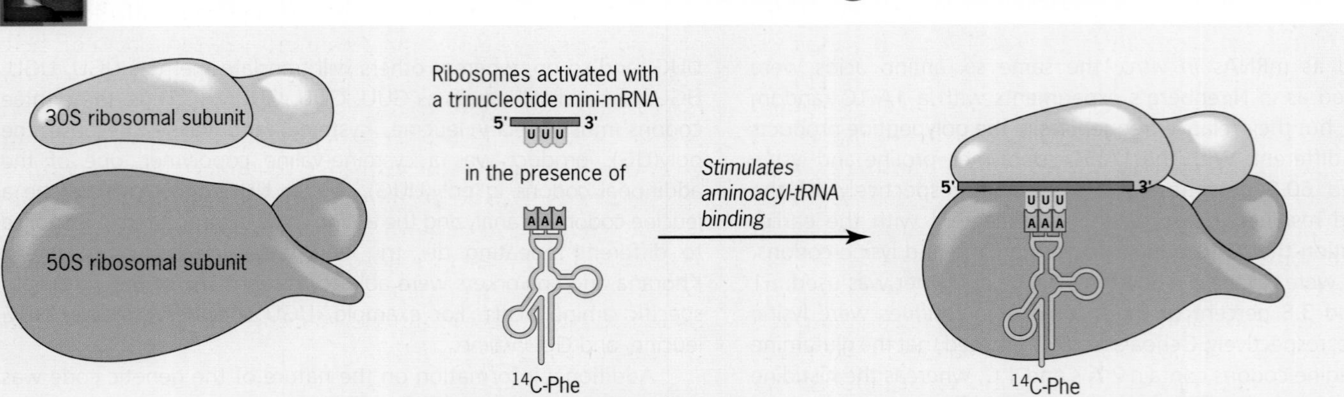

Figure 1 ▶ Stimulation of aminoacyl-tRNA binding to ribosomes by synthetic trinucleotide mini-mRNAs. The results of these trinucleotide-activated ribosome binding assays helped scientists crack the genetic code.

researchers who discovered them), result in the synthesis of truncated polypeptides. Mutations that produce chain-termination triplets within genes have come to be known as **nonsense mutations,** in contrast to **missense mutations,** which change a triplet so that it specifies a different amino acid. A gene that contains a missense mutation encodes a complete polypeptide, but with an amino acid substitution in the polypeptide gene product. A nonsense mutation results in a truncated polypeptide, with the length of the chain depending on the position of the mutation within the gene. Nonsense mutations frequently result from single base-pair substitutions, as illustrated in **FIGURE 12.22*a***. The polypeptide fragments produced from genes containing nonsense mutations (**FIGURE 12.22*b***) often are completely nonfunctional.

Suppression of nonsense mutations has been shown to result from mutations in tRNA genes that cause the mutant tRNAs to recognize the termination (UAG, UAA, or UGA) codons, albeit with varying efficiencies. These mutant tRNAs are referred to as **suppressor tRNAs.** When the *amber* (UAG) suppressor tRNA produced by the *amber su3* mutation in *E. coli* was sequenced, it was found to have an altered anticodon. This particular *amber* suppressor mutation occurs in the tRNA^{Tyr2} gene (one of two tyrosine tRNA genes in *E. coli*). The anticodon of the wild-type (nonsuppressor) tRNA^{Tyr2} was shown to be 5′-G′UA-3′ (where G′ is a derivative of guanine). The anticodon of the mutant (suppressor) tRNA^{Tyr2} is 5′-CUA-3′. Because of the single-base substitution, the anticodon of the suppressor tRNA^{Tyr2} base-pairs with the 5′-UAG-3′ *amber* codon (recall that base-pairing always involves strands of opposite polarity); that is,

mRNA: 5′-UAG-3′ (codon)
tRNA: 3′-AUC-5′ (anticodon)

Thus, suppressor tRNAs allow complete polypeptides to be synthesized from mRNAs containing termination codons within genes (**FIGURE 12.22*c***). Such polypeptides will be functional if the amino acid inserted by the suppressor tRNA does not significantly alter the protein's chemical properties.

KEY POINTS

- The wobble hypothesis explains how a single tRNA can respond to two or more codons.
- Some suppressor mutations alter the anticodons of tRNAs so that the mutant tRNAs recognize chain-termination codons and insert amino acids in response to their presence in mRNA molecules.

▶ *In Vivo* Confirmation of the Nature of the Genetic Code

Comparisons of the nucleotide sequences of genes and the amino acid sequences of their polypeptide products have shown that the codon assignments deduced from *in vitro* studies are utilized during protein synthesis in living cells.

The codon assignments shown in **TABLE 12.1** were initially based on the results of *in vitro* translation and trinucleotide-stimulated aminoacyl-tRNA binding studies. The source of these assignments raised an obvious question. Are the assignments based on *in vitro* experiments valid *in vivo*? The recent genome-sequencing projects have demonstrated that these codon assignments are correct for protein synthesis *in vivo* for many, if not all, species. When the amino acid substitutions that result from mutations induced by chemical mutagens with specific effects (Chapter 13) were determined by amino acid sequencing, the substitutions were almost always consistent with the codon assignments given in **TABLE 12.1** and the known effect of the mutagen.

More convincingly, when the nucleotide sequences of genes or mRNAs were determined and compared with the amino acid sequences of the polypeptides encoded by those genes or mRNAs, the observed correlations were those predicted from the codon assignments shown in **TABLE 12.1.** This relationship was first demonstrated for bacteriophage MS2 by Walter Fiers and colleagues in 1972. Fiers and coworkers determined the amino acid sequence of the phage MS2 coat protein and the nucleotide sequence of the MS2 gene that encodes the coat protein (see **FIGURE 14.8**). Phage MS2 stores its genetic information in single-stranded RNA; its chromosome is equivalent to an mRNA molecule in organisms with DNA genomes. Fiers and colleagues then compared the nucleotide sequence of the coat protein gene with the amino acid sequence of the coat polypeptide. The amino acid sequence of the coat protein was exactly the sequence predicted from the nucleotide sequence of the coat protein gene and the codon assignments shown in **TABLE 12.1.**

In the last two decades, similar comparisons of the nucleotide sequences of genes and the amino acid sequences of their polypeptide products have clearly established that the codon assignments shown in **TABLE 12.1** are indeed valid *in vivo*. When the nucleotide sequence of the normal human β-globin gene was determined, the sequence predicted the sequence of the 146 amino acids in the human β-globin polypeptide, including the glutamic acid at position six. When the sequence of the sickle-cell allele was determined, it predicted the presence of valine at position six. Indeed, scientists now understand the molecular basis of the symptoms of sickle-cell anemia described by Herrick and Irons in 1910.

KEY POINTS

- Comparisons of the nucleotide sequences of genes with the amino acid sequences of their polypeptide products have verified the codon assignments deduced from *in vitro* studies.

Basic Exercises

ILLUSTRATE BASIC GENETIC ANALYSIS

1. The human β-globin polypeptide is 146 amino acids long. How long is the coding portion of the human β-globin mRNA?

Answer: Each amino acid is specified by a codon containing three nucleotides. Therefore, the 146 amino acids in β-globin will be specified by 438 (146 × 3) nucleotides. However, a termination codon must be present at the end of the coding sequence, bringing the length to 438 + 3 = 441 nucleotides. In the case of β-globin and many other proteins, the amino-terminal methionine (specified by the initiation codon AUG) is removed from the β-globin during synthesis. Adding the initiation codon increases the coding sequence of the β-globin mRNA to 444 nucleotides (441 + 3).

2. If the coding segment of an mRNA with the sequence 5′-AUGUUUCCCAAAGGG-3′ is translated, what amino acid sequence will be produced?

Answer: (Amino-terminus)-methionine-phenylalanine-proline-lysine-glycine-(carboxyl-terminus). The amino acid sequence is deduced using the genetic code shown in **TABLE 12.1.** AUG is the methionine initiation codon followed by the phenylalanine codon UUU, the proline codon CCC, the lysine codon AAA, and the glycine codon GGG.

3. If a coding segment of the template strand of a gene (DNA) has the sequence 3′-TACAAAGGGTTTCCC-5′, what amino acid sequence will be produced if it is transcribed and translated?

Answer: The mRNA sequence produced by transcription of this segment of the gene will be 5′-AUGUUUCCCAAAGGG-3′. Note that this mRNA has the same nucleotide sequence as the one discussed in Exercise 2. Thus, it will produce the same peptide when translated: NH_2-Met-Phe-Pro-Lys-Gly-COOH.

4. What sequence of nucleotide pairs in a gene in *Drosophila* will encode the amino acid sequence methionine-tryptophan (reading from the amino terminus to the carboxyl terminus)?

Answer: The codons for methionine and tryptophan are AUG and UGG, respectively. Thus, the nucleotide sequence in the mRNA specifying the dipeptide sequence methionine-tryptophan must be 5′-AUGUGG-3′. The template DNA strand must be complementary and antiparallel to the mRNA sequence (3′-TACACC-5′), and the other strand of DNA must be complementary to the template strand. Therefore, the sequence of base pairs in the gene must be:

5′-ATGTGG-3′
3′-TACACC-5′

5. A wild-type gene contains the trinucleotide-pair sequence:

5′-GAG-3′
3′-CTC-5′

This triplet specifies the amino acid glutamic acid. If the second base pair in this gene segment were to change from A:T to T:A, yielding the following DNA sequence:

5′-GTG-3′
3′-CAC-5′

would it still encode glutamic acid?

Answer: No, it would now specify the amino acid valine. The codon for glutamic acid is 5′-GAG-3′, which tells us that the bottom strand of DNA is the template strand. Transcription of the wild-type gene yields the mRNA sequence 5′-GAG-3′, which is a glutamic acid codon. Transcription of the altered gene produces the mRNA sequence 5′-GUG-3′, which is a valine codon. Indeed, this is exactly the same nucleotide-pair change that gave rise to the altered hemoglobin in Herrick's sickle-cell anemia patient, discussed at the beginning of this chapter. See Figure 1.9 for further details.

▶ Testing Your Knowledge

INTEGRATE DIFFERENT CONCEPTS AND TECHNIQUES

1. The average mass of the 20 common amino acids is about 137 daltons. Estimate the approximate length of an mRNA molecule that encodes a polypeptide with a mass of 65,760 daltons. Assume that the polypeptide contains equal amounts of all 20 amino acids.

Answer: Based on this assumption, the polypeptide would contain about 480 amino acids (65,760 daltons/137 daltons per amino acid). Since each codon contains three nucleotides, the coding region of the mRNA would have to be 1440 nucleotides long (480 amino acids $\times$ 3 nucleotides per amino acid).

2. The antibiotic streptomycin kills sensitive *E. coli* by inhibiting the binding of $tRNA_f^{Met}$ to the *P* site of the ribosome and by causing misreading of codons in mRNA. In sensitive bacteria, streptomycin is bound by protein S12 in the 30S subunit of the ribosome. Resistance to streptomycin can result from a mutation in the gene-encoding protein S12 so that the altered protein will no longer bind the antibiotic. In 1964, Luigi Gorini and Eva Kataja isolated mutants of *E. coli* that grew on minimal medium supplemented with either the amino acid arginine or streptomycin. That is, in the absence of streptomycin, the mutants behaved like typical arginine-requiring bacteria. However, in the absence of arginine, they were streptomycin-dependent conditional-lethal mutants. That is, they grew in the presence of streptomycin but not in the absence of streptomycin. Explain the results obtained by Gorini and Kataja.

Answer: The streptomycin-dependent conditional-lethal mutants isolated by Gorini and Kataja contained missense mutations in genes encoding arginine biosynthetic enzymes. If arginine was present in the medium, these enzymes were unessential. However, these enzymes were required for growth in the absence of arginine (one of the 20 amino acids required for protein synthesis).

Streptomycin causes misreading of mRNA codons in bacteria. This misreading allowed the codons that contained the missense mutations to be translated ambiguously—with the wrong amino acids incorporated—when the antibiotic was present. When streptomycin was present in the mutant bacteria, an amino acid occasionally would be inserted (at the site of the mutation) that resulted in an active enzyme, which, in turn, allowed the cells to grow, albeit slowly. In the absence of streptomycin, no misreading occurred, and all of the mutant polypeptides were inactive.

▶ Questions and Problems

ENHANCE UNDERSTANDING AND DEVELOP ANALYTICAL SKILLS

12.1 Identify three different types of RNA that are involved in translation and list the characteristics and functions of each.

12.2 (a) Where in the cells of higher organisms do ribosomes originate? (b) Where in the cells are ribosomes most active in protein synthesis?

12.3 In a general way, describe the molecular organization of proteins and distinguish proteins from DNA, chemically and functionally. Why is the synthesis of proteins of particular interest to geneticists?

12.4 At what locations in the cell does protein synthesis occur?

12.5 Outline the process of aminoacyl-tRNA formation.

12.6 GO The thymine analog 5-bromouracil is a chemical mutagen that induces single base-pair substitutions in DNA called transitions (substitutions of one purine for another purine and one pyrimidine for another pyrimidine). Using the known nature of the genetic code (**TABLE 12.1**), which of the following amino acid substitutions should you expect to be induced by 5-bromouracil with the highest frequency:

(a) Met → Val;
(b) Met → Leu;
(c) Lys → Thr;
(d) Lys → Gln;
(e) Pro → Arg; or
(f) Pro → Gln?
Why?

12.7 In what sense and to what extent is the genetic code (a) degenerate, (b) ordered, and (c) universal?

12.8 (a) How is messenger RNA related to polysome formation? (b) How does rRNA differ from mRNA and tRNA in specificity?

(c) How does the tRNA molecule differ from that of DNA and mRNA in size and helical arrangement?

12.9 Of what significance is the wobble hypothesis?

12.10 How is translation (a) initiated and (b) terminated?

12.11 The bases A, G, U, C, I (inosine) all occur at the 5′ positions of anticodons in tRNAs.

(a) Which base can pair with three different bases at the 3′ positions of codons in mRNA?
(b) What is the minimum number of tRNAs required to recognize all codons of amino acids specified by codons with complete degeneracy?

12.12 GO If the average molecular mass of an amino acid is assumed to be 100 daltons, about how many nucleotides will be present in an mRNA coding sequence specifying a single polypeptide with a molecular mass of 350,000 daltons?

12.13 (a) Why is the genetic code a triplet code instead of a singlet or doublet code? (b) How many different amino acids are specified by the genetic code? (c) How many different amino acid sequences are possible in a polypeptide 146 amino acids long?

12.14 What is the minimum number of tRNAs required to recognize the six codons specifying the amino acid leucine?

12.15 What are the basic differences between translation in prokaryotes and in eukaryotes?

12.16 GO Assume that in the year 2025, the first expedition of humans to Mars discovers several Martian life forms thriving in hydrothermal vents that exist below the planet's surface. Several teams of molecular biologists extract proteins and nucleic acids from these organisms and make some momentous discoveries. Their first discovery is that the proteins in Martian life forms contain only 12 different amino acids instead of the 20 present in life forms on earth. Their second discovery is that the DNA and RNA in these organisms have only two different nucleotides instead of the four nucleotides present in living organisms on earth. (a) Assuming that transcription and translation work similarly in Martians and earthlings, what is the minimum number of nucleotides that must be present in the Martian codon to specify all the amino acids in Martians? (b) Assuming that the Martian code proposed above has translational start-and-stop signals, would you expect the Martian genetic code to be degenerate like the genetic code used on earth?

12.17 An *E. coli* gene has been isolated and shown to be 68 nm long. What is the maximum number of amino acids that this gene could encode?

12.18 What is the function of each of the following components of the protein-synthesizing apparatus?

(a) aminoacyl-tRNA synthetase
(b) release factor 1
(c) peptidyl transferase
(d) initiation factors
(e) elongation factor G

12.19 The human α-globin chain is 141 amino acids long. How many nucleotides in mRNA are required to encode human α-globin?

12.20 (a) What is the difference between a nonsense mutation and a missense mutation? (b) Are nonsense or missense mutations more frequent in living organisms? (c) Why?

12.21 (a) In what ways does the order in the genetic code minimize mutational lethality? (b) Why do base-pair changes that cause the substitution of a leucine for a valine in the polypeptide gene product seldom produce an mutant phenotype?

12.22 What are the functions of the *A*, *P*, and *E* aminoacyl-tRNA binding sites on the ribosome?

12.23 (a) In what ways are ribosomes and spliceosomes similar? (b) In what ways are they different?

12.24 (a) What is the function of the Shine-Dalgarno sequence in prokaryotic mRNAs? (b) What effect does the deletion of the Shine-Dalgarno sequence from a mRNA have on its translation?

12.25 Alan Garen extensively studied a particular nonsense (chain-termination) mutation in the alkaline phosphatase gene of *E. coli*. This mutation resulted in the termination of the alkaline phosphatase polypeptide chain at a position where the amino acid tryptophan occurred in the wild-type polypeptide. Garen induced revertants (in this case, mutations altering the same codon) of this mutant with chemical mutagens that induced single base-pair substitutions and sequenced the polypeptides in the revertants. Seven different types of revertants were found, each with a different amino acid at the tryptophan position of the wild-type polypeptide (termination position of the mutant polypeptide fragment). The amino acids present at this position in the various revertants included tryptophan, serine, tyrosine, leucine, glutamic acid, glutamine, and lysine. Did the nonsense mutation studied by Garen contain a UAG, a UAA, or a UGA chain termination mutation? Explain the basis of your deduction.

12.26 If you were to (1) purify cysteine transfer RNA and charge it with labeled cysteine (that is, activate it by attaching ^{3}H-labeled cysteine), (2) use Raney nickel (a highly reducing nickel powder) to convert the cysteine to alanine still attached to the cysteyl-specific transfer RNA, and (3) place the alanine-charged cysteyl transfer RNA into an *in vitro* protein-synthesizing system activated with poly(UG) templates that normally stimulate the incorporation of cysteine, but not alanine, into polypeptide chains (that is, when you use the normal alanine-charged alanyl and cysteine-charged cysteyl transfer RNAs), what result would you expect?

12.27 The following DNA sequence occurs in a bacterium (the promoter sequence is located to the left but is not shown).

↓

5′-CAATCATGGACTGCCATGCTTCATATGAATAGTTGACAT-3′
3′-GTTAGTACCTGACGGTACGAAGTATACTTATCAACTGTA-5′

(a) What is the ribonucleotide sequence of the mRNA molecule that is transcribed from the template strand of this piece of DNA? Assume that both translational start and termination codons are present.
(b) What is the amino acid sequence of the polypetide encoded by this mRNA?

(c) If the nucleotide indicated by the arrow undergoes a mutation that causes this C:G base pair to be deleted, what will be the polypeptide encoded by the mutant gene?

12.28 The 5′ terminus of a human mRNA has the following sequence:

5′cap-GAAGAGACAAGGTCAUGGCCAUAUGCU-UGUUCCAAUCGUUAGCUGCGCAGGAUCGC-CCUGGG......3′

When this mRNA is translated, what amino acid sequence will be specified by this portion of the mRNA?

12.29 A partial (5′ subterminal) nucleotide sequence of a prokaryotic mRNA is as follows:

5′-.....AGGAGGCUCGAACAUGUCAAUAUGCUU-GUUCCAAUCGUUAGCUGCGCAGGACCGUC-CCGGA......3′

When this mRNA is translated, what amino acid sequence will be specified by this portion of the mRNA?

12.30 GO The following DNA sequence occurs in the nontemplate strand of a structural gene in a bacterium (the promoter sequence is located to the left but is not shown):

↓
5′-GAATGTCAGAACTGCCATGCTTCATATGAATAGACCTCTAG-3′

(a) What is the ribonucleotide sequence of the mRNA molecule that is transcribed from this piece of DNA?
(b) What is the amino acid sequence of the polypeptide encoded by this mRNA?
(c) If the nucleotide indicated by the arrow undergoes a mutation that changes T to A, what will be the resulting amino acid sequence following transcription and translation?

↓
5′-CAATCATGGACTGCCATGCTTCATATGAATAGTTGACAT-3′
3′-GTTAGTACCTGACGGTACGAAGTATACTTATCAACTGTA-5′

(a) What is the ribonucleotide sequence of the mRNA molecule that is transcribed from the template strand of this piece of DNA? Assume that both translational start and termination codons are present.
(b) What is the amino acid sequence of the polypetide encoded by this mRNA?
(c) If the nucleotide indicated by the arrow undergoes a mutation that causes this C:G base pair to be deleted, what will be the polypeptide encoded by the mutant gene?

▶ Genomics on the Web

at http://www.ncbi.nlm.nih.gov/

The genetic code is nearly, but not completely, universal. As discussed in this chapter, some codons have different meanings in the mitochondria of humans and other mammals. Even in nuclear transcripts, rare differences in codon meaning occur in some protozoa. Indeed, there is another exception to the universality of the genetic code that we have not discussed. There is one more amino acid—sometimes called the 21st amino acid—that is incorporated into polypeptides during translation, and the mechanism by which this happens is fascinating.

1. What are selenoproteins? What is the structure of the amino acid selenocysteine? What role does selenium play in the function of selenoproteins? Are selenoproteins present in bacteria? archaea? eukaryotes?

2. What codon(s) specify selenocysteine? How are these codons diverted from their more common meaning to specifying the incorporation of selenocysteine during translation? Is the mechanism the same in bacteria, archaea, and eukaryotes? If not, what are the differences?

3. How many different families of selenoproteins and selenoprotein genes are present in the NCBI databases? What are some of the functions of selenoproteins? What is Keshan disease in humans, and what does it have to do with selenoproteins?

Hint: At the NCBI web site, perform a search of all databases for selenoproteins, and start by examining the information available in the online books. They will provide answers to the basic questions asked above. For protein and gene families, start with the HomoloGene search results. Then, try proteins, genes, OMIM (The Online Mendelian Inheritance in Man), and, if you want to become an expert on the subject, PubMed Central.

Chapter 13
Mutation, DNA Repair, and Recombination

CHAPTER OUTLINE

- Mutation: Source of the Genetic Variability Required for Evolution
- Mutation: Basic Features of the Process
- Mutation: Phenotypic Effects
- The Molecular Basis of Mutation
- Screening Chemicals for Mutagenicity: The Ames Test
- DNA Repair Mechanisms
- Inherited Human Diseases with Defects in DNA Repair
- DNA Recombination Mechanisms

Children playing outdoors. The child in the white coveralls has Xeroderma pigmentosum, an autosomal recessive disorder characterized by acute sensitivity to sunlight. He must avoid exposure to sunlight to prevent skin cancer.

Xeroderma Pigmentosum: Defective Repair of Damaged DNA in Humans

The sun shone brightly on a midsummer day—a perfect day for most children to spend at the playground. All of Nathan's friends were dressed in shorts and tee shirts. As Nathan prepared to join his friends, he pulled on full-length pants, a jacket with a hood and face mask, and gloves. Then he applied a thick layer of sunscreen to his forehead. Whereas his friends enjoy playing in the sunshine, Nathan lives in constant fear of the effects of sunlight. Nathan was born with the inherited disorder xeroderma pigmentosum, an autosomal recessive trait that affects about one out of 250,000 children. Nathan's skin cells are extremely sensitive to ultraviolet radiation—the high-energy rays of sunlight. Ultraviolet light causes chemical changes in the DNA in Nathan's skin cells, changes that lead not only to intense freckling but also to skin cancer.

Nathan's friends gave little thought to playing in the sun; sunburn was their only major concern. Their skin cells contain enzymes that correct the changes in DNA resulting from exposure to ultraviolet light. However, Nathan's skin cells are lacking one of the enzymes required to repair ultraviolet light-induced alterations in the structure of DNA. Xeroderma pigmentosum can result from inherited defects in any of nine different human genes. Moreover, other inherited disorders are known to result from the failure to repair DNA damaged by other physical and chemical agents. The life-threatening consequences of these inherited defects in the DNA repair enzymes dramatically emphasize their importance.

Given the key role that DNA plays in living organisms, the evolution of mechanisms to protect its integrity would seem inevitable. Indeed, as we discuss in this chapter, living cells contain numerous enzymes that constantly scan DNA to search for damaged or incorrectly paired nucleotides. When detected, these defects are corrected by a small army of DNA repair enzymes, each having evolved to combat a particular type of damage. In this chapter, we examine the types of changes that occur in DNA, the processes by which these alterations are corrected, and the related processes of recombination between homologous DNA molecules.

Mutation: Source of the Genetic Variability Required for Evolution

Mutations—inherited changes in the genetic material—provide new genetic variation that allows organisms to evolve.

We know from preceding chapters that inheritance is based on genes that are transmitted from parents to offspring during reproduction and that the genes store genetic information encoded in the sequences of nucleotide pairs in DNA or nucleotides in RNA. We have examined how this genetic information is accurately duplicated during the semiconservative replication of DNA. This accurate replication was shown to depend in part on proofreading activities built into the DNA polymerases that catalyze DNA synthesis. Thus, mechanisms have evolved to facilitate the faithful transmission of genetic information from cell to cell and ultimately from generation to generation. Nevertheless, mistakes in the genetic material do occur. Such heritable changes in the genetic material are called mutations.

The term **mutation** refers to both (1) the change in the genetic material and (2) the process by which the change occurs. An organism that exhibits a novel phenotype resulting from a mutation is called a **mutant.** Used in its broad historical sense, mutation refers to any sudden, heritable change in the genotype of a cell or an organism. However, changes in the genotype, and thus in the phenotype, of an organism that result from recombination events that produce new combinations of preexisting genetic variation must be carefully distinguished from changes caused by new mutations. Both events sometimes give rise to new phenotypes at very low frequencies. Mutational changes in the genotype of an organism include changes in chromosome number and structure (Chapter 6), as well as changes in the structures of individual genes. Mutations that involve changes at specific sites in a gene are referred to as **point mutations.** They include the substitution of one base pair for another or the insertion or deletion of one or a few nucleotide pairs at a specific site in a gene. Today, the term *mutation* sometimes is used in a narrow sense to refer only to changes in the structures of individual genes. In this chapter, we explore the process of mutation as defined in the narrow sense.

Mutation is the ultimate source of all genetic variation; it provides the raw material for evolution. Recombination mechanisms rearrange genetic variability into new combinations, and natural or artificial selection preserves the combinations best adapted to the existing environmental conditions or desired by the plant or animal breeder. Without mutation, all genes would exist in only one form. Alleles would not exist, and classical genetic analysis would not be possible. Most important, populations of organisms would not be able to evolve and adapt to environmental changes. Some level of mutation is essential to provide new genetic variability and allow organisms to adapt to new environments. At the same time, if mutations occurred too frequently, they would disrupt the faithful transfer of genetic information from generation to generation. Moreover, most mutations with easily detected phenotypic effects are deleterious to the organisms in which they occur. As we would expect, the rate of mutation is influenced by genetic factors, and mechanisms have evolved that regulate the level of mutation that occurs under various environmental conditions.

KEY POINT

- Mutations are heritable changes in the genetic material that provide the raw material for evolution.

Mutation: Basic Features of the Process

Mutations occur in all organisms from viruses to humans. They can occur spontaneously or be induced by mutagenic agents. Mutation is usually a random, nonadaptive process.

Mutations occur in all genes of all living organisms. These mutations provide new genetic variability that allows organisms to adapt to environmental changes. Thus, mutations have been, and continue to be, essential to the evolutionary process. Before we discuss specific phenotypic effects of mutations and the mechanisms by which various types of mutation occur, we will consider some of the basic features of this important process.

MUTATION: SOMATIC OR GERMINAL

A mutation may occur in any cell and at any stage in the development of a multicellular organism. The immediate effects of the mutation and its ability to produce a phenotypic change are determined by its dominance, the type of cell in which it occurs, and the time at which it takes place during the life cycle of the organism. In higher animals, the germ-line cells that give rise to the gametes separate from other cell lineages early in development (Chapter 2). All nongerm-line cells are somatic cells. **Germinal mutations** are those that occur in germ-line cells, whereas **somatic mutations** occur in somatic cells. If a mutation occurs in a somatic cell, the resulting mutant phenotype will occur only in the descendants of that cell. The mutation will not be transmitted through the gametes to the progeny. The Delicious apple (**FIGURE 13.1**) and the navel orange are

Figure 13.1 ▶ The original Delicious apple was the result of a somatic mutation. It has subsequently been modified by the selection of additional somatic mutations.

examples of mutant phenotypes that resulted from mutations occurring in somatic cells. The Delicious apple was discovered in 1881 by Jessie Hiatt, an Iowa farmer. It has subsequently been modified by the selection of additional somatic mutations. The fruit trees in which the original mutations occurred were somatic mosaics. Fortunately, vegetative propagation was feasible for both the Delicious apple and the navel orange, and today numerous progeny from grafts and buds have perpetuated the original mutations.

If dominant mutations occur in germ-line cells, their effects may be expressed immediately in progeny. If the mutations are recessive, their effects are often obscured in diploids. Germinal mutations may occur at any stage in the reproductive cycle of the organism. If the mutation arises in a gamete, only a single member of the progeny is likely to have the mutant gene. If a mutation occurs in a primordial germ-line cell of the testis or ovary, several gametes may receive the mutant gene, enhancing its potential for perpetuation. Thus, the dominance of a mutant allele and the stage in the reproductive cycle at which a mutation occurs are major factors in determining the likelihood that the mutant allele will be manifested in an organism.

The earliest recorded dominant germinal mutation in domestic animals was that observed by Seth Wright in 1791 on his farm by the Charles River in Dover, Massachusetts. Among his flock of sheep, Wright noticed a peculiar male lamb with unusually short legs. It occurred to him that it would be an advantage to have a whole flock of these short-legged sheep, which could not jump over the low stone fences in his New England neighborhood. Wright used the new short-legged ram to breed his ewes in the next season. Two of their lambs had short legs. Short-legged sheep were then bred together, and a line was developed in which the new trait was expressed in all individuals.

MUTATION: SPONTANEOUS OR INDUCED

When a new mutation—such as the one that produced Wright's short-legged sheep—occurs, is it caused by some agent in the environment or does it result from an inherent process in living organisms? **Spontaneous mutations** are those that occur without a known cause. They may truly be spontaneous, resulting from a low level of inherent metabolic errors, or they may actually be caused by unknown agents present in the environment. **Induced mutations** are those resulting from exposure of organisms to physical and chemical agents that cause changes in DNA (or RNA in some viruses). Such agents are called **mutagens;** they include ionizing irradiation, ultraviolet light, and a wide variety of chemicals.

Operationally, it is impossible to prove that a particular mutation occurred spontaneously or was induced by a mutagenic agent. Geneticists must restrict such distinctions to the population level. If the mutation rate is increased a hundredfold by treatment of a population with a mutagen, an average of 99 of every 100 mutations present in the population will have been induced by the mutagen. Researchers can thus make valid comparisons between spontaneous and induced mutations statistically by comparing populations exposed to a mutagenic agent with control populations that have not been exposed to the mutagen.

Spontaneous mutations occur infrequently, although the observed frequencies vary from gene to gene and from organism to organism. Measurements of spontaneous mutation frequencies for various genes of phage and bacteria range from about 10^{-8} to 10^{-10} detectable mutations per nucleotide pair per generation. For eukaryotes, estimates of mutation rates range from about 10^{-7} to 10^{-9} detectable mutations per nucleotide pair per generation (considering only those genes for which extensive data are available). In comparing mutation rates per nucleotide with mutation rates per gene, the coding region of the average gene is usually assumed to be 1000 nucleotide pairs in length. Thus, the mutation rate per gene varies from about 10^{-4} to 10^{-7} per generation.

Treatment with mutagenic agents can increase mutation frequencies by orders of magnitude. The mutation frequency per gene in bacteria and viruses can be increased to over 1 percent by treatment with potent chemical mutagens. That is, over 1 percent of the genes of the treated organisms will contain a mutation, or, stated differently, over 1 percent of the phage or bacteria in the population will have a mutation in a given gene.

MUTATION: USUALLY A RANDOM, NONADAPTIVE PROCESS

The rats in many cities are no longer affected by the anticoagulants that have traditionally been used as rodent poisons. Many cockroach populations are insensitive to chlordane, the poison used to control them in the 1950s. Housefly populations often exhibit high levels of resistance to many insecticides. More and more pathogenic microorganisms are becoming resistant to antibiotics developed to control them. The introduction of these pesticides and antibiotics by humans produced new environments for these organisms. Mutations producing resistance to these pesticides and antibiotics occurred; the sensitive organisms were killed; and the mutants multiplied to produce new resistant populations. Many such cases of evolution via mutation and natural selection are well documented. These examples raise a basic question about the nature of mutation. Is mutation a purely random event in which the environmental stress merely preserves preexisting mutations? Or is mutation directed by the environmental stress? For example, if you cut off the tails of mice for many generations, will you eventually produce a strain of tailless mice? Despite the beliefs of Jean Lamarck and Trofim Lysenko, who believed in the inheritance of "acquired traits"—traits imposed on organisms by environmental factors—the answer is no; the mice will continue to be born with tails.

Today, it is hard to understand how Lysenko could have sold his belief in Lamarckism—the inheritance of acquired traits—to those in power in the Soviet Union from 1937 through 1964. However, disproving Lamarckism was not an easy task, especially in the case of microorganisms, where even small cultures often contain billions of organisms. As an example, let us consider a population of bacteria such as *E. coli* growing in a streptomycin-free environment. When exposed to streptomycin, most of the bacteria will be killed by the antibiotic. However, if the population is large enough, it will soon give rise to a streptomycin-resistant culture in which all the cells are resistant to the antibiotic. Does streptomycin simply select rare, randomly occurring mutants that preexist in the population, or do all of the cells have some low probability of developing resistance in response to the presence of streptomycin? How can geneticists distinguish between these two possibilities? Resistance to streptomycin can only be detected by treating the culture with the antibiotic. How, then, can a geneticist determine whether resistant bacteria are present prior to exposure to streptomycin, or are induced by the presence of the antibiotic?

In 1952, Joshua and Esther Lederberg developed an important new technique called **replica plating.** This technique allowed them to demonstrate the presence of antibiotic-resistant mutants in bacterial cultures prior to exposure to the antibiotic (**FIGURE 13.2**). The Lederbergs first diluted the bacterial cultures, spread the bacteria on the surface of semisolid nutrient

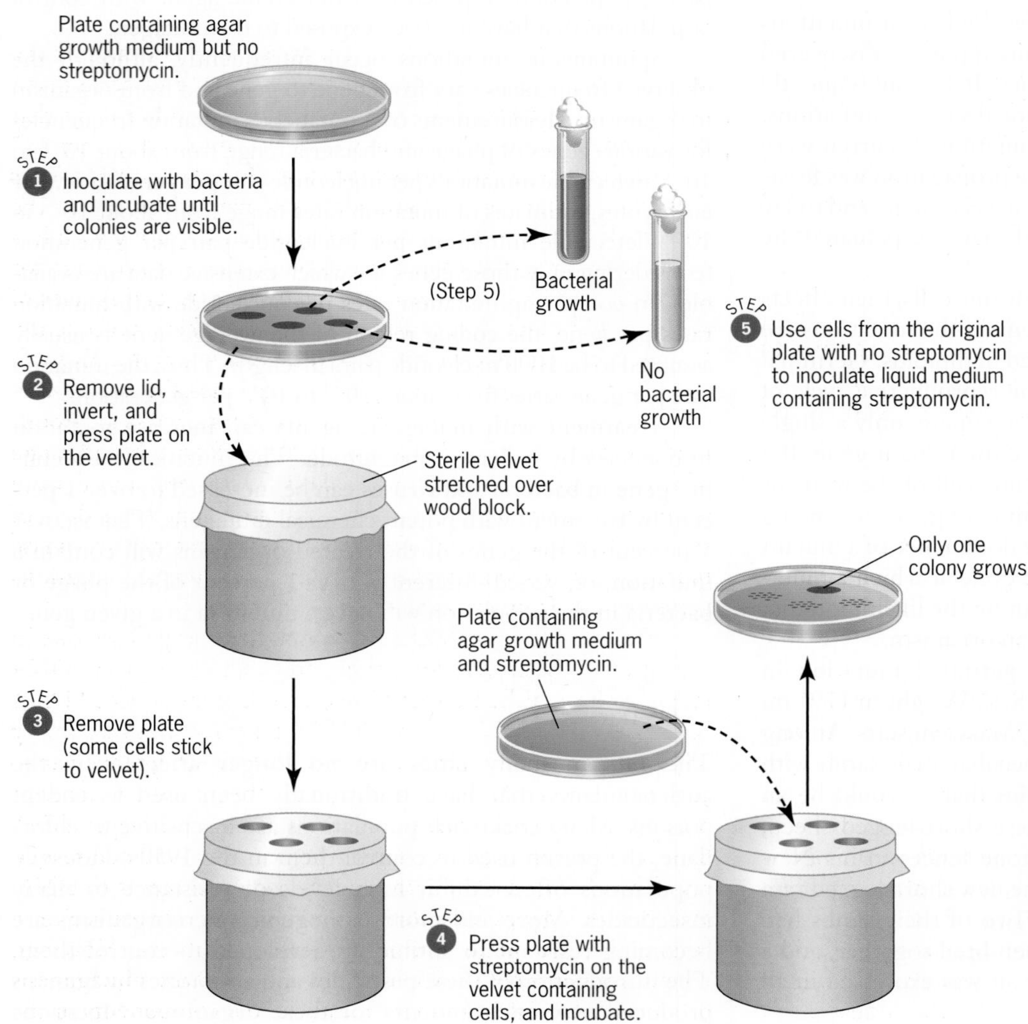
Plate containing agar growth medium but no streptomycin.
STEP 1 Inoculate with bacteria and incubate until colonies are visible.
(Step 5)
Bacterial growth
No bacterial growth
STEP 5 Use cells from the original plate with no streptomycin to inoculate liquid medium containing streptomycin.
STEP 2 Remove lid, invert, and press plate on the velvet.
Sterile velvet stretched over wood block.
Only one colony grows.
Plate containing agar growth medium and streptomycin.
STEP 3 Remove plate (some cells stick to velvet).
STEP 4 Press plate with streptomycin on the velvet containing cells, and incubate.

Figure 13.2 ▶ Joshua and Esther Lederberg's use of replica plating to demonstrate the random or nondirected nature of mutation. For simplicity, only four colonies are shown on each plate, and only two are tested for streptomycin resistance in step 5. Actually, each plate would contain about 200 colonies, and many plates would be used to find an adequate number of mutant colonies.

agar medium in petri dishes, and incubated the plates until each bacterium had produced a visible colony on the surface of the agar. They next inverted each plate and pressed it onto sterile velvet placed over a wood block. Some of the cells from each colony stuck to the velvet. They then gently pressed a sterile plate of nutrient agar medium containing streptomycin onto the velvet. They repeated this replica-plating procedure with many plates, each containing about 200 bacterial colonies. After they incubated the selective plates (those containing streptomycin) overnight, rare streptomycin-resistant colonies had formed.

The Lederbergs subsequently tested the colonies on the nonselective plates (those not containing streptomycin) for their ability to grow on medium containing streptomycin. Their results were definitive. The colonies that grew on the selective replica plates almost always contained streptomycin-resistant cells, whereas those that did not grow on the selective medium seldom contained any resistant cells (**FIGURE 13.2**).

If a mutation that makes a bacterium resistant to streptomycin occurs at an early stage in the growth of a colony, the resistant cell will divide and produce two, then four, then eight, and eventually a large number of resistant bacteria. Thus, if mutation is a randomly occurring, nonadaptive process, many of the colonies that form on the nonselective plates will contain more than one antibiotic-resistant bacterium and will give rise to resistant cultures when tested for growth on selective media. However, if mutation is adaptive and the mutations to streptomycin resistance occur only after exposure to the antibiotic, then the colonies on the nonselective plates that gave rise to resistant colonies on the selective plates after replica plating would be no more likely to contain streptomycin-resistant cells than the other colonies on the nonselective plates.

Thus, by using their replica-plating technique, the Lederbergs demonstrated the existence of streptomycin-resistant mutants in a population of bacteria prior to their exposure to the antibiotic. Their results, along with those of many other experiments, have shown that environmental stress does not direct or cause genetic changes as Lysenko believed; it simply selects rare preexisting mutations that result in phenotypes better adapted to the new environment.

ADAPTIVE, OR STATIONARY-PHASE, MUTAGENESIS IN BACTERIA

An **adaptive mutation** provides a selective advantage to the mutant organism when grown in the environment in which it originated. The formation of adaptive mutations in bacteria is perhaps more appropriately called **stationary-phase mutagenesis** because it occurs when populations of bacteria quit growing—enter the stationary phase—due to starvation or some other environmental stress (**FIGURE 13.3**). Bacteria in stationary phase either cease dividing or die at the same rate as new cells are produced. The adaptive mutations in these stationary-phase cells are produced along with other randomly occurring mutations (nonadaptive mutations); that is, they result from a stress-induced increase in the mutation rate. This increased mutagenesis, in turn, is caused by the induction of error-prone DNA repair processes (see the section "DNA Repair Mechanisms" later in this chapter).

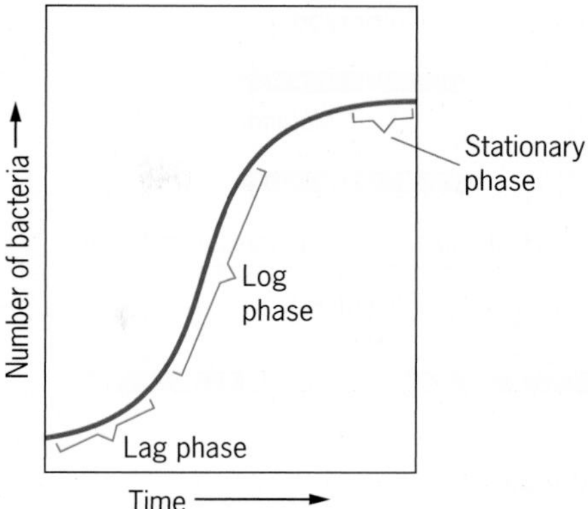

Figure 13.3 ▶ A bacterial growth curve. Transfer of nongrowing bacteria to a new nutrient medium results in three successive stages of growth: lag phase, log phase, and stationary phase. During the lag phase, the bacteria adjust their metabolism to facilitate growth on the new medium. Once this adjustment has been made, exponential growth commences and continues throughout the log phase. When one or more required nutrients in the medium becomes growth-limiting, growth slows and eventually stops in the stationary phase.

We should emphasize that "adaptive" does not mean that a specific type of mutation that improves growth under a given environmental stress has been preferentially induced by that environment stress; the latter would be "directed mutation," for which there is no valid evidence. Rather, "adaptive" means that some mutations that provide a selective advantage to the bacteria have occurred in response to the stress, along with lots of other nonadaptive mutations. Indeed, there is evidence for adaptive, or stationary-phase, mutagenesis in several species of bacteria. However, most of the research on this phenomenon has been carried out with *E. coli* and *Bacillus subtilis*. Even in these two species, stationary-phase mutagenesis is not universal; it occurs in some strains and not in others. One study examined 787 strains of *E. coli* that were isolated from a broad range of natural environments worldwide. When these strains were tested for stationary-phase mutagenesis during seven days of starvation, 80 percent exhibited some level of stationary-phase mutagenesis. However, the frequency of induced mutations varied widely from strain to strain.

When bacteria such as *E. coli* enter stationary phase due to starvation or other environmental stress, an error-prone DNA repair pathway called the SOS response is induced (see "DNA Repair Mechanisms"). During the SOS response, a large number of genes that encode proteins involved in DNA metabolism—replication, recombination, and repair—are turned on. Some of the induced proteins are involved in the repair of damaged DNA by recombination; others are error-prone DNA polymerases (IV and V in *E. coli*) that replicate past damaged segments of DNA, and, in so doing, produce mutations. The SOS response is clearly responsible for some stationary-phase mutations. However, recent studies have shown that stationary-phase mutagenesis occurs in bacterial strains that lack essential components of the

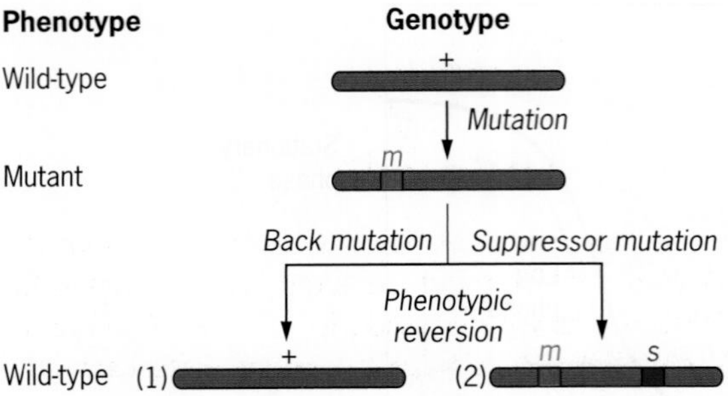

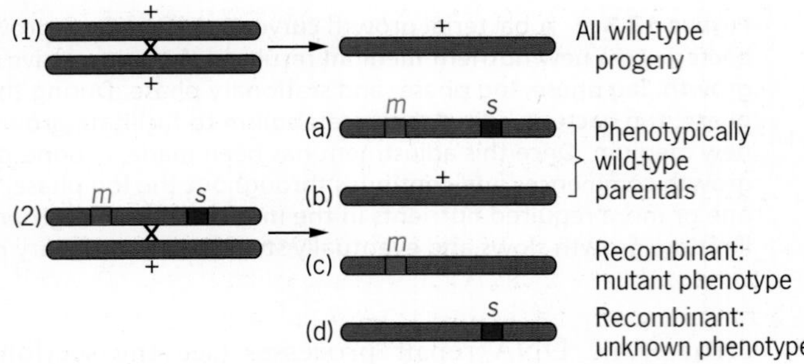

Figure 13.4 ▶ Restoration of the original wild-type phenotype of an organism may occur by (1) back mutation or (2) suppressor mutation (shown on the same chromosome for simplicity). Some mutants can revert to the wild-type phenotype by both mechanisms. Revertants of the two types can be distinguished by backcrosses to the original wild-type. If back mutation has occurred, all backcross progeny will be wild-type. If a suppressor mutation is responsible, some of the backcross progeny will have the mutant phenotype (2c).

SOS response system. Still other studies have provided evidence of multiple pathways of stationary-phase mutagenesis.

At present, we can conclude that adaptive, or stationary-phase, mutagenesis is widespread in bacteria and contributes to their ability to adapt to diverse environments. However, the molecular mechanisms by which stationary-phase mutagenesis occurs need further study, as does the question of whether the phenomenon occurs in other species. Although there is evidence of stationary-phase mutagenesis in yeast, and error-prone DNA polymerases are present in eukaryotes, the possibility that similar processes occur in eukaryotes remains largely unstudied.

MUTATION: A REVERSIBLE PROCESS

As we discussed earlier, a mutation in a wild-type gene can produce a mutant allele that results in an abnormal phenotype. However, the mutant allele can also mutate back to a form that restores the wild-type phenotype. That is, mutation is a reversible process.

The mutation of a wild-type gene to a form that results in a mutant phenotype is referred to as *forward mutation*. However, sometimes the designation of the wild-type and mutant phenotypes is quite arbitrary. They may simply represent two different, but normal, phenotypes. For example, geneticists consider the alleles for brown and blue eye color in humans both to be wild-type. However, in a population composed almost entirely of brown-eyed individuals, the allele for blue eyes might be thought of as a mutant allele. When a second mutation restores the original phenotype lost because of an earlier mutation, the process is called **reversion** or **reverse mutation.** Reversion may occur in two different ways: (1) by **back mutation,** a second mutation at the same site in the gene as the original mutation, restoring the wild-type nucleotide sequence, or (2) by the occurrence of a **suppressor mutation,** a second mutation at a different location in the genome, which compensates for the effects of the first mutation (**FIGURE 13.4**). Back mutation restores the original wild-type nucleotide sequence of the gene, whereas a suppressor mutation does not. Suppressor mutations may occur at distinct sites in the same gene as the original mutation or in different genes, even on different chromosomes. Some mutations revert primarily by back mutation, whereas others do so almost exclusively through the occurrence of suppressor mutations. Thus, in genetic studies, researchers often must distinguish between these two possibilities by backcrossing the phenotypic revertant with the original wild-type organism. If the wild-type phenotype is restored by a suppressor mutation, the original mutation will still be present and can be separated from the suppressor mutation by recombination (**FIGURE 13.4**). If the wild-type phenotype is restored by back mutation, all of the progeny of the backcross will be wild-type.

KEY POINTS

- Mutations occur in both germ-line and somatic cells, but only germ-line mutations are transmitted to progeny.
- Mutations can occur spontaneously or be induced by mutagenic agents in the environment.
- Mutation usually is a nonadaptive process in which an environmental stress simply selects organisms with preexisting, randomly occurring mutations.
- Adaptive, or stationary-phase, mutagenesis occurs in bacteria that have been exposed to an environmental stress such as starvation.
- Restoration of the wild-type phenotype in a mutant organism can result from either back mutation or a suppressor mutation.

Mutation: Phenotypic Effects

The effects of mutations on phenotype range from no observable change to lethality.

The effects of mutations on phenotype range from alterations so minor that they can be detected only by special genetic or biochemical techniques, to gross modifications of morphology, to lethals. A gene is a sequence of nucleotide pairs that usually encodes a specific polypeptide. Any mutation occurring within a given gene will thus produce a new allele of that gene. Genes containing mutations with no effect on phenotype or small effects that can be recognized only by special techniques are called **isoalleles.** Other mutations produce **null alleles** that result in no gene product or totally nonfunctional gene products. If mutations of the latter type occur in genes that are required for the growth of the organism, individuals that are homozygous for the mutation will not survive. Such mutations are called **recessive lethals.**

Mutations can be either recessive or dominant. In monoploid organisms such as viruses and bacteria, both recessive and dominant mutations can be recognized by their effect on the phenotype of the organism in which they occur. In diploid organisms such as fruit flies and humans, recessive mutations will alter the phenotype only when present in the homozygous condition. Thus, in diploids, most recessive mutations will not be recognized at the time of their occurrence because they will be present in the heterozygous state. X-linked recessive mutations are an exception; they will be expressed in the hemizygous state in the heterogametic sex (for example, males in humans and fruit flies; females in birds). X-linked recessive lethal mutations will alter the sex ratio of offspring because hemizygous individuals that carry the lethal will not survive (**FIGURE 13.5**).

MUTATIONS WITH PHENOTYPIC EFFECTS: USUALLY DELETERIOUS AND RECESSIVE

Most of the thousands of mutations that have been identified and studied by geneticists are deleterious and recessive. This result is to be expected if we consider what is known about the genetic control of metabolism and the techniques available for identifying mutations. As we discussed in Chapter 4, metabolism occurs by sequences of chemical reactions, with each step catalyzed by a specific enzyme encoded by one or more genes. Mutations in these genes frequently produce blocks in metabolic pathways (**FIGURE 13.6**). These blocks occur because alterations in the base-pair sequences of genes often cause changes in the amino acid sequences of polypeptides (**FIGURE 13.7**), which may result in nonfunctional products (**FIGURE 13.6**). Indeed, this is the most commonly observed effect of easily detected mutations. Given a wild-type allele encoding an active enzyme and mutant alleles encoding less active or totally inactive enzymes, it is apparent why most of the observed mutations would be recessive. If a cell contains both active and inactive forms of a given enzyme, the active form usually will catalyze the reaction in question. Therefore, the allele specifying the active product usually will be dominant, and the allele encoding the inactive product will be recessive (Chapter 4).

Because of the degeneracy and order in the genetic code (Chapter 12), many mutations have no effect on the phenotype of the organism; they are called **neutral mutations.** But why should most mutations with phenotypically recognizable effects result in decreased gene-product activity or no gene-product

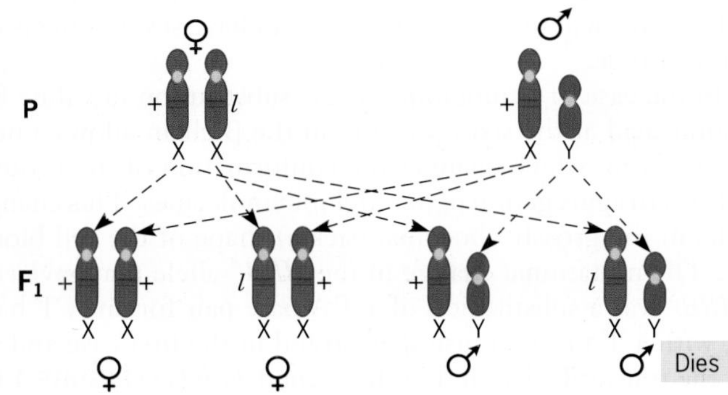

Figure 13.5 ► Alteration of the sex ratio by an X-linked recessive lethal mutation. Females heterozygous for an X-linked recessive lethal will produce female and male progeny in a 2:1 ratio.

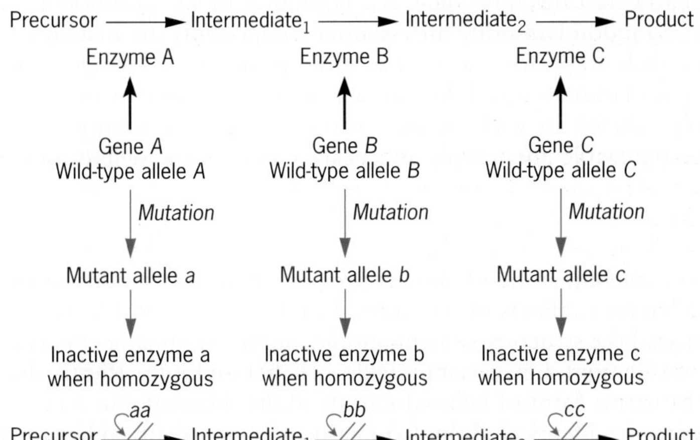

Figure 13.6 ► Recessive mutant alleles often result in blocks in metabolic pathways. The pathways can be only a few steps long, as diagrammed here, or many steps long. The wild-type allele of each gene usually encodes a functional enzyme that catalyzes the appropriate reaction. Most mutations that occur in wild-type genes result in altered forms of the enzyme with reduced or no activity. In the homozygous state, mutant alleles that produce inactive products cause metabolic blocks (–/→) owing to the lack of the required enzyme activity.

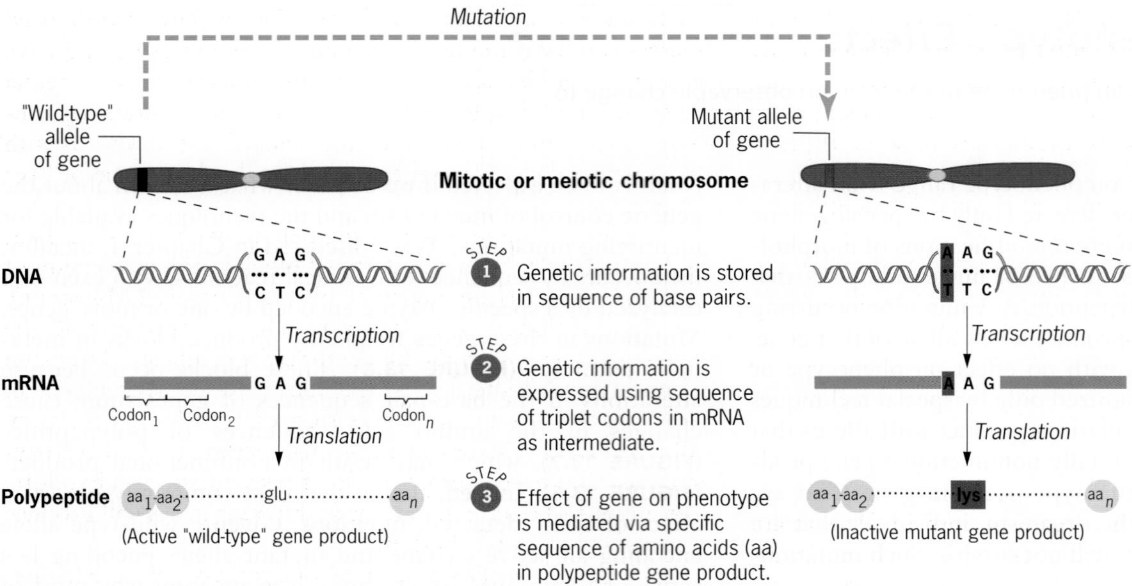

Figure 13.7 ▶ Overview of the mutation process and the expression of wild-type and mutant alleles. Mutations alter the sequences of nucleotide pairs in genes, which, in turn, cause changes in the amino acid sequences of the polypeptides encoded by these genes. A G:C base pair (top, left) has mutated to an A:T base pair (top, right). This mutation changes one mRNA codon from GAG to AAG and one amino acid in the polypeptide product from glutamic acid (glu) to lysine (lys). Such changes often yield nonfunctional gene products.

activity? A wild-type allele of a gene encoding a wild-type enzyme or structural protein will have been selected for optimal activity during the course of evolution. Thus mutations, which cause random changes in the highly adapted amino acid sequences, usually will produce less active or totally inactive products. You can make an analogy with any complex, carefully engineered machine such as a computer or an automobile. If you randomly modify an essential component, the machine is unlikely to perform as well as it did prior to the change. This view of mutation and the interaction between mutant and wild-type alleles fits with the observation that most mutations with recognizable phenotypic effects are recessive and deleterious.

EFFECTS OF MUTATIONS IN HUMAN GLOBIN GENES

Mutant human hemoglobins provide good illustrations of the deleterious effects of mutation. In Chapters 1 and 12, we discussed the structure of hemoglobin and the traumatic effects of one hemoglobin variant, sickle-cell hemoglobin. Recall that the major form of hemoglobin in adults (hemoglobin A) contains two identical **alpha (α) chains** and two identical **beta (β) chains.** Each α polypeptide consists of a specific sequence of 141 amino acids, whereas each β chain is 146 amino acids long. Because of similarities in their amino acid sequences, all the globin chains (and, thus, their structural genes) are believed to have evolved from a common progenitor.

Many different variants of adult hemoglobin have been identified in human populations, and several of them have severe phenotypic effects. Many of the variants were initially detected by their altered electrophoretic behavior (movement in an electric field due to charge differences—see Chapter 9). The hemoglobin variants provide an excellent illustration of the effects of mutation on the structures and functions of gene products and, ultimately, on the phenotypes of the affected individuals.

When the amino acid sequences of the β chains of hemoglobin A and the hemoglobin in patients with sickle-cell anemia (hemoglobin S) were determined and compared, hemoglobin S was found to differ from hemoglobin A at only one position. The sixth amino acid from the amino terminus of the β chain of hemoglobin A is glutamic acid (a negatively charged amino acid). The β chain of hemoglobin S contains valine (no charge at neutral pH) at that position. The α chains of hemoglobin A and hemoglobin S are identical. Thus, the change of a single amino acid in one polypeptide can have severe effects on the phenotype.

In the case of hemoglobin S, the substitution of valine for glutamic acid at the sixth position in the β chain allows a new bond to form, which changes the conformation of the protein and leads to aggregation of hemoglobin molecules. This change results in the grossly abnormal (sickle) shape of the red blood cells. The mutational change in the HBB^A allele that gave rise to HBB^S was a substitution of a T:A base pair for an A:T base pair, with a T in the transcribed strand in the first case and an A in the transcribed strand in the second case (see **FIGURE 1.9**). This A:T → T:A base-pair substitution was first predicted from protein sequence data and the known codon assignments, and was later verified by sequencing the HBB^A and HBB^S alleles.

Over 100 hemoglobin variants with amino acid changes in the β chain are known (see the Genomics on the Web questions at the end of the chapter). Most of them differ from the normal β chain of hemoglobin A by a single amino acid substitution. Some differ by two amino acids. Numerous variants of the α polypeptide also have been identified.

The hemoglobin examples show that mutation is a process in which changes in gene structure, often changes in one or a few base pairs, can cause changes in the amino acid sequences of the polypeptide gene products. These alterations in protein structure, in turn, cause changes in the phenotype that are recognized as mutant.

MUTATION IN HUMANS: BLOCKS IN METABOLIC PATHWAYS

In Chapter 4, we discussed the genetic control of metabolic pathways, in which each step in a pathway is catalyzed by an enzyme encoded by one or more genes. When mutations occur in such genes, they often cause metabolic blocks (see **FIGURE 13.6**) that lead to abnormal phenotypes. This picture of the genetic control of metabolism is valid for all living organisms, including humans (see Focus on Tay-Sachs Disease, a Childhood Tragedy).

We can illustrate the effects of mutations on human metabolism by considering virtually any metabolic pathway. However, the metabolism of the aromatic amino acids phenylalanine and tyrosine provides an especially good example because some of the early studies of mutations in humans revealed blocks in this pathway (see A Milestone in Genetics: Garrod's Inborn Errors of Metabolism in Chapter 4). Phenylalanine and tyrosine are essential amino acids required for protein synthesis; they are not synthesized *de novo* in humans as they are in microorganisms. Thus, both amino acids must be obtained from dietary proteins.

The best-known inherited defect in phenylalanine-tyrosine metabolism is phenylketonuria, which is caused by the absence of phenylalanine hydroxylase, the enzyme that converts phenylalanine to tyrosine. Newborns with phenylketonuria, an autosomal recessive disease, develop severe mental retardation if not placed on a diet low in phenylalanine (see Chapter 4 Milestone). The first inherited disorder in the phenylalanine-tyrosine metabolic pathway to be studied in humans was alkaptonuria, which is caused by autosomal recessive mutations that inactivate the enzyme homogentisic acid oxidase. Alkaptonuria played an important role in the evolution of the concept of the gene (see Chapter 4 Milestone).

Two other inherited disorders are caused by mutations in genes encoding enzymes required for the catabolism of tyrosine; both are inherited as autosomal recessives. Tyrosinosis and tyrosinemia result from the lack of the enzymes tyrosine transaminase and *p*-hydroxyphenylpyruvic acid oxidase, respectively. Both enzymes are required to degrade tyrosine to CO_2 and H_2O. Tyrosinosis is very rare; only a few cases have been studied. Individuals with tyrosinosis show pronounced increases in tyrosine levels in their blood and urine and have various congenital abnormalities. Individuals with tyrosinemia have elevated levels of both tyrosine and *p*-hydroxyphenylpyruvic acid in their blood and urine. Most newborns with tyrosinemia die within six months after birth because of liver failure.

Albinism, the absence of pigmentation in the skin, hair, and eyes, results from a mutational block in the conversion of tyrosine to the dark pigment melanin (see Chapter 4 Milestone). One type of albinism is caused by the absence of tyrosinase, the enzyme that catalyzes the first step in the synthesis of melanin from tyrosine. Other types of albinism result from blocks in subsequent steps in the conversion of tyrosine to melanin. Albinism is inherited as an autosomal recessive trait; heterozygotes usually have normal levels of pigmentation. Therefore, two albinos who have mutations in different genes will produce normally pigmented children.

Thus, studies of a single metabolic pathway, phenylalanine-tyrosine metabolism, have revealed five different inherited disorders, all caused by mutations in genes that control steps in this pathway. Similar examples of the genetic control of metabolism can be obtained by examining essentially any other metabolic pathway in humans.

CONDITIONAL LETHAL MUTATIONS: POWERFUL TOOLS FOR GENETIC STUDIES

Of all the mutations—from isoalleles to lethals—**conditional lethal mutations** are the most useful for genetic studies. These are mutations that are (1) lethal in one environment, the *restrictive condition*, but are (2) viable in a second environment, the *permissive condition*. Conditional lethal mutations allow geneticists to identify and study mutations in essential genes that result in complete loss of gene-product activity even in haploid organisms. Mutants carrying conditional lethals can be propagated under permissive conditions, and information about the functions of the gene products can be inferred by studying the consequences of their absence under the restrictive conditions. Conditional lethal mutations have been used to investigate a vast array of biological processes from development to photosynthesis.

The three major classes of mutants with conditional lethal phenotypes are (1) auxotrophic mutants, (2) temperature-sensitive mutants, and (3) suppressor-sensitive mutants. **Auxotrophs** are mutants that are unable to synthesize an essential metabolite (amino acid, purine, pyrimidine, vitamin, and so forth) that is synthesized by wild-type or *prototrophic* organisms of the same species. The auxotrophs will grow and reproduce when the metabolite is supplied in the medium (the permissive condition); they will not grow when the essential metabolite is absent (the restrictive condition). **Temperature-sensitive mutants** will grow at one temperature but not at another. Most temperature-sensitive mutants are heat-sensitive; however, some are cold-sensitive. The temperature sensitivity usually

▶ FOCUS ON **Tay-Sachs Disease, a Childhood Tragedy**

Of all the inherited human disorders, one of the most tragic is Tay-Sachs disease. Infants homozygous for the mutant gene that causes Tay-Sachs disease are normal at birth. However, within a few months, they become hypersensitive to loud noises and develop a cherry-red spot on the retina of the eye. These early symptoms of the disease often go undetected by parents and physicians. At six months to one year after birth, Tay-Sachs children begin to undergo progressive neurological degeneration that rapidly leads to mental retardation, blindness, deafness, and general loss of control of body functions. By two years of age, they are usually totally paralyzed and develop chronic respiratory infections. Death commonly occurs at three to four years of age.

Although the molecular defect responsible for Tay-Sachs disease is known, there is no effective treatment for the disorder. The only positive aspect of Tay-Sachs disease is that it is rare in most populations. However, this is of little comfort to the Ashkenazi Jewish people of Central Europe and their descendants. Tay-Sachs disease occurs in about 1 of 3600 of their children, and about 1 of 30 adults in these Jewish populations carries the mutant gene in the heterozygous state. If two individuals from these populations marry, the chance that both will carry the mutant gene is about 1 in 1000 (0.033×0.033); if both are carriers, on average, one-fourth of their children will be homozygous for the mutant gene and develop Tay-Sachs disease.

The mutation that causes Tay-Sachs disease is located in the *HEXA* gene, which encodes the enzyme hexosaminidase A. This enzyme acts on a complex lipid called ganglioside G_{M2}, cleaving it into a smaller ganglioside (G_{M3}) and *N*-acetyl-*D*-galactosamine, as shown in **FIGURE 1.**

The function of ganglioside G_{M2} is to coat nerve cells, insulating them from events occurring in neighboring cells and thus speeding up the transmission of nerve impulses. In the absence of the enzyme that breaks it down, ganglioside G_{M2} accumulates and literally smothers nerve cells. This buildup of complex lipids on neurons blocks their action, leading to deterioration of the nervous system and eventually to paralysis.

Although Tay-Sachs disease was described by Warren Tay in 1881 and the biochemical basis has been known for over 20 years, there is still no effective treatment of this tragic disorder. Whereas some inherited disorders can be treated by enzyme therapy—by supplying

Ganglioside G_{M2}: *N*-acetyl-*D*-galactosamine–β-1,4,-galactose-β-1,4-glucose-β-1,1-ceramide
3
α–2
N-acetylneuraminic acid

Hexosaminidase A ⇸ Tay-Sachs disease

N-acetyl-*D*-galactosamine
+
Ganglioside G_{M3}:
Galactose-β-1,4-glucose-β-1,1-ceramide
3
α–2
N-acetylneuraminic acid

Figure 1 ▶ **The metabolic defect in humans with Tay-Sachs disease.**

the missing enzyme to patients—this approach is not feasible with Tay-Sachs disease because the enzyme will not penetrate the barrier separating brain cells from the circulatory system. Moreover, somatic-cell gene therapy—providing functional copies of the defective gene to somatic cells (Chapter 16)—is not possible at present because there is no established procedure for introducing genes into neurons. Indeed, scientists still don't know which nerve cells are responsible for the neurological degeneration that occurs in children with the disease.

Tay-Sachs disease can be detected prenatally by amniocentesis (Chapter 6), and this procedure has been used extensively to diagnose the disorder. Recently, a sensitive DNA test has been developed that allows scientists to detect the mutant gene that causes Tay-Sachs disease in DNA isolated from a single cell (Chapter 17). This DNA test has been used to screen eight-cell pre-embryos produced by *in vitro* fertilization for the Tay-Sachs mutation. One cell is used for the DNA test, and the other seven cells retain the capacity to develop into a normal embryo when implanted into the uterus of the mother. Only embryos that test normal—those not homozygous for the deadly Tay-Sachs gene—are implanted. This procedure allows parents who are both carriers of the mutant gene to have children without worrying about the birth of a child with Tay-Sachs disease.

results from increased heat or cold lability of the mutant gene product—for example, an enzyme that is active at low temperature but partially or totally inactive at higher temperatures. Occasionally, only the synthesis of the gene product is sensitive to temperature, and once synthesized, the mutant gene product may be as stable as the wild-type gene product. **Suppressor-sensitive mutants** are viable when a second genetic factor, a suppressor, is present, but they are nonviable in the absence of the suppressor. The suppressor gene may correct or compensate for the defect in phenotype that is caused by the suppressor-sensitive mutation, or it may cause the gene product altered by the mutation to be nonessential. We have discussed one class of suppressor-sensitive mutations, the *amber* mutations, in Chapter 12.

Now, let's briefly consider how conditional lethal mutations can be used to investigate biological processes—to dissect

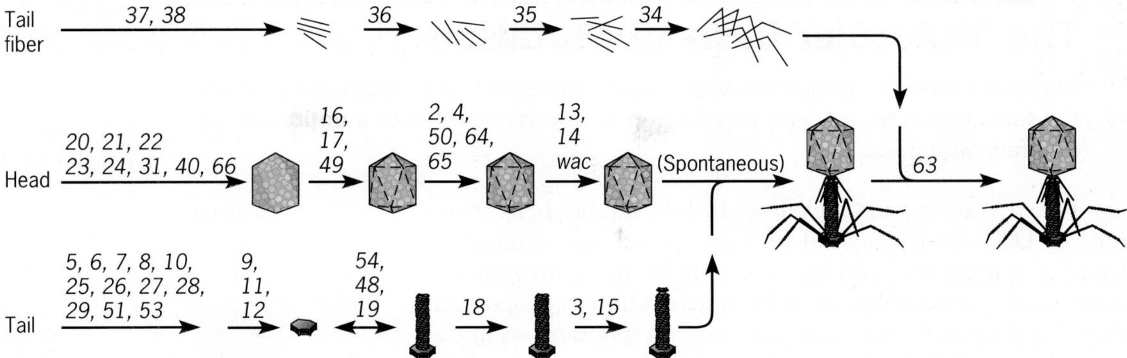

Figure 13.8 ▶ Abbreviated map of morphogenesis in bacteriophage T4. The head, the tail, and the tail fibers are produced via separate branches of the pathway and are then joined in the final stages of morphogenesis. The numbers identify the T4 genes whose products are required at each step in the pathway. The *wac* gene encodes the six "whiskers" and collar at the base of the head. The sequences of early steps in head and tail formation are known but are omitted here to keep the diagram concise.

biological processes into their individual parts or steps. Let's begin with a simple biosynthetic pathway:

$$\begin{array}{ccccc} & \text{Gene A} & & \text{Gene B} & \\ & \downarrow & & \downarrow & \\ & \text{Enzyme A} & & \text{Enzyme B} & \\ \text{Precursor X} & \longrightarrow & \text{Intermediate Y} & \longrightarrow & \text{Product Z} \end{array}$$

Intermediate Y is produced from precursor X by the action of enzyme A, the product of gene *A*, but intermediate Y may be rapidly converted to product Z by enzyme B, the product of gene *B*. If so, intermediate Y may be present in minute quantities and be difficult to isolate and characterize. However, in a mutant organism that has a mutation in gene *B*, resulting in the synthesis of either an inactive form of enzyme B or no enzyme B, intermediate Y may accumulate to much higher concentrations, facilitating its isolation and characterization. Similarly, a mutation in gene *A* may aid in the identification of precursor X. In this way, the sequence of steps in a given metabolic pathway can often be determined.

Morphogenesis in living organisms occurs in part by the sequential addition of proteins to macromolecular structures to produce the final three-dimensional conformations, and the sequence of protein additions can often be determined by isolating and studying mutant organisms with defects in the genes encoding the proteins involved. Because an appropriate mutation will eliminate the activity of a single polypeptide, mutations provide a powerful tool with which to dissect biological processes—to break the processes down into individual steps.

The resolving power of mutational dissection of biological processes has been elegantly documented by the research of Robert Edgar, Jonathan King, William Wood, and colleagues, who worked out the complete pathway of morphogenesis for bacteriophage T4. This complex process involves the products of about 50 of the roughly 200 genes in the T4 genome. Each gene encodes a structural protein of the virus or an enzyme that catalyzes one or more steps in the morphogenetic pathway. By (1) isolating mutant strains of phage T4 with temperature-sensitive and suppressor-sensitive conditional lethal mutations in each of the approximately 50 genes, and (2) using electron microscopy and biochemical techniques to analyze the structures that accumulate when these mutant strains are grown under the restrictive conditions, Edgar, King, Wood, and coworkers established the complete pathway of phage T4 morphogenesis (**FIGURE 13.8**).

Many other biological processes also have been successfully dissected by mutational studies. Examples include the photosynthetic electron transport chains in plants and pathways of nitrogen fixation in bacteria. Currently, mutational dissection is yielding new insights into the processes of differentiation and development in higher plants and animals (Chapter 21). Researchers are also using mutations to dissect behavior and learning in *Drosophila*. In principle, scientists should be able to use mutations to dissect any biological process. Every gene can mutate to a nonfunctional state. Thus, mutational dissection of biological processes is limited only by the ingenuity of researchers in identifying mutations of the desired types.

KEY POINTS

- The effects of mutations on the phenotypes of living organisms range from minor to lethal changes.
- Most mutations exert their effects on the phenotype by altering the amino acid sequences of polypeptides, the primary gene products.
- The mutant polypeptides, in turn, cause blocks in metabolic pathways.
- Conditional lethal mutations provide powerful tools with which to dissect biological processes.

The Molecular Basis of Mutation

Mutations alter the nucleotide sequences of genes in several ways, for example, the substitution of one base pair for another or the deletion or addition of one or a few base pairs.

When Watson and Crick described the double-helix structure of DNA and proposed its semiconservative replication based on specific base-pairing to account for the accurate transmission of genetic information from generation to generation, they also proposed a mechanism to explain spontaneous mutation. Watson and Crick pointed out that the structures of the bases in DNA are not static. Hydrogen atoms can move from one position in a purine or pyrimidine to another position—for example, from an amino group to a ring nitrogen. Such chemical fluctuations are called **tautomeric shifts.** Although tautomeric shifts are rare, they may be of considerable importance in DNA metabolism because some alter the pairing potential of the bases. The nucleotide structures that we discussed in Chapter 9 are the common, more stable forms, in which adenine always pairs with thymine and guanine always pairs with cytosine. The more stable keto forms of thymine and guanine and the amino forms of adenine and cytosine may infrequently undergo tautomeric shifts to less stable enol and imino forms, respectively (**FIGURE 13.9**). The bases would be expected to exist in their less stable tautomeric forms for only short periods of time. However, if a base existed in the rare form at the moment that it was being replicated or being incorporated into a nascent DNA chain, a mutation would result. When the bases are present in their rare imino or enol states, they can form adenine-cytosine and guanine-thymine base pairs (**FIGURE 13.10*a***). The net effect of such an event, and the subsequent replication required to segregate the mismatched base pair, is an A:T to G:C or a G:C to A:T base-pair substitution (**FIGURE 13.10*b***).

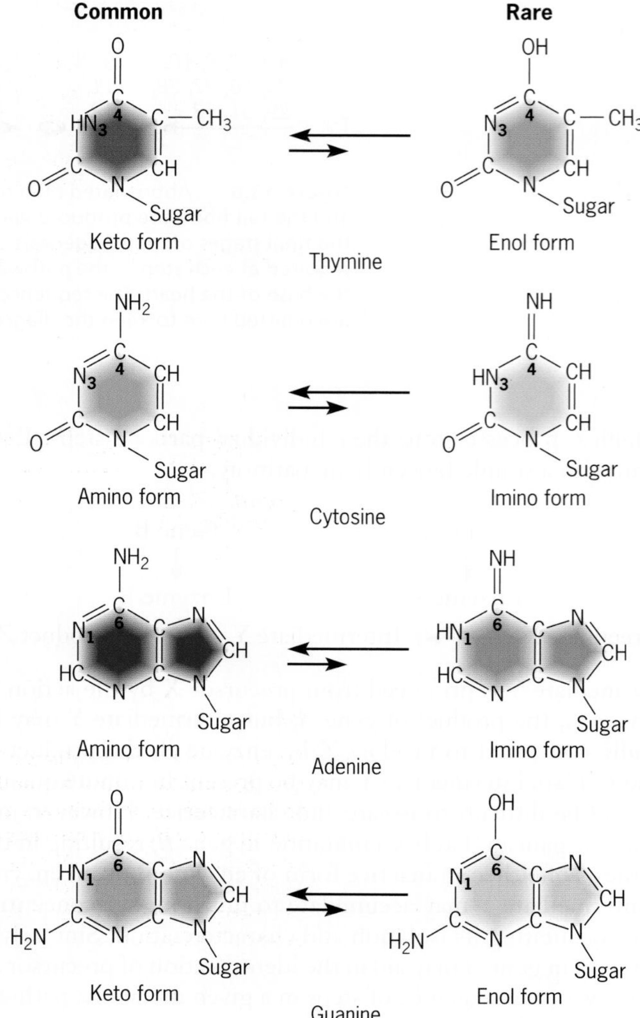

Figure 13.9 ► Tautomeric forms of the four common bases in DNA. The shifts of hydrogen atoms between the number 3 and number 4 positions of the pyrimidines and between the number 1 and number 6 positions of the purines change their base-pairing potential.

Mutations resulting from tautomeric shifts in the bases of DNA involve the replacement of a purine in one strand of DNA with the other purine and the replacement of a pyrimidine in the complementary strand with the other pyrimidine. Such base-pair substitutions are called **transitions.** Base-pair substitutions involving the replacement of a purine with a pyrimidine and vice versa are called **transversions.** There are three substitutions—one transition and two transversions—possible for every base pair. A total of four different transitions and eight different transversions are possible (**FIGURE 13.11*a***). Another type of point mutation involves the addition or deletion of one or a few base pairs. Base-pair additions and deletions are collectively referred to as **frameshift mutations** because they alter the reading frame of all base-pair triplets (DNA triplets that specify codons in mRNA and amino acids in the polypeptide gene product) in the gene that are distal to the site at which the mutation occurs (**FIGURE 13.11*b***).

All three types of point mutations—transitions, transversions, and frameshift mutations—are present among spontaneously occurring mutations. A surprisingly large proportion of the spontaneous mutations that have been studied are single base-pair additions and deletions rather than base-pair substitutions. These frameshift mutations almost always result in the synthesis of nonfunctional protein gene products.

Although much remains to be learned about the causes, molecular mechanisms, and frequency of spontaneously occurring mutations, three major factors are (1) the accuracy of the DNA replication machinery, (2) the efficiency of the mechanisms that have evolved for the repair of damaged DNA, and (3) the degree of exposure to mutagenic agents present in the

Hydrogen-bonded A:C and G:T base pairs that form when cytosine and guanine are in their rare imino and enol tautomeric forms.

Rare A:C base pair

Cytosine (rare imino form)

Adenine (common amino form)

Rare G:T base pair

Thymine (common keto form)

Guanine (rare enol form)

(*a*)

Mechanism by which tautomeric shifts in the bases in DNA cause mutations.

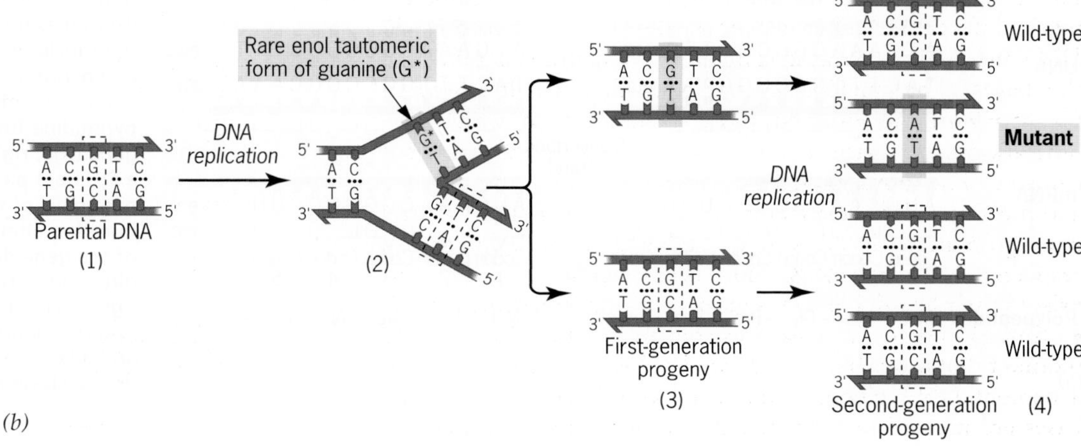

(*b*)

Figure 13.10 ▶ The effects of tautomeric shifts in the nucleotides in DNA on (*a*) base-pairing and (*b*) mutation. Rare A:C and G:T base pairs like those shown in (*a*) also form when thymine and adenine are in their rare enol and imino forms, respectively. (*b*) A guanine (1) undergoes a tautomeric shift to its rare enol form (G*) at the time of replication (2). In its enol form, guanine pairs with thymine (2). During the subsequent replication (3 to 4), the guanine shifts back to its more stable keto form. The thymine incorporated opposite the enol form of guanine (2) directs the incorporation of adenine during the next replication (3 to 4). The net result is a G:C to A:T base-pair substitution.

environment. Perturbations of the DNA replication apparatus or DNA repair systems, both of which are under genetic control, have been shown to cause large increases in mutation rates.

INDUCED MUTATIONS

Many naturally occurring mutations were identified and studied by the early geneticists. However, the science of genetics changed dramatically in 1927 when Hermann J. Muller discovered that X rays induced mutations in *Drosophila*. The ability to induce mutations opened the door to a completely new approach to genetic analysis. Geneticists could now induce mutations in genes of interest and then study the effects of the missing gene products. We discuss Muller's ingenious demonstration of X-ray-induced mutations on the X chromosome of *Drosophila* in A Milestone in Genetics: Muller Demonstrates That X Rays Are Mutagenic. Muller was awarded the Nobel Prize in Physiology or Medicine in 1946 for this breakthrough. Subsequent work showed that X rays are mutagenic to all organisms and that many other agents—physical, chemical, and transposable genetic elements—are similarly mutagenic.

X rays have many effects on living tissues. Therefore, X-ray-induced mutations provide little information about the molecular mechanisms by which mutations are produced. The discovery of chemical mutagens with specific effects on DNA has led to a better understanding of mutation at the molecular level.

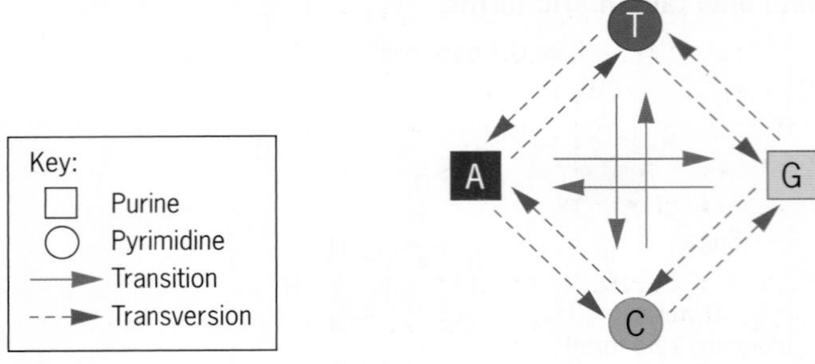

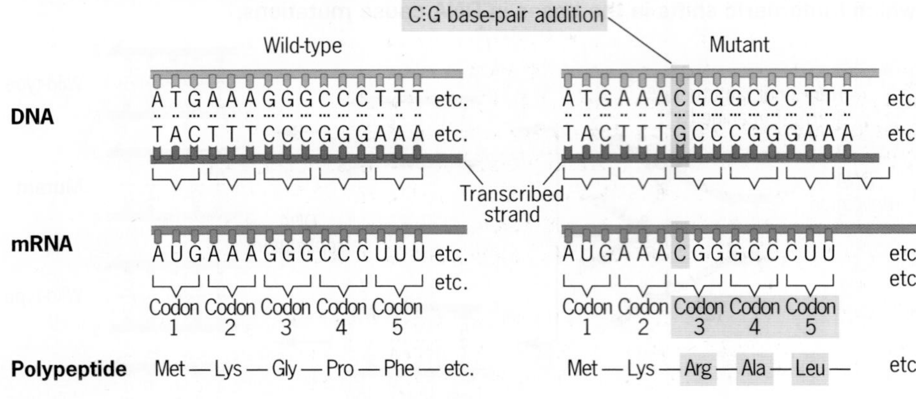

Figure 13.11 ▶ Types of point mutations that occur in DNA: (a) base substitutions and (b) frameshift mutations. (*a*) The base substitutions include four transitions (purine for purine and pyrimidine for pyrimidine; green arrows) and eight transversions (purine for pyrimidine and pyrimidine for purine; blue arrows). (*b*) A mutant gene (top, right) was produced by the insertion of a C:G base pair between the sixth and seventh base pairs of the wild-type gene (top, left). This insertion alters the reading frame of that portion of the gene distal to the mutation, relative to the direction of transcription and translation (left to right, as diagrammed). The shift in reading frame, in turn, changes all of the codons in the mRNA and all of the amino acids in the polypeptide specified by base-pair triplets distal to the mutation.

Mustard gas (sulfur mustard) was the first chemical shown to be mutagenic. Charlotte Auerbach and her associates discovered the mutagenic effects of mustard gas and related compounds during World War II. However, because of the potential use of mustard gas in chemical warfare, the British government placed their results on the classified list. Thus, Auerbach and coworkers could neither publish their results nor discuss them with other geneticists until the war ended. The compounds that they studied are examples of a large class of chemical mutagens that transfer alkyl (CH_3^-, $CH_3CH_2^-$, and so forth) groups to the bases in DNA; thus, they are called alkylating agents. Like X rays, mustard gas has many effects on DNA. In the 1950s, chemical mutagens that have specific effects on DNA were discovered (**FIGURE 13.12**).

MUTATIONS INDUCED BY CHEMICALS

Chemical mutagens can be divided into two groups: (1) those that are mutagenic to both replicating and nonreplicating DNA, such as the alkylating agents and nitrous acid; and (2) those that are mutagenic only to replicating DNA, such as base analogs—purines and pyrimidines with structures similar to the normal bases in DNA. The base analogs must be incorporated into DNA chains in the place of normal bases during replication in order to exert their mutagenic effects. The second group of mutagens also includes the acridine dyes, which intercalate into DNA and increase the probability of mistakes during replication.

The mutagenic **base analogs** have structures similar to the normal bases and are incorporated into DNA during replication. However, their structures are sufficiently different from the normal bases in DNA that they increase the frequency of mispairing, and thus mutation, during replication. The two most commonly used base analogs are 5-bromouracil and 2-aminopurine. The pyrimidine 5-bromouracil is a thymine analog; the bromine at the 5 position is similar in several respects to the methyl (—CH_3) group at the 5 position in thymine. However, the bromine at this position changes the charge distribution and increases the frequency of tautomeric shifts (see **FIGURE 13.9**). In its more stable keto form, 5-bromouracil pairs with adenine. After a tautomeric shift to its enol form, 5-bromouracil pairs with guanine (**FIGURE 13.13**). The mutagenic effect of 5-bromouracil is the same as that predicated for tautomeric shifts in normal bases (see **FIGURE 13.10*b***), namely, transitions.

If 5-bromouracil is present in its less frequent enol form as a nucleoside triphosphate at the time of its incorporation into a nascent strand of DNA, it will be incorporated opposite guanine

Alkylating agents

$Cl—CH_2—CH_2—S—CH_2—CH_2—Cl$
Di-(2-chloroethyl) sulfide
(Mustard gas)

$CH_3—CH_2—O—SO_2—CH_3$
Ethyl methane sulfonate
(EMS)

$CH_3—CH_2—O—SO_2—CH_2—CH_3$
Ethyl ethane sulfonate
(EES)

(*a*)

Base analogs

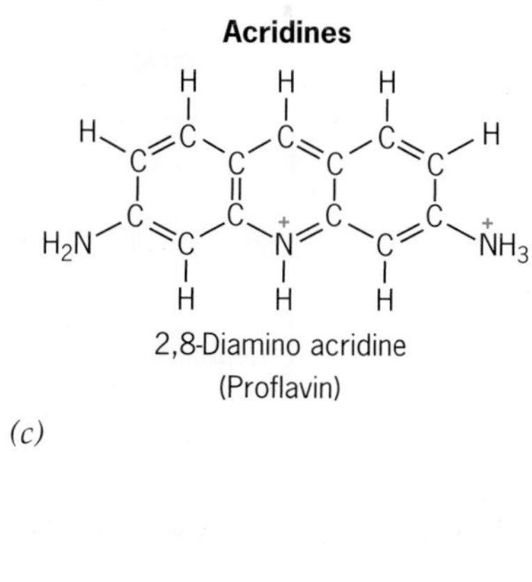

5-Bromouracil
(5-BU)

2-Aminopurine
(2-AP)

(*b*)

Acridines

2,8-Diamino acridine
(Proflavin)

(*c*)

Deaminating agent

HNO_2
Nitrous acid

(*d*)

Hydroxylating agent

NH_2OH
Hydroxylamine

(*e*)

Figure 13.12 ▶ Some potent chemical mutagens.

in the template strand and cause a G:C → A:T transition (**FIGURE 13.14*a***). If, however, 5-bromouracil is incorporated in its more frequent keto form opposite adenine (in place of thymine) and undergoes a tautomeric shift to its enol form during a subsequent replication, it will cause an A:T → G:C transition (**FIGURE 13.14*b***). Thus, 5-bromouracil induces transitions in both directions, A:T ↔ G:C. An important consequence of the bidirectionality of 5-bromouracil-induced transitions is that mutations originally induced with this thymine analog can also be induced to mutate back to the wild-type with 5-bromouracil. 2-Aminopurine acts in a similar manner but is incorporated in place of adenine or guanine.

Nitrous acid (HNO_2) is a potent mutagen that acts on either replicating or nonreplicating DNA. Nitrous acid causes oxidative deamination of the amino groups in adenine, guanine, and cytosine. This reaction converts the amino groups to keto groups and changes the hydrogen-bonding potential of the modified bases (**FIGURE 13.15**). Adenine is deaminated to hypoxanthine, which base-pairs with cytosine rather than thymine. Cytosine is converted to uracil, which base-pairs with adenine instead of guanine. Deamination of guanine produces xanthine, but xanthine—just like guanine—base-pairs with cytosine. Thus, the deamination of guanine is not mutagenic. Because the deamination of adenine results in A:T → G:C transitions, and the deamination of cytosine produces G:C → A:T transitions, nitrous acid induces transitions in both directions, A:T ↔ G:C (see Focus on Problem Solving: Predicting

5-Bromouracil : adenine base pair.

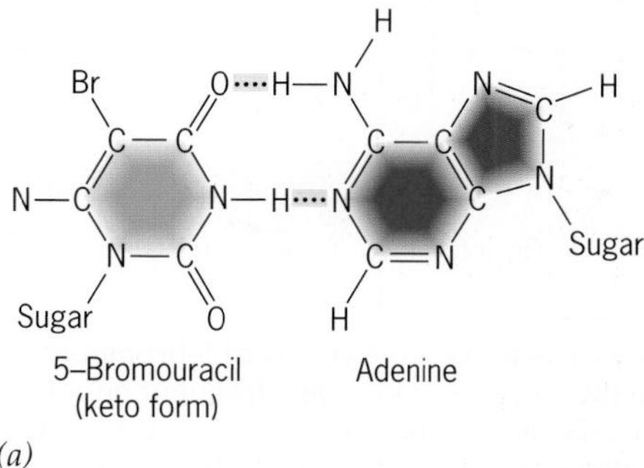

5-Bromouracil
(keto form)

Adenine

(*a*)

5-Bromouracil : guanine base pair.

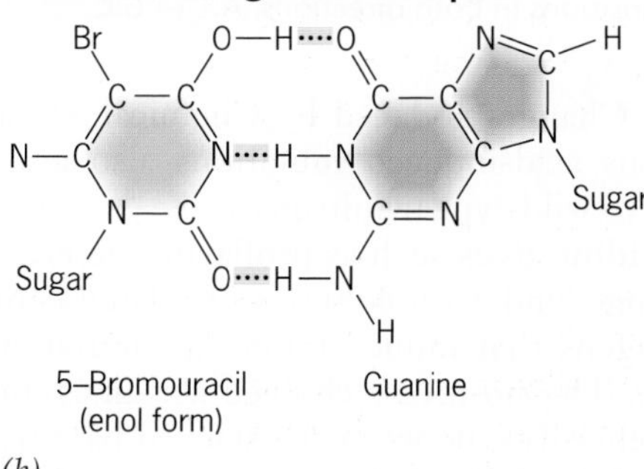

5-Bromouracil
(enol form)

Guanine

(*b*)

Figure 13.13 ▶ Base-pairing between 5-bromouracil and (*a*) adenine or (*b*) guanine.

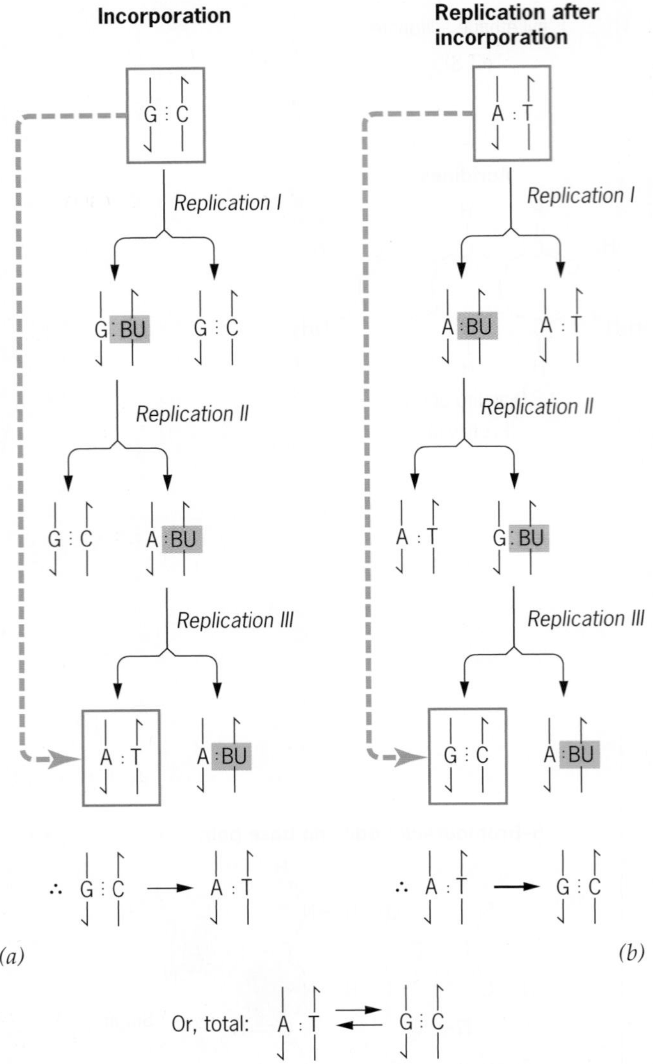

Figure 13.14 ▶ The mutagenic effects of 5-bromouracil. (*a*) When 5-bromouracil (BU) is present in its less frequent enol form (orange) at the time of incorporation into DNA, it induces G:C → A:T transitions. (*b*) When 5-bromouracil is incorporated into DNA in its more common keto form (blue) and shifts to its enol form during a subsequent replication, it induces A:T → G:C transitions. Thus, 5-bromouracil can induce transitions in both directions, A:T ↔ G:C.

Amino Acid Changes Induced by Chemical Mutagens). As a result, nitrous acid-induced mutations also are induced to mutate back to wild-type by nitrous acid.

The **acridine dyes** such as proflavin (see **FIGURE 13.12*c***), acridine orange, and a whole series of related compounds are potent mutagens that induce frameshift mutations (see **FIGURE 13.11*b***). The positively charged acridines intercalate, or sandwich themselves, between the stacked base pairs in DNA (**FIGURE 13.16**). In so doing, they increase the rigidity and alter the conformation of the double helix, causing slight bends or kinks in the molecule. When DNA molecules containing intercalated acridines replicate, additions and deletions of one to a few base pairs occur. As we might expect, these small additions and deletions, usually of a single base pair, result in altered reading frames for the portion of the gene distal to the mutation (see **FIGURE 13.11b**). Thus, acridine-induced mutations usually result in nonfunctional gene products.

Alkylating agents are chemicals that donate alkyl groups to other molecules. They include nitrogen mustard, and methyl and ethyl methane sulfonate (MMS and EMS) (see **FIGURE 13.12*a***)—chemicals that have multiple effects on DNA. Alkylating agents induce all types of mutations, including transitions, transversions, frameshifts, and even chromosome aberrations, with relative frequencies that depend on the reactivity of the agent involved. One mechanism of mutagenesis by alkylating agents involves the transfer of methyl or ethyl groups to the bases, resulting in altered base-pairing potentials. For example, EMS causes ethylation of the bases in DNA at the 7-*N* and the 6-*O* positions. When 7-ethylguanine is produced, it base-pairs with thymine to cause G:C → A:T transitions. Other base alkylation products activate error-prone DNA repair processes that introduce transitions, transversions, and frameshift mutations during the repair process. Some alkylating agents, particularly difunctional alkylating agents (those with two reactive alkyl groups), cross-link DNA strands or molecules and induce chromosome breaks, which result in various kinds of chromosomal aberrations (Chapter 6). Alkylating agents as a class therefore exhibit less specific mutagenic effects than do base analogs, nitrous acid, or acridines.

In contrast to most alkylating agents, the **hydroxylating agent** hydroxylamine (NH_2OH) has a specific mutagenic effect. It induces only G:C → A:T transitions. When DNA is treated with hydroxylamine, the amino group of cytosine is hydroxylated. The resulting hydroxylaminocytosine base-pairs with adenine, leading to G:C → A:T transitions. Because of its specificity, hydroxylamine has been very useful in classifying transition mutations. Mutations that are induced to revert to wild-type by nitrous acid or base analogs, and therefore were originally caused by transitions, can be divided into two classes on the basis of their revertibility with hydroxylamine. (1) Those with an A:T base pair at the mutant site will not be induced to revert by hydroxylamine. (2) Those with a G:C base pair at the mutant site will be induced to revert by hydroxylamine. Thus, hydroxylamine can be used to determine whether a particular mutation was an A:T → G:C or a G:C → A:T transition.

MUTATIONS INDUCED BY RADIATION

The portion of the electromagnetic spectrum (**FIGURE 13.17**) with wavelengths shorter and of higher energy than visible light is subdivided into **ionizing radiation** (X rays, gamma rays, and cosmic rays) and **nonionizing radiation** (ultraviolet light). Ionizing radiations are of high energy and are useful for medical diagnosis because they penetrate living tissues for substantial distances. In the process, these high-energy rays collide with atoms and cause the release of electrons, creating positively

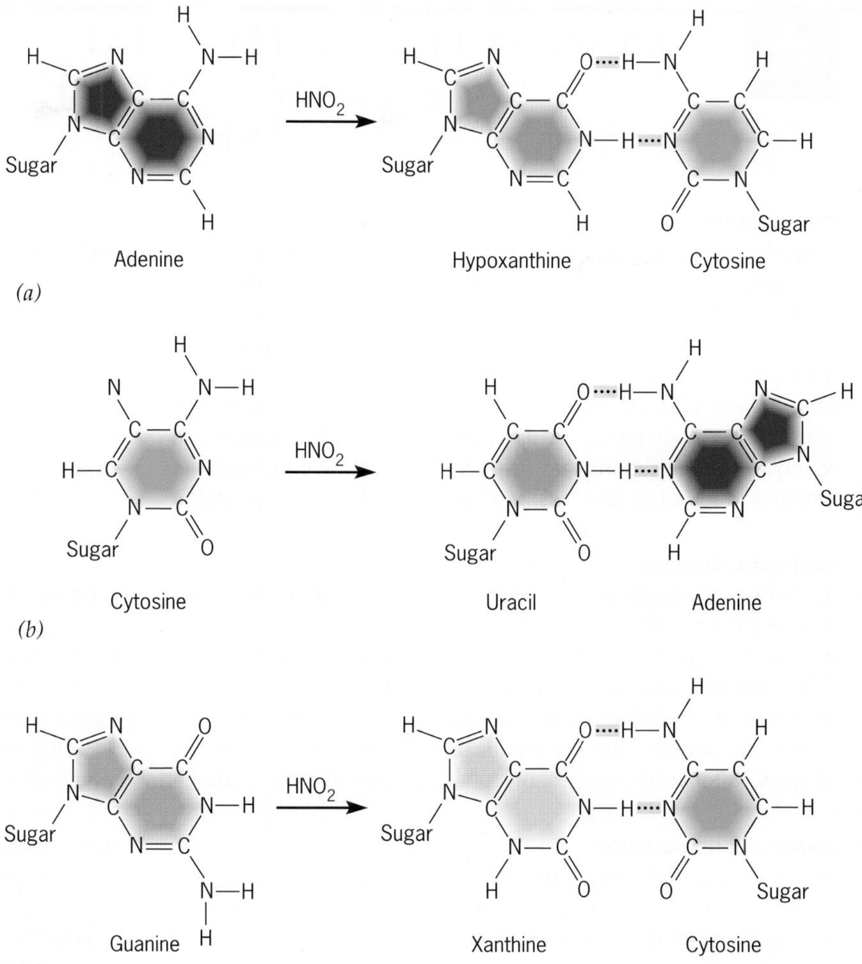

Figure 13.15 ▶ Nitrous acid induces mutations by oxidative deamination of the bases in DNA. Nitrous acid converts (*a*) adenine to hypoxanthine, causing A:T → G:C transitions; (*b*) cytosine to uracil, causing G:C → A:T transitions; and (*c*) guanine to xanthine, which is not mutagenic. Together, the effects of nitrous acid on adenine and cytosine explain its ability to induce transitions in both directions, A:T ↔ G:C.

charged free radicals or ions. The ions, in turn, collide with other molecules and cause the release of additional electrons. The result is that a cone of ions is formed along the track of each high-energy ray as it passes through living tissues. This process of ionization is induced by machine-produced X rays,

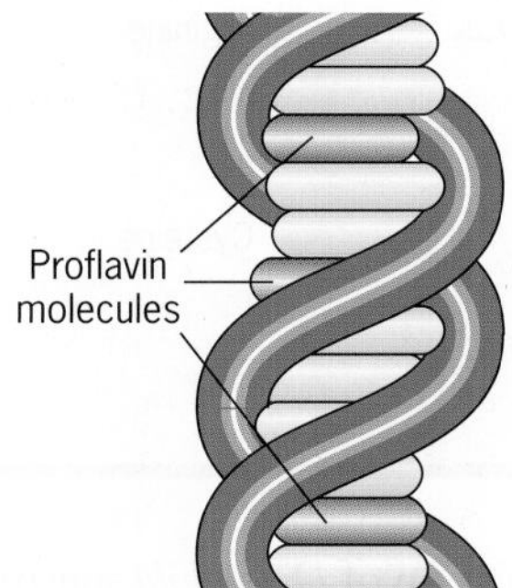

Figure 13.16 ▶ Intercalation of proflavin into the DNA double helix. X-ray diffraction studies have shown that these positively charged acridine dyes become sandwiched between the stacked base pairs.

protons, and neutrons, as well as by the alpha, beta, and gamma rays released by radioactive isotopes such as ^{32}P, ^{35}S, and the uranium-238 used in nuclear reactors.

Ultraviolet rays, having lower energy than ionizing radiations, penetrate only the surface layer of cells in higher plants and animals and do not cause ionizations. Ultraviolet rays dissipate their energy to the atoms they encounter, raising the electrons in the outer orbitals to higher energy levels, a state referred to as *excitation*. Molecules containing atoms in either ionic forms or excited states are chemically more reactive than those containing atoms in their normal stable states. The increased reactivity of atoms present in DNA molecules is responsible for most of the mutagenicity of ionizing radiation and ultraviolet light.

X rays and other forms of ionizing radiation are quantitated in **roentgen (r)** units, which are measures of the number of ionizations per unit volume under a standard set of conditions. Specifically, one roentgen unit is a quantity of ionizing radiation that produces 2.083×10^9 ion pairs in one cubic centimeter of air at 0°C and a pressure of 760 mm of mercury. Note that the dosage of irradiation in roentgen units does not involve a time scale. The same dosage may be obtained by a low intensity of

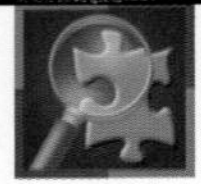

▶ FOCUS ON PROBLEM SOLVING
Predicting Amino Acid Changes Induced by Chemical Mutagens

THE PROBLEM

You are given the nature of the genetic code in Table 12.1. As is illustrated in Figure 13.15, the chemical nitrous acid deaminates adenine, cytosine, and guanine (adenine → hypoxanthine, which base-pairs with cytosine; cytosine → uracil, which base-pairs with adenine; and guanine → xanthine, which base-pairs with cytosine). If you treat a population of nonreplicating tobacco mosaic viruses (TMV) with nitrous acid, would you expect the nitrous acid to induce any mutations that result in the substitution of another amino acid for a histidine (His) residue in a wild-type polypeptide? that is, polypeptide : aa_1......histidine..........................aa_n

? ↓ Nitrous acid

aa_1...... aa_x (not histidine)............aa_n

If so, what amino acid(s) and by what mechanism(s)? If not, why not?

FACTS AND CONCEPTS

1. TMV stores its genetic information in single-stranded RNA that is equivalent to mRNA.

2. The TMV genomic RNA replicates like DNA via a complementary (base-paired) double-stranded intermediate.

3. Although the tobacco mosaic viruses are not replicating at the time of treatment with nitrous acid, they will subsequently be allowed to replicate by infecting tobacco leaves in order to determine whether or not any mutations of the indicated type were induced by treatment with nitrous acid.

4. The histidine codons are CAU and CAC. Therefore, the TMV genome (RNA) contains one of these sequences at all sites specifying histidine in the polypeptides encoded by TMV.

5. The adenines and cytosines in the TMV genome are potential targets of nitrous acid-induced mutation.

ANALYSIS AND SOLUTION

When nitrous acid deaminates adenine and cytosine, it produces hypoxanthine and uracil, respectively. During subsequent replication of the modified TMV RNAs, hypoxanthine pairs with cytosine and uracil pairs with adenine. As a result, some of the A's and C's in TMV RNA will be converted to G's and U's. The deamination of these bases results in tyrosine, arginine, and cysteine codons in the TMV genomes produced by the semiconservative replication of the mutagenized viral RNA. Thus, nitrous acid mutagenesis will lead to the replacement of some histidines in wild-type TMV proteins with tyrosines, arginines, and cysteines in mutant proteins, as shown in the following diagram.

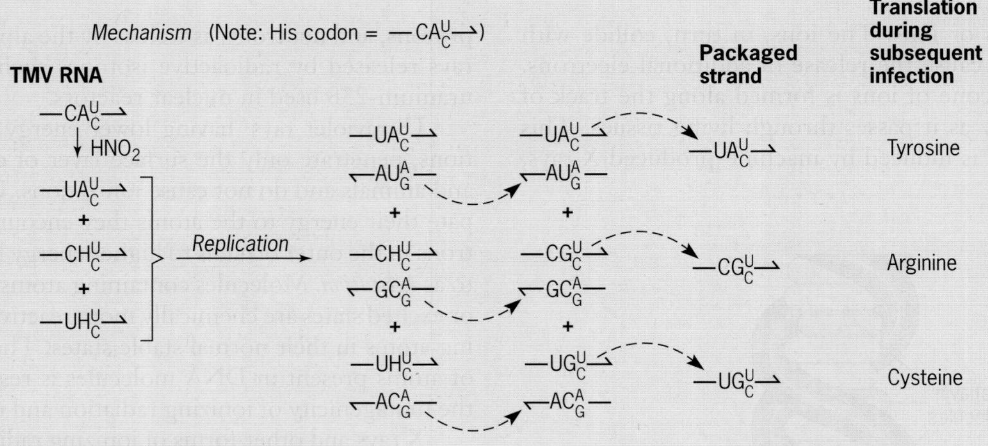

For further discussion go to your *WileyPLUS* course.

irradiation over a long period of time or a high intensity of irradiation for a short period of time. This point is important because in most studies the frequency of induced point mutations is directly proportional to the dosage of irradiation (**FIGURE 13.18**). For example, X-irradiation of *Drosophila* sperm causes an approximately 3 percent increase in mutation rate for each 1000 r increase in irradiation dosage. This linear relationship shows that the induction of mutations by X rays exhibits single-hit

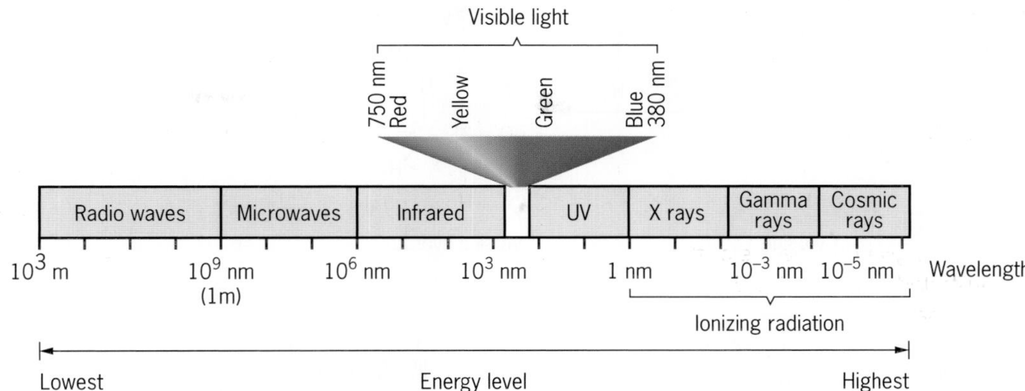

Figure 13.17 ▶ The electromagnetic spectrum.

kinetics, which means that each mutation results from a single ionization event. That is, every ionization has a fixed probability of inducing a mutation under a standard set of conditions.

What is a safe level of irradiation? The development and use of the atomic bomb and the accidents at nuclear power plants have generated concern about exposure to ionizing radiations. The linear relationship between mutation rate and radiation dosage indicates that there is no safe level of irradiation. Rather, the results indicate that the higher the dosage of irradiation, the higher the mutation rate, and the lower the dosage, the lower the mutation rate. Even very low levels of irradiation have certain low, but real, probabilities of inducing mutations.

In *Drosophila* sperm, chronic irradiation (low levels of irradiation over long periods of time) is as effective in inducing mutations as acute irradiation (the same total dosage of irradiation administered at high intensity for short periods of time). However, in mice, chronic irradiation results in fewer mutations than the same dosage of acute irradiation. Moreover, when mice are treated with intermittent doses of irradiation, the mutation frequency is slightly lower than when they are treated with the same total amount of irradiation in a continuous dose. The differential response of fruit flies and mammals to chronic irradiation is thought to result from differences in the efficiency with which these species repair irradiation-induced damage in DNA. Repair mechanisms may exist in the spermatogonia and oocytes of mammals that do not function in *Drosophila* sperm. Nevertheless, we should emphasize that all of these irradiation treatments are mutagenic, albeit to different degrees, in both *Drosophila* and mammals.

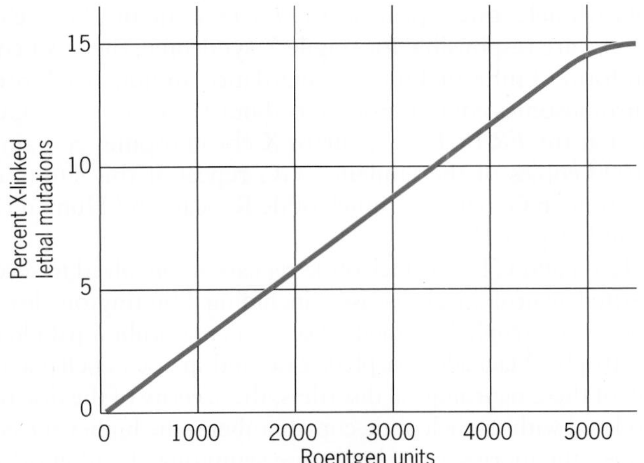

Figure 13.18 ▶ Relationship between irradiation dosage and mutation frequency in *Drosophila*.

Ionizing radiation also induces gross changes in chromosome structure, including deletions, duplications, inversions, and translocations (Chapter 6). These chromosome aberrations result from radiation-induced breaks in chromosomes. Because these aberrations require two chromosomal breaks, they exhibit two-hit kinetics rather than the single-hit kinetics observed for point mutations.

Ultraviolet (UV) radiation does not possess sufficient energy to induce ionizations. However, it is readily absorbed by many organic molecules such as the purines and pyrimidines in DNA, which then enter a more reactive or excited state. UV rays penetrate tissue only slightly. Thus, in multicellular organisms, only the epidermal layer of cells usually is exposed to the effects of UV. However, ultraviolet light is a potent mutagen for unicellular organisms. The maximum absorption of UV by DNA is at a wavelength of 254 nm. Maximum mutagenicity also occurs at 254 nm, suggesting that the UV-induced mutation process is mediated directly by the absorption of UV by purines and pyrimidines. *In vitro* studies show that the pyrimidines absorb strongly at 254 nm and, as a result, become very reactive. Two major products of UV absorption by pyrimidines (thymine and cytosine) are pyrimidine hydrates and pyrimidine dimers (**FIGURE 13.19**). Thymine dimers cause mutations in two ways. (1) Dimers perturb the structure of DNA double helices and interfere with accurate DNA replication. (2) Errors occur during the cellular processes that repair defects in DNA, such as UV-induced thymine dimers (see the section "DNA Repair Mechanisms" later in this chapter).

MUTATIONS INDUCED BY TRANSPOSABLE GENETIC ELEMENTS

Living organisms contain remarkable DNA elements that can move from one site in the genome to another site. These **transposons,** or transposable genetic elements, are the subject

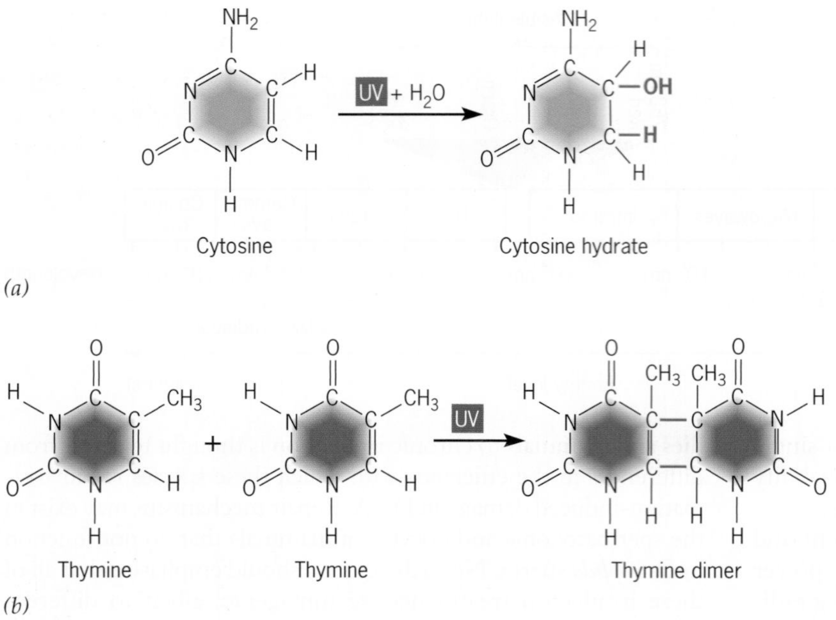

Figure 13.19 ▶ Pyrimidine photoproducts of UV irradiation. (*a*) Hydrolysis of cytosine to a hydrate form that may cause mispairing of bases during replication. (*b*) Cross-linking of adjacent thymine molecules to form thymine dimers, which block DNA replication.

of Chapter 18. The insertion of a transposon into a gene will often render the gene nonfunctional (**FIGURE 13.20**). If the gene encodes an important product, a mutant phenotype is likely to result. Geneticists now know that many of the classical mutants of maize, *Drosophila*, *E. coli*, and other organisms were caused by the insertion of transposable genetic elements into important genes (see **FIGURE 18.21**). Indeed, Mendel's *wrinkled* allele in the pea (Chapter 3) and the first mutation (w^1) causing white eyes in *Drosophila* (Chapter 5) both resulted from the insertion of transposable elements. See Chapter 18 for additional details about the mechanisms by which transposons move and, in the process, produce mutations.

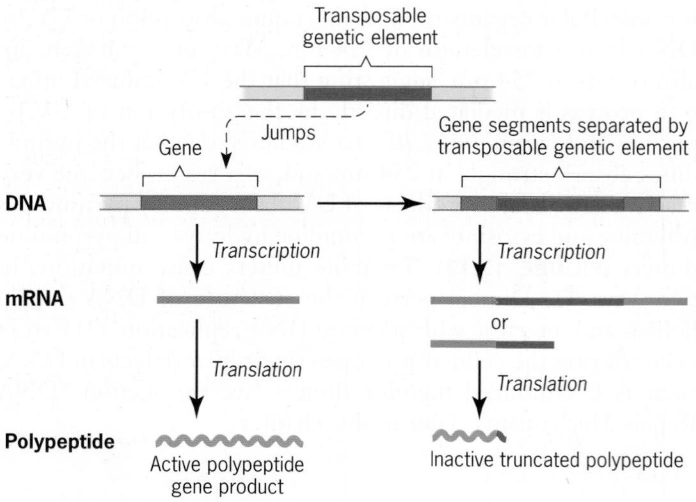

Figure 13.20 ▶ Mechanism of transposon-induced mutation. The insertion of a transposable genetic element (red) into a wild-type gene (left) will usually render the gene nonfunctional (right). A truncated gene product usually results from transcription- or translation-termination signals, or both, located within the transposon.

EXPANDING TRINUCLEOTIDE REPEATS AND INHERITED HUMAN DISEASES

All of the types of mutations discussed in the preceding sections of this chapter occur in humans. In addition, another type of mutation occurs that is associated with human diseases. Repeated sequences of one to six nucleotide pairs are known as **simple tandem repeats.** Such repeats are dispersed throughout the human genome. Repeats of three nucleotide pairs, **trinucleotide repeats,** can increase in copy number and cause inherited diseases in humans. Several trinucleotides have been shown to undergo such increases in copy number. Expanded CGG trinucleotide repeats at the *FRAXA* site on the X chromosome are responsible for fragile X syndrome, the most common form of inherited mental retardation in humans. Normal X chromosomes contain from 6 to about 50 copies of the CGG repeat at the *FRAXA* site. Mutant X chromosomes contain up to 1000 copies of the tandem CGG repeat at this site (see A Milestone in Genetics: Trinucleotide Repeats and Human Disease in Chapter 17).

CAG and CTG trinucleotide repeats are involved in several inherited neurological diseases, including Huntington disease, myotonic dystrophy, Kennedy disease, dentatorubral pallidoluysian atrophy, Machado-Joseph disease, and spinocerebellar ataxia. In all of these neurological disorders, the severity of the disease is correlated with trinucleotide copy number—the higher the copy number, the more severe the disease symptoms. In addition, the expanded trinucleotides associated with these diseases are unstable in somatic cells and between generations. This instability gives rise to the phenomenon of *anticipation*, which is the increasing severity of the disease or earlier age of onset that occurs in successive generations as the trinucleotide copy number increases. The mechanism of trinucleotide expansion is unknown.

KEY POINTS

- Mutations are induced by chemicals, ionizing irradiation, ultraviolet light, and endogenous transposable genetic elements.
- Point mutations are of three types: (1) transitions—purine for purine and pyrimidine for pyrimidine substitutions; (2) transversions—purine for pyrimidine and pyrimidine for purine substitutions; and (3) frameshift mutations—additions or deletions of one or two nucleotide pairs, which alter the reading frame of the gene distal to the site of the mutation.
- Several inherited human diseases are caused by expanded trinucleotide repeats.

Screening Chemicals for Mutagenicity: The Ames Test

The Ames test provides a simple and inexpensive method for detecting the mutagenicity of chemicals.

Mutagenic agents are also **carcinogens**; that is, they induce cancers. The one characteristic that the hundreds of types of cancer have in common is that the malignant cells continue to divide after cell division would have stopped in normal cells. Of course, cell division, like all other biological processes, is under genetic control. Specific genes encode products that regulate cell division in response to intracellular, intercellular, and environmental signals. When these genes mutate to nonfunctional states, uncontrolled cell division sometimes results. Clearly, we wish to avoid being exposed to mutagenic and carcinogenic agents. However, our technological society depends on the extensive use of chemicals in both industry and agriculture. Hundreds of new chemicals are produced each year, and the mutagenicity and carcinogenicity of these chemicals need to be evaluated before their use becomes widespread.

Traditionally, the carcinogenicity of chemicals has been tested on rodents, usually newborn mice. These studies involve feeding or injecting the substance being tested and subsequently examining the animals for tumors. Mutagenicity tests have been done in a similar fashion. However, because mutation is a low-frequency event and because maintaining large populations of mice is an expensive undertaking, the tests have been relatively insensitive; that is, low levels of mutagenicity could not be detected.

Bruce Ames and his associates developed sensitive techniques that allow the mutagenicity of large numbers of chemicals to be tested quickly at relatively low cost. Ames and coworkers constructed auxotrophic strains of the bacterium *Salmonella typhimurium* carrying various types of mutations—transitions, transversions, and frameshifts—in genes required for the biosynthesis of the amino acid histidine. They monitored the reversion of these auxotrophic mutants to prototrophy by placing a known number of mutant bacteria on medium lacking histidine and scoring the number of colonies produced by prototrophic revertants. Because some chemicals are mutagenic only to replicating DNA, they added a small amount of histidine—enough to allow a few cell divisions but not the formation of visible colonies—to the medium. They measured the mutagenicity of a chemical by comparing the frequency of reversion in its presence with the spontaneous reversion frequency (**FIGURE 13.21**). They assessed its ability to induce different types of mutations by using a set of tester strains that carry different types of mutations— one strain with a transition, one with a frameshift mutation, and so forth.

Over a period of several years during which they tested thousands of different chemicals, Ames and his colleagues observed a greater than 90 percent correlation between the mutagenicity and the carcinogenicity of the substances tested. Initially, they found several potent carcinogens to be nonmutagenic to the tester strains. Subsequently, they discovered that many of these carcinogens are metabolized to strongly mutagenic derivatives in eukaryotic cells. Thus, Ames and his associates added a rat liver extract to their assay systems in an attempt to detect the mutagenicity of metabolic derivatives of the substances being tested. Coupling of the rat liver activation system to the microbial mutagenicity tests expanded the utility of the system considerably. For example, nitrates (found in charred meats) are not themselves mutagenic or carcinogenic. However, in eukaryotic cells, nitrates are converted to nitrosamines, which are highly mutagenic and carcinogenic. Ames's mutagenicity tests demonstrated the presence of frameshift mutagens in several components of chemically fractionated cigarette smoke condensates. In some cases, activation by the liver extract preparation was required for mutagenicity; in other cases, activation was not required. The Ames test provides a rapid, inexpensive, and sensitive procedure for testing the mutagenicity of chemicals. Since mutagenic chemicals are also carcinogens, the Ames test can be used to identify chemicals that have a high likelihood of being carcinogenic.

KEY POINT

- Bruce Ames and coworkers developed an inexpensive and sensitive method for testing the mutagenicity of chemicals with histidine auxotrophic mutants of *Salmonella*.

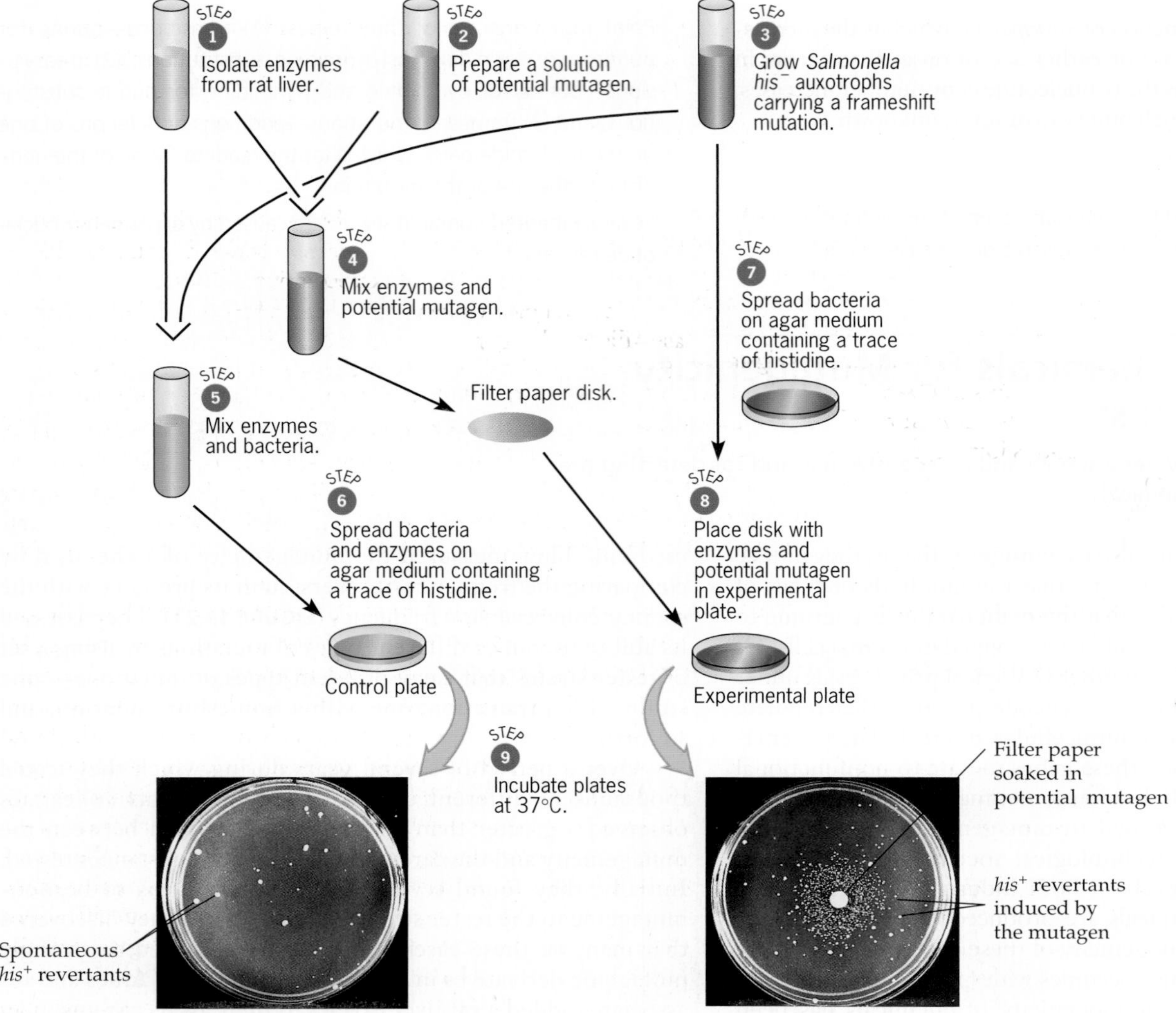

Figure 13.21 ▶ The Ames test for mutagenicity. The medium in each petri dish contains a trace of histidine and a known number of his^- cells of a specific *Salmonella typhimurium* "tester strain" harboring a frameshift mutation. The control plate shown on the left provides an estimate of the frequency of spontaneous reversion of this particular tester strain. The experimental plate on the right shows the frequency of reversion induced by the potential mutagen, in this case, the carcinogen 2-aminofluorene.

DNA Repair Mechanisms

Living organisms contain many enzymes that scan their DNA for damage and initiate repair processes when damage is detected.

The multiplicity of repair mechanisms that have evolved in organisms ranging from bacteria to humans emphatically documents the importance of keeping mutation at a tolerable level. For example, *E. coli* cells possess five well-characterized mechanisms for the repair of defects in DNA: (1) light-dependent repair or photoreactivation, (2) excision repair, (3) mismatch repair, (4) postreplication repair, and (5) the error-prone repair system (SOS response). Moreover, there are at least two different types of excision repair, and the excision repair pathways can be initiated by several different enzymes, each acting on a specific kind of damage in DNA. Mammals seem to possess all of the repair mechanisms found in *E. coli* except photoreactivation. Because most mammalian cells do not have access to light, photoreactivation would be of relatively little value to them.

The importance of DNA repair pathways to human health is clear. Inherited disorders such as xeroderma pigmentosum, which was discussed at the beginning of this chapter, vividly document the serious consequences of defects in DNA repair. We discuss some of these inherited disorders in a subsequent section of this chapter.

LIGHT-DEPENDENT REPAIR

Light-dependent repair or **photoreactivation** of DNA in bacteria is carried out by a light-activated enzyme called **DNA photolyase**. When DNA is exposed to ultraviolet light, thymine dimers are produced by covalent cross-linkages between adjacent thymine residues (see **FIGURE 13.19*b***). DNA photolyase recognizes and binds to thymine dimers in DNA, and uses light energy to cleave the covalent cross-links (**FIGURE 13.22**). Photolyase will bind to thymine dimers in DNA in the dark, but it cannot catalyze cleavage of the bonds joining the thymine moieties without energy derived from visible light, specifically light within the blue region of the spectrum. Photolyase also splits cytosine dimers and cytosine-thymine dimers. Thus, when

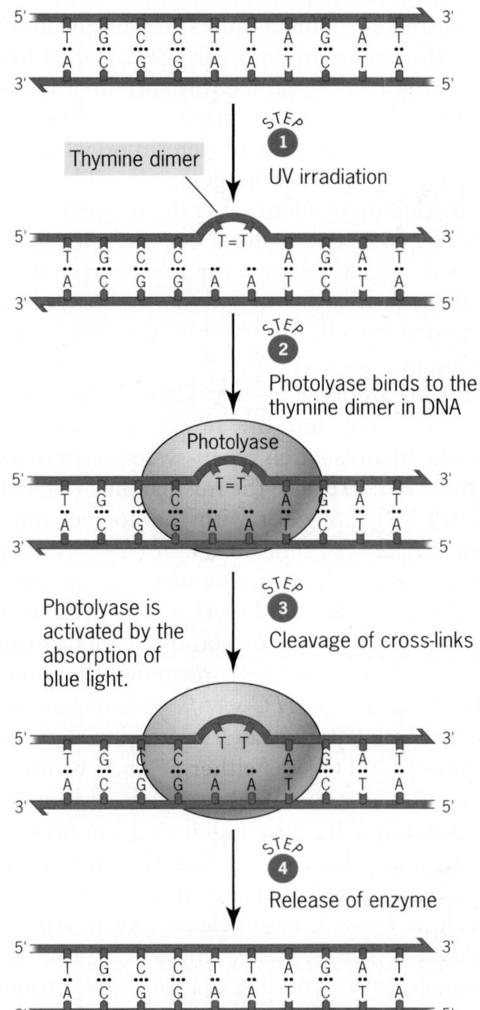

Figure 13.22 ▶ Cleavage of thymine dimer cross-links by light-activated photolyase. The arrows indicate the opposite polarity of the complementary strands of DNA.

ultraviolet light is used to induce mutations in bacteria, the irradiated cells are grown in the dark for a few generations to maximize the mutation frequency.

EXCISION REPAIR

Excision repair of damaged DNA involves at least three steps. In step 1, a DNA repair endonuclease or endonuclease-containing enzyme complex recognizes, binds to, and excises the damaged base or bases in DNA. In step 2, a DNA polymerase fills in the gap by using the undamaged complementary strand of DNA as template. In step 3, the enzyme DNA ligase seals the break left by DNA polymerase to complete the repair process. There are two major types of excision repair: **base excision repair** systems remove abnormal or chemically modified bases from DNA, whereas **nucleotide excision repair** pathways remove larger defects like thymine dimers. Both excision pathways are operative in the dark, and both occur by very similar mechanisms in *E. coli* and humans.

Base excision repair (**FIGURE 13.23**) can be initiated by any of a group of enzymes called DNA glycosylases that recognize abnormal bases in DNA. Each glycosylase recognizes a specific type of altered base, such as deaminated bases, oxidized bases, and so on (step 2). The glycosylases cleave the glycosidic bond between the abnormal base and 2-deoxyribose, creating apurinic or apyrimidinic sites (AP sites) with missing bases (step 3). AP sites are recognized by enzymes called AP endonucleases, which act together with phosphodiesterases to excise the sugar-phosphate groups at these sites (step 4). DNA polymerase then replaces the missing nucleotide according to the specifications of the complementary strand (step 5), and DNA ligase seals the nick (step 6).

Nucleotide excision repair removes larger lesions like thymine dimers and bases with bulky side-groups from DNA. In nucleotide excision repair, a unique excision nuclease activity produces cuts on either side of the damaged nucleotide(s) and excises an oligonucleotide containing the damaged base(s). This nuclease is called an **excinuclease** to distinguish it from the endonucleases and exonucleases that play other roles in DNA metabolism.

The *E. coli* nucleotide excision repair pathway is shown in **FIGURE 13.24**. In *E. coli*, excinuclease activity requires the products of three genes, *uvrA*, *uvrB*, and *uvrC* (designated *uvr* for *UV r*epair). A trimeric protein containing two UvrA polypeptides and one UvrB polypeptide recognizes the defect in DNA, binds to it, and uses energy from ATP to bend the DNA at the damaged site. The UvrA dimer is then released, and the UvrC protein binds to the UvrB/DNA complex. The UvrB protein cleaves the fifth phosphodiester bond from the damaged nucleotide(s) on the 3′ side, and the UvrC protein hydrolyzes the eighth phosphodiester linkage from the damage on the 5′ side. The *uvrD* gene product, DNA helicase II, releases the excised dodecamer. In the last two steps of the pathway, DNA polymerase I fills in the gap, and DNA ligase seals the remaining nick in the DNA molecule.

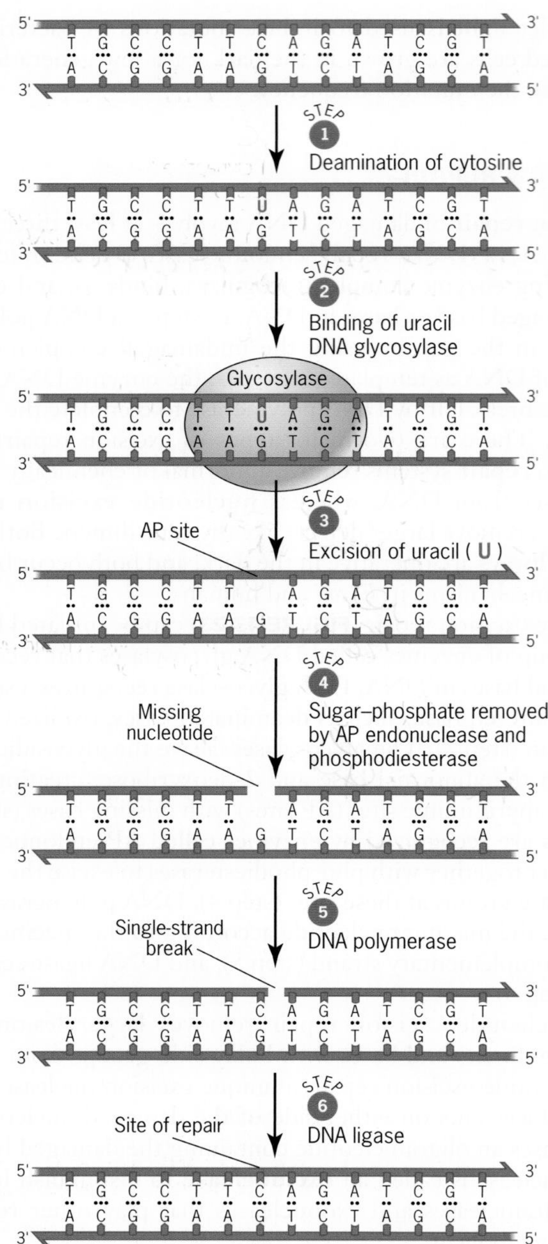

Figure 13.23 ▶ Repair of DNA by the base excision pathway. Base excision repair may be initiated by any one of several different DNA glycosylases. In the example shown, uracil DNA glycosylase starts the repair process.

Nucleotide excision repair in humans occurs through a pathway similar to the one in *E. coli*, but it involves about four times as many proteins. In humans, the excinuclease activity contains 15 polypeptides. Protein XPA (for *x*eroderma *p*igmentosum protein *A*) recognizes and binds to the damaged nucleotide(s) in DNA. It then recruits the other proteins required for excinuclease activity. In humans, the excised oligomer is 24 to 32 nucleotides long rather than the dodecamer removed in *E. coli*. The gap is filled in by either DNA polymerase δ or ε in humans, and DNA ligase completes the job.

OTHER DNA REPAIR MECHANISMS

During the last few years, research on DNA repair mechanisms has demonstrated the presence of an army of DNA repair enzymes that constantly scan DNA for damage ranging from the presence of thymine dimers induced by ultraviolet light to modifications too diverse and numerous to describe here. New results of this work have shown that several previously unknown DNA polymerases play critical roles in various DNA repair processes. Detailed discussions of these important DNA repair processes are beyond the scope of this text. Nevertheless, the importance of these repair mechanisms cannot be overstated. What is more important to the survival of a species than maintaining the integrity of its genetic blueprint?

In Chapter 10, we examined the mechanism by which the $3' \rightarrow 5'$ exonuclease activity built into DNA polymerases proofreads DNA strands during their synthesis, removing any mismatched nucleotides at the $3'$ termini of growing strands. Another postreplication DNA repair pathway, **mismatch repair,** provides a backup to this replicative proofreading by correcting mismatched nucleotides remaining in DNA after replication. Mismatches often involve the normal four bases in DNA. For example, a T may be mispaired with a G. Because both T and G are normal components of DNA, mismatch repair systems need some way to determine whether the T or the G is the correct base at the given site. The repair system makes this distinction by identifying the template strand, which contains the original nucleotide sequence, and the newly synthesized strand, which contains the misincorporated base (the error). In bacteria, this distinction can be made based on the pattern of methylation in newly replicated DNA. In *E. coli*, the A in GATC sequences is methylated subsequent to its synthesis. Thus, a time interval occurs during which the template strand is methylated, and the newly synthesized strand is unmethylated. The mismatch repair system uses this difference in methylation state to excise the mismatched nucleotide in the nascent strand and replace it with the correct nucleotide by using the methylated parental strand of DNA as template.

In *E. coli*, mismatch repair requires the products of four genes, *mutH*, *mutL*, *mutS*, and *mutU* (=*uvrD*). The MutS protein recognizes mismatches and binds to them to initiate the repair process. MutH and MutL proteins then join the complex. MutH contains a *GATC-specific endonuclease activity* that cleaves the unmethylated strand at hemimethylated (that is, half methylated) GATC sites either $5'$ or $3'$ to the mismatch. The incision sites may be 1000 exonucleotide pairs or more from the mismatch. The subsequent excision process requires MutS, MutL, DNA helicase II (MutU), and an appropriate exonuclease. If the incision occurs at a GATC sequence $5'$ to the mismatch, a $5' \rightarrow 3'$ exonuclease like *E. coli* exonuclease VII is required. If the incision occurs $3'$ to the mismatch, a $3' \rightarrow 5'$exonuclease activity like that of *E. coli* exonuclease I is needed. After the excision process has removed the mismatched nucleotide from the unmethylated strand, DNA polymerase III fills in the large—up to 1000 bp—gap, and DNA ligase seals the nick.

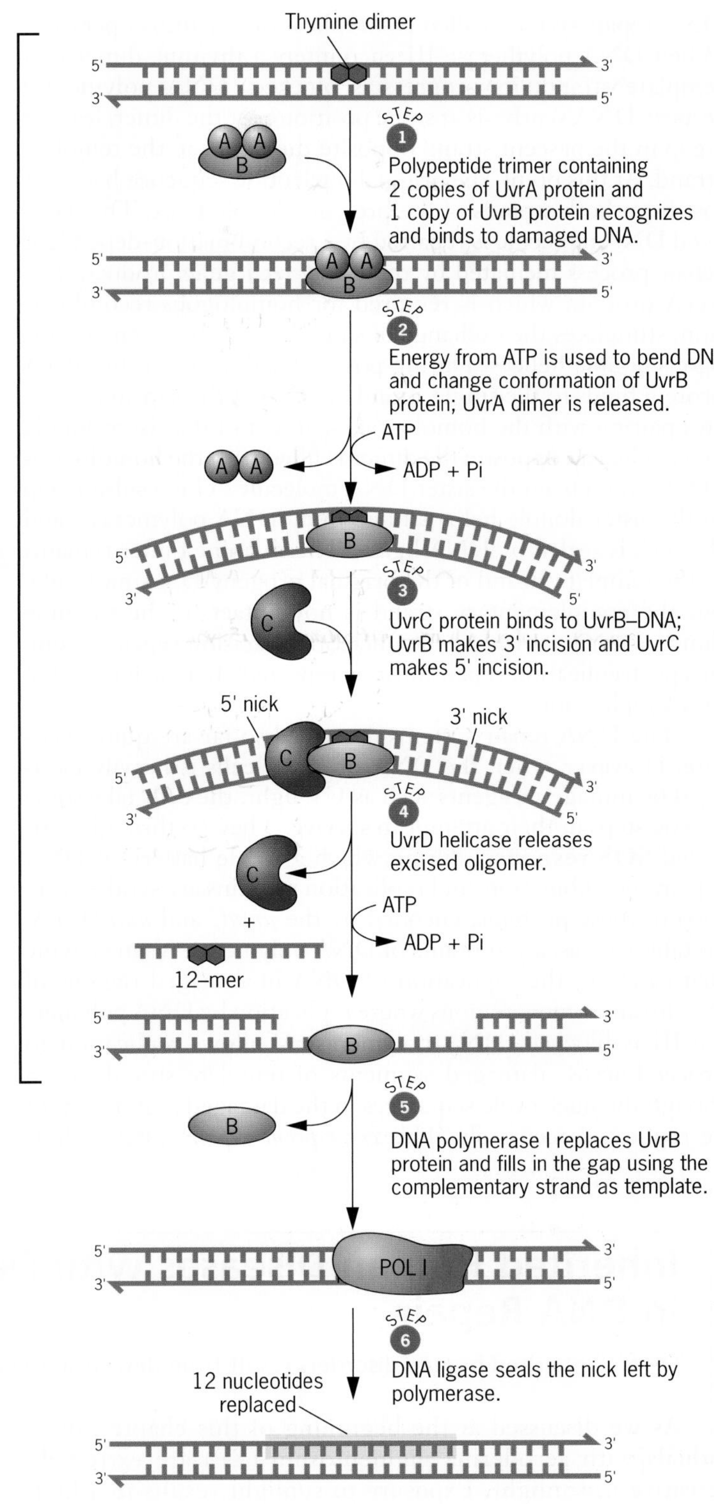

Figure 13.24 ▶ Repair of DNA by the nucleotide excision pathway in *E. coli*. The excinuclease (excision nuclease) activity requires the products of three genes—*uvrA, uvrB,* and *uvrC*. Nucleotide excision occurs by a similar pathway in humans, except that many more proteins are involved and a 24- to-32-nucleotide-long oligomer is excised.

Homologues of the *E. coli* MutS and MutL proteins have been identified in fungi, plants, and mammals—an indication that similar mismatch repair pathways occur in eukaryotes. In fact, mismatch excision has been demonstrated *in vitro* with nuclear extracts prepared from human cells. Thus, mismatch repair is probably a universal or nearly universal mechanism for safeguarding the integrity of genetic information stored in double-stranded DNA.

In *E. coli*, light-dependent repair, excision repair, and mismatch repair can be eliminated by mutations in the *phr* (*ph*otor-eactivation), *uvr*, and *mut* genes, respectively. In mutants deficient in more than one of these repair mechanisms, still another

DNA repair system, called postreplication repair, is operative. When DNA polymerase III encounters a thymine dimer in a template strand, its progress is blocked. DNA polymerase restarts DNA synthesis at some position past the dimer, leaving a gap in the nascent strand opposite the dimer in the template strand. At this point, the original nucleotide sequence has been lost from both strands of the progeny double helix. The damaged DNA molecule is repaired by a recombination-dependent repair process mediated by the *E. coli recA* gene product. The RecA protein, which is required for homologous recombination, stimulates the exchange of single strands between homologous double helices. During postreplication repair, the RecA protein binds to the single strand of DNA at the gap and mediates pairing with the homologous segment of the sister double helix. The gap opposite the dimer is filled with the homologous DNA strand from the sister DNA molecule. The resulting gap in the sister double helix is filled in by DNA polymerase, and the nick is sealed by DNA ligase. The thymine dimer remains in the template strand of the original progeny DNA molecule, but the complementary strand is now intact. If the thymine dimer is not removed by the nucleotide excision repair system, this postreplication repair must be repeated after each round of DNA replication.

The DNA repair systems described so far are quite accurate. However, when the DNA of *E. coli* cells is heavily damaged by mutagenic agents such as UV light, the cells take some drastic steps in their attempt to survive. They go through a so-called **SOS response**, during which a whole battery of DNA repair, recombination, and replication proteins are synthesized. Two of these proteins, encoded by the *umuC* and *umuD* (*UV mu*table) genes, are subunits of DNA polymerase V, an enzyme that catalyzes the replication of DNA in damaged regions of the chromosome—regions where replication by DNA polymerase III is blocked. DNA polymerase V allows replication to proceed across damaged segments of template strands, even though the nucleotide sequences in the damaged region cannot be replicated accurately. This *error-prone repair* system eliminates gaps in the newly synthesized strands opposite damaged nucleotides in the template strands but, in so doing, increases the frequency of replication errors.

The mechanism by which the SOS system is induced by DNA damage has been worked out in considerable detail. Two key regulatory proteins—LexA and RecA—control the SOS response. Both are synthesized at low background levels in the cell in the absence of damaged DNA. Under this condition, LexA binds to the DNA regions that regulate the transcription of the genes that are induced during the SOS response and keep their expression levels low. When cells are exposed to ultraviolet light or other agents that cause DNA damage, the RecA protein binds to single-stranded regions of DNA caused by the inability of DNA polymerase III to replicate the damaged regions. The interaction of RecA with DNA activates RecA, which then stimulates LexA to inactivate itself by self-cleavage. With LexA inactive, the level of expression of the SOS genes—including *recA*, *lexA*, *umuC*, *umuD*, and others—increases and the error-prone repair system is activated.

The SOS response appears to be a somewhat desperate and risky attempt to escape the lethal effects of heavily damaged DNA. When the error-prone repair system is operative, mutation rates increase sharply.

Recent research on DNA repair mechanisms indicates that many new repair processes remain to be elucidated. During the last few years, several new DNA polymerases that have unique roles in DNA repair have been characterized. The results of these studies suggest that we have much to learn about the mechanisms that safeguard the integrity of our genetic information.

KEY POINTS

- Multiple DNA repair systems have evolved to safeguard the integrity of genetic information in living organisms.
- Each repair pathway corrects a specific type of damage in DNA.

Inherited Human Diseases with Defects in DNA Repair

Several inherited human disorders result from defects in DNA repair pathways.

As we discussed at the beginning of this chapter, individuals with xeroderma pigmentosum (XP) are extremely sensitive to sunlight. Exposure to sunlight results in a high frequency of skin cancer in XP patients (**FIGURE 13.25**). The cells of individuals with XP are deficient in the repair of UV-induced damage to DNA, such as thymine dimers. The XP syndrome can result from defects in any of at least eight different genes. The products of seven of these genes, *XPA*, *XPB*, *XPC*, *XPD*, *XPE*, *XPF*, and *XPG*, are required for nucleotide excision repair (**TABLE 13.1**). They have been purified and shown to be essential for excinuclease activity. Since excinuclease activity in humans requires 15 polypeptides, the list of *XP* genes will probably expand in the future. Two other human disorders, Cockayne syndrome and trichothiodystrophy, also result from defects in nucleotide excision repair. Individuals with Cockayne syndrome exhibit retarded growth and mental skills, but not increased rates of skin cancer. Patients with trichothiodystrophy have short stature, brittle hair, and scaly skin; they also have underdeveloped mental abilities. Individuals with either Cockayne

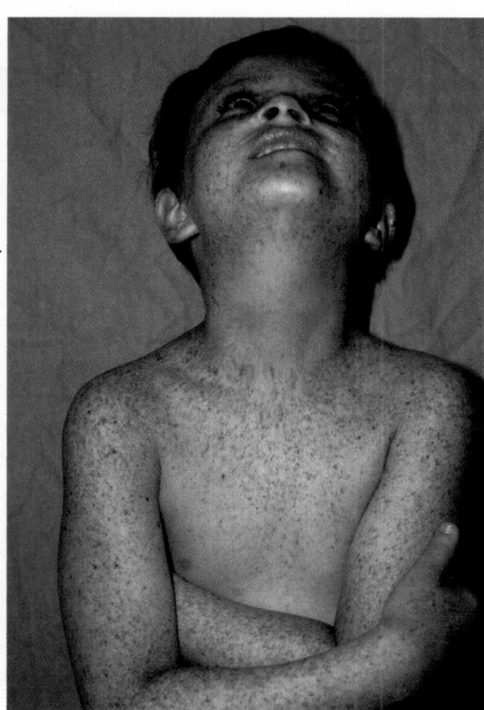

Figure 13.25 ► Phenotypic effects of the inherited disease xeroderma pigmentosum. Individuals with this autosomal recessive disease develop skin cancer after exposure to sunlight.

syndrome or trichothiodystrophy are defective in a type of excision repair that is coupled to transcription. However, details of this transcription-coupled repair process are still being worked out.

In addition to the damage to skin cells, some individuals with XP develop neurological abnormalities, which appear to result from the premature death of nerve cells. This effect on the very long-lived nerve cells may have interesting implications with respect to the causes of aging. One theory is that aging results from the accumulation of somatic mutations. If so, a defective repair system would be expected to speed up the aging process, and this appears to be the case with the nerve cells of XP patients. However, at present, there is little evidence linking somatic mutation to senescence.

Ataxia-telangiectasia, Fanconi anemia, Bloom syndrome, Werner syndrome, Rothmund-Thomson syndrome, and Nijmegan breakage syndrome are six other inherited diseases in humans associated with known defects in DNA metabolism. All six disorders exhibit autosomal recessive patterns of inheritance, and all result in a high risk of malignancy, especially leukemia in the case of ataxia-telangiectasia and Fanconi anemia. Cells of patients with ataxia-telangiectasia exhibit an abnormal sensitivity to ionizing radiation, suggesting a defect in the repair of radiation-induced DNA damage. Cells of individuals with Fanconi anemia are impaired in the removal of DNA interstrand cross-links, such as those formed by the antibiotic mitomycin C. Individuals with Bloom syndrome and Nijmegan breakage syndrome exhibit a high frequency of chromosome breaks that results in chromosome aberrations (Chapter 6) and sister chromatid exchanges. Ataxia-telangiectasia is caused by defects in a kinase involved in the control of the cell cycle, and Bloom syndrome, Werner syndrome, and Rothmund-Thomson syndrome result from alterations in specific DNA helicases (members of the RecQ family of helicases). **TABLE 13.1** lists some of the better known human diseases resulting from inherited defects in DNA repair pathways. The demonstration that the onset of hereditary nonpolypoid colorectal cancer is associated with a defect in the repair of mismatched base pairs in DNA hints that similar defects in other DNA repair pathways may be involved in the development of specific types of human cancer (Chapter 22).

KEY POINTS

- The importance of DNA repair pathways is documented convincingly by inherited human disorders that result from defects in DNA repair.
- Certain types of cancer are also associated with defects in DNA repair pathways.

DNA Recombination Mechanisms

Recombination between homologous DNA molecules involves the activity of numerous enzymes that cleave, unwind, stimulate single-strand invasions of double helices, repair, and join strands of DNA.

In Chapter 7, we discussed the main features of recombination between homologous chromosomes, but we did not consider the molecular details of the process. Because many of the gene products involved in the repair of damaged DNA also are required for recombination between homologous chromosomes, or crossing over, we will now examine some of the molecular aspects of this important process. Moreover, recombination usually, perhaps always, involves some DNA repair synthesis. Thus, much of the information discussed in the preceding sections is relevant to the process of recombination.

In eukaryotes, crossing over is associated with the formation of the synaptonemal complex, which forms during prophase of the first meiotic division. This structure is composed primarily of proteins and RNA. For unknown reasons, crossing over occurs only rarely in male *Drosophila*. (Crossing over does occur in both sexes of most species; the near absence of crossing over

▶ **TABLE 13.1**

Inherited Human Diseaes Caused by Defects in a DNA Repair

Inherited Disorder	Gene	Chromosome	Function of Product	Major Symptoms
1. Xeroderma pigmentosum	*XPA*	9	DNA-damage-recognition protein	UV sensitivity, early onset skin cancers, neurological disorders
	XPB	2	$3' \rightarrow 5'$ helicase	
	XPC	3	DNA-damage-recognition protein	
	XPD	19	$5' \rightarrow 3'$ helicase	
	XPE	11	DNA-damage-recognition protein	
	XPF	16	Nuclease, 3′ incision	
	XPG	13	Nuclease, 5′ incision	
	XPV	6	Translesion DNA polymerase η	
2. Cockayne syndrome	*CSA*	5	DNA excision repair protein	UV sensitivity, neurological and developmental disorders, premature aging
	CSB	10	DNA excision repair protein	
3. Trichothiodystrophy	*TTDA*	6	Basal transcription factor IIH	UV sensitivity, neurological disorders, mental retardation
4. Ataxia-telangiectasia	*ATM*	11	Serine/threonine kinase	Radiation sensitivity, chromosome instability, early onset progressive neurodegeneration, cancer prone
5. Fanconi anemia	*FA* (8 genes, *A-H,* on 5 different chromosomes)			Sensitivity to DNA-cross-linking agents, chromosome instability, cancer prone
6. Bloom syndrome	*BLM*	15	BLM RecQ helicase	Chromosome instability, mental retardation, cancer prone
7. Werner syndrome	*WRN*	8	WRN RecQ helicase	Chromosome instability, progressive neurodegeneration, cancer prone
8. Rothmund-Thomson syndrome	*RECQL4*	8	RecQ helicase L4	Chromosome instability, mental retardation, cancer prone
9. Nijmegan breakage syndrome	*NBSI*	8	DNA-double-strand-break-recognition protein	Chromosome instability, microcephaly (small cranium), cancer prone

in the heterogametic sex is unique to *Drosophila* and a few other species.) Of interest here is the fact that no synaptonemal complex is present during the first meiotic division of spermatogenesis in *Drosophila*. In addition, mutations in the *c3G* gene of *Drosophila* eliminate both crossing over and the formation of the synaptonemal complex in females. Thus, crossing over and the formation of the synaptonemal complex appear to be linked. A small amount of DNA synthesis occurs during the formation of the synaptonemal complex, and this DNA synthesis is probably involved in synapsis and crossing over.

RECOMBINATION: CLEAVAGE AND REJOINING OF DNA MOLECULES

In Chapter 7, we discussed the experiment of Creighton and McClintock showing that crossing over occurs by breakage of parental chromosomes and rejoining of the parts in new combinations. Evidence demonstrating that recombination occurs by breakage and rejoining has also been obtained by autoradiography and other techniques. Indeed, the main features of the process of recombination are now well established, even though specific details remain to be elucidated.

Much of what we know about the molecular details of crossing over is based on the study of *recombination-deficient mutants* of *E. coli* and *S. cerevisiae*. Biochemical studies of these mutants have shown that they are deficient in various enzymes and other proteins required for recombination. Together, the results of genetic and biochemical studies have provided a fairly complete picture of recombination at the molecular level.

Many of the currently popular models of crossing over were derived from a model proposed by Robin Holliday in 1964. Holliday's model was one of the first that explained most of the genetic data available at the time by a mechanism involving the breakage, reunion, and repair of DNA molecules. An updated version of the Holliday model is shown in **FIGURE 13.26**. This mechanism, like many others that have been invoked, begins when an endonuclease cleaves single strands of each of the two parental DNA molecules (breakage). Segments of the single strands on one side of each cut are then displaced from their complementary strands with the aid of DNA helicases and single-strand binding proteins. The helicases unwind the two strands of DNA in the region adjacent to single-strand incisions. In *E. coli*, the *RecBCD complex* contains both an endonuclease activity that makes single-strand breaks in DNA and a DNA helicase activity that unwinds the complementary strands of DNA in the region adjacent to each nick.

The displaced single strands then exchange pairing partners, base-pairing with the intact complementary strands of the

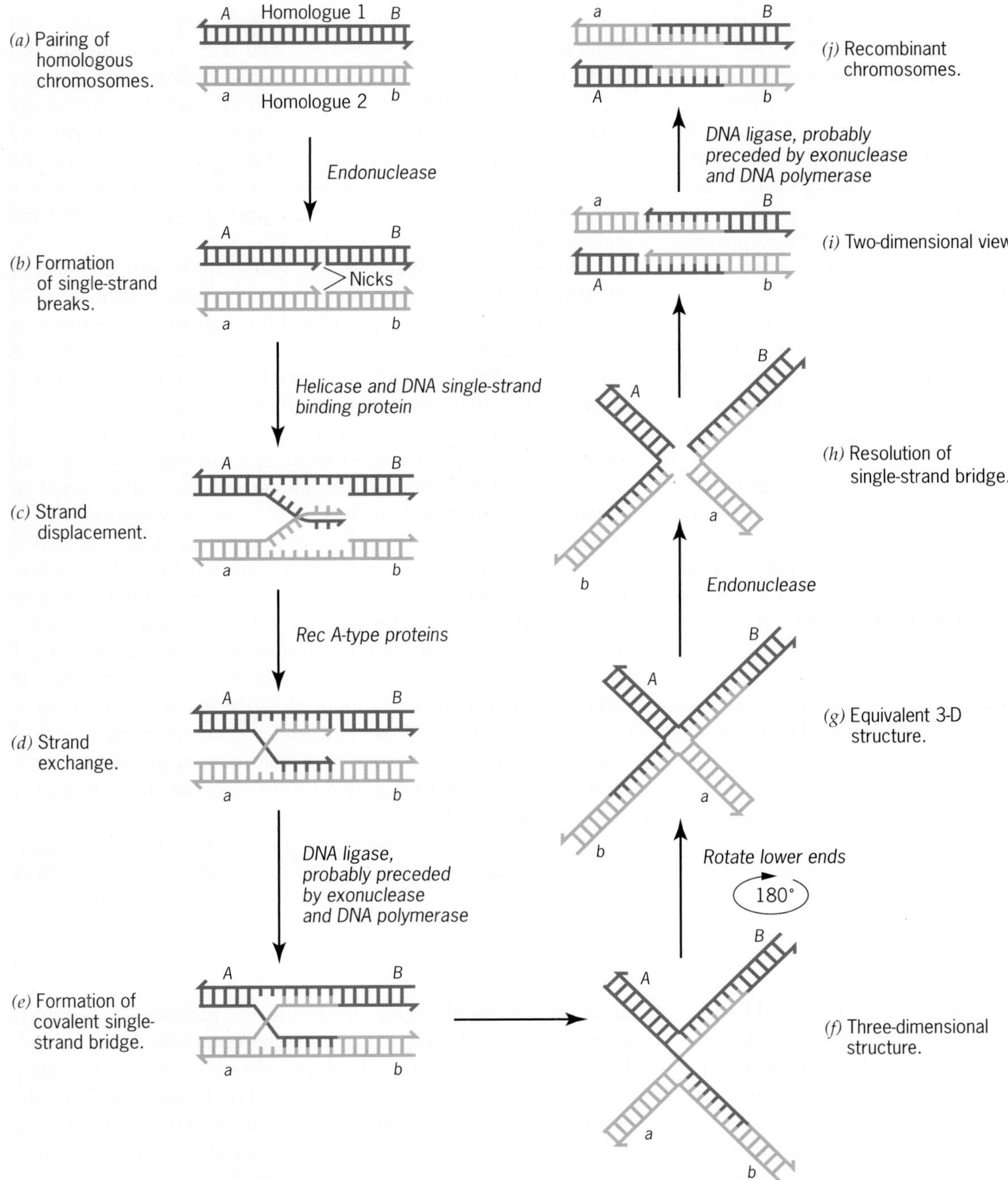

Figure 13.26 ▶ A mechanism for recombination between homologous DNA molecules. The pathway shown is based on the model originally proposed by Robin Holliday in 1964.

▶ A MILESTONE IN GENETICS: **Muller Demonstrates That X Rays Are Mutagenic**

In the late 1880's, Hugo de Vries observed the sudden appearance of new phenotypes in the evening primroses growing in his experimental garden. He called the novel plants "sports" and gave them Latin names because he thought they were new species. In 1901, de Vries introduced the term *mutation* to describe the origin of these "sports."[1] As it turned out, de Vries's "sports" were the result of gross changes in chromosome structure—translocations (see Chapter 6).

New phenotypes resulting from mutations were described and shown to be heritable by many of the early plant breeders. However, the work on *Drosophila* in Thomas Hunt Morgan's laboratory provided the most extensive evidence for the effects of new mutant alleles. The first quantitative measurements of mutation frequency were made in 1919 by Hermann J. Muller and Edgar Altenburg.[2] These researchers studied the frequency of X-linked lethals in *Drosophila.* The preeminent breakthrough in mutation research occurred in 1927, when Muller demonstrated that the treatment of *Drosophila* sperm with X rays sharply increased the frequency of X-linked recessive lethals.[3] (X rays are a form of electromagnetic radiation with shorter wavelengths and higher energy than visible light; see **FIGURE 13.17**.) Muller's study was the first demonstration that mutation could be induced by an external factor. In 1946, he received the Nobel Prize in Physiology or Medicine for this important discovery.

Muller's unambiguous demonstration of the mutagenicity of X rays became possible because he developed a simple and accurate technique that could be used to identify lethal mutations on the X chromosome of *Drosophila.* This technique, called the ***ClB method***, is performed with females heterozygous for a normal X chromosome and an altered X chromosome—the ***ClB chromosome***—that Muller constructed specifically for use in his experiment.

The *ClB* chromosome has three essential components. (1) *C*, for crossover suppressor, refers to a long inversion that suppresses recombination between the *ClB* chromosome and the structurally normal X chromosome in heterozygous females. The inversion does not prevent crossing over between the two chromosomes, but causes progeny carrying recombinant X chromosomes produced by crossing over between the two chromosomes to abort because of duplications and deficiencies (see Chapter 6). (2) *l* refers to a recessive lethal mutation on the *ClB* chromosome. Homozygous females and hemizygous males carrying this X-linked lethal mutation are not viable. (3) *B* refers to a mutation that causes the bar-eye phenotype, a condition in which the large compound eyes of wild-type flies are reduced in size to narrow, bar-shaped eyes. Because *B* is partially dominant, females heterozygous for the *ClB* chromosome can be readily identified. Both the recessive lethal (*l*) and the bar-eye mutation (*B*) are located within the inverted segment of the *ClB* chromosome.

Muller irradiated male flies and mated them with *ClB* females (**FIGURE 1**). All the bar-eyed daughters of this mating carried the *ClB* chromosome of the female parent and the irradiated X chromosome of the male parent. Because the entire population of reproductive cells of the males was irradiated, each bar-eyed daughter carried a potentially mutated X chromosome. These bar-eyed daughters were then mated individually (in separate cultures) with wild-type males. If the irradiated X chromosome carried by a bar-eyed daughter had acquired an X-linked lethal, all the progeny of the mating would be female. Males hemizygous for the *ClB* chromosome would die because of the recessive lethal (*l*) this chromosome carries; in addition, males hemizygous for the irradiated X chromosome would die if a recessive lethal had been induced on it. Matings of bar-eyed daughters carrying an irradiated X chromosome in which no lethal mutation had been induced would produce female and male progeny in a ratio of 2:1 (only the males with the *ClB* chromosome will die). With the *ClB* technique, detecting newly induced recessive, X-linked lethals is unambiguous and error free; it involves nothing more complex than scoring for the presence or absence of male progeny. By this procedure, Muller was able to demonstrate a 150-fold increase in the frequency of X-linked lethals after treating male flies with X rays.

After Muller's pioneering experiments with *Drosophila*, other researchers soon demonstrated that X rays are mutagenic in other

[1]de Vries, H. 1901. *Die Mutationstheorie,* Vol. 1, Viet, Leipzig.

[2]Muller, H. J., and E. Altenburg. 1919. The rate of change of hereditary factors in *Drosophila. Proc. Soc. Exp. Biol. Med.* 17: 10–14.

[3]Muller, H. J. 1927. Artificial transmutation of the gene. *Science* 66: 84–87.

homologous chromosomes. This process is stimulated by proteins like the *E. coli* RecA protein. RecA-type proteins have been characterized in many species, both prokaryotic and eukaryotic. RecA protein and its homologues stimulate *single-strand assimilation*, a process by which a single strand of DNA displaces its homologue in a DNA double helix. RecA-type proteins promote reciprocal exchanges of DNA single strands between two DNA double helices in two steps. In the first step, a single strand of one double helix is assimilated by a second, homologous double helix, displacing the identical or homologous strand and base-pairing with the complementary strand. In the second step, the displaced single strand is similarly assimilated by the first double helix. The RecA protein mediates these exchanges by binding to the unpaired strand of DNA, aiding in the search for a homologous DNA sequence, and, once a homologous double helix is found, promoting the replacement of one strand with the unpaired strand. If complementary sequences already exist as single strands, the presence

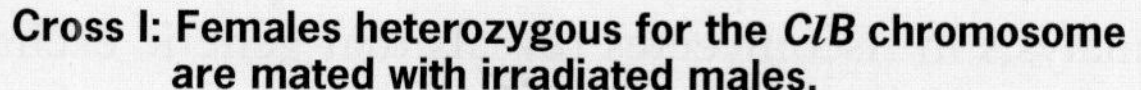

Cross I: Females heterozygous for the ClB chromosome are mated with irradiated males.

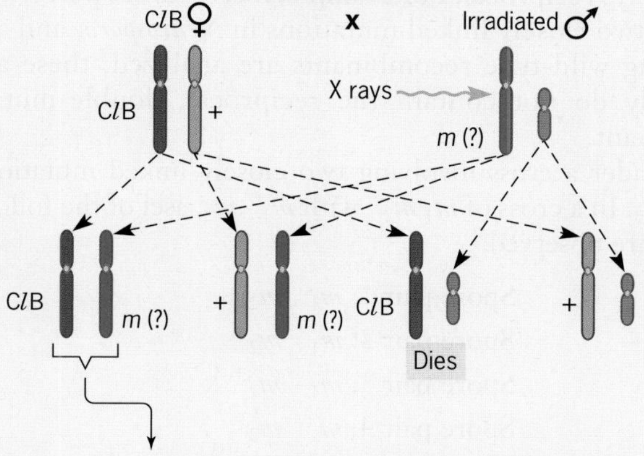

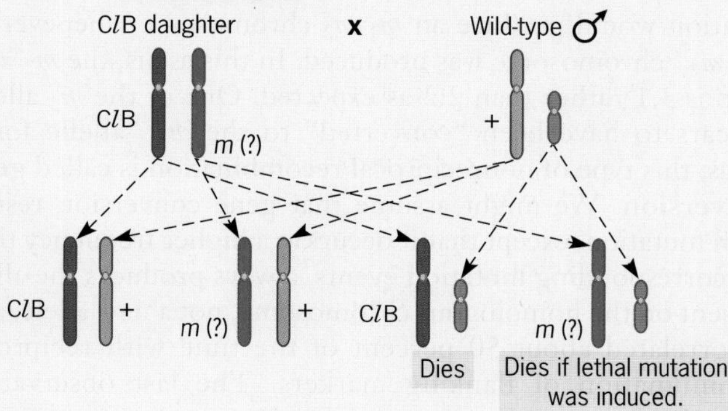

Figure 1 ▶ The *ClB* technique used by Muller to detect X-linked recessive lethal mutations in *Drosophila*. The mating shown in cross II will produce only female progeny if an X-linked recessive lethal is present on the irradiated X chromosome. One-third of the progeny produced from cross II will be males if there is no recessive lethal on the irradiated X chromosome. Thus, scoring for lethal mutations simply involves screening the progeny of cross II for the presence or absence of males.

organisms, including plants, other animals, and microbes. Moreover, other types of high-energy electromagnetic radiation and many chemicals were soon shown to be potent mutagens. The ability to induce mutations in genes contributed immensely to progress in genetics. It allowed researchers to induce mutations in genes of interest and to "knock out" their functions. The mutant organisms could then be studied to gain information about the function of the wild-type gene product. This approach—mutational dissection of biological processes—has proven to be a powerful tool in the analysis of many biological processes.

QUESTIONS FOR DISCUSSION

1. Muller demonstrated that X rays induce mutations in *Drosophila*. Subsequent studies in many laboratories have shown that X rays are mutagenic in all living organisms, including humans. However, X rays also provide an important diagnostic tool in medicine and dentistry. When you have diagnostic X rays performed, the medical professional provides a lead apron to prevent the irradiation of your reproductive cells. This safeguards them and minimizes the chance of new mutations that could be transmitted to your offspring. Still, X rays induce mutations in somatic cells, and some of these mutations increase the chance of cancer in the future. What regulations apply to the use of X rays for diagnostic purposes? What factors should be considered in establishing these regulations?
2. In addition to X rays, other forms of high-energy radiation such as gamma rays, cosmic rays, and the particles released by the decay of radioactive chemicals used in nuclear bombs and nuclear reactors are highly mutagenic. In 1986, the potential dangers of nuclear reactors became reality when the nuclear power plant near Chernobyl in Ukraine exploded and contaminated the surrounding countryside. Although nuclear reactors are an economical source of electricity, the potential for accidents like the one at Chernobyl remains. What are the pros and cons of using nuclear reactors to provide economical and "ecologically friendly" energy?

of RecA protein increases the rate of renaturation by over 50-fold.

The cleaved strands are then covalently joined in new combinations (reunion) by DNA ligase. If the original breaks in the two strands do not occur at exactly the same site in the two homologues, some tailoring will be required before DNA ligase can catalyze the reunion step. This tailoring involves the excision of nucleotides by an exonuclease and repair synthesis by a DNA polymerase. The sequence of events described so far will produce *X*-shaped recombination intermediates called *chi forms*, which have been observed by electron microscopy in several species (**FIGURE 13.27**). The chi forms are resolved by enzyme-catalyzed breakage and rejoining of the complementary DNA strands to produce two recombinant DNA molecules. In *E. coli*, chi structures can be resolved by the product of either the *recG* gene or the *ruvC* gene (*r*epair of *UV*-induced damage). Each gene encodes an endonuclease that catalyzes the cleavage of single strands at chi junctions (see **FIGURE 13.26**).

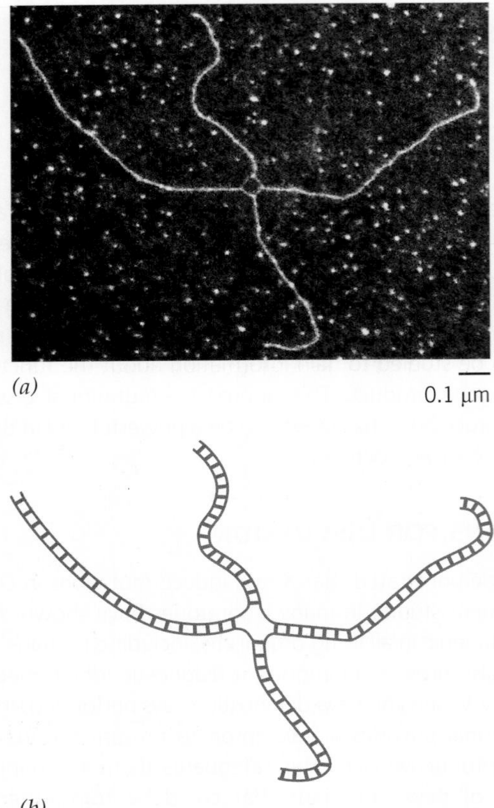

Figure 13.27 ▶ Electron micrograph (*a*) and diagram (*b*) of an X-shaped recombination intermediate or chi structure.

A substantial body of evidence indicates that homologous recombination occurs by more than one mechanism—probably by several different mechanisms. In *S. cerevisiae*, the ends of DNA molecules produced by double-strand breaks are highly recombinogenic. This fact and other evidence suggest that recombination in yeast often involves a double-strand break in one of the parental double helices. Thus, in 1983, Jack Szostak, Franklin Stahl, and colleagues proposed a *double-strand break model* of crossing over. According to their model, recombination involves a double-strand break in one of the parental double helices, not just single-strand breaks as in the Holliday model. The initial breaks are then enlarged to gaps in both strands. The two single-stranded termini produced at the double-stranded gap of the broken double helix invade the intact double helix and displace segments of the homologous strand in this region. The gaps are then filled in by repair synthesis. This process yields two homologous chromosomes joined by two single-strand bridges. The bridges are resolved by endonucleolytic cleavage, just as in the Holliday model. Both the double-strand-break model and the Holliday model nicely explain the production of chromosomes that are recombinant for genetic markers flanking the region in which the crossover occurs.

GENE CONVERSION: DNA REPAIR SYNTHESIS ASSOCIATED WITH RECOMBINATION

Up to this point, we have discussed only recombination events that can be explained by breakage of homologous chromatids and the reciprocal exchange of parts. However, tetrad analysis in Ascomycetes reveals that genetic exchange is not always reciprocal. For example, if crosses are performed between two closely linked mutations in *Neurospora*, and asci containing wild-type recombinants are analyzed, these asci frequently do not contain the reciprocal, double-mutant recombinant.

Consider a cross involving two closely linked mutations, m_1 and m_2. In a cross of $m_1\ m_2^+$ with $m_1^+\ m_2$, asci of the following type are observed:

Spore pair 1: $m_1^+\ m_2$
Spore pair 2: $m_1^+\ m_2^+$
Spore pair 3: $m_1\ \ m_2^+$
Spore pair 4: $m_1\ \ m_2^+$

Wild-type $m_1^+ m_2^+$ spores are present, but the $m_1\ m_2$ double-mutant spores are not present in the ascus. Reciprocal recombination would produce an $m_1\ m_2$ chromosome whenever an $m_1^+ m_2^+$ chromosome was produced. In this ascus, the $m_2^+ : m_2$ ratio is 3:1 rather than 2:2 as expected. One of the m_2 alleles appears to have been "converted" to the m_2^+ allelic form. Thus, this type of nonreciprocal recombination is called **gene conversion**. We might assume that gene conversion results from mutation, except that it occurs at a higher frequency than the corresponding mutation events, always produces the allele present on the homologous chromosome, not a new allele, and is correlated about 50 percent of the time with reciprocal recombination of flanking markers. The last observation strongly suggests that gene conversion results from events that occur during crossing over. Indeed, gene conversion is now believed to result from DNA repair synthesis associated with the breakage, excision, and reunion events of crossing over.

With closely linked markers, gene conversion occurs more frequently than reciprocal recombination. In one study of the *his1* gene of yeast, 980 of 1081 asci containing his^+ recombinants exhibited gene conversion, whereas only 101 showed classical reciprocal recombination.

The most striking feature of gene conversion is that the input 1:1 allele ratio is not maintained. This can be explained easily if short segments of parental DNA are degraded and then resynthesized with template strands provided by DNA carrying the other allele. Given the mechanisms of excision repair discussed earlier in this chapter, the Holliday model of crossing over explains gene conversion for genetic markers located in the immediate vicinity of the crossover. In **FIGURE 13.26*d–i***, there is a segment of DNA between the *A* and *B* loci where complementary strands of DNA from the two homologous chromosomes are base-paired. If a third pair of alleles located within this segment were segregating in the cross, mismatches in the two double helices would be present. DNA molecules containing such mismatches, or different alleles in the two complementary strands of a double helix, are called **heteroduplexes**. Such heteroduplex molecules occur as intermediates in the process of recombination.

If **FIGURE 13.26e** were modified to include a third pair of alleles, and the other two chromatids were added, the tetrad would have the following composition:

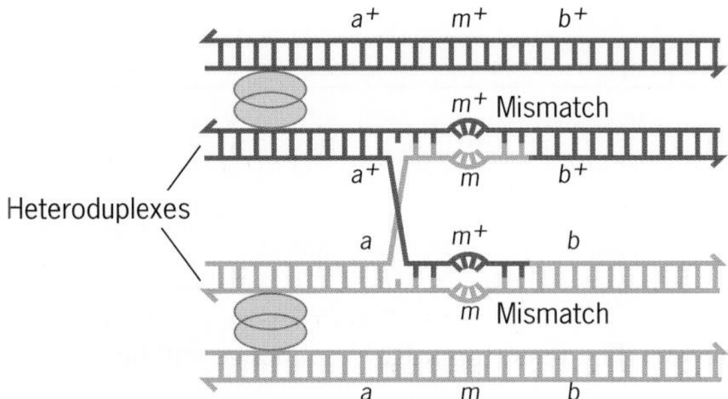

If the mismatches are resolved by excision repair (see **FIGURE 13.24**), in which the m strands are excised and resynthesized with the complementary m^+ strands as templates, the following tetrad will result:

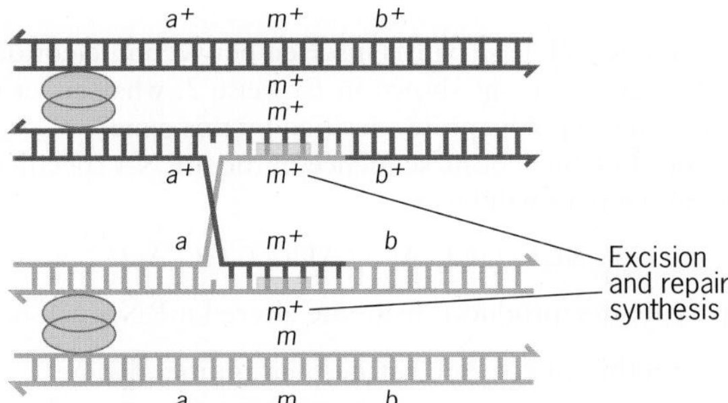

As a result of semiconservative DNA replication during the subsequent mitotic division, this tetrad will yield an ascus containing six m^+ ascospores and two m ascospores, the 3:1 gene conversion ratio.

Suppose that only one of the two mismatches in the tetrad just described is repaired prior to the mitotic division. In this case, the semiconservative replication of the remaining heteroduplex will yield one m^+ homoduplex and one m homoduplex, and the resulting ascus will contain a $5m^+{:}3m$ ratio of ascospores. Such 5:3 gene conversion ratios do occur. They result from postmeiotic (mitotic) segregation of unrepaired heteroduplexes.

Gene conversion is associated with the reciprocal recombination of flanking markers approximately 50 percent of the time. This correlation is nicely explained by the Holliday model of recombination presented in **FIGURE 13.26**. If the two recombinant chromatids of the tetrad just diagrammed are drawn in a form equivalent to that shown in **FIGURE 13.26g**, the association of gene conversion with reciprocal recombination of flanking markers can easily be explained (**FIGURE 13.28**).

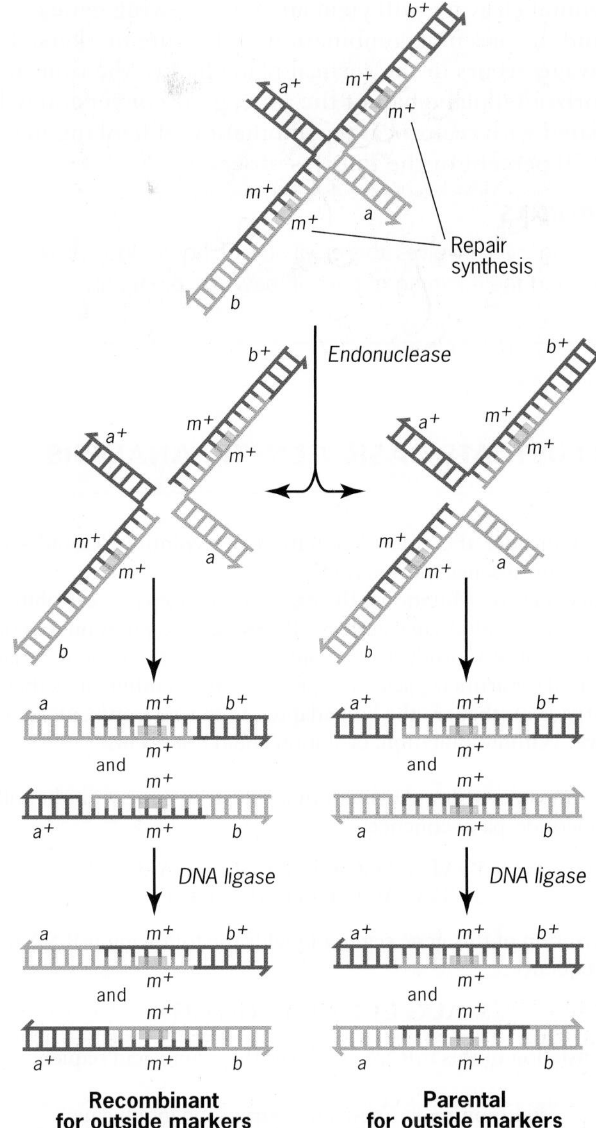

Figure 13.28 ▶ Formation of either the recombinant (bottom left) or parental (bottom right) combinations of flanking markers in association with gene conversion. The recombination intermediate at the top is equivalent to that illustrated in Figure 13.29*g*, but shows the mismatch-repaired chromatids of the tetrad diagrammed in the text. This tetrad produces an ascus showing 3 m^+ to 1 m gene conversion. Cleavage of the single-strand bridge in the vertical plane (left) produces the recombinant ($a^+ b$ and $a\ b^+$) arrangement of flanking markers, whereas cleavage in the horizontal plane yields the parental ($a^+ b^+$ and $a\ b$) arrangement of the flanking markers.

The single-strand bridge connecting the two chromatids must be resolved by endonucleolytic cleavage to complete the recombination process. This cleavage may occur either horizontally or vertically on the chi form drawn in **FIGURE 13.28**. Vertical cleavage will yield an ascus showing both gene conversion and reciprocal recombination of flanking markers.

Horizontal cleavage will yield an ascus showing gene conversion and the parental combination of flanking markers. Thus, if cleavage occurs in the vertical plane half of the time and in the horizontal plane half of the time, gene conversion will be associated with reciprocal recombination of flanking markers about 50 percent of the time, as observed.

KEY POINTS

- Crossing over involves the breakage of homologous DNA molecules and the rejoining of parts in new combinations.
- When genetic markers are closely linked, nonreciprocal recombination, or gene conversion, often occurs, yielding 3:1 ratios of the segregating alleles.
- Gene conversion results from DNA repair synthesis that occurs during the recombination process.

▶ Basic Exercises

ILLUSTRATE BASIC GENETIC ANALYSIS

1. Consider the role of mutation in evolution. Could species evolve in the absence of mutation?

Answer: No. Mutation is the essential first step in the evolutionary process; it is the ultimate source of all new genetic variation. Recombination mechanisms produce new combinations of this genetic variation, and natural (or artificial) selection preserves the combinations that produce organisms that are the best adapted to the environments in which they live. Without mutation, evolution could not occur.

2. Consider a short segment of a wild-type gene with the following nucleotide-pair sequence:

5′-ATG TCC GCA TGG GGA-3′
3′-TAC AGG CGT ACC CCT-5′

Transcription of this gene segment yields the following mRNA nucleotide sequence:

5′-AUG UCC GCA UGG GGA-3′

and translation of this mRNA produces the amino acid sequence:

methionine-serine-alanine-tryptophan, glycine

If a single nucleotide-pair substitution occurs in this gene, changing the G:C at position 7 to A:T, what effect will this mutation have on the polypeptide produced by this gene?

Answer: The mRNA produced by the gene segment with the mutation will now be:

5′-AUG UCC ACA UGG GGA-3′

and will encode the amino acid sequence:

methionine-serine-threonine-tryptophan-glycine

Note that the third amino acid of the mutant polypeptide is threonine instead of alanine as in the wild-type polypeptide. Thus, this base-pair substitution, like most base-pair substitutions, results in a single amino acid substitution in the polypeptide encoded by the gene.

3. If a single nucleotide-pair substitution occurs in the gene segment shown in Exercise 2, changing the G:C at position 12 to A:T, what effect will this mutation have on the polypeptide produced by this gene?

Answer: The resulting mRNA sequence will be:

5′-AUG UCC GCA UGA GGA-3′
(UGA: termination codon)

with the fourth codon changed from UGG, a tryptophan codon, to UGA, one of the three chain-termination codons. As a result, the mutant polypeptide will be prematurely terminated at this position, yielding a truncated protein.

4. If a single A:T base pair is inserted between nucleotide pairs 6 and 7 in the gene segment shown in Exercise 2, what effect will this change have on the polypeptide specified by this gene?

Answer: The nucleotide sequence of the mRNA specified by the mutant gene segment will be:

5′-AUG UCC AGC AUG GGG A-3′

and the polypeptide produced from the altered mRNA will be:

methionine-serine-serine-methionine-glycine
(serine-methionine-glycine: altered amino acid sequence)

The base-pair insertion will alter the reading frame of the mRNA (trinucleotides read as codons) distal to the site of the mutation. As a result, all of the amino acids specified by codons downstream from the site of the insertion will be changed, producing an abnormal (usually nonfunctional) protein. In many cases, an insertion will shift a termination codon into the proper reading frame for translation, causing a truncated polypeptide to be produced.

5. If the two DNA molecules shown below, where the arrowhead indicates the 3′-end of each strand, undergo crossing over by breakage and reunion, will both of the recombinants shown be produced with equal frequency?

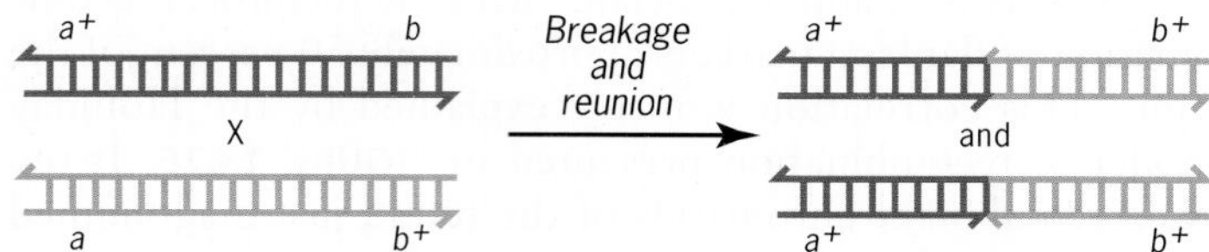

Answer: No. During recombination, only DNA strands with the same polarity can be joined. The second recombinant will not be produced.

► Testing Your Knowledge

INTEGRATE DIFFERENT CONCEPTS AND TECHNIQUES

1. Charles Yanofsky isolated a large number of auxotrophic mutants of *E. coli* that could grow only on medium containing the amino acid tryptophan. How could such mutants be identified? If a specific tryptophan auxotroph resulted from a nitrous acid-induced mutation, could it be induced to revert back to prototrophy by treatment with 5-bromouracil (5–BU)?

Answer: The culture of mutagenized bacteria must be grown in medium containing tryptophan so that the desired mutants can survive and reproduce. The bacteria should then be diluted, plated on agar medium containing tryptophan, and incubated until visible colonies are produced. The colonies are next transferred to plates lacking tryptophan by the replica-plating technique developed by the Lederbergs (see **FIGURE 13.2**). The desired tryptophan auxotrophs will grow on the plates containing tryptophan, but not on the replica plates lacking tryptophan. Because nitrous acid and 5–BU produce transition mutations in both directions, A:T ↔ G:C, any mutation induced with nitrous acid should be induced to back-mutate with 5–BU.

2. Assume that you recently discovered a new species of bacteria and named it *Escherichia mutaphilium*. During the last year, you have been studying the *mutA* gene and its polypeptide product, the enzyme trinucleotide mutagenase, in this bacterium. *E. mutaphilium* has been shown to use the established, nearly universal genetic code and to behave like *Escherichia coli* in all other respects relevant to molecular genetics.

The sixth amino acid from the amino terminus of the wild-type trinucleotide mutagenase is histidine, and the wild-type *mutA* gene has the triplet nucleotide-pair sequence

3′-GTA-5′
5′-CAT-3′

at the position corresponding to the sixth amino acid of the gene product. Seven independently isolated mutants with single nucleotide-pair substitutions within this triplet have also been characterized. Furthermore, the mutant trinucleotide mutagenases have all been purified and sequenced. All seven are different: they contain, respectively, glutamine, tyrosine, asparagine, aspartic acid, arginine, proline, and leucine as the sixth amino acid from the amino terminus.

Mutants *mutA1*, *mutA2*, and *mutA3* will not recombine with each other, but each will recombine with each of the other four mutants (*mutA4*, *mutA5*, *mutA6*, and *mutA7*) to yield true wild-type recombinants. Similarly, mutants *A4*, *A5*, and *A6* will not recombine with each other but will each yield true wild-type recombinants in crosses with each of the other four mutants. Finally, crosses between *mutA1* and *mutA7* yield about twice as many true wild-type recombinants as do crosses between *mutA6* and *mutA7*.

Mutants *A1* and *A6* are induced to back-mutate to wild-type by treatment with 5-bromouracil (5-BU), whereas mutants *A2*, *A3*, *A4*, *A5*, and *A7* are not induced to back-mutate by treatment with 5–BU. Mutants *A2* and *A4* grow slowly on minimal medium, whereas mutants *A3* and *A5* carry null mutations (producing completely inactive gene products) and are incapable of growth on minimal medium. This difference has been used to select for mutation events from genotypes *mutA3* and *mutA5* to genotypes *mutA2* and *mutA4*. Mutants *A3* and *A5* can be induced to mutate to *A2* and *A4*, respectively, by treatment with 5-bromouracil or hydroxylamine. However, mutant *A3* cannot be induced to mutate to *A4*, nor *A5* to *A2*, by treatment with either mutagen.

Use the information given above and the nature of the genetic code (**TABLE 12.1**) to deduce which mutant allele specifies the mutant polypeptide with each of the seven different amino acid substitutions at position 6 of trinucleotide mutagenase, and describe the rationale behind each of your deductions.

Answer: The following deductions can be made from the information given.

(a) The wild-type His codon must be CAU based on the nucleotide-pair sequence of the gene.

(b) The codons for the seven amino acids found at position 6 in the mutant polypeptides must be connected to CAU by a single-base change because the mutants were all derived from wild-type by a single nucleotide-pair substitution. Thus, the degeneracy of the genetic code is not a factor in deducing specific codon assignments.

(c) Because of the nature of the genetic code—specifically the degeneracy at the third (3′) position in each codon, there are three possible amino acid substitutions due to single-base substitutions (caused by single base-pair substitutions in DNA) at each of the first two positions (the 5′ base and the middle base), but only one possible amino acid change due to a single-base change at position 3 (the 3′ base in the codon). For ease of discussion, the three nucleotide-pair positions in the triplet under consideration will be referred to as position 1 (corresponding to the 5′ base in the codon), position 2 (the middle nucleotide pair), and position 3 (corresponding to the 3′ base in the mRNA codon).

(d) Since *A1*, *A2*, and *A3* do not recombine with each other, they must all result from base-pair substitutions at the same position in the triplet, at either position 1 or position 2. The same is true for *A4*, *A5*, and *A6*. Since *A7* recombines with each of the other six mutant alleles, it must result from the single base-pair substitution at position 3 that leads to an amino acid change.

(e) The only amino acid with codons connected to the His codon CAU by single-base changes at position 3 is Gln (codons CAA and CAG). Thus, the *mutA7* polypeptide must have glutamine as the sixth amino acid.

(f) Since *mutA7* (the third position substitution) yields about twice as many wild-type recombinants in crosses with *mutA1* as in crosses with *mutA6*, the *A1* substitution must be at position 1 and the *mutA6* substitution must be at position 2. Combined with (d) above, this places the *A2* and *A3* substitutions at position 1 and the *A4* and *A5* substitutions at position 2.

(g) Since *mutA1* and *mutA6* are induced to revert to wild-type by 5–BU, they must be connected to the triplet of nucleotide pairs encoding His by transition mutations—that is,

$$(\textit{mutA3})\ \frac{\text{ATA}}{\text{TAT}} \xleftrightarrow{\text{5–BU}} \frac{\text{GTA}}{\text{CAT}} \xleftrightarrow{\text{5–BU}} \frac{\text{GCA}}{\text{CGT}}\ (\textit{mutA6})$$

(h) Since *mutA3* and *mutA5* are induced to mutate to *mutA2* and *mutA4*, respectively, by hydroxylamine, *A3* must be connected to *A2*, and *A5* to *A4*, specifically by G:C → A:T transitions—that is,

$$(\textit{mutA3})\ \frac{\text{CTA}}{\text{GAT}} \xrightarrow{\text{HA}} \frac{\text{TTA}}{\text{AAT}}\ (\textit{mutA2})$$

and

$$(\textit{mutA5})\ \frac{\text{GGA}}{\text{CCT}} \xrightarrow{\text{HA}} \frac{\text{GAA}}{\text{CTT}}\ (\textit{mutA4})$$

Collectively, these deductions establish that the following relationships between the amino acids, codons, and nucleotide-pair triplets are present at the position of interest in the trinucleotide mutagenase polypeptides, mRNAs, and genes in the seven different mutants:

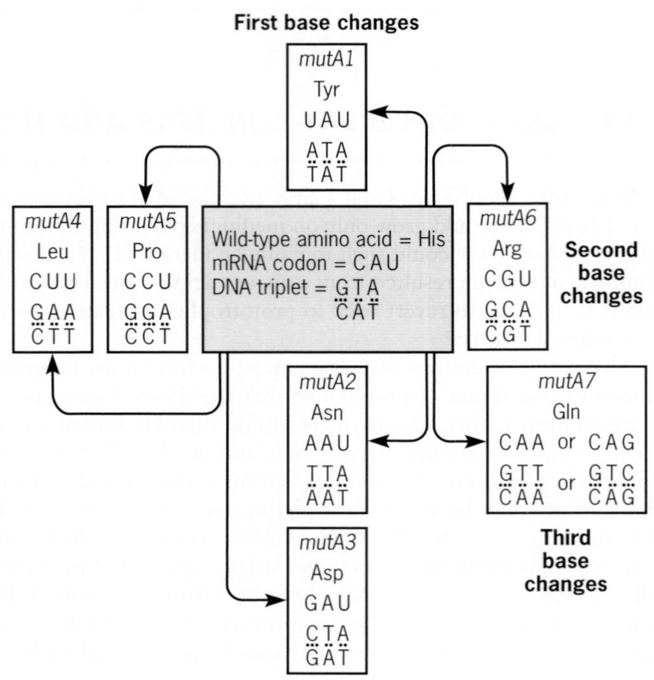

▸ Questions and Problems

ENHANCE UNDERSTANDING AND DEVELOP ANALYTICAL SKILLS

13.1 Published spontaneous mutation rates for humans are generally higher than those for bacteria. Does this indicate that individual genes of humans mutate more frequently than those of bacteria? Explain.

13.2 How can mutations in bacteria causing resistance to a particular drug be detected? How can it be determined whether a particular drug causes mutations or merely identifies mutations already present in the organisms under investigation?

13.3 Identify the following point mutations represented in DNA and in RNA as (1) transitions, (2) transversions, or (3) reading frameshifts. (a) A to G; (b) C to T; (c) C to G; (d) T to A; (e) UAU ACC UAU to UAU AAC CUA; (f) UUG CUA AUA to UUG CUG AUA.

13.4 GO Of all possible missense mutations that can occur in a segment of DNA encoding the amino acid tryptophan, what is the ratio of transversions to transitions if all single base-pair substitutions occur at the same frequency?

13.5 Both lethal and visible mutations are expected to occur in fruit flies that are subjected to irradiation. Outline a method for detecting (a) X-linked lethals and (b) X-linked visible mutations in irradiated *Drosophila*.

13.6 Products resulting from somatic mutations, such as the navel orange and the Delicious apple, have become widespread in citrus groves and apple orchards. However, traits resulting from somatic mutations are seldom maintained in animals. Why?

13.7 If a single short-legged sheep should occur in a flock, suggest experiments to determine whether the short legs are the result of a mutation or an environmental effect. If due to a mutation, how can one determine whether the mutation is dominant or recessive?

13.8 A precancerous condition (intestinal polyposis) in a particular human family group is determined by a single dominant gene. Among the descendants of one woman who died with cancer of the colon, 12 people have died with the same type of cancer and 5 now have intestinal polyposis. All other branches of the large kindred have been carefully examined, and no cases have been found. Suggest an explanation for the origin of the defective gene.

13.9 Juvenile muscular dystrophy in humans depends on an X-linked recessive gene. In an intensive study, 33 cases were found in a population of some 800,000 people. The investigators were confident that they had found all cases that were well enough advanced to be detected at the time the study was made. The symptoms of the disease were expressed only in males. Most of those with the disease died at an early age, and none lived beyond 21 years of age. Usually, only one case was detected in a family, but sometimes two or three cases occurred in the same family. Suggest an explanation for the sporadic occurrence of the disease and the tendency for the gene to persist in the population.

13.10 The bacteriophage T4 genome contains about 50 percent A:T base pairs and 50 percent G:C base pairs. The base analog 2-aminopurine induces A:T → G:C and G:C → A:T base-pair substitutions by undergoing tautomeric shifts. Hydroxylamine is a mutagenic chemical that reacts specifically with cytosine and induces only G:C → A:T substitutions. If a large number of independent mutations were produced in bacteriophage T4 by treatment with 2-aminopurine, what percentage of these mutations should you expect to be induced to mutate back to the wild-type genotype by treatment with hydroxylamine?

13.11 A single mutation blocks the conversion of phenylalanine to tyrosine. (a) Is the mutant gene expected to be pleiotropic? (b) Explain.

Use the known codon-amino acid assignments given in Chapter 12 to work the following problem.

13.12 Would such mutations occur if a nonreplicating suspension of MS2 phage was treated with 5-bromouracil?

13.13 Assuming that the β-globin chain and the α-globin chain shared a common ancestor, what mechanisms might explain the differences that now exist in these two chains? What changes in DNA and mRNA codons would account for the differences that have resulted in unlike amino acids at corresponding positions

13.14 A mutator gene *Dt* in maize increases the rate at which the gene for colorless aleurone (*a*) mutates to the dominant allele (*A*), which yields colored aleurone. When reciprocal crosses were made (i.e., seed parent *dt/dt, a/a* × *Dt/Dt, a/a* and seed parent *Dt/Dt, a/a* × *dt/dt, a/a*), the cross with *Dt/Dt* seed parents produced three times as many dots per kernel as the reciprocal cross. Explain these results.

13.15 One stock of fruit flies was treated with 1000 roentgens (r) of X rays. The X-ray treatment increased the mutation rate of a particular gene by 2 percent. What percentage increases in the mutation rate of this gene would be expected if this stock of flies was treated with X-ray doses of 3000 r, 3500 r, and 5000 r?

13.16 In a given strain of bacteria, all of the cells are usually killed when a specific concentration of streptomycin is present in the medium. Mutations that confer resistance to streptomycin occur. The streptomycin-resistant mutants are of two types: some can live with or without streptomycin; others cannot survive unless this drug is present in the medium. Given a streptomycin-sensitive strain of this species, outline an experimental procedure by which streptomycin-resistant strains of the two types could be established.

13.17 If CTT is a DNA triplet (transcribed strand of DNA) specifying glutamic acid, what DNA and mRNA base triplet alterations could account for valine and lysine in position 6 of the β-globin chain?

13.18 Why does the frequency of chromosome breaks induced by X rays vary with the total dosage and not with the rate at which it is delivered?

13.19 A cross was performed in *Neurospora crassa* between a strain of mating type *A* and genotype $x^+ m^+ z$ and a strain of mating type *a* and genotype $x\ m\ z^+$. Genes *x*, *m*, and *z* are closely linked and present in the order *x*–*m*–*z* on the chromosome. An ascus produced from this cross contained two copies ("identical twins") of each of the four products of meiosis. If the genotypes of the four products of meiosis showed that gene conversion had occurred at the *m* locus and that reciprocal recombination had occurred at the *x* and *z* loci, what might the genotypes of the four products look like? In the parentheses below, write the genotypes of the four haploid products of meiosis in an ascus showing gene conversion at the *m* locus and reciprocal recombination of the flanking markers (at the *x* and *z* loci).

Ascus Spore Pairs

1–2	3–4	5–6	7–8
( )	( )	( )	( )

13.20 GO You are screening three new pesticides for potential mutagenicity using the Ames test. Two *his*⁻ strains resulting from either a frameshift or a transition mutation were used and produced the following results (number of revertant colonies):

Strain 1	Transition Mutant Control (no chemical)	Transition Mutant + Chemical	Transition Mutant + Chemical + Rat Liver Enzymes
Pesticide #1	21	150	17
Pesticide #2	18	25	13
Pesticide #3	25	300	250

Strain 2	Frameshift Mutant Control (no chemical)	Frameshift Mutant + Chemical	Frameshift Mutant + Chemical + Rat Liver Enzymes
Pesticide #1	5	3	7
Pesticide #2	7	7	120
Pesticide #3	6	11	5

What type of mutations, if any, do the three pesticides induce?

13.21 Are mutational changes induced by nitrous acid more likely to be transitions or transversions?

13.22 How does nitrous acid induce mutations? What specific end results might be expected in DNA and mRNA from the treatment of viruses with nitrous acid?

13.23 How does the action and mutagenic effect of 5-bromouracil differ from that of nitrous acid?

13.24 GO One person was in an accident and received 50 roentgens (r) of X rays at one time. Another person received 5 r in each of 20 treatments. Assuming no intensity effect, what proportionate number of mutations would be expected in each person?

13.25 Seymour Benzer and Ernst Freese compared spontaneous and 5-bromouracil-induced mutants in the *rII* gene of the bacteriophage T4; the mutagen increased the mutation rate ($rII^+ \rightarrow rII$) several hundred times above the spontaneous mutation rate. Almost all (98 percent) of the 5-bromouracil-induced mutants could be induced to revert to wild-type (*rII* → *rII*$^+$) by 5-bromouracil treatment, but only 14 percent of the spontaneous mutants could be induced to revert to wild-type by this treatment. Discuss the reason for this result.

13.26 Sydney Brenner and A. O. W. Stretton found that nonsense mutations did not terminate polypeptide synthesis in the *rII* gene of the bacteriophage T4 when these mutations were located within a DNA sequence interval in which a single nucleotide insertion had been made on one end and a single nucleotide deletion had been made on the other. How can this finding be explained?

13.27 A reactor overheats and produces radioactive tritium (H^3), radioactive iodine (I^{131}), and radioactive xenon (Xn^{133}). Why should we be more concerned about radioactive iodine than the other two radioactive isotopes?

13.28 How do acridine-induced changes in DNA result in inactive proteins?

Use the known codon-amino acid assignments given in Chapter 12 to work the following problems.

13.29 Recall that nitrous acid deaminates adenine, cytosine, and guanine (adenine → hypoxanthine, which base-pairs with cytosine; cytosine → uracil, which base-pairs with adenine; and guanine → xanthine, which base-pairs with cytosine). Would you expect nitrous acid to induce any mutations that result in the substitution of another amino acid for a glycine residue in a wild-type polypeptide (i. e., glycine → another amino acid) if the mutagenesis were carried out on a suspension of mature (nonreplicating) T4 bacteriophage? (*Note*: After the mutagenic treatment of the phage suspension, the nitrous acid is removed. The treated phage are then allowed to infect *E. coli* cells to express any induced mutations.) If so, by what mechanism? If not, why not?

13.30 GO Bacteriophage MS2 carries its genetic information in RNA. Its chromosome is analogous to a polygenic molecule of mRNA in organisms that store their genetic information in DNA. The MS2 minichromosome encodes 4 polypeptides (i.e., it has four genes). One of these four genes encodes the MS2 coat protein, a polypeptide 129 amino acids long. The entire nucleotide sequence in the RNA of MS2 is known. Codon 112 of the coat protein gene is CUA, which specifies the amino acid leucine. If you were to treat a replicating population of bacteriophage MS2 with the mutagen 5-bromouracil, what amino acid substitutions would you expect to be induced at position 112 of the MS2 coat protein (i.e., Leu → other amino acid)? (*Note*: Bacteriophage MS2 RNA replicates using a complementary strand of RNA and base-pairing like DNA.)

13.31 Would the different amino acid substitutions induced by 5-bromouracil at position 112 of the coat polypeptide that you indicated in Problem 13.30 be expected to occur with equal frequency? If so, why? If not, why not? Which one(s), if any, would occur more frequently?

13.32 Keeping in mind the known nature of the genetic code, the information given about phage MS2 in Problem 13.32, and the information you have learned about nitrous acid in Problem 13.30, would you expect nitrous acid to induce any mutations that would result in amino acid substitutions of the type glycine → another amino acid if the mutagenesis were carried out on a suspension of mature (nonreplicating) MS2 bacteriophage? If so, by what mechanism? If not, why not?

13.33 Mutations in the genes encoding the α and β subunits of hemoglobin lead to blood diseases such as thalassemias and sickle-cell anemia. You have found a family in China in which some members suffer from a new genetic form of anemia. The DNA sequences at the 5′ end of the nontemplate strand of the normal and mutant DNA encoding the α subunit of hemoglobin are as follows:

Normal 5′-ACGTTATGCCGTACTGCCAGCTAACT-
GCTAAAGAACAATTA.......-3′

Mutant 5′-ACGTTATGCCCGTACTGCCAGCTAACT-
GCTAAAGAACAATTA.......-3′

(a) What type of mutation is present in the mutant hemoglobin gene?
(b) What are the codons in the translated portion of the mRNA transcribed from the normal and mutant genes?
(c) What are the amino acid sequences of the normal and mutant polypeptides?

13.34 Which of the following amino acid substitutions should you expect to be induced by 5-bromouracil with the highest frequency? (a) Met → Leu; (b) Met → Thr; (c) Lys → Thr; (d) Lys → Gln; (e) Pro → Arg; or (f) Pro → Gln? Why?

13.35 The wild-type sequence of part of a protein is

NH_2-Trp-Trp-Trp-Met-Arg-Glu-Trp-Thr-Met

Each mutant in the following table differs from wild-type by a single point mutation. Using this information, determine the mRNA sequence coding for the wild-type polypeptide. If there is more than one possible nucleotide, list all possibilities.

Mutant	Amino Acid Sequence of Polypeptide
1	Trp-Trp-Trp Met
2	Trp-Trp-Trp-Met-Arg-Asp-Trp-Thr-Met
3	Trp-Trp-Trp-Met-Arg-Lys-Trp-Thr-Met
4	Trp-Trp-Trp-Met-Arg-Glu-Trp-Met-Met

13.36 Acridine dyes such as proflavin are known to induce primarily single base-pair additions and deletions. Suppose that the wild-type nucleotide sequence in the mRNA produced from a gene is

5′-AUGCCCUUUGGGAAAGGGUUUCCCUAA-3′

Also, assume that a mutation is induced within this gene by proflavin, and, subsequently, a revertant of this mutation is similarly induced, with proflavin and shown to result from a second-site suppressor mutation within the same gene. If the amino acid sequence of the polypeptide encoded by this gene in the revertant (double mutant) strain is

NH_2-Met-Pro-Phe-Gly-Glu-Arg-Phe-Pro-COOH

what would be the most likely nucleotide sequence in the mRNA of this gene in the revertant (double mutant)?

13.37 Would you expect nitrous acid to induce a higher frequency of Tyr → Ser or Tyr → Cys substitutions? Why?

► Genomics on the Web

at http://www.ncbi.nlm.nih.gov/

Sickle-cell anemia is caused by a single base-pair substitution in the human β-globin gene. This mutation changes the sixth amino acid in the mature polypeptide from glutamic acid to valine (see **FIGURE 1.9**). This single amino acid change, in turn, causes all the symptoms of this painful and eventually fatal disease.

1. What other mutations in the human β-globin gene have changed the glutamic acid at position 6 to some other amino acid? What are these hemoglobin variants called? Are there β-globin variants with an amino acid substitution at position 6 and another amino acid substitution elsewhere in the polypeptide?

2. Proline is present at position 5 in normal human β-globin. What amino acid substitutions have occurred at this position in mutant β-globins? How about the glutamic acid present at position 7? Are there mutations that change this amino acid to something else?

3. Mutations have been documented at a large number of the 146 base-pair triplets (specifying mRNA codons) in the human β-globin gene. How many of these triplets have mutated to produce an amino acid substitution in the polypeptide?

4. What genes are located next to the β-globin gene on human chromosome 11? What are the functions of the delta-, gamma A-, gamma G-, and epsilon-globin genes? Is there any significance to their arrangement on the chromosome?

Hint*:* At the NCBI web site, search all databases with the query "beta-globin variants." Start with the results in OMIM (Online Medical Inheritance in Man), click HBB (online symbol for the human β-globin gene), on the left bar, click HbVar for a list of all the human β-globin variants characterized to date, then return to the HBB page and click "Gene Map" for a list of the genes next to *HBB*.

Chapter 14
Definitions of the Gene

CHAPTER OUTLINE

- Evolution of the Concept of the Gene: Summary
- Evolution of the Concept of the Gene: Function
- Evolution of the Concept of the Gene: Structure
- A Genetic Definition of the Gene
- The *rII* Locus of Bacteriophage T4
- Genes-within-Genes in Bacteriophage ΦX174
- Complex Gene–Protein Relationships

Albino and normally pigmented male red deer.

What Is Life?

In an article about genetic research in the years just before the discovery of the structure of DNA, James Watson wrote:

> *As an undergraduate at Chicago, I had already decided to go into genetics even though my formal training in it was negligible, with most of my course work reflecting a boyhood interest in natural history. Population genetics at first intrigued me, but from the moment I read Schrödinger's "What is Life?"*[1] *I became polarized toward finding out the secret of the gene.*[2]

Watson is not the only scientist to have been influenced by Erwin Schrödinger's little book. *What Is Life* was written by this Austrian physicist, famous for his work in quantum mechanics, to explore fundamental biological questions. Among other things, it included a discussion of how irradiation of living matter might be used to determine the size of a gene. Schrödinger knew of Muller's demonstration that X rays induce mutations, and he was intrigued by fellow physicist Max Delbrück's ideas about the underlying mechanism. Delbrück had collaborated with the Russian geneticist Nikolaj V. Timofeeff-Ressovsky to analyze the effects of radiation and to develop a "quantum mechanical" model of a gene. The discussion of this work in *What Is Life?* was read by other physicists, and some of them, with curiosity aroused, decided to abandon physics and pursue careers in genetics. One of these "turncoats" was a young American named Seymour Benzer. In a tribute to Max Delbrück (and, indirectly, to Erwin Schrödinger), Benzer wrote:

> *Delbrück first entered my life in the form of the chapter heading "Delbrück's Model" in Schrödinger's book "What is Life?" I read that book at an impressionable age, while still a graduate student in pre-transistor solid state physics at Purdue University. . . . Thus, I was suddenly plunged into the biology business.*[3]

For 10 years, the "biology business" into which Benzer plunged himself involved the exhaustive—and some would say exhausting—analysis of two genes in the chromosome of bacteriophage T4. Benzer's study of these genes produced the most detailed genetic maps ever constructed. And, as we shall see in this chapter, it also helped geneticists clarify what they mean by the word "gene." Thus, Schrödinger's book, entitled with the question "What is life?" inspired work that ultimately came to answer another fundamental question, "What is a gene?"

[1]Schrödinger, E. 1945. *What Is Life? The Physical Aspect of the Living Cell.* Cambridge University Press, London.

[2]Watson, J. D. 1966. Growing up in the phage group. In *Phage and the Origins of Molecular Biology* (J. Cairns, G. S. Stent, and J. D. Watson, eds.), pp. 239–245. Cold Spring Harbor Laboratory Press, Cold Spring Harbor, New York.

[3]Benzer, S. 1966. Adventures in the *rII* region. In *Phage and the Origins of Molecular Biology* (J. Cairns, G. S. Stent, and J. D. Watson, eds.), pp. 157–165. Cold Spring Harbor Laboratory Press, Cold Spring Harbor, New York.

The gene is to genetics what the atom is to chemistry. Thus, throughout this text we have focused our attention on the gene and the alternate forms of genes (alleles). In the preceding chapters, we have examined the patterns of transmission of independently assorting and linked genes, the chromosomal location of genes, the chemical composition of genes and chromosomes, the mechanism of replication, mutational events in genes, and the mechanisms by which genes exert their effects on the phenotype of the organism. What is this unit of genetic information that we call the gene? As we will see, the concept of a gene is not static; it has evolved through several phases since Wilhelm Johannsen introduced the term in 1909, and it will undoubtedly evolve through additional refinements in the future.

The gene was first defined as the unit of genetic information that controls a specific aspect of the phenotype. Such a description, though accurate, does not provide a precise, unambiguous definition that can be used to identify a gene in molecular terms. At a more fundamental level, the gene has been defined as the unit of genetic information that specifies the synthesis of one polypeptide. How can the gene be defined operationally? An *operational definition* spells out an operation or experiment that can be carried out to define or delimit something. In this chapter, we focus on the **complementation test** as an operational definition of the gene, the unit of genetic information that controls the synthesis of one polypeptide or one structural RNA molecule. We will also consider the limitations of the complementation test, along with the unique structural features of selected genes.

Evolution of the Concept of the Gene: Summary

Our concept of the gene has undergone many refinements since Mendel's discovery of "unit factors" in 1866.

Before discussing evidence supporting the various concepts, let's summarize the important stages in the evolution of the gene concept. The gene theory of inheritance began with the publication of Mendel's classic paper in 1866 but did not become an accepted part of scientific knowledge until after the rediscovery of Mendel's work in 1900. Mendel's gene (not so named) was the "character" or "unit factor" that controlled one specific phenotypic trait such as flower color in peas. At the time of the rediscovery of Mendel's work, the English physician Sir Archibald E. Garrod was studying several inherited diseases in humans. Garrod first recognized that homozygosity for recessive mutant alleles sometimes causes defects in the normal processes of metabolism. His concept of the gene is probably stated most accurately as one mutant gene–one metabolic block, which about 40 years later was refined to the one gene–one enzyme concept by George W. Beadle and Edward L. Tatum. Because many enzymes contain two or more different polypeptides, each encoded by a separate gene, the one gene–one enzyme concept subsequently was modified to one gene–one polypeptide. For cases where the final product of a gene is an RNA molecule, the concept of the gene can be modified to one gene–one transcript.

Prior to 1940, genes were considered analogous to beads on a string; recombination occurred between, but not within, genes. The gene was both the basic functional unit, which controlled one phenotypic trait, and the elementary structural unit, which could not be subdivided by recombination or mutation. Clarence Oliver's 1940 report that recombination had occurred within the *lozenge* gene of *Drosophila* stimulated both excitement and much debate about its significance. When the debate ended, the nucleotide pair had replaced the gene as the basic unit of structure, the unit of genetic material not subdivisible by mutation or recombination.

In the early 1940s, Edward B. Lewis developed the complementation, or *cis-trans*, test for functional allelism in *Drosophila*. This test subsequently was exploited by Seymour Benzer to define the gene experimentally in bacteriophage T4. The complementation test provides an operational definition of the gene. This test allows a geneticist to determine whether two independent mutations are located in the same gene or in two different genes. Moreover, the gene as defined by the complementation test is usually a good fit to the one gene–one polypeptide concept.

In the 1960s, elegant experiments by Charles Yanofsky, Sydney Brenner, and their collaborators showed that the gene and its polypeptide product were colinear structures, with a direct correlation between the sequence of nucleotide pairs in the gene and the sequence of amino acids in the polypeptide. However, this simple concept of the gene as a continuous sequence of nucleotide pairs specifying a colinear sequence of amino acids in the polypeptide gene product was short-lived. Overlapping genes and genes-within-genes were discovered in the late 1960s, and the sequences of eukaryotic genes were shown in the late 1970s to be interrupted by intron sequences. Moreover, some genes, for example, genes encoding immunoglobulins, were shown to be stored in germ-line chromosomes as short "gene segments," which are assembled into mature, functional genes during development.

Thus, the definition of the gene needs to remain somewhat pliable if it is to encompass all of the different structure–function relationships that occur in living organisms. Here, we

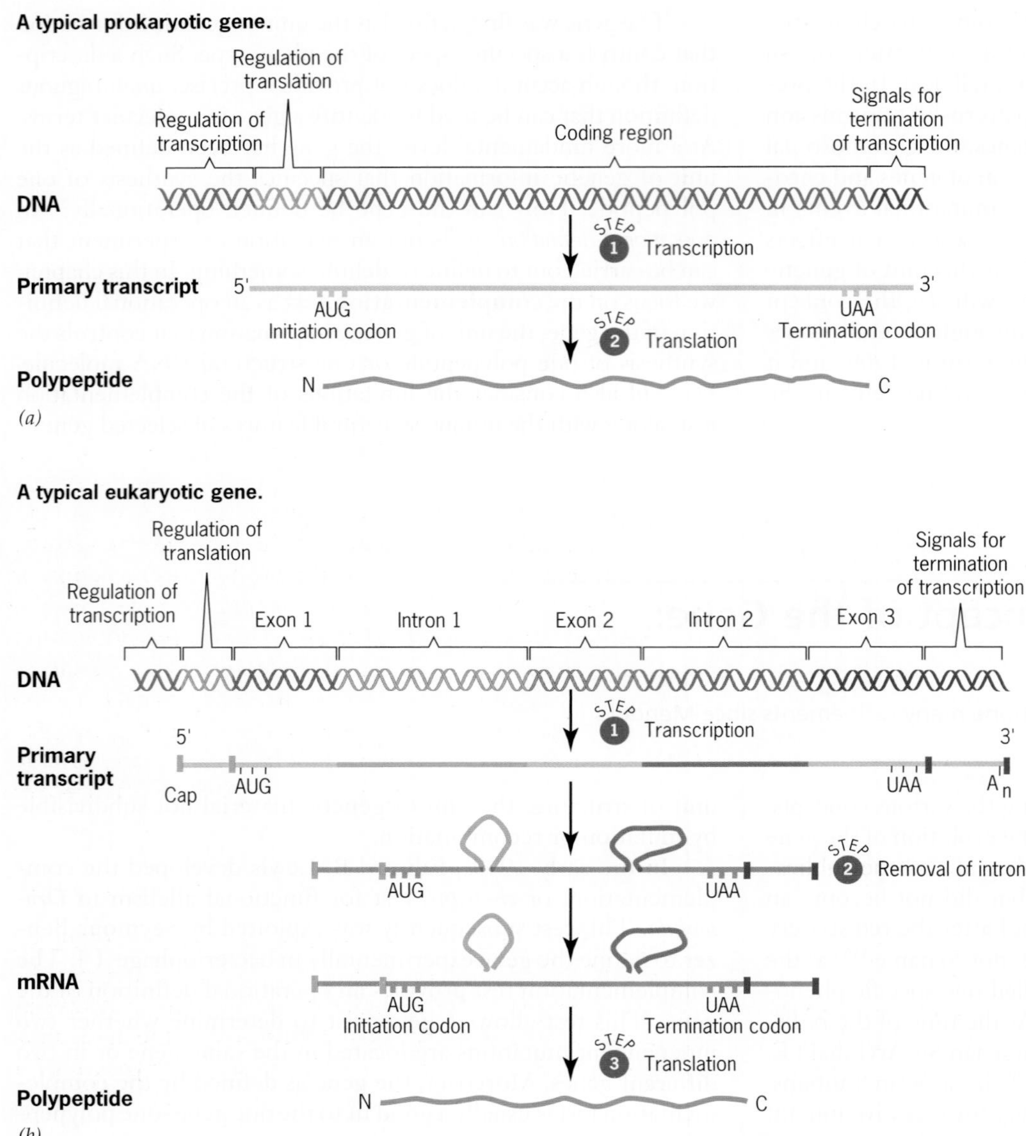

Figure 14.1 ▶ Gene structure. Prokaryotic genes usually contain uninterrupted coding sequences (*a*), whereas the coding sequences of eukaryotic genes are commonly interrupted by noncoding sequences (*b*).

define the gene as the unit of genetic information that controls the synthesis of one polypeptide or one structural RNA molecule. As just defined, the gene can be identified operationally by the complementation test. As such, the gene includes the 5′ and 3′ noncoding regions that are involved in regulating the transcription and translation of the gene and all introns within the gene (**FIGURE 14.1**). The structural gene refers to the portion that is transcribed to produce the RNA product. In the case of overlapping genes, this definition requires that some nucleotide-pair sequences be considered components of two or more genes. For those cases where exons are spliced together in various combinations to make related but different proteins, the gene may be defined as a DNA sequence that is a single unit of transcription and encodes a set of closely related polypeptides, sometimes called "protein isoforms." In germ-line chromosomes, the DNA sequences that encode segments of antibody chains probably should be called "gene segments" because this genetic information is not organized into units that fit any of the standard definitions of the gene (Chapter 21).

KEY POINTS

- The concept of the gene has undergone many refinements since the rediscovery of Mendel's work in 1900.
- Most genes encode one polypeptide and can be operationally defined by the complementation test.

Evolution of the Concept of the Gene: Function

Our concept of the gene has evolved from Mendel's "unit factor" controlling one phenotypic trait to the unit of genetic material specifying one polypeptide and operationally defined by the complementation test.

In the preceding section, we briefly examined the evolution of the concept of the gene, the basic unit of genetic information. That summary included both functional and structural aspects of the gene. Now, let's take a closer look at the gene as a unit of function.

MENDEL: CONSTANT FACTORS CONTROLLING PHENOTYPIC TRAITS

The law of combination of different characters, which governs the development of the hybrids, finds therefore its foundation and explanation in the principle enunciated, that the hybrids produce egg cells and pollen cells which in equal numbers represent all constant forms which result from the combinations of the characters brought together in fertilization. (Mendel, 1866; translation by William Bateson)

Mendel's characters or factors, which are now called genes, controlled specific phenotypic traits such as flower color, seed color, and seed shape. They were the basic units of function, the units of genetic information that governed one specific aspect of the phenotype. This definition of the gene as the basic unit of function is accepted by most geneticists. There has been no change in the concept of the gene as the basic unit of function since its discovery by Mendel in 1866. However, the elucidation of the chemical nature of the genetic material raised questions about the structure of the gene, and our understanding of the molecular structure of this basic unit of function has undergone several refinements since the rediscovery of Mendel's work in 1900.

If we examine what is known about how genes control phenotypic traits, the need for a more precise definition of the gene will be obvious. The pathway by which a gene exerts its effect on the phenotype of an organism is often very complex (**FIGURE 14.2**). Several genes may have similar effects on the same phenotypic trait, making it difficult to sort out the effects of individual genes. All the genes of an organism are located in the same nuclei, and they do not all function independently. The phenotype of an organism is the product of the action of all the genes acting within the restrictions imposed by the environment. Each gene also has an effect on the population to which the organism carrying the gene belongs. Ultimately, each gene has a potential effect, small though it may be, on the cumulative phenotype of the biosphere, for each gene may affect the ability of the organism, or the population, or the species to compete for an ecological niche in the biosphere (Chapter 25).

GARROD: ONE MUTANT GENE–ONE METABOLIC BLOCK

At the time of the rediscovery of Mendel's work in 1900, Sir Archibald Garrod was studying several congenital metabolic diseases in humans. One of these was the inherited disease **alkaptonuria**, which is easily detected because of the blackening of the urine upon exposure to air. The substance responsible for

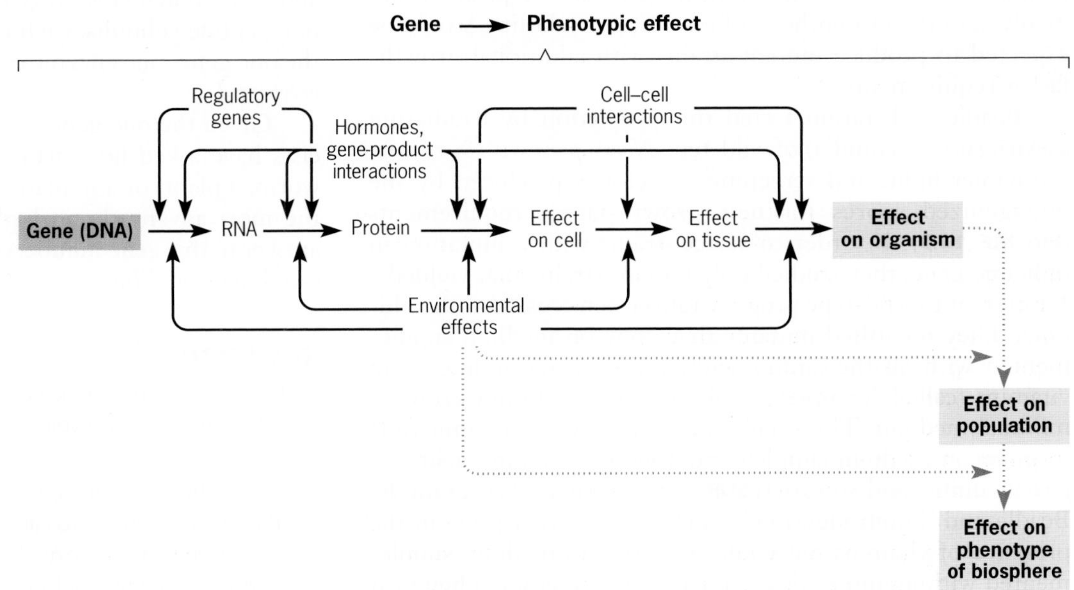

Figure 14.2 ► The complex pathway by which a gene exerts its effect on the phenotype of an organism, a population, or the biosphere.

this blackening is alkapton (or homogentisic acid), an intermediate in the degradation of the aromatic amino acids tyrosine and phenylalanine (see Figure 2 in A Milestone in Genetics: "Garrod's Inborn Errors of Metabolism" in Chapter 4). Garrod believed that the presence of homogentisic acid in the urine was due to a block in the normal pathway of metabolism of this compound. Moreover, on the basis of the family pedigree studies, Garrod proposed that alkaptonuria was inherited as a single recessive gene. The results of Garrod's studies of alkaptonuria and a few other congenital diseases in humans, such as albinism, were presented in detail in his book *Inborn Errors of Metabolism*. Although the details of the biochemical pathway affected by the recessive mutations that cause alkaptonuria were not worked out until many years later, Garrod clearly understood the relationship between genes and metabolism. His concept might be best stated as *one mutant gene–one metabolic block*.

BEADLE AND TATUM: ONE GENE–ONE ENZYME

During the late 1930s, George Beadle and Boris Ephrussi performed pioneering experiments on *Drosophila* eye color mutants. They identified genes that are required for the synthesis of specific eye pigments, indicating that enzyme-catalyzed metabolic pathways are under genetic control. Their results motivated Beadle to search for the ideal organism to use in extending this work. He chose the salmon-colored bread mold *Neurospora crassa* because it can grow on medium containing only (1) inorganic salts, (2) a simple sugar, and (3) one vitamin, biotin. *Neurospora* growth medium containing only these components is called "minimal medium." Beadle and his new collaborator, Edward Tatum, reasoned that *Neurospora* must be capable of synthesizing all the other essential metabolites, such as the purines, pyrimidines, amino acids, and other vitamins, *de novo*. Furthermore, they reasoned that the biosynthesis of these growth factors must be under genetic control. If so, mutations in genes whose products are involved in the biosynthesis of essential metabolites would be expected to produce mutant strains with additional growth-factor requirements.

Beadle and Tatum tested this prediction by irradiating asexual spores (conidia) of wild-type *Neurospora* with X rays or ultraviolet light, and screening the clones produced by the mutagenized spores for new growth-factor requirements (**FIGURE 14.3**). In order to select strains with a mutation in only one gene, they studied only mutant strains that yielded a 1:1 mutant to wild-type progeny ratio when crossed with wild-type. They identified mutants that grew on medium supplemented with all the amino acids, purines, pyrimidines, and vitamins (called "complete medium") but could not grow on minimal medium. They analyzed the ability of these mutants to grow on medium supplemented with just amino acids, or just vitamins, and so on (**FIGURE 14.3**, step 2). For example, Beadle and Tatum identified mutant strains that grew in the presence of vitamins but could not grow in medium supplemented with amino acids or other growth factors. They next investigated the ability of these vitamin-requiring strains to grow on media supplemented with each of the vitamins separately (**FIGURE 14.3** step 3).

In this way, Beadle and Tatum demonstrated that each mutation resulted in a requirement for one growth factor. By correlating their genetic analyses with biochemical studies of the mutant strains, they demonstrated in several cases that one mutation resulted in the loss of one enzyme activity. This work, for which Beadle and Tatum received a Nobel Prize in 1958, was soon verified by similar studies of many other organisms in many laboratories. The *one gene–one enzyme* concept thus became a central tenet of molecular genetics.

Appropriately, in his Nobel Prize acceptance speech, Beadle stated:

> *In this long, roundabout way, we had discovered what Garrod had seen so clearly so many years before. By now we knew of his work and were aware that we had added little if anything new in principle. . . . Thus we were able to demonstrate that what Garrod had shown for a few genes and a few chemical reactions in man was true for many genes and many reactions in* Neurospora.

ONE GENE–ONE POLYPEPTIDE

Subsequent to the work of Beadle and Tatum, many enzymes and structural proteins were shown to be heteromultimeric—that is, to contain two or more different polypeptide chains, with each polypeptide encoded by a separate gene. For example, in *E. coli*, the enzyme tryptophan synthetase is a heterotetramer composed of two α polypeptides encoded by the *trpA* gene and two β polypeptides encoded by the *trpB* gene. Similarly, the hemoglobins, which transport oxygen from our lungs to all other tissues of our bodies, are tetrameric proteins that contain two α-globin chains and two β-globin chains, as well as four oxygen-binding heme groups (see **FIGURE 12.3**). Other enzymes, for example, *E. coli* DNA polymerase III (Chapter 10) and RNA polymerase II (Chapter 11), contain many different polypeptide subunits, each encoded by a separate gene. Thus, the one gene–one enzyme concept was modified to *one gene–one polypeptide*.

Given the one gene–one polypeptide relationship, geneticists have asked how many genes it takes to produce a fly, a worm, a plant, or a human. With humankind's view of itself as the most advanced and highly evolved species on earth, the answer to this gene numbers question has been a major surprise (see Focus on The Human Genome: How Many Genes?).

KEY POINTS

- The existence of a basic genetic element—the gene—that controlled a specific phenotypic trait was established by Mendel's work in 1866.
- During the last century, the concept of the gene evolved from the unit that can mutate to cause a specific block in metabolism, to the unit specifying one enzyme, to the sequence of nucleotide pairs in DNA encoding one polypeptide chain or one RNA molecule.

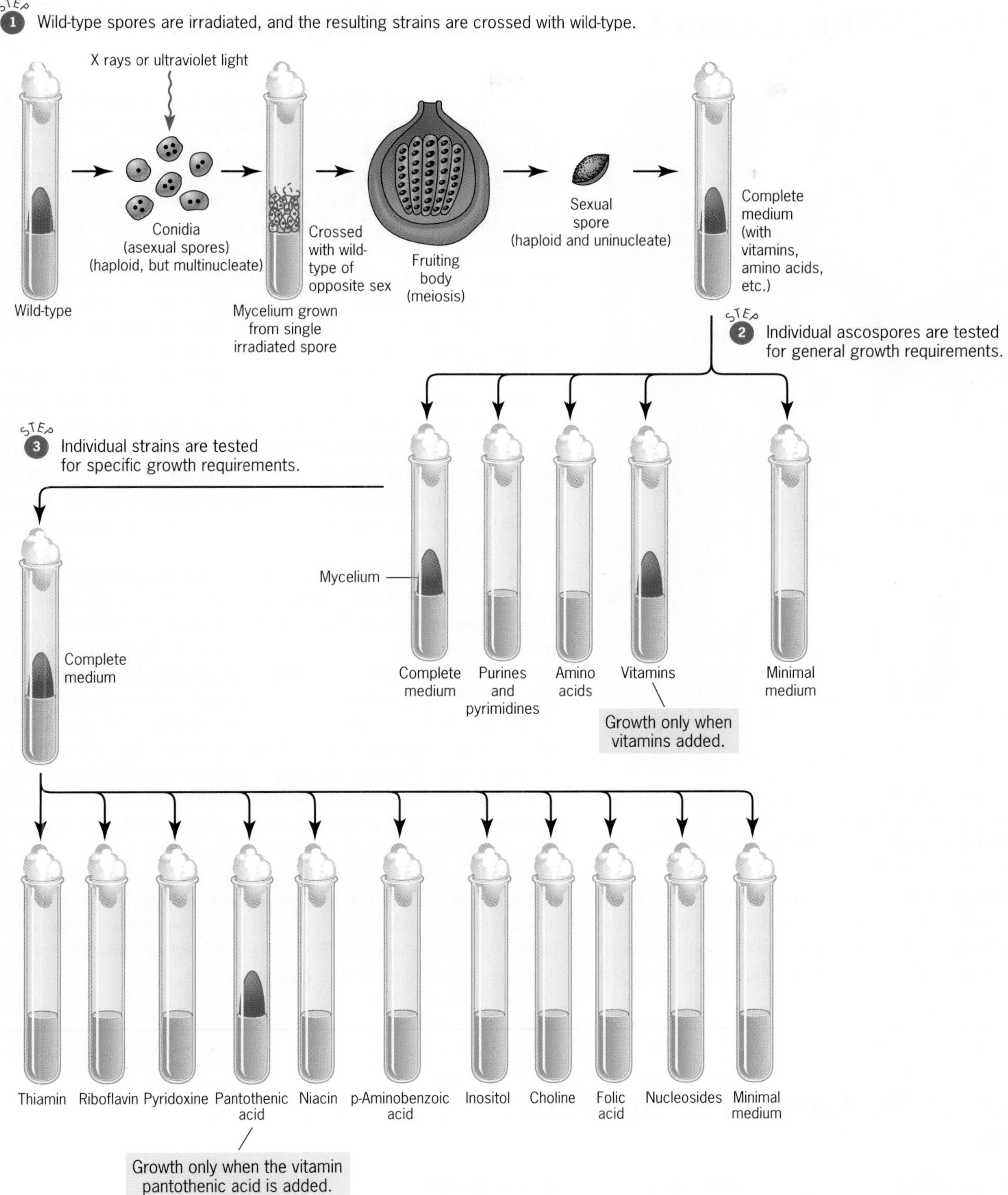

Figure 14.3 ▶ Diagram of Beadle and Tatum's experiment with *Neurospora* that led to the one gene–one enzyme hypothesis.

▶ FOCUS ON The Human Genome: How Many Genes?

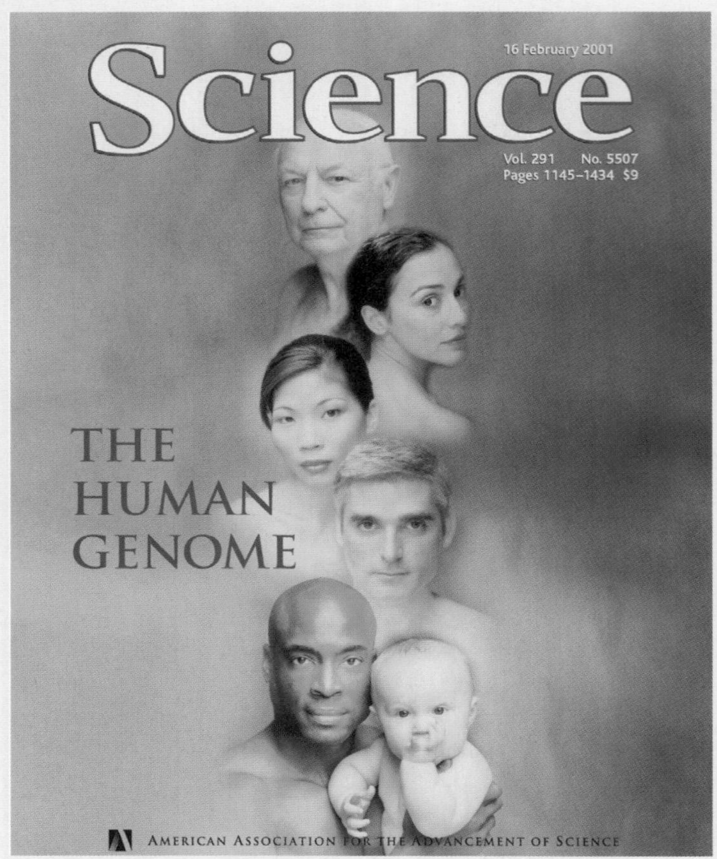

Figure 1 ▶ The cover of *Science* highlighting the publication of one of the two first drafts of the nucleotide sequence of the human genome. The other first draft was published in *Nature* the same week.

How many genes are needed to direct the growth and development of a human? Prior to the release of the first drafts of the human genome sequence in February 2001 by both private and public human genome sequencing teams (respectively, Celera, Inc. and the International Human Genome Sequencing Consortium) (**FIGURE 1**), most estimates of the number of human genes were in the 70,000 to 100,000 range, with some as high as 140,000. After all, earlier genome sequencing projects had shown that the yeast *S. cerevisiae* has about 6000 genes, the fruit fly *D. melanogaster* about 13,000 genes, the tiny worm *C. elegans* approximately 19,000 genes, and the small plant *A. thaliana* about 26,000 genes. Surely, we humans are much more complex than any of the species mentioned above. Right? If it takes 26,000 genes to "build" a small plant, it seemed reasonable that it should take at least twice that many to provide enough genetic information to produce a human.

Thus, the results of the two sequencing groups were somewhat shocking. Scientists at Celera identified only 26,588 protein-encoding sequences in their draft of the human genome, with weak evidence for up to 12,000 additional genes. The International Consortium's analysis of their data led them to predict a total of 30,000 to 35,000 genes in the human genome. If they are correct, the human genome contains only a few thousand more genes than the genome of the plant *Arabidopsis*.

As it turns out, the International Consortium's estimate of 30,000 to 35,000 genes in the human genome apparently was too high. They have subsequently published (*Nature*, October 21, 2004) a "nearly complete" sequence—only 341 gaps—that includes 99 percent of the euchromatin (gene-rich chromatin) in the genome. Surprisingly, as the accuracy of the sequence data has improved, the predicted gene number has decreased. In this latest report, the Consortium estimates that the human genome contains only 20,000 to 25,000 genes that encode proteins. Perhaps we are not as complex and sophisticated as we thought.

However, Eric Lander, lead author on the International Consortium's initial human genome paper, cautions that (1) genes expressed at very low levels in specialized cells may well have escaped detection by the gene-prediction software used in these studies, and (2) human genes give rise to multiple proteins by alternate pathways of exon splicing more often than do the genes of flies, worms, and plants. (See the subsection "Alternate Pathways of Transcript Splicing: Protein Isoforms" later in this chapter.)

Evolution of the Concept of the Gene: Structure

Our concept of the gene has evolved from the "bead-on-a-string" not divisible by mutation or recombination to the sequence of nucleotide pairs encoding one polypeptide or, in some cases, one RNA molecule.

In the preceding section, we examined the evolution of the concept of the gene as the basic functional component of the genetic material. In this section, we examine the gene from a structural perspective. What is the structure of the gene? Do all genes have the same structure?

THE PRE-1940 BEADS-ON-A-STRING CONCEPT

Prior to 1940, the genes in a chromosome were considered analogous to beads on a string. Recombination was believed to occur only between the beads or genes, not within genes. The gene was believed to be indivisible. According to this beads-on-a-string concept, the gene was the basic unit of genetic information defined by three criteria: (1) function, (2) recombination, and (3) mutation. More specifically, the gene was

1. *The unit of function*, the unit of genetic material that controlled the inheritance of one "character" or one attribute of phenotype.
2. *The unit of structure*, operationally defined in two ways:

   a. *By recombination:* as the unit of genetic information not subdivisible by recombination.
   b. *By mutation:* as the smallest unit of genetic material capable of independent mutation.

Geneticists initially thought that all three criteria defined the same basic unit of inheritance, namely, the gene.

Geneticists now know that these criteria define two different units of inheritance. According to the current molecular concept, the gene is the unit of function, the unit of genetic information controlling the synthesis of one polypeptide chain or, in some cases, one RNA molecule. The unit of structure is simply the structural unit in DNA, the nucleotide pair. Because it does not make sense to call each nucleotide pair a gene, geneticists have focused on the original definition of the gene as the unit of function and have discarded the beads-on-a-string view that the gene is not subdivisible by recombination or mutation. This is clearly appropriate since the emphasis in Mendel's work was on the factor (or gene, as it is now called) controlling one phenotypic characteristic.

DISCOVERY OF RECOMBINATION WITHIN THE GENE

In 1940, Clarence P. Oliver published the first evidence indicating that recombination could occur within a gene. Oliver was studying mutations at the *lozenge* locus on the X chromosome of *Drosophila melanogaster*. Two mutations, lz^s ("spectacle" eye) and lz^g ("glassy" eye), were thought to be alleles—that is, different forms of the same gene. The data available prior to 1940 indicated that they mapped at the same locus on the X chromosome. They had similar effects on the phenotype of the eye, and heterozygous lz^s/lz^g females had lozenge rather than wild-type eyes. However, when lz^s/lz^g females were crossed with either lz^s or lz^g males and large numbers of progeny were examined, wild-type progeny occurred with a frequency of about 0.2 percent.

These rare wild-type progeny could be explained by reversion of either the lz^s or the lz^g mutation. But there were two strong arguments against the reversion explanation. (1) The frequency of reversion of lz^s or lz^g to wild-type in hemizygous lozenge males was much lower than 0.2 percent. (2) When the female lz^s/lz^g heterozygotes carried genetic markers bracketing the *lozenge* locus, the rare progeny with wild-type eyes always carried an X chromosome with lz^+ that was flanked by recombinant outside markers. Moreover, the same combination of outside markers always occurred, as though the sites of lz^s and lz^g were fixed relative to each other and crossing over was occurring between them. If the lz^s/lz^g heterozygous female carried X chromosomes of the type

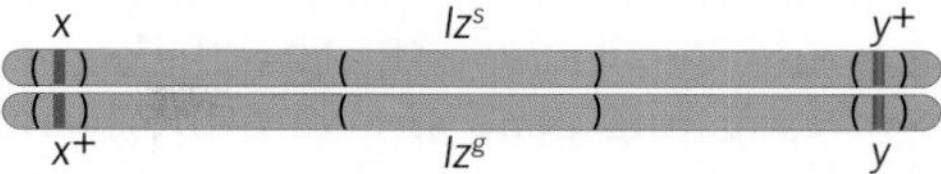

the rare progeny with wild-type eyes all (with one exception) contained an X chromosome with the following composition:

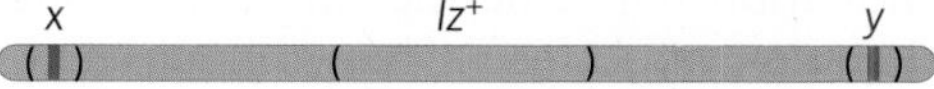

Among progeny of these matings, the reciprocal combination of outside markers (x^+y^+) never appeared in combination with lz^+. This result strongly suggested that the lz^s and lz^g mutations were located at distinct sites in the *lozenge* locus and that the lz^+ chromosome was produced by crossing over between the two sites, as shown in the following diagram.

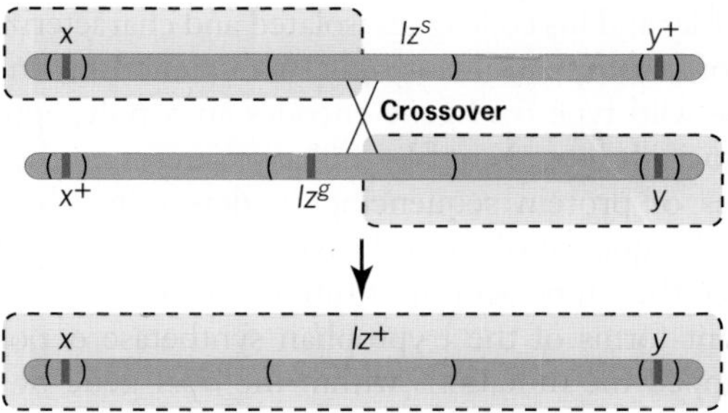

Definitive evidence for the involvement of recombination required the recovery and identification of the lz^s-lz^g double mutant with the reciprocal combination of outside markers—that is,

Oliver was not able to identify this double mutant because it could not be distinguished from the parental single-mutant phenotypes. The identification of both products, the wild-type and double-mutant chromosomes, produced by crossing over within the *lozenge* gene, was first accomplished by Melvin M. Green, one of Oliver's students.

The results of these pioneering studies first indicated that the gene was more complex than a bead on a string. They showed that the gene was divisible, containing sites that were separable by crossing over. Oliver's and Green's results were the first step toward the present concept of the gene as a long sequence of nucleotide pairs, capable of mutating and recombining at many different sites along its length.

RECOMBINATION BETWEEN ADJACENT NUCLEOTIDE PAIRS

The results obtained by Oliver and Green in their studies of *Drosophila* genes all indicated that mutable sites separable by recombination can exist within a single gene. Seymour Benzer extended this picture of the gene by demonstrating the existence of 199 distinct mutation sites that were separable by recombination within the *rIIA* gene of bacteriophage T4 (see The *rII* Locus of Bacteriophage T4). Benzer's picture of the gene as a sequence of nucleotide pairs capable of mutating at many distinct sites was soon verified by the results of many researchers investigating gene structure in several different organisms, both prokaryotes and eukaryotes. Given this information about the structure of genes and the known structure of DNA, it followed that the smallest unit of genetic material capable of mutation might be the single nucleotide pair and that recombination might occur between adjacent nucleotide pairs, whether between or within genes. Recombination between adjacent nucleotide pairs of a gene was first demonstrated by Charles Yanofsky in his studies of the *trpA* gene encoding the α polypeptide of tryptophan synthetase in *E. coli.* This enzyme, a tetramer containing two α polypeptides and two β polypeptides, catalyzes the final step in the biosynthesis of the amino acid tryptophan.

Yanofsky and his colleagues isolated and characterized a large number of tryptophan auxotrophs with mutations in the *trpA* gene. The wild-type *trpA* gene encodes an α polypeptide that is 268 amino acids long. Yanofsky and associates used the laborious techniques of protein sequencing to determine the complete amino acid sequence of the wild-type α polypeptide. They also determined the amino acid substitutions that had occurred in several mutant forms of the tryptophan synthetase α polypeptide. They mapped the mutations within the *trpA* gene by two- and three-factor crosses, and they compared the map positions with the locations of the amino acid substitutions in the mutant polypeptides.

Yanofsky and coworkers showed that recombination can occur between mutations that alter the same amino acid. Mutations *trpA23* and *trpA46* both result in the substitution of another amino acid (arginine in the case of *A23*, glutamic acid in the case of *A46*) for the glycine present at position 211 of the wild-type tryptophan synthetase α polypeptide. However, these two mutations occur at different mutable sites; that is, the *A23* and *A46* sites are separable by recombination. Yanofsky and colleagues determined the amino acids present at position 211 of the α polypeptide in other mutants as well as in revertants and partial revertants of the *trpA23* and *trpA46* mutants. By using this information and the known codon assignments, they were able to determine which of the glycine, arginine, and glutamic acid codons were present in the *trpA* mRNA at the position encoding amino acid residue 211 of the α polypeptides present in *trp*$^+$, *trpA23*, and *trpA46* cells, respectively (**FIGURE 14.4**).

Once the specific codons in mRNA are known, the corresponding base-pair sequences in the structural gene from which the mRNA is transcribed are also known. One strand of DNA will be complementary to the mRNA, and the second strand will be complementary to the first strand. Therefore, Yanofsky's data demonstrated that the mutational events that produced the *trpA23* and *trpA46* alleles were G:C to A:T transitions at adjacent nucleotide pairs. The *trp*$^+$ cells produced by recombination between chromosomes carrying mutations *A23* and *A46* demonstrated that recombination had occurred between adjacent nucleotide pairs, as shown in **FIGURE 14.5.** Yanofsky's results showed that the unit of genetic material not divisible by recombination is the single nucleotide pair.

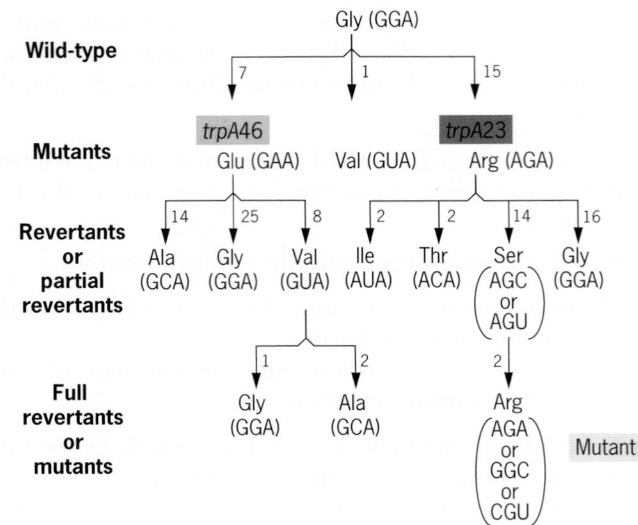

Figure 14.4 ▶ Pedigree of amino acid residue 211 (from the NH_2 terminus) of the α polypeptide of tryptophan synthetase of *E. coli*. Amino acid 211 is altered in *trpA23* and *trpA46* mutants (see Figure 14.7). The triplet codons shown in parentheses are the only codons specific to the indicated amino acids that will permit all of the observed amino acid replacements to occur by single base-pair substitutions. The number beside each arrow indicates the number of times that particular substitution was observed.

COLINEARITY BETWEEN THE CODING SEQUENCE OF A GENE AND ITS POLYPEPTIDE PRODUCT

The genetic information is stored in linear sequences of nucleotide pairs in DNA (or nucleotides in RNA, in some cases). Transcription and translation convert this genetic information into linear sequences of amino acids in polypeptides, which function as the key intermediaries in the genetic control of the phenotype.

It is now known that the nucleotide-pair sequences of the coding regions of the structural genes and the amino acid sequences of the polypeptides that they encode are **colinear**: the first three base pairs of the coding sequence of a gene specify the first amino acid of the polypeptide, the next three base pairs (four to six) specify the second amino acid, and so on, in a systematic way (**FIGURE 14.6*a***). It is also known that most of the genes in higher eukaryotes are interrupted by introns (Chapter 11). However, the presence of introns in genes does not invalidate the concept of colinearity. The presence of introns in genes simply means that there is no direct correlation in physical distances between the positions of base-pair coding triplets in a gene and the positions of amino acids in the polypeptide specified by that gene (**FIGURE 14.6*b***).

The first strong evidence for colinearity between a gene and its polypeptide product resulted from studies by Charles Yanofsky and colleagues on the *E. coli* gene that encodes the α subunit of the enzyme tryptophan synthetase. As mentioned earlier, this

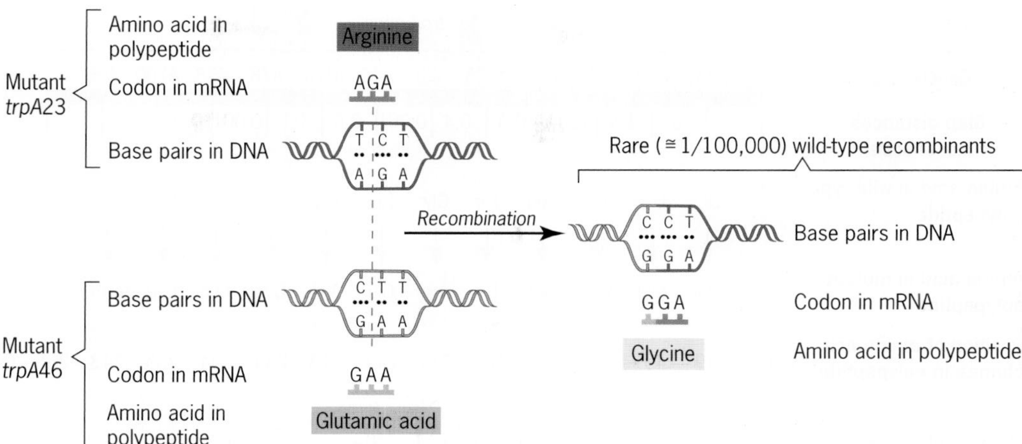

Figure 14.5 ► Recombination between mutations at adjacent nucleotide pairs in the *trpA* gene of *E. coli*. Mutations *A23* and *A46* both result in an amino acid substitution at position 211 of the tryptophan synthetase α polypeptide. Wild-type *E. coli* has a glycine residue at this position of the α polypeptide. *A23* causes a glycine to arginine substitution; *A46* causes a glycine to glutamic acid substitution (see Figure 14.4).

enzyme contains two α polypeptides encoded by the *trpA* gene and two β polypeptides encoded by the *trpB* gene. Yanofsky and coworkers performed a detailed genetic analysis of mutations in the *trpA* gene and correlated the genetic data with biochemical data on the sequences of the wild-type and mutant tryptophan synthetase α polypeptides. They demonstrated that there was a direct correlation between the map positions of mutations in the *trpA* gene and the positions of the resultant amino acid substitutions in the tryptophan synthetase α polypeptide (**FIGURE 14.7**).

About the same time, Sydney Brenner and associates demonstrated a similar colinearity between the positions of mutations in the gene of bacteriophage T4 that encodes the major structural protein of the phage head and the positions in the polypeptide affected by these mutations. Brenner and colleagues studied *amber*

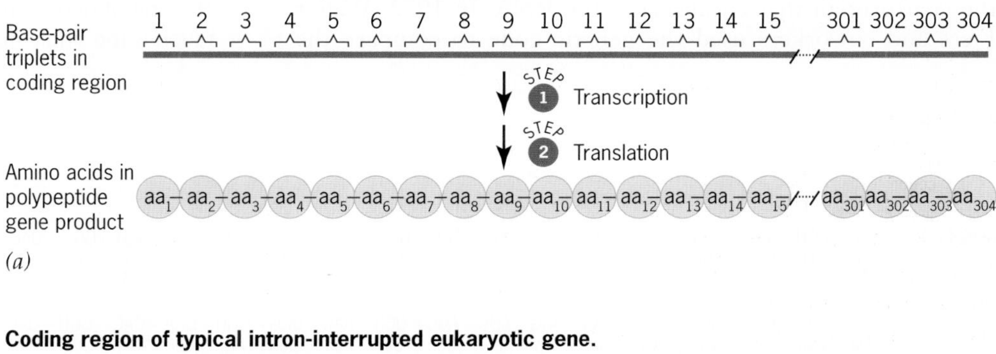

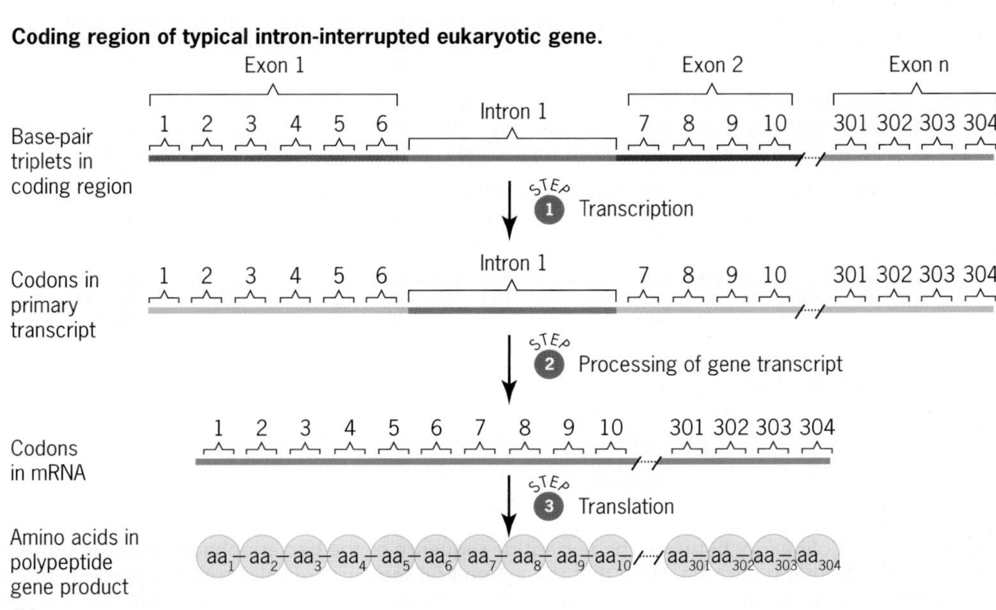

Figure 14.6 ► Colinearity between the coding regions of genes and their polypeptide products.

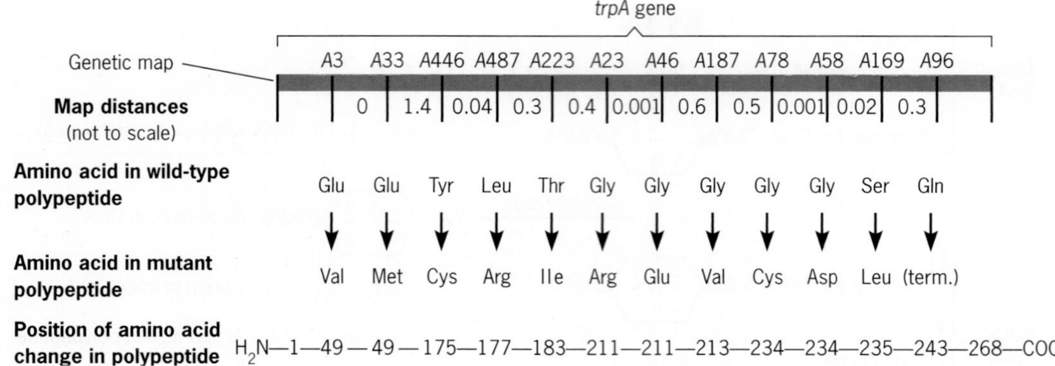

Figure 14.7 ▶ Colinearity between the *E. coli trpA* gene and its polypeptide product, the α polypeptide of tryptophan synthetase. The map positions of mutations in the *trpA* gene are shown at the top, and the locations of the amino acid substitutions produced by these mutations are shown below the map.

(UAG chain-termination) mutations and demonstrated a direct correlation between the length of the polypeptide fragment produced and the position of the mutation within the gene.

Definitive evidence for colinearity has been provided by direct comparisons of the nucleotide sequences of genes and the amino acid sequences of their polypeptide products. One of the first cases where the amino acid sequence of a polypeptide and the nucleotide sequence of the gene encoding it were both determined experimentally involved the coat protein of bacteriophage MS2 and the gene that encodes it. This small virus has an RNA genome that encodes only four proteins, one being the coat protein that encapsulates the RNA. In 1972, Walter Fiers and coworkers used the genetic code to compare the nucleotide sequence of the coat protein gene with the amino acid sequence of the coat polypeptide and found that the sequences exhibited perfect colinearity (**FIGURE 14.8**). Since then, similar results have established colinearity between many genes and their protein products in organisms ranging from viruses to humans.

KEY POINTS

- The concept of the gene has evolved from a bead on a string, not divisible by recombination or mutation, to a sequence of nucleotide pairs in DNA encoding one polypeptide chain.
- The unit of genetic material not divisible by recombination or mutation is the single nucleotide pair.

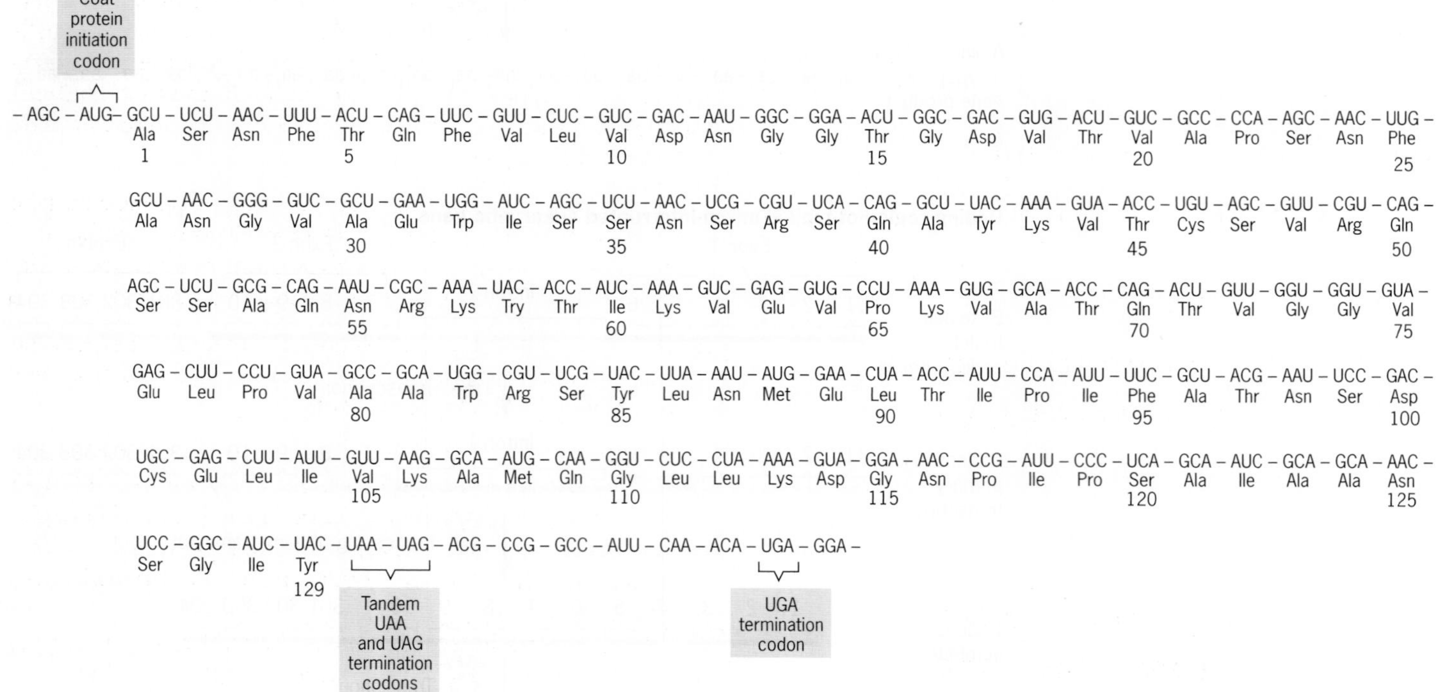

Figure 14.8 ▶ Colinearity between the nucleotide sequence of the bacteriophage MS2 coat protein gene and the amino acid sequence of the coat polypeptide that it encodes. Note that the amino acid sequence of this protein is precisely that predicted from the nucleotide sequence based on the genetic code. In addition, note that all three termination codons are present between the coat gene and the gene downstream from it on the MS2 chromosome.

A Genetic Definition of the Gene

The complementation or *trans* test provides an operational definition of the gene.

With the emergence of the one gene–one polypeptide concept, scientists could define the gene biochemically, but they had no genetic tool to use in determining whether two mutations were in the same or different genes. This deficiency was resolved when Edward Lewis developed the complementation test for functional allelism in the 1940s. We discuss Lewis's experiments in A Milestone in Genetics: Lewis's *Cis-Trans* Position Effect.

Before we discuss the complementation test in the next section and Lewis's experiments in A Milestone in Genetics, we need to introduce some new terms. A double heterozygote, which carries two mutations and their wild-type alleles, that is, m_1 and m_1^+ plus m_2 and m_2^+, can exist in either of two arrangements (**FIGURE 14.9**). When the two mutations are on the same chromosome, the arrangement is called the **coupling** or ***cis* configuration;** a heterozygote with this genotype is called a ***cis* heterozygote** (**FIGURE 14.9*a***). When the two mutations are on different chromosomes, the arrangement is called the **repulsion** or ***trans* configuration.** An organism with this genotype is a ***trans* heterozygote** (**FIGURE 14.9*b***).

THE COMPLEMENTATION TEST AS AN OPERATIONAL DEFINITION OF THE GENE

Lewis's discovery of the *cis-trans* position effects with *Star-asteroid* and w^a–w (see A Milestone in Genetics in this chapter) led to the development of the **complementation test** or ***trans* test** for functional allelism. The complementation test allows geneticists to determine whether mutations that produce the same or similar phenotypes are in the same gene or in different genes. They must test mutations pairwise by determining the phenotypes of *trans* heterozygotes. That is, they must construct *trans* heterozygotes with each pair of mutations to be analyzed and determine whether these heterozygotes have mutant or wild-type phenotypes.

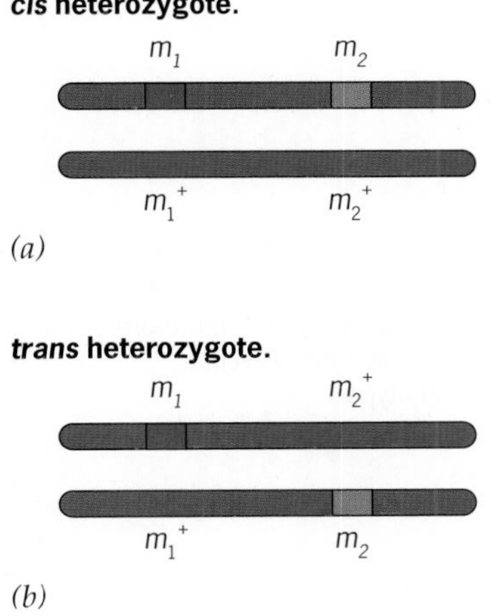

Figure 14.9 ▶ The arrangement of genetic markers in *cis* and *trans* heterozygotes.

Ideally, the complementation or *trans* test should be done in conjunction with the ***cis* test**—a control that is often omitted. *Cis* tests are performed by constructing *cis* heterozygotes for each pair of mutations to be analyzed and determining whether they have mutant or wild-type phenotypes. Together, the complementation or *trans* test and the *cis* test are referred to as the ***cis-trans* test.** Each *cis* heterozygote, which contains one wild-type chromosome, should have the wild-type phenotype whether the mutations are in the same gene or in two different genes (**FIGURE 14.10**). Indeed, the *cis* heterozygote must have the wild-type phenotype for the results of the *trans* test to be valid. If the *cis* heterozygote has the mutant phenotype, the *trans* test cannot be used to determine whether the two mutations are in the same gene. Thus, the *trans* test cannot be used with dominant mutations. Because *cis* heterozygotes contain one chromosome with wild-type copies of all relevant genes, these genes should specify functional products and a wild-type phenotype.

Whether two mutations are in the same gene or two different genes is determined by the results of the complementation or *trans* test. With diploid organisms, the *trans* heterozygote is produced simply by crossing organisms that are homozygous for each of the mutations of interest. With viruses, *trans* heterozygotes are produced by simultaneously infecting host cells with two different mutants. Regardless of how the two mutations are placed in a common protoplasm in the *trans* configuration, the results of the *trans* or complementation test provide the same information.

1. If the *trans* heterozygote has the mutant phenotype (the phenotype of organisms or cells homozygous for either one of the two mutations), then the two mutations are in the same unit of function, the same gene (**FIGURE 14.11*a***).
2. If the *trans* heterozygote has the wild-type phenotype, then the two mutations are in two different units of function, two different genes (**FIGURE 14.11*b***).

When the two mutations present in a *trans* heterozygote are both in the same gene, as shown for mutations m_1 and m_2 in **FIGURE 14.11*a***, both chromosomes will carry defective copies of that gene. As a result, the *trans* heterozygote will contain only nonfunctional products of the gene involved and will have a mutant phenotype.

When a *trans* heterozygote has the wild-type phenotype, the two mutations are said to exhibit complementation or to complement each other and are located in different genes. In the example illustrated in **FIGURE 14.11*b***, the chromosome carrying mutation m_1 in gene *1* has a wild-type copy of gene *2*, which specifies functional gene *2* product, and the chromosome

cis heterozygote: mutations in one gene.

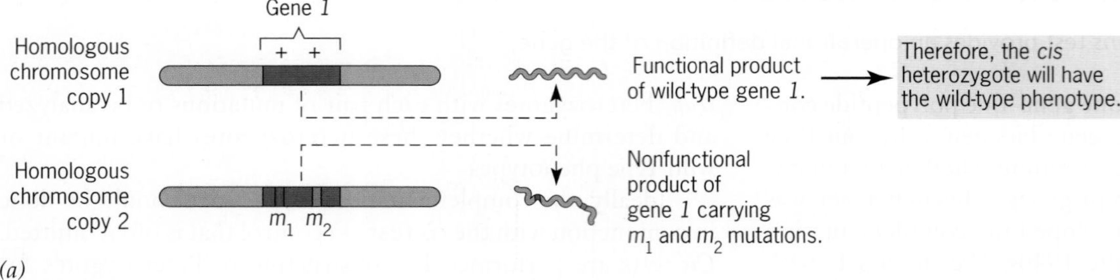

cis heterozygote: mutations in two different genes.

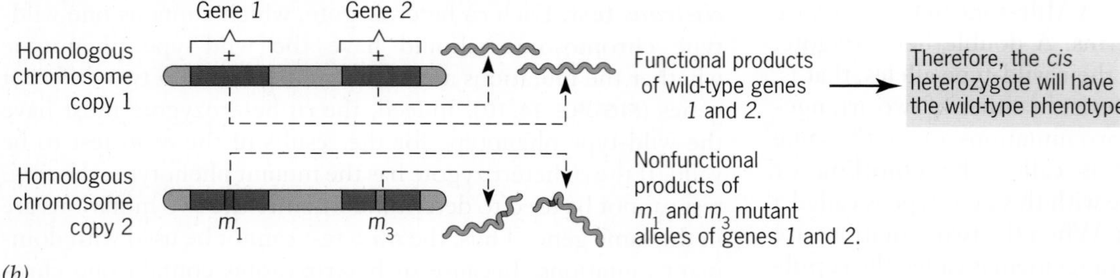

Figure 14.10 ▶ The *cis* test. The *cis* heterozygote should have the wild-type phenotype whether the mutations are in the same gene (*a*) or in two different genes (*b*).

carrying mutation m_3 in gene *2* has a wild-type copy of gene *1*, which encodes functional gene *1* product. Thus, the *trans* heterozygote shown contains functional products of both genes and has a wild-type phenotype.

Only the complementation test, or the *trans* part of the *cis-trans* test, is included in most genetic analyses. Constructing a chromosome that carries both mutations for the *cis* test is often difficult, especially with eukaryotes. Moreover, the *cis* hetero-

trans heterozygote: mutations in one gene.

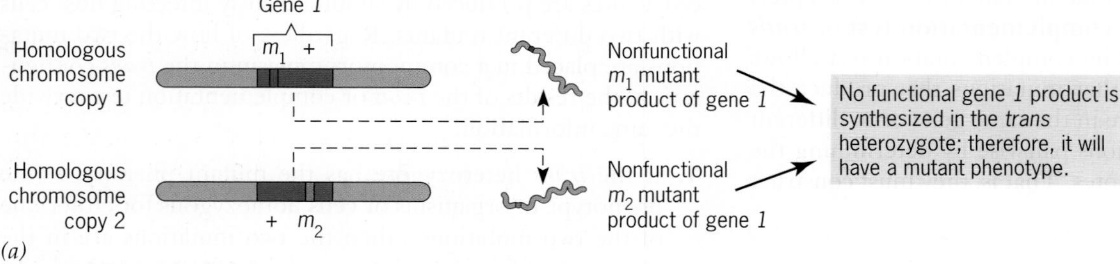

trans heterozygote: mutations in two different genes.

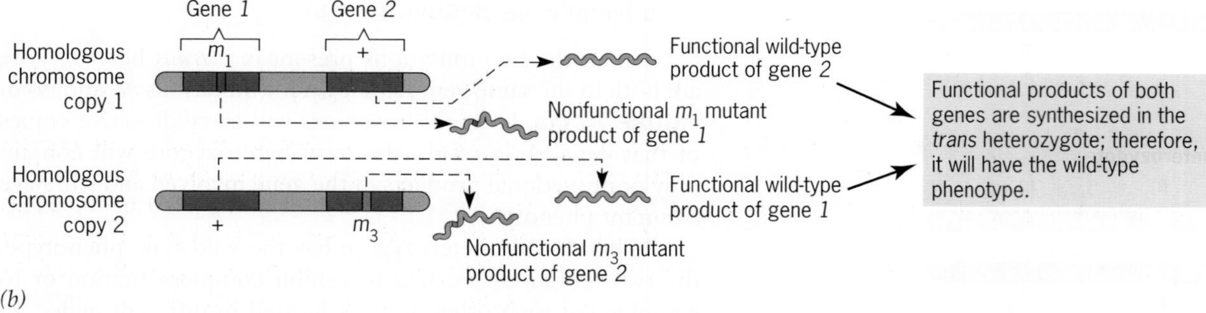

Figure 14.11 ▶ The *trans* test. The *trans* heterozygote should have (*a*) the mutant phenotype if the two mutations are in the same gene and (*b*) the wild-type phenotype if the mutations are in two different genes.

zygotes almost always have wild-type phenotypes if the mutations being analyzed are recessive. Thus, in most instances, the results of complementation or *trans* tests can be interpreted correctly without carrying out the laborious *cis* tests.

Seymour Benzer introduced the term **cistron** to refer to the unit of function operationally defined by the *cis-trans* test. However, today, most geneticists consider the terms *gene* and *cistron* to be synonyms. Thus, we will use gene, rather than cistron, throughout this text.

The gene is operationally defined as the unit of function by the complementation or *trans* test, which is used to determine whether mutations are in the same gene or different genes. The complementation test is one of the three basic tools of genetics; the other two are recombination (Chapter 7) and mutation (Chapter 13).

Note that the information provided by complementation tests is totally distinct from that obtained from recombination analyses. The results of complementation tests indicate whether mutations are allelic, whereas the results of recombination analyses indicate whether mutations are linked and, if so, provide estimates of how far apart they are on a chromosome. Nevertheless, complementation and recombination are sometimes misunderstood. Thus, we will contrast complementation and recombination by illustrating both phenomena with the same three mutations in bacteriophage T4, which we discussed in Chapter 8. Because of the simple structure of the virus and the direct relationship between specific gene products and phenotypes, the phage T4 system provides an excellent visualization of the difference between complementation and recombination.

We discussed the *amber* mutations of bacteriophage T4 in Chapter 12 (see **FIGURE 12.22**). *Amber* mutations produce translation–termination triplets within the coding regions of genes. As a result, the products of the mutant genes are truncated polypeptides, which are almost always totally nonfunctional. Therefore, complementation tests performed with *amber* mutations are usually unambiguous. When *amber* mutations occur in essential genes, the mutant phenotype is lethality—that is, no progeny are produced when a restrictive host cell is infected. The wild-type phenotype is a normal yield (about 300 phage per cell) of progeny phage in each infected restrictive host cell. In Chapter 13, we called such mutations conditional lethals and emphasized their utility in genetic analysis. With conditional lethals, the distinction between the mutant and wild-type phenotypes is maximal: lethality versus normal growth.

Two of the three *amber* mutations that we will consider (*am*B17 and *am*H32) are in gene *23*, which encodes the major structural protein of the phage head; the other mutation (*am*E18) is in gene *18*, which specifies the major structural protein of the phage tail. We can see from **FIGURE 14.12** why complementation occurs between mutations *am*B17 (head gene) and *am*E18 (tail gene), and why complementation does not occur between mutations *am*B17 and *am*H32 (both in the head gene). Complementation is the result of the functionality of the gene *products* specified by chromosomes carrying two different mutations when they are present in the same protoplasm. Complementation does not depend on recombination of the two chromosomes. *Complementation, or the lack of it, is assessed by the phenotype (wild-type or mutant) of each* trans *heterozygote*. You can test your understanding of the complementation test by answering the questions posed in Focus on Problem Solving: Assigning Mutations to Genes.

FIGURE 14.13 illustrates the occurrence of recombination between the *amber* mutations used to illustrate complementation in **FIGURE 14.12**. We discussed the basic features of crosses between different mutant T4 phage in Chapter 8. Recall that recombination of phage genes occurs by a process analogous to crossing over in eukaryotes, with linkage distances measured in map units, just as in eukaryotes. Recombination frequencies are measured by infecting permissive host cells with two mutants so that the mutant chromosomes can replicate and participate in crossing over. Then, the progeny are screened for wild-type recombinants by plating them on lawns of restrictive host cells (*E. coli* cells in which only the wild-type phage can grow). In the example shown in **FIGURE 14.13**, recombination is observed in both crosses: (1) *am*B17 × *am*E18, mutations in two different genes, and (2) *am*B17 × *am*H32, mutations in the same gene. The only difference is that more recombinants are produced in cross 1, which involves two *amber* mutations that are relatively far apart on the phage T4 chromosome, than in cross 2, which involves two mutations located near one another in the same gene. *Recombination involves the actual breakage of chromosomes and reunion of parts to produce wild-type and double-mutant chromosomes.*

Structural allelism is the occurrence of two or more different mutations at the same site and is determined by the recombination test. Two mutations that do not recombine are structurally allelic; the mutations either occur at the same site or overlap a common site. *Functional allelism* is determined by the complementation test as just described; two mutations that do not complement are in the same unit of function, the same gene. Mutations that are both structurally and functionally allelic are called **homoalleles;** they do not complement or recombine with each other. Mutant homoalleles have defects at the same site or overlap a common site in the same gene. Mutations that are functionally allelic, but structurally nonallelic, are called **heteroalleles;** they recombine with each other but do not complement one another. Mutant heteroalleles occur at different sites but within the same gene.

INTRAGENIC COMPLEMENTATION

The results of complementation tests are usually unambiguous when mutations that result in the synthesis of no gene product, partial gene products, or totally defective gene products are used—for example, deletions of segments of genes, frameshift mutations, or polypeptide chain-terminating mutations. Of course, the mutations must be recessive. When mutations causing amino acid substitutions are used, the results of complementation tests are sometimes ambiguous because of the occurrence of a phenomenon called **intragenic complementation.**

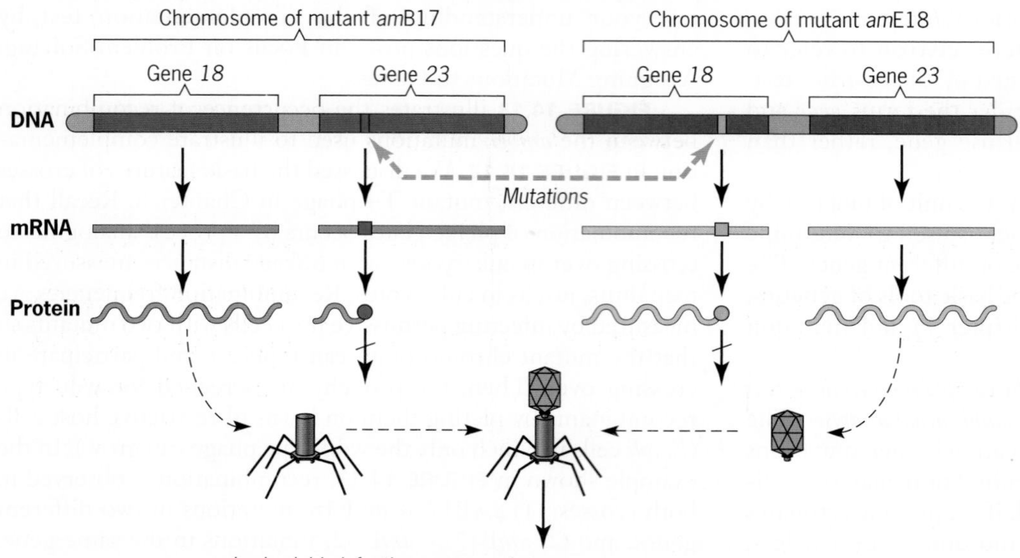

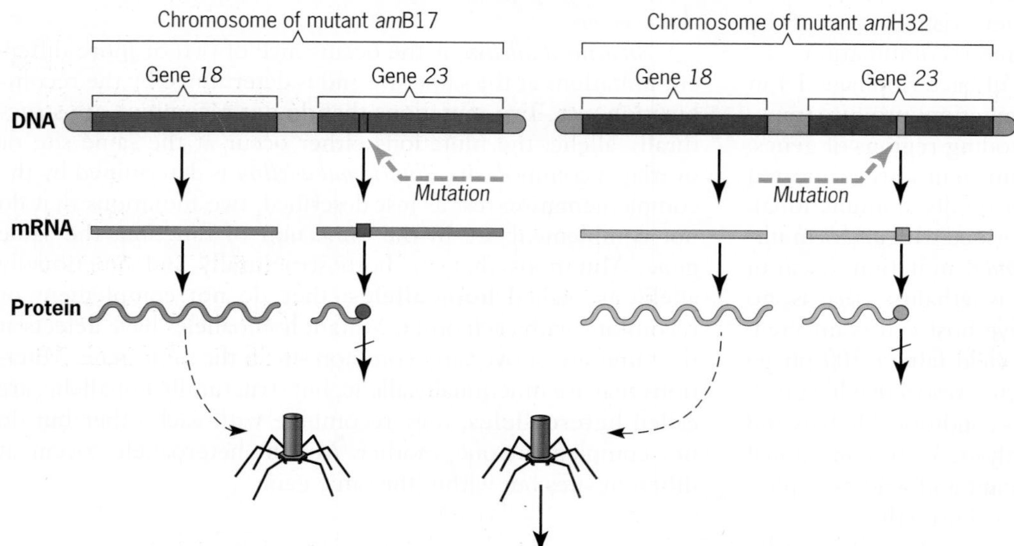

Figure 14.12 ▶ Complementation and noncomplementation in *trans* heterozygotes. (*a*) Complementation between mutation *am*B17 in gene *23*, which encodes the major structural protein of the phage T4 head, and mutation *am*E18 in gene *18*, which encodes the major structural protein of the phage tail. Phage heads and tails are both synthesized in the cell, with the result that infective progeny phage are produced. (*b*) When the *trans* heterozygote contains two mutations (*am*B17 and *am*H32) in gene *23*, no heads are produced, and no infective progeny phage can be assembled. Compare with Figure 14.13.

The functional forms of some proteins are dimers or higher multimers consisting of two or more polypeptides. These polypeptides may be either homologous, the products of a single gene, or nonhomologous, the products of two or more distinct genes. When the active form of the protein contains two or more homologous polypeptides (it may or may not also contain nonhomologous polypeptides), intragenic complementation may occur. *Inter*genic complementation (discussed in the preceding section) and *intra*genic complementation (described in this section) are two distinct phenomena.

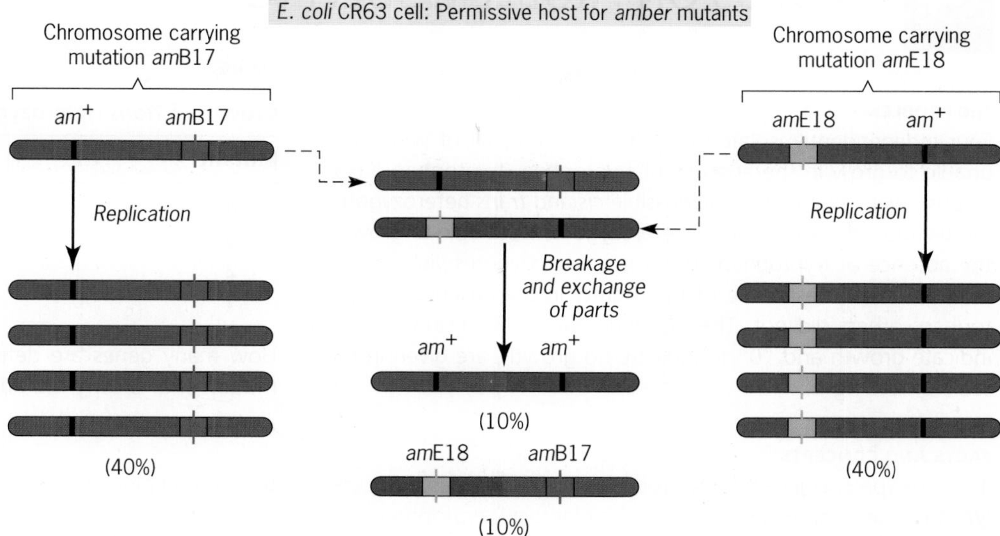

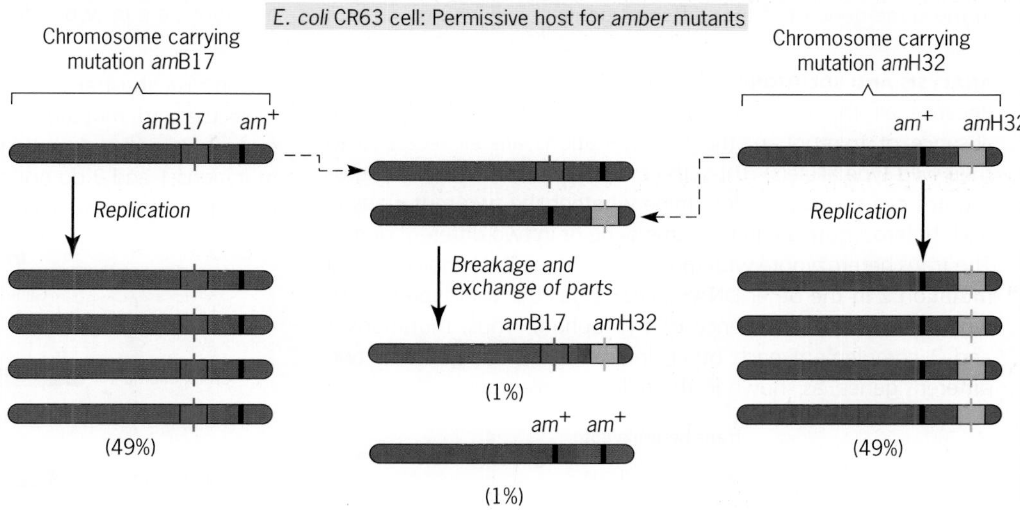

Figure 14.13 ▶ Recombination between (*a*) the complementing mutations *am*B17 (gene *23*) and *am*E18 (gene *18*), and (*b*) the noncomplementing mutations *am*B17 and *am*H32 (both in gene *23*). Recombination occurs in both cases; however, fewer recombinants are produced in cells infected with *am*B17 and *am*H32 because the two mutations are located closer together on the phage T4 chromosome. Compare with Figure 14.12.

Let us consider an enzyme that functions as a homodimer, that is, a protein containing two copies of a specific gene product (**FIGURE 14.14**). In organisms that are homozygous for the wild-type allele of the gene, all the protein dimers will contain identical wild-type polypeptides. Similarly, organisms that are homozygous for any mutation in the gene will contain dimers with two mutant polypeptides. An organism that is heterozygous for two different mutations in the gene will produce some dimers that contain the two different mutant polypeptides. We call these heterodimers. Such heterodimers may have partial or complete (wild-type) function. If they do have partial or complete function, intragenic complementation will occur, and the *trans* heterozygote will have a wild-type phenotype or a phenotype intermediate between mutant and wild-type

▶ FOCUS ON PROBLEM SOLVING
Assigning Mutations to Genes

THE PROBLEM

Four independently isolated mutants of *E. coli,* all of which are unable to grow in the absence of tryptophan (tryptophan auxotrophs), were examined in all possible *cis* and *trans* heterozygotes (partial diploids). All of the *cis* heterozygotes were able to grow in the absence of tryptophan. The *trans* heterozygotes yielded two different responses: some of them grew in the absence of tryptophan; others did not. The experimental results, using "+" to indicate growth and "0" to indicate no growth, are given in the table on the right.

Growth of *Trans* Heterozygotes on Medium Lacking Tryptophan

Mutant	1	2	3	4
4	+	0	+	0
3	0	+	0	
2	+	0		
1	0			

How many genes are defined by these four mutations? Which mutant strains carry mutations in the same gene(s)?

FACTS AND CONCEPTS

1. Wild-type *E. coli* cells (prototrophs) encode enzymes that catalyze the biosynthesis of the essential amino acid tryptophan.

2. Tryptophan auxotrophs are unable to synthesize their own tryptophan.

3. Tryptophan auxotrophs grow normally in the presence of tryptophan; however, they cannot grow in its absence.

4. The complementation test (the phenotype of a *trans* heterozygote) often can be used to determine whether two mutations are in the same gene or in two different genes.

5. The complementation test yields unambiguous results only if the *cis* heterozygote (see **FIGURE 14.10**) containing the same two mutations has the wild-type phenotype. Therefore, complementation tests are informative only when the mutant alleles are recessive to the wild-type allele(s).

6. If a *trans* heterozygote has a mutant phenotype, the two mutations are in the same gene (see **FIGURE 14.11*a***).

7. If a *trans* heterozygote has the wild-type phenotype, the two mutations are in two different genes (see **FIGURE 14.11*b***).

ANALYSIS AND SOLUTION

Because all the *cis* heterozygotes were able to grow in the absence of tryptophan, the four mutations are all recessive to their wild-type allele(s). Thus, the phenotypes of the *trans* heterozygotes can be used to determine whether the two mutations in each heterozygote are in the same gene or in two different genes. The *trans* heterozygote with mutation 1 in one DNA molecule and mutation 2 in the other DNA molecule has the wild-type phenotype (growth in the absence of tryptophan). Thus, mutations 1 and 2 complement each other, indicating that they are in two different genes, as shown in the following diagram.

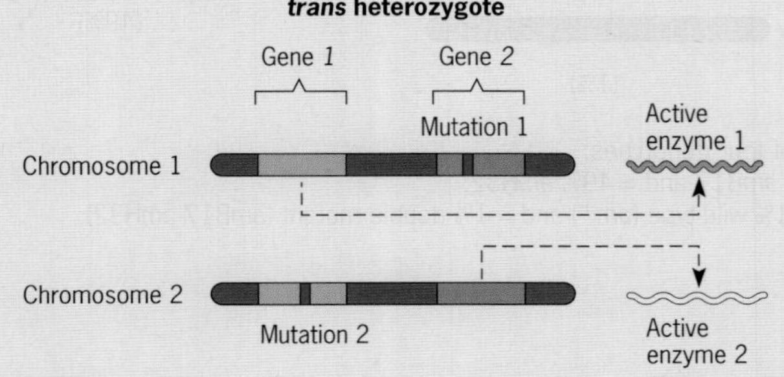

Mutations 1 and 2 complement each other, and the *trans* heterozygote has the wild-type phenotype. Therefore, mutations 1 and 2 are in different genes.

Likewise, the *trans* heterozygotes containing (i) mutations 1 and 4, (ii) mutations 2 and 3, and (iii) mutations 3 and 4 all exhibited the wild-type phenotype, indicating that the mutations are in different genes.

In contrast, the *trans* heterozygote with mutation 1 in one DNA molecule and mutation 3 in the other DNA molecule has the mutant phenotype (no growth in the absence of tryptophan). Thus, mutations 1 and 3 do not complement each other, indicating that they are in the same gene, as shown in the following diagram.

***trans* heterozygote**

Gene *1* Gene *2*

Mutation 1

Chromosome 1

No active enzyme 2

Chromosome 2

Mutation 3

Mutations 1 and 3 do NOT complement each other, and the *trans* heterozygote has the mutant phenotype because no active enzyme 2 is present. Therefore, mutations 1 and 3 are in the same gene.

The same result was obtained with mutations 2 and 4; they do not complement each other and are located in the same gene. Collectively, the results of the six complementation tests show that the four mutations define two different genes, with mutations 1 and 3 in one gene and mutations 2 and 4 in another gene. There are five genes in *E. coli* that encode tryptophan biosynthetic enzymes (see **FIGURE 19.11**); however, the four mutations analyzed in this problem are located in just two of these genes.

For further discussion go to your *WileyPLUS* course.

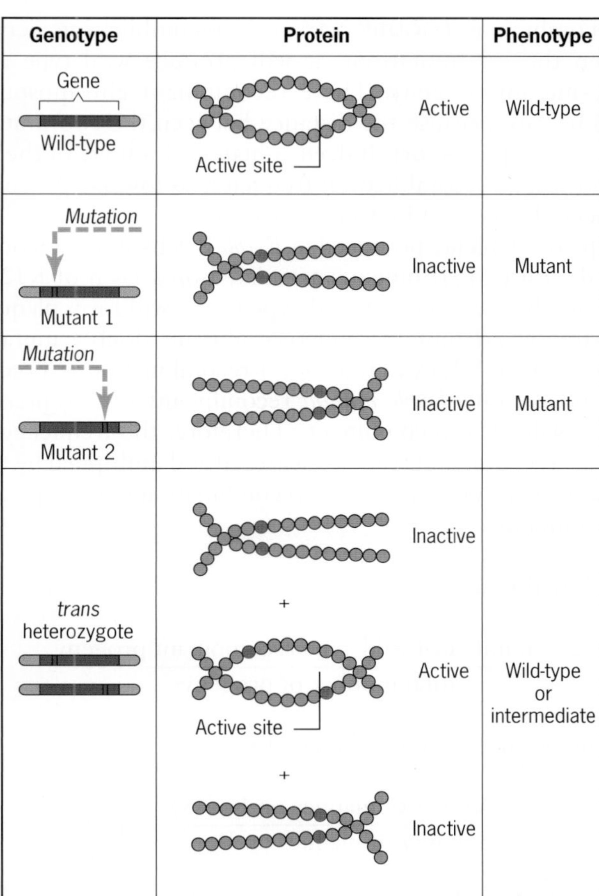

Figure 14.14 ▶ *Intra*genic complementation sometimes occurs when the active form of an enzyme or structural protein is a multimer that contains at least two copies of any one gene product. Here, the functional form of the enzyme is a dimer composed of two polypeptides encoded by one gene. The amino acids altered by the mutations are shown as red circles in the polypeptide chains; they disrupt the normal three-dimensional structure of the protein.

(**FIGURE 14.14**, bottom). In the case of noncomplementing mutations in a gene encoding a multimeric protein, the heteromultimers are nonfunctional, just like the mutant homomultimers (protein multimers composed of two or more identical mutant polypeptides).

LIMITATIONS ON THE USE OF THE COMPLEMENTATION TEST

The complementation test has been very useful in operationally delimiting genes. Usually, two or more mutations that produce the same phenotype can be assigned to one or more genes based on the results of complementation tests. However, in some cases, the results of complementation tests cannot be used to delimit genes. As previously mentioned, *complementation tests are not informative in studies of dominant or codominant mutations or in cases where intragenic complementation occurs*. In addition, complementation tests are sometimes uninformative because of epistatic interactions between the mutant gene products. If the *cis* test is done, such interactions are readily detected because the *cis* heterozygotes will have mutant phenotypes rather than the required wild-type phenotype.

Another limitation of the complementation test is encountered in working with so-called polar mutations. A **polar mutation** is a mutation that not only results in a defective product of the gene in which it is located but also interferes with the expression of one or more adjacent genes. The adjacent genes are always located on one side of the gene carrying the mutation (thus the term *polar mutation*). Such polar mutations are frequently observed in prokaryotes in coordinately regulated sets of genes called operons (Chapter 19). They usually are mutations resulting in polypeptide chain-termination signals (nucleotide-pair triplets yielding UAA, UAG, and UGA codons in mRNA) within genes. These polar mutations interfere with the expression of genes located downstream of the mutant gene. As a result, polar mutations fail to complement mutations in genes subject to the polar effect. Thus, the results of complementation tests performed with polar mutations are often ambiguous.

KEY POINTS

- The complementation or *trans* test provides an operational definition of the gene; it is used to determine whether mutations are in the same gene or different genes.
- Intragenic complementation may occur when a protein is a multimer containing at least two copies of one gene product.

The *rII* Locus of Bacteriophage T4

Benzer's map of the *rII* locus of phage T4 is the most detailed genetic map constructed to date.

The most detailed genetic map ever constructed is that of the *rII* locus of bacteriophage T4. We discussed the life cycle of phage T4 and the procedure used to perform crosses between phage mutants in Chapter 8. We also discussed rapid lysis (*r*) mutants that alter the morphology of the plaques produced when the phage are grown on a lawn of *E. coli* cells on agar medium (see **FIGURE 8.7**).

THE *rII* MUTANTS ARE CONDITIONAL LETHALS

In the discussion of the *r* mutants in Chapter 8, we did not tell you that T4 *r* mutants of one type (the *rII* mutants) are conditional lethal mutants, like the *amber* mutants that we discussed in Chapter 12. (Mutants at two other T4 loci—*rI* and *rIII*—are not conditional lethals.) The *rII* mutants can grow on certain strains of *E. coli*, such as B. They cannot grow—are lethal—on other *E. coli* strains, such as K12(λ)—K12 cells that contain λ prophages. This conditional lethality of the *rII* mutants provides a powerful tool for selecting rare wild-type (r^+) recombinants produced in genetic crosses.

Seymour Benzer and coworkers exploited the conditional lethality of the *rII* mutants to construct an amazingly detailed map of the *rII* locus. They mapped 2400 *rII* mutants to 304 sites separable by recombination in two contiguous genes in a small region of the bacteriophage T4 chromosome. Four additional sites were identified in another study, bringing the total number of sites to 308. Benzer's experiments extended the results of earlier studies on *Drosophila* by Oliver, Green, Lewis, and others, which showed that the gene is divisible by both mutation and recombination.

Because of their unique plaque morphology (see **FIGURE 8.7**), *r* mutants are easy to distinguish from wild-type phage. The *r* mutants then have to be tested for their ability to grow on *E. coli* K12(λ) to eliminate *rI* and *rIII* mutants. Only the *rII* mutants fail to grow on *E. coli* K12(λ). Thus, it was quite easy for Benzer and colleagues to isolate a large number of *rII* mutants.

COMPLEMENTATION TESTS SHOW THAT THE *rII* LOCUS CONTAINS TWO GENES

Given a large collection of *rII* mutants, Benzer asked how many genes they defined. To answer this question, he performed complementation tests. He simultaneously infected *E. coli* K12(λ) cells (the restrictive host) with two different *rII* mutants and asked whether the infected cells (*trans* heterozygotes) had the mutant—lethality, or no progeny—phenotype or the wild-type—300 progeny per infected cell—phenotype (**FIGURE 14.15**). The results of the complementation tests showed that all of Benzer's *rII* mutants contained mutations in one of two contiguous genes, designated *rIIA* and *rIIB*. In a few cases, the mutations extended into both genes.

Once Benzer had identified mutations in the two genes, he used them as reference mutants for the two genes. He infected *E. coli* K12(λ) cells with each new *rII* mutant and (1) an *rIIA* reference mutant and (2) an *rIIB* reference mutant. Because the *rII* locus contains only two genes, two complementation tests were sufficient to assign each new mutation to either the *rIIA* or the *rIIB* gene. If the two *rII* mutants contained mutations in the same gene (that is, both were *rIIA* mutants or both were *rIIB* mutants), they failed to grow on *E. coli* K12(λ); if they carried mutations in different genes (one was *rIIA* and the other was *rIIB*), they grew on *E. coli* K12(λ).

MAPPING *rII* MUTATIONS BY TWO-FACTOR CROSSES

Crosses between *rII* mutants were performed by simultaneously infecting *E. coli* B cells—the permissive host—with two different *rII* mutants and screening the progeny for wild-type (r^+) recombinants (**FIGURE 14.16**). If recombination occurs between the two mutations, it will produce wild-type and double-mutant progeny. These recombinant chromosomes should be rare because the distance between the two mutant sites must be quite short. Indeed, a major advantage of the *rII* system is that a recombination frequency as low as 10^{-6} could have been detected in Benzer's experiments.

If progeny phage produced in *E. coli* B cells simultaneously infected with two *rII* mutants are plated on a lawn of K12(λ) cells, only the recombinant wild-type phage will form plaques. The parental *rII* mutants and recombinant double mutants cannot grow on K12(λ). However, if recombination is a reciprocal process, one *double-mutant* recombinant will be present for each wild-type recombinant. Therefore, the frequency of wild-type recombinants can be measured and multiplied by two to obtain the total frequency of recombination (wild-type and double mutants).

Recombination frequency =

$$\frac{2 \times \text{number of wild-type recombinant progeny}}{\text{total number of progeny}}$$

and map distance (% recombinants) =

$$\frac{2 \times \text{number of plaques on K12}(\lambda)}{\text{number of plaques on B}} \times 100$$

To calculate the frequency of recombinant progeny, the total number of progeny phage produced per unit volume of infected cells must be determined. The total number of progeny phage is calculated by diluting the lysate appropriately and plating a sample of the diluted lysate on a lawn of *E. coli* B cells. All of the progeny phage (parental and recombinant) produced in a cross between two *rII* mutants can grow on *E. coli* B cells. If 250 plaques are counted in a sample diluted 10^6-fold, then the original sample contained 2.5×10^8 phage per unit volume (250×10^6). If 300 recombinant phage (wild-type and double mutant) are present in a 10^4 dilution of the lysate, the recombination frequency between the two mutant sites is $3 \times 10^6/2.5 \times 10^8 = 0.012$, and the map distance is 1.2 centiMorgans.

DELETION MAPPING

Benzer characterized about 60 independent *rII* mutations by the two-factor crosses described above. However, the procedures were labor-intensive. To map all 2400 of his mutations by two- and three-factor crosses would have taken many years. Thus, Benzer devised a new and more efficient mapping procedure—**deletion mapping**—and used it to complete the map of the *rII* locus in just a few years. Benzer demonstrated that some of his *rII* mutants contained deletions of all or part of the *rII* region. A phage that carries a deletion cannot mutate back to wild-type, nor can it recombine with another phage that has a mutation in

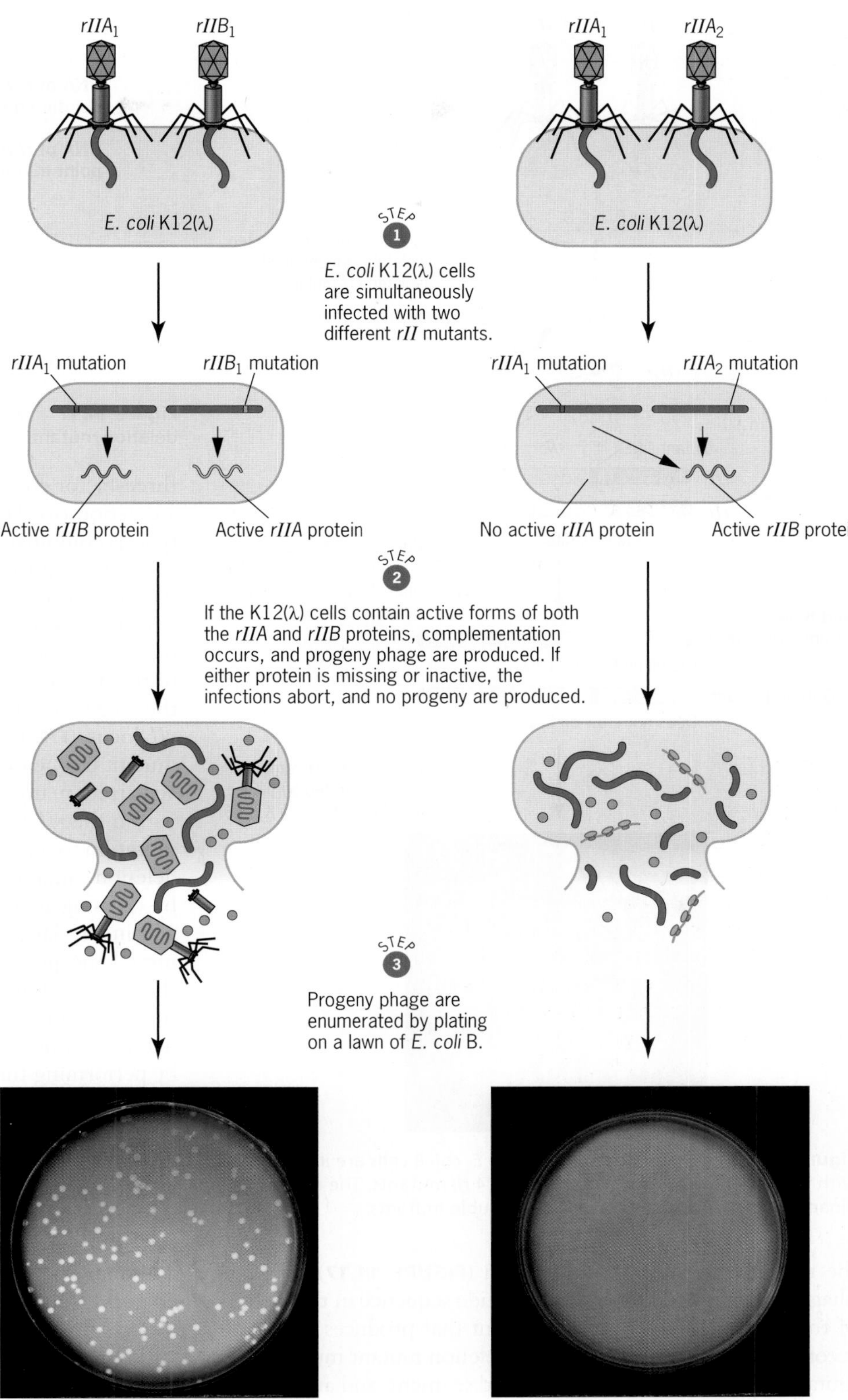

Figure 14.15 ► Complementation tests between *rII* mutants of phage T4. *E. coli* K12(λ) cells are simultaneously infected with two different *rII* mutants. (*a*) If the two mutations are in different genes, complementation will occur, and progeny phage will be produced. (*b*) If the two mutations are in the same gene, they will not complement each other, and no progeny will be produced.

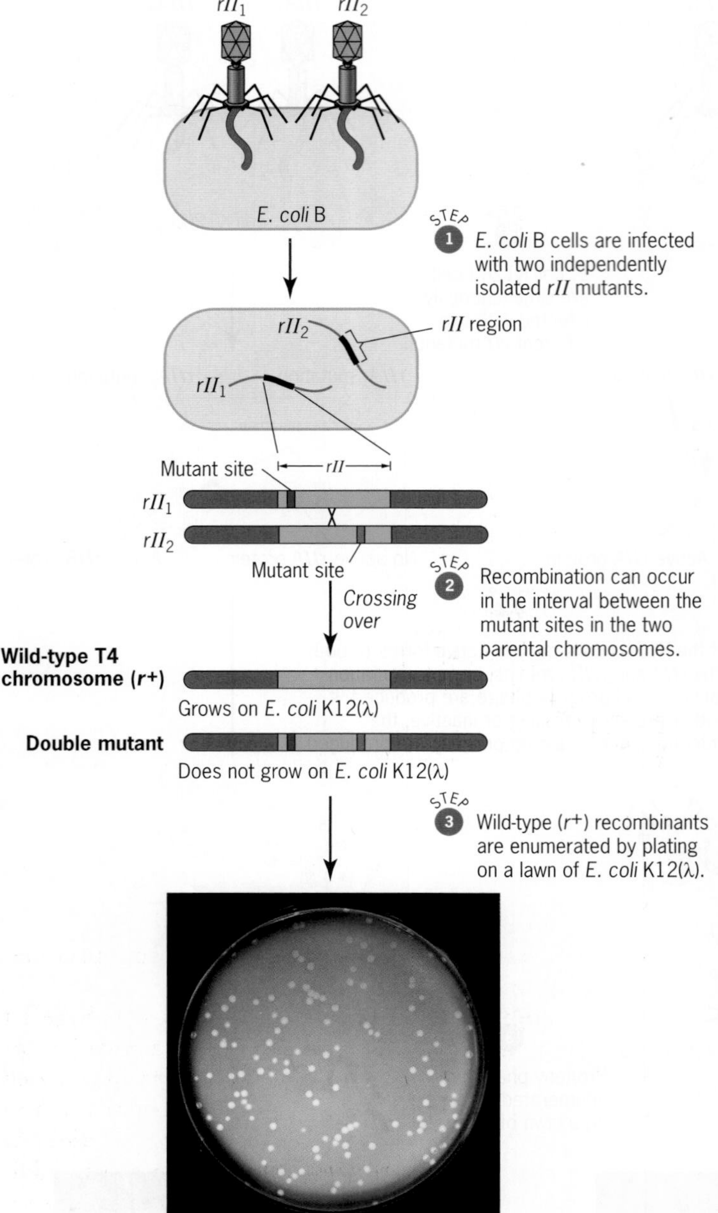

Figure 14.16 ▶ A genetic cross in which *E. coli* B cells are infected with two independently isolated phage T4 *rII* mutants. The recombinant progeny are wild-type (r^+) and double mutants.

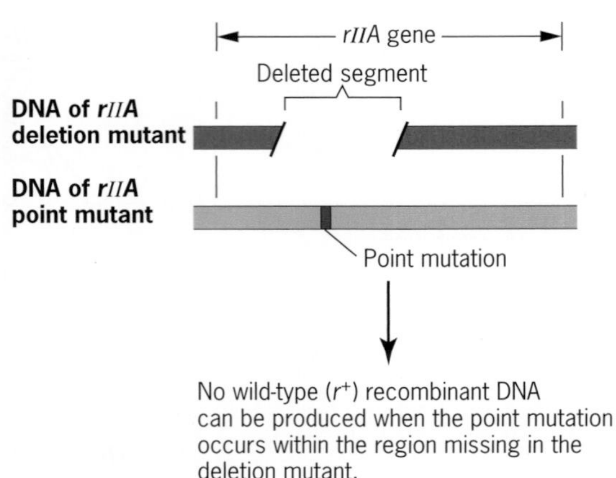

Figure 14.17 ▶ A cross between an *rIIA* point mutant and an *rIIA* deletion mutant.

the region removed by the deletion (**FIGURE 14.17**). Neither phage will have the wild-type nucleotide sequence in this region of the gene. Therefore, an *rII* mutant that produces wild-type recombinants when crossed with a deletion mutant must carry a mutation that maps *outside* the deleted segment, and any mutant that does not produce wild-type recombinants in the cross must carry a mutation that maps *within* the deleted area.

Before he could use deletions in his short-cut mapping procedure, Benzer first had to determine the sizes of the deleted regions. He accomplished this by crossing the deletions with point mutations that had been mapped by standard two- and three-factor crosses (Chapter 7). His rationale was simple: (1) if a deletion overlapped the site of the point mutation, no wild-type recombinants could be produced because both mutants had defects at the same site in the *rII* region, or (2) if a deletion did not overlap the site of the defect in the point mutant, then wild-type recombinant progeny would be produced. Benzer used this procedure to determine the sizes and the end points of the deletions in these mutants. Seven large deletions (called the "big seven") that were missing overlapping segments of the *rII* locus (**FIGURE 14.18**) were used to map each new point mutant to one of seven intervals of the locus. Benzer also characterized many smaller deletions that defined 47 intervals within the *rII* region (**FIGURE 14.18**). After he mapped a point mutation to one of the large intervals defined by the "big seven" deletion mutants, he mapped it to one of the 47 small intervals by crossing it with each of the deletions that had end points within the large interval (**FIGURE 14.19**). Benzer then determined the precise location of the point mutation within the short interval by crossing it with each of the other point mutations that mapped to the same interval. He determined the definitive order of the point mutations within each small interval by performing three-factor crosses as described in Chapter 7.

Let's consider an example. When point mutation *r*548 was crossed with the "big seven"—*r*1272, *r*1241, *r*J3, *r*PT1, *r*PB242, *r*A105, and *r*638—it recombined with *r*A105 and *r*638, but not with the other five large deletions. These results showed that the mutation in *r*548 overlapped with deletions *r*1272, *r*1241, *r*J3, *r*PT1, and *r*PB242, but not with deletions *r*A105 and *r*638, and demonstrated that *r*548 is located in interval A5 (**FIGURE 14.19*b***). Point mutation *r*548 was then crossed with deletions *r*1605, *r*1589, *r*PB230, and *r*1993—deletions with endpoints within the A5 interval. It recombined with *r*PB230 and *r*1993 but not with *r*1605 and *r*1589. These results demonstrated that *r*548 is located in subinterval A5c2, the segment of interval A5 missing in deletions *r*1605 and *r*1589 but present in deletions *r*PB230 and *r*1993 (**FIGURE 14.19*c***). Benzer then only had to cross *r*548 with the other mutations that mapped in interval A5 to complete the fine structure map of this segment of the *rIIA* gene (**FIGURE 14.19*d***).

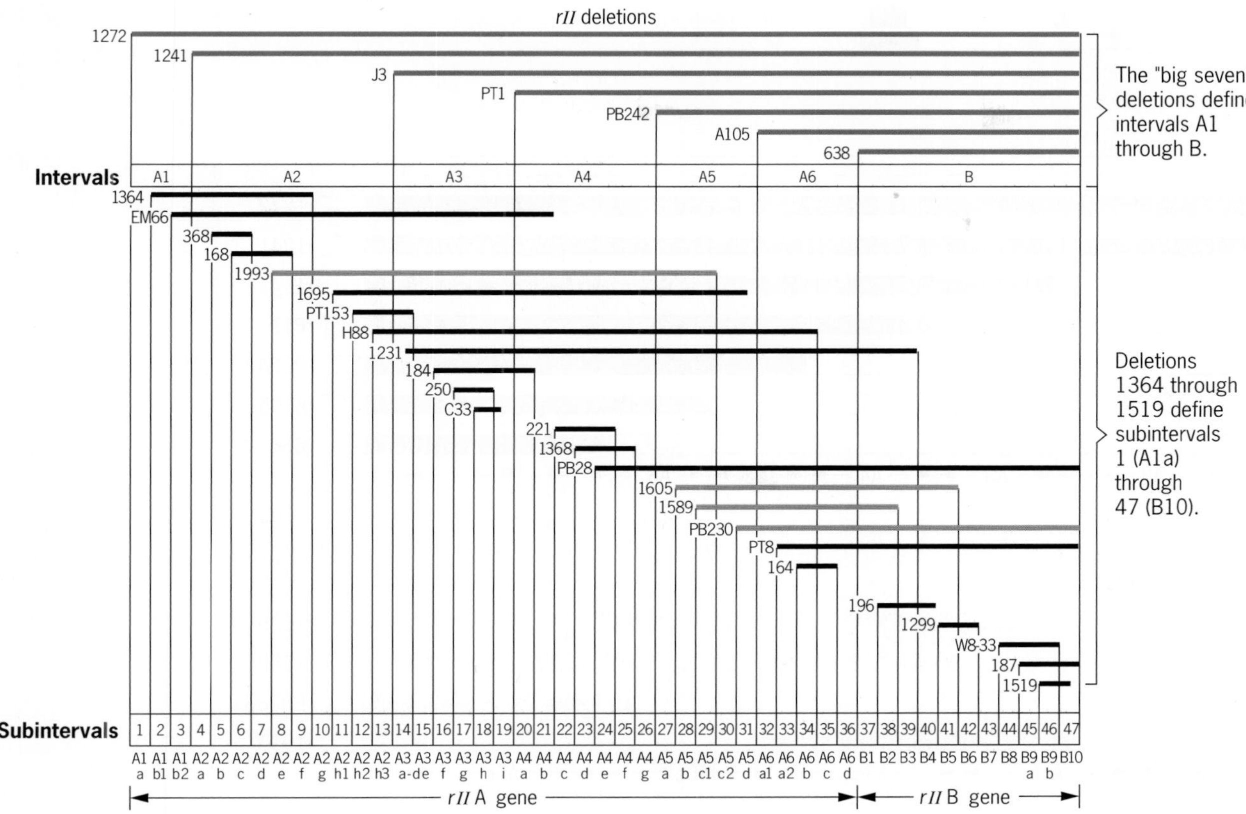

Figure 14.18 ▶ Benzer's phage T4 *rII* deletion mutants. The deletions shown in orange define the four subintervals of interval A5 (see Figure 14.19).

THE *rII* LOCUS: MANY SITES OF MUTATION IN TWO ADJACENT GENES

Benzer used the procedures described above to map 2400 *rII* mutants to 308 distinct sites in the *rIIA* and *rIIB* genes (**FIGURE 14.20**). The smallest recombination frequency that Benzer observed was 0.02 percent, which corresponds to about 2.3 base pairs. Thus, Benzer's results suggested, and later experiments proved, that recombination can occur between adjacent nucleotide pairs. How large is the *rII* locus? How many base pairs are present in each gene? Now that the T4 genome has been sequenced, we know that the coding regions of the *rIIA* and *rIIB* genes are 2175 and 936 nucleotide pairs, respectively, in length. Thus, the *rIIA* gene is over twice the size of the *rIIB* gene, in good agreement with Benzer's genetic map (**FIGURE 14.20**).

Interestingly, the *rII* mutants were not randomly distributed over the 308 sites. Some *rII* sites, called mutational *hot spots*, mutated more often than others; indeed, over 500 spontaneous mutations occurred at one hot spot in the *rIIB* gene (**FIGURE 14.20**). We now know that these hot spots are sites where a particular base pair is repeated several times. The wild-type nucleotide-pair sequence at the *rIIB* hot spot is six tandem A:T base pairs. The mutations at this site are of two types: 5 A:T base pairs and 7 A:T base pairs. Apparently, DNA polymerase tends to "stutter" during the replication of a tandem array of A:T base pairs—sometimes deleting a base pair and sometimes inserting an extra base pair.

The existence of a detailed map of the *rII* locus of phage T4 provided researchers with opportunities to explore other important questions in genetics. For example, the *rII* system was used to study the mechanisms by which various chemicals induce mutations (Chapter 13). The first evidence that the genetic code was a triplet code resulted from studies of single base-pair addition and deletion mutants (Chapter 12). In addition, crosses involving *rII* mutants provided researchers with important information about recombination mechanisms. Indeed, Benzer's analysis of the T4 *rII* locus was a true tour de force; it yielded an amazingly detailed picture of the structure of the gene.

KEY POINTS

- Benzer mapped over 2400 mutations to 308 distinct sites in the two genes at the *rII* locus of T4.
- Benzer's results showed that the base pair was the smallest unit altered by mutation and that recombination occurred between adjacent base pairs.
- Overlapping deletions were used to speed up the mapping of large numbers of mutations.
- Benzer's results solidified the concept of the gene as a unit of function that is divisible by mutation and recombination.

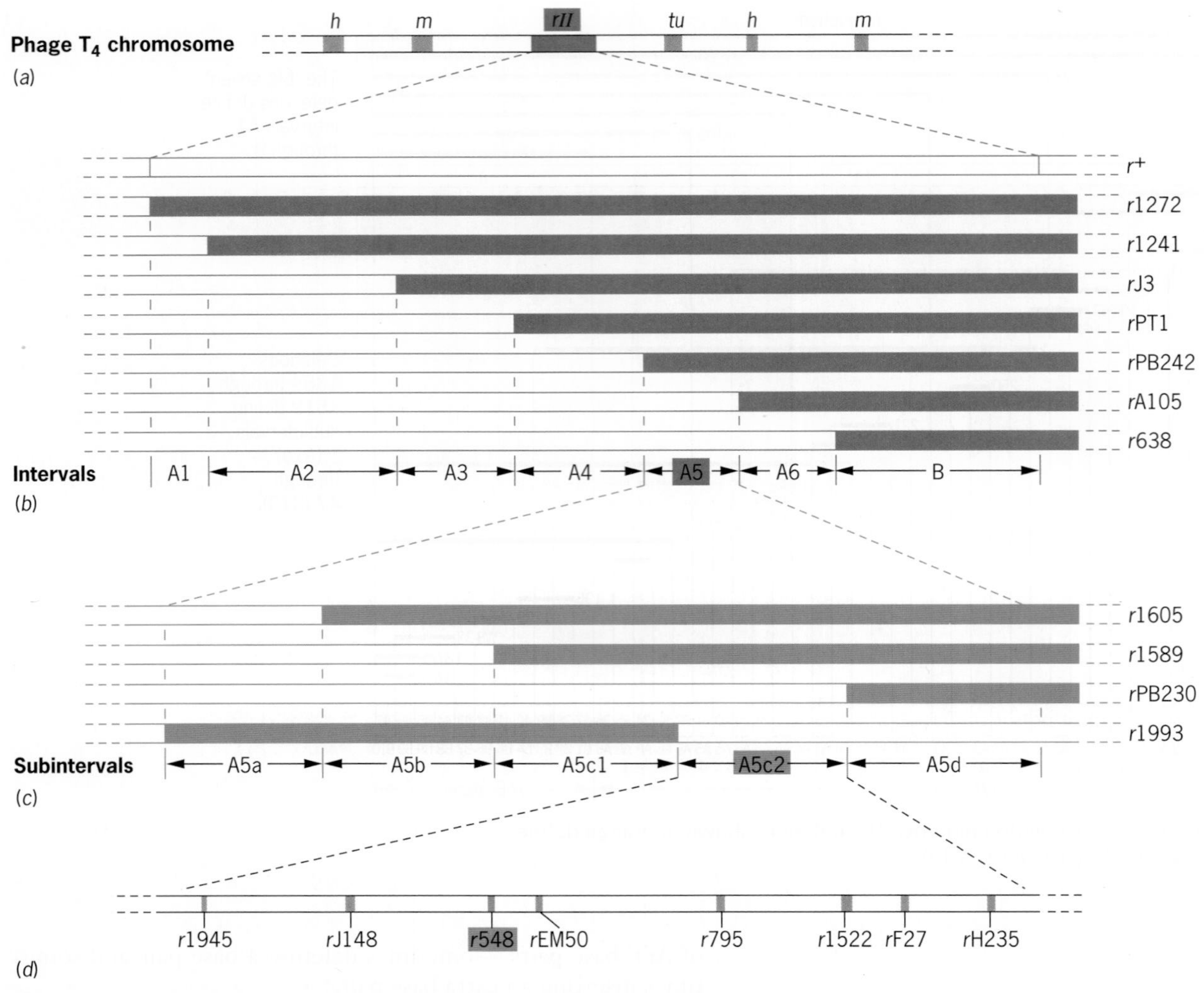

Figure 14.19 ▶ Benzer's short-cut deletion mapping procedure. Mutant *r*548 is used to illustrate how an *rII* point mutation is mapped by this procedure (see text). (*a*) Map of a short region of the phage T4 chromosome spanning the *rII* locus. (*b*) The "big seven" deletions were used to map *r*548 to interval A5. (*c*) Mutant *r*548 was next mapped to the A5c2 subinterval based on the results of crosses to deletions with end points in the A5 interval. (*d*) The map of *r*548 and the other mutations in the A5c2 subinterval is based on the results of standard two- and three-factor crosses.

Genes-within-Genes in Bacteriophage ΦX174

The genome of phage ΦX174 contains genes that overlap one another and genes that lie entirely within the coding sequences of other genes.

Bacteriophage ΦX174 contains a circular, single-stranded DNA molecule, 5386 nucleotides in length. The results of early studies indicated that the ΦX174 genome contained nine genes; however, more recent results have shown that the ΦX174 genome encodes 11 different proteins, which collectively contain over 2300 amino acids. However, the genetic code is a triplet code; that is, a sequence of three nucleotides specifies each amino acid (Chapter 12). If one assumes that all 5386 nucleotides of the ΦX174 genome encode amino acids, the maximum number of amino acids that it could specify would be 5386/3 or 1795, over 500 fewer amino acid residues than are present in the 11 proteins. Thus, the question was how the ΦX174 genome could encode these 11 proteins.

This enigma was not solved until the ΦX174 genome was sequenced. When the sequence of the ΦX174 genome was compared with the sequences of some of the gene products, it became clear that the ΦX174 genome contains **overlapping genes** and **genes-within-genes.** Genes overlap when different

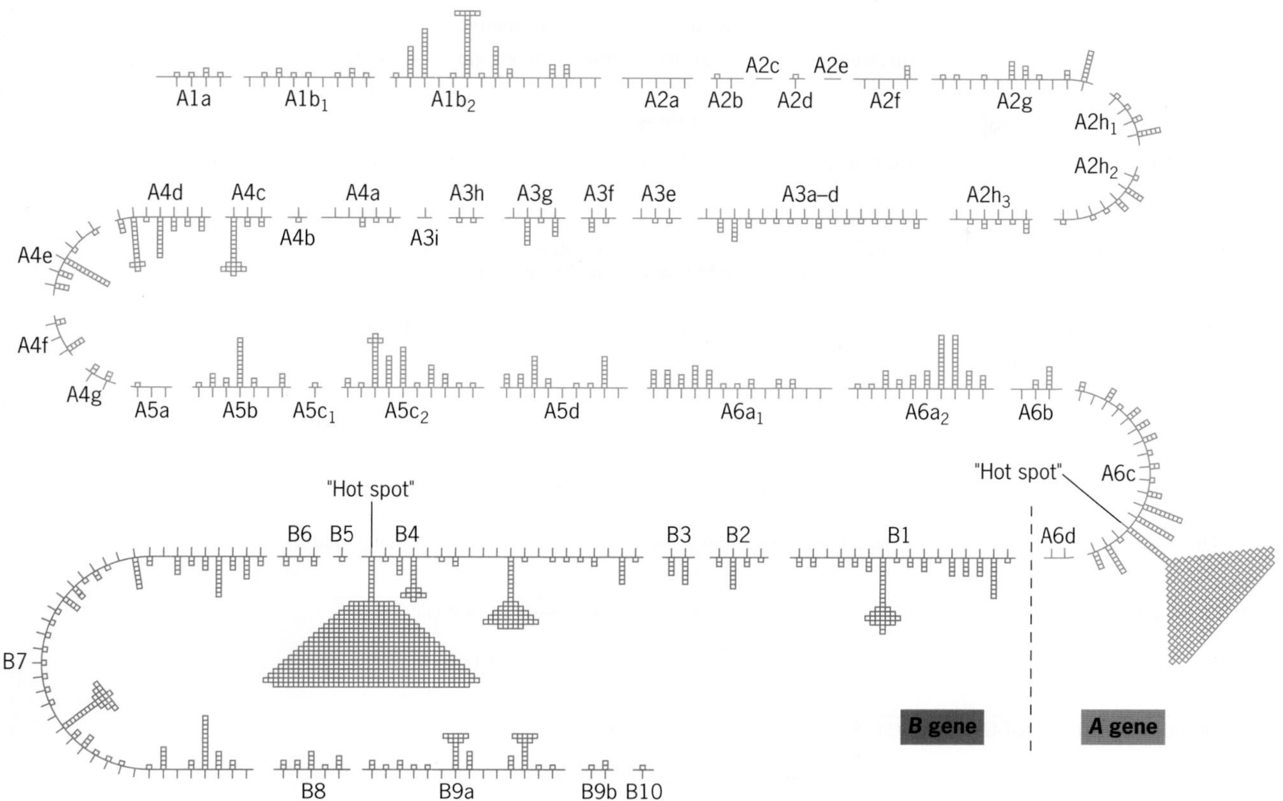

Figure 14.20 ▶ The genetic map of the *rII* locus of phage T4. The *rII* locus contains two genes, *rIIA* and *rIIB*. Each square represents the independent occurrence of a spontaneous mutation at the indicated site. Vertical lines at which no squares occur represent positions defined by mutations induced by mutagenic agents. Some sites, called hot spots, mutate at very high frequencies.

reading frames of the same DNA sequence encode different proteins. Some segments of the ΦX174 DNA molecule specify two or three different amino acid sequences by being translated in two or all three reading frames (**FIGURE 14.21**). For example, the *B* gene is located entirely within the *A* gene, and the *E* gene is located entirely within the *D* gene. Use of the two different reading frames of a single DNA sequence to encode two different proteins is illustrated for the *D* and *E* genes in **FIGURE 14.22**. A key feature of these overlapping genes is the presence of the initiation triplet TAC (specifying an AUG initiation codon in the mRNA) within the coding sequence of another gene. The initiation codon specifies N-formylmethionine (f-Met), the first amino acid incorporated into polypeptides during translation (Chapter 12).

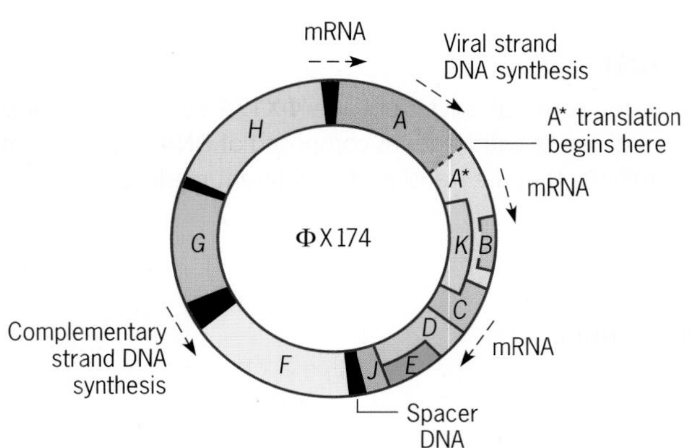

Figure 14.21 ▶ Physical map of the phage ΦX174 genome showing the overlapping genes and start and stop points for transcription. The black regions are nontranslated spacers.

In addition to the two cases of a gene-within-a-gene discussed above, the nucleotide sequence of the *K* gene overlaps the DNA sequences that encode polypeptides A and A*. These polypeptides are specified by the same reading frame of the *A* gene, but with the initiation of translation occurring at different sites (see **FIGURE 14.21**). As in the two cases of genes-within-genes, the A* and K polypeptides are specified by different reading frames. Thus, even though the same nucleotides are used, the amino acid sequences of the A* and K polypeptides are different.

Genes-within-genes and overlapping genes allow phage ΦX174 to make maximum use of its 5386 nucleotides. More proteins can be specified by small genomes if two or all three reading frames specify polypeptides. However, such overlapping coding sequences cannot evolve independently. Indeed, mutations in overlapping genes will often alter the structure of more than one protein, and, in some cases, these mutations will

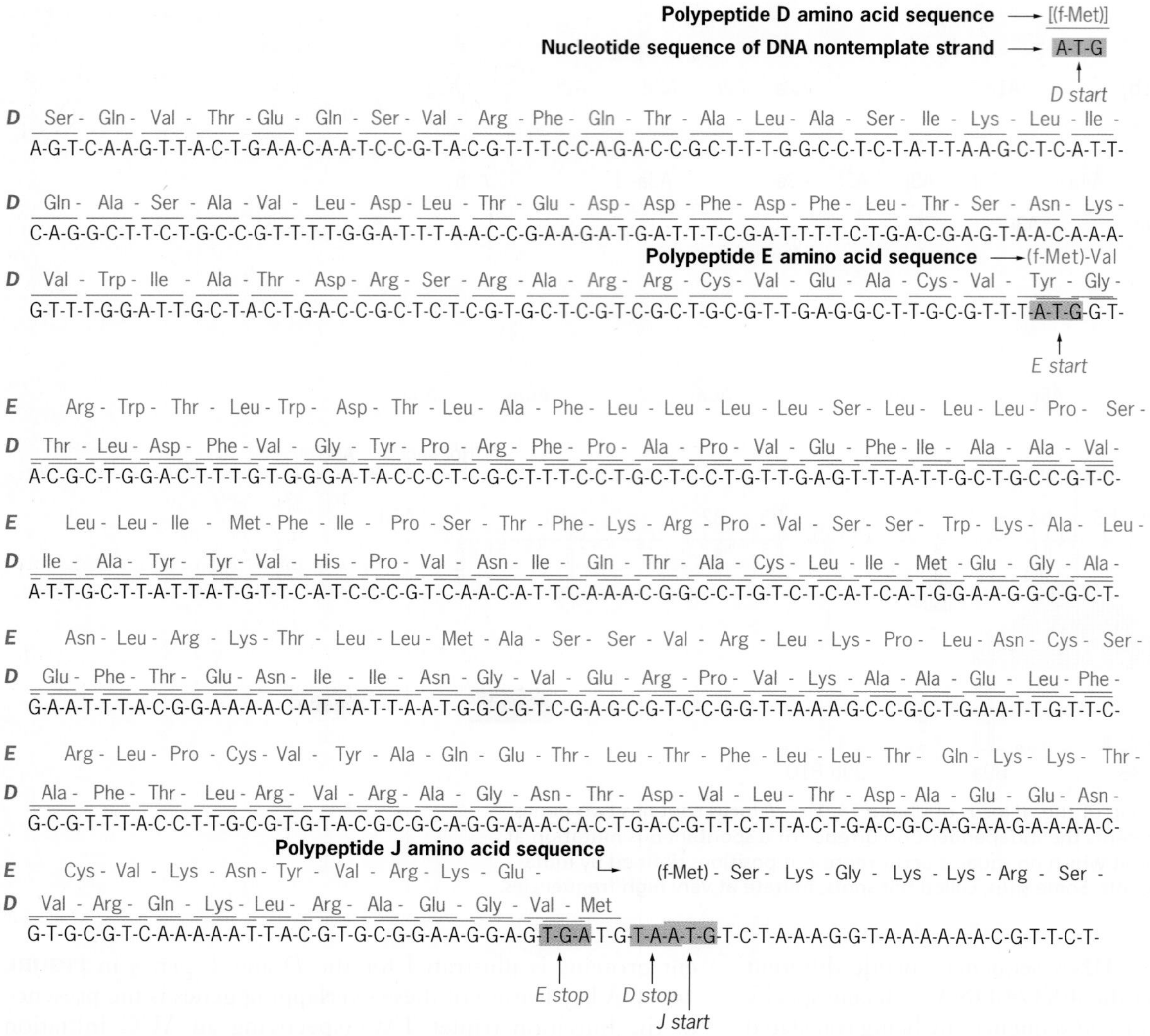

Polypeptide D amino acid sequence → [(f-Met)]
Nucleotide sequence of DNA nontemplate strand → A-T-G
D start

D Ser - Gln - Val - Thr - Glu - Gln - Ser - Val - Arg - Phe - Gln - Thr - Ala - Leu - Ala - Ser - Ile - Lys - Leu - Ile -
A-G-T-C-A-A-G-T-T-A-C-T-G-A-A-C-A-A-T-C-C-G-T-A-C-G-T-T-T-C-C-A-G-A-C-C-G-C-T-T-T-G-G-C-C-T-C-T-A-T-T-A-A-G-C-T-C-A-T-T-

D Gln - Ala - Ser - Ala - Val - Leu - Asp - Leu - Thr - Glu - Asp - Asp - Phe - Asp - Phe - Leu - Thr - Ser - Asn - Lys -
C-A-G-G-C-T-T-C-T-G-C-C-G-T-T-T-T-G-G-A-T-T-T-A-A-C-C-G-A-A-G-A-T-G-A-T-T-T-C-G-A-T-T-T-T-C-T-G-A-C-G-A-G-T-A-A-C-A-A-A-
Polypeptide E amino acid sequence → (f-Met)-Val
D Val - Trp - Ile - Ala - Thr - Asp - Arg - Ser - Arg - Ala - Arg - Arg - Cys - Val - Glu - Ala - Cys - Val - Tyr - Gly -
G-T-T-T-G-G-A-T-T-G-C-T-A-C-T-G-A-C-C-G-C-T-C-T-C-G-T-G-C-T-C-G-T-C-G-C-T-G-C-G-T-T-G-A-G-G-C-T-T-G-C-G-T-T-T-A-T-G-G-T-
E start

E Arg - Trp - Thr - Leu - Trp - Asp - Thr - Leu - Ala - Phe - Leu - Leu - Leu - Leu - Ser - Leu - Leu - Leu - Pro - Ser -
D Thr - Leu - Asp - Phe - Val - Gly - Tyr - Pro - Arg - Phe - Pro - Ala - Pro - Val - Glu - Phe - Ile - Ala - Ala - Val -
A-C-G-C-T-G-G-A-C-T-T-T-G-T-G-G-G-A-T-A-C-C-C-T-C-G-C-T-T-T-C-C-T-G-C-T-C-C-T-G-T-T-G-A-G-T-T-T-A-T-T-G-C-T-G-C-C-G-T-C-

E Leu - Leu - Ile - Met - Phe - Ile - Pro - Ser - Thr - Phe - Lys - Arg - Pro - Val - Ser - Ser - Trp - Lys - Ala - Leu -
D Ile - Ala - Tyr - Tyr - Val - His - Pro - Val - Asn - Ile - Gln - Thr - Ala - Cys - Leu - Ile - Met - Glu - Gly - Ala -
A-T-T-G-C-T-T-A-T-T-A-T-G-T-T-C-A-T-C-C-C-G-T-C-A-A-C-A-T-T-C-A-A-A-C-G-G-C-C-T-G-T-C-T-C-A-T-C-A-T-G-G-A-A-G-G-C-G-C-T-

E Asn - Leu - Arg - Lys - Thr - Leu - Leu - Met - Ala - Ser - Ser - Val - Arg - Leu - Lys - Pro - Leu - Asn - Cys - Ser -
D Glu - Phe - Thr - Glu - Asn - Ile - Ile - Asn - Gly - Val - Glu - Arg - Pro - Val - Lys - Ala - Ala - Glu - Leu - Phe -
G-A-A-T-T-T-A-C-G-G-A-A-A-A-C-A-T-T-A-T-T-A-A-T-G-G-C-G-T-C-G-A-G-C-G-T-C-C-G-G-T-T-A-A-A-G-C-C-G-C-T-G-A-A-T-T-G-T-T-C-

E Arg - Leu - Pro - Cys - Val - Tyr - Ala - Gln - Glu - Thr - Leu - Thr - Phe - Leu - Leu - Thr - Gln - Lys - Lys - Thr -
D Ala - Phe - Thr - Leu - Arg - Val - Arg - Ala - Gly - Asn - Thr - Asp - Val - Leu - Thr - Asp - Ala - Glu - Glu - Asn -
G-C-G-T-T-T-A-C-C-T-T-G-C-G-T-G-T-A-C-G-C-G-C-A-G-G-A-A-A-C-A-C-T-G-A-C-G-T-T-C-T-T-A-C-T-G-A-C-G-C-A-G-A-A-G-A-A-A-A-C-
Polypeptide J amino acid sequence
E Cys - Val - Lys - Asn - Tyr - Val - Arg - Lys - Glu - → (f-Met) - Ser - Lys - Gly - Lys - Lys - Arg - Ser -
D Val - Arg - Gln - Lys - Leu - Arg - Ala - Glu - Gly - Val - Met
G-T-G-C-G-T-C-A-A-A-A-A-T-T-A-C-G-T-G-C-G-G-A-A-G-G-A-G-T-G-A-T-G-T-A-A-T-G-T-C-T-A-A-A-G-G-T-A-A-A-A-A-A-C-G-T-T-C-T-
E stop *D stop* *J start*

Figure 14.22 ▶ Overlapping genes in phage ΦX174. Gene *E* is located entirely within gene *D*, but the two genes are translated using different reading frames. The reading frame of the *D* gene and the correlated amino acid sequence of the *D* polypeptide are shown in red above the nucleotide sequence. The reading frame of the *E* gene and the correlated amino acid sequence of the *E* polypeptide are shown in green. Note that the translation initiation triplet (ATG) of the *J* gene also overlaps the termination triplet (TAA) of the *D* gene.

probably have unfavorable effects on one or more of the gene products. Nevertheless, overlapping genes have evolved in phage ΦX174 and other viruses where small genome size seems to provide a selective advantage.

KEY POINT

- The chromosome of bacteriophage ΦX174 contains overlapping genes and genes-within-genes composed of DNA sequences specifying mRNAs that are translated in different reading frames.

Complex Gene–Protein Relationships

In some cases the relationship between DNA coding sequences and polypeptide products is complex, involving alternate pathways of transcript splicing or assembly of genes from gene segments during development.

Most prokaryotic genes consist of continuous sequences of nucleotide pairs, which specify colinear sequences of amino acids in the polypeptide gene products. As we discussed in Chapter 11, most eukaryotic genes are split into expressed sequences (exons) and intervening sequences (introns). However, because the spliceosomes usually excise introns from primary transcripts by *cis*-splicing mechanisms (processes that join exons from the same RNA molecule), the presence of introns in genes does not invalidate the complementation test as an operational definition of the gene. Nevertheless, in some cases, transcripts of split genes may undergo several different types of splicing, making the relationships between genes and proteins more complex than the usual one gene–one polypeptide relationship. In other cases, expressed genes are assembled from "gene pieces" during the development of the specialized cells in which they are expressed.

ALTERNATE PATHWAYS OF TRANSCRIPT SPLICING: PROTEIN ISOFORMS

Many interrupted eukaryotic genes, such as the mammalian β-globin genes and the chicken ovalbumin and 1α2 collagen genes discussed in Chapter 11, each encode a single polypeptide chain with a specific function. In these cases, the mRNA produced from the gene contains all the exons joined together in the same order as they occur in the gene. However, the transcripts of some interrupted genes undergo alternate pathways of transcript splicing; that is, different exons of a gene may be joined to produce a related set of mRNAs encoding a small family of closely related polypeptides called **protein isoforms** (**FIGURE 14.23**). In some cases, alternate splicing changes the reading frame of distal exons producing very different proteins. Note, however, that even in cases where alternate splicing occurs, colinearity between the genes and their polypeptide products is maintained.

The alternate splicing pathways are often tissue-specific, producing related proteins that carry out similar, but not necessarily identical, functions in different types of cells (Chapter 20). The mammalian tropomyosin genes provide striking examples of genes that each produce a family of protein isoforms. Tropomyosins are proteins involved in the regulation of muscle contraction in animals. Because the various organs of an animal contain different muscle types, all of which need to be regulated, the availability of a family of related tropomyosins might be beneficial. In any case, one mouse tropomyosin gene is known to produce at least 10 different tropomyosin polypeptides as a result of alternate pathways of transcript splicing. Genes of this type obviously do not fit the one gene–one polypeptide concept very well. For such genes, where alternate splicing pathways give rise to two or more different polypeptides, the gene can be defined as a DNA sequence that is a single unit of transcription and encodes a set of protein isoforms.

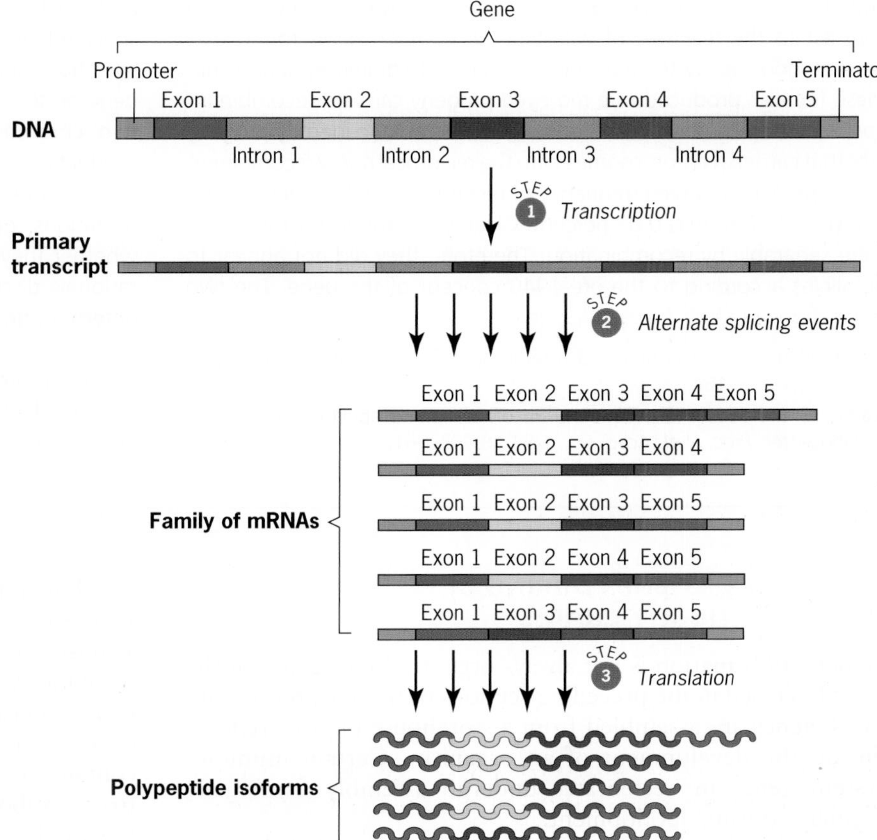

Figure 14.23 ▶ A single gene may produce a family of closely related polypeptides by using alternate pathways of exon splicing.

▶ A MILESTONE IN GENETICS: Lewis's *Cis-Trans* Position Effect

For many years, the gene was construed to be the unit of function controlling one phenotypic trait and the unit of structure not divisible by mutation or recombination. This view of the gene began to change in 1940 with the work of Clarence P. Oliver, a professor of zoology at the University of Minnesota, and his students Edward Lewis and Melvin Green. Lewis went on to share the 1995 Nobel Prize in Physiology or Medicine for his later work on the genetic control of development in *Drosophila*. However, it was his early work started while he was an undergraduate in Oliver's lab that contributed in a major way to our concept of the gene.

In 1942, Lewis reported the results of a study with mutants at the *Star-asteroid* locus (named after two mutations that result in small, rough-surfaced eyes) in *Drosophila*. He observed that flies with the genotype *S* ast^+/S^+*ast* had a more extreme mutant phenotype than flies with the genotype *S ast*/S^+ast^+, even though the two genotypes carried the same mutant alleles, *S* (*Star*) and *ast* (*asteroid*).[1] In the genotype with the more extreme phenotype, the two mutant alleles were on different chromosomes, whereas in the genotype with the less extreme phenotype, they were on the same chromosome. Lewis's observations were difficult to interpret because the *Star* mutation is partially dominant. More clear-cut results were obtained when he performed similar experiments with the recessive eye color mutations *white* (*w*) and *apricot* (*apr*).[2]

Fruit flies that are homozygous for the X-linked mutations *apr* and *w* have apricot-colored eyes and white eyes, respectively, in contrast to the red eyes of wild-type *Drosophila*. Lewis found that heterozygous *apr*/*w* females had light apricot-colored eyes and that these females produced rare red-eyed progeny carrying recombinant $apr^+ w^+$ chromosomes. In addition, Lewis was able to identify progeny flies that carried X chromosomes with the reciprocal *apr w* recombinant genotype. The observed frequency of recombination between the *apr* and *w* mutations was 0.03 percent. Clearly, the *apr* and *w* mutations were separable by recombination. Therefore, they did not appear to be alleles according to the pre-1940 concept of the gene. The two mutations seemed to be in the same unit of function, but they were in two different units of structure.

When Lewis produced flies with the genotype *apr w*/$apr^+ w^+$ he found that they had red eyes just like wild-type flies (**FIGURE 1*a***). When he constructed flies of genotype *apr* w^+/apr^+*w* he found that they had light apricot-colored eyes (**FIGURE 1*b***). Both genotypes contained the same mutant and wild-type genetic information but in different arrangements. When we observe different phenotypes in organisms that contain the same genetic markers, but in different arrangements, we say that the markers exhibit a **position effect**. The type of position effect that Lewis observed is called a ***cis-trans* position effect**.

In the *apr w*/$apr^+ w^+$ flies, the two mutations were present on the same chromosome, and the wild-type alleles were present on the other chromosome—that is, the flies were *cis* heterozygotes. In the *apr* w^+/apr^+*w* flies, the two mutations were on different chromosomes—that is, the flies were *trans* heterozygotes. The *cis* heterozygotes had wild-type eyes; the *trans* heterozygotes had mutant eyes. This is precisely the result that would be expected if *apr* and *w* were mutations at different sites in the same gene, the unit of genetic information encoding a single polypeptide (**FIGURE 1**). Thus, *apr* and *w* are considered to be alleles of the same gene, and *apr* is now designated w^a in recognition of this relationship. If *apr* and *w* had been mutations in two different genes, both the *cis* and the *trans* heterozygotes should have expressed the wild-type phenotype—red eyes. In the *trans* heterozygote, the apr^+ gene product would be specified by the apr^+ gene on the chromosome carrying the *w* mutation, and the w^+ gene product would be specified by the w^+ gene on the chromosome harboring the *apr* mutation. Together, the two chromosomes would produce both of the wild-type gene products.

To determine how many genes are identified by a collection of mutations, geneticists can construct *trans* heterozygotes, observe whether they have mutant or wild-type phenotypes, and, using the rationale discussed above, assign the mutations to genes. If a *trans* heterozygote has a mutant phenotype, the two mutations involved are in the same gene. If it has the wild-type phenotype, the two mutations are in different genes. This *trans*-heterozygote test is now usually called the complementation test because mutants in different genes "complement each other" when present in *trans* heterozygotes.

[1]Lewis, E. B. 1942. The Star and asteroid loci in *Drosophila melanogaster*. *Genetics* 27:153–154.

[2]Lewis, E. B. 1952. The pseudoallelism of white and apricot in *Drosophila melanogaster*. *Proc. Natl. Acad. Sci., USA* 38:953–961.

ASSEMBLY OF GENES DURING DEVELOPMENT: HUMAN ANTIBODY CHAINS

Genetic information is not always organized into genes of the type described in the preceding sections of this chapter. In rare cases, genes are assembled from a storehouse of *gene segments* during the development of an organism. Certain immune-system genes in vertebrates undergo assembly from gene segments during development.

The immune system of vertebrate animals depends on the synthesis of proteins called **antibodies** to provide protection against infections by viruses, bacteria, toxins, and other foreign substances. Each antibody contains four polypeptides, two identical heavy chains and two identical light chains. The light chains are of two types: kappa and lambda. Each antibody chain contains a variable region, which exhibits extensive diversity from antibody to antibody, and a constant region, which is largely the same in all antibodies. In germ-line chromosomes,

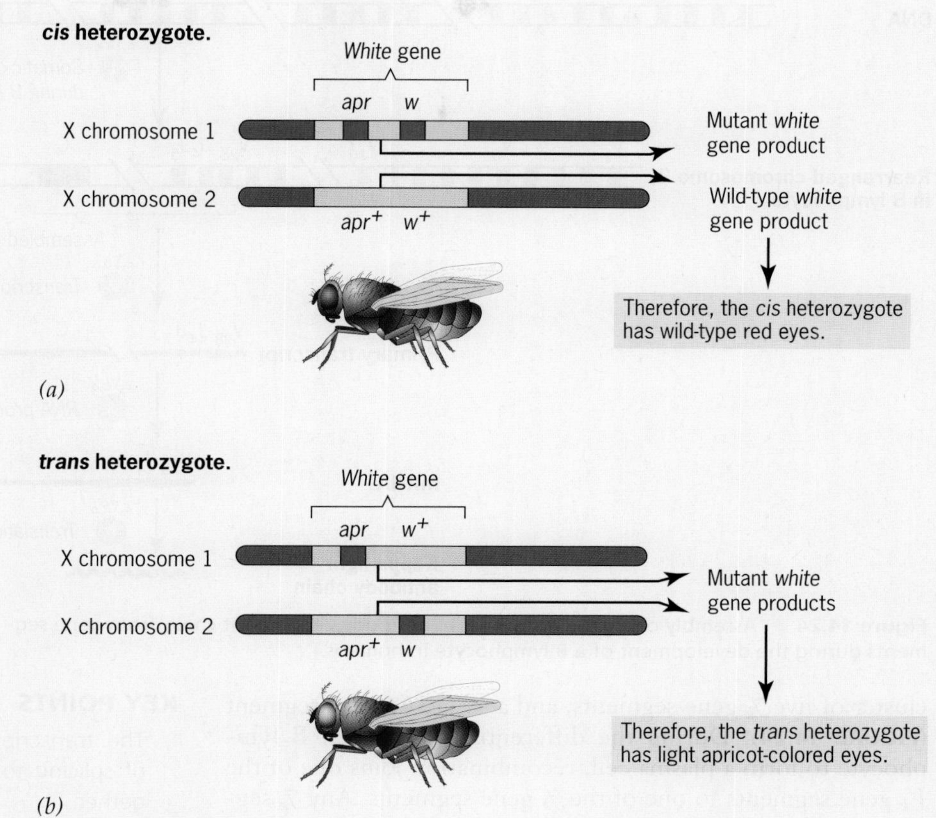

Figure 1 ► The *cis-trans* position effect observed by Edward Lewis with the *apr* and *w* mutations of *Drosophila*.

The *trans* or complementation test has proven to be a powerful tool for use in identifying genes as units of function.

QUESTIONS FOR DISCUSSION

1. Although the *trans* or complementation test has been invaluable in defining the genes of experimental organisms, it has been of only limited value in human genetics. Why? Can you think of ways of performing complementation tests to define human genes without performing crosses between different people?

2. When asked about the formative experiences in his education, Edward Lewis said: "As an undergraduate at the University of Minnesota, I was greatly indebted to Professor C. P. Oliver (a student of H. J. Muller) who gave me a desk in his laboratory so I could carry on my *Drosophila* experiments." How important is hands-on research experience to the education of undergraduates majoring in biology? Should participation in undergraduate research be required of all college biology majors? all college science majors? all college students?

the DNA sequences encoding these antibody chains are present in gene segments, and the gene segments are joined together to produce genes during the differentiation of antibody-producing B-lymphocytes from progenitor cells. Although B-lymphocytes do not secrete the antibodies that they produce, they terminally differentiate into plasma cells, which do.

To illustrate this process of gene assembly during development, let us briefly consider the DNA sequences encoding kappa light chains in humans. (We discuss this topic in more detail in Chapter 21.) A kappa light chain gene is assembled from three gene segments: V_k (**V** for variable region), J_k (**J** for joining segment), and C_k (**C** for constant region), during B lymphocyte development. Together, the V_k and J_k gene segments encode the variable region of the kappa light chain, whereas the C_k gene segment encodes the constant region. No functional $V_kJ_kC_k$ kappa light chain gene is present in any human germ-line chromosome. Instead, human chromosome 2 contains a cluster of 76 V_k gene segments (40 functional), another

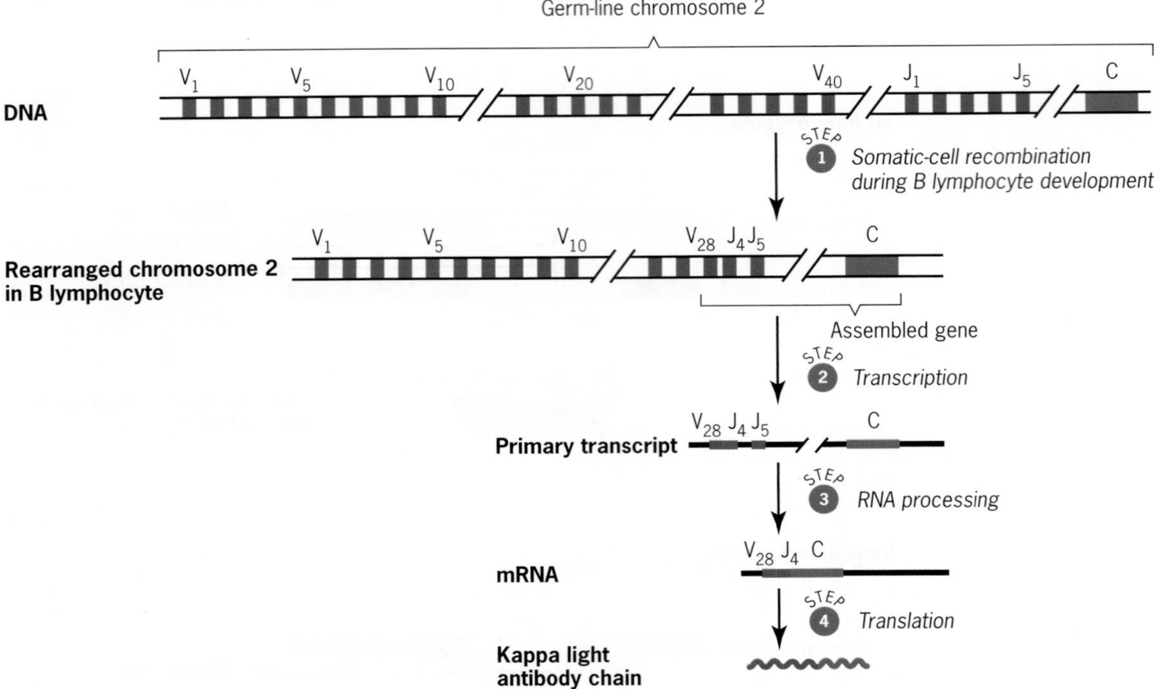

Figure 14.24 ▶ Assembly of a gene encoding an antibody kappa light chain from gene segments during the development of a B lymphocyte in humans.

cluster of five J_k gene segments, and a single C_k gene segment (**FIGURE 14.24**). During the differentiation of each B lymphocyte to form a plasma cell, recombination joins one of the V_k gene segments to one of the J_k gene segments. Any J_k segments remaining between the newly formed V_kJ_k exon and the C_k gene segment become part of an intron that is removed during the processing of the primary transcript. Similar somatic recombination events are responsible for the assembly of the genes encoding antibody heavy chains, lambda light chains, and T-lymphocyte receptor proteins.

KEY POINTS

- The transcripts of some genes undergo alternate pathways of splicing to produce mRNAs with different exons joined together.
- Translation of these mRNAs commonly produces closely related polypeptides called protein isoforms.
- Other genes, such as those encoding antibody chains, are assembled from gene segments during development by regulated processes of somatic recombination.

▶ Basic Exercises

ILLUSTRATE BASIC GENETIC ANALYSIS

1. How has our concept of the gene changed since its discovery by Mendel in 1866?

Answer: Each of Mendel's "constant factors" (genes) controlled one phenotypic trait (for example, red vs. white flowers in peas). In 1908, Sir Archibald E. Garrod extended the concept of the gene to one mutant gene–one metabolic block. In 1941, George Beadle and Edward Tatum rediscovered Garrod's work and refined his concept to one gene–one enzyme. When some enzymes were found to contain two or more different polypeptides, the concept was further refined to one gene–one polypeptide. In some cases, alternate pathways of transcript splicing produce a family of related proteins (protein isoforms) from a single gene. In a few cases, genes are assembled from "gene segments" by recombination events that occur during the differentiation of specialized cells.

2. How does one produce a *trans* heterozygote in *Drosophila*?

Answer: A *trans* heterozygote is produced by crossing flies homozygous for each of the mutations. As an example, an $m_1m_2^+/m_1^+m_2$ heterozygote would be produced by crossing flies homozygous for the m_1 mutation with flies homozygous for the m_2 mutation.

3. How does a *cis* heterozygote differ from a *trans* heterozygote?

Answer: A *cis* heterozygote has both mutations on the same chromosome, with the wild-type alleles on the other chromosome, as shown below, whereas a *trans* heterozygote has one mutant allele and one wild-type allele on each chromosome, as in Exercise 2.

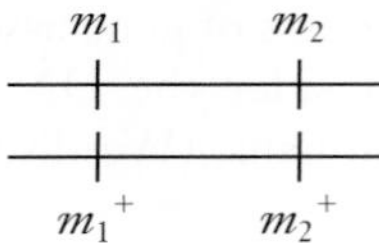

4. Two X-linked mutations in *Drosophila*, *white ivory* and *white apricot*, cause the eyes to be ivory- and apricot-colored, respectively. When hemizygous *white ivory* males are crossed with homozygous *white apricot* females, the female progeny (*trans* heterozygotes) have light-apricot colored eyes, whereas wild-type *Drosophila* have dark-red eyes. Are the *white ivory* and *white apricot* mutations in the same gene or in two different genes?

Answer: When the *trans* heterozygote has a mutant phenotype, the two mutations are in the same gene (see **FIGURE 14.11*a***).

5. Two X-linked mutations in *Drosophila*, *white carrot* and *carnation*, cause the eyes to be light carrot- and carnation-colored, respectively. When hemizygous *carrot* males are crossed with homozygous *carnation* females, the female progeny (*trans* heterozygotes) have dark-red (wild-type) eyes. Are the *carrot* and *carnation* mutations in the same gene or in two different genes?

Answer: When the *trans* heterozygote has the wild-type phenotype, the two mutations are in two different genes (see **FIGURE 14.11*b***).

▶ Testing Your Knowledge

INTEGRATE DIFFERENT CONCEPTS AND TECHNIQUES

1. In *Drosophila*, *white*, *white cherry*, and *vermilion* are all sex-linked mutations affecting eye color. All three mutations are recessive to their wild-type allele(s) for red eyes. A white-eyed female crossed with a vermilion-eyed male produces white-eyed male offspring and red-eyed (wild-type) female offspring. A white-eyed female crossed with a white cherry-eyed male produces white-eyed sons and light cherry-eyed daughters. Do these results indicate whether or not any of the three mutations affecting eye color are located in the same gene? If so, which mutations?

Answer: The complementation test for allelism involves placing mutations pairwise in a common protoplasm in the *trans* configuration and determining whether the resulting *trans* heterozygotes have mutant or wild-type phenotypes. If the two mutations are in the same gene, both copies of the gene in the *trans* heterozygote will produce defective gene products, resulting in a mutant phenotype (see **FIGURE 14.11*a***). However, if the two mutations are in different genes, the two mutations will complement each other, because the wild-type copies of each gene will produce functional gene products (see **FIGURE 14.11*b***). When complementation occurs, the *trans* heterozygote will have the wild-type phenotype. Thus, the complementation test allows one to determine whether any two recessive mutations are located in the same gene or in different genes.

If the *trans* heterozygote has the mutant phenotype, the two mutations are in the same gene. If the *trans* heterozygote has the wild-type phenotype, the two mutations are in two different genes. Because the mutations of interest are sex-linked, all the male progeny will have the same phenotype as the female parent. They are hemizygous, with one X chromosome obtained from their mother. In contrast, the female progeny are *trans* heterozygotes. In the cross between the white-eyed female and the vermilion-eyed male, the female progeny have red eyes, the wild-type phenotype. Thus, the *white* and *vermilion* mutations are in different genes, as illustrated in the following diagram:

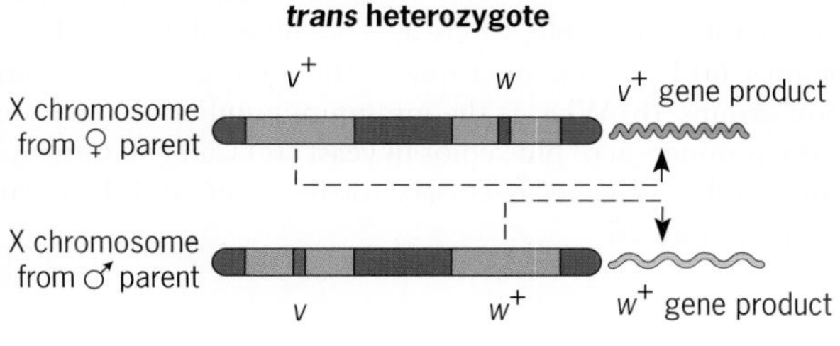

Complementation yields wild-type phenotype; both v^+ and w^+ gene products are produced in the *trans* heterozygote.

In the cross between a white-eyed female and a white cherry-eyed male, the female progeny have light cherry-colored eyes (a mutant phenotype), not wild-type red eyes as in the first cross. Since the *trans* heterozygote has a mutant phenotype, the two mutations, *white* and *white cherry*, are in the same gene:

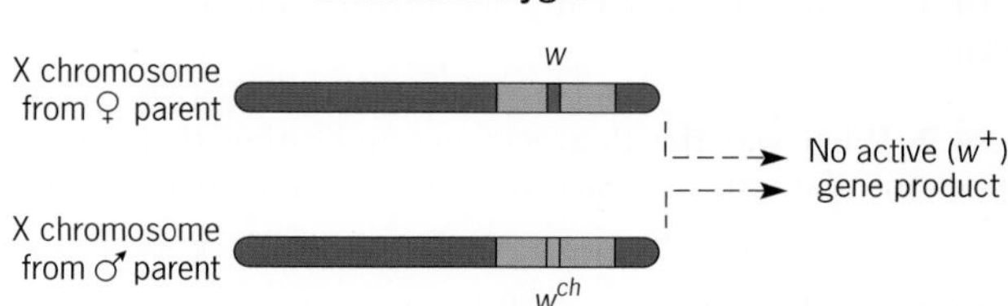

No w^+ gene product; therefore, mutant phenotype.

2. Suppressor-sensitive (*sus*) mutants of bacteriophage ϕ29 can grow on *Bacillus subtilis* strain L15 but cannot grow (that is, are lethal) on *B. subtilis* strain 12A. Wild-type (sus^+) ϕ29 phage can reproduce on both strains, L15 and 12A. Thus, the ϕ29 *sus* mutants are conditional lethal mutants like the *amber* mutants of bacteriophage T4 (see **FIGURE 12.22**). Seven different *sus* mutants of phage ϕ29 were analyzed for complementation by simultaneously infecting the restrictive host (*B. subtilis* strain 12A) with each possible pair of mutants. Single infections with each of the mutants and with wild-type ϕ29 were also done as controls. The results of these complementation or *trans* tests and the controls are given as progeny phage per infected cell in the accompanying table. Several infections performed with wild-type ϕ29 phage yielded 300 to 400 progeny phage per infected cell. The results of the *cis* controls are not given, but assume that all of the *cis* heterozygote controls yielded over 300 progeny phage per infected cell. Also assume that no intragenic complementation occurs between any of the *sus* mutants studied.

Phage ϕ29 Progeny per Infected Bacterium

Mutant	1	2	3	4	5	6	7
7	365	384	344	371	347	333	0.01
6	341	301	351	369	329	0.1	
5	386	326	322	0.04	<0.01		
4	327	398	374	0.06			
3	354	387	<0.01				
2	0.01	<0.01					
1	0.02						

(a) Based on these data, how many genes are identified by the seven *sus* mutants? (b) Which *sus* mutations are located in the same gene(s)?

Answer: The seven *sus* mutants yielded <0.01 to 0.1 progeny phage per infected *B. subtilis* strain 12A cell; those data define the mutant phenotype (basically no progeny). Infections of strain 12A cells with wild-type ϕ29 produced 300 to 400 progeny phage per infected cell, defining the wild-type phenotype. We then examine the phenotypes of the *trans* heterozygotes to determine whether any of the mutations are located in the same gene(s). In each case, we must ask whether the *trans* heterozygote has the mutant or the wild-type phenotype. If a *trans* heterozygote has the mutant phenotype, the two *sus* mutations are in the same gene. If it has the wild-type phenotype, the two *sus* mutations are in different genes. If you are unsure of why this is true, review **FIGURE 14.11**. Of the 21 *trans* heterozygotes examined, 19 exhibited the wild-type phenotype, indicating that in each case the two mutations are in different genes. Two *trans* heterozygotes—(1) *sus*1 on one chromosome and *sus*2 on a second chromosome and (2) *sus*4 on one chromosome and *sus*5 on the other—had the mutant phenotype. Therefore, (*a*) the seven *sus* mutations are located in five different genes, with (*b*) mutations *sus*1 and *sus*2 in one gene and mutations *sus*4 and *sus*5 in another gene.

▸ Questions and Problems

ENHANCE UNDERSTANDING AND DEVELOP ANALYTICAL SKILLS

14.1 What is the currently accepted operational definition of the gene?

14.2 What was the first evidence indicating that the unit of function and the unit of structure of genetic material were not the same?

14.3 Of what value are conditional lethal mutations for genetic fine structure analysis?

14.4 Seven mutants of *Neurospora* are unable to grow on minimal medium unless it is supplemented with one or more of the metabolites A through G. On the basis of the data given below (where "+" indicates growth and "0" indicates no growth), draw a biochemical pathway for the synthesis of these seven substances and show where the mutants are blocked in the pathway.

Growth in the Presence of Metabolite(s)

Mutant	A	B	C	D	E	F	G	B+C	D+G	B+C+E
1	+	0	0	0	0	0	0	0	+	+
2	0	+	0	0	0	0	0	+	0	+
3	0	0	+	0	0	0	0	+	0	+
4	0	0	0	+	+	0	0	0	+	+
5	0	0	0	0	+	0	0	0	0	+
6	+	0	0	0	0	+	0	0	+	+
7	0	0	0	0	0	0	+	+	+	+

14.5 Assume that the mutants described in Problem 14.8 yielded the following results. How many genes would they have defined? Which mutations would have been in the same gene(s)?

Growth of *trans* Heterozygotes (without Histidine)

Mutant	1	2	3	4	5	6	7	8
8	+	+	+	+	+	+	0	0
7	+	+	+	+	+	+	0	
6	+	+	+	+	0	0		
5	+	+	+	+	0			
4	+	+	0	0				
3	+	+	0					
2	0	0						
1	0							

14.6 Based on our current concept of the gene, (a) what is the smallest unit of genetic material that can be changed by mutation, and (b) what is the smallest region of genetic material in which recombination can occur?

14.7 A researcher is interested in delineating the pathway for the production of blue colony color in yeast. She finds that it is possible to mate two mutant strains of yeast with gray colonies and produce diploid yeast cells that produce blue colonies. She then proceeds to mate nine different yeast mutants (*A* through *I*) with gray colonies in all possible pairwise combinations and inspects the diploid progeny for colony color. The results of her complementation analysis are presented in the following table, where a "+" = blue colonies and a "–" = gray colonies. (a) Place the mutations in the nine strains into complementation groups. (b) What is the minimum number of genes necessary for the production of blue color in yeast? (c) Can you estimate the maximum number of genes necessary for this trait, and if so, what is the maximum number?

	A	*B*	*C*	*D*	*E*	*F*	*G*	*H*	*I*
A	–	–	+	+	+	–	–	–	–
B	–	–	+	+	+	–	–	–	–
C	+	+	–	–	+	+	+	+	+
D	+	+	–	–	+	+	+	+	+
E	+	+	+	+	–	+	+	+	+
F	–	–	+	+	+	–	–	–	–
G	–	–	+	+	+	–	–	–	–
H	–	–	+	+	+	–	–	–	–
I	–	–	+	+	+	–	–	–	–

14.8 Two different inbred varieties of a particular plant species have white flowers. All other varieties of this species have red flowers. What experiments might be done to obtain evidence to determine whether the difference in flower color in these varieties is the result of different alleles of a single gene or the result of genetic variation in two or more genes?

14.9 Four mutant strains of *Neurospora* require one or more of the supplemented metabolites A through D to grow. From the data below (where "+" = growth; "0" = no growth), draw the metabolic pathway for the synthesis of these metabolites and show where the mutant strains (1, 2, 3, and 4) are blocked.

	Metabolite			
Mutant	**A**	**B**	**C**	**D**
1	+	0	0	0
2	+	+	+	+
3	+	0	+	0
4	+	+	+	0

14.10 Eight independently isolated mutants of *E. coli*, all of which are unable to grow in the absence of histidine (his^-), were examined in all possible *cis* and *trans* heterozygotes (partial diploids). All of the *cis* heterozygotes were able to grow in the absence of histidine. The *trans* heterozygotes yielded two different responses: some of them grew in the absence of histidine; others did not. The experimental results, using "+" to indicate growth and "0" to indicate no growth, are given in the accompanying table. How many genes are defined by these eight mutations? Which mutant strains carry mutations in the same gene(s)?

Growth of *trans* Heterozygotes (without Histidine)

Mutant	**1**	**2**	**3**	**4**	**5**	**6**	**7**	**8**
8	0	0	0	0	0	0	+	0
7	+	+	+	+	+	+	0	
6	0	0	0	0	0	0		
5	0	0	0	0	0			
4	0	0	0	0				
3	0	0	0					
2	0	0						
1	0							

14.11 What is the difference between a pair of homoalleles and a pair of heteroalleles?

14.12 Considering only base-pair substitutions, how many different mutant homoalleles can occur at one site in a gene?

14.13 The *amber* mutants of phage T4 are conditional lethal mutants. They grow on *E. coli* strain CR63 but are lethal on *E. coli* strain B. An *amber* mutant almost never exhibits *intra*genic complementation with any other *amber* mutant; for this problem, assume that no *intra*genic complementation occurs between any of the mutants involved. The following results were obtained when eight *amber* mutants were analyzed for complementation by infecting the restrictive host (*E. coli* strain B) with each possible pair of mutants. The results of mixed infections by pairs of mutants are shown as "0" if no progeny are produced and as "+" if progeny phage resulted from the infection with that particular pair of mutants.

Mutant	**1**	**2**	**3**	**4**	**5**	**6**	**7**	**8**
8	+	+	+	+	+	+	0	0
7	+	+	+	+	+	+	0	
6	+	+	+	+	+	0		
5	0	+	0	+	0			
4	+	+	+	0				
3	0	+	0					
2	+	0						
1	0							

(a) These data indicate that the eight *amber* mutations are located in how many different genes?
(b) Which mutations are located in the same gene or genes?

14.14 What determines the maximum number of different alleles that can exist for a given gene?

14.15. Are the following statements concerning the genetic element referred to as the gene true or false?
(a) The classical (pre-1940) conception of the gene was that it was (1) a unit of physiological function or expression, (2) the smallest unit that could undergo mutation, and (3) a unit not subdivisible by recombination.
(b) In bacteria, the *cis-trans* test provides an operational definition by which we usually can identify a gene as the unit that specifies one mRNA molecule.
(c) Our present knowledge of the structure of the gene indicates that the units defined by criteria (2) and (3) in statement (a) above are both equivalent to a single nucleotide pair.
(d) Studies in the 1940s demonstrated the existence of heteroalleles, clearly indicating that many mutations that were allelic by the functional criterion could be separated by recombination, and thereby indicating that the units of function, mutation, and recombination are not equivalent.
(e) Homoalleles are functionally and structurally allelic; heteroalleles are functionally allelic but structurally nonallelic.

14.16 GO Suppressor-sensitive (*sus*) mutants of bacteriophage lambda can grow on *E. coli* strain C600 but cannot grow (that is, are lethal) on *E. coli* strain W3350. In other words, *sus* mutants are conditional-lethal mutants. Seven *sus* mutants were analyzed for complementation by simultaneously infecting the restrictive host (*E. coli* strain W3350) with each possible pair of mutants. Single infections with each mutant and with wild-type lambda were also done as controls. The results of these complementation or *trans* tests and the controls are given as progeny per infected cell in the accompanying table. Several infections with wild-type lambda yielded 120 to 150 progeny phage per infected cell, and all of the *cis* heterozygotes yielded over 100 progeny phage per infected cell. Assume that no intragenic complementation occurs between any of these *sus* mutants.

Lambda Progeny per Infected Cell

Mutant	1	2	3	4	5	6	7
7	0.01	129	0.01	151	130	125	0.01
6	128	150	170	0.06	0.1	0.1	
5	121	123	131	0.05	<0.01		
4	143	119	130	0.06			
3	<0.01	150	<0.01				
2	180	<0.01					
1	0.02						

(a) Based on the above data, how many genes are defined by the seven *sus* mutants?
(b) Which *sus* mutations are located in the same gene(s)?

14.17 Both *temperature-sensitive* (*ts*) mutant alleles and *amber* (*am*) mutant alleles have been identified and studied for many of the genes of bacteriophage T4. Different *ts* mutations within the same gene are frequently found to complement each other, whereas different *am* mutations within the same gene practically never complement one another. Why is this difference to be expected?

14.18 The *rosy* (*ry*) gene of *Drosophila* encodes the enzyme xanthine dehydrogenase; the active form of xanthine dehydrogenase is a dimer containing two copies of the *rosy* gene product. Mutations *ry2* and *ry42* are both located within the region of the *rosy* gene that encodes the *rosy* polypeptide gene product. However, *ry2/ry42 trans* heterozygotes have wild-type eye color. How can the observed complementation between *ry2* and *ry42* be explained given that these two mutations are located in the same gene?

14.19 The recessive mutations *b* (*black*) and *e* (*ebony*) in *Drosophila* both produce flies with black bodies rather than gray bodies like wild-type flies. Mapping studies showed that *b* is located on chromosome 2, whereas *e* is on chromosome 3. When homozygous *b/b* flies are crossed with homozygous *e/e* flies, the heterozygous *b/e* progeny have gray bodies. The observed complementation indicates that the two mutations are in two different genes. Was it necessary to perform a complementation test to conclude that the *b* and *e* mutations were located in two different genes? If so, why? If not, why not?

14.20 GO Suppressor-sensitive (*sus*) mutants of bacteriophage ϕ29 can grow on *Bacillus subtilis* strain L15 but cannot grow (that is, are lethal) on *B. subtilis* strain 12A. Wild-type (sus^+) ϕ29 phage can reproduce on both strains, L15 and 12A. Thus, the ϕ29 *sus* mutants are conditional lethal mutants like the *amber* mutants of bacteriophage T4. Seven different *sus* mutants of phage ϕ29 were analyzed for complementation by simultaneously infecting the restrictive host (*B. subtilis* strain 12A) with each possible pair of mutants. Single infections with each of the mutants and with wild-type ϕ29 were also done as controls. The results of these complementation or *trans* tests and the controls are given as progeny phage per infected cell in the accompanying table. Several infections performed with wild-type ϕ29 phage yielded 300 to 400 progeny phage per infected cell, and all of the *cis* heterozygote controls yielded over 300 progeny phage per infected cell. Assume that no intragenic complementation occurs between any of the *sus* mutants studied.

Phage ϕ29 Progeny per Infected Bacterium

Mutant	1	2	3	4	5	6	7
7	0.01	384	0.01	380	334	330	0.01
6	345	290	353	0.06	321	0.1	
5	390	323	319	371	<0.01		
4	320	401	377	0.06			
3	<0.01	380	<0.01				
2	348	<0.01					
1	0.02						

(a) Based on these data, how many genes are identified by the seven *sus* mutants?
(b) Which *sus* mutations are located in the same gene(s)?

14.21 Is the number of potential alleles of a gene directly related to the number of nucleotide pairs in the gene? Is such a relationship more likely to occur in prokaryotes or in eukaryotes? Why?

14.22 Based on the information provided in **FIGURE 14.7,** (a) are mutations *trpA3* and *trpA33* heteroalleles or homoalleles? (b) Are mutations *trpA78* and *trpA58* heteroalleles or homoalleles?

14.23 Assume that the mutants described in Problem 14.24 had yielded the following results.

Phage ϕ29 Progeny per Infected Bacterium

Mutant	1	2	3	4	5	6	7
7	0.01	0.01	0.01	0.03	<0.01	0.01	0.01
6	0.08	0.09	0.05	0.06	0.1	0.1	
5	0.02	<0.01	<0.01	0.04	<0.01		
4	0.05	0.06	0.03	0.06			
3	<0.01	<0.01	<0.01				
2	0.01	<0.01					
1	0.02						

(a) How many genes would they have defined?
(b) Which mutations would have been in the same gene(s)?

14.24 Why was it necessary to modify Beadle and Tatum's one gene–one enzyme concept of the gene to one gene–one polypeptide?

14.25 In their analysis of gene function, Beadle and Tatum used *Neurospora* as an experimental organism, whereas Garrod had studied gene function in humans. What advantages does *Neurospora* have over humans for such studies?

14.26 Arthur Chovnick and colleagues have mapped a large number of recessive mutations that produce fruit flies with rose-colored eyes. They also have performed complementation tests on these *ry* (*rosy*) mutations. Heterozygotes that carried mutations *ry42* and *ry406* in the *trans* configuration had wild-type eyes, whereas *trans* heterozygotes that harbored *ry5* and *ry41* had rose-colored eyes. The results of two- and three-factor crosses unambiguously demonstrated that mutations *ry42* and *ry406* both map between mutations *ry5* and *ry41*. How can these results be explained?

14.27 Based on the information given in **FIGURE 14.7**, what is the maximum number of nucleotide pairs separating mutations *trpA*78 and *trpA*58?

14.28 GO Professor Jennifer Ross Mendel has characterized five *E. coli* F-strains, each harboring a different deletion in the *glnA* gene, which encodes the enzyme glutamine synthetase. Glutamine synthetase is an important enzyme in organisms from bacteria to plants and animals; it catalyzes the final step in the biosynthesis of the amino acid glutamine. The shaded boxes in the following diagram show the relative locations and sizes of the five deletions (A, B, C, D, and E) characterized by Dr. Mendel:

glnA Gene

Deletion
A
B
C
D
E

Dr. Mendel has also induced several $glnA^-$ point mutations by treatment of an Hfr strain with nitrous acid. When the professor crossed seven of her nitrous acid-induced mutant strains with each of the five deletion strains, the following results were obtained, where a "+" indicates the formation of $glnA^+$ recombinants and a "0" indicates that no $glnA^+$ recombinants were produced:

	Point Mutants						
Deletion	**1**	**2**	**3**	**4**	**5**	**6**	**7**
A	0	0	0	0	0	0	+
B	0	0	0	0	0	+	+
C	+	+	0	0	+	+	+
D	+	0	0	0	0	0	+
E	+	+	+	0	0	0	+

(a) What is the linear order of the six-point mutations that can be ordered based on these data?
(b) Which of the seven point mutations cannot be ordered relative to the other mutations studied on the basis of these data?

14.29 How can a deletion mutant be distinguished from a point mutant?

14.30 The horizontal lines in the following topological map represent the relative positions and extents of five deletions (R, S, T, U, and V) in the *rII* region of T4 phage.

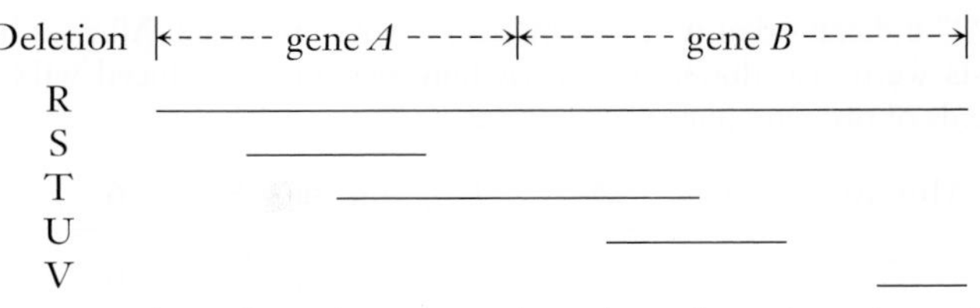

Segments | A1 | A2 | A3 | A4 | B1 | B2 | B3 | B4 | B5 |

(a) In which gene does a mutation occur if it complements with U but does not complement with S?
(b) Where is a mutation located if it does not complement with S or V?
(c) Where is a mutation located if it complements with T?
(d) If a mutation is in gene *B*, will it complement with S or V, both, or neither?
(e) Can a mutation complement with V but nor with U?
(f) Give the most probable position of a point mutation if it recombines with deletions S and V but not with the other three deletions.
(g) Give the most probable position of a point mutation if it does not complement V, does not recombine with R, but does recombine with the other four deletions.

14.31 The following deletion map of the *rII* locus in phage T4 shows the extent of deletions present in T4 strains d1, d2, d3, d4, d5, and d6. Map the *rII* point mutations *r*41, *r*42, *r*43, *r*44, *r*45, and *r*46 relative to the deletions.

d1
d2
d3
d4
d5
d6

Segments | A1 | A2 | A3 | A4 | A5 | A6 | A7 |

	Deletion Mutants					
Point Mutants	**d1**	**d2**	**d3**	**d4**	**d5**	**d6**
*r*41	0	0	+	+	+	0
*r*42	0	0	0	0	+	0
*r*43	0	0	0	+	0	+
*r*44	0	0	0	0	0	+
*r*45	0	0	+	0	+	0
*r*46	0	+	0	+	0	+

("+" = r^+ recombinants produced; "0" = no r^+ recombinants produced)

14.32 The *loz* (*lethal on Z*) mutants of bacteriophage X are conditional lethal mutants that can grow on *E. coli* strain Y but cannot grow on *E. coli* strain Z. The results shown in the following table were obtained when seven *loz* mutants were analyzed for complementation by infecting *E. coli* strain Z with each possible pair of mutants. A "+" indicates that progeny phage were produced in the infected cells, and a "0" indicates that no progeny phage were produced. All possible *cis* tests were also done, and all *cis* heterozygotes produced wild-type yields of progeny phage.

Mutant	1	2	3	4	5	6	7
7	+	+	0	+	0	0	0
6	+	+	+	+	+	0	
5	+	+	0	+	0		
4	0	0	+	0			
3	+	+	0				
2	0	0					
1	0						

Given that intragenic complementation does not occur between any of the seven *loz* mutants analyzed here, (a) propose three plausible explanations for the apparently anomalous complementation behavior of *loz* mutant number 7. (b) What simple genetic experiments can be used to distinguish between the three possible explanations? (c) Explain why specific outcomes of the proposed experiments will distinguish between the three possible explanations.

14.33 In *Drosophila*, *car* (*carnation*) and *g* (*garnet*) are X-linked mutations that produce brown eyes, in contrast to the dark-red eyes of wild-type flies. The *g* and *car* mutations map at positions 44.4 and 62.5, respectively, on the linkage map of the X chromosome. Is a complementation test needed to determine whether these two mutations are in the same gene or two different genes? If so, why? If not, why not?

14.34 (a) What are the genetic implications of overlapping genes and genes-within-genes? (b) What is the maximum number of different amino acid sequences that can be produced from the same segment of a single strand of DNA? (c) From the two strands of a DNA double helix? (d) What restrictions are imposed on the evolution of two polypeptides specified by two different reading frames of the same DNA sequence?

14.35 Professor Francine H. Crick has characterized five *E. coli* F^- strains, each harboring a different deletion in the *lacZ* gene, which encodes the enzyme β-galactosidase. β-galactosidase catalyzes the first step in the catabolism of lactose, cleaving it into glucose and galactose.

The shaded boxes in the following diagram show the relative locations and sizes of the five deletions (A, B, C, D, and E) characterized by Dr. Crick:

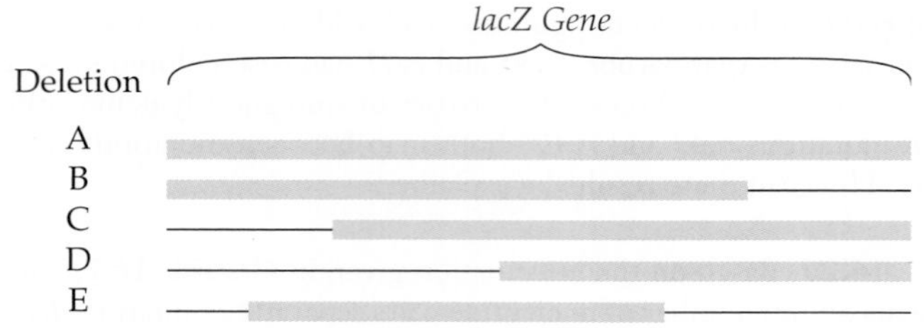

Dr. Crick has also induced several *lacZ* point mutations by treatment of an Hfr strain with nitrous acid. When the professor crossed eight of her nitrous acid-induced mutant strains with each of the five deletion strains, the following results were obtained, where a "+" indicates the formation of $lacZ^+$ recombinants and a "0" indicates that no $lacZ^+$ recombinants were produced:

	Point Mutants							
Deletion	**1**	**2**	**3**	**4**	**5**	**6**	**7**	**8**
A	+	0	0	0	0	+	0	0
B	+	0	0	0	+	+	0	0
C	+	0	+	0	0	+	0	+
D	+	0	+	0	0	+	+	+
E	+	+	+	0	+	+	0	0

(a) What is the linear order of the point mutations that can be ordered on the basis of the above data?

(b) Which of the eight point mutations cannot be ordered relative to the other mutations studied on the basis of these data?

14.36 Tropomyosins are proteins that mediate the interactions between actin and troponin and regulate muscle contractions. In *Drosophila*, six different tropomyosins that have some amino acid sequences in common, but differ in other sequences, are encoded by two tropomyosin genes (*TmI* and *TmII*). How can two genes encode six different polypeptides?

▶ Genomics on the Web

at http://www.ncbi.nlm.nih.gov/

1. In this chapter, we have discussed the use of the *cis-trans* or complementation test to define the basic unit of function in genetics—the GENE. With the development of new technologies that allow us to sequence entire genomes, new strategies have been developed that allow scientists to identify genes based on conserved structural features—a topic we will take up in Chapter 15. When these strategies are used to predict the number of genes in humans, related mammals, and other vertebrates, all of these species seem to have a similar number of genes. What are the current estimates of the number of genes in the genomes of *Homo sapiens* (humans), *Pan troglodytes* (chimps), *Canis lupis familiaris* (dogs), *Bos taurus* (cattle), *Mus musculus* (mice), *Rattus norvegicus* (rats), *Gallus gallus* (chickens), and *Danio rerio* (zebra fish)?

2. Estimates of gene number are very similar in all of these vertebrates, indicating that their genomes have been conserved during the course of evolution. Let's examine this conservation further by comparing the structures of important genes and gene products in these species. Actually, let's focus on one very important gene and its product, namely, β-globin. Hemoglobin is required for the transport of oxygen to the cells of all vertebrates. As discussed in Chapter 12, hemogobin is a tetramer composed of two α-globin chains and two β-globin chains joined to an oxygen-transporting heme group. When hemoglobin does not function properly as, for example, in individuals with sickle-cell disease, hemolytic anemia occurs. This fatal disorder results from a single base-pair substitution in the gene encoding β-globin (Chapter 12). How similar are the amino acid sequences of the β-globins in the vertebrate species discussed in question 1? If you compare the amino acid sequences of the β-globins of these species, which species appear to be the most closely related? When you compare the β-globins of the other vertebrates to human β-globin, which species appears to be our closest relative? Which species is the most distantly related to humans?

Hint: At the NCBI web site, go to Entrez Home and click on HomoloGene (eukaryotic homology groups), then search for "beta hemoglobin," and examine HomoloGene: 68066 (β-globins in these species). Scroll down to "Protein Alignments" and click on "Show Pairwise Alignment Scores." Compare the *Homo sapiens* β-globin amino acid sequence with that of each of the other species by clicking on "Blast."

Chapter 15 The Techniques of Molecular Genetics

CHAPTER OUTLINE

- Basic Techniques Used to Identify, Amplify, and Clone Genes
- Construction and Screening of DNA Libraries
- Rapid Site-Specific Mutagenesis Using PCR
- The Molecular Analysis of DNA, RNA, and Protein
- The Molecular Analysis of Genes and Chromosomes

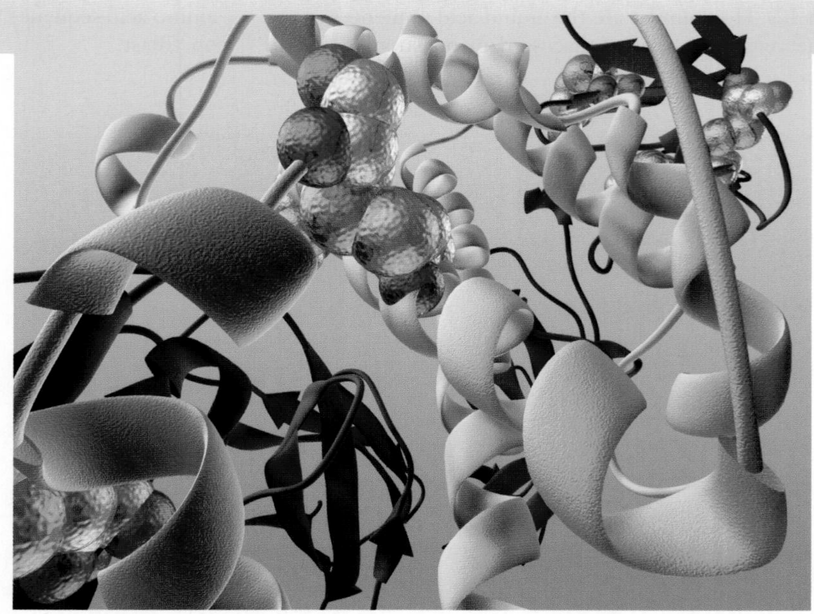

Molecular structure of human growth hormone bound to its receptor (dark brown ribbons) based on X-ray crystallographic data.

Treatment of Pituitary Dwarfism with Human Growth Hormone

Kathy was a typical child in most respects—happy, playful, a bit mischievous, and intelligent. Indeed, the only thing unusual about Kathy was her small stature. She was born with pituitary dwarfism, which results from a deficiency of human growth hormone (hGH). Kathy seemed destined to remain abnormally small throughout her life. Then, at age 10, Kathy began receiving treatments of hGH synthesized in bacteria. She grew 5 inches during her first year of treatments. By continuing to receive hGH during maturation, Kathy reached the short end of the normal height distribution for adults. Without these treatments, she would have remained abnormally small in stature.

The hGH that allowed Kathy to grow to near-normal size was one of the first products of genetic engineering, the use of designed or modified genes to synthesize desired products. hGH was initially produced in *E. coli* cells harboring a modified gene composed of the coding sequence for hGH fused to synthetic bacterial regulatory elements. This chimeric gene was constructed *in vitro* and introduced into *E. coli* by transformation. In 1985, hGH produced in *E. coli* became the second pharmaceutical product of genetic engineering to be approved for use in humans by the U.S. Food and Drug Administration. Human insulin, which was the first such product, was approved in 1982.

How do scientists construct a gene that will produce hGH or human insulin in *E. coli*? They accomplish this feat by combining the coding sequence of the human growth hormone or human insulin gene with regulatory sequences that will ensure its expression in *E. coli* cells. Once they have pieced together the gene in the test tube, they must introduce it into living bacteria so that it can be expressed. In the past, the synthesis of human proteins in bacteria seemed like science fiction. Today, human proteins are routinely produced in bacteria or eukaryotic cells growing in culture. In this chapter, we focus on the powerful tools of molecular genetics that allow researchers to construct genes from components derived from different species and to express these novel genes in both bacteria and eukaryotic cells.

Much of what we know about the structure of genes has been obtained by molecular studies of genes and chromosomes made possible by the development of **recombinant DNA technologies.** Recombinant DNA approaches begin with the **cloning** of specific genes. The cloning of a gene involves its isolation, its insertion into a small self-replicating genetic element such as a plasmid or viral chromosome, and its amplification during the replication of the plasmid or viral chromosome in an appropriate host cell (usually an *E. coli* cell). The small self-replicating genetic elements used to clone genes are called **cloning vectors.** Gene cloning—the isolation and amplification of a given gene—should not be confused with the cloning of organisms—the production of an organism, such as the lamb named Dolly, from a single cell obtained from an adult organism.

The isolation and cloning of a specific gene is a complex process. However, after a gene has been cloned, it can be subjected to a whole array of manipulations that allow investigations of gene structure–function relationships. Usually, a cloned gene is sequenced; that is, the nucleotide-pair sequence of the gene is determined. If the function of the gene is unknown, its nucleotide sequence can be compared with thousands of gene sequences stored in three large computer gene banks—one in Germany, a second in Japan, and a third in the United States (see Focus on GenBank in Chapter 16). Sometimes the function of a gene can be deduced based on its similarity to other genes whose functions are known. Given the nucleotide sequence of a gene and knowledge of the genetic code, the amino acid sequence of the polypeptide encoded by the gene can be predicted. The predicted amino acid sequence of the polypeptide can then be searched for amino acid sequences that may provide clues about its function. Nucleic acid and protein sequence databases have become important resources for research in molecular genetics, and they will become increasingly important in both basic biological research and the diverse applications of this research (Chapters 16 and 17).

Basic Techniques Used to Identify, Amplify, and Clone Genes

Recombinant DNA, gene cloning, and DNA amplification techniques allow scientists to isolate and characterize essentially any DNA sequence from any organism.

The haploid genome of a mammal contains about 3×10^9 nucleotide pairs. If the combined exons of the average gene are 3000 nucleotide pairs long (many are larger), the coding region of the gene will represent one of a million such sequences in the genome. Although most of the DNA in mammalian genomes does not consist of genes, still, isolating any one gene is like searching for the proverbial needle in a haystack. Most techniques used in the analysis of genes and other DNA sequences require that the sequence be available in significant quantities in pure or essentially pure form. How can one identify the segment of a DNA molecule that carries a single gene and isolate enough of this sequence in pure form to permit molecular analyses of its structure and function?

The development of recombinant DNA and gene-cloning technologies has provided molecular geneticists with methods by which genes or other segments of large chromosomes can be isolated, replicated, and studied by nucleic acid sequencing techniques, electron microscopy, and other analytical techniques. Indeed, genes or other DNA sequences can be amplified by two distinct approaches—one with amplification of the sequence occurring *in vivo* and the other *in vitro*. The second approach can only be used when short nucleotide sequences on either side of the DNA sequence of interest are known.

In the first approach, a minichromosome carrying the gene of interest is produced in the test tube and is then introduced into an appropriate host cell. This gene-cloning procedure involves two essential steps: (1) the incorporation of the gene of interest into a small self-replicating chromosome (*in vitro*) and (2) the amplification of the recombinant minichromosome by its replication in an appropriate host cell (*in vivo*). Step 1 involves the joining of two or more different DNA molecules *in vitro* to produce **recombinant DNA molecules,** for example, a human gene inserted into an *E. coli* plasmid or other self-replicating minichromosome. Step 2 is really the gene-cloning event in which the recombinant DNA molecule is replicated or "cloned" to produce many identical copies for subsequent biochemical analysis. In step 2, the recombinant minichromosome is introduced into *E. coli* cells where it replicates to produce many copies of the recombinant DNA molecule. Although the entire procedure is often referred to as the recombinant DNA or gene-cloning technique, these terms actually refer to two separate steps in the process.

In the second approach, short DNA strands that are complementary to DNA sequences on either side of the gene or DNA sequence of interest are synthesized and used to initiate its amplification *in vitro* by a special (heat-stable) DNA polymerase. This procedure—called the polymerase chain reaction (PCR)—is an extremely powerful gene-amplification tool. The amplified products can then be analyzed and sequenced, and, if desired, they can be inserted into cloning vectors and replicated *in vivo* for additional studies. Amplification of a DNA sequence by PCR frequently eliminates the need to clone the sequence by replication *in vivo*. Thus, procedures involving the amplification of DNA sequences by PCR have commonly replaced earlier *in vivo* amplification protocols. However, PCR can only be used when nucleotide sequences flanking the gene or DNA sequence of interest are known.

THE DISCOVERY OF RESTRICTION ENDONUCLEASES

The ability to clone and sequence essentially any gene or other DNA sequence of interest from any species depends on a special class of enzymes called **restriction endonucleases** (from the Greek term *éndon* meaning "within"; endonucleases make internal cuts in DNA molecules). Many endonucleases make random cuts in DNA, but the restriction endonucleases are site-specific, and Type II restriction enzymes cleave DNA molecules only at specific nucleotide sequences called **restriction sites.** Type II restriction enzymes cleave DNA at these sites regardless of the source of the DNA. Different restriction endonucleases are produced by different microorganisms and recognize different nucleotide sequences in DNA (**TABLE 15.1**). The restriction endonucleases are named by using the first letter of the genus and the first two letters of the species that produces the enzyme. If an enzyme is produced only by a specific strain, a letter designating the strain is appended to the name. The first restriction enzyme identified from a bacterial strain is designated I, the second II, and so on. Thus, restriction endonuclease *Eco*RI is produced by *Escherichia coli* strain *R*Y13. About 400 different restriction enzymes have been characterized and purified; thus, restriction endonucleases that cleave DNA molecules at many different DNA sequences are available.

Restriction endonucleases were discovered in 1970 by Hamilton Smith and Daniel Nathans (see A Milestone in Genetics: Restriction Endonucleases). Smith and Nathans shared the 1986 Nobel Prize in Physiology or Medicine with Werner Arber, who carried out pioneering research that led to the discovery of restriction enzymes. The biological function of restriction endonucleases is to protect the genetic material of bacteria from "invasion" by foreign DNAs, such as DNA molecules from another species or viral DNAs. As a result, restriction endonucleases are sometimes referred to as the immune systems of prokaryotes.

All cleavage sites in the DNA of an organism must be protected from cleavage by the organism's own restriction endonucleases; otherwise the organism would commit suicide by degrading its own DNA. In many cases, this protection of endogenous cleavage sites is accomplished by **methylation** of one or more nucleotides in each nucleotide sequence that is recognized by the organism's own restriction endonuclease (**FIGURE 15.1**). Methylation occurs rapidly after replication,

▶ **TABLE 15.1**

Recognition Sequences and Cleavage Sites of Representative Restriction Endonucleases

Enzyme	Source	Recognition Sequence[a] and Cleavage Sites[b]	Number of Recognition Sequences per Chromosome of: Phage λ	SV40 Virus
*Eco*RI	*Escherichia coli* strain RY13	G↓AA·TTC CTT·AA↑G	5	1
*Hin*cII	*Haemophilus influenzae* strain R$_c$	GTPy↓·PuAC CAPu·↑PyTG	34	7
*Hin*dIII	*Haemophilus influenzae* strain R$_d$	A↓AG·CTT TTC·GA↑A	6	6
*Hpa*II	*Haemophilus parainfluenzae*	C↓C·GG GG·C↑C	750	1
*Alu*I	*Arthrobacter luteus*	AG↓·CT TC·↑GA	143	34

[a]The axis of dyad symmetry in each palindromic recognition sequence is indicated by the red dot; the DNA sequences are the same reading in opposite directions from this point and switching the top and bottom strands to correct for their opposite polarity. Pu indicates that either purine (adenine or guanine) may be present at this position; Py indicates that either pyrimidine (thymine or cytosine) may be present.

[b]The position of each bond cleaved is indicated by an arrow. Note that with some restriction endonucleases the cuts are staggered (at different positions in the two complementary strands).

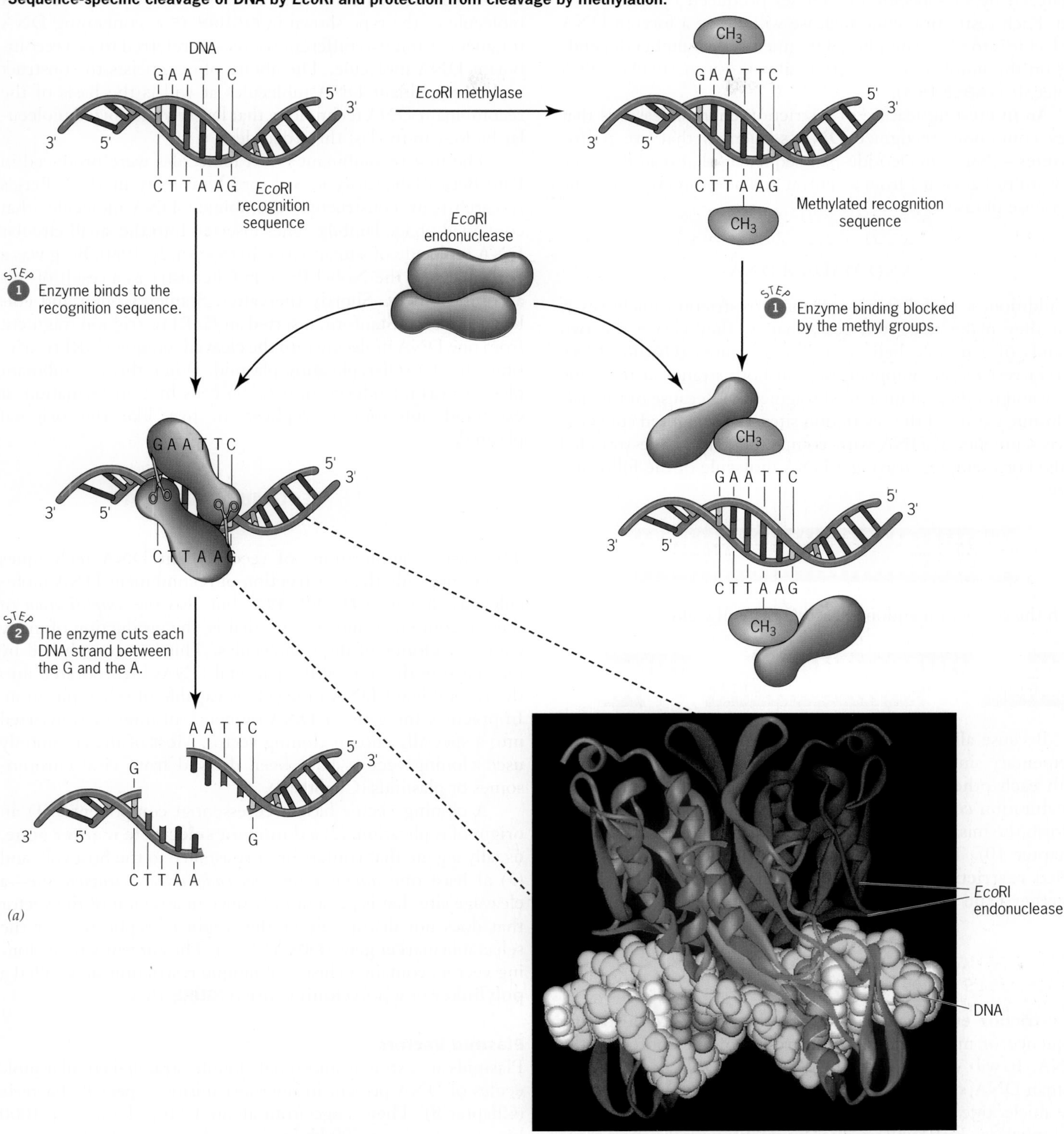

(b) **Structure of an *Eco*RI-DNA complex based on X-ray diffraction data.**

Figure 15.1 ▶ The *Eco*RI restriction-modification system. (a) Cleavage of the unmethylated *Eco*RI recognition sequence by *Eco*RI restriction endonuclease and protection of the recognition sequence from cleavage by methylation catalyzed by the *Eco*RI methylase. (*b*) Diagram of the structure of the *Eco*RI-DNA complex based on X-ray diffraction data. The two subunits of the *Eco*RI endonuclease are shown in red and purple, respectively.

catalyzed by site-specific methylases produced by the organism. Each restriction endonuclease will cleave a foreign DNA molecule into a fixed number of fragments, the number depending on the number of restriction sites in the particular DNA molecule (**TABLE 15.1**).

An interesting feature of restriction endonucleases is that they commonly recognize DNA sequences that are **palindromes**—that is, nucleotide-pair sequences that read the same forward or backward from a central axis of symmetry, as in the nonsense phrase

←——— ———→
AND MADAM DNA

In addition, a useful feature of many restriction nucleases is that they make staggered cuts; that is, they cleave the two strands of a double helix at different points (**FIGURE 15.1**). (Other restriction endonucleases cut both strands at the same place and produce blunt-ended fragments.) Because of the palindromic nature of the restriction sites, the staggered cuts produce segments of DNA with complementary single-stranded ends. For example, cleaving a DNA molecule of the following type:

with the restriction endonuclease *Eco*RI will yield

5′ G + AATTC G + AATTC 3′
3′ CTTAA G CTTAA G 5′

Because all the resulting DNA fragments will have complementary single-stranded termini, they will hydrogen bond with each other and can be rejoined under the appropriate renaturation conditions by using the enzyme **DNA ligase** to reform the missing phosphodiester linkages in each strand (see Chapter 10). Thus, DNA molecules can be cut into pieces, called **restriction fragments,** and the pieces can be joined together again with DNA ligase, almost at will.

THE PRODUCTION OF RECOMBINANT DNA MOLECULES *In Vitro*

A restriction endonuclease catalyzes the cleavage of a specific sequence of nucleotide pairs regardless of the source of the DNA. It will cleave phage DNA, *E. coli* DNA, corn DNA, human DNA, or any other DNA, as long as the DNA contains the nucleotide sequence that it recognizes. Thus, restriction endonuclease *Eco*RI will produce fragments with the same complementary single-stranded ends, 5′-AATT-3′, regardless of the source of DNA, and two *Eco*RI fragments can be covalently fused regardless of their origin; that is, an *Eco*RI fragment from human DNA can be joined to an *Eco*RI fragment from *E. coli* DNA just as easily as two *Eco*RI fragments from *E. coli* DNA or two *Eco*RI fragments from human DNA can be joined. A DNA molecule of the type shown in **FIGURE 15.2,** containing DNA fragments from two different sources, is referred to as a recombinant DNA molecule. The ability of geneticists to construct such recombinant DNA molecules at will is the basis of the recombinant DNA technology that has revolutionized molecular biology in the last three decades.

The first recombinant DNA molecules were produced in Paul Berg's laboratory at Stanford University in 1972. Berg's research team constructed recombinant DNA molecules that contained phage lambda genes inserted into the small circular DNA molecule of simian virus 40 (SV40). In 1980, Berg was a corecipient of the Nobel Prize in Chemistry as a result of this accomplishment. Shortly thereafter, Stanley Cohen and colleagues, also at Stanford, inserted an *Eco*RI restriction fragment from one DNA molecule into the cleaved, unique *Eco*RI restriction site of a self-replicating plasmid. When this recombinant plasmid was introduced into *E. coli* cells by transformation, it exhibited autonomous replication, just like the original plasmid.

AMPLIFICATION OF RECOMBINANT DNA MOLECULES IN CLONING VECTORS

The various applications of recombinant DNA techniques require not only the construction of recombinant DNA molecules, as shown in **FIGURE 15.2,** but also the *amplification* of these recombinant molecules—that is, the production of many copies or **clones** of these molecules. This is accomplished by making sure that one of the parental DNAs incorporated into the recombinant DNA molecule is capable of self-replication. In practice, the gene or DNA sequence of interest is inserted into a specially chosen cloning vector. Most of the commonly used cloning vectors have been derived from viral chromosomes or plasmids (Chapter 8).

A cloning vector has three essential components: (1) an origin of replication, (2) a **dominant selectable marker gene,** usually a gene that confers drug resistance to the host cell, and (3) at least one *unique restriction endonuclease cleavage site*—a cleavage site that is present only once in a region of the vector that does not disrupt either the origin of replication or the selectable marker gene (**FIGURE 15.3**). The currently used cloning vectors contain a cluster of unique restriction sites called a **polylinker** or a **polycloning site** (**FIGURE 15.4**).

Plasmid Vectors

Plasmids are extrachromosomal, double-stranded circular molecules of DNA present in microorganisms, especially bacteria (Chapter 8). They range from about 1 kb (1 kilobase = 1000 base pairs) to over 200 kb in size, and many replicate autonomously. Many plasmids carry antibiotic-resistance genes, which are ideal selectable markers. Plasmid pBR322 was one of the first widely used cloning vectors; it contains both ampicillin- and tetracycline-resistance genes and a number of unique restriction enzyme cleavage sites. Many of the cloning vectors

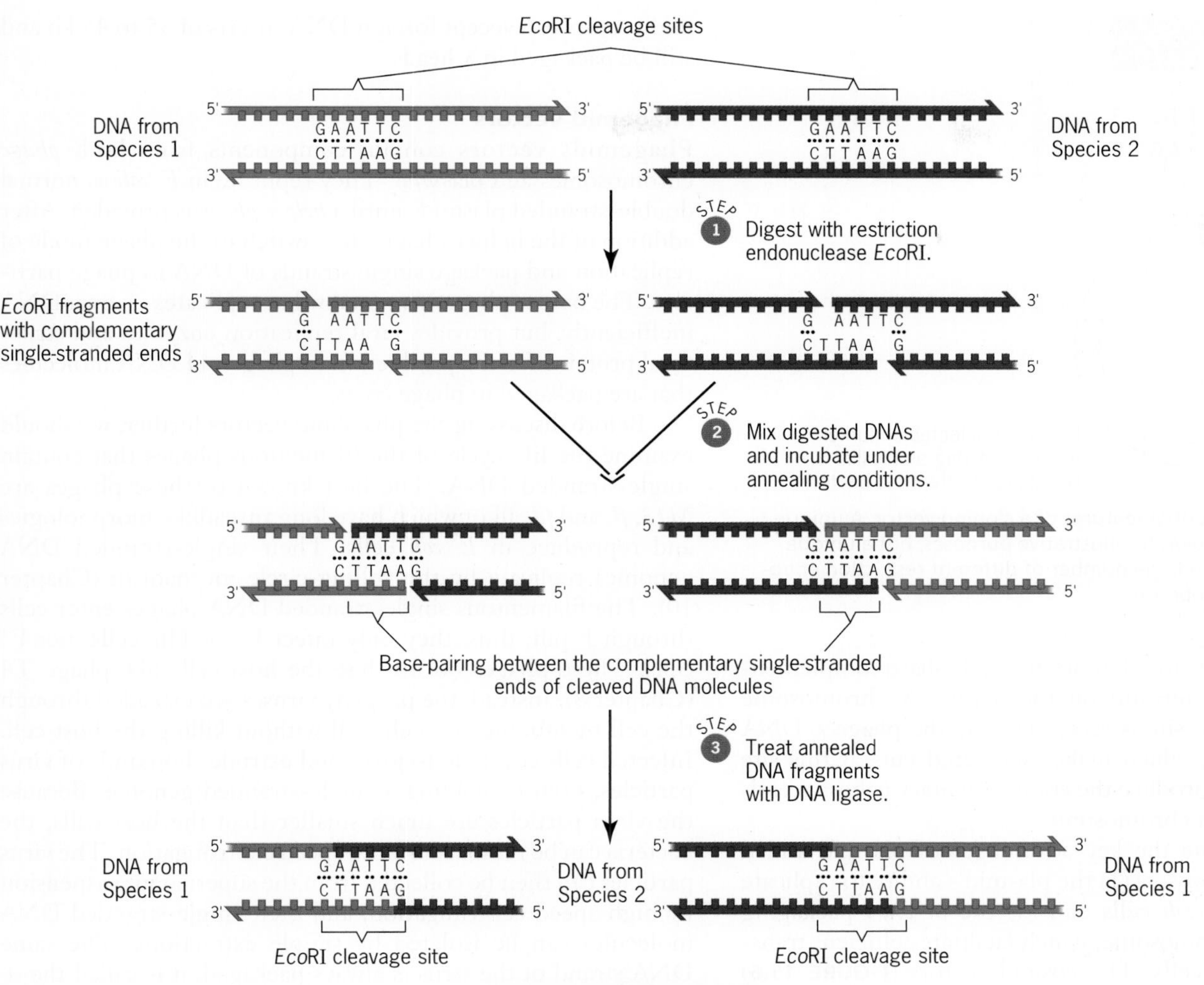

Figure 15.2 ▶ The construction of recombinant DNA molecules *in vitro*. DNA molecules isolated from two different species are cleaved with a restriction enzyme, mixed under annealing conditions, and covalently joined by treatment with DNA ligase. The DNA molecules can be obtained from any species—animal, plant, or microbe. The digestion of DNA with the restriction enzyme *Eco*RI produces the same complementary single-stranded 5′-AATT-3′ ends regardless of the source of the DNA.

in use today were derived, at least in part, from plasmid pBR322.

Bacteriophage Vectors

Most bacteriophage cloning vectors have been constructed from the phage λ chromosome. The complete 48,502 nucleotide-pair sequence of the wild-type lambda genome was determined in 1982. The central one-third of the λ chromosome contains genes that are required for lysogeny (Chapter 8) but not for lytic growth. Thus, the central part (about 15 kb in length) of the λ chromosome can be excised with restriction enzymes and replaced with foreign DNA (**FIGURE 15.5**). The resulting recombinant DNA molecules can be packaged in phage heads *in vitro*. The phage particles can inject the recombinant DNA molecules into *E. coli* cells, where they will replicate and produce clones of the recombinant DNA molecules. DNA molecules that are too large or too small cannot be packaged in the lambda head; only molecules 45 to 50 kb in size are packaged. As a result, the lambda cloning vectors can only accommodate inserts of 10 to 15 kb.

Cosmid Vectors

Some eukaryotic genes are much larger than 15 kb in size and cannot be cloned intact in either plasmid or lambda cloning vectors. For this and other reasons, scientists have developed vectors that can accommodate larger DNA insertions. The first such vectors, called **cosmids** (for λ *cos* site and plas*mid*), were hybrids between plasmids and the phage λ chromosome. *Cos*

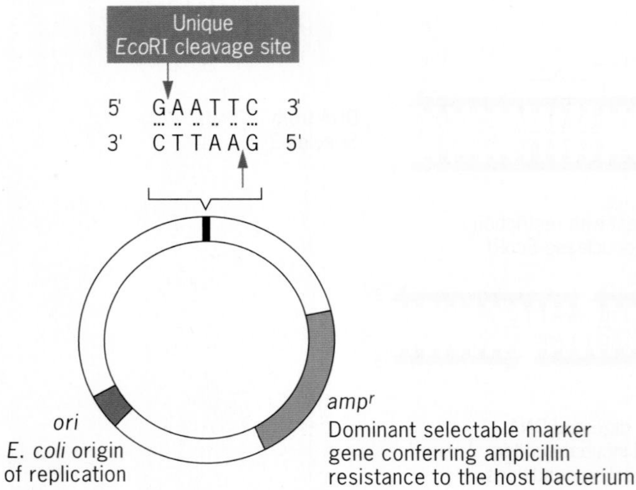

Figure 15.3 ▶ The essential features of a cloning vector. A unique *Eco*RI cleavage site is shown for illustrative purposes, but a unique cleavage site for any of a large number of different restriction endonucleases is just as appropriate.

stands for *co*hesive *s*ite, in reference to the 12-base complementary single-stranded termini on the mature λ chromosome (Chapter 10). The *cos* site is recognized by the phage λ DNA packaging apparatus, which makes staggered cuts at this site during packaging to produce the complementary cohesive ends of the mature lambda chromosome.

Cosmids combine the key advantages of plasmid and λ phage vectors: they possess (1) the plasmid's ability to replicate autonomously in *E. coli* cells and (2) the *in vitro* packaging capacity of the λ chromosome, which facilitates efficient transformation of *E. coli* cells. The cosmid vectors (**FIGURE 15.6**) carry the origin of replication and an antibiotic-resistance gene of the plasmid plus the λ *cos* site, which is needed to package DNA in λ heads. Because none of the λ genes is present, cosmid vectors can accept foreign DNA inserts of 35 to 45 kb and still be packaged in λ heads.

Phagemid Vectors

Phagemids vectors contain components from both *phage* chromosomes and plas*mids*. They replicate in *E. coli* as normal double-stranded plasmids until a *helper phage* is provided. After addition of the helper phage, they switch to the phage mode of replication and package single strands of DNA in phage particles. The helper phage is a mutant that replicates its own DNA inefficiently, but provides viral replication enzymes and structural proteins for the production of phagemid DNA molecules that are packaged in phage coats.

Before discussing the phagemid vectors further, we should examine the life cycle of the filamentous phages that contain single-stranded DNA. The best known of these phages are *M13*, *fl*, and *fd*, all of which have long threadlike morphologies and reproduce in *E. coli* cells. Their single-stranded DNA genomes replicate by the rolling-circle mechanism (Chapter 10). The filamentous single-stranded DNA phages enter cells through F pili; thus, they only infect F^+ or Hfr cells, not F^- cells. These phages do not lyse the host cells like phage T4 (Chapter 8). Instead, the progeny viruses are extruded through the cell membrane and cell wall without killing the host cell. Infected cells continue to grow and extrude thousands of virus particles, each containing a single-stranded genome. Because the virus particles are much smaller than the host cells, the bacteria can be removed by low-speed centrifugation. The virus particles can then be collected from the supernatant suspension by high-speed centrifugation, and their single-stranded DNA molecules can be isolated by simple extractions. The same DNA strand of the virus is always packaged; it is called the + strand, and its complement is the − strand. The packaged + strand has the same sense as mRNA; its nucleotide triplets correspond to mRNA codons, but with T in place of U.

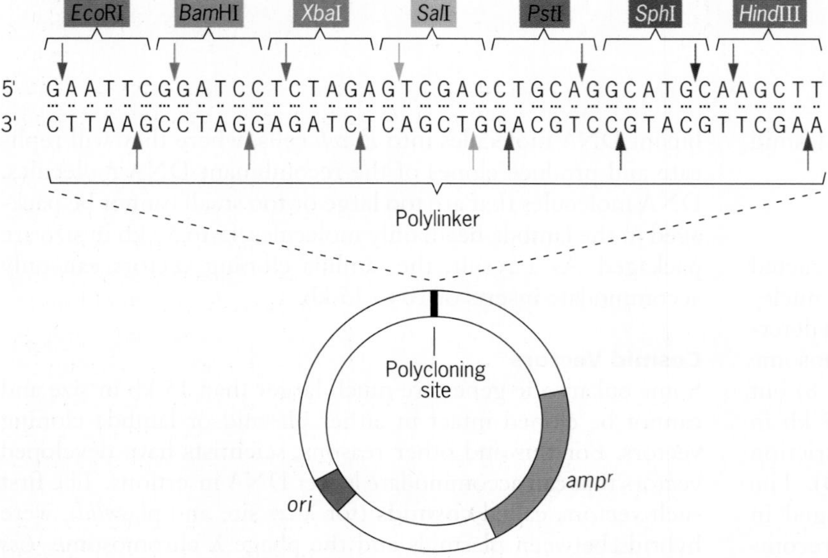

Figure 15.4 ▶ Structure of a polycloning site in a cloning vector. The phosphodiester bond cleaved by each restriction endonuclease is indicated by an arrow. The restriction endonucleases shown are produced by the following microorganisms: *Eco*RI, *Escherichia coli* strain RY13; *Bam*HI, *Bacillus amyloliquefaciens* strain H; *Xba*I, *Xanthomonas badrii; Sal*I, *Streptomyces albus; Pst*I, *Providencia stuartii; Sph*I, *Streptomyces phaeochromogenes;* and *Hin*dIII, *Haemophilus influenzae* strain R_d.

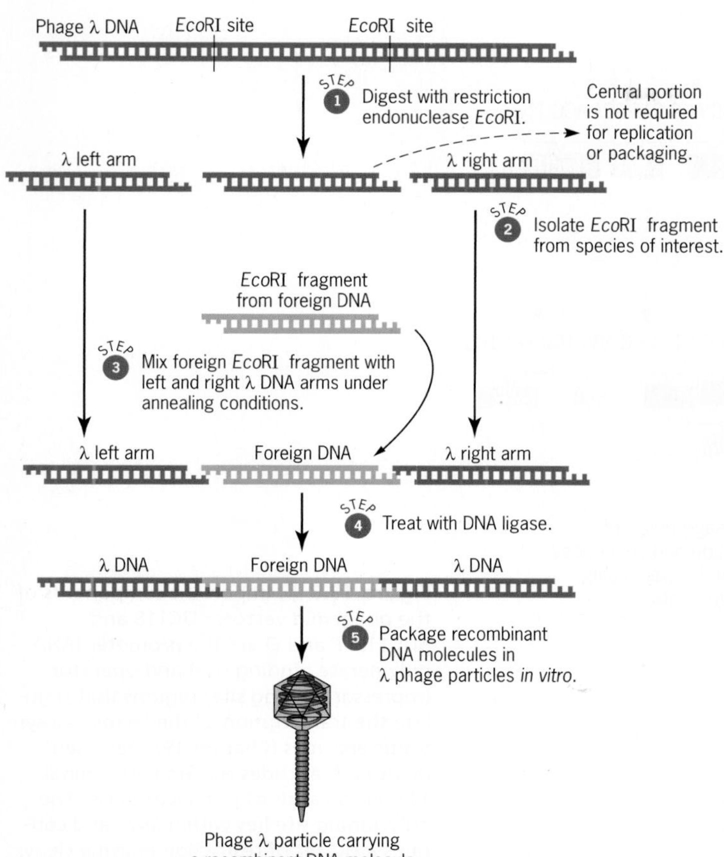

Figure 15.5 ► Strategy employed in using phage λ as a cloning vector.

The major features of the M13 life cycle are shown in **FIGURE 15.7.** Note that the packaging of single strands of phage DNA in progeny phage provides a neat biological purification of single-stranded DNA.

The phagemid vectors *pUC118* and *pUC119* are virtually identical, but they contain the polycloning region in opposite orientations (turned end-for-end) relative to the rest of the genes of the vector (**FIGURE 15.8**). Thus, if a foreign DNA is inserted into a specific restriction site in both vectors, one

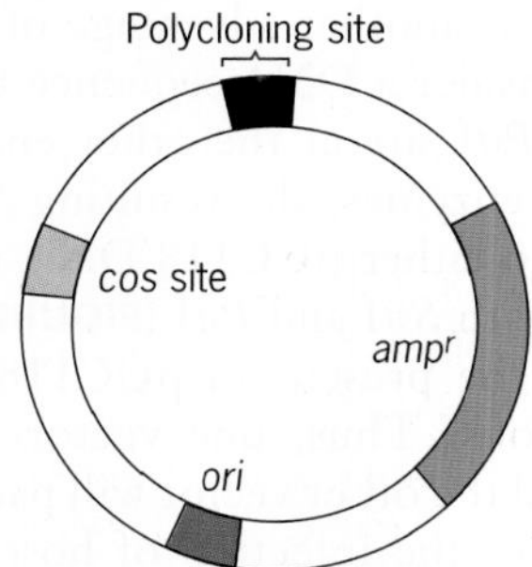

Figure 15.6 ► Structure of a typical cosmid cloning vector. Cosmids combine important features of phage λ and plasmid cloning vectors (see text).

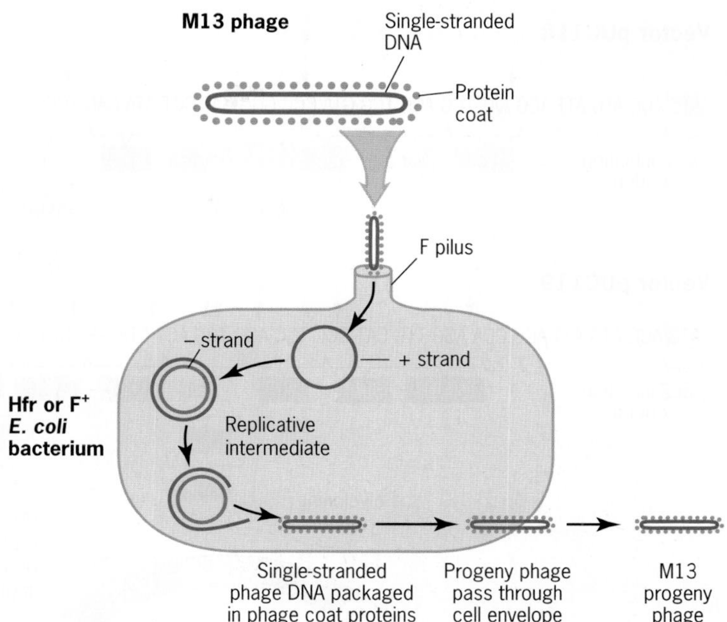

Figure 15.7 ► The life cycle of bacteriophage M13.

vector will package one strand of the foreign DNA, and the other vector will package the complementary strand. These vectors were designated *pUC* for *p*lasmid and *U*niversity of *C*alifornia, where they were constructed. Vectors pUC118 and pUC119 contain the origin of replication from phage M13. Phagemids pUC118 and pUC119 can replicate either (1) as double-stranded plasmids in the absence of helper phage (**FIGURE 15.9*a***) or (2) as single-stranded DNAs, which are packaged in M13 phage coats and extruded from the cell, in the presence of helper phage (**FIGURE 15.9*b***). In the absence of helper phage, replication is controlled by the plasmid origin of replication. In the presence of helper phage, replication is directed by the phage M13 origin of replication.

The utility of the pUC vectors is greatly enhanced by a simple color test that allows cells harboring plasmids with foreign DNA inserts to be distinguished from those containing plasmids with no insert. The basis of this color indicator test is the functional inactivation of the 5′ segment of the *E. coli lacZ* gene, which is present in the vector, by the insertion of foreign DNA into the polycloning region.

The *E. coli lacZ* gene encodes β-galactosidase, the enzyme that cleaves lactose into glucose and galactose. This is the first step in the catabolism of lactose in *E. coli* (Chapter 19). The presence of β-galactosidase in cells can be monitored on the basis of its ability to cleave the substrate 5-bromo-4-chloro-3-indolyl-β-D-galactoside (usually called X-gal) to galactose and 5-bromo-4-chloroindigo. X-gal is colorless; 5-bromo-4-chloroindigo is blue. Thus, cells containing active β-galactosidase produce blue colonies on agar medium containing X-gal. Cells lacking β-galactosidase activity produce white colonies on X-gal plates (**FIGURE 15.10**).

The molecular basis of the β-galactosidase activity that provides the color indicator test for pUC vectors is somewhat

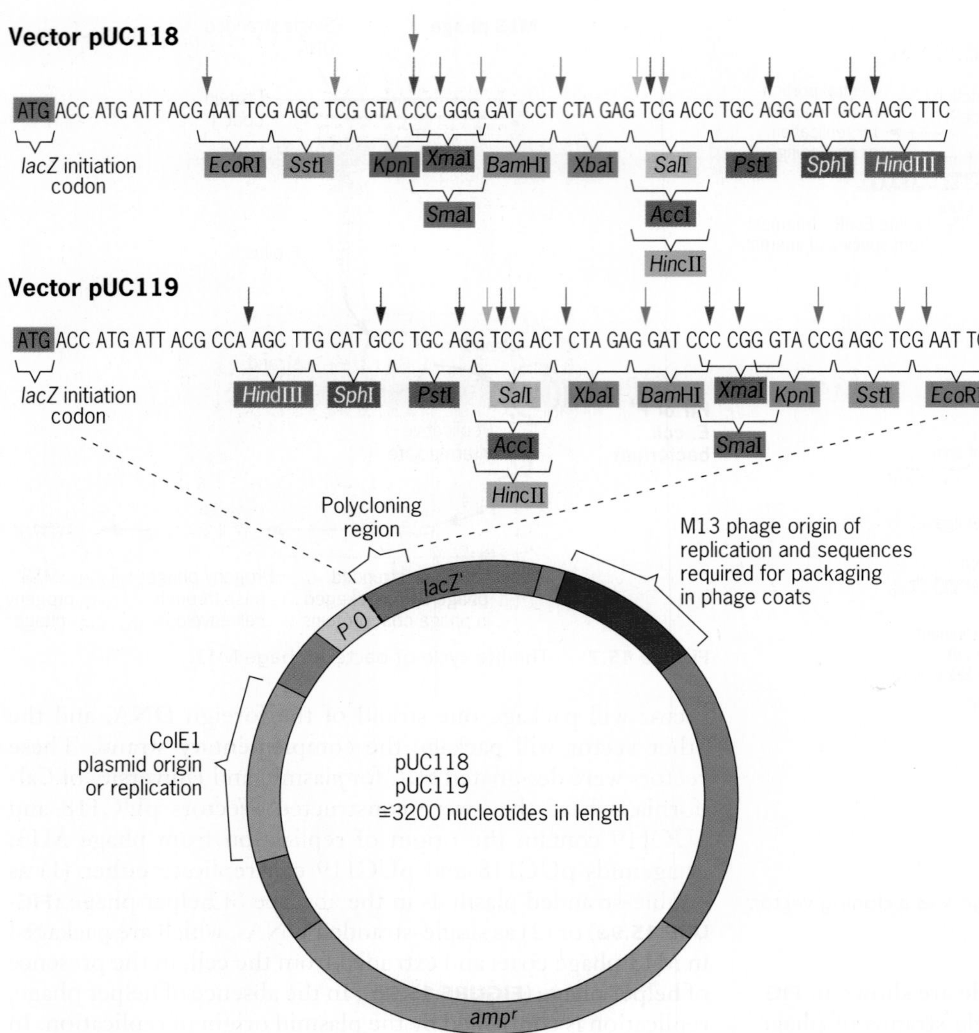

Figure 15.8 ▶ Important components of the phagemid vectors pUC118 and pUC119. *P* and *O* are the promoter (RNA polymerase binding site) and operator (repressor binding site) regions that regulate the transcription of the lactose biosynthetic enzymes (Chapter 19). Gene segment *lacZ′* encodes the amino terminal 147 amino acids of β-galactosidase. The polycloning site lies within *lacZ′* and contains a cluster of restriction enzyme cleavage sites (top). The ColE1 plasmid origin controls replication in the absence of helper phage, whereas the M13 origin controls replication in the presence of helper phage. The arrows denote sites of cleavage by the respective restriction enzymes.

more complex. The *lacZ* gene of *E. coli* is over 3000 nucleotide pairs long, and placing the entire gene in the plasmid would make the vector larger than desired. The pUC vectors contain only a small part of the *lacZ* gene. This *lacZ′* gene segment encodes only the amino terminal portion of β-galactosidase. However, the presence of a functional copy of the *lacZ′* gene segment can be detected because of a unique type of intragenic complementation. When a functional copy of the *lacZ′* gene segment on the pUC plasmid is present in a cell that contains a particular *lacZ* mutant allele on the chromosome or on an F′ plasmid, the two defective *lacZ* sequences yield polypeptides that together have β-galactosidase activity. The mutant allele, designated *lacZ*Δ*M15*, synthesizes a lac protein that lacks amino acids 11 through 14 from the amino terminus. The absence of these amino acids prevents the mutant polypeptides from interacting to produce the active tetrameric form of the enzyme.

The presence of the amino terminal fragment (the first 147 amino acids) of the lacZ polypeptide encoded by the *lacZ′* gene fragment on pUC plasmids facilitates tetramer formation by the ΔM15 deletion polypeptides. This yields active β-galactosidase, which permits the Xgal color test to be utilized without placing the entire *lacZ* gene in the pUC vectors.

The ability to perform *directional cloning*, the insertion of a foreign DNA into the polycloning region in a predetermined orientation, is another advantage of the pUC118-119 vector system. Consider a DNA sequence that has a *Sst*I site at one end and a *Pst*I site at the other end. If this DNA is cleaved with both enzymes, the resulting *Sst*I-*Pst*I fragment can be inserted into either pUC118 DNA or pUC119 DNA that has been cut with *Sst*I and *Pst*I (**FIGURE 15.8**). This *Sst*I-*Pst*I fragment will be present in pUC118 and pUC119 in opposite orientations. Thus, one vector will package one strand of DNA, and the other vector will package the complementary strand after the infection of host cells with helper phage (**FIGURE 15.9*b***). By using both vectors, each of the two complementary single strands of the cloned DNA can be isolated in large quantities.

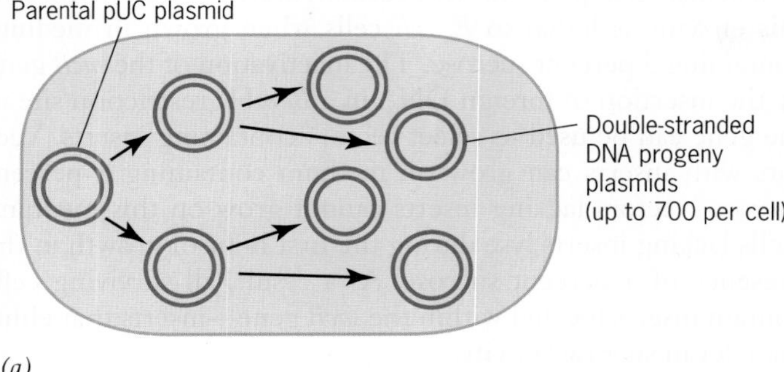

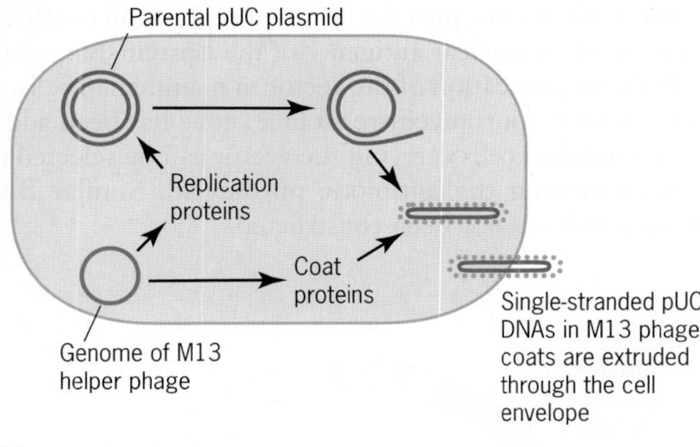

Figure 15.9 ▶ The plasmid and phage modes of replication of the pUC118 and pUC119 phagemid vectors. (*a*) In the absence of M13 helper phage, replication is controlled by an origin of replication derived from plasmid ColE1. (*b*) In bacteria infected with an M13 helper phage, rolling-circle replication is controlled by the M13 origin of replication.

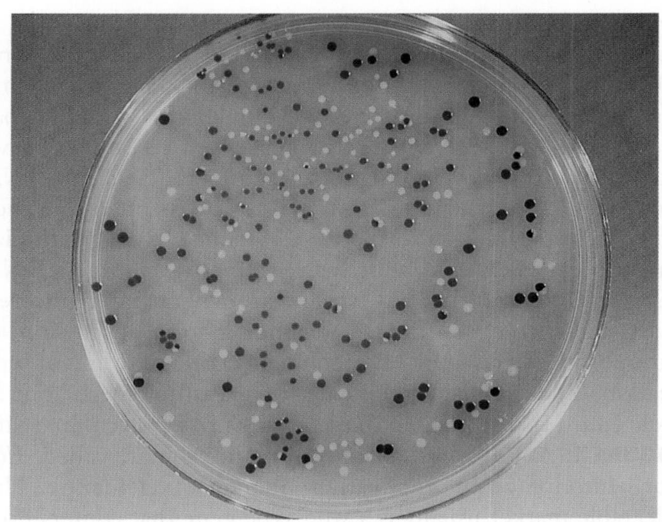

Figure 15.10 ▶ Photograph illustrating the use of X-gal to identify *E. coli* colonies containing (blue) or lacking (white) β-galactosidase activity. In this case, the cells in the white colonies harbor pUC119 plasmids with foreign DNA fragments inserted into the polycloning site, and the cells in the blue colonies contain pUC119 plasmids with no insert.

Eukaryotic and Shuttle Vectors

The plasmid, phage λ, and cosmid cloning vectors all replicate in *E. coli* cells. Because various taxonomic groups utilize distinct origins of replication and regulatory signals, different cloning vectors must be used in different species. Thus, special cloning vectors have been developed that can replicate in other prokaryotes and in eukaryotic organisms. Many unique cloning vectors are available for use in *S. cerevisiae*, *D. melanogaster*, mammals, plants, and other species.

E. coli is the host cell of choice for many of the manipulations performed on cloned DNAs. Thus, some of the most useful cloning vectors are **shuttle vectors** that can replicate in both *E. coli* and another species, such as a eukaryotic organism. Shuttle vectors designed for use in the yeast *S. cerevisiae* contain both *E. coli* and *S. cerevisiae* origins of replication and selectable marker genes, along with a polycloning site (**FIGURE 15.11**). Such shuttle vectors are extremely useful for genetic studies. A yeast gene can be cloned in a shuttle vector, subjected to site-specific mutagenesis (directed changes in nucleotide sequences) in *E. coli*, and then moved back to yeast to examine the effects of the induced modifications. Similar shuttle vectors are available for use in *E. coli* and various animal systems.

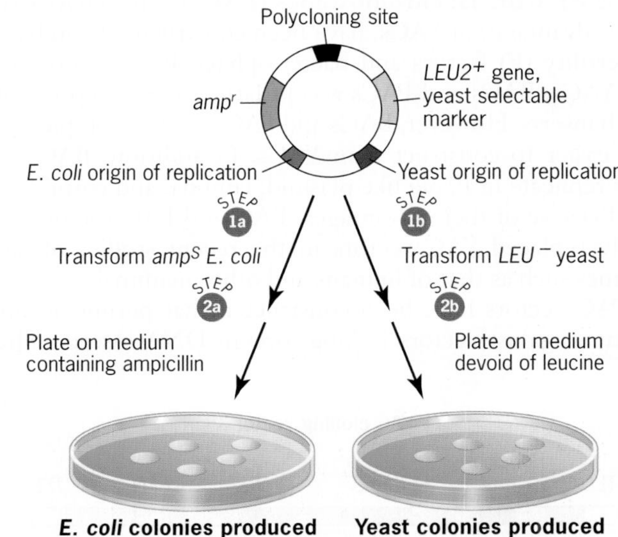

Figure 15.11 ▶ Basic structure and utility of an *E. coli*–yeast shuttle vector, which can replicate in both *E. coli* and *S. cerevisiae*. Shuttle vectors allow investigators to move genes back and forth between two organisms and study the function of these genes in both hosts.

Artificial Chromosomes

Some eukaryotic genes are very large. For example, the gene for human dystrophin (a protein that links filaments to membranes in muscle cells) is over 2000 kb in length. With the goal of cloning large segments of chromosomes, researchers have worked on the development of vectors that accept DNA sequences longer than the 35- to 45-kb inserts of cosmid vectors. This research led to the development of **yeast artificial chromosomes (YACs),** which can accommodate foreign DNA inserts of 200 to 500 kb. YAC vectors are genetically engineered yeast minichromosomes (see Chapter 9 for a discussion of chromosome structure). They contain (1) a yeast origin of replication, (2) a yeast centromere, (3) two yeast telomeres, one at each end, (4) a selectable marker, and (5) a polycloning site (**FIGURE 15.12**). In yeast, origins of replication are called *ARS* (for *A*utonomously *R*eplicating *S*equence) elements. The selectable marker usually is a wild-type gene that confers prototrophy on the host cell. For example, *URA3*$^+$ can be used as the selectable marker in transformations of *URA3*$^-$ auxotrophs growing on medium lacking uracil.

YAC cloning vectors are especially valuable in investigations such as the Human Genome Project, which was designed to map, clone, and sequence large eukaryotic genomes (Chapter 16). The large DNA inserts make it much easier to identify and characterize sets of clones that collectively span an entire genome. If the average insert in cosmid vectors is 40 kb and the average insert in YAC vectors is 200 kb, five cosmid clones are needed to cover the region represented by one YAC clone. Because genome mapping projects require the isolation of overlapping clones, this difference in insert size translates into more than a fivefold difference in the effort required to obtain a complete physical map.

Bacterial artificial chromosomes (BACs) and bacteriophage **P1 artificial chromosomes (PACs),** which have many of the advantages of YACs, have been constructed from bacterial fertility (F) factors and bacteriophage P1 chromosomes. Like YACs, BACs and PACs accept large inserts—up to 150–300 kb inserts. However, BACs and PACs are less complex and, thus, easier to construct than YACs. In addition, BACs and PACs replicate in *E. coli* like plasmid, lambda, and cosmid vectors. Because of their advantages, BAC and PAC vectors have largely replaced YAC vectors in the recent studies of large genomes such as that of humans and other mammals.

PAC vectors have been constructed that permit negative selection against vectors lacking foreign DNA inserts. These PAC vectors contain the *sacB* gene of *Bacillus subtilis*. This gene encodes the enzyme levan sucrase, which catalyzes the transfer of fructose groups to various carbohydrates. The presence of this enzyme is lethal to *E. coli* cells when grown in medium containing 5 percent sucrose. The inactivation of the *sacB* gene by the insertion of foreign DNA in a *Bam*HI restriction site in the gene can be used to select vectors containing inserts. Vectors with inserts can grow on medium containing 5 percent sucrose; vectors lacking inserts cannot grow on this medium. Cells lacking inserts lyse during the first hour of growth in the presence of 5 percent sucrose. As a result, all surviving cells contain inserts located within the *sacB* gene—inserts that eliminate levan sucrase activity.

During the last few years, PAC and BAC vectors have been modified so that they can replicate both in *E. coli* and in mammalian cells. The structure of one of these vectors is shown in **FIGURE 15.13.** This shuttle vector, pJCPAC-Mam1, contains the *sacB* gene, which allows for positive selection of cells carrying vectors with inserts, plus the origin of replication (*oriP*) and the gene encoding nuclear antigen 1 of the Epstein-Barr virus, which facilitate replication of the vector in mammalian cells. In addition, the *pur*^{r} (puromycin-resistance) gene has been added so that mammalian cells carrying the vector can be selected on medium containing the antibiotic puromycin. Similar BAC shuttle vectors have also been constructed.

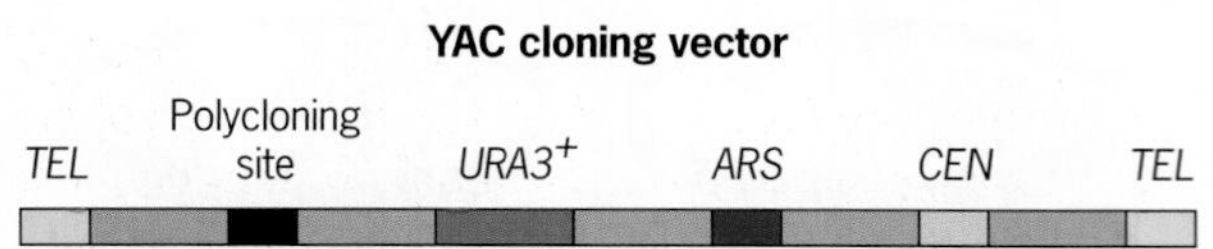

Figure 15.12 ▶ Structure of a YAC cloning vector. The components are: (1) *ARS,* autonomously replicating sequence (a yeast origin of replication), (2) *CEN,* a yeast centromere, (3) *TEL,* a yeast telomere, (4) *URA3*$^+$, a wild-type gene required for the biosynthesis of uracil, and (5) the polycloning site.

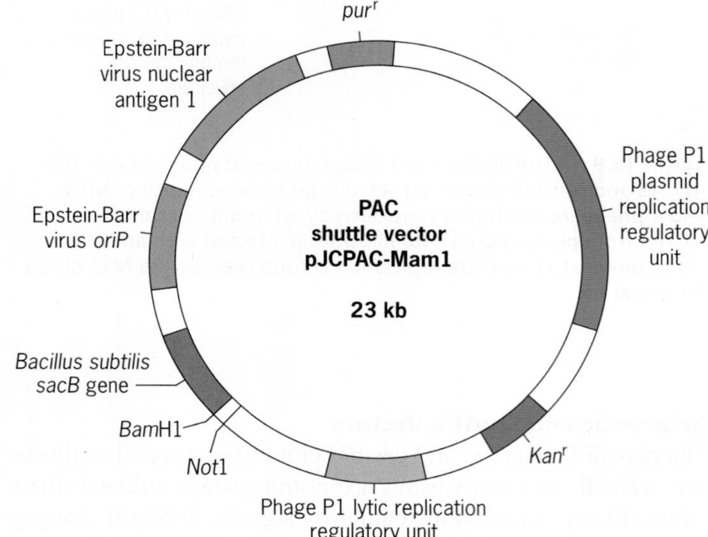

Figure 15.13 ▶ Structure of the PAC mammalian shuttle vector pJCPAC-Mam1. The vector can replicate in either *E. coli* or mammalian cells. It can replicate in *E. coli* at low copy number under the control of the bacteriophage P1 plasmid replication unit or be amplified by inducing the phage P1 lytic replication unit (under the control of the *lac* inducible promoter; see Chapter 19). It can replicate in mammalian cells by using the origin of replication (*oriP*) and nuclear antigen 1 of the Epstein-Barr virus. Genes *kan*^{r} and *pur*^{r} provide dominant selectable markers for use in *E. coli* and mammalian cells, respectively. The *sacB* gene (derived from *Bacillus subtilis*) is used for negative selection against vectors lacking DNA inserts (see text for details). *Bam*HI and *Not*I are cleavage sites for these two restriction endonucleases.

AMPLIFICATION OF DNA SEQUENCES BY THE POLYMERASE CHAIN REACTION (PCR)

Today, we have complete or nearly complete nucleotide sequences of many genomes, including the human genome. The availability of these sequences in GenBank and other databases allows researchers to isolate genes or other DNA sequences of interest without using cloning vectors or host cells. The amplification of the DNA sequence is performed entirely *in vitro*, and the sequence can be amplified a millionfold or more in just a few hours. All that is required to use this procedure is knowledge of short nucleotide sequences flanking the sequence of interest. This *in vitro* amplification of genes and other DNA sequences is accomplished by the **polymerase chain reaction** (usually referred to as **PCR**). PCR involves using synthetic oligonucleotides complementary to known sequences flanking the sequence of interest to prime enzymatic amplification of the intervening segment of DNA in the test tube. The PCR procedure for amplifying DNA sequences was developed by Kary Mullis, who received the 1993 Nobel Prize in Chemistry for this work.

The PCR procedure involves three steps, each repeated many times (**FIGURE 15.14**). In step 1, the genomic DNA containing the sequence to be amplified is denatured by heating to 92–95°C for about 30 seconds. In step 2, the denatured DNA is annealed to an excess of the synthetic oligonucleotide primers by incubating them together at 50–60°C for 30 seconds. The ideal annealing temperature depends on the base composition of the primer. In step 3, DNA polymerase is used to replicate the DNA segment between the sites complementary to the oligonucleotide primers. The primer provides the free 3′-OH required for covalent extension, and the denatured genomic DNA provides the required template function (Chapter 10). Polymerization is usually carried out at 70–72°C for 1.5 minutes. The products of the first cycle of replication are then denatured, annealed to oligonucleotide primers, and replicated again with DNA polymerase. The procedure is repeated many times until the desired level of amplication is achieved. Note that *amplification occurs exponentially*. One DNA double helix will yield 2 double helices after one cycle of replication, 4 after two cycles, 8 after three cycles, 16 after four cycles, 1024 after ten cycles, and so on. After 30 cycles of amplification, more than a billion copies of the DNA sequence will have been produced.

Initially, PCR was performed with DNA polymerase I of *E. coli* as the replicase. Because this enzyme is heat-inactivated during the denaturation step, fresh enzyme had to be added at step 3 of each cycle. A major improvement in PCR amplification of DNA came with the discovery of a heat-stable DNA polymerase in the thermophilic bacterium, *Thermus aquaticus*. This polymerase, called ***Taq* polymerase** (*T. aquaticus* polymerase), remains active during the heat denaturation step. As a result, polymerase does not have to be added after each cycle of denaturation. Instead, excess *Taq* polymerase and oligonucleotide primers can be added at the start of the PCR process, and amplification cycles can be carried out by sequential alterations in temperature. PCR machines or thermal cyclers change the temperature automatically and hold large numbers of samples, making PCR amplification of specific DNA sequences a relatively simple task.

One disadvantage of PCR is that errors are introduced into the amplified DNA copies at low but significant frequencies. Unlike most DNA polymerases, *Taq* polymerase does not contain a built-in 3′ → 5′ proofreading activity, and, consequently, it produces a higher than normal frequency of replication errors. If an incorrect nucleotide is incorporated during an early PCR cycle, it will be amplified just like any other nucleotide in the DNA sequence. When high fidelity is required, PCR is performed using heat-stable polymerases—such as *Pfu* (from *Pyrococcus furiosus*) or *Tli* (from *Thermococcus litoralis*)—that possess 3′ → 5′ proofreading activity. A second disadvantage of *Taq* polymerase is that it amplifies long tracts of DNA—greater than a few thousand nucleotide pairs—inefficiently. If long segments of DNA need to be amplified, the more processive *Tfl* polymerase from *Thermus flavus* is used in place of *taq* polymerase. *Tfl* polymerase will amplify DNA fragments up to about 35 kb in length. Fragments longer than 35 kb cannot be efficiently amplified by PCR.

PCR technologies provide shortcuts for many applications that require large amounts of a specific DNA sequence. These procedures permit scientists to obtain definitive structural data on genes and DNA sequences when very small amounts of DNA are available. One important application occurs in the diagnosis of inherited human diseases, especially in cases of prenatal diagnosis, where limited amounts of fetal DNA are available. A second major application occurs in forensic cases involving the identification of individuals by using DNA isolated from very small tissue samples. Few criteria can provide more definitive evidence of identity than DNA sequences. By using PCR amplification, DNA sequences can be obtained from minute amounts of DNA isolated from a few drops of blood, semen, or even individual human hairs. Thus, *PCR DNA fingerprinting* experiments play important roles in legal cases involving uncertain identity. Some of the applications of PCR are discussed in Chapter 17.

KEY POINTS

- The discovery of restriction endonucleases—enzymes that recognize and cleave DNA in a sequence-specific manner—allowed scientists to produce recombinant DNA molecules *in vitro*.
- DNA sequences can be inserted into small, self-replicating DNA molecules called cloning vectors and amplified by replication *in vivo* after being introduced into living cells by transformation.
- A variety of cloning vectors have been constructed, each with advantages for certain research purposes.
- The polymerase chain reaction (PCR) can be used to amplify specific DNA sequences *in vitro*.

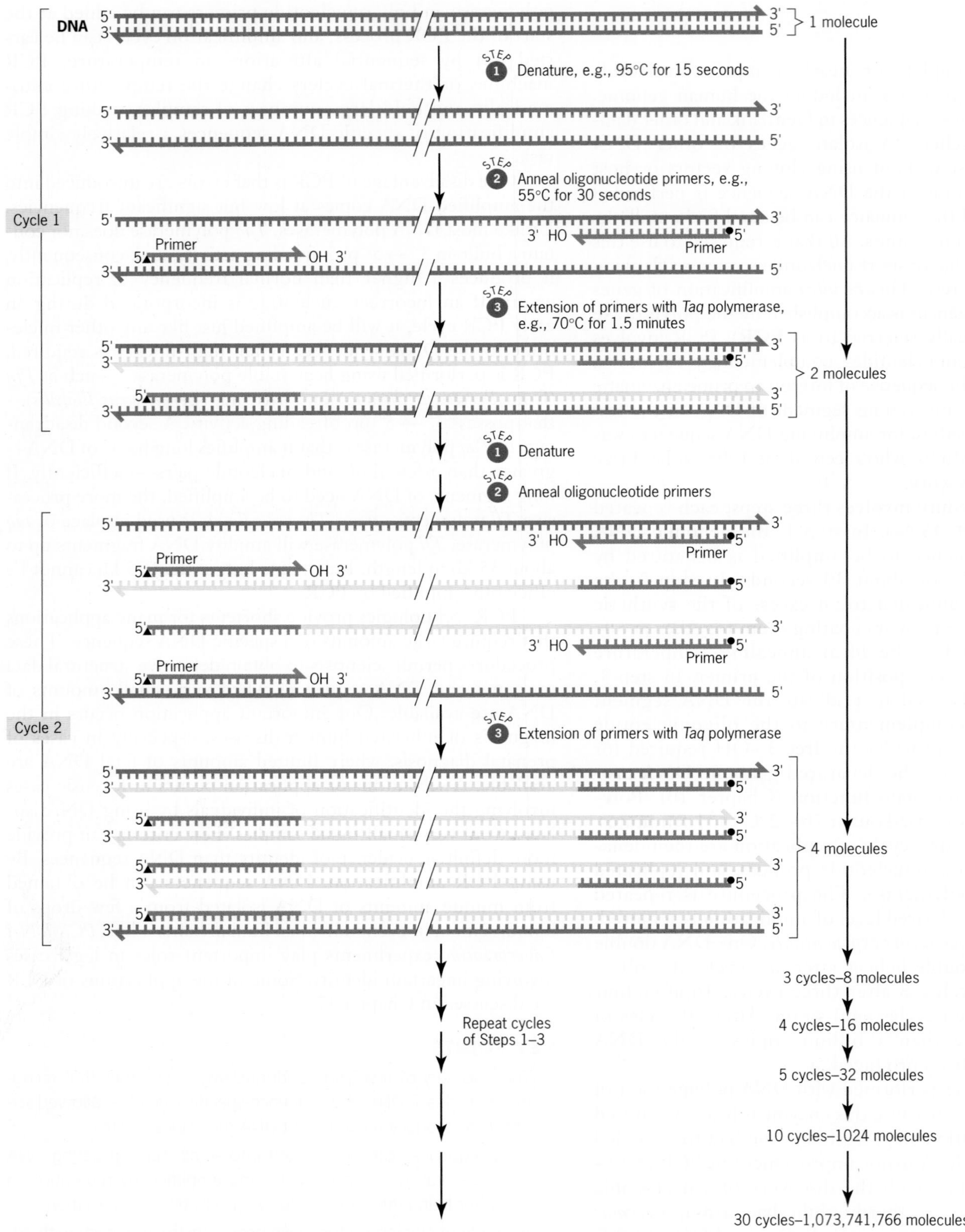

Figure 15.14 ▶ The use of PCR to amplify DNA molecules *in vitro*. Each cycle of amplification involves three steps: (1) denaturation of the genomic DNA being analyzed, (2) annealing of the denatured DNA to chemically synthesized oligonucleotide primers with sequences complementary to sites on opposite sides of the DNA region of interest, and (3) enzymatic replication of the region of interest by *Taq* polymerase.

Construction and Screening of DNA Libraries

DNA libraries can be constructed and screened for genes and other sequences of interest.

The first step in cloning a gene from an organism usually involves the construction of a **genomic DNA library**—a set of DNA clones collectively containing the entire genome. Sometimes, individual chromosomes of an organism are isolated by a procedure that sorts chromosomes based on size and DNA content. The DNAs from the isolated chromosomes are then used to construct chromosome-specific DNA libraries. The availability of chromosome-specific DNA libraries facilitates the search for a gene that is known to reside on a particular chromosome, especially for organisms like humans with large genomes. After their construction, libraries are amplified by replication and used to identify individual genes or DNA sequences of interest to the researcher.

An alternative approach to gene cloning restricts the search for a gene to DNA sequences that are transcribed into mRNA copies. The RNA retroviruses (Chapter 18) encode an enzyme called reverse transcriptase, which catalyzes the synthesis of DNA molecules complementary to single-stranded RNA templates. These DNA molecules are called **complementary DNAs (cDNAs).** They can be converted to double-stranded cDNA molecules with DNA polymerases (Chapter 10), and the double-stranded cDNAs can be cloned in plasmid or phage λ vectors. By starting with mRNA, geneticists are able to construct **cDNA libraries** that contain only the coding regions of the expressed genes of an organism.

CONSTRUCTION OF GENOMIC LIBRARIES

Genomic DNA libraries are usually prepared by isolating total DNA from an organism, digesting the DNA with a restriction endonuclease, and inserting the restriction fragments into an appropriate cloning vector. Two different procedures are used to insert the DNA fragments into the cloning vector. If the restriction enzyme that is used makes staggered cuts in DNA, producing complementary single-stranded ends, the restriction fragments can be ligated directly into vector DNA molecules cut with the same enzyme (**FIGURE 15.15**). An advantage of this procedure is that the foreign DNA inserts can be precisely excised from the vector DNA by cleavage with the restriction endonuclease used to prepare the genomic DNA fragments for cloning.

If the restriction enzyme cuts both strands of DNA at the same position, producing blunt ends, complementary single-stranded tails must be added to the DNA fragments *in vitro.* This is accomplished by using the enzyme **terminal transferase** to add nucleotides to the 3′ termini of the DNA strands after the 5′ ends are cut back with phage λ exonuclease. Usually, poly(A) tails are added to the cleaved vector DNA, and poly(T) tails are added to the genomic DNA fragments, or vice versa. Then, the T-tailed genomic DNA fragments are inserted into the A-tailed vector DNA molecules with DNA ligase. Since the T and A tails will not always be the same length, the *E. coli* enzymes exonuclease III and DNA polymerase I are used to cut back overhangs and fill in gaps, respectively. DNA ligase will only seal nicks between adjacent nucleotides; it will not add nucleotides if gaps are present.

Once the genomic DNA fragments are ligated into vector DNA, the recombinant DNA molecules must be introduced into host cells for amplification by replication *in vivo.* This step usually involves transforming antibiotic-sensitive recipient cells under conditions where a single recombinant DNA molecule is introduced per cell (for most cells) (Chapter 8). When *E. coli* is used, the bacteria must first be made permeable to DNA by treatment with chemicals or a short pulse of electricity. Transformed cells are then selected by growing the cells under conditions where the selectable marker gene of the vector is essential for growth.

A good genomic DNA library contains essentially all of the DNA sequences in the genome of interest. For large genomes, complete libraries contain hundreds of thousands of different recombinant clones.

CONSTRUCTION OF cDNA LIBRARIES

Most of the DNA sequences present in the large genomes of higher animals and plants do not encode proteins. Thus, expressed DNA sequences can be identified more easily by working with complementary DNA (cDNA) libraries. Because most mRNA molecules contain 3′ poly(A) tails, poly(T) oligomers can be used to prime the synthesis of complementary DNA strands by reverse transcriptase (**FIGURE 15.16**). Then, the RNA–DNA duplexes are converted to double-stranded DNA molecules by the combined activities of ribonuclease H, DNA polymerase I, and DNA ligase. Ribonuclease H degrades the RNA template strand, and short RNA fragments produced during degradation serve as primers for DNA synthesis. DNA polymerase I catalyzes the synthesis of the second DNA strand and replaces RNA primers with DNA strands, and DNA ligase seals the remaining single-strand breaks in the double-stranded DNA molecules. These double-stranded cDNAs can be inserted into plasmid or phage λ cloning vectors by adding complementary single-stranded tails to the cDNAs and vectors as described above for blunt-ended restriction fragments.

SCREENING DNA LIBRARIES FOR GENES OF INTEREST

The genomes of higher plants and animals are very large. For example, the human genome contains 3×10^9 nucleotide pairs. Thus, searching genomic DNA or cDNA libraries of

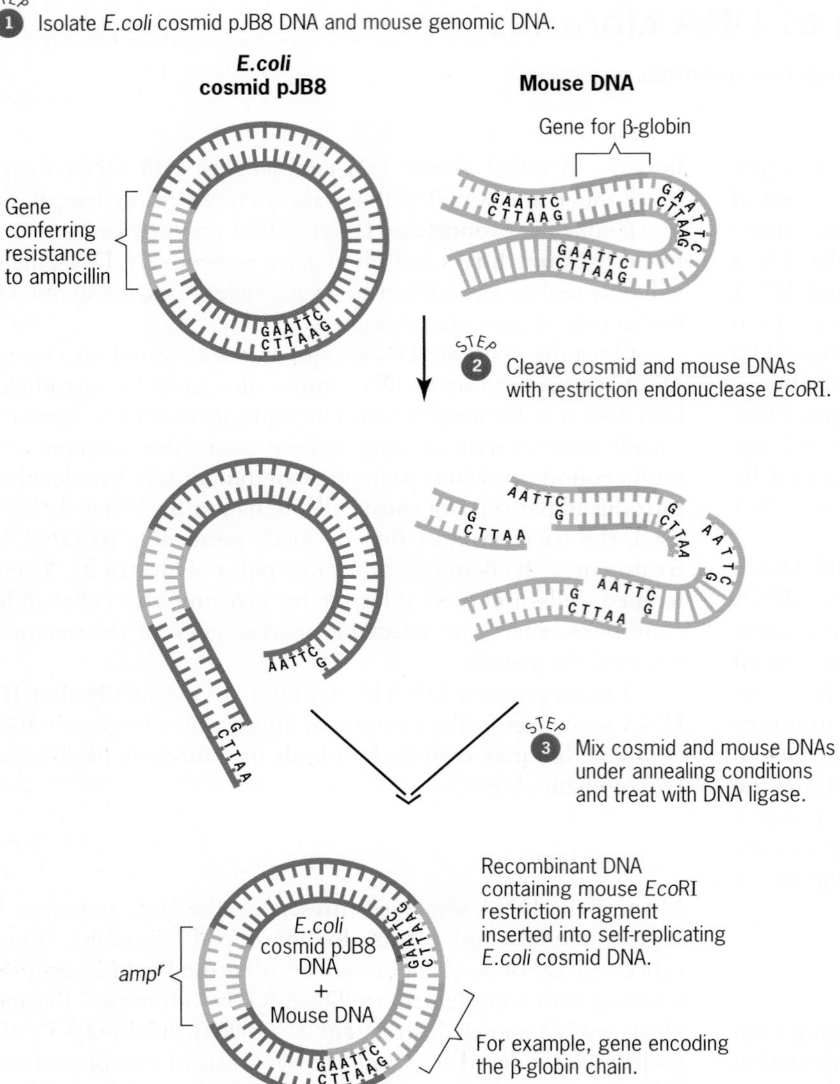

Figure 15.15 ▶ Procedure used to clone DNA restriction fragments with complementary single-stranded ends.

multicellular eukaryotes for a specific gene or other DNA sequence of interest requires the identification of a single DNA sequence in a library that contains a million or more different sequences. The most powerful screening procedure is genetic selection: searching for a DNA sequence in the library that can restore the wild-type phenotype to a mutant organism. When genetic selection cannot be employed, more laborious molecular screens must be carried out. Molecular screens usually involve the use of DNA or RNA sequences as hybridization probes or the use of antibodies to identify gene products synthesized by cDNA clones.

Genetic Selection

The simplest procedure for identifying a clone of interest is **genetic selection.** For example, the *Salmonella typhimurium* gene that confers resistance to penicillin can be easily cloned. A genomic library is constructed from the DNA of a *pen*^{r} strain of *S. typhimurium*. Penicillin-sensitive *E. coli* cells are transformed with the recombinant DNA clones in the library and are plated on medium containing penicillin. Only the transformed cells harboring the *pen*^{r} gene will be able to grow in the presence of penicillin.

When mutations are available in the gene of interest, genetic selection can be based on the ability of the wild-type allele of a gene to restore the normal phenotype to a mutant organism. Although this type of selection is called *complementation screening*, it really depends on the dominance of wild-type alleles over mutant alleles that encode inactive products. For example, the genes of *S. cerevisiae* that encode histidine-biosynthetic enzymes were cloned by transforming *E. coli* histidine auxotrophs with yeast cDNA clones and selecting transformed cells that could grow on histidine-free medium. Indeed, many plant and animal genes have been identified based on their ability to complement mutations in *E. coli* or yeast.

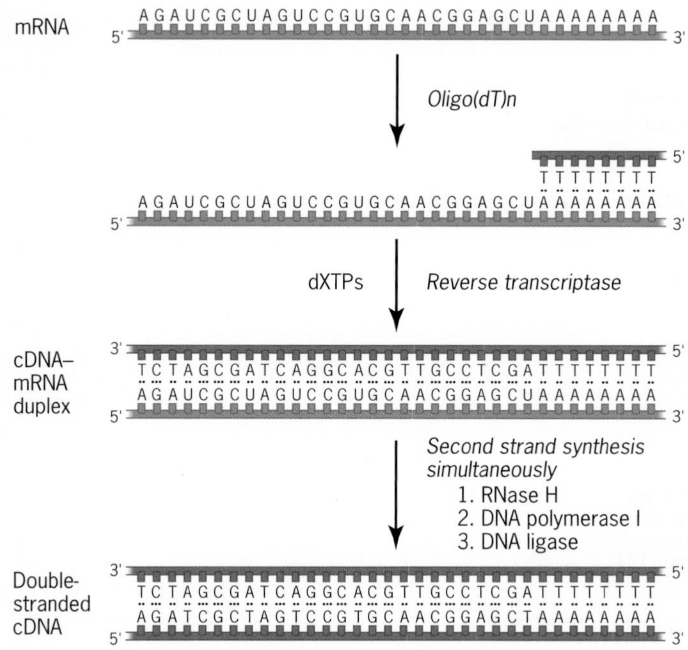

Figure 15.16 ► The synthesis of double-stranded cDNAs from mRNA molecules.

Complementation screening has limitations. Eukaryotic genes contain introns, which must be spliced out of gene transcripts prior to their translation. Because *E. coli* cells do not possess the machinery required to excise introns from eukaryotic genes, complementation screening of eukaryotic clones in *E. coli* is restricted to cDNAs, from which the intron sequences have already been excised. In addition, the complementation screening procedure depends on the correct transcription of the cloned gene in the new host. Eukaryotes have signals that regulate gene expression that are different from those in prokaryotes; therefore, the complementation approach is more likely to work with prokaryotic genes in prokaryotic organisms, and eukaryotic genes in eukaryotic organisms. For this reason, researchers often use *S. cerevisiae* to screen eukaryotic DNA libraries by the complementation procedure.

Molecular Hybridization

The first eukaryotic DNA sequences to be cloned were genes that are highly expressed in specialized cells. These genes included the mammalian α- and β-globin genes and the chicken ovalbumin gene. Red blood cells are highly specialized for the synthesis and storage of hemoglobin. Over 90 percent of the protein molecules synthesized in red blood cells during their period of maximal biosynthetic activity are globin chains. Similarly, ovalbumin is a major product of chicken oviduct cells. As a result, RNA transcripts of the globin and ovalbumin genes can be easily isolated from reticulocytes and oviduct cells, respectively. These RNA transcripts can be employed to synthesize radioactive cDNAs, which, in turn, can be used to screen genomic DNA libraries by ***in situ*** **colony** or **plaque hybridization** (**FIGURE 15.17**). Colony hybridization is used with libraries constructed in plasmid and cosmid vectors; plaque hybridization is used with libraries in phage lambda vectors. We shall focus on *in situ* colony hybridization here, but the two procedures are virtually identical.

The colony hybridization screening procedure involves transfer of the colonies formed by transformed cells onto nylon membranes, hybridization with a labeled DNA or RNA probe, and autoradiography (**FIGURE 15.17**). The labeled DNA or RNA is employed as a probe for hybridization (see Chapter 9) to denatured DNA from colonies grown on the nylon membranes. The DNA from the lysed cells is bound to the membranes before hybridization so that it won't come off during subsequent steps in the procedure. After time is allowed for hybridization between complementary strands of DNA, the membranes are washed with buffered salt solutions to remove nonhybridized cDNA and are then exposed to X-ray film to detect the presence of radioactivity on the membrane. Only colonies that contain DNA sequences complementary to the radioactive cDNA will yield radioactive spots on the autoradiographs (**FIGURE 15.17**). The locations of the radioactive spots are used to identify colonies that contain the desired sequence on the original replicated plates. These colonies are used to purify DNA clones harboring the gene or DNA sequence of interest.

KEY POINTS

- DNA libraries can be constructed that contain complete sets of genomic DNA sequences or DNA copies (cDNAs) of mRNAs in an organism.
- Specific genes or other DNA sequences can be isolated from DNA libraries by genetic complementation or by hybridization to labeled nucleic acid probes containing sequences of known function.

Rapid Site-Specific Mutagenesis Using PCR

PCR can be used to produce mutations of interest quickly and efficiently.

New information about the structures of eukaryotic genes and chromosomes has accumulated at an unprecedented rate since the development of recombinant DNA technologies. Researchers can now clone and characterize any gene or other DNA sequence of interest from virtually any organism. These cloned genes can be modified *in vitro* and reintroduced into the same or different species for further study. Scientists can produce new restriction enzyme cleavage sites in DNA molecules or change a particular codon in a gene to a different codon or even alter numerous codons in a gene. In short, scientists can

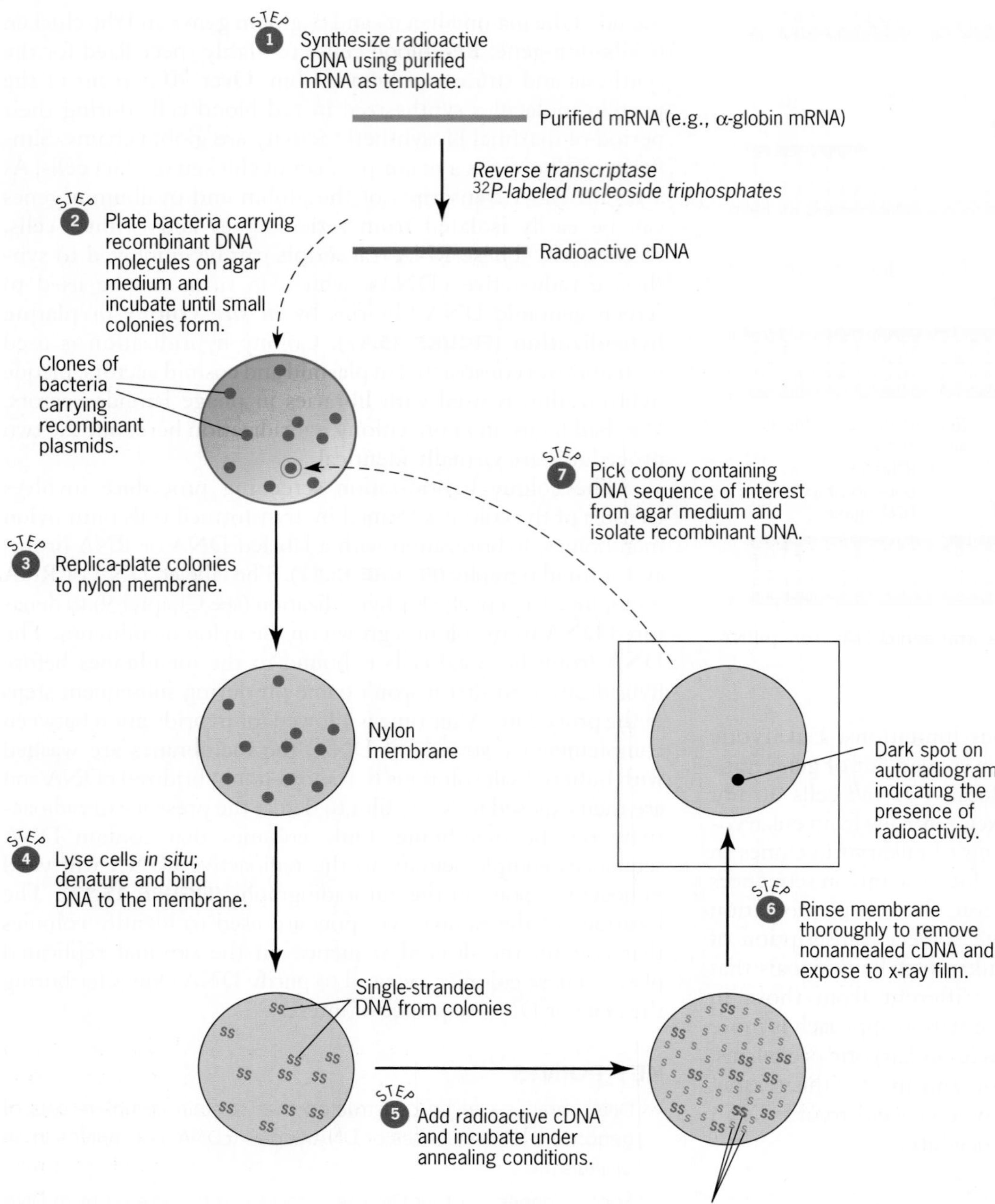
STEP 1 Synthesize radioactive cDNA using purified mRNA as template.
Purified mRNA (e.g., α-globin mRNA)
Reverse transcriptase
32P-labeled nucleoside triphosphates
Radioactive cDNA
STEP 2 Plate bacteria carrying recombinant DNA molecules on agar medium and incubate until small colonies form.
Clones of bacteria carrying recombinant plasmids.
STEP 7 Pick colony containing DNA sequence of interest from agar medium and isolate recombinant DNA.
STEP 3 Replica-plate colonies to nylon membrane.
Nylon membrane
Dark spot on autoradiogram indicating the presence of radioactivity.
STEP 4 Lyse cells in situ; denature and bind DNA to the membrane.
STEP 6 Rinse membrane thoroughly to remove nonannealed cDNA and expose to x-ray film.
Single-stranded DNA from colonies
STEP 5 Add radioactive cDNA and incubate under annealing conditions.
Radioactive cDNA

Figure 15.17 ▶ Screening DNA libraries by colony hybridization. A radioactive cDNA is employed as a hybridization probe. See text for details.

now dissect genes at the nucleotide level by changing one nucleotide at a time and examining the effect of each change on the function of the gene or its product. Such nucleotide changes are accomplished by a procedure called **site-specific mutagenesis,** or, more precisely, **oligonucleotide-directed site-specific mutagenesis.** Site-specific mutagenesis can be performed by many different procedures. In this section, we will examine one of the ways that PCR can be used to produce site-specific mutations. This protocol can be carried out on genes or other DNA sequences amplified in cloning vectors or amplified *in vitro* by PCR. The site-specific mutagenesis is accomplished by using a mutagenic PCR primer as illustrated in **FIGURE 15.18.**

The mutagenic primer is an oligonucleotide 12 to 15 nucleotides in length, which is largely complementary to one strand of the DNA sequence of interest, but which contains one or more noncomplementary or "mismatched" bases. The mismatched bases will provide the desired mutant sequence. The synthetic oligonucleotide primer is used in conjunction with another primer complementary to the other strand of the DNA sequence of interest. The two primers are used to

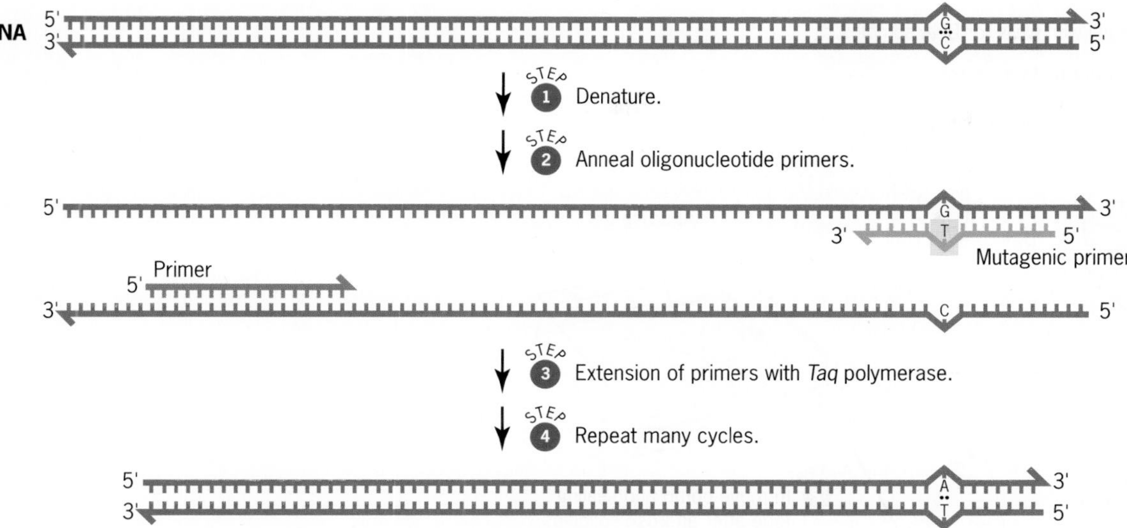

Figure 15.18 ▶ Site-specific mutagenesis performed with mutagenic PCR primers.

amplify the intervening DNA sequence as in standard PCR (see **FIGURE 15.14**). After many cycles of amplification by PCR, the PCR products will consist almost entirely of mutant DNA fragments. The protocol shown in **FIGURE 15.18** can also be used with additional flanking PCR primers to incorporate the induced mutation into larger fragments of a gene or chromosome.

KEY POINT

▶ Site-specific mutagenesis can be performed efficiently by PCR.

The Molecular Analysis of DNA, RNA, and Protein

DNA, RNA, or protein molecules can be separated by gel electrophoresis, transferred to membranes, and analyzed by various procedures.

The development of recombinant DNA techniques has spawned many new approaches to the analysis of genes and gene products. Questions that were totally unapproachable just 20 years ago can now be investigated with relative ease. Geneticists can isolate and characterize essentially any gene from any organism; however, the isolation of genes from large eukaryotic genomes is sometimes a long and laborious process (Chapter 17). Once a gene has been cloned, its expression can be investigated in even the most complex organisms such as humans.

Is a particular gene expressed in the kidney, the liver, bone cells, hair follicles, erythrocytes, or lymphocytes? Is this gene expressed throughout the development of the organism or only during certain stages of development? Is a mutant allele of this gene similarly expressed, spatially and temporally, during development? Or does the mutant allele have an altered pattern of expression? If the latter, is this altered pattern of expression responsible for an inherited syndrome or disease? These questions and many others can now be routinely investigated using well-established methodologies.

A comprehensive discussion of the techniques used to investigate gene structure and function is beyond the scope of this text. However, let's consider some of the most important methods used to investigate the structure of genes (DNA), their transcripts (RNA), and their final products (usually proteins).

ANALYSIS OF DNAs BY SOUTHERN BLOT HYBRIDIZATIONS

Gel electrophoresis is a powerful tool for the separation of macromolecules with different sizes and charges. DNA molecules have an essentially constant charge per unit mass; thus, they separate in agarose and acrylamide gels almost entirely on the basis of size or conformation. Agarose or acrylamide gels act as molecular sieves, retarding the passage of large molecules more than small molecules. Agarose gels are better sieves for large molecules (larger than a few hundred nucleotides); acrylamide gels are better for separating small DNA molecules.

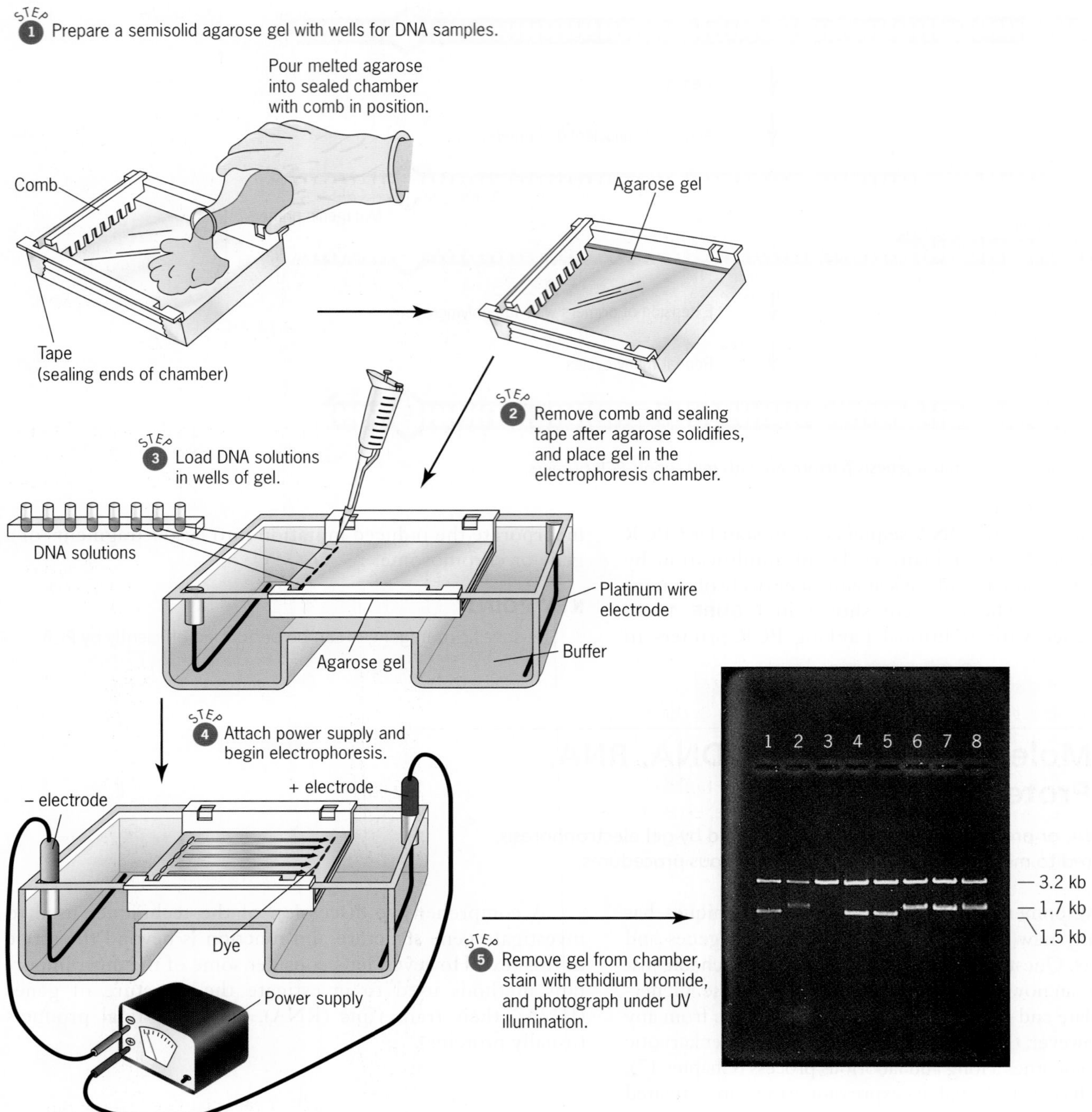

Figure 15.19 ▶ The separation of DNA molecules by agarose gel electrophoresis. The DNAs are dissolved in loading buffer with density greater than that of the electrophoresis buffer so that DNA samples settle to the bottoms of the wells, rather than diffusing into the electrophoresis buffer. The loading buffer also contains a dye to monitor the rate of migration of molecules through the gel. Ethidium bromide binds to DNA and fluoresces when illuminated with ultraviolet light. In the photograph shown, lane 3 contained *Eco*RI-cut pUC119 DNA; the other lanes contained *Eco*RI-cut pUC119 DNAs carrying maize glutamine synthetase cDNA inserts.

FIGURE 15.19 illustrates the separation of DNA restriction fragments by agarose gel electrophoresis. The procedures used to separate RNA and protein molecules are largely the same in principle but involve slightly different techniques because of the unique properties of each class of macromolecule (see Variation in Protein Structure in Chapter 25).

In 1975, E. M. Southern published an important new procedure that allowed investigators to identify the locations of

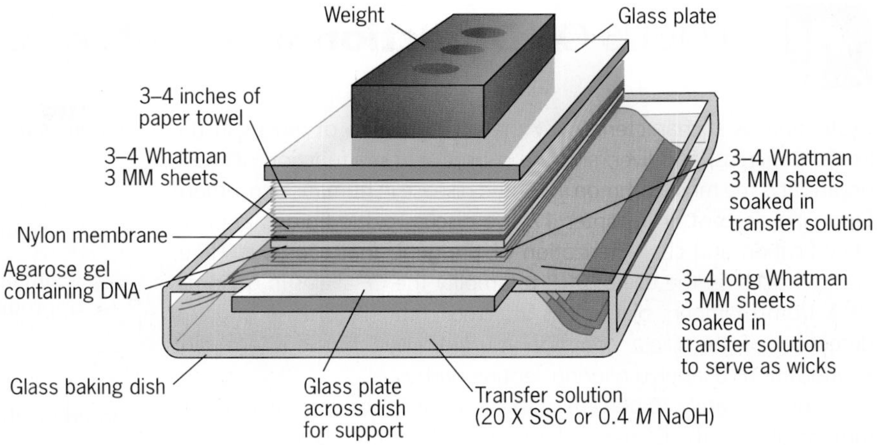

Weight
Glass plate
3–4 inches of paper towel
3–4 Whatman 3 MM sheets
Nylon membrane
Agarose gel containing DNA
Glass baking dish
Glass plate across dish for support
Transfer solution (20 X SSC or 0.4 *M* NaOH)
3–4 long Whatman 3 MM sheets soaked in transfer solution to serve as wicks
3–4 Whatman 3 MM sheets soaked in transfer solution

Figure 15.20 ▶ Procedure used to transfer DNAs separated by gel electrophoresis to nylon membranes. The transfer solution carries the DNA from the gel to the membrane as the dry paper towels on top draw the salt solution from the reservoir through the gel to the towels. The DNA binds to the membrane on contact. The membrane with the DNA bound to it is dried and baked under vacuum to affix the DNA firmly prior to hybridization. SSC is a solution containing sodium chloride and sodium citrate.

genes and other DNA sequences on restriction fragments separated by gel electrophoresis. The essential feature of this technique is the transfer of the DNA molecules that have been separated by gel electrophoresis onto nitrocellulose or nylon membranes (**FIGURE 15.20**). Such transfers of DNA to membranes are called Southern blots after the scientist who developed the technique. The DNA is denatured either prior to or during transfer by placing the gel in an alkaline solution. After transfer, the DNA is immobilized on the membrane by drying or UV irradiation. A radioactive DNA probe containing the sequence of interest is then hybridized (Chapter 9) with the immobilized DNA on the membrane. The probe will hybridize only with DNA molecules that contain a nucleotide sequence complementary to the sequence of the probe. Nonhybridized probe is then washed off the membrane, and the washed membrane is exposed to X-ray film to detect the presence of the radioactivity. After the film is developed, the dark bands show the positions of DNA sequences that have hybridized with the probe (**FIGURE 15.21**).

The ability to transfer DNA molecules that have been separated by gel electrophoresis to nylon membranes for hybridization studies and other types of analyses has proven to be extremely useful (see Focus on Detection of a Mutant Gene Causing Cystic Fibrosis).

ANALYSIS OF RNAs BY NORTHERN BLOT HYBRIDIZATIONS

If DNA molecules can be transferred from agarose gels to nylon membranes for hybridization studies, we might expect that RNA molecules separated by agarose gel electrophoresis could be similarly transferred and analyzed. Indeed, such RNA transfers are used routinely in genetics laboratories. RNA blots are called **northern blots** in recognition of the fact that the procedure is analogous to the Southern blotting technique, but with RNA molecules being separated and transferred to a membrane. As we will discuss in the next section, this terminology has been extended to the transfer of proteins from gels to membranes, a procedure called western blotting.

The northern blot procedure is essentially identical to that used for Southern blot transfers (**FIGURE 15.20**). However, RNA molecules are very sensitive to degradation by RNases.

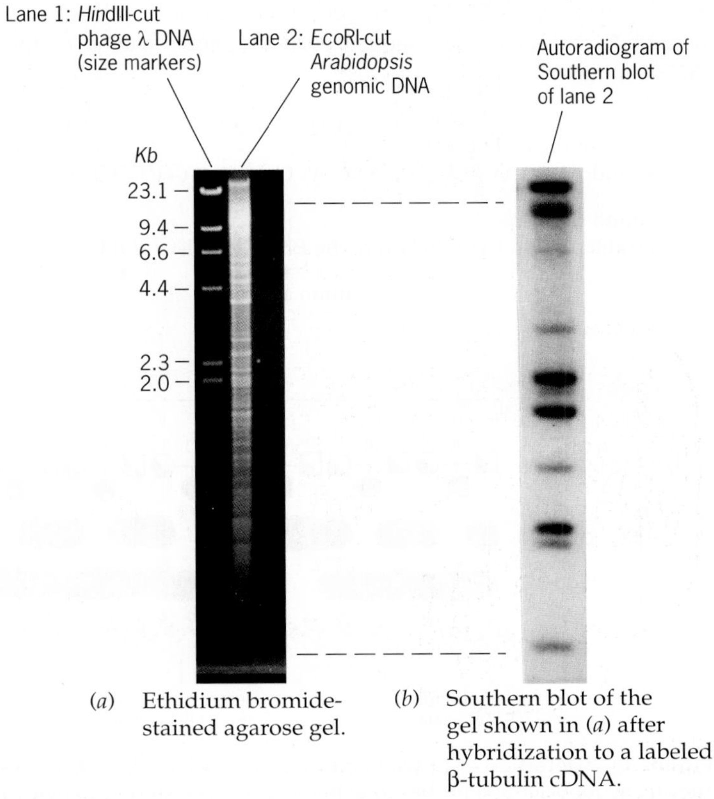

Figure 15.21 ▶ Identification of genomic restriction fragments harboring specific DNA sequences by the Southern blot hybridization procedure. (*a*) Photograph of an ethidium bromide-stained agarose gel containing phage λ DNA digested with *Hin*dIII (left lane), and *Arabidopsis thaliana* DNA digested with *Eco*RI (right lane). The λ DNA digest provides size markers. The *A. thaliana* DNA digest was transferred to a nylon membrane by the Southern procedure (**FIGURE 15.20**) and hybridized to a radioactive DNA fragment of a cloned β-tubulin gene. The resulting Southern blot is shown in (*b*); nine different *Eco*RI fragments hybridized with the β-tubulin probe.

▶ FOCUS ON **Detection of a Mutant Gene Causing Cystic Fibrosis**

Cystic fibrosis is characterized by the accumulation of mucus in the lungs, pancreas, and liver, and the subsequent malfunction of these organs. It is the most common inherited disease in humans of northern European descent. In Chapter 17, we discuss cystic fibrosis and the identification and characterization of the gene that causes it. Here, we will focus on the use of PCR to amplify the *CF* alleles in genomic DNA from members of families afflicted with this disease and the detection of the most common mutant allele by Southern blot hybridization to labeled oligonucleotide probes.

Approximately 70 percent of the cases of CF result from a specific mutant allele of the *CF* gene. This mutant allele, *CF*Δ*F508,* contains a three-base deletion that eliminates a phenylalanine residue at position 508 in the polypeptide product. Because the nucleotide sequence of the *CF* gene is known and since the *CF*Δ*F508* allele differs from the wild-type allele by three base pairs, it was possible to design oligonucleotide probes that hybridize specifically with the wild-type *CF* allele or the Δ*F508* allele under the appropriate conditions.

The wild-type *CF* gene and gene product have the following nucleotide and amino acid sequences in the region altered by the Δ*F508* mutation:

deleted in Δ*F508* (over TTT)

bases in the coding strand: 5′-AAA GAA AAT ATC ATC TTT GGT GTT-3′

amino acids in product: NH$_2$-Lys Glu Asn Ile Ile Phe Gly Val-COOH

amino acid 508 (under Phe)

whereas the Δ*F508* allele and product have these sequences:

deletion

bases in the coding strand: 5′-AAA GAA AAT ATC AT. . .T GGT GTT-3′

amino acids in product: NH$_2$-Lys Glu Asn Ile Ile — Gly Val-COOH

Phe absent

Based on these nucleotide sequences, Lap-Chee Tsui and colleagues synthesized oligonucleotides spanning this region of the mutant and wild-type alleles of the *CF* gene and tested their specificity. They demonstrated that at 37°C under a standard set of conditions, one oligonucleotide probe (oligo-N: 3′-CTTTTATAGTAGAAACCAC-5′) hybridized only with the wild-type allele, whereas another (oligo-ΔF: 3′-TTCTTTTATAGTA . . . ACCACAA-5′) hybridized only with the Δ*F508* allele. Their results showed that the oligo-ΔF probe could be used to detect the Δ*F508* allele in either the homozygous or heterozygous state. When Tsui and coworkers used these allele-specific oligonucleotide probes to analyze CF patients and their parents for the presence of the Δ*F508* mutation, they found that many of the patients were homozygous for this mutation, whereas most of their parents were heterozygous, as is expected. Some of their results are shown in **FIGURE 1.**

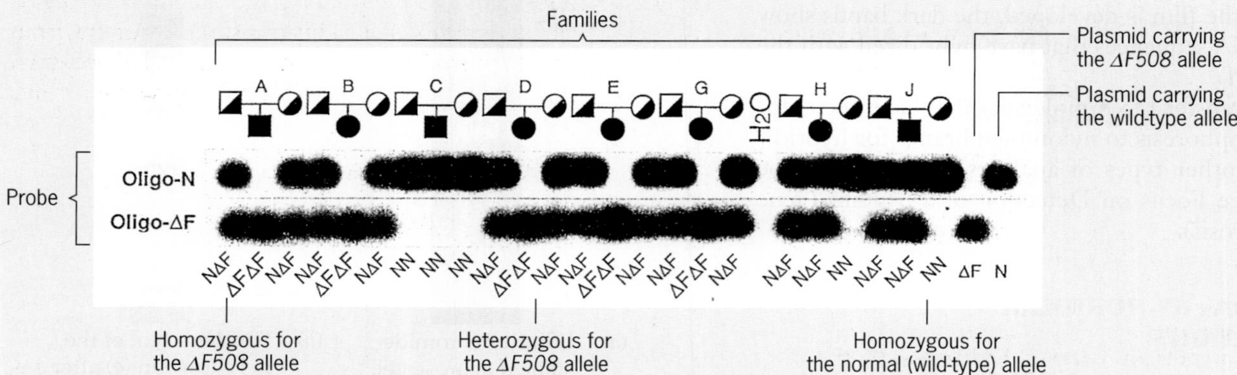

Figure 1 ▶ Detection of *CF* wild-type and Δ*F508* alleles by hybridization of labeled allele-specific oligonucleotide probes to genomic DNAs transferred to nylon membranes by the Southern blotting procedure (**FIGURE 15.20**). PCR was used to amplify the *CF* loci in genomic DNAs isolated from individual family members. The PCR products were separated by gel electrophoresis, transferred to membranes, denatured, and hybridized to the radioactive oligonucleotide probes (described above). Duplicate Southern blots were prepared; one blot was hybridized to the probe specific for the wild-type *CF* allele (top lane), and the other was hybridized to the probe specific for the Δ*F508* allele (bottom lane). The family pedigrees shown at the top represent offspring with *CF* and their heterozygous parents. Note that the Δ*F508* allele is present in families A, B, D, E, and G. Family C carries a different *CF* allele, and families H and J have one parent with the Δ*F508* allele and the other parent with a different *CF* allele.

Thus, care must be taken to prevent contamination of materials with these extremely stable enzymes. Furthermore, most RNA molecules contain considerable secondary structure and must therefore be kept denatured during electrophoresis in order to separate them on the basis of size. Denaturation is accomplished by adding formaldehyde or some other chemical denaturant to the buffer used for electrophoresis. After transfer to an appropriate membrane, the RNA blot is hybridized to either RNA or DNA probes just as with a Southern blot.

Northern blot hybridizations (**FIGURE 15.22**) are extremely helpful in studies of gene expression. They can be used to determine when and where a particular gene is expressed. However, we must remember that northern blot hybridizations only measure the accumulation of RNA transcripts. They provide no information about why the observed accumulation has occurred. Changes in transcript levels may be due to changes in the rate of transcription or to changes in the rate of transcript degradation. More sophisticated procedures must be used to distinguish between these possibilities.

ANALYSIS OF RNAs BY REVERSE TRANSCRIPTASE-PCR (RT-PCR)

The enzyme reverse transcriptase catalyzes the synthesis of DNA strands that are complementary to RNA templates. It can be used *in vitro* to synthesize DNAs that are complementary to RNA template strands. The resulting DNA strands can then be converted to double-stranded DNA by several different procedures (for example, see Figure 15.16), including the use of a second primer and the heat-stable *Taq* DNA polymerase. The resulting DNA molecules can then be amplified in standard PCR (see Amplification of DNA by the Polymerase Chain Reaction [PCR] in this chapter).

The first strand of DNA, often called a **cDNA** because it is complementary to the mRNA under study, may be synthesized by using an oligo(dT) primer that will anneal to the 3′-poly(A) tails of all mRNAs, or by using gene-specific primers (sequences complementary to the RNA molecule of interest). Gene-specific oligonucleotide primers are usually chosen to anneal to sequences in the 3′-noncoding regions of the mRNAs. **FIGURE 15.23** illustrates how such primers can be used in RT-PCR to amplify a specific gene transcript. The products of these amplifications are analyzed by gel electrophoresis. Wherever a product appears in the gel, the investigator knows that the sample from which it was generated contained the mRNA under study. This procedure is therefore a quick and easy way of ascertaining whether or not a particular gene is being transcribed.

Many modifications of the RT-PCR procedure have been developed, with a major emphasis on making it more quantitative. For example, known amounts of the RNA under study can be analyzed to determine the relationship between RNA input and DNA output. By knowing this relationship, an investigator can use the quantity of DNA generated by an experimental sample to extrapolate back to the amount of RNA that was initially present in that sample.

ANALYSIS OF PROTEINS BY WESTERN BLOT TECHNIQUES

Polyacrylamide gel electrophoresis is an important tool for the separation and characterization of proteins. Because many functional proteins are composed of two or more subunits, individual polypeptides are separated by electrophoresis in the presence of the detergent sodium dodecyl sulfate (SDS), which denatures the proteins. After electrophoresis, the proteins are detected by staining with Coomassie blue or silver stain. However, the separated polypeptides also can be transferred from the gel to a nitrocellulose membrane, and individual proteins can be detected with antibodies. This transfer of proteins from acrylamide gels to nitrocellulose membranes, called **western blotting,** is performed by using an electric current to move

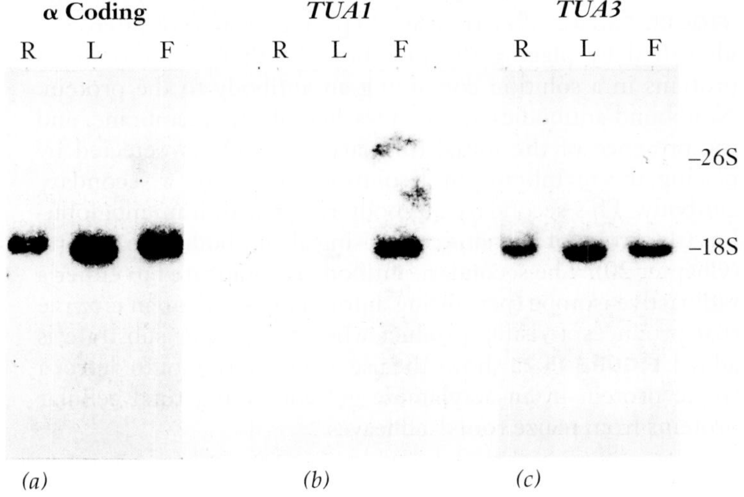

Figure 15.22 ▶ Typical northern blot hybridization data. Total RNAs were isolated from roots (R), leaves (L), and flowers (F) of *A. thaliana* plants, separated by agarose gel electrophoresis, and then transferred to nylon membranes. The autoradiogram shown in (*a*) is of a blot that was hybridized to a radioactive probe containing an α-tubulin coding sequence. This probe hybridizes to the transcripts of all six α-tubulin genes in *A. thaliana*. The autoradiograms shown in (*b*) and (*c*) are of RNA blots that were hybridized to DNA probes specific for the α1- and α3-tubulin genes (*TUA1* and *TUA3*, respectively). The results show that the α3-tubulin transcript is present in all organs analyzed, whereas the α1-tubulin transcript is present only in flowers. The 18S and 26S ribosomal RNAs provide size markers. Their positions were determined from a photograph of the ethidium-bromide stained gel prior to transfer of the RNAs to the nylon membrane.

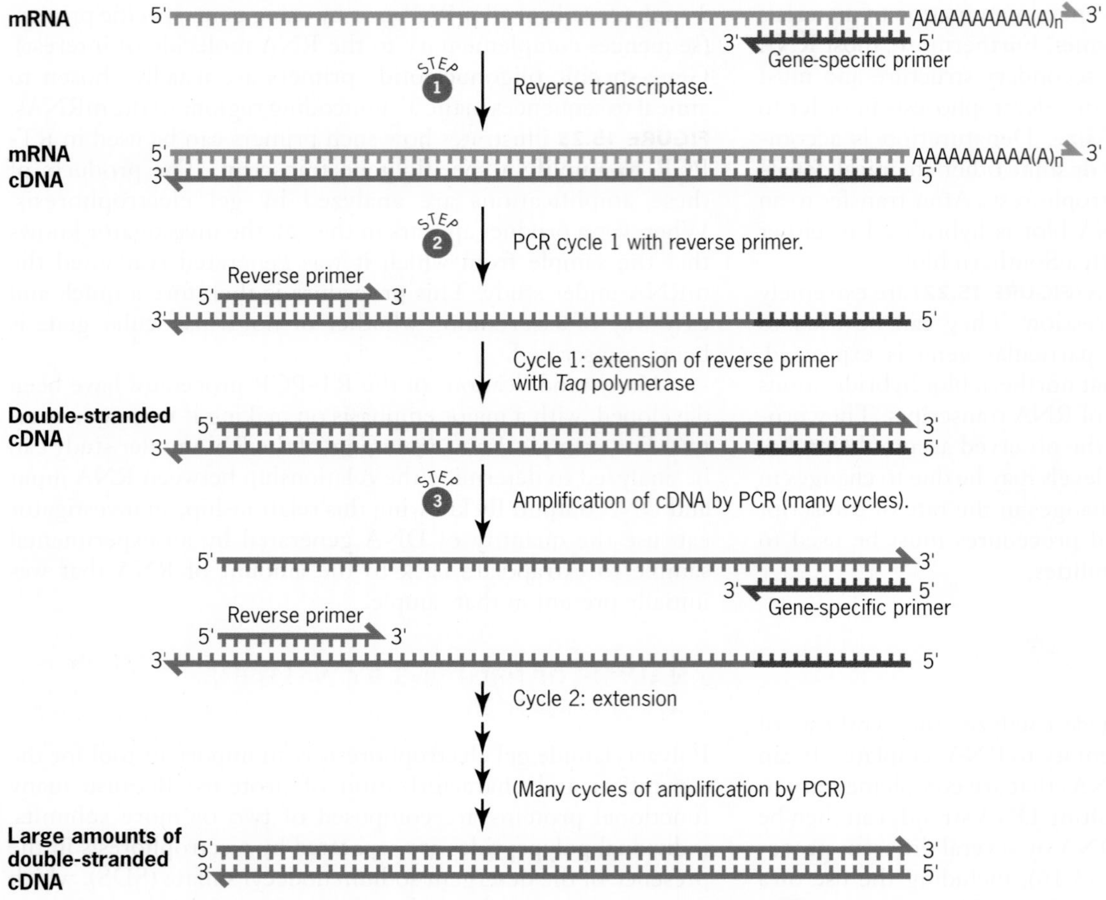

Figure 15.23 ▶ Detection and amplification of RNAs by reverse transcriptase PCR (RT-PCR). Specific gene transcripts are amplified by first using reverse transcriptase to synthesize a single-stranded DNA that is complementary to the mRNA of interest. The synthesis is initiated with a gene-specific oligonucleotide primer (a primer that will only anneal to the mRNA of interest). The complementary DNA strand is then synthesized by using a reverse primer and *Taq* polymerase. Large quantities of double-stranded cDNA are subsequently synthesized by standard PCR reactions in the presence of both the gene-specific and reverse PCR primers.

the proteins from the gel to the surface of the membrane (**FIGURE 15.24**). After transfer, a specific protein of interest is identified by placing the membrane with the immobilized proteins in a solution containing an antibody to the protein. Nonbound antibodies are then washed off the membrane, and the presence of the initial (primary) antibody is detected by placing the membrane in a solution containing a secondary antibody. This secondary antibody reacts with immunoglobulins (the group of proteins comprising all antibodies) in general (Chapter 20). The secondary antibody is conjugated to either a radioactive isotope (permitting autoradiography) or an enzyme that produces a visible product when the proper substrate is added. **FIGURE 15.25** shows the use of a western blot to detect a single protein in an acrylamide gel containing total cellular proteins from maize roots and leaves.

KEY POINTS

- DNA restriction fragments and other small DNA molecules can be separated by agarose or acrylamide gel electrophoresis and transferred to nylon membranes to produce DNA gel blots called Southern blots.
- The DNAs on Southern blots can be hybridized to labeled DNA probes to detect sequences of interest by autoradiography.
- When RNA molecules are separated by gel electrophoresis and transferred to membranes for analysis, the resulting RNA gel blots are called northern blots.
- RNA molecules can be detected and analyzed by reverse transcriptase-PCR (RT-PCR).
- When proteins are transferred from gels to membranes and detected with antibodies, the products are called western blots.

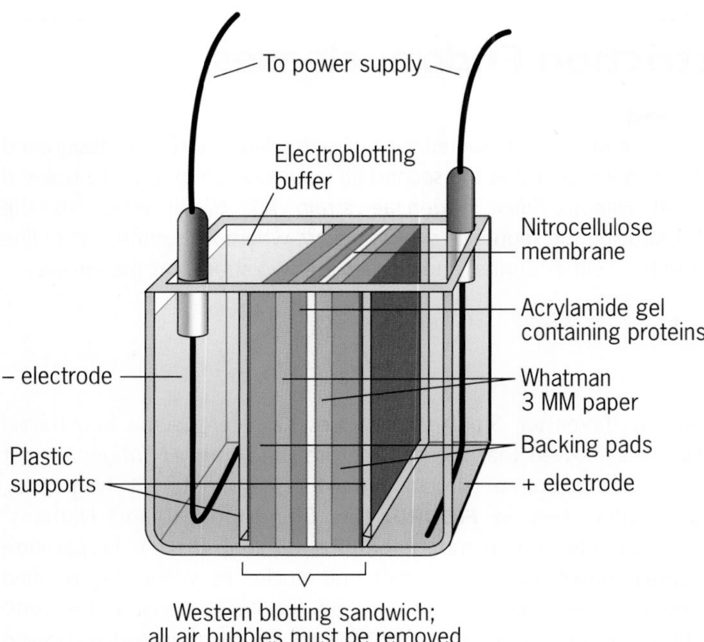

Figure 15.24 ▶ A typical western blotting or electroblotting apparatus. An electric current is used to transfer the proteins from a gel to a nitrocellulose membrane placed next to it in the blotting sandwich. All other components of the sandwich function to provide gentle but firm support; tight contact between the gel and the membrane is essential for good transfer.

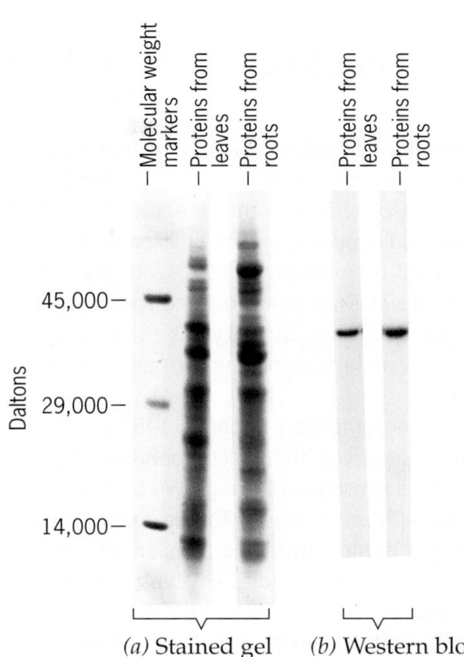

Figure 15.25 ▶ The use of western blots to identify individual proteins separated by polyacrylamide gel electrophoresis. (*a*) Proteins isolated from roots or leaves of maize were separated by polyacrylamide gel electrophoresis and stained with Coomassie blue. (*b*) The chloroplastic form of glutamine synthetase was identified by western blot analysis of the gel shown in (*a*).

The Molecular Analysis of Genes and Chromosomes

The sites at which restriction enzymes cleave DNA molecules can be used to construct physical maps of the molecules; however, nucleotide sequences provide the ultimate physical maps of DNA molecules.

Recombinant DNA techniques allow geneticists to determine the structure of genes, chromosomes, and entire genomes. Indeed, molecular geneticists are constructing detailed genetic and physical maps of the genomes of many organisms (Chapter 16).

The ultimate physical map of a genetic element is its nucleotide sequence, and the complete nucleotide sequences of the genomes of many viruses, bacteria, mitochondria, chloroplasts, and of a few eukaryotic organisms have already been determined. In addition, in October 2004 the International Human Genome Sequencing Consortium published a "nearly complete" sequence of the human genome. This sequence contains only 341 gaps and covers 99 percent of the gene-rich chromatin in the human genome (Chapter 16). In the following sections, we discuss the construction of restriction enzyme cleavage site maps of genes and chromosomes and the determination of DNA sequences.

PHYSICAL MAPS OF DNA MOLECULES BASED ON RESTRICTION ENZYME CLEAVAGE SITES

Most restriction endonucleases cleave DNA molecules in a site-specific manner (see **TABLE 15.1**). As a result, they can be used to generate **physical maps** of chromosomes that are of great value in assisting researchers in isolating DNA fragments carrying genes or other DNA sequences of interest. The sizes of the restriction fragments can be determined by polyacrylamide or agarose gel electrophoresis (**FIGURE 15.19**). Because of the nucleotide subunit structure of DNA, with one phosphate group per nucleotide, DNA has an essentially constant charge per unit of mass. Thus, the rates of migration of DNA fragments during electrophoresis provide accurate estimates of their lengths, with the rate of migration inversely proportional to length.

The procedure that is used to map the restriction enzyme cleavage sites is illustrated in **FIGURE 15.26**. The sizes of DNA

▶ A MILESTONE IN GENETICS: **Restriction Endonucleases**

Prior to 1970, the idea of isolating a specific human gene was pure science fiction. The human genome contains 3 billion nucleotide pairs, and trying to isolate a gene that is a few thousand nucleotide-pairs long would require finding one of a million such sequences. That perspective changed dramatically in 1970 with the identification of an enzyme that cleaves DNA in a sequence-specific manner. Indeed, the discovery of this class of enzymes—the type II restriction endonucleases—paved the way for what we now know as recombinant DNA technology.

The restriction enzyme story began in 1952 when Salvador Luria and Mary Human observed that bacteriophage T2 grown on one strain of *E. coli* could subsequently be grown on that strain but could not be grown on a different strain.[1] The next year, Giuseppe Bertani and Jean Weigle demonstrated a similar phenomenon using bacteriophage λ.[2] These two studies established that phage growth was restricted by some *E. coli* strains, but that growth became possible if the phage DNA was modified by something in these strains. In 1968 Stuart Linn and Werner Arber elucidated the mechanism underlying this restriction/modification system.[3] Linn and Arber showed that the *E. coli* cells they were studying contained two enzymes, one that methylated a specific DNA sequence in the phage DNA, and another that cleaved the phage DNA when this target sequence was unmethylated. However, the phage DNA was not cleaved at the target sequence, but rather, at a random sequence some distance from it. Once cleaved, the phage DNA could not replicate; thus, the cleaving enzyme had the effect of restricting phage growth in the *E. coli* cells. Such enzymes are now called restriction endonucleases. Phage can grow in restrictive *E. coli* cells only if the phage DNA has been modified by the addition of methyl groups to some of the nucleotides within the target sequence. This modification protects the phage DNA from cleavage by the restriction endonuclease and therefore allows the phage to replicate.

Enzymatic cleavage of DNA at a distance from a target sequence that is recognized by the enzyme is characteristic of type I and type III restriction endonucleases. In contrast, type II restriction endonucleases cut DNA molecules within the target sequence; thus, their cutting action is sequence-specific. The first type II restriction endonuclease was isolated in 1970 from the bacterium *Haemophilus influenzae* by Hamilton Smith and his students Kent Wilcox and Thomas Kelly, Jr.[4]

The enzyme characterized in Smith's laboratory was designated *Hin*dII because it was the second (II) restriction endonuclease isolated from *Haemophilus influenzae* strain R_d. *Hin*dII binds to the hexanucleotide sequence shown below, where Py is either pyrmidine and Pu is either purine, and cleaves the two strands at the arrows.

↓
5′–GTPyPuAC–3′
3′–CAPuPyTG–5′
↑

Shortly thereafter, Smith's colleagues Kathleen Danna and Daniel Nathans showed that the enzyme prepared from *H. influenzae* cut the chromosome of simian virus 40 (SV40) into 11 fragments, designated A–K in **FIGURE 1*a*.**[5] Ching-Juh Lai and Nathans[6] subsequently demonstrated that the original enzyme preparation contained two distinct restriction endonucleases. When they purified one of these enzymes, now called *Hin*dIII, it cleaved the SV40 chromosome into six fragments (**FIGURE 1*b***). Further studies showed that purified *Hin*dII cuts the SV40 chromosome into seven fragments; however, two of them are too small to be detected by the electrophoretic procedure used to separate the DNA fragments shown in **FIGURE 1*a*.** The 11 fragments observed by Danna and Nathans, therefore, were produced by the combined action of *Hin*dII and *Hin*dIII. *Hin*dIII is now known to recognize the sequence shown below and cut it in a staggered fashion (arrows) to produce single-stranded complementary overhangs.

↓
5′–AAGCTT–3′
3′–TTCGAA–5′
↑

Arber, Nathans, and Smith shared the 1978 Nobel Prize in Physiology or Medicine for their research on restriction endonucleases. Their work led to the discovery of a large number of type II restriction enzymes, each recognizing a specific sequence of nucleotide pairs and cleaving DNA into discrete fragments.

Today, over 400 different restriction enzymes have been purified and characterized. Scientists use these enzymes to cleave DNA molecules from diverse organisms—from viruses to elephants—into discrete fragments, which can then be cloned, sequenced, and recombined in various ways. Indeed, restriction endonucleases were largely responsible for the explosion of new knowledge in biology during the last decades of the twentieth century. They made the recombinant DNA revolution possible.

[1]Luria, S. E., and M. L. Human. 1952. A nonhereditary, host-induced variation of bacterial viruses. *J. Bacteriol.* 64:557–569.

[2]Bertani, G., and J. J. Weigle. 1953. Host controlled variation in bacterial viruses. *J. Bacteriol.* 65:113–121.

[3]Linn, S., and W. Arber. 1968. Host specificity of DNA produced by *Escherichia coli*, X. *In vitro* restriction of phage fd replicative form. *Proc. Natl. Acad. Sci. U.S.A.* 59:1300–1306.

[4]Smith, H. O., and K. W. Wilcox. 1970. A restriction enzyme from *Haemophilus influenzae*, I. Purification and general properties. *J. Mol. Biol.* 51: 379–391. Kelly, Jr., T. J., and H. O. Smith. 1970. A restriction enzyme from *Haemophilus influenzae*, II. Base sequence of the recognition site. *J. Mol. Biol.* 51:393–409.

[5]Danna, K., and D. Nathans. 1971. Specific cleavage of simian virus 40 DNA by restriction endonuclease of *Haemophilus influenzae. Proc. Natl. Acad. Sci. U.S.A.* 68:2913–2917.

[6]Lai, C-J., and D. Nathans. 1974. Deletion mutants of simian virus 40 generated by enzymatic excision of DNA segments from the viral genome. *J. Mol. Biol.* 89:179–193.

Autoradiogram showing the DNA fragments produced by cleavage of the SV40 chromosome with restriction endonucleases *Hind*II and *Hind*III and then separation of the fragments by polyacrylamide gel electrophoresis.

Location of the *Hind*II and *Hind*III cleavage sites on the circular chromosome of SV40.

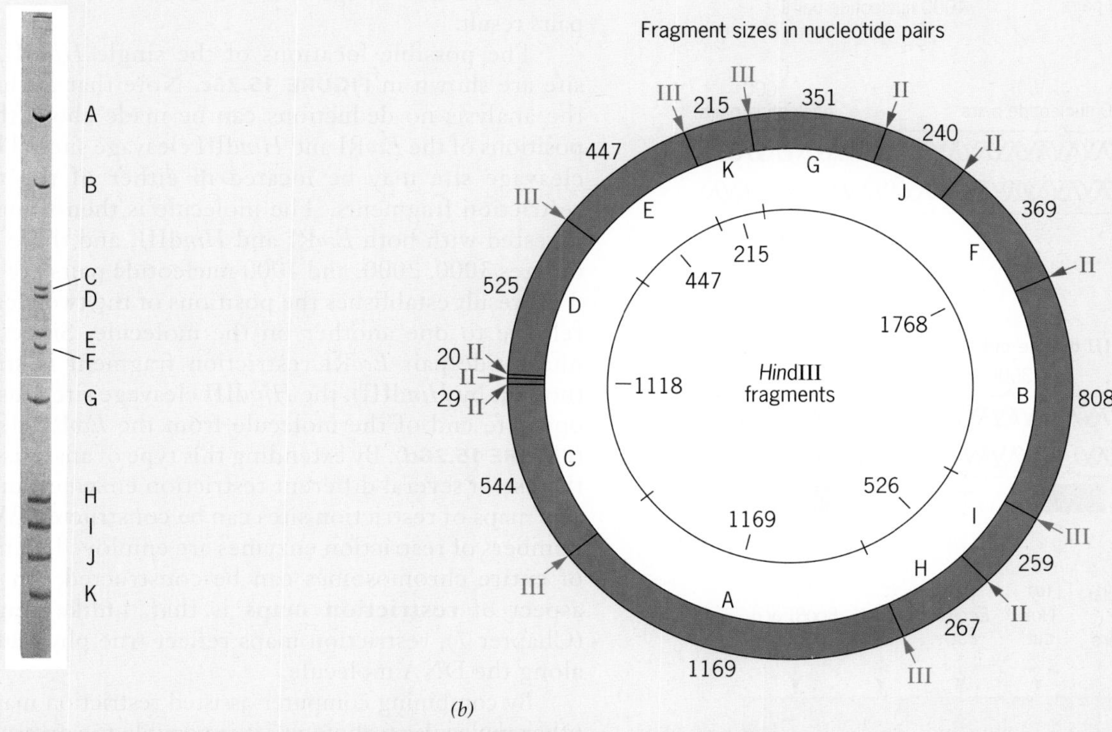

Figure 1 ▶ *Hind*II and *Hind*III cleavage sites in the SV40 chromosome. (*a*) Autoradiogram in Danna and Nathans' 1971 paper showing the fragments—after their separation by gel electrophoresis—produced by cleavage of SV40 DNA with both enzymes. The SV40 DNA was labeled with ^{14}C-thymidine prior to cleavage to allow detection of the fragments by autoradiography. (*b*) A map of the *Hind*II and *Hind*III restriction sites in the SV40 chromosome. The outer circle shows the locations of *Hind*II (red arrows) and *Hind*III (green arrows) sites on the SV40 chromosomes. The positions of fragments A through K shown in (a) and the sizes of these fragments are also shown. The inner circle shows the sites of cleavage by purified *Hind*III and the sizes of the *Hind*III restriction fragments. The sizes of the restriction fragments were calculated based on the nucleotide sequence of SV40 in GenBank.

QUESTIONS FOR DISCUSSION

1. The discovery of restriction endonucleases has allowed scientists to join DNA molecules from one species with DNA molecules from a different species—to produce "recombinant DNA" molecules—in the test tube. These molecules can then be introduced into living cells, where they can be expressed. Some people reject the use of recombinant DNA technology to modify living organisms because it is "unnatural." Others say that recombinant DNA molecules are formed at a low frequency in nature and that scientists are just speeding up natural processes. What is your position on this issue? Why?

2. Restriction enzymes are extremely important tools in genetics today. The discovery of these enzymes emerged from investigations on how *E. coli* cells restrict bacteriophage infections and on how they modify bacteriophage DNA by methylating specific sequences. These phenomena—restriction and modification—seem specialized and obscure compared to the technological significance of restriction enzymes. Would you say that the restriction enzyme story demonstrates the importance of basic biological research?

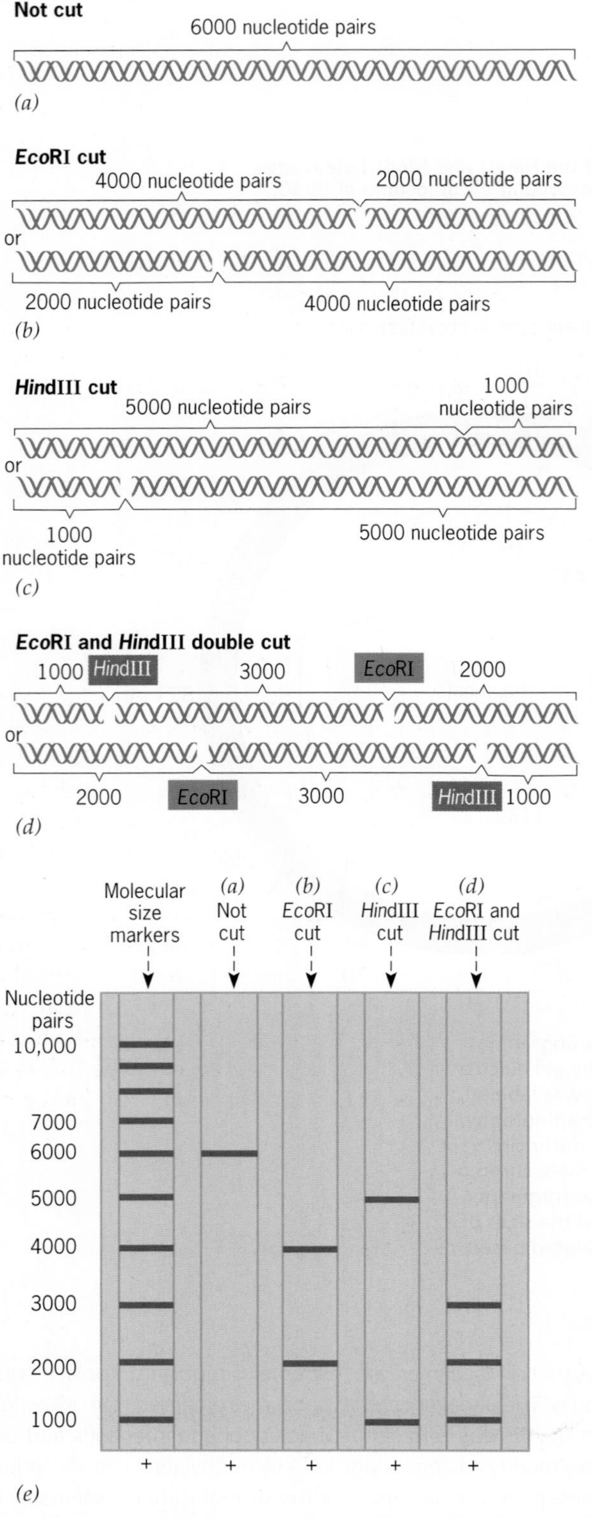

Figure 15.26 ▶ Procedure used to map restriction enzyme cleavage sites in DNA molecules. (*a–d*) Structures of the DNA molecule or of restriction fragments of the molecule either (*a*) uncut or cut with (*b*) *Eco*RI, (*c*) *Hin*dIII, or (*d*) *Eco*RI and *Hin*dIII. (*e*) The separation of these DNA molecules and fragments by agarose gel electrophoresis. The left lane on the gel contains a set of molecular size markers, a set of DNA molecules of size 1000 nucleotide pairs, and multiples thereof.

restriction fragments are estimated by using a set of DNA markers of known size. In **FIGURE 15.26**, a set of DNA molecules that differ in length by 1000 nucleotide pairs are used as size markers. Consider a DNA molecule approximately 6000 nucleotide pairs (6 kb) in length. When the 6-kb DNA molecule is cut with *Eco*RI, two fragments of sizes 4000 and 2000 nucleotide pairs are produced. The possible positions of the single *Eco*RI cleavage site in the molecule are shown in **FIGURE 15.26*b***. When the same DNA molecule is cleaved with *Hin*dIII, two fragments of sizes 5000 and 1000 nucleotide pairs result.

The possible locations of the single *Hin*dIII cleavage site are shown in **FIGURE 15.26*c***. Note that at this stage of the analysis no deductions can be made about the relative positions of the *Eco*RI and *Hin*dIII cleavage sites. The *Hin*dIII cleavage site may be located in either of the two *Eco*RI restriction fragments. The molecule is then simultaneously digested with both *Eco*RI and *Hin*dIII, and three fragments of sizes 3000, 2000, and 1000 nucleotide pairs are produced. This result establishes the positions of the two cleavage sites relative to one another on the molecule. Since the 2000-nucleotide-pair *Eco*RI restriction fragment is still present (not cut by *Hin*dIII), the *Hin*dIII cleavage site must be at the opposite end of the molecule from the *Eco*RI cleavage site (**FIGURE 15.26*d***). By extending this type of analysis to include the use of several different restriction enzymes, more extensive maps of restriction sites can be constructed. When large numbers of restriction enzymes are employed, detailed maps of entire chromosomes can be constructed. An important aspect of **restriction maps** is that, unlike genetic maps (Chapter 7), restriction maps reflect true physical distances along the DNA molecule.

By combining computer-assisted restriction mapping with other molecular techniques, it is possible to construct physical maps of entire genomes. The first multicellular eukaryote for which this was accomplished is *Caenorhabditis elegans*, a worm that is important for studies on the genetic control of development (Chapter 21). Moreover, the physical map of the *C. elegans* genome has been correlated with its genetic map. Thus, when an interesting new mutation is identified in *C. elegans*, its position on the genetic map often can be used to obtain clones of the wild-type gene from a large international *C. elegans* clone bank.

NUCLEOTIDE SEQUENCES OF GENES AND CHROMOSOMES

The ultimate fine structure map of a specific gene or chromosome is its nucleotide-pair sequence, complete with a chart of all nucleotide-pair changes that alter the function of that gene or chromosome. Prior to 1975, the thought of trying to sequence entire chromosomes was barely conceivable—at best, a laborious task requiring years of work. By late 1976, however, the entire 5386-nucleotide-long chromosome of phage ΦX174 had been sequenced. Today, sequencing is a routine laboratory procedure. The complete, or nearly complete, nucleotide sequences of the genomes of over 2000 viruses, over 1000 plasmids, about 1500 chloroplasts and mitochondria, over 700 bacteria and archaea, and about 30 eukaryotes are now known. In addition, the sequencing of another 200 eukaryotic genomes is underway, and the sequence of 99 percent of the euchromatin in the human genome is known (Chapter 16).

Our initial ability to sequence essentially any DNA molecule was the result of four major developments. The most important breakthrough was the discovery of restriction enzymes and their use in preparing homogeneous samples of specific segments of chromosomes. Another major advance was the improvement of gel electrophoresis procedures to the point where DNA fragments that differ in length by a single nucleotide could be resolved. Gene-cloning techniques to facilitate the preparation of large quantities of a particular DNA molecule were also important. Finally, researchers invented two different procedures by which the nucleotide sequences of DNA molecules can be determined.

Both DNA sequencing procedures depend on the generation of a population of DNA fragments that all have one end in common (all end at exactly the same nucleotide) and terminate at all possible positions (every consecutive nucleotide) at the other end. The common end is the 5′-terminus of the sequencing primer. The 3′-terminus of the primer contains a free —OH, which is the site of chain extension by DNA polymerase. These fragments are then separated on the basis of chain length by polyacrylamide gel electrophoresis. In both cases, four separate biochemical reactions are carried out simultaneously, each of which generates a set of fragments terminating at one of the four bases (A, G, C, or T) in DNA.

The first procedure, called the Maxam and Gilbert procedure after Allan Maxam and Walter Gilbert who invented it, uses four different chemical reactions to cleave DNA chains specifically at As, Gs, Cs, or Cs + Ts. This procedure is no longer used. The second procedure, developed by Fred Sanger and colleagues, uses *in vitro* DNA synthesis in the presence of radioactive nucleotides and specific chain-terminators to generate four populations of radioactively labeled fragments that end at As, Gs, Cs, and Ts, respectively. We will discuss the Sanger procedure.

2′,3′-Dideoxyribonucleoside triphosphates (**FIGURE 15.27**) are the chain-terminators most frequently used in the Sanger sequencing procedure. Recall that DNA polymerases have an absolute requirement for a free 3′–OH on the DNA primer strand (Chapter 10). If a 2′,3′-dideoxynucleotide is added to the end of a chain, it will block subsequent extension of that chain since the 2′,-3′ dideoxynucleotides have no 3′–OH. By using (1) 2′,3′-dideoxythymidine triphosphate (ddTTP), (2) 2′,3′-dideoxycytidine triphosphate (ddCTP), (3) 2′,3′-dideoxyadenosine triphosphate (ddATP), and (4) 2′,3′-dideoxyguanosine triphosphate (ddGTP) as chain-terminators in four separate DNA synthesis reactions, four populations of fragments can be generated, and each population will contain chains that all terminate with the same base (T, C, A, or G) (**FIGURE 15.28**).

In a given reaction, the ratio of dXTP:ddXTP (where X can be any one of the four bases) is kept at approximately 100:1, so that the probability of termination at a given X in the nascent chain is about 1/100. This yields a population of fragments terminating at all potential (X) termination sites within a distance of a few hundred nucleotides from the original primer terminus.

After the DNA fragments generated in the four parallel reactions are released from the template strands by denaturation, they are separated by polyacrylamide gel electrophoresis and their positions in the gel are detected by autoradiography. The bands on the autoradiograms correspond to radioactive chains of different lengths; they produce a "ladder" defining the nucleotide sequence of the longest chain that has been synthesized (**FIGURE 15.29**).

Base
3'–OH required
for chain elongation
Normal DNA precursor
2'-deoxyribonucleoside triphosphate

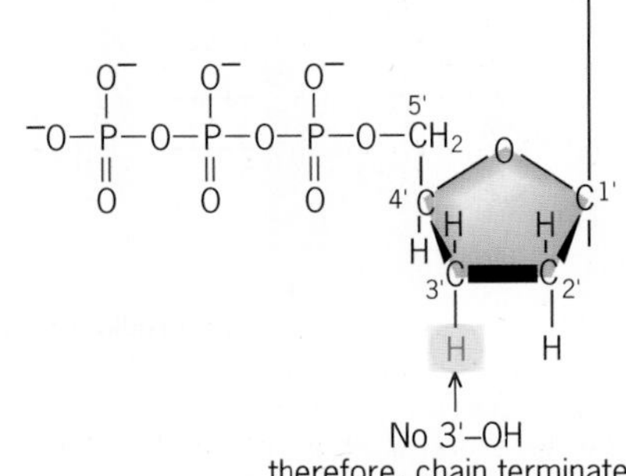

Chain-termination precursor
2',3'-dideoxyribonucleoside triphosphate

Figure 15.27 ▶ Comparison of the structures of the normal DNA precursor 2′-deoxyribonucleoside triphosphate and the chain-terminator 2′,3′-dideoxyribonucleoside triphosphate used in DNA sequencing reactions.

The shortest fragment will migrate the greatest distance and give rise to the band nearest the anode (the positive electrode). Each successive band will contain chains that are one nucleotide longer than the chains in the preceding band of the ladder. The 3′-terminal nucleotide of the chain in each band will be the dideoxynucleotide chain-terminator present in the reaction mixture (1, 2, 3, or 4) in which that specific chain was produced (**FIGURE 15.28**). By reading the ladder produced by autoradiography of the polyacrylamide gels used to separate the fragments generated in each of the four parallel reactions, the complete nucleotide sequence of a DNA chain can be determined. This is illustrated in **FIGURE 15.28** for a hypothetical nucleotide sequence. An autoradiogram of an actual dideoxynucleotide chain-terminator sequencing gel is shown in **FIGURE 15.29.** Under optimal conditions, long sequences of several hundred nucleotides can be determined from a single sequencing gel.

Today, all large-scale DNA sequencing is done with automated DNA sequencing machines that use the dideoxy chain-termination procedure described above, but with modifications. Different DNA sequencing machines use slightly different protocols. However, the major differences between the slab-gel sequencing procedure illustrated in **FIGURE 15.28** and

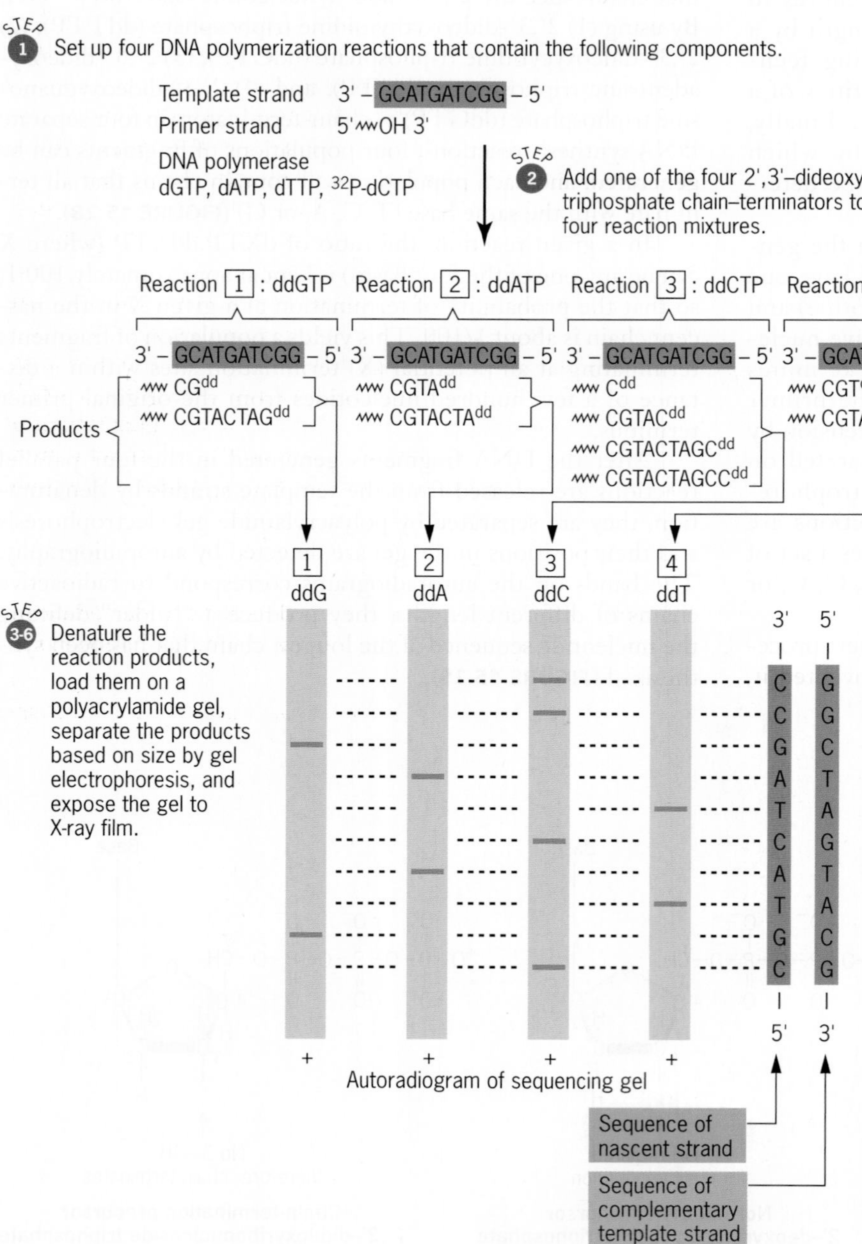

Figure 15.28 ▶ Sequencing DNA by the 2′,3′-dideoxynucleoside triphosphate chain-termination procedure. Four reactions are carried out in parallel, each of which contains one of the four 2′,3′-dideoxy chain-terminators: ddGTP, ddATP, ddCTP, and ddTTP. All four reaction mixtures contain the components required for DNA synthesis *in vitro*, including a primer strand that anneals to the template strand. The primer strand determines the common 5′-end of all products; it has a free 3′-OH and is extended in the 5′ → 3′ direction by DNA polymerase (see Chapter 10). One radioactive DNA precursor (^{32}P-dCTP here) is present in each reaction so that the products can be detected by autoradiography. The products of the four reactions are separated by polyacrylamide gel electrophoresis, and the positions of the nascent DNA chains in the gel are determined by autoradiography. Because the shortest chain migrates the greatest distance, the nucleotide sequence of the nascent chains (shown in red at the right of the autoradiogram) is obtained by reading the gel from the bottom (anode) to the top (cathode).

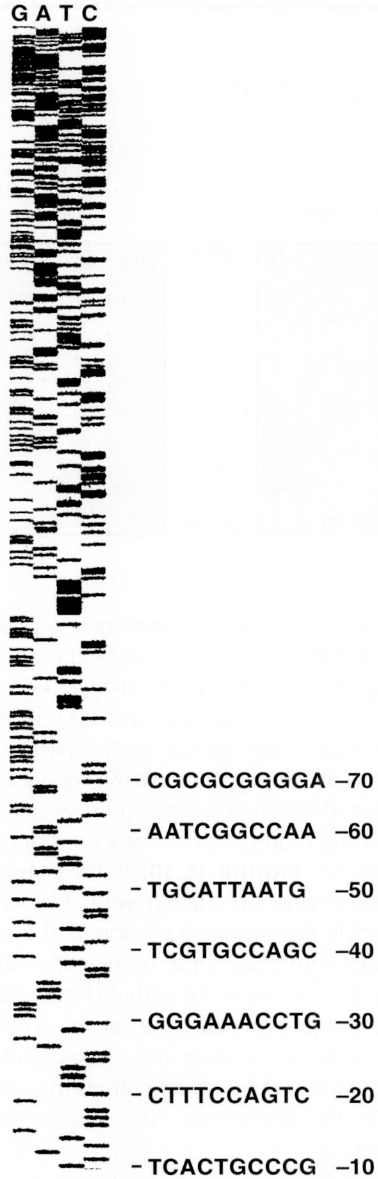

Figure 15.29 ► Photograph of an autoradiograph of a 2′,3′-dideoxy-nucleotide chain-terminator sequencing gel. The sequence defined by the lower portion of the gel is shown on the right.

automated DNA sequencing are (1) the use of fluorescent dyes, rather than radioactive isotopes, to detect DNA chains; (2) the separation of the products of all four dideoxy chain-terminator reactions by electrophoresis through a single gel or capillary tube; (3) the use of photocells to detect the fluorescence of the dyes as they pass through the gel or capillary tube; and (4) the direct transfer of the output of the photocell to a computer, which automatically analyzes, records, and prints out the results.

A different fluorescent dye is used to label the products of each of the four dideoxy chain-terminator sequencing reactions. As a result, the products of the four reactions can be distinguished by their fluorescence as they pass through a gel or capillary tube. The fluorescent dyes can be coupled to the primers used in the sequencing reactions or coupled directly to the dideoxynucleoside triphosphate chain-terminators. **FIGURE 15.30** compares the automated DNA sequencing procedure with the slab-gel procedure described in **FIGURE 15.28** and shows a computer printout of the results of automated sequencing of a short segment of DNA. See Focus on Problem Solving: Determining the Nucleotide Sequences of Genetic Elements to test your understanding of the slab-gel and automated DNA sequencing procedures.

The automated DNA sequencing machines can perform 96 capillary electrophoresis separations simultaneously, and sample loading, electrophoresis, data collection, and data analysis are all fully automated. If operated continuously, a single 96-capillary machine can crank out over 100,000 nucleotides of sequence per day. Although that seems like a lot, remember that the human genome contains three billion nucleotide pairs!

KEY POINTS

- Detailed physical maps of DNA molecules can be prepared by identifying the sites that are cleaved by various restriction endonucleases.
- The nucleotide sequences of DNA molecules provide the ultimate physical maps of genes and chromosomes.

▶ FOCUS ON PROBLEM SOLVING

Determining the Nucleotide Sequences of Genetic Elements

THE PROBLEM

Ten micrograms of a decanucleotide-pair *Hpa*I restriction fragment were isolated from the double-stranded DNA chromosome of a small virus. Octanucleotide poly-A tails were then added to the 3′-ends of both strands using terminal transferase and dATP; that is,

5′-X X X X X X X X X X-3′
3′-X′X′X′X′X′X′X′X′X′X′-5′

↓ terminal transferase, dATP

5′-X X X X X X X X X X A A A A A A A A-3′
3′-A A A A A A A A X′X′X′X′X′X′X′X′X′X′-5′

where X and X′ can be any of the four standard nucleotides, but X′ is always complementary to X.

The two complementary strands (Watson strand and Crick strand) were then separated and sequenced by the 2′,3′-dideoxyribonucleoside triphosphate chain-termination method.

1. The first set of sequencing reactions was primed using radioactive (^{32}P-labeled) synthetic poly-T octamers; that is,

Watson strand:

3′-A A A A A A A A X′X′X′X′X′X′X′X′X′X′-5′
5′-^{32}P-T T T T T T T T -OH

Crick strand:

5′-X X X X X X X X X X A A A A A A A A-3′
HO-T T T T T T T T-^{32}P-5′

The four dideoxy nucleoside triphosphate sequencing reactions—(1) ddTTP, (2) ddCTP, (3) ddATP, and (4) ddGTP—were carried out using DNA polymerase for both strands. Each reaction mixture was applied to a lane in a polyacrylamide gel, fractionated by electrophoresis, and autoradiographed. The autoradiogram of the sequencing gel for one of the strands (Watson) is shown on the left of the following diagram. Draw the banding pattern that would be expected on the autoradiogram of the gel for the complementary strand (Crick strand) on the right in the diagram.

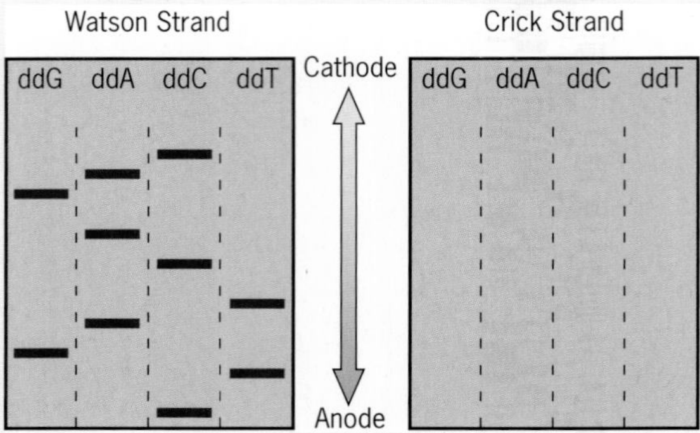

2. The second set of sequencing reactions was primed using nonradioactive synthetic poly-T octamers, and the reactions were performed using fluorescent dyes to detect the nascent DNA chains present in each of the four dideoxy (ddX) chain-termination reactions. The products were subsequently separated by capillary gel electrophoresis using automated DNA sequencing machines. The dyes fluoresce at different wavelengths, which are recorded by a photocell as the products of the reactions are separated in the capillary tubes (see **FIGURE 15.30**). In the standard sequencing reactions, the chains terminating with ddG fluoresce dark blue (appear black in the computer printouts), those terminating with ddC fluoresce light blue, those terminating with ddA fluoresce green, and those terminating with ddT fluoresce red. What would the computer printout look like for the sequencing reactions performed using as template the Crick strand of the *Hpa*I restriction fragment described above? Draw the expected computer printout in the box below. (Use the format shown in **FIGURE 15.30.**)

FACTS AND CONCEPTS

1. All DNA polymerases have an absolute requirement for a free 3'-hydroxyl on the end of the primer strand that will be extended by DNA polymerization reactions.
2. All DNA synthesis occurs 5′ to 3′; that is, all synthesis occurs by the addition of nucleotides to the 3′-end of the primer strand.
3. The addition of a 2′,3′-dideoxyribonucleoside monophosphate to the 3′ end of a primer strand will block its extension.
4. Polyacrylamide gel electrophoresis separates DNA strands on the basis of size and conformation.
5. DNA chains have a constant charge per unit mass; that is, they have one negative charge per nucleotide.
6. Because of their constant charge per unit mass, polynucleotide chains can be separated based on their size (length in nucleotides or nucleotide pairs).
7. Linear DNA molecules that differ in length by one nucleotide can be separated by polyacrylamide gel electrophoresis for chains up to a few hundred nucleotides long.
8. The shortest chains will migrate the largest distance during gel electrophoresis.
9. The two strands of a Watson-Crick double helix have opposite chemical polarity; if one strand has 5′ to 3′ polarity, the complementary strand has 3′ to 5′ polarity.

ANALYSIS AND SOLUTION

1. Because all DNA synthesis occurs by the addition of nucleotides to the 3'-OH terminus of the primer strand, all synthesis occurs in the 5′ → 3′ direction. Therefore, the sequence of the nascent DNA chain synthesized with the Watson strand as template is read 5' to 3' from the bottom to the top of the gel. The shortest nascent DNA fragment shown was present in the ddC reaction; it terminates with ddC, which means there was a G at this position in the template strand. Reading the ladder of bands from the bottom to the top of the gel reveals that the sequence of the nascent strand is 5′-CTGATCAGAC-3′. Therefore, the sequence of the complementary Watson template strand is 5′-GTCTGATCAG-3′. Now, if the Crick strand is used as the template strand in the sequencing reactions, the nascent strand will have the sequence of the Watson strand, so the ladder of bands on the audoradiogram will be as shown below. The sequence of the nascent strand will be 5′-GTCTGATCAG-3′ reading the ladder of bands from the bottom to the top of the gel, and the sequence of the complementary Crick template strand will be 5′-CTGATCAGAC-3′.

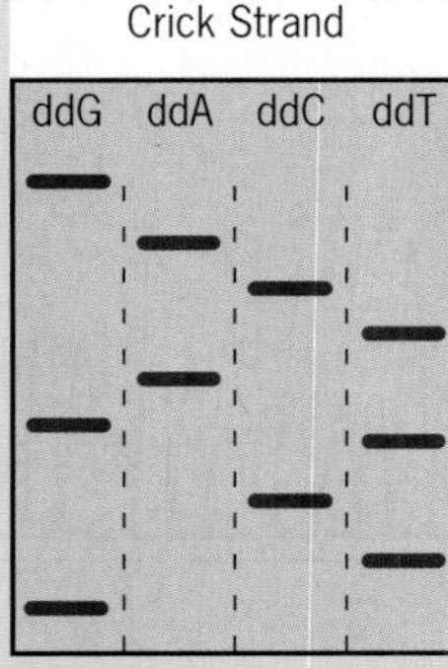

2. If the sequencing reactions are performed in a machine using the Crick strand as template and the products are analyzed by capillary gel electrophoresis, the predicted printout will be as shown below. The sequence of the nascent chain will be 5′-GTCTGATCAG-3′, reading the printout from left to right.

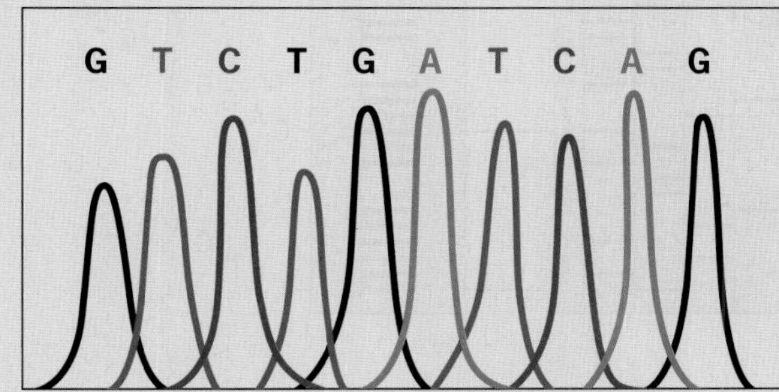

For further discussion go to your *WileyPLUS* course.

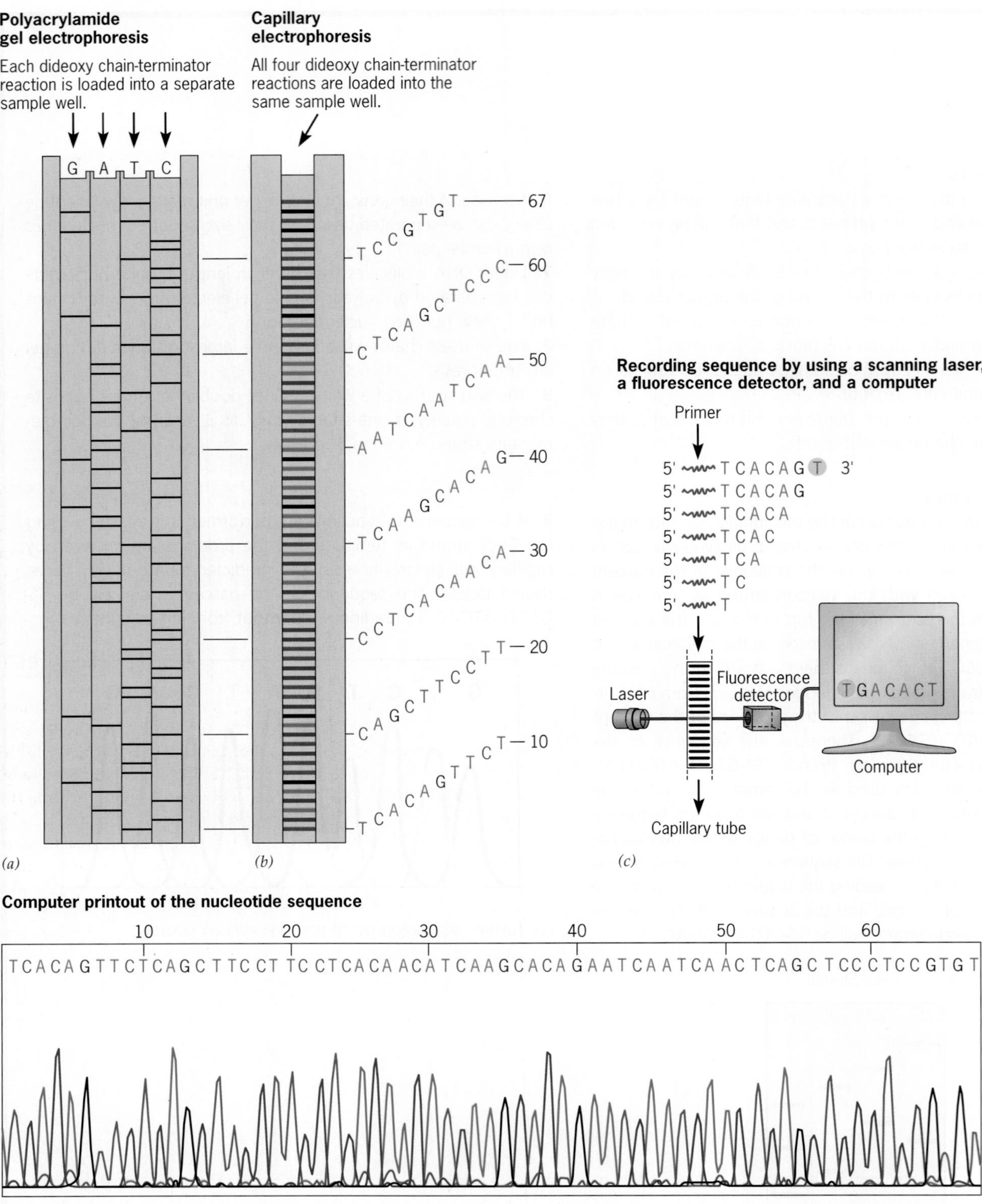

Figure 15.30 ▶ Comparison of the standard slab-gel (*a*) and automated capillary gel (*b*) DNA sequencing procedures. (*c*) In the automated sequencing procedure, the products in the four dideoxy chain-terminator reactions are labeled with four dyes that fluoresce at different wavelengths. The products of all four reactions are denatured, loaded into the same well, and separated based on length by capillary gel electrophoresis. The fluorescence of each dye is detected by a photocell (fluorescence detector) as the nascent DNA chains move through a laser beam and is recorded by a computer. (*d*) A computer printout of the results of an automated DNA sequencing run, showing the nucleotide sequence of a short segment of DNA.

► Basic Exercises

ILLUSTRATE BASIC GENETIC ANALYSIS

1. What is a recombinant DNA molecule?

Answer: A recombinant DNA molecule is constructed *in vitro* from portions of two different DNA molecules, often DNA molecules from two different species.

2. What are restriction endonucleases?

Answer: Restriction endonucleases are enzymes that cleave DNA molecules in a sequence-specific manner such that all of the fragments produced have the same nucleotide sequences at their ends. Many restriction enzymes make staggered cuts in palindromic DNA sequences, yielding fragments with complementary single-stranded termini, as shown here.

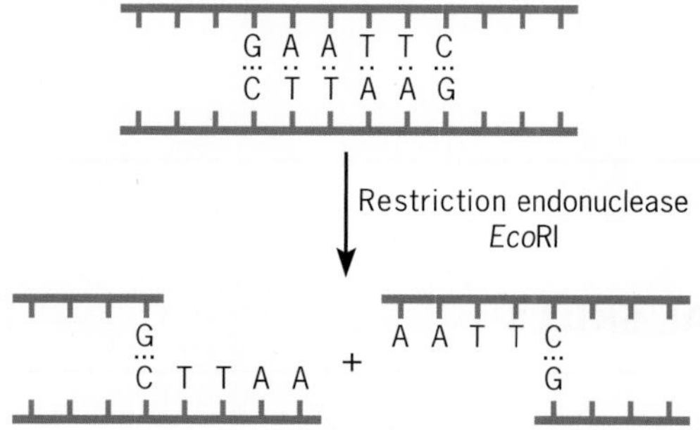

3. How are restriction endonucleases used to construct recombinant DNA molecules *in vitro*?

Answer: If DNA molecules from two different sources (perhaps different species) are both digested with a restriction endonuclease that recognizes a palindromic DNA sequence and makes staggered cuts in the two strands, the resulting fragments will have complementary single-stranded ends. If these DNA fragments are mixed, the complementary ends will pair, and the addition of DNA ligase will produce recombinant DNA molecules, as shown upper right.

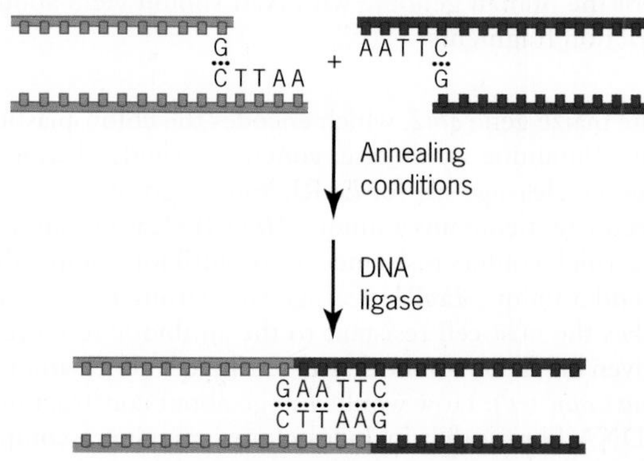

Exercise 3.

4. Why is the polymerase chain reaction (PCR) such a powerful tool for use in analyses of DNA?

Answer: Because PCR amplifies DNA sequences exponentially, large quantities of specific sequences can be obtained starting with just one or a few molecules. If one begins with a single molecule of DNA, 10 cycles of replication will yield 1024 DNA double helices, and 20 cycles will yield 1,048,576.

5. How are 2′,3′-dideoxyribonucleoside triphosphates used in DNA sequencing protocols?

Answer: The 2′,3′-dideoxyribonucleoside triphosphates function as specific terminators of DNA synthesis. When a 2′,3′-dideoxyribonucleoside monophosphate is added to the end of a nascent DNA chain, that chain can no longer be extended by DNA polymerase because of the absence of the 3′-OH required for chain extension. By using the appropriate ratios of 2′-deoxyribonucleoside triphosphates to 2′,3′-dideoxyribonucleoside triphosphates in DNA synthesis reactions *in vitro*, DNA chains are produced that terminate at all possible nucleotide positions. Separation of these nascent DNA chains by gel electrophoresis and detection of their positions in the gel with radioactive nucleotides or fluorescent dyes are then used to determine their nucleotide sequences (see **FIGURES 15.28–15.30**).

► Testing Your Knowledge

INTEGRATE DIFFERENT CONCEPTS AND TECHNIQUES

1. The human genome (haploid) contains about 3×10^9 nucleotide pairs of DNA. If you digest a preparation of human DNA with *Not*I, a restriction endonuclease that recognizes and cleaves the octameric sequence 5′-GCGGCCGC-3′, how many different restriction fragments would you expect to produce? Assume that the four bases (G, C, A, and T) are equally prevalent and randomly distributed in the human genome.

Answer: Assuming that the four bases are present in equal amounts and are randomly distributed, the chance of a specific nucleotide occurring at a given site is 1/4. The chance of a specific dinucleotide sequence (e.g., AG) occurring is $1/4 \times 1/4 = (1/4)^2$, and the probability of a specific octanucleotide sequence is $(1/4)^8$ or 1/65,536. Therefore, *Not*I will cleave such DNA molecules an average of once in every 65,536 nucleotide pairs. If a DNA molecule is cleaved at *n* sites,

$n + 1$ fragments will result. A genome of 3×10^9 nucleotide pairs should contain about 45,776 ($3 \times 10^9/65{,}536$) *Not*I cleavage sites. If the entire human genome consisted of a single molecule of DNA, *Not*I would cleave it into 45,776 + 1 fragments. Given that these cleavage sites are distributed on 24 different chromosomes, complete digestion of the human genome with *Not*I should yield about 45,776 + 24 restriction fragments.

2. The maize gene *gln2*, which encodes the chloroplastic form of the enzyme glutamine synthetase, contains a single cleavage site for *Hin*dIII, but no cleavage site for *Eco*RI. You are given an *E. coli* plasmid cloning vector that contains a unique *Hin*dIII cleavage site within the gene amp^r, which confers resistance to the antibiotic ampicillin on the host cell, and a unique *Eco*RI cleavage site within a second gene tet^r, which makes the host cell resistant to the antibiotic tetracycline. You are also given an *E. coli* strain that is sensitive to both ampicillin and tetracycline (amp^s tet^s). How would you go about constructing a maize genomic DNA library that includes clones carrying a complete *gln2* gene?

Answer: Maize genomic DNA should be purified and digested with *Eco*RI. Vector DNA should be similarly purified and digested with *Eco*RI. The maize *Eco*RI restriction fragments and the *Eco*RI-cut plasmid DNA molecules will now have complementary single-stranded ends (5′-AATT-3′). The maize restriction fragments should next be mixed with the *Eco*RI-cut plasmid molecules and covalently inserted into the linearized vector molecules in an ATP-dependent reaction catalyzed by DNA ligase. The ligation reaction will produce circular recombinant plasmids, some of which will contain maize *Eco*RI fragment inserts. Insertion of maize DNA fragments into the *Eco*RI site of the plasmid disrupts the tet^r gene so that the resulting recombinant plasmids will no longer confer tetracycline resistance to host cells. amp^s tet^s *E. coli* cells should then be transformed with the recombinant plasmid DNAs, and the cells should be plated on medium containing ampicillin to select for transformed cells harboring plasmids. The majority of the cells will not be transformed and, thus, will not grow in the presence of ampicillin. The cells that grow on ampicillin-containing medium should be pooled and frozen at -80°C in 20 percent glycerol. This collection of cells harboring different *Eco*RI fragments of the maize genome represents a clone library that should contain clones with an intact *gln2* gene since this gene contains no *Eco*RI cleavage site. Note that the *Hin*dIII site of the vector could be used to construct a similar maize genomic *Hin*dIII fragment library, but such a library would not contain intact *gln2* genes because of the *Hin*dIII cleavage site in *gln2*.

▶ Questions and Problems

ENHANCE UNDERSTANDING AND DEVELOP ANALYTICAL SKILLS

15.1 (a) In what ways is the introduction of recombinant DNA molecules into host cells similar to mutation? (b) In what ways is it different?

15.2 In what ways do restriction endonucleases differ from other endonucleases?

15.3 Restriction endonucleases are invaluable tools for biologists. However, genes encoding restriction enzymes obviously did not evolve to provide tools for scientists. Of what possible value are restriction endonucleases to the microorganisms that produce them?

15.4 What determines the sites at which DNA molecules will be cleaved by a restriction endonuclease?

15.5 If the sequence of base pairs along a DNA molecule occurs strictly at random, what is the expected frequency of a specific restriction enzyme recognition sequence of length (a) 3 and (b) 4 base pairs?

15.6 GO Following are four different single strands of DNA. Which of these, in their double-stranded form, would you expect to be cleaved by a restriction endonuclease?

(a) ACTCCAGAATTCACTCCG
(b) GCCTCATTCGAAGCCTGA
(c) CTCGCCAATTGACTCGTC
(d) ACTCCACTCCCGACTCCA

15.7 One of the procedures for cloning foreign DNA segments takes advantage of restriction endonucleases like *Hin*dIII (see **TABLE 15.1**) that produce complementary single-stranded ends. These enzymes produce identical complementary ends on cleaved foreign DNAs and on the vector DNAs into which the foreign DNAs are inserted. What major advantage does this cloning strategy have over procedures that use terminal transferase to synthesize complementary single-stranded ends on foreign DNAs and vector DNAs *in vitro?*

15.8 You are working as part of a research team studying the structure and function of a particular gene. Your job is to clone the gene. A restriction map is available for the region of the chromosome in which the gene is located; the map is as follows:

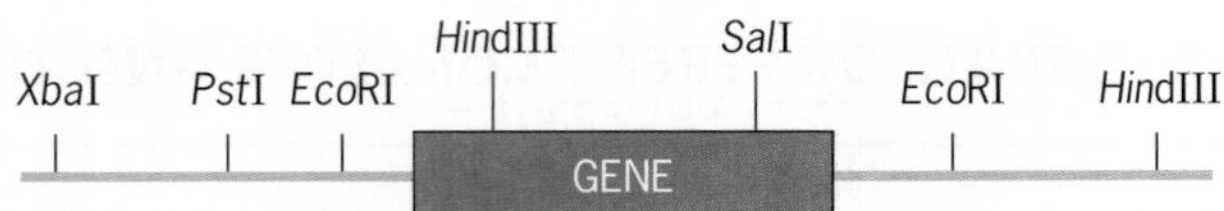

Your first task is to prepare a genomic DNA library that contains clones carrying the entire gene. Describe how you would prepare such a library in plasmid vector pUC118 (see Figure 15.8), indicating which restriction enzymes, media, and host cells you would use.

15.9 Genomic clones of the chloroplastic glutamine synthetase gene (*gln2*) of maize are cleaved into two fragments by digestion with restriction endonuclease *Hin*dIII, whereas full-length maize *gln2* cDNA clones are not cut by *Hin*dIII. Explain these results.

15.10 Most of the genes of plants and animals that were cloned soon after the development of recombinant DNA technologies were genes encoding products that are synthesized in large quantities in specialized cells. For example, about 90 percent of the protein synthesized in mature red blood cells of mammals consists of α- and β-globin chains, and the globin genes were among the first mammalian genes cloned. Why were genes of this type so prevalent among the first eukaryotic genes that were cloned?

15.11 You are studying a gene in *E. coli* that is expressed at 45°C but not at 37°C. You have shown that the regulation of this gene occurs at the level of transcription. In addition, you have isolated a protein required for induction of transcription of the gene at high temperature and have demonstrated that it binds to a specific octameric nucleotide-pair sequence upstream (5′) from the gene. You have cloned the complete gene plus the upstream regulatory sequence in a plasmid, and you have available an *E. coli* mutant carrying a deletion of the entire gene. You now want to determine which of the nucleotide pairs in the 5′ octameric protein binding site are involved in the interaction with (binding of) the regulatory protein. How could you determine this experimentally?

15.12 GO In the following illustration, the upper line shows a gene composed of segments A–D. The lower circle shows a mutant version of this gene, consisting of two fused pieces (A′-B′, C′-D′), carried on a plasmid. You attempt a targeted mutagenesis of a diploid cell by transforming cells with the cloned mutant gene. The following diagram shows the desired pairing of the plasmid and chromosome just prior to recombination.

A x B C Probe x D
A′ x B′ C′ x D′
Plasmid

You prepare DNA from the cells, digest it with an enzyme that cuts at x, and hybridize the cleaved DNA with the probe shown above. The following diagram shows a Southern blot of possible results. (a) Which lane shows fragments produced from DNA in the cell before transformation? (b) Which lane shows fragments produced from DNA in the cell in which the anticipated targeted mutagenesis occurred? (c) Which of these blot patterns might be expected if two crossovers occurred, one between A and B, and the other between C and D?

1 2 3 4 5

15.13 Compare the nucleotide-pair sequences of genomic DNA clones and cDNA clones of specific genes of higher plants and animals. What is the most frequent difference that you would observe?

15.14 What major advantage does the polymerase chain reaction (PCR) have over other methods for analyzing nucleic acid structure and function?

15.15 Almost all the sophisticated cloning vectors—phages, plasmids, or phage-plasmid hybrids—in use today have one component (in addition to an origin of replication) in common. What is this component, and what is its function?

15.16 GO The following drawing shows a restriction map of a segment of a DNA molecule. Eco refers to locations where the restriction endonuclease *Eco*RI cuts the DNA, and Pst refers to locations where the restriction enzyme *Pst*I cuts the DNA. Potential restriction sites are numbered 1–6. Distances between restriction sites are shown on the bottom scale in base pairs (bp). The thick line represents the part of the molecule that has homology with a probe.

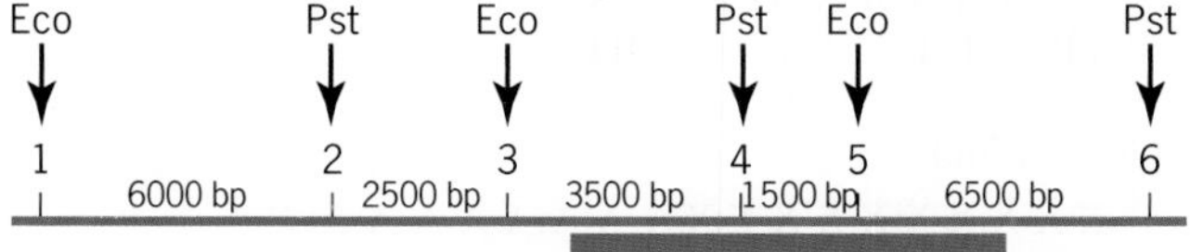

(a) Assume that individual 1 has restriction sites 1 through 6. If DNA is digested with *Pst*I, what are the expected sizes of the DNA fragments that will hybridize with the probe?
(b) Assume that individual 2 has a mutation that eliminates site 4. If DNA is digested with *Pst*I, what are the expected sizes of the DNA fragments that will hybridize with the probe?
(c) Assume that individual 3 has a mutation that eliminates site 5. If the DNA is digested with *Pst*I, what are the expected sizes of the DNA fragments that will hybridize with the probe?
(d) If the DNA of individual 1 is digested with both *Pst*I and *Eco*RI, what are the expected sizes of the DNA fragments that will hybridize with the probe?
(e) If the DNA of individual 3 is digested with both *Pst*I and *Eco*RI, what are the expected sizes of the DNA fragments that will hybridize with the probe?

15.17 Certain types of molecular analyses are facilitated by the availability of large quantities of pure single-stranded DNA from a normally double-stranded DNA molecule. However, the separation and purification of the two single strands of a given double helix are usually very difficult to accomplish by standard biochemical techniques. How have molecular biologists taken advantage of a natural biological mechanism for the purification of DNA single strands?

15.18 An important tool of molecular biologists involves the transfer of proteins that have been separated by gel electrophoresis to nitrocellulose membranes and the detection of specific proteins on the membranes by using antibodies and coupled radioactive labels or coupled enzymatic reactions. When this procedure is used, the resulting display of the visualized protein bands is called a western blot. What is the significance of the name "western blot"?

15.19 Ten micrograms of a decanucleotide-pair *Hpa*I restriction fragment were isolated from the double-stranded DNA chromosome of a small virus. Octanucleotide poly(A) tails were then added to the 3′-ends of both strands using terminal transferase and dATP; that is,

5′-X X X X X X X X X X-3′
3′-X′X′X′X′X′X′X′X′X′X′-5′
↓ terminal transferase, dATP
5-X X X X X X X X X X A A A A A A A A-3′
3′-A A A A A A A A X′X′X′X′X′X′X′X′X′X′-5′

where X and X′ can be any of the four standard nucleotides, but X′ is always complementary to X.

The two complementary strands ("Watson" strand and "Crick" strand) were then separated and sequenced by the 2′,3′-dideoxyribonucleoside triphosphate chain-termination method. The reactions were all primed using a radioactive (^{32}P-labeled) synthetic poly(T) octamer; that is,

Watson strand

3′-A A A A A A A A X′X′X′X′X′X′X′X′X′X′-5′
5′-^{32}P-T T T T T T T T-OH

Crick strand

5′-X X X X X X X X X X A A A A A A A A-3′
HO-T T T T T T T T-^{32}P-5′

The usual four parallel reactions: (1) ddTTP, (2) ddCTP, (3) ddATP, and (4) ddGTP (plus DNA polymerase and all other substrates and required components) were carried out for both strands. Each reaction mixture was applied to a lane in a polyacrylamide gel, fractionated by electrophoresis, and autoradiographed. The autoradiogram of the sequencing gel for one of the strands is shown on the left of the diagram. Draw the banding pattern that would be expected on the autoradiogram of the gel for the complementary strand on the right in the diagram at upper right.

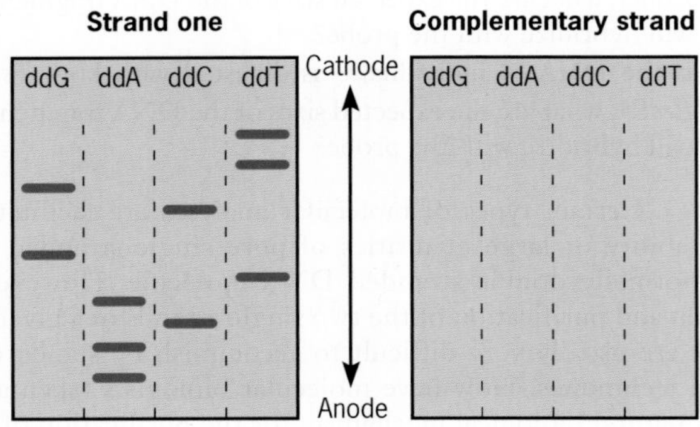

Problem 15.19

15.20 The automated DNA sequencing machines utilize fluorescent dyes to detect the nascent DNA chains present in each of the four dideoxy (ddX) chain-termination reactions. The dyes fluoresce at different wavelengths, which are recorded by a photocell as the products of the reactions are separated based on length by capillary gel electrophoresis (see **FIGURE 15.30**). In the standard sequencing reactions, the chains terminating with ddG fluoresce dark blue (appears black in print out below), those terminating with ddC fluoresce light blue, those terminating with ddA fluoresce green, and those terminating with ddT fluoresce red. The computer printout for the sequence of a short segment of DNA is as follows:

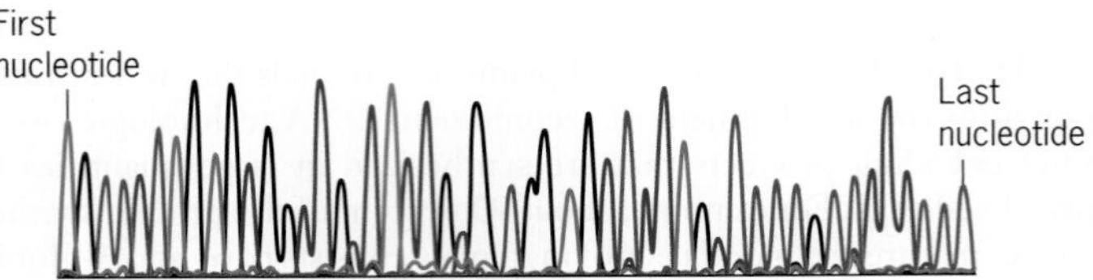

What is the nucleotide sequence of the nascent strand of DNA? What is the nucleotide sequence of the DNA template strand?

15.21 (a) What experimental procedure is carried out in Southern, northern, and western blot analyses? (b) What is the major difference between Southern, northern, and western blot analyses?

15.22 A DNA molecule is subjected to single and double digestions with restriction enzymes, and the products are separated by gel electrophoresis. The results are given at upper right (fragment sizes are in kb):

*Eco*RI	*Eco*RI and *Hin*dIII	*Hin*dIII	*Bam*HI	*Eco*RI and *Bam*HI	*Hin*dIII and *Bam*HI
8	5	12	6	6	6
4	4		6	4	5
	3			2	1

Draw the restriction map of this DNA molecule.

15.23 The cystic fibrosis (*CF*) gene (location: chromosome 7, region q31) has been cloned and sequenced, and studies of CF patients have shown that about 70 percent of them are homozygous for a mutant *CF* allele that has a specific three-nucleotide-pair deletion (equivalent to one codon). This deletion results in the loss of a phenylalanine residue at position 508 in the predicted CF gene product. Assume that you are a genetic counselor responsible for advising families with CF in their pedigrees regarding the risk of CF among their offspring. How might you screen putative CF patients and their parents and relatives for the presence of the *CFΔF508* mutant gene? What would the detection of this mutant gene in a family allow you to say about the chances that CF will occur again in the family?

15.24 Cereal grains are major food sources for humans and other animals in many regions of the world. However, most cereal grains contain inadequate supplies of certain of the amino acids that are essential for monogastric animals such as humans. For example, corn contains insufficient amounts of lysine, tryptophan, and threonine. Thus, a major goal of plant geneticists is to produce corn varieties with increased kernel lysine content. As a prerequisite to the engineering of high-lysine corn, molecular biologists need more basic information about the regulation of the biosynthesis and the activity of the enzymes involved in the synthesis of lysine. The first step in the anabolic pathway unique to the biosynthesis of lysine is catalyzed by the enzyme dihydrodipicolinate synthase. Assume that you have recently been hired by a major U.S. plant research institute and that you have been asked to isolate a clone of the nucleic acid sequence encoding dihydrodipicolinate synthase in maize. Briefly describe four different approaches you might take in attempting to isolate such a clone and include at least one genetic approach.

15.25 You are studying a circular plasmid DNA molecule of size 10.5 kilobase pairs (kb). When you digest this plasmid with restriction endonucleases *Bam*HI, *Eco*RI, and *Hin*dIII, singly and in all possible combinations, you obtain linear restriction fragments of the following sizes:

Enzymes	Fragment Sizes (in kb)
*Bam*HI	7.3, 3.2
*Eco*RI	10.5
*Hin*dIII	5.1, 3.4, 2.0
*Bam*HI + *Eco*RI	6.7, 3.2, 0.6
*Bam*HI + *Hin*dIII	4.6, 2.7, 2.0, 0.7, 0.5
*Eco*RI + *Hin*dIII	4.0, 3.4, 2.0, 1.1
*Bam*HI + *Eco*RI + *Hin*dIII	4.0, 2.7, 2.0, 0.7, 0.6, 0.5

Draw a restriction map for the plasmid that fits your data.

15.26 The following diagram is a segment of an autoradiogram of a dideoxy sequencing gel. The origin is at the top. What is the sequence (including polarity) of the single-stranded DNA that served as a template for generating this pattern? (Use all of the bands on the gel.)

15.27 A linear DNA molecule is subjected to single and double digestions with restriction endonucleases, and the following results are obtained:

Enzymes	Fragment Sizes (in kb)
*Eco*RI	2.9, 4.5, 7.4, 8.0
*Hin*dIII	3.9, 6.0, 12.9
*Eco*RI and *Hin*dIII	1.0, 2.0, 2.9, 3.5, 6.0, 7.4

Draw the restriction map defined by these data.

▶ Genomics on the Web

at http://www.ncbi.nlm.nih.gov/

In this chapter, we have discussed a DNA test for one of the most prevalent mutant alleles that causes cystic fibrosis, and in Chapter 17 (**FIGURE 17.2**), we will examine a DNA test for mutant genes that result in Huntington's disease.

1. Are DNA tests available for mutant genes that cause other inherited human diseases? If so, what are some of the diseases for which DNA tests are currently available?

2. What are some of the molecular techniques used in these DNA tests? Gel electrophoresis? PCR? Southern blots?

3. How reliable are these tests? Can they be performed on fetal cells obtained by amniocentesis? On single cells obtained from eight-cell pre-embryos?

Hint: At the NCBI web site, go to Human Genome Resources, and then to OMIM (Online Mendelian Inheritance in Man), and search for "DNA tests for mutant alleles." Also visit http://www.genetests.org/ for information on 607 laboratories providing tests for 1,549 different human genetic diseases.

Chapter 16 Genomics

CHAPTER OUTLINE

- Genomics: An Overview
- Correlated Genetic, Cytological, and Physical Maps of Chromosomes
- Map Position-Based Cloning of Genes
- The Human Genome Project
- RNA and Protein Assays of Genome Function
- Comparative Genomics

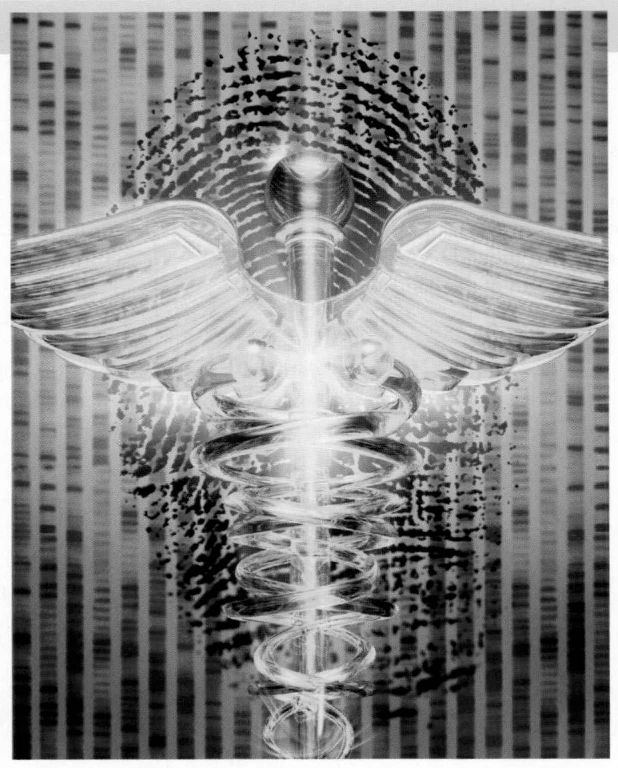

A symbol of the medical profession superimposed on a thumb print and a DNA sequencing gel.

Human Gene Prospecting in Iceland

A major search for human genes with important pharmacological value is taking place in a somewhat unexpected location—the remote island of Iceland. Because of their history and geographical isolation, the 270,000 people of Iceland provide a unique resource for genetic studies. They are the genetically quite homogeneous descendants of Vikings who settled on this island more than 1100 years ago. This homogeneity has been enhanced by two "genetic bottlenecks" during which the population of Iceland was sharply reduced. During the fifteenth century, the population plummeted from about 70,000 to around 25,000 when bubonic plague ravaged the island. During the 1700s, the population dropped below 50,000 on three occasions because of famine and disease caused in part by the eruption of the volcano Hekla. Thus, the human gene pool of Iceland is much more homogeneous than the gene pools of most other human populations. In addition, Iceland's national health service has kept superb family medical records since 1915.

In 1997, Kari Stefansson, a Harvard geneticist, recognized the uniqueness of Iceland's human gene pool and family records. He returned to his homeland to launch a private company, deCODE Genetics, with the goal of identifying human genes that would lead to the development of new pharmaceutical drugs and diagnostic tests. In 1998, the government of Iceland granted deCODE an exclusive license to construct and analyze a genetic database derived from the country's health records. The company's first success was the identification of the *familial essential tremor* gene—a gene associated with shakiness in the elderly. Scientists at deCode have also identified genes involved in susceptibility to heart disease, stroke, hypertension, Alzheimer's disease, and prostate cancer.

Based on these results, deCODE Genetics negotiated a $200 million contract with the Swiss pharmaceutical giant Hoffmann-LaRoche, which will give the Swiss firm exclusive rights to any drugs or diagnostic products resulting from the work of deCODE scientists. To the people of Iceland, the contract specifies that Hoffmann-LaRoche must provide free of charge all drugs, diagnostic tests, and other products resulting from this research. Therefore, at least in this one case, the people who are providing the genetic data and the DNA samples for analysis will personally benefit from the results of research by a private company.

The deCODE saga has not progressed without controversy, however. The key issue in the ongoing debate in Iceland is the question of *presumed consent* versus *informed consent.*

According to the legislation passed by the parliament of Iceland, an individual's health records are automatically added to the database unless she or he specifically requests that the records be excluded. This law established the practice of "presumed consent." The general practice in the field of medicine is one of "informed consent," where an individual's records are considered private unless the individual specifically signs an informed consent document stating that they may be released. Many citizens and doctors in Iceland object to the presumed consent law regulating deCODE, claiming that it is an invasion of their privacy. Some doctors are refusing to submit their patients' health records to deCODE. Nevertheless, Iceland's law establishing the practice of presumed consent is very clear, and the deCODE database continues to expand—as does the debate on the procedures for collecting and using genetic data from human populations.

Gregor Mendel studied the effects of seven genes on traits in peas, but he studied no more than three genes in any one cross. Today's geneticists can study the expression of all the genes—the entire genome—of an organism in a single experiment. As of May 2008, the complete nucleotide sequences of the genomes of over 2000 viruses, defective viruses, and viroids, 1325 plasmids, 1373 mitochondria, 131 chloroplasts, 109 archaea, 687 true bacteria, and 23 eukaryotes had been determined. In addition, the genomes of another 208 eukaryotes have been sequenced, and the sequences are currently being assembled into complete genome sequences. Sequencing projects for another 233 eukaryotic genomes are underway, and the complete genome sequences of two individuals—Craig Venter and James Watson—are now available. Some scientists are predicting that it may be possible to sequence an entire human genome for as little as $1000 in the not-too-distant future.

The list of eukaryotes whose genomes have been sequenced includes important model organisms in genetics: baker's yeast *Saccharomyces cerevisae*, the fruit fly *Drosophila melanogaster*, and the plant *Arabidopsis thaliana*. It also includes the protozoan *Plasmodium falciparum*, which causes the most dangerous form of malaria, and the mosquito *Anopheles gambiae*, which is the host organism most responsible for the spread of this disease. The silkworm (*Bombyx mori*), an economically important insect, is on the list, and so are several vertebrates: the mouse (*Mus musculus*), the Norwegian rat (*Rattus norvegicus*), the Red Jungle Fowl—an ancestor of domestic chickens (*Gallus gallus*), the puffer fish (*Fugu rubripes*), and our own species (*Homo sapiens*).

One of the original goals of the Human Genome Project was to determine the complete nucleotide sequence of the human genome by the year 2005. As it turned out, two first drafts of the sequence—one by a public consortium and the other by a private company—were published in February 2001 (see A Milestone in Genetics: Two Drafts of the Sequence of the Human Genome at the end of this chapter). Indeed, a nearly complete sequence of the human genome comprising 99 percent of the euchromatic DNA was released in October 2004, a full year ahead of the original goal.

The improvements in DNA sequencing technology that occurred during the last two decades of the twentieth century have allowed researchers to collect large amounts of sequence data. However, sequencing was not always so easy. It took Robert Holley, 1968 Noble Prize recipient, several years to determine the 77-nucleotide sequence of the alanine tRNA from yeast. A few of the major advances in sequencing technology, as well as some of the landmarks in the study of genomes, are highlighted in **FIGURE 16.1.**

In the present era, vast amounts of sequence data accumulate daily. Most of these data are the results of research projects funded by government agencies—the National Institutes of Health (NIH), the National Science Foundation (NSF), and the Department of Energy (DOE) in the United States—and comparable agencies in other countries. Thus, these data are public information and are available to anyone who wants to use them. Making the sequences public has been accomplished by establishing sequence databases that are available free on the web at http://www.ncbi.nlm.nih.gov/entrez/query.fcgi (see the Focus on GenBank).

Of course, just making the databases available is not enough. We must be able to extract information from them—that is, to "mine" the databases—and then analyze the extracted information efficiently and accurately. This process requires computer software that can search the vast DNA sequences in genomes of interest. The need for such software has spawned a new scientific discipline called bioinformatics. Mathematicians, computer scientists, and molecular biologists who work in this discipline develop computer-search algorithms that can extract information from DNA and protein sequence data.

The availability of entire genome sequences has opened the door to bioinformatic analyses and to functional studies of the genes contained in these sequences. Microarrays—including the so-called gene chips—allow scientists to investigate the expression of all the genes in an organism simultaneously (see RNA and Protein Assays of Genome Function in this chapter). Other procedures use known nucleotide sequences to dissect metabolic pathways by "knocking out" or turning off the expression of genes (see Reverse Genetics in Chapter 17).

In this chapter, we will discuss some of the tools and techniques that are used to study the structure and function of genomes, we will examine the spectacular progress of the Human Genome Project, and we will see how comparisons of genomes can contribute to our understanding of evolution. In the following chapter, we will examine other technical advances—DNA prints, human gene therapy, and the production of transgenic microorganisms, plants, and animals. We will also see how geneticists have identified the defective genes that are responsible for two tragic human conditions, Huntington's disease and cystic fibrosis. The procedures used to identify these genes have become methodological paradigms for the identification of many other disease-related genes in humans.

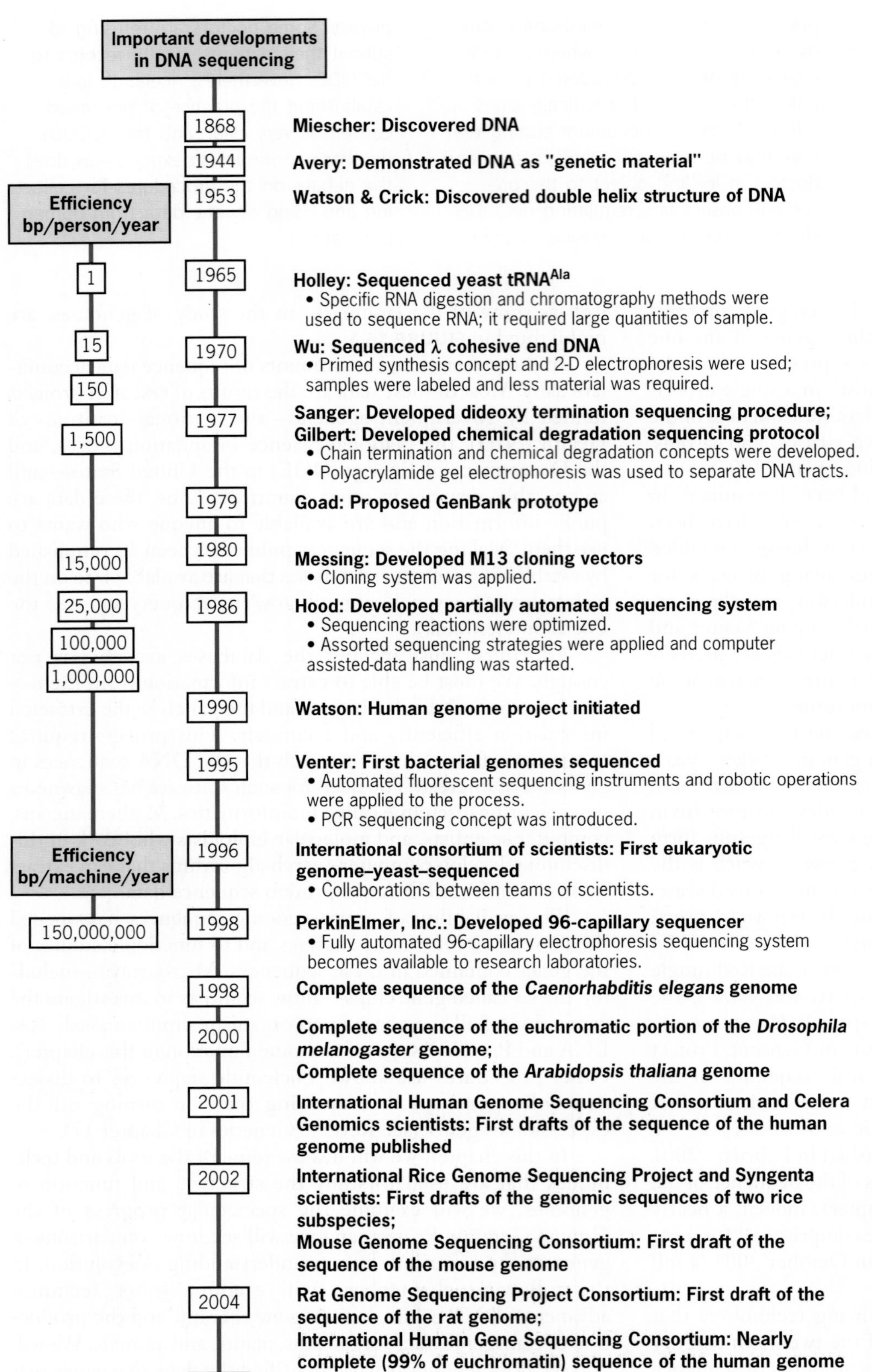

Important developments in DNA sequencing
Efficiency bp/person/year
1
15
150
1,500
15,000
25,000
100,000
1,000,000
Efficiency bp/machine/year
150,000,000
1868
Miescher: Discovered DNA
1944
Avery: Demonstrated DNA as "genetic material"
1953
Watson & Crick: Discovered double helix structure of DNA
1965
Holley: Sequenced yeast tRNA^Ala
• Specific RNA digestion and chromatography methods were used to sequence RNA; it required large quantities of sample.
1970
Wu: Sequenced λ cohesive end DNA
• Primed synthesis concept and 2-D electrophoresis were used; samples were labeled and less material was required.
1977
Sanger: Developed dideoxy termination sequencing procedure;
Gilbert: Developed chemical degradation sequencing protocol
• Chain termination and chemical degradation concepts were developed.
• Polyacrylamide gel electrophoresis was used to separate DNA tracts.
1979
Goad: Proposed GenBank prototype
1980
Messing: Developed M13 cloning vectors
• Cloning system was applied.
1986
Hood: Developed partially automated sequencing system
• Sequencing reactions were optimized.
• Assorted sequencing strategies were applied and computer assisted-data handling was started.
1990
Watson: Human genome project initiated
1995
Venter: First bacterial genomes sequenced
• Automated fluorescent sequencing instruments and robotic operations were applied to the process.
• PCR sequencing concept was introduced.
1996
International consortium of scientists: First eukaryotic genome–yeast–sequenced
• Collaborations between teams of scientists.
1998
PerkinElmer, Inc.: Developed 96-capillary sequencer
• Fully automated 96-capillary electrophoresis sequencing system becomes available to research laboratories.
1998
Complete sequence of the Caenorhabditis elegans genome
2000
Complete sequence of the euchromatic portion of the Drosophila melanogaster genome;
Complete sequence of the Arabidopsis thaliana genome
2001
International Human Genome Sequencing Consortium and Celera Genomics scientists: First drafts of the sequence of the human genome published
2002
International Rice Genome Sequencing Project and Syngenta scientists: First drafts of the genomic sequences of two rice subspecies;
Mouse Genome Sequencing Consortium: First draft of the sequence of the mouse genome
2004
Rat Genome Sequencing Project Consortium: First draft of the sequence of the rat genome;
International Human Gene Sequencing Consortium: Nearly complete (99% of euchromatin) sequence of the human genome
2005

Figure 16.1 ▶ Advances in DNA sequencing efficiency, some of the technological developments that enhanced the productivity of sequencers, and some landmarks in DNA sequencing. Initially, all the steps in DNA sequencing were performed manually, making it a very labor-intensive process. However, fully automated sequencing machines have now replaced human sequencers, greatly increasing efficiency.

▶ FOCUS ON GenBank

In 1979 Walter Goad, a physicist working at the Los Alamos National Laboratory (LANL) in New Mexico, came up with the idea for a database that would contain all available DNA sequences. From 1982 until 1992, Goad and his colleagues incorporated sequences into the database—now named GenBank—and maintained it at LANL. Today this database is maintained by the National Center for Biotechnology Information (NCBI), which is part of the National Library of Medicine (NLM) at the National Institutes of Health (NIH) in Bethesda, Maryland. The content of the database has grown enormously since Goad and his colleagues created it (**FIGURE 1**). At the end of 1982, GenBank contained 680,338 nucleotide pairs of sequenced DNA, but by April 2008, it contained almost 90 billion nucleotide pairs.

Databases comparable to GenBank have also been established in Europe and Japan. The European Molecular Biology Laboratory (EMBL) Data Library was set up in Germany in 1980, and the DNA DataBank of Japan (DDBJ) was established in 1984. GenBank, EMBL, and DDBJ subsequently joined forces and formed the International Nucleotide Sequence Database Collaboration, which allows researchers to search all three databases simultaneously.

The development of search and retrieval programs that screen databases for sequences similar to input sequences has provided scientists with a major research tool. In particular, NCBI's ENTREZ retrieval system has proven invaluable. This system was first distributed on CD-ROM in 1992, then as a network version in 1993, and finally was made available—free—on the Internet in 1994 (URL: http://www.ncbi.nlm.nih.gov/entrez). The amount of searchable and retrievable information available at the Entrez web site has increased every year. It not only encompasses DNA and protein sequence databases, but it also includes a huge bibliographic database called PubMed that covers most of the journals in medicine and biology. Today, you can search all these databases simultaneously by using NCBI's global cross-database search engine, and the search page will give you the number of items found (that is, the "hits") in each database.

A discussion of all the databases that can be searched with Entrez is far beyond the scope of this textbook. You are encouraged to visit the site and explore some of its databases. They include the PubMed and DNA databases mentioned above, and databases of protein sequences, three-dimensional macromolecular structures, cancer chromosomes and genes, expressed sequences, single-nucleotide polymorphisms, whole-genome sequences, and many more.

Let's perform one Entrez search to illustrate how it works. Assume that you have just determined the nucleotide sequence of a segment of DNA from an organism of interest, and you want to know if that DNA has already been sequenced or if it is similar to sequences in any of the current databases. One of the quickest ways to obtain this information is to perform a BLAST (Basic Local Alignment Search Tool) search with your sequence as the input, or query, sequence. Let's start at the NCBI home page: http://www.ncbi.nlm.nih.gov/. First, select "BLAST" along the top tool bar. Then, under "Nucleotide" at the top left, click on "Quickly search for highly similar sequences (megablast)," and a window similar to that shown in **FIGURE 2*a*** will appear.

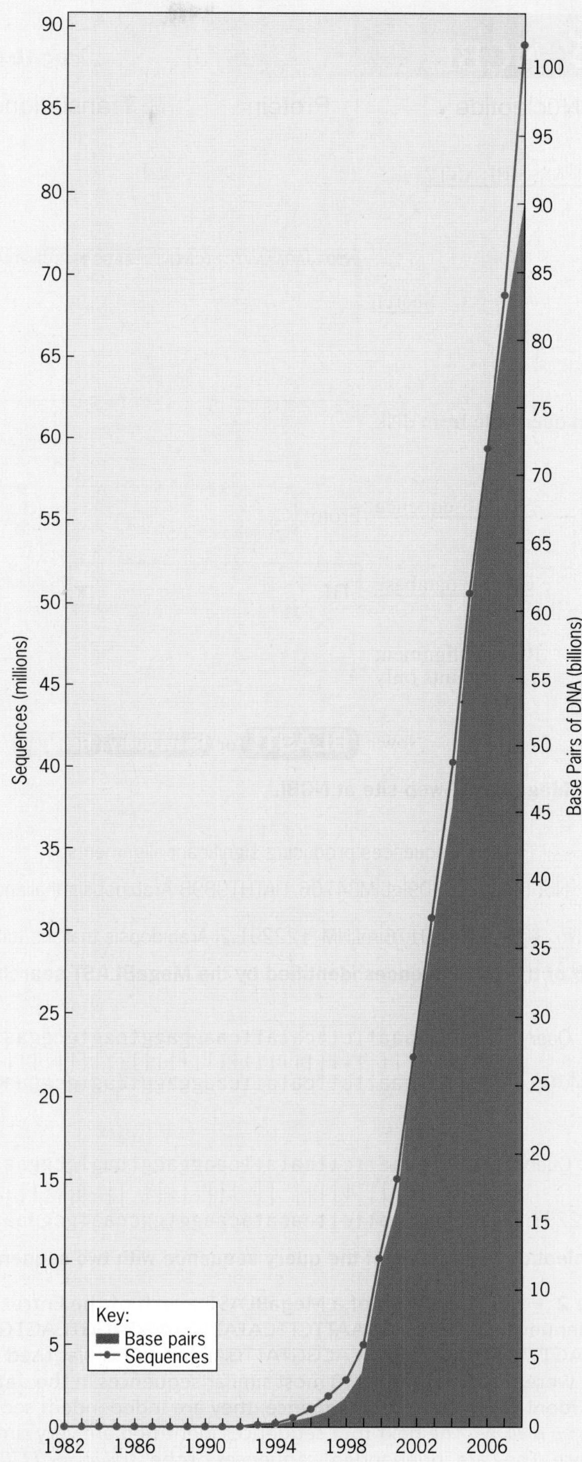

Figure 1 ▶ Growth of GenBank from its origin in 1982 to April 15, 2008. The left and right ordinates show the size of the collection in number of DNA sequences (red) and number of nucleotide pairs (blue), respectively. The number of different sequences grew from 606 at the end of 1982 to over 100 million in April 2008.

▶ FOCUS ON GenBank *(continued...)*

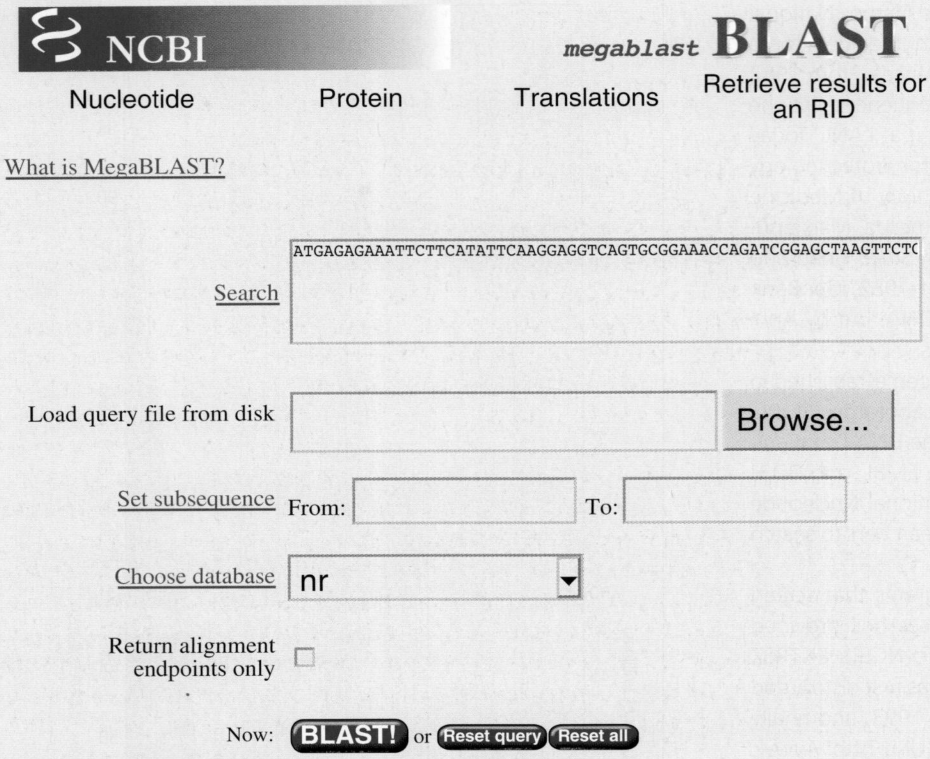

(a) **The MegaBLAST web site at NCBI.**

	Sequences producing significant alignments:	Score (bits)	E value
No. 5:	gi|166909|gb|M84706.1|ATHTUB9B Arabidopsis thaliana beta-9 tubul. . .	192	8e–47
No. 10:	gi|30689176|ref|NM_122291.2| Arabidopsis thaliana tubulin beta-8. . .	110	6e–22

(b) **Two of the 10 sequences identified by the MegaBLAST search.**

```
Query:      atgagagaaattcttcatattcaaggaggtcagtgcggaaaccagatcggagctaagttctgggaagttatttgcggcgagcacggtattgatcaaaccg 100
            ||||||||||||||||||||||||||||||||||||||||||||||||||||||||||||||||||||||||||||||||||||||||||||||||||||
M84706.1    atgagagaaattcttcatattcaaggaggtcagtgcggaaaccagatcggagctaagttctgggaagttatttgcggcgagcacggtattgatcaaaccg 227

Query:      atgagagaaattcttcatattcaaggaggtcagtgcggaaaccagatcggagctaagttctgggaagtta-tttgcggcgagcacggtattgat 93
            ||| |||| ||||||| || ||||| || ||||||||||||||||||||||||||||||||||||||| |||||| |||||||| || |||
NM_122291.2 atgcgagagattcttcacatacaaggtggccaatgcggaaaccagatcggagctaagttctgggaagt-agtttgcgccgagcacgggatcgat 172
```

(c) **Nucleotide alignments of the query sequence with two sequences identified by the MegaBLAST search.**

Figure 2 ▶ *(a)* Illustration of a MegaBLAST search of the Entrez "Nucleotide" databases. The 100-nucleotide sequence: ATGAGAGAAATTCTTCATATTCAAGGAGGTCAGTGCGGAAACCAGATCGGAGCTAAGTTCT-GGGAAGTTATTTGCGGCGAGCACGGTATTGATCAAACCG was used as the query sequence. The search results were restricted to the 10 most similar sequences in the databases. The first six sequences identified are all identical to the query sequence; they are independent sequences of the β9-tubulin gene (*TUB9*) of *Arabidopsis thaliana*. The next four sequences identified are very similar, but not identical to the query sequence. They are independent sequences of the *Arabidopsis TUB8* gene, which encodes the closely related β8-tubulin. *(b)* The two distinct sequences identified by the MegaBLAST search. The higher the score, the more closely related the sequences. *(c)* Alignments of the query sequence with each of the sequences identified in the MegaBLAST search. Note the perfect identity of the sequence M84706.1 (*TUB9*) with the query sequence, and the close similarity of NM_122291.2 (*TUB8*) with the query sequence. *TUB8* and *TUB9* encode very similar β-tubulins.

► FOCUS ON GenBank

Suppose that your nucleotide sequence is as follows:

5′-ATGAGAGAAATTCTTCATATTCAAGGAGGTCAGTGCGGAAACC AGATCGGAGCTAAGTTCTGGGAAGTTATTTGCGGCGAGCACGGT ATTGATCAAACCG-3′

Either type or paste this query sequence in the box at the top of the BLAST page. Before you click the "BLAST!" button, scroll down to "Format" and make a few changes to decrease the amount of information that you will receive. Change "alignment" from "HTML" to "Plain text," and reduce both the number of "descriptions" and the number of "alignments" to 10. Lastly, change the "alignment view" to "Pairwise." Now, click the "BLAST!" button. Your results should appear in less than a minute. They should include a list of 10 "Sequences producing significant alignments" (descriptions of sequences 5 and 10 are shown in **FIGURE 2*b***) and the alignment of each sequence with your query sequence (alignments are shown for numbers 5 and 10 in **FIGURE 2*c***).

The 10 most similar sequences in the databases actually include only two distinct sequences. The first six sequences are all independently obtained sequences of the same gene; the next four are independent sequences of a second, closely related gene. Note in **FIGURE 2c** that the query sequence is a perfect match with sequence 5 and that it differs from sequence 10 at 12 nucleotide positions. Sequences 5 and 10 are members of a gene family that encodes a set of very closely related proteins with the same or very similar functions.

Suppose you want to know more about the sequences identified in your search. Let's go to the NCBI home page: http://www.ncbi.nlm.nih.gov/entrez/. Scroll to the bottom of the page, and click on "Go" for a search across all databases. In the "Search across databases" box, type the accession number for sequence 5—"M84706"—and click "Go." The search will take about a second, and the summary page will show two matches in "PubMed" and one each in "PubMed Central," "Nucleotide," and "Gene." Click "Nucleotide" first, and you will see that the sequence is part of the beta-9 tubulin gene of *Arabidopsis thaliana.* (β-Tubulin is one of the structural proteins of anaphase spindle fibers.) Next, click on "M84706," and the resulting page will give a description of the gene, including the complete nucleotide sequence.

Now, let's return to the cross-database search summary page, and click on "PubMed" to see the matches in the literature. It will show you two references. Click on the second one: "Snustad *et al.*," and the title page and abstract of this paper will appear in the window. Click on "Free full text article," and a PDF version of the paper will be downloaded to your computer. This reference contains the first report of the sequences of the nine β-tubulin genes of *Arabidopsis*.

This brief exploration of the Entrez web site illustrates the power and convenience of the software and the databases now available to analyze DNA sequences. Without these tools, geneticists would be hard-pressed to make much sense out of the vast number of DNA sequences currently available.

Genomics: An Overview

Genomics is the subdiscipline of genetics that focuses on the structure and function of entire genomes.

Geneticists have used the term *genome* for over seven decades to refer to one complete copy of the genetic information or one complete set of chromosomes (monoploid or haploid) of an organism. In contrast, **genomics** is a relatively new term. The word "genomics" appears to have been coined by Thomas Roderick in 1986 to refer to the genetics subdiscipline of mapping, sequencing, and analyzing the functions of entire genomes, and to serve as the name of a new journal—*Genomics*—dedicated to the communication of new information in this subdiscipline.

As more detailed maps and sequences of genomes became available, the genomics subdiscipline was divided into *structural genomics*—the study of the genome structure; *functional genomics*—the study of the genome function; and **comparative genomics**—the study of genome evolution. Functional genomics includes analyses of the **transcriptome,** the complete set of RNAs transcribed from a genome, and the **proteome,** the complete set of proteins encoded by a genome. Indeed, functional genomics has spawned an entirely new discipline, **proteomics,** which has as its goal the determination of the structures and functions of all the proteins in an organism.

Whereas structural genomics is quite advanced—given the availability of complete nucleotide sequences for many organisms—functional genomics has just entered an explosive growth phase. New array hybridization and gene-chip technologies allow researchers to monitor the expression of entire genomes—all the genes in an organism—at various stages of growth and development or in response to environmental changes. These powerful new tools promise to provide a wealth of information about genes and how they interact with each other and with the environment.

KEY POINT

- Genomics is the subdiscipline of genetics devoted to the mapping, sequencing, and functional and comparative analyses of genomes.

Correlated Genetic, Cytological, and Physical Maps of Chromosomes

The chromosomal locations of genes and other molecular markers can be mapped based on recombination frequencies, positions relative to cytological features, or physical distances.

The ability of scientists to identify and isolate genes based on information about their location in the genome was one of the first major contributions of genomics research. In principle, this approach, called *positional cloning*, can be used to identify and clone any gene with a known phenotypic effect in any species. Positional cloning has been used extensively in many species, including humans. Indeed, in Chapter 17, we will consider the use of positional cloning to identify the human genes responsible for Huntington's disease and cystic fibrosis.

Because the utility of positional cloning depends on the availability of detailed maps of the regions of the chromosomes where the genes of interest reside, major efforts have focused on the development of detailed maps of the human genome and the genomes of important model organisms such as *D. melanogaster*, *C. elegans*, and *A. thaliana*. The goal of this research is to construct correlated genetic and physical maps with markers distributed at relatively short intervals throughout the genome. In the case of the human and *Drosophila* genomes, the genetic and physical maps can also be correlated with cytological maps (banding patterns) of the chromosomes (**FIGURE 16.2**). We will discuss the construction of these maps in the following sections of this chapter.

Recall that genetic maps (**FIGURE 16.2**, left) are constructed from recombination frequencies, with 1 centiMorgan (cM) equal to the distance that yields an average frequency of recombination of 1 percent (Chapter 7). Genetic maps with markers spaced at short intervals—high-density genetic maps—are often constructed by using molecular markers such as restriction fragments of different lengths (*restriction fragment-length polymorphisms*, or *RFLPs*). Cytological maps (**FIGURE 16.2**, center) are based on the banding patterns of chromosomes observed with the microscope after treatment with various stains (Chapter 6). **Physical maps** (**FIGURE 16.2,**

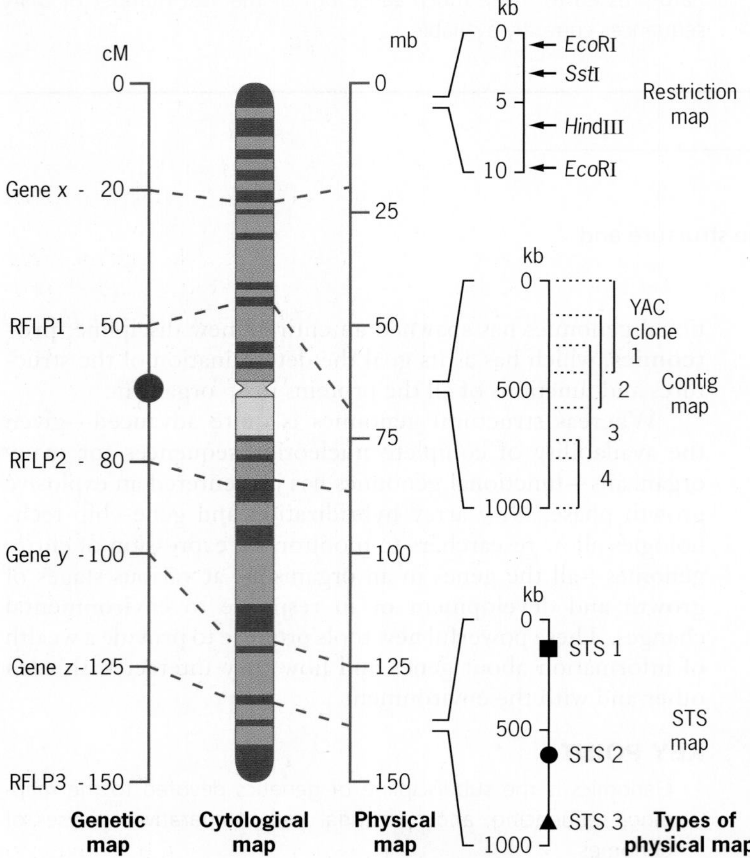

Figure 16.2 ▶ Correlation of the genetic, cytological, and physical maps of a chromosome. Genetic map distances are based on crossing-over frequencies and are measured in percentage recombination, or centiMorgans (cM), whereas physical distances are measured in kilobase pairs (kb) or megabase pairs (mb). Restriction maps, contig maps, and STS (sequence-tagged site) maps are described in the text.

right), such as the restriction maps (**FIGURE 16.2**, top right) discussed in Chapter 15, are based on the molecular distances—base pairs (bp), kilobases (kb, 1000 bp), and megabases (mb, 1 million bp)—separating sites on the giant DNA molecules present in chromosomes. Physical maps often contain the locations of overlapping genomic clones or *contigs* (**FIGUROE 16.2**, right center) and unique nucleotide sequences called *sequence-tagged sites*, or *STSs* (**FIGURE 16.2**, bottom right).

Physical maps of a chromosome can be correlated with the genetic and cytological maps in several ways. Genes that have been cloned can be positioned on the cytological map by *in situ* hybridization (Chapter 9). Correlations between the genetic and physical maps can be established by locating clones of genetically mapped genes or RFLPs on the physical map. Markers that are mapped both genetically and physically are called *anchor markers;* they anchor the physical map to the genetic map and vice versa. Physical maps of chromosomes can also be correlated with genetic and cytological maps by using (1) PCR (see Figure 15.14) to amplify short—usually 200 to 500 bp—unique genomic DNA sequences, (2) Southern blots to relate these sequences to overlapping clones on physical maps, and (3) *in situ* hybridization to determine their chromosomal locations (cytological map positions). These short, unique anchor sequences are called *sequence-tagged sites (STSs).* Another approach uses short cDNA sequences (DNA copies of mRNAs), or *expressed-sequence tags (ESTs)*, as hybridization probes to anchor physical maps to RFLP maps (genetic maps) and cytological maps.

Physical distances do not correlate directly with genetic map distances because recombination frequencies are not always proportional to molecular distances. However, the two are often reasonably well correlated in euchromatic regions of chromosomes. In humans, 1 cM is equivalent, on average, to about 1 mb of DNA.

RESTRICTION FRAGMENT-LENGTH POLYMORPHISM (RFLP) AND MICROSATELLITE MAPS

When mutations change the nucleotide sequences in restriction enzyme cleavage sites, the enzymes no longer recognize them (**FIGURE 16.3*a***). Other mutations may create new restriction sites. These mutations result in variations in the lengths of the DNA fragments produced by digestion with various restriction enzymes (**FIGURE 16.3*b***). Such **restriction fragment-length polymorphisms,** or **RFLPs,** have proven invaluable in constructing detailed genetic maps for use in positional cloning. The RFLPs are mapped just like other genetic markers; they segregate in crosses as though they were codominant alleles.

The DNAs of different geographical isolates, different ecotypes (strains adapted to different environmental conditions), and different inbred lines of a species contain many RFLPs that can be used to construct detailed genetic maps. Indeed, the DNAs of different individuals—even relatives—often exhibit RFLPs. Some RFLPs can be visualized directly when the fragments in DNA digests are separated by agarose gel electrophoresis, stained with ethidium bromide, and viewed under ultraviolet light. Other RFLPs can be detected only by using specific cDNA or genomic clones as radioactive hybridization probes on genomic Southern blots (**FIGURE 16.3*b***). The RFLPs themselves are the phenotypes used to classify the progeny of crosses as parental or recombinant. RFLPs segregate as codominant markers in crosses, with the restriction fragments from both of the homologous chromosomes visible in gels or detected on autoradiograms of Southern blots produced from the gels.

RFLP markers have proven especially valuable in mapping the chromosomes of humans, where researchers must rely on the segregation of spontaneously occurring mutant alleles in families to estimate map distances. Pedigree-based mapping of this type is done by comparing the probabilities that the genetic markers segregating in the pedigree are unlinked or linked by various map distances. In 1992, geneticists used this procedure to construct an early map of about 2000 RFLPs on the 24 human chromosomes. **FIGURE 16.4** shows the correlation between an RFLP map and the cytological map of human chromosome 1.

In humans, the most useful RFLPs involve short sequences that are present as tandem repeats. The number of copies of each sequence present at a given site on a chromosome is highly variable. These sites, called **variable number tandem repeats,** or **VNTRs,** are therefore highly polymorphic. VNTRs vary in fragment length not because of differences in the positions of restriction enzyme cleavage sites, but because of differences in the number of copies of the repeated sequence between the restriction sites.

Microsatellites are another class of polymorphisms that have proven extremely valuable in constructing high-density maps of eukaryotic chromosomes. Microsatellites are polymorphic tandem repeats of sequences only two to five nucleotide pairs long. In humans, microsatellite sequences composed of polymorphic tandem repeats of the dinucleotide sequence AC/TG (AC in one strand; TG in the complementary strand) provide especially useful markers. In 1996, a group of French and Canadian researchers published a comprehensive map of 5264 AC/TG microsatellites in the human genome. These microsatellites defined 2335 sites with an average distance of 1.6 cM or about 1.6 mb between adjacent markers.

By 1997, a large international consortium had used RFLPs to map over 16,000 human genes (ESTs and cloned genes) and had integrated their map with the physical map of the human genome. In this collaborative study, over 20,000 STSs were mapped to 16,354 distinct loci. These genetic maps composed primarily of RFLP markers have made it possible to identify and characterize mutant genes that are responsible for many human diseases (Chapter 17).

CYTOGENETIC MAPS

In some species, genes and clones can be positioned on the cytological maps of the chromosomes by *in situ* hybridization

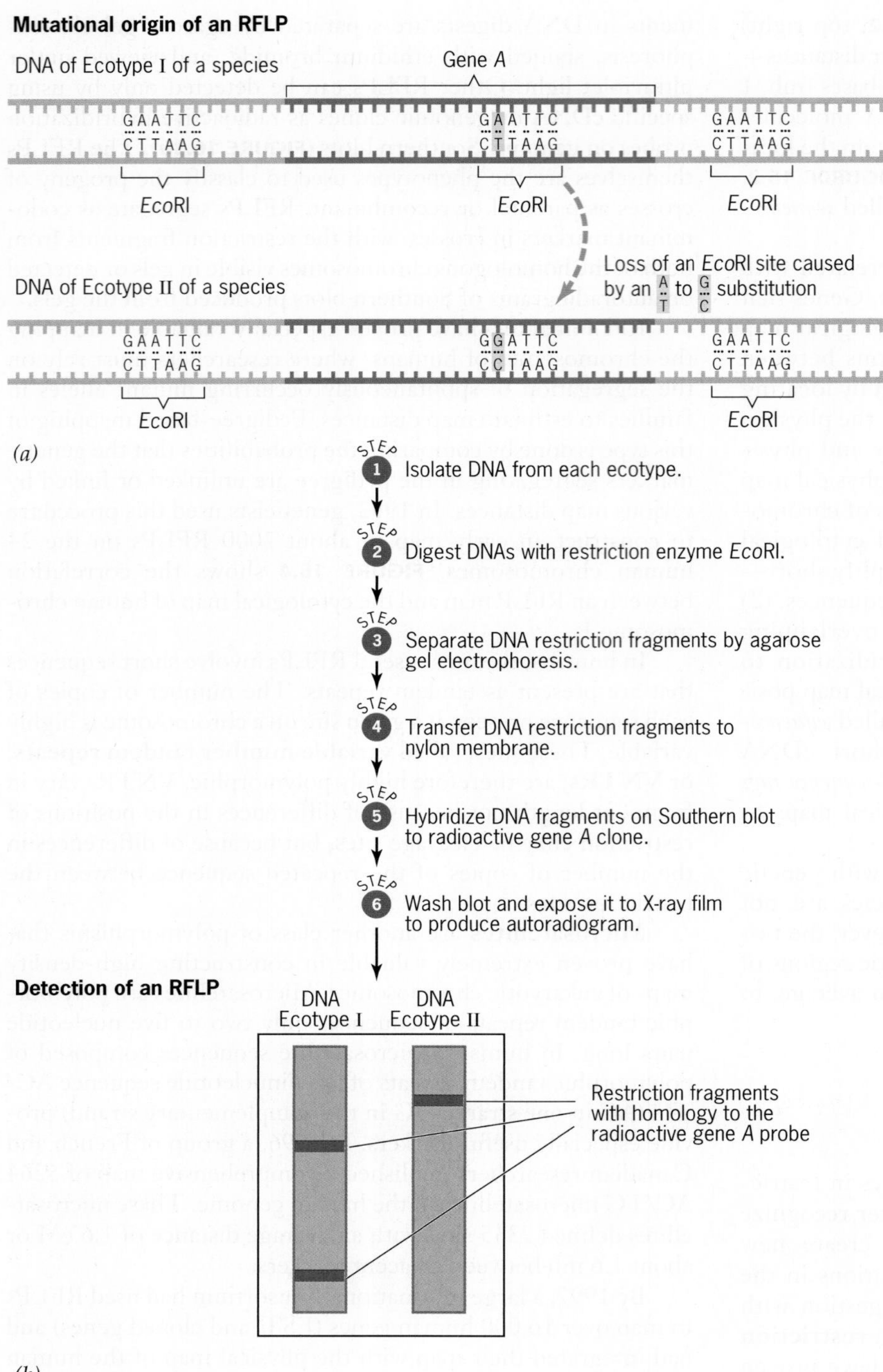
Mutational origin of an RFLP
DNA of Ecotype I of a species
Gene A
GAATTC
CTTAAG
EcoRI
EcoRI
EcoRI
Loss of an EcoRI site caused by an A:T to G:C substitution
DNA of Ecotype II of a species
GGATTC
CCTAAG
(a)
STEP 1 Isolate DNA from each ecotype.
STEP 2 Digest DNAs with restriction enzyme EcoRI.
STEP 3 Separate DNA restriction fragments by agarose gel electrophoresis.
STEP 4 Transfer DNA restriction fragments to nylon membrane.
STEP 5 Hybridize DNA fragments on Southern blot to radioactive gene A clone.
STEP 6 Wash blot and expose it to X-ray film to produce autoradiogram.
Detection of an RFLP
DNA Ecotype I
DNA Ecotype II
Restriction fragments with homology to the radioactive gene A probe
(b)

Figure 16.3 ▶ The mutational origin (*a*) and detection (*b*) of RFLPs in different ecotypes of a species. In the example shown, an A:T → G:C base-pair substitution results in the loss of the central *Eco*RI recognition sequence present in gene *A* of the DNA of ecotype I. This mutation might have occurred in an ecotype II ancestor during the early stages of its divergence from ecotype I.

(Chapter 9). For example, in *Drosophila*, the banding patterns of the giant polytene chromosomes in the salivary glands (Chapter 6) provide high-resolution maps of the chromosomes (see **FIGURE 6.11**). Thus, a clone of unknown genetic content can be positioned on the cytological map with considerable precision. In mammals, including humans, fluorescent *in situ* hybridization (FISH; see Focus on *in situ* Hybridization in Chapter 9) can be used to position clones on chromosomes stained by any of several chromosome-banding protocols. **FIGURE 1** in Focus on *in situ* Hybridization in Chapter 9 illustrates how FISH can be used to determine the chromosomal location of specific DNA sequences on human chromosomes. If RFLPs can be

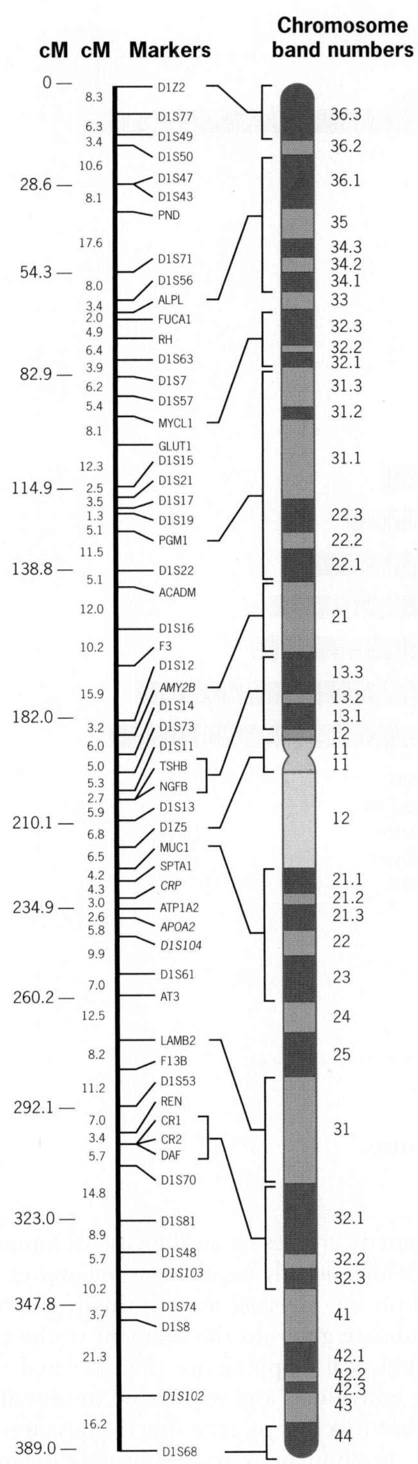

Figure 16.4 ► Correlation of the RFLP map (left) and the cytological map (right) of human chromosome 1. Molecular markers and a few genes are shown in the center. Distances are in centiMorgans (cM), with the uppermost marker set at position 0 on the left and distances between adjacent markers shown in the second column from the left. The brackets on the left of the cytological map show the chromosomal locations of the indicated genes and molecular markers.

identified that overlap with these sequences, they can be used as STS sites that anchor genetic maps of chromosomes to cytological maps, producing cytogenetic maps. If the sequences can be positioned on the physical maps by Southern blot hybridization experiments (Chapter 15), they can also be used to tie the physical maps to both the genetic and cytological maps of the chromosomes.

PHYSICAL MAPS AND CLONE BANKS

The RFLP mapping procedure has been used to construct detailed genetic maps of chromosomes, which, in turn, have made positional cloning feasible. These genetic maps have been supplemented with physical maps of chromosomes. By isolating and preparing restriction maps of large numbers of genomic clones, overlapping clones can be identified and used to construct physical maps of chromosomes and even entire genomes. In principle, this procedure is simple (**FIGURE 16.5**). However, in practice, it is a formidable task, especially for large genomes. The restriction maps of large genomic clones in YAC, PAC, or BAC vectors (Chapter 15) are analyzed by computer and organized in overlapping sets of clones called **contigs.** As more data are added, adjacent contigs are joined; when the physical map of a genome is complete, each chromosome is represented by a single contig map.

The construction of physical maps of entire genomes requires that vast amounts of data be searched for overlaps. Nevertheless, detailed physical maps are available for several genomes, including the human, *C. elegans*, *D. melanogaster*, and *A. thaliana* genomes. These physical maps are being used to prepare clone banks that contain catalogued clones collectively spanning entire chromosomes. Thus, if a researcher needs a clone of a particular gene or segment of a chromosome, that clone may already have been catalogued in the clone bank, or clone library, and be available on request. Obviously, the availability of such clone banks and the correlated physical maps of entire genomes are dramatically accelerating genetic research. Indeed, searching for a specific gene with and without the aid of a physical map is like searching for a book in a huge library with and without a computer catalog giving the locations of the books in the library.

KEY POINTS

- Genetic maps of chromosomes are based on recombination frequencies between markers.
- Cytogenetic maps are based on the location of markers within, or near, cytological features of chromosomes observed by microscopy.
- Physical maps of chromosomes are based on distances in base pairs, kilobase pairs, or megabase pairs separating markers.
- High-density maps that integrate the genetic, cytological, and physical maps of chromosomes have been constructed for many chromosomes, including all of the human chromosomes.

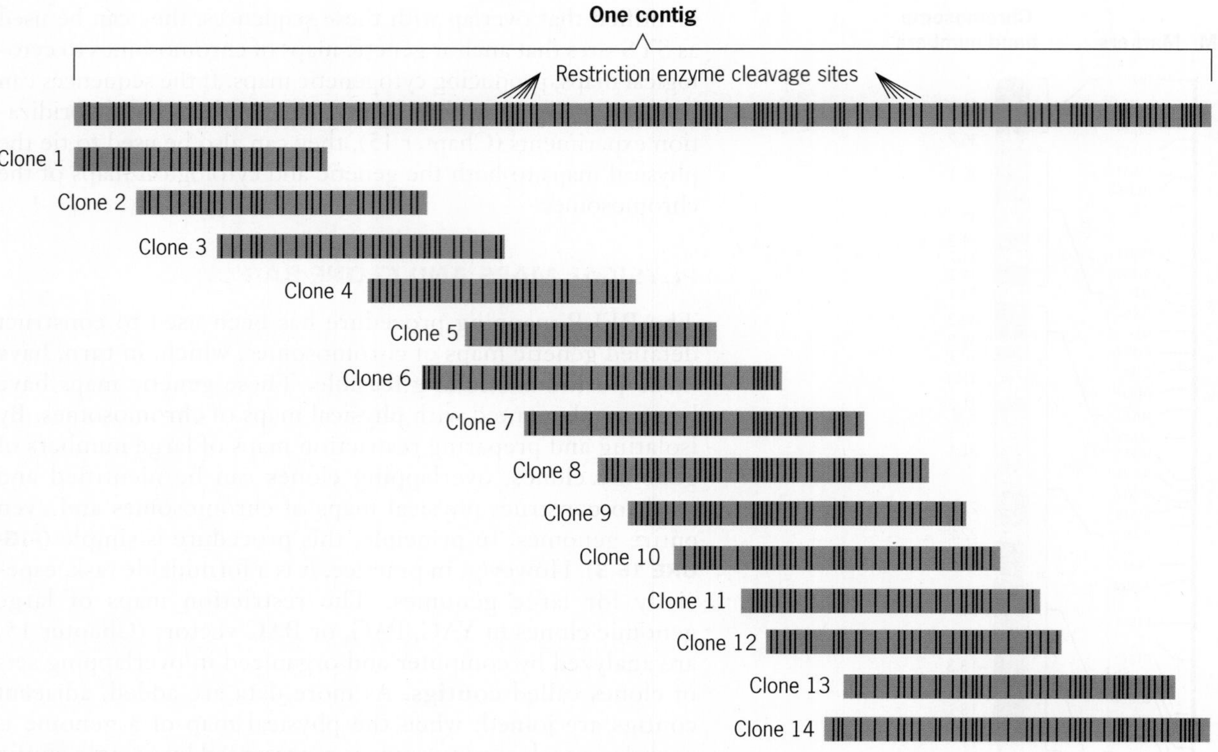

Figure 16.5 ▶ A contig map produced from overlapping genomic clones. Large—200 to 500 kb—genomic clones, such as those present in YAC, PAC, or BAC vectors (Chapter 15), are used to construct contig maps. Restriction maps of individual clones are prepared and searched by computer for overlaps. Overlapping clones are then organized into contig maps like the one shown here. When the physical map of a genome is complete, each chromosome will be represented by a single contig map.

Map Position-Based Cloning of Genes

Detailed genetic, cytogenetic, and physical maps of chromosomes allow scientists to isolate genes by chromosome walks and chromosome jumps.

The first eukaryotic genes to be cloned were genes that are expressed at very high levels in specialized tissues or cells. For example, about 90 percent of the protein synthesized in mammalian reticulocytes is hemoglobin. Thus, α- and β-globin mRNAs could be easily isolated from reticulocytes and used to prepare radioactive cDNA probes for genomic library screens. However, most genes are not expressed at such high levels in specialized cells. How, then, are genes that are expressed at moderate or low levels cloned? One important approach is to map the gene precisely and to search for a clone of the gene by using procedures that depend on its location in the genome. This approach, called **positional cloning,** can be used to identify any gene, given an adequate map of the region of the chromosome in which it is located.

The steps in positional cloning are illustrated in **FIGURE 16.6.** The gene is first mapped to a specific region of a given chromosome by genetic crosses or, in the case of humans, by pedigree analysis, which usually requires large families. The gene is next localized on the physical map of this region of the chromosome. Candidate genes in the segment of the chromosome identified by physical mapping are then isolated from mutant and wild-type individuals and sequenced to identify mutations that would result in a loss of gene function. As we shall discuss in Chapter 17, the human genes responsible for inherited disorders such as Huntington's disease and cystic fibrosis have been identified by using the positional cloning approach. In species where transformation is possible, copies of the wild-type alleles of candidate genes are introduced into mutant organisms to determine whether the wild-type genes will restore the wild phenotype. Restoration of the wild phenotype to a mutant organism provides strong evidence that the introduced wild-type gene is the gene of interest.

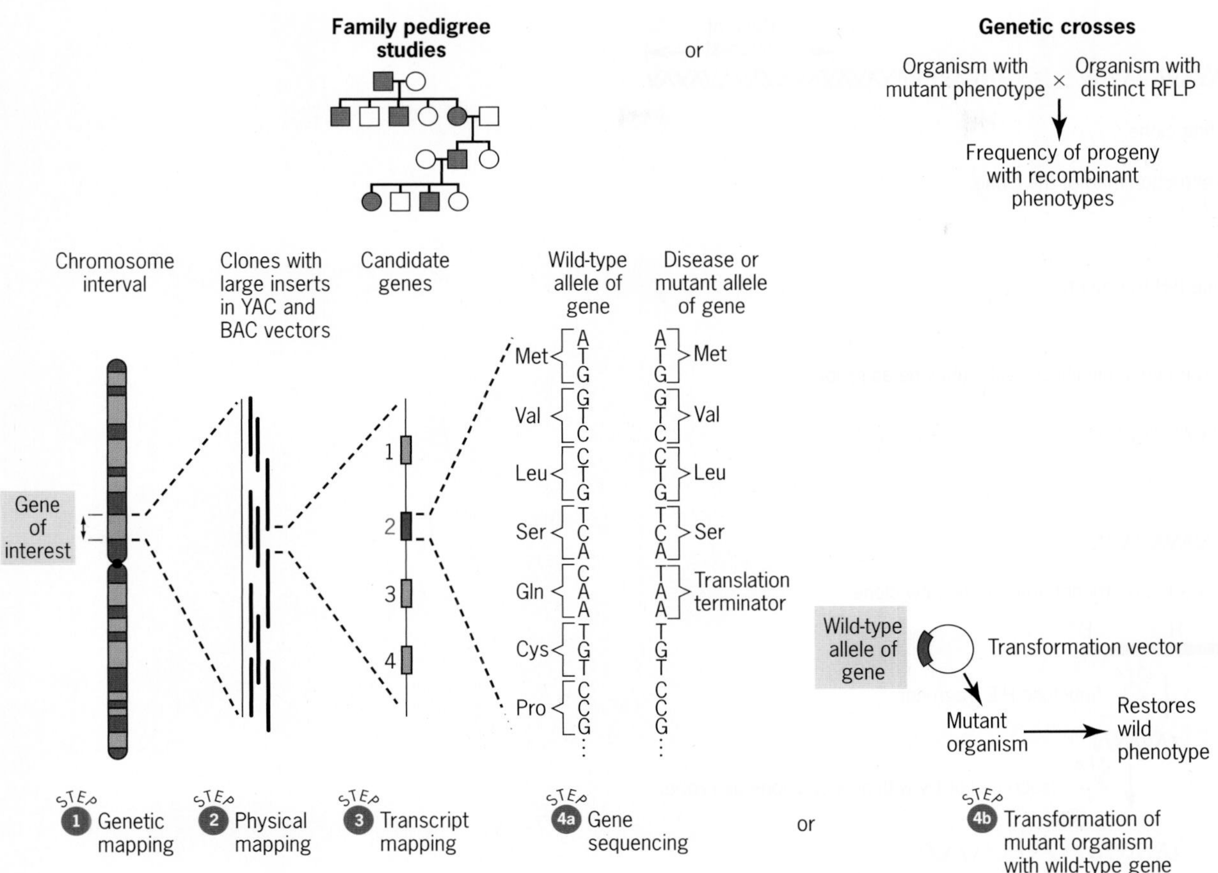

Figure 16.6 ▶ Steps involved in the positional cloning of genes. In humans, genetic mapping must be done by pedigree analysis, and candidate genes must be screened by sequencing wild-type and mutant alleles (step 4a). In other species, the gene of interest is mapped by appropriate genetic crosses, and the candidate genes are screened by transforming the wild-type alleles into mutant organisms and determining whether or not they restore the wild-type phenotype (step 4b).

CHROMOSOME WALKS

Positional cloning is accomplished by mapping the gene of interest, identifying an RFLP or other molecular marker near the gene, and then "walking" or "jumping" along the chromosome until the gene is reached.

Chromosome walks are initiated by the selection of a molecular marker (RFLP or known gene clone) close to the gene of interest and the use of this clone as a hybridization probe to screen a genomic library for overlapping sequences. Restriction maps are constructed for the overlapping clones identified in the library screen, and the restriction fragment farthest from the original probe is used to screen a second genomic library constructed by using a different restriction enzyme or to rescreen a library prepared from a partial digest of genomic DNA. Repeating this procedure several times and isolating a series of overlapping genomic clones allow a researcher to walk along the chromosome to the gene of interest (**FIGURE 16.7**). Without information about the orientation of the starting clone on the linkage map, the initial walk will have to proceed in both directions until another RFLP is identified and it is determined whether the new RFLP is closer to or farther away from the gene of interest than is the starting RFLP.

Verification that a clone of the gene of interest has been isolated is accomplished in various ways. In experimental organisms such as *Drosophila* and *Arabidopsis*, verification is achieved by introducing the wild-type allele of the gene into a mutant organism and showing that it restores the wild-type phenotype. In humans, verification usually involves determining the nucleotide sequences of the wild-type gene and several mutant alleles and showing that the coding sequences of the mutant genes are defective and unable to produce functional gene products.

Chromosome walking is very difficult in species with large genomes (the walk is usually too far) and an abundance of dispersed repetitive DNA (each repeated sequence is a potential roadblock). Chromosome walking is easier in organisms such as *A. thaliana* and *C. elegans*, which have small genomes and little dispersed-repetitive DNA.

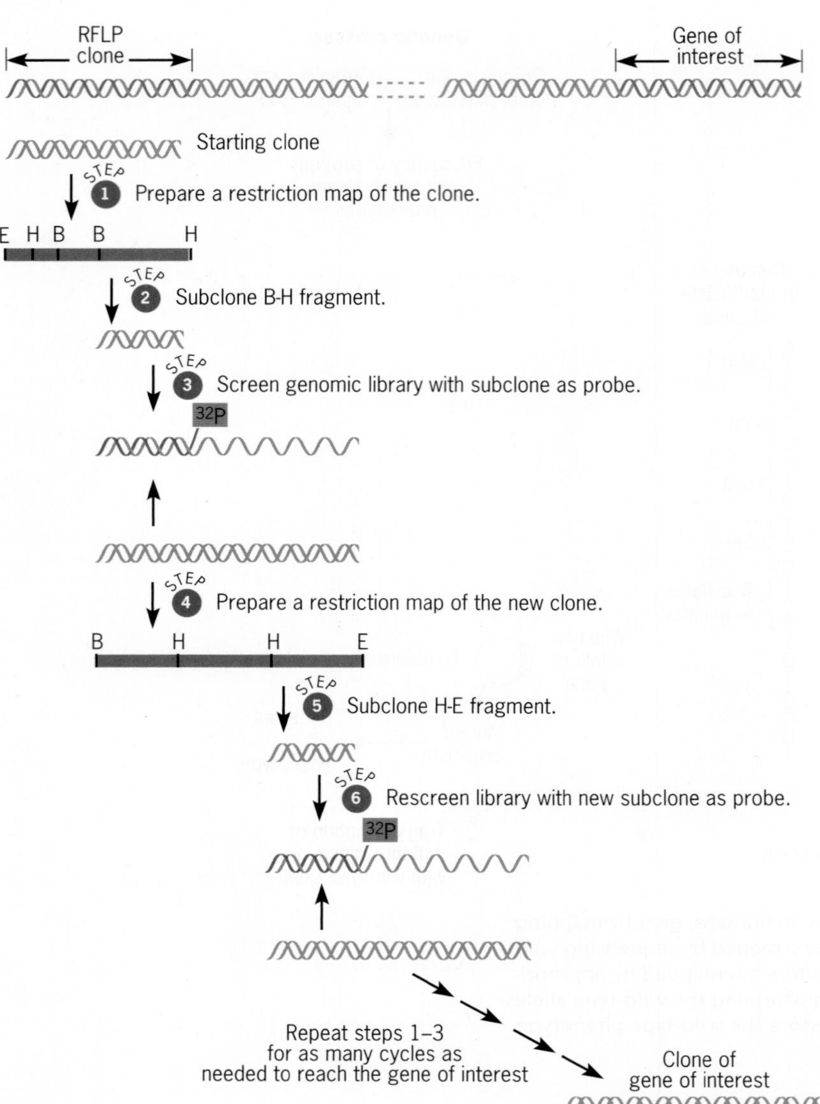

Figure 16.7 ▶ Positional cloning of a gene by chromosome walking. A chromosome walk starts with the identification of a molecular marker (such as the RFLP shown at top) close to the gene of interest and proceeds by repeating steps 1 through 3 as many times as is required to reach the gene of interest (bottom).

CHROMOSOME JUMPS

When the distance from the closest molecular marker to the gene of interest is large, a technique called **chromosome jumping** can be used to speed up an otherwise long walk. Each jump can cover a distance of 100 kb or more. The chromosome jumping procedure is illustrated in **FIGURE 16.8.** Like a walk, a jump is initiated by using a molecular probe such as an RFLP as a starting point. However, with chromosome jumps, large DNA fragments are prepared by partial digestion of genomic DNA with a restriction endonuclease. The large genomic fragments are then circularized with DNA ligase. A second restriction endonuclease is used to excise the junction fragment from the circular molecule. This junction fragment will contain both ends of the long fragment; it can be identified by hybridizing the DNA fragments on Southern blots to the initial molecular probe. A restriction map of the junction fragment is prepared, and a restriction fragment that corresponds to the distal end of the long genomic fragment is cloned and used to initiate a chromosome walk or a second chromosome jump. Chromosome jumping has proven especially useful in work with large genomes such as the human genome. Chromosome jumps played a key role in identifying the human cystic fibrosis gene (Chapter 17).

KEY POINTS

- Detailed genetic, cytogenetic, and physical maps of chromosomes permit researchers to isolate genes based on their location in the genome.
- If a molecular marker such as a restriction fragment-length polymorphism (RFLP) maps close to a gene, the gene can usually be isolated by chromosome walks or chromosome jumps.

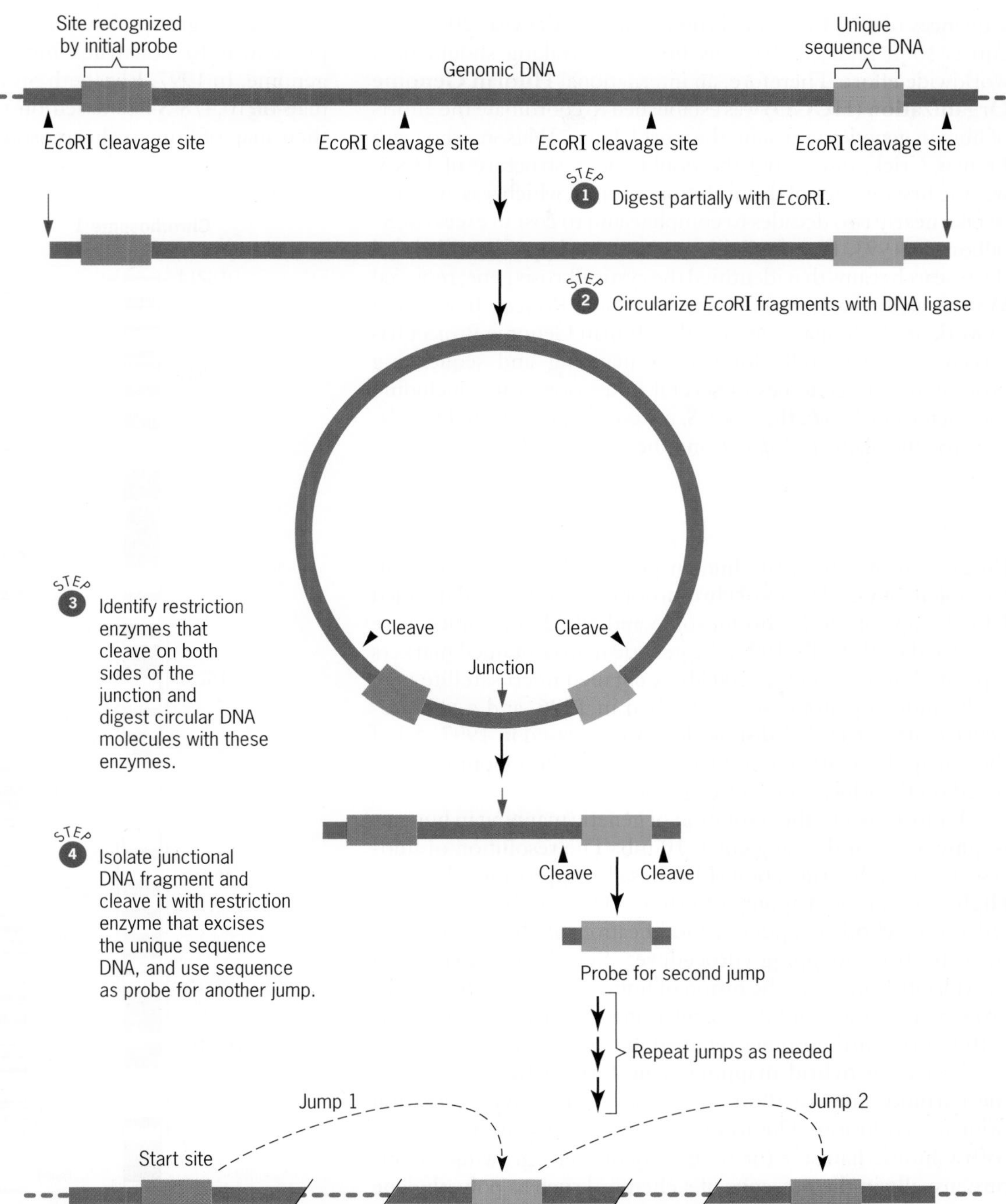
Site recognized by initial probe
Genomic DNA
Unique sequence DNA
EcoRI cleavage site
EcoRI cleavage site
EcoRI cleavage site
EcoRI cleavage site
STEP 1 Digest partially with EcoRI.
STEP 2 Circularize EcoRI fragments with DNA ligase
STEP 3 Identify restriction enzymes that cleave on both sides of the junction and digest circular DNA molecules with these enzymes.
Cleave
Cleave
Junction
STEP 4 Isolate junctional DNA fragment and cleave it with restriction enzyme that excises the unique sequence DNA, and use sequence as probe for another jump.
Cleave
Cleave
Probe for second jump
Repeat jumps as needed
Jump 1
Jump 2
Start site

Figure 16.8 ▶ Chromosome jumping as a shortcut method for replacing long chromosome walks. This procedure can also be used to jump over repetitive DNA sequences that block chromosome walks.

The Human Genome Project

Detailed genetic, cytogenetic, and physical maps are available for all 24 human chromosomes, and complete, or nearly complete, nucleotide sequences are available for the genomes of many species, including *Homo sapiens*.

As the recombinant DNA, gene cloning, and DNA sequencing technologies improved in the 1970s and early 1980s, scientists began discussing the possibility of sequencing all 3×10^9 nucleotide pairs in the human genome. These discussions led to the launching of the **Human Genome Project** in 1990. The initial goals of the Human Genome Project were (1) to map all of the human genes, (2) to construct a detailed physical map of the entire human genome, and (3) to determine the nucleotide sequences of all 24 human chromosomes by the year 2005. Scientists soon realized that this huge undertaking should be a worldwide effort. Therefore, an international **Human Genome Organization (HUGO)** was established to coordinate the efforts of human geneticists around the world. James Watson, who, with Francis Crick, discovered the double-helix structure of DNA, was the first director of this ambitious project, which was expected to take nearly two decades to complete and to cost in excess of $3 billion. In 1993, Francis Collins, who, with Lap-Chee Tsui, led the research teams that identified the cystic fibrosis gene, replaced Watson as director of the Human Genome Project. In addition to work on the human genome, the Human Genome Project has served as an umbrella for similar mapping and sequencing projects on the genomes of several other organisms, including the bacterium *E. coli*, the yeast *S. cerevisiae*, the fruit fly *D. melanogaster*, the plant *A. thaliana*, and the worm *C. elegans*.

MAPPING THE HUMAN GENOME

Progress in mapping the human genome has been excellent. Complete physical maps of chromosomes Y and 21 and detailed RFLP maps of the X chromosome and all 22 autosomes were published in 1992. By 1995, the genetic map contained markers separated by, on average, 200 kb. A detailed microsatellite map of the human genome was published in 1996, and a comprehensive map of 16,354 distinct loci was released in 1997. All of these maps have proven invaluable to researchers cloning genes based on their locations in the genome.

Unfortunately, the resolution of genetic mapping in humans is quite low—in the range of 1–10 mb. The resolution of fluorescent *in situ* hybridization (FISH) is also approximately 1 mb. Higher resolution mapping (down to 50 kb) can be achieved by radiation hybrid mapping, a modification of the somatic-cell hybridization mapping procedure. Standard somatic-cell hybridization involves the fusion of human cells and rodent cells growing in culture and the correlation of human gene products with human chromosomes retained in the hybrid cells.

Radiation hybrid mapping is performed by fragmenting the chromosomes of the human cells with heavy irradiation prior to cell fusion. The irradiated human cells are then fused with Chinese hamster (or other rodent) cells growing in culture, usually in the presence of a chemical such as polyethylene glycol to increase the efficiency of cell fusion. The human–Chinese hamster somatic-cell hybrids are then identified by growth in an appropriate selection medium.

Many of the human chromosome fragments become integrated into the Chinese hamster chromosomes during this process and are transmitted to progeny cells just like the normal genes in the Chinese hamster chromosomes. The polymerase chain reaction (PCR; Chapter 15) is then used to screen a large panel of the selected hybrid cells for the presence of human genetic markers. Chromosome maps are constructed based on the assumption that the probability of an X-ray-induced break between two markers is directly proportional to the distances separating them in chromosomal DNA.

Several groups have used the radiation hybrid mapping procedure to construct high-density maps of the human genome. In 1997, Elizabeth Stewart and coworkers published a map of 10,478 STSs based on radiation hybrid mapping data; their map of human chromosome 1 is shown in **FIGURE 16.9.**

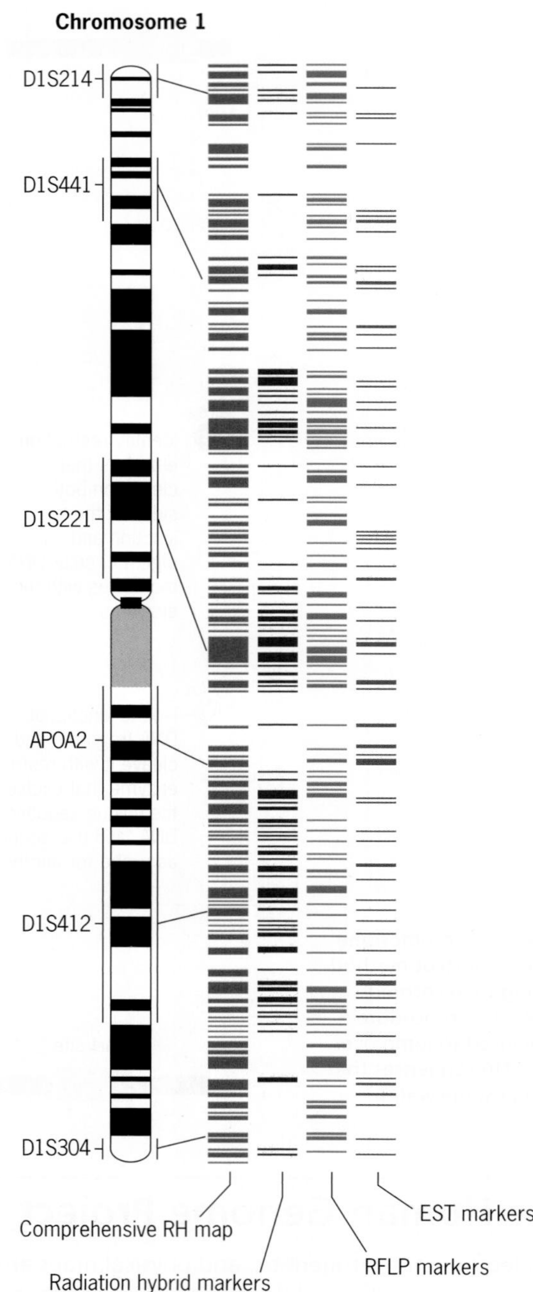

Figure 16.9 ▶ A high-resolution map of human chromosome 1. The cytogenetic map of chromosome 1 is shown on the left, along with the locations of six anchor markers. To the right of the cytogenetic map are four genetic maps that show the locations of the comprehensive radiation hybrid markers (red lines), the high-confidence radiation hybrid markers (blue lines), the RFLP markers (green lines), and the ESTs (purple lines).

SEQUENCING THE HUMAN GENOME

Whereas the gene-mapping work advanced quickly, progress toward sequencing the human genome initially lagged behind schedule. However, that all changed rapidly beginning in 1998. During May of 1998, J. Craig Venter announced that he had formed a private company, Celera Genomics, with the goal of sequencing the human genome in just three years. (For details, see A Milestone in Genetics: Two Drafts of the Sequence of the Human Genome.) Shortly thereafter, the leaders of the public Human Genome Project's sequencing laboratories announced that they had revised their schedule and planned to complete the sequence of the human genome by 2003—two years earlier than originally proposed. From this point in time, everything accelerated.

The complete sequence of the first human chromosome—small chromosome 22—was published in December 1999. The complete sequence of human chromosome 21 followed in May 2000. Then, with the intervention of the White House, Venter, of Celera Genomics, and Francis Collins, director of the public Human Genome Project, agreed to publish first drafts of the sequence of the human genome at the same time. The Celera and public sequences were both published in February 2001. **FIGURE 16.10** shows an annotated, sequence-based map of a

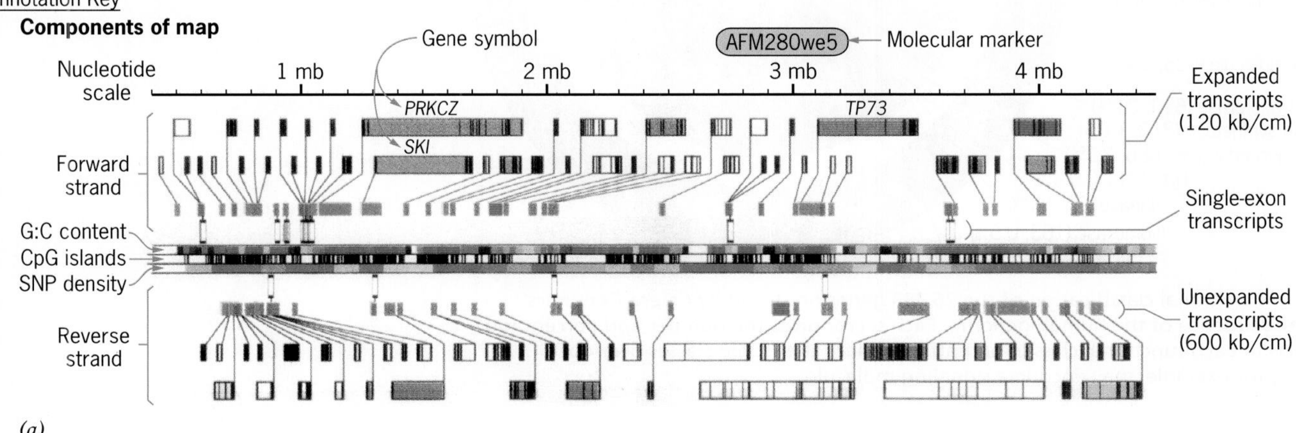

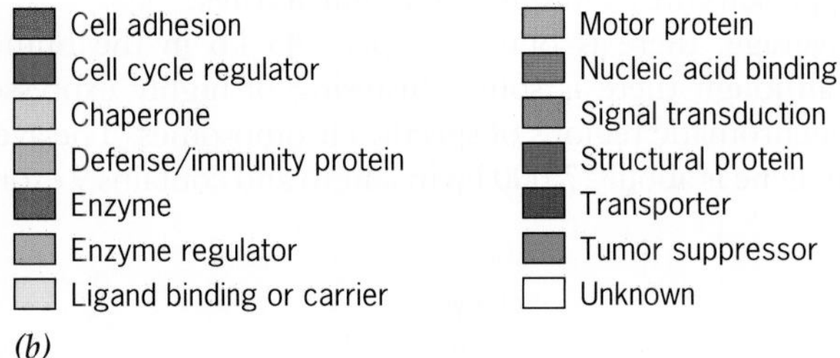

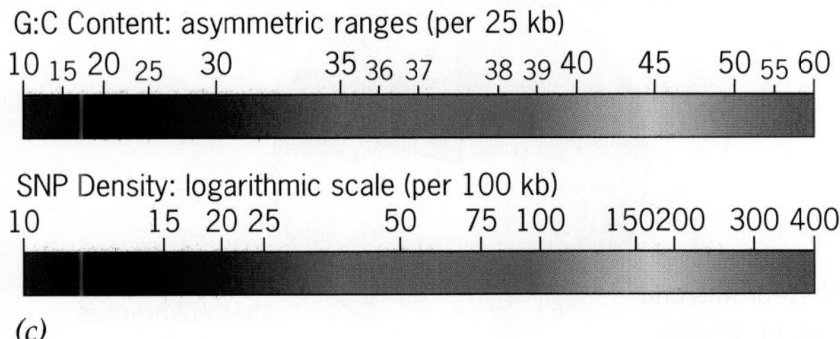

Figure 16.10 ▶ Annotated, sequence-based map of a 4-mb segment of DNA at the tip of human chromosome 1, assembled by researchers at Celera Genomics. (*a*) The top line gives distances in mb. The next three panels show predicted transcripts from one strand of DNA (the "forward strand"), whereas the bottom three panels show transcripts specified by the other strand of DNA (the "reverse strand"). The middle three panels give the G:C content, the positions of CpG islands, which occur upstream of genes, and the density of single nucleotide polymorphisms (SNPs), respectively. (*b*) The color code for gene-product functions, and (*c*) the color codes for G:C content and SNP density.

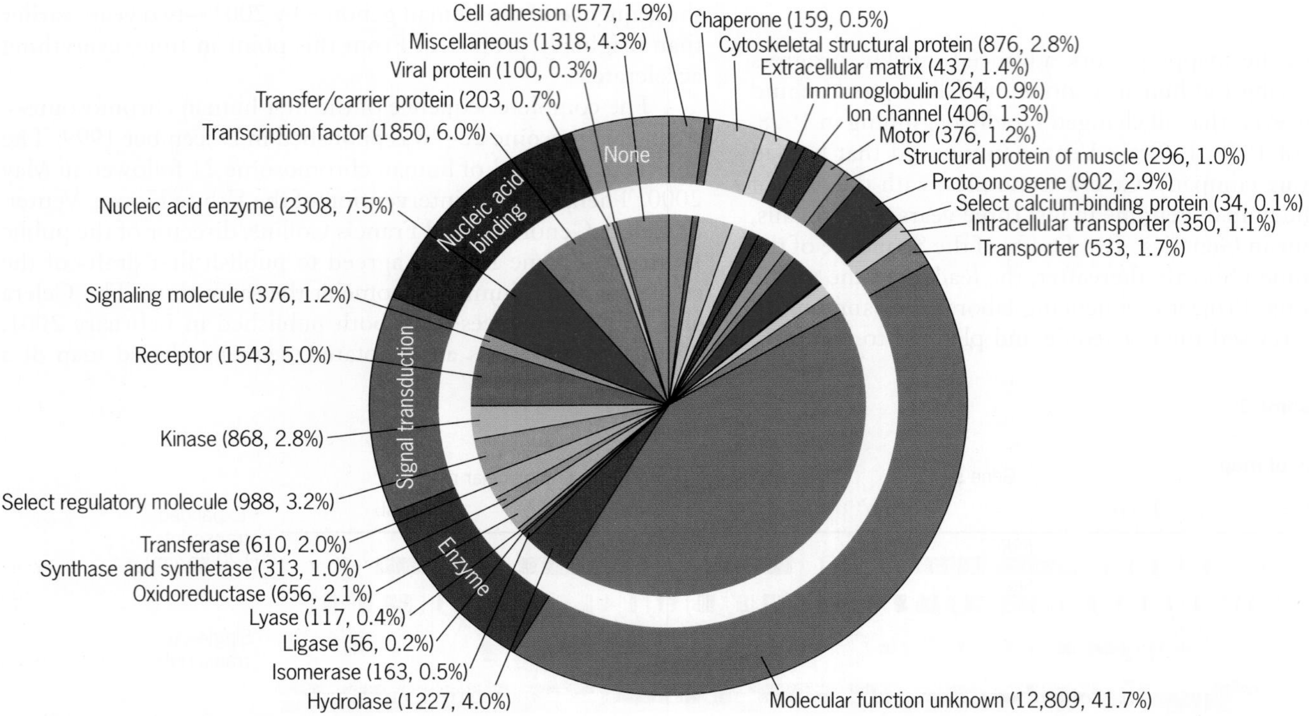

Figure 16.11 ▶ Functional classification of the 26,383 genes predicted by Celera Genomics' first draft of the sequence of the human genome. Each sector gives the number and percentage of gene products in each functional class in parentheses. Note that some classes overlap: a proto-oncogene, for example, may encode a signaling molecule.

4-mb segment at the tip of the short arm of human chromosome 1. This map illustrates the positions and orientations of known and predicted genes in one small portion of the human genome. For similar maps of the entire human genome, see the February 15, 2001, issue of *Nature* and the February 16, 2001, issue of *Science*.

The amount of information in these first drafts of the human genome was overwhelming, including the sequence of over 2650 megabase pairs of DNA (over 2,650,000,000 bp). The human genome is more than 25 times the size of the previously sequenced *Drosophila* and *Arabidopsis* genomes, and it is eight times the sum of all previously sequenced genomes.

The sequence of the human genome provided one surprise: there appeared to be only about 30,000 to 35,000 genes rather than the estimated 50,000 to 120,000 genes suggested by earlier studies (see Focus on the Human Genome: How Many Genes? in Chapter 14). The distribution of functions for the 26,383 genes predicted by the Celera sequence is shown in **FIGURE 16.11.** About 60 percent of the predicted proteins have similarities with proteins of other species whose genomes have been sequenced (**FIGURE 16.12**). Over 40 percent of the predicted human proteins share similarities with *Drosophila* and *C. elegans* proteins. The picture is quite different for families of closely related proteins, which tend to perform important basic cellular functions. Only 94 of 1278 protein families predicted by the sequence of the human genome are specific to vertebrates. The rest have evolved from domains of proteins in distant ancestors, including prokaryotes and unicellular eukaryotes.

On average, there is one gene per 145 kb in the human genome, although there is some clustering of highly expressed genes in euchromatic regions of specific chromosomes. The average human gene is about 27,000 bp in length and contains 9 exons.

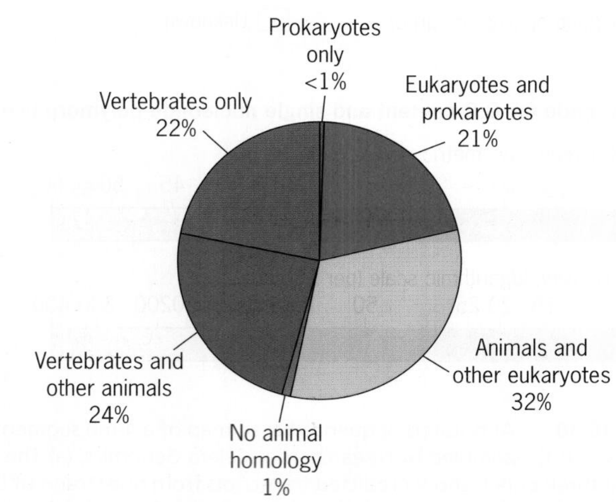

Figure 16.12 ▶ Pie chart showing homology of predicted human proteins to proteins of other species for those where homologues were detected by computer searches of the public databases.

Exons make up only 1.1 percent of the genome, whereas introns make up 24 percent, with 75 percent of the genome being intergenic DNA. Of the intergenic DNA, at least 44 percent is derived from transposable genetic elements (see Chapter 18 for details).

The two first drafts of the sequence of the human genome were incomplete, containing over 100,000 gaps. Therefore, the International Human Genome Sequencing Consortium continued to work on filling in these gaps and completing the sequence. By October 2004, they had reduced the number of gaps to 341 and had completed the sequence of 99 percent of the euchromatic DNA in the human genome. Surprisingly, the estimated number of genes in the genome had decreased again—to just 22,287 protein-coding genes—in the more complete sequence. There are, of course, other genes that specify RNA products—rRNAs, tRNAs, snRNAs, and miRNAs—that will increase the total number of genes significantly.

With the development of new sequencing technology that enables large genomes like the human genome to be sequenced quickly and at much lower costs, it has become feasible to sequence many individual human genomes. Indeed, two individuals—James D. Watson and J. Craig Venter—have had their genomes sequenced. Moreover, Jeffrey M. Kidd and coworkers recently mapped and sequenced structural variation in eight human genomes of diverse origin. The individuals selected for study were of African, Asian, and European ancestry. The researchers focused on changes in the genomes in the range of 1 kb to 1 mb and documented a large amount of structural diversity—especially deletions, inversions, and insertions. They discovered 525 new insertion sequences that are not present in Celera's or HUGO's human genome sequences. The results of the study by Kidd and colleagues indicate that a large amount of genetic variation occurs among genomes. Just how much variation will become clear as the sequences of more individual genomes become available. Would you like to know the sequence of your genome? How much would you be willing to pay to have your genome sequenced?

The wealth of information provided by the nearly complete sequence of the human genome is just beginning to be exploited. Given that only about 1.1 percent of the genome encodes amino acid sequences in polypeptides, the big question is what are the functions of the rest of the components of the human genome. Francis Collins and other leaders of the Sequencing Consortium are already focusing on this question. They have organized a new consortium, ENCODE (*ENC*yclopedia *O*f *D*NA *E*lements), whose goal is to identify all of the nongenic functional elements in the human genome. These elements will include regulatory sequences such as promoters, enhancers, silencers, sites of methylation, and acetylation, and other factors involved in the control of chromatin structure and gene expression (**FIGURE 16.13**). The identification and functional characterization of these elements may prove to be a larger challenge than the sequencing of the human genome.

Another international consortium—the *Hu*man *P*roteome *O*rganization (HUPO)—has been formed with the goal of determining the structures and functions of all the proteins encoded by the human genome. Despite the wealth of data provided by the sequence of the human genome, the functional dissection of the genome is just beginning. We still have a long way to go before we will really understand the structure and function of the 3×10^9 nucleotide pairs in the human genome.

The availability of the sequence of the human genome will raise a whole new set of questions about the proper use of this new knowledge. Many of these questions focus on an individual's right to privacy. For example, if a mutation that causes a late-onset disorder such as Huntington's disease (Chapter 17) is discovered in a family, who should have access to this information? If such information were available to the public, widespread discrimination might occur. Employers might not hire members of the family, and medical schools might not admit talented young scholars to their M.D. programs (see A Milestone in Genetics: Mapping the Gene for Huntington's Disease in Chapter 7). Will insurance companies provide health and life insurance to someone who carries a mutant gene that increases the risk of cancer or leads to a late-onset disorder like Huntington's disease? If they do, will the insurance be affordable, or will it be priced beyond the reach of all but the wealthy? Given the recent increase in the amount of genetic information available, it seems clear that laws protecting the privacy of this information will be needed in the future. Indeed, in the United States, the Genetic Information Nondiscrimination Act (GINA) has just been passed (May 2008) and is poised to become law. It will protect individuals from discrimination by employers and insurance companies based on genetics and DNA tests.

THE HUMAN HAPMAP PROJECT

Human genomes contain a large amount of genetic variation. We discussed gross changes—deletions, duplications, inversions, and translocations—in the structures of genomes in Chapter 6. In the preceding section, we discussed changes of intermediate size—deletions, insertions, and inversions in the range of 1 kb to 1 mb—in eight human genomes. Small changes—insertions or deletions of one or a few nucleotide pairs—are even more frequent. The most common changes in human genomes are single nucleotide-pair substitutions, for example, A:T to G:C or G:C to A:T substitutions (Chapter 13). Base-pair substitutions of this type have produced a large number of ***s*ingle-*n*ucleotide *p*olymorphism*s*** (**SNPs,** pronounced "snips") in human genomes. Most of these SNPs are not located in the coding regions of genes and do not result in mutant phenotypes. When the nucleotide sequences of the same chromosomes of two individuals are compared, one SNP is present, on average, in every 1200 nucleotide pairs.

SNPs can be detected in human genomes by the microarray hybridization or "gene-chip" technology described in a later subsection of this chapter (see Array Hybridizations and Gene Chips). In brief, hybridization probes can be synthesized that can detect single-nucleotide differences in DNA molecules. If a DNA molecule matches a probe exactly, it will bind

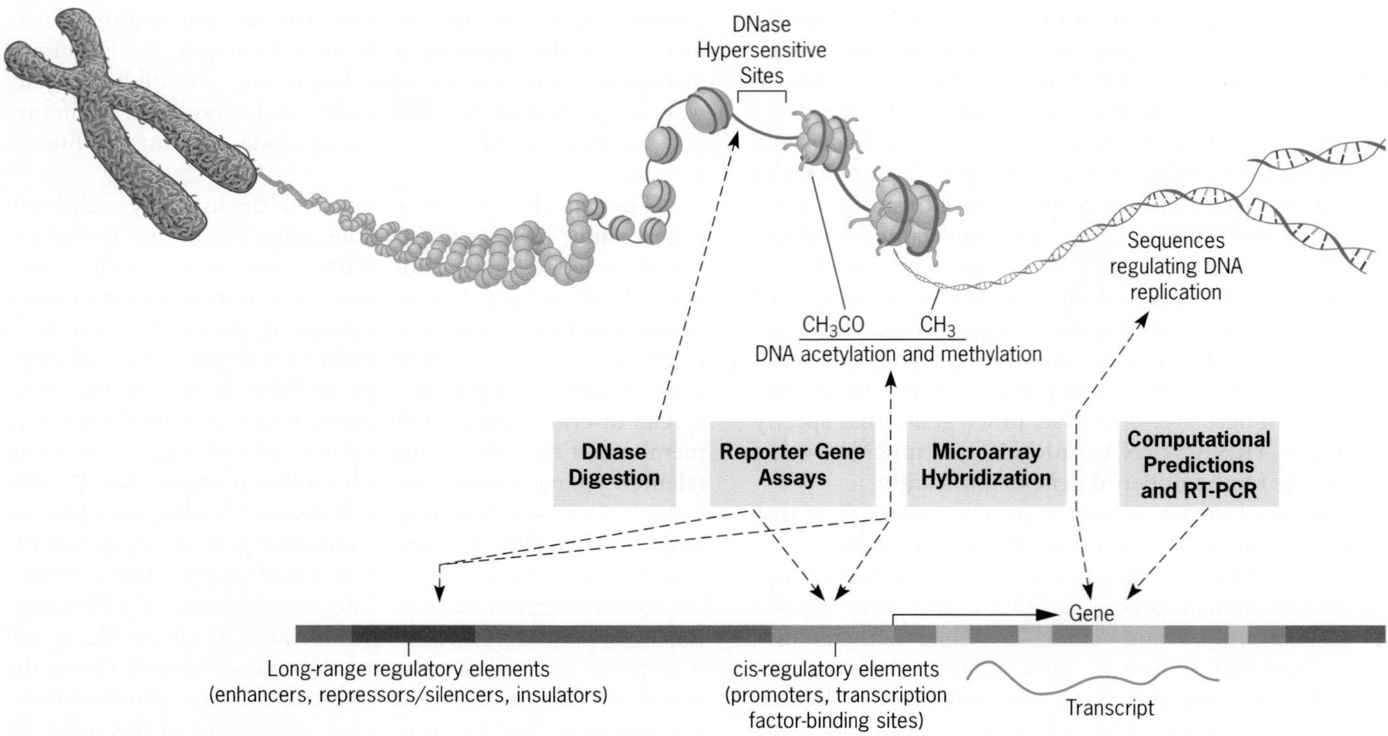

Figure 16.13 ▶ The goal of the ENCODE (*ENC*yclopedia *O*f *DNA* *E*lements) Project Consortium is to identify the nongenic functional elements in the human genome. The elements will include regulatory sequences such as promoters, enhancers, silencers, repressor-binding sites, transcription factor-binding sites, and sites of chemical modifications such as acetylation and methylation. They will also include sequences that alter chromatin structure by interacting with DNA-binding proteins and the histones that package DNA into nucleosomes. Some of these elements will alter chromatin structure producing DNase hypersensitive sites (characteristic of chromatin that is transcriptionally active—see Chapter 20). Tools to be used in these studies will include reporter gene assays and microarray hybridizations (discussed in subsequent sections of this chapter) and reverse-transcript PCR (RT-PCR), polymerase chain reactions using RNAs as templates to identify transcribed regions of the genome.

to that probe; if it does not match exactly, it will not bind. Thus, if a segment of DNA from one individual has an A:T base pair at a specific position, and the corresponding segment of DNA from another individual has a G:C base pair at this position, it is possible to distinguish these two individuals genetically by hybridizing their DNA to probes that will bind to one or the other of the two DNA segments. These and thousands of other diagnostic probes can be arrayed systematically on a silicon wafer (see **FIGURE 16.17**) to screen for single-nucleotide differences in genomic DNA collected from a sample of individuals. Usually the DNA from each individual is amplified by PCR using primers that flank genomic regions of interest, and the amplified DNA is labeled in some way before hybridizing it with the diagnostic array of probes. In a study conducted at Perlegen Sciences, Inc., researchers used this microarray technology to determine the genotypes of 71 people at more than 1.5 million sites in the human genome—a stupendous accomplishment! By studying these polymorphisms in different subpopulations, it may be possible to trace important genetic events in the evolutionary history of our species and to predict a person's susceptibility to diseases like cancer and heart disease.

Individual SNPs may be present in one human population and absent from another. When present, they may vary in frequency from one population to another. Most SNPs present in human populations were produced by a single mutation in one individual that subsequently spread through the population. Each SNP is associated with other SNPs that were present on the ancestral chromosome at the time that the mutation generating the SNP occurred. SNPs that are closely linked tend to be passed on to progeny as a unit because there is little chance for a crossover to recombine them. The SNPs on a chromosome or a segment of a chromosome that tend to be inherited together define a genetic unit called a **haplotype** (**FIGURE 16.14**). Of course, mutation will modify haplotypes, and crossing over will generate new haplotypes during the course of evolution.

Because of their frequency and distribution throughout the human genome, SNPs have proven to be valuable genetic markers. The study of haplotypes defined by SNPs is providing important information about the relationships among different ethnic groups and about human evolution (see Chapter 25). The study of SNPs and haplotypes is also helping researchers identify genes that are involved in susceptibility to diseases such

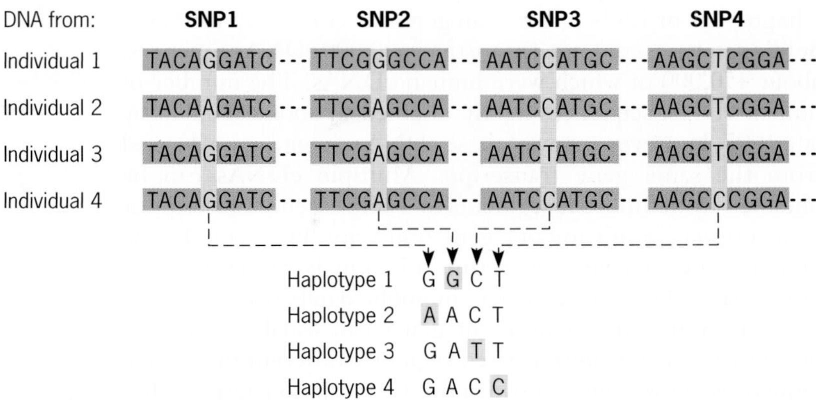

Figure 16.14 ► Haplotypes are sets of linked SNPs and other genetic markers that tend to be inherited as a unit.

as breast cancer, glaucoma, amyotrophic lateral sclerosis (ALS, also known as Lou Gehrig's disease), and rheumatoid arthritis. The strategy in these studies has been to determine the SNP genotypes of large samples of people and then search for associations between the SNPs (or the haplotypes defined by linked SNPs) and particular diseases. Once an association has been found, the SNP or haplotype can be used to help predict the risk that an individual will develop the disease, and in favorable cases, it may help to identify the actual disease-causing gene.

Because of the value of SNP haplotypes in studying ancestry and evolution in human populations and in finding disease associations, researchers from around the world have initiated the International HapMap Project. The goal of this collaborative enterprise is to to identify and map SNPs using DNA samples from many different human populations. The data collected by the Project are being made available as a resource for all genomic researchers.

KEY POINTS

- Researchers collaborating on the Human Genome Project have constructed detailed maps of all 24 human chromosomes.
- Other participants of the Human Genome Project have determined the complete, or nearly complete, nucleotide sequences of the genomes of several important model organisms.
- A nearly complete sequence of the euchromatic DNA in the human genome was released in October 2004.
- Scientists throughout the world have initiated the International Human HapMap Project with the goal of characterizing the similarities and differences in human genomes worldwide.

RNA and Protein Assays of Genome Function

The availability of the nucleotide sequences of entire genomes has led to the development of microarray, gene-chip, and reporter gene technologies that permit researchers to study the expression of all the genes of an organism simultaneously.

Knowing the complete sequence of the human genome will help identify genes responsible for human diseases and should lead to successful gene therapies for some of these diseases. However, it will not tell us what these genes do or how they control biological processes. Indeed, by itself, the nucleotide sequence of a gene, a chromosome, or an entire genome is uninformative. Only when supplemented with information about their functions do sequences become truly meaningful. Thus, information about the functions of nucleotide sequences must still be obtained by traditional genetic studies and by molecular analyses. If geneticists want to understand the genetic control of the growth and development of a mature human from a single fertilized egg (Chapter 21), they will need to know much more than the sequence of the human genome. But the availability of the ultimate map of the human genome, its nucleotide sequence, will certainly accelerate progress toward understanding the programs of gene expression that control morphogenesis. Indeed, the development of new technologies such as "gene-chip" hybridizations is designed to take advantage of the availability of the sequences of complete genomes (see the later subsection, Array Hybridizations and Gene Chips).

EXPRESSED SEQUENCES

In large eukaryotic genomes, only a small proportion of the DNA encodes proteins. In the yeast *S. cerevisiae*, almost 70 percent of the genome encodes proteins, and there is one gene for every 2 kb of sequence. In humans, only about 1 percent of the genome encodes amino acid sequences, and there is one gene for every 130 kb of sequence. Thus, in order to focus on the protein-coding content of genomes, many scientists have analyzed cDNA clones (DNAs complementary to RNA molecules;

Chapter 15) or ESTs rather than genomic clones. By 1996, the public databases contained more than 600,000 cDNA sequences, about 450,000 of which were human cDNAs. The number of human cDNA sequences nearly doubled to about 800,000 by late 1997. However, many of these cDNA sequences are derived from the same gene transcripts. Multiple cDNAs can be obtained from different segments of a single gene transcript or from alternative splicing of a gene transcript. For example, the human gene that encodes serum albumin is represented by more than 1300 EST sequences in public databases.

The transcripts of different genes can usually be recognized by distinct 3′ untranslated regions—different nucleotide sequences in the region between the 3′ translation-termination codon and the 3′ terminus of the transcript. When sequence comparisons were made between the regions of cDNAs corresponding to the 3′ untranslated regions of the transcripts, the cDNAs were grouped into 49,625 clusters (with 97 percent sequence identity within clusters). Prior to the publication of the two drafts of the human genome, the number of clusters was considered a good estimate of the number of distinct genes. Of the sequence clusters, 4563 corresponded to known human genes. One problem with estimating gene numbers from EST sequences is that the ESTs may be derived from nonoverlapping regions of a gene transcript. In any case, it now seems clear that human gene number estimates based on EST databases gave gene numbers that were too high.

ARRAY HYBRIDIZATIONS AND GENE CHIPS

Given the sequence of an entire genome, geneticists can immediately begin to study the expression of every gene in the organism. Oligonucleotide hybridization probes can be synthesized that are complementary to segments of the transcripts of every gene, or PCR can be used to make millions of copies of each gene in a genome. Thus, scientists can monitor changes in total genome expression over time, throughout development, or in response to changes in the environment. Such knowledge should prove invaluable in understanding human diseases such as cancer and, perhaps, even the aging process.

In Chapter 15, we discussed the use of northern blot hybridizations—RNAs separated by gel electrophoresis and transferred to membranes for hybridization to radioactive probes—to study gene expression. To investigate the expression of entire genomes or large sets of genes, scientists basically reverse the northern blot procedure and perform **dot blot hybridizations.** In dot blot hybridizations, the gene-specific nucleotide sequences or probes are applied and bound to membranes in specific patterns, or arrays (**FIGURE 16.15**), and these probe arrays (called dot blots) are hybridized to radioactive, or fluorescent, RNA or cDNA preparations (**FIGURE 16.16*a*** and ***b***). The amounts of RNA or cDNA hybridized to each probe on a dot blot, or hybridization array, can be measured by scanning the blot with an imaging system that measures the amount of radioactivity, or fluorescence, and analyzing the results with computer programs that compare the signals with those produced by known control probes and RNAs or cDNAs. **FIGURE 16.16*c*** and ***d*** show the results of scanning and computer analysis of the autoradiograms shown in **FIGURE 16.16*a*** and ***b***.

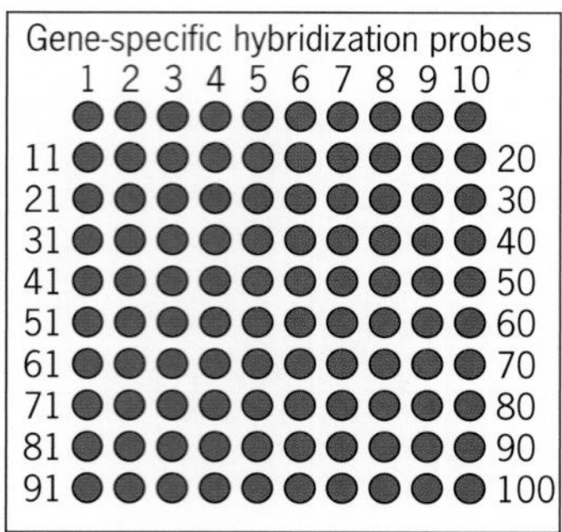

Nylon membrane containing an array of 100 probes

Figure 16.15 ▶ Dot blot or array hybridization analysis of gene expression. Gene-specific hybridization probes are applied to a membrane with the aid of a vacuum filtration apparatus. The membrane is then placed in hybridization solution containing RNAs or cDNAs labeled with a radioactive isotope or a fluorescent dye. After hybridization, the labeled RNA bound to each probe is either visualized by autoradiography (see **FIGURE 16.16**) or quantitated by densitometry or by scanners that measure the amount of radioactivity or fluorescence.

The hybridization arrays shown in **FIGURE 16.16** contain 588 human EST probes. However, new technologies now allow scientists to produce microarrays that contain thousands of hybridization probes on a single membrane or other solid support. The gene chips shown in **FIGURE 16.17** can array over 10,000 oligonucleotide probes on a silicon wafer of only a few square centimeters. Microarrays of oligonucleotide or cDNA probes are produced in several ways: (1) microsynthesis of oligonucleotides *in situ* (on the chip), (2) spotting prefabricated oligonucleotide probes on solid supports, and (3) spotting DNA fragments or cDNAs on the supports. The probes on the microarrays are then hybridized to labeled (usually with a fluorescent tag) RNA or cDNA samples. The amount of RNA or cDNA hybridized to each probe in a microarray is quantified by using sophisticated scanners with micrometer resolution and appropriate computer software.

The genome-sequencing projects and the microarray hybridization technologies have spawned the new subdiscipline of functional genomics, which focuses on the expression of entire genomes. However, some geneticists have argued that this has been the goal of the science of genetics since its inception. As knowledge in the field has advanced, geneticists have been able to study the expression of more and more genes. Now, for the first time, they can study the expression of all the genes of an organism simultaneously. Probe microarrays are available that allow geneticists to analyze the expression of the nearly 6000 genes of budding yeast, and DNA chips that permit scientists to study the expression of the approximately 14,000 genes of *Drosophila melanogaster*, the roughly 26,000 genes of *Arabidopsis thaliana*, and the approximately 23,000 human genes are now available to the research community. The ability to analyze the expression of entire genomes will enhance the current explosion of new information in biology and will eventually

Autoradiograms

Untreated cancer cells

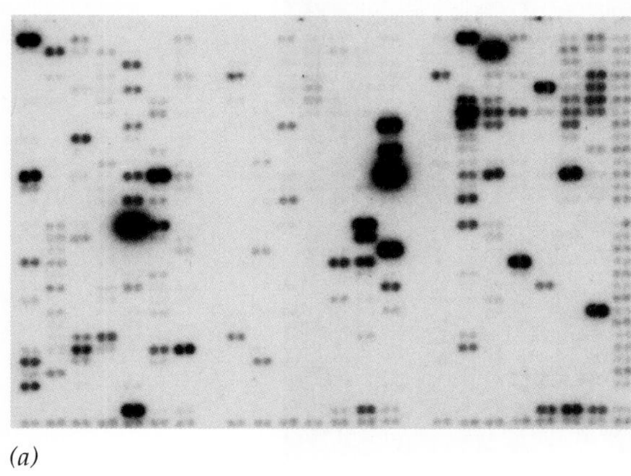

(*a*)

Cancer cells treated with chemotherapeutic agent

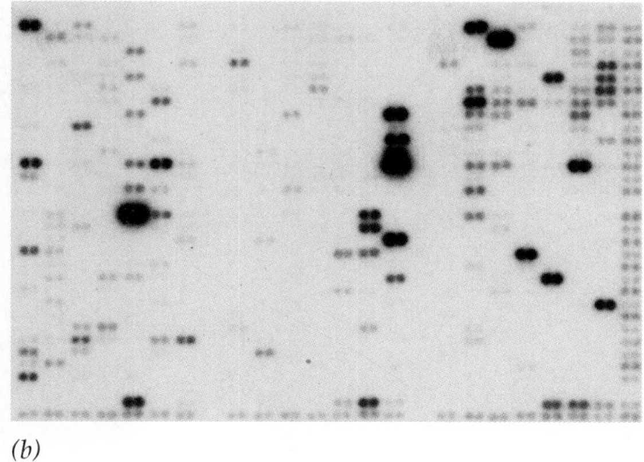

(*b*)

Computer analysis of autoradiogram in ***(a)***

(*c*)

Computer analysis of autoradiogram in ***(b)***

(*d*)

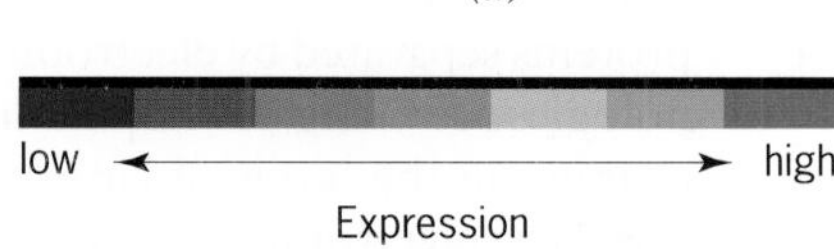

Figure 16.16 ▶ The use of dot blot, or array, hybridizations to investigate changes in levels of gene expression. The autoradiograms (top) show the results of hybridizations of arrays of 588 human ESTs to ^{32}P-labeled cDNAs prepared from poly(A) RNAs isolated from human cancer cells growing in culture, either (*a*) untreated or (*b*) treated with a chemotherapeutic agent. The photographs shown in (*c*) and (*d*) were produced by using a scanner to measure the intensities of the hybridization signals on the autoradiograms in (*a*) and (*b*), respectively, and converting them to visual images with appropriate computer software. Changes in levels of gene expression induced by the chemotherapeutic agent can be detected by comparing the two arrays.

lead to an understanding of the normal process of human development and the causes of at least some human diseases.

USE OF GREEN FLUORESCENT PROTEIN AS A REPORTER OF PROTEIN SYNTHESIS

Array hybridizations and gene chips can be used to determine whether genes are transcribed, but they provide no information about the translation of the gene transcripts. Thus, biologists often use antibodies to detect the protein products of genes of interest. Western blots are used to detect proteins separated by electrophoresis (Chapter 15), and antibodies coupled to fluorescent compounds are used to detect the location of proteins *in vivo*. However, both of these approaches provide only a single time-point assay of a protein in a cell, tissue, or organism.

The discovery of a naturally occurring fluorescent protein, the **green fluorescent protein (GFP)** of the jellyfish *Aequorea victoria*, has provided a powerful tool that can be used to study gene expression at the protein level. GFP is now being used to monitor the synthesis and localization of specific proteins in a wide variety of living cells. These studies entail constructing fusion genes that contain the nucleotide sequence

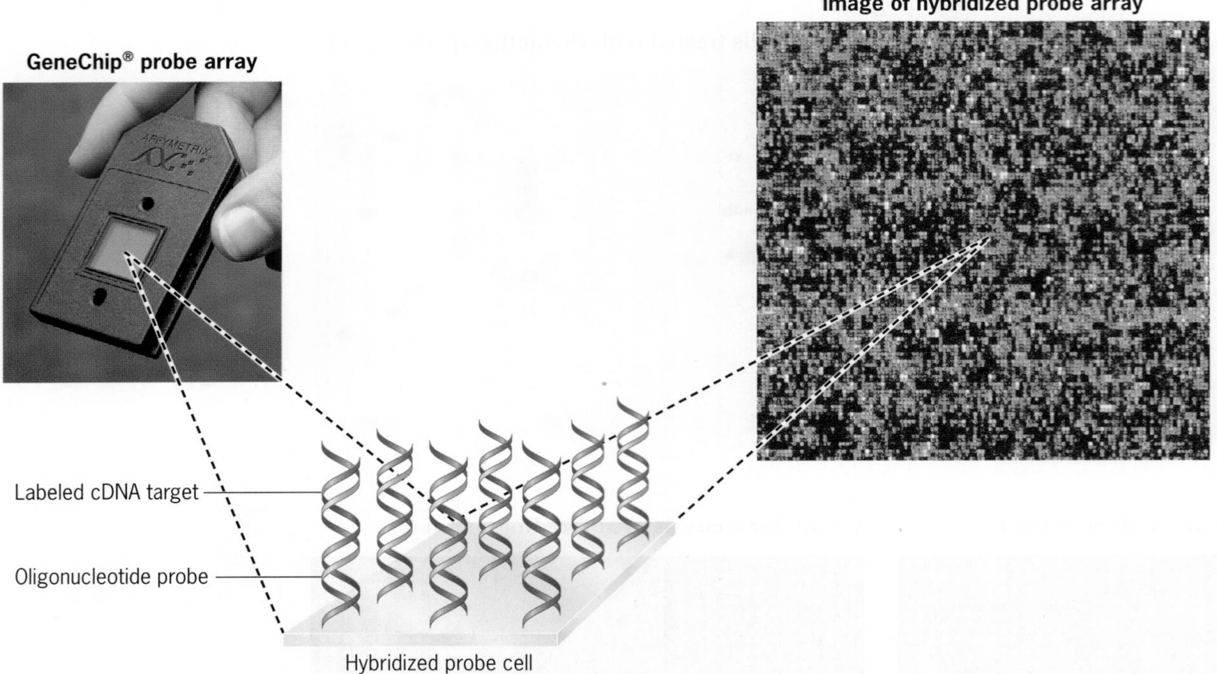

Figure 16.17 ▶ Photograph of a gene chip (top left) and an illustration of the use of the gene chip to analyze the expression of all the genes of an organism simultaneously. These gene chips contain thousands of oligonucleotide hybridization probes. Gene chips permit researchers to detect the transcripts of thousands of genes in one experiment.

encoding GFP, coupled in frame to the nucleotide sequence encoding the protein of interest; introducing the chimeric gene into cells by transformation; and studying the fluorescence of the fusion protein in transgenic cells exposed to blue or UV light (**FIGURE 16.18**). Because GFP is a small protein, it can often be coupled to proteins without interfering with their activity or interaction with other cellular components.

As the name implies, GFP fluoresces bright green when exposed to blue or ultraviolet light. The chromophore of GFP is produced by the post-translational cyclization and oxidation of an encoded serine/tyrosine/glycine tripeptide. This chromophore is largely protected from ion and solvent effects by encasement in a barrel-like fold of the mature protein. Unlike other bioluminescent proteins, GFP does not require the addition of substrates, cofactors, or any other substances to fluoresce—only exposure to blue or UV light. Thus, GFP can be used to study gene expression in living cells and to study protein localization and movement in cells over time. By mutagenizing the GFP gene, molecular biologists have produced variant forms of GFP that emit blue or yellow light, variants that fluoresce up to 35 times more intensely than the wild-type GFP, and variants whose fluorescence depends on the pH of the microenvironment. These GFP variants can be used to study the synthesis and intracellular localization of two or more proteins simultaneously (**FIGURE 16.18*d***).

Some geneticists are using GFP fusions to study changes in the expression of all the genes encoding proteins that are involved in a particular metabolic pathway in response to treatment of cells or tissues with a specific drug or potentially therapeutic agent. They construct an entire set of chimeric genes containing the GFP coding region fused in frame to the coding regions of other genes, introduce them into host cells, and monitor their expression by quantifying the fluorescence of fusion proteins separated by electrophoresis or other techniques. Technologies are being developed that will allow scientists to observe changes in the levels of large arrays of GFP fusion proteins by capillary electrophoresis (electrophoresis performed in small capillary tubes), monitored by sensitive microphotodetectors and sophisticated computer software. Indeed, in the not-too-distant future, functional genomics may involve the use of DNA chips to detect gene transcripts and "protein chips" to detect the polypeptides encoded by these transcripts.

KEY POINTS

- Once the complete nucleotide sequence of a genome has been determined, scientists can study the temporal and spatial patterns of expression of all the genes of the organism.
- Microarrays of gene-specific hybridization probes on gene chips allow researchers to study the transcription of thousands of genes simultaneously.
- Chimeric genes that contain the coding region of the green fluorescent protein of the jellyfish fused with the coding regions of the genes of experimental organisms can be used to study the localization of proteins in living cells.

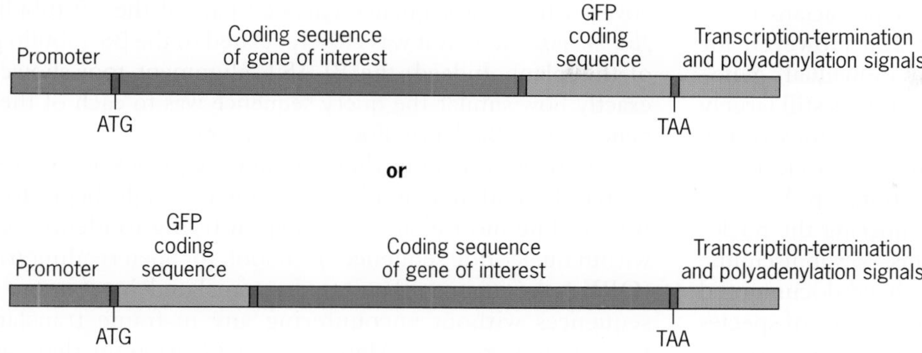

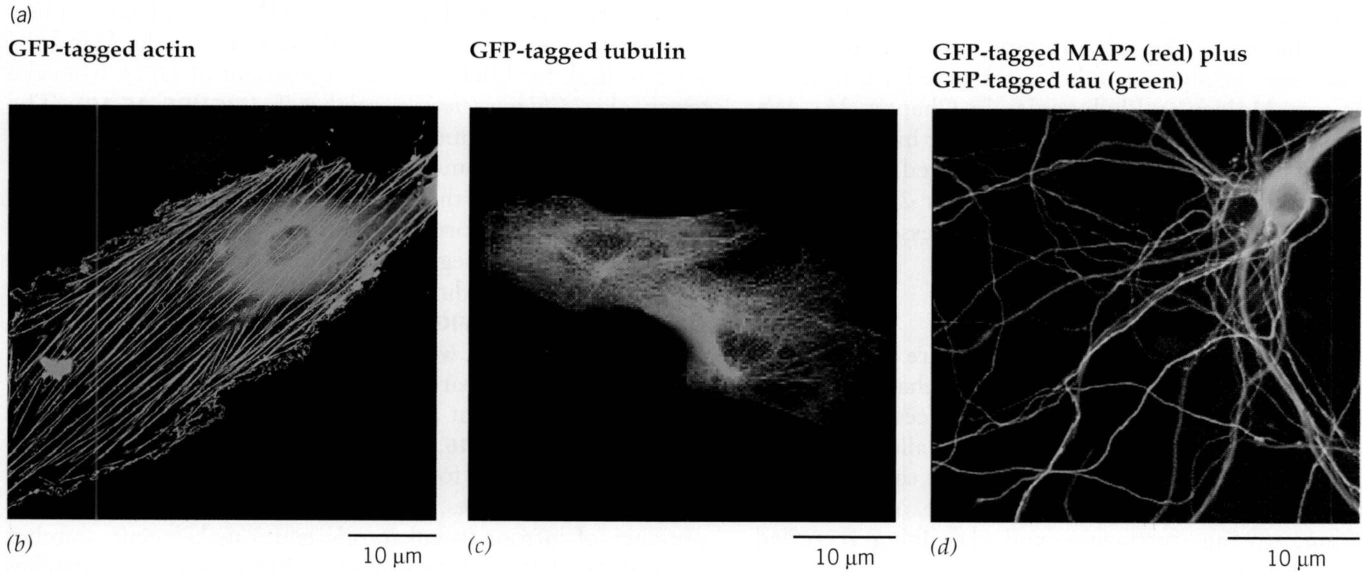

Figure 16.18 ► Use of the green fluorescent protein (GFP) of the jellyfish to study protein localization in living cells. (*a*) Structure of GFP fusion genes. The GFP coding sequence may be placed at either end of the gene of interest or at internal positions. (*b–d*) Immunofluorescence localization of GFP-tagged proteins: (*b*) smooth-muscle actin in a fibroblast cell; (*c*) the microtubule structural protein tubulin in Chinese hamster ovary cells; and (*d*) double labeling of two microtubule-binding proteins, MAP2 labeled with blue light-emitting GFP and tau labeled with green light-emitting GFP, in a rat neuron. With the light filters used for microscopy, MAP2 and tau appear red and green, respectively.

Comparative Genomics

Comparisons of the nucleotide sequences of the genomes of organisms have resulted in a better understanding of taxonomic relationships and of the changes responsible for the evolution of species from common ancestors.

Now that we know the complete nucleotide sequences of the genomes of over 2000 viruses, over 700 archaea and eubacteria, and 23 eukaryotes (plus another 208 currently being assembled)—over 89 billion nucleotide pairs of DNA in total—how can we use this information? By itself, the sequence of a DNA molecule is totally uninformative. In genetics, we use mutational dissection to determine the functions of various units of DNA. But how can we extract information from the vast amounts of sequence information in the databanks? In this section, we will briefly discuss a few of the tools that are used to "mine" information from DNA sequences—those currently available and the vast number of new sequences that are accumulating daily.

We have been promised that the sequence of the human genome will lead to new approaches in the practice of medicine and that treatments will be tailored to the genotype of each individual. Women with a mutation in either the *BRCA1*

(*BR*east *CA*ncer *1*) or the *BRCA2* gene, for example, will probably be monitored for breast cancer by their physicians more vigilantly than other women (see Chapter 22). However, for the most part, medical treatment based on an individual's genotype, other than for classical hereditary disorders, is still largely futuristic. So, what were the immediate contributions of the genomics era? The answer to this question seems clear—an enhanced knowledge of the evolutionary relationships between species and other taxonomic groups. By comparing the nucleotide sequences of the genomes of organisms—a subdiscipline called **comparative genomics**—scientists have documented many of the changes responsible for the divergence of species from common ancestors. Comparative genomics is a powerful new tool for studies of evolution. Evolutionary "trees"—phylogenies that show the relationships between species or other taxonomic groups—can be constructed from DNA sequences (see Molecular Phylogenies in Chapter 25). We will briefly examine some of the changes that have occurred in the genomes of the cereal grasses and selected mammals in the last two sections of this chapter. Chapter 25 presents a more detailed discussion of evolutionary processes.

BIOINFORMATICS

Knowledge of the nucleotide sequences of entire genomes has provided a wealth of information and a new challenge. How can we extract information from these sequences? This challenge has spawned a new scientific discipline called **bioinformatics** (**bio**logy + **informatics**). You all know that biology is the study of life. Informatics is the science of gathering, manipulating, storing, retrieving, and classifying recorded information. Bioinformatics involves doing all these things with biological information—most notably DNA and protein sequences. A comprehensive discussion of bioinformatics is beyond the scope of this textbook. Nevertheless, let's briefly examine some of the tools used to study the nucleotide sequences of DNA and the amino acid sequences of proteins.

Suppose that you have just sequenced a DNA restriction fragment isolated from your favorite organism. How would you begin to analyze the function of this DNA molecule? First, you might ask whether anyone had previously sequenced this DNA or similar DNA molecules. To answer this question, you need a computer program that can search large DNA databases for similar sequences. Software programs designed to search sequence databases were first developed in the 1980s, and today there are programs designed to do almost anything that you can imagine. Some of the more popular programs were developed by the *G*enetics *C*omputer *G*roup (GCG) at the University of Wisconsin. We will use a couple of their programs to illustrate how nucleotide sequences can be studied. Earlier in this chapter (see Focus on GenBank), we discussed the use of the Entrez web site (http://www.ncbi.nlm.nih.gov/entrez) to search for DNA sequences similar to a query sequence by using megaBLAST software. The other program that is used for rapidly searching huge databases is called FASTA. In the search that we performed in the Focus on GenBank exercise, we discovered that our sequence encoded part of the β9-tubulin of *Arabidopsis* and that it was closely related to the β8-tubulin gene of this plant. Indeed, the BLAST alignment tool showed us exactly how similar the query sequence was to each of the two genes in GenBank (see Focus **FIGURE 2c**).

Now, let's assume that our new sequence is not represented in GenBank, and let's ask how we might begin to analyze it. The most elementary step in trying to identify genes within nucleotide sequences is to look for *o*pen *r*eading *f*rames (ORFs)—sequences that can be translated into amino acid sequences without encountering any in-frame translation-termination codons. "Map" is a GCG program that can be used to translate double-stranded DNA in all six reading frames (three in each strand of DNA). Let's use the Map software to look for ORFs in a short segment of DNA from the green alga *Chlamydomonas reinhardti* (**FIGURE 16.19**). The standard three-letter amino acid abbreviations (see **TABLE 12.1**) are quite cumbersome when dealing with large protein databases; therefore, the single-letter code (see **FIGURE 12.1**) is preferred in bioinformatic analyses.

When the short segment of *Chlamydomonas* DNA is translated in all three reading frames, only reading frame 5 lacks termination codons (**FIGURE 16.19**). Thus, any ORF spanning this segment of DNA would have to be in reading frame 5. Of course, if we were searching for genes, we would usually be looking for ORFs that are much longer than the short DNA sequence in **FIGURE 16.19**. The software developed to search nucleotide sequences for genes can also screen for promoters, ribosome-binding sites, and other conserved sequences. The presence of introns in eukaryotic genes makes gene searches more difficult than in prokaryotes. Eukaryotic gene searches involve scanning for intron splice sites in addition to ORFs and other regulatory sequences. Today, the gene-search programs are tailored to individual species by factoring in features of their genomes—for instance, base composition, codon usage, and preferences for certain sequences in regulatory elements. The only way to be absolutely certain that a predicted gene is real and that its introns have been identified correctly is to isolate and sequence a full-length cDNA clone and compare its sequence with the sequence of a genomic clone. There are currently dozens of gene-prediction programs; they go by names like GRAIL (*G*ene *R*ecognition and *A*nalysis *I*nternet *L*ink), GeneMark (one of the first used to search for genes in *E. coli*), GeneScan, GeneFinder, and so on.

One common feature of eukaryotic genomes is the presence of gene families—sets of genes that encode very similar proteins (often called isoforms). These proteins usually have redundant or overlapping functions. In order to compare all of the genes in a gene family, bioinformaticians have developed programs that will align multiple nucleotide or amino acid sequences, allowing a direct visual comparison of all members of a gene or protein family. Multiple alignments are especially useful in identifying conserved DNA sequences that are important regulatory elements, such as protein-binding sites. They

DNA sequence

```
                  5' ggcgtgtattaaattgggtaactcttcattgcgtggtgttgtagatagatgagggatggc 3'
Nucleotide pair: 1   ---------+---------+---------+---------+---------+---------+  60
                  3' ccgcacataatttaacccattgagaagtaacgcaccacaacatctatctactccctaccg 5'
```

Translation products

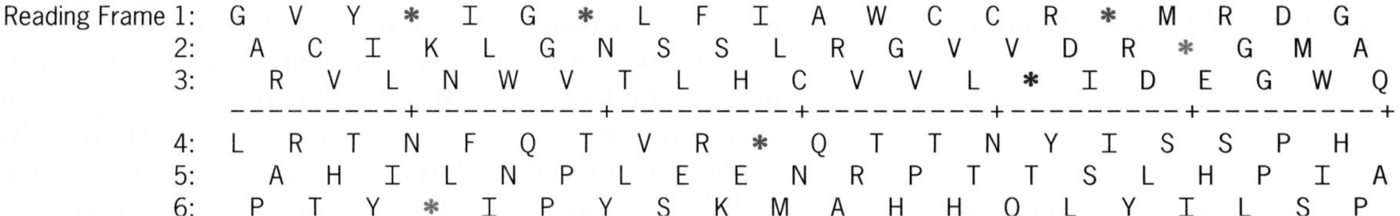

Reading Frame 1: G V Y * I G * L F I A W C C R * M R D G
2: A C I K L G N S S L R G V V D R * G M A
3: R V L N W V T L H C V V L * I D E G W Q
4: L R T N F Q T V R * Q T T N Y I S S P H
5: A H I L N P L E E N R P T T S L H P I A
6: P T Y * I P Y S K M A H H Q L Y I L S P

Figure 16.19 ▶ Illustration of the use of the Wisconsin GCG "Map" program to identify ORFs by translating a 60-base-pair segment of DNA from *C. reinhardtii* in all six reading frames. Note the presence of translation-termination codons in all reading frames except number 5. Thus, it is the only reading frame that could be part of a larger ORF. Recall that translation is always 5′ → 3′; thus, the amino-termini are on the left for translation products 1–3 and on the right for products 4–6. The translation termination sites are designated by asterisks. They correspond to the termination triplets shown underlined in red, green, or blue, depending on the reading frame.

are also useful in identifying important, and thus conserved, functional domains within proteins. **FIGURE 16.20** shows the alignment of the amino-terminal regions of the eight β-tubulins of *Arabidopsis thaliana*. Note that within the 60 or 61 amino acids shown there are five regions of four or more amino acids that are conserved in all eight proteins.

Genes with very similar nucleotide sequences, such as the nine genes encoding the β-tubulins shown in **FIGURE 16.20**, often—but not always—owe their similarity to having evolved from a common ancestral gene. Such genes are said to be **homologous;** note that "similar" and "homologous" are not synonyms. The nine β-tubulin genes of *Arabidopsis* are homologous;

```
Tubulin     Amino acid                                                                Amino acid
β6:         1  MREILHIQGGQCGNQIGsKFWEVVCdEHGIDpTGRYvG nsDLQLERVNVYYNEASCGRyV   60
β8:         1  MREILHIQGGQCGNQIGAKFWEVVCAEHGIDsTGRYqG enDLQLERVNVYYNEASCGRFV   60
β2 and β3:  1  MREILHIQGGQCGNQIGAKFWEVVCAEHGIDpTGRYtGD SDLQLERiNVYYNEASCGRFV   60
β7:         1  MREILHIQGGQCGNQIGSKFWEVVnIEHGIDqTGRYvGD SeLQLERvNVYYNEASCGRYV   60
β5:         1  MREILHIQGGQCGNQIGSKFWEVICDEHGIDsTGRYsGDtADLQLERINVYYNEASGGRYV   61
β1:         1  MREILHvQGGQCGNQIGSKFWEVICDEHGvDpTGRYnGDSADLQLERINVYYNEASGGRYV   61
β4:         1  MREILHIQGGQCGNQIGAKFWEVICDEHGIDbTGQYvGDS pLQLERIdVYFNEASGGKYV   60
β9:         1  MREILHIQGGQCGNQIGAKFWEVICgEHGIDqTGQscGD tDLQLERINVYFNEASGGKYV   60
Conserved seqs.  1      2       3                       4        5
```

Figure 16.20 ▶ Illustration of a multi-sequence protein alignment using the single-letter amino acid code (see **FIGURE 12.1**). The alignment—generated with the GCG "PileUp" program—compares the amino-terminal regions of the eight β-tubulins of *Arabidopsis*, which are encoded by nine genes. Genes *TUB2* (*TU*bulin *B*eta number *2*) and *TUB3* encode identical products (the β2- and β3-tubulins): these two genes differ only at positions corresponding to the degenerate third bases of codons. The most similar sequences are grouped together. The single-letter codes for amino acids are in upper case when adjacent sequences are identical and are in lower case when adjacent sequences are different.

that is, they are **homologues.** Indeed, two of the genes were produced by a recent (on the evolutionary timescale) gene-duplication event; they differ at only 30 base pairs (corresponding to degenerate bases in codons) and encode the same polypeptide. These genes are also called **paralogous** genes or **paralogues**—homologous genes within a species. Homologous genes present in different species are called **orthologous** genes or **orthologues.** The tubulin genes of *Arabidopsis* and *Chlamydomonas* are orthologues.

PROKARYOTIC GENOMES

In 1995, *Haemophilus influenzae* became the first cellular organism to have its genome sequenced in its entirety. By May 2008, the complete sequences of the genomes of 109 archaea and 687 bacteria were available in the public databases, and that number will be considerably larger by the time this book is in print. The genomes range in size from 490,885 bp for *Nanoarchaeum equitans*—an obligate symbiont—to 580,076 bp for *Mycoplasma genitalium*—thought to have the smallest genome of any nonsymbiotic bacterium; to 4,403,837 bp for *Mycobacterium tuberculosis*—the cause of more human deaths than any other infectious bacterium; to 4,639,675 bp for *Escherichia coli* strain K12—the best-known cellular microorganism; to 9,105,828 bp for *Bradyrhizobium japonicum*—a soil bacterium capable of colonizing plant root nodules. The size and predicted gene content of selected prokaryotic genomes are shown in **TABLE 16.1.**

Before discussing bacterial genomes in more detail, we need to emphasize that the sizes of genomes are highly variable even within species, varying from one strain to another. Indeed, studies of different isolates of *E. coli*, *Prochlorococcus marinus*, and *Streptococcus coelicolor* have documented variations in genome size of up to a million nucleotide pairs between different strains of the same species.

The genomes of the archaea are of particular interest because of the ability of these organisms to survive in extremely harsh environments. Their genomes should assist us in determining how they have adapted to these extreme conditions.

TABLE 16.1

Size and Gene Content of Selected Prokaryotic Genomes

Species	Genome Size in Nucleotide Pairs	Predicted Number of Genes
Archaea		
Archaeoglobus fulgidus	2,178,400	2,486
Methanosarcina acetivorans	5,751,492	4,721
Nanoarchaeum equitans	490,885	582
Pyrococcus furiosus	1,908,256	2,228
Sulfolobus solfataricus	2,992,245	3,033
Thermoplasma volcanium	1,584,804	1,548
Eubacteria		
Bacillus subtilis	4,214,630	4,225
Bordetella parapertussis	4,773,551	4,467
Bradyrhizobium japonicum	9,105,828	8,373
Buchnera aphidicola	615,980	550
Chlamydia pneumoniae, strain AR39	1,229,853	1,167
Escherichia coli, strain K12 MG1655	4,639,675	4,467
Escherichia coli, strain O157 EDL933	5,528,970	5,463
Haemophilus influenzae Rd KW20	1,830,138	1,789
Legionella pneumophila, strain Paris	3,503,610	3,136
Mycobacterium tuberculosis, strain CDC	4,403,837	4,293
Mycobacterium genitalium	580,076	525
Neisseria meningitidis Z2491	2,184,406	2,208
Pseudomonas syringae strain DC3000	6,397,126	5,660
Rickettsia typhi	1,111,496	919
Salmonella typhimurium	4,857,432	4,622
Staphylococcus aureus, strain MW2	2,820,462	2,712
Streptomyces coelicolor	8,667,507	7,912
Ureaplasma parum ATCC 700970	751,719	653
Yersinia pestis, strain KIM	4,600,755	4,240

Data are from the NCBI web site (http://www.ncbi.nim.nih.gov/Genomes/ as of May 2008.

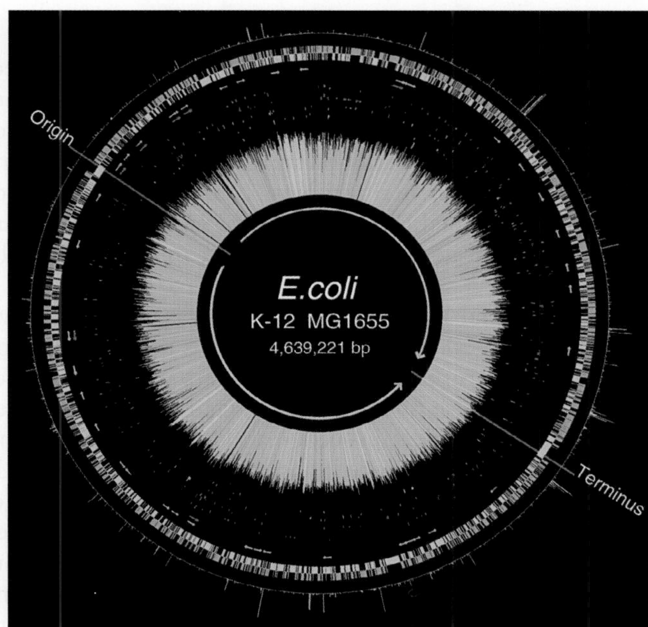

Key :
➡ tRNA genes
➡ rRNA genes
— origin and terminus of replication

Figure 16.21 ▶ Sequence-based map of chromosome of *Escherichia coli*. The blue arrows mark the halves of the chromosome traversed by the replication forks. The outer concentric circle gives the positions of genes encoding proteins similar to bacteriophage proteins. The second concentric circle shows the location of genes that are transcribed clockwise (gold) from one strand or counterclockwise (yellow) from the complementary strand. The sunburst in the center is a histogram in which the length of each ray is proportional to the randomness of codon usage within each coding sequence.

The *Sulfolobus* species, for example, can grow at temperatures up to 85°C (185°F) in highly acidic (pH = 1) soil. Some *Geobacter* species can grow at temperatures up to 121°C—well above the boiling point of water, which allows them to live in the hydrothermal vents at the bottom of the ocean. Other archaea—the halophiles—have adapted to high-salt environments such as the Great Salt Lake in Utah. By determining how the genomes of these organisms have changed to allow them to survive in such extreme environments, we will develop a better understanding of the process of evolution and life on our planet.

Of the bacterial genomes sequenced to date, the sequence of the *E. coli* strain K12 genome (**FIGURE 16.21**) has undoubtedly caused the most excitement among geneticists. *E. coli* is the most studied and best understood cellular organism on our planet. Geneticists, biochemists, and molecular biologists have utilized *E. coli* as the preferred model organism for decades. Most of what is known about bacterial genetics was learned from research on *E. coli*. Thus, the publication of the complete sequence of the *E. coli* genome in 1997 was a significant milestone in the history of genetics.

The *E. coli* genome contains 4467 putative protein-coding genes. Of these predicted genes, about one-third are well-studied genes encoding known products, whereas 38 percent are of unknown function. **FIGURE 16.22** shows a classification of the genes according to gene-product function. The average distance between genes (size of intergenic regions) in the *E. coli* genome is 118 bp. Known and putative genes specifying proteins and stable RNAs make up 87.8 percent and 0.8 percent of the genome, respectively, and noncoding repetitive elements account for 0.7 percent of the genome. Thus, 10.7 percent of the genome must involve regulatory sequences, sequences with other unknown functions, and perhaps some nonfunctional nucleotide sequences.

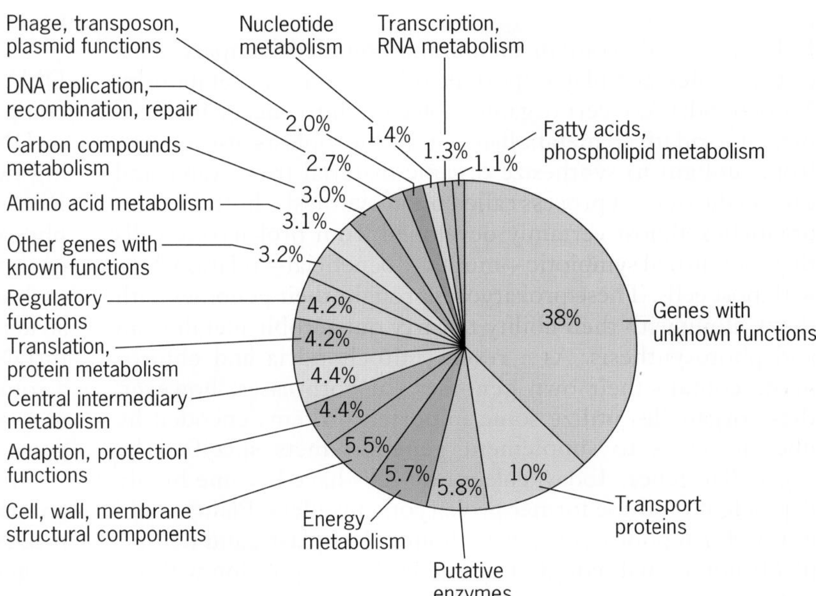

Figure 16.22 ▶ Classification of the 4288 known and putative genes of *E. coli* according to function. The largest class of known genes encodes proteins involved in transport. However, note that 38 percent of the genes encode proteins of unknown function.

Other bacterial genomes exhibit the same gene density as *E. coli*—about one gene per kb of DNA. This contrasts sharply with the variation in gene density observed in the genomes of eukaryotes, where larger genomes have very low gene densities.

The genomes of *M. tuberculosis* (tuberculosis), *Mycobacterium leprae* (leprosy), *Corynebacterium diphtheriae* (diphtheria), *Legionella pneumophila* (Legionnaire's disease), *Bacillus anthracis* (anthrax), *Yersinia pestis* (bubonic plague), and other infectious bacteria are also of great interest because of the pathogenicity of these organisms and the hope that a complete understanding of their metabolism will suggest ways to prevent these often fatal diseases. The need for new ways to combat bacterial pathogens has been enhanced by the recent evolution of antibiotic-resistant strains of these infectious bacteria (Chapter 8).

The genome of *M. genitalium* is of special interest because it may approximate the "minimal gene set" for a cellular organism—the smallest set of genes that will allow a cell to reproduce. The genome of *M. genitalium* contains only 525 predicted genes, in contrast to the 1789 predicted genes of *H. influenzae* and the 4467 predicted genes of *E. coli* strain K12. The parasitic life cycle of *M. genitalium* has apparently allowed it to shed genes encoding proteins with functions provided by its host—structural components of cell walls and enzymes involved in the biosynthesis of metabolites provided by the host cell. By comparing the 525 genes of *M. genitalium* with those of other bacteria, and using information about the functions of these genes in other bacteria, researchers have estimated that the minimal number of genes required for the reproduction of a cellular organism is somewhere between 265 and 350. Of course, this estimate is highly speculative.

THE GENOMES OF CHLOROPLASTS AND MITOCHONDRIA

Eukaryotic cells contain membrane-bounded compartments or organelles that play important roles in energy metabolism. Mitochondria convert organic molecules into energy by aerobic, or oxidative, metabolism, and chloroplasts use energy from sunlight to synthesize organic material from water and carbon dioxide—a process called photosynthesis. Both of these organelles almost certainly developed from prokaryotic cells that established symbiotic—mutually beneficial—relationships with host cells. These prokaryotes brought their genomes with them, along with their ability to carry out aerobic metabolism and photosynthesis. As a result, mitochondria and chloroplasts contain their own genomes. In both cases, however, these organelles utilize some imported proteins encoded by nuclear genes to supplement gene products specified by organellar genes. Today, eukaryotic cells have become highly dependent on these former prokaryotic invaders. Plants could not perform photosynthesis without chloroplasts, and neither plants nor animals could carry out aerobic respiration without mitochondria.

MITOCHONDRIAL GENOMES

Mitochondrial genetic systems consist of DNA and the molecular machinery needed to replicate and express the genes contained in this DNA. This machinery includes the macromolecules needed for transcription and translation. Mitochondria even possess their own ribosomes. Many of these macromolecules are encoded by mitochondrial genes, but some are encoded by nuclear genes and are therefore imported from the cytosol.

Mitochondrial DNA, or **mtDNA** as it is usually abbreviated, was discovered in the 1960s, initially through electron micrographs that revealed DNA-like fibers within the mitochondria. Later, these fibers were extracted and characterized by physical and chemical procedures. The advent of recombinant DNA techniques made it possible to analyze mtDNA in great detail. In fact, the complete nucleotide sequences of mtDNA molecules from many different species have now been determined (**TABLE 16.2**). Mitochondrial DNA molecules vary enormously in size, from about 6 kb in the malaria-causing parasite *Plasmodium* to 2500 kb in some of the flowering plants. Each mitochondrion appears to contain several copies of the DNA, and because each cell usually has many mitochondria, the number of mtDNA molecules per cell can be very large. A vertebrate oocyte, for example, may contain as many as 10^8 copies of the mtDNA. Somatic cells, however, have fewer copies, perhaps less than 1000.

Most mtDNA molecules are circular, but in some species, such as the alga *Chlamydomonas reinhardtii* and the ciliate *Paramecium aurelia*, they are linear. The circular mtDNA molecules, which have been studied the most thoroughly, appear to be organized in many different ways. In the vertebrates 37 distinct genes are packed into a 16- to 17-kb circle leaving little or no space between genes. In some of the flowering plants an unknown number of genes are dispersed over a very large circular DNA molecule hundreds or thousands of kilobases in size.

The structure of mtDNA molecules has been studied by DNA sequencing (**TABLE 16.2**). Animal mtDNA is small and compact (**FIGURE 16.23**). In human beings, for example, the mtDNA is 16,571 base pairs long and contains 37 genes, including two that encode ribosomal RNAs, 22 that encode transfer RNAs, and 13 that encode polypeptides involved in oxidative phosphorylation, the process that mitochondria use to recruit energy. In mice, cattle, and frogs, the mtDNA is similar to that of human beings—an indication of a basic conservation of structure within the vertebrate subphylum. Invertebrate mtDNA is about the same size as vertebrate mtDNA, but it has a somewhat different genetic organization. These differences seem to have been caused by structural rearrangements of the genes within the circular mtDNA molecule.

In fungi, the mtDNA is considerably larger than it is in animals (**TABLE 16.2**). Yeast, for example, possesses circular mtDNA molecules 78 kb long. These molecules contain at least 33 genes, including 2 that encode ribosomal RNAs, 23 to 25 that encode transfer RNAs, 1 that encodes a ribosomal protein,

▶ **TABLE 16.2**

Size and Gene Content of Selected Mitochondrial and Chloroplast Genomes

Species	Common Name	Genome Size in Nucleotide Pairs	Predicted Number of Genes
Mitochondrial Genomes			
Apis mellifera	honeybee	16,343	13
Arabidopsis thaliana	mouse ear cress	366,924	57
Caenorhabditis elegans	roundworm	13,794	12
Candida glabrata	yeast (infectious)	20,063	37
Chlamydomonas reinhardtii	green alga	15,758	25
Drosophila melanogaster	fruit fly	19,517	37
Danio rerio	zebra fish	16,596	37
Homo sapiens	human	16,571	37
Mus musculus	mouse	16,299	37
Oryza sativa	Indica rice	491,515	96
Plasmodium falciparum	malaria protozoan	5,967	3
Rattus norvegicus	Norwegian rat	16,313	37
Saccharomyces cerevisae	baker's yeast	85,779	43
Zea mays subsp. *mays*	corn	569,630	218
Chloroplast Genomes			
Arabidopsis thaliana	mouse ear cress	154,478	129
Chlamydomonas reinhardtii	green alga	203,828	109
Marchantia polymorpha	liverwort	121,024	134
Oryza sativa	Japonica rice	134,525	159
Zea mays subsp. *mays*	corn	140,384	158

Data are from the NCBI web site (http://www.ncbi.nim.nih.gov/Genomes/) as of May 15, 2008.

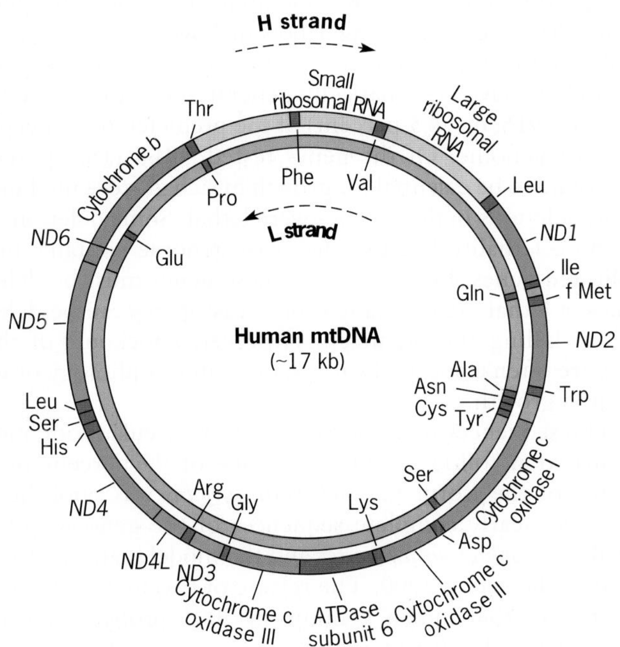

Figure 16.23 ▶ Map of the human mitochondrial genome *ND1–6* are genes encoding subunits of the enzyme NADH reductase; the tRNA genes in the mtDNA are indicated by abbreviations for the amino acids. Arrows show the direction of transcription. Genes on the inner circle are transcribed from the L (light) strand of the DNA, whereas genes on the outer circle are transcribed from the H (heavy) strand of the DNA.

and 7 that encode different polypeptides involved in oxidative phosphorylation. The yeast mtDNA is larger than animal mtDNA because several of its genes contain introns and there are long noncoding sequences between some of the genes. Animal mtDNA does not contain introns.

Plant mtDNA is much larger than the mtDNA of other organisms (**TABLE 16.2**). It is also more variable in structure. One of the first plant mtDNAs to be sequenced is from the liverwort, *Marchantia polymorpha*. The mtDNA from this primitive, nonvascular plant is a 186-kb circular molecule with 94 substantial open reading frames (ORFs), some corresponding to known genes and others having still unassigned genetic functions. The latter ORFs are therefore called *URFs*, for *u*nassigned *r*eading *f*rames. Thirty-two distinct introns have been found in the *Marchantia* mtDNA, accounting for about 20 percent of the molecule. In vascular plants, the mtDNA is larger than it is in *Marchantia;* for example, it is a 570-kb circular molecule in maize and a 300-kb circle in the watermelon. Higher plant mtDNA molecules contain many noncoding sequences, including some that are duplicated. Physical mapping of mitochondrial genes in plants has shown that they are located in different positions on the mtDNA circles of different species, even when the species are fairly closely related. This implies that the mtDNA of higher plants has undergone many genetic rearrangements during its evolution.

Most—perhaps all—mitochondrial gene products function solely within the mitochondrion. However, they do not function alone. Many nuclear gene products are imported to augment or facilitate their function. Mitochondrial ribosomes, for example, are constructed with ribosomal RNA transcribed from mitochondrial genes and with ribosomal proteins encoded by nuclear genes. The ribosomal proteins are synthesized in the cytosol and imported into the mitochondria for assembly into ribosomes.

Many of the polypeptides needed for aerobic metabolism are also synthesized in the cytosol. These include subunits of several proteins involved in oxidative phosphorylation—for example, the ATPase that is responsible for binding the energy of aerobic metabolism into ATP. However, because some of the subunits of this protein are synthesized in the mitochondria, the complete protein is actually a mixture of nuclear and mitochondrial gene products. This dual composition suggests that the nuclear and mitochondrial genetic systems are coordinated in some way so that equivalent amounts of their products are made; possible molecular mechanisms for this coordination are currently under investigation.

CHLOROPLAST GENOMES

Chloroplasts are specialized forms of a general class of plant organelles called **plastids.** Botanists distinguish among several kinds of plastids, including chromoplasts (plastids containing pigments), amyloplasts (plastids containing starch), and elaioplasts (plastids containing oil or lipid). All three types seem to develop from small membrane-bounded organelles called proplastids, and, within a particular plant species, all seem to contain the same DNA. This DNA is generally referred to as **chloroplast DNA,** abbreviated simply as **cpDNA.**

In higher plants, cpDNAs typically range from 120 to 160 kb in size, and in algae, from 85 to 292 kb (**TABLE 16.2**). In a few species of green algae in the genus *Acetabularia*, the cpDNA is much larger, about 2000 kb. Among the species of plants whose chloroplast DNA has been at least partially characterized, the cpDNA seems to be organized as a closed circular molecule. However, in some species, especially those with large cpDNAs, a linear arrangement cannot be ruled out.

The number of cpDNA molecules in a cell depends on two factors: the number of chloroplasts and the number of cpDNA molecules within each chloroplast. For example, in the unicellular alga *Chlamydomonas reinhardtii*, there is only one chloroplast per cell, and it contains about 100 copies of the cpDNA. In *Euglena gracilis*, another unicellular organism, there are about 15 chloroplasts per cell, and each contains about 40 copies of the cpDNA.

All cpDNA molecules carry basically the same set of genes, but in different species these genes are arranged in different ways. The basic gene set includes genes for ribosomal RNAs, transfer RNAs, some ribosomal proteins, various polypeptide components of the photosystems that are involved in capturing solar energy, the catalytically active subunit of the enzyme ribulose 1,5-bisphosphate carboxylase, and four subunits of a chloroplast-specific RNA polymerase. More than 100 cpDNA molecules have been sequenced in their entirety (see **TABLE 16.2** for a few examples). Two of the first cpDNAs sequenced were from the liverwort, *Marchantia polymorpha* (**FIGURE 16.24**), and from the tobacco plant, *Nicotiana tabacum*. The tobacco cpDNA is larger (155,844 bp) and contains about 150 genes. The best estimate for the gene number in the liverwort cpDNA (121,024 bp) is 134. Most cpDNAs have a pair of large inverted repeats that contain the genes for the ribosomal RNAs. These repeats range anywhere from 10 to 76 kb in length and are variously located in different cpDNA molecules.

As mentioned earlier, the development of functional chloroplasts depends on the expression of both nuclear and chloroplast genes. The nuclear genes are transcribed in the nucleus and translated in the cytosol. The products of nuclear genes that function in the chloroplast must be imported from the cytosol. Once imported, these proteins must act in concert with cpDNA-encoded proteins. Functional chloroplasts thus depend on the coordinated activities of both nuclear and chloroplast gene products.

EUKARYOTIC GENOMES

Baker's yeast, *Saccharomyces cerevisiae*, was the first eukaryotic organism to have its entire genome sequenced. The complete 12,068-kb sequence of the *S. cerevisiae* genome was assembled in 1996 through an international collaboration of about 600 scientists working in Europe, North America, and Japan. The yeast genome contains 5885 potential protein-coding genes, about 140 genes specifying ribosomal RNAs, 40 genes for small nuclear RNA molecules, and over 200 tRNA genes. Researchers have systematically generated deletions of essentially all (5916, or 96.5 percent) of the predicted 6268 genes in the yeast genome. Of the genes tested, 1105 (18.7 percent) were found to be essential for growth in rich glucose medium—that is, deletions in these genes were lethal. Some deletions did not cause lethality because the yeast genome contains many duplicated genes. Both copies of these genes must be deleted to have a lethal effect. Many other yeast genes can be deleted without killing the organism. However, knockouts of these genes are often associated with changes in morphology or with impaired growth.

The sequences of the genomes of other eukaryotic model systems soon followed. The sequence of 99 percent of the genome of the worm *Caenorhabditis elegans* was published in 1998, and nearly complete sequences of the genomes of the fruit fly *Drosophila melanogaster* and the model plant *Arabidopsis thaliana* followed in 2000. The release of two first drafts of the sequence of the human genome in 2001 probably received more coverage by the international news media than any other event in the history of biology. The publication of the sequences of the genomes of several other eukaryotes followed, and sequencing projects are underway for many other species. As mentioned earlier, a nearly complete sequence of the human genome was released in 2004.

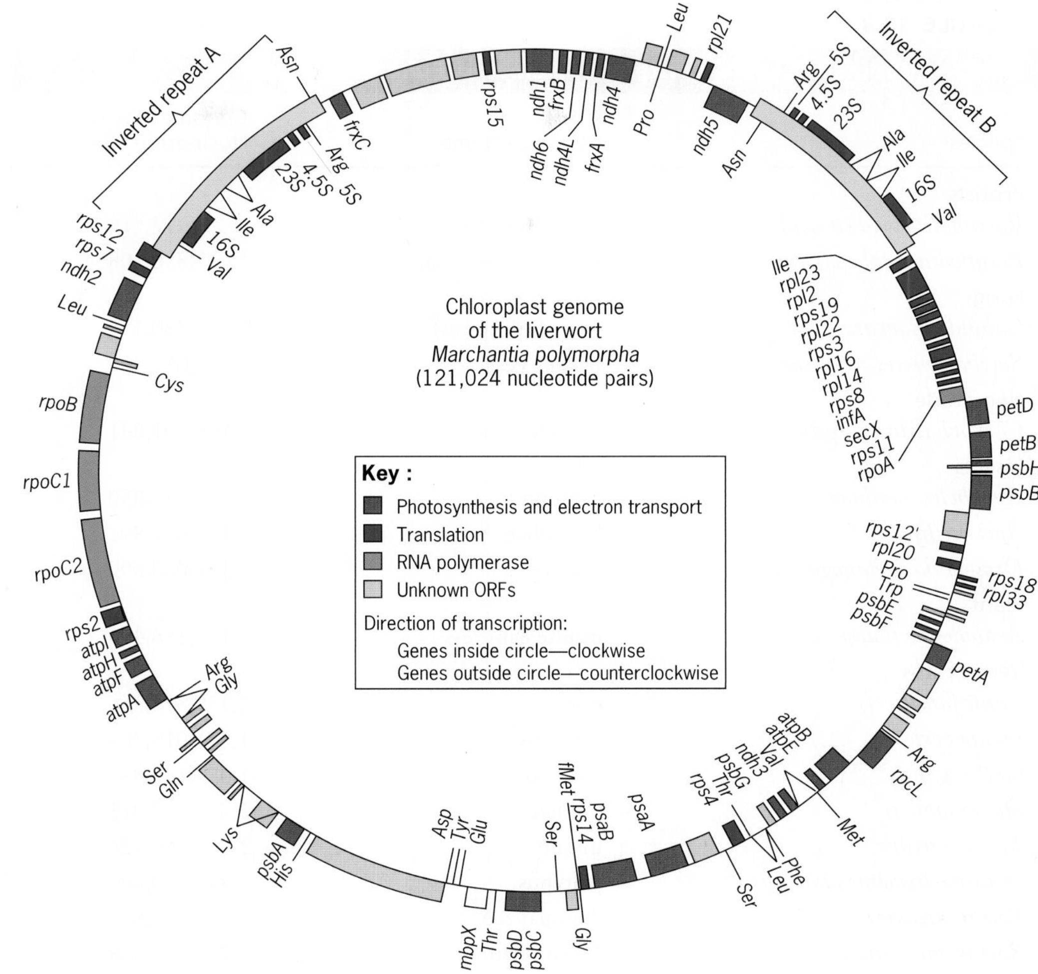

Figure 16.24 ► Genetic organization of the chloroplast genome in the liverwort *Marchantia polymorpha*. Symbols: *rpo*, RNA polymerase; *rps*, ribosomal proteins of small subunit; *rpl* and *secX*, ribosomal proteins of large subunit; *4.5S, 5S, 16S, 23S*, rRNAs of the indicated size; *rbs*, ribulose bisphosphate carboxylase; *psa*, photosystem I; *psb*, photosystem II; *pet*, cytochrome b/f complex; *atp*, ATP synthesis; *infA*, initiation factor A; *frx*, iron-sulfur proteins; *ndh*, putative NADH reductase; *mpb*, chloroplast permease; tRNA genes are indicated by abbreviations for the amino acids.

What have we learned from all these sequences? In contrast to the genomes of archaea and eubacteria, gene density varies widely among different eukaryotic species, ranging from one gene per 1900 bp in baker's yeast to one gene per 127,900 bp (145,000 bp if the unsequenced heterochromatin is included) in humans. Genome size and gene content are shown for the genomes of selected eukaryotes in **TABLE 16.3.** The genomes of the single-celled eukaryotes are like yeast, with one gene for every 1000 to 2000 bp. Gene density decreases to one gene per 4000 to 5000 bp for *Arabidopsis* and *C. elegans*, to one gene per 9500 bp in *D. melanogaster*, to one gene per 15,000 bp in the pufferfish, and is the lowest in mammals at one gene for every 115,000 to 129,000 bp. The observed decrease in gene density with increased developmental complexity raises questions about the functions of the noncoding DNA. As mentioned earlier (see **FIGURE 16.13**), an international *ENC*yclopedia *O*f *DNA E*lements (ENCODE) Project Consortium was recently organized to investigate this material.

One reason for the decreased gene density in the larger eukaryotic genomes is that these genomes contain considerable amounts of repetitive DNA (see Chapter 9). Baker's yeast contains very little repetitive DNA, although about 30 percent of its genes are duplicated. By contrast, the genomes of multicellular eukaryotes contain lots of repetitive DNA, and the amount of this material is, in most cases, directly related to genome size. For example, only about 10 percent of the small genome of *C. elegans* consists of moderately repetitive DNA, whereas about 45 percent of the large genomes of mammals is composed of moderately repeated DNA sequences. Most of these moderately repetitive sequences are derived from transposable genetic elements (see Chapter 18).

Highly repetitive DNA is also more abundant in the larger genomes; however, there is no direct correlation between the amount of highly repetitive DNA and genome size. Indeed, closely related species sometimes differ significantly in the amount of highly repetitive DNA. For example, 18 percent of the genome of *D. melanogaster* consists of highly repetitive DNA, whereas 45 percent of the DNA in *D. virilis* is highly repetitive. Much of the highly repetitive DNA in most species, including humans, is present in the regions of chromosomes that flank the

▶ **TABLE 16.3**

Size and Predicted Gene Content of Selected Eukaryotic Genomes

Species	Common Name	Genome Size in Nucleotide Pairs	Predicted Number of Genes*	Gene Density (bp/gene)†
Protists				
Encephalitozoon cuniculi	microsporidian	2,497,519	2,029	1,200
Plasmodium falciparum	malaria protozoan	22,820,308	5,361	4,300
Fungi				
Candida glabrata	yeast (infectious)	12,280,357	5,272	2,500
Saccharomyces cerevisae	baker's yeast	12,057,909	6,268	1,900
Nematode				
Caenorhabditis elegans	roundworm	100,291,841	20,516	4,900
Insects				
Anopheles gambiae	mosquito	278,253,050	14,707	18,900
Apis mellifera	honeybee	197,657,892	29,832	6,600
Drosophila melanogaster	fruit fly	131,000,899	13,792	9,500
Plant				
Arabidopsis thaliana	mouse ear cress	119,186,496	28,152	4,200
Vertebrates				
Canis familiaris	dog	2,359,826,366	18,201	129,700
Danio rerio	zebra fish	1,571,018,465	23,524	66,800
Gallus gallus	chicken	1,054,180,845	17,709	59,600
Homo sapiens	human	2,851,330,913	22,287	127,900
Mus musculus	mouse	2,932,368,526	25,396	115,500
Ornithorhynchus anatinus	platypus	1,840,000,000	18,527	99,500
Pan troglodytes	chimpanzee	2,928,563,828	21,098	139,000
Rattus norvegicus	Norwegian rat	2,571,104,688	22,159	115,800
Takifugu rubripes	Japanese pufferfish	329,140,338	20,796	15,800
Tetraodon nigroviridis	spotted pufferfish	402,240,326	27,918	14,400

Data are from the NCBI web site (http://www.ncbi.nim.nih.gov/Genomes/), the Ensembl web site (http://www.ensembl.org), or the CBS Genome Atlas Database (http://www.cbs.dtu.dk/services/GenomeAtlas) as of May 15, 2008.

*Gene numbers are Ensembl predictions (minus pseudogenes when data are available) except for the honeybee (a Genscan prediction) and the spotted pufferfish (a Genoscope prediction). Note that the latter programs yield higher estimates than Ensembl.

†Values are rounded to the nearest 100 bp.

centromeres (centromeric heterochromatin) and in the telomeres. This DNA is difficult to sequence. In fact, most of the unsequenced DNA in the human genome—472 million base pairs—consists of highly repetitive sequences, and 24 of the gaps in the human genome sequence correspond to blocks of centromeric heterochromatin in the 24 chromosomes.

Introns are a significant component of eukaryotic DNA, and they are more prevalent and longer in the larger eukaryotic genomes. Intergenic regions are also longer in the larger eukaryotic genomes. In contrast, the number of distinct protein domains—functional regions of proteins—encoded by genes does not seem to vary much with genome size. The predicted numbers of protein domains encoded by the *A. thaliana*, *D. melanogaster*, and human genomes are 1012, 1035, and 1262, respectively. However, humans and other vertebrates make greater use of alternate pathways of transcript splicing (see Chapter 20) to shuffle these domains into more combinations, increasing protein diversity.

Genomics research has shown that even distantly related species have many of the same genes. For example, 18 percent of the genes in *Arabidopsis* and 50 percent of the genes in *Drosophila* have human homologues. Among closely related species, the proportion of homologous genes is even greater. For example, 99 percent of the genes in mice are homologous to human genes. Furthermore, in closely related species, entire chromosomes often show similar arrangements of genes. We will briefly examine these kinds of similarities in the last two sections of this chapter.

GENOME EVOLUTION IN THE CEREAL GRASSES

The cereal crops provide much of the food for humans and their domestic livestock, which provide additional human foodstuff. Therefore, enhanced cereal productivity is an important component of the effort to feed the constantly expanding human population on our planet. Increased knowledge of the genomes of these agronomic species is key to achieving this objective. Recent comparative analyses of the genomes of several cereal species indicate that much of the information obtained by mapping and sequencing the smallest of these genomes—the 400-mb genome of rice—will be directly applicable to other cereal grasses because of the conservation of genome structure in these species.

When Graham Moore and colleagues compared the high-density genetic maps of the chromosomes of several cereal grasses, they discovered that despite large differences in genome size and chromosome number, the linkage relationships of blocks of unique DNA sequences and known genes were remarkably conserved. In contrast, the quantities and locations of repetitive DNA sequences were highly variable.

The striking conservation of genome structure in the cereal grasses is illustrated most clearly by drawing the rice genome as a circular array and aligning the conserved blocks of genes in the other species with the rice genome (**FIGURE 16.25**). The circular display of the cereal genomes does not imply any circularity of ancestral chromosomes; it simply permits maximal alignment of homologous blocks of genes. The alignment also emphasizes the presence of duplicate copies of each block of genes in the maize genome, indicating that maize has evolved from a tetraploid ancestor. Interestingly, one set of genes is present largely in the small chromosomes of maize, and the second set is present primarily in the large chromosomes.

The conserved structures of the cereal grass genomes should assist plant breeders in their attempts to produce varieties with increased yield, pest resistance, drought tolerance, and other desired traits. Because of the conserved genome structure, information obtained from sequencing the relatively small rice genome should be more easily applied to breeding and genetic engineering projects on the other cereal crop species.

GENOME EVOLUTION IN MAMMALS

Mammalian genomes exhibit conservation of chromosome structure similar to that observed in the cereal grasses. Although genetic mapping is being done in over 200 mammalian species, high-density maps are currently available for only human, mouse, dog, rat, and a few agriculturally important farm animals such as swine and cattle. These detailed chromosome maps can be used to demonstrate the conserved linkage relationships of genes in species where such maps are available. For other species, a procedure called **chromosome painting** has

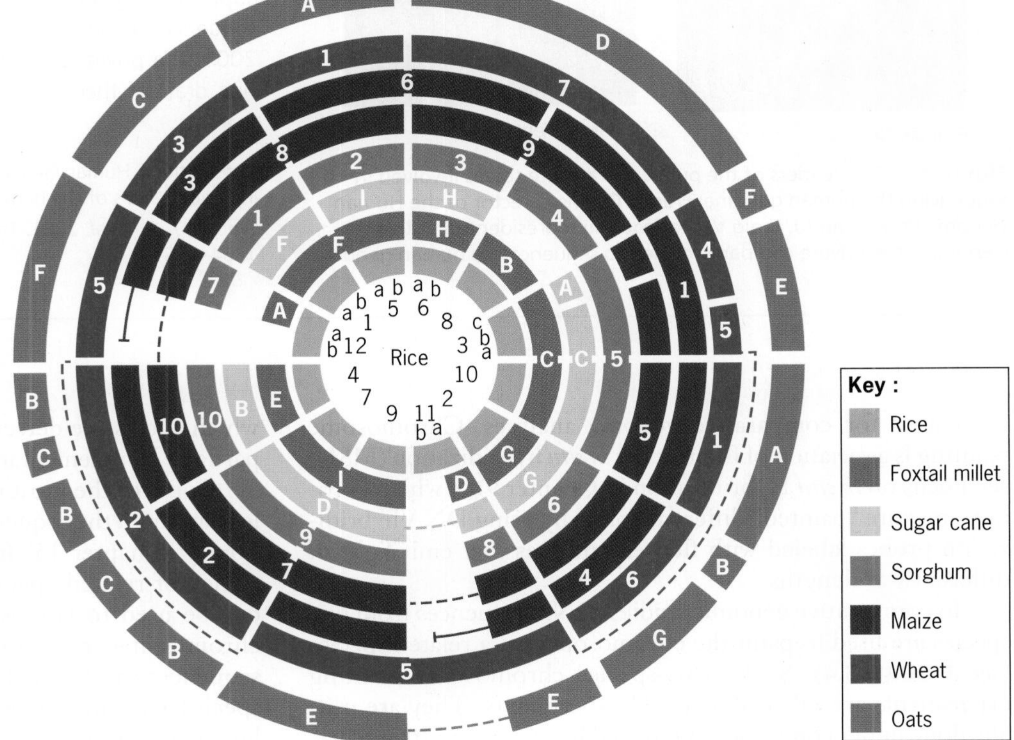

Figure 16.25 ▶ Simplified comparative map of the genomes of seven cereal grasses. Chromosomes and segments of chromosomes (denoted by capital letters) of the various cereal grasses are aligned with the chromosomes of rice, the grass species with the smallest genome (center). The maize genome has two similar copies of each block of genes and thus occupies two rings of the circle. The outer dashed lines connect adjacent segments of wheat chromosomes. Similar segments of chromosomes in the oats genome are not connected by dashed lines for the sake of simplicity.

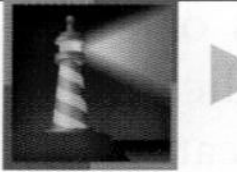

▶ A MILESTONE IN GENETICS: **Two Drafts of the Sequence of the Human Genome**

The Human Genome Project was launched in 1990, with the goal of sequencing the entire human genome—all 3 billion nucleotide pairs—by 2005, at an estimated cost of $3 billion. During 1997 and 1998, eight research teams in the United States and a team at the Sanger Centre in the United Kingdom geared up to begin high-output sequencing of their allotted portions of the genome. The international project was headed by Francis Collins (**FIGURE 1**), one of the leaders of the research team that identified the mutant gene responsible for cystic fibrosis (Chapter 17). Then, in May 1998, J. Craig Venter (**FIGURE 1**), whose Institute for Genomic Research in Rockville, Maryland, had sequenced several bacterial genomes, stunned the scientific world. He announced that he had teamed up with PerkinElmer Corporation of Norwalk, Connecticut, and had formed a new company, Celera Genomics, which would sequence the human genome in just three years and at a cost of only $300 million. Clearly, the race to identify, sequence, and patent important human genes was underway.

Francis Collins

J. Craig Venter

Figure 1 ▶ The leaders of the public and private teams involved in sequencing the human genome: Francis Collins, director of the Human Genome Project, and J. Craig Venter, founding president of Celera Genomics, the private company formed to sequence the human genome.

Venter's proposal was based on two key developments: (1) a sequencing strategy called **whole-genome shotgun sequencing** and (2) the production of faster, *fully automated sequencing machines* by PerkinElmer's Applied Biosystems division. Instead of marching down a chromosome, sequencing carefully mapped genomic clones, whole-genome sequencing involves chopping the entire genome into small fragments, sequencing just the ends of the fragments, and then using supercomputers to assemble a complete sequence by finding overlaps among the fragments. This procedure had worked well for the much smaller bacterial genomes, but some sequencing experts were skeptical about its feasibility with the large genomes of eukaryotes. Whereas bacterial genomes contain little repetitive DNA, eukaryotic genomes contain large amounts of highly repetitive DNA. These highly repetitive sequences might match too many fragments and therefore stymie the computer's sequence assembly process.

The DNA sequencing machine developed by PerkinElmer could separate the products of 96 dideoxy chain-terminator reactions (Chapter 15) simultaneously and could operate unattended for 24 hours. All functions, including gel loading, sample loading, electrophoresis, data collection, and data analysis, were fully automated. The sequencers were capable of performing about 10 automated sequencing runs per day. On good days, the 230 sequencing machines at Celera cranked out over 100 million nucleotides of sequence—a dramatic increase over earlier sequencing technologies.

Clearly, a little competition had a significant effect. On February 15, 2001, the International Human Genome Sequencing Consortium—supported by public funding—published their first draft of the sequence of the human genome in *Nature*.[1] Then, on February 16, 2001, the privately funded Celera Genomics group published their first draft of the human genome in *Science*.[2] The two drafts were

[1]International Human Genome Sequencing Consortium. 2001. Initial sequencing and analysis of the human genome. *Nature* 409:860–921.

[2]Venter, J. C., *et al.* 2001. The sequence of the human genome. *Science* 291: 1304–1351.

been used for comparative genome analyses. Chromosome painting is a variation of fluorescent *in situ* hybridization (FISH; see Focus on *in situ* Hybridization in Chapter 9) in which chromosomes are "painted" different colors by using DNA hybridization probes labeled with fluorescent dyes that emit light of different wavelengths.

In comparative genomic studies, DNA sequences from one species are used to paint the chromosomes of a related species (see **FIGURE 6.4**). Such cross-species chromosome painting experiments are called Zoo-FISH experiments. They are usually done under conditions of reduced hybridization stringency, which allows the detection of cross-hybridization between the partially complementary strands of homologous genes.

Some of the most interesting chromosome painting studies have used the sequences in a chromosome-specific genomic library (Chapter 15) from one species to paint the chromosomes of related species. Indeed, different fluorescent dyes can be used to label sequences from two or more different chromosomes of one species, and these fluorescence-tagged sequences can be used to "paint" the chromosomes of related species. Chromosome-specific libraries are available for all 24 human chromosomes, and sequences from these libraries have amazingly consistent (**TABLE 1**). The Consortium's draft covered about 92 percent of the genome, and Celera's draft covered about 95 percent of the euchromatic DNA in the genome. The only major surprise was that the genome was found to contain fewer genes—about 30,000—than previously estimated—50,000 to 100,000.

The Consortium continued to work on a more complete sequence of the human genome, and in October 2004, released a nearly complete sequence of the genome (covering approximately 99 percent of the euchromatic DNA).[3] Once again, the surprise was that the estimated number of genes was even smaller—only 20,000 to 25,000 genes that encode protein products.

Now that a nearly complete sequence of the genome is in hand, the next questions to be addressed are: "What are the functions of its genes?" and "What nongenic elements influence the expression of these genes?" With the goal of answering the first question, scientists formed an International Human Proteome Organization (HUPO). Their objective is to determine the structures and functions of all of the proteins encoded by the human genome. Other scientists are focused on the second question. They have organized the ENCODE (*ENC*yclopedia *O*f *D*NA *E*lements) Consortium and will try to identify functional elements other than genes in the human genome.

[3]International Human Genome Sequencing Consortium. 2004. Finishing the euchromatic sequence of the human genome. *Nature* 431: 931–945.

QUESTIONS FOR DISCUSSION

1. Patents are granted to protect the investments of individuals and companies in acquiring new methods and products. However, there has been considerable debate as to whether genes should be subject to patent protection. Do you think that human genes should be patentable? If so, should the patent be granted to the person who provided the DNA or to the person or company who sequenced the gene? What are some of the pros and cons to awarding patents on human genes?
2. While working on their draft of the sequence of the human genome, scientists at Celera Genomics had access to the sequence data generated by the International Human Genome Sequencing Consortium. The Consortium's data were deposited in the GenBank, EMBL, and DDBJ databases with free access to everyone. In contrast, Celera's data were not available to the Consortium scientists. What are the advantages of making sequence data available free to everyone? What are the disadvantages of doing so, especially when the data are generated by research at a private company?

TABLE 1

Comparisons of the Two First Drafts of the Human Genome and the More Recent Sequence Covering 99 Percent of the Euchromatic DNA

Draft	Nucleotide Pairs Sequenced	Estimated Number of Gaps	Estimated Number of Genes
HGP Consortium 2001	2,692,900,000	147,821	31,778
Celera Genomics 2001	2,653,979,733	105,264	26,588
HGP Consortium 2004	2,851,330,913	341	22,287

been used to "paint" the chromosomes of several related species, including most of the primates and a few more distantly related mammals. Upon reviewing comparative linkage and chromosome painting data, Bhanu Chowdhary and colleagues concluded that the evolution of mammalian chromosomes involved three classes of conserved synteny. (Synteny is the presence of genes on the same chromosome.) The three classes are (1) conservation of entire chromosomes, (2) conservation of large segments of chromosomes, and (3) the joining of segments of different chromosomes to produce new synteny. Each type of chromosome evolution is illustrated in **FIGURE 16.26.**

The synteny of genes on human chromosome 17 is conserved on chromosome 12 of the pig and chromosome 19 of cattle (**FIGURE 16.26*a***). Similar patterns of conservation of human chromosomes 13 and 20 have been observed in these species.

Human chromosome 2 provides an example of the second class of conserved synteny, the conservation of large segments of chromosomes (**FIGURE 16.26*b***). Major portions of the long and short arms of human chromosome 2 are conserved on two chromosomes of the pig and cattle (also in the horse and cat—not shown). The genes on human chromosome 2 are more dispersed on the chromosomes of Indian muntjac and mouse (not shown). Human chromosomes 4, 5, 6, 9, and 11 also have large chromosome segments conserved in the other mammals studied.

The third pattern of conserved synteny—the joining of segments of chromosomes to produce new synteny—is illustrated by human chromosomes 3 and 21 (**FIGURE 16.26c**). Chromosome 13 of the pig appears to contain the genes of both human chromosomes arranged as if they had simply fused together. In cattle, one block of genes present on human chromosome 3 is present on a separate chromosome (22), whereas most of the rest of the genes from human chromosome 3 are on a chromosome (1) with the genes from human chromosome 21. Other examples of this pattern of conserved synteny involve human chromosomes 12 and 22, 14 and 15, and 16 and 19.

Despite the extensive conservation of blocks of genes illustrated by the examples discussed above, more detailed comparisons of chromosome structure in mammals clearly show that numerous chromosome rearrangements have occurred during the evolution of even closely related species. A comparison of human chromosomes with the chromosomes of the white-cheeked gibbon, *Hylobates concolor*, revealed that at least 21 translocations have occurred during the evolution of these two species from their common ancestor. Inversions and other intrachromosomal rearrangements are especially prevalent in the genomes of closely related species.

KEY POINTS

- Comparative genomics—comparing the nucleotide sequences of genomes—has provided new information about the relationships between various taxonomic groups.

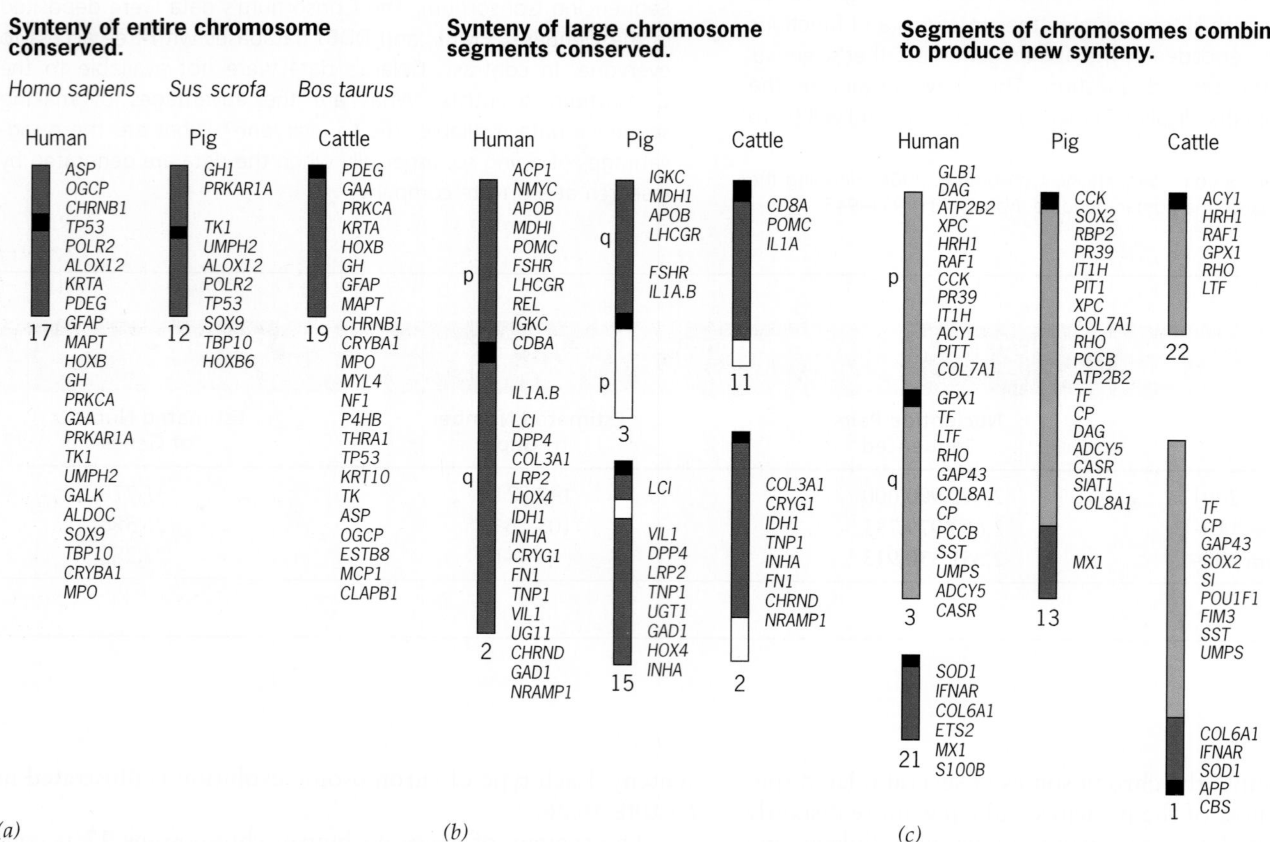

Figure 16.26 ▶ Chromosome evolution in mammals. Examples of three classes of conserved synteny (conserved blocks of linked genes) are illustrated. (*a*) Human chromosome 17 is an example of the conservation of entire chromosomes. Note that even with conserved synteny inversions occur, changing the order of some genes. (*b*) Human chromosome 2 provides an example of the conservation of large chromosomal segments. (c) Human chromosomes 3 and 21 illustrate the formation of new synteny by chromosome fusion. The chromosomal locations of a few genes are shown to the right of the chromosomes. Fewer genes have been mapped in pigs than in cattle or humans. Species-specific chromosome designators are given below the chromosomes.

► FOCUS ON PROBLEM SOLVING
Using Bioinformatics to Investigate DNA Sequences

THE PROBLEM

You have decided to follow the lead of Craig Venter and James Watson and have your genome sequenced. The first 100 nucleotides had the sequence acatttgctt ctgacacaac tgtgttcact agcaacctca aacagacacc atggtgcatc tgactcctga ggagaagtct gccgttactg ccctgtgggg. What is the function of this DNA? On what chromosome is it located? Is the sequence unique, or are similar sequences present elsewhere in your genome? Is this sequence present in the genomes of other species?

FACTS AND CONCEPTS

1. The entire human genome—excluding some regions of highly repetitive DNA in heterochromatin—has been sequenced and the sequences have been deposited in GenBank.
2. The sequences of the genomes of several other mammals including our closest relative—the chimpanzee—are also available in GenBank.
3. The NCBI web site (http://www.ncbi.nlm.nih.gov/) contains bioinformatic tools that can be used to search GenBank for specific DNA sequences and/or for the proteins encoded by these sequences.
4. The BLAST (Basic Local Alignment Search Tool) software allows you to search through specific genome sequences or all of the sequences in GenBank for similar sequences.
5. The NCBI web site can also be searched for publications that report the results of studies on specific DNA sequences and their products.

ANALYSIS AND SOLUTION

A BLAST search of the "Human genomic + transcript" sequences in the GenBank nucleotide database informs us that the 100-nucleotide sequence is part of the human β-globin gene *(HBB)* on chromosome 11. The 100-nucleotide sequence is identical to the sequence of one strand of the human *HBB* gene. The sequence is also very similar (93 percent identical) to the sequence of the human δ-globin gene located adjacent to the β-globin gene. A BLAST search of all NCBI Genomes (Chromosomes) shows that the sequence differs at only 1 nucleotide from the homologous sequence on chromosome 11 of the chimpanzee (*Pan troglodytes*) and at only 7 nucleotides from the homologous sequence on chromosome 14 of the rhesus monkey (*Macaca mulatta*). Clearly, the sequences of the β-globin genes are highly conserved in all primates. Indeed, a more detailed analysis would show that the β-globin genes of all vertebrates are highly conserved.

For further discussion go to your *WileyPLUS* course.

- Bioinformatics is the science of storing, comparing, and extracting information from biological systems, especially DNA and protein sequences.
- The nucleotide sequences of diverse prokaryotic organisms have provided insights into their adaptation to unique environmental niches.
- Mitochondrial genomes are usually circular and range in size from 6 kb to 2500 kb, whereas chloroplast genomes can be either circular or linear and are typically 120 to 292 kb in size with more than 100 genes.
- As eukaryotic organisms have increased in complexity, the proportion of their genomes that encodes proteins has decreased, and the function of most of the noncoding DNA is unknown.
- Comparative genomics has revealed a remarkable conservation of synteny in related eukaryotic species, such as mammals and the cereal grasses.

► Basic Exercises

ILLUSTRATE BASIC GENETIC ANALYSIS

1. What is a genetic map?
Answer: A genetic map shows the positions of genes and other markers such as restriction fragment-length polymorphisms (RFLPs) on a chromosome based on recombination frequencies.

2. What is a cytological map?
Answer: A cytological map shows the positions of genes and other genetic markers relative to the banding patterns of chromosomes.

3. What is a physical map of a DNA molecule or chromosome?

Answer: A physical map of a DNA molecule or chromosome gives the positions of genes or other markers based on the actual distances in base pairs (bp), kilobase pairs (kb), or megabase pairs (mb) separating them. Restriction maps, contig maps, and sequence-tagged site (STS) maps are examples of physical maps.

4. How can genetic maps, cytological maps, and physical maps of chromosomes be correlated?

Answer: If a gene is cloned and positioned on all three maps, it provides an anchor marker that can be used to relate the genetic, cytological, and physical maps to each other. All three types of maps are colinear arrays showing the locations of nucleotide sequences on the chromosome. They differ in the units that are used to assign the positions of markers along the linear arrays.

5. How can the map position of a gene on a chromosome be used to identify and clone the gene?

Answer: Once a gene has been positioned on the genetic, cytological, or physical map of a chromosome, molecular markers such as RFLPs close to the gene can be used to initiate chromosome walks and jumps starting at the linked marker and progressing along the chromosome to the position of the gene of interest. The identity of the gene must be established by transforming a mutant organism with a wild-type copy of the gene and showing that it restores the wild-type phenotype or, in humans, by comparing the nucleotide sequences of the gene in a number of affected and nonaffected individuals (see **FIGURE 16.6**).

▶ Testing Your Knowledge

INTEGRATE DIFFERENT CONCEPTS AND TECHNIQUES

1. Best disease is a form of blindness in humans that develops gradually in adults. It is caused by an autosomal dominant mutation on chromosome 11. Nine RFLPs, designated 1 through 9, map on chromosome 11 in numerical order. The polymorphisms at each site are designated by superscripts 0 through N, where $N + 1$ is the number of polymorphisms present at a site in the family represented by the following pedigree. DNA was obtained from each member of the family, digested with the appropriate restriction enzyme, subjected to gel electrophoresis, transferred to a nylon membrane by Southern blotting, denatured, and hybridized to radioactive probes that detect all the RFLPs. After hybridization, the membranes were exposed to X-ray film, and the autoradiograms were used to determine which RFLP was present in each member of the family. The results are shown in the following pedigree. Circles represent females; squares represent males; red symbols indicate individuals with Best disease. Which RFLP site is closest to the mutation that causes Best disease? Which allele of this RFLP is present on the chromosome that carries the Best disease mutation?

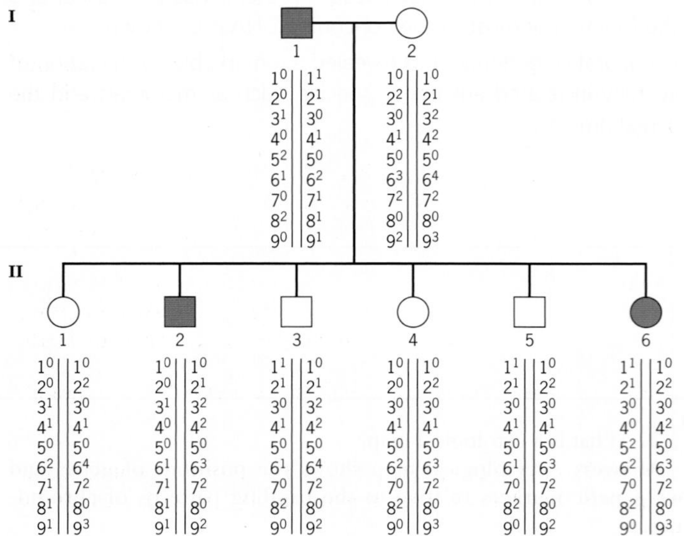

Answer: RFLP site 4 is closest to the Best disease mutation, which is present on the copy of chromosome 11 carrying the 4^0 allele of the polymorphism. Of the polymorphisms on chromosome 11, only the 4^0 allele is present in all three family members with Best disease and absent from all five members with normal vision.

2. Eleven genomic clones, each containing DNA from chromosome 4 of *Drosophila melanogaster*, were tested for cross-hybridization in all pairwise combinations. The clones are designated A through K, and the results of the hybridizations are shown in the following table. A plus sign indicates that hybridization occurred; a minus sign indicates that no hybridization was observed.

	A	B	C	D	E	F	G	H	I	J	K
K:	–	–	–	–	+	–	–	–	–	–	+
J:	–	–	+	+	–	+	–	+	–	+	
I:	–	+	–	–	+	–	–	–	+		
H:	–	–	–	–	–	+	–	+			
G:	+	–	+	–	–	–	+				
F:	–	–	–	+	–	+					
E:	–	–	–	–	+						
D:	–	–	+	+							
C:	–	–	+								
B:	–	+									
A:	+										

Based on the hybridization results shown in the table, how many contigs do these clones define? Draw the contig map(s) defined by these data.

Answer: Maps of the two contigs defined by the 11 mutations are as follows:

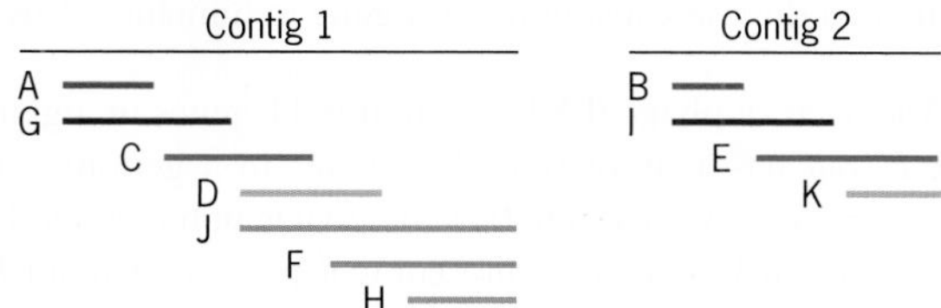

▶ Questions and Problems

ENHANCE UNDERSTANDING AND DEVELOP ANALYTICAL SKILLS

16.1 What is a contig? an RFLP? a VNTR? an STS? an EST? How is each of these used in the construction of chromosome maps?

16.2 GO The following is a Southern blot of *Eco*RI-digested DNA of rye plants from two different inbred lines, A and B. Developed autoradiogram I below shows the bands resulting from probing the blot with ^{32}P-labeled cDNA1. Autoradiogram II below shows the same Southern blot after it was stripped of probe and reprobed with ^{32}P-labeled cDNA2.

	I: A	I: B		II: A	II: B
a1	—	—	b1	—	
			b2	—	
a2		—			
a3		—	b3		—
a4	—				

(a) Which bands would you expect to see in the autoradiogram of a similarly probed Southern blot prepared using *Eco*RI-digested DNA from F_1 hybrid plants produced by crossing the two inbred lines? (b) What can you conclude about the gene(s) represented by band a1 on blot I in the two inbreds? (c) The F_1 plants were crossed to plants possessing only bands a1, a4, and b3. DNA was isolated from several individual progeny and digested with *Eco*RI. The resulting DNA fragments were separated by gel electrophoresis, transferred to a nylon membrane, and hybridized with radioactive cDNA1 and cDNA2 probes. The following table summarizes the bands present in autoradiograms obtained using DNA from individual progeny. Interpret these data. Do the data provide evidence for RFLPs? at how many loci? Are any of the RFLPs linked? If so, what are the linkage distances defined by the data?

	Bands Present						
Plant No.	**a1**	**a2**	**a3**	**a4**	**b1**	**b2**	**b3**
1	+	+	+	+			+
2	+	+	+	+			+
3	+	+	+	+			+
4	+	+	+	+			+
5	+	+	+	+	+	+	+
6	+			+	+	+	+
7	+			+	+	+	+
8	+			+	+	+	+
9	+			+	+	+	+
10	+			+			+

16.3 Distinguish between a genetic map, a cytogenetic map, and a physical map. How can each of these types of maps be used to identify a gene by positional cloning?

16.4 What is the difference between chromosome walking and chromosome jumping? Why must chromosome jumps, rather than just chromosome walks, sometimes be used to identify a gene of interest?

16.5 VNTRs and microsatellites are specific classes of polymorphisms. What is the difference between a VNTR and a microsatellite? Why would it be inappropriate to call most VNTRs microsatellites?

16.6 What are microsatellites? Why are they called microsatellites?

16.7 Why is the resolution of radiation hybrid mapping of human genes higher than the resolution of standard somatic-cell hybrid mapping?

16.8 GO An RFLP and a mutant allele that causes albinism in humans cannot be shown to be separated by recombination based on pedigree analysis or by radiation hybrid mapping. Do these observations mean that the RFLP occurs within or overlaps the gene harboring the mutation that causes albinism? If so, why? If not, why not?

16.9 As part of the Human Genome Mapping Project, you are trying to clone a gene involved in colon cancer. Your first step is to localize the gene using RFLP markers. In the following table, RFLP loci are defined by STS (sequence-tagged site) number (for example, STS1), and the gene for colon cancer is designated *C*.

Loci	% Recombination	Loci	% Recombination
C, STS1	50	STS1, STS5	10
C, STS2	15	STS2, STS3	30
C, STS3	15	STS2, STS4	14
C, STS4	1	STS2, STS5	50
C, STS5	40	STS3, STS4	16
STS1, STS2	50	STS3, STS5	25
STS1, STS3	35	STS4, STS5	41
STS1, STS4	50		

(a) Given the percentage recombination between different RFLP loci and the gene for colon cancer shown in the table, draw a genetic map showing the order and genetic distances between adjacent RFLP markers and the gene for colon cancer. (b) Given that the human genome contains approximately 3.3×10^9 base pairs of DNA and that the human genetic map contains approximately 3300 centiMorgans, approximately how many base pairs of DNA are located along the stretch of chromosome defined by this RFLP map? (*Hint:* First figure how many base pairs of DNA are present per cM in the human genome.) (c) How many base pairs of DNA are present in the region between the colon cancer gene and the nearest STS?

16.10 To be useful as a genetic marker for positional cloning of a mutant allele that causes an inherited abnormality in humans, an RFLP must be present on the same homologue as the mutation. Why?

16.11 You have cloned a previously unknown human gene. What procedure will allow you to position this gene on the cytological map of the human genome without performing any pedigree analyses? Describe how you would carry out this procedure.

16.12 GO Eight human–Chinese hamster radiation hybrids were tested for the presence of six human ESTs designated A through F. The results are shown in the following table, where a plus indicates that a marker was present and a minus indicates that it was absent.

Marker	Radiation hybrid 1	2	3	4	5	6	7	8
A	−	+	−	−	+	+	−	+
B	+	−	+	−	−	+	−	+
C	−	+	+	+	−	−	−	+
D	+	−	+	+	−	−	+	−
E	+	−	+	+	−	−	+	−
F	−	+	−	+	+	+	+	−

Based on these data, do any of the ESTs appear to be closely linked? Which ones? What would be needed for you to be more certain of your answer?

16.13 Which type of molecular marker, RFLP or EST, is most likely to mark a disease-causing mutant gene in humans? Why?

16.14 Bacteriophage ΦX174 contains 11 genes in a genome of 5386 bp; *E. coli* has a predicted 4288 genes in a genome of about 4.639 kb; *S. cerevisiae* has about 6000 genes in a genome of size 12.1 mb; *C. elegans* has about 19,000 genes present in a genome of about 100 mb; and *H. sapiens* has an estimated 22,000 genes in its 3000-mb genome. Which genome has the highest gene density? the lowest gene density? Does there appear to be any correlation between gene density and developmental complexity? If so, describe the correlation.

16.15 What advantages do gene chips have over traditional dot blot hybridization techniques?

16.16 An RFLP and a mutation that causes deafness in humans both map to the same location on the same chromosome. How can you determine whether or not the RFLP overlaps with the gene containing the deafness mutation?

16.17 Five human genomic DNA clones present in YAC vectors were tested by hybridization for the presence of six sequence-tagged sites designated STS1 through STS6. The results are given in the following table; a plus indicates the presence of the STS, and a minus indicates the absence of the STS.

YAC clone	STS 1	2	3	4	5	6
A	+	−	+	+	−	−
B	+	−	−	−	+	−
C	−	−	+	+	−	+
D	−	+	−	−	+	−
E	−	−	+	−	−	+

(a) What is the order of the STS sites on the chromosome?
(b) Draw the contig map defined by these data.

16.18 What major advantage does the green fluorescent protein of the jellyfish have over other methods for studying protein synthesis and localization?

16.19 A contig map of one segment of chromosome 3 of *Arabidopsis* is as follows. (a) If an EST hybridizes with genomic clones C, D, and E, but not with the other clones, in which segment of chromosome 3 is the EST located? (b) If a clone of gene *ARA* hybridizes only with genomic clones C and D, in which chromosome segment is the gene located? (c) If a restriction fragment hybridizes with only one of the genomic clones shown above, in which chromosome segment(s) could the fragment be located?

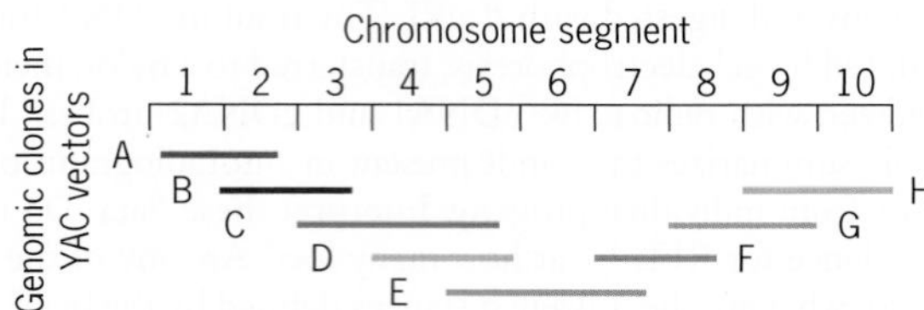

16.20 Of the cereal grass species, only maize contains two copies of each block of linked genes. What does this duplication of sets of maize genes indicate about the origin of this agronomically important species?

16.21 The sequence of a gene in *Drosophila melanogaster* that encodes a histone H2A polypeptide is:

aagtagtcgaaaccgaattccgtagaaacaactcgcacgctccggtttcgtgtt-
gcaacaaaataggcattcccatcgcggcagttagaatcaccgagtgcccagagtcacgt-
tcgtaagcaggcgcagtttacaggcagcagaaaaatcgattgaacagaaatggct-
ggcggtaaagcaggcaaggattcgggcaaggccaaggcgaaggcggtatcgcgttc-
cgcgcgcgcgggtcttcagttccccgtgggtcgcatccatcgtcatctcaagagccg-
cactacgtcacatggacgcgtcggagccactgcagccgtgtactccgctgccatatt-
ggaatacctgaccgccgaggtcctggagttggcaggcaacgcatcgaaggactt-
gaaagtgaaacgtatcactcctcgccacttacagctcgccattcgcggagacgag-
gagctggacagcctgatcaaggcaaccatcgctggtggcggtgtcattccgcacata-
cacaagtcgctgatcggcaaaaaggaggaaacggtgcaggatccgcagcggaag-
ggcaacgtcattctgtcgcaggcctactaagccagtcggcaatcggacgccttcgaaa-
catgcaacactaatgtttaattcagatttcagcagagacaagctaaaacaccgacgagtt-
gtaatcatttctgtgcgccagcatatatttcttatatacaacgtaatacataattatgtaat-
tctagcatctccccaacactcacatacatacaaacaaaaaatacaaacacacaaaacg-
tatttacccgcacgcatccttggcgaggttgagtatgaaacaaaaacaaaacttaatt-
tagagcaaagtaattacacgaataaatttaataaaaaaaactataataaaaacgcc.

Let's use the translation software available on the Internet at http://www.expasy.org/tools/dna.html to translate this gene in all six possible reading frames and see which reading frame specifies histone H2A. Just type or paste the DNA sequence in the "ExPASy Translate Tool" box, and click "TRANSLATE SEQUENCE." The results will show the products of translation in all six reading frames, with **Met**'s and **Stop**'s boldfaced to highlight potential open reading frames. Which reading frame specifies histone H2A?

16.22 At the beginning of this chapter, we discussed the unique gene pool in Iceland and research at deCODE Genetics focused on the identification of genes associated with human diseases. One of the genes studied by deCODE scientists encodes 5-lipoxygenase activating protein (FLAP), which is involved in the synthesis of signaling molecules in white blood cells. Use PubMed's Entrez web site (URL: http://www.ncbi.nlm.nih.gov/entrez) to search for information about deCODE's work on the gene encoding FLAP. *Hint:* Shorten your search by using a two-part query—"5-lipoxygenase activating protein" and "deCODE." (The two components of the query must be separated by a comma followed by a space.) What human diseases are affected by different alleles of the human gene encoding this protein?

16.23 Assume that you have just sequenced a small fragment of DNA that you had cloned. The nucleotide sequence of this segment of DNA is as follows.

aagtagtcgaaaccgaattccgtagaaacaactcgcacgctccggtttcgtgtt-
gcaacaaaataggcattcccatcgcggcagttagaatcaccgagtgcccagagtcacgt-
tcgtaagcaggcgcagtttacaggcagcagaaaaatcgattgaacagaaatggct-
ggcggtaaagcaggcaaggattcgggcaaggccaaggcgaaggcggtatcgcg-
ttccgcgcgcgcggg

In an attempt to learn something about the identity or possible function of this DNA sequence, you decide to perform a BLAST (megablast) search on PubMed's Entrez web site (URL: http://www.ncbi.nlm.nih.gov/entrez). Paste or type this sequence into the query sequence box and make the same changes in Format that were suggested in Focus on GenBank so as to limit the amount of information retrieved. Run the search and examine the sequences most closely related to your query sequence. Are they coding sequences? What proteins do they encode? The first sequence listed is NM_079795.2. Go to the Entrez "Search across databases" tool (click the box at the bottom of the Entrez home page), type or paste NM_079795.2 in as the query, and click "Go." You will get results ("hits") in three databases: nucleotide, HomoloGene, and UniGene. Examine the information in all three databases and see what you can learn about the function of your DNA or the protein that it encodes. What have you learned about your DNA sequence from these searches? Repeat the megablast search with only half of your sequence as the query sequence. Do you still identify the same sequences in the databases? If you use one-fourth of your sequence as a query, do you still retrieve the same sequences? What is the shortest DNA sequence that you can use as a query and still identify the same sequences in the databanks?

16.24 PubMed's Entrez web site (URL: http://www.ncbi.nlm.nih.gov/entrez) can also be used to search for protein sequences. Instead of performing a "megablast" search with a nucleic acid query, one performs a "blastp" with a polypeptide (amino acid sequence) query. Assume that you have the following partial sequence of a polypeptide:

GYDVEKNNSRIKLGLKSLVSKGILVQTKGTGASGS-
FKLNKKAASGEAKPQAKKAGAAKA

Go to the Entrez web site and click protein in the top tool bar, then click BLAST on the left. Then click on protein-protein BLAST (blastp) and enter your query sequence in the box at the top. Scroll down to "Format" and change from "HTML" to "Plain text." Also change the number of "Descriptions" and "Alignments" both to 10. Then click "BLAST!" and "Format" on the subsequent page. Your results should be available in 10 to 15 seconds. The first sequence listed should be identical to your query sequence; the second sequence should differ from your query by a single amino acid. What is the identity of your query sequence?

▶ Genomics on the Web

at http://www.ncbi.nlm.nih.gov/

The chimpanzee, *Pan troglodytes*, is our closest living relative. Humans and chimps evolved from a common ancestor that lived approximately 6 million years ago.

1. How similar are the chimpanzee and human genomes?

2. If you compare some important proteins—for example, the α- and β-globins—of humans and chimps, how similar are their amino acid sequences?

3. If you compare the nucleotide sequences of the genes encoding the α- and β-globins, how similar are they?

4. Are the amino acid sequences of the proteins or the nucleotide sequences of the genes more similar? Why might this be expected?

5. Given the striking similarities between the human and chimpanzee genomes, what kinds of differences do you think are likely to explain the behavioral differences between humans and chimps?

Hint: At the NCBI web site, go to Genome Biology, then under Genome Resources, Genome Projects Database, click on Mammals and go to *Pan troglodytes*. To compare sequences, go back to Genome Biology and click on HomoloGene. Search using *HBB* (the gene symbol for β-hemoglobin), and click on 1. HomoloGene:68066. Scroll down to Show Pairwise Alignments and do a BLAST comparison of the *Pan* and *Homo* β-globins. To compare the nucleotide sequences of the genes, return to the Entrez home page and perform the search in the Nucleotide database and carry out a similar BLAST search using the nucleotide sequence that you found as a query sequence.

Chapter 17
Applications of Molecular Genetics

CHAPTER OUTLINE

- **Use of Recombinant DNA Technology to Identify Human Genes**
- **Molecular Diagnosis of Human Diseases**
- **Human Gene Therapy**
- **DNA Fingerprints**
- **Production of Eukaryotic Proteins in Bacteria**
- **Transgenic Animals and Plants**
- **Reverse Genetics: Dissecting Biological Processes By Inhibiting Gene Expression**

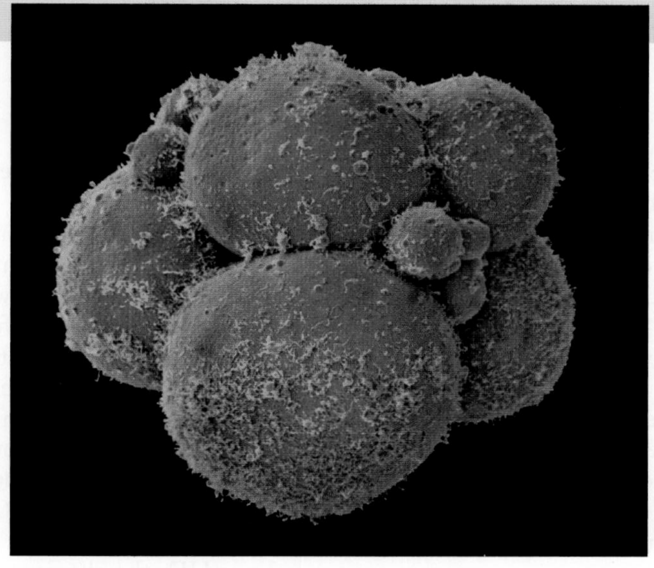

An eight-cell human pre-embryo.

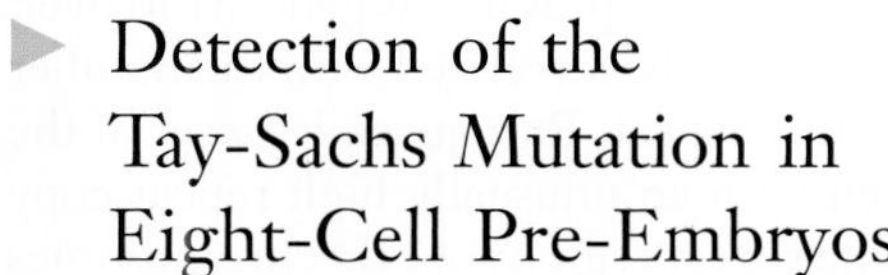

Detection of the Tay-Sachs Mutation in Eight-Cell Pre-Embryos

Brittany was conceived *in vitro* from ova and sperm obtained from her mother and father. Three days after conception, one of the eight cells of the pre-embryo from which Brittany developed was used to conduct DNA tests for a mutant gene that causes Tay-Sachs disease. The results of the tests were negative, and the pre-embryo was implanted in the womb of Brittany's mother. Nine months later, Brittany was delivered with no concern that she might someday die from Tay-Sachs disease as her older sister had. The removal of one cell from an eight-cell pre-embryo does not affect its subsequent development.

Tay-Sachs disease (see Focus on Tay-Sachs Disease, a Childhood Tragedy in Chapter 13) is a lethal autosomal recessive disorder. Infants with Tay-Sachs are normal at birth but undergo rapid neurological degeneration leading to blindness, paralysis, mental retardation, and death at three to four years of age. The degeneration results from the absence of an enzyme called hexosaminidase A, which catalyzes the first step in the breakdown of a complex lipid named ganglioside G_{M2}. In the absence of hexosaminidase A, ganglioside G_{M2} accumulates in neurons and causes progressive degeneration of the central nervous system.

Both of Brittany's parents are heterozygous for the mutant gene that causes Tay-Sachs disease. They knew that if they conceived another child, it would have a 1/4 chance of suffering from Tay-Sachs disease, as did their first daughter. Although they longed for a child, Brittany's parents were not willing to risk having another infant with Tay-Sachs disease. Tay-Sachs disease can be detected prenatally; however, because of religious convictions, the couple had ruled out the abortion of an affected fetus. *In vitro* fertilization and a DNA test for the mutant hexosaminidase A gene that both parents carried allowed confirmation that Brittany was not homozygous for the mutant gene before implantation at the pre-embryo stage. To her parents, who suffered through the degeneration and death of their first child, Brittany is a priceless treasure.

The DNA test for Tay-Sachs disease became possible when the genetic defect that causes this disease was identified. In this chapter, we will consider some of the applications of the recombinant DNA and genomic technologies described in the preceding two chapters. These technologies are, without doubt, the most powerful tools ever developed in the field of biology.

Just as geneticists now know the complete pathway of morphogenesis of bacteriophage T4 (Chapter 13), in the future they will know the complete pathway of morphogenesis of a yeast cell, a fruit fly, an *Arabidopsis* plant, or, indeed, even a human being. Moreover, at some point, biologists will understand the molecular basis of learning and memory and will know what molecular events underlie the aging process. Most important, they will understand the complex mechanisms that regulate cell division in humans and should be able to use this knowledge to prevent or cure at least some types of human cancer and life-threatening viral infections.

Use of Recombinant DNA Technology to Identify Human Genes

The mutant genes that cause Huntington's disease and cystic fibrosis were identified by positional cloning.

Recombinant DNA techniques have revolutionized the search for defective genes that cause human diseases. Indeed, numerous major "disease genes" have already been identified by positional cloning (Chapter 16). In addition, the mutations responsible for the diseases have been determined by comparing the nucleotide sequences of wild-type and mutant alleles of the genes. The coding sequences of the wild-type alleles have been translated by computer to predict the amino acid sequences of the gene products. Oligopeptides have been synthesized based on the predicted amino acid sequences and used to produce antibodies, which, in turn, have been used to localize the gene products and to investigate their functions *in vivo*. The results of these studies will allow future treatment of some of these diseases by gene therapy.

HUNTINGTON'S DISEASE

Huntington's disease (HD) is a genetic disorder caused by an autosomal dominant mutation, which occurs in about one of every 10,000 individuals of European descent. Individuals with HD undergo progressive degeneration of the central nervous system, usually beginning at age 30 to 50 years and terminating in death 10 to 15 years later. To date, HD is untreatable. However, identification of the gene and the mutational defect responsible for HD has kindled hope for an effective treatment in the future. Because of the late age of onset of the disease, most HD patients already have children before the disease symptoms appear. Since the disorder is caused by a dominant mutation, each child of a heterozygous HD patient has a 50 percent chance of being afflicted with the disease. These children observe the degeneration and death of their HD parent, knowing that they have a 50:50 chance of suffering the same fate.

The HD gene was one of the first human genes shown to be tightly linked to an RFLP. In 1983, James Gusella, Nancy Wexler, and coworkers demonstrated that the HD gene cosegregated with an RFLP that mapped near the end of the short arm of chromosome 4. They based their findings largely on data from studies of two large families, one in Venezuela and one in the United States. Subsequent research showed that the linkage was about 96 percent complete; 4 percent of the offspring of HD heterozygotes were recombinant for the RFLP and the HD allele. Given this early localization of the HD gene to a relatively short segment of chromosome 4, some geneticists predicted that the HD gene would soon be cloned and characterized. However, the task was more difficult than anticipated and took a full 10 years to accomplish.

By using positional cloning procedures, Gusella, Wexler, and coworkers identified a gene, first called *IT15* (for *I*nteresting *T*ranscript number *15*) and subsequently named *huntingtin*, that spans about 210 kb near the end of the short arm of chromosome 4 (**FIGURE 17.1**). This gene contains a trinucleotide repeat, $(CAG)_n$, which is present in 11 to 34 copies on each chromosome 4 of healthy individuals. In individuals with HD, the chromosome carrying the HD mutation contains 42 to 100 or more copies of the CAG repeat in this gene. Moreover, the age of onset of HD is inversely correlated with the number of copies of the trinucleotide repeat. Rare juvenile onset of the disease occurs in children with an unusually high repeat copy number. The trinucleotide repeat regions of HD chromosomes are unstable, with repeat numbers often expanding and sometimes contracting between generations. Gusella, Wexler, and collaborators detected expanded CAG repeat regions in chromosomes from 72 different families with HD, leaving little doubt that they had identified the correct gene.

The *huntingtin* gene is expressed in many different cell types, producing a large 10- to 11-kb mRNA. The coding region of the *huntingtin* mRNA predicts a protein 3144 amino acids in length. Unfortunately, the predicted amino acid sequence of the huntingtin protein has provided little information about its function. The dominance of the HD mutation indicates that the mutant protein causes the disease.

The expanded CAG repeat region in the mutant *huntingtin* gene encodes an abnormally long polyglutamine region near the amino terminus of the protein. The elongated polyglutamine region fosters protein-protein interactions that lead to the accumulation of aggregates of the huntingtin protein in brain cells. These protein aggregates are thought to cause the clinical symptoms of HD, and current approaches to treatment involve attempts to disrupt or eliminate these protein aggregates.

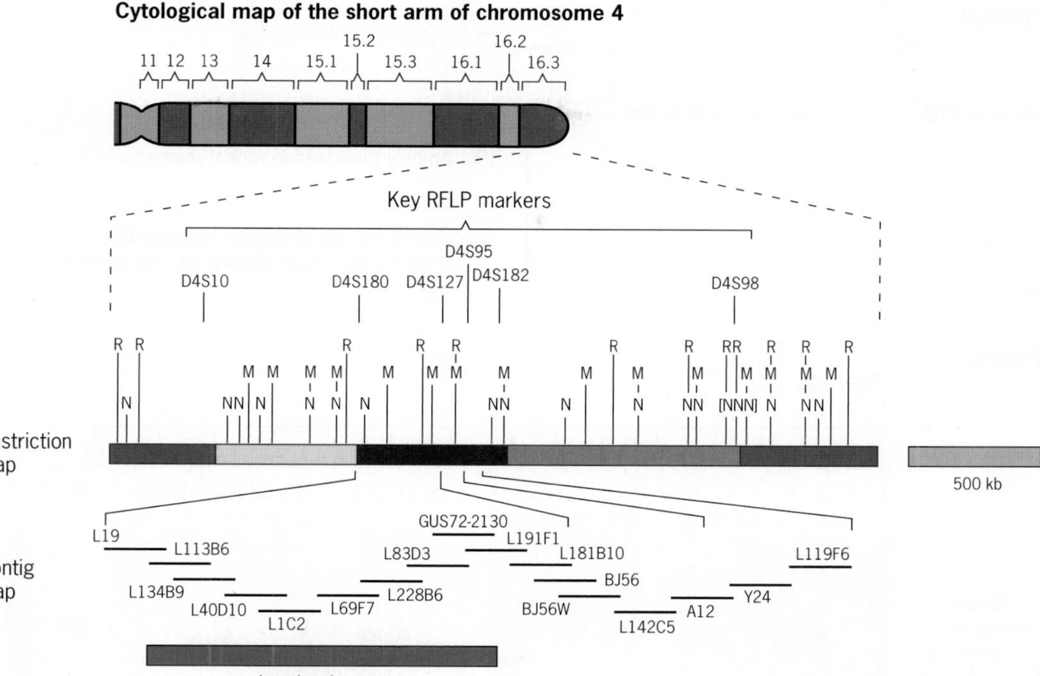

Figure 17.1 ► Identification of the gene responsible for Huntington's disease by positional cloning. The cytological map of the short arm of chromosome 4 is shown at the top. The RFLP markers, restriction map, and contig map used to locate the *huntingtin* gene are shown below the cytological map. M, N, and R represent *Mlu*I, *Not*I, and *Nru*I restriction sites, respectively.

HD was the fourth human disease to be associated with an unstable trinucleotide repeat. In 1991, fragile X syndrome—the most common form of mental retardation in humans—was the first human disorder to be associated with an expanded trinucleotide repeat. We discuss the discovery of the nature of the mutation responsible for the fragile X syndrome in A Milestone in Genetics: Trinucleotide Repeats and Human Disease. Shortly thereafter, myotonic dystrophy and spinobulbar muscular atrophy (both diseases associated with loss of muscle control) were shown to result from expanded trinucleotide repeats. By 2004, over 40 different human disorders—many associated with neurodegenerative abnormalities—had been shown to result from expanded trinucleotide repeats. They include several types of spinocerebellar ataxia, dentatorubro-pallidoluysian atrophy (Haw River syndrome), Friedreich ataxia, and fragile X syndrome. The high frequency of human disorders caused by the expansion of trinucleotide repeats indicates that this may be a common mutational event in our species.

Although the identification of the genetic defect, the expanded trinucleotide repeat in the *huntingtin* gene, has not led to a treatment of the disorder, it has provided a simple and accurate DNA test for the HD mutation (**FIGURE 17.2**). Once the nucleotide sequences of the *huntingtin* gene on either side of the trinucleotide repeat region were known, oligonucleotide primers could be synthesized and used to amplify the region by PCR, and the number of CAG repeats could be determined by polyacrylamide gel electrophoresis. Thus, individuals at risk of carrying the mutant *huntingtin* gene can easily be tested for its presence. Because the PCR procedure requires little DNA, the test for HD also can be performed prenatally on fetal cells obtained by amniocentesis or chorionic biopsy (see Focus on Amniocentesis and Chorionic Biopsy in Chapter 6). Test your understanding of the DNA test for HD by devising a DNA test for the mutant alleles containing expanded CGG trinucleotide repeats that cause the most common form of mental retardation in humans (see Focus on Problem Solving: Testing for Mutant Alleles that Cause Fragile X Mental Retardation).

Given the availability of the DNA test for the HD mutation, individuals who are at risk of transmitting the defective gene to their children can determine whether they carry it before starting a family. Each person with a heterozygous parent has a 50 percent chance of not carrying the defective gene. If the test is negative, she or he can begin a family with no concern about transmitting the mutation. If the test is positive, the fetus can be tested prenatally, or the couple can consider *in vitro* fertilization, as did the parents we discussed at the beginning of this chapter. If the eight-cell pre-embryo tests negative for the HD mutation, it can be implanted in the mother's uterus with the knowledge that it carries two normal copies of the *huntingtin* gene. If used conscientiously, the DNA test for the HD mutation should diminish human suffering from this dreaded disease.

CYSTIC FIBROSIS

Cystic fibrosis (CF) is one of the most common inherited diseases in humans, affecting 1 in 2000 newborns of northern European heritage. CF is inherited as an autosomal recessive mutation, and the frequency of heterozygotes is estimated to be about 1 in 25 in Caucasian populations. In the United States alone, over 30,000 people suffer from this devastating disease.

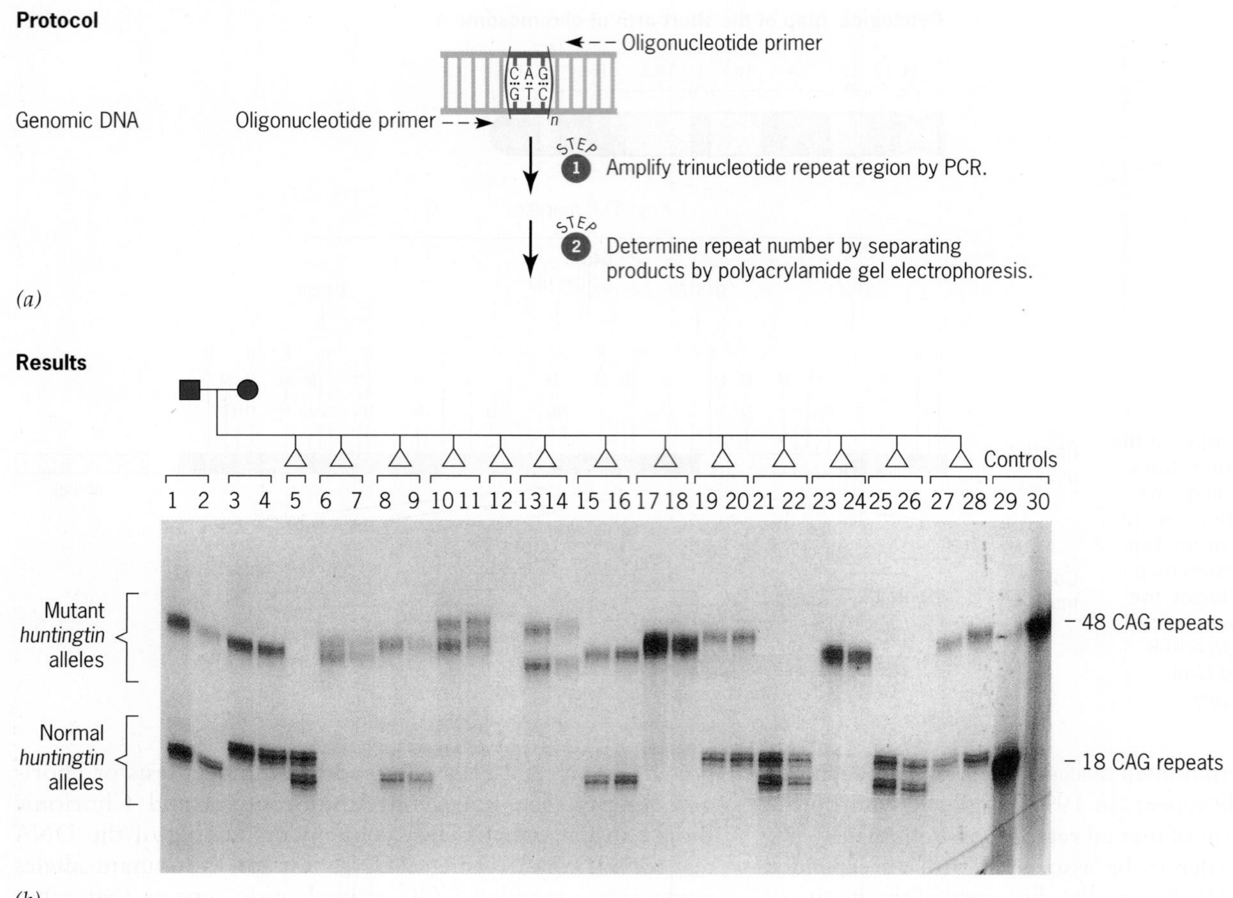

Figure 17.2 ▶ Testing for the expanded trinucleotide repeat regions (*a*) in the *huntingtin* gene that are responsible for Huntington's disease by PCR. The results shown in (*b*) are from a Venezuelan family in which the parents are heterozygous for the same mutant *huntingtin* allele. The order of birth of the children has been changed, and their sex is not given to assure anonymity. Most individuals were tested twice to minimize errors.

One easily diagnosed symptom of CF is excessively salty sweat, a largely benign effect of the mutant gene. Other symptoms are anything but benign. The lungs, pancreas, and liver become clogged with a thick mucus, which results in chronic infections and the eventual malfunction of these vital organs. In addition, mucus often builds up in the digestive tract, causing individuals to be malnourished no matter how much they eat. Lung infections are recurrent, and patients often die from pneumonia or other infections of the respiratory system. In 1940, the average life expectancy for a newborn with CF was less than two years. With improved methods of treatment, life expectancy has gradually increased. Today, the life expectancy for someone with CF is about 32 years, but the quality of life is poor.

Identification of the *CF* gene is one of the major successes of positional cloning (Chapter 16). Biochemical analyses of cells from CF patients had failed to identify any specific metabolic defect or mutant gene product. Then, in 1989, Francis Collins and Lap-Chee Tsui and their coworkers identified the *CF* gene and characterized some of the mutations that cause this tragic disease. The cloning and sequencing of the *CF* gene quickly led to the identification of its product, which in turn has suggested approaches to clinical treatment of the disease and hope for successful gene therapy in the future.

The *CF* gene was first mapped to the long arm of chromosome 7 by its cosegregation with RFLPs. Further RFLP mapping localized the gene to a 500-kb region of chromosome 7. The two RFLP markers closest to the *CF* gene were then used to initiate chromosome walks (see **FIGURE 16.7**) and jumps (see **FIGURE 16.8**) and to begin construction of a detailed physical map of the region (**FIGURE 17.3**). Three kinds of information were used to narrow the search for the *CF* gene.

1. Human genes are often preceded by clusters of cytosines and guanines called **CpG islands** (Chapter 20). Three such clusters are present just upstream from the *CF* gene (**FIGURE 17.3**).
2. Important coding sequences usually are conserved in related species. When exon sequences from the *CF* gene were used to probe Southern blots containing restriction fragments from human, mouse, hamster, and bovine genomic DNAs (often called *zoo blots*), the exons were found to be highly conserved.
3. As previously mentioned, CF is known to be associated with abnormal mucus in the lungs, pancreas, and sweat glands. A

FOCUS ON PROBLEM SOLVING

Testing for Mutant Alleles that Cause Fragile X Mental Retardation

THE PROBLEM

The most common inherited type of mental retardation in humans is caused by expanded CGG trinucleotide repeats in the FMR-1 (for fragile X mental retardation gene 1) gene. See A Milestone in Genetics: Trinucleotide Repeats and Human Disease later in this chapter for details. Design a DNA test for the presence of FMR-1 mutant alleles. How will the results of the test tell you whether an individual is homozygous or heterozygous for the mutant allele when present?

FACTS AND CONCEPTS

1. Normal individuals usually have 6 to 59 copies of the CGG trinucleotide present in the region between the promoter and the translation start site of the FMR-1 gene.
2. Individuals with fragile X syndrome usually have more than 200 copies of this trinucleotide.
3. The entire euchromatic portion of the human genome has been sequenced. Thus, the sequence of the FMR-1 gene and the genomic sequences flanking it are known.
4. PCR can be used to amplify the region of the FMR-1 gene that contains the CGG trinucleotide repeats.
5. Agarose gel electrophoresis can be used to determine the sizes of DNA molecules.

ANALYSIS AND SOLUTION

1. Synthesize forward and reverse oligonucleotide PCR primers (see **FIGURE 15.14**) that are complementary to sequences flanking the trinucleotide repeat region of the FMR-1 gene.

2. Use these primers to amplify the trinucleotide repeat region in genomic DNA samples from the individuals to be tested. Genomic DNAs from individuals with a known number of CGG trinucleotide repeats—both normal and expanded—should be included as controls.

3. Use agarose gel electrophoresis to determine the sizes of the amplified DNAs (see **FIGURE 15.19**). The controls will serve as size markers in this analysis.

4. DNA samples from individuals who are heterozygous for normal and expanded FMR-1 alleles will yield two amplified DNA fragments—a smaller fragment containing 6 to 59 copies of the repeat and a larger fragment containing more than 200 copies of the repeat. DNA samples from individuals who are homozygous for an FMR-1 allele will yield one amplified DNA fragment—small if two normal alleles are present, larger if two mutant alleles are present.

For further discussion go to your *WileyPLUS* course.

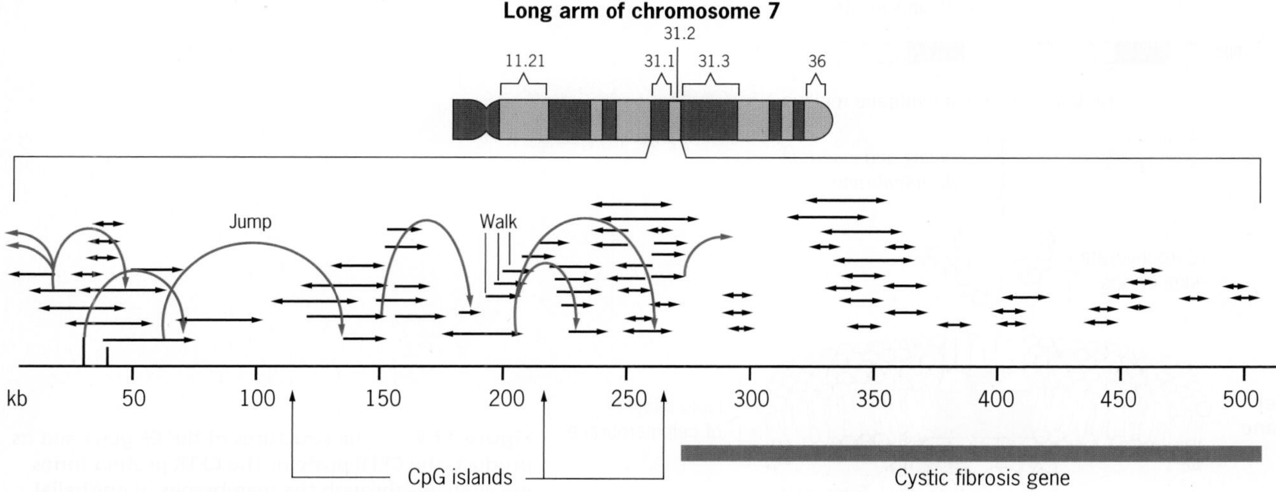

Figure 17.3 ▶ The sequence of chromosome walks and jumps used to locate and characterize the cystic fibrosis gene. The positions of CpG islands used as landmarks in locating the 5′ end of the gene are also shown.

cDNA library was prepared from mRNA isolated from sweat gland cells growing in culture and screened by colony hybridization using exon probes from the *CF* gene (candidate *CF* gene at the time).

Use of the sweat gland cDNA library proved to be critical in identifying the *CF* gene because northern blot experiments subsequently showed that this gene is expressed only in epithelial cells of the lungs, pancreas, salivary glands, sweat glands, intestine, and reproductive tract. Thus, cDNA clones of the *CF* gene would not have been identified using cDNA libraries prepared from other tissues and organs. The northern blot results also showed that the putative *CF* gene is expressed in the appropriate tissues.

Identification of a candidate gene as a disease gene hinges on comparisons of normal and mutant alleles from several different families. CF is unusual in that 70 percent of the mutant alleles contain the same three-base deletion, *ΔF508*, which eliminates the phenylalanine residue at position 508 in the *CF* gene product. Unlike the *huntingtin* gene, the nucleotide sequence of the *CF* gene proved very informative. The gene is huge, spanning 250 kb and containing 24 exons (**FIGURE 17.4**). The CF mRNA is about 6.5 kb in length and encodes a protein of 1480 amino acids. A computer search of the protein databanks quickly showed that the *CF* gene product is similar to several ion channel proteins, which form pores between cells through which ions pass. The *CF* gene product, called the *cystic fibrosis transmembrane conductance regulator*, or *CFTR protein*, forms ion channels (**FIGURE 17.4**) through the membranes of cells that line the respiratory tract, pancreas, sweat glands, intestine, and other organs and regulates the flow of salts and water in and out of these cells. Because the mutant CFTR protein does not function properly in CF patients, salt accumulates in epithelial cells and mucus builds up on the surfaces of these cells.

The presence of mucus on the lining of the respiratory tract leads to chronic, progressive infections by *Pseudomonas aeruginosa*, *Staphylococcus aureus*, and related bacteria. These infections, in turn, frequently result in respiratory failure and death. However, the mutations in the *CF* gene are pleiotropic; they cause a

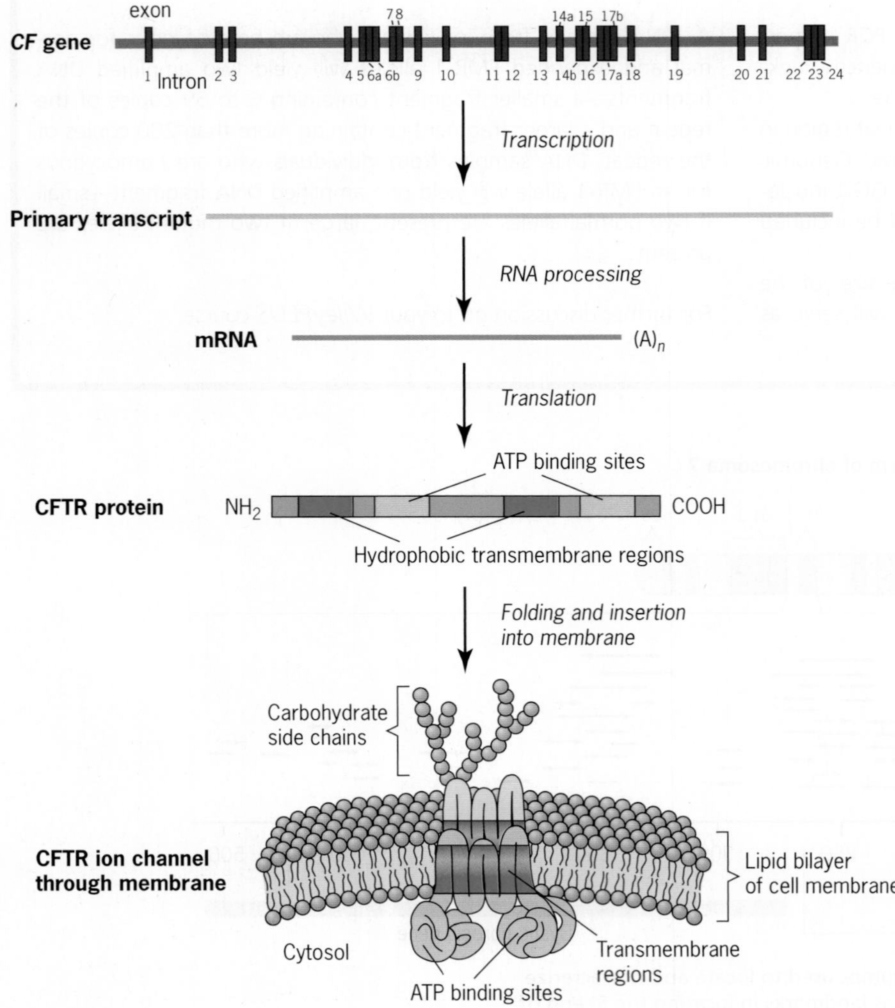

Figure 17.4 ► The structures of the *CF* gene and its product, the CFTR protein. The CFTR protein forms ion channels through the membranes of epithelial cells of the lungs, intestine, pancreas, sweat glands, and some other organs.

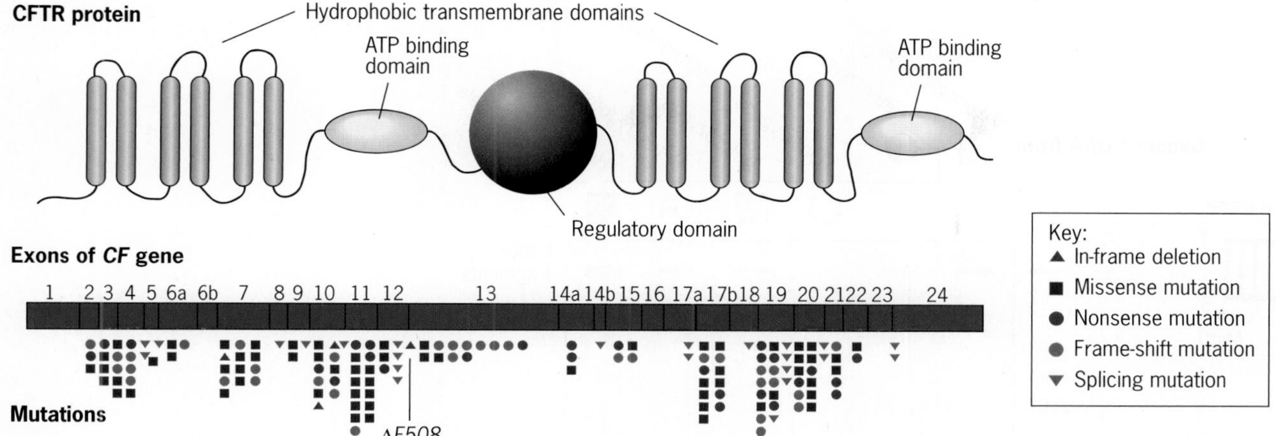

Figure 17.5 ▶ Mutations in the *CF* gene that cause cystic fibrosis. The distribution and classification of the mutations that cause cystic fibrosis are shown below the exons of the *CF* gene. A schematic diagram of the CFTR protein is shown above the exon map to illustrate the domains of the protein that are altered by the mutations. About 70 percent of all cases of CF result from mutation *ΔF508*, which deletes the phenylalanine present at position 508 of the normal CFTR protein.

number of distinct phenotypic effects. Malfunctions of the pancreas, liver, bones, and intestinal tract are common in individuals with CF. Although CFTR forms chloride channels (**FIGURE 17.4**), it also regulates the activity of several other transport systems such as potassium and sodium channels. Some work suggests that CFTR may play a role in regulating lipid metabolism and transport. CFTR interacts with a number of other proteins and undergoes phosphorylation/dephosphorylation by kinases and phosphatases. Thus, CFTR should be considered multifunctional. Indeed, some of the symptoms of CF may result from the loss of CFTR functions other than the chloride channels.

Although 70 percent of the of CF cases are due to the *ΔF508* trinucleotide deletion, over 170 different CF mutations have been identified (**FIGURE 17.5**). About 20 of these mutations are quite common; others are rare, and many have been identified in only one individual. Several of these mutations can be detected by DNA screens such as the test for the *ΔF508* deletion illustrated in the Focus on Detection of a Mutant Gene Causing Cystic Fibrosis in Chapter 15. These tests can be performed on fetal cells obtained by amniocentesis or chorionic biopsy. They have also been done successfully on eight-cell pre-implantation embryos produced by *in vitro* fertilization. The diversity of the mutations that cause CF (see **FIGURE 17.5**) makes it very difficult to devise DNA tests for all of the mutant CF alleles.

KEY POINTS

- The mutant genes responsible for Huntington's disease and cystic fibrosis were identified by positional cloning.
- The nucleotide sequences of the *huntingtin* and *CF* genes were used to predict the amino acid sequences of their polypeptide products and to obtain information about the functions of the gene products.
- The characterization of the *huntingtin* and *CF* genes has led to the development of DNA tests that detect some of the mutations that cause Huntington's disease and cystic fibrosis.

Molecular Diagnosis of Human Diseases

Mutant genes that cause inherited human diseases can often be detected by tests performed on genomic DNA.

Once the gene responsible for a human disease has been cloned and sequenced and the nature of the mutations that cause the disorder is known, a molecular test for the mutant alleles usually can be designed. These tests can be performed on small amounts of DNA by using PCR to amplify the DNA segment of interest (**FIGURE 15.14**). Thus, they can be performed prenatally on fetal cells obtained by amniocentesis or chorionic biopsy, or even on a single cell from a pre-embryo produced by *in vitro* fertilization.

Some molecular diagnoses involve simply testing for the presence or absence of a specific restriction enzyme cleavage site in DNA. For example, the mutation that causes sickle-cell anemia (Chapter 1) removes a cleavage site for the restriction enzyme *Mst*II (**FIGURE 17.6**). The HBB^S (sickle-cell) allele can be distinguished from the normal β-globin allele (HBB^A) by amplifying part of the β-globin gene by PCR, cutting the amplified DNA with *Mst*II, separating the resulting restriction fragments by agarose gel electrophoresis, preparing a Southern

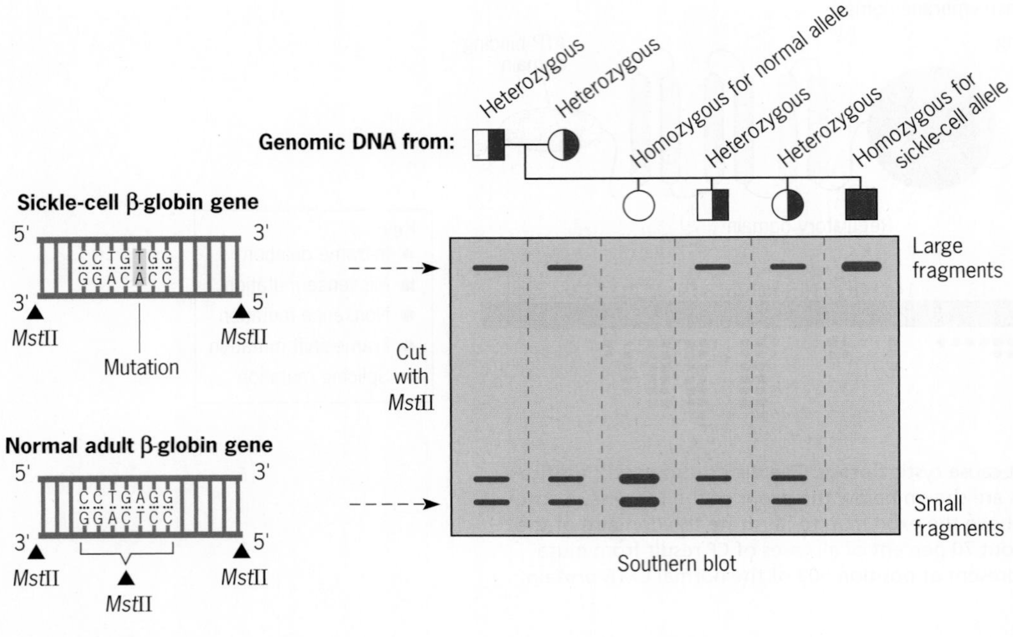

Figure 17.6 ▶ Detection of the sickle-cell hemoglobin mutation by Southern blot analysis of genomic DNAs cut with restriction enzyme *Mst*II.

blot of the separated fragments, and hybridizing the DNA on the blot to a probe spanning the site of the mutation. The probe will hybridize with two small fragments from the normal β-globin gene, but with only one fragment from the sickle-cell β-globin gene (**FIGURE 17.6**). Thus, this test allows for the detection of heterozygotes as well as homozygotes.

Alternatively, PCR primers can be synthesized that are complementary to DNA sequences flanking the sickle-cell mutation in the $Hb_{\beta}{}^{S}$ gene and used to amplify this segment from genomic DNA. The amplified DNA can be treated with *Mst*II and the products of the reaction examined by agarose gel electrophoresis to see whether or not the amplified fragment was cleaved by *Mst*II.

For inherited disorders such as Huntington's disease and fragile X syndrome, which result from expanded trinucleotide repeat regions in genes, PCR and Southern blots can be used to detect the mutant alleles. The DNA test for the *huntingtin* gene is illustrated in **FIGURE 17.2**. Other types of mutations can be detected by using allele-specific oligonucleotides to probe genomic Southern blots. This procedure is illustrated for the Δ*F508* mutation in the *CF* gene—the most frequent cause of cystic fibrosis—in Focus on Detection of a Mutant Gene Causing Cystic Fibrosis in Chapter 15. Indeed, once the mutations responsible for a disease have been characterized, the development of DNA tests to detect the most common ones is usually routine. Clearly, the availability of diagnostic tests for mutations that cause human diseases has contributed greatly to the field of genetic counseling, providing invaluable information to families in which the genetic defects occur.

KEY POINTS

- Mutant genes that are responsible for inherited human disorders can often be diagnosed by DNA tests.
- The results of DNA tests for mutant genes that cause inherited diseases allow genetic counselors to inform families of the risks of having affected children.

Human Gene Therapy

Gene therapy—introducing functional copies of a gene into an individual with two defective copies of the gene—is a potential tool for treating inherited human diseases.

Of the approximately 5000 inherited human diseases catalogued to date, only a few are currently treatable. For many of these diseases, the missing or defective gene product cannot be supplied exogenously, as insulin is supplied to diabetics. Most enzymes are unstable and cannot be delivered in functional form to their sites of action in the body, at least not in a form that provides for long-term activity. Cell membranes are impermeable to large macromolecules such as proteins; thus, enzymes must be synthesized in the cells where they are needed. Therefore, treatment of inherited diseases is largely restricted to those cases where the missing metabolite is a small molecule that can be distributed to the appropriate tissues of the body through the circulatory system, or the symptoms can be controlled by modifying the individual's diet. For many other inherited diseases, **gene therapy** offers the most promising approach to successful treatment. Gene therapy involves adding a normal (wild-type) copy of a gene to the genome of an individual carrying defective copies of the gene. A gene that has been introduced into a cell or organism is called a **transgene** (for *trans*ferred *gene*) to distinguish it from endogenous genes, and the organism carrying the introduced gene is said to be **transgenic.** If gene therapy is successful, the transgene will synthesize the missing gene product and restore the normal phenotype.

Before considering specific examples, we need to discuss two types of gene therapy: **somatic-cell** or **nonheritable gene therapy,** and **germ-line** or **heritable gene therapy.** In higher animals such as humans, the reproductive or germ-line cells are produced by a cell lineage separate from all somatic-cell lineages. Thus, somatic-cell gene therapy will treat the disease symptoms of the individual but will not cure the disease. That is, the defective gene(s) will still be present in the germ-line cells of the patient after somatic-cell gene therapy and may be transmitted to his or her children. All of the gene-therapy treatments of human diseases that we will discuss here are somatic-cell gene therapies. Germ-line gene therapy is being performed on mice and other animals, but not on humans.

The distinction between somatic-cell and germ-line gene therapy is important when we discuss humans. The frequently expressed concerns about humankind's "tinkering with nature" or "playing God" apply to germ-line gene transfers, not to somatic-cell gene therapy. Major moral and ethical considerations are involved in any decision to perform germ-line modifications of human genes. In contrast, somatic-cell gene therapy is no different from enzyme (gene-product) therapy or cell, tissue, and organ transplants. In transplants, entire organs, with all the foreign genes present in the genome of every cell in the organ, are implanted in patients. In current somatic-cell gene therapies, some of the patient's own cells are removed, repaired, and reimplanted in the patient. Thus, somatic-cell gene therapy is less complex and less life-threatening for an individual than an organ transplant.

To perform somatic-cell gene therapy, wild-type genes must be introduced into and expressed in cells homozygous or hemizygous for a mutant allele of the gene. In principle, the wild-type gene could be delivered to the mutant cells by any of several different procedures. Most commonly, viruses are used to carry the wild-type gene into cells. In the case of retroviral vectors, the wild-type transgene is integrated—along with the retroviral DNA—into the DNA of the host cell. Thus, when retroviral vectors are used, the transgene is transmitted to all progeny cells in the affected cell lineage.

With other viral vectors, such as those derived from adenoviruses, the transgenes are present only transiently in host cells, because the genomes of these viruses replicate autonomously and persist only until the immune system eliminates the viruses along with the infected cells. The advantage of these vectors over retroviral vectors is that no potentially harmful mutations are induced during the integration step (Chapter 13). However, they have two major disadvantages: (1) transgene expression is transient, lasting only as long as the viral infection persists, and (2) most humans exhibit strong immune responses to these viruses, presumably because of prior exposure to the same or closely related viruses. For example, in early attempts to treat cystic fibrosis by somatic-cell gene therapy, an adenoviral vector carrying the *CF* gene was inhaled by patients, with the hope that lung cells would become infected and synthesize enough of the *CF* gene product to alleviate some of the symptoms of the disease. Unfortunately, these treatments proved ineffective, at least in part because of rapid immune responses to these viruses in the individuals receiving the treatments.

With diseases such as cystic fibrosis, where effective gene therapy will require long-term transgene expression, the standard adenovirus vectors probably will not work. Because transgene expression is transient, the treatments will need to be repeated periodically. However, given that secondary immune responses are very rapid and efficient, subsequent treatments with the same viral vector probably will be ineffective.

Human gene therapy is performed under strict guidelines developed by the National Institutes of Health (NIH). Each proposed gene-therapy procedure is scrutinized by review committees at both the local (institution or medical center) and national (NIH) levels. Several requirements must be fulfilled before a gene-therapy procedure will be approved:

1. The gene must be cloned and well characterized; that is, it must be available in pure form.
2. An effective method must be available for delivering the gene into the desired tissue(s) or cells.
3. The risks of gene therapy to the patient must have been carefully evaluated and shown to be minimal.
4. The disease must not be treatable by other strategies.
5. Data must be available from preliminary experiments with animal models or human cells and must indicate that the proposed gene therapy should be effective.

A gene-therapy proposal will not be approved by the local and national review committees until they are convinced that all of the above conditions have been fulfilled. Moreover, with the unfortunate death in September 1999 of Jesse Gelsinger, an 18-year-old with ornithine transcarbamylase deficiency, due to a severe immune reaction to the adenovirus vector used in his experimental gene therapy, the review committees are being especially cautious in their evaluation of gene-therapy proposals.

The first use of gene therapy in humans occurred in 1990, when a four-year-old girl with **adenosine deaminase-deficient severe combined immunodeficiency disease (ADA^- SCID)** received her first transgene treatment. Several other ADA^-

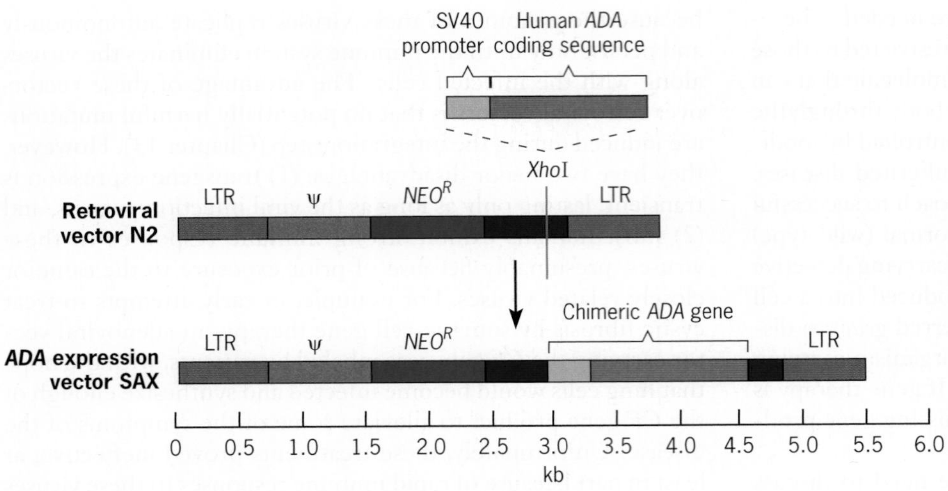

Figure 17.7 ► Construction of the human adenosine deaminase (*ADA*) gene-transfer vector SAX (for simian virus 40, *ADA*, *Xho*I site). In this vector, expression of the human *ADA* coding sequence is controlled by a strong promoter obtained from simian virus 40. The NEO^R gene makes cells resistant to the antibiotics neomycin, kanamycin, and G418. LTR refers to the long terminal repeats from the Moloney murine leukemia virus; ψ is a DNA sequence required to package vector DNA in virions.

SCID patients have subsequently been treated, along with a few individuals with other inherited disorders. SCID is a rare autosomal disease of the immune system. Individuals with SCID have essentially no immune system, so that even minor infections can become serious and often fatal. Some SCID patients lack an enzyme called adenosine deaminase (ADA). In the absence of this enzyme, toxic levels of the phosphorylated form of its substrate, deoxyadenosine, accumulate in T lymphocytes (white blood cells essential to an immune response) and kill them. T lymphocytes stimulate cells called B lymphocytes to develop into antibody-producing plasma cells. Thus, in the absence of T lymphocytes, no immune response is possible, and newborns with ADA^- SCID seldom live more than a few years. Four factors made ADA^- SCID a good candidate for somatic-cell gene therapy. First, the *ADA* gene was one of the first human disease genes to be cloned and characterized. Second, white blood cells can easily be obtained from ADA^- SCID patients and reintroduced after functional copies of the *ADA* gene are added. Third, even a small amount of functional ADA will restore partial immune function. Fourth, the overproduction of ADA does not appear to have toxic effects on the patient.

As a prerequisite to gene therapy, a retroviral gene-transfer vector was constructed with the human *ADA* gene under the control of a strong viral promoter (**FIGURE 17.7**), and expression of the chimeric gene was demonstrated in white blood cells infected with the retroviral vector. The actual treatment of ADA^- SCID by gene therapy involves four sequential steps: (1) isolation of white blood cells from the patient, (2) introduction of functional copies of the *ADA* gene into these cells, (3) demonstration of transgene expression in these cells, and (4) infusion of the transgenic cells back into the patient (**FIGURE 17.8**).

Although the number of ADA^- SCID patients who have been treated by gene therapy is still small, and these patients have continued to receive treatments with purified ADA, the therapy has resulted in improved immune function. Most are attending normal schools, and some have survived infectious diseases such as chicken pox. It was anticipated that the effects of gene therapy using white blood cells would be short-lived because of the limited life span of these cells. Thus, repeated infusions of white blood cells carrying functional *ADA* genes have been performed on the patients. The major problem encountered in attempts to treat ADA^- SCID patients by gene therapy has not been the short life span of the white blood cells, but the transient expression of the introduced *ADA* transgenes. Within a few weeks after treatment, transcription of the *ADA* transgene has subsided. The activity of the viral promoters that control this transcription has apparently been arrested, indicating that additional research is needed to understand why transgenes are often silenced (expression turned off) in their new hosts.

To avoid the limitations resulting from the short life span of white blood cells, the bone marrow stem cells that give rise to white blood cells could be used to treat immune disorders such as ADA^- SCID. The modified stem cells should continually produce T lymphocytes with the *ADA* transgene and could provide a permanent or long-term treatment of the disease. Indeed, stem-cell gene therapy was first used to treat two infants with ADA^- SCID in 1993, and this procedure will undoubtedly be the method of choice in the future.

During the year 2000, British and French physicians performed what at the time appeared to be the first successful somatic-cell gene-therapy treatment of individuals with a fatal inherited disease. They treated boys with a type of SCID similar to the ADA^- SCID previously discussed but caused by mutations in a gene on the X chromosome. This X-linked SCID results from the loss or inactivation of the γ subunit of the interleukin-2 receptor. Interleukin-2 is a signaling molecule required for the development of cells of the immune system. However, the γ polypeptide of the interleukin-2 receptor is also a component of several other lymphocyte-specific growth factors. Collectively, they stimulate the development of B and T lymphocytes—cells required for the production of antibody-producing plasma cells and killer T cells, respectively. In the absence of the γ polypeptide, an individual has no functional immune system and seldom survives for more than a few years.

Like the individuals with ADA^- SCID, boys with X-linked SCID seemed to be good candidates for treatment by somatic-

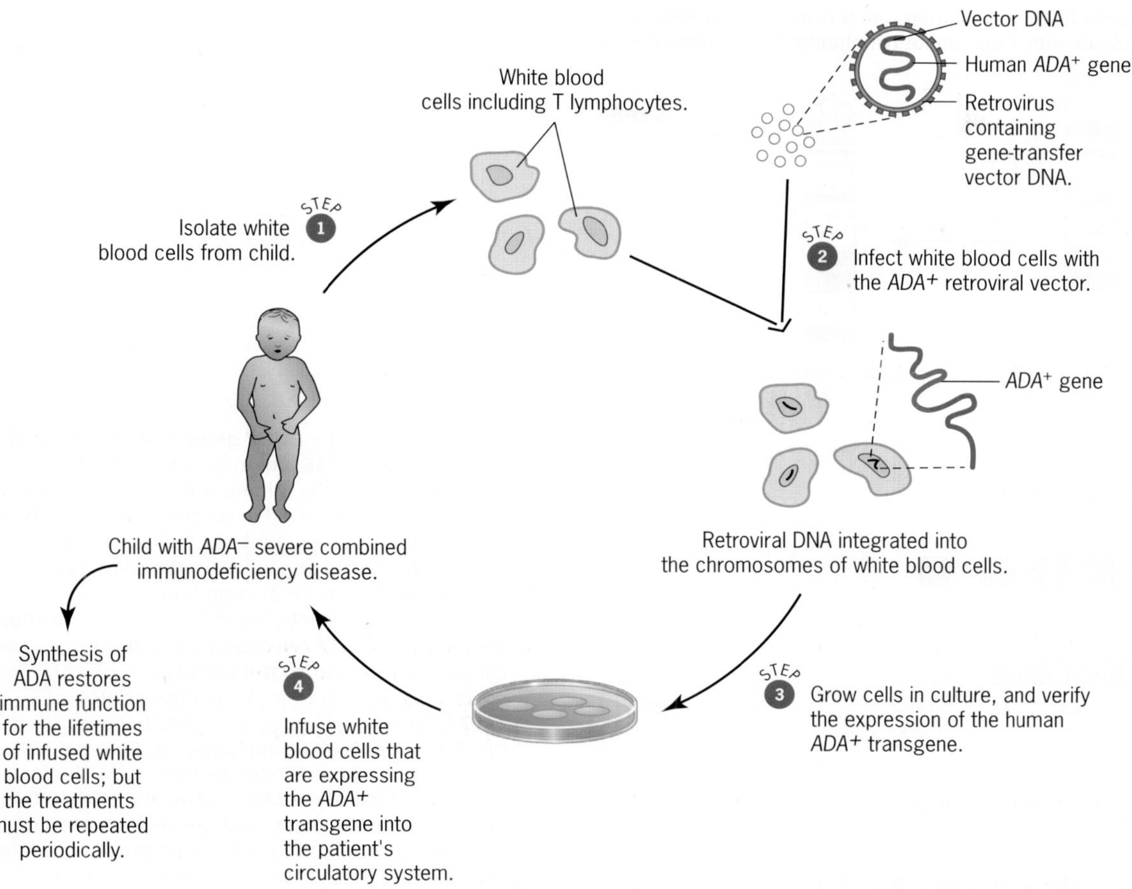

Figure 17.8 ▶ Treatment of adenosine deaminase-deficient severe combined immunodeficiency disease (ADA⁻ SCID) by somatic-cell gene therapy. ADA⁻ SCID results from the loss or lack of activity of the enzyme adenosine deaminase (ADA). Gene therapy is performed by isolating white blood cells from the patient, introducing a wild-type copy of the *ADA* gene into these cells with a retroviral vector, verifying the expression of the transgene in cultured cells, and infusing the transformed white blood cells back into the patient.

cell gene therapy. Thus, the gene encoding the γ subunit of the human interleukin-2 receptor was cloned, inserted into a retroviral vector, introduced into hematopoietic stem cells (precursors to cells of the circulatory system) isolated from patients with X-linked SCID, and checked for gene expression while the cells were still growing in culture medium. After verifying expression of the gene (designated *IL2Rγc* for *i*nter*l*eukin-*2* *r*eceptor γ *c*ommon), the stem cells were transfused back into the SCID patients from whom they had been isolated. During the next two years, 14 boys with X-linked SCID were treated. In all 14 cases, gene therapy cured the immunodeficiency, resulting in normal T-cell levels within a few months after treatment. Thus, for two years, everything indicated that the gene therapy had been a major success. Then one of the boys developed acute T-cell leukemia, a cancer of the white blood cells. Later, the same T-cell leukemia was detected in three more of the gene-therapy patients. Clearly, something had gone wrong.

One advantage of retroviral vectors is that they insert themselves into the chromosomes of host cells and, therefore, are transmitted to progeny cells during cell division. However, like transposable elements, they can cause mutation by inserting themselves into genes of host cells (see **FIGURE 13.20**). In addition, some retroviral DNAs up-regulate the expression of genes close to their sites of integration, and the vector (derived from components of the Moloney murine leukemia virus) used to introduce the *IL2Rγc* gene into X-linked SCID patients was of this type.

When the location of the viral DNA carrying the *IL2Rγc* gene was determined in the first two boys who developed leukemia, the vector was found in the same gene in both cases (**FIGURE 17.9**). The retroviral DNA had integrated into a gene that was known to be associated with T-cell acute lymphoblastic leukemia (T-cell ALL) in individuals carrying a unique translocation chromosome. The translocation fused the *TCRβ* (*T*-*c*ell *r*eceptor β subunit) gene on chromosome 7 with the 5′ region of the *LMO2* (LIM-only) gene on chromosome 11 (**FIGURE 17.9*a*** and ***b***). *LMO2* encodes a protein that is essential for the formation of certain transcription factor complexes. The expression of *LMO2* is normally down-regulated during the development of T cells. When it is overexpressed in T cells, it

Identification of the *LMO2* oncogene by its association with a translocation between chromosomes 7 and 11 in individuals with T-cell acute lymphoblastic leukemia (T-cell ALL).

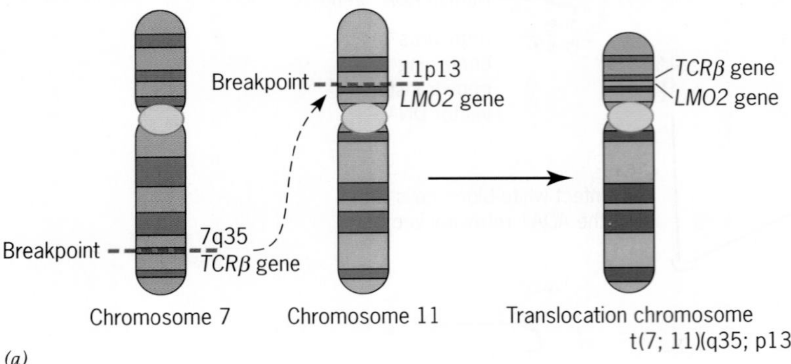

(a)

Structure of the normal *LMO2* gene and mRNA.

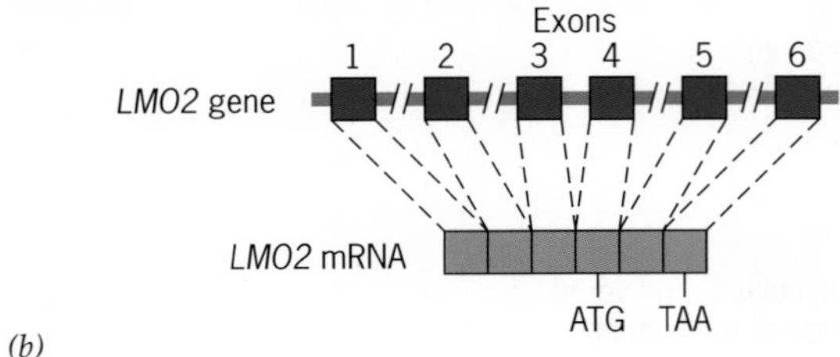

(b)

Site of retroviral DNA insertion in boy 1 with leukemia.

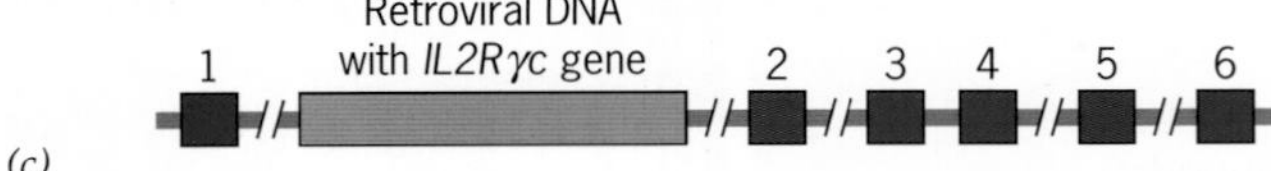

(c)

Site of retroviral DNA insertion in boy 2 with leukemia.

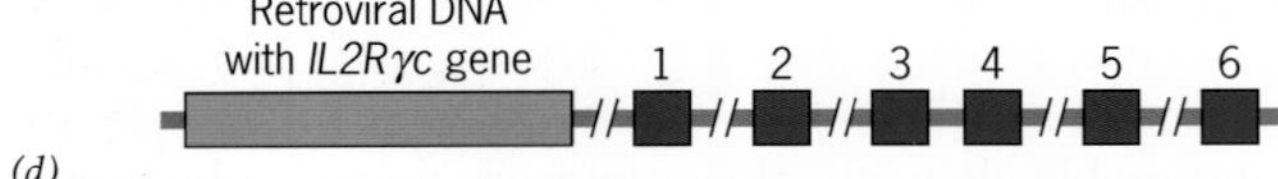

(d)

Figure 17.9 ▶ The *LMO2* gene (*LIM*-only gene 2) encodes a small (156 amino acids) protein that functions as a bridge joining different transcription factors. It was identified in studies of individuals with T-cell acute lymphoblastic leukemia (T-cell ALL, a cancer affecting white blood cells). (*a*) In these patients, a translocation had occurred between chromosomes 7 and 11. This translocation moved the *TCR*β (*T*-cell *r*eceptor β subunit) gene on chromosome 7 next to the *LMO2* gene on chromosome 11 and resulted in the overexpression of *LMO2*. When overexpressed, *LMO2* behaves as an oncogene (cancer-causing gene; see Chapter 22) in a pathway leading to T-cell leukemia. (*b*) The structures of the wild-type *LMO2* gene and mRNA. The gene contains six exons, with exons 4–6 encoding the *LMO2* protein; the positions of the ATG translation start (ATG) and termination (TAA) sites are shown. In the two gene-therapy patients who developed leukemia, the retroviral vector carrying the *IL2R*γ*c* gene (encoding *i*nter*l*eukin-*2* *r*eceptor *g*amma *c*ommon subunit) integrated either (*c*) between exons 1 and 2 (boy 1) or (*d*) upstream of exon 1 (boy 2). In both cases, the integrated retroviral DNA up-regulated the expression of the *LMO2* gene in white blood cells, where its expression is normally down-regulated.

stimulates cell division. As such, *LMO2* is classified as a proto-oncogene, a gene that can become a cancer-causing oncogene by mutation or altered expression (see Chapter 22). Indeed, *LMO2* is overexpressed in the T cells of individuals with acute leukemia resulting from the translocation shown in **FIGURE 17.9*a*** and from the integration of the retroviral DNA carrying the *IL2R*γ*c* gene in X-linked SCID patients treated by gene therapy (**FIGURE 17.9*c*** and ***d***).

Scientists have known for many years that there was some risk that the retroviral vectors used in gene therapy might cause mutations by integrating within genes. However, the risk was thought to be small. If a vector integrated at random into the human genome (3×10^9 nucleotide pairs), the chance that the vector would insert into a specific gene would be about 1 in a million. However, retroviral vectors are known to insert preferentially into expressed genes. Given that there are about 25,000 genes in the human genome, even if all insertions were into genes, the random insertion of vectors into genes would hit a given gene with a probability of about 1 in 25,000. Obviously, with 2 out of 15 insertions occurring within the *LMO2* gene, insertions are not occurring at random. Instead, this particular vector exhibits a strong tendency to insert into the *LMO2* gene. Indeed, recent evidence has shown that another of the boys treated for X-linked SCID with this vector has the *IL2R*γ*c* gene/vector DNA inserted near the *LMO2* gene. At present, he exhibits no symptoms of leukemia.

Clearly, we still have a lot to learn before gene therapy can be used as an effective treatment of inherited human disorders. We need safer vectors, and we need to learn how to regulate the expression of the genes in these vectors. How long will it take to develop effective and safe gene-therapy protocols? We do not have an answer to that question; however, we can predict that there will be a time when gene therapy is used routinely and safely in the treatment of inherited human diseases.

Current somatic-cell gene-therapy protocols are **gene-addition** procedures; they simply add functional copies of the gene that is defective in the patient to the genomes of recipient cells. They do not replace the defective gene with a functional gene. In fact, the introduced genes are inserted at random or nearly random sites in the chromosomes of the host cells. The ideal gene-therapy protocol would replace the defective gene with a functional gene. **Gene replacements** would be mediated by homologous recombination and would place the introduced gene at its normal location in the host genome. In humans, gene replacements are usually referred to as *targeted gene transfers*. Oliver Smithies and coworkers first used homologous recombination to target DNA sequences to the β-globin locus of human tissue-culture cells in 1985. However, the frequency of the targeted gene-transfer was very low (about 10^{-5}). Since then, Smithies, Mario Capecchi, and others have developed improved gene-targeting vectors and selection strategies. As a result, more efficient targeted gene replacements are possible, and cells with the desired gene replacement can be identified more easily. In the future, targeted gene replacements will probably become the method of choice for somatic-cell gene therapy treatment of human diseases.

KEY POINTS

- Gene therapy involves the addition of a normal (wild-type) copy of a gene to the genome of an individual who carries defective copies of the gene.
- Although somatic-cell gene therapy effectively restored immunological function in boys with X-linked severe combined immunodeficiency disease, four of the boys subsequently developed leukemia or leukemia-like disorders.
- Somatic-cell gene therapy holds promise for the treatment of many inherited human diseases; however, the results to date have been disappointing.

DNA Fingerprints

DNA fingerprints—recorded patterns of DNA polymorphisms—provide strong evidence of an individual's identity or nonidentity.

Fingerprints have played a central role in human identity cases for decades. Indeed, fingerprints often provide the key evidence that places a suspect at a crime scene. The use of fingerprints in forensic cases is based on the premise that no two individuals will have identical prints. Similarly, no two individuals, except for identical twins, will have genomes with the same nucleotide sequences. The human genome contains 3×10^9 nucleotide pairs; each site is occupied by one of the four base pairs in DNA. Many base-pair substitutions are silent; they occur in unessential noncoding sequences or at genomic positions corresponding to the third base of codons and do not change the amino acid sequences of the gene products because of the degeneracy in the code. Therefore, such nucleotide-pair substitutions accumulate in genomes during the course of evolution. In addition, duplications and deletions of DNA sequences and other genome rearrangements contribute to the evolutionary divergence of genomes. Indeed, recent evidence has demonstrated that the human genome contains large families of DNA polymorphisms of many different types, polymorphisms that can provide valuable evidence in cases of uncertain identity. These polymorphisms can be used to produce **DNA fingerprints,** specific banding patterns on Southern blots of genomic DNA cleaved with a specific restriction enzyme and hybridized to appropriate DNA probes.

The power and utility of the DNA fingerprinting procedure in personal identity cases are obvious to anyone familiar with molecular genetics and the techniques utilized in the production of DNA prints. The controversies regarding the use of DNA fingerprints in forensic cases relate to the competency of the research laboratories involved, the probability of human error in producing prints, and the methods for calculating the probability that two individuals have identical fingerprints. To make accurate estimates of the likelihood of identical prints, researchers must have reliable information about the frequency of the polymorphisms in the population in question. For example, if inbreeding (matings between related individuals) is common in the population, the probability of identical fingerprints will increase. Thus, accurate estimates of the probability that two individuals will have matching fingerprints requires reliable information about the frequencies of the polymorphisms in the relevant population. Data obtained from one population should never be extrapolated to another population because different polymorphism frequencies may be present in different populations.

DNA fingerprinting provides a powerful forensic tool if used properly. The DNA prints can be prepared from minute amounts of blood, semen, hair bulbs, or other cells. The DNA is extracted from these cells, amplified by PCR, and analyzed with carefully chosen DNA probes by the Southern blot procedure (Chapter 15). Indeed, fingerprinting can sometimes be done with DNA isolated from preserved tissues of individuals long after their deaths. As previously mentioned, the human genome contains numerous short DNA sequences that are present as tandem repeats of varied lengths at several chromosomal locations. These variable number tandem repeats, or VNTRs, are important components of DNA fingerprints (**FIGURE 17.10**). Although DNA prints are applicable in all cases of questionable identity, they have proven especially useful in paternity and forensic cases.

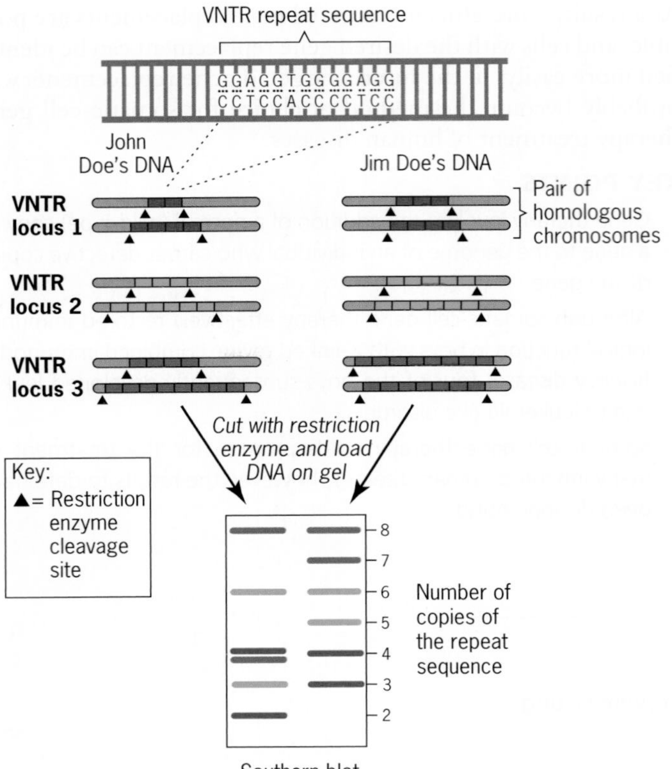

Figure 17.10 ▶ Simplified diagram of the use of variable number tandem repeats in preparing DNA fingerprints.

PATERNITY TESTS

In the past, cases of uncertain paternity often have been decided by comparing the blood types of the child, the mother, and possible fathers. Blood-type data can be used to prove that men with particular blood types could not have fathered the child. Unfortunately, these blood-type comparisons contribute little toward a positive identification of the father. In contrast, DNA fingerprints not only exclude misidentified fathers, but also come close to providing a positive identification of the true father. DNA samples are obtained from cells of the child, the mother, and possible fathers, and DNA fingerprints are prepared as described in **FIGURE 17.10**. When the fingerprints are compared, all the bands in the child's DNA print should be present in the combined DNA prints of the parents. For each pair of homologous chromosomes, the child will have received one from each parent. Thus, approximately half of the bands in the child's DNA print will result from DNA sequences inherited from the mother, and the other half from DNA sequences inherited from the father.

FIGURE 17.11 shows the DNA fingerprints of a child, the mother, and two men suspected of being the child's father. In this case, the DNA prints indicate that the second father candidate is probably the child's biological father. The accuracy of DNA fingerprints in identifying child–parent relationships can be enhanced by increasing the number of hybridization probes used in the analysis. With the use of more probes, more polymorphisms can be surveyed and a larger proportion of the genomes of the child and parents can be compared; the result is a more reliable identification.

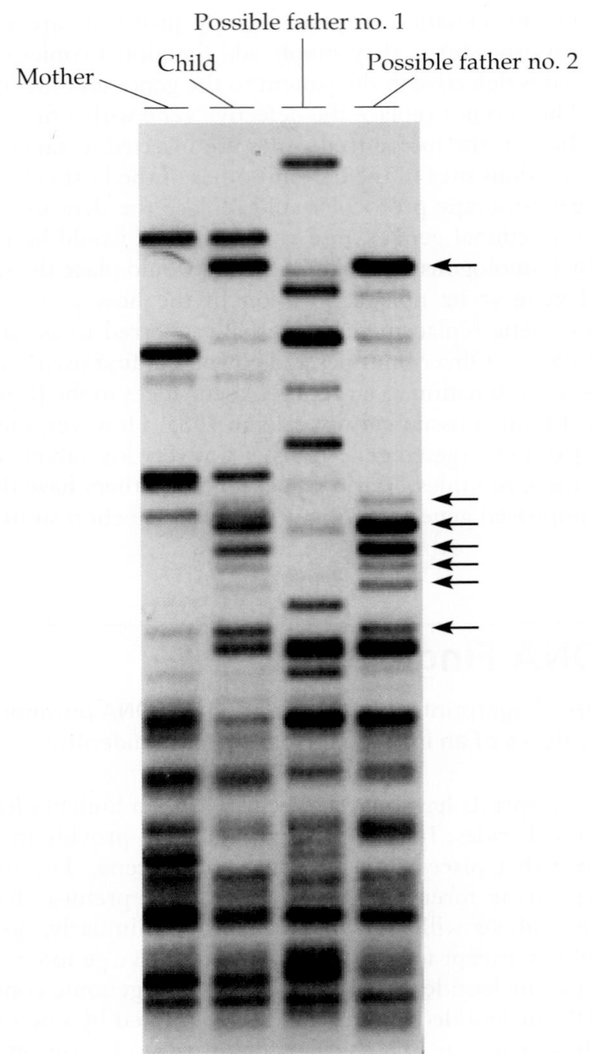

Figure 17.11 ▶ DNA fingerprints of a mother, her child, and two men, each of whom claimed to be the child's father. Arrows mark bands that identify male no. 2 as the biological father.

FORENSIC APPLICATIONS

DNA fingerprints were first used as evidence in a criminal case in 1988. In 1987, a Florida judge denied the prosecutor's request to present statistical interpretations of DNA evidence against an accused rapist. After a mistrial, the suspect was released. Three months later, he was again in court, accused of another rape. This time the judge allowed the prosecutor to present a statistical analysis of the data, based on appropriate population surveys. The analysis showed that the DNA fingerprint prepared from semen recovered from the victim had a probability of about one in 10 billion of matching the DNA fingerprint of the suspect purely by chance. This time the suspect was convicted.

There can be no question about the value of DNA prints in forensic cases of this type when good tissue or cell samples are obtained from the scene of the crime. If performed carefully by trained scientists and interpreted conservatively using valid population-based data on the distributions of the polymorphisms involved, DNA fingerprints can provide a much-needed and powerful tool in the ongoing fight against crime.

FIGURE 17.12 illustrates one type of DNA fingerprints used in forensic cases, namely, VNTR prints. The DNA fingerprint prepared from the bloodstain at the crime scene matches the DNA print from suspect 1, but not the prints from the other two suspects. Of course, these matching DNA prints by themselves do not prove that suspect 1 committed the crime, but, if combined with additional DNA prints and supporting evidence, they can provide strong indications that suspect 1 was at the scene of the crime. Perhaps more importantly, these prints clearly show that the blood cells in the stain were not from either of the other two suspects. Thus, DNA fingerprints have proven invaluable in reducing the frequency of wrongful convictions.

By combining VNTR fingerprints with prints produced using other types of DNA probes, the possibility that DNA fingerprints from two individuals will match just by chance can virtually be eliminated. The rationale behind using DNA fingerprints to identify individuals is that each person's DNA has a unique nucleotide sequence. Despite the human population explosion, there are far more possible combinations of the four base pairs in a human genome of 3×10^9 base pairs than there are humans on planet Earth. Thus, except for identical twins, no two humans should have identical genomes. DNA fingerprints provide a tool by which these differences can be detected and recorded, just as fingerprints have been recorded for decades.

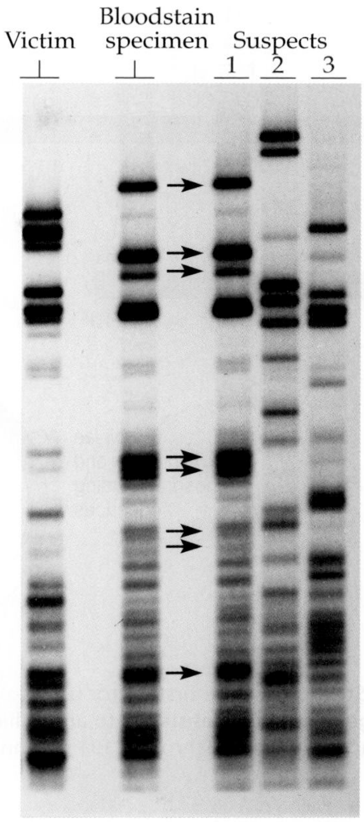

Figure 17.12 ▶ DNA fingerprints prepared from DNA isolated from a bloodstain at the site of a crime and from blood obtained from three individuals suspected of committing the crime. The arrows denote DNA fragments from suspect 1—not present in suspects 2 or 3—that match those from the bloodstain at the crime scene.

KEY POINTS

- ▶ DNA fingerprints detect and record polymorphisms in the genomes of individuals.
- ▶ DNA prints provide strong evidence of individual identity, evidence that may be extremely valuable in paternity and forensic cases.

Production of Eukaryotic Proteins in Bacteria

Human insulin, human growth hormone, and other valuable eukaryotic proteins can be produced economically in genetically engineered bacteria.

For decades, microorganisms have been used to produce important products for humans. We are all aware of the impact of antibiotics on human health; fewer of us are aware of their economic importance. The wholesale market value of antibiotics in the United States is over $2 billion annually. Microbes also play important roles in the production of many other materials, for example, antifungal drugs, amino acids, and vitamins. Today, because of genetic engineering, bacteria are being used in the production of important eukaryotic proteins such as human insulin, human growth hormone, and the entire family of human interferons. In addition, genetically engineered microbes are being used to synthesize valuable enzymes and other organic molecules and to provide metabolic machinery for the detoxification of pollutants and the conversion of biomass to combustible compounds.

HUMAN GROWTH HORMONE

In 1982, human insulin became the first commercial success of the new recombinant DNA technologies in the field of pharmaceuticals. Since then, several other human proteins with medicinal value have been synthesized in bacteria. Some of the first human proteins to be produced in microorganisms were blood-clotting factor VIII (lacking in individuals with one type

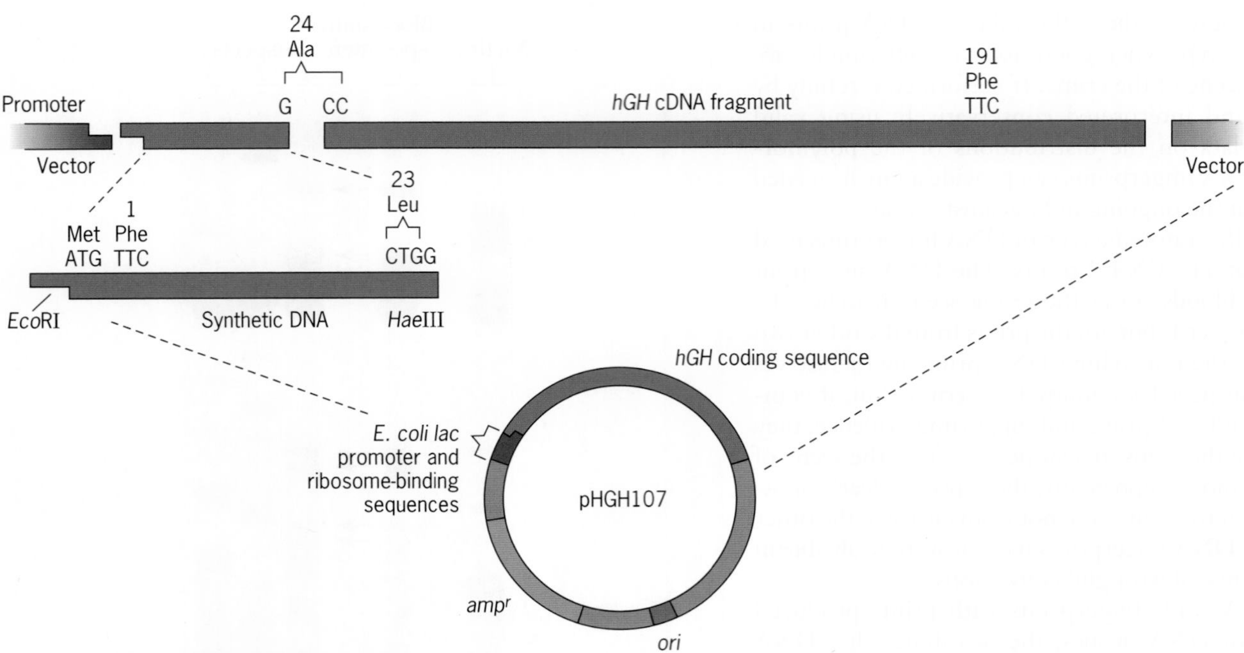

Figure 17.13 ▶ Structure of the first vector used to produce human growth hormone (hGH) in *E. coli*. The *amp*^{r} gene provides resistance to ampicillin; *ori* is the plasmid's origin of replication. The amino acids are numbered 1 through 191 beginning at the amino terminus.

of hemophilia), plasminogen activator (a protein that disperses blood clots), and human growth hormone (a protein deficient in certain types of dwarfism). As an example, let's examine the synthesis of **human growth hormone (hGH)** in *E. coli*. hGH, which is required for normal growth, is a single polypeptide chain 191 amino acids in length. In contrast to insulin, porcine and bovine pituitary growth hormones do not work in humans. Only growth hormones from humans or from closely related primates will function in humans. Thus, prior to 1985, the major source of growth hormone suitable for treatment of humans was from human cadavers.

To obtain expression in *E. coli*, the hGH coding sequence must be placed under the control of *E. coli* regulatory elements. Therefore, the hGH coding sequence was joined to the promoter and ribosome-binding sequences of the *E. coli lac* operon (a set of genes encoding proteins required for growth on the sugar lactose; see Chapter 19). To accomplish this, a *Hae*III cleavage site in the nucleotide-pair triplet specifying codon 24 of hGH was used to fuse a synthetic DNA sequence encoding amino acids 1–23 to a partial cDNA sequence encoding amino acids 24–191. This unit was then inserted into a plasmid carrying the *lac* regulatory signals and introduced into *E. coli* by transformation. The structure of the first plasmid used to produce hGH in *E. coli* is shown in **FIGURE 17.13.**

The hGH produced in *E. coli* in these first experiments contained methionine at the amino terminus (the methionine specified by the ATG initiator codon). Native hGH has an amino-terminal phenylalanine: a methionine is initially present but is then enzymatically removed. *E. coli* also removes many amino-terminal methionine residues post-translationally. However, the excision of the terminal methionine is sequence-dependent, and *E. coli* cells do not excise the amino-terminal methionine residue from hGH. Nevertheless, the hGH synthesized in *E. coli* was found to be fully active in humans despite the presence of the extra amino acid. More recently, a DNA sequence encoding a signal peptide (the amino acid sequence required for transport of proteins across membranes) has been added to an *HGH* gene construct similar to the one shown in **FIGURE 17.13.** With the signal sequence added, hGH is both secreted and correctly processed; that is, the methionine residue is removed with the rest of the signal peptide during the transport of the primary translation product across the membrane. This product is identical to native hGH. In 1985, hGH became the second genetically engineered pharmaceutical to be approved for use in humans by the U.S. Food and Drug Administration. Human insulin produced in *E. coli* had been approved for use by diabetics in 1982.

PROTEINS WITH INDUSTRIAL APPLICATIONS

Some enzymes with important industrial applications have been manufactured for many years by using microorganisms to carry out their synthesis. For example, proteases have been produced from *Bacillus licheniformis* and other bacteria. These proteases have been employed extensively as cleaning aids in detergents and in smaller amounts as meat tenderizers and as digestive aids in animal feeds. Amylases have been widely used to break down complex carbohydrates such as starch to glucose. The glucose is then converted to fructose with the enzyme glucose isomerase, and this fructose is used as a food sweetener. The amylases and glucose isomerase are all manufactured by microbiological processes.

The protein rennin is used in making cheeses. Prior to the advent of genetic engineering, rennin was extracted from the fourth stomach of cattle. Genetically engineered bacteria are now used for the commercial production of rennin. These examples are all proteins that have had important industrial applications for some time. In the future, we can expect many additional enzymes to be manufactured and used in industrial applications because of the ease of producing these proteins by means of recombinant microorganisms (or by transgenic plants and animals; see the next section).

KEY POINTS

- Valuable proteins that could be isolated from eukaryotes only in small amounts and at great expense can now be produced in large quantities in genetically engineered bacteria.
- Proteins such as human insulin and human growth hormone are valuable pharmaceuticals used to treat diabetes and pituitary dwarfism, respectively.

Transgenic Animals and Plants

Synthetic, modified, or other foreign genes can be introduced into animals and plants, and the resulting transgenic organisms can be used to study the functions of the genes, for example, by insertional mutagenesis, to produce novel products, or to serve as animal models for studies of inherited human diseases.

Although a complete discussion of the methods used to produce transgenic animals and plants is beyond the scope of this book, let's examine a couple of the commonly used procedures, and some of the initial applications of recombinant DNA technologies in animal and plant breeding.

TRANSGENIC ANIMALS: MICROINJECTION OF DNA INTO FERTILIZED EGGS AND TRANSFECTION OF EMBRYONIC STEM CELLS

Many different animals have been modified by the introduction of foreign DNA. The mouse, however, has been studied more than any other vertebrate, and we will restrict our discussion of the techniques used to produce transgenic animals to those used with mice. There are two general methods of introducing transgenes into mouse chromosomes. One relies on the injection of DNA into fertilized eggs or embryos, and the other involves the transfection of embryonic stem cells growing in culture.

The first transgenic mice were produced by microinjection of DNA into fertilized eggs. Indeed, this procedure has been used almost exclusively to produce transgenic pigs, sheep, cattle, and other domestic animals. Prior to the microinjection of DNA, the eggs are surgically removed from the female parent and are fertilized *in vitro*. The DNA is then microinjected into the male pronucleus (the haploid nucleus contributed by the sperm, prior to nuclear fusion) of the fertilized egg through a very fine-tipped glass needle (**FIGURE 17.14**). Usually, several hundred to several thousand copies of the gene of interest are injected into each egg, and multiple integrations often occur. Surprisingly, when multiple copies do integrate into the genome, they usually do so as tandem, head-to-tail arrays at a single chromosomal site. The integration of injected DNA molecules appears to occur at random sites in the genome.

Because the DNA is injected into the fertilized egg, integration of the injected DNA molecules usually occurs early during embryonic development. As a result, some germ-line cells may carry the transgene. As would be expected, the animals that develop from the injected eggs—called the G_0 *generation*—are almost always genetic mosaics, with some somatic cells carrying the transgene and others not carrying it. The initial (G_0) transgenic animals must be mated and G_1 progeny produced to obtain animals in which all cells carry the transgene. In most

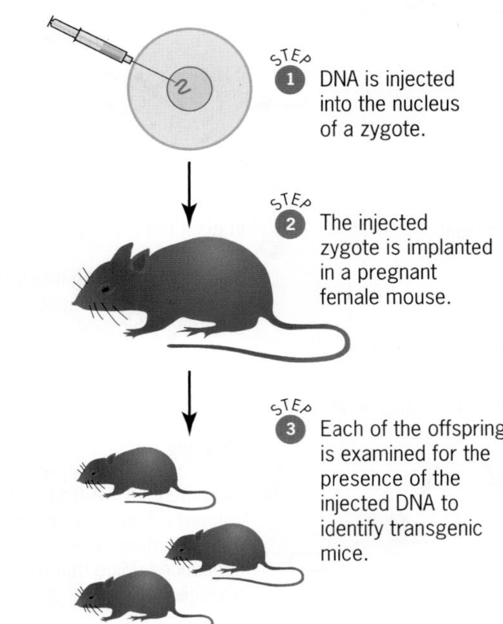

STEP 1 DNA is injected into the nucleus of a zygote.
STEP 2 The injected zygote is implanted in a pregnant female mouse.
STEP 3 Each of the offspring is examined for the presence of the injected DNA to identify transgenic mice.

Figure 17.14 ► The production of transgenic mice by injecting DNA into eggs and implanting them into females to complete their development.

of the cases where their inheritance has been studied, the transgenes were transmitted to the progeny in a stable fashion.

The other procedure that is now widely used to produce transgenic mice relies on the injection or transfection of DNA into large populations of cultured cells that were derived from very young mouse embryos (**FIGURE 17.15**). These **embryonic stem cells** (or **ES cells**) come from the inner cell mass, a group of cells found in the blastula stage of mouse embryos. Such cells can be cultured *in vitro*, transfected or injected with DNA, and then introduced into other developing mouse embryos. By chance, some of the introduced ES cells may contribute to the formation of adult tissues, so that when the mouse is born, it may consist of a mixture of two types of cells, its own and those derived from the cultured (and potentially transfected) ES cells. Such mice are called **chimeras.** If the ES cells happened to contribute to the chimera's germ line, the introduced foreign DNA has a chance of being transmitted to the next generation. Breeding a chimeric mouse may therefore establish a transgenic strain.

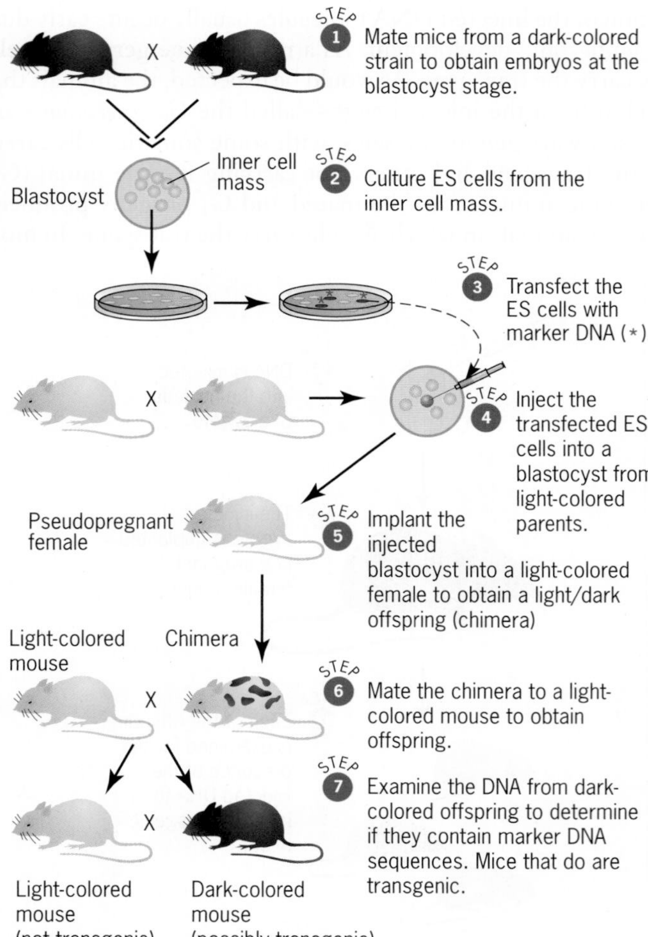
STEP 1 Mate mice from a dark-colored strain to obtain embryos at the blastocyst stage.
X
Blastocyst
Inner cell mass
STEP 2 Culture ES cells from the inner cell mass.
STEP 3 Transfect the ES cells with marker DNA (*).
X
STEP 4 Inject the transfected ES cells into a blastocyst from light-colored parents.
Pseudopregnant female
STEP 5 Implant the injected blastocyst into a light-colored female to obtain a light/dark offspring (chimera)
Light-colored mouse
Chimera
X
STEP 6 Mate the chimera to a light-colored mouse to obtain offspring.
STEP 7 Examine the DNA from dark-colored offspring to determine if they contain marker DNA sequences. Mice that do are transgenic.
X
Light-colored mouse (not transgenic)
Dark-colored mouse (possibly transgenic)

Figure 17.15 ▶ The production of transgenic mice by embryonic stem (ES) cell technology.

Transgenic mice are produced routinely in laboratories throughout the world, with thousands of transgenic strains having been created. They provide valuable tools for the study of gene expression in mammals and an excellent model system with which to test various gene-transfer vectors and methodologies for possible use in humans. In most cases, the transgenes show normal patterns of inheritance, indicating that they have been integrated into the host genome. We discuss one important application of this technology later in this chapter (see the section Knockout Mutations in the Mouse).

One of the first experiments with transgenic mice showed that growth rate could be increased when rat, bovine, or human growth hormone genes were expressed in the mice (**FIGURE 17.16**). This prompted animal breeders to ask whether the introduction of either (1) extra copies of the homologous (same-species) growth hormone gene or (2) copies of heterologous growth hormone genes from related species might result in domestic animals with enhanced growth rates. Transgenic pigs were produced with the hope that enhanced growth hormone levels might result in leaner pigs with improved meat quality and with faster growth. Other scientists introduced growth hormone transgenes into fish and chickens with similar objectives.

The experiments with transgenic pigs indicated that growth rates were increased, but not on standard diets, only on high-protein diets. However, the transgenic pigs were found to be leaner than the controls, as was expected, because growth hormone favors the synthesis of proteins instead of fats. Unfortunately, the transgenic pigs also exhibited several undesirable side effects of the higher growth hormone levels. Most notably, the female transgenic pigs were sterile. In addition, transgenic animals of both sexes were lethargic with weak muscles and were

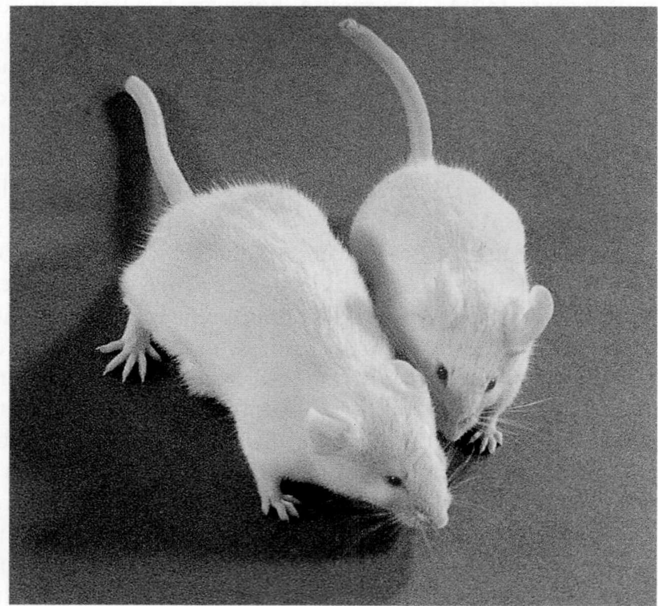

Figure 17.16 ▶ The transgenic mouse on the left, which carries a chimeric human growth hormone gene, is about twice the size of the control mouse on the right.

highly susceptible to arthritis and ulcers. Although scientists are hopeful that ways can be found to overcome these side effects, the initial results suggest that attempts to improve growth rates and enhance meat quality in domestic animals by increasing growth hormone levels with transgenes may not be very effective.

Other transgenic animals are being tested for resistance to viral infections. Avian leukosis virus (ALV) is a major viral pathogen of chickens, causing losses to the poultry industry of $50 million to $100 million per year. Obviously, the availability of an ALV-resistant strain of chickens would be of major commercial value. Therefore, researchers have produced transgenic chickens that carry a defective ALV genome. These chickens produce viral RNA and the viral envelope protein, but no progeny viruses. Most importantly, they are resistant to infection by ALV. The synthesis of large amounts of the retroviral envelope protein somehow blocks the reproductive cycle of intact, pathogenic ALV viruses. That ALV resistance has been transmitted to several generations of progeny indicates that the trait is stable. These encouraging results suggest that the introduction of defective viral genomes into domestic animals may be a useful tool in the production of virus-resistant genotypes.

Another potentially important use of transgenic animals is for the production and secretion of valuable proteins in milk. Many native human proteins contain carbohydrate or lipid side groups that are added post-translationally. Bacteria do not contain the enzymes that catalyze the addition of these moieties to nascent proteins. In such cases, recombinant bacteria cannot be used to synthesize the final product; they will synthesize the polypeptide only in its unmodified form. For this reason, some researchers have begun to explore alternative methods for producing valuable human proteins, especially glycoproteins and lipoproteins. Indeed, mouse and hamster cells growing in culture are now commonly used for the production of human proteins with medicinal applications.

TRANSGENIC PLANTS: THE TI PLASMID OF *AGROBACTERIUM TUMEFACIENS*

Plant breeders have modified plants genetically for decades. Today, however, plant breeders can directly modify the DNA of plants, and they can quickly add genes from other species to plant genomes by recombinant DNA techniques. Indeed, transgenic plants can be produced by several different procedures. One widely used procedure, called **microprojectile bombardment,** involves shooting DNA-coated tungsten or gold particles into plant cells. Another procedure, called **electroporation,** uses a short burst of electricity to get the DNA into cells. However, the most widely used method of generating transgenic plants, at least in dicots, is ***Agrobacterium tumefaciens*-mediated transformation.** *A. tumefaciens* is a soil bacterium that has evolved a natural genetic engineering system; it contains a segment of DNA that is transferred from the bacterium to plant cells.

An important feature of plant cells is their **totipotency**—that is, the ability of a single cell to produce all the differentiated cells of the mature plant. Many differentiated plant cells are able to dedifferentiate to the embryonic state and subsequently to redifferentiate to new cell types. Thus, there is no separation of germ-line cells from somatic cells as in higher animals. This totipotency of plant cells is a major advantage for genetic engineering because it permits the regeneration of entire plants from individual modified somatic cells.

A. tumefaciens is the causative agent of crown gall disease of dicotyledonous plants. The name refers to the galls or tumors that often form at the crown (junction between the root and the stem) of infected plants (**FIGURE 17.17*a***). Because the crown of the plant is usually located at the soil surface, it is here that a plant is most likely to be wounded (for example, from a soil abrasion as it blows in a strong wind) and infected by a soil bacterium such as *A. tumefaciens*. However, *A. tumefaciens* can infect a plant and induce a tumor at any wound site (**FIGURE 17.17*b***). After the infection of a wound site by *A. tumefaciens*, two key events occur: (1) the plant cells begin to proliferate and form tumors, and (2) they begin to synthesize an arginine derivative called an opine. The opine synthesized is usually either nopaline or octopine depending on the strain of *A. tumefaciens*. These opines are catabolized and used as energy sources by the infecting

Formation of a crown gall at the soil surface.

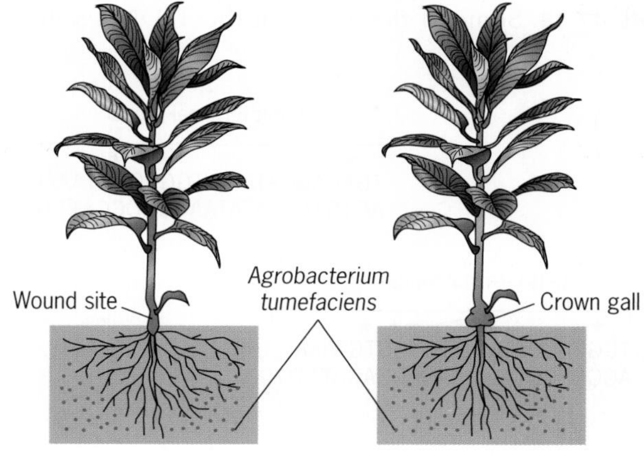

(a)

Formation of tumors, or galls, at wound sites on a leaf.

(b)

Figure 17.17 ► The formation of tumors, or "galls," on plants at wound sites of (*a*) a plant growing in soil or (*b*) a leaf infected by *Agrobacterium tumefaciens*.

bacteria. *A. tumefaciens* strains that induce the synthesis of nopaline can grow on nopaline, but not on octopine, and vice versa. Clearly, an interesting interrelationship has evolved between *A. tumefaciens* strains and their plant hosts. *A. tumefaciens* is able to divert the metabolic resources of the host plant to the synthesis of opines, which are of no apparent benefit to the plant but which provide sustenance to the bacterium.

The ability of *A. tumefaciens* to induce crown galls in plants is controlled by genetic information carried on a large (about 200,000 nucleotide pairs) plasmid called the **Ti plasmid** for its *t*umor-*i*nducing capacity. Two components of the Ti plasmid, the **T-DNA** and the ***vir* region,** are essential for the transformation of plant cells. During the transformation process, the T-DNA (for *T*ransferred *DNA*) is excised from the Ti plasmid, transferred to a plant cell, and integrated (covalently inserted) into the DNA of the plant cell. The available data indicate that integration of the T-DNA occurs at random chromosomal sites; moreover, in some cases, multiple T-DNA integration events occur in the same cell. In nopaline-type Ti plasmids, the T-DNA is a 23,000-nucleotide-pair segment that carries 13 known genes. In octopine-type Ti plasmids, there are two separate T-DNA segments. For the sake of brevity, we consider only nopaline-type Ti plasmids in the subsequent discussion.

The structure of a typical nopaline Ti plasmid is shown in **FIGURE 17.18.** Some of the genes on the T-DNA segment of

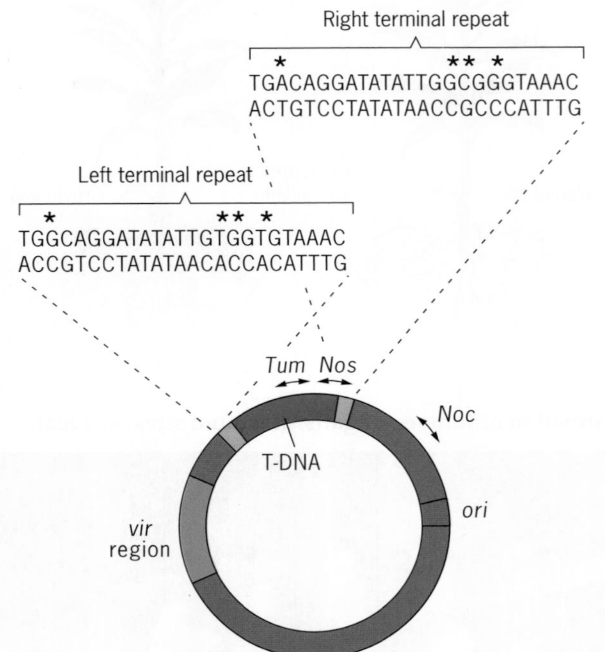

Figure 17.18 ▶ Structure of the nopaline Ti plasmid pTi C58, showing selected components. The Ti plasmid is 210 kb in size. Symbols used are: *ori*, origin of replication; *Tum*, genes responsible for tumor formation; *Nos*, genes involved in nopaline biosynthesis; *Noc*, genes involved in the catabolism of nopaline; and *vir*, virulence genes required for T-DNA transfer. The nucleotide-pair sequences of the left and right terminal repeats are shown at the top; the asterisks mark the four base pairs that differ in the two border sequences.

the Ti plasmid encode enzymes that catalyze the synthesis of phytohormones (the auxin indoleacetic acid and the cytokinin isopentenyl adenosine). These phytohormones are responsible for the tumorous growth of cells in crown galls. The T-DNA region is bordered by 25-nucleotide-pair imperfect repeats, one of which must be present in *cis* for T-DNA excision and transfer. The deletion of the right border sequence completely blocks the transfer of T-DNA to plant cells.

The *vir* (for virulence) region of the Ti plasmid contains the genes required for the T-DNA transfer process. These genes encode the DNA processing enzymes required for excision, transfer, and integration of the T-DNA segment during the transformation process. The *vir* genes can supply the functions needed for T-DNA transfer when located either *cis* or *trans* to the T-DNA. They are expressed at very low levels in *A. tumefaciens* cells growing in soil. However, exposure of the bacteria to wounded plant cells or exudates from plant cells induces enhanced levels of expression of the *vir* genes. This induction process is very slow for bacteria, taking 10 to 15 hours to reach maximum levels of expression. Phenolic compounds such as acetosyringone act as inducers of the *vir* genes, and transformation rates can often be increased by adding these inducers to plant cells inoculated with *Agrobacterium*. The transformation of plant cells by the Ti plasmid of *A. tumefaciens* occurs as illustrated in **FIGURE 17.19.**

Once it had been established that the T-DNA region of the Ti plasmid of *A. tumefaciens* is transferred to plant cells and becomes integrated in plant chromosomes, the potential use of *Agrobacterium* in plant genetic engineering was obvious. Foreign genes could be inserted into the T-DNA and then transferred to the plant with the rest of the T-DNA. This procedure works very well. The problem is that the transformed plant cells carrying wild-type T-DNAs lose their normal control of cell division and form tumors. This feature of T-DNA renders wild-type Ti plasmids incompatible with the goals of most gene-transfer experiments. Fortunately, this problem was solved with the identification of the genes in the T-DNA that are responsible for tumor formation (see **FIGURE 17.18**). The deletion of one or more of these genes produces a disarmed Ti plasmid. Unfortunately, the deletion of the tumor-causing genes also makes it difficult to identify plant cells that have received the disarmed T-DNA. Thus, a way to identify plant cells transformed with disarmed Ti plasmids—ideally, a good selectable marker gene located within the T-DNA region of the disarmed Ti plasmid—is needed.

A good selectable marker gene is one that will provide resistance to a drug, an antibiotic, or another agent that arrests the growth of normal plant cells. The selective agent should inhibit the growth of plant cells or kill them slowly. Agents that kill cells rapidly result in the release of phenolic compounds and other substances that are toxic to the growth of the remaining, otherwise resistant cells. The antibiotic kanamycin has been the most widely used selective agent in plants.

The *kan^r* gene from the *E. coli* transposon Tn*5* has been used extensively as a selectable marker in plants; it encodes an

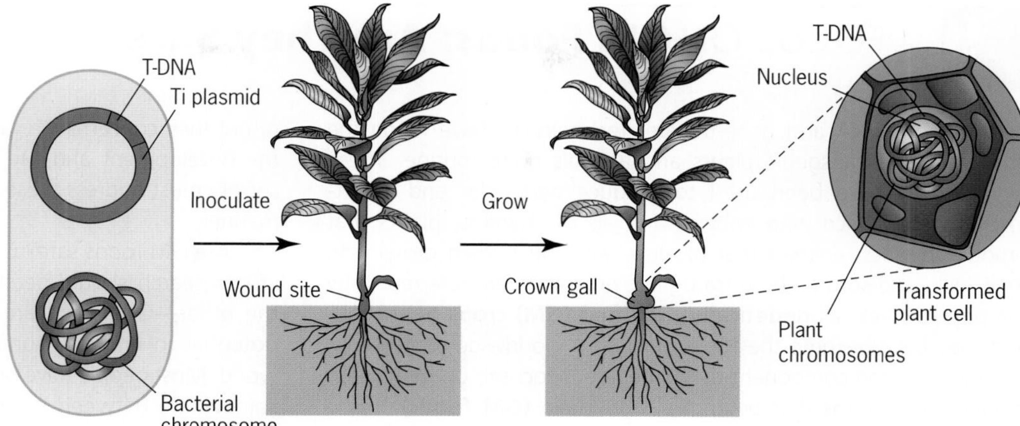

Figure 17.19 Transformation of plant cells by *Agrobacterium tumefaciens* harboring a wild-type Ti plasmid. Plant cells in the tumor contain the T-DNA segment of the Ti plasmid integrated into chromosomal DNA.

enzyme called neomycin phosphotransferase type II (NPTII). NPTII is one of several prokaryotic enzymes that detoxify the kanamycin family of aminoglycoside antibiotics by phosphorylating them. Because the promoter sequences and transcription-termination signals are different in bacteria and plants, the native Tn*5* *kan*^{r} gene cannot be used in plants. Instead, the NPTII coding sequence must be provided with a plant promoter (5′ to the coding sequence) and plant termination and polyadenylation signals (3′ to the coding sequence). Such constructions with prokaryotic coding sequences flanked by eukaryotic regulatory sequences are called **chimeric selectable marker genes.**

Regulatory sequences from several different plant genes have been used to construct chimeric marker genes. One widely used chimeric selectable marker gene contains the cauliflower mosaic virus (CaMV) 35S (transcript size) promoter, the NPTII coding sequence, and the Ti nopaline synthase (*nos*) termination sequence; this chimeric gene is usually symbolized 35S/NPTII/*nos*. The Ti vectors used to transfer genes into plants have the tumor-inducing genes of the plasmid replaced with a chimeric selectable marker gene such as 35S/NPTII/*nos*. A large number of sophisticated Ti plasmid gene-transfer vectors are now used routinely to transfer genes into plants.

The powerful new tools that permit plant and animal breeders to produce transgenic plants and animals with relative ease have a vast array of applications. In Chapter 1, we discussed the production of corn borer-resistant corn. The most widely used transgenes are those that produce herbicide resistance in agronomic crops. With the development of these and other genetically modified plants and animals have come questions about their safety. Indeed, an intense and highly emotional controversy about the safety of genetically modified (GM) crops and other foods is occurring in many parts of the world (see Focus on GM Foods: Are They Safe?).

KEY POINTS

- DNA sequences of interest can now be introduced into most plant and animal species.
- The resulting transgenic organisms provide valuable resources for studies of gene function and biological processes.
- The Ti plasmid of *Agrobacterium tumefaciens* is an important tool for transferring genes into plants.

Reverse Genetics: Dissecting Biological Processes by Inhibiting Gene Expression

Reverse genetic approaches make use of known nucleotide sequences to devise procedures for inhibiting the expression of specific genes.

The explosion of new information in biology during the twentieth century resulted in part because of the application of genetic approaches to the dissection of biological processes (see Chapter 13). The classical genetic approach was to identify organisms with abnormal phenotypes and to characterize the mutant genes responsible for these phenotypes. Comparative molecular studies were then performed on mutant and wild-type organisms to determine the effects of the mutations. These studies identified genes encoding products that were involved in the biological processes under investigation. In some cases, the results of these studies allowed biologists to determine the precise sequence of events or pathway by which a process

▶ FOCUS ON GM Foods: Are They Safe?

Recombinant DNA and gene-transfer technologies have made the production of transgenic plants and animals quite routine. These technologies have been used to produce herbicide- and insect-resistant crops, rice with enhanced levels of vitamins, plants that produce plastics, animals that produce valuable human proteins in their milk, and so on. These transgenic organisms are referred to by the popular press as **genetically modified (GM) crops** or animals, and they have become the focus of intense worldwide controversy. The most heated component of the debate is focused on genetically modified organisms that enter the food chain **(GM foods).** Some social and environmental groups have fueled this controversy with labels such as "Frankenstein food," "genetic pollution," and "mutant crops." Greenpeace activists have attempted to block the international transport of genetically modified crops, advertising their activities with huge banners reading "Floodgate: Genetic Pollution." Whether or not their concerns are valid, they have created major opposition to the development and use of GM foods, especially in Europe, and society must address these concerns in a careful and comprehensive manner.

Are GM foods safe? Opinions range from one extreme—all GM crop research should be stopped and all GM foods destroyed—to the other—GM foods are completely safe and have phenomenal potential to improve human nutrition and health throughout the world. Most people have opinions about GM foods that are probably somewhere between these extremes. Whereas proponents of GM foods emphasize their numerous potential applications, opponents argue that some GM foods may contain toxic and allergenic substances and that the cultivation of GM crops will lead to a decrease in biodiversity. Some of the potential risks and benefits of GM crops are listed in **TABLE 1.**

▶ **TABLE 1**

Potential Risks and Benefits of GM Crops

Potential Risks	Potential Benefits
1. Production of toxic or allergenic products.	1. Foods with enhanced nutritional quality, such as added vitamins and minerals.
2. Unexpected deleterious effects of "unnatural" products.	2. Production of pharmaceuticals and edible vaccines.
3. "Escape" of transgenes by transfer to related species, with deleterious effects.	3. Crops with built-in resistance to insects, herbicides, and diseases.
4. Pollination of "organic" crops, reducing their value.	4. Delayed fruit ripening, extending shelf life.
5. Harmful effects on honey production by bees.	5. Production of biodegradable plastics in crops.
6. Harmful effects of transgene products on other species, such as monarch butterflies.	6. Biological control of weeds, resulting in decreased use of chemical herbicides and pesticides.
7. Reduction in biodiversity.	7. Synthesis of products that are valuable raw materials for the production of paper, detergents, lubricants, and the like.
8. Economic losses due to international boycotts of GM products.	8. Genetic control of flowering time, seed content, and other important traits of important crops.
9. Increased dependence of farmers on large agrochemical companies.	

occurs. The complete pathway of morphogenesis in bacteriophage T4 (see **FIGURE 13.8**) provided early documentation of the power of the mutational dissection approach.

During the last couple of decades, the nucleotide sequences of genes and entire genomes have become available. Today, we often know the nucleotide sequence of a gene before we know its function. This knowledge has led to new approaches to the genetic dissection of biological processes, approaches collectively called **reverse genetics.** Reverse genetic approaches use the nucleotide sequences of genes to devise procedures for either isolating null mutations in them or shutting off their expression. The function of a specific gene often can be deduced by studying organisms lacking any functional product of the gene. In the following sections of this chapter, we examine four different reverse genetic approaches: antisense RNA, foreign DNA insertions producing "knockout" mutations in mice, T-DNA and transposon insertions in plants, and RNA interference.

ANTISENSE RNA

The first reverse genetic procedure for shutting off the expression of specific genes was the use of **antisense RNAs.** The antisense RNA method involves the synthesis of RNA molecules that are complementary to the mRNA molecules pro-

A 1999 paper in *Nature* entitled "Transgenic Pollen Harms Monarch Larvae" by John Losey and colleagues led to major concerns about the possible ecological effects of GM crops. They fed monarch butterfly larvae (**FIGURE 1**) milkweed leaves coated with pollen from corn plants containing a gene encoding an insecticidal protein (*bt* toxin) from the bacterium *Bacillus thuringiensis* (see **FIGURE 1.14**). Many of the larvae died, resulting in concern about the effects of the 26 million acres of *bt* corn growing in the United States at the time. Subsequent studies demonstrated that this concern was probably ill founded for several reasons: (1) corn pollen does not adhere well to milkweed leaves; (2) the amount of corn pollen present on milkweed leaves near *bt* corn fields is much lower than the amount used in Losey's study; (3) the most widely grown lines of *bt* corn have very low levels of the toxin in their pollen; and (4) in the field, most corn pollen is released before monarchs lay their eggs. Subsequently, Losey demonstrated that monarch butterflies avoid leaves coated with *bt* corn pollen when laying eggs. In addition, the commonly used alternative method of controlling insect pests—spraying with chemical insecticides—is almost certainly more detrimental to monarchs than *bt* corn. Nevertheless, public concern has been heavily influenced by Losey's original publication, with less publicity given to subsequent studies.

The positive aspect of the monarch butterfly saga is that it heightened concern about the possible ecological effects of GM crops, which in turn stimulated important discussions and research. Perhaps the most valid concern about GM crops is whether they will decrease biodiversity. Will the widespread growth of herbicide-resistant crops and control of weeds with a specific herbicide eliminate some of the native plants in the ecosystem? Will it be detrimental to wildlife, especially birds? Numerous studies are being planned to address these questions. However, even these studies, which were designed to obtain more information about the effects of GM crops on biodiversity, are being criticized as potentially damaging by Greenpeace and other environmental groups.

Figure 1 ▶ A larva of the monarch butterfly feeding on a milkweed leaf.

Clearly, the emotional debate over GM foods is as much about cultural and societal values as it is about the science it involves. The concerns about GM foods must be dealt with by considering all of these factors. In New Zealand, a commission on GM crops has attempted to obtain and evaluate opinions from all sectors of society (for details, see http://www.gmcommission.govt.nz). The work of this commission provides an excellent model for what needs to be done on a worldwide scale. Documents submitted to the commission were largely opposed to GM foods (65 percent strongly opposed). However, a subsequent large survey of public opinion found that only 2 percent of the respondents listed GM foods as one of their concerns. These results suggest that those who are opposed to GM foods are very proactive in expressing their opposition, whereas the average citizen is less concerned about the potential risks associated with GM foods. However, two things are certain: the debate will continue, and the final outcome cannot be predicted.

duced by transcription of a given gene. The normal mRNA of a gene is said to be sense because it carries the codons that are read during translation to produce the specified sequence of amino acids in the polypeptide gene product. Normally, the complement to the mRNA sense strand will not contain a sequence of codons that can be translated to produce a functional protein; thus, this complementary strand is called antisense RNA. In addition, an antisense RNA usually will not contain the regulatory sequences required for translation. When antisense RNA molecules are present in the same cell with sense (mRNA) molecules from a gene, the antisense RNA and mRNA molecules will anneal to form duplex RNA molecules, which will either be degraded or translationally repressed.

The antisense RNA approach has proven useful in dissecting pathways of gene expression in plants, animals, and microorganisms. It has also been a commercial success. The first genetically engineered plant product approved for human consumption, the FlavrSavr™ tomato, was introduced in supermarkets in California and Illinois in 1994. The FlavrSavr™ tomato, which remains firm longer during the ripening process, was produced by using antisense RNA to decrease the rate of expression of an endogenous gene encoding an enzyme called polygalacturonase to 10 percent of the normal level. Polygalacturonase is an enzyme that breaks down cell walls and causes softening in tomatoes as they ripen.

As most of us know from first-hand experience, vine-ripened tomatoes are much more flavorful than those that are picked green and allowed to ripen en route to the marketplace. However, vine-ripened tomatoes are too soft to survive the handling required during transport; they bruise too easily. The Flavr-Savr™ tomato has partially solved this problem. FlavrSavr™ tomatoes remain firm further into the ripening process, allowing them to remain on the vines longer before they are picked and shipped to market.

The simplest way to produce the antisense RNA of a gene within a cell or an organism is to (1) clone the gene of interest, (2) separate the coding sequence of the gene from its promoter by cleavage of the DNA with an appropriate restriction enzyme, (3) ligate the coding sequence to its promoter in the inverse orientation (flipped end-for-end), and (4) introduce the promoter/inverted coding sequence construct or **antisense gene** into the host cell or organism by transformation. The effect of a gene specifying antisense RNA is illustrated in **FIGURE 17.20.** The transcription of the antisense gene will produce antisense RNA transcripts. These antisense RNAs will hybridize with mRNA (sense RNA) molecules in cells and either mediate their degradation or block their translation by the RNA interference pathway (Chapter 20). As a result, the antisense RNAs will block the expression of genes specifying complementary mRNA. Although antisense RNAs block the expression of many genes, they have little effect on the expression of other genes. Thus, the use of antisense RNAs for reverse genetic analyses has been largely replaced by approaches using T-DNA and transposon insertional mutagenesis or RNA interference.

KNOCKOUT MUTATIONS IN THE MOUSE

We discussed the procedures used to generate transgenic mice in an earlier section of this chapter (see **FIGURES 17.14** and **17.15**). Normally, the transgenes are inserted into the genome at random sites. However, if the injected or transfected DNA contains a sequence homologous to a sequence in the mouse genome, it will sometimes be inserted into that sequence by homologous recombination. The insertion of this foreign DNA into a gene will disrupt or "knock out" the function of the gene just like the insertion of a transposable genetic element (see **FIGURE 13.20**). Indeed, this approach has been used to generate knockout mutations in hundreds of mouse genes.

The first step in the production of mice carrying a knockout mutation in a gene of interest is to construct a gene-targeting vector, a vector with the potential to undergo homologous recombination with one of the chromosomal copies of the gene and, in so doing, insert foreign DNA into the gene and disrupt its function. A gene (*neo*^{r}) that confers resistance to the antibiotic neomycin is inserted into a cloned copy of the gene of interest, splitting it into two parts and making it nonfunctional (**FIGURE 17.21,** step 1). The presence of the *neo*^{r} gene in the vector will allow neomycin to be used to eliminate cells not carrying an integrated copy of the gene-targeting vector or the *neo*^{r} gene. The segments of the gene retained on either side of the inserted *neo*^{r} gene provide sites of homology for recombination with chromosomal copies of the gene. The thymidine kinase gene (tk^{HSV}) from herpes simplex virus is inserted into the cloning vector (**FIGURE 17.21,** step 2) for subsequent use in eliminating transgenic mouse cells resulting from the random integration of the vector. The thymidine kinase from herpes simplex virus (HSV) phosphorylates the drug gancyclovir, and when this phosphorylated nucleotide-analog is incorporated into DNA, it kills the host cell. In the absence of the HSV thymidine kinase, gancyclovir is harmless to the host cell.

The next step is to transfect embryonic stem (ES) cells (from dark-colored mice) growing in culture with linear copies of the gene-targeting vector (**FIGURE 17.21,** step 3) and

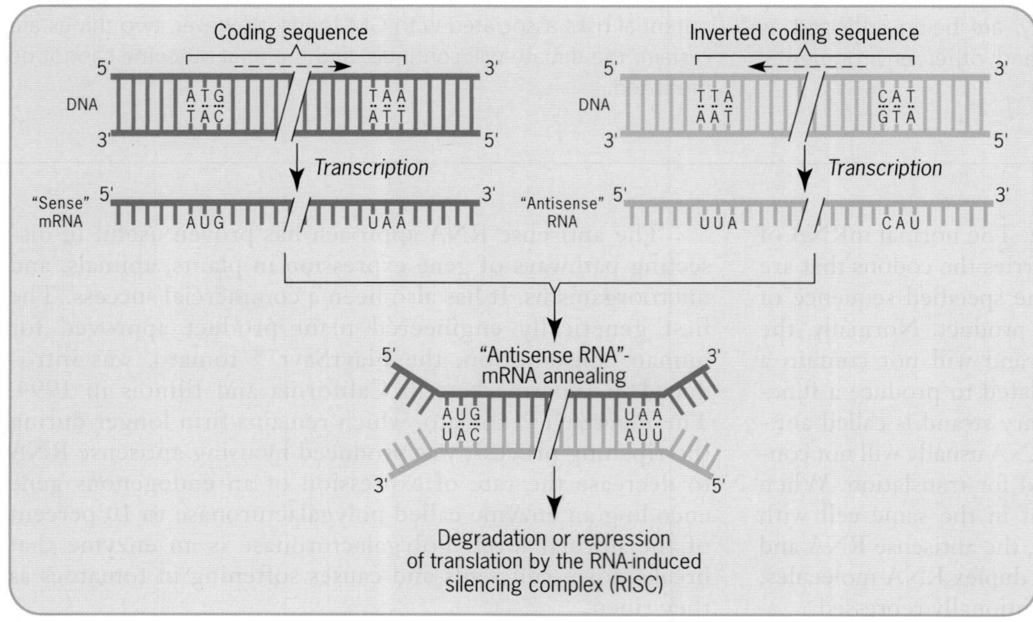

Figure 17.20 ▶ The antisense RNA procedure for blocking or reducing the level of expression of specific genes. The double-stranded RNA produced when antisense RNA hybridizes with mRNA (sense RNA) is targeted for degradation or translational repression by the *RNA-induced* silencing complex (RISC) as discussed in Chapter 20.

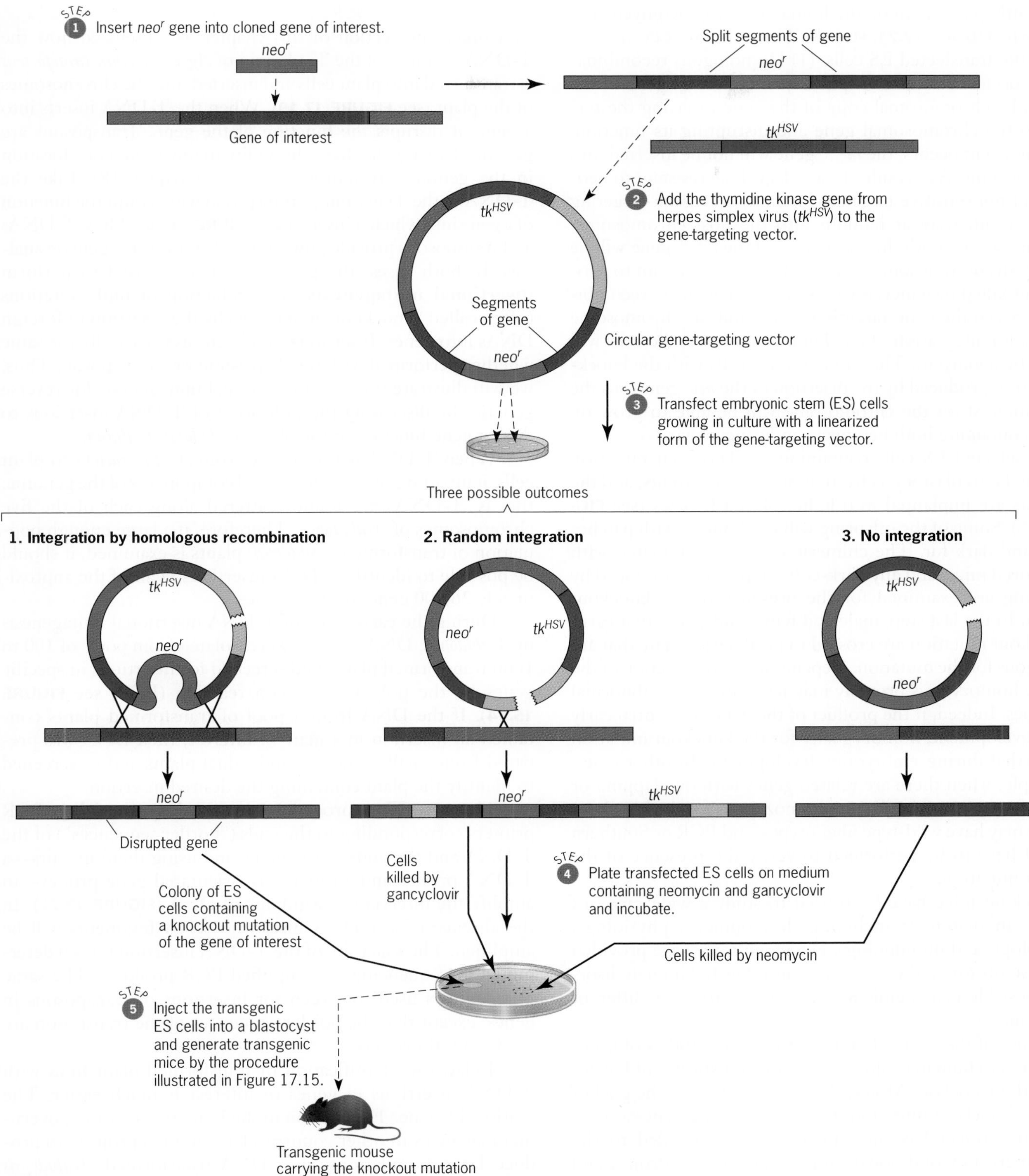

Figure 17.21 ► The generation of knockout mutations in the mouse by homologous recombination between gene-targeting vectors and chromosomal genes in transfected embryonic stem (ES) cells. The procedure used to produce transgenic mice from transgenic ES cells growing in culture is illustrated in **FIGURE 17.15.** The neo^r gene confers mouse cells with resistance to the antibiotic neomycin, and the tk^{HSV} gene makes them sensitive to the nucleotide-analog gancyclovir. See text for additional details.

subsequently plate them on medium containing neomycin and gancyclovir (**FIGURE 17.21,** step 4). Three different events will occur in the transfected ES cells. (1) Homologous recombination may occur between the split sequences of the gene in the vector and a chromosomal copy of the gene inserting the *neo*^{r} gene into the chromosomal gene and disrupting its function. When this event occurs, the tk^{HSV} gene will not be inserted into the chromosome. As a result, these cells will be resistant to neomycin, but not sensitive to gancyclovir. (2) The gene-targeting vector may integrate at random into the host chromosome. When this occurs, both the *neo*^{r} gene and the tk^{HSV} gene will be present in the chromosome. These cells will be resistant to neomycin, but killed by gancyclovir. (3) There may be no recombination between the gene-targeting vector and the chromosome and, thus, no integration of any kind. In this case, the cells will be killed by neomycin. Thus, only the ES cells with the knockout mutation produced by the insertion of the *neo*^{r} gene into the gene of interest on the chromosome will be able to grow on medium containing both neomycin and gancyclovir.

The selected ES cells containing the knockout mutation are injected into blastocysts from light-colored parents, and the blastocysts are implanted into light-colored females (see **FIGURE 17.15**). Some of the offspring will be chimeric with patches of light and dark fur. The chimeric offspring are mated with light-colored mice, and any dark-colored progeny produced by this mating are examined for the presence of the knockout mutation. In the last step, male and female offspring that carry the knockout mutation are crossed to produce progeny that are homozygous for the mutation. Depending on the function of the gene, the homozygous progeny may have normal or abnormal phenotypes. Indeed, if the product of the gene is essential early during development, homozygosity for the knockout mutation will be lethal during embryonic development. In other cases, for example, when there are related genes with overlapping or identical functions, mice that are homozygous for the knockout mutation may have wild-type phenotypes, and PCR or Southern blots will have to be performed to verify the presence of the knockout mutation.

Knockout mice have been used to study a wide range of processes in mammals including development, physiology, neurobiology, and immunology. Knockout mice have provided model systems for studies of numerous inherited human disorders from sickle-cell anemia to heart disease to many different types of cancer.

Because of the value of knockout mice for studies of processes related to human health, the National Institutes of Health initiated the Knockout Mouse Project in 2006 with the goal of producing knockout mutations in as many mouse genes as possible. This project has subsequently been expanded to the North American Conditional Mouse Mutagenesis Project and is working together with the European Conditional Mouse Mutagenesis Project to produce at least one knockout mutation in each of the over 20,000 genes in the mouse genome. All of the knockout strains produced by this collaborative effort are being made available to researchers throughout the world.

T-DNA AND TRANSPOSON INSERTIONS

In a preceding section of this chapter, we discussed how the T-DNA segment of the Ti plasmid of *Agrobacterium tumefaciens* is transferred into plant cells and inserted into the chromosomes of the plant (see **FIGURE 17.19**). When the T-DNA inserts into a gene, it disrupts the function of the gene. Transposons are genetic elements that have the ability to move from one location in the genome to another location (Chapter 18). Like the T-DNA of the Ti plasmid, a transposon will disrupt the function of a gene into which it inserts (see **FIGURE 13.20**). Thus, T-DNAs and transposons provide powerful tools for reverse genetic analysis. In both cases, the genetic element is used to perform **insertional mutagenesis**—the induction of null mutations (often called "knockout" mutations) by the insertion of foreign DNAs into genes. Insertional mutagenesis is basically the same whether performed with the Ti plasmid or a transposon. Thus, we will illustrate the use of insertional mutagenesis for reverse genetics by discussing the utilization of T-DNA insertions to dissect gene function in the plant *Arabidopsis thaliana*.

When T-DNA is transferred from *A. tumefaciens* to plant cells, it integrates into essentially all components of the genome; that is, T-DNAs are found scattered along each of the five chromosomes of *Arabidopsis*. Therefore, if a large enough population of transformed *Arabidopsis* plants is examined, it should be possible to identify T-DNA insertions in all of the approximately 26,000 genes of this species.

During the early years of T-DNA insertional mutagenesis in *Arabidopsis*, DNA samples were isolated from pools of 100 to 1000 transformed plants and screened for insertions in specific genes by the polymerase chain reaction (PCR; see **FIGURE 15.14**). If the DNA from a pool of transformed plants contained an insertion in a gene of interest, then DNA was prepared from smaller pools or individual plants and re-screened to identify the plant containing the desired insertion.

The screening procedure involved synthesizing PCR primers corresponding to the ends ("border sequences") of the T-DNA and the ends of the gene, and using them in pairs—a T-DNA primer and an inversely orientated gene primer—to amplify the intervening segments of DNA (**FIGURE 17.22**). In the absence of a T-DNA insert, no DNA fragments will be amplified. The exact site of the T-DNA insertion is then determined by sequencing the amplified PCR products. The same procedure is used to screen for insertions of transposons in genes, except that the border sequences of the transposon are used as PCR primers.

Today, the identification of transformed plant lines with T-DNA insertions in genes of interest is much easier. The National Science Foundation in the United States and government agencies in other countries have funded projects to produce large populations of T-DNA-transformed *Arabidopsis* plants. The nucleotide sequences of the DNAs flanking the T-DNA insertions have been determined and used to identify the sites of the insertions. Because the sequence of the *Arabidopsis* genome is known (Chapter 16), the flanking DNA sequences can be used to precisely map each insertion. Indeed,

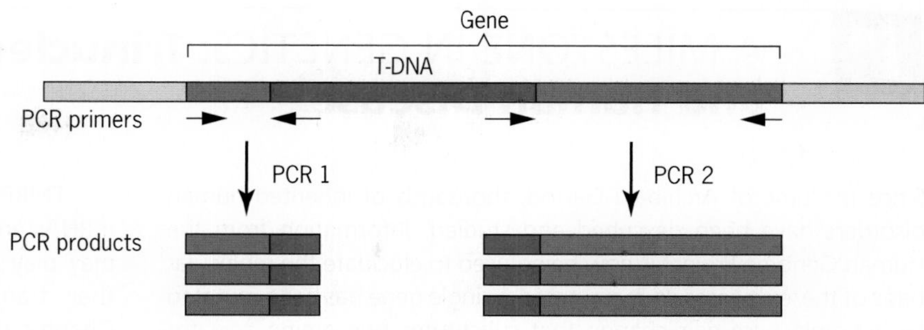

Figure 17.22 ► Identification of T-DNA insertions within genes by PCR. PCR primers that contain the 5′ → 3′ nucleotide sequences at the ends of the T-DNA (called border sequences) and the ends of the gene of interest are used in PCR to amplify the segments of DNA between the T-DNA insert and gene termini. Because the T-DNA can insert in either orientation (the one shown and flipped end-for-end), four PCR amplifications are usually performed—the two shown and two with the right and left T-DNA border primers switched. The precise location of the T-DNA insert is determined by sequencing the DNA fragments amplified by PCR.

as of January 2005, approximately 336,000 T-DNA and transposon insertions had been mapped in the *Arabidopsis* genome, and the map of these insertions was made available to everyone by placing it on the web. Researchers at the Salk Institute in La Jolla, California, have integrated the map of T-DNA insertions from their laboratory with the maps of T-DNA and transposon insertions characterized by other research groups. Their sequence-based map of these insertions is available at http://signal.salk.edu/cgi-bin/tdnaexpress; an abbreviated version of their map of the tip of chromosome 1 is shown in **FIGURE 17.23.**

SIGnAL "T-DNA Express" Arabidopsis Gene Mapping Tool (Dec.20, 2004)

Arabidopsis thaliana chromosome 1, nucleotide pairs 1 through 10,001.

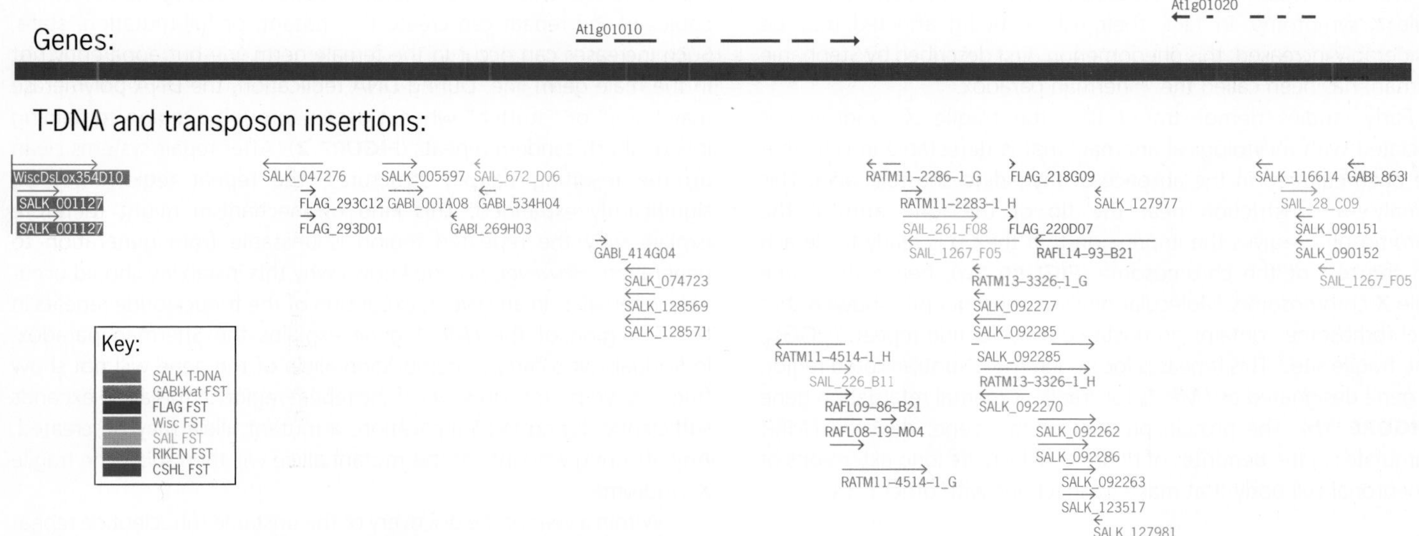

Figure 17.23 ► Map of T-DNA and transposon insertions in the 10 kb region at the tip of chromosome 1 in *Arabidopsis*. The positions of the flanking sequence tags (FSTs) are shown as arrows below the chromosome (dark blue box). The data shown are from the SIGnAL (*S*alk *I*nstitute *G*enomic *A*nalysis *L*aboratory) web site, http://signal.salk.edu/cgi-bin/tdnaexpress. The two genes (At1g01010 and At1g01020) in this region of chromosome 1 have unknown functions. The T-DNA and transposon insertion lines are from the Salk Institute (Salk T-DNA), the German collection (GABI-Kat), the French collection (FLAG), the University of Wisconsin (Wisc), the *S*yngenta *A*rabidopsis *I*nsertion *L*ibrary (SAIL), the Riken BioResource Center in Japan, and the *C*old *S*pring *H*arbor *L*aboratory (CSHL) collection.

▶ A MILESTONE IN GENETICS: **Trinucleotide Repeats and Human Disease**

Since the time of Archibald Garrod, thousands of inherited human disorders have been described and studied. Information from the Human Genome Project is now being used to elucidate the molecular basis of these diseases. In many cases, a single gene has been mutated by a simple base-pair change that substitutes one amino acid for another in the gene's polypeptide product. In other cases, the underlying mutation has a more dramatic effect; it may prematurely terminate polypeptide synthesis, or it may throw off the reading frame so that a garbled polypeptide is produced. As this molecular detective work has progressed, geneticists have learned that several dozen inherited diseases arise from instabilities in the sequence composition of genes. Most often these instabilities involve regions in which a unit of three adjacent nucleotides—a trinucleotide—is tandemly repeated many times.

The fragile X syndrome was the first human disorder to be connected to an unstable trinucleotide repeat. Individuals affected with this disorder show significant mental impairment; they may also exhibit facial and behavioral abnormalities. The fragile X syndrome occurs in about 1 in 4000 males and in about 1 in 7000 females; after Down syndrome, it is the most frequent cause of inherited mental retardation in humans. Pedigree studies indicate that the fragile X syndrome is caused by a dominant, X-linked mutation that is incompletely penetrant. About 20 percent of hemizygous males and about 30 percent of heterozygous females do not show symptoms. However, descendants from an asymptomatic individual may show fragile X symptoms. In fact, their risk of being affected may be considerably increased; this phenomenon, first described by Stephanie Sherman, has been called the "Sherman paradox."[1]

Early studies demonstrated that the fragile X syndrome is associated with a cytological anomaly that is detectable in cells that have been cultured in the absence of thymidine and folic acid. This anomaly—a constriction near the tip of the long arm of the X chromosome—gives the impression that the tip is ready to detach from the rest of the chromosome (**FIGURE 1*a***), hence the name fragile X chromosome. Molecular analysis subsequently showed that this chromosome contains an unstable trinucleotide repeat, $(CGG)_n$, at the fragile site.[2] This repeat is located in the 5′-untranslated region of a gene designated as *FMR-1*, for *f*ragile X *m*ental *r*etardation gene ***1*** (**FIGURE 1*b***). The protein product of this gene, denoted FMRP, accumulates in the dendrites of neurons, which are long extensions of the neuronal cell body that make connections with other cells.

FMRP is an RNA-binding protein. It is found in complexes with mRNAs and other components of the translation apparatus, and it may play a role in transporting mRNA molecules or in regulating their translation, possibly by interacting with micro RNAs (see Chapter 20). Individuals with the fragile X syndrome apparently do not make this protein. Thus, a lack of *FMR-1* gene expression seems to be the cause of the fragile X syndrome. How is the loss of FMRP expression related to the unstable trinucleotide repeat in the 5′-region of the *FMR-1* gene? Normal—that is, expressed—*FMR-1* genes contain 6 to 59 copies of this repeat. By contrast, abnormal—that is, unexpressed—*FMR-1* genes, which are found in people who have the fragile X syndrome, contain more than 200 copies. Somehow an increase in the number of trinucleotide repeats interferes with the expression of the *FMR-1* gene. One hypothesis currently under investigation is that the increased number of repeats leads to chemical modification of the DNA in the promoter of the *FMR-1* gene. This promoter is highly methylated in individuals who have the fragile X syndrome. Various studies have shown that hypermethylation of DNA, especially in and around promoters, silences gene expression (see Chapter 20).

What about cases in which the trinucleotide repeat number is between 60 and 200? Individuals with this number of repeats usually do not show the fragile X syndrome; however, their descendants often do. *FMR-1* genes with 60 to 200 copies of the trinucleotide repeat are said to be in a premutation state. Increasing the number of copies of the repeat can create the mutant, or full-mutation, state. Such increases can occur in the female germ line but apparently not in the male germ line. During DNA replication, the DNA polymerase may "slip," or "stutter," when it passes through a region containing lots of short, tandem repeats (**FIGURE 2**). After repair systems clean up the resulting hairpin structures, the repeat region may be significantly expanded. This kind of mechanism might therefore explain why the repeated region is unstable from generation to generation. However, no one knows why this instability should occur only in females. In any event, expansion of the trinucleotide repeats in the 5′ region of the *FMR-1* gene explains the Sherman paradox. Individuals who carry a premutation allele of the gene will not show fragile X symptoms. However, if the repeat region in this allele expands sufficiently during DNA replication, a mutant allele may be created. Any offspring who inherit this mutant allele will then show the fragile X syndrome.

Within a year of the discovery of the unstable trinucleotide repeat in the *FMR-1* gene, another neurodegenerative disorder, spinobulbar muscular atrophy (also known as Kennedy's disease), was linked to an unstable trinucleotide repeat, this time $(CAG)_n$.[3] Many other neurodegerative disorders have since been shown to result from

[1]Sherman, S. L., N. E. Morton, P. A. Jacobs, and G. Turner. 1984. The marker (X) syndrome: a cytogenetic and genetic analysis. *Annals of Human Genetics* 48:21–37.

[2]Fu, Y-H., D. P. A. Kuhl, A. Pizzuti, M. Pieretti, J. S. Sutcliffe, S. Richards, A. J. M. H. Verkerk, J. J. A. Holden, R. G. Fenwick, Jr., S. T. Warren, B. A. Oostra, D. L. Nelson, and C. T. Caskey. 1991. Variation of the CGG repeat at the fragile X site results in genetic instability: resolution of the Sherman paradox. *Cell* 67:1047–1058.

[3]La Spada, A. R., E. M. Wilson, D. B. Lubahn, A. E. Harding, and H. Fischbeck. 1991. Androgen receptor gene mutations in X-linked spinal and bulbar muscular atrophy. *Nature* 352:77–79.

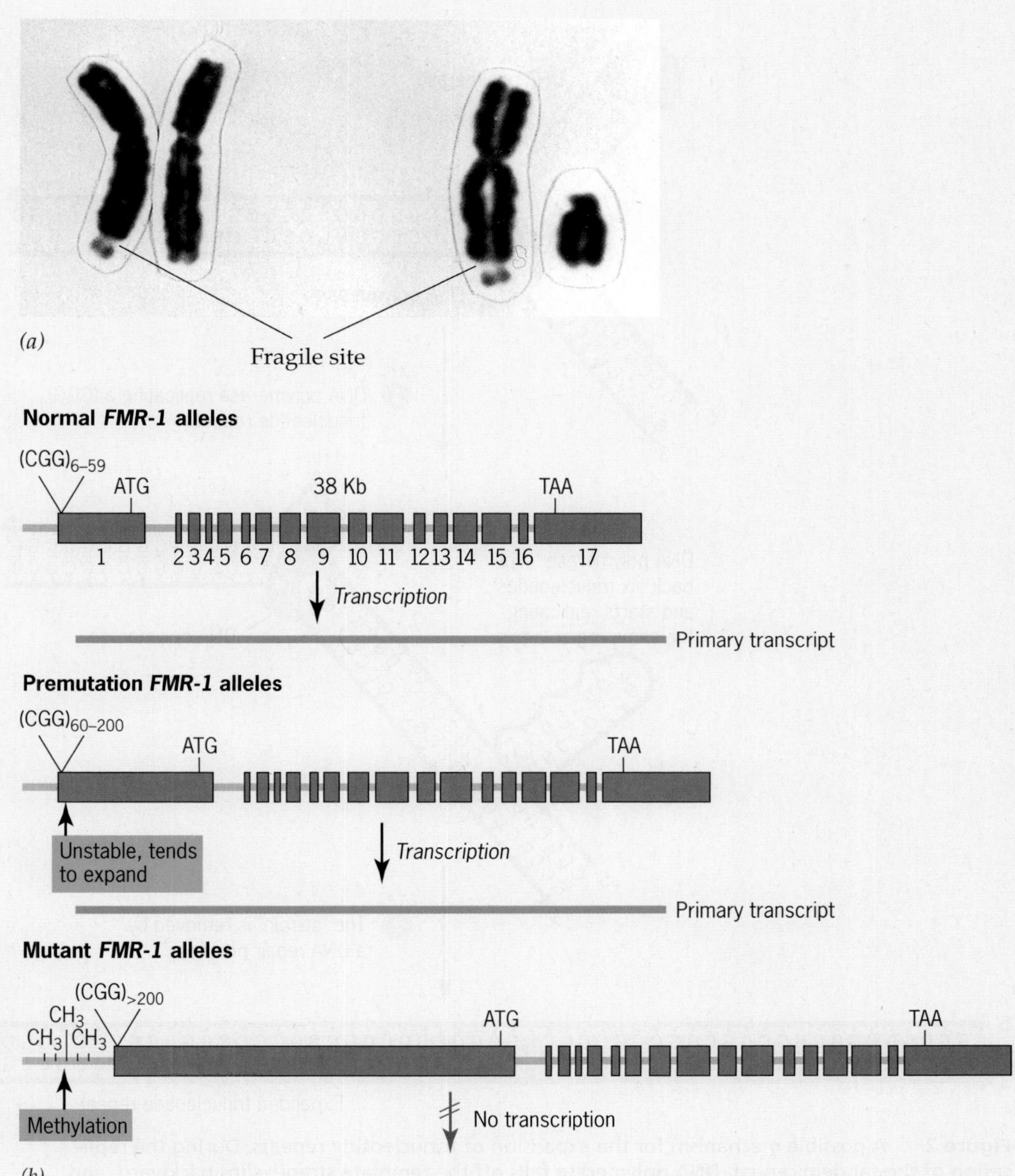

Figure 1 ▶ (*a*) The fragile X and a normal X chromosome from a female (left), and the fragile X and a normal Y chromosome from a male (right). (*b*) The location and number of CGG trinucleotide repeats in normal (top), premutation (center), and mutant alleles (bottom) of the *FMR-1* gene. The promoters of the mutant alleles are heavily methylated, which blocks transcription.

expanded trinucleotide repeats. The best known of these is Huntington's disease. Mutations involving unstable trinucleotide repeats therefore seem to be a significant type of genetic defect in our species.

QUESTIONS FOR DISCUSSION

1. The presence of the expanded trinucleotide repeat in a gene can be detected by a simple DNA test (see **FIGURE 17.2**). If you thought you might be carrying a premutation allele of the *FMR-1* gene, would you have the test done? If so, with whom would you share the results?
2. What might be the evolutionary significance of trinucleotide repeats? How might they influence the composition and structure of genomes?

▶ A MILESTONE IN GENETICS: Trinucleotide Repeats and Human Disease *(continued...)*

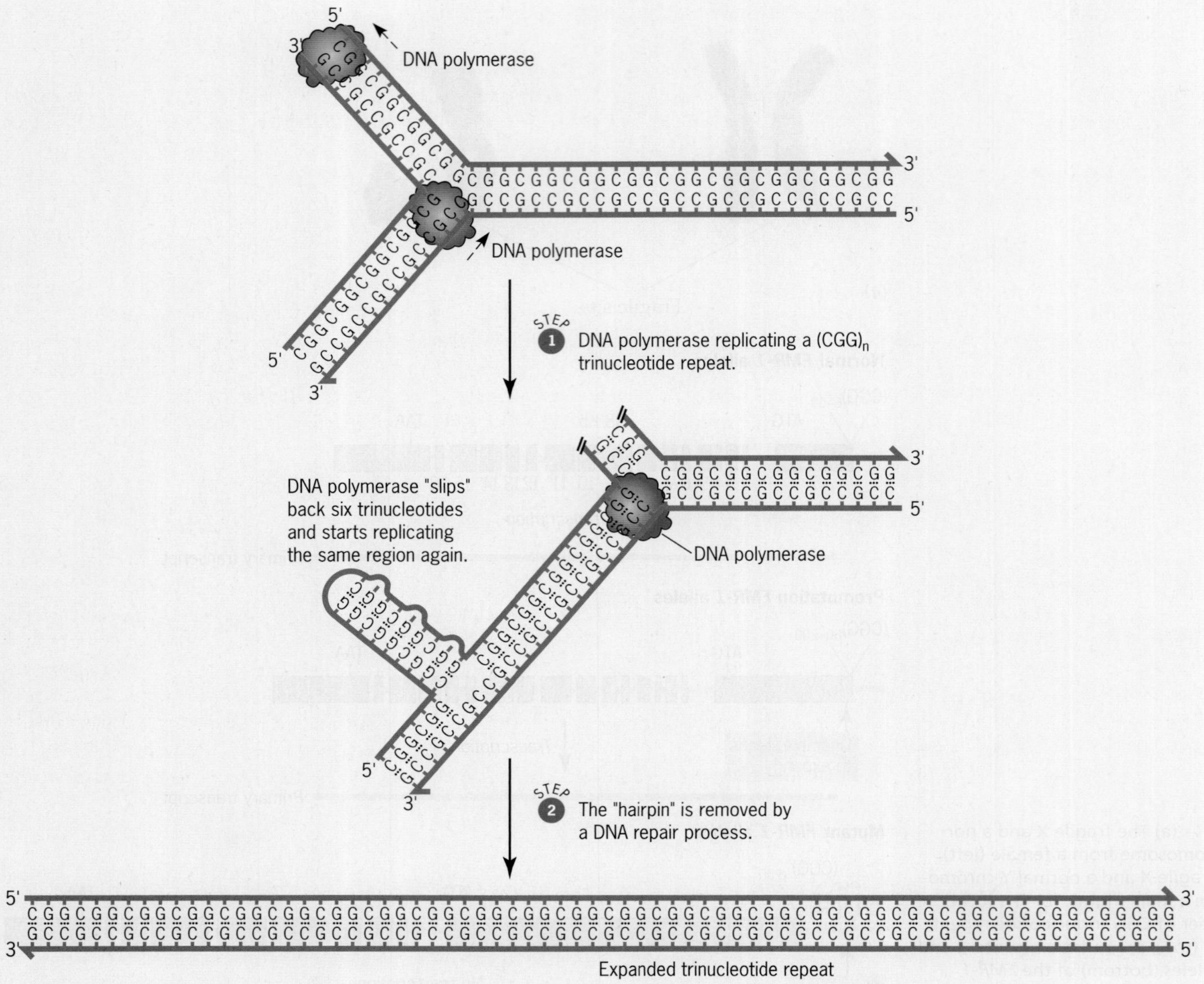

DNA polymerase

DNA polymerase

STEP 1 DNA polymerase replicating a $(CGG)_n$ trinucleotide repeat.

DNA polymerase "slips" back six trinucleotides and starts replicating the same region again.

DNA polymerase

STEP 2 The "hairpin" is removed by a DNA repair process.

Expanded trinucleotide repeat

Figure 2 ▶ A possible mechanism for the expansion of trinucleotide repeats. During the replication of the tandem repeat, DNA polymerase falls off the template strand, slips backward, and then reinitiates synthesis in a previously replicated region. The hairpin formed as a result of the slippage is recognized as a defect by a DNA repair enzyme, which initiates the repair process. A DNA polymerase involved in the repair pathway catalyzes the synthesis of a strand complementary to the unfolded hairpin, producing an expanded trinucleotide repeat region.

Seeds of transgenic plants from the Salk lines and many of the other collections are available upon request from the *Arabidopsis Biological Resource Center* (ABRC) at Ohio State University. In addition, seeds of T-DNA and transposon insertion lines characterized at the *Versailles Genomic Resource Center* (VGRC) in France, the *Nottingham Arabidopsis Stock Centre* (NASC) in Germany, and the Riken BioResource Center in Japan are also available to the *Arabidopsis* research community. Seeds of desired lines can be ordered online from the ABRC web site http://www.arabidopsis.org/abrc. In addition, links to other sites with information about the *Arabidopsis* genome are available at http://www.arabidopsis.org, the web site of *T*he *Arabidopsis* *I*nformation *R*esource (TAIR). Therefore, if someone is interested in the function of a particular *Arabidopsis* gene, she or he can search the Salk web site for T-DNA and transposon insertions in that gene; once the insertions have been identified, seeds carrying the desired insertion mutations can be ordered online. These large collections of insertional mutations have proven to be invaluable resources for studies of gene function in this model plant.

RNA INTERFERENCE

Although its effects were first observed in petunias a few years earlier, the discovery of the third reverse genetics approach—**RNA interference (RNAi)**—is usually credited to the work of Andrew Fire, Craig Mello, and colleagues, published in 1998. Indeed, Fire and Mello shared the 2006 Noble Prize in Physiology or Medicine in recognition of this work. When they injected double-stranded RNA (dsRNA) into *C. elegans*, it "interfered with" (or shut off) the expression of genes containing the same nucleotide sequence (see A Milestone in Genetics: The Discovery of RNA Interference in Chapter 20). During the last few years, RNAi has moved to the cutting edge in molecular biology. We now know that double-stranded RNA (dsRNA) plays important roles in preventing viral infections, in combating the expansion of populations of transposable genetic elements, and in regulating gene expression (see Chapter 20). Indeed, RNAi is not only at the cutting edge of molecular biology, but it has enormous potential in the fight against human diseases. In this chapter, however, we will focus on the use of RNAi as a tool for reverse genetics, a tool with which to study gene function and dissect biological processes.

RNAi is used extensively to silence genes—turn down or turn off their expression—in *C. elegans*, *D. melanogaster*, and many plants. It has potential uses in all species, including humans. RNAi can be carried out in several different ways. The common feature in all RNAi experiments is the presence of dsRNA carrying at least a portion of the nucleotide sequence of the gene that one wishes to silence in the organism or cells under investigation. Two different approaches are frequently used to achieve this goal. In one approach, the dsRNA is synthesized *in vitro* and microinjected into the organism (**FIGURE 17.24*a***). In the second approach, a gene expression cassette is constructed that carries two copies of at least a portion of the gene of interest in inverse orientations and is introduced into the organism by transformation or transfection (**FIGURE 17.24*b***). When the introduced transgene is transcribed, it produces an RNA molecule that is self-complementary and forms a partially double-stranded stem-and-loop, or "hairpin" structure. In both cases, the dsRNAs stimulate RNA-induced gene silencing. The dsRNAs are ultimately bound by the RNA-induced silencing complex (RISC) and either degraded or translationally repressed, depending on the organism and cells involved (see Chapter 20 for details).

RNAi is quite easy to perform in *C. elegans*; these little worms can be microinjected with the dsRNA, soaked in media containing the dsRNA, or fed bacteria synthesizing the dsRNA of interest. All three procedures lead to effective gene silencing in *C. elegans*.

The sequence of 99 percent of the genome of *C. elegans* was published in December 1998. Within two years, collaborative research groups in Great Britain, Germany, Switzerland, and Canada had used RNAi to systematically silence more than 90 percent of the 2769 predicted genes on chromosome I and more than 96 percent of the 2300 predicted genes on chromosome III of *C. elegans*. These studies provided new information about the functions of over 400 genes. Clearly, RNAi is a powerful tool with which to dissect biological processes. RNAi makes use of natural pathways involved in the regulation of gene expression. There are hundreds of genes in plant and animal genomes that encode **microRNAs,** which form dsRNAs *in vivo*. At present, we know the regulatory functions of only a few of these microRNAs (see Chapter 20); however, the functions of the rest of the microRNAs are the subject of many ongoing investigations.

Can RNAi be used to inhibit the reproduction of viruses such as the human immunodeficiency virus (HIV) or to down-regulate the expression of oncogenes (cancer-causing genes)? We don't know the answer to that question. However, we do know that the business world is excited about the potential therapeutic applications of RNAi. Not only are the big pharmaceutical firms investing heavily in RNAi technology, but a plethora of start-up companies have been formed specifically to exploit RNAi for commercial goals. Whether or not the RNAi technologies will live up to expectations remains to be seen.

KEY POINTS

- Reverse genetic approaches use known nucleotide sequences to devise procedures for isolating null mutations of genes or inhibiting gene expression.
- Antisense RNA, which is complementary to sense RNA (mRNA), can be used to shut off or reduce the expression of individual genes.
- Knockout mutations of genes in the mouse can be produced by inserting foreign DNAs into chromosomal genes by homologous recombination.
- T-DNA or transposon insertions provide a source of null mutations of genes.
- RNA interference—blocking gene expression with double-stranded RNA—can be used to dissect biological processes by inhibiting the functions of specific genes.

Initiation of RNAi by synthesis and injection of dsRNA.

STEP 1 Double-stranded RNA containing the desired sequence is synthesized *in vitro.*

5' 3'
3' 5'
dsRNA

STEP 2 The dsRNA is microinjected into the organism.

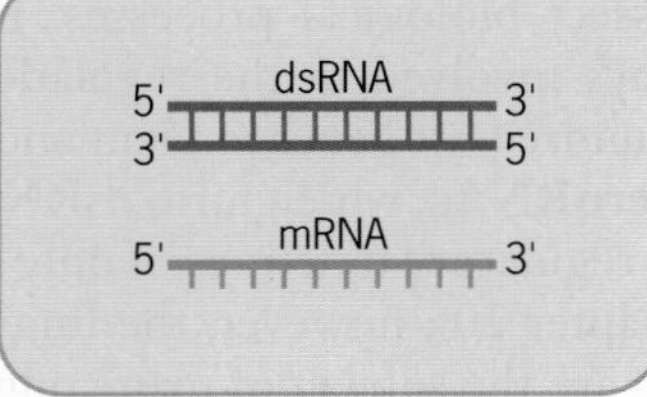

STEP 3 Degradation of mRNA or repression of translation by the RNA-induced silencing complex (RISC).

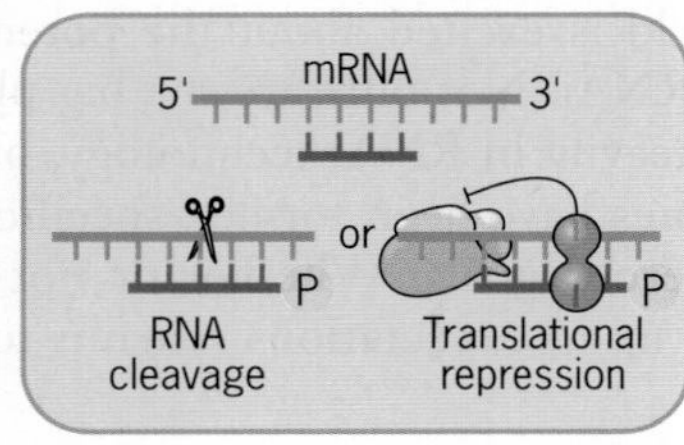

(a)

Initiation of RNAi by introducing a transgene encoding self-complementary RNA.

STEP 1 A gene-expression cassette carrying two copies of the desired sequence in inverse orientations is introduced into the genome.

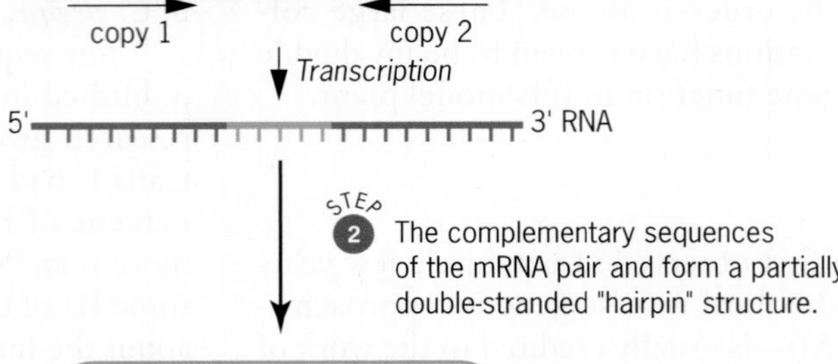

STEP 2 The complementary sequences of the mRNA pair and form a partially double-stranded "hairpin" structure.

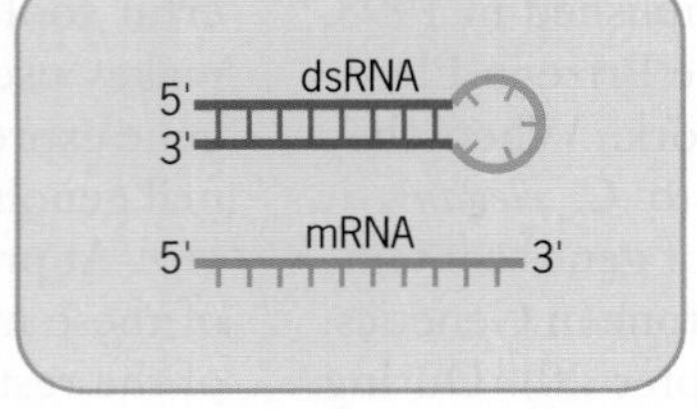

STEP 3 Degradation of mRNA or repression of translation by the RNA-induced silencing complex (RISC).

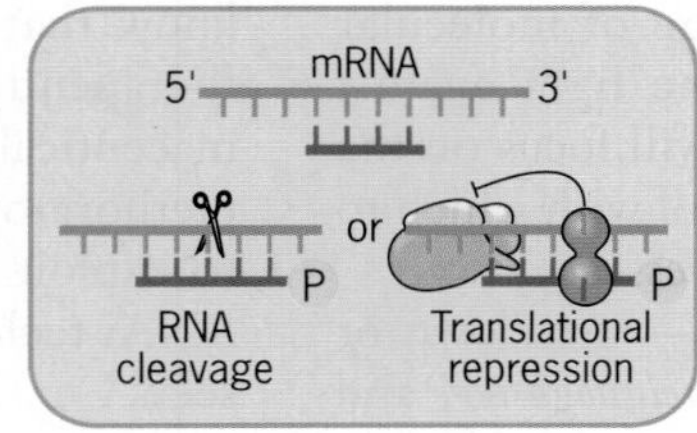

(b)

Figure 17.24 ▸ Two procedures for initiating RNAi with double-stranded RNA (dsRNA). (*a*) A dsRNA molecule containing a portion of the nucleotide sequence of the gene to be silenced is synthesized *in vitro* and injected into the organism. (*b*) A gene expression cassette containing two copies of a segment of the gene in inverse orientations is constructed and introduced into the organism under investigation. The self-complementary RNA transcript forms a partially double-stranded RNA hairpin. In both cases, the dsRNA initiates silencing of the targeted gene via the RNA-induced silencing complex (RISC) pathway, which results in the degradation of the targeted mRNA or repression of its translation (see Chapter 20 for details).

Basic Exercises

ILLUSTRATE BASIC GENETIC ANALYSIS

1. How were restriction fragment-length polymorphisms (RFLPs) used in the search for the mutant gene that causes Huntington's disease (HD)?

Answer: The HD research teams screened members of two large families for linkage between RFLPs and the HD gene. They found an RFLP on chromosome 4 that was tightly linked to the HD gene (4 percent recombination).

2. Once tight linkage had been established between the HD gene and the RFLP on chromosome 4, what was the research teams' next step in their search for the HD gene?

Answer: They next prepared a detailed restriction map of this region (spanning 500 kb) of chromosome 4 (see **FIGURE 17.1**).

3. How did the research teams identify candidate genes within the mapped region of chromosome 4?

Answer: They used cDNA clones to identify the coding segments or exons of genes in the region and to screen genomic libraries for clones overlapping the exons. The sequences of the cDNAs and genomic DNAs were then compared to deduce the exon-intron structures of genes in the mapped region.

4. How did the HD research teams determine which of the candidate genes was the HD gene?

Answer: They sequenced the candidate genes of individuals with HD and nonaffected members of their families and looked for structural abnormalities in the genes of affected individuals. Their results showed that one gene, now called the *huntingtin* gene, contains a trinucleotide repeat, $(CAG)_n$, which was present in 11 to 34 copies in nonaffected individuals and in 42 to over 100 copies in affected individuals. They identified this expanded trinucleotide repeat in the *huntingtin* alleles of affected members of 72 different families, leaving little doubt that *huntingtin* is the HD gene.

5. Of what value is knowledge of the nucleotide sequence of the *huntingtin* gene to genetic counselors?

Answer: Knowing the nucleotide sequence of the *huntingtin* gene has provided counselors with a simple and accurate diagnostic test for the presence of mutant alleles of the gene. Oligonucleotide primers to sequences flanking the trinucleotide repeat region of the gene can be used to amplify this segment of the gene, and the number of trinucleotide repeats can be determined by polyacrylamide gel electrophoresis (see **FIGURE 17.2**). As a result, individuals at risk of transmitting the mutant gene can be tested for its presence before starting a family. If the mutant gene is present in one of the parents, fetal cells or even a single cell from an eight-cell pre-implantation embryo can be tested for its presence. Thus, genetic counselors are able to provide families at risk for the disorder with accurate information regarding the presence of the gene in individuals planning families, in fetal cells, and even in eight-cell embryos.

Testing Your Knowledge

INTEGRATE DIFFERENT CONCEPTS AND TECHNIQUES

1. Spinocerebellar ataxia (type 1) is a progressive neurological disease with onset typically occurring between ages 30 and 50. The neurodegeneration results from the selective loss of specific neurons. Although it is not understood why selective neuronal death occurs, it is known that the disease is caused by the expansion of a CAG trinucleotide repeat, with normal alleles containing about 28 copies and mutant alleles harboring 43 to 81 copies of the trinucleotide. Given the nucleotide sequences on either side of the repeat region, how would you test for the presence of the expanded trinucleotide repeat region responsible for type 1 spinocerebellar ataxia?

Answer: The DNA test for spinocerebellar ataxia (type 1) would be similar to the test for the *huntingtin* allele described in **FIGURE 17.2.** You would first make PCR primers corresponding to DNA sequences on either side of the CAG repeat region. These primers would be used to amplify the desired CAG repeat region from genomic DNA of the individual being tested by PCR. Then, the sizes of the trinucleotide repeat regions would be determined by measuring the sizes of the PCR products by gel electrophoresis. Any gene with fewer than 30 copies of the CAG repeat would be considered a normal allele, whereas the presence of a gene with 40 or more copies of the trinucleotide would be diagnostic of the mutant alleles that cause spinocerebellar ataxia.

2. Assume that you have just performed the DNA test for spinocerebellar ataxia on a 25-year-old woman whose mother died from the disease. The results came back positive for the ataxia mutation. The woman and her husband long for their own biological children, but do not want to risk transmitting the defective gene to any of these children. What are their options?

Answer: Their options will depend on their religious and moral convictions. One possibility involves the use of amniocentesis or chorionic biopsy to obtain fetal cells early in pregnancy, performing the DNA test for the expanded trinucleotide region responsible for spinocerebellar ataxia on the fetal cells, and allowing the pregnancy to continue only if the defective gene is not present. Another possibility is the use of *in vitro* fertilization. The ataxia DNA test is then performed on a cell from an eight-cell pre-embryo, and the pre-embryo is implanted only if the test for the defective ataxia gene is negative. A third option may be available in the future, namely, an effective method of treating the disease prior to the onset of neurodegeneration, perhaps by gene-replacement therapy.

▶ Questions and Problems

ENHANCE UNDERSTANDING AND DEVELOP ANALYTICAL SKILLS

17.1 How was the nucleotide sequence of the *CF* gene used to obtain information about the structure and function of its gene product?

17.2 In humans, the absence of an enzyme called purine nucleoside phosphorylase (PNP) results in a severe T-cell immunodeficiency similar to that of severe combined immunodeficiency disease (SCID). PNP deficiency exhibits an autosomal recessive pattern of inheritance, and the gene encoding human PNP has been cloned and sequenced. Would PNP deficiency be a good candidate for treatment by gene therapy? Design a procedure for the treatment of PNP deficiency by somatic-cell gene therapy.

17.3 Myotonic dystrophy (MD), occurring in about 1 of 8000 individuals, is the most common form of muscular dystrophy in adults. The disease, which is characterized by progressive muscle degeneration, is caused by a dominant mutant gene that contains an expanded CAG repeat region. Wild-type alleles of the *MD* gene contain 5 to 30 copies of the trinucleotide. Mutant *MD* alleles contain 50 to over 2000 copies of the CAG repeat. The complete nucleotide sequence of the *MD* gene is available. Design a diagnostic test for the mutant gene responsible for myotonic dystrophy that can be carried out using genomic DNA from newborns, fetal cells obtained by amniocentesis, and single cells from eight-cell pre-embryos produced by *in vitro* fertilization.

17.4 Why is the mutant gene that causes Huntington's disease called *huntingtin?* Why might this gene be renamed in the future?

17.5 Human proteins can now be produced in bacteria such as *E. coli*. However, one cannot simply introduce a human gene into *E. coli* and expect it to be expressed. What steps must be taken to construct an *E. coli* strain that will produce a mammalian protein such as human growth hormone?

17.6 How might the characterization of the *CF* gene and its product lead to the treatment of cystic fibrosis by somatic-cell gene therapy? What obstacles must be overcome before cystic fibrosis can be treated successfully by gene therapy?

17.7 What are CpG islands? Of what value are CpG islands in positional cloning of human genes?

17.8 GO A group of bodies are found buried in a forest. The police suspect that they may include the missing Jones family (two parents and two children). They extract DNA from bones and examine (using PCR) genes *A* and *B*, which are known to contain tandem triplet repeats of variable length. They also analyze DNA from two other men. The results are shown below where the numbers indicate the number of copies of a tandem repeat in a particular allele; for example, male 1 has one allele with 8 and another allele with 9 copies of a tandem repeat in gene *A*.

	Gene A	Gene B
male 1	8/9	5/7
male 2	6/8	5/5
male 3	7/10	7/7
woman	8/8	3/5
child 1	7/8	5/7
child 2	8/8	3/7

Could the woman have been the mother of both children? Why or why not? Which man, if any, could have been the father of child 1?

17.9 DNA fingerprints have played central roles in many recent rape and murder trials. What is a DNA fingerprint? What roles do DNA prints play in these forensic cases? In some cases, geneticists have been concerned that DNA fingerprint data were being used improperly. What were some of their concerns, and how can these concerns be properly addressed?

17.10 You have constructed a synthetic gene that encodes an enzyme that degrades the herbicide glyphosate. You wish to introduce your synthetic gene into *Arabidopsis* plants and test the transgenic plants for resistance to glyphosate. How could you produce a transgenic *Arabidopsis* plant harboring your synthetic gene by *A. tumefaciens*-mediated transformation?

17.11 A human VNTR locus contains a tandem repeat $(TAA)_n$, where n may be between 5 and 15. How many alleles of this locus would you expect to find in the human population?

17.12 GO The DNA fingerprints shown below were prepared using genomic DNA from blood cells obtained from a woman, her daughter, and three men who all claim to be the girl's father. Based on the DNA prints, what can be determined about paternity in this case?

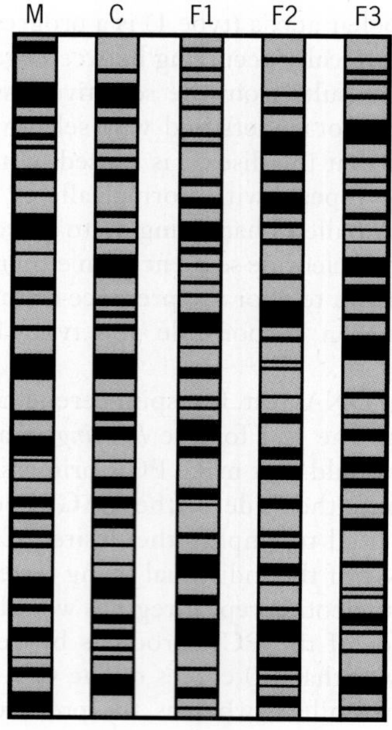

17.13 The generation of transgenic plants using *A. tumefaciens*-mediated transformation often results in multiple sites of insertion. These sites frequently vary in the level of transgene expression. What approaches could you use to determine whether or not transgenic plants carry more than one transgene and, if so, where the transgenes are inserted into chromosomes?

17.14 What is antisense RNA? Of what use is antisense RNA in genetic research? in agriculture? How can a cell or an organism that produces a specific antisense RNA be produced?

17.15 Most forensic experts agree that fingerprinting of DNA from blood samples obtained at crime scenes and on personal items can provide convincing evidence for murder convictions. However, the defense attorneys sometimes argue successfully that sloppiness in handling blood samples results in contamination of the samples. What problems would contamination of blood samples present in the interpretation of DNA fingerprints? Would you expect such errors to lead to the conviction of an innocent person or the acquittal of a guilty person?

17.16 Richard Meagher and coworkers have cloned a family of 10 genes that encode actins (a major component of the cytoskeleton) in *Arabidopsis thaliana*. The 10 actin gene products are similar, often differing by just a few amino acids. Thus, the coding sequences of the 10 genes are also very similar, so that the coding region of one gene will cross-hybridize with the coding regions of the other nine genes. In contrast, the noncoding regions of the 10 genes are quite divergent. Meagher has hypothesized that the 10 actin genes exhibit quite different temporal and spatial patterns of expression. You have been hired by Meagher to test this hypothesis. Design experiments that will allow you to determine the temporal and spatial pattern of expression of each of the 10 actin genes in *Arabidopsis*.

17.17 Transgenic mice are now routinely produced and studied in research laboratories throughout the world. How are transgenic mice produced? What kinds of information can be obtained from studies performed on transgenic mice? Does this information have any importance to the practice of medicine? If so, what?

17.18 Disarmed retroviral vectors can be used to introduce genes into higher animals, including humans. What advantages do retroviral vectors have over other kinds of gene-transfer vectors? What disadvantages?

17.19 The first transgenic mice resulted from microinjecting fertilized eggs with vector DNA similar to that diagrammed in **FIGURE 17.13** except that it contained a promoter for the mammalian metallothionein gene linked to the *HGH* gene. The resulting transgenic mice showed elevated levels of HGH in tissues of organs other than the pituitary gland—for example, in heart, lung, and liver—and the pituitary gland underwent atrophy. How might the production of HGH in transgenic animals be better regulated, with expression restricted to the pituitary gland?

17.20 How do insertional mutagenesis approaches differ from other reverse genetic approaches?

17.21 In what ways are antisense RNA and RNAi gene-silencing procedures the same? different?

17.22 We discussed the unfortunate effects of insertional mutagenesis in the four boys who developed leukemia after treatment of X-linked severe combined immunodeficiency disease by gene therapy (see **FIGURE 17.9**). How might this consequence of gene therapy be avoided in the future? Do you believe that the use of somatic-cell gene therapy to treat human diseases can ever be made 100 percent risk free? Why? Why not?

17.23 Insertional mutagenesis is a powerful tool in both plants and animals. However, when performing large-scale insertional mutagenesis, what major advantage do plants have over animals?

17.24 Let's check the Salk Institute's Genome Analysis Laboratory web site (http://signal.salk.edu/cgi-bin/tdnaexpress) to see if any of their T-DNA lines have insertions in the gene shown in the previous question. At the SIGnAL web site, scroll down to "Blast" and paste or type the sequence in the box. The resulting map will show the location of mapped T-DNA insertions relative to the location of the gene (green rectangle at the top). The blue arrows at the top right will let you focus on just the short region containing the gene or relatively long regions of chromosome 4 of *Arabidopsis*. Are there any T-DNA insertions in the gene in question? Near the gene?

17.25 One strand of a gene in *Arabidopsis thaliana* has the following nucleotide sequence:

atgagtgacgggaggaggaagaagagcgtgaacggaggtgcaccggcg
caaacaatcttggatgatcggagatctagtcttccggaagttgaagcttctccaccggct
gggaaacgagctgttatcaagagtgccgatatgaaagatgatatgcaaaag
gaagctatcgaaatcgccatctccgcgtttgagaagtacagtgtggagaaggatat
agctgagaatataaagaaggagtttgacaagaaacatggtgctacttggcattgcatt
gttggtcgcaactttggttcttatgtaacgcatgagacaaaccatttcgtttacttctacctc
gaccagaaagctgtgctgctcttcaagtcgggttaa

The function(s) of this gene is still uncertain. (a) How might you use antisense RNA to study its function(s)? (b) How might insertional mutagenesis be used to investigate the function(s) of the gene? (c) Design an experiment using RNA interference to probe the function(s) of the gene.

17.26 How do the reverse genetic approaches used to dissect biological processes differ from classical genetic approaches?

▶ Genomics on the Web

at http://www.ncbi.nlm.nih.gov/

Muscular dystrophy is a group of human disorders that involve progressive muscle weakness and loss of muscle cells.

1. How many different types of inherited muscular dystrophy have been characterized in humans to date?

2. What are the chromosomal locations of the defective genes responsible for the different forms of muscular dystrophy?

3. Duchenne and Becker muscular dystrophy both result from mutations in a gene on the X chromosome that encodes a protein called dystrophin. How do these two types of muscular dystrophy differ?

4. The dystrophin (*DMD*) gene has been cloned and sequenced. What are the unique features of this gene and the protein that it encodes? What obstacles do they present for the treatment of Duchenne and Becker muscular dystrophy by gene therapy?

5. Are gene tests for Duchenne muscular dystrophy available? How are they performed?

Hint: At the ncbi web site, click on OMIM (Online Mendelian Inheritance in Man) and search using "muscular dystrophy" as the query. From the resulting list of muscular dystrophies, note the chromosomal locations of the genes responsible for the various forms of the disorder to estimate the number of different genes involved. For information on Duchenne and Becker muscular dystrophy, click on #310200, "Muscular Dystrophy, Duchenne Type," and #300376, "Muscular Dystrophy, Becker Type." For information on gene tests, click on "Gene Tests" and follow the links provided to the different types of tests available.

Chapter 18
Transposable Genetic Elements

CHAPTER OUTLINE

- **Transposable Elements: An Overview**
- **Transposable Elements in Bacteria**
- **Cut-and-Paste transposons in Eukaryotes**
- **Retroviruses and Retrotransposons**
- **Transposable Elements in Humans**
- **The Genetic and Evolutionary Significance of Transposable Elements**

Color variation among kernels of maize. Studies of the genetic basis of this variation led to the discovery of transposable elements.

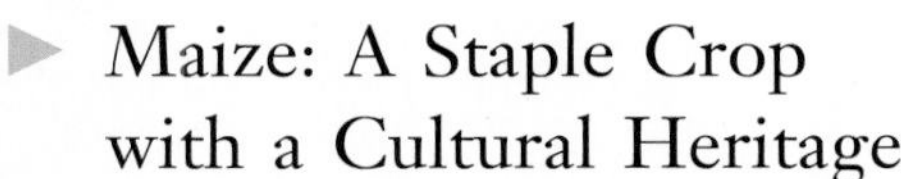

Maize: A Staple Crop with a Cultural Heritage

Maize is one of the world's most important crop plants. The cultivation of maize began at least 5000 years ago in Central America. By the time Christopher Columbus arrived in the New World, maize cultivation had spread north to Canada and south to Argentina. The native peoples of North and South America developed many different varieties of maize, each adapted to particular conditions. They developed varieties that had colorful kernels—red, blue, yellow, white, and purple—and associated each color with a special aesthetic or religious value. To the peoples of the American Southwest, for example, blue maize is considered sacred, and each of the four cardinal directions of the compass is represented by a particular maize color. Some groups consider kernels with stripes and spots to be signs of strength and vigor.

The colorful patterns that we see on maize ears also have an important scientific significance. Modern research has shown that the stripes and spots on maize kernels are the result of a genetic phenomenon called *transposition*. Within the maize genome—indeed, within the genomes of most organisms—geneticists have found DNA sequences that can move from one position to another. These *transposable elements*—or, more simply, *transposons*—constitute an appreciable fraction of the genome. In maize, for example, they account for more than half of all the DNA. When transposable elements move from one location to another, they may break chromosomes or mutate genes. Thus, these elements have a profound genetic significance.

Transposable Elements: An Overview

Transposable elements—transposons—are found in the genomes of many kinds of organisms; they are structurally and functionally diverse.

The world of transposable elements is rich with diversity. Many different kinds of elements have been identified in an assortment of organisms, including bacteria, fungi, protists, plants, and animals. These elements are prominent components of genomes—for example, more than 40 percent of the human genome—and they clearly have roles in shaping the structure of chromosomes and in modulating the expressions of genes. In this chapter we shall explore the structural and behavioral diversity of different types of transposable elements, and we shall investigate their genetic and evolutionary significance.

Although each kind of transposable element has its own special characteristics, most of them can be classified into one of three categories based on how they transpose (**TABLE 18.1**). In the first category, transposition is accomplished by excising an element from its position in a chromosome and inserting it into another position. The excision and insertion events are catalyzed by an enzyme called the transposase, which is usually encoded by the element itself. Geneticists refer to this mechanism as *cut-and-paste transposition* because the element is physically cut out of one site in a chromosome and pasted into a new site, which may even be on a different chromosome. We shall refer to the elements in this category as **cut-and-paste transposons.**

In the second category, transposition is accomplished through a process that involves replication of the transposable element's DNA. A transposase encoded by the element mediates an interaction between the element and a potential insertion site. During this interaction, the element is replicated and one copy of it is inserted at the new site; one copy also remains at the original site. Because there is a net gain of one copy of the element, geneticists refer to this mechanism as *replicative transposition*. We shall refer to the elements in the category as **replicative transposons.**

In the third category, transposition is accomplished through a process that involves the insertion of copies of an element that were synthesized from the element's RNA. An enzyme called reverse transcriptase uses the element's RNA as a template to synthesize DNA molecules, which are then inserted into new chromosomal sites. Because this mechanism reverses the usual direction in which genetic information flows in cells—that is, it flows from RNA to DNA

▶ TABLE 18.1

Categorization of Transposable Elements by Transposition Mechanism

Category	Examples	Host Organism
I. Cut-and-paste transposons	IS elements (e.g., IS*50*)	Bacteria
	Composite transposons (e.g., Tn*5*)	Bacteria
	Ac/*Ds* elements	Maize
	P elements	*Drosophila*
	mariner elements	*Drosophila*
	hobo elements	*Drosophila*
	Tc1 elements	Nematodes
II. Replicative transposons	Tn*3* elements	Bacteria
III. Retrotransposons		
A. Retroviruslike elements	Ty*1*	Yeast
(also called long terminal repeat, or LTR, retrotransposons)	*copia*	*Drosophila*
B. Retroposons	*F*, *G*, and *I* elements	*Drosophila*
	Telomere-specific retroposons (*HeT-A*, *TART*)	*Drosophila*
	LINEs (e.g., *L1*)	Humans
	SINEs (e.g., *Alu*)	Humans

instead of from DNA to RNA—geneticists refer to it as *retrotransposition*. We shall refer to the elements in this category as **retrotransposons**. Some of the elements that transpose in this way are related to the retroviruses; consequently, they are called *retroviruslike elements*. Other elements that engage in retrotransposition are simply called *retroposons*.

We shall encounter many different transposable elements in this chapter, each with its own peculiar story. **TABLE 18.1** categorizes these elements according to their transposition mechanisms. The cut-and-paste transposons are found in both prokaryotes and eukarotes. The replicative transposons are found only in prokaryotes, and the retrotransposons are found only in eukaryotes.

KEY POINTS

- A cut-and-paste transposon is excised from one genomic position and inserted into another by an enzyme, the transposase, which is usually encoded by the transposon itself.
- A replicative transposon is copied during the process of transposition.
- A retrotransposon produces RNA molecules that are reverse-transcribed into DNA molecules; these DNA molecules are subsequently inserted into new genomic positions.

Transposable Elements in Bacteria

Bacterial transposons move within and between the bacterial chromosome and various types of plasmids.

Although transposable elements were originally discovered in eukaryotes, bacterial transposons were the first to be studied at the molecular level. There are three main types: the insertion sequences, or IS elements, the composite transposons, and the Tn*3* elements. These three types of transposons differ in size and structure. The IS elements are the simplest, containing only genes that encode proteins involved in transposition. The composite transposons and Tn*3* elements are more complex, containing some genes that encode products unrelated to the transposition process.

IS ELEMENTS

The simplest bacterial transposons are the **insertion sequences,** or **IS elements,** so named because they can insert at many different sites in bacterial chromosomes and plasmids. IS elements were first detected in certain *lac*$^-$ mutations of *E. coli*. These mutations had the unusual property of reverting to wild-type at a high rate. Molecular analyses revealed that these unstable mutations possessed extra DNA in or near the *lac* genes. When DNA from the wild-type revertants of these mutations was compared with that from the mutations themselves, it was found that the extra DNA had been lost. Thus, these genetically unstable mutations were caused by DNA sequences that had inserted into *E. coli* genes, and reversion to wild-type was caused by excision of these sequences. Similar insertion sequences have been found in many other bacterial species.

IS elements are compactly organized. Typically, they consist of fewer than 2500 nucleotide pairs and contain only genes whose products are involved in promoting or regulating transposition. Many distinct types of IS elements have been identified. The smallest, IS*1*, is 768 nucleotide pairs long. Each type of IS element is demarcated by short identical, or nearly identical, sequences at its ends (**FIGURE 18.1**). Because these terminal sequences are always in inverted orientation with respect to each other, they are called **inverted terminal repeats.** Their lengths range from 9 to 40 nucleotide pairs. Inverted terminal repeats are characteristic of most—but not all—types of transposons. When nucleotides in these repeats are mutated, the transposon usually loses its ability to move. These mutations therefore demonstrate that inverted terminal repeats play an important role in the transposition process.

At least some IS elements encode a protein that is needed for transposition. This protein, called **transposase,** seems to bind at or near the ends of the element, where it cuts both strands of the DNA. Cleavage of the DNA at these sites excises the element from the chromosome or plasmid, so that it can be inserted at a new position in the same or a different DNA molecule. IS elements are therefore cut-and-paste transposons. When IS elements insert into chromosomes or plasmids, they create a duplication of part of the DNA sequence at the site of the insertion. One copy of the duplication is located on each side of the element. These short (2 to 13 nucleotide pairs), directly repeated sequences, called **target site duplications,**

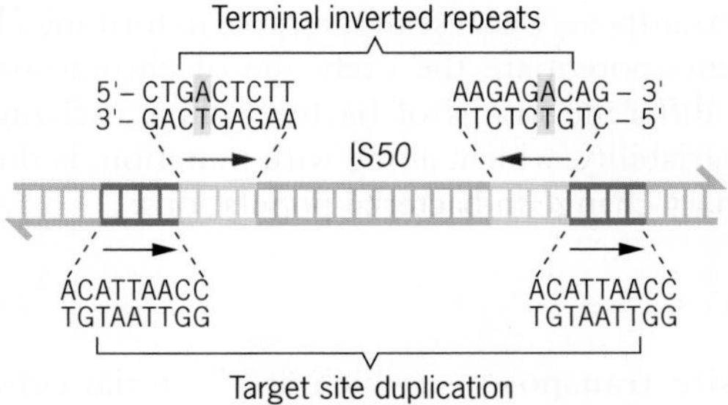

Figure 18.1 ▶ Structure of an inserted IS*50* element showing its terminal inverted repeats and target site duplication. The terminal inverted repeats are imperfect because the fourth nucleotide pair from each end is different.

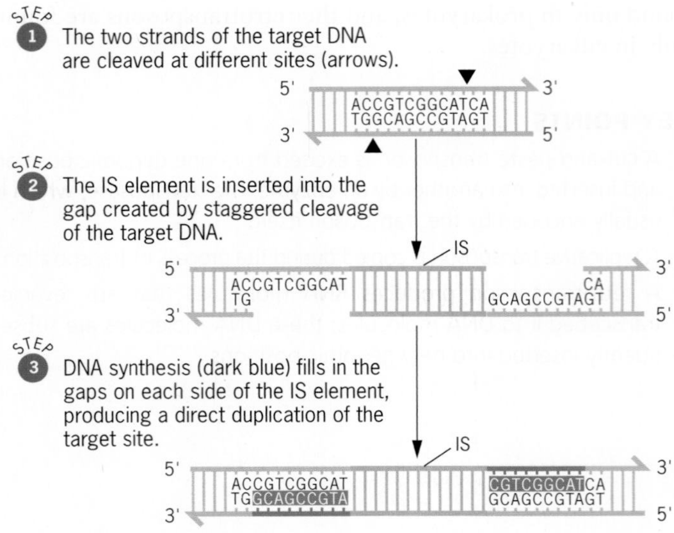

Figure 18.2 ▶ Production of target site duplications by the insertion of an IS element.

are thought to arise from staggered cleavage of the double-stranded DNA molecule (**FIGURE 18.2**).

A bacterial chromosome may contain several copies of a particular type of IS element. For example, 6 to 10 copies of IS*1* are found in the *E. coli* chromosome. Plasmids may also contain IS elements. The F plasmid, for example, typically has at least two different IS elements, IS*2* and IS*3*. When a particular IS element resides in both a plasmid and a chromosome, it creates the opportunity for homologous recombination between different DNA molecules. Such recombination appears to be responsible for the integration of the F plasmid into the *E. coli* chromosome (Chapter 8).

Both the *E. coli* chromosome and the F plasmid are circular DNA molecules. When the plasmid and the chromosome recombine in a region of homology—for example, in an IS element common to both of them—the smaller plasmid is integrated into the larger chromosome, creating a single circular molecule. Such integration events produce Hfr strains capable of transferring their chromosomes during conjugation. These strains vary in the integration site of the F plasmid because the IS elements that mediate recombination occupy different chromosomal positions in different *E. coli* strains—a result of their ability to transpose. Because of their role in forming Hfr strains, IS elements potentiate the exchange of chromosomal genes between different strains of bacteria. This exchange creates genetic variability, which, along with mutation, is the basis for evolutionary change in bacterial populations.

COMPOSITE TRANSPOSONS

Composite transposons, which are bacterial cut-and-paste transposons denoted by the symbol Tn, are created when two IS elements insert near each other. The region between them can then be transposed by the joint action of the flanking elements. In effect, two IS elements "capture" a DNA sequence that is otherwise immobile and endow it with the ability to move. **FIGURE 18.3** gives three examples. In Tn*9*, the flanking IS elements are in the same orientation with respect to each other, whereas in Tn*5* and Tn*10*, the orientation is inverted. The region between the IS elements in each of these transposons contains genes that have nothing to do with transposition. In fact, in all three transposons, the genes between the flanking IS elements confer resistance to antibiotics.

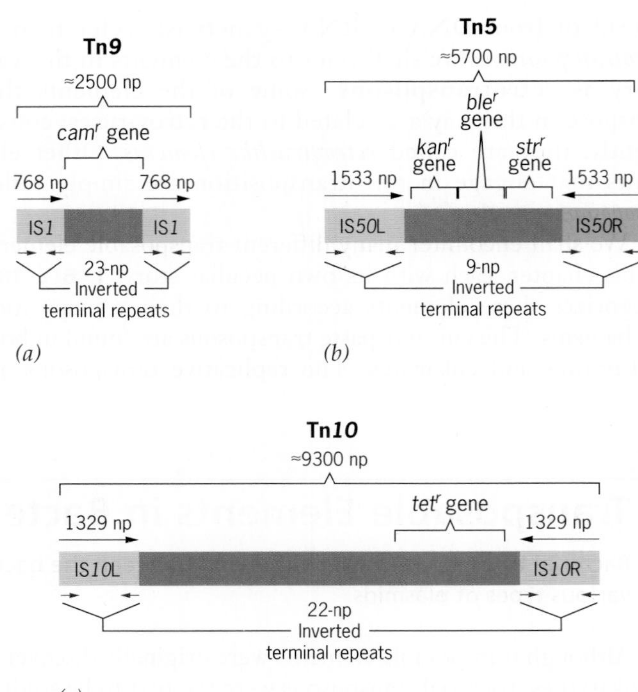

Figure 18.3 ▶ Genetic organization of composite transposons. The orientation and length (in nucleotide pairs, np) of the constituent sequences are indicated. (*a*) Tn*9* consists of two IS*1* elements flanking a gene for chloramphenicol resistance. (*b*) Tn*5* consists of two IS*50* elements flanking genes for kanamycin, bleomycin, and streptomycin resistance. (c) Tn*10* consists of two IS*10* elements flanking a gene for tetracycline resistance.

Sometimes the flanking IS elements in a composite transposon are not quite identical. For instance, in Tn*5*, the element on the right, called IS*50*R, is capable of producing a transposase to stimulate transposition, but the element on the left, called IS*50*L, is not. This difference is due to a change in a single nucleotide pair that prevents IS*50*L from specifying the active transposase.

Tn*5* also illustrates another feature of the composite transposons: their movement is regulated (**FIGURE 18.4**). When a bacterial cell is infected with a nonlytic bacteriophage that carries Tn*5* on its chromosome, the frequency of Tn*5* transposition is dramatically reduced if the infected cell already carries a copy of Tn*5*. This reduction implies that the resident transposon inhibits the transposition of an incoming transposon, possibly by synthesizing a repressor. Analyses by Michael Syvanen, William Reznikoff, and their colleagues have shown that this hypothesis is correct. The IS*50*R element of Tn*5*

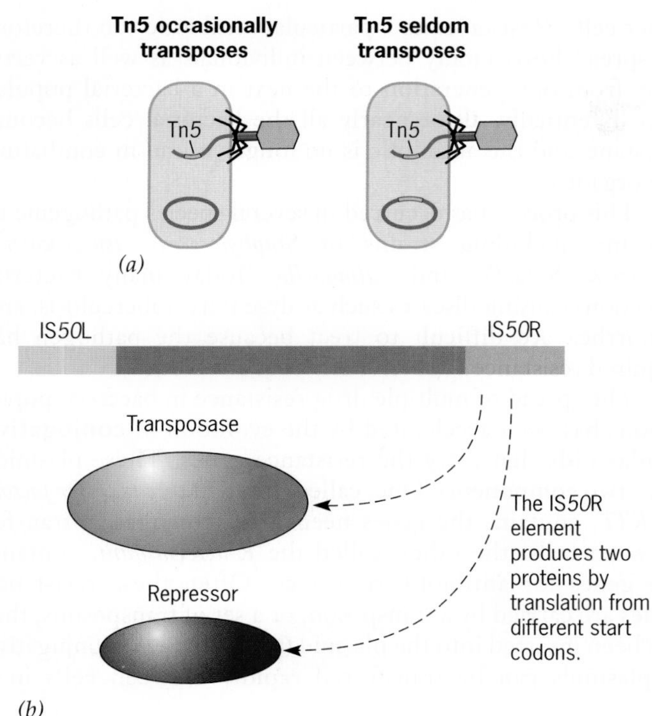

Figure 18.4 ▶ Regulation of Tn*5*. (*a*) Infection of *E. coli* cells with bacteriophage carrying Tn*5*. Cells that already possess a copy of Tn*5* repress transposition. (*b*) Genetic basis of Tn*5* regulation. One of the proteins produced by IS*50*R is a transposase that catalyzes transposition, but the other is a repressor that inhibits transposition. The effect of the repressor usually prevails.

actually produces two proteins. One, the transposase, catalyzes transposition, whereas the other, a truncated transposase created by translation from a start codon within the transposase gene, prevents transposition. Because the shorter protein, called the Tn*5* repressor, is the more abundant, Tn*5* transposition tends to be repressed.

Tn*3* ELEMENTS

The elements in this group of transposons are larger than the IS elements and usually contain genes that are not necessary for transposition—a feature that is also characteristic of the composite transposons. However, unlike the composite transposons, the Tn*3* elements do not have IS elements at each of their ends. Instead, the Tn*3* elements have simple inverted repeats 38 to 40 nucleotide pairs long at their termini. The Tn*3* elements also produce target site duplications when they insert into DNA.

The genetic organization of the element known specifically as Tn*3* is shown in **FIGURE 18.5**. There are three genes, *tnpA*, *tnpR*, and *bla*, encoding, respectively, a transposase, a resolvase/repressor, and an enzyme called beta lactamase. The beta lactamase confers resistance to the antibiotic ampicillin, and the other two proteins play important roles in transposition.

The transposition of Tn*3* occurs in two stages (**FIGURE 18.6**). First, the transposase mediates the fusion of two circular molecules, one carrying Tn*3* and the other not. The resulting

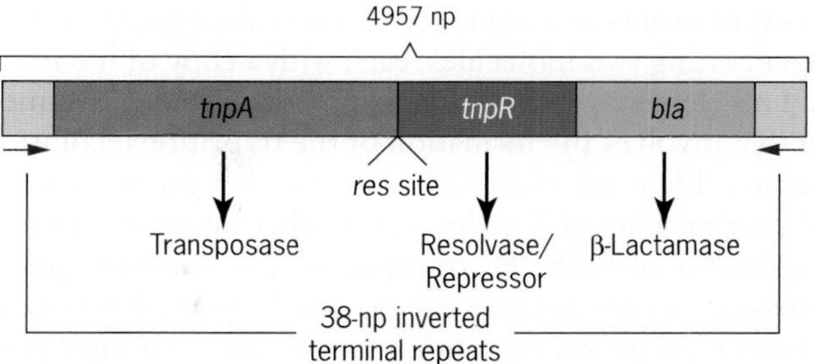

Figure 18.5 ▶ Genetic organization of Tn*3*. Lengths of DNA sequences are given in nucleotide pairs (np).

structure is called a **cointegrate.** During this process, the transposon is replicated, and one copy is inserted at each junction in the cointegrate; the two Tn*3* elements in the cointegrate are oriented in the same direction. In the second stage of transposition, the *tnpR*-encoded resolvase mediates a site-specific recombination event between the two Tn*3* elements.

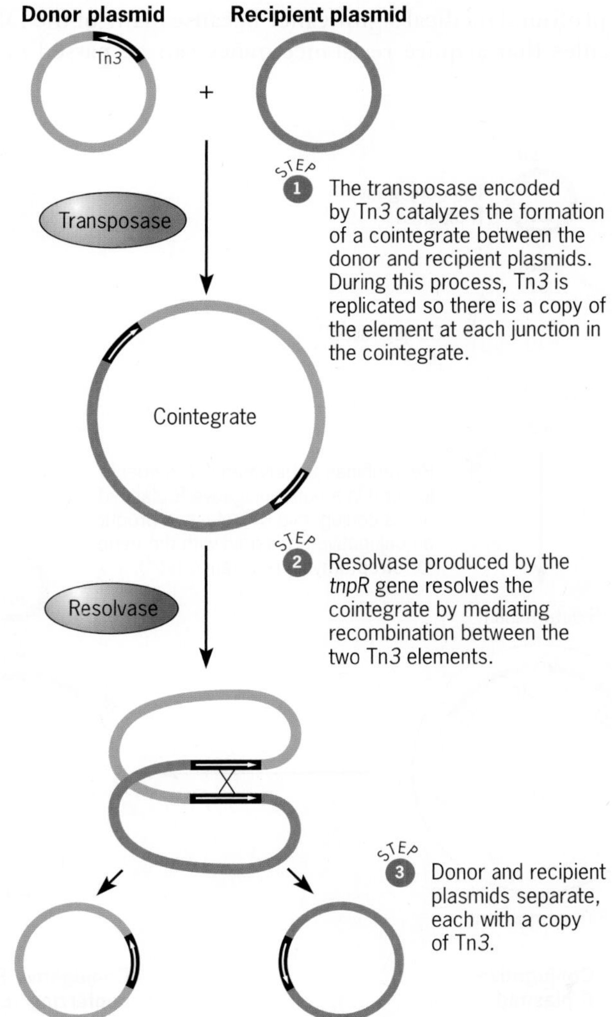

Figure 18.6 ▶ Transposition of Tn*3* via the formation of a cointegrate.

This event occurs at a sequence in Tn*3* called *res*, the *resolution site*, generating two molecules, each with a copy of the transposon. Tn*3* elements therefore transpose by a replicative mechanism that involves the formation of the transitional cointegrate structure. They are classified as replicative transposons. The *tnpR* gene product of Tn*3* has yet another function—to repress the synthesis of both the transposase and resolvase proteins. Repression occurs because the *res* site is located between the *tnpA* and *tnpR* genes. By binding to this site, the tnpR protein interferes with the transcription of both genes, leaving their products in chronic short supply. As a result, the Tn*3* element tends to remain immobile.

THE MEDICAL SIGNIFICANCE OF BACTERIAL TRANSPOSONS

Many bacterial transposons carry genes for antibiotic resistance. Consequently, it is a relatively simple matter for these genes to move from one DNA molecule to another—for instance, from a chromosome to a plasmid. This genetic flux has a profound medical significance because many of the DNA molecules that acquire resistance genes can be passed on to other cells. Resistance to a particular antibiotic can therefore be spread horizontally between individuals as well as vertically from one generation to the next in a bacterial population. Eventually, all or nearly all the bacterial cells become resistant, and the antibiotic is no longer useful in combating the organism.

This process has occurred in several species pathogenic to humans, including strains of *Staphylococcus*, *Enterococcus*, *Neisseria*, *Shigella*, and *Salmonella*. Today many bacterial infections causing diseases such as dysentery, tuberculosis, and gonorrhea are difficult to treat because the pathogen has acquired resistance to several different antibiotics.

The spread of multiple drug resistance in bacterial populations has been accelerated by the evolution of **conjugative R plasmids** that carry the resistance genes. These plasmids have two components. One, called the *resistance transfer factor*, or *RTF*, contains the genes needed for conjugative transfer between cells; the other, called the *R-determinant*, contains the genes for antibiotic resistance. Often these resistance genes are carried by a transposon, or a set of transposons, that has been inserted into the plasmid (**FIGURE 18.7**). Conjugative R plasmids can be transferred rapidly between cells in a

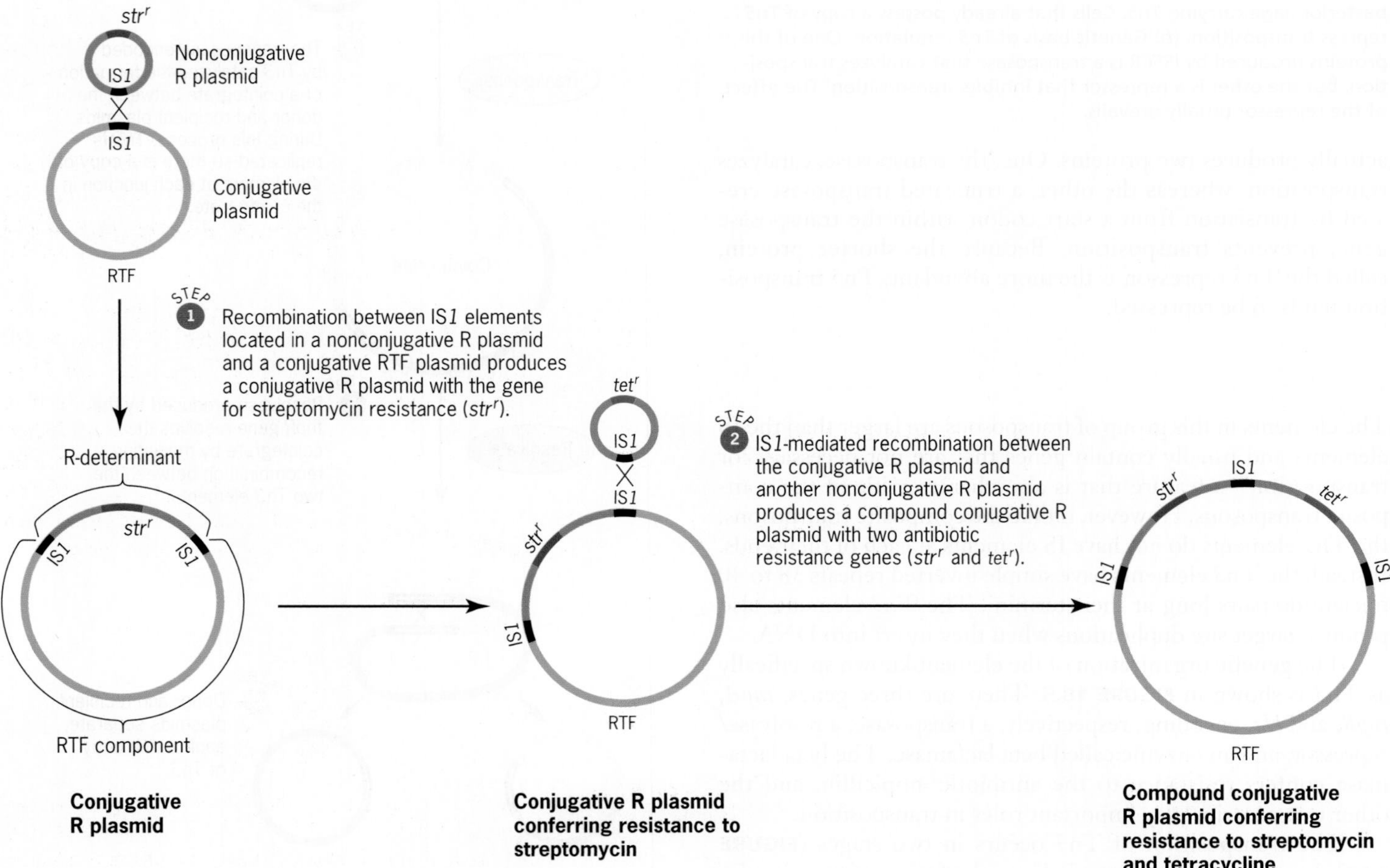

Figure 18.7 ▶ Evolution of conjugative plasmids carrying genes for antibiotic resistance.

bacterial population. Conjugative plasmids can, in fact, be passed from one species to another, even between quite dissimilar cell types—for example, between a coccus and a bacillus. Thus, once multiple drug resistance has evolved in a part of the microbial kingdom, it can spread to other parts with relative ease.

KEY POINTS

- Insertion sequences (IS elements) are cut-and-paste transposons that reside in bacterial chromosomes and plasmids.
- Composite transposons consist of two IS elements flanking a region that contains one or more genes for antibiotic resistance.
- Tn*3* is a replicative transposon that transposes by temporarily fusing DNA molecules into a cointegrate; when the cointegrate is resolved, each of the constituent DNA molecules emerges with a copy of Tn*3*.
- Bacterial transposons are demarcated by inverted terminal repeats; when they insert into a DNA molecule, they create a duplication at the insertion site.
- Conjugative plasmids can move transposons that contain genes for antibiotic resistance from one bacterial cell to another.

Cut-and-Paste Transposons in Eukaryotes

Transposable elements were discovered by analyzing genetic instabilities in maize; genetic analyses have also revealed transposable elements in *Drosophila*.

Geneticists have found many different types of transposons in eukaryotes. These elements vary in size, structure, and behavior. Some are abundant in the genome, others rare. In the following sections, we discuss a few of the eukaryotic transposons that move by a cut-and-paste mechanism. All these elements have inverted repeats at their termini and create target site duplications when they insert into DNA molecules. Some encode a transposase that catalyzes the movement of the element from one position to another.

Ac AND *Ds* ELEMENTS IN MAIZE

The *Ac* and *Ds* elements in maize were discovered through the pioneering work of Barbara McClintock (see the Focus on Barbara McClintock, the Discoverer of Transposable Elements). Through genetic analysis, McClintock showed that the activities of these elements are responsible for the striping and spotting of maize kernels. Many years later, Nina Federoff, Joachim Messing, Peter Starlinger, Heinz Saedler, Susan Wessler, and their colleagues isolated the elements and determined their molecular structure.

McClintock discovered the *Ac* and *Ds* elements by studying chromosome breakage. She used genetic markers that controlled the color of maize kernels to detect the breakage events. When a particular marker was lost, McClintock inferred that the chromosome segment on which it was located had also been lost, an indication that a breakage event had occurred. The loss of a marker was detected by a change in the color of the aleurone, the outermost layer of the triploid endosperm of maize kernels.

In one set of experiments, the genetic marker that McClintock followed was an allele of the *C* locus on the short arm of chromosome 9. Because this allele, C^I, is a dominant inhibitor of aleurone coloration, any kernel possessing it is colorless. McClintock fertilized *CC* ears with pollen from C^IC^I tassels, producing kernels in which the endosperm was C^ICC. (The triploid endosperm receives two alleles from the female parent and one from the male parent.) Although McClintock found that most of these kernels were colorless, as expected, some showed patches of brownish-purple pigment (**FIGURE 18.8**). McClintock guessed that in such mosaics, the inhibitory C^I allele had been lost sometime during endosperm development, leading to a clone of tissue that was able to make pigment. The genotype in such a clone would be *-CC*, where the dash indicates the missing C^I allele.

The mechanism that McClintock proposed to explain the loss of the C^I allele is diagrammed in **FIGURE 18.9**. A break at the site labeled by the arrow detaches a segment of the chromosome from its centromere, creating an acentric fragment. Such a fragment tends to be lost during cell division; thus, all the descendants of this cell will lack part of the paternally derived chromosome. Because the lost fragment carries the C^I allele, none of the cells in this clone is inhibited from forming pigment. If any of them produces a part of the aleurone, a patch of purple tissue will appear, creating a mosaic kernel similar to the one shown in **FIGURE 18.8**.

McClintock found that the breakage responsible for these mosaic kernels occurred at a particular site on chromosome 9. She named the factor that produced these breaks ***Ds***, for **Dissociation**. However, by itself, this factor was unable to induce chromosome breakage. In fact, McClintock found that *Ds* had to be stimulated by another factor, called ***Ac***, for **Activator**. The *Ac* factor was present in some maize stocks but absent in others. When different stocks were crossed, *Ac* could be combined with *Ds* to create the condition that led to chromosome breakage.

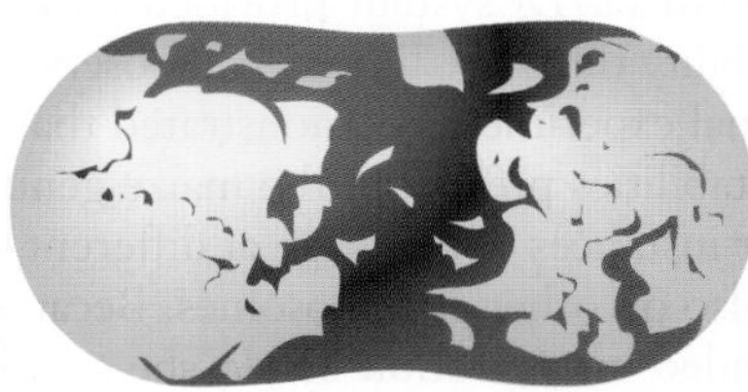

Figure 18.8 ► Maize kernel (top view) showing loss of the C^I allele for the inhibition of pigmentation in the aleurone. The brownish purple patches are *-CC*, whereas the yellow patches are C^ICC.

► FOCUS ON Barbara McClintock, the Discoverer of Transposable Elements

Scientific advances can come through the unflagging persistence of a single individual, someone who focuses on a question or problem and researches it for many years. Barbara McClintock was such an individual (**FIGURE 1**). Her long life was devoted to the study of maize genetics. Born in New England in 1902, she grew up in New York City and attended Cornell University, first as an undergraduate, then as a graduate student. After receiving her Ph.D., McClintock remained at Cornell for several years, collaborating with an illustrious group of maize geneticists, including George Beadle, R. A. Emerson, Charles Burnham, Marcus Rhoades, and Lowell Randolph. Together, these researchers developed the materials and methods of maize genetics into a rich intellectual discipline. McClintock played an important role in this work. Because of her skill in cytological analysis, she succeeded in identifying each of the 10 maize chromosomes and was able to connect them with linkage groups. She also did pioneering work on the mechanism of crossing over and on the origin of the nucleolus. However, McClintock's most notable achievement was to elucidate the genetic properties of transposable elements. Most of her research on this subject was done during the many years she spent in the Genetics Department of the Carnegie Institution at Cold Spring Harbor, New York. There she found the freedom to pursue her studies of transposon-mediated mutation and chromosome breakage.

Figure 1 ► Barbara McClintock.

McClintock published her first report about transposable elements in 1948. Several other reports followed, including a major paper in the 1951 *Cold Spring Harbor Symposium on Quantitative Biology*. For many reasons, the ideas that she espoused in these papers were not well received. The concept of transposition contradicted the established view that genes occupied fixed positions on chromosomes. McClintock's data were complex, and she had difficulty communicating them to her colleagues. In addition, transposition did not seem to be a general phenomenon. Although there was little doubt that it occurred in maize, no one had seen it in other organisms. This situation changed in the 1960s and 1970s, when transposition was discovered in bacteria and *Drosophila*. At last the scientific world awoke to the broad significance of McClintock's ideas.

McClintock was highly respected by her colleagues. In 1944, at the relatively young age of 42, she was admitted to the prestigious National Academy of Sciences of the United States. In 1945 she was elected president of the Genetics Society of America, and in 1970 she was awarded the National Medal of Science. Her Nobel Prize came in 1983, 35 years after her first publication on transposable elements. The day the prize was announced, she was out in the woods collecting mushrooms. During her life, McClintock was a model of "adamant individuality and self-containment."[1] She died in 1992, a few months after her ninetieth birthday.

[1]Federoff, N. V. 1994. Barbara McClintock (June 16, 1902–September 2, 1992). *Genetics* 136:1–10.

This two-factor *Ac*/*Ds* system provided an explanation for the genetic instability that McClintock had observed on chromosome 9. Additional experiments demonstrated that this was only one of many instabilities present in the maize genome. McClintock found other instances of breakage at different sites on chromosome 9 and also on other chromosomes. Because breakage at these sites depended on activation by *Ac*, she concluded that *Ds* factors were also involved. To explain all these observations, McClintock proposed that *Ds* could exist at many different sites in the genome and that it could move from one site to another.

This explanation has been borne out by subsequent analyses. The *Ac* and *Ds* elements belong to a family of transposons. These elements are structurally related to each other and can insert at many different sites on the chromosomes. Multiple copies of the *Ac* and *Ds* elements are often present in the maize genome. Through genetic analysis, McClintock demonstrated that both *Ac* and *Ds* can move. When one of these elements inserts in or near a gene, McClintock found that the gene's function is altered—sometimes completely abolished. Thus, *Ac* and *Ds* can induce mutations by inserting into genes. To empha-

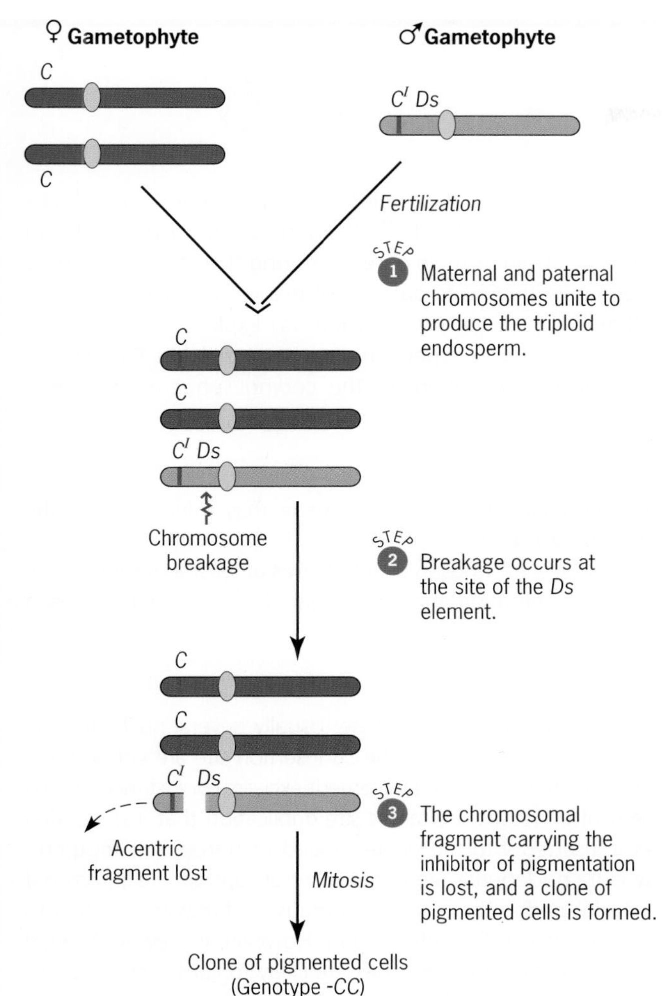

Figure 18.9 ► Chromosome breakage caused by the transposable element *Ds* in maize. The allele *C* on the short arm of chromosome 9 produces normal pigmentation in the aleurone; the allele C^I inhibits this pigmentation.

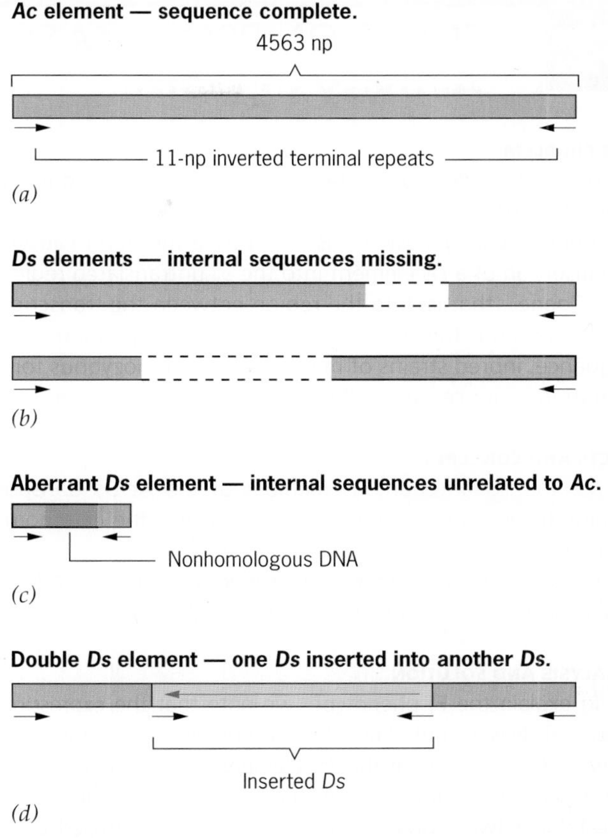

Figure 18.10 ► Structural organization of the members of the *Ac/Ds* family of transposable elements in maize. The terminal inverted repeats (short arrows underneath) and DNA sequence lengths (in nucleotide pairs, np) are indicated.

size this effect on gene expression, McClintock called the *Ac* and *Ds* transposons **controlling elements.**

DNA sequencing has shown that *Ac* elements consist of 4563 nucleotide pairs bounded by inverted repeats that are 11 nucleotide pairs long (**FIGURE 18.10*a***); these inverted terminal repeats are essential for transposition. Each *Ac* element is also flanked by direct repeats 8 nucleotide pairs long. Because the direct repeats are created at the time the element is inserted into the chromosome, they are target site duplications, not integral parts of the element.

Unlike *Ac*, *Ds* elements are structurally heterogeneous. They all possess the same inverted terminal repeats as *Ac* elements, demonstrating that they belong to the same transposon family, but their internal sequences vary. Some *Ds* elements appear to have been derived from *Ac* elements by the loss of internal sequences (**FIGURE 18.10*b***). The deletions in these elements may have been caused by incomplete DNA synthesis during replication or transposition. Other *Ds* elements contain non-*Ac* DNA between their inverted terminal repeats (**FIGURE 18.10*c***). These unusual members of the *Ac/Ds* family are called *aberrant Ds* elements. A third class of *Ds* elements is characterized by a peculiar piggybacking arrangement (**FIGURE 18.10*d***); one *Ds* element is inserted into another but in an inverted orientation. These so-called *double Ds* elements were apparently responsible for the chromosome breakage that McClintock observed in her experiments.

The activities of the *Ac/Ds* elements—excision and transposition, and all their genetic correlates, including mutation and chromosome breakage—are caused by a transposase encoded by the *Ac* elements. The *Ac* transposase apparently interacts with sequences at or near the ends of *Ac* and *Ds* elements, catalyzing their movement. Deletions or mutations in the gene that encodes the transposase abolish this catalytic function. Thus *Ds* elements, which have such lesions, cannot activate themselves. However, they can be activated if a transposase-producing *Ac* element is present somewhere in the genome. The transposase made by this element can diffuse through the nucleus, bind to a *Ds* element, and activate it. The *Ac* transposase is, therefore, a *trans*-acting protein.

Transposons related to the *Ac/Ds* elements have been found in other species, including animals. Perhaps the best-studied of

▸ FOCUS ON PROBLEM SOLVING Analyzing Transposon Activity in Maize

THE PROBLEM

In maize, the wild-type allele of the *C* gene is needed for dark coloration of the aleurone in kernels. Without this allele, the aluerone is pale yellow. c^{Ds} is a recessive mutation caused by the insertion of a *Ds* element into the 5′ untranslated region of the *C* gene—that is, into the region between the transcription start site and the first codon in the polypeptide coding sequence. Inbred strains of maize that are homozygous for this mutation produce pale yellow kernels, just like inbred strains that are homozygous for a deletion of the *C* gene ($c^{\triangle}$). A maize breeder crosses an inbred $c^{Ds}c^{Ds}$ strain as female parent to an inbred $c^{\triangle}c^{\triangle}$ strain as male parent. Among the kernels in the F_1, he sees many that have patches of brownish purple tissue on an otherwise pale yellow aleurone. (a) Explain the F_1 phenotype. (b) Would you expect this phenotype if the *Ds* element were inserted somewhere in the coding sequence of the *C* gene?

FACTS AND CONCEPTS

1. *Ds,* the nonautonomous member of the *Ac/Ds* transposon family, moves only in the presence of *Ac,* the autonomous member.

2. The 5′ untranslated region of a gene does not contain codons for amino acids in the polypeptide specified by the gene.

3. A transposon insertion into a gene may interfere with the gene's expression.

4. Excision of a transposon usually leaves at least a portion of the target site duplication that was created when the transposon inserted.

ANALYSIS AND SOLUTION

a. To explain the F_1 phenotype, we note that the expression of the c^{Ds} allele is disrupted by a *Ds* insertion into the 5′ untranslated region of the *C* gene. If this *Ds* element were to be excised, the gene's expression might be restored. When the maize breeder crossed the two inbred strains, he unwittingly crossed a strain with a *Ds* insertion in the *C* gene to a strain that carried a cryptic *Ac* element. The triploid aleurone in the F_1 kernels must have been $c^{Ds}c^{Ds}c^{\triangle}$ *(Ac).* The two copies of the c^{Ds} allele were derived from the female parent and the single copy of the $c^{\triangle}$ deletion allele and the single copy of *Ac* were derived from the male parent. In this hybrid genotype, *Ac* can activate the *Ds* element, causing it to excise from the *C* gene. Because the element was inserted into noncoding DNA, its excision is expected to restore *C* gene expression. Therefore, if cells in which such excisions occur give rise to aleurone tissue, that tissue will be brownish purple in an otherwise pale yellow kernel.

b. *Ds* excisions are seldom precise. Usually, several nucleotides in the gene's sequence around the *Ds* insertion site are either duplicated or deleted when the *Ds* element excises. For instance, the *Ds* element often leaves the target site duplication that it generated when it inserted into the gene—a kind of transposon footprint. These extra nucleotides are not likely to disrupt gene expression if they are located in the gene's 5′ untranslated region, which does not contain any coding information. However, if they are located in the gene's coding region, they are likely to cause serious problems. They could alter the length or composition of the polypeptide encoded by the gene. Thus, excising a *Ds* element from the coding sequence of the *C* gene is not likely to restore that gene's function. With such a *Ds* insertion, we would not expect to see patches of brownish purple tissue in the F_1 kernels.

For further discussion go to your *WileyPLUS* course.

these elements is one called *hobo*, whimsically named for its ability to transpose. The *hobo* element is found in some species of *Drosophila*. To explore other genetic effects of *Ac*/*Ds* elements, work through the Focus on Problem Solving: Analyzing Transposon Activity in Maize.

P ELEMENTS AND HYBRID DYSGENESIS IN *DROSOPHILA*

Some of the most extensive research on transposable elements has focused on the *P* elements of *Drosophila melanogaster*. These transposons were identified through the cooperation of geneticists working in several different laboratories. In 1977 Margaret and James Kidwell, working in Rhode Island, and John Sved, working in Australia, discovered that crosses between certain strains of *Drosophila* produce hybrids with an assortment of aberrant traits, including frequent mutation, chromosome breakage, and sterility. The term **hybrid dysgenesis,** derived from Greek roots meaning "a deterioration in quality," was used to denote this syndrome of abnormalities.

Kidwell and her colleagues found that they could classify *Drosophila* strains into two main types based on whether or not they produce dysgenic hybrids in testcrosses. The two types of strains are denoted M and P. Only crosses between M and P strains produce dysgenic hybrids, and they do so only if the male in the cross is from the P strain. Crosses between two different P strains, or between two different M strains, produce hybrids that are normal. We can summarize the phenotypes of the hybrid offspring from these different crosses in a simple table:

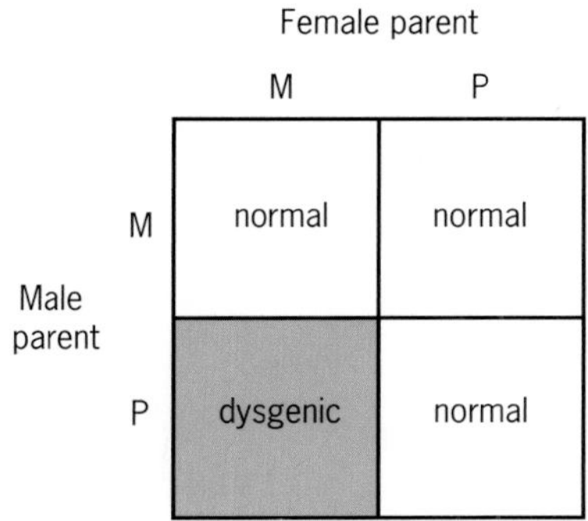

To Kidwell and her colleagues, these findings suggested that the chromosomes of P strains carry genetic factors that are activated when they enter eggs made by M females, and that once activated, these factors induce mutations and chromosome breakage. Inspired by this work, William Engels, a graduate student at the University of Wisconsin, began to study mutations induced in dysgenic hybrids. In 1979 Engels found a particular mutation that reverted to wild-type at a high rate. This instability, which is reminiscent of the behavior of IS-induced mutations in *E. coli*, strongly suggested that a transposable element was involved.

The discovery by Michael Simmons and Johng Lim of dysgenesis-induced mutations in the *white* gene allowed the transposon hypothesis to be tested. In 1980, Simmons and Lim, working in Minnesota and Wisconsin, respectively, sent the newly discovered *white* mutations to Paul Bingham, a geneticist in North Carolina. Bingham and his collaborator, Gerald Rubin, a geneticist in Maryland, had just finished isolating DNA from the *white* gene. Using this DNA as a probe, Bingham and Rubin were able to isolate DNA from the mutant *white* alleles and compare it to the wild-type *white* DNA. In each mutation, they found that a small element had been inserted into the coding region of the *white* gene. Additional experiments demonstrated that these elements are present in multiple copies and at different locations in the genomes of P strains; however, they are completely absent from the genomes of M strains. Geneticists therefore began calling these P strain–specific transposons ***P* elements.**

DNA sequence analysis has shown that *P* elements vary in size. The largest elements are 2907 nucleotide pairs long, including terminal inverted repeats of 31 nucleotide pairs. These *complete P* elements carry a gene that encodes a transposase. When the P transposase binds near the ends of a complete *P* element, it can move that element to a new location in the genome. *Incomplete P* elements (**FIGURE 18.11**) lack the ability to produce the transposase because some of their internal sequences are deleted; however, they do possess the terminal and subterminal sequences that bind the transposase. Consequently, these elements can be mobilized if a transposase-producing complete element is present somewhere in the genome.

Surveys of natural populations of *Drosophila* conducted by Dominique Anxolabéhère and his colleagues in France have demonstrated that there is considerable variation in the number

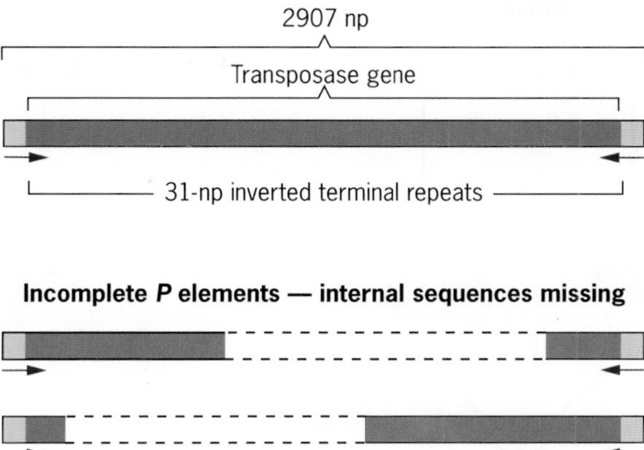

Figure 18.11 ► Structure of *P* elements in *Drosophila* showing orientations and lengths (in nucleotide pairs, np) of DNA sequences.

of *P* elements in the chromosomes. Some flies have as many as 50, whereas others have only a few. Perhaps the most surprising discovery is that flies derived from strains captured before 1950 have no *P* elements at all. Margaret Kidwell has suggested that these "empty" strains represent the primitive condition and that *P* elements have invaded natural populations of *Drosophila* during recent times. Curiously, the closest relatives of *D. melanogaster* have preserved the "empty" condition, but other, more distantly related species have acquired *P* elements. It is not possible to say how these species acquired their *P* elements, but one possibility is that the elements were carried into the genome by viruses that naturally infect *Drosophila.* Such a process would be analogous to the transduction of *E. coli* cells by a bacteriophage that carried an IS element.

Populations of *Drosophila* that possess *P* elements have evolved mechanisms to regulate their movement. In some strains, this regulation depends on **cytotype,** a cellular condition that is transmitted maternally through the egg cytoplasm. The P cytotype represses *P* element movement, and the M cytotype permits it. P cytotype is characteristic of strains that have *P* elements on their chromosomes, whereas M cytotype is characteristic of strains that lack *P* elements. When *P* elements are combined with the M cytotype by making appropriate crosses, they are induced to transpose.

The maternal transmission of cytotype can be seen in the offspring of reciprocal crosses between P and M cytotype strains (**FIGURE 18.12**). There are two crosses: (1) P cytotype female × M cytotype male, and (2) P cytotype male × M cytotype female. The offspring from both crosses inherit *P* elements on their chromosomes and are, in fact, genotypically identical; however, only those from the second cross allow *P* movement. This difference between the reciprocal crosses indicates that the condition that permits or represses *P* movement is transmitted only by the female parent, presumably in the egg cytoplasm. In Cross 1, the females transmit the P cytotype to their progeny, which then represses *P* movement, whereas in Cross 2, the females transmit the M cytotype, which allows *P* movement. Recent research

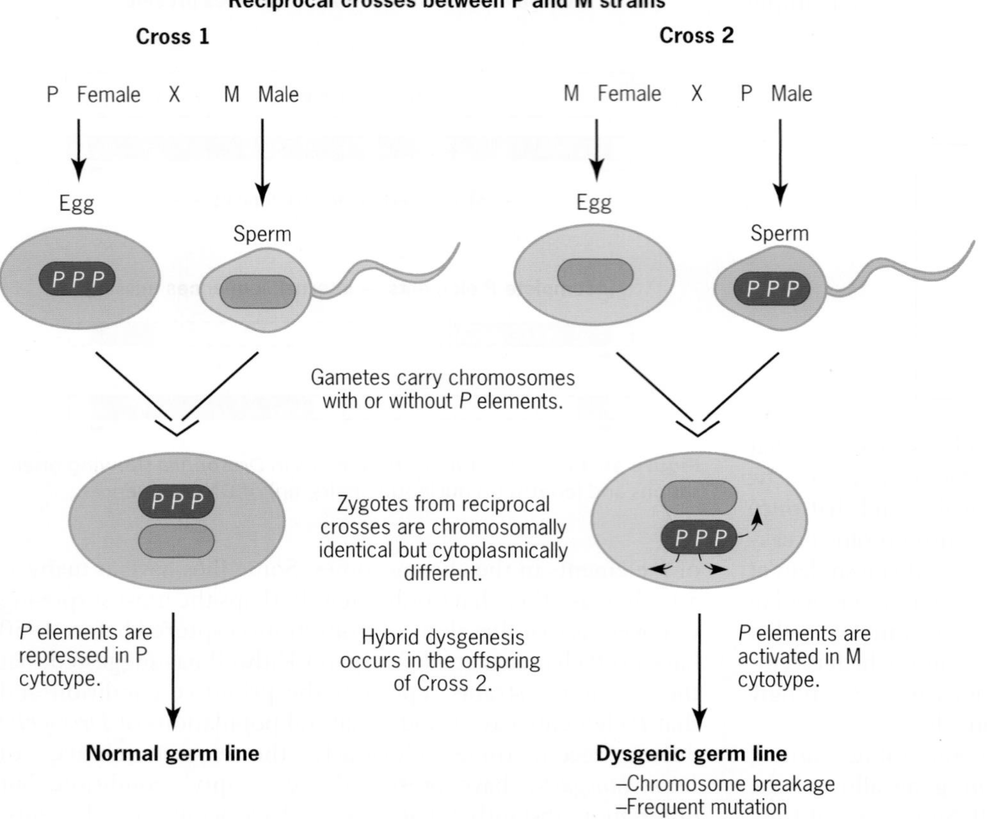

Figure 18.12 ▶ *P* element–mediated hybrid dysgenesis in *Drosophila.* Cytotype is inherited maternally. The P cytotype (dark beige) represses *P* element movement, whereas the M cytotype (light beige) permits it. Dysgenesis occurs only in the M cytotype hybrids of Cross 2.

suggests that repression by the P cytotype involves small RNA molecules derived from certain *P* elements. These RNA molecules appear to target *P* transposase mRNA for destruction, thereby preventing the transposase from being synthesized. We shall investigate the targeting of mRNAs by small RNAs—a phenomenon called RNA interference—in Chapter 20. It may be a general mechanism for the regulation of many types of transposable elements in a wide variety of organisms. Given the damage that can be caused by extensive *P* element movement, it may seem surprising that P male × M female crosses produce any viable progeny at all. Donald Rio, Frank Laski, and Gerald Rubin have shown that these progeny are healthy because *P* elements move only in the germ line. In the somatic tissues, where *P* element movement would cause serious problems, there is little, if any, transposition because the P transposase is not synthesized there. The metabolic block occurs at the level of RNA splicing; one of the introns remains in the transcript of the transposase gene, creating a stop codon that prematurely terminates translation. Thus, instead of making the transposase, somatic cells synthesize a shorter protein that does not have the transposase's catalytic function. Without this function, *P* elements are unable to move, and the fly is protected from massive damage in its somatic tissues.

Geneticists routinely use hybrid dysgenesis to obtain *P* insertion mutations in the laboratory. Dysgenic hybrids produced by crossing P males with M females are mated to recover mutations that have occurred in their germ lines. These mutations are detected by appropriate techniques, such as Muller's *ClB* crossing scheme to identify X-linked lethals (Chapter 13). Hybrid dysgenesis has an advantage over traditional methods of inducing mutations because a gene that has been mutated by the insertion of a transposable element is "tagged" with a known DNA sequence. The transposon tag may subsequently be used to isolate the gene from a large, heterogeneous mixture of DNA. Mutagenesis by **transposon tagging** is therefore a standard technique in molecular genetics. A Milestone in Genetics: Transformation of *Drosophila* with *P* Elements discusses another important use of *P* elements in genetics research.

MARINER, AN ANCIENT AND WIDESPREAD TRANSPOSON

Two of the closest relatives of *Drosophila melanogaster*, *D. simulans* and *D. mauritiana*, possess a small transposon called *mariner* (1286 nucleotide pairs long with 28-np inverted terminal repeats). Although *mariner* is not present in *D. melanogaster*, similar transposons are found in many other insects, in nematodes, fungi, and even humans. This widespread distribution suggests that *mariner* elements are ancient, dating from the earliest evolutionary times.

Sequence analysis of *mariners* from many different organisms has suggested that these elements are occasionally transferred horizontally between species. Hugh Robertson has shown that *mariner*-like elements are present in several different orders of insects. Curiously, the *mariners* in distantly related species are sometimes more similar to each other than the *mariners* in closely related species. For example, the *mariner* elements in the Mediterranean fruit fly are very similar to those in the honeybee; yet these two species are separated by hundreds of millions of years of evolutionary time. A plausible explanation is that the *mariners* in these two species are actually recent invaders from some outside source, perhaps another insect. It is not known how these elements could have been transferred between species, but one possibility is that they hitchhiked in the genome of a virus with a wide host range. During infection in one host species, the virus may have acquired the transposon; then during infection in another species, the transposon may have jumped from the virus into the host genome.

DNA sequencing has also identified another group of transposons related to the *mariner* elements. These are slightly larger, 1.6 to 1.7 kilobases long, with inverted repeats at their termini. Their repeat sequences typically range from 54 to 234 nucleotide pairs. The best-studied transposon in this group, called *Tc1*, is found in the nematode *Caenorhabditis elegans*. Other *Tc1*-like elements have been found in vertebrates. The *Tc1* and *mariner* elements therefore belong to a transposon superfamily of ancient origin.

KEY POINTS

- The maize transposable element *Ds*, discovered because of its ability to break chromosomes, is activated by another transposable element, *Ac*, which encodes a transposase.
- Transposable *P* elements are responsible for hybrid dysgenesis, a syndrome of germ-line abnormalities that occurs in the offspring of crosses between P and M strains of *Drosophila*.
- The *mariner* transposons, which were originally discovered in certain species of *Drosophila*, are widespread in the animal kingdom and appear to have spread horizontally among different species.

Retroviruses and Retrotransposons

Retroviruses and related transposable elements utilize the enzyme reverse transcriptase to copy RNA into DNA. The DNA copies of these entities are subsequently inserted at different positions in genomic DNA.

In addition to cut-and-paste transposons such as *Ac*, *P*, and *mariner*, eukaryotic genomes contain transposable elements whose movement depends on the reverse transcription of RNA into DNA. This reversal in the flow of genetic information has led geneticists to call these elements **retrotransposons,** from a Latin prefix meaning "backward." Reverse transcription also plays a crucial role in the life cycles of some viruses. The genomes of these viruses are composed of single-stranded RNA. When one of these viruses infects a cell, its RNA is copied into double-stranded DNA. Because the genetic information moves from RNA to DNA, these viruses are called **retroviruses.** We shall begin our investigation of retrotransposons with a discussion of the retroviruses. Later, we shall delve into the two main classes of retrotransposons. One class is composed of elements similar to the DNA forms of the retroviruses. The other class is composed of elements whose DNA has been copied from polyadenylated RNA.

RETROVIRUSES

The retroviruses were discovered by studying the causes of certain types of tumors in chickens, cats, and mice. In each case, an RNA virus was implicated in the production of the tumor. We shall explore how these viruses cause tumors in Chapter 22. Here, we focus on the key features of their life cycles. An important advance in understanding these life cycles came in 1970 when David Baltimore, Howard Temin, and Satoshi Mizutani discovered an RNA-dependent DNA polymerase—that is, a **reverse transcriptase,** which allows these viruses to copy RNA into DNA. This discovery initiated research on the process of reverse transcription, and provided a glimpse into what might be called the "retro-world"—that vast collection of DNA sequences derived from reverse transcription. We now know that reverse transcription is responsible for populating genomes with many kinds of DNA sequences, including, of course, the retroviruses. The discovery of reverse transcriptase therefore opened a view onto a component of genomes that had previously been unexplored.

Many different types of retroviruses have been isolated and identified. However, the epitome is the **human immunodeficiency virus, HIV,** which causes **acquired immune deficiency syndrome,** or **AIDS,** a disease that now affects tens of millions of people. AIDS was first detected in the last quarter of the twentieth century. It is a serious disease of the immune system. As it progresses, a person loses the ability to fight off infections by an assortment of pathogens, including organisms that are normally benign. Without treatment, infected individuals succumb to these infections, and eventually they die. AIDS is transmitted from one individual to another through bodily fluids such as blood or semen that have been contaminated with HIV. The initial symptoms of the disease are flu-like. Infected individuals experience aches, fever, and fatigue. After a few weeks, these symptoms abate and health is seemingly restored. This asymptomatic state may last several years. However, the virus continues to multiply and spread through the body, targeting specialized cells that play important roles in the immune system. Eventually, these cells are so depleted by the killing action of the virus that the immune system fails and opportunistic pathogens assert themselves. Many types of illnesses, such as pneumonia, may ensue. AIDS is a major cause of death among subpopulations in many countries—for example, among intravenous drug users and sex industry workers—and in sub-Saharan Africa, it is a major cause of death in the population at large.

Because of its lethality and pandemic status, HIV/AIDS has been the focus of an enormous amount of research. One outcome of this effort has been a detailed understanding of HIV's

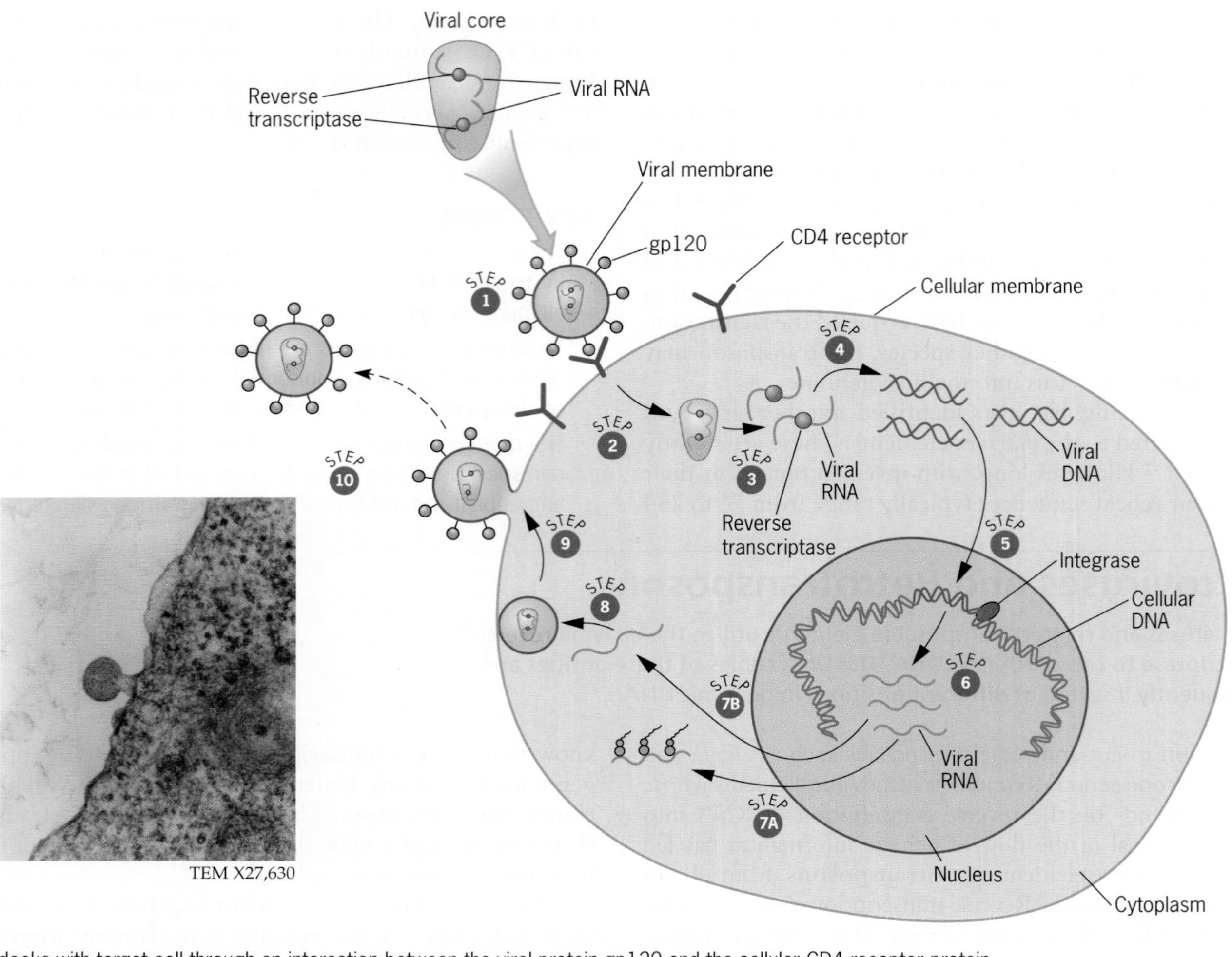

STEP 1 HIV docks with target cell through an interaction between the viral protein gp120 and the cellular CD4 receptor protein.

STEP 2 The viral and cellular membranes fuse, allowing the viral core to enter the cell.

STEP 3 RNA and associated proteins are released from the viral core.

STEP 4 Reverse transcriptase catalyzes the synthesis of double-stranded viral DNA from single-stranded viral RNA in the cytoplasm.

STEP 5 Integrase catalyzes the insertion of viral DNA into cellular DNA in the nucleus.

STEP 6 Cellular RNA polymerase transcribes viral DNA into viral RNA.

STEP 7a Some viral RNA serves as mRNA for the synthesis of viral proteins.

STEP 7b Some viral RNA forms the genomes of progeny viruses.

STEP 8 Progeny virus particles are assembled near the cellular membrane.

STEP 9 Progeny virus particles are extruded from the cell by budding.

STEP 10 Progeny virus particles are free to infect other cells.

Figure 18.13 ▶ The HIV life cycle. The inset shows a virus particle budding from a cell.

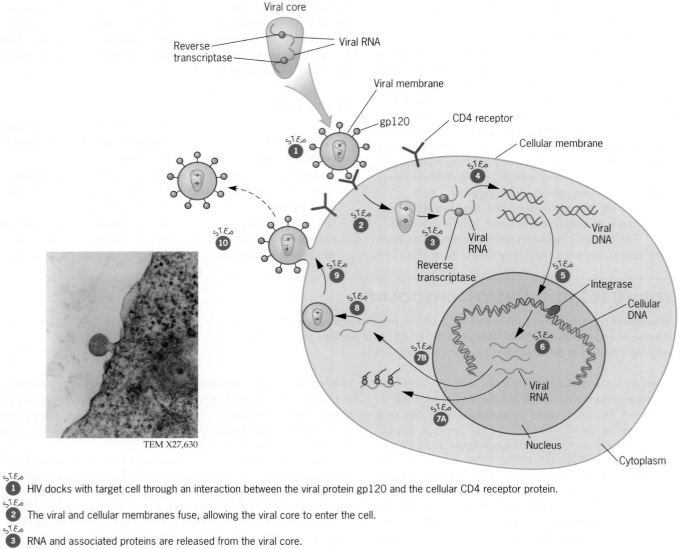

life cycle (**FIGURE 18.13**). The roughly spherical virus enters a host cell by interacting with specific receptor proteins, called CD4 receptors, which are located on the cell's surface. This interaction is mediated by a glycoprotein (a protein to which sugars have been attached) called gp120, which is embedded in the lipid membrane that surrounds the viral particle. Once gp120 has "docked" with the CD4 receptor, the viral and cellular membranes fuse and the viral particle is admitted to the cell. Inside the cell, the lipid membrane and the protein coat that surround the virus particle are removed, and materials within the virus's core are released into the cell's cytoplasm. This core contains two identical single-stranded RNA molecules—the virus's genome—and a small number of proteins that facilitate replication of the genome, including two molecules of the viral reverse transcriptase, one bound to each strand of viral RNA.

HIV's reverse transcriptase—and other reverse transcriptases as well—converts single-stranded RNA into double-stranded DNA. The resulting double-stranded DNA molecules are then inserted at random positions in the chromosomes of the infected cell, in effect populating that cell's genome with many copies of the viral genome. These copies can then be transcribed by the cell's ordinary RNA polymerases to produce a large amount of viral RNA, which serves to direct the synthesis of viral proteins and also provides genomic RNA for the assembly of new viral particles. These particles are extruded from the cell by a process of budding through the cell's membrane. The extruded particles may then infect other cells by interacting with the CD4 receptors on their surfaces. In this way, HIV's genetic material is replicated and disseminated through a population of susceptible immune cells.

The HIV genome, slightly more than 10 kilobases long, contains several genes. Three of these genes, denoted *gag*, *pol*, and *env*, are found in all other retroviruses. The *gag* gene encodes proteins of the viral particle; the *pol* gene encodes the reverse transcriptase and another enzyme called integrase, which catalyzes the insertion of the DNA form of the HIV genome into the chromosomes of a host cell; and the *env* gene encodes the glycoproteins that are embedded in the virus's lipid envelope.

Let's now take a closer look at replication of the HIV genome (**FIGURE 18.14**). This process, catalyzed by reverse transcriptase, begins with the synthesis of a single DNA strand complementary to the single-stranded RNA of the viral genome. It is primed by a tRNA that is complementary to a sequence called PBS (*p*rimer *b*inding *s*ite) situated to the left of center in the HIV RNA (step 1 in **FIGURE 18.14**). This tRNA is packaged prehybridized to the PBS in the HIV core. After reverse transcriptase catalyzes the synthesis of the 3′ end of the viral DNA, ribonuclease H (RNase H) degrades the genomic RNA in the RNA-DNA duplex (step 2). This degradation leaves the repeated (R) sequence of the nascent DNA strand free to hybridize with the R sequence at the 3′ end of the HIV RNA. The net result is that the R region of the nascent DNA strand "jumps" from the 5′ end of the HIV RNA to the 3′ end of the HIV RNA (step 3). Reverse transcriptase next extends the DNA copy by using the 5′ region of the HIV RNA as template (step 4).

In step 5, RNaseH degrades all the RNA in the RNA-DNA duplex except a small region, the polypurine tract, which is composed mostly of the purines adenine and guanine. This polypurine tract is used to prime second-strand DNA synthesis of the 3′ half of the HIV genome (step 6). After the tRNA and the genomic RNA present in the RNA-DNA duplexes are removed (step 7), a second DNA "jump" occurs during which the PBS at the 5′ end of the second DNA strand hybridizes with the complementary PBS at the 5′ end of the first DNA strand (step 8). The 3′-hydroxyl termini of the two DNA strands are then used to prime DNA synthesis to complete the synthesis of double-stranded HIV DNA (step 9). Note that the conversion of the viral RNA to viral DNA produces signature sequences at both ends of the DNA molecule. These sequences, called **long terminal repeats (LTRs),** are required for integration of the viral genome into the DNA of the host cell.

Integration (**FIGURE 18.15**) of the viral DNA is catalyzed by the enzyme integrase, which has endonuclease activity. Integrase first produces recessed 3′ ends in the HIV DNA by making single-stranded cuts near the ends of both LTRs (step 1). These recessed ends are next used for integrase-catalyzed attacks on phosphodiester bonds in a target sequence in the DNA of the host cell. This process results in the formation of new phosphodiester linkages between the 3′ ends of the HIV DNA and 5′ phosphates in the host DNA (step 2). In the final stage of integration, DNA repair enzymes of the host cell fill in the single-strand gaps to produce an HIV DNA genome covalently inserted into the chromosomal DNA of the host cell (step 3). Notice that the target sequence at the site of integration is duplicated in the process. The integrated HIV genome thereafter becomes a permanent part of the host cell genome, replicating just like any other segment of the host DNA.

Integrated retroviruses of many different types are present in vertebrate genomes, including our own. Because these retroviruses are replicated along with the rest of the DNA, they are transmitted to daughter cells during division, and if they are integrated in germ-line cells, they are also passed on to the next generation through the gametes. Geneticists call the heritable DNA sequences that are derived from the reverse transcription and integration of viral genomes *endogenous retroviruses*. For the most part, these sequences have lost their ability to produce infectious viral particles; they are, therefore, innocuous remnants of ancient viral infections. HIV is not an endogenous retrovirus, but if it should lose its lethal potential and be transmitted in integrated form through the germ line, it could become one.

We now turn our attention to two classes of retrotransposons: the retroviruslike elements, which resemble the integrated forms of retroviruses, and the retroposons, which are DNA copies of polyadenylated RNA.

RETROVIRUSLIKE ELEMENTS

Retroviruslike elements are found in many different eukaryotes, including yeast, plants, and animals. Despite differences in size and nucleotide sequence, they all have the same basic

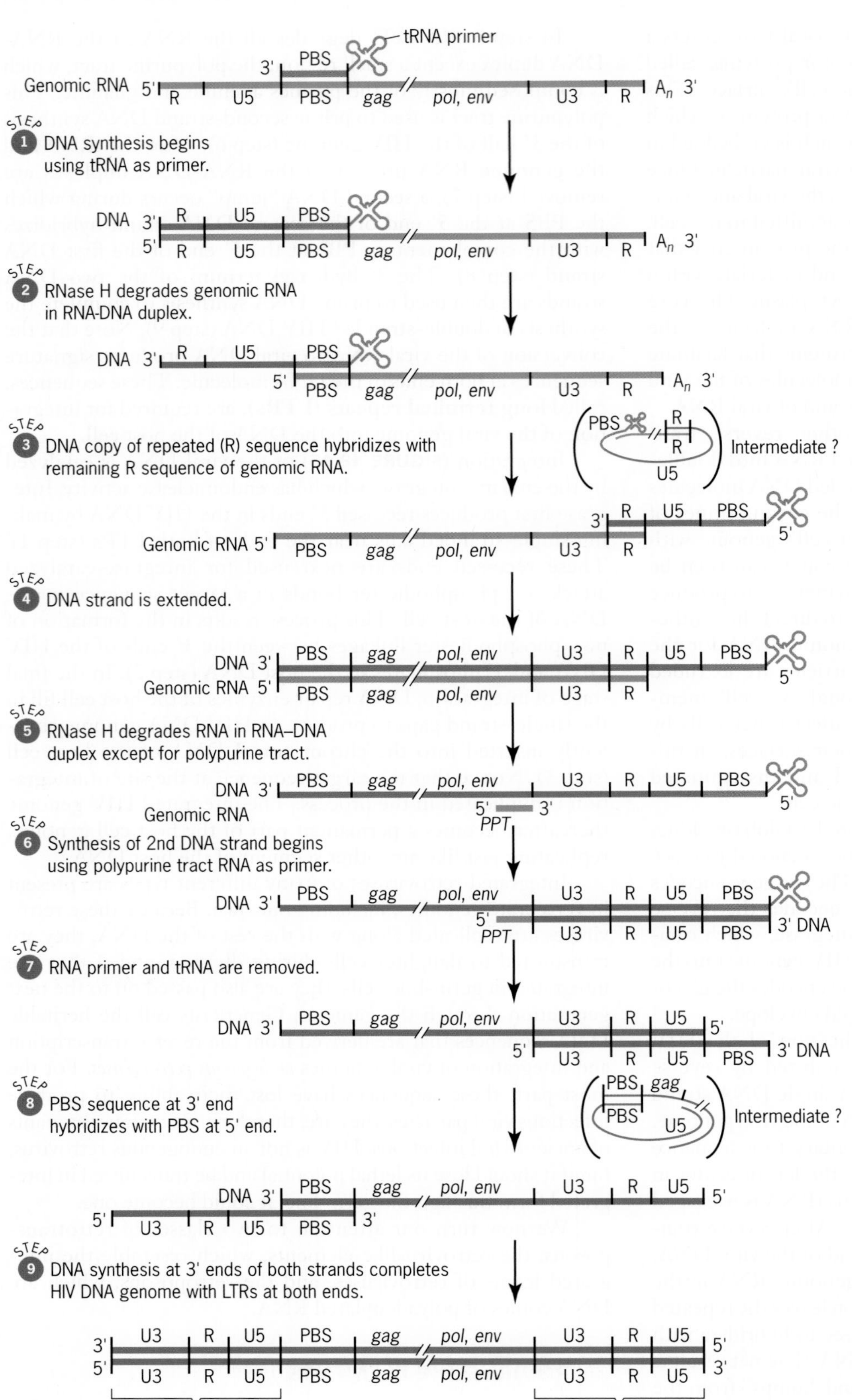

Figure 18.14 ▶ Conversion of HIV genomic RNA into double-stranded DNA. R, *r*epeated sequence; U5, *u*nique sequence near *5'* terminus; U3, *u*nique sequence near *3'* terminus; PBS, *p*rimer *b*inding *s*ite; A_n poly(*A*) tail; *gag*, *pol*, and *env*, sequences encoding HIV proteins; PPT, polypurine tract rich in adenine and guanine; LTR, *l*ong *t*erminal *r*epeat.

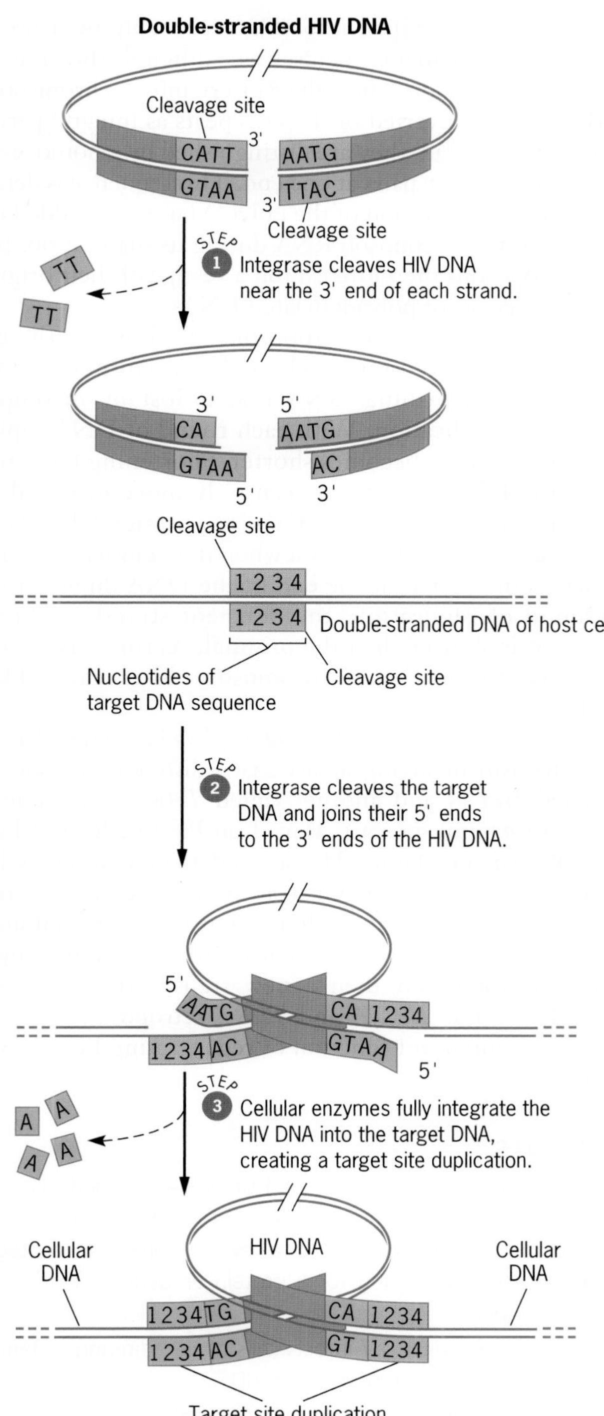

Figure 18.15 ▶ Integration of the HIV double-stranded DNA into the chromosomal DNA of the host cell.

structure: a central coding region flanked by long terminal repeats, or LTRs, which are oriented in the same direction. The repeated sequences are typically a few hundred nucleotide pairs long. Each LTR is, in turn, usually bounded by short, inverted repeats like those found in other types of transposons. Because of their characteristic LTRs, the retroviruslike elements are sometimes called *LTR retrotransposons.*

The coding region of a retroviruslike element contains a small number of genes, usually only two. These are homologous to the *gag* and *pol* genes found in retroviruses; *gag* encodes a structural protein of the virus capsule, and *pol* encodes a reverse transcriptase/integrase protein. The retroviruses have a third gene, *env*, which encodes a protein component of the virus envelope. In the retroviruslike elements, the gag and pol proteins play important roles in the transposition process.

One of the best-studied retroviruslike elements is the Ty*1* transposon from the yeast *Saccharomyces cerevisiae* (**FIGURE 18.16*a***). This element is about 5.9 kilobase pairs long; its LTRs are about 340 base pairs long, and it creates a 5-bp target site duplication upon insertion into a chromosome. Most yeast strains have about 35 copies of the Ty*1* element; sometimes they also contain LTRs that have been detached from Ty*1* elements. These solo LTRs, or delta sequences as they are sometimes called, are apparently formed by recombination between the LTRs of complete Ty*1* elements (**FIGURE 18.16*b***). The recombination event puts the central coding region and a portion of each LTR onto a circular molecule. When the circle leaves the chromosome, the remaining portions of the LTRs fuse, creating the solo delta sequence. The released circular molecule is then lost.

Ty*1* elements have only two genes, *TyA* and *TyB*, which are homologous to the *gag* and *pol* genes of the retroviruses. Biochemical studies have shown that the products of these two genes can form viruslike particles in the cytoplasm of yeast cells. The transposition of Ty*1* elements involves reverse

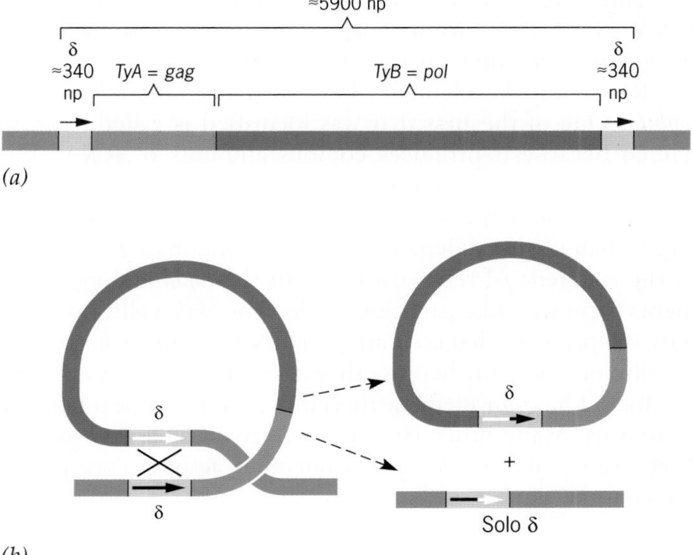

Figure 18.16 ▶ The retroviruslike element Ty*1* of yeast. (*a*) Genetic organization of the yeast Ty*1* element, showing the long terminal repeat sequences (LTRs, denoted by the Greek letter delta) and the two genes (*TyA* and *TyB*). Lengths of sequences are in nucleotide pairs (np). (*b*) Formation of a solo delta sequence by homologous recombination between the delta sequences at the ends of the element.

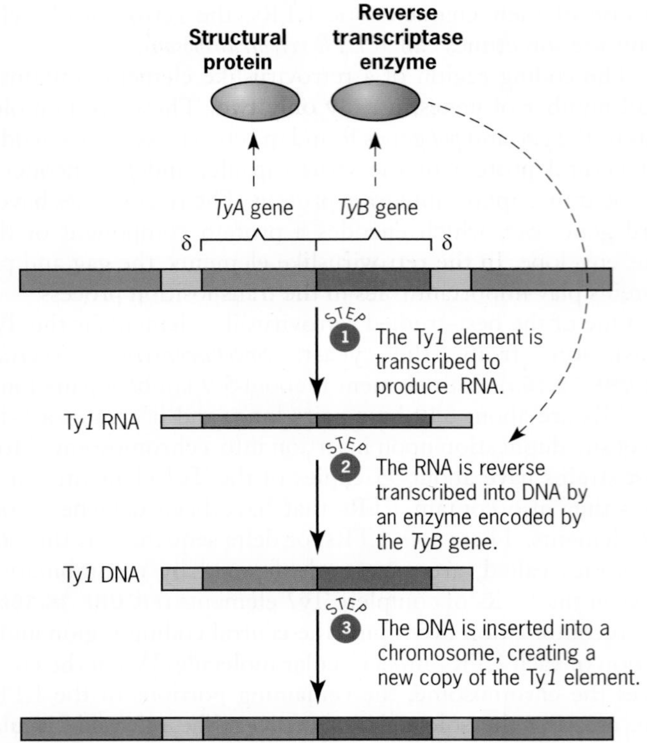

Figure 18.17 ▶ Transposition of the yeast Ty*1* element.

transcription of RNA (**FIGURE 18.17**). After the RNA is synthesized from Ty*1* DNA, a reverse transcriptase encoded by the *TyB* gene uses it as a template to make double-stranded DNA, probably in the viruslike particles. Then the newly synthesized DNA is transported to the nucleus and inserted somewhere in the genome, creating a new Ty*1* element.

Retroviruslike elements have also been found in *Drosophila*. One of the first that was identified is called *copia*, so named because it produces copious amounts of RNA. The *copia* element is structurally similar to the Ty*1* element of yeast. The *gypsy* element, another *Drosophila* retrotransposon, is larger than the *copia* element because it contains a gene similar to the *env* gene of retroviruses. Both the *copia* and *gypsy* elements form viruslike particles inside *Drosophila* cells; however, only the particles that contain *gypsy* RNA can move across cell membranes, possibly because they also contain *gypsy*'s *env* gene product. The *gypsy* element therefore appears to be a genuine retrovirus. Many other families of retroviruslike transposons have been found in *Drosophila*, but their activities are poorly understood.

RETROPOSONS

The **retroposons,** or non-LTR retrotransposons, are a large and widely distributed class of retrotransposons, including the *F*, *G*, and *I* elements of *Drosophila* and several types of elements in mammals. These elements move through an RNA molecule that is reverse transcribed into DNA, probably by a protein encoded by the elements themselves. Although they create a target site duplication when they insert into a chromosome, they do not have inverted or direct repeats as integral parts of their termini. Instead, they are distinguished by a homogenous sequence of A:T base pairs at one end. This sequence is derived from reverse transcription of the poly(A) tail that is added near the 3′ end of the retroposon RNA during its maturation. Integrated retroposons therefore exhibit a vestige of their origin as reverse transcripts of polyadenylated RNAs.

In *Drosophila*, special retroposons are found at the ends (telomeres) of chromosomes, where they perform the critical function of replenishing DNA that is lost by incomplete chromosome replication. With each round of DNA replication, a chromosome becomes shorter. Shortening takes place because the DNA polymerase can only move in one direction, adding nucleotides to the 3′ end of a primer (Chapter 10). Usually, the primer is RNA, and when it is removed, a single-stranded region is left at the end of the DNA duplex. In the next round of replication, the deficient strand produces a duplex that is shorter than the original. As this process continues, cycle after cycle, the chromosome loses material from its end.

To counterbalance this loss, *Drosophila* has evolved a curious mechanism involving at least two different retroposons, one called *HeT-A* and another called *TART* (for *t*elomere-*a*ssociated *r*etro*t*ransposon). Mary Lou Pardue, Robert Levis, Harald Biessmann, James Mason, and their colleagues have shown that these two elements transpose preferentially to the ends of chromosomes, extending them by several kilobases. Eventually, the transposed sequences are lost by incomplete DNA replication, but then a new transposition occurs to restore them. The *HeT-A* and *TART* retroposons therefore perform the important function of regenerating lost chromosome ends.

KEY POINTS

- Retrovirus genomes are composed of single-stranded RNA comprising at least three genes: *gag* (coding for structural proteins of the viral particle), *pol* (coding for a reverse transcriptase/integrase protein), and *env* (coding for a protein embedded in the virus's lipid envelope).
- The human retrovirus HIV infects cells of the immune system and causes the life-threatening disease AIDS.
- Retroviruslike elements possess genes homologous to *gag* and *pol*, but not to *env*.
- Retroviruslike elements and the DNA forms of retroviruses inserted in cellular chromosomes are demarcated by long terminal repeat (LTR) sequences.
- Retroposons lack LTRs; however, at one end they have a sequence of A:T base pairs derived from the reverse transcription of a poly(A) tail attached to the retroposon's RNA.
- The retroposons *HeT-A* and *TART* are components of the ends of *Drosophila* chromosomes.

Transposable Elements in Humans

The human genome is populated by a diverse array of transposable elements that collectively account for 44 percent of all human DNA.

With the sequencing of the human genome, it is now possible to assess the significance of transposable elements in our own species. At least 44 percent of human DNA is derived from transposable elements, including retroviruslike elements (8 percent of the sequenced genome), retroposons (33 percent), and several families of elements that transpose by a cut-and-paste mechanism (3 percent).

The principal transposable element is a retroposon called *L1*. This element belongs to a class of sequences known as the **long interspersed nuclear elements,** or **LINEs.** Complete *L1* elements are about 6 kb long, they have an internal promoter that is recognized by RNA polymerase II, and they have two open reading frames, ORF1, which encodes a nucleic acid-binding protein, and ORF2, which encodes a protein with endonuclease and reverse transcriptase activities. The human genome contains between 3000 and 5000 complete *L1* elements. In addition, it contains more than 500,000 *L1* elements that are truncated at their 5′ ends; these incomplete *L1* elements are transpositionally inactive. Each *L1* element in the genome, whether complete or incomplete, is usually flanked by a short target site duplication.

L1 elements are authentic retroposons. Their transposition involves the transcription of a complete *L1* element into RNA and the reverse transcription of this RNA into DNA (**FIGURE 18.18**). Both processes take place in the nucleus. However, before the *L1* RNA is reverse transcribed, it journeys to the cytoplasm where it is translated into polypeptides that apparently remain associated with it when it returns to the nucleus. The polypeptide encoded by ORF2 possesses an endonuclease function that catalyzes cleavage of one strand of the DNA duplex at a prospective insertion site in a chromosome. The exposed 3′ end of this cleaved DNA strand then serves as a primer for DNA synthesis using the *L1* RNA as a template and the reverse transcriptase activity provided by the ORF2 polypeptide. In this way, an *L1* DNA sequence is synthesized at the point in the chromosome where the ORF2 polypeptide has introduced a single-strand nick. The newly synthesized *L1* DNA is subsequently made double-stranded by further DNA synthesis, and the double-stranded product is then covalently integrated into the chromosome, thereby creating a new copy of the *L1* element in the genome. Sometimes the 5′ region of the *L1* RNA is not copied into DNA. When this happens, the resulting *L1* insertion will lack 5′ sequences—that is, it will be an incomplete *L1* element.

Only a small number of the complete *L1* elements in the human genome appear to be transpositionally active. Transposed copies of these elements have been discovered through the analysis of individuals with genetic diseases such as hemophilia and muscular dystrophy. Two other types of LINE sequences, *L2* (315,000 copies) and *L3* (37,000 copies), are found in the human genome; however, neither of these elements is transpositionally active.

The **short interspersed nuclear elements,** or **SINES,** are the second most abundant class of transposable elements in the human genome. These elements are typically less than 400 base pairs long and do not encode proteins. Like all retroposons, they have a sequence of A:T base pairs at one end. SINEs transpose through a process that involves reverse transcription of an RNA that has been transcribed from an internal promoter. Although the details of the transposition process are not well understood, it seems that the reverse transcriptase needed for the synthesis of DNA from the SINE RNA is furnished by a LINE-type element. Thus, the SINEs depend on the LINEs to multiply and insert within the genome. In this sense, they can be considered as retroposons that are parasites on the functionally autonomous and authentic retroposons. The human genome contains three families of SINEs, the *Alu*, *MIR*, and *Ther2/MIR3* elements. However, only the *Alu* elements—named for an enzyme that recognizes a specific nucleotide sequence within them—are transpositionally active.

The human genome possesses more than 400,000 sequences that are derived from retroviruslike elements. Most of these sequences are solo LTRs like the delta sequences found in the yeast genome. Although more than 100 different families of retroviruslike elements have been identified in human DNA, only a few appear to have been transpositionally active in recent evolutionary history. Like the inactive LINEs and SINEs, nearly all of the human retroviruslike sequences are genetic fossils left over from a time when they were actively transposing.

Cut-and-paste transposons are a small component of the human genome. Genome sequencing has identified two elements that are distantly related to the *Ac/Ds* elements of maize, as well as a few elements that are members of the *Tc1/mariner* superfamily. All the available evidence indicates that these types of transposons have been transpositionally inactive for many millions of years.

KEY POINTS

- The human genome contains four basic types of transposable elements: LINEs, SINEs, retroviruslike elements, and cut-and-paste transposons.
- The *L1* LINE and the *Alu* SINE are transpositionally active; other human transposons appear to be inactive.

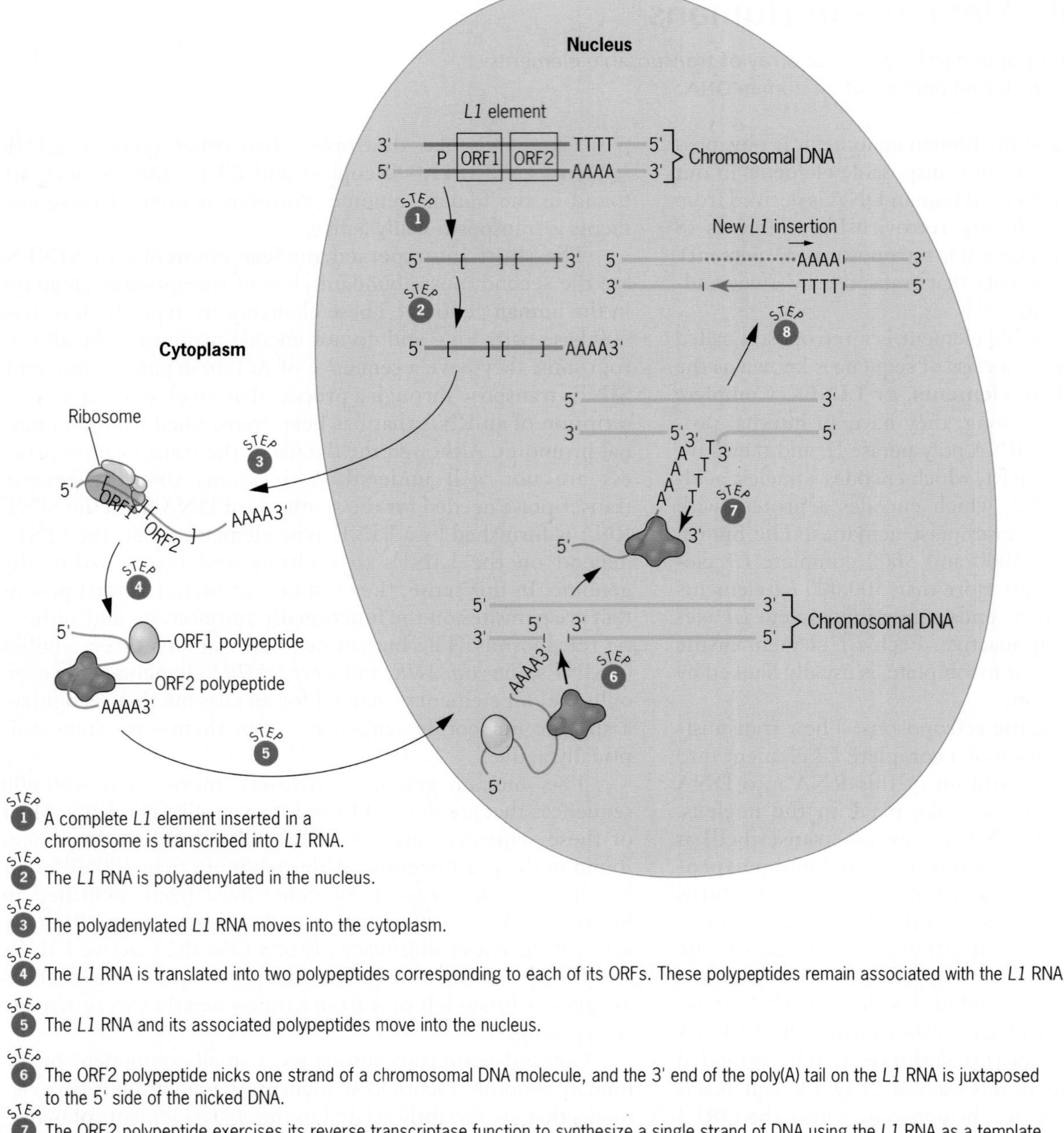

STEP 1 A complete *L1* element inserted in a chromosome is transcribed into *L1* RNA.

STEP 2 The *L1* RNA is polyadenylated in the nucleus.

STEP 3 The polyadenylated *L1* RNA moves into the cytoplasm.

STEP 4 The *L1* RNA is translated into two polypeptides corresponding to each of its ORFs. These polypeptides remain associated with the *L1* RNA.

STEP 5 The *L1* RNA and its associated polypeptides move into the nucleus.

STEP 6 The ORF2 polypeptide nicks one strand of a chromosomal DNA molecule, and the 3' end of the poly(A) tail on the *L1* RNA is juxtaposed to the 5' side of the nicked DNA.

STEP 7 The ORF2 polypeptide exercises its reverse transcriptase function to synthesize a single strand of DNA using the *L1* RNA as a template. The 3' end of the nicked chromosomal DNA serves as the primer for this DNA synthesis.

STEP 8 The newly synthesized single strand of DNA swings into place between the two sides of the nicked chromosomal DNA. Simultaneously, the *L1* RNA is eliminated, and the other strand of chromosomal DNA is nicked to allow for synthesis of a second strand of DNA (dotted line), complementary to the *L1* sequence, in the direction indicated by the thin arrow. All the nicks are repaired to link the newly inserted *L1* element to the chromosomal DNA.

Figure 18.18 ▶ Hypothesized mechanism for transposition of *L1* elements in the human genome. The approximately 6-kb *L1* element contains two open reading frames, ORF1 and ORF2, transcribed from a common promoter (P). The polypeptide encoded by ORF1 remains associated with the *L1* RNA and may be responsible for returning the RNA to the nucleus. The polypeptide encoded by ORF2 has at least two catalytic functions. First, it is capable of cleaving DNA strands; thus, it is an endonuclease. Second, it is capable of synthesizing DNA from an RNA template; thus, it is a reverse transcriptase. The size of the new *L1* insertion will depend on how far the reverse transcriptase travels along the *L1* RNA template. If it fails to reach the 5′ end, the insertion will be incomplete. Incomplete insertions usually have nonfunctional promoters and therefore cannot produce *L1* RNA for future transpositions.

The Genetic and Evolutionary Significance of Transposable Elements

Transposable elements have shaped genomes during the course of evolution.

Transposable elements are widespread and probably ancient components of the genome. In some species, they constitute an appreciable fraction of the total DNA. Mobile DNA is therefore an important component of the genome. It also contributes significantly to the total mutation rate. In *Drosophila*, for example, perhaps half the mutations that occur spontaneously are caused by transposable element insertions. This ability to inflict mutational damage raises questions about the evolutionary status of transposable elements. Do they perform any useful function, or are they merely genetic parasites, causing mutations as they wander about the genome? Where did transposable elements come from? What mechanisms have evolved to control and limit their movement?

TRANSPOSONS AND GENOME ORGANIZATION

Certain chromosome regions are especially rich in transposon sequences. In maize, transposable elements are concentrated in the DNA between genes; collectively, these elements account for more than half the DNA in the maize genome. In *Drosophila*, transposons are concentrated in the centric heterochromatin and in the heterochromatin abutting the euchromatin of each chromosome arm. However, many of these transposons have mutated to the point where they cannot be mobilized; genetically, they are the equivalent of "dead." Heterochromatin therefore seems to be a kind of graveyard filled with degenerate transposable elements.

Transposable elements are also found in the euchromatin, where they are dispersed at many different sites. In *Drosophila*, more than 90 distinct families of transposable elements have been identified. The number of members in each family ranges from a few to one or two hundred, but most families seem to have between 20 and 80 members.

There is some evidence, especially from cytological studies of *Drosophila* by Johng Lim, that transposable elements play a role in the evolution of chromosome structure. Several *Drosophila* transposons have been implicated in the formation of chromosome rearrangements, and a few seem to rearrange chromosomes at high frequencies. One possible mechanism is crossing over between homologous transposons located at different positions in a chromosome (**FIGURE 18.19**). If two transposons in opposite orientation pair and cross over, the segment between them will be inverted. If two transposons in the same orientation pair and cross over, the segment between them will be deleted. These events are examples of *ectopic intrachromosomal exchanges*—that is, exchanges between

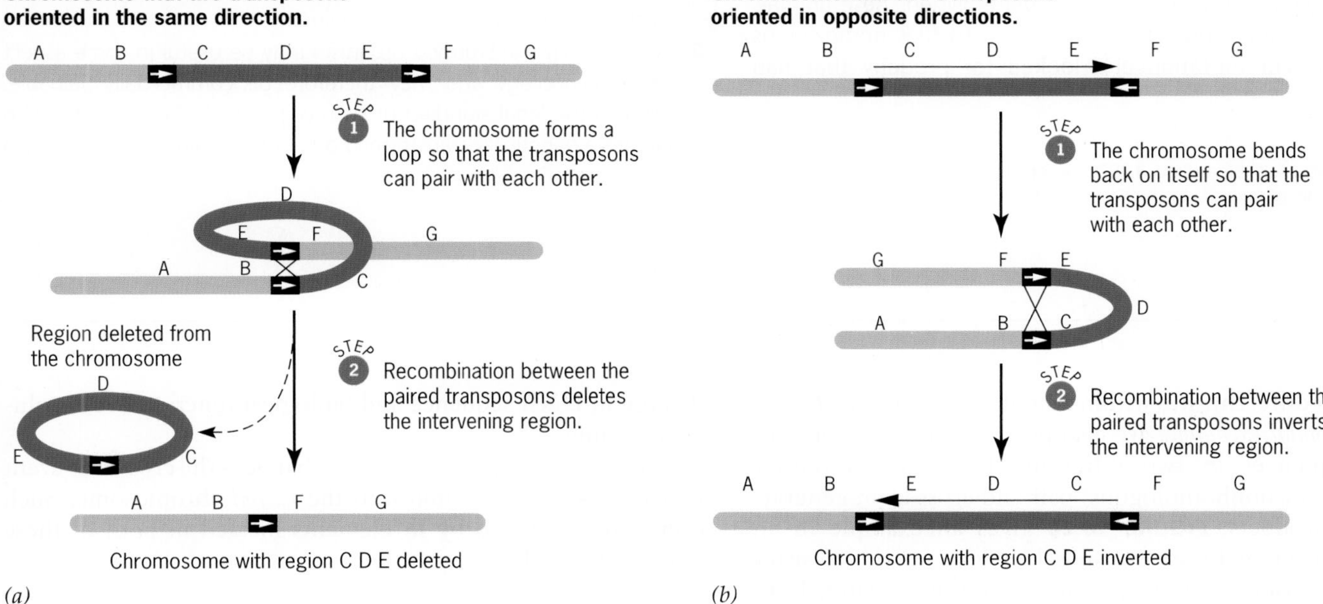

Chromosome with two transposons oriented in the same direction.
A B C D E F G
STEP 1 The chromosome forms a loop so that the transposons can pair with each other.
D
E F G
A B C
Region deleted from the chromosome
D
E C
STEP 2 Recombination between the paired transposons deletes the intervening region.
A B F G
Chromosome with region C D E deleted
(a)
Chromosome with two transposons oriented in opposite directions.
A B C D E F G
STEP 1 The chromosome bends back on itself so that the transposons can pair with each other.
G F E
D
A B C
STEP 2 Recombination between the paired transposons inverts the intervening region.
A B E D C F G
Chromosome with region C D E inverted
(b)

Figure 18.19 ▶ Transposon-mediated chromosome rearrangements. The transposons are shown as white arrows. (*a*) Formation of a deletion by intrachromosomal recombination between two transposons in the same orientation. (*b*) Formation of an inversion by intrachromosomal recombination between two transposons in opposite orientations.

▶ A MILESTONE IN GENETICS: **Transformation of *Drosophila* with *P* Elements**

In the 1940s, Oswald Avery and colleagues discovered that bacteria could be genetically altered by treating them with isolated DNA. In this process, a DNA fragment enters a cell and is physically recombined into the chromosome. The recombinant cell can then be cultured to produce a strain of genetically transformed organisms. Avery's discovery raised the hope that someday it would be possible to alter the genomes of eukaryotic organisms by inserting specific DNA fragments into them. When applied to human beings, such a procedure might provide a way of correcting genetic diseases.

For nearly 40 years, the experimental production of genetic transformants was limited to microorganisms. Many researchers attempted to transform higher eukaryotes, but none succeeded. This string of failures was broken in 1982, when Gerald Rubin and Allan Spradling produced the first genetically transformed *Drosophila*.[1]

Rubin and Spradling used transposable *P* elements to insert purified DNA into living *Drosophila* embryos. First, they constructed two bacterial plasmids that contained *P* elements. One plasmid contained a complete *P* element capable of producing the *P* transposase *in vivo*. The other contained an incomplete *P* element into which a gene for wild-type eye color had been inserted. Next, Rubin and Spradling injected a mixture of the two plasmids into *Drosophila* embryos that were homozygous for a recessive mutation of the eye color gene. They hoped that transposase produced by the complete *P* element would catalyze the incomplete *P* element to jump from its plasmid into the chromosomes of the injected animals, carrying with it the eye color gene. When these animals matured, Rubin and Spradling mated them to flies homozygous for the eye color mutation and looked for progeny that had wild-type eyes. They found many, indicating that the wild-type gene carried by the incomplete *P* element had been successfully incorporated into the genomes of some of the injected embryos. In effect, Rubin and Spradling had corrected the mutant eye color phenotype by inserting a copy of the wild-type gene into the fly genome.

The technique that Rubin and Spradling developed is now routinely used to transform *Drosophila* with isolated DNA (**FIGURE 1**). An incomplete *P* element serves as the *transformation vector,* and a complete *P* element serves as the source of the transposase that is needed to insert the vector into the chromosomes of an injected embryo. The term *vector* comes from the Latin word for "carrier"; it is used in this context because the incomplete *P* element *carries* a fragment of DNA into the genome. Practically any DNA sequence can be placed into the vector and ultimately inserted into the animal. Genes from organisms as diverse as bacteria and humans have successfully been incorporated into *Drosophila* chromosomes. Similar techniques are now being developed for the genetic transformation of other organisms, including our own species.

QUESTIONS FOR DISCUSSION

1. Genetic transformation with transposable element vectors has allowed scientists to create strains of *Drosophila* and other species that contain "foreign" genes—genes derived from entirely different organisms. What safety issues are associated with the creation of these genetically transformed strains?
2. A genetically transformed organism may be useful in some aspect of biotechnology and may therefore be commercially valuable. What is the legal standing of such organisms? Can a person who creates a genetically transformed organism obtain a patent on such an organism?

[1]Rubin, G. M. and A. C. Spardling, 1982. Genetic transformation of *Drosophila* with transposable element vectors, *Science* 218:348–353.

sequences at different sites within a single chromosome. *Ectopic interchromosomal exchanges* are also possible. In these types of events, sequences in two different chromosomes (either homologous or nonhomologous) pair and cross over, generating novel products. **FIGURE 18.20** gives an example of an ectopic exchange between transposable elements in two sister chromatids. Notice that one product from this exchange lacks the segment between the two transposons, whereas the other contains a duplication of this segment. If a gene is located within the segment flanked by the two transposons, this process can create a duplicate copy of the gene, which may then diverge in DNA sequence and biological function over evolutionary time.

Another example of ectopic exchange is the crossover event that inserts the F plasmid into the *E. coli* chromosome. Such events are mediated by IS elements located in both of these circular molecules.

TRANSPOSONS AND MUTATION

Transposable elements are responsible for mutations in a wide variety of organisms. **FIGURE 18.21** shows some of the transpo-

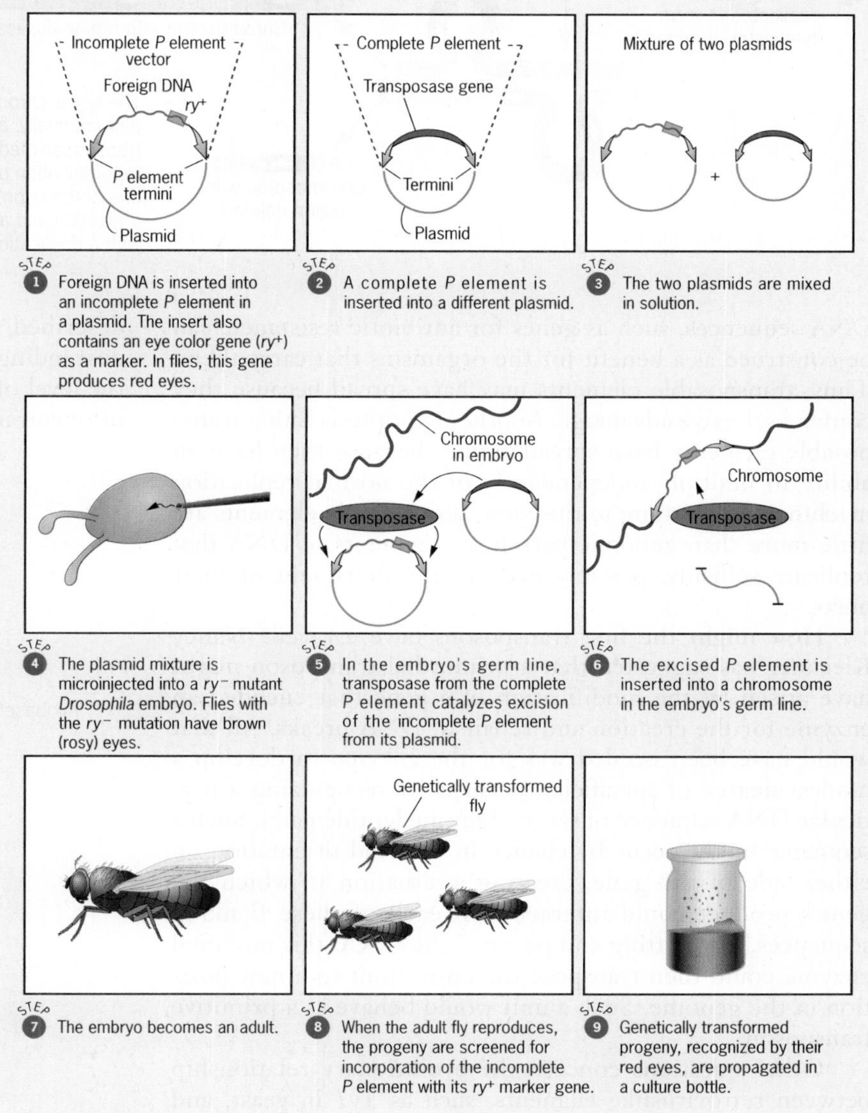

Incomplete P element vector
Foreign DNA
ry+
P element termini
Plasmid
Complete P element
Transposase gene
Termini
Plasmid
Mixture of two plasmids
+
STEP 1 Foreign DNA is inserted into an incomplete P element in a plasmid. The insert also contains an eye color gene (ry+) as a marker. In flies, this gene produces red eyes.
STEP 2 A complete P element is inserted into a different plasmid.
STEP 3 The two plasmids are mixed in solution.
Chromosome in embryo
Transposase
Chromosome
Transposase
STEP 4 The plasmid mixture is microinjected into a ry− mutant Drosophila embryo. Flies with the ry− mutation have brown (rosy) eyes.
STEP 5 In the embryo's germ line, transposase from the complete P element catalyzes excision of the incomplete P element from its plasmid.
STEP 6 The excised P element is inserted into a chromosome in the embryo's germ line.
Genetically transformed fly
STEP 7 The embryo becomes an adult.
STEP 8 When the adult fly reproduces, the progeny are screened for incorporation of the incomplete P element with its ry+ marker gene.
STEP 9 Genetically transformed progeny, recognized by their red eyes, are propagated in a culture bottle.

Figure 1 ▶ Genetic transformation of *Drosophila* using *P* element vectors. Foreign DNA inserted between *P* element termini is integrated into the genome through action of a transposase encoded by the complete *P* element. Flies with this DNA in their genomes can be propagated in laboratory cultures.

son insertions that have been found in different mutant alleles of the *Drosophila white* gene. These include several types of elements: *P*, retroviruslike elements, and retroposons. Some of these elements are inserted in exons, others in introns, and still others in regulatory DNA upstream of the actual gene. The very first mutant allele of *white*, w^1, discovered by T. H. Morgan, resulted from a transposon insertion.

Although transposon insertions are common in stocks of mutant organisms, the occurrence of new insertion mutations is a rare event. This suggests that the movement of many transposon families is regulated. When this regulation is upset, a burst of transposition may occur, causing many mutations simultaneously. This is apparently what happens when *P* elements are mobilized in dysgenic hybrids of *Drosophila.*

EVOLUTIONARY ISSUES CONCERNING TRANSPOSABLE ELEMENTS

The widespread distribution of transposable elements suggests that they have played a role in evolution. One hypothesis is that these elements are nature's tools for genetic engineering. Their ability to copy, transpose, and rearrange other

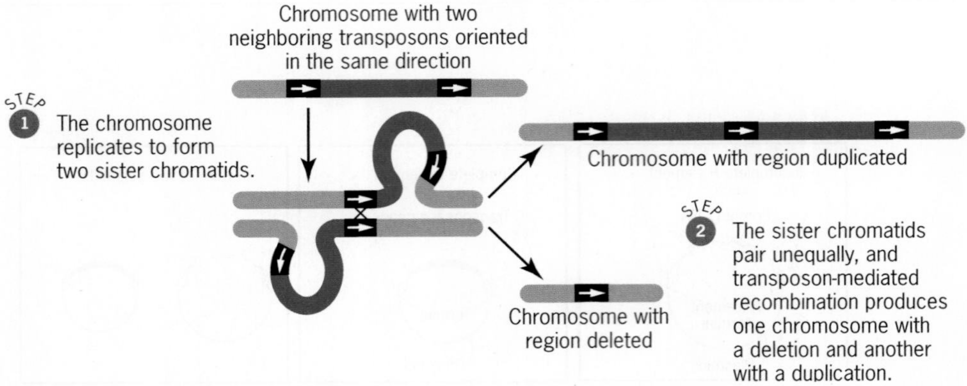

Figure 18.20 ▶ Origin of a gene duplication by transposon-mediated unequal crossing over between sister chromatids.

DNA sequences, such as genes for antibiotic resistance, can be construed as a benefit for the organisms that carry them. Thus, transposable elements may have spread because they confer a selective advantage. Another hypothesis is that transposable elements have spread simply because they have an ability to multiply independently of the normal replication machinery. According to this view, transposable elements are little more than genomic parasites—segments of DNA that replicate selfishly, possibly even to the detriment of their hosts.

How might the first transposons have evolved? Nancy Kleckner has suggested that a primordial transposon might have arisen by the modification of a gene that encoded an enzyme for the creation and repair of DNA breaks. All that would have been needed was for the enzyme to develop a modest degree of specificity, perhaps by recognizing a particular DNA sequence of six or eight nucleotide pairs. Such a sequence might occur by chance in inverted orientation on either side of the gene, creating a situation in which the gene's product could interact with each of these flanking sequences. By "cutting and pasting" the DNA, this modified enzyme could then transpose the entire unit to a new position in the genome. Such a unit would behave as a primitive transposon.

Other questions concern the evolutionary relationship between retroviruslike elements, such as Ty*1* in yeast, and full-fledged retroviruses. Collectively, these entities have been referred to as **retroelements.** Alan Kingsman and Susan Kingsman have proposed that retroviruses have developed from the simpler retrotransposons by the addition of a gene (called *env*) that synthesizes a membrane protein. With this addition, the retroelement could produce a particle capable of escaping from one cell and entering another one. Such a particle would be infectious and would therefore provide the retroelement with the opportunity to transpose between genomes as well as within them. Of course, the situation could be reversed—a retrovirus could lose its ability to escape from a cell and become trapped inside. Such a mutant virus would be reduced to the status of a retrotransposon, capable of moving within cells but not between them. Using a musical metaphor, the Nobel Prize–winning virologist Howard Temin once described these contrasting scenarios as ascending and descending evolutionary scales. Retrotransposons can rise to the level of retroviruses, and retroviruses can fall to the level of retrotransposons.

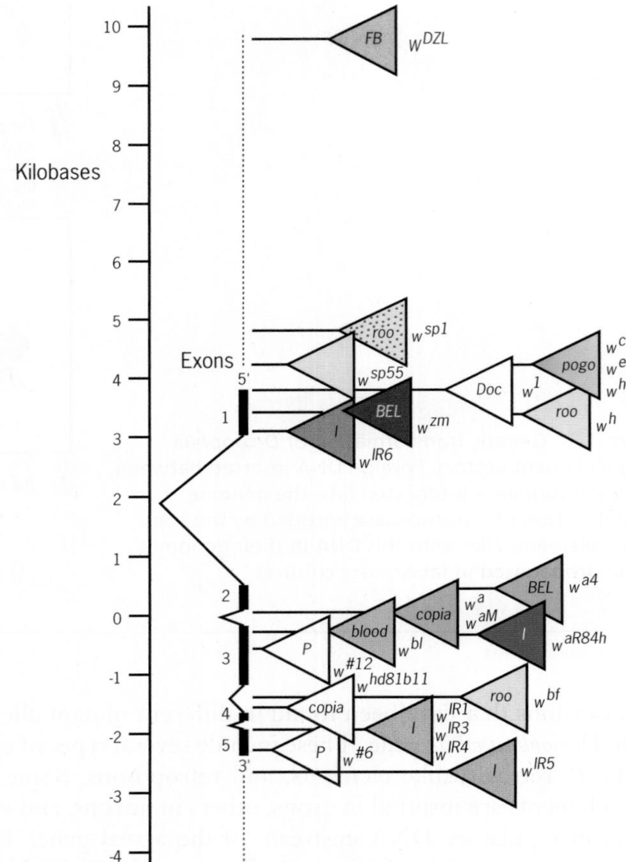

Figure 18.21 ▶ Transposon-insertion mutations in the *white* gene of *Drosophila.* Distance along the molecular map is given in kilobase pairs, with the zero coordinate arbitrarily positioned in the second intron of the gene at the site of the *copia* insertion in the w^a mutation. Each triangle represents a different transposon insertion in the *white* gene. Colors indicate the mutant eye colors.

KEY POINTS

- Transposons are found in the genomes of many types of organisms.
- Transposons are a major cause of mutations and chromosome rearrangements.
- Some transposons confer a selective advantage on the organisms that carry them; others are genetic parasites.

▶ Basic Exercises

ILLUSTRATE BASIC GENETIC ANALYSIS

1. Sketch a bacterial *IS* element inserted in a circular plasmid. Indicate the positions of (a) the transposase gene, (b) the inverted terminal repeats, and (c) the target site duplication.

Answer:

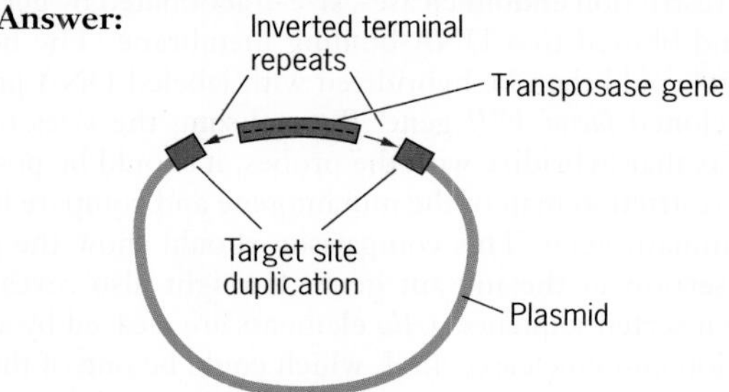

2. What factor must be present in maize to mobilize a *Ds* element inserted in a chromosome arm?

Answer: A *Ds* element is mobilized when the transposase encoded by an *Ac* element acts on it. An *Ac* element must therefore be present somewhere in the maize genome.

3. A geneticist has two strains of *Drosophila*. One, a long-standing laboratory stock with white eyes, is devoid of *P* elements; the other, recently derived from wild-type flies collected in a fruit market, has *P* elements in its genome. Which of the following crosses would be expected to produce dysgenic hybrid offspring: (a) white females × wild-type males, (b) white males × wild-type females, (c) white females × white males, (d) wild-type females × wild-type males?

Answer: (a) white females × wild-type males. The white females lack *P* elements in their genomes, and they have the M cytotype, which is the cellular condition that would permit *P* elements to become active if they were present. This condition is transmitted to the offspring through the egg. The wild-type males have *P* elements in their genomes, and they also have the P cytotype, which represses the activity of these elements. However, the P cytotype cannot be transmitted through the sperm. Thus, when the wild-type males are crossed to the white females, the offspring inherit *P* elements from their fathers and the M cytotype from their mothers. This combination of factors allows the paternally inherited *P* elements to become active in the germline tissues of the offspring, and hybrid dysgenesis ensues.

4. What are the similarities and differences among retroviruses, retroviruslike elements, and retroposons?

Answer: All three types of retroelements use reverse transcription to insert DNA copies of their RNA into new sites in the cell's genome. Furthermore, the enzyme (reverse transcriptase) that catalyzes reverse transcription is encoded by each type of element. For retroviruses and retroviruslike elements, reverse transcription of the RNA occurs in the cytoplasm, whereas for retroposons, it occurs in the nucleus. Retroviruses and retroviruslike elements encode another protein that functions in the assembly of virus or viruslike particles in the cytoplasm. Retroposons encode a different protein that appears to bind to the retroposon RNA and convey it into the nucleus. Retroviral RNA is packaged into viral particles, which can exit from the cell. This exit capability requires a protein encoded by the *env* gene in the viral genome. Because neither retroviruslike elements nor retroposons carry an *env* gene, their RNA cannot be packaged for exit from the cell. Retroviruses are infectious; retroviruslike elements and retroposons are not.

5. What transposable element is most abundant in the human genome?

Answer: The LINE known as *L1* is the most abundant human transposon. It accounts for about 17 percent of all human DNA.

6. How could two transposons in the same family cause deletion of DNA between them on a chromosome?

Answer: The two transposons would have to be in the same orientation. Ectopic pairing between the transposons followed by recombination would excise the chromosomal material between them. See Figure 18.19*a*.

▶ Testing Your Knowledge

INTEGRATE DIFFERENT CONCEPTS AND TECHNIQUES

1. A copy of the wild-type *white* gene (w^+) from *Drosophila* was inserted in the middle of an incomplete *P* element contained within a plasmid. The plasmid was mixed with another plasmid that contained a complete *P* element, and the mixture was carefully injected into *Drosophila* embryos homozygous for a null mutation (w^-) of the *white* gene. The adults that developed from these injected embryos all had white eyes, but when they were mated to uninjected white flies, some of their progeny had red eyes. Explain the origin of these red-eyed progeny.

Answer: The complete *P* element in one of the plasmids would produce the P transposase, the enzyme that catalyzes *P* element transposition, in the germ lines of the injected embryos. The incomplete *P* element in the other plasmid would be a target for this transposase. If this incomplete *P* element were mobilized by the transposase to jump from its plasmid into the chromosomes of the injected embryo, the fly that developed from this embryo would carry a copy of the wild-type *white* gene in its germ line. (*P* element movement is limited to the germ line; therefore, the incomplete *P* element would not jump into the chromosomes of the somatic cells, such as those that eventually form the eye.) Such a genetically transformed fly would, in effect, have the germ-line genotype w^-/w^-; $P(w^+)$ or w^-/Y; $P(w^+)$, where $P(w^+)$ denotes the incomplete *P* element that contains the w^+ gene. This element could be inserted on any of the chromosomes. If the transformed fly were mated to an uninjected white fly, some of its offspring would inherit the $P(w^+)$ insertion, which, because it carries a wild-type *white* gene, would cause red eyes to develop. The red-eyed progeny are therefore the result of genetic transformation of a mutant white fly by the w^+ gene within the incomplete *P* element.

2. The *Alu* element is one of the SINEs in the human genome. Each *Alu* retroposon is about 300 base pairs long—not long enough to encode a reverse transcriptase that could catalyze the conversion of *Alu* RNA into *Alu* DNA during the process of retrotransposition. In spite of this deficiency, the *Alu* elements have accumulated to such an extent that they constitute 11 percent of human DNA—over 1 million copies. How might this dramatic expansion of *Alu* elements have occurred during the evolutionary history of the human lineage without an *Alu*-encoded reverse transcriptase?

Answer: The *Alu* elements may have "borrowed" the services of a reverse transcriptase encoded by a different retroposon such as the *L1* element, which is large enough to encode a reverse transcriptase and at least one other polypeptide. If *L1*-encoded reverse transcriptase, or the reverse transcriptase encoded by some other retroelement—perhaps another LINE—can bind to *Alu* RNA, then it is conceivable that the reverse transcriptase could use the *Alu* RNA to synthesize *Alu* DNA, which could subsequently be integrated into chromosomal DNA. Repetition of this process over evolutionary time could explain the accumulation of so many copies of the *Alu* element in the human genome.

3. What techniques could be used to demonstrate that a mutation in a man with hemophilia is due to the insertion of an *Alu* element into the coding sequence of the X-linked gene for factor VIII, which is one of the proteins needed for efficient blood clotting in humans?

Answer: A molecular geneticist would have several ways of showing that the mutant gene for hemophilia is due to an *Alu* insertion in the gene's coding sequence. One technique is genomic Southern blotting. Genomic DNA from the hemophiliac could be digested with different restriction endonucleases, size-fractionated by gel electrophoresis, and blotted to a DNA-binding membrane. The bound DNA fragments could then be hybridized with labeled DNA probes made from a cloned *factor VIII* gene. By analyzing the sizes of the DNA fragments that hybridize with the probes, it should be possible to construct a restriction map of the mutant gene and compare it to a map of a nonmutant gene. This comparison should show the presence of an insertion in the mutant gene. It might also reveal the identity of the inserted sequence. (*Alu* elements are cleaved by a particular restriction endonuclease, *Alu* I, which could be one of the enzymes used in the analysis.) A simpler technique is to amplify portions of the coding sequence of the *factor VIII* gene by using the polymerase chain reaction (PCR). Pairs of primers positioned appropriately down the length of the coding sequence could be used in a series of amplification reactions, each of which would be seeded with template DNA from the hemophiliac. Each pair of primers would be expected to amplify a segment of the *factor VIII* gene. The sizes of the PCR products could then be determined by gel electrophoresis. An *Alu* insertion in a particular segment of the gene would increase the size of that segment by about 300 base pairs. The putative *Alu* insertion could be identified definitively by sequencing the DNA of the larger-than-normal PCR product.

Questions and Problems

ENHANCE UNDERSTANDING AND DEVELOP ANALYTICAL SKILLS

18.1 One strain of *E. coli* is resistant to the antibiotic streptomycin, and another strain is resistant to the antibiotic ampicillin. The two strains were cultured together and then plated on selective medium containing streptomycin and ampicillin. Several colonies appeared, indicating that cells had acquired resistance to both antibiotics. Suggest a mechanism to explain the acquisition of double resistance.

18.2 What distinguishes IS and Tn*3* elements in bacteria?

18.3 Which of the following pairs of DNA sequences could qualify as the terminal repeats of a bacterial IS element? Explain.
(a) 5′-GAATCCGCA-3′ and 5′-TGCGGATTC-3′
(b) 5′-GAATCCGCA-3′ and 5′-GAATCCGCA-3′
(c) 5′-GAATCCGCA-3′ and 5′-CTTAGGCGT-3′
(d) 5′-GAATCCGCA-3′ and 5′-ACGCCTAAG-3′

18.4 Which of the following pairs of DNA sequences could qualify as target site duplications at the point of an IS*50* insertion? Explain.
(a) 5′-AATTCGCGT-3′ and 5′-TGCGCTTAA-3′
(b) 5′-AATTCGCGT-3′ and 5′-ACGCGAATT-3′
(c) 5′-AATTCGCGT-3′ and 5′-TTAAGCGCA-3′
(d) 5′-AATTCGCGT-3′ and 5′-AATTCGCGT-3′

18.5 By chance, an IS*1* element has inserted near an IS*2* element in the *E. coli* chromosome. The gene between them, sug^+, confers the ability to metabolize certain sugars. Will the unit IS*1* sug^+ IS*2* behave as a composite transposon? Explain.

18.6 What enzymes are necessary for replicative transposition of Tn*3*? What are their respective functions?

18.7 The circular order of genes on the *E. coli* chromosome is **A B C D E F G H**, with the * indicating that the ends of the chromosome are attached to each other. Two copies of an IS element are located in this chromosome, one between genes *C* and *D*, and the other between genes *D* and *E*. A single copy of this element is also present in the F plasmid. Two Hfr strains were obtained by selecting for integration of the F plasmid into the chromosome. During conjugation, one strain transfers the chromosomal genes in the order *D E F G H A B C*, whereas the other transfers them in the order *D C B A H G F E*. Explain the origin of these two Hfr strains. Why do they transfer genes in different orders? Does the order of transfer reveal anything about the orientation of the IS elements in the *E. coli* chromosome?

18.8 The composite transposon Tn*5* consists of two IS*50* elements, one on either side of a group of three genes for antibiotic resistance. The entire unit IS*50*L kan^r ble^r str^r IS*50*R can transpose to a new location in the *E. coli* chromosome. However, of the two IS*50* elements in this transposon, only IS*50*R produces the catalytically active transposase. Would you expect IS*50*R to be able to be excised from the Tn*5* composite transposon and insert elsewhere in the chromosome? Would you expect IS*50*L to be able to do this?

18.9 What is the medical significance of bacterial transposons?

18.10 If DNA from a *P* element insertion mutation of the *Drosophila white* gene and DNA from a wild-type *white* gene were purified, denatured, mixed with each other, renatured, and then viewed with an electron microscope, what would the hybrid DNA molecules look like?

18.11 Would a Tn*3* element with a frameshift mutation early in the *tnpA* gene be able to form a cointegrate? Would a Tn*3* element with a frameshift mutation early in the *tnpR* gene be able to form a cointegrate?

18.12 The X-linked *singed* locus is one of several in *Drosophila* that controls the formation of bristles on the adult cuticle. Males that are hemizygous for a mutant *singed* allele have bent, twisted bristles that are often much reduced in size. Several *P* element insertion mutations of the *singed* locus have been characterized, and some have been shown to revert to the wild-type allele by excision of the inserted element. What conditions must be present to allow such reversions to occur?

18.13 In maize, the recessive allele *bz* (*bronze*) produces a lighter color in the aleurone than does the dominant allele, *Bz*. Ears on a homozygous *bz/bz* plant were fertilized by pollen from a homozygous *Bz/Bz* plant. The resulting cobs contained kernels that were uniformly dark except for a few on which light spots occurred. Suggest an explanation.

18.14 (a) What are retroviruslike elements? (b) Give examples of retroviruslike elements in yeast and *Drosophila*. (c) Describe how retroviruslike elements transpose. (d) After a retroviruslike element has been inserted into a chromosome, is it ever expected to be excised?

18.15 In homozygous condition, a deletion mutation of the *c* locus, c^n, produces colorless (white) kernels in maize; the dominant wild-type allele, *C*, causes the kernels to be purple. A newly identified recessive mutation of the *c* locus, c^m, has the same phenotype as the deletion mutation (white kernels), but when c^mc^m and c^nc^n plants are crossed, they produce white kernels with purple stripes. If it is known that the c^nc^n plants harbor *Ac* elements, what is the most likely explanation for the c^m mutation?

18.16 Would you ever expect the genes in a retrotransposon to possess introns? Explain.

18.17 When complete *P* elements are injected into embryos from an M strain, they transpose into the chromosomes of the germ line, and progeny reared from these embryos can be used to establish new P strains. However, when complete *P* elements are injected into embryos from insects that lack these elements, such as mosquitoes, they do not transpose into the chromosomes of the germ line. What does this failure to insert in the chromosomes of other insects indicate about the nature of *P* element transposition?

18.18 In maize, the *O2* gene, located on chromosome 7, controls the texture of the endosperm, and the *C* gene, located on chromosome 9, controls its color. The gene on chromosome 7 has two alleles, a recessive, *o2*, which causes the endosperm to be soft, and a dominant, *O2*, which causes it to be hard. The gene on chromosome 9 also has two alleles, a recessive, *c*, which allows the endosperm to be colored, and a dominant, C^I, which inhibits coloration. In one homozygous C^I strain, a *Ds* element is inserted on chromosome 9 between the *C* gene and the centromere. This element can be activated by introducing an *Ac* element by appropriate crosses. Activation of *Ds* causes the C^I allele to be lost by chromosome breakage. In $C^I/c/c$ kernels, such loss produces patches of colored tissue in an otherwise colorless background (see Figures 18.8 and 18.9). A geneticist crosses a strain with the genotype *o2/o2*; C^I *Ds*/C^I *Ds* to a strain with the genotype *O2/o2*; *c/c*. The latter strain also carries an *Ac* element somewhere in the genome. Among the offspring, only those with hard endosperm show patches of colored tissue. What does this tell you about the location of the *Ac* element in the *O2/o2*; *c/c* strain?

18.19 Sometimes solo copies of the LTR of a retrotransposon called *gypsy* are found in *Drosophila* chromosomes. How might these solo LTRs originate?

18.20 Approximately half of all spontaneous mutations in *Drosophila* are caused by transposable element insertions. In human beings, however, the accumulated evidence suggests that the vast majority of spontaneous mutations are *not* caused by transposon insertions. Propose a hypothesis to explain this difference.

18.21 Suggest a method to determine whether the *TART* retroposon is situated at the telomeres of each of the chromosomes in the *Drosophila* genome.

18.22 GO It has been proposed that the *hobo* transposable elements in *Drosophila* mediate intrachromosomal recombination—that is, two *hobo* elements on the same chromosome pair and recombine with each other. What would such a recombination event produce if the *hobo* elements were oriented in the same direction on the chromosome? What if they were oriented in opposite directions?

18.23 What evidence suggests that some transposable elements are not simply genetic parasites?

18.24 The human genome contains about 5000 "processed pseudogenes," which are derived from the insertion of DNA copies of mRNA molecules derived from many different genes. Predict the structure of these pseudogenes. Would each type of processed pseudogene be expected to found a new family of retrotransposons within the human genome? Would the copy number of each type of processed pseudogene be expected to increase significantly over evolutionary time, as the copy number of the *Alu* family has? Explain your answers.

18.25 Z. Ivics, Z. Izsvák, and P. B. Hackett have "resurrected" a nonmobile member of the *Tc1/mariner* family of transposable elements isolated from the DNA of salmon. These researchers altered 12 codons within the coding sequence of the transposase gene of the salmon element to restore the catalytic function of its transposase. The altered element, called *Sleeping Beauty*, is being tested as an agent for the genetic transformation of vertebrates such as mice and zebra fish (and possibly humans). Suppose that you have a bacterial plasmid containing the gene for green fluorescent protein (*gfp*) inserted between the ends of a *Sleeping Beauty* element. How would you go about obtaining mice or zebra fish that express the *gfp* gene?

▶ Genomics on the Web

at http://www.ncbi.nlm.nih.gov/

Here is a path to the sequence of a complete *P* element inserted in genomic DNA of *Drosophila melanogaster*:

Genomic biology → Insects → Drosophila melanogaster → Related Resources: Flybase → Files → Transposons (Dmel) → P-element → Sequence Accession → X06779.

1. Click on "repeat region (direct repeat)" to find the target site duplication created when this *P* element inserted into the genome. Note the length of the duplication and its sequence.

2. Click on "repeat region (P element)" to find the 2907 base-pair sequence of the *P* element itself; the first and last 31 base pairs are the terminal inverted repeats. Note the sequence.

3. Copy the first line of nucleotides in the entire sequence (genomic DNA plus inserted *P* element) and use the BLAST function under the Tools tab on the Flybase web site to locate this sequence in the *D. melanogaster* genome (be sure to delete the spaces between segments of 10 nucleotides when you carry out your search). What chromosome is the insertion on and what gene is it near? What phenotype is associated with mutations in this gene? Would the *P* element insertion be expected to cause a mutant phenotype?

Chapter 19
Regulation of Gene Expression in Prokaryotes and Their Viruses

CHAPTER OUTLINE

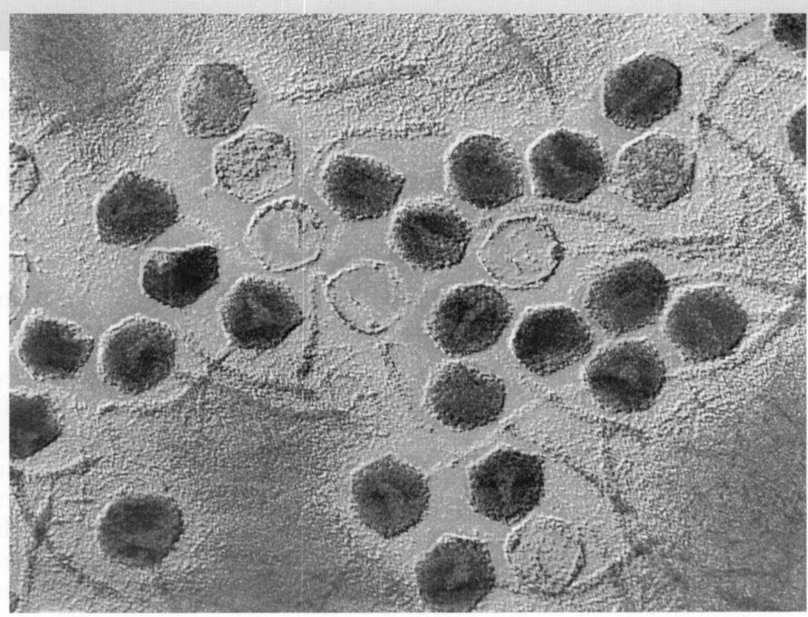

Scanning electron micrograph of bacteriophage *lambda.*

D'Hérelle's Dream of Treating Dysentery in Humans by Phage Therapy

In 1910, the French-Canadian microbiologist Felix d'Hérelle was in Mexico investigating a bacterial disease that was killing entire populations of locusts. The infected locusts developed severe diarrhea, excreting almost pure suspensions of bacilli prior to death. When he studied the bacteria in the feces of the locusts, d'Hérelle observed circular clear spots in the bacterial cultures grown on agar. However, when he examined the material in the clear spots microscopically, he could not see anything. In 1915, d'Hérelle returned to the Pasteur Institute in Paris, where he studied an epidemic of bacterial dysentery that was raging through army units stationed in France. He once again observed clear spots in lawns of bacteria. In addition, he demonstrated that whatever was killing the *Shigella*—a bacterium that causes dysentery in humans—could pass through a porcelain filter that retained all known bacteria. In 1917, d'Hérelle published his results and named the submicroscopic bacteriocidal agents bacteriophages (from the Greek for "bacteria-devouring"). About the same time, an English medical bacteriologist, Frederick W. Twort, discovered a similar submicroscopic agent that killed micrococci. Unfortunately, Twort's research was soon interrupted by a call to serve in the Royal Army Medical Corps in World War I.

Meanwhile, d'Hérelle continued to study the submicroscopic agents that killed *Shigella*. He provided the following account of one of his experiments: ". . . . in a flash I had understood: what caused my clear spots was in fact an invisible microbe, a filtrable virus, but a virus parasitic on bacteria. . . . 'If this is true, the same thing has probably occurred during the night in the sick man, who yesterday was in serious condition. In his intestine, as in my test tube, the dysentery bacilli will have dissolved away under the action of their parasite. He should now be cured.' I dashed to the hospital. In fact, during the night, his condition had greatly improved and convalescence was beginning" (d'Hérelle, F. 1949. The Bacteriophage. *Science News* 14:44–59).

Indeed, d'Hérelle became obsessed with his belief that human diseases caused by bacteria could be treated, perhaps even eradicated, by bacteriophage therapy. Unfortunately, it was soon demonstrated that this simple form of bacteriophage therapy is not effective in treating bacterial infections because, too frequently, the bacteria mutate to phage-resistant forms. Nevertheless, d'Hérelle's work set the stage for research that would eventually produce a whole new field—microbial genetics—and yield insights into the mechanisms by which gene expression is regulated. In this chapter, we examine some of these mechanisms.

Microorganisms exhibit remarkable capacities to adapt to diverse environmental conditions. This adaptability depends in part on their ability to turn on and turn off the expression of specific sets of genes in response to changes in the environment. The expression of particular genes is turned on when the products of these genes are needed for growth. Their expression is turned off when the gene products are no longer needed. The synthesis of gene transcripts and translation products requires the expenditure of considerable energy. By turning off the expression of genes when their products are not needed, an organism can save energy and can utilize the conserved energy to synthesize products that maximize growth rate. What, then, are the mechanisms by which microorganisms regulate gene expression in response to changes in the environment?

Gene expression in prokaryotes is regulated at several different levels: transcription, mRNA processing, mRNA turnover, translation, and post-translation (**FIGURE 19.1**). However, the regulatory mechanisms with the largest effects on phenotype act at the level of transcription.

Based on what is known about the regulation of transcription, the various regulatory mechanisms seem to fit into two general categories:

1. *Mechanisms that involve the rapid turn-on and turn-off of gene expression in response to environmental changes.* Regulatory mechanisms of this type are important in microorganisms because of the frequent exposure of these organisms to sudden changes in environment. They provide microorganisms with considerable "plasticity," an ability to adjust their metabolic processes rapidly in order to achieve maximal growth and reproduction under a wide range of environmental conditions.
2. *Mechanisms referred to as preprogrammed circuits or cascades of gene expression.* In these cases, some event triggers the expression of one set of genes. The product(s) of one or more of these genes functions by turning off the transcription of the first set of genes or turning on the transcription of a second set of genes. Then, one or more of the products of the second set acts by turning on a third set, and so on. In these cases, the sequential expression of genes is genetically preprogrammed, and the genes cannot usually be turned on out of sequence. Such preprogrammed sequences of gene expression are well documented in prokaryotes and the viruses that attack them.

Constitutive, Inducible, and Repressible Gene Expression

Genes that specify cellular components that perform housekeeping functions—for example, the ribosomal RNAs and proteins involved in protein synthesis—are expressed constitutively. Other genes often are expressed only when their products are required for growth.

Certain gene products—such as tRNA molecules, rRNA molecules, ribosomal proteins, RNA polymerase subunits, and enzymes catalyzing metabolic processes that are frequently referred to as cellular "housekeeping" functions—are essential components of almost all living cells. Genes that specify products of this type are continually being expressed in most cells. Such genes are said to be expressed constitutively and are referred to as **constitutive genes.**

Other gene products are needed for cell growth only under certain environmental conditions. Constitutive synthesis of such gene products would be wasteful, using energy that could otherwise be utilized for more rapid growth. The evolution of regulatory mechanisms that provide for the synthesis of such gene products only when and where they are needed would clearly endow the organisms that possess these regulatory mechanisms with a selective advantage over organisms that lack

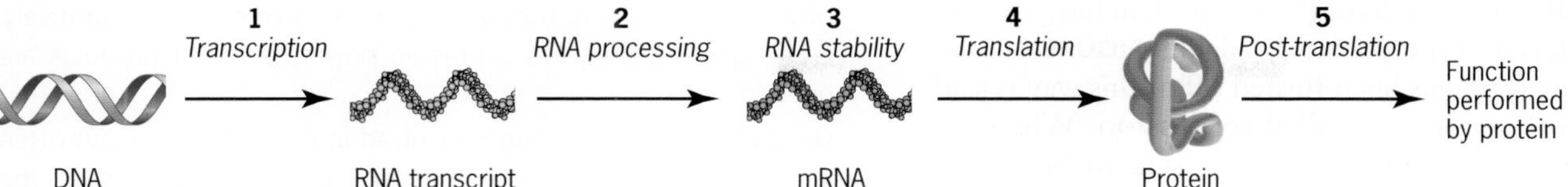

Figure 19.1 ▶ An abbreviated pathway of gene expression, showing five important levels of regulation in prokaryotes.

them. This undoubtedly explains why currently existing organisms, including bacteria and viruses, exhibit highly efficient mechanisms for the control of gene expression.

Escherichia coli and most other bacteria are capable of growth using any one of several carbohydrates—for example, glucose, sucrose, galactose, arabinose, and lactose—as an energy source. If glucose is present in the environment, it will be preferentially metabolized by *E. coli* cells. However, in the absence of glucose, *E. coli* cells can grow very well on other carbohydrates. Cells growing in medium containing the sugar lactose, for example, as the sole carbon source synthesize two enzymes, β-galactosidase and β-galactoside permease, which are uniquely required for the catabolism of lactose. β-Galactoside permease pumps lactose into the cell, where β-galactosidase cleaves it into glucose and galactose. Neither of these enzymes is of any use to *E. coli* cells if no lactose is available to them. The synthesis of these two enzymes requires considerable energy (in the form of ATP and GTP; see Chapters 11 and 12). Thus, *E. coli* cells have evolved a regulatory mechanism by which the synthesis of these lactose-catabolizing enzymes is turned on in the presence of lactose and turned off in its absence.

In natural environments (intestinal tracts and sewers), *E. coli* cells probably encounter an absence of glucose and the presence of lactose relatively infrequently. Therefore, the *E. coli* genes encoding the enzymes involved in lactose utilization are probably turned off most of the time. If cells growing on a carbohydrate other than lactose are transferred to medium containing lactose as the only carbon source, they quickly begin to synthesize the enzymes required for lactose utilization (**FIGURE 19.2*a***). This process of turning on the expression of genes in response to a substance in the environment is called **induction.** Genes whose expression is regulated in this manner are called **inducible genes;** their products, if enzymes, are called **inducible enzymes.**

Enzymes that are involved in **catabolic** (degradative) **pathways,** such as in lactose, galactose, or arabinose utilization, are characteristically inducible. As we discuss below, induction occurs at the level of transcription. Induction alters the rate of enzyme synthesis, not the activity of existing enzyme molecules. Induction should not be confused with enzyme activation, which occurs when the binding of a small molecule to an enzyme increases the activity of the enzyme, but does not affect its rate of synthesis.

Bacteria can synthesize most of the organic molecules required for growth, such as amino acids, purines, pyrimidines, and vitamins. For example, the *E. coli* genome contains five genes encoding enzymes that catalyze steps in the biosynthesis of tryptophan. These five genes must be expressed in *E. coli* cells growing in an environment devoid of tryptophan in order to provide adequate amounts of this amino acid for ongoing protein synthesis.

When *E. coli* cells are present in an environment containing enough tryptophan to support optimal growth, the continued

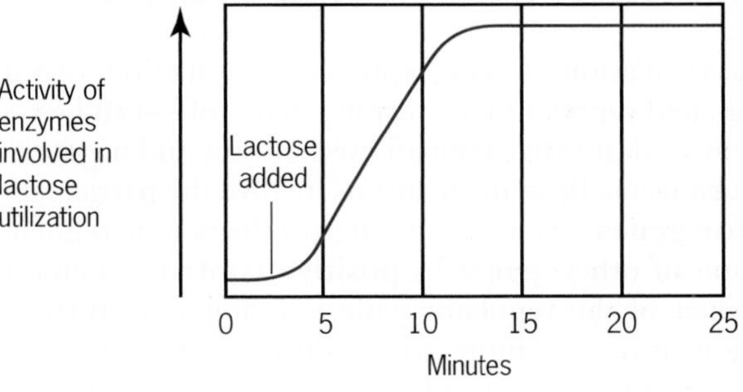

(*a*)

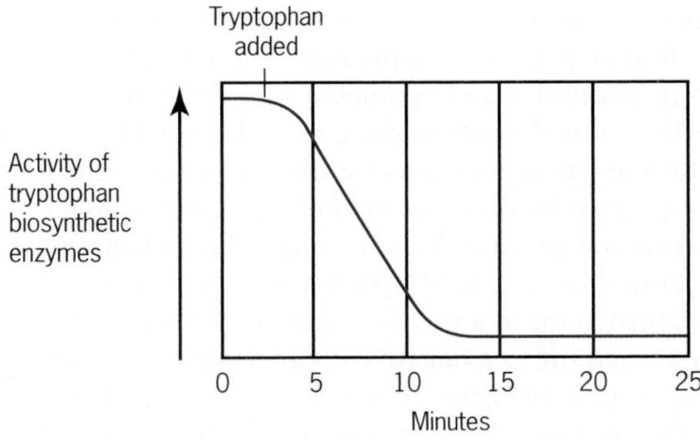

(*b*)

Figure 19.2 ▶ (*a*) Induction of the synthesis of enzymes required for the utilization of lactose as an energy source and (*b*) repression of the synthesis of the enzymes required for the biosynthesis of tryptophan, both in *E. coli*. Note that low levels of enzyme synthesis occur whether the metabolites are present or absent.

synthesis of the tryptophan biosynthetic enzymes would be a waste of energy. Thus, a regulatory mechanism has evolved in *E. coli* that turns off the synthesis of the tryptophan biosynthetic enzymes when external tryptophan is available (**FIGURE 19.2*b***). A gene whose expression has been turned off in this way is said to be "repressed"; the process is called **repression.** When the expression of this gene is turned on, it is said to be "derepressed"; such a response is called **derepression.**

Enzymes that are components of **anabolic** (biosynthetic) **pathways** often are repressible. Repression, like induction, occurs at the level of transcription. Repression should not be confused with feedback inhibition, which occurs when the product of a biosynthetic pathway binds to and inhibits the activity of the first enzyme in the pathway, but does not affect the synthesis of the enzyme.

KEY POINTS

- In prokaryotes, genes that specify housekeeping functions such as rRNAs, tRNAs, and ribosomal proteins are expressed constitutively. Other genes usually are expressed only when their products are needed.
- Genes that encode enzymes involved in catabolic pathways often are expressed only in the presence of the substrates of the enzymes; their expression is inducible.
- Genes that encode enzymes involved in anabolic pathways usually are turned off in the presence of the end product of the pathway; their expression is repressible.
- Although gene expression can be regulated at many levels, transcriptional regulation is the most common.

Positive and Negative Control of Gene Expression

In some cases, the product of a regulatory gene is required to initiate the expression of one or more genes. In other cases, the product of a regulatory gene is required to turn off the expression of one or more genes.

The regulation of gene expression—induction, or turning genes on, and repression, or turning genes off—can be accomplished by both positive control mechanisms and negative control mechanisms. Both mechanisms involve the participation of **regulator genes**—genes encoding products that regulate the expression of other genes. In **positive control mechanisms,** the product of the regulator gene is required to turn on the expression of one or more structural genes (genes specifying the amino acid sequences of enzymes or structural proteins), whereas in **negative control mechanisms,** the product of the regulator gene is necessary to shut off the expression of structural genes. Positive and negative regulation are illustrated for both inducible and repressible systems in **FIGURE 19.3**.

Recall that a gene is expressed when RNA polymerase binds to its promoter and synthesizes an RNA transcript that contains the coding region of the gene (Chapter 11). The product of the regulator gene acts by binding to a site called the regulator protein binding site (*RPBS*) adjacent to the promoter of the structural gene(s). When the product of the regulator gene is bound at the *RPBS*, transcription of the structural gene(s) is turned on in a positive control system (**FIGURE 19.3**, right) or turned off in a negative control system (**FIGURE 19.3**, left). The regulator gene products are called **activators**—because they activate gene expression—in positive control systems, and **repressors**—because they repress gene expression—in negative control systems. Whether or not a regulator protein can bind to the *RPBS* depends on the presence or absence of **effector molecules** in the cell. The effectors are usually small molecules such as amino acids, sugars, and similar metabolites. The effector molecules involved in induction of gene expression are called **inducers;** those involved in repression of gene expression are called **co-repressors.**

The effector molecules (inducers and co-repressors) bind to regulator gene products (activators and repressors) and cause changes in the three-dimensional structures of these proteins. Conformational changes in protein structure resulting from the binding of small molecules are called **allosteric transitions**. Conformational changes in proteins frequently result in alterations in their activity. In the case of activators and repressors, allosteric transitions caused by the binding of effector molecules usually alter their ability to bind to regulator protein-binding sites adjacent to the promoters of the structural genes they control.

In a negative, inducible control mechanism (**FIGURE 19.3*a***, left), the free repressor binds to the *RPBS* and prevents the transcription of the structural gene(s) in the absence of inducer. When inducer is present, it is bound by the repressor, and the repressor/inducer complex cannot bind to the *RPBS*. With no repressor bound to the *RPBS*, RNA polymerase binds to the promoter and transcribes the structural gene(s). In a positive, inducible control mechanism (**FIGURE 19.3*a***, right), the activator cannot bind to the *RPBS* unless inducer is present, and RNA polymerase cannot transcribe the structural gene(s) unless the activator/inducer complex is bound to the *RPBS*. Thus, transcription of the structural genes is turned on only in the presence of inducer.

In a negative, repressible regulatory mechanism (**FIGURE 19.3*b***, left), transcription of the structural gene(s) occurs in the absence of the co-repressor, but not in its presence. When the repressor/co-repressor complex is bound to the *RPBS*, it

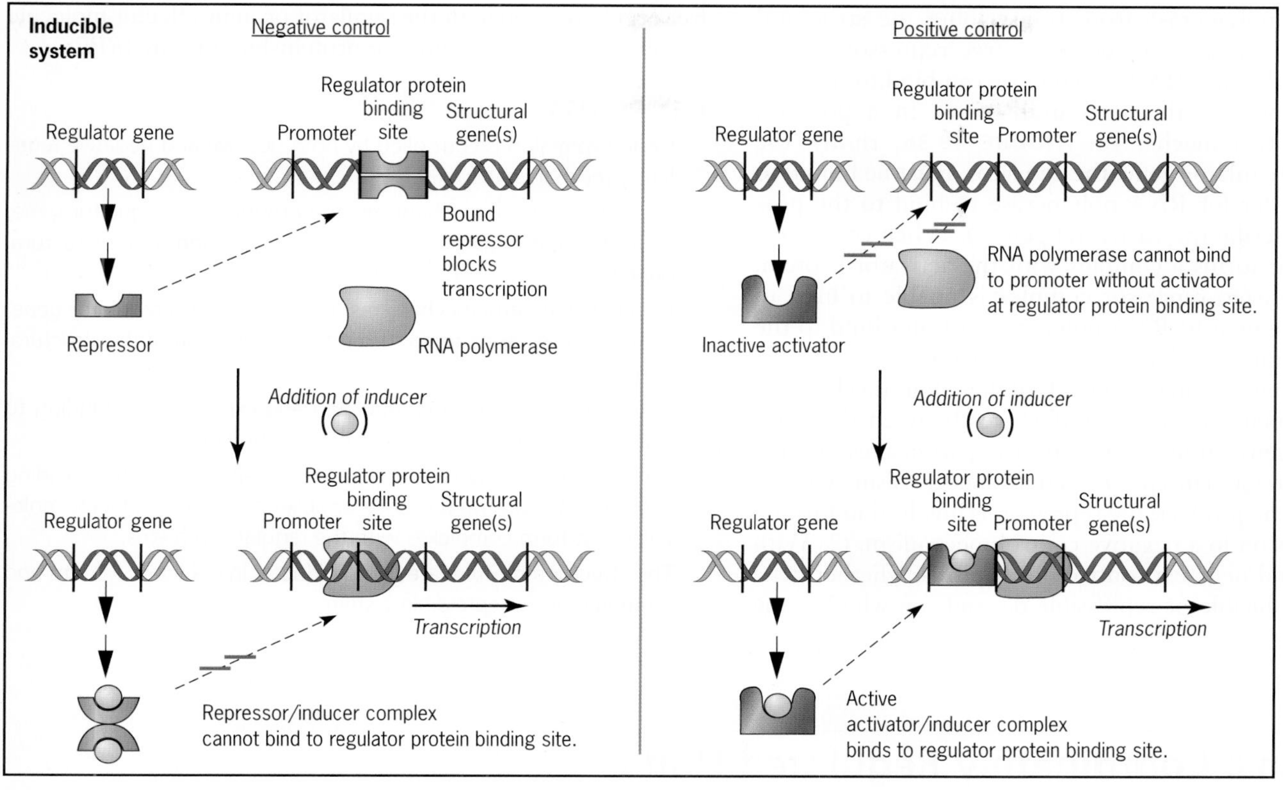

(a)

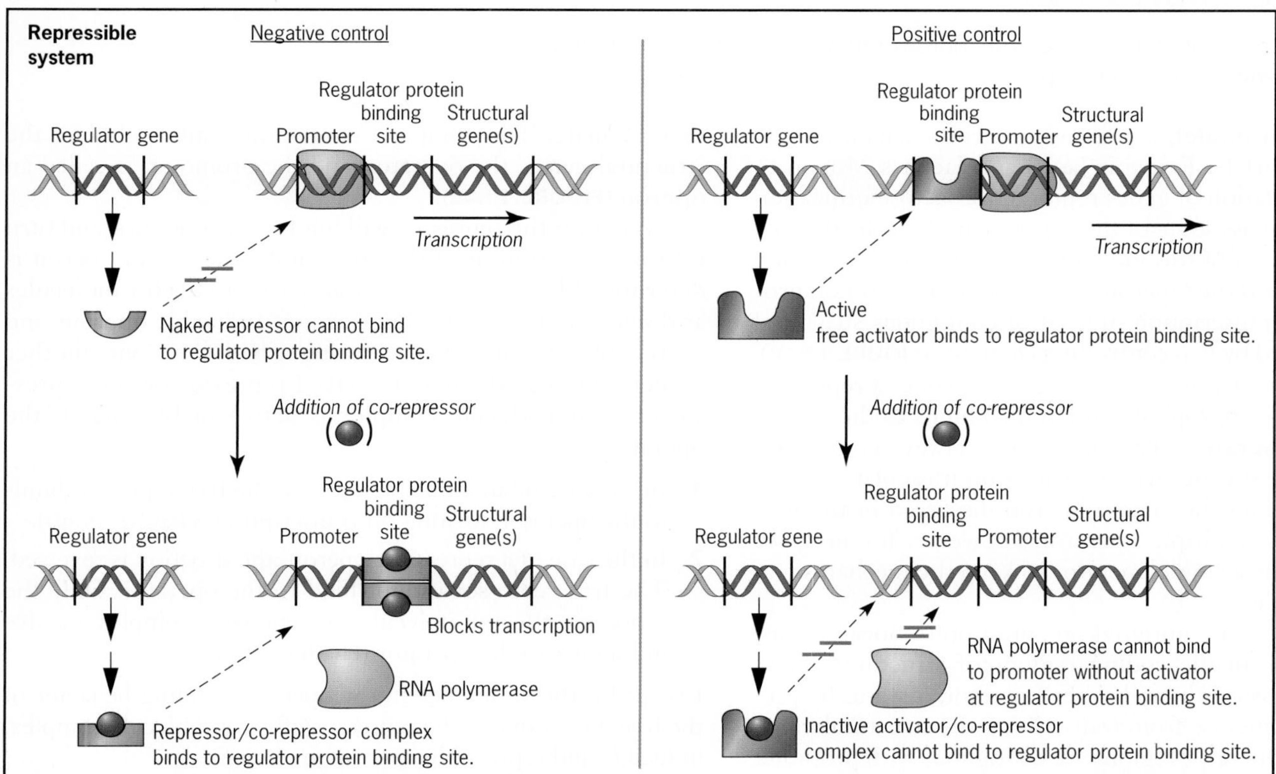

(b)

Figure 19.3 ► Negative and positive control of inducible (*a*) and repressible (*b*) gene expression. The regulator gene product is required to turn on gene expression in positive control circuits and to turn off gene expression in negative control systems.

prevents RNA polymerase from transcribing the structural genes. In the absence of co-repressor, free repressor cannot bind to the *RPBS*; thus, RNA polymerase can bind to the promoter and transcribe the structural genes. In a positive, repressible control mechanism (**FIGURE 19.3*b***, right), the product of the regulator gene, the activator, must be bound to the *RPBS* in order for RNA polymerase to bind to the promoter and transcribe the structural gene(s). When co-repressor is present, it forms a complex with the activator protein, and this activator/co-repressor complex is unable to bind to the *RPBS*; consequently, RNA polymerase cannot bind to the promoter and transcribe the structural gene(s).

In order to understand the details of these four mechanisms of regulation, focus on the key differences between them. (1) The regulator gene product, the activator, participates in turning on gene expression in a positive control mechanism, whereas the regulator gene product, the repressor, is involved in turning off gene expression in a negative control mechanism. (2) With both positive and negative control mechanisms, whether gene expression is inducible or repressible depends on whether the free regulator protein or the regulator protein/effector molecule complex binds to the regulator protein-binding site (*RPBS*).

KEY POINTS

- Gene expression is controlled by both positive and negative regulatory mechanisms.
- In positive control mechanisms, the product of a regulator gene, an activator, is required to turn on the expression of the structural gene(s).
- In negative control mechanisms, the product of a regulator gene, a repressor, is necessary to turn off the expression of the structural gene(s).
- Activators and repressors regulate gene expression by binding to sites adjacent to the promoters of structural genes.
- Whether or not the regulator proteins can bind to their binding sites depends on the presence or absence of small effector molecules that form complexes with the regulator proteins.
- The effector molecules are called inducers in inducible systems and co-repressors in repressible systems.

Operons: Coordinately Regulated Units of Gene Expression

In prokaryotes, genes with related functions often are present in coordinately regulated genetic units called operons.

The **operon model,** a negative control mechanism, was developed in 1961 by François Jacob and Jacques Monod to explain the regulation of genes required for lactose utilization in *E. coli*. We discuss some of the experimental results that led to the development of this model in A Milestone in Genetics: Jacob, Monod, and the Operon Model. Jacob and Monod proposed that the transcription of a set of contiguous structural genes is regulated by two controlling elements (**FIGURE 19.4*a***). One of the elements, the repressor gene, encodes a repressor, which (under the appropriate conditions) binds to the second element, the **operator.** The operator is always contiguous with the structural genes whose expression it regulates. Some operons—including the lactose operon discussed in the next section—contain multiple operators; however, for now, we will consider only a single operator to keep the mechanism as simple as possible.

Transcription is initiated at promoters located just upstream (5′) from the coding regions of structural genes. When the repressor is bound to the operator, it sterically prevents RNA polymerase from transcribing the structural genes in the operon. Operator regions are contiguous with promoter regions; sometimes operators and promoters even overlap, sharing a short DNA sequence. Operator regions are often located between the promoters and the structural genes that they regulate. The complete contiguous unit, including the structural genes, the operator, and the promoter, is called an **operon** (**FIGURE 19.4*a***).

Whether the repressor will bind to the operator and turn off the transcription of the structural genes in an operon is determined by the presence or absence of effector molecules as discussed in the preceding section. Inducible operons and repressible operons can be distinguished from one another by determining whether the naked repressor or the repressor/effector molecule complex is active in binding to the operator.

1. In the case of an inducible operon, the free repressor binds to the operator, turning off transcription (**FIGURE 19.4*b***).
2. In the case of a repressible operon, the situation is reversed. The free repressor cannot bind to the operator. Only the repressor/effector molecule (co-repressor) complex is active in binding to the operator (**FIGURE 19.4*c***).

Except for this difference in the operator-binding behavior of the free repressor and the repressor/effector molecule complex, inducible and repressible operons are identical.

A single mRNA transcript carries the coding information of an entire operon. Thus, the mRNAs of operons consisting of more than one structural gene are multigenic. For example, the

The operon: components

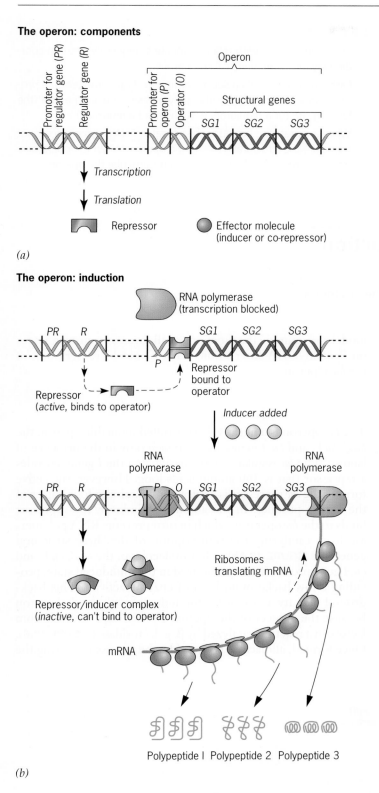

Promoter for regulator gene (PR)
Regulator gene (R)
Operon
Promoter for operon (P)
Operator (O)
Structural genes
SG1 SG2 SG3
Transcription
Translation
Repressor
Effector molecule (inducer or co-repressor)

(a)

The operon: induction

RNA polymerase (transcription blocked)
PR R SG1 SG2 SG3
P
Repressor bound to operator
Repressor (*active*, binds to operator)
Inducer added
RNA polymerase
RNA polymerase
PR R P O SG1 SG2 SG3
Ribosomes translating mRNA
Repressor/inducer complex (*inactive*, can't bind to operator)
mRNA
Polypeptide I Polypeptide 2 Polypeptide 3

(b)

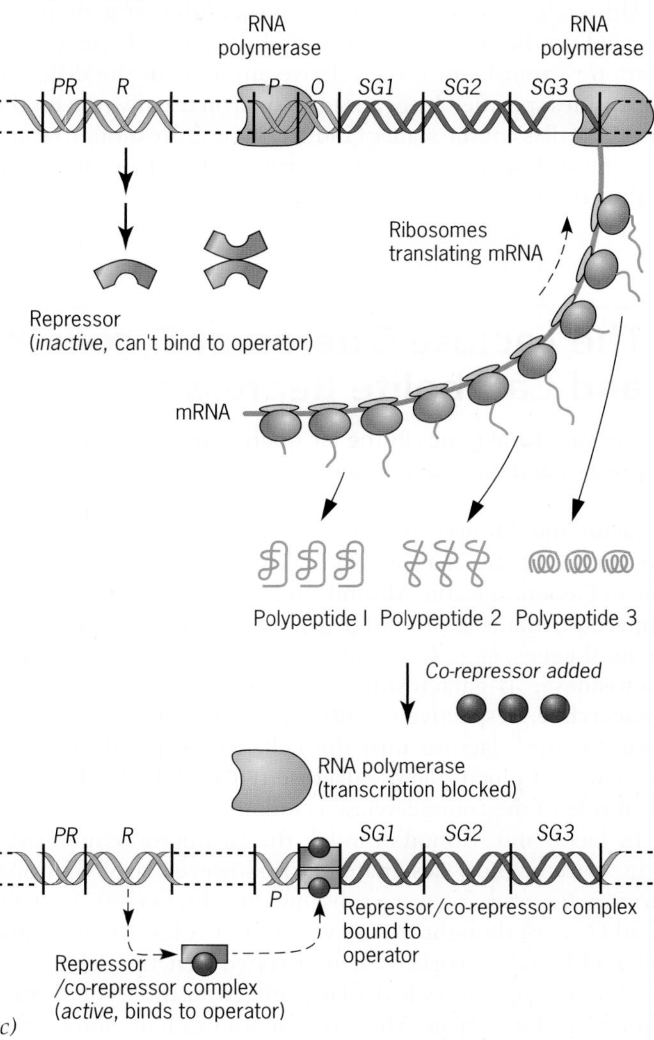

(c)

Figure 19.4 ▶ Regulation of gene expression by the operon mechanism. (*a*) Components of an operon: one or more structural genes (three, *SG1*, *SG2*, and *SG3*, are shown) and the adjoining operator (*O*) and promoter (*P*) sequences. The transcription of the regulator gene (*R*) is initiated by RNA polymerase, which binds to its promoter (*PR*). When repressor is bound to the operator, it sterically prevents RNA polymerase from initiating transcription of the structural genes. The difference between an inducible operon (*b*) and a repressible operon (*c*) is that free repressor binds to the operator of an inducible operon, whereas the repressor/effector molecule complex binds to the operator of a repressible operon. Thus, an inducible operon is turned off in the absence of the effector (inducer) molecule, and a repressible operon is turned on in the absence of the effector (co-repressor) molecule.

tryptophan operon mRNA of *E. coli* contains the coding sequences of five different genes. Because they are co-transcribed, all structural genes in an operon are coordinately expressed.

Although the molar quantities of the different gene products need not be the same (because of different efficiencies of initiation of translation), the relative amounts of the different polypeptides specified by genes in an operon usually remain the same, regardless of the state of induction or repression. In some cases, the differential use of transcription termination signals can alter the amounts of gene products synthesized.

KEY POINTS

- In bacteria, genes with related functions frequently occur in coordinately regulated units called operons.
- Each operon contains a set of contiguous structural genes, a promoter (the binding site for RNA polymerase), and an operator (the binding site for a regulatory protein called a repressor).
- When a repressor is bound to the operator, RNA polymerase cannot transcribe the structural genes in the operon. When the operator is free of repressor, RNA polymerase can transcribe the operon.

The Lactose Operon in *E. coli*: Induction and Catabolite Repression

The structural genes in the *lac* operon are transcribed only when lactose is present and glucose is absent.

Jacob and Monod proposed the operon model based on their studies of the lactose (*lac*) operon in *E. coli* (see A Milestone in Genetics: Jacob, Monod, and the Operon Model). The *lac* operon contains a promoter (*P*), an operator (*O*), and three structural genes, *lacZ*, *lacY*, and *lacA*, encoding the enzymes β-galactosidase, β-galactoside permease, and β-galactoside transacetylase, respectively (**FIGURE 19.5**). β-galactoside permease "pumps" lactose into the cell, where β-galactosidase cleaves it into glucose and galactose (**FIGURE 19.6**). The biological role of the transacetylase is unknown.

In Jacob and Monod's model, the *lac* operon contained a single operator (now designated O_1). However, two additional operators (O_2 and O_3) were subsequently discovered. Initially, O_2 and O_3 were thought to play very minor roles. Then, Benno Müller-Hill and coworkers demonstrated that the deletion of both "minor" operators had a large effect on the level of transcription of the operon. More recent studies have shown that efficient repression of the *lac* operon requires the major operator (O_1) and at least one of the minor operators (O_2 or O_3). Nevertheless, we will first develop Jacob and Monod's model of the *lac* operon with a single operator. Then, we will extend the model and examine the functions of all three operators in Focus on Protein–DNA Interactions That Control Transcription of the *lac* Operon.

INDUCTION

The *lac* operon is a negatively controlled inducible operon; the *lacZ*, *lacY*, and *lacA* genes are expressed only in the presence of lactose. The *lac* regulator gene, designated the *I* gene, encodes a repressor that is 360 amino acids long. However, the active form of the *lac* repressor is a tetramer containing four copies of the *I* gene product. In the absence of inducer, the repressor binds to the *lac* operator, which in turn prevents RNA polymerase from catalyzing the transcription of the three structural genes (see **FIGURE 19.4b**). A few molecules of the *lacZ*, *lacY*, and *lacA* gene products are synthesized in the uninduced state, providing a low background level of enzyme activity. This background activity is essential for induction of the *lac* operon because the inducer of the operon, allolactose, is derived from lactose in a reaction catalyzed by β-galactosidase (**FIGURE 19.6**). Once formed, allolactose is bound by the repressor, causing the

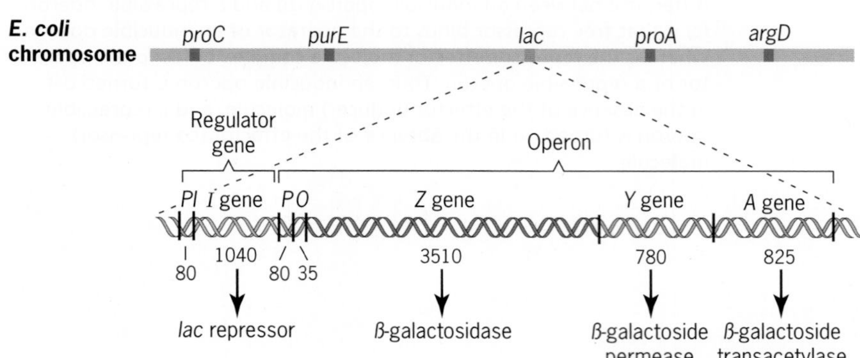

Figure 19.5 ▶ The *lac* operon of *E. coli*. The *lac* operon consists of three structural genes, *Z*, *Y*, and *A*, plus the promoter (*P*) and operator (*Q*) regions adjoining the *Z* gene. The regulator gene (*I*) is contiguous with the operon in the case of *lac*. The regulator gene has its own promoter (*PI*). The numbers below the various genetic elements indicate their sizes in nucleotide pairs.

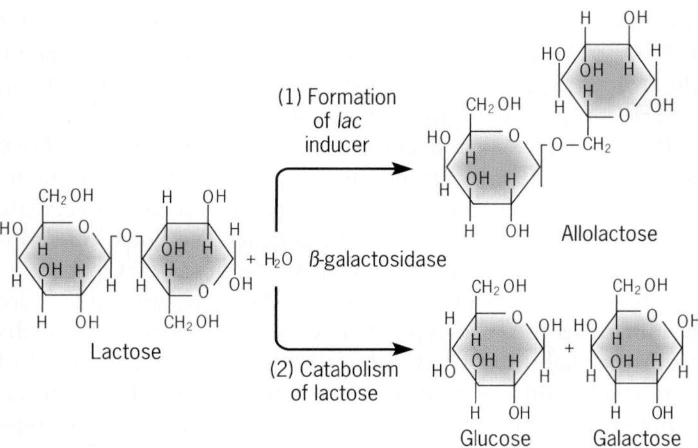

Figure 19.6 ▶ Two physiologically important reactions catalyzed by β-galactosidase: (1) conversion of lactose to the *lac* operon inducer allolactose, and (2) cleavage of lactose to produce the monosaccharides glucose and galactose.

release of the repressor from the operator. In this way, allolactose induces the transcription of the *lacZ*, *lacY*, and *lacA* structural genes (see **FIGURE 19.4*b***).

The *lacI* gene, the *lac* operator, and the *lac* promoter were all initially identified genetically by the isolation of mutant strains that exhibited altered expression of the *lac* operon genes. Mutations in the *I* gene and the operator frequently result in constitutive synthesis of the *lac* gene products. These mutations are designated I^- and O^c, respectively. The I^- and O^c constitutive mutations can be distinguished not only by map position, but also by their behavior in partial diploids in which they are located in *cis* and *trans* configurations relative to mutations in *lac* structural genes (**TABLE 19.1**). Recall that partial diploids can be constructed using fertility (F) factors that carry chromosomal genes—F′ factors (Chapter 8). F′ factors that carry the *lac* operon have been used to study the interactions between the various components of the operon.

Like monoploid wild-type ($I^+P^+O^+Z^+Y^+A^+$) cells, partial diploids (also called "merozygotes") of genotype F′ $I^+P^+O^+Z^+A^+/I^+P^+O^+Z^-Y^-A^-$ or of genotype F′ $I^+P^+O^+Z^-Y^-A^-/I^+P^+O^+Z^+Y^+A^+$ are inducible for the utilization of lactose as a carbon source. The wild-type alleles (Z^+, Y^+, and A^+) of the three structural genes are dominant to their mutant alleles (Z^-, Y^-, and A^-). This dominance is expected because the wild-type alleles produce functional enzymes, whereas the mutant alleles produce no enzymes or defective (inactive) enzymes. Partial diploids of genotype $I^+P^+O^+Z^+Y^+A^+/I^-P^+O^+Z^+Y^+A^+$ (I^+/I^-) are also inducible for the synthesis of the three enzymes specified by the *lac* operon. Thus, I^+ is dominant to I^- as expected, because I^+ encodes a functional repressor molecule and its I^- allele specifies an inactive repressor. The dominance of I^+ over I^- also indicates that the repressor is diffusible, because the repressor produced by the $lacI^+$ allele on one chromosome can turn off the *lac* structural genes on both operons in the cell (**FIGURE 19.7*a***).

Like wild-type cells, partial diploids of genotype F′ $I^+P^+O^+Z^+Y^+A^+/I^-P^+O^+Z^-Y^-A^-$ or genotype F′ $I^+P^+O^+Z^-Y^-A^-/I^-P^+O^+Z^+Y^+A^+$ are inducible for β-galactosidase, β-galactoside permease, and β-galactoside transacetylase. The inducibility of these genotypes demonstrates that the *lac* repressor (I^+ gene product) controls the expression of structural genes located either *cis* (**FIGURE 19.7*b***) or *trans* (**FIGURE 19.7*c***) to the $lacI^+$ allele.

The operator constitutive (O^c) mutations act only in *cis*; that is, O^c mutations affect the expression of only those structural genes located on the same chromosome. The *cis*-acting nature of *O* mutations is logical given the function of the operator. *O* mutations should not act in *trans* if the operator is the binding site for the repressor; as such, the operator does not encode any product, diffusible or otherwise. A regulator gene should act in *trans* only if it specifies a diffusible product.

▶ **TABLE 19.1**

Phenotypic Effects of Mutations in the Repressor Gene (*I*) and the Operator (*O*) Region of the *lac* Operon

	β-Galactosidase Activity[a]		β-Galactoside Permease Activity[a]		
Genotype	With Lactose	Without Lactose	With Lactose	Without Lactose	Deduction
$I^+P^+O^+Z^+Y^+$	100 units	1 unit	100 units	1 unit	Wild-type is inducible
$I^+P^+O^+Z^+Y^+$/ F′ $I^+P^+O^+Z^-Y^-$	100 units	1 unit	100 units	1 unit	Z^+ is dominant to Z^-; Y^+ is dominant to Y^-
$I^+P^+O^+Z^+Y^+$/ F′ $I^+P^+O^+Z^+Y^+$	200 units	2 units	200 units	2 units	Activity depends on gene dosage
$I^-P^+O^+Z^+Y^+$	100 units	100 units	100 units	100 units	$lacI^-$ mutants are constitutive
$I^+P^+O^+Z^+Y^+$/ F′ $I^-P^+O^+Z^+Y^+$	200 units	2 units	200 units	2 units	I^+ is dominant to I^-
$I^+P^+O^cZ^+Y^+$	100 units	100 units	100 units	100 units	$lacO^c$ mutants are constitutive
$I^+P^+O^cZ^+Y^-$/ F′ $I^+P^+O^+Z^-Y^+$	100 units	100 units	100 units	1 unit	O^c and O^+ are *cis*-acting regulators

[a]Activity levels in wild-type bacteria have been set at 100 units for both β-galactosidase (the product of gene *Z*) and β-galactoside permease (the product of gene *Y*). The *A* gene and its product β-galactoside transacetylase are not shown for the sake of brevity.

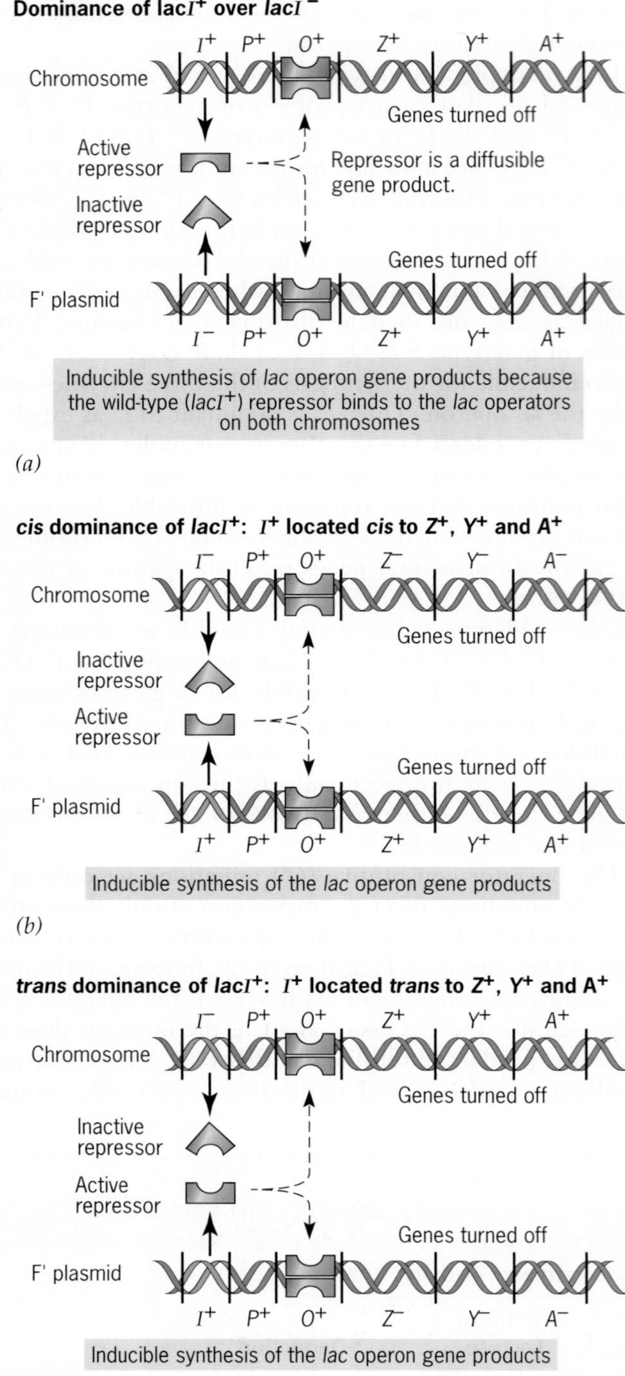

Figure 19.7 ▶ Studies of *E. coli* partial diploids have shown that the *lacI*$^+$ gene is dominant to *lacI*$^-$ alleles (*a*) and controls *lac* operators located either *cis* (*b*) or *trans* (*c*) to itself. These effects demonstrate that the *lacI* gene product is diffusible. Although the functional form of the *lac* repressor is a tetramer, the two molecules at the back of the tetramer are not shown for the sake of simplicity.

Therefore, a partial diploid of genotype $F'\ I^+P^+O^c\ Z^-Y^-A^-/I^+P^+O^+Z^+Y^+A^+$ is inducible for the three enzymes specified by the structural genes of the *lac* operon (**TABLE 19.2, FIGURE 19.8*a***), whereas a partial diploid of genotype $F'\ I^+P^+O^c\ Z^+Y^+A^+/I^+P^+O^+Z^-Y^-A^-$ synthesizes these enzymes constitutively (**TABLE 19.2, FIGURE 19.8*b***). Once you are confident that you understand how the components of the operon interact to regulate the transcription of the *lac* structural genes, see Focus on Problem Solving: Testing Your Understanding of the *lac* Operon.

Some of the *I* gene mutations, those designated I^{-d}, are dominant to the wild-type allele (I^+). This dominance results from the inability of heteromultimers (proteins composed of two or more different forms of a polypeptide; recall that the *lac* repressor functions as a tetramer) that contain both wild-type and mutant polypeptides to bind to the operator. Other *I* gene mutations, those designated I^s (*s* for superrepressed), cause the *lac* operon to be uninducible. In strains carrying these I^s mutations, the *lac* structural genes can usually be induced to some degree with high concentrations of inducer, but they are not induced at normal concentrations of inducer. When studied *in vitro*, the mutant I^s polypeptides form tetramers that bind to *lac* operator DNA. However, they either do not bind inducer or exhibit a low affinity for inducer. Thus, the I^s mutations alter the inducer binding site of the *lac* repressor.

Promoter mutations do not change the inducibility of the *lac* operon. Instead, they modify the levels of gene expression in the induced and uninduced state by changing the frequency of initiation of *lac* operon transcription—that is, the efficiency of RNA polymerase binding.

The *lac* promoter actually contains two separate components: (1) the RNA polymerase binding site and (2) a binding site for another protein called *c*atabolite *a*ctivator *p*rotein (abbreviated CAP) that prevents the *lac* operon from being induced in the presence of glucose. This second control circuit, which we consider next, assures the preferential utilization of glucose as an energy source when it is available.

CATABOLITE REPRESSION

The presence of glucose has long been known to prevent the induction of the *lac* operon, as well as other operons controlling enzymes involved in carbohydrate catabolism. This phenomenon, called **catabolite repression** (or the **glucose effect**), assures that glucose is metabolized when present, in preference to other, less efficient, energy sources.

The catabolite repression of the *lac* operon and several other operons is mediated by a regulatory protein called **CAP** (for ***c*atabolite *a*ctivator *p*rotein**) and a small effector molecule called **cyclic AMP** (adenosine-3′, 5′-monophosphate; abbreviated cAMP) (**FIGURE 19.9**). Because CAP binds cAMP when this mononucleotide is present at sufficient concentrations, it is sometimes called the cyclic AMP receptor protein.

The *lac* promoter contains two separate binding sites, one for RNA polymerase and one for the CAP/cAMP complex (**FIGURE 19.10**). The CAP/cAMP complex must be present at

TABLE 19.2

The *lac* Repressor Gene (I) Acts both *Cis* and *Trans*; the *lac* Operator Acts only in the *Cis* Configuration

Genotype	β-Galactosidase Activity[a] With Lactose	β-Galactosidase Activity[a] Without Lactose	β-Galactoside Permease Activity[a] With Lactose	β-Galactoside Permease Activity[a] Without Lactose	Deduction
$I^+P^+O^+Z^+Y^+$	100 units	1 unit	100 units	1 unit	Wild-type is inducible
$I^+P^+O^+\,Z^+Y^+/\,F'\,\boxed{I^-}P^+O^+\boxed{Z^-Y^-}$	100 units	1 unit	100 units	1 unit	
$\boxed{I^-}P^+O^+Z^+Y^+/\,F'\,I^+P^+O^+\boxed{Z^-Y^-}$	100 units	1 unit	100 units	1 unit	I^+ acts both *cis* and *trans*
$I^+P^+O^+Z^+Y^+/\,F'\,I^+P^+\boxed{O^cZ^-Y^-}$	100 units	1 unit	100 units	1 unit	O^+ acts only in *cis*
$I^+P^+O^+\boxed{Z^-Y^-}/\,F'\,I^+P^+\boxed{O^c}\,Z^+Y^+$	100 units	100 units	100 units	100 units	$\boxed{O^c}$ acts only in *cis*

[a]Activity levels in wild-type bacteria have been set at 100 units for both β-galactosidase (the product of gene *Z*) and β-galactoside permease (the product of gene *Y*). The *A* gene and its product β-galactoside transacetylase are not shown for the sake of brevity.

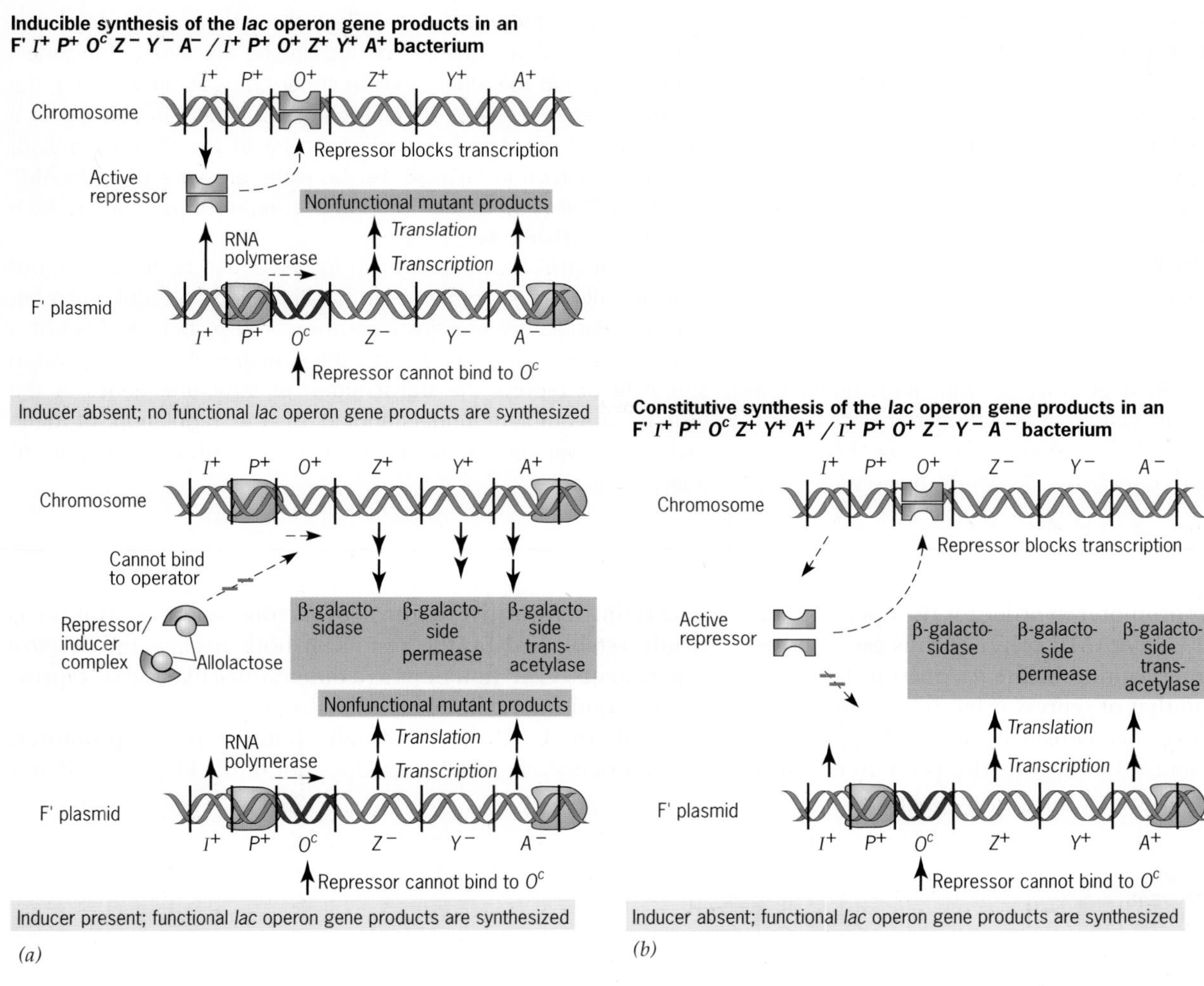

Figure 19.8 ► Studies of *E.coli* partial diploids have shown that the operator acts only in the *cis* configuration. The synthesis of functional β-galactosidase, β-galactoside permease, and β-galactoside transacetylase is (*a*) inducible in a partial diploid of genotype F′ $I^+P^+O^cZ^-Y^-A^-/I^+P^+O^+Z^+Y^+A^+$ and (b) constitutive in a partial diploid of genotype F′ $I^+P^+O^cZ^+Y^+A^+/I^+P^+O^+Z^-Y^-A^-$. These results demonstrate that the operator (*O*) is *cis*-acting; that is, it regulates only those structural genes located on the same chromosome.

▶ FOCUS ON **Protein–DNA Interactions That Control Transcription of the *lac* Operon**

Transcription of the structural genes in the *lac* operon is regulated by sequence-specific interactions between proteins and their DNA binding sites. One such interaction is the binding of RNA polymerase to its binding site in the promoter (discussed in Chapter 11). A second is the binding of CAP/cAMP to its binding site in the *lac* promoter (discussed in the preceding section). A third is the binding of the *lac* repressor to the operator (actually, operators).

Let's first examine the binding of CAP/cAMP to its binding site in the lac promoter. CAP/cAMP controls catabolite repression; the binding of CAP/cAMP to the promoter is required for efficient induction of the *lac* operon. How does the binding of CAP/cAMP stimulate transcription of the *lac* structural genes? RNA polymerase cannot bind efficiently to its binding site in the *lac* promoter unless CAP/cAMP is already bound. When CAP/cAMP binds to DNA, it bends the DNA (**FIGURE 1*a***). X-ray studies show that the DNA is bent as it is wrapped on the surface of the CAP/cAMP complex (**FIGURE 1*b***). Recall that the CAP/cAMP and RNA polymerase binding sites are adjacent to one another in the *lac* promoter (**FIGURE 19.10**). Presumably, the bending of the DNA by CAP/cAMP promotes a more open site for RNA polymerase and thus enhanced binding and transcription of the structural genes. However, there is also evidence for contact between RNA polymerase and CAP/cAMP, so the complete picture may be more complex than just the bending of the DNA.

Next, let's examine the binding of the *lac* repressor to the operators, which prevents RNA polymerase from transcribing the structural genes in the operon. Originally, the *lac* operon was thought to contain a single operator. Later, it was shown to contain three operators, the primary operator—O_1—and two secondary operators—O_2 and O_3. O_1 is the original operator; it is located between the promoter and the *Z* gene (see **FIGURE 19.10**). O_2 is located downstream from O_1 within the *Z* gene, and O_3 is located upstream of the promoter (**FIGURE 1c**). Maximum repression requires all three operators: however, strong repression occurs as long as O_1 and either O_2 or O_3 are present. Why are two of the operators required for efficient repression? To answer this question, we need to look at the sequence-specific binding of the repressor to the operators.

The active form of the *lac* repressor is a tetramer containing four copies of the product of the *lacI* gene. X-ray studies of the structures formed by the *lac* repressor and 21-bp-synthetic binding sites showed that each tetrameric repressor binds two operator sequences simultaneously (**FIGURE 1*d***). In effect, the tetramer consists of two dimers, each with a sequence-specific binding site. One of the dimers binds to O_1, and the other binds to either O_2 or O_3. In so doing, the repressor bends the DNA forming either a hairpin (O_1 and O_2) or a loop (O_1 and O_3). The proposed structure of the O_1–O_3–repressor complex is shown in **FIGURE 1e.** Note the presence of CAP/cAMP within the DNA loop formed when the *lac* repressor is bound to both O_1 and O_3 (**FIGURE 1e**).

Similar DNA loops are known to be formed by the binding of protein activators and repressors of other operons in *E. coli* and other bacteria. Regulatory proteins have the ability to bind to DNA in a sequence-specific manner, to alter the structure of the DNA, and to stimulate or repress the transcription of structural genes in the vicinity. A complete understanding of the regulation of gene expression will require detailed knowledge of these important interactions.

its binding site in the *lac* promoter in order for the operon to be induced normally. The CAP/cAMP complex thus exerts positive control over the transcription of the *lac* operon. It has an effect exactly opposite to that of repressor binding to an operator. Although the precise mechanism by which CAP/cAMP stimulates RNA polymerase binding to the promoter is still uncertain, its positive control of *lac* operon transcription is firmly established by the results of both *in vivo* and *in vitro* experiments. CAP functions as a dimer; thus, like the *lac* repressor, it is multimeric in its functional state.

Only the CAP/cAMP complex binds to the *lac* promoter; in the absence of cAMP, CAP does not bind. Thus, cAMP acts

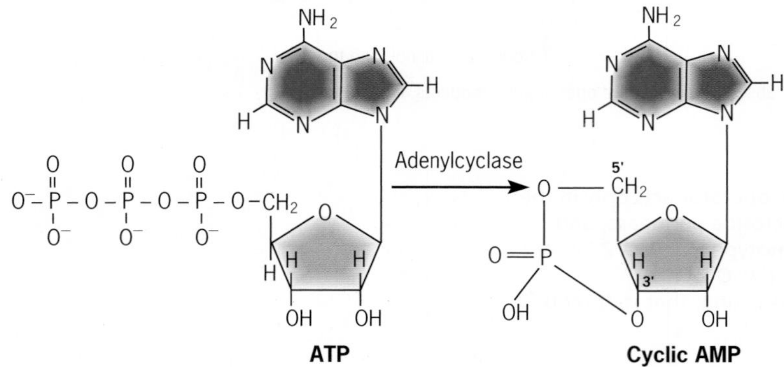

Figure 19.9 ▶ The adenylcyclase-catalyzed synthesis of cyclic AMP (cAMP) from ATP.

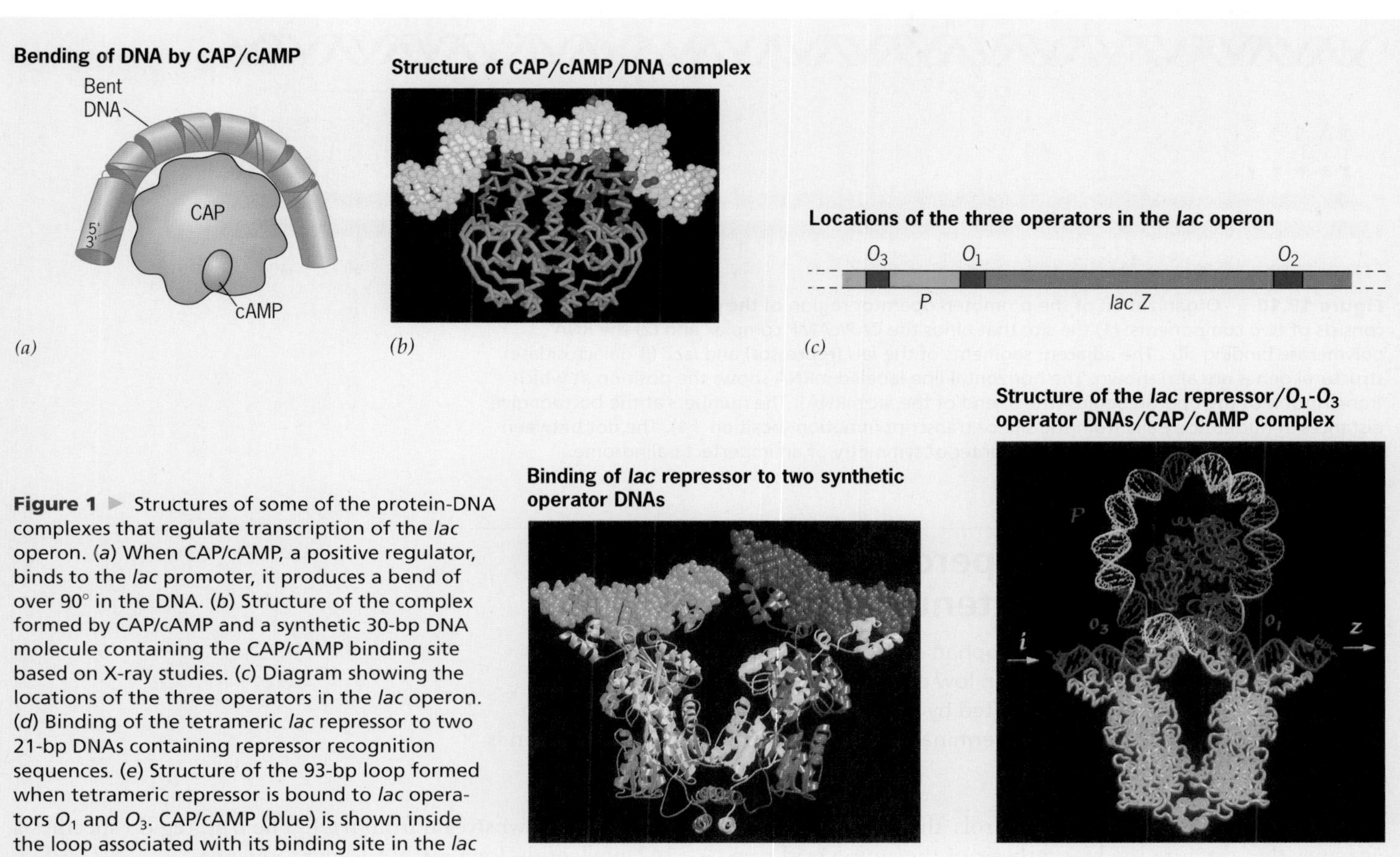

Figure 1 ▶ Structures of some of the protein-DNA complexes that regulate transcription of the *lac* operon. (*a*) When CAP/cAMP, a positive regulator, binds to the *lac* promoter, it produces a bend of over 90° in the DNA. (*b*) Structure of the complex formed by CAP/cAMP and a synthetic 30-bp DNA molecule containing the CAP/cAMP binding site based on X-ray studies. (*c*) Diagram showing the locations of the three operators in the *lac* operon. (*d*) Binding of the tetrameric *lac* repressor to two 21-bp DNAs containing repressor recognition sequences. (e) Structure of the 93-bp loop formed when tetrameric repressor is bound to *lac* operators O_1 and O_3. CAP/cAMP (blue) is shown inside the loop associated with its binding site in the *lac* promoter.

as the effector molecule, determining the effect of CAP on *lac* operon transcription. The intracellular cAMP concentration is sensitive to the presence or absence of glucose. High concentrations of glucose cause sharp decreases in the intracellular concentration of cAMP. Glucose prevents the activation of adenylcyclase, the enzyme that catalyzes the formation of cAMP from ATP. Thus, the presence of glucose results in a decrease in the intracellular concentration of cAMP. In the presence of a low concentration of cAMP, CAP cannot bind to the *lac* operon promoter. In turn, RNA polymerase cannot bind efficiently to the *lac* promoter in the absence of bound CAP/cAMP. Thus, in the presence of glucose, *lac* operon transcription never exceeds 2 percent of the induced rate observed in the absence of glucose. By similar mechanisms, CAP and cAMP keep the arabinose (*ara*) and galactose (*gal*) operons of *E. coli* from being induced in the presence of glucose.

The complete nucleotide-pair sequence of the *lac* operon regulatory region is shown in **FIGURE 19.10**. Comparative nucleotide-sequence studies of mutant and wild-type promoters and operators, in addition to *in vitro* CAP/cAMP, RNA polymerase, and repressor binding studies and X-ray crystallographic data, have provided detailed information about the nature of these important **sequence-specific protein–nucleic acid interactions.** Some of these interactions are examined more closely in Focus on Protein–DNA Interactions That Control Transcription of the *lac* Operon.

KEY POINTS

- The *E. coli lac* operon is a negative inducible and catabolite repressible system; the three structural genes in the *lac* operon are transcribed at high levels only in the presence of lactose and the absence of glucose.
- In the absence of lactose, the *lac* repressor binds to the *lac* operator and prevents RNA polymerase from initiating transcription of the operon.
- Catabolite repression keeps operons such as *lac* encoding enzymes involved in carbohydrate catabolism from being induced in the presence of glucose, the preferred energy source.

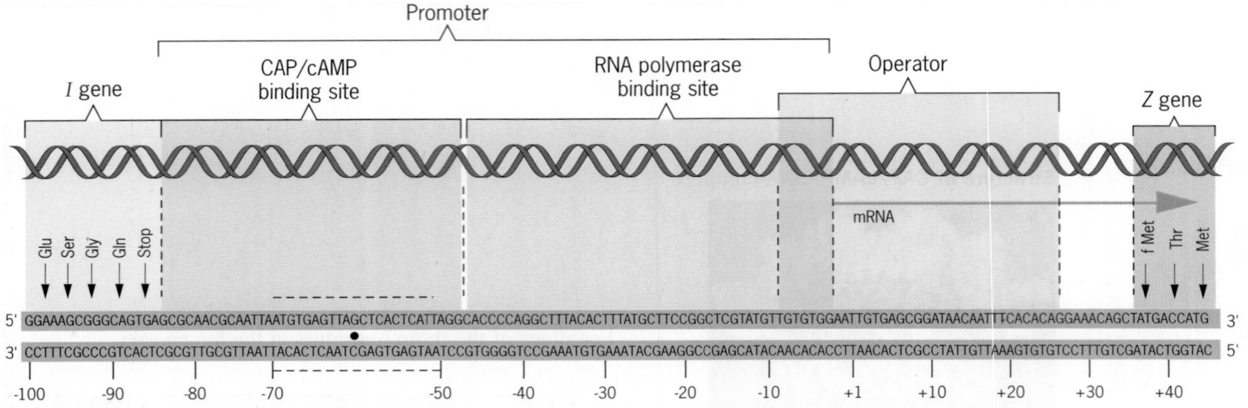

Figure 19.10 ▶ Organization of the promoter–operator region of the *lac* operon. The promoter consists of two components: (1) the site that binds the CAP/cAMP complex and (2) the RNA polymerase binding site. The adjacent segments of the *lacI* (repressor) and *lacZ* (β-galactosidase) structural genes are also shown. The horizontal line labeled mRNA shows the position at which transcription of the operon begins (the 5′-end of the *lac* mRNA). The numbers at the bottom give distances in nucleotide pairs from the site of transcript initiation (position +1). The dot between the two nucleotide strands indicates the center of symmetry of an imperfect palindrome.

The Tryptophan Operon in *E. coli*: Repression and Attenuation

The structural genes in the tryptophan operon are transcribed only when tryptophan is absent or present in low concentrations. The expression of the genes in the *trp* operon is regulated by repression of transcriptional initiation and by attenuation (premature termination) of transcription when tryptophan is prevalent in the environment.

The *trp* operon of *E. coli* controls the synthesis of the enzymes that catalyze the biosynthesis of the amino acid tryptophan. The functions of the five structural genes and the adjacent regulatory sequences of the *trp* operon have been analyzed in detail by Charles Yanofsky and colleagues. The five structural genes encode enzymes that convert chorismic acid to tryptophan. The expression of the *trp* operon is regulated at two levels: repression, which controls the initiation of transcription, and attenuation, which governs the frequency of premature transcript termination. We will discuss these regulatory mechanisms in the following two sections.

REPRESSION

The *trp* operon of *E. coli* is a negative repressible operon. The organization of the *trp* operon and the pathway of biosynthesis of tryptophan are shown in **FIGURE 19.11**. The *trpR* gene, which encodes the *trp* repressor, is not closely linked to the *trp* operon. The operator (*O*) region of the *trp* operon lies within the primary promoter (P_1) region. There is also a weak promoter (P_2) at the operator-distal end of the *trpD* gene. The P_2 promoter increases the basal level of transcription of the *trpC*, *trpB*, and *trpA* genes. Two transcription termination sequences (*t* and *t*′) are located downstream from *trpA*. The *trpL* region specifies a 162-nucleotide-long mRNA leader sequence.

The regulation of transcription of the *trp* operon is diagrammed in **FIGURE 19.4c**. In the absence of tryptophan (the co-repressor), RNA polymerase binds to the promoter region and transcribes the structural genes of the operon. In the presence of tryptophan, the co-repressor/repressor complex binds to the operator region and prevents RNA polymerase from initiating transcription of the genes in the operon.

The rate of transcription of the *trp* operon in the derepressed state (absence of tryptophan) is 70 times the rate that occurs in the repressed state (presence of tryptophan). In *trpR* mutants, which lack a functional repressor, the rate of synthesis of the tryptophan biosynthetic enzymes is still reduced about tenfold by the addition of tryptophan to the medium. This additional reduction in *trp* operon expression is caused by attenuation, which is discussed next.

ATTENUATION

Deletions that remove part of the *trpL* region (**FIGURE 19.11**) result in increased rates of expression of the *trp* operon. However, these deletions have no effect on the repressibility of the

FOCUS ON PROBLEM SOLVING

Testing Your Understanding of the *lac* Operon

THE PROBLEM

The following table gives the relative activities of the enzymes β-galactosidase and β-galactoside permease in cells with different genotypes at the *lac* locus in *E. coli*. The induced level of activity of each enzyme in wild-type cells was arbitrarily set at 100 units, and all other enzyme levels were measured relative to the levels observed in these cells. Based on the data given in the table for genotypes 1 through 4, fill in the levels of activity that would be expected for genotype 5 in the spaces (parentheses) provided.

	β-Galactosidase		β-Galactoside Permease	
Genotype	**−inducer**	**+inducer**	**−inducer**	**+inducer**
1. $I^+O^+Z^+Y^+$	0.2	100	0.2	100
2. $I^-O^+Z^+Y^+$	100	100	100	100
3. $I^+O^cZ^+Y^+$	75	100	75	100
4. $I^-O^+Z^+Y^-$ / F′ $I^-O^+Z^+Y^+$	200	200	100	100
5. $I^-O^cZ^-Y^+$ / F′ $I^+O^+Z^+Y^+$	( )	( )	( )	( )

FACTS AND CONCEPTS

1. The *lacZ* and lac*Y* genes encode the enzymes β-galactosidase and β-galactoside permease, respectively. β-galactoside permease transports lactose into cells where β-galactosidase cleaves it into glucose and galactose. The *lacZ*$^+$ and *lacY*$^+$ alleles of these genes encode functional enzymes, whereas the *lacZ*$^-$ and *lacY*$^-$ alleles encode nonfunctional gene products.

2. In wild-type *E. coli* cells, the *lacZ*$^+$ and *lacY*$^+$ genes are transcribed only in the presence of lactose. Their transcription is repressed in the absence of lactose when β-galactosidase and β-galactoside permease have nothing to catabolize or transport. Their transcription is induced when lactose is added to the medium (see **FIGURE 19.4*b***).

3. Constitutive mutants of *E. coli* synthesize β-galactosidase and β-galactoside permease continually whether or not lactose is present. These constitutive mutations are of two types and map at two distinct sites in and near the *lac* operon. Some of the constitutive mutations—*lacI*$^-$ mutations—map in the gene that encodes the lac repressor; others—*lacO*^{c} mutations—map in the operator region—the site where the *lac* repressor binds.

4. The *lac* repressor (*lacI*$^+$ gene product) binds to the *lac* operator (*O*) and prevents RNA polymerase from binding to the *lac* promoter and transcribing the genes in the *lac* operon (see **FIGURE 19.4**). The *lacI*$^-$ mutant alleles encode inactive repressors that cannot bind to the *lac* operator. Allele *lacI*$^+$ is dominant to *lacI*$^-$.

5. The *lac* repressor is a diffusible protein; thus, *lacI*$^+$ regulates the expression of *lac* operon genes located both *cis* (on the same chromosome) and *trans* (on a different chromosome) to it. Regulatory elements of this type are said to be *cis*- and *trans*-acting.

6. The wild-type operator (O^+) contains a nucleotide sequence that functions as a binding site for the *lac* repressor. Operator-constitutive (O^c) mutants contain an operator with an altered nucleotide sequence (often a deletion) to which the *lac* repressor either no longer binds or binds inefficiently. Because *lacO*$^+$ and *lacO*^{c} operators only regulate the expression of *lac* genes on the same chromosome, they are called *cis*-acting regulators.

7. The amount of β-galactosidase and β-galactoside permease synthesized in a cell depends on the number of functional copies of the *lacZ*$^+$ and *lacY*$^+$ genes in the cell.

ANALYSIS AND SOLUTION

1. The data given for genotype 1 ($I^+O^+Z^+Y^+$ = wild-type) show that these cells synthesize 0.2 unit of each enzyme in the absence of lactose and 100 units in the presence of lactose.

2. The data for genotype 2 ($I^-O^+Z^+Y^+$ = repressor mutant) show that in the absence of a functional repressor cells synthesize 100 units of each enzyme whether lactose is present or absent.

3. The operator-constitutive mutant (genotype 3) makes 75 units of each enzyme in the absence of lactose and 100 units in the presence of lactose. There is some binding of the *lac* repressor to the *lac* operator in the absence of lactose. When lactose is present, that binding no longer occurs, and synthesis of the *lac* enzymes increases to the fully induced level (100 units).

4. Genotype 4 (the partial diploid $I^-O^+Z^+Y^-$/F′ $I^-O^+Z^+Y^+$) shows the effect of gene dosage. Cells make twice as much enzyme when two copies of a wild-type gene are present.

5. Genotype 5 ($I^-O^cZ^-Y^+$/F′ $I^+O^+Z^+Y^+$) is a partial diploid with two copies of the *lac* operon. It has two copies of Y^+, but only one copy of Z^+. It has an I^+ allele on the F′, so functional repressor will be present in the cells. Transcription of chromosomal genes will be controlled by O^c, whereas transcription of genes on the F′ will be controlled by O^+. All of the β-galactosidase will be produced by the Z^+ allele on the F′; there is a Z^- mutation on the chromosome. The F′ contains a wild-type *lac* operon, so 0.2 unit of β-galactosidase will be synthesized in the absence of lactose, and 100 units will be synthesized in the presence of lactose. In the case of β-galactoside permease, the contributions of both copies of the Y^+ gene must be considered and combined to calculate the total amount of the enzyme per cell. In the absence of lactose, 75 units will be produced from the chromosomal copy of the Y^+ gene and 0.2 unit from the copy on the F′, for a total of 75.2 units. In the presence of lactose, 100 units will be made from each copy of the Y^+ gene, for a total of 200 units.

For further discussion go to your *WileyPLUS* course.

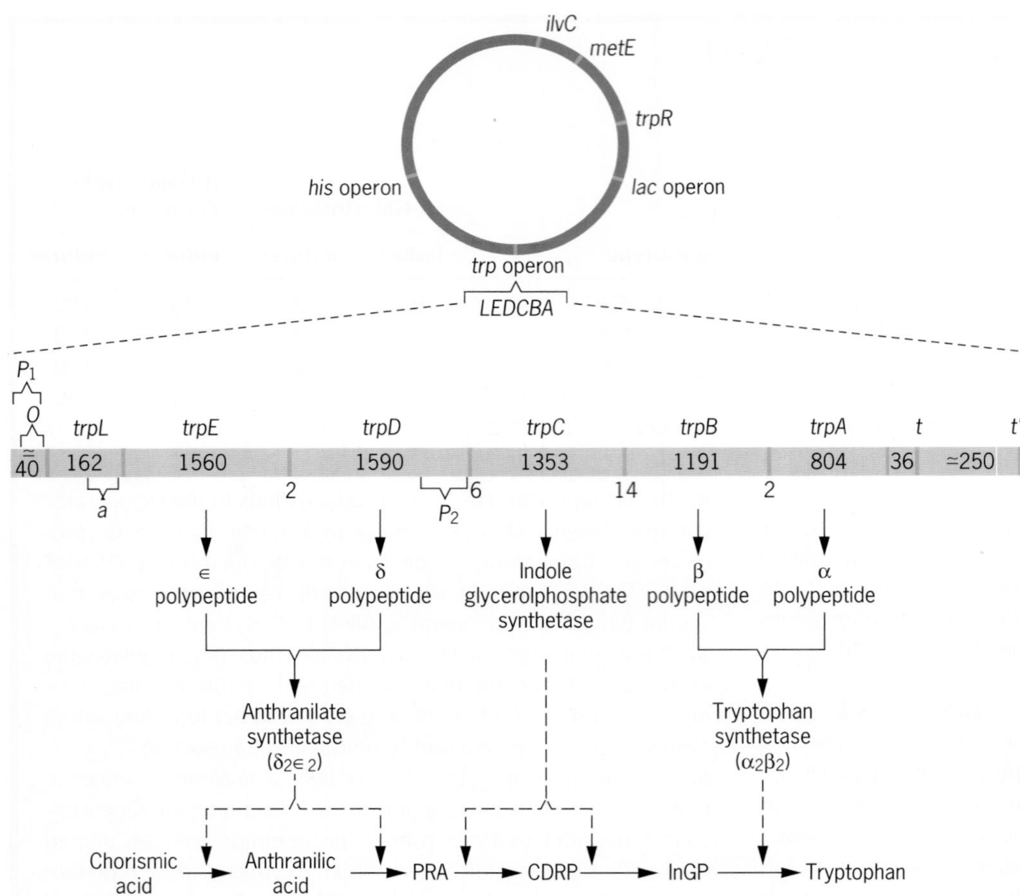

Figure 19.11 ▶ Organization of the *trp* (tryptophan) operon in *E. coli*. The *trp* operon contains five structural genes that encode enzymes involved in the biosynthesis of tryptophan, as shown at the bottom, and the *trpL* regulatory region. The length of each gene or region is given in nucleotide pairs; the intergenic distances are shown below the gene sequence. Key: PRA, phosphoribosyl anthranilate; CDRP, carboxyphenylamino-deoxyribulose phosphate; InGP, indole-glycerol phosphate.

trp operon; that is, repression and derepression occur just as in $trpL^+$ strains. These results indicate that the synthesis of the tryptophan biosynthetic enzymes is regulated at a second level by a mechanism that is independent of repression/derepression and requires nucleotide sequences present in the *trpL* region of the *trp* operon.

This second level of regulation of the *trp* operon is called **attenuation,** and the sequence within *trpL* that controls this phenomenon is called the **attenuator** (**FIGURE 19.12*a***). Attenuation occurs by control of the termination of transcription at a site near the end of the mRNA leader sequence. This "premature" termination of *trp* operon transcription occurs only in the presence of tryptophan-charged $tRNA^{Trp}$. When this premature termination or attenuation occurs, a truncated (140 nucleotides) *trp* transcript is produced.

The attenuator region has a nucleotide-pair sequence essentially identical to the *transcription-termination signals* found at the ends of most bacterial operons. These termination signals contain a G:C-rich palindrome followed by several A:T base pairs. Transcription of these termination signals yields a nascent RNA with the potential to form a hydrogen-bonded hairpin structure followed by several uracils. When a nascent transcript forms this hairpin structure, it causes a conformational change in the associated RNA polymerase, resulting in termination of transcription within the following, more weakly hydrogen-bonded (A:U)*n* region of DNA-RNA base-pairing.

The nucleotide sequence of the attenuator therefore explains its ability to terminate *trp* operon transcription prematurely. But how can this be regulated by the presence or absence of tryptophan?

First, recall that transcription and translation are coupled in prokaryotes; that is, ribosomes begin translating mRNAs while they are still being synthesized. Thus, events that occur during translation may also affect transcription.

Second, note that the 162-nucleotide-long leader sequence of the *trp* operon mRNA contains sequences that can base-pair to form alternate stem-and-loop or hairpin structures (**FIGURE 19.12*b***). The four leader regions that can base-pair to form these structures are: (1) nucleotides 60–68, (2) nucleotides 75–83, (3) nucleotides 110–121, and (4) nucleotides 126–134. The actual lengths of these regions involved in base-pairing vary depending on which regions pair. The nucleotide sequences of these four regions are such that region 1 can base-pair with region 2, region 2 can pair with region 3, and region 3 can pair with region 4. Region 2 can base-pair with either region 1 or region 3, but, obviously, it can pair with only one of these regions at any given time. Thus, there are two possible secondary structures for the *trp* leader sequence: (1)

Regulatory components of the *trpL* region

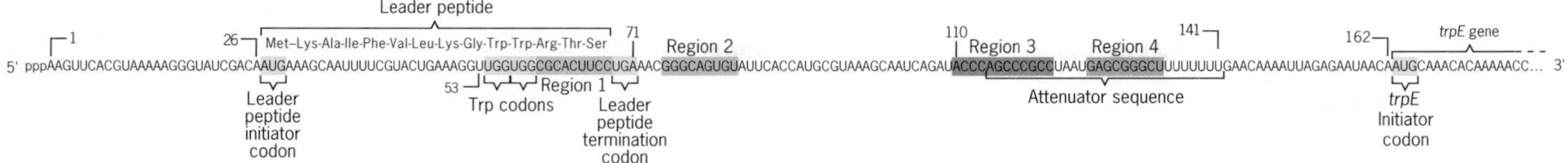

(a)

Alternate secondary structures formed by the *trpL* transcript

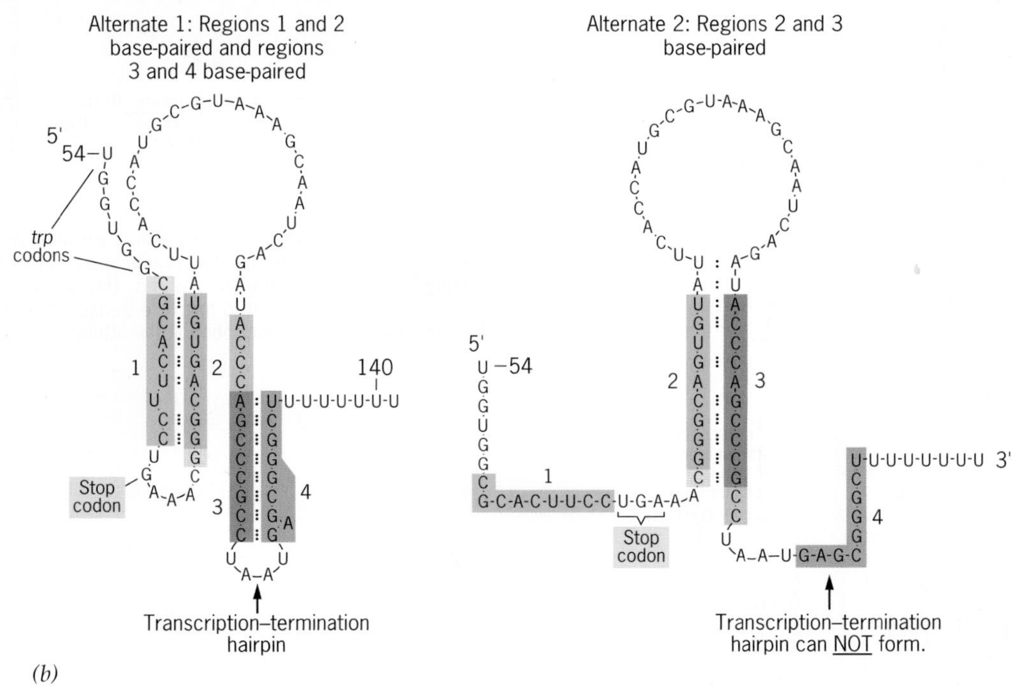

(b)

Figure 19.12 ▶ Sequences in the leader region of the *trp* mRNA responsible for attenuation. (*a*) The *trpL* sequence, highlighting the sequence encoding the leader peptide, the two tandem tryptophan codons responsible for the control of attenuation by tryptophan, and the four regions (shaded) that form the stem-and-loop or hairpin structures shown in (*b*). (*b*) Alternate secondary structures formed by the *trpL* mRNA—either (1) region 1 will pair with region 2 and region 3 with region 4, forming a transcription-termination hairpin, or (2) region 2 will base-pair with region 3, preventing region 3 from pairing with region 4. The concentration of tryptophan in the cell determines which of these structures will form during the transcription of the *trp* operon.

region 1 paired with region 2 and region 3 paired with region 4 or (2) region 2 paired with region 3, leaving regions 1 and 4 unpaired. The pairing of regions 3 and 4 produces the previously mentioned transcription-termination hairpin. If region 3 is base-paired with region 2, it cannot pair with region 4, and the transcription-termination hairpin cannot form. As you have probably guessed by now, the presence or absence of tryptophan determines which of these alternative structures will form.

Third, note that the leader sequence contains an AUG translation-initiation codon, followed by 13 codons for amino acids, followed in turn by a UGA translation-termination codon (**FIGURE 19.12*a***). In addition, the *trp* leader sequence contains an efficient ribosome-binding site located in the appropriate position for the initiation of translation at the leader AUG initiation codon. All the available evidence indicates that a 14-amino-acid "leader peptide" is synthesized as diagrammed in **FIGURE 19.12*a***.

The normal *trp* operon transcription-termination hairpin is shown in **FIGURE 19.13*a***, and the proposed mechanism of attenuation of *trp* operon transcription is diagrammed in **FIGURE 19.13*b*** and ***c***. The leader peptide contains two contiguous tryptophan residues. The two Trp codons are positioned such that in low concentrations of tryptophan (and thus low concentrations of Trp-tRNA^{Trp}), the ribosome will stall before it encounters the base-paired structure formed by leader regions

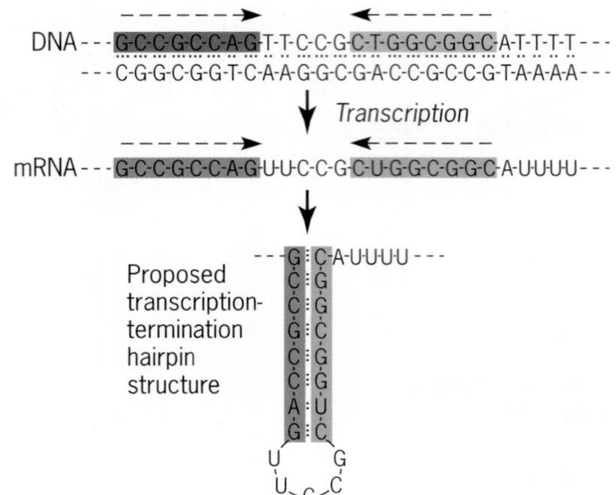

(*a*) Structure of *trp* operon transcription-termination sequence *t* and formation of the transcription-termination hairpin.

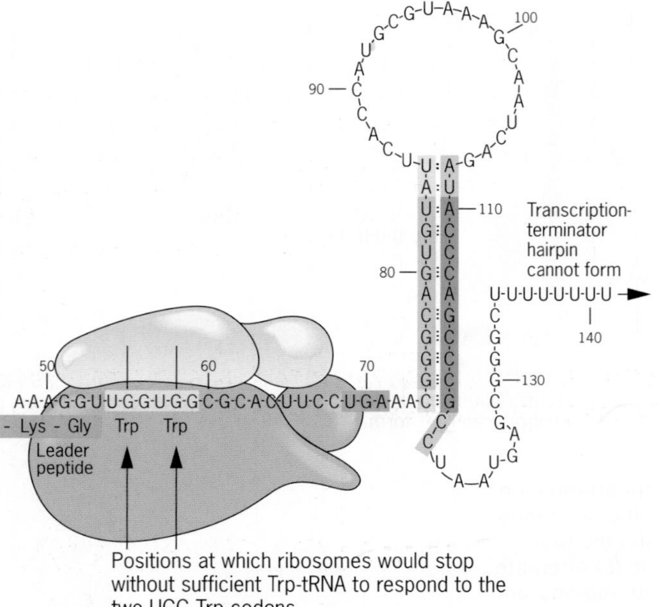

(*b*) With low levels of tryptophan, translation of the leader sequence stalls at one of the Trp codons. This stalling allows leader regions 2 and 3 to pair, which prevents region 3 from pairing with region 4 to form the transcription-termination hairpin. Thus transcription proceeds through the entire *trp* operon.

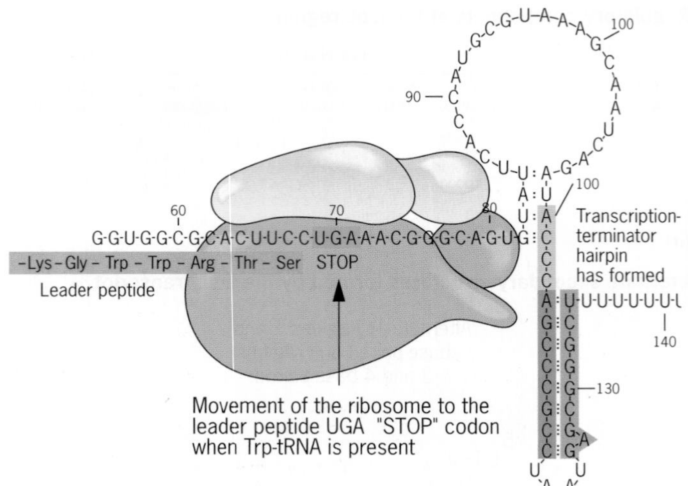

(*c*) In the presence of sufficient tryptophan, translation proceeds past the Trp codons to the termination codon and disrupts the base-pairing between leader regions 2 and 3. This process leaves region 3 free to pair with region 4 to form the transcription-termination hairpin, which stops transcription at the attenuator sequence.

Figure 19.13 ▶ Control of the *trp* operon by attenuation. (*a*) The transcription-termination signal in *E. coli* contains a region of dyad symmetry (arrows) that results in mRNA sequences that can form hairpin structures. (*b*) In low concentrations of tryptophan, transcription proceeds past the attenuator sequence through the entire *trp* operon. (c) In the presence of sufficient tryptophan, transcription frequently terminates at the attenuator sequence.

2 and 3 (**FIGURE 19.13*b***). Because the pairing of regions 2 and 3 precludes the formation of the transcription-termination hairpin by the base-pairing of regions 3 and 4, transcription will continue past the attenuator into the *trpE* gene in the absence of tryptophan.

In the presence of sufficient tryptophan, the ribosome can translate past the Trp codons to the leader-peptide termination codon. In the process, it will disrupt the base-pairing between leader regions 2 and 3. This disruption leaves region 3 free to pair with region 4, forming the transcription-termination hairpin (**FIGURE 19.13*c***). Thus, in the presence of sufficient tryptophan, transcription frequently (about 90 percent of the time) terminates at the attenuator, reducing the amount of mRNA for the *trp* structural genes.

The transcription of the *trp* operon can be regulated over a range of almost 700-fold by the combined effects of repression (up to 70-fold) and attenuation (up to 10-fold).

Regulation of transcription by attenuation is not unique to the *trp* operon. Five other operons (*thr*, *ilv*, *leu*, *phe*, and *his*) are known to be regulated by attenuation. The *his* operon, which for many years was thought to be repressible, is now believed to be regulated entirely by attenuation. Although minor details vary from operon to operon, the main features of attenuation are the same for all six operons.

KEY POINTS

- The *E. coli trp* operon is a negative repressible system; transcription of the five structural genes in the *trp* operon is repressed in the presence of significant concentrations of tryptophan.
- Operons such as *trp* that encode enzymes involved in amino acid biosynthetic pathways often are controlled by a second regulatory mechanism called attenuation.
- Attenuation occurs by the premature termination of transcription at a site in the mRNA leader sequence (the sequence 5′ to the coding region) when tryptophan is prevalent in the environment in which the bacteria are growing.

Bacteriophage Lambda: Repression of Lambda Lytic Pathway Genes During Lysogeny

When temperate bacteriophages, such as phage λ, enter the lysogenic pathway, during which they covalently insert their chromosome into the chromosome of the host, their lytic genes must be kept repressed (turned off).

When a temperate bacteriophage such as lambda (λ) infects a bacterium, it can follow either of two pathways of development (see **FIGURE 8.5**). It can either (1) enter the lytic cycle, during which it reproduces and lyses the host cell, just like a virulent phage (Chapter 8; see **FIGURE 8.3**), or (2) enter the lysogenic pathway (Chapter 8; see **FIGURE 8.6**), during which its chromosome is inserted into the chromosome of the host and thereafter is replicated like any other segment of that chromosome. When integrated in the chromosome of the host cell, the phage chromosome is called a *prophage*. In a lysogenic bacterium, the genes of the prophage that encode products involved in the lytic pathway must not be expressed. The prophage genes specifying enzymes involved in the replication of phage DNA, structural proteins required for phage morphogenesis, and the

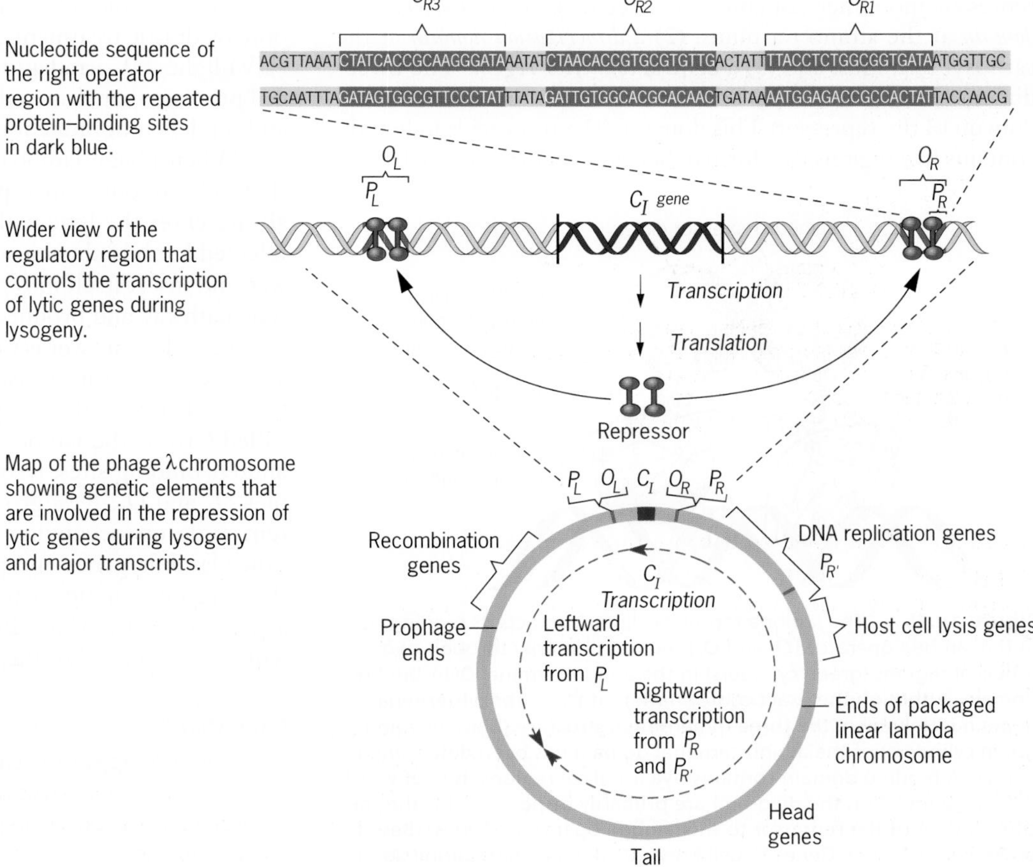

Figure 19.14 ▶ Repression of the lambda lytic genes in a lysogenic *E. coli* cell. Transcription of the lytic genes is repressed by the binding of the lambda repressor to two operator sequences (O_L and O_R), which regulate leftward and rightward transcription of the lambda chromosome (bottom). The lambda repressor is encoded by the C_I gene (center) and represses the synthesis of the major transcripts by binding to triplicate protein-binding sites in O_L and O_R. The arrows show the relative sizes and directions of synthesis of the major lambda transcripts.

lysozyme that catalyzes cell lysis must be kept turned off to maintain a stable lysogenic state.

The lytic pathway genes of the prophage are kept turned off in lysogenic cells by a simple repressor-operator-promoter circuit, much like the negative regulatory circuits of bacterial operons. Let's examine the mechanism by which this repression of lytic gene expression occurs. The C_I gene of phage λ encodes a repressor, a well-characterized protein with a weight of 27,000 daltons. The C_I repressor, as a dimer, binds to two operator regions that control transcription of the λ genes involved in lytic growth (**FIGURE 19.14**). These two operator regions, designated O_L (for transcription in the *l*eftward direction) and O_R (for transcription in the *r*ightward direction), overlap with promoter sequences at which RNA polymerase binds to initiate transcription of the genes controlling lytic development. With repressor bound to the two operators, RNA polymerase cannot bind to the two promoters and cannot initiate transcription. In this way, the lambda lytic genes continue to be repressed, allowing the dormant prophage to be transmitted from parental host cells to progeny cells generation after generation.

Each lambda operator contains three repressor binding sites, each 17 nucleotide pairs long, with similar but not identical nucleotide sequences. Each repressor binding site has partial twofold symmetry around the central base pair. This symmetry plays an important role in the interaction of repressor dimers with these operator binding sites (**FIGURE 19.15.**)

The lambda repressor has been studied in great detail. The repressor monomer contains three regions: (1) a *DNA-binding domain* at the amino terminus, (2) a *dimerization domain* at the carboxyl terminus, and (3) a central *connector region*. The three-dimensional structure is known only for the DNA-binding domain of the repressor. This domain, 92 amino acids in length, contains five regions that form α-helical structures; the α-helices are numbered 1 through 5, starting from the NH_3 terminus. Two of the α-helical regions, numbers 2 and 3, are largely responsible for the DNA-binding specificity of the repressor. When a repressor dimer makes contact with its DNA binding site (**FIGURE 19.15**), the two region 3 helices lie almost entirely within adjacent major grooves on one face of the λ operator DNA. Each monomer contacts one side (half-site) of the palindromic nucleotide-pair sequence in each repressor binding site. The two region 2 helices lie across the major grooves occupied by the region 3 helices.

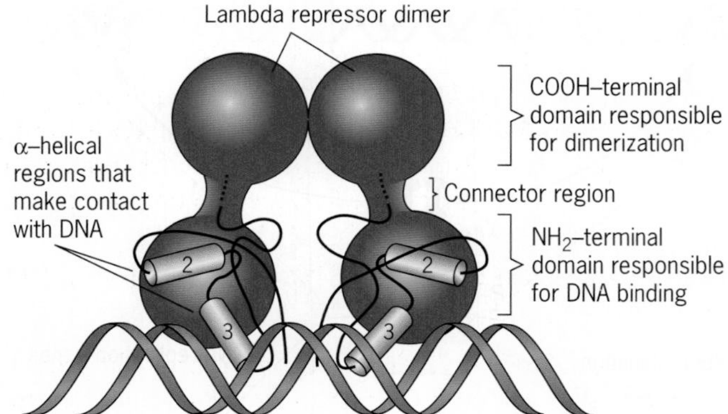

Figure 19.15 ▶ The lambda repressor dimer contacts its binding sites in the lambda operator (O_L and O_R) regions primarily through two α-helical regions (green cylinders) in the amino-terminal DNA-binding domain. Although the exact conformation of the carboxyl-terminal domain is unknown, the three-dimensional structure (black line and green cylinders) of the amino-terminal domain has been determined. This DNA-binding domain contains five α-helical regions, two of which (labeled 2 and 3 on the diagram) are primarily responsible for the specific binding of the repressor to the lambda operator regions. (Based on Benjamin Lewin, *Genes V,* Cell Press, 1994. Used with permission.)

Because the lambda repressor has the highest affinity for sites O_{L1} and O_{R1}, the first repressor dimers usually bind at these two sites. The λ repressor exhibits cooperative binding to (1) sites O_{L1} and O_{L2}, and (2) sites O_{R1} and O_{R2}. Thus, the presence of a dimer at site O_{L1} or O_{R1} increases the affinity of repressor for sites O_{L2} or O_{R2}, respectively. This cooperativity does not extend to O_{L3} or O_{R3}; at normal intracellular concentrations of repressor, only sites O_{L1}, O_{L2}, O_{R1}, and O_{R2} are occupied by repressor. When repressor dimers are bound to O_{R1}–O_{R2} and O_{L1}–O_{L2}, RNA polymerase cannot bind to P_R and P_L and therefore cannot initiate transcription. Thus, the λ genes encoding functions involved in lytic development are maintained in a repressed state.

Lambda lysogeny is quite stable. The lambda lytic pathway genes are tightly repressed in a lysogenic cell; thus, lytic-function gene products are not produced by transcription of prophage genes. In addition, spontaneous switches from lysogeny to lytic development are rare under normal growth conditions. Populations of lysogenic cells can be induced to enter the lytic pathway only by drastic treatments, such as by irradiation with ultraviolet (UV) light. UV irradiation of lambda lysogens activates a host cell protease that cleaves the connector region of the λ repressor and renders it nonfunctional.

When phage lambda injects its DNA into an *E. coli* cell, that cell contains no repressor. Why then does lambda not always enter the lytic pathway? What determines whether the injected lambda chromosome will (1) enter the lysogenic pathway and insert itself into the host chromosome, or (2) enter the lytic pathway and produce progeny phage at the expense of the host cell? The answer is that an intricate and fascinating genetic switch controls this decision. The decision hinges on a delicate balance between the λ repressor and another lambda protein called Cro. If the lambda repressor ends up occupying the O_L and O_R binding sites, lysogeny will occur. If Cro protein occupies O_L and O_R, lytic growth will occur. However, which protein ends up occupying these operators involves numerous other lambda genes and proteins. For a detailed discussion of this elegant genetic switch, see *A Genetic Switch: Gene Control and Phage* λ by Mark Ptashne, one of the researchers who worked out several of the details of this regulatory mechanism.

KEY POINT

- When a temperate bacteriophage, such as phage λ, enters the lysogenic state, the lytic genes of the prophage are kept turned off by a repressor–operator–promoter circuit similar to those of bacterial operons.

Temporal Sequences of Gene Expression During Phage Infection

When bacteriophage infect bacterial cells, lytic growth is governed by preprogrammed sequences of viral gene expression.

Regulation of gene expression during the life cycles of virulent bacteriophages is quite different from the reversible on-off switches characteristic of bacterial operons. In phage-infected bacteria, the viral genes are expressed in genetically preprogrammed sequences or cascades. Although different bacterial viruses exhibit variations of the specific mechanisms involved, a common picture emerges. One set of phage genes, usually called early genes, is expressed immediately after infection. The product(s) of one or more of the early genes is (are) responsible for turning on the expression of the next set of genes and turning off the expression of the early genes, and so on. Two to four sets of genes are usually involved, depending on the virus. In all cases studied so far, the regulation of sequential gene expression during phage infection occurs primarily at the level of transcription.

We will illustrate these temporal sequences of gene expression in phage-infected bacteria by examining the life cycle of bacteriophage SP01, a virus that infects *Bacillus subtilis*. Phage SP01 contains three sets of genes, called early, middle, and late genes, in reference to their time of expression during the phage reproductive cycle (**FIGURE 19.16**). The phage SP01 early genes are transcribed by the *B. subtilis* RNA polymerase. One of the early gene products is a polypeptide that binds to the host cell's RNA polymerase, changing its specificity so that the modified RNA polymerase transcribes the middle genes of SP01. Two of the products of the middle genes are, in turn, polypeptides that associate with the *B. subtilis* RNA polymerase, further changing its specificity so that it then transcribes the SP01 late genes.

KEY POINTS

- Preprogrammed temporal sequences of viral gene expression occur in bacteriophage-infected cells.
- The first viral genes expressed in an infected cell are transcribed by unmodified bacterial RNA polymerase.
- Subsequent sets of expressed viral genes are transcribed either by an RNA polymerase encoded by the phage genome or by bacterial RNA polymerase modified by the addition of viral protein(s).

Translational Control of Gene Expression

The regulation of gene expression is often fine-tuned by modulating either the frequency of initiation of translation or the rate of polypeptide chain elongation.

Although gene expression in prokaryotes is regulated predominantly at the level of transcription, fine-tuning often occurs at the level of translation. In prokaryotes, mRNA molecules are frequently multigenic, carrying the coding sequences of several genes. For example, the *E. coli lac* operon mRNA harbors nucleotide sequences encoding β-galactosidase, β-galactoside permease, and β-galactoside transacetylase. Thus, the three genes encoding these proteins must be turned on and turned off together at the transcription level because the genes are co-transcribed. Nevertheless, the three gene products are not synthesized in equal amounts. An *E. coli* cell that is growing on rich medium with lactose as the sole carbon source contains about 3000 molecules of β-galactosidase, 1500 molecules of β-galactoside permease, and 600 molecules of β-galactoside transacetylase. Clearly, the different molar quantities of these proteins per cell must be controlled posttranscriptionally.

Remember that transcription, translation, and mRNA degradation are coupled in prokaryotes; an mRNA molecule usually is involved in all three processes at any given time. Thus, gene products may be produced in different amounts from the same transcript by several mechanisms.

1. *Unequal efficiencies of translational initiation* are known to occur at the ATG start codons of different genes.
2. *Altered efficiencies of ribosome movement* through intergenic regions of a transcript are quite common. Decreased translation rates often result from hairpins or other forms of secondary structure that impede ribosome migration along the mRNA molecule.
3. *Differential rates of degradation* of specific regions of mRNA molecules also occur.

The synthesis of the ribosomal proteins of *E. coli* provides several well-documented examples of translational regulation of gene expression. *E. coli* cells that are growing rapidly under optimal conditions need more ribosomes for protein synthesis than those that are growing slowly under adverse conditions. Recall (Chapter 12) that the *E. coli* ribosome contains three RNA molecules and 52 polypeptides. The syntheses of these structural components must be coordinated to assure proper stoichiometry for ribosome assembly. Thus, regulatory mechanisms have evolved that assure that ribosomal RNAs and proteins are synthesized in the appropriate

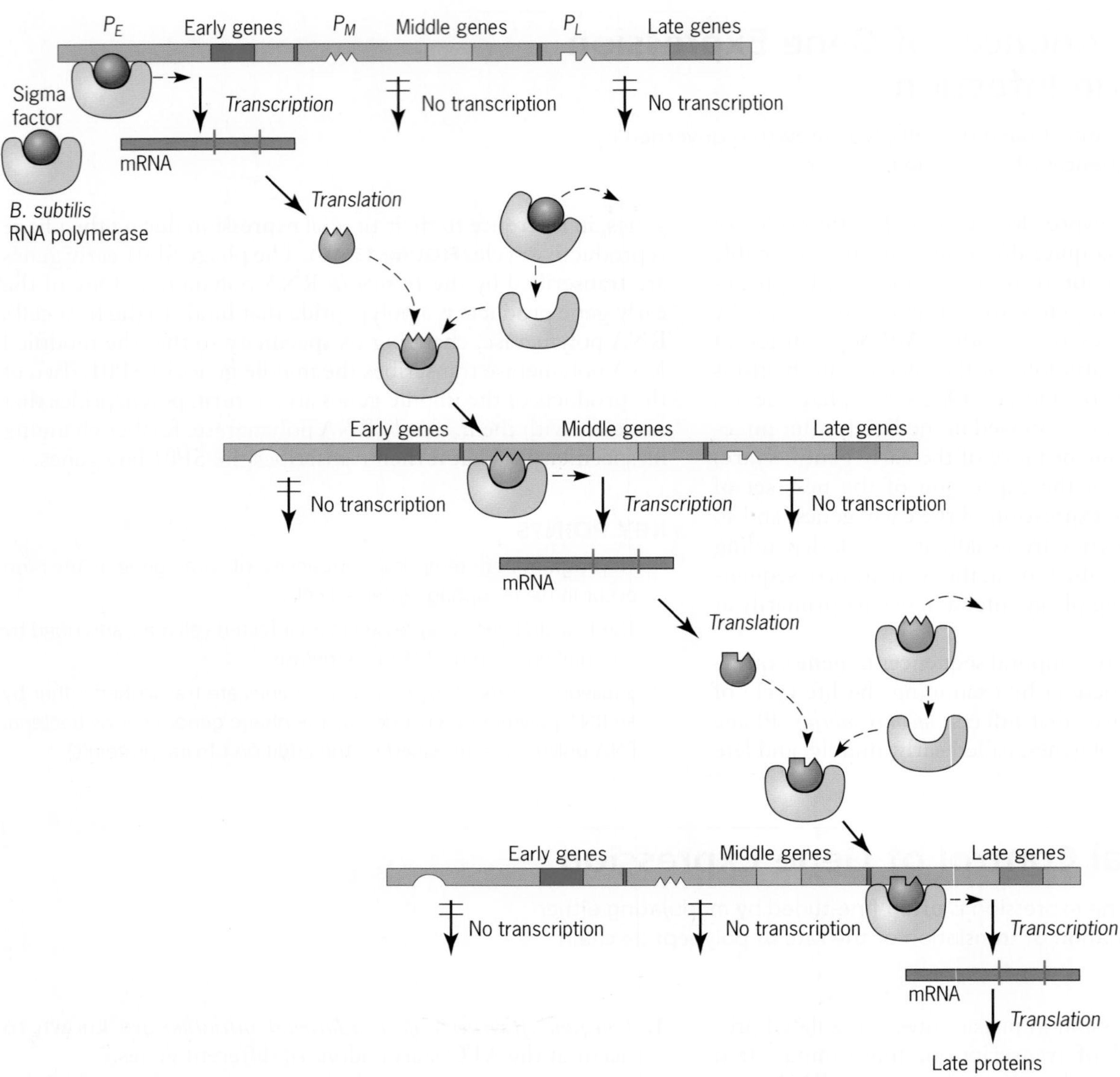

Figure 19.16 ▶ The *Bacillus subtilis* phage SP01 regulatory cascade. Early genes are transcribed by the RNA polymerase of the host cell (top). Middle and late genes are transcribed by RNA polymerases in which the host σ factor has been replaced by one or two SP01 proteins, respectively (middle and bottom). P_E, P_M, and P_L are the promoters for early, middle, and late transcripts, respectively.

amounts, and some of these regulatory mechanisms act at the level of translation.

Most of the *E. coli* genes encoding ribosomal proteins are located in clusters, and the genes in each cluster are co-transcribed. All but one of the ribosomal proteins are utilized in equimolar amounts during ribosome assembly. Moreover, the synthesis of the ribosomal proteins must be coordinated with the synthesis of the three ribosomal RNAs. This coordination between ribosomal protein synthesis and ribosomal RNA synthesis occurs at the level of translation.

The ribosomal protein S10 (for *s*mall subunit protein number *10*) gene cluster provides a good illustration of how this regulation works. The *S10* transcriptional unit contains 11 coordinately regulated genes, all encoding ribosomal proteins (**FIGURE 19.17*a***). In the case of the *S10* transcriptional unit and at least five other such units containing ribosomal protein genes, the product of one of the genes inhibits the translation of the multigenic transcript. Thus, the gene encoding the regulatory protein is itself one of the regulated genes, so that the regulatory gene is negatively self-regulated. In the *S10* transcriptional unit, the regulatory gene is *rplD* (for *r*ibosomal *p*rotein *l*arge subunit gene *D*), which encodes protein L4 (*l*arge subunit protein number *4*). When free rRNAs are present in the cell, nascent L4 protein binds to the RNA (**FIGURE 19.17*b***) and is assembled into ribosomes. In the absence of rRNA, the L4 protein binds to the 5′ end of the *S10* transcriptional unit mRNA and inhibits its

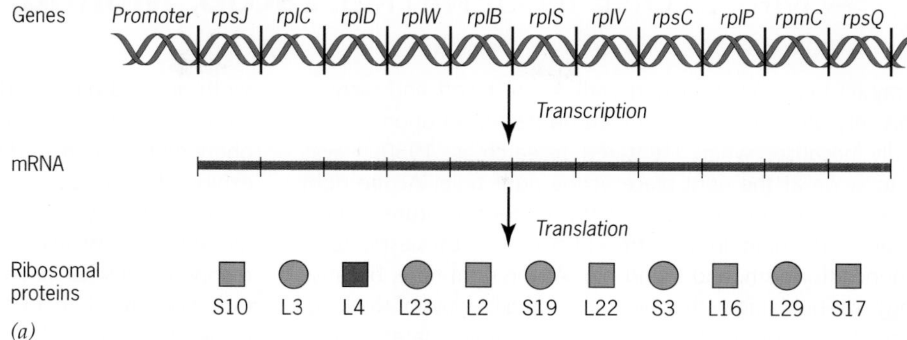

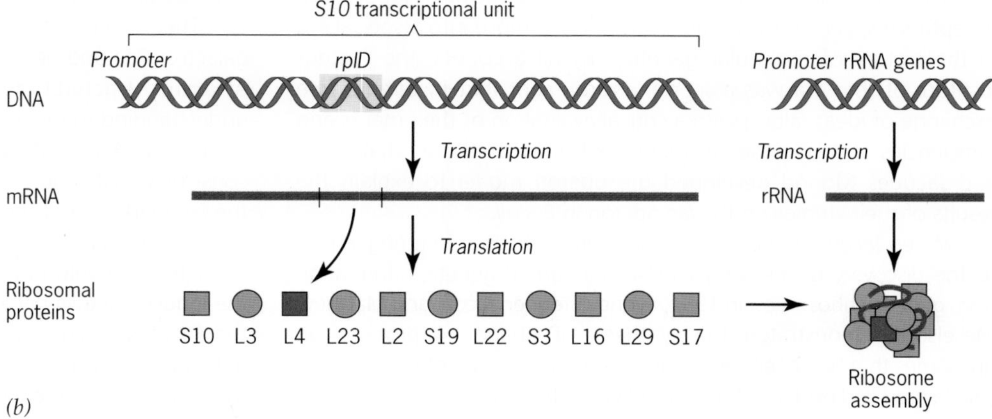

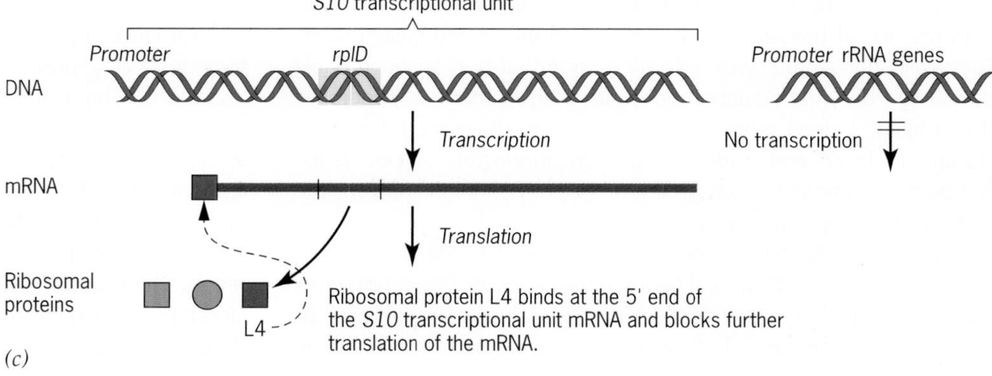

Figure 19.17 ► Organization (*a*) and translational regulation (*b* and *c*) of the *E. coli S10* transcriptional unit, which contains 11 genes encoding ribosomal proteins. Translation of the mRNA of the *S10* transcriptional unit is regulated by ribosomal protein L4, which binds to a nucleotide sequence near the 5′ end of the *S10* transcript. (*b*) In the presence of ribosomal RNA, the ribosomal proteins interact with the rRNA in the assembly of ribosomes. (*c*) In the absence of ribosomal RNA, ribosomal protein L4 binds near the 5′ end of the *S10* transcript and blocks its translation.

translation (**FIGURE 19.17*c***). These events prevent the synthesis of ribosomal proteins that cannot be used by the cell.

The inhibition of translation of an mRNA molecule by one of the products that it encodes is common in both prokaryotes and eukaryotes. This mechanism is referred to as **negative self-regulation** or **negative autogenous regulation.** When the gene product is a structural component of the cell or of some organelle within the cell, the autoregulation often is carried out by the free monomers present in the cell.

KEY POINTS

- Regulatory fine-tuning frequently occurs at the level of translation by modulation of the rate of polypeptide chain initiation or chain elongation.
- Sometimes gene expression is regulated by the differential degradation of specific regions of polygenic mRNAs.

▶ A MILESTONE IN GENETICS: **Jacob, Monod, and the Operon Model**

"If I find myself here today, sharing with André Lwoff and Jacques Monod this very great honor which is being bestowed upon us, it is undoubtedly because, when I entered research in 1950, I was fortunate to arrive at the right place at the right time. At the right place, because there, in the attics of the Pasteur Institute, a new discipline was emerging in an atmosphere of enthusiasm, lucid criticism, nonconformism, and friendship. At the right time, because then biology was bubbling with activity, changing its ways of thinking, discovering in microorganisms a new and simple material, and drawing closer to physics and chemistry. A rare moment, in which ignorance could become a virtue."[1]

This introductory statement in François Jacob's Nobel Prize acceptance speech provides an informative glimpse into a special era in the history of molecular genetics. By all accounts, the Pasteur Institute of that era was a special environment that fostered a free exchange of ideas, along with a critical evaluation of their merits and deficiencies. It was in the "attics of the Pasteur Institute" that Jacob and Jacques Monod developed the operon model to explain the results of their studies on the *lac* operon in *E. coli*.[2]

When Jacob mentioned "the right time," he was referring in part to the discovery of messenger RNA that carried genetic information from genes to ribosomes. In 1961, Sydney Brenner, Jacob, and Matthew Meselson demonstrated the synthesis of phage T4 proteins on ribosomes that had been present in *E. coli* cells prior to infection.[3] Their results provided evidence for a short-lived RNA molecule that mediated the synthesis of proteins. Indeed, the operon model would not have made sense in the absence of this unstable RNA intermediary.

When Jacob and Monod began their study of lactose utilization in *E. coli,* they observed that β-galactosidase was synthesized only in the presence of the substrate lactose or closely related galactosides. Their colleague André Lwoff informed them that they were probably observing "enzymatic adaptation" and referred them to papers on the subject. Indeed, this phenomenon was discovered by Émile Duclaux in 1899[4] and studied extensively during the next 60 years. Although the changes in enzyme levels were well characterized, the underlying mechanism was unknown.

Jacob and Monod isolated mutant strains of *E. coli* with defects in the two important structural genes at the *lac* locus. These mutations were designated z^- (loss of β-galactosidase) and y^- (loss of β-galactoside permease).[5] They also isolated mutants in which the synthesis of β-galactosidase and permease occurred continuously—regardless of whether lactose was present. They called this phenomenon *constitutive synthesis* and referred to strains that exhibited it as *constitutive mutants*. Wild-type strains exhibited inducible (i^+) synthesis of the *lac* enzymes; thus, the constitutive mutant strains were designated i^-. All the various types of *lac* mutants mapped to a small region of the *E. coli* chromosome. The z^- and y^- mutations resulted in the loss of β-galactosidase and permease, respectively, and had no effect on the inducibility of the enzymes. The z^- mutants had normal permease activity, and the y^- mutants had normal β-galactosidase activity. Both the z^- and y^- mutations appeared to be defects in the structural genes encoding these enzymes.

The i^- mutants exhibited constitutive synthesis of both β-galactosidase and permease. The i^- mutations altered the expression of both structural genes. One of the first breakthroughs in understanding these constitutive mutants resulted from collaborative work with Arthur Pardee, a postdoctoral fellow at the time. He constructed Hfr and F^- strains that carried various combinations of the *lac* markers and then performed mating experiments with these strains. One mating was between a z^+i^+ Hfr strain and a z^-i^- F^- strain. The Hfr cells could not synthesize β-galactosidase in the absence of the inducer lactose because some type of repressor was present, and the F^- cells could not synthesize this enzyme because they had a defect in the structural gene. However, during mating, the z^+ genes from the Hfr cells were expressed immediately in the transient partial diploids formed by z^+ gene transfer to the F^- cells (**FIGURE 1**).[6] Because other experiments had shown that i^+ is dominant to i^-, this result demonstrated that the i^+ vs. i^- phenotype depends on the state of the cytoplasm.

Although the transient partial diploids yielded informative results, their instability prevented a thorough analysis of the various *lac* mutations. The instability problem was soon eliminated by the discovery of F factors that carried segments of the *E. coli* chromosome (F′ factors; see Chapter 8).[7] Jacob and Monod used F′ factors carrying *lac* genes to produce stable partial diploids that contained *lac* mutations in many different combinations. One example is discussed here.

In addition to the i^- constitutive mutations mentioned above, Jacob and Monod identified constitutive mutations that mapped closer to the *z* gene than the i^- mutations. Unlike the i^- mutants, these constitutive mutants—designated o^c mutants for operator constitutive—were dominant to their wild-type (inducible) allele; that is, the *lac* structural genes were expressed constitutively in $o^c\ z^+y^+/$

[1]Jacob, F. 1965. Genetics of the bacterial cell. In *Noble Lectures in Molecular Biology, 1933–1975,* pp. 219–244, Elsevier, New York.

[2]Jacob, F., and J. Monod. 1961. Genetic regulatory mechanisms in the synthesis of proteins. *J. Mol. Biol.* 3:318–356.

[3]Brenner, S., F. Jacob, and M. Meselson. 1961. An unstable intermediate carrying information from genes to ribosomes for protein synthesis. *Nature* 190:576–580.

[4]Duclaux, E. 1899. *Traité de Microbiologie*. Masson et Cie, Paris.

[5]Because Jacob and Monod's original model is reproduced in Milestone Figure 2, we will use their lowercase gene symbols here rather than the currently used uppercase symbols (see text).

[6]Pardee, A. B., F. Jacob, and J. Monod. 1959. The genetic control and cytoplasmic expression of "inducibility" in the synthesis of β-galactosidase by *E. coli*. *J. Mol. Biol.* 1:165–178.

[7]Jacob, F., and E. A. Adelberg. 1959. Transfert de caractéres génétiques par incorporation au facteur sexuel d'*Escherichia coli*. *Comptes Rendus des Séances de L'Académie des Sciences* 249:189–191. English translation published in *Papers on Bacterial Genetics* (E. A. Adelberg, ed.), 1960. Little, Brown, Boston.

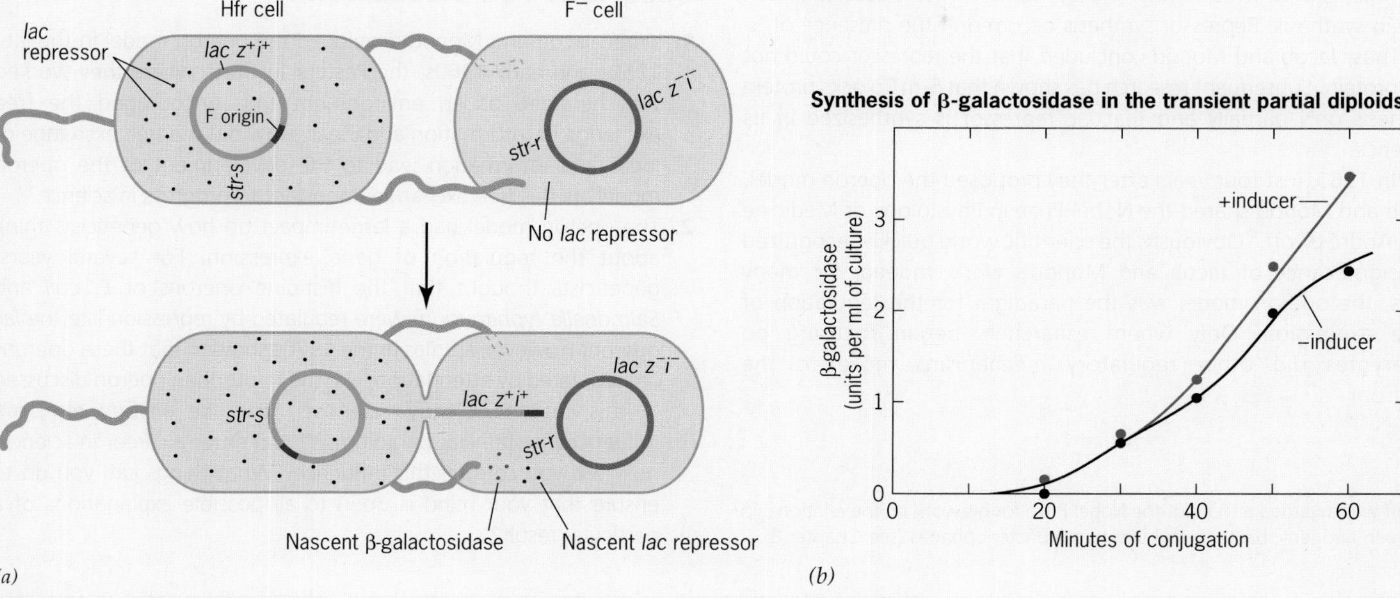

Figure 1 ▶ Synthesis of β-galactosidase in the transient partial diploids produced in matings between *lac* z^+i^+ Hfr cells and z^-i^- F^- cells. (*a*) Diagram of the mating. (*b*) Synthesis of β-galactosidase in the partial diploids in the presence and absence of inducer.

$o^+z^+y^+$ partial diploids. Actually, o^c alleles are *cis*-dominant; they only affect the expression of genes located *cis*—on the same chromosome—to themselves. Thus, a partial diploid of the type

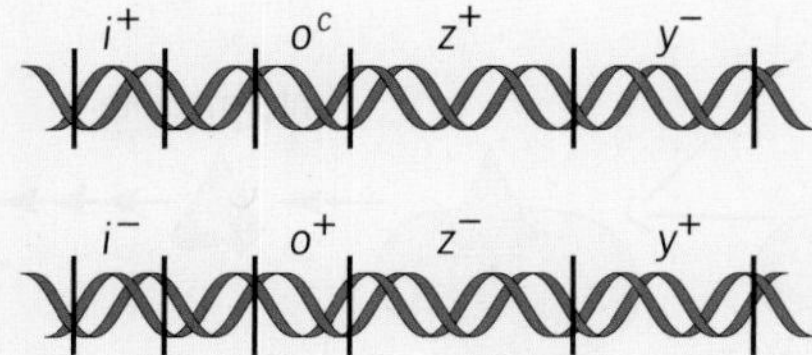

synthesizes β-galactosidase (*z* gene product) constitutively and β-galactoside permease (*y* gene product) inducibly. This result indicated to Jacob and Monod that the i^+ regulator gene acts both *cis* and *trans* and encodes a diffusible product. Even though the i^+ gene is on the top chromosome in the diagram, its product—a repressor—diffuses throughout the cell, binds to o^+ on the bottom chromosome, and prevents the transcription of the genes on that chromosome when the inducer lactose is absent. In contrast, the operator does **not** make a diffusible product; it is the binding site for the repressor. Thus, o^c and o^+ only affect transcription of structural genes located *cis* to themselves. In the partial dipoid above, o^c controls *lac* z^+ expression and o^+ controls *lac* y^+ expression. Thus, Jacob and Monod's operon model distinguished between *cis*- and *trans*-acting regulatory elements.

The clarity and accuracy of Jacob and Monod's operon model was quite amazing. Their original model is reproduced in **FIGURE 2**. At the time, they did not know whether repression occurred at the level of transcription (Model I, shown in **FIGURE 2**) or at the level of translation (Model II, not shown). They also concluded incorrectly that the repressor was RNA, not protein. This conclusion was based on

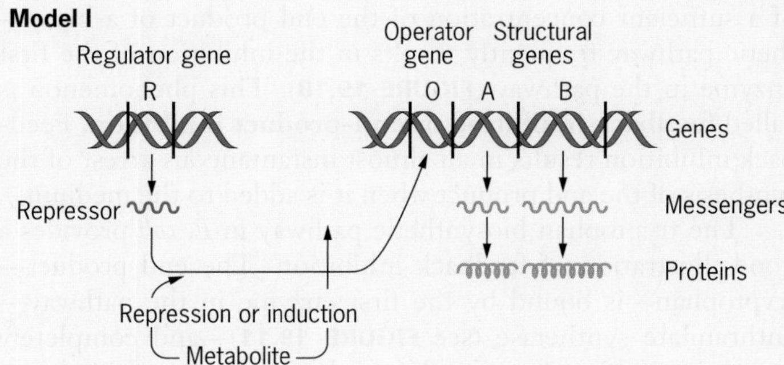

Figure 2 ▶ Diagram of the operon model as it appeared in Jacob and Monod's 1961 paper. In Model II (not shown), the repressor interacted with the mRNA and prevented its translation. In their paper, Jacob and Monod referred to the operator as a gene, which we would not do today.

A MILESTONE IN GENETICS: **Jacob, Monod, and the Operon Model *(continued...)***

experiments in which 5-methyl tryptophan (5-mT) was used to inhibit protein synthesis. Repressor synthesis occurred in the presence of 5-mT. Thus, Jacob and Monod concluded that the repressor could not be a protein. Subsequent research has shown that 5-mT blocks protein synthesis only partially and that *lac* repressor is synthesized in its presence.

In 1965, just four years after they proposed the operon model, Jacob and Monod shared the Nobel Prize in Physiology or Medicine with André Lwoff.[8] Obviously, the scientific world quickly recognized the significance of Jacob and Monod's work. Indeed, for many years, the operon model was the paradigm for the regulation of gene expression. Only when researchers began focusing on eukaryotes did other regulatory mechanisms move to the forefront.

[8]Lwoff was awarded a share of the Nobel Prize for his work on the relationship between lysogenic bacteria and temperate bacteriophages (see Chapter 8).

QUESTIONS FOR DISCUSSION

1. When Jacob and Monod developed the operon model in the late 1950s and early 1960s, the Pasteur Institute where they worked was heralded as an environment that encouraged the free exchange of information and ideas. How did the free exchange of ideas and information lead to the development of the operon model? Is this free exchange important to progress in science?
2. The operon model had a large impact on how geneticists think about the regulation of gene expression. For several years, geneticists thought that the histidine operons of *E. coli* and *Salmonella typhimurium* were regulated by repression like the *lac* operon. However, studies in the 1970s showed that these operons are regulated by attenuation, like the tryptophan operon discussed in this chapter. Clearly, scientists, just like anyone else, are influenced by familiar paradigms. If you enter a career in science, how will you counter this influence? What things can you do to ensure that your mind is open to all possible explanations of a particular result?

Post-Translational Regulatory Mechanisms

Feedback inhibition occurs when the product of a biosynthetic pathway inhibits the activity of the first enzyme in the pathway, rapidly shutting off the synthesis of the product.

Earlier in this chapter, we discussed the mechanism by which the transcription of bacterial genes encoding enzymes in a biosynthetic pathway is repressed when the product of the pathway is present in the medium in which the cells are growing. A second, and more rapid, regulatory fine-tuning of metabolism often occurs at the level of enzyme activity. The presence of a sufficient concentration of the end product of a biosynthetic pathway frequently results in the inhibition of the first enzyme in the pathway (**FIGURE 19.18**). This phenomenon is called **feedback inhibition** or **end-product inhibition.** Feedback inhibition results in an almost instantaneous arrest of the synthesis of the end product when it is added to the medium.

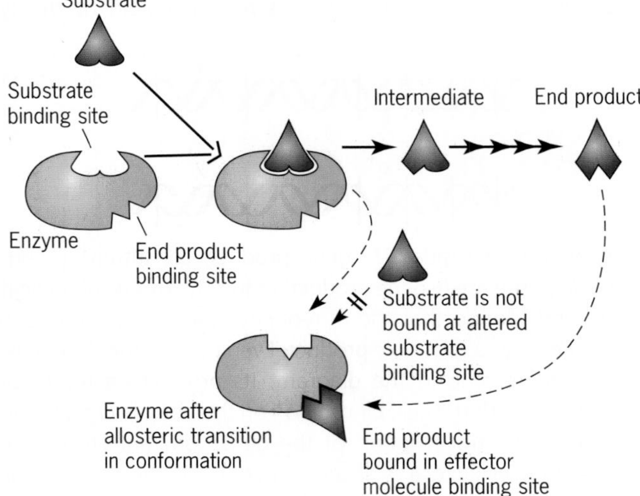

Figure 19.18 ▶ Feedback inhibition of gene-product activity. The end product of a biosynthetic pathway often binds to and arrests the activity of the first enzyme in the pathway, quickly blocking the synthesis of the end product.

The tryptophan biosynthetic pathway in *E. coli* provides a good illustration of feedback inhibition. The end product—tryptophan—is bound by the first enzyme in the pathway—anthranilate synthetase (see **FIGURE 19.11**)—and completely arrests its activity, stopping the synthesis of tryptophan almost immediately.

Feedback inhibition-sensitive enzymes contain an end-product binding site (or sites) in addition to the substrate binding site (or sites). In the case of multimeric enzymes, the *end-product* or *regulatory binding site* often is on a subunit (polypeptide) different from that of the substrate site. Upon binding the end product, such enzymes undergo allosteric transitions that reduce their affinity for their substrates. Proteins that undergo such conformational changes are referred to as allosteric proteins. Many, perhaps most, enzymes undergo allosteric transitions of some kind.

Allosteric transitions also appear to be responsible for enzyme activation, which often occurs when an enzyme binds one or more of its substrates or some other small molecule. Some enzymes exhibit a broad spectrum of activation and inhibition by many different effector molecules. An example is the enzyme glutamine synthetase, which catalyzes the final step in the biosynthesis of the amino acid glutamine. Glutamine synthetase is a complex multimeric enzyme in both prokaryotes and eukaryotes. The glutamine synthetase of *E. coli* has been shown to respond, either by activation or inhibition, to 16 different metabolites, presumably through allosteric transitions.

KEY POINTS

- Feedback inhibition occurs when the product of a biosynthetic pathway inhibits the activity of the first enzyme in the pathway, rapidly arresting the biosynthesis of the product.
- Enzyme activation occurs when a substrate or other effector molecule enhances the activity of an enzyme, increasing the rate of synthesis of the product of the biosynthetic pathway.

► Basic Exercises

ILLUSTRATE BASIC GENETIC ANALYSIS

1. How can positive and negative regulatory mechanisms be distinguished?

Answer: Mutations in regulator genes that yield nonfunctional products will have very different effects in positive and negative control systems. In positive control circuits, such mutations will make it impossible to turn on the expression of the regulated genes, whereas in negative control circuits, these mutations will make it impossible to turn off the expression of the regulated genes.

2. How can inducible and repressible operons be distinguished?

Answer: In the absence of the effector molecule, inducible operons will be turned off, whereas repressible operons will be turned on.

3. How can *cis*- and *trans*-acting regulatory elements be distinguished?

Answer: They can be distinguished by constructing partial diploids in which the regulatory elements are positioned (1) *cis* to the regulated genes and (2) *trans* to the regulated genes. A *cis*-acting element will only influence the expression of the genes when present in the *cis* configuration, whereas a *trans*-acting element will exert its effect in either the *cis* or *trans* configuration (compare **FIGURES 19.7** and **19.8**).

4. What is attenuation, and how does it work?

Answer: Attenuation is a mechanism for regulating gene expression by the premature termination of transcription in the leader region of a transcript. In the case of the tryptophan (*trp*) operon of *E. coli*, for example, the presence or absence of the end product, tryptophan, determines whether or not attenuation occurs. The leader region of the mRNA has sequences that can base-pair to form alternative hairpin structures, one of which is a typical transcription-termination signal. Whether or not this hairpin forms depends on the translation of a leader peptide containing two tryptophan residues. When low levels of tryptophan are present, translation stops at the Trp codons, which prevents the formation of the transcription-termination hairpin (see **FIGURE 19.13*b***). When sufficient tryptophan is present, translation proceeds past the Trp codons to the translation-termination codon, disrupting the first hairpin. This, in turn, allows the transcription-termination hairpin to form and attenuation (termination of transcription at the attenuator) to occur (see **FIGURE 19.13*c***). Attenuation decreases the synthesis of the tryptophan biosynthetic enzymes tenfold. Attenuation is possible in prokaryotes because transcription and translation are coupled, so events occurring during translation can affect transcription.

5. How are the lytic genes of the λ prophage kept turned off in lysogenic cells?

Answer: Transcription of the λ genes is controlled by two operator-promoter sites: P_LO_L and P_RO_R (see **FIGURE 19.14**). The C_1 gene of λ encodes a repressor that binds to O_L and O_R. When the λ repressor is bound to O_L and O_R, RNA polymerase cannot bind to P_L and P_R and initiate transcription (see **FIGURES 19.14** and **19.15**). Thus, the lytic genes on the λ prophage are kept in a repressed state by a repressor-promoter-operator circuit similar to the ones that regulate transcription in operons of bacteria.

► Testing Your Knowledge

INTEGRATE DIFFERENT CONCEPTS AND TECHNIQUES

1. The operon model for the regulation of enzyme synthesis concerned in lactose utilization by *E. coli* includes a regulator gene (*I*), an operator region (*O*), a structural gene (*Z*) for the enzyme β-galactosidase, and another structural gene (*Y*) for β-galactoside permease. β-Galactoside permease transports lactose into the bacterium, where β-galactosidase cleaves it into galactose and glucose.

Mutations in the *lac* operon have the following effects: Z^- and Y^- mutant strains are unable to make functional β-galactosidase and β-galactoside permease, respectively, whereas I^- and O^c mutant strains synthesize the *lac* operon gene products constitutively. The following figure shows a partially diploid strain of *E. coli* that carries two copies of the *lac* operon. On the diagram, fill in a genotype that will result in the constitutive synthesis of β-galactosidase and the inducible synthesis of β-galactoside permease by this partial diploid.

I O Z Y

I O Z Y

Answer: Several different genotypes will produce β-galactosidase constitutively and β-galactoside permease inducibly. They must meet two key requirements: (1) the cell must contain at least one copy of the I^+ gene, which encodes the repressor, and (2) the Z^+ gene and an O^c mutation must be on the same chromosome because the operator acts only in *cis*; that is, it only affects the expression of genes on the same chromosome. In contrast, the cell can be either homozygous or heterozygous for the I^+ gene, and, if heterozygous, I^+ may be on either chromosome because I^+ is dominant to I^- and I^+ acts in both the *cis* and *trans* arrangement. One possible genotype is given in the following diagram. How many other genotypes can you devise that will synthesize β-galactosidase constitutively and β-galactoside permease inducibly?

I^+ O^c Z^+ Y^-

I^- O^+ Z^- Y^+

2. Wild-type *E. coli* cells have been growing exponentially in culture medium containing very low concentrations of tryptophan for 20 minutes when someone adds a large amount of tryptophan to the culture medium. What physiological changes will occur in these cells after the addition of tryptophan?

Answer: (a) The first thing that will happen is that tryptophan will be bound by the first enzyme—anthranilate synthetase—in the tryptophan biosynthetic pathway, inhibiting the activity of the enzyme and arresting the synthesis of tryptophan almost immediately. This regulatory mechanism is called feedback inhibition (see **FIGURE 19.18**). (b) The second thing that will happen is that the high concentration of this amino acid will decrease the rates of synthesis of the tryptophan biosynthetic enzymes by the premature termination—attenuation—of transcription of the genes in the tryptophan operon (see **FIGURES 19.12** and **19.13**). (c) The third thing that will happen is that the high concentration of tryptophan will lead to the repression of transcription of the *trp* operon, further decreasing the rates of synthesis of the tryptophan biosynthetic enzymes (see **FIGURE 19.4c**). Working in concert, feedback inhibition, attenuation, and repression/derepression quickly and rather precisely adjust the rates of synthesis of metabolites such as tryptophan in bacteria in response to changes in environmental conditions.

► Questions and Problems

ENHANCE UNDERSTANDING AND DEVELOP ANALYTICAL SKILLS

19.1 In the lactose operon of *E. coli*, what is the function of each of the following genes or sites: (a) regulator, (b) operator, (c) promoter, (d) structural gene *Z*, and (e) structural gene *Y*?

19.2 What would be the result of inactivation by mutation of the following genes or sites in the *E. coli* lactose operon: (a) regulator, (b) operator, (c) promoter, (d) structural gene *Z*, and (e) structural gene *Y*?

19.3 How can inducible and repressible enzymes of microorganisms be distinguished?

19.4 Distinguish between (a) repression and (b) feedback inhibition caused by the end product of a biosynthetic pathway. How do these two regulatory phenomena complement each other to provide for the efficient regulation of metabolism?

19.5 Write the partial diploid genotype for a strain that will (a) produce β-galactosidase constitutively and permease inducibly and (b) produce β-galactosidase constitutively but not permease either constitutively or inducibly, even though a Y^+ gene is known to be present.

19.6 GO Groups of alleles associated with the lactose operon are as follows (in order of dominance for each allelic series): repressor I^s (superrepressor), I^+ (inducible), and I^- (constitutive); operator, O^c (constitutive, cis-dominant), and 0^+ (inducible, cis-dominant); structural Z^+ and Y^+. For each of the following partial diploids indicate whether enzyme synthesis is constitutive or inducible:

(a) $I^+O^+Z^+Y^+/I^+O^+Z^+Y^+$,
(b) $I^+O^+Z^+Y^+/I^+O^cZ^+Y^+$,
(c) $I^+O^cZ^+Y^+/I^+O^cZ^+Y^+$,
(d) $I^+O^+Z^+Y^+/I^-O^+Z^+Y^+$,
(e) $I^-O^+Z^+Y^+/I^-O^+Z^+Y^+$. Why?

19.7 Groups of alleles associated with the lactose operon are as follows (in order of dominance for each allelic series): repressor, I^s (superrepressor), I^+ (inducible), and I^- (constitutive); operator, O^c (constitutive, *cis*-dominant) and O^+ (inducible, *cis*-dominant); structural, Z^+ and Y^+. (*a*) Which of the following genotypes will produce β-galactosidase and β-galactoside permease if lactose is present: (1) $I^+O^+Z^+Y^+$, (2) $I^-O^cZ^+Y^+$, (3) $I^sO^cZ^+Y^+$, (4) $I^sO^+Z^+Y^+$, and (5) $I^-O^+Z^+Y^+$? (b) Which of the above genotypes will produce β-galactosidase and β-galactoside permease if lactose is absent? Why?

19.8 GO Assume that you have discovered a new strain of *E. coli* that has a mutation in the *lac* operator region that causes the wild-type repressor protein to bind irreversibly to the operator. You have named this operator mutant O^{sb} for "*s*uper*b*inding" operator. (a) What phenotype would a partial diploid of genotype $I^+O^{sb}Z^-Y^+/I^+O^+Z^+Y^-$ have with respect to the synthesis of the enzymes β-galactosidase and β-galactoside permease? (b) Does your new O^{sb} mutation exhibit *cis* or *trans* dominance in its effects on the regulation of the *lac* operon?

19.9 Of what biological significance is the phenomenon of catabolite repression?

19.10 Would it be possible to isolate *E. coli* mutants in which the transcription of the *lac* operon is not sensitive to catabolite repression? If so, in what genes might the mutations be located?

19.11 Is the CAP–cAMP effect on the transcription of the *lac* operon an example of positive or negative regulation? Why?

19.12 As a genetics historian, you are repeating some of the classic experiments conducted by Jacob and Monod with the lactose operon in *E. coli*. You use an F′ plasmid to construct several *E. coli* strains that are partially diploid for the *lac* operon. You construct strains with the following genotypes: (1) $I^+O^cZ^+Y^-/I^+O^+Z^-Y^+$, (2) $I^+O^cZ^-Y^+/I^+O^+Z^+Y^-$, (3) $I^-O^+Z^+Y^+/I^+O^+Z^-Y^-$, (4) $I^sO^+Z^-Y^-/I^+O^+Z^+Y^+$, and (5) $I^+O^cZ^+Y^+/I^sO^+Z^-Y^+$. (a) Which of these strains will produce functional β-galactosidase in both the presence and absence of lactose? (b) Which of these strains will exhibit constitutive synthesis of functional β-galactoside permease? (c) Which of these strains will express both gene *Z* and gene *Y* constitutively and will produce functional products (β-galactosidase and β-galactoside permease) of both genes? (d) Which of these strains will show *cis* dominance of *lac* operon regulatory elements? (e) Which of these strains will exhibit *trans* dominance of *lac* operon regulatory elements?

19.13 Using examples, distinguish between negative regulatory mechanisms and positive regulatory mechanisms.

19.14 How might the concentration of glucose in the medium in which an *E. coli* cell is growing regulate the intracellular level of cyclic AMP?

19.15 Constitutive mutations produce elevated enzyme levels at all times; they may be of two types: O^c or I^-. Assume that all other DNA present is wild-type. Outline how the two constitutive mutants can be distinguished with respect to (a) map position, (b) regulation of enzyme levels in O^c/O^+ versus I^-/I^+ partial diploids, and (c) the position of the structural genes affected by an O^c mutation versus the genes affected by an I^- mutation in a partial diploid.

19.16 GO The following table gives the relative activities of the enzymes β-galactosidase and β-galactoside permease in cells with different genotypes at the *lac* locus in *E. coli*. The level of activity of each enzyme in wild-type *E. coli* not carrying F′s was arbitrarily set at 100; all other values are relative to the observed levels of activity in these wild-type bacteria. Based on the data given in the table for genotypes 1 through 4, fill in the levels of enzyme activity that would be expected for the fifth genotype.

	β-Galactoside		β-Galactosidase Permease	
Genotype	**−Inducer**	**+Inducer**	**−Inducer**	**+Inducer**
1. $I^+O^+Z^+Y^+$	0.2	100	0.2	100
2. $I^-O^+Z^+Y^+$	100	100	100	100
3. $I^+O^cZ^+Y^+$	20	100	20	100
4. $I^-O^+Z^+Y^-/F'\ I^-O^+Z^+Y^+$	200	200	100	100
5. $I^-O^cZ^-Y^+/F'\ I^+O^+Z^+Y^+$	——	——	——	——

19.17 By what mechanism does the presence of tryptophan in the medium in which *E. coli* cells are growing result in premature termination or attenuation of transcription of the *trp* operon?

19.18 What effect will deletion of the *trpL* region of the *trp* operon have on the rates of synthesis of the enzymes encoded by the five genes in the *trp* operon in *E. coli* cells growing in the presence of tryptophan?

19.19 What is a prophage? In what ways does the phage λ prophage differ from the λ chromosome present during lytic infections?

19.20 Suppose that you used site-specific mutagenesis to modify the *trpL* sequence such that the two UGG Trp codons at positions 54–56 and 57–60 (see **FIGURE 19.12**) in the mRNA leader sequence were changed to GGG Gly codons. Will attenuation of the *trp* operon still be regulated by the presence or absence of tryptophan in the medium in which the *E. coli* cells are growing?

19.21 The rate of transcription of the *trp* operon in *E. coli* is controlled by both (1) repression/derepression and (2) attenuation. By what mechanisms do these two regulatory processes modulate *trp* operon transcript levels?

19.22 Would attenuation of the type that regulates the level of *trp* transcripts in *E. coli* be likely to occur in eukaryotic organisms?

19.23 A lambda phage has a UAG chain-termination mutation in the middle of the C_I gene. What phenotype will result from this mutation?

19.24 What is the major structural difference between the operator region in the *lac* operon of *E. coli* and the phage λ operators O_L and O_R?

▶ Genomics on the Web

at http://www.ncbi.nlm.nih.gov/

The *E. coli* catabolite activator protein (CAP) plays an important regulatory role by preventing the induction of the *lac* operon in the presence of high concentrations of glucose, which is a more efficient energy source than lactose. High concentrations of glucose prevent the activation of the enzyme adenylcyclase, which catayzes the synthesis of cyclic AMP (cAMP) from ATP. CAP must form a complex with cAMP in order to bind to the *lac* promoter and, in turn, stimulate the binding of RNA polymerase.

Without CAP-cAMP bound to the promoter, transcription of the *lac* operon never exceeds 2 percent of the level observed in the absence of glucose. The CAP-cAMP complex has the same effect on the *gal* operon, the *ara* operon, and several other operons. It serves as a global regulator of catabolic pathways in bacteria. This phenomenon—catabolite repression or the "glucose effect"— involves specific interactions between DNA-binding domains of the CAP-cAMP complex and nucleotide sequences in bacterial promoters.

1. What kinds of interactions are involved in the binding of CAP-cAMP to DNA?

2. What is the three-dimensional structure of CAP-cAMP?

3. What are the three-dimensional structures of CAP-cAMP-DNA complexes?

4. Does the binding of CAP-cAMP have any effect on DNA structure?

5. Does CAP share any three-dimensional structural domains with other DNA-binding proteins?

Hint: At the NCBI web site, click on "Molecular databases," scroll down and click on "Structure (MMDB = *M*olecular *M*odeling Data*b*ase), and search using "CAP-cAMP" as a query. Click on "1O3T," "Crystal Structures of CAP-DNA Complexes," "1G6N, 2.1 Angstrom Structure of CAP-cAMP," and others, to view three-dimensional models of these molecular interactions.

Chapter 20
Regulation of Gene Expression in Eukaryotes

CHAPTER OUTLINE

- **Ways of Regulating Eukaryotic Gene Expression: An Overview**
- **Induction of Transcriptional Activity by Environmental and Biological Factors**
- **Molecular Control of Transcription in Eukaryotes**
- **Posttranscriptional Regulation of Gene Expression by RNA Interference**
- **Gene Expression and Chromosome Organization**
- **Activation and Inactivation of Whole Chromosomes**

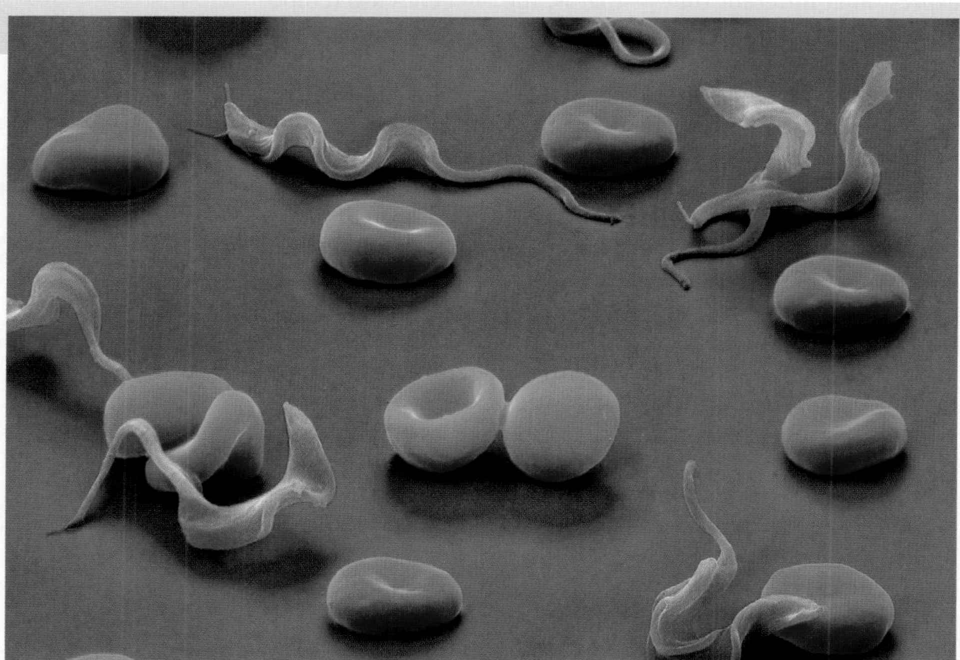

Trypanosomes among red blood cells.

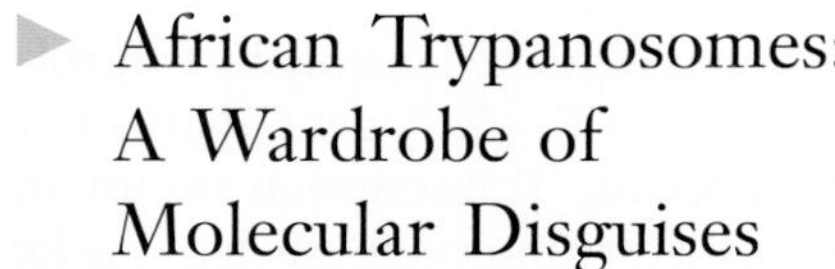

African Trypanosomes: A Wardrobe of Molecular Disguises

Near the end of the nineteenth century, David Bruce, a surgeon in the British Medical Service, summarized his observations and experiments on a disease of wild and domesticated animals in southern Africa. The disease, called nagana from a Zulu word meaning "loss of spirit," is characterized by fever, swelling, lethargy, and emaciation. Bruce recognized that nagana is transmitted by the tsetse, a type of biting fly common in the open spaces of the African scrub plain. Furthermore, his examination of diseased animals led him to conclude that the causative agent is a flagellated, unicellular protozoan that is injected into the animal's blood when the tsetse bites. This blood parasite, a type of trypanosome, is now called *Trypanosoma brucei* in Bruce's honor. Humans can also be infected with tsetse-borne trypanosomes, whereupon they develop the debilitating illness known as African sleeping sickness.

In both animals and humans, trypanosome infections last a long time. This is remarkable because, in the blood, trypanosomes are subjected to repeated attacks by the immune system. With each immune attack, most of the trypanosomes are destroyed; however, a few always survive to repopulate the blood and maintain the infection. The key to this resurgence is the trypanosome's ability to change the protein that coats its surface. Each trypanosome is covered with about 10 million molecules of a single glycoprotein. When the immune system recognizes this protein coat, the infecting trypanosome is in trouble; immune cells will trap and destroy it. However, before all the trypanosomes in the animal are completely wiped out, a few manage to change their surface glycoprotein to one that the immune system does not immediately recognize. These altered trypanosomes escape destruction and proliferate. Eventually, the immune system will learn to recognize them too, but in the meantime another group of altered trypanosomes arises to keep the infection going. The seemingly endless supply of molecular disguises available to trypanosomes is due to a large array of genes that encode the variant surface glycoproteins (VSGs) coating these organisms. At any one time, only one of these genes is expressed; all the others are silent. However, during the course of an infection the identity of the expressed gene changes. With each change, the trypanosomes acquire a new surface protein and manage to stay one step ahead of the animal's immune defenses. Thus, the infection is maintained for weeks or even months until, through exhaustion, the animal dies.

Ways of Regulating Eukaryotic Gene Expression: An Overview

Eukaryotic gene expression can be regulated at the transcriptional, processing, or translational level.

DIMENSIONS OF EUKARYOTIC GENE REGULATION

The story of how trypanosomes evade attacks by the immune system is a story about gene regulation. Different *vsg* genes are expressed at different times—that is, the *vsg* genes are temporally regulated. Among eukaryotes, especially multicellular organisms like ourselves, genes are also regulated in a spatial dimension. Multicellular organisms contain many different cell types organized into tissues and organs. A particular gene might be expressed in blood cells, but never in nerve cells. Another gene might have just the opposite expression profile. The regulation that creates such differences in gene expression underlies the anatomical and physiological complexity of multicellular eukaryotes.

As in prokaryotes, the expression of genes in eukaryotes involves the transcription of DNA into RNA and the subsequent translation of that RNA into polypeptides. However, prior to translation, most eukaryotic RNA is "processed." During processing, the RNA is capped at its 5′ end, polyadenylated at its 3′ end, and altered internally by losing its noncoding intron sequences (see Chapter 11). Prokaryotic RNAs typically do not undergo these terminal and internal modifications.

Gene expression in eukaryotes is more complicated than it is in prokaryotes because eukaryotic cells are compartmentalized by an elaborate system of membranes. This compartmentalization subdivides the cells into separate organelles, the most conspicuous one being the nucleus; eukaryotic cells also possess mitochondria, chloroplasts (if they are plant cells), and an endoplasmic reticulum. Each of these organelles performs a different function. The nucleus stores the genetic material, the mitochondria and chloroplasts recruit energy, and the reticulum transports materials within the cell.

The subdivision of eukaryotic cells into organelles physically separates the events of gene expression. The primary event, transcription of DNA into RNA, occurs in the nucleus. RNA transcripts are also modified in the nucleus by capping, polyadenylation, and the removal of introns. The resulting messenger RNAs are then exported to the cytoplasm where they become associated with ribosomes, many of which are located on the membranes of the endoplasmic reticulum. Once associated with ribosomes, these mRNAs are translated into polypeptides. This physical separation of the events of gene expression makes it possible for regulation to occur in different places (**FIGURE 20.1**). Regulation can occur in the nucleus at either the DNA or RNA level, or in the cytoplasm at either the RNA or polypeptide level.

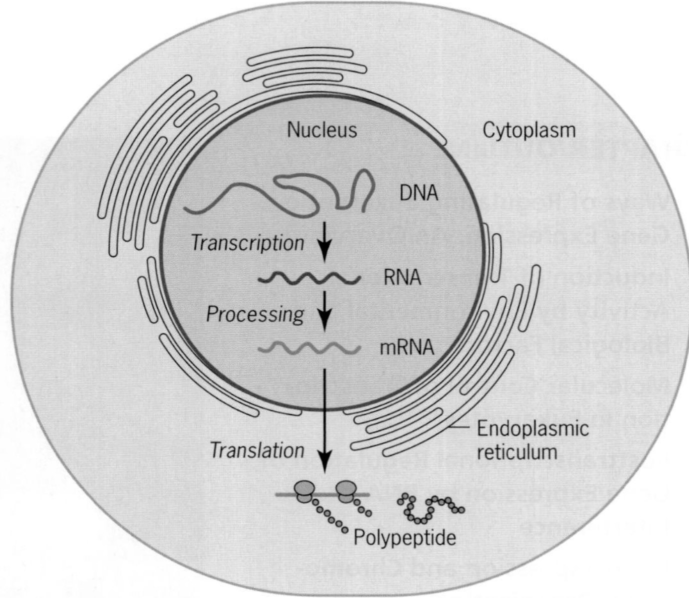

Figure 20.1 ▶ Eukaryotic gene expression showing the stages at which expression can be regulated: transcription, processing, and translation.

CONTROLLED TRANSCRIPTION OF DNA

In prokaryotes, gene expression is regulated mainly by controlling the transcription of DNA into RNA. A gene that is not transcribed is simply not expressed. Transcription occurs in prokaryotes when negative regulatory molecules such as the *lac* repressor protein have been removed from the vicinity of a gene and positive regulatory molecules such as the catabolite activator protein (CAP)/cyclic AMP complex have bound to it (Chapter 19). These protein-DNA interactions control whether or not a gene is accessible to RNA polymerase. Furthermore, the mechanisms that have evolved to control transcription in these organisms respond quickly to environmental changes. As we discussed in Chapter 19, this hair-trigger control is an efficient strategy for prokaryotic survival.

The control of transcription is more complex in eukaryotes than it is in prokaryotes. One reason is that genes are sequestered in the nucleus. Before environmental signals can have any effect on the level of transcription, they must be transmitted from the cell surface, where they are usually received, through the cytoplasm and the nuclear membrane, and onto the chromosomes. Eukaryotic cells therefore need fairly elaborate internal signaling systems to control the transcription of DNA. Another complicating factor is that many eukaryotes are multicellular. Environmental cues may have to pass through layers of cells in order to have an impact on the transcription of genes in a particular tissue. Intercellular communication is therefore an important aspect of eukaryotic transcriptional regulation.

As in prokaryotes, eukaryotic transcriptional regulation is mediated by protein-DNA interactions. Positive and negative regulator proteins bind to specific regions of the DNA and stimulate or inhibit transcription. As a group, these proteins are called **transcription factors.** Many different types have been identified, and most seem to have characteristic domains that allow them to interact with DNA. The structure of these proteins, and the nature of their interactions with DNA, will be discussed in a later section.

ALTERNATE SPLICING OF RNA

Most eukaryotic genes possess introns, noncoding regions that interrupt the sequence that specifies the amino acids of a polypeptide. Each intron must be removed from the RNA transcript of a gene in order for the coding sequence to be expressed properly. As we discussed in Chapter 11, this process involves the precise joining of the coding sequences, or exons, into a messenger RNA. The formation of the mRNA is mediated by tiny nuclear organelles called spliceosomes.

Genes with multiple introns present a curious problem to the RNA splicing machinery. These introns can be removed separately or in combination, depending on how the splicing machinery interacts with the RNA. If two successive introns are removed together, the exon between them will also be removed. Thus, the splicing machinery has the opportunity to modify the coding sequence of an RNA by deleting some of its exons. This phenomenon of splicing an RNA transcript in different ways is apparently a way of economizing on genetic information. Instead of duplicating genes, or pieces of genes, the *alternate splicing* of transcripts makes it possible for a single gene to encode different polypeptides.

One example of alternate splicing occurs during the expression of the gene for troponin T, a protein found in the skeletal muscles of vertebrates; the size of this protein ranges from about 150 to 250 amino acids. In the rat, the troponin T gene is more than 16 kb long and contains 18 different exons (**FIGURE 20.2**). Transcripts of this gene are spliced in different ways to create a large array of mRNAs. When these are translated, many different troponin T polypeptides are produced. All these polypeptides share amino acids from exons 1–3, 9–15, and 18. However, the regions encoded by exons 4–8 may be present or absent, depending on the splicing pattern, and apparently in any combination. Additional variation is provided by the presence or absence of regions encoded by exons 16 and 17; if 16 is present, 17 is not, and vice versa. These different forms of troponin T presumably function in slightly different ways within the muscles, contributing to the variability of muscle cell action.

Another example of alternate splicing occurs during the expression of genes involved in sex determination in *Drosophila*.

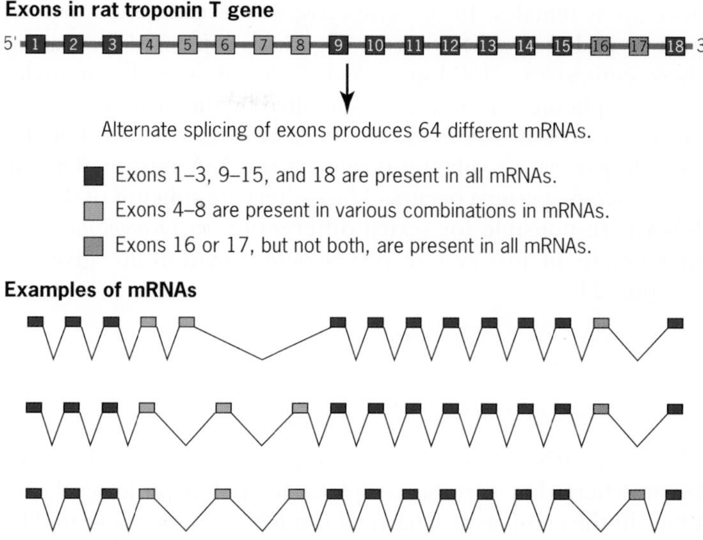

Figure 20.2 ▶ Alternate splicing of transcripts from the rat troponin T gene. Only 3 of the possible 64 different mRNAs are shown.

The master regulator of the sex-determination process is the X-linked *Sex-lethal (Sxl)* gene. In chromosomal (XX) females, the transcript of this gene is spliced to produce an mRNA that encodes a regulatory protein. In chromosomal (XY) males, the *Sxl* transcript is alternately spliced to include an exon with a stop codon; thus, when this RNA is translated, it generates a short polypeptide without regulatory function (**FIGURE 20.3**). In XX embryos, where the regulatory *Sxl* protein is present, a particular set of genes is expressed that causes these embryos to

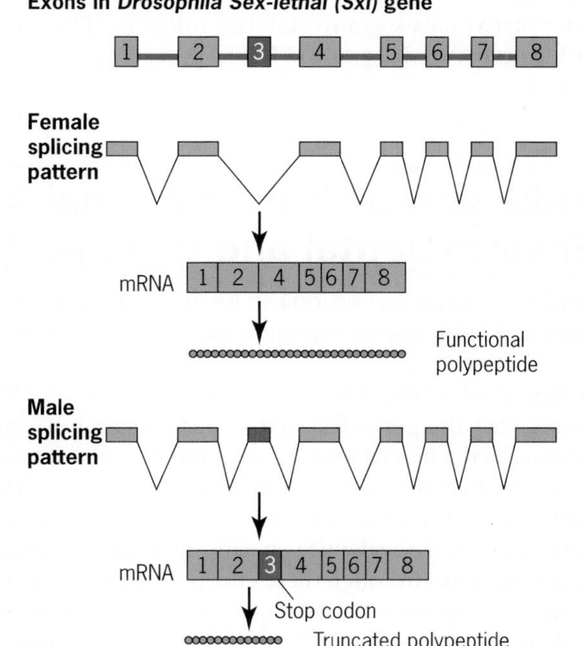

Figure 20.3 ▶ Alternate splicing of transcripts from the *Sex-lethal* gene in male and female *Drosophila*.

develop as females. In XY embryos, where the regulatory *Sxl* protein is absent, a different set of genes is expressed that causes these embryos to develop as males. Female-specific or male-specific splicing is ultimately controlled by the ratio of X chromosomes to autosomes. If the X:A ratio is 1.0, the female-specific pattern of splicing occurs; if the X:A ratio is 0.5, the male-specific pattern occurs. Thus, alternate splicing of the *Sxl* RNA is responsible for sexual differention in *Drosophila.* Further details of this system of sex determination are given in Chapter 21.

CYTOPLASMIC CONTROL OF MESSENGER RNA STABILITY

Messenger RNAs are exported from the nucleus to the cytoplasm where they serve as templates for polypeptide synthesis. Once in the cytoplasm, a particular mRNA can be translated by several ribosomes that move along it in sequential fashion. This translational assembly line continues until the mRNA is degraded. Messenger RNA degradation is therefore another control point in the overall process of gene expression. Long-lived mRNAs can support multiple rounds of polypeptide synthesis, whereas short-lived mRNAs cannot.

An mRNA that is rapidly degraded must be replenished by additional transcription; otherwise, the polypeptide it encodes will cease to be synthesized. This cessation of polypeptide synthesis may, of course, be part of a developmental program. Once the polypeptide has had its effect, it may no longer be needed; in fact, its continued synthesis may be harmful. In such cases, rapid degradation of the mRNA would be a reasonable way of preventing undesired polypeptide synthesis.

Messenger RNA longevity can be influenced by several factors. Poly(A) tails seem to stabilize mRNAs. The sequence of the 3′ untranslated region (3′ UTR) preceding a poly(A) tail also seems to affect mRNA stability. Several short-lived mRNAs have the sequence AUUUA repeated several times in their 3′ untranslated regions. When this sequence is artificially transferred to the 3′ untranslated region of more stable mRNAs, they, too, become unstable. Chemical factors, such as hormones, may also affect mRNA stability. In the toad *Xenopus laevis*, the *vitellogenin* gene is transcriptionally activated by the steroid hormone estrogen. However, in addition to inducing transcription of this gene, estrogen also increases the longevity of its mRNA.

Recent research has revealed that the stability of mRNAs and the translation of mRNAs into polypeptides are also regulated by small, noncoding RNA molecules called small interfering RNAs (siRNAs) or microRNAs (miRNAs). These regulatory RNA molecules, which are between 21 and 28 nucleotides long, are produced from larger, double-stranded RNAs in a wide variety of eukaryotic organisms, including fungi, plants, and animals. Short interfering and microRNAs base pair with sequences in specific mRNAs; once paired, they either cause the mRNA to be cleaved and subsequently degraded, or they prevent the mRNA from being translated into a polypeptide. In plants, these small RNA molecules provide a critical defense against infection by RNA viruses, and in both plants and animals they regulate the expression of genes involved in maturation and development. We shall discuss them in more detail later in this chapter.

KEY POINTS

- Proteins called transcription factors interact with DNA to control the transcription of eukaryotic genes.
- Eukaryotic gene transcripts may be alternately spliced to produce messenger RNAs that encode distinct, but related, polypeptides.
- The stability of eukaryotic messenger RNAs can influence the level of polypeptide synthesis.

Induction of Transcriptional Activity by Environmental and Biological Factors

Eukaryotic gene expression can be induced by environmental factors such as heat and light, and by signaling molecules such as hormones and growth factors.

In their study of the *lactose* operon in *E. coli*, Jacob and Monod discovered that the genes for lactose metabolism were specifically transcribed when lactose was given to the cells. Thus, they demonstrated that lactose was an **inducer** of gene transcription. Following in the footsteps of Jacob and Monod, many researchers have attempted to identify specific inducers of eukaryotic gene transcription. Although these efforts have met with considerable success, the overall extent to which eukaryotic genes are induced by environmental and nutritional factors seems to be less than it is in prokaryotes. Here we will consider three examples of inducible gene expression in eukaryotes. The first two involve induction by environmental factors—temperature and light—and the third involves induction by that special group of signaling molecules called hormones.

TEMPERATURE: THE HEAT-SHOCK GENES

When organisms are subjected to the stress of high temperature, they respond by synthesizing a group of proteins that help to stabilize the internal cellular environment. These *heat-shock proteins*, found in both prokaryotes and eukaryotes, are among the most conserved polypeptides known. Comparisons of the amino acid sequences of heat-shock proteins from organisms as diverse as *E. coli* and *Drosophila* show that they are 40 to 50 percent identical—a remarkable finding considering the length of evolutionary time separating these organisms.

The expression of the heat-shock proteins is regulated at the transcriptional level; that is, heat stress specifically induces the transcription of the genes encoding these proteins (**FIGURE 20.4**). In *Drosophila*, for example, one of the heat-shock proteins called HSP70 (for *h*eat-*s*hock *p*rotein, molecular weight 70 kilodaltons) is encoded by a family of genes located in two nearby clusters on one of the autosomes. Altogether, there are five to six copies of these *hsp70* genes in the two clusters. When the temperature exceeds 33°C, as it does on hot summer days, each of the genes is transcribed into RNA, which is then processed and translated to produce HSP70 polypeptides. This heat-induced transcription of the *hsp70* genes is mediated by a polypeptide called the heat-shock transcription factor, or HSTF, which is present in the nuclei of *Drosophila* cells. When *Drosophila* are heat stressed, the HSTF is chemically altered by phosphorylation. In this altered state, it binds specifically to nucleotide sequences upstream of the *hsp70* genes and makes the genes more accessible to RNA polymerase II, the enzyme that transcribes most protein-encoding genes. The transcription of the *hsp70* genes is then vigorously stimulated. The sequences to which the phosphorylated HSTF binds are called *heat-shock response elements (HSEs).*

LIGHT: THE RIBULOSE 1,5-BISPHOSPHATE CARBOXYLASE GENES IN PLANTS

One of the most abundant proteins on earth is ribulose 1,5-bisphosphate carboxylase (RBC), an enzyme that plays a critical role in photosynthesis in green plants. Through the work of this enzyme, carbon dioxide is incorporated into molecules of sugar, which are then metabolized to provide energy for cells. This process ultimately depends on the ability of plants to absorb light energy. Without light, the entire process comes to a halt, and there is no need for enzymes such as RBC. It is therefore no surprise that the production of RBC is specifically induced when plants are exposed to light.

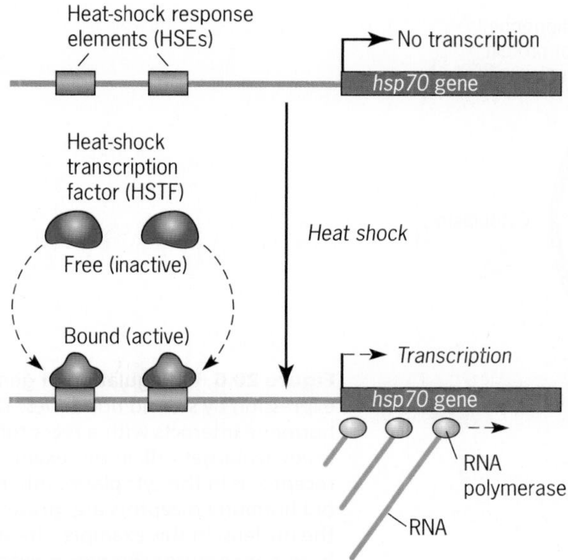

Figure 20.4 ▶ Induction of transcription from the *Drosophila hsp*70 gene by heat shock. The HSEs are located between 40 and 90 base pairs upstream of the transcription initiation site (bent arrow).

RBC is a complex enzyme consisting of large and small subunits, each encoded by different genes. In some plant species, both genes are located in the DNA of the chloroplast, but in others, the gene for the small subunit is located in the nuclear DNA and the gene for the large subunit is located in the chloroplast DNA. The expression of the nuclear gene for the small subunit (denoted *rbcS*) has been analyzed in several different plant species. These studies have shown that *rbcS* is vigorously transcribed after plants have been exposed to light (**FIGURE 20.5**).

The light-inducible transcription of *rbcS* is only partly understood. One important event is the absorbtion of light by a cytoplasmic protein called phytochrome. A light-absorbing molecule called a chromophore is attached to each molecule of phytochrome. The absorption of light by the chromophore apparently causes a conformational change in the phytochrome polypeptide, which then triggers changes in other proteins. Although the details of subsequent events are sketchy, it seems that eventually some of these proteins bind to regions upstream of the *rbcS* gene and stimulate its transcription. This response is quick and vigorous; a substantial amount of *rbcS* RNA is generated after exposure to light, providing numerous templates for the synthesis of the small subunit of RBC. A similar process may be involved in the production of the large subunit. Exposure to light therefore induces the production of one of the key enzymes needed for photosynthesis.

SIGNAL MOLECULES: GENES THAT RESPOND TO HORMONES

In multicellular eukaryotes, one type of cell can signal another by secreting a **hormone.** Hormones circulate through the body, make contact with their target cells, and then initiate a

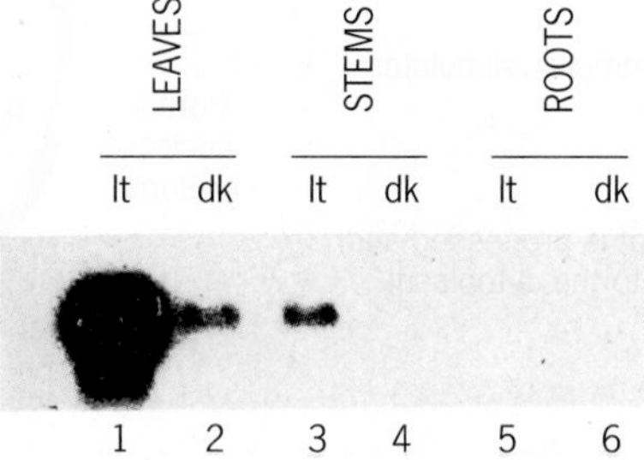

Figure 20.5 ▶ Light-induced transcription of the gene for the small subunit of ribulose 1,5-bisphosphate carboxylase. Plants were either exposed to light (lt) or kept in the dark (dk), and RNA was extracted from different plant tissues. The extracted RNA was analyzed by northern blotting using a radioactively labeled probe for the *rbcS* gene. The resulting autoradiogram shows that *rbcS* RNA is produced in leaves and stems, especially after exposure to light.

series of events that regulate the expression of particular genes. In animals there are two general classes of hormones. The first class, the *steroid hormones*, are small, lipid-soluble molecules derived from cholesterol. Because of their lipid nature, they have little or no trouble passing through cell membranes. Examples are estrogen and progesterone, which play important roles in female reproductive cycles, testosterone, a hormone of male differentiation and behavior, the glucocorticoids, which are involved in regulating blood sugar levels, and ecdysone, a hormone that controls developmental events in insects. Once these hormones have entered a cell, they interact with cytoplasmic or nuclear proteins called *hormone receptors*. The receptor/hormone complex that is formed then interacts with the DNA where it acts as a transcription factor to regulate the expression of certain genes (**FIGURE 20.6**).

The second class of hormones, the *peptide hormones*, are linear chains of amino acids. Like all other polypeptides, these molecules are encoded by genes. Examples are insulin, which regulates blood sugar levels, somatotropin, which is a growth hormone, and prolactin, which targets tissues in the breasts of female mammals. Because peptide hormones are typically too large to pass freely through cell membranes, the signals they convey must be transmitted to the interior of cells by *membrane-bound receptor proteins* (**FIGURE 20.7**). When a peptide hormone interacts with its receptor, it causes a conformational change in the receptor that eventually leads to changes in other proteins inside the cell. Through a cascade of such changes, the hormonal signal is transmitted through the cytoplasm of the cell and into the nucleus, where it ultimately has the effect of regulating the expression of specific genes. This process of transmitting the hormonal signal through the cell and into the nucleus is called **signal transduction**.

Hormone-induced gene expression is mediated by specific sequences in the DNA. These sequences, called *hormone response elements (HREs)*, are analogous to the heat-shock response elements discussed earlier. They are situated near the genes they regulate and serve to bind specific proteins, which then act as transcription factors. With steroid hormones such as estrogen, the HREs are bound by the hormone/receptor complex, which then stimulates transcription. The vigor of this transcriptional response depends on the number of HREs present. When there are multiple response elements, hormone/receptor complexes bind cooperatively with each other, significantly increasing the rate of transcription; that is, a gene with two response elements

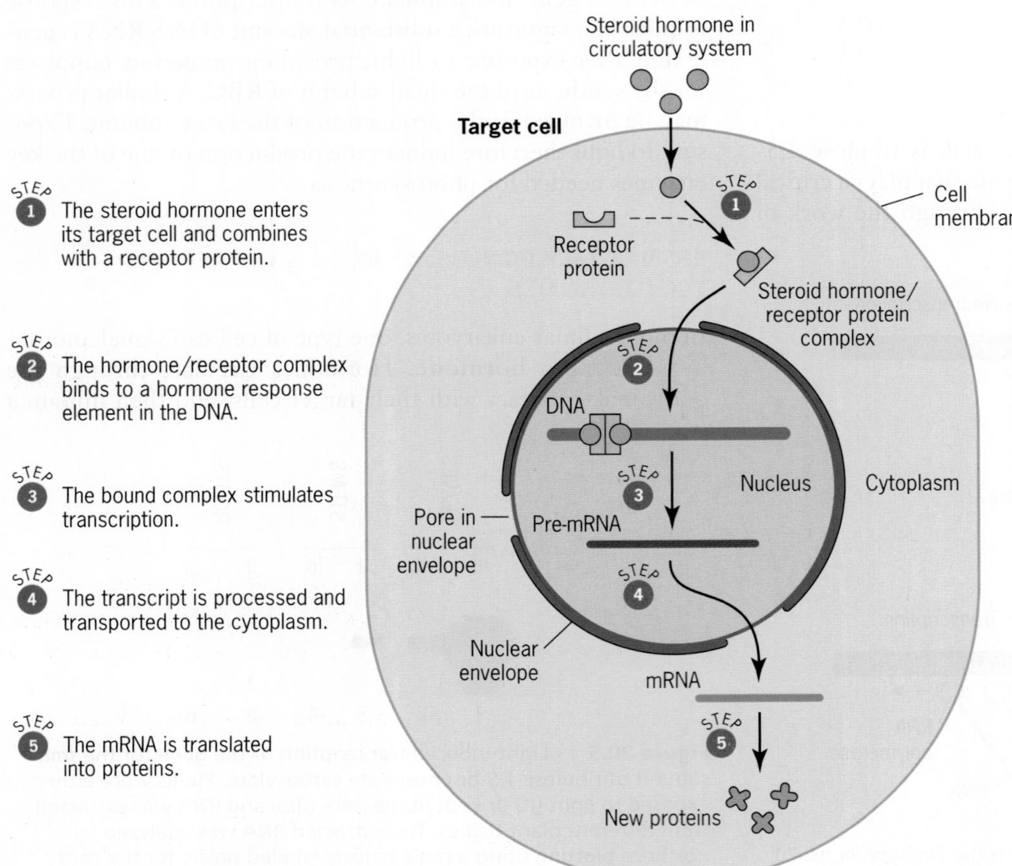
Steroid hormone in circulatory system
Target cell
STEP 1 The steroid hormone enters its target cell and combines with a receptor protein.
Cell membrane
Receptor protein
Steroid hormone/ receptor protein complex
STEP 2 The hormone/receptor complex binds to a hormone response element in the DNA.
DNA
STEP 3 The bound complex stimulates transcription.
Nucleus
Cytoplasm
Pore in nuclear envelope
Pre-mRNA
STEP 4 The transcript is processed and transported to the cytoplasm.
Nuclear envelope
mRNA
STEP 5 The mRNA is translated into proteins.
New proteins

Figure 20.6 ▶ Regulation of gene expression by steroid hormones. The hormone interacts with a receptor inside its target cell. In this example the receptor is in the cytoplasm; other steroid hormone receptors are located in the nucleus. In this example, the steroid/hormone receptor complex moves into the nucleus where it activates the transcription of particular genes.

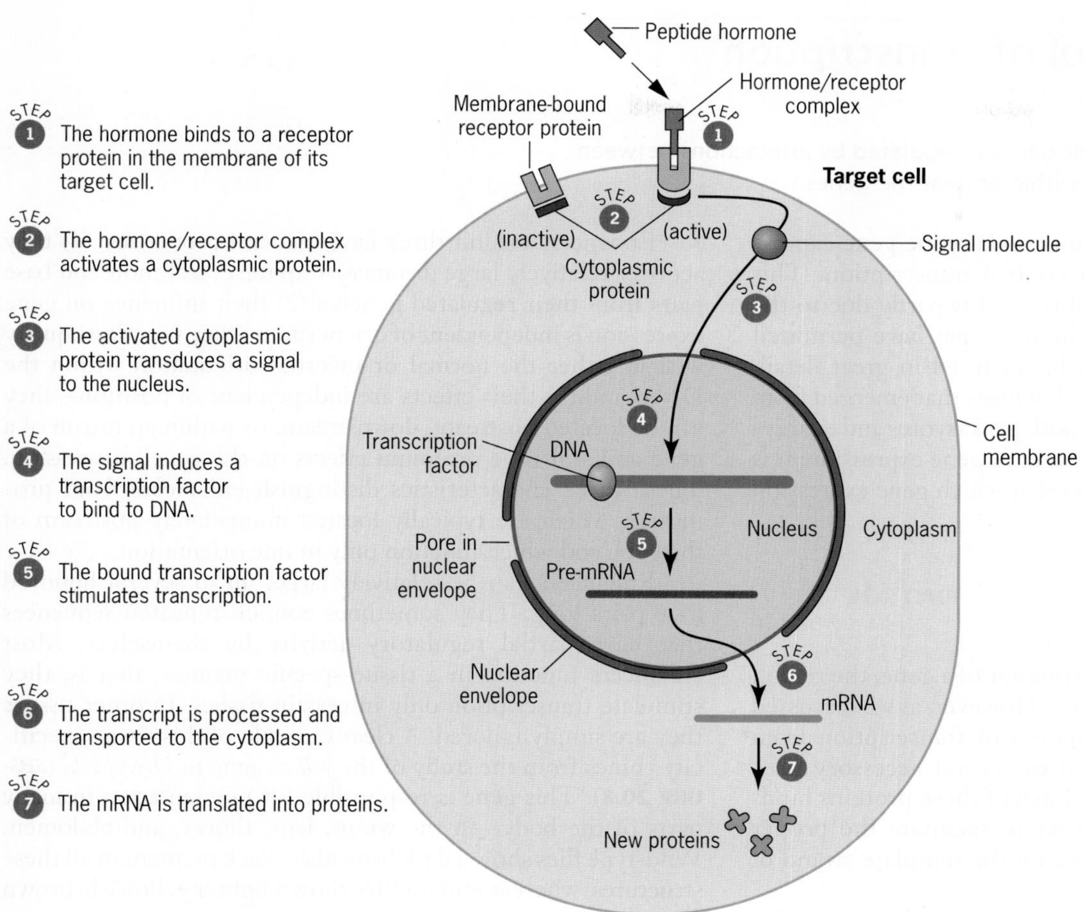

STEP 1 The hormone binds to a receptor protein in the membrane of its target cell.
STEP 2 The hormone/receptor complex activates a cytoplasmic protein.
STEP 3 The activated cytoplasmic protein transduces a signal to the nucleus.
STEP 4 The signal induces a transcription factor to bind to DNA.
STEP 5 The bound transcription factor stimulates transcription.
STEP 6 The transcript is processed and transported to the cytoplasm.
STEP 7 The mRNA is translated into proteins.
Peptide hormone
Hormone/receptor complex
Membrane-bound receptor protein
Target cell
(inactive)
(active)
Cytoplasmic protein
Signal molecule
Transcription factor
DNA
Cell membrane
Pore in nuclear envelope
Nucleus
Cytoplasm
Pre-mRNA
Nuclear envelope
mRNA
New proteins

Figure 20.7 ▶ Regulation of gene expression by peptide hormones. The hormone (an extracellular signal) interacts with a receptor in the membrane of its target cell. The resulting hormone/receptor complex activates a cytoplasmic protein that triggers a cascade of intracellular changes. These changes transmit the signal into the nucleus, where a transcription factor stimulates the expression of particular genes.

is transcribed more than twice as vigorously as a gene with only one. With peptide hormones, the receptor usually remains in the cell membrane, even after it has formed a complex with the hormone. The hormonal signal is therefore conveyed to the nucleus by other proteins, some of which bind to sequences near the genes that are regulated by the hormone. These proteins then act as transcription factors to control the expression of the genes.

Transcriptional activity can be induced by many other kinds of proteins that are not hormones in the classical sense—that is, not produced by a particular gland or organ. These include a variety of secreted, circulating molecules such as nerve growth factor, epidermal growth factor, and platelet-derived growth factor, and other noncirculating molecules associated with cell surfaces or with the matrix between cells. Although each of these proteins has its own peculiarities, the general mechanism whereby they induce transcription resembles that of the peptide hormones. An interaction between the signaling protein and a membrane-bound receptor initiates a chain of events inside the cell that ultimately results in specific transcription factors binding to particular genes, which are then transcribed.

KEY POINTS

- Transcription of the *hsp70* genes in response to increased temperature is mediated by a heat-shock transcription factor.
- Transcription of the gene for the photosynthetic enzyme ribulose 1,5-bisphosphate carboxylase (RBC) is induced by exposure to light.
- Steroid hormones and their receptor proteins form complexes that act as transcription factors to regulate the expression of specific genes.
- Peptide hormones interact with membrane-bound receptor proteins to activate a signaling system that regulates the expression of specific genes.

Molecular Control of Transcription in Eukaryotes

The transcription of eukaryotic genes is regulated by interactions between proteins and DNA sequences within or near the genes.

Much of the current research on eukaryotic gene expression focuses on the factors that control transcription. This heavy emphasis on transcriptional control is partly due to the development of experimental techniques that have permitted this aspect of gene regulation to be analyzed in great detail. However, it is also due to the appeal of ideas that emerged from the study of prokaryotic genes. In both prokaryotes and eukaryotes, transcription is the primary event in gene expression; it is therefore the most fundamental level at which gene expression can be controlled.

DNA SEQUENCES INVOLVED IN THE CONTROL OF TRANSCRIPTION

Transcription is initiated in the promoter of a gene, the region recognized by the RNA polymerase. However, as we discussed in Chapter 11, the accurate initiation of transcription from eukaryotic gene promoters requires several accessory proteins, or *basal transcription factors*. Each of these proteins binds to a sequence within the promoter to facilitate the proper alignment of the RNA polymerase on the template strand of the DNA.

The transcription of eukaryotic genes is also controlled by a variety of *special transcription factors*, such as those involved in the regulation of the heat-, light-, and hormone-inducible genes we have discussed. These factors bind to response elements, or, more generally, to sequences called **enhancers** located in the vicinity of a gene. The special transcription factors that bind to these enhancers may interact with the basal transcription factors and the RNA polymerase, which bind to the promoter of a gene. The interactions that take place among the special transcription factors, the basal transcription factors, and the RNA polymerase regulate the transcriptional activity of a gene.

Enhancers exhibit three fairly general properties: (1) they act over relatively large distances—up to several thousand base pairs from their regulated gene(s); (2) their influence on gene expression is independent of orientation—they function equally well in either the normal or inverted orientation within the DNA; and (3) their effects are independent of position—they can be located upstream, downstream, or within an intron of a gene and still have profound effects on the gene's expression. These three characteristics distinguish enhancers from promoters, which are typically located immediately upstream of the gene and which function only in one orientation.

Enhancers can be relatively large, up to several hundred base pairs long. They sometimes contain repeated sequences that have partial regulatory activity by themselves. Most enhancers function in a tissue-specific manner; that is, they stimulate transcription only in certain tissues. In other tissues they are simply ignored. A clear example of this tissue specificity comes from the study of the *yellow* gene in *Drosophila* (**FIGURE 20.8**). This gene is responsible for pigmentation in many parts of the body—in the wings, legs, thorax, and abdomen. Wild-type flies show a dark brownish-black pigment in all these structures, whereas mutant flies show a lighter yellowish-brown pigment. However, in some mutants, there is a mosaic pattern of pigmentation, brownish-black in some tissues and yellowish-brown in others. These mosaic patterns are due to mutations that alter the transcription of the *yellow* gene in some tissues but not in others. Pamela Geyer and Victor Corces have shown that the *yellow* gene is regulated by several enhancers, some of which are located within an intron, and that each enhancer activates transcription in a different tissue. If, for example, the enhancer for expression in the wing is mutated, the bristles on the wings are yellowish-brown instead of brownish-black. The battery of enhancers associated with the *yellow* gene allows its expression to be controlled in a tissue-specific way. To see

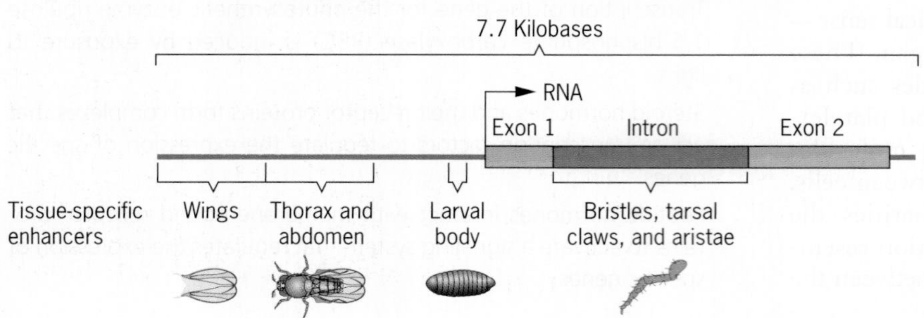

Figure 20.8 ▶ The tissue-specific enhancers of the *Drosophila yellow* gene.

another way of studying enhancers, work through the Focus on Problem Solving: Defining the Sequences Required for a Gene's Expression.

One of the first enhancers to be studied extensively is located in the chromosome of the eukaryotic virus SV40. This virus infects monkey cells and has been widely used in biological research. Its 5.2-kb circular chromosome contains a single prominent enhancer about 220 base pairs long (**FIGURE 20.9*a*, *b***). The enhancer regulates the transcription of two groups of genes on the virus chromosome. One group, situated to the right of the enhancer, is transcribed early during infection, and the other group, situated to the left, is transcribed later. The SV40 enhancer contains two 72-bp direct repeats, either of which is sufficient for enhancer function. It can be inverted or moved to different locations on the SV40 chromosome and it still retains its regulatory ability. Furthermore, if it is inserted upstream or downstream from other eukaryotic genes, it stimulates their transcription. These effects are mediated by proteins that bind to the enhancer. Curiously, examining SV40 chromosomes with the electron microscope shows that the enhancer region is not wrapped around nucleosomes (**FIGURE 20.9*c***). A plausible interpretation is that enhancer-binding transcription factors prevent nucleosome formation.

How do enhancers influence the transcription of genes? The results of many studies indicate that the proteins that bind to enhancers influence the activity of the proteins that bind to

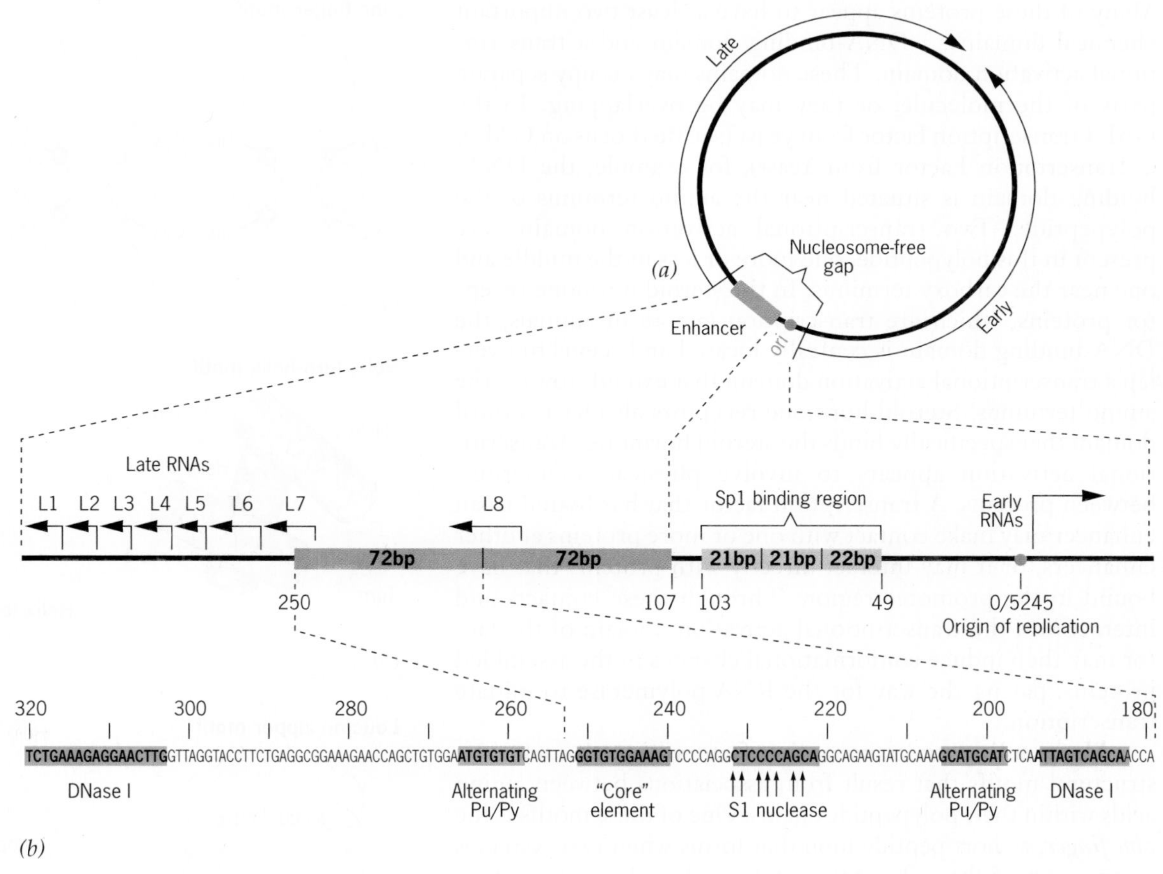

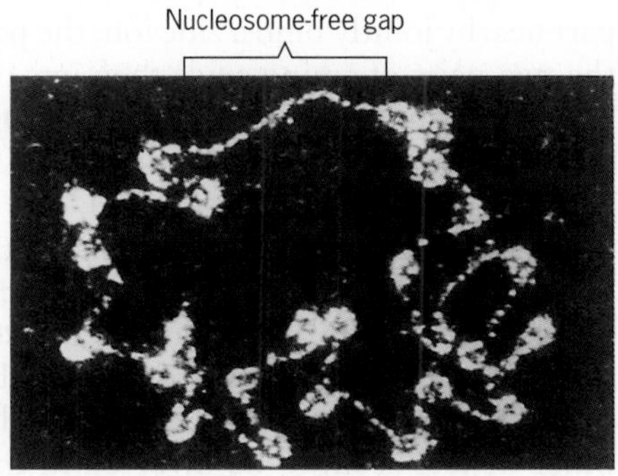

Figure 20.9 ► Structure of the enhancer of the simian virus 40 (SV40). (*a*) Diagram of the SV40 chromosome showing the location of the enhancer and the origin of replication (*ori*). (*b*) Diagram showing the components of the enhancer. Each of the two 72-base-pair repeats contains regions that are sensitive to DNase I and S1 nuclease digestion, regions with alternating purines (Pu) and pyrimidines (Py), and a core-enhancer element. A 64-base-pair element that binds the transcription factor SP1 is adjacent to one of the repeats. The origin of replication is at coordinate 0/5245 on the circular chromosome. (*c*) Electron micrograph of an SV40 chromosome showing nucleosomes everywhere except in the vicinity of the enhancer.

promoters, including the basal transcription factors and the RNA polymerase. The two types of proteins are brought into physical contact by a multimeric complex consisting of at least 20 different proteins. This *mediator complex* appears to bend the DNA in such a way that the proteins bound to an enhancer are juxtaposed to those bound at the promoter. In this way, then, proteins bound to the enhancer exert control over transcription, which is initiated at the promoter.

PROTEINS INVOLVED IN THE CONTROL OF TRANSCRIPTION: TRANSCRIPTION FACTORS

Research over the last two decades has identified a large number of eukaryotic proteins that stimulate transcription. Many of these proteins appear to have at least two important chemical domains: a DNA-binding domain and a transcriptional activation domain. These domains may occupy separate parts of the molecule, or they may be overlapping. In the GAL4 transcription factor from yeast (see the Focus on GAL4, a Transcription Factor from Yeast), for example, the DNA-binding domain is situated near the amino terminus of the polypeptide. Two transcriptional activation domains are present in this polypeptide, one more or less in the middle and one near the carboxy terminus. In the steroid hormone receptor proteins, which are transcription factors in animals, the DNA-binding domain is centrally located and seems to overlap a transcriptional activation domain that extends toward the amino terminus. Steroid hormone receptors also have a third domain that specifically binds the steroid hormone. Transcriptional activation appears to involve physical interactions between proteins. A transcription factor that has bound to an enhancer may make contact with one or more proteins at other enhancers, or it may interact directly with proteins that have bound in the promoter region. Through these contacts and interactions, the transcriptional activation domain of the factor may then induce conformational changes in the assembled proteins, paving the way for the RNA polymerase to initiate transcription.

Many eukaryotic transcription factors have characteristic structural motifs that result from associations between amino acids within their polypeptide chains. One of these motifs is the *zinc finger*, a short peptide loop that forms when two cysteines in one part of the polypeptide and two histidines in another part nearby jointly bind a zinc ion; the peptide segment between the two pairs of amino acids then juts out from the main body of the protein as a kind of finger (**FIGURE 20.10*a***). Mutational analysis has demonstrated that these fingers play important roles in DNA binding.

A second motif in many transcription factors is the *helix-turn-helix*, a stretch of three short helices of amino acids separated from each other by turns (**FIGURE 20.10*b***). Genetic and biochemical analyses have shown that the helical segment closest to the carboxy terminus is required for DNA binding; the other helices seem to be involved in the formation of protein dimers. In many transcription factors, the helix-turn-helix motif coincides with a highly conserved region of approximately 60 amino acids called the *homeodomain*, so named because it occurs in proteins encoded by the homeotic genes of *Drosophila*. Classical analyses have demonstrated that mutations in these genes alter the developmental fates of groups of cells (Chapter 21). Thus, for example, mutations in the *Antennapedia* gene can cause antennae to develop as legs. This bizarre phenotype is an example of a homeotic transformation—the substitution of one body part for another during the developmental process. Molecular analyses of the homeotic genes in *Drosophila* have demonstrated that each encodes a protein with a homeodomain and that these proteins can bind to DNA. The

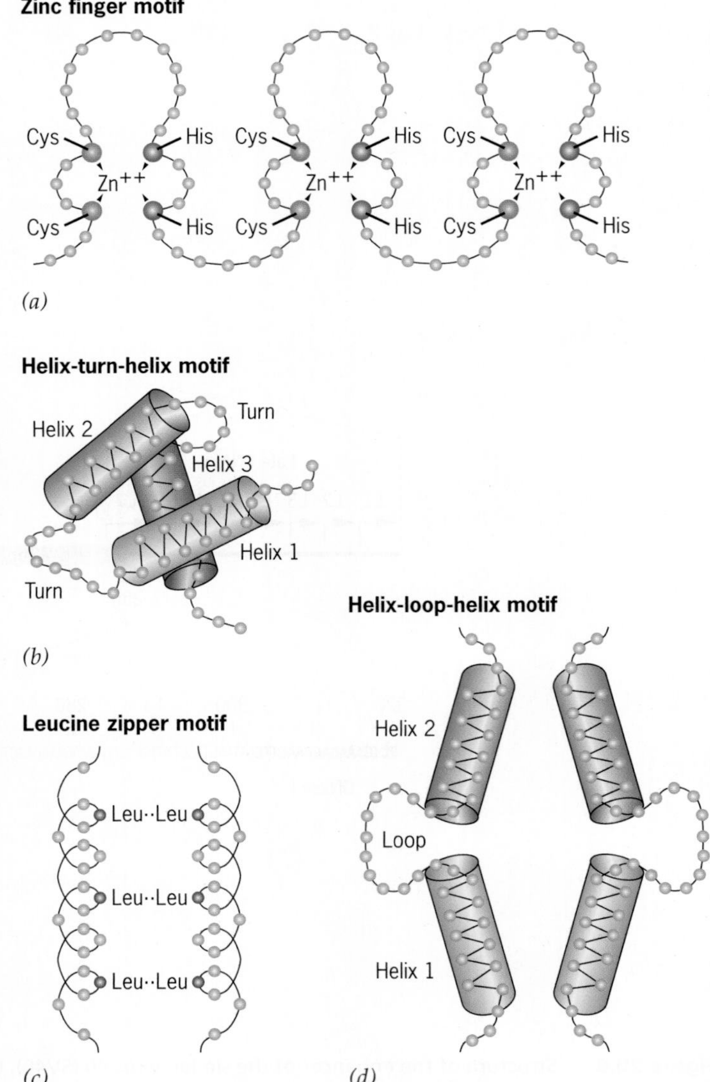

Figure 20.10 ► Structural motifs within different types of transcription factors. (*a*) Zinc-finger motifs in the mammalian transcription factor SP1. (*b*) Helix-turn-helix motif in a homeodomain transcription factor. (*c*) A leucine zipper motif that allows two polypeptides to dimerize and then bind to DNA. (*d*) A helix-loop-helix motif that allows two polypeptides to dimerize and then bind to DNA.

► FOCUS ON PROBLEM SOLVING

Defining the Sequences Required for a Gene's Expression

THE PROBLEM

The tubulins are important proteins of the cytoskeleton in eukaryotes. In *Arabidopsis thaliana* the tubulin encoded by the *TUA1* gene is expressed primarily in mature anthers within the flowers. To determine the sequences responsible for this tissue-specific expression, 533 base pairs of DNA upstream of the *TUA1* transcription start site plus the first 56 base pairs of the 5′ untranslated region of the *TUA1* gene were fused to the coding sequence of the β-glucuronidase (GUS) gene from *E. coli*. β-glucuronidase catalyzes the conversion of a colorless substance called X-gluc into a dark blue pigment. Thus, the appearance of blue pigment in X-gluc-treated material is an indication that the *GUS* gene is being expressed. When this assay was applied to *Arabidopsis* plants that had been genetically transformed with the *GUS* gene fused behind the upstream sequences of the *TUA1* gene, the anthers turned dark blue; all the other tissues remained colorless (see **FIGURE 1**). Further tests demonstrated that pollen within the anthers was expressing the *GUS* transgene—that is, when treated with X-gluc, the pollen grains turned blue. The entire experiment was then repeated using progressively shorter segments from the *TUA1* upstream sequences to drive expression of the *GUS* gene. From the results shown in **FIGURE 2**, what part of the upstream region is required for expression of the *TUA1* gene?

Figure 1 ► Expression of the β-glucuronidase (GUS) gene from *E. coli* in a transgenic *Arabidopsis* flower. The *GUS* gene has been fused to sequences from the upstream region of the *Arabidopsis TUA1* gene. Its expression is detected by treating the flower with X-gluc, a colorless substrate of GUS that turns blue when it is metabolized. In this flower, the *GUS* gene is expressed exclusively in the anthers.

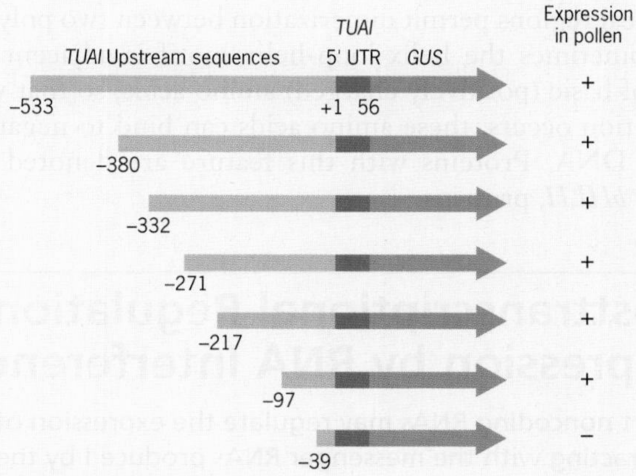

Figure 2 ► Expression of *TUA1/GUS* transgenes in *Arabidopsis* pollen. Progressively shorter segments from the upstream region of the *TUA1* gene and a short sequence from the 5′ untranslated region (UTR) of this gene have been fused to the coding sequences of the *GUS* gene from *E. coli*. +1 is the transcription start site of the *TUA1* gene. Nucleotides to the left of this site are indicated by negative numbers. GUS activity in transgenic pollen is indicated by a plus sign; no GUS activity is indicated by a minus sign. For further details see Carpenter, J., S. E. Ploense, D. P. Snustad, and C. D. Silflow, 1992. Preferential expression of an α-tubulin gene of Arabidopsis in pollen. *The Plant Cell* 4:557–571.

FACTS AND CONCEPTS

1. The region upstream of a gene's transcription start site contains the gene's promoter.

2. This region may also contain enhancers that regulate the gene's expression in a spatially or temporally specific way.

3. The 5′ untranslated region of a gene lies between the transcription start site and the translation start site.

4. *E. coli* genes such as *GUS* can be expressed in eukaryotes such as *Arabidopsis* if they are fused to eukaryotic promoters.

ANALYSIS AND SOLUTION

In this series of experiments, *GUS* is a "reporter" that tells us if the upstream sequences from the *TUA1* gene are able to drive gene expression. All but the smallest of the upstream sequences can function as a successful driver. Thus, there must be a sequence between base pairs –97 and –39 in the upstream sequence of *TUA1* that is critical for the gene's expression. Without this sequence, the *TUA1* gene cannot be expressed. Furthermore, this sequence is sufficient to drive *TUA1* expression in mature pollen. Thus, it functions as an enhancer controlling the tissue-specific expression of the *TUA1* gene.

For further discussion go to your *WileyPLUS* course.

homeodomain proteins stimulate the transcription of particular genes in a spatially and temporally specific manner during development. Homeodomain proteins have also been identified in other organisms, including human beings, where they may play important roles as transcription factors.

A third structural motif found in transcription factors is the *leucine zipper*, a stretch of amino acids with a leucine at every seventh position (**FIGURE 20.10*c***). Polypeptides with this feature can form dimers by interactions between the leucines in each of their zipper regions. Usually, the zipper sequence is adjacent to a positively charged stretch of amino acids. When two zippers interact, these charged regions splay out in opposite directions, forming a surface that can bind to negatively charged DNA.

A fourth structural motif found in some transcription factors is the *helix-loop-helix*, a stretch of two helical regions of amino acids separated by a nonhelical loop (**FIGURE 20.10*d***). The helical regions permit dimerization between two polypeptides. Sometimes the helix-loop-helix motif is adjacent to a stretch of basic (positively charged) amino acids, so that when dimerization occurs, these amino acids can bind to negatively charged DNA. Proteins with this feature are denoted *basic HLH*, or *bHLH*, proteins.

Transcription factors with dimerization motifs such as the leucine zipper or the helix-loop-helix could, in principle, combine with polypeptides like themselves to form homodimers, or they could combine with different polypeptides to form heterodimers. This second possibility suggests a way in which complex patterns of gene expression can be achieved. The transcription of a gene in a particular tissue might depend on activation by a heterodimer, which could form only if its constituent polypeptides were synthesized in that tissue. Moreover, these two polypeptides would have to be present in the correct amounts to favor the formation of the heterodimer over the corresponding homodimers. Subtle modulations in gene expression might therefore be achieved by shifting the concentrations of the two components of a heterodimer.

KEY POINTS

- Enhancers act in an orientation-independent manner over considerable distances to regulate transcription from a gene's promoter.
- Transcription factors recognize and bind to specific DNA sequences within enhancers.
- Transcription factors possess characteristic structural motifs such as the zinc finger, the helix-turn-helix, the leucine zipper, and the helix-loop-helix.

Posttranscriptional Regulation of Gene Expression by RNA Interference

Short noncoding RNAs may regulate the expression of eukaryotic genes by interacting with the messenger RNAs produced by these genes.

Although a great deal of eukaryotic gene regulation occurs at the transcriptional level, recent research has demonstrated that posttranscriptional mechanisms also play important roles in regulating the expression of eukaryotic genes. Some of these mechanisms involve small, noncoding RNAs. By base pairing with target sequences in messenger RNA molecules, these small RNAs interfere with gene expression. Hence, this type of posttranscriptional gene regulation is called **RNA interference**, often abbreviated as **RNAi.** Most types of eukaryotic organisms are capable of RNAi. Among the model genetic organisms, this phenomenon has been well studied in the nematode *Caenorhabditis elegans*, in *Drosophila*, and in *Arabidopsis*. It also exists in mammals, including human beings. Curiously, RNAi is not present in the budding yeast *Saccharomyces cerevisiae*, although it is present in the fission yeast *Schizosaccharomyces pombe*. As we shall see, the widespread capacity of eukaryotic organisms to regulate gene expression by RNAi has allowed geneticists to analyze the functions of genes in organisms that are not amenable to standard genetic approaches.

RNAi PATHWAYS

The phenomenon of RNA interference, which is summarized in **FIGURE 20.11**, involves small RNA molecules called short interfering RNAs (siRNAs) or microRNAs (miRNAs). These molecules, 21 to 28 base pairs long, are produced from larger, double-stranded RNA molecules by the enzymatic action of proteins that are double-stranded RNA-specific endonucleases. Because these endonucleases "dice" large RNA into small pieces, they are called *Dicer* enzymes. The nematode *Caenorhabditis elegans* produces a single kind of Dicer enzyme; *Drosophila* produces two different Dicer enzymes; and *Arabidopsis* produces at least three. In *C. elegans* and *Drosophila*, these enzymes act in the cytoplasm; in *Arabidopsis*, they probably act in the nucleus. The siRNAs and miRNAs produced by Dicer activity are base-paired throughout their lengths except at their 3′ ends, where two nucleotides are unpaired.

In the cytoplasm, siRNAs and miRNAs become incorporated into ribonucleoprotein particles. The double-stranded siRNA or miRNA in these particles is unwound, and one of its strands is preferentially eliminated. The surviving single strand of RNA is then able to interact with specific messenger RNA molecules. This interaction is mediated by base pairing between the single strand of RNA in the RNA-protein complex and a complementary sequence in the messenger RNA molecule. Because this interaction prevents the expression of the gene that produced the mRNA, the RNA-protein particle is called an **RNA-Induced Silencing Complex (RISC).**

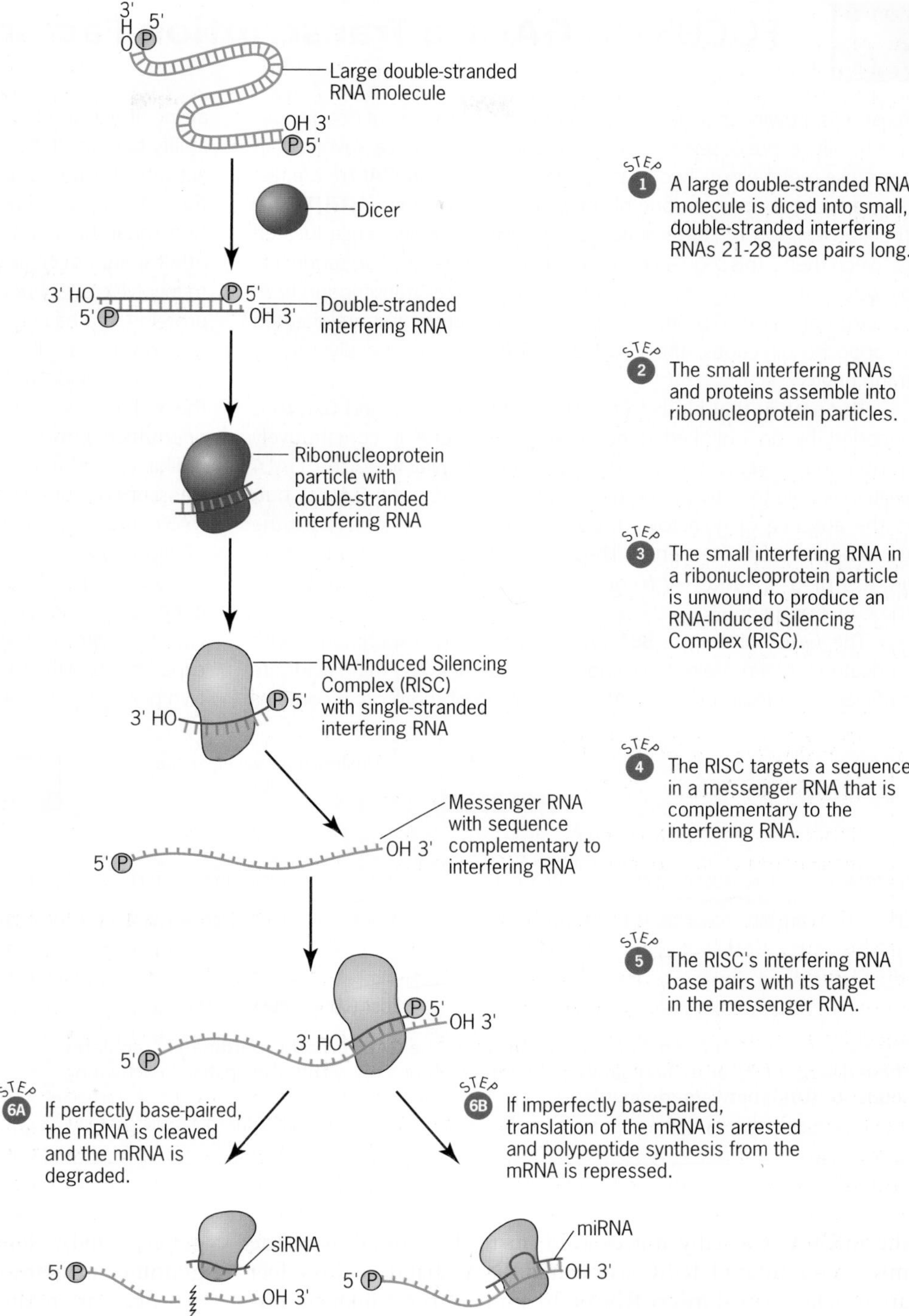

Figure 20.11 ▶ Summary of events involved in RNA interference pathways.

RISCs from different organisms vary in size and composition. However, they all contain at least one molecule from the whimsically named Argonaute family of proteins. The function of these proteins is not fully understood. Whenever the base pairing between the RNA within the RISC and the target sequence in the mRNA is perfect or nearly so, the RISC cleaves the target mRNA in the middle of the base-paired region. The cleaved mRNA is then degraded. The endonuclease responsible for cleaving the mRNA is currently unknown; however, it may be an Argonaute protein. After cleavage, the RISC may associate with another molecule of mRNA and induce its cleavage. Because a RISC may be used repeatedly without losing its ability to target and cleave mRNA, it behaves as a catalyst. RISC-associated RNAs that result in mRNA cleavage are usually termed **short interfering RNAs (siRNAs).** Whenever the RNA within the RISC pairs imperfectly with its target sequence,

▶ FOCUS ON **GAL4, a Transcription Factor from Yeast**

Yeast cells growing in a medium that contains galactose but not glucose specifically express several genes whose products are involved in galactose metabolism. Two of these genes, *GAL1* and *GAL10,* are less than a kilobase apart on one of the yeast chromosomes (**FIGURE 1**). The transcription of these genes is controlled by a sequence located between them. This sequence, called the *upstream activating sequence,* or *UAS,* is the binding site for a protein that activates transcription from the *GAL1* and *GAL10* promoters. Because these promoters are oriented in opposite directions, the *GAL1* and *GAL10* genes are divergently transcribed.

The protein that binds to the UAS between *GAL1* and *GAL10* is encoded by an unlinked gene, *GAL4.* This gene is constitutively transcribed in yeast cells. In the presence of galactose, the GAL4 protein is able to activate the transcription of *GAL1* and *GAL10,* but, in the absence of galactose, the GAL4 protein cannot activate the transcription of these genes. Thus galactose can be considered an inducer of *GAL1* and *GAL10* gene expression—through an effect of the GAL4 protein.

The GAL4 protein is 881 amino acids long. Experiments with truncated GAL4 proteins have shown that the first 73 amino acids are sufficient for binding to the UAS. This portion of the protein contains a zinc finger motif, a structural feature frequently associated with the ability to bind DNA. Once the GAL4 protein has bound to DNA, it can activate the transcription of nearby genes. Studies with portions of the GAL4 protein have shown that this activating property resides in two separate domains. One spans amino acids 148–196, and the other spans amino acids 768–881. Transcriptional activation is thought to involve contact between these parts of the GAL4 protein and other proteins that have bound to the gene's promoter, including, possibly, RNA polymerase II.

The UAS located between the *GAL1* and *GAL10* genes resembles the enhancers found in higher eukaryotes. It functions in either orientation and at some distance from the genes it regulates; however, unlike an enhancer, it cannot function downstream of a gene's transcription initiation site. Molecular analysis has revealed that the UAS contains four GAL4 binding sites. Each is an imperfect palindrome of eight base pairs centered on an A:T base pair.

When the UAS is artificially inserted upstream of a gene, that gene comes under the control of the GAL4 protein. This phenomenon was first demonstrated by Hitoshi Kakidani and Mark Ptashne in experiments with Chinese hamster cells that were transfected with two plasmids, one carrying a version of the *GAL4* gene and the other

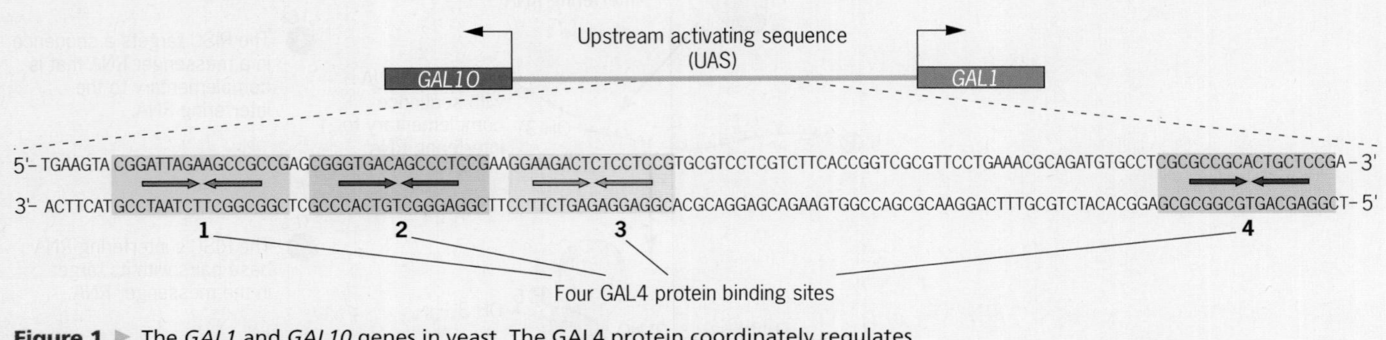

Figure 1 ▶ The *GAL1* and *GAL10* genes in yeast. The GAL4 protein coordinately regulates these divergently transcribed genes by binding to four sites within the upstream activating sequence (UAS) between them.

the mRNA is usually not cleaved; instead, translation of the mRNA is inhibited. RISC-associated RNAs that have this effect are usually termed **microRNAs.** In animals, the sequences targeted by these RISCs are found in the 3′ untranslated regions of mRNA molecules, and often these sequences are present several times within the 3′ untranslated region (UTR). In plants, the sequences targeted by RISCs are usually located within the coding region of the mRNA, or within the mRNA's 5′ UTR.

SOURCES OF SHORT INTERFERING RNAs AND MICRORNAs

Some of the small RNA molecules that induce RNAi are derived from the transcripts of microRNA genes. These genes, usually denoted by the symbol *mir*, are found in the genomes of many kinds of eukaryotes; about 100 *mir* genes are present in the *C. elegans* and *Drosophila* genomes, and about 250 are present in vertebrate genomes. Initially, a few of these genes were identified through analysis of mutations that altered the regulation of other genes. When the *mir* genes defined by these mutations were analyzed at the molecular level, they were found to have little or no protein-coding potential. Instead, they possessed a peculiar structure. Each of them contained a short stretch of nucleotides repeated in opposite orientations around a short intervening segment of DNA. When transcribed, this inverted repeat structure generates an RNA that can fold back on itself to form a short double-stranded stem at the base of a single-stranded loop carrying a "reporter" gene located downstream of the UAS.[1] The reporter gene encoded an enzyme (chloramphenicol acetyltransferase, CAT) for which a simple biochemical assay was available. Kakidani and Ptashne set out to determine whether the GAL4 protein expressed from one of the plasmids could stimulate the expression of the *CAT* reporter gene on the other plasmid by binding to the UAS near it. Their results (**FIGURE 2**) showed that plasmids carrying an intact *GAL4* gene, or a *GAL4* gene encoding the UAS-binding domain and at least one of the transcriptional activation domains of the protein, were able to stimulate expression of the *CAT* reporter gene. However, a plasmid with a *GAL4* gene encoding only the UAS-binding domain could not stimulate *CAT* gene expression. Thus, both the UAS-binding and transcriptional activation domains of the GAL4 protein are necessary for biological function.

The discovery that the GAL4 protein can regulate the expression of a gene near a UAS has become the basis of a procedure to study the effects of eukaryotic gene expression *in vivo.* In this procedure, a UAS is inserted upstream of a gene's transcription initiation site, and the resulting fusion gene is introduced into the cells of an organism—either transiently by transfection or permanently by transformation. A functional *GAL4* gene is then introduced into the cells to stimulate the expression of the fusion gene. This procedure provides a way to induce the expression of the fusion gene and thereby to ascertain the significance of that gene's function in a particular population of cells.

[1]Kakidani, H., and M. Ptashne. 1988. GAL4 activates gene expression in mammalian cells. *Cell* 52:161–167.

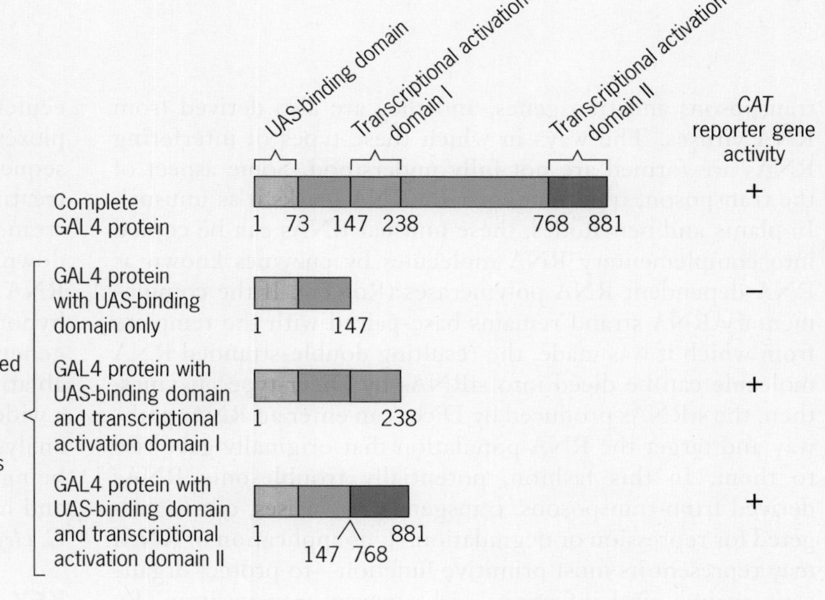

Figure 2 ► Results of the experiments of Kakidani and Ptashne on GAL4 regulation of a *CAT* reporter gene fused to a UAS. The complete GAL4 protein and two deleted proteins are capable of stimulating expression of the reporter gene. All three of these GAL4 proteins contain the UAS-binding domain and at least one of the transcriptional activation domains. A truncated GAL4 protein containing only the UAS-binding domain does not stimulate reporter gene expression.

(**FIGURE 20.12*a***). An enzyme called Drosha recognizes this stem-loop region and excises it from the primary transcript of the *mir* gene. The liberated stem-loop is then exported to the cytoplasm where it is cleaved by Dicer to form an miRNA. In *C. elegans*, where this process was discovered, Dicer removes the loop and trims the stem to a length of 22 nucleotides on each of its strands. After maturing in a RISC, the miRNA—now single-stranded—can target a sequence in the mRNA produced by another gene. **FIGURE 20.12*b*** shows base pairing between the miRNA from the *C. elegans mir* gene *lin-4* and one of this miRNA's targets in the 3′ UTR of the mRNA from a protein-coding gene, *lin-14*. Through this base pairing, the *lin-4* miRNA represses translation of the *lin-14* mRNA.

Since the discovery of these mutationally defined *mir* genes, many other *mir* genes have been found by using computer programs to screen the genomic DNA sequences of *C. elegans*, *Drosophila*, and other model organisms for the characteristic inverted repeat structure. Many of the candidate *mir* genes identified by this computer-based genomic approach have been verified by detecting miRNAs derived from these genes in cell extracts. Genes whose mRNAs contain sequences targeted by miRNAs are also being identified by a combination of computer-based analysis and *in vivo* experimentation. Many of these genes encode transcription factors or other developmentally significant proteins.

Some of the RNAs that induce RNAi are derived from the transcription of other elements in the genome such as

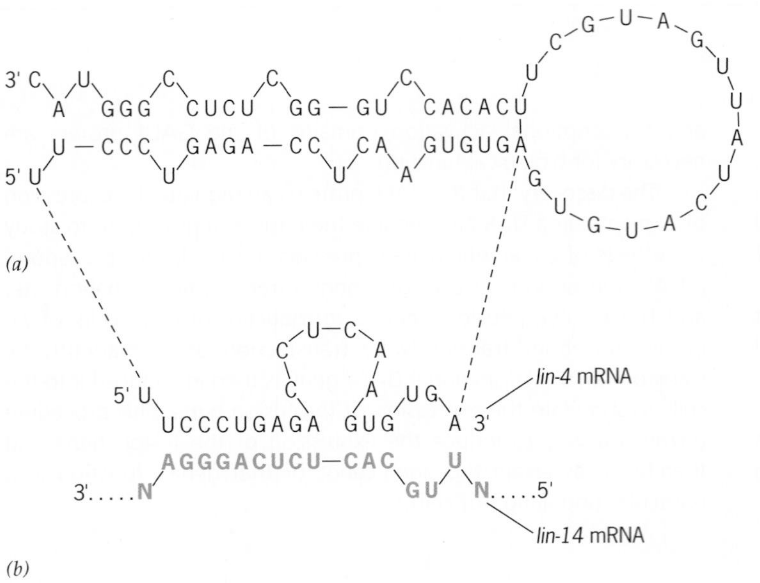

Figure 20.12 ▶ Regulation of gene expression by RNA interference. (*a*) Stem-loop structure of a transcript from the *C. elegans* microRNA gene *lin-4*. (*b*) Base-pairing between the microRNA derived from the *lin-4* transcript and a sequence in the 3′ untranslated region of the *lin-14* messenger RNA.

transposons and transgenes, and they are also derived from RNA viruses. The ways in which these types of interfering RNAs are formed are not fully understood. Some aspect of the transposon, transgene, or viral RNA marks it as unusual. In plants and nematodes, these unusual RNAs can be copied into complementary RNA molecules by enzymes known as RNA-dependent RNA polymerases (RdRPs). If the complementary RNA strand remains base-paired with the template from which it was made, the resulting double-stranded RNA molecule can be diced into siRNAs by Dicer-type enzymes; then, the siRNAs produced by Dicer can enter an RNAi pathway and target the RNA population that originally gave rise to them. In this fashion, potentially troublesome RNAs derived from transposons, transgenes, or viruses, can be targeted for repression or degradation. This application of RNAi may represent its most primitive function—to protect organisms against viral infections and runaway transposition. By contrast, the intricate miRNA-based systems for gene regulation evident in organisms like *C. elegans* seem to represent highly evolved applications of RNAi.

Researchers have discovered that RNAi can also be induced by double-stranded RNA that has been prepared *in vitro* by transcription from cloned genes or gene segments (see A Milestone in Genetics: The Discovery of RNA Interference later in this chapter). The DNA is transcribed in both directions by inserting it between promoters in opposite orientations in a suitable cloning vector, or by inserting inverted copies of the DNA downstream of a single promoter (see Chapter 17). Double-stranded RNA molecules derived from the transcripts of such clones can be transfected into cultured cells; they can also be injected into living organisms. Once inside cells, the double-stranded RNA enters an RNAi pathway. It is diced into siRNA molecules, which are then incorporated into RNA-protein complexes and targeted to mRNAs containing complementary sequences. The targeted mRNAs are usually degraded. Thus, treating cells or organisms with a particular type of double-stranded RNA has the effect of knocking out or knocking down the expression of the gene that corresponds to that RNA. It is therefore equivalent to inducing an amorphic or hypomorphic mutation in the gene. Using this approach, geneticists have been able to study the consequences of ablating or attenuating the expression of particular genes in a wide variety of organisms, including some in which genetic analysis is difficult, slow, or impossible. Thus, RNAi is now being used to analyze the function of genes in fish, rodents, and humans, as well as in simpler model organisms such as *C. elegans*, *Drosophila*, and *Arabidopsis*.

KEY POINTS

- Short interfering RNAs and microRNAs are produced from larger double-stranded precursors by the action of Dicer-type endonucleases.
- In RNA-Induced Silencing Complexes (RISCs), siRNAs and miRNAs become single-stranded so they can target complementary sequences in messenger RNA molecules.
- Messenger RNA that has been targeted by siRNA is cleaved, and mRNA that has been targeted by miRNA is prevented from serving as a template for polypeptide synthesis.
- Hundreds of genes for miRNAs are present in eukaryotic genomes.
- Transposons and transgenes may stimulate the synthesis of siRNAs.
- RNA interference is used as a research tool to knock out or knock down the expression of genes in cells and whole organisms.

Gene Expression and Chromosome Organization

Many features of chromosomes influence the transcription of genes.

The primary event in gene expression is transcription of DNA into RNA. For this to occur, the DNA must be accessible to RNA polymerase and an assortment of other proteins that help to initiate transcription. If the DNA is tightly bound by histones or other kinds of "packaging" proteins, it may be too condensed to allow transcription. Chromosomal position may also affect the transcription of genes because genes that have been transposed to different sites within a chromosome often show altered expression. These alterations may be caused by enhancers at the new chromosomal site or by other aspects of chromosome structure. All these phenomena indicate that gene expression is influenced by chromosome organization.

TRANSCRIPTION IN LAMPBRUSH CHROMOSOME LOOPS

Few if any of the genes in highly condensed chromosomes, such as those found at metaphase of mitosis or in mature sperm, are transcribed. In order to be transcribed, the chromatin must be "open" to the transcriptional apparatus. Some of the first evidence supporting this idea came from the cytological localization of transcription in the meiotic chromosomes of amphibian oocytes. These very large (400 to 800 μm long), duplicated chromosomes—called lampbrush chromosomes—consist of a highly condensed axis surrounded by numerous pairs of lateral loops (Chapter 9).

The lateral loops in lampbrush chromosomes are regions of intense transcriptional activity (**FIGURE 20.13**). Researchers have demonstrated this activity by pulse-labeling amphibian oocytes with ^{3}H-uridine and then examining the chromosomes by autoradiography. The ^{3}H-uridine is incorporated into newly synthesized RNA. Autoradiography of pulse-labeled oocytes reveals that the radioactive uridine is localized around the lateral loops of the lampbrush chromosomes rather than around the condensed axes; thus, the more loosely organized loops are actively involved in RNA synthesis.

TRANSCRIPTION IN POLYTENE CHROMOSOME PUFFS

Additional cytological evidence for transcription in "open" chromosome regions comes from the study of polytene chromosomes in *Drosophila* and other Dipteran insects. As discussed in Chapter 6, these chromosomes consist of hundreds of sister chromatids aligned side by side, creating a large, thick cable that is longitudinally differentiated into light and dark bands. For many years, the bands, called **chromomeres,** have been thought to have a functional significance. One reason for this speculation is that during the course of development, particular bands expand into diffuse, less densely staining structures called **puffs** (**FIGURE 20.14**). *In situ* hybridization experiments have shown that puffs contain genes that are actively transcribed. In these experiments, a radioactively labeled RNA or DNA that

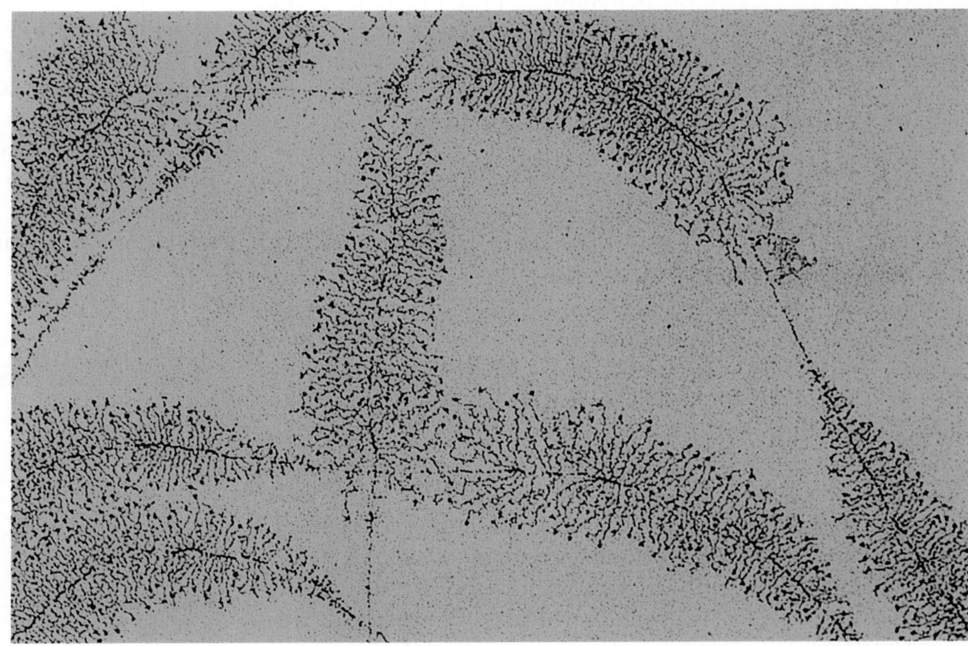

Figure 20.13 ▶ Photomicrograph showing transcription from the DNA loops of amphibian lampbrush chromosomes.

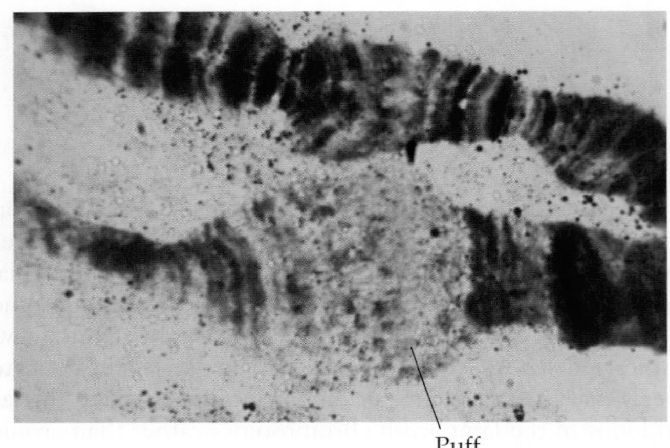

Figure 20.14 ▶ A puff in the polytene chromosomes of *Drosophila*.

contains a specific sequence is denatured and hybridized with denatured RNA or DNA in chromosomes that have been prepared for cytological analysis. The labeled RNA or DNA probe binds to its complementary sequence in the chromosomes—that is, it binds *in situ;* the location of the bound probe is then revealed by autoradiography. Application of this technique shows that puffs in polytene chromosomes contain DNA sequences that are actively transcribed into RNA. Over time, particular puffs regress and new ones form at different sites in the chromosomes. This temporal sequence of puffing is controlled by **ecdysone,** a steroid hormone that appears in pulses at different times during development. A plausible interpretation is that ecdysone induces transcription in puffed bands. Puffing may also be induced by other factors, for example, heat shock. When *Drosophila* larvae are incubated at temperatures greater than 33°C, several new puffs appear in the polytene chromosomes and old ones regress. The new puffs correspond to the loci for the heat-shock genes. These cytological studies with *Drosophila* therefore show that diffuse, expanded chromosomal regions are sites of intense RNA synthesis.

MOLECULAR ORGANIZATION OF TRANSCRIPTIONALLY ACTIVE DNA

The findings with lampbrush and polytene chromosomes raise questions about the molecular organization of transcriptionally active DNA. Is this DNA actually more "open" than nontranscribed DNA? This question has been answered by measuring the sensitivity of DNA in chromatin to the action of pancreatic deoxyribonuclease I (DNase I), an enzyme that cleaves DNA molecules and degrades them into their constituent nucleotides. In 1976, Mark Groudine and Harold Weintraub demonstrated that transcriptionally active DNA is more sensitive to DNase I than nontranscribed DNA. Groudine and Weintraub extracted chromatin from chicken red blood cells and partially digested it with DNase I. Then they probed the residual chromatin material for sequences of two genes, β-globin, which is actively transcribed in red blood cells, and ovalbumin, which is not. They found that over 50 percent of the β-globin DNA had been digested by the DNase I enzyme, compared with only 10 percent of the ovalbumin DNA. These results strongly implied that the actively transcribed gene was more "open" to nuclease attack. Subsequent research has shown that the nuclease sensitivity of transcriptionally active genes depends on at least two small nonhistone proteins, HMG14 and HMG17 (HMG for *h*igh *m*obility *g*roup, because they have high mobility during gel electrophoresis). When these proteins are removed from active chromatin, nuclease sensitivity is lost; when they are added again, it is restored.

The treatment of isolated chromatin with a very low concentration of DNase I causes the DNA to be cleaved at a few specific sites, appropriately called *DNase I hypersensitive sites.* Some of these sites have been shown to lie upstream of transcriptionally active genes, in either promoter or enhancer regions. The functional significance of these hypersensitive sites is still unclear, but some evidence suggests that they may mark regions in which the DNA is locally unwound, perhaps because transcription has begun.

In the case of the human genes for β-globin, several DNase I hypersensitive sites are located in a 15-kb-long *locus control region (LCR)* upstream of the genes themselves (**FIGURE 20.15**). The human β-globin genes reside in a cluster spanning 28 kilobases on chromosome 11. Each of the genes in the cluster is a duplicate of an ancestral β-globin gene. Over evolutionary time, the individual genes in the cluster have diverged from one another by random mutation so that today, each one of them encodes a slightly different polypeptide. In one of the genes, a chain-terminating mutation has abolished the ability to make a polypeptide. Such noncoding genes are called pseudogenes, and they are usually denoted by the Greek letter psi (ψ)—thus, the ψ β gene in this cluster. The human β-globin genes are spatially and temporally regulated. In fact, a remarkable feature of this gene cluster is that its members are expressed at different times during development. The ε gene is expressed in the embryo, the two γ genes are expressed in the fetus, and the δ and β genes are expressed in infants and adults. This sequential activation of genes from one side to the other in the cluster is apparently related to the need to produce slightly different kinds of hemoglobin during the course of human development.

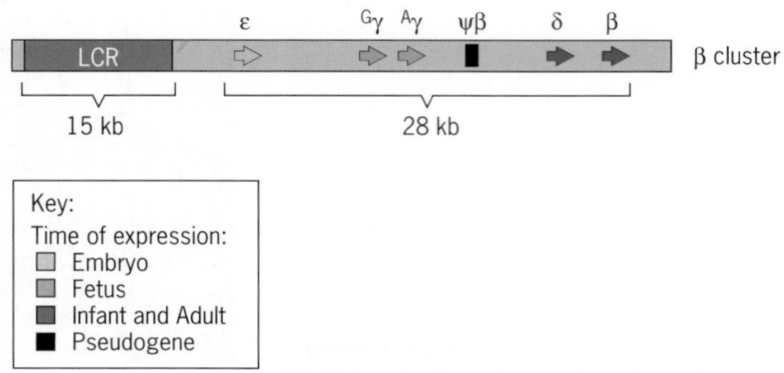

Figure 20.15 ▶ The β-globin gene cluster on human chromosome 11.

Embryo, fetus, and infant have different oxygen requirements, different circulatory systems, and different physical environments. The temporal switching in β-globin gene expression is apparently an adaptation to this changing array of conditions.

The LCR of the β-globin gene cluster contains binding sites for transcription factors that preactivate the individual genes for transcription. Preactivation is detected by an increase in the sensitivity of the DNA within the LCR to digestion with low concentrations of DNase I. Transcription of the β-globin genes appears to require this preactivation and is stimulated by transcription factors that bind to specific enhancers in the β-globin gene complex. However, the tissue and temporal specificity of β-globin gene expression depends on sequences embedded in the LCR. Studies with transgenic mice indicate that the LCR is not simply a large collection of enhancers that exert control over the various β-globin genes. The LCR must be situated upstream of the β-globin genes and in its natural orientation in order to control gene expression properly. That is, it functions in an orientation-dependent manner. Enhancers typically function in an orientation-independent manner and in different positions relative to a gene's promoter. The LCR has one other feature that distinguishes it from simple enhancers: it can control β-globin gene expression when the entire gene cluster (LCR plus β-globin genes) is inserted in a different chromosomal position. Enhancers, by contrast, often fail to function when they and their associated genes are transposed to a different chromosomal location. Thus, the LCR seems to insulate the β-globin genes from the influence of the chromatin around them.

CHROMATIN REMODELING

Experiments that assess the sensitivity of DNA to digestion with DNase I have established that transcribed DNA is more accessible to nuclease attack than nontranscribed DNA. Is transcribed DNA packaged in nucleosomes? If it is, what structural changes occur in the nucleosomes during transcription? Are the nucleosomes "opened" and "closed" as the RNA polymerase passes along the DNA template? Efforts to answer these questions have involved a combination of genetic and biochemical approaches that have demonstrated that transcribed DNA is indeed packaged into nucleosomes. However, in transcribed DNA, the nucleosomes are altered by multiprotein complexes that ultimately facilitate the action of the RNA polymerase. This alteration of nucleosomes in preparation for transcription is called **chromatin remodeling.**

Two general types of chromatin-remodeling complexes have been identified. One type is composed of enzymes that transfer acetyl groups to the amino acid lysine at specific positions in the histones of the nucleosomes. As a class, these enzymes are called histone acetyl transferases (HATs). Numerous studies have shown that acetylation of histones is correlated with increased gene expression, perhaps because the addition of the acetyl groups loosens the association between the DNA and the histone octamers in the nucleosomes. Kinases—enzymes that transfer phosphate groups to molecules—may also play a role along with these chromatin-remodeling complexes. It is known, for example, that acetylation of lysine-14 in histone H4 is often preceded by phosphorylation of serine-10 in that molecule. Together, these two modifications of histone H4 seem to "open" the chromatin for increased transcriptional activity.

Another type of chromatin-remodeling complex disrupts nucleosome structure in the vicinity of a gene's promoter. The most intensively studied of these complexes is the SWI/SNF complex found in baker's yeast. This complex is named for the two types of mutations (*sw*itching-*i*nhibited and *s*ucrose *n*on*f*ermenter) that led to the discovery of its constituent proteins. Related complexes have been found in the cells of other organisms, including humans. The SWI/SNF complex consists of at least eight proteins. It regulates transcription by sliding histone octamers along the associated DNA in nucleosomes; it can also transfer these octamers to other locations on a DNA molecule. The nucleosome shifting catalyzed by the SWI/SNF complex apparently gives transcription factors access to the DNA. These factors then stimulate a gene's expression.

We have discussed chromatin remodeling from the point of view of gene activation. However, active chromatin can also be remodeled into inactive chromatin. This reverse remodeling seems to involve two biochemical modifications to the histones in nucleosomes: deacetylation, catalyzed by the histone deacetylases (HDACs), and methylation, catalyzed by the histone methyl transferases (HMTs). Some of the nucleotides in the DNA may also be methylated by a group of enzymes called the DNA methyl transferases (DNMTs). Chromatin that has been subjected to these modifications tends to be transcriptionally silent. The subject of DNA methylation is explored more fully in a later section.

EUCHROMATIN AND HETEROCHROMATIN

Variation in the density of chromatin within the nuclei of cells leads to differential staining of sections of chromosomes. The deeply staining material is called **heterochromatin,** and its lightly staining counterpart is called **euchromatin.** What, if any, is the functional significance of these different types of chromatin?

A combination of genetic and molecular analyses has shown that the vast majority of eukaryotic genes are located in euchromatin. Moreover, when euchromatic genes are artificially transposed to a heterochromatic environment, they tend to function abnormally, and, in some cases, not to function at all. This impaired ability to function can create a mixture of normal and mutant characteristics in the same individual, a condition referred to as **position-effect variegation.** This term is used because the variability in the phenotype is caused by changing the position of the euchromatic gene, specifically by relocating it to the heterochromatin. Many examples of position-effect variegation have been discovered in *Drosophila,* usually in association with inversions or translocations that move a euchromatic gene into the heterochromatin. The *white mottled* allele, w^{m4}, is a good example. In this case, a wild-type allele of the *white* gene has been relocated by an inversion, with one break near the *white* locus and the other in the basal heterochromatin of the X chromosome. This rearrangement interferes with the normal expression of the *white* gene and causes a mottled-eye phenotype. Apparently, the euchromatic *white* gene cannot function well in a heterochromatic environment. This and other examples have led to the view that heterochromatin represses gene function, perhaps because it is condensed into a form that is not accessible to the transcriptional machinery. Ongoing research is attempting to identify the proteins that might be involved in this condensation process.

One research strategy has been to identify genes that, when mutant, suppress or enhance position-effect variegation. Screens for such mutations have used the variegating w^{m4} allele of the *white* locus in *Drosophila* as a convenient assay. When w^{m4} is combined with a *su*ppressor of position-effect *var*iegation—that is, with a *su(var)* mutation, the eye color of the fly will be wild type, or nearly so; when w^{m4} is combined with an *e*nhancer of position-effect *var*iegation—that is, with an *e(var)* mutation, the eye color will be more mottled than it is with w^{m4} alone. **FIGURE 20.16** shows the effect of a dominant suppressor of position-effect variegation, $Su(var)205^4$, on the w^{m4} phenotype. Molecular analysis has demonstrated that this suppressor mutation impairs the production of a nonhistone chromosomal protein that associates predominantly with the heterochromatin. Flies heterozygous for the $Su(var)205^4$ mutation suppress the w^{m4} phenotype; however, flies that are homozygous for it die early in development. Thus, the protein encoded by the wild-type *Su(var)205* gene, called heterochromatin protein 1, or simply HP1, is essential for life. Perhaps a mutation in one copy of the *Su(var)205* gene reduces the amount of HP1 enough to relax the heterochromatin but not enough to unravel it. In this relaxed state, a euchromatic gene such as w^{m4} that has come under the repressive influence of the heterochromatin is expressed at normal levels; consequently, the phenotype that it controls ceases to variegate.

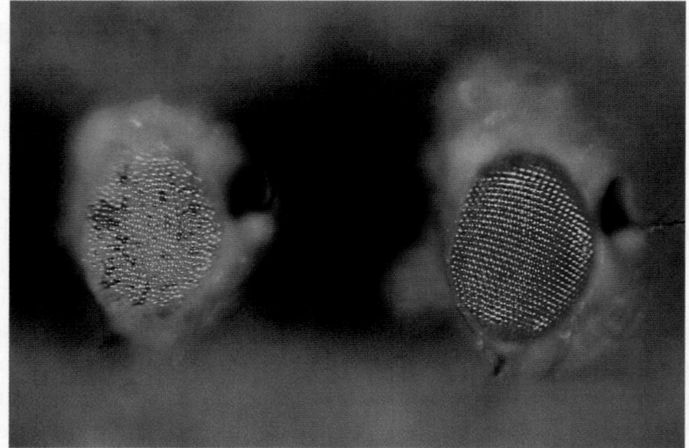

Figure 20.16 ▶ Effect of a suppressor of variegation mutation on eye color in *Drosophila*. Left: Eye from a fly carrying the w^{m4} allele. Right: Eye from a fly carrying w^{m4} along with a dominant suppressor of variegation, *Su(var)205*^{4}.

Approximately 30 percent of *Drosophila* DNA is heterochromatic. Most of this DNA is located around the centromeres of the chromosomes. However, some of it is interspersed in the euchromatin, and some is located at the ends, or telomeres, of the chromosomes. Thus, in *Drosophila*, three types of heterochromatin are present: *centric* heterochromatin (around the centromeres), *intercalary* heterochromatin (scattered in the euchromatin), and *telomeric* heterochromatin (at the ends of the chromosomes).

Most of the heterochromatic regions in the *Drosophila* genome are rich in repetitive DNA sequences. Both highly repetitive and moderately repetitive sequences can be found in these regions, and the moderately repetitive sequences comprise many types of transposable elements, especially inactive retrotransposons (Chapter 18). The heterochromatic regions of the *Drosophila* genome also contain some expressed genes. Several hundred genes for ribosomal RNA are located in the centric heterochromatin of the X and Y chromosomes, and about 20 protein-encoding genes are known to reside in other heterochromatic regions of the *Drosophila* genome. Thus, heterochromatin is not always associated with complete genetic inactivity. Curiously, when a heterochromatic gene is moved into a euchromatic environment by a chromosome rearrangement, its expression tends to variegate. Furthermore, this variegation tends to be enhanced by the *su(var)* mutations that suppress the variegation of a euchromatic gene that has been placed in a heterochromatic environment. The fact that heterochromatic genes exhibit position effects opposite to those of euchromatic genes indicates that they are specially adapted to function in a heterochromatic environment. However, the nature of this adaptation is currently unknown.

GENE SILENCING

Although heterochromatin has the ability to repress euchromatic genes that have been placed near it by a chromosome rearrangement, this feature of chromatin is not the only one that prevents gene expression in eukaryotes. Studies with many different organisms indicate that other aspects of chromatin organization, some possibly related to the protein components of heterochromatin, control gene activity. We have already seen that transcription factors, acting in concert with particular enhancer and promoter sequences, turn genes on in a spatially and temporally specific manner. However, at a particular time in a given eukaryotic cell, most genes are not being transcribed. Thus, the rule, rather than the exception, is that genes are transcriptionally silent. What features of chromatin organization bring about this widespread gene silencing?

Silencing by the Polycomb Group Proteins

In *Drosophila*, studies of genes that control the formation of particular body parts have identified proteins that are involved in gene silencing. These proteins are encoded by a special class of genes known as the *Polycomb group*, abbreviated *PcG*. This class is named for *Polycomb*, a gene defined by a dominant mutation that causes extra bristles to form in clusters on the legs. These clusters, known as sex combs, are normally found on the forelegs of male *Drosophila;* however, in males that carry a *Polycomb* mutation, they are found on the middle and hind legs as well. *Polycomb* mutations also cause other phenotypic anomalies, including the transformation of antennae into rudimentary legs.

Polycomb and the other *PcG* genes encode proteins that regulate the expression of transcription factors involved in *Drosophila* development. These regulatory proteins appear to form large multiprotein complexes that associate with specific regions of the genome—in particular, with the loci that encode these developmentally important transcription factors. When and how they do this remain a mystery. However, several facts are known. First, the PcG proteins associate with chromosomal regions that contain special DNA sequences known as *Polycomb response elements* (*PREs*). Individually, most of the PcG proteins are not able to bind to the PRE DNA itself; however, they may be able to bind as a complex, or they may interact with other types of proteins that can bind to the PRE DNA. Second, the PcG proteins repress the genes located near the PREs. Whether or not the PcG proteins associate with a particular PRE and repress the genes in its neighborhood depends on factors present in the animal very early in its embryonic development. Third, once established, the association between the PcG proteins and a PRE is maintained faithfully throughout development. Thus, the repressed state of the genes near a PRE is transmitted from a cell to its daughters through each mitotic division. This form of gene silencing therefore assures that clones of cells develop consistently into particular body parts. When the PcG mechanism of gene silencing is compromised by mutations in any of the *PcG* genes, some body parts develop inappropriately, as, for example, with the *Polycomb* mutant itself, where antennae may develop as legs, and middle and hind legs may develop as forelegs.

Silencing of Mating Type Cassettes

The phenomenon of gene silencing has also been studied in yeast where one of the most informative phenotypes is mating type. The haploid yeast *Saccharomyces cerevisiae* has two possible mating types, "a" and α. The type that is expressed by a particular cell is controlled by the mating type locus (*MAT*) situated on chromosome 3. Genetic information in this locus specifies the mating type as either "a" or α. Thus, the *MAT* locus has two alleles, *MATa* and *MAT*α. Colonies grown from single cells all express the same mating type information. However, at a frequency that depends on the particular yeast strain, the mating type can switch from "a" to α, or from α to "a."

The basis for this switching is now well understood. At some distance on either side of the *MAT* locus there are DNA segments called *silent mating type cassettes* that contain extra genetic information for mating type (**FIGURE 20.17**). The cassette on the left, denoted *HML*α, contains information for the α mating type, and the cassette on the right, denoted *HMRa*, contains information for the "a" mating type. However, a gene-silencing mechanism prevents either of these cassettes from being expressed. Thus, the mating type of a cell is exclusively controlled by the genetic information at the *MAT* locus. When a yeast cell expresses an enzyme called the HO endonuclease, DNA in the *MAT* locus can be replaced by DNA from one of the flanking cassettes. If the *MAT* locus carries the *a* allele, the silent α allele located in the cassette to its left can be copied

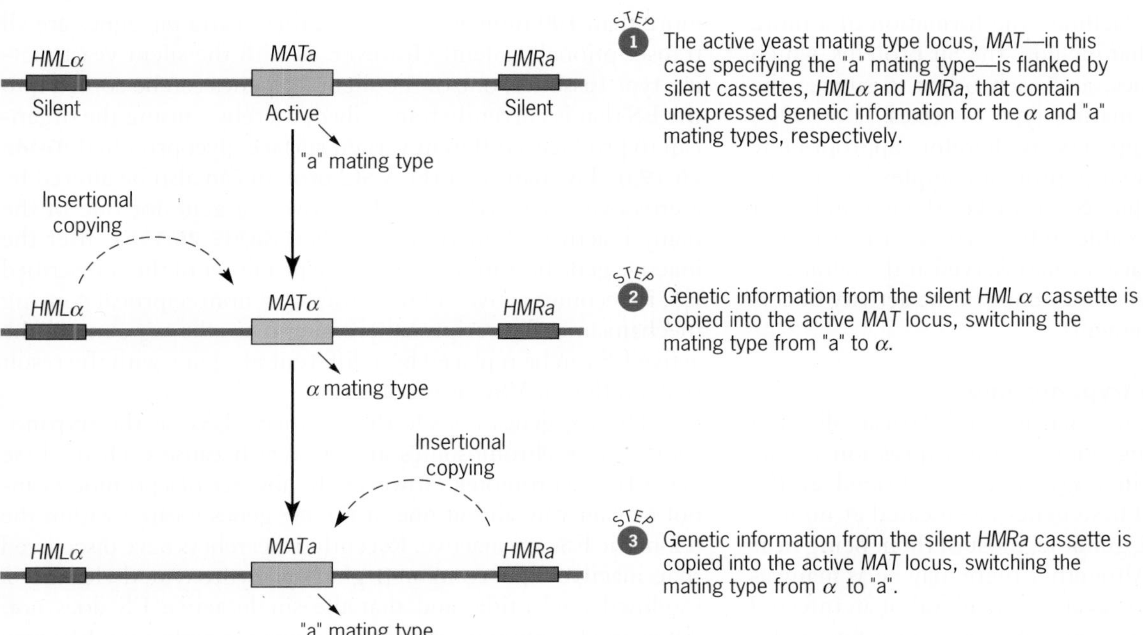
HMLα
MATa
HMRa
Silent
Active
Silent
"a" mating type
Insertional copying
HMLα
MATα
HMRa
α mating type
Insertional copying
HMLα
MATa
HMRa
"a" mating type
STEP 1 The active yeast mating type locus, MAT—in this case specifying the "a" mating type—is flanked by silent cassettes, HMLα and HMRa, that contain unexpressed genetic information for the α and "a" mating types, respectively.
STEP 2 Genetic information from the silent HMLα cassette is copied into the active MAT locus, switching the mating type from "a" to α.
STEP 3 Genetic information from the silent HMRa cassette is copied into the active MAT locus, switching the mating type from α to "a".

Figure 20.17 ▶ The process of mating type switching in yeast.

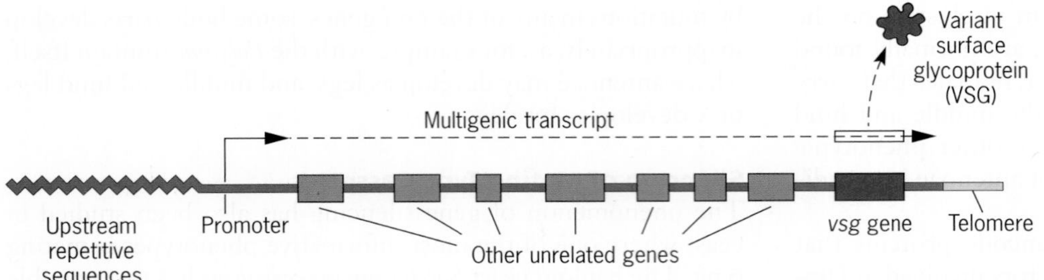

Figure 20.18 ▶ Structure of an expression site (ES) containing an active *vsg* gene near the end of one of the chromosomes in *Trypanosoma brucei.*

into the *MAT* locus and subsequently expressed. This DNA replacement will cause the mating type of the cell to switch from "a" to α. Conversely, if the *MAT* locus carries the α allele, the silent *a* allele in the cassette to its right can be copied into the *MAT* locus and then expressed, switching the mating type from α to "a."

An important feature of the entire yeast mating type system is that a particular yeast cell expresses the genetic information for only one of the two possible mating types. Thus, the cassettes that flank the expressed *MAT* locus must be silenced. Recent research has demonstrated that the silencing of these cassettes involves several proteins, including one called the repressor/activator protein, abbreviated Rap1p; several silent information regulator proteins, abbreviated Sir(*n*)p, where the numeral *n* identifies the specific protein; and several proteins in the origin recognition complex, abbreviated ORC—a complex that is also involved in DNA replication. Experimental evidence indicates that histone H4 is involved in silencing the left and right mating type cassettes. Although the detailed mechanism remains to be elucidated, it appears that the ORC proteins and Rap1p interact with specific DNA sequences situated alongside the mating type loci in the left and right cassettes. These bound proteins then facilitate the formation of a more elaborate protein complex that includes the Sir proteins, which interact with Rap1p, histones, and each other to silence the genes in the left and right mating type cassettes. The DNA sequences that mediate this process are therefore appropriately called **silencers.** When this large protein complex has assembled around the silencer sequences, it makes the left and right mating type cassettes inaccessible to the enzymes that catalyze transcription. Silencing has also been observed at the telomeres of yeast chromosomes, where it may control the activity of certain types of transposable elements.

Silencing of *vsg* Genes in Trypanosomes

In the trypanosomes that cause nagana and African sleeping sickness, a form of silencing controls the expression of the variant surface glycoprotein (*vsg*) genes mentioned at the beginning of this chapter. These genes are located at numerous sites in the trypanosome genome, most of them being near the ends of chromosomes. Altogether there may be as many as 1000 different *vsg* genes. However, in the blood of an infected animal, trypanosomes express only one of the *vsg* genes, and it is always located near a chromosome end. All the other *vsg* genes, including all the *vsg* genes that are located near telomeres, are silenced. The single expressed *vsg* gene resides in a multigenic unit controlled by a promoter about 40 to 50 kb away (**FIGURE 20.18**); other unrelated genes lie between the promoter and the *vsg* gene. This entire multigenic unit is called an *expression site (ES).*

Expression sites appear to exist at the ends of each of the trypanosome's 24 large chromosomes; however, at any one time, only one of them is transcribed. An as-yet-unknown mechanism silences the other potential expression sites, thereby restricting the infecting trypanosomes to the production of a single type of variant surface glycoprotein (VSG). During the course of an infection, the ES that is transcribed may change. The active ES may be silenced and a previously inactive ES become active, leading to the production of a different VSG. Such changes provide a way for the trypanosomes to alter their surface coats and evade destruction by the infected animal's immune system.

The trypanosome's surface coat can be changed in two other ways. In addition to the *vsg* genes located at the telomeres of the large chromosomes, trypanosomes possess numerous other *vsg* genes situated at internal loci and at the telomeres of more than 100 minichromosomes; these extra *vsg* genes are all transcriptionally silent. However, as with the silent yeast mating type cassettes, one of the silent *vsg* genes can be copied into the ES that is currently transcribed, thereby causing the organism to produce a different variant surface glycoprotein (**FIGURE 20.19*a***). The nature of the VSG protein can also be altered by a crossover that exchanges the active *vsg* gene for one of the many inactive telomeric *vsg* genes (**FIGURE 20.19*b***); after the inactive gene becomes linked to the promoter of the transcribed ES, it becomes active. Thus, either by a nonreciprocal copying mechanism or by reciprocal crossing over, the *vsg* gene in the active ES can be replaced by a different *vsg* gene, with the result that a different VSG is produced.

The *vsg* genes outside the telomeric ESs on the trypanosome's large chromosomes are inactive because each of these genes lacks a promoter. However, the absence of a promoter cannot explain why all but one of the *vsg* genes located within the telomeric ESs are inactive. Recently, researchers have discovered that inactive ESs contain β-glucosyl-hydroxymethyluracil, a modified nucleotide, and that the single active ES does not. Thus, modification of nucleotide sequences within the ESs may play a role in telomeric *vsg* gene silencing.

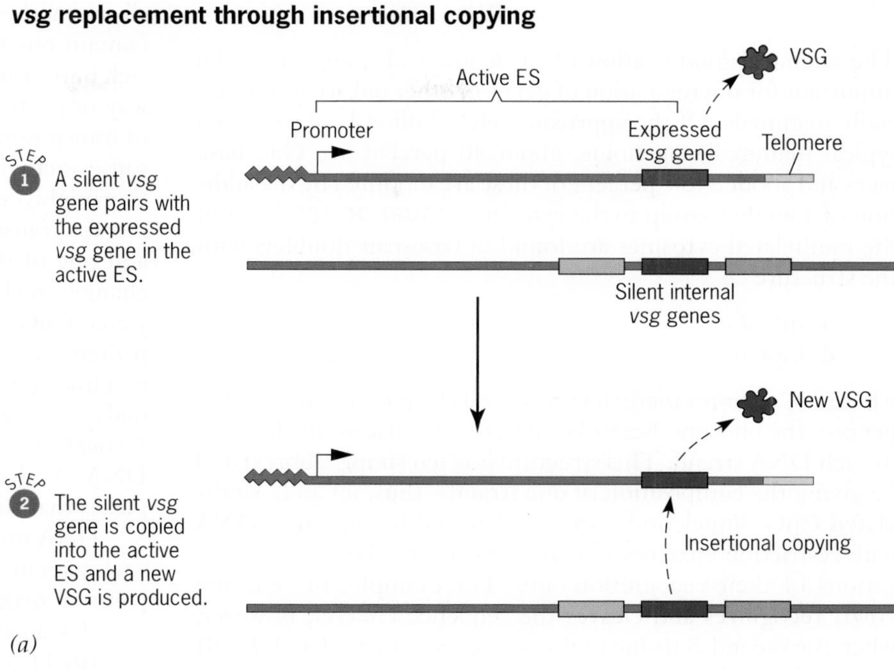
vsg replacement through insertional copying
VSG
Active ES
Promoter
Expressed
vsg gene
Telomere
STEP 1
A silent vsg gene pairs with the expressed vsg gene in the active ES.
Silent internal
vsg genes
New VSG
STEP 2
The silent vsg gene is copied into the active ES and a new VSG is produced.
Insertional copying
(a)

vsg replacement through crossing-over

STEP 1 A silent *vsg* gene pairs with the expressed *vsg* gene in the active ES. Crossing over occurs between the paired genes and the upstream promoter in the active ES.

VSG

Active ES

Promoter

Expressed *vsg* gene

Crossing over

Silent telomeric *vsg* gene

New VSG

STEP 2 The *vsg* genes are switched, and a new VSG is produced by expression of the formerly silent *vsg* gene.

(b)

Figure 20.19 ▶ Changing the expression of *vsg* genes in trypanosomes. (*a*) Replacement of the telomeric *vsg* gene in an active trypanosome ES by a silent internal *vsg* gene encoding a different variant surface glycoprotein (VSG). The replacement mechanism excises the currently active *vsg* gene and inserts a copy of the silent *vsg* gene into the vacant site within the active ES. At the end of this process, a new variant surface glycoprotein is produced. (*b*) Replacement of the telomeric *vsg* gene in an active trypanosome ES by a silent telomeric *vsg* gene encoding a different VSG. The replacement mechanism involves crossing over between the paired *vsg* genes and the promoter of the ES.

RNAi-Mediated Silencing of Transcription

We have already seen that RNA interference can impact gene expression by causing the degradation of mRNAs or by arresting the translation of these molecules. Thus, RNAi provides a mechanism for posttranscriptional gene silencing. Recent research has implicated RNAi in the transcriptional silencing of genes as well. Transgenes and transposons, especially those located in heterochromatic regions of a genome, may be silenced through the influence of short interfering RNAs. The principal evidence for this transcriptional silencing comes from studies of transposons in *C. elegans*, transgenes in *Drosophila*, and the mating type loci in *Schizosaccharomyces pombe*. In *Drosophila* and *S. pombe*, RNAi pathways appear to be involved in the formation of heterochromatin around the centromeres of chromosomes. However, many other factors are also involved, and they may, in fact, play the major roles in this form of transcriptional silencing.

DNA METHYLATION AND IMPRINTING

The chemical modification of nucleotides also appears to be important for the regulation of genes in other eukaryotes, especially mammals. Of the approximately 3 billion base pairs in a typical mammalian genome, about 40 percent are G:C base pairs and about 2 to 7 percent of these are modified by the addition of a methyl group to the cytosine (**FIGURE 20.20**). Most of the methylated cytosines are found in base-pair doublets with the structure

5′ mCpG 3′
3′ GpCm 5′

where mC denotes methylcytosine and the p between C and G denotes the phosphodiester bond between adjacent nucleotides in each DNA strand. This structure is often simply abbreviated by giving the composition of one strand—thus, mCpG. Methylated CpG dinucleotides can be detected by digesting DNA with restriction enzymes that are sensitive to chemical modifications of their recognition sites. For example, the enzyme *Hpa*II recognizes and cleaves the sequence CCGG; however, when the second cytosine in this sequence is methylated, *Hpa*II cannot cleave the sequence. Thus, methylated and unmethylated DNAs give different patterns of restriction fragments when they are digested with this enzyme.

CpG dinucleotides occur less often than expected in mammalian genomes, probably because they have been mutated into TpG dinucleotides over the course of evolution. Moreover, the distribution of CpG dinucleotides is uneven, with numerous short segments of DNA having a much higher density of CpG dinucleotides than other regions of the genome. These CpG-rich segments, usually about 1 to 2 kb long, are called **CpG islands.** In the human genome, there are about 30,000 such islands, most being situated near transcription start sites. Molecular analysis has demonstrated that the cytosines in these islands are rarely, if ever, methylated, and that this un- or undermethylated state is conducive to transcription. Thus, DNA in the vicinity of a CpG island is hypersensitive to digestion with DNase I, and its nucleosomes are usually somewhat different than nucleosomes elsewhere in the genome—typically, there is less histone H1, and some of the core histones are acetylated.

Where methylated DNA is found, it is associated with transcriptional repression. This is most dramatically seen in female mammals where the inactive X chromosome is extensively methylated. Regions of the mammalian genome that contain repetitive sequences, including those regions that are rich in transposable elements, are also methylated, perhaps as a way of protecting the organism against the deleterious effects of transposon expression and movement. The mechanisms that cause methylated DNA to be transcriptionally silent are not thoroughly understood; however, at least two proteins that repress transcription are known to bind to methylated DNA, and one of these, denoted MeCP2, has been shown to cause changes in chromatin structure. Thus, it is possible that methylated CpG dinucleotides bind specific proteins and that these proteins form a complex that prevents the transcription of neighboring genes. In this respect, methylated DNA in mammals may act like the Polycomb response elements in *Drosophila*—an organism that has little, if any, methylated DNA. As discussed earlier, these elements bind protein complexes that silence nearby genes.

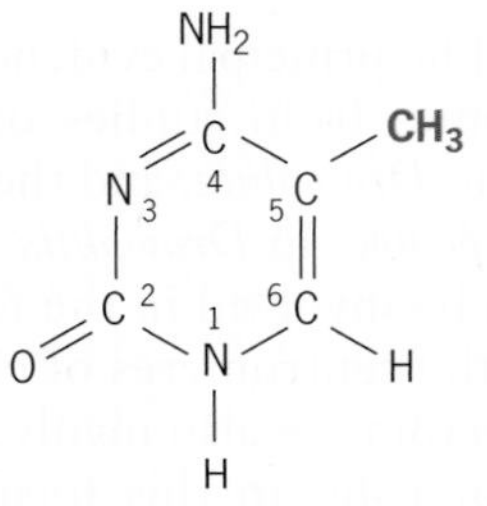

DNA methylation in mammals is also responsible for unusual cases in which the expression of a gene is controlled by its parental origin. For example, in mice, the *Igf 2* gene, which encodes an *i*nsulin-like *g*rowth *f*actor, is expressed when it is inherited from the father but not from the mother. By contrast, a gene known as *H19* is expressed when it is inherited from the mother but not from the father. Whenever the expression of a gene is conditioned by its parental origin, geneticists say that the gene has been **imprinted**—a term intended to convey the idea that the gene has been marked in some way so that it "remembers" which parent it came from.

Recent molecular analysis has demonstrated that the mark that conditions the expression of a gene is methylation of one or more CpG dinucleotides in the gene's vicinity. These methylated dinucleotides are initially formed in the parental germ line (**FIGURE 20.21**). Thus, for example, the *Igf 2* gene is methylated in the female germ line but not in the male germ line. At fertilization, a methylated, maternally contributed *Igf 2* gene is combined with an unmethylated, paternally contributed *Igf 2* gene. During embryogenesis, the methylated and unmethylated states are preserved each time the genes replicate. Because a methylated gene is silent, only the paternally contributed *Igf 2* gene is expressed in the developing animal. Exactly the opposite happens with the *H19* gene, which is methylated in the male germ line but not in the female germ line. More than 20 different imprinted genes have been identified in mice and humans. For each, the methylation imprint is established in the parental germ line. However, a methylated gene that was inherited from one sex can be unmethylated when it passes through an offspring of the opposite sex. Thus, the methylation imprints are reset each generation, depending on the sex of the animal. The fact that some genes are methylated in one sex but not in the other implies that sex-specific factors control the methylation machinery.

Figure 20.20 ▶ The structure of 5-methylcytosine.

GENE AMPLIFICATION

Sometimes, gene expression is facilitated by an increase in gene number. This process, called **gene amplification,** is designed to increase the number of DNA templates for RNA synthesis.

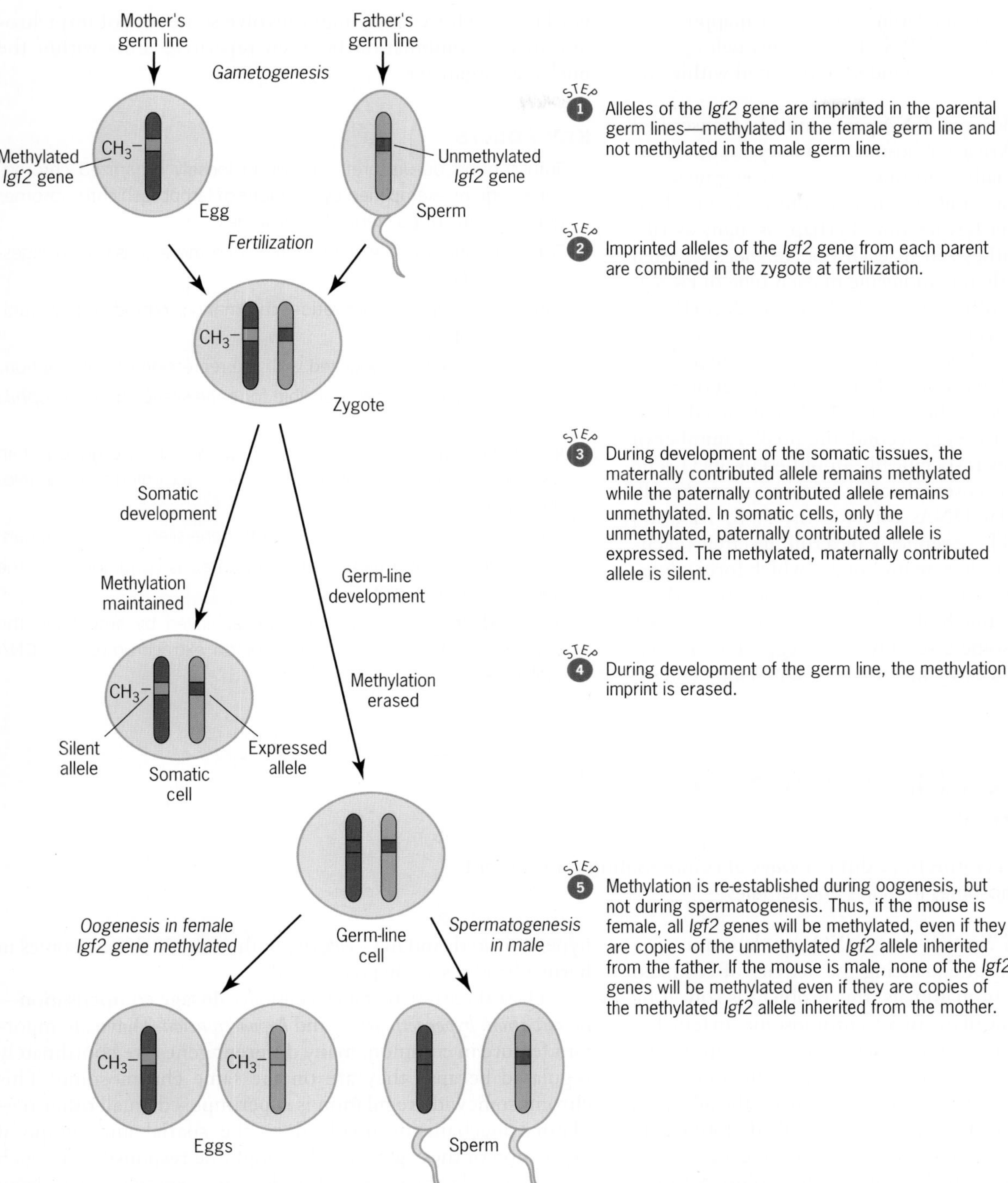
Mother's germ line
Father's germ line
Gametogenesis
Methylated Igf2 gene
CH_3–
Egg
Unmethylated Igf2 gene
Sperm
Fertilization
CH_3–
Zygote
Somatic development
Methylation maintained
Germ-line development
Methylation erased
CH_3–
Silent allele
Somatic cell
Expressed allele
Oogenesis in female Igf2 gene methylated
Germ-line cell
Spermatogenesis in male
CH_3–
CH_3–
Eggs
Sperm
STEP 1 Alleles of the Igf2 gene are imprinted in the parental germ lines—methylated in the female germ line and not methylated in the male germ line.
STEP 2 Imprinted alleles of the Igf2 gene from each parent are combined in the zygote at fertilization.
STEP 3 During development of the somatic tissues, the maternally contributed allele remains methylated while the paternally contributed allele remains unmethylated. In somatic cells, only the unmethylated, paternally contributed allele is expressed. The methylated, maternally contributed allele is silent.
STEP 4 During development of the germ line, the methylation imprint is erased.
STEP 5 Methylation is re-established during oogenesis, but not during spermatogenesis. Thus, if the mouse is female, all Igf2 genes will be methylated, even if they are copies of the unmethylated Igf2 allele inherited from the father. If the mouse is male, none of the Igf2 genes will be methylated even if they are copies of the methylated Igf2 allele inherited from the mother.

Figure 20.21 ▶ Methylation and imprinting of the *Igf2* gene in mice. The gene is methylated in females but not in males.

Perhaps the most dramatic example of expression-related gene amplification involves the ribosomal RNA genes in amphibian oocytes. These genes are needed to produce structural components of the ribosomes. In eukaryotes, there are four main types of rRNA genes: the 5S gene, which encodes a 120-base rRNA; the 5.8S gene, which encodes a 160-base rRNA; the 18S gene, which encodes a 1.8-kb rRNA; and the 28S gene, which encodes a 4.7-kb rRNA. In the genome of the toad *Xenopus laevis*, thousands of 5S genes are scattered over all the chromosomes, but there are far fewer copies of the 5.8S, 18S, and 28S genes, perhaps only 800 to 1000 per diploid cell. These latter genes are all concentrated at a single site called the **nucleolar organizer,** so named because it forms the nucleolus. Molecular mapping has shown that the 5.8S, 18S, and 28S rRNA genes belong to a single transcriptional unit that is tandemly repeated within the nucleolar organizer.

All four types of ribosomal RNA genes must be transcribed to produce rRNAs for *Xenopus* ribosomes. In oocytes, this transcription must be especially vigorous because the egg needs to accumulate enough ribosomes to support the early development of the embryo after fertilization. Perhaps as many as 10^{12} ribosomes must be synthesized in each oocyte. Because each ribosome contains exactly one molecule of each type of rRNA, 10^{12} molecules of each of the 5S, 5.8S, 18S, and 28S rRNAs must be synthesized. This requirement is fulfilled in two ways. First, *Xenopus* has about 24,000 5S genes, including many that are specifically activated in oocytes. This enormous set of genes is therefore able to generate the 5S rRNAs that are needed for ribosome production in the egg. Second, the smaller number of 5.8S, 18S, and 28S genes is specifically amplified in oocytes by the creation of extrachromosomal copies of these genes. Small, covalently closed circular DNAs carrying the 5.8S, 18S, and 28S genes are formed. These replicate by a rolling-circle mechanism (Chapter 10) to produce many copies, which form supernumerary nucleoli within the oocyte. Transcription from these circular DNAs provides much of the 5.8S, 18S, and 28S rRNA that is needed for ribosome assembly in the egg. The mechanism that generates these extrachromosomal DNA molecules is not known. However, it might involve some sort of intrachromosomal recombination between repetitive units within the nucleolar organizer.

KEY POINTS

- Transcription occurs preferentially in loosely organized chromosome regions exemplified by the loops of lampbrush chromosomes and the puffs of polytene chromosomes.
- Transcriptionally active DNA tends to be more sensitive to digestion with DNase I.
- During transcriptional activation, chromatin is remodeled by multiprotein complexes.
- Heterochromatin is associated with the repression of transcription.
- Protein complexes are responsible for gene silencing in *Drosophila* and yeast.
- In trypanosomes, only one telomeric variant surface glycoprotein (*vsg*) gene is expressed; all other *vsg* genes, including those at telomeres, are silent.
- Methylation of DNA is associated with gene silencing in mammals.
- The expression of a gene that is imprinted is conditioned by the gene's parental origin.
- Increased gene expression may be achieved by amplifying the DNA, either within chromosomes or on extrachromosomal DNA molecules.

Activation and Inactivation of Whole Chromosomes

Mammals, flies, and worms have distinct ways of compensating for different dosages of X chromosomes in males and females.

Organisms with an XX/XY or XX/XO sex-determination system face the problem of equalizing the activity of X-linked genes in the two sexes. In mammals, this problem is solved by randomly inactivating one of the two X chromosomes in females; each female therefore has the same number of transcriptionally active X-linked genes as a male. In *Drosophila*, neither of the two X chromosomes in a female is inactivated; instead, the genes on the single X chromosome in a male are transcribed more vigorously to bring their output in line with that of the genes on the two X chromosomes in a female. Still another solution to the problem of unequal numbers of X-linked genes has been found in the nematode *Caenorhabiditis elegans*. In this organism, XX individuals are hermaphrodites (they function as both male and female), and XO individuals are males. X-linked transcriptional activity is equalized in these two genotypes by partial repression of the genes on both of the X chromosomes in the hermaphrodites. Therefore, mammals, flies, and worms have solved the problem of X-linked gene dosage in different ways (**FIGURE 20.22**). In mammals, one of the X chromosomes in females is inactivated; in *Drosophila*, the single X chromosome in males is hyperactivated; and in *C. elegans*, both of the X chromosomes in hermaphrodites are hypoactivated.

These three different mechanisms of **dosage compensation**—*inactivation*, *hyperactivation*, and *hypoactivation*—have an important feature in common: many different genes are coordinately regulated because they are on the same chromosome. This chromosomewide regulation is superimposed on all other regulatory mechanisms involved in the spatial and temporal expression of these genes. What might be responsible for such a global regulatory system? For decades, geneticists have been trying to elucidate the molecular basis of dosage compensation. The working hypothesis has been that some factor or factors bind specifically to the X chromosome and alter its transcriptional activities. Recent discoveries indicate that this idea is correct.

INACTIVATION OF X CHROMOSOMES IN MAMMALS

In mammals, X chromosome inactivation begins at a particular site called the *X inactivation center (XIC)* and then spreads in

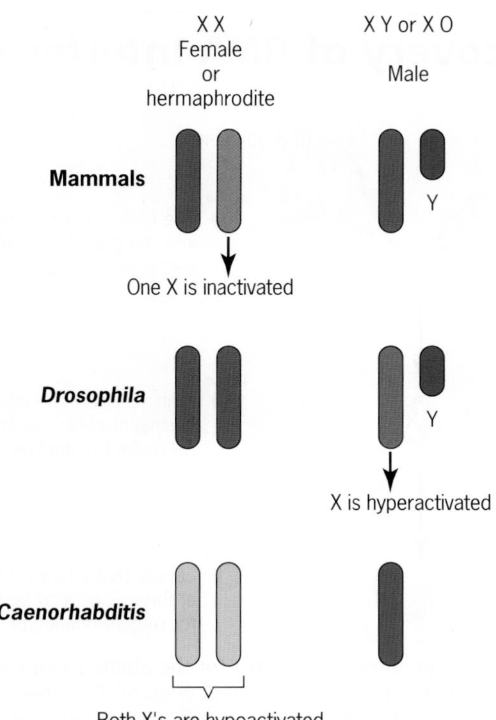

Figure 20.22 ▶ Three mechanisms of dosage compensation for X-linked genes: inactivation, hyperactivation, and hypoactivation.

opposite directions toward the ends of the chromosome. Curiously, not all genes on an inactivated X chromosome are transcriptionally silent. One that remains active is called *XIST* (for *X inactive specific transcript*); this gene is located within the XIC (**FIGURE 20.23**). Recent research has shown that, in human beings, the *XIST* gene encodes a 17-kb transcript devoid of any significant open reading frames. It therefore seems unlikely that the *XIST* gene codes for a protein. Instead, the RNA itself

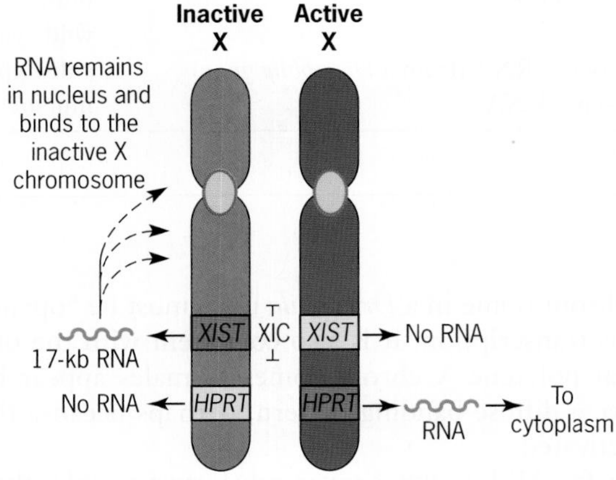

Figure 20.23 ▶ Expression of the *XIST* gene in the inactive X chromosome of human females. For comparison, the expression of the *HPRT* gene on the active X chromosome is shown. This gene encodes hypoxanthine phosphoribosyl transferase, an enzyme that plays a role in the metabolism of purines.

is probably the functional product of the *XIST* gene. Though polyadenylated, this RNA is restricted to the nucleus and is specifically localized to inactivated X chromosomes; it does not appear to be associated with active X chromosomes in either males or females.

In mice, where fairly detailed experimental analysis has been possible, researchers have found that the homologue of the human *XIST* gene is transcribed during the early stages of embryonic development at a low level from both of the X chromosomes that are present in females. The transcripts from each of a female mouse's *Xist* genes are unstable and remain closely associated with their respective genes. As development proceeds, the transcripts from one of the genes stabilize and eventually envelop the entire X chromosome on which that gene is located; the transcripts from the other *Xist* gene disintegrate, and further transcription from that gene is repressed by methylation of nucleotides in the gene's promoter. Thus, in the female mouse, one X chromosome—the one whose *Xist* gene continues to be transcribed—becomes coated with *Xist* RNA and the other does not. The choice of the chromosome that becomes coated is apparently random. Although the coating mechanism is not yet understood, the consequence of coating is clear: most of the genes on the coated chromosome are repressed, and that chromosome becomes the inactive X chromosome. In the mammalian dosage compensation system, therefore, the X chromosome that remains active is, paradoxically, the one that represses its *Xist* gene.

Inactive X chromosomes are readily identified in mammalian cells. During interphase, they condense into a darkly staining mass associated with the nuclear membrane. This mass, the Barr body, decondenses during S phase to allow the inactive X chromosome to be replicated. However, because decondensation takes some time, the inactive X replicates later than the rest of the chromosomes. Inactive X chromosomes must therefore have a very different chromatin structure than that of other chromosomes. This difference is partly determined by the kinds of histones associated with the DNA. One of the four core histones, H4, can be chemically modified by the addition of acetyl groups to any of several lysines in the polypeptide chain. Acetylated H4 is associated with all the chromosomes in the human genome. However, on the inactive X it seems to be restricted to three fairly narrow bands, each corresponding to a region that contains some active genes. Acetylated H4 is also depleted in areas of heterochromatin on the other chromosomes. These findings suggest that the depletion of acetylated H4 is a key feature of the inactive X chromosome.

HYPERACTIVATION OF X CHROMOSOMES IN *DROSOPHILA*

In *Drosophila*, dosage compensation requires the protein products of at least five different genes. Null mutations in these genes result in male-specific lethality because the single X chromosome in males is not hyperactivated. Mutant males usually die during the late larval or early pupal stages. These

▶ A MILESTONE IN GENETICS: The Discovery of RNA Interference

Sometimes a scientific discovery emerges from an effort to explain a paradox. The discovery that double-stranded RNA induces a pathway that interferes with gene expression is a case in point. In 1995, Su Guo and Kenneth Kemphues published an analysis of *par-1*, a maternally acting gene of *C. elegans*.[1] The product of this gene helps to establish anterior-posterior polarity in *C. elegans* embryos. Hermaphrodites homozygous for a *par-1* null mutation produce embryos that arrest at an early stage of development. These embryos have too many pharyngeal muscle cells, and they completely lack intestinal cells. Consequently, they die. Guo and Kemphues had isolated a cDNA, designated ZC22, which they thought had come from the *par-1* gene. To test the authenticity of this cDNA, they injected antisense RNA made from ZC22 into the gonads of wild-type hermaphrodites and scored the resulting embryos for the *par-1* mutant phenotype (**FIGURE 1**).

The experiment was a success (see **TABLE 1**). Antisense RNA made from ZC22 caused the embryos to look like they had come from homozygous mutant *par-1* parents. These embryos had too many pharyngeal muscle cells, they lacked intestinal cells, and 52 percent of them died. Guo and Kemphues did not observe the *par-1* mutant phenotype when they injected hermaphrodites with a solution of plain water, or when they injected them with solutions of antisense RNA made from other genes. Thus, they were confident that the cDNA they had tested had come from the *par-1* gene.

Guo and Kemphues performed one more control experiment, however. They injected sense RNA made from ZC22 into wild-type hermaphrodites and examined the resulting embryos. To their surprise, they found that these embryos had the *par-1* mutant phenotype and that 54 percent of them died. Thus, while striving to be thorough, Guo and Kemphues had stumbled onto a paradox. They found that injections with ZC22 antisense RNA produced the *par-1* mutant phenotype, as expected. But so did injections with ZC22 sense RNA. The effect of the antisense RNA could be explained by standard genetic theory—base-pairing between injected antisense RNA and *par-1* messenger RNA would block the translation of *par-1* messenger RNA into protein—but the effect of the sense RNA could not be readily explained.

[1]Guo, S., and K. J. Kemphues. 1995. *par-1*, a gene required for establishing polarity in *C. elegans* embryos, encodes a putative Ser/Thr kinase that is asymmetrically distributed. *Cell* 81:611–620.

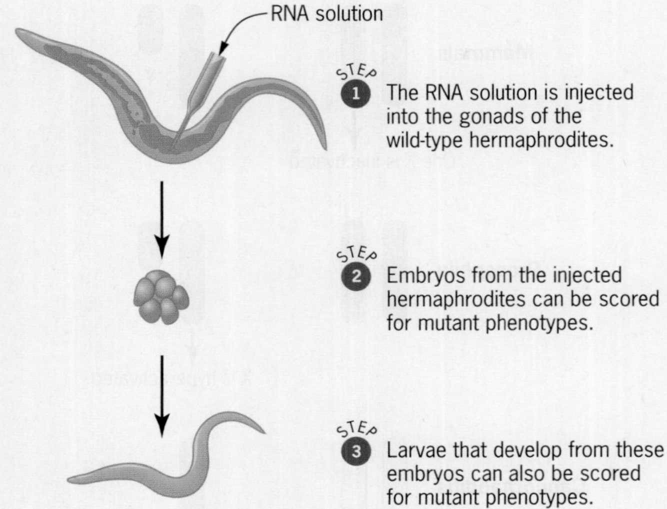

Figure 1 ▶ Scheme for testing RNA for the ability to interfere with the expression of a wild-type gene in *C. elegans*. The phenotypic effect of the test solution is assayed in embryos or larvae that develop from the injected animals.

▶ **TABLE 1**

Mutant (*par-1*) and Wild Phenotypes in Embryos from Wild-type *C. elegans* Hermaphrodites Injected by Guo and Kemphues

Injection Solution	Embryonic Phenotype
ZC22 antisense RNA	mutant
Water	wild-type
TS antisense RNA (from a *Drosophila* gene)	wild-type
ZC22 sense RNA	mutant

dosage compensation genes are therefore called male-specific lethal (*msl*) loci, and their products are called the MSL proteins. Antibodies prepared against these proteins have been used as probes to localize the proteins inside cells. The remarkable finding is that each of the MSL proteins binds specifically to the X chromosome in males (**FIGURE 20.24**). One of these proteins, the product of the *msl* gene called *maleless* (*mle*), is homologous to DNA and RNA helicases, which are enzymes capable of unwinding nucleic acids. The putative helicase function of the Maleless protein is consistent with the idea that the X chromosome in a *Drosophila* male must be "opened" for vigorous transcription. It is also consistent with the observation that polytene X chromosomes in males appear bloated and have a diffuse banding pattern, perhaps because they are hyperactivated.

The five MSL proteins apparently form a complex that binds to the single X chromosome in *Drosophila* males. Recent studies indicate that this MSL protein complex also contains specific types of RNA. Two different RNAs, *roX1* and *roX2* (for *R*NA *on* the *X* chromosome), co-localize on the male's X chromosome

▶ **TABLE 2**

Mutant (*unc-22*) and Wild Phenotypes in Larvae from Wild-type *C. elegans* Hermaphrodites Injected by Fire and Colleagues

Injection Solution	Larval Phenotype
unc-22 antisense RNA	wild-type
unc-22 sense RNA	wild-type
unc-22 antisense RNA+ *unc-22* sense RNA	mutant

In 1998, researchers discovered an explanation for the paradoxical effect of the sense RNA. Andrew Fire, SiQun Xu, Mary Montgomery, Steven Kostas, Samuel Driver, and Craig Mello published a paper describing the effects of single- and double-stranded RNA preparations on the expression of different *C. elegans* genes.[2] One set of experiments focused on *unc-22*, a gene that encodes a myofilament protein found in muscle cells (see **TABLE 2**). When *unc-22* expression is impaired, *C. elegans* exhibits a twitching behavior. Because this phenotype is easily scored, it provides a convenient assay to gauge the effect of injected RNA on *unc-22* expression. Fire and his colleagues looked for the twitching phenotype in the offspring of wild-type worms that had been injected with single-stranded—either sense or antisense—or double-stranded *unc-22* RNA; the double-stranded RNA was obtained by mixing the sense and antisense RNAs together.

The results of these experiments were definitive. Neither type of single-stranded RNA, either sense or antisense, had a significant effect on the phenotype of the offspring of the injected worms. However, double-stranded RNA caused these offspring to be twitchers—that is, it mimicked the phenotype of an *unc-22* mutant. Injected double-stranded *unc-22* RNA had therefore interfered with the expression of the endogenous *unc-22* gene.

Subsequent work has shown that this type of interference involves the enzyme Dicer, which cleaves the injected double-stranded RNA into short segments, and a riboprotein complex called RISC, which uses these segments as guides to target complementary messenger RNAs for degradation. The experiments of Fire and colleagues, which were originally designed to resolve the paradox found by Guo and Kemphues, wound up uncovering a new phenomenon: RNA interference.

But why did the sense RNA that Guo and Kemphues injected impair gene expression? Most likely, it was not pure sense RNA but a mixture of sense and antisense molecules. Double-stranded RNA is a highly efficient inducer of RNA interference. Even a few double-stranded molecules will trigger the interference pathway. If the sense RNA that Guo and Kemphues used was contaminated with antisense RNA, then some double-stranded RNA molecules would have formed in solution. Worms injected with this solution would respond appropriately by targeting the endogenous messenger RNA for degradation. The sense RNA would appear to have interfered with the expression of the target gene. The paradoxical effect of the sense RNA is therefore most likely to have been caused by antisense contaminants in the injection solution.

QUESTIONS FOR DISCUSSION

1. How has double-stranded RNA been used to study gene function in model organisms such as *C. elegans* and *Drosophila*? How has it been used to study gene function in organisms that are not amenable to standard genetic analysis? Does double-stranded RNA have potential as a therapeutic agent to ameliorate or cure human diseases?
2. RNA interference has become a useful tool in genetic research. However, it existed long before geneticists discovered it. Why do organisms have RNA interference pathways? When might RNA interference pathways have evolved? Have some organisms lost the capacity for RNA interference?

[2]Fire, A., S. Xu, M. K. Montgomery, S. A. Kostas, S. E. Driver, and C. C. Mello. 1998. Potent and specific genetic interference by double-stranded RNA in *Caenorhabditis elegans*. *Nature* 391:806–811.

with the MSL proteins. Both of these RNAs are transcribed from X-linked genes, and both are spliced and polyadenylated. However, they do not contain long open reading frames and are therefore probably not translated into proteins. Like the *Xist* RNAs in mammals, the *roX* RNAs in *Drosophila* seem to be the functional end-products of the genes that encode them. Curiously, the expression of the *roX* genes depends on the proper functioning of the *msl* genes. Furthermore, this expression is turned off when *Sxl*, the master regulatory gene of the *Drosophila* dosage compensation and sex-determination systems, is turned on, as it naturally is in *Drosophila* females. The *roX* RNAs are therefore produced only in *Drosophila* males.

The current model proposes that the MSL proteins form a complex that is joined by the *roX* RNAs. This complex then binds to 30 to 40 sites along the male's X chromosome, including the loci that contain the two *roX* genes. From each of these entry sites, the MSL/*roX* complex spreads bidirectionally until it reaches all the genes on the male's X chromosome that need to be hyperactivated. The process of hyperactivation may involve chromatin remodeling by the MSL/*roX* complex. One

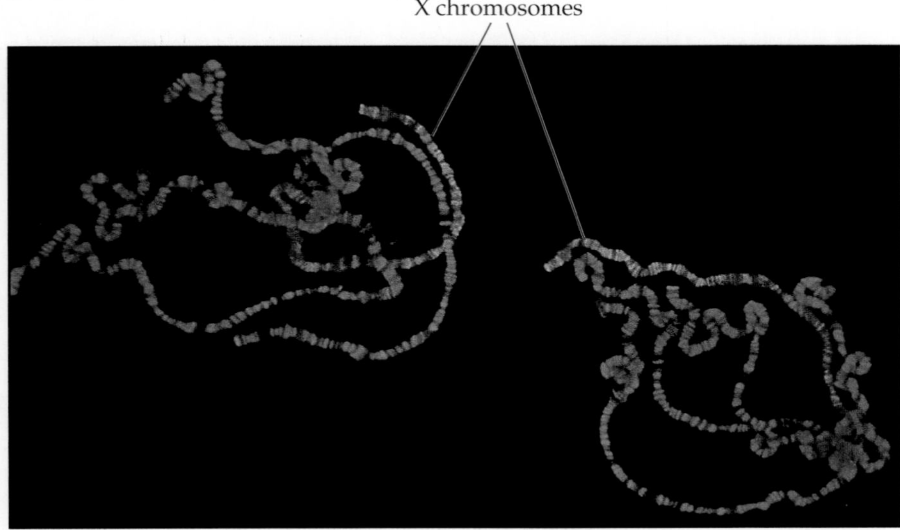

Figure 20.24 ▶ Binding of the protein product of the *Drosophila mle* gene to the single X chromosome in males.

of the MSL proteins is a histone acetyl transferase, and a particular acetylated version of histone H4 is exclusively associated with hyperactivated X chromosomes.

HYPOACTIVATION OF X CHROMOSOMES IN *CAENORHABDITIS*

In *C. elegans*, dosage compensation involves the partial repression of X-linked genes in the somatic cells of hermaphrodites. The mechanism is not fully understood, but the products of several genes are involved. Like the MSL proteins in *Drosophila*, the proteins encoded by these genes bind specifically to the X chromosome. However, unlike the situation in *Drosophila*, they bind only when two X chromosomes are present. The proteins apparently do not bind to the single X chromosome in males, nor do they bind to any of the autosomes in either males or hermaphrodites. Dosage compensation in *C. elegans* therefore seems to involve a mechanism exactly opposite to the one in *Drosophila*. A protein complex binds to the X chromosomes and represses rather than enhances transcription.

KEY POINTS

- Inactivation of an X chromosome in XX female mammals is mediated by a noncoding RNA transcribed from the *XIST* gene on that chromosome.
- Hyperactivation of the single X chromosome in male *Drosophila* is mediated by an RNA-protein complex that binds to many sites on that chromosome and stimulates the transcription of its genes.
- Hypoactivation of the two X chromosomes in *C. elegans* hermaphrodites is mediated by proteins that bind to these chromosomes and reduce the transcription of their genes.

▶ Basic Exercises

ILLUSTRATE BASIC GENETIC ANALYSIS

1. Arrange the following events in chronological order, beginning with the earliest: (a) splicing of an RNA molecule, (b) migration of an mRNA molecule into the cytoplasm, (c) transcription of a gene, (d) degradation of an mRNA molecule, (e) polypeptide synthesis.

Answer: c-a-b-e-d.

2. What factors induce the expression of the *hsp70* gene in *Drosophila* and the ribulose bisphosphate carboxylase (*RBC*) gene in plants?

Answer: The *hsp70* gene is induced by heat stress; the *RBC* gene is induced by exposure to light.

3. Indicate whether each of the following phenomena related to the regulation of gene expression occurs in the nucleus or the cytoplasm of a eukaryotic cell.

(a) Stimulation of gene expression by a transcription factor.
(b) Alternate splicing of the primary transcript of a gene.
(c) Polyadenylation of a gene's primary transcript.
(d) Translation of a messenger RNA.
(e) Inhibition of translation by a microRNA binding to a messenger RNA.
(f) Degradation of a messenger RNA induced by a short interfering RNA.
(g) Binding of a peptide hormone to its receptor.
(h) Binding of a steroid hormone to its receptor.
(i) Silencing of gene expression by heterochromatin.
(j) Whole chromosome inactivation.

Answer: Item (h) may take place in the cytoplasm or the nucleus, depending on the particular steroid hormone. Items (a), (b), (c), (i), and (j) take place in the nucleus. All other items take place in the cytoplasm.

4. Indicate whether the following are associated with gene activity or inactivity: (a) Polycomb group protein, (b) DNA methylation, (c) histone acetylation, (d) histone methylation, (e) heterochromatin, (f) locus control region, (g) GAL4 protein, (h) DNase I sensitivity.

Answer: (a) Inactivity, (b) inactivity, (c) activity, (d) inactivity, (e) inactivity, (f) activity, (g) activity, (h) activity.

5. What are some differences between euchromatin and heterochromatin?

Answer: Heterochromatin stains darkly throughout the cell cycle; euchromatin does not stain darkly during interphase. Heterochromatin is rich in repeated DNA sequences and in transposable elements; euchromatin may contain repeated sequences and transposons, but usually not to the extent that heterochromatin does. Heterochromatin has few protein-coding genes; euchromatin has many protein-coding genes.

6. How is the level of X-linked gene expression equalized in the two sexes of (a) humans, (b) flies, (c) worms?

Answer: (a) In humans, one of the two X chromosomes in females is randomly inactivated. (b) In flies, the single X chromosome in males is hyperactivated. (c) In worms, the two X chromosomes in hermaphrodites are hypoactivated.

▶ Testing Your Knowledge

INTEGRATE DIFFERENT CONCEPTS AND TECHNIQUES

1. The bacterial *lacZ* gene for β-galactosidase was inserted into a transposable *P* element from *Drosophila* (Chapter 18) so that it could be transcribed from the *P* element promoter. This fusion gene was then injected into the germ line of a *Drosophila* embryo along with an enzyme that catalyzes the transposition of *P* elements. During development, the modified *P* element became inserted into the chromosomes of some of the germ-line cells. Progeny from this injected animal were then individually mated to flies from a standard laboratory stock to establish strains that carried the *P/lacZ* fusion gene in their genomes. Three of these strains were analyzed for *lacZ* expression by staining dissected tissues from adult flies with X-gal, a chromogenic substrate that turns blue in the presence of β-galactosidase. In the first strain, only the eyes stained blue, in the second, only the intestines stained blue, and in the third, all the tissues stained blue. How do you explain these results?

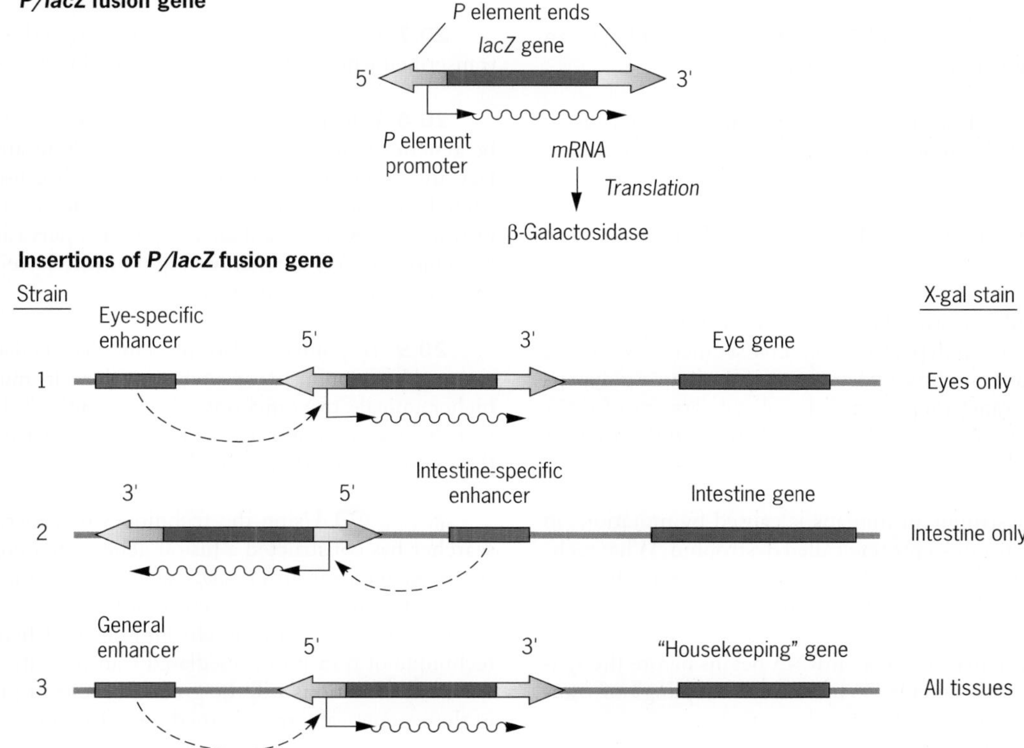

Answer: The three strains evidently carried different insertions of the *P/lacZ* fusion gene (see accompanying diagram). In each strain, the expression of the *P/lacZ* fusion gene must have come under the influence of a different regulatory sequence, or enhancer, capable of interacting with the *P* promoter and initiating transcription into the *lacZ* gene. In the first strain, the modified *P* element must have inserted near an eye-specific enhancer, which would drive transcription only in eye tissue. In the second strain, it must have inserted near an enhancer that drives transcription in the intestinal cells, and in the third strain, it must have inserted near an enhancer that drives transcription in all, or nearly all, cells, regardless of tissue affiliation. Presumably each of these different enhancers lies near a gene that would normally be expressed under its control. For example, the eye-specific enhancer would be near a gene needed for some aspect of eye function or development. These results show that random insertions of the *P/lacZ* fusion gene can be used to identify different types of enhancers and, through them, the genes they control. These fusion gene insertions are therefore often called *enhancer traps*.

2. In their seminal paper on RNA interference, Andrew Fire and coworkers (1998 *Nature* 391: 806–811) describe the results of experiments in which RNA derived from the *mex-3* gene was injected into *C. elegans* hermaphrodites. Embryos obtained from these injected hermaphrodites were analyzed by *in situ* hybridization using probes for *mex-3* RNA. The probes were designed to bind to *mex-3* messenger RNA, which normally accumulates in the gonads of hermaphrodites and in their embryos. Binding of the probe molecules to mRNA in the embryos is easily detected if the probe molecules have been labeled. When Fire and his colleagues performed these *in situ* hybridization experiments, they found that embryos from worms that had been injected with double-stranded *mex-3* RNA were not labeled by the probe molecules, whereas embryos from worms that had been injected with single-stranded RNA complementary to *mex-3* mRNA—that is, with antisense *mex-3* RNA—were labeled, though not quite as intensively as embryos from worms that had not been injected at all. What do these results indicate about the efficacy of double-stranded versus single-stranded antisense RNA to silence gene expression?

Answer: The results of these *in situ* hybridization experiments indicate that double-stranded RNA is a strong silencer of *mex-3* gene expression in *C. elegans* embryos. By contrast, single-stranded antisense RNA barely has an effect on *mex-3* gene expression. The embryos from worms injected with double-stranded *mex-3* RNA did not carry any detectable *mex-3* messenger RNA. The absence of *mex-3* messenger RNA in these embryos is the result of RNA interference induced by the injected double-stranded RNA. The embryos from worms injected with single-stranded antisense *mex-3* RNA did carry some *mex-3* messenger RNA. Thus, single-stranded antisense *mex-3* RNA is not as effective as double-stranded *mex-3* RNA in the induction of RNAi.

▶ Questions and Problems

ENHANCE UNDERSTANDING AND DEVELOP ANALYTICAL SKILLS

20.1 How could you use the polytene chromosomes of Dipteran insects to study the regulation of transcription?

20.2 Why do steroid hormones interact with receptors inside the cell, whereas peptide hormones interact with receptors on the cell surface?

20.3 Operons are common in bacteria but not in eukaryotes. Suggest a reason why.

20.4 A polypeptide consists of three separate segments of amino acids, A—B—C. Another polypeptide contains segments A and C, but not segment B. How might you determine if these two polypeptides are produced by translating alternately spliced versions of RNA from a single gene, or by translating mRNA from two different genes?

20.5 Muscular dystrophy in humans is caused by mutations in an X-linked gene that encodes a protein called dystrophin. What techniques could you use to determine if this gene is active in different types of cells, say skin cells, nerve cells, and muscle cells?

20.6 In bacteria, translation of an mRNA begins before the synthesis of that mRNA is completed. Why is this "coupling" of transcription and translation not possible in eukaryotes?

20.7 What techniques could be used to show that a plant gene is transcribed when the plant is illuminated with light?

20.8 When introns were first discovered, they were thought to be genetic "junk"—that is, sequences without any useful function. In fact, they appeared to be worse than junk because they actually interrupted the coding sequences of genes. However, among eukaryotes, introns are pervasive and anything that is pervasive in biology usually has a function. What function might introns have? What benefit might they confer on an organism?

20.9 Tropomyosins are proteins that mediate the interaction of actin and troponin, two proteins involved in muscle contractions. In higher animals, tropomyosins exist as a family of closely related proteins that share some amino acid sequences but differ in others. Explain how these proteins could be created from the transcript of a single gene.

20.10 GO Using the techniques of genetic engineering, a researcher has constructed a fusion gene containing the heat-shock response elements from a *Drosophila hsp70* gene and the coding region of a jellyfish gene (*gfp*) for green fluorescent protein. This fusion gene has been inserted into the chromosomes of living *Drosophila* by the technique of transposon-mediated transformation (Chapter 18). Under what conditions will the green fluorescent protein be synthesized in these genetically transformed flies? Explain.

20.11 Suppose that the segment of the *hsp70* gene that was used to make the *hsp70/gfp* fusion in the preceding problem had mutations in each of its heat-shock response elements. Would the green fluorescent protein encoded by this fusion gene be synthesized in genetically transformed flies?

20.12 In *Drosophila*, expression of the *yellow* gene is needed for the formation of dark pigment in many different tissues; without this expression, a tissue appears yellow in color. In the wings, the expression of the *yellow* gene is controlled by an enhancer located upstream of the gene's transcription initiation site. In the tarsal claws, expression is controlled by an enhancer located within the gene's only intron. Suppose that by genetic engineering, the wing enhancer is placed within the intron and the claw enhancer is placed upstream of the transcription initiation site. Would a fly that carried this modified *yellow* gene in place of its natural *yellow* gene have darkly pigmented wings and claws? Explain.

20.13 Why does it make sense for the photosynthetic enzyme ribulose 1,5-bisphosphate carboxylase (RBC) to be synthesized specifically when plants are exposed to light?

20.14 GO The alternately spliced forms of the RNA from the *Drosophila doublesex* gene encode proteins that are needed to block the development of one or the other set of sexual characteristics. The protein that is made in female animals blocks the development of male characteristics, and the protein that is made in male animals blocks the development of female characteristics. Predict the phenotype of XX and XY animals homozygous for a null mutation in the *doublesex* gene.

20.15 The protein product of the *Drosophila Sex-lethal* gene is needed in females but not in males. Predict the phenotype of XY *Drosophila* that are hemizygous for a null mutation in the X-linked *Sex-lethal* gene.

20.16 Suppose that the LCR of the β-globin gene cluster was deleted from one of the two chromosomes 11 in a man. What disease might this deletion cause?

20.17 A researcher suspects that a 550 bp-long intron contains an enhancer that drives expression of an *Arabidopsis* gene specifically in root tip tissue. Outline an experiment to test this hypothesis.

20.18 RNA interference has been implicated in the regulation of transposable elements. In *Drosophila*, two of the key proteins involved in RNA interference are encoded by the genes *aubergine* and *piwi*. Flies that are homozygous for mutant alleles of these genes are lethal or sterile, but flies that are heterozygous for them are viable and fertile. Suppose that you have strains of *Drosophila* that are heterozygous for *aubergine* or *piwi* mutant alleles. Why might the genomic mutation rate in these mutant strains be greater than the genomic mutation rate in a wild-type strain?

20.19 In *Drosophila* larvae, the single X chromosome in males appears diffuse and bloated in the polytene cells of the salivary gland. Is this observation compatible with the idea that X-linked genes are hyperactivated in *Drosophila* males?

20.20 In trypanosomes, what mechanisms lead to the expression of a different variant surface glycoprotein?

20.21 Would double-stranded RNA derived from an intron be able to induce RNA interference?

20.22 What is the nature of each of the following classes of enzymes? What does each type of enzyme do to chromatin? (a) HATs, (b) HDACs, (c) HMTs.

20.23 Suppose female mice homozygous for the *a* allele of the *Igf2* gene are crossed to male mice homozygous for the *b* allele of this gene. Which of these two alleles will be expressed in the F_1 progeny?

20.24 GO A researcher hypothesizes that in mice gene A is actively transcribed in liver cells, whereas gene B is actively transcribed in brain cells. Suppose that the hypothesis is correct and that gene A is actively transcribed in liver cells whereas gene B is actively transcribed in brain cells. The researcher now extracts equivalent amounts of chromatin from liver and brain tissues and treats these extracts separately with DNase I for a limited period of time. If the DNA that remains after the treatments is then fractionated by gel electrophoresis, transferred to a membrane by Southern blotting, and hybridized with a radioactively labeled probe specific for gene A, which sample (liver or brain) will be expected to show the greater signal on the autoradiogram? Explain your answer.

20.25 A researcher hypothesizes that in mice gene *A* is actively transcribed in liver cells, whereas gene *B* is actively transcribed in brain cells. Describe procedures that would allow the researcher to test this hypothesis.

20.26 Suppose that a woman carries an X chromosome in which the *XIST* locus has been deleted. The woman's other X chromosome has an intact *XIST* locus. What pattern of X-inactivation would be observed throughout the woman's body?

20.27 Why do null mutations in the *mle* gene in *Drosophila* have no effect in females?

► Genomics on the Web

at http://www.ncbi.nlm.nih.gov/

The human β-globin genes are located in a cluster on the short arm of chromosome 11.

1. Search for the namesake gene of the cluster, the adult β-globin gene, in the human genome database. What is the official symbol of this gene? How many exons does it contain?

2. Use the Map Viewer function to locate the β-globin gene cluster on the ideogram of chromosome 11. In what cytological band does it reside? Is it closer to the telomere of the short arm or to the centromere?

3. Use the Sequence Viewer to inspect the adult β-globin gene in detail. Is the gene transcribed toward the centromere or toward the telomere? How long is the transcript of the gene? How long is the mature mRNA? How many amino acids does the mRNA specify? What are the first three amino acids, and what codons specify them?

4. Bring up the text sequence of the adult β-globin gene by clicking the ATGC button on the Sequence Viewer page. Locate the initiation codon for methionine in the first exon. Because the sequence in the window is that of the template strand of the DNA, this codon reads 5′-CAT-3′ from left to right on the screen.

5. GATA1 and MyoD are two transcription factors that recognize short sequences in mammalian genomes. The sequence recognized by GATA1 is 5′-TGATAG-3′ and the sequence recognized by MyoD is 5′-CAAATG-3′. Copy the sequence of the transcribed portion of the adult β-globin gene into a text file and scan it for each of these recognition sequences. Where are they located? Which of these two transcription factors might be involved in regulating the expression of the adult β-globin gene?

Chapter 21
The Genetic Control of Animal Development

CHAPTER OUTLINE

- Model Organisms for the Genetic Analysis of Development
- Genetic Analysis of Developmental Pathways
- Maternal Gene Activity in Development
- Zygotic Gene Activity in Development
- Genetic Analysis of Development in Vertebrates

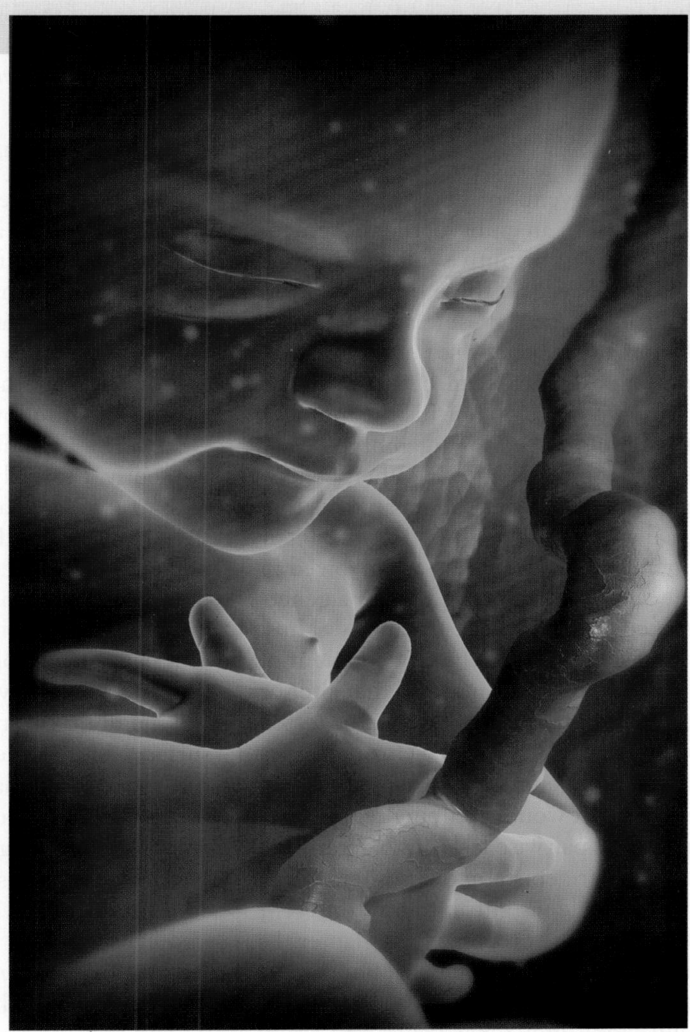

Human fetus late during development.

Stem Cell Therapy

Stem cells are in the news. Scientists are discussing their possible uses, and all sorts of people—politicians, religious leaders, journalists, victims of illnesses such as Parkinson's disease, diabetes, and arthritis, and even Hollywood celebrities—are joining the conversation. Though nondescript themselves, stem cells have the ability to produce offspring that can differentiate into special cell types, like muscle fiber, lymphocyte, neuron, or bone cell. They might, therefore, be used to regenerate worn-out tissues, to replace lost organs or body parts, to correct injuries, or to alleviate biochemical deficits. These prospects point out the importance of understanding how different types of cells acquire their specialized functions, and how, in a multicelled organism, they form tissues and organs in an orderly manner over time. In other words, they point out the importance of understanding the process of development—from fertilized egg to embryo to adult. The possibility for stem cell therapy also raises important ethical questions. Must the stem cells be derived by destroying embryos? Should embryonic life be sacrificed to prolong and enhance adult life? Is it acceptable to produce embryos merely to harvest stem cells for therapeutic purposes? Around the world, people and their governments are debating these questions, while scientists continue to explore the properties of stem cells and how they might be used.

Model Organisms for the Genetic Analysis of Development

Drosophila and *C. elegans* have been the premier model organisms for the genetic analysis of animal development.

The development of a multicelled animal from a fertilized egg demonstrates the power of controlled gene expression. Genes must be expressed carefully over time to bring about the specialization of cells, the orderly assembly of these cells into tissues and organs, and the formation of the animal's body. The process of animal development therefore depends on the faithful execution of a genetic program encoded in the animal's DNA. It should come as no surprise, therefore, that genetics has contributed greatly to our understanding of this process.

Classical studies from anatomy and embryology provided detailed observations about the events of development—the division of the fertilized egg to form an embryo, the movement of cells within the embryo to form primitive tissues, and the subsequent differentiation of cells within these tissues to form different organs. For practical reasons, these classical studies focused on a few kinds of animals, especially sea urchins, frogs, and chickens. The eggs of such animals can be manipulated experimentally, and their embryos develop outside the mother's body. Embryologists could therefore see how an embryo developed in response to an experimental treatment. When geneticists began to study development, they focused on animals that were easy to breed, especially *Drosophila* and *C. elegans*. Their objective was to identify genes whose products are involved in important developmental events. The standard way for a geneticist to achieve such an objective is to collect mutations. Thus, for example, if a geneticist wishes to study the development of *Drosophila* wings, he or she would collect mutations that alter or prevent wing formation. These mutations would then be tested for allelism with one another and mapped on the chromosomes to define and position the relevant genetic loci. Once these loci have been identified, the geneticist would combine representative mutations from each locus in pairwise fashion to determine whether some of the mutations are epistatic over others. Such epistasis testing can provide valuable insights into how different genes contribute to a developmental process (see Chapter 4). Finally, to investigate the molecular basis of gene action and to elucidate the role that each gene's product plays in development, the geneticist would clone individual genes and study them with the full panoply of techniques now available—sequencing, RNA and protein blotting, RT-PCR, fluorescent labeling, the production of transgenics and so on (see Chapters 15 and 17).

Using this general strategy, geneticists have learned a great deal about the way that development proceeds in *Drosophila* and *C. elegans*. Much is now known about how cells become specialized, how tissues and organs form, and how the body plan is laid out. This knowledge has also provided an intellectual framework to guide the study of development in other animals, including vertebrates such as the mouse and the zebra fish. The study of these model vertebrates has, in turn, provided many insights into the process of development in human beings. However, before exploring any of these topics, we need to discuss some of the basic features of development in the two premier models for studying the genetic control of development, *Drosophila* and *C. elegans*.

FEATURES OF *DROSOPHILA* DEVELOPMENT

Adult *Drosophila* develop from ellipsoidal eggs about 1 mm long and 0.5 mm wide at their maximum diameter (**FIGURE 21.1*a***). Each egg is surrounded by a *chorion*, a tough shell-like structure that is made of materials synthesized by somatic cells in the ovary. The anterior end of the egg is distinguished by two filaments that help to bring oxygen into the egg. Sperm enter the egg through another anterior structure, the *micropyle*. The cell divisions that follow fertilization are rapid—so rapid that there is no time for membranes to form between daughter cells. Consequently, the early *Drosophila* embryo is actually a single cell with many identical nuclei; such a cell is called a *syncytium* (**FIGURE 21.1*b***). After division cycle 9 within the syncytium, the 512 nuclei that have been created migrate to the cytoplasmic membrane on the periphery of the embryo, where they continue to divide four more times. In addition, a few of the nuclei migrate to the posterior pole of the embyro. At division cycle 13, all the nuclei in the syncytium become separated from each other by cell membranes, creating a single layer of cells on the embryo's surface. This single layer, called the *cellular blastoderm*, will give rise to all the somatic tissues of the animal. Cellularization of the nuclei at the posterior pole creates the *pole cells*, which give rise to the adult germ line. Thus, at this very early stage of development, the somatic and germ-cell lineages of the future adult have already been separated.

It takes about a day for the *Drosophila* embryo to develop into a wormlike *larva*. This larva hatches by chewing its way through the egg shell and begins feeding voraciously. It sheds its skin twice to accommodate increases in size and then, about five days after hatching, becomes immobile and hardens its skin, forming a *pupa*. During the next four days, many of the larval tissues are destroyed, and flat packets of cells that were sequestered during the larval stages expand and differentiate into adult structures such as antennae, eyes, wings, and legs. Because an adult insect is called an *imago*, these packets are referred to as *imaginal discs*; see **FIGURE 2.17** for their locations in the larva. When this anatomical reorganization is completed, a radically different animal emerges from the pupal casing—one that can fly and reproduce!

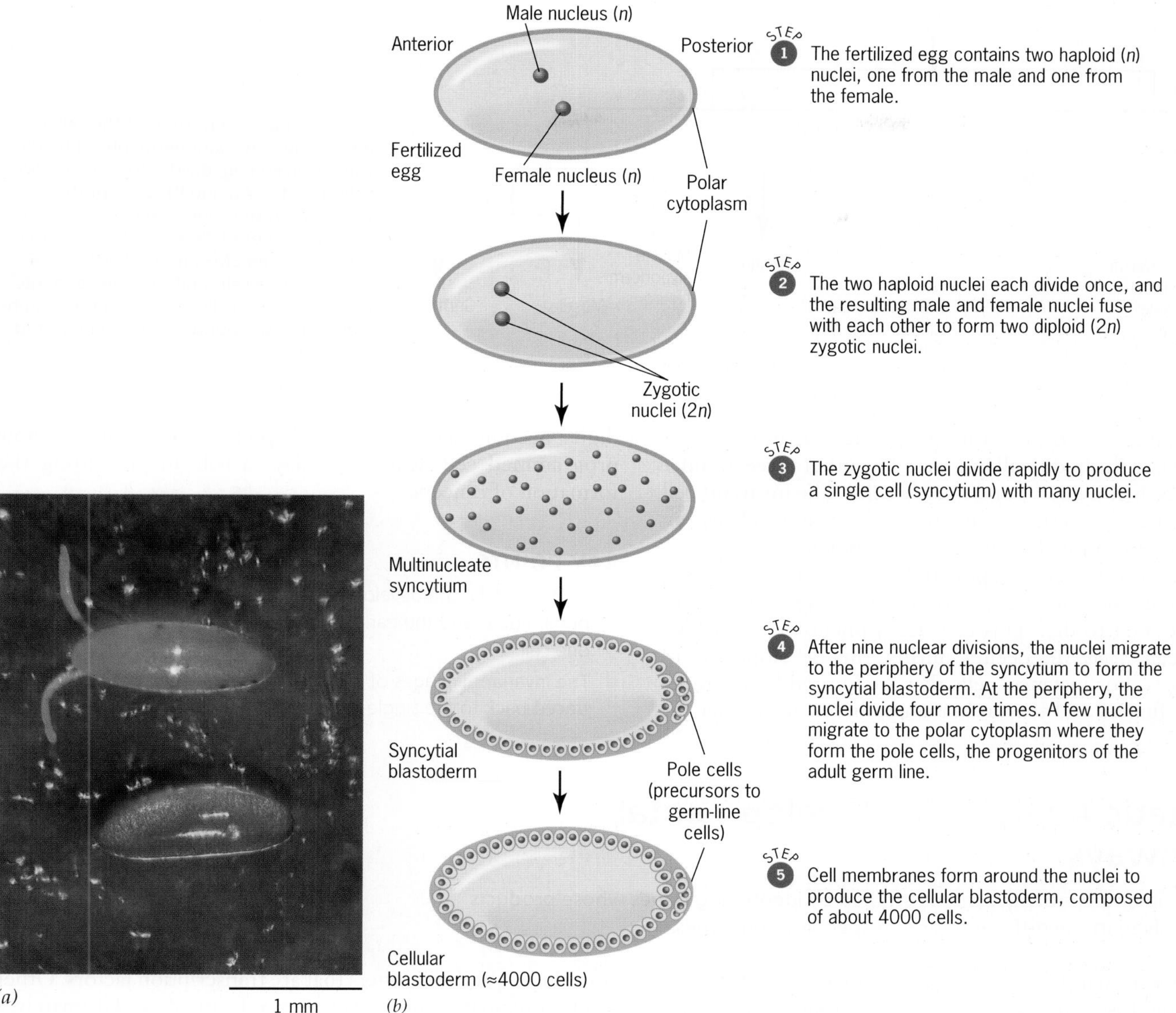

Male nucleus (n)
Anterior
Posterior
STEP 1 The fertilized egg contains two haploid (n) nuclei, one from the male and one from the female.
Fertilized egg
Female nucleus (n)
Polar cytoplasm
STEP 2 The two haploid nuclei each divide once, and the resulting male and female nuclei fuse with each other to form two diploid (2n) zygotic nuclei.
Zygotic nuclei (2n)
STEP 3 The zygotic nuclei divide rapidly to produce a single cell (syncytium) with many nuclei.
Multinucleate syncytium
STEP 4 After nine nuclear divisions, the nuclei migrate to the periphery of the syncytium to form the syncytial blastoderm. At the periphery, the nuclei divide four more times. A few nuclei migrate to the polar cytoplasm where they form the pole cells, the progenitors of the adult germ line.
Syncytial blastoderm
Pole cells (precursors to germ-line cells)
STEP 5 Cell membranes form around the nuclei to produce the cellular blastoderm, composed of about 4000 cells.
Cellular blastoderm (≈4000 cells)
(a)
1 mm
(b)

Figure 21.1 ▶ Basic features of *Drosophila* development. (*a*) Photograph of *Drosophila* eggs, with (top) and without (bottom) the surrounding chorion. (*b*) Early embryonic development in *Drosophila*.

FEATURES OF *C. ELEGANS* DEVELOPMENT

Adult *C. elegans* are about 1 mm long, roughly the same size as a *Drosophila* egg. This small organism completes its life cycle in about three days. *C. elegans* is a hermaphroditic species; see **FIGURE 2.18** for diagrams of the hermaphrodite and male forms of the animal. An individual hermaphrodite can produce both sperm and eggs, and self-fertilization is possible. To the researcher, this self-fertilization is a convenient way of making recessive mutations homozygous. Hermaphrodites have two X chromosomes and five pairs of autosomes. Occasionally, animals with a single X chromosome and five pairs of autosomes are produced by meiotic nondisjunction; these animals are males, capable of making sperm but not eggs. XO males can be crossed with XX hermaphrodites to carry out standard genetic procedures such as recombination mapping and complementation testing.

C. elegans is a transparent animal, affording researchers the opportunity to observe each cell in the course of development. John Sulston and coworkers have been able to trace the lineage of all the cells in the adult worm from the single-celled zygote (**FIGURE 21.2**). This work has shown that the events of *C. elegans* development are essentially invariant. Each adult hermaphrodite consists of 959 somatic nuclei (some cells are multinucleate) as well as an indeterminant number of germ cells. A *C. elegans* zygote goes through a series of asymmetric cell divisions to produce six "founder" cells. One of these

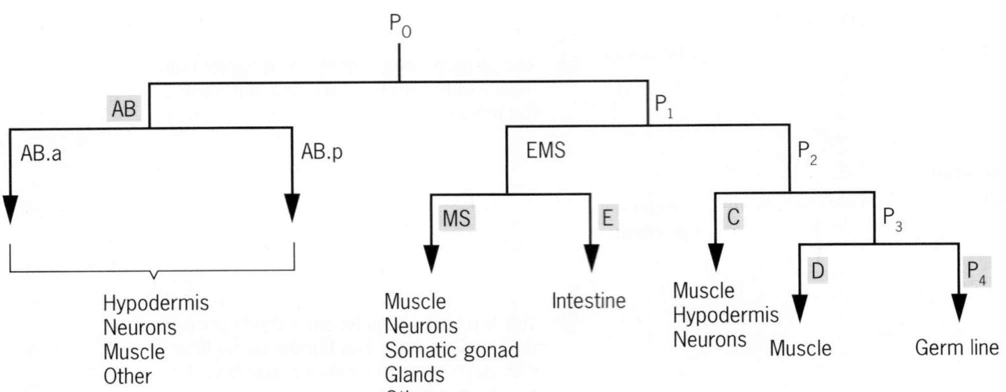

Figure 21.2 ▶ A portion of the cell lineage of the *C. elegans* hermaphrodite. P0, the fertilized egg, divides to produce two cells denoted AB and P1. Each of these then divides to produce two pairs of cells, one pair denoted AB.a and AB.p and the other denoted EMS and P2. Further divisions of these cells and their descendants produce all the cells of the adult hermaphrodite. The six founder cells are indicated in yellow.

eventually gives rise to the entire germ line, one to the intestine, and one to the body wall muscles. The other three founder cells produce mainly nerve and muscle cells. The invariant cell lineages that form the tissues of the adult make *C. elegans* an excellent organism for the study of development.

Another feature of development that is evident in *C. elegans* is that some cells do not survive to the adult stage. These cells are programmed to die. This cell death phenomenon, which scientists call *apoptosis*, occurs during the development of many animals. For instance, cells that are located between the developing fingers and toes in human embryos die so that the digits become separated. In Chapter 22, we shall see how programmed cell death also plays a role in preventing the formation of cancers.

KEY POINTS

- ▶ In *Drosophila,* the developmental sequence is egg, embryo, larva, pupa, adult, and the early embryo is a syncytium—many nuclei in one cell.
- ▶ The invariant lineages of all the cells in an adult *C. elegans* can be traced back to the single-celled zygote.

Genetic Analysis of Developmental Pathways

Developmental pathways can be studied by identifying genes whose products are involved in the differentiation of specific phenotypes.

The notion that phenotypes result from a series of steps in a pathway and that genes control each of these steps emerged in the 1930s and 1940s from a combination of genetic and biochemical investigations. By that time, it had already been established that cells carry out a multitude of biochemical reactions, each catalyzed by an enzyme, and that several reactions can be linked to form a pathway (see Chapter 14). Genetics provided the key insight that each enzyme in a pathway is encoded by a gene. A mutation in one of these genes could therefore inactivate an essential enzyme, blocking the entire pathway and causing a mutant phenotype. Geneticists quickly realized that by studying such mutations, they could identify each step in the pathway. Furthermore, by analyzing pairs of mutations, they could sometimes arrange the steps in the correct temporal order. This genetic dissection of biochemical pathways was soon extended to other biological processes, including development.

A developmental pathway consists of the events involved in the differentiation of tissues and organs. Different gene products participate in these events, including some that are signal molecules, some that are signal receptors, others that are signal transducers, and still others that are transcription factors. Other kinds of regulatory proteins may also be involved. Ultimately, a pathway generates the components that form the structures of particular tissues and organs. It therefore creates a phenotype.

The components of a pathway are causally ordered by the way in which they bring about the phenotype. Gene *A* may produce a secreted protein that acts as a signal to stimulate the transcription of gene *X*, whose product is a component of a differentiated—that is, structurally and functionally specialized—cell. This stimulation may be mediated by other gene products, including, for example, a membrane-bound receptor for the A protein, intracellular proteins that are activated by this receptor, and one or more transcription factors that respond to this activation by binding to enhancers near gene *X* to induce its expression. The overall structure of the pathway is

Gene *A* →	Gene *R* →	Gene *C* →	Gene *T* →	Gene *X*
↓	↓	↓	↓	↓
Secreted signal protein A	Membrane-bound receptor protein	Cytoplasmic protein	Transcription factor	Protein X in a differentiated cell

In this developmental pathway, the different proteins are ordered by a causal chain that ultimately produces a protein characteristic of a particular differentiated cell. The arrows between the genes in the pathway indicate the order in which the gene products act to bring about a differentiated state.

As an example of how geneticists have analyzed developmental pathways, let's examine the processes of sexual differentiation in *Drosophila* and *C. elegans*. The pathways for sex determination in these two organisms have been studied thoroughly. They control the differentiation of the somatic tissues into male or female structures. Other, less well understood, pathways govern the development of the tissues within the male and female germ lines. As it turns out, the somatic sex-determination pathways in *Drosophila* and *C. elegans* involve entirely different molecular mechanisms. In *Drosophila*, the key genes encode proteins that regulate RNA splicing, whereas in *C. elegans*, they encode signaling molecules, their receptors, and transcription factors. In both animals, however, the sex-determinaton pathway responds to the same fundamental signal, which is the ratio of X chromosomes to autosomes. When this ratio is 1.0 or greater, the pathway produces a female phenotype (or a hermaphrodite phenotype in *C. elegans*), and when it is 0.5 or less, it produces a male phenotype.

SEX DETERMINATION IN *DROSOPHILA*

The sex-determination pathway in *Drosophila* has three components: (1) a system to ascertain the X:A ratio very early in the embryo; (2) a system to convert this ratio into a developmental signal; and (3) a system to respond to this signal by producing either male or female structures.

Ascertaining the X:A Ratio

The system to ascertain the X:A ratio involves interactions between maternally synthesized proteins that have been deposited in the egg cytoplasm, and embryonically synthesized proteins that are encoded by several X-linked genes (**FIGURE 21.3**). Because of strict dose-dependence, these latter proteins are twice as abundant in XX embryos as in XY embryos, and therefore provide a means of "counting" the number of X chromosomes

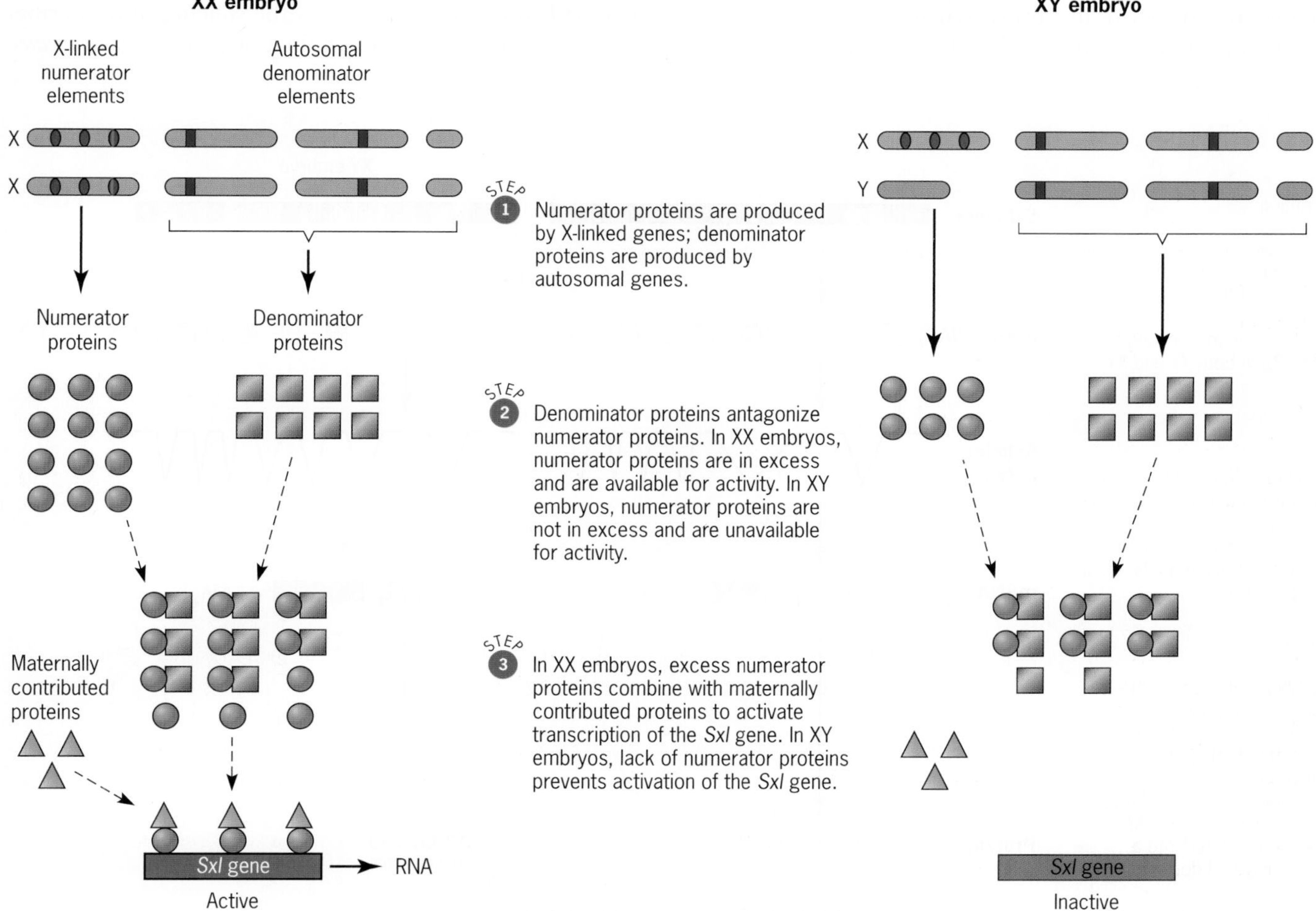

Figure 21.3 ▶ Ascertainment of the X:A ratio by numerator and denominator elements in *Drosophila*. The ratio of X chromosomes to sets of autosomes is ascertained by interactions between the protein products of these genes.

present. Because the genes that encode these proteins affect the numerator of the X:A ratio, they are called *numerator elements*. Other genes located on the autosomes affect the denominator of the X:A ratio. These so-called *denominator elements* encode proteins that antagonize the products of the numerator elements. Consequently, as the dosage of denominator elements is increased, the "perceived" dosage of numerator elements is decreased, and the number of X chromosomes present in the genotype is underestimated. This process occurs in *Drosophila* with two X chromosomes and three pairs of autosomes (genotype XX; AAA); such flies develop as intersexes rather than as females. The system for ascertaining the X:A ratio in *Drosophila* is therefore based on antagonism between X-linked (numerator) and autosomal (denominator) gene products.

Generating a Developmental Signal

Once the X:A ratio has been ascertained, it is converted into a molecular signal that controls the expression of the X-linked *Sex-lethal* (*Sxl*) gene, the master regulator of the sex-determination pathway (**FIGURE 21.4**). Early in development, this signal activates transcription of the *Sxl* gene from P_E, the gene's "early" promoter, but only in XX embryos. The early transcripts from this promoter are processed and translated to produce functional Sex-lethal protein, denoted SXL. After only a few division cycles, transcription from the P_E promoter is replaced by transcription from another promoter, P_M, the so-called maintenance promoter of the *Sxl* gene. Curiously, transcription from the P_M promoter is also initiated in XY embryos. However, the transcripts from P_M are correctly processed only if SXL protein is present. Consequently, in XY embryos, where this protein has not yet been made, the *Sxl* transcripts are alternately spliced to include an exon with a stop codon, and when these alternately spliced transcripts are translated, they generate a short polypeptide without regulatory function. Thus, alternate splicing of the *Sxl* transcripts in XY embryos does not lead to the production of functional SXL protein, and in the absence of this protein, these embryos develop as males. In XX embryos, where SXL protein was initially made in response to the X:A signal, *Sxl* transcripts from the P_M promoter are spliced to encode more SXL protein. This expression pattern is maintained because the SXL protein regulates the splicing of transcripts from the *Sxl* gene in such a way that they encode the full-length SXL protein. In XX embryos, this protein is therefore a positive regulator of its own synthesis—a curious feedback mechanism that maintains the expression of the SXL protein in XX embryos and prevents its expression in XY embryos.

The SXL protein also regulates the splicing of transcripts from another gene in the sex-determination pathway, *transformer*

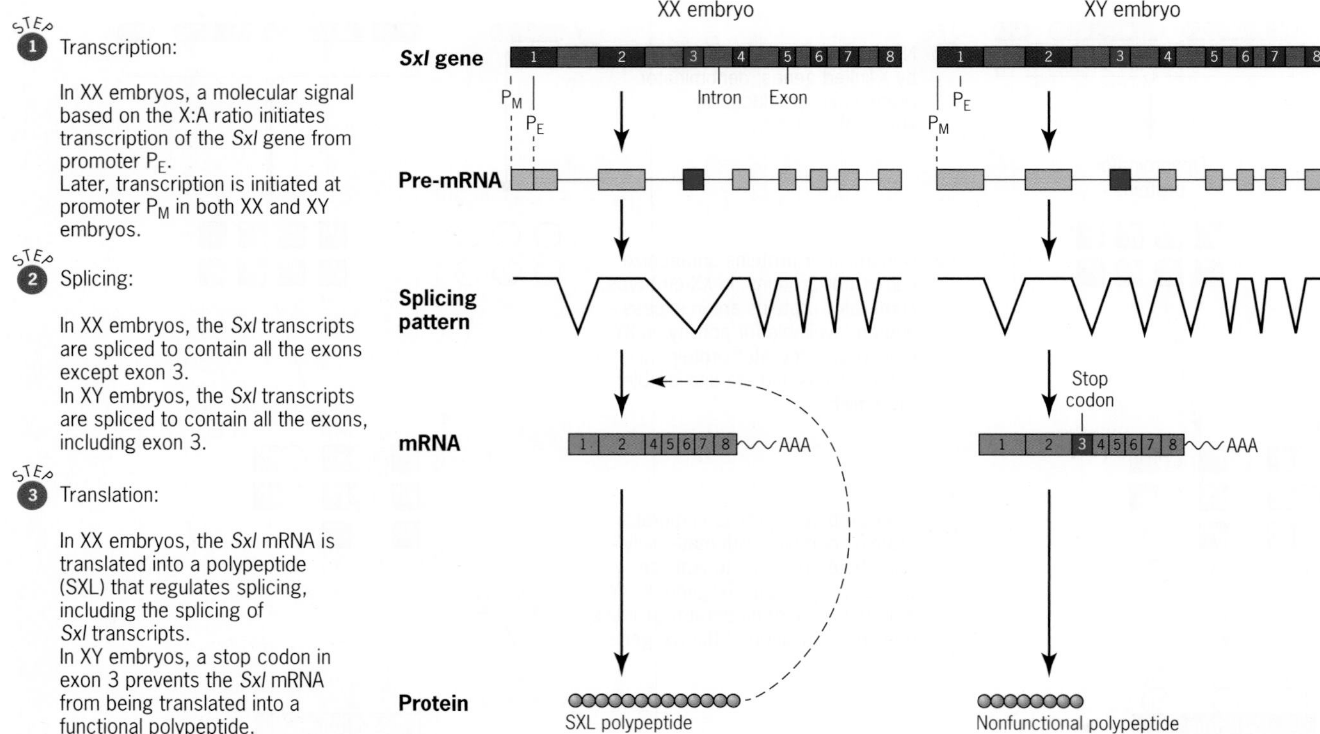

Figure 21.4 ▶ Sex-specific expression of the *Sex-lethal* (*Sxl*) gene in *Drosophila*. Although this gene is transcribed in both XX and XY embryos, alternate splicing of its RNA limits the synthesis of the SXL protein to XX embryos, which develop as females. The absence of SXL protein in XY embryos causes them to develop as males.

(*tra*) (**FIGURE 21.5**) These transcripts can be processed in two different ways. In chromosomal males, where the SXL protein is absent, the splicing apparatus always leaves a stop codon in the second exon of the *tra* RNA. Thus, when spliced *tra* RNA is translated, it generates a truncated (and nonfunctional) polypeptide. In females, where the SXL protein is present, this premature stop codon is removed by alternate splicing in at least some of the transcripts. Thus, when they are translated, some functional *transformer* protein (denoted TRA) is produced. The SXL protein therefore allows the synthesis of functional TRA protein in XX embryos but not in XY embryos.

Differentiating in Response to the Signal

The TRA protein also turns out to be a regulator of RNA processing. Along with TRA2, a protein encoded by the *transformer2* (*tra2*) gene, it controls the expression of *doublesex* (*dsx*), an autosomal gene that can produce two different proteins through alternate splicing of its RNA. In XX embryos, where the TRA protein is present, *dsx* transcripts are processed to encode a DSX protein that represses the genes required for male development. Therefore, such embryos develop into females. In XY embryos, where the TRA protein is absent, *dsx* transcripts are processed to encode a DSX protein that represses the genes required for female development. Consequently, such embryos develop into males. The TRA and TRA2 proteins also control the processing of transcripts from another autosomal gene called *fruitless* (*fru*). The Focus on *fruitless* describes how this gene's role in sexual differentiation has been elucidated.

Mutations in the Sex-determination Genes

Mutations have been obtained in each of the genes of the *Drosophila* sex-determination pathway (**TABLE 21.1**). Loss-of-function mutations in *Sxl* prevent SXL protein from being made in females. Homozygous mutants would therefore develop into males; however, they die as embryos. This embryonic death is not due to the incipient sexual transformation but rather to an abnormality in the dosage compensation system (Chapter 20). In addition to regulating the sex-determination pathway, *Sxl* also regulates dosage compensation. Although the mechanism is not fully understood, it apparently prevents the hyperactivation of X-linked genes in XX animals. When this hyperactivation occurs, as it does in homozygous *Sxl* mutants, XX embryos die because there is too much X-linked gene expression. However, in XY animals, loss-of-function mutations in the *Sxl* gene have no effect, which is consistent with the fact that the SXL protein is normally not made in males.

Loss-of-function mutations in *transformer* and *transformer2* have the same phenotype: both XX and XY animals develop into males. The sexual transformation in XX animals demonstrates that both the tra^+ and $tra2^+$ genes are needed for female development; however, they are perfectly dispensable for male development. Loss-of-function mutations in the *dsx* gene cause both XX and XY embryos to develop into intersexes. The intersexual

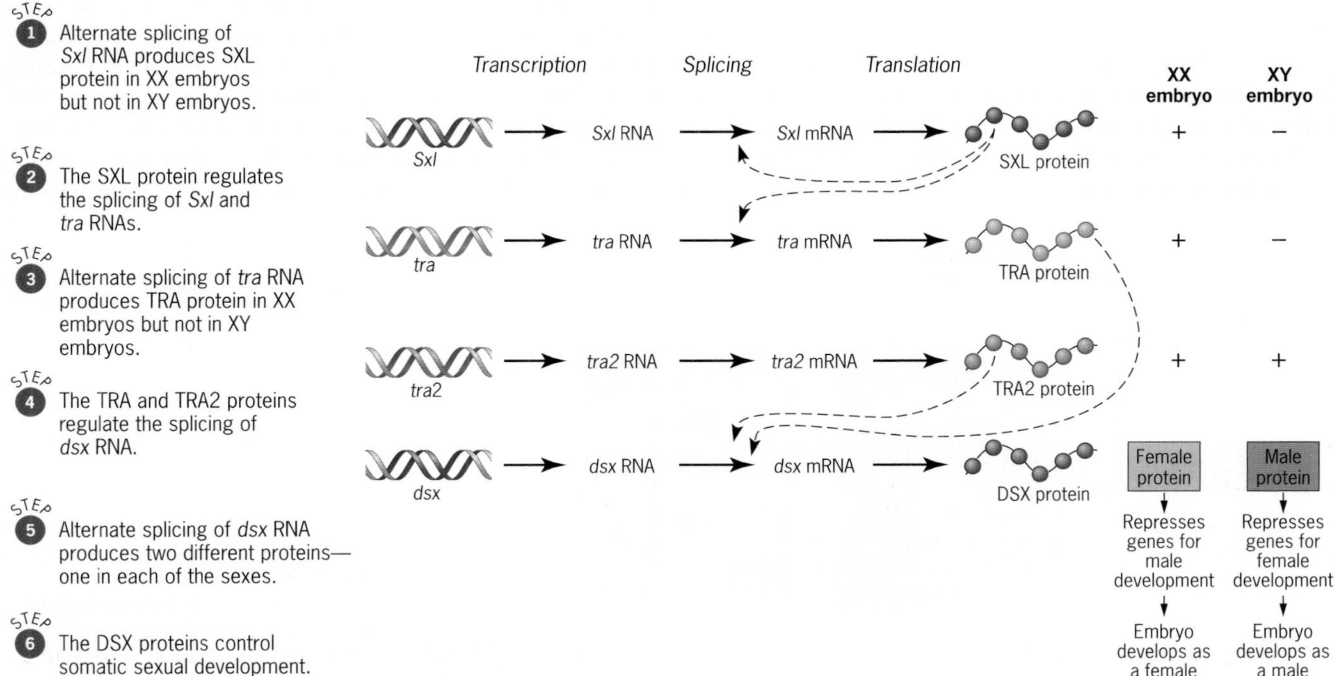
STEP 1 Alternate splicing of Sxl RNA produces SXL protein in XX embryos but not in XY embryos.
STEP 2 The SXL protein regulates the splicing of Sxl and tra RNAs.
STEP 3 Alternate splicing of tra RNA produces TRA protein in XX embryos but not in XY embryos.
STEP 4 The TRA and TRA2 proteins regulate the splicing of dsx RNA.
STEP 5 Alternate splicing of dsx RNA produces two different proteins—one in each of the sexes.
STEP 6 The DSX proteins control somatic sexual development.
Transcription
Splicing
Translation
XX embryo
XY embryo
Sxl
Sxl RNA
Sxl mRNA
SXL protein
+
−
tra
tra RNA
tra mRNA
TRA protein
+
−
tra2
tra2 RNA
tra2 mRNA
TRA2 protein
+
+
dsx
dsx RNA
dsx mRNA
DSX protein
Female protein
Male protein
Represses genes for male development
Represses genes for female development
Embryo develops as a female
Embryo develops as a male

Figure 21.5 ▶ Regulation of sex determination in *Drosophila* by the *Sex-lethal* (*Sxl*) gene. The *Sxl* gene regulates the expression of the *transformer* (*tra*) gene, which, in turn, regulates the expression of the *doublesex* (*dsx*) gene; the *transformer2* (*tra2*) gene also participates in the regulation of *dsx*. The + and − signs indicate the presence or absence of the various proteins.

▶ **TABLE 21.1**

Phenotypes of Loss-of-Function Mutations in Sex-Determination Genes in *Drosophila melanogaster* and *Caenorhabditis elegans*[a]

Gene	XX Mutant Phenotype	XY (or XO) Mutant Phenotype
Drosophila melanogaster		
Numerator gene	lethal	no effect
Denominator gene	no effect	reduced viability
Sxl	lethal	no effect
tra	male	no effect
tra2	male	sterile male
dsx	sterile intersex	sterile intersex
Caenorhabditis elegans		
xol-1	no effect	lethal
sdc-1	masculinized	no effect
sdc-2	masculinized	no effect
sdc-3	no sex-determination effect	no effect
her-1	no effect	fertile hermaphrodite
tra-2	male	no effect
fem-1	female	female
fem-2	female	female
fem-3	female	female
tra-1	male	minor effects in gonad

[a]*Source:* Parkhurst, S. M., and P. M. Meneely. 1994. *Science* 264:924–932.

phenotype appears because *both* the male and the female developmental pathways are activated in *dsx* mutants.

SEX DETERMINATION IN *C. ELEGANS*

The somatic sex-determination pathway in *C. elegans* involves at least 10 different genes (**FIGURE 21.6**). As in *Drosophila*, mutations in these genes alter sexual development (**TABLE 21.1**). For example, loss-of-function mutations in the two *transformer* genes *tra-1* and *tra-2* (not to be confused with the *tra* and *tra2* genes of *Drosophila*) cause XX animals to develop into males, and loss-of-function mutations in the *hermaphrodite* gene *her-1* cause XO animals to develop into hermaphrodites. These phenotypic alterations show that the *tra-1* and *tra-2* gene products are needed for normal hermaphrodite development and that the *her-1* gene product is needed for normal male development. Loss-of-function mutations in three other genes, *fem-1*, *fem-2*, and *fem-3* (for feminization), cause XO animals to

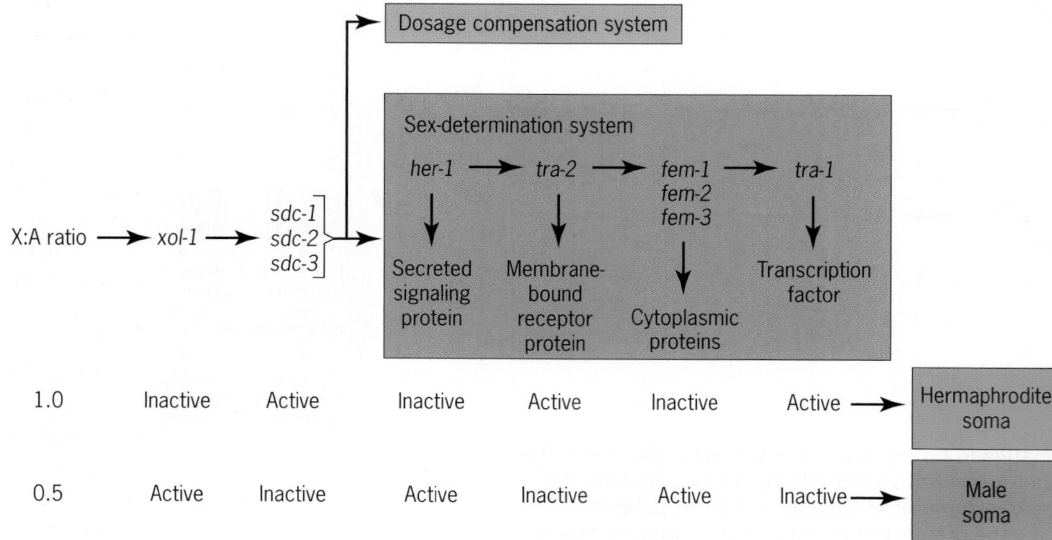

Figure 21.6 ▶ Developmental pathway for sex determination in *C. elegans.* Differentiation of the animal as a male or as a hermaphrodite depends on a cascade of gene expression, which in turn depends on the X:A ratio. Dosage compensation also depends on the X:A ratio.

develop into spermless females. This feminized phenotype shows that the *fem* gene products are needed for normal male development.

Researchers have been able to arrange these six genes in a developmental pathway. The first gene, *her-1*, encodes a secreted protein that is a signaling molecule. The next gene, *tra-2*, encodes a membrane-bound protein, which functions as a receptor for the *her-1* signaling protein. The products of the *fem* genes are cytoplasmic proteins that transduce the *her-1* signal, and the last gene in the pathway, *tra-1*, encodes a zinc-finger-type transcription factor, which regulates the genes involved in sexual differentiation.

In *C. elegans*, the sex-determination pathway seems to involve a series of negative regulators of gene expression. In XO animals, the secreted *her-1* gene product apparently interacts with the *tra-2* gene product, causing it to become inactive. This interaction allows the three *fem* gene products to be activated, and they collectively inactivate the *tra-1* gene product, which is a positive regulator of female differentiation. Because the animal cannot develop as a hermaphrodite without active *tra-1* protein, it develops as a male. In XX animals, the *her-1* protein is not made; thus, its receptor, the *tra-2* protein, remains active. Active *tra-2* protein causes the *fem* gene products to be inactivated, which in turn allows the *tra-1* protein to stimulate female differentiation. The animal therefore develops as a hermaphrodite.

Sexual development in *C. elegans* fundamentally depends on the X:A ratio, just as it does in *Drosophila*. Somehow this ratio is converted into a molecular signal that controls sexual differentiation. This same signal also controls the phenomenon of dosage compensation, which in *C. elegans* involves hypoactivation of the two X chromosomes in the hermaphrodite (Chapter 20). The signal from the X:A ratio is funneled into the sex-determination and dosage-compensation pathways through a short pathway involving at least four genes. One of these, *xol-1*, is required in males but not in hermaphrodites. Loss-of-function mutations in *xol-1* cause XO animals to die, hence the name of the gene, *XO-lethal*. Three other genes, *sdc-1*, *sdc-2*, and *sdc-3* (for *s*ex determination and *d*osage *c*ompensation), are negatively regulated by *xol-1*. Loss-of-function mutations in these genes either kill XX animals or turn them into males. Thus, the *sdc* genes are needed in hermaphrodites but not in males.

KEY POINTS

- In *Drosophila* the pathway that controls sexual differentiation involves some genes that ascertain the X:A ratio, some that convert this ratio into a developmental signal, and others that respond to the signal by producing either male or female structures.
- The *Sex-lethal* (*Sxl*) gene plays a key role in *Drosophila* sexual development by regulating the splicing of its own transcript and that of another gene (*tra*).
- In *C. elegans*, the sexual-differentiation pathway involves genes that encode signaling proteins, their receptors, signal transducers, and transcription factors.

Maternal Gene Activity in Development

Materials transported into the egg during oogenesis play a major role in embryonic development.

Important events occur in animal development even before an egg is fertilized. At this time, nutritive and determinative materials are transported into the egg from surrounding cells, laying up food stores and organizing the egg for its subsequent development—the molecular equivalent of a mother's love. These materials are generated by the expression of genes in the female reproductive system, some being expressed in somatic reproductive tissues and others only in germ-line tissues. Collectively, these genes help to form eggs that can develop into embryos after fertilization. In some species, these maternal gene products lay out the basic body plan of the embryo, distinguishing head from tail and back from belly. These maternally supplied materials therefore establish a molecular coordinate system to guide an embryo's development.

MATERNAL-EFFECT GENES

Mutations in genes that contribute to the formation of healthy eggs may have no effect on the viability or appearance of the female making those eggs. Instead, their effects may be seen only in the next generation. Such mutations are called **maternal-effect mutations** because the mutant phenotype in the offspring is caused by a mutant genotype in its mother.

Genes identified by such mutations are called **maternal-effect genes.** The *dorsal* (*dl*) gene in *Drosophila* is a good example (**FIGURE 21.7**). Matings between flies homozygous for recessive mutations in this gene produce inviable progeny. This lethal effect is strictly maternal. A cross between homozygous mutant females and homozygous wild-type males produces inviable progeny, but the reciprocal cross (homozygous mutant males × homozygous wild-type females) produces viable progeny. The lethal effect of the *dorsal* mutation is therefore manifested only if females are homozygous for it. The male genotype is irrelevant.

Molecular characterization of the *dorsal* gene has revealed the basis for this maternal effect. The *dorsal* gene encodes a transcription factor that is produced during oogenesis and stored in the egg. Early in development, this transcription

▶ FOCUS ON *fruitless*

In 1963 at the annual meeting of the American Society of Zoologists, Kulbir Gill reported the discovery of a new male-sterile mutation in *Drosophila*. Males homozygous for this mutation did not attempt to mate with females even though they courted them. Even more remarkably, the mutant males courted other males, both mutant and wild-type. This behavior was dramatically seen when Gill placed a group of mutant males together in a culture. The males formed long chains or circles, with each male in the chain courting the male in front of it (**FIGURE 1**). However, the male-sterile mutation apparently had no effects in females. Homozygous mutant females mated readily with wild-type males and produced offspring.

For many years, the male-sterile mutation that Gill discovered was little more than a laboratory curiosity. Then, in the late 1970s and early 1980s, Jeffrey Hall and coworkers carried out experiments to map the mutation and to characterize its phenotypic effects more fully. The mutation was also given a name, *fruitless* (*fru*), in recognition of its male-sterile effect.

In the 1990s, researchers from several different laboratories embarked on a cooperative effort to clone the *fru* gene. Their work, published in 1996 with Lisa Ryner as the principal author,[1] established that *fru* encodes a zinc-finger-type transcription factor that regulates the genes for male sexual behavior—indeed, male sexual preference—in *Drosophila*.

Earlier genetic experiments by Barbara Taylor had suggested that *fru* might control a branch in the *Drosophila* sex-determination pathway (**FIGURE 22.9**). The TRA and TRA2 proteins regulate the processing of *dsx* RNA. Alternate splicing of *dsx* transcripts in males and females produces different forms of a transcription factor (DSX-female and DSX-male) that sex-specifically regulates the genes involved in sexual differentiation. However, Taylor observed that the formation of a particular abdominal muscle in males was not controlled by the *dsx* gene. This finding implied that the sex-determination pathway had a branch, regulated not by *dsx* but by another as yet unidentified gene, and that this unknown gene might, like *dsx*, encode sex-specific transcription factors. Furthermore, it seemed likely that the unknown gene would itself be regulated posttranscriptionally by the splicing-factor proteins TRA and TRA2.

The TRA and TRA2 proteins bind to an RNA sequence 13 nucleotides long. This sequence is repeated multiple times in transcripts from the *dsx* gene. Thus, when Ryner and her colleagues began their efforts to clone *fru*, they searched for other genes that possessed the 13-nucleotide TRA/TRA2 RNA-binding sites. Their search led them to a locus on chromosome 3 where the *fru* gene had been mapped. Cloning and analysis of the DNA at this locus revealed the molecular structure of the *fru* gene. It contains three copies of the 13-nucleotide TRA/TRA2 RNA-binding sequence, and, as expected, its expression is regulated at the splicing level by the TRA and TRA2 proteins.

[1]Ryner, L. C., S. F. Goodwin, D. H. Castrillon, A. Anand, A. Villella, B. S. Baker, J. C. Hall, B. J. Taylor, and S. A. Wasserman. 1996. Control of male sexual behavior and sexual orientation in *Drosophila* by the *fruitless* gene. *Cell* 87:1079–1089.

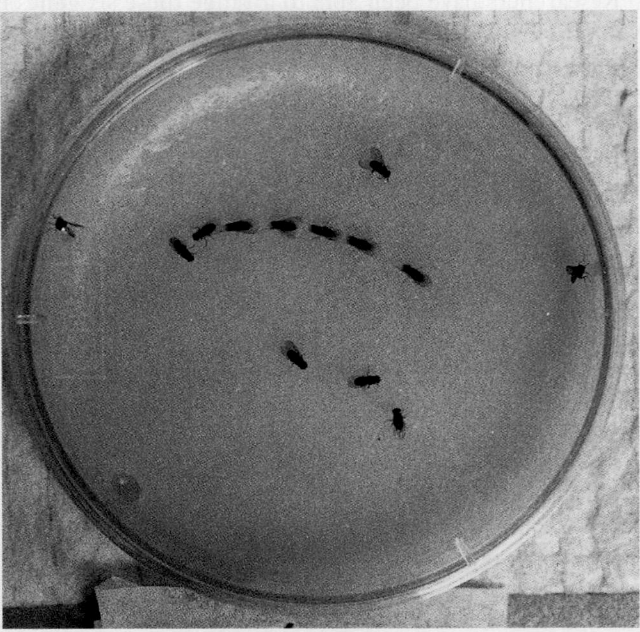

Figure 1 ▶ *Drosophila* males homozygous for the *fruitless* (*fru*) mutation courting each other. Each male in the chain is courting the male in front of it.

The pattern of *fru* expression is complex; there are multiple promoters, and the transcripts from the gene are alternately spliced in males and females. In males, at least one of *fru*'s products is a transcription factor; it is unclear if a different version of this transcription factor is made in females. *In situ* RNA hybridization experiments indicate that *fru* expression is limited to a relatively small number of neurons in the fly's central nervous system, some of which have been implicated in the control of male sexual behavior.

The discovery of *Drosophila*'s *fru* gene raises the possibility that sexual orientation is under genetic control in other species, including our own. Indeed, Dean Hamer and his associates have suggested that a gene in the long arm of the X chromosome contributes to male homosexuality in humans.[2] However, the nature of this gene—if it even exists[3]—is unknown.

[2]Hamer, D. H., S. Hu, V. L. Magnuson, N. Hu, and A.M.L. Pattatucci. 1993. A linkage between DNA markers on the X chromosome and male sexual orientation. *Science* 261:321–327.

[3]Rice, G., C. Anderson, N. Risch, and G. Ebers. 1999. Male homosexuality: absence of linkage to microsatellite markers at Xq28. *Science* 284:665–667.

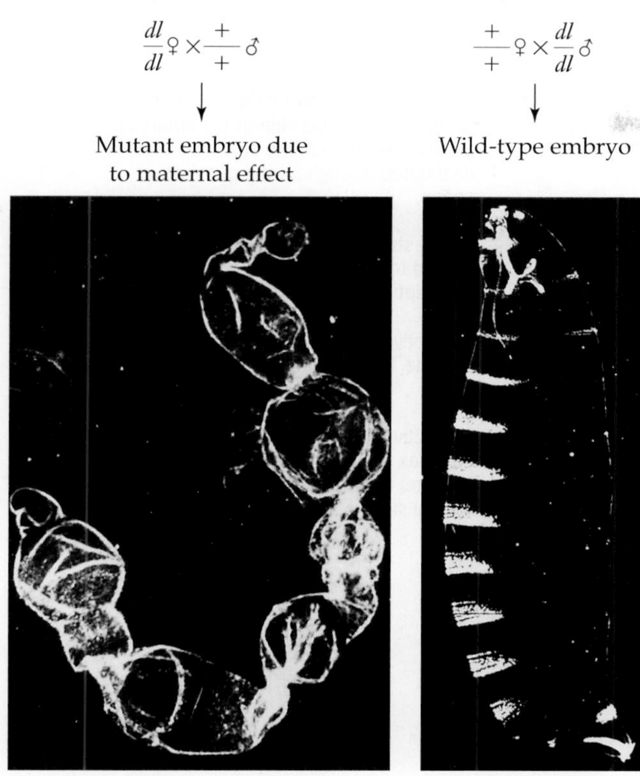
dl/dl ♀ × +/+ ♂
Mutant embryo due to maternal effect
+/+ ♀ × dl/dl ♂
Wild-type embryo

Figure 21.7 ► The maternal effect of a mutation in the *dorsal* (*dl*) gene of *Drosophila.* The mutant phenotype is an embryo that lacks ventral tissues; that is, it is dorsalized.

DETERMINATION OF THE DORSAL-VENTRAL AND ANTERIOR-POSTERIOR AXES IN *DROSOPHILA* EMBRYOS

Animals with bilateral symmetry have two primary body axes, one distinguishing back from belly (dorsal from ventral) and the other distinguishing head from tail (anterior from posterior). Both of these axes are established very early in development, in some species even before fertilization. In *Drosophila,* the processes of axis formation have been dissected genetically by collecting mutations that affect early embryonic development.

In the 1970s and 1980s, massive searches for such mutations were carried out by Christiane Nüsslein-Volhard, Eric Wieschaus, Trudi Schüpbach, Gerd Jürgens, and others. These researchers used chemical mutagens to induce mutations in each of the *Drosophila* chromosomes. Many mutations were identified, including maternal-effect lethals in genes such as *dorsal.* Molecular and genetic analyses of these mutations have provided considerable insight into the events of early *Drosophila* development.

Formation of the Dorsal-Ventral Axis

Differentiation of a *Drosophila* embryo along the dorsal-ventral axis hinges on the action of the transcription factor encoded by the *dorsal* gene (**FIGURE 21.8**). This protein is synthesized maternally and stored in the cytoplasm of the egg. At the time of blastoderm formation, the dorsal protein enters the nuclei on the ventral side of the embryo, inducing the transcription of two genes called *twist* and *snail* (whimsically named for their mutant phenotypes). In these same nuclei, it represses the genes *zerknüllt* (from the German for "crumpled") and *decapentaplegic* (from the Greek words for "15" and "stroke"). The selective induction and repression of genes cause the ventral cells to differentiate into a primitive embryonic layer of tissue called the mesoderm. On the opposite side of the embryo, where the dorsal protein is excluded from the nuclei, *twist* and *snail* are not induced and *zerknüllt* and *decapentaplegic* are not repressed. Consequently, these cells differentiate into a different primitive tissue, the embryonic epidermis. The entrance of the dorsal factor plays an important role in the differentiation of the dorsal and ventral parts of the embryo. When it is missing, the ventral parts incorrectly differentiate as if they were on the dorsal side, creating an embryo with two dorsal surfaces. This lethal condition cannot be prevented by a wild-type *dorsal* allele inherited from the father because *dorsal* is not transcribed in the embryo. Expression of the *dorsal* gene is, in fact, limited to the female germ line. Mutations in *dorsal* are therefore strict maternal-effect lethals.

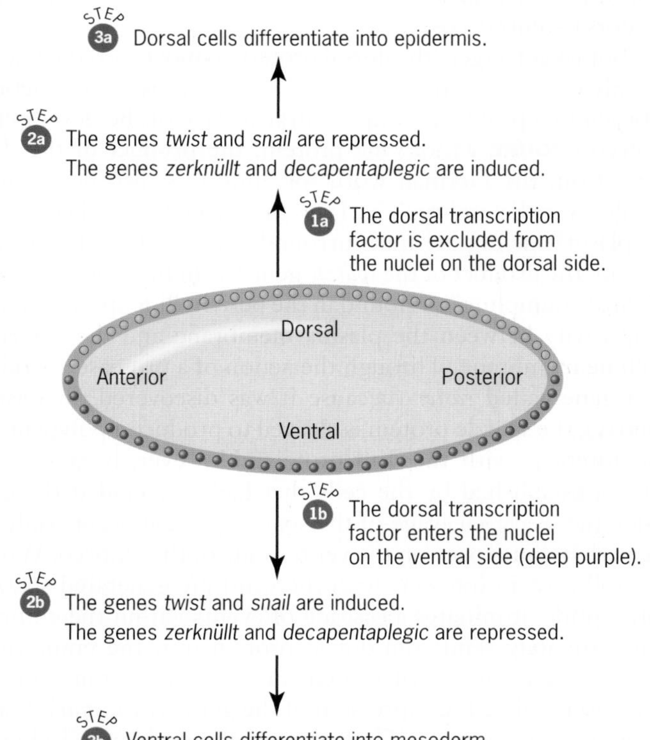
STEP 3a Dorsal cells differentiate into epidermis.
STEP 2a The genes twist and snail are repressed.
The genes zerknüllt and decapentaplegic are induced.
STEP 1a The dorsal transcription factor is excluded from the nuclei on the dorsal side.
Dorsal
Anterior
Posterior
Ventral
STEP 1b The dorsal transcription factor enters the nuclei on the ventral side (deep purple).
STEP 2b The genes twist and snail are induced.
The genes zerknüllt and decapentaplegic are repressed.
STEP 3b Ventral cells differentiate into mesoderm.

Figure 21.8 ► Determination of the dorsal-ventral axis in *Drosophila* by the dorsal protein. This protein is a transcription factor that acts only in the nuclei on the ventral side of the embryo. The genes *twist, snail, zerknüllt,* and *decapentaplegic* are regulated by dorsal protein.

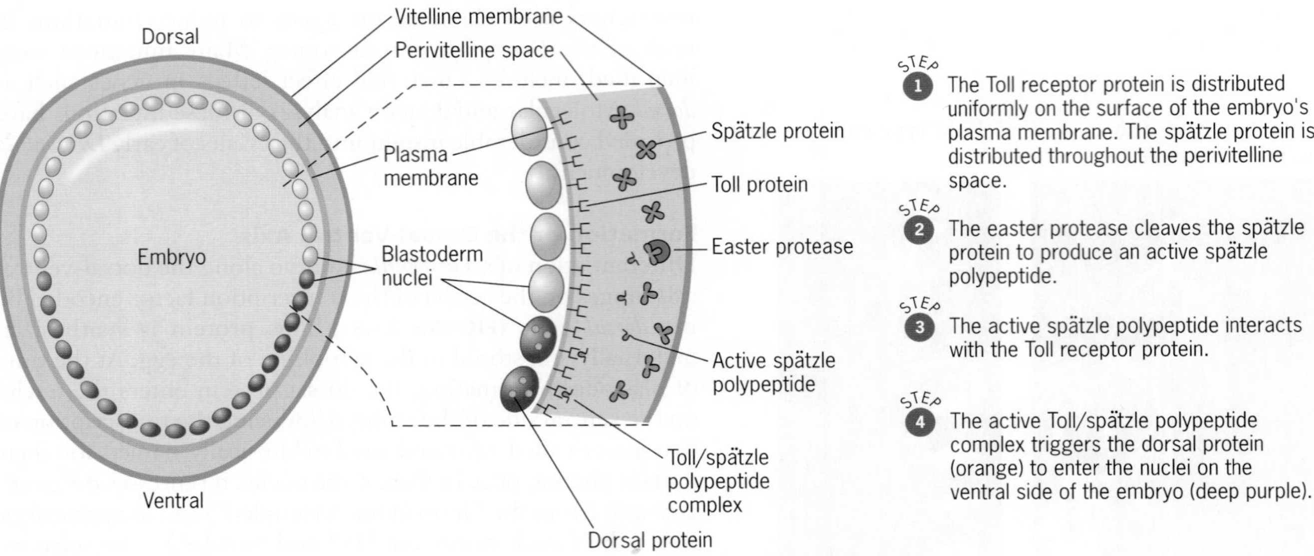

Dorsal
Vitelline membrane
Perivitelline space
Plasma membrane
Embryo
Blastoderm nuclei
Ventral
Spätzle protein
Toll protein
Easter protease
Active spätzle polypeptide
Toll/spätzle polypeptide complex
Dorsal protein
STEP 1 The Toll receptor protein is distributed uniformly on the surface of the embryo's plasma membrane. The spätzle protein is distributed throughout the perivitelline space.
STEP 2 The easter protease cleaves the spätzle protein to produce an active spätzle polypeptide.
STEP 3 The active spätzle polypeptide interacts with the Toll receptor protein.
STEP 4 The active Toll/spätzle polypeptide complex triggers the dorsal protein (orange) to enter the nuclei on the ventral side of the embryo (deep purple).

Figure 21.9 ▶ Differentiation of the dorsal-ventral axis in a *Drosophila* embryo. The cross section shows the interaction between the membrane-bound Toll receptor protein and a polypeptide from the spätzle protein that induces differentiation along the dorsal-ventral axis. Formation of the interacting spätzle polypeptide occurs in the space between the plasma membrane and the vitelline membrane on the ventral side of the embryo.

transcription factor into the ventral nuclei and its exclusion from the dorsal nuclei therefore initiate differentiation along the dorsal-ventral axis.

But what triggers the dorsal protein to move into the nuclei on only one side of the embryo? The answer is an interaction between two proteins on the ventral surface of the developing embryo (**FIGURE 21.9**). One protein, the product of the *Toll* gene (from the German word for "tuft"), is distributed uniformly over the embryo's surface; this protein is embedded in the plasma membrane that surrounds the embryo. The other protein, the product of the *spätzle* gene (from the German word for "little dumpling"), is found in the perivitelline space, a fluid-filled cavity between the plasma membrane and the external vitelline membrane. Through the action of a protease encoded by a gene called *easter* (because it was discovered on Easter Sunday), the spätzle protein is cleaved to produce a polypeptide that interacts with the Toll protein. However, because of a pattern established by the cells that had surrounded the egg inside the ovary, cleavage of the spätzle protein occurs only in the perivitelline space on the ventral side of the embryo. When the Toll protein interacts with the ventrally generated spätzle polypeptide, it initiates a cascade of events within the embryo that ultimately sends the dorsal protein into the embryonic nuclei. There the dorsal protein functions as a transcription factor to regulate the expression of the genes *twist*, *snail*, *decapentaplegic*, and *zerknüllt*. Thus, the membrane-bound Toll protein acts as a receptor for the determinative spätzle polypeptide, and the physical interaction between these two molecules acts as a signal to trigger a genetic program for the differentiation of the embryo along its dorsal-ventral axis.

Formation of the Anterior-Posterior Axis

The anterior-posterior axis in *Drosophila* is created by the regional synthesis of transcription factors encoded by the *hunchback* and *caudal* genes (**FIGURE 21.10**). These two genes are transcribed in the nurse cells of the maternal germ line. These special cells support the growth and development of the oocyte.The maternal transcripts of the *hunchback* and *caudal* genes are then carried from the nurse cells into the oocyte where they become uniformly distributed in the cytoplasm. However, both types of transcripts are translated in different parts of the embryo. The hunchback RNA is translated only in the anterior part, and the caudal RNA is translated only in the posterior part. This differential translation produces concentration gradients of the proteins encoded by these two genes; hunchback protein is concentrated in the anterior part of the embryo, and caudal protein is concentrated in the posterior part. These two proteins then function to activate or repress transcription of the genes whose products are involved in the differentiation of the embryo along its anterior-posterior axis.

What limits the translation of hunchback RNA to the anterior part of the embryo and caudal RNA to the posterior part? It turns out that two maternally supplied RNAs are involved, one transcribed from the *bicoid* gene and the other from the *nanos* gene. Both of these RNAs are synthesized in the nurse cells of the maternal germ line and are then transported into the oocyte. The bicoid RNA becomes anchored at the anterior end of the developing oocyte, and the nanos RNA becomes anchored at the posterior end. After fertilization, each type of RNA is translated locally, and the resulting

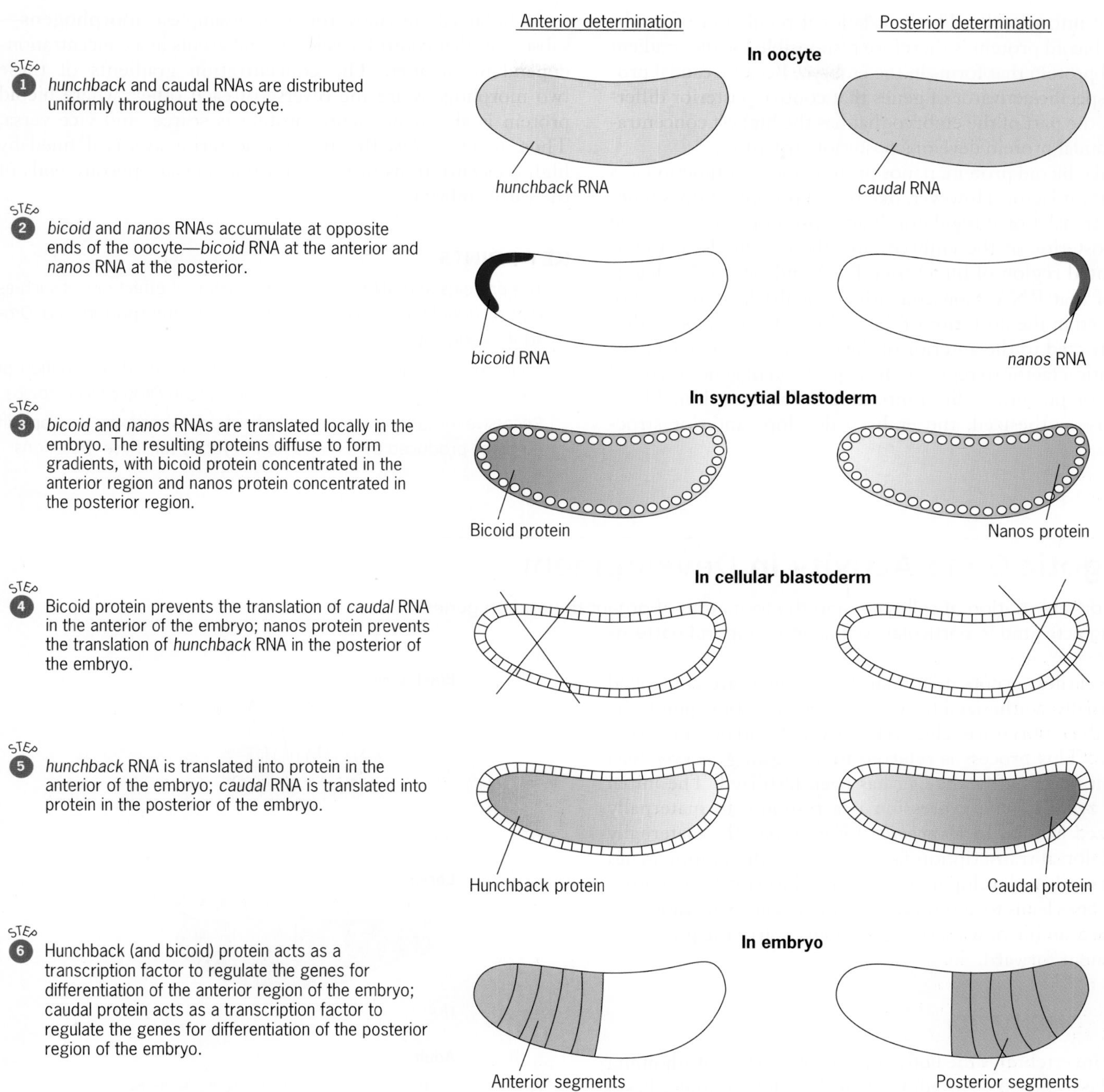

Anterior determination
Posterior determination
In oocyte
STEP 1 hunchback and caudal RNAs are distributed uniformly throughout the oocyte.
hunchback RNA
caudal RNA
STEP 2 bicoid and nanos RNAs accumulate at opposite ends of the oocyte—bicoid RNA at the anterior and nanos RNA at the posterior.
bicoid RNA
nanos RNA
In syncytial blastoderm
STEP 3 bicoid and nanos RNAs are translated locally in the embryo. The resulting proteins diffuse to form gradients, with bicoid protein concentrated in the anterior region and nanos protein concentrated in the posterior region.
Bicoid protein
Nanos protein
In cellular blastoderm
STEP 4 Bicoid protein prevents the translation of caudal RNA in the anterior of the embryo; nanos protein prevents the translation of hunchback RNA in the posterior of the embryo.
STEP 5 hunchback RNA is translated into protein in the anterior of the embryo; caudal RNA is translated into protein in the posterior of the embryo.
Hunchback protein
Caudal protein
In embryo
STEP 6 Hunchback (and bicoid) protein acts as a transcription factor to regulate the genes for differentiation of the anterior region of the embryo; caudal protein acts as a transcription factor to regulate the genes for differentiation of the posterior region of the embryo.
Anterior segments
Posterior segments

Figure 21.10 ▶ Determination of the anterior-posterior axis in *Drosophila* by maternally supplied RNAs. These RNAs come from the *hunchback, caudal, bicoid,* and *nanos* genes. For each oocyte or embryo, anterior is at the left and posterior is at the right.

protein products diffuse through the embryo to form concentration gradients; bicoid protein is concentrated at the anterior end, and nanos protein is concentrated at the posterior end.

The bicoid protein has two functions. First, it acts as a transcription factor to stimulate the synthesis of RNAs from several genes, including *hunchback*. These RNAs are then translated into proteins that control the formation of the anterior structures of the embryo. Second, bicoid protein prevents the translation of caudal RNA by binding to sequences in the 3′ untranslated region of that RNA. Thus, wherever bicoid protein is abundant (that is, in the anterior of the embryo), caudal RNA is not translated into protein. Conversely, wherever bicoid protein is scarce (that is, in the posterior of the embryo), caudal RNA is translated into protein. The translational regulation of caudal RNA by bicoid protein is therefore responsible for the gradient of caudal protein that forms in the embryo. Because caudal protein is a specific activator of genes that control posterior differentation, the part of the embryo that has the highest concentration of caudal protein develops posterior structures.

Unlike bicoid protein, nanos protein does not function as a transcription factor. However, like bicoid protein, it does function as a translational regulator. Nanos protein is concentrated in the posterior of the embryo, and there it binds to the 3′ untranslated region of hunchback RNA and causes the degradation of that RNA. Consequently, hunchback protein is not synthesized in the posterior of the embryo. Instead, its synthesis is restricted to the anterior of the embryo where it acts as a transcription factor to regulate the expression of genes involved in anterior-posterior differentiation. Wherever hunchback protein is synthesized, the embryo develops anterior structures.

The bicoid and nanos proteins are examples of **morphogens**—substances that control developmental events in a concentration-dependent manner. The concentration gradients of these two morphogens are the reverse of each other; where bicoid protein is abundant, nanos protein is scarce, and vice versa. Thus, in *Drosophila* the anterior-posterior axis is defined by high concentrations of these morphogens at opposite ends of the early embryo.

KEY POINTS

- The proteins and RNAs encoded by maternal-effect genes such as *dorsal, hunchback, bicoid,* and *nanos* are transported into *Drosophila* eggs during oogenesis.
- Maternal-effect gene products are involved in the determination of the dorsal-ventral and anterior-posterior axes in *Drosophila* embryos.
- Recessive mutations in maternal-effect genes are expressed only in embryos produced by females homozygous for these mutations.

Zygotic Gene Activity in Development

The differentiation of cell types and the formation of organs depend on genes being activated in particular spatial and temporal patterns.

The earliest events in animal development are controlled by maternally synthesized factors. However, at some point, the genes in the embryo are selectively activated, and new materials are made. This process is referred to as *zygotic gene expression* because it occurs after the egg has been fertilized. The initial wave of zygotic gene expression is a response to maternally synthesized factors. In *Drosophila*, for example, the maternally supplied dorsal transcription factor activates the zygotic genes *twist* and *snail.* As development proceeds, the activation of other zygotic genes leads to complex cascades of gene expression. We shall now examine how these zygotic genes carry the process of development forward.

BODY SEGMENTATION

In many invertebrates the body consists of an array of adjoining units called *segments.* An adult *Drosophila*, for example, has a head, three distinct thoracic segments, and eight abdominal segments. Within the thorax and abdomen, each segment can be identified by coloration, bristle pattern, and the kinds of appendages attached to it. These segments can also be identified in the embryo and the larva (**FIGURE 21.11**). In vertebrates, a segmental pattern is not so evident in the adult, but it can be recognized in the embryo from the way that nerve fibers grow from the central nervous system, from the formation of branchial arches in the head, and from the organization of muscle masses along the anterior-posterior axis. Later in development, these features are modified, and the original segmental pattern becomes obscured. Nonetheless, in both vertebrates

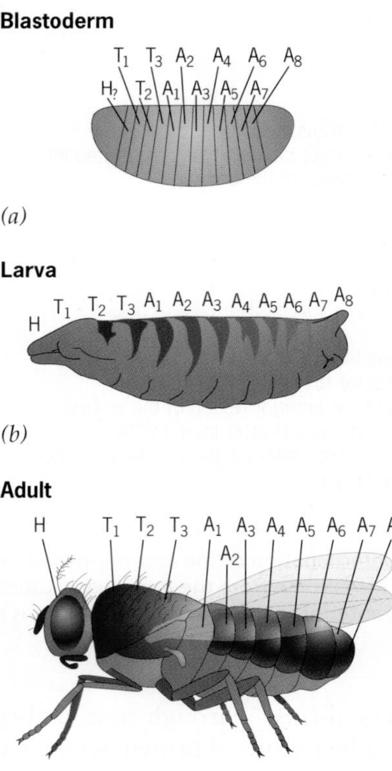

Figure 21.11 ▶ Segmentation in *Drosophila* at the (*a*) blastoderm, (*b*) larval, and (*c*) adult stages of development. Although segments are not visible in the blastoderm, its cells are already committed to form segments as shown; H, head segment; T, thoracic segment; A, abdominal segment.

and many invertebrates, segmentation is a key aspect of the overall body plan.

Homeotic Genes

Interest in the genetic control of segmentation began with the discovery of mutations that transform one segment into another. The first such mutation was found in *Drosophila* in 1915 by Calvin Bridges. He named it *bithorax* (*bx*) because it affected two thoracic segments. In this mutant, the third thoracic segment was transformed, albeit weakly, into the second, creating a fly with a small pair of rudimentary wings in place of the small balancing structures called halteres (**FIGURE 21.12**). Later, other segment-transforming mutations were found in *Drosophila*—for example, *Antennapedia* (*Antp*), a mutant that partially transforms the antennae on the head into legs, which characteristically grow from the thorax. These mutations have come to be called **homeotic mutations** because they cause one body part to look like another. The word homeotic comes from William Bateson, who coined the term *homeosis* to refer to cases in which "something has been changed into the likeness of something else." Like so many other words Bateson coined, this one has become a standard term in the modern genetics vocabulary.

The bithorax and Antennapedia phenotypes result from mutations in **homeotic genes.** Several such genes have now been identified in *Drosophila*, where they form two large clusters on one of the autosomes (**FIGURE 21.13**). The *bithorax complex*, usually denoted *BX-C*, consists of three genes: *Ultrabithorax* (*Ubx*), *abdominal-A* (*abd-A*), and *Abdominal-B (Abd-B);* the *Antennapedia complex*, denoted *ANT-C*, consists of five genes: *labial (lab)*, *proboscipedia (pb)*, *Deformed (Dfd)*, *Sex combs reduced (Scr)*, and *Antennapedia (Antp)*. Molecular analysis of these genes has demonstrated that they all encode helix-turn-helix transcription factors with a conserved region of 60 amino acids. This region, called the **homeodomain,** is involved in DNA binding.

The BX-C was the first of the two homeotic gene complexes to be dissected genetically. Analysis of this complex began in the late 1940s with the work of Edward Lewis. By studying mutations in the BX-C, Lewis showed that the wild-type function of each part of the complex is restricted to a specific region in the developing animal. Molecular analyses later

Figure 21.12 ▶ The phenotype of a *bithorax* mutation in *Drosophila.*

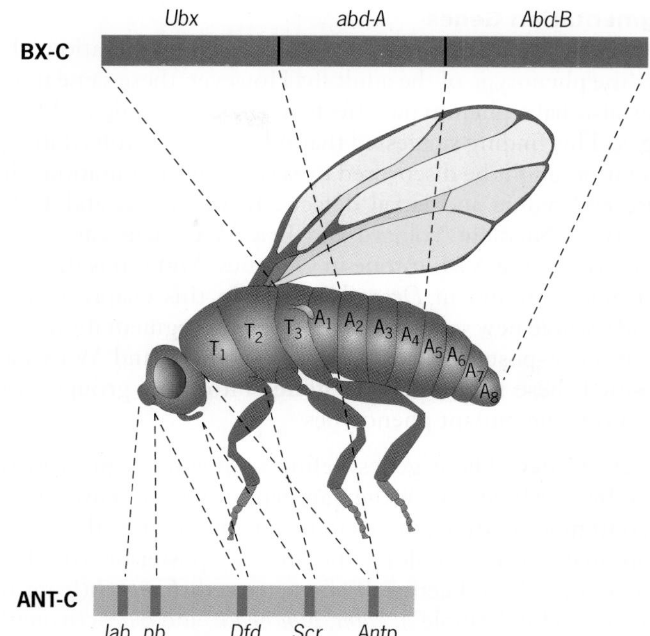

Figure 21.13 ▶ The homeotic genes in the bithorax complex (BX-C) and Antennapedia complex (ANT-C) of *Drosophila.* The body regions in which each gene is expressed are indicated.

reinforced and refined this conclusion. Study of the ANT-C began in the 1970s, principally through the work of Thomas Kaufman, Matthew Scott, and their collaborators. Through a combination of genetic and molecular analyses, these investigators showed that the genes of the ANT-C are also expressed in a regionally specific fashion. However, the ANT-C genes are expressed more anteriorly than the BX-C genes. Curiously, the pattern of expression of the ANT-C and BX-C genes along the anterior-posterior axis corresponds exactly to the order of the genes along the chromosome (**FIGURE 21.13**); it is not yet clear why this is so. The developmental pathway that each cell takes seems to depend simply on the set of homeotic genes that are expressed within it. Because the homeotic genes play such a key role in selecting the segmental identities of individual cells, they are often called **selector genes.**

The proteins encoded by the homeotic genes are homeodomain transcription factors. These proteins bind to regulatory sequences in the DNA, including some within the bithorax and Antennapedia complexes themselves. For example, the UBX and ANTP proteins bind to a sequence within the promoter of the *Ubx* gene—a suggestion that the homeotic genes can regulate themselves and each other. Other gene targets of the homeodomain transcription factors have been identified, including some that encode other types of transcription factors. The homeotic genes therefore seem to control a regulatory cascade of target genes, which in turn act to determine the segmental identities of individual cells. However, the homeotic genes do not stand at the top of this regulatory cascade. Their activities are controlled by another group of genes expressed earlier in development.

Segmentation Genes

Most of the homeotic genes were identified by mutations that alter the phenotype of the adult fly. However, these same mutations also have phenotypic effects in the embryonic and larval stages. This finding suggested that other genes involved in segmentation might be discovered by screening for mutations that cause embryonic and larval defects. In the 1970s and 1980s, Christiane Nüsslein-Volhard and Eric Wieschaus carried out such screens (see A Milestone in Genetics: Mutations that Disrupt Segmentation in *Drosophila* later in this chapter). They found a whole new set of genes required for segmentation along the anterior-posterior axis. Nüsslein-Volhard and Wieschaus classified these **segmentation genes** into three groups based on embryonic mutant phenotypes.

1. ***Gap Genes.*** These genes define segmental regions in the embryo. Mutations in the gap genes cause an entire set of contiguous body segments to be missing; that is, they create an anatomical gap along the anterior-posterior axis. Four gap genes have been well characterized: *Krüppel* (from the German for "cripple"), *giant*, *hunchback*, and *knirps* (from the German for "dwarf"). Each gene is expressed in characteristic regions in the early embryo under the control of the maternal-effect genes *bicoid* and *nanos*. The gap genes encode transcription factors.
2. ***Pair-Rule Genes.*** These genes define a pattern of segments within the embryo. The pair-rule genes are regulated by the gap genes and are expressed in seven alternating bands, or stripes, along the anterior-posterior axis, in effect dividing the embryo into 14 distinct zones, or *parasegments* (**FIGURE 21.14**). Some of the mutations in pair-rule genes produce embryos with only half as many parasegments as wild-type have. In each mutant, every other parasegment is missing, although the missing parasegments are not the same in different pair-rule mutants. Examples of pair-rule genes are *fushi tarazu* (from the Japanese for "something missing") and *even-skipped*. In *fushi tarazu* mutants, each of the odd-numbered parasegments is missing; in *even-skipped* mutants, each of the even-numbered parasegments is missing. The pair-rule genes also encode transcription factors.
3. ***Segment-Polarity Genes.*** These genes define the anterior and posterior compartments of individual segments along the anterior-posterior axis. Mutations in **segment-polarity genes** cause part of each segment to be replaced by a mirror-image copy of an adjoining half-segment. For example, mutations in the segment-polarity gene *gooseberry* cause the posterior half of each segment to be replaced by a mirror-image copy of the adjacent anterior half-segment. Many of the segment-polarity genes are expressed in 14 narrow bands along the anterior-posterior axis. Thus, they refine the segmental pattern established by the pair-rule genes. Two of the best-studied segment-polarity genes are *engrailed* and *wingless; engrailed* encodes a transcription factor, and *wingless* encodes a signaling molecule.

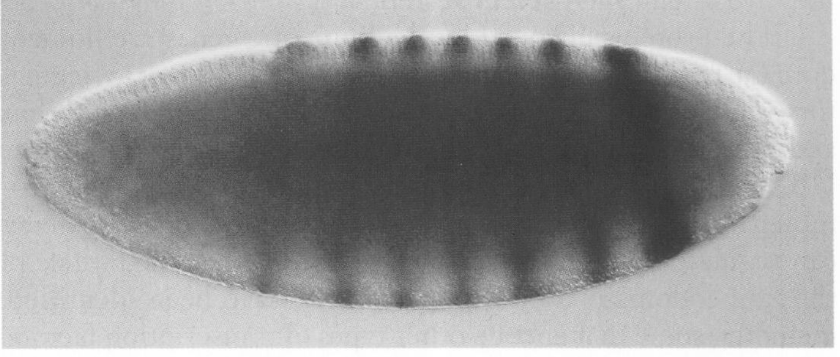

Figure 21.14 ▶ The seven-stripe pattern of RNA expression of the pair-rule gene *fushi tarazu* (*ftz*) in a *Drosophila* blastoderm embryo. The RNA was detected by *in situ* hybridization with a *ftz*-specific probe. Anterior is at the left; dorsal is at the top. Other pair-rule genes show a different seven-stripe pattern.

These three groups of genes form a regulatory hierarchy (**FIGURE 21.15**). The gap genes, which are regionally activated by the maternal-effect genes, regulate the expression of the pair-rule genes, which in turn regulate the expression of the segment-polarity genes. Concurrent with this process, the homeotic genes are activated under the control of the gap and pair-rule genes to give unique identities to the segments that form along the anterior-posterior axis. Interactions among the products of all these genes then refine and stabilize the segmental boundaries. In this way, the *Drosophila* embryo is progressively subdivided into smaller and smaller developmental units.

ORGAN FORMATION

When many different types of cells are organized for a specific purpose, they form an organ. The heart, stomach, kidney, liver, and eye are all examples of organs. One of the remarkable features of an organ is that it forms in a specific part of the body. The development of a heart in the head or an eye in the thorax of a fly, for example, would be extremely abnormal, and we would wonder what had gone wrong. Anatomically correct organ formation is obviously under tight genetic control.

Geneticists have obtained insights into the nature of this control from the study of another gene in *Drosophila*. This gene is called *eyeless* after the phenotype of flies that are mutant for it (**FIGURE 21.16**). The wild-type *eyeless* gene encodes a homeodomain transcription factor whose action switches on a developmental pathway that involves several thousand genes. Initially, several subordinate regulatory genes are activated. Their products then trigger a cascade of events that create specific cell types within the developing eye.

The role of the *eyeless* gene has been demonstrated by expressing it in tissues that normally do not form eyes (**FIGURE 21.17**). Walter Gehring and colleagues did this by creating transgenic flies in which the *eyeless* gene was fused to a promoter that could be activated in specific tissues. Activation of this promoter caused transcription of the *eyeless* gene outside its normal domain of expression. This, in turn, caused eyes to form in

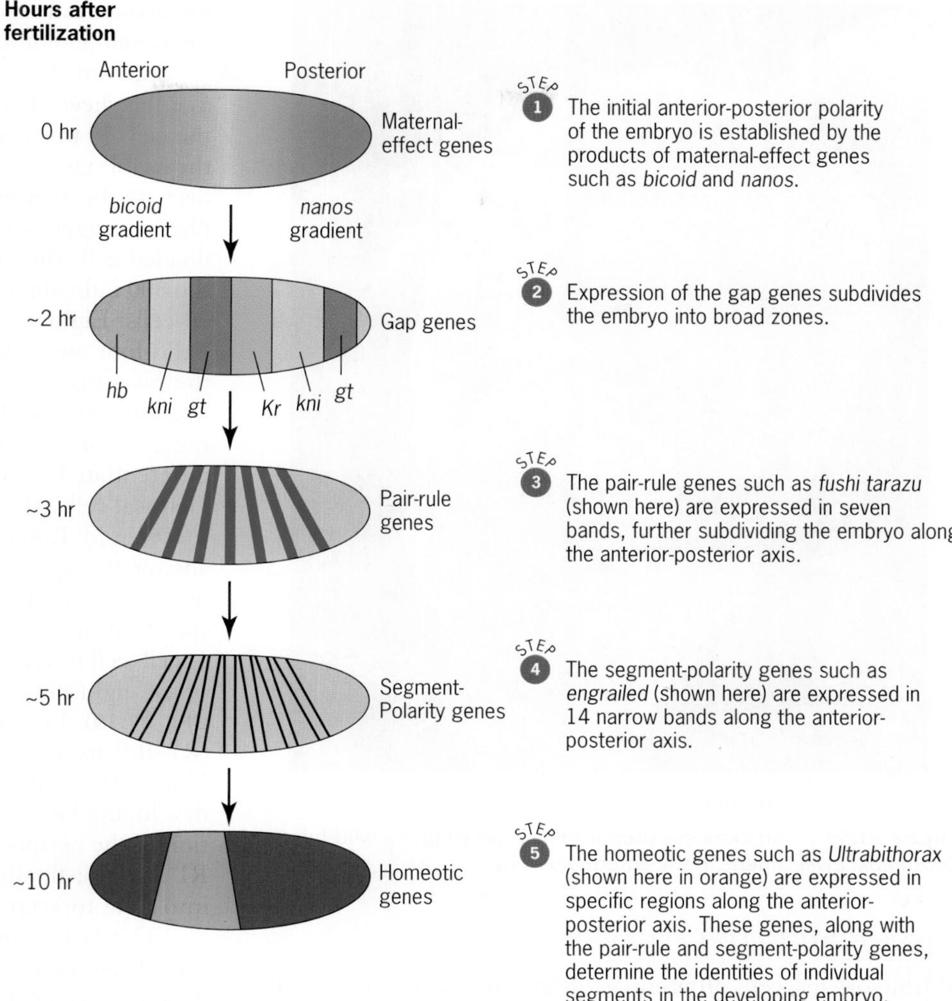
Hours after fertilization
Anterior
Posterior
0 hr
Maternal-effect genes
bicoid gradient
nanos gradient
~2 hr
Gap genes
hb
kni
gt
Kr
kni
gt
~3 hr
Pair-rule genes
~5 hr
Segment-Polarity genes
~10 hr
Homeotic genes
STEP 1 The initial anterior-posterior polarity of the embryo is established by the products of maternal-effect genes such as bicoid and nanos.
STEP 2 Expression of the gap genes subdivides the embryo into broad zones.
STEP 3 The pair-rule genes such as fushi tarazu (shown here) are expressed in seven bands, further subdividing the embryo along the anterior-posterior axis.
STEP 4 The segment-polarity genes such as engrailed (shown here) are expressed in 14 narrow bands along the anterior-posterior axis.
STEP 5 The homeotic genes such as Ultrabithorax (shown here in orange) are expressed in specific regions along the anterior-posterior axis. These genes, along with the pair-rule and segment-polarity genes, determine the identities of individual segments in the developing embryo.

Figure 21.15 ► Cascade of gene expression to produce segmentation in *Drosophila* embryos.

unorthodox places such as wings, legs, and antennae. These extra (or ectopic) eyes were anatomically well developed and functional; in fact, their photoreceptors responded to light.

An even more remarkable finding is that a mammalian homologue of the *eyeless* gene, called *Pax6*, also produces these extra eyes when it is inserted into *Drosophila* chromosomes. Gehring and coworkers used the mouse homologue of *eyeless* to transform *Drosophila* and got the same result as they did with the *eyeless* gene itself. This showed that the mouse gene, which also encodes a homeodomain protein, is functionally equivalent to the *Drosophila* gene; that is, it regulates the pathway for eye development. However, when the mouse gene is put into *Drosophila*, it produces *Drosophila* eyes, not mouse eyes. *Drosophila* eyes develop because the genes that respond to the regulatory command of the inserted mouse gene are normal *Drosophila* genes, which must, of course, specify the formation of a *Drosophila* eye. In mice, mutations in the homologue of the *eyeless* gene reduce the size of the eyes; for that reason, the mutant phenotype is called *Small eye*. A homologue of *eyeless* and *Small eye* has also been found in human beings. Mutations in this gene cause a syndrome of eye defects called *aniridia* in which the iris is reduced or missing.

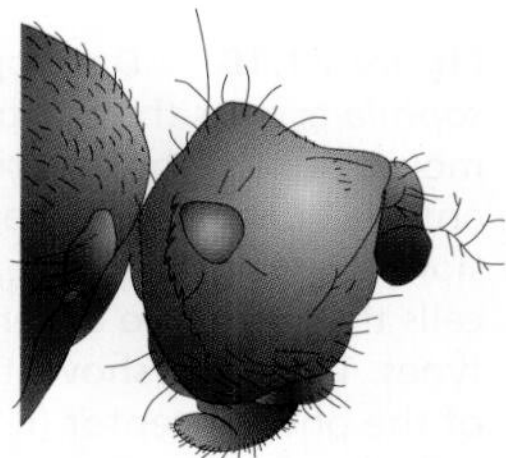

Figure 21.16 ► The phenotype of an *eyeless* mutant in *Drosophila.*

The discovery of homologous genes that control eye development in different organisms has profound evolutionary implications. It suggests that the function of these genes is very ancient, dating back to the common ancestor of flies and mammals. Perhaps the eyes in this ancestral organism were nothing more than a cluster of light-sensitive cells organized through the regulatory effects of a primitive *eyeless* gene. Over evolutionary time, this gene continued to regulate the increasingly more complicated process of eye development, so that today, eyes as different as those in insects and those in mammals are still formed under its control.

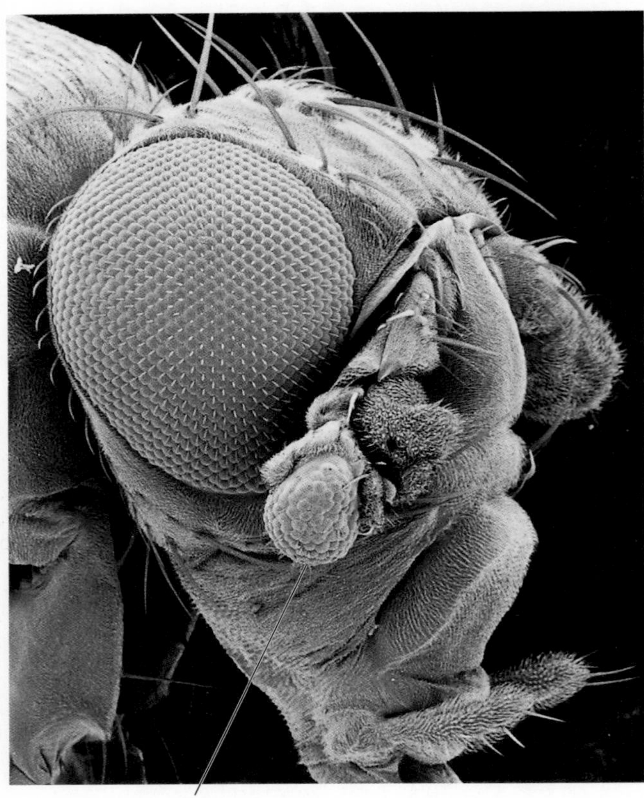

Figure 21.17 ▶ An extra eye produced by expressing the wild-type *Drosophila eyeless* gene in the antenna of a fly.

SPECIFICATION OF CELL TYPES

Within organs, cells differentiate in specific ways. For example, some cells become neurons whereas others become neuronal support cells. The mechanisms that regulate this differentiation have been analyzed by studying very simple situations involving a few distinct cell types. One such situation occurs in the development of the *Drosophila* eye (**FIGURE 21.18**).

Each of the large compound eyes in *Drosophila* originates as a flat sheet of cells in one of the imaginal discs. Initially, all the cells in this epithelial sheet look the same, but late during the larval stage, a furrow forms near the posterior margin of the disc. As this furrow moves in the anterior direction across the disc, it triggers a wave of cell divisions in its wake. The newly divided cells then differentiate into specific cell types to form the 800 individual facets of the adult eye. Each facet consists of 20 cells. Eight are photoreceptor neurons designed to absorb light; four are cone cells that secrete a lens to focus light into the photoreceptors; six are sheath cells to provide insulation and support; and the two remaining cells form sensory hairs on the eye's surface. Thus, a highly patterned array of intricately differentiated facets develops from what had been a flat sheet of identical cells. What brings this transformation about?

Gerald Rubin and his collaborators have attempted to answer this question by collecting mutations that disrupt eye development. Their research has led to the idea that the specification of cell types within each facet depends on a series of cell–cell interactions. This is illustrated in the differentiation of the eight photoreceptor cells, denoted R1, R2, . . . R8 (**FIGURE 21.19**). In a fully formed facet, six of the photoreceptors (R1–R6) are arranged in a circle around the other two (R7, R8). One of the central cells, R8, is the first to differentiate in the developing facet. Its appearance is followed by the differentiation of the peripheral cells R2 and R5, then by R3 and R4, and R1 and R6; finally, the second central cell, R7, differentiates into a photoreceptor.

This last event has been studied in great detail. Rubin and his colleagues have shown that the differentiation of the R7 cell depends on reception of a signal from the already differentiated R8 cell. To receive this signal, the R7 cell must synthesize a specific receptor, a membrane-bound protein encoded by a

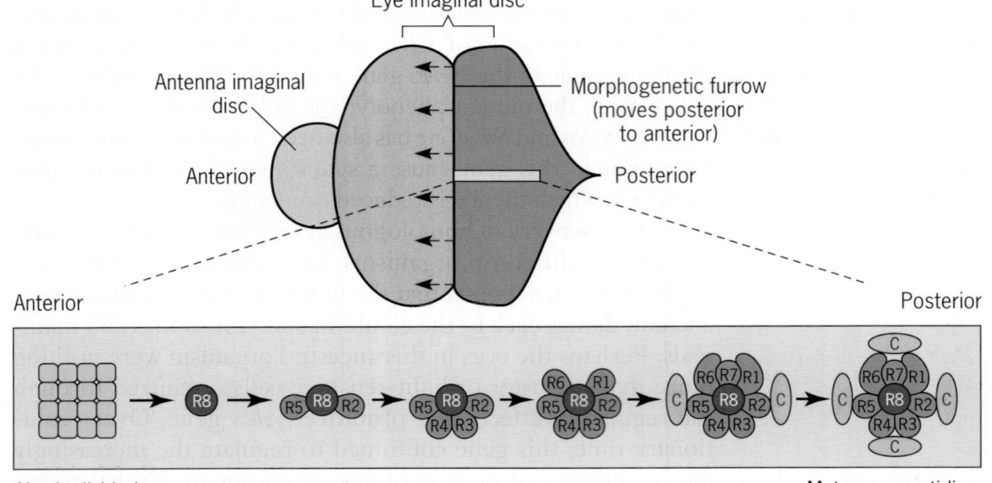

Figure 21.18 ▶ Development of the *Drosophila* eye. As the morphogenetic furrow moves toward the anterior of the eye-antenna imaginal disc, a wave of cell divisions follows in its wake. The newly divided cells then begin to differentiate into specific types. The insert shows the differentiation of the photoreceptor (R1–R8) and cone (C) cells that form each ommatidium (facet) of the compound eye.

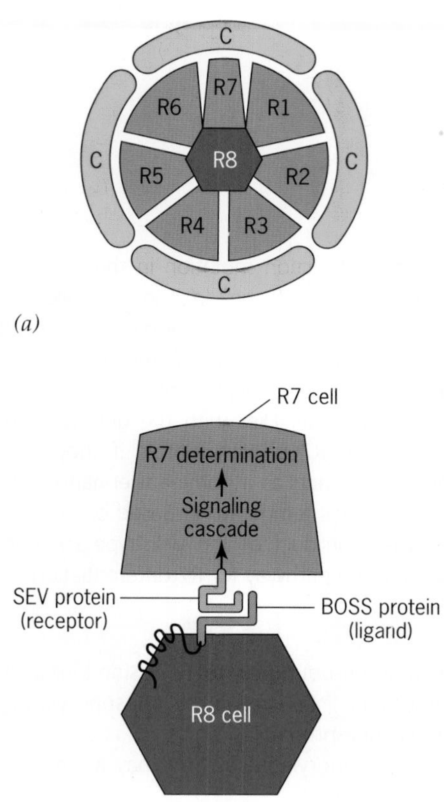

Figure 21.19 ▶ Determination of the R7 photoreceptor of an ommatidium (facet) in the *Drosophila* compound eye. (*a*) Arrangement of the eight photoreceptors (1–8) and four cone cells (C) in an ommatidium. (*b*) Signaling between the differentiated R8 cell and the presumptive R7 cell. The bride of sevenless (BOSS) protein on the R8 cell is the ligand for the sevenless (SEV) receptor protein on the surface of the R7 cell. Activation of this receptor initiates a signaling cascade within the R7 cell that induces it to differentiate.

gene called *sevenless* (*sev*). Mutations in this gene abolish the function of the receptor and prevent the R7 cell from differentiating as a neuron; instead, it differentiates as a cone cell. The signal for the R7 receptor is produced by a gene called *bride of sevenless* (*boss*) and is specifically expressed on the surface of the R8 cell. Contact between the differentiated R8 cell and the undifferentiated R7 cell allows the R8 signal, or **ligand** as it is technically called, to interact with the R7 receptor and activate it. This activation induces a cascade of changes within the R7 cell that ultimately prompt it to differentiate as a light-receiving neuron. This differentiation is presumably mediated by one or more transcription factors acting on genes within the R7 nucleus. Thus, the signal from the R8 cell is "transduced" into the R7 nucleus, where it alters the pattern of gene expression. The analysis of eye development in *Drosophila* therefore shows that **induction,** the process of determining the fate of an undifferentiated cell by a signal from a differentiated cell, can play an important role in the specification of cell types.

The protein encoded by the *sev* gene is a tyrosine kinase—that is, a protein that phosphorylates tyrosine residues in other proteins. Once the SEV protein has been activated by contact with the BOSS ligand, it phosphorylates other proteins inside the R7 cell. These intracellular proteins are downstream effectors of the BOSS signal. Ultimately, they activate transcription factors to stimulate the expression of the genes that are involved in the differentiation of the R7 cell as a photoreceptor. To explore the BOSS-SEV signaling pathway further, work through the Focus on Problem Solving: Dissecting Pathways for Cell Differentiation Using Mutations.

KEY POINTS

- The zygotic genes are activated after fertilization in response to maternal gene products.
- In *Drosophila,* the products of the segmentation genes regulate the subdivision of the embryo into a series of segments along the anterior-posterior axis.
- The identity of each body segment is determined by the products of genes in the bithorax and Antennapedia homeotic gene complexes.
- The formation of an organ may depend on the product of a master regulatory gene, such as the *eyeless* gene in *Drosophila*.
- In *Drosophila* specific cell types differentiate after segmental identities have been established.
- Differentiation events may involve a signal produced by one cell and a receptor produced by another cell.

Genetic Analysis of Development in Vertebrates

Geneticists can study development in vertebrates by applying knowledge gained from the study of model invertebrates, by analyzing mutations and phenocopies of mutant genes in model vertebrates such as mice and zebra fish, and by examining the differentiation of stem cells.

Much of the knowledge about the genetic control of development comes from the study of *Drosophila* and *C. elegans,* two model invertebrates. Geneticists would like to apply and extend this knowledge to other groups of animals, in particular, to vertebrates. The ultimate goal would be to learn about the genetic control of development in our own species. One

▶ FOCUS ON PROBLEM SOLVING

Dissecting Pathways for Cell Differentiation Using Mutations

THE PROBLEM

In *Drosophila,* the interaction between the SEV and BOSS proteins signals R7 cells to differentiate as photoreceptors in the ommatidia of the compound eyes; when this interaction does not occur, the R7 cells differentiate as cone cells. Neither the SEV nor the BOSS proteins appear to be needed for any other developmental event in the fly. (*a*) Predict the phenotypes of flies that are homozygous for recessive, loss-of-function mutations in either the *sev* or the *boss* genes. (*b*) Predict the phenotype of a fly that is heterozygous for a dominant, gain-of-function mutation that constitutively activates the SEV protein. (*c*) Suppose that one copy of this dominant, gain-of-function *sev* mutation was introduced into a fly that was homozygous for a recessive, loss-of-function mutation in the *boss* gene. What would the phenotype of that fly be? (*d*) The sev^{B4} allele is temperature-sensitive; at 22.7°C, flies that are homozygous for it develop normal R7 photoreceptors, but at 24.3°C, they fail to develop these photoreceptors. sos^{2A} is a recessive, loss-of-function mutation in the *son of sevenless* (*sos*) gene. Flies with the genotype sev^{B4}/sev^{B4}; $sos^{2A}/+$ fail to develop R7 photoreceptors if they are raised at 22.7°C. Therefore sos^{2A} acts as a dominant enhancer of the sev^{B4} mutant phenotype at this temperature. Based on this observation, where is the protein product of the wild-type *sos* gene—called SOS—likely to act in the pathway for R7 differentiation?

FACTS AND CONCEPTS

1. A loss-of-function mutation in a gene abolishes the function of that gene's protein product.

2. A gain-of-function mutation in a gene endows that gene's product with a new function.

3. A protein that is constitutively active carries out its function all the time.

4. The protein encoded by a temperature-sensitive mutation functions at one temperature but fails to function, or functions poorly, at another (usually higher) temperature.

5. SEV is a membrane-bound protein tyrosine kinase that transduces the extracellular BOSS signal by phosphorylating tyrosine residues in intracellular proteins.

6. The proteins phosphorylated by SEV may act as downstream effectors of the BOSS signal.

7. Downstream effector proteins ultimately activate the genes needed for R7 differentiation.

ANALYSIS AND SOLUTION

This problem focuses on a developmental event in the *Drosophila* eye—differentiation of the R7 photoreceptor cell. A key step in the process that leads to this event is signaling between the BOSS ligand molecule, which is located in the membrane of the already differentiated R8 cell, and the SEV receptor, which is located in the membrane of the still undifferentiated R7 cell (see **FIGURE 21.19**). The failure of either protein to function will prevent the signal from "going through." **a.** Recessive, loss-of-function mutations in either the *sev* and or *boss* genes will therefore lead to flies that do not have R7 photoreceptors in the ommatidia of their eyes. **b.** However, a dominant, gain-of-function mutation that constitutively activates the SEV protein would be expected to lead to R7 differentiation. **c.** Furthermore, this differentiation would be expected to occur even if the fly is homozygous for a recessive, loss-of-function mutation in the *boss* gene, because with a constitutively activated SEV protein, BOSS function is irrelevant. **d.** If the SEV protein is activated—either by the BOSS ligand or by a gain-of-function mutation in the *sev* gene, a faulty effector protein could stop it from inducing the R7 cell to differentiate. The SOS protein is likely to be a downstream effector in the pathway for R7 differentiation because when it is depleted by mutating one copy of the *sos* gene, flies that have a partially functional SEV protein show a mutant phenotype—that is, transmission of the developmental signal through SEV and its downstream effector proteins is weakened.

For further discussion go to your *WileyPLUS* course.

strategy for achieving this goal is to use the information obtained from the study of invertebrate genes to identify developmentally significant genes in vertebrates. Another is to study model vertebrate species with techniques similar to those that are being used in *Drosophila* and *C. elegans.*

VERTEBRATE HOMOLOGUES OF INVERTEBRATE GENES

Once a gene has been isolated and sequenced, researchers can screen DNA sequence databases for homologous genes in other organisms. If the gene's sequences have been reasonably well conserved over evolutionary time, this procedure works even for distantly related species. Thus, it has been possible to identify genes from various vertebrate species that are homologous to genes from *Drosophila* and *C. elegans*. The identification of a vertebrate gene then makes many kinds of experimental analyses possible, including assays for the gene's expression at both the RNA and protein levels.

One of the most dramatic applications of this approach has shown that vertebrates contain homologues of the homeotic genes of *Drosophila*. These so-called *Hox* genes were initially identified by probing Southern blots of mouse and human genomic DNA with segments of the *Drosophila* homeotic genes. Subsequently, the cross-hybridizing DNA fragments were cloned, mapped with restriction enzymes, and sequenced. The results of all these analyses have established that mice, humans, and many other vertebrates so far examined have 38 *Hox* genes in their genomes (**FIGURE 21.20**). These genes are usually organized in four clusters (*a*, *b*, *c*, and *d*), each about 120 kb long; in mice and humans, each cluster is located on a different chromosome. It seems that the four *Hox* gene clusters were created by the quadruplication of a primordial cluster very early in the evolution of the vertebrates, probably 500 to 600 million years ago.

The molecular analysis of the vertebrate *Hox* genes has revealed striking structural and functional similarities to the homeotic genes of the bithorax and Antennapedia complexes in *Drosophila*. These latter genes actually seem to have been members of a larger cluster, called HOM-C, which was split by a chromosome rearrangement during the evolution of the flies. In other types of insects such as the flour beetle, *Tribolium castaneum*, the bithorax and Antennapedia complexes are united in a single cluster. Comparison between vertebrates and invertebrates shows that the basic organization of the HOM/*Hox* genes has been preserved during evolution. The structural homologues of the ANT-C genes are at one end of each vertebrate *Hox* gene cluster, and the structural homologues of the BX-C genes are at the other end. Moreover, within each cluster, the physical order of the *Hox* genes corresponds to the position of their expression along the anterior-posterior axis of the embryo, just as it does for the homeotic genes in *Drosophila* (**FIGURE 21.21**). With one exception (the *Deformed* gene in *Drosophila*), all the HOM and *Hox* genes are transcribed in the same direction, with expression proceeding from one end of a cluster to the other end, both spatially (anterior to posterior in the embryo) and temporally (early to late in development). This phenomenon, called *colinearity* (not to be confused with the same word, which was used in Chapter 14 to describe the linear sequence of sites in a gene and its polypeptide product), suggests that the HOM and *Hox* genes provide a common molecular mechanism for establishing the identities of specific regions in many different types of embryos.

THE MOUSE: RANDOM INSERTION MUTATIONS AND GENE-SPECIFIC KNOCKOUT MUTATIONS

The genetic control of development cannot be studied in vertebrates with the same thoroughness as it can in invertebrates such as *Drosophila*. Obviously, there are technical and logistical constraints. Vertebrates have comparatively long life cycles, their husbandry is expensive, and it is difficult to obtain and analyze mutant strains, especially those with a developmental significance. In spite of these shortcomings, geneticists have been able to make headway in the genetic analysis of development in two complex vertebrates—the mouse and the zebra fish.

Over 500 loci responsible for genetic diseases have been identified in the mouse, and some are involved in developmental processes. Most of these loci were discovered through

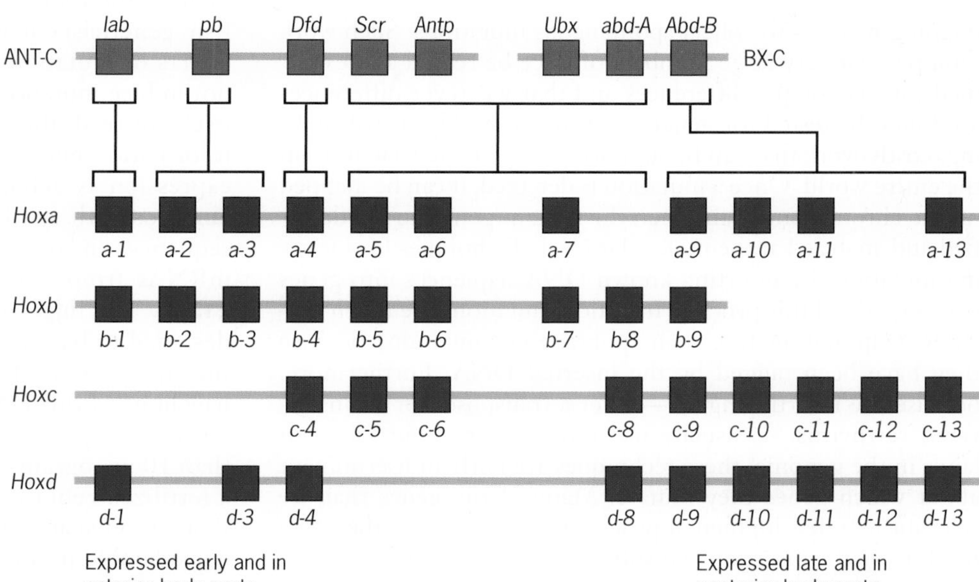

Figure 21.20 ▶ Organization and expression of the mammalian *Hox* genes homologous to the *Drosophila* genes in the BX-C and ANT-C. Homologies are indicated by brackets. All the genes except *Dfd* are transcribed from right to left. The time and anatomical location of expression are indicated.

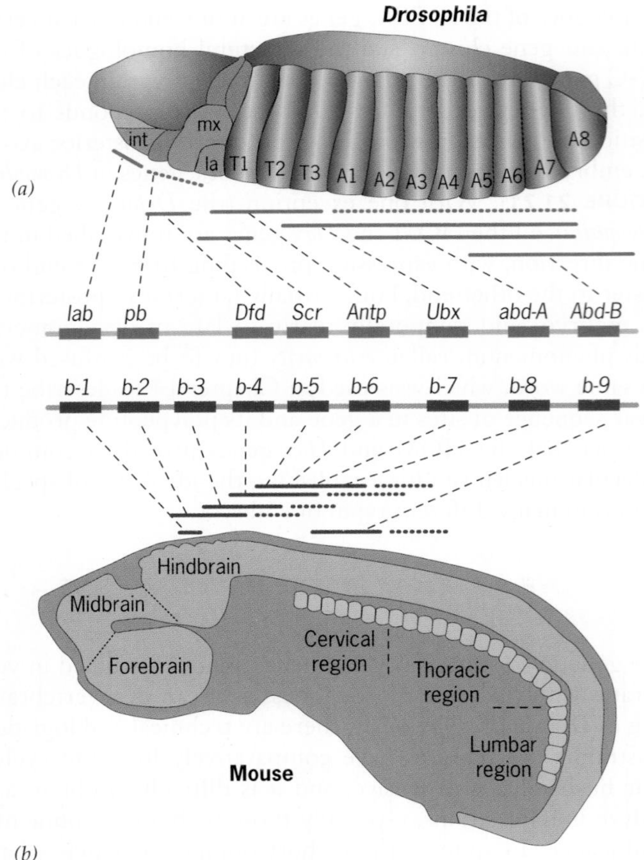

Figure 21.21 ▶ Expression patterns of *Drosophila* and mouse homeotic genes. (*a*) A 10-hour-old *Drosophila* embryo showing the approximate expression of the homeotic genes in the epidermis. In the head, int, mx, and la designate intercalary, maxillary, and labial segments, respectively. Thoracic and abdominal segments are indicated as T1–T3, and A1–A8. (*b*) A 12-day-old mouse embryo showing the approximate expression of the *Hoxb* genes in the central nervous system. The dotted lines indicate extension of the expression toward the posterior.

ongoing projects to collect spontaneous mutations. Such work requires that very large numbers of mice be reared and examined for phenotypic differences and that whatever differences are found be tested for genetic transmission. This is painstaking, costly work that can be supported only at a few facilities in the entire world. Once a mutation is detected, it can be mapped on the chromosomes, and then the mutant gene can be identified and analyzed at the molecular level. Techniques for inducing mutations by inserting known DNA sequences into genes have expedited this process. Insertion mutations are much easier to map and analyze than spontaneous mutations because they have been tagged by the inserted DNA. Furthermore, because the inserting agent—either a transposon or an inactivated retrovirus—is usually not too specific about where it lands in the genome, these techniques are fairly indiscriminant about which genes they mutate. Many of the genes that are relevant to a developmental process under study can therefore be "hit" by the insertion and subsequently identified.

Mouse geneticists have also invented procedures to mutate specific genes. In these procedures, which are discussed in Chapter 17, the integrity of a gene is disrupted by an insertion that is specifically targetted to that gene. Such a disruption, called a **knockout mutation,** can help a researcher determine what role the normal gene plays during development. For example, mice that are homozygous for a knockout mutation in the *Hoxc-8* gene develop an extra pair of ribs posterior to the normal set of ribs; they also have clenched toes on their forepaws. The extra-rib phenotype in these mutant mice is reminiscent of the segmental transformations that are seen with homeotic mutations in *Drosophila*. Thus, the mouse's *Hoxc-8* gene appears to be involved in establishing the identities of tissues along the anterior-posterior axis and also within the digits.

The genetic analysis of development in mice is providing clues about development in our own species. For example, mutations in at least two different mouse genes mimic the development of abnormal left-right asymmetries in human beings. Normally, humans, mice, and other vertebrates exhibit structures that are asymmetric along the left-right body axis. The heart tube always loops to the right, and the liver, stomach, and other viscera are shifted either to the left or right away from the body's midline. In mutant individuals, these characteristic asymmetries are not seen, perhaps because of a defect in the mechanisms that establish the basic body plan. Studying these types of mutants in the mouse may therefore help to elucidate how the organs are positioned in human beings.

THE ZEBRA FISH: MORPHOLINO KNOCKDOWN MUTATIONS

As with the mouse, ambitious mutagenesis programs have been implemented to define genes important for development in the zebra fish. However, to speed up progress, zebra fish geneticists have developed procedures to mimic the effects of mutations in specific genes. These procedures do not induce mutations—that is, they do not change the DNA itself. Instead, they block or attenuate the expression of the information encoded in the DNA. This disruption of gene expression is achieved by injecting developing zebra fish embryos with synthetic molecules that can bind to specific sequences in messenger RNA molecules and prevent these mRNAs from being translated into polypeptides (**FIGURE 21.22**). The injected material is a stable DNA analog that is designed to base pair specifically to the 5′ region, including the start codon, of an mRNA. Because the sugar-phosphate backbone of this DNA analog is built from morpholine moieties, the injected molecule is called a **morpholino.** Fewer than 10 nanograms of a morpholino injected into the yolk of a fertilized zebra fish egg or early embryo (containing fewer than 16 cells) are effective in disrupting gene expression for a considerable period of time. The animals that develop from

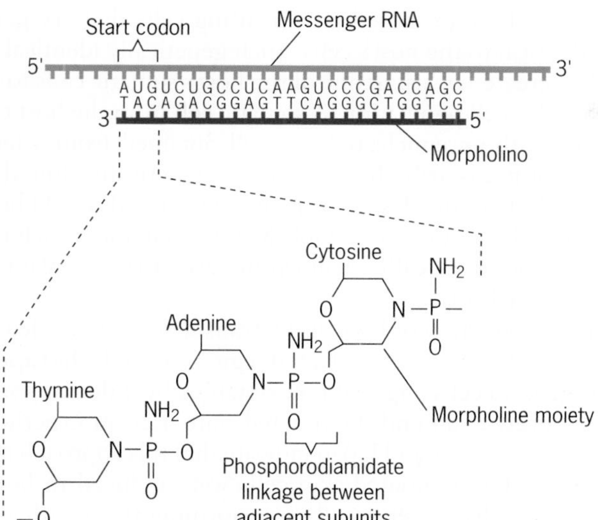

Figure 21.22 ▶ A morpholino for the zebra fish *no tail* gene base paired with a sequence around the start codon of the messenger RNA from this gene. Morpholinos are typically 20 to 25 subunits long. Each subunit consists of a morpholine moiety bonded to a nitrogenous base (A, T, U, G, or C). The subunits are linked to each other by phosphorodiaminate bonds instead of by phosphodiester bonds, as they are in DNA and RNA.

these injected eggs or embryos have phenotypes like those of animals that carry *bona fide* mutations in the targeted gene. Thus, they are **phenocopies** of the mutant phenotype. With the completion of the zebra fish genome sequence, geneticists can now use morpholinos to knock down gene expression and ascertain whether a particular gene product has a significant role in development. For example, Benjamin Feldman and Derek Stemple used a morpholino to knock down expression of the *no tail* (*ntl*) gene. This gene is homologous to the mouse *brachyury* gene, also known as the *tail-length* (*T*) gene (see Chapter 4). Feldman and Stemple found that injection of just a few nanograms of a morpholino designed to base pair to the start codon of the *ntl* gene and to 22 nucleotides adjacent and downstream of this codon (see **FIGURE 21.22**) caused embryos to have short tails. This morpholino-induced, or morphant, phenotype was a mimic of the mutant phenotype caused by an actual mutation in the *ntl* gene.

The morpholino technique has been used to knock down gene expression in the embryos of other vertebrates, including the toad *Xenopus laevis*, the mouse, and the chicken. It is an antisense technique that has developed in parallel with the RNA interference techniques discussed in Chapters 17 and 20. However, unlike RNA interference, morpholinos do not cause the degradation of messenger RNA. Instead, they work by inhibiting the start of translation, probably by impairing the binding of ribosomes to the messenger RNA to which they are complementary. Also, morpholinos are single-stranded molecules; by contrast, the molecules that induce RNA interference are double-stranded.

STUDIES WITH MAMMALIAN STEM CELLS

The terminally differentiated cells in the human body—lymphocytes, neurons, muscle fibers, and so on—usually do not divide. When these types of cells are lost through death, they must be replenished or the tissue they belong to will atrophy. Replenishment occurs when unspecialized cells present in the tissue divide to produce cells that subsequently differentiate into the specialized cell type. These unspecialized precursors of specialized cells are called **stem cells.** For example, the marrow in a human femur contains undifferentiated cells that can replenish various types of blood cells. These hematopoietic stem cells keep the circulatory system supplied with lymphocytes, erythrocytes, and platelets. The tissues in some organs such as the heart appear to have very few stem cells; consequently, their ability to regenerate lost or damaged material is limited. Other tissues, such as the gut lining and the skin, have large populations of stem cells that vigorously replace differentiated cells as they are lost. Because these types of stem cells are found in developed organisms, they are called adult stem cells.

Stem cells are also found in developing organisms. In fact, during the earliest stages of development, all or most of the cells have the properties of stem cells. Cells taken from a mouse embryo, for example, can be cultured *in vitro* and subsequently transplanted into another mouse embryo, where they will divide and ultimately contribute to the formation of many kinds of tissues and organs. Embryonic stem cells therefore have tremendous developmental potential; that is, they are **pluripotent**—able to develop in many ways.

No matter if they are derived from embryonic or adult tissue, stem cells provide an opportunity to study the mechanisms involved in the differentiation of special cell types. Stem cells can be obtained from a variety of mammals, including mice, monkeys, and humans. They can be cultured *in vitro* and examined for differentiation while growing there or after being transplanted into a host organism. While in culture, stem cells can be treated in various ways to ascertain what triggers their development in a specific direction. Molecular techniques, including gene-chip technologies, allow researchers to determine which genes the cells are expressing as their developmental programs unfold.

Because embryonic stem (ES) cells have the greatest developmental potential, they are ideally suited for this kind of analysis. These cells are usually derived from the inner cell mass of embryos that had been created by *in vitro* fertilization. Cells isolated from this mass are plated on a layer of mitotically inactive "feeder cells," which provide growth factors to stimulate division. For mouse ES cells growing in culture, the doubling time is about 12 hours; for human ES cells, it is about 36 hours. After the isolated embryonic cells have grown for a while on the feeder cells, they are dissociated and replated to establish clonal stem cell populations, which may then be frozen for long-term storage. A clonal cell population is one that has come from a single progenitor cell.

ES cells begin to differentiate when they are transferred from feeder cell cultures to suspension cultures supplied with an appropriate medium. Under these conditions they form **embryoid bodies,** which are multicellular aggregates consisting of differentiated and undifferentiated cells. For some species, the embryoid bodies resemble early embryos. The cells in these bodies may differentiate into the types of specialized cells that are derived from each of the three primary tissue layers—ectoderm, mesoderm, and endoderm. For example, they may form neurons, which are derived from ectoderm; smooth muscle cells or rhythmically contracting cardiac cells, which are derived from mesoderm; or pancreatic islet cells, which are derived from endoderm. By observing this process in different cell lines—for instance, in lines in which particular genes have been mutated—it may be possible to dissect the genetic network of interactions involved in the differentiation of various cell types.

The issue of procuring and analyzing human ES cells is, of course, controversial. The human ES cell lines now in use were derived from embryos that were donated by people who had sought medical help to have children through *in vitro* fertilization. Typically, many more embryos are created through this process than are eventually used to produce children. A couple may then decide to donate its unused embryos for research purposes. The derivation of ES cells from such embryos necessarily requires that the embryos be destroyed. Some people view the destruction of early embryos as an acceptable practice; others consider it immoral. The controversy surrounding this practice has caused many governments to withhold or restrict financial support for research on human embryonic stem cells. In the United States, for example, federal government support is provided only to projects using human stem cell lines established before August 9, 2001. Funds for projects using lines established after that date must come from other sources.

The debate on funding for human embryonic stem cell research has been intensified by the prospect of using human ES cells to cure diseases that result from the loss of specific cell types, such as diabetes mellitus (in which the pancreatic islet cells have been lost) and Parkinson's disease (in which certain types of neurons in a particular region of the brain have been lost). ES cell therapy has also been proposed to treat disabilities such as those resulting from spinal cord damage. The idea is to transplant cells derived from ES cells into the diseased or injured tissue and allow these cells to regenerate the lost or damaged parts of the tissue. Experiments with mice and rats suggest that this strategy might work in humans. However, many technical problems have yet to be solved. For instance, it is not yet possible to obtain pure cultures of a particular differentiated cell type. When human ES cells develop in culture, they differentiate into many kinds of cells; isolating one kind—say, for example, cardiac cells—is a formidable technical challenge.

The proponents of human stem cell therapy also have to solve other kinds of problems. Cells derived from an *in vitro* culture might divide uncontrollably and form tumors upon being transplanted into a host, or they might be wiped out by the host's immune system. To circumvent the latter problem, researchers have proposed transplanting cells that are genetically identical to the host's cells. Such genetically identical cells could be created by using one of the host's somatic cells to generate the ES cell population. A somatic cell from the host could be fused with an enucleated egg cell obtained from a female donor (not necessarily the host). If the genetically altered egg, which is diploid, divides to form an embryo, cells could be isolated from that embryo to establish an ES cell line, which could then provide genetically identical material for transplantation back into the host.

The production of ES cells by transferring the nucleus of a somatic cell into an enucleated egg is called **therapeutic cloning.** Stem cells might also be obtained by inducing somatic cells to revert to an undifferentiated state. Recent experiments in the United States and Japan indicate that this approach might be feasible. Differentiated skin cells were induced to become pluripotent cells by genetically transforming them with a mixture of four cloned genes. However, some of the genes that were used in these experiments are associated with tumor formation when they are expressed inappropriately. Thus, more research is needed before induced pluripotent cells can be used in stem cell therapy.

REPRODUCTIVE CLONING

Therapeutic cloning is different from **reproductive cloning,** which aims to produce a complete individual by transferring a somatic-cell nucleus from a donor into an enucleated egg and then allowing that egg to develop into a genetically identical copy of the donor. In 1997, researchers at the Roslin Institute in Scotland produced the first cloned mammal—a sheep named Dolly (see the opening essay in Chapter 2). Dolly was created by replacing the nucleus of an egg with the nucleus from a cell that had been taken from the udder of an adult female sheep. The transplanted nucleus evidently contained all the genetic information needed to direct Dolly's development even though it came from a differentiated cell. Since the creation of Dolly, scientists have produced many other animals by reproductive cloning—mice, cats, cows, and goats. Differentiated cells therefore seem to have the genetic potential to direct development.

However, animals produced by reproductive cloning sometimes have developmental abnormalities and shortened life spans. Frequently, they fail to thrive. This lack of vigor suggests that the somatic nuclei used in reproductive cloning are different from the zygotic nuclei produced by ordinary fertilization. Perhaps the somatic nuclei have accumulated mutations, or perhaps they have undergone changes associated with genetic imprinting or chromosome inactivation—methylation of some nucleotides, actetylation of histones, and so on. Such changes would have to be reversed for a somatic nucleus to function as a zygotic nucleus. Because of the problems encountered in the reproductive cloning of animals, the international scientific community does not consider reproductive cloning of humans to be safe. Consequently, there is widespread agreement that the reproductive cloning of humans should not be attempted.

GENETIC CHANGES IN THE DIFFERENTIATION OF VERTEBRATE IMMUNE CELLS

Although evidence from reproductive cloning suggests that differentiated cells may have the same DNA content as a fertilized egg, we know of some types of differentiated vertebrate cells that do not. These cells are components of the system that protects animals against infection by viruses, bacteria, fungi, and protists—the immune system.

In mammals, where most of the research has been focused, the immune system comprises several distinct types of cells, all derived from stem cells that reside in the bone marrow. These stem cells divide to produce more of their own kind, as well as precursors of specialized immune cells. Two important classes of specialized immune cells participate directly in the fight against invading pathogens. The plasma B cells produce and secrete proteins called **immunoglobulins,** also known as **antibodies,** and the killer T cells produce proteins that project from their surfaces and act as receptors for a variety of substances. Both the B-cell antibodies and the T-cell receptors are able to recognize other molecules—for example, the foreign materials introduced by a pathogen—through a lock-and-key mechanism. The foreign molecule, called an **antigen,** is the key that fits precisely into the lock formed by the B-cell antibody or the T-cell receptor (**FIGURE 21.23**). This specificity of fit is the basis of an animal's ability to defend itself against pathogens. However, because there are many different potential pathogens, an animal must be able to produce many different types of antibodies and T-cell receptors in order to ward off infection.

Antibodies and T-cell receptors are proteins, and proteins are encoded by genes. Therefore, to produce the large array of antibodies and T-cell receptors needed to counter all possible pathogens, it would seem that an animal would have to possess an enormous number of genes—too many to fit even in a large genome such as our own. This predicament perplexed geneticists for years. In the last quarter of the twentieth century, however, researchers discovered how an animal could produce a large number of different antibodies and T-cell receptors by recombining small genetic elements into functional genes. The coding potential achieved by this combinatorial shuffling of gene segments is truly astounding. With a modest amount of DNA dedicated to immune system functions, an animal can produce hundreds of thousands, if not millions, of antibodies and T-cell receptors, each with a different ability to lock on to a foreign molecule from an invading organism.

To see how this recombination system works, we'll focus on the production of antibodies. Each antibody is a tetramer composed of four polypeptides, two identical *light chains* and two identical *heavy chains,* joined by disulfide bonds (**FIGURE 21.24**). The light chains are about 220 amino acids long, and the heavy chains are about 445 amino acids long. Every chain, light or heavy, has an amino-terminal *variable region,* within which the amino acid sequence varies among the different kinds of antibodies that an animal produces, and a carboxy-terminal *constant region,* within which the amino acid sequence is the same for all antibodies of a particular class.

The light and heavy chains of an antibody are encoded by different loci in the genome. In humans, there are two light chain loci, the kappa (κ) locus on chromosome 2 and the lambda (λ) locus on chromosome 22, and there is one heavy chain locus, located on chromosome 14. Each of these loci consists of a long array of gene segments. We'll focus on the kappa locus to see how these segments are organized and how they are recombined into coherent coding sequences to produce different polypeptides.

A kappa polypeptide is encoded by three types of gene segments:

1. An $L_\kappa V_\kappa$ gene segment, which encodes a *l*eader peptide and the amino-terminal 95 amino acids of the *v*ariable region of the kappa light chain; the leader peptide is removed from the kappa light chain by cleavage after it guides the nascent polypeptide through the membrane of the endoplasmic reticulum in an antibody-synthesizing plasma cell.

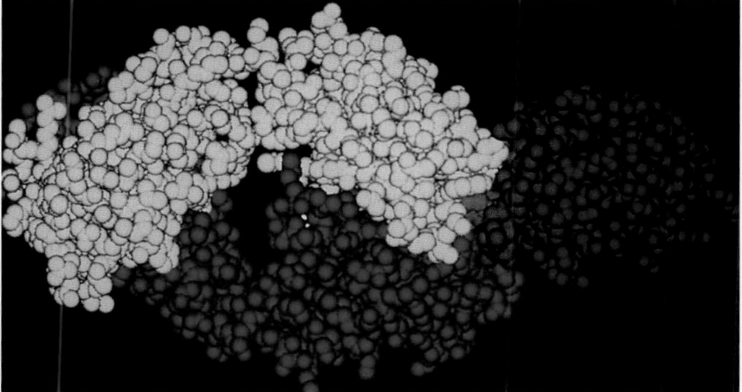

Figure 21.23 ► The three-dimensional structure of an antigen-antibody complex. Only one of the two antigen-binding sites of a typical antibody is shown. The antigen (green) is the enzyme lysozyme. The antigen-binding site of the antibody is formed by the amino-terminal portions of a light chain (yellow) and a heavy chain (blue). A glutamine residue that protrudes from lysozyme where the antibody binds is shown in red. The structure is based on X-ray diffraction data.

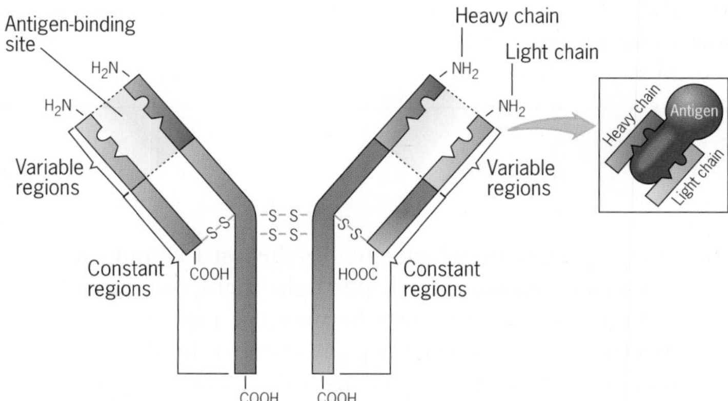

Figure 21.24 ► Structure of an antibody molecule. The inset shows the lock-and-key interaction between the antibody and the antigen that it recognizes.

▶ A MILESTONE IN GENETICS: Mutations that Disrupt Segmentation in *Drosophila*

The bodies of many types of animals are constructed from segments. For example, insects possess three main body parts—a head, a thorax, and an abdomen—and the thorax and abdomen are built from smaller, segmental units. Even vertebrates exhibit a segmented anatomy, especially during the early stages of development when discrete masses of tissue called somites form at regular intervals along the backbone. Segmentation therefore seems to be a standard feature in the blueprints of many animals.

Anatomists and embryologists have learned about the ways in which segments form and align by dissecting developing animals. Geneticists, too, have learned about segmentation by performing dissections. However, their dissections do not use scalpels or knives; rather, they employ mutations to define and analyze the genes that are involved in the segmentation process.

The genetic dissection of segmentation took a great step forward in 1980 when Christiane Nüsslein-Volhard and Eric Wieschaus published a paper describing mutations that alter the number, size, and composition of segments in *Drosophila* embryos and larvae.[1] Nüsslein-Volhard and Wieschaus knew that the spatial organization of *Drosophila* embryos is influenced by the products of maternal-effect genes. These gene products are deposited in the egg, and after fertilization, they guide the development of the embryo along both its anterior-posterior and dorsal-ventral axes. Nüsslein-Volhard and Wieschaus also knew that the later stages of *Drosophila* development involve the products of homeotic genes such as *Ultrabithorax*. These gene products specify the fates of individual segments in the body. Some regions differentiate into thoracic segments, and others differentiate into abdominal segments. Against this background, Nüsslein-Volhard and Wieschaus recognized that there must be an additional class of gene products that act after the maternally supplied gene products have established the basic body plan, but before the homeotic gene products have fleshed it out. They sought to define these gene products by screening for mutations that disrupt segmentation in *Drosophila* embryos and larvae.

Because such mutations would act as recessive lethals, Nüsslein-Volhard and Wieschaus devised a scheme to induce them by treating adult *Drosophila* with the mutagen ethyl methanesulfonate. Mutagenized chromosomes were recovered in heterozygous condition the next generation, and then, through a series of crosses involving balancer chromosomes, they were made homozygous to determine if new recessive lethal mutations had been induced. In one experiment in which 5800 mutagenized chromosomes were screened, 4500 were found to carry new recessive lethal mutations. Once identified, these mutant chromosomes were maintained in heterozygous condition in balanced stocks.

Balanced lethal stocks segregate recessive lethal homozygotes every generation (**FIGURE 1**). If the lethal mutation in a stock acts

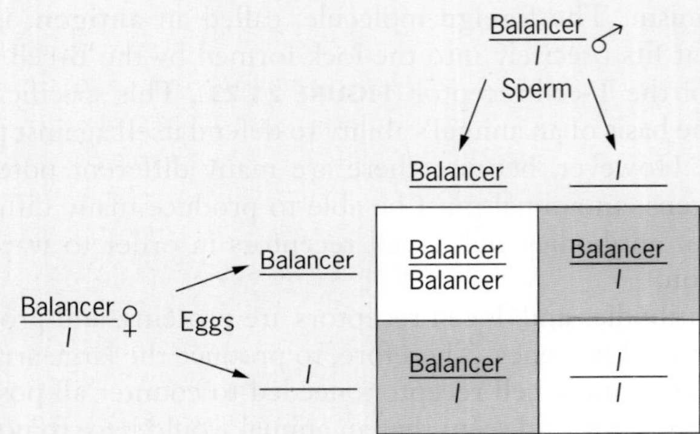

Figure 1 ▶ Segregation of embryos homozygous for a newly induced recessive lethal mutation (*l*) in a balanced *Drosophila* stock. The balancer chromosome carries a dominant marker and multiple inversions to suppress recombination between it and the chromosome with the lethal mutation. Most balancer chromosomes also carry a recessive lethal mutation that is not allelic to the lethal mutation on the other chromosome. Thus, the offspring that are homozygous for the balancer chromosome as well as those that are homozygous for the newly induced lethal mutation die during development.

[1] Nüsslein-Volhard, C., and E. Wieschaus. 1980. Mutations affecting segment number and polarity in *Drosophila*. *Nature* 287:795–801.

2. A J_κ gene segment, which encodes the last 13 amino acids of the variable region of the kappa light chain; the symbol J_κ is used for this gene segment because the peptide it encodes *j*oins the amino-terminal peptide encoded by the $L_\kappa V_\kappa$ segment to a carboxy-terminal peptide encoded by the next type of gene segment.
3. A C_κ gene segment, which encodes the *c*onstant region of the kappa light chain.

In humans, the kappa locus contains 76 $L_\kappa V_\kappa$ gene segments (although only 40 are functional), five J_κ gene segments, and a single C_κ gene segment. The J_κ gene segments are located between the $L_\kappa V_\kappa$ gene segments and the C_κ gene segment. In germ-line cells, the five J_κ segments are separated from the $L_\kappa V_\kappa$ segments by a long noncoding sequence, and from the C_κ gene segment by another noncoding sequence approximately 2 kb long (**FIGURE 21.25**). During the development of a par-

Figure 2 ▶ *Drosophila* embryos showing the phenotypes of mutations in three different segmentation genes: *knirps* (a gap gene), *even-skipped* (a pair-rule gene), and *gooseberry* (a segment-polarity gene). The colored areas are bands of denticles (short, hairlike projections from the skin of the embryo) associated with the anterior portion of each thoracic and abdominal segment in the wild-type embryo. These denticle bands are a means of identifying segments and portions of segments in developing embryos. In the knirps mutant, the denticle pattern indicates that much of the material in adjacent abdominal segments is missing. In the even-skipped mutant, it shows that throughout the body alternating segments are missing, and in the gooseberry mutant, it shows that the posterior portion of each segment has been deleted and replaced with a mirror-image copy of the anterior portion, which is marked by the denticles; the mirror-image copies of the anterior portion of each segment are shown in orange in the thorax and in gray in the abdomen. *Source:* After Nüsslein-Volhard & Wieschaus, 1980. *Nature* 287: 795–801. Figure 1, p. 796.

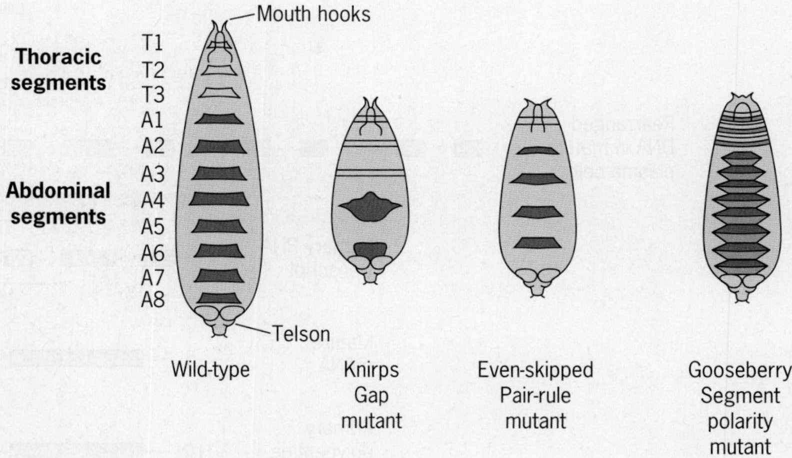

early in development, embryos homozygous for it will not hatch from the egg. Thus, to search for mutations affecting early development, Nüsslein-Volhard and Wieschaus examined unhatched embryos in their collection of stocks. They looked especially hard for embryos in which the segmental pattern was disrupted.

This painstaking search was successful. Nüsslein-Volhard and Wieschaus identified segmentation-disrupting mutations in 15 different genes. Mutations in three of the genes created segmental gaps in the embryos, mutations in six of the genes deleted alternating parts of the segmental pattern, and mutations in six of the genes deleted parts of segments and replaced these parts with duplicates of other parts (**FIGURE 2**). The gap genes, pair-rule genes, and segment-polarity genes were thereby defined. Subsequent work has shown that the products of these genes play key roles in establishing the segmental pattern in *Drosophila* embryos. Homologues of these gene products also function in the embryological development of other organisms.

For this path-breaking work, Nüsslein-Volhard and Wieschaus were awarded the Nobel Prize in Physiology or Medicine in 1995. Edward Lewis also shared this Prize for his pioneering work on the homeotic genes of *Drosophila*.

QUESTIONS FOR DISCUSSION

1. Nüsslein-Volhard and Wieschaus looked for mutant phenotypes in unhatched embryos. Before Nüsslein-Volhard and Wieschaus began their work, most *Drosophila* geneticists looked for mutant phenotypes in adults. What does this tell us about the importance of "viewpoint" or "plane of vision" in the search for new knowledge?
2. Once the segmentation genes had been defined by mutations, they had to be characterized at the molecular level by cloning, analysis of expression, and so on; once characterized at the molecular level, probes made from these genes could be used to study homologous genes in other organisms. What issues involving the segmentation genes would be interesting to study in other segmented organisms such as crustaceans and flatworms, and in nonsegmented organisms such as jellyfish and starfish?

ticular B cell, the kappa light chain gene that will be expressed is assembled from one $L_\kappa V_\kappa$ segment, one J_κ segment, and the single C_κ segment by a process of somatic recombination. Any one of the 40 functional $L_\kappa V_\kappa$ gene segments can be joined with any one of the five J_κ segments in this process; the DNA between the joined segments is simply deleted (**FIGURE 21.26**). The joining event is mediated by sites called recombination signal sequences (RSS), which are adjacent to each of the gene segments. These sites are composed of 7- or 9-base pair-long repeats separated by 12- or 23-base pair-long spacers. The repeats within the RSS immediately downstream of an $L_\kappa V_\kappa$ gene segment are complementary to the repeats within the RSS immediately upstream of a J_κ gene segment. When these repeats pair, a protein complex can catalyze recombination between them, joining the $L_\kappa V_\kappa$ segment to the J_κ segment. The *r*ecombination *a*ctivating *g*ene proteins 1 and 2 (RAG1

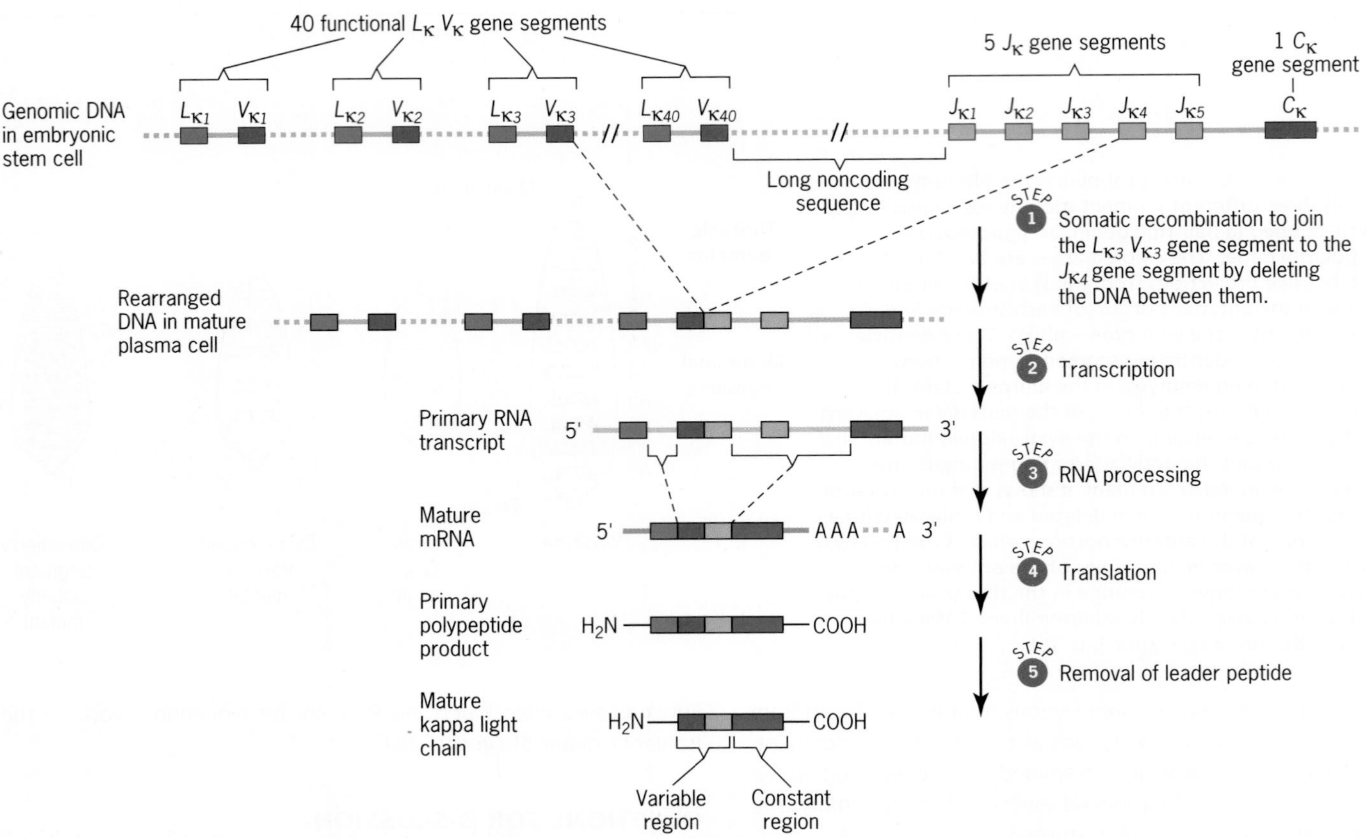

Figure 21.25 ► The genetic control of human antibody kappa light chains. Each kappa light chain is encoded by a gene assembled from different types of gene segments within the immunoglobulin kappa locus (*IGK*) on chromosome 2. This assembly occurs during the differentiation of the plasma B cells of the immune system.

and RAG2) are important components of this complex; together, they control the specificity of the recombination event.

The $L_κ V_κ J_κ$ fusion that is produced by this recombination event encodes the variable portion of the kappa light chain. The entire DNA sequence—$L_κ V_κ J_κ$ -noncoding stretch-$C_κ$—in the rearranged kappa locus is then transcribed. The noncoding sequence between the fused $L_κ V_κ J_κ$ segments and the $C_κ$ segment is removed during RNA processing, just as are the introns of other genes, and the resulting mRNA is translated into a polypeptide. The amino-terminal leader peptide is cleaved from this polypeptide to create the finished kappa light chain. The total number of functional kappa light chains that can be produced by this mechanism is 40 (the number of functional $L_κ V_κ$ gene segments) × 5 (the number of $J_κ$ gene segments) × 1 (the number of $C_κ$ gene segments) = 200. In a similar manner, recombination of gene segments can create 120 different lambda light chains and 6600 different heavy chains. The combinatorial assembly of all these chains then makes it possible for a human to produce 320 (200 + 120) × 6600 = 2,112,000 different antibodies. However, the actual number of different antibodies is even greater because of slight variations in the sites where the recombination events take place, and because of hypermutability in the sequences that encode the variable regions of the antibody chains. All these events occur independently in the precursors of the plasma B cells. Thus, as these cells differentiate, each one acquires the ability to produce a different antibody.

The first definitive experimental evidence for the rearrangement of DNA sequences during immune cell differentiation was published in 1976 by Nobumichi Horzumi and Susumu Tonegawa. These researchers demonstrated that the DNA sequences encoding the variable and constant regions of kappa light chains were present on the same *Bam*HI restriction fragment in the genomic DNA of antibody-producing cells, but were on different *Bam*HI fragments in the genomic DNA of embryonic cells (**FIGURE 21.27**). Tonegawa and colleagues subsequently showed that the DNA sequences encoding lamba light chains and heavy

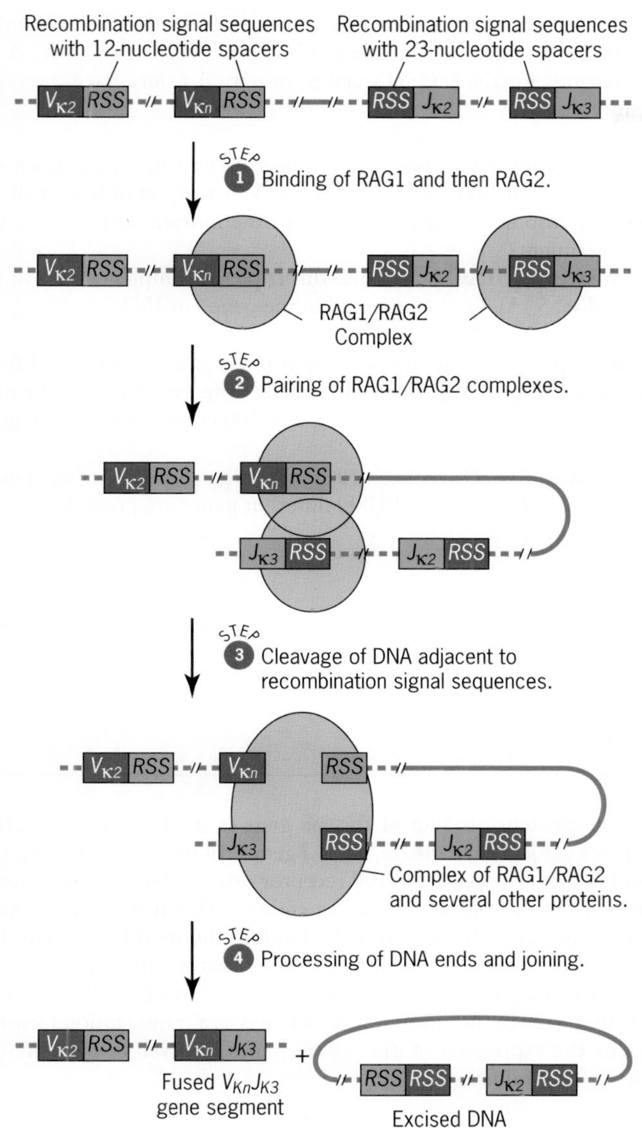

Figure 21.26 ► Simplified model of V_κ-J_κ joining. The joining process is mediated by the specific binding of RAG1 and RAG2 to the recombination signal sequences (RSS) adjacent to the V_κ and J_κ gene segments. The RSS adjacent to each V_κ segment contains 12-nucleotide spacers; those adjacent to J_κ segments contain 23-nucleotide spacers. The RAG1/RAG2 complex catalyzes recombination only when one RSS contains a 12-nucleotide spacer and the other RSS contains a 23-nucleotide spacer.

DNA from embryo cells

*Bam*HI *Bam*HI *Bam*HI *Bam*HI

5' V C 3'

(*a*)

DNA from antibody-producing cell

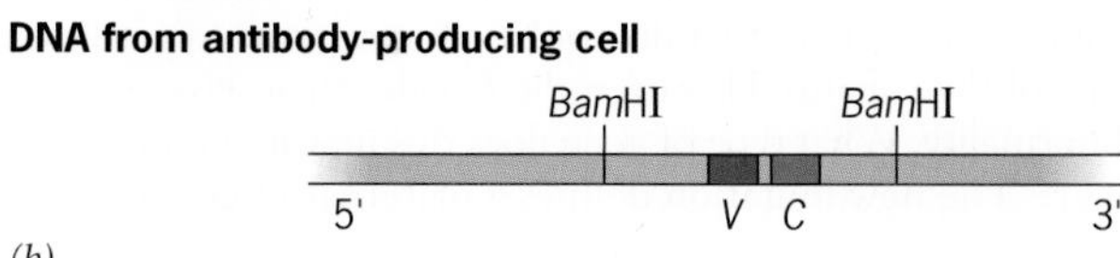

(*b*)

Figure 21.27 ► Evidence for DNA rearrangement during immune cell differentiation. (*a*) In embryonic cells, the sequences that encode the variable (V) and constant (C) regions of kappa light chains are in different *Bam*HI restriction fragments. (*b*) In antibody producing cells, these sequences are close together in the same *Bam*HI fragment.

chains was also rearranged in antibody-producing cells. For this work, Tonegawa received the Nobel Prize in Physiology or Medicine in 1987.

KEY POINTS

- Many vertebrate genes—for example, the *Hox* genes—have been identified by homology with genes isolated from model organisms such as *Drosophila* and *C. elegans*.
- Among vertebrates, the mouse and the zebra fish provide opportunities to study mutations that affect development.
- Injecting antisense morpholinos into vertebrate eggs or embryos can disrupt gene expression and produce mutant phenocopies that reveal the developmental significance of the targeted gene.
- Mammalian stem cells, especially those derived from embryos, can be cultured *in vitro* to study the mechanisms that underlie differentiation.
- Animals produced by reproductive cloning suggest that differentiated cells have the same genetic potential as the zygote.
- Recombination between gene segments during immune cell differentiation creates the sequences that encode the light and heavy chains of antibodies.

► Basic Exercises

ILLUSTRATE BASIC GENETIC ANALYSIS

1. Arrange the following developmental stages in *Drosophila melanogaster* in chronological order from earliest to latest: pupa, blastoderm, zygote, unfertilized egg, larva, adult.

Answer: unfertilized egg, zygote, blastoderm, larva, pupa, adult.

2. In *C. elegans* with an X:A ratio of 0.5, what signal transduction pathway brings about the male phenotype?

Answer: The her-1 protein is a secreted signal that is expressed when the X:A ratio is 0.5. This signal interacts with the tra-2 membrane-bound receptor protein and inactivates it. When the tra-2 protein is inactivated, the cytoplasmic proteins encoded by the three *fem* genes are activated. Their activity enables the tra-1 protein to act as a transcription factor to stimulate the expression of genes involved in male development.

3. *Drosophila* females homozygous for a newly discovered recessive, autosomal mutation lay eggs that do not hatch into larvae, regardless of the genotype of their mates. However, the females themselves show no obvious abnormality. What type of gene does this new mutation define?

Answer: The new mutation defines a maternal-effect gene.

4. Predict the eye phenotype of a fly homozygous for a recessive loss-of-function mutation in the *sevenless* gene. Would a fly homozygous for a recessive loss-of-function mutation in the *bride of sevenless* gene have the same phenotype?

Answer: A fly homozygous for the *sevenless* mutation would not develop the R7 photoreceptor in each of the ommatidia in its compound eyes. The *sevenless* gene encodes the membrane-bound receptor for the extracellular ligand that triggers the R7 cell to differentiate; the ligand is encoded by the *bride of sevenless* gene. A fly homozygous for the *bride of sevenless* mutation would have the same phenotype.

5. A geneticist wishes to use a morpholino to knock down expression of a gene in zebra fish embryos. What triplet of bases will the geneticist absolutely want to include in the morpholino to carry out this experiment?

Answer: 3′ TAC 5′, because this triplet is complementary to the start codon, 5′ AUG 3′, contained in the gene's mRNA.

6. Suppose that an antibody light chain gene is assembled from three different gene segments. How many different chains can be produced if the genome contains 5, 20, and 200 copies of the three gene segments?

Answer: If each gene is assembled using one copy of each gene segment, $5 \times 20 \times 200 = 20{,}000$ different genes are possible.

▶ Testing Your Knowledge

INTEGRATE DIFFERENT CONCEPTS AND TECHNIQUES

1. The protein product of the *dorsal* (*dl*) gene in *Drosophila* has been called a ventral morphogen—that is, a substance that brings about the formation of ventral structures in the embryo by virtue of its high concentration in the nuclei on the ventral side of the blastoderm. However, the dorsal protein can enter these ventral nuclei only if a receptor on the embryo's ventral surface has been activated. This receptor is encoded by the *Toll* (*Tl*) gene. The extracellular ligand for the Toll receptor is encoded by the *spätzle* (*spz*) gene. This ligand can exist in two states, "native" and "modified," and the modified state is needed for activation of the Toll receptor. The products of three genes, *snake* (*snk*), *easter* (*ea*), and *gastrulation defective* (*gd*), are required to convert the native ligand into the modified ligand. All three of these gene products are serine proteases, proteins capable of cleaving other proteins at certain serines in the polypeptide chain. Using these facts, diagram the developmental pathway that ultimately causes the dorsal protein to induce the formation of ventral structures in the *Drosophila* embryo.

Answer: Here is one representation.

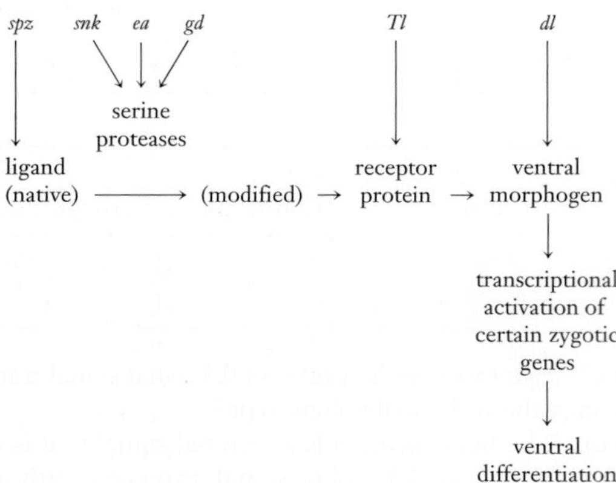

The protein product of the *spz* gene is modified by the serine proteases made by the *snk*, *ea*, and *gd* genes. In its modified form, this ligand is able to activate the Toll receptor protein, but the activation is restricted to the ventral side of the embryo. When the Toll receptor has been activated (presumably, by binding the modified spätzle ligand), it transduces a signal into the cytoplasm of the embryo. This signal ultimately causes the dorsal protein to move into the nuclei on the ventral side of the embryo, where it acts as a transcription factor to regulate the expression of the zygotic genes involved in the differentiation of ventral fates.

2. Considering the pathway described above, what would be the phenotypes of recessive loss-of-function mutations in the *spz* and *Tl* genes?

Answer: For reference, we should note that loss-of-function mutations in *dl* are maternal effect lethals; that is, embryos from *dl*/*dl* mothers die during development. When these dying embryos are examined, they are found to lack ventral structures. Geneticists say that they are "dorsalized." This peculiar phenotype is due to the failure of the dorsal transcription factor to induce appropriate development in the ventral nuclei of the embryo. In the absence of this induction, the ventral cells differentiate as if they were on the dorsal side of the embryo. Mutations in *spz* and *Tl* might be expected to have the same phenotypic effect because they would block steps in the pathway that ultimately causes the dorsal protein to induce ventral differentiation. Recessive mutations in *spz* and *Tl* are therefore maternal-effect lethals. Females homozygous for these mutations produce dorsalized embryos that die during development.

Questions and Problems

ENHANCE UNDERSTANDING AND DEVELOP ANALYTICAL SKILLS

21.1 Predict the phenotype of a fruit fly that develops from an embryo in which the posterior pole cells had been destroyed by a laser beam.

21.2 Why is the early *Drosophila* embryo a syncytium?

21.3 In *C. elegans*, the cell lineage that unfolds during development is essentially invariant. Why is this an advantage to someone studying development?

21.4 GO In *C. elegans*, homozygosity for a *dumpy* (*dpy*) mutation causes the animal to be shorter than wild-type. If a hermaphrodite heterozygous for such a mutation were self-fertilized, what fraction of its progeny would be dumpy?

21.5 In *Drosophila*, what larval tissues produce the external organs of the adult?

21.6 In early *C. elegans* development, the EMS cell divides into two cells, MS and E. MS subsequently develops into muscle cells, neurons, and some other cells while E develops into the cells of the intestine. What do you think would happen if the E cell were destroyed by a finely focused laser beam shortly after it was formed? What would happen if material extracted from the MS cell of one embryo were injected into the E cell of another embryo?

21.7 A researcher wants to perform a complementation test between two independently isolated *dumpy* mutations in *C. elegans*. Both mutations are autosomal recessives. In addition to being homozygous for the *dumpy* mutation, one strain is homozygous for an X-linked recessive mutation (*unc*) that causes the worms to be uncoordinated. The other strain is homozygous for the *dumpy* mutation alone. How should the complementation test be performed?

21.8 Diagram the sex-determination pathway in *Drosophila*.

21.9 A *Drosophila* researcher has discovered an autosomal recessive mutation that causes XX animals to develop into sterile males. Propose a scheme to find out if the mutation is an allele of either the *tra* or *tra2* genes.

21.10 Predict the sexual phenotype (anatomical and behavioral) of the following mutant genotypes in *Drosophila:* (a) XY; *tra fru/tra fru;* (b) XX; *tra fru/tra fru;* (c) XY; *dsx/dsx; fru/fru;* (d) XX; *dsx/dsx; fru/fru*.

21.11 What phenotype would you expect in an XX *Drosophila* that is homozygous for loss-of-function mutations in both the *tra* and *dsx* genes? Explain.

21.12 Loss-of-function mutations in the *Drosophila Sxl* gene cause females to die, but they have no effect in males. What phenotype would a gain-of-function mutation most likely have?

21.13 Predict the phenotype of XO and XX *C. elegans* that are homozygous for loss-of-function mutations in both the *tra-1* and *fem-1* genes.

21.14 Triploid *Drosophila* with two X chromosomes develop as intersexes. Suggest a mechanism to explain this phenotype.

21.15 Why do women, but not men, who are homozygous for the mutant allele that causes phenylketonuria produce children that are physically and mentally retarded?

21.16 Loss-of-function mutations in the *C. elegans xol-1* gene cause males to die, whereas loss-of-function mutations in the *Drosophila Sxl* gene cause females to die. What is the reason for these opposite effects?

21.17 In *Drosophila*, recessive mutations in the dorsal-ventral axis gene *dorsal* (*dl*) cause a dorsalized phenotype in embryos produced by *dl/dl* mothers; that is, no ventral structures develop. Predict the phenotype of embryos produced by females homozygous for a recessive mutation in the anterior-posterior axis gene *nanos*.

21.18 GO A researcher is trying to clone the *dpy-3* gene from *C. elegans*. Recessive mutations in this gene cause worms to be shorter than normal, a phenotype referred to as "dumpy." The researcher has mapped the *dpy-3* gene relative to other genes on one of the *C. elegans* chromosomes and, using the map data, has obtained cosmid clones with wild-type DNA from the vicinity of the *dpy-3* gene. To determine which clone contains the *dyp-3* gene, the researcher has injected each of them into the gonads of hermaphrodites homozygous for a recessive mutation in *dpy-3*. She has then scored the transgenic progeny derived from these injected hermaphrodites for the dumpy phenotype. From the results shown below, where on the molecular map should the researcher place the *dpy-3* gene?

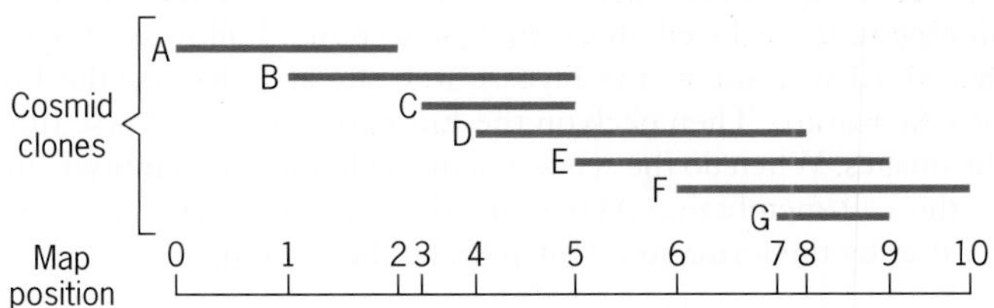

Cosmid Clone	Phenotype of Transgenic Progeny
A	dumpy
B	dumpy
C	dumpy
D	wild-type
E	wild-type
F	dumpy
G	dumpy

21.19 A researcher is planning to collect mutations in the gap genes, which control the first steps in the segmentation of *Drosophila* embryos. What phenotype should the researcher look for in this search for gap gene mutations?

21.20 A researcher is planning to collect mutations in maternal-effect genes that control the earliest events in *Drosophila* development. What phenotype should the researcher look for in this search for maternal-effect mutations?

21.21 Diagram a pathway that shows the contributions of the *sevenless* (*sev*) and *bride of sevenless* (*boss*) genes to the differentiation of the R7 photoreceptor in the ommatidia of *Drosophila* eyes. Where would *eyeless* (*ey*) fit in this pathway?

21.22 GO Like *dorsal*, *bicoid* is a strict maternal-effect gene in *Drosophila;* that is, it has no zygotic expression. Recessive mutations in *bicoid* (*bcd*) cause embryonic death by preventing the formation of anterior structures. Predict the phenotypes of (a) *bcd/bcd* animals produced by mating heterozygous males and females; (b) *bcd/bcd* animals produced by mating *bcd/bcd* females with *bcd/+* males; (c) *bcd/+* animals produced by mating *bcd/bcd* females with *bcd/+* males; (d) *bcd/bcd* animals produced by mating *bcd/+* females with *bcd/bcd* males; (e) *bcd/+* animals produced by mating *bcd/+* females with *bcd/bcd* males.

21.23 How might you show that two mouse *Hox* genes are expressed in different tissues and at different times during development?

21.24 With reference to **FIGURE 21.27**, what technique could you use to demonstrate that the V and C regions of the kappa light chain locus reside in different *Bam*HI restriction fragments in genomic DNA taken from embryos and in the same *Bam*HI fragment in genomic DNA taken from a clone of differentiated antibody-producing plasma cells? If this technique were applied to genomic DNA taken from differentiated fibroblasts (skin cells), what result would you expect?

21.25 Assume that an animal is capable of producing 100 million different antibodies and that each antibody contains a light chain 200 amino acids long and a heavy chain 500 amino acids long. How much genomic DNA would be needed to accommodate the coding sequences of these genes?

21.26 The methylation of DNA, the acetylation of histones, and the packaging of DNA into chromatin by certain kinds of proteins are sometimes referred to as epigenetic modifications of the DNA. These modifications portend difficulties for reproductive cloning. Do they also portend difficulties for therapeutic cloning and for the use of stem cells to treat diseases or injuries that involve the loss of specific cell types?

21.27 Each $L_\kappa V_\kappa$ gene sement in the kappa light chain locus on chromosome 2 consists of two coding exons, one for the leader peptide and one for the variable portion of the kappa light chain. Would you expect to find a stop codon at the end of the coding sequence in the second (V_κ) exon?

21.28 When the mouse *Pax6* gene, which is homologous to the *Drosophila eyeless* gene, is expressed in *Drosophila*, it produces extra compound eyes with ommatidia, just like normal *Drosophila* eyes. If the *Drosophila eyeless* gene were introduced into mice and expressed there, what effect would you expect? Explain.

▶ Genomics on the Web

at http://www.ncbi.nlm.nih.gov/

1. Images showing the anatomy and developmental stages of *Drosophila* are archived on the Flybase web site. Follow the links from the NCBI web site to the Flybase web site and click on the Image-Browse feature. Then click on the Embryo icon and browse through the images. When do the syncytial nuclei in the early embryo migrate to the cell membrane? When are these nuclei separated from one another by the formation of membranes between them?

2. The Flybase web site also has movies of *Drosophila* development. Click on the Movies icon and explore embryogenesis by looking at the film that shows the cell migration events called gastrulation from a lateral perspective—that is, from a side view. Then look at the film that shows gastrulation in an embryo that is homozygous for a mutation in the pair-rule gene *fushi tarazu* (*ftz*). Describe what is abnormal in the *ftz* embryo.

3. *C. elegans* researchers have compiled a wealth of information about the anatomy and development of their model organism in a database called Wormatlas. Follow the links on the NCBI web site to the Wormatlas. View the videos of living *C. elegans*. Then click on the Lineage Tree icon to bring up and view a diagram showing when each of the cells in an adult *C. elegans* is formed. Zoom in on the embryonic lineage to determine when the E and C cells are born. An X in the lineage marks a cell that succumbs to apoptosis. In which of these sublineages—E or C—does apoptosis occur?

4. Follow the links on the NCBI web site to ZFIN, a web site for zebra fish researchers. Click on Anatomy and Resources to bring up a page that provides access to still images and movies of zebra fish development. Why, in the early stages of embryogenesis, is cell division confined to the dorsal part of each image?

Chapter 22 The Genetic Basis of Cancer

CHAPTER OUTLINE

- **Cancer: A Genetic Disease**
- **Oncogenes**
- **Tumor Suppressor Genes**
- **Genetic Pathways to Cancer**

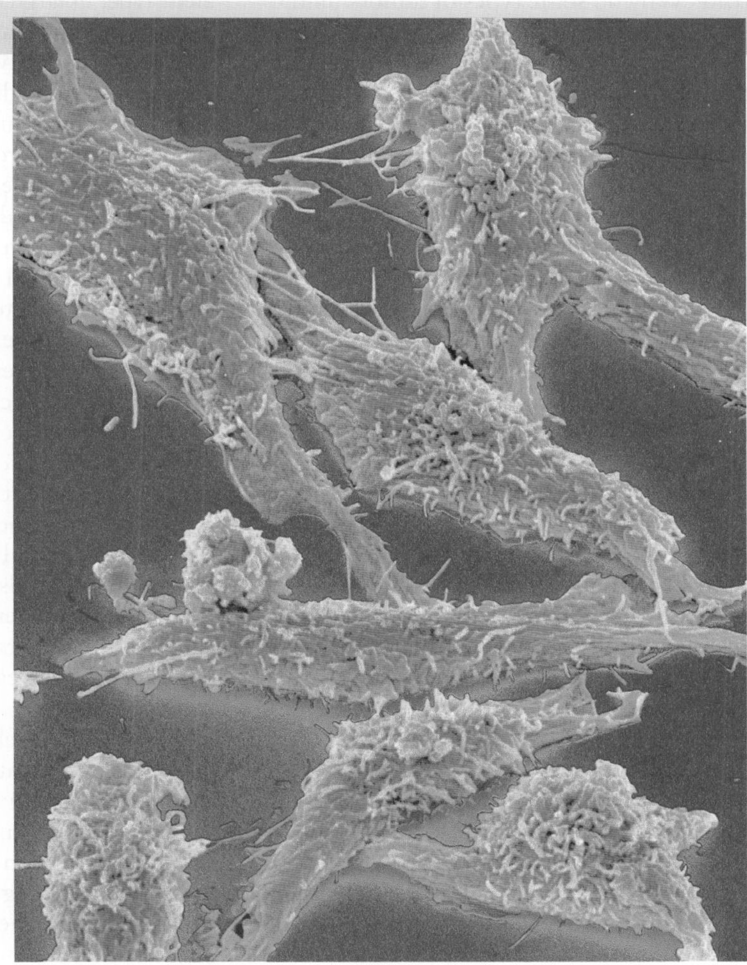

Colored scanning electron migrograph of cancer cells from a human colon.

A Molecular Family Connection

When Allison Romano started looking at colleges and universities, she wanted to find a school where she could study genetics in depth, maybe even do some hands-on research. Her plans were, in a sense, genetically motivated. At age 12 she was diagnosed with a tumor on one of her adrenal glands. This tumor was removed surgically, and after a lengthy convalescence, Allison returned to seventh grade, healthy and happy, and imbued with an interest in learning about the disease that had afflicted her. In high school, the courses Allison took reinforced this interest. She read a lot and met several students who enjoyed studying biology. Then another adrenal tumor appeared, but this time not in Allison. Rather, the tumor was found in her father. Louis Romano's tumor—the size of a golf ball—was successfully removed, and Louis recovered fully. After this incident, the oncologist suspected that both Louis and Allison had developed adrenal tumors—a rare form of cancer called pheochromocytoma—because they carried a mutation in the *VHL* gene, located in the short arm of chromosome 3. Published research had shown that such mutations are sometimes associated with this type of cancer. The oncologist therefore sent DNA samples from Louis and Allison to a genetics laboratory. DNA tests showed that both Louis and Allison were heterozygous for a mutant *VHL* allele. At nucleotide 490 in the *VHL* gene, a G:C base pair had been changed into an A:T base pair, causing serine to be substituted for glycine at position 93 in the polypeptide encoded by the gene. When Allison learned of this result, she resolved to study genetics. Her older sister, who showed no sign of pheochromocytoma, asked to be tested for the mutant allele and was found to have it. Her doctor then advised her to have regular screenings for any sign of pheochromocytoma. Louis Romano's two siblings—both asymptomatic—were also informed about the *VHL* mutation, but neither of them opted for testing. Allison subsequently majored in biology at a large university and worked for two semesters in a cancer genetics lab. Her project, on the identification of cancer-related genes in mice, was presented as a poster at the university's annual undergraduate research symposium, where her father and sister could see how she had found purpose in their family's molecular connection.

Cancer: A Genetic Disease

Mutations in genes that control cell growth and division are responsible for cancer.

Cancerous tumors kill several hundred thousand Americans every year. What causes tumors to form, and what causes some of them to spread? Why do some types of tumors tend to be found in families? Is the tendency to develop cancer inherited? Do environmental factors contribute to the development of cancer? In recent years, these and other questions have stimulated an enormous amount of research on the basic biology of cancer. Although many details are still unclear, the fundamental finding is that cancers result from genetic malfunctions. In some instances, these malfunctions may be triggered or exacerbated by environmental factors such as diet, excessive exposure to sunlight, or chemical pollutants. Cancers arise when critical genes are mutated. These mutations can cause biochemical processes to go awry and lead to the unregulated proliferation of cells. Without regulation, cancer cells divide ceaselessly, piling up on top of each other to form tumors. When cells detach from a tumor and invade the surrounding tissues, the tumor is *malignant*. When the cells do not invade the surrounding tissues, the tumor is *benign*. Malignant tumors may spread to other locations in the body, forming secondary tumors. This process is called **metastasis,** from Greek words meaning to "change state." In both benign and malignant tumors, something has gone wrong with the systems that control cell division. Researchers have now firmly established that this loss of control is due to underlying genetic defects.

THE MANY FORMS OF CANCER

Cancer is not a single disease, but rather a group of diseases. Cancers can originate in many different tissues of the body. Some grow aggressively, others more slowly. Some types of cancer can be stopped by appropriate medical treatment; others cannot. **FIGURE 22.1** shows the frequencies of new cases of different types of cancer in the United States, as well as the number of fatalities attributed to each type. Lung cancer is the most prevalent type, in large measure due to the effects of cigarette smoking. Breast cancer and prostate cancer are also fairly common.

The most prevalent types of cancer are derived from cell populations that divide actively, for example, from epithelial cells in the intestines, lungs, or prostate gland. Rarer forms of cancer develop from cell populations that typically do not divide, for example, from differentiated muscle or nerve cells.

Although the death rate from cancer is still high, enormous progress has been made in detecting and treating different types of cancer. The techniques of molecular genetics have enabled scientists to characterize cancers in ways that were not previously possible, and they have allowed them to devise new strategies for cancer therapy. There is little doubt that the large investment in basic cancer research is paying off.

Cancer cells can be obtained for experimental study by removing tissue from a tumor and dissociating it into its constituent cells. With appropriate nutrients, these dissociated tumor cells can be cultured *in vitro*, sometimes indefinitely. Cancer cells can also be derived from cultures of normal cells by treating the cells with agents that induce the cancerous state. Radiation, mutagenic chemicals, and certain types of viruses can irreversibly transform normal cells into cancerous cells. The agents that cause this type of transformation are called **carcinogens.**

The abiding characteristic of all cancer cells is that their growth is unregulated. When normal cells are cultured *in vitro*, they form a single cell layer—a monolayer—on the surface of the culture medium. Cancer cells, by contrast, overgrow each other, piling up on the surface of the culture medium to form masses. This unregulated pileup occurs because cancer cells do not respond to the chemical signals that inhibit cell division

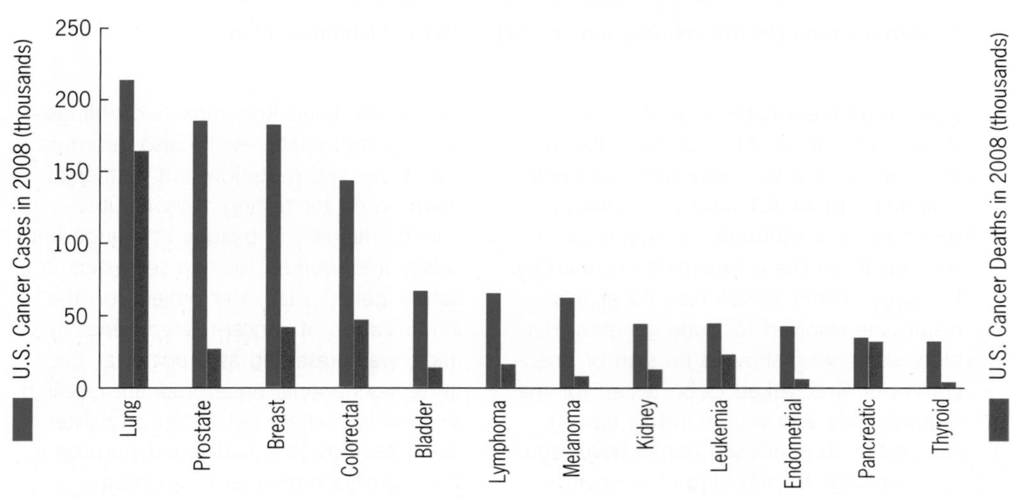

Figure 22.1 ▶ Estimated number of new cases and deaths from specific types of cancer in the United States in 2008.

and because they cannot form stable associations with their neighbors.

The external abnormalities that are apparent in a culture of cancer cells are correlated with profound intracellular abnormalities. Cancer cells often have a disorganized cytoskeleton, they may synthesize unusual proteins and display them on their surfaces, and they frequently have abnormal chromosome numbers—that is, they are aneuploid.

CANCER AND THE CELL CYCLE

The cell cycle consists of periods of growth, DNA synthesis, and division. The length of this cycle and the duration of each of its components are controlled by external and internal chemical signals. The transition from each phase of the cycle requires the integration of specific chemical signals and precise responses to these signals. If the signals are incorrectly sensed or if the cell is not properly prepared to respond, the cell could become cancerous.

The current view of cell-cycle control is that transitions between different phases of the cycle (G_1, S, G_2, and M; see Chapter 2) are regulated at "checkpoints." A **checkpoint** is a mechanism that halts progression through the cycle until a critical process such as DNA synthesis is completed, or until damaged DNA is repaired. When a checkpoint is satisfied, the cell cycle can progress. Two types of proteins play important roles in this progression: the *cyclins* and the *cyclin-dependent kinases*, often abbreviated CDKs. Complexes formed between the cyclins and the CDKs cause the cell cycle to advance.

The CDKs are the catalytically active components of the cell-cycling mechanism. These proteins regulate the activities of other proteins by transferring phosphate groups to them. However, the phosphorylation activity of the CDKs depends on the presence of the cyclins. The cyclins enable the CDKs to carry out their function by forming cyclin/CDK complexes. When the cyclins are absent, these complexes cannot form, and the CDKs are inactive. Cell cycling therefore requires the alternate formation and degradation of cyclin/CDK complexes.

One of the most important cell-cycle checkpoints, called *START*, is in mid-G_1 (**FIGURE 22.2**). The cell receives both external and internal signals at this checkpoint to determine when it is appropriate to move into the S phase. This checkpoint is regulated by D-type cyclins in conjunction with CDK_4. If a cell is driven past the *START* checkpoint by the cyclin D/CDK_4 complex, it becomes committed to another round of DNA replication. Inhibitory proteins with the capability of sensing problems late in the G_1 phase, such as low levels of nutrients or DNA damage, can put a brake on the cyclin/CDK complex and prevent the cell from entering the S phase. In the absence of such problems, the cyclin D/CDK_4 complex drives the cell through the end of the G1 phase and into the S phase, thereby initiating the DNA replication that is a prelude to cell division.

In tumor cells, checkpoints in the cell cycle are typically deregulated. This deregulation is due to genetic defects in the machinery that alternately raise and lower the abundance of the

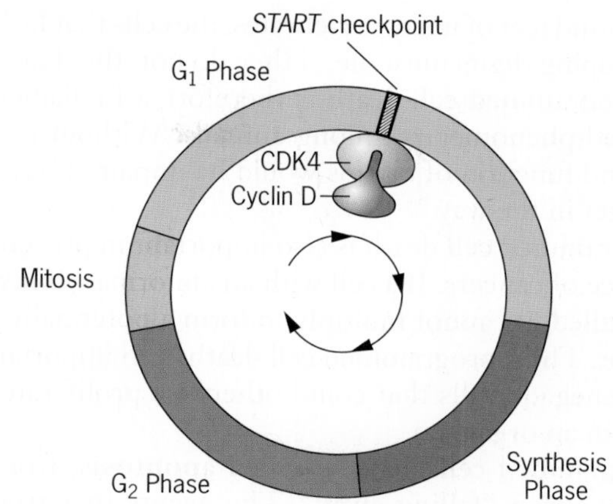

Figure 22.2 ▶ A schematic view of the *START* checkpoint in the mammalian cell cycle. Passage through the checkpoint depends on the activity of the cyclin D/CDK4 protein complex.

cyclin/CDK complexes. For example, the genes encoding the cyclins or the CDKs may be mutated, or the genes encoding the proteins that respond to specific cyclin/CDK complexes or that regulate the abundance of these complexes may be mutated. Many different types of genetic defects can deregulate the cell cycle, with the ultimate consequence that the cells may become cancerous.

Cells in which the *START* checkpoint is dysfunctional are especially prone to become cancerous. The *START* checkpoint controls entry into the S phase of the cell cycle. If DNA within a cell has been damaged, it is important that entry into the S phase be delayed to allow for the damaged DNA to be repaired. Otherwise, the damaged DNA will be replicated and transmitted to all the cell's descendants. Normal cells are programmed to pause at the *START* checkpoint to ensure that repair is completed before DNA replication commences. By contrast, cells in which the *START* checkpoint is dysfunctional move into S phase without repairing their damaged DNA. Over a series of cell cycles, mutations that result from the replication of unrepaired DNA may accumulate and cause further deregulation of the cell cycle. A clone of cells with a dysfunctional *START* checkpoint may therefore become aggressively cancerous.

CANCER AND PROGRAMMED CELL DEATH

Every cancer involves the accumulation of unwanted cells. In many animals, superfluous cells can be disposed of by mechanisms that are programmed into the cells themselves. This programmed cell death was originally discovered in studies with the nematode *Caenorhabditis elegans*. This tiny roundworm loses some of the cells that accumulate during the 10 or so cycles of division that occur during its development from a fertilized egg. Genetic analyses by Robert Horvitz and his colleagues demonstrated that the loss of these cells does not occur in certain mutant strains of *C. elegans*. Thus, cell death is part of the normal developmental program in this animal—and in others too, for we know, for example, that during the development of the hands and feet of many vertebrates, the cells that lie between the developing digits must die; if they do not, the digits remain fused. Programmed cell death is therefore a fundamental and widespread phenomenon among animals. Without it, the formation and function of organs would be impaired by cells that simply "get in the way."

Programmed cell death is also important in preventing the occurrence of cancers. If a cell with an abnormal ability to replicate is killed, it cannot multiply to form a potentially dangerous tumor. Thus, programmed cell death is an important check against renegade cells that could otherwise proliferate uncontrollably in an organism.

Programmed cell death is called **apoptosis,** from Greek roots that mean "falling away." The events that trigger cell death are only partially understood; we shall investigate some of them later in this chapter. However, the actual killing events are known in some detail. A family of proteolytic enzymes called *caspases* plays a crucial role in the cell death phenomenon. The caspases remove small parts of other proteins by cleaving peptide bonds. Through this enzymatic trimming, the target proteins are inactivated. The caspases attack many different kinds of proteins, including the lamins, which make up the inner lining of the nuclear envelope, and several components of the cytoskeleton. The collective impact of this proteolytic cleavage is that cells in which it occurs lose their integrity; their chromatin becomes fragmented, blebs of cytoplasm form at their surfaces, and they begin to shrink. Cells undergoing this kind of disintegration are usually engulfed by phagocytes, which are scavenger cells of the immune system, and are then destroyed. If the apoptotic mechanism has been impaired or inactivated, a cell that should otherwise be killed can survive and proliferate. Such a cell has the potential to form a clone that could become cancerous if it acquires the ability to divide uncontrollably.

A GENETIC BASIS FOR CANCER

The recent great advances in understanding cancer have come through application of molecular genetic techniques. However, before these techniques were available to researchers, there was strong evidence that the underlying causes of cancer are genetic. First, it was known that the cancerous state is clonally inherited. When cancer cells are grown in culture, their descendants are all cancerous. The cancerous condition is therefore transmitted from each cell to its daughters at the time of division—a phenomenon indicating that cancer has a genetic basis. Second, it was known that certain types of viruses can induce the formation of tumors in experimental animals. The induction of cancer by viruses implies that the proteins encoded by viral genes are involved in the production of the cancerous state. Third, it was known that cancer can be induced by agents capable of causing mutations. Mutagenic chemicals and ionizing radiation had been shown to induce tumors in experimental animals. In addition, a wealth of epidemiological data had implicated these agents as the causes of cancer in human beings. Fourth, it was known that certain types of cancer tend to run in families. In particular, susceptibility to retinoblastoma, a rare cancer of the eye, and susceptibility to some forms of colon cancer appeared to be inherited as simple dominant conditions, albeit with incomplete penetrance and variable expressivity. Because susceptibility to these special types of cancer is inherited, it seemed plausible that all cancers might have their basis in genetic defects—either inherited mutations or mutations acquired somatically during a person's lifetime. Finally, it was known that certain types of white blood cell cancers (leukemias and lymphomas) are associated with particular chromosomal aberrations. Collectively, these diverse observations strongly suggested that cancer is caused by genetic malfunctions.

In the 1980s, when molecular genetic techniques were first used to study cancer cells, researchers discovered that the cancerous state is, indeed, traceable to specific genetic defects. Typically, however, not one but several such defects are required to convert a normal cell into a cancerous cell. Cancer researchers have identified two broad classes of genes that, when mutated, can contribute to the development of a cancerous state. In one of these classes, mutant genes actively promote cell division; in the other class, mutant genes fail to repress cell division. Genes in the first class are called **oncogenes,** from the Greek word for "tumor." Genes in the second class are called **tumor suppressor genes.** In the sections that follow, we discuss the discovery, characteristics, and significance of each of these classes of cancer-related genes.

KEY POINTS

- Cancer is a group of diseases in which the cellular cycle of growth and division is unregulated.
- Cancers may develop if the mechanism for programmed cell death (apoptosis) is impaired.
- Cancers are due to the occurrence of mutations in genes whose protein products are involved in the control of the cell cycle.

Oncogenes

Many cancers involve the overexpression of certain genes or the abnormal activity of their mutant protein products.

Oncogenes comprise a diverse group of genes whose products play important roles in the regulation of biochemical activities within cells, including those activities related to cell division. These genes were first discovered in the genomes of RNA viruses that are capable of inducing tumors in vertebrate hosts. Later, cellular counterparts of these viral oncogenes were discovered in many different organisms, ranging from *Drosophila* to humans.

TUMOR-INDUCING RETROVIRUSES AND VIRAL ONCOGENES

Fundamental insights into the genetic basis of cancer have come from the study of tumor-inducing viruses. Many of these viruses have a genome composed of RNA instead of DNA. After entering a cell, the viral RNA is used as a template to synthesize complementary DNA, which is then inserted at one or more positions in the cell's chromosomes. The synthesis of DNA from RNA is catalyzed by the viral enzyme reverse transcriptase. This reversal of the normal flow of genetic information from DNA to RNA has prompted biologists to call these pathogens **retroviruses** (see Chapter 18).

The first tumor-inducing virus was discovered in 1910 by Peyton Rous; it caused a special kind of tumor, or sarcoma, in the connective tissue of chickens and has since been called the Rous sarcoma virus. Modern research has shown that the RNA genome of this retrovirus contains four genes: *gag*, which encodes the capsid protein of the virion; *pol*, which encodes the reverse transcriptase; *env*, which encodes a protein of the viral envelope; and *v-src*, which encodes a protein kinase that inserts into the plasma membranes of infected cells. The distinguishing feature of a kinase is that it can phosphorylate other proteins. Of these four genes, only the *v-src* gene is responsible for the virus's ability to form tumors. A virus in which the *v-src* gene has been deleted is infectious but unable to induce tumors. Genes such as *v-src* that cause cancer are called oncogenes.

Studies with other tumor-inducing retroviruses have uncovered at least 20 different viral oncogenes, usually denoted *v-onc* (**TABLE 22.1**). Some of these oncogenes are related to cellular genes that encode growth factors. For example, *v-sis*, an oncogene from the simian sarcoma virus, codes for a version of the platelet-derived growth factor (PDGF). PDGF is normally produced by platelet cells to promote wound healing, which it does by stimulating the growth of cells at the wound site. Simian sarcoma viruses carrying the *v-sis* gene induce tumors in monkeys. They also transform cultured cells to a cancerous state, presumably by producing large amounts of the *v-sis* version of PDGF, which then causes uncontrolled cell growth.

Other viral oncogenes encode proteins that are similar to growth-factor and hormone receptors. For example, the *v-erbB* gene from the avian erythroblastosis virus encodes a protein much like the cellular receptor for epidermal growth factor (EGF), and the *v-fms* gene from the feline sarcoma virus encodes a protein much like the receptor for the cellular CSF-1 (colony-stimulating factor 1) growth factor. Both of these growth-factor receptors are transmembrane proteins with a growth-factor-binding domain on the outside of the cell and a protein kinase domain on the inside. This domain on the inside enables the protein to phosphorylate certain amino acids, usually tyrosine, in other proteins that interact with it.

▶ **TABLE 22.1**

Retroviral Oncogenes

Oncogene	Host Virus	Species	Function of Gene Product
abl	Abelson murine leukemia virus	Mouse	Tyrosine-specific protein kinase
erbA	Avian erythroblastosis virus	Chicken	Analog of thyroid hormone receptor
erbB	Avian erythroblastosis virus	Chicken	Truncated version of epidermal growth-factor (EGF) receptor
fes	ST feline sarcoma virus	Cat	Tyrosine-specific protein kinase
fgr	Gardner-Rasheed feline sarcoma virus	Cat	Tyrosine-specific protein kinase
fms	McDonough feline sarcoma virus	Cat	Analog of colony stimulating growth-factor (CSF-1) receptor
fos	FJB osteosarcoma virus	Mouse	Transcriptional activator protein
fps	Fuginami sarcoma virus	Chicken	Tyrosine-specific protein kinase
jun	Avian sarcoma virus 17	Chicken	Transcriptional activator protein
mil (*mht*)	MH2 virus	Chicken	Serine/threonine protein kinase
mos	Moloney sarcoma virus	Mouse	Serine/threonine protein kinase
myb	Avian myeloblastosis virus	Chicken	Transcription factor
myc	MC29 myelocytomatosis virus	Chicken	Transcription factor
raf	3611 murine sarcoma virus	Mouse	Serine/threonine protein kinase
H-ras	Harvey murine sarcoma virus	Rat	GTP-binding protein
K-ras	Kirsten murine sarcoma virus	Rat	GTP-binding protein
rel	Reticuloendotheliosis virus	Turkey	Transcription factor
ros	URII avian sarcoma virus	Chicken	Tyrosine-specific protein kinase
sis	Simian sarcoma virus	Monkey	Analog of platelet-derived growth factor (PDGF)
src	Rous sarcoma virus	Chicken	Tyrosine-specific protein kinase
yes	Y73 sarcoma virus	Chicken	Tyrosine-specific protein kinase

Many viral oncogenes, including *v-src*, encode tyrosine kinases that do not span the plasma membrane. Instead, these proteins are situated on the inner face of the plasma membrane, where they perform their phosphorylating function. The various *v-ras* oncogenes encode proteins that bind GTP, much like the cellular G proteins, which play an important role in regulating the level of cyclic AMP.

Another group of viral oncogenes encode proteins that apparently function as transcription factors. These include the *v-jun*, *v-fos*, *v-erbA*, and *v-myc* genes, each carried by a different retrovirus. These proteins are homologous to cellular proteins that bind to DNA and regulate transcription.

Each type of viral oncogene therefore appears to encode a protein that could theoretically play a role in regulating the expression of cellular genes, including those involved in the processes of growth and division. Some of these proteins may act as signals to stimulate certain types of cellular activity; others may act as receptors to pick up these signals or as intracellular agents to convey them from the plasma membrane to the nucleus; yet another category of viral oncogene proteins may act as transcription factors to stimulate gene expression.

CELLULAR HOMOLOGUES OF VIRAL ONCOGENES: THE PROTO-ONCOGENES

The proteins encoded by viral oncogenes are similar to cellular proteins with important regulatory functions. Many of these cellular proteins were actually identified by isolating the cellular homologue of the viral oncogene. For example, the cellular homologue of the *v-src* gene was obtained by screening a genomic DNA library made from uninfected chicken cells. For this screening, the *v-src* gene was used as a hybridization probe to detect recombinant DNA clones that could base-pair with it. Analysis of these clones established that chicken cells contain a gene that is similar to *v-src*—indeed, that is related to it in an evolutionary sense. However, this gene is not associated with an integrated sarcoma provirus, and it differs from the *v-src* gene in a very important respect: it contains introns (**FIGURE 22.3**). There are, in fact, 11 introns in the chicken homologue of *v-src*, compared to zero in the *v-src* gene itself. This startling discovery suggested that perhaps *v-src* had evolved from a normal cellular gene and that, concomitantly, it had lost its introns.

The cellular homologues of viral oncogenes are called **proto-oncogenes,** or sometimes, *normal cellular oncogenes,*

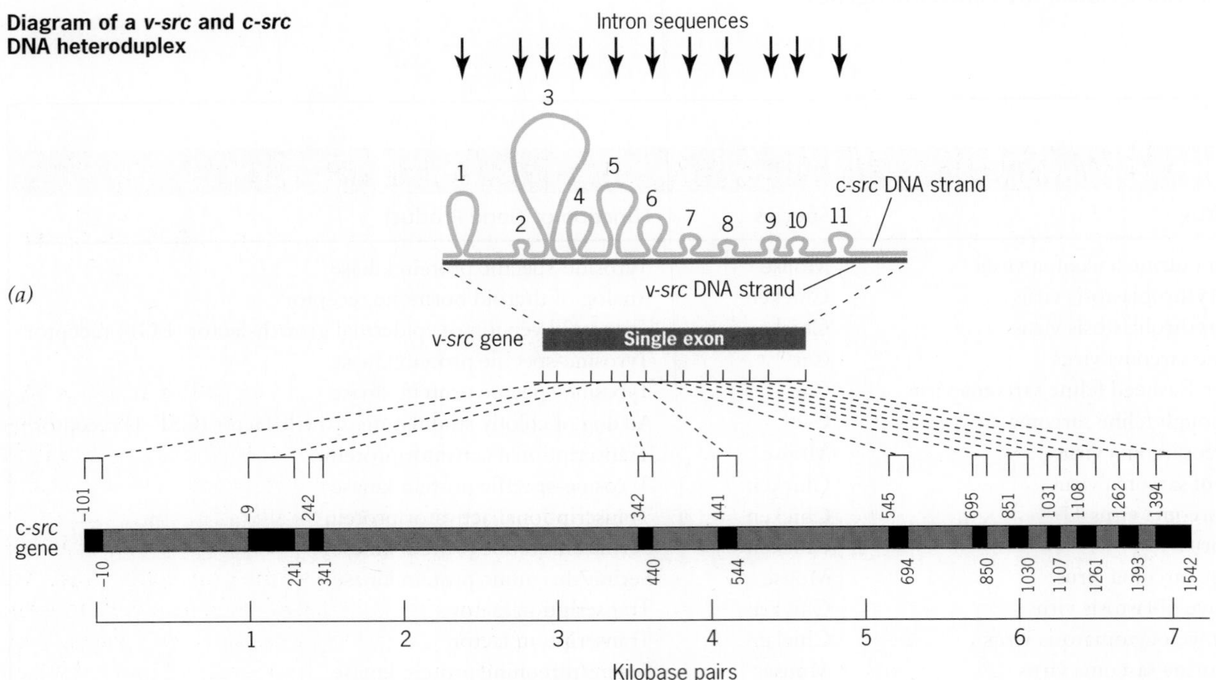

Figure 22.3 ▶ Structures of the *v-src* and *c-src* genes. (*a*) Diagram of the predicted DNA heteroduplex formed by hybridization of a strand from the *c-src* gene (top) and a partially complementary strand from the *v-src* gene (bottom). The introns (numbered 1–11) form single-stranded loops. (*b*) Schematic comparison of these two genes, with exons shown in black. The coordinate system for the exons in the *c-src* gene is based on the first nucleotide in the coding sequence (position 1). The first exon (position −101 to −10) is in the 5′— leader of the gene.

denoted *c-onc*. The cellular homologue of *v-src* is therefore *c-src*. The coding sequences of these two genes are very similar, differing only in 18 nucleotides; *v-src* encodes a protein of 526 amino acids, and *c-src* encodes a protein of 533 amino acids. By using *v-onc* genes as probes, other *c-onc* genes have been isolated from many different organisms, including human beings. As a rule, these cellular oncogenes show considerable conservation in structure. *Drosophila*, for example, carries very similar homologues of the vertebrate cellular oncogenes *c-abl*, *c-erbB*, *c-fps*, *c-raf*, *c-ras*, and *c-myb*. The similarity of oncogenes from different species strongly suggests that the proteins they encode are involved in important cellular functions.

Why do *c-onc*s have introns whereas *v-onc*s do not? The most plausible answer is that *v-onc*s were derived from *c-onc*s by the insertion of a fully processed *c-onc* mRNA into the genome of a retrovirus. A virion that packaged such a recombinant molecule would then be able to transduce the *c-onc* gene whenever it infected another cell. During infection, the recombinant RNA would be reverse-transcribed into DNA and then integrated into the cell's chromosomes. What could be of greater value to a virus than to have a new gene that stimulates increased growth of its host, while its integrated genome goes along for the ride?

In many cases, the acquisition of an oncogene by a retrovirus has been accompanied by the loss of some viral genetic material. Because this lost material is needed for viral replication, these oncogenic viruses are able to reproduce only if a helper virus is present. In this respect, they resemble the defective transducing bacteriophages we discussed in Chapter 8.

Why do *v-onc*s induce tumors, whereas normal *c-onc*s do not? In some cases it appears that the viral oncogene produces much more protein than its cellular counterpart, perhaps because it has been transcriptionally activated by enhancers embedded in the viral genome. In chicken tumor cells, for example, the *v-src* gene produces 100 times as much tyrosine kinase as the *c-src* gene. This vast oversupply of the kinase evidently upsets the delicate signaling mechanisms that control cell division, causing unregulated growth. Other *v-onc* genes may induce tumors by expressing their proteins at inappropriate times, or by expressing altered—that is, mutant—forms of these proteins.

MUTANT CELLULAR ONCOGENES AND CANCER

The products of the *c-onc*s play key roles in regulating cellular activities. Consequently, a mutation in one of these genes can upset the biochemical balance within a cell and put it on the track to becoming cancerous. Studies of many different types of human cancer have demonstrated that mutant cellular oncogenes are associated with the development of a cancerous state.

The first evidence linking cancer to a mutant *c-onc* came from the study of a human bladder cancer. The mutation responsible for this bladder cancer was isolated by Robert

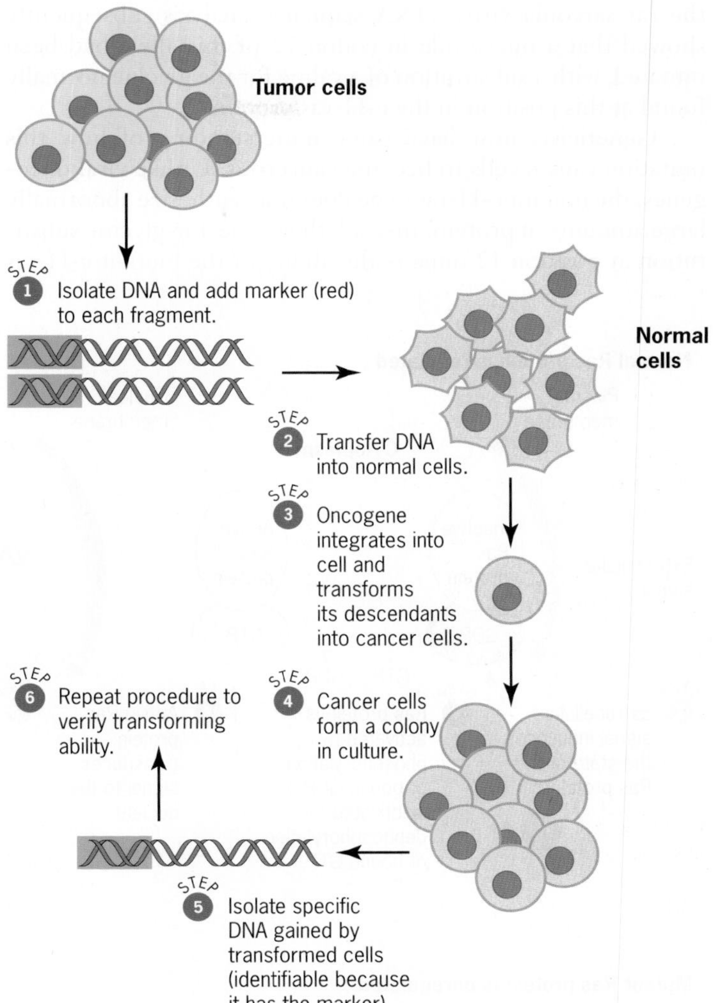

Figure 22.4 ▶ The transfection test to identify DNA sequences capable of transforming normal cells into cancer cells.

Weinberg and colleagues using a transfection test (**FIGURE 22.4**). DNA was extracted from the cancerous tissue and fragmented into small pieces; then each of these pieces was joined to a segment of bacterial DNA, which served as a molecular marker. The marked DNA fragments were then introduced, or transfected, into cells growing in culture to determine if any of them could transform the cells into a cancerous state. This state could be recognized by the tendency of the cancer cells to form small clumps, or foci, when grown on soft agar plates. The DNA from such cells was extracted and screened to see if it carried the molecular marker that was linked to the original transfecting fragments. If it did, this DNA was retested for its ability to induce the cancerous state. After several tests, Weinberg's research team identified a DNA fragment from the original bladder cancer that reproducibly transformed cultured cells into cancer cells. This fragment carried an allele of the *c-H-ras* oncogene, a homologue of an oncogene in the Harvey strain of the rat sarcoma virus. DNA sequence analysis subsequently showed that a nucleotide in codon 12 of this allele had been mutated, with a substitution of a valine for the glycine normally found at this position in the c-H-ras protein.

Geneticists now have some understanding of how this mutation causes cells to become cancerous. Unlike viral oncogenes, the mutant c-H-ras gene does not synthesize abnormally large amounts of protein. Instead, the valine-for-glycine substitution at position 12 impairs the ability of the mutant c-H-ras protein to hydrolyze one of its substrates, guanosine triphosphate (GTP). Because of this impairment, the mutant protein is kept in an active signaling mode, transmitting information that ultimately stimulates the cells to divide in an uncontrolled way (**FIGURE 22.5**).

Mutant versions of the *c-ras* oncogenes have now been found in a large number of different human tumors, including lung, colon, mammary, prostate, and bladder tumors, as well as neuroblastomas (nerve cell cancers), fibrosarcomas (cancers of

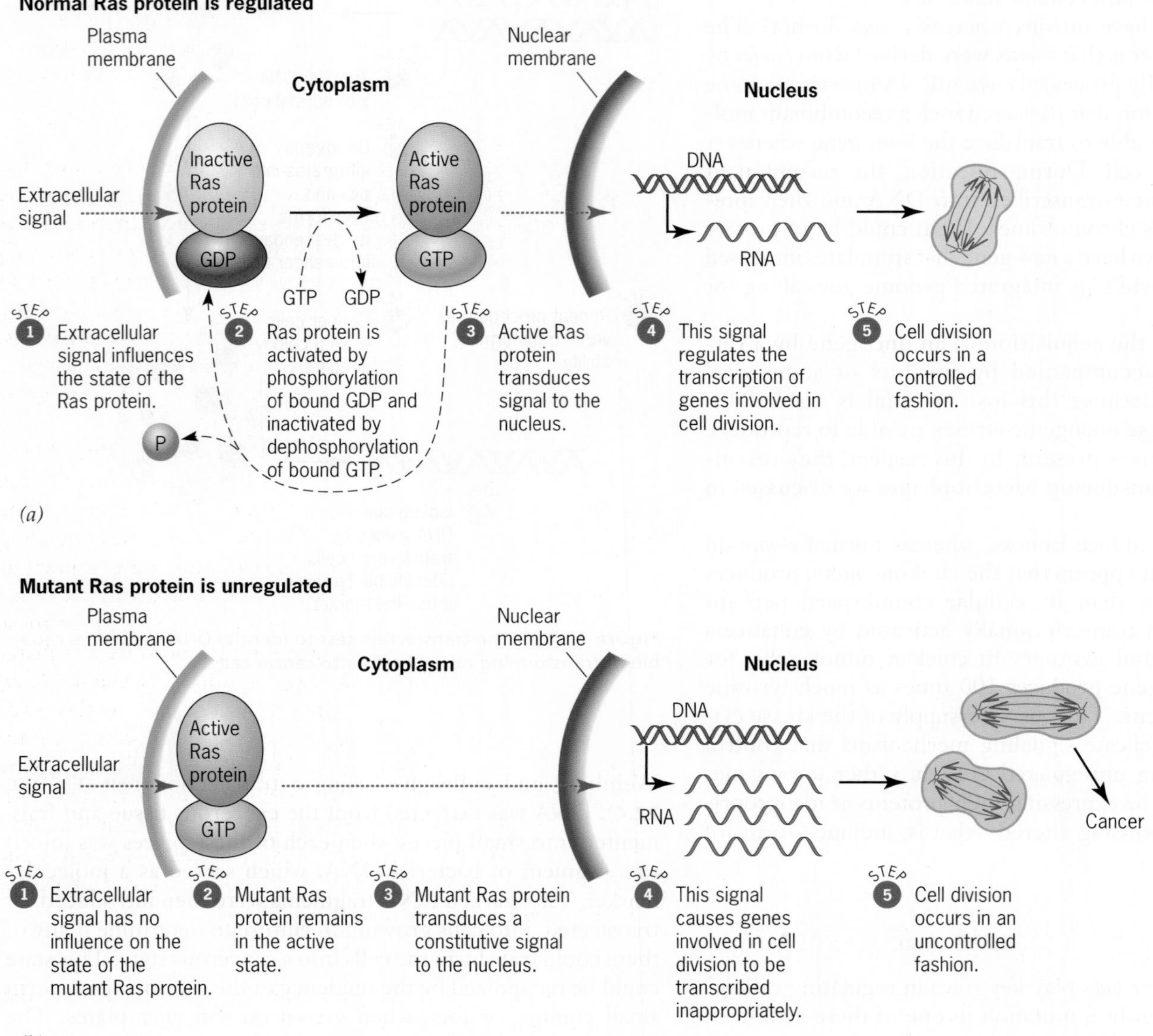

Figure 22.5 ▶ Ras protein signaling and cancer. (*a*) The normal protein product of the *ras* gene alternates between inactive and active states, depending on whether it is bound to GDP or GTP. Extracellular signals such as growth factors stimulate the conversion of inactive Ras to active Ras. Through active Ras, these signals are transmitted to other proteins and eventually to the nucleus, where they induce the expression of genes involved in cell division. Because this signaling is intermittent and regulated, cell division occurs in a controlled manner. (*b*) Mutant Ras proteins exist mainly in the active state. These proteins transmit their signals more or less constantly, leading to uncontrolled cell division, the hallmark of cancer.

the connective tissues), and teratocarcinomas (cancers that contain different embryonic cell types). In all cases, the mutations involve amino acid changes in one of three positions—12, 59, or 61. Each of these amino acid changes impairs the ability of the mutant Ras protein to switch out of its active signaling mode. These types of mutations therefore stimulate cells to grow and divide.

In these types of cancer, only one of the two copies of the *c-ras* gene has been mutated. The single mutant allele is dominant in its ability to bring about the cancerous state. Mutations in *c-ras* and other cellular oncogenes that lead to cancer in this way are therefore *dominant activators* of uncontrolled cell growth.

Dominant activating mutations in cellular oncogenes are seldom inherited through the germ line; rather, the vast majority of them occur spontaneously in the soma during the course of cell division. Because the number of cell divisions in a human life is very large—more than 10^{16}—thousands of potentially oncogenic mutations are bound to occur, and if each one functioned as a dominant activator of uncontrolled cell growth, the development of a tumor would be inevitable. However, many people lead long lives without developing tumors. The explanation for this paradox is that each individual oncogene mutation is, by itself, seldom able to induce a cancerous state. However, when several different growth-regulating genes have been mutated, the cell cannot compensate for their separate effects, its growth becomes unregulated, and cancer ensues. In many tumors, at least one of these deleterious mutations is in a cellular oncogene. Thus, this group of genes plays an important role in the etiology of human cancer.

CHROMOSOME REARRANGEMENTS AND CANCER

Certain types of human cancer are associated with chromosome rearrangements. For example, chronic myelogenous leukemia (CML) is associated with an aberration of chromosome 22. This abnormal chromosome was originally discovered in the city of Philadelphia and thus is called the *Philadelphia chromosome*. Initially, it was thought to have a simple deletion in its long arm; however, subsequent analysis using molecular techniques has shown that the Philadelphia chromosome is actually the result of a reciprocal translocation between chromosomes 9 and 22. (For a general discussion of translocations, see Chapter 6.) In the Philadelphia translocation, the tip of the long arm of chromosome 9 has been joined to the body of chromosome 22, and the distal portion of the long arm of chromosome 22 has been joined to the body of chromosome 9 (**FIGURE 22.6**). The translocation breakpoint on chromosome 9 is in the *c-abl* oncogene, which encodes a tyrosine kinase, and the breakpoint on chromosome 22 is in a gene called *bcr*. Through the translocation, the *bcr* and *c-abl* genes have been physically joined, creating a fusion gene whose polypeptide product has the amino terminus of the Bcr protein and the carboxy terminus of the Abl protein. Although it is not understood precisely why, this fusion polypeptide causes white blood cells to become cancerous. The mechanism may involve the tyrosine kinase activity of the c-Abl protein, which is tightly controlled in normal cells but is deregulated in cells that produce the fusion polypeptide. In effect, the tyrosine kinase function of the c-Abl protein has been constitutively activated by the *bcr/c-abl* gene fusion. This fusion is therefore a dominant activator of the c-Abl tyrosine kinase. Deregulation of the c-Abl tyrosine kinase leads to abnormal phosphorylation of other proteins, including some that are involved in controlling the cell cycle. In their phosphorylated state, these proteins cause cells to grow and divide uncontrollably.

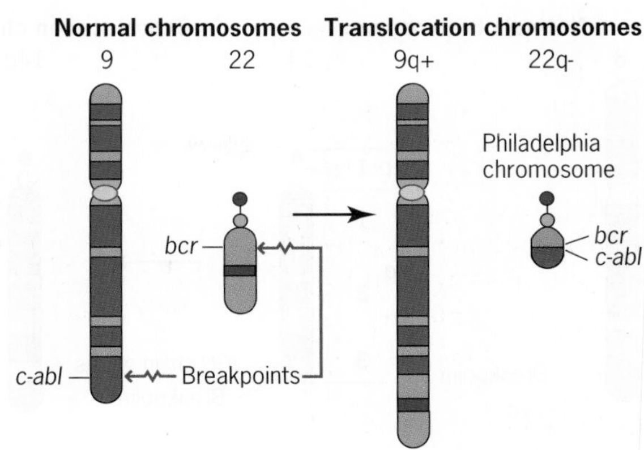

Figure 22.6 ▶ The reciprocal translocation involved in the Philadelphia chromosome associated with chronic myelogenous leukemia.

Burkitt's lymphoma is another example of a white blood cell cancer associated with reciprocal translocations. These translocations invariably involve chromosome 8 and one of the three chromosomes (2, 14, and 22) that carry genes encoding the polypeptides that form immunoglobulins (also known as antibodies; see Chapter 21). Translocations involving chromosomes 8 and 14 are the most common (**FIGURE 22.7**). In these translocations, the *c-myc* oncogene on chromosome 8 is juxtaposed to the genes for the immunoglobulin heavy chains (*IGH*) on chromosome 14. This rearrangement results in the overexpression of the *c-myc* oncogene in cells that produce immunoglobulin heavy chains—that is, in the B cells of the immune system. The *c-myc* gene encodes a transcription factor that activates genes involved in promoting cell division. Consequently, the overexpression of *c-myc* that occurs in cells that carry the *IGH/c-myc* fusion created by the t8;14 translocation causes those cells to become cancerous.

KEY POINTS

- Some viruses carry genes (oncogenes) that can induce the formation of tumors in animals.
- Viral oncogenes are homologous to cellular genes (proto-oncogenes), which can induce tumors when they are overexpressed or when they are mutated to produce abnormally active protein products.
- Mutations in proto-oncogenes actively promote cell proliferation.
- Some cancers are associated with chromosome rearrangements that enhance the expression of proto-oncogenes or that alter the nature of their protein products.

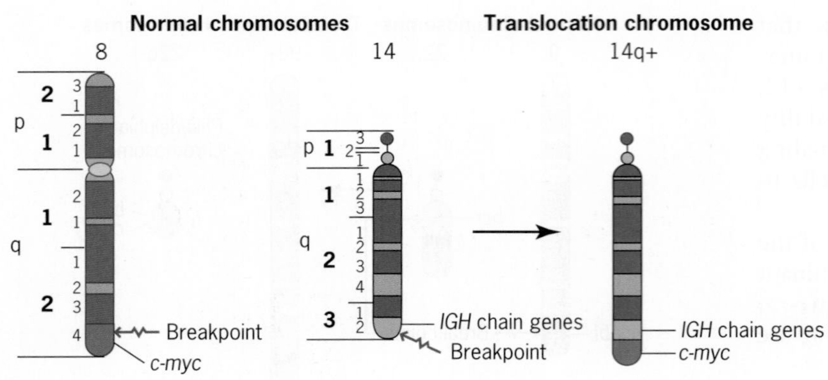

Figure 22.7 ▶ A reciprocal translocation involved in Burkitt's lymphoma. Only the translocation chromosome (14q+) that carries both the *c-myc* oncogene and the immunoglobulin heavy chain genes (*IGH*) is shown.

Tumor Suppressor Genes

Many cancers involve the inactivation of genes whose products play important roles in regulating the cell cycle.

The normal alleles of genes such as *c-ras* and *c-myc* produce proteins that regulate the cell cycle. When these genes are overexpressed, or when they produce proteins that function as dominant activators, the cell is predisposed to become cancerous. However, the full development of a cancerous state usually requires additional mutations, and typically these mutations affect genes that are normally involved in the restraint of cell growth. These mutations therefore define a second class of cancer-related genes—the anti-oncogenes, or, as they are more often called, the tumor suppressor genes.

INHERITED CANCERS AND KNUDSON'S TWO-HIT HYPOTHESIS

Many of the tumor suppressor genes were initially discovered through the analysis of rare cancers in which a predisposition to develop the cancer follows a dominant pattern of inheritance. This predisposition is due to heterozygosity for an inherited loss-of-function mutation in the tumor suppressor gene. A cancer develops only if a second mutation occurs in the somatic cells and if this mutation knocks out the function of the wild-type allele of the tumor suppressor gene. Thus, development of the cancer requires two loss-of-function mutations—that is, two inactivating "hits," one in each of the two copies of the tumor suppressor gene.

In 1971 Alfred Knudson proposed this explanation for the occurrence of *retinoblastoma*, a rare childhood cancer of the eye. In most human populations, the incidence of retinoblastoma is about 5 in 100,000 children. Pedigree analysis indicates that approximately 40 percent of the cases involve an inherited mutation that predisposes the individual to develop the cancer. The other 60 percent of the cases cannot be traced to a specific inherited mutation. These noninherited cases are said to be *sporadic*. On the basis of statistical analyses, Knudson proposed that both the inherited and sporadic cases of retinoblastoma occur because the two copies of a particular gene have been inactivated (**FIGURE 22.8**). In the inherited cases, one of the inactivating mutations has been transmitted through the germ line, and the other occurs during the development of the somatic tissues of the eye. In the sporadic cases, both of the inactivating mutations occur during eye development. Thus, in either type of retinoblastoma, two mutational "hits" are required to knock out a gene that normally functions to suppress tumor formation in the eye.

Subsequent research findings have verified the correctness of Knudson's two-hit hypothesis. First, several cases of retinoblastoma were found to be associated with a small deletion in the long arm of chromosome 13. The gene that normally prevents retinoblastoma—symbolized *RB*—must therefore be located in the region defined by this deletion. More refined cytogenetic mapping subsequently placed the *RB* gene in locus 13q14.2. Second, positional cloning techniques were used to isolate a candidate *RB* gene. Once isolated, the gene's structure, sequence, and expression patterns were determined. Third, the structure of the candidate gene was examined in cells taken from tumorous eye tissue. As predicted by Knudson's two-hit hypothesis, both copies of this gene were inactivated in retinoblastoma cells. Thus, the candidate gene appeared to be the authentic *RB* gene. Finally, cell culture experiments demonstrated that a cDNA from the wild-type allele of the candidate gene could revert the cancerous properties of cultured tumor cells. These cancer reversion experiments proved beyond a doubt that the candidate gene was the authentic *RB* tumor suppressor gene. The protein product of this gene—denoted pRB—was subsequently found to be a ubiquitously expressed protein that interacts with a family of transcription factors involved in regulating the cell cycle.

Knudson's two-hit hypothesis has since been applied to other inherited cancers, including Wilms' tumor, Li-Fraumeni syndrome, neurofibromatosis, von Hippel-Lindau disease, and

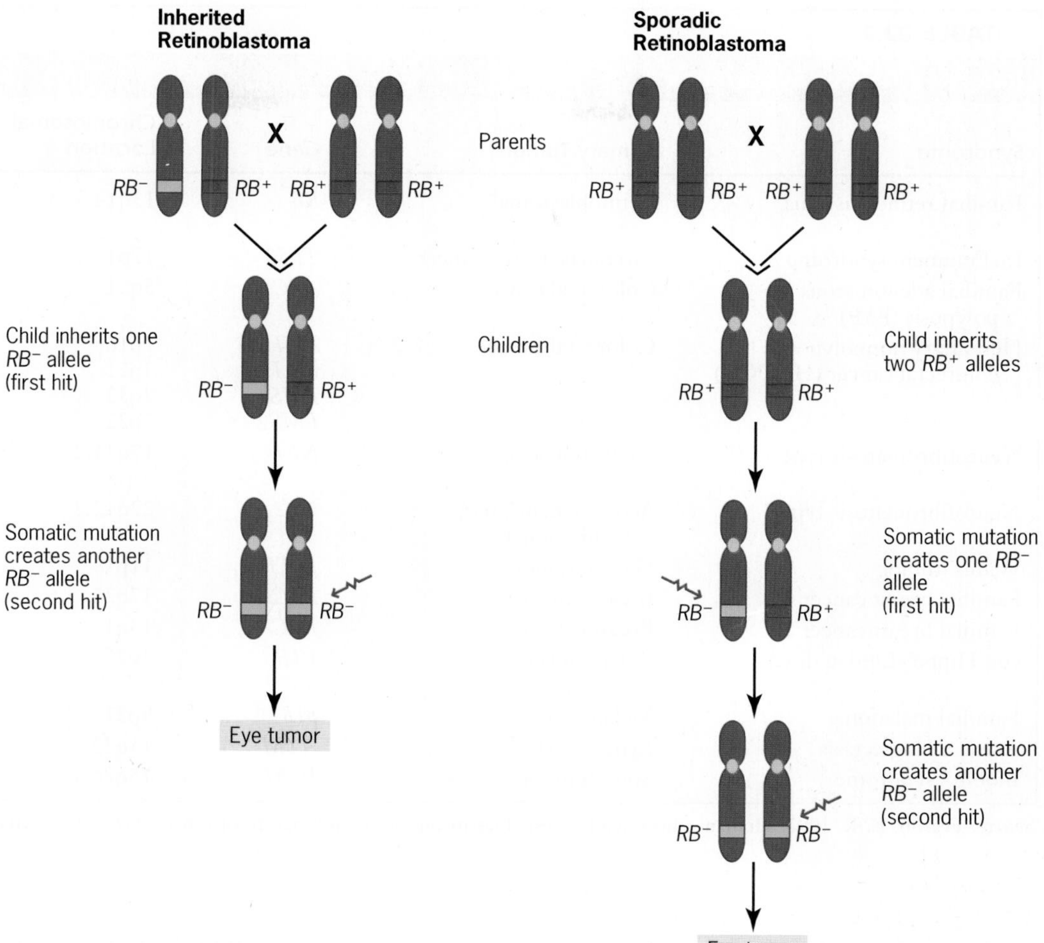

Figure 22.8 ► Knudson's two-hit hypothesis to explain the occurrence of inherited and sporadic cases of retinoblastoma. Two inactivating mutations are required to eliminate the function of the *RB* gene.

certain types of colon and breast cancer (**TABLE 22.2**). In each case, a different tumor suppressor gene is involved. For example, in Wilms' tumor, a cancer of the urogenital system, the relevant tumor suppressor gene is the *WT1* gene located in the short arm of chromosome 11; in neurofibromatosis, a disease characterized by benign tumors and skin lesions, it is the *NF1* gene located in the long arm of chromosome 17; and in familial adenomatous polyposis, a condition characterized by the occurrence of numerous tumors in the colon, it is the *APC* gene located in the long arm of chromosome 5. Like retinoblastoma, these three diseases are rare, and only a fraction of the observed cases involve an inherited mutation in the relevant tumor suppressor gene. The other cases are caused either by two independent somatic mutations in that gene or by mutations in other, as-yet-unidentified tumor suppressor genes.

To explore the genetic dimensions of the two-hit hypothesis, work through the Focus on Problem Solving: Estimating Mutation Rates in Retinoblastoma.

CELLULAR ROLES OF TUMOR SUPPRESSOR PROTEINS

Only about 1 percent of all cancers are hereditary. However, more than 20 different inherited cancer syndromes have been identified, and in nearly all of them the underlying defect is in a tumor suppressor gene rather than in an oncogene. The proteins encoded by these tumor suppressor genes function in a diverse array of cellular processes, including division, differentiation, programmed cell death, and DNA repair. In the following sections, we discuss some of the tumor suppressor proteins that have been studied intensively.

pRB

Recent research has revealed that the RB tumor suppressor protein plays a key role in regulation of the cell cycle. Although the *RB* gene was discovered through its association with retinoblastoma, mutations in this gene are also associated with other types of cancer, including small-cell lung carcinomas, osteosarcomas, and bladder, cervical, and prostate carcinomas. Furthermore, mice that are homozygous for an *RB* knockout mutation die during embryonic development. Thus, the RB gene product is essential for life.

The RB gene product, symbolized pRB, is a 105-kilodalton nuclear protein that is involved in cell-cycle regulation. Two genes homologous to *RB* have been found in mammalian genomes, and their protein products, p107 and p130 (each named

► **TABLE 22.2**

Inherited Cancer Syndromes

Syndrome	Primary Tumor	Gene	Chromosomal Location	Proposed Protein Function
Familial retinoblastoma	Retinoblastoma	*RB*	13q14.3	Cell cycle and transcriptional regulation
Li-Fraumeni syndrome	Sarcomas, breast cancer	*TP53*	17p13.1	Transcription factor
Familial adenomatous polyposis (FAP)	Colorectal cancer	*APC*	5q21	Regulation of β-catenin
Hereditary nonpolyposis colorectal cancer (HNPCC)	Colorectal cancer	*MSH2* *MLH1* *PMS1* *PMS2*	2p16 3p21 2q32 7p22	DNA mismatch repair
Neurofibromatosis type 1	Neurofibromas	*NF1*	17q11.2	Regulation of Ras-mediated signaling
Neurofibromatosis type 2	Acoustic neuromas, meningiomas	*NF2*	22q12.2	Linkage of membrane proteins to cytoskeleton
Wilms' tumor	Wilms' tumor	*WT1*	11p13	Transcriptional repressor
Familial breast cancer 1	Breast cancer	*BRCA1*	17q21	DNA repair
Familial breast cancer 2	Breast cancer	*BRCA2*	13q12	DNA repair
von Hippel-Lindau disease	Renal cancer	*VHL*	3p25	Regulation of transcriptional elongation
Familial melanoma	Melanoma	*p16*	9p21	Inhibitor of CDKs
Ataxia telangiectasia	Lymphoma	*ATM*	11q22	DNA repair
Bloom's syndrome	Solid tumors	*BLM*	15q26.1	DNA helicase

Source: Fearon, E. R. 1997. Human cancer syndromes: clues to the origin and nature of cancer. *Science* 278:1043–1050.

for its mass in kilodaltons), may also play key roles in cell-cycle regulation. No human tumors are known to have inactivating mutations in either of these two genes, and mice homozygous for a knockout mutation in either of them do not show abnormal phenotypes. However, mice that are homozygous for knockout mutations in both of these genes die shortly after birth. Thus, together the p107 and p130 members of the RB family of proteins are involved in important cellular processes.

Molecular and biochemical analyses have elucidated the role of pRB in cell-cycle regulation (**FIGURE 22.9**). Early in the G_1 phase of the cell cycle, pRB binds to the E2F proteins, a family of transcription factors that control the expression of several genes whose products move the cell through its cycle. When E2F transcription factors are bound to pRB, they cannot bind to specific enhancer sequences in their target genes. Consequently, the cell-cycle factors encoded by these genes are not produced, and the machinery for DNA synthesis and cell division remains quiescent. Later in G_1, pRB is phosphorylated through the action of cyclin-dependent kinases. In this changed state, pRB releases the E2F transcription factors that have bound to it. These released transcription factors are then free to activate their target genes, which encode proteins that induce the cell to progress through S phase and into mitosis. After mitosis, pRB is dephosphorylated, and each of the daughter cells enters the quiescent phase of a new cell cycle.

This orderly and rhythmic progression through the cell cycle is disrupted in cancer cells. In many types of cancer—not just retinoblastoma—both copies of the *RB* gene have been inactivated, either by deletions or by mutations that impair or abolish the ability of the RB protein to bind E2F transcription factors. The inability of pRB to bind to these transcription factors leaves them free to activate their target genes, thereby setting in motion the machinery for DNA synthesis and cell division. In effect, one of the natural brakes on the process of cell division has been released. In the absence of this brake, cells have a tendency to move through their cycle quickly. If other cell-cycle brakes fail, the cells divide ceaselessly to form tumors.

p53

The 53-kilodalton tumor suppressor protein p53 was discovered through its role in the induction of cancers by certain DNA viruses. This protein is encoded by a tumor suppressor gene called *TP53*. Inherited mutations in *TP53* are associated with the Li-Fraumeni syndrome, a rare dominant condition in which any of several different types of cancer may develop. Somatic mutations that inactivate both copies of the *TP53* gene are also associated with a variety of cancers. In fact, such mutations are found in a majority of all human tumors. Loss of p53 function is therefore a key step in carcinogenesis.

▶ FOCUS ON PROBLEM SOLVING
Estimating Mutation Rates in Retinoblastoma

THE PROBLEM

Alfred Knudson based his two-hit hypothesis of cancer on a statistical analysis of retinoblastoma. Patients with retinoblastoma (RB) may have tumors in one eye (unilateral RB) or in both eyes (bilateral RB), and within each eye, there may be more than one tumor. Among patients who had inherited an *RB* gene mutation from a parent, Knudson found that the average total number of tumors that formed was 3. Furthermore, he estimated that the total number of retinoblasts—the cells that form the embryonic retina—was about 2 million in each eye.

a. If each tumor in this group of patients is due to the occurrence of another *RB* gene mutation within the first two years of life—the second hit in Knudson's hypothesis—what is the somatic mutation rate for the *RB* gene per year?

b. Given this mutation rate, and assuming that mutations occur independently, would you ever expect a sporadic case of retinoblastoma—one in which both *RB* gene mutations occur after birth—to have the bilateral form of the disease?

c. Would you ever expect to find sporadic cases of unilateral retinoblastoma with more than one tumor in the affected eye?

d. Assuming that tumors form independently, what is the chance that a patient with three tumors will have all three tumors in one eye?

FACTS AND CONCEPTS

1. Retinoblastoma occurs when both *RB* genes have been inactivated by mutations.

2. One of these inactivating mutations may be inherited from a parent.

3. Sporadic cases of retinoblastoma occur when both of the inactivating mutations arise during eye development.

4. When two events are independent, we multiply their probabilities to obtain the probability that they will both occur.

5. When two events are mutually exclusive, we add their probabilities to obtain the probability that either one or the other will occur.

ANALYSIS AND SOLUTION

a. To estimate the somatic mutation rate, we need to count the number of mutational events in comparison to the total number of chances for such events. The average number of tumors (3) is an estimate of the average number of mutational events. The number of chances for such events is a function of the total number of genes that can mutate to produce a tumor: 1 RB^+ gene per cell in a patient that has already inherited one RB^- mutation from a parent $\times$ 2 $\times$ 10^6 cells per eye $\times$2 eyes per patient $= 4 \times 10^6$ chances for a mutational event. Thus, the mutation rate is $3/(4 \times 10^6) = 7.5 \times 10^{-7}$ mutations, or, on an annualized basis, 7.5×10^{-7} mutations/2 years $= 3.7 \times 10^{-7}$ mutations/year.

b. We would not expect sporadic cases to show bilateral retinoblastoma. In these cases, both of the mutations that cause a cell to form a tumor must occur somatically. The probability of two mutations occurring in the same cell is $(3.7 \times 10^{-7})^2 = 1.4 \times 10^{-13}$, and the probability of this double mutational event happening in two different cells, one in each eye, is $(1.4 \times 10^{-13})^2 = 2 \times 10^{-26}$, a very small number.

c. The same reasoning as in (b) applies, except that we do not specify that the two mutant cells arise in different eyes. The probability of more than one tumor forming in an individual that has not inherited a preexisting *RB* gene mutation is exceedingly small.

d. If three tumors form, the chance that they will all be in the right eye is $(1/2)^3$. Here the exponent is the number of tumors, and 1/2 is the chance that a particular tumor will be in one eye rather than the other. However, because there are two eyes, the total probability that all three tumors will be in one eye is $2 \times (1/2)^3 = 1/4$.

For further discussion go to your *WileyPLUS* course.

The p53 protein is a 393-amino-acid-long transcription factor that consists of three distinct domains: an N-terminal transcription-activation domain (TAD), a central DNA-binding core domain (DBD), and a C-terminal homo-oligomerization domain (OD) (**FIGURE 22.10*a***). Most of the mutations that inactivate p53 are located in the DBD. These mutations evidently impair or abolish the ability of p53 to bind to specific DNA sequences that are embedded in its target genes, thereby preventing the transcriptional activation of these genes. Thus, mutations in the DBD are typically recessive loss-of-function mutations. Other types of mutations are found in the OD portion of the polypeptide. Molecules of p53 with these types of mutations dimerize with wild-type p53 polypeptides and prevent the wild-type polypeptides from functioning as transcriptional activators. Thus, mutations in the OD have a *dominant negative* effect on p53 function.

The p53 protein plays a key role in cellular responses to stress (**FIGURE 22.10*b***). In normal cells the level of p53 is low, but when the cells are treated with a DNA-damaging agent such as radiation, the level of p53 increases dramatically. This

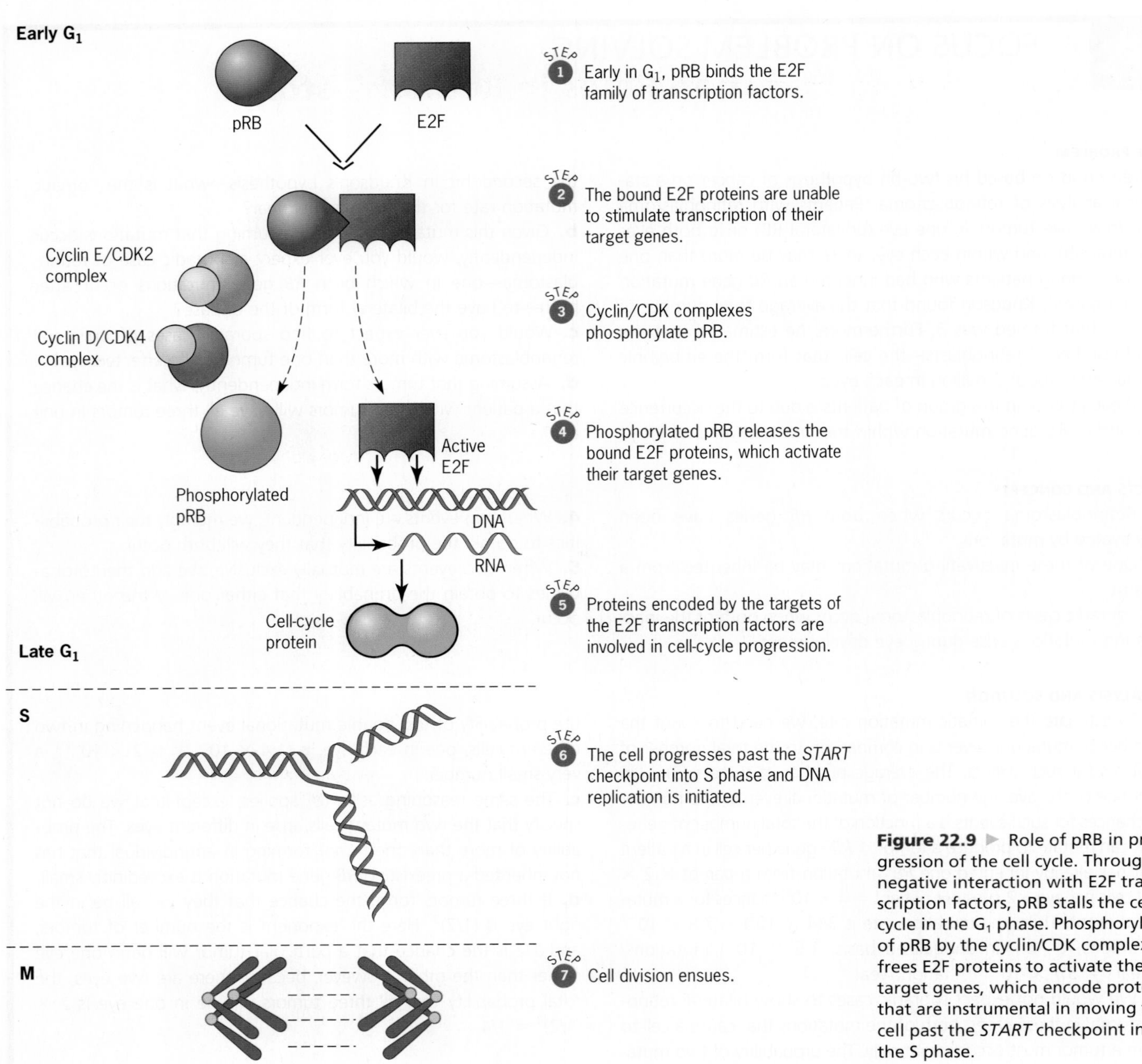

Figure 22.9 ▶ Role of pRB in progression of the cell cycle. Through its negative interaction with E2F transcription factors, pRB stalls the cell cycle in the G_1 phase. Phosphorylation of pRB by the cyclin/CDK complexes frees E2F proteins to activate their target genes, which encode proteins that are instrumental in moving the cell past the *START* checkpoint into the S phase.

response to DNA damage is mediated by a pathway that decreases the degradation of p53. In response to DNA damage, p53 is phosphorylated, converting it into a stable and active form. Once activated, p53 either stimulates the transcription of genes whose products arrest the cell cycle, thereby allowing the damaged DNA to be repaired, or it activates another set of genes whose products ultimately cause the damaged cell to die.

One prominent factor in the response that arrests the cell cycle is p21, a protein encoded by a gene that is activated by the p53 transcription factor. The p21 protein is an inhibitor of cyclin/CDK protein complexes. When p21 is synthesized in response to cell stress, the cyclin/CDK complexes are inactivated and the cell cycle is arrested. During this timeout, the cell's damaged DNA can be repaired. Thus, p53 is responsible for activating a brake on the cell cycle, and this brake allows the cell to maintain its genetic integrity. Cells that lack functional p53 have difficulty applying this brake. If these cells progress through the cell cycle and proceed into subsequent divisions, additional mutations that cause them to be unregulated may accumulate. Mutational inactivation of p53 is therefore often a key step in the pathway to cancer.

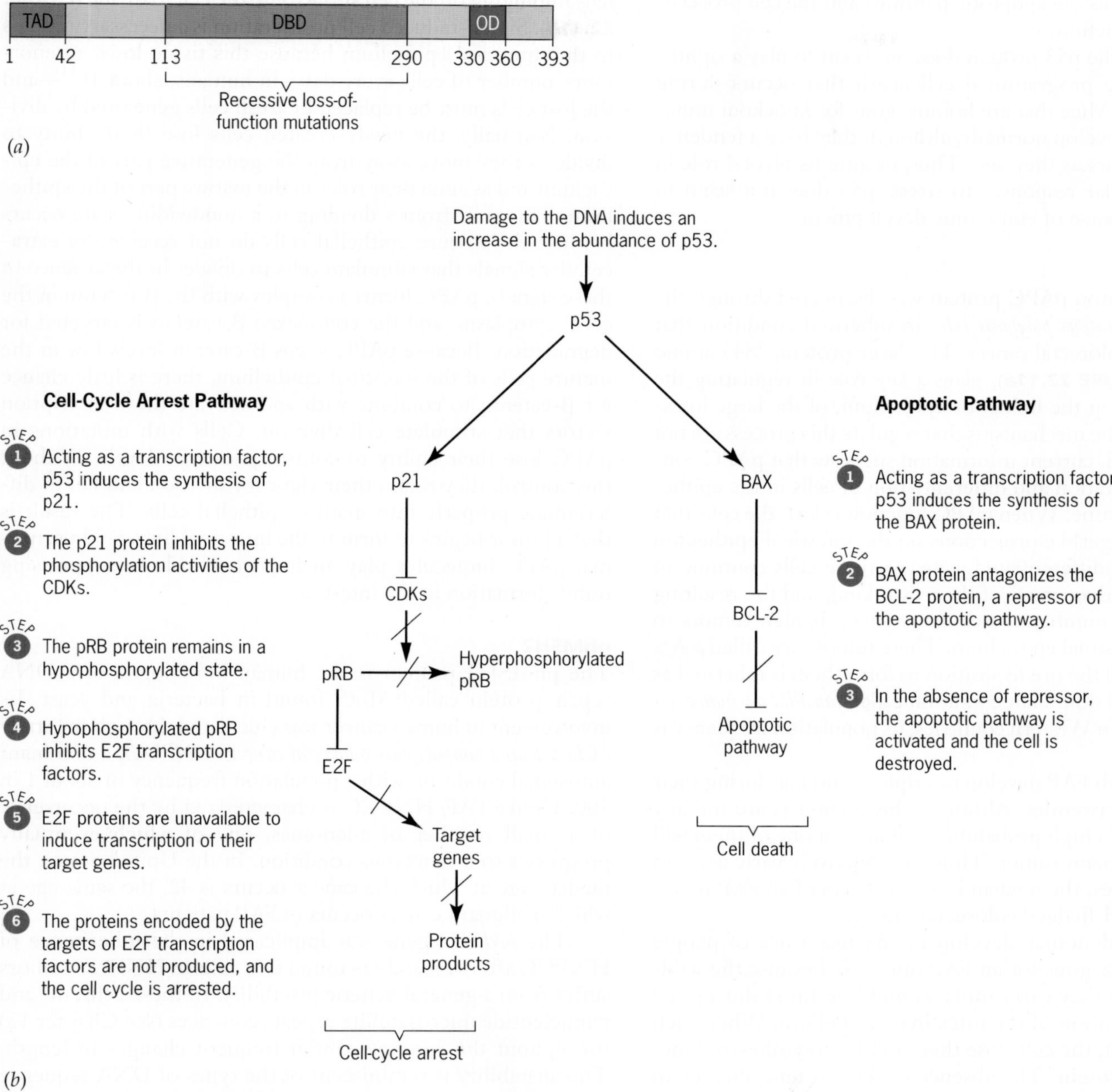

Figure 22.10 ▶ *(a)* Principal domains within p53. TAD = transcription-activation domain; DBD = DNA-binding domain; OD = oligomerization domain. The numbers refer to amino acid positions in the polypeptide. *(b)* Role of p53 in the cellular response to DNA damage. Two response pathways have been identified. Within each pathway, a pointed arrow indicates a positive influence or a directional change (e.g., a protein is synthesized or phosphorylated, a protein catalyzes a reaction, or a gene is expressed), and a blunted arrow indicates a negative influence (e.g., repression of protein synthesis or protein activity, or repression of a pathway). A slash through an arrow indicates that the influence—positive or negative—is blocked.

The p53 protein can also mediate another response to cell stress. Instead of orchestrating efforts to repair damage within a cell, p53 may trigger a suicidal response in which the damaged cell is programmed for destruction. The way in which p53 programs cell death is not well understood. One mechanism seems to involve the protein product of the *BAX* gene. The BAX protein is an antagonist of another protein called BCL-2, which normally suppresses the apoptotic, or cell-death, pathway.

When the *BAX* gene is activated by p53, its protein product releases the BCL-2 protein from its suppressing mode. This release then opens the apoptotic pathway, and the cell proceeds to its own destruction.

Curiously, the p53 protein does not seem to play a significant role in the programmed cell death that occurs during embryogenesis. Mice that are homozygous for knockout mutations in *TP53* develop normally, although they have a tendency to develop tumors as they age. Thus, despite its pivotal role in regulating cellular responses to stress, p53 does not seem to influence the course of embryonic development.

pAPC

The 310-kilodalton pAPC protein was discovered through the study of *adenomatous polyposis coli*, an inherited condition that often leads to colorectal cancer. This large protein, 2843 amino acids long (**FIGURE 22.11*a***), plays a key role in regulating the renewal of cells in the lining, or epithelium, of the large intestine. Although the mechanisms that regulate this process are not fully understood, current information suggests that pAPC controls the proliferation and differentiation of cells in the epithelium of the intestine. When pAPC function is lost, the cells that generate the fingerlike projections on the intestinal epithelium remain in an undifferentiated state. As these cells continue to divide, they produce more of their own kind, and the resulting increase in cell number causes many small, benign tumors to form in the intestinal epithelium. These tumors are called *polyps* or *adenomas*, and the predisposition to form them is inherited as a rare autosomal dominant condition called *familial adenomatous polyposis (FAP)*. In Western countries, its population frequency is about 1 in 7000.

Patients with FAP develop multiple adenomas during their teens and early twenties. Although the adenomas are initially benign, there is a high probability that at least one of them will become a malignant tumor. Thus, at a relatively early age—in the United States, the median is 42—carriers of an FAP mutation develop full-fledged colorectal cancer.

Multiple adenomas develop in the intestines of people who are heterozygous for an FAP mutation because the wild-type *APC* allele they carry mutates multiple times during the natural regeneration of the intestinal epithelium. When such mutations occur, the cells lose their ability to synthesize functional pAPC protein. The absence of this protein releases an important brake on cell proliferation, and cell division proceeds unchecked. Thus, the formation of numerous benign tumors in the intestines of FAP heterozygotes results from the independent occurrence of second mutational "hits" in the cells of the intestinal epithelium. Individuals who do not carry an FAP mutation seldom form multiple adenomas. However, they may produce one or a few adenomas if by chance both of their *APC* genes are inactivated by somatic mutations.

The pAPC protein appears to regulate cell division through its ability to bind β-catenin, a protein that is present inside cells. β-catenin naturally binds to other proteins as well, including certain transcription factors that stimulate the expression of genes whose protein products promote cell division. The interactions with these transcription factors are favored when signals impinging on the cell surface cue the cell to divide (**FIGURE 22.11*b***). Signal-induced cell proliferation is a necessary process in the intestinal epithelium because this tissue loses an enormous number of cells every day—in humans, about 10^{11}—and the lost cells must be replaced by fresh cells generated by division. Normally, the newly created cells lose their ability to divide as they move away from the generative part of the epithelium and assume their roles in the mature part of the epithelium. This shift from a dividing to a nondividing state occurs because the mature epithelial cells do not receive the extracellular signals that stimulate cells to divide. In the absence of these signals, pAPC forms a complex with the β-catenin in the cells' cytoplasm, and the complexed β-catenin is targeted for degradation. Because pAPC keeps β-catenin levels low in the mature cells of the intestinal epithelium, there is little chance for β-catenin to combine with and activate the transcription factors that stimulate cell division. Cells with mutations in pAPC lose their ability to control β-catenin levels. Without this control, they retain their vigor for division and fail to differentiate properly into mature epithelial cells. The result is that a tumor begins to form in the intestinal lining. Thus, normal pAPC molecules play an important role in suppressing tumor formation in the intestine.

phMSH2

The phMSH2 protein is the human homologue of a DNA repair protein called MutS found in bacteria and yeast. Its involvement in human cancer was elucidated through the study of *hereditary nonpolyposis colorectal cancer (HNPCC)*, a dominant autosomal condition with a population frequency of about 1 in 500. Unlike FAP, HNPCC is characterized by the occurrence of a small number of adenomas, one of which eventually progresses to a cancerous condition. In the United States, the median age at which the cancer occurs is 42, the same age at which malignant cancer occurs in FAP patients.

The *hMSH2* gene was implicated in the inheritance of HNPCC after researchers found that cells in HNPCC tumors suffer from a general genetic instability. In these cells, di- and trinucleotide microsatellite repeat sequences (see Chapter 13) throughout the genome exhibit frequent changes in length. This instability is reminiscent of the types of DNA sequence changes observed in bacteria with mutations in the genes that control DNA mismatch repair (see Chapter 13). The human homologue of one of these bacterial genes maps to the short arm of chromosome 2, a chromosome that had previously been implicated in HNPCC by linkage analysis. Sequence analysis of this gene—denoted *hMSH2*—indicated that it was inactivated in tumors removed from some HNPCC patients. Thus, loss of *hMSH2* function was causally connected to the genome-wide instability observed in HNPCC tumors. Subsequent analysis has demonstrated that germ-line mutations in *hMSH2*, or in three other human homologues of bacterial mismatch repair genes, account for the inherited cases of HNPCC.

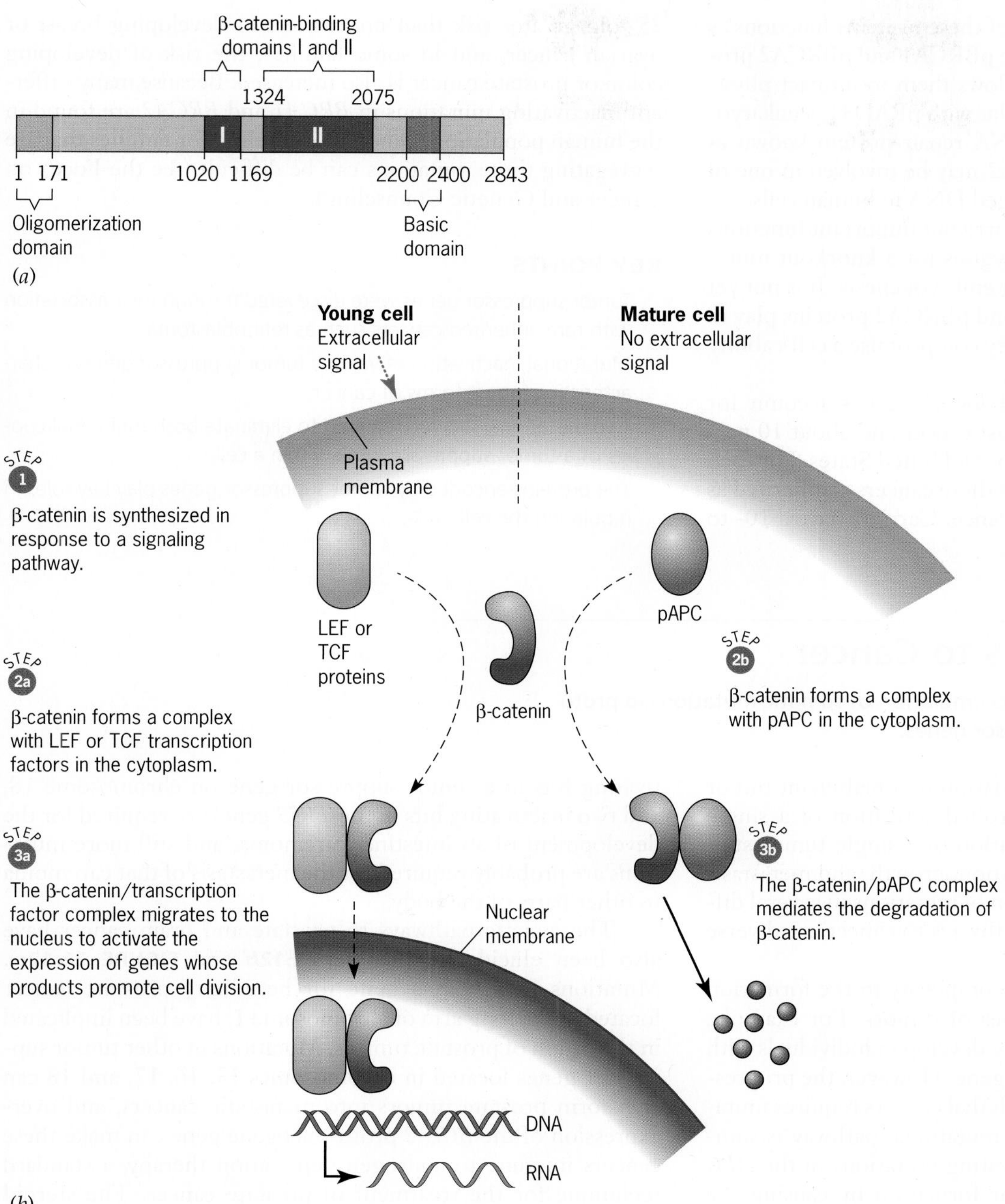
β-catenin-binding
domains I and II
1324
2075
I
II
1 171
1020 1169
2200 2400 2843
Oligomerization
domain
Basic
domain
(a)
Young cell
Extracellular
signal
Mature cell
No extracellular
signal
Plasma
membrane
STEP 1
β-catenin is synthesized in
response to a signaling
pathway.
LEF or
TCF
proteins
β-catenin
pAPC
STEP 2a
β-catenin forms a complex
with LEF or TCF transcription
factors in the cytoplasm.
STEP 2b
β-catenin forms a complex
with pAPC in the cytoplasm.
STEP 3a
The β-catenin/transcription
factor complex migrates to the
nucleus to activate the
expression of genes whose
products promote cell division.
STEP 3b
The β-catenin/pAPC complex
mediates the degradation of
β-catenin.
Nuclear
membrane
DNA
RNA
(b)

Figure 22.11 ▶ (*a*) Principal domains within pAPC. The numbers refer to amino acid positions in the polypeptide. (*b*) Role of pAPC in cell-cycle control. The pAPC protein influences progression through the cell cycle by interacting with β-catenin, a protein that can activate LEF or TCF transcription factors. In young cells (steps 2a, 3a), an extracellular signal activates these transcription factors and cell division is stimulated. In mature cells (steps 2b, 3b), interactions between pAPC and β-catenin prevent the transcription factors from being activated and cell division is inhibited.

pBRCA1 and pBRCA2

Mutant versions of the tumor suppressor genes *BRCA1* and *BRCA2* genes have been implicated in hereditary breast and ovarian cancer. *BRCA1* was mapped to chromosome 17 in 1990 and isolated in 1994 (see A Milestone in Genetics: The Identification of the *BRCA1* Gene later in this chapter), and *BRCA2* was mapped to chromosome 13 in 1994 and isolated in 1995. Both genes encode large proteins; pBRCA1 is a 220-kilodalton polypeptide, and pBRCA2 is a 384-kilodalton polypeptide. Cellular and biochemical studies have shown that each of these proteins is located within the nuclei of normal cells and that each contains a putative transcriptional activation domain.

However, it is not known if either of these proteins functions as a bona fide transcription factor. The pBRCA1 and pBRCA2 proteins also contain a domain that allows them to interact physically with other proteins, in particular with pRAD51, a eukaryotic homologue of the bacterial DNA repair protein known as RecA. Thus, pBRCA1 and pBRCA2 may be involved in one of the many systems that repair damaged DNA in human cells.

Both pBRCA1 and pBRCA2 carry out important functions within cells. Mice that are homozygous for a knockout mutation in either gene die early during embryogenesis. It is not yet clear what roles mutant pBRCA1 and pBRCA2 proteins play in human carcinogenesis. Possibly they compromise a cell's ability to detect or repair damaged DNA.

Mutations in the *BRCA1* and *BRCA2* genes account for about 7 percent of all cases of breast cancer and about 10 percent of all cases of ovarian cancer in the United States. For each gene, the predisposition to develop these cancers is inherited as a dominant allele with high penetrance. Carriers have a 10- to 25-fold greater risk than noncarriers of developing breast or ovarian cancer, and in some families, the risk of developing colon or prostate cancer is also increased. Because many different inactivating mutations in *BRCA1* and *BRCA2* are found in the human population, genetics counseling for families that are segregating these mutations can be difficult (see the Focus on Cancer and Genetic Counseling).

KEY POINTS

- Tumor suppressor genes were discovered through their association with rare, inherited cancers such as retinoblastoma.
- Mutational inactivation of various tumor suppressor genes is characteristic of most forms of cancer.
- Two mutational hits are required to eliminate both functional copies of a tumor suppressor gene within a cell.
- The proteins encoded by tumor suppressor genes play key roles in regulating the cell cycle.

Genetic Pathways to Cancer

Cancers develop through an accumulation of somatic mutations in proto-oncogenes and tumor suppressor genes.

In most cancer cases, the formation of a malignant tumor is not attributable to the uncontrolled activation of a single proto-oncogene or to the inactivation of a single tumor suppressor gene. Rather, tumor formation, growth, and metastasis usually depend on the accumulation of mutations in several different genes. Thus, the genetic pathways to cancer are diverse and complex.

We can see this diversity and complexity in the formation and development of different types of tumors. For example, benign tumors of the large intestine develop in individuals with inactivating mutations in the *APC* gene. However, the progression of these tumors to potentially lethal cancers requires mutations in several other genes. This mutational pathway is summarized in **FIGURE 22.12*a***. Inactivating mutations in the *APC* gene initiate the process of tumor formation by causing the development of abnormal tissues within the intestinal epithelium. These abnormal tissues contain dysplastic cells—cells with unusual shapes and enlarged nuclei—that may grow into early-stage adenomas. If the *K-ras* proto-oncogene is activated in one of these adenomas, that adenoma may grow and develop more fully. Inactivating mutations in any of several tumor suppressor genes located in the long arm of chromosome 18 may then induce the adenoma to progress further, and inactivating mutations in the *TP53* tumor suppressor gene on chromosome 17 may transform it into a vigorously growing carcinoma. Additional tumor suppressor gene mutations may allow carcinoma cells to break away and invade other tissues. Thus, no less than seven independent mutations (two inactivating hits in the *APC* gene, one activating mutation in the *K-ras* gene, two inactivating hits in a tumor suppressor gene on chromosome 18, and two inactivating hits in the *TP53* gene) are required for the development of an intestinal carcinoma, and still more mutations are probably required for the metastasis of that carcinoma to other parts of the body.

The genetic pathways to prostate and brain cancer have also been elucidated (**FIGURE 22.12*b*** and **FIGURE 22.12*c***). Mutations in *HPC1*, a gene for hereditary prostate cancer located in the long arm of chromosome 1, have been implicated in the origin of prostate tumors. Mutations in other tumor suppressor genes located in chromosomes 13, 16, 17, and 18 can transform prostate tumors into metastatic cancers, and overexpression of the *BCL-2* proto-oncogene gene can make these cancers immune to androgen deprivation therapy, a standard technique for the treatment of prostate cancer. The steroid hormone androgen is required for the proliferation of cells in the prostate epithelium. In the absence of androgen, these cells are programmed to die. However, prostate tumor cells may acquire the ability to survive in the absence of androgen, probably because an excess of the *BCL-2* gene product represses the programmed cell death pathway. Prostate cancers that have progressed to the stage of androgen independence are almost always fatal.

Glioblastomas are tumors that develop from glial cells in the brain. Primary glioblastomas usually occur in older individuals; they grow rapidly and are almost always fatal. Secondary glioblastomas usually occur in children and young adults; they develop from preexisting, low-grade tumors called astrocytomas and grow slowly—a feature that makes them more

▶ FOCUS ON **Cancer and Genetic Counseling**[1]

The identification of inherited mutations in tumor suppressor genes has opened a new era in genetics counseling. The carriers of such mutations are often at high risk to develop potentially life-threatening tumors, sometimes at a relatively early age. If molecular tests reveal that an individual carries a mutant tumor suppressor gene, medical treatment can be given to reduce the chance that he or she will develop a lethal cancer. For example, a child who carries a mutation in the *APC* gene could be checked periodically by endoscopy and suspicious lesions in the intestine could be removed, or a woman who carries a mutation in either of the *BRCA* genes could undergo a prophylactic mastectomy (removal of the breasts) or oophorectomy (removal of the ovaries).

A negative result from a test for a mutant tumor suppressor gene would, of course, be a cause for celebration—at least to the extent that the test can be trusted. For a large gene with many different mutant alleles segregating in the population, it is difficult to design a cost-effective test to detect mutations located anywhere in the gene. Typically, these tests are based on the polymerase chain reaction, and most of them are designed to detect specific mutant alleles. An individual who is at risk to carry a mutant tumor suppressor gene can be tested for the known mutations—at least the most frequent ones. However, a negative result is not definitive because that individual could carry a "*private*" mutation—that is, one that has not previously been identified in the population.

The existence of private alleles makes counseling for inherited cancers difficult. For example, over 300 different mutations have been identified in the *BRCA1* gene, and about 50 percent of them are private. If an individual with a family history of breast cancer comes to a genetic counselor for evaluation, which mutations should the counselor look for? Sometimes data from other family members or information collected from the individual's ethnic group can provide clues. If other individuals in the family have been found to carry a particular mutant allele, then the counselor should test for that allele. If certain mutant alleles are characteristic of the individual's ethnic group, then the counselor should test for them. In Ashkenazi Jewish populations, for example, some *BRCA1* and *BRCA2* mutant alleles have frequencies as high as 2.5 percent. By comparison, the combined frequency of all mutant alleles in non-Jewish Caucasian populations is only 0.1 percent. Thus, an Ashkenazi Jew at risk for inherited breast or ovarian cancer should be tested for the mutant alleles that are likely to be segregating in Ashkenazi Jewish families.

Genetic testing for mutant tumor suppressor genes raises a host of psychological issues. In cases where therapeutic medical treatment is not available, an individual might choose not to be tested because the psychological burden of living with the knowledge that one carries a potentially lethal mutant gene could be overwhelming. Knowledge that one is a carrier might be expected to influence career plans and decisions about marriage and child-bearing. The prospect of an early death might dissuade an individual from seeking permanent commitments—to a spouse, to children, or to a vocation—and the chance of transmitting a mutant allele to children might deter the individual from reproducing. Knowledge that one is a carrier can also influence other people—family members, friends, and coworkers. A young daughter whose mother has tested positively for a *BRCA1* mutation must herself begin to grapple with the prospect of being a carrier, and a husband whose wife carries a *BRCA1* mutation must share in the decision of whether or not she should undergo a prophylactic oophorectomy and preclude the couple from ever having children of their own.

Testing for mutant tumor suppressor genes raises many ethical issues. To whom should the test results be revealed? the patient? the patient's family? parents? children? employer? landlord? insurance agent? What measures should society take to safeguard the privacy of genetic test results? What policies should governments adopt to protect individuals from discrimination on the basis of their genotypes? How should insurance and employment policies be modified? Should the reproductive rights of individuals who carry harmful mutations be limited? As with any technological advance, the ability to detect mutations in tumor suppressor genes leaves us with many questions about how we should proceed. Currently, the answers to these questions are far from clear.

[1]Ponder, Bruce. 1997. Genetic testing for cancer risk. *Science* 278:1050–1054.

treatable than primary glioblastomas. Two proto-oncogenes have been implicated in the formation of glioblastomas. The gene for the epidermal growth factor receptor (EGFR) is frequently overexpressed in primary glioblastomas, and the gene for the platelet-derived growth factor (PDGF) protein is often overexpressed in secondary glioblastomas. Mutations in a variety of tumor suppressor genes, including *TP53*, *RB*, and *NF2*, have been implicated in the formation of glioblastomas, mainly in the development of secondary glioblastomas from low-grade astrocytomas. Many other tumor suppressor genes appear to be involved in the formation of glioblastomas, but they have not yet been identified.

Douglas Hanahan and Robert Weinberg have proposed six hallmarks of the pathways leading to malignant cancer:

1. *Cancer cells acquire self-sufficiency in the signaling processes that stimulate division and growth.* This self-sufficiency may arise from changes in the extracellular factors that cue cells to divide, or from changes in any part of the system that transduces these cues or translates their instructions into action

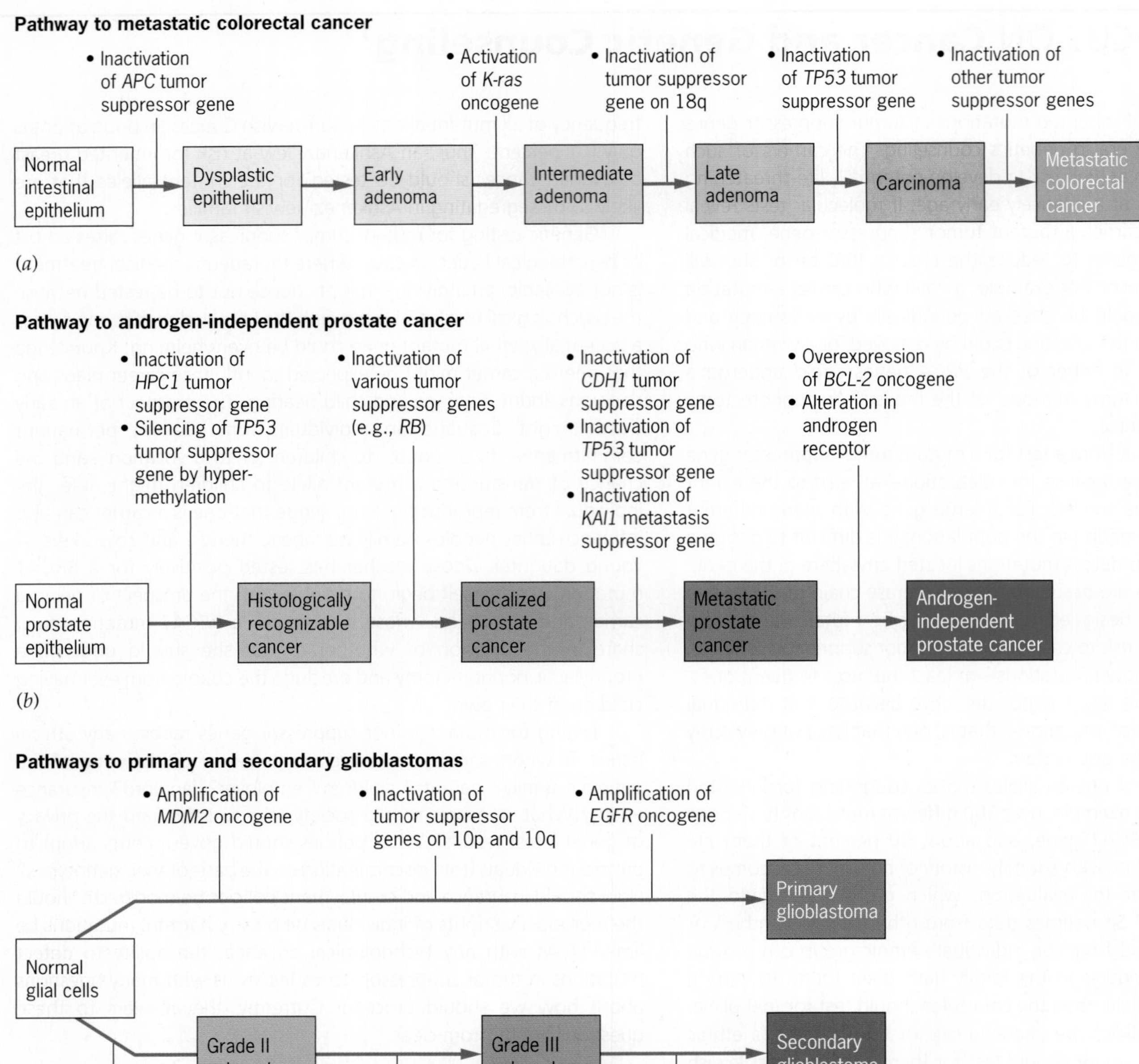

Figure 22.12 ▶ Genetic pathways to cancer.

inside the cell. In the most extreme case, self-sufficiency occurs when cells respond to growth factors that they themselves produce, thereby creating a positive feedback loop that stimulates ceaseless cell division.

2. *Cancer cells are abnormally insensitive to signals that inhibit growth.* Cell division is stimulated by a variety of biochemical signals; however, other signals inhibit cell division. In normal cells, these countervailing factors balance each other

▶ A MILESTONE IN GENETICS: The Identification of the *BRCA1* Gene

In 1994, research groups from the National Institute of Environmental Health Sciences in Research Triangle, North Carolina, McGill University in Montreal, Canada, the University of Utah in Salt Lake City, Myriad Genetics, a company based in Salt Lake City, and Eli Lily, the pharmaceutical giant based in Indianapolis, Indiana—45 people in all—published an article reporting the identification of a gene involved in hereditary breast and ovarian cancer.[1]

This gene had been long sought. In the early 1980s, evidence began to suggest that one or more genes predispose women to develop breast or ovarian cancer. Often these cancers appear early in adult life, and often they are fatal. Analysis of pedigrees showing the inheritance of these cancers was complicated by variation in the age of onset, by incomplete penetrance, and by the sex-limited nature of the trait. Nevertheless, in 1990 one of the cancer-predisposing genes was mapped to chromosome 17. Further progress refined the location of this gene, designated *BRCA1*, to the middle of the chromosome's long arm. Eventually researchers were able to situate *BRCA1* within a region containing roughly 600 kilobases of DNA. Then the problem was to identify which segment of the 600-kb-long region was actually the *BRCA1* gene.

To tackle this problem, the consortium of research groups collected clones of DNA from within the candidate region. All sorts of clones were obtained—large clones propagated in yeast artificial chromosomes (YACs) and bacterial artificial chromosomes (BACs), and smaller clones propagated in bacteriophages and cosmids. The collection of clones enabled the researchers to construct a physical map of the candidate region. Points along this map were then analyzed for expression as gene transcripts. A total of 65 expressed sequences were identified. DNA corresponding to the coding portions of these transcripts was analyzed in the hope of finding mutations in individuals who were thought to carry a predisposing allele of the breast cancer gene. The evidence from several putative carriers indicated that *BRCA1* is in the middle of the 600-kb region. It comprises 22 exons spread over 100 kb, and its 7.8-kb-long mRNA encodes a polypeptide of 1863 amino acids.

The researchers found four mutations within the candidate coding sequence: (1) an 11-base pair deletion in exon 2 that causes a reading frame shift, (2) a nonsense mutation in exon 11 that causes premature termination of polypeptide synthesis, (3) a 1-base pair insertion in exon 20 that causes a reading frame shift, and (4) a missense mutation in exon 21 that substitutes arginine for methionine in the protein. In addition, they found a case in which the candidate gene's transcript was absent, possibly because a regulatory mutation had occurred in some noncoding element. Each of these five mutations was from a different family.

Putative carriers from three other families did not show mutations in the candidate gene's coding sequence. However, in these families, the evidence for linkage between the cancer-predisposing gene and molecular markers in the long arm of chromosome 17 was weak. Thus, they may represent cases in which a gene other than *BRCA1* is involved. Indeed, at the end of their paper, the researchers noted that in one of these families, the predisposition to develop breast cancer might be linked to *BRCA2*, a gene on a different chromosome. Alternately, these three families might have cancer-predisposing mutations in the noncoding elements of *BRCA1*.

Since 1994 many more mutations have been identified in the *BRCA1* gene. The discovery of these mutations has made it possible to develop DNA tests that can help to identify women who, because of their genotypes, are at high risk to develop breast and ovarian cancer. Such tests represent the first steps in combating this deadly disease.

QUESTIONS FOR DISCUSSION

1. The *BRCA1* gene was identified by a consortium of researchers who were supported by a combination of public and private funds. One of the principal investigators, Mark Skolnick, was both a professor at the University of Utah, where he received government grants, and an officer in a biotech company, Myriad Genetics, which has subsequently developed and marketed tests for mutant *BRCA1* alleles. What ethical issues are involved when research is jointly funded by government grants and company money, especially when the company ultimately stands to profit from the research?
2. A woman who tests positive for a predisposing mutant allele of the *BRCA1* gene has a high probability (>80 percent) of developing breast or ovarian cancer—and of dying from one of these types of cancer. What measures are available to prevent these cancers from occurring? If you were such a woman, what would you do?

[1]Miki, Y., et al. 1994. A strong candidate for the breast and ovarian cancer susceptibility gene *BRCA1*. *Science* 266:66–71.

with the result that growth occurs in a regulated manner. In cancer cells, growth is unregulated because the stimulatory signals have the upper hand. During the progression to malignancy, cancer cells lose their ability to respond appropriately to signals that inhibit growth. For example, cells in intestinal adenomas often no longer respond to TGFβ, a protein that instructs pRB to block progression through the cell cycle. When this block fails, the cells advance from G_1 into S, replicate their DNA, and divide. These cells are then on their way to forming a malignant tumor.

3. *Cancer cells can evade programmed cell death.* As we have seen, p53 plays a key role in protecting an organism from the accumulation of damaged cells that could endanger its life. Through mechanisms that are still incompletely understood, p53 sends damaged cells into an autodestruct pathway that clears them from the organism. When p53 malfunctions, this autodestruct pathway is blocked, and the damaged cells survive and multiply. Such cells are likely to produce descendants that are even more abnormal than they are. Consequently, lineages derived from damaged cells are prone to advance to a cancerous state. The ability to evade programmed cell death is therefore a key characteristic in the progression to malignant cancer.

4. *Cancer cells acquire limitless replicative potential.* Normal cells are able to divide around 60 to 70 times. This limitation arises from the minute, but inexorable, loss of DNA from the ends of chromosomes every time the DNA is replicated (Chapter 10). The cumulative effect of this loss enforces a finite reproductive ability on every cell lineage. Cells that go past the reproductive limit become genetically unstable and die. Cancer cells manage to transcend this limit by replenishing their lost DNA. They do so by increasing the activity of the enzyme telomerase, which adds DNA sequences to the ends of chromosomes. When cells have acquired limitless replicative potential by overcoming the loss of DNA at the ends of chromosomes, they are said to be *immortalized.*

5. *Cancer cells develop ways to nourish themselves.* Any tissue in a complex, multicellular organism needs a vascular system to bring nutrients to it. In humans and other vertebrate animals, the circulatory system provides this function. The cells in premalignant tumors fail to grow aggressively because they are not directly fed by the circulatory system. However, when blood vessels are induced to grow among these cells—through a process called **angiogenesis**—the tumor is nourished and can then expand. Thus, a key step in the progression to malignant cancer is the induction of blood vessel growth by the cells of the tumor. Many factors that induce or inhibit angiogenesis are known. In normal tissues, these factors are kept in balance so that blood vessels grow appropriately in the body; in cancerous tissues, the balance is tipped in favor of the inducing factors, which act to stimulate blood vessel development. Once capillaries have grown into a tumor, a reliable means of nourishment is at hand. The tumor can then feed itself and grow to a size where it becomes a danger to the organism.

6. *Cancer cells acquire the ability to invade other tissues and colonize them.* More than 90 percent of all cancer deaths are caused by metastasis of the cancer to other parts of the body. When tumors metastasize, the cancer cells detach from the primary tumor and travel through the bloodstream to another location, where they establish a new, lasting, and, in the end, lethal, relationship with the surrounding cells. Profound changes must take place on the surfaces of the cancer cells for this process to occur. When it does, secondary tumors may develop in tissues far removed from the primary tumor. Cancers that have spread in this fashion are extremely difficult to control and eradicate. Metastasis is therefore the most serious occurrence in the progression of a cancer.

Numerous studies have established that somatic mutation is the basis for the development and progression of all types of cancer. As a cancer progresses on the pathway to malignancy, its cells become increasingly unregulated. Mutations accumulate, and whole chromosomes or chromosome segments may be lost. This genetic instability increases the likelihood that the cancer will develop each of the hallmarks discussed above.

Because of the importance of somatic mutations in the etiology of cancer, factors that increase the mutation rate are bound to increase the incidence of cancer. Today many countries maintain research programs to identify mutagenic and carcinogenic agents (see Chapter 13 for a discussion of the Ames test to identify chemical mutagens). When such agents are identified, public health authorities devise policies to minimize human exposure to them. However, no environment is carcinogen-free, and human behaviors that contribute to the risk of cancer such as smoking, excessive exposure to sunlight, and consumption of fatty foods that contain little fiber are difficult to change. Understanding of the processes that cause cancer has advanced significantly during the last decade. In the next decade, we can expect this understanding to lead to more effective strategies for cancer prevention and treatment.

KEY POINTS

- Different types of cancer are associated with mutations in different genes.
- Cancer cells may stimulate their own growth and division.
- Cancer cells do not respond to factors that inhibit cell growth.
- Cancer cells can evade the natural mechanisms that kill abnormal cells.
- Immortalized cancer cells can divide endlessly.
- Tumors can expand when they induce the in-growth of blood vessels to nourish their cells.
- Metastatic cancer cells can invade other tissues and colonize them.

▶ Basic Exercises

ILLUSTRATE BASIC GENETIC ANALYSIS

1. Which cell-cycle checkpoint prevents a cell from replicating damaged DNA?

Answer: The *START* checkpoint in mid-G_1 of the cell cycle.

2. (a) In which class of genes do dominant gain-of-function mutations cause cancer? (b) In which class of genes do recessive loss-of-function mutations cause cancer?

Answer: (a) Oncogenes. (b) Tumor suppressor genes.

3. Why do some chromosomal rearrangements lead to cancer?

Answer: The breakpoints of these rearrangements often juxtapose a cellular oncogene to a promoter that stimulates the vigorous expression of the oncogene. Overexpression of the gene product can lead to excessive cell division and growth.

4. Intestinal cancer occurs in individuals with inactivating mutations in the *APC* gene. Explain how it might also occur in individuals with mutations in the β-catenin gene.

Answer: A mutation that specifically prevented β-catenin from binding to pAPC might lead to cancer. β-catenin that cannot bind to pAPC would be available to bind to the transcription factors that stimulate the expression of genes whose products promote cell division and growth.

5. Which tumor suppressor gene is most frequently mutated in human cancers?

Answer: *TP53*, the gene that encodes p53.

▶ Testing Your Knowledge

INTEGRATE DIFFERENT CONCEPTS AND TECHNIQUES

1. An oncogene within the genome of a retrovirus has a high probability of causing cancer, but an oncogene in its normal chromosomal position does not. If these two oncogenes encode exactly the same polypeptide, how can we explain their different properties?

Answer: There are at least three possibilities. One (*a*) is that the virus simply adds extra copies of the oncogene to the cell and that collectively these produce too much of the polypeptide. An excess of polypeptide might cause uncontrolled cell division, that is, cancer.

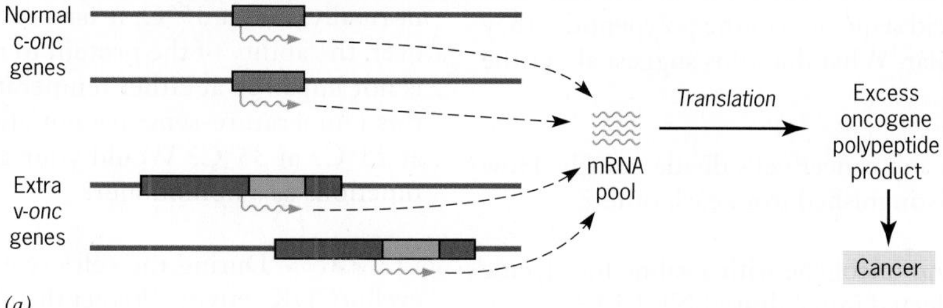
The virus adds extra copies of the oncogene to the cell
Normal c-onc genes
Extra v-onc genes
mRNA pool
Translation
Excess oncogene polypeptide product
Cancer
(*a*)

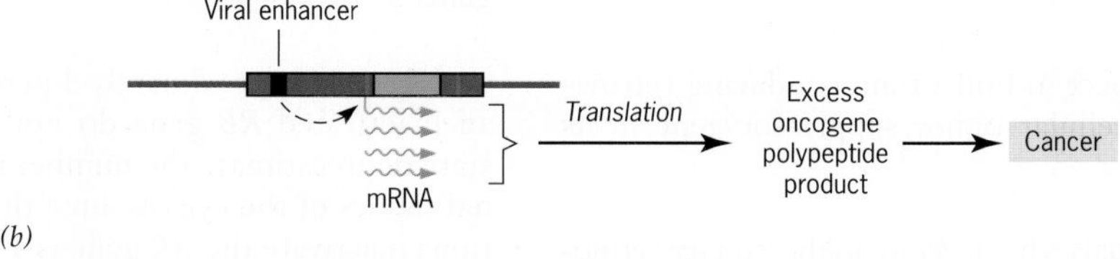
The viral oncogene is expressed inappropriately under the control of a viral enhancer
Viral enhancer
mRNA
Translation
Excess oncogene polypeptide product
Cancer
(*b*)

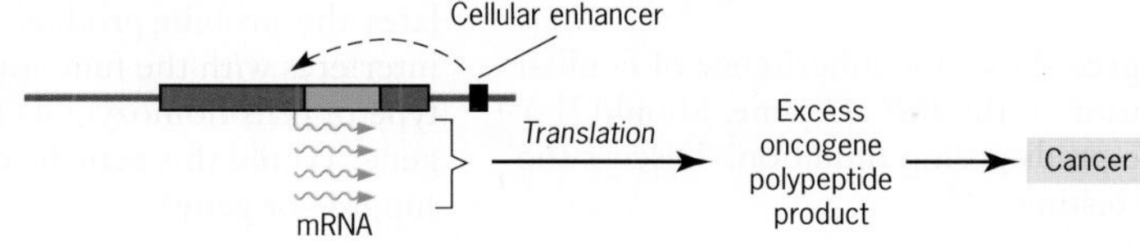
The viral oncogene is expressed inappropriately under the control of a cellular enhancer
Cellular enhancer
mRNA
Translation
Excess oncogene polypeptide product
Cancer
(*c*)

Another possibility (*b*) is that the viral oncogene is expressed inappropriately under the control of enhancers in the viral DNA. These enhancers might trigger the oncogene to be expressed at the wrong time or to be overexpressed constitutively. In either case, the polypeptide would be inappropriately produced and might thereby upset the normal controls on cell division. A third possibility (*c*) is that integration of the virus into the chromosomes of the infected cell might put the viral oncogene in the vicinity of an enhancer in the chromosomal DNA and that this enhancer might elicit inappropriate expression. All three explanations stress the idea that the expression of an oncogene must be correctly regulated. Misexpression or overexpression could lead to uncontrolled cell division.

▶ Questions and Problems

ENHANCE UNDERSTANDING AND DEVELOP ANALYTICAL SKILLS

22.1 Most cancer cells are aneuploid. Suggest how aneuploidy might contribute to deregulation of the cell cycle.

22.2 How might the absence of introns in a retroviral oncogene explain that gene's overexpression in the tissues of an infected animal?

22.3 Many cancers seem to involve environmental factors. Why, then, is cancer called a genetic disease?

22.4 GO A mutation in the *ras* cellular oncogene can cause cancer when it is in heterozygous condition, but a mutation in the *RB* tumor suppressor gene can cause cancer only when it is in homozygous condition. What does this difference between dominant and recessive mutations imply about the roles that the *ras* and *RB* gene products play in normal cellular activities?

22.5 When cellular oncogenes are isolated from different animals and compared, the amino acid sequences of the polypeptides they encode are found to be very similar. What does this suggest about the functions of these polypeptides?

22.6 Both embryonic cells and cancer cells divide quickly. How can these two types of cells be distinguished from each other?

22.7 When a mutant *c-H-ras* oncogene with a valine for glycine substitution in codon 12 is transfected into cultured NIH 3T3 cells, it transforms those cells into cancer cells. When the same mutant oncogene is transfected into cultured embryonic cells, it does not transform them. Why?

22.8 Would you ever expect to find a tumor-inducing retrovirus that carried a processed cellular tumor suppressor gene in its genome?

22.9 Explain why individuals who develop nonhereditary retinoblastoma usually have tumors in only one eye, whereas individuals with hereditary retinoblastoma usually develop tumors in both eyes.

22.10 The following pedigree shows the inheritance of familial ovarian cancer caused by a mutation in the *BRCA1* gene. Should II-1 be tested for the presence of the predisposing mutation? Discuss the advantages and disadvantages of testing.

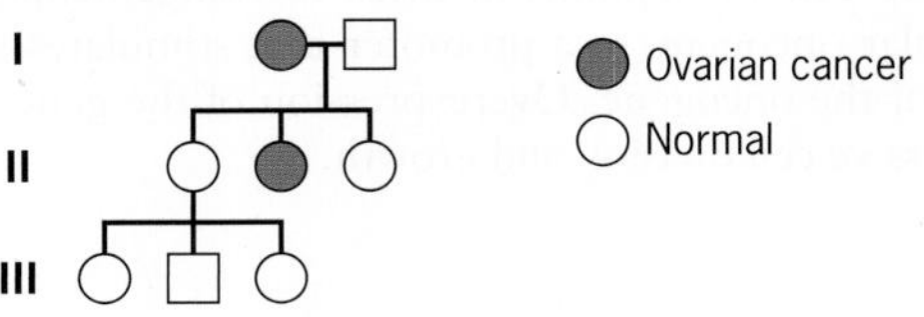

22.11 Inherited cancers like retinoblastoma show a dominant pattern of inheritance. However, the underlying genetic defect is a recessive loss-of-function mutation—often the result of a deletion. How can the dominant pattern of inheritance be reconciled with the recessive nature of the mutation?

22.12 A particular E2F transcription factor recognizes the sequence TTTCGCGC in the promoter of its target gene. A temperature-sensitive mutation in the gene encoding this E2F transcription factor alters the ability of its protein product to activate transcription; at 25°C the mutant protein activates transcription normally, but at 35°C, it fails to activate transcription at all. However, the ability of the protein to recognize its target DNA sequence is not impaired at either temperature. Would cells heterozygous for this temperature-sensitive mutation be expected to divide normally at 25°C? at 35°C? Would your answers change if the E2F protein functions as a homodimer?

22.13 During the cell cycle, the p16 protein is an inhibitor of cyclin/CDK activity. Predict the phenotype of cells homozygous for a loss-of-function mutation in the gene that encodes p16. Would this gene be classified as a proto-oncogene or as a tumor suppressor gene?

22.14 Approximately 4 percent of the individuals who inherit an inactivated *RB* gene do not develop retinoblastoma. Use this statistic to estimate the number of cell divisions that form the retinal tissues of the eye. Assume that the rate at which somatic mutations inactivate the *RB* gene is 1 mutation per 10^8 cell divisions.

22.15 The protein product of the *BAX* gene negatively regulates the protein product of the *BCL-2* gene—that is, BAX protein interferes with the function of the BCL-2 protein. Predict the phenotype of cells homozygous for a loss-of-function mutation in the *BAX* gene. Would this gene be classified as a proto-oncogene or as a tumor suppressor gene?

22.16 Mice homozygous for a knockout mutation of the *TP53* gene are viable. Would they be expected to be more or less sensitive to the killing effects of ionizing radiation?

22.17 Suppose that a cell is heterozygous for a mutation that caused p53 to bind tightly and constitutively to the DNA of its target genes. How would this mutation affect the cell cycle? Would such a cell be expected to be more or less sensitive to the effects of ionizing radiation?

22.18 GO Mice that are heterozygous for a knockout mutation in the *RB* gene develop pituitary and thyroid tumors. Mice that are homozygous for this mutation die during embryonic development. Mice that are homozygous for a knockout mutation in the gene encoding the p130 homologue of RB and heterozygous for a knockout mutation in the gene encoding the p107 homologue of RB do not have a tendency to develop tumors. However, homozygotes for knockout mutations in both of these genes die during embryonic development. What do these findings suggest about the roles of the *RB*, *p139*, and *p107* genes in embryos and adults?

22.19 Would cancer-causing mutations of the *APC* gene be expected to increase or decrease the ability of pAPC to bind β-catenin?

22.20 The *BCL-2* gene encodes a protein that represses the pathway for programmed cell death. Predict the phenotype of cells heterozygous for a dominant activating mutation in this gene. Would the *BCL-2* gene be classified as a proto-oncogene or as a tumor suppressor gene?

22.21 It has been demonstrated that individuals with diets poor in fiber and rich in fatty foods have an increased risk to develop colorectal cancer. Fiber-poor, fat-rich diets may irritate the epithelial lining of the large intestine. How could such irritation contribute to the increased risk for colorectal cancer?

22.22 Cancer cells frequently are homozygous for loss-of-function mutations in the *TP53* gene, and many of these mutations map in the portion of *TP53* that encodes the DNA-binding domain of p53. Explain how these mutations contribute to the cancerous phenotype of the cells.

22.23 The p21 protein is strongly expressed in cells that have been irradiated. Researchers have thought that this strong expression is elicited by transcriptional activation of the *p21* gene by the p53 protein acting as a transcription factor. Does this hypothesis fit with the observation that p21 expression is induced by radiation treatment in mice homozygous for a knockout mutation in the *TP53* gene? Explain.

22.24 Messenger RNA from the *KAI1* gene is strongly expressed in normal prostate tissues but weakly expressed in cell lines derived from metastatic prostate cancers. What does this finding suggest about the role of the KAI1 gene product in the etiology of prostate cancer?

Genomics on the Web

at http://www.ncbi.nlm.nih.gov/

The von Hippel-Lindau syndrome is characterized by the occurrence of cancer in the kidney. Often the *VHL* tumor suppressor gene has been mutated in this type of cancer.

1. Search the NCBI databases for information on the *VHL* gene. Where is it located in the genome? How long is its polypeptide product? Are different isoforms of the VHL protein created by alternate splicing?

2. The VHL protein physically interacts with other proteins inside cells. One interactant is the von Hippel-Lindau binding protein, VBP1. Search the databases for the gene encoding this protein. Where is this gene located? How long is the VBP1 polypeptide? How is this polypeptide thought to function inside cells?

3. The VHL protein plays a role in biochemical pathways inside cells. Find the Pathways section on the *VHL* page and click on KEGG pathway: renal cell carcinoma to see where the VHL protein functions. What is its role in renal cells? With what proteins does it interact?

4. Homologues of the *VHL* gene exist in the genomes of the rat and mouse. Use the Map Viewer function under the Homology section on the *VHL* page to locate these homologues. What chromosomes are they on? Is the region around these homologues similar in all three organisms—rat, mouse, and human? What does the structure of this chromosomal region in these three organisms suggest about the evolutionary process?

Chapter 23
Inheritance of Complex Traits

CHAPTER OUTLINE

- Complex Traits
- Statistics of Quantitative Genetics
- Analysis of Quantitative Traits
- Correlations Between Relatives
- Quantitative Genetics of Human Behavioral Traits

Cardiovascular Disease: A Combination of Genetic and Environmental Factors

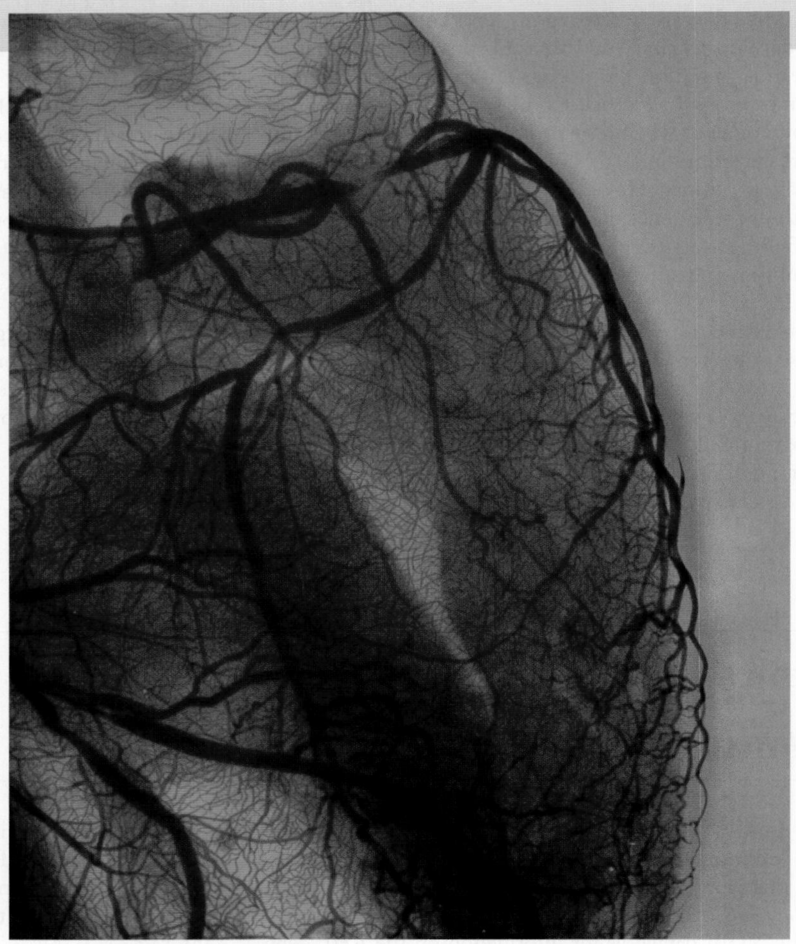

Color-enhanced angiogram of the heart showing blockage of one of the coronary arteries (upper center). Total blockage can lead to a heart attack.

Near the end of December, Paul Reston, a 47-year-old biology teacher in a suburban high school outside Pittsburgh, Pennsylvania, was spending his Saturday morning grading examinations. He was somewhat tired that day and felt a bit of stomach distress. He also had a slight pain in his left arm and shoulder. These symptoms had persisted for a few days. At first, Mr. Reston thought he had a mild case of the flu, but the arm and shoulder pain suggested another possibility: that he was having a heart attack. This possibility seemed more real when he remembered that his father had died from a sudden heart attack many years earlier at the relatively young age of 45. After a telephone conversation with a nurse in his health care clinic, Mr. Reston had his son drive him to a nearby hospital, where he spent two hours in the emergency room. The attending physician gave Mr. Reston a battery of tests to evaluate his condition. His heartbeat was regular, his blood pressure was normal, and an electrocardiogram revealed no abnormalities. Biochemical tests for telltale signs of heart damage were also negative. In addition, except for a family history of heart disease, Mr. Reston did not present other major risk factors. He was not overweight, he did not smoke, and he exercised regularly. The physician released Mr. Reston but advised him to return to the hospital the following week for a cardiac stress test. The following Monday, he was tested for heart function while running on a treadmill. The test results were good. Based on his performance, the supervising cardiologist concluded that Mr. Reston had less than a 1 percent chance of suffering a fatal heart attack.

In spite of his family history of heart disease, Mr. Reston's risk to develop this disease was low. The cardiologist explained that heart disease is a complex trait influenced by many factors: diet, physical activity, and smoking, for example, as well as a fairly large number of genes. Because Mr. Reston's father had succumbed to a heart attack, Mr. Reston may have inherited genes that put him at risk. However, the cardiologist emphasized that heart disease is not inherited as a simple Mendelian trait; rather, it involves the interplay of many different genetic and environmental factors.

Complex Traits

Breeding experiments and comparisons between relatives reveal that complex phenotypes may be influenced by a combination of genetic and environmental factors.

Many traits, such as disease susceptibility, body size, and various aspects of behavior, do not show simple patterns of inheritance. Nonetheless, we know that genes influence these types of traits. One indication is that genetically related individuals resemble one another. We see these resemblances between siblings, between parents and offspring, and sometimes between more distant relatives. The extreme case is monozygotic twins—twins that have developed from a single fertilized egg. Such twins are often strikingly similar, in behavior as well as in appearance. Another indication for a genetic influence is that these types of traits respond to selective breeding. In agriculture, crops and livestock have been shaped by propagating individuals with desirable features—greater protein content, reduced body fat, greater productivity, resistance to disease, and so forth. This ability to change phenotypes through selective breeding indicates that the traits have a genetic basis. Usually, however, this genetic basis is complex. Several to many genes are involved, and their individual effects are difficult to discern through conventional genetic analysis. Consequently, other techniques are needed to study the inheritance of complex traits.

QUANTIFYING COMPLEX TRAITS

Many complex traits vary continuously in a population. One phenotype seems to blend imperceptibly into the next. Examples are body size, height, weight, enzyme activity, blood pressure, and reproductive ability. The phenotypic variation in these types of traits can be quantified by measuring the trait in a sample of individuals from the population. We might, for example, capture mice in a barn and weigh each of them, or we might collect corncobs from a field and count the number of kernels on each. With such a quantitative approach, the phenotype of every individual in the sample is reduced to a number. These numbers can be analyzed with a variety of statistical techniques, enabling us to study the trait and, ultimately, to investigate its genetic basis. Traits that are amenable to this kind of treatment are called **quantitative traits.** Their essential characteristic is that they can be measured.

GENETIC AND ENVIRONMENTAL FACTORS INFLUENCE QUANTITATIVE TRAITS

The Danish biologist Wilhelm Johannsen was one of the first people to show that variation in a quantitative trait is due to a combination of genetic and environmental factors. Johannsen studied the weight of seeds from the broad bean, *Phaseolus vulgaris*. Among the plants available to him, seed weight varied from 150 mg to 900 mg. Johannsen established lines from individual seeds across this range and maintained each line by self-fertilization for several generations. The seeds from each of these "pure" lines tended to resemble the seed from which they were founded. This ability to establish lines of beans with characteristically different seed weights indicated that some variation in this trait is due to genetic differences. However, Johannsen observed that seed weight also varied within each of the pure lines. This residual variation was not likely to be due to genetic differences because each line had been systematically inbred to make it homozygous for its genes. Rather, it must have been due to variation in uncontrolled factors in the environment. Johannsen's work, published in 1903 and 1909, therefore led to the realization that phenotypic variation in a quantitative trait has two components—one genetic, the other environmental.

MULTIPLE GENES INFLUENCE QUANTITATIVE TRAITS

Another Scandinavian, Herman Nilsson-Ehle, provided evidence that the genetic component of this variation could involve the contributions of several different genes. Nilsson-Ehle studied color variation in wheat grains. When he crossed a white-grained variety with a dark red-grained variety, he obtained an F_1 with an intermediate red phenotype (**FIGURE 23.1**). Self-fertilization of the F_1 produced an F_2 with seven distinct classes, ranging from white to dark red. The number of F_2 classes and the phenotypic ratio that Nilsson-Ehle observed suggested that three independently assorting genes were involved in the determination of grain color. Nilsson-Ehle hypothesized that each gene had two alleles, one causing red grain color and the other white grain color, and that the alleles for red grain color contributed to pigment intensity in an additive fashion. Based on this hypothesis, the genotype of the white-grained parent could be represented as *aa bb cc*, and the genotype of the red-grained parent could be represented as *AA BB CC*. The F_1 genotype would be *Aa Bb Cc*, and the F_2 would contain an array of genotypes that would differ in the number of pigment-contributing alleles present. Each phenotypic class in the F_2 would carry a different number of these pigment-contributing alleles. The white class, for example, would carry none, the intermediate red class would carry three, and the dark red class would carry six. Nilsson-Ehle's work, published in 1909, showed that a complex inheritance pattern could be explained by the segregation and assortment of multiple genes.

The American geneticist Edward M. East extended Nilsson-Ehle's studies to a trait that did not show simple Mendelian ratios in the F_2. East studied the length of the corolla in tobacco flowers (**FIGURE 23.2*a***). In one pure line, the corolla length averaged 41 mm; in another, it averaged 93 mm. Within each pure line, East observed some phenotypic variation—presumably the result of environmental influences (**FIGURE 23.2*b***). By crossing

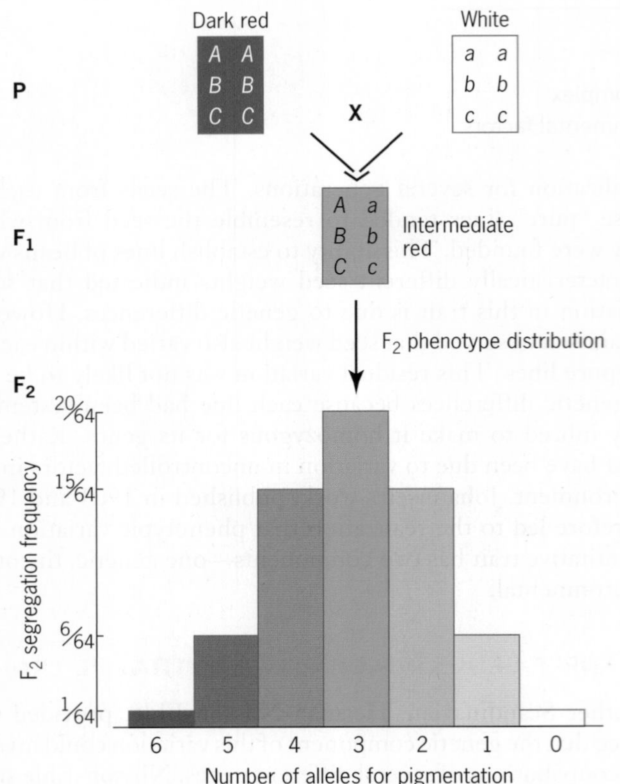

Figure 23.1 ▶ Inheritance of grain color in wheat. Three independently assorting genes (*A*, *B*, and *C*) are assumed to control grain color. Each gene has two alleles. The alleles that contribute additively to pigmentation are represented by uppercase letters.

the two lines, East obtained an F_1 that had intermediate corolla length and approximately the same amount of variation that he had seen within each of the parental strains. When East intercrossed the F_1 plants, he obtained an F_2 with about the same corolla length, on average, that he saw in the F_1; however, the F_2 plants were much more variable than the F_1. This variability was due to two sources: (1) the segregation and independent assortment of different pairs of alleles controlling corolla length, and (2) environmental factors. East inbred some of the F_2 plants to produce an F_3 and observed less variation within the different F_3 lines than in the F_2. The reduced amount of variation within the F_3 lines was presumably due to the segregation of fewer allelic differences. Thus, the complex inheritance pattern that East observed with corolla length could be explained by a combination of genetic segregation and environmental influences.

How many genes were involved in determining corolla length in East's strains of tobacco? We can make a crude guess by comparing the F_2 plants with each of the inbred parental strains. Let's suppose that the strain with the shorter corollas was homozygous for one set of alleles and that the strain with the longer corollas was homozygous for another set of alleles. Furthermore, let's suppose that the long-corolla alleles act additively, that all length-controlling genes assort independently, and that each gene makes an equal contribution to the phenotype. If corolla length were determined by one gene, with alleles *a* (for short corolla) and *A* (for long corolla), we would expect 1/4 of the F_2 plants to have short corollas (like the short parental strain) and 1/4 to have long corollas (like the long parental strain). If two genes determined corolla length, we would expect 1/16 of the F_2 plants to resemble the short-corolla parent and 1/16 to resemble the long-corolla parent. If three genes were involved, the frequency of each parental type in the F_2 would be 1/64, and if four genes were involved, it would be 1/256. With five genes, the parental frequencies in the F_2 would each be 1/1024. East studied 444 F_2 plants and failed to find even one with either of the parental phenotypes. This failure would seem to rule out the hypothesis of four or fewer

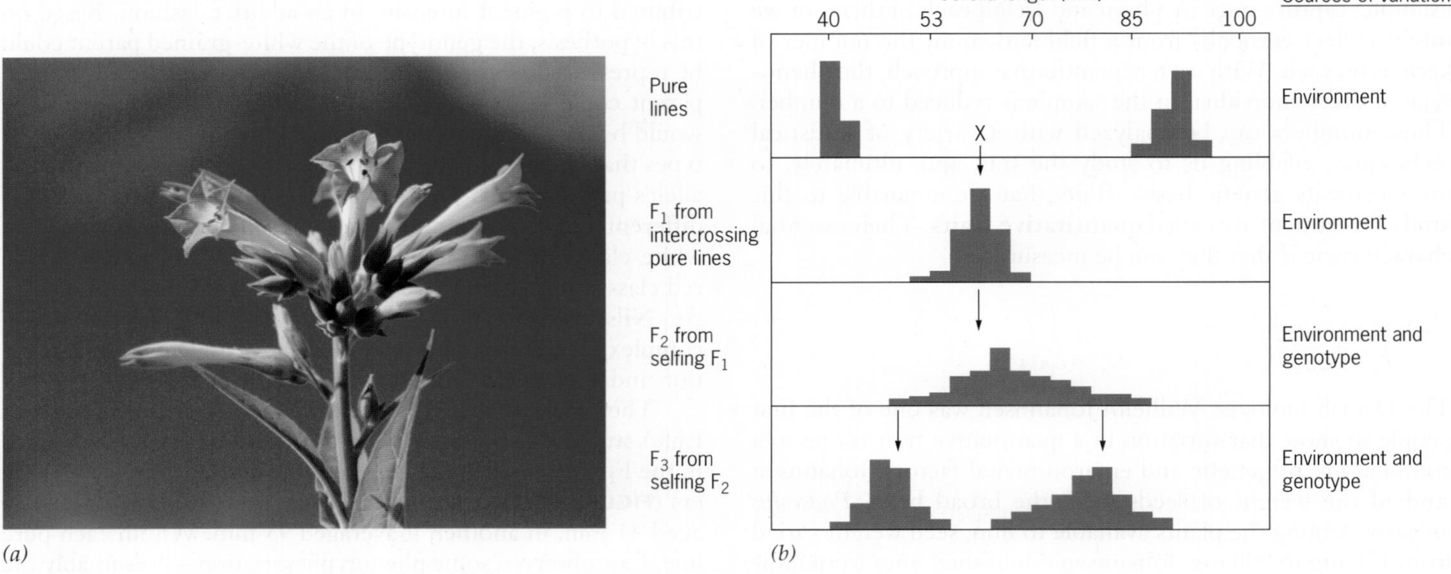

Figure 23.2 ▶ Corolla length as a quantitative trait. (*a*) Tobacco flowers showing the long corolla. (*b*) Inheritance of corolla length in tobacco. At least five genes appear to be involved.

genes controlling corolla length. Thus, we can conclude that at least five genes are responsible for the difference in corolla length between East's two inbred strains.

THRESHOLD TRAITS

Continuously varying traits such as bean size, grain color, and corolla length are controlled by multiple factors, both genetic and environmental. Geneticists have found that some traits that do not vary continuously in the population also appear to be influenced by multiple factors. For example, many people develop heart disease in their fifth or sixth decade of life. Heart disease is not a quantitative trait in the usual sense; individuals either have it or they don't. However, many factors predispose an individual to develop heart disease: body weight, amount of exercise, diet, blood cholesterol level, whether or not the individual smokes, and the presence of heart disease in close relatives such as parents or siblings. These underlying risk factors contribute to a variable called the *liability*. Geneticists theorize that when the liability exceeds a certain level, or threshold, the trait appears. This type of trait is therefore called a **threshold trait** (**FIGURE 23.3**).

In humans, the evidence that threshold traits are influenced by genetic factors comes from comparisons between relatives, especially twins. Occasionally a fertilized human egg splits and forms two genetically identical zygotes. The individuals who develop from these zygotes are referred to as one-egg, or **monozygotic (MZ),** twins; they share 100 percent of their genes. More frequently, two independently fertilized eggs develop at the same time in the mother's womb. These two-egg, or **dizygotic (DZ),** twins are as closely related as ordinary siblings; thus, they share 50 percent of their genes. Because of their genetic identity, we would expect MZ twins to be phenotypically more similar than DZ twins.

Similarity with respect to a threshold trait is assessed by determining the **concordance rate**—the fraction of twin pairs in which both twins show the trait among pairs in which at least one of them does. For cleft lip, a congenital condition due to an error in embryological development, the concordance rate has been estimated to be about 40 percent for MZ twins and about 4 percent for DZ twins. The much greater concordance rate for MZ twins strongly suggests that genetic factors influence an individual's likelihood of being born with cleft lip. Mental illnesses such as schizophrenia and bipolar disorder can also be regarded as threshold traits. For schizophrenia, the concordance rate ranges from 30 to 60 percent for MZ twins and from 6 to 18 percent for DZ twins; for bipolar disorder, the concordance rate is 70–80 percent for MZ twins and about 20 percent for DZ twins. Thus, twin studies suggest that both of these mental illnesses are influenced by genetic factors.

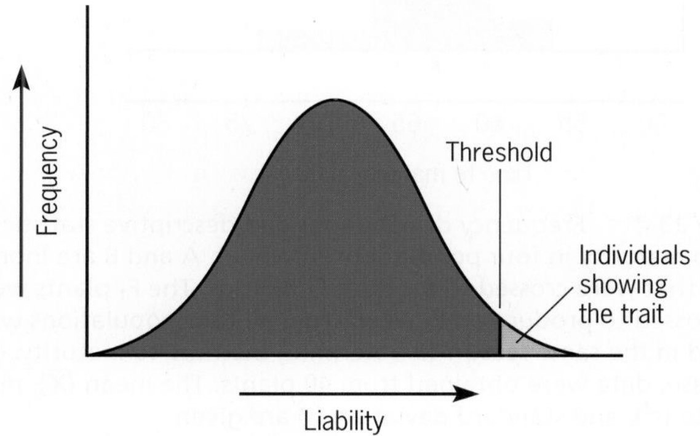

Figure 23.3 ▶ A model for expression of a threshold trait. When the underlying variable, the liability, reaches a threshold value, the trait is expressed. This variable is assumed to be continuously distributed in the population.

KEY POINTS

- Resemblances between relatives and responses to selective breeding indicate that complex traits have a genetic basis.
- Some complex traits can be quantified to permit genetic analysis.
- Many genetic and environmental factors influence the variation observed in quantitative traits.
- Phenotypic segregations may provide a way to estimate the number of genes that influence a quantitative trait.
- Traits that are manifested when an underlying continuous variable (the liability) reaches a threshold value may be influenced by genetic factors.
- In humans, evidence that a threshold trait has a genetic basis comes from studies with twins.
- The concordance rate is the fraction of twin pairs in which both twins show a trait among pairs in which at least one of them does.

Statistics of Quantitative Genetics

The frequency distributions of quantitative traits can be characterized by summary statistics.

The hallmark of quantitative traits is that they vary continuously in a population of individuals. This type of variation poses a formidable problem for the geneticist. Segregation ratios are difficult, if not impossible, to discern because the number of phenotypes is large and one phenotype blends imperceptibly into the next. For quantitatively varying traits, routine genetic analyses of the sort that we have done with eye color in *Drosophila* and with human disorders such as albinism are out of the question. For these types of traits we must resort to a different kind of analysis, one that is based on statistical descriptions of the phenotype in a population. In the sections that follow, we introduce the basic statistical concepts that are needed for this type of analysis.

FREQUENCY DISTRIBUTIONS

The first step in the study of any quantitative trait is to collect measurements of the trait from individuals in a population. Usually, only a small fraction of all the individuals in the population can be measured. We call this group the **sample.** The data from the sample can be presented graphically as a **frequency distribution.** In the graph the horizontal or x-axis measures values of the trait. This axis is divided into regular intervals that allow each individual in the population to be categorized for the trait. Thus, each observation in the sample can be placed into one of the intervals on the x-axis. The vertical or y-axis measures the frequency of the observations within each interval.

FIGURE 23.4 shows frequency distributions that were obtained in a genetic study of wheat. The investigators measured the time that wheat takes to mature. Four different populations of wheat were grown in test plots in the same season, and 40 plants from each population were monitored until the heads of grain matured. The time to maturity for each plant was recorded in days. Two of the populations (A and B) were inbred strains, and one was an F_1 produced by crossing these two strains. The fourth population was an F_2 produced by intercrossing the F_1 plants.

The two parental strains A and B were highly inbred varieties that were completely or almost completely homozygous. As the frequency distributions indicate, strain A matured quickly and strain B matured slowly. The lack of phenotypic overlap between the samples from these two strains demonstrates their genetic distinctiveness. Apparently, strains A and B were homozygous for different alleles of genes controlling maturation time. Within each strain, however, there was still some phenotypic variation, presumably the result of microenvironmental differences within the test plots.

The distributions of the F_1 and F_2 samples indicate that these populations had intermediate maturation times. Their intermediate position on the x-axis suggests that the alleles controlling maturation time contribute additively to the trait. Notice that the distribution of the F_2 sample is considerably broader than that of the F_1. The additional variability seen in the F_2 population reflects the genetic segregation that occurred when the F_1 plants reproduced. We now explore ways in which quantitative geneticists summarize the data in a frequency distribution.

THE MEAN AND THE MODAL CLASS

The essential characteristics of a frequency distribution can be summarized by simple statistics calculated from the data. One of these summary statistics is called the **mean** or average. It gives us the "center" of the distribution—the "typical" value.

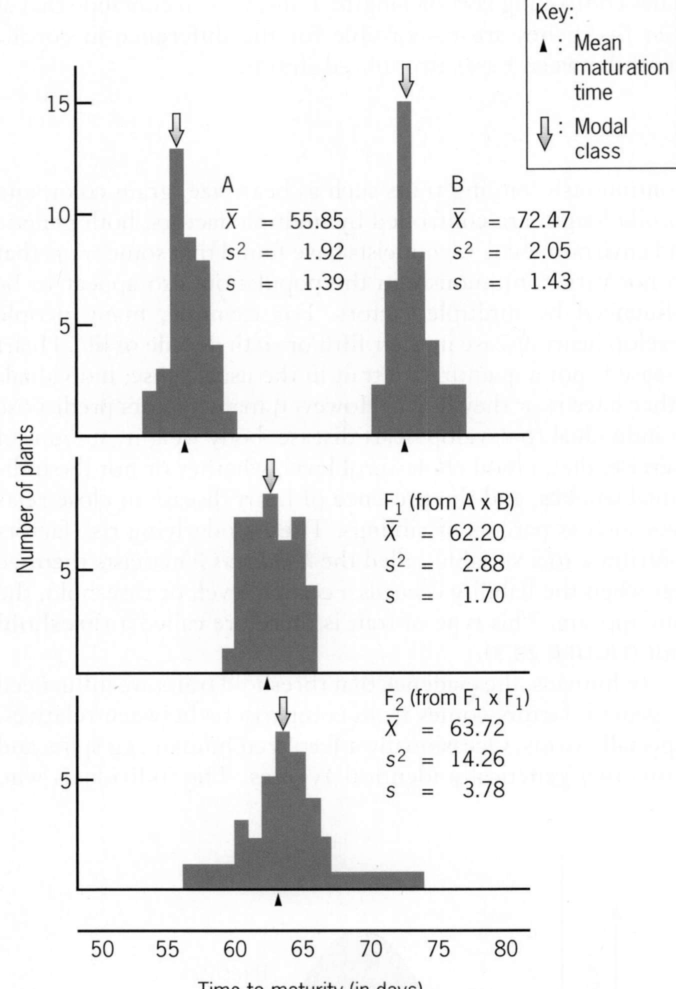

Figure 23.4 ▶ Frequency distributions and descriptive statistics of time to maturity in four populations of wheat. A and B are inbred strains that were crossed to produce F_1 hybrids. The F_1 plants were intercrossed to produce an F_2. Seed from all four populations was planted in the same season to determine the time to maturity. In each case, data were obtained from 40 plants. The mean ($\overline{X}$), mode, variance (s^2), and standard deviation (s) are given.

We calculate the sample mean ($\overline{X}$) by summing all the data in the sample and dividing by the total number of observations (n). In mathematical notation, the mean is:

$$\overline{X} = (\Sigma X_k)/n$$

The Greek letter Σ in this formula is a mathematical shorthand for the sum of all the individual measurements in the sample; thus, $\Sigma X_k = (X_1 + X_2 + X_3 + \ldots X_n)$, where X_k represents the kth of the n individual observations. In **FIGURE 23.4** the positions of the sample means are indicated by triangles beneath the distributions; the numerical values of these means are given on the right. The means of the F_1 and F_2 samples are 62.20 and 63.72 days, respectively; both are a little less than the average of the means of the two inbred parental strains (64.16 days).

The **modal class** in a sample is the class that contains the most observations. Like the mean, it also captures the "center" of the distribution. In **FIGURE 23.4** the modal classes are indicated by short arrows. We see that in each of the distributions the mean is within or very close to the modal class. This coincidence reflects the symmetry of the distributions; in each case, roughly equal numbers of observations are above and below the mean and the modal class. Not all distributions have this feature. Some are skewed, with most of the observations clustered at one end and only a few at the other end, forming a long tail. Statisticians have developed an extensive theory about a particular type of symmetrical distribution called a *normal distribution* (**FIGURE 23.5**). In this bell-shaped distribution, the mean and the modal class are located exactly in the center. Often distributions of sample data approximate the shape of a normal distribution. Thus, we can apply the extensive theory about normal distributions to analyze such data.

THE VARIANCE AND THE STANDARD DEVIATION

The data in a frequency distribution could be dispersed, or they could be clustered. To measure the spread of data in a frequency distribution, we use a statistic called the **variance.** Data that are widely dispersed produce a large value for the variance, whereas data that are tightly clustered produce a small value. The sample variance, denoted s^2, is calculated from the formula

$$s^2 = \Sigma(X_k - \overline{X})^2/(n - 1)$$

In this formula, $(X_k - \overline{X})^2$ is the squared difference between the kth observation and the sample mean (often called the *squared deviation from the mean*), and the Greek letter Σ indicates that all such squared deviations are summed. The sum of the squared deviations is averaged by dividing by $n - 1$. (For technical reasons, the divisor is one less than the sample size.)

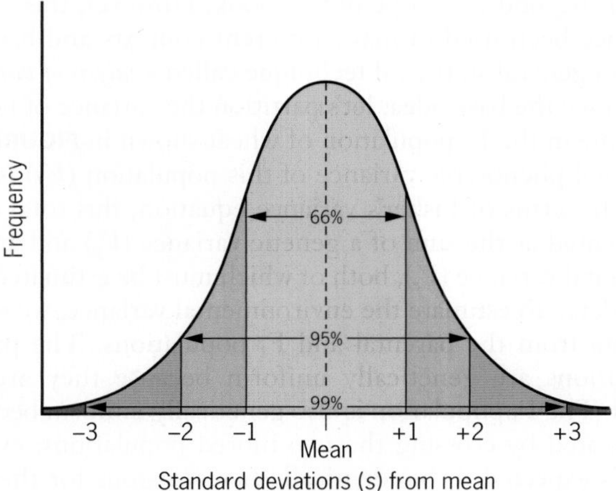

Figure 23.5 ▶ A normal frequency distribution showing the percentage of measurements within 1, 2, and 3 standard deviations of the mean.

We should note two features of the variance. First, it measures the dispersion of the data around the *mean*. When we calculate the variance, we take the mean to be the central value of the distribution and find the difference between it and each of the observations in the sample. Second, the variance is always positive. When we calculate the variance, we square the difference between each observation and the mean, and then sum the squared differences. Because each of the squared differences is positive, the variance, calculated by summing these squared differences, is also positive.

Although the variance has desirable mathematical properties, it is difficult to interpret because the units of measurement are squared (for example, $s^2 = 2.88$ days^2). Consequently, another statistic, called the **standard deviation,** is often used to describe the variability of a sample. The standard deviation (s) is the square root of the sample variance

$$s = \sqrt{s^2}$$

This statistic is easier to interpret than the variance because it is expressed in the same units as the original measurements.

The variances and standard deviations of the four wheat populations are given in **FIGURE 23.4**. The F_2 population has the greatest variance and standard deviation, no doubt because it is segregating for genes that control maturation time. In the F_2 plants, both genetic and environmental differences produce the observed variability. In the other populations, most if not all of the observed variation is due to environmental factors alone. Each of the two parental strains is highly inbred and is therefore expected to be homozygous for most of its genes. The F_1 plants are heterozygous for the alleles that are different in the two parental strains, but they all have the same genotype. Thus, in neither the parental strains nor the F_1 do we expect to find much genetic variation among plants. In a later section we will see how to estimate that part of the variance in a quantitative trait that is due to genetic differences among individuals in a population.

As mentioned above, the distribution of a quantitative trait often looks like a normal distribution. The shape and the position of a normal distribution are completely specified by its mean and standard deviation. Thus, if we know only the mean and standard deviation of a quantitative trait, and assume that the trait is normally distributed, we can construct the approximate shape of the trait's distribution. In this distribution, 66 percent of the measurements will lie within one standard deviation of the mean, 95 percent will lie within two standard deviations of the mean, and 99 percent will lie within three standard deviations of the mean (**FIGURE 23.5**).

KEY POINTS

- The mean ($\overline{X} = (\Sigma X_k)/n$) and modal class point to the center of a frequency distribution.
- The variance ($s^2 = \Sigma(X_k - \overline{X})^2) / (n - 1)$ and standard deviation $s = \sqrt{s^2}$ are statistics that indicate the extent to which data are scattered around the mean in a frequency distribution.

Analysis of Quantitative Traits

Quantitative geneticists focus their analyses on phenotypic variability as measured by the variance.

In this section we will see how statistics are used in the genetic analysis of quantitative traits. The thrust of the analysis is to partition the observed variation in the trait into genetic and environmental components, and then to use the genetic component to make predictions about the phenotypes of the offspring of particular crosses.

THE MULTIPLE FACTOR HYPOTHESIS

The key idea in quantitative genetics is that traits are controlled by many different factors in the environment and in the genotype. This **Multiple Factor Hypothesis** emerged in the second decade of the twentieth century through the experimental investigations of E. M. East, W. Johannsen, H. Nilsson-Ehle, and others. However, it was a theoretician, R. A. Fisher, who crystallized the Multiple Factor Hypothesis into its modern form. Fisher did this work during World War I while he was teaching school in Great Britain. His theoretical analysis was published in 1918, the year the war ended.

Fisher hypothesized that a particular value of a quantitative trait, *T*, is the result of the combined influence of genetic and environmental factors. He represented the effects of these factors as deviations from the overall population mean:

$$T = \mu + g + e$$

In this equation, the Greek letter μ represents the population mean, *g* represents the deviation from the mean that is due to genetic factors, and *e* represents the deviation from the mean that is due to environmental factors. In Fisher's scheme, the position of a particular value of the trait, *T*, in the population depends on the genetic and environmental factors that have affected it (**FIGURE 23.6**). Some factors produce large values of *T*, and some produce small values of *T*. For each individual, these factors are different. Furthermore, Fisher emphasized that a multitude of factors are involved. He hypothesized that many genes contribute to a quantitative trait, and he assumed that many aspects of the environment also make contributions. Today we say that a trait that is controlled by many genes is **polygenic.**

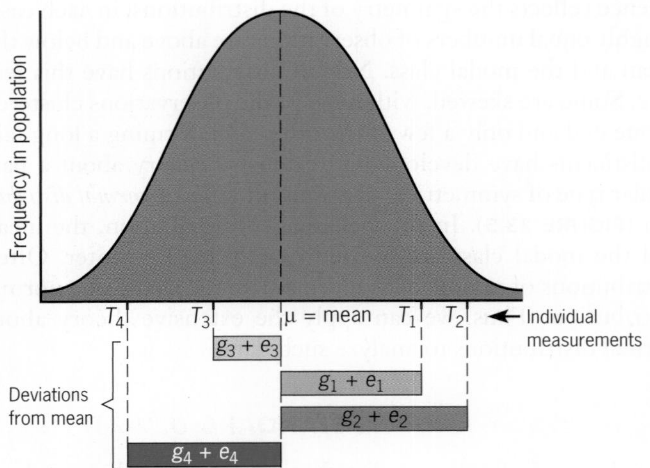

Figure 23.6 ▶ Quantitative phenotypes and the deviations of individual measurements from the population mean. Each individual's deviation is hypothesized to consist of a deviation due to its genotype (*g*) and a deviation due to its environment (*e*).

PARTITIONING THE PHENOTYPIC VARIANCE

With these simple ideas, Fisher was able to develop a procedure to analyze the variability of a quantitative trait in terms of the contributing genetic and environmental factors. To measure the variability of the trait, he focused on the statistic we have called the variance. Specifically, he discovered how to split the overall variance of the trait into two component variances, one measuring the effects of genetic differences among individuals and the other measuring the effects of environmental differences. Thus, in Fisher's analysis, the variance of a quantitative trait, symbolized V_T, is equal to the sum of a *genetic variance*, symbolized V_g, and an *environmental variance*, symbolized V_e:

$$V_T = V_g + V_e$$

In this variance equation, the variance of the quantitative trait, V_T, is often referred to as the *total phenotypic variance*.

A discussion of Fisher's method of splitting the total phenotypic variance into its genetic and environmental components is beyond the scope of this book. However, this method has since been used in many different contexts and has given rise to a general statistical technique called *analysis of variance*.

To see the basic idea, let's partition the variance of maturation time in the F_2 population of wheat shown in **FIGURE 23.4**. The total phenotypic variance of this population (V_T) is 14.26 days^2. In terms of Fisher's variance equation, this total can be represented as the sum of a genetic variance (V_g) and an environmental variance (V_e), both of which must be estimated using other data. To estimate the environmental variance, we can use the data from the parental and F_1 populations. The parental populations are genetically uniform because they are both inbred. The F_1 population is also genetically uniform because it was created by crossing the two inbred populations; every F_1 plant is expected to be identically heterozygous for the genes that differ in the inbred parental populations. Because of this genetic uniformity, the variability that we see in each of these three populations must reflect differences due to environmental effects. To obtain a representative value for V_e, we can average the variances of these groups:

$$\begin{aligned} V_e &= (V_A + V_B + V_{F1})/3 \\ &= (1.92 \text{ days}^2 + 2.05 \text{ days}^2 + 2.88 \text{ days}^2)/3 \\ &= 2.28 \text{ days}^2 \end{aligned}$$

With this estimate of the environmental variance, we can now estimate V_g by subtraction from the total variance V_T:

$$\begin{aligned} V_g &= V_T - V_e \\ &= 14.26 \text{ days}^2 - 2.28 \text{ days}^2 \\ &= 11.98 \text{ days}^2 \end{aligned}$$

Thus, the total phenotypic variance for maturation time in the F_2 wheat population has been split into two components:

$$\begin{aligned} V_T &= V_g + V_e \\ 14.26 \text{ days}^2 &= 11.98 \text{ days}^2 + 2.28 \text{ days}^2 \end{aligned}$$

From this partition, we see that most of the variance in maturation time in the F_2 wheat population is due to genetic differences among the individuals. This genetic variability arose from the segregation and assortment of genes when the F_1 plants reproduced. These plants were heterozygous for the genes that differed in the parental populations. When they reproduced, segregation and assortment produced an array of genotypes—three distinct genotypes for each heterozygous gene. The variation that we see in the F_2 is due primarily to phenotypic differences among these genotypes.

BROAD-SENSE HERITABILITY

Often it is informative to calculate the proportion of the total phenotypic variance that is due to genetic differences among individuals in a population. This proportion is called the **broad-sense heritability,** symbolized H^2. In terms of Fisher's variance components,

$$\begin{aligned} H^2 &= V_g / V_T \\ &= V_g / (V_g + V_e) \end{aligned}$$

The symbol for the broad-sense heritability, H^2, is written with the exponent 2 to remind us that this statistic is calculated from variances, which are squared quantities.

Because of the way it is calculated, the broad-sense heritability must lie between 0 and 1. If it is close to 0, little of the observed variability in the population is attributable to genetic differences among individuals. If it is close to 1, most of the observed variability is attributable to genetic differences. The broad-sense heritability therefore summarizes the relative contributions of genetic and environmental factors to the observed variability in a population. However, it is important to note that this statistic is population-specific. For a given trait, different populations may have different values of the broad-sense heritability. Thus, the broad-sense heritability of one population cannot automatically be assumed to represent the broad-sense heritability of another population.

In the F_2 wheat population, $H^2 = 11.98/14.26 = 0.84$. This result tells us that in this population 84 percent of the observed variability in wheat maturation time is due to genetic differences among individuals. However, it does not tell us what these differences are. The genetic variance upon which the broad-sense heritability depends includes all the factors that cause genotypes to have different phenotypes: the effects of individual alleles, the dominance relationships between alleles, and the epistatic interactions among different genes. In Chapter 4 we saw how these factors influence phenotypes. In the next two sections, we will see that by breaking out these components of genetic variability and by focusing on the component that involves the effects of individual alleles, we can predict the phenotypes of offspring from the phenotypes of their parents.

NARROW-SENSE HERITABILITY

The ability to make predictions in quantitative genetics depends on the amount of genetic variation that is due to the effects of individual alleles. Genetic variation that is due to the effects of dominance and epistasis has little predictive power.

To see how dominance limits the ability to make predictions, consider the ABO blood types in humans (**TABLE 4.1** in Chapter 4). This trait is determined strictly by the genotype; environmental variation has essentially no effect on the phenotype. However, because of dominance, two individuals with the same phenotype can have different genotypes. For example, a person with type A blood could be either I^AI^A or I^Ai. If two people with type A blood produce a child, we cannot predict precisely what phenotype the child will have. It could be either type A or type O, depending on the genotypes of the parents; however, we know that it will not have type B or type AB blood. Thus, although we can make some kind of prediction about the child's phenotype, dominance prevents us from making a precise prediction.

Our ability to make predictions about an offspring's phenotype is improved in situations where the genotypes are not confused by dominance. Consider, for example, the inheritance of flower color in the snapdragon, *Antirrhinum majus*. Flowers in this plant are white, red, or pink, depending on the genotype (**FIGURE 4.1** in Chapter 4). As with the ABO blood types, variation in flower color has essentially no environmental component; all the variance is the result of genetic differences. However, for the flower color trait, the genotype of an individual is not obscured by the complete dominance of one allele over the other. A plant with two *w* alleles has white flowers, a plant with one *w* allele and one *W* allele has pink flowers, and a plant with two *W* alleles has red flowers. In this system, the phenotype depends simply on the number of *W* alleles present; each *W* allele intensifies the color by a fixed amount. Thus, we can say that the color-determining alleles contribute to the phenotype in a strictly additive fashion. This kind of allele action improves our ability to make predictions in crosses between different plants. A mating between two red plants produces only red offspring; a mating between two white plants produces only white offspring; and a mating between red and white plants produces only pink offspring. The only uncertainty is in a cross between two heterozygotes, and in this case the uncertainty is due to Mendelian segregation, not to dominance.

Quantitative geneticists distinguish between genetic variance that is due to alleles that act additively (such as those in the flower color example just discussed) and genetic variance that is due to dominance. These different variance components are symbolized as:

V_a = additive genetic variance
V_d = dominance variance

In addition, geneticists define a third variance component that measures variation due to epistatic interactions between alleles of different genes:

$$V_i = \text{epistatic variance}$$

Epistatic interactions, like dominance, are of little help in predicting phenotypes. Altogether, these three variance components constitute the total genetic variance:

$$V_g = V_a + V_d + V_i$$

If we recall that $V_T = V_g + V_e$, we can express the total phenotypic variance as the sum of four components:

$$V_T = V_a + V_d + V_i + V_e$$

Of these four variance components, only the additive genetic variance, V_a, is useful in predicting the phenotypes of offspring from the phenotypes of their parents. This variance, as a fraction of the total phenotypic variance, is called the **narrow-sense heritability,** symbolized h^2. Thus,

$$h^2 = V_a / V_T$$

Like the broad-sense heritability, h^2 lies between 0 and 1. The closer it is to one, the greater is the proportion of the total phenotypic variance that is additive genetic variance, and the greater is our ability to predict an offspring's phenotype. **TABLE 23.1** gives some estimates for the narrow-sense heritability for several traits. Human stature is highly heritable, but litter size in pigs is not. Thus, if we knew the parental phenotypes, we would be better able to predict the height of a human's offspring than the litter size of a pig's offspring.

▶ **TABLE 23.1**

Estimates of Narrow-Sense Heritability (h^2) for Quantitative Traits

Trait	h^2
Stature in human beings	0.65
Milk yield in dairy cattle	0.35
Litter size in pigs	0.05
Egg production in poultry	0.10
Tail length in mice	0.40
Body size in *Drosophila*	0.40

Source: D. S. Falconer. 1981. *Introduction to Quantitative Genetics*, 2nd ed., p. 51. Longman, London.

PREDICTING PHENOTYPES

To gain insight into the meaning of the narrow-sense heritability, let's consider the situation diagrammed in **FIGURE 23.7**. Michael (M) and Frances (F) have taken a standardized intelligence test, and their Intelligence Quotients (IQs) have been determined. Michael's score is 110 and Frances' score is 120. The mean IQ score in the population is 100. Michael and Frances had an infant son Oswald (O), who was given up for adoption when he was born, and the adoptive parents wish to predict Oswald's IQ. If IQ had no genetic component, our best estimate for Oswald's IQ would be 100, the mean of the population. We have no way of predicting what kind of home environment Oswald will receive and therefore cannot predict what

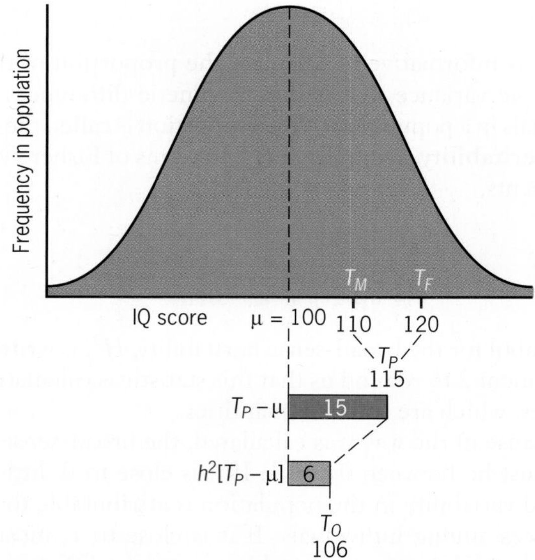

Figure 23.7 ▶ Predicting an offspring's phenotype based on the phenotypes of its parents and the narrow-sense heritability of the trait. Only a portion of the deviation of the midparent (T_P) value from the population mean is heritable. The magnitude of this portion is determined by the narrow-sense heritability.

kind of nongenetic factors will influence his mental development. Nor can we use the IQs of Michael and Frances to predict anything about Oswald's IQ, since, by assumption, the genes they gave to him would have nothing to do with mental development. However, several studies have indicated that variation in IQ scores does have a genetic component. In fact, the narrow-sense heritability of IQ has been estimated to be about 0.4—that is, about 40 percent of the observed variation in IQ scores is due to the additive effects of alleles. Can we use this statistic along with the parental IQs to predict Oswald's IQ score?

Let's symbolize the IQs of Oswald, Michael, and Frances as T_O, T_M, and T_F, respectively, and let's symbolize the population mean as μ. The best prediction for Oswald's IQ is

$$T_O = \mu + h^2[(T_M + T_F)/2 - \mu]$$

The expression with parentheses, $(T_M + T_F)/2$, is usually called the *midparent value*. It is the average of the phenotypes of the two parents. If we denote the midparent value with the symbol T_P, the prediction equation for Oswald's phenotype simplifies to

$$T_O = \mu + h^2[T_P - \mu]$$

The expression in brackets, $[T_P - \mu]$, is the difference between the midparent value and the mean of the population. The product of this difference and the narrow-sense heritability is the predicted deviation of the offspring's phenotype from the mean of the population. In effect, the narrow-sense heritability translates the difference between the midparent value and the mean of the population into a "heritable" difference that we can expect to see in the offspring. By adding this heritable difference to the mean, we can predict the offspring's phenotype.

Let's now substitute the known quantities for each of the terms in the prediction equation: $\mu = 100$, $T_P = (110 + 120)/2 = 115$, and $h^2 = 0.4$. Thus, the predicted value of T_O is

$$\begin{aligned} T_O &= 100 + (0.4)[115 - 100] \\ &= 106 \end{aligned}$$

This result tells us that Oswald's IQ is expected to be between the midparent value (115) and the mean of the population (100). In fact, it is at a point 40 percent of the distance between the population mean and the midparent value. This 40 percent corresponds to the narrow-sense heritability (0.4). If the narrow-sense heritability of IQ were greater than 0.4, the predicted value of Oswald's IQ would be closer to the midparent value. For a perfectly heritable trait, $h^2 = 1$, and the predicted value of the offspring's phenotype would equal the average of the two parents' phenotypes. Thus, the narrow-sense heritability is a critical statistic. It tells us how closely the offspring will resemble the average of their parents. We should emphasize, however, that the IQ score we have calculated for Oswald is a predicted value—not one we know for certain. If we were to look at thousands of couples, each having a midparent IQ value of 115, the IQs of their children would be expected to form a frequency distribution. The mean IQ of this distribution would be 106; however, most children would have higher or lower IQs—some even higher than the IQs of either parent and some even lower than the population mean of 100. The variability in this distribution comes from Mendelian segregation of the alleles that influence IQ and from factors in the environment. If, for example, Oswald were raised in a home with little or no intellectual stimulation, with a poor diet, and with other unfavorable conditions, his IQ might turn out to be considerably lower than 106. Conversely, in a nurturing home environment, Oswald's IQ might turn out to be much greater than 106. We have predicted Oswald's IQ to be 106; however, we should keep in mind that this number is a prediction, not an absolutely determined value.

ARTIFICIAL SELECTION

In addition to predicting an offspring's phenotype, the narrow-sense heritability has another use: to predict the outcome of a program of selective breeding in a population. The ideas are summarized in **FIGURE 23.8**, which shows the frequency distri-

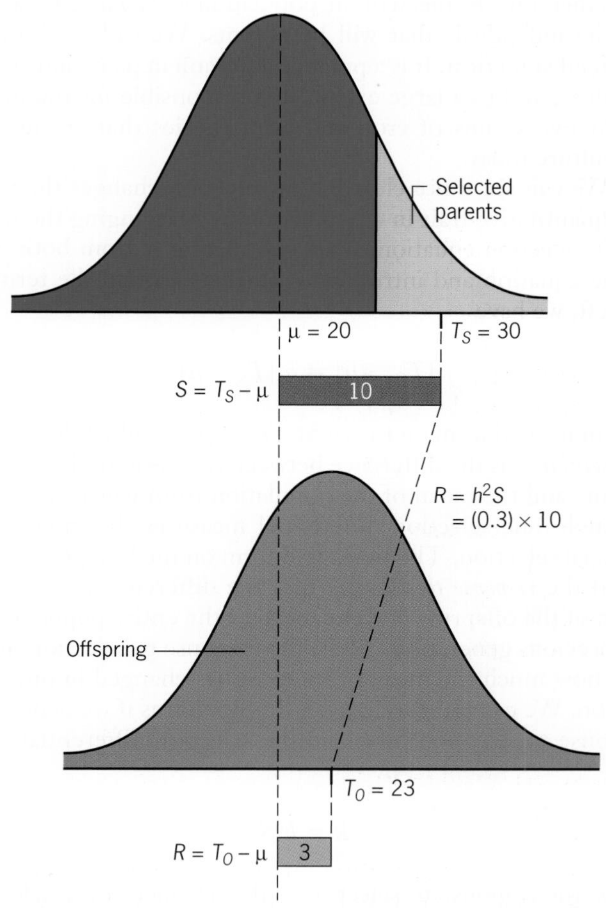

Figure 23.8 ▶ The process of artificial selection. The selection differential (*S*) is the difference between the mean of the selected parents and the mean of the population. The response to selection (*R*) is the difference between the mean of the offspring and the mean of the overall population that included their parents. The ratio *R*/*S* equals the narrow-sense heritability.

butions of a quantitative trait among parents and their offspring. In the parental generation, the mean value of the trait is 20. To form the next generation, we select the individuals in the upper tail of the distribution to be parents; let's suppose that the mean of these selected individuals is 30 units. Can we predict the mean value of the trait in the offspring of these selected parents? The answer is yes, providing we know the narrow-sense heritability of the trait. The prediction equation is

$$T_O = \mu + h^2[T_S - \mu]$$

where T_O is the mean of the offspring, μ is the mean of the overall population, T_S is the mean of the selected parents, and h^2 is the narrow-sense heritability. Notice that this equation is the same as the prediction equation for the phenotype of a single offspring, except that T_S has been substituted for T_P. In effect, we have adapted the single-offspring prediction equation to a situation in which many parents (albeit *selected* parents) produce a whole group of offspring, which then forms the population in the next generation. Thus, the new equation allows us to predict how the mean of the population will change by selecting the individuals that will be parents. We call this process **artificial selection.** It is a practice common in plant and animal breeding, and to a large extent, it is responsible for the highly productive strains of crop and farm species that are used in agriculture today.

We can see more clearly how selection changes the mean of a quantitative trait in a population by rearranging the terms in the selection equation. After subtracting μ from both sides of the equation and introducing brackets around the term on the left, we have

$$[T_O - \mu] = h^2[T_S - \mu]$$

The bracketed term on the right, $[T_S - \mu]$, is called the *selection differential;* it is the difference between the mean of the selected parents and the mean of the population from which they were selected. The selection differential measures the intensity of artificial selection. The bracketed term on the left, $[T_O - \mu]$, is called the *response to selection;* it is the difference between the mean of the offspring and the mean of the entire population in the previous generation. Thus, the response to selection measures how much the mean of the trait has changed in one generation. We can put this in even simpler terms if we denote the response to selection by R and the selection differential by S; then

$$R = h^2S$$

Thus, the response to selection is the product of the selection differential and the narrow-sense heritability. Let's now return to our example; $\mu = 20$, $T_S = 30$, and let's suppose that $h^2 = 0.3$. With these values, $S = 10$ and $R = (0.3) \times 10 = 3$; thus, $T_O = 20 + 3 = 23$. If the selection process were repeated generation after generation, we would expect the mean of the population to increase incrementally. The Focus on Artificial Selection shows how this is accomplished in practice.

Now let's suppose that we select for a change in another trait whose narrow-sense heritability is unknown. For this trait, the mean of the population is 100 and the mean of the selected parents is 120. Among the offspring of these parents, we find that the mean is 104. What is the narrow-sense heritability? From the equation for the response to selection, we see that $R/S = h^2$, and in this example, $R = 104 - 100 = 4$, and $S = 120 - 100 = 20$. Thus, $R/S = 4/20 = 0.2 = h^2$, the narrow-sense heritability. From this example, we see that the response to an artificial selection experiment can be used to estimate the narrow-sense heritability.

QUANTITATIVE TRAIT LOCI

Statistical analysis has been a mainstay of quantitative genetics since Fisher's 1918 paper. With this type of analysis, quantitative geneticists have studied many different traits in many different organisms, and recently they have developed techniques to identify individual genes that influence complex traits. A gene's position in a chromosome is called a *locus* (plural, loci), and the locus for a gene that influences a quantitative trait is called a **quantitative trait locus**—abbreviated QT locus, or more simply QTL.

Modern molecular techniques have made it possible to search genomes for QT loci. These loci have been identified and mapped on specific chromosomes in model laboratory organisms such as the fruit fly and the mouse, in agriculturally significant plants such as corn and rice, in livestock such as pigs and cows, and in our own species. The traits that have been studied include bristle number in the fruit fly, obesity in the mouse, crop yield in rice and corn, milk production in dairy cattle, fatness and growth rate in pigs, and susceptibility to illnesses such as diabetes, cancer, cardiovascular disease, and schizophrenia in human beings.

To illustrate the methods used to identify QT loci in organisms where breeding experiments are possible, let's consider a study on fruit weight in tomatoes conducted by Steven Tanksley and colleagues. Cultivated tomatoes belong to the species *Lycopersicon esculentum*. There are many different varieties, and in each variety the fruits have a characteristic size, shape, and color (**FIGURE 23.9**). All these varieties were derived by artificial selection from wild tomatoes, which are native to South America. *L. pimpinellifolium*, which has small, berry-like fruits, is thought to be the genetic ancestor of cultivated tomatoes. A fruit from *L. pimpinellifolium* weighs about one gram, whereas a fruit from the cultivated variety Giant Heirloom weighs as much as 1000 grams—a dramatic indication of the power of artificial selection.

Tanksley and colleagues began their efforts to identify the loci responsible for variation in tomato fruit weight by constructing detailed molecular maps for each of the tomato's 12 chromosomes. They exploited the fact that *L. pimpinellifolium* and *L. esculentum* differ in the sites where restriction enzymes

▶ FOCUS ON **Artificial Selection**

Artificial selection is a standard practice to improve crop plants and livestock. However, improvement is usually slow because the generation time of agriculturally significant species is typically measured in years rather than weeks or months. To study the efficacy of artificial selection, Franklin Enfield and his colleagues carried out extensive experiments with a laboratory animal, the flour beetle, *Tribolium castaneum* (**FIGURE 1**). In these experiments, Enfield selected for increased body size. He measured the weight of the animals at the pupal stage and selected the heaviest pupae to be the parents of the next generation. This process was continued for 125 generations. At the start of the experiment, the weight of the individual pupae ranged from 1800 to 3000 μg, the mean was 2400 μg, and the variance was 40,000 μg^2. After 125 generations of selection, the mean pupa weight had increased to 5800 μg, more than twice the mean of the starting population (**FIGURE 2*a***). Moreover, none of the individuals in the selected population was as small as the largest individuals in the original starting population (**FIGURE 2*b***). This complete lack of overlap in the frequency distributions indicates that the genetic makeup of the population had been radically altered.

To achieve this stunning result, Enfield used a selection differential of 200 μg in each generation. Initially, the narrow-sense heritability for pupa weight was estimated to be about 0.3; thus, the predicted response to selection was 0.3×200 μg $=$ 60 μg per generation (see Figure 23.8). For the first 40 generations, this was approximately what Enfield observed. However, the cumulative response during this time was 2000 μg, a little less than the 2400 μg that was expected (60 μg/generation $\times$ 40 generations). This discrepancy was due to factors that reduced the selection efficiency, including such things as infertility among the selected individuals. Thus, although the narrow-sense heritability is a reasonably good predictor of the response to selection over a few generations, in the long term it tends to overestimate this response.

The later generations of Enfield's project dramatically demonstrate this point. Between generations 40 and 125, the cumulative response was 1400 μg, which, though impressive, is much less than the expected response of 5100 μg (60 μg/generation $\times$ 85 generations). Enfield checked narrow-sense heritability in these later generations to see if any of the additive genetic variance had been lost during the long selective process. To his surprise, he found that h^2 was only slightly changed, indicating that the population still retained selectable genetic variability. A more detailed analysis demonstrated that during these generations, the efficiency of selection was severely reduced by a negative correlation between size and reproductive ability. (After a certain point, the larger the beetle, the less reproductively successful it is.) This reduced the effective selection differential and made it difficult to select for further increases in size. In fact, when selection was relaxed in generation 50 and again in generation 110, mean pupa weight began dropping back to a nearly normal value. This result suggests that a force of natural selection strongly opposed the artificial selection that was practiced in the main experiment. Enfield's attempts to increase pupa weight beyond 5800 μg failed, not for any lack of genetic variability, but simply because artificial selection had exceeded the natural limits.

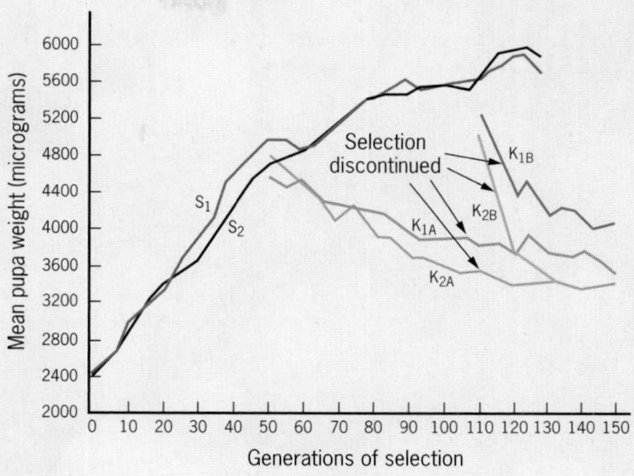

(*a*)

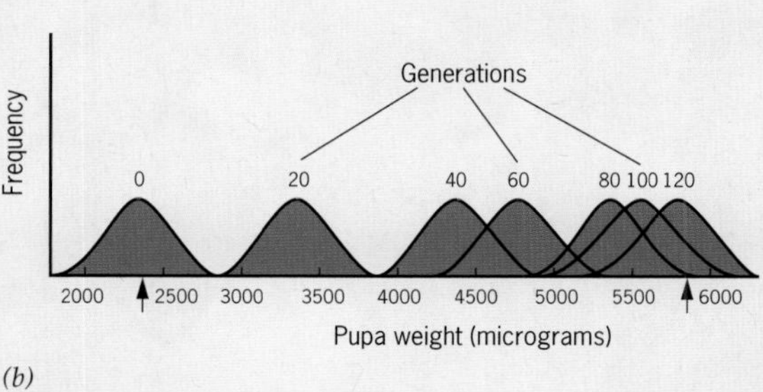

(*b*)

Figure 2 ▶ Enfield's experiments with the flour beetle, *Tribolium*. (*a*) Artificial selection for increased size as measured by pupa weight. The curves S_1 and S_2 show the response to selection in two replicate populations. The curves K_{1A}, K_{2A}, K_{1B}, and K_{2B} show what happened when artificial selection was discontinued in subpopulations that were established from the selected populations. (*b*) Frequency distributions of pupa weight in *Tribolium* populations selected for increased size. The shape of the distributions is only approximate. The means at generations 0 and 120 are indicated by arrows.

Figure 1 ▶ Contrasting sizes of flour beetles. The smaller beetle is from a standard laboratory stock. The larger beetle shows the result of over 120 generations of selection for increased body size.

Figure 23.9 ▶ Variation in fruit size, shape, and color in tomatoes.

cleave genomic DNA. For example, *Eco*RI may cleave at a particular site in the DNA of *L. pimpinellifolium*, but not cleave at this site in the DNA of *L. esculentum* because the *Eco*RI recognition sequence there (GAATTC) had mutated. Differences of this sort create *restriction fragment-length polymorphisms* (RFLPs) that can be analyzed by Southern blotting (see Figure 16.3 and the associated discussion in Chapter 16). Tanksley and colleagues catalogued a large number of RFLPs in the tomato genome and then positioned them on the genetic maps of the chromosomes by observing the frequency of recombination in hybrids created by crossing the two different species. In effect, they treated the RFLPs as molecular genetic markers and performed recombination experiments similar to the ones using phenotypic markers that we discussed in Chapter 7. Altogether, 88 RFLP loci were positioned on the maps of the tomato chromosomes. Then Tanksley and Zachary Lippman carried out an experiment to determine which of these loci was associated with differences in fruit weight. The experimental procedures are outlined in **FIGURE 23.10**.

L. pimpinellifolium plants were crossed to the Giant Heirloom variety of *L. esculentum*, and a single F_1 plant was self-fertilized to produce F_2 progeny. At each stage in the experiment, the fruits produced by each plant were weighed. The parental strains differed dramatically in fruit weight: 1 gram for *L. pimpinellifolium* and 500 grams for *L. esculentum*. The fruit of the F_1 plant averaged 10.5 grams, and the fruit of the 188 F_2 plants that were generated averaged 11.1 grams. However, among the F_2 plants, fruit weight varied considerably, with some plants bearing fruit that averaged more than 20 grams. This variation is due to the segregation of genes affecting fruit weight. To locate these genes—or QT loci—on the genetic map, Tanksley and Lippman determined the RFLP genotypes of the F_2 plants. DNA was extracted from individual plants, digested with restriction enzymes, and analyzed by Southern blotting to determine what RFLP markers were present. For a particular RFLP locus, an F_2 plant could be homozygous for the marker from *L. pimpinellifolium*, it could be homozygous for the marker from *L. esculentum*, or it could be heterozygous—that is, carry a marker from each species. We can designate these genotypes as *LP/LP*, *LE/LE*, and *LP/LE*, respectively. Each F_2 plant was genotyped for the *LP* and *LE* markers at each of the 88 RFLP loci—a heroic undertaking.

Then Tanksley and Lippman studied the relationship between the genotypes at each *RFLP* locus and fruit weight. For example, at the *TG167* RFLP locus on chromosome 2, they found that plants that were homozygous for the *LP* marker had fruits that weighed 8.4 grams, that plants that were heterozygous for the *LP* and *LE* markers had fruits that weighed 10.0 grams, and that plants that were homozygous for the *LE* marker had fruits that weighed 17.5 grams. Thus, at this RFLP locus it seems that the *LE* marker is associated with increased fruit weight, which suggests that in *L. esculentum* there is an allele for increased fruit weight somewhere near the *TG167* locus. However, we cannot conclude that the allele for increased fruit weight is actually at the *TG167* locus—only that it is nearby. Thus, this analysis points to the existence of a QTL affecting fruit weight near *TG167* on chromosome 2. Tanksley and Lippman designated this QTL as *fw2.2*.

After examining the relationship between fruit weight and the genotypes at all the other RFLP loci, Tanksley and Lippman concluded that there are five additional fruit weight loci, including one more on chromosome 2, two on chromosome 1, and one each on chromosomes 3 and 11 (**FIGURE 23.11**). More detailed mapping studies ultimately allowed Tanksley and colleagues to pinpoint the *fw2.2* QTL and show that it is a single gene, *ORFX*. This gene is expressed early in floral development and is structurally similar to the human *c-ras* oncogene. Thus, its product might be involved in signal transduction within cells (see Chapters 21 and 22). To delve deeper into the analysis of QT loci in the tomato, work through the Focus on Problem Solving: Detecting Dominance at a QTL.

Tanksley's research shows that identifying and mapping QT loci can be an elaborate and time-consuming enterprise. Fortunately, newer technologies such as gene chips that detect

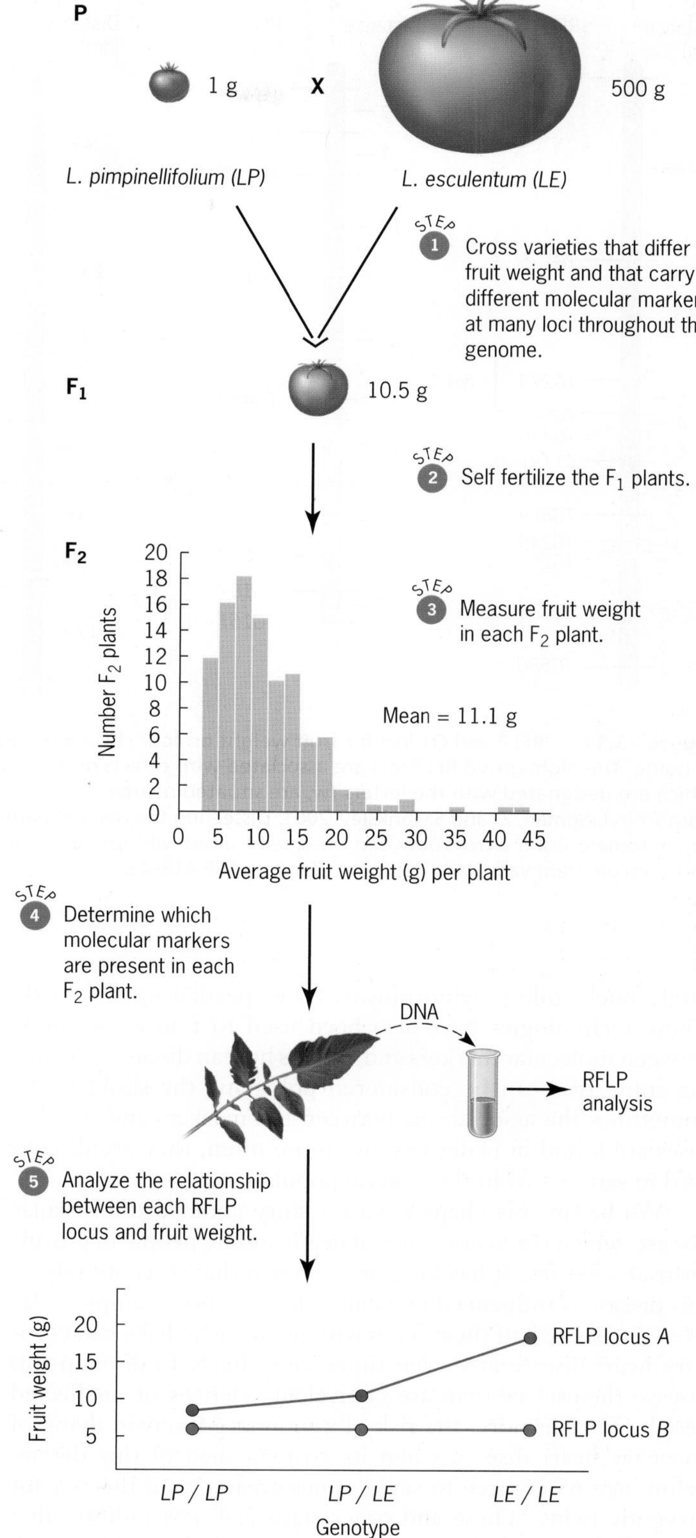

Figure 23.10 ▶ Methods to identify QT loci for fruit weight in tomatoes. Two different species of tomatoes were crossed to produce an F_1 plant, which was self-fertilized to produce many F_2 plants, each of which was characterized for the quantitative trait fruit weight and a battery of loci whose alleles are defined by restriction fragment-length polymorphisms (RFLPs). The resulting data were analyzed to determine if fruit weight was related to the genotypes at any of the RFLP loci. The *LP* allele is derived from *L. pimpinellifolium,* and the *LE* allele is derived from *L. esculentum*. For one RFLP locus (*A*), the *LE* allele increases fruit weight when it is homozygous. For the other RFLP locus (*B*), the *LE* allele has no effect on fruit weight. A QTL for fruit weight therefore appears to be located near RFLP locus A.
Data from Lippman, Z. and S. Tanksley. 2001. Dissecting the genetic pathway to extreme fruit size in tomato using a cross between the small-fruited wild species *Lycopersicon pimpinellifolium* and *L. esculentum* var. Giant Heirloom. *Genetics* 158:413–422.

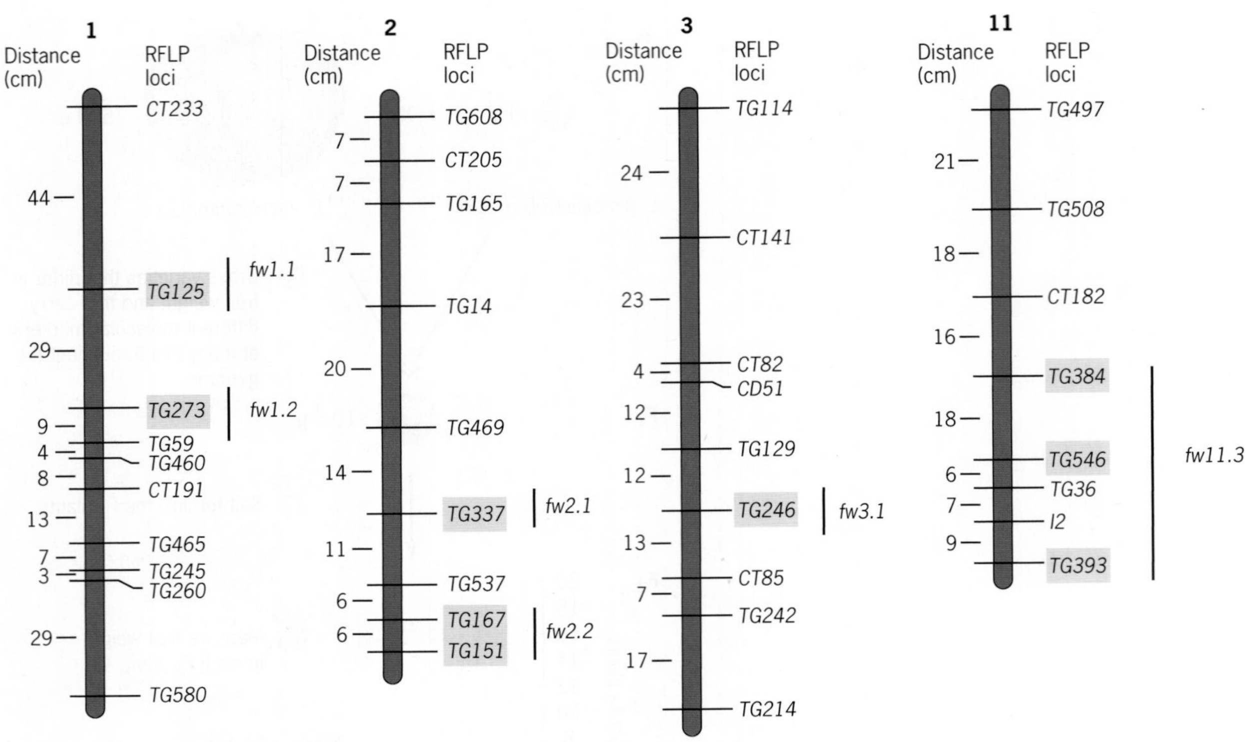

Figure 23.11 ▶ RFLP and QT loci for fruit weight on four chromosomes in the tomato genome. The highlighted RFLP loci are associated with effects on fruit weight. The QT loci, which are designated with the letters *fw*, are situated nearby.
Data from Lippman, Z. and S. Tanksley. 2001. Dissecting the genetic pathway to extreme fruit size in tomato using a cross between the small-fruited wild species *Lycopersicon pimpinellifolium* and *L. esculentum* var. Giant Heirloom. *Genetics* 158:413–422.

single nucleotide polymorphisms have speeded up the work. These technologies have also been used to find associations between molecular markers and various human diseases, including some that can be considered polygenic threshold traits. Sometimes the associations between the markers and the diseases are found in pedigrees, but more often, they are discovered in samples from the general population.

We began this chapter with a story about cardiovascular disease, which is a major cause of death among people in postindustrial societies. It has long been known that susceptibility to this disease is influenced by genetic factors. For example, relatives who share half their genes with people who have had coronary heart disease are seven times more likely to develop this disease themselves than are equivalent relatives of unaffected people. Furthermore, the risk of a monozygotic twin dying of coronary heart disease when its co-twin died of this disease before age 65 is three to seven times greater than the risk for dizygotic twins. These and other statistical data indicate that susceptibility to cardiovascular disease is under genetic control. Current research is focusing on efforts to identify specific genes that contribute to variation in the factors that put people at risk to develop this disease. These factors include plasma cholesterol level, obesity, blood pressure, high- and low-density lipoprotein levels, and triglyceride level. **TABLE 23.2** lists some of the QT loci that have been identified in these efforts.

KEY POINTS

- The total phenotypic variance can be partitioned into genetic and environmental components: $V_T = V_g + V_e$.
- The phenotypic variance in a population that is genetically uniform estimates V_e.
- The broad-sense heritability is the proportion of the total phenotypic variance that is genetic variance: $H^2 = V_g / V_T$.
- The genetic variance can be subdivided into additive genetic, dominance, and epistatic variances: $V_g = V_a + V_d + V_i$.
- The narrow-sense heritability is the proportion of the total phenotypic variance that is due to the additive effects of alleles: $h^2 = V_a / V_T$.
- The narrow-sense heritability is used to predict the phenotypes of offspring (T_O) given the average phenotype of the parents (T_P) and the mean phenotype in the population (μ) from which the parents came: $T_O = \mu + h^2 (T_P - \mu)$.
- The response to artificial selection can be predicted from the narrow-sense heritability and the selection differential: $R = h^2S$.
- By using molecular markers, geneticists are able to identify and map quantitative trait loci.

TABLE 23.2

Quantitative Trait Loci That Contribute to Variation in Risk Factors for Cardiovascular Disease

Locus	Gene Product	Chromosome	Risk Factor
AGT	Angiotensin	1	Blood pressure
APOA-1	Apolipoprotein A1	11	HDL[a] cholesterol
APOA-2	Apolipoprotein A2	1	HDL cholesterol
APOA-4	Apolipoprotein A4	11	HDL cholesterol, triglycerides
APOB	Apolipoprotein B	2	LDL[b] cholesterol
APOC-3	Apolipoprotein C3	11	Triglycerides
APOE	Apolipoprotein E	19	LDL cholesterol, triglycerides
CETP	Cholesterol ester transfer protein	16	HDL cholesterol
DCP	Dipeptidyl carboxypeptidase	17	HDL cholesterol, blood pressure
FGA/B	Fibrinogen A and B	4	Fibrinogen
HRG	Histidine-rich glycoprotein	3	Histidine-rich glycoprotein
LDLR	Low-density lipoprotein receptor	19	LDL cholesterol
LPA	Lipoprotein (a)	6	HDL cholesterol, triglycerides
LPL	Lipoprotein lipase	8	Triglycerides
PLAT	Plasminogen activator tissue-type	8	Tissue plasminogen activator level
PLANH1	Plasminogen activator inhibitor-1	7	PAI-1 level

Source: G. P. Vogler et al. 1997. Genetics and behavioral medicine: risk factors for cardiovascular disease. *Behavioral Medicine* 22:141–149.

[a]High-density lipoprotein.

[b]Low-density lipoprotein.

Correlations Between Relatives

Quantitative analyses of the resemblance between relatives can provide estimates of broad- and narrow-sense heritabilities.

Much of classical genetic analysis involves comparisons between relatives—parents and offspring, siblings, half siblings, and so forth. The usual procedure is to follow a particular trait through a series of crosses or to trace it through a collection of pedigrees. By analyzing the data, it is possible to discern whether or not the trait has a genetic basis. If it does, further work may allow the researcher to identify the gene or genes involved—to locate these genes on chromosomes and, ultimately, to analyze them at the molecular level. For complex traits that involve many genes and that are also influenced by a host of environmental factors, this type of analysis is extremely difficult. Nevertheless, comparisons between relatives can provide useful information about the underlying genetic variation in the trait.

CORRELATING QUANTITATIVE PHENOTYPES BETWEEN RELATIVES

Relatives often have similar phenotypes for a quantitative trait. As an example, let's consider data on the heights of monozygotic twins. **FIGURE 23.12*a*** shows such data, with each twin pair represented as a point in a graph. The height of one member of each pair is plotted on the horizontal or x-axis, and the height of its co-twin is plotted on the vertical or y-axis. From the graph it is clear that monozygotic twins are remarkably similar with respect to height. When one twin is short, the other tends to be short too; when one twin is tall, the other also tends to be tall. We refer to this pattern of resemblance as a positive correlation, and we summarize it quantitatively by calculating a statistic called the *correlation coefficient*, usually symbolized by the letter r. Let's denote the height of the twin plotted on the x-axis by the letter X and that of its co-twin plotted on the y-axis by the letter Y; then the correlation coefficient for all the twin pairs in the graph is calculated from the expression

$$r = \Sigma[(X_k - \overline{X})(Y_k - \overline{Y})] / [(n - 1) s_X s_Y]$$

In this formula, $\overline{X}$ and $\overline{Y}$ are the sample means of the twins plotted on the x- and y-axes, s_X and s_Y are the respective sample standard deviations, and n is the number of twin pairs. The Greek letter Σ indicates a summation on the index k over all the twin pairs. This formula provides researchers with a way of

▶ FOCUS ON PROBLEM SOLVING
Detecting Dominance at a QTL

THE PROBLEM

a. FIGURE 23.10 shows how Zachary Lippman and Steven Tanksley identified QT loci for fruit weight in tomatoes. The parents in the initial cross differed dramatically in the average weights of their fruits—1 gram versus 500 grams. The F_1 fruits averaged 10.5 grams, and the F_2 fruits averaged 11.1 grams. Why do these data indicate that dominance plays a role in determining fruit weight in tomatoes?

b. Lippman and Tanksley identified six QT loci affecting fruit weight. One locus, *fw11.3*, was located near the RFLP locus *TG36* on chromosome 11. Another locus, *fw2.2*, was located near the RFLP locus *TG167* on chromosome 2. When F_2 plants were genotyped for these two loci, Lippman and Tanksley found the following relationship between the genotypes and average fruit weight (all values in grams)[1]:

		Genotype of F_2 Plants		
QTL	**RFLP Locus**	***LP/LP***	***LP/LE***	***LE/LE***
fw11.3	*TG36*	6.2	12.2	20.0
fw2.2	*TG167*	8.4	10.0	17.5

Which QTL shows dominance for the trait fruit weight? Which of the alleles, *LE* or *LP*, is dominant?

FACTS AND CONCEPTS

1. When alleles act additively, the phenotype of the heterozygote is midway between the phenotypes of the two homozygotes.

2. To a quantitative geneticist, dominance exists when the alleles are not acting in a strictly additive fashion. Dominance is, therefore, a deviation from strict additivity.

3. For a single locus acting on a trait, dominance is indicated when the phenotype of the heterozygote is not midway between the phenotypes of the two homozygotes.

4. For many loci acting on a trait, dominance is indicated when the phenotype of the F_1 is not midway between the phenotypes of the two parents.

ANALYSIS AND SOLUTION

a. The mean fruit weights in the P, F_1, and F_2 generations indicate that dominance plays a role in determining this quantitative trait. The F_1 and F_2 averages are much closer to the fruit weight of *L. pimpinellifolium* than *L. esculentum*. This skew is clear evidence for dominance.

b. The *fw2.2* QTL shows dominance, whereas the *fw11.3* QTL does not. For *fw2.2*, the heterozygote's phenotype is close to the phenotype of the *LP/LP* homozygote, not midway between the two homozygotes. This observation indicates that the *LP* allele of the *fw2.2* QTL is partially dominant over the *LE* allele. By contrast, the phenotype of the heterozygote at the *fw11.3* QTL is nearly midway between that of the two homozygotes. Thus, the alleles of this locus appear to act more or less additively to determine fruit weight.

For further discussion go to your *WileyPLUS* course.

[1]Data from Lippman, Z. and S. Tanksley. 2001. Dissecting the genetic pathway to extreme fruit size in tomato using a cross between the small-fruited wild species *Lycopersicon pimpinellifolium* and *L. esculentum* var. Giant Heirloom. *Genetics* 158:413–422.

assigning a numerical score to a set of paired measurements such as the heights of twins in the graph. The value of the correlation coefficient can range from −1 to +1, with −1 indicating a perfect negative correlation between the *X*'s and the *Y*'s (high values on one axis consistently paired with low values on the other axis) and +1 indicating a perfect positive correlation. When the correlation coefficient is zero, we say that the measurements are uncorrelated. This type of situation is illustrated in **FIGURE 23.12*b***, where there is no consistent relationship between the values plotted on the *x*- and *y*-axes. For the twin data in **FIGURE 23.9*a***, the correlation coefficient is +0.84, which is very close to +1. Thus, monozygotic twins show a strong positive correlation with respect to height.

Correlation coefficients can be calculated for all sorts of quantitative phenotypes—height, weight, IQ score, and so forth. Furthermore, these coefficients can be calculated using data from different types of relatives—for example, from pairs of twins, pairs of full siblings, pairs of half siblings, and pairs of first cousins. We can also calculate correlation coefficients using data from unrelated individuals—for example, from pairs of college roommates. If some of the variation in a quantitative trait is due to genetic differences among individuals, we would expect the value of the correlation coefficient to increase with the closeness

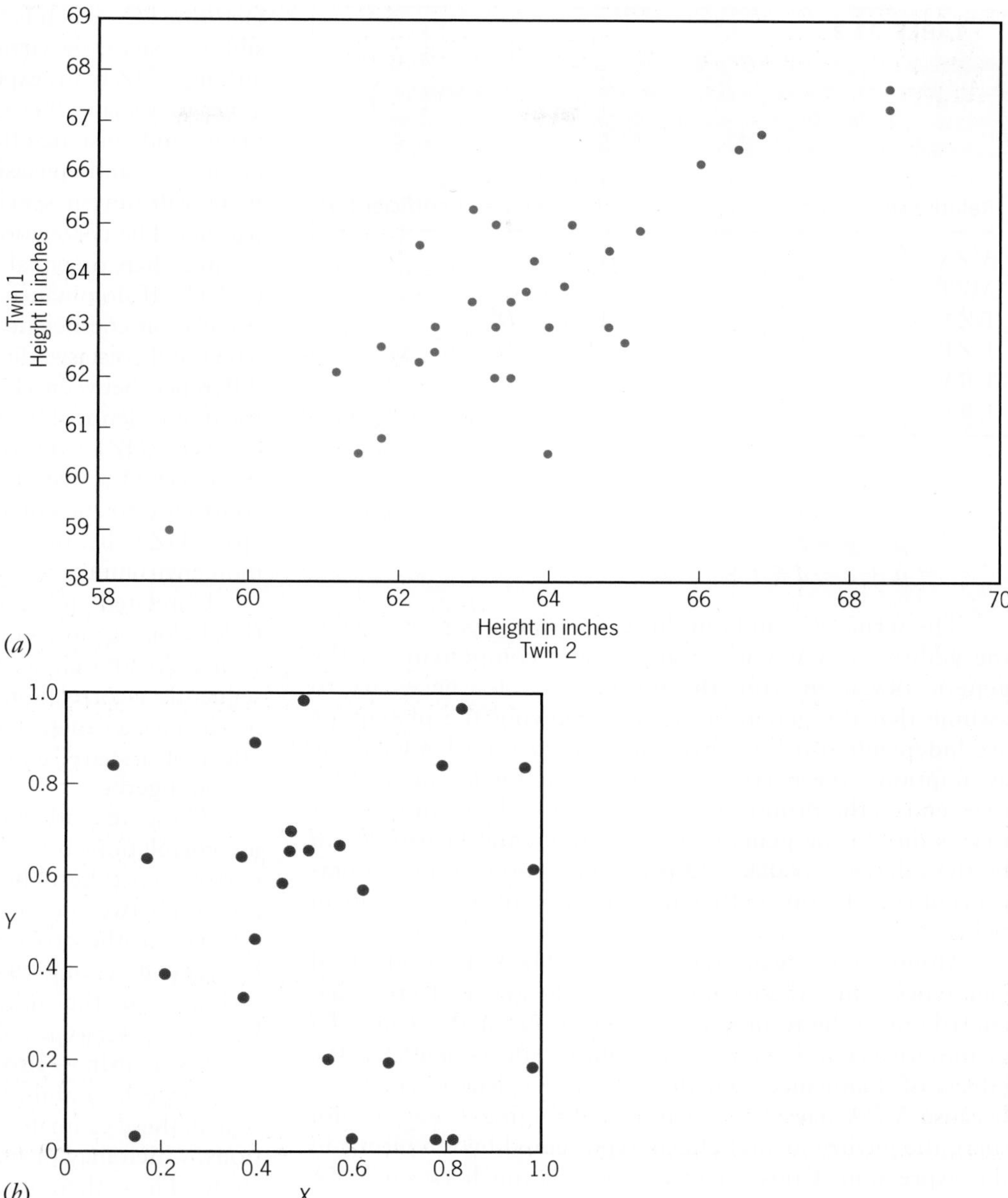

Figure 23.12 ▶ Correlations between paired data points. (*a*) Positive correlation for height between monozygotic twins (data courtesy of Thomas Bouchard, University of Minnesota). (*b*) A set of paired data in which the correlation coefficient is close to zero.

of the genetic relationship. Thus, monozygotic twins, who share 100 percent of their genes, should be more strongly correlated than first cousins, who share 12.5 percent of their genes.

INTERPRETING CORRELATIONS BETWEEN RELATIVES

We have already seen that variation in a quantitative trait can be partitioned into genetic and environmental components. The broad-sense heritability (H^2) is the proportion of the phenotypic variance that is due to genetic variation in a population, and the narrow-sense heritability (h^2) is the proportion of the phenotypic variance that is due to additive genetic variation in a population. If dominance and epistasis influence a trait, we expect the broad-sense heritability to be greater than the narrow-sense heritability. If these factors do not influence a trait, then the broad-sense heritability and the narrow-sense heritability are equivalent.

Correlation coefficients calculated by the formula given in the previous section can be interpreted in terms of broad- and narrow-sense heritabilties. Geneticists have analyzed the relationships among these quantities, beginning with the pioneering work of R. A. Fisher. This analysis assumes that T, the value of a trait in an individual, is equal to the mean of the population (μ) plus genetic (g) and environmental (e) deviations from the mean:

▶ **TABLE 23.3**

Theoretical Values of Correlation Coefficients for MZ and DZ Twins and Unrelated Individuals Reared Together or Apart

Relationship	Theoretical Value of Correlation Coefficient (r)
MZA	H^2
MZT	$H^2 + C^2$
DZA	$(1/2)h^2 + D^2$
DZT	$(1/2)h^2 + D^2 + C^2$
URA	0
URT	C^2

$$T = \mu + g + e$$
$$= \mu + a + d + i + e$$

The terms a, d, and i in this expression are, respectively, the additive, dominance, and epistatic components of the genetic deviation from the mean. It is also necessary to assume that the genetic factors influencing the phenotype are independent of the environmental factors. Under these assumptions, the correlation coefficient for a pair of relatives equals the proportion of the total variance in the trait that is due to the genetic and environmental factors *shared* by the relatives. **TABLE 23.3** presents theoretical interpretations of correlation coefficients for different types of human twins.

Monozygotic twins reared apart (MZA) have identical genotypes. Thus, these twins share all the genetic factors that contribute to the term g in the expression for the value of a quantitative trait, including the additive effects of alleles, the effects of dominance, and the effects of epistasis. However, because MZA have had separate upbringings, they do not share the environmental effects represented by the term e in the expression. Consequently, a correlation between MZA depends only on their identical genotypes. In the theory of quantitative genetics, this correlation equals the proportion of the total phenotypic variance that is due to genetic differences among the twin pairs—that is, it equals the broad-sense heritability, H^2.

Monozygotic twins reared together (MZT) have a common environment as well as identical genotypes. A correlation between them therefore equals the proportion of the total variance that is due to shared gentoypes (H^2), plus the proportion that is due to shared environmental factors. This latter component, which is denoted by the term C^2 in **TABLE 23.3**, is called the **environmentality.**

Dizygotic (DZ) twins are as closely related as ordinary siblings. To interpret a correlation coefficient between DZ twins, we must therefore discount its genetic component by a factor of 1/2, which is the fraction of genes that DZ twins (or siblings) share by virtue of common ancestry. Furthermore, although DZ twins experience the same additive effects of the genes they share, they experience only some of the same dominance and epistatic effects. This diminished similarity due to dominance and epistasis reflects the low probability that DZ twins will inherit specific combinations of alleles from their parents. The correlation coefficient for DZ twins is therefore greater than or equal to $(1/2)h^2$, but less than or equal to $(1/2)H^2$. If dominance and epistasis are negligible, then the correlation coefficient equals $(1/2)h^2$. If there is some dominance and epistasis, then it equals $(1/2)h^2$ plus a fraction of the difference between $(1/2)H^2$ and $(1/2)h^2$. In **TABLE 23.3**, this fraction is denoted by the term D^2. For dizygotic twins reared together (DZT) the correlation coefficient will also include the effect of a shared environment (C^2). This effect will not contribute to the correlation between dizygotic twins reared apart (DZA) because these types of twins do not share a common environment.

Unrelated individuals reared apart (URA) or together (URT, for example unrelated children adopted into the same family) do not share genes by virtue of common ancestry. Consequently, a correlation between these types of individuals does not involve a genetic component. However, it does involve the effect of a shared environment (C^2) if the individuals were reared together.

These and other theoretical results allow geneticists to use correlations between relatives to estimate the broad- and narrow-sense heritabilities for quantitative traits. The correlation between monozygotic twins reared apart provides an estimate of the broad-sense heritability, and the correlation between dizygotic twins reared apart provides a maximal estimate of the narrow-sense heritability. Correlations between other types of relatives—full siblings, half-siblings, and first cousins—also provide maximal estimates of the narrow-sense heritability. It should be emphasized, however, that all these estimates depend on several simplifying assumptions, which may or may not be met in the population under study. Thus, their interpretation is subject to considerable uncertainty.

KEY POINTS

- The correlation coefficient summarizes the degree of association between paired measurements, X_k and Y_k: $r = \Sigma\left[(X_k - \overline{X})(Y_k - \overline{Y})\right] / \left[(n-1)\, s_X s_Y\right]$.
- A correlation coefficient can be used to estimate the proportion of the total variance in a quantitative trait that is due to genetic and environmental factors shared by relatives.
- The correlation between monozygotic twins reared apart provides an estimate of the broad-sense heritability.
- The correlation between dizygotic twins reared apart provides a maximum estimate of the narrow-sense heritability.

Quantitative Genetics of Human Behavioral Traits

Quantitative genetics theory has been used to assess the heritability of intelligence and personality traits in humans.

Animals exhibit a wide range of behaviors associated with feeding, courtship, reproduction, and a host of other activities. The genetic determinants of these behaviors are only now beginning to be identified through experimental work. Studies with mutant strains of worms, fruit flies, and mice have revealed several genes that influence behavior. Research on human beings has also indicated that behavior is affected by genetic factors. For example, people with Huntington's disease gradually lose motor control and mental function; as the disease progresses, they may become depressed, even psychotic. Huntington's disease is due to a dominant mutation that is manifested in adults, usually after age 30. At present, there is no cure. Phenylketonuria is another human genetic condition with a behavioral phenotype. People with this disease accumulate toxic metabolites in their nervous tissues, including the brain. Without treatment—which involves restricting the amount of phenylalanine consumed in the food—individuals with this disorder fail to develop normal mental abilities. Still another example of how the genotype can influence behavior is Down syndrome, a condition that arises from the presence of an extra chromosome 21. People with this condition have below-normal mental abilities, and if they survive to middle age, they invariably develop Alzheimer's disease, a form of dementia that also occurs in chromosomally normal individuals, although at a much lower rate and usually much later in life. People with Alzheimer's disease gradually, but inexorably, lose their memories and intellectual functions; they become progressively more forgetful and disoriented, and need to be monitored constantly to prevent them from hurting themselves or others. Researchers now believe that Alzheimer's disease may be caused by extra copies or mutant alleles of a gene located on chromosome 21. Mutant alleles of other genes may also lead to Alzheimer's disease.

Conditions such as Huntington's disease, phenylketonuria, and Down syndrome indicate that genetic factors can influence human behavior. However, these conditions do not offer much insight into the nature of the behavioral differences that we see in the general population. Does genetic variation account for some of these differences, and if it does, what proportion of the overall variability is due to genetic factors? These provocative questions fall within the purview of quantitative genetics. In the following sections, we apply quantitative genetics theory to the study of two complex human behavioral traits, intelligence and personality.

INTELLIGENCE

The term *intelligence* refers to an assortment of mental abilities, including verbal and mathematical skills, memory and recall, reasoning and problem solving, discrimination of different objects, and spatial perception. For more than a century, psychologists have tried to characterize and quantify these abilities by administering intelligence tests. The tests—and many different ones have been used—attempt to measure general reasoning ability. The score that an individual makes on one of these tests is converted into an *intelligence quotient*, or *IQ*, which is scaled so that the mean of the population is 100 and the standard deviation is 15. Although there is considerable debate about what an IQ score actually measures—is it a true reflection of a person's intelligence?—these scores have been used to assess whether variation in mental abilities has a genetic component. Some of the most revealing data have come from studies of monozygotic and dizygotic twins.

For IQ test scores, the correlation coefficients of MZ twins, reared together or apart, are very high—in the range of 0.7–0.8 (**TABLE 23.4**). By comparison, the correlation coefficients of DZ twins tend to be lower—presumably because they share only half their genes, and the correlation coefficients for unrelated individuals reared together are essentially zero. Such analyses strongly suggest that whatever an IQ test measures, it has a large genetic component. This conclusion is supported by other correlation analyses. For example, the IQs of adopted children are more strongly correlated with the IQs of their biological parents than with those of their adoptive parents. Thus, in the determination of IQ, the biological (that is, genetic) link between parents and children seems to be more influential than the environmental one.

TABLE 23.4

Correlation Coefficients for IQ Test Scores for MZ and DZ Twins, Reared Together or Apart[a]

Study	MZT	MZA	DZT	DZA
Newman et al. 1937		0.71		
Juel-Nielsen 1980		0.69		
Shields 1962		0.75		
Bouchard et al. 1990	0.83	0.75		
Pedersen et al. 1992	0.80	0.78	0.22	0.32
Newman et al. 1998				0.47
Average	0.82	0.75	0.22	0.38

[a]Data and references from Bouchard, T. J. 1998. Genetic and environmental influences on adult intelligence and special mental abilities. *Human Biol.* 70:257–279. By permission of the Wayne State University Press.

▶ A MILESTONE IN GENETICS: The Minnesota Study of Twins Reared Apart

Professor Thomas Bouchard, Jr., and his colleagues at the University of Minnesota consider twins who were separated early in life and reared apart to be "a fascinating experiment of nature."[1] These researchers have studied such twins to investigate the extent to which genes influence variation in human behavior. Their project, the Minnesota Study of Twins Reared Apart, began in 1979 and continues to provide information about the influence of the genome on an assortment of behavioral traits, including intelligence, personality, vocational interests, and social attitudes.

Because very few twins are separated after birth and reared in different households, the Minnesota Study of Twins Reared Apart is one of the few attempts to use such twins to disentangle genetic and environmental components of human phenotypic variation. In this project, reared-apart twins are identified through adoption records. Those twins who agree to become part of the study undergo extensive medical and psychological assessments, including IQ tests, life history interviews, personality tests, and psychiatric interviews. So far, more than 130 reared-apart twin pairs have been studied. They include both monozygotic and dizygotic twins.

Analysis of the data on these twins has led Bouchard and his colleagues to conclude that "(i) genetic factors exert a pronounced and pervasive influence on behavioral variability, and (ii) the effect of being reared in the same home is negligible for many psychological traits."[2] These conclusions run counter to the conventional wisdom about traits such as IQ and personality test scores. For decades, psychologists and social scientists have assumed that genetic variation contributed little to observed variation in these traits. However, Bouchard and colleagues estimate that the broad-sense heritability for personality test scores is around 0.5, and for IQ, it may be as high as 0.7. Both estimates come from correlations between monozygotic twins reared apart.

These conclusions should not be accepted uncritically. The reared-apart twins in the study shared the same prenatal environment, and some of them remained together for as long as four years before being placed in different homes. In addition, some of these twins were in contact with each other at various times after their separation. A portion of the similarity between the twins could therefore be due to shared experiences—that is, environmental effects—rather than to shared genes. However, monozygotic twins reared together show about the same level of phenotypic similarity as monozygotic twins reared apart. If a shared upbringing accounts for some of the similarity between monozygotic twins reared together, these twins might be expected to be more similar than monozygotic twins reared apart. Another complicating factor is that reared-apart twins might have been placed in similar homes. Nonrandom placement of separated twins could make them appear more alike than they would otherwise be. Despite these concerns, the twin correlation data seem to suggest that a significant portion of the variation in human behavioral traits is attributable to genetic differences in the population. Data from correlations between adopted children and their biological parents also support this view.

Thus, Bouchard and colleagues write:

> Our findings support and extend those from many family, twin, and adoption studies, a broad consilience of findings leading to the following generalization: For almost every behavioral trait so far investigated, from reaction time to religiosity, an important fraction of the variation among people turns out to be associated with genetic variation.[3]

Establishing that human populations are genetically variable is the first step in attempting to identify specific genes that influence behavior. Data from the Human Genome Project will facilitate this process; however, finding these genes will not be an easy task.

QUESTIONS FOR DISCUSSION

1. What are the implications of genetic variability for behavioral traits in human populations? How might the discovery of genes that influence behavior affect societal concerns such as education and justice?
2. In the first half of the twentieth century, the eugenics movement based its political program on the idea that genes influence behavior. Individuals deemed to have abnormal or undesirable behaviors were sterilized so that they could not pass on their genes. What is the legacy of the eugenics movement? How has it affected attitudes toward research on human behavior genetics today?

[1] Bouchard, T. J., Jr., D. T. Lykken, M. McGue, N. L. Segal, and A. Tellegen. 1990. Sources of human psychological differences: The Minnesota Study of Twins Reared Apart. *Science* 250:223–250.

[2] Ibid.

[3] Ibid.

What fraction of the variation among IQ scores is attributable to genetic differences among people? The most direct estimate comes from the correlation coefficient for MZ twins reared apart. Observed values of this correlation coefficient are around 0.7; thus, as much as 70 percent of the variation in IQ scores is attributable to genetic variability in the population. This estimate of the broad-sense heritability implies that, with respect to intelligence (as measured by IQ), people differ one from another more because of genetic factors than because of environmental factors.

PERSONALITY

Personality traits, like intelligence, can be assessed by testing. Psychologists use many different tests, some to measure personality characteristics and others to measure vocational and social interests. The results of these tests tend to be less reliable than those of IQ tests. Nevertheless, they quantify aspects of human personality in ways that allow them to be analyzed for genetic influences.

Perhaps the most thorough genetic analysis of personality in the general population has come from the Minnesota Study of Twins Reared Apart, a long-term research project carried out at the University of Minnesota. (See A Milestone in Genetics: The Minnesota Study of Twins Reared Apart.) The results from this project suggest that genetic differences explain a significant fraction of the overall variation in human personality, perhaps as much as 50 percent (**TABLE 23.5**). The correlation coefficient for the personality and psychological interest test scores of MZ twins reared apart ranges from 0.39 to 0.50. Thus, the broad-sense heritability for these traits is reasonably high. Additional insight into the genetic control of personality has come from studying conditions such as manic depression, schizophrenia, and alcoholism. The occurrence of these traits in the members of MZ and DZ twin pairs has been estimated, and the general finding is that MZ twins are more similar than DZ twins. Thus, for example, among male MZ twin pairs with one member identified as alcoholic, the co-twin is alcoholic 41 percent of the time. By contrast, a male DZ co-twin is alcoholic only 22 percent of the time. The greater concordance for alcoholism between MZ twins suggests that this trait is influenced by genetic factors.

TABLE 23.5

Mean Correlation Coefficients for MZ Twins Reared Together or Apart Who Were Evaluated for Personality Traits, Psychological Interests, and Social Attitudes as Part of the Minnesota Study of Twins Reared Apart[a]

Test Instrument	MZT	MZA
Personality traits		
Multidimensional Personality Questionnaire	0.49	0.50
California Psychological Inventory	0.49	0.48
Psychological interests		
Strong Campbell Interest Inventory	0.48	0.39
Jackson Vocational Interest Survey	NA	0.43
Minnesota Occupational Interest Scales	0.49	0.40
Social attitudes		
Religiosity Scales	0.51	0.49
Nonreligious Social Attitude Items	0.28	0.34
MPQ Traditionalism Scale	0.50	0.53

[a]Abstracted with permission from Bouchard et al. 1990. *Science* 250:223–228. Copyright 1990 American Association for the Advancement of Science.

KEY POINTS

- Studying monozygotic and dizygotic twins, reared together or apart, has been useful in assessing the extent to which genes influence behavior in the general human population.
- The broad-sense heritability for intelligence, as measured by IQ tests, is estimated to be 70 percent.
- The broad-sense heritability for personality traits is estimated to be between 34 and 50 percent.

Basic Exercises

ILLUSTRATE BASIC GENETIC ANALYSIS

1. In a plant species, stalk height is determined by four independently assorting genes, *A*, *B*, *C*, and *D*, each segregating two alleles; with each gene one allele, denoted by the superscript zero, adds nothing to the basic stalk height of 10 centimeters, whereas the other allele, denoted by the superscript one, adds one centimeter to the basic stalk height. If all the alleles of these genes act additively to determine stalk height, (a) what is the phenotype of a plant with the genotype $A^0A^1B^0B^1C^0C^1D^0D^1$, and (b) if this plant is selfed, what fraction of its offspring will be 10 centimeters tall?

Answer: (a) The phenotype of the quadruple heterozygote should be the basic height (10 cm) plus the contributions of each of the one-superscript alleles (4 cm)—that is, 14 cm. (b) Among the progeny of the selfed plant, only those that are homozygous for all the zero-superscript alleles will manifest the basic phenotype of 10 cm. These quadruple zero-homozygotes will have a frequency of $(1/4)^4 = 1/256$.

2. For schizophrenia, the concordance for monozygotic twins is 60 percent, and for dizygotic twins it is 10 percent. Do these facts argue that schizophrenia is a threshold trait with a genetic basis?

Answer: The greater concordance for monozygotic twins, which are genetically identical, does argue that schizophrenia is a threshold trait with a genetic basis. The lower concordance for dizygotic twins presumably reflects the fact that they share only 50 percent of their genes.

3. Which of the two frequency distributions shown below has (a) the greater mean, (b) the greater variance, (c) the greater standard deviation?

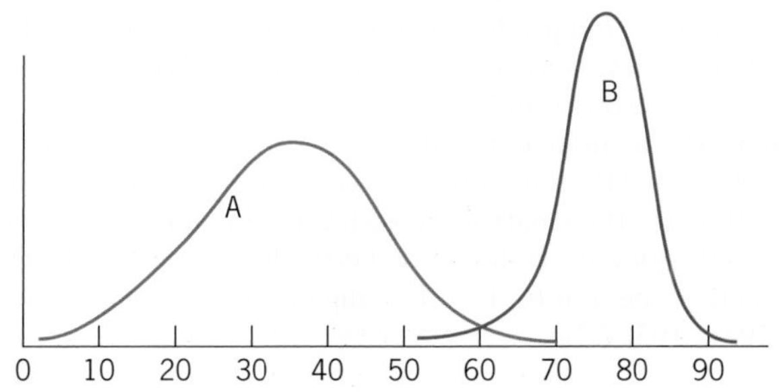

Answer: Distribution B has the greater mean. Distribution A has the greater variance and standard deviation.

4. Two phenotypically different highly inbred strains, P_1 and P_2, were crossed to produce an F_1 population, which was intercrossed to produce an F_2 population. In which strain or population is the genetic variance for a quantitative trait expected to be greater than zero?

Answer: The genetic variance is expected to be greater than zero in the F_2 population because it is segregating for the genetic differences introduced by the initial cross between P_1 and P_2. The inbred strains themselves as well as the F_1 population created by crossing them are expected to have little, if any, genetic variability. Thus, in each of these populations the genetic variance should be essentially zero.

5. Distinguish between the broad- and narrow-sense heritabilities.

Answer: The broad-sense heritability includes all the genetic variance as a fraction of the total phenotypic variance. The narrow-sense heritability includes only the additive genetic variance as a fraction of the total phenotypic variance.

6. Suppose that the correlation coefficient for height between human DZ twins reared apart is 0.30. What does this correlation suggest about the value of the narrow-sense heritability for height in this population?

Answer: Theoretically, the correlation coefficient for DZ twins reared apart estimates $(1/2)h^2 + D^2$, where D^2 reflects correlations due to dominance and epistasis. If we assume that neither dominance nor epistasis causes variation in this trait, then the correlation coefficient estimates $(1/2)h^2$. Thus, if we double the correlation coefficient, we obtain a maximum estimate of the narrow-sense heritability; $h^2 < 2 \times 0.30 = 0.60$.

▶ Testing Your Knowledge

INTEGRATE DIFFERENT CONCEPTS AND TECHNIQUES

1. A group of researchers studied variation in the number of abdominal bristles in female *Drosophila*. Two inbred strains that differed in bristle number were crossed to produce F_1 hybrids. The variance in bristle number among the F_1 flies was 3.33. These F_1 flies were intercrossed with one another to produce an F_2 population, in which the variance in bristle number was 5.44. Estimate the broad-sense heritability for bristle number in the F_2 population.

Answer: Because the F_1 flies were produced by crossing two inbred strains, they are genetically uniform. The variance observed among these flies therefore estimates the environmental variance, V_e. The variance observed among the F_2 flies, V_T, is the sum of the genetic variance, V_g, and the environmental variance, V_e. Thus, we can estimate V_g by subtracting the variance observed in the F_1 flies from that observed in the F_2 flies: $V_g = V_T - V_e = 5.44 - 3.33 = 2.11$. The broad-sense heritability, which is defined as V_g/V_T, is therefore $2.11/5.44 = 0.37$.

2. The mean value of a trait is 100 units, and the narrow-sense heritability is 0.3. A male and a female measuring 130 and 90 units, respectively, mate and produce a large number of offspring, which are reared in randomized environments. What is the expected value of the trait among these offspring?

Answer: The midparent value (the average of the two parents) is $(130 + 90)/2 = 110$. This value deviates from the population mean (100) by 10 units. If the narrow-sense heritability for the trait is 0.3, 30 percent of this deviation should be heritable. Consequently, the predicted value of the trait for the offspring of these two parents is $100 + (0.3 \times 10) = 103$.

3. In a study of MZ and DZ twins, reared together and apart, a group of Swedish researchers obtained the following correlation coefficients for IQ test scores: MZT, 0.80; MZA, 0.78; DZT, 0.22; DZA, 0.32. What do these correlations suggest about the extent to which variation in IQ scores is attributable to genetic variation? Are the results internally consistent?

Answer: The correlation for MZ twins reared apart, 0.78, implies that 78 percent of the population's variability in IQ is due to genetic variation—that is, the broad-sense heritability is 0.78. The slightly higher correlation for MZ twins reared together reinforces this conclusion and suggests that the effect of a common environment on the correlation for IQ is negligible. Thus, common environmental influences seem to account for a very small percentage of the overall variation in IQ within the population. The correlations for the DZ twins are generally in agreement with this view, but there is one inconsistency: the correlation for DZ twins reared together is less than that for DZ twins reared apart. We might have expected the correlation for DZ twins reared together to be as great as or greater than the correlation for DZ twins reared together. This inconsistency is probably due to sampling error. If we accept the correlation for DZ twins reared apart at face value, then doubling it should provide a maximal estimate of the narrow-sense heritability; $2 \times 0.32 = 0.64$. The fact that this estimate is less than the broad-sense heritability estimated from the correlation between MZ twins reared apart (0.78) suggests (albeit not too strongly given all the statistical uncertainties associated with these data) that some of the genetic variation in IQ is due to nonadditive genetic factors such as dominance and epistasis.

▶ Questions and Problems

ENHANCE UNDERSTANDING AND DEVELOP ANALYTICAL SKILLS

23.1 Assume that size in rabbits is determined by genes with equal and additive effects. From a total of 2012 F_2 progeny from crosses between true-breeding large and small varieties, eight rabbits were as small as the small variety and eight were as large as the large variety. How many size-determining genes were segregating in these crosses?

23.2 GO The height of the seed head in wheat at maturity is determined by several genes. In one variety, the head is just 7 inches above the ground; in another, it is 34 inches above the ground. Plants from the 7-inch variety were crossed to plants from the 34-inch variety. Among the F_1, the seed head was 20 inches above the ground. After self-fertilization, the F_1 plants produced an F_2 population in which 7-inch and 34-inch plants each appeared with a frequency of 1/256. (a) How many genes are involved in the determination of seed head height in these strains of wheat? (b) How much does each allele of these genes contribute to seed head height? (c) If a 20-inch F_1 plant were crossed to a 7-inch plant, how often would you expect 14-inch wheat to occur in the progeny?

23.3 If heart disease is considered to be a threshold trait, what genetic and environmental factors might contribute to the underlying liability for a person to develop this disease?

23.4 A sample of 10 plants from a population was measured in inches as follows: 17, 21, 20, 22, 20, 21, 20, 22, 19, and 23. Calculate (a) the mean, (b) the variance, and (c) the standard deviation.

23.5 For alcoholism, the concordance rate for monozygotic twins is 58 percent, whereas for dizygotic twins, it is 24 percent. Do these data suggest that alcoholism has a genetic basis?

23.6 GO A wheat variety with red kernels (genotype *A′*(*A′ B′B′*) was crossed with a variety with white kernels (genotype *AA BB*). The F_1 were intercrossed to produce an F_2. If each primed allele increases the amount of pigment in the kernel by an equal amount, what phenotypes will be expected in the F_2? Assuming that the *A* and *B* loci assort independently, what will the phenotypic frequencies be?

23.7 Measurements on ear length were obtained from three populations of corn—two inbred varieties and a randomly pollinated population derived from a cross between the two inbred strains. The phenotypic variances were 9.2 cm^2 and 9.6 cm^2 for the two inbred varieties and 26.4 cm^2 for the randomly pollinated population. Estimate the broad-sense heritability of ear length for these populations.

23.8 A researcher has been studying kernel number on ears of corn. In one highly inbred strain, the variance for kernel number is 357. Within this strain, what is the broad-sense heritability for kernel number?

23.9 Quantitative geneticists use the variance as a measure of scatter in a sample of data; they calculate this statistic by averaging the squared deviations between each measurement and the sample mean. Why don't they simply measure the scatter by computing the average of the deviations without bothering to square them?

23.10 Figure 23.4 summarizes data on maturation time in populations of wheat. Do these data provide any insight as to whether or not this trait is influenced by dominance? Explain.

23.11 The narrow-sense heritability for abdominal bristle number in a population of *Drosophila* is 0.3. The mean bristle number is 12. A male with 10 bristles is mated to a female with 20 bristles, and a large number of progeny are scored for bristle number. What is the expected number of bristles among these progeny?

23.12 GO Leo's IQ is 84 and Julie's IQ is 113. The mean IQ in the population is 100. Assume that the narrow-sense heritability for IQ is 0.35. What is the expected IQ of Leo and Julie's first child?

23.13 A fish breeder wishes to increase the rate of growth in a stock by selecting for increased length at six weeks after hatching. The mean length of six-week-old fingerlings is currently 10 cm. Adult fish that had a mean length of 15 cm at six weeks of age were used to produce a new generation of fingerlings. Among these, the mean length was 12.5 cm. Estimate the narrow-sense heritability of fingerling length at six weeks of age and advise the breeder about the feasibility of the plan to increase growth rate.

23.14 A breeder is trying to decrease the maturation time in a population of sunflowers. In this population, the mean time to flowering is 120 days. Plants with a mean flowering time of only 90 days were used to produce the next generation. If the narrow-sense heritability for flowering time is 0.3, what will the average time to flowering be in the next generation?

23.15 A quantitative geneticist claims that the narrow-sense heritability for body mass in human beings is 0.7, while the broad-sense heritability is only 0.3. Why must there be an error?

23.16 The mean value of a trait is 100 units, and the narrow-sense heritability is 0.5. A male and a female measuring 122 and 128 units, respectively, mate and produce a large number of offspring, which are reared in an average environment. What is the expected value of the trait among these offspring?

23.17 On the basis of the observed correlations for personality traits shown in Table 23.5, what can you say about the value of the environmentality (C^2 in Table 23.3)?

23.18 GO A selection differential of 30 μg per generation was used in an experiment to select for increased pupa weight in *Tribolium*. The narrow-sense heritability for pupa weight was estimated to be 0.2. If the mean pupa weight was initially 3000 μg and selection was practiced for 10 generations, what was the mean pupa weight expected to become?

23.19 One way to estimate a maximum value for the narrow-sense heritability is to calculate the correlation between half-siblings that have been reared apart and divide it by the fraction of genes that half-siblings share by virtue of common ancestry. A study of human half-siblings found that the correlation coefficient for height was 0.14. From this result, what is the maximum value of the narrow-sense heritability for height in this population?

▸ Genomics on the Web

at http://www.ncbi.nlm.nih.gov/

1. QTL mapping has been carried out for many organisms, including crop plants such as rice and maize. Follow the links to the *Zea mays* page, and then under Related Resources, go to the web site of "Gramene," a resource for comparative grass genomics. Explore the rice and maize QTL data by clicking on the links under Traits. For 100-grain weight in rice, how many QT loci have been mapped? How many of rice's 12 chromosomes contain at least one of these QT loci? For kernel row number in maize, how many QT loci have been mapped? On how many of maize's 10 chromosomes do these QT loci lie?

2. With many people now living into their seventh and eighth decades of life, Alzheimer's disease has become more frequent. Geneticists have found variants at several loci that seem to predispose people to develop this condition. These loci include *APOE*, *APP*, *PSEN1*, and *PSEN2*. Use the search function on the *Homo sapiens* web page to locate each of these loci in the human genome. On what chromosomes do they reside? Click on each locus to bring up a summary about the gene. How are the gene products thought to function in the etiology of Alzheimer's disease?

Chapter 24
Population Genetics

CHAPTER OUTLINE

- The Theory of Allele Frequencies
- Natural Selection
- Random Genetic Drift
- Populations in Genetic Equilibrium

Pitcairn Island in the south Pacific.

A Remote Colony

In September 1787, Lieutenant William Bligh and a crew of 45 men set sail from England aboard the ship H.M.S. *Bounty*. Their destination was the Pacific island of Tahiti, where they were to collect breadfruit tree saplings for transplantation to the Caribbean island of Jamaica. Because their passage around Cape Horn was blocked by ferociously bad weather, they sailed to Tahiti by crossing the south Atlantic, rounding the Cape of Good Hope, and then traversing the southern Indian Ocean and the western Pacific. Their voyage was long and difficult. When they finally reached Tahiti, they relaxed there and enjoyed the hospitality of the local people. After collecting the breadfruit saplings, Bligh and his crew departed Tahiti on April 6, 1789, bound for the Caribbean. Barely three weeks into the voyage, the crew mutinied. Led by Bligh's friend and chief subordinate Fletcher Christian, the mutineers put Bligh and his supporters into the ship's launch and set them adrift in the lonely waters of the south Pacific. Eventually Bligh and his men reached civilization. The mutineers initially returned to Tahiti, where some decided to stay, but nine of them, including Fletcher Christian, resolved to find another place to live. Along with a group of Polynesians—six men, twelve women, and a baby—they set sail in the *Bounty,* and on January 15, 1790, landed on Pitcairn Island, an uninhabited speck of land 1350 miles from Tahiti. Pitcairn Island had been discovered decades earlier, but because cartographers had put it in the wrong place on their charts, it held promise as a refuge for the mutineers. On January 23, 1790, Fletcher Christian and his followers burned the *Bounty* and set about establishing their new home.

Life on Pitcairn Island was not easy. The men fought over land and women, and the women murdered some of the men. In 1808, the island was visited by an American whaling ship, which found that only one of the original mutineers was still alive. British ships subsequently stopped at the island, and in 1838, Pitcairn Island was formally incorporated into the British Empire. By 1855 the population of the colony had increased to nearly 200, which was more than it could sustain, and in 1856 all the people were moved to Norfolk Island, a former British penal colony 3500 miles away. Two years later, 17 of the former inhabitants returned to Pitcairn Island to reestablish the colony, which has survived for nearly 150 years and today is home to about 50 people, all descendants of the original settlers.

The population on Pitcairn Island is the result of mixing two different groups of people, Britons and Polynesians. The offspring of the original settlers received genes from each of these groups, and when they reproduced, some of these genes were transmitted to their offspring and ultimately to the current members of the population. Which of the founding genes were passed down through time? How did factors such as the health, vigor, and reproductive ability of the people, and the ways in which they chose mates, influence the pathways of genetic descent? Did any of the genes mutate as they were transmitted through time? How did migration to and from the island affect its genetic composition? Has the island's genetic diversity increased, decreased, or remained the same? What is the significance of the population's size? Has the genetic composition of the population changed over time—that is, has it evolved?

These and other questions about the genetic makeup and history of the people on Pitcairn Island fall within the purview of *population genetics*, a discipline that studies genes in groups of individuals. Population genetics examines allelic variation among individuals, the transmission of allelic variants from parents to offspring generation after generation, and the temporal changes that occur in the genetic makeup of a population because of systematic and random evolutionary forces. In this chapter, we shall investigate how these forces—mutation, migration, selection, and random genetic drift—shape the genetic composition of a population. We begin with an introduction to the basic methods of population genetic analysis. As we shall see, these methods focus on the frequencies of the alleles that are present in the members of the population.

The Theory of Allele Frequencies

When the members of a population mate randomly, it is easy to predict the frequencies of the genotypes from the frequencies of their constituent alleles.

The theory of population genetics is a theory of allele frequencies. Each gene in the genome exists in different allelic states, and, if we focus on a particular gene, a diploid individual is either a homozygote or a heterozygote. Within a population of individuals, we can calculate the frequencies of the different types of homozygotes and heterozygotes of a gene, and from these frequencies we can estimate the frequency of each of the gene's alleles. These calculations are the foundation for population genetics theory.

ESTIMATING ALLELE FREQUENCIES

Because an entire population is usually too large to study, we resort to analyzing a representative sample of individuals from it. **TABLE 24.1** presents data from a sample of people who were tested for the M-N blood types. These blood types are determined by two alleles of a gene on chromosome 4: L^M, which produces the M blood type, and L^N, which produces the N blood type (see Chapter 4). People who are L^ML^N heterozygotes have the MN blood type.

TABLE 24.1

Frequency of the M-N Blood Types in a Sample of 6129 Individuals

Blood Type	Genotype	Number of Individuals
M	L^ML^M	1787
MN	L^ML^N	3039
N	L^NL^N	1303

To estimate the frequencies of the L^M and L^N alleles, we simply calculate the incidence of each allele among all the alleles sampled:

1. Because each individual in the sample carries two alleles of the blood-type locus, the total number of alleles in the sample is two times the sample size: $2 \times 6129 = 12{,}258$.
2. The frequency of the L^M allele is two times the number of L^ML^M homozygotes plus the number of L^ML^N heterozygotes, all divided by the total number of alleles sampled: $[(2 \times 1787) + 3039]/12{,}258 = 0.5395$.
3. The frequency of the L^N allele is two times the number of L^NL^N homozygotes plus the number of L^ML^N heterozygotes, all divided by the total number of alleles sampled: $[(2 \times 1303) + 3039]/12{,}258 = 0.4605$.

Thus, letting p represent the frequency of the L^M allele and letting q represent the frequency of the L^N allele, we estimate that in the population from which the sample was taken, $p = 0.5395$ and $q = 0.4605$. Furthermore, because L^M and L^N represent 100 percent of the alleles of this particular gene, $p + q = 1$.

When directly counting the number of alleles in a sample is not possible because one of the alleles is dominant, we cannot use this method to estimate the allele frequencies. However, another method, discussed in a later section, does provide these estimates.

When the gene under study is X-linked, we only need to count the different alleles in males. For example, in a sample of 200 men, 24 have X-linked color blindness and all the others have normal color vision. Assuming that each color-blind man is hemizygous for the same mutant allele, we estimate the frequency of that allele to be 24/200 = 0.12 and the frequency of the normal allele to be 1 − 0.12 = 0.88.

In these examples, each of the alleles has a reasonably high frequency—one that can be estimated reliably with a sample of moderate size. However, some alleles have frequencies of 0.01 or less, and estimating their frequencies, or even detecting them, requires a large sample. Whenever the second most frequent allele of a gene has a frequency greater than 0.01, we refer to the situation as a genetic **polymorphism.** Later in this chapter we shall discuss the evolutionary forces that maintain genetic polymorphisms in nature.

RELATING GENOTYPE FREQUENCIES TO ALLELE FREQUENCIES: THE HARDY–WEINBERG PRINCIPLE

Do the estimated allele frequencies have any predictive power? Can we use them to predict the frequencies of genotypes? In the first decade of the twentieth century, these questions were posed independently by G. H. Hardy, a British mathematician, and by Wilhelm Weinberg, a German physician. In 1908 Hardy and Weinberg each published papers describing a mathematical relationship between allele frequencies and genotype frequencies. This relationship, now called the **Hardy–Weinberg principle,** allows us to predict a population's genotype frequencies from its allele frequencies.

Let's suppose that in a population a particular gene is segregating two alleles, *A* and *a*, and that the frequency of *A* is p and that of *a* is q. If we assume that the members of the population mate randomly, then the diploid genotypes of the next generation will be formed by the random union of haploid eggs and haploid sperm (**FIGURE 24.1**). The probability that an egg (or sperm) carries *A* is p, and the probability that it carries *a* is q. Thus, the probability of producing an *AA* homozygote in the population is simply $p \times p = p^2$, and the probability of producing an *aa* homozygote is $q \times q = q^2$. For the *Aa* heterozygotes, there are two possibilities: An *A* sperm can unite with an *a* egg, or an *a* sperm can unite with an *A* egg. Each of these events occurs with probability $p \times q$, and because they are equally likely, the total probability of forming an *Aa* zygote is $2pq$. Thus, on the assumption of random mating, the predicted frequencies of the three genotypes in the population are:

Genotype	Frequency
AA	p^2
Aa	$2pq$
aa	q^2

These predicted frequencies can be obtained by expanding the binomial expression $(p + q)^2 = p^2 + 2pq + q^2$. Population geneticists refer to them as the Hardy–Weinberg genotype frequencies.

The key assumption underlying the Hardy–Weinberg principle is that the members of the population mate at random with respect to the gene under study. This assumption means that the adults of the population essentially form a pool of gam-

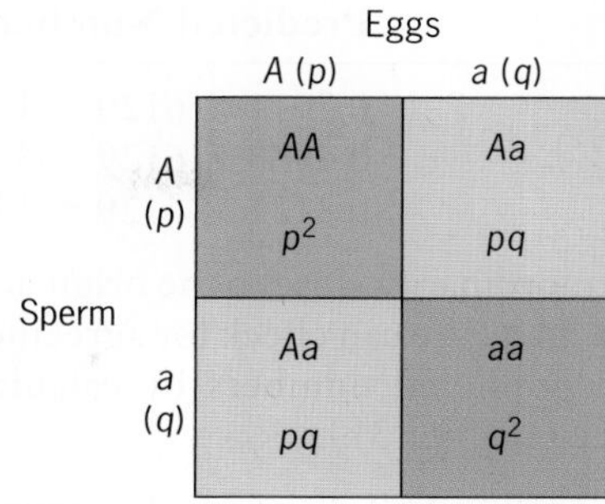

Figure 24.1 ▶ Punnett square showing the Hardy–Weinberg principle.

etes that, at fertilization, combine randomly to produce the zygotes of the next generation. If these zygotes have equal chances of surviving to the adult stage, then the genotype frequencies created at the time of fertilization will be preserved, and when the next generation reproduces, these frequencies will once again appear in the offspring. Thus, with random mating and no differential survival or reproduction among the members of the population, the Hardy–Weinberg genotype frequencies—and, of course, the underlying allele frequencies—persist generation after generation. This condition is referred to as the *Hardy–Weinberg equilibrium.* Later in this chapter we shall consider forces that upset this equilibrium by altering allele frequencies; these forces—mutation, migration, natural selection, and random genetic drift—play key roles in the evolutionary process.

APPLICATIONS OF THE HARDY–WEINBERG PRINCIPLE

The intellectual roots of the Hardy–Weinberg principle are discussed in A Milestone in Genetics later in this chapter. Here, let's return to the M-N blood type example to see how the Hardy–Weinberg principle applies to a real population. From the sample data given in **TABLE 24.1**, the frequency of the L^M allele was estimated to be $p = 0.5395$ and the frequency of the L^N allele was estimated to be $q = 0.4605$. With the Hardy–Weinberg principle, we can now use these frequencies to predict the genotype frequencies of the M-N blood type gene:

Genotype	Hardy–Weinberg Frequency
L^ML^M	$p^2 = (0.5395)^2 = 0.2911$
L^ML^N	$2pq = 2\,(0.5395)\,(0.4605) = 0.4968$
L^NL^N	$q^2 = (0.4605)^2 = 0.2121$

Do these predictions fit with the original data from which the two allele frequencies were estimated? To answer this question, we must compare the observed genotype numbers with numbers predicted by the Hardy–Weinberg principle. We obtain these predicted numbers by multiplying the Hardy–Weinberg frequencies by the size of the sample taken from the population:

Genotype	Predicted Number
$L^M L^M$	$0.2911 \times 6129 = 1784.2$
$L^M L^N$	$0.4968 \times 6129 = 3044.8$
$L^N L^N$	$0.2121 \times 6129 = 1300.0$

The results are extraordinarily close to the original sample data presented in **TABLE 24.1**. We can check for agreement between the observed and predicted numbers by calculating a chi-square statistic (see Chapter 3):

$$\chi^2 = \frac{(1787 - 1784.2)^2}{1784.2} + \frac{(3039 - 3044.8)^2}{3044.8} + \frac{(1303 - 1300.0)^2}{1300.0}$$
$$= 0.223$$

This chi-square statistic has $3 - 2 = 1$ degree of freedom because (1) the sum of the three predicted numbers is fixed by the sample size, and because (2) the allele frequency p was estimated directly from the sample data. (The frequency q can be estimated indirectly as $1 - p$ and therefore does not reduce the degrees of freedom any further.) The critical value for a chi-square statistic with one degree of freedom is 3.841 (see **TABLE 3.2**), which is much greater than the observed value. Consequently, we conclude that the predicted genotype frequencies are in agreement with the observed frequencies in the sample, and furthermore, we infer that in the population from which the sample was obtained, the M-N genotypes are in Hardy–Weinberg proportions—a finding that is not too surprising given that marriage is usually not based on blood type.

The preceding analysis indicates how we can use the Hardy–Weinberg principle to predict genotype frequencies from allele frequencies. Can we turn the Hardy–Weinberg principle around and use it to predict allele frequencies from genotype frequencies? For example, in the United States, the incidence of the recessive metabolic disorder phenylketonuria (PKU) is about 0.0001. Does this statistic allow us to calculate the frequency of the mutant allele that causes PKU?

We cannot proceed as before by counting the different types of alleles, mutant and normal, that are present in the population because heterozygotes and normal homozygotes are phenotypically indistinguishable. Instead, we must proceed by applying the Hardy–Weinberg principle in reverse to estimate the mutant allele frequency. The incidence of PKU, 0.0001, represents the frequency of mutant homozygotes in the population. Under the assumption of random mating, these individuals should occur with a frequency equal to the square of the mutant allele frequency. Denoting this allele frequency by q, we have

$$q^2 = 0.0001$$
$$q = \sqrt{0.0001} = 0.01$$

Thus, 1 percent of the alleles in the population are estimated to be mutant. Using the Hardy–Weinberg principle in the usual way, we can then predict the frequency of people in the population who are heterozygous carriers of the mutant allele:

$$\text{Carrier frequency} = 2pq = 2(0.99)(0.01) = 0.0198$$

Thus, approximately 2 percent of the population are predicted to be carriers.

The Hardy–Weinberg principle also applies to X-linked genes and to genes with multiple alleles. For an X-linked gene such as the one that controls color vision, the allele frequencies are estimated from the frequencies of the genotypes in males, and the frequencies of the genotypes in females are obtained by applying the Hardy–Weinberg principle to these estimated allele frequencies. (We assume, of course, that the allele frequencies are the same in the two sexes.) If the frequency of the allele for normal color vision (C) is $p = 0.88$ (taken from our earlier example) and the frequency of the allele for color blindness (c) is $q = 0.12$, then, under the assumptions of random mating and equal allele frequencies in the two sexes, we have:

Sex	Genotype	Frequency	Phenotype
Males	C	$p = 0.88$	Normal vision
	c	$q = 0.12$	Color blind
Females	CC	$p^2 = 0.77$	Normal vision
	Cc	$2pq = 0.21$	Normal vision
	cc	$q^2 = 0.02$	Color blind

For genes with multiple alleles, the Hardy–Weinberg genotype proportions are obtained by expanding a multinomial expression. For example, the A–B–O blood types are determined by three alleles I^A, I^B, and i. If the frequencies of these are p, q, and r, respectively, then the frequencies of the six different genotypes in the A–B–O blood-typing system are obtained by expanding the trinomial $(p + q + r) = p^2 + q^2 + r^2 + 2pq + 2qr + 2pr$:

Blood Type	Genotype	Frequency
A	$I^A I^A$	p^2
	$I^A i$	$2pr$
B	$I^B I^B$	q^2
	$I^B i$	$2qr$
AB	$I^A I^B$	$2pq$
O	ii	r^2

EXCEPTIONS TO THE HARDY–WEINBERG PRINCIPLE

There are many reasons why the Hardy–Weinberg principle might not apply to a particular population. Mating might not be random, the members of the population carrying different alleles might not have equal chances of surviving and reproducing, the population might be subdivided into partially isolated units, or it might be an amalgam of different populations that have come together recently by migration. We now briefly consider each of these exceptions to the Hardy–Weinberg principle.

1. *Nonrandom mating*. Random mating is the key assumption underlying the Hardy–Weinberg principle. If mating is not random, the simple relationship between allele frequencies and genotype frequencies breaks down. There are two ways

in which the members of a population might mate nonrandomly. First, they might mate with each other because they are genetically related—for example, because they are siblings or first cousins. We refer to this type of nonrandom mating as *consanguineous mating* (see Chapter 4). Second, the members of the population might mate with each other because they are phenotypically similar—for example, because the mates have the same stature or skin color. We refer to this type of nonrandom mating as *assortative mating*.

Consanguineous mating and assortative mating have the same qualitative effect; they reduce the frequency of heterozygotes and increase the frequency of homozygotes compared to the Hardy–Weinberg genotype frequencies. For the case of consanguineous mating, we can quantify this effect by using the inbreeding coefficient, F (see Chapter 4). Let's suppose that a gene has two alleles, A and a, with respective frequencies p and q, and that the population in which the gene is segregating has reached a level of inbreeding measured by F. (Recall from Chapter 4 that the range of F is between 0 and 1, with 0 corresponding to no inbreeding and 1 corresponding to complete inbreeding.) The genotype frequencies in this population are given by the following formulas:

Genotype	Frequency with Consanguineous Mating
AA	$p^2 + pqF$
Aa	$2pq - 2pqF$
aa	$q^2 + pqF$

From these formulas, it is clear that the frequencies of the two homozygotes have increased compared to the Hardy–Weinberg frequencies and that the frequency of the heterozygotes has decreased compared to the Hardy–Weinberg frequency. Notice that for each homozygote, the increase in frequency is exactly half the decrease in the frequency of the heterozygotes. Furthermore, each change in genotype frequency is directly proportional to the inbreeding coefficient. For a population that is completely inbred, $F = 1$, and the genotype frequencies become:

Genotype	Frequency with $F = 1$
AA	p
Aa	0
Aa	q

With assortative mating, mathematical expressions for the genotype frequencies are more complicated than those for consanguineous mating and are beyond the scope of this book. However, assortative mating has the same general effect as consanguineous mating: it increases the frequency of homozygotes and decreases the frequency of heterozygotes. These changes occur because phenotypically similar individuals tend to have similar genotypes. Thus, when such individuals mate, they tend to produce more homozygous offspring than do randomly mated individuals.

2. *Unequal survival.* If zygotes produced by random mating have different survival rates, we would not expect the genotype frequencies of the individuals that develop from these zygotes to conform to the Hardy-Weinberg predictions. For example, consider a randomly mating population of *Drosophila* that is segregating two alleles, A_1 and A_2, of an autosomal gene. A sample of 200 adults from this population yielded the following data:

Genotype	Observed Number	Expected Number
A_1A_1	26	46.1
A_1A_2	140	99.8
A_2A_2	34	54.1

The expected numbers were obtained by estimating the frequencies of the two alleles among the flies in the sample; the frequency of the A_1 allele is $(2 \times 26 + 140)/(2 \times 200) = 0.48$, and the frequency of the A_2 allele is $1 - 0.48 = 0.52$. Then the Hardy–Weinberg formulas were applied to these estimated frequencies. Obviously, the expected numbers are not in agreement with the observed numbers, which show an excess of heterozygotes and a dearth of both types of homozygotes. Here the disagreement is so obvious that a chi-square calculation to test the goodness of fit between the observed and expected numbers is unnecessary. The explanation for the disagreement probably lies with differential survival of the three genotypes during development from the zygote to the adult stage. The A_1A_2 heterozygotes survive better than either of the two homozygotes. Unequal survival rates can therefore lead to genotype frequencies that deviate from the Hardy–Weinberg predictions.

3. *Population subdivision.* When a population is a single interbreeding unit, it is said to be **panmictic. Panmixis** (the noun) implies that any member of the population is able to mate with any other member—that is, there are no geographical or ecological barriers to mating in the population. In nature, however, populations are often subdivided. We can think of fish living in a group of lakes that are intermittently connected by rivers, or of birds living on a chain of islands in an archipelago. Such populations are structured by geographical and ecological features that might be correlated with genetic differences. For example, the fish in one lake might have a high frequency of allele A, while those in another lake might have a low frequency of this allele. Although the genotype frequencies might conform to Hardy–Weinberg predictions within each lake, across the entire range of the fish population, they will not. Geographical subdivision makes the population genetically inhomogeneous, and such inhomogeneity violates a tacit assumption of the Hardy–Weinberg principle: that allele frequencies are uniform throughout the population.

4. *Migration.* When individuals move from one territory to another, they carry their genes with them. The introduction of genes by recent migrants can alter allele and genotype frequencies within a population and disrupt the state of

Hardy–Weinberg equilibrium. As an example, let's consider the situation in **FIGURE 24.2**. Two populations of equal size are separated by a geographical barrier. In population I the frequencies of *A* and *a* are both 0.5, whereas in population II the frequency of *A* is 0.8 and that of *a* is 0.2. With random mating within each population, the Hardy–Weinberg principle predicts that the two populations will have different genotype frequencies (see **FIGURE 24.2**).

Let's suppose that the geographical barrier between the populations breaks down and that the two populations merge completely. In the merged population, the allele frequencies will be the simple averages of the frequencies of the separate populations; the frequency of *A* will be $(0.5 + 0.8)/2 = 0.65$, and the frequency of *a* will be $(0.5 + 0.2)/2 = 0.35$. Moreover, the genotype frequencies in the merged population will be the simple averages of the genotype frequencies in the separate populations: the frequency of *AA* will be $(0.25 + 0.64)/2 = 0.445$, that of *Aa* will be $(0.50 + 0.32)/2 = 0.410$, and that of *aa* will be $(0.25 + 0.04)/2 = 0.145$. Notice, however, that these observed genotype frequencies are not equal to the frequencies predicted by the Hardy–Weinberg principle: $(0.65)^2 = 0.422$ for *AA*, $2(0.65)(0.35) = 0.455$ for *Aa*, and $(0.35)^2 = 0.123$ for *aa*. The reason for this discrepancy is that the observed genotype frequencies were not created by random mating within the entire merged population. Rather, they were created by amalgamating genotype frequencies from separate randomly mating populations. Thus, the merger of two randomly mating populations does not produce a population with Hardy–Weinberg genotype frequencies. However, if the merged population mates randomly for just one generation, Hardy–Weinberg genotype frequencies will be established, and the allele frequencies of the merged population will allow prediction of these genotype frequencies. This example demonstrates that merging randomly mating populations temporarily upsets Hardy–Weinberg equilibrium. The migration of individuals from one population to another also causes a temporary upset in Hardy–Weinberg equilibrium. However, if a population that has received migrants mates randomly for just one generation, Hardy–Weinberg equilibrium will be restored.

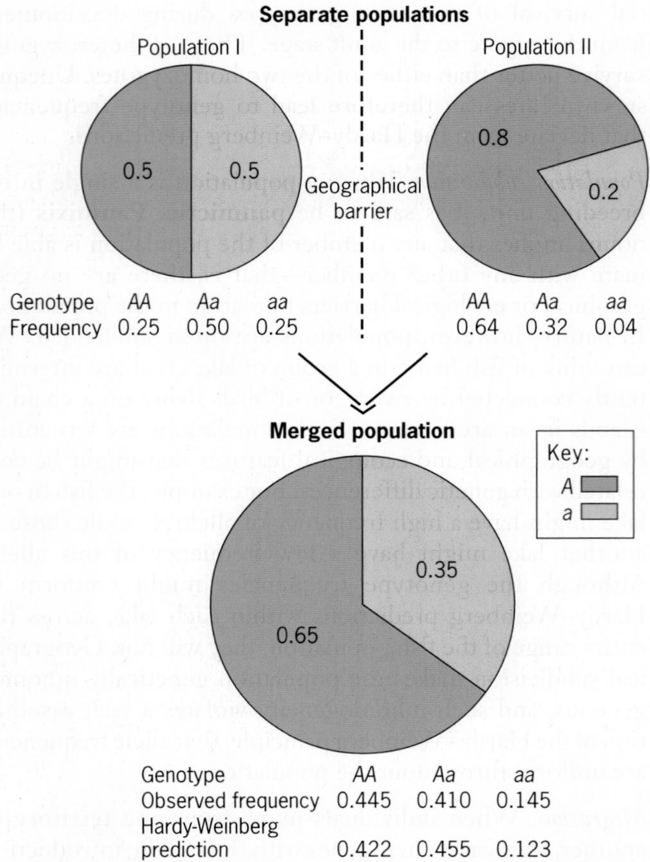

Figure 24.2 ▶ Effects of population merger on allele and genotype frequencies.

USING ALLELE FREQUENCIES IN GENETIC COUNSELING

Genetic counselors sometimes use allele frequency data in conjunction with pedigree analysis to calculate the risk that an individual will develop a genetic disease. A simple case is shown in **FIGURE 24.3**. The man and woman in generation I have had three children, the last of whom suffered from Tay-Sachs disease, which is caused by an autosomal recessive mutation (*ts*) with a frequency approaching 0.017 in certain populations. Assuming that the frequency of the mutant allele is 0.017 in II-1's ethnic group, her chance of being a carrier (*TS ts*) is obtained by using the Hardy–Weinberg principle: $2\ (0.017)\ (0.983) = 0.033$, which is approximately 1/30. The chance that her husband (II-2) is a carrier is determined by analyzing the pedigree. Because II-4 died of Tay-Sachs disease, we know that both I-1 and I-2 were heterozygous for the mutant allele. Either of them could have transmitted this allele to II-2. However, both of

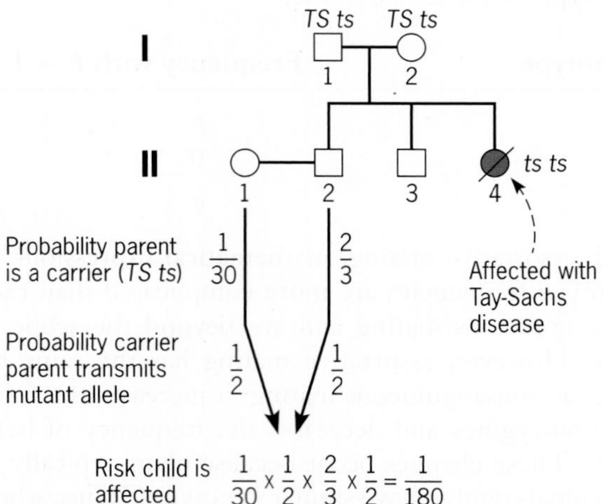

Figure 24.3 ▶ Pedigree analysis using population data to calculate the risk for Tay-Sachs disease in a child.

them did not transmit it to him because II-2 does not have the disease. Thus, the chance that II-2 is a carrier of the mutant allele is 2/3. To calculate the risk that II-1 and II-2 will have a child with Tay-Sachs disease, we combine the probabilities that each parent is a carrier (1/30 for II-1 and 2/3 for II-2) with the probability that if they are carriers, they will both transmit the mutant allele to their offspring ((1/2) × (1/2) = 1/4). Thus, the risk for the child to have Tay-Sachs disease is (1/30) × (2/3) × (1/4) = 1/180 = 0.006, which is 20 times the risk for a random child in a population where the mutant allele frequency is 0.017.

KEY POINTS

- Allele frequencies can be estimated by enumerating the genotypes in a sample from a population.
- Under the assumption of random mating, the Hardy–Weinberg principle allows genotype frequencies for autosomal and X-linked genes to be predicted from allele frequencies.
- The Hardy–Weinberg principle does not apply to populations with consanguineous or assortative mating, unequal survival among genotypes, geographic subdivision, or migration.
- The Hardy–Weinberg principle is useful in genetic counseling.

Natural Selection

Allele frequencies change systematically in populations because of differential survival and reproduction among genotypes.

Charles Darwin described the key force that drives evolutionary change in populations. He argued that organisms produce more offspring than the environment can support and that a struggle for survival ensues. In the face of this competition, the organisms that survive and reproduce transmit to their offspring traits that favor survival and reproduction. After many generations of such competition, traits associated with strong competitive ability become prevalent in the population, and traits associated with weak competitive ability disappear. Selection for survival and reproduction in the face of competition is therefore the mechanism that changes the physical and behavioral characteristics of a species. Darwin called this process **natural selection.**

NATURAL SELECTION AT THE LEVEL OF THE GENE

To put the mechanism of natural selection into a genetic context, we must recognize that the ability to survive and reproduce is a phenotype—arguably the most important phenotype of all—and that it is determined, at least partly, by genes. Geneticists refer to this ability to survive and reproduce as **fitness,** a quantitative variable they usually symbolize by the letter w. Each member of a population has its own fitness value: 0 if it dies or fails to reproduce, 1 if it survives and produces 1 offspring, 2 if it survives and produces 2 offspring, and so forth. The average of all these values is the average fitness of the population, usually symbolized $\overline{w}$.

For a population with a stable size, the average fitness is 1; each individual in such a population produces, on average, one offspring. Of course, some individuals will produce more than one offspring, and some will not produce any offspring at all. However, when the population size is not changing, the average number of offspring (that is, the average fitness) is 1. In a declining population, the average number of offspring is less than 1, and in a growing population it is greater than 1 (**FIGURE 24.4**).

To see how fitness differences among individuals lead to change in the characteristics of a population, let's assume that fitness is determined by a single gene segregating two alleles, A and a, in a particular species of insect. Furthermore, let's assume that allele A causes the insects to be dark in color, that allele a causes them to be light in color, and that A is completely dominant to a. In a forest habitat, where plant growth is luxuriant, the dark form of the insect survives better than the light form. Consequently, the fitnesses of genotypes AA and Aa are greater than the fitness of genotype aa. By contrast, in open fields, where plant growth is scarce, the light form of the insect survives better than the dark form, and the fitness relationships are reversed.

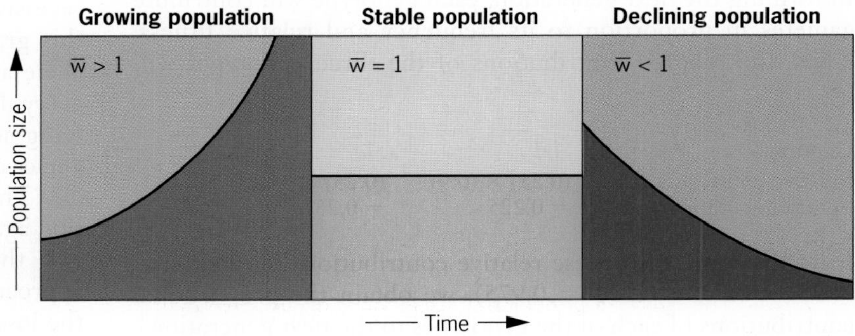

Figure 24.4 ► Significance of average fitness ($\overline{w}$) for population size as a function of time. Population size grows, is stable, or declines depending on the value of the average fitness.

We can express these relationships mathematically by applying the concept of **relative fitness.** In each of the two environments, we arbitrarily define the fitness of the competitively superior genotype(s) to be equal to 1 and express the fitness of the inferior genotype(s) as a deviation from 1. This fitness deviation, symbolized by the letter s, is called the **selection coefficient;** it measures the intensity of natural selection acting on the genotypes in the population. We can summarize the fitness relationships among the three insect genotypes in each of the two habitats in the following table:

Genotype:	*AA*	*Aa*	*aa*
Phenotype:	dark	dark	light
Relative fitness in forest habitat:	1	1	$1 - s_1$
Relative fitness in field habitat:	$1 - s_2$	$1 - s_2$	1

These relative fitnesses tell us nothing about the absolute reproductive abilities of the different genotypes in the two habitats. However, they do tell us how well each genotype competes with the other genotypes within a particular environment. Thus, for example, we know that *aa* is a weaker competitor than either *AA* or *Aa* in the forest habitat. How much weaker depends, of course, on the actual value of the selection coefficient, s_1. If $s_1 = 1$, then *aa* is effectively a lethal genotype (its relative fitness is 0), and we would expect natural selection to reduce the frequency of the *a* allele from the population. If s_1 were much smaller, say only 0.01, natural selection would still reduce the frequency of the *a* allele, but it would do so very slowly.

To see the effect of natural selection on allele frequencies, let's focus on an insect population in the forest habitat. We shall assume that initially the frequency of *A* is $p = 0.5$, that the frequency of *a* is $q = 0.5$, and that $s_1 = 0.1$. Furthermore, let's assume that the population mates randomly and that the genotypes are present in Hardy–Weinberg frequencies at fertilization each generation. (Differential survival among the genotypes will change these frequencies as the insects mature.) Under these assumptions, the initial genetic composition of the population is:

Genotype:	*AA*	*Aa*	*aa*
Relative fitness:	1	1	$1 - 0.1 = 0.9$
Frequency: (at fertilization)	$q^2 = 0.25$	$p^2 = 0.25$	$2pq = 0.50$

In forming the next generation, each genotype will contribute gametes in proportion to its frequency and relative fitness. Thus, the relative contributions of the three genotypes will be:

Genotype:	*AA*	*Aa*	*aa*
Relative contribution to next generation:	$(0.25) \times (0.9)$ $= 0.225$	$(0.25) \times 1$ $= 0.25$	$(0.50) \times 1$ $= 0.50$

If we divide each of these relative contributions by their sum $(0.25 + 0.50 + 0.225 = 0.975)$, we obtain the proportional contributions of each of the genotypes to the next generation:

Genotype:	*AA*	*Aa*	*aa*
Proportional contribution to next generation:	0.513	0.231	0.256

From these numbers we can calculate the frequency of the *a* allele after one generation of selection simply by noting that all the genes transmitted by the *aa* homozygotes are *a* and that half the genes transmitted by the *Aa* heterozygotes are *a*. In the next generation, the frequency of *a*, symbolized q', will be

$$q' = 0.231 + (1/2)(0.513) = 0.487$$

which is slightly less than the starting frequency of 0.5. Thus, in the forest habitat, natural selection, acting through the lower fitness of the *aa* homozygotes, has decreased the frequency of *a* from 0.5 to 0.487. In every subsequent generation, the frequency of *a* will be reduced slightly because of selection against the *aa* homozygotes, and eventually, this allele will be eliminated from the population altogether. **FIGURE 24.5*a*** shows how natural selection will drive the *a* allele to extinction.

In the field habitat, *aa* homozygotes are selectively superior to the other two genotypes. Thus, starting with $q = 0.5$, Hardy–Weinberg genotype frequencies, and the selection coefficient $s_2 = 0.1$, we have:

Genotype:	*AA*	*Aa*	*aa*
Relative fitness:	$1 - 0.1 = 0.9$	$1 - 0.1 = 0.9$	1
Frequency:	0.25	0.50	0.25

After one generation of selection in the field habitat, the frequency of *a* will be $q' = 0.513$, which is slightly greater than the starting frequency. Every generation afterward, the frequency of *a* will rise, and eventually it will equal 1, at which point it is said that the allele has been fixed in the population. **FIGURE 24.5*b*** shows the selection-driven path toward fixation of *a*.

These two scenarios illustrate selection for or against a recessive allele. In the forest habitat, the recessive allele *a* is deleterious in homozygous condition and selection acts against it. In the field habitat, *a* is selectively favored over the dominant allele *A*, which is deleterious in both homozygous and heterozygous condition.

Notice that selection *for* a recessive allele—and therefore against a harmful dominant allele—is more effective than selection *against* a recessive allele. The curve in **FIGURE 24.5*b*** shows the time course of selection in favor of a recessive allele. This curve rises steeply to the top of the graph, at which point the recessive allele is fixed in the population. The process shown in this graph efficiently changes the frequency of the recessive allele and rather quickly gets it to a final value of 1, because every dominant allele in the population is exposed to the purifying action of selection. By virtue of their dominance, these alleles cannot "hide out" in heterozygous condition.

The curve in **FIGURE 24.5a** shows the time course of selection against a recessive allele. This curve changes more gradually than the curve in **FIGURE 24.5*b*** and asymptotically approaches a limit at the bottom of the graph, which represents the loss of the recessive allele. Selection is less effective in this

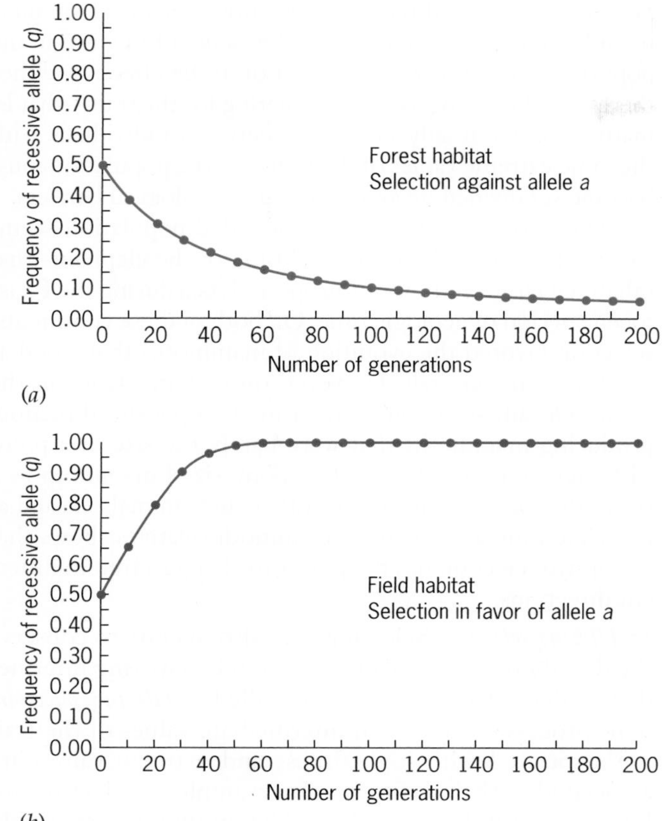

Figure 24.5 ▶ (*a*) Selection against the recessive allele *a* in the forest habitat. (*b*) Selection in favor of the recessive allele *a* in the field habitat.

Figure 24.6 ▶ (*a*) The dark form of the peppered moth on tree bark covered with lichens. (*b*) The light form of the peppered moth on tree bark covered with soot from industrial pollution.

case because it can only act against the recessive allele when it is homozygous. Once the recessive allele has been reduced in frequency, recessive homozygotes will be rare; most of the surviving recessive alleles will therefore be found in heterozygotes, where they are immune from the purifying effect of selection. By comparing the two graphs in **FIGURE 24.5**, we see that a harmful recessive allele can linger in a population much longer than a harmful dominant allele.

Studies of the moth *Biston betularia*, an inhabitant of wooded areas in Great Britain, have shown that selection of the type we have been discussing does operate to change allele frequencies in nature. This species, commonly known as the peppered moth, exists in two color forms, light and dark (**FIGURE 24.6**); the light form is homozygous for a recessive allele *c*, and the dark form carries a dominant allele *C*. From 1850 onward, the frequency of the dark form increased in certain areas of England, particularly in the industrialized Midlands section of the country. Around the heavily industrialized cities of Manchester and Birmingham, for example, the frequency of the dark form increased from 1 to 90 percent. This dramatic increase has been attributed to selection against the light form in the soot-polluted landscapes of industrialized areas. In recent times, the level of pollution has abated considerably and the light form of the moth has made a comeback, although not quite to its preindustrial frequencies. Whatever processes have been at work against the light form of the moth appear to have been reversed by environmental restoration in this region of England.

NATURAL SELECTION AT THE LEVEL OF THE PHENOTYPE

Although the example of *Biston betularia* shows that fitness can be dramatically influenced by different alleles of a single gene, in most circumstances it is influenced by the alleles of many genes. Typically, fitness depends on sets of genes that control quantitative traits such as body size, disease susceptibility, and fecundity. Thus, we would expect natural selection to affect the statistical distributions of these kinds of traits within a population. We now consider three ways in which selection can affect the distribution of a quantitative trait (**FIGURE 24.7**).

1. *Directional selection*. Selection that favors values of a trait at one end of its distribution is *directional selection*. This type of situation commonly occurs in agriculture where plant and animal breeders practice artificial selection to improve traits such as crop yield, nutritional content, and egg production (see Chapter 23). In nature, directional selection may occur when a deterioriating environment steadily challenges the population to adapt. R. A. Fisher, who studied this situation theoretically, concluded that directional selection increases

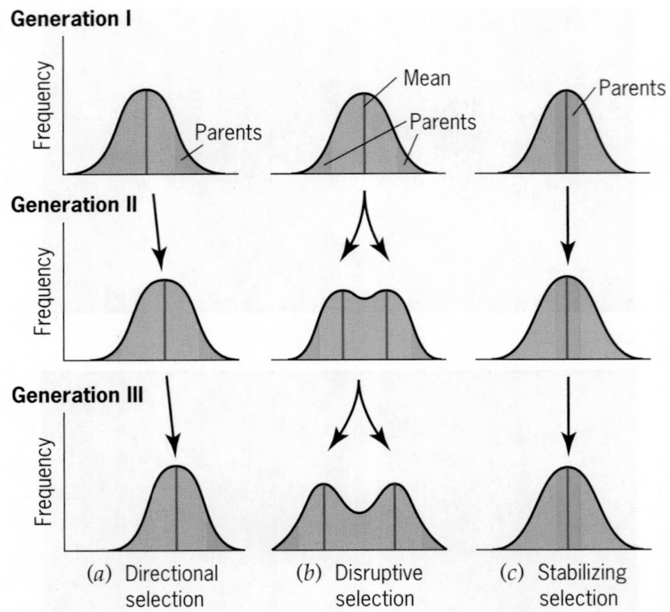

Figure 24.7 ▶ The effects of directional, disruptive, and stabilizing selection on the frequency distribution of a quantitative trait. The mean of the trait is indicated by a red line.

the average fitness of a population at a rate that is proportional to the additive genetic variance for fitness—a principle that he immodestly called the "fundamental theorem of natural selection." Although a discussion of Fisher's theorem is beyond the scope of this textbook, it should be noted that the theorem parallels the principle that the response to artificial selection depends on the proportion of variance in a trait that is additive genetic variance—that is, it depends on the narrow-sense heritability (see Chapter 23). Thus, the rate at which natural selection can change fitness in a population is a function of the narrow-sense heritability for fitness.

Evolutionary biologists have discovered many examples of directional selection. The increase in body size of the horse during the last 40 million years, the development of extravagant body ornaments such as antlers in deer and feathers in birds, and the increase in brain size in our own species all probably involved directional selection.

2. *Disruptive selection*. Selection that favors extreme values of a trait at the expense of intermediate values is *disruptive selection*. It is, in effect, directional selection that works simultaneously to increase and decrease the values of the trait in the population. For this type of selection to be effective, either there must be strong assortative mating for the trait—that is, matings preferentially take place between individuals with the same extreme values of the trait—or the population must become subdivided by geographical or ecological barriers.

   Disruptive selection in a subdivided population seems to have played a role in the evolution of the elephants that inhabited large sections of Europe and Asia during the Pleistocene period of geologic time. On both of these continents, selection favored the evolution of mammoths that stood as much as 5 meters tall. However, on certain islands in the Arctic Ocean, selection worked in the opposite direction, producing mammoths that were barely the size of a pony. Although some would regard a pony-sized mammoth as a contradiction in terms, the fact that such animals existed at the same time as their truly mammoth relatives shows that disruptive selection has the power to change a trait in different directions.

3. *Stabilizing selection*. Selection may also operate to conserve the distribution of a quantitative trait by favoring intermediate values. Such a process is called *stabilizing selection*. This process occurs when intermediate values of the trait are associated with high fitness and extreme values are associated with low fitness. An example is selection for birth weight in human babies. The optimum birth weight is around 8 pounds. Babies that deviate significantly from this weight are less likely to survive; larger ones may be injured during birth, and smaller ones are more likely to die after birth.

KEY POINTS

- Natural selection occurs when genotypes differ in the ability to survive and reproduce—that is, when they differ in fitness.
- The intensity of natural selection is quantified by the selection coefficient.
- At the level of the gene, natural selection changes the frequencies of alleles in populations.
- At the level of the phenotype, natural selection influences the distributions of quantitative traits.
- Natural selection may be directional, disruptive, or stabilizing.

Random Genetic Drift

Allele frequencies change unpredictably in populations because of uncertainties during reproduction.

In his book *The Origin of Species*, Darwin emphasized the role of natural selection as a systematic force in evolution. However, he also recognized that evolution is affected by random processes. New mutants appear unpredictably in populations. Thus, mutation, the ultimate source of all genetic variability, is a random process that profoundly affects evolution; without mutation, evolution could not occur. Darwin also recognized that inheritance (which he did not understand) is unpredictable. Traits are inherited, but offspring are not exact replicas of their parents; there is always some unpredictability in the transmission of a trait from one generation to the next. In the twentieth century, after Mendel's principles were rediscovered, the evolutionary implications of this unpredictability were investigated by Sewall Wright, R. A. Fisher, and Motoo Kimura. From their theoretical analyses, it is clear that the randomness associated with the Mendelian mechanism profoundly affects the evolutionary process. In the following sections, we explore how the uncertainties of genetic transmission can lead to random changes in allele frequencies—a phenomenon called **random genetic drift.**

RANDOM CHANGES IN ALLELE FREQUENCIES

To investigate how the uncertainties associated with the Mendelian mechanism can lead to random changes in allele frequencies, let's consider a mating between two heterozygotes, *Cc* × *Cc*, that produces two offspring, which is the number expected if each individual in the population replaces itself (**FIGURE 24.8**). We can enumerate the possible genotypes of the two offspring and compute the probability associated with each of the possible combinations by using the methods discussed in Chapter 3. For example, the probability that the first offspring is *CC* is 1/4 and the probability that the second offspring is *CC* is also 1/4; thus, the probability that both offspring are *CC* is (1/4) × (1/4) = 1/16. The probability that one of the offspring is *CC* and the other is *Cc* is (1/4) × (1/2) × 2 (because there are two possible birth orders: *CC* then *Cc*, or *Cc* then *CC*); thus, the probability of observing the genotypic combination *CC* and *Cc* in the two offspring is 1/4. The entire probability distribution for the various genotypic combinations of offspring is given in **FIGURE 24.8**. This figure also gives the frequency of the *c* allele associated with each combination.

Among the parents, the frequency of *c* is 0.5. This frequency is the most probable frequency for *c* among the two offspring. In fact, the probability that the frequency of *c* will not change between parents and offspring is 6/16. However, there is an appreciable chance that the frequency of *c* will increase or decrease among the offspring simply because of the uncertainties associated with the Mendelian mechanism. The chance that the frequency of *c* will increase is 5/16, and the chance that it will decrease is also 5/16. Thus, the chance that the frequency of *c* will change in one direction or the other, 5/16 + 5/16 = 10/16, is actually greater than the chance that it will remain the same.

This situation illustrates the phenomenon of random genetic drift. For every pair of parents in the population that is segregating different alleles of a gene, there is a chance that the Mendelian mechanism will lead to changes in the frequencies of those alleles. When these random changes are summed over all pairs of parents, there may be aggregate changes in the allele frequencies. Thus, the genetic composition of the population can change even without the force of natural selection.

Random genetic drift is essentially the result of a gene sampling process that occurs when organisms reproduce. This sampling process has two components. First, the alleles of segregating genes are randomly incorporated into gametes. The offspring produced by a heterozygote with genotype *Cc* inherit either allele *C* or allele *c*, each with probability 1/2. Thus, in segregating individuals, there is always uncertainty as to which allele a given offspring will receive. Second, there is random variation in the number of offspring that a parent produces. Some parents produce many offspring, some produce a few, and some produce none. Although part of this variation may be due to intrinsic fitness differences among the members of the population, part of it may be due to purely random factors—accidental deaths, bad weather, environmental catastrophes. This purely random variation in reproductive output compounds the randomness associated with Mendelian segregation, and the net result is random change in allele frequencies.

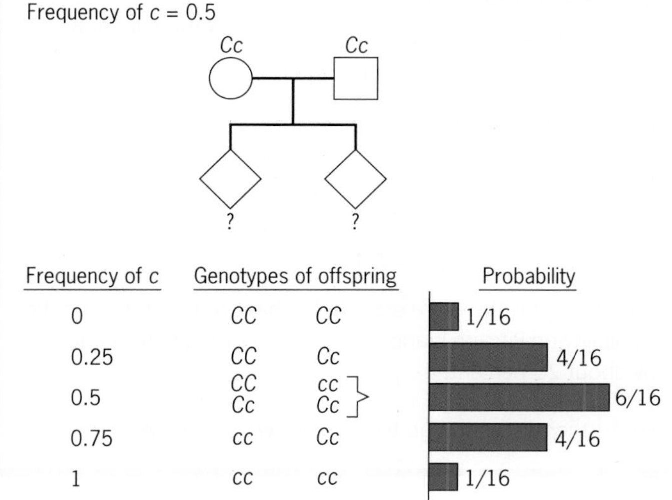

Figure 24.8 ▶ Probabilities associated with possible frequencies of the allele *c* among the two children of heterozygous parents.

THE EFFECTS OF POPULATION SIZE

A population's susceptibility to random genetic drift depends on its size. In large populations, the effect of genetic drift is minimal, whereas in small ones, it may be the primary evolutionary force. Geneticists gauge the effect of population size by monitoring the frequency of heterozygotes over time. Let's focus, once again, on alleles *C* and *c*, with respective frequencies p and q, and let's assume that neither allele has any effects on fitness; that is, *C* and *c* are selectively neutral. Furthermore, let's assume that the population mates randomly and that in any given generation, the genotypes are present in Hardy–Weinberg proportions.

In a very large population—essentially infinite in size—the frequencies of *C* and *c* will be constant, and the frequency of the heterozygotes that carry these two alleles will be $2pq$. In a small population of finite size N, the allele frequencies will change randomly as a result of genetic drift. Because of these changes, the frequency of heterozygotes, often called the **heterozygosity,** will also change. To express the magnitude of this change over one generation, let's define the current frequency of heterozygotes as H and the frequency of heterozygotes in the next generation as H'. Then the mathematical relationship between H' and H is

$$H' = \left(1 - \frac{1}{2N}\right) H$$

This equation tells us that in one generation, random genetic drift causes the heterozygosity to decline by a factor of $1/2N$. In a total of t generations, we would expect the heterozygosity to decline to a level given by the equation

$$H_t = \left(1 - \frac{1}{2N}\right)^t H$$

This equation enables us to see the cumulative effect of random genetic drift over many generations. In each generation, the heterozygosity is expected to decline by a factor of $1/2N$; over many generations, the heterozygosity will eventually be reduced to 0, at which point all genetic variability in the population will be lost. At this point the population will possess only one allele of the gene, and either $p = 1$ and $q = 0$, or $p = 0$ and $q = 1$. Thus, through random changes in allele frequencies, drift steadily erodes the genetic variability of a population, ultimately leading to the fixation and loss of alleles. It is important to recognize that this process depends critically on the population size (**FIGURE 24.9**). Small populations are the most sensitive to the variability-reducing effects of drift. Large populations are less sensitive. To see how drift might have reduced genetic variability in the population of Pitcairn Island described at the beginning of this chapter, work through the Focus on Problem Solving: Applying Genetic Drift to Pitcairn Island.

If selectively neutral alleles of the sort we have been discussing are ultimately destined for fixation or loss, can we determine the probabilities that are associated with these two ultimate outcomes? Let's suppose that at the current time, the frequency of C is p and that of c is q. Then, as long as the alleles

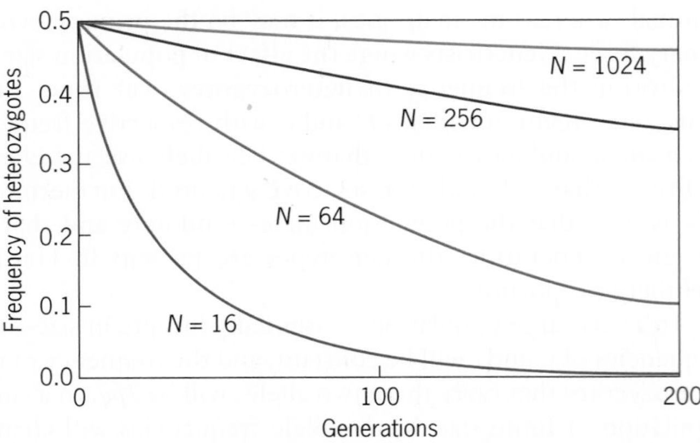

Figure 24.9 ▶ Decline in the frequency of heterozygotes due to random genetic drift in populations of different size N. The populations begin with $p = q = 0.5$.

▶ FOCUS ON PROBLEM SOLVING
Applying Genetic Drift to Pitcairn Island

THE PROBLEM

When Fletcher Christian and his fellow mutineers on the H.M.S. *Bounty* settled on Pitcairn Island, they didn't realize that they were beginning a genetic experiment. The founding group of men and women brought a finite sample of genes to the island—a sample from two larger populations, Britain and Polynesia. From its beginning in 1790, the Pitcairn Island colony has essentially been a closed system. Some people have left the island, but very few have migrated to it. Most of the alleles that are present on the island today are copies of alleles that were brought there by the colony's founders. Of course, not every allele that was present at the founding is present today. Some alleles were lost through the death or infertility of their carriers. Others have been lost through genetic drift. Let's suppose that the average population size of Pitcairn Island has been 20 and that when the colony was founded, H (the heterozygosity) was 0.20. Let's also suppose that 10 generations have elapsed since the founding of the colony. What is the expected value of H today?

FACTS AND CONCEPTS

1. The heterozygosity is a measure of genetic variability in a population.

2. In a population of size N, genetic drift is expected to reduce the heterozygosity by a factor of $1/2N$ each generation.

3. The loss in variability is cumulative; after t generations, the heterozygosity is given by $H_t = (1 - 1/2N)^t H$.

ANALYSIS AND SOLUTION

To predict the value of H today, we can use the equation

$$H_t = (1 - 1/2N)^t H$$

with $t = 10$, $N = 20$, and $H = 0.20$

$$\begin{aligned} H_{10} &= (1 - 1/2N)^{10} H \\ &= (1 - 1/40)^{10}\,(0.20) \\ &= (0.78)(0.20) \\ &= 0.15 \end{aligned}$$

Genetic drift is therefore expected to have reduced the genetic variability on Pitcairn Island, as measured by the heterozygosity, by about 25 percent.

For further discussion go to your *WileyPLUS* course.

are selectively neutral and the population mates randomly, the probability that a particular allele will ultimately be fixed in the population is its current frequency—p for allele C and q for allele c—and the probability that the allele will ultimately be lost from the population is 1 minus its current frequency; that is, $1 - p$ for allele C and $1 - q$ for allele c. Thus, when random genetic drift is the driving force in evolution, we can assign specific probabilities to the possible evolutionary outcomes and, remarkably, these probabilities are independent of population size.

KEY POINTS

- Genetic drift, the random change of allele frequencies in populations, is due to uncertainties in Mendelian segregation and to unpredictable variation in the number of offspring.
- In diploid organisms, the rate at which genetic variability is lost by random genetic drift is $1/2N$, where N is the population size.
- Small populations are more susceptible to drift than large ones.
- Drift ultimately leads to the fixation of one allele at a locus and the loss of all other alleles; the probability that an allele will ultimately be fixed is equal to its current frequency in the population.

Populations in Genetic Equilibrium

The evolutionary forces of mutation, selection, and drift may oppose each other to create a dynamic equilibrium in which allele frequencies no longer change.

In a randomly mating population without selection or drift to change allele frequencies, and without migration or mutation to introduce new alleles, the Hardy–Weinberg genotype frequencies persist indefinitely. Such an idealized population is in a state of genetic equilibrium. In reality, the situation is much more complicated; selection and drift, migration and mutation are almost always at work changing the population's genetic composition. However, these evolutionary forces may act in contrary ways to create a *dynamic equilibrium* in which there is no net change in allele frequencies. This type of equilibrium differs fundamentally from the equilibrium of the ideal Hardy–Weinberg population. In a dynamic equilibrium, the population simultaneously tends to change in opposite directions, but these opposing tendencies cancel each other and bring the population to a point of balance. In the ideal Hardy–Weinberg equilibrium, the population does not change because there are no evolutionary forces at work. We now explore how opposing evolutionary forces can create a dynamic equilibrium within a population.

BALANCING SELECTION

One type of dynamic equilibrium arises when selection favors the heterozygotes at the expense of each type of homozygote in the population. In this situation, called *balancing selection* or *heterozygote advantage*, we can assign the relative fitness of the heterozygotes to be 1 and the relative fitnesses of the two types of homozygotes to be less than 1:

Genotype:	*AA*	*Aa*	*aa*
Relative fitness:	$1 - s$	1	$1 - t$

In this formulation, the terms $1 - s$ and $1 - t$ contain selection coefficients that are assumed to lie between 0 and 1. Thus, each of the homozygotes has a lower fitness than the heterozygotes. The superiority of the heterozygotes is sometimes referred to as *overdominance*.

In cases of heterozygote advantage, selection tends to eliminate both the A and a alleles through its effects on the homozygotes, but it also preserves these alleles through its effects on the heterozygotes. At some point these opposing tendencies balance each other, and a dynamic equilibrium is established. To determine the frequencies of the two alleles at the point of equilibrium, we must derive an equation that describes the process of selection, and then solve this equation for the allele frequencies when the opposing selective forces are in balance—that is, when the allele frequencies are no longer changing (**TABLE 24.2**). At the balance point, the frequency of A is $p = t/(s + t)$, and the frequency of a is $q = s/(s + t)$.

As an example, let's suppose that the AA homozygotes are lethal ($s = 1$) and that the aa homozygotes are 50 percent as fit as the heterozygotes ($t = 0.5$). Under these assumptions, the population will establish a dynamic equilibrium when $p = 0.5/(0.5 + 1) = 1/3$ and $q = 1/(0.5 + 1) = 2/3$. Both alleles will be maintained at appreciable frequencies by selection in favor of the heterozygotes—a condition known as a **balanced polymorphism.**

TABLE 24.2

Calculating Equilibrium Allele Frequencies with Balancing Selection

Genotypes:	*AA*	*Aa*	*aa*
Relative fitnesses:	$1 - s$	1	$1 - t$
Frequencies:	p^2	$2pq$	q^2

Average relative fitness: $\overline{w} = p^2 \times (1 - s) + 2pq \times 1 + q^2 (1 - t)$

Frequency of A in the next generation after selection:

$$p' = [p^2(1 - s) + (1/2)2pq]/\overline{w} = p(1 - sp)/\overline{w}$$

Change in frequency of A due to selection:

$$\Delta p = p' - p = pq(tq - sp)/\overline{w}$$

At equilibrium, $\Delta p = 0$: $p = t/(s + t)$ and $q = s/(s + t)$

In humans, the disease sickle-cell anemia is associated with a balanced polymorphism. Individuals with this disease are homozygous for a mutant allele of the β-globin gene, denoted HBB^S, and they suffer from a severe form of anemia in which the hemoglobin molecules crystallize in the blood. This crystallization causes the red blood cells to assume a characteristic sickle shape. Because sickle-cell anemia is usually fatal without medical treatment, the fitness of HBB^SHBB^S homozygotes has historically been 0. However, in some parts of the world, particularly in tropical Africa, the frequency of the HBB^S allele is as high as 0.2. With such harmful effects, why does the HBB^S allele remain in the population at all?

The answer is that there is moderate selection against homozygotes that carry the wild-type allele HBB^A. These homozygotes are less fit than the HBB^SHBB^A heterozygotes because they are more susceptible to infection by the parasites that cause malaria, a fitness-reducing disease that is widespread in regions where the frequency of the HBB^S allele is high (**FIGURE 24.10**). We can schematize this situation by assigning relative fitnesses to each of the genotypes of the β-globin gene:

Genotype:	HBB^SHBB^S	HBB^SHBB^A	HBB^AHBB^A
Relative fitness:	$1 - s$	1	$1 - t$

If we assume that the equilibrium frequency of HBB^S is $p = 0.1$—a typical value in West Africa—and if we note that $s = 1$ because the HBB^SHBB^S homozygotes die, we can estimate the intensity of selection against the HBB^AHBB^A homozygotes because of their greater susceptibility to malaria:

$$p = t / (s + t)$$
$$0.1 = t / (1 + t)$$
$$t = (0.1) / (0.9) = 0.11$$

This result tells us that the HBB^AHBB^A homozygotes are about 11 percent less fit than the HBB^SHBB^A heterozygotes. Thus, the selective inferiority of the HBB^SHBB^S and HBB^AHBB^A homozygotes compared to the heterozygotes creates a balanced polymorphism in which both alleles of the β-globin gene are maintained in the population.

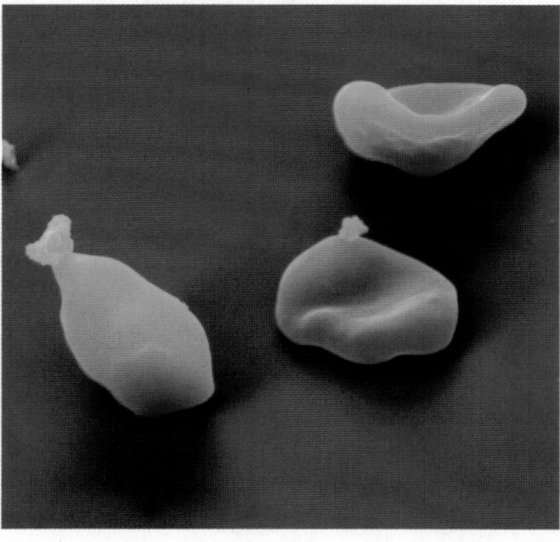

Figure 24.10 ▶ The malaria parasite *Plasmodium falciparum* (yellow) emerging from red blood cells that it had infected.

Various other mutant *HBB* alleles are found at appreciable frequencies in tropical and subtropical regions of the world in which *falciparum* malaria is—or was—endemic. It is plausible that these alleles have also been maintained in human populations by balancing selection.

MUTATION–SELECTION BALANCE

Another type of dynamic equilibrium is created when selection eliminates deleterious alleles that are produced by recurrent mutation. For example, let's consider the case of a deleterious recessive allele a that is produced by mutation of the wild-type allele A at rate u. A typical value for u is 3×10^{-6} mutations per generation. Even though this rate is very low, over time, the mutant allele will accumulate in the population, and, because it is recessive, it can be carried in heterozygous condition without having any harmful effects. At some point, however, the mutant allele will become frequent enough for aa homozygotes to appear in the population, and these will be subject to the force of selection in proportion to their frequency and the value of the selection coefficient s. Selection against these homozygotes will counteract the force of mutation, which introduces the mutant allele into the population.

If we assume that the population mates randomly, and if we denote the frequency of A as p and that of a as q, then we can summarize the situation as follows:

Mutation:		Selection:		
produces a		eliminates a		
$A \rightarrow a$	Genotype:	AA	Aa	aa
rate $= u$	Relative fitness:	1	1	$1 - s$
	Frequency:	p^2	$2pq$	q^2

Mutation introduces mutant alleles into the population at rate u, and selection eliminates them at rate sq^2 (**FIGURE 24.11**). When these two processes are in balance, a dynamic equilibrium will be established. We can calculate the frequency of the mutant allele at the equilibrium created by mutation–selection balance by equating the rate of mutation to the rate of elimination by selection:

$$u = sq^2$$

Thus, after solving for q, we obtain

$$q = \sqrt{u/s}$$

For a mutant allele that is lethal in homozygous condition, $s = 1$, and the equilibrium frequency of the mutant allele is simply the square root of the mutation rate. If we use the value for u that was given above, then for a recessive lethal allele the equilibrium frequency is $q = 0.0017$. If the mutant allele is not

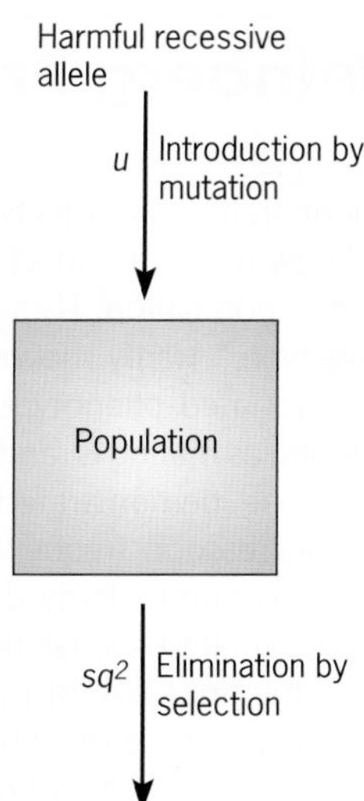

Figure 24.11 ▶ Mutation–selection balance for a deleterious recessive allele with frequency q. Genetic equilibrium is reached when the introduction of the allele into the population by mutation at rate u is balanced by the elimination of the allele by selection with intensity s against the recessive homozygotes.

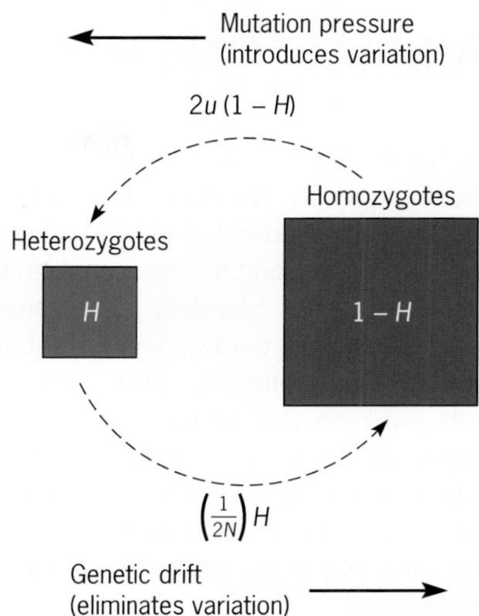

Figure 24.12 ▶ Mutation–drift balance for variability as measured by the frequency of heterozygotes H in a population of size N. An equilibrium frequency of heterozygotes is reached when the introduction of variability by mutation at rate u is balanced by the elimination of variability by genetic drift at rate $\frac{1}{2N}$.

completely lethal in homozygous condition, then the equilibrium frequency will be higher than 0.0017 by a factor that depends on $1/\sqrt{s}$. For example, if s is 0.1, then at equilibrium the frequency of this slightly deleterious allele will be $q = 0.0055$, or 3.2 times greater than the equilibrium frequency of a recessive lethal allele.

Studies with natural populations of *Drosophila* have indicated that lethal alleles are less frequent than the preceding calculations predict. The discrepancy between the observed and predicted frequencies has been attributed to partial dominance of the mutant alleles—that is, these alleles are not completely recessive. Natural selection appears to act against deleterious alleles in heterozygous condition as well as in homozygous condition. Thus, the equilibrium frequencies of these alleles are lower than we would otherwise predict. Selection that acts against mutant alleles in homozygous or heterozygous condition is sometimes called *purifying selection*.

MUTATION–DRIFT BALANCE

We have already seen that random genetic drift eliminates variability from a population. Without any counteracting force, this process would eventually make all populations completely homozygous. However, mutation replenishes the variability that is lost by drift. At some point, the opposing forces of mutation and genetic drift come into balance and a dynamic equilibrium is established.

Previously, we saw that genetic variability can be quantified by calculating the frequency of heterozygotes in a population—a statistic called the heterozygosity, which is symbolized by the letter H. The frequency of homozygotes in a population—often called the *homozygosity*—is equal to $1 - H$. Over time, genetic drift decreases H and increases $1 - H$, and mutation does just the opposite (**FIGURE 24.12**). Let's assume that each new mutation is selectively neutral. In a randomly mating population of size N, the rate at which drift decreases H is $(\frac{1}{2N})$ H (see the earlier section, The Effects of Population Size). The rate at which mutation increases H is proportional to the frequency of the homozygotes in the population $(1 - H)$ and the probability that one of the two alleles in a particular homozygote mutates to a different allele, thereby converting that homozygote into a heterozygote. This probability is simply the mutation rate u for each of the two alleles in the homozygote; thus, the total probability of mutation converting a particular homozygote into a heterozygote is $2u$. The rate at which mutation increases H in a population is therefore equal to $2u(1 - H)$.

When the opposing forces of mutation and drift come into balance, the population will achieve an equilibrium level of variability denoted by $\hat{H}$. We can calculate this equilibrium value of H by equating the rate at which mutation increases H to the rate at which drift decreases it:

$$2u(1 - H) = \left(\frac{1}{2N}\right)H$$

By solving for H, we obtain the equilibrium heterozygosity at the point of mutation–drift balance:

$$\hat{H} = 4Nu \,/\, (4Nu + 1)$$

Thus, the equilibrium level of variability (as measured by the heterozygosity) is a function of the population size and the mutation rate.

If we assume that the mutation rate is $u = 1 \times 10^{-6}$, we can plot $\hat{H}$ for different values of N (**FIGURE 24.13**). For $N < 10{,}000$,

▶ A MILESTONE IN GENETICS: **The Hardy–Weinberg Principle**

The modern science of genetics was born in 1866 when Gregor Mendel published his paper on inheritance in peas. Because Mendel's paper appeared in an obscure journal, it initially had no impact. Thirty-four years elapsed before the world finally recognized the significance of Mendel's discoveries. After Mendel's ideas came to light, the science of genetics developed quickly. Various subdisciplines were born—for example, biochemical genetics, which started with Archibald Garrod's work on the inborn errors of metabolism, and *Drosophila* genetics, which started when T. H. Morgan found the white-eye mutant in one of his laboratory cultures. Population genetics also began about this time. In fact, we can date its birth to 1908, the year in which the constancy of genotype frequencies under random mating was first described.

The distribution of genotype frequencies was explored in two articles, one published in the high-profile American journal *Science* and the other in the annual volume of the Society for Natural History in Württemberg, Germany—a publication that was not too widely read. G. H. Hardy, an eminent British mathematician, was the author of the *Science* paper. Wilhelm Weinberg, a German physician, was the author of the paper published in the Württemberg annual. Hardy and Weinberg arrived at their conclusions independently, and today, we refer to their discovery about genotype frequencies as the Hardy–Weinberg principle: If *A* and *a* are alleles with frequencies *p* and *q*, respectively, then in a large population with random mating and without selection, the frequencies of the three genotypes are p^2 (*AA*), $2pq$ (*Aa*), and q^2 (*aa*). Furthermore, these frequencies will persist generation after generation—that is, the population will remain in a state of genetic equilibrium.

Hardy was prompted to write his short paper[1] describing this principle in response to remarks made by Udny Yule, who had suggested that under Mendelism a dominant trait should eventually be expressed in three-fourths of the members of a population. Yule had a particular dominant trait, brachydactyly or short fingers, in mind, and he obviously knew that brachydactyly is not manifested in three-fourths of the human population. Using "a little mathematics of the multiplication-table type," Hardy showed that the frequencies of genotypes and their associated phenotypes are stable from one generation to the next as long as mating is random and the population is reasonably large. Thus, he demonstrated that a trait such as brachydactyly should not increase in frequency simply because it is dominant, as Yule had conjectured. Hardy did note that from one generation to the next there might be small fluctuations in genotype frequencies on account of the finite size of the population. Thus, he anticipated the concept of genetic drift, which was analyzed two decades later by Sewall Wright and R. A. Fisher.

Hardy regarded his paper as utterly trivial and may have published it in an American journal to minimize the chance that his British colleagues would see it. He was "a pure mathematician's pure mathematician. He abhorred any 'practical' mathematics. For him, pure mathematics was beautiful and useless, while useful mathematics was dull and ugly. It must have embarrassed him that his mathematically most trivial paper is not only far and away his most widely known, but has been of such distastefully practical value."[2]

Weinberg wrote his paper[3] to investigate whether or not the tendency for women to produce twins is determined by a Mendelian factor. His interest in twinning is not surprising because as a physician he attended at more than 3500 births. Despite a busy medical practice, he had time to read and think about heredity, and he looked for ways to apply Mendelian concepts to human traits. The title of his article translates as "On the Demonstration of Inheritance in Humans."

Weinberg recognized that pedigree analysis provides one way to study human heredity; however, the trait that interested him—

[1]Hardy, G. H. 1908. Mendelian proportions in a mixed population. *Science* 28: 49–50.

[2]Crow, J. F. 1988. Eighty years ago. The beginnings of population genetics. *Genetics* 119:473–476.

[3]Weinberg, W. 1908. Über den Nachweis der Vererbung beim Menschen. *Jahreshefte Vereins für vaterländische Naturkunde in Württemberg* 64:369–382.

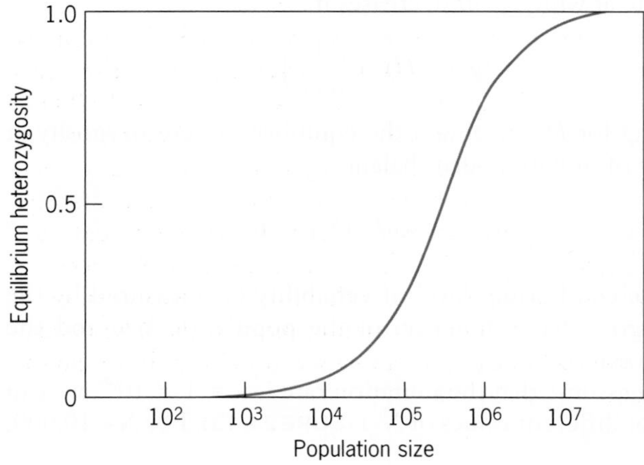

Figure 24.13 ▶ Equilibrium frequency of heterozygotes (heterozygosity) under mutation–drift balance as a function of genetically effective population size. The mutation rate is assumed to be 10^{-6}.

twinning—was not amenable to conventional pedigree analysis. As an alternative, he took a broader approach, which involved ascertaining the frequencies of traits (and their underlying Mendelian determinants) in whole populations. One fact suggested to him that twinning is heritable. The frequency of dizygotic twins varies among different ethnic groups—for example, it is higher among Germans than among Italians. The mathematical relationship between allele frequencies and genotype frequencies that we now know as the Hardy-Weinberg principle provided Weinberg with a theoretical foundation on which to build a methodology for his genetic studies. However, his analysis was not definitive; the best he could do was to suggest that twinning in humans is due to a recessive allele—clearly an oversimplification for such a complex trait.

Until the 1940s, the Hardy–Weinberg principle was known as Hardy's law in the English-speaking world. Curt Stern, a geneticist who fled Nazi Germany to work in America, added Weinberg's name to the law by publishing a note in *Science* in 1943.[4] Stern translated the relevant passages of Weinberg's 1908 paper and offered an explanation for why the paper had been largely ignored:

> While Weinberg's paper, like Mendel's, appeared in an obscure journal, its failure to be recognized can not be ascribed to this fact alone. His later contributions dealing with extensions of the statistical treatment of the genetics of populations are found in the "regular" journals. These papers have received some attention and in them Weinberg refers to his 1908 pioneer work. However, both Weinberg and Hardy were ahead of contemporary thought and similar problems were not generally considered for at least eight years. At that time perhaps Hardy's name and the prominent place of his publication both helped to leave Weinberg's contribution neglected.[5]

[4]Stern, C. 1943. The Hardy–Weinberg law. *Science* 97:137–138.

[5]Ibid., p. 138.

Stern also made a proposal, which the scientific community has accepted:

> Hardy as a mathematician did not follow up his discovery by any further consideration of its genetic implications. Weinberg in 1909 reformulated his theorem in terms valid for multiple alleles—at a time when no case of multiple alleles had been discovered in man. He also for the first time investigated polyhybrid populations and recognized their essentially different method of attaining equilibrium. Considering these facts it seems a matter of justice to attach the names of both the discoverers to the population formula.[6]

So today we have the *Hardy–Weinberg* principle to recognize the two 1908 papers—scientific twins, if you like—that marked the beginning of population genetics.

QUESTIONS FOR DISCUSSION

1. Stern's proposal to credit Weinberg along with Hardy for discovering the genotype frequency formula was, in his words, "a matter of justice." But by the time that Stern made his proposal, Weinberg was already dead. Thus, like Mendel, Weinberg received credit for his discovery posthumously. We do not know if Weinberg was bothered by the fact that during his lifetime the formula was known simply as Hardy's law. However, scientists—like most other people—generally like to receive credit for their work. Can you think of other instances in which a scientist's work was recognized belatedly or posthumously? Can you think of cases in which the wrong person was credited with a scientific discovery?
2. Because his focus was on human heredity, Weinberg could not avail himself of the experimental techniques possible with plants and animals. Instead, he chose a "population approach" to the questions that interested him. How is the population approach relevant to issues in genetics today?

[6]Ibid.

the equilibrium frequency of heterozygotes in the population is quite low; thus, drift dominates over mutation in small populations. For N equal to $1/u$, the reciprocal of the mutation rate, the equilibrium frequency of heterozygotes is 0.8, and for even greater values of N, the frequency of heterozygotes increases asymptotically toward 1. Thus, in large populations, mutation dominates over drift; every mutational event creates a new allele, and each new allele contributes to the heterozygosity because the large size of the population protects the allele from being lost by random genetic drift.

Values of $\hat{H}$ in natural populations vary among species. In the African cheetah, for example, $\hat{H}$ is 1 percent or less among a sample of loci, suggesting that over evolutionary time, population size in this species has been small. In humans, $\hat{H}$ is estimated to be about 12 percent, suggesting that over evolutionary time population size has averaged about 30,000 to 40,000 individuals. Estimates of population size that are derived from heterozygosity data are typically much smaller than estimates obtained from census data. The reason for this discrepancy is that the estimates based on heterozygosity data are *genetically effective* population sizes—sizes that take into account restrictions on mating and reproduction, as well as temporal fluctuations in the number of mating individuals. The genetically effective size of a population is almost always less than the census size of a population.

KEY POINTS

- Selection involving heterozygote superiority (balancing selection) creates a dynamic equilibrium in which different alleles are retained in a population despite their being harmful in homozygotes.
- In humans sickle-cell anemia is associated with balancing selection at the locus for β-globin.
- Selection against a deleterious recessive allele that is replenished in the population by mutation leads to a dynamic equilibrium in which the frequency of the recessive allele is a simple function of the mutation rate and the selection coefficient: $q = \sqrt{us}$.
- A population's acquisition of selectively neutral alleles through mutation is balanced by the loss of these alleles through genetic drift. At equilibrium, the frequency of heterozygotes involving these alleles is a function of the population's size and the mutation rate: $H = 4Nu/(4Nu + 1)$.

▶ Basic Exercises

ILLUSTRATE BASIC GENETIC ANALYSIS

1. Calculate the allele frequencies from the following population data:

Genotype	Number
AA	68
Aa	42
aa	24
Total	134

Answer: The frequency of the *A* allele, *p*, is $[(2 \times 68) + 42]/(2 \times 134) = 0.664$. The frequency of the *a* allele, *q*, is $[(2 \times 24) + 42]/(2 \times 134) = 0.336$.

2. Predict the Hardy–Weinberg genotype frequencies using the allele frequencies calculated in Exercise 1. Are these frequencies in agreement with the observed frequencies?

Answer: The basic calculations are summarized in the following table:

Genotype	Obs. No.	H-W Frequency	Exp. No.	Obs.−Exp. No.
AA	68	$p^2 = 0.441$	59.1	8.9
Aa	42	$2pq = 0.446$	59.8	−17.8
aa	24	$q^2 = 0.113$	15.1	8.9

To test for agreement between the observed and expected numbers, we calculate a χ^2 test statistic with 1 degree of freedom: $\chi^2 = \Sigma(\text{Obs} - \text{Exp})^2/\text{Exp} = 12.0$, which exceeds the critical value for this test statistic. Thus, we reject the hypothesis that the genotype frequencies calculated from the Hardy–Weinberg principle agree with the observed frequencies. Evidently, the population is not in Hardy–Weinberg equilibrium.

3. In a population that has been mating randomly for many generations, two phenotypes are segregating; one is due to a dominant allele *G*, the other to a recessive allele *g*. The frequencies of the dominant and recessive phenotypes are 0.7975 and 0.2025, respectively. Estimate the frequencies of the dominant and recessive alleles.

Answer: The frequency of the dominant phenotype represents the sum of two Hardy–Weinberg genotype frequencies: p^2 (*GG*) + $2pq$ (*Gg*). The frequency of the recessive phenotype represents just one Hardy–Weinberg genotype frequency, q^2 (*gg*). To estimate the frequency of the recessive allele, we take the square root of the observed frequency of the recessive phenotype: $q = \sqrt{0.2025} = 0.45$. The frequency of the dominant allele is obtained by subtraction: $p = 1 - q = 0.55$.

4. A gene with two alleles is segregating in a population. The fitness of the recessive homozygotes is 90 percent that of the heterozygotes and the dominant homozygotes. What is the value of the selection coefficient that measures the intensity of natural selection against the recessive allele?

Answer: Using *s* to represent the selection coefficient, the fitness scheme is

Genotype	Relative Fitness
AA	1
Aa	1
aa	$1 - s$

Because the recessive homozygotes are 90 percent as fit as either of the other genotypes, the expression $1 - s = 0.9$; thus, $s = 0.1$.

5. Suppose that the alleles of the *T* gene are selectively neutral. In a population of 50 individuals, currently 34 are heterozygotes. Predict the frequency of heterozygotes in this population 10 generations in the future. Assume that the population size is constant and that mating is completely random (including the possibility of self-fertilization).

Answer: For a selectively neutral gene, evolution occurs by random genetic drift. The governing equation is $H_t = (1 - \frac{1}{2N})^t H$, where H_t is the frequency of heterozygotes *t* generations in the future, *N* is the population size, and *H* is the frequency of heterozygotes now. From the data given in the problem, $N = 50$, $H = 34/50 = 0.68$, and $t = 10$. Thus, $H_t = (0.99)^{10} \times (0.68) = 0.615$.

6. Purifying selection eliminates deleterious alleles from a population, but recurrent mutation replenishes them. Suppose that recessive lethal alleles of the *B* gene are created at the rate of 2×10^{-6} per generation. What is the expected frequency of lethal alleles in a population in mutation–selection equilibrium?

Answer: The frequency of lethal alleles is given by the equation $q = \sqrt{u/s}$ where u is the mutation rate (from dominant normal allele to recessive lethal allele) and s is the intensity of selection against the deleterious allele (in this case, $s = 1$). Thus, the expected frequency of lethal alleles in the population is $q = \sqrt{2 \times 10^{-6}} = 0.0014$.

► Testing Your Knowledge

INTEGRATE DIFFERENT CONCEPTS AND TECHNIQUES

1. The A–B–O blood types of 1000 people from an isolated village were determined to obtain the following data:

Blood Type	Number of People
A	42
B	672
AB	36
O	250

Estimate the frequencies of the I^A, I^B, and i alleles of the A–B–O blood group gene from these data.

Answer: Let's symbolize the frequencies of the I^A, I^B, and i alleles of the I gene as p, q, and r, respectively, and let's assume that the genotypes of this gene are in Hardy–Weinberg proportions. We begin by estimating r, the frequency of the i allele. To obtain this estimate, we note that the frequency of the O blood type, which is $250/1000 = 0.25$ in the data, should correspond to the Hardy–Weinberg frequency of the ii genotype, r^2. Thus, if we use the Hardy–Weinberg principle in reverse, we can estimate the frequency of the i allele as $r = \sqrt{0.250} = 0.500$.

To estimate p, the frequency of the I^A allele, we note that $(p + r)^2 = p^2 + 2pr + r^2$ corresponds to the combined frequencies of the A ($p^2 + 2pr$) and O (r^2) blood types. From the data, these combined frequencies are estimated to be $(42 + 250)/1000 = 0.292$. If we set $(p + r)^2 = 0.292$ and take the square root, we obtain $p + r = 0.540$; then, by subtracting r, we can estimate the frequency of the I^A allele as $p = 0.540 - 0.500 = 0.040$. To estimate q, the frequency of the I^B allele, we note that $p + q + r = 1$. Thus, $q = 1 - p - r = 1 - 0.040 - 0.500 = 0.460$.

2. A man and a woman who both have normal color vision have had three children, including a male who is color blind. The incidence of color-blind males in the population from which this couple came is 0.30, which is unusually high for X-linked color blindness. If the color-blind male marries a female with normal color vision, what is the chance that their first child will be color blind?

Answer: Clearly, the risk that the couple will have a color-blind child depends on the female's genotype. If the female is heterozygous for the allele for color blindness, she has a probability of 1/2 of transmitting this allele to her first child. The male will transmit either an X chromosome, which carries the mutant allele, or a Y chromosome; in either case, the female's contribution to the zygote will be determinative. To obtain the probability that the female is heterozygous for the mutant allele, we note that the incidence of color blindness among males in the population is 0.30; this number provides an estimate of the frequency of the mutant allele, q, in the population. Furthermore, because $q = 0.30$, the frequency of the wild-type allele, p, is $1 - q = 0.70$. If the genotypes in the population are in Hardy–Weinberg proportions, then the frequency of heterozygous females is $2pq = 2 \times (0.7) \times (0.3) = 0.42$. However, among females who have normal color vision, the frequency of heterozygotes is greater because homozygous mutant females have been excluded from the total. To adjust for this effect, we calculate the ratio of heterozygotes to wild-type homozygotes plus heterozygotes and specifically exclude the mutant homozygotes—that is, we compute $2pq/(p^2 + 2pq) = 2pq/[p(p + 2q)] = 2q/(p + q + q) = 2q/(1 + q)$. Substituting $q = 0.3$ into the last expression, we estimate the frequency of heterozygotes among females with normal color vision (wild-type homozygotes plus heterozygotes) to be $2 \times (0.3)/(1 + 0.3) = 0.46$. This number is the chance that the female in question is a heterozygous carrier of the mutant allele. The probability that her first child will be color blind is the chance that she is a carrier (0.46) times the chance that she will transmit the mutant allele to her child (1/2); thus, the risk for the child to be color blind is $(0.46) \times (1/2) = 0.23$.

3. The HBB^S allele responsible for sickle-cell anemia is maintained in many human populations because in heterozygous condition it confers some resistance to infection by malaria parasites; however, in homozygous condition, this allele is essentially lethal. Thus, as malaria is eradicated we might expect the HBB^S allele to disappear from human populations. If the normal allele HBB^A mutates to HBB^S at a rate of 10^{-8} per generation, what ultimate frequency would you predict for the HBB^S allele in a malaria-free world?

Answer: In a malaria-free world, the advantage of maintaining the HBB^S allele in a balanced polymorphism would disappear. HBB^S HBB^A heterozygotes would have the same fitness as Hb^AHb^A homozygotes, and HBB^SHBB^S homozygotes would continue to have very low fitness—essentially zero compared to the other two genotypes. Under these circumstances, the frequency of the HBB^S allele (q) would be determined by a balance between selection against it in homozygous condition (selection coefficient $s = 1$) and introduction into the population by mutation at rate $u = 10^{-8}$ per generation. The equilibrium frequency of the Hb^S allele would be $q = \sqrt{u/s} = 0.0001$ a thousandfold less than its current frequency in malaria-infested regions of the world.

▶ Questions and Problems

ENHANCE UNDERSTANDING AND DEVELOP ANALYTICAL SKILLS

24.1 The incidence of recessive albinism is 0.008 in a human population. If mating for this trait is random in the population, what is the frequency of the recessive allele?

24.2 GO A gene has three alleles, A_1, A_2, and A_3, with frequencies 0.5, 0.3, 0.2, respectively. If mating is random, predict the combined frequency of all the heterozygotes in the population.

24.3 The following data for the M-N blood types were obtained from native villages in Central and North America:

Group	Sample Size	M	MN	N
Central American	98	55	35	8
North American	240	68	55	117

Calculate the frequencies of the L^M and L^N alleles for the two groups.

24.4 In *Drosophila* the ruby eye phenotype is caused by a recessive, X-linked mutant allele. The wild-type eye color is red. A laboratory population of *Drosophila* is started with 30 percent ruby-eyed females, 30 percent homozygous red-eyed females, 5 percent ruby-eyed males, and 35 percent red-eyed males. (a) If this population mates randomly for one generation, what is the expected frequency of ruby-eyed males and females? (b) What is the frequency of the recessive allele in each of the sexes?

24.5 Hemophilia is caused by an X-linked recessive allele. In a particular population, the frequency of males with hemophilia is 1/4000. What is the expected frequency of females with hemophilia?

24.6 The frequency of an allele in a large randomly mating population is 0.8. What is the frequency of heterozygous carriers?

24.7 Human beings carrying the dominant allele T can taste the substance phenylthiocarbamide (PTC). In a population in which the frequency of this allele is 0.4, what is the probability that a particular taster is homozygous?

24.8 In a sample from an African population, the frequencies of the L^M and L^N alleles were 0.68 and 0.32, respectively. If the population mates randomly with respect to the M-N blood types, what are the expected frequencies of the M, MN, and N phenotypes?

24.9 What frequencies of alleles A and a in a randomly mating population maximize the frequency of heterozygotes?

24.10 A population of Hawaiian *Drosophila* is segregating two alleles, P^1 and P^2, of the phosphoglucose isomerase (*PGI*) gene. In a sample of 100 flies from this population, 35 were P^1P^1 homozygotes, 40 were P^1P^2 heterozygotes, and 25 were P^2P^2 homozygotes. (a) What are the frequencies of the P^1 and P^2 alleles in this sample? (b) Perform a chi-square test to determine if the genotypes in the sample are in Hardy–Weinberg proportions. (c) Assuming that the sample is representative of the population, how many generations of random mating would be required to establish Hardy–Weinberg proportions in the population?

24.11 A trait determined by an X-linked dominant allele shows 100 percent penetrance and is expressed in 36 percent of the females in a population. Assuming that the population is in Hardy–Weinberg equilibrium, what proportion of the males in this population express the trait?

24.12 A phenotypically normal couple has had one normal child and a child with cystic fibrosis, an autosomal recessive disease. The incidence of cystic fibrosis in the population from which this couple came is 1/1000. If their normal child eventually marries a phenotypically normal person from the same population, what is the risk that the newlyweds will produce a child with cystic fibrosis?

24.13 In a large population that reproduces by random mating, the frequencies of the genotypes *GG*, *Gg*, and *gg* are 0.04, 0.32, and 0.64, respectively. Assume that a change in the climate induces the population to reproduce exclusively by self-fertilization. Predict the frequencies of the genotypes in this population after many generations of self-fertilization.

24.14 In an isolated population, the frequencies of the I^A, I^B, and i alleles of the A–B–O blood-type gene are, respectively, 0.14, 0.28, and 0.58. If the genotypes of the A–B–O blood type gene are in Hardy–Weinberg proportions, what fraction of the people who have type A blood in this population are expected to be homozygous for the I^A allele?

24.15 Each of two isolated populations is in Hardy–Weinberg equilibrium with the following genotype frequencies:

Genotype	*AA*	*Aa*	*aa*
Frequency in Population 1:	0.04	0.32	0.64
Frequency in Population 2:	0.64	0.32	0.04

(a) If the populations are equal in size and they merge to form a single large population, predict the allele and genotype frequencies in the large population immediately after merger.
(b) If the merged population reproduces by random mating, predict the genotype frequencies in the next generation.
(c) If the merged population continues to reproduce by random mating, will these genotype frequencies remain constant?

24.16 GO The frequencies of the alleles A and a are 0.4 and 0.6, respectively, in a particular plant population. After many generations of random mating, the population goes through one cycle of self-fertilization. What is the expected frequency of heterozygotes in the progeny of the self-fertilized plants?

24.17 In a survey of moths collected from a natural population, a researcher found 51 dark specimens and 49 light specimens. The dark moths carry a dominant allele, and the light moths are homozygous for a recessive allele. If the population is in Hardy–Weinberg equilibrium, what is the estimated frequency of the recessive allele in the population? How many of the dark moths in the sample are likely to be homozygous for the dominant allele?

24.18 In a large randomly mating population, 0.84 of the individuals express the phenotype of the dominant allele A and 0.16 express the phenotype of the recessive allele a. (a) What is the frequency of the dominant allele? (b) If the aa homozygotes are 5 percent less fit than the other two genotypes, what will the frequency of A be in the next generation?

24.19 The frequency of newborn infants homozygous for a recessive lethal allele is about 1 in 25,000. What is the expected frequency of carriers of this allele in the population?

24.20 A population consists of 25 percent tall individuals (genotype TT), 25 percent short individuals (genotype tt), and 50 percent individuals of intermediate height (genotype Tt). Predict the ultimate phenotypic and genotypic composition of the population if, generation after generation, mating is strictly assortative (that is, tall individuals mate with tall individuals, short individuals mate with short individuals, and intermediate individuals mate with intermediate individuals).

24.21 A population is segregating three alleles, A_1, A_2, and A_3, with frequencies 0.2, 0.5, and 0.3, respectively. If these alleles are selectively neutral, what is the probability that A_2 will ultimately be fixed by genetic drift? What is the probability that A_3 will ultimately be lost by genetic drift?

24.22. GO Mice with the genotype Hh are twice as fit as either of the homozygotes HH and hh. With random mating, what is the expected frequency of the h allele when the mouse population reaches a dynamic equilibrium because of balancing selection?

24.23 Because individuals with cystic fibrosis die before they can reproduce, the coefficient of selection against them is $s = 1$. Assume that heterozygous carriers of the recessive mutant allele responsible for this disease are as fit as wild-type homozygotes and that the population frequency of the mutant allele is 0.02. (a) Predict the incidence of cystic fibrosis in the population after one generation of selection. (b) Explain why the incidence of cystic fibrosis hardly changes even with $s = 1$.

24.24 For each set of relative fitnesses for the genotypes AA, Aa, and aa, explain how selection is operating. Assume that $0 < t < s < 1$.

	AA	Aa	aa
Case 1	1	1	$1 - s$
Case 2	$1 - s$	$1 - s$	1
Case 3	1	$1 - t$	$1 - s$
Case 4	$1 - s$	1	$1 - t$

24.25 In some regions of West Africa, the frequency of the HBB^S allele is 0.2. If this frequency is the result of a dynamic equilibrium due to the superior fitness of HBB^SHBB^A heterozygotes, and if HBB^SHBB^S homozygotes are essentially lethal, what is the intensity of selection against the HBB^AHBB^A homozygotes?

24.26 Individuals with the genotype bb are 20 percent less fit than individuals with the genotypes BB or Bb. If B mutates to b at a rate of 10^{-6} per generation, what is the expected frequency of the allele b when the population reaches mutation–selection equilibrium?

24.27 A completely recessive allele g is lethal in homozygous condition. If the dominant allele G mutates to g at a rate of 10^{-6} per generation, what is the expected frequency of the lethal allele when the population reaches mutation–selection equilibrium?

▶ Genomics on the Web

at http://www.ncbi.nlm.nih.gov/

The mutant allele that causes sickle-cell anemia is prevalent in areas where people have a high probability of contracting malaria, which is caused by a parasite transmitted by mosquitoes. Click on the links for Malaria and Mosquito on the Genomic biology page to find information on the malaria parasite *Plasmodium falciparum* and on the mosquito vector *Anopheles gambiae*.

1. How large is the *Plasmodium* genome? How many chromosomes does it comprise? How large is the *Anopheles* genome? How many chromosomes does it comprise? Have the genomes of these organisms been sequenced completely?

2. On the *Plasmodium* web page, click on the overview link to bring up a page with summary information on this parasite. Under related resources, click on WHO/Malaria info to bring up a page with links to information about various aspects of malaria. How widespread is the disease? How is it being treated today? How is the *Plasmodium* parasite transmitted from one person to another?

Chapter 25
Evolutionary Genetics

"Where do we come from? What are we? Where are we going?" An 1897 painting by the French artist Paul Gauguin.

CHAPTER OUTLINE

- The Emergence of Evolutionary Theory
- Genetic Variation in Natural Populations
- Molecular Evolution
- Speciation
- Human Evolution

D'ou venons nous? Que sommes nous? Ou allons nous?

In 1897 in Tahiti, the French artist Paul Gauguin created an enormous painting with a provocative title: "Where do we come from? What are we? Where are we going?" The painting, now on display in the Boston Museum of Fine Arts, shows a group of Polynesian people, both young and old, reclining, sitting, walking, and eating in a strangely colored landscape. The figures are forlorn and abstracted, and a few of them seem to stare interrogatively at the viewer, posing, as it were, those three haunting questions that Gauguin inscribed in the painting's margin. This melancholy canvas, created near the end of Gauguin's life, seems to depict the artist's personal search for answers to some of life's deep questions. However, it is more than the statement of an individual who sought inspiration, freedom, and fulfillment in the South Seas. Gauguin's painting reflects a universal quest for what it means to be human. During the nineteenth century, people began to see this issue in a new light, especially with the emergence of evolutionary theory. Charles Darwin's *The Origin of Species,* first published in 1859, advanced the ideas that species are not fixed and that populations of organisms change over time. In a later book, *The Descent of Man,* Darwin proposed that the human species was also subject to evolutionary forces. Darwin's ideas have troubled many people.

The Emergence of Evolutionary Theory

The theory of evolution, initially enunciated by Charles Darwin, is based on genetic principles.

The publication of *The Origin of Species* in 1859 provoked a storm of controversy—not because the idea that species evolve was new, but rather because Darwin made the case for it so well. Darwin's book was cogently written and rich in evidence. He argued that species change gradually over long periods of time. Some species split into two or more separate species; other species become extinct. Darwin's ideas were unsettling to many people who held to the notion that each species was divinely created and that except for trivial variations among individuals, species do not change—that is, they are immutable. Darwin's book contested this view. Although he did not say much about the origin of the first organisms on earth, he argued that during millions of years they had changed and diversified to produce the plethora of species now alive. Furthermore, Darwin argued that this change and diversification, what he called "divergence of character," was the result of purely natural processes.

DARWIN'S THEORY OF EVOLUTION

Darwin proposed that a species changes as a result of generations of competition among individuals. Within a species individuals vary with respect to heritable characteristics that influence the ability to survive and reproduce. Individuals that possess these characteristics will, on average, have more offspring than individuals that do not possess them. Because of this unequal contribution to the next generation, the characteristics that enhance survival and reproduction will tend to become more frequent within the species. Over many generations, this process, which Darwin called *natural selection*, changes the characteristics of the species—that is, the species evolves. In his book, Darwin summarized his thoughts about evolution by natural selection:

> Again, it may be asked, how is it that varieties, which I have called incipient species, become ultimately converted into good and distinct species which in most cases obviously differ from each other far more than do the varieties of the same species? How do those groups of species, which constitute what are called distinct genera, and which differ from each other more than do the species of the same genus, arise? All these results . . . follow from the struggle for life. Owing to this struggle, variations, however slight and from whatever cause proceeding, if they be in any degree profitable to the individuals of a species, in their infinitely complex relations to other organic beings and to their physical conditions of life, will tend to the preservation of such individuals, and will generally be inherited by the offspring. The offspring, also, will thus have a better chance of surviving, for, of the many individuals of any species which are periodically born, but a small number can survive. I have called this principle, by which each slight variation, if useful, is preserved, by the term Natural Selection.

Darwin hypothesized that selection was the driving force of evolution in nature because he was powerfully aware of how artificial selection has changed the characteristics of domesticated species. He recognized the impact that artificial selection has had in creating different breeds of cattle, dogs, and fowl (**FIGURE 25.1**); he also knew of its role in shaping horticultural and agricultural varieties of plants.

Darwin was also a first-rate naturalist. As a young man, he served for five years on the British survey ship H.M.S. *Beagle*. The *Beagle* departed from England in 1831, traveled to South America, and returned to England in 1836. The lengthy sojourn along the coast of South America afforded Darwin many opportunities to observe plants, animals, and geological formations. For example, on the Galapagos Islands off the coast of Ecuador he observed several species of birds that were different from each other in appearance and behavior, but that he subsequently

Cocker spaniel

English bulldog

Golden Laced Wyandotte

Black Belgian Bantam

Figure 25.1 ▶ Variation among breeds of dogs and chickens.

Warbler finch (*Geospiza olivace*)

Common cactus finch (*G. scandens*)

Medium ground finch (*G. fortis*)

Figure 25.2 ▶ Finches on the Galapagos Islands.

recognized were related to each other and to birds on the South American mainland (**FIGURE 25.2**). From these and other observations, Darwin was led to the view that species are not fixed entities. Rather, he inferred that they change over time, and that some—exemplified by the fossils he saw during his travels—even became extinct.

Darwin spent more than 20 years analyzing and interpreting the data that he collected on the voyage of the *Beagle*. In addition, at his country estate in Kent, England, he performed experiments with a variety of plants and domesticated animals. The observations that he made in this experimental work, along with his extensive reading and analysis of the data that he collected on the *Beagle*'s voyage, gave Darwin the insights that eventually led to the publication of *The Origin of Species*.

EVOLUTIONARY GENETICS

Darwin's theory of evolution had one major gap. It offered no explanation for the origin of variation among individuals, and it could not explain how particular variants are inherited. Eventually, Darwin did propose a theory of inheritance based on the transmission of acquired characteristics. However, his theory was flawed. Biologists who were attracted to Darwin's ideas on evolution struggled, as he did, to explain how the variants that natural selection favors are transmitted from parents to offspring. In 1900 the rediscovery of Mendel's principles provided the long-sought-after explanation: traits are determined by genes, which segregate different alleles, and genes are transmitted to the offspring in gametes produced by their parents. The analysis of genetic transmission in experimental crosses and pedigrees quickly gave rise to a new type of analysis that involved whole populations. The discipline of evolutionary genetics was born, and by 1930, especially through the contributions of Sewall Wright, R. A. Fisher, and J. B. S. Haldane (**FIGURE 25.3**), it had become the foundation for Darwinian theory.

KEY POINTS

- Charles Darwin formulated a theory in which species evolve through natural selection.
- After the rediscovery of Mendel's work, Darwin's ideas became grounded on Mendelian principles of inheritance.

Sewall Wright

R. A. Fisher

J. B. S. Haldane

Figure 25.3 ▶ The founders of evolutionary genetics theory.

Genetic Variation in Natural Populations

Many different experimental approaches provide information about genetic variation in populations of organisms.

Darwin's *The Origin of Species* begins with a discussion of variation. Without variation, populations cannot evolve. Soon after Mendel's principles were rediscovered, biologists began to document genetic variation in natural populations. Initially, these efforts focused on conspicuous features of the phenotype—pigmentation, size, and so forth. Later, they emphasized characteristics that are more directly related to chromosomes and genes. In the following sections, we discuss variation at the phenotypic, chromosomal, and molecular levels.

VARIATION IN PHENOTYPES

Naturalists have described phenotypic variation within many species. For example, land snails have different colored bands on their shells, squirrels and other small mammals have different coat colors, and butterflies and moths have different patterns in their wings (**FIGURE 25.4**). In the plant kingdom, phenotypic variation may be manifested by different kinds of flowers. All these sorts of phenotypic differences may have a genetic basis.

Brown-banded snail (*Liguus fasciatus*)

Yellow-banded snail

Gray squirrel (*Sciurus carolinensis*)

Albino squirrel

Yellow tiger swallowtail (*Papilio glaucus*)

Black tiger swallowtail

Figure 25.4 ▶ Naturally occurring phenotypic variation in land snails, squirrels, and butterflies.

However, to elucidate the underlying genetic factors, it is necessary to bring the organisms into the laboratory and cross them with one another. Unfortunately, for many organisms this approach is not feasible. Thus, geneticists have tended to focus their investigations of naturally occurring phenotypic variation on organisms that can be reared and bred in the laboratory.

Some of the classic studies were carried out in Russia, where researchers collected *Drosophila* from natural populations, inbred them, and examined the progeny for characteristics associated with mutant genes—for example, white eyes (instead of red eyes) and yellow bodies (instead of gray bodies). This work documented the presence of mutant alleles in natural populations, but none of the mutant alleles was common enough to be considered part of a polymorphism. However, significant polymorphisms have been found in research with other species. We have seen, for example, that light and dark forms of the peppered moth exist in Great Britain, and that these forms are due to different alleles of a single gene (Chapter 24). We have also seen that the status of this polymorphism has changed since the nineteenth century, with the dark form rising in frequency in the heavily industrialized areas of England. In the plant kingdom, the tiny annual *Linanthus parryae*, known familiarly as the desert snow, produces either white or blue flowers throughout its habitat on the western and southern margins of the Mojave Desert in North America (**FIGURE 25.5**). Plants with blue flowers carry a dominant allele, and plants with white flowers are homozygous for a recessive allele. The blue-flower phenotype ranges from 1 to 80 percent, depending on location. Thus, this color polymorphism varies spatially among *Linanthus* populations.

Humans are also polymorphic. Pedigree analysis and population sampling have enabled researchers to identify many human polymorphisms. The classic data come from the study of blood types, which are determined by antigens on the surfaces of cells. The alleles that encode these antigens are often polymorphic. For example, the Duffy blood-typing system identifies two antigens, each encoded by a different allele of a gene on chromosome 1. The two Duffy alleles, denoted Fy^a and Fy^b, have different frequencies among different populations (**TABLE 25.1**). In England, both Fy^a and Fy^b are common, but in Korea, only Fy^a is common, and in southern Africa, only Fy^b is common. Thus, the status of the Duffy polymorphism varies among human ethnic groups.

Figure 25.5 ▶ Color variation in the desert snow (*Linanthus parryae*).

▶ **TABLE 25.1**

Frequencies of Alleles of the Duffy Blood Group Locus in Different Human Populations

Allele	Korea	South Africa	England
Fy^a	0.995	0.060	0.421
Fy^b	0.005	0.940	0.579

Source: Data from Cavalli-Sforza, L. L., and A. W. F. Edwards. 1967. Phylogenetic analysis: models and estimation procedures. *Evolution* 21:550–570.

VARIATION IN CHROMOSOME STRUCTURE

Phenotypic variation can be a reflection of underlying genetic variability. Is there a way to detect variability by looking at the genetic material itself? The polytene chromosomes from the salivary glands of *Drosophila* larvae afford researchers an unparalleled opportunity to look for variation in chromosome structure. Flies captured in the wild can be brought into the laboratory and bred to produce larvae, which can then be examined for alterations in the banding patterns of their polytene chromosomes. For more than 25 years, Theodosius Dobzhansky and his collaborators performed this type of analysis on several species of *Drosophila* native to North and South America. The most thorough studies involved three closely related species, *D. pseudoobscura*, *D. persimilis*, and *D. miranda*, which are found in western North America. *D. pseudoobscura* and *D. persimilis* are sibling species—morphologically similar and able to be interbred in the laboratory; however, hybrid males from crosses between *D. pseudoobscura* and *D. persimilis* are sterile, and hybrid females from these crosses have severely reduced fertility. *D. miranda* is less closely related to the other two species. It rarely produces hybrids in crosses with either of them, and when hybrids are produced, they are always sterile.

Dobzhansky and his collaborators identified many different arrangements of the banding patterns in the polytene chromosomes of these species. Each arrangement consists of one or more inversions of the most common banding pattern. For example, in the third chromosome of *D. pseudoobscura*, they found 17 different arrangements in natural populations. The Standard banding pattern, denoted ST, was most frequent in populations along the coast of California and in northern Mexico; in these areas 48 to 58 percent of all third chromosomes in the sample of captured flies showed the ST banding pattern. Different arrangements predominated in other areas. For example, the arrangement known as Arrowhead (AR) was found in 88 percent of chromosomes sampled from Arizona, Utah, and Nevada, and the arrangement known as Pike's Peak (PP) was found in 71 percent of chromosomes sampled from Texas. Repeated sampling of selected populations established that the frequencies of the arrangements changed seasonally. For example, at Piñon Flats, California, the frequency of the ST arrangement declined from greater than 50 percent in March to around 30 percent in June. This shift in frequency was observed in each of several years in which samples were collected. In addition, Dobzhansky and his coworkers observed long-term changes in the frequencies of arrangements in some populations. At Lone Pine, California, for instance, the ST arrangement increased from a frequency of 21 percent in 1938 to 65 percent in 1963. These researchers also performed laboratory experiments to measure the competitive abilities of flies carrying different chromosome arrangements. Their experiments suggested that balancing selection plays an important role in maintaining these chromosomal polymorphisms in nature.

VARIATION IN PROTEIN STRUCTURE

In 1966 R. C. Lewontin, J. L. Hubby, and H. Harris initiated a new era in the study of genetic variation in natural populations when they applied the technique of gel electrophoresis to detect amino acid differences in proteins. Lewontin and Hubby studied protein variation in *Drosophila*, and Harris studied it in humans. Their technique proved to be so successful that it was quickly applied to study genetic variation in all sorts of organisms, including creatures as diverse as starfish, wild oats, and spittle bugs.

Gel electrophoresis is a sieving technique that separates macromolecules on the basis of size and charge (see Chapter 15). The sieving agent is a thin, rectangular gel made from a polymer such as starch or polyacrylamide. Samples containing the macromolecules of interest are deposited in wells formed in a line across the gel. The gel is immersed in a solution of buffer formulated to conduct electricity, and the ends of the tank containing the buffer are connected to a power source. When the power is turned on, an electric field is created in the buffer tank. In this field, macromolecules migrate through the gel at a rate that depends on their size and charge. Smaller, more highly charged molecules migrate faster than larger, less highly charged molecules. Proteins that differ in size and charge can therefore be separated from each other by moving them through the gel. After sufficient time has elapsed, the power is turned off and the gel is treated to reveal how far each protein has migrated. The treatment may involve a reagent that stains the proteins, or, if the proteins are enzymes, it may involve a substrate whose chemical change is coupled to the production of a characteristic color.

As an example, consider the staining pattern shown in **FIGURE 25.6**. The proteins in tissue samples from 10 different *Trillium pusillum* plants were separated by electrophoresis in a gel, and the gel was stained to detect the enzyme isocitrate dehydrogenase. This enzyme occurs naturally as a dimer—that is, it consists of two polypeptide chains encoded by a particular locus in the plant's genome. The uppermost and lowermost bands in the gel represent, respectively, slow- and fast-migrating forms of the enzyme. Within each of these bands, the two subunits of the enzyme are identical. The subunit in the fast-moving form of the enzyme is encoded by one allele of the gene for isocitrate dehydrogenase, and the subunit in the slow-moving form of the enzyme is encoded by another allele of this gene. Samples 1, 2, 8, and 10 in the photograph came from plants that were homozygous for the allele encoding the slow-moving subunit. All the other samples came from plants that were heterozygous for this allele and the allele for the fast-moving subunit. These heterozygous plants produce three types of isocitrate dehydrogenase. One type consists of two identical fast-moving subunits, the second type consists of two identical slow-moving subunits, and the third type consists of a fast-moving subunit joined with a slow-moving subunit—that is, a "hybrid" enzyme—which has an intermediate electrophoretic mobility and therefore appears as a band in the middle of the gel.

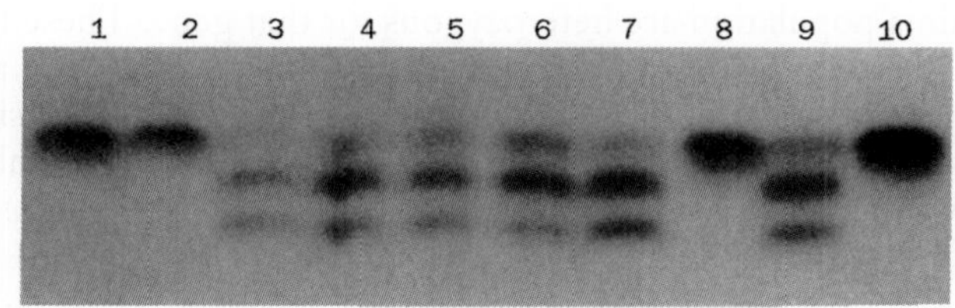

Figure 25.6 ▶ Photograph of an electrophoretic gel stained to detect the enzyme isocitrate dehydrogenase in samples extracted from *Trillium pusillum* plants.

The fast- and slow-moving forms of isocitrate dehydrogenase are examples of **allozymes**—forms of an enzyme encoded by different alleles of a gene. Allozymes differ from each other by one or more amino acids in their overall sequence. Protein gel electrophoresis therefore allows a researcher to detect variation at the level of gene products—that is, as amino acid differences in polypeptide chains. This technique can be applied to proteins extracted from almost any kind of organism, including those that are not amenable to genetic analysis in the laboratory. In addition, because protein extracts are easy to obtain, it provides a way to survey genetic variation in large samples of individuals from different populations. Protein gel electrophoresis therefore allows researchers to investigate the spatial and temporal dimensions of genetic variation in nature.

Proteins that exhibit electrophoretic variation are said to be polymorphic if at least two of the variants have frequencies greater than 1 percent in the population. Proteins that do not exhibit electrophoretic variation are said to be *monomorphic*. In humans, approximately one-third of soluble enzymes are polymorphic.

Protein gel electrophoresis provided the first extensive evidence of genetic variation at the molecular level. In many species one-fourth to one-third of all genes that encode soluble proteins exhibit electrophoretic polymorphisms, and for a given polymorphic gene, about 12 to 15 percent of individuals within a population are heterozygous for that gene. These two statistics—the proportion of genes that are polymorphic and the proportion of individuals that are heterozygous—are simple and convenient measures of the amount of genetic variability within a population.

VARIATION IN NUCLEOTIDE SEQUENCES

Protein gel electrophoresis provides researchers with a way to survey genetic variation in natural populations. For nearly two decades after it was first introduced, this technique was the mainstay of experimental population genetics. However, even during this time geneticists recognized that it had some significant shortcomings. First, it works best with soluble enzymes. Nonsoluble, hydrophobic proteins, which are often associated with membraneous organelles, cannot readily be analyzed. Second, it focuses on gene products rather than on the genes themselves. The noncoding components of genes—promoters, introns, enhancers, and the like—cannot be analyzed by protein gel electrophoresis. Third, it tells us nothing about variation in the nongenic portion of a genome, which, for complex eukaryotes, comprises most of the DNA. In humans, for example, 78 percent of the DNA is nongenic. To address these shortcomings, researchers have developed techniques to analyze variation in the DNA itself.

DNA sequencing provides the ultimate data on genetic variation. Any sequence—coding, noncoding, genic, nongenic—can be analyzed. The first efforts to study genetic variation by DNA sequencing used material that had been cloned from the genomes of different individuals. Each clone was obtained by virtue of its ability to hybridize with a specific DNA probe. The clones were then sequenced, and the sequences were compared to identify differences along their lengths.

As an example of this type of analysis, consider the results of a study of sequence variability in the gene for alcohol dehydrogenase, *Adh*, in *Drosophila melanogaster* performed by Martin Kreitman. Eleven cloned *Adh* genes from different populations were sequenced to obtain the data for the study. The *Adh* gene consists of four exons and three introns. Transcription of the *Adh* gene can be initiated from either of two promoters—one that functions in the adult and another that functions in the larva. The adult promoter is located upstream of the larval promoter. Thus, adult transcripts of the *Adh* gene contain all four exons and all three introns, whereas larval transcripts contain only the last three exons and the last two introns. The coding sequences of the *Adh* gene begin in the second exon; therefore, all the coding sequences are present in the larval transcript as well as in the adult transcript. Kreitman catalogued the differences among the *Adh* genes that he sequenced (**FIGURE 25.7**). Altogether, 43 nucleotide positions were polymorphic. The majority of the polymorphisms were in noncoding regions of the *Adh* gene—in introns, or in the 3′ and 5′ untranslated regions—and in the DNA flanking the gene. Some polymorphisms were also found within the gene's coding sequences; however, only one of these polymorphisms caused an amino acid difference in the Adh polypeptide. This difference, a lysine versus a threonine at position 192, alters the mobility of the Adh protein during gel electrophoresis; the polypeptide with lysine moves faster than the one with threonine. All the other nucleotide differences in the coding sequence of the *Adh* gene have no effect on the amino acid sequence of the polypeptide. Geneticists refer to them as *silent polymorphisms*; they arise from

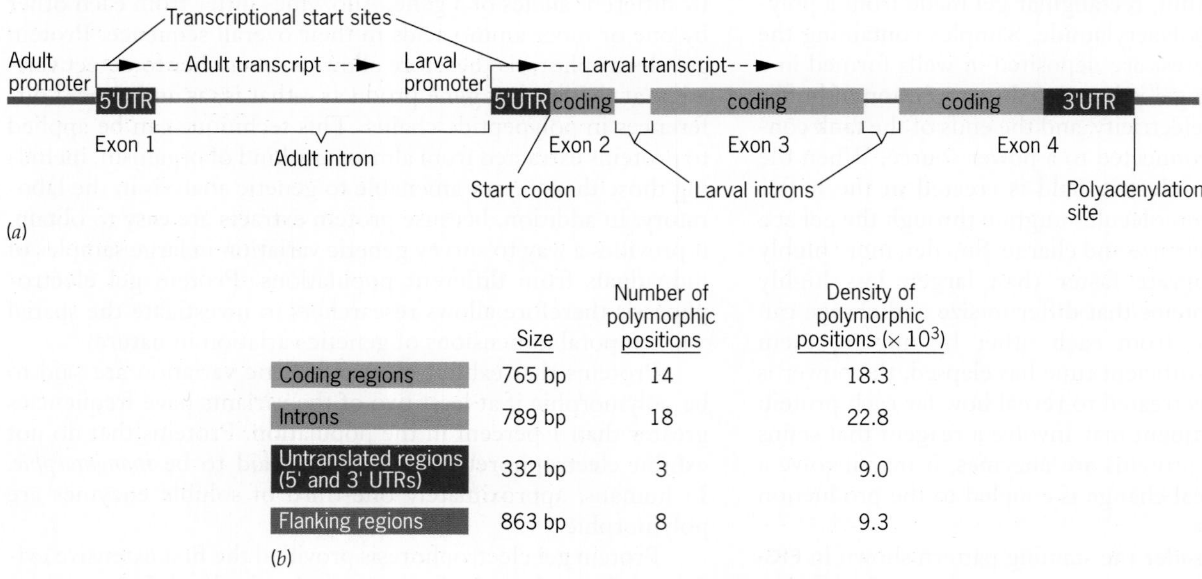

	Size	Number of polymorphic positions	Density of polymorphic positions ($\times 10^3$)
Coding regions	765 bp	14	18.3
Introns	789 bp	18	22.8
Untranslated regions (5′ and 3′ UTRs)	332 bp	3	9.0
Flanking regions	863 bp	8	9.3

Figure 25.7 ▶ (*a*) Molecular structure of the *Alcohol dehydrogenase* (*Adh*) gene in *Drosophila melanogaster*. (*b*) DNA sequence polymorphisms in different regions of the *Adh* gene. Data from Kreitman, M. 1983. Nucleotide polymorphism at the alcohol dehydrogenase locus of *Drosophila melanogaster. Nature* 304:412–417.

the degeneracy of the genetic code—that is, more than one codon being able to specify the incorporation of a particular amino acid into a polypeptide.

Today, obtaining DNA sequence data to study naturally occurring genetic variation is not nearly as difficult as it used to be. Particular regions of the genome can be amplified by PCR, and the resulting DNA products can be sequenced by machine. Sophisticated computer programs can then be used to analyze the sequence data and identify variation among individuals. This technique permits researchers to assess the level of variation in functionally different regions of DNA—for instance, in exons compared to introns.

Gene chip technologies (see Chapter 16) provide another means of documenting variation at the DNA level. These technologies allow researchers to screen genomic DNA for single-nucleotide polymorphisms (SNPs), which are found every 1–2 kb. Many different genomic DNA samples can be analyzed in parallel, and a great many SNPs can be detected on a single chip.

KEY POINTS

- Genetic variation in natural populations can be detected at the phenotypic, chromosomal, and molecular levels.
- Classic studies established the existence of genetic polymorphisms for conspicuous phenotypic traits, such as flower color, and for blood types.
- Polymorphisms in chromosome structure have been documented in various species of *Drosophila* by analyzing banding patterns in the polytene chromosomes.
- Polymorphisms in polypeptide structure have been detected by using the technique of protein gel electrophoresis.
- Polymorphisms in DNA structure have been detected by sequencing cloned or PCR-amplified DNA, and by using diagnostic gene chips.

Molecular Evolution

DNA and protein sequences provide information on the phylogenetic relationships among different organisms, and on their evolutionary history.

The ability to clone, amplify, manipulate, and sequence DNA molecules from any type of organism has had an enormous impact on the study of evolution. In *The Origin of Species*, Darwin repeatedly referred to evolution as a process of "descent with modification." His focus was on the traits of organisms, which are passed on more or less faithfully to their offspring every generation, but which also undergo modifications as the organisms adapt to changing environmental conditions. Today, knowing that heredity depends on the sequence of nucleotides in DNA, we understand the molecular basis of Darwin's concept. DNA molecules are passed from parents to offspring generation after generation. However, this process of genetic transmission is not perfect. Mutations occur, and when they do, modified DNA molecules are transmitted to the offspring. Over long periods of time, mutations accumulate and the DNA sequence is changed; segments of DNA molecules may also be duplicated or rearranged. This process of molecular evolution must underlie the evolution of organisms that Darwin wrote about.

MOLECULES AS "DOCUMENTS OF EVOLUTIONARY HISTORY"

One body of evidence that led Darwin to propose that species evolve came from the study of rocks in the ground. Fossils—the mineralized remains of animals and plants long since dead—were avidly collected in Darwin's day. These unusual rocks were curios for display in Victorian drawing rooms, but they were also evidence of organisms that had once lived on the earth. From the detailed study of fossils naturalists could reconstruct, at least crudely, what ancient organisms looked like and how they might have behaved. Comparisons between living organisms and the fossilized remains of extinct organisms stimulated speculation about the origin of species. Thus, with the perspective gained from studying fossils, naturalists began to think about life in historical terms.

DNA molecules, like fossils, contain information about life's history. The DNA molecules in creatures today are derived from their ancestors—parents, grandparents, and so on, going back in time to the very first organisms. Each DNA molecule is the end result of a long historical process involving mutation, recombination, selection, and genetic drift. In metaphorical terms, the sequence of nucleotides in a DNA molecule is the current version of an ancient text that, in the course of being copied generation after generation, has been altered (mutated), cut and pasted (recombined), preserved for its value (selected), and randomly disseminated (subjected to drift). Emile Zuckerkandl, one of the pioneers in the study of molecular evolution, put it this way: DNA molecules are "documents of evolutionary history."

So, too, are protein molecules. Polypeptides are encoded by genes, which are segments of DNA molecules. As the genes evolve, so do the proteins they encode. Geneticists can therefore investigate evolution at the molecular level either by studying nucleotide sequences in DNA or amino acid sequences in proteins.

The analysis of DNA and protein sequences has several advantages over more traditional methods of studying evolution based on comparative anatomy, physiology, and embryology. First, DNA and protein sequences follow simple rules of heredity. By contrast, anatomical, physiological, and embryological traits are subject to all the vicissitudes of complex heredity (see Chapter 23). Second, molecular sequence data are easy to obtain, and they are also amenable to quantitative analyses framed in the context of evolutionary genetics theory. The interpretation of these analyses is usually much more straightforward than the interpretation of analyses based on morphological data. Third, molecular sequence data allow researchers to investigate evolutionary relationships among organisms that are phenotypically very dissimilar. For instance, DNA and protein sequences from bacteria, yeast, protozoa, and humans can be compared to study the evolutionary relationships among them.

One problem with the molecular approach to evolution is that researchers usually cannot obtain DNA or protein sequence data from extinct organisms. In a few exceptional cases, such data have been obtained from fossils. However, in none of these cases was the fossilized specimen more than a few tens of thousands of years old. Thus, truly ancient organisms are beyond the reach of any molecular investigation. Another problem is that it is not always clear how molecular sequence data bear on questions about evolution at the phenotypic level.

MOLECULAR PHYLOGENIES

The evolutionary relationships among organisms are summarized in diagrams called **phylogenetic trees,** or more simply, **phylogenies.** These trees may only show the relationships among the organisms, or they may superimpose the relationships on a time line to indicate how each of the organisms evolved. A phylogeny that only shows the relationships is an *unrooted tree*, whereas one that shows their derivation is a *rooted tree* (**FIGURE 25.8**). In both rooted and unrooted trees, the lineages bifurcate to produce branches. The branches at the tips of the tree—called terminal branches—lead to the organisms that are under study. Each bifurcation in a tree represents a common ancestor of the organisms farther out in the tree.

In molecular analyses of evolutionary relationships, the organisms are represented by DNA or protein sequences. Some analyses are based on a single gene or gene product. Other analyses combine data obtained by sequencing different genes or gene products. Sometimes the analyses utilize nongenic DNA sequences to ascertain the relationships among organisms.

The descendants of an ancestral DNA or protein sequence are said to be *homologous*, even if they have diverged significantly from the ancestor and are different from each other. Two sequences that come to resemble each other even though they

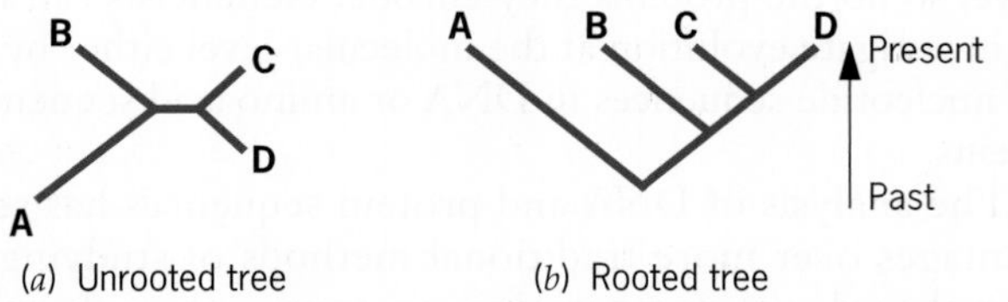

Figure 25.8 ▶ Difference between unrooted (*a*) and rooted (*b*) phylogenetic trees.

are derived from entirely different ancestral sequences are said to be *analogous*. The construction of phylogenetic trees should always be based on the analysis of homologous sequences.

Many methods are now available to construct phylogenetic trees from DNA or protein sequence data. These methods usually have four features in common: (1) aligning the sequences to allow comparisons among them; (2) ascertaining the amount of similarity (or difference) between any two sequences; (3) grouping the sequences on the basis of similarity; and (4) placing the sequences at the tips of a tree.

As an example, consider the sequence data summarized in **FIGURE 25.9*a***. A researcher has sequenced a gene from four populations of an organism, denoted 1, 2, 3, and 4. This gene consists of two exons separated by an intron; a 5′ untranslated region (UTR) is included in the first exon, and a 3′ UTR is included in the second exon. When we align the four sequences of this gene, we find that two of them have a transposable element insertion in the 3′ UTR, and three of them have lost a short sequence within the intron. We also see that each of the four sequences has at least one feature that uniquely distinguishes it from all the other sequences—a G:C base pair near the start of the coding sequence in exon I in sequence 1, an A:T base pair near the start of exon II in sequence 2, a C:G base pair in the 5′ UTR and a T:A base pair near the end of the coding sequence in exon II in sequence 3, and an A:T base pair in the 5′ UTR, a C:G base pair in the middle of exon II, and a G:C base pair in the 3′ UTR in sequence 4. Based on these similarities and differences, how are the four sequences related?

If we assume that all four sequences were derived from a common ancestral sequence, we can construct a rooted tree that shows their evolutionary relationships (**FIGURE 25.9*b***). Sequences 1 and 2 are the most similar—they both have the transposon insertion in the 3′ UTR, and they are both missing the A:T base pair within the intron. Sequence 3 also lacks the A:T base pair within the intron. Because of these similarities, we could place sequences 1 and 2 close together in the tree—on branches diverging from a common point—and we could place sequence 3 on another branch nearby. Sequence 4 has none of the features found in the other sequences. Thus, because it is the most different in the sample—the outlier—we could place it on a branch that diverged from the other lineages at the root of the tree.

The tree we have constructed is one of 15 possible rooted trees showing the evolutionary relationships among four sequences (**FIGURE 25.9*c***). Twelve of the 15 possible rooted trees have the same basic structure as the tree we have built: a bifurcation at the root of the tree, producing two main branches, and then two more bifurcations in one (but not the other) of these branches. Three of the 15 possible rooted trees have a different basic structure: a bifurcation at the root of the tree and then another bifurcation in each of the main branches. Thus, in this example we have two classes of trees. Each class is characterized by a distinct basic structure—what tree builders call a *topology*.

Which tree is the best one to summarize the relationships among the sequences? We can answer this question only if we

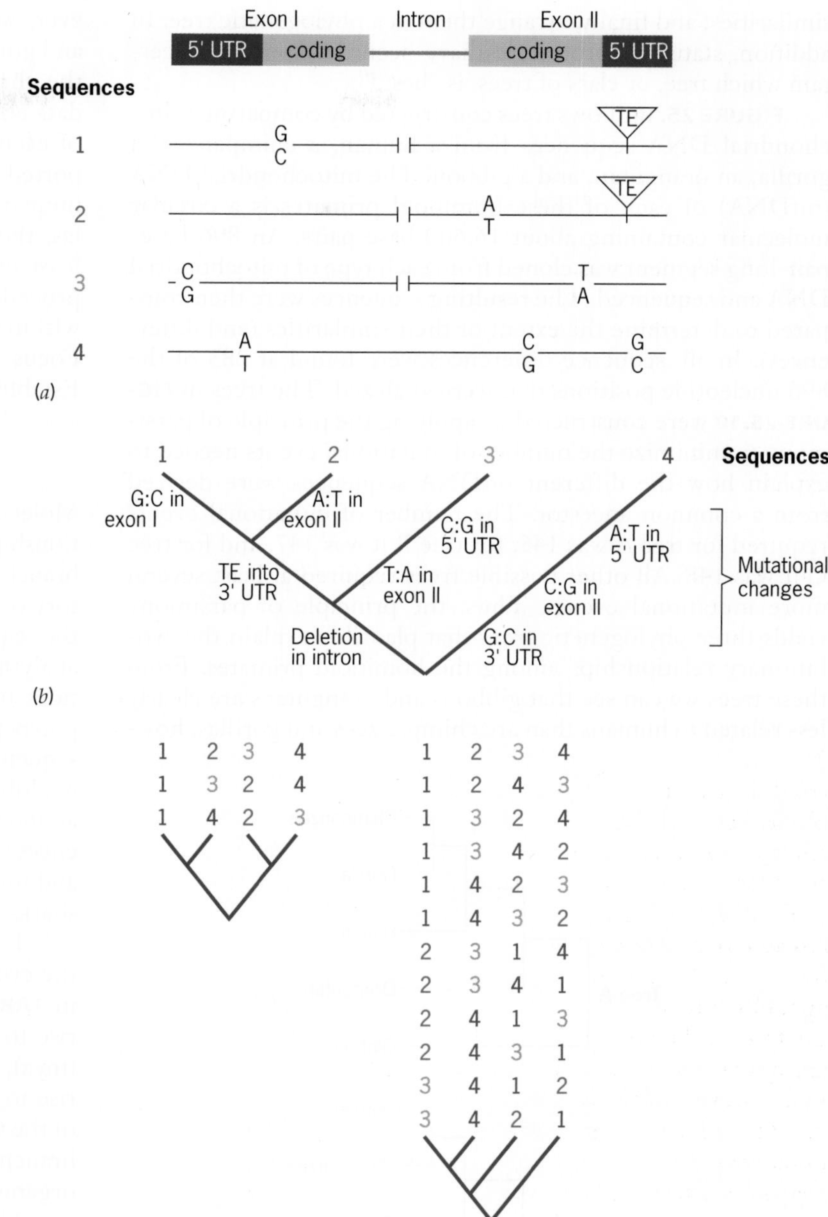

Figure 25.9 ▶ (*a*) Differences among the sequences of a gene from four populations (1, 2, 3, and 4). The structure of the gene is shown at the top. Base-pair differences (G:C, A:T, C:G, and T:A), a deletion (gap), and a transposable element insertion (TE) in the otherwise identical sequences from the four populations are shown below. (*b*) Rooted phylogenetic tree showing the evolutionary relationships among the four sequences. The mutational changes needed to account for the differences among the sequences are shown on the branches of the tree. (*c*) The 15 possible rooted trees showing how four sequences are related. Three of the possible trees have one topology; 12 of the possible trees have another topology.

have a criterion with which to judge the merit of a tree. Many tree builders have adopted the *principle of parsimony* as the criterion for judgment. By this they mean that the best tree is the one that requires the fewest mutational changes to explain the evolution of all the tree's sequences from a common ancestor. **FIGURE 25.9*b*** details the mutational changes that are required by the tree we have constructed: a single transposon insertion, a single base-pair deletion, and seven independent base-pair substitutions. The total of nine mutations is less than or equal to the total that would be required to account for the differences among the sequences using any other tree. For example, if sequences 1 and 3 were switched in position, 10 separate mutational changes would be needed to explain the origin of the four sequences from a common ancestor. The principle of parsimony comes from William of Occam, a fourteenth-century British scholar, who argued against multiplying hypotheses to explain phenomena.

Obviously, the number of possible trees will increase if more sequences are included in the analysis. For 6 sequences, there are 945 possible rooted trees; for 10 sequences, there are 34,459,425. Constructing trees with numerous sequences is therefore a daunting task. Fortunately, computer algorithms have been developed to align sequences, compare them to determine their similarities, cluster them by virtue of these similarities, and finally, arrange them in a phylogenetic tree. In addition, statistical procedures have been developed to ascertain which tree, or class of trees, is "best."

FIGURE 25.10 shows trees constructed by comparing mitochondrial DNA sequences from a human, a chimpanzee, a gorilla, an orangutan, and a gibbon. The mitochondrial DNA (mtDNA) of each of these hominoid primates is a circular molecular containing about 16,600 base pairs. An 896-base-pair-long segment was cloned from each type of mitochondrial DNA and sequenced. The resulting sequences were then compared to determine the extent of their similarities (and differences). In all, sequence differences were found at 283 of the 896 nucleotide positions that were analyzed. The trees in **FIGURE 25.10** were constructed by applying the principle of parsimony to minimize the number of mutational events needed to explain how the different mtDNA sequences were derived from a common ancestor. The number of mutational events required for tree A was 145, for tree B it was 147, and for tree C it was 148. All other possible trees required at least several more mutational events. Thus, the principle of parsimony yields three phylogenetic trees that plausibly explain the evolutionary relationships among the hominoid primates. From these trees we can see that gibbons and orangutans are clearly less related to humans than are chimpanzees and gorillas; however, we cannot readily discern how humans, chimpanzees, and gorillas are related to each other. Trees A, B, and C present the three possibilities. A more sophisticated analysis of the data employing statistical techniques that estimate the lengths of each of the branches favors tree B. This tree is also supported by the analysis of other types of DNA sequences. Thus, humans are more closely related to chimpanzees than to gorillas, they are less closely related to orangutans, and they are least closely related to gibbons. To apply the tree-building procedure to mtDNA sequences from different individuals within a particular human ethnic group, work through the Focus on Problem Solving: Using Mitochondrial DNA to Establish a Phylogeny.

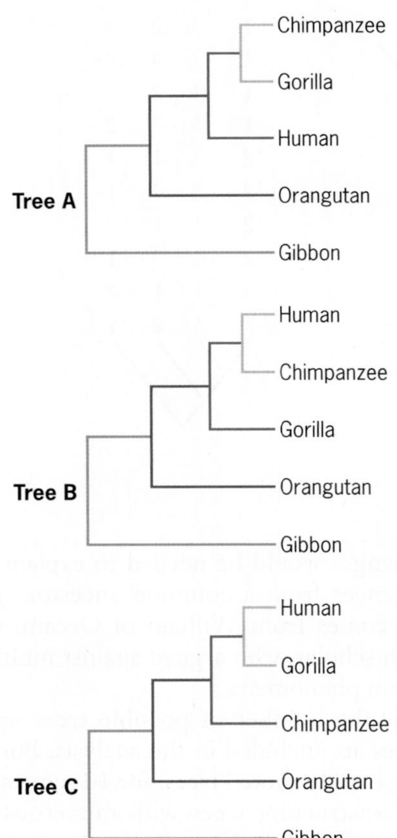

Figure 25.10 ▶ Phylogenetic trees of hominids constructed from the analysis of an 896-base-pair-long sequence of mitochondrial DNA.

RATES OF MOLECULAR EVOLUTION

Molecular phylogenetic trees tell us about the evolutionary relationships among DNA or protein sequences. If we can link the branch points of a tree to specific times in the evolutionary history of the sequences, then we can determine the rate at which the sequences have been evolving. As an example of this kind of analysis, consider α-globin, which is one of two kinds of polypeptides found in the blood protein hemoglobin. The α-globin polypeptide consists of 141 amino acids. We can compare the sequence of α-globin from one organism with the sequence of α-globin from another organism and count the number of amino acids that differ between the two sequences. Such differences are tabulated in **TABLE 25.2**. The α-globins from humans and mice have the fewest differences (16); those from carp and sharks have the greatest number of differences (85).

The fossil record provides information about key events in the evolutionary history of the six types of organisms included in **TABLE 25.2**. For instance, the evolutionary lines that gave rise to humans and mice diverged about 80 million years ago (mya), near the end of the Mesozoic Era, and the lines that gave rise to carp and sharks diverged at least 440 mya near the end of the Ordovician period in the Paleozoic Era. These and other branch points in the evolutionary history of the six different organisms are depicted in **FIGURE 25.11**.

The tree in **FIGURE 25.11** was constructed by using evidence from the fossil record. However, its topology is consistent with the molecular data presented in **TABLE 25.2**. Humans and mice show the fewest amino acid differences in α-globin,

TABLE 25.2

Number of Dissimilar Amino Acids in the α-Globins of Representative Vertebrates

	Mouse	Chicken	Newt	Carp	Shark
Human 1	6	35	62	68	79
Mouse		39	63	68	79
Chicken			63	72	83
Newt				74	84
Carp					85

FOCUS ON PROBLEM SOLVING
Using Mitochondrial DNA to Establish a Phylogeny

THE PROBLEM

Derbeneva *et al.* (2002 *Am. J. Hum. Genet.* 71:415-421) sequenced the mitochondrial DNA (mtDNA) obtained from a sample of 30 Aleut people from the Commander Islands, the westernmost islands of the Aleutian chain. Nine distinct types of mtDNA, each denoted by a Roman numeral, were identified. Each entry in the following table indicates the position of a nucleotide that differs from the nucleotide found in Type VIII, which was used as a standard. These differences are consistent across the various types—that is, the different nucleotide at position 9667 in Type I is the same as the different nucleotide at position 9667 in Type II. From these data, construct a diagram that shows the phylogenetic relationships among the different types of Aleut mtDNA.

Type	Number in Sample	Nucleotide Positions Different from Type VIII
I	13	8910, 9667
II	4	6554, 8639, 8910, 9667, 16311
III	3	8910, 9667, 16519
IV	3	8910, 9667, 11062
V	1	8460, 8910, 9667
VI	1	5081, 8910, 9667
VII	1	8910, 9667, 10695, 11113
VIII	3	Standard
IX	1	16092

FACTS AND CONCEPTS

1. Human mtDNA is a circular molecule consisting of around 16,570 nucleotide pairs (see Chapter 16).

2. When two mtDNA sequences are compared, each single base-pair difference represents a mutation.

3. Phylogenetic trees are constructed by grouping the most similar DNA sequences near one another and by minimizing the number of mutations needed to explain the differences among all the DNA sequences.

ANALYSIS AND SOLUTION

The standard type of Aleut mtDNA differs from type IX at one nucleotide position (16092). It differs from all the other types at two or more nucleotide positions—positions 8910 and 9667 in Type I, and these two positions plus at least one other position in all the other types. The standard type is therefore one mutational step removed from Type IX, and it is two mutational steps removed from Type I, but in a different direction. All the other types are one or more mutational step removed from Type I. We can summarize the relationships among the types in a phylogenetic diagram, which, however, does not tell us which mtDNA is ancestral to the others—that is, it is an unrooted tree:

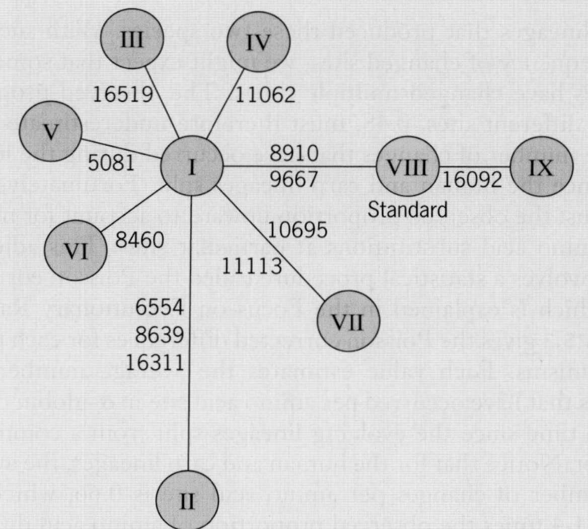

For further discussion go to your *WileyPLUS* course.

and they are closest together—that is, separated by the shortest evolutionary time—in **FIGURE 25.11**. The chicken's α-globin is the next closest to the α-globins of the two mammals, followed by the newt's, the carp's, and the shark's. The extent to which the amino acid sequences of these six organisms differ can be used to estimate the rate at which α-globin has been evolving.

To obtain this rate, we first need to determine the average number of amino acid changes that have occurred since any two of the lineages split from a common ancestor. We can start with the two most closely related organisms, humans and mice, which are different in 16 of the 141 amino acid sites in α-globin. The proportion of different sites in the α-globins of these two species is therefore $16/141 = 0.11$, which we can also interpret as the average number of differences per amino acid site. Now consider two very distantly related organisms, humans and carp. The α-globins of these two organisms differ in 68 of the 141 amino acid sites; thus, the proportion of different sites is $68/141 = 0.48$—that is, almost half the sites have changed during the evolution

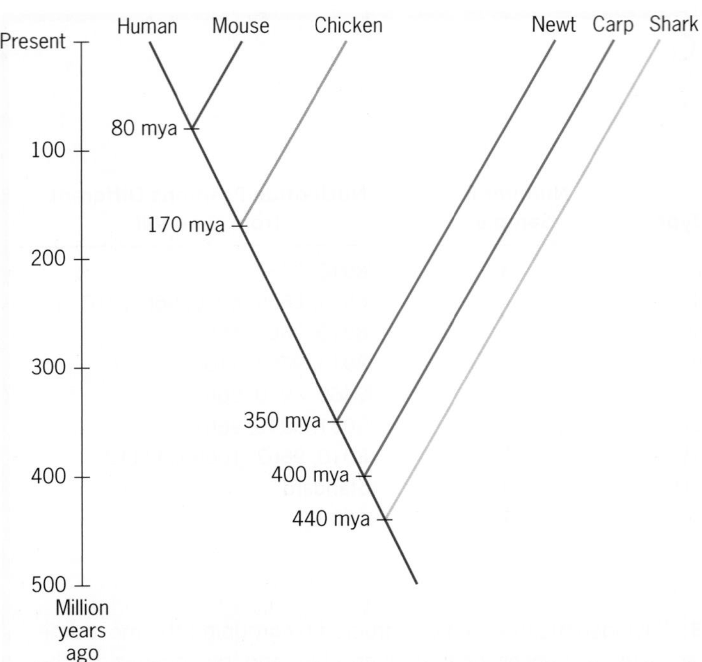

Figure 25.11 ▶ Phylogeny of representative vertebrates constructed from the fossil record.

▶ **TABLE 25.3**

Poisson-corrected Average Number of Amino Acid Differences per Site in the α-Globins of Representative Vertebrates and Associated Evolutionary Rates[a]

	Mouse	Chicken	Newt	Carp	Shark
Human	0.12	0.28	0.58	0.66	0.82
	0.74	0.84	0.83	0.82	0.93
Mouse		0.33	0.59	0.66	0.82
		0.95	0.85	0.82	0.93
Chicken			0.59	0.72	0.89
			0.85	0.89	1.01
Newt				0.74	0.91
				0.93	1.03
Carp					0.92
					1.05

[a]The top number is the average number of amino acid differences between the α-globins of the two organisms. The bottom number is the annualized rate of amino acid substitution per site during the evolution of the α-globins in the lineages that produced these organisms ($\times 10^9$ years).

of the lineages that produced these two species. With such a high frequency of changed sites, we might expect that some of the sites have changed multiple times. The observed proportion of different sites, 0.48, must therefore underestimate the average number of changes that have occurred during the long time since the human and carp lineages split. Fortunately, we can adjust the observed proportion upward to account for multiple amino acid substitutions at particular sites. This adjustment involves a statistical procedure called the Poisson correction, which is explained in the Focus on Evolutionary Rates. **TABLE 25.3** gives the Poisson corrected differences for each pair of organisms. Each value estimates the average number of changes that have occurred per amino acid site in α-globin during the time since the evolving lineages split from a common ancestor. Notice that for the human and carp lineages, the average number of changes per amino acid site is 0.66, which is almost 1.4 times the observed proportion of amino acid differences between the human and carp α-globins.

With the average number of changes per amino acid site for each pair of organisms, we can now calculate the rate at which α-globin has evolved. This rate is the average number of changes per amino acid site divided by the total time that the two lineages have been evolving. For example, the lineages that produced humans and mice split from a common ancestor 80 mya. The total time that these lineages have been evolving is therefore 2×80 million years = 160 my. If we divide the average number of amino acid changes per site by this length of time, we obtain an estimate of the evolutionary rate of α-globin in the human–mouse lineages. Using the Poisson-corrected average number of amino acid changes per site from **TABLE 25.3**, we find that the average number of amino acid changes per site during the total evolutionary time is 0.12 amino acid changes per site/160 my = 0.74×10^{-9} amino acid changes per site/year. From this rate and all the other rates presented in **TABLE 25.3**, we see that α-globin has been evolving at a little less than one amino acid change per site every billion years.

THE MOLECULAR CLOCK

The values calculated for each pair of organisms in **TABLE 25.3** imply that α-globin has evolved at more or less the same rate in all the evolutionary lineages analyzed. This apparent constancy of rate has been observed for other proteins as well. To evolutionary biologists, it suggests that amino acid substitutions occur in clocklike fashion over time. Thus, they sometimes metaphorically speak of the evolutionary process as one that follows a *molecular clock*. Extensive analyses have indicated that the rate of molecular evolution actually varies somewhat among different lineages. We see a hint of this variation in the data in **TABLE 25.3**, where the calculated rate of evolution in the mammalian lineages is slightly less than the rates in the other lineages. Therefore, a universal molecular clock probably does not keep the same time in all evolving lines. However, within some lines, local clocks may be operating—that is, within them the rate of molecular evolutionary change is approximately constant.

Wherever a molecular clock is operating, we can use the estimated rate of evolution to calculate the time since two lineages diverged from a common ancestor. For example, human and kangaroo α-globins differ in 26 of 141 amino acid sites. The proportion of different sites is therefore 0.18. Using the Poisson correction, we calculate that the average number of substitutions that have occurred per site in the evolution of these lineages is 0.20. How long ago did these lineages split? The estimated evolutionary rate of α-globin for mammals is 0.74 substitutions per site every billion years. If we assume that this rate is characteristic of all mammals, including the marsupials, then the total evolutionary time since the human and kangaroo lineages split is $0.20/0.74 \times 10^{-9} = 270$ my. Because this time span must be apportioned equally between the two lineages, the time since the human and kangaroo lineages diverged from a common ancestor is 270 my/2 = 135 my.

Calculations based on the assumption of a molecular clock can be very helpful in estimating when, in historical time, lineages diverged from a common ancestor. This approach has been used to date events in the evolution of our own species, for which fossil evidence is scarce. For instance, the lines that gave rise to humans and chimpanzees are estimated to have diverged between 5 and 6 mya.

VARIATION IN EVOLUTIONARY RATES

The α-globin polypeptide seems to be evolving at a rate of slightly less than one amino acid substitution per site every billion years. Do other proteins evolve at this rate too? Extensive analyses have shown that some do, but others evolve either faster or slower. The observed rates of amino acid sequence evolution range over three orders of magnitude. At the extremes, fibrinopeptide, which is derived from a protein involved in blood clotting, evolves at a rate of greater than 8 amino acid substitutions per site every billion years, whereas the histones, which interact intimately with DNA, evolve at a rate of only 0.01 amino acid substitutions per site every billion years. We can also see variation in evolutionary rates within some polypeptides. For example, amino acids on the surface of α-globin change at a rate of about 1.3 substitutions per site every billion years, whereas amino acids in the interior of the molecule change at a rate of only 0.17 substitutions per site every billion years.

Preproinsulin, the precursor of the peptide hormone insulin, provides another example of intramolecular variation in evolutionary rate. This polypeptide consists of four segments. The first segment is a signal peptide, the second and fourth segments form the active insulin molecule, and the third segment is a peptide bridge that initially links the two active segments. When active insulin is formed, this bridge segment is deleted and the two active segments are joined together covalently by disulfide bonds. The signal and bridge segments evolve at a rate of slightly more than one amino acid substitution per site every billion years; however, the two active segments evolve at a rate of only 0.2 substitutions per site every billion years. Thus, within the preproinsulin polypeptide, the evolutionary rate varies significantly.

What might explain the observed variation in evolutionary rates? Geneticists hypothesize that in more rapidly evolving proteins, the exact amino acid sequence is not as important as it is in more slowly evolving proteins. They speculate that in some proteins, amino acid changes can occur with relative impunity, whereas in others, they are rigorously selected against. According to this view, the rate of evolution depends on the degree to which the amino acid sequence of a protein is constrained by selection to preserve that protein's function. Slowly evolving proteins are more constrained than rapidly evolving proteins. Variation in evolutionary rates is therefore explained by the amount of *functional constraint* on the amino acid sequence. This idea also applies to parts of proteins. For example, the specific amino acids at or near the active sites of enzymes might be expected to be more rigorously constrained by selection than amino acids that simply take up space, such as those in the bridge segment of preproinsulin, which is discarded during the formation of the active insulin molecule. Thus, functionally more important proteins, or parts of proteins, evolve more slowly than functionally less important ones.

Variation in the evolutionary rate is also seen when DNA sequences are examined. The DNA sequences in pseudogenes—duplicated genes that do not encode functional products because they have sustained one or more lesions such as frameshifting or nonsense mutations—have the highest evolutionary rates. For example, the evolutionary rate of the ψα1 pseudogene of α-globin is 5.1 nucleotide substitutions per site every billion years. By contrast, nucleotides in the first or second positions of codons in a functional α-globin gene evolve at the rate of 0.7 nucleotide substitutions per site every billion years. This sevenfold difference in the evolutionary rate can be explained by the concept of functional constraint. The nucleotides in a pseudogene are not constrained by selection because the function of the pseudogene has already been destroyed. However, the nucleotides in the first and second positions of a codon in a functional gene are constrained because changing them will almost always change the amino acid specified by that codon. Some of these changes will be conservative in the sense that the new amino acid will be structurally and functionally like the original amino acid. For example, if the first nucleotide in the codon CTT mutates to A, the amino acid specified by this codon will change from leucine to isoleucine. These two amino acids have similar properties. However, other substitutions in this codon may cause a nonconservative change in the amino acid sequence. For instance, if CTT mutates to TTT, the amino acid specified by the codon will change from leucine to phenylalanine, which has very different chemical properties.

Nucleotides in the third position of codons within functional genes present a special—and interesting—case. These nucleotides evolve much faster than nucleotides in either the first or the second position. In α-globin, for example, the observed evolutionary rate for third position nucleotides is 2.6 substitutions per site every billion years—about half the evolutionary rate of nucleotides in a pseudogene. This rather high rate of nucleotide substitution can be explained only if nucleotides in the third position are less constrained by selection than nucleotides in either the first or the second position. To see that this is the case, we must remember that the genetic

▶ FOCUS ON **Evolutionary Rates**

Nucleotide and amino acid sequences are the fundamental data for the study of molecular evolution. Once homologous sequences from different organisms have been aligned, we can ascertain how many positions in the molecules are the same or different; then, with the help of fossil data on the history of the organisms, we can estimate the rate of molecular evolution.

The simplest case is when we compare the amino acid sequences of two homologous polypeptides. Consider, for example, the two polypeptides shown in **FIGURE 1**. In three of the four positions in these two polypeptides, the amino acids are identical; in the remaining position, they are different—glycine in polypeptide A and serine in polypeptide B. This single amino acid difference indicates that at least one amino acid substitution occurred during the evolution of the two polypeptides. The ancestral amino acid might have been serine, in which case the glycine in polypeptide A represents a substitution event, or the ancestral amino acid might have been glycine, in which case the serine in polypeptide B represents a substitution event.

However, the history of these polypeptides might have been more complicated. The ancestral amino acid at the variable position might have been something other than serine or glycine—say, for example, arginine. In this case, both of the descendant polypeptides must have sustained amino acid substitutions during their evolution. Thus, the minimum number of amino acid substitutions is two. We say "minimum" because multiple substitutions might have occurred at the variable position in either of the descendant polypeptides

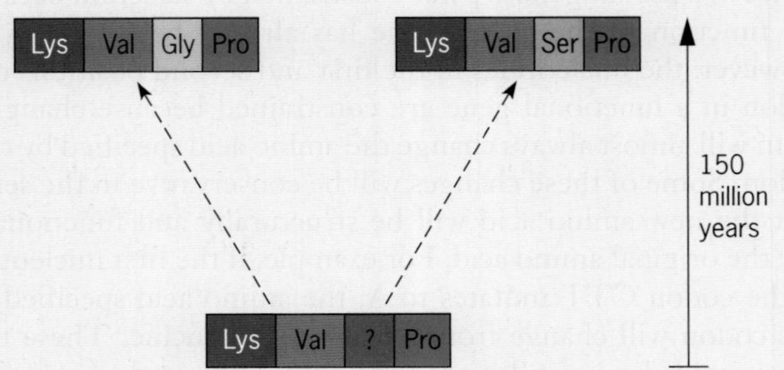

Figure 1 ▶ Comparison of two homologous polypeptides that have evolved independently for 150 million years.

during their evolution. Thus, by focusing on amino acid differences at corresponding positions in homologous polypeptides, we cannot count the actual number of amino acid substitutions that have taken place. All we can say is that *at least* one such substitution has occurred. This uncertainty poses a problem for estimating the rate of molecular evolution, which, after all, is the total number of amino acid substitutions that have occurred divided by the total time the polypeptides have been evolving.

To get around this problem we focus—paradoxically—on the amino acids that are the same in the two polypeptides. These amino acids have presumably not changed in either polypeptide since the two evolving lineages diverged from a common ancestor. Thus, they provide information about the probability that an amino acid substitution does *not* occur during the course of evolution. If we can estimate this probability, then we can turn the situation around and estimate the probability that a substitution does occur, and from it we can obtain the evolutionary rate.

Suppose that S is the proportion of amino acids that are the same in two polypeptides—in our example, $S = 0.75$—and suppose that v is the probability that an amino acid substitution occurs at a site in either polypeptide during one year of evolutionary time—that is, v is the yearly rate of amino acid substitution per site in these polypeptides. By defining v in this way, $1 - v$ is the probability that an amino acid substitution does not occur at a site in any one year of evolutionary time.

From the fossil record we can determine when the two lineages carrying these polypeptides diverged from a common ancestor. For the polypeptides in **FIGURE 1**, this divergence occurred 150 million years ago. In general, if the time since divergence from the common ancestor is T years, then the total evolutionary time for the two lineages is $T + T = 2T$ years. This sum represents the total number of yearly opportunities for an amino acid substitution to occur at a particular site in the evolving polypeptides. It also represents the total number of yearly opportunities for a substitution *not* to occur at this site. Thus, at the end of the evolutionary process, the probability that an amino acid substitution has not occurred at a particular site in either of the polypeptides is the product of all the individual, independent chances for it not to occur, which equals $(1 - v)^{2T}$. To say

code is degenerate. Many amino acids are specified by more than one codon. For example, proline is specified by four different codons: CCT, CCC, CCA, and CCG. As long as the first two nucleotides in a codon are both C, any nucleotide can be present in the third position and the codon will specify proline—that is, the third nucleotide position is fourfold degenerate. Changing the last nucleotide in a proline codon—for example, changing the T in CCT to C to create the codon CCC—should therefore be inconsequential for the structure and function of the polypeptide encoded by a gene. However, changing either the first or second nucleotide in CCT to any other nucleotide will change the amino acid specified by the codon. The first two positions in the CCT codon are therefore more constrained than the third position.

About half of all codons are fourfold degenerate in the third nucleotide position. A majority of all the other codons are twofold degenerate in this position—that is, either of two nucleotides in the third position will specify the same amino it another way, the probability that corresponding amino acids in the two polypeptides have remained the same during the evolutionary process is the probability that neither of them has changed in any one year, which is $(1 - v)^{2T}$. We can estimate this probability by the proportion of amino acids that are currently the same in the two polypeptides—that is, by S. Thus,

$$S = (1 - v)^{2T}$$

To solve for v, the yearly rate of amino acid substitution per site, we take the natural logarithm of both sides of the equation.

$$\ln S = \ln (1 - v)^{2T}$$
$$\ln S = 2T \ln (1 - v)$$

Because v is a very small number, in fact quite close to zero, $\ln (1 - v)$ is approximately equal to $-v$ (the logarithm curve is nearly linear when the argument of the logarithm function is close to 1). Thus,

$$\ln S = -2Tv$$

which implies that

$$v = (-\ln S)/2T$$

With this formula we can estimate the rate of molecular evolution of two homologous polypeptides by (1) calculating the proportion of sites in them that are the same, (2) taking the natural logarithm of this proportion, and then (3) dividing by the total elapsed evolutionary time. In our example, $S = 0.75$ and $2T = 300$ million years; thus, v is $[-\ln(0.75)]/300 = 0.97$ amino acid substitutions per site every billion years.

As discussed above, some of the amino acid sites that are different in two polypeptides have changed once, others have changed twice, and still others have changed multiple times during the evolutionary process. The quantity $2Tv$ is the average number of amino acid substitutions that have occurred per site during the evolution of the polypeptides. If we assume that amino acid substitutions occur randomly and independently throughout time, then we can use this average to calculate the probability that a site has changed a specified number of times. The calculation uses the formula for a probability distribution that is widely used by scientists. It is called the **Poisson probability distribution.** In the context of molecular evolution, the Poisson formula is

$$\text{Probability of } n \text{ changes occurring at an amino acid site} = e^{-2Tv}(2Tv)^n/n!$$

The average number of amino acid substitutions that have occurred per site ($2Tv$) appears twice in the numerator of this formula—as the exponent of the first term and as the argument of the power function in the second term. Thus, it is the key parameter of the Poisson formula.

In our example, $2Tv$ is estimated from $-\ln S = -\ln(0.75)$ to be 0.29 amino acid substitutions per site. This estimate is slightly greater than the proportion of amino acids that are different in the two polypeptides ($1 - S = 0.25$) because it takes into account the possibility that multiple substitutions may have occurred at individual amino acid sites. We say that $2Tv$ is the Poisson-corrected number of amino acid differences between the two polypeptides.

With an estimate of $2Tv$, we can use the Poisson formula to calculate the probability that a particular amino acid site has changed exactly once, twice, and so on.

$$\text{Probability of 1 change} = e^{-2Tv}(2Tv) = 0.22$$
$$\text{Probability of 2 changes} = e^{-2Tv}(2Tv)^2/2 = 0.03$$

The probability that no changes have occurred is

$$\text{Probability of 0 changes} = e^{-2Tv} = 0.75$$

In this example, the probability that more than two changes have occurred is negligible. However, if the Poisson parameter $2Tv$ were greater, multiple changes would have some chance of occurring. For example, if $2Tv = 0.7$, the probability for three changes at a site is 0.03, and the probability for four changes is 0.005.

Statistical procedures analogous to the Poisson correction have been developed to estimate evolutionary rates from comparisons of homologous DNA sequences. However, these procedures are more complicated because the identity of a nucleotide in two DNA sequences does not necessarily imply that this nucleotide remained unchanged during the evolution of these sequences. Methods to deal with this issue can be found in specialized texts on the subject of molecular evolution.

acid. This high level of degeneracy accounts for the faster evolutionary rate of third position nucleotides.

A nucleotide substitution that does not change the amino acid specified by a codon is called a *synonymous* substitution. A nucleotide substitution that does change the amino acid specified by a codon is called a *nonsynonymous* substitution. A wealth of DNA sequence data has now established that synonymous substitutions occur more frequently than nonsynonymous substitutions in evolving lineages.

We also see variation in the evolutionary rates of nucleotides in the noncoding portions of genes (**FIGURE 25.12**). Nucleotides in introns evolve more rapidly than nucleotides in 5′ and 3′ untranslated regions. The different evolutionary rates observed for these types of noncoding sequences presumably reflect variation in the functional constraints on them. In general, these types of sequences do not evolve as fast as pseudogenes, nor do they evolve as slowly as nucleotides in the first or second positions of codons. Rather, they show intermediate evolutionary rates.

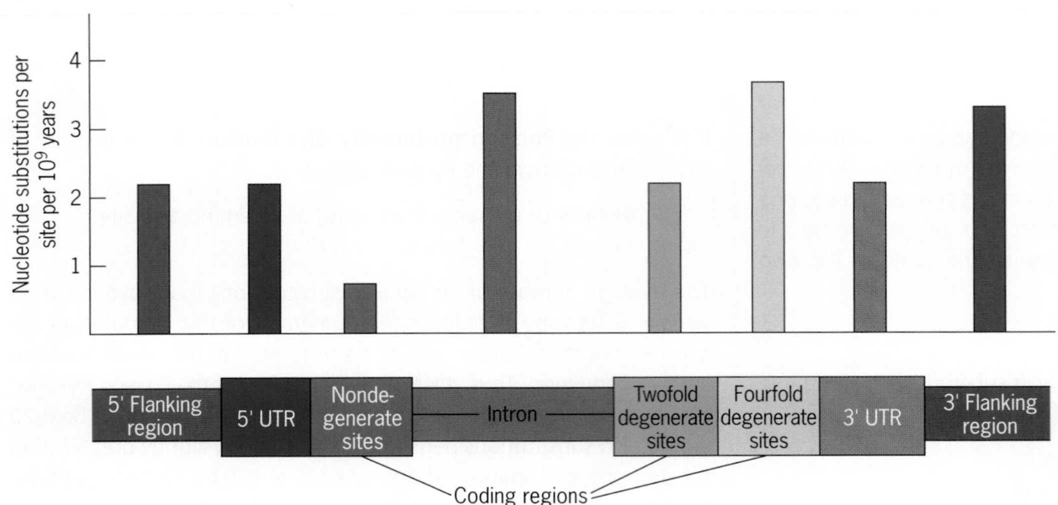

Figure 25.12 ▶ Variation in evolutionary rates among different parts of genes.

THE NEUTRAL THEORY OF MOLECULAR EVOLUTION

Evolutionary geneticists have developed a theory—called the Neutral Theory—to explain the evolution of DNA and protein sequences. It focuses on three processes: mutation, purifying selection, and random genetic drift.

Mutation is at the root of all nucleotide and amino acid substitutions that occur during evolution. Without mutation, DNA and protein molecules could not evolve. Experimentally determined mutation rates are on the order of 10^{-9}–10^{-8} events per nucleotide each generation. These rates reflect the effects of polymerase errors and chemical damage to DNA. They would surely be higher if cells were not equipped with an assortment of mechanisms to prevent replication errors and to repair damaged DNA.

Some of the mutations that occur spontaneously improve the fitness of organisms—that is, they are beneficial mutations that might, over time, spread through a population and become fixed. Other mutations depress fitness and are eliminated from a population by the force of *purifying selection*. Because each gene is already the end result of a long evolutionary process, it is improbable that very many new mutations will improve a gene's function. Many mutations, like random changes in a piece of complex machinery, are likely to impair function. However, some mutations may have little or no effect on fitness. Geneticists say that such mutations are *selectively neutral*. We could easily imagine that synonymous nucleotide substitutions in the third positions of codons might be selectively neutral, as might any type of nucleotide substitution in a pseudogene, which has already been impaired by a previous mutation. Conservative amino acid substitutions in proteins might also be selectively neutral.

The fate of a selectively neutral mutation depends completely on *random genetic drift*. Most selectively neutral mutations are lost from a population shortly after they first appear. A very small fraction of them survive for a few generations, and an even smaller fraction ultimately spread throughout the population and become fixed. What is the rate at which selectively neutral mutations are fixed in a population? The answer is surprisingly simple: the rate of fixation is the rate at which genes mutate to selectively neutral alleles.

To see that the rate of fixation is equal to the neutral mutation rate, suppose that the population size is N and that each generation a fraction u of the $2N$ copies of a gene mutate to selectively neutral alleles. Let's also assume that each new mutant is unique. This is not an unreasonable assumption given the large number of nucleotides that can mutate within a gene, and the extremely low probability that exactly the same nucleotide change will occur more than once. Each generation the number of selectively neutral mutations that appear in the population will be $2Nu$. Because each mutant is unique, its frequency will be $1/2N$. With random genetic drift, the probability that a neutral allele will ultimately be fixed is the allele's current frequency (see Chapter 24). Thus, the probability that a particular new mutation will ultimately be fixed is $1/2N$. The rate at which selectively neutral mutants of any sort are fixed is simply the number of such mutants that appear each generation ($2Nu$) multiplied by the probability that any one of them will ultimately be fixed ($1/2N$). Thus, the rate of fixation of selectively neutral mutants is

$$2Nu \times (1/2N) = u$$

This simple result is the cornerstone of the Neutral Theory. It says that for selectively neutral mutations, the rate of molecular evolution is equal to the rate at which these mutations occur in the population. The rate of evolution does not depend on population size, the efficiency of selection, or peculiarities of the mating system. Furthermore, if the neutral mutation rate is constant, then nucleotide and amino acid substitutions, which are due to mutations, should occur in clocklike fashion in all evolving lineages. Thus, the Neutral Theory explains why substitution rates are approximately constant in many lineages over long periods of evolutionary time.

It also explains why substitution rates differ among proteins and DNA regions. For some positions within a sequence, all or nearly all mutations will be selectively neutral—for example, the nucleotides in a pseudogene or in the third position of a codon that is fourfold degenerate. For other positions, a smaller fraction of all mutations will be selectively neutral, and for some positions, almost no mutations will be selectively neutral. Thus, the Neutral Theory explains the variation in evolutionary rates that is observed among proteins and DNA regions by invoking differences in functional constraints. The highest rates are observed in molecules or in portions of molecules that are not constrained by selection to preserve a function—that is, in molecules in which mutational changes have little or no effect on function. The lowest evolutionary rates are observed in molecules where selection pressure is strongest.

The Neutral Theory has had an enormous impact on the study of evolution at the molecular level. We discuss its intellectual roots in A Milestone in Genetics: The Neutral Theory of Molecular Evolution later in this chapter.

MOLECULAR EVOLUTION AND PHENOTYPIC EVOLUTION

By definition, the Neutral Theory has nothing to say about the evolution of traits that are adaptive. The giraffe's long neck, the elephant's trunk, and the camel's hump are all adaptations that enhance fitness. So, too, is the large, highly convoluted brain of a human being. Darwin emphasized that adaptations such as these evolve because natural selection favors them. In the classic Darwinian sense, then, evolution implies positive selection for something, not just negative selection against deleterious mutants, or, as the Neutral Theory assumes, no selection at all. In addition, Darwin recognized a connection between the evolution of adaptations and the diversification of organisms. As organisms adapt to what Darwin called "the conditions of life," they become different from one another. The phenotypic changes that occur during this process produce new varieties and eventually new species.

The evolution of adaptations and the diversification of organisms must ultimately be due to change at the molecular level. However, change at the molecular level is not a guarantee that phenotypic evolution will occur. Crocodiles, sharks, and horseshoe crabs (**FIGURE 25.13**) have all accumulated amino acid and nucleotide changes at rates similar to highly diversified groups of animals such as birds, mammals, and insects. Yet, to judge from the fossil record, these types of organisms have changed very little in phenotype since they first appeared hundreds of millions of years ago. "Living fossils" therefore seem to have roughly the same rate of molecular evolution as organisms that have diverged extensively at the phenotypic level. This observation suggests that many nucleotide and amino acid substitutions have little to do with phenotypic evolution.

What sorts of genetic changes might be responsible for the evolution of novel phenotypes? Some possible answers are coming from the genome sequencing projects and from ongoing studies in developmental genetics. One observation is that genes often become duplicated during evolution and that the duplicates sometimes acquire different functions. The classic example of *gene duplication* comes from the study of the globin genes in animals (**FIGURE 25.14**). Today we find two classes of globin genes—those encoding components of hemoglobin, which carries oxygen in the blood, and those encoding myoglobin, which stores oxygen in muscle. These functionally different classes of genes are derived from a primordial globin gene, which was duplicated about 800 mya, long before the diversification of the animals at the start of the Paleozoic Era. The hemoglobin genes have, in turn, been duplicated several times during the evolution of the vertebrates. As best we can tell, the α- and β-globin genes were created by a duplication more than 450 mya in the evolutionary line that produced the jawed fishes. Jawless fish—the lampreys and their kin—have only one kind of hemoglobin gene (α); sharks and bony fish have at least two. About 300 to 350 mya, the α- and β-globin genes were separated from each other and took up residence on different chromosomes. Each of these genes subsequently underwent several duplication events to produce clusters of α- and β-globin genes. In humans, for example, seven α-globin genes are clustered together on chromosome 16, and six β-globin genes are clustered together on chromosome 11. Three of the α-globin genes and one of the α-globin genes in humans are nonfunctional pseudogenes. The other globin genes in these clusters encode different, but related, polypeptides that carry oxygen in the blood at different times during life. Some of these polypeptides function only in the embryo, others only

(a) Nile crocodile (*Crocodylus niloticus*)

(b) Great white shark (*Carcharodon carcharias*)

(c) Horseshoe crab (*Limulus polyphemus*)

Figure 25.13 ► Some organisms considered to be "living fossils."

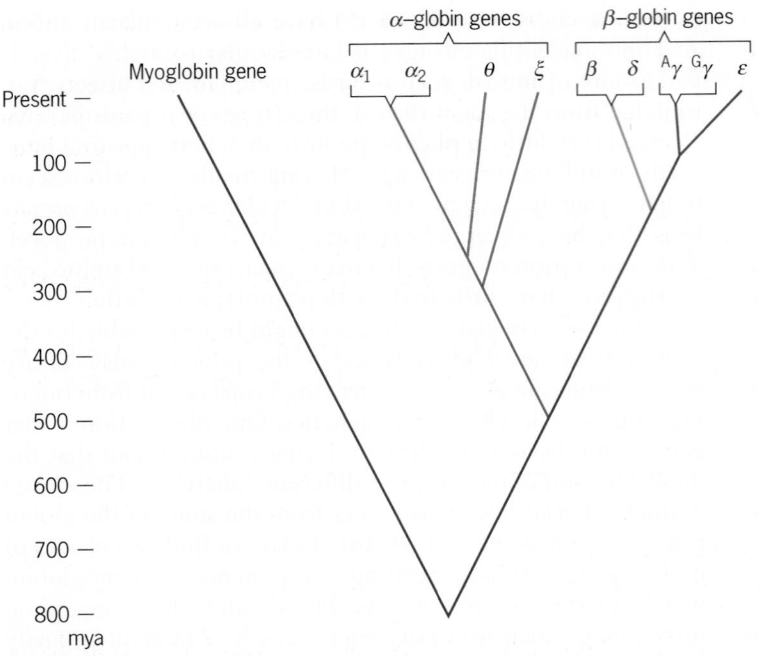

Figure 25.14 ▶ Role of gene duplication in the evolution of the globin genes.

in the fetus, and still others only in the adult (see Chapter 20). Thus, these families of hemoglobin genes indicate that duplicated genes can acquire different functions.

Another phenomenon that might help to explain phenotypic evolution is that portions of genes may be duplicated and recombined with other genes. Eukaryotic genes are segmented into exons and introns. Shortly after this segmented structure was discovered, Walter Gilbert speculated that each exon in a gene encodes a separate functional domain in the gene's polypeptide product. He further speculated that exons from one gene could be combined with exons from another gene to create a coding sequence that would specify a protein with some of the properties of each of the original gene products. Thus, he proposed that novel proteins could be created by combining exons in modular fashion—a process now called *exon shuffling* (**FIGURE 25.15**). DNA sequencing studies have provided evidence for Gilbert's hypothesis. For example, tissue plasminogen activator (TPA), a protein involved in the breakup of blood clots, is encoded by a gene that seems to have acquired exons from several different sources. One exon comes from the gene for fibronectin, another from the gene for epidermal growth factor, two exons come from the gene for plasminogen, and one comes from a gene that encodes a protease. Altogether, then, at least four genes have contributed exons to the formation of the TPA gene. The recombination of evolutionarily proven exons provides almost limitless possibilities to form mosaic proteins. Mixing and matching exons, and the polypeptide domains they encode, may be an important process in evolution, and it may partly explain why eukaryotes are anatomically, physiologically, and behaviorally so diverse.

In addition to gene duplication and exon shuffling, evolutionary diversification seems to have benefited from spatial and temporal changes in the expression of genes, especially those whose products regulate the expression of other genes. For example, the homeobox genes play important roles in the formation of animal bodies along an anterior–posterior axis; these genes encode transcription factors. Changing the time or place in which specific homeobox genes are expressed may profoundly change the appearance of the animal. In *Drosophila*, where the homeobox genes have been studied thoroughly, it is clear that altering the pattern of expression of one or a few of these genes can produce a fly with four wings instead of two, or

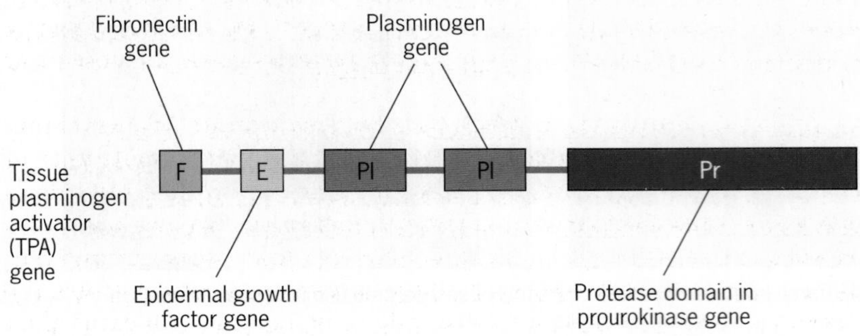

Figure 25.15 ▶ Exon shuffling exemplified by the gene for tissue plasminogen activator (TPA). Exons from at least four different genes have been recombined to produce the TPA gene.

a fly with extra appendages on either the head or the thorax. With these kinds of observations in the laboratory, it is not hard to imagine that similar kinds of changes might have occurred in nature during the course of evolution.

KEY POINTS

- Phylogenetic trees based on the comparison of DNA and protein sequences show the evolutionary relationships among organisms.
- The rate of molecular evolution can be determined by calculating the average number of amino acid or nucleotide changes that have occurred per site in a molecule since two or more evolving lineages diverged from a common ancestor.
- The near uniformity of the rate of molecular evolution in different lineages is metaphorically described as a "molecular clock."
- The rate of evolution varies among different protein and DNA sequences and appears to depend on the extent to which these sequences are constrained by natural selection to preserve their function.
- Selectively neutral mutations are fixed in a population at a rate equal to the neutral mutation rate.
- Gene duplication and exon shuffling have played important roles in evolution.
- Changes in the spatial and temporal aspects of gene regulation may have contributed to the rapid evolution of some types of organisms.

Speciation

Species arise when a population of organisms splits into genetically distinct groups that can no longer interbreed with each other.

Biologists have named and described a large number of plant, animal, and microbial species. Many more species have yet to be identified. Where did all this diversity come from? How is it maintained? Why are different species distinct from one another? What factors contribute to the formation of species? Charles Darwin raised these kinds of questions nearly 150 years ago when he wrote *The Origin of Species*. Today, biologists continue to grapple with them as they address the central problem of evolutionary genetics—the problem of speciation.

WHAT IS A SPECIES?

The term *species* is usually applied to a group of organisms that share certain characteristics. However, species have been defined in different ways. In classical taxonomy, a species is defined exclusively on the basis of phenotypic characteristics. If the characteristics of two groups of organisms are sufficiently different, then the groups are considered to be separate species. This approach to defining a species relies on careful observation of the organisms, either as specimens in a zoo, arboretum, herbarium, or museum collection, or, better still, as the inhabitants of a natural environment where their behavior as well as their morphology can be studied. This approach to defining a species also relies on the expertise of the taxonomist, who must decide if groups of organisms are sufficiently different to warrant their classification as separate species. Thus, it is a subjective approach that may lead to different classifications in the hands of different people.

In evolutionary genetics, a species is defined on the basis of a shared gene pool. A group of interbreeding, or potentially interbreeding, organisms that does not exchange genes with other such groups is considered to be a species. Evolutionary geneticists say that each species is *reproductively isolated* from every other species. This approach to defining a species relies on a researcher's ability to determine whether groups of organisms exchange genes in nature. If they do, they are classified as a single species; if they do not, they are classified as separate species. The genetic approach to defining species therefore involves an objective assessment of whether or not groups of organisms are reproductively isolated from each other.

These two ways of defining species are not always in agreement. Organisms may be reproductively isolated, but they may not be distinguished by easily recognized phenotypic characteristics. In taxonomy, such organisms would be regarded as a single species, whereas in evolutionary genetics, they would be regarded as separate species. Conversely, organisms may have different phenotypic characteristics, but they may not be reproductively isolated. A taxonomist would regard such organisms as separate species, whereas an evolutionary geneticist would regard them as a single species. When it is possible to determine whether different organisms are reproductively isolated, we can apply the genetic definition of species. However, when such determinations are not possible—as, for example, with fossilized organisms—we are limited to the taxonomic definition of species.

Reproductive isolation is the key to the genetic definition of a species. Groups of organisms that inhabit the same territory can be reproductively isolated from each other by different mechanisms. *Prezygotic isolating mechanisms* prevent the members of different groups from producing hybrid offspring. *Postzygotic isolating mechanisms* prevent any hybrid offspring that are produced from passing on their genes to subsequent generations.

Prezygotic isolating mechanisms operate by preventing matings between individuals from different populations of organisms, or by preventing the gametes of these individuals from uniting to form zygotes. For example, two populations of organisms that inhabit the same area might seek out different habitats within that area. If the habitat preference is strong, the two populations will have little or no contact with each other.

Ecological isolation based on habitat preference can therefore prevent the populations from producing hybrid zygotes. Temporal or behavioral factors can also bring about reproductive isolation between populations of organisms. For instance, the organisms might become sexually mature at different times, or they might have different courting rituals. If ecological, temporal, and behavioral isolating mechanisms fail to prevent mating between different organisms, then anatomical or chemical incompatibilities in their reproductive organs or gametes might prevent them from producing hybrid zygotes. The organisms might be unable to copulate successfully, or to exchange pollen, or their sperm or pollen might die in the reproductive tissues of their mates. Any of these prezygotic isolating mechanisms could prevent genes from being exchanged between populations occupying the same territory.

Postzygotic isolating mechanisms operate after hybrid zygotes have been formed, either by reducing hybrid viability or by impairing hybrid fertility. The zygotes from matings between different organisms might not survive, or they might not reach sexual maturity. If they do reach sexual maturity, they might not produce functional gametes. Any of these circumstances could prevent populations of organisms that live in the same territory from exchanging genes.

MODES OF SPECIATION

The key event in speciation is the splitting of a population of organisms into one or more subpopulations that become reproductively isolated from each other. The most straightforward way for this event to happen is for the subpopulations to become geographically separated so that they evolve independently—that is, geographical barriers keep the subpopulations apart so that they accumulate their own sets of genetic changes over time (**FIGURE 25.16**). Then, if the subpopulations are reunited by the disappearance of the geographical barriers, the genetic changes they have accumulated may make them reproductively isolated from each other. For example, one subpopulation may have evolved a preference for a particular food source and another subpopulation may have evolved a preference for a different food source. When the two subpopulations are rejoined in the same territory, their distinctive food preferences may limit contact between them to such an extent that interpopulational matings never occur. Another possibility is that during the time the subpopulations were separated, they may have evolved different physiological processes or mating habits. When the subpopulations are reunited, they may not be able to mate with each other, or if they can mate with each other, their hybrids may not be viable or fertile. The process whereby subpopulations evolve reproductive isolation while they are geographically separated is called **allopatric speciation** (from Greek roots meaning "in other villages").

It is conceivable that subpopulations might evolve reproductive isolation without being separated geographically (**FIGURE 25.17**). Perhaps the subpopulations become ecologically specialized so that they evolve more or less independently, or perhaps their members mate assortatively so that there is little or no genetic exchange between the subpopulations. The process of evolving reproductive isolation between subpopulations that exist in the same territory is called **sympatric speciation** (from Greek roots meaning "in the same villages").

Because the evolution of reproductive isolation may require hundreds of thousands of years, it is not easily studied. Most investigations of speciation are done *post factum*—that is, after the species have already formed. Based on data collected from the species, researchers attempt to determine how and why they became reproductively isolated from each other.

One issue in these studies is whether the species evolved allopatrically or sympatrically. Did they develop reproductive

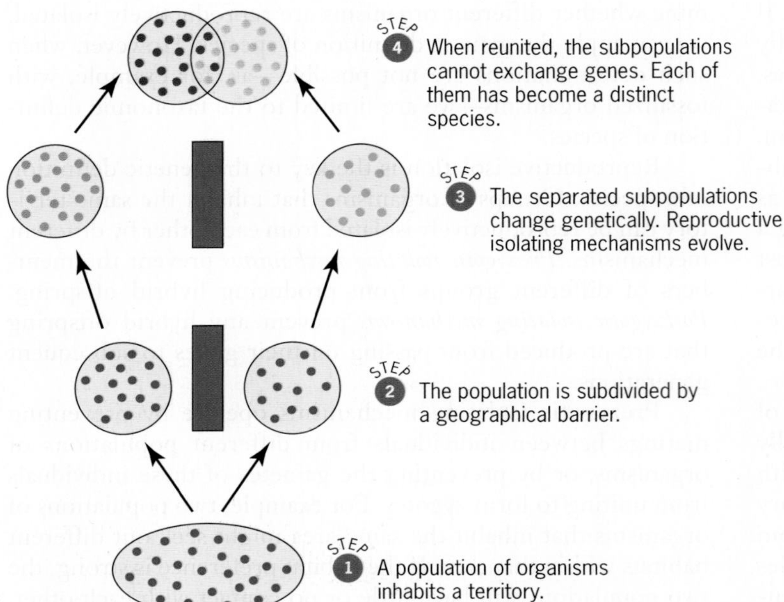

STEP 4 When reunited, the subpopulations cannot exchange genes. Each of them has become a distinct species.
STEP 3 The separated subpopulations change genetically. Reproductive isolating mechanisms evolve.
STEP 2 The population is subdivided by a geographical barrier.
STEP 1 A population of organisms inhabits a territory.

Figure 25.16 ▶ The process of allopatric speciation.

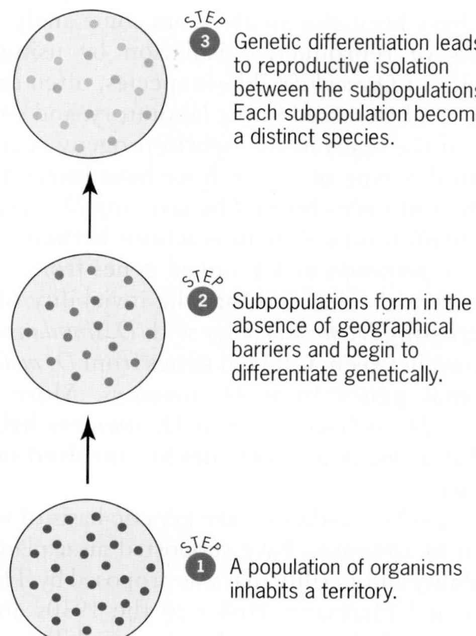

STEP 3 Genetic differentiation leads to reproductive isolation between the subpopulations. Each subpopulation becomes a distinct species.
STEP 2 Subpopulations form in the absence of geographical barriers and begin to differentiate genetically.
STEP 1 A population of organisms inhabits a territory.

Figure 25.17 ► The process of sympatric speciation.

Figure 25.18 ► Four species of *Drosophila* from the Hawaiian Islands. Starting at the upper left and moving clockwise: *D. heteroneura, D. grimshawi, D. ornata,* and *D. differens*. These and hundreds of other *Drosophila* species have evolved during the last few million years on the Hawaiian Islands, which are far removed from other land masses in and around the Pacific Ocean.

isolating mechanisms while they were geographically separated, or did they develop these mechanisms while they inhabited the same territory? This question cannot be answered with certainty. However, most evolutionary geneticists are inclined toward the view that allopatric speciation is more prevalent than sympatric speciation, if only because allopatric speciation is a more straightforward process. For example, imagine that a small number of organisms migrate to a remote oceanic island where they found a population that evolves independently of the main population on the nearest continent. The island population may change significantly over time and eventually become reproductively isolated from its closest relatives on the continent. This scenario—which is allopatric speciation pure and simple—may have played out many times on oceanic islands (**FIGURE 25.18**). Indeed, Darwin proposed it as an explanation for the species of plants and animals he observed on the Galapagos Islands off the west coast of South America. It is not too hard to imagine other types of geographic separation that would permit allopatric speciation to occur. Deserts and mountain ranges can subdivide continents; reductions in rainfall can isolate lakes and river systems; land masses can rise up to separate oceans. Populations that are subdivided by these kinds of barriers have the potential to evolve into distinct, reproductively isolated species (**FIGURE 25.19**).

(*a*)

(*b*)

Figure 25.19 ► Species of grosbeaks that may have arisen by allopatric speciation. (*a*) The black-headed grosbeak (*Pheucticus melanocephalus*) found in the western United States. (*b*) The rose-breasted grosbeak (*P. ludoviciannus*) found in the eastern United States.

Although allopatric speciation may have been the prevalent mode in creating the species that exist today, there is evidence that sympatric speciation has also contributed to species diversity. The strongest case for sympatric speciation comes from the study of cichlid fish in two small crater lakes located in west central Africa. Today these lakes are isolated from other significant bodies of water. However, in the relatively recent past they were apparently colonized by cichlids from the surrounding river systems. These colonists then evolved into the groups of species now present in the lakes. Analysis of mitochondrial DNA sequences indicates that the cichlid species within each lake are derived from a common ancestor and that they are more closely related to each other than to the cichlid species found in the surrounding river systems (**FIGURE 25.20**). There are no obvious geographic barriers within these lakes. Their shorelines are regular, and they do not seem to have been subdivided during their history. Thus, it appears that the crater-lake cichlids evolved into different species sympatrically.

Cichlid fish inhabit many of the lakes and rivers in tropical Africa, especially the East African Great Lakes—Lake Victoria, Lake Malawi, and Lake Tanganyika, where over 1500 cichlid species have been identified. The apparent sympatric speciation of cichlids in the small crater lakes of west central Africa raises the possibility that some of the species in these large lakes may also have originated sympatrically. More research is needed to determine how the Great Lake cichlids evolved.

THE GENETICS OF SPECIATION

One of the fundamental questions in evolutionary biology is what types of genetic changes bring about reproductive isolation between populations of organisms. This question is difficult to address because if the populations are reproductively isolated, they either cannot be crossed, or if they can be crossed, their hybrids are sterile or unviable. Fortunately, researchers have been able to carry out some analyses of the genetic basis of reproductive isolation by using closely related species of *Drosophila*. These species, often called sibling species, can be crossed in the laboratory, and sometimes at least one of the sexes in the hybrid progeny is viable and fertile. From this type of research we have learned that the sterility of hybrid males formed by crossing *D. pseudoobscura* with *D. persimilis* is caused by interactions between X-linked genes from *D. persimilis* and Y-linked genes from *D. pseudoobscura*. We have also learned that the inviability of hybrids formed by crossing *D. melanogaster* with *D. simulans* is caused by interactions between X-linked genes from *D. melanogaster* and autosomal genes from *D. simulans*. More detailed research with *D. melanogaster* and *D. simulans* hybrids has indicated that as many as 200 genes are involved in causing this inviability.

These and other studies on the genetic basis of reproductive isolation in *Drosophila* have supported an explanation for hybrid inviability and sterility that was proposed by Theodosius Dobzhanzky and Hermann Muller in the 1930s and 1940s. Dobzhansky and Muller argued that the inviability or sterility of hybrids is due to incompatibilities between genes that evolved separately in different populations. For example, if a population with the genotype $A_1A_1\ B_1B_1$ becomes subdivided by geographical barriers, one of its subpopulations might fix a new mutant allele of the A locus to become $A_2A_2\ B_1B_1$, and another subpopulation might fix a new mutant allele of the B locus to become $A_1A_1\ B_2B_2$ (**FIGURE 25.21**). If these subpopulations rejoin after the disappearance of the geographical barriers that separated them, they might cross to produce $A_1A_2\ B_1B_2$ hybrids. Such crosses would bring the A_2 and B_2 alleles together in the same genotype for the first time. Because these alleles evolved in separate populations, they might interact unfavorably with each other and cause death or sterility. Thus, the independent evolution of genes in separated populations might produce genotypes that, when hybridized, lead to inviable or sterile offspring. In other words, the biochemical, cellular,

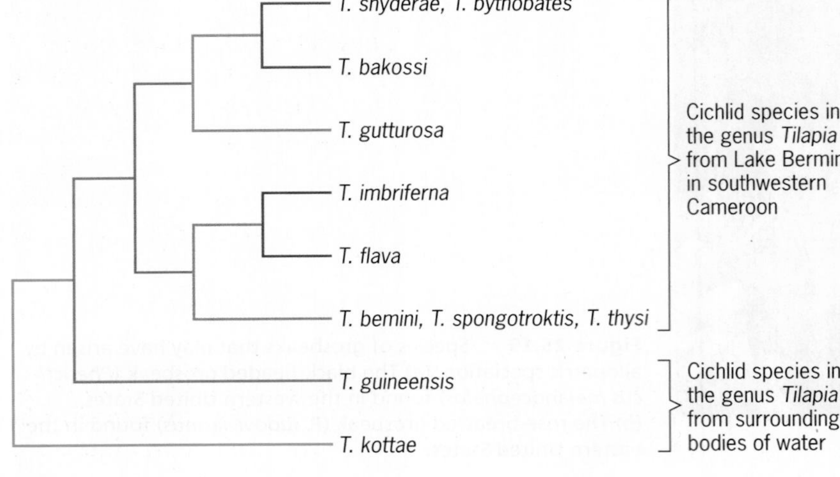

Figure 25.20 ▶ Phylogenetic tree based on mitochondrial DNA sequences showing the relationships among cichlid fish in the genus *Tilapia* from Lake Bermin, an isolated crater lake, and from surrounding bodies of water in Cameroon.

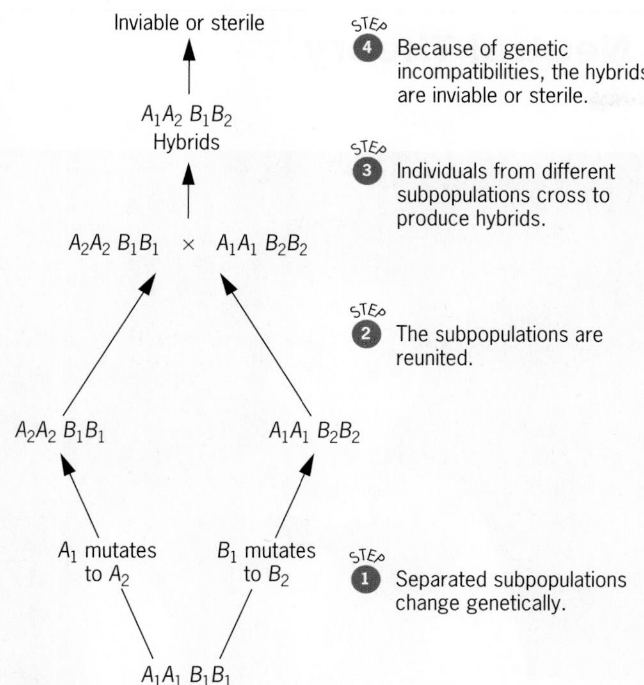
Inviable or sterile
STEP 4 Because of genetic incompatibilities, the hybrids are inviable or sterile.
$A_1A_2\ B_1B_2$
Hybrids
STEP 3 Individuals from different subpopulations cross to produce hybrids.
$A_2A_2\ B_1B_1 \times A_1A_1\ B_2B_2$
STEP 2 The subpopulations are reunited.
$A_2A_2\ B_1B_1$
$A_1A_1\ B_2B_2$
A_1 mutates to A_2
B_1 mutates to B_2
STEP 1 Separated subpopulations change genetically.
$A_1A_1\ B_1B_1$

Figure 25.21 ▶ The Dobzhansky-Muller model for the evolution of reproductive isolation between populations.

or developmental programs encoded by these genes are simply incompatible with each other.

Sometimes the reproductive isolation between populations is due to chromosomal differences. Hybrids formed by crossing organisms that have different karyotypes may be sterile because the chromosomes from the two karyotypes cannot disjoin and segregate properly during meiosis. Meiotic irregularities account for the sterility of some polyploidy plant species. These plants persist only because they can reproduce asexually. Meiotic irregularities also explain why fertile allopolyploids are reproductively isolated from the diploid species that hybridized to produce them.

KEY POINTS

- In evolutionary genetics, a species is a group of populations that share a common gene pool.
- The development of reproductive isolation between populations is the key event in the speciation process.
- Speciation may occur when the populations are geographically separated (allopatric) or when they coexist in the same territory (sympatric).
- The inviability or sterility of interspecific hybrids may be due to incompatibilities among genes that changed while the species were evolving.

Human Evolution

Fossil evidence and DNA sequence analysis have provided information about the origin of modern humans.

When Darwin proposed his theory of evolution in 1859, and later, when he suggested that human beings had evolved from more primitive organisms, he provoked a great controversy. The idea that organisms evolve, and more specifically, that humans evolve, has troubled many people. In the ensuing 150 years, much has been learned about the course of human evolution. Paleontologists have analyzed the fossilized remains of organisms that are likely to have been the ancestors of modern humans, and geneticists have analyzed DNA sequence data in order to study the relationships among humans and their closest nonhuman relatives, the great apes. In the following sections, we discuss some of these analyses.

HUMANS AND THE GREAT APES

Several morphological features distinguish human beings from chimpanzees and gorillas. The apes have larger canine and incisor teeth than modern humans, and their jaws are larger and heavier. Ape brains are smaller than human brains, and the point where the ape brain attaches to the spinal cord is placed farther to the back of the skull than it is in humans. The shape and proportions of an ape's body are also different from those of a human. In an ape, the body's trunk widens toward the base, whereas in a human, it tends to have the same width from the shoulders to the waist. The legs of an ape are proportionately shorter than those of a human, and the pelvis is not constructed to accommodate a regular upright stance. Although apes can walk upright on two legs, they cannot do so for long periods of time. By contrast, humans are exclusively bipedal—except, of course, in early childhood. The hands and feet of apes also differ from those of humans. Apes do not have opposable thumbs, and their feet do not provide the support that is needed for bipedal locomotion.

Despite all these morphological differences, the DNA of apes and humans is remarkably similar. When the genomes of chimpanzees and humans are compared, they are found to be more than 99 percent identical. This high degree of identity implies that chimpanzees and humans are quite closely related, and suggests that they diverged from a common ancestor rather recently in evolutionary time, perhaps 5 to 6 million years ago. The other great ape species, the gorilla, appears to be less closely related to humans than the chimpanzee is.

HUMAN EVOLUTION IN THE FOSSIL RECORD

Though rare, fossils have provided important information about human evolution (**FIGURE 25.22**). The oldest fossils that appear to be strictly within the human evolutionary line come

▶ A MILESTONE IN GENETICS: The Neutral Theory of Molecular Evolution

In 1968, molecular genetics was in its infancy. Efficient techniques for sequencing DNA were still a decade in the future. No one knew about introns or pseudogenes, or about the ubiquity and diversity of transposons. The determination of subunit sequences in complex macromolecules was limited to a handful of proteins such as hemoglobin, myoglobin, and cytochrome c, and the analysis of these sequences was performed by eye and hand, not by sophisticated algorithms programmed into a desktop computer. Studies of evolution focused on phenotypic changes and emphasized the Darwinian theory of "selection for desirable characteristics and advantageous genes."[1] There was not much opportunity to study evolution at the molecular level, and little thought was given to the possibility that molecules might evolve through mutation and random genetic drift rather than through positive Darwinian selection. Thus, when the Japanese population geneticist Motoo Kimura (**FIGURE 1**) published a paper entitled "Evolutionary Rate at the Molecular Level" with a summary that read, "Calculating the rate of evolution in terms of nucleotide substitutions seems to give a value so high that many of the mutations involved must be neutral ones,"[2] people took note.

Kimura had few data on which to base his calculation of the evolutionary rate. Because DNA sequences were not available, he used the sequences of proteins—α- and β-globins, cytochrome c, and the enzyme triosephosphate dehydrogenase. Most of the sequences he analyzed were from mammals. Averaging over the four types of proteins, Kimura concluded that the rate of molecular evolution is "approximately one [amino acid] substitution in 28 $\times$ 10^6 yr for a polypeptide chain consisting of 100 amino-acids."[3] He then tried to convert this rate into a rate of nucleotide substitution. To this end, he assumed that a haploid mammalian genome contains 4 $\times$ 10^9 base pairs of DNA, that all the DNA encodes protein, and that each amino acid substitution detected at the protein level is equivalent to 1.2 nucleotide substitutions in the DNA. This last assumption recognizes that some—Kimura estimated about 20 percent—of all nucleotide substitutions do not change an amino acid in the protein; that is, they are synonymous substitutions. The calculation on the next page is a summary of Kimura's conversion of the rate of protein evolution into a rate of DNA evolution.

Figure 1 ▶ Motoo Kimura.

The converted evolutionary rate, 0.57 base-pair substitutions in a genome per year, implies that one substitution occurs every 1.8 years—an extraordinarily high evolutionary rate, so high, in fact, that it made Kimura uncomfortable.

Kimura's discomfort came from his inability to reconcile the estimated genomic rate of nucleotide substitution with a prediction that J. B. S. Haldane had made a decade earlier. Haldane was interested in calculating what he called the "cost of natural selection," a measure of the evolutionary impact on a population of substituting a favorable allele for a deleterious allele. During the selection process, less fit

[1]King, J. L., and T. H. Jukes. 1969. Non-Darwinian evolution. *Science* 164: 788–798.

[2]Kimura, M. 1968. Evolutionary rate at the molecular level. *Nature* 217: 624–626.

[3]Ibid., p. 625.

from East Africa where they were formed 4 to 5 million years ago. These first humanlike—that is, *hominin*—creatures have been given the name *Ardipithecus ramidus*. Later in the fossil record, 3 to 4 million years ago, another hominin creature appeared. This organism, known as *Australopithecus afarensis*, probably stood 1 to 1.5 m tall and walked upright, at least for short distances. The fossil known as Lucy is a specimen of *Australopithecus afarensis*.

The first organisms to be classified in the same genus as *Homo sapiens* appeared 2 to 2.5 million years ago. Two species have been named, *H. rudolfensis* and *H. habilis*. Both of these "early *Homo*" species have many apelike features; however, compared to *Australopithecus*, the opening for the spinal cord is closer to the middle of the skull, and the skull itself is reduced in length and increased in width—all hominin characteristics. Nevertheless, many paleontologists have questioned the inclusion

$$\frac{\left(\frac{\text{1 a.a. subst.}}{\text{polypeptide}}\right)}{28 \times 10^6\ \text{yr}} \times \left(\frac{\text{1 polypeptide}}{\text{100 a.a.}}\right) \times \left(\frac{\text{1.2 bp subst.}}{\text{1 a.a. subst.}}\right) \times \left(\frac{\text{1 a.a.}}{\text{3 bp}}\right) \times \left(\frac{4 \times 10^9\ \text{bp}}{\text{genome}}\right) = 0.57\ \text{bp subst/genome/yr}$$

Rate of protein evolution (first term); Correction for size of polypeptide (second term); Correction for synonymous substitutions (third term); Correction for size of a codon (fourth term); Genome size (fifth term); Rate of DNA evolution (result).

individuals must die or fail to reproduce. Thus, each allele substitution entails a cost to the population. Haldane recognized that there must be a limit to the number of substitutions that can be carried out simultaneously. If too many substitutions occur at the same time, the deaths and reproductive failures they entail will so deplete the population that it will not be able to survive. This phenomenon is well known in plant and animal breeding, where selection for several traits simultaneously decimates the breeding stock. Haldane concluded that the cost imposed by natural selection must limit the rate of evolution to roughly one allele substitution every 300 generations. Thus, for an organism with a generation time of three to four years, the evolutionary rate is approximately one substitution every 1000 years—about 500 times less than the evolutionary rate estimated by Kimura. Faced with the enormous discrepancy between the estimated and "permissible" evolutionary rates, Kimura concluded that molecular evolution is not greatly driven by positive selection; rather, he said, it is driven by the fixation of selectively neutral mutations by random genetic drift. Thus, the Neutral Theory of Molecular Evolution was born.

In 1969, Jack Lester King and Thomas H. Jukes amplified Kimura's ideas in a lengthy article entitled "Non-Darwinian Evolution."[4] After analyzing variation in the amino acid sequences of an assortment of proteins, King and Jukes wrote:

> From these considerations it is not difficult to conclude that the stream of spontaneous alterations in DNA, continuously fed into the genetic pool, should include far more acceptable changes that are neutral than changes that are adaptive. Protein molecules are subjected to incessant probing as a result of point mutations and other DNA alterations. The genome becomes virtually saturated with such changes as are not thrown off through natural selection. We conclude that most proteins contain regions where substitutions of many amino acids can be made without producing appreciable changes in protein function. The principal evidence for this is the astounding variability in primary structure of homologous proteins from various species, and the rapid rate at which molecular changes accumulate in evolution.[5]

The article by King and Jukes attracted plenty of attention. Although data were still hard to come by, many geneticists took up the study of molecular evolution. Within a decade DNA sequencing came on stream, and the data drought ended. Today, there is no shortage of DNA data, or of people to study it, and the Neutral Theory has become a mainstay in our efforts to understand evolution at the molecular level.

QUESTIONS FOR DISCUSSION

1. Kimura's estimate of the rate of nucleotide substitution made assumptions about the amount of DNA in a haploid genome and about the proportion of DNA that codes for protein. In light of information now available, how would you refine his estimate?

2. Why is the Neutral Theory of Molecular Evolution non-Darwinian?

[4]King, and Jukes, Non-Darwinian evolution.

[5]Ibid., p. 797.

of these two species within the genus *Homo*, and there is some sentiment to reclassify them in the genus *Australopithecus*.

Between 1.9 and 1.5 million years ago, another hominin appeared in the fossil record. This creature, called *Homo ergaster*, had a body shape and limb proportions like those of modern humans, and its teeth and jaws were also human in structure. Thus, *H. ergaster* is the first hominin that can confidently be placed within the genus *Homo*.

All the early hominin fossils come from Africa. The first hominin species to produce fossils outside of Africa was *Homo erectus*. These fossils, formed about 1 million years ago, have been found in China and Indonesia. Thus, *H. erectus* was widespread and probably gave rise to archaic populations of humans in Europe, Asia, and Africa. The best known of the archaic humans were the Neanderthals, a species that evolved in Europe and the Near East several hundred thousand years ago. Analysis

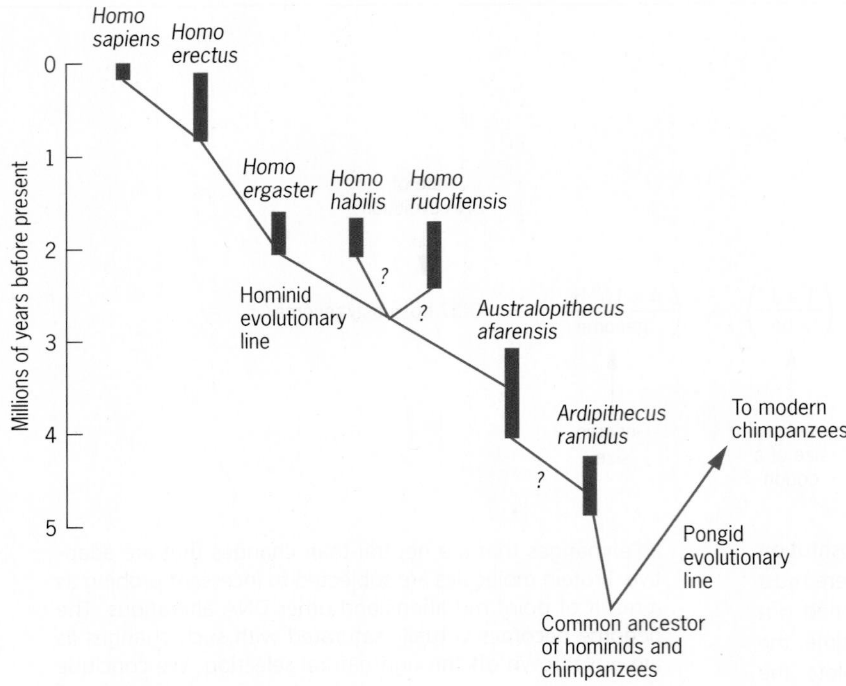

Figure 25.22 ▶ Ancestors of human beings that have been discovered through fossil evidence. The hominid evolutionary line leads from the common ancestor of humans and chimpanzees to modern humans (*Homo sapiens*). The pongid evolutionary line leads from this common ancestor to modern chimpanzees. Uncertainties in the hominid evolutionary line are indicated by question marks.

of DNA from fossilized Neanderthal bones suggests that the Neanderthals did not contribute genetic material to the modern human species, *H. sapiens*. In fact, they may have been competitors with the populations that evolved into modern humans, and eventually lost out in the competition and became extinct.

Modern humans may have evolved simultaneously in Europe, Asia, and Africa from the archaic human populations that existed on each of those continents, or they may have evolved on one continent—probably Africa—and subsequently spread to the others. The first of these ideas about the origin of modern humans is called the *Multiregional Hypothesis*, and the second is called the *Out-of-Africa Hypothesis*. According to the Multiregional Hypothesis, the evolution of modern humans involved archaic populations in different regions of the Old World. According to the Out-of-Africa Hypothesis, it involved only populations in Africa, which then migrated to each of the other continents and replaced the archaic human populations that were living there. Fossil evidence cannot discriminate between these two hypotheses. However, genetic evidence obtained by studying DNA sequences in living human beings has provided ways of testing them.

DNA SEQUENCE VARIATION AND HUMAN ORIGINS

Genetic data allow researchers to study human evolution by investigating the relationships among extant human populations. Populations that are closely related share genetic properties that distantly related populations do not. Thus, by analyzing variation in genes, gene products, and DNA sequences, it is possible to determine the relatedness of different racial and ethnic groups, and to arrange them in a phylogenetic tree. Genetic analysis also permits researchers to decipher key events in human evolutionary history.

Many types of genetic variation have been used to study human evolution: blood group and allozyme polymorphisms, and variation in the composition of DNA sequences themselves. Both nuclear and mitochondrial genetic variation has been investigated. The nuclear genome contains the preponderance of human polymorphisms, but the mitochondrial genome has the unique property of being transmitted exclusively through females. Variation in mitochondrial DNA therefore provides a way of tracing maternal lineages in human evolutionary history.

Compared to other species, the human species is genetically rather uniform. At the nucleotide level, humans have about one-fourth the genetic variation of chimpanzees and about one-tenth that of *Drosophila*. Furthermore, most of the genetic variation in the human species—perhaps 85 to 95 percent of it—is within rather than between populations.

The relative absence of genetic variation in human populations implies that during its evolutionary history, the genetically effective size of the human population was small—between 10,000 and 100,000 individuals. The census size may have been larger (today, it certainly is), but the mating system, various constraints on reproduction, and bottlenecks in size caused by famine, disease, or weather-related catastrophes apparently conspired to keep the effective population size under 100,000. In such a population, random genetic drift dominates over mutation to determine the equilibrium level of variability for selectively neutral alleles (see Chapter 24).

When different human populations are analyzed for genetic variation, those in Africa are found to have more variation than

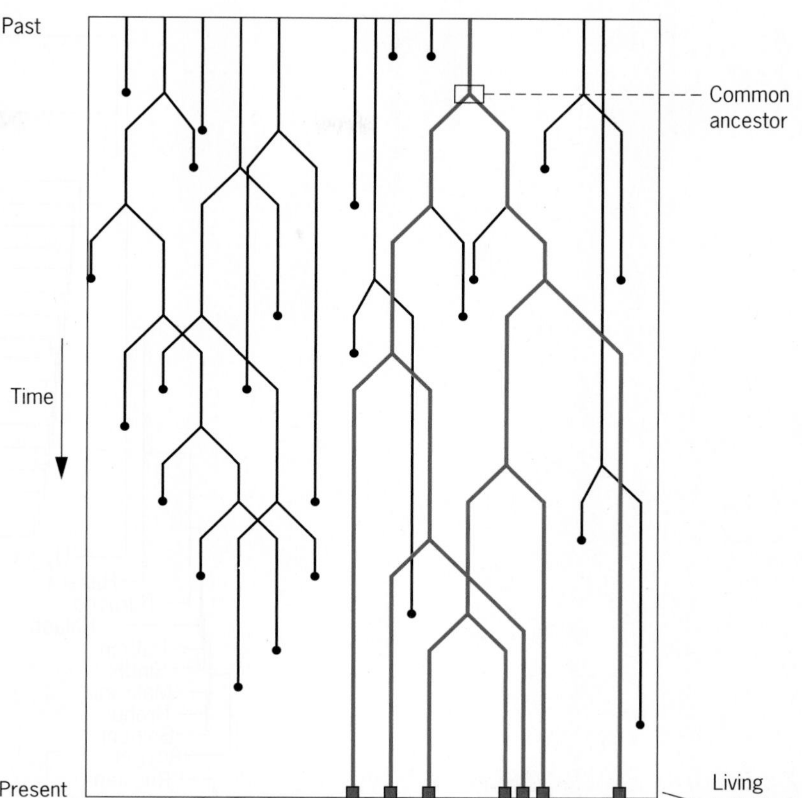

Figure 25.23 ▶ A coalescence process. If the lineages of the DNA sequences found in living individuals are traced back into the past, they coalesce in a common ancestor. These lineages are highlighted in red in the time line. Other DNA sequences from the past are not represented in living individuals; the time each became extinct is indicated by a dot.

those in other continents. The greater accumulation of genetic variation in African populations suggests that these populations are the oldest—an idea that is consistent with the Out-of-Africa Hypothesis of human origins. Fairly strong evidence for this hypothesis has come from studies of mitochondrial DNA sequences from different human populations. By analyzing sequences from living individuals, it is possible to work back to the ancestral sequence from which all the existing sequences could have sprung. This ancestral sequence represents the point at which the lineages of the living individuals coalesce into one individual, the common ancestor of them all (**FIGURE 25.23**). Then, by counting the number of mutations that occurred between the ancestral DNA sequence and the current sequences, and by dividing this number by the known mutation rate, it is possible to calculate the time that has elapsed since the common ancestor existed.

When this type of analysis is performed on mitochondrial DNA sequences, the elapsed time between the present and the time when the common ancestor lived is estimated to be 100,000 to 200,000 years. Analyses of DNA sequences on the Y chromosome, which is transmitted exclusively through males, yield a similar estimate. Thus, the coalescent principle suggests that all modern humans are descended from maternal and paternal common ancestors who lived between 100,000 and 200,000 years ago. This result does not imply, however, that these common ancestors were the only two people alive at that remote time. Certainly many others were alive too. Their genetic lineages—mitochondrial in the case of females and Y chromosomal in the case of males—simply became extinct. With the coalescent method, current DNA sequences can be traced back to the individuals whose mitochondrial or Y chromosomal lineages were lucky enough to survive and spread through the species, modified, of course, by the random process of mutation.

These analyses of mitochondrial and Y-linked DNA sequences have now been supplemented with analyses of autosomal DNA. One recent study analyzed single-nucleotide polymorphisms in more than nine hundred human genomes from 51 different populations all over the world (**FIGURE 25.24**). The results indicate that the modern human species is relatively young and that it originated in the archaic human populations of Africa. From Africa, humans migrated to Asia and Europe, and later to Australia and the Americas, ultimately becoming the dominant species on the earth.

KEY POINTS

- Fossil evidence indicates that the remote ancestors of human beings evolved in Africa, beginning about 4 to 5 million years ago.
- Genetic evidence indicates that modern human populations may have emerged from Africa about 100,000 to 200,000 years ago and subsequently spread to other continents.

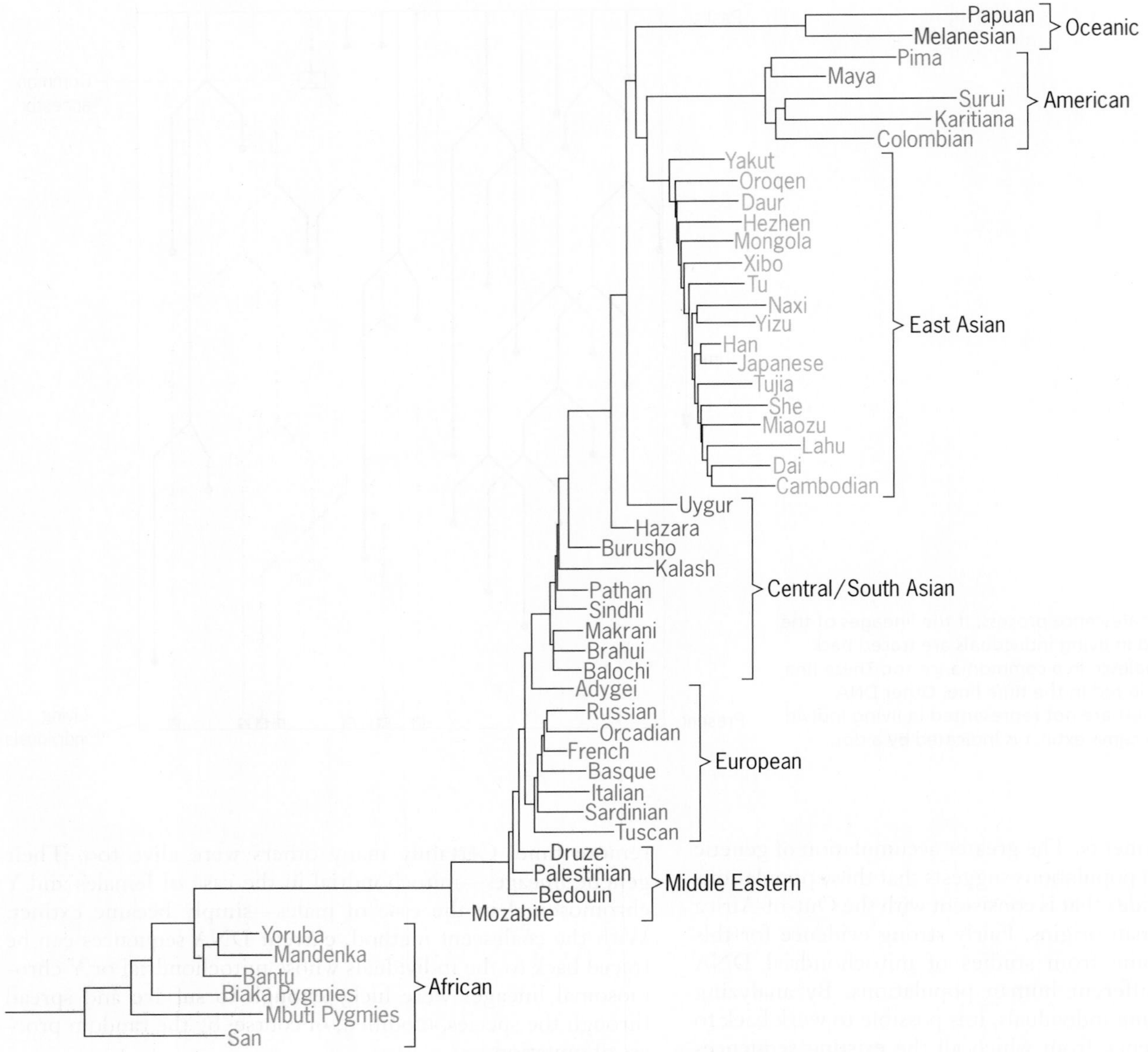

Figure 25.24 ▶ Phylogeny of human populations based on the analysis of single-nucleotide polymorphisms.

▶ Basic Exercises

ILLUSTRATE BASIC GENETIC ANALYSIS

1. Nevin Aspinwall investigated the frequencies of electrophoretically distinguishable alleles of the gene encoding alpha-glycerophosphate dehydrogenase (α-GPDH) in the pink salmon (*Onchorhynchus gorbuscha*) in rivers along the northwest coast of North America, from Alaska to Washington State (1974, *Evolution* 28:295–305). Fast, slow, and hybrid allozymes of α-GPDH were detected in this study; the fast and slow allozymes were each encoded by different alleles of the gene, and the hybrid allozyme was produced in fish heterozygous for these alleles. In the sample from Dungeness River, Washington, Aspinwall observed 32 fish with the slow allozyme, 6 with the hybrid allozyme, and 1 with the fast allozyme. What are the frequencies of the "fast" and "slow" alleles of the α-*GPDH* gene in the sample from this locality?

Answer: In the Dungeness River sample of 39 fish, each with two copies of the α-*GPDH* gene, the frequency of the fast allele is $(2 \times 1 + 6)/(2 \times 39) = 0.10$, and the frequency of the slow allele is $1 - 0.10 = 0.90$.

2. How many distinct rooted, bifurcating phylogenetic trees could show the evolutionary relationships among three different organisms.

Answer: If we denote the organisms as A, B, and C, three distinct rooted, bifurcating phylogenetic trees could show the evolutionary relationships among them:

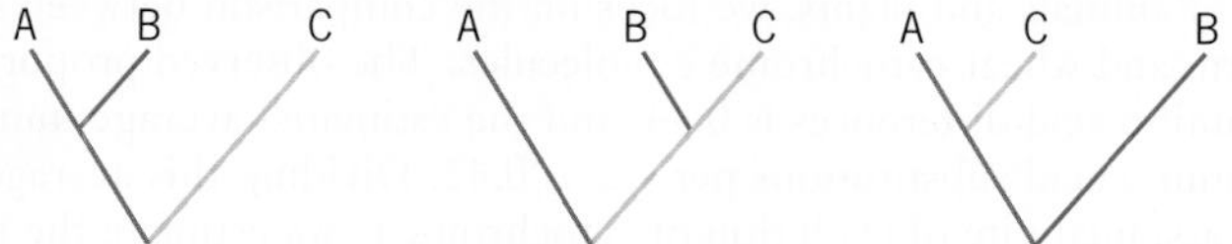

3. Human and horse α-globin polypeptides differ in 18 of 141 amino acid positions. On average, how many amino acid substitutions have occurred per site in this polypeptide since the human and horse lineages diverged from a common ancestor? If the evolutionary rate for α-globin among mammals has been 0.74 substitutions per site every billion years, how much time has elapsed since the common ancestor of humans and horses existed?

Answer: Human and horse α-globin differ in 18/141 = 0.128 of their amino acids. To obtain the average number of amino acid substitutions that have occurred per site since the human and the horse α-globins began evolving independently, we use the Poisson correction: $-\ln(1 - 0.128) = 0.136$ amino acid substitutions per site. Then, to calculate the total time that has elapsed since the common ancestor of humans and horses, we divide 0.136 amino acid substitutions per site by the estimated evolutionary rate for mammals (0.74 amino acid substitutions per site every billion years): 0.136/0.74 = 184 million years. This span of time must be divided equally between the human and horse lineages to obtain the time since their common ancestor existed: 184 million years/2 = 92 million years.

4. Under the Neutral Theory of Molecular Evolution, what is the rate at which selectively neutral mutations are fixed in a population by random genetic drift?

Answer: The rate of fixation of selectively neutral mutations is simply the rate at which these mutations occur.

5. What is the genetic definition of a species?

Answer: A species is a population that is reproductively isolated from all other populations—that is, it cannot exchange genes with other populations.

Testing Your Knowledge

INTEGRATE DIFFERENT CONCEPTS AND TECHNIQUES

1. In his study of allozyme polymorphism in populations of pink salmon in rivers from Alaska to Washington State, Nevin Aspinwall collected data from mature salmon captured in 1969, 1970, and 1971. Salmon are born in rivers, and after about nine months, they migrate into the ocean, where they increase in size. When they reach two years of age, the salmon return to the river of their birth to spawn, and then they die. Because of this two-year life cycle, Pacific salmon are split into odd- and even-year populations that do not interbreed. Aspinwall found that among the salmon captured in odd years, 870 were homozygous for the slow allele of α-*GPDH*, 17 were homozygous for the fast allele, and 231 were heterozygous. Among the salmon captured in the even year, 649 were homozygous for the slow allele, 45 were homozygous for the fast allele, and 309 were heterozygous. What is interesting about these data?

Answer: From Aspinwall's summary data, we can calculate the frequencies of the fast allele of the α-*GPDH* gene in the odd- and even-year populations. In the odd-year population, the frequency is $(2 \times 17 + 231)/(2 \times 1118) = 0.119$, and in the even-year population, it is $(2 \times 45 + 309)/(2 \times 1003) = 0.199$. Thus, the frequency of the fast allele in the even-year population is almost twice the corresponding frequency in the odd-year population. Because the two salmon populations inhabit the same territory, they are presumably subject to the same selection pressures. Thus, the observed difference in allele frequency between these populations suggests that they have diverged by random genetic drift.

2. The following table shows the number of amino acid differences among molecules of cytochrome c.

	Tuna	Silkworm	Wheat
Human	20	26	35
Tuna		27	40
Silkworm			37

If the number of amino acid sites that can be compared among these molecules is 110, what is the average number of amino acid substitutions that have occurred per site during the evolution of each pair of organisms? What is the rate at which cytochrome c has evolved among the vertebrates? If the evolutionary rate among the vertebrates can be applied to other branches of cytochrome c's phylogeny, how long ago did the insect and fish lineages diverge from a common ancestor? How long ago did the animal and plant lineages diverge from a common ancestor?

Answer: To estimate the average number of amino acid substitutions per site, we first compute the proportion of amino acid differences for each pair of organisms by dividing the observed number of differences by 110, which is the total number of sites in the cytochrome c molecule. Then we use the Poisson correction to calculate the average number of amino acid substitutions per site. If d is the proportion of amino acid differences between the cytochrome c molecules of two organisms, then the average number of substitutions per site is obtained from the formula $-\ln(1 - d)$. In the following table, the proportion of amino acid differences is given in black, and the average number of amino acid substitutions per site is given in red:

	Tuna	Silkworm	Wheat
Human	0.18	0.24	0.32
	0.20	0.27	0.38
Tuna		0.24	0.36
		0.28	0.45
Silkworm			0.34
			0.42

To calculate the rate of evolution among the vertebrates, we focus on the comparison between human and tuna cytochrome c molecules. The observed proportion of amino acid differences is 0.18, and the estimated average number of substitutions per site is slightly higher, 0.20. From the fossil record, the fish (represented by the tuna) and the tetrapod (represented by the human being) lineages are estimated to have split about 400 million years ago (mya). The total elapsed evolutionary time in these lineages is therefore 2 × 400 my = 800 my. We obtain the rate of amino acid substitution per site in cytochrome c by dividing the average number of amino acid substitutions per site by the total elapsed evolutionary time: 0.20 amino acid substitutions per site/800 my = 0.25 amino acid substitutions per site every billion years.

If we assume that this rate holds throughout the phylogeny of tetrapods, fish, insects, and plants—that is, if we assume a molecular clock—then we can calculate the time that has passed since the fish and insect lineages and the animal and plant lineages split from common ancestors. For the common ancestor of fish and insects, we focus on the comparison between tuna and silkworm cytochrome c molecules. The observed proportion of amino acid differences is 0.24, and the estimated average number of amino acid substitutions per site is 0.28. Dividing this average by the estimated rate of evolution of cytochrome c (0.25 amino acid substitutions per site every billion years), we obtain the total elapsed evolutionary time: 0.28 substitutions per site/0.25 substitutions per site every billion years = 1.1 billion years. We must apportion this time equally between the fish and insect lineages. Thus, the time since they diverged from a common ancestor is calculated to be 550 million years. For the ancestor of animals and plants, we focus on the comparison between silkworm and wheat cytochrome c molecules. The observed proportion of amino acid differences is 0.34, and the estimated average number of amino acid substitutions per site is 0.42. Dividing this average by the assumed rate of evolution of cytochrome c, we estimate the total elapsed evolutionary time to be 1.68 billion years. The time since these two lineages diverged from a common ancestor is therefore 840 million years.

3. In an extensive analysis of single-nucleotide polymorphisms (SNPs) among three groups of Americans, David Hinds and his collaborators (2005, *Science* 307:1072–1079) found that among 1,586,383 SNPs examined by microarray technology, 93.5 percent were segregating in a sample of 23 African Americans, 81.1 percent were segregating in a sample of 24 European Americans, and 73.6 percent were segregating in a sample of 24 Han Chinese Americans. What do these data indicate about genetic diversity among these three groups, and how do they fit with current ideas about human evolutionary history?

Answer: If we use the percentage of SNPs segregating in a population as an indicator of its genetic diversity, then clearly the African American group is the most diverse of the three groups studied. The fact that African Americans are the most diverse also fits with the idea that modern humans originated in Africa. African populations, being the oldest among all populations of modern humans, have had the longest time to accumulate genetic variants.

▶ Questions and Problems

ENHANCE UNDERSTANDING AND DEVELOP ANALYTICAL SKILLS

25.1 Using the data in Table 25.1, and assuming that mating is random with respect to the blood type, predict the frequencies of the three genotypes of the Duffy blood-type locus in a South African and an English population.

25.2 Theodosius Dobzahnsky and his collaborators studied chromosomal polymorphisms in *Drosophila pseudoobscura* and its sister species in the western United States. In one study of polymorphisms in chromosome III of *D. pseudoobscura* sampled from populations at different locations in the Yosemite region of the Sierra Nevada, Dobzhansky (1948, *Genetics* 33:158–176) recorded the following frequencies of the Standard (ST) banding pattern:

Location	Frequency ST	Elevation (in feet)
Jacksonville	0.46	850
Lost Claim	0.41	3,000
Mather	0.32	4,600
Aspen	0.26	6,200
Porcupine	0.14	8,000
Tuolumne	0.11	8,600
Timberline	0.10	9,900
Lyell Base	0.10	10,500

What is interesting about these data?

25.3 What was some of the evidence that led Charles Darwin to argue that species change over time?

25.4 Darwin stressed that species evolve by natural selection. What was the main gap in his theory?

25.5 DNA and protein molecules are "documents of evolutionary history." Why aren't complex carbohydrate molecules such as starch, cellulose, and glycogen considered "documents of evolutionary history"?

25.6 During the early evolutionary history of the vertebrates, a primordial globin gene was duplicated to form the α- and β-globin genes. The rate of evolution of the polypeptides encoded by these duplicate genes has been estimated to be about 0.9 amino acid substitutions per site every billion years. By comparing the human α- and β-globins, the average number of amino acid substitutions per site has been estimated to be 0.800. From this estimate, calculate when the duplication event that produced the α- and β-globin genes must have occurred.

25.7 The heme group in hemoglobin is held in place by histidines in the globin polypeptides. All vertebrate globins possess these histidines. Explain this observation in terms of the Neutral Theory of molecular evolution.

25.8 A researcher has been studying genetic variation in fish populations by using PCR to amplify trinucleotide repeats at a particular site on a chromosome (see chapter 17). The diagram below shows the gel-fractionated products of amplifications with DNA samples from 10 different fish. How many distinct alleles of this trinucleotide repeat locus are evident in the gel?

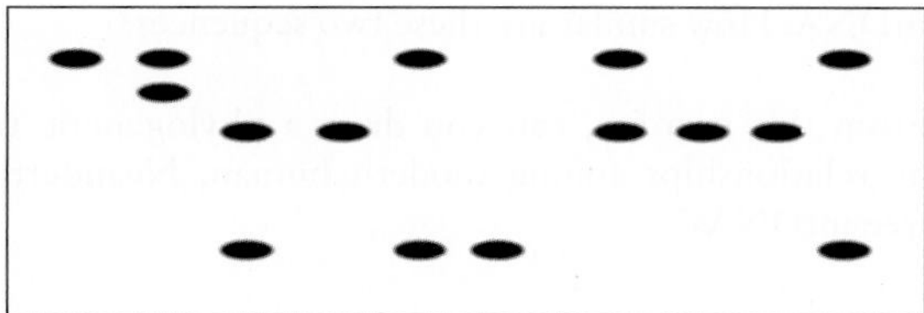

25.9 Ribonuclease, a protein that degrades RNA, is 124 amino acids long. A comparison between the amino acid sequences of cow and rat ribonucleases reveals 45 differences. What is the average number of amino acid substitutions that have occurred per site in these two evolutionary lineages? If the cow and the rat lineages diverged from a common ancestor 90 million years ago, what is the rate of ribonuclease evolution?

25.10 The coding sequence of the alcohol dehydrogenase (*Adh*) gene of *Drosophila melanogaster* consists of 765 nucleotides (255 codons); 192 of these nucleotides are functionally silent—that is, they can be changed without changing an amino acid in the Adh polypeptide. In a study of genetic variation in the *Adh* gene, Martin Kreitman observed that 13 of the 192 silent nucleotides were polymorphic. If the same level of polymorphisms existed among the nonsilent nucleotides of the *Adh* gene, how many amino acid polymorphisms would Kreitman have observed in the populations he studied?

25.11 Within the coding region of a gene, where would you most likely find silent polymorphisms?

25.12 Using the DNA sequences in **FIGURE 25.9*a***, compute the minimum number of mutations required to explain the derivation of the four sequences (1, 2, 3, and 4) in the following phylogenetic trees:

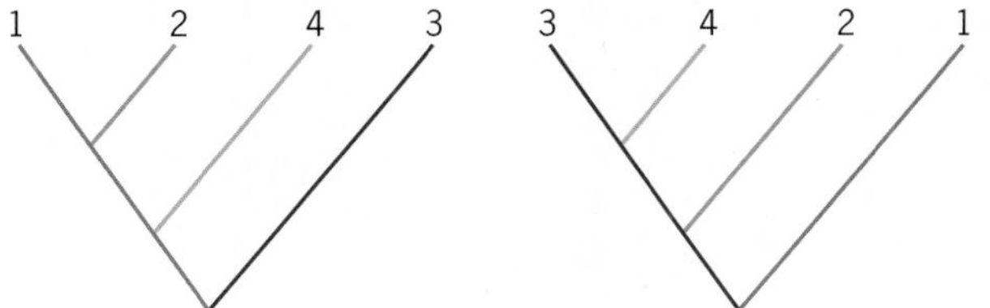

How do the results compare with the number of mutations required by the most parsimonious tree given in **FIGURE 25.9*b***?

25.13 If the evolutionary rate of amino acid substitution in a protein is *K*, what is the average length of time between successive amino acid substitutions in this protein?

25.14 A geneticist has studied the sequence of a gene in each of three species, A, B, and C. Species A and species B are sister species; species C is more distantly related. The geneticist has calculated the ratio of nonsynonymous (NS) to synonymous (S) nucleotide substitutions in the coding region of the gene in two ways—first, by comparing the gene sequences of species A and C, and second, by comparing the gene sequences of species B and C. The NS:S ratio for the comparison of species B and C is six times greater than it is for the comparison of species A and C. What might this difference in the NS:S ratios suggest?

25.15 How might you explain the thousandfold difference in the evolutionary rates of fibrinopeptide and histone 3?

25.16 If a randomly mating population is segregating *n* selectively neutral alleles of a gene and each allele has the same frequency, what is the frequency of all the homozygotes in the population?

25.17 Dispersed, repetitive sequences such as transposable elements may have played a role in duplicating short regions in a genome. Can you suggest a mechanism? (*Hint*: see Chapter 18.)

25.18 Distinguish between allopatric and sympatric modes of speciation.

25.19 *Drosophila mauritiana* inhabits the island of Mauritius in the Indian Ocean. *Drosophila simulans*, a close relative, is widely distributed throughout the world. What experimental tests would you perform to determine if *D. mauritiana* and *D. simulans* are genetically different species?

25.20 A segment of DNA in an individual may differ at several nucleotide positions from a corresponding DNA segment in another individual. For instance, one individual may have the sequence ...A...G...C... and another individual may have the sequence ...T...A...A.... These two DNA segments differ in three nucleotide positions. Because the nucleotides within each segment are tightly linked, they will tend to be inherited together as a unit, that is, without being scrambled by recombination. We call such heritable units DNA haplotypes. Through sampling and DNA sequencing, researchers can determine which DNA haplotypes are present in a particular population. When this kind of analysis is performed on human populations by sequencing, for example, a segment of mitochondrial DNA, it is found that samples from Africa exhibit more haplotype diversity than samples from other continents. What does this observation tell us about human evolution?

25.21 The *prune* gene (symbol *pn*) is X-linked in *Drosophila melanogaster*. Mutant alleles of this gene cause the eyes to be brown instead of red. A dominant mutant allele of another gene located on a large autosome causes hemizygous or homozygous *pn* flies to die; this dominant mutant allele is therefore called *Killer of prune* (symbol *Kpn*). How could mutants such as these play a role in the evolution of reproductive isolation between populations?

25.22 Exon shuffling is a mechanism that combines exons from different sources into a coherent sequence that can encode a composite protein—one that contains peptides from each of the contributing exons. Alternate splicing is a mechanism that allows exons to be deleted during the expression of a gene; the mRNAs produced by alternate splicing may encode different, but related, polypeptides (see Chapter 20). What bearing do these two mechanisms have on the number of genes in a eukaryotic genome? Do these mechanisms help to explain why the gene number in the nematode *Caenorabditis elegans* is not too different from the gene number in *Homo sapiens*?

▶ Genomics on the Web

at http://www.ncbi.nlm.nih.gov/

1. Search GenBank for AY149291, a 357-bp fragment of mitochondrial DNA (mtDNA) obtained from a Neanderthal fossil found in Germany. Use the BLAST tool to find the homologous DNA sequence in the mtDNA of modern humans. What are the coordinates of the modern human DNA sequence? How similar is the Neanderthal sequence to the modern human sequence?

2. Now use the BLAST tool to find the homologous DNA sequence in the mtDNA of chimpanzee (*Pan troglodytes*). Click on the first item in the list of results to see the comparison of the Neanderthal and chimpanzee mtDNA sequences. How similar are these two sequences?

3. Now search GenBank for AF347015, the complete sequence of the mtDNA of a modern human. When this sequence appears, copy the part of it that corresponds to the 357-bp fragment of Neanderthal mtDNA into a text file, delete the numbers and spaces from the copied text, and then use the resulting sequence in BLAST to compare this region in modern human mtDNA to the homologous region in chimpanzee mtDNA. How similar are these two sequences?

4. From this exercise, can you draw a phylogenetic tree that shows the relationships among modern human, Neanderthal, and chimpanzee mtDNAs?

Photo Credits

Chapter 1 Opener: Hybrid Medical Animation/Photo Researchers, Inc. Fig. 1.1: James King-Holmes//Photo Researchers, Inc. Fig. 1.3: Perrin Pierre/©Corbis. Fig. 1.5: Scott Sinklier/AGStock USA/Photo Researchers, Inc. Fig. 1.9: Stan Flegler/Visuals Unlimited. Fig. 1.10*a*: Courtesy of the American Museum of Natural History. Fig. 1.10*b*: Hulton Archive/Getty Images. Fig. 1.12*a*: Courtesy American Angus Association. Fig. 1.12*b*: Courtesy Crown Prince 920, www.towildoaksfarms.com. Fig. 1.12*c*: Courtesy American Simmental Association. Fig. 1.12*d*: Courtesy American International Charolais Association. Fig. 1.13*a,b*: Courtesy John Doebley, Genetics, University Wisconsin. Fig. 1.14*a,b*: Courtesy Mycogen. Fig. 1.15: James King-Holmes/Photo Researchers, Inc. ***A Milestone in Genetics:*** Fig. 1: Arthur Foster/Hulton Archive/Getty Images

Chapter 2 Opener (top): Najiah Feanny/Corbis SABA. Opener (bottom): Getty Images. Fig. 2.1*a*: Biophoto Associates/Photo Researchers, Inc. Fig. 2.1*b*: Kwangshin Kim/Photo Researchers, Inc. Fig. 2.1*c*: Eye of Science/Photo Researchers, Inc. Fig. 2.1*d*: Dr. Torsten Wittmann/Photo Researchers, Inc. Fig. 2.3*a*: Dr. Gopal Murti/Photo Researchers, Inc. Fig. 2.3*b*: Alfred Pasieka/Photo Researchers, Inc. Fig. 2.5*a*: Courtesy Conley L. Rieder, from *J. Cell Biol.* 1212:364, 1993. Reproduced with permission of Rockefeller University Press. Fig. 2.5*b*: From Jerome B. Rattner and Stephanie G. Phillips, *J. Cell Biol.* 57:363, 1973. Reproduced with permission of Rockefeller University Press. Fig. 2.6: Andrew S. Bajer. Fig. 2.7*a*: David M. Phillips/Visuals Unlimited. Fig. 2.7*b*: Cabisco/Visuals Unlimited. Fig. 2.8: L. Willatt/Photo Researchers, Inc. Fig. 2.10*a-m*: Clare A. Hasenkampf/Biological Photo Service. Fig. 2.11*a*: From M. Westergaard and D. von Wettstein, *Annual Review of Genetics*, Vol. 6; ©1972 by Annual Reviews, Inc. Original photograph courtesy of D. von Wettstein. Fig. 2.12*b*: From Bernard John, *Meiosis*, Cambridge University Press, 1990. Fig. 2.16: SPL/Photo Researchers, Inc. Fig. 2.17: Carolina Biological Supply/Phototake. Fig. 2.18*a,b*: From J. E. Sulston and H. R. Horvitz, *Develop. Biol.* 56:110–156, 1977. Photograph courtesy of Dr. John Sulston. Fig. 2.19*a*: TUNS/ArcoImages/Peter Arnold, Inc. Fig. 2.19*b*: Science Source/Photo Researchers, Inc. Fig. 2.20*a*: Mark Smith/Photo Researchers, Inc. Fig. 2.20*b*: Courtesy GloFishTM www.glofish.com, Yorktown Technologies, L.P. ***A Milestone in Genetics:*** Nancy Kedersha/Photo Researchers, Inc.

Chapter 3 Opener: Mitch Hrdlicka/Photodisc/Getty Images. Fig. 3.3: Martin Shields/Photo Researchers, Inc. Fig. 3.4: Martin Shields/Photo Researchers, Inc. Fig. 3.5: Martin Shields/Photo Researchers, Inc. Fig. 3.10*a*: H. Reinhard/ArcoImages/Peter Arnold, Inc. Fig. 3.11: Martin Shields/Photo Researchers, Inc.

Chapter 4 Opener: The Irish Image Collection/©Corbis. Fig. 4.10*a*: Courtesy Lester V. Bergman & Associates, Inc. Fig. 4.12*a-d*: Courtesy Ralph Somes. Fig. 4.15*a*: TH Foto-Werbung/Science Photo Library/Photo Researchers, Inc. Fig. 4.15*a*: (inset photo) ©TH Foto/Alamy. Fig. 4.18*a,b*: Photo by Leah Sandall, University of Nebraska-Lincoln, Plant and Soil Science Library, http://plantandsoil.unl.edu. ***A Milestone in Genetics:*** Courtesy National Portrait Gallery.

Chapter 5 Opener: Jan Van Der Knokke/Foto Natura/Getty Images. Fig. 5.9*a*: Corbis-Bettmann. ***Focus on:*** Dr. Dennis Kunkel/Phototake. Fig. 5.17*b*: Grant Heilman Photography. Fig. 5.18: Michael Abbey/Photo Researchers, Inc. ***A Milestone in Genetics:*** Columbia University, Columbiana Collection.

Chapter 6 Opener: Peter Turner/Getty Images. Fig. 6.2: From C. G. Vosa, 1971.The quinacrine-fluorescence patterns of the chromosomes of *Albium carinatum*, *Chromosoma* 33:382–385, Fig 1. Fig. 6.3: From R. M. Patterson & J. C. Petricciani, 1973. A comparison of prophase & metaphase G-bands in the muntjak. *J. of Heredity* 64 (2):80–82. Fig. 2*a*. Fig. 6.4: From Weinberg and Stanyon, 1995. *Current Opinions In Genetics & Development*. 5: 792–797. Fig. 6.5: Phototake. Fig. 6.7*a*: C. G. Maxwell/Photo Researchers, Inc. Fig. 6.7*b*: Eric Crichton//Bruce Coleman, Inc. Fig. 6.7*c*: Ken Wagner/Phototake. Fig. 6.7*d*: M. Timothy O'Keefe//Bruce Coleman, Inc. Fig. 6.11: Courtesy Todd R. Laverty, HHMI/Berkeley Drosophila Genome Project. Fig. 6.14*a*: Hattie Young/Photo Researchers, Inc. Fig. 6.14*b*: Courtesy of R. M. Fineman. ***Focus on:*** Yoav Levy/Phototake. Fig. 6.17: Courtesy University of Utah Cytogenetics Laboratory. Fig. 6.18: Courtesy J. K. Lim. ***A Milestone in Genetics:*** From Tijo, J. H. and A. Levan, The chromsome number of man. *Hereditas* 42:1–6, p. 2, Fig. 1*a* (1956). Courtesy Hereditas.

Chapter 7 Opener: ©Miss Gracie B/Alamy Fig. 7.9: Courtesy Bernard John. Fig. 7.29: Science Source/Photo Researchers, Inc. ***A Milestone in Genetics:*** Courtesy of Nancy Wexler.

Chapter 8 Opener: Dennis Kunkel/Visuals Unlimited. Fig. 8.1*a*: Harold Fisher/Visuals Unlimited. Fig. 8.2: A. K. Kleinschmidt, *Biophys. Acta.* 61, 1962, p. 861, Fig. 1, Elsevier Publishing Company. Fig. 8.4*a*: Courtesy A. F. Howatson, From *Gene Expression*, Vol. 3, John Wiley & Sons, 1977. Fig. 8.7*a,b*: Photographs by R. Wagner and D. P. Snustad. Fig. 8.8: From G.S. Stent, *Molecular Biology of Bacterial Viruses*. W.H. Freeman and Company. 1963. Fig. 8.10: From G. S. Stent, *Molecular Biology of Bacterial Viruses*. W.H. Freeman and Company. 1963. Fig. 8.14: Christine Case/Visuals Unlimited. Fig. 8.18: From O. T. Avery, C. M. MacLeod, and M. McCarty, *J. Exp. Med.* 79:153, 1944; by copyright permission of Rockefeller University Press. Fig. 8.21: T. F. Anderson, E. Wollman, and F. Jacob, from *Annals Institute Pasteur*, Masson and Cie, Editors.

Chapter 9 Opener: Phototake, Inc/Alamy. Fig. 9.8: Omikron/Science Source/Photo Researchers, Inc. ***A Milestone in Genetics:*** (center right) Corbis-Bettmann/(bottom left) Courtesy National Institute of Health/(center left and bottom right) A. Barrington Brown/Photo Researchers, Inc. Fig. 9.12: Courtesy Dr. Svend Freytag. Fig. 9.18*a*: Omikron/Science Source/Photo Researchers, Inc. Fig. 9.19: From Vollrath & Davis, *Nucleic Acid Res.* 15:7876 (1987) Oxford University Press. Photo courtesy Dr. Douglas Vollrath. Fig. 9.20: From R. Kavenoff, L. C. Klotz, & B. H. Zimm, *Cold Spring Harbor Symp. Quant. Biol.* 38:1–8, 1973. © 1973 by Cold Spring Harbor Laboratory. Original photo courtesy R. Kavenoff. Fig. 9.21*a*: Courtesy Victoria Foe. Fig. 9.23*a,b*: Reprinted with permission from Karolyn Luger, et al., *Nature* 389:251m 1997, courtesy of Timothy J. Richmond. Fig. 9.23*c*: Courtesy of D. Rhodes. © Copyright 1997 by Macmillan Magazines Limited. Fig. 9.24: Gunter F. Bahr/Biological Photo Service. Fig. 9.25*a*: Courtesy Barbara Hamkalo. Fig. 9.25*b*: Reprinted from Bednar et al., 1998 *Proc. Natl. Acad. Sci. USA* 95: 4173-14178 with permission, Photos courtesy Dr. Christopher L. Woodcock, University of Massachusetts, Amherst. Courtesy Christopher L. Woodcock. Fig. 9.26: From J. R. Paulson & U. K. Laemmli, *Cell* 12: 817–828, 1977. Copyright 1977, MIT; published by MIT Press. Original photo courtesy U.K. Laemmli. Fig. 9.30: From Huntington F. Willard, *Trends Genetics* 6:414, 1990. ***Focus on:*** (top) Courtesy Dr. Mary Lou Pardue. MIT./From J. Meyne, in R. P. Wagner, *Chromosomes: A Synthesis*, Copyright © 1993. Reprinted with permissions of Wiley-Liss, Inc., a subsidiary of John Wiley & Sons, Inc.

Chapter 10 Opener: Courtesy Dr. Karl Fredga, Department of Genetics, Uppsala University, Sweden. Fig. 10.2*a,b*: From J. H. Taylor, *The Replication and Organization of DNA in Chromosomes*, Molecular Genetics, Part I, J. H. Taylor (ed), Academic Press, New York, 1963. Fig. 10.3*a*: J. Cairns, Imperial Cancer Research Fund, London, England. Reproduced with permission from *Cold Spring Harbor Sympos. Quant. Biol.* 28:43, 1963. Original photo courtesy John Cairns. Fig. 10.7*c*: Photo reproduced with permission from M. Schnös and R. B. Inman, *J. Mol. Biol.* 51: 61–73,

1970. © 1970 by Academic Press, Inc. (London), Ltd. Fig. 10.9*a,b*: From J. Wolfson, D. Dressler, and M. Magazin, *Proceedings of the National Academy of Sciences, U.S.* 69:499, 1972. Original micrographs courtesy D. Dressler. Fig. 10.13*a,b*: From S. Doublie, S. Tabor, A. M. Long, C. C. Richardson, and T. Ellenberger, Nature 391:251–258. Fig. 10.14*a*: From X. P. Kong et al., *Cell* 69:425–437, 1992. © Cell Press. Original photograph courtesy Dr. John Kuriyan, Howard Hughes Medical Institute, Rockefeller University. Fig. 10.28: From K. Koths and D. Dressler, Harvard Medical School. Fig. 10.29*a,b*: Reproduced with permission from J. A. Huberman and A. D. Riggs, *Journal of Molecular Biology* 32:327–341, 1968. © 1968 by Academic Press. Original photographs courtesy J. A. Huberman. Fig. 10.31*a*: Courtesy of Steven L. McKnight and Oscar L. Miller, Jr. Fig. 10.33: Gerald Herbert/ © AP/Wide World Photos.

Chapter 11 Opener: Courtesy M. Jurica. Fig. 11.4*a,b*: From D. Prescott, Cellular Sites of RNA Synthesis, *Prog. Nucleic Acid Res. Mol. Biol.* 3:33–57, 1964. Fig. 11.12: From O. L. Miller, Jr., B. A. Hamkalo, and C. A. Thomas, Jr., *Science* 169:392–395, 1970. Copyright © 1970 by the American Association for the Advancement of Science. Original micrograph courtesy O. L. Miller, Jr. Fig. 11.16: Courtesy Dr. Seth Darst. Fig. 11.22: Courtesy Jack Griffith, Lineberger Comprehensive Cancer Center, University of North Carolina at Chapel Hill. ***A Milestone in Genetics:*** S. M. Berget, C. Moore, P. A. Sharp, 1977. Spliced segments at the 5 termminnus off adenovirus 2 late RNA, *Proc. Natl. Acad. Sci. USA* 74:3171–3175. From Tilghman et al., *Proc. Natl. Acad. Sci. USA* 75:1309–1313. Courtesy Pierre Chambon.

Chapter 12 Opener: Science Photo Library/Photo Researchers, Inc. Fig. 12.7: E. Gwyn Jordan, *Molecular Biology*, Kings College. Fig. 12.8: O. L. Miller, B. R. Beatty, D. W. Fawcett/Visuals Unlimited. Fig. 12.11*a*: From S. H. Kim, F. L. Suddath, G. J. Quiqley, A. McPherson, J. L. Sussman, A. H. J. Wang, N. C. Seeman, and A. Rich, *Science* 185: 435–440, 1974 by the American Association for the Advancement of Science. Original photo courtesy S. H. Kim. Fig. 12.13*b*: Courtesy Dr. Joachim Frank. From Frank, et al. 1995. *Biochemistry and Cell Biology* 73:357. Fig. 12.14*a-c*: Reproduced with permission of M. Yusupov et al., *Science* 292:823–826, 2001. Courtesy Albion Baucom. Fig. 12.18: From S. L. McKnight, N. L. Sullivan, and O. L. Miller, Jr., *Prog. Nucleic Acid Res. Mol. Biol.* 19:313–318, 1976. Micrograph courtesy S. L. McKnight and O. L. Miller, Jr., University of Virginia.

Chapter 13 Opener: Sarah Leen/NG Image Collection. Fig. 13.1: © Corbis. Fig. 13.21: From B. N. Ames, J. McCann, and E. Yamasaki, *Mutat. Res.* 31:347, 1975. Photograph courtesy B. N. Ames. Fig. 13.25: Ken Greer/ Visuals Unlimited. Fig. 13.27*a*: Courtesy of H. Potter and D. Dressler, Harvard Medical School.

Chapter 14 Opener: Mellor Images/ Alamy. ***Focus on:*** *Science* Vol. 291, No. 5507, 16 February 2001. Photos courtesy Ann Elliott Cutting/© 2001 American Association for the Advancement of Science. Fig. 14.15*a,b*: Courtesy D. Peter Snustad. Fig. 14.16: Bruce Iverson Photomicrography.

Chapter 15 Opener: Kenneth Eward/ Photo Researchers, Inc. Fig. 15.1*b*: Based on an X-ray structure by John Rosenberg, U. of Pittsburgh. Fig. 15.1: Courtesy S. Kopczak and D. P. Snustad, University of Minnesota. Fig. 15.19: Courtesy D. Peter Snustad. Fig. 15.21*a*: Photographs courtesy of S. R. Ludwig and D. P. Snustad, University of Minnesota. Fig. 15.21*b*: Courtesy D. Peter Snustad. ***Focus on:*** From Kerem, et al. (1989), *Science* 245:1073–1080. ***A Milestone in Genetics:*** Danna & Nathans 1971, *Proc. Natl. Acad. Sci. USA* 68:2913–2917, Fig. 3. Courtesy Kathleen Danna. Fig. 15.22*a,b,c*: Courtesy of S. R. Ludwig and D. P. Snustad, University of Minnesota. Fig. 15.25*a,b*: Courtesy of M. G. Li and D. P. Snustad, University of Minnesota. Fig. 15.29: From *DNA Sequencing*, 2nd edition, Amersham Life Science. Fig. 12, pg. 16.

Chapter 16 Opener: Steve Hunt/ Getty Images. Fig. 16.16*a-d*: Courtesy CLONTECH Laboratories, Inc. Fig. 16.17: Courtesy Affymetrix, Inc. Fig. 16.17: Affymetrix. Fig. 16.18*b-d*: Ludin & Matus, 1998. *Trends in Cell Biology*, 8:72. Fig. 16.21: Courtesy Frederick Blattner. ***A Milestone in Genetics:*** UN (left) Courtesy of the National Institutes of Health/UN (right) Celera Genomics Corp.

Chapter 17 Opener: Dr. Yorgos Nikas/Photo Researchers, Inc. Fig. 17.2*b*: From *Cell* 72:971–983, Fig. 7, March 26,1993, Copyright © 1993 Cell Press. Fig. 17.11: Courtesy of Cellmark Diagnostics, Inc., Germantown, Maryland. Fig. 17.12: Courtesy of Cellmark Diagnostics, Germantown, Maryland. Fig. 17.16: Dr. R. L. Brinster, School of Veterinary Medicine, University of Pennsylvania. Fig. 17.17*b*: Photograph courtesy of G. Panzour and A. Das, University of Minnesota. ***A Milestone in Genetics:*** Scott Camazine/Photo Researchers, Inc. From Richards and Sutherland, *Trends in Genetics*, vol. 8 (7), p. 249, 1992. Photo courtesy Grant Sutherland. Fig. 17.23: Courtesy National Science Foundation.

Chapter 18 Opener: Jane Grushow/Grant Heilman Photography. ***Focus on:*** Used with permission of Marjorie M. Bhavnani, photo from Cold Spring Harbor Laboratory Archives. Fig. 18.13: Dennis Kunkel/Getty Images.

Chapter 19 Opener: CNRI/Photo Researchers, Inc. ***Focus on:*** Fig. (top) Courtesy T. A. Steitz, Yale University. Fig. (center left) Courtesy Ponzy Lu and Mitchell Lewis, University of Pennsylania. Lewis et al. 1966. *Science*, 271:1247–1254, Fig. 6. (center right) Courtesy Ponzy Lu and Mitchell Lewis, University of Pennsylania, *Science*, 271:1247–1254, Fig. 11.

Chapter 20 Opener: Oliver Meckes/ Photo Researchers, Inc. Fig. 20.5: From Gloria Coruzzi, Richard Broglie, Carol Edwards and Nam-Hai Chua, 1984 Tissue-specific and light-regulated expression of a pea nuclear gene encoding the small subunit of ribulose-1, 5-bisphosphate carboxylase. *EMBO J.* 3: 1671–1679). Fig. 20.9*c*: Reproduced with permission from E. Serfling, M. Jasin, and W. Schaffner, *Trends in Genet.* 1:224–230, 1985. Fig. 20.13: O. L. Miller/Visuals Unlimited. Fig. 20.14: Jack M. Bostrack/Visuals Unlimited. Fig. 20.16: Courtesy of Joseph Fong, University of Minnesota. Fig. 20.24: Courtesy Dr. Mitzi Kuroda, from M. I. Kuroda et al., *Cell* 66:935–947, 1991, Fig. 6. Photograph courtesy of Dr. Mitzi Kuroda. ***Focus on Problem Solving:*** Courtesy D. Peter Snustad.

Chapter 21 Opener: 3D4Medical.com/Getty Images. Fig. 21.1*a*: Courtesy Kevin Haley, University of Minnesota. ***Focus on:*** Courtesy Jeffrey Hall. Fig. 21.7: Daniel St. Johnston and Christiane Nüsslein-Vohard, *Cell* 68:201–219, 1992. Photograph courtesy of Christiane Nüsslein-Vohard. Fig. 21.14: Courtesy of Matthew Scott, Howard Hughes Medical Institute. Fig. 21.17: Courtesy Walter Gehring, Universitüt Basel, Switzerland. Fig. 21.23: From Amit et al., *Science* 233: 747, Copyright © 1986 the American Association for the Advancement of Science. Photograph courtesy of R. J. Poljak.

Chapter 22 Opener: Susumu Nishinaga/Photo Researchers, Inc.

Chapter 23 Opener: SPL/Photo Researchers, Inc. Fig. 23.2*a*: TH Foto-Werbung/SPL/Photo Researchers, Inc. ***Focus on:*** Courtesy of Franklin D. Enfield, Department of Genetics and Cell Biology, University of Minnesota. Fig. 23.9: D. Cavagnaro/ Visuals Unlimited.

Chapter 24 Opener: Wolfgang Kaehler/© Corbis. Fig. 24.6*a*: Courtesy of Professor Lawrence Cook, University of Manchester. Fig. 24.6*b*: Professor Lawrence Cook, University of Manchester. Fig. 24.10: Eye of Science/Photo Researchers, Inc.

Chapter 25 Opener: Museum of Fine Arts, Boston/Superstock. Fig. 25.1: (far left) Herbert Spichtinger/ zefa/© Corbis. Fig. 25.1: (left) Greg Stott/Masterfile. Fig. 25.1: (center) Courtesy Poultry World. Fig. 25.1: (right) Courtesy Poultry World. Fig. 25.2: (left) Tui De Roy/Minden Pictures, Inc. Fig. 25.2: (center) Frans Lanting/Minden Pictures, Inc. Fig. 25.2: (right) Tui De Roy/Minden Pictures, Inc. Fig. 25.3: (left) Science Photo Library/Photo Researchers, Inc. Fig. 25.3: (center) © Corbis. Fig. 25.3: (right) Science Photo Library/ Photo Researchers, Inc. Fig. 25.4: (top left) J. H. Robinson/Photo Researchers, Inc. Fig. 25.4: (top right) J. H. Robinson/Photo Researchers, Inc. Fig. 25.4: (center left) Alvin E. Staffan/Photo Researchers, Inc. Fig. 25.4: (center right) Gregory K. Scott/Photo Researchers, Inc. Fig. 25.4: (bottom left) John Kaprielian/Photo Researchers, Inc. Fig. 25.4: (bottom right) Millard Sharp/Photo Researchers, Inc. Fig. 25.5: Barbara J. Collins. Fig. 25.6: Data courtesy of Paul R. Cabe, University of Minnesota. From Gardner et al. *Principles of Genetics*, 8e., 1991, Fig. 22.1. Fig. 25.13*a*: Taxi/Getty Images. Fig. 25.13*b*: Mike Parry/Minden Pictures, Inc. Fig. 25.13*c*: Rich Reid/National Geographic Society. Fig. 25.18: Couretesy Kenneth Y. Kaneshiro, University of Hawaii. Fig. 25.19*a*: John A. L. Cooke/ Animals Animals/ Earth Scenes. Fig. 25.19*b*: Marie Read/Animals Animals/Earth Scenes. ***A Milestone in Genetics:*** Courtesy American Philosophical Society.

Illustration Credits

Chapter 3 Fig. 3.15: H. T. Lynch, R. Fusaro, and J. F. Lynch. 1997. Cancer genetics in the new era of molecular biology. *NY Acad. Sci.* 833:1.

Chapter 4 Fig. 4.10*b*: *Principles of Human Genetics*, 3/e by Curt Stern, © 1973 by W. H. Freeman and Company. Used with permission. Fig. 4.17: Nance, W. E., Jackson, C. E., and Witkop, C. J., Jr. 1970. *American Journal of Human Genetics* 22:579–586. Used with permission of the University of Chicago Press.

Chapter 8 Fig. 8.4*b*: After Pfashne *A Genetic Switch* 2e. Cell and BSP Press. Blackwell.

Chapter 9 Fig. 9.27: After Figure 1 in The ENCODE Project Consortium. *Science* 306:636–640, Oct. 22, 2004.

Chapter 10 Fig. 10.25: Adapted from *DNA Replication* by Kornberg and Baker © 1992 by W.H. Freeman and Company. Used with permission.

Chapter 11 Fig. 11.21: Reprinted with permission of *Nature* from Zang, N. J., Grabowski, P. J., and Cech, T. R. 1983. *Nature* 301: 578–583. Copyright 1983 Macmillan Magazines Limited.

Chapter 12 Figs. 12.3: Alberts, B., et al., *Molecular Biology of the Cell, 3/e*, page 114. New York: Garland Publishing Inc., 1994. Fig. 12.4: Figure from *Biology*, Second Edition by Claude A. Villee, Eldra Pearl Solomon, Charles E. Martin, Diana W. Martin, Linda R. Berg, and P. William Davis, copyright © 1989 by Saunders College Publishing, reproduced by permission of Harcourt Brace & Company. Fig. 12.10: Reprinted with permission from Holley, R. W., et al., 1965. *Science* 147: 1462–1465. Copyright 1965 American Association for the Advancement of Science.

Chapter 15 Fig. 15.1*b*: X-ray drawing by John Rosenberg, Univ. of Pittsburgh.

Chapter 16 Fig. 16.1: After Messing, J. W., and Llaca, V. 1998. *Proceedings of the National Academy of Sciences* 95:2017. Copyright 1998 National Academy of Sciences, U.S.A. Fig. 16.4: Based on data from NIH/CEPH Collaboration Mapping Group, 1992. *Science* 258: 67–86. Fig. 16.9: From Stewart, E. A., et al. 1997. *Genome Research* 7:422–433. Figs. 16.10 and 16.11: Venter et al. (2000). *Science* 291: 1304–1351. Fig. 16.12: Data from the International Human Genome Sequencing Consortium. (2001). *Nature* 409:860–921. Fig. 16.13: After Figure 1 in The ENCODE Project Consortium. *Science* 306:636–640, Oct. 22, 2004. Fig. 16.22: Data used with permission from Blattner, F. R., et al. 1997. *Science* 277:1453. Table 4. Copyright 1997 American Association for the Advancement of Science. Fig. 16.25: After Moore, et al. 1995. *Current Opinion Genetics & Development* 5:537 and Gale, M. D., and Devos, K. M. 1998. *Proceedings of the National Academy of Sciences* 95:1971–1973. Fig. 16.26: Chowdhary B. P., et al., 1998 *Genome Research* 8: 577–584. ***Focus On:*** Data from the GenBank website. http://www.ncbi.hihigov/Genbank/genbankstats.html.

Chapter 17 Fig. 17.1: From Huntington's Disease Collaborative Research Group. 1993. *Cell* 72:971–983. Copyright Cell Press. Fig. 17.3: From Marx, J. L. 1989. *Science* 245:923–925. Reprinted with permission of Dr. Lapchee Tsui, The Hospital for Sick Children, Toronto, Canada. Figs. 17.4, 17.5: Reprinted by permission from Collins, F. S., 1992. *Science* 256:774–779. Copyright 1992 American Association for Advancement of Science. Fig. 17.9: McCormack and Rabbitts. 2004. *New England Journal of Medicine.* 350:913–922 (Feb. 26, 2004 #9).

Chapter 18 Fig. 18.19: From Lim, J. K., and Simmons, M. J. 1994. *BioEssays* 16:269–275. © ICSN Press.

Chapter 19 Fig. 19.15: Lewin, B. 1994. *Genes V.* Copyright Cell Press. Fig. 19.17: Jinks-Robertson, S., and Nomura, M. 1982. *Escherichia coli and Salmonella typhimurium, Cellular and Molecular Biology* 2:1358–1385. ***A Milestone in Genetics:*** Fig. 1*b*: Data from Pardee, Jacob and Minod. 1959. *J Molecular Biology*. 1:165–178. ***Focus on*** Fig. 1*a*: Based on Gartenberg and Crothers. 1988. *Nature* 333:824–829. (Fig. 5).

Chapter 20 Fig. 20.4: *Molecular Cell Biology* by Darnell, Lodish and Baltimore. © 1996, 1990, 1985 by Scientific American Books. Used with permission by W. H. Freeman and Company.

Chapter 21 Fig. 21.20: Reprinted from *Trends in Genetics* 10, Duboule and Morata, pp. 358–364, copyright 1994, with permission from Elsevier Science.

Chapter 22 Fig. 22.12*a*: From Kinzler, K. W., and Vogelstein, B. 1996. *Cell* 87:159–170. Copyright Cell Press. Fig. 22.12*b*: Based on Latil, A., and Lidereau, R. 1998. *Virchow's Archive* 432:389–406. Fig. 22.12*c*: From Ng, H. K., and Lam, P. V. P. 1998 *Pathology* 30:196–202. Carfax Publishing, 11 New Feter Lane, London, EC4P 4EE.

Chapter 25 Fig. 25.22: Based on Wood, B. 1996. *BioEssays* 18:945–954. Copyright © 1996 John Wiley & Sons, Inc. Reprinted with permission of Wiley-Liss, Inc., a division of John Wiley & Sons, Inc.

Glossary

This glossary provides an introduction to some basic and recurring terms in the text. Names of chemical compounds, definitions of specialized terms, and variants of basic names have been omitted from the glossary but are given in the index. Please locate terms that are not in the glossary by referring to the index.

Abscissa. The horizontal scale on a graph.

Acentric chromosome. Chromosome fragment lacking a centromere.

Acridine dyes. A class of positively charged polycyclic molecules that intercalate into DNA and induce frameshift mutations.

Acrocentric. A modifying term for a chromosome or chromatid that has its centromere near the end.

Acrosome. An apical organelle in the head of the sperm.

Activator (of gene expression). Regulator gene products that turn on, or activate, the expression of other genes.

***Activator* (*Ac*).** A transposable element in maize that encodes a transacting transposase capable of catalyzing the movement of *Ac* elements and other members of the *Ac/Ds* family.

Adaptation. Adjustment of an organism or a population to an environment.

Adaptive mutation. A mutation that provides a selective advantage to the mutant organism when grown in the environment in which it originated.

ADA-SCID (adenosine deaminase-deficient severe combined immunodeficiency disease). An autosomal recessive disorder in humans caused by a lack of the enzyme adenosine deaminase, which catalyzes the breakdown of deoxyadenosine. In the absence of this enzyme, toxic derivatives of this nucleoside accumulate and kill cells required for normal immune responses to infections.

Additive allelic effects. Genetic factors that raise or lower the value of a phenotype on a linear scale of measurement.

Additive genetic variance. The portion of the total phenotypic variance in a quantitive trait that is due to the additive effects of alleles.

Adenine. A purine base found in RNA and DNA.

A-DNA. A right-handed DNA double helix that has 11 base pairs per turn. DNA exists in this form when partially dehydrated.

***Agrobacterium tumefaciens*-mediated transformation.** A naturally occurring process of DNA transfer from the bacterium *A. tumefaciens* to plants.

AIDS (acquired immunodeficiency syndrome). The usually fatal human disease in which the immune system is destroyed by the human immunodeficiency virus (HIV).

Albinism. Absence of pigment in skin, hair, and eyes of an animal. Absence of chlorophyll in plants.

Aleurone. The outermost layer of the endosperm in a seed.

Alkaptonuria. An inherited metabolic disorder. Alkaptonurics excrete excessive amounts of homogentisic acid (alkapton) in the urine.

Alkylating agents. Chemicals that transfer alkyl (methyl, ethyl, and so on) groups to the bases in DNA.

Allele (allelomorph; *adj.*, allelic, allelomorphic). One of a pair, or series, of alternative forms of a gene that occur at a given locus in a chromosome. Alleles are symbolized with the same basic symbol (for example, *D* for tall peas and *d* for dwarf). (See **Multiple alleles.**)

Allele frequency. The proportion of one allele relative to all alleles at a locus in a population.

Allopatric speciation. Speciation occurring at least in part because of geographic isolation.

Allopolyploid. A polyploid having chromosome sets from different species; a polyploid containing genetically different chromosome sets derived from two or more species.

Allosteric transition. A reversible interaction of a small molecule with a protein molecule that causes a change in the shape of the protein and a consequent alteration of the interaction of that protein with a third molecule.

Allotetraploid. An organism with four genomes derived from hybridization of different species. Usually, in forms that become established, two of the four genomes are from one species and two are from another species.

Allozyme. A variant of an enzyme detected by electrophoresis.

Amino acid. Any one of a class of organic compounds containing an amino (NH_2) group and a carboxyl (COOH) group. Amino acids are the building blocks of proteins. Alanine, proline, threonine, histidine, lysine, glutamine, phenylalanine, tryptophan, valine, arginine, tyrosine, and leucine are among the common amino acids.

Aminoacyl (*A*) site. The ribosome binding site that contains the incoming aminoacyl-tRNA.

Aminoacyl-tRNA synthetases. Enzymes that catalyze the formation of high energy bonds between amino acids and tRNA molecules.

Amniocentesis. A procedure for obtaining amniotic fluid from a pregnant woman. Chemical contents of the fluid are studied directly for the diagnosis of some diseases. Cells are cultured, and metaphase chromosomes are examined for irregularities (for example, trisomy).

Amnion. The thin membrane that lines the fluid-filled sac in which the embryo develops in higher vertebrates.

Amniotic fluid. Liquid contents of the amniotic sac of higher vertebrates containing cells of the embryo (not of the mother). Both fluid and cells are used for diagnosis of genetic abnormalities of the embryo or fetus.

Amorph. A mutation that obliterates gene function; a null mutation.

Amphidiploid. A species or type of plant derived from doubling the chromosomes in the F_1 hybrid of two species; an allopolyploid. In an amphidiploid the two species are known, whereas in other allopolyploids they may not be known.

Anabolic pathway. A pathway by which a metabolite is synthesized; a biosynthetic pathway.

Anaphase. The stage of mitosis or meiosis during which the daughter chromosomes pass from the equatorial plate to opposite poles of the cell (toward the ends of the spindle). Anaphase follows metaphase and precedes telophase.

Anchor gene. A gene that has been positioned on both the physical map and the genetic map of a chromosome.

Androgen. A male hormone that controls sexual activity in vertebrate animals.

Anemia. Abnormal condition characterized by pallor, weakness, and breathlessness, resulting from a deficiency of hemoglobin or a reduced number of red blood cells.

Aneuploid. An organism or cell having a chromosome number that is not an exact multiple of the monoploid (n) with one genome, that is, hyperploid, higher (for example, $2n + 1$), or hypoploid, lower (for example, $2n - 1$). Also applied to cases where part of a chromosome is duplicated or deficient.

Anther. The organ in flowers that produces pollen.

Antibody. Substance in a tissue or fluid of the body that acts in antagonism to a foreign substance (antigen).

Antibody-mediated (humoral) immune response. The synthesis of antibodies by plasma cells in response to an encounter of the cells of the immune system with a foreign immunogen.

Anticodon. Three bases in a transfer RNA molecule that are complementary to the three bases of a specific codon in messenger RNA.

Antigen. A substance, usually a protein, that is bound by an antibody or a T-cell receptor when introduced into a vertebrate organism (cf. **Immunogen**).

Antisense gene. A gene that produces a transcript that is complementary to pre-mRNA or mRNA of a normal gene (usually constructed by inverting the coding region relative to promoter).

Antisense RNA. RNA that is complementary to the pre-mRNA or mRNA produced from a gene.

Apomixis. An asexual method of reproduction involving the production of unreduced (usually diploid) eggs, which then develop without fertilization.

Apoptosis. A phenomenon in which eukaryotic cells die because of genetically programmed events within those cells.

Artificial selection. The practice of choosing individuals from a population for reproduction, usually because these individuals possess one or more desirable traits.

Ascospore. One of the spores contained in the ascus of certain fungi such as *Neurospora*.

Ascus (*pl.*, **asci**). Reproductive sac in the sexual stage of a type of fungi (Ascomycetes) in which ascospores are produced.

Asexual reproduction. Any process of reproduction that does not involve the formation and union of gametes from the different sexes or mating types.

Assortative mating. Mating in which the partners are chosen because they are phenotypically similar.

Asynapsis. The failure or partial failure in the pairing of homologous chromosomes during the meiotic prophase.

ATP. Adenosine triphosphate: an energy-rich compound that promotes certain activities in the cell.

Attenuation. A mechanism for controlling gene expression in prokaryotes that involves premature termination of transcription.

Attenuator. A nucleotide sequence in the 5′ region of a prokaryotic gene (or in its RNA) that causes premature termination of transcription, possibly by forming a secondary structure.

Autocatalytic reaction. A reaction catalyzed by a substrate without the involvement of any other catalytic agent.

Autoimmune diseases. Disorders in which the immune systems of affected individuals produce antibodies against self antigens—antigens synthesized in their own cells.

Autonomous. A term applied to any biological unit that can function on its own, that is, without the help of another unit. For example, a transposable element that encodes an enzyme for its own transposition (cf. **Nonautonomous**).

Autopolyploid. A polyploid that has multiple and identical or nearly identical sets of chromosomes (genomes). A polyploid species with genomes derived from the same original species.

Autoradiograph. A record or photograph prepared by labeling a substance such as DNA with a radioactive material such as tritiated thymidine and allowing the image produced by radioactive decay to develop on a film over a period of time.

Autosome. Any chromosome that is not a sex chromosome.

Auxotroph. A mutant microorganism (for example, bacterium or yeast) that will not grow on a minimal medium but that requires the addition of some compound such as an amino acid or a vitamin.

Backcross. The cross of an F_1 hybrid to one of the parental types. The offspring of such a cross are referred to as the backcross generation or backcross progeny. (See **Testcross.**)

Back mutation. A second mutation at the same site in a gene as the original mutation, which restores the wild-type nucleotide sequence.

BACs (bacterial artificial chromosomes). Cloning vectors constructed from bacterial fertility (F) factors; like YAC vectors, they accept large inserts of size 200 to 500 kb.

Bacteriophage. A virus that attacks bacteria. Such viruses are called bacteriophages because they destroy their bacterial hosts.

Balanced lethal. Lethal mutations in different genes on the same pair of chromosomes that remain in repulsion because of close linkage or crossover suppression. In a closed population, only the trans-heterozygotes (l_1 + / + l_2) for the lethal mutations survive.

Balanced polymorphism. Two or more types of individuals maintained in the same breeding population by a selection mechanism.

Balancer chromosome. In *Drosophila* genetics, a dominantly marked, multiply inverted chromosome that suppresses recombination with a homologous chromosome that is structurally normal.

Barr body. A condensed mass of chromatin found in the nuclei of placental mammals that contains one or more X chromosomes; named for its discoverer, Murray Barr.

Basal body. Small granule to which a cilium or flagellum is attached.

Basal transcription factors. Proteins required for the initiation of transcription in eukaryotes.

Base analogs. Unnatural purine or pyrimidine bases that differ slightly from the normal bases and that can be incorporated into nucleic acids. They are often mutagenic.

Base excision repair. The removal of abnormal or chemically modified bases from DNA.

Base substitution. A single base change in a DNA molecule. (See **Transition; Transversion.**)

B-DNA. Double-stranded DNA that exists as a right-handed helix with 10.4 base pairs per turn; the conformation of DNA when present in aqueous solutions containing low salt concentrations.

Binomial coefficient. The term that gives the number of ways of obtaining the two possible outcomes in an experiment in which only two outcomes are possible.

Binomial expansion. Exponential multiplication of an expression consisting of two terms connected by a plus (+) or minus (−) sign, such as $(a + b)^n$.

Binomial probability. The frequency associated with the occurrence of an outcome in an experiment which has only two possible outcomes, such as head or tail in coin tossing.

Bioinformatics. The study of genetic and other biological information using computer and statistical techniques.

Biometry. Application of statistical methods to the study of biological problems.

Bivalent. A pair of synapsed or associated homologous chromosomes that have undergone the duplication process to form a group of four chromatids.

Blastomere. Any one of the cells formed from the first few cleavages in animal development.

Blastula. In animals, an early embryo form that follows the morula stage; typically, a single-layered sheet or ball of cells.

B lymphocytes (B cells). An important class of cells that mature in bone marrow and are largely responsible for the antibody-mediated or humoral immune response; they give rise to the antibody-producing plasma cells and some other cells of the immune system.

Breeding value. In quantitative genetics, the part of the deviation of an individual phenotype from the population mean that is due to the additive effects of alleles.

Broad-sense heritability. In quantitative genetics, the proportion of the total phenotypic variance that is due to genetic factors.

CAAT box. A conserved nucleotide sequence in eukaryotic promoters involved in the initiation of transcription.

5′ CAP (mRNA). The 7-methy guanosine cap that is added to most eukaryotic mRNAs posttranscriptionally.

Carbohydrate. A molecule consisting of carbon, hydrogen and oxygen in the proportions 1:2:1; a molecule of sugar or a macromolecule composed of sugar subunits.

Carcinogen. An agent capable of inducing cancer in an organism.

Carrier. An individual who carries a recessive allele that is not expressed (that is, is obscured by a dominant allele).

Catabolic pathway. A pathway by which an organic molecule is degraded in order to obtain energy for growth and other cellular processes; degradative pathway.

Catabolite activator protein (CAP). A positive regulatory protein that in the presence of cyclic AMP (cAMP) binds to the promoter regions of operons and stimulates their transcription. CAP/cAMP assures that glucose is used as a carbon source when present rather than less efficient energy sources such as lactose, arabinose, and other sugars. When glucose is present, it prevents the synthesis of cAMP and thus the activation of transcription by CAP/cAMP.

Catabolite repression. Glucose-mediated reduction in the rates of transcription of operons that specify enzymes involved in catabolic pathways (such as the *lac* operon).

cDNA (complementary DNA). A DNA molecule synthesized *in vitro* from an RNA template.

cDNA library. A collection of cDNA clones containing copies of the RNAs isolated from an organism or a specific tissue or cell type of an organism.

Cell cycle. The cyclical events that occur during the divisions of mitotic cells. The cell cycle oscillates between mitosis and the interphase, which is divided into G_1, S, and G_2.

Cellular immune response. See **T cell-mediated immune response.**

CentiMorgan. See **Crossover unit.**

Centriole. An organelle in many animal cells that appears to be involved in the formation of the spindle during mitosis.

Centromere. Spindle-fiber attachment region of a chromosome.

Centrosome. A barrel-shaped organelle associated with the mitotic spindle in animal cells.

Chain-termination codon. A codon that specifies polypeptide chain termination rather than the incorporation of an amino acid. There are three (UAA, UAG, and UGA), and they are recognized by protein release factors rather than tRNAs.

Chaperone. A protein that helps nascent polypeptides fold into their proper three-dimensional structures.

Character (*contraction of the word* characteristic). One of the many details of structure, form, substance, or function that make up an individual organism.

Checkpoint. A mechanism that halts progression through the eukaryotic cell cycle.

Chemotaxis. Attraction or repulsion of organisms by a diffusing substance.

Chiasma (*pl.*, Chiasmata). A visible change of partners in two of a group of four chromatids during the first meiotic prophase. In the diplotene stage of meiosis, the four chromatids of a bivalent are associated in pairs, but in such a way that one part of two chromatids is exchanged. This point of "change of partner" is the chiasma.

Chimera (animal). Individual derived from two embryos by experimental intervention.

Chimera (plant). Part of a plant with a genetically different constitution as compared with other parts of the same plant. It may result from different zygotes that grow together or from artificial fusion (grafting); it may either be pernical, with parallel layers of genetically different tissues, or sectorial.

Chimeric selectable marker gene. A gene constructed using DNA sequences from two or more sources that allows a cell or organism to survive under conditions where it would otherwise die.

Chi-square. A statistic used to test the goodness of fit of data to the predictions of an hypothesis.

Chloroplast. A green organelle in the cytoplasm of plants that contains chlorophyll and in which starch is synthesized. A mode of cytoplasmic inheritance, independent of nuclear genes, has been associated with these cytoplasmic organelles.

Chorionic biopsy. A procedure in which cells are taken from an embryo for the purpose of genetic testing.

Chromatid. In mitosis or meiosis, one of the two identical strands resulting from self-duplication of a chromosome.

Chromatin. The complex of DNA and proteins in eukaryotic chromosomes; originally named because of the readiness with which it stains with certain dyes.

Chromatin fibers. A basic organizational unit of eukaryotic chromosomes that consists of DNA and associated proteins assembled into strands of average diameter 30 nm.

Chromatin remodeling. The alteration of the structure of DNA and its associated protein molecules, especially histones, by a protein complex; this remodeling often involves the chemical modification of the histones.

Chromatography. A method for separating and identifying the components from mixtures of molecules having similar chemical and physical properties.

Chromocenter. Body produced by fusion of the heterochromatic regions of the chromosomes in the polytene tissues (for example, the salivary glands) of certain *Diptera*.

Chromomeres. Small bodies that are identified by their characteristic size and linear arrangement along a chromosome.

Chromonema (*pl.*, chromonemata). An optically single thread forming an axial structure within each chromosome.

Chromosome aberration. Abnormal structure or number of chromosomes; includes deficiency, duplication, inversion, translocation, aneuploidy, polyploidy, or any other change from the normal pattern.

Chromosome banding. Staining of chromosomes in such a way that light and dark areas occur along the length of the chromosomes. Lateral comparisons identify pairs. Each human chromosome can be identified by its banding pattern.

Chromosome jumping. A procedure that uses large DNA fragments to move discontinuously along a chromosome from one site to another site. (See **Positional cloning.**)

Chromosome painting. The study of the organization and evolution of chromosomes by *in situ* hybridization using DNA probes labeled with fluorescent dyes that emit light at different wavelengths.

Chromosomes. Darkly staining nucleoprotein bodies that are observed in cells during division. Each chromosome carries a linear array of genes.

Chromosome theory of heredity. The theory that chromosomes carry the genetic information and that their behavior during meiosis provides the physical basis for the segregation and independent assortment of genes.

Chromosome walking. A procedure that uses overlapping clones to move sequentially down a chromosome from one site to another site. (See **Positional cloning.**)

Cilium (*pl.*, cilia; *adj.*, ciliate). Hairlike locomotor structure on certain cells; a locomotor structure on a ciliate protozoan.

Circularly permuted chromosomes. Linear chromosomes with nucleotide sequences that are circular permutations of one another. Cleaving a circular molecule of DNA at random points around the circle will generate a population of circularly permuted linear molecules.

***cis*-acting sequence.** A nucleotide sequence that only affects the expression of genes located on the same chromosome, that is, *cis* to itself.

***cis* configuration.** See **Coupling.**

***cis* heterozygote.** A heterozygote that contains two mutations arranged in the *cis* configuration—for example, a^+b^+ / $a\ b$.

***cis-trans* position effect.** The occurrence of different phenotypes when two mutations are present in *cis*- and *trans*-heterozygotes.

***cis-trans* test.** The construction and analysis of *cis* and *trans* heterozygotes of pairs of mutations to determine whether the mutations are in the same gene or in two different genes. For the test to be informative, the *cis* heterozygote must have the wild-type phenotype. If this condition is met, the two mutations are in the same gene if the *trans* heterozygote has the mutant phenotype and are in two different genes if the *trans* heterozygote has the mutant phenotype.

Cistron. See **Gene.**

***ClB* chromosome.** An X chromosome in *Drosophila* that carries a mutation causing bar-shaped eyes and a recessive lethal mutation within a large inversion.

***ClB* method.** The use of a special X chromosome in *Drosophila* that carries a mutation causing bar-shaped eyes and a recessive lethal mutation within a long inversion to detect new recessive X-linked lethal mutations. H. J. Muller used this chromosome to demonstrate that X rays are mutagenic. See **ClB chromosome.**

Clone. All the individuals derived by vegetative propagation from a single original individual. In molecular biology, a population of identical DNA molecules all carrying a particular DNA sequence from another organism.

Cloning (gene). The production of many copies of a gene or specific DNA sequence.

Cloning vector. A small, self-replicating DNA molecule—usually a plasmid or viral chromosome—into which foreign DNAs are inserted in the process of cloning genes or other DNA sequences of interest.

Codominant alleles. Alleles that produce independent effects when heterozygous.

Codon. A set of three adjacent nucleotides in an mRNA molecule that specifies the incorporation of an amino acid into a polypeptide chain or that signals the end of polypeptide synthesis. Codons with the latter function are called termination codons.

Coefficient. A number expressing the amount of some change or effect under certain conditions (for example, the coefficient of inbreeding).

Coefficient of relationship. The fraction of genes two individuals share by virtue of common ancestry.

Coenzyme. A substance necessary for the activity of an enzyme.

Coincidence. The ratio of the observed frequency of double crossovers to the expected frequency, where the expected frequency is calculated by assuming that the two crossover events occur independently of each other.

Cointegrate. A DNA molecule formed by the fusion of two different DNA molecules, usually mediated by a transposable element.

Colchicine. An alkaloid derived from the autumn crocus that is used as an agent to arrest spindle formation and interrupt mitosis.

Colinearity. A relationship in which the units in one molecule occur in the same sequence as the units in another molecule which they specify; for example, the nucleotides in a gene are colinear with the amino acids in the polypeptide encoded by that gene.

Colony. A compact collection of cells produced by the division of a single progenitor cell.

Comparative genomics. The branch of genomics that compares the structure and function of the genomes of different species.

Competence. Ability of a bacterial cell to incorporate DNA and become genetically transformed.

Complementarity. The relationship between the two strands of a double helix of DNA. Thymine in one strand pairs with adenine in the other strand, and cytosine in one strand pairs with guanine in the other strand.

Complementation test (*trans* test). Introduction of two recessive mutations into the same cell to determine whether they are alleles of the same gene, that is, whether they affect the same genetic function. If the mutations are allelic, the genotype $m_1\ +/+\ m_2$ will exhibit a mutant phenotype, whereas if they are nonallelic, it will exhibit the wild phenotype.

Composite transposon. A transposable element formed when two identical or nearly identical transposons insert on either side of a nontransposable segment of DNA—for example, the bacterial transposon Tn*5*.

Compound chromosome. A chromosome formed by the union of two separate chromosomes from the same pair, as in attached-X chromosomes or attached X-Y chromosomes.

Concatamer. A long DNA molecule that contains many copies of a viral chromosome or other DNA molecule joined end-to-end.

Concordance rate. Among pairs of items identified because one member of the pair has a particular trait, the frequency with which the other member of the pair has the same trait.

Conditional lethal mutation. A mutation that is lethal under one set of environmental conditions—the restrictive conditions—but is viable under another set of environmental conditions—the permissive conditions.

Conidium (*pl.*, conidia). An asexual spore produced by a specialized hypha in certain fungi.

Conjugation. Union of sex cells (gametes) or unicellular organisms during fertilization; in *Escherichia coli*, a one-way transfer of genetic material from a donor ("male" cell) to a recipient ("female" cell).

Consanguineous mating. A mating between relatives.

Consanguinity. Relationship due to descent from a common ancestor.

Consensus sequence. The nucleotide sequence that is present in the majority of genetic signals or elements that perform a specific function.

Constitutive enzyme. An enzyme that is synthesized continually regardless of growth conditions (cf. **Inducible enzyme** and **Repressible enzyme**).

Constitutive gene. A gene that is continually expressed in all cells of an organism.

Contig. A set of overlapping clones that provide a physical map of a portion of a chromosome.

Continuous variation. Variation not represented by distinct classes. Individuals grade into each other, and measurement data are required for analysis (cf. **Discontinuous variation**). Multiple genes are usually responsible for this type of variation.

Controlling element. In maize, a transposable element such as *Ac* or *Ds* that is capable of influencing the expression of a nearby gene.

Coordinate repression. Correlated regulation of the structural genes in an operon by a molecule that interacts with the operator sequence.

Copolymers. Mixtures consisting of more than one monomer; for example, polymers of two kinds of organic bases such as uracil and cytosine (poly-UC) have been combined for studies of the genetic code.

Co-repressor. An effector molecule that forms a complex with a repressor and turns off the expression of a gene or set of genes.

Correlation. A statistical association between variables.

Cosmids. Cloning vectors that are hybrids between phage λ chromosomes and plasmids; they contain λ ***cos*** sites and plasmid origins of replication.

Coupling (*cis* configuration). The condition in which a double heterozygote has received two linked mutations from one parent and their wild-type alleles from the other parent (for example, *a b*/*a b* × + +/+ + produces *a b*/+ + (cf. **Repulsion**).

Covalent bond. A bond in which an electron pair is equally shared by protons in two adjacent atoms.

Covariance. A measure of the statistical association between variables.

cpDNA. The DNA of plant plastids, including chloroplasts.

CpG islands. Clusters of cytosines and guanines that often occur upstream of human genes.

Cri-du-chat syndrome. A condition produced when a small region in the short arm of one human chromosome 5 is deleted.

Critical value. The threshold value of a statistic that marks off a fraction of the statistic's frequency distribution. A sample statistic greater than this critical value warrants rejection of the hypothesis being tested.

Crossbreeding. Mating between members of different races or species.

Crossing over. A process in which chromosomes exchange material through the breakage and reunion of their DNA molecules. (See **Recombination.**)

Crossover unit. A measure of distance on genetic maps that is based on the average number of crossing-over events that take place during meiosis. A map interval that is one crossover unit in length (sometimes called a centiMorgan) implies that only one in every hundred chromatids recovered from meiosis will have undergone a crossing-over event in this interval.

Cut-and-paste transposon. A transposable element that is excised from one position in the genome and inserted into another position through the action of a transposon-encoded enzyme called the transposase.

Cyclic AMP. Adenosine-3′, 5′-monophosphate, a small molecule that must be bound by the catabolite activator protein (CAP) in order for the complex (CAP/cAMP) to bind to the promoters of operons and stimulate transcription.

Cystic fibrosis. An autosomal recessive disorder in humans characterized by clogging of the lungs, pancreas, and liver with mucus and, as a result, chronic infections. The average life expectancy of an individual with cystic fibrosis is about 35 years.

Cytogenetics. Area of biology concerned with chromosomes and their implications in genetics.

Cytokinesis. Cytoplasmic division and other changes exclusive of nuclear division that are a part of mitosis or meiosis.

Cytological map. A diagram of a chromosome based on differential staining—the "banding pattern"—along its length.

Cytology. The study of the structure and function of cells.

Cytoplasm. The protoplasm of a cell outside the nucleus in which cell organelles (mitochondria, plastids, and the like) reside; all living parts of the cell except the nucleus.

Cytoplasmic inheritance. Hereditary transmission dependent on the cytoplasm or structures in the cytoplasm rather than the nuclear genes; extrachromosomal inheritance. Example: Plastid characteristics in plants may be inherited by a mechanism independent of nuclear genes.

Cytosine. A pyrimidine base found in RNA and DNA.

Cytoskeleton. A complex system of fibers and filaments that provides support for cells and that is involved in moving the components of cells throughout the cytoplasm.

Cytotype. A maternally inherited cellular condition in *Drosophila* that regulates the activity of transposable P elements.

Dalton. The mass of a hydrogen atom.

Daughter cell. A product of cell division.

Deficiency (deletion). Absence of a segment of a chromosome, reducing the number of loci.

Degeneracy (of the genetic code). The specification of an amino acid by more than one codon.

Degrees of freedom. An index associated with the frequency distribution of a test statistic calculated from sample data.

Deletion mapping. A short-cut mapping procedure using a set of overlapping deletions to map point mutations in a gene or on a chromosome.

Deme. A local population of organisms.

Denaturation. Loss of native configuration of a macromolecule, usually accompanied by loss of biological activity. Denatured proteins often unfold their polypeptide chains and express changed properties of solubility.

de novo. Arising anew, afresh, once more.

Deoxyribonuclease (DNase). Any enzyme that hydrolyzes DNA.

Deoxyribonucleic acid. See **DNA.**

Derepression. The process of turning on the expression of a gene or set of genes whose expression has been repressed (turned off).

Determination. Process by which undifferentiated cells in an embryo become committed to develop into specific cell types, such as neuron, fibroblast, and muscle cell.

Deviation. As used in statistics, a departure from an expected value.

Diakinesis. A stage of meiosis just before metaphase I in which the bivalents are shortened and thickened.

Dicentric chromosome. One chromosome having two centromeres.

Dicot. A plant with two cotyledons, or seed leaves.

2′, 3′-Dideoxyribonucleoside triphosphates (ddNTPs). Chain-terminating DNA precursors (nucleoside triphosphates) with a hydrogen (H) linked to the 3′ carbon in place of the hydroxyl (OH) group in normal DNA precursors (2′-deoxyribonucleotide triphosphates); ddNTPs are used in DNA sequencing reactions.

Differentiation. A process in which unspecialized cells develop characteristic structures and functions.

Dihybrid. An individual that is heterozygous for two pairs of alleles; the progeny of a cross between homozygous parents differing in two respects.

Dimer. A compound having the same percentage composition as another but twice the molecular weight; one formed by polymerization.

Dimorphism. Two different forms in a group as determined by such characteristics as sex, size, or coloration.

Diploid. An organism or cell with two sets of chromosomes ($2n$) or two genomes. Somatic tissues of higher plants and animals are ordinarily diploid in chromosome constitution in contrast with the haploid (monoploid) gametes.

Diplonema (*adj.*, **diplotene**). That stage in prophase of meiosis I following the pachytene stage, but preceding diakinesis, in which the chromosomes of bivalents separate from each other at and around their centromeres.

Discontinuous variation. Phenotypic variability involving distinct classes such as red versus white, tall versus dwarf (cf. **Continuous variation**).

Discordant. Members of a pair showing different, rather than similar, characteristics.

Disjunction. Separation of homologous chromosomes during anaphase of mitotic or meiotic divisions. (See **Nondisjunction.**)

Disome. See **Monosomic.**

***Dissociation* (*Ds*).** A transposable element in maize, originally detected as an agent that mediates chromosome breakage in response to the effect of *Activator (Ac)*, another transposable element.

Ditype. In fungi, a tetrad that contains two kinds of meiotic products (spores) (for example, 2 *AB* and 2 *ab*).

Dizygotic twins. Two-egg or fraternal twins.

DNA. Deoxyribonucleic acid; the information-carrying genetic material that comprises the genes. DNA is a macromolecule composed of a long chain of deoxyribonucleotides joined by phosphodiester linkages. Each deoxyribonucleotide contains a phosphate group, the five-carbon sugar 2-deoxyribose, and a nitrogen-containing base.

DNA fingerprint. The banding pattern on a Southern blot of an individual's genomic DNA that has been cleaved with a restriction enzyme(s) and hybridized to an appropriate nucleic acid probe(s).

DNA gyrase. An enzyme in bacteria that catalyzes the formation of negative supercoils in DNA.

DNA helicase. An enzyme that catalyzes the unwinding of the complementary strands of a DNA double helix.

DNA ligase. An enzyme that catalyzes covalent closure of nicks in DNA double helices.

DNA polymerase. An enzyme that catalyzes the synthesis of DNA.

DNA primase. An enzyme that catalyzes the synthesis of short strands of RNA that initiate the synthesis of DNA strands.

DNA repair enzymes. Enzymes that catalyze the repair of damaged DNA.

DNA topoisomerase. An enzyme that catalyzes the introduction or removal of supercoils from DNA.

Dominance. A condition in which one member of an allele pair is manifested to the exclusion of the other.

Dominant selectable marker gene. A gene that allows the host cell to survive under conditions where it would otherwise die.

Donor cell. A bacterium that donates DNA to another (recipient) cell during recombination in bacteria (cf. **Recipient cell**).

Dosage compensation. A phenomenon in which the activity of a gene is increased or decreased according to the number of copies of that gene in the cell.

Dot blot hydridization. Hybridization of a labeled DNA or RNA probe to multiple DNA sequences fixed to a membrane in a specific pattern.

Double helix. A DNA molecule composed of two complementary strands.

Downstream sequence. A sequence in a unit of transcription that follows (is located 3′ to) the transcription start site. The nucleotide pair in DNA corresponding to the nucleotide at the 5′ end of the transcript (RNA) is designated +1. The following nucleotide pair is designated +2. All of the following (+) nucleotide sequences are downstream sequences (cf. **Upstream sequence**).

Drift. See **Random genetic drift.**

Duplication. The occurrence of a segment more than once in the same chromosome or genome; also, the multiplication of cells.

Ecdysone. A hormone that influences development in insects.

Eclosion. Emergence of an adult insect from the pupal stage.

Ecotype. A population or strain of organisms that is adapted to a particular habitat.

Ectopic. A term used to describe a phenomenon that occurs in an abnormal place.

Effector molecule. A molecule that influences the behavior of a regulatory molecule, such as a repressor protein, thereby influencing gene expression.

Egg (ovum). A germ cell produced by a female organism.

Electrophoresis. The migration of suspended particles in an electric field.

Electroporation. A process whereby cell membranes are made permeable to DNA by applying an intense electric current.

Elongation (of DNA, RNA, or protein synthesis). The incorporation of the second and subsequent subunits (nucleotides or amino acids) during the synthesis of a macromolecule (DNA, RNA, or polypeptide).

Elongation factors. Soluble proteins that are required for polypeptide chain elongation.

Embryo. An organism in the early stages of development; in humans, the first two months in the uterus.

Embryonic stem cells. Cells present in embryos that can differentiate into many different types of tissues and/or organs.

Embryo sac. A large thin-walled space within the ovule of the seed plant in which the egg and, after fertilization, the embryo develop; the mature female gametophyte in higher plants.

Endogenote. The part of the bacterial chromosome that is homologous to a genome fragment (exogenote) transferred from the donor to the recipient cell in the formation of a merozygote.

Endomitosis. Duplication of chromosomes without division of the nucleus, resulting in increased chromosome number within a cell. Chromosome strands separate, but the cell does not divide.

Endonuclease. An enzyme that breaks strands of DNA at internal positions; some are involved in recombination of DNA.

Endoplasmic reticulum. Network of membranes in the cytoplasm to which ribosomes adhere.

Endopolyploidy. A state in which the cells of a diploid organism contain multiples of the diploid chromosome number (that is, $4n$, $8n$, and so on).

Endosperm. Nutritive tissue that develops in the embryo sac of most angiosperms. It usually forms after the fertilization of the two fused primary endosperm nuclei of the embryo sac with one of the two male gamete nuclei. In most diploid plants, the endosperm is triploid ($3n$).

Endosymbiosis. A mutually beneficial relationship in which one organism lives inside another organism.

End-product inhibition. See **Feedback inhibition.**

Enhancer. A substance or object that increases a chemical activity or a physiological process; a major or modifier gene that increases a physiological process; a DNA sequence that influences transcription of a nearby gene.

Environment. The aggregate of all the external conditions and influences affecting the life and development of an organism.

Environmentality. The proportion of the total phenotypic variance in a quantitative trait that is due to the effects of a shared environment.

Enzyme. A protein that accelerates a specific chemical reaction in a living system.

Epigenetic. A term referring to the nongenetic causes of a phenotype.

Episome. A genetic element that may be present or absent in different cells and that may be inserted in a chromosome or independent in the cytoplasm (for example, the fertility factor (F) in *Escherichia coli*).

Epistasis. Interactions between products of nonallelic genes. Genes suppressed are said to be hypostatic. Dominance is associated with members of allelic pairs, whereas epistasis is interaction among products of nonalleles.

Equational division. Mitotic-type division that is usually the second division in the meiotic sequence; somatic mitosis and the nonreductional division of meiosis.

Equatorial plate. The figure formed by the chromosomes in the center (equatorial plane) of the spindle in mitosis.

Equilibrium. A state of dynamical systems in which there is no net change.

Equilibrium density gradient centrifugation. A procedure used to separate macromolecules based on their density (mass per unit volume).

Estrogen. Female hormone or estrus-producing compound.

ESTs (expressed sequence tags). Short cDNA sequences that are used to link physical maps and genetic (RFLP) maps.

Euchromatin. Genetic material that is not stained so intensely by certain dyes during interphase and that comprises many different kinds of genes (cf. **Heterochromatin**).

Eugenics. The application of the principles of genetics to the improvement of humankind.

Eukaryote. A member of the large group of organisms that have nuclei enclosed by a membrane within their cells (cf. **Prokaryote**).

Euploid. An organism or cell having a chromosome number that is an exact multiple of the monoploid (n) or haploid number. Terms used to identify different levels in an euploid series are diploid, triploid, tetraploid, and so on (cf. **Aneuploid**).

Excinuclease. The endonuclease-containing protein complex that excises a segment of damaged DNA during excision repair.

Excision repair. DNA repair processes that involve the removal of the damaged segment of DNA and its replacement by the synthesis of a new strand using the complementary strand of DNA as template.

Exit (*E*) site. The ribosome binding site that contains the free tRNA prior to its release.

Exogenote. Chromosomal fragment homologous to an endogenote and donated to a merozygote.

Exon amplification. A procedure that is used to identify coding regions (exons) that are flanked by 5′ and 3′ intron splice sites.

Exons. The segments of a eukaryotic gene that correspond to the sequences in the final processed RNA transcript of that gene.

Exonuclease. An enzyme that digests DNA or RNA, beginning at the ends of strands.

Expressivity. Degree of expression of a trait controlled by a gene. A particular gene may produce different degrees of expression in different individuals.

Extrachromosomal. Structures that are not part of the chromosomes; DNA units in the cytoplasm that control cytoplasmic inheritance.

F_1. The first filial generation; the first generation of descent from a given mating.

F_2. The second filial generation produced by crossing *inter se* or by self-pollinating the F_1. The inbred "grandchildren" of a given mating, but in controlled genetic experimentation, self-fertilization of the F_1 (or equivalent) is implied.

F^+ cell. A bacterium that contains an autonomous fertility (F) factor. (See **F factor.**)

F factor. A bacterial episome that confers the ability to function as a genetic donor ("male") in conjugation; the fertility factor in bacteria.

Feedback inhibition. The accumulated end product of a biochemical pathway stops synthesis of that product. A late metabolite of a synthetic pathway regulates synthesis at an earlier step of the pathway.

Fertilization. The fusion of a male gamete (sperm) with a female gamete (egg) to form a zygote.

Fetus. Prenatal stage of a viviparous animal between the embryonic stage and the time of birth; in humans, the final seven months before birth.

Filial. See F_1 and F_2.

Fission. A mode of cell division among the prokaryotes in which the genetic material of the mother cell is first duplicated and then apportioned equally to the two daughter cells.

Fitness. The number of offspring left by an individual, often compared with the average of the population or with some other standard, such as the number left by a particular genotype.

Fixation. An event that occurs when all the alleles at a locus except one are eliminated from a population. The remaining allele, with frequency 100 percent, is said to have been fixed.

Flagellum (*pl.*, **flagella;** *adj.* **flagellate**). A whiplike organelle of locomotion in certain cells; locomotor structures in flagellate protozoa.

Fluorescence *in situ* hybridization (FISH). *In situ* hybridization performed using a DNA or RNA probe coupled to a fluorescent dye.

Folded genome. The condensed intracellular state of the DNA in the nucleoid of a bacterium. The DNA is segregated into domains, and each domain is independently negatively supercoiled.

Forced cloning. The insertion of a foreign DNA into a cloning vector in a predetermined orientation.

Founder principle. The possibility that a new, small, isolated population may diverge genetically because the founding individuals are a random sample from a large, main population.

Frameshift mutation. A mutation that changes the reading frame of an mRNA, either by inserting or deleting nucleotides.

Frequency distribution. A graph showing either the relative or absolute incidence of classes in a population. The classes may be defined by either a discrete or a continuous variable; in the latter case, each class represents a different interval on the scale of measurement.

Fusion protein. A polypeptide made from a recombinant gene that contains portions of two or more different genes. The different genes are joined so that their coding sequences are in the same reading frame.

Gain-of-function mutation. A mutation that endows a gene product with a new function.

Gall. A tumorous growth in plants.

Gamete. A mature male or female reproductive cell (sperm or egg).

Gametogenesis. The formation of gametes.

Gametophyte. That phase of the plant life cycle that bears the gametes; the cells have *n* chromosomes.

Gametophytic incompatibility. A botanical phenomenon controlled by the complex *S* locus in which a pollen grain cannot fertilize an ovule produced by a plant that carries the same *S* allele as the pollen grain. For example, S_1 pollen cannot fertilize an ovule made by an S_1/S_2 plant.

Gap gene. A gene that controls the formation of adjacent segments in the body of *Drosophila*.

Gastrula. An early animal embryo consisting of two layers of cells; an embryological stage following the blastula.

GenBank. The DNA sequence databank maintained by the National Center for Biotechnology Information at the National Institutes of Health in the United States. Similar databanks are maintained in Europe (the European Molecular Biology Laboratory Data Library) and Japan (the DNA DataBank of Japan).

Gene. A hereditary determinant of a specific biological function; a unit of inheritance (DNA) located in a fixed position on a chromosome; a segment of DNA encoding one polypeptide and defined operationally by the *cis-trans* or complementation test.

Gene addition. The addition of a functional copy of a gene to the genome of an organism.

Gene amplification. A phenomenon whereby the DNA of a specific gene or set of genes is replicated independently of the rest of the genome to increase the number of gene copies.

Gene chip. A small silicon wafer or other solid support containing a large number of oligonucleotide or cDNA hybridization probes arranged on its surface in a specific pattern, or microarray.

Gene cloning. The incorporation of a gene of interest into a self-replicating DNA molecule and the amplification of the resulting recombinant DNA molecule in an appropriate host cell.

Gene conversion. A process, often associated with recombination, during which one allele is replicated at the expense of another, leading to non-Mendelian segregation ratios. In whole tetrads, for example, the ratio may be 6:2 or 5:3 instead of the expected 4:4.

Gene expression. The process by which genes produce RNAs and proteins and exert their effects on the phenotype of an organism.

Gene flow. The spread of genes from one breeding population to another by migration, possibly leading to allele frequency changes.

Gene pool. The sum total of all different alleles in the breeding members of a population at a given time.

Generalized transduction. Recombination in bacteria mediated by a bacteriophage that can transfer any bacterial gene of the donor cell to a recipient cell (cf. **Specialized transduction**).

Gene replacement. The incorporation of a transgene into a chromosome at its normal location by homologous recombination, thus replacing the copy of the gene originally present at the locus.

Genes-within-genes. Genes that are located entirely within other genes, but with their reading frames offset by one or two nucleotides.

Gene therapy. The treatment of inherited diseases by introducing wild-type copies of the defective gene causing the disorder into the cells of affected individuals. If reproductive cells are modified, the procedure is called *germ-line* or *heritable gene therapy*. If cells other than reproductive cells are modified, the procedure is called *somatic-cell* or *noninheritable gene therapy*.

Genetically modified (GM) plant or animal. A plant or animal that has been altered by the introduction of DNA from a different species.

Genetic code. The set of 64 nucleotide triplets that specify the 20 amino acids and polypeptide chain initiation and termination.

Genetic drift. See **Random genetic drift.**

Genetic equilibrium. Condition in a group of interbreeding organisms in which the allele frequencies remain constant over time.

Genetic map. A diagram of a chromosome with distances based on recombination frequencies—centiMorgans.

Genetics. The science of heredity and variation.

Genetic selection. The exposure of a cell or an organism to environmental conditions in which it can survive only if it carries a specific gene or genetic element.

Genome. A complete set (*n*) of chromosomes (hence, of genes) inherited as a unit from one parent.

Genomic DNA library. A collection of clones containing the genomic DNA sequences of an organism.

Genomics. The study of the structure and function of entire genomes.

Genotype. The genetic constitution (gene makeup) of an organism (cf. **Phenotype**).

Germ cell. A reproductive cell capable when mature of being fertilized and reproducing an entire organism (cf. **Somatic cell**).

Germinal mutation. A mutation that occurs in the reproductive cells (germ-line cells) of the body and is transmitted to progeny (cf. **Somatic mutation).**

Germ line. The tissue that ultimately produces the gametes.

Germ-line (heritable) gene therapy. Treatment of an inherited disorder by adding functional (wild-type) copies of a gene to reproductive (germ-line) cells of an individual carrying defective copies of that gene (cf. **Somatic-cell [nonheritable] gene therapy).**

Germ plasm. The hereditary material transmitted to the offspring through the germ cells.

Globulins. Common proteins in the blood that are insoluble in water and soluble in salt solutions. Alpha, beta, and gamma globulins can be distinguished in human blood serum. Gamma globulins are important in developing immunity to diseases.

Glucocorticoid. A steroid hormone that regulates gene expression in higher animals.

Golgi complex. A membranous system within cells that is involved in the secretion of cellular substances.

Gonad. A sexual organ (that is, ovary or testis) that produces gametes.

Green fluorescent protein (GFP). A naturally occurring fluorescent protein synthesized by the jellyfish *Aequorea victoria*.

Guanine. A purine base found in DNA and RNA.

Guide RNAs. RNA molecules that contain sequences that function as templates during RNA editing.

Gynandromorph. An individual in which one part of the body is female and another part is male; a sex mosaic.

Haploid (monoploid). An organism or cell having only one complete set (n) of chromosomes or one genome.

Haplotype. A set of linked genetic variants, especially single nucleotide polymorphisms (SNPs), on a chromosome.

Haptoglobin. A serum protein, alpha globulin, in the blood.

Hardy-Weinberg Principle. Mathematical relationship that allows the frequencies of genotypes in a population to be predicted from their constituent allele frequencies; a consequence of random mating.

Helix. Any structure with a spiral shape. The Watson and Crick model of DNA is in the form of a double helix.

Helper T cells. T cells that respond to an antigen displayed by a macrophage by stimulating B and T lymphocytes to develop into antibody-producing plasma cells and killer T cells, respectively.

Hemizygote. An individual that carries one copy of a chromosome or gene, as in sex linkage or as a result of deletion.

Hemoglobin. Conjugated protein compound containing iron, located in erythrocytes of vertebrates; important in the transportation of oxygen to the cells of the body.

Hemolymph. The mixture of blood and other fluids in the body cavity of an invertebrate.

Hemophilia. A bleeder's disease; tendency to bleed freely from even a slight wound; hereditary condition dependent on a sex-linked recessive gene.

Heredity. Resemblance among individuals related by descent; transmission of traits from parents to offspring.

Heritability. Degree to which a given trait is controlled by inheritance. (See **Broad-sense heritability** and **Narrow-sense heritability**.)

Hermaphrodite. An individual with both male and female reproductive organs.

Heteroalleles. Mutations that are functionally allelic but structurally nonallelic; mutations at different sites in a gene.

Heterochromatin. Chromatin that stains darkly even during interphase, often containing repetitive DNA with few genes.

Heteroduplex. A double-stranded nucleic acid containing one or more mismatched (noncomplementary) base pairs.

Heterogametic sex. Producing unlike gametes with regard to the sex chromosomes. In humans, the XY male is heterogametic, and the XX female is homogametic.

Heterogeneous nuclear RNA (hnRNA). The population of primary transcripts in the nucleus of a eukaryotic cell.

Heterokaryon. A cell containing two or more different nuclei.

Heterologous chromosome. A chromosome that contains a different set of genes than the chromosome to which it is compared.

Heteroplasmy. A cellular condition in which two genetically different types of an organelle are present (cf. **Homoplasmy**).

Heteropyknosis (*adj.*, **heteropyknotic).** Property of certain chromosomes, or of their parts, to remain more dense and to stain more intensely than other chromosomes or parts during the cell cycle.

Heterosis. Superiority of heterozygous genotypes in respect to one or more traits in comparison with corresponding homozygotes.

Heterozygosity. The proportion of heterozygous individuals in a population; used as a measure of genetic variability.

Heterozygote (adj., **heterozygous).** An organism with unlike members of any given pair or series of alleles that consequently produces unlike gametes.

Hfr. High-frequency recombination strain of *Escherichia coli;* in such strains, the F episome is integrated into the bacterial chromosome.

Histones. Group of proteins rich in basic amino acids. They function in the coiling of DNA in chromosomes and in the regulation of gene activity.

HIV (human immunodeficiency virus). The retrovirus that causes AIDS in humans.

Holoenzyme. The form of a multimeric enzyme in which all of the component polypeptides are present.

Homeobox. A DNA sequence found in several genes that are involved in the specification of organs in different body parts in animals; characteristic of genes that influence segmentation in animals. The homeobox corresponds to an amino acid sequence in the polypeptide encoded by these genes; this sequence is called the homeodomain.

Homeodomain. See **Homeobox.**

Homeotic genes. A group of genes whose products control formation of the body of an embryo by regulating the expression of other genes in segmental regions along the anterior-posterior axis.

Homeotic mutation. A mutation that causes a body part to develop in an inappropriate position in an organism; for example, a mutation in *Drosophila* that causes legs to develop on the head in the place of antennae.

Homoalleles. Mutations that are both functionally and structurally allelic; mutations at the same site in the same gene.

Homogametic sex. Producing like gametes with regard to the sex chromosomes (cf. **Heterogametic sex**).

Homologous chromosomes. Chromosomes that occur in pairs and are generally similar in size and shape, one having come from the male parent and the other from the female parent. Such chromosomes contain the same array of genes.

Homologous genes. Genes that have evolved from a common ancestral gene (cf. **Orthologous genes; Paralogous genes).**

Homologues. See **Homologous chromosomes; Homologous genes.**

Homoplasmy. A cellular condition in which all copies of an organelle are genetically identical (cf. **Heteroplasmy**).

Homozygote (*adj.*, homozygous). An individual in which the two copies of a gene are the same allele.

Hormone. An organic product of cells of one part of the body that is transported by the body fluids to another part where it influences activity or serves as a coordinating agent.

Human Genome Organization (HUGO). An international group of scientists formed to coordinate the sequencing and mapping of the human genome.

Human Genome Project. A huge international effort to map and sequence the entire human genome.

Human growth hormone. A signaling polypeptide required for normal growth in humans; it is deficient in individuals with certain types of dwarfism.

Human immunodeficiency virus (HIV). The retrovirus that causes acquired immune deficiency syndrome (AIDS) in humans.

Humoral immune response. See **Antibody-mediated (humoral) immune response.**

Huntington's disease. A late-onset (age 30 to 50 years) neurodegenerative disorder in humans caused by an autosomal dominant mutation. The genetic defect is an expanded $(CAG)_n$ trinucleotide repeat that encodes an abnormally long polyglutamine region near the amino terminus of the *huntingtin* gene product.

Hybrid. An offspring of homozygous parents differing in one or more genes; more generally, an offspring of a cross between unrelated strains.

Hybrid dysgenesis. In *Drosophila*, a syndrome of abnormal germ-line traits, including mutation, chromosome breakage, and sterility, which results from transposable element activity.

Hybridization. Interbreeding of species, races, varieties, and so on, among plants or animals; a process of forming a hybrid by cross pollination of plants or by mating animals of different types.

Hybrid vigor (heterosis). Unusual growth, strength, and health of heterozygous hybrids derived from two less vigorous homozygous parents.

Hydrogen bonds. Weak interactions between electronegative atoms and hydrogen atoms (electropositive) that are linked to other electronegative atoms.

Hydrophobic interactions. Association of nonpolar groups with each other when present in aqueous solutions because of their insolubility in water.

Hyperploid. A genetic condition in which a chromosome or a segment of a chromosome is overrepresented in the genotype (cf. **Hypoploid**).

Hypersensitive sites. Regions in the DNA that are highly susceptible to digestion with endonucleases.

Hypomorph. A mutation that reduces but does not completely abolish gene expression.

Hypoploid. A genetic condition in which a chromosome or segment of a chromosome is underrepresented in the genotype (cf. **Hyperploid**).

Hypostasis. See **Epistasis.**

Hypothesis. In science, a statement about how a phenomenon can be explained.

Imaginal disc. A mass of cells in the larvae of *Drosophila* and other holometabolous insects that give rise to particular adult organs such as antennae, eyes, and wings.

Immunoglobulin. See **Globulin.**

Imprinting. A process that alters the state of a gene without altering its nucleotide sequence; often associated with methylation of specific nucleotides in the gene. The altered state is established in the germ line and is transmitted to the offspring where it may persist throughout the offspring's life. A gene that has been altered in this way is said to have been imprinted.

***in situ*.** From the Latin, meaning in the natural place; refers to experimental treatments performed on cells or tissue rather than on extracts from them.

***in situ* colony or plaque hybridization.** A procedure for screening colonies or plaques growing on plates or membranes for the presence of specific DNA sequences by the hybridization of nucleic acid probes to the DNA molecules present in these colonies or plaques.

***in situ* hybridization.** A method for determining the location of specific DNA sequences in chromosomes by hybridizing labeled DNA or RNA to denatured DNA in chromosome preparations and visualizing the hybridized probe by autoradiography or fluorescence microscopy.

Intein. A short amino acid sequence in a primary translation product that can excise itself from the polypeptide.

***in vitro*.** From the Latin meaning "within glass"; biological processes made to occur experimentally outside the organism in a test tube or other container.

***in vivo*.** From the Latin meaning "within the living organism."

Inbred line. A strain produced by many generations of systematic inbreeding, for example, by repeated self-fertilization or by repeated full-sib mating.

Inbreeding. Matings between related individuals.

Inbreeding coefficient. The probability that two alleles in an individual are identical to each other by descent from a common ancestor.

Inbreeding depression. The observation that inbred lines are weaker than noninbred lines.

Incomplete dominance. Expression of two alleles in a heterozygote that allows the heterozygote to be distinguished from either of its homozygous parents.

Independent assortment. The random distribution of alleles to the gametes that occurs when genes are located in different chromosomes. The distribution of one pair of alleles is independent of other genes located in nonhomologous chromosomes.

Induced mutation. A mutation that results from the exposure of an organism to a chemical or physical agent that causes changes in the structure of DNA or RNA (cf. **Spontaneous mutation**).

Inducer. A substance of low molecular weight that is bound by a repressor to produce a complex that can no longer bind to the operator; thus, the presence of the inducer turns on the expression of the gene(s) controlled by the operator.

Inducible enzyme. An enzyme that is synthesized only in the presence of the substrate that acts as an inducer.

Inducible gene. A gene that is expressed only in the presence of a specific metabolite, the inducer.

Induction. The process of turning on the expression of a gene or set of genes by an inducer.

Inhibitor. Any substance or object that retards a chemical reaction; a major or modifier gene that interferes with a reaction.

Initiation (of DNA, RNA, or protein synthesis). The incorporation of the first subunit (nucleotide or amino acid) during the synthesis of a macromolecule (DNA, RNA, or polypeptide).

Initiation factors. Soluble proteins required for the initiation of translation.

Insertional mutation. A mutation caused by the insertion of foreign DNA such as a transposable element or the T-DNA of the Ti plasmid of *Agrobacterium tumefaciens*.

Interaction. In statistics, an effect that cannot be explained by the additive action of contributing factors; a departure from strict additivity.

Intercalating agent. A chemical capable of inserting between adjacent base pairs in a DNA molecule.

Intercross. A cross between the F_1 hybrids derived from a cross between two parental strains.

Interference. Crossing over at one point that reduces the chance of another crossover nearby; detected by studying the pattern of crossing over with three or more linked genes.

Interphase. The stage in the cell cycle when the cell is not dividing; the metabolic stage during which DNA replication occurs; the stage following telophase of one division and extending to the beginning of prophase in the next division.

Intersex. An organism displaying secondary sexual characters intermediate between male and female; a type that shows some phenotypic characteristics of both males and females.

Intragenic complementation. Complementation that occurs between two mutant alleles of a gene; common only when the product of the gene functions as a homomultimer.

Introns. Intervening sequences of DNA bases within eukaryotic genes that are not represented in the mature RNA transcript because they are spliced out of the primary RNA transcript.

Invariant. Constant, unchanging, usually referring to the portion of a molecule that is the same across species.

Inversion. A rearrangement that reverses the order of a linear array of genes in a chromosome.

Inverted repeat. A sequence present twice in a DNA molecule but in reverse orientation.

Ionic bonds. Attractions between oppositely charged chemical groups.

Ionizing radiation. The portion of the electromagnetic spectrum that results in the production of positive and negative charges (ion pairs) in molecules. X rays and gamma rays are examples of ionizing radiation (cf. **Nonionizing radiation**).

IS element (insertion sequence). A short (800–1400 nucleotide pairs) DNA sequence found in bacteria that is capable of transposing to a new genomic location; other DNA sequences that are bounded by IS elements may also be transposed.

Isoalleles. Different forms of a gene that produce the same phenotype or very similar phenotypes.

Isochromosome. A chromosome with two identical arms and identical genes. The arms are mirror images of each other.

Isoform. A member of a family of closely related proteins—proteins that have some amino acid sequences in common and some different.

Isogenic stocks. Strains of organisms that are genetically uniform; completely homozygous.

Kappa chain. One of two classes of antibody light chains (cf. **Lambda chain**).

Karyotype. The chromosome constitution of a cell or an individual; chromosomes arranged in order of length and according to position of centromere; also, the abbreviated formula for the chromosome constitution, such as 47, XX + 21 for human trisomy-21.

Killer T cells (cytotoxic T cells). T cells that carry T-cell receptors and kill cells displaying the recognized antigens.

Kinetics. A dynamic process involving motion.

Kinetochore. A proteinaceous structure associated with the centromere of a chromosome during eukaryotic cell division; the point at which microtubules attach to move the chromosome through the division process.

Kinetosome. Granular body at the base of a flagellum or a cilium.

Klinefelter syndrome. A condition produced when two X chromosomes and one Y chromosome are present in the human karyotype.

Knockout mutation. A mutation that completely abolishes a gene's function.

Lagging strand. The strand of DNA that is synthesized discontinuously during replication.

Lambda chain. One of two classes of antibody light chains (cf. **Kappa chain**).

Lamella. A double-membrane structure, plate, or vesicle that is formed by two membranes lying parallel to each other.

Lampbrush chromosomes. Large diplotene chromosomes present in oocyte nuclei, particularly conspicuous in amphibians. These chromosomes have extended regions called loops, which are active sites of transcription.

Leader sequence. The segment of an mRNA molecule from the 5′ terminus to the translation initiation codon.

Leading strand. The strand of DNA that is synthesized continuously during replication.

Leptonema (*adj.*, **leptotene).** Stage in meiosis immediately preceding synapsis in which the chromosomes appear as single, fine, threadlike structures (but they are really double because DNA replication has already taken place).

Lethal mutation. A mutation that renders an organism or a cell possessing it inviable.

Ligand. A molecule that can bind to another molecule in or on cells.

Ligase. An enzyme that joins the ends of two strands of nucleic acid.

Ligation. The joining of two or more DNA molecules by covalent bonds.

Lipid. A molecule composed of fatty acids and triglycerides.

LINEs (long interspersed nuclear elements). Families of long (average length is 6500 bp) moderately repetitive transposable elements in eukaryotes.

Linkage. A relationship among genes in the same chromosome. Such genes tend to be inherited together.

Linkage equilibrium. A state in which the alleles of linked loci are randomized with respect to each other on the chromosomes of a population.

Linkage map. A linear or circular diagram that shows the relative positions of genes on a chromosome as determined by genetic analysis.

Locus (*pl.*, **loci).** A fixed position on a chromosome that is occupied by a given gene or one of its alleles.

Lod score. The logarithm of the ratio of odds for pedigree data, where the odds are calculated under the assumption of linkage with a specified frequency of recombination and under the assumption of no linkage, that is, 50 percent recombination.

Loss-of-function mutation. A mutation that impairs or abolishes gene expression or the function of a gene product.

LTR (long terminal repeat). A DNA sequence present at each end of a retrotransposon.

Lymphocyte. A general class of white blood cells that are important components of the immune system of vertebrate animals.

Lysis. Bursting of a cell by the destruction of the cell membrane following infection by a virus.

Lysogenic bacteria. Those harboring temperate bacteriophages.

Lysosome. A small, membrane-bound cellular organelle that contains enzymes dedicated to the degradation of macromolecules.

Lytic phage. See **Virulent phage.**

Macromolecule. A large molecule; term used to identify molecules of proteins and nucleic acids.

Map unit. See **Crossover unit.**

Mass selection. As practiced in plant and animal breeding, the choosing of individuals for reproduction from the entire population on the basis of the individuals' phenotypes rather than on the phenotypes of their relatives.

Maternal effect. Trait controlled by a gene of the mother but expressed in the progeny.

Maternal-effect gene. A gene whose product acts in the offspring of the female who carries the gene.

Maternal-effect mutation. A mutation that causes a mutant phenotype in the offspring of a female that carries the mutation; however, the female herself may not show the mutant phenotype.

Maternal inheritance. Inheritance controlled by extrachromosomal (that is, cytoplasmic) factors that are transmitted through the egg.

Maturation. The formation of gametes or spores.

Mean. The arithmetic average; the sum of all measurements or values in a sample divided by the sample size.

Median. In a set of measurements, the central value above and below which there are an equal number of measurements.

Megaspore. The single large cell produced at the end of meiosis in the female reproductive tissues of plants.

Meiosis. The process by which the chromosome number of a reproductive cell becomes reduced to half the diploid ($2n$) or somatic number; results in the formation of gametes in animals or of spores in plants; important source of variability through recombination.

Meiotic drive. Any mechanism that causes alleles to be recovered unequally in the gametes of a heterozygote.

Melanin. Brown or black pigment.

Membrane. A macromolecular structure composed of lipids and proteins that surrounds a cell or certain of the organelles within a cell, such as the mitochondria and chloroplasts; also, a component of the endoplasmic reticulum within cells.

Memory cells. Long-lived B and T cells that mediate rapid secondary immune responses to a previously encountered antigen.

Mendelian population. A natural interbreeding unit of sexually reproducing plants or animals sharing a common gene pool.

Merozygote. Partial zygote produced by a process of partial genetic exchange, such as transformation in bacteria. An exogenote may be introduced into a bacterial cell in the formation of a merozygote.

Mesoderm. The middle germ layer that forms in the early animal embryo and gives rise to such parts as bone and connective tissue.

Messenger RNA (mRNA). RNA that carries information necessary for protein synthesis from the DNA to the ribosomes.

Metabolic cell. A cell that is not dividing.

Metabolism. Sum total of all chemical processes in living cells by which energy is provided and used.

Metacentric chromosome. A chromosome with the centromere near the middle and two arms of about equal length.

Metafemale (superfemale). In *Drosophila*, abnormal female, usually sterile, with an excess of X chromosomes compared with sets of autosomes (for example, XXX; AA).

Metaphase. That stage of cell division in which the chromosomes are most discrete and arranged in an equatorial plate; stage following prophase and preceding anaphase.

Metastasis. The spread of cancer cells to previously unaffected organs.

Methylation (of DNA and RNA). The addition of a methyl ($-CH_3$) group(s) to one or more of the nucleotides in a nucleic acid.

Microprojectile bombardment. A procedure for transforming plant cells by shooting DNA-coated tungsten or gold particles into the cells.

MicroRNA. See **Short interfering RNA.**

Microsatellites. DNA sequences composed of polymorphic tandem repeats of two to five nucleotide pairs.

Microspore. One of the four end products of meiosis in the male reproductive tissues of plants.

Microtubule Organizing Center (MTOC). A region in a eukaryotic cell that generates the microtubules used during cell division. In animal cells, the MTOC is associated with distinct organelles called centrosomes.

Microtubules. Hollow filaments in the cytoplasm making up a part of the locomotor apparatus of a motile cell; component of the mitotic spindle.

Midparent value. In quantitative genetics, the average of the phenotypes of two mates.

Mismatch repair. DNA repair processes that correct base pairs that are not properly hydrogen-bonded.

Missense mutation. A mutation that changes a codon specifying an amino acid to a codon specifying a different amino acid.

Mitochondria. Organelles in the cytoplasm of plant and animal cells where oxidative phosphorylation takes place to produce ATP.

Mitosis. Disjunction of duplicated chromosomes and division of the cytoplasm to produce two genetically identical daughter cells.

Modal class. In a frequency distribution, the class having the greatest frequency.

Model. A mathematical description of a biological phenomenon.

Modifier (modifying gene). A gene that affects the expression of some other gene.

Monohybrid. An offspring of two homozygous parents that differ from one another by the alleles present at only one gene locus.

Monohybrid cross. A cross between parents differing in only one trait or in which only one trait is being considered.

Monomer. A single molecular entity that may combine with others to form more complex structures.

Monomorphism. The absence of genetic variability **(cf. Polymorphism)**.

Monoploid. Organism or cell having a single set of chromosomes or one genome (chromosome number n).

Monosomic. A diploid organism lacking one chromosome of its proper complement ($2n-1$); an aneuploid. Monosome refers to the single chromosome, disome to two chromosomes of a kind, and trisome to three chromosomes of a kind.

Monozygotic twins. One-egg or identical twins.

Morphogen. A substance that stimulates the development of form or structure in an organism.

Morpholino. A synthetic molecule analogous to a short single strand of DNA in which the deoxyribose moiety has been replaced by a morpholine moiety in each of the subunits.

Morphology. Study of the form of an organism; developmental history of visible structures and the comparative relation of similar structures in different organisms.

Mosaic. An organism or part of an organism that is composed of cells of different genotypes.

mtDNA. The DNA of mitochondria.

Multifactorial trait. A trait determined by a combination of several genetic and environmental factors.

Multigene family. A group of genes that are similar in nucleotide sequence or that produce polypeptides with similar amino acid sequences.

Multiple alleles. A condition in which a particular gene occurs in three or more allelic forms in a population of organisms.

Multiple Factor Hypothesis. A theory advanced by R. A. Fisher and others to explain variation in complex phenotypes such as height, weight, and disease susceptibility.

Mutable genes. Genes with an unusually high mutation rate.

Mutagen. An environmental agent, either physical or chemical, that is capable of inducing mutations.

Mutant. A cell or individual organism that shows a change brought about by a mutation; a changed gene.

Mutation. A change in the DNA at a particular locus in an organism. The term is used loosely to include point mutations involving a single gene change as well as a chromosomal change.

Mutation pressure. A constant mutation rate that adds mutant genes to a population; repeated occurrences of mutations in a population.

Mycelium (*pl.*, mycelia). Threadlike filament making up the vegetative portion of thallus fungi.

Narrow-sense heritability. In quantitative genetics, the proportion of the phenotypic variance that is due to the additive effects of alleles.

Natural selection. Differential survival and reproduction in nature that favors individuals that are better adapted to their environment; elimination of less fit organisms.

Negative autogenous regulation (negative self-regulation). Inhibition of the expression of a gene or set of coordinately regulated genes by the product of the gene or the product of one of the genes.

Negative control system. A mechanism in which the regulatory protein(s) is required to turn off gene expression.

Negative supercoiling. The formation of coiled tertiary structures in double-stranded DNA molecules with fixed (not free to rotate) ends when the molecules are underwound.

Neutral mutation. A mutation that changes the nucleotide sequence of a gene but has no effect on the fitness of the organism.

Neutral theory. The theory that the evolution of traits with little or no effect on fitness is a random process involving mutation and genetic drift.

Nick-translation. A procedure for labeling DNA by nicking it with an endonuclease and "translating" the nick along the DNA molecule in the presence of labeled deoxyribonucleoside triphosphates by the concerted action of the 5′ —> 3′ exonuclease and 5′ —> 3′ polymerase activities of *E. coli* DNA polymerase I.

Nitrous acid. HNO_2, a potent chemical mutagen.

Nonautonomous. A term referring to biological units that cannot function by themselves; such units require the assistance of another unit, or "helper" (cf. **Autonomous**).

Nondisjunction. Failure of disjunction or separation of homologous chromosomes in mitosis or meiosis, resulting in too many chromosomes in some daughter cells and too few in others. Examples: In meiosis, both members of a pair of chromosomes go to one pole so that the other pole does not receive either of them; in mitosis, both sister chromatids go to the same pole.

Nonhistone chromosomal proteins. All of the proteins in chromosomes except the histones.

Nonionizing radiation. The portion of the electromagnetic spectrum that does not lead to the production of positive and negative charges (ion pairs) in molecules. Visible and ultraviolet light are examples of nonionizing radiation (cf. **Ionizing radiation**).

Nonpolypoid Colorectal Cancer. A form of cancer found in the lower digestive tract, sometimes inherited as a dominant condition.

Nonsense mutation. A mutation that changes a codon specifying an amino acid to a termination codon.

Nonsynonymous substitution. A base-pair change in a codon that alters the amino acid specified by the codon.

Nontemplate strand. In transcription, the nontranscribed strand of DNA. It will have the same sequence as the RNA transcript, except that T is present at positions where U is present in the RNA transcript.

Northern blot. The transfer of RNA molecules from an electrophoretic gel to a cellulose or nylon membrane by capillary action.

Nuclease. An enzyme that catalyzes the degradation of nucleic acids.

Nucleic acid. A macromolecule composed of phosphoric acid, pentose sugar, and organic bases; DNA and RNA.

Nucleolar Organizer (NO). A chromosomal segment containing genes that control the synthesis of ribosomal RNA, located at the secondary constriction of some chromosomes.

Nucleolus. An RNA-rich, spherical sack in the nucleus of metabolic cells; associated with the nucleolar organizer; storage place for ribosomes and ribosome precursors.

Nucleoprotein. Conjugated protein composed of nucleic acid and protein; the material of which the chromosomes are made.

Nucleoside. An organic compound consisting of a base covalently linked to ribose or deoxyribose.

Nucleosome. Spherical subunits of eukaryotic chromatin that are composed of a core particle consisting of an octamer of histones and 146 nucleotide pairs.

Nucleotide. A subunit of DNA and RNA molecules containing a phosphate group, a sugar, and a nitrogen-containing organic base.

Nucleotide excision repair. The removal of relatively large defects such as thymine dimers in DNA via the excision of a segment of the DNA strand spanning the defect and repair synthesis by a DNA polymerase using the complementary strand as template.

Nucleus. The part of a eukaryotic cell that contains the chromosomes; separated from the cytoplasm by a membrane.

Null allele. A mutant form of a gene that either produces no product or produces a totally nonfunctional product.

Nullisomic. An otherwise diploid cell or organism lacking both members of a chromosome pair (chromosome formula $2n - 2$).

Null mutation. A mutation that abolishes the expression of a gene. (See **Amorph.**)

Octoploid. Cell or organism with eight genomes or sets of chromosomes (chromosome number $8n$).

Oligonucleotide-directed site-specific mutagenesis. A procedure by which a specific nucleotide sequence can be changed to another predetermined sequence.

Oncogene. A gene that can cause cancerous transformation in animal cells growing in culture and tumor formation in animals themselves; a gene that promotes cell division.

Oocyte. The egg-mother cell; the cell that undergoes two meiotic divisions (oogenesis) to form the egg cell. Primary oocyte—before completion of the first meiotic division; secondary oocyte—after completion of the first meiotic division.

Oogenesis. The formation of the egg or ovum in animals.

Oogonium (*pl.*, oogonia). A germ cell of the female animal before meiosis begins.

Open Reading Frame (ORF). A DNA segment containing the sequences required to encode a polypeptide. The RNA transcript of an ORF begins with a translation start codon, followed by a sequence of codons specifying amino acids, and ending with a translation stop codon. An ORF is presumed, but not known, to encode a polypeptide.

Operational definition. An operation or procedure that can be carried out to define or delimit something.

Operator. A part of an operon that controls the expression of one or more structural genes by serving as the binding site for one or more regulatory proteins.

Operon. A group of genes making up a regulatory or control unit. The unit includes an operator, a promoter, and structural genes.

Operon model. The negative control mechanism proposed by Jacob and Monod in 1961 to explain the coordinate regulation of co-transcribed sets of structural genes. The mechanism involves a regulator gene encoding a repressor that controls transcription of the set of genes by binding to an operator region and blocking transcription by RNA polymerase.

Order (in the genetic code). There are two types of order in the genetic code: (1) multiple codons for a given amino acid usually differ only at the third position, and (2) the codons for amino acids with similar chemical properties are closely related.

Ordered tetrad. A set of spores formed by meiosis in certain fungi in which the placement of the spores in the ascus reveals the way in which the chromosomes segregated during meiosis.

Ordinate. The vertical axis in a graph.

Organelle. Specialized part of a cell with a particular function or functions (for example, the cilium of a protozoan).

Organizer. An inductor; a chemical substance in a living system that determines the fate in development of certain cells or groups of cells.

Origin of replication. The site or nucleotide sequence on a chromosome or DNA molecule at which replication is initiated.

Orthologous genes. Homologous genes present in different species (cf. **Homologous genes**).

Orthologues. See **Orthologous genes.**

Outbreeding. Mating of unrelated individuals.

Ovary. The swollen part of the pistil of a plant flower that contains the ovules; the female reproductive organ or gonad in animals.

Overdominance. A condition in which heterozygotes are superior (on some scale of measurement) to either of the associated homozygotes.

Overlapping genes. Genes with overlapping coding sequences.

Ovule. The macrosporangium of a flowering plant that becomes the seed. It includes the nucellus and the integuments.

P. Symbol for the parental generation or parents of a given individual.

Pachynema (*adj.*, pachytene). A mid-prophase stage in meiosis immediately following zygonema and preceding diplonema. In favorable microscopic preparations, the chromosomes are visible as long, paired threads. Rarely, four chromatids are detectable.

Pair-rule gene. A gene that influences the formation of body segments in *Drosophila.*

Palindrome. A segment of DNA in which the base-pair sequence reads the same in both directions from a central point of symmetry.

Panmictic population. A population in which mating occurs at random.

Panmixis. Random mating in a population.

Paracentric inversion. An inversion that is entirely within one arm of a chromosome and does not include the centromere.

Paralogous genes. Homologous genes present within a species (cf. **Homologous genes**).

Paralogues. See **Paralogous genes.**

Parameter. A value or constant based on an entire population (cf. **Statistic**).

Parental. Pertaining to the founding strains used in a cross; having the characteristics of these strains. In a series of crosses, the parental generation is symbolized as **P**.

Parental ditype. In tetrad analysis, the occurrence of the two parental genotypes in the tetrad of meiotic products.

Parthenogenesis. The development of a new individual from an egg without fertilization.

Paternal. Pertaining to the father.

Pathogen. An organism that causes a disease.

Pattern baldness. A hereditary form of baldness in which the thinning of the hair begins on the crown of the head.

PCR. See **Polymerase chain reaction.**

Pedigree. A table, chart, or diagram representing the ancestry of an individual.

***P* element.** A transposable element in *Drosophila* that, when activated, causes hybrid dysgenesis.

Penetrance. The percentage of individuals that show a particular phenotype among those capable of showing it.

Peptide. A compound containing amino acids; a breakdown or buildup unit in protein metabolism.

Peptide bond. A chemical bond holding amino acid subunits together in proteins.

Peptidyl (*P*) site. The ribosome binding site that contains the tRNA to which the growing polypeptide chain is attached.

Peptidyl transferase. An enzyme activity—built into the large subunit of the ribo- some—that catalyzes the formation of peptide bonds between amino acids during translation.

Pericentric inversion. An inversion including the centromere, hence involving both arms of a chromosome.

Peroxisome. A subcellular organelle that contains enzymes involved in the degradation of fatty acids and amino acids.

Petite mutant. A respiration-deficient yeast mutant that produces small colonies when grown on glucose-containing medium.

Phage. See **Bacteriophage.**

Phagemids. Cloning vectors that contain components derived from both phage chromosomes and plasmids.

Phagocytes. Immune system cells that ingest and destroy viruses, bacteria, fungi, and other foreign substances or cells.

Phenocopy. An organism whose phenotype (but not genotype) has been changed by the environment to resemble the phenotype of a different (mutant) organism.

Phenotype. The observable characteristics of an organism.

Phenylalanine. See **Amino acid.**

Phenylketonuria. Metabolic disorder resulting in mental retardation; transmitted as a Mendelian recessive and treated in early childhood by special diet.

Photoreactivation. A DNA repair process that is light-dependent.

Phylogeny. A diagram showing the evolutionary relationships among a group of organisms; an evolutionary tree.

Physical map. A diagram of a chromosome or DNA molecule with distances given in base pairs, kilobases, or megabases.

Pistil. The centrally located organ in flowers that contains the ovary.

Plaque. Clear area on an otherwise opaque culture plate of bacteria where the bacteria have been killed by a virus.

Plasma cells. Antibody-producing white blood cells devived from B lymphocytes.

Plasmid. An extrachromosomal hereditary determinant that exists in an autonomous state and is transferred independently of chromosomes.

Plastid. A cytoplasmic body found in the cells of plants and some protozoa. Chloroplasts, for example, produce chlorophyll that is involved in photosynthesis.

Pleiotropy (*adj.*, **Pleiotropic).** Condition in which a single gene influences more than one trait.

Pluripotent. An adjective applied to cells that have the potential to differentiate into many different cell types.

Point mutations. Changes that occur at specific sites in genes. They include nucleotide-pair substitutions and the insertion or deletion of one or a few nucleotide pairs.

Polar bodies. In female animals, the smaller cells produced at meiosis that do not develop into egg cells. The first polar body is produced at division I and may not go through division II. The second polar body is produced at division II.

Polar mutation. A mutation that influences the functioning of genes that are downstream in the same transcription unit.

Pole cells. A group of cells in the posterior of *Drosophila* embryos that are precursors to the adult germ line.

Polyadenylation. The addition of poly(A) tails to eukaryotic gene transcripts (RNAs).

Poly(A) polymerase. An enzyme that adds the poly(A) tails to the 3′ termini of eukaryotic gene transcripts (RNAs).

Poly(A) tail (mRNA). A polyadenosine tract 20 to 200 nucleotides long that is added to the 3′ ends of most eukaryotic mRNAs posttranscriptionally.

Polycloning site. See **Polylinker.**

Polydactyly. The occurrence of more than the usual number of fingers or toes.

Polygene (*adj.*, **polygenic).** One of many genes involved in quantitative inheritance.

Polylinker (polycloning site). A segment of DNA that contains a set of unique restriction enzyme cleavage sites.

Polymer. A compound composed of many smaller subunits; results from the process of polymerization.

Polymerase. An enzyme that catalyzes the formation of DNA or RNA.

Polymerase chain reaction (PCR). A procedure involving multiple cycles of denaturation, hybridization to oligonucleotide primers, and polynucleotide synthesis that amplifies a particular DNA sequence.

Polymerization. Chemical union of two or more molecules of the same kind to form a new compound having the same elements in the same proportions but a higher molecular weight and different physical properties.

Polymorphism. The existence of two or more variants in a population of individuals, with at least two of the variants having frequencies greater than 1 percent.

Polynucleotide. A linear sequence of joined nucleotides in DNA or RNA.

Polypeptide. A linear molecule with two or more amino acids and one or more peptide groups. They are called dipeptides, tripeptides, and so on, according to the number of amino acids present.

Polyploid. An organism with more than two sets of chromosomes ($2n$ diploid) or genomes—for example, triploid ($3n$), tetraploid ($4n$), pentaploid ($5n$), hexaploid ($6n$), heptaploid ($7n$), octoploid ($8n$).

Polysaccharide capsules. Carbohydrate coverings with antigenic specificity that are present on some types of bacteria.

Polytene chromosomes. Giant chromosomes produced by interphase replication without division and consisting of many identical chromatids arranged side by side in a cablelike pattern.

Population. Entire group of organisms of one kind; an interbreeding group of plants or animals; the extensive group from which a sample might be taken.

Population (effective). Breeding members of the population.

Population genetics. The branch of genetics that deals with frequencies of alleles and genotypes in breeding populations.

Positional cloning. The isolation of a clone of a gene or other DNA sequence based on its map position in the genome.

Position effect. A difference in phenotype that is dependent on the position of a gene or group of genes, often caused by heterochromatin that is nearby.

Positive control system. A mechanism in which the regulatory protein(s) is required to turn on gene expression.

Postreplication repair. A recombination-dependent mechanism for repairing damaged DNA.

Primary transcript. The RNA molecule produced by transcription prior to any posttranscriptional modifications; also called a pre-mRNA in eukaryotes.

Primer. A short nucleotide sequence with a reactive 3′ OH that can initiate DNA synthesis along a template.

Primosome. A protein replication complex that catalyzes the initiation of Okazaki fragments during discontinuous synthesis. It contains DNA primase and DNA helicase activities.

Prion. A small infectious (disease-causing) protein molecule.

Probability. The frequency of occurrence of an event.

Proband. The individual in a family in whom an inherited trait is first identified.

Progeny testing. The practice of ascertaining the genotype of an individual by mating it to an individual of known genotype and examining the progeny.

Progerias. Inherited diseases characterized by premature aging.

Prokaryote. A member of a large group of organisms (including bacteria and bluegreen algae) that lack true nuclei in their cells and that do not undergo mitosis.

Promoter. A nucleotide sequence to which RNA polymerase binds and initiates transcription; also, a chemical substance that enhances the transformation of benign cells into cancerous cells.

Proofreading. The enzymatic scanning of DNA for structural defects such as mismatched base pairs.

Prophage (provirus). The genome of a temperate bacteriophage integrated into the chromosome of a lysogenic bacterium and replicated along with the host chromosome.

Prophase. The stage of mitosis between interphase and metaphase. During this phase, the centriole divides and the two daughter centrioles move apart. Each sister DNA strand from interphase replication becomes coiled, and the chromosome is longitudinally double except in the region of the centromere. Each partially separated chromosome is called a chromatid. The two chromatids of a chromosome are sister chromatids.

Protamines. Small basic proteins that replace the histones in the chromosomes of some sperm cells.

Protease. Any enzyme that hydrolyzes proteins.

Protein. A macromolecule composed of one to several polypeptides. Each polypeptide consists of a chain of amino acids linked together by peptide bonds.

Proteome. The complete set of proteins encoded by a genome.

Proteomics. The science focused on determining the structures and functions of all the proteins produced by living organisms.

Proto-oncogene. A normal cellular gene that can be changed to an oncogene by mutation.

Protoplast. A plant or bacterial cell from which the wall has been removed.

Prototroph. An organism such as a bacterium that will grow on a minimal medium.

Provirus. A viral chromosome that has integrated into a host—either prokaryotic or eukaryotic—genome (cf. **Prophage**).

Pseudoautosomal gene. A gene located on both the X and Y chromosomes.

Pseudogene. An inactive but stable component of a genome resembling a gene; apparently derived from active genes by mutation.

Pulsed-field gel electrophoresis. A procedure used to separate very large DNA molecules by alternating the direction of electric currents across a semisolid gel in a pulsed manner.

Punctuated equilibrium. The occurrence of speciation events in bursts separated by long intervals of species stability.

Purine. A double-ring nitrogen-containing base present in nucleic acids; adenine and guanine are the two purines present in most DNA and RNA molecules.

Pyrimidine. A single-ring nitrogen-containing base present in nucleic acids; cytosine and thymine are commonly present in DNA, whereas uracil usually replaces thymine in RNA.

Quantitative inheritance. Inheritance of measurable traits (height, weight, color intensity) that depend on the cumulative action of many genes, each producing a small effect on the phenotype.

Quantitative trait loci (QTL). Two or more genes that affect a single quantitative trait.

Quantitative traits. Phenotypes that can be measured, such as height, weight, and growth rate.

Quantum speciation. The formation of a new species in one or a few generations by selection and genetic drift.

Race. A distinguishable group of organisms of a particular species.

Radiation hybrid mapping. The use of human-rodent hybrid cells containing fragments of human chromosomes (produced by irradiation) fused to rodent chromosomes to determine the linkage relationships of human genes.

Radioactive isotope. An unstable isotope (form of an atom) that emits ionizing radiation.

Random genetic drift. Changes in allele frequency in small breeding populations due to chance fluctuations.

Reading frame. The series of nucleotide triplets that are sequentially positioned in the *A* site of the ribosome during translation of an mRNA; also, the sequence of nucleotide-pair triplets in DNA that correspond to these codons in mRNA.

Receptor. A molecule that can accept the binding of a ligand.

Recessive. A term applied to one member of an allelic pair lacking the ability to manifest itself when the other or dominant member is present.

Recessive lethal mutation. A mutant form of a gene that results in the death of an organism that is homozygous for it.

Recipient cell. A bacterium that receives DNA from another (donor) cell during recombination in bacteria (cf. **Donor cell**).

Reciprocal crosses. Crosses between different strains with the sexes reversed; for example, female A × male B and male A × female B are reciprocal crosses.

Recognition sequence (-35 sequence). A nucleotide sequence (consensus TTGACA) in prokaryotic promoters to which the sigma factor of RNA polymerase binds during the initiation of transcription.

Recombinant DNA molecule. A DNA molecule constructed *in vitro* by joining all or parts of two different DNA molecules.

Recombination. The production of gene combinations not found in the parents by the assortment of nonhomologous chromosomes and crossing over between homologous chromosomes during meiosis. For linked genes, the frequency of recombination can be used to estimate the genetic map distance; however, high frequencies (approaching 50 percent) do not yield accurate estimates.

Reduction division. Phase of meiosis in which the maternal and paternal chromosomes of the bivalent separate (cf. **Equational division**).

Regulator gene. A gene that controls the rate of expression of another gene or genes. Example: The *lacI* gene produces a protein that controls the expression of the structural genes of the *lac* operon in *Escherichia coli*.

Release factors. Soluble proteins that recognize termination codons in mRNAs and terminate translation in response to these codons.

Renaturation. The restoration of a molecule to its native form. In nucleic acid biochemistry, this term usually refers to the formation of a double-stranded helix from complementary single-stranded molecules.

Repetitive DNA. DNA sequences that are present in a genome in multiple copies—sometimes a million times or more.

Replica plating. A procedure for duplicating the bacterial colonies growing on agar medium in one petri plate to agar medium in another petri plate.

Replication. A duplication process that is accomplished by copying from a template (for example, reproduction at the level of DNA).

Replicative transposons. A transposable element that is replicated during the transposition process. Tn*3* in *E. coli* is an example.

Replicon. A unit of replication. In bacteria, replicons are associated with segments of the cell membrane that control replication and coordinate it with cell division.

Replisome. The complete replication apparatus—present at a replication fork—that carries out the semiconservative replication of DNA.

Repressible enzyme. An enzyme whose synthesis is diminished by a regulatory molecule.

Repression. The process of turning off the expression of a gene or set of genes in response to some signal.

Repressor. A protein that binds to DNA and turns off gene expression.

Repressor gene. A gene that encodes a repressor.

Reproductive cloning. A process in which the nucleus of an egg cell is replaced with the nucleus of a cell from a developed organism with the purpose of producing a new organism genetically identical to the donor.

Repulsion (*trans* configuration). The condition in which a double heterozygote has received a mutant and a wild-type allele from each parent; for example, $a + /a + \times + b/ + b$ produces $a + / + b$ (cf. **Coupling**).

Resistance factor. A plasmid that confers antibiotic resistance to a bacterium.

Restitution nucleus. A nucleus with unreduced or doubled chromosome number that results from the failure of a meiotic or mitotic division.

Restriction endonuclease. See **Restriction enzyme.**

Restriction enzyme. An endonuclease that recognizes a specific short sequence in DNA and cleaves the DNA molecule at or near that site.

Restriction fragment. A fragment of DNA produced by cleaving a DNA molecule with one or more restriction endonucleases.

Restriction fragment-length polymorphism (RFLP). The existence of two or more genetic variants detectable by visualizing fragments of genomic DNA that were obtained by digesting the DNA with a restriction enzyme. Usually the DNA fragments are fractionated by electrophoresis, transferred to a membrane by Southern blotting, and then visualized by hybridizing the membrane with a labeled DNA probe.

Restriction map. A linear or circular physical map of a DNA molecule showing the sites that are cleaved by different restriction enzymes.

Restriction site. A DNA sequence that is cleaved by a restriction enzyme.

Reticulocyte. A young red blood cell.

Retroelement. Any of the integrated retroviruses or the transposable elements that resemble them.

Retroposon. A transposable element that creates new copies via reverse transcription of RNA into DNA but that lacks the long terminal repeat sequences.

Retrotransposon. A transposable element that creates new copies by reverse transcription of RNA into DNA.

Retrovirus. A virus that stores its genetic information in RNA and replicates by using reverse transcriptase to synthesize a DNA copy of its RNA genome.

Retroviruslike element. A type of retrotransposon that resembles the integrated form of a retrovirus.

Reverse genetics. Genetic approaches that use the nucleotide sequence of a gene to devise procedures for isolating mutations in the gene or shutting off its expression.

Reverse transcriptase. An enzyme that catalyzes the synthesis of DNA using an RNA template.

Reversion (reverse mutation). Restitution of a mutant gene to the wild-type condition, or at least to a form that gives the wild phenotype; more generally, the appearance of a trait expressed by a remote ancestor.

RFLP (restriction fragment-length polymorphism). A genetic difference among individuals that is detected by comparing DNA fragments released by digestion with one or more restriction enzymes.

Rh factor. Antigen on the red blood corpuscles of certain people. A pregnant Rh negative woman carrying an Rh positive child may develop antibodies, causing the child to develop a hemolytic disease.

Ribonuclease. Any enzyme that hydrolyzes RNA.

Ribonucleic acid. See **RNA.**

Ribosomal RNAs (rRNAs). The RNA molecules that are structural components of ribosomes.

Ribosome. Cytoplasmic organelle on which proteins are synthesized.

R-loops. Single-stranded DNA regions in RNA-DNA hybrids formed *in vitro* under conditions where RNA-DNA duplexes are more stable than DNA-DNA duplexes.

RNA. Ribonucleic acid; the information-carrying material in some viruses; more generally, a molecule derived from DNA by transcription that may carry information (messenger or mRNA), provide subcellular structure (ribosomal or rRNA), transport amino acids (transfer or tRNA), or facilitate the biochemical modification of itself or other RNA molecules.

RNA editing. Post-transcriptional processes that alter the information encoded in gene transcripts (RNAs).

RNA interference. A phenomenon in which double-stranded RNA prevents the expression of a gene homologous to at least part of the RNA.

RNA polymerase. An enzyme that catalyzes the synthesis of RNA.

RNA primer. A short (10 to 60 nucleotides) segment of RNA that is used to initiate the synthesis of a new strand of DNA; synthesized by the enzyme DNA primase.

Robertsonian translocation. A rearrangement in which two nonhomologous chromosomes have fused to form a single chromosome, usually with the loss of some material from the ends of each of the fusing chromosomes.

Roentgen (r). Unit of ionizing radiation.

Sample. A group of items selected to represent a large population.

Satellite DNA. A component of the genome that can be isolated from the rest of the DNA by density gradient centrifugation. Usually, it consists of short, highly repetitious sequences.

Scaffold. The central core structure of condensed chromosomes. The scaffold is composed of nonhistone chromosomal proteins.

SCID (severe combined immunodeficiency syndrome). A group of diseases characterized by the inability to mount an immune response, either humoral or cellular.

Secondary oocyte. See **Oocyte.**

Secondary spermatocyte. See **Spermatocyte.**

Segmentation genes. A group of genes that control the early development of *Drosophila* embryos. Their products define segments along the anterior-posterior axis.

Segment-polarity genes. A group of genes whose products define the anterior and posterior compartments in each of the segments that form along the anterior-posterior axis of *Drosophila* embryos.

Segregation. The separation of paternal and maternal chromosomes from each other at meiosis; the separation of alleles from each other in heterozygotes; the occurrence of different phenotypes among offspring, resulting from chromosome or allele separation in their heterozygous parents; Mendel's first principle of inheritance.

Selection. Differential survival and reproduction among genotypes; the most important of the factors that change allele frequencies in large populations.

Selection coefficient. A number that measures the fitness of a genotype relative to a standard.

Selection differential. In plant and animal breeding, the difference between the mean of the individuals selected to be parents and the mean of the overall population.

Selection pressure. Effectiveness of differential survival and reproduction in changing the frequency of alleles in a population.

Selection response. In plant and animal breeding, the difference between the mean of the individuals selected to be parents and the mean of their offspring.

Selector gene. A gene that influences the development of specific body segments in *Drosophila;* a homeotic gene.

Self-fertilization. The process by which pollen of a given plant fertilizes the ovules of the same plant. Plants fertilized in this way are said to have been selfed. An analogous process occurs in some animals, such as nematodes and molluscs.

Semiconservative replication. Replication of DNA by a mechanism in which the parental strands are conserved (remain intact) and serve as templates for the synthesis of new complementary strands.

Semisterility. A condition of only partial fertility in plant zygotes (for example, maize); usually associated with translocations.

Sense RNA. A primary transcript or mRNA that contains a coding region (contiguous sequence of codons) that is translated to produce a polypeptide.

Sense strand (of RNA). See **Sense RNA.**

Serology (*adj.*, **serological).** The study of interactions between antigens and antibodies.

Sex chromosomes. Chromosomes that are connected with the determination of sex.

Sexduction. The incorporation of bacterial genes into F factors and their subsequent transfer by conjugation to a recipient cell.

Sex factor. A bacterial episome (for example, the F plasmid in *E. coli*) that enables the cell to be a donor of genetic material. The sex factor may be propagated in the cytoplasm, or it may be integrated into the bacterial chromosome.

Sex-influenced dominance. A dominant expression that depends on the sex of the individual. For example, horns in some breeds of sheep are dominant in males and recessive in females.

Sex-limited. Expression of a trait in only one sex. Examples: milk production in mammals; horns in Rambouillet sheep; egg production in chickens.

Sex linkage. Association or linkage of a hereditary trait with sex; the gene is in a sex chromosome, usually the X; often used synonymously with X-linkage.

Sex mosaic. See **Gynandromorph.**

Sexual reproduction. Reproduction involving the formation of mature germ cells (that is, eggs and sperm).

Shine-Dalgarno sequence. A conserved sequence in prokaryotic mRNAs that is complementary to a sequence near the 5′ terminus of the 16S ribosomal RNA and is involved in the initiation of translation.

Short interfering RNA (siRNA). Double-stranded RNA molecules 21–28 base pairs long that mediate the phenomenon of RNA interference; also known as microRNA molecules.

Shuttle vector. A plasmid capable of replicating in two different organisms, such as yeast and *E. coli.*

Sib-mating (crossing of siblings). Matings involving two individuals of the same parentage; brother-sister matings.

Sigma factor. The subunit of prokaryotic RNA polymerases that is responsible for the initiation of transcription at specific initiation sequences.

Signal transduction. The process whereby a molecular signal such as a hormone is passed internally within a cell by a system of molecules to effect a change in the cell's state.

Silencer. A DNA sequence that helps to reduce or shut off the expression of a nearby gene.

Silent polymorphism. A variant in DNA that does not alter the amino acid sequence of a protein.

SINEs (short interspersed nuclear elements). Families of short (150 to 300 bp), moderately repetitive transposable elements of eukaryotes. The best known SINE family is the Alu family in humans.

Single nucleotide polymorphism (SNP). A single base pair in the DNA that varies in a population.

Single-strand DNA-binding protein. A protein that coats DNA single strands, keeping them in an extended state.

Sister chromatid. One of the products of chromosome duplication.

Site-specific mutagenesis. See **Oligonucleotide-directed site-specific mutagenesis.**

Small nuclear ribonucleoproteins (snRNPs). RNA-protein complexes that are components of spliceosomes.

Small nuclear RNAs (snRNAs). Small RNA molecules that are located in the nuclei of eukaryotic cells; most snRNAs are components of the spliceosomes that excise introns from pre-mRNAs.

Somatic cell. A cell that is a component of the body, in contrast with a germ cell that is capable, when fertilized, of reproducing the organism.

Somatic-cell (nonheritable) gene therapy. Treatment of an inherited disorder by adding functional (wild-type) copies of a gene to nongerm-line cells of an individual carrying defective copies of that gene (cf. **Germ-line [heritable] gene therapy).**

Somatic hypermutation. A high frequency of mutation that occurs in the gene segments encoding the variable regions of antibodies during the differentiation of B lymphocytes into antibody-producing plasma cells.

Somatic mutation. A mutation that occurs in the nonreproductive cells (somatic cells) of the body and is not transmitted to progeny (cf. **Germinal mutation).**

SOS response. The synthesis of a whole set of DNA repair, recombination, and replication proteins in bacteria containing severely damaged DNA (for example, following exposure to UV light).

Southern blot. The transfer of DNA fragments from an electrophoretic gel to a cellulose or nylon membrane by capillary action.

Specialized transduction. Recombination in bacteria mediated by a bacteriophage that can only transfer genes in a small segment of the chromosome of the donor cell to a recipient cell (cf. **Generalized transduction**).

Species. Interbreeding, natural populations that are reproductively isolated from other such groups.

Sperm (abbreviation of spermatozoon, *pl.*, **spermatozoa).** A mature male germ cell.

Spermatids. The four cells formed by the meiotic divisions in spermatogenesis. Spermatids become mature spermatozoa or sperm.

Spermatocyte (sperm mother cell). The cell that undergoes two meiotic divisions (spermatogenesis) to form four spermatids; the *primary* spermatocyte before completion of the first meiotic division; the *secondary* spermatocyte after completion of the first meiotic division.

Spermatogenesis. The process by which maturation of the gametes (sperm) of the male takes place.

Spermatogonium (*pl.*, **spermatogonia).** Primordial male germ cell that may divide by mitosis to produce more spermatogonia. A spermatogonium may enter a growth phase and give rise to a primary spermatocyte.

Spermiogenesis. Formation of sperm from spermatids; the part of spermatogenesis that follows the meiotic divisions of spermatocytes.

Spheroplast. A plant or bacterial cell from which the wall has been removed. See **Protoplast.**

Spindle. A system of microtubules that distributes duplicated chromosomes equally and exactly to each of the daughters of a dividing eukaryotic cell.

Spliceosome. The RNA/protein complex that excises introns from primary transcripts of nuclear genes in eukaryotes.

Splicing. The process that covalently joins exon sequences of RNA and eliminates the intervening intron sequences.

Spontaneous mutation. A mutation that occurs without a known cause (cf. **Induced mutation).**

Sporophyte. The diploid generation in the life cycle of a plant that produces haploid spores by meiosis.

SRY. A Y-linked gene in humans and other mammals encoding a protein, the testis-determining factor, which plays a key role in male development.

Stamen. The elongated structure that bears the anthers in flowering plants.

Standard deviation. A measure of variability in a set of data; the square root of the variance.

Standard error. A measure of variation among a population of means.

Stationary-phase mutation. See **Adaptive mutation.**

Statistic. A value based on a sample or samples of a population from which estimates of a population value or parameter may be obtained.

Stem cell. A cell with the ability to proliferate extensively and whose offspring can differentiate into specialized cell types.

Sterility. Inability to produce offspring.

Structural gene. A gene that specifies the synthesis of a polypeptide.

STSs (sequence-tagged sites). Short, unique DNA sequences (usually 200 to 500 bp) that are amplified by PCR and used to link physical maps and genetic maps.

Subspecies. One of two or more morphologically or geographically distinct but interbreeding populations of a species.

Supercoil. A DNA molecule that contains extra twists as a result of overwinding (positive supercoils) or underwinding (negative supercoils).

Suppressor mutation. A mutation that partially or completely cancels the phenotypic effect of another mutation.

Suppressor-sensitive mutant. An organism that can grow when a second genetic factor—a suppressor—is present, but not in the absence of this factor.

Symbiont. An organism living in intimate association with another, dissimilar organism.

Sympatric speciation. The formation of new species by populations that inhabit the same or overlapping geographic regions.

Synapsis. The pairing of homologous chromosomes in the meiotic prophase.

Synaptoinemal complex. A ribbonlike structure formed between synapsed homologues at the end of the first meiotic prophase, binding the chromatids along their length and facilitating chromatid exchange.

Syndrome. A group of symptoms that occur together and represent a particular disease.

Synkaryon. A nucleus formed by the fusion of nuclei from two different somatic cells during somatic-cell hybridization.

Synonymous substitution. A base-pair change in a codon that does not alter the amino acid specified by the codon.

Synteny. The occurrence of two loci on the same chromosome, without regard to the distance between them.

TACTAAC box. A conserved DNA sequence located about 30 nucleotides upstream of the 3′ splice site of introns in nuclear genes of eukaryotes. The 3′ adenine in this sequence is involved in intron excision.

Taq polymerase. A heat-stable DNA polymerase isolated from the thermophilic bacterium *Thermus aquaticus.*

Target site duplication. A sequence of DNA that is duplicated when a transposable element inserts; usually found at each end of the insertion.

TATAAT sequence (-10 sequence). An AT-rich sequence in prokaryotic promoters that facilitates the localized unwinding of DNA and the initiation of RNA synthesis.

TATA box. A conserved promoter sequence that determines the transcription start site.

Tautomeric shift. The transfer of a hydrogen atom from one position in an organic molecule to another position.

Tay-Sachs disease. A lethal autosomal recessive disorder in humans characterized by neurological degeneration and death in early childhood. The disease is caused by the absence of an enzyme called hexosaminidase A.

T cell-mediated (cellular) immune response. The synthesis of antigen-specific T-cell receptors and the development of killer T cells in response to an encounter of immune system cells with a foreign immunogen.

T-cell receptor. An antigen-binding protein that is located on the surfaces of killer T cells and mediates the cellular immune response of mammals. The genes that encode T-cell antigens are assembled from gene segments by somatic recombination processes that occur during T lymphocyte differentiation.

T-DNA. The segment of DNA in the Ti plasmid of *Agrobacterium tumefaciens* that is transferred to plant cells and inserted into the chromosomes of the plant.

Telomerase. An enzyme that adds telomere sequences to the ends of eukaryotic chromosomes.

Telomere. The unique structure found at the ends of eukaryotic chromosomes.

Telophase. The last stage in each mitotic or meiotic division in which the chromosomes are assembled at the poles of the division spindle.

Temperate phage. A phage (virus) that invades but may not destroy (lyse) the host (bacterial cell) (cf. **Virulent phage**). However, it may subsequently enter the lytic cycle.

Temperature-sensitive mutant. An organism that can grow at one temperature but not at another.

Template. A pattern or mold. DNA stores coded information and acts as a model or template from which information is copied into complementary strands of DNA or transcribed into messenger RNA.

Template strand. In transcription, the DNA strand that is copied to produce a complementary strand of RNA.

Terminalization. Repelling movement of the centromeres of bivalents in the diplotene stages of the meiotic prophase that tends to move the visible chiasmata toward the ends of the bivalents.

Terminally redundant chromosome. A linear chromosome that contains the same nucleotide sequence at both ends.

Terminal transferase. An enzyme that adds nucleotides to the 3′ termini of DNA molecules.

Termination (of DNA, RNA, or protein synthesis). The release of a complete macromolecule (DNA, RNA, or polypeptide) after the incorporation of the final subunit (nucleotide or amino acid).

Termination signal. In transcription, a nucleotide sequence that specifies RNA chain termination.

Testcross. Backcross to the recessive parental type, or a cross between genetically unknown individuals with a fully recessive tester to determine whether an individual in question is heterozygous or homozygous for a certain allele. It is also used as a test for linkage.

Testis-determining factor (TDF). A protein produced early in the development of male mammals that stimulates the differentiation of the testes from the embryonic gonads.

Tetrad. The four cells arising from the second meiotic division in plants (pollen tetrads) or fungi (ascospores). The term is also used to identify the quadruple group of chromatids that is formed by the association of duplicated homologous chromosomes during meiosis.

Tetrad analysis. The determination of genotypes of the four products of a single meiotic event and the statistical analysis of their frequencies.

Tetraploid. An organism whose cells contain four haploid ($4n$) sets of chromosomes or genomes.

Tetrasomic (*noun*, tetrasome). Pertaining to a nucleus or an organism with four members of one of its chromosomes whereas the remainder of its chromosome complement is diploid (chromosome formula: $2n + 2$).

Tetratype. In fungi, a tetrad of spores that contains four different types; for example, *AB*, *aB*, *Ab*, and *ab*.

TFIIX (Transcription Factor X for RNA polymerase II). A protein required for the initiation of transcription by RNA polymerase II in eukaryotes; X represents any one of several different factors designated A through F.

Therapeutic cloning. A process in which the nucleus of an egg cell is replaced with the nucleus of a donor cell (possibly differentiated) to produce a population of stem cells that have the same genotype as the donor cell. These stem cells could then be used to replace lost cells in the donor organism.

Threshold trait. A trait that is manifested discontinuously but that is a function of underlying continuous genetic and environmental variation.

Thymine. A pyrimidine base found in DNA. The other three organic bases—adenine, cytosine, and guanine—are found in both RNA and DNA, but in RNA, thymine is replaced by uracil.

Ti plasmid. The large plasmid in *Agrobacterium tumefaciens*. It is responsible for the induction of tumors in plants with crown gall disease and is an important vector for transferring genes into plants, especially dicots.

T lymphocytes (T cells). Cells that differentiate in the thymus gland and are primarily responsible for the T cell-mediated or cellular immune response.

Topoisomerase. An enzyme that introduces or removes supercoils from DNA.

Totipotent cell (or nucleus). An undifferentiated cell (or nucleus) such as a blastomere that when isolated or suitably transplanted can develop into a complete embryo.

Trafficking. The movement of materials through the cytoplasm of a cell, usually guided by membranes, vesicles, and components of the cytoskeleton.

***trans*-acting.** A term describing substances that are diffusable and that can affect spatially separated entities within cells.

***trans* configuration.** See **Repulsion.**

Transcript. The RNA molecule produced by transcription of a gene.

Transcription. Process through which RNA is formed along a DNA template. The enzyme RNA polymerase catalyzes the formation of RNA from ribonucleoside triphosphates.

Transcriptional antiterminator. A protein that prevents RNA polymerase from terminating transcription at specific transcription-termination sequences.

Transcription factor. A protein that regulates the transcription of genes.

Transcription unit. A segment of DNA that contains transcription initiation and termination signals and is transcribed into one RNA molecule.

Transcriptome. The complete set of RNAs transcribed from a genome.

Transduction (t). Genetic recombination in bacteria mediated by bacteriophage. Abortive t: Bacterial DNA is injected by a phage into a bacterium, but it does not replicate. Generalized t: Any bacterial gene may be transferred by a phage to a recipient bacterium. Restricted t: Transfer of bacterial DNA by a temperate phage is restricted to only one site on the bacterial chromosome.

Transfection. The uptake of DNA by a eukaryotic cell, followed by the incorporation of genetic markers present in the DNA into the cell's genome.

Transfer RNAs (tRNAs). RNAs that transport amino acids to the ribosomes, where the amino acids are assembled into proteins.

Transformation (cancerous). The conversion of eukaryotic cells growing in culture to a state of uncontrolled cell growth (similar to tumor cell growth).

Transformation (genetic). Genetic alteration of an organism brought about by the incorporation of foreign DNA into cells.

Transgene. A foreign or modified gene that has been introduced into an organism.

Transgenic. A term applied to organisms that have been altered by introducing DNA molecules into them.

Transgressive variation. The appearance in the F_2 (or later) generation of individuals showing more extreme development of a trait than in either of the original parents.

***trans* heterozygote.** A heterozygote that contains two mutations arranged in the *trans* configuration—for example, $a\ b^+ / a^+ b$.

Transition. A mutation caused by the substitution of one purine by another purine or one pyrimidine by another pyrimidine in DNA or RNA.

Translation. Protein (polypeptide) synthesis directed by a specific messenger RNA; occurs on ribosomes.

Translocation. Change in position of a segment of a chromosome to another part of the same chromosome or to a different chromosome.

Transposable genetic element. A DNA element that can move from one location in the genome to another.

Transposase. An enzyme that catalyzes the movement of a DNA sequence to a different site in a DNA molecule.

Transposons. DNA elements that can move ("transpose") from one position in a DNA molecule to another.

Transposon tagging. The insertion of a transposable element into or nearby a gene, thereby marking that gene with a known DNA sequence.

Transversion. A mutation caused by the substitution of a purine for a pyrimidine or a pyrimidine for a purine in DNA or RNA.

Trihybrid. The offspring from homozygous parents differing in three pairs of genes.

Trinucleotide repeats. Tandem repeats of three nucleotides that are present in many human genes. In several cases, these trinucleotide repeats have undergone expansions in copy number that have resulted in inherited diseases.

Trisomic. An otherwise diploid cell or organism that has an extra chromosome of one pair (chromosome formula: $2n + 1$).

Trivalent. An association between three chromosomes during meiosis.

Tubulin. The major protein component of the microtubules of eukaryotic cells.

Tumor suppressor gene. A gene whose product is involved in the repression of cell division.

Ultraviolet (UV) radiation. The portion of the electromagnetic spectrum—wavelengths from about 1 to 350 nm—between ionizing radiation and visible light. UV is absorbed by DNA and is highly mutagenic to unicellular organisms and to the epidermal cells of multicellular organisms.

Unequal crossing over. Crossing over between repeated DNA sequences that have paired out of register, creating duplicated and deficient products.

Univalent. An unpaired chromosome at meiosis.

Universality (of the genetic code). The codons have the same meaning, with minor exceptions, in virtually all species.

Unordered tetrad. A set of four end products of meiosis not ordered with respect to the way in which they were produced in the two meiotic divisions.

Upstream sequence. A sequence in a unit of transcription that precedes (is located 5′ to) the transcription start site. The nucleotide pair in DNA corresponding to the nucleotide at the 5' end of the transcript (RNA) is designated +1. The preceding nucleotide pair is designated −1. All preceding (−) nucleotide sequences are upstream sequences (cf. **Downstream sequence**).

Uracil. A pyrimidine base found in RNA but not in DNA. In DNA, uracil is replaced by thymine.

Van der Waals interactions. Weak attractions between atoms placed in close proximity.

Variance. A measure of variation in a population; the square of the standard deviation.

Variation. In biology, the occurrence of differences among individuals.

Vector. A plasmid or viral chromosome that may be used to construct recombinant DNA molecules for introduction into living cells.

Velocity density gradient centrifugation. A procedure used to separate macromolecules based on their rate of movement through a density gradient.

Whole-genome shotgun sequencing. An approach to sequencing genomes that involves randomly cleaving the entire genome into small fragments, sequencing the ends of these fragments, and using supercomputers to assemble the complete sequence by aligning overlapping sequences.

Viability. The capability to live and develop normally.

Viroid. A small infectious (disease-causing) agent consisting of a naked circular molecule of RNA.

***vir* region (of Ti plasmid).** The region of the Ti plasmid of *Agrobacterium tumefaciens* that contains genes encoding products required for the transfer of the T-DNA from the bacterium to plant cells.

Virulent phage. A phage (virus) that destroys the host (bacterial) cell (cf. **Temperate phage**).

Viscoelastometry. A method to study the physical properties of molecules in solution.

VNTR (variable number tandem repeat). A short DNA sequence that is present as tandem repeats and in highly variable copy number.

Western blot. The transfer of proteins from an electrophoretic gel to a cellulose or nylon membrane by means of an electric force.

Wild type. The customary phenotype or standard for comparison.

Wobble hypothesis. Hypothesis to explain how one tRNA may recognize two codons. The first two bases of the mRNA codon and anticodon pair properly, but the third base in the anticodon has some play (or wobble) that permits it to pair with more than one base.

X chromosome. A chromosome associated with sex determination. In most animals, the female has two, and the male has one X chromosome.

YACs (yeast artificial chromosomes). Linear cloning vectors constructed from essential elements of yeast chromosomes. They can accommodate foreign DNA inserts of 200 to 500 kb in size.

Y chromosome. The partner of the X chromosome in the male of many animal species.

Z-DNA. A left-handed double helix that forms in G:C-rich DNA molecules. The Z refers to the zig-zagged paths of the sugar-phosphate backbones in this form of DNA.

Zygonema (*adj*., zygotene). Stage in meiosis during which synapsis occurs; after the leptotene stage and before the pachytene stage in the meiotic prophase.

Zygote. The cell produced by the union of two mature sex cells (gametes) in reproduction; also used in genetics to designate the individual developing from such a cell.

Answers to Odd-Numbered Questions and Problems

CHAPTER 1

1.1 The bases present in DNA are adenine, thymine, guanine, and cytosine; the bases present in RNA are adenine, uracil, guanine, and cytosine. The sugar in DNA is deoxyribose; the sugar in RNA is ribose.

1.3 Mendel postulated transmissible factors —genes—to explain the inheritance of traits. He discovered that genes exist in different forms, which we now call alleles. Each organism carries two copies of each gene. During reproduction, one of the gene copies is randomly incorporated into each gamete. When the male and female gametes unite at fertilization, the gene copy number is restored to two. Different alleles may coexist in an organism. During the production of gametes, they separate from each other without having been altered by coexistence.

1.5 AUCGAAUCA

1.7 ATTCGGTGC

1.9 The two mutant forms of the β-globin gene are properly described as alleles. Because neither of the mutant alleles can specify a "normal" polypeptide, an individual who carries each of them would probably suffer from anemia.

1.11 Sometimes DNA is synthesized from RNA in a process called reverse transcription. This process plays an important role in the life cycles of some viruses.

CHAPTER 2

2.1 Prokaryotic chromosomes are typically (but not always) smaller than eukaryotic chromosomes; in addition, prokaryotic chromosomes are circular, whereas eukaryotic chromosomes are linear. For example, the circular chromosome of *E. coli*, a prokaryote, is about 1.4 mm in circumference. By contrast, a linear human chromosome may be 10 to 30 cm long. Prokaryotic chromosomes also have a comparatively simple composition: DNA, some RNA, and some protein. Eukaryotic chromosomes are more complex: DNA, some RNA, and lots of protein.

2.3 In a eukaryotic cell the many chromosomes are contained within a membrane-bounded structure called the nucleus; the chromosomes of prokaryotic cells are not contained within a special subcellular compartment. Eukaryotic cells usually possess a well-developed internal system of membranes and they also have membrane-bounded subcellular organelles such as mitochondria and chloroplasts; prokaryotic cells do not typically have a system of internal membranes (although some do), nor do they possess membrane-bounded organelles.

2.5 (1) Anaphase: (f), (h); (2) metaphase: (e), (i); (3) prophase: (b), (c), (d); (4) telophase: (a), (g).

2.7 Interphase typically lasts longer than M phase. During interphase, DNA must be synthesized to replicate all the chromosomes. Other materials must also be synthesized to prepare for the upcoming cell division.

2.9 Crossing over occurs *after* chromosomes have duplicated in cells going through meiosis.

2.11 Chromosome disjunction occurs during anaphase I. Chromatid disjunction occurs during anaphase II.

2.13 Chromosomes 11 and 16 would not be expected to pair with each other during meiosis; these chromosomes are heterologues, not homologues.

2.15 Yeast chromosomes are, on average, smaller than *E. coli* chromosomes. Thus, some eukaryotic chromosomes are smaller than some prokaryotic chromosomes.

2.17 Among eukaryotes, there doesn't seem to be a clear relationship between genome size and gene number. For example, humans, with 3.2 billion base pairs of genomic DNA, have between 20,000 and 25,000 genes, and *Arabidopsis* plants, with about 150 million base pairs of genomic DNA, have roughly the same number of genes as humans. However, among prokaryotes, gene number is rather tightly correlated with genome size, probably because there is so little nongenic DNA.

2.19 One of the pollen nuclei fuses with the egg nucleus in the female gametophyte to form the zygote, which then develops into an embryo and ultimately into a sporophyte. The other genetically functional pollen nucleus fuses with two nuclei in the female gametophyte to form a triploid nucleus, which then develops into a triploid tissue, the endosperm; this tissue nourishes the developing plant embryo.

2.21 (a) 7, (b) 7, (c) 14.

CHAPTER 3

3.1 (a) All tall; (b) 3/4 tall, 1/4 dwarf; (c) all tall; (d) 1/2 tall, 1/2 dwarf.

3.3 (a) Checkered, red (*CC BB*) × plain, brown (*cc bb*) → F_1 all checkered, red (*Cc Bb*); (b) F_2 progeny: 9/16 checkered, red (*C- B-*), 3/16 plain, red (*cc B-*), 3/16 checkered, brown (*C- bb*), 1/16 plain, brown (*cc bb*).

3.5

F_1 gametes	F_2 genotypes	F_2 phenotypes
(a) 2	3	2
(b) 2 × 2 = 4	3 × 3 = 9	2 × 2 = 4
(c) 2 × 2 × 2 = 8	3 × 3 × 3 = 27	2 × 2 × 2 = 8
(d) 2^n	3^n	2^n, where *n* is the number of genes

3.7 Among the F_2 progeny with long, black fur, the genotypic ratio is 1 *BB RR*: 2 *BB Rr*: 2 *Bb RR*: 4 *Bb Rr*; thus, 1/9 of the rabbits with long, black fur are homozygous for both genes.

3.9 Half the children from *Aa* × *aa* matings would not have albinism. In a family of three children, the chance that 2 will be unaffected and 1 affected is $3 \times (1/2)^1 \times (1/2)^2 = 3/8$.

3.11 $(1/2)^4 = 1/16$

3.13 (10/32) + (10/32) + (5/32) + (1/32) = 26/32

3.15 Man (*Cc ff*) × woman (*cc Ff*). (a) *cc ff*, (1/2) × (1/2) = 1/4; (b) *Cc ff*, (1/2) × (1/2) = 1/4; (c) *cc Ff*, (1/2) × (1/2) = 1/4; (d) *Cc Ff*, (1/2) × (1/2) = 1/4.

3.17 (a) 1, reject; (b) 2, reject; (c) 3, accept; (d) 3, accept.

3.19 $\chi^2 = (30 - 25)^2/25 + (20 - 25)^2/25 = 2$, which is less than 3.84, the 5 percent critical value for a chi-square statistic with one degree of freedom; consequently, the observed segregation ratio is consistent with the expected ratio of 1:1.

3.21 (a) (1/2) × (1/4) = 1/8; (b) (1/2) × (1/2) × (1/4) = 1/16; (c) (2/3) × (1/4) = 1/6; (d) (2/3) × (1/2) × (1/2) × (1/4) = 1/24

3.23 For III-1 × III-2, the chance of an affected child is 1/2. For IV-2 × IV-3, the chance is zero.

3.25 1/2.

CHAPTER 4

4.1

Parents	Offspring
(a) yellow × yellow	2 yellow: 1 light belly
(b) yellow × light belly	2 yellow: 1 light belly: 1 black and tan
(c) black and tan × yellow	2 yellow: 1 black and tan: 1 black
(d) light belly × light belly	all light belly
(e) light belly × yellow	1 yellow: 1 light belly
(f) agouti × black and tan	1 agouti: 1 black and tan
(g) black and tan × black	1 black and tan: 1 black
(h) yellow × agouti	1 yellow: 1 light belly
(i) yellow × yellow	2 yellow: 1 light belly

4.3 MN.

4.5 No. The woman is I^AI^B. One man could be either I^AI^A or I^Ai; the other must be ii. Either could be the father of the child.

4.7 (a) 1 A: 1 B; (b) all B; (c) 1 A: 1 B: 1 AB: 1 O; (d) 1 A: 1 O

4.9 The woman is I^AI^B L^ML^M; the man is *ii* L^ML^N; the blood types of the children will be A and M, A and MN, B and M, B and MN, all equally likely.

4.11 The mother is *Bb* and the father is *bb*. The chance that a daughter is *Bb* is 1/2. (a) The chance that the daughter will have a bald son is (1/2) × (1/2) = 1/4. (b) The chance that the daughter will have a bald daughter is 1/4.

4.13 The individuals III-4 and III-5 must be homozygous for recessive mutations in different genes; that is, one is *aa BB* and the other is *AA bb*; none of their children is deaf because all of them are heterozygous for both genes (*Aa Bb*).

4.15 12/16 white, 3/16 yellow, 1/16 green.

4.17 (a) 3/4 walnut, 1/4 rose; (b) 1/2 walnut, 1/2 pea; (c) 3/8 walnut, 3/8 rose, 1/8 pea, 1/8 single; (d) 1/2 rose, 1/2 single.

4.19 (a) Because the F_2 segregation is approximately 9 red: 7 white, flower color is due to epistasis between two independently assorting genes: red = *A-B-* and white = *aa B-*, *A-bb*, or *aa bb*. (b) colorless precursor—*A*→ colorless product—*B*→ red pigment.

4.21 9/16 dark red (wild-type), 3/16 brownish-purple, 3/16 bright red, 1/16 white.

4.23 The pedigree is shown below:

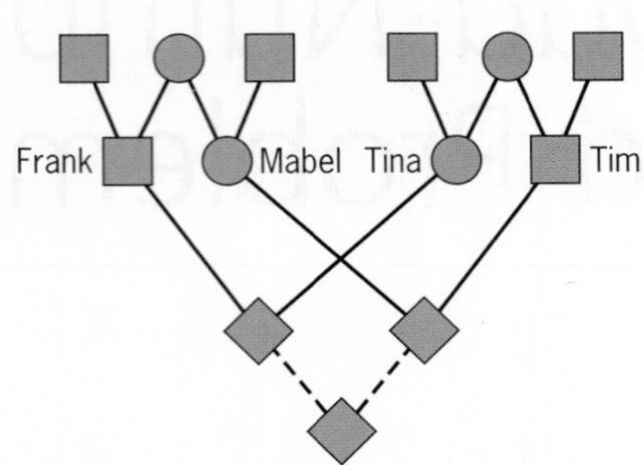

The coefficient of relationship between the offspring of the two couples is obtained by calculating the inbreeding coefficient of the imaginary child from a mating between these offspring and multiplying by 2: $[(1/2)^5 \times 2] \times 2 = 1/8$.

4.25 $F_A = (1/2)^5 = 1/32$; $F_B = 2 \times (1/2)^6 = 1/32$; $F_C = 2 \times (1/2)^7 = 1/64$.

4.27 (a) purple × red; (b) proportion white (*aa*) = 1/4; (c) proportion red (*A- B- C- dd*) = (3/4)(3/4)(3/4)(1/2) = 27/128, proportion white (*aa*) = 1/4 = 32/128, proportion blue (*A- B- cc Dd*) = (3/4)(3/4)(1/4)(1/2) = 9/128.

4.29 The mean ear length for randomly mated maize is 24 cm and that for maize from one generation of self-fertilization is 20 cm. The inbreeding coefficient of the offspring of one generation of self-fertilization is 1/2, and the inbreeding coefficient of the offspring of two generations of self-fertilization is (1/2)(1 + 1/2) = 3/4. Mean ear length (Y) is expected to decline linearly with inbreeding according to the equation $Y = 24 - b\,F_1$ where b is the slope of the line. The value of b can be determined from the two values of Y that are given. The difference between these two values (4 cm) corresponds to an increase in F from 0 to 1/2. Thus, $b = 4/(1/2) = 8$ cm, and for $F = 3/4$, the predicted mean ear length is $Y = 24 - 8 \times (3/4) = 18$.

CHAPTER 5

5.1 All the daughters will be green and all the sons will be rosy.

5.3 XX is female, XY is male, XXY is female, XXX is female (but barely viable), XO is male (but sterile).

5.5 The male-determining sperm carries a Y chromosome; the female-determining sperm carries an X chromosome.

5.7 Each of the rare vermilion daughters must have resulted from the union of an X(*v*) X(*v*) egg with a Y-bearing sperm. The diplo-X eggs must have originated through nondisjunction of the X chromosomes during oogenesis in the mother. However, we cannot determine if the nondisjunction occurred in the first or the second meiotic division.

5.9 1/8

5.11 The risk for the child is P(mother is *C/c*) × P(mother transmits *c*) × P(child is male) = (1/2) × (1/2) × (1/2) = 1/8; if the couple has already had a child with color blindness, P(mother is *C/c*) = 1, and the risk for each subsequent child is 1/4.

5.13 Male.

5.15 (a) Female; (b) intersex; (c) intersex; (d) male: (e) female; (f) male.

5.17 Female.

5.19 *Drosophila* does not achieve dosage compensation by inactivating one of the X chromosomes in females.

5.21

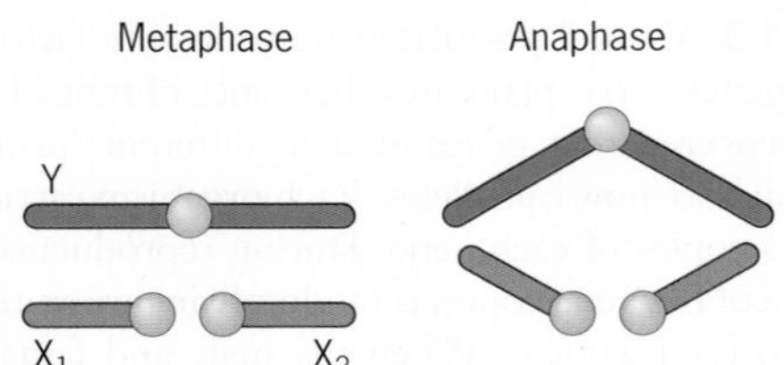

5.23 Eye color in canaries is due to a gene on the *Z* chromosome, which is present in two copies in males and one copy in females. The allele for pink color at hatching (*p*) is recessive to the allele for black color at hatching (*P*). There is no eye color gene on the other sex chromosome (*W*), which is present in one copy in females and absent in males. The parental birds were genotypically *p/W* (cinnamon females) and *P/P* (green males). Their F_1 sons were genotypically *p/P* (with black eyes at hatching). When these sons were crossed to green females (genotype *P/W*), they produced F_2 progeny that sorted into three categories: males with black eyes at hatching (*P/-*, half the total progeny), females with black eyes at hatching (*P/W*, a fourth of the total progeny), and females with pink eyes at hatching (*p/W*, a fourth of the total progeny). When these sons were crossed to cinnamon females (genotype *p/W*), they produced F_2 progeny that sorted into four equally frequent categories: males with black eyes at hatching (genotype *P/p*), males with pink eyes at hatching (genotype *p/p*), females with black eyes at hatching (genotype *P/W*), and females with pink eyes at hatching (genotype *p/W*).

CHAPTER 6

6.1 In allotetraploids, each member of the different sets of chromosomes can pair with a homologous partner during prophase I and then disjoin during anaphase I. In triploids, disjunction is irregular because homologous chromosomes associate during prophase I by forming either bivalents and univalents or by forming trivalents.

6.3 XX is female, XY is male, XO is female (but sterile), XXX is female, XXY is male (but sterile), XYY is male.

6.5 The fly is a gynandromorph, that is, a sexual mosaic. The yellow tissue is X(*y*)/O and the gray tissue is X(*y*)/X(+). This mosaicism must have arisen through loss of the X chromosome that carried the wild-type allele, presumably during one of the early embryonic cleavage divisions.

6.7 The fertile plant is an allotetraploid with 7 pairs of chromosomes from species A and 9 pairs of chromosomes from species B; the total number of chromosomes is $(2 \times 7) + (2 \times 9) = 32$.

6.9 Nondisjunction must have occurred in the mother. The color blind woman with Turner syndrome was produced by the union of an X-bearing sperm, which carried the mutant allele for color blindness, and a nullo-x egg.

6.11 XYY men would produce more children with sex chromosome abnormalities because their three sex chromosomes will disjoin irregularly during meiosis. This irregular disjunction will produce a variety of aneuploid gametes, including the XY, YY, XYY, and nullo sex chromosome constitutions.

6.13 (a) Deletion

3 4
1 2 5 6 7 8
1 2 3 4 5 6 7 8

(b) Duplication:

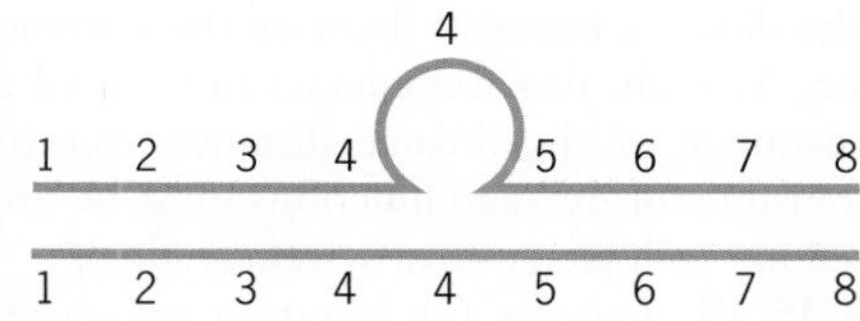

(c) A terminal inversion:

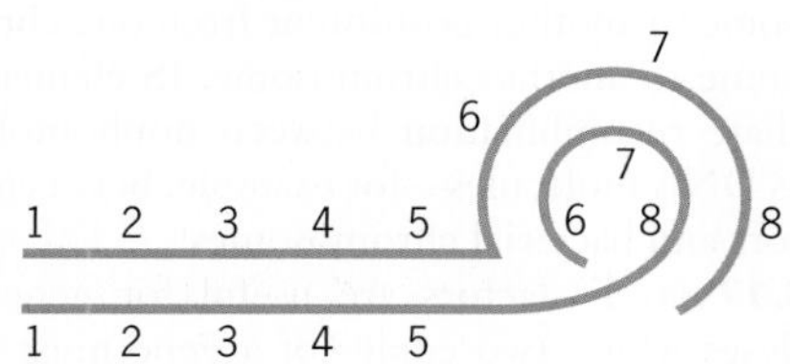

6.15

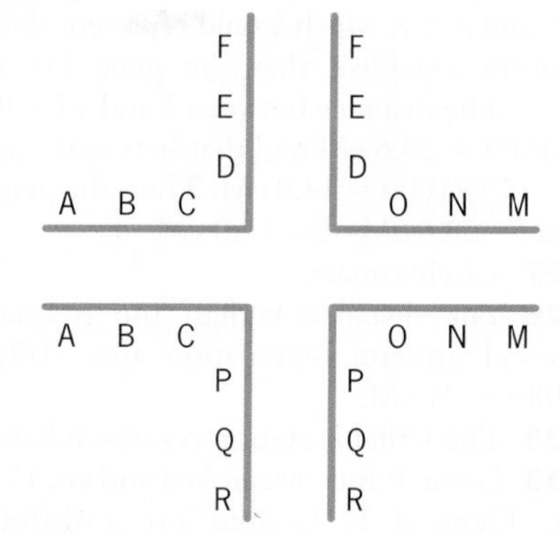

6.17 All the daughters will be yellow-bodied and all the sons will be white-eyed.

6.19 The boy carries a translocation between chromosome 21 and another chromosome, say chromosome 14. He also carries a normal chromosome 21 and a normal chromosome 14. The boy's sister carries the translocation, one normal chromosome 14, and two normal copies of chromosome 21.

6.21 The three populations are related by a series of inversions:

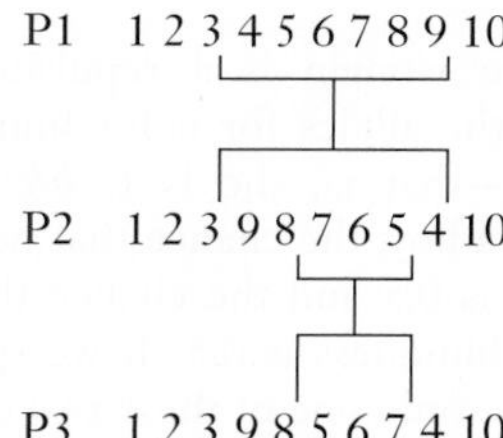

6.23 The mother is heterozygous for a reciprocal translocation between the long arms of the large and small chromosomes; a piece from the long arm of the large chromosome has been broken off and attached to the long arm of the short chromosome. The child has inherited the rearranged large chromosome and the normal small chromosome from the mother. Thus, because the rearranged large chromosome is deficient for some of its genes, the child is hypoploid.

6.25 XX zygotes will develop into males because one of their X chromosomes carries the *SRY* gene that was translocated from the Y chromosome. XY zygotes will develop into females because their Y chromosome has lost the *SRY* gene.

CHAPTER 7

7.1 A two-strand double crossover must have occurred.

7.3 (a) Cross: $a^+b^+/a^+b^+ \times a\,b/a\,b$ Gametes: a^+b^+ from one parent, $a\,b$ from the other F_1: $a^+b^+/a\,b$; (b) 40% a^+b^+, 40% $a\,b$, 10% a^+b, 10% $a\,b^+$; (c) F_2 from testcross: 40% $a^+b^+/a\,b$, 40% $a\,b/a\,b$, 10% $a^+b/a\,b$, 10% $a\,b^+/a\,b$; (d) Coupling linkage phase; (e) F_2 from intercross:

		Sperm			
		40% a^+b^+	40% $a\,b$	10% a^+b	10% $a\,b^+$
Eggs	40% a^+b^+	16% a^+b^+/a^+b^+	16% $a^+b^+/a\,b$	4% a^+b^+/a^+b	4% $a^+b^+/a\,b^+$
	40% $a\,b$	16% $a\,b/a^+b^+$	16% $a\,b/a\,b$	4% $a\,b/a^+b$	4% $a\,b/a\,b^+$
	10% a^+b	4% a^+b/a^+b^+	4% $a^+b/a\,b$	1% a^+b/a^+b	1% $a^+b/a\,b^+$
	10% $a\,b^+$	4% $a\,b^+/a^+b^+$	4% $a\,b^+/a\,b$	1% $a\,b^+/a^+b$	1% $a\,b^+/a\,b^+$

Summary of phenotypes:

a^+ and b^+	66%
a^+ and b	9%
a and b^+	9%
a and b	16%

7.5 No. The genes *a* and *d* could be very far apart on the same chromosome—so far apart that they recombine freely, that is, 50 percent of the time.

7.7 Coupling heterozygotes $a^+b^+/a\,b$ would produce the following gametes: 35% a^+b^+, 35% $a\,b$, 15% a^+b, 15% $a\,b^+$; repulsion heterozygotes $a^+b/a\,b^+$ would produce the following gametes: 35% a^+b, 35% $a\,b^+$, 15% a^+b^+, 15% $a\,b$. In each case, the frequencies of the testcross progeny would correspond to the frequencies of the gametes.

7.9 Yes. Recombination frequency = $(39 + 42)/(140 + 39 + 42 + 147) = 0.171$. Cross:

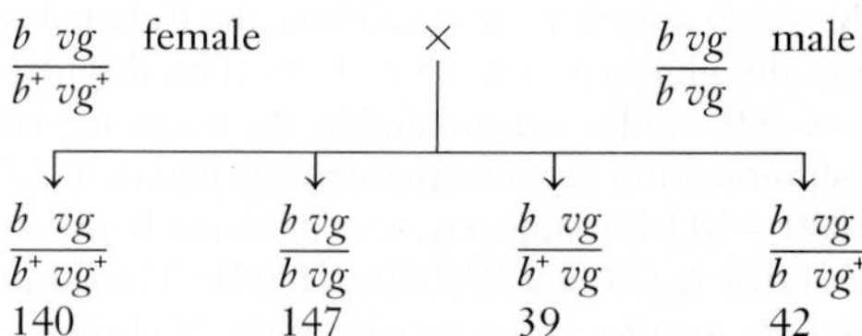

7.11 Yes. Recombination frequency is estimated by the frequency of black offspring among the colored offspring: $21/(58 + 21) = 0.27$. Cross:

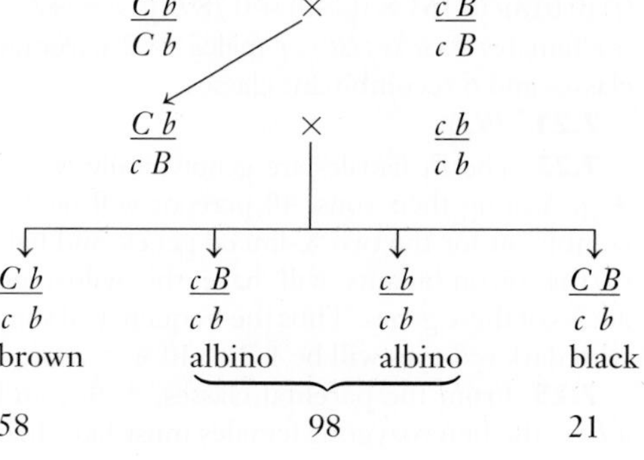

7.13 (a) The F_1 females, which are *sr* e^+/ sr^+ *e*, produce four types of gametes: 46% *sr* e^+, 46% sr^+ *e*, 4% *sr e*, 4% sr^+ e^+. (b) The F_1 males, which have the same genotype as the F_1 females, produce two types of gametes: 50% *sr* e^+, 50% sr^+ *e*; remember, there is no crossing over in Drosophila males. (c) 46% striped, gray; 46% unstriped, ebony; 4% striped, ebony; 4% unstriped, gray. (d) The offspring from the intercross can be obtained from the following table.

		Sperm	
		sr e^+ 0.50	sr^+ *e* 0.50
Eggs	*sr* e^+ 0.46	*sr* e^+/*sr* e^+ 0.23	*sr* e^+/sr^+ *e* 0.23
	sr^+ *e* 0.46	sr^+ *e*/*sr* e^+ 0.23	sr^+ *e*/sr^+ *e* 0.23
	sr e 0.04	*sr e*/*sr* e^+ 0.002	*sr e*/sr^+ *e* 0.002
	sr^+ e^+ 0.04	sr^+ e^+/*sr* e^+ 0.002	sr^+ e^+/sr^+ *e* 0.002

Summary of phenotypes	
striped, gray	0.25
unstriped, gray	0.50
striped, ebony	0
unstriped, ebony	0.25

7.15 (a) The F_1 females, which are *cn* vg^+/ cn^+*vg*, produce four types of gametes: 45% *cn* vg^+, 45% cn^+*vg*, 5% cn^+vg^+, 5% *cn vg*. (b) 45% cinnabar eyes, normal wings; 45% reddish-brown eyes, vestigial wings; 5% reddish-brown eyes, normal wings; 5% cinnabar eyes, vestigial wings.

7.17 The double crossover classes, which are the two that were not observed, establish that the gene order is *y*—*w*—*ec*. Thus, the F_1 females had the genotype *y w ec*/+ + +. The distance between *y* and *w* is estimated by the frequency of recombination between these two genes: (8 + 7)/ 1000 = 0.015; similarly, the distance between *w* and *ec* is (18 + 23)/1000 = 0.041. Thus, the genetic map for this segment of the X chromosome is *y*—1.5 cM—*w*—4.1 cM—*ec*.

7.19 (a) Two of the classes (the parental types) vastly outnumber the other six classes (recombinant types); (b) *st* + +/+ *ss e*; (c) *st*—*ss*—*e*; (d) [(145 + 122) × 1 + (18) × 2]/1000 = 30.3 cM; (e) (122 + 18)/1000 = 14.0 cM; (f) (0.018)/ (0.163 × 0.140) = 0.789. (g) *st* + +/ + *ss e* females × *st ss e*/*st ss e* males → 2 parental classes and 6 recombinant classes.

7.21 $(P/2)^2$

7.23 The F_1 females are genotypically *pn* +/ + *g*. Among their sons, 40 percent will be recombinant for the two X-linked genes, and half of the recombinants will have the wild-type alleles of these genes. Thus the frequency of sons with dark red eyes will be 1/2 × 40% = 20%.

7.25 From the parental classes, + + *c* and *a b* +, the heterozygous females must have had the genotype + + *c*/*a b* +. The missing classes, + *b* + and *a* + *c*, which would represent double crossovers, establish that the gene order is *b*—*a*—*c*. The distance between *b* and *a* is (96 + 110)/1000 = 20.6 cM and that between *a* and *c* is (65 + 75)/1000 = 14.0 cM. Thus, the genetic map is *b*—20.6 cM—*a*—14.0 cM—*c*.

7.27 5.4 chiasmata.

7.29 The distance is half the frequency of second division segregation asci: (1/2) × (84/200) = 21 cM.

7.31 The lethal mutation resides in band 7.

7.33 Genes *b* and *c* are linked and are 17 cM apart. Gene *a* is located on a different chromosome.

7.35 (a) 2 crossovers; (b) one exchange between *x* and *y*, the other between *y* and *z*; (c) 4-strand double crossover.

7.37 A two-strand double crossover within the inversion; the exchange points of the double crossover must lie between the genetic markers and the inversion breakpoints.

7.39 II-1 has the genotype *C h*/*c H*, that is, she is a repulsion heterozygote for the alleles for color blindness (*c*) and hemophilia (*h*). None of her children are recombinant for these alleles.

7.41 The woman is a repulsion heterozygote for the alleles for color blindness and hemophilia—that is, she is *C h*/*c H*. If the woman has a boy, the chance that he will have hemophilia is 0.5 and the chance that he will have color blindness is 0.5. If we specify that the boy have only one of these two conditions, then the chance that he will have color blindness is 0.45. The reason is that the boy will inherit a nonrecombinant X chromosome with a probability of 0.9, and half the nonrecombinant X chromosomes will carry the mutant allele for color blindness and the other half will carry the mutant allele for hemophilia. The chance that the boy will have both conditions is 0.05 and the chance that he will have neither condition is 0.05. The reason is that the boy will inherit a recombinant X chromosome with a probability of 0.1, and half the recombinant X chromosomes will carry both mutant alleles and the other half will carry neither mutant allele.

CHAPTER 8

8.1 Bacteriophage T4 is a virulent phage. When it infects a host cell, it reproduces and kills the host cell in the process. Bacteriophage lambda can reproduce and kill the host bacterium—the lytic response—just like phage T4, or it can insert its chromosome into the chromosome of the host and remain there in a dormant state—the lysogenic response.

8.3 Linkage maps are based on population averages. With circularly permuted chromosomes, genes that are close together on some chromosomes are at opposite ends of other chromosomes. At the population level, these circularly permuted chromosomes yield a circular genetic map.

8.5 Viruses reproduce and transmit their genes to progeny viruses. They utilize energy provided by host cells and respond to environmental and cellular signals like other living organisms. However, viruses are obligate parasites; they can reproduce only in appropriate host cells.

8.7 The *a*, *b*, and *c* mutations are closely linked and in the order *b*—*a*—*c* on the chromosome.

8.9 The insertion of the phage λ chromosome into the host chromosome is a site-specific recombination process catalyzed by an enzyme that recognizes specific sequences in the λ and *E. coli* chromosomes. Crossing over between homologous chromosomes is not sequence-specific. It can occur at many sites along the two chromosomes.

8.11 Perform two experiments: (1) determine whether the process is sensitive to DNase, and (2) determine whether cell contact is required for the process to take place. The cell contact requirement can be tested by a U-tube experiment (see Fig. 8.17). If the process is sensitive to DNase, it is similar to transformation. If cell contact is required, it is similar to conjugation. If it is neither sensitive to DNase nor requires cell contact, it is similar to transduction.

8.13 Cotransduction refers to the simultaneous transduction of two different genetic markers to a single recipient cell. Since bacteriophage particles can package only 1/100 to 1/50 of the total bacterial chromosome, only markers that are relatively closely linked can be cotransduced. The frequency of cotransduction of any two markers will be an inverse function of the distance between them on the chromosome. As such, this frequency can be used as an estimate of the linkage distance. Specific cotransduction-linkage functions must be prepared for each phage-host system studied.

8.15 IS elements (or insertion sequences) are short (800–1400 nucleotide pairs) DNA sequences that are transposable—that is, capable of moving from one position in a chromosome to another position or from one chromosome to another chromosome. IS elements mediate recombination between nonhomologous DNA molecules—for example, between F factors and bacterial chromosomes.

8.17 (a) F′ factors are useful for genetic analyses where two copies of a gene must be pre-sent in the same cell, for example, in determining dominance relationships. (b) F′ factors are formed by abnormal excision of F factors from Hfr chromosomes (see Fig. 8.31). (c) By the conjugative transfer of an F′ factor from a donor cell to a recipient (F^-) cell.

8.19

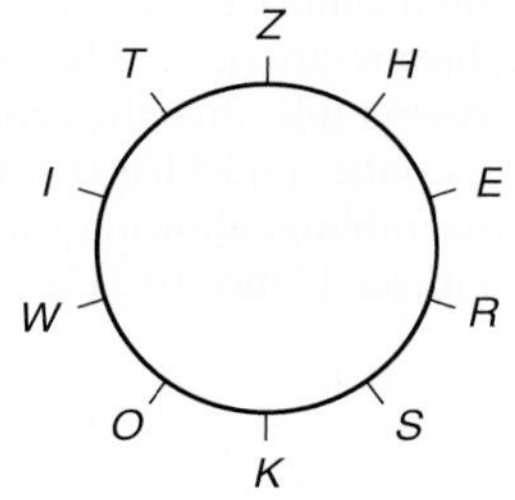

8.21 *pro—pur—his.*

8.23

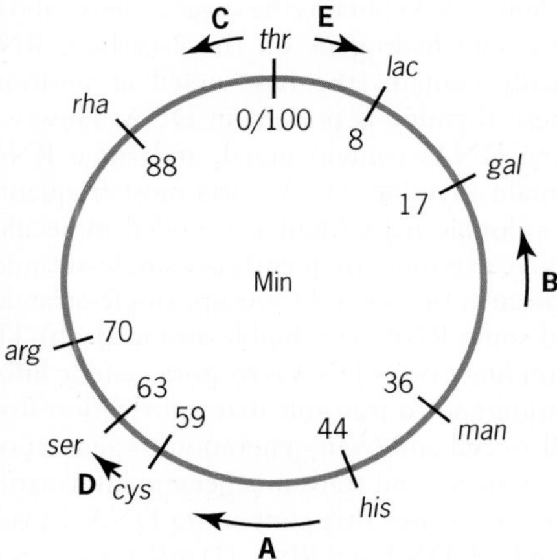

CHAPTER 9

9.1 (a) The objective was to determine whether the genetic material was DNA or protein. (b) By labeling phosphorus, a constituent of DNA, and sulfur, a constituent of protein, in a virus, it was possible to demonstrate that only the labeled phosphorus was introduced into the host cell during the viral reproductive cycle. The DNA was enough to produce new phages. (c) Therefore DNA, not protein, is the genetic material.

9.3 (a) Griffith's *in vivo* experiments demonstrated the occurrence of transformation in pneumococcus. They provided no indication as to the molecular basis of the transformation phenomenon. Avery and colleagues carried out *in vitro* experiments, employing biochemical analyses to demonstrate that transformation was mediated by DNA. (b) Griffith showed that a transforming substance existed; Avery et al. defined it as DNA. (c) Griffith's experiments did not include any attempt to characterize the substance responsible for transformation. Avery et al. isolated DNA in "pure" form and demonstrated that it could mediate transformation.

9.5 (a) The RNA genome of a virus is packaged inside a protein coat, whereas a viroid is composed of naked RNA. (b) A prion is composed of protein, whereas a viroid is RNA. (c) Viroids and prions are both infectious; they can maintain the phenotype they produce from generation to generation.

9.7 DNA has one atom less of oxygen than RNA in the sugar part of the molecule. In DNA, thymine replaces the uracil that is present in RNA. (In certain bacteriophages, DNA-containing uracil is present.) DNA is most frequently double-stranded, but bacteriophages such as ΦX174 contain single-stranded DNA. RNA is most frequently single-stranded. Some viruses, such as the Reoviruses, however, contain double-stranded RNA chromosomes.

9.9 No. TMV RNA is single-stranded. Thus the base-pair stoichiometry of DNA does not apply.

9.11 (a) A multistranded, spiral structure was suggested by the X-ray diffraction patterns. A double-stranded helix with specific base-pairing nicely fits the 1:1 stoichiometry observed for A:T and G:C in DNA. (b) Use of the known hydrogen-bonding potential of the bases provided a means of holding the two complementary strands in a stable configuration in such a double helix.

9.13 (a) false; (b) false; (c) true; (d) true; (e) true; (f) true; (g) false; (h) true; (i) true; (j) false; (k) true; (l) false; (m) true.

9.15 The B form of DNA helix is that proposed by Watson and Crick and is the conformation that DNA takes under physiological conditions. It is a right-handed double helical coil with 10 bases per turn of the helix and a diameter of 1.9 nm. It has a major and a minor groove. Z-DNA is left-handed, has 12 bases per turn, a single deep groove, and is 1.8 nm in diameter. Its sugar-phosphate backbone takes a zigzagged path, and it is G:C rich. A-DNA is a right-handed helix with 11 base pairs per turn. It is a shorter, thicker double helix with a diameter of 0.23 nm and has a narrow, deep major groove and a broad, shallow minor groove. A-DNA forms *in vitro* under high salt concentrations or in a partially dehydrated state.

9.17 (a) 85.4° C; (b) about 15%

9.19 The satellite DNA fragments would renature much more rapidly than the main-band DNA fragments. In *D. virilus* satellite DNAs, all three have repeating heptanucleotide-pair sequences. Thus essentially every 40 nucleotide-long (average) single-stranded fragment from one strand will have a sequence complementary (in part) with every single-stranded fragment from the complementary strand. Many of the nucleotide-pair sequences in main-band DNA will be unique sequences (present only once in the genome).

9.21 Interphase. Chromosomes are for the most part metabolically inactive (exhibiting little transcription) during the various stages of condensation in mitosis and meiosis.

9.23 (1) The nucleosome level; the core containing an octamer of histones plus 146 nucleotide pairs of DNA arranged as $1\frac{3}{4}$ turns of a supercoil (see Figure 9.22), yielding an approximately 10-nm diameter spherical body; or juxtaposed, a roughly 10-nm diameter fiber. (2) The 30-nm fiber observed in condensed mitotic and meiotic chromosomes; it appears to be formed by coiling or folding the 10-nm nucleosome fiber. (3) The highly condensed mitotic and meiotic chromosomes (for example, metaphase chromosomes); the tight folding or coiling maintained by a "scaffold" composed of nonhistone chromosomal proteins (see Figure 9.26).

9.25 Viscoelastometry is a procedure used to measure the viscosity of molecules in solution. In addition, viscoelastometric methods can be used to estimate the sizes of the largest DNA molecules present in aqueous solutions. By using viscoelastometry, scientists have obtained evidence which indicates that all of the DNA present in chromosomes of eukaryotes exists as giant, "chromosome-size" DNA molecules (one huge DNA molecule per chromosome). These data eliminated early models of chromosome structure with multiple DNA molecules joined end-to-end by protein or RNA "linkers."

9.27 The locations of the satellite DNA sequence in the chromosomes can be determined by *in situ* hybridization (see Focus on *In Situ* Hybridization).

9.29 Purified DNA from Type III cells was shown to be sufficient to transform Type II cells. This occurred in the absence of any dead Type III cells.

CHAPTER 10

10.1 If nascent DNA is labeled by exposure to ^{3}H-thymidine for very short periods of time, continuous replication predicts that the label would be incorporated into chromosome-sized DNA molecules, whereas discontinuous replication predicts that the label would first appear in small pieces of nascent DNA (prior to covalent joining, catalyzed by polynucleotide ligase).

10.3 (a) Both $3' \rightarrow 5'$ and $5' \rightarrow 3'$ exonuclease activities. (b) The $3' \rightarrow 5'$ exonuclease "proofreads" the nascent DNA strand during its synthesis. If a mismatched base pair occurs at the 3′-OH end of the primer, the 3′ → 5′ exonuclease removes the incorrect terminal nucleotide before polymerization proceeds again. The 5′ → 3′ exonuclease is responsible for the removal of RNA primers during DNA replication and functions in pathways involved in the repair of damaged DNA (see Chapter 13). (c) Yes, both exonuclease activities appear to be very important. Without the 3′ → 5′ proofreading activity during replication, an intolerable mutation frequency would occur. The 5′ → 3′ exonuclease activity is essential to the survival of the cell. Conditional mutations that alter the 5′ → 3′ exonuclease activity of DNA polymerase I are lethal to the cell under conditions where the exonuclease is nonfunctional.

10.5 One-half of the DNA molecules fully heavy (^{15}N in both strands); the other half of the molecules "hybrid" (^{15}N in one strand, ^{14}N in the complementary strand).

10.7 Current evidence suggests that polymerases α, δ, and/or ε are required for the replication of nuclear DNA. Polymerase δ and/or ε are thought to catalyze the continuous synthesis of the leading strand, and polymerase α is believed to function as a primase in the discontinuous synthesis of the lagging strand. Polymerase γ catalyzes replication of organellar chromosomes. Polymerases β, ζ, η, θ, ι, κ, λ, μ and σ function in various DNA repair pathways (see Chapter 13).

10.9

Two

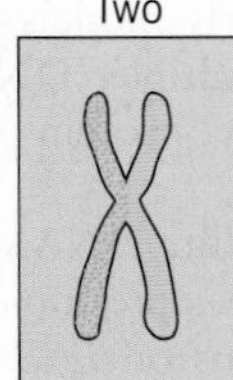

Plus two

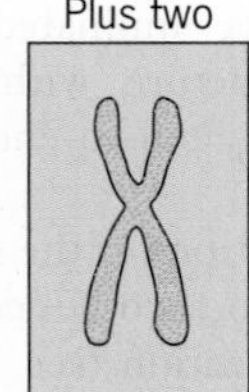

For both the large and small chromosomes

10.11 Sucrose velocity density gradient centrifugation is the standard technique for separating DNA molecules in this size range. Pulsed-field gel electrophoresis (Chapter 9) could also be used.

10.13 In eukaryotes, the rate of DNA synthesis at each replication fork is about 2500 to 3000 nucleotide pairs per minute. Large eukaryotic chromosomes often contain 10^7 to 10^8 nucleotide pairs. A single replication fork could not replicate the giant DNA molecule in one of these large chromosomes fast enough to permit the observed cell generation times.

10.15 No. *E. coli* strains carrying *polA* mutations that eliminate the 3′ → 5′ exonuclease activity of DNA polymerase I will exhibit unusually high mutation rates.

10.17 (a) DNA gyrase; (b) primase; (c) the 5′ → 3′ exonuclease activity of DNA polymerase I; (d) the 5′ → 3′ polymerase activity of DNA polymerase III; (e) the 3′ → 5′ exonuclease activity of DNA polymerase III.

10.19 (a) Rolling-circle replication begins when an endonuclease cleaves one strand of a circular DNA double helix. This cleavage produces a free 3′-OH on one end of the cut strand, allowing it to function as a primer. (b) The discontinuous synthesis of the lagging strand requires the *de novo* initiation of each Okazaki fragment, which requires DNA primase activity.

10.21 DNA helicase unwinds the DNA double helix, and single-strand DNA-binding protein coats the unwound strands, keeping them in an extended state. DNA gyrase catalyzes the formation of negative supercoiling in *E. coli* DNA, and this negative supercoiling behind the replication forks is thought to drive the unwinding process because superhelical tension is reduced by unwinding the complementary strands.

10.23 DnaA protein initiates the formation of the replication bubble by binding to the 9-bp repeats of *OriC*. DnaA protein is known to be required for the initiation process because bacteria with temperature-sensitive mutations in the *dnaA* gene cannot initiate DNA replication at restrictive temperatures.

10.25 (1) DNA replication usually occurs continuously in rapidly growing prokaryotic cells but is restricted to the S phase of the cell cycle in eukaryotes. (2) Most eukaryotic chromosomes contain multiple origins of replication, whereas most prokaryotic chromosomes contain a single origin of replication. (3) Prokaryotes utilize two catalytic complexes that contain the same DNA polymerase to replicate the leading and lagging strands, whereas eukaryotes utilize two or three distinct DNA polymerases for leading and lagging strand synthesis. (4) Replication of eukaryotic chromosomes requires the partial disassembly and reassembly of nucleosomes as replisomes move along parental DNA molecules. In prokaryotes, replication probably involves a similar partial disassembly/reassembly of nucleosome-like structures. (5) Most prokaryotic chromosomes are circular and thus have no ends. Most eukaryotic chromosomes are linear and have unique termini called telomeres that are added to replicating DNA molecules by a unique, RNA-containing enzyme called telomerase.

10.27 The chromosomes of haploid yeast cells that carry the *est1* mutation become shorter during each cell division. Eventually, chromosome instability results from the complete loss of telomeres, and cell death occurs because of the deletion of essential genes near the ends of chromosomes.

10.29 Nucleosomes and replisomes are both large macromolecular structures, and the packaging of eukaryotic DNA into nucleosomes raises the question of how a replisome can move past a nucleosome and replicate the DNA in the nucleosome in the process. The most obvious solution to this problem would be to completely or partially disassemble the nucleosome to allow the replisome to pass. The nucleosome would then reassemble after the replisome had passed. One popular model has the nucleosome partially disassembling, allowing the replisome to move past it (see Figure 10.31*b*).

CHAPTER 11

11.1 3′—TACGT—5′

11.3 (a) RNA contains the sugar ribose, which has an hydroxyl (OH) group on the 2-carbon; DNA contains the sugar 2-deoxyribose, with only hydrogens on the 2-carbon. RNA usually contains the base uracil at positions where thymine is present in DNA. However, some DNAs contain uracil, and some RNAs contain thymine. DNA exists most frequently as a double helix (double-stranded molecule); RNA exists more frequently as a single-stranded molecule; but some DNAs are single-stranded and some RNAs are double-stranded. (b) The main function of DNA is to store genetic information and to transmit that information from cell to cell and from generation to generation. RNA stores and transmits genetic information in some viruses that contain no DNA. In cells with both DNA and RNA: (1) mRNA acts as an intermediary in protein synthesis, carrying the information from DNA in the chromosomes to the ribosomes (sites at which proteins are synthesized). (2) tRNAs carry amino acids to the ribosomes and function in codon recognition during the synthesis of polypeptides. (3) rRNA molecules are essential components of the ribosomes. (4) snRNAs are important components of spliceosomes, and (5) miRNAs play key roles in regulating gene expression (see Chapter 20). (c) DNA is located primarily in the chromosomes (with some in cytoplasmic organelles, such as mitochondria and chloroplasts), whereas RNA is located throughout cells.

11.5 Both prokaryotic and eukaryotic organisms contain messenger RNAs, transfer RNAs, and ribosomal RNAs. In addition, eukaryotes contain small nuclear RNAs and micro RNAs. Messenger RNA molecules carry genetic information from the chromosomes (where the information is stored) to the ribosomes in the cytoplasm (where the information is expressed during protein synthesis). The linear sequence of triplet codons in an mRNA molecule specifies the linear sequence of amino acids in the polypeptides produced during translation of that mRNA. Transfer RNA molecules are small (about 80 nucleotides long) molecules that carry amino acids to the ribosomes and provide the codon-recognition specificity during translation. Ribosomal RNA molecules provide part of the structure and function of ribosomes; they represent an important part of the machinery required for the synthesis of polypeptides. Small nuclear RNAs are structural components of spliceosomes, which excise noncoding intron sequences from nuclear gene transcripts. Micro RNAs are involved in the regulation of gene expression.

11.7 Protein synthesis occurs on ribosomes. In eukaryotes, most of the ribosomes are located in the cytoplasm and are attached to the extensive membranous network of endoplasmic reticulum. Some protein synthesis also occurs in cytoplasmic organelles such as chloroplasts and mitochondria.

11.9 The introns of protein-encoding nuclear genes of higher eukaryotes almost invariably begin (5′) with GT and end (3′) with AG. In addition, the 3′ subterminal A in the "TACTAAC box" is completely conserved; this A is involved in bond formation during intron excision.

11.11 "Self-splicing" of RNA precursors demonstrates that RNA molecules can also contain catalytic sites; this property is not restricted to proteins.

11.13

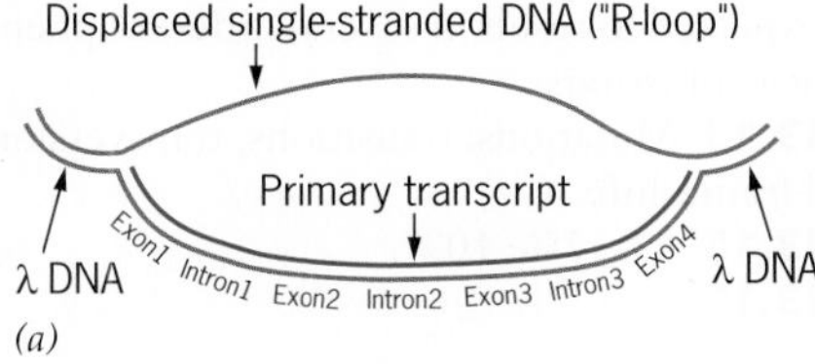

(a)

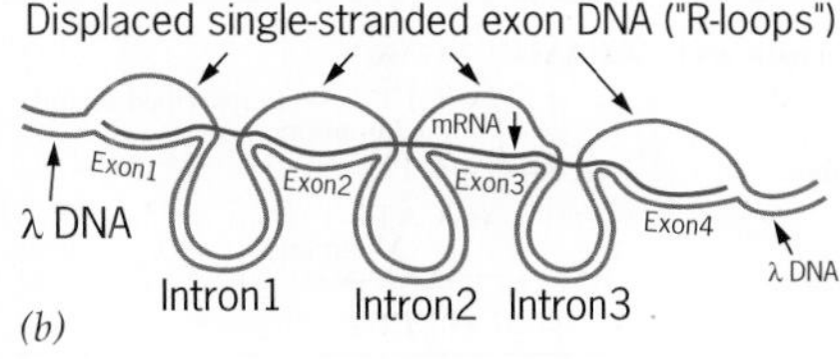

(b)

11.15 (a) Sequence 5. It contains the conserved intron sequences: a 5′ GU, a 3′ AG, and a UACUAAC internal sequence providing a potential bonding site for intron excision. Sequence 4 has a 5′ GU and a 3′ AG, but contains no internal A for the bonding site during intron excision. (b) 5′—UAGUCUCAA—3′; the putative intron from the 5′ GU through the 3′ AG has been removed.

11.17 Assuming that there is a CAAT box located upstream from the TATA box shown in this segment of DNA, the nucleotide sequence of the transcript will be 5′-ACCCGACAU-AGCUACGAUGACGAUA-3′.

11.19 DNA, RNA, and protein synthesis all involve the synthesis of long chains of repeating subunits. All three processes can be divided into three stages: chain initiation, chain elongation, and chain termination.

11.21 In eukaryotes, the genetic information is stored in DNA in the nucleus, whereas proteins are synthesized on ribosomes in the cytoplasm. How could the genes, which are separated from the sites of protein synthesis by a double-membrane—the nuclear envelope, direct the synthesis of polypeptides without some kind of intermediary to carry the specifications for the polypeptides from the nucleus to the cytoplasm? Researchers first used labeled RNA and protein precursors and autoradiography (see Figure 11.4) to demonstrate that RNA synthesis and protein synthesis occurred in the nucleus and the cytoplasm, respectively.

11.23 According to the central dogma, genetic information is stored in DNA and is transferred from DNA to RNA to protein during gene expression. RNA tumor viruses store their genetic information in RNA, and that information is copied into DNA by the enzyme reverse transcriptase after a virus infects a host cell. Thus the discovery of RNA tumor viruses or retroviruses—retro for backwards flow of genetic information—provided an exception to the central dogma.

11.25 TATA and CAAT boxes. The TATA and CAAT boxes are usually centered at positions -30 and -80, respectively, relative to the startpoint (+1) of transcription. The TATA box is responsible for positioning the transcription startpoint; it is the binding site for the first basal transcription factor that interacts with the promoter. The CAAT box enhances the efficiency of transcriptional initiation.

11.27 The primary transcripts of eukaryotes undergo more extensive post-transcriptional processing than those of prokaryotes. Thus the largest differences between mRNAs and primary transcripts occur in eukaryotes. Transcript processing is usually restricted to the excision of terminal sequences in prokaryotes. In contrast, eukaryotic transcripts are usually modified by: (1) the excision of intron sequences; (2) the addition of 7-methyl guanosine caps to the 5′ termini; (3) the addition of poly(A) tails to the 3′ termini. In addition, the sequences of some eukaryotic transcripts are modified by RNA editing processes.

11.29 Some individuals with systemic lupus erythematosus produce antibodies that react with proteins in snRNPs. These antibodies have been used to immunoprecipitate snRNPs, facilitating their purification and subsequent characterization.

11.31 RNA editing sometimes leads to the synthesis of two or more distinct polypeptides from a single mRNA.

11.33 The first preparation of RNA polymerase is probably lacking the sigma subunit and, as a result, initiates the synthesis of RNA chains at random sites along both strands of the *argH* DNA. The second preparation probably contains the sigma subunit and initiates RNA chains only at the site used *in vivo*, which is governed by the position of the -10 and -35 sequences of the promoter.

11.35 This zygote will probably be nonviable because the gene product is essential and the elimination of the 5' splice site will almost certainly result in the production of a nonfunctional gene product.

CHAPTER 12

12.1 Messenger RNA molecules carry genetic information from the chromosomes (where the information is stored) to the ribosomes in the cytoplasm (where the information is expressed during protein synthesis). The linear sequence of triplet codons in an mRNA molecule specifies the linear sequence of amino acids in the polypeptide(s) produced during translation of that mRNA. Transfer RNA molecules are small (about 80 nucleotides long) molecules that carry amino acids to the ribosomes and provide the codon-recognition specificity during translation. Ribosomal RNA molecules provide part of the structure and function of ribosomes; they represent an important part of the machinery required for the synthesis of polypeptides.

12.3 Proteins are long chainlike molecules made up of amino acids linked together by peptide bonds. Proteins are composed of carbon, hydrogen, nitrogen, oxygen, and usually sulfur. They provide the enzymatic capacity and much of the structure of living organisms. DNA is composed of phosphate, the pentose sugar 2-deoxyribose, and four nitrogen-containing organic bases (adenine, cytosine, guanine, and thymine). DNA stores and transmits the genetic information in most living organisms. Protein synthesis is of particular interest to geneticists because proteins are the primary gene products—the key intermediates through which genes control the phenotypes of living organisms.

12.5 A specific aminoacyl-tRNA synthetase catalyzes the formation of an amino acid-AMP complex from the appropriate amino acid and ATP (with the release of pyrophosphate). The same enzyme then catalyzes the formation of the aminoacyl-tRNA complex, with the release of AMP. The amino acid-AMP and aminoacyl-tRNA linkages are both high-energy phosphate bonds.

12.7 (a) The genetic code is degenerate in that all but 2 of the 20 amino acids are specified by two or more codons. Some amino acids are specified by six different codons. The degeneracy occurs largely at the third or 3′ base of the codons. "Partial degeneracy" occurs where the third base of the codon may be either of the two purines or either of the two pyrimidines and the codon still specifies the same amino acid. "Complete degeneracy" occurs where the third base of the codon may be any one of the four bases and the codon still specifies the same amino acid. (b) The code is ordered in the sense that related codons (codons that differ by a single base change) specify chemically similar amino acids. For example, the codons CUU, AUU, and GUU specify the structurally related amino acids, leucine, isoleucine, and valine, respectively. (c) The code appears to be almost completely universal. Known exceptions to universality include strains carrying suppressor mutations that alter the reading of certain codons (with low efficiencies in most cases) and the use of UGA as a tryptophan codon in yeast and human mitochondria.

12.9 Crick's wobble hypothesis explains how the anticodon of a given tRNA can base-pair with two or three different mRNA codons. Crick proposed that the base-pairing between the 5′ base of the anticodon in tRNA and the 3′ base of the codon in mRNA was less stringent than normal and thus allowed some "wobble" at this site. As a result, a single tRNA often recognizes two or three of the related codons specifying a given amino acid (see Table 12.2).

12.11 (a) Inosine. (b) Two.

12.13 (a) Singlet and doublet codes provide a maximum of 4 and $(4)^2$ or 16 codons, respectively. Thus neither code would be able to specify all 20 amino acids. (b) 20. (c) $(20)^{146}$.

12.15 Translation occurs by very similar mechanisms in prokaryotes and eukaryotes; however, there are some differences. (1) In prokaryotes, the initiation of translation involves base pairing between a conserved sequence (AGGAGG)—the Shine-Dalgarno box—in mRNA and a complementary sequence near the 3′ end of the 16S rRNA. In eukaryotes, the initiation complex forms at the 5′ end of the transcript when a cap-binding protein interacts with the 7-methyl guanosine on the mRNA. The complex then scans the mRNA processively and initiates translation (with a few exceptions) at the AUG closest to the 5′ terminus. (2) In prokaryotes, the amino group of the initiator methionyl-$tRNA_i^{Met}$ is formylated; in eukaryotes, the amino group of methionyl-$tRNA_i^{Met}$ is not formylated. (3) In prokaryotes, two soluble protein release factors (RFs) are required for chain termination. RF-1 terminates polypeptides in response to UAA and UAG condons; RF-2 terminates chains in response to UAA and UGA codons. In eukaryotes, one release factor responds to all three termination codons.

12.17 Assuming 0.34 nm per nucleotide pair in B-DNA, a gene 68 nm long would contain 200 nucleotide pairs. Given the triplet code, this gene would contain 200/3 = 66.7 triplets, one of which must specify chain termination. Disregarding the partial triplet, this gene could encode a maximum of 65 amino acids.

12.19 426 nucleotides—3 × 141 = 423 specifying amino acids plus three (one codon) specifying chain termination.

12.21 (a) Related codons often specify the same or very similar amino acids. As a result, single base-pair substitutions frequently result in the synthesis of identical proteins (degeneracy) or proteins with amino acid substitutions involving very similar amino acids. (b) Leucine and valine have very similar structures and chemical properties; both have nonpolar side groups and fold into essentially the same three-dimensional structures when present in polypeptides. Thus, substitutions of leucine for valine or valine for leucine seldom alter the function of a protein.

12.23 (a) Ribosomes and spliceosomes both play essential roles in gene expression, and both are complex macromolecular structures composed of RNA and protein molecules. (b) Ribosomes are located in the cytoplasm; spliceosomes in the nucleus. Ribosomes are larger and more complex than spliceosomes.

12.25 (UAG). This is the only nonsense codon that is related to tryptophan, serine, tyrosine, leucine, glutamic acid, glutamine, and lysine codons by a single base-pair substitution in each case.

12.27 (a) 5′-CAAUCAUGGACUGCCAUGCUUCAUAUGAAUAGUUGACAU-3′. (b) NH_2-fMet-Asp-Cys-His-Ala-Ser-Tyr-Glu-COOH. (c) NH_2-fMet-Asp-Cys-Met-Leu-His-Met-Asn-Ser-COOH.

12.29 Met-Ser-lle-Cys-Leu-Phe-Gln-Ser-Leu-Ala-Ala-Gln-Asp-Arg-Pro-Gly

CHAPTER 13

13.1 Probably not. A human is larger than a bacterium, with more cells and a longer life span. If mutation frequencies are calculated in terms of cell generations, the rates for human cells and bacterial cells are similar.

13.3 (a) Transition, (b) transition, (c) transversion, (d) transversion, (e) frameshift, (f) transition.

13.5 (a) *ClB* method, (b) attached-X method (see Chapter 6).

13.7 The sheep with short legs could be mated to unrelated animals with long legs. If the trait is expressed in the first generation, it could be presumed to be inherited and to depend on a dominant gene. On the other hand, if it does not appear in the first generation, F_1 sheep could be crossed back to the short-legged parent. If the trait is expressed in one-half of the backcross progeny, it might be presumed to be inherited as a simple recessive. If two short-legged sheep of different sex could be obtained, they could be mated repeatedly to test the hypothesis of dominance. In the event that the trait is not transmitted to the progeny that result from these matings, it might be considered to be environmental or dependent on some complex genetic mechanism that could not be identified by the simple test used in the experiments.

13.9 The X-linked gene is carried by mothers, and the disease is expressed in half of their sons. Such a disease is difficult to follow in pedigree studies because of the recessive nature of the gene, the tendency for the expression to skip generations in a family line, and the loss of the males who carry the gene. One explanation for the sporadic occurrence and tendency for the gene to persist is that, by mutation, new defective genes are constantly being added to the load already present in the population.

13.11 (a) Yes. (b) A block would result in the accumulation of phenylalanine and a decrease in the amount of tyrosine, which would be expected to result in several different phenotypic expressions.

13.13 Mutations: transitions, transversions, and frameshifts.

13.15 6%; 7%; 10%.

13.17

Amino Acid	mRNA	DNA
Glumatic acid	—GAA→	—GAA→ ←CTT— ← Transcribed strand ↓ Mutation
Valine	—GUA→	—GTA→ ←CAT— Mutation
Lysine	—AAA→	—AAA→ ←TTT—

13.19 $(x^+ m^+ z)(x^+ m^+ z^+)(x m^+ z)(x m z^+)$ or equivalent.

13.21 Transitions.

13.23 Nitrous acid acts as a mutagen on either replicating or nonreplicating DNA and produces transitions from A to G or C to T, whereas 5-bromouracil does not affect nonreplicating DNA but acts during the replication process causing GC ↔ AT transitions. 5-Bromouracil must be incorporated into DNA during the replication process in order to induce mispairing of bases and thus mutations.

13.25 5-BU causes GC ↔ AT transitions. 5-BU can, therefore, revert almost all of the mutations that it induces by enhancing the transition event that is the reverse of the one that produced the mutation. In contrast, the spontaneous mutations will include transversions, frameshifts, deletions, and other types of mutations, including transitions. Only the spontaneous transitions will show enhanced reversion after treatment with 5-BU.

13.27 Radioactive iodine is concentrated by living organisms and food chains.

13.29 Yes:

DNA:	←GGX— —CCX′→	$\xrightarrow{HNO_2}$	←GGX— —UCX′→
	↓		↓
mRNA:	GGX		AGX
	↓		↓
Polypeptide:	Gly		Ser or Arg (depending on X)

or

DNA:	←GGX— —CCX′→	$\xrightarrow{HNO_2}$	←GGX— —CUX′→
	↓		↓
mRNA:	GGX		GAX
	↓		↓
Polypeptide:	Gly		Asp or Glu (depending on X)

or

DNA:	←GGX— —CCX′→	$\xrightarrow{HNO_2}$	←GGX— —UUX′→
	↓		↓
mRNA:	GGX		AAX
	↓		↓
Polypeptide:	Gly		Asn or Lys (depending on X)

Note: The X at the third position in each codon in mRNA and in each triplet of base pairs in DNA refers to the fact that there is complete degeneracy at the third base in the glycine codon. Any base may be present in the codon, and it will still specify glycine.

13.31 No. Leucine → proline would occur more frequently. Leu (CUA) —5-BU→ Pro (CCA) occurs by a single base-pair transition, whereas Leu (CUA) —5-BU→ Ser (UCA) requires two base-pair transitions. Recall that 5-bromouracil (5-BU) induces only transitions (see Figure 13.14).

13.33 (a) Frameshift due to the insertion of C at the 9th, 10th, or 11th nucleotide from the 5′ end; (b) normal: 5′-AUGCCGUACUGCCAGCUAACUGCUAAAGAACAAUUA-3′. mutant: 5′-AUGCCCGUACUGCCAGCUAACUGCUAAAGAACAAUUA-3′. (c) normal: NH_2-Met-Pro-Tyr-Cys-Gln-Leu-Thr-Ala-Lys-Glu-Gln-Leu. mutant: NH_2-Met-Pro-Val-Leu-Pro-Ala-Asn-Cys.

13.35 5′-UGG-UGG-UGG-AUG-CGA or AGA-GAA or GAG-UGG-AUG-3′

13.37 Tyr → Cys substitutions; Tyr to Cys requires a transition, which is induced by nitrous acid. Tyr to Ser would require a transversion, and nitrous acid is not expected to induce transversions.

CHAPTER 14

14.1 The *cis-trans* test, which defines the unit of genetic material specifying the amino acid sequence of one polypeptide.

14.3 They provide powerful selective sieves for identifying rare recombinants. This is accomplished by using the restrictive environmental conditions to select wild-type recombinant progeny from crosses between pairs of conditional lethal mutants.

14.5 Four genes; mutations 1 and 2 in one gene; mutations 3 and 4 in a second gene; mutations 5 and 6 in a third gene; mutations 7 and 8 in a fourth gene.

14.7 (a) (A, B, F, G, H, I) and (C, D) and (E); (b) Blue colony requires at least three genes; (c) The maximum number of genes for blue color cannot be estimated from the above data.

14.9 Precursor → D → B → C → A
2 4 3 1

14.11 Homoalleles are structurally and functionally allelic; they are not separable by recombination. Heteroalleles are functionally allelic (based on *cis-trans* tests) but are structurally nonallelic (based on recombination tests). Heteroalleles thus result from mutations occurring at different sites within a gene.

14.13 (a) Five genes. (b) Mutations 1, 3, and 5 are in one gene; mutations 7 and 8 are in a second gene; mutations 2, 4, and 6 identify genes 3, 4, and 5, respectively.

14.15 (a) True. (b) False. (c) True. (d) True. (e) True.

14.17 *am* mutations result in UAG chain-termination codons within the coding sequence of the mRNA product of a gene; they thus produce truncated polypeptide gene-products. Since all *am* mutant alleles of a gene will produce polypeptides lacking the COOH-terminus, they would not be expected to exhibit intragenic complementation except in very rare cases. In contrast, most *ts* mutations are caused by missense mutations that change the amino acid sequence of the polypeptide gene-product making it more heat-labile. However, most *ts* mutant alleles produce a complete, although altered, gene-product. As a result, *ts* mutant alleles often exhibit intragenic complementation when the active form of the protein gene-product is a homomultimer. For this reason, *cis-trans* tests carried out with *am* mutants, not *ts* mutants, have been used whenever possible to define the genes of phage T4 operationally.

14.19 No. Because the two mutations map to different chromosomes, they could not be located within the same gene, at least, based on our current concept of the gene. However, if two transcripts are spliced in *trans*, it is possible for two parts of a gene—a nucleotide sequence encoding one polypeptide—to be located on two different chromosomes. Remember that our concept of the gene has evolved considerably since it was introduced by Mendel in 1865, and it will undoubtedly continue to evolve in the future.

14.21 It depends on how you define alleles. If every variation in nucleotide sequence is considered to be a different allele, even if the gene product and the phenotype of the organism carrying the mutation are unchanged, then the number of alleles will be directly related to gene size. However, if the nucleotide sequence change must produce an altered gene product or phenotype before it is considered a distinct allele, then there will be a positive correlation, but not a direct relationship, between the number of alleles of a gene and its size in nucleotide pairs. The relationship is more likely to occur in prokaryotes where most genes lack introns. In eukaryotic genes, nucleotide sequence changes within introns usually are neutral; that is, they do not affect the activity of the gene product or the phenotype of the organism. Thus, in the case of eukaryotic genes with introns, there may be no correlation between gene size and number of alleles producing altered phenotypes.

14.23 One gene; all seven mutations are in the same gene.

14.25 *Neurospora* has many advantages over humans as an experimental organism. The most important advantages are the ability to grow organisms under carefully controlled conditions, to enhance mutation frequency by treatment with mutagenic agents, and to perform controlled crosses for genetic analysis.

14.27 One. The *trpA*58 and *trpA*78 mutations alter the same codon. If they alter the 5′ and 3′ nucleotide pairs of the triplet, they will be separated by the middle nucleotide pair of the triplet specifying one mRNA codon.

14.29 The most reliable way to determine if mutation is a deletion is to perform a reverse mutation analysis. Deletions do not revert to wild type, but point mutations do.

14.31

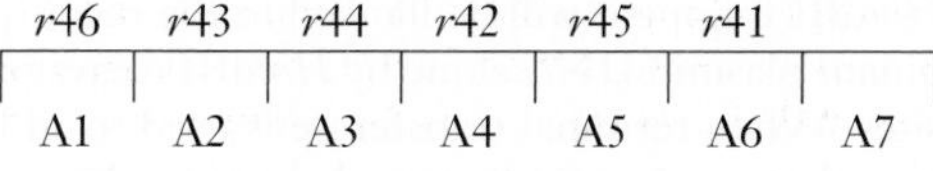

14.33 No. These mutations are located far apart on the X chromosome. They are separated by millions of nucleotide pairs and many other genes, and it is virtually impossible for mutations located that far apart to be part of the same gene. For these mutations to be located in the same gene, all of the intervening genes would have to be part of a huge intron or some type of *trans* splicing would have to occur.

14.35 (a) The linear order of the six mutations that can be ordered based on the data given is 3-8-7-4-2-5. (b) Mutations 1 and 6 cannot be ordered based on the data provided.

CHAPTER 15

15.1 (a) Both introduce new genetic variability into the cell. In both cases, only one gene or a small segment of DNA representing a small fraction of the total genome is changed or added to the genome. The vast majority of the genes of the organism remain the same. (b) The introduction of recombinant DNA molecules, if they come from a very different species, is more likely to result in a novel, functional gene product in the cell, if the introduced gene (or genes) is capable of being expressed in the foreign protoplasm. The introduction of recombinant DNA molecules is more analogous to duplication mutations (see Chapter 6) than to other types of mutations.

15.3 Restriction endonucleases are believed to provide a kind of primitive immune system to the microorganisms that produce them—protecting their genetic material from "invasion" by foreign DNAs from viruses or other pathogens or just DNA in the environment that might be taken up by the microorganism. Obviously, these microorganisms do not have a sophisticated immune system like that of higher animals (Chapter 21).

15.5 (a) $(1/4)^3 = 1/64$; (b) $(1/4)^4 = 1/256$

15.7 A foreign DNA cloned using an enzyme that produces single-stranded complementary ends can always be excised from the cloning vector by cleavage with the same restriction enzyme that was originally used to clone it. For example, if a *Hin*dIII fragment from the human genome is cloned into *Hin*dIII-cleaved pUC119, the human *Hin*dIII fragment can be excised from a plasmid DNA preparation of this clone by cleavage with restriction endonuclease *Hin*dIII. The human *Hin*dIII fragment will be flanked in the recombinant plasmid DNA clone by *Hin*dIII cleavage sites. When terminal transferase is used to add complementary single-stranded ends during cloning, the original restriction endonuclease cleavage sites are destroyed. Thus, the restriction enzyme used to generate the fragment for cloning cannot be used to excise the original fragments from the cloning vector.

15.9 The maize *gln2* gene contains many introns, and one of the introns contains a *Hin*dIII cleavage site. The intron sequences (and thus the *Hin*dIII cleavage site) are not present in mRNA sequences and thus are also not present in full-length *gln2* cDNA clones.

15.11 By oligonucleotide-directed site-specific mutagenesis, you can change each of the eight nucleotide pairs in the protein binding site to each of the other three possible nucleotide pairs. You can then examine the binding of the regulatory protein to each of the mutant sequences (24 single nucleotide-pair changes are possible at the eight positions), and you can examine the ability of each of the mutant sequences to mediate induction of transcription of the gene at 45°C *in vivo*. You could study the binding of the protein using synthetic oligonucleotide sequences more easily, but that approach would not let you study induction of transcription of the gene *in vivo*. The site-specific mutagenesis approach will allow you to carry out a saturation mutational analysis of the octameric regulatory protein binding sequence.

15.13 Most genes of higher plants and animals contain noncoding intron sequences. These intron sequences will be present in genomic clones, but not in cDNA clones, because cDNAs are synthesized using mRNA templates and intron sequences are removed during the processing of the primary transcripts to produce mature mRNAs.

15.15 All modern cloning vectors contain a "polycloning site"—a cluster of cleavage sites for a number of different restriction endonucleases in a nonessential region of the vector into which the foreign DNA can be inserted. In general, the greater the complexity of the polycloning site—that is, the more restriction endonuclease cleavage sites that are present—the greater the utility of the vector for cloning a wide variety of different restriction fragments. For example, see the polycloning site present in pUC118 and pUC119 shown in Figure 15.8.

15.17 Some bacterial viruses package single-stranded DNA rather than double-stranded DNA. These bacteriophages employ a double-stranded DNA intermediate for replication, but then switch to the production of single-stranded DNAs for packaging during phage maturation. When a foreign DNA segment is cloned into the replicative form of the chromosome of one of these phages, only one of the two strands gets packaged during the subsequent maturation processes. In the case of the filamentous single-stranded DNA phages such as M13 and f2, the progeny phage particles are simply extruded through the cell membrane and wall without lysing the host bacterium. Because the phage are so small relative to the size of the bacteria, the phage particles can be separated from the bacteria by a simple low-speed centrifugation step. The bacteria form a pellet at the bottom of the centrifuge tube; the phage remain in the supernatant suspension. The bacterial pellets are discarded, and the phage particles can be collected by high-speed centrifugation or by precipitation with polyethylene glycol and low-speed centrifugation. Pure single-stranded DNA can then be separated from the phage proteins by phenol-chloroform extractions. Many cloning vectors like pUC118 and pUC119 (see Figure 15.8) are phage-plasmid hybrids. They have both plasmid and M13 phage origins of replication as well as M13 phage packaging signals so that they can replicate either as plasmids or as phages and can package single-stranded phage DNA when in the phage mode of replication. The switch from the plasmid mode of replication to the phage mode is accomplished by superinfecting the host bacteria with a mutant "helper" phage.

15.19

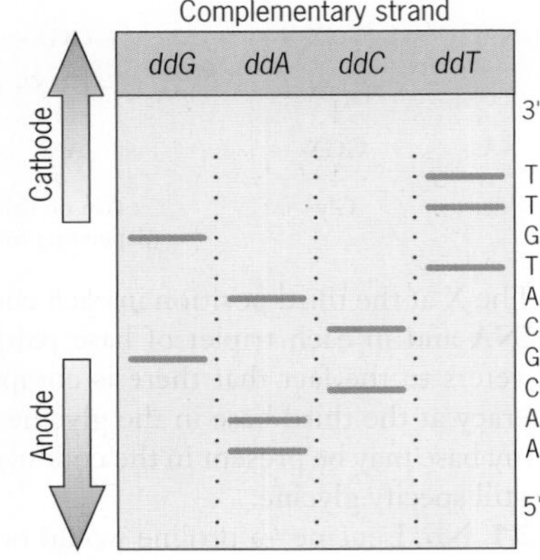

15.21 (a) Southern, northern, and western blot procedures all share one common step, namely, the transfer of macromolecules (DNAs, RNAs, and proteins, respectively) that have been separated by gel electrophoresis to a solid support—usually a nitrocellulose or nylon membrane—for further analysis. (b) The major difference between these techniques is the class of macromolecules that are separated during the electrophoresis step: DNA for Southern blots, RNA for northern blots, and protein for western blots.

15.23 Because the nucleotide-pair sequences of both the normal *CF* gene and the *CF*Δ*508* mutant gene are known, labeled oligonucleotides can be synthesized and used as hybridization probes to detect the presence of each allele (normal and $\Delta 508$). Under high-stringency hybridization conditions, each probe will hybridize only with the *CF* allele that exhibits perfect complementarity to itself. Since the sequences of the *CF* gene flanking the $\Delta 508$ site are known, oligonucleotide PCR primers can be synthesized and used to amplify this segment of the DNA obtained from small tissue explants of putative CF patients and their relatives by PCR. The amplified DNAs can then be separated by agarose gel electrophoresis, transferred to nylon membranes, and hybridized to the respective labeled oligonucleotide probes, and the presence of each *CF* allele can be detected by autoradiography. For a demonstration of the utility of this procedure, see B. Kerem et al. "Identification of the cystic fibrosis gene: Genetic analysis." *Science 245*: 1073–1080, 1989. Kerem and coworkers used two synthetic oligonucleotide probes (oligo-N = 3′-CTTTTATAGTAGAAACCAC-5′ and oligo-ΔF = 3′-TTCTTTTATAGTA—ACCACAA-5′; the dash indicates the deleted nucleotides in the *CF*Δ*508* mutant allele) to analyze the DNA of CF patients and their parents. For confirmed CF families, the results of these Southern blot hybridizations with the oligo-N (normal) and oligo-ΔF (*CF*Δ*508*) labeled probes were often as follows:

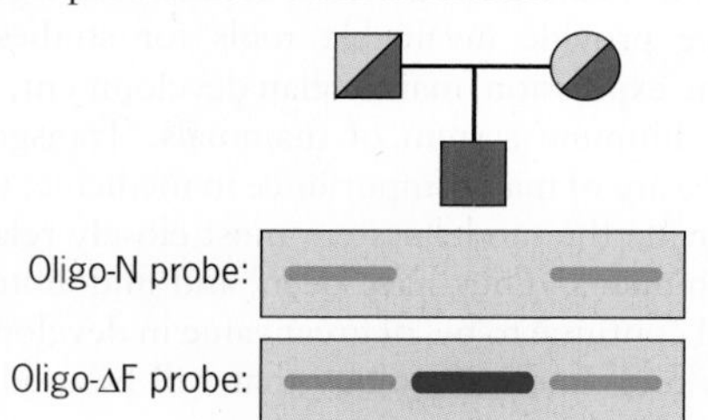

Both parents were heterozygous for the normal *CF* allele and the mutant *CF*Δ*508* allele as would be expected for a rare recessive trait, and the CF patient was homozygous for the *CF*Δ*508* allele. In such families, one-fourth of the children would be expected to be homozygous for the $\Delta 508$ mutant allele and exhibit the symptoms of CF, whereas three-fourths would be normal (not have CF). However, two-thirds of these normal children would be expected to be heterozygous and transmit the allele to their children. Only one-fourth of the children of this family would be homozygous for the normal *CF* allele and have no chance of transmitting the mutant *CF* gene to their offspring. Note that the screening procedure described here can be used to determine which of the normal children are carriers of the *CF*Δ*508* allele: that is, the mutant gene can be detected in heterozygotes as well as homozygotes.

15.25 There are two possible restriction maps for these data as shown below:

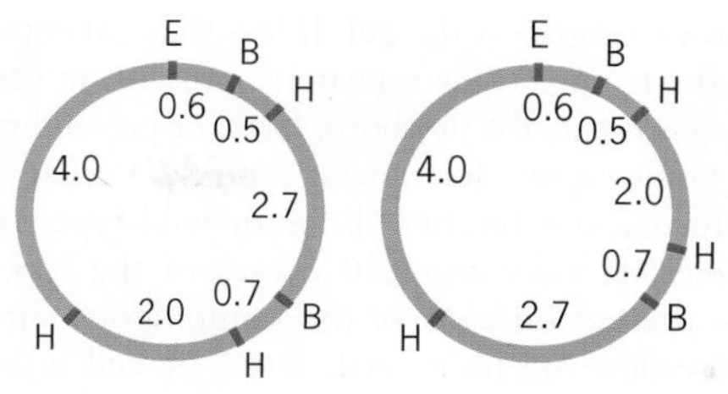

Restriction enzyme cleavage sites for *Bam*HI, *Eco*RI, and *Hin*dIII are denoted by B, E, and H, respectively. The numbers give distances in kilobase pairs.

15.27

H H
6.0 | 2.0 7.4 3.5 | 1.0 2.9
E E E

CHAPTER 16

16.1 A contig (*contig*uous clones) is a physical map of a chromosome or part of a chromosome prepared from a set of overlapping genomic DNA clones. An RFLP (*r*estriction *f*ragment *l*ength *p*olymorphism) is a variation in the length of a specific restriction fragment excised from a chromosome by digestion with one or more restriction endonucleases. A VNTR (*v*ariable *n*umber *t*andem *r*epeat) is a short DNA sequence that is present in the genome as tandem repeats and in highly variable copy number. An STS (*s*equence *t*agged *s*ite) is a unique DNA sequence that has been mapped to a specific site on a chromosome. An EST (*e*xpressed *s*equence *t*ag) is a cDNA sequence—a genomic sequence that is transcribed. Contig maps permit researchers to obtain clones harboring genes of interest directly from DNA Stock Centers—to "clone by phone." RFLPs are used to construct the high density genetic maps that are needed for positional cloning. VNTRs are especially valuable RFLPs that are used to identify multiple sites in genomes. STSs and ESTs provide molecular probes that can be used to initiate chromosome walks to nearby genes of interest.

16.3 Genetic map distances are determined by crossover frequencies. Cytogenetic maps are based on chromosome morphology or physical features of chromosomes. Physical maps are based on actual physical distances—the number of nucleotide pairs (0.34 nm per bp)—separating genetic markers. If a gene or other DNA sequence of interest is shown to be located near a mutant gene, a specific knob on a chromosome, or a particular DNA restriction fragment, that genetic or physical marker (mutation, knob, or restriction fragment) can be used to initiate a chromosome walk (see Figure 16.7).

16.5 VNTRs and microsatellites are both polymorphic tandem repeats of relatively short nucleotide sequences. Microsatellites, however, are repeats of sequences only 2–5 nucleotide pairs in length.

16.7 The resolution of radiation hybrid mapping is higher than that of standard somatic-cell hybrid mapping because the frequency of recombination is greatly increased in radiation hybrids by using X rays to fragment the human chromosomes prior to cell fusion. The rationale of radiation hybrid mapping is that the probability of breaking a DNA molecule in the region between the two genes and thus separating them is directly proportional to the physical distance (number of base pairs) between them.

16.9 (a)

10 cM	25 cM	15 cM	1 cM	14 cM
STS1 — STS5	STS5 — STS3	STS3 — *C*	*C* — STS4	STS4 — STS2

(b) 3.3×10^9 bp/3.3×10^3 cM = 1×10^6 bp/cM. The total map length is 65 cM, which equates to about 65×10^6 or 65 million bp. (c) The cancer gene (*C*) and STS4 are separated by 1 cM or about one million base pairs.

16.11 With a clone of the gene available, fluorescent *in situ* hybridization (FISH) can be used to determine which human chromosome carries the gene and to localize the gene on the chromosome. Single-stranded copies of the clone are coupled to a fluorescent probe and hybridized to denatured DNA in chromosomes spread on a slide (see Focus on *In Situ* Hybridization in Chapter 9). After hybridization and removal on nonhybridized probe, fluorescence microscopy; is performed to determine the location of the fluorescent probe (see Figure 1*d* in Focus on *In Situ* Hybridization in Chapter 9).

16.13 An EST is more likely than an RFLP to occur in a disease-causing human gene. ESTs all correspond to expressed sequences in a genome. RFLPs occur throughout a genome, in both expressed and unexpressed sequences. Because less than 5 percent of the human genome encodes proteins, most RFLPs occur in noncoding DNA.

16.15 The advantage that gene chips have over traditional dot blot hybridization methods is that a single gene chip can be used to quantify thousands of distinct nucleotide sequences simultaneously. The hybridization probes are arranged on gene chips in microarrays on a solid surface, whereas traditional dot blots were macroarrays of a few to, at most, a few hundred sequences. The gene-chip technology allows researchers to investigate the levels of expression of large numbers of genes more efficiently than was possible using traditional dot blot techniques.

16.17 (a) Order of STS sites: 2-5-1-4-3-6.

(b) STS markers: 2 5 1 4 3 6

A———

YAC clones B——— ——— C

D——— ——— E

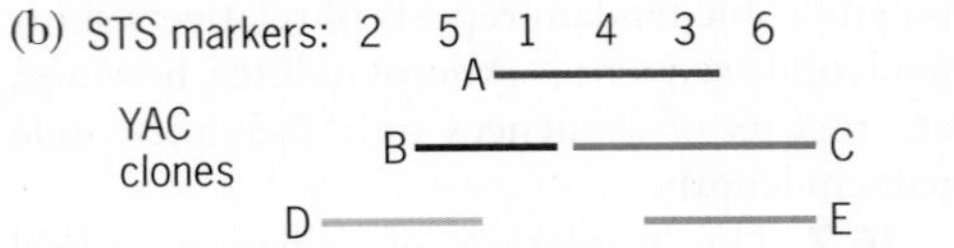

16.19 (a) Segment 5; (b) segment 4; (c) segment 1, 6, or 10.

16.21 Reading frame 5′ → 3′ number 1 has a large open reading frame with a methionine codon near the 5′ end. You can verify that this is the correct reading frame by using the predicted translation product as a query to search one of the protein databases (see Question 16.26).

16.23 All of the sequences identified by the megablast search encode histone H2a proteins. The query sequence is identical to the coding sequence of the *Drosophila melanogaster* histone *H2aV* gene (a member of the gene family encoding histone H2a proteins). The query sequence encodes a *Drosophila* histone H2a polypeptide designated variant V. The same databank sequences are identified when one-half or one-fourth of the given nucleotide sequence is used as the query in the megablast search. Query sequences as short as 15 to 20 nucleotides can be used to identify the *Drosophila* gene encoding the histone H2a variant. However, the results will vary depending on the specific nucleotide sequence used as the query sequence.

CHAPTER 17

17.1 The *CF* gene was identified by map position-based cloning, and the nucleotide sequences of *CF* cDNAs were used to predict the amino acid sequence of the *CF* gene product. A computer search of the protein data banks revealed that the *CF* gene product was similar to several ion channel proteins. This result focused the attention of scientists studying cystic fibrosis on proteins involved in the transport of salts between cells and led to the discovery that the *CF* gene product was a transmembrane conductance regulator—now called the CFTR protein.

17.3 Oligonucleotide primers complementary to DNA sequences on both sides (upstream and downstream) of the CAG repeat region in the *MD* gene can be synthesized and used to amplify the repeat region by PCR. One primer must be complementary to an upstream region of the template strand, and the other primer must be complementary to a downstream region of the nontemplate strand. After amplification, the size(s) of the CAG repeat regions can be determined by gel electrophoresis (see Figure 17.2). Trinucleotide repeat lengths can be measured by including repeat regions of known length on the gel. If less than 30 copies of the trinucleotide repeat are present on each chromosome, the newborn, fetus, or pre-embryo is homozygous for a wild-type *MD* allele or heterozygous for two different wild-type *MD* alleles. If more than 50 copies of the repeat are present on each of the homologous chromosomes, the individual, fetus, or cell is homozygous for a dominant mutant *MD* allele or heterozygous for two different mutant alleles. If one chromosome contains less than 30 copies of the CAG repeat and the homologous chromosome contains more than 50 copies, the newborn, fetus, or pre-embryo is heterozygous, carrying one wild-type *MD* allele and one mutant *MD* allele.

17.5 The transcription initiation and termination and translation initiation signals or eukaryotes differ from those of prokaryotes such as *E. coli*. Therefore, to produce a human protein in *E. coli*, the coding sequence of the human gene must be joined to appropriate *E. coli* regulatory signals—promoter, transcription terminator, and translation initiator sequences. Moreover, if the gene contains introns, they must be removed or the coding sequence of a cDNA must be used, because *E. coli* does not possess the spliceosomes required for the excision of introns from nuclear gene transcripts. In addition, many eukaryotic proteins undergo post-translational processing events that are not carried out in prokaryotic cells. Such proteins are more easily produced in transgenic eukaryotic cells growing in culture.

17.7 CpG islands are clusters of cytosines and guanines that are often located just upstream (5′) from the coding regions of human genes. Their presence in nucleotide sequences can provide hints as to the location of genes in human chromosomes.

17.9 DNA fingerprints are the specific patterns of bands present on Southern blots of genomic DNAs that have been digested with particular restriction enzymes and hybridized to appropriate DNA probes such as VNTR sequences. DNA fingerprints, like epidermal fingerprints, are used as evidence for identity or nonidentity in forensic cases. Geneticists have expressed concerns about the statistical uses of fingerprint data. In particular, they have questioned some of the methods used to calculate the probability that DNA from someone other than the suspect could have produced the observed fingerprints. This concern is based in part on the lack of adequate databases for various human subpopulations and the lack of precise information about the amount of variability in DNA fingerprints for individuals of different ethnic backgrounds. These concerns can best be addressed by the acquisition of data on fingerprint variability in different subpopulations and ethnic groups.

17.11 Eleven, ranging in multiples of 3, from 15 to 45 nucleotides long.

17.13 Probing Southern blots of restriction enzyme-digested DNA of the transgenic plants with ^{32}P-labeled transgene may provide evidence of multiple insertions, but would not reveal the genomic location of the inserts. Fluorescence *in situ* hybridization (FISH) is a powerful procedure for determining the genomic location of gene inserts. FISH is used to visualize the location of transgenes in chromosomes.

17.15 Contamination of blood samples would introduce more variability into the fingerprint. This would lead to a lack of allelic matching of fingerprints obtained from the blood samples and from the defendant. Mixing errors would be expected only to lead to the acquittal of a guilty person and not to the conviction of an innocent person. Only the mislabeling of samples could implicate someone who is innocent.

17.17 Transgenic mice are usually produced by microinjecting the genes of interest into pronuclei of fertilized eggs or by infecting pre-implantation embryos with retroviral vectors containing the genes of interest. Transgenic mice provide invaluable tools for studies of gene expression, mammalian development, and the immune system of mammals. Transgenic mice are of major importance in medicine; they provide the model system most closely related to humans. They have been, and undoubtedly will continue to be, of great value in developing the tools and technology that will be used for human gene therapy in the future.

17.19 The vector described contains the HGH gene; however, it does not contain a mammalian HGH-promoter that will regulate the expression of the transgene in the appropriate tissues. Construction of vectors containing a properly positioned mammalian HGH-promoter sequence should result in transgenic mice in which HGH synthesis is restricted to the pituitary gland.

17.21 Both antisense RNA and RNAi make use of RNA strands that are complementary to the mRNA target strand to block its translation (compare Figures 17.20 and 17.24). Antisense RNA uses single stranded RNA that is complementary to the mRNA. RNAi involves the use of double-stranded RNA, where one strand is complementary to the mRNA and the other strand is equivalent to the mRNA, to silence the expression of the target gene. RNAi makes use of the RNA-induced silencing complex (RISC) to block gene expression.

17.23 Plants have an advantage over animals in that once insertional mutations are induced they can be stored for long periods of time and distributed to researchers as dormant seeds.

17.25 (a) You can invert the coding sequence of a gene, introduce it into *Arabidopsis* plants by transformation with *Agrobacterium tumefaciens*, and examine the effects of the antisense RNA on the phenotypes of the transgenic plants. (b) You would first want to check the Salk Institute's Genome Analysis Laboratory web site to see if a T-DNA of transposon insertion has already been identified in this gene (see Question 17.28). If so, you can simply order seeds of the transgenic line from the *Arabidopsis* Biological Resource Center at Ohio State University. If no insertion is available in the gene, you can determine where it maps in the genome and use transposons that preferentially jump to nearby sites to identify a new insertional mutation (see http://www.arabidopsis.org/abrc/ima/jsp). (c) You can construct a gene that has sense and antisense sequences transcribed to a single mRNA molecule (see Figure 17.24*b*). The transcript will form a partially base-paired hairpin that will enter the RISC silencing pathway and block the expression of the gene.

CHAPTER 18

18.1 Resistance for the second antibiotic was acquired by conjugative gene transfer between the two types of cells.

18.3 The pair in (a) are inverted repeats and could therefore qualify.

18.5 No. IS*1* and IS*2* are mobilized by different transposases.

18.7 In the first strain, the F factor integrated into the chromosome by recombination with the IS element between genes *C* and *D*. In the second strain, it integrated by recombination with the IS element between genes *D* and *E*. The two strains transfer their genes in different orders because the two chromosomal IS elements are in opposite orientation.

18.9 Many bacterial transposons carry genes for antibiotic resistance and it is relatively simple for these genes to move from one DNA molecule to another. DNA molecules that acquire resistance genes can be passed to other cells in a bacterial population, both vertically (by descent) and horizontally (by conjugative transfer). Over time, continued exposure to an antibiotic will select for cells that have acquired a gene for resistance to that antibiotic. The antibiotic will therefore no longer be useful in combating these bacteria.

18.11 The *tnpA* mutation: no; the *tnpR* mutation: yes.

18.13 The paternally inherited *Bz* allele was inactivated by a transposable element insertion.

18.15 The c^m mutation is due to a *Ds* or an *Ac* insertion.

18.17 Factors made by the host's genome are required.

18.19 Through crossing over between the LTRs of a *gypsy* element.

18.21 *In situ* hybridization to polytene chromosomes using a *TART* probe (see Chapter 9).

18.23 *TART* and *HeT-A* replenish the ends of *Drosophila* chromosomes.

18.25 The *Sleeping Beauty* element could be used as a transformation vector in vertebrates much like the *P* element has been used in *Drosophila*. The *gfp* gene could be inserted between the ends of the *Sleeping Beauty* element and injected into eggs or embryos along with an intact *Sleeping Beauty* element capable of encoding the element's transposase. If the transposase that is produced in the injected egg or embryo acts on the element that contains the *gfp* gene, it might cause the latter to be inserted into genomic DNA. Then, if the egg or embryo develops into an adult, that adult can be bred to determine if a *Sleeping Beauty/gfp* transgene is transmitted to the next generation. In this way, it would be possible to obtain strains of mice or zebra fish that express the *gfp* gene.

CHAPTER 19

19.1

Gene or Regulatory Element	Function
(a) Regulator gene	Codes for repressor
(b) Operator	Binding site of repressor
(c) Promoter	Binding site of RNA polymerase and CAP-cAMP complex
(d) Structural gene *Z*	Encodes β-galactosidase
(e) Structural gene *Y*	Encodes β-galactoside permease

19.3 By studying the synthesis or lack of synthesis of the enzyme in cells grown on chemically defined media. If the enzyme is synthesized only in the presence of a certain metabolite or a particular set of metabolites, it is probably inducible. If it is synthesized in the absence but not in the presence of a particular metabolite or group of metabolites, it is probably repressible.

19.5

$$\text{(a)}\ \frac{I^+O^cZ^+Y^-}{I^+O^+Z^-Y^+};$$

$$\text{(b)}\ \frac{I^sO^cZ^+Y^-}{I^sO^+Z^-Y^+}$$

19.7 (a) 1, 2, 3, and 5; (b) 2, 3, and 5.

19.9 Catabolite repression has apparently evolved to assure the use of glucose as a carbon source when this carbohydrate is available, rather than less efficient energy sources.

19.11 Positive regulation; the CAP-cAMP complex has a positive effect on the expression of the *lac* operon. It functions in turning on the transcription of the structural genes in the operon.

19.13 Negative regulatory mechanisms such as that involving the repressor in the lactose operon block the transcription of the structural genes of the operon, whereas positive mechanisms such as the CAP-cAMP complex in the *lac* operon promote the transcription of the structural genes of the operon.

19.15 (a) The O^c mutations map very close to the *Z* structural gene; I^- mutations map slightly farther from the structural gene (but still very close by; see Figure 19.5). (b) An $I^+O^+Z^+Y^+/I^+O^cZ^+Y^+$ partial diploid would exhibit constitutive synthesis of β-galactosidase and β-galactoside permease, whereas an $I^+O^+Z^+Y^+/I^-O^+Z^+Y^+$ partial diploid would be inducible for the synthesis of these enzymes. (c) The O^c mutation is *cis*-dominant; the I^- mutation is *trans*-recessive.

19.17 First, remember that transcription and translation are coupled in prokaryotes. When tryptophan is present in cells, tryptophan-charged $tRNA^{Trp}$ is produced. This allows translation of the *trp* leader sequence through the two UGG Trp codons to the *trp* leader sequence UGA termination codon. This translation of the *trp* leader region prevents base-pairing between the partially complementary mRNA leader sequences 75–83 and 110–121 (see Figure 19.13*b*), which in turn permits formation of the transcription-termination "hairpin" involving leader sequences 110–121 and 126–134 (see Figure 19.13*c*).

19.19 A prophage is a chromosome of a bacteriophage after it has become integrated into the chromosome of the host bacterium. The lambda prophage is present as a transcriptionally inactive linear structure in the *E. coli* chromosome. During lytic infections, the lambda chromosome begins its reproductive cycle by replicating as a transcriptionally active circular DNA molecule.

19.21 Repression/derepression of the *trp* operon occurs at the level of transcription initiation, modulating the frequency at which RNA polymerase initiates transcription from the *trp* operon promoters. Attenuation modulates *trp* transcript levels by altering the frequency of termination of transcription within the *trp* operon leader region (*trpL*).

19.23 The mutation will produce the socalled

clear-plaque phenotype. In the absence of functional repressor, lambda phage can only reproduce lytically. Lysogeny requires the lambda repressor to be present to keep the lytic genes of the lambda prophage repressed. Thus all infected cells will be killed and lysed during the lambda lytic cycle, resulting in clear plaques rather than the turbid plaques formed by wild-type lambda.

CHAPTER 20

20.1 By monitoring puffs in response to environmental signals, such as heat shock, or to hormonal signals.

20.3 In multicellular eukaryotes, the environment of an individual cell is relatively stable. There is no need to respond quickly to changes in the external environment. In addition, the development of a multicellular organism involves complex regulatory hierarchies composed of hundreds of different genes. The expression of these genes is regulated spatially and temporally, often through intricate intercellular signaling processes.

20.5 Activity of the *dystrophin* gene could be assessed by blotting RNA extracted from the different types of cells and hybridizing it with a probe from the gene (northern blotting); or the RNA could be reverse transcribed into cDNA using one or more primers specific to the *dystrophin* gene and the resulting cDNA could be amplified by the polymerase chain reaction (RT-PCR). Another technique would be to hybridize *dystrophin* RNA *in situ*—that is, in the cells themselves—with a probe from the gene. It would also be possible to check each cell type for production of dystrophin protein by using anti-dystrophin antibodies to analyze proteins from the different cell types on western blots, or to analyze the proteins in the cells themselves—that is, *in situ*.

20.7 Northern blotting of RNA extracted from plants grown with and without light, or PCR amplification of cDNA made by reverse transcribing these same RNA extracts.

20.9 By alternate splicing of the transcript.

20.11 Probably not unless the promoter of the *gfp* gene is recognized and transcribed by the Drosophila RNA polymerase independently of the heat shock response elements.

20.13 This enzyme plays an important role in photosynthesis, a light-dependent process. Thus, it makes sense that its production should be triggered by exposure to light.

20.15 The flies would be phenotypically normal males.

20.17 The intron could be placed in a GUS expression vector, which could then be inserted into *Arabidopsis* plants. If the intron contains an enhancer that drives gene expression in root tips, transgenic plants should show GUS expression in their root tips. See the Focus on Problem Solving in Chapter 20 for an example of this type of analysis.

20.19 Yes.

20.21 Short interfering RNAs target messenger RNA molecules, which are devoid of introns. Thus, if siRNA were made from double-stranded RNA derived from an intron, it would be ineffective against an mRNA target.

20.23 The paternally contributed allele (*b*) will be expressed in the F_1 progeny.

20.25 RNA could be isolated from liver and brain tissue. Northern blotting or RT-PCR with this RNA could then establish which of the genes (A or B) is transcribed in which tissue. For northern blotting, the RNA samples would be fractionated in a denaturing gel and blotted to a membrane, and then the RNA on the membrane would be hybridized with gene-specific probes, first for one gene, then for the other (or the researcher could prepare two separate blots and hybridize each one with a different probe). For RT-PCR, the RNA samples would be reverse transcribed into cDNA using primers specific for each gene; then the cDNA molecules would be amplified by standard PCR and the products of the amplifications would be fractionated by gel electrophoresis to determine which gene's RNA was present in the original samples.

20.27 The *mle* gene is not functional in females.

CHAPTER 21

21.1 The fly will be sterile because the posterior pole cells form the germ line in adults of both sexes.

21.3 The invariant cell lineage of *C. elegans* affords researchers the opportunity to follow the development of tissues and organs over time; they can observe cell division, cell movement, and cell differentiation—all leading to the formation of the adult animal.

21.5 Imaginal discs.

21.7 Mate homozygous *unc* hermaphrodites that carry one of the *dumpy* mutations with males that carry the other *dumpy* mutation and score the non-uncoordinated (*unc*/+) F_1 hermaphrodites for the dumpy phenotype. If these worms are dumpy, the two mutations are allelic; if they are wild-type, the two mutations are not allelic.

21.9 Cross heterozygous carriers separately with *tra*/+ and *tra2*/+ flies. If any of the XX progeny of these matings are transformed into sterile males, the new mutation is an allele of the *tra* or *tra2* tester.

21.11 Intersex. The *dsx* mutation is epistatic to the *tra* mutation.

21.13 XX, male phenotype; XO, female phenotype.

21.15 In homozygous condition, the mutation that causes phenylketonuria has a maternal effect. Women homozygous for this mutation influence the development of their children *in utero*.

21.17 Some structures fail to develop in the posterior portion of the embryo.

21.19 Screen for lethal mutations that prevent regions of the embryo from developing normally.

21.21 *ey* → *boss* → *sev* → R7 differentiation

21.23 Northern blotting of RNA extracted from the tissues at different times during development. Hybridize the blot with gene-specific probes.

21.25 If each antibody consists of one kind of light chain and one kind of heavy chain, and if light and heavy chains can combine freely, the potential to produce 100 million different antibodies implies the existence of 10,000 light chain genes and 10,000 heavy chain genes (10,000 × 10,000 = 100 million). If each light chain is 200 amino acids long, each light chain gene must comprise 3 × 200 = 600 nucleotides because each amino acid is specified by a triplet of nucleotides; similarly, each heavy chain gene must comprise at 3 × 500 = 1500 nucleotides. Therefore, the genome must contain 10,000 × 600 = 6 million nucleotides devoted to light chain production and 10,000 × 1500 = 15 million nucleotides devoted to heavy chain production. Altogether, then, the genome must contain 21 million nucleotides dedicated to encoding the amino acids of the various antibody chains.

21.27 No, because the V_k coding sequence must be joined to the coding sequence of the constant region to encode a complete kappa light chain.

CHAPTER 22

22.1 Aneuploidy might involve the loss of functional copies of tumor suppressor genes, or it might involve the inappropriate duplication of proto-oncogenes. Loss of tumor suppressor genes would remove natural brakes on cell division, and duplication of proto-oncogenes would increase the abundance of factors that promote cell division.

22.3 Cancer has been called a genetic disease because it results from mutations of genes that regulate cell growth and division. Nonhereditary forms of cancer result from mutations in somatic cells. These mutations, however, can be induced by environmental factors including tobacco smoke, chemical pollutants, ionizing radiation, and UV light. Hereditary forms of cancer also frequently involve the occurrence of environmentally-induced somatic mutations.

22.5 The products of these genes play important roles in cell activities.

22.7 The cultured NIH 3T3 cells probably carry other mutations that predispose them to become cancerous; transfection of such cells with a mutant *c-H-ras* oncogene may be the last step in the process of transforming the cells into cancer cells. Cultured embryonic cells probably do not carry the predisposing mutations needed for them to become cancerous; thus, when they are transfected with the mutant *c-H-ras* oncogene, they continue to divide normally.

22.9 Retinoblastoma results from homozygosity for a loss-of-function (recessive) allele. The sporadic occurrence of retinoblastoma requires two mutations of this gene in the same cell or cell lineage. Therefore retinoblastoma is rare among individuals who, at conception, are homozygous for the wild-type allele of the *RB* gene. For such individuals, we would expect the frequency of tumors in both eyes to be the square of the frequency of tumors in one eye. Individuals who are heterozygous for a mutant *RB* allele require only one somatic mutation to occur for them to develop retinoblastoma. Because there are millions of cells in each retina, there is a high probability that this somatic mutation will occur in at least one cell in each eye, causing both eyes to develop tumors.

22.11 At the cellular level, loss-of-function mutations in the *RB* are recessive; a cell that is heterozygous for such a mutation divides normally. However, when a second mutation occurs, that cell becomes cancerous. If the first *RB* mutation was inherited, there is a high probability that the individual carrying this mutation will develop retinoblastoma because a second mutation can occur any time during the formation of the retinas in either eye. Thus, the individual is predisposed to develop retinoblastoma, and it is this predisposition that shows a dominant pattern of inheritance.

22.13 Cells homozygous for a loss-of-function mutation in the *p16* gene might be expected to divide in an uncontrolled fashion because the p16 protein would not be able to inhibit cyclin-CDK activity during the cell cycle. The *p16* gene would therefore be classified as a tumor suppressor gene.

22.15 Cells homozygous for a loss-of-function mutation in the *BAX* gene would be unable to prevent repression of the programmed cell death pathway by the BCL-2 gene product. Consequently, these cells would be unable to execute that pathway in response to DNA damage induced by radiation treatment. Such cells would continue to divide and accumulate mutations; ultimately, they would have a good chance of becoming cancerous. The *BAX* gene would therefore be classified as a tumor suppressor gene.

22.17 If a cell were heterozygous for a mutation that caused p53 to bind tightly and constitutively to the DNA of its target genes, its growth and division might be retarded, or it might be induced to undergo apoptosis. Such a cell would be expected to be more sensitive to the effects of ionizing radiation because radiation increases the expression of p53, and in this case, the p53 would be predisposed to activate its target genes, causing the cell to respond vigorously to the radiation treatment.

22.19 They would probably decrease the ability of pAPC to bind β-catenin.

22.21 The increased irritation to the intestinal epithelium caused by a fiber-poor, fat-rich diet would be expected to increase the need for cell division in this tissue (to replace the cells that were lost because of the irritation), with a corresponding increase in the opportunity for the occurrence of cancer-causing mutations.

22.23 No. Apparently there is another pathway—one not mediated by p53—that leads to the activation of the *p21* gene.

CHAPTER 23

23.1 Because 8/2012 is approximately 1/256 $= (1/4)^4$, it appears that four size-determining genes were segregating in the crosses.

23.3 Some of the genes implicated in heart disease are listed in Table 23.2. Environmental factors might include diet, amount of exercise, and whether or not the person smokes.

23.5 The concordance for monozygotic twins is almost twice as great as that for dizygotic twins. Monozygotic twins share twice as many genes as dizygotic twins. The data strongly suggest that alcoholism has a genetic basis.

23.7 V_e is estimated by the average of the variances of the inbreds: 9.4 cm^2. V_g is estimated by the difference between the variances of the randomly pollinated population and the inbreds: $(26.4 - 9.4) = 7.4$ cm^2. The broad-sense heritability is $H^2 = V_g/V_t = 7.4/26.4 = 0.64$.

23.9 Because $\Sigma(X_i - \text{mean}) = 0$

23.11 $(15 - 12)(0.3) + 12 = 12.9$ bristles.

23.13 $h^2 = R/S = (12.5 - 10)/(15 - 10) = 0.5$; selection for increased growth rate should be effective.

23.15 Broad-sense heritability must be greater than narrow-sense heritability because $H^2 = V_g/V_t < V_a/V_t = h^2$.

23.17 The correlations for MZT are not much different from those for MZA. Evidently, for these personality traits, the environmentality (C^2 in Table 23.3) is negligible.

23.19 Half-siblings share 25 percent of their genes. The maximum value for h^2 is therefore $0.14/0.25 = 0.56$.

CHAPTER 24

24.1 $q^2 = 0.008$; $q = 0.089$.

24.3 Frequency of L^M in Central American population: $p = (2 \times 55 + 35)/(2 \times 98) = 0.74$; $q = 0.26$. Frequency of L^M in North American population: $p = (2 \times 68 + 55)/(2 \times 240) = 0.40$; $q = 0.60$.

24.5 $(0.00025)^2 = 6.25 \times 10^{-8}$.

24.7 Frequency of tasters (genotypes *TT* and *Tt*): $(0.4)^2 + 2(0.4)(0.6) = 0.64$. Frequency of *TT* tasters among all tasters: $(0.4)^2/(0.64) = 0.25$.

24.9 Frequency of heterozygotes $= H = 2pq = 2p(1 - p)$. Using calculus, take the derivative of H and set the result to zero to solve for the value of p that maximizes H: $dH/dp = 2 - 4p = 0$ implies that $p = 2/4 = 0.5$.

24.11 In females, the frequency of the dominant phenotype is 0.36. The frequency of the recessive phenotype is $0.64 = q^2$; thus, $q = 0.8$ and $p = 0.2$. The frequency of the dominant phenotype in males is therefore $p = 0.2$.

24.13 Ultimate frequency of *GG* is 0.2; ultimate frequency of *gg* is 0.8.

24.15 (a) Frequency of *A* in merged population is 0.5, and that of *a* is also 0.5; (b) 0.25 (*AA*), 0.50 (*Aa*), and 0.25 (*aa*); (c) Frequencies in (b) will persist.

24.17 Under the assumption that the population is in Hardy–Weinberg equilibrium, the frequency of the allele for light coloration is the square root of the frequency of recessive homozygotes. Thus, $q = \sqrt{0.49} = 0.7$, and the frequency of the allele for dark color is $1 - q = p = 0.3$. From $p^2 = 0.09$, we estimate that $0.09 \times 100 = 9$ of the dark moths in the sample are homozygous for the dominant allele.

24.19 $q^2 = 4 \times 10^{-5}$; thus $q = 6.3 \times 10^{-3}$ and $2pq = 0.0126$.

24.21 Probability of ultimate fixation of A_2 is 0.5; probability of ultimate loss of A_3 is $1 - 0.3 = 0.7$.

24.23 Use the following scheme:

Genotype	*CC*	*Cc*	*cc*
Hardy–Weinberg frequency	$(0.98)^2 = 0.9604$	$2(0.98)(0.02) = 0.0392$	$(0.02)^2 = 0.0004$
Relative fitness	1	1	0
Relative contribution to next generation	$(0.9604) \times 1$	$(0.0392) \times 1$	0
Proportional contribution	$0.9604/0.9996 = 0.9608$	$0.0392/0.9996 = 0.0392$	

New frequency of the allele for cystic fibrosis = (0.5)(0.0392) = 0.0196; thus, the incidence of the disease will be $(0.0196)^2 = 0.00038$, which is very slightly less than the incidence in the previous generation.

24.25 $p = 0.2$; at equilibrium, $p = t/(s + t)$. Because $s = 1$, we can solve for t; $t = 0.25$.

24.27 At mutation-selection equilibrium $q = \sqrt{u/s} = \sqrt{10^{-6}/1} = 0.001$.

CHAPTER 25

25.1 The frequency of the *a* allele is 0.06 in the South African population and 0.42 in the English population. The predicted genotype frequencies under the assumption of random mating are:

Genotype	South Africa	England
aa	$(0.06)^2 = 0.004$	$(0.42)^2 = 0.18$
ab	$2(0.06)(0.94) = 0.11$	$2(0.42)(0.58) = 0.49$
bb	$(0.94)^2 = 0.88$	$(0.58)^2 = 0.33$

25.3 Among other things, Darwin observed species on islands that were different from each other and from continental species, but that were still similar enough to indicate that they were related. He also observed variation within species, especially within domesticated breeds, and saw how the characteristics of an organism could be changed by selective breeding. His observations of fossilized organisms indicated that some species have become extinct.

25.5 Complex carbohydrates are not "documents of evolutionary history" because, although they are polymers, they are typically made of one subunit incorporated repetitiously into a chain. Such a polymer has little or no "information content." Thus, there is little or no opportunity to distinguish a complex carbohydrate obtained from two different organisms. Moreover, complex carbohydrates are not part of the genetic machinery; their formation is ultimately specified by the action of enzymes, which are gene products, but they themselves are not genetic material or the products of the genetic material.

25.7 The histidines are rigorously conserved because they perform an important function—anchoring the heme group in hemoglobin. Because these amino acids are strongly constrained by natural selection, they do not evolve by mutation and random genetic drift.

25.9 Estimate the average number of substitutions per site in the ribonuclease molecule as $-\ln(S)$, where $S = (124 - 45)/124 = 0.64$, the proportion of amino acids that are the same in the rat and cow molecules. The average number of substitutions per site since the cow and rat lineages diverged from a common ancestor is therefore 0.45. The evolutionary rate in the cow and rat lineages is $0.45/(2 \times 90$ million years) = 2.5 substitutions per site every billion years.

25.11 In the third position of some of the codons. Due to the degeneracy of the genetic code, different codons can specify the same amino acid. The degeneracy is most pronounced in the third position of many codons, where different nucleotides can be present without changing the amino acid that is specified.

25.13 The reciprocal of the rate, that is, $1/K$.

25.15 The protein with the higher evolutionary rate is not as constrained by natural selection as the protein with the lower evolutionary rate.

25.17 Repetitive sequences that are near each other can mediate displaced pairing during meiosis. Exchange involving the displaced sequences can duplicate the region between them. See Figure 18.20.

25.19 Cross *D. mauritiana* with *D. simulans* and determine if these two species are reproductively isolated. For instance, can they produce offspring? If they can, are the offspring fertile?

25.21 The *Kpn-pn* interaction is an example of the kind of negative epistasis that might prevent populations that have evolved separately from merging into one panmictic population. The *Kpn* mutation would have evolved in one population and the *pn* mutation in another, geographically separate population. When the populations merge, the two mutations can be brought into the same fly by interbreeding. If the combination of these mutations is lethal, then the previously separate populations will not be able to exchange genes, that is, they will be reproductively isolated.

Index